Operations Research

APPLICATIONS AND ALGORITHMS

Operations Research

APPLICATIONS

AND ALGORITHMS

FOURTH EDITION

Wayne L. Winston

INDIANA UNIVERSITY

WITH CASES BY
Jeffrey B. Goldberg
UNIVERSITY OF ARIZONA

BROOKS/COLE
CENGAGE Learning™

Australia • Brazil • Japan • Korea • Mexico • Singapore • Spain • United Kingdom • United States

Operations Research: Applications and Algorithms, Fourth Edition
Wayne L. Winston, Jeffrey B. Goldberg

Publisher: Curt Hinrichs

Assistant Editor: Ann Day

Editorial Assistant: Katherine Brayton

Technology Project Manager: Burke Taft

Marketing Manager: Joseph Rogove

Marketing Assistant: Jessica Perry

Advertising Project Manager: Tami Strang

Print/Media Buyer: Jessica Reed

Permissions Editor: Bob Kauser

Production Project Manager: Hal Humphrey

Production Service: Hoyt Publishing Services

Text Designer: Kaelin Chappell

Copy Editors: David Hoyt and Erica Lee

Illustrator: Electronic Illustrators Group

Cover Designer: Lisa Langhoff

Cover Image: Getty Images, photographer Nick Koudis

Compositor: ATLIS Graphics

For product information and technology assistance, contact us at
Cengage Learning Customer & Sales Support, 1-800-354-9706
For permission to use material from this text or product,
submit all requests online at **cengage.com/permissions**
Further permissions questions can be emailed to
permission request@cengage.com

Library of Congress Control Number: 2003105883

Student Edition with InfoTrac College Edition:

ISBN-13: 978-0-534-38058-8

ISBN-10: 0-534-38058-1

Student Edition without InfoTrac College Edition:

ISBN-13: 978-0-534-42358-2

ISBN-10: 0-534-42358-2

Brooks/Cole
10 Davis Drive
Belmont, CA 94002
USA

Cengage Learning is a leading provider of customized learning solutions with office locations around the globe, including Singapore, the United Kingdom, Australia, Mexico, Brazil, and Japan. Locate your local office at: **international.cengage.com/region**

Cengage Learning products are represented in Canada by Nelson Education, Ltd.

For your course and learning solutions, visit **www.cengage.com**

Purchase any of our products at your local college store or at our preferred online store **www.ichapters.com**

Printed in The United States of America
9 10 11 12 12

Brief Contents

Contents

Appendix 3: Answers to Selected Problems 1370

Preface

In recent years, operations research software has become widely available. Its use is illustrated throughout this book. Like most tools, however, it is of little value unless the user understands its application and purpose. Users must ensure that the mathematical input accurately reflects the real-life problems to be solved and that the numerical results are correctly applied to solve them. With this in mind, this book emphasizes model formulation and model building as well as the interpretation of software output.

Intended Audience and Prerequisites

This book is intended as an advanced beginning or intermediate text in operations research or management science. The following groups can benefit from using it.

- Undergraduate majors in information systems or decision sciences in business, operations research, management science, industrial engineering, mathematics, or agricultural/resource economics.

- MBA students or masters students in public administration enrolled in an applications-oriented operations research or management science course.

- Graduate students who need an overview of the major topics in operations research and management science.

- Practitioners who need a comprehensive reference.

For courses specializing in deterministic models or in probabilistic models of operations research, or for those wishing to cover state-of-the-art methods of operations research (OR), the publisher offers split volumes of this text that feature additional coverage.

Introduction to Mathematical Programming (Operations Research: Volume One—ISBN 0-534-35964-7) includes Chapters 1 through 10, 11, and 14 of *Operations Research,* along with three unique chapters covering recent developments in mathematical programming. Unique topics include heuristic methods,

artificial intelligence, genetic algorithms, simulated annealing, Tabu search, and neural networks.

Introduction to Probability Models (Operations Research: Volume Two—ISBN 0-534-40572-X) includes *OR* Chapters 12, 13, and 15 through 24, plus three additional chapters on financial engineering topics. Topics include option pricing, real options, the scenario approach to portfolio optimization, stochastic calculus, and stochastic control.

Operations Research is designed for students who have had some calculus, matrix algebra, and an introductory statistics course. A formal course in probability theory is not required. Chapter 2 provides a review of matrix algebra, and Chapter 12 reviews the probability and calculus required for the rest of the book.

Features

The following features help to make this text reader-friendly.

- The book is completely self-contained, with all the necessary mathematical background reviewed in Chapters 2 and 12. Each chapter is designed to be modular, so the book can be tailored to the needs of a course. Additionally, each section of the book is written to be as self-contained as possible; instructors can be extremely flexible in designing a course. The Instructor's Notes identify which portions of the book must be covered as prerequisites to each section.

- To provide immediate feedback to students, problems are placed at the end of sections, and most chapters conclude with review problems. There are approximately 1,500 problems, grouped by level of difficulty: Group A for practice of basic techniques, Group B for underlying concepts, and Group C for mastering the theory independently.

- The book avoids excessive theoretical exercises in favor of applied word problems. Many problems are based on published applications. The exposition takes great pain, by means of several examples in each chapter, to guide the student step by step through even the most complex topics.

- To help students review for exams, most chapters have a summary of concepts and formulas. Answers to selected problems appear in an appendix. A *Student Solutions Manual* is available, providing worked-out solutions to selected problems. The *Student Solutions Manual* may be purchased separately or packaged with the text at a nominal additional price.

- Instructors who adopt this text in their courses may receive the *Instructor's Suite CD-ROM.* This CD contains complete solutions to every problem in the text, PowerPoint slides, and Instructor's Notes.

- The book is accompanied by a CD containing special versions of LINDO, LINGO, Premium Solver, Process Model, and @Risk.

- The text contains instruction for using the software contained on the CD. All of the files needed for examples and exercises are also included on the CD.

Coverage and Organization

The linear programming section of the book is completely self-contained; all necessary mathematical background is given in Chapter 2. Students who are familiar with matrix multiplication should have no problems with Chapters 2–11. Portions of the remaining chapters require rudimentary knowledge of calculus and probability equivalent to that obtained from a one-semester calculus course and a one-semester statistics course. All topics in calculus and probability used in Chapters 13–24 are reviewed in Chapter 12.

Since not all students need a full-blown theoretical treatment of sensitivity analysis, there are two chapters on the topic. Chapter 5 is an applied approach to sensitivity analysis, emphasizing the interpretation of computer output. Chapter 6 contains a full discussion of sensitivity analysis, duality, and the dual simplex method. The instructor should cover Chapter 5 or Chapter 6, but not both. Classes emphasizing model building and model formulation skills should cover Chapter 5. Those paying close attention to the algorithms of mathematical programming (particularly classes in which students will go on to further study in operations research) should study Chapter 6. If Chapter 5 rather than Chapter 6 is covered, then Chapter 2 may be omitted.

Changes to the Fourth Edition

The fourth edition of *Operations Research* contains many substantial changes. Most significant is the inclusion of Process Model (Chapter 22) to perform queuing simulations and @Risk (Chapter 23) to perform spreadsheet-based simulations. Other major changes include the following.

- Over 200 new problems have been added.

- Microsoft Excel is featured. All Lotus spreadsheets appearing in the previous edition have been converted to Excel.

- There is more discussion of optimization with spreadsheets. The method of solving optimization problems with spreadsheets has been changed from What's Best to the Excel Solver.

- Discussion of important Excel functions such as MMULT, OFFSET, MINVERSE, and NPV has been added.

- Chapter 4 includes more extensive instruction in the use of LINDO and LINGO.

- Chapter 4 includes more discussion of the geometry of LPs.

- Chapter 11 contains new applications of nonlinear programming to pricing problems.

- Eleven new cases involving mathematical programming are included. Professor Jeff Goldberg of the University of Arizona wrote the cases.

- Chapter 12 contains a discussion of Excel's normal distribution functions and z-transforms.

- Chapter 13 covers the applications of prospect theory and framing effects in decision making.

- Chapter 15 discusses power-of-two inventory policies and multiple-product EOQ models.

- Chapter 20 now covers computing Poisson and exponential probabilities with Excel, Buzen's method for closed queuing networks, approximations for $G/G/s$ queuing systems, the use of data tables in queuing optimization, and computing transient probabilities for queuing systems.

- Chapter 22 shows how to use the powerful, user-friendly simulation package Process Model to simulate queuing systems.

- Chapter 23 deals with the Excel add-in @Risk, for Monte Carlo simulation. Application areas include capital budgeting, project management, and reliability.

- In Chapter 24, Excel data tables and the OFFSET

function are used to optimize the number of periods in a moving-average forecast.

Use of the Computer

In deference to the virtually universal usage of Excel, this software is featured throughout the book when appropriate. When Excel's native capabilities are limited, the text discusses add-in software that builds on the capabilities of Excel, or uses stand-alone software.

The CD accompanying the book contains several valuable software packages.

- **LINDO** and **LINGO.** These easy-to-use linear and nonlinear programming software packages are provided by Lindo Systems, Inc.
- **Premium Solver for Education.** Generously provided by Frontline Systems (the developers of Microsoft Excel's Solver), Premium Solver provides evolutionary solving techniques utilized in nonlinear optimization problems.
- **@Risk.** A professional Monte Carlo simulation add-in for Excel by Palisade Corporation.
- **Process Model.** This discrete-event simulation software is easy to learn and use. It is illustrated in Chapter 22. Process Model is provided by Process Model Inc.

Software illustrations, with all the necessary step-by-step instructions, appear at the ends of sections, to provide maximum flexibility to instructors who wish to employ different software packages in their courses.

Acknowledgments

Many people have played significant roles in the development of the fourth edition. My views on teaching operations research were greatly influenced by the many excellent teachers I have had, including Gordon Bradley, Eric Denardo, George Fishman, Gordon Kaufman, Richard Larson, John Little, Robert Mifflin, Martin Shubik, Matthew Sobel, Arthur Veinott, Jr., Harvey Wagner, and Ward Whitt. In particular, I would like to acknowledge Professor Wagner's *Principles of Operations Research,* which taught me more about operations research than any other single book.

I am grateful to the following:

- Linus Schrage, Mark Wiley, and Kevin Cunningham of LINDO Systems, for their help in including LINDO and LINGO.

- Dan Fylstra and Edwin Straver of Frontline Systems, for allowing me to include the Premium Solver.
- Sam McLafferty of Palisade, for allowing me to include @Risk.
- Matthew Jorgensen at Process Model Inc., for allowing me to include Process Model.

Thanks go to all the people at Brooks/Cole who worked on the book, especially our editor, Curt Hinrichs, for his unflagging editoral support.

I greatly appreciate the editing and production skills of David Hoyt, and the fine typesetting of ATLIS Graphics.

Thanks to the 146 survey respondents who provided valuable feedback on the book and the course and its emerging needs. They include Nikolaos Adamou, University of Athens & Sage Graduate School; Jeffrey Adler, Rensselaer Polytechnic Institute; Victor K. Akatsa, Chicago State University; Steven Andelin, Kutztown University of Pennsylvania; Badiollah R. Asrabadi, Nicholls State University; Rhonda Aull-Hyde, University of Delaware; Jonathan Bard, Mechanical Engineering; John Barnes, Virginia Commonwealth University; Harold P. Benson, University of Florida; Elinor Berger, Columbus College; Richard H. Bernhard, North Carolina State University; R. L. Bulfin, Auburn University; Laura Burke, Lehigh University; Jonathan Caulkins, Carnegie Mellon University; Beth Chance, University of the Pacific; Alan Chesen, Wright State University; Young Chun, Louisiana State University; Chia-Shin Chung, Cleveland State University; Ken Currie, Tennessee Technological University; Ani Dasgupta, Pennsylvania State University; Nirmil Devi, Embry-Riddle Aeronautical University; James Falk, George Washington University; Kambiz Farahmand, Texas A & M University–Kingsville; Yahya Fathi, North Carolina State University; Steve Fisk, Bowdoin College; William P. Fox, United States Military Academy; Michael C. Fu, University of Maryland; Saul I. Gass, University of Maryland; Ronald Gathro, Western New England College; Perakis Georgia, Massachusetts Institute of Technology; Alan Goldberg, California State University–Hayward; Jerold Griggs, University of South Carolina; David Grimmett, Austin Peay State University; Melike Baykal Gursoy, Rutgers University; Jorge Haddock, Rensselaer Polytechnic Institute; Jane Hagstrom, University of Illinois; Carl Harris, George Mason University; Miriam Heller, University of Houston; Sundresh S. Heragu, Rensselaer Polytechnic Insitute; Rebecca E. Hill, Rochester Institute of Technology; David Holdsworth, Alaska Pacific University; Elained Hubbard, Kennesaw State College; Robert Hull, Western Illinois University; Jeffrey Jarrett, University of Rhode Island; David Kaufman, University of Massachusetts; Davook Khalili, San Jose University; Morton Klein, Columbia University; S. Kumar, Rochester Institute of Technology; David Larsen, University of New Orleans; Mark Lawley, University of Alabama; Kenneth D. Lawrence, New Jersey Institute of Technology; Andreas Lazari, Valdosta State University; Jon Lee, University of Kentucky; Luanne Lohr, University of Georgia; Joseph Malkovitch, York College; Masud Mansuri, California State University–Fresno; Steven C. McKelvey, Saint Olaf College; Ojjat Mehri, Youngstown State University; Robert Mifflin, Washington State University; Katya Mints, Columbia University; Rafael Moras, St. Mary's University; James G. Morris, University of Wisconsin, Madison; Frederic Murphy, Temple University; David Olson, Texas A & M University; Mufit Ozden, Miami University; R. Gary Parker, Georgia Institute of Technology; Barry Pasternack, California State University, Fullerton; Walter M. Patterson, Lander University; James E Pratt, Cornell University; B. Madhu Rao, Bowling Green State University; T. E. S. Raghavan, University of Illinois–Chicago; Gary Reeves, University of South Carolina; Gaspard Rizzuto, University of Southwestern; David Ronen, University of Missouri–St. Louis; Paul Savory, University of Nebraska, Lincoln; Jon Schlosser, New Mexico Highlands University; Delray Schultz, Millersville University; Richard Serfozo, Georgia Technological Institute; Morteza Shafi-Mousavi, Indiana University–South Bend; Dooyoung Shin, Mankato State University; Ronald L. Shubert, Elizabethtown College; Joel Sobel, University of California, San Diego; Manbir S. Sodhi, University of Rhode Island; Ariela Sofer, George Mason University; Toni M. Somers, Wayne State University; Robert Stark, University of Delaware; Joseph A. Svestka, Cleveland State University; Alexander Sze, Concordia College; Roman Sznajder, University of Maryland–Baltimore County; Bijan Vasigh, Embry-Riddle Aeronautical University; John H. Vande Vate, Georgia Technological University; Richard G. Vinson, University of South Alabama; Jin Wang, Valdodsta State University; Zhongxian Wang, Montclair State University; Robert C. Williams, Alfred University; Arthur Neal Willoughby, Morgan State; Shmuel Yahalom, SUNY–Maritime College; James Yates, University of Central Oklahoma; Bill Yurcik, University of Pittsburgh.

Thanks also go to reviewers of the previous editions: Esther Arkin, Sant Arora, Harold Benson, Warren J. Boe, Bruce Bowerman, James W. Chrissis, Jerald Dauer, S. Selcuk Erenguc, Yahya Fathi, Robert Freund, Irwin Greenberg, Rebecca E. Hill, John Hooker, Sidney Lawrence, Patrick Lee, Edward Minieka, James G. Morris, Joel A. Nachlas, David L. Olson, Sudhakar Pandit, David W. Pentico, Bruce Pollack-Johnson, Michael Richey, Gary D. Scudder, Lawrence Seiford, Michael Sinchcomb, and Paul Stiebitz.

I retain responsibility for all errors and would love to hear from users of the book. I can be reached at

Indiana University
Department of Operations and Decision Technology
Kelley School of Business
Room 570
Bloomington, IN 47405

Wayne Winston (Winston@indiana.edu)

About the Author

Wayne Winston

Wayne L. Winston is Professor of Operations and Decision Technologies in the Kelley School of Business at Indiana University, where he has taught since 1975. Wayne received his B.S. degree in mathematics from MIT and his Ph.D. degree in operations research from Yale. He has written the successful textbooks *Operations Research: Applications and Algorithms; Introduction to Mathematical Programming; Simulation Modeling with @Risk; Practical Management Science;* and *Financial Models Using Simulation and Optimization.* Wayne has published over 20 articles in leading journals and has won many teaching awards, including the school-wide MBA award four times. His current interest is in showing how spreadsheet models can be used to solve business problems in all disciplines, particularly in finance and marketing.

Wayne enjoys swimming and basketball, and his passion for trivia won him an appearance several years ago on the television game show *Jeopardy,* where he won two games. He is married to the lovely and talented Vivian. They have two children, Gregory and Jennifer.

1

An Introduction to Model Building

1.1 An Introduction to Modeling

Operations research (often referred to as **management science**) is simply a scientific approach to decision making that seeks to best design and operate a system, usually under conditions requiring the allocation of scarce resources.

By a **system,** we mean an organization of interdependent components that work together to accomplish the goal of the system. For example, Ford Motor Company is a system whose goal consists of maximizing the profit that can be earned by producing quality vehicles.

The term *operations research* was coined during World War II when British military leaders asked scientists and engineers to analyze several military problems such as the deployment of radar and the management of convoy, bombing, antisubmarine, and mining operations.

The scientific approach to decision making usually involves the use of one or more **mathematical models.** A mathematical model is a mathematical representation of an actual situation that may be used to make better decisions or simply to understand the actual situation better. The following example should clarify many of the key terms used to describe mathematical models.

EXAMPLE 1 | **Maximizing Wozac Yield**

Eli Daisy produces Wozac in huge batches by heating a chemical mixture in a pressurized container. Each time a batch is processed, a different amount of Wozac is produced. The amount produced is the *process yield* (measured in pounds). Daisy is interested in understanding the factors that influence the yield of the Wozac production process. Describe a model-building process for this situation.

Solution Daisy is first interested in determining the factors that influence the yield of the process. This would be referred to as a *descriptive model,* because it describes the behavior of the actual yield as a function of various factors. Daisy might determine (using regression methods discussed in Chapter 24) that the following factors influence yield:

- container volume in liters (V)
- container pressure in milliliters (P)
- container temperature in degrees Celsius (T)
- chemical composition of the processed mixture

If we let A, B, and C be percentage of mixture made up of chemicals A, B, and C, then Daisy might find, for example, that

$$(1)\ \text{yield} = 300 + .8V + .01P + .06T + .001T*P - .01T^2 - .001P^2$$
$$+ 11.7A + 9.4B + 16.4C + 19A*B + 11.4A*C - 9.6B*C$$

To determine this relationship, the yield of the process would have to be measured for many different combinations of the previously listed factors. Knowledge of this equation would enable Daisy to describe the yield of the production process once volume, pressure, temperature, and chemical composition were known.

Prescriptive or Optimization Models

Most of the models discussed in this book will be **prescriptive** or **optimization** models. A prescriptive model "prescribes" behavior for an organization that will enable it to best meet its goal(s). The components of a prescriptive model include

- objective function(s)
- decision variables
- constraints

In short, an optimization model seeks to find values of the decision variables that optimize (maximize or minimize) an objective function among the set of all values for the decision variables that satisfy the given constraints.

The Objective Function

Naturally, Daisy would like to maximize the yield of the process. In most models, there will be a function we wish to maximize or minimize. This function is called the model's *objective function.* Of course, to maximize the process yield we need to find the values of V, P, T, A, B, and C that make (1) as large as possible.

In many situations, an organization may have more than one objective. For example, in assigning students to the two high schools in Bloomington, Indiana, the Monroe County School Board stated that the assignment of students involved the following objectives:

- Equalize the number of students at the two high schools.
- Minimize the average distance students travel to school.
- Have a diverse student body at both high schools.

Multiple objective decision-making problems are discussed in Sections 4.14 and 11.13.

The Decision Variables

The variables whose values are under our control and influence the performance of the system are called *decision variables*. In our example, V, P, T, A, B, and C are decision variables. Most of this book will be devoted to a discussion of how to determine the value of decision variables that maximize (sometimes minimize) an objective function.

Constraints

In most situations, only certain values of decision variables are possible. For example, certain volume, pressure, and temperature combinations might be unsafe. Also, A B, and C must be nonnegative numbers that add to 1. Restrictions on the values of decision variables are called *constraints*. Suppose the following:

- Volume must be between 1 and 5 liters.
- Pressure must be between 200 and 400 milliliters.
- Temperature must be between 100 and 200 degrees Celsius.
- Mixture must be made up entirely of A, B, and C.
- For the drug to properly perform, only half the mixture at most can be product A.

These constraints can be expressed mathematically by the following constraints:

$$V \leq 5$$
$$V \geq 1$$
$$P \leq 400$$
$$P \geq 200$$
$$T \leq 200$$
$$T \geq 100$$
$$A \geq 0$$
$$B \geq 0$$
$$A + B + C = 1$$
$$A \leq 5$$

The Complete Optimization Model

After letting z represent the value of the objective function, our entire optimization model may be written as follows:

Maximize $z = 300 + .8V + .01P + .06T + .001T*P - .01T^2 - .001P^2$
$$+ 11.7A + 9.4B + 16.4C + 19A*B + 11.4A*C - 9.6B*C$$

Subject to (s.t.)

$$V \leq 5$$
$$V \geq 1$$
$$P \leq 400$$
$$P \geq 200$$
$$T \leq 200$$
$$T \geq 100$$
$$A \geq 0$$
$$B \geq 0$$
$$C \geq 0$$
$$A + B + C = 1$$
$$A \leq 5$$

Any specification of the decision variables that satisfies all of the model's constraints is said to be in the **feasible region.** For example, V = 2, P = 300, T = 150, A = .4, B = .3, and C = .1 is in the feasible region. An **optimal solution** to an optimization model is any point in the feasible region that optimizes (in this case, *maximizes*) the objective function. Using the LINGO package that comes with this book, it can be determined that the optimal solution to this model is V = 5, P = 200, T = 100, A = .294, B = 0, C = .706, and $z = 183.38$. Thus, a maximum yield of 183.38 pounds can be obtained with a 5-liter container, pressure of 200 milliliters, temperature of 100 degrees Celsius, and 29% A and 71% C. This means no other feasible combination of decision variables can obtain a yield exceeding 183.38 pounds.

Static and Dynamic Models

A **static model** is one in which the decision variables do not involve sequences of decisions over multiple periods. A **dynamic model** is a model in which the decision variables *do* involve sequences of decisions over multiple periods. Basically, in a static model we solve a "one-shot" problem whose solutions prescribe optimal values of decision variables at all points in time. Example 1 is an example of a static model; the optimal solution will tell Daisy how to maximize yield at all points in time.

For an example of a dynamic model, consider a company (call it Sailco) that must determine how to minimize the cost of meeting (on time) the demand for sailboats during the next year. Clearly Sailco's must determine how many sailboats it will produce during each of the next four quarters. Sailco's decisions involve decisions made over multiple periods, hence a model of Sailco's problem (see Section 3.10) would be a dynamic model.

Linear and Nonlinear Models

Suppose that whenever decision variables appear in the objective function and in the constraints of an optimization model, the decision variables are always multiplied by constants and added together. Such a model is a **linear model.** If an optimization model is not linear, then it is a **nonlinear model.** In the constraints of Example 1, the decision variables are always multiplied by constants and added together. Thus, Example 1's constraints pass the test for a linear model. However, in the objective function for Example 1, the terms $.001T*P$, $-.01T^2$, $19A*B$, $11.4A*C$, and $-9.6B*C$ make the model nonlinear. In general, nonlinear models are much harder to solve than linear models. We will discuss linear models in Chapters 2 through 10. Nonlinear models will be discussed in Chapter 11.

Integer and Noninteger Models

If one or more decision variables must be integer, then we say that an optimization model is an **integer model.** If all the decision variables are free to assume fractional values, then the optimization model is a **noninteger** model. Clearly, volume, temperature, pressure, and percentage composition of our inputs may all assume fractional values. Thus, Example 1 is a noninteger model. If the decision variables in a model represent the number of workers starting work during each shift at a fast-food restaurant, then clearly we have an integer model. Integer models are much harder to solve than nonlinear models. They will be discussed in detail in Chapter 9.

Deterministic and Stochastic Models

Suppose that for any value of the decision variables, the value of the objective function and whether or not the constraints are satisfied is known with certainty. We then have a **deterministic model.** If this is not the case, then we have a **stochastic model.** All models in the first 12 chapters will be deterministic models. Stochastic models are covered in Chapters 13, 16, 17, and 19–24.

If we view Example 1 as a deterministic model, then we are making the (unrealistic) assumption that for given values of V, P, T, A, B, and C, the process yield will always be the same. This is highly unlikely. We can view (1) as a representation of the *average* yield of the process for given values of the decision variables. Then our objective is to find values of the decision variables that maximize the average yield of the process.

We can often gain useful insights into optimal decisions by using a deterministic model in a situation where a stochastic model is more appropriate. Consider Sailco's problem of minimizing the cost of meeting the demand (on time) for sailboats. The uncertainty about future demand for sailboats implies that for a given production schedule, we do not know whether demand is met on time. This leads us to believe that a stochastic model is needed to model Sailco's situation. We will see in Section 3.10, however, that we can develop a deterministic model for this situation that yields good decisions for Sailco.

1.2 The Seven-Step Model-Building Process

When operations research is used to solve an organization's problem, the following seven-step model-building procedure should be followed:

Step 1: Formulate the Problem   The operations researcher first defines the organization's problem. Defining the problem includes specifying the organization's objectives and the parts of the organization that must be studied before the problem can be solved. In Example 1, the problem was to determine how to maximize the yield from a batch of Wozac.

Step 2: Observe the System   Next, the operations researcher collects data to estimate the value of parameters that affect the organization's problem. These estimates are used to develop (in step 3) and evaluate (in step 4) a mathematical model of the organization's problem. For example, in Example 1, data would be collected in an attempt to determine how the values of T, P, V, A, B, and C influence process yield.

Step 3: Formulate a Mathematical Model of the Problem   In this step, the operations researcher develops a mathematical model of the problem. In this book, we will describe many mathematical techniques that can be used to model systems. For Example 1, our optimization model would be the result of step 3.

Step 4: Verify the Model and Use the Model for Prediction   The operations researcher now tries to determine if the mathematical model developed in step 3 is an accurate representation of reality. For example, to validate our model, we might check and see if (1) accurately represents yield for values of the decision variables that were not used to estimate (1). Even if a model is valid for the current situation, we must be aware of blindly applying it. For example, if the government placed new restrictions on Wozac, then we might have to add new constraints to our model, and the yield of the process [and Equation (1)] might change.

Step 5: Select a Suitable Alternative   Given a model and a set of alternatives, the operations researcher now chooses the alternative that best meets the organization's objectives. (There may be more than one!) For instance, our model enabled us to determine that yield was maximized with V = 5, P = 200, T = 100, A = .294, B = 0, C = .706, and z = 183.38.

Step 6: Present the Results and Conclusion of the Study to the Organization   In this step, the operations researcher presents the model and recommendation from step 5 to the decision-making individual or group. In some situations, one might present several alternatives and let the organization choose the one that best meets its needs. After presenting the results of the operations research study, the analyst may find that the organization does not approve of the recommendation. This may result from incorrect definition of the organization's problems or from failure to involve the decision maker from the start of the project. In this case, the operations researcher should return to step 1, 2, or 3.

Step 7: Implement and Evaluate Recommendations   If the organization has accepted the study, then the analyst aids in implementing the recommendations. The system must be constantly monitored (and updated dynamically as the environment changes) to ensure that the recommendations enable the organization to meet its objectives.

In what follows, we discuss three successful management science applications. We will give a detailed (but nonquantitative) description of each application. We will tie our discussion of each application to the seven-step model-building process described in Section 1.2.

1.3 CITGO Petroleum

Klingman et al. (1987) applied a variety of management-science techniques to CITGO Petroleum. Their work saved the company an estimated $70 million per year. CITGO is an oil-refining and -marketing company that was purchased by Southland Corporation (the owners of the 7-Eleven stores). We will focus on two aspects of the CITGO team's work:

1   a mathematical model to optimize operation of CITGO's refineries, and

2   a mathematical model—supply distribution marketing (SDM) system—that was used to develop an 11-week supply, distribution, and marketing plan for the entire business.

Optimizing Refinery Operations

Step 1   Klingman et al. wanted to minimize the cost of operating CITGO's refineries.

Step 2   The Lake Charles, Louisiana, refinery was closely observed in an attempt to estimate key relationships such as:

1   How the cost of producing each of CITGO's products (motor fuel, no. 2 fuel oil, turbine fuel, naptha, and several blended motor fuels) depends on the inputs used to produce each product.

2   The amount of energy needed to produce each product. This required the installation of a new metering system.

3   The yield associated with each input–output combination. For example, if 1 gallon of crude oil would yield .52 gallons of motor fuel, then the yield would equal 52%.

4   To reduce maintenance costs, data were collected on parts inventories and equipment breakdowns. Obtaining accurate data required the installation of a new database-management system and integrated maintenance-information system. A process control system was also installed to accurately monitor the inputs and resources used to manufacture each product.

Step 3   Using linear programming (LP), a model was developed to optimize refinery operations. The model determines the cost-minimizing method for mixing or blending together inputs to produce desired outputs. The model contains **constraints** that ensure that inputs are blended so that each output is of the desired quality. Blending constraints are discussed in Section 3.8. The model ensures that plant capacities are not exceeded and allows for the fact that each refinery may carry an inventory of each end product. Sections 3.10 and 4.12 discuss inventory constraints.

Step 4 To validate the model, inputs and outputs from the Lake Charles refinery were collected for one month. Given the actual inputs used at the refinery during that month, the actual outputs were compared to those predicted by the model. After extensive changes, the model's predicted outputs were close to the actual outputs.

Step 5 Running the LP yielded a daily strategy for running the refinery. For instance, the model might, say, produce 400,000 gallons of turbine fuel using 300,000 gallons of crude 1 and 200,000 gallons of crude 2.

Steps 6 and 7 Once the database and process control were in place, the model was used to guide day-to-day refinery operations. CITGO estimated that the overall benefits of the refinery system exceeded $50 million annually.

The Supply Distribution Marketing (SDM) System

Step 1 CITGO wanted a mathematical model that could be used to make supply, distribution, and marketing decisions such as:

1 Where should crude oil be purchased?

2 Where should products be sold?

3 What price should be charged for products?

4 How much of each product should be held in inventory?

The goal, of course, was to maximize the profitability associated with these decisions.

Step 2 A database that kept track of sales, inventory, trades, and exchanges of all refined products was installed. Also, regression analysis (see Chapter 24) was used to develop forecasts for wholesale prices and wholesale demand for each CITGO product.

Steps 3 and 5 A minimum-cost network flow model (MCNFM) (see Section 7.4) is used to determine an 11-week supply, marketing, and distribution strategy. The model makes all decisions mentioned in step 1. A typical model run that involved 3,000 equations and 15,000 decision variables required only 30 seconds on an IBM 4381.

Step 4 The forecasting modules are continuously evaluated to ensure that they continue to give accurate forecasts.

Steps 6 and 7 Implementing the SDM required several organizational changes. A new vice-president was appointed to coordinate the operation of the SDM and LP refinery model. The product supply and product scheduling departments were combined to improve communication and information flow.

1.4 San Francisco Police Department Scheduling

Taylor and Huxley (1989) developed a police patrol scheduling system (PPSS). All San Francisco (SF) police precincts use PPSS to schedule their officers. It is estimated that PPSS saves the SF police more than $5 million annually. Other cities such as Virginia

Beach, Virginia, and Richmond, California, have also adopted PPSS. Following our seven-step model-building procedure, here is a description of PPSS.

Step 1 The SFPD wanted a method to schedule patrol officers in each precinct that would quickly produce (in less than one hour) a schedule and graphically display it. The program should first determine the personnel requirements for each hour of the week. For example, 38 officers might be needed between 1 A.M. and 2 A.M. Sunday but only 14 officers might be needed from 4 A.M. to 5 A.M. Sunday. Officers should then be scheduled to minimize the sum over each hour of the week of the shortages and surpluses relative to the needed number of officers. For example, if 20 officers were assigned to the midnight to 8 A.M. Sunday shift, we would have a shortage of $38 - 20 = 18$ officers from 1 to 2 A.M. and a surplus of $20 - 14 = 6$ officers from 4 to 5 A.M. A secondary criterion was to minimize the maximum shortage because a shortage of 10 officers during a single hour is far more serious than a shortage of one officer during 10 different hours. The SFPD also wanted a scheduling system that precinct captains could easily fine-tune to produce the optimal schedule.

Step 2 The SFPD had a sophisticated computer-aided dispatch (CAD) system to keep track of all calls for police help, police travel time, police response time, and so on. SFPD had a standard percentage of time that administrators felt each officer should be busy. Using CAD, it is easy to determine the number of workers needed each hour. Suppose, for example, an officer should be busy 80% of the time and CAD indicates that 30.4 hours of work come in from 4 to 5 A.M. Sunday. Then we need 38 officers from 4 to 5 A.M. on Sunday [.8*(38) = 30.4 hours].

Step 3 An LP model was formulated (see Section 3.5 for a discussion of scheduling models). As discussed in step 1, the primary objective was to minimize the sum of hourly shortages and surpluses. At first, schedulers assumed that officers worked five consecutive days for eight hours a day (this was the policy prior to PPSS) and that there were three shift starting times (say, 6 A.M., 2 P.M., and 10 A.M.). The constraints in the PPSS model reflected the limited number of officers available and the relationship of the number of officers working each hour to the shortages and surpluses for that hour. Then PPSS would produce a schedule that would tell the precinct captain how many officers should start work at each possible shift time. For example, PPSS might say that 20 officers should start work at 6 A.M. Monday (working 6 A.M.–2 P.M. Monday–Friday) and 30 officers should start work at 2 P.M. Saturday (working 2 P.M.–10 P.M. Saturday–Wednesday). The fact that the number of officers assigned to a start time must be an integer made it far more difficult to find an optimal schedule. (Problems in which decision variables must be integers are discussed in Chapter 9.)

Step 4 Before implementing PPSS, the SFPD tested the PPSS schedules against manually created schedules. PPSS produced an approximately 50% reduction in both surpluses and shortages. This convinced the department to implement PPSS.

Step 5 Given the starting times for shifts and the type of work schedule [four consecutive days for 10 hours per day (the 4/10 schedule) or five consecutive days for eight hours per day (the 5/8 schedule)], PPSS can produce a schedule that minimizes the sum of shortages and surpluses. More important, PPSS can be used to experiment with shift times and work rules. Using PPSS, it was found that if only three shift times are allowed, then a 5/8 schedule was superior to a 4/10 schedule. If, however, five shift times were allowed, then a 4/10 schedule was found to be superior. This finding was of critical importance because police officers had wanted to switch to a 4/10 schedule for years. The city had resisted 4/10 schedules because they appeared to reduce productivity. PPSS showed that 4/10 schedules need not reduce productivity. After the introduction of PPSS, the SFPD went to 4/10 schedules and *improved productivity!* PPSS also enables the department to experiment with a mix of one-officer and two-officer patrol cars.

Steps 6 and 7 It is estimated that PPSS created an extra 170,000 productive hours per year, thereby saving the city of San Francisco $5.2 million per year. Ninety-six percent of all workers preferred PPSS generated schedules to manually generated schedules. PPSS enabled SFPD to make strategic changes (such as adopting the 4/10 schedule), which made officers happier and increased productivity. Response times to calls improved by 20% after PPSS was adopted.

A major reason for the success of PPSS was that the system allowed precinct captains to fine-tune the computer-generated schedule and obtain a new schedule in less than one minute. For example, precinct captains could easily add or delete officers and add or delete shifts and quickly see how these changes modified the master schedule.

1.5 GE Capital

GE Capital provides credit card service to 50 million accounts. The average total outstanding balance exceeds $12 billion. GE Capital, led by Makuch et al. (1989), developed the PAYMENT system to reduce delinquent accounts and the cost of collecting from delinquent accounts.

Step 1 At any one time, GE Capital has more than $1 billion in delinquent accounts. The company spends $100 million per year processing these accounts. Each day, workers contact more than 200,000 delinquent credit card holders with letters, messages, or live calls. The company's goal was to reduce delinquent accounts and the cost of processing them. To do this, GE Capital needed to come up with a method of assigning scarce labor resources to delinquent accounts. For example, PAYMENT determines which delinquent accounts receive live phone calls and which delinquent accounts receive no contact.

Step 2 The key to modeling delinquent accounts is the concept of a **delinquency movement matrix (DMM).** The DMM determines how the probability of the payment on a delinquent account during the current month depends on the following factors: size of unpaid balance (either <$300 or ≥$300), action taken (no action, live phone call, taped message, letters), and a performance score (high, medium, or low). The higher the performance score associated with a delinquent account, the more likely the account is to be collected. Table 1 lists the probabilities for a $250 account that is two months delinquent, has a high performance score, and is contacted with a phone message.

TABLE 1
Sample Entries in DMM

Event	Probability
Account completely paid	.30
One month is paid	.40
Nothing is paid	.30

Because GE Capital has millions of delinquent accounts, there is ample data to accurately estimate the DMM. For example, suppose there were 10,000 two-month delinquent accounts with balances under $300 that have a high performance score and are contacted with phone messages. If 3,000 of those accounts were completely paid off during the current month, then we would estimate the probability of an account being completely paid off during the current month as $3,000/10,000 = .30$.

Step 3 GE Capital developed a linear optimization model. The objective function for the PAYMENT model was to maximize the expected delinquent accounts collected during the next six months. The decision variables represented the fraction of each type of delinquent account (accounts are classified by payment balance, performance score, and months delinquent) that experienced each type of contact (no action, live phone call, taped message, or letter). The constraints in the PAYMENT model ensure that available resources are not overused. Constraints also relate the number of each type of delinquent account present in, say, January to the number of delinquent accounts of each type present during the next month (February). This **dynamic** aspect of the PAYMENT model is crucial to its success. Without this aspect, the model would simply "skim" the accounts that are easiest to collect each month. This would result in few collections during later months.

Step 4 PAYMENT was piloted on a $62 million portfolio for a single department store. GE Capital managers came up with their own strategies for allocating resources (collectively called CHAMPION). The store's delinquent accounts were randomly assigned to the CHAMPION and PAYMENT strategies. PAYMENT used more live phone calls and more "no action" than the CHAMPION strategies. PAYMENT also collected $180,000 per month more than any of the CHAMPION strategies, a 5% to 7% improvement. Note that using more of the no-action strategy certainly leads to a long-run increase in customer goodwill!

Step 5 As described in step 3, for each type of account, PAYMENT tells the credit managers the fraction that should receive each type of contact. For example, for three-month delinquent accounts with a small (<$300) unpaid balance and high performance score, PAYMENT might prescribe 30% no action, 20% letters, 30% phone messages, and 20% live phone calls.

Steps 6 and 7 PAYMENT was next applied to the 18 million accounts of the $4.6 billion Montgomery-Ward department store portfolio. Comparing the collection results to the same time period a year earlier, it was found that PAYMENT increased collections by $1.6 million per month (more than $19 million per year). This is actually a conservative estimate of the benefit obtained from PAYMENT, because PAYMENT was first applied to the Montgomery-Ward portfolio during the depths of a recession—and a recession makes it much more difficult to collect delinquent accounts.

Overall, GE Capital estimates that PAYMENT increased collections by $37 million per year and used fewer resources than previous strategies.

REFERENCES

Klingman, D., N. Phillips, D. Steiger, and W. Young, "The Successful Deployment of Management Science Throughout Citgo Corporation," *Interfaces* 17 (1987, no. 1):4–25.

Makuch, W., J. Dodge, J. Ecker, D. Granfors, and G. Hahn, "Managing Consumer Credit Delinquency in the US Economy: A Multi-Billion Dollar Management Science Application," *Interfaces* 22 (1992, no. 1):90–109.

Taylor, P., and S. Huxley, "A Break from Tradition for the San Francisco Police: Patrol Officer Scheduling Using an Optimization-Based Decision Support Tool," *Interfaces* 19 (1989, no. 1):4–24.

2 Basic Linear Algebra

In this chapter, we study the topics in linear algebra that will be needed in the rest of the book. We begin by discussing the building blocks of linear algebra: matrices and vectors. Then we use our knowledge of matrices and vectors to develop a systematic procedure (the Gauss–Jordan method) for solving linear equations, which we then use to invert matrices. We close the chapter with an introduction to determinants.

The material covered in this chapter will be used in our study of linear and nonlinear programming.

2.1 Matrices and Vectors

Matrices

DEFINITION ■ A **matrix** is any rectangular array of numbers. ■

For example,

$$\begin{bmatrix} 1 & 2 \\ 3 & 4 \end{bmatrix}, \quad \begin{bmatrix} 1 & 2 & 3 \\ 4 & 5 & 6 \end{bmatrix}, \quad \begin{bmatrix} 1 \\ -2 \end{bmatrix}, \quad [2 \quad 1]$$

are all matrices.

If a matrix A has m rows and n columns, we call A an $m \times n$ matrix. We refer to $m \times n$ as the **order** of the matrix. A typical $m \times n$ matrix A may be written as

$$A = \begin{bmatrix} a_{11} & a_{12} & \cdots & a_{1n} \\ a_{21} & a_{22} & \cdots & a_{2n} \\ \vdots & \vdots & & \vdots \\ a_{m1} & a_{m2} & \cdots & a_{mn} \end{bmatrix}$$

DEFINITION ■ The number in the ith row and jth column of A is called the **ijth element** of A and is written a_{ij}. ■

For example, if

$$A = \begin{bmatrix} 1 & 2 & 3 \\ 4 & 5 & 6 \\ 7 & 8 & 9 \end{bmatrix}$$

then $a_{11} = 1$, $a_{23} = 6$, and $a_{31} = 7$.

Sometimes we will use the notation $A = [a_{ij}]$ to indicate that A is the matrix whose ijth element is a_{ij}.

Two matrices $A = [a_{ij}]$ and $B = [b_{ij}]$ are **equal** if and only if A and B are of the same order and for all i and j, $a_{ij} = b_{ij}$. ■

For example, if

$$A = \begin{bmatrix} 1 & 2 \\ 3 & 4 \end{bmatrix} \quad \text{and} \quad B = \begin{bmatrix} x & y \\ w & z \end{bmatrix}$$

then $A = B$ if and only if $x = 1$, $y = 2$, $w = 3$, and $z = 4$.

Vectors

Any matrix with only one column (that is, any $m \times 1$ matrix) may be thought of as a **column vector.** The number of rows in a column vector is the **dimension** of the column vector. Thus,

$$\begin{bmatrix} 1 \\ 2 \end{bmatrix}$$

may be thought of as a 2×1 matrix or a two-dimensional column vector. R^m will denote the set of all m-dimensional column vectors.

In analogous fashion, we can think of any vector with only one row (a $1 \times n$ matrix as a **row vector.** The dimension of a row vector is the number of columns in the vector. Thus, [9 2 3] may be viewed as a 1×3 matrix or a three-dimensional row vector. In this book, vectors appear in boldface type: for instance, vector **v.** An m-dimensional vector (either row or column) in which all elements equal zero is called a **zero vector** (written **0**). Thus,

$$[0 \quad 0] \quad \text{and} \quad \begin{bmatrix} 0 \\ 0 \end{bmatrix}$$

are two-dimensional zero vectors.

Any m-dimensional vector corresponds to a directed line segment in the m-dimensional plane. For example, in the two-dimensional plane, the vector

$$\mathbf{u} = \begin{bmatrix} 1 \\ 2 \end{bmatrix}$$

corresponds to the line segment joining the point

$$\begin{bmatrix} 0 \\ 0 \end{bmatrix}$$

to the point

$$\begin{bmatrix} 1 \\ 2 \end{bmatrix}$$

The directed line segments corresponding to

$$\mathbf{u} = \begin{bmatrix} 1 \\ 2 \end{bmatrix}, \quad \mathbf{v} = \begin{bmatrix} 1 \\ -3 \end{bmatrix}, \quad \mathbf{w} = \begin{bmatrix} -1 \\ -2 \end{bmatrix}$$

are drawn in Figure 1.

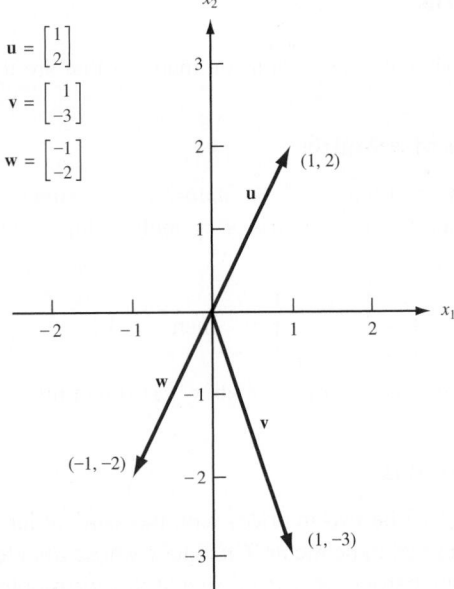

FIGURE 1
Vectors Are Directed
Line Segments

The Scalar Product of Two Vectors

An important result of multiplying two vectors is the *scalar product*. To define the scalar product of two vectors, suppose we have a row vector $\mathbf{u} = [u_1 \quad u_2 \quad \cdots \quad u_n]$ and a column vector

$$\mathbf{v} = \begin{bmatrix} v_1 \\ v_2 \\ \vdots \\ v_n \end{bmatrix}$$

of the same dimension. The **scalar product** of $\mathbf{u}$ and $\mathbf{v}$ (written $\mathbf{u} \cdot \mathbf{v}$) is the number $u_1v_1 + u_2v_2 + \cdots + u_nv_n$.

For the scalar product of two vectors to be defined, the first vector must be a row vector and the second vector must be a column vector. For example, if

$$\mathbf{u} = [1 \quad 2 \quad 3] \qquad \text{and} \qquad \mathbf{v} = \begin{bmatrix} 2 \\ 1 \\ 2 \end{bmatrix}$$

then $\mathbf{u} \cdot \mathbf{v} = 1(2) + 2(1) + 3(2) = 10$. By these rules for computing a scalar product, if

$$\mathbf{u} = \begin{bmatrix} 1 \\ 2 \end{bmatrix} \qquad \text{and} \qquad \mathbf{v} = [2 \quad 3]$$

then $\mathbf{u} \cdot \mathbf{v}$ is not defined. Also, if

$$\mathbf{u} = [1 \quad 2 \quad 3] \qquad \text{and} \qquad \mathbf{v} = \begin{bmatrix} 3 \\ 4 \end{bmatrix}$$

then $\mathbf{u} \cdot \mathbf{v}$ is not defined because the vectors are of two different dimensions.

Note that two vectors are perpendicular if and only if their scalar product equals 0. Thus, the vectors $[1 \quad -1]$ and $[1 \quad 1]$ are perpendicular.

We note that $\mathbf{u} \cdot \mathbf{v} = \|\mathbf{u}\| \, \|\mathbf{v}\| \cos \theta$, where $\|\mathbf{u}\|$ is the length of the vector $\mathbf{u}$ and θ is the angle between the vectors $\mathbf{u}$ and $\mathbf{v}$.

Matrix Operations

We now describe the arithmetic operations on matrices that are used later in this book.

The Scalar Multiple of a Matrix

Given any matrix A and any number c (a *number* is sometimes referred to as a *scalar*), the matrix cA is obtained from the matrix A by multiplying each element of A by c. For example,

$$\text{if} \quad A = \begin{bmatrix} 1 & 2 \\ -1 & 0 \end{bmatrix}, \quad \text{then} \quad 3A = \begin{bmatrix} 3 & 6 \\ -3 & 0 \end{bmatrix}$$

For $c = -1$, scalar multiplication of the matrix A is sometimes written as $-A$.

Addition of Two Matrices

Let $A = [a_{ij}]$ and $B = [b_{ij}]$ be two matrices with the same order (say, $m \times n$). Then the matrix $C = A + B$ is defined to be the $m \times n$ matrix whose ijth element is $a_{ij} + b_{ij}$. Thus, to obtain the sum of two matrices A and B, we add the corresponding elements of A and B. For example, if

$$A = \begin{bmatrix} 1 & 2 & 3 \\ 0 & -1 & 1 \end{bmatrix} \quad \text{and} \quad B = \begin{bmatrix} -1 & -2 & -3 \\ 2 & 1 & -1 \end{bmatrix}$$

then

$$A + B = \begin{bmatrix} 1-1 & 2-2 & 3-3 \\ 0+2 & -1+1 & 1-1 \end{bmatrix} = \begin{bmatrix} 0 & 0 & 0 \\ 2 & 0 & 0 \end{bmatrix}.$$

This rule for matrix addition may be used to add vectors of the same dimension. For example, if $\mathbf{u} = [1 \quad 2]$ and $\mathbf{v} = [2 \quad 1]$, then $\mathbf{u} + \mathbf{v} = [1 + 2 \quad 2 + 1] = [3 \quad 3]$. Vectors may be added geometrically by the parallelogram law (see Figure 2).

We can use scalar multiplication and the addition of matrices to define the concept of a line segment. A glance at Figure 1 should convince you that any point u in the m-dimensional plane corresponds to the m-dimensional vector $\mathbf{u}$ formed by joining the origin to the point u. For any two points u and v in the m-dimensional plane, the **line segment** joining u and v (called the line segment uv) is the set of all points in the m-dimensional plane that correspond to the vectors $c\mathbf{u} + (1 - c)\mathbf{v}$, where $0 \leq c \leq 1$ (Figure 3). For example, if $u = (1, 2)$ and $v = (2, 1)$, then the line segment uv consists

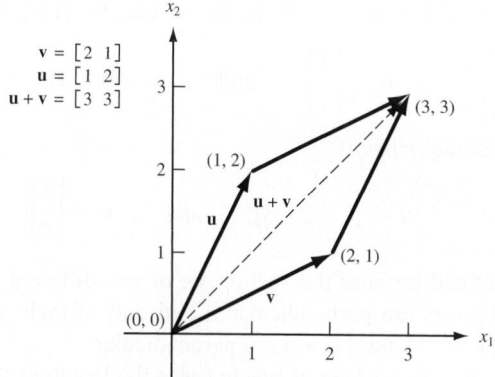

FIGURE 2
Addition of Vectors

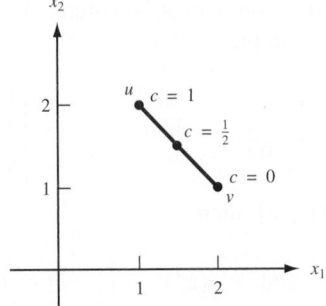

FIGURE 3
Line Segment Joining
$u = (1, 2)$ and
$v = (2, 1)$

of the points corresponding to the vectors $c[1 \quad 2] + (1 - c)[2 \quad 1] = [2 - c \quad 1 + c]$, where $0 \leq c \leq 1$. For $c = 0$ and $c = 1$, we obtain the endpoints of the line segment uv; for $c = \frac{1}{2}$, we obtain the midpoint $(0.5\mathbf{u} + 0.5\mathbf{v})$ of the line segment uv.

Using the parallelogram law, the line segment uv may also be viewed as the points corresponding to the vectors $\mathbf{u} + c(\mathbf{v} - \mathbf{u})$, where $0 \leq c \leq 1$ (Figure 4). Observe that for $c = 0$, we obtain the vector $\mathbf{u}$ (corresponding to point u), and for $c = 1$, we obtain the vector $\mathbf{v}$ (corresponding to point v).

The Transpose of a Matrix

Given any $m \times n$ matrix

$$A = \begin{bmatrix} a_{11} & a_{12} & \cdots & a_{1n} \\ a_{21} & a_{22} & \cdots & a_{2n} \\ \vdots & \vdots & & \vdots \\ a_{m1} & a_{m2} & \cdots & a_{mn} \end{bmatrix}$$

the **transpose** of A (written A^T) is the $n \times m$ matrix

$$A^T = \begin{bmatrix} a_{11} & a_{21} & \cdots & a_{m1} \\ a_{12} & a_{22} & \cdots & a_{m2} \\ \vdots & \vdots & & \vdots \\ a_{1n} & a_{2n} & \cdots & a_{mn} \end{bmatrix}$$

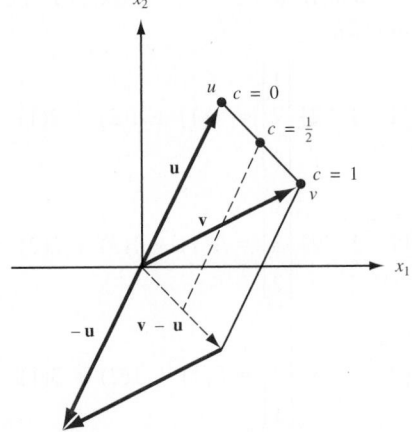

FIGURE 4
Representation of Line
Segment uv

Thus, A^T is obtained from A by letting row 1 of A be column 1 of A^T, letting row 2 of A be column 2 of A^T, and so on. For example,

$$\text{if} \quad A = \begin{bmatrix} 1 & 2 & 3 \\ 4 & 5 & 6 \end{bmatrix}, \quad \text{then} \quad A^T = \begin{bmatrix} 1 & 4 \\ 2 & 5 \\ 3 & 6 \end{bmatrix}$$

Observe that $(A^T)^T = A$. Let $B = [1 \quad 2]$; then

$$B^T = \begin{bmatrix} 1 \\ 2 \end{bmatrix} \quad \text{and} \quad (B^T)^T = [1 \quad 2] = B$$

As indicated by these two examples, for any matrix A, $(A^T)^T = A$.

Matrix Multiplication

Given two matrices A and B, the matrix product of A and B (written AB) is defined if and only if

$$\text{Number of columns in } A = \text{number of rows in } B \tag{1}$$

For the moment, assume that for some positive integer r, A has r columns and B has r rows. Then for some m and n, A is an $m \times r$ matrix and B is an $r \times n$ matrix.

DEFINITION ■ The **matrix product** $C = AB$ of A and B is the $m \times n$ matrix C whose ijth element is determined as follows:

ijth element of C = scalar product of row i of $A \times$ column j of B ■ (2)

If Equation (1) is satisfied, then each row of A and each column of B will have the same number of elements. Also, if (1) is satisfied, then the scalar product in Equation (2) will be defined. The product matrix $C = AB$ will have the same number of rows as A and the same number of columns as B.

EXAMPLE 1 **Matrix Multiplication**

Compute $C = AB$ for

$$A = \begin{bmatrix} 1 & 1 & 2 \\ 2 & 1 & 3 \end{bmatrix} \quad \text{and} \quad B = \begin{bmatrix} 1 & 1 \\ 2 & 3 \\ 1 & 2 \end{bmatrix}$$

Solution Because A is a 2×3 matrix and B is a 3×2 matrix, AB is defined, and C will be a 2×2 matrix. From Equation (2),

$$c_{11} = [1 \quad 1 \quad 2] \begin{bmatrix} 1 \\ 2 \\ 1 \end{bmatrix} = 1(1) + 1(2) + 2(1) = 5$$

$$c_{12} = [1 \quad 1 \quad 2] \begin{bmatrix} 1 \\ 3 \\ 2 \end{bmatrix} = 1(1) + 1(3) + 2(2) = 8$$

$$c_{21} = [2 \quad 1 \quad 3] \begin{bmatrix} 1 \\ 2 \\ 1 \end{bmatrix} = 2(1) + 1(2) + 3(1) = 7$$

$$c_{22} = [2 \quad 1 \quad 3] \begin{bmatrix} 1 \\ 3 \\ 2 \end{bmatrix} = 2(1) + 1(3) + 3(2) = 11$$

$$C = AB = \begin{bmatrix} 5 & 8 \\ 7 & 11 \end{bmatrix}$$

EXAMPLE 2   Column Vector Times Row Vector

Find AB for

$$A = \begin{bmatrix} 3 \\ 4 \end{bmatrix} \quad \text{and} \quad B = [1 \quad 2]$$

Solution   Because A has one column and B has one row, $C = AB$ will exist. From Equation (2), we know that C is a 2×2 matrix with

$$c_{11} = 3(1) = 3 \qquad c_{21} = 4(1) = 4$$
$$c_{12} = 3(2) = 6 \qquad c_{22} = 4(2) = 8$$

Thus,

$$C = \begin{bmatrix} 3 & 6 \\ 4 & 8 \end{bmatrix}$$

EXAMPLE 3   Row Vector Times Column Vector

Compute $D = BA$ for the A and B of Example 2.

Solution   In this case, D will be a 1×1 matrix (or a scalar). From Equation (2),

$$d_{11} = [1 \quad 2] \begin{bmatrix} 3 \\ 4 \end{bmatrix} = 1(3) + 2(4) = 11$$

Thus, $D = [11]$. In this example, matrix multiplication is equivalent to scalar multiplication of a row and column vector.

Recall that if you multiply two real numbers a and b, then $ab = ba$. This is called the *commutative property of multiplication*. Examples 2 and 3 show that for matrix multiplication, it may be that $AB \neq BA$. Matrix multiplication is not necessarily commutative. (In some cases, however, $AB = BA$ will hold.)

EXAMPLE 4   Undefined Matrix Product

Show that AB is undefined if

$$A = \begin{bmatrix} 1 & 2 \\ 3 & 4 \end{bmatrix} \quad \text{and} \quad B = \begin{bmatrix} 1 & 1 \\ 0 & 1 \\ 1 & 2 \end{bmatrix}$$

Solution   This follows because A has two columns and B has three rows. Thus, Equation (1) is not satisfied.

TABLE **1**

Gallons of Crude Oil Required to Produce 1 Gallon
of Gasoline

Crude Oil	Premium Unleaded	Regular Unleaded	Regular Leaded
1	$\frac{3}{4}$	$\frac{2}{3}$	$\frac{1}{4}$
2	$\frac{1}{4}$	$\frac{1}{3}$	$\frac{3}{4}$

Many computations that commonly occur in operations research (and other branches of mathematics) can be concisely expressed by using matrix multiplication. To illustrate this, suppose an oil company manufactures three types of gasoline: premium unleaded, regular unleaded, and regular leaded. These gasolines are produced by mixing two types of crude oil: crude oil 1 and crude oil 2. The number of gallons of crude oil required to manufacture 1 gallon of gasoline is given in Table 1.

From this information, we can find the amount of each type of crude oil needed to manufacture a given amount of gasoline. For example, if the company wants to produce 10 gallons of premium unleaded, 6 gallons of regular unleaded, and 5 gallons of regular leaded, then the company's crude oil requirements would be

$$\text{Crude 1 required} = \left(\tfrac{3}{4}\right)(10) + \left(\tfrac{2}{3}\right)(6) + \left(\tfrac{1}{4}\right)5 = 12.75 \text{ gallons}$$
$$\text{Crude 2 required} = \left(\tfrac{1}{4}\right)(10) + \left(\tfrac{1}{3}\right)(6) + \left(\tfrac{3}{4}\right)5 = 8.25 \text{ gallons}$$

More generally, we define

$$p_U = \text{gallons of premium unleaded produced}$$
$$r_U = \text{gallons of regular unleaded produced}$$
$$r_L = \text{gallons of regular leaded produced}$$
$$c_1 = \text{gallons of crude 1 required}$$
$$c_2 = \text{gallons of crude 2 required}$$

Then the relationship between these variables may be expressed by

$$c_1 = \left(\tfrac{3}{4}\right)p_U + \left(\tfrac{2}{3}\right)r_U + \left(\tfrac{1}{4}\right)r_L$$
$$c_2 = \left(\tfrac{1}{4}\right)p_U + \left(\tfrac{1}{3}\right)r_U + \left(\tfrac{3}{4}\right)r_L$$

Using matrix multiplication, these relationships may be expressed by

$$\begin{bmatrix} c_1 \\ c_2 \end{bmatrix} = \begin{bmatrix} \frac{3}{4} & \frac{2}{3} & \frac{1}{4} \\ \frac{1}{4} & \frac{1}{3} & \frac{3}{4} \end{bmatrix} \begin{bmatrix} p_U \\ r_U \\ r_L \end{bmatrix}$$

Properties of Matrix Multiplication

To close this section, we discuss some important properties of matrix multiplication. In what follows, we assume that all matrix products are defined.

1 Row i of AB = (row i of A)B. To illustrate this property, let

$$A = \begin{bmatrix} 1 & 1 & 2 \\ 2 & 1 & 3 \end{bmatrix} \quad \text{and} \quad B = \begin{bmatrix} 1 & 1 \\ 2 & 3 \\ 1 & 2 \end{bmatrix}$$

Then row 2 of the 2×2 matrix AB is equal to

$$[2 \quad 1 \quad 3] \begin{bmatrix} 1 & 1 \\ 2 & 3 \\ 1 & 2 \end{bmatrix} = [7 \quad 11]$$

This answer agrees with Example 1.

2 Column j of $AB = A$(column j of B). Thus, for A and B as given, the first column of AB is

$$\begin{bmatrix} 1 & 1 & 2 \\ 2 & 1 & 3 \end{bmatrix} \begin{bmatrix} 1 \\ 2 \\ 1 \end{bmatrix} = \begin{bmatrix} 5 \\ 7 \end{bmatrix}$$

Properties 1 and 2 are helpful when you need to compute only *part* of the matrix AB.

3 Matrix multiplication is associative. That is, $A(BC) = (AB)C$. To illustrate, let

$$A = [1 \quad 2], \qquad B = \begin{bmatrix} 2 & 3 \\ 4 & 5 \end{bmatrix}, \qquad C = \begin{bmatrix} 2 \\ 1 \end{bmatrix}$$

Then $AB = [10 \quad 13]$ and $(AB)C = 10(2) + 13(1) = [33]$.

On the other hand,

$$BC = \begin{bmatrix} 7 \\ 13 \end{bmatrix}$$

so $A(BC) = 1(7) + 2(13) = [33]$. In this case, $A(BC) = (AB)C$ does hold.

4 Matrix multiplication is distributive. That is, $A(B + C) = AB + AC$ and $(B + C)D = BD + CD$.

Matrix Multiplication with Excel

Using the Excel MMULT function, it is easy to multiply matrices. To illustrate, let's use Excel to find the matrix product AB that we found in Example 1 (see Figure 5 and file Mmult.xls). We proceed as follows:

Step 1 Enter A and B in D2:F3 and D5:E7, respectively.

Step 2 Select the range (D9:E10) in which the product AB will be computed.

Step 3 In the upper left-hand corner (D9) of the selected range, type the formula

$$= \text{MMULT(D2:F3,D5:E7)}$$

Then hit **Control Shift Enter** (not just Enter), and the desired matrix product will be computed. Note that MMULT is an *array* function and not an ordinary spreadsheet function. This explains why we must preselect the range for AB and use Control Shift Enter.

Mmult.xls

	A	B	C	D	E	F
1	MatrixMultiplication					
2				1	1	2
3			A	2	1	3
4						
5			B	1	1	
6				2	3	
7				1	2	
8						
9				5	8	
10			C	7	11	
11						

FIGURE 5

PROBLEMS

Group A

1 For $A = \begin{bmatrix} 1 & 2 & 3 \\ 4 & 5 & 6 \\ 7 & 8 & 9 \end{bmatrix}$ and $B = \begin{bmatrix} 1 & 2 \\ 0 & -1 \\ 1 & 2 \end{bmatrix}$, find:

a $-A$ **b** $3A$ **c** $A + 2B$

d A^T **e** B^T **f** AB

g BA

2 Only three brands of beer (beer 1, beer 2, and beer 3) are available for sale in Metropolis. From time to time, people try one or another of these brands. Suppose that at the beginning of each month, people change the beer they are drinking according to the following rules:

30% of the people who prefer beer 1 switch to beer 2.

20% of the people who prefer beer 1 switch to beer 3.

30% of the people who prefer beer 2 switch to beer 3.

30% of the people who prefer beer 3 switch to beer 2.

10% of the people who prefer beer 3 switch to beer 1.

For $i = 1, 2, 3$, let x_i be the number who prefer beer i at the beginning of this month and y_i be the number who prefer beer i at the beginning of next month. Use matrix multiplication to relate the following:

$$\begin{bmatrix} y_1 \\ y_2 \\ y_3 \end{bmatrix} \qquad \begin{bmatrix} x_1 \\ x_2 \\ x_3 \end{bmatrix}$$

Group B

3 Prove that matrix multiplication is associative.

4 Show that for any two matrices A and B, $(AB)^T = B^T A^T$.

5 An $n \times n$ matrix A is symmetric if $A = A^T$.

a Show that for any $n \times n$ matrix, AA^T is a symmetric matrix.

b Show that for any $n \times n$ matrix A, $(A + A^T)$ is a symmetric matrix.

6 Suppose that A and B are both $n \times n$ matrices. Show that computing the matrix product AB requires n^3 multiplications and $n^3 - n^2$ additions.

7 The **trace of a matrix** is the sum of its diagonal elements.

a For any two matrices A and B, show that trace $(A + B) = $ trace $A + $ trace B.

b For any two matrices A and B for which the products AB and BA are defined, show that trace $AB = $ trace BA.

2.2 Matrices and Systems of Linear Equations

Consider a system of linear equations given by

$$\begin{aligned}
a_{11}x_1 + a_{12}x_2 + \cdots + a_{1n}x_n &= b_1 \\
a_{21}x_1 + a_{22}x_2 + \cdots + a_{2n}x_n &= b_2 \\
\vdots \qquad \vdots \qquad\qquad \vdots \qquad & \\
a_{m1}x_1 + a_{m2}x_2 + \cdots + a_{mn}x_n &= b_m
\end{aligned} \tag{3}$$

In Equation (3), $x_1, x_2, \ldots, x_n$ are referred to as **variables,** or unknowns, and the a_{ij}'s and b_i's are **constants.** A set of equations such as (3) is called a linear system of m equations in n variables.

DEFINITION ■ A **solution** to a linear system of m equations in n unknowns is a set of values for the unknowns that satisfies each of the system's m equations. ■

To understand linear programming, we need to know a great deal about the properties of solutions to linear equation systems. With this in mind, we will devote much effort to studying such systems.

We denote a possible solution to Equation (3) by an n-dimensional column vector **x,** in which the ith element of **x** is the value of x_i. The following example illustrates the concept of a solution to a linear system.

EXAMPLE 5   Solution to Linear System

Show that

$$\mathbf{x} = \begin{bmatrix} 1 \\ 2 \end{bmatrix}$$

is a solution to the linear system

$$x_1 + 2x_2 = 5$$
$$2x_1 - x_2 = 0$$

(4)

and that

$$\mathbf{x} = \begin{bmatrix} 3 \\ 1 \end{bmatrix}$$

is not a solution to linear system (4).

Solution   To show that

$$\mathbf{x} = \begin{bmatrix} 1 \\ 2 \end{bmatrix}$$

is a solution to Equation (4), we substitute $x_1 = 1$ and $x_2 = 2$ in both equations and check that they are satisfied: $1 + 2(2) = 5$ and $2(1) - 2 = 0$.

The vector

$$\mathbf{x} = \begin{bmatrix} 3 \\ 1 \end{bmatrix}$$

is not a solution to (4), because $x_1 = 3$ and $x_2 = 1$ fail to satisfy $2x_1 - x_2 = 0$.

Using matrices can greatly simplify the statement and solution of a system of linear equations. To show how matrices can be used to compactly represent Equation (3), let

$$A = \begin{bmatrix} a_{11} & a_{12} & \cdots & a_{1n} \\ a_{21} & a_{22} & \cdots & a_{2n} \\ \vdots & \vdots & & \vdots \\ a_{m1} & a_{m2} & \cdots & a_{mn} \end{bmatrix}, \quad \mathbf{x} = \begin{bmatrix} x_1 \\ x_2 \\ \vdots \\ x_n \end{bmatrix}, \quad \mathbf{b} = \begin{bmatrix} b_1 \\ b_2 \\ \vdots \\ b_m \end{bmatrix}$$

Then (3) may be written as

$$A\mathbf{x} = \mathbf{b}$$

(5)

Observe that both sides of Equation (5) will be $m \times 1$ matrices (or $m \times 1$ column vectors). For the matrix $A\mathbf{x}$ to equal the matrix $\mathbf{b}$ (or for the vector $A\mathbf{x}$ to equal the vector $\mathbf{b}$), their corresponding elements must be equal. The first element of $A\mathbf{x}$ is the scalar product of row 1 of A with $\mathbf{x}$. This may be written as

$$[a_{11} \quad a_{12} \quad \cdots \quad a_{1n}] \begin{bmatrix} x_1 \\ x_2 \\ \vdots \\ x_n \end{bmatrix} = a_{11}x_1 + a_{12}x_2 + \cdots + a_{1n}x_n$$

This must equal the first element of $\mathbf{b}$ (which is b_1). Thus, (5) implies that $a_{11}x_1 + a_{12}x_2 + \cdots + a_{1n}x_n = b_1$. This is the first equation of (3). Similarly, (5) implies that the scalar product of row i of A with $\mathbf{x}$ must equal b_i, and this is just the ith equation of (3). Our discussion shows that (3) and (5) are two different ways of writing the same linear system. We call (5) the **matrix representation** of (3). For example, the matrix representation of (4) is

$$\begin{bmatrix} 1 & 2 \\ 2 & -1 \end{bmatrix} \begin{bmatrix} x_1 \\ x_2 \end{bmatrix} = \begin{bmatrix} 5 \\ 0 \end{bmatrix}$$

Sometimes we abbreviate (5) by writing

$$A|\mathbf{b} \tag{6}$$

If A is an $m \times n$ matrix, it is assumed that the variables in (6) are $x_1, x_2, \ldots, x_n$. Then (6) is still another representation of (3). For instance, the matrix

$$\begin{bmatrix} 1 & 2 & 3 & 2 \\ 0 & 1 & 2 & 3 \\ 1 & 1 & 1 & 1 \end{bmatrix}$$

represents the system of equations

$$x_1 + 2x_2 + 3x_3 = 2$$
$$x_2 + 2x_3 = 3$$
$$x_1 + x_2 + x_3 = 1$$

PROBLEM

Group A

1 Use matrices to represent the following system of equations in two different ways:

$$x_1 - x_2 = 4$$
$$2x_1 + x_2 = 6$$
$$x_1 + 3x_2 = 8$$

2.3 The Gauss–Jordan Method for Solving Systems of Linear Equations

We develop in this section an efficient method (the Gauss–Jordan method) for solving a system of linear equations. Using the Gauss–Jordan method, we show that any system of linear equations must satisfy one of the following three cases:

Case 1 The system has no solution.

Case 2 The system has a unique solution.

Case 3 The system has an infinite number of solutions.

The Gauss–Jordan method is also important because many of the manipulations used in this method are used when solving linear programming problems by the simplex algorithm (see Chapter 4).

Elementary Row Operations

Before studying the Gauss–Jordan method, we need to define the concept of an **elementary row operation** (ERO). An ERO transforms a given matrix A into a new matrix A' via one of the following operations.

Type 1 ERO

A' is obtained by multiplying any row of A by a nonzero scalar. For example, if

$$A = \begin{bmatrix} 1 & 2 & 3 & 4 \\ 1 & 3 & 5 & 6 \\ 0 & 1 & 2 & 3 \end{bmatrix}$$

then a Type 1 ERO that multiplies row 2 of A by 3 would yield

$$A' = \begin{bmatrix} 1 & 2 & 3 & 4 \\ 3 & 9 & 15 & 18 \\ 0 & 1 & 2 & 3 \end{bmatrix}$$

Type 2 ERO

Begin by multiplying any row of A (say, row i) by a nonzero scalar c. For some $j \neq i$, let row j of $A' = c(\text{row } i \text{ of } A) + \text{row } j \text{ of } A$, and let the other rows of A' be the same as the rows of A.

For example, we might multiply row 2 of A by 4 and replace row 3 of A by $4(\text{row } 2$ of $A) + \text{row } 3$ of A. Then row 3 of A' becomes

$$4 \begin{bmatrix} 1 & 3 & 5 & 6 \end{bmatrix} + \begin{bmatrix} 0 & 1 & 2 & 3 \end{bmatrix} = \begin{bmatrix} 4 & 13 & 22 & 27 \end{bmatrix}$$

and

$$A' = \begin{bmatrix} 1 & 2 & 3 & 4 \\ 1 & 3 & 5 & 6 \\ 4 & 13 & 22 & 27 \end{bmatrix}$$

Type 3 ERO

Interchange any two rows of A. For instance, if we interchange rows 1 and 3 of A, we obtain

$$A' = \begin{bmatrix} 0 & 1 & 2 & 3 \\ 1 & 3 & 5 & 6 \\ 1 & 2 & 3 & 4 \end{bmatrix}$$

Type 1 and Type 2 EROs formalize the operations used to solve a linear equation system. To solve the system of equations

$$\begin{aligned} x_1 + x_2 &= 2 \\ 2x_1 + 4x_2 &= 7 \end{aligned} \tag{7}$$

we might proceed as follows. First replace the second equation in (7) by $-2(\text{first equation in (7)}) + \text{second equation in (7)}$. This yields the following linear system:

$$\begin{aligned} x_1 + x_2 &= 2 \\ 2x_2 &= 3 \end{aligned} \tag{7.1}$$

Then multiply the second equation in (7.1) by $\frac{1}{2}$, yielding the system

$$\begin{aligned} x_1 + x_2 &= 2 \\ x_2 &= \tfrac{3}{2} \end{aligned} \tag{7.2}$$

Finally, replace the first equation in (7.2) by $-1[\text{second equation in (7.2)}] + \text{first equation in (7.2)}$. This yields the system

$$x_1 = \frac{1}{2}$$
$$x_2 = \frac{3}{2}$$
(7.3)

System (7.3) has the unique solution $x_1 = \frac{1}{2}$ and $x_2 = \frac{3}{2}$. The systems (7), (7.1), (7.2), and (7.3) are *equivalent* in that they have the same set of solutions. This means that $x_1 = \frac{1}{2}$ and $x_2 = \frac{3}{2}$ is also the unique solution to the original system, (7).

If we view (7) in the augmented matrix form $(A|\mathbf{b})$, we see that the steps used to solve (7) may be seen as Type 1 and Type 2 EROs applied to $A|\mathbf{b}$. Begin with the augmented matrix version of (7):

$$\begin{bmatrix} 1 & 1 & | & 2 \\ 2 & 4 & | & 7 \end{bmatrix}$$
(7')

Now perform a Type 2 ERO by replacing row 2 of (7') by -2(row 1 of (7')) + row 2 of (7'). The result is

$$\begin{bmatrix} 1 & 1 & | & 2 \\ 0 & 2 & | & 3 \end{bmatrix}$$
(7.1')

which corresponds to (7.1). Next, we multiply row 2 of (7.1') by $\frac{1}{2}$ (a Type 1 ERO), resulting in

$$\begin{bmatrix} 1 & 1 & | & 2 \\ 0 & 1 & | & \frac{3}{2} \end{bmatrix}$$
(7.2')

which corresponds to (7.2). Finally, perform a Type 2 ERO by replacing row 1 of (7.2') by -1(row 2 of (7.2')) + row 1 of (7.2'). The result is

$$\begin{bmatrix} 1 & 0 & | & \frac{1}{2} \\ 0 & 1 & | & \frac{3}{2} \end{bmatrix}$$
(7.3')

which corresponds to (7.3). Translating (7.3') back into a linear system, we obtain the system $x_1 = \frac{1}{2}$ and $x_2 = \frac{3}{2}$, which is identical to (7.3).

Finding a Solution by the Gauss–Jordan Method

The discussion in the previous section indicates that if the matrix $A'|\mathbf{b}'$ is obtained from $A|\mathbf{b}$ via an ERO, the systems $A\mathbf{x} = \mathbf{b}$ and $A'\mathbf{x} = \mathbf{b}'$ are equivalent. Thus, any sequence of EROs performed on the augmented matrix $A|\mathbf{b}$ corresponding to the system $A\mathbf{x} = \mathbf{b}$ will yield an equivalent linear system.

The Gauss–Jordan method solves a linear equation system by utilizing EROs in a systematic fashion. We illustrate the method by finding the solution to the following linear system:

$$\begin{aligned} 2x_1 + 2x_2 + \ x_3 &= 9 \\ 2x_1 - \ x_2 + 2x_3 &= 6 \\ x_1 - \ x_2 + 2x_3 &= 5 \end{aligned}$$
(8)

The augmented matrix representation is

$$A|\mathbf{b} = \begin{bmatrix} 2 & 2 & 1 & | & 9 \\ 2 & -1 & 2 & | & 6 \\ 1 & -1 & 2 & | & 5 \end{bmatrix}$$
(8')

Suppose that by performing a sequence of EROs on (8') we could transform (8') into

$$\begin{bmatrix} 1 & 0 & 0 & | & 1 \\ 0 & 1 & 0 & | & 2 \\ 0 & 0 & 1 & | & 3 \end{bmatrix} \qquad (9')$$

We note that the result obtained by performing an ERO on a system of equations can also be obtained by multiplying both sides of the matrix representation of the system of equations by a particular matrix. This explains why EROs do not change the set of solutions to a system of equations.

Matrix (9') corresponds to the following linear system:

$$\begin{aligned} x_1 \quad\;\; &= 1 \\ x_2 \;\; &= 2 \qquad\qquad (9) \\ x_3 &= 3 \end{aligned}$$

System (9) has the unique solution $x_1 = 1$, $x_2 = 2$, $x_3 = 3$. Because (9') was obtained from (8') by a sequence of EROs, we know that (8) and (9) are equivalent linear systems. Thus, $x_1 = 1$, $x_2 = 2$, $x_3 = 3$ must also be the unique solution to (8). We now show how we can use EROs to transform a relatively complicated system such as (8) into a relatively simple system like (9). This is the essence of the Gauss–Jordan method.

We begin by using EROs to transform the first column of (8') into

$$\begin{bmatrix} 1 \\ 0 \\ 0 \end{bmatrix}$$

Then we use EROs to transform the second column of the resulting matrix into

$$\begin{bmatrix} 0 \\ 1 \\ 0 \end{bmatrix}$$

Finally, we use EROs to transform the third column of the resulting matrix into

$$\begin{bmatrix} 0 \\ 0 \\ 1 \end{bmatrix}$$

As a final result, we will have obtained (9'). We now use the Gauss–Jordan method to solve (8). We begin by using a Type 1 ERO to change the element of (8') in the first row and first column into a 1. Then we add multiples of row 1 to row 2 and then to row 3 (these are Type 2 EROs). The purpose of these Type 2 EROs is to put zeros in the rest of the first column. The following sequence of EROs will accomplish these goals.

Step 1 Multiply row 1 of (8') by $\frac{1}{2}$. This Type 1 ERO yields

$$A_1|\mathbf{b}_1 = \begin{bmatrix} 1 & 1 & \frac{1}{2} & | & \frac{9}{2} \\ 2 & -1 & 2 & | & 6 \\ 1 & -1 & 2 & | & 5 \end{bmatrix}$$

Step 2 Replace row 2 of $A_1|\mathbf{b}_1$ by $-2(\text{row } 1 \text{ of } A_1|\mathbf{b}_1) + \text{row } 2 \text{ of } A_1|\mathbf{b}_1$. The result of this Type 2 ERO is

$$A_2|\mathbf{b}_2 = \begin{bmatrix} 1 & 1 & \frac{1}{2} & | & \frac{9}{2} \\ 0 & -3 & 1 & | & -3 \\ 1 & -1 & 2 & | & 5 \end{bmatrix}$$

Step 3   Replace row 3 of $A_2|\mathbf{b}_2$ by -1(row 1 of $A_2|\mathbf{b}_2$ + row 3 of $A_2|\mathbf{b}_2$). The result of this Type 2 ERO is

$$A_3|\mathbf{b}_3 = \begin{bmatrix} 1 & 1 & \frac{1}{2} & \Big| & \frac{9}{2} \\ 0 & -3 & 1 & \Big| & -3 \\ 0 & -2 & \frac{3}{2} & \Big| & \frac{1}{2} \end{bmatrix}$$

The first column of $(8')$ has now been transformed into

$$\begin{bmatrix} 1 \\ 0 \\ 0 \end{bmatrix}$$

By our procedure, we have made sure that the variable x_1 occurs in only a single equation and in that equation has a coefficient of 1. We now transform the second column of $A_3|\mathbf{b}_3$ into

$$\begin{bmatrix} 0 \\ 1 \\ 0 \end{bmatrix}$$

We begin by using a Type 1 ERO to create a 1 in row 2 and column 2 of $A_3|\mathbf{b}_3$. Then we use the resulting row 2 to perform the Type 2 EROs that are needed to put zeros in the rest of column 2. Steps 4–6 accomplish these goals.

Step 4   Multiply row 2 of $A_3|\mathbf{b}_3$ by $-\frac{1}{3}$. The result of this Type 1 ERO is

$$A_4|\mathbf{b}_4 = \begin{bmatrix} 1 & 1 & \frac{1}{2} & \Big| & \frac{9}{2} \\ 0 & 1 & -\frac{1}{3} & \Big| & 1 \\ 0 & -2 & \frac{3}{2} & \Big| & \frac{1}{2} \end{bmatrix}$$

Step 5   Replace row 1 of $A_4|\mathbf{b}_4$ by -1(row 2 of $A_4|\mathbf{b}_4$) + row 1 of $A_4|\mathbf{b}_4$. The result of this Type 2 ERO is

$$A_5|\mathbf{b}_5 = \begin{bmatrix} 1 & 0 & \frac{5}{6} & \Big| & \frac{7}{2} \\ 0 & 1 & -\frac{1}{3} & \Big| & 1 \\ 0 & -2 & \frac{3}{2} & \Big| & \frac{1}{2} \end{bmatrix}$$

Step 6   Replace row 3 of $A_5|\mathbf{b}_5$ by 2(row 2 of $A_5|\mathbf{b}_5$) + row 3 of $A_5|\mathbf{b}_5$. The result of this Type 2 ERO is

$$A_6|\mathbf{b}_6 = \begin{bmatrix} 1 & 0 & \frac{5}{6} & \Big| & \frac{7}{2} \\ 0 & 1 & -\frac{1}{3} & \Big| & 1 \\ 0 & 0 & \frac{5}{6} & \Big| & \frac{5}{2} \end{bmatrix}$$

Column 2 has now been transformed into

$$\begin{bmatrix} 0 \\ 1 \\ 0 \end{bmatrix}$$

Observe that our transformation of column 2 did not change column 1.

To complete the Gauss–Jordan procedure, we must transform the third column of $A_6|\mathbf{b}_6$ into

$$\begin{bmatrix} 0 \\ 0 \\ 1 \end{bmatrix}$$

We first use a Type 1 ERO to create a 1 in the third row and third column of $A_6|\mathbf{b}_6$. Then we use Type 2 EROs to put zeros in the rest of column 3. Steps 7–9 accomplish these goals.

Step 7 Multiply row 3 of $A_6|\mathbf{b}_6$ by $\frac{6}{5}$. The result of this Type 1 ERO is

$$A_7|\mathbf{b}_7 = \begin{bmatrix} 1 & 0 & \frac{5}{6} & \Big| & \frac{7}{2} \\ 0 & 1 & -\frac{1}{3} & \Big| & 1 \\ 0 & 0 & 3 & \Big| & 3 \end{bmatrix}$$

Step 8 Replace row 1 of $A_7|\mathbf{b}_7$ by $-\frac{5}{6}$(row 3 of $A_7|\mathbf{b}_7$) + row 1 of $A_7|\mathbf{b}_7$. The result of this Type 2 ERO is

$$A_8|\mathbf{b}_8 = \begin{bmatrix} 1 & 0 & 0 & \Big| & 1 \\ 0 & 1 & -\frac{1}{3} & \Big| & 1 \\ 0 & 0 & 1 & \Big| & 3 \end{bmatrix}$$

Step 9 Replace row 2 of $A_8|\mathbf{b}_8$ by $\frac{1}{3}$(row 3 of $A_8|\mathbf{b}_8$) + row 2 of $A_8|\mathbf{b}_8$. The result of this Type 2 ERO is

$$A_9|\mathbf{b}_9 = \begin{bmatrix} 1 & 0 & 0 & \Big| & 1 \\ 0 & 1 & 0 & \Big| & 2 \\ 0 & 0 & 1 & \Big| & 3 \end{bmatrix}$$

$A_9|\mathbf{b}_9$ represents the system of equations

$$\begin{aligned} x_1 \qquad &= 1 \\ x_2 \quad &= 2 \\ x_3 &= 3 \end{aligned} \tag{9}$$

Thus, (9) has the unique solution $x_1 = 1$, $x_2 = 2$, $x_3 = 3$. Because (9) was obtained from (8) via EROs, the unique solution to (8) must also be $x_1 = 1$, $x_2 = 2$, $x_3 = 3$.

The reader might be wondering why we defined Type 3 EROs (interchanging of rows). To see why a Type 3 ERO might be useful, suppose you want to solve

$$\begin{aligned} 2x_2 + x_3 &= 6 \\ x_1 + x_2 - x_3 &= 2 \\ 2x_1 + x_2 + x_3 &= 4 \end{aligned} \tag{10}$$

To solve (10) by the Gauss–Jordan method, first form the augmented matrix

$$A|\mathbf{b} = \begin{bmatrix} 0 & 2 & 1 & \Big| & 6 \\ 1 & 1 & -1 & \Big| & 2 \\ 2 & 1 & 1 & \Big| & 4 \end{bmatrix}$$

The 0 in row 1 and column 1 means that a Type 1 ERO cannot be used to create a 1 in row 1 and column 1. If, however, we interchange rows 1 and 2 (a Type 3 ERO), we obtain

$$\begin{bmatrix} 1 & 1 & -1 & \Big| & 2 \\ 0 & 2 & 1 & \Big| & 6 \\ 2 & 1 & 1 & \Big| & 4 \end{bmatrix} \tag{10'}$$

Now we may proceed as usual with the Gauss–Jordan method.

Special Cases: No Solution or an Infinite Number of Solutions

Some linear systems have no solution, and some have an infinite number of solutions. The following two examples illustrate how the Gauss–Jordan method can be used to recognize these cases.

EXAMPLE 6   Linear System with No Solution

Find all solutions to the following linear system:

$$x_1 + 2x_2 = 3$$
$$2x_1 + 4x_2 = 4$$

(11)

Solution   We apply the Gauss–Jordan method to the matrix

$$A|\mathbf{b} = \begin{bmatrix} 1 & 2 & | & 3 \\ 2 & 4 & | & 4 \end{bmatrix}$$

We begin by replacing row 2 of $A|\mathbf{b}$ by -2(row 1 of $A|\mathbf{b}$) + row 2 of $A|\mathbf{b}$. The result of this Type 2 ERO is

$$\begin{bmatrix} 1 & 2 & | & 3 \\ 0 & 0 & | & -2 \end{bmatrix}$$

(12)

We would now like to transform the second column of (12) into

$$\begin{bmatrix} 0 \\ 1 \end{bmatrix}$$

but this is not possible. System (12) is equivalent to the following system of equations:

$$x_1 + 2x_2 = 3$$
$$0x_1 + 0x_2 = -2$$

(12′)

Whatever values we give to x_1 and x_2, the second equation in (12′) can never be satisfied. Thus, (12′) has no solution. Because (12′) was obtained from (11) by use of EROs, (11) also has no solution.

Example 6 illustrates the following idea: *If you apply the Gauss–Jordan method to a linear system and obtain a row of the form* $[0 \quad 0 \quad \cdots \quad 0|c]$ $(c \neq 0)$, *then the original linear system has no solution.*

EXAMPLE 7   Linear System with Infinite Number of Solutions

Apply the Gauss–Jordan method to the following linear system:

$$x_1 + x_2 \qquad = 1$$
$$x_2 + x_3 = 3$$
$$x_1 + 2x_2 + x_3 = 4$$

(13)

Solution   The augmented matrix form of (13) is

$$A|\mathbf{b} = \begin{bmatrix} 1 & 1 & 0 & | & 1 \\ 0 & 1 & 1 & | & 3 \\ 1 & 2 & 1 & | & 4 \end{bmatrix}$$

We begin by replacing row 3 (because the row 2, column 1 value is already 0) of $A|\mathbf{b}$ by -1(row 1 of $A|\mathbf{b}$) + row 3 of $A|\mathbf{b}$. The result of this Type 2 ERO is

$$A_1|\mathbf{b}_1 = \begin{bmatrix} 1 & 1 & 0 & | & 1 \\ 0 & 1 & 1 & | & 3 \\ 0 & 1 & 1 & | & 3 \end{bmatrix} \tag{14}$$

Next we replace row 1 of $A_1|\mathbf{b}_1$ by -1(row 2 of $A_1|\mathbf{b}_1$) + row 1 of $A_1|\mathbf{b}_1$. The result of this Type 2 ERO is

$$A_2|\mathbf{b}_2 = \begin{bmatrix} 1 & 0 & -1 & | & -2 \\ 0 & 1 & 1 & | & 3 \\ 0 & 1 & 1 & | & 3 \end{bmatrix}$$

Now we replace row 3 of $A_2|\mathbf{b}_2$ by -1(row 2 of $A_2|\mathbf{b}_2$) + row 3 of $A_2|\mathbf{b}_2$. The result of this Type 2 ERO is

$$A_3|\mathbf{b}_3 = \begin{bmatrix} 1 & 0 & -1 & | & -2 \\ 0 & 1 & 1 & | & 3 \\ 0 & 0 & 0 & | & 0 \end{bmatrix}$$

We would now like to transform the third column of $A_3|\mathbf{b}_3$ into

$$\begin{bmatrix} 0 \\ 0 \\ 1 \end{bmatrix}$$

but this is not possible. The linear system corresponding to $A_3|\mathbf{b}_3$ is

$$x_1 \quad\quad - \quad x_3 = -2 \tag{14.1}$$
$$x_2 + \quad x_3 = 3 \tag{14.2}$$
$$0x_1 + 0x_2 + 0x_3 = 0 \tag{14.3}$$

Suppose we assign an arbitrary value k to x_3. Then (14.1) will be satisfied if $x_1 - k = -2$, or $x_1 = k - 2$. Similarly, (14.2) will be satisfied if $x_2 + k = 3$, or $x_2 = 3 - k$. Of course, (14.3) will be satisfied for any values of x_1, x_2, and x_3. Thus, for any number k, $x_1 = k - 2$, $x_2 = 3 - k$, $x_3 = k$ is a solution to (14). Thus, (14) has an infinite number of solutions (one for each number k). Because (14) was obtained from (13) via EROs, (13) also has an infinite number of solutions. A more formal characterization of linear systems that have an infinite number of solutions will be given after the following summary of the Gauss–Jordan method.

Summary of the Gauss–Jordan Method

Step 1 To solve $A\mathbf{x} = \mathbf{b}$, write down the augmented matrix $A|\mathbf{b}$.

Step 2 At any stage, define a current row, current column, and current entry (the entry in the current row and column). Begin with row 1 as the current row, column 1 as the current column, and a_{11} as the current entry. **(a)** If a_{11} (the current entry) is nonzero, then use EROs to transform column 1 (the current column) to

$$\begin{bmatrix} 1 \\ 0 \\ \vdots \\ 0 \end{bmatrix}$$

Then obtain the new current row, column, and entry by moving down one row and one column to the right, and go to step 3. (**b**) If a_{11} (the current entry) equals 0, then do a Type 3 ERO involving the current row and any row that contains a nonzero number in the current column. Use EROs to transform column 1 to

$$\begin{bmatrix} 1 \\ 0 \\ \vdots \\ 0 \end{bmatrix}$$

Then obtain the new current row, column, and entry by moving down one row and one column to the right. Go to step 3. (**c**) If there are no nonzero numbers in the first column, then obtain a new current column and entry by moving one column to the right. Then go to step 3.

Step 3 (**a**) If the new current entry is nonzero, then use EROs to transform it to 1 and the rest of the current column's entries to 0. When finished, obtain the new current row, column, and entry. If this is impossible, then stop. Otherwise, repeat step 3. (**b**) If the current entry is 0, then do a Type 3 ERO with the current row and any row that contains a nonzero number in the current column. Then use EROs to transform that current entry to 1 and the rest of the current column's entries to 0. When finished, obtain the new current row, column, and entry. If this is impossible, then stop. Otherwise, repeat step 3. (**c**) If the current column has no nonzero numbers below the current row, then obtain the new current column and entry, and repeat step 3. If it is impossible, then stop.

This procedure may require "passing over" one or more columns without transforming them (see Problem 8).

Step 4 Write down the system of equations $A'\mathbf{x} = \mathbf{b}'$ that corresponds to the matrix $A'|\mathbf{b}'$ obtained when step 3 is completed. Then $A'\mathbf{x} = \mathbf{b}'$ will have the same set of solutions as $A\mathbf{x} = \mathbf{b}$.

Basic Variables and Solutions to Linear Equation Systems

To describe the set of solutions to $A'\mathbf{x} = \mathbf{b}'$ (and $A\mathbf{x} = \mathbf{b}$), we need to define the concepts of basic and nonbasic variables.

DEFINITION ■
> After the Gauss–Jordan method has been applied to any linear system, a variable that appears with a coefficient of 1 in a single equation and a coefficient of 0 in all other equations is called a **basic variable** (BV). ■
>
> Any variable that is not a basic variable is called a **nonbasic variable** (NBV). ■

Let BV be the set of basic variables for $A'\mathbf{x} = \mathbf{b}'$ and NBV be the set of nonbasic variables for $A'\mathbf{x} = \mathbf{b}'$. The character of the solutions to $A'\mathbf{x} = \mathbf{b}'$ depends on which of the following cases occurs.

Case 1 $A'\mathbf{x} = \mathbf{b}'$ has at least one row of form $[0 \quad 0 \quad \cdots \quad 0|c]$ $(c \neq 0)$. Then $A\mathbf{x} = \mathbf{b}$ has no solution (recall Example 6). As an example of Case 1, suppose that when the Gauss–Jordan method is applied to the system $A\mathbf{x} = \mathbf{b}$, the following matrix is obtained:

$$A'|\mathbf{b}' = \begin{bmatrix} 1 & 0 & 0 & 1 & | & 1 \\ 0 & 1 & 0 & 2 & | & 1 \\ 0 & 0 & 1 & 3 & | & -1 \\ 0 & 0 & 0 & 0 & | & 0 \\ 0 & 0 & 0 & 0 & | & 2 \end{bmatrix}$$

In this case, $A'\mathbf{x} = \mathbf{b}'$ (and $A\mathbf{x} = \mathbf{b}$) has no solution.

Case 2 Suppose that Case 1 does not apply and NBV, the set of nonbasic variables, is empty. Then $A'\mathbf{x} = \mathbf{b}'$ (and $A\mathbf{x} = \mathbf{b}$) will have a unique solution. To illustrate this, we recall that in solving

$$\begin{aligned} 2x_1 + 2x_2 + x_3 &= 9 \\ 2x_1 - x_2 + 2x_3 &= 6 \\ x_1 - x_2 + 2x_3 &= 5 \end{aligned}$$

the Gauss–Jordan method yielded

$$A'|\mathbf{b}' = \begin{bmatrix} 1 & 0 & 0 & | & 1 \\ 0 & 1 & 0 & | & 2 \\ 0 & 0 & 1 & | & 3 \end{bmatrix}$$

In this case, BV $= \{x_1, x_2, x_3\}$ and NBV is empty. Then the unique solution to $A'\mathbf{x} = \mathbf{b}'$ (and $A\mathbf{x} = \mathbf{b}$) is $x_1 = 1$, $x_2 = 2$, $x_3 = 3$.

Case 3 Suppose that Case 1 does not apply and NBV is nonempty. Then $A'\mathbf{x} = \mathbf{b}'$ (and $A\mathbf{x} = \mathbf{b}$) will have an infinite number of solutions. To obtain these, first assign each nonbasic variable an arbitrary value. Then solve for the value of each basic variable in terms of the nonbasic variables. For example, suppose

$$A'|\mathbf{b}' = \begin{bmatrix} 1 & 0 & 0 & 1 & 1 & | & 3 \\ 0 & 1 & 0 & 2 & 0 & | & 2 \\ 0 & 0 & 1 & 0 & 1 & | & 1 \\ 0 & 0 & 0 & 0 & 0 & | & 0 \end{bmatrix} \tag{15}$$

Because Case 1 does not apply, and BV $= \{x_1, x_2, x_3\}$ and NBV $= \{x_4, x_5\}$, we have an example of Case 3: $A'\mathbf{x} = \mathbf{b}'$ (and $A\mathbf{x} = \mathbf{b}$) will have an infinite number of solutions. To see what these solutions look like, write down $A'\mathbf{x} = \mathbf{b}'$:

$$\begin{aligned} x_1 + x_4 + x_5 &= 3 \tag{15.1} \\ x_2 + 2x_4 &= 2 \tag{15.2} \\ x_3 + x_5 &= 1 \tag{15.3} \\ 0x_1 + 0x_2 + 0x_3 + 0x_4 + 0x_5 &= 0 \tag{15.4} \end{aligned}$$

Now assign the nonbasic variables (x_4 and x_5) arbitrary values c and k, with $x_4 = c$ and $x_5 = k$. From (15.1), we find that $x_1 = 3 - c - k$. From (15.2), we find that $x_2 = 2 - 2c$. From (15.3), we find that $x_3 = 1 - k$. Because (15.4) holds for all values of the variables, $x_1 = 3 - c - k$, $x_2 = 2 - 2c$, $x_3 = 1 - k$, $x_4 = c$, and $x_5 = k$ will, for any values of c and k, be a solution to $A'\mathbf{x} = \mathbf{b}'$ (and $A\mathbf{x} = \mathbf{b}$).

Our discussion of the Gauss–Jordan method is summarized in Figure 6. We have devoted so much time to the Gauss–Jordan method because, in our study of linear programming, examples of Case 3 (linear systems with an infinite number of solutions) will occur repeatedly. Because the end result of the Gauss–Jordan method must always be one of Cases 1–3, we have shown that any linear system will have no solution, a unique solution, or an infinite number of solutions.

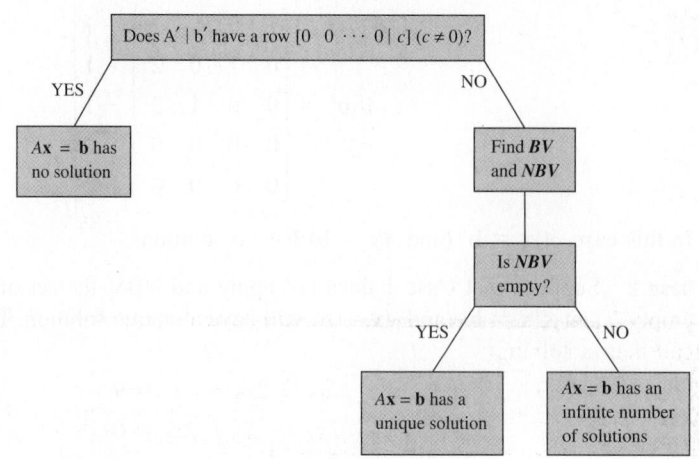

FIGURE 6
Description of
Gauss–Jordan Method
for Solving Linear
Equations

PROBLEMS

Group A

Use the Gauss–Jordan method to determine whether each of the following linear systems has no solution, a unique solution, or an infinite number of solutions. Indicate the solutions (if any exist).

1 $x_1 + x_2 \qquad + x_4 = 3$
$\qquad x_2 + x_3 \qquad = 4$
$x_1 + 2x_2 + x_3 + x_4 = 8$

2 $x_1 + x_2 + x_3 = 4$
$x_1 + 2x_2 \qquad = 6$

3 $x_1 + x_2 = 1$
$2x_1 + x_2 = 3$
$3x_1 + 2x_2 = 4$

4 $2x_1 - x_2 + x_3 + x_4 = 6$
$x_1 + x_2 + x_3 \qquad = 4$

5 $x_1 \qquad + \qquad x_4 = 5$
$\qquad x_2 \qquad + 2x_4 = 5$
$\qquad x_3 + 0.5x_4 = 1$
$\qquad 2x_3 + \qquad x_4 = 3$

6 $\qquad 2x_2 + 2x_3 = 4$
$x_1 + 2x_2 + x_3 = 4$
$\qquad x_2 - x_3 = 0$

7 $x_1 + x_2 \qquad = 2$
$\qquad -x_2 + 2x_3 = 3$
$\qquad x_2 + x_3 = 3$

8 $x_1 + x_2 + x_3 \qquad = 1$
$\qquad x_2 + 2x_3 + x_4 = 2$
$\qquad x_4 = 3$

Group B

9   Suppose that a linear system $A\mathbf{x} = \mathbf{b}$ has more variables than equations. Show that $A\mathbf{x} = \mathbf{b}$ cannot have a unique solution.

2.4 Linear Independence and Linear Dependence[†]

In this section, we discuss the concepts of a linearly independent set of vectors, a linearly dependent set of vectors, and the rank of a matrix. These concepts will be useful in our study of matrix inverses.

Before defining a linearly independent set of vectors, we need to define a linear combination of a set of vectors. Let $V = \{\mathbf{v}_1, \mathbf{v}_2, \ldots, \mathbf{v}_k\}$ be a set of row vectors all of which have the same dimension.

[†]This section covers topics that may be omitted with no loss of continuity.

A **linear combination** of the vectors in V is any vector of the form $c_1\mathbf{v}_1 + c_2\mathbf{v}_2$ $\cdots + c_k\mathbf{v}_k$, where $c_1, c_2, \ldots, c_k$ are arbitrary scalars. ■

For example, if $V = \{[1 \quad 2], [2 \quad 1]\}$, then

$$2\mathbf{v}_1 - \mathbf{v}_2 = 2([1 \quad 2]) - [2 \quad 1] = [0 \quad 3]$$
$$\mathbf{v}_1 + 3\mathbf{v}_2 = [1 \quad 2] + 3([2 \quad 1]) = [7 \quad 5]$$
$$0\mathbf{v}_1 + 3\mathbf{v}_2 = [0 \quad 0] + 3([2 \quad 1]) = [6 \quad 3]$$

are linear combinations of vectors in V. The foregoing definition may also be applied to a set of column vectors.

Suppose we are given a set $V = \{\mathbf{v}_1, \mathbf{v}_2, \ldots, \mathbf{v}_k\}$ of m-dimensional row vectors. Let $\mathbf{0} = [0 \quad 0 \quad \cdots \quad 0]$ be the m-dimensional $\mathbf{0}$ vector. To determine whether V is a linearly independent set of vectors, we try to find a linear combination of the vectors in V that adds up to $\mathbf{0}$. Clearly, $0\mathbf{v}_1 + 0\mathbf{v}_2 + \cdots + 0\mathbf{v}_k$ is a linear combination of vectors in V that adds up to $\mathbf{0}$. We call the linear combination of vectors in V for which $c_1 = c_2 = \cdots = c_k = 0$ the *trivial* linear combination of vectors in V. We may now define linearly independent and linearly dependent sets of vectors.

DEFINITION ■ A set V of m-dimensional vectors is **linearly independent** if the only linear combination of vectors in V that equals $\mathbf{0}$ is the trivial linear combination. ■

A set V of m-dimensional vectors is **linearly dependent** if there is a nontrivial linear combination of the vectors in V that adds up to $\mathbf{0}$. ■

The following examples should clarify these definitions.

EXAMPLE 8　0 Vector Makes Set LD

Show that any set of vectors containing the $\mathbf{0}$ vector is a linearly dependent set.

Solution To illustrate, we show that if $V = \{[0 \quad 0], [1 \quad 0], [0 \quad 1]\}$, then V is linearly dependent, because if, say, $c_1 \neq 0$, then $c_1([0 \quad 0]) + 0([1 \quad 0]) + 0([0 \quad 1]) = [0 \quad 0]$. Thus, there is a nontrivial linear combination of vectors in V that adds up to $\mathbf{0}$.

EXAMPLE 9　LI Set of Vectors

Show that the set of vectors $V = \{[1 \quad 0], [0 \quad 1]\}$ is a linearly independent set of vectors.

Solution We try to find a nontrivial linear combination of the vectors in V that yields $\mathbf{0}$. This requires that we find scalars c_1 and c_2 (at least one of which is nonzero) satisfying $c_1([1 \quad 0]) + c_2([0 \quad 1]) = [0 \quad 0]$. Thus, c_1 and c_2 must satisfy $[c_1 \quad c_2] = [0 \quad 0]$. This implies $c_1 = c_2 = 0$. The only linear combination of vectors in V that yields $\mathbf{0}$ is the trivial linear combination. Therefore, V is a linearly independent set of vectors.

EXAMPLE 10　LD Set of Vectors

Show that $V = \{[1 \quad 2], [2 \quad 4]\}$ is a linearly dependent set of vectors.

Solution Because $2([1 \quad 2]) - 1([2 \quad 4]) = [0 \quad 0]$, there is a nontrivial linear combination with $c_1 = 2$ and $c_2 = -1$ that yields $\mathbf{0}$. Thus, V is a linearly dependent set of vectors.

Intuitively, what does it mean for a set of vectors to be linearly dependent? To understand the concept of linear dependence, observe that a set of vectors V is linearly dependent (as

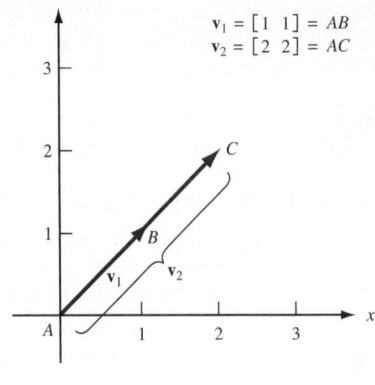

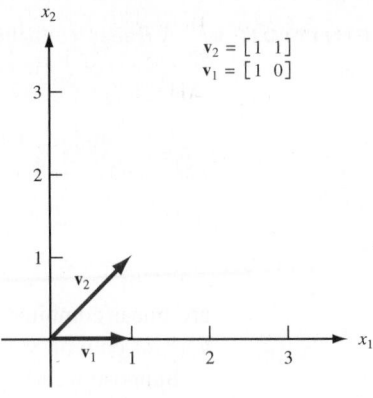

FIGURE 7
(a) Two Linearly Dependent Vectors (b) Two Linearly Independent Vectors

a

b

long as $\mathbf{0}$ is not in V) if and only if some vector in V can be written as a nontrivial linear combination of other vectors in V (see Problem 9 at the end of this section). For instance, in Example 10, $[2 \quad 4] = 2([1 \quad 2])$. Thus, if a set of vectors V is linearly dependent, the vectors in V are, in some way, not all "different" vectors. By "different" we mean that the direction specified by any vector in V cannot be expressed by adding together multiples of other vectors in V. For example, in two dimensions it can be shown that two vectors are linearly dependent if and only if they lie on the same line (see Figure 7).

The Rank of a Matrix

The Gauss–Jordan method can be used to determine whether a set of vectors is linearly independent or linearly dependent. Before describing how this is done, we define the concept of the rank of a matrix.

Let A be any $m \times n$ matrix, and denote the rows of A by $\mathbf{r}_1, \mathbf{r}_2, \ldots, \mathbf{r}_m$. Also define $R = \{\mathbf{r}_1, \mathbf{r}_2, \ldots, \mathbf{r}_m\}$.

DEFINITION ■ The **rank** of A is the number of vectors in the largest linearly independent subset of R. ■

The following three examples illustrate the concept of rank.

EXAMPLE 11 **Matrix with 0 Rank**

Show that rank $A = 0$ for the following matrix:

$$A = \begin{bmatrix} 0 & 0 \\ 0 & 0 \end{bmatrix}$$

Solution For the set of vectors $R = \{[0 \quad 0], [0, \quad 0]\}$, it is impossible to choose a subset of R that is linearly independent (recall Example 8).

EXAMPLE 12 **Matrix with Rank of 1**

Show that rank $A = 1$ for the following matrix:

$$A = \begin{bmatrix} 1 & 1 \\ 2 & 2 \end{bmatrix}$$

Solution Here $R = \{[1 \quad 1], [2 \quad 2]\}$. The set $\{[1 \quad 1]\}$ is a linearly independent subset of R, so rank A must be at least 1. If we try to find two linearly independent vectors in R, we fail because $2([1 \quad 1]) - [2 \quad 2] = [0 \quad 0]$. This means that rank A cannot be 2. Thus, rank A must equal 1.

EXAMPLE 13 **Matrix with Rank of 2**

Show that rank $A = 2$ for the following matrix:

$$A = \begin{bmatrix} 1 & 0 \\ 0 & 1 \end{bmatrix}$$

Solution Here $R = \{[1 \quad 0], [0 \quad 1]\}$. From Example 9, we know that R is a linearly independent set of vectors. Thus, rank $A = 2$.

To find the rank of a given matrix A, simply apply the Gauss–Jordan method to the matrix A. Let the final result be the matrix $\bar{A}$. It can be shown that performing a sequence of EROs on a matrix does not change the rank of the matrix. This implies that rank $A = $ rank $\bar{A}$. It is also apparent that the rank of $\bar{A}$ will be the number of nonzero rows in $\bar{A}$. Combining these facts, we find that rank $A = $ rank $\bar{A} = $ number of nonzero rows in $\bar{A}$.

EXAMPLE 14 **Using Gauss–Jordan Method to Find Rank of Matrix**

Find

$$\text{rank } A = \begin{bmatrix} 1 & 0 & 0 \\ 0 & 2 & 1 \\ 0 & 2 & 3 \end{bmatrix}$$

Solution The Gauss–Jordan method yields the following sequence of matrices:

$$A = \begin{bmatrix} 1 & 0 & 0 \\ 0 & 2 & 1 \\ 0 & 2 & 3 \end{bmatrix} \rightarrow \begin{bmatrix} 1 & 0 & 0 \\ 0 & 1 & \frac{1}{2} \\ 0 & 2 & 3 \end{bmatrix} \rightarrow \begin{bmatrix} 1 & 0 & 0 \\ 0 & 1 & \frac{1}{2} \\ 0 & 0 & 2 \end{bmatrix} \rightarrow \begin{bmatrix} 1 & 0 & 0 \\ 0 & 1 & \frac{1}{2} \\ 0 & 0 & 1 \end{bmatrix} \rightarrow \begin{bmatrix} 1 & 0 & 0 \\ 0 & 1 & 0 \\ 0 & 0 & 1 \end{bmatrix}$$
$$= \bar{A}$$

Thus, rank $A = $ rank $\bar{A} = 3$.

How to Tell Whether a Set of Vectors Is Linearly Independent

We now describe a method for determining whether a set of vectors $V = \{\mathbf{v}_1, \mathbf{v}_2, \ldots, \mathbf{v}_m\}$ is linearly independent.

Form the matrix A whose ith row is $\mathbf{v}_i$. A will have m rows. If rank $A = m$, then V is a linearly independent set of vectors, whereas if rank $A < m$, then V is a linearly dependent set of vectors.

EXAMPLE 15 **A Linearly Dependent Set of Vectors**

Determine whether $V = \{[1 \quad 0 \quad 0], [0 \quad 1 \quad 0], [1 \quad 1 \quad 0]\}$ is a linearly independent set of vectors.

Solution The Gauss–Jordan method yields the following sequence of matrices:

$$A = \begin{bmatrix} 1 & 0 & 0 \\ 0 & 1 & 0 \\ 0 & 1 & 0 \end{bmatrix} \rightarrow \begin{bmatrix} 1 & 0 & 0 \\ 0 & 1 & 0 \\ 0 & 1 & 0 \end{bmatrix} \rightarrow \begin{bmatrix} 1 & 0 & 0 \\ 0 & 1 & 0 \\ 0 & 0 & 0 \end{bmatrix} = \bar{A}$$

Thus, rank A = rank $\bar{A}$ = 2 < 3. This shows that V is a linearly dependent set of vectors. In fact, the EROs used to transform A to $\bar{A}$ can be used to show that [1 1 0] = [1 0 0] + [0 1 0]. This equation also shows that V is a linearly dependent set of vectors.

PROBLEMS

Group A

Determine if each of the following sets of vectors is linearly independent or linearly dependent.

1 $V = \{[1 \quad 0 \quad 1], [1 \quad 2 \quad 1], [2 \quad 2 \quad 2]\}$

2 $V = \{[2 \quad 1 \quad 0], [1 \quad 2 \quad 0], [3 \quad 3 \quad 1]\}$

3 $V = \{[2 \quad 1], [1 \quad 2]\}$

4 $V = \{[2 \quad 0], [3 \quad 0]\}$

5 $V = \left\{ \begin{bmatrix} 1 \\ 2 \\ 3 \end{bmatrix}, \begin{bmatrix} 4 \\ 5 \\ 6 \end{bmatrix}, \begin{bmatrix} 5 \\ 7 \\ 9 \end{bmatrix} \right\}$

6 $V = \left\{ \begin{bmatrix} 1 \\ 0 \\ 0 \end{bmatrix}, \begin{bmatrix} 0 \\ 2 \\ 1 \end{bmatrix}, \begin{bmatrix} 1 \\ 0 \\ 1 \end{bmatrix} \right\}$

Group B

7 Show that the linear system $A\mathbf{x} = \mathbf{b}$ has a solution if and only if $\mathbf{b}$ can be written as a linear combination of the columns of A.

8 Suppose there is a collection of three or more two-dimensional vectors. Provide an argument showing that the collection must be linearly dependent.

9 Show that a set of vectors V (not containing the **0** vector) is linearly dependent if and only if there exists some vector in V that can be written as a nontrivial linear combination of other vectors in V.

2.5 The Inverse of a Matrix

To solve a single linear equation such as $4x = 3$, we simply multiply both sides of the equation by the multiplicative inverse of 4, which is 4^{-1}, or $\frac{1}{4}$. This yields $4^{-1}(4x) = (4^{-1})3$, or $x = \frac{3}{4}$. (Of course, this method fails to work for the equation $0x = 3$, because zero has no multiplicative inverse.) In this section, we develop a generalization of this technique that can be used to solve "square" (number of equations = number of unknowns) linear systems. We begin with some preliminary definitions.

DEFINITION ■ A **square matrix** is any matrix that has an equal number of rows and columns. ■

The **diagonal elements** of a square matrix are those elements a_{ij} such that $i = j$. ■

A square matrix for which all diagonal elements are equal to 1 and all nondiagonal elements are equal to 0 is called an **identity matrix.** ■

The $m \times m$ identity matrix will be written as I_m. Thus,

$$I_2 = \begin{bmatrix} 1 & 0 \\ 0 & 1 \end{bmatrix}, \qquad I_3 = \begin{bmatrix} 1 & 0 & 0 \\ 0 & 1 & 0 \\ 0 & 0 & 1 \end{bmatrix}, \qquad \cdots$$

If the multiplications $I_m A$ and AI_m are defined, it is easy to show that $I_m A = AI_m = A$. Thus, just as the number 1 serves as the unit element for multiplication of real numbers, I_m serves as the unit element for multiplication of matrices.

Recall that $\frac{1}{4}$ is the multiplicative inverse of 4. This is because $4(\frac{1}{4}) = (\frac{1}{4})4 = 1$. This motivates the following definition of the inverse of a matrix.

<table>
<tr><td>DEFINITION ■</td><td>

For a given $m \times m$ matrix A, the $m \times m$ matrix B is the **inverse** of A if
$$BA = AB = I_m \tag{16}$$
(It can be shown that if $BA = I_m$ or $AB = I_m$, then the other quantity will also equal I_m.) ■

</td></tr>
</table>

Some square matrices do not have inverses. If there does exist an $m \times m$ matrix B that satisfies Equation (16), then we write $B = A^{-1}$. For example, if

$$A = \begin{bmatrix} 2 & 0 & -1 \\ 3 & 1 & 2 \\ -1 & 0 & 1 \end{bmatrix}$$

the reader can verify that

$$\begin{bmatrix} 2 & 0 & -1 \\ 3 & 1 & 2 \\ -1 & 0 & 1 \end{bmatrix} \begin{bmatrix} 1 & 0 & 1 \\ -5 & 1 & -7 \\ 1 & 0 & 2 \end{bmatrix} = \begin{bmatrix} 1 & 0 & 0 \\ 0 & 1 & 0 \\ 0 & 0 & 1 \end{bmatrix}$$

and

$$\begin{bmatrix} 1 & 0 & 1 \\ -5 & 1 & -7 \\ 1 & 0 & 2 \end{bmatrix} \begin{bmatrix} 2 & 0 & -1 \\ 3 & 1 & 2 \\ -1 & 0 & 1 \end{bmatrix} = \begin{bmatrix} 1 & 0 & 0 \\ 0 & 1 & 0 \\ 0 & 0 & 1 \end{bmatrix}$$

Thus,

$$A^{-1} = \begin{bmatrix} 1 & 0 & 1 \\ -5 & 1 & -7 \\ 1 & 0 & 2 \end{bmatrix}$$

To see why we are interested in the concept of a matrix inverse, suppose we want to solve a linear system $A\mathbf{x} = \mathbf{b}$ that has m equations and m unknowns. Suppose that A^{-1} exists. Multiplying both sides of $A\mathbf{x} = \mathbf{b}$ by A^{-1}, we see that any solution of $A\mathbf{x} = \mathbf{b}$ must also satisfy $A^{-1}(A\mathbf{x}) = A^{-1}\mathbf{b}$. Using the associative law and the definition of a matrix inverse, we obtain

$$(A^{-1}A)\mathbf{x} = A^{-1}\mathbf{b}$$
$$\text{or} \qquad I_m\mathbf{x} = A^{-1}\mathbf{b}$$
$$\text{or} \qquad \mathbf{x} = A^{-1}\mathbf{b}$$

This shows that knowing A^{-1} enables us to find the unique solution to a square linear system. This is the analog of solving $4x = 3$ by multiplying both sides of the equation by 4^{-1}.

The Gauss–Jordan method may be used to find A^{-1} (or to show that A^{-1} does not exist). To illustrate how we can use the Gauss–Jordan method to invert a matrix, suppose we want to find A^{-1} for

$$A = \begin{bmatrix} 2 & 5 \\ 1 & 3 \end{bmatrix}$$

This requires that we find a matrix

$$\begin{bmatrix} a & b \\ c & d \end{bmatrix} = A^{-1}$$

that satisfies

$$\begin{bmatrix} 2 & 5 \\ 1 & 3 \end{bmatrix}\begin{bmatrix} a & b \\ c & d \end{bmatrix} = \begin{bmatrix} 1 & 0 \\ 0 & 1 \end{bmatrix} \tag{17}$$

From Equation (17), we obtain the following pair of simultaneous equations that must be satisfied by a, b, c, and d:

$$\begin{bmatrix} 2 & 5 \\ 1 & 3 \end{bmatrix}\begin{bmatrix} a \\ c \end{bmatrix} = \begin{bmatrix} 1 \\ 0 \end{bmatrix}; \qquad \begin{bmatrix} 2 & 5 \\ 1 & 3 \end{bmatrix}\begin{bmatrix} b \\ d \end{bmatrix} = \begin{bmatrix} 0 \\ 1 \end{bmatrix}$$

Thus, to find

$$\begin{bmatrix} a \\ c \end{bmatrix}$$

(the first column of A^{-1}), we can apply the Gauss–Jordan method to the augmented matrix

$$\left[\begin{array}{cc|c} 2 & 5 & 1 \\ 1 & 3 & 0 \end{array}\right]$$

Once EROs have transformed

$$\begin{bmatrix} 2 & 5 \\ 1 & 3 \end{bmatrix}$$

to I_2,

$$\begin{bmatrix} 1 \\ 0 \end{bmatrix}$$

will have been transformed into the first column of A^{-1}. To determine

$$\begin{bmatrix} b \\ d \end{bmatrix}$$

(the second column of A^{-1}), we apply EROs to the augmented matrix

$$\left[\begin{array}{cc|c} 2 & 5 & 0 \\ 1 & 3 & 1 \end{array}\right]$$

When

$$\begin{bmatrix} 2 & 5 \\ 1 & 3 \end{bmatrix}$$

has been transformed into I_2,

$$\begin{bmatrix} 0 \\ 1 \end{bmatrix}$$

will have been transformed into the second column of A^{-1}. Thus, to find each column of A^{-1}, we must perform a sequence of EROs that transform

$$\begin{bmatrix} 2 & 5 \\ 1 & 3 \end{bmatrix}$$

into I_2. This suggests that we can find A^{-1} by applying EROs to the 2×4 matrix

$$A|I_2 = \begin{bmatrix} 2 & 5 & | & 1 & 0 \\ 1 & 3 & | & 0 & 1 \end{bmatrix}$$

When

$$\begin{bmatrix} 2 & 5 \\ 1 & 3 \end{bmatrix}$$

has been transformed to I_2,

$$\begin{bmatrix} 1 \\ 0 \end{bmatrix}$$

will have been transformed into the first column of A^{-1}, and

$$\begin{bmatrix} 0 \\ 1 \end{bmatrix}$$

will have been transformed into the second column of A^{-1}. Thus, *as A is transformed into* I_2, I_2 *is transformed into* A^{-1}. The computations to determine A^{-1} follow.

Step 1 Multiply row 1 of $A|I_2$ by $\frac{1}{2}$. This yields

$$A'|I_2' = \begin{bmatrix} 1 & \frac{5}{2} & | & \frac{1}{2} & 0 \\ 1 & 3 & | & 0 & 1 \end{bmatrix}$$

Step 2 Replace row 2 of $A'|I_2'$ by -1(row 1 of $A'|I_2'$) + row 2 of $A'|I_2'$. This yields

$$A''|I_2'' = \begin{bmatrix} 1 & \frac{5}{2} & | & \frac{1}{2} & 0 \\ 0 & \frac{1}{2} & | & -\frac{1}{2} & 1 \end{bmatrix}$$

Step 3 Multiply row 2 of $A''|I_2''$ by 2. This yields

$$A'''|I_2''' = \begin{bmatrix} 1 & \frac{5}{2} & | & \frac{1}{2} & 0 \\ 0 & 1 & | & -1 & 2 \end{bmatrix}$$

Step 4 Replace row 1 of $A'''|I_2'''$ by $-\frac{5}{2}$(row 2 of $A'''|I_2'''$) + row 1 of $A'''|I_2'''$. This yields

$$\begin{bmatrix} 1 & 0 & | & 3 & -5 \\ 0 & 1 & | & -1 & 2 \end{bmatrix}$$

Because A has been transformed into I_2, I_2 will have been transformed into A^{-1}. Hence,

$$A^{-1} = \begin{bmatrix} 3 & -5 \\ -1 & 2 \end{bmatrix}$$

The reader should verify that $AA^{-1} = A^{-1}A = I_2$.

A Matrix May Not Have an Inverse

Some matrices do not have inverses. To illustrate, let

$$A = \begin{bmatrix} 1 & 2 \\ 2 & 4 \end{bmatrix} \quad \text{and} \quad A^{-1} = \begin{bmatrix} e & f \\ g & h \end{bmatrix} \tag{18}$$

To find A^{-1} we must solve the following pair of simultaneous equations:

$$\begin{bmatrix} 1 & 2 \\ 2 & 4 \end{bmatrix}\begin{bmatrix} e \\ g \end{bmatrix} = \begin{bmatrix} 1 \\ 0 \end{bmatrix} \tag{18.1}$$

$$\begin{bmatrix} 1 & 2 \\ 2 & 4 \end{bmatrix}\begin{bmatrix} f \\ h \end{bmatrix} = \begin{bmatrix} 0 \\ 1 \end{bmatrix} \tag{18.2}$$

When we try to solve (18.1) by the Gauss–Jordan method, we find that

$$\left[\begin{array}{cc|c} 1 & 2 & 1 \\ 2 & 4 & 0 \end{array}\right]$$

is transformed into

$$\left[\begin{array}{cc|c} 1 & 2 & 1 \\ 0 & 0 & -2 \end{array}\right]$$

This indicates that (18.1) has no solution, and A^{-1} cannot exist.

Observe that (18.1) fails to have a solution, because the Gauss–Jordan method transforms A into a matrix with a row of zeros on the bottom. This can only happen if rank $A < 2$. If $m \times m$ matrix A has rank $A < m$, then A^{-1} will not exist.

The Gauss–Jordan Method for Inverting an $m \times m$ Matrix A

Step 1   Write down the $m \times 2m$ matrix $A|I_m$.

Step 1   Use EROs to transform $A|I_m$ into $I_m|B$. This will be possible only if rank $A = m$. In this case, $B = A^{-1}$. If rank $A < m$, then A has no inverse.

Using Matrix Inverses to Solve Linear Systems

As previously stated, matrix inverses can be used to solve a linear system $A\mathbf{x} = \mathbf{b}$ in which the number of variables and equations are equal. Simply multiply both sides of $A\mathbf{x} = \mathbf{b}$ by A^{-1} to obtain the solution $\mathbf{x} = A^{-1}\mathbf{b}$. For example, to solve

$$\begin{aligned} 2x_1 + 5x_2 &= 7 \\ x_1 + 3x_2 &= 4 \end{aligned} \tag{19}$$

write the matrix representation of (19):

$$\begin{bmatrix} 2 & 5 \\ 1 & 3 \end{bmatrix}\begin{bmatrix} x_1 \\ x_2 \end{bmatrix} = \begin{bmatrix} 7 \\ 4 \end{bmatrix} \tag{20}$$

Let

$$A = \begin{bmatrix} 2 & 5 \\ 1 & 3 \end{bmatrix}$$

We found in the previous illustration that

$$A^{-1} = \begin{bmatrix} 3 & -5 \\ -1 & 2 \end{bmatrix}$$

	A	B	C	D	E	F	G	H
1		Inverting						
2		a						
3		Matrix			2	0	-1	
4				A	3	1	2	
5					-1	0	1	
6								
7					1	0	1	
8				A^{-1}	-5	1	-7	
9					1	0	2	

FIGURE 8

Multiplying both sides of (20) by A^{-1}, we obtain

$$\begin{bmatrix} 3 & -5 \\ -1 & 2 \end{bmatrix}\begin{bmatrix} 2 & 5 \\ 1 & 3 \end{bmatrix}\begin{bmatrix} x_1 \\ x_2 \end{bmatrix} = \begin{bmatrix} 3 & -5 \\ -1 & 2 \end{bmatrix}\begin{bmatrix} 7 \\ 4 \end{bmatrix}$$

$$\begin{bmatrix} x_1 \\ x_2 \end{bmatrix} = \begin{bmatrix} 1 \\ 1 \end{bmatrix}$$

Thus, $x_1 = 1$, $x_2 = 1$ is the unique solution to system (19).

Inverting Matrices with Excel

Minverse.xls

The Excel =MINVERSE command makes it easy to invert a matrix. See Figure 8 and file Minverse.xls. Suppose we want to invert the matrix

$$A = \begin{bmatrix} 2 & 0 & -1 \\ 3 & 1 & 2 \\ -1 & 0 & 1 \end{bmatrix}$$

Simply enter the matrix in E3:G5 and select the range (we chose E7:G9) where you want A^{-1} to be computed. In the upper left-hand corner of the range E7:G9 (cell E7), we enter the formula

$$= \text{MINVERSE(E3:G5)}$$

and select **Control Shift Enter.** This enters an array function that computes A^{-1} in the range E7:G9. You cannot edit part of an array function, so if you want to delete A^{-1}, you must delete the entire range where A^{-1} is present.

PROBLEMS

Group A

Find A^{-1} (if it exists) for the following matrices:

1 $\begin{bmatrix} 1 & 3 \\ 2 & 5 \end{bmatrix}$

2 $\begin{bmatrix} 1 & 0 & 1 \\ 4 & 1 & -2 \\ 3 & 1 & -1 \end{bmatrix}$

3 $\begin{bmatrix} 1 & 0 & 1 \\ 1 & 1 & 1 \\ 2 & 1 & 2 \end{bmatrix}$

4 $\begin{bmatrix} 1 & 2 & 1 \\ 1 & 2 & 0 \\ 2 & 4 & 1 \end{bmatrix}$

5 Use the answer to Problem 1 to solve the following linear system:

$$x_1 + 3x_2 = 4$$
$$2x_1 + 5x_2 = 7$$

6 Use the answer to Problem 2 to solve the following linear system:

$$x_1 + \qquad x_3 = 4$$
$$4x_1 + x_2 - 2x_3 = 0$$
$$3x_1 + x_2 - \quad x_3 = 2$$

Group B

7 Show that a square matrix has an inverse if and only if its rows form a linearly independent set of vectors.

8 Consider a square matrix B whose inverse is given by B^{-1}.

a In terms of B^{-1}, what is the inverse of the matrix $100B$?

b Let B' be the matrix obtained from B by doubling every entry in row 1 of B. Explain how we could obtain the inverse of B' from B^{-1}.

c Let B' be the matrix obtained from B by doubling every entry in column 1 of B. Explain how we could obtain the inverse of B' from B^{-1}.

9 Suppose that A and B both have inverses. Find the inverse of the matrix AB.

10 Suppose A has an inverse. Show that $(A^T)^{-1} = (A^{-1})^T$. (*Hint:* Use the fact that $AA^{-1} = I$, and take the transpose of both sides.)

11 A square matrix A is *orthogonal* if $AA^T = I$. What properties must be possessed by the columns of an orthogonal matrix?

2.6 Determinants

Associated with any square matrix A is a number called the *determinant* of A (often abbreviated as det A or $|A|$). Knowing how to compute the determinant of a square matrix will be useful in our study of nonlinear programming.

For a 1×1 matrix $A = [a_{11}]$,

$$\det A = a_{11} \tag{21}$$

For a 2×2 matrix

$$A = \begin{bmatrix} a_{11} & a_{12} \\ a_{21} & a_{22} \end{bmatrix} \tag{22}$$

$$\det A = a_{11}a_{22} - a_{21}a_{12}$$

For example,

$$\det \begin{bmatrix} 2 & 4 \\ 3 & 5 \end{bmatrix} = 2(5) - 3(4) = -2$$

Before we learn how to compute det A for larger square matrices, we need to define the concept of the *minor* of a matrix.

DEFINITION ■ If A is an $m \times m$ matrix, then for any values of i and j, the ijth **minor** of A (written A_{ij}) is the $(m-1) \times (m-1)$ submatrix of A obtained by deleting row i and column j of A. ■

For example,

$$\text{if} \quad A = \begin{bmatrix} 1 & 2 & 3 \\ 4 & 5 & 6 \\ 7 & 8 & 9 \end{bmatrix}, \quad \text{then} \quad A_{12} = \begin{bmatrix} 4 & 6 \\ 7 & 9 \end{bmatrix} \quad \text{and} \quad A_{32} = \begin{bmatrix} 1 & 3 \\ 4 & 6 \end{bmatrix}$$

Let A be any $m \times m$ matrix. We may write A as

$$A = \begin{bmatrix} a_{11} & a_{12} & \cdots & a_{1n} \\ a_{21} & a_{22} & \cdots & a_{2n} \\ \vdots & \vdots & & \vdots \\ a_{m1} & a_{m2} & \cdots & a_{mn} \end{bmatrix}$$

To compute det A, pick any value of i ($i = 1, 2, \ldots, m$) and compute det A:

$$\det A = (-1)^{i+1}a_{i1}(\det A_{i1}) + (-1)^{i+2}a_{i2}(\det A_{i2}) + \cdots + (-1)^{i+m}a_{im}(\det A_{im}) \tag{23}$$

Formula (23) is called the expansion of det A by the cofactors of row i. The virtue of (23) is that it reduces the computation of det A for an $m \times m$ matrix to computations involving only $(m - 1) \times (m - 1)$ matrices. Apply (23) until det A can be expressed in terms of 2×2 matrices. Then use Equation (22) to find the determinants of the relevant 2×2 matrices.

To illustrate the use of (23), we find det A for

$$A = \begin{bmatrix} 1 & 2 & 3 \\ 4 & 5 & 6 \\ 7 & 8 & 9 \end{bmatrix}$$

We expand det A by using row 1 cofactors. Notice that $a_{11} = 1$, $a_{12} = 2$, and $a_{13} = 3$. Also

$$A_{11} = \begin{bmatrix} 5 & 6 \\ 8 & 9 \end{bmatrix}$$

so by (22), det $A_{11} = 5(9) - 8(6) = -3$;

$$A_{12} = \begin{bmatrix} 4 & 6 \\ 7 & 9 \end{bmatrix}$$

so by (22), det $A_{12} = 4(9) - 7(6) = -6$; and

$$A_{13} = \begin{bmatrix} 4 & 5 \\ 7 & 8 \end{bmatrix}$$

so by (22), det $A_{13} = 4(8) - 7(5) = -3$. Then by (23),

$$\det A = (-1)^{1+1} a_{11}(\det A_{11}) + (-1)^{1+2} a_{12}(\det A_{12}) + (-1)^{1+3} a_{13}(\det A_{13})$$
$$= (1)(1)(-3) + (-1)(2)(-6) + (1)(3)(-3) = -3 + 12 - 9 = 0$$

The interested reader may verify that expansion of det A by either row 2 or row 3 cofactors also yields det $A = 0$.

We close our discussion of determinants by noting that they can be used to invert square matrices and to solve linear equation systems. Because we already have learned to use the Gauss–Jordan method to invert matrices and to solve linear equation systems, we will not discuss these uses of determinants.

PROBLEMS

Group A

1 Verify that det $\begin{bmatrix} 1 & 2 & 3 \\ 4 & 5 & 6 \\ 7 & 8 & 9 \end{bmatrix} = 0$ by using expansions by row 2 and row 3 cofactors.

2 Find det $\begin{bmatrix} 1 & 0 & 0 & 0 \\ 0 & 2 & 0 & 0 \\ 0 & 0 & 3 & 0 \\ 0 & 0 & 0 & 5 \end{bmatrix}$

3 A matrix is said to be upper triangular if for $i > j$, $a_{ij} = 0$. Show that the determinant of any upper triangular 3×3 matrix is equal to the product of the matrix's diagonal elements. (This result is true for any upper triangular matrix.)

Group B

4 **a** Show that for any 1×1 and 3×3 matrix, det $-A = -\det A$.

b Show that for any 2×2 and 4×4 matrix, det $-A = \det A$.

c Generalize the results of parts (a) and (b).

Matrices

A **matrix** is any rectangular array of numbers. For the matrix A, we let a_{ij} represent the element of A in row i and column j.

A matrix with only one row or one column may be thought of as a **vector**. Vectors appear in boldface type (**v**). Given a row vector $\mathbf{u} = [u_1 \quad u_2 \quad \cdots \quad u_n]$ and a column

$$\mathbf{v} = \begin{bmatrix} v_1 \\ v_2 \\ \vdots \\ v_n \end{bmatrix}$$

of the same dimension, the **scalar product** of **u** and **v** (written $\mathbf{u} \cdot \mathbf{v}$) is the number $u_1v_1 + u_2v_2 + \cdots + u_nv_n$.

Given two matrices A and B, the **matrix product** of A and B (written AB) is defined if and only if the number of columns in A = the number of rows in B. Suppose this is the case and A has m rows and B has n columns. Then the matrix product $C = AB$ of A and B is the $m \times n$ matrix C whose ijth element is determined as follows: The ijth element of C = the scalar product of row i of A with column j of B.

Matrices and Linear Equations

The **linear equation system**

$$\begin{aligned}
a_{11}x_1 + a_{12}x_2 + \cdots + a_{1n}x_n &= b_1 \\
a_{21}x_1 + a_{22}x_2 + \cdots + a_{2n}x_n &= b_2 \\
\vdots \qquad \vdots \qquad\qquad \vdots \enspace &= \enspace \vdots \\
a_{m1}x_1 + a_{m2}x_2 + \cdots + a_{mn}x_n &= b_m
\end{aligned}$$

may be written as $A\mathbf{x} = \mathbf{b}$ or $A|\mathbf{b}$, where

$$A = \begin{bmatrix} a_{11} & a_{12} & \cdots & a_{1n} \\ a_{21} & a_{22} & \cdots & a_{2n} \\ \vdots & \vdots & & \vdots \\ a_{m1} & a_{m2} & \dots & a_{mn} \end{bmatrix}, \quad \mathbf{x} = \begin{bmatrix} x_1 \\ x_2 \\ \vdots \\ x_n \end{bmatrix}, \quad \mathbf{b} = \begin{bmatrix} b_1 \\ b_2 \\ \vdots \\ b_m \end{bmatrix}$$

The Gauss–Jordan Method

Using **elementary row operations** (EROs), we may solve any linear equation system. From a matrix A, an ERO yields a new matrix A' via one of three procedures.

Type 1 ERO

Obtain A' by multiplying any row of A by a nonzero scalar.

Type 2 ERO

Multiply any row of A (say, row i) by a nonzero scalar c. For some $j \neq i$, let row j of $A' = c(\text{row } i \text{ of } A) + \text{row } j$ of A, and let the other rows of A' be the same as the rows of A.

Type 3 ERO

Interchange any two rows of A.

The Gauss–Jordan method uses EROs to solve linear equation systems, as shown in the following steps.

Step 1   To solve $A\mathbf{x} = \mathbf{b}$, write down the augmented matrix $A|\mathbf{b}$.

Step 2   Begin with row 1 as the current row, column 1 as the current column, and a_{11} as the current entry. (**a**) If a_{11} (the current entry) is nonzero, then use EROs to transform column 1 (the current column) to

$$\begin{bmatrix} 1 \\ 0 \\ \vdots \\ 0 \end{bmatrix}$$

Then obtain the new current row, column, and entry by moving down one row and one column to the right, and go to step 3. (**b**) If a_{11} (the current entry) equals 0, then do a Type 3 ERO switch with any row with a nonzero value in the same column. Use EROs to transform column 1 to

$$\begin{bmatrix} 1 \\ 0 \\ \vdots \\ 0 \end{bmatrix}$$

and proceed to step 3 after moving into a new current row, column, and entry. (**c**) If there are no nonzero numbers in the first column, then proceed to a new current column and entry. Then go to step 3.

Step 3   (**a**) If the current entry is nonzero, use EROs to transform it to 1 and the rest of the current column's entries to 0. Obtain the new current row, column, and entry. If this is impossible, then stop. Otherwise, repeat step 3. (**b**) If the current entry is 0, then do a Type 3 ERO switch with any row with a nonzero value in the same column. Transform the column using EROs and move to the next current entry. If this is impossible, then stop. Otherwise, repeat step 3. (**c**) If the current column has no nonzero numbers below the current row, then obtain the new current column and entry, and repeat step 3. If it is impossible, then stop.

This procedure may require "passing over" one or more columns without transforming them.

Step 4   Write down the system of equations $A'\mathbf{x} = \mathbf{b}'$ that corresponds to the matrix $A'|\mathbf{b}'$ obtained when step 3 is completed. Then $A'\mathbf{x} = \mathbf{b}'$ will have the same set of solutions as $A\mathbf{x} = \mathbf{b}$.

To describe the set of solutions to $A'\mathbf{x} = \mathbf{b}'$ (and $A\mathbf{x} = \mathbf{b}$), we define the concepts of basic and nonbasic variables. After the Gauss–Jordan method has been applied to any linear system, a variable that appears with a coefficient of 1 in a single equation and a coefficient of 0 in all other equations is called a **basic variable.** Any variable that is not a basic variable is called a **nonbasic variable.**

Let BV be the set of basic variables for $A'\mathbf{x} = \mathbf{b}'$ and NBV be the set of nonbasic variables for $A'\mathbf{x} = \mathbf{b}'$.

Case 1 $A'\mathbf{x} = \mathbf{b}'$ contains at least one row of the form $[0 \quad 0 \quad \cdots \quad 0|c](c \neq 0)$. In this case, $A\mathbf{x} = \mathbf{b}$ has no solution.

Case 2 If Case 1 does not apply and NBV, the set of nonbasic variables, is empty, then $A\mathbf{x} = \mathbf{b}$ will have a unique solution.

Case 3 If Case 1 does not hold and NBV is nonempty, then $A\mathbf{x} = \mathbf{b}$ will have an infinite number of solutions.

Linear Independence, Linear Dependence, and the Rank of a Matrix

A set V of m-dimensional vectors is **linearly independent** if the only linear combination of vectors in V that equals $\mathbf{0}$ is the trivial linear combination. A set V of m-dimensional vectors is **linearly dependent** if there is a nontrivial linear combination of the vectors in V that adds to $\mathbf{0}$.

Let A be any $m \times n$ matrix, and denote the rows of A by $\mathbf{r}_1, \mathbf{r}_2, \ldots, \mathbf{r}_m$. Also define $R = \{\mathbf{r}_1, \mathbf{r}_2, \ldots, \mathbf{r}_m\}$. The **rank** of A is the number of vectors in the largest linearly independent subset of R. To find the rank of a given matrix A, apply the Gauss–Jordan method to the matrix A. Let the final result be the matrix $\bar{A}$. Then rank A = rank $\bar{A}$ = number of nonzero rows in $\bar{A}$.

To determine if a set of vectors $V = \{\mathbf{v}_1, \mathbf{v}_2, \ldots, \mathbf{v}_m\}$ is linearly dependent, form the matrix A whose ith row is $\mathbf{v}_i$. A will have m rows. If rank $A = m$, then V is a linearly independent set of vectors; if rank $A < m$, then V is a linearly dependent set of vectors.

Inverse of a Matrix

For a given square ($m \times m$) matrix A, if $AB = BA = I_m$, then B is the **inverse** of A (written $B = A^{-1}$). The Gauss–Jordan method for inverting an $m \times m$ matrix A to get A^{-1} is as follows:

Step 1 Write down the $m \times 2m$ matrix $A|I_m$.

Step 2 Use EROs to transform $A|I_m$ into $I_m|B$. This will only be possible if rank $A = m$. In this case, $B = A^{-1}$. If rank $A < m$, then A has no inverse.

Determinants

Associated with any square ($m \times m$) matrix A is a number called the **determinant** of A (written det A or $|A|$). For a 1×1 matrix, det $A = a_{11}$. For a 2×2 matrix, det $A = a_{11}a_{22} - a_{21}a_{12}$. For a general $m \times m$ matrix, we can find det A by repeated application of the following formula (valid for $i = 1, 2, \ldots, m$):

$$\det A = (-1)^{i+1}a_{i1}(\det A_{i1}) + (-1)^{i+2}a_{i2}(\det A_{i2}) + \cdots + (-1)^{i+m}a_{im}(\det A_{im})$$

Here A_{ij} is the ijth **minor** of A, which is the $(m-1) \times (m-1)$ matrix obtained from A after deleting the ith row and jth column of A.

REVIEW PROBLEMS

Group A

1 Find all solutions to the following linear system:

$$
\begin{aligned}
x_1 + x_2 \quad\;\; &= 2 \\
x_2 + x_3 &= 3 \\
x_1 + 2x_2 + x_3 &= 5
\end{aligned}
$$

2 Find the inverse of the matrix $\begin{bmatrix} 0 & 3 \\ 2 & 1 \end{bmatrix}$.

3 Each year, 20% of all untenured State University faculty become tenured, 5% quit, and 75% remain untenured. Each year, 90% of all tenured S.U. faculty remain tenured and 10% quit. Let U_t be the number of untenured S.U. faculty at the beginning of year t, and T_t the tenured number.

Use matrix multiplication to relate the vector $\begin{bmatrix} U_{t+1} \\ T_{t+1} \end{bmatrix}$ to the vector $\begin{bmatrix} U_t \\ T_t \end{bmatrix}$.

4 Use the Gauss–Jordan method to determine all solutions to the following linear system:

$$
\begin{aligned}
2x_1 + 3x_2 &= 3 \\
x_1 + x_2 &= 1 \\
x_1 + 2x_2 &= 2
\end{aligned}
$$

5 Find the inverse of the matrix $\begin{bmatrix} 0 & 2 \\ 1 & 3 \end{bmatrix}$.

6 The grades of two students during their last semester at S.U. are shown in Table 2.

Courses 1 and 2 are four-credit courses, and courses 3 and 4 are three-credit courses. Let GPA_i be the semester grade point average for student i. Use matrix multiplication to express the vector $\begin{bmatrix} GPA_1 \\ GPA_2 \end{bmatrix}$ in terms of the information given in the problem.

7 Use the Gauss–Jordan method to find all solutions to the following linear system:

$$
\begin{aligned}
2x_1 + x_2 &= 3 \\
3x_1 + x_2 &= 4 \\
x_1 - x_2 &= 0
\end{aligned}
$$

8 Find the inverse of the matrix $\begin{bmatrix} 2 & 3 \\ 3 & 5 \end{bmatrix}$.

9 Let C_t = number of children in Indiana at the beginning of year t, and A_t = number of adults in Indiana at the beginning of year t. During any given year, 5% of all children become adults, and 1% of all children die. Also, during any given year, 3% of all adults die. Use matrix multiplication to express the vector $\begin{bmatrix} C_{t+1} \\ A_{t+1} \end{bmatrix}$ in terms of $\begin{bmatrix} C_t \\ A_t \end{bmatrix}$.

10 Use the Gauss–Jordan method to find all solutions to the following linear equation system:

$$
\begin{aligned}
x_1 \quad\;\; - x_3 &= 4 \\
x_2 + x_3 &= 2 \\
x_1 + x_2 \quad\;\; &= 5
\end{aligned}
$$

11 Use the Gauss–Jordan method to find the inverse of the matrix $\begin{bmatrix} 1 & 0 & 2 \\ 0 & 1 & 0 \\ 0 & 1 & 1 \end{bmatrix}$.

12 During any given year, 10% of all rural residents move to the city, and 20% of all city residents move to a rural area (all other people stay put!). Let R_t be the number of rural residents at the beginning of year t, and C_t be the number of city residents at the beginning of year t. Use matrix multiplication to relate the vector $\begin{bmatrix} R_{t+1} \\ C_{t+1} \end{bmatrix}$ to the vector $\begin{bmatrix} R_t \\ C_t \end{bmatrix}$.

13 Determine whether the set $V = \{[1 \quad 2 \quad 1], [2 \quad 0 \quad 0]\}$ is a linearly independent set of vectors.

14 Determine whether the set $V = \{[1 \quad 0 \quad 0], [0 \quad 1 \quad 0], [-1 \quad -1 \quad 0]\}$ is a linearly independent set of vectors.

15 Let $A = \begin{bmatrix} a & 0 & 0 & 0 \\ 0 & b & 0 & 0 \\ 0 & 0 & c & 0 \\ 0 & 0 & 0 & d \end{bmatrix}$.

  a For what values of a, b, c, and d will A^{-1} exist?

  b If A^{-1} exists, then find it.

16 Show that the following linear system has an infinite number of solutions:

$$
\begin{bmatrix} 1 & 1 & 0 & 0 \\ 0 & 0 & 1 & 1 \\ 1 & 0 & 1 & 0 \\ 0 & 1 & 0 & 1 \end{bmatrix} \begin{bmatrix} x_1 \\ x_2 \\ x_3 \\ x_4 \end{bmatrix} = \begin{bmatrix} 2 \\ 3 \\ 4 \\ 1 \end{bmatrix}
$$

17 Before paying employee bonuses and state and federal taxes, a company earns profits of $60,000. The company pays employees a bonus equal to 5% of after-tax profits. State tax is 5% of profits (after bonuses are paid). Finally, federal tax is 40% of profits (after bonuses and state tax are paid). Determine a linear equation system to find the amounts paid in bonuses, state tax, and federal tax.

18 Find the determinant of the matrix $A = \begin{bmatrix} 2 & 4 & 6 \\ 1 & 0 & 0 \\ 0 & 0 & 1 \end{bmatrix}$.

19 Show that any 2×2 matrix A that does not have an inverse will have det $A = 0$.

TABLE 2

Student	Course			
	1	2	3	4
1	3.6	3.8	2.6	3.4
2	2.7	3.1	2.9	3.6

Group B

20 Let A be an $m \times m$ matrix.

a Show that if rank $A = m$, then $A\mathbf{x} = \mathbf{0}$ has a unique solution. What is the unique solution?

b Show that if rank $A < m$, then $A\mathbf{x} = \mathbf{0}$ has an infinite number of solutions.

21 Consider the following linear system:

$$[x_1 \quad x_2 \quad \cdots \quad x_n] = [x_1 \quad x_2 \quad \cdots \quad x_n]P$$

where

$$P = \begin{bmatrix} p_{11} & p_{12} & \cdots & p_{1n} \\ p_{21} & p_{22} & \cdots & p_{2n} \\ \vdots & \vdots & & \vdots \\ p_{n1} & p_{n2} & \cdots & p_{nn} \end{bmatrix}$$

If the sum of each row of the P matrix equals 1, then use Problem 20 to show that this linear system has an infinite number of solutions.

22[†] The national economy of Seriland manufactures three products: steel, cars, and machines. (1) To produce \$1 of steel requires 30¢ of steel, 15¢ of cars, and 40¢ of machines. (2) To produce \$1 of cars requires 45¢ of steel, 20¢ of cars, and 10¢ of machines. (3) To produce \$1 of machines requires 40¢ of steel, 10¢ of cars, and 45¢ of machines. During the coming year, Seriland wants to consume d_s dollars of steel, d_c dollars of cars, and d_m dollars of machinery.

For the coming year, let

s = dollar value of steel produced

c = dollar value of cars produced

m = dollar value of machines produced

Define A to be the 3×3 matrix whose ijth element is the dollar value of product i required to produce \$1 of product j (steel = product 1, cars = product 2, machinery = product 3).

a Determine A.

b Show that

$$\begin{bmatrix} s \\ c \\ m \end{bmatrix} = A\begin{bmatrix} s \\ c \\ m \end{bmatrix} + \begin{bmatrix} d_s \\ d_c \\ d_m \end{bmatrix} \qquad \text{(24)}$$

(*Hint:* Observe that the value of next year's steel production = (next year's consumer steel demand) + (steel needed to make next year's steel) + (steel needed to make next year's cars) + (steel needed to make next year's machines). This should give you the general idea.)

c Show that Equation (24) may be rewritten as

$$(I - A)\begin{bmatrix} s \\ c \\ m \end{bmatrix} = \begin{bmatrix} d_s \\ d_c \\ d_m \end{bmatrix}$$

d Given values for d_s, d_c, and d_m, describe how you can use $(I - A)^{-1}$ to determine if Seriland can meet next year's consumer demand.

e Suppose next year's demand for steel increases by \$1. This will increase the value of the steel, cars, and machines that must be produced next year. In terms of $(I - A)^{-1}$, determine the change in next year's production requirements.

REFERENCES

The following references contain more advanced discussions of linear algebra. To understand the theory of linear and nonlinear programming, master at least one of these books:

Dantzig, G. *Linear Programming and Extensions.* Princeton, N.J.: Princeton University Press, 1963.

Hadley, G. *Linear Algebra.* Reading, Mass.: Addison-Wesley, 1961.

Strang, G. *Linear Algebra and Its Applications,* 3d ed. Orlando, Fla.: Academic Press, 1988.

Leontief, W. *Input–Output Economics.* New York: Oxford University Press, 1966.

Teichroew, D. *An Introduction to Management Science: Deterministic Models.* New York: Wiley, 1964. A more extensive discussion of linear algebra than this chapter gives (at a comparable level of difficulty).

[†]Based on Leontief (1966). See references at end of chapter.

3 Introduction to Linear Programming

Linear programming (LP) is a tool for solving optimization problems. In 1947, George Dantzig developed an efficient method, the simplex algorithm, for solving linear programming problems (also called LP). Since the development of the simplex algorithm, LP has been used to solve optimization problems in industries as diverse as banking, education, forestry, petroleum, and trucking. In a survey of Fortune 500 firms, 85% of the respondents said they had used linear programming. As a measure of the importance of linear programming in operations research, approximately 70% of this book will be devoted to linear programming and related optimization techniques.

In Section 3.1, we begin our study of linear programming by describing the general characteristics shared by all linear programming problems. In Sections 3.2 and 3.3, we learn how to solve graphically those linear programming problems that involve only two variables. Solving these simple LPs will give us useful insights for solving more complex LPs. The remainder of the chapter explains how to formulate linear programming models of real-life situations.

3.1 What Is a Linear Programming Problem?

In this section, we introduce linear programming and define important terms that are used to describe linear programming problems.

EXAMPLE 1 | Giapetto's Woodcarving

Giapetto's Woodcarving, Inc., manufactures two types of wooden toys: soldiers and trains. A soldier sells for $27 and uses $10 worth of raw materials. Each soldier that is manufactured increases Giapetto's variable labor and overhead costs by $14. A train sells for $21 and uses $9 worth of raw materials. Each train built increases Giapetto's variable labor and overhead costs by $10. The manufacture of wooden soldiers and trains requires two types of skilled labor: carpentry and finishing. A soldier requires 2 hours of finishing labor and 1 hour of carpentry labor. A train requires 1 hour of finishing and 1 hour of carpentry labor. Each week, Giapetto can obtain all the needed raw material but only 100 finishing hours and 80 carpentry hours. Demand for trains is unlimited, but at most 40 soldiers are bought each week. Giapetto wants to maximize weekly profit (revenues − costs). Formulate a mathematical model of Giapetto's situation that can be used to maximize Giapetto's weekly profit.

Solution In developing the Giapetto model, we explore characteristics shared by all linear programming problems.

Decision Variables We begin by defining the relevant **decision variables.** In any linear programming model, the decision variables should completely describe the decisions to be made (in this case, by Giapetto). Clearly, Giapetto must decide how many soldiers and trains should be manufactured each week. With this in mind, we define

$$x_1 = \text{number of soldiers produced each week}$$

$$x_2 = \text{number of trains produced each week}$$

Objective Function   In any linear programming problem, the decision maker wants to maximize (usually revenue or profit) or minimize (usually costs) some function of the decision variables. The function to be maximized or minimized is called the **objective function.** For the Giapetto problem, we note that *fixed costs* (such as rent and insurance) do not depend on the values of x_1 and x_2. Thus, Giapetto can concentrate on maximizing (weekly revenues) − (raw material purchase costs) − (other variable costs).

Giapetto's weekly revenues and costs can be expressed in terms of the decision variables x_1 and x_2. It would be foolish for Giapetto to manufacture more soldiers than can be sold, so we assume that all toys produced will be sold. Then

$$\text{Weekly revenues} = \text{weekly revenues from soldiers}$$
$$+ \text{ weekly revenues from trains}$$
$$= \left(\frac{\text{dollars}}{\text{soldier}}\right)\left(\frac{\text{soldiers}}{\text{week}}\right) + \left(\frac{\text{dollars}}{\text{train}}\right)\left(\frac{\text{trains}}{\text{week}}\right)$$
$$= 27x_1 + 21x_2$$

Also,

$$\text{Weekly raw material costs} = 10x_1 + 9x_2$$
$$\text{Other weekly variable costs} = 14x_1 + 10x_2$$

Then Giapetto wants to maximize

$$(27x_1 + 21x_2) - (10x_1 + 9x_2) - (14x_1 + 10x_2) = 3x_1 + 2x_2$$

Another way to see that Giapetto wants to maximize $3x_1 + 2x_2$ is to note that

$$\text{Weekly revenues} = \text{weekly contribution to profit from soldiers}$$
$$- \text{ weekly nonfixed costs} + \text{weekly contribution to profit from trains}$$
$$= \left(\frac{\text{contribution to profit}}{\text{soldier}}\right)\left(\frac{\text{soldiers}}{\text{week}}\right)$$
$$+ \left(\frac{\text{contribution to profit}}{\text{train}}\right)\left(\frac{\text{trains}}{\text{week}}\right)$$

Also,

$$\frac{\text{Contribution to profit}}{\text{Soldier}} = 27 - 10 - 14 = 3$$
$$\frac{\text{Contribution to profit}}{\text{Train}} = 21 - 9 - 10 = 2$$

Then, as before, we obtain

$$\text{Weekly revenues} - \text{weekly nonfixed costs} = 3x_1 + 2x_2$$

Thus, Giapetto's objective is to choose x_1 and x_2 to maximize $3x_1 + 2x_2$. We use the variable z to denote the objective function value of any LP. Giapetto's objective function is

$$\text{Maximize } z = 3x_1 + 2x_2 \tag{1}$$

(In the future, we will abbreviate "maximize" by *max* and "minimize" by *min.*) The coefficient of a variable in the objective function is called the **objective function coefficient** of the variable. For example, the objective function coefficient for x_1 is 3, and the objective function coefficient for x_2 is 2. In this example (and in many other problems), the objective function coefficient for each variable is simply the contribution of the variable to the company's profit.

Constraints As x_1 and x_2 increase, Giapetto's objective function grows larger. This means that if Giapetto were free to choose any values for x_1 and x_2, the company could make an arbitrarily large profit by choosing x_1 and x_2 to be very large. Unfortunately, the values of x_1 and x_2 are limited by the following three restrictions (often called **constraints**):

Constraint 1 Each week, no more than 100 hours of finishing time may be used.

Constraint 2 Each week, no more than 80 hours of carpentry time may be used.

Constraint 3 Because of limited demand, at most 40 soldiers should be produced each week.

The amount of raw material available is assumed to be unlimited, so no restrictions have been placed on this.

The next step in formulating a mathematical model of the Giapetto problem is to express Constraints 1–3 in terms of the decision variables x_1 and x_2. To express Constraint 1 in terms of x_1 and x_2, note that

$$\frac{\text{Total finishing hrs.}}{\text{Week}} = \left(\frac{\text{finishing hrs.}}{\text{soldier}}\right)\left(\frac{\text{soldiers made}}{\text{week}}\right)$$
$$+ \left(\frac{\text{finishing hrs.}}{\text{train}}\right)\left(\frac{\text{trains made}}{\text{week}}\right)$$
$$= 2(x_1) + 1(x_2) = 2x_1 + x_2$$

Now Constraint 1 may be expressed by

$$2x_1 + x_2 \leq 100 \tag{2}$$

Note that the units of each term in (2) are finishing hours per week. *For a constraint to be reasonable, all terms in the constraint must have the same units.* Otherwise one is adding apples and oranges, and the constraint won't have any meaning.

To express Constraint 2 in terms of x_1 and x_2, note that

$$\frac{\text{Total carpentry hrs.}}{\text{Week}} = \left(\frac{\text{carpentry hrs.}}{\text{solider}}\right)\left(\frac{\text{soldiers}}{\text{week}}\right)$$
$$+ \left(\frac{\text{carpentry hrs.}}{\text{train}}\right)\left(\frac{\text{trains}}{\text{week}}\right)$$

$$= 1(x_1) + 1(x_2) = x_1 + x_2$$

Then Constraint 2 may be written as

$$x_1 + x_2 \leq 80 \tag{3}$$

Again, note that the units of each term in (3) are the same (in this case, carpentry hours per week).

Finally, we express the fact that at most 40 soldiers per week can be sold by limiting the weekly production of soldiers to at most 40 soldiers. This yields the following constraint:

$$x_1 \leq 40 \tag{4}$$

Thus (2)–(4) express Constraints 1–3 in terms of the decision variables; they are called the *constraints* for the Giapetto linear programming problem. The coefficients of the decision variables in the constraints are called **technological coefficients.** This is because the technological coefficients often reflect the technology used to produce different products. For example, the technological coefficient of x_2 in (3) is 1, indicating that a soldier requires 1 carpentry hour. The number on the right-hand side of each constraint is called the constraint's **right-hand side** (or **rhs**). Often the rhs of a constraint represents the quantity of a resource that is available.

Sign Restrictions To complete the formulation of a linear programming problem, the following question must be answered for each decision variable: Can the decision variable only assume nonnegative values, or is the decision variable allowed to assume both positive and negative values?

If a decision variable x_i can only assume nonnegative values, then we add the **sign restriction** $x_i \geq 0$. If a variable x_i can assume both positive and negative (or zero) values, then we say that x_i is **unrestricted in sign** (often abbreviated **urs**). For the Giapetto problem, it is clear that $x_1 \geq 0$ and $x_2 \geq 0$. In other problems, however, some variables may be urs. For example, if x_i represented a firm's cash balance, then x_i could be considered negative if the firm owed more money than it had on hand. In this case, it would be appropriate to classify x_i as urs. Other uses of urs variables are discussed in Section 4.12.

Combining the sign restrictions $x_1 \geq 0$ and $x_2 \geq 0$ with the objective function (1) and Constraints (2)–(4) yields the following optimization model:

$$\max z = 3x_1 + 2x_2 \quad \text{(Objective function)} \tag{1}$$

subject to (s.t.)

$$2x_1 + x_2 \leq 100 \quad \text{(Finishing constraint)} \tag{2}$$

$$x_1 + x_2 \leq 80 \quad \text{(Carpentry constraint)} \tag{3}$$

$$x_1 \leq 40 \quad \text{(Constraint on demand for soldiers)} \tag{4}$$

$$x_1 \geq 0 \quad \text{(Sign restriction)}^\dagger \tag{5}$$

$$x_2 \geq 0 \quad \text{(Sign restriction)} \tag{6}$$

"Subject to" (s.t.) means that the values of the decision variables x_1 and x_2 must satisfy all constraints and all sign restrictions.

Before formally defining a linear programming problem, we define the concepts of linear function and linear inequality.

DEFINITION ■ A function $f(x_1, x_2, \ldots, x_n)$ of $x_1, x_2, \ldots, x_n$ is a **linear function** if and only if for some set of constants $c_1, c_2, \ldots, c_n$, $f(x_1, x_2, \ldots, x_n) = c_1x_1 + c_2x_2 + \cdots + c_nx_n$. ■

For example, $f(x_1, x_2) = 2x_1 + x_2$ is a linear function of x_1 and x_2, but $f(x_1, x_2) = x_1^2x_2$ is not a linear function of x_1 and x_2.

DEFINITION ■ For any linear function $f(x_1, x_2, \ldots, x_n)$ and any number b, the inequalities $f(x_1, x_2, \ldots, x_n) \leq b$ and $f(x_1, x_2, \ldots, x_n) \geq b$ are **linear inequalities.** ■

Thus, $2x_1 + 3x_2 \leq 3$ and $2x_1 + x_2 \geq 3$ are linear inequalities, but $x_1^2x_2 \geq 3$ is not a linear inequality.

---

†The sign restrictions do constrain the values of the decision variables, but we choose to consider the sign restrictions as being separate from the constraints. The reason for this will become apparent when we study the simplex algorithm in Chapter 4.

A **linear programming problem** (LP) is an optimization problem for which we do the following:

1 We attempt to maximize (or minimize) a *linear* function of the decision variables. The function that is to be maximized or minimized is called the *objective function.*

2 The values of the decision variables must satisfy a set of *constraints.* Each constraint must be a linear equation or linear inequality.

3 A *sign restriction* is associated with each variable. For any variable x_i, the sign restriction specifies that x_i must be either nonnegative ($x_i \geq 0$) or unrestricted in sign (urs). ■

Because Giapetto's objective function is a linear function of x_1 and x_2, and all of Giapetto's constraints are linear inequalities, the Giapetto problem is a linear programming problem. Note that the Giapetto problem is typical of a wide class of linear programming problems in which a decision maker's goal is to maximize profit subject to limited resources.

The Proportionality and Additivity Assumptions

The fact that the objective function for an LP must be a linear function of the decision variables has two implications.

1 The contribution of the objective function from each decision variable is proportional to the value of the decision variable. For example, the contribution to the objective function from making four soldiers ($4 \times 3 = \$12$) is exactly four times the contribution to the objective function from making one soldier (\$3).

2 The contribution to the objective function for any variable is independent of the values of the other decision variables. For example, no matter what the value of x_2, the manufacture of x_1 soldiers will always contribute $3x_1$ dollars to the objective function.

Analogously, the fact that each LP constraint must be a linear inequality or linear equation has two implications.

1 The contribution of each variable to the left-hand side of each constraint is proportional to the value of the variable. For example, it takes exactly three times as many finishing hours ($2 \times 3 = 6$ finishing hours) to manufacture three soldiers as it takes to manufacture one soldier (2 finishing hours).

2 The contribution of a variable to the left-hand side of each constraint is independent of the values of the variable. For example, no matter what the value of x_1, the manufacture of x_2 trains uses x_2 finishing hours and x_2 carpentry hours.

The first implication given in each list is called the **Proportionality Assumption of Linear Programming.** Implication 2 of the first list implies that the value of the objective function is the sum of the contributions from individual variables, and implication 2 of the second list implies that the left-hand side of each constraint is the sum of the contributions from each variable. For this reason, the second implication in each list is called the **Additivity Assumption of Linear Programming.**

For an LP to be an appropriate representation of a real-life situation, the decision variables must satisfy both the Proportionality and Additivity Assumptions. Two other assumptions must also be satisfied before an LP can appropriately represent a real situation: the Divisibility and Certainty Assumptions.

The Divisibility Assumption

The **Divisibility Assumption** requires that each decision variable be allowed to assume fractional values. For example, in the Giapetto problem, the Divisibility Assumption implies that it is acceptable to produce 1.5 soldiers or 1.63 trains. Because Giapetto cannot actually produce a fractional number of trains or soldiers, the Divisibility Assumption is not satisfied in the Giapetto problem. A linear programming problem in which some or all of the variables must be nonnegative integers is called an **integer programming problem.** The solution of integer programming problems is discussed in Chapter 9.

In many situations where divisibility is not present, rounding off each variable in the optimal LP solution to an integer may yield a reasonable solution. Suppose the optimal solution to an LP stated that an auto company should manufacture 150,000.4 compact cars during the current year. In this case, you could tell the auto company to manufacture 150,000 or 150,001 compact cars and be fairly confident that this would reasonably approximate an optimal production plan. On the other hand, if the number of missile sites that the United States should use were a variable in an LP and the optimal LP solution said that 0.4 missile sites should be built, it would make a big difference whether we rounded the number of missile sites down to 0 or up to 1. In this situation, the integer programming methods of Chapter 9 would have to be used, because the number of missile sites is definitely not divisible.

The Certainty Assumption

The **Certainty Assumption** is that each parameter (objective function coefficient, right-hand side, and technological coefficient) is known with certainty. If we were unsure of the exact amount of carpentry and finishing hours required to build a train, the Certainty Assumption would be violated.

Feasible Region and Optimal Solution

Two of the most basic concepts associated with a linear programming problem are feasible region and optimal solution. For defining these concepts, we use the term *point* to mean a specification of the value for each decision variable.

DEFINITION ∎ The **feasible region** for an LP is the set of all points that satisfies all the LP's constraints and sign restrictions. ∎

For example, in the Giapetto problem, the point ($x_1 = 40$, $x_2 = 20$) is in the feasible region. Note that $x_1 = 40$ and $x_2 = 20$ satisfy the constraints (2)–(4) and the sign restrictions (5)–(6):

Constraint (2), $2x_1 + x_2 \leq 100$, is satisfied, because $2(40) + 20 \leq 100$.

Constraint (3), $x_1 + x_2 \leq 80$, is satisfied, because $40 + 20 \leq 80$.

Constraint (4), $x_1 \leq 40$, is satisfied, because $40 \leq 40$.

Restriction (5), $x_1 \geq 0$, is satisfied, because $40 \geq 0$.

Restriction (6), $x_2 \geq 0$, is satisfied, because $20 \geq 0$.

On the other hand, the point ($x_1 = 15$, $x_2 = 70$) is not in the feasible region, because even though $x_1 = 15$ and $x_2 = 70$ satisfy (2), (4), (5), and (6), they fail to satisfy (3): $15 + 70$ is not less than or equal to 80. Any point that is not in an LP's feasible region is said to be an **infeasible point.** As another example of an infeasible point, consider ($x_1 = 40$, $x_2 = -20$). Although this point satisfies all the constraints and the sign restriction (5), it is infeasible because it fails to satisfy the sign restriction (6), $x_2 \geq 0$. The feasible region for the Giapetto problem is the set of possible production plans that Giapetto must consider in searching for the optimal production plan.

DEFINITION ■ For a maximization problem, an **optimal solution** to an LP is a point in the feasible region with the largest objective function value. Similarly, for a minimization problem, an optimal solution is a point in the feasible region with the smallest objective function value. ■

Most LPs have only one optimal solution. However, some LPs have no optimal solution, and some LPs have an infinite number of solutions (these situations are discussed in Section 3.3). In Section 3.2, we show that the unique optimal solution to the Giapetto problem is ($x_1 = 20$, $x_2 = 60$). This solution yields an objective function value of

$$z = 3x_1 + 2x_2 = 3(20) + 2(60) = \$180$$

When we say that ($x_1 = 20$, $x_2 = 60$) is the optimal solution to the Giapetto problem, we are saying that no point in the feasible region has an objective function value that exceeds 180. Giapetto can maximize profit by building 20 soldiers and 60 trains each week. If Giapetto were to produce 20 soldiers and 60 trains each week, the weekly profit would be $180 less weekly fixed costs. For example, if Giapetto's only fixed cost were rent of $100 per week, then weekly profit would be $180 - 100 = \$80$ per week.

PROBLEMS

Group A

1 Farmer Jones must determine how many acres of corn and wheat to plant this year. An acre of wheat yields 25 bushels of wheat and requires 10 hours of labor per week. An acre of corn yields 10 bushels of corn and requires 4 hours of labor per week. All wheat can be sold at $4 a bushel, and all corn can be sold at $3 a bushel. Seven acres of land and 40 hours per week of labor are available. Government regulations require that at least 30 bushels of corn be produced during the current year. Let $x_1 =$ number of acres of corn planted, and $x_2 =$ number of acres of wheat planted. Using these decision variables, formulate an LP whose solution will tell Farmer Jones how to maximize the total revenue from wheat and corn.

2 Answer these questions about Problem 1.
  a Is ($x_1 = 2$, $x_2 = 3$) in the feasible region?
  b Is ($x_1 = 4$, $x_2 = 3$) in the feasible region?
  c Is ($x_1 = 2$, $x_2 = -1$) in the feasible region?
  d Is ($x_1 = 3$, $x_2 = 2$) in the feasible region?

3 Using the variables $x_1 =$ number of bushels of corn produced and $x_2 =$ number of bushels of wheat produced, reformulate Farmer Jones's LP.

4 Truckco manufactures two types of trucks: 1 and 2. Each truck must go through the painting shop and assembly shop. If the painting shop were completely devoted to painting Type 1 trucks, then 800 per day could be painted; if the painting shop were completely devoted to painting Type 2 trucks, then 700 per day could be painted. If the assembly shop were completely devoted to assembling truck 1 engines, then 1,500 per day could be assembled; if the assembly shop were completely devoted to assembling truck 2 engines, then 1,200 per day could be assembled. Each Type 1 truck contributes $300 to profit; each Type 2 truck contributes $500. Formulate an LP that will maximize Truckco's profit.

Group B

5 Why don't we allow an LP to have $<$ or $>$ constraints?

3.2 The Graphical Solution of Two-Variable Linear Programming Problems

Any LP with only two variables can be solved graphically. We always label the variables x_1 and x_2 and the coordinate axes the x_1 and x_2 axes. Suppose we want to graph the set of points that satisfies

$$2x_1 + 3x_2 \leq 6 \tag{7}$$

The same set of points (x_1, x_2) satisfies

$$3x_2 \leq 6 - 2x_1$$

This last inequality may be rewritten as

$$x_2 \leq \tfrac{1}{3}(6 - 2x_1) = 2 - \tfrac{2}{3}x_1 \tag{8}$$

Because moving downward on the graph decreases x_2 (see Figure 1), the set of points that satisfies (8) and (7) lies on or below the line $x_2 = 2 - \tfrac{2}{3}x_1$. This set of points is indicated by darker shading in Figure 1. Note, however, that $x_2 = 2 - \tfrac{2}{3}x_1$, $3x_2 = 6 - 2x_1$, and $2x_1 + 3x_2 = 6$ are all the same line. This means that the set of points satisfying (7) lies on or below the line $2x_1 + 3x_2 = 6$. Similarly, the set of points satisfying $2x_1 + 3x_2 \geq 6$ lies on or above the line $2x_1 + 3x_2 = 6$. (These points are shown by lighter shading in Figure 1.)

Consider a linear inequality constraint of the form $f(x_1, x_2) \geq b$ or $f(x_1, x_2) \leq b$. In general, it can be shown that in two dimensions, the set of points that satisfies a linear inequality includes the points on the line $f(x_1, x_2) = b$, defining the inequality plus all points on one side of the line.

There is an easy way to determine the side of the line for which an inequality such as $f(x_1, x_2) \leq b$ or $f(x_1, x_2) \geq b$ is satisfied. Just choose any point P that does not satisfy the line $f(x_1, x_2) = b$. Determine whether P satisfies the inequality. If it does, then all points on the *same* side as P of $f(x_1, x_2) = b$ will satisfy the inequality. If P does not satisfy the inequality, then all points on the *other* side of $f(x_1, x_2) = b$, which does not contain P, will satisfy the inequality. For example, to determine whether $2x_1 + 3x_2 \geq 6$ is satisfied by points above or below the line $2x_1 + 3x_2 = 6$, we note that $(0, 0)$ does not satisfy $2x_1 + 3x_2 \geq 6$. Because $(0, 0)$ is *below* the line $2x_1 + 3x_2 = 6$, the set of points satisfying $2x_1 + 3x_2 \geq 6$ includes the line $2x_1 + 3x_2 = 6$ and the points *above* the line $2x_1 + 3x_2 = 6$. This agrees with Figure 1.

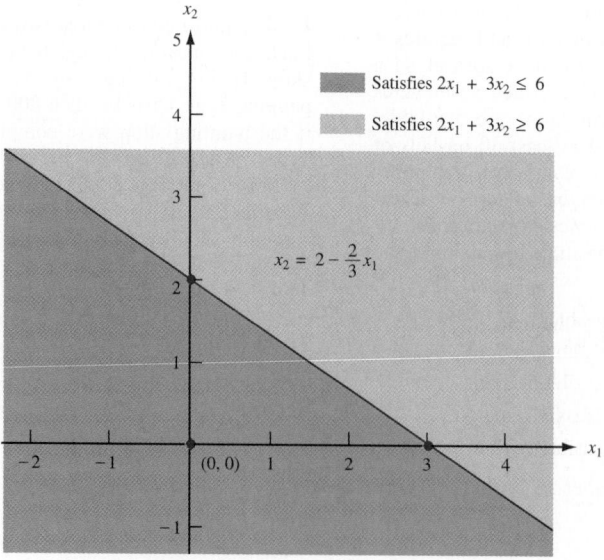

FIGURE 1
Graphing a Linear Inequality

Finding the Feasible Solution

We illustrate how to solve two-variable LPs graphically by solving the Giapetto problem. To begin, we graphically determine the feasible region for Giapetto's problem. The feasible region for the Giapetto problem is the set of all points (x_1, x_2) satisfying

$$2x_1 + x_2 \leq 100 \qquad \text{(Constraints)} \qquad (2)$$
$$x_1 + x_2 \leq 80 \qquad (3)$$
$$x_1 \qquad \leq 40 \qquad (4)$$
$$x_1 \qquad \geq 0 \qquad \text{(Sign restrictions)} \qquad (5)$$
$$x_2 \geq 0 \qquad (6)$$

For a point (x_1, x_2) to be in the feasible region, (x_1, x_2) must satisfy *all* the inequalities (2)–(6). Note that the only points satisfying (5) and (6) lie in the first quadrant of the x_1–x_2 plane. This is indicated in Figure 2 by the arrows pointing to the right from the x_2 axis and upward from the x_1 axis. Thus, any point that is outside the first quadrant cannot be in the feasible region. This means that the feasible region will be the set of points in the first quadrant that satisfies (2)–(4).

Our method for determining the set of points that satisfies a linear inequality will also identify those that meet (2)–(4). From Figure 2, we see that (2) is satisfied by all points below or on the line AB (AB is the line $2x_1 + x_2 = 100$). Inequality (3) is satisfied by all points on or below the line CD (CD is the line $x_1 + x_2 = 80$). Finally, (4) is satisfied by all points on or to the left of line EF (EF is the line $x_1 = 40$). The side of a line that satisfies an inequality is indicated by the direction of the arrows in Figure 2.

From Figure 2, we see that the set of points in the first quadrant that satisfies (2), (3), and (4) is bounded by the five-sided polygon $DGFEH$. Any point on this polygon or in its interior is in the feasible region. Any other point fails to satisfy at least one of the inequalities (2)–(6). For example, the point (40, 30) lies outside $DGFEH$ because it is above the line segment AB. Thus (40, 30) is infeasible, because it fails to satisfy (2).

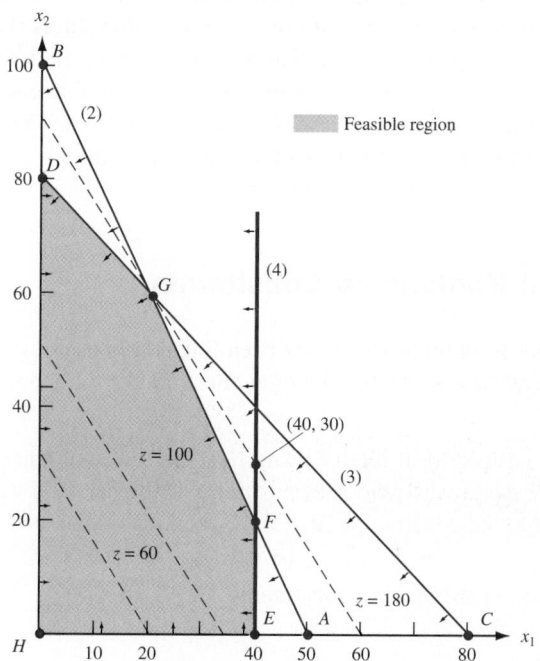

FIGURE 2
Graphical Solution of Giapetto Problem

An easy way to find the feasible region is to determine the set of infeasible points. Note that all points above line AB in Figure 2 are infeasible, because they fail to satisfy (2). Similarly, all points above CD are infeasible, because they fail to satisfy (3). Also, all points to the right of the vertical line EF are infeasible, because they fail to satisfy (4). After these points are eliminated from consideration, we are left with the feasible region ($DGFEH$).

Finding the Optimal Solution

Having identified the feasible region for the Giapetto problem, we now search for the optimal solution, which will be the point in the feasible region with the largest value of $z = 3x_1 + 2x_2$. To find the optimal solution, we need to graph a line on which all points have the same z-value. In a max problem, such a line is called an **isoprofit line** (in a min problem, an **isocost line**). To draw an isoprofit line, choose any point in the feasible region and calculate its z-value. Let us choose $(20, 0)$. For $(20, 0)$, $z = 3(20) + 2(0) = 60$. Thus, $(20, 0)$ lies on the isoprofit line $z = 3x_1 + 2x_2 = 60$. Rewriting $3x_1 + 2x_2 = 60$ as $x_2 = 30 - \frac{3}{2}x_1$, we see that the isoprofit line $3x_1 + 2x_2 = 60$ has a slope of $-\frac{3}{2}$. Because all isoprofit lines are of the form $3x_1 + 2x_2 = $ constant, all isoprofit lines have the same slope. *This means that once we have drawn one isoprofit line, we can find all other isoprofit lines by moving parallel to the isoprofit line we have drawn.*

It is now clear how to find the optimal solution to a two-variable LP. After you have drawn a single isoprofit line, generate other isoprofit lines by moving parallel to the drawn isoprofit line in a direction that increases z (for a max problem). After a point, the isoprofit lines will no longer intersect the feasible region. The last isoprofit line intersecting (touching) the feasible region defines the largest z-value of any point in the feasible region and indicates the optimal solution to the LP. In our problem, the objective function $z = 3x_1 + 2x_2$ will increase if we move in a direction for which both x_1 and x_2 increase. Thus, we construct additional isoprofit lines by moving parallel to $3x_1 + 2x_2 = 60$ in a northeast direction (upward and to the right). From Figure 2, we see that the isoprofit line passing through point G is the last isoprofit line to intersect the feasible region. Thus, G is the point in the feasible region with the largest z-value and is therefore the optimal solution to the Giapetto problem. Note that point G is where the lines $2x_1 + x_2 = 100$ and $x_1 + x_2 = 80$ intersect. Solving these two equations simultaneously, we find that ($x_1 = 20$, $x_2 = 60$) is the optimal solution to the Giapetto problem. The optimal value of z may be found by substituting these values of x_1 and x_2 into the objective function. Thus, the optimal value of z is $z = 3(20) + 2(60) = 180$.

Binding and Nonbinding Constraints

Once the optimal solution to an LP has been found, it is useful (see Chapters 5 and 6) to classify each constraint as being a binding constraint or a nonbinding constraint.

DEFINITION ■ A constraint is **binding** if the left-hand side and the right-hand side of the constraint are equal when the optimal values of the decision variables are substituted into the constraint. ■

Thus, (2) and (3) are binding constraints.

Because $x_1 = 20$ is less than 40, (4) is a nonbinding constraint.

Convex Sets, Extreme Points, and LP

The feasible region for the Giapetto problem is an example of a convex set.

Figure 3 gives four illustrations of this definition. In Figures 3a and 3b, each line segment joining two points in S contains only points in S. Thus, in both these figures, S is convex. In Figures 3c and 3d, S is not convex. In each figure, points A and B are in S, but there are points on the line segment AB that are not contained in S. In our study of linear programming, a certain type of point in a convex set (called an *extreme point*) is of great interest.

For example, in Figure 3a, each point on the circumference of the circle is an extreme point of the circle. In Figure 3b, points A, B, C, and D are extreme points of S. Although point E is on the boundary of S in Figure 3b, E is not an extreme point of S. This is because E lies on the line segment AB (AB lies completely in S), and E is not an endpoint of the line segment AB. Extreme points are sometimes called **corner points,** because if the set S is a polygon, the extreme points of S will be the vertices, or corners, of the polygon.

The feasible region for the Giapetto problem is a convex set. This is no accident: It can be shown that the feasible region for any LP will be a convex set. From Figure 2, we see that the extreme points of the feasible region are simply points D, F, E, G, and H. It can be shown that the feasible region for any LP has only a finite number of extreme points. Also note that the optimal solution to the Giapetto problem (point G) is an extreme point of the feasible region. It can be shown that *any LP that has an optimal solution has an extreme point that is optimal.* This result is very important, because it reduces the set of points that yield an optimal solution from the entire feasible region (which generally contains an *infinite* number of points) to the set of extreme points (a *finite* set).

S = shaded area

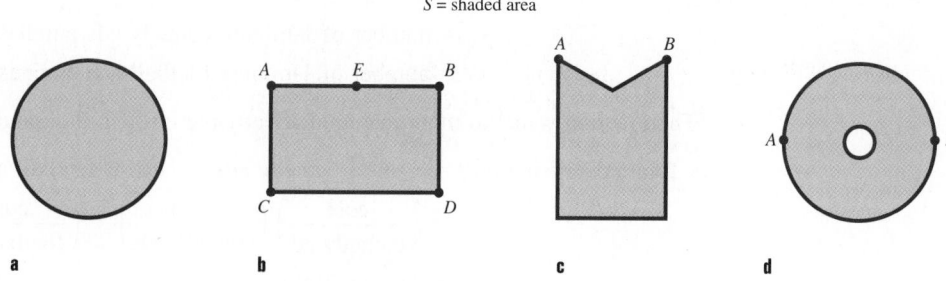

FIGURE 3
Convex and Nonconvex Sets

a   b   c   d

For the Giapetto problem, it is easy to see why the optimal solution must be an extreme point of the feasible region. We note that z increases as we move isoprofit lines in a northeast direction, so the largest z-value in the feasible region must occur at some point P that has no points in the feasible region northeast of P. This means that the optimal solution must lie somewhere on the boundary of the feasible region $DGFEH$. The LP must have an extreme point that is optimal, because for any line segment on the boundary of the feasible region, the largest z-value on that line segment must be assumed at one of the endpoints of the line segment.

To see this, look at the line segment FG in Figure 2. FG is part of the line $2x_1 + x_2 = 100$ and has a slope of -2. If we move along FG and decrease x_1 by 1, then x_2 will increase by 2, and the value of z changes as follows: $3x_1$ goes down by $3(1) = 3$, and $2x_2$ goes up by $2(2) = 4$. Thus, in total, z increases by $4 - 3 = 1$. This means that moving along FG in a direction of decreasing x_1 increases z. Thus, the value of z at point G must exceed the value of z at any other point on the line segment FG.

A similar argument shows that for any objective function, the maximum value of z on a given line segment must occur at an endpoint of the line segment. Therefore, for any LP, the largest z-value in the feasible region must be attained at an endpoint of one of the line segments forming the boundary of the feasible region. In short, one of the extreme points of the feasible region must be optimal. (To test your understanding, show that if Giapetto's objective function were $z = 6x_1 + x_2$, point F would be optimal, whereas if Giapetto's objective function were $z = x_1 + 6x_2$, point D would be optimal.)

Our proof that an LP always has an optimal extreme point depended heavily on the fact that both the objective function and the constraints were linear functions. In Chapter 11, we show that for an optimization problem in which the objective function or some of the constraints are not linear, the optimal solution to the optimization problem may not occur at an extreme point.

The Graphical Solution of Minimization Problems

EXAMPLE 2   Dorian Auto

Dorian Auto manufactures luxury cars and trucks. The company believes that its most likely customers are high-income women and men. To reach these groups, Dorian Auto has embarked on an ambitious TV advertising campaign and has decided to purchase 1-minute commercial spots on two types of programs: comedy shows and football games. Each comedy commercial is seen by 7 million high-income women and 2 million high-income men. Each football commercial is seen by 2 million high-income women and 12 million high-income men. A 1-minute comedy ad costs \$50,000, and a 1-minute football ad costs \$100,000. Dorian would like the commercials to be seen by at least 28 million high-income women and 24 million high-income men. Use linear programming to determine how Dorian Auto can meet its advertising requirements at minimum cost.

Solution   Dorian must decide how many comedy and football ads should be purchased, so the decision variables are

$$x_1 = \text{number of 1-minute comedy ads purchased}$$
$$x_2 = \text{number of 1-minute football ads purchased}$$

Then Dorian wants to minimize total advertising cost (in thousands of dollars).

Total advertising cost = cost of comedy ads + cost of football ads

$$= \left(\frac{\text{cost}}{\text{comedy ad}}\right)\left(\frac{\text{total}}{\text{comedy ads}}\right) + \left(\frac{\text{cost}}{\text{football ad}}\right)\left(\frac{\text{total}}{\text{football ads}}\right)$$
$$= 50x_1 + 100x_2$$

Thus, Dorian's objective function is

$$\min z = 50x_1 + 100x_2 \tag{9}$$

Dorian faces the following constraints:

Constraint 1 Commercials must reach at least 28 million high-income women.

Constraint 2 Commercials must reach at least 24 million high-income men.

To express Constraints 1 and 2 in terms of x_1 and x_2, let HIW stand for high-income women viewers and HIM stand for high-income men viewers (in millions).

$$\text{HIW} = \left(\frac{\text{HIW}}{\text{comedy ad}}\right)\left(\begin{array}{c}\text{total}\\\text{comedy ads}\end{array}\right) + \left(\frac{\text{HIW}}{\text{football ad}}\right)\left(\begin{array}{c}\text{total}\\\text{football ads}\end{array}\right)$$
$$= 7x_1 + 2x_2$$
$$\text{HIM} = \left(\frac{\text{HIM}}{\text{comedy ad}}\right)\left(\begin{array}{c}\text{total}\\\text{comedy ads}\end{array}\right) + \left(\frac{\text{HIM}}{\text{football ad}}\right)\left(\begin{array}{c}\text{total}\\\text{football ads}\end{array}\right)$$
$$= 2x_1 + 12x_2$$

Constraint 1 may now be expressed as

$$7x_1 + 2x_2 \geq 28 \tag{10}$$

and Constraint 2 may be expressed as

$$2x_1 + 12x_2 \geq 24 \tag{11}$$

The sign restrictions $x_1 \geq 0$ and $x_2 \geq 0$ are necessary, so the Dorian LP is given by:

$$\min z = 50x_1 + 100x_2$$
$$\text{s.t.} \quad 7x_1 + 2x_2 \geq 28 \quad \text{(HIW)}$$
$$2x_1 + 12x_2 \geq 24 \quad \text{(HIM)}$$
$$x_1, x_2 \geq 0$$

This problem is typical of a wide range of LP applications in which a decision maker wants to minimize the cost of meeting a certain set of requirements. To solve this LP graphically, we begin by graphing the feasible region (Figure 4). Note that (10) is satisfied by points on or above the line AB (AB is part of the line $7x_1 + 2x_2 = 28$) and that

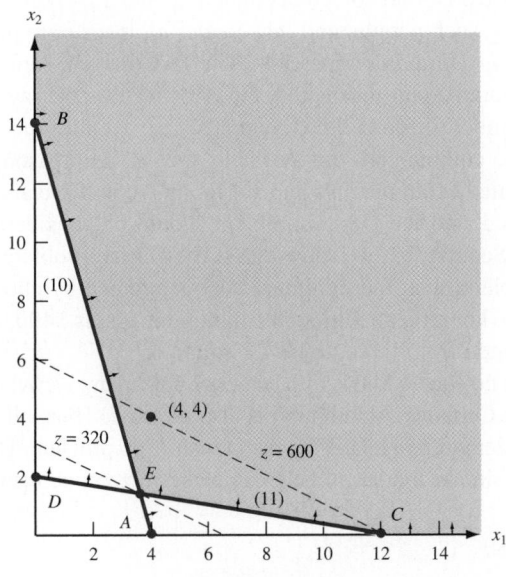

FIGURE 4
Graphical Solution of
Dorian Problem

(11) is satisfied by the points on or above the line CD (CD is part of the line $2x_1 + 12x_2 = 24$). From Figure 4, we see that the only first-quadrant points satisfying both (10) and (11) are the points in the shaded region bounded by the x_1 axis, CEB, and the x_2 axis.

Like the Giapetto problem, the Dorian problem has a convex feasible region, but the feasible region for Dorian, unlike Giapetto's, contains points for which the value of at least one variable can assume arbitrarily large values. Such a feasible region is called an **unbounded feasible region.**

Because Dorian wants to minimize total advertising cost, the optimal solution to the problem is the point in the feasible region with the *smallest z*-value. To find the optimal solution, we need to draw an *isocost line* that intersects the feasible region. An isocost line is any line on which all points have the same z-value (or same cost). We arbitrarily choose the isocost line passing through the point ($x_1 = 4$, $x_2 = 4$). For this point, $z = 50(4) + 100(4) = 600$, and we graph the isocost line $z = 50x_1 + 100x_2 = 600$.

We consider lines parallel to the isocost line $50x_1 + 100x_2 = 600$ in the direction of decreasing z (southwest). The last point in the feasible region that intersects an isocost line will be the point in the feasible region having the *smallest z*-value. From Figure 4, we see that point E has the smallest z-value of any point in the feasible region; this is the optimal solution to the Dorian problem. Note that point E is where the lines $7x_1 + 2x_2 = 28$ and $2x_1 + 12x_2 = 24$ intersect. Simultaneously solving these equations yields the optimal solution ($x_1 = 3.6$, $x_2 = 1.4$). The optimal z-value can then be found by substituting these values of x_1 and x_2 into the objective function. Thus, the optimal z-value is $z = 50(3.6) + 100(1.4) = 320 = \$320,000$. Because at point E both the HIW and HIM constraints are satisfied with equality, both constraints are binding.

Does the Dorian model meet the four assumptions of linear programming outlined in Section 3.1?

For the Proportionality Assumption to be valid, each extra comedy commercial must add exactly 7 million HIW and 2 million HIM. This contradicts empirical evidence, which indicates that after a certain point advertising yields diminishing returns. After, say, 500 auto commercials have been aired, most people have probably seen one, so it does little good to air more commercials. Thus, the Proportionality Assumption is violated.

We used the Additivity Assumption to justify writing (total HIW viewers) = (HIW viewers from comedy ads) + (HIW viewers from football ads). In reality, many of the same people will see a Dorian comedy commercial and a Dorian football commercial. We are double-counting such people, and this creates an inaccurate picture of the total number of people seeing Dorian commercials. The fact that the same person may see more than one type of commercial means that the effectiveness of, say, a comedy commercial depends on the number of football commercials. This violates the Additivity Assumption.

If only 1-minute commercials are available, then it is unreasonable to say that Dorian should buy 3.6 comedy commercials and 1.4 football commercials, so the Divisibility Assumption is violated, and the Dorian problem should be considered an integer programming problem. In Section 9.3, we show that if the Dorian problem is solved as an integer programming problem, then the minimum cost is attained by choosing ($x_1 = 6$, $x_2 = 1$) or ($x_1 = 4$, $x_2 = 2$). For either solution, the minimum cost is \$400,000. This is 25% higher than the cost obtained from the optimal LP solution.

Because there is no way to know with certainty how many viewers are added by each type of commercial, the Certainty Assumption is also violated. Thus, all the assumptions of linear programming seem to be violated by the Dorian Auto problem. Despite these drawbacks, analysts have used similar models to help companies determine their optimal media mix.[†]

[†]Lilien and Kotler (1983).

PROBLEMS

Group A

1 Graphically solve Problem 1 of Section 3.1.

2 Graphically solve Problem 4 of Section 3.1.

3 Leary Chemical manufactures three chemicals: A, B, and C. These chemicals are produced via two production processes: 1 and 2. Running process 1 for an hour costs $4 and yields 3 units of A, 1 of B, and 1 of C. Running process 2 for an hour costs $1 and produces 1 unit of A and 1 of B. To meet customer demands, at least 10 units of A, 5 of B, and 3 of C must be produced daily. Graphically determine a daily production plan that minimizes the cost of meeting Leary Chemical's daily demands.

4 For each of the following, determine the direction in which the objective function increases:

   a $z = 4x_1 - x_2$

   b $z = -x_1 + 2x_2$

   c $z = -x_1 - 3x_2$

5 Furnco manufactures desks and chairs. Each desk uses 4 units of wood, and each chair uses 3. A desk contributes $40 to profit, and a chair contributes $25. Marketing restrictions require that the number of chairs produced be at least twice the number of desks produced. If 20 units of wood are available, formulate an LP to maximize Furnco's profit. Then graphically solve the LP.

6 Farmer Jane owns 45 acres of land. She is going to plant each with wheat or corn. Each acre planted with wheat yields $200 profit; each with corn yields $300 profit. The labor and fertilizer used for each acre are given in Table 1. One hundred workers and 120 tons of fertilizer are available. Use linear programming to determine how Jane can maximize profits from her land.

TABLE 1

	Wheat	Corn
Labor	3 workers	2 workers
Fertilizer	2 tons	4 tons

3.3 Special Cases

The Giapetto and Dorian problems each had a unique optimal solution. In this section, we encounter three types of LPs that do not have unique optimal solutions.

1 Some LPs have an infinite number of optimal solutions (*alternative* or *multiple optimal solutions*).

2 Some LPs have no feasible solutions (*infeasible* LPs).

3 Some LPs are *unbounded:* There are points in the feasible region with arbitrarily large (in a max problem) z-values.

Alternative or Multiple Optimal Solutions

EXAMPLE 3   **Alternative Optimal Solutions**

An auto company manufactures cars and trucks. Each vehicle must be processed in the paint shop and body assembly shop. If the paint shop were only painting trucks, then 40 per day could be painted. If the paint shop were only painting cars, then 60 per day could be painted. If the body shop were only producing cars, then it could process 50 per day. If the body shop were only producing trucks, then it could process 50 per day. Each truck contributes $300 to profit, and each car contributes $200 to profit. Use linear programming to determine a daily production schedule that will maximize the company's profits.

Solution   The company must decide how many cars and trucks should be produced daily. This leads us to define the following decision variables:

$$x_1 = \text{number of trucks produced daily}$$
$$x_2 = \text{number of cars produced daily}$$

The company's daily profit (in hundreds of dollars) is $3x_1 + 2x_2$, so the company's objective function may be written as

$$\max z = 3x_1 + 2x_2 \qquad (12)$$

The company's two constraints are the following:

Constraint 1 The fraction of the day during which the paint shop is busy is less than or equal to 1.

Constraint 2 The fraction of the day during which the body shop is busy is less than or equal to 1.

We have

$$\text{Fraction of day paint shop works on trucks} = \left(\frac{\text{fraction of day}}{\text{truck}}\right)\left(\frac{\text{trucks}}{\text{day}}\right)$$

$$= \tfrac{1}{40}\, x_1$$

Fraction of day paint shop works on cars $= \tfrac{1}{60}\, x_2$

Fraction of day body shop works on trucks $= \tfrac{1}{50}\, x_1$

Fraction of day body shop works on cars $= \tfrac{1}{50}\, x_2$

Thus, Constraint 1 may be expressed by

$$\frac{1}{40}\, x_1 + \frac{1}{60}\, x_2 \le 1 \qquad \text{(Paint shop constraint)} \qquad (13)$$

and Constraint 2 may be expressed by

$$\frac{1}{50}\, x_1 + \frac{1}{50}\, x_2 \le 1 \qquad \text{(Body shop constraint)} \qquad (14)$$

Because $x_1 \ge 0$ and $x_2 \ge 0$ must hold, the relevant LP is

$$\max z = 3x_1 + 2x_2 \qquad (12)$$

$$\text{s.t.} \quad \frac{1}{40}\, x_1 + \frac{1}{60}\, x_2 \le 1 \qquad (13)$$

$$\frac{1}{50}\, x_1 + \frac{1}{50}\, x_2 \le 1 \qquad (14)$$

$$x_1, x_2 \ge 0$$

The feasible region for this LP is the shaded region in Figure 5 bounded by $AEDF$.[†]

For our isoprofit line, we choose the line passing through the point (20, 0). Because (20, 0) has a z-value of $3(20) + 2(0) = 60$, this yields the isoprofit line $z = 3x_1 + 2x_2 = 60$. Examining lines parallel to this isoprofit line in the direction of increasing z (northeast), we find that the last "point" in the feasible region to intersect an isoprofit line is the *entire* line segment AE. This means that any point on the line segment AE is optimal. We can use any point on AE to determine the optimal z-value. For example, point A, (40, 0), gives $z = 3(40) = 120$.

In summary, the auto company's LP has an infinite number of optimal solutions, or *multiple* or *alternative optimal solutions*. This is indicated by the fact that as an isoprofit

[†]Constraint (13) is satisfied by all points on or below AB (AB is $\tfrac{1}{40}\, x_1 + \tfrac{1}{60}\, x_2 = 1$), and (14) is satisfied by all points on or below CD (CD is $\tfrac{1}{50}\, x_1 + \tfrac{1}{50}\, x_2 = 1$).

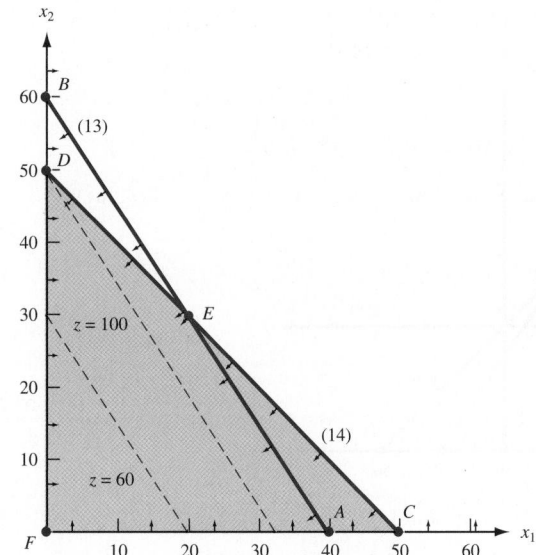

FIGURE 5
Graphical Solution of
Example 3

line leaves the feasible region, it will intersect an entire line segment corresponding to the binding constraint (in this case, AE).

From our current example, it seems reasonable (and can be shown to be true) that if two points (A and E here) are optimal, then *any* point on the line segment joining these two points will also be optimal.

If an alternative optimum occurs, then the decision maker can use a secondary criterion to choose between optimal solutions. The auto company's managers might prefer point A because it would simplify their business (and still allow them to maximize profits) by allowing them to produce only one type of product (trucks).

The technique of **goal programming** (see Section 4.14) is often used to choose among alternative optimal solutions.

Infeasible LP

It is possible for an LP's feasible region to be empty (contain no points), resulting in an *infeasible* LP. Because the optimal solution to an LP is the best point in the feasible region, an infeasible LP has no optimal solution.

EXAMPLE 4   **Infeasible LP**

Suppose that auto dealers require that the auto company in Example 3 produce at least 30 trucks and 20 cars. Find the optimal solution to the new LP.

Solution   After adding the constraints $x_1 \geq 30$ and $x_2 \geq 20$ to the LP of Example 3, we obtain the following LP:

$$\max z = 3x_1 + 2x_2$$

$$\text{s.t.} \quad \frac{1}{40} x_1 + \frac{1}{60} x_2 \leq 1 \tag{15}$$

$$\frac{1}{50} x_1 + \frac{1}{50} x_2 \leq 1 \tag{16}$$

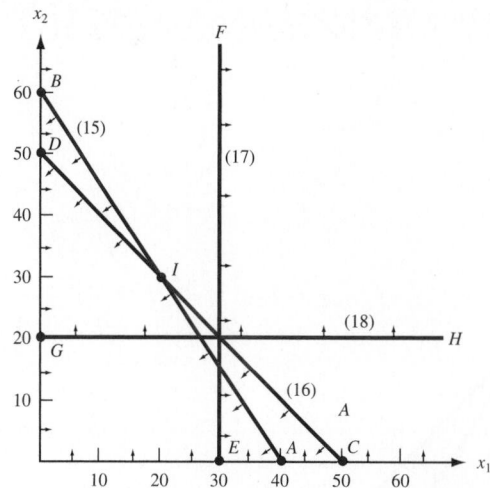

FIGURE 6
An Empty Feasible Region (Infeasible LP)

$$x_1 \qquad \geq 30 \qquad \qquad \text{(17)}$$
$$x_2 \geq 20 \qquad \qquad \text{(18)}$$
$$x_1, x_2 \geq 0$$

The graph of the feasible region for this LP is Figure 6.

Constraint (15) is satisfied by all points on or below AB (AB is $\frac{1}{40}x_1 + \frac{1}{60}x_2 = 1$).

Constraint (16) is satisfied by all points on or below CD (CD is $\frac{1}{50}x_1 + \frac{1}{50}x_2 = 1$).

Constraint (17) is satisfied by all points on or to the right of EF (EF is $x_1 = 30$).

Constraint (18) is satisfied by all points on or above GH (GH is $x_2 = 20$).

From Figure 6 it is clear that no point satisfies all of (15)–(18). This means that Example 4 has an empty feasible region and is an infeasible LP.

In Example 4, the LP is infeasible because producing 30 trucks and 20 cars requires more paint shop time than is available.

Unbounded LP

Our next special LP is an *unbounded* LP. For a max problem, an unbounded LP occurs if it is possible to find points in the feasible region with arbitrarily large z-values, which corresponds to a decision maker earning arbitrarily large revenues or profits. This would indicate that an unbounded optimal solution should not occur in a correctly formulated LP. Thus, if the reader ever solves an LP on the computer and finds that the LP is unbounded, then an error has probably been made in formulating the LP or in inputting the LP into the computer.

For a minimization problem, an LP is unbounded if there are points in the feasible region with arbitrarily small z-values. When graphically solving an LP, we can spot an unbounded LP as follows: A max problem is unbounded if, when we move parallel to our original isoprofit line in the direction of increasing z, we never entirely leave the feasible region. A minimization problem is unbounded if we never leave the feasible region when moving in the direction of decreasing z.

EXAMPLE 5 **Unbounded LP**

Graphically solve the following LP:

$$\max z = 2x_1 - x_2$$
$$\text{s.t.} \quad x_1 - x_2 \le 1 \tag{19}$$
$$2x_1 + x_2 \ge 6 \tag{20}$$
$$x_1, x_2 \ge 0$$

Solution From Figure 7, we see that (19) is satisfied by all points on or above AB (AB is the line $x_1 - x_2 = 1$). Also, (20) is satisfied by all points on or above CD (CD is $2x_1 + x_2 = 6$). Thus, the feasible region for Example 5 is the (shaded) unbounded region in Figure 7, which is bounded only by the x_2 axis, line segment DE, and the part of line AB beginning at E. To find the optimal solution, we draw the isoprofit line passing through $(2, 0)$. This isoprofit line has $z = 2x_1 - x_2 = 2(2) - 0 = 4$. The direction of increasing z is to the southeast (this makes x_1 larger and x_2 smaller). Moving parallel to $z = 2x_1 - x_2$ in a southeast direction, we see that any isoprofit line we draw will intersect the feasible region. (This is because any isoprofit line is steeper than the line $x_1 - x_2 = 1$.)

Thus, there are points in the feasible region that have arbitrarily large z-values. For example, if we wanted to find a point in the feasible region that had $z \ge 1,000,000$, we could choose any point in the feasible region that is southeast of the isoprofit line $z = 1,000,000$.

From the discussion in the last two sections, we see that every LP with two variables must fall into one of the following four cases:

Case 1 The LP has a unique optimal solution.

Case 2 The LP has alternative or multiple optimal solutions: Two or more extreme points are optimal, and the LP will have an infinite number of optimal solutions.

Case 3 The LP is infeasible: The feasible region contains no points.

Case 4 The LP is unbounded: There are points in the feasible region with arbitrarily large z-values (max problem) or arbitrarily small z-values (min problem).

In Chapter 4, we show that every LP (not just LPs with two variables) must fall into one of Cases 1–4.

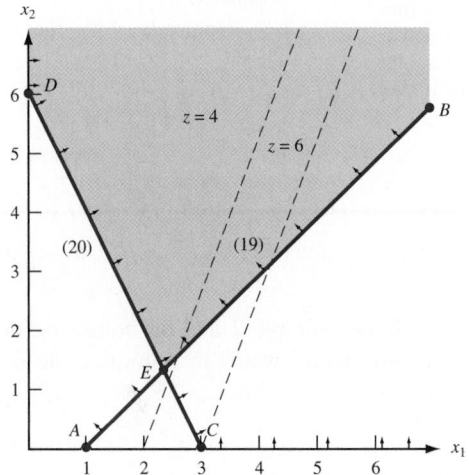

FIGURE 7
An Unbounded LP

In the rest of this chapter, we lead the reader through the formulation of several more complicated linear programming models. The most important step in formulating an LP model is the proper choice of decision variables. If the decision variables have been properly chosen, the objective function and constraints should follow without much difficulty. Trouble in determining an LP's objective function and constraints is usually the result of an incorrect choice of decision variables.

PROBLEMS

Group A

Identify which of Cases 1–4 apply to each of the following LPs:

1
$$\max z = x_1 + x_2$$
$$\text{s.t.} \quad x_1 + x_2 \le 4$$
$$x_1 - x_2 \ge 5$$
$$x_1, x_2 \ge 0$$

2
$$\max z = 4x_1 + x_2$$
$$\text{s.t.} \quad 8x_1 + 2x_2 \le 16$$
$$5x_1 + 2x_2 \le 12$$
$$x_1, x_2 \ge 0$$

3
$$\max z = -x_1 + 3x_2$$
$$\text{s.t.} \quad x_1 - x_2 \le 4$$
$$x_1 + 2x_2 \ge 4$$
$$x_1, x_2 \ge 0$$

4
$$\max z = 3x_1 + x_2$$
$$\text{s.t.} \quad 2x_1 + x_2 \le 6$$
$$x_1 + 3x_2 \le 9$$
$$x_1, x_2 \ge 0$$

5 True or false: For an LP to be unbounded, the LP's feasible region must be unbounded.

6 True or false: Every LP with an unbounded feasible region has an unbounded optimal solution.

7 If an LP's feasible region is not unbounded, we say the LP's feasible region is bounded. Suppose an LP has a bounded feasible region. Explain why you can find the optimal solution to the LP (without an isoprofit or isocost line) by simply checking the z-values at each of the feasible region's extreme points. Why might this method fail if the LP's feasible region is unbounded?

8 Graphically find all optimal solutions to the following LP:
$$\min z = x_1 - x_2$$
$$\text{s.t.} \quad x_1 + x_2 \le 6$$
$$x_1 - x_2 \ge 0$$
$$x_2 - x_1 \ge 3$$
$$x_1, x_2 \ge 0$$

9 Graphically determine two optimal solutions to the following LP:
$$\min z = 3x_1 + 5x_2$$
$$\text{s.t.} \quad 3x_1 + 2x_2 \ge 36$$
$$3x_1 + 5x_2 \ge 45$$
$$x_1, x_2 \ge 0$$

Group B

10 Money manager Boris Milkem deals with French currency (the franc) and American currency (the dollar). At 12 midnight, he can buy francs by paying .25 dollars per franc and dollars by paying 3 francs per dollar. Let $x_1 =$ number of dollars bought (by paying francs) and $x_2 =$ number of francs bought (by paying dollars). Assume that both types of transactions take place simultaneously, and the only constraint is that at 12:01 A.M. Boris must have a nonnegative number of francs and dollars.

a Formulate an LP that enables Boris to maximize the number of dollars he has after all transactions are completed.

b Graphically solve the LP and comment on the answer.

3.4 A Diet Problem

Many LP formulations (such as Example 2 and the following diet problem) arise from situations in which a decision maker wants to minimize the cost of meeting a set of requirements.

EXAMPLE 6 **Diet Problem**

My diet requires that all the food I eat come from one of the four "basic food groups" (chocolate cake, ice cream, soda, and cheesecake). At present, the following four foods are available for consumption: brownies, chocolate ice cream, cola, and pineapple cheesecake. Each brownie costs 50¢, each scoop of chocolate ice cream costs 20¢, each bottle of cola costs 30¢, and each piece of pineapple cheesecake costs 80¢. Each day, I must ingest at least 500 calories, 6 oz of chocolate, 10 oz of sugar, and 8 oz of fat. The nutritional content per unit of each food is shown in Table 2. Formulate a linear programming model that can be used to satisfy my daily nutritional requirements at minimum cost.

Solution As always, we begin by determining the decisions that must be made by the decision maker: how much of each type of food should be eaten daily. Thus, we define the decision variables:

$$x_1 = \text{number of brownies eaten daily}$$
$$x_2 = \text{number of scoops of chocolate ice cream eaten daily}$$
$$x_3 = \text{bottles of cola drunk daily}$$
$$x_4 = \text{pieces of pineapple cheesecake eaten daily}$$

My objective is to minimize the cost of my diet. The total cost of any diet may be determined from the following relation: (total cost of diet) = (cost of brownies) + (cost of ice cream) + (cost of cola) + (cost of cheesecake). To evaluate the total cost of a diet, note that, for example,

$$\text{Cost of cola} = \left(\frac{\text{cost}}{\text{bottle of cola}}\right)\left(\frac{\text{bottles of}}{\text{cola drunk}}\right) = 30x_3$$

Applying this to the other three foods, we have (in cents)

$$\text{Total cost of diet} = 50x_1 + 20x_2 + 30x_3 + 80x_4$$

Thus, the objective function is

$$\min z = 50x_1 + 20x_2 + 30x_3 + 80x_4$$

The decision variables must satisfy the following four constraints:

Constraint 1 Daily calorie intake must be at least 500 calories.

Constraint 2 Daily chocolate intake must be at least 6 oz.

Constraint 3 Daily sugar intake must be at least 10 oz.

Constraint 4 Daily fat intake must be at least 8 oz.

TABLE 2
Nutritional Values for Diet

Type of Food	Calories	Chocolate (Ounces)	Sugar (Ounces)	Fat (Ounces)
Brownie	400	3	2	2
Chocolate ice cream (1 scoop)	200	2	2	4
Cola (1 bottle)	150	0	4	1
Pineapple cheesecake (1 piece)	500	0	4	5

To express Constraint 1 in terms of the decision variables, note that (daily calorie intake) = (calories in brownies) + (calories in chocolate ice cream) + (calories in cola) + (calories in pineapple cheesecake).

The calories in the brownies consumed can be determined from

$$\text{Calories in brownies} = \left(\frac{\text{calories}}{\text{brownie}}\right)\left(\begin{array}{c}\text{brownies}\\\text{eaten}\end{array}\right) = 400x_1$$

Applying similar reasoning to the other three foods shows that

$$\text{Daily calorie intake} = 400x_1 + 200x_2 + 150x_3 + 500x_4$$

Constraint 1 may be expressed by

$$400x_1 + 200x_2 + 150x_3 + 500x_4 \geq 500 \qquad \text{(Calorie constraint)} \qquad (21)$$

Constraint 2 may be expressed by

$$3x_1 + 2x_2 \geq 6 \qquad \text{(Chocolate constraint)} \qquad (22)$$

Constraint 3 may be expressed by

$$2x_1 + 2x_2 + 4x_3 + 4x_4 \geq 10 \qquad \text{(Sugar constraint)} \qquad (23)$$

Constraint 4 may be expressed by

$$2x_1 + 4x_2 + x_3 + 5x_4 \geq 8 \qquad \text{(Fat constraint)} \qquad (24)$$

Finally, the sign restrictions $x_i \geq 0$ ($i = 1, 2, 3, 4$) must hold.

Combining the objective function, constraints (21)–(24), and the sign restrictions yields the following:

$$\min z = 50x_1 + 20x_2 + 30x_3 + 80x_4$$

s.t.						
$400x_1 +$	$200x_2 +$	$150x_3 +$	$500x_4 \geq 500$	(Calorie constraint)	(21)
$3x_1 +$	$2x_2$		≥ 6	(Chocolate constraint)	(22)
$2x_1 +$	$2x_2 +$	$4x_3 +$	$4x_4 \geq 10$	(Sugar constraint)	(23)
$2x_1 +$	$4x_2 +$	$x_3 +$	$5x_4 \geq 8$	(Fat constraint)	(24)
	$x_i \geq 0$ ($i = 1, 2, 3, 4$)			(Sign restrictions)	

The optimal solution to this LP is $x_1 = x_4 = 0$, $x_2 = 3$, $x_3 = 1$, $z = 90$. Thus, the minimum-cost diet incurs a daily cost of 90¢ by eating three scoops of chocolate ice cream and drinking one bottle of cola. The optimal z-value may be obtained by substituting the optimal value of the decision variables into the objective function. This yields a total cost of $z = 3(20) + 1(30) = 90$¢. The optimal diet provides

$$200(3) + 150(1) = 750 \text{ calories}$$
$$2(3) = 6 \text{ oz of chocolate}$$
$$2(3) + 4(1) = 10 \text{ oz of sugar}$$
$$4(3) + 1(1) = 13 \text{ oz of fat}$$

Thus, the chocolate and sugar constraints are binding, but the calories and fat constraints are nonbinding.

A version of the diet problem with a more realistic list of foods and nutritional requirements was one of the first LPs to be solved by computer. Stigler (1945) proposed a diet problem in which 77 types of food were available and 10 nutritional requirements (vitamin A, vitamin C, and so on) had to be satisfied. When solved by computer, the optimal solution yielded a diet consisting of corn meal, wheat flour, evaporated milk, peanut butter, lard, beef, liver, potatoes, spinach, and cabbage. Although such a diet is clearly high in vital nutrients, few people would be satisfied with it because it does not seem to meet a minimum standard of tastiness (and Stigler required that the same diet be eaten each day). The optimal solution to any LP model will reflect only those aspects of reality that are captured by the objective function and constraints. Stigler's (and our) formulation of the diet problem did not reflect people's desire for a tasty and varied diet. Integer programming has been used to plan institutional menus for a weekly or monthly period.[†] Menu-planning models do contain constraints that reflect tastiness and variety requirements.

PROBLEMS

Group A

1 There are three factories on the Momiss River (1, 2, and 3). Each emits two types of pollutants (1 and 2) into the river. If the waste from each factory is processed, the pollution in the river can be reduced. It costs $15 to process a ton of factory 1 waste, and each ton processed reduces the amount of pollutant 1 by 0.10 ton and the amount of pollutant 2 by 0.45 ton. It costs $10 to process a ton of factory 2 waste, and each ton processed will reduce the amount of pollutant 1 by 0.20 ton and the amount of pollutant 2 by 0.25 ton. It costs $20 to process a ton of factory 3 waste, and each ton processed will reduce the amount of pollutant 1 by 0.40 ton and the amount of pollutant 2 by 0.30 ton. The state wants to reduce the amount of pollutant 1 in the river by at least 30 tons and the amount of pollutant 2 in the river by at least 40 tons. Formulate an LP that will minimize the cost of reducing pollution by the desired amounts. Do you think that the LP assumptions (Proportionality, Additivity, Divisibility, and Certainty) are reasonable for this problem?

2[‡] U.S. Labs manufactures mechanical heart valves from the heart valves of pigs. Different heart operations require valves of different sizes. U.S. Labs purchases pig valves from three different suppliers. The cost and size mix of the valves purchased from each supplier are given in Table 3. Each month, U.S. Labs places one order with each supplier. At least 500 large, 300 medium, and 300 small valves must be purchased each month. Because of limited availability of pig valves, at most 700 valves per month can be purchased from each supplier. Formulate an LP that can be used to minimize the cost of acquiring the needed valves.

3 Peg and Al Fundy have a limited food budget, so Peg is trying to feed the family as cheaply as possible. However, she still wants to make sure her family members meet their daily nutritional requirements. Peg can buy two foods. Food

TABLE 3

Supplier	Cost Per Value ($)	Percent Large	Percent Medium	Percent Small
1	5	40	40	20
2	4	30	35	35
3	3	20	20	60

1 sells for $7 per pound, and each pound contains 3 units of vitamin A and 1 unit of vitamin C. Food 2 sells for $1 per pound, and each pound contains 1 unit of each vitamin. Each day, the family needs at least 12 units of vitamin A and 6 units of vitamin C.

a Verify that Peg should purchase 12 units of food 2 each day and thus oversatisfy the vitamin C requirement by 6 units.

b Al has put his foot down and demanded that Peg fulfill the family's daily nutritional requirement exactly by obtaining precisely 12 units of vitamin A and 6 units of vitamin C. The optimal solution to the new problem will involve ingesting less vitamin C, but it will be more expensive. Why?

4 Goldilocks needs to find at least 12 lb of gold and at least 18 lb of silver to pay the monthly rent. There are two mines in which Goldilocks can find gold and silver. Each day that Goldilocks spends in mine 1, she finds 2 lb of gold and 2 lb of silver. Each day that Goldilocks spends in mine 2, she finds 1 lb of gold and 3 lb of silver. Formulate an LP to help Goldilocks meet her requirements while spending as little time as possible in the mines. Graphically solve the LP.

[†]Balintfy (1976).
[‡]Based on Hilal and Erickson (1981).

3.5 A Work-Scheduling Problem

Many applications of linear programming involve determining the minimum-cost method for satisfying workforce requirements. The following example illustrates the basic features common to many of these applications.

EXAMPLE 7   Post Office Problem

A post office requires different numbers of full-time employees on different days of the week. The number of full-time employees required on each day is given in Table 4. Union rules state that each full-time employee must work five consecutive days and then receive two days off. For example, an employee who works Monday to Friday must be off on Saturday and Sunday. The post office wants to meet its daily requirements using only full-time employees. Formulate an LP that the post office can use to minimize the number of full-time employees who must be hired.

Solution   Before giving the correct formulation of this problem, let's begin by discussing an *incorrect* solution. Many students begin by defining x_i to be the number of employees working on day i (day 1 = Monday, day 2 = Tuesday, and so on). Then they reason that (number of full-time employees) = (number of employees working on Monday) + (number of employees working on Tuesday) + $\cdots$ + (number of employees working on Sunday). This reasoning leads to the following objective function:

$$\min z = x_1 + x_2 + \cdots + x_6 + x_7$$

To ensure that the post office has enough full-time employees working on each day, they add the constraints $x_i \geq$ (number of employees required on day i). For example, for Monday add the constraint $x_1 \geq 17$. Adding the sign restrictions $x_i \geq 0$ ($i = 1, 2, \ldots, 7$) yields the following LP:

$$\min z = x_1 + x_2 + x_3 + x_4 + x_5 + x_6 + x_7$$
$$
\begin{aligned}
\text{s.t.} \quad x_1 & \geq 17 \\
x_2 & \geq 13 \\
x_3 & \geq 15 \\
x_4 & \geq 19 \\
x_5 & \geq 14 \\
x_6 & \geq 16 \\
x_7 & \geq 11 \\
x_i \geq 0 \quad (i & = 1, 2, \ldots, 7)
\end{aligned}
$$

There are at least two flaws in this formulation. First, the objective function is *not* the number of full-time post office employees. The current objective function counts each employee five times, not once. For example, each employee who starts work on Monday works Monday to Friday and is included in x_1, x_2, x_3, x_4, and x_5. Second, the variables x_1, $x_2, \ldots, x_7$ are interrelated, and the interrelation between the variables is not captured by the current set of constraints. For example, some of the people who are working on Monday (the x_1 people) will be working on Tuesday. This means that x_1 and x_2 are interrelated, but our constraints do not indicate that the value of x_1 has any effect on the value of x_2.

The key to correctly formulating this problem is to realize that the post office's primary decision is not how many people are working each day but rather how many people *begin* work on each day of the week. With this in mind, we define

TABLE 4
Requirements for Post Office

Day	Number of Full-time Employees Required
1 = Monday	17
2 = Tuesday	13
3 = Wednesday	15
4 = Thursday	19
5 = Friday	14
6 = Saturday	16
7 = Sunday	11

$$x_i = \text{number of employees beginning work on day } i$$

For example, x_1 is the number of people beginning work on Monday (these people work Monday to Friday). With the variables properly defined, it is easy to determine the correct objective function and constraints. To determine the objective function, note that (number of full-time employees) = (number of employees who start work on Monday) + (number of employees who start work on Tuesday) $+\cdots+$ (number of employees who start work on Sunday). Because each employee begins work on exactly one day of the week, this expression does not double-count employees. Thus, when we correctly define the variables, the objective function is

$$\min z = x_1 + x_2 + x_3 + x_4 + x_5 + x_6 + x_7$$

The post office must ensure that enough employees are working on each day of the week. For example, at least 17 employees must be working on Monday. Who is working on Monday? Everybody except the employees who begin work on Tuesday or on Wednesday (they get, respectively, Sunday and Monday, and Monday and Tuesday off). This means that the number of employees working on Monday is $x_1 + x_4 + x_5 + x_6 + x_7$. To ensure that at least 17 employees are working on Monday, we require that the constraint

$$x_1 + x_4 + x_5 + x_6 + x_7 \geq 17$$

be satisfied. Adding similar constraints for the other six days of the week and the sign restrictions $x_i \geq 0$ ($i = 1, 2, \ldots, 7$) yields the following formulation of the post office's problem:

$$
\begin{aligned}
\min z = x_1 + x_2 + x_3 + x_4 + x_5 + x_6 + x_7 & \\
\text{s.t.} \quad x_1 \qquad\qquad + x_4 + x_5 + x_6 + x_7 &\geq 17 \quad \text{(Monday constraint)} \\
x_1 + x_2 \qquad\qquad + x_5 + x_6 + x_7 &\geq 13 \quad \text{(Tuesday constraint)} \\
x_1 + x_2 + x_3 \qquad\qquad + x_6 + x_7 &\geq 15 \quad \text{(Wednesday constraint)} \\
x_1 + x_2 + x_3 + x_4 \qquad\qquad + x_7 &\geq 19 \quad \text{(Thursday constraint)} \\
x_1 + x_2 + x_3 + x_4 + x_5 \qquad\qquad &\geq 14 \quad \text{(Friday constraint)} \\
x_2 + x_3 + x_4 + x_5 + x_6 \qquad &\geq 16 \quad \text{(Saturday constraint)} \\
x_3 + x_4 + x_5 + x_6 + x_7 &\geq 11 \quad \text{(Sunday constraint)} \\
x_i \geq 0 \quad (i = 1, 2, \ldots, 7) & \qquad\quad \text{(Sign restrictions)}
\end{aligned}
$$

The optimal solution to this LP is $z = \frac{67}{3}$, $x_1 = \frac{4}{3}$, $x_2 = \frac{10}{3}$, $x_3 = 2$, $x_4 = \frac{22}{3}$, $x_5 = 0$, $x_6 = \frac{10}{3}$, $x_7 = 5$. Because we are only allowing full-time employees, however, the variables must be integers, and the Divisibility Assumption is not satisfied. To find a reasonable answer in which all variables are integers, we could try to round the fractional variables up, yielding the feasible solution $z = 25$, $x_1 = 2$, $x_2 = 4$, $x_3 = 2$, $x_4 = 8$, $x_5 = 0$, $x_6 = 4$, $x_7 = 5$. It turns out, however, that integer programming can be used to show that an optimal solution to the post office problem is $z = 23$, $x_1 = 4$, $x_2 = 4$, $x_3 = 2$, $x_4 = 6$, $x_5 = 0$, $x_6 = 4$, $x_7 = 3$. Notice that there is no way that the optimal linear programming solution could have been rounded to obtain the optimal all-integer solution.

Baker (1974) has developed an efficient technique (that does not use linear programming) to determine the minimum number of employees required when each worker receives two consecutive days off.

If you solve this problem using LINDO, LINGO, or the Excel Solver, you may get a different workforce schedule that uses 23 employees. This shows that Example 7 has alternative optimal solutions.

Creating a Fair Schedule for Employees

The optimal solution we found requires 4 workers to start on Monday, 4 on Tuesday, 2 on Wednesday, 6 on Thursday, 4 on Saturday, and 3 on Sunday. The workers who start on Saturday will be unhappy because they never receive a weekend day off. By rotating the schedules of the employees over a 23-week period, a fairer schedule can be obtained. To see how this is done, consider the following schedule:

- weeks 1–4: start on Monday
- weeks 5–8: start on Tuesday
- weeks 9–10: start on Wednesday
- weeks 11–16: start on Thursday
- weeks 17–20: start on Saturday
- weeks 21–23: start on Sunday

Employee 1 follows this schedule for a 23-week period. Employee 2 starts with week 2 of this schedule (starting on Monday for 3 weeks, then on Tuesday for 4 weeks, and closing with 3 weeks starting on Sunday and 1 week on Monday). We continue in this fashion to generate a 23-week schedule for each employee. For example, employee 13 will have the following schedule:

- weeks 1–4: start on Thursday
- weeks 5–8: start on Saturday
- weeks 9–11: start on Sunday
- weeks 12–15: start on Monday
- weeks 16–19: start on Tuesday
- weeks 20–21: start on Wednesday
- weeks 22–23 start on Thursday

This method of scheduling treats each employee equally.

Modeling Issues

1 This example is a **static scheduling problem,** because we assume that the post office faces the same schedule each week. In reality, demands change over time, workers take vacations in the summer, and so on, so the post office does not face the same situation each week. A **dynamic scheduling problem** will be discussed in Section 3.12.

2 If you wanted to set up a weekly scheduling model for a supermarket or a fast-food restaurant, the number of variables could be very large and the computer might have difficulty finding an exact solution. In this case, **heuristic methods** can be used to find a good solution to the problem. See Love and Hoey (1990) for an example of scheduling a fast-food restaurant.

3 Our model can easily be expanded to handle part-time employees, the use of overtime, and alternative objective functions such as maximizing the number of weekend days off. (See Problems 1, 3, and 4.)

4 How did we determine the number of workers needed each day? Perhaps the post office wants to have enough employees to ensure that 95% of all letters are sorted within an hour. To determine the number of employees needed to provide adequate service, the post office would use queuing theory, which is discussed in *Stochastic Models in Operations Research: Applications and Algorithms;* and forecasting, which is discussed in Chapter 14 of this book.

Real-World Application

Krajewski, Ritzman, and McKenzie (1980) used LP to schedule clerks who processed checks at the Ohio National Bank. Their model determined the minimum-cost combination of part-time employees, full-time employees, and overtime labor needed to process each day's checks by the end of the workday (10 P.M.). The major input to their model was a forecast of the number of checks arriving at the bank each hour. This forecast was produced using multiple regression (see *Stochastic Models in Operations Research: Applications and Algorithms*). The major output of the LP was a work schedule. For example, the LP might suggest that 2 full-time employees work daily from 11 A.M. to 8 P.M., 33 part-time employees work every day from 6 P.M. to 10 P.M., and 27 part-time employees work from 6 P.M. to 10 P.M. on Monday, Tuesday, and Friday.

PROBLEMS

Group A

1 In the post office example, suppose that each full-time employee works 8 hours per day. Thus, Monday's requirement of 17 workers may be viewed as a requirement of $8(17) = 136$ hours. The post office may meet its daily labor requirements by using both full-time and part-time employees. During each week, a full-time employee works 8 hours a day for five consecutive days, and a part-time employee works 4 hours a day for five consecutive days. A full-time employee costs the post office $15 per hour, whereas a part-time employee (with reduced fringe benefits) costs the post office only $10 per hour. Union requirements limit part-time labor to 25% of weekly labor requirements. Formulate an LP to minimize the post office's weekly labor costs.

2 During each 4-hour period, the Smalltown police force requires the following number of on-duty police officers: 12 midnight to 4 A.M.—8; 4 to 8 A.M.—7; 8 A.M. to 12 noon—6; 12 noon to 4 P.M.—6; 4 to 8 P.M.—5; 8 P.M. to 12 midnight—4. Each police officer works two consecutive 4-hour shifts. Formulate an LP that can be used to minimize the number of police officers needed to meet Smalltown's daily requirements.

Group B

3 Suppose that the post office can force employees to work one day of overtime each week. For example, an employee whose regular shift is Monday to Friday can also be required to work on Saturday. Each employee is paid $50 a day for each of the first five days worked during a week and $62 for the overtime day (if any). Formulate an LP whose solution will enable the post office to minimize the cost of meeting its weekly work requirements.

4 Suppose the post office had 25 full-time employees and was not allowed to hire or fire any employees. Formulate an LP that could be used to schedule the employees in order to maximize the number of weekend days off received by the employees.

5 Each day, workers at the Gotham City Police Department work two 6-hour shifts chosen from 12 A.M. to 6 A.M., 6 A.M. to 12 P.M., 12 P.M. to 6 P.M., and 6 P.M. to 12 A.M. The following number of workers are needed during each shift: 12 A.M. to 6 A.M.—15 workers; 6 A.M. to 12 P.M.—5 workers; 12 P.M. to 6 P.M.—12 workers; 6 P.M. to 12 A.M.—6 workers. Workers whose two shifts are consecutive are paid $12 per hour; workers whose shifts are not consecutive are paid $18 per hour. Formulate an LP that can be used to minimize the cost of meeting the daily workforce demands of the Gotham City Police Department.

6 During each 6-hour period of the day, the Bloomington Police Department needs at least the number of policemen shown in Table 5. Policemen can be hired to work either 12 consecutive hours or 18 consecutive hours. Policemen are paid $4 per hour for each of the first 12 hours a day they work and are paid $6 per hour for each of the next 6 hours they work in a day. Formulate an LP that can be used to minimize the cost of meeting Bloomington's daily police requirements.

7 Each hour from 10 A.M. to 7 P.M., Bank One receives checks and must process them. Its goal is to process all the checks the same day they are received. The bank has 13 check-processing machines, each of which can process up to 500 checks per hour. It takes one worker to operate each machine. Bank One hires both full-time and part-time workers. Full-time workers work 10 A.M.–6 P.M., 11 A.M.– 7 P.M., or Noon–8 P.M. and are paid $160 per day. Part-time workers work either 2 P.M.–7 P.M. or 3 P.M.–8 P.M. and are paid $75 per day. The number of checks received each hour is given in Table 6. In the interest of maintaining continuity, Bank One believes it must have at least three full-time workers under contract. Develop a cost-minimizing work schedule that processes all checks by 8 P.M.

TABLE 5

Time Period	Number of Policemen Required
12 A.M.–6 A.M.	12
6 A.M.–12 P.M.	8
12 P.M.–6 P.M.	6
6 P.M.–12 A.M.	15

TABLE 6

Time	Checks Received
10 A.M.	5,000
11 A.M.	4,000
Noon	3,000
1 P.M.	4,000
2 P.M.	2,500
3 P.M.	3,000
4 P.M.	4,000
5 P.M.	4,500
6 P.M.	3,500
7 P.M.	3,000

3.6 A Capital Budgeting Problem

In this section (and in Sections 3.7 and 3.11), we discuss how linear programming can be used to determine optimal financial decisions. This section considers a simple capital budgeting model.[†]

We first explain briefly the concept of net present value (NPV), which can be used to compare the desirability of different investments. Time 0 is the present.

Suppose investment 1 requires a cash outlay of $10,000 at time 0 and a cash outlay of $14,000 two years from now and yields a cash flow of $24,000 one year from now. Investment 2 requires a $6,000 cash outlay at time 0 and a $1,000 cash outlay two years from now and yields a cash flow of $8,000 one year from now. Which investment would you prefer?

Investment 1 has a net cash flow of

$$-10{,}000 + 24{,}000 - 14{,}000 = \$0$$

and investment 2 has a net cash flow of

$$-6{,}000 + 8{,}000 - 1{,}000 = \$1{,}000$$

On the basis of net cash flow, investment 2 is superior to investment 1. When we compare investments on the basis of net cash flow, we are assuming that a dollar received at

[†]This section is based on Weingartner (1963).

any point in time has the same value. This is not true! Suppose that there exists an investment (such as a money market fund) for which $1 invested at a given time will yield (with certainty) $(1 + r)$ one year later. We call r the *annual interest rate*. Because $1 now can be transformed into $(1 + r)$ one year from now, we may write

$$\$1 \text{ now} = \$(1 + r) \text{ one year from now}$$

Applying this reasoning to the $(1 + r)$ obtained one year from now shows that

$$\$1 \text{ now} = \$(1 + r) \text{ one year from now} = \$(1 + r)^2 \text{ two years from now}$$

and

$$\$1 \text{ now} = \$(1 + r)^k \, k \text{ years from now}$$

Dividing both sides of this equality by $(1 + r)^k$ shows that

$$\$1 \text{ received } k \text{ years from now} = \$(1 + r)^{-k} \text{ now}$$

In other words, a dollar received k years from now is equivalent to receiving $\$(1 + r)^{-k}$ now.

We can use this idea to express all cash flows in terms of time 0 dollars (this process is called *discounting cash flows to time* 0). Using discounting, we can determine the total value (in time 0 dollars) of the cash flows for any investment. The total value (in time 0 dollars) of the cash flows for any investment is called the **net present value, or NPV,** of the investment. The NPV of an investment is the amount by which the investment will increase the firm's value (as expressed in time 0 dollars).

Assuming that $r = 0.20$, we can compute the NPV for investments 1 and 2.

$$\text{NPV of investment } 1 = -10{,}000 + \frac{24{,}000}{1 + 0.20} - \frac{14{,}000}{(1 + 0.20)^2}$$

$$= \$277.78$$

This means that if a firm invested in investment 1, then the value of the firm (in time 0 dollars) would increase by $277.78. For investment 2,

$$\text{NPV of investment } 2 = -6{,}000 + \frac{8{,}000}{1 + 0.20} - \frac{1{,}000}{(1 + 0.20)^2}$$

$$= -\$27.78$$

If a firm invested in investment 2, then the value of the firm (in time 0 dollars) would be reduced by $27.78.

Thus, the NPV concept says that investment 1 is superior to investment 2. This conclusion is contrary to the one reached by comparing the net cash flows of the two investments. Note that the comparison between investments often depends on the value of r. For example, the reader is asked to show in Problem 1 at the end of this section that for $r = 0.02$, investment 2 has a higher NPV than investment 1. Of course, our analysis assumes that the future cash flows of an investment are known with certainty.

Computing NPV with Excel

If we receive a cash flow of c_t in t years from now ($t = 1, 2, \ldots T$) and we discount cash flows at a rate r, then the NPV of our cash flows is given by

$$\sum_{t=1}^{t=T} \frac{c_t}{(1 + r)^t}$$

The basic idea is that \$1 today equals $\$(1 + r)$ a year from now, so

$$\frac{1}{1 + r} \text{ today} = \$1 \text{ a year from now}$$

The Excel function = NPV makes this computation easy. The syntax is

$$= \text{NPV } (r, \text{ range of cash flows})$$

The formula assumes that cash flows occur at the end of the year.

Projects with NPV > 0 add value to the company, while projects with negative NPV reduce the company's value.

NPV.xls

We illustrate the computation of NPV in the file NPV.xls.

EXAMPLE 8   **Computing NPV**

For a discount rate of 15%, consider a project with the cash flows shown in Figure 8.

a Compute project NPV if cash flows are at the end of the year.

b Compute project NPV if cash flows are at the beginning of the year.

c Compute project NPV if cash flows are at the middle of the year.

Solution **a** We enter in cell C7 the formula

$$= \text{NPV}(C1, C4:I4)$$

and obtain \$375.06.

b Because all cash flows are received a year earlier, we multiply each cash flow's value by $(1 + 1.15)$, so the answer is obtained in C8 with formula

$$= (1 + C1) \cdot C7$$

NPV is now larger: \$431.32.

We checked this in cell D8 with the formula

$$= C4 + \text{NPV}(C1, D4:I4)$$

c Because all cash flows are received six months earlier, we multiply each cash flow's value by $\sqrt{1.15}$. NPV is now computed in C9 with the formula

$$= (1.15)\text{\textasciicircum}0.5 \cdot C7$$

Now NPV is \$402.21.

	A	B	C	D	E	F	G	H	I
1		dr	0.15						
2									
3		Time	1	2	3	4	5	6	7
4			-400	200	600	-900	1000	250	230
5									
6									
7	end of year	end of yr.	\$375.06						
8	beginning of yr.	beg. of yr.	\$431.32	\$431.32					
9	middle of year	middle of yr.	\$402.21						

FIGURE 8

The XNPV Function

Often cash flows occur at irregular intervals. This makes it difficult to compute the NPV of these cash flows. Fortunately, the Excel XNPV function makes computing NPVs of irregularly timed cash flows a snap. To use the XNPV function, you must first have added the Analysis Toolpak. To do this, select Tools Add-Ins and check the Analysis Toolpak and Analysis Tookpak VBA boxes. Here is an example of XNPV in action.

| EXAMPLE 9 | Finding NPV of Nonperiodic Cash Flows |

Suppose on April 8, 2001, we paid out $900. Then we receive

- $300 on 8/15/01
- $400 on 1/15/02
- $200 on 6/25/02
- $100 on 7/03/03.

If the annual discount rate is 10%, what is the NPV of these cash flows?

Solution We enter the dates (in Excel date format) in D3:D7 and the cash flows in E3:E7 (see Figure 9). Entering the formula

$$= XNPV(A9,E3:E7,D3:D7)$$

in cell D11 computes the project's NPV in terms of April 8, 2001, dollars because that is the first date chronologically. What Excel did was as follows:

1 Compute the number of years after April 8, 2001, that each date occurred. (We did this in column F). For example, August 15, 2001, is .3534 years after April 8.

2 Then discount cash flows at a rate $\left(\dfrac{1}{1 + \text{rate}}\right)^{\text{years after}}$. For example, the August 15, 2001, cash flow is discounted by $\left(\dfrac{1}{1 + .1}\right)^{.3534} = .967$.

3 We obtained Excel dates in serial number form by changing format to General.

If you want the XNPV function to determine a project's NPV in today's dollars, insert a $0 cash flow on today's date and include this row in the XNPV calculation. Excel will then return the project's NPV as of today's date.

	A	B	C	D	E	F	G
1							
2	XNPV Function		Code	Date	Cash Flow	Time	df
3			36989.00	4/8/01	-900		1
4			37118.00	8/15/01	300	0.353425	0.966876
5			37271.00	1/15/02	400	0.772603	0.929009
6			37432.00	6/25/02	200	1.213699	0.890762
7			37805.00	7/3/03	100	2.235616	0.808094
8	Rate						
9		0.1					
10				XNPV	Direct		
11				20.62822	20.628217		
12							
13				XIRR			
14				12.97%			

FIGURE 9

Example of XNPV Function

With this background information, we are ready to explain how linear programming can be applied to problems in which limited investment funds must be allocated to investment projects. Such problems are called **capital budgeting problems.**

EXAMPLE 10 **Project Selection**

Star Oil Company is considering five different investment opportunities. The cash outflows and net present values (in millions of dollars) are given in Table 7. Star Oil has $40 million available for investment now (time 0); it estimates that one year from now (time 1) $20 million will be available for investment. Star Oil may purchase any fraction of each investment. In this case, the cash outflows and NPV are adjusted accordingly. For example, if Star Oil purchases one-fifth of investment 3, then a cash outflow of $\frac{1}{5}(5) = \$1$ million would be required at time 0, and a cash outflow of $\frac{1}{5}(5) = \$1$ million would be required at time 1. The one-fifth share of investment 3 would yield an NPV of $\frac{1}{5}(16) = \$3.2$ million. Star Oil wants to maximize the NPV that can be obtained by investing in investments 1–5. Formulate an LP that will help achieve this goal. Assume that any funds left over at time 0 cannot be used at time 1.

Solution Star Oil must determine what fraction of each investment to purchase. We define

$$x_i = \text{fraction of investment } i \text{ purchased by Star Oil} \quad (i = 1, 2, 3, 4, 5)$$

Star's goal is to maximize the NPV earned from investments. Now, (total NPV) = (NPV earned from investment 1) + (NPV earned from investment 2) + ··· + (NPV earned from investment 5). Note that

NPV from investment 1 = (NPV from investment 1)(fraction of investment 1 purchased)

$$= 13x_1$$

Applying analogous reasoning to investments 2–5 shows that Star Oil wants to maximize

$$z = 13x_1 + 16x_2 + 16x_3 + 14x_4 + 39x_5 \tag{25}$$

Star Oil's constraints may be expressed as follows:

Constraint 1 Star cannot invest more than $40 million at time 0.

Constraint 2 Star cannot invest more than $20 million at time 1.

Constraint 3 Star cannot purchase more than 100% of investment i ($i = 1, 2, 3, 4, 5$).

To express Constraint 1 mathematically, note that (dollars invested at time 0) = (dollars invested in investment 1 at time 0) + (dollars invested in investment 2 at time 0) + ··· + (dollars invested in investment 5 at time 0). Also, in millions of dollars,

$$\begin{pmatrix} \text{Dollars invested in investment 1} \\ \text{at time 0} \end{pmatrix} = \begin{pmatrix} \text{dollars required for} \\ \text{investment 1 at time 0} \end{pmatrix} \begin{pmatrix} \text{fraction of} \\ \text{investment 1 purchased} \end{pmatrix}$$

$$= 11x_1$$

TABLE 7
Cash Flows and Net Present Value for Investments in Capital Budgeting

	Investment ($)				
	1	2	3	4	5
Time 0 cash outflow	11	53	5	5	29
Time 1 cash outflow	3	6	5	1	34
NPV	13	16	16	14	39

Similarly, for investments 2–5,

$$\text{Dollars invested at time } 0 = 11x_1 + 53x_2 + 5x_3 + 5x_4 + 29x_5$$

Then Constraint 1 reduces to

$$11x_1 + 53x_2 + 5x_3 + 5x_4 + 29x_5 \leq 40 \qquad \text{(Time 0 constraint)} \qquad \textbf{(26)}$$

Constraint 2 reduces to

$$3x_1 + 6x_2 + 5x_3 + x_4 + 34x_5 \leq 20 \qquad \text{(Time 1 constraint)} \qquad \textbf{(27)}$$

Constraints 3–7 may be represented by

$$x_i \leq 1 \qquad (i = 1, 2, 3, 4, 5) \qquad \textbf{(28-32)}$$

Combining (26)–(32) with the sign restrictions $x_i \geq 0$ ($i = 1, 2, 3, 4, 5$) yields the following LP:

$$\max z = 13x_1 + 16x_2 + 16x_3 + 14x_4 + 39x_5$$

$$\begin{aligned}
\text{s.t.} \quad & 11x_1 + 53x_2 + 5x_3 + 5x_4 + 29x_5 \leq 40 \qquad \text{(Time 0 constraint)} \\
& 3x_1 + 6x_2 + 5x_3 + x_4 + 34x_5 \leq 20 \qquad \text{(Time 1 constraint)} \\
& x_1 \qquad\qquad\qquad\qquad\qquad\quad \leq 1 \\
& \quad x_2 \qquad\qquad\qquad\qquad\quad \leq 1 \\
& \qquad x_3 \qquad\qquad\qquad\quad \leq 1 \\
& \qquad\quad x_4 \qquad\qquad\quad \leq 1 \\
& \qquad\qquad\quad x_5 \leq 1 \\
& x_i \geq 0 \qquad (i = 1, 2, 3, 4, 5)
\end{aligned}$$

The optimal solution to this LP is $x_1 = x_3 = x_4 = 1$, $x_2 = 0.201$, $x_5 = 0.288$, $z = 57.449$. Star Oil should purchase 100% of investments 1, 3, and 4; 20.1% of investment 2; and 28.8% of investment 5. A total NPV of $57,449,000 will be obtained from these investments.

It is often impossible to purchase only a fraction of an investment without sacrificing the investment's favorable cash flows. Suppose it costs $12 million to drill an oil well just deep enough to locate a $30-million gusher. If there were a sole investor in this project who invested $6 million to undertake half of the project, then he or she would lose the entire investment and receive no positive cash flows. Because, in this example, reducing the money invested by 50% reduces the return by more than 50%, this situation would violate the Proportionality Assumption.

In many capital budgeting problems, it is unreasonable to allow the x_i to be fractions: Each x_i should be restricted to 0 (not investing at all in investment i) or 1 (purchasing all of investment i). Thus, many capital budgeting problems violate the Divisibility Assumption.

A capital budgeting model that allows each x_i to be only 0 or 1 is discussed in Section 9.2.

PROBLEMS

Group A

1 Show that if $r = 0.02$, investment 2 has a larger NPV than investment 1.

2 Two investments with varying cash flows (in thousands of dollars) are available, as shown in Table 8. At time 0, $10,000 is available for investment, and at time 1, $7,000 is available. Assuming that $r = 0.10$, set up an LP whose solution maximizes the NPV obtained from these investments. Graphically find the optimal solution to the LP.

TABLE 8

	Cash Flow (in $ Thousands) at Time			
Investment	0	1	2	3
1	−6	−5	7	9
2	−8	−3	9	7

(Assume that any fraction of an investment may be purchased.)

3 Suppose that r, the annual interest rate, is 0.20, and that all money in the bank earns 20% interest each year (that is, after being in the bank for one year, $1 will increase to $1.20). If we place $100 in the bank for one year, what is the NPV of this transaction?

4 A company has nine projects under consideration. The NPV added by each project and the capital required by each project during the next two years is given in Table 9. All figures are in millions. For example, Project 1 will add $14 million in NPV and require expenditures of $12 million during year 1 and $3 million during year 2. Fifty million is available for projects during year 1 and $20 million is available during year 2. Assuming we may undertake a fraction of each project, how can we maximize NPV?

Group B

5[†] Finco must determine how much investment and debt to undertake during the next year. Each dollar invested reduces the NPV of the company by 10¢, and each dollar of debt increases the NPV by 50¢ (due to deductibility of interest payments). Finco can invest at most $1 million during the coming year. Debt can be at most 40% of investment. Finco now has $800,000 in cash available. All investment must be paid for from current cash or borrowed money. Set up an LP whose solution will tell Finco how to maximize its NPV. Then graphically solve the LP.

TABLE 9

	Project								
	1	2	3	4	5	6	7	8	9
Year 1 Outflow	12	54	6	6	30	6	48	36	18
Year 2 Outflow	3	7	6	2	35	6	4	3	3
NPV	14	17	17	15	40	12	14	10	12

3.7 Short-Term Financial Planning[‡]

LP models can often be used to aid a firm in short- or long-term financial planning (also see Section 3.11). Here we consider a simple example that illustrates how linear programming can be used to aid a corporation's short-term financial planning.[§]

EXAMPLE 11	Short-Term Financial Planning

Semicond is a small electronics company that manufactures tape recorders and radios. The per-unit labor costs, raw material costs, and selling price of each product are given in Table 10. On December 1, 2002, Semicond has available raw material that is sufficient to manufacture 100 tape recorders and 100 radios. On the same date, the company's balance sheet is as shown in Table 11, and Semicond's asset–liability ratio (called the current ratio) is 20,000/10,000 = 2.

Semicond must determine how many tape recorders and radios should be produced during December. Demand is large enough to ensure that all goods produced will be sold. All sales are on credit, however, and payment for goods produced in December will not

---

[†]Based on Myers and Pogue (1974).
[‡]This section covers material that may be omitted with no loss of continuity.
[§]This section is based on an example in Neave and Wiginton (1981).

TABLE 10

Cost Information for Semicond

	Tape Recorder	Radio
Selling price	$100	$90
Labor cost	$ 50	$35
Raw material cost	$ 30	$40

TABLE 11

Balance Sheet for Semicond

	Assets	Liabilities
Cash	$10,000	
Accounts receivable[§]	$ 3,000	
Inventory outstanding[¶]	$ 7,000	
Bank loan		$10,000

[§]Accounts receivable is money owed to Semicond by customers who have previously purchased Semicond products.

[¶]Value of December 1, 2002, inventory = 30(100) + 40(100) = $7,000.

be received until February 1, 2003. During December, Semicond will collect $2,000 in accounts receivable, and Semicond must pay off $1,000 of the outstanding loan and a monthly rent of $1,000. On January 1, 2003, Semicond will receive a shipment of raw material worth $2,000, which will be paid for on February 1, 2003. Semicond's management has decided that the cash balance on January 1, 2003, must be at least $4,000. Also, Semicond's bank requires that the current ratio at the beginning of January be at least 2. To maximize the contribution to profit from December production, (revenues to be received) − (variable production costs), what should Semicond produce during December?

Solution Semicond must determine how many tape recorders and radios should be produced during December. Thus, we define

x_1 = number of tape recorders produced during December

x_2 = number of radios produced during December

To express Semicond's objective function, note that

$$\frac{\text{Contribution to profit}}{\text{Tape recorder}} = 100 - 50 - 30 = \$20$$

$$\frac{\text{Contribution to profit}}{\text{Radio}} = 90 - 35 - 40 = \$15$$

As in the Giapetto example, this leads to the objective function

$$\max z = 20x_1 + 15x_2 \tag{33}$$

Semicond faces the following constraints:

Constraint 1 Because of limited availability of raw material, at most 100 tape recorders can be produced during December.

Constraint 2 Because of limited availability of raw material, at most 100 radios can be produced during December.

Constraint 3 Cash on hand on January 1, 2002, must be at least $4,000.

Constraint 4 (January 1 assets)/(January 1 liabilities) ≥ 2 must hold.

Constraint 1 is described by

$$x_1 \leq 100 \tag{34}$$

Constraint 2 is described by

$$x_2 \leq 100 \tag{35}$$

To express Constraint 3, note that

$$
\begin{aligned}
\text{January 1 cash on hand} = {}& \text{December 1 cash on hand} \\
& + \text{accounts receivable collected during December} \\
& - \text{portion of loan repaid during December} \\
& - \text{December rent} - \text{December labor costs} \\
= {}& 10{,}000 + 2{,}000 - 1{,}000 - 1{,}000 - 50x_1 - 35x_2 \\
= {}& 10{,}000 - 50x_1 - 35x_2
\end{aligned}
$$

Now Constraint 3 may be written as

$$10{,}000 - 50x_1 - 35x_2 \geq 4{,}000 \tag{36'}$$

Most computer codes require each LP constraint to be expressed in a form in which all variables are on the left-hand side and the constant is on the right-hand side. Thus, for computer solution, we should write (36′) as

$$50x_1 + 35x_2 \leq 6{,}000 \tag{36}$$

To express Constraint 4, we need to determine Semicond's January 1 cash position, accounts receivable, inventory position, and liabilities in terms of x_1 and x_2. We have already shown that

$$\text{January 1 cash position} = 10{,}000 - 50x_1 - 35x_2$$

Then

$$
\begin{aligned}
\text{January 1 accounts receivable} = {}& \text{December 1 accounts receivable} \\
& + \text{accounts receivable from December sales} \\
& - \text{accounts receivable collected during December} \\
= {}& 3{,}000 + 100x_1 + 90x_2 - 2000 \\
= {}& 1{,}000 + 100x_1 + 90x_2
\end{aligned}
$$

It now follows that

$$
\begin{aligned}
\text{Value of January 1 inventory} = {}& \text{value of December 1 inventory} \\
& - \text{value of inventory used in December} \\
& + \text{value of inventory received on January 1} \\
= {}& 7{,}000 - (30x_1 + 40x_2) + 2{,}000 \\
= {}& 9{,}000 - 30x_1 - 40x_2
\end{aligned}
$$

We can now compute the January 1 asset position:

$$
\begin{aligned}
\text{January 1 asset position} = {}& \text{January 1 cash position} + \text{January 1 accounts receivable} \\
& + \text{January 1 inventory position} \\
= {}& (10{,}000 - 50x_1 - 35x_2) + (1{,}000 + 100x_1 + 90x_2) \\
& + (9{,}000 - 30x_1 - 40x_2) \\
= {}& 20{,}000 + 20x_1 + 15x_2
\end{aligned}
$$

Finally,

$$\begin{aligned} \text{January 1 liabilities} &= \text{December 1 liabilities} - \text{December loan payment} \\ &\quad + \text{amount due on January 1 inventory shipment} \\ &= 10{,}000 - 1{,}000 + 2{,}000 \\ &= \$11{,}000 \end{aligned}$$

Constraint 4 may now be written as

$$\frac{20{,}000 + 20x_1 + 15x_2}{11{,}000} \geq 2$$

Multiplying both sides of this inequality by 11,000 yields

$$20{,}000 + 20x_1 + 15x_2 \geq 22{,}000$$

Putting this in a form appropriate for computer input, we obtain

$$20x_1 + 15x_2 \geq 2{,}000 \tag{37}$$

Combining (33)–(37) with the sign restrictions $x_1 \geq 0$ and $x_2 \geq 0$ yields the following LP:

$$\max z = 20x_1 + 15x_2$$

$$\begin{array}{llll}
\text{s.t.} & x_1 & \leq 100 & \text{(Tape recorder constraint)} \\
& x_2 \leq 100 & & \text{(Radio constraint)} \\
& 50x_1 + 35x_2 \leq 6{,}000 & & \text{(Cash position constraint)} \\
& 20x_1 + 15x_2 \geq 2{,}000 & & \text{(Current ratio constraint)} \\
& x_1, x_2 \geq 0 & & \text{(Sign restrictions)}
\end{array}$$

When solved graphically (or by computer), the following optimal solution is obtained: $z = 2{,}500$, $x_1 = 50$, $x_2 = 100$. Thus, Semicond can maximize the contribution of December's production to profits by manufacturing 50 tape recorders and 100 radios. This will contribute $20(50) + 15(100) = \$2{,}500$ to profits.

PROBLEMS

Group A

1 Graphically solve the Semicond problem.

2 Suppose that the January 1 inventory shipment had been valued at $7,000. Show that Semicond's LP is now infeasible.

3.8 Blending Problems

Situations in which various inputs must be blended in some desired proportion to produce goods for sale are often amenable to linear programming analysis. Such problems are called **blending problems.** The following list gives some situations in which linear programming has been used to solve blending problems.

1 Blending various types of crude oils to produce different types of gasoline and other outputs (such as heating oil)

2 Blending various chemicals to produce other chemicals

3 Blending various types of metal alloys to produce various types of steels

4 Blending various livestock feeds in an attempt to produce a minimum-cost feed mixture for cattle

5 Mixing various ores to obtain ore of a specified quality

6 Mixing various ingredients (meat, filler, water, and so on) to produce a product like bologna

7 Mixing various types of papers to produce recycled paper of varying quality

The following example illustrates the key ideas that are used in formulating LP models of blending problems.

EXAMPLE 12   Oil Blending

Sunco Oil manufactures three types of gasoline (gas 1, gas 2, and gas 3). Each type is produced by blending three types of crude oil (crude 1, crude 2, and crude 3). The sales price per barrel of gasoline and the purchase price per barrel of crude oil are given in Table 12. Sunco can purchase up to 5,000 barrels of each type of crude oil daily.

The three types of gasoline differ in their octane rating and sulfur content. The crude oil blended to form gas 1 must have an average octane rating of at least 10 and contain at most 1% sulfur. The crude oil blended to form gas 2 must have an average octane rating of at least 8 and contain at most 2% sulfur. The crude oil blended to form gas 3 must have an octane rating of at least 6 and contain at most 1% sulfur. The octane rating and the sulfur content of the three types of oil are given in Table 13. It costs $4 to transform one barrel of oil into one barrel of gasoline, and Sunco's refinery can produce up to 14,000 barrels of gasoline daily.

Sunco's customers require the following amounts of each gasoline: gas 1—3,000 barrels per day; gas 2—2,000 barrels per day; gas 3—1,000 barrels per day. The company considers it an obligation to meet these demands. Sunco also has the option of advertising to stimulate demand for its products. Each dollar spent daily in advertising a particular type of gas increases the daily demand for that type of gas by 10 barrels. For example, if Sunco decides to spend $20 daily in advertising gas 2, then the daily demand for gas 2 will increase by 20(10) = 200 barrels. Formulate an LP that will enable Sunco to maximize daily profits (profits = revenues − costs).

Solution   Sunco must make two types of decisions: first, how much money should be spent in advertising each type of gas, and second, how to blend each type of gasoline from the three types of crude oil available. For example, Sunco must decide how many barrels of crude 1 should be used to produce gas 1. We define the decision variables

a_i = dollars spent daily on advertising gas i ($i = 1, 2, 3$)

x_{ij} = barrels of crude oil i used daily to produce gas j ($i = 1, 2, 3; j = 1, 2, 3$)

For example, x_{21} is the number of barrels of crude 2 used each day to produce gas 1.

TABLE 12
Gas and Crude Oil Prices for Blending

Gas	Sales Price per Barrel ($)	Crude	Purchase Price per Barrel ($)
1	70	1	45
2	60	2	35
3	50	3	25

TABLE **13**
Octane Ratings and Sulfur Requirements
for Blending

Crude	Octane Rating	Sulfur Content (%)
1	12	0.5
2	6	2.0
3	8	3.0

Knowledge of these variables is sufficient to determine Sunco's objective function and constraints, but before we do this, we note that the definition of the decision variables implies that

$$x_{11} + x_{12} + x_{13} = \text{barrels of crude 1 used daily}$$
$$x_{21} + x_{22} + x_{23} = \text{barrels of crude 2 used daily} \qquad \textbf{(38)}$$
$$x_{31} + x_{32} + x_{33} = \text{barrels of crude 3 used daily}$$

$$x_{11} + x_{21} + x_{31} = \text{barrels of gas 1 produced daily}$$
$$x_{12} + x_{22} + x_{32} = \text{barrels of gas 2 produced daily} \qquad \textbf{(39)}$$
$$x_{13} + x_{23} + x_{33} = \text{barrels of gas 3 produced daily}$$

To simplify matters, let's assume that gasoline cannot be stored, so it must be sold on the day it is produced. This implies that for $i = 1, 2, 3$, the amount of gas i produced daily should equal the daily demand for gas i. Suppose that the amount of gas i produced daily exceeded the daily demand. Then we would have incurred unnecessary purchasing and production costs. On the other hand, if the amount of gas i produced daily is less than the daily demand for gas i, then we are failing to meet mandatory demands or incurring unnecessary advertising costs.

We are now ready to determine Sunco's objective function and constraints. We begin with Sunco's objective function. From (39),

$$\text{Daily revenues from gas sales} = 70(x_{11} + x_{21} + x_{31}) + 60(x_{12} + x_{22} + x_{32})$$
$$+ 50(x_{13} + x_{23} + x_{33})$$

From (38),

$$\text{Daily cost of purchasing crude oil} = 45(x_{11} + x_{12} + x_{13}) + 35(x_{21} + x_{22} + x_{23})$$
$$+ 25(x_{31} + x_{32} + x_{33})$$

Also,

Daily advertising costs $= a_1 + a_2 + a_3$

Daily production costs $= 4(x_{11} + x_{12} + x_{13} + x_{21} + x_{22} + x_{23} + x_{31} + x_{32} + x_{33})$

Then,

$$\text{Daily profit} = \text{daily revenue from gas sales}$$
$$- \text{daily cost of purchasing crude oil}$$
$$- \text{daily advertising costs} - \text{daily production costs}$$
$$= (70 - 45 - 4)x_{11} + (60 - 45 - 4)x_{12} + (50 - 45 - 4)x_{13}$$
$$+ (70 - 35 - 4)x_{21} + (60 - 35 - 4)x_{22} + (50 - 35 - 4)x_{23}$$
$$+ (70 - 25 - 4)x_{31} + (60 - 25 - 4)x_{32}$$
$$+ (50 - 25 - 4)x_{33} - a_1 - a_2 - a_3$$

Thus, Sunco's goal is to maximize

$$z = 21x_{11} + 11x_{12} + x_{13} + 31x_{21} + 21x_{22} + 11x_{23} + 41x_{31}$$
$$+ 31x_{32} + 21x_{33} - a_1 - a_2 - a_3 \tag{40}$$

Regarding Sunco's constraints, we see that the following 13 constraints must be satisfied:

Constraint 1 Gas 1 produced daily should equal its daily demand.

Constraint 2 Gas 2 produced daily should equal its daily demand.

Constraint 3 Gas 3 produced daily should equal its daily demand.

Constraint 4 At most 5,000 barrels of crude 1 can be purchased daily.

Constraint 5 At most 5,000 barrels of crude 2 can be purchased daily.

Constraint 6 At most 5,000 barrels of crude 3 can be purchased daily.

Constraint 7 Because of limited refinery capacity, at most 14,000 barrels of gasoline can be produced daily.

Constraint 8 Crude oil blended to make gas 1 must have an average octane level of at least 10.

Constraint 9 Crude oil blended to make gas 2 must have an average octane level of at least 8.

Constraint 10 Crude oil blended to make gas 3 must have an average octane level of at least 6.

Constraint 11 Crude oil blended to make gas 1 must contain at most 1% sulfur.

Constraint 12 Crude oil blended to make gas 2 must contain at most 2% sulfur.

Constraint 13 Crude oil blended to make gas 3 must contain at most 1% sulfur.

To express Constraint 1 in terms of decision variables, note that

$$\text{Daily demand for gas 1} = 3,000 + \text{gas 1 demand generated by advertising}$$

$$\text{Gas 1 demand generated by advertising} = \left(\frac{\text{gas 1 demand}}{\text{dollar spent}}\right)\left(\frac{\text{dollars}}{\text{spent}}\right)$$
$$= 10a_1{}^\dagger$$

Thus, daily demand for gas 1 $= 3,000 + 10a_1$. Constraint 1 may now be written as

$$x_{11} + x_{21} + x_{31} = 3,000 + 10a_1 \tag{41'}$$

which we rewrite as

$$x_{11} + x_{21} + x_{31} - 10a_1 = 3,000 \tag{41}$$

Constraint 2 is expressed by

$$x_{12} + x_{22} + x_{32} - 10a_2 = 2,000 \tag{42}$$

---

†Many students believe that gas 1 demand generated by advertising should be written as $\frac{1}{10}a_1$. Analyzing the units of this term will show that this is not correct. $\frac{1}{10}$ has units of dollars spent per barrel of demand, and a_1 has units of dollars spent. Thus, the term $\frac{1}{10}a_1$ would have units of (dollars spent)^2 per barrel of demand. This cannot be correct!

Constraint 3 is expressed by

$$x_{13} + x_{23} + x_{33} - 10a_3 = 1{,}000 \tag{43}$$

From (38), Constraint 4 reduces to

$$x_{11} + x_{12} + x_{13} \leq 5{,}000 \tag{44}$$

Constraint 5 reduces to

$$x_{21} + x_{22} + x_{23} \leq 5{,}000 \tag{45}$$

Constraint 6 reduces to

$$x_{31} + x_{32} + x_{33} \leq 5{,}000 \tag{46}$$

Note that

$$\text{Total gas produced} = \text{gas 1 produced} + \text{gas 2 produced} + \text{gas 3 produced}$$
$$= (x_{11} + x_{21} + x_{31}) + (x_{12} + x_{22} + x_{32}) + (x_{13} + x_{23} + x_{33})$$

Then Constraint 7 becomes

$$x_{11} + x_{21} + x_{31} + x_{12} + x_{22} + x_{32} + x_{13} + x_{23} + x_{33} \leq 14{,}000 \tag{47}$$

To express Constraints 8–10, we must be able to determine the "average" octane level in a mixture of different types of crude oil. We assume that the octane levels of different crudes blend linearly. For example, if we blend two barrels of crude 1, three barrels of crude 2, and one barrel of crude 3, the average octane level in this mixture would be

$$\frac{\text{Total octane value in mixture}}{\text{Number of barrels in mixture}} = \frac{12(2) + 6(3) + 8(1)}{2 + 3 + 1} = \frac{50}{6} = 8\frac{1}{3}$$

Generalizing, we can express Constraint 8 by

$$\frac{\text{Total octane value in gas 1}}{\text{Gas 1 in mixture}} = \frac{12x_{11} + 6x_{21} + 8x_{31}}{x_{11} + x_{21} + x_{31}} \geq 10 \tag{48'}$$

Unfortunately, (48') is not a linear inequality. To transform (48') into a linear inequality, all we have to do is multiply both sides by the denominator of the left-hand side. The resulting inequality is

$$12x_{11} + 6x_{21} + 8x_{31} \geq 10(x_{11} + x_{21} + x_{31})$$

which may be rewritten as

$$2x_{11} - 4x_{21} - 2x_{31} \geq 0 \tag{48}$$

Similarly, Constraint 9 yields

$$\frac{12x_{12} + 6x_{22} + 8x_{32}}{x_{12} + x_{22} + x_{32}} \geq 8$$

Multiplying both sides of this inequality by $x_{12} + x_{22} + x_{32}$ and simplifying yields

$$4x_{12} - 2x_{22} \geq 0 \tag{49}$$

Because each type of crude oil has an octane level of 6 or higher, whatever we blend to manufacture gas 3 will have an average octane level of at least 6. This means that any values of the variables will satisfy Constraint 10. To verify this, we may express Constraint 10 by

$$\frac{12x_{13} + 6x_{23} + 8x_{33}}{x_{13} + x_{23} + x_{33}} \geq 6$$

Multiplying both sides of this inequality by $x_{13} + x_{23} + x_{33}$ and simplifying, we obtain

$$6x_{13} + 2x_{33} \geq 0 \tag{50}$$

Because $x_{13} \geq 0$ and $x_{33} \geq 0$ are always satisfied, (50) will automatically be satisfied and thus need not be included in the model. A constraint such as (50) that is implied by other constraints in the model is said to be a **redundant constraint** and need not be included in the formulation.

Constraint 11 may be written as

$$\frac{\text{Total sulfur in gas 1 mixture}}{\text{Number of barrels in gas 1 mixture}} \leq 0.01$$

Then, using the percentages of sulfur in each type of oil, we see that

Total sulfur in gas 1 mixture = Sulfur in oil 1 used for gas 1
$$+ \text{ sulfur in oil 2 used for gas 1}$$
$$+ \text{ sulfur in oil 3 used for gas 1}$$
$$= 0.005x_{11} + 0.02x_{21} + 0.03x_{31}$$

Constraint 11 may now be written as

$$\frac{0.005x_{11} + 0.02x_{21} + 0.03x_{31}}{x_{11} + x_{21} + x_{31}} \leq 0.01$$

Again, this is not a linear inequality, but we can multiply both sides of the inequality by $x_{11} + x_{21} + x_{31}$ and simplify, obtaining

$$-0.005x_{11} + 0.01x_{21} + 0.02x_{31} \leq 0 \tag{51}$$

Similarly, Constraint 12 is equivalent to

$$\frac{0.005x_{12} + 0.02x_{22} + 0.03x_{32}}{x_{12} + x_{22} + x_{32}} \leq 0.02$$

Multiplying both sides of this inequality by $x_{12} + x_{22} + x_{32}$ and simplifying yields

$$-0.015x_{12} + 0.01x_{32} \leq 0 \tag{52}$$

Finally, Constraint 13 is equivalent to

$$\frac{0.005x_{13} + 0.02x_{23} + 0.03x_{33}}{x_{13} + x_{23} + x_{33}} \leq 0.01$$

Multiplying both sides of this inequality by $x_{13} + x_{23} + x_{33}$ and simplifying yields the LP constraint

$$-0.005x_{13} + 0.01x_{23} + 0.02x_{33} \leq 0 \tag{53}$$

Combining (40)–(53), except the redundant constraint (50), with the sign restrictions $x_{ij} \geq 0$ and $a_i \geq 0$ yields an LP that may be expressed in tabular form (see Table 14). In Table 14, the first row (max) represents the objective function, the second row represents the first constraint, and so on. When solved on a computer, an optimal solution to Sunco's LP is found to be

$$z = 287{,}500$$

$$x_{11} = 2222.22 \quad x_{12} = 2111.11 \quad x_{13} = 666.67$$
$$x_{21} = 444.44 \quad x_{22} = 4222.22 \quad x_{23} = 333.34$$
$$x_{31} = 333.33 \quad x_{32} = 3166.67 \quad x_{33} = 0$$
$$a_1 = 0 \quad a_2 = 750 \quad a_3 = 0$$

TABLE 14
Objective Function and Constraints for Blending

x_{11}	x_{12}	x_{13}	x_{21}	x_{22}	x_{23}	x_{31}	x_{32}	x_{33}	a_1	a_2	a_3	
21	11	1	31	21	11	41	31	21	−1	−1	−1	(max)
1	0	0	1	0	0	1	0	0	−10	0	0	= 3,000
0	1	0	0	1	0	0	1	0	0	−10	0	= 2,000
0	0	1	0	0	1	0	0	1	0	0	−10	= 1,000
1	1	1	0	0	0	0	0	0	0	0	0	≤ 5,000
0	0	0	1	1	1	0	0	0	0	0	0	≤ 5,000
0	0	0	0	0	0	1	1	1	0	0	0	≤ 5,000
1	1	1	1	1	1	1	1	1	0	0	0	≤ 14,000
2	0	0	−4	0	0	−2	0	0	0	0	0	≥ 0
0	4	0	0	−2	0	0	0	0	0	0	0	≥ 0
−0.005	0	0	0.01	0	0	0.02	0	0	0	0	0	≤ 0
0	−0.015	0	0	0	0	0	0.01	0	0	0	0	≤ 0
0	0	−0.005	0	0	0.01	0	0	0.02	0	0	0	≤ 0

Thus, Sunco should produce $x_{11} + x_{21} + x_{31} = 3,000$ barrels of gas 1, using 2222.22 barrels of crude 1, 444.44 barrels of crude 2, and 333.33 barrels of crude 3. The firm should produce $x_{12} + x_{22} + x_{32} = 9,500$ barrels of gas 2, using 2,111.11 barrels of crude 1, 4222.22 barrels of crude 2, and 3,166.67 barrels of crude 3. Sunco should also produce $x_{13} + x_{23} + x_{33} = 1,000$ barrels of gas 3, using 666.67 barrels of crude 1 and 333.34 barrels of crude 2. The firm should also spend \$750 on advertising gas 2. Sunco will earn a profit of \$287,500.

Observe that although gas 1 appears to be most profitable, we stimulate demand for gas 2, not gas 1. The reason for this is that given the quality (with respect to octane level and sulfur content) of the available crude, it is difficult to produce gas 1. Therefore, Sunco can make more money by producing more of the lower-quality gas 2 than by producing extra quantities of gas 1.

Modeling Issues

1 We have assumed that the quality level of a mixture is a **linear** function of each input used in the mixture. For example, we have assumed that if gas 3 is made with $\frac{2}{3}$ crude 1 and $\frac{1}{3}$ crude 2, then octane level for gas 3 $= \left(\frac{2}{3}\right) \cdot$ (octane level for crude 1) $+ \left(\frac{1}{3}\right) \cdot$ (octane level for crude 2). If the octane level of a gas is not a linear function of the fraction of each input used to produce the gas, then we no longer have a linear programming problem; we have a **nonlinear programming** problem. For example, let $g_{i3} =$ fraction of gas 3 made with oil i. Suppose that the octane level for gas 3 is given by gas 3 octane level $= g_{13}{}^{.5} \cdot$ (oil 1 octane level) $+ g_{23}{}^{.4} \cdot$ (oil 2 octane level) $+ g_{33}{}^{.3} \cdot$ (oil 3 octane level). Then we do not have an LP problem. The reason for this is that the octane level of gas 3 is not a linear function of g_{13}, g_{23}, and g_{33}. We discuss nonlinear programming in Chapter 11.

2 In reality, a company using a blending model would run the model periodically (each day, say) and set production on the basis of the current inventory of inputs and current demand forecasts. Then the forecast levels and input levels would be updated, and the model would be run again to determine the next day's production.

Real-World Applications

Blending at Texaco

Texaco (see Dewitt et al., 1980) uses a nonlinear programming model (OMEGA) to plan and schedule its blending applications. The company's model is nonlinear because blend volatilities and octanes are nonlinear functions of the amount of each input used to produce a particular gasoline.

Blending in the Steel Industry

Fabian (1958) describes a complex LP model that can be used to optimize the production of iron and steel. For each product produced there are several blending constraints. For example, basic pig iron must contain at most 1.5% silicon, at most .05% sulphur, between .11% and .90% phosphorus, between .4% and 2% manganese, and between 4.1% and 4.4% carbon. See Problem 6 (in the Review Problems section) for a simple example of blending in the steel industry.

Blending in the Oil Industry

Many oil companies use LP to optimize their refinery operations. Problem 14 contains an example (based on Magoulas and Marinos-Kouris [1988]) of a blending model that can be used to maximize a refinery's profit.

PROBLEMS

Group A

1 You have decided to enter the candy business. You are considering producing two types of candies: Slugger Candy and Easy Out Candy, both of which consist solely of sugar, nuts, and chocolate. At present, you have in stock 100 oz of sugar, 20 oz of nuts, and 30 oz of chocolate. The mixture used to make Easy Out Candy must contain at least 20% nuts. The mixture used to make Slugger Candy must contain at least 10% nuts and 10% chocolate. Each ounce of Easy Out Candy can be sold for 25¢ , and each ounce of Slugger Candy for 20¢. Formulate an LP that will enable you to maximize your revenue from candy sales.

2 O.J. Juice Company sells bags of oranges and cartons of orange juice. O.J. grades oranges on a scale of 1 (poor) to 10 (excellent). O.J. now has on hand 100,000 lb of grade 9 oranges and 120,000 lb of grade 6 oranges. The average quality of oranges sold in bags must be at least 7, and the average quality of the oranges used to produce orange juice must be at least 8. Each pound of oranges that is used for juice yields a revenue of $1.50 and incurs a variable cost (consisting of labor costs, variable overhead costs, inventory costs, and so on) of $1.05. Each pound of oranges sold in bags yields a revenue of 50¢ and incurs a variable cost of 20¢. Formulate an LP to help O.J. maximize profit.

3 A bank is attempting to determine where its assets should be invested during the current year. At present, $500,000 is available for investment in bonds, home loans, auto loans, and personal loans. The annual rate of return on each type of investment is known to be: bonds, 10%; home loans, 16%; auto loans, 13%; personal loans, 20%. To ensure that the bank's portfolio is not too risky, the bank's investment manager has placed the following three restrictions on the bank's portfolio:

a The amount invested in personal loans cannot exceed the amount invested in bonds.

b The amount invested in home loans cannot exceed the amount invested in auto loans.

c No more than 25% of the total amount invested may be in personal loans.

The bank's objective is to maximize the annual return on its investment portfolio. Formulate an LP that will enable the bank to meet this goal.

4 Young MBA Erica Cudahy may invest up to $1,000. She can invest her money in stocks and loans. Each dollar invested in stocks yields 10¢ profit, and each dollar invested in a loan yields 15¢ profit. At least 30% of all money invested must be in stocks, and at least $400 must be in loans. Formulate an LP that can be used to maximize total profit earned from Erica's investment. Then graphically solve the LP.

5 Chandler Oil Company has 5,000 barrels of oil 1 and 10,000 barrels of oil 2. The company sells two products: gasoline and heating oil. Both products are produced by combining oil 1 and oil 2. The quality level of each oil is as follows: oil 1—10; oil 2—5. Gasoline must have an average quality level of at least 8, and heating oil at least 6. Demand for each product must be created by advertising. Each dollar spent advertising gasoline creates 5 barrels of demand and each spent on heating oil creates 10 barrels of demand. Gasoline is sold for $25 per barrel, heating oil for $20. Formulate an LP to help Chandler maximize profit. Assume that no oil of either type can be purchased.

6 Bullco blends silicon and nitrogen to produce two types of fertilizers. Fertilizer 1 must be at least 40% nitrogen and sells for $70/lb. Fertilizer 2 must be at least 70% silicon and sells for $40/lb. Bullco can purchase up to 80 lb of nitrogen at $15/lb and up to 100 lb of silicon at $10/lb. Assuming that all fertilizer produced can be sold, formulate an LP to help Bullco maximize profits.

7 Eli Daisy uses chemicals 1 and 2 to produce two drugs. Drug 1 must be at least 70% chemical 1, and drug 2 must be at least 60% chemical 2. Up to 40 oz of drug 1 can be sold at $6 per oz; up to 30 oz of drug 2 can be sold at $5 per oz. Up to 45 oz of chemical 1 can be purchased at $6 per oz, and up to 40 oz of chemical 2 can be purchased at $4 per oz. Formulate an LP that can be used to maximize Daisy's profits.

8 Highland's TV-Radio Store must determine how many TVs and radios to keep in stock. A TV requires 10 sq ft of floorspace, whereas a radio requires 4 sq ft; 200 sq ft of floorspace is available. A TV will earn Highland $60 in profits, and a radio will earn $20. The store stocks only TVs and radios. Marketing requirements dictate that at least 60% of all appliances in stock be radios. Finally, a TV ties up $200 in capital, and a radio, $50. Highland wants to have at most $3,000 worth of capital tied up at any time. Formulate an LP that can be used to maximize Highland's profit.

9 Linear programming models are used by many Wall Street firms to select a desirable bond portfolio. The following is a simplified version of such a model. Solodrex is considering investing in four bonds; $1,000,000 is available for investment. The expected annual return, the worst-case annual return on each bond, and the "duration" of each bond are given in Table 15. The duration of a bond is a measure of the bond's sensitivity to interest rates. Solodrex wants to maximize the expected return from its bond investments, subject to three constraints.

Constraint 1 The worst-case return of the bond portfolio must be at least 8%.
Constraint 2 The average duration of the portfolio must be at most 6. For example, a portfolio that invested $600,000

in bond 1 and $400,000 in bond 4 would have an average duration of

$$\frac{600,000(3) + 400,000(9)}{1,000,000} = 5.4$$

Constraint 3 Because of diversification requirements, at most 40% of the total amount invested can be invested in a single bond.

Formulate an LP that will enable Solodrex to maximize the expected return on its investment.

10 Coalco produces coal at three mines and ships it to four customers. The cost per ton of producing coal, the ash and sulfur content (per ton) of the coal, and the production capacity (in tons) for each mine are given in Table 16. The number of tons of coal demanded by each customer are given in Table 17.
   The cost (in dollars) of shipping a ton of coal from a mine to each customer is given in Table 18. It is required that the total amount of coal shipped contain at most 5% ash and at most 4% sulfur. Formulate an LP that minimizes the cost of meeting customer demands.

11 Eli Daisy produces the drug Rozac from four chemicals. Today they must produce 1,000 lb of the drug. The three active ingredients in Rozac are A, B, and C. By weight, at least 8% of Rozac must consist of A, at least 4% of B, and at least 2% of C. The cost per pound of each chemical and the amount of each active ingredient in 1 lb of each chemical are given in Table 19.
   It is necessary that at least 100 lb of chemical 2 be used. Formulate an LP whose solution would determine the cheapest way of producing today's batch of Rozac.

TABLE 16

Mine	Production Cost ($)	Capacity	Ash Content (Tons)	Sulfur Content (Tons)
1	50	120	.08	.05
2	55	100	.06	.04
3	62	140	.04	.03

TABLE 17

Customer 1	Customer 2	Customer 3	Customer 4
80	70	60	40

TABLE 15

Bond	Expected Return (%)	Worst-Case Return (%)	Duration
1	13	6%	3
2	8	8%	4
3	12	10%	7
4	14	9%	9

TABLE 18

	Customer			
Mine	1	2	3	4
1	4	6	8	12
2	9	6	7	11
3	8	12	3	5

TABLE **19**

Chemical	Cost ($ per Lb)	A	B	C
1	8	.03	.02	.01
2	10	.06	.04	.01
3	11	.10	.03	.04
4	14	.12	.09	.04

12 (A spreadsheet might be helpful on this problem.) The *risk index* of an investment can be obtained from return on investment (ROI) by taking the percentage of change in the value of the investment (in absolute terms) for each year, and averaging them.

Suppose you are trying to determine what percentage of your money should be invested in T-bills, gold, and stocks. In Table 20 (or File Inv68.xls) you are given the annual returns (change in value) for these investments for the years 1968–1988. Let the risk index of a portfolio be the weighted (according to the fraction of your money assigned to each investment) average of the risk index of each individual investment. Suppose that the amount of each investment must be between 20% and 50% of the total invested. You would like the risk index of your portfolio to equal .15, and your goal is to maximize the expected return on your portfolio. Formulate an LP whose solution will maximize the expected return on your portfolio, subject to the given constraints. Use the average return earned by each investment during the years 1968–1988 as your estimate of expected return.[†]

Group B

13 The owner of Sunco does not believe that our LP optimal solution will maximize daily profit. He reasons, "We have 14,000 barrels of daily refinery capacity, but your optimal solution produces only 13,500 barrels. Therefore, it cannot be maximizing profit." How would you respond?

14 Oilco produces two products: regular and premium gasoline. Each product contains .15 gram of lead per liter. The two products are produced from six inputs: reformate, fluid catalytic cracker gasoline (FCG), isomerate (ISO), polymer (POL), MTBE (MTB), and butane (BUT). Each input has four attributes:

Attribute 1 Research octane number (RON)
Attribute 2 RVP
Attribute 3 ASTM volatility at 70°C
Attribute 4 ASTM volatility at 130°C

The attributes and daily availability (in liters) of each input are given in Table 21.

The requirements for each output are given in Table 22.

The daily demand (in thousands of liters) for each product must be met, but more can be produced if desired. The RON and ASTM requirements are minimums. Regular gasoline sells for 29.49¢/liter, premium gasoline for 31.43¢. Before being ready for sale, .15 gram/liter of lead must be re-moved from each product. The cost of removing .1 gram/liter is 8.5¢. At most 38% of each type of gasoline can consist of FCG. Formulate and solve an LP whose solution will tell Oilco how to maximize their daily profit.[‡]

TABLE **20**

Year	Stocks	Gold	T-Bills
1968	11	11	5
1969	−9	8	7
1970	4	−14	7
1971	14	14	4
1972	19	44	4
1973	−15	66	7
1974	−27	64	8
1975	37	0	6
1976	24	−22	5
1977	−7	18	5
1978	7	31	7
1979	19	59	10
1980	33	99	11
1981	−5	−25	15
1982	22	4	11
1983	23	−11	9
1984	6	−15	10
1985	32	−12	8
1986	19	16	6
1987	5	22	5
1988	17	−2	6

TABLE **21**

	Availability	RON	RVP	ASTM(70)	ASTM(130)
Reformate	15,572	98.9	7.66	−5	46
FCG	15,434	93.2	9.78	57	103
ISO	6,709	86.1	29.52	107	100
POL	1,190	97	14.51	7	73
MTB	748	117	13.45	98	100
BUT	Unlimited	98	166.99	130	100

TABLE **22**

	Demand	RON	RVP	ASTM(70)	ASTM(130)
Regular	9.8	90	21.18	10	50
Premium	30	96	21.18	10	50

[†]Based on Chandy (1987).

[‡]Based on Magoulas and Marinos-Kouris (1988).

3.9 Production Process Models

We now explain how to formulate an LP model of a simple production process.[†] The key step is to determine how the outputs from a later stage of the process are related to the outputs from an earlier stage.

EXAMPLE 13   **Brute Production Process**

Rylon Corporation manufactures Brute and Chanelle perfumes. The raw material needed to manufacture each type of perfume can be purchased for $3 per pound. Processing 1 lb of raw material requires 1 hour of laboratory time. Each pound of processed raw material yields 3 oz of Regular Brute Perfume and 4 oz of Regular Chanelle Perfume. Regular Brute can be sold for $7/oz and Regular Chanelle for $6/oz. Rylon also has the option of further processing Regular Brute and Regular Chanelle to produce Luxury Brute, sold at $18/oz, and Luxury Chanelle, sold at $14/oz. Each ounce of Regular Brute processed further requires an additional 3 hours of laboratory time and $4 processing cost and yields 1 oz of Luxury Brute. Each ounce of Regular Chanelle processed further requires an additional 2 hours of laboratory time and $4 processing cost and yields 1 oz of Luxury Chanelle. Each year, Rylon has 6,000 hours of laboratory time available and can purchase up to 4,000 lb of raw material. Formulate an LP that can be used to determine how Rylon can maximize profits. Assume that the cost of the laboratory hours is a fixed cost.

Solution   Rylon must determine how much raw material to purchase and how much of each type of perfume should be produced. We therefore define our decision variables to be

$$x_1 = \text{number of ounces of Regular Brute sold annually}$$
$$x_2 = \text{number of ounces of Luxury Brute sold annually}$$
$$x_3 = \text{number of ounces of Regular Chanelle sold annually}$$
$$x_4 = \text{number of ounces of Luxury Chanelle sold annually}$$
$$x_5 = \text{number of pounds of raw material purchased annually}$$

Rylon wants to maximize

$$\text{Contribution to profit} = \text{revenues from perfume sales} - \text{processing costs}$$
$$- \text{costs of purchasing raw material}$$
$$= 7x_1 + 18x_2 + 6x_3 + 14x_4 - (4x_2 + 4x_4) - 3x_5$$
$$= 7x_1 + 14x_2 + 6x_3 + 10x_4 - 3x_5$$

Thus, Rylon's objective function may be written as

$$\max z = 7x_1 + 14x_2 + 6x_3 + 10x_4 - 3x_5 \tag{54}$$

Rylon faces the following constraints:

Constraint 1   No more than 4,000 lb of raw material can be purchased annually.

Constraint 2   No more than 6,000 hours of laboratory time can be used each year.

Constraint 1 is expressed by

$$x_5 \leq 4,000 \tag{55}$$

[†]This section is based on Hartley (1971).

To express Constraint 2, note that

Total lab time used annually = time used annually to process raw material

+ time used annually to process Luxury Brute

+ time used annually to process Luxury Chanelle

$= x_5 + 3x_2 + 2x_4$

Then Constraint 2 becomes

$$3x_2 + 2x_4 + x_5 \leq 6{,}000 \tag{56}$$

After adding the sign restrictions $x_i \geq 0$ ($i = 1, 2, 3, 4, 5$), many students claim that Rylon should solve the following LP:

$$\max z = 7x_1 + 14x_2 + 6x_3 + 10x_4 - 3x_5$$
$$\text{s.t.} \qquad\qquad\quad x_5 \leq 4{,}000$$
$$3x_2 + 2x_4 + x_5 \leq 6{,}000$$
$$x_i \geq 0 \quad (i = 1, 2, 3, 4, 5)$$

This formulation is incorrect. Observe that the variables x_1 and x_3 do not appear in any of the constraints. This means that any point with $x_2 = x_4 = x_5 = 0$ and x_1 and x_3 very large is in the feasible region. Points with x_1 and x_3 large can yield arbitrarily large profits. Thus, this LP is unbounded. Our mistake is that the current formulation does not indicate that the amount of raw material purchased determines the amount of Brute and Chanelle that is available for sale or further processing. More specifically, from Figure 10 (and the fact that 1 oz of processed Brute yields exactly 1 oz of Luxury Brute), it follows that

$$\begin{array}{c} \text{Ounces of Regular Brute Sold} \\ \text{+ ounces of Luxury Brute sold} \end{array} = \left(\frac{\text{ounces of Brute produced}}{\text{pound of raw material}} \right) \left(\begin{array}{c} \text{pounds of raw} \\ \text{material purchased} \end{array} \right)$$

$$= 3x_5$$

This relation is reflected in the constraint

$$x_1 + x_2 = 3x_5 \qquad \text{or} \qquad x_1 + x_2 - 3x_5 = 0 \tag{57}$$

Similarly, from Figure 10 it is clear that

Ounces of Regular Chanelle sold + ounces of Luxury Chanelle sold $= 4x_5$

This relation yields the constraint

$$x_3 + x_4 = 4x_5 \qquad \text{or} \qquad x_3 + x_4 - 4x_5 = 0 \tag{58}$$

Constraints (57) and (58) relate several decision variables. Students often omit constraints of this type. As this problem shows, leaving out even one constraint may very well

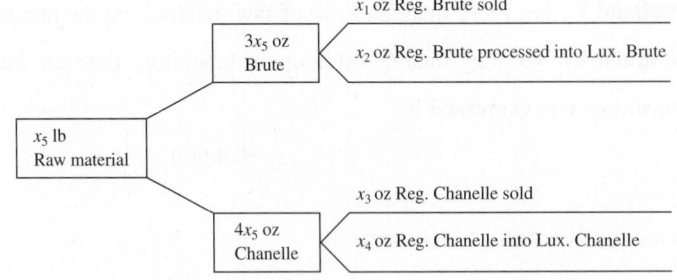

FIGURE 10
Production Process for
Brute and Chanelle

lead to an unacceptable answer (such as an unbounded LP). If we combine (53)–(58) with the usual sign restrictions, we obtain the *correct* LP formulation.

$$\max z = 7x_1 + 14x_2 + 6x_3 + 10x_4 - 3x_5$$

$$\text{s.t.} \qquad\qquad\qquad\qquad\qquad\qquad x_5 \leq 4{,}000$$

$$3x_2 \qquad + 2x_4 + x_5 \leq 6{,}000$$

$$x_1 + x_2 \qquad\qquad - 3x_5 = 0$$

$$x_3 + x_4 - 4x_5 = 0$$

$$x_i \geq 0 \qquad (i = 1, 2, 3, 4, 5)$$

The optimal solution is $z = 172{,}666.667$, $x_1 = 11{,}333.333$ oz, $x_2 = 666.667$ oz, $x_3 = 16{,}000$ oz, $x_4 = 0$, and $x_5 = 4{,}000$ lb. Thus, Rylon should purchase all 4,000 lb of available raw material and produce 11,333.333 oz of Regular Brute, 666.667 oz of Luxury Brute, and 16,000 oz of Regular Chanelle. This production plan will contribute $172,666.667 to Rylon's profits. In this problem, a fractional number of ounces seems reasonable, so the Divisibility Assumption holds.

We close our discussion of the Rylon problem by discussing an error that is made by many students. They reason that

$$1 \text{ lb raw material} = 3 \text{ oz Brute} + 4 \text{ oz Chanelle}$$

Because $x_1 + x_2 = $ total ounces of Brute produced, and $x_3 + x_4 = $ total ounces of Chanelle produced, students conclude that

$$x_5 = 3(x_1 + x_2) + 4(x_3 + x_4) \tag{59}$$

This equation might make sense as a statement for a computer program; in a sense, the variable x_5 is replaced by the right side of (59). As an LP constraint, however, (59) makes no sense. To see this, note that the left side has the units "pounds of raw material," and the term $3x_1$ on the right side has the units

$$\left(\frac{\text{Ounces of Brute}}{\text{Pounds of raw material}} \right) (\text{ounces of Brute})$$

Because some of the terms do not have the same units, (59) cannot be correct. *If there are doubts about a constraint, then make sure that all terms in the constraint have the same units.* This will avoid many formulation errors. (Of course, even if the units on both sides of a constraint are the same, the constraint may still be wrong.)

PROBLEMS

Group A

1 Sunco Oil has three different processes that can be used to manufacture various types of gasoline. Each process involves blending oils in the company's catalytic cracker. Running process 1 for an hour costs $5 and requires 2 barrels of crude oil 1 and 3 barrels of crude oil 2. The output from running process 1 for an hour is 2 barrels of gas 1 and 1 barrel of gas 2. Running process 2 for an hour costs $4 and requires 1 barrel of crude 1 and 3 barrels of crude 2. The output from running process 2 for an hour is 3 barrels of gas 2. Running process 3 for an hour costs $1 and requires 2 barrels of crude 2 and 3 barrels of gas 2. The output from running process 3 for an hour is 2 barrels of gas 3. Each week, 200 barrels of crude 1, at $2/barrel, and 300 barrels of crude 2, at $3/barrel, may be purchased. All gas produced can be sold at the following per-barrel prices: gas 1, $9; gas 2, $10; gas 3, $24. Formulate an LP whose solution will maximize revenues less costs. Assume that only 100 hours of time on the catalytic cracker are available each week.

2 Furnco manufactures tables and chairs. A table requires 40 board ft of wood, and a chair requires 30 board ft of wood. Wood may be purchased at a cost of $1 per board ft, and 40,000 board ft of wood are available for purchase. It takes 2 hours of skilled labor to manufacture an unfinished table or an unfinished chair. Three more hours of skilled labor will turn an unfinished table into a finished table, and 2 more hours of skilled labor will turn an unfinished chair into a finished chair. A total of 6,000 hours of skilled labor are available (and have already been paid for). All furniture produced can be sold at the following unit prices: unfinished table, $70; finished table, $140; unfinished chair, $60; finished chair, $110. Formulate an LP that will maximize the contribution to profit from manufacturing tables and chairs.

3 Suppose that in Example 11, 1 lb of raw material could be used to produce either 3 oz of Brute *or* 4 oz of Chanelle. How would this change the formulation?

4 Chemco produces three products: 1, 2, and 3. Each pound of raw material costs $25. It undergoes processing and yields 3 oz of product 1 and 1 oz of product 2. It costs $1 and takes 2 hours of labor to process each pound of raw material. Each ounce of product 1 can be used in one of three ways.

It can be sold for $10/oz.

It can be processed into 1 oz of product 2. This requires 2 hours of labor and costs $1.

It can be processed into 1 oz of product 3. This requires 3 hours of labor and costs $2.

Each ounce of product 2 can be used in one of two ways.

It can be sold for $20/oz.

It can be processed into 1 oz of product 3. This requires 1 hour of labor and costs $6.

Product 3 is sold for $30/oz. The maximum number of ounces of each product that can be sold is given in Table 23. A maximum of 25,000 hours of labor are available. Determine how Chemco can maximize profit.

TABLE 23

Product	Oz
1	5,000
2	5,000
3	3,000

Group B

5 A company produces A, B, and C and can sell these products in unlimited quantities at the following unit prices: A, $10; B, $56; C, $100. Producing a unit of A requires 1 hour of labor; a unit of B, 2 hours of labor plus 2 units of A; and a unit of C, 3 hours of labor plus 1 unit of B. Any A that is used to produce B cannot be sold. Similarly, any B that is used to produce C cannot be sold. A total of 40 hours of labor are available. Formulate an LP to maximize the company's revenues.

6 Daisy Drugs manufactures two drugs: 1 and 2. The drugs are produced by blending two chemicals: 1 and 2. By weight, drug 1 must contain at least 65% chemical 1, and drug 2 must contain at least 55% chemical 1. Drug 1 sells for $6/oz, and drug 2 sells for $4/oz. Chemicals 1 and 2 can be produced by one of two production processes. Running process 1 for an hour requires 3 oz of raw material and 2 hours skilled labor and yields 3 oz of each chemical. Running process 2 for an hour requires 2 oz of raw material and 3 hours of skilled labor and yields 3 oz of chemical 1 and 1 oz of chemical 2. A total of 120 hours of skilled labor and 100 oz of raw material are available. Formulate an LP that can be used to maximize Daisy's sales revenues.

7[†] Lizzie's Dairy produces cream cheese and cottage cheese. Milk and cream are blended to produce these two products. Both high-fat and low-fat milk can be used to produce cream cheese and cottage cheese. High-fat milk is 60% fat; low-fat milk is 30% fat. The milk used to produce cream cheese must average at least 50% fat and that for cottage cheese, at least 35% fat. At least 40% (by weight) of the inputs to cream cheese and at least 20% (by weight) of the inputs to cottage cheese must be cream. Both cottage cheese and cream cheese are produced by putting milk and cream through the cheese machine. It costs 40¢ to process 1 lb of inputs into a pound of cream cheese. It costs 40¢ to produce 1 lb of cottage cheese, but every pound of input for cottage cheese yields 0.9 lb of cottage cheese and 0.1 lb of waste. Cream can be produced by evaporating high-fat and low-fat milk. It costs 40¢ to evaporate 1 lb of high-fat milk. Each pound of high-fat milk that is evaporated yields 0.6 lb of cream. It costs 40¢ to evaporate 1 lb of low-fat milk. Each pound of low-fat milk that is evaporated yields 0.3 lb of cream. Each day, up to 3,000 lb of input may be sent through the cheese machine. Each day, at least 1,000 lb of cottage cheese and 1,000 lb of cream cheese must be produced. Up to 1,500 lb of cream cheese and 2,000 lb of cottage cheese can be sold each day. Cottage cheese is sold for $1.20/lb and cream cheese for $1.50/lb. High-fat milk is purchased for 80¢/lb and low-fat milk for 40¢/lb. The evaporator can process at most 2,000 lb of milk daily. Formulate an LP that can be used to maximize Lizzie's daily profit.

8 A company produces six products in the following fashion. Each unit of raw material purchased yields four units of product 1, two units of product 2, and one unit of product 3. Up to 1,200 units of product 1 can be sold, and up to 300 units of product 2 can be sold. Each unit of product 1 can be sold or processed further. Each unit of product 1 that is processed yields a unit of product 4. Demand for products 3 and 4 is unlimited. Each unit of product 2 can be sold or processed further. Each unit of product 2 that is processed further yields 0.8 unit of product 5 and 0.3 unit of product 6. Up to 1,000 units of product 5 can be sold, and up to 800 units of product 6 can be sold. Up to 3,000 units of raw material can be purchased at $6 per unit. Leftover units of products 5 and 6 must be destroyed. It costs $4 to destroy each leftover unit of product 5 and $3

[†]Based on Sullivan and Secrest (1985).

TABLE 24

Product	Sales Price ($)	Production Cost ($)
1	7	4
2	6	4
3	4	2
4	3	1
5	20	5
6	35	5

TABLE 26

Grade	1	2	3	4	5	6
Amount produced	2	3	1	1.5	2	3

TABLE 27

Grade	1	2	3	4	5	6
Maximum demand	20	30	40	35	25	50

to destroy each leftover unit of product 6. Ignoring raw material purchase costs, the per-unit sales price and production costs for each product are shown in Table 24. Formulate an LP whose solution will yield a profit-maximizing production schedule.

9 Each week Chemco can purchase unlimited quantities of raw material at $6/lb. Each pound of purchased raw material can be used to produce either input 1 or input 2. Each pound of raw material can yield 2 oz of input 1, requiring 2 hours of processing time and incurring $2 in processing costs. Each pound of raw material can yield 3 oz of input 2, requiring 2 hours of processing time and incurring $4 in processing costs.

Two production processes are available. It takes 2 hours to run process 1, requiring 2 oz of input 1 and 1 oz of input 2. It costs $1 to run process 1. Each time process 1 is run 1 oz of product A and 1 oz of liquid waste are produced. Each time process 2 is run requires 3 hours of processing time, 2 oz of input 2 and 1 oz of input 1. Process 2 yields 1 oz of product B and .8 oz of liquid waste. Process 2 incurs $8 in costs.

Chemco can dispose of liquid waste in the Port Charles River or use the waste to produce product C or product D. Government regulations limit the amount of waste Chemco is allowed to dump into the river to 1,000 oz/week. One ounce of product C costs $4 to produce and sells for $11. One hour of processing time, 2 oz of input 1, and .8 oz of liquid waste are needed to produce an ounce of product C. One unit of product D costs $5 to produce and sells for $7. One hour of processing time, 2 oz of input 2, and 1.2 oz of liquid waste are needed to produce an ounce of product D.

At most 5,000 oz of product A and 5,000 oz of product B can be sold each week, but weekly demand for products C and D is unlimited. Product A sells for $18/oz and product B sells for $24/oz. Each week 6,000 hours of processing time is available. Formulate an LP whose solution will tell Chemco how to maximize weekly profit.

10 LIMECO owns a lime factory and sells six grades of lime (grades 1 through 6). The sales price per pound is given in Table 25. Lime is produced by kilns. If a kiln is run for an 8-hour shift, the amounts (in pounds) of each grade

of lime given in Table 26 are produced. It costs $150 to run a kiln for an 8-hour shift. Each day the factory believes it can sell up to the amounts (in pounds) of lime given in Table 27.

Lime that is produced by the kiln may be reprocessed by using any one of the five processes described in Table 28.

For example, at a cost of $1/lb, a pound of grade 4 lime may be transformed into .5 lb of grade 5 lime and .5 lb of grade 6 lime.

Any extra lime leftover at the end of each day must be disposed of, with the disposal costs (per pound) given in Table 29.

Formulate an LP whose solution will tell LIMECO how to maximize their daily profit.

11 Chemco produces three products: A, B, and C. They can sell up to 30 pounds of each product at the following prices (per pound): product A, $10; product B, $12; product C, $20. Chemco purchases raw material at $5/lb. Each pound of raw material can be used to produce *either* 1 lb of A or 1 lb of B. For a cost of $3/lb processed, product A can be converted to .6 lb of product B and .4 lb of product C. For a cost of $2/lb processed, product B can be converted to .8 lb of product C. Formulate an LP whose solution will tell Chemco how to maximize their profit.

12 Chemco produces 3 chemicals: B, C, and D. They begin by purchasing chemical A for a cost of $6/100 liters. For an

TABLE 28

Input (1 Lb)	Output	Cost ($ per Lb of Input)
Grade 1	.3 lb Grade 3	
	.2 lb Grade 4	2
	.3 lb Grade 5	
	.2 lb Grade 6	
Grade 2	1 lb Grade 6	1
Grade 3	.8 lb Grade 4	1
Grade 4	.5 lb Grade 5	1
	.5 lb Grade 6	
Grade 5	.9 lb Grade 6	2

TABLE 25

Grade	1	2	3	4	5	6
Price($)	12	14	10	18	20	25

TABLE 29

Grade	1	2	3	4	5	6
Cost of Disposition ($)	3	2	3	2	4	2

additional cost of $3 and the use of 3 hours of skilled labor, 100 liters of A can be transformed into 40 liters of C and 60 liters of B. Chemical C can either be sold or processed further. It costs $1 and takes 1 hour of skilled labor to process 100 liters of C into 60 liters of D and 40 liters of B. For each chemical the sales price per 100 liters and the maximum amount (in 100s of liters) that can be sold are given in Table 30.

A maximum of 200 labor hours are available. Formulate an LP whose solution will tell Chemco how to maximize their profit.

13 Carrington Oil produces two types of gasoline, gas 1 and gas 2, from two types of crude oil, crude 1 and crude 2. Gas 1 is allowed to contain up to 4% impurities, and gas 2 is allowed to contain up to 3% impurities. Gas 1 sells for $8 per barrel, whereas gas 2 sells for $12 per barrel. Up to 4,200 barrels of gas 1 and up to 4,300 barrels of gas 2 can be sold. The cost per barrel of each crude, availability, and the level of impurities in each crude are as shown in Table 31. Before blending the crude oil into gas, any amount of each crude can be "purified" for a cost of $0.50 per barrel. Purification eliminates half the impurities in the crude oil. Determine how to maximize profit.

14 You have been put in charge of the Melrose oil refinery. The refinery produces gas and heating oil from crude oil. Gas sells for $8 per barrel and must have an average "grade level" of at least 9. Heating oil sells for $6 a barrel and must

have an average grade level of at least 7. At most, 2,000 barrels of gas and 600 barrels of heating oil can be sold. Incoming crude can be processed by one of three methods. The per barrel yield and per barrel cost of each processing method are shown in Table 32. For example, if we refine 1 barrel of incoming crude by method 1, it costs us $3.40 and yields .2 barrels of grade 6, .2 barrels of grade 8, and .6 barrels of grade 10.

Before being processed into gas and heating oil, processed grades 6 and 8 may be sent through the catalytic cracker to improve their quality. For $1.30 per barrel, a barrel of grade 6 may be "cracked" into a barrel of grade 8. For $2 per barrel, a barrel of grade 8 may be cracked into a barrel of grade 10. Any leftover processed or cracked oil that cannot be used for heating oil or gas must be disposed of at a cost of $0.20 per barrel. Determine how to maximize the refinery's profit.

TABLE 31

Oil	Cost per Barrel ($)	Impurity Level (%)	Availability (Barrels)
Crude 1	6	10%	5,000
Crude 2	8	2%	4,500

TABLE 30

	B	C	D
Price ($)	12	16	26
Maximum demand	30	60	40

TABLE 32

Method	Grade 6	Grade 8	Grade 10	Cost ($)
1	.2	.2	.6	3.40
2	.3	.3	.4	3.00
3	.4	.4	.2	2.60

3.10 Using Linear Programming to Solve Multiperiod Decision Problems: An Inventory Model

Up to this point, all the LP formulations we have discussed are examples of *static,* or *one-period, models.* In a static model, we assume that all decisions are made at a single point in time. The rest of the examples in this chapter show how linear programming can be used to determine optimal decisions in **multiperiod,** or **dynamic, models.** Dynamic models arise when the decision maker makes decisions at more than one point in time. In a dynamic model, decisions made during the current period influence decisions made during future periods. For example, consider a company that must determine how many units of a product should be produced during each month. If it produced a large number of units during the current month, this would reduce the number of units that should be produced during future months. The examples discussed in Sections 3.10–3.12 illustrate how earlier decisions affect later decisions. We will return to dynamic decision models when we study dynamic programming in Chapters 18 and 19.

EXAMPLE 14   **Sailco Inventory**

Sailco Corporation must determine how many sailboats should be produced during each of the next four quarters (one quarter = three months). The demand during each of the next four quarters is as follows: first quarter, 40 sailboats; second quarter, 60 sailboats; third quarter, 75 sailboats; fourth quarter, 25 sailboats. Sailco must meet demands on time. At the beginning of the first quarter, Sailco has an inventory of 10 sailboats. At the beginning of each quarter, Sailco must decide how many sailboats should be produced during that quarter. For simplicity, we assume that sailboats manufactured during a quarter can be used to meet demand for that quarter. During each quarter, Sailco can produce up to 40 sailboats with regular-time labor at a total cost of $400 per sailboat. By having employees work overtime during a quarter, Sailco can produce additional sailboats with overtime labor at a total cost of $450 per sailboat.

At the end of each quarter (after production has occurred and the current quarter's demand has been satisfied), a carrying or holding cost of $20 per sailboat is incurred. Use linear programming to determine a production schedule to minimize the sum of production and inventory costs during the next four quarters.

Solution   For each quarter, Sailco must determine the number of sailboats that should be produced by regular-time and by overtime labor. Thus, we define the following decision variables:

x_t = number of sailboats produced by regular-time labor (at $400/boat)
during quarter t ($t = 1, 2, 3, 4$)

y_t = number of sailboats produced by overtime labor (at $450/boat)
during quarter t ($t = 1, 2, 3, 4$)

It is convenient to define decision variables for the inventory (number of sailboats on hand) at the end of each quarter:

i_t = number of sailboats on hand at end of quarter t ($t = 1, 2, 3, 4$)

Sailco's total cost may be determined from

Total cost = cost of producing regular-time boats

+ cost of producing overtime boats + inventory costs

= $400(x_1 + x_2 + x_3 + x_4) + 450(y_1 + y_2 + y_3 + y_4)$

+ $20(i_1 + i_2 + i_3 + i_4)$

Thus, Sailco's objective function is

$$\min z = 400x_1 + 400x_2 + 400x_3 + 400x_4 + 450y_1 + 450y_2$$
$$+ 450y_3 + 450y_4 + 20i_1 + 20i_2 + 20i_3 + 20i_4 \tag{60}$$

Before determining Sailco's constraints, we make two observations that will aid in formulating multiperiod production-scheduling models.

For quarter t,

Inventory at end of quarter t = inventory at end of quarter $(t - 1)$

+ quarter t production − quarter t demand

This relation plays a key role in formulating almost all multiperiod production-scheduling models. If we let d_t be the demand during period t (thus, $d_1 = 40$, $d_2 = 60$, $d_3 = 75$, and $d_4 = 25$), our observation may be expressed in the following compact form:

$$i_t = i_{t-1} + (x_t + y_t) - d_t \qquad (t = 1, 2, 3, 4) \tag{61}$$

In (61), i_0 = inventory at end of quarter 0 = inventory at beginning of quarter 1 = 10. For example, if we had 20 sailboats on hand at the end of quarter 2 ($i_2 = 20$) and produced 65 sailboats during quarter 3 (this means $x_3 + y_3 = 65$), what would be our ending third-quarter inventory? Simply the number of sailboats on hand at the end of quarter 2 plus the sailboats produced during quarter 3, less quarter 3's demand of 75. In this case, $i_3 = 20 + 65 - 75 = 10$, which agrees with (61). Equation (61) relates decision variables associated with different time periods. In formulating any multiperiod LP model, the hardest step is usually finding the relation (such as (61)) that relates decision variables from different periods.

We also note that quarter t's demand will be met on time if and only if (sometimes written *iff*) $i_t \geq 0$. To see this, observe that $i_{t-1} + (x_t + y_t)$ is available to meet period t's demand, so that period t's demand will be met if and only if

$$i_{t-1} + (x_t + y_t) \geq d_t \quad \text{or} \quad i_t = i_{t-1} + (x_t + y_t) - d_t \geq 0$$

This means that the sign restrictions $i_t \geq 0$ ($t = 1, 2, 3, 4$) will ensure that each quarter's demand will be met on time.

We can now determine Sailco's constraints. First, we use the following four constraints to ensure that each period's regular-time production will not exceed 40: $x_1, x_2, x_3, x_4 \leq 40$. Then we add constraints of the form (61) for each time period ($t = 1, 2, 3, 4$). This yields the following four constraints:

$$i_1 = 10 + x_1 + y_1 - 40 \qquad i_2 = i_1 + x_2 + y_2 - 60$$
$$i_3 = i_2 + x_3 + y_3 - 75 \qquad i_4 = i_3 + x_4 + y_4 - 25$$

Adding the sign restrictions $x_t \geq 0$ (to rule out negative production levels) and $i_t \geq 0$ (to ensure that each period's demand is met on time) yields the following formulation:

$$\min z = 400x_1 + 400x_2 + 400x_3 + 400x_4 + 450y_1 + 450y_2 + 450y_3 + 450y_4$$
$$+ 20i_1 + 20i_2 + 20i_3 + 20i_4$$

s.t. $\quad x_1 \leq 40, \qquad x_2 \leq 40, \qquad x_3 \leq 40, \qquad x_4 \leq 40$

$\quad\quad i_1 = 10 + x_1 + y_1 - 40, \qquad i_2 = i_1 + x_2 + y_2 - 60$

$\quad\quad i_3 = i_2 + x_3 + y_3 - 75, \qquad i_4 = i_3 + x_4 + y_4 - 25$

$\quad\quad i_t \geq 0, \qquad y_t \geq 0, \qquad \text{and} \qquad x_t \geq 0 \qquad (t = 1, 2, 3, 4)$

The optimal solution to this problem is $z = 78{,}450$; $x_1 = x_2 = x_3 = 40$; $x_4 = 25$; $y_1 = 0$; $y_2 = 10$; $y_3 = 35$; $y_4 = 0$; $i_1 = 10$; $i_2 = i_3 = i_4 = 0$. Thus, the minimum total cost that Sailco can incur is $78,450. To incur this cost, Sailco should produce 40 sailboats with regular-time labor during quarters 1–3 and 25 sailboats with regular-time labor during quarter 4. Sailco should also produce 10 sailboats with overtime labor during quarter 2 and 35 sailboats with overtime labor during quarter 3. Inventory costs will be incurred only during quarter 1.

Some readers might worry that our formulation allows Sailco to use overtime production during quarter t even if period t's regular production is less than 40. True, our formulation does not make such a schedule infeasible, but any production plan that had $y_t > 0$ and $x_t < 40$ could not be optimal. For example, consider the following two production schedules:

Production schedule A $= x_1 = x_2 = x_3 = 40$; $\quad x_4 = 25$;

$\quad\quad\quad\quad\quad\quad\quad\quad\quad y_2 = 10$; $\quad\quad y_3 = 25$; $\quad\quad y_4 = 0$

Production schedule B $= x_1 = 40$; $\quad\quad x_2 = 30$; $\quad\quad x_3 = 30$; $\quad\quad x_4 = 25$;

$\quad\quad\quad\quad\quad\quad\quad\quad\quad y_2 = 20$; $\quad\quad y_3 = 35$; $\quad\quad y_4 = 0$

Schedules A and B both have the same production level during each period. This means that both schedules will have identical inventory costs. Also, both schedules are feasible, but schedule B incurs more overtime costs than schedule A. Thus, in minimizing costs, schedule B (or any schedule having $y_t > 0$ and $x_t < 40$) would never be chosen.

In reality, an LP such as Example 14 would be implemented by using a **rolling horizon,** which works in the following fashion. After solving Example 14, Sailco would implement only the quarter 1 production strategy (produce 40 boats with regular-time labor). Then the company would observe quarter 1's actual demand. Suppose quarter 1's actual demand is 35 boats. Then quarter 2 begins with an inventory of $10 + 40 - 35 = 15$ boats. We now make a forecast for quarter 5 demand (suppose the forecast is 36). Next determine production for quarter 2 by solving an LP in which quarter 2 is the first quarter, quarter 5 is the final quarter, and beginning inventory is 15 boats. Then quarter 2's production would be determined by solving the following LP:

$$\min z = 400(x_2 + x_3 + x_4 + x_5) + 450(y_2 + y_3 + y_4 + y_5) + 20(i_2 + i_3 + i_4 + i_5)$$

$$\text{s.t.} \quad x_2 \leq 40, \quad x_3 \leq 40, \quad x_4 \leq 40, \quad x_5 \leq 40$$

$$i_2 = 15 + x_2 + y_2 - 60, \quad i_3 = i_2 + x_3 + y_3 - 75$$

$$i_4 = i_3 + x_4 + y_4 - 25, \quad i_5 = i_4 + x_5 + y_5 - 36$$

$$i_t \geq 0, \quad y_t \geq 0, \quad \text{and} \quad x_t \geq 0 \quad (t = 2, 3, 4, 5)$$

Here, $x_5 =$ quarter 5's regular-time production, $y_5 =$ quarter 5's overtime production, and $i_5 =$ quarter 5's ending inventory. The optimal values of x_2 and y_2 for this LP are then used to determine quarter 2's production. Thus, each quarter, an LP (with a planning horizon of four quarters) is solved to determine the current quarter's production. Then current demand is observed, demand is forecasted for the next four quarters, and the process repeats itself. This technique of "rolling planning horizon" is the method by which most dynamic or multiperiod LP models are implemented in real-world applications.

Our formulation of the Sailco problem has several other limitations.

1 Production cost may not be a linear function of the quantity produced. This would violate the Proportionality Assumption. We discuss how to deal with this problem in Chapters 9 and 13.

2 Future demands may not be known with certainty. In this situation, the Certainty Assumption is violated.

3 We have required Sailco to meet all demands on time. Often companies can meet demands during later periods but are assessed a penalty cost for demands that are not met on time. For example, if demand is not met on time, then customer displeasure may result in a loss of future revenues. If demand can be met during later periods, then we say that demands can be **backlogged.** Our current LP formulation can be modified to incorporate backlogging (see Problem 1 of Section 4.12).

4 We have ignored the fact that quarter-to-quarter variations in the quantity produced may result in extra costs (called **production-smoothing costs.**) For example, if we increase production a great deal from one quarter to the next, this will probably require the costly training of new workers. On the other hand, if production is greatly decreased from one quarter to the next, extra costs resulting from laying off workers may be incurred. In Section 4.12, we modify the present model to account for smoothing costs.

5 If any sailboats are left at the end of the last quarter, we have assigned them a value of zero. This is clearly unrealistic. In any inventory model with a finite horizon, the inventory left at the end of the last period should be assigned a **salvage value** that is indicative of the worth of the final period's inventory. For example, if Sailco feels that each sailboat left at the end of quarter 4 is worth \$400, then a term $-400i_4$ (measuring the worth of quarter 4's inventory) should be added to the objective function.

PROBLEMS

Group A

1 A customer requires during the next four months, respectively, 50, 65, 100, and 70 units of a commodity (no backlogging is allowed). Production costs are $5, $8, $4, and $7 per unit during these months. The storage cost from one month to the next is $2 per unit (assessed on ending inventory). It is estimated that each unit on hand at the end of month 4 could be sold for $6. Formulate an LP that will minimize the net cost incurred in meeting the demands of the next four months.

2 A company faces the following demands during the next three periods: period 1, 20 units; period 2, 10 units; period 3, 15 units. The unit production cost during each period is as follows: period 1—$13; period 2—$14; period 3—$15. A holding cost of $2 per unit is assessed against each period's ending inventory. At the beginning of period 1, the company has 5 units on hand.

In reality, not all goods produced during a month can be used to meet the current month's demand. To model this fact, we assume that only one half of the goods produced during a period can be used to meet the current period's demands. Formulate an LP to minimize the cost of meeting the demand for the next three periods. (*Hint:* Constraints such as $i_1 = x_1 + 5 - 20$ are certainly needed. Unlike our example, however, the constraint $i_1 \geq 0$ will not ensure that period 1's demand is met. For example, if $x_1 = 20$, then $i_1 \geq 0$ will hold, but because only $\frac{1}{2}(20) = 10$ units of period 1 production can be used to meet period 1's demand, $x_1 = 20$ would not be feasible. Try to think of a type of constraint that will ensure that what is available to meet each period's demand is at least as large as that period's demand.)

Group B

3 James Beerd bakes cheesecakes and Black Forest cakes. During any month, he can bake at most 65 cakes. The costs per cake and the demands for cakes, which must be met on time, are listed in Table 33. It costs 50¢ to hold a cheesecake, and 40¢ to hold a Black Forest cake, in inventory for a month. Formulate an LP to minimize the total cost of meeting the next three months' demands.

4 A manufacturing company produces two types of products: A and B. The company has agreed to deliver the products on the schedule shown in Table 34. The company has two assembly lines, 1 and 2, with the available production hours shown in Table 35. The production rates for each assembly line and product combination, in terms of

TABLE 34

Date	A	B
March 31	5,000	2,000
April 30	8,000	4,000

TABLE 35

Month	Production Hours Available	
	Line 1	Line 2
March	800	2,000
April	400	1,200

TABLE 36

Product	Production Rate	
	Line 1	Line 2
A	0.15	0.16
B	0.12	0.14

hours per product, are shown in Table 36. It takes 0.15 hour to manufacture 1 unit of product A on line 1, and so on. It costs $5 per hour of line time to produce any product. The inventory carrying cost per month for each product is 20¢ per unit (charged on each month's ending inventory). Currently, there are 500 units of A and 750 units of B in inventory. Management would like at least 1,000 units of each product in inventory at the end of April. Formulate an LP to determine the production schedule that minimizes the total cost incurred in meeting demands on time.

5 During the next two months, General Cars must meet (on time) the following demands for trucks and cars: month 1—400 trucks, 800 cars; month 2—300 trucks, 300 cars. During each month, at most 1,000 vehicles can be produced. Each truck uses 2 tons of steel, and each car uses 1 ton of steel. During month 1, steel costs $400 per ton; during month 2, steel costs $600 per ton. At most, 1,500 tons of steel may be purchased each month (steel may only be used

TABLE 33

Item	Month 1		Month 2		Month 3	
	Demand	Cost/Cake ($)	Demand	Cost/Cake ($)	Demand	Cost/Cake ($)
Cheesecake	40	3.00	30	3.40	20	3.80
Black Forest	20	2.50	30	2.80	10	3.40

during the month in which it is purchased). At the beginning of month 1, 100 trucks and 200 cars are in inventory. At the end of each month, a holding cost of $150 per vehicle is assessed. Each car gets 20 mpg, and each truck gets 10 mpg. During each month, the vehicles produced by the company must average at least 16 mpg. Formulate an LP to meet the demand and mileage requirements at minimum cost (include steel costs and holding costs).

6 Gandhi Clothing Company produces shirts and pants. Each shirt requires 2 sq yd of cloth, each pair of pants, 3. During the next two months, the following demands for shirts and pants must be met (on time): month 1—10 shirts, 15 pairs of pants; month 2—12 shirts, 14 pairs of pants. During each month, the following resources are available: month 1—90 sq yd of cloth; month 2—60 sq yd. (Cloth that is available during month 1 may, if unused during month 1, be used during month 2.)

During each month, it costs $4 to make an article of clothing with regular-time labor and $8 with overtime labor. During each month, a total of at most 25 articles of clothing may be produced with regular-time labor, and an unlimited number of articles of clothing may be produced with overtime labor. At the end of each month, a holding cost of $3 per article of clothing is assessed. Formulate an LP that can be used to meet demands for the next two months (on time) at minimum cost. Assume that at the beginning of month 1, 1 shirt and 2 pairs of pants are available.

7 Each year, Paynothing Shoes faces demands (which must be met on time) for pairs of shoes as shown in Table 37. Workers work three consecutive quarters and then receive one quarter off. For example, a worker may work during quarters 3 and 4 of one year and quarter 1 of the next year. During a quarter in which a worker works, he or she can produce up to 50 pairs of shoes. Each worker is paid $500 per quarter. At the end of each quarter, a holding cost of $50 per pair of shoes is assessed. Formulate an LP that can be used to minimize the cost per year (labor + holding) of meeting the demands for shoes. To simplify matters, assume

TABLE **37**

Quarter 1	Quarter 2	Quarter 3	Quarter 4
600	300	800	100

that at the end of each year, the ending inventory is zero. (*Hint:* It is allowable to assume that a given worker will get the same quarter off during each year.)

8 A company must meet (on time) the following demands: quarter 1—30 units; quarter 2—20 units; quarter 3—40 units. Each quarter, up to 27 units can be produced with regular-time labor, at a cost of $40 per unit. During each quarter, an unlimited number of units can be produced with overtime labor, at a cost of $60 per unit. Of all units produced, 20% are unsuitable and cannot be used to meet demand. Also, at the end of each quarter, 10% of all units on hand spoil and cannot be used to meet any future demands. After each quarter's demand is satisfied and spoilage is accounted for, a cost of $15 per unit is assessed against the quarter's ending inventory. Formulate an LP that can be used to minimize the total cost of meeting the next three quarters' demands. Assume that 20 usable units are available at the beginning of quarter 1.

9 Donovan Enterprises produces electric mixers. During the next four quarters, the following demands for mixers must be met on time: quarter 1—4,000; quarter 2—2,000; quarter 3—3,000; quarter 4—10,000. Each of Donovan's workers works three quarters of the year and gets one quarter off. Thus, a worker may work during quarters 1, 2, and 4 and get quarter 3 off. Each worker is paid $30,000 per year and (if working) can produce up to 500 mixers during a quarter. At the end of each quarter, Donovan incurs a holding cost of $30 per mixer on each mixer in inventory. Formulate an LP to help Donovan minimize the cost (labor and inventory) of meeting the next year's demand (on time). At the beginning of quarter 1, 600 mixers are available.

3.11 Multiperiod Financial Models

The following example illustrates how linear programming can be used to model multi-period cash management problems. The key is to determine the relations of cash on hand during different periods.

EXAMPLE 15 **Finco Multiperiod Investment**

Finco Investment Corporation must determine investment strategy for the firm during the next three years. Currently (time 0), $100,000 is available for investment. Investments A, B, C, D, and E are available. The cash flow associated with investing $1 in each investment is given in Table 38.

For example, $1 invested in investment B requires a $1 cash outflow at time 1 and returns 50¢ at time 2 and $1 at time 3. To ensure that the company's portfolio is diversified, Finco requires that at most $75,000 be placed in any single investment. In addition to investments A–E, Finco can earn interest at 8% per year by keeping uninvested cash in

TABLE 38

	Cash Flow (\$) at Time*			
	0	**1**	**2**	**3**
A	−1	+0.50	+1	0
B	0	−1	+0.50	+1
C	−1	+1.2	0	0
D	−1	0	0	+1.9
E	0	0	−1	+1.5

*Note: Time 0 = present; time 1 = 1 year from now; time 2 = 2 years from now; time 3 = 3 years from now.

money market funds. Returns from investments may be immediately reinvested. For example, the positive cash flow received from investment C at time 1 may immediately be reinvested in investment B. Finco cannot borrow funds, so the cash available for investment at any time is limited to cash on hand. Formulate an LP that will maximize cash on hand at time 3.

Solution Finco must decide how much money should be placed in each investment (including money market funds). Thus, we define the following decision variables:

A = dollars invested in investment A

B = dollars invested in investment B

C = dollars invested in investment C

D = dollars invested in investment D

E = dollars invested in investment E

S_t = dollars invested in money market funds at time t ($t = 0, 1, 2$)

Finco wants to maximize cash on hand at time 3. At time 3, Finco's cash on hand will be the sum of all cash inflows at time 3. From the description of investments A–E and the fact that from time 2 to time 3, S_2 will increase to $1.08S_2$,

$$\text{Time 3 cash on hand} = B + 1.9D + 1.5E + 1.08S_2$$

Thus, Finco's objective function is

$$\max z = B + 1.9D + 1.5E + 1.08S_2 \tag{62}$$

In multiperiod financial models, the following type of constraint is usually used to relate decision variables from different periods:

Cash available at time t = cash invested at time t

       + uninvested cash at time t that is carried over to time $t + 1$

If we classify money market funds as investments, we see that

$$\text{Cash available at time } t = \text{cash invested at time } t \tag{63}$$

Because investments A, C, D, and S_0 are available at time 0, and \$100,000 is available at time 0, (63) for time 0 becomes

$$100,000 = A + C + D + S_0 \tag{64}$$

At time 1, $0.5A + 1.2C + 1.08S_0$ is available for investment, and investments B and S_1 are available. Then for $t = 1$, (63) becomes

$$0.5A + 1.2C + 1.08S_0 = B + S_1 \tag{65}$$

At time 2, $A + 0.5B + 1.08S_1$ is available for investment, and investments E and S_2 are available. Thus, for $t = 2$, (63) reduces to

$$A + 0.5B + 1.08S_1 = E + S_2 \tag{66}$$

Let's not forget that at most \$75,000 can be placed in any of investments A–E. To take care of this, we add the constraints

$$A \leq 75,000 \tag{67}$$
$$B \leq 75,000 \tag{68}$$
$$C \leq 75,000 \tag{69}$$
$$D \leq 75,000 \tag{70}$$
$$E \leq 75,000 \tag{71}$$

Combining (62) and (64)–(71) with the sign restrictions (all variables ≥ 0) yields the following LP:

$$\max z = B + 1.9D + 1.5E + 1.08S_2$$
$$\text{s.t.} \quad A + C + D + S_0 = 100,000$$
$$0.5A + 1.2C + 1.08S_0 = B + S_1$$
$$A + 0.5B + 1.08S_1 = E + S_2$$
$$A \leq 75,000$$
$$B \leq 75,000$$
$$C \leq 75,000$$
$$D \leq 75,000$$
$$E \leq 75,000$$
$$A, B, C, D, E, S_0, S_1, S_2 \geq 0$$

We find the optimal solution to be $z = 218,500$, $A = 60,000$, $B = 30,000$, $D = 40,000$, $E = 75,000$, $C = S_0 = S_1 = S_2 = 0$. Thus, Finco should not invest in money market funds. At time 0, Finco should invest \$60,000 in A and \$40,000 in D. Then, at time 1, the \$30,000 cash inflow from A should be invested in B. Finally, at time 2, the \$60,000 cash inflow from A and the \$15,000 cash inflow from B should be invested in E. At time 3, Finco's \$100,000 will have grown to \$218,500.

You might wonder how our formulation ensures that Finco never invests more money at any time than the firm has available. This is ensured by the fact that each variable S_i must be nonnegative. For example, $S_0 \geq 0$ is equivalent to $100,000 - A - C - D \geq 0$, which ensures that at most \$100,000 will be invested at time 0.

Real-World Application

Using LP to Optimize Bond Portfolios

Many Wall Street firms buy and sell bonds. Rohn (1987) discusses a bond selection model that maximizes profit from bond purchases and sales subject to constraints that minimize the firm's risk exposure. See Problem 4 for a simplified version of this model.

PROBLEMS

Group A

1 A consultant to Finco claims that Finco's cash on hand at time 3 is the sum of the cash inflows from all investments, not just those investments yielding a cash inflow at time 3. Thus, the consultant claims that Finco's objective function should be

$$\max z = 1.5A + 1.5B + 1.2C + 1.9D + 1.5E$$
$$+ 1.08S_0 + 1.08S_1 + 1.08S_2$$

Explain why the consultant is incorrect.

2 Show that Finco's objective function may also be written as

$$\max z = 100,000 + 0.5A + 0.5B + 0.2C + 0.9D + 0.5E$$
$$+ 0.08S_0 + 0.08S_1 + 0.08S_2$$

3 At time 0, we have $10,000. Investments A and B are available; their cash flows are shown in Table 39. Assume that any money not invested in A or B earns *no* interest. Formulate an LP that will maximize cash on hand at time 3. Can you guess the optimal solution to this problem?

Group B

4[†] Broker Steve Johnson is currently trying to maximize his profit in the bond market. Four bonds are available for purchase and sale, with the bid and ask price of each bond as shown in Table 40. Steve can buy up to 1,000 units of each bond at the ask price or sell up to 1,000 units of each bond at the bid price. During each of the next three years, the person who sells a bond will pay the owner of the bond the cash payments shown in Table 41.

Steve's goal is to maximize his revenue from selling bonds less his payment for buying bonds, subject to the constraint that after each year's payments are received, his current cash position (due only to cash payments from bonds and not purchases or sale of bonds) is nonnegative. Assume

TABLE 39

Time	A	B
0	−$1	$0
1	$0.2	−$1
2	$1.5	$0
3	$0	$1.0

TABLE 40

Bond	Bid Price	Ask Price
1	980	990
2	970	985
3	960	972
4	940	954

[†]Based on Rohn (1987).

TABLE 41

Year	Bond 1	Bond 2	Bond 3	Bond 4
1	100	80	70	60
2	110	90	80	50
3	1,100	1,120	1,090	1,110

TABLE 42

Month	Cash Flow	Month	Cash Flow
January	−12	July	−7
February	−10	August	−2
March	−8	September	15
April	−10	October	12
May	−4	November	−7
June	5	December	45

that cash payments are discounted, with a payment of $1 one year from now being equivalent to a payment of 90¢ now. Formulate an LP to maximize net profit from buying and selling bonds, subject to the arbitrage constraints previously described. Why do you think we limit the number of units of each bond that can be bought or sold?

5 A small toy store, Toyco projects the monthly cash flows (in thousands of dollars) in Table 42 during the year 2003. A negative cash flow means that cash outflows exceed cash inflows to the business. To pay its bills, Toyco will need to borrow money early in the year. Money can be borrowed in two ways:

a Taking out a long-term one-year loan in January. Interest of 1% is charged each month, and the loan must be paid back at the end of December.

b Each month money can be borrowed from a short-term bank line of credit. Here, a monthly interest rate of 1.5% is charged. All short-term loans must be paid off at the end of December.

At the end of each month, excess cash earns 0.4% interest. Formulate an LP whose solution will help Toyco maximize its cash position at the beginning of January, 2004.

6 Consider Problem 5 with the following modification: Each month Toyco can delay payments on some or all of the cash owed for the current month. This is called "stretching payments." Payments may be stretched for only one month, and a 1% penalty is charged on the amount stretched. Thus, if it stretches payments on $10,000 cash owed in January, then it must pay 10,000(1.01) = $10,100 in February. With this modification, formulate an LP that would help Toyco maximize its cash on hand at the beginning of January 1, 2004.

7 Suppose we are borrowing $1,000 at 12% annual interest with 60 monthly payments. Assume equal payments are made at the end of month 1, month 2, . . . month 60. We know that entering into Excel the function

$$= PMT(.01, 60, 1,000)$$

would yield the monthly payment ($22.24).

It is instructive to use LP to determine the montly payment. Let p be the (unknown) monthly payment. Each month we owe $.01 \cdot$ (our current unpaid balance) in interest. The remainder of our monthly payment is used to reduce the unpaid balance. For example, suppose we paid $30 each month. At the beginning of month 1, our unpaid balance is $1,000. Of our month 1 payment, $10 goes to interest and $20 to paying off the unpaid balance. Then we would begin month 2 with an unpaid balance of $980. The trick is to use LP to determine the monthly payment that will pay off the loan at the end of month 60.

8 You are a CFA (chartered financial analyst). Madonna has come to you because she needs help paying off her credit card bills. She owes the amounts on her credit cards shown in Table 43. Madonna is willing to allocate up to $5,000 per month to pay off these credit cards. All cards must be paid off within 36 months. Madonna's goal is to minimize the total of all her payments. To solve this problem, you must understand how interest on a loan works. To illustrate, suppose Madonna pays $5,000 on Saks during month 1. Then her Saks balance at the beginning of month 2 is

$$20,000 - (5,000 - .005(20,000))$$

This follows because during month 1 Madonna incurs $.005(20,000)$ in interest charges on her Saks card. Help Madonna solve her problems!

9 Winstonco is considering investing in three projects. If we fully invest in a project, the realized cash flows (in millions of dollars) will be as shown in Table 44. For example, project 1 requires cash outflow of $3 million today

TABLE 43

Card	Balance ($)	Monthly Rate (%)
Saks Fifth Avenue	20,000	.5
Bloomingdale's	50,000	1
Macys	40,000	1.5

TABLE 44

Time (Years)	Cash Flow		
	Project 1	Project 2	Project 3
0	−3	−2	−2
.5	−1	−.5	−2
1	+1.8	1.5	−1.8
1.5	1.4	1.5	1
2	1.8	1.5	1
2.5	1.8	.2	1
3	5.5	−1	6

and returns $5.5 million 3 years from now. Today we have $2 million in cash. At each time point (0, .5, 1, 1.5, 2, and 2.5 years from today) we may, if desired, borrow up to $2 million at 3.5% (per 6 months) interest. Leftover cash earns 3% (per 6 months) interest. For example, if after borrowing and investing at time 0 we have $1 million we would receive $30,000 in interest at time .5 years. Winstonco's goal is to maximize cash on hand after it accounts for time 3 cash flows. What investment and borrowing strategy should be used? Remember that we may invest in a fraction of a project. For example, if we invest in .5 of project 3, then we have cash outflows of −$1 million at time 0 and .5.

3.12 Multiperiod Work Scheduling

In Section 3.5, we saw that linear programming could be used to schedule employees in a static environment where demand did not change over time. The following example (a modified version of a problem from Wagner [1975]) shows how LP can be used to schedule employee training when a firm faces demand that changes over time.

EXAMPLE 16 | **Multiperiod Work Scheduling**

CSL is a chain of computer service stores. The number of hours of skilled repair time that CSL requires during the next five months is as follows:

Month 1 (January): 6,000 hours

Month 2 (February): 7,000 hours

Month 3 (March): 8,000 hours

Month 4 (April): 9,500 hours

Month 5 (May): 11,000 hours

At the beginning of January, 50 skilled technicians work for CSL. Each skilled technician can work up to 160 hours per month. To meet future demands, new technicians must be trained. It takes one month to train a new technician. During the month of training, a trainee must be supervised for 50 hours by an experienced technician. Each experienced technician is paid $2,000 a month (even if he or she does not work the full 160 hours). During the month of training, a trainee is paid $1,000 a month. At the end of each month, 5% of CSL's experienced technicians quit to join Plum Computers. Formulate an LP whose solution will enable CSL to minimize the labor cost incurred in meeting the service requirements for the next five months.

Solution CSL must determine the number of technicians who should be trained during month t ($t = 1, 2, 3, 4, 5$). Thus, we define

$$x_t = \text{number of technicians trained during month } t \quad (t = 1, 2, 3, 4, 5)$$

CSL wants to minimize total labor cost during the next five months. Note that

Total labor cost = cost of paying trainees + cost of paying experienced technicians

To express the cost of paying experienced technicians, we need to define, for $t = 1, 2, 3, 4, 5$,

$$y_t = \text{number of experienced technicians at the beginning of month } t$$

Then

$$\text{Total labor cost} = (1{,}000x_1 + 1{,}000x_2 + 1{,}000x_3 + 1{,}000x_4 + 1{,}000x_5)$$
$$+ (2{,}000y_1 + 2000y_2 + 2{,}000y_3 + 2{,}000y_4 + 2{,}000y_5)$$

Thus, CSL's objective function is

$$\min z = 1{,}000x_1 + 1{,}000x_2 + 1{,}000x_3 + 1{,}000x_4 + 1{,}000x_5$$
$$+ 2{,}000y_1 + 2{,}000y_2 + 2{,}000y_3 + 2{,}000y_4 + 2{,}000y_5$$

What constraints does CSL face? Note that we are given $y_1 = 50$, and that for $t = 1, 2, 3, 4, 5$, CSL must ensure that

Number of available technician hours during month t

$$\geq \text{Number of technician hours required during month } t \quad \textbf{(72)}$$

Because each trainee requires 50 hours of experienced technician time, and each skilled technician is available for 160 hours per month,

Number of available technician hours during month $t = 160y_t - 50x_t$

Now (72) yields the following five constraints:

$$160y_1 - 50x_1 \geq 6{,}000 \qquad \text{(month 1 constraint)}$$
$$160y_2 - 50x_2 \geq 7{,}000 \qquad \text{(month 2 constraint)}$$
$$160y_3 - 50x_3 \geq 8{,}000 \qquad \text{(month 3 constraint)}$$
$$160y_4 - 50x_4 \geq 9{,}500 \qquad \text{(month 4 constraint)}$$
$$160y_5 - 50x_5 \geq 11{,}000 \qquad \text{(month 5 constraint)}$$

As in the other multiperiod formulations, we need constraints that relate variables from different periods. In the CSL problem, it is important to realize that the number of skilled technicians available at the beginning of any month is determined by the number of skilled technicians available during the previous month and the number of technicians trained during the previous month:

Experienced technicians available at beginning of month t = Experienced technicians available at beginning of month $(t - 1)$ (73)

+ technicians trained during month $(t - 1)$

− experienced technicians who quit during month $(t - 1)$

For example, for February, (73) yields

$$y_2 = y_1 + x_1 - 0.05y_1 \quad \text{or} \quad y_2 = 0.95y_1 + x_1$$

Similarly, for March, (73) yields

$$y_3 = 0.95y_2 + x_2$$

and for April,

$$y_4 = 0.95y_3 + x_3$$

and for May,

$$y_5 = 0.95y_4 + x_4$$

Adding the sign restrictions $x_t \geq 0$ and $y_t \geq 0$ ($t = 1, 2, 3, 4, 5$), we obtain the following LP:

$$\min z = 1{,}000x_1 + 1{,}000x_2 + 1{,}000x_3 + 1{,}000x_4 + 1{,}000x_5$$
$$+ 2{,}000y_1 + 2{,}000y_2 + 2{,}000y_3 + 2{,}000y_4 + 2{,}000y_5$$

s.t.
$$160y_1 - 50x_1 \geq 6{,}000 \qquad y_1 = 50$$
$$160y_2 - 50x_2 \geq 7{,}000 \qquad 0.95y_1 + x_1 = y_2$$
$$160y_3 - 50x_3 \geq 8{,}000 \qquad 0.95y_2 + x_2 = y_3$$
$$160y_4 - 50x_4 \geq 9{,}500 \qquad 0.95y_3 + x_3 = y_4$$
$$160y_5 - 50x_5 \geq 11{,}000 \qquad 0.95y_4 + x_4 = y_5$$
$$x_t, y_t \geq 0 \qquad (t = 1, 2, 3, 4, 5)$$

The optimal solution is $z = 593{,}777$; $x_1 = 0$; $x_2 = 8.45$; $x_3 = 11.45$; $x_4 = 9.52$; $x_5 = 0$; $y_1 = 50$; $y_2 = 47.5$; $y_3 = 53.58$; $y_4 = 62.34$; and $y_5 = 68.75$.

In reality, the y_t's must be integers, so our solution is difficult to interpret. The problem with our formulation is that assuming that exactly 5% of the employees quit each month can cause the number of employees to change from an integer during one month to a fraction during the next month. We might want to assume that the number of employees quitting each month is the integer closest to 5% of the total workforce, but then we do not have a linear programming problem!

PROBLEMS

Group A

1 If $y_1 = 38$, then what would be the optimal solution to CSL's problem?

2 An insurance company believes that it will require the following numbers of personal computers during the next six months: January, 9; February, 5; March, 7; April, 9; May, 10; June, 5. Computers can be rented for a period of one, two, or three months at the following unit rates: one-month rate, $200; two-month rate, $350; three-month rate, $450. Formulate an LP that can be used to minimize the cost of renting the required computers. You may assume that if a machine is rented for a period of time extending beyond June, the cost of the rental should be prorated. For example, if a computer is rented for three months at the beginning of May, then a rental fee of $\frac{2}{3}(450) = \$300$, not $450, should be assessed in the objective function.

3 The IRS has determined that during each of the next 12 months it will need the number of supercomputers given in Table 45. To meet these requirements, the IRS rents

TABLE 45

Month	Computer Requirements
1	800
2	1,000
3	600
4	500
5	1,200
6	400
7	800
8	600
9	400
10	500
11	800
12	600

TABLE 46

Month	Selling Price ($)	Purchase Price ($)
1	3	8
2	6	8
3	7	2
4	1	3
5	4	4
6	5	3
7	5	3
8	1	2
9	3	5
10	2	5

supercomputers for a period of one, two, or three months. It costs $100 to rent a supercomputer for one month, $180 for two months, and $250 for three months. At the beginning of month 1, the IRS has no supercomputers. Determine the rental plan that meets the next 12 months' requirements at minimum cost. *Note:* You may assume that fractional rentals are okay, so if your solution says to rent 140.6 computers for one month we can round this up or down (to 141 or 140) without having much effect on the total cost.

Group B

4 You own a wheat warehouse with a capacity of 20,000 bushels. At the beginning of month 1, you have 6,000 bushels

of wheat. Each month, wheat can be bought and sold at the price per 1000 bushels given in Table 46.

The sequence of events during each month is as follows:

a You observe your initial stock of wheat.

b You can sell any amount of wheat up to your initial stock at the current month's selling price.

c You can buy (at the current month's buying price) as much wheat as you want, subject to the warehouse size limitation.

Your goal is to formulate an LP that can be used to determine how to maximize the profit earned over the next 10 months.

S U M M A R Y   **Linear Programming Definitions**

A **linear programming problem (LP)** consists of three parts:

1 A linear function (the **objective function**) of decision variables (say, $x_1, x_2, \ldots, x_n$) that is to be maximized or minimized.

2 A set of **constraints** (each of which must be a linear equality or linear inequality) that restrict the values that may be assumed by the decision variables.

3 The **sign restrictions,** which specify for each decision variable x_j either (1) variable x_j must be nonnegative—$x_j \geq 0$; or (2) variable x_j may be positive, zero, or negative—x_j is **unrestricted in sign (urs).**

The coefficient of a variable in the objective function is the variable's **objective function coefficient.** The coefficient of a variable in a constraint is a **technological coefficient.** The right-hand side of each constraint is called a **right-hand side (rhs).**

A *point* is simply a specification of the values of each decision variable. The **feasible region** of an LP consists of all points satisfying the LP's constraints and sign restrictions. Any point in the feasible region that has the largest z-value of all points in the feasible region (for a max problem) is an **optimal solution** to the LP. An LP may have no optimal solution, one optimal solution, or an infinite number of optimal solutions.

A constraint in an LP is **binding** if the left-hand side and the right-hand side are equal when the values of the variables in the optimal solution are substituted into the constraint.

Graphical Solution of Linear Programming Problems

The feasible region for any LP is a **convex set.** If an LP has an optimal solution, there is an extreme (or corner) point of the feasible region that is an optimal solution to the LP.

We may graphically solve an LP (max problem) with two decision variables as follows:

Step 1   Graph the feasible region.

Step 2   Draw an isoprofit line.

Step 3   Move parallel to the isoprofit line in the direction of increasing z. The last point in the feasible region that contacts an isoprofit line is an optimal solution to the LP.

LP Solutions: Four Cases

When an LP is solved, one of the following four cases will occur:

Case 1   The LP has a unique solution.

Case 2   The LP has more than one (actually an infinite number of) optimal solutions. This is the case of **alternative optimal solutions.** Graphically, we recognize this case when the isoprofit line last hits an entire line segment before leaving the feasible region.

Case 3   The LP is **infeasible** (it has no feasible solution). This means that the feasible region contains no points.

Case 4   The LP is unbounded. This means (in a max problem) that there are points in the feasible region with arbitrarily large z-values. Graphically, we recognize this case by the fact that when we move parallel to an isoprofit line in the direction of increasing z, we never lose contact with the LP's feasible region.

Formulating LPs

The most important step in formulating most LPs is to determine the decision variables correctly.

In any constraint, the terms must have the same units. For example, one term cannot have the units "pounds of raw material" while another term has the units "ounces of raw material."

REVIEW PROBLEMS

Group A

1   Bloomington Breweries produces beer and ale. Beer sells for $5 per barrel, and ale sells for $2 per barrel. Producing a barrel of beer requires 5 lb of corn and 2 lb of hops. Producing a barrel of ale requires 2 lb of corn and 1 lb of hops. Sixty pounds of corn and 25 lb of hops are available. Formulate an LP that can be used to maximize revenue. Solve the LP graphically.

2   Farmer Jones bakes two types of cake (chocolate and vanilla) to supplement his income. Each chocolate cake can be sold for $1, and each vanilla cake can be sold for 50¢. Each chocolate cake requires 20 minutes of baking time and uses 4 eggs. Each vanilla cake requires 40 minutes of baking time and uses 1 egg. Eight hours of baking time and 30 eggs are available. Formulate an LP to maximize Farmer Jones's revenue, then graphically solve the LP. (A fractional number of cakes is okay.)

3 I now have $100. The following investments are available during the next three years:

Investment A Every dollar invested now yields $0.10 a year from now and $1.30 three years from now.
Investment B Every dollar invested now yields $0.20 a year from now and $1.10 two years from now.
Investment C Every dollar invested a year from now yields $1.50 three years from now.

During each year, uninvested cash can be placed in money market funds, which yield 6% interest per year. At most $50 may be placed in each of investments A, B, and C. Formulate an LP to maximize my cash on hand three years from now.

4 Sunco processes oil into aviation fuel and heating oil. It costs $40 to purchase each 1,000 barrels of oil, which is then distilled and yields 500 barrels of aviation fuel and 500 barrels of heating oil. Output from the distillation may be sold directly or processed in the catalytic cracker. If sold after distillation without further processing, aviation fuel sells for $60 per 1,000 barrels, and heating oil sells for $40 per 1,000 barrels. It takes 1 hour to process 1,000 barrels of aviation fuel in the catalytic cracker, and these 1,000 barrels can be sold for $130. It takes 45 minutes to process 1,000 barrels of heating oil in the cracker, and these 1,000 barrels can be sold for $90. Each day, at most 20,000 barrels of oil can be purchased, and 8 hours of cracker time are available. Formulate an LP to maximize Sunco's profits.

5 Finco has the following investments available:

Investment A For each dollar invested at time 0, we receive $0.10 at time 1 and $1.30 at time 2. (Time 0 = now; time 1 = one year from now; and so on.)
Investment B For each dollar invested at time 1, we receive $1.60 at time 2.
Investment C For each dollar invested at time 2, we receive $1.20 at time 3.

At any time, leftover cash may be invested in T-bills, which pay 10% per year. At time 0, we have $100. At most, $50 can be invested in each of investments A, B, and C. Formulate an LP that can be used to maximize Finco's cash on hand at time 3.

6 All steel manufactured by Steelco must meet the following requirements: 3.2–3.5% carbon; 1.8–2.5% silicon; 0.9–1.2% nickel; tensile strength of at least 45,000 pounds per square inch (psi). Steelco manufactures steel by combining two alloys. The cost and properties of each alloy are given in Table 47. Assume that the tensile strength of a mixture of the two alloys can be determined by averaging that of the alloys that are mixed together. For example, a one-ton mixture that is 40% alloy 1 and 60% alloy 2 has a tensile strength of 0.4(42,000) + 0.6(50,000). Use linear programming to determine how to minimize the cost of producing a ton of steel.

7 Steelco manufactures two types of steel at three different steel mills. During a given month, each steel mill has 200 hours of blast furnace time available. Because of differences in the furnaces at each mill, the time and cost to produce a ton of steel differs for each mill. The time and cost for each mill are shown in Table 48. Each month, Steelco must manufacture at least 500 tons of steel 1 and 600 tons of steel 2. Formulate an LP to minimize the cost of manufacturing the desired steel.

8[†] Walnut Orchard has two farms that grow wheat and corn. Because of differing soil conditions, there are differences in the yields and costs of growing crops on the two farms. The yields and costs are shown in Table 49. Each farm has 100 acres available for cultivation; 11,000 bushels of wheat and 7,000 bushels of corn must be grown. Determine a planting plan that will minimize the cost of meeting these demands. How could an extension of this model be used to allocate crop production efficiently throughout a nation?

9 Candy Kane Cosmetics (CKC) produces Leslie Perfume, which requires chemicals and labor. Two production processes are available: Process 1 transforms 1 unit of labor and 2 units of chemicals into 3 oz of perfume. Process 2 transforms 2 units of labor and 3 units of chemicals into 5 oz of perfume. It costs CKC $3 to purchase a unit of labor and $2 to purchase a unit of chemicals. Each year, up to 20,000 units of labor and 35,000 units of chemicals can be purchased. In the absence of advertising, CKC believes it can sell 1,000 oz of perfume. To stimulate demand for

TABLE 48
Producing a Ton of Steel

Mill	Steel 1 Cost	Steel 1 Time (Minutes)	Steel 2 Cost	Steel 2 Time (Minutes)
1	$10	20	$11	22
2	$12	24	$ 9	18
3	$14	28	$10	30

TABLE 49

	Farm 1	Farm 2
Corn yield/acre (bushels)	500	650
Cost/acre of corn ($)	100	120
Wheat yield/acre (bushels)	400	350
Cost/acre of wheat ($)	90	80

[†]Based on Heady and Egbert (1964).

TABLE 47

	Alloy 1	Alloy 2
Cost per ton ($)	$190	$200
Percent silicon	2	2.5
Percent nickel	1	1.5
Percent carbon	3	4
Tensile strength (psi)	42,000	50,000

Leslie, CKC can hire the lovely model Jenny Nelson. Jenny is paid $100/hour. Each hour Jenny works for the company is estimated to increase the demand for Leslie Perfume by 200 oz. Each ounce of Leslie Perfume sells for $5. Use linear programming to determine how CKC can maximize profits.

10 Carco has a $150,000 advertising budget. To increase automobile sales, the firm is considering advertising in newspapers and on television. The more Carco uses a particular medium, the less effective is each additional ad. Table 50 shows the number of new customers reached by each ad. Each newspaper ad costs $1,000, and each television ad costs $10,000. At most, 30 newspaper ads and 15 television ads can be placed. How can Carco maximize the number of new customers created by advertising?

11 Sunco Oil has refineries in Los Angeles and Chicago. The Los Angeles refinery can refine up to 2 million barrels of oil per year, and the Chicago refinery up to 3 million. Once refined, oil is shipped to two distribution points: Houston and New York City. Sunco estimates that each distribution point can sell up to 5 million barrels per year. Because of differences in shipping and refining costs, the profit earned (in dollars) per million barrels of oil shipped depends on where the oil was refined and on the point of distribution (see Table 51). Sunco is considering expanding the capacity of each refinery. Each million barrels of annual refining capacity that is added will cost $120,000 for the Los Angeles refinery and $150,000 for the Chicago refinery. Use linear programming to determine how Sunco can maximize its profits less expansion costs over a ten-year period.

12 For a telephone survey, a marketing research group needs to contact at least 150 wives, 120 husbands, 100 single adult males, and 110 single adult females. It costs $2 to make a daytime call and (because of higher labor costs) $5 to make an evening call. Table 52 lists the results. Because of limited staff, at most half of all phone calls can be evening calls. Formulate an LP to minimize the cost of completing the survey.

TABLE 50

	Number of Ads	New Customers
Newspaper	1–10	900
	11–20	600
	21–30	300
Television	1–5	10,000
	6–10	5,000
	11–15	2,000

TABLE 51

From	Profit per Million Barrels ($)	
	To Houston	To New York
Los Angeles	20,000	15,000
Chicago	18,000	17,000

TABLE 52

Person Responding	Percent of Daytime Calls	Percent of Evening Calls
Wife	30	30
Husband	10	30
Single male	10	15
Single female	10	20
None	40	5

13 Feedco produces two types of cattle feed, both consisting totally of wheat and alfalfa. Feed 1 must contain at least 80% wheat, and feed 2 must contain at least 60% alfalfa. Feed 1 sells for $1.50/lb, and feed 2 sells for $1.30/lb. Feedco can purchase up to 1,000 lb of wheat at 50¢/lb and up to 800 lb of alfalfa at 40¢/lb. Demand for each type of feed is unlimited. Formulate an LP to maximize Feedco's profit.

14 Feedco (see Problem 13) has decided to give its customer (assume it has only one customer) a quantity discount. If the customer purchases more than 300 lb of feed 1, each pound over the first 300 lb will sell for only $1.25/lb. Similarly, if the customer purchases more than 300 pounds of feed 2, each pound over the first 300 lb will sell for $1.00/lb. Modify the LP of Problem 13 to account for the presence of quantity discounts. (*Hint:* Define variables for the feed sold at each price.)

15 Chemco produces two chemicals: A and B. These chemicals are produced via two manufacturing processes. Process 1 requires 2 hours of labor and 1 lb of raw material to produce 2 oz of A and 1 oz of B. Process 2 requires 3 hours of labor and 2 lb of raw material to produce 3 oz of A and 2 oz of B. Sixty hours of labor and 40 lb of raw material are available. Demand for A is unlimited, but only 20 oz of B can be sold. A sells for $16/oz, and B sells for $14/oz. Any B that is unsold must be disposed of at a cost of $2/oz. Formulate an LP to maximize Chemco's revenue less disposal costs.

16 Suppose that in the CSL computer example of Section 3.12, it takes two months to train a technician and that during the second month of training, each trainee requires 10 hours of experienced technician time. Modify the formulation in the text to account for these changes.

17 Furnco manufactures tables and chairs. Each table and chair must be made entirely out of oak or entirely out of pine. A total of 150 board ft of oak and 210 board ft of pine are available. A table requires either 17 board ft of oak or 30 board ft of pine, and a chair requires either 5 board ft of oak or 13 board ft of pine. Each table can be sold for $40, and each chair for $15. Formulate an LP that can be used to maximize revenue.

18[†] The city of Busville contains three school districts. The number of minority and nonminority students in each district is given in Table 53. Of all students, 25% ($\frac{200}{800}$) are minority students.

[†]Based on Franklin and Koenigsberg (1973).

TABLE 53

District	Minority Students	Nonminority Students
1	50	200
2	50	250
3	100	150

TABLE 54

District	Cooley High	Walt Whitman High
1	1	2
2	2	1
3	1	1

The local court has decided that both of the town's two high schools (Cooley High and Walt Whitman High) must have approximately the same percentage of minority students (within ±5%) as the entire town. The distances (in miles) between the school districts and the high schools are given in Table 54. Each high school must have an enrollment of 300–500 students. Use linear programming to determine an assignment of students to schools that minimizes the total distance students must travel to school.

19[†] Brady Corporation produces cabinets. Each week, it requires 90,000 cu ft of processed lumber. The company may obtain lumber in two ways. First, it may purchase lumber from an outside supplier and then dry it in the supplier's kiln. Second, it may chop down logs on its own land, cut them into lumber at its sawmill, and finally dry the lumber in its own kiln. Brady can purchase grade 1 or grade 2 lumber. Grade 1 lumber costs $3 per cu ft and when dried yields 0.7 cu ft of useful lumber. Grade 2 lumber costs $7 per cubic foot and when dried yields 0.9 cu ft of useful lumber. It costs the company $3 to chop down a log. After being cut and dried, a log yields 0.8 cu ft of lumber. Brady incurs costs of $4 per cu ft of lumber dried. It costs $2.50 per cu ft of logs sent through the sawmill. Each week, the sawmill can process up to 35,000 cu ft of lumber. Each week, up to 40,000 cu ft of grade 1 lumber and up to 60,000 cu ft of grade 2 lumber can be purchased. Each week, 40 hours of time are available for drying lumber. The time it takes to dry 1 cu ft of grade 1 lumber, grade 2 lumber, or logs is as follows: grade 1—2 seconds; grade 2—0.8 second; log—1.3 seconds. Formulate an LP to help Brady minimize the weekly cost of meeting the demand for processed lumber.

20[‡] The Canadian Parks Commission controls two tracts of land. Tract 1 consists of 300 acres and tract 2, 100 acres. Each acre of tract 1 can be used for spruce trees or hunting, or both. Each acre of tract 2 can be used for spruce trees or camping, or both. The capital (in hundreds of dollars) and labor (in worker-days) required to maintain one acre of each tract, and the profit (in thousands of dollars) per acre for each possible use of land are given in Table 55. Capital of $150,000 and 200 man-days of labor are available. How should the land be allocated to various uses to maximize profit received from the two tracts?

21[§] Chandler Enterprises produces two competing products: A and B. The company wants to sell these products to two groups of customers: group 1 and group 2. The value each customer places on a unit of A and B is as shown in Table 56. Each customer will buy either product A or product B, but not both. A customer is willing to buy product A if she believes that

Value of product A − price of product A

≥ Value of product B − price of product B

and

Value of product A − price of product A ≥ 0

A customer is willing to buy product B if she believes that

Value of product B − price of product B

≥ value of product A − price of product A

and

Value of product B − price of product B ≥ 0

Group 1 has 1,000 members, and group 2 has 1,500 members. Chandler wants to set prices for each product that ensure that group 1 members purchase product A and group 2 members purchase product B. Formulate an LP that will help Chandler maximize revenues.

22[¶] Alden Enterprises produces two products. Each product can be produced on one of two machines. The length of time needed to produce each product (in hours) on each machine is as shown in Table 57. Each month, 500 hours of time are available on each machine. Each month, customers are willing to buy up to the quantities of each product at the

TABLE 55

Tract	Capital	Labor	Profit
1 Spruce	3	0.1	0.2
1 Hunting	3	0.2	0.4
1 Both	4	0.2	0.5
2 Spruce	1	0.05	0.06
2 Camping	30	5	0.09
2 Both	10	1.01	1.1

TABLE 56

	Group 1 Customer	Group 2 Customer
Value of A to	$10	$12
Value of B to	$8	$15

[§]Based on Dobson and Kalish (1988).
[¶]Based on Jain, Stott, and Vasold (1978).

[†]Based on Carino and Lenoir (1988).
[‡]Based on Cheung and Auger (1976).

TABLE 57

Product	Machine 1	Machine 2
1	4	3
2	7	4

TABLE 58

Product	Demands		Prices	
	Month 1	Month 2	Month 1	Month 2
1	100	190	$55	$12
2	140	130	$65	$32

prices given in Table 58. The company's goal is to maximize the revenue obtained from selling units during the next two months. Formulate an LP to help meet this goal.

23 Kiriakis Electronics produces three products. Each product must be processed on each of three types of machines. When a machine is in use, it must be operated by a worker. The time (in hours) required to process each product on each machine and the profit associated with each product are shown in Table 59. At present, five type 1 machines, three type 2 machines, and four type 3 machines are available. The company has 10 workers available and must determine how many workers to assign to each machine. The plant is open 40 hours per week, and each worker works 35 hours per week. Formulate an LP that will enable Kiriakis to assign workers to machines in a way that maximizes weekly profits. (*Note:* A worker need not spend the entire work week operating a single machine.)

24 Gotham City Hospital serves cases from four diagnostic-related groups (DRGs). The profit contribution, diagnostic service use (in hours), bed-day use (in days), nursing care use (in hours), and drug use (in dollars) are

TABLE 59

	Product 1	Product 2	Product 3
Machine 1	2	3	4
Machine 2	3	5	6
Machine 3	4	7	9
Profit ($)	6	8	10

TABLE 60

DRG	Profit	Diagnostic Services	Bed-Day	Nursing Use	Drugs
1	2,000	7	5	30	800
2	1,500	4	2	10	500
3	500	2	1	5	150
4	300	1	0	1	50

given in Table 60. The hospital now has available each week 570 hours of diagnostic services, 1,000 bed-days, 50,000 nursing hours, and $50,000 worth of drugs. To meet the community's minimum health care demands at least 10 DRG1, 15 DRG2, 40 DRG3, and 160 DRG4 cases must be handled each week. Use LP to determine the hospital's optimal mix of DRGs.[†]

25 Oliver Winery produces four award-winning wines in Bloomington, Indiana. The profit contribution, labor hours, and tank usage (in hours) per gallon for each type of wine are given in Table 61. By law, at most 100,000 gallons of wine can be produced each year. A maximum of 12,000 labor hours and 32,000 tank hours are available annually. Each gallon of wine 1 spends an average of $\frac{1}{3}$ year in inventory; wine 2, an average of 1 year; wine 3, an average of 2 years; wine 4, an average of 3.333 years. The winery's warehouse can handle an average inventory level of 50,000 gallons. Determine how much of each type of wine should be produced annually to maximize Oliver Winery's profit.

26 Graphically solve the following LP:

$$\min z = 5x_1 + x_2$$
$$\text{s.t.} \quad 2x_1 + x_2 \geq 6$$
$$x_1 + x_2 \geq 4$$
$$2x_1 + 10x_2 \geq 20$$
$$x_1, x_2 \geq 0$$

27 Grummins Engine produces diesel trucks. New government emission standards have dictated that the average pollution emissions of all trucks produced in the next three years cannot exceed 10 grams per truck. Grummins produces two types of trucks. Each type 1 truck sells for $20,000, costs $15,000 to manufacture, and emits 15 grams of pollution. Each type 2 truck sells for $17,000, costs $14,000 to manufacture, and emits 5 grams of pollution. Production capacity limits total truck production during each year to at most 320 trucks. Grummins knows that the maximum number of each truck type that can be sold during each of the next three years is given in Table 62.

Thus, *at most*, 300 type 1 trucks can be sold during year 3. Demand may be met from previous production or the current year's production. It costs $2,000 to hold 1 truck (of any type) in inventory for one year. Formulate an LP to help Grummins maximize its profit during the next three years.

TABLE 61

Wine	Profit ($)	Labor (Hr)	Tank (Hr)
1	6	.2	.5
2	12	.3	.5
3	20	.3	1
4	30	.5	1.5

[†]Based on Robbins and Tuntiwonpiboon (1989).

TABLE 62
Maximum Demand for Trucks

Year	Type 1	Type 2
1	100	200
2	200	100
3	300	150

TABLE 63

Shift	Hourly Salary	Defects (per Capacitor)	Price
8 A.M.–4 P.M.	$12	4	$18
4 P.M.–Midnight	$16	3	$22
Midnight–8 A.M.	$20	2	$24

28 Describe all optimal solutions to the following LP:

$$\min z = 4x_1 + x_2$$
$$\text{s.t.} \quad 3x_1 + x_2 \geq 6$$
$$4x_1 + x_2 \geq 12$$
$$x_1 \geq 2$$
$$x_1, x_2 \geq 0$$

29 Juiceco manufactures two products: premium orange juice and regular orange juice. Both products are made by combining two types of oranges: grade 6 and grade 3. The oranges in premium juice must have an average grade of at least 5, those in regular juice, at least 4. During each of the next two months Juiceco can sell up to 1,000 gallons of premium juice and up to 2,000 gallons of regular juice. Premium juice sells for $1.00 per gallon, while regular juice sells for 80¢ per gallon. At the beginning of month 1, Juiceco has 3,000 gallons of grade 6 oranges and 2,000 gallons of grade 3 oranges. At the beginning of month 2, Juiceco may purchase additional grade 3 oranges for 40¢ per gallon and additional grade 6 oranges for 60¢ per gallon. Juice spoils at the end of the month, so it makes no sense to make extra juice during month 1 in the hopes of using it to meet month 2 demand. Oranges left at the end of month 1 may be used to produce juice for month 2. At the end of month 1 a holding cost of 5¢ is assessed against each gallon of leftover grade 3 oranges, and 10¢ against each gallon of leftover grade 6 oranges. In addition to the cost of the oranges, it costs 10¢ to produce each gallon of (regular or premium) juice. Formulate an LP that could be used to maximize the profit (revenues − costs) earned by Juiceco during the next two months.

30 Graphically solve the following linear programming problem:

$$\max z = 5x_1 - x_2$$
$$\text{s.t.} \quad 2x_1 + 3x_2 \geq 12$$
$$x_1 - 3x_2 \geq 0$$
$$x_1 \geq 0, x_2 \geq 0$$

31 Graphically find all solutions to the following LP:

$$\min z = x_1 - 2x_2$$
$$\text{s.t.} \quad x_1 \geq 4$$
$$x_1 + x_2 \geq 8$$
$$x_1 - x_2 \leq 6$$
$$x_1, x_2 \geq 0$$

32 Each day Eastinghouse produces capacitors during three shifts: 8 A.M.–4 P.M., 4 P.M.–midnight, midnight–8 A.M. The hourly salary paid to the employees on each shift, the price charged for each capacitor made during each shift, and

the number of defects in each capacitor produced during a given shift are shown in Table 63. Each of the company's 25 workers can be assigned to one of the three shifts. A worker produces 10 capacitors during a shift, but because of machinery limitations, no more than 10 workers can be assigned to any shift. Each day, at most 250 capacitors can be sold, and the average number of defects per capacitor for the day's production cannot exceed three. Formulate an LP to maximize Eastinghouse's daily profit (sales revenue − labor cost).

33 Graphically find all solutions to the following LP:

$$\max z = 4x_1 + x_2$$
$$\text{s.t.} \quad 8x_1 + 2x_2 \leq 16$$
$$x_1 + x_2 \leq 12$$
$$x_1, x_2 \geq 0$$

34 During the next three months Airco must meet (on time) the following demands for air conditioners: month 1, 300; month 2, 400; month 3, 500. Air conditioners can be produced in either New York or Los Angeles. It takes 1.5 hours of skilled labor to produce an air conditioner in Los Angeles, and 2 hours in New York. It costs $400 to produce an air conditioner in Los Angeles, and $350 in New York. During each month, each city has 420 hours of skilled labor available. It costs $100 to hold an air conditioner in inventory for a month. At the beginning of month 1, Airco has 200 air conditioners in stock. Formulate an LP whose solution will tell Airco how to minimize the cost of meeting air conditioner demands for the next three months.

35 Formulate the following as a linear programming problem: A greenhouse operator plans to bid for the job of providing flowers for city parks. He will use tulips, daffodils, and flowering shrubs in three types of layouts. A Type 1 layout uses 30 tulips, 20 daffodils, and 4 flowering shrubs. A Type 2 layout uses 10 tulips, 40 daffodils, and 3 flowering shrubs. A Type 3 layout uses 20 tulips, 50 daffodils, and 2 flowering shrubs. The net profit is $50 for each Type 1 layout, $30 for each Type 2 layout, and $60 for each Type 3 layout. He has 1,000 tulips, 800 daffodils, and 100 flowering shrubs. How many layouts of each type should be used to yield maximum profit?

36 Explain how your formulation in Problem 35 changes if both of the following conditions are added:

a The number of Type 1 layouts cannot exceed the number of Type 2 layouts.

b There must be at least five layouts of each type.

37 Graphically solve the following LP problem:

$$\min z = 6x_1 + 2x_2$$
$$\text{s.t.} \quad 3x_1 + 2x_2 \geq 12$$
$$2x_1 + 4x_2 \geq 12$$
$$x_2 \geq 1$$
$$x_1, x_2 \geq 0$$

38 We produce two products: product 1 and product 2 on two machines (machine 1 and machine 2). The number of hours of machine time and labor depends on the machine and the product as shown in Table 64.

The cost of producing a unit of each product is shown in Table 65.

The number of labor hours and machine time available this month are in Table 66.

This month, at least 200 units of product 1 and at least 240 units of product 2 must be produced. Also, at least half of product 1 must be made on machine 1, and at least half of product 2 must be made on machine 2. Determine how we can minimize the cost of meeting our monthly demands.

39 Carrotco manufactures two products: 1 and 2. Each unit of each product must be processed on machine 1 and machine 2 and uses raw material 1 and raw material 2. The resource usage is as in Table 67.

TABLE 64

	Product 1 Machine 1	Product 2 Machine 1	Product 1 Machine 2	Product 2 Machine 2
Machine time	0.7	0.75	0.8	0.9
Labor	0.75	0.75	1.2	1

TABLE 65

Product 1 Machine 1	Product 2 Machine 1	Product 1 Machine 2	Product 2 Machine 2
$1.50	$0.40	$2.20	$4.00

TABLE 66

Resource	Hours Available
Machine 1	200
Machine 2	200
Labor	400

TABLE 67

	Product 1	Product 2
Machine 1	0.6	0.4
Machine 2	0.4	0.3
Raw material 1	2	1
Raw material 2	1	2

TABLE 68

	Product 1	Product 2
Demand	400	300
Sales Price	30	35

Thus, producing one unit of product 1 uses .6 unit of machine 1 time, .4 unit of machine 2 time, 2 units of raw material 1, and 1 unit of raw material 2. The sales price per unit and demand for each product are in Table 68.

It costs $4 to purchase each unit of raw material 1 and $5 to produce each unit of raw material 2. Unlimited amounts of raw material can be purchased. Two hundred units of machine 1 time and 300 units of machine 2 time are available. Determine how Carrotco can maximize its profit.

40 A company assembles two products: A and B. Product A sells for $11 per unit, and product B sells for $23 per unit. A unit of product A requires 2 hours on assembly line 1 and 1 unit of raw material. A unit of product B requires 2 units of raw material, 1 unit of A, and 2 hours on line 2. For line 1, 1,300 hours of time are available and 500 hours of time are available on line 2. A unit of raw material may be bought (for $5 a unit) or produced (at no cost) by using 2 hours of time on line 1. Determine how to maximize profit.

41 Ann and Ben are getting divorced and want to determine how to divide their joint property: retirement account, home, summer cottage, investments, and miscellaneous assets. To begin, Ann and Ben are told to allocate 100 total points to the assets. Their allocation is as shown in Table 69.

Assuming that all assets are divisible (that is, a fraction of each asset may be given to each person), how should the assets be allocated? Two criteria should govern the asset allocation:

Criteria 1 Each person should end up with the same number of points. This prevents Ann from envying Ben and Ben from envying Ann.

Criteria 2 The total number of points received by Ann and Ben should be maximized.

If assets could not be split between people, what problem arises?

42 Eli Daisy manufactures two drugs in Los Angeles and Indianapolis. The cost of manufacturing a pound of each drug is shown in Table 70.

TABLE 69

	Points	
Item	Ann's	Ben's
Retirement account	50	40
Home	20	30
Summer cottage	15	10
Investments	10	10
Miscellaneous	5	10

TABLE 70

City	Drug 1 Cost ($)	Drug 2 Cost ($)
Indianapolis	4.10	4.50
Los Angeles	4.00	5.20

TABLE 71

City	Drug 1 Time (Hr)	Drug 2 Time (Hr)
Indianapolis	.2	.3
Los Angeles	.24	.33

The machine time (in hours) required to produce a pound of each drug at each city is as in Table 71.

Daisy needs to produce at least 1,000 pounds of drug 1 and 2,000 pounds of drug 2 per week. The company has 500 hours per week of machine time in Indianapolis and 400 hours per week of machine time in Los Angeles. Determine how Lilly can minimize the cost of producing the needed drugs.

43 Daisy also produces Wozac in New York and Chicago. Each month, it can produce up to 30 units in New York and up to 35 units in Chicago. The cost of producing a unit each month at each location is shown in Table 72.

The customer demands shown in Table 73 must be met on time.

The cost of holding a unit in inventory (measured against ending inventory) is shown in Table 74.

TABLE 72

	Cost ($)	
Month	New York	Chicago
1	8.62	8.40
2	8.70	8.75
3	8.90	9.00

TABLE 73

Month	Demand (Units)
1	50
2	60
3	40

TABLE 74

Month	Holding Cost ($)
1	0.26
2	0.12
3	0.12

At the beginning of month 1, we have 10 units of Wozac in inventory. Determine a cost-minimizing schedule for the next three months.

44 You have been put in charge of the Dawson Creek oil refinery. The refinery produces gas and heating oil from crude oil. Gas sells for $11 per barrel and must have an average grade level of at least 9. Heating oil sells for $6 a barrel and must have an average grade level of at least 7. At most, 2,000 barrels of gas and 600 barrels of heating oil can be sold.

Incoming crude can be processed by one of three methods. The per barrel yield and per barrel cost of each processing method are shown in Table 75.

For example, if we refine one barrel of incoming crude by method 1, it costs us $3.40 and yields .2 barrels of grade 6, .2 barrels of grade 8, and .6 barrels of grade 10. These costs include the costs of buying the crude oil.

Before being processed into gas and heating oil, grades 6 and 8 may be sent through the catalytic cracker to improve their quality. For $1 per barrel, one barrel of grade 6 can be "cracked" into a barrel of grade 8. For $1.50 per barrel, a barrel of grade 8 can be cracked into a barrel of grade 10. Determine how to maximize the refinery's profit.

45 Currently we own 100 shares each of stocks 1 through 10. The original price we paid for these stocks, today's price, and the expected price in one year for each stock is shown in Table 76.

We need money today and are going to sell some of our stocks. The tax rate on capital gains is 30%. If we sell 50 shares of stock 1, then we must pay tax of $.3 \cdot 50(30 - 20) = \$150$. We must also pay transaction costs of 1% on each transaction. Thus, our sale of 50 shares of stock 1 would incur transaction costs of $.01 \cdot 50 \cdot 30 = \$15$. After taxes and transaction costs, we must be left with $30,000 from our stock sales. Our goal is to maximize the expected (before-tax) value in one year of our remaining stock. What stocks should we sell? Assume it is all right to sell a fractional share of stock.

Group B

46 Gotham City National Bank is open Monday–Friday from 9 A.M. to 5 P.M. From past experience, the bank knows that it needs the number of tellers shown in Table 77. The bank hires two types of tellers. Full-time tellers work 9–5 five days a week, except for 1 hour off for lunch. (The bank determines when a full-time employee takes lunch hour, but each teller must go between noon and 1 P.M. or between 1 P.M. and 2 P.M.) Full-time employees are paid (including fringe benefits) $8/hour (this includes payment for lunch hour). The bank may also hire part-time tellers. Each part-

TABLE 75

| Method | Grade | | | Cost ($ per Barrel) |
	6	8	10	
1	.2	.3	.5	3.40
2	.3	.4	.3	3.00
3	.4	.4	.2	2.60

TABLE **76**

Stock	Shares Owned	Price ($)		
		Purchase	Current	In One Year
1	100	20	30	36
2	100	25	34	39
3	100	30	43	42
4	100	35	47	45
5	100	40	49	51
6	100	45	53	55
7	100	50	60	63
8	100	55	62	64
9	100	60	64	66
10	100	65	66	70
Tax rate (%)	0.3			
Transaction cost (%)	0.01			

TABLE **77**

Time Period	Tellers Required
9–10	4
10–11	3
11–Noon	4
Noon–1	6
1–2	5
2–3	6
3–4	8
4–5	8

time teller must work exactly 3 consecutive hours each day. A part-time teller is paid $5/hour (and receives no fringe benefits). To maintain adequate quality of service, the bank has decided that at most five part-time tellers can be hired. Formulate an LP to meet the teller requirements at minimum cost. Solve the LP on a computer. Experiment with the LP answer to determine an employment policy that comes close to minimizing labor cost.

47[†]   The Gotham City Police Department employs 30 police officers. Each officer works 5 days per week. The crime rate fluctuates with the day of the week, so the number of police officers required each day depends on which day of the week it is: Saturday, 28; Sunday, 18; Monday, 18; Tuesday, 24; Wednesday, 25; Thursday, 16; Friday, 21. The police department wants to schedule police officers to minimize the number whose days off are not consecutive. Formulate an LP that will accomplish this goal. (*Hint:* Have a constraint for each day of the week that ensures that the proper number of officers are *not* working on the given day.)

[†]Based on Rothstein (1973).

48[‡]Alexis Cornby makes her living buying and selling corn. On January 1, she has 50 tons of corn and $1,000. On the first day of each month Alexis can buy corn at the following prices per ton: January, $300; February, $350; March, $400; April, $500. On the last day of each month, Alexis can sell corn at the following prices per ton: January, $250; February, $400; March, $350; April, $550. Alexis stores her corn in a warehouse that can hold at most 100 tons of corn. She must be able to pay cash for all corn at the time of purchase. Use linear programming to determine how Alexis can maximize her cash on hand at the end of April.

49[§]At the beginning of month 1, Finco has $400 in cash. At the beginning of months 1, 2, 3, and 4, Finco receives certain revenues, after which it pays bills (see Table 78). Any money left over may be invested for one month at the interest rate of 0.1% per month; for two months at 0.5% per month; for three months at 1% per month; or for four months at 2% per month. Use linear programming to determine an investment strategy that maximizes cash on hand at the beginning of month 5.

50   City 1 produces 500 tons of waste per day, and city 2 produces 400 tons of waste per day. Waste must be incinerated at incinerator 1 or 2, and each incinerator can process up to 500 tons of waste per day. The cost to incinerate waste is $40/ton at incinerator 1 and $30/ton at 2.

TABLE **78**

Month	Revenues ($)	Bills ($)
1	400	600
2	800	500
3	300	500
4	300	250

[‡]Based on Charnes and Cooper (1955).
[§]Based on Robichek, Teichroew, and Jones (1965).

Incineration reduces each ton of waste to 0.2 tons of debris, which must be dumped at one of two landfills. Each landfill can receive at most 200 tons of debris per day. It costs $3 per mile to transport a ton of material (either debris or waste). Distances (in miles) between locations are shown in Table 79. Formulate an LP that can be used to minimize the total cost of disposing of the waste of both cities.

51[†] Silicon Valley Corporation (Silvco) manufactures transistors. An important aspect of the manufacture of transistors is the melting of the element germanium (a major component of a transistor) in a furnace. Unfortunately, the melting process yields germanium of highly variable quality.

Two methods can be used to melt germanium; method 1 costs $50 per transistor, and method 2 costs $70 per transistor. The qualities of germanium obtained by methods 1 and 2 are shown in Table 80. Silvco can refire melted germanium in an attempt to improve its quality. It costs $25 to refire the melted germanium for one transistor. The results of the refiring process are shown in Table 81. Silvco has sufficient furnace capacity to melt or refire germanium for at most 20,000 transistors per month. Silvco's monthly demands are for 1,000 grade 4 transistors, 2,000 grade 3 transistors, 3,000 grade 2 transistors, and 3,000 grade 1 transistors. Use linear programming to minimize the cost of producing the needed transistors.

TABLE 79

City	Incinerator	
	1	2
1	30	5
2	36	42

Incinerator	Landfill	
	1	2
1	5	8
2	9	6

TABLE 80

Grade of[‡] Melted Germanium	Percent Yielded by Melting	
	Method 1	Method 2
Defective	30	20
1	30	20
2	20	25
3	15	20
4	5	15

[‡]*Note:* Grade 1 is poor; grade 4 is excellent. The quality of the germanium dictates the quality of the manufactured transistor.

[†]Based on Smith (1965).

TABLE 81

Refired Grade of Germanium	Percent Yielded by Refiring			
	Defective	Grade 1	Grade 2	Grade 3
Defective	30	0	0	0
1	25	30	0	0
2	15	30	40	0
3	20	20	30	50
4	10	20	30	50

TABLE 82

Input	Cost ($)	Pulp Content (%)
Box board	5	15
Tissue paper	6	20
Newsprint	8	30
Book paper	10	40

52[‡] A paper-recycling plant processes box board, tissue paper, newsprint, and book paper into pulp that can be used to produce three grades of recycled paper (grades 1, 2, and 3). The prices per ton and the pulp contents of the four inputs are shown in Table 82. Two methods, de-inking and asphalt dispersion, can be used to process the four inputs into pulp. It costs $20 to de-ink a ton of any input. The process of de-inking removes 10% of the input's pulp, leaving 90% of the original pulp. It costs $15 to apply asphalt dispersion to a ton of material. The asphalt dispersion process removes 20% of the input's pulp. At most, 3,000 tons of input can be run through the asphalt dispersion process or the de-inking process. Grade 1 paper can only be produced with newsprint or book paper pulp; grade 2 paper, only with book paper, tissue paper, or box board pulp; and grade 3 paper, only with newsprint, tissue paper, or box board pulp. To meet its current demands, the company needs 500 tons of pulp for grade 1 paper, 500 tons of pulp for grade 2 paper, and 600 tons of pulp for grade 3 paper. Formulate an LP to minimize the cost of meeting the demands for pulp.

53 Turkeyco produces two types of turkey cutlets for sale to fast-food restaurants. Each type of cutlet consists of white meat and dark meat. Cutlet 1 sells for $4/lb and must consist of at least 70% white meat. Cutlet 2 sells for $3/lb and must consist of at least 60% white meat. At most, 50 lb of cutlet 1 and 30 lb of cutlet 2 can be sold. The two types of turkey used to manufacture the cutlets are purchased from the GobbleGobble Turkey Farm. Each type 1 turkey costs $10 and yields 5 lb of white meat and 2 lb of dark meat. Each type 2 turkey costs $8 and yields 3 lb of white meat and 3 lb of dark meat. Formulate an LP to maximize Turkeyco's profit.

54 Priceler manufactures sedans and wagons. The number of vehicles that can be sold each of the next three months

[‡]Based on Glassey and Gupta (1975).

TABLE 83

Month	Sedans	Wagons
1	1,100	600
2	1,500	700
3	1,200	50

are listed in Table 83. Each sedan sells for $8,000, and each wagon sells for $9,000. It costs $6,000 to produce a sedan and $7,500 to produce a wagon. To hold a vehicle in inventory for one month costs $150 per sedan and $200 per wagon. During each month, at most 1,500 vehicles can be produced. Production line restrictions dictate that during month 1 at least two-thirds of all cars produced must be sedans. At the beginning of month 1, 200 sedans and 100 wagons are available. Formulate an LP that can be used to maximize Priceler's profit during the next three months.

55 The production-line employees at Grummins Engine work four days a week, 10 hours a day. Each day of the week, (at least) the following numbers of line employees are needed: Monday–Friday, 7 employees; Saturday and Sunday, 3 employees. Grummins has 11 production-line employees. Formulate an LP that can be used to maximize the number of consecutive days off received by the employees. For example, a worker who gets Sunday, Monday, and Wednesday off receives two consecutive days off.

56 Bank 24 is open 24 hours per day. Tellers work two consecutive 6-hour shifts and are paid $10 per hour. The possible shifts are as follows: midnight–6 A.M., 6 A.M.–noon, noon–6 P.M., 6 P.M.–midnight. During each shift, the following numbers of customers enter the bank: midnight–6 A.M., 100; 6 A.M.–noon, 200; noon–6 P.M., 300; 6 P.M.–midnight, 200. Each teller can serve up to 50 customers per shift. To model a cost for customer impatience, we assume that any customer who is present at the end of a shift "costs" the bank $5. We assume that by midnight of each day, all customers must be served, so each day's midnight–6 A.M. shift begins with 0 customers in the bank. Formulate an LP that can be used to minimize the sum of the bank's labor and customer impatience costs.

57[†] Transeast Airlines flies planes on the following route: L.A.–Houston–N.Y.–Miami–L.A. The length (in miles) of each segment of this trip is as follows: L.A.–Houston, 1,500 miles; Houston–N.Y., 1,700 miles; N.Y.–Miami, 1,300 miles; Miami–L.A., 2,700 miles. At each stop, the plane may purchase up to 10,000 gallons of fuel. The price of fuel at each city is as follows: L.A., 88¢; Houston, 15¢; N.Y., $1.05; Miami, 95¢. The plane's fuel tank can hold at most 12,000 gallons. To allow for the possibility of circling over a landing site, we require that the ending fuel level for each leg of the flight be at least 600 gallons. The number of gallons used per mile on each leg of the flight is

1 + (average fuel level on leg of flight/2,000)

[†]Based on Darnell and Loflin (1977).

To simplify matters, assume that the average fuel level on any leg of the flight is

(Fuel level at start of leg) + (fuel level at end of leg)
                                2

Formulate an LP that can be used to minimize the fuel cost incurred in completing the schedule.

58[‡] To process income tax forms, the IRS first sends each form through the data preparation (DP) department, where information is coded for computer entry. Then the form is sent to data entry (DE), where it is entered into the computer. During the next three weeks, the following number of forms will arrive: week 1, 40,000; week 2, 30,000; week 3, 60,000. The IRS meets the crunch by hiring employees who work 40 hours per week and are paid $200 per week. Data preparation of a form requires 15 minutes, and data entry of a form requires 10 minutes. Each week, an employee is assigned to either data entry or data preparation. The IRS must complete processing of all forms by the end of week 5 and wants to minimize the cost of accomplishing this goal. Formulate an LP that will determine how many workers should be working each week and how the workers should be assigned over the next five weeks.

59 In the electrical circuit of Figure 11, I_t = current (in amperes) flowing through resistor t, V_t = voltage drop (in volts) across resistor t, and R_t = resistance (in ohms) of resistor t. Kirchoff's Voltage and Current Laws imply that $V_1 = V_2 = V_3$ and $I_1 + I_2 + I_3 = I_4$. The power dissipated by the current flowing through resistor t is $I_t^2 R_t$. Ohm's Law implies that $V_t = I_t R_t$. The two parts of this problem should be solved independently.

a Suppose you are told that $I_1 = 4$, $I_2 = 6$, $I_3 = 8$, and $I_4 = 18$ are required. Also, the voltage drop across each resistor must be between 2 and 10 volts. Choose the R_t's to minimize the total dissipated power. Formulate an LP whose solution will solve your problem.

b Suppose you are told that $V_1 = 6$, $V_2 = 6$, $V_3 = 6$, and $V_4 = 4$ are required. Also, the current flowing through each resistor must be between 2 and 6 amperes. Choose the R_t's to minimize the total dissipated power. Formulate an LP whose solution will solve your problem. (*Hint:* Let $\frac{1}{R_t}$ ($t = 1, 2, 3, 4$) be your decision variables.)

60 The mayor of Llanview is trying to determine the number of judges needed to handle the judicial caseload.

FIGURE 11

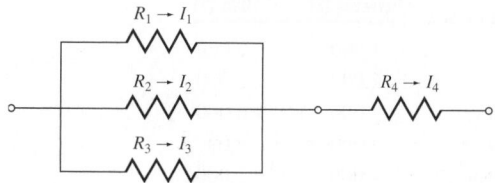

[‡]Based on Lanzenauer et al. (1987).

TABLE 84

Month	Hours
January	400
February	300
March	200
April	600
May	800
June	300
July	200
August	400
September	300
October	200
November	100
December	300

During each month of the year it is estimated that the number of judicial hours needed is as given in Table 84.

a Each judge works all 12 months and can handle as many as 120 hours per month of casework. To avoid creating a backlog, all cases must be handled by the end of December. Formulate an LP whose solution will determine how many judges Llanview needs.

b If each judge received one month of vacation each year, how would your answer change?

Group C

61[†] E.J. Korvair Department Store has $1,000 in available cash. At the beginning of each of the next six months, E.J. will receive revenues and pay bills as shown in Table 85. It is clear that E.J. will have a short-term cash flow problem until the store receives revenues from the Christmas shopping season. To solve this problem, E.J. must borrow money.

At the beginning of July, E.J. may take out a six-month loan. Any money borrowed for a six-month period must be paid back at the end of December along with 9% interest (early payback does not reduce the interest cost of the loan). E.J. may also meet cash needs through month-to-month borrowing. Any money borrowed for a one-month period incurs an interest cost of 4% per month. Use linear programming to determine how E.J. can minimize the cost of paying its bills on time.

62[‡] Olé Oil produces three products: heating oil, gasoline, and jet fuel. The average octane levels must be at least 4.5 for heating oil, 8.5 for gas, and 7.0 for jet fuel. To produce these products Olé purchases two types of oil: crude 1 (at $12 per barrel) and crude 2 (at $10 per barrel). Each day, at most 10,000 barrels of each type of oil can be purchased.

Before crude can be used to produce products for sale, it must be distilled. Each day, at most 15,000 barrels of oil can be distilled. It costs 10¢ to distill a barrel of oil. The result of distillation is as follows: (1) Each barrel of crude 1 yields 0.6 barrel of naphtha, 0.3 barrel of distilled 1, and 0.1 barrel of distilled 2. (2) Each barrel of crude 2 yields 0.4 barrel of naphtha, 0.2 barrel of distilled 1, and 0.4 barrel of distilled 2. Distilled naphtha can be used only to produce gasoline or jet fuel. Distilled oil can be used to produce heating oil or it can be sent through the catalytic cracker (at a cost of 15¢ per barrel). Each day, at most 5,000 barrels of distilled oil can be sent through the cracker. Each barrel of distilled 1 sent through the cracker yields 0.8 barrel of cracked 1 and 0.2 barrel of cracked 2. Each barrel of distilled 2 sent through the cracker yields 0.7 barrel of cracked 1 and 0.3 barrel of cracked 2. Cracked oil can be used to produce gasoline and jet fuel but not to produce heating oil.

The octane level of each type of oil is as follows: naphtha, 8; distilled 1, 4; distilled 2, 5; cracked 1, 9; cracked 2, 6.

All heating oil produced can be sold at $14 per barrel; all gasoline produced, $18 per barrel; and all jet fuel produced, $16 per barrel. Marketing considerations dictate that at least 3,000 barrels of each product must be produced daily. Formulate an LP to maximize Olé's daily profit.

63 Donald Rump is the international funds manager for Countribank. Each day Donald's job is to determine how the bank's current holdings of dollars, pounds, marks, and yen should be adjusted to meet the day's currency needs. Today the exchange rates between the various currencies are given in Table 86. For example, one dollar can be converted to .58928 pounds, or one pound can be converted to 1.697 dollars.

At the beginning of the day, Countribank has the currency holdings given in Table 87.

At the end of the day, Countribank must have at least the amounts of each currency given in Table 88.

Donald's goal is to each day transfer funds in a way that makes currency holdings satisfy the previously listed mini-

TABLE 85

Month	Revenues ($)	Bills ($)
July	1,000	5,000
August	2,000	5,000
September	2,000	6,000
October	4,000	2,000
November	7,000	2,000
December	9,000	1,000

TABLE 86

	To			
From	Dollars	Pounds	Marks	Yen
Dollars	1	.58928	1.743	138.3
Pounds	1.697	1	2.9579	234.7
Marks	.57372	.33808	1	79.346
Yen	.007233	.00426	.0126	1

[†]Based on Robichek, Teichroew, and Jones (1965).

[‡]Based on Garvin et al. (1957).

TABLE **87**	
Currency	Units (in Billions)
Dollars	8
Pounds	1
Marks	8
Yen	0

TABLE **88**	
Currency	Units (in Billions)
Dollars	6
Pounds	3
Marks	1
Yen	10

mums, and maximizes the dollar value of the currency holdings at the end of the day.

To figure out the dollar value of, say, one pound, average the two conversion rates. Thus, one pound is worth approximately

$$\frac{1.697 + (1/.58928)}{2} = 1.696993 \text{ dollars}$$

REFERENCES

Each of the following seven books is a cornucopia of interesting LP formulations:

Bradley, S., A. Hax, and T. Magnanti. *Applied Mathematical Programming.* Reading, Mass.: Addison-Wesley, 1977.

Lawrence, K., and S. Zanakis. *Production Planning and Scheduling: Mathematical Programming Applications.* Atlanta, Ga: Industrial Engineering and Management Press, 1984.

Murty, K. *Operations Research: Deterministic Optimization Models.* Saddle River, N.J.: Prentice-Hall, 1995.

Schrage, L. *Linear Integer and Quadratic Programming With LINDO.* Palo Alto, Calif.: Scientific Press, 1986.

Shapiro, J. *Optimization Models for Planning and Allocation: Text and Cases in Mathematical Programming.* New York: Wiley, 1984.

Wagner, H. *Principles of Operations Research,* 2d ed. Englewood Cliffs, N.J.: Prentice Hall, 1975.

Williams, H. *Model Building in Mathematical Programming,* 2d ed. New York: Wiley, 1999.

Baker, K. "Scheduling a Full-Time Work Force to Meet Cyclic Staffing Requirements," *Management Science* 20(1974):1561–1568. Presents a method (other than LP) for scheduling personnel to meet cyclic workforce requirements.

Balintfy, J. "A Mathematical Programming System for Food Management Applications," *Interfaces* 6(no. 1, pt 2, 1976):13–31. Discusses menu planning models.

Carino, H., and C. Lenoir. "Optimizing Wood Procurement in Cabinet Manufacturing," *Interfaces* 18(no. 2, 1988):11–19.

Chandy, K. "Pricing in the Government Bond Market," *Interfaces* 16(1986):65–71.

Charnes, A., and W. Cooper. "Generalization of the Warehousing Model," *Operational Research Quarterly* 6(1955):131–172.

Cheung, H., and J. Auger. "Linear Programming and Land Use Allocation," *Socio-Economic Planning Science* 10(1976):43–45.

Darnell, W., and C. Loflin. "National Airlines Fuel Management and Allocation Model," *Interfaces* 7(no. 3, 1977):1–15.

Dobson, G., and S. Kalish. "Positioning and Pricing a Product Line," *Marketing Science* 7(1988):107–126.

Fabian, T. "A Linear Programming Model of Integrated Iron and Steel Production," *Management Science,* 4(1958):415–449.

Forgionne, G. "Corporate MS Activities: An Update," *Interfaces* 13(1983):20–23. Concerns the fraction of large firms using linear programming (and other operations research techniques).

Franklin, A., and E. Koenigsberg. "Computed School Assignments in a Large District," *Operations Research* 21(1973):413–426.

Garvin, W., et al. "Applications of Linear Programming in the Oil Industry," *Management Science* 3(1957): 407–430.

Glassey, R., and V. Gupta. "An LP Analysis of Paper Recycling." In *Studies in Linear Programming,* ed. H. Salkin and J. Saha. New York: North-Holland, 1975.

Hartley, R. "Decision Making When Joint Products Are Involved," *Accounting Review* (1971):746–755.

Heady, E., and A. Egbert. "Regional Planning of Efficient Agricultural Patterns," *Econometrica* 32(1964):374–386.

Hilal, S., and W. Erickson. "Matching Supplies to Save Lives: Linear Programming the Production of Heart Valves," *Interfaces* 11(1981):48–56.

Jain, S., K. Stott, and E. Vasold. "Orderbook Balancing Using a Combination of LP and Heuristic Techniques," *Interfaces* 9(no. 1, 1978):55–67.

Krajewski, L., L. Ritzman, and P. McKenzie. "Shift Scheduling in Banking Operations: A Case Application," *Interfaces,* 10(no. 2, 1980):1–8.

Love, R., and J. Hoey, "Management Science Improves Fast Food Operations," *Interfaces,* 20(no. 2, 1990): 21–29.

Magoulas, K., and D. Marinos-Kouris. "Gasoline Blending LP," *Oil and Gas Journal* (July 18, 1988):44–48.

Moondra, S. "An LP Model for Workforce Scheduling in Banks," *Journal of Bank Research* (1976).

Myers, S., and C. Pogue. "A Programming Approach to Corporate Financial Management," *Journal of Finance* 29(1974):579–599.

Neave, E., and J. Wiginton. *Financial Management: Theory and Strategies.* Englewood Cliffs, N.J.: Prentice Hall, 1981.

Robbins, W., and N. Tuntiwonpiboon. "Linear Programming a Useful Tool in Case-Mix Management," *HealthCare Financial Management* (1989):114–117.

Robichek, A., D. Teichroew, and M. Jones. "Optimal Short-Term Financing Decisions," *Management Science* 12(1965):1–36.

Rohn, E. "A New LP Approach to Bond Portfolio Management," *Journal of Financial and Quantitative Analysis* 22(1987):439–467.

Rothstein, M. "Hospital Manpower Shift Scheduling by Mathematical Programming," *Health Services Research* (1973).

Smith, S. "Planning Transistor Production by Linear Programming," *Operations Research* 13(1965): 132–139.

Stigler, G. "The Cost of Subsistence," *Journal of Farm Economics* 27(1945). Discusses the diet problem.

Sullivan, R., and S. Secrest. "A Simple Optimization DSS for Production Planning at Dairyman's Cooperative Creamery Association," *Interfaces* 15(no. 5, 1985): 46–54.

Weingartner, H. *Mathematical Programming and the Analysis of Capital Budgeting.* Englewood Cliffs, N.J.: Prentice Hall, 1963.

4

The Simplex Algorithm and Goal Programming

In Chapter 3, we saw how to solve two-variable linear programming problems graphically. Unfortunately, most real-life LPs have many variables, so a method is needed to solve LPs with more than two variables. We devote most of this chapter to a discussion of the simplex algorithm, which is used to solve even very large LPs. In many industrial applications, the simplex algorithm is used to solve LPs with thousands of constraints and variables.

In this chapter, we explain how the simplex algorithm can be used to find optimal solutions to LPs. We also detail how two state-of-the-art computer packages (LINDO and LINGO) can be used to solve LPs. Briefly, we also discuss Karmarkar's pioneering approach for solving LPs. We close the chapter with an introduction to goal programming, which enables the decision maker to consider more than one objective function.

4.1 How to Convert an LP to Standard Form

We have seen that an LP can have both equality and inequality constraints. It also can have variables that are required to be nonnegative as well as those allowed to be unrestricted in sign (urs). Before the simplex algorithm can be used to solve an LP, the LP must be converted into an equivalent problem in which all constraints are equations and all variables are nonnegative. An LP in this form is said to be in **standard form.**[†]

To convert an LP into standard form, each inequality constraint must be replaced by an equality constraint. We illustrate this procedure using the following problem.

EXAMPLE 1 — Leather Limited

Leather Limited manufactures two types of belts: the deluxe model and the regular model. Each type requires 1 sq yd of leather. A regular belt requires 1 hour of skilled labor, and a deluxe belt requires 2 hours. Each week, 40 sq yd of leather and 60 hours of skilled labor are available. Each regular belt contributes $3 to profit and each deluxe belt, $4. If we define

$$x_1 = \text{number of deluxe belts produced weekly}$$
$$x_2 = \text{number of regular belts produced weekly}$$

[†]Throughout the first part of the chapter we assume that all variables must be nonnegative (≥ 0). The conversion of urs variables to nonnegative variables is discussed in Section 4.12.

the appropriate LP is

$$\max z = 4x_1 + 3x_2 \qquad \text{(LP 1)}$$

$$\text{s.t.} \qquad x_1 + x_2 \leq 40 \qquad \text{(Leather constraint)} \qquad (1)$$

$$2x_1 + x_2 \leq 60 \qquad \text{(Labor constraint)} \qquad (2)$$

$$x_1, x_2 \geq 0$$

How can we convert (1) and (2) to equality constraints? We define for each $\leq$ constraint a **slack variable** s_i(s_i = slack variable for ith constraint), which is the amount of the resource unused in the ith constraint. Because $x_1 + x_2$ sq yd of leather are being used, and 40 sq yd are available, we define s_1 by

$$s_1 = 40 - x_1 - x_2 \qquad \text{or} \qquad x_1 + x_2 + s_1 = 40$$

Similarly, we define s_2 by

$$s_2 = 60 - 2x_1 - x_2 \qquad \text{or} \qquad 2x_1 + x_2 + s_2 = 60$$

Observe that a point (x_1, x_2) satisfies the ith constraint if and only if $s_i \geq 0$. For example, $x_1 = 15$, $x_2 = 20$ satisfies (1) because $s_1 = 40 - 15 - 20 = 5 \geq 0$.

Intuitively, (1) is satisfied by the point (15, 20), because $s_1 = 5$ sq yd of leather are unused. Similarly, (15, 20) satisfies (2), because $s_2 = 60 - 2(15) - 20 = 10$ labor hours are unused. Finally, note that the point $x_1 = x_2 = 25$ fails to satisfy (2), because $s_2 = 60 - 2(25) - 25 = -15$ indicates that (25, 25) uses more labor than is available.

In summary, to convert (1) to an equality constraint, we replace (1) by $s_1 = 40 - x_1 - x_2$ (or $x_1 + x_2 + s_1 = 40$) and $s_1 \geq 0$. To convert (2) to an equality constraint, we replace (2) by $s_2 = 60 - 2x_1 - x_2$ (or $2x_1 + x_2 + s_2 = 60$) and $s_2 \geq 0$. This converts LP 1 to

$$\max z = 4x_1 + 3x_2$$

$$\text{s.t.} \qquad x_1 + x_2 + s_1 \qquad = 40$$

$$2x_1 + x_2 \qquad + s_2 = 60 \qquad \text{(LP 1$'$)}$$

$$x_1, x_2, s_1, s_2 \geq 0$$

Note that LP $1'$ is in standard form. In summary, *if constraint i of an LP is a $\leq$ constraint, then we convert it to an equality constraint by adding a slack variable s_i to the ith constraint and adding the sign restriction $s_i \geq 0$.*

To illustrate how a $\geq$ constraint can be converted to an equality constraint, let's consider the diet problem of Section 3.4.

$$\min z = 50x_1 + 20x_2 + 30x_3 + 80x_4$$

$$\text{s.t.} \quad 400x_1 + 200x_2 + 150x_3 + 500x_4 \geq 500 \qquad \text{(Calorie constraint)} \qquad (3)$$

$$3x_1 + 2x_2 \qquad\qquad\qquad \geq 6 \qquad \text{(Chocolate constraint)} \qquad (4)$$

$$2x_1 + 2x_2 + 4x_3 + 4x_4 \geq 10 \qquad \text{(Sugar constraint)} \qquad (5)$$

$$2x_1 + 4x_2 + x_3 + 5x_4 \geq 8 \qquad \text{(Fat constraint)} \qquad (6)$$

$$x_1, x_2, x_3, x_4 \geq 0$$

To convert the ith $\geq$ constraint to an equality constraint, we define an **excess variable** (sometimes called a surplus variable) e_i. (e_i will always be the excess variable for the ith constraint.) We define e_i to be the amount by which the ith constraint is oversatisfied. Thus, for the diet problem,

$$e_1 = 400x_1 + 200x_2 + 150x_3 + 500x_4 - 500, \qquad \text{or} \qquad \text{(3')}$$
$$400x_1 + 200x_2 + 150x_3 + 500x_4 - e_1 = 500$$

$$e_2 = 3x_1 + 2x_2 - 6, \qquad \text{or} \qquad 3x_1 + 2x_2 - e_2 = 6 \qquad \text{(4')}$$
$$e_3 = 2x_1 + 2x_2 + 4x_3 + 4x_4 - 10, \qquad \text{or} \qquad 2x_1 + 2x_2 + 4x_3 + 4x_4 - e_3 = 10 \quad \text{(5')}$$
$$e_4 = 2x_1 + 4x_2 + x_3 + 5x_4 - 8, \qquad \text{or} \qquad 2x_1 + 4x_2 + x_3 + 5x_4 - e_4 = 8 \qquad \text{(6')}$$

A point (x_1, x_2, x_3, x_4) satisfies the ith $\geq$ constraint if and only if e_i is nonnegative. For example, from (4'), $e_2 \geq 0$ if and only if $3x_1 + 2x_2 \geq 6$. For a numerical example, consider the point $x_1 = 2$, $x_3 = 4$, $x_2 = x_4 = 0$, which satisfies all four of the diet problem's constraints. For this point,

$$e_1 = 400(2) + 150(4) - 500 = 900 \geq 0$$
$$e_2 = 3(2) - 6 = 0 \geq 0$$
$$e_3 = 2(2) + 4(4) - 10 = 10 \geq 0$$
$$e_4 = 2(2) + 4 - 8 = 0 \geq 0$$

As another example, consider $x_1 = x_2 = 1$, $x_3 = x_4 = 0$. This point is infeasible; it violates the chocolate, sugar, and fat constraints. The infeasibility of this point is indicated by

$$e_2 = 3(1) + 2(1) - 6 = -1 < 0$$
$$e_3 = 2(1) + 2(1) - 10 = -6 < 0$$
$$e_4 = 2(1) + 4(1) - 8 = -2 < 0$$

Thus, to transform the diet problem into standard form, replace (3) by (3'); (4) by (4'); (5) by (5'); and (6) by (6'). We must also add the sign restrictions $e_i \geq 0$ ($i = 1, 2, 3, 4$). The resulting LP is in standard form and may be written as

$$\min z = 50x_1 + 20x_2 + 30x_3 + 80x_4$$

$$
\begin{aligned}
\text{s.t.} \quad & 400x_1 + 200x_2 + 150x_3 + 500x_4 - e_1 && = 500 \\
& 3x_1 + 2x_2 && - e_2 && = 6 \\
& 2x_1 + 2x_2 + 4x_3 + 4x_4 && - e_3 && = 10 \\
& 2x_1 + 4x_2 + x_3 + 5x_4 && - e_4 && = 8 \\
& x_i, e_i \geq 0 \qquad (i = 1, 2, 3, 4)
\end{aligned}
$$

In summary, *if the ith constraint of an LP is a $\geq$ constraint, then it can be converted to an equality constraint by subtracting an excess variable e_i from the ith constraint and adding the sign restriction $e_i \geq 0$.*

If an LP has both $\leq$ and $\geq$ constraints, then simply apply the procedures we have described to the individual constraints. As an example, let's convert the short-term financial planning model of Section 3.7 to standard form. Recall that the original LP was

$$\max z = 20x_1 + 15x_2$$

$$
\begin{aligned}
\text{s.t.} \quad & x_1 && \leq 100 \\
& x_2 \leq 100 \\
& 50x_1 + 35x_2 \leq 6{,}000 \\
& 20x_1 + 15x_2 \geq 2{,}000 \\
& x_1, x_2 \geq 0
\end{aligned}
$$

Following the procedures described previously, we transform this LP into standard form by adding slack variables s_1, s_2, and s_3, respectively, to the first three constraints and subtracting an excess variable e_4 from the fourth constraint. Then we add the sign restrictions $s_1 \geq 0$, $s_2 \geq 0$, $s_3 \geq 0$, and $e_4 \geq 0$. This yields the following LP in standard form:

$$\max z = 20x_1 + 15x_2$$

$$\begin{aligned}
\text{s.t.} \quad x_1 + + s_1 &= 100 \\
x_2 + s_2 &= 100 \\
50x_1 + 35x_2 + s_3 &= 6{,}000 \\
20x_1 + 15x_2 - e_4 &= 2{,}000 \\
\end{aligned}$$

$$x_i \geq 0 \quad (i = 1, 2); \qquad s_i \geq 0 \quad (i = 1, 2, 3); \qquad e_4 \geq 0$$

Of course, we could easily have labeled the excess variable for the fourth constraint e_1 (because it is the first excess variable). We chose to call it e_4 rather than e_1 to indicate that e_4 is the excess variable for the fourth constraint.

PROBLEMS

Group A

1 Convert the Giapetto problem (Example 1 in Chapter 3) to standard form.

2 Convert the Dorian problem (Example 2 in Chapter 3) to standard form.

3 Convert the following LP to standard form:

$$\min z = 3x_1 + x_2$$

$$\begin{aligned}
\text{s.t.} \quad x_1 &\geq 3 \\
x_1 + x_2 &\leq 4 \\
2x_1 - x_2 &= 3 \\
x_1, x_2 &\geq 0
\end{aligned}$$

4.2 Preview of the Simplex Algorithm

Suppose we have converted an LP with m constraints into standard form. Assuming that the standard form contains n variables (labeled for convenience $x_1, x_2, \ldots, x_n$), the standard form for such an LP is

$$\max z = c_1x_1 + c_2x_2 + \cdots + c_nx_n$$

(or min)

$$\begin{aligned}
\text{s.t.} \quad a_{11}x_1 + a_{12}x_2 + \cdots + a_{1n}x_n &= b_1 \\
a_{21}x_1 + a_{22}x_2 + \cdots + a_{2n}x_n &= b_2 \\
\vdots \qquad \vdots \qquad\qquad \vdots \\
a_{m1}x_1 + a_{m2}x_2 + \cdots + a_{mn}x_n &= b_m \\
x_i \geq 0 \quad (i = 1, 2, \ldots, n)
\end{aligned} \tag{7}$$

If we define

$$A = \begin{bmatrix} a_{11} & a_{12} & \cdots & a_{1n} \\ a_{21} & a_{22} & \cdots & a_{2n} \\ \vdots & \vdots & & \vdots \\ a_{m1} & a_{m2} & \cdots & a_{mn} \end{bmatrix}$$

and

$$\mathbf{x} = \begin{bmatrix} x_1 \\ x_2 \\ \vdots \\ x_n \end{bmatrix}, \qquad \mathbf{b} = \begin{bmatrix} b_1 \\ b_2 \\ \vdots \\ b_m \end{bmatrix}$$

the constraints for (7) may be written as the system of equations $A\mathbf{x} = \mathbf{b}$. Before proceeding further with our discussion of the simplex algorithm, we must define the concept of a basic solution to a linear system.

Basic and Nonbasic Variables

Consider a system $A\mathbf{x} = \mathbf{b}$ of m linear equations in n variables (assume $n \geq m$).

DEFINITION ■ A basic solution to $A\mathbf{x} = \mathbf{b}$ is obtained by setting $n - m$ variables equal to 0 and solving for the values of the remaining m variables. This assumes that setting the $n - m$ variables equal to 0 yields unique values for the remaining m variables or, equivalently, the columns for the remaining m variables are linearly independent. ■

To find a basic solution to $A\mathbf{x} = \mathbf{b}$, we choose a set of $n - m$ variables (the **nonbasic variables,** or **NBV**) and set each of these variables equal to 0. Then we solve for the values of the remaining $n - (n - m) = m$ variables (the **basic variables,** or **BV**) that satisfy $A\mathbf{x} = \mathbf{b}$.

Of course, the different choices of nonbasic variables will lead to different basic solutions. To illustrate, we find all the basic solutions to the following system of two equations in three variables:

$$\begin{aligned} x_1 + x_2 \qquad &= 3 \\ -x_2 + x_3 &= -1 \end{aligned} \qquad (8)$$

We begin by choosing a set of $3 - 2 = 1$ (3 variables, 2 equations) nonbasic variables. For example, if NBV $= \{x_3\}$, then BV $= \{x_1, x_2\}$. We obtain the values of the basic variables by setting $x_3 = 0$ and solving

$$\begin{aligned} x_1 + x_2 &= 3 \\ -x_2 &= -1 \end{aligned}$$

We find that $x_1 = 2, x_2 = 1$. Thus, $x_1 = 2, x_2 = 1, x_3 = 0$ is a basic solution to (8). However, if we choose NBV $= \{x_1\}$ and BV $= \{x_2, x_3\}$, we obtain the basic solution $x_1 = 0$, $x_2 = 3, x_3 = 2$. If we choose NBV $= \{x_2\}$, we obtain the basic solution $x_1 = 3, x_2 = 0$, $x_3 = -1$. The reader should verify these results.

Some sets of m variables do not yield a basic solution. For example, consider the following linear system:

$$\begin{aligned} x_1 + 2x_2 + x_3 &= 1 \\ 2x_1 + 4x_2 + x_3 &= 3 \end{aligned}$$

If we choose NBV $= \{x_3\}$ and BV $= \{x_1, x_2\}$, the corresponding basic solution would be obtained by solving

$$\begin{aligned} x_1 + 2x_2 &= 1 \\ 2x_1 + 4x_2 &= 3 \end{aligned}$$

Because this system has no solution, there is no basic solution corresponding to BV = $\{x_1, x_2\}$.

Feasible Solutions

A certain subset of the basic solutions to the constraints $A\mathbf{x} = \mathbf{b}$ of an LP plays an important role in the theory of linear programming.

DEFINITION ■ Any basic solution to (7) in which all variables are nonnegative is a **basic feasible solution** (or **bfs**). ■

Thus, for an LP with the constraints given by (8), the basic solutions $x_1 = 2$, $x_2 = 1$, $x_3 = 0$, and $x_1 = 0$, $x_2 = 3$, $x_3 = 2$ are basic *feasible* solutions, but the basic solution $x_1 = 3$, $x_2 = 0$, $x_3 = -1$ fails to be a basic solution (because $x_3 < 0$).

In the rest of this section, we assume that all LPs are in standard form. Recall from Section 3.2 that the feasible region for any LP is a convex set. Let S be the feasible region for an LP in standard form. Recall that a point P is an extreme point of S if all line segments that contain P and are completely contained in S have P as an endpoint. It turns out that the extreme points of an LP's feasible region and the LP's basic feasible solutions are actually one and the same. More formally,

THEOREM **1**

A point in the feasible region of an LP is an extreme point if and only if it is a basic feasible solution to the LP.

See Luenburger (1984) for a proof of Theorem 1.

To illustrate the correspondence between extreme points and basic feasible solutions outlined in Theorem 1, let's look at the Leather Limited example of Section 4.1. Recall that the LP was

$$\max z = 4x_1 + 3x_2$$
$$\text{s.t.} \quad x_1 + x_2 \le 40 \qquad \text{(LP 1)}$$
$$2x_1 + x_2 \le 60 \qquad \text{(1)}$$
$$x_1, x_2 \ge 0 \qquad \text{(2)}$$

By adding slack variables s_1 and s_2, respectively, to (1) and (2), we obtain LP 1 in standard form:

$$\max z = 4x_1 + 3x_2$$
$$\text{s.t.} \quad x_1 + x_2 + s_1 \qquad = 40$$
$$2x_1 + x_2 \qquad + s_2 = 60 \qquad \text{(LP 1')}$$
$$x_1, x_2, s_1, s_2 \ge 0$$

The feasible region for the Leather Limited problem is graphed in Figure 1. Both inequalities are satisfied: (1) by all points below or on the line $AB(x_1 + x_2 = 40)$, and (2) by all points on or below the line $CD(2x_1 + x_2 = 60)$. Thus, the feasible region for LP 1 is the shaded region bounded by the quadrilateral *BECF*. The extreme points of the feasible region are $B = (0, 40)$, $C = (30, 0)$, $E = (20, 20)$, and $F = (0, 0)$.

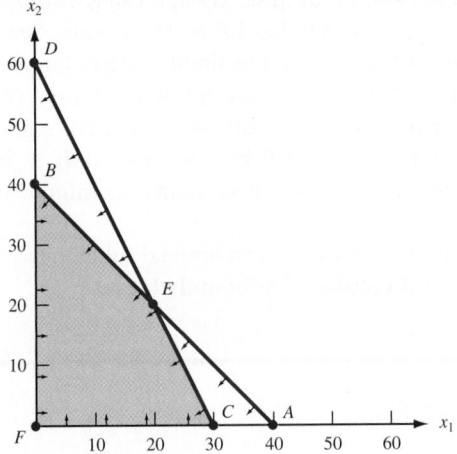

FIGURE 1
Feasible Region for
Leather Limited

Table 1 shows the correspondence between the basic feasible solutions to LP 1′ and the extreme points of the feasible region for LP 1. This example should make it clear that the basic feasible solutions to the standard form of an LP correspond in a natural fashion to the LP's extreme points.

In the context of the Leather Limited example, it is easy to show why any bfs is an extreme point. The converse is harder! We now show that for the LL problem, any bfs is an extreme point. Any point in the feasible region for LL may be specified as a four-dimensional column vector with the four elements of the vector denoting x_1, x_2, s_1, and s_2, respectively. Consider the bfs B with BV $= \{x_2, s_2\}$. If B is not an extreme point, then there exists two distinct feasible points v_1 and v_2 and non-negative numbers σ_1 and σ_2 satisfying $0 < \sigma_i < 1$ and $\sigma_1 + \sigma_2 = 1$ such that

$$\begin{bmatrix} 0 \\ 40 \\ 0 \\ 20 \end{bmatrix} = \sigma_1 v_1 + \sigma_2 v_2$$

Clearly, both v_1 and v_2 must both have $x_1 = s = 0$. But because v_1 and v_2 are both feasible, the values of x_2 and s_2 for both v_1 and v_2 can be determined by solving $x_2 = 40$ and $x_2 + s_2 = 60$. These equations have a unique solution (because columns corresponding to basic variables x_2 and s_2 are linearly independent). This shows that $v_1 = v_2$, so B is indeed an extreme point.

TABLE 1
Correspondence between Basic Feasible Solutions and Corner Points for Leather Limited

Basic Variables	Nonbasic Variables	Basic Feasible Solution	Corresponds to Corner Point
x_1, x_2	s_1, s_2	$s_1 = s_2 = 0, x_1 = x_2 = 20$	E
x_1, s_1	x_2, s_2	$x_2 = s_2 = 0, x_1 = 30, s_1 = 10$	C
x_1, s_2	x_2, s_1	$x_2 = s_1 = 0, x_1 = 40, s_2 = -20$	Not a bfs because $s_2 < 0$
x_2, s_1	x_1, s_2	$x_1 = s_2 = 0, s_1 = -20, x_2 = 60$	Not a bfs because $s_1 < 0$
x_2, s_2	x_1, s_1	$x_1 = s_1 = 0, x_2 = 40, s_2 = 20$	B
s_1, s_2	x_1, x_2	$x_1 = x_2 = 0, s_1 = 40, s_2 = 60$	F

We note that more than one set of basic variables may correspond to a given extreme point. If this is the case, then we say the LP is **degenerate.** See Section 4.11 for a discussion of the impact of degeneracy on the simplex algorithm.

We will soon see that if an LP has an optimal solution, then it has a bfs that is optimal. This is important because any LP has only a finite number of bfs's. Thus we can find the optimal solution to an LP by *searching only a finite number of points.* Because the feasible region for any LP contains an infinite number of points, this helps us a lot!

Before explaining why any LP that has an optimal solution has an optimal bfs, we need to define the concept of a **direction of unboundedness.**

4.3 Direction of Unboundedness

Consider an LP in standard form with feasible region S and constraints $A\mathbf{x} = \mathbf{b}$ and $\mathbf{x} \geq \mathbf{0}$. Assuming that our LP has n variables, $\mathbf{0}$ represents an n-dimensional column vector consisting of all 0's. A nonzero vector $\mathbf{d}$ is a **direction of unboundedness** if for all $\mathbf{x} \in S$ and any $c \geq 0$, $x + c\mathbf{d} \in S$. In short, if we are in the LP's feasible region, then we can move as far as we want in the direction $\mathbf{d}$ and remain in the feasible region. Figure 2 displays the feasible region for the Dorian Auto example (Example 2 of Chapter 3). In standard form, the Dorian example is

$$\min z = 50x_1 + 100x_2$$
$$7x_1 + 2x_2 - e_1 = 28$$
$$2x_1 + 12x_2 - e_2 = 24$$
$$x_1, x_2, e_1, e_2 \geq 0$$

Looking at Figure 2 it is clear that if we start at any feasible point and move up and to the right at a 45-degree angle, we will remain in the feasible region. This means that

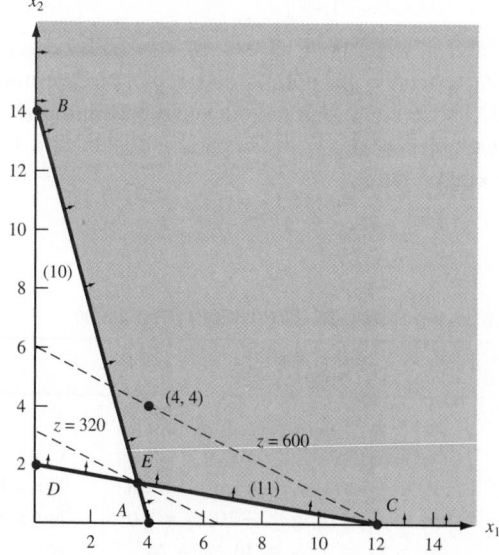

FIGURE 2
Graphical Solution of
Dorian Problem

$$d = \begin{bmatrix} 1 \\ 1 \\ 9 \\ 14 \end{bmatrix}$$

is a direction of unboundedness for this LP. It is easy to show (see Problem 6) that $\mathbf{d}$ is a direction of unboundedness if and only if $A\mathbf{d} = 0$ and $\mathbf{d} \geq \mathbf{0}$.

The following Representation Theorem [for a proof, see Nash and Sofer (1996)] is the key insight needed to show why any LP with an optimal solution has an optimal bfs.

THEOREM 2

Consider an LP in standard form, having bfs $\mathbf{b}_1, \mathbf{b}_2, \ldots, \mathbf{b}_k$. Any point $\mathbf{x}$ in the LP's feasible region may be written in the form

$$\mathbf{x} = \mathbf{d} + \sum_{i=1}^{i=k} \sigma_i \mathbf{b}_i$$

where $\mathbf{d}$ is $\mathbf{0}$ or a direction of unboundedness and $\sum_{i=1}^{i=k} \sigma_i = 1$ and $\sigma_i \geq 0$.

If the LP's feasible region is bounded, then $\mathbf{d} = \mathbf{0}$, and we may write $\mathbf{x} = \sum_{i=1}^{i=k} \sigma_i \mathbf{b}_i$, where the σ_i are nonnegative weights adding to 1. In this case, we see that any feasible $\mathbf{x}$ may be written as a **convex combination** of the LP's bfs. We now give two illustrations of Theorem 2.

Consider the Leather Limited example. The feasible region is bounded. To illustrate Theorem 2, we can write the point $G = (20, 10)$ (G is not a bfs!) in Figure 3 as a convex combination of the LP's bfs. Note from Figure 3 that point G may be written as $\frac{1}{6}F + \frac{5}{6}H$ [here $H = (24, 12)$]. Then note that point H may be written as $.6E + .4C$. Putting these two relationships together, we may write point G as $\frac{1}{6}F + \frac{5}{6}(.6E + .4C) = \frac{1}{6}F + \frac{1}{2}E + \frac{1}{3}C$. This expresses point G as a convex combination of the LP's extreme points.

To illustrate Theorem 2 for an unbounded LP, let's consider Example 2 of Chapter 3 (the Dorian example; see Figure 4) and try to express the point $F = (14, 4)$ in the representation given in Theorem 2. Recall that in standard form the constraints for the Dorian example are given by

$$7x_1 + 2x_2 - e_1 = 28$$
$$2x_1 + 12x_2 - e_2 = 24$$

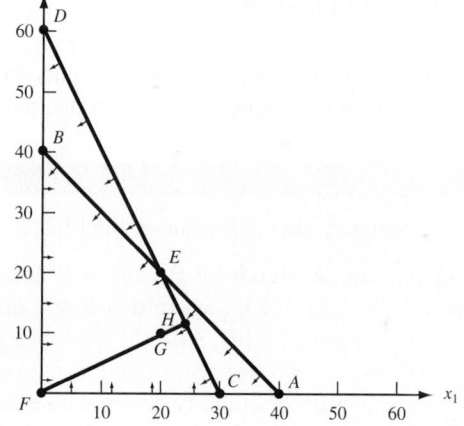

FIGURE 3
Writing (20, 10) as a
Convex Combination
of bfs

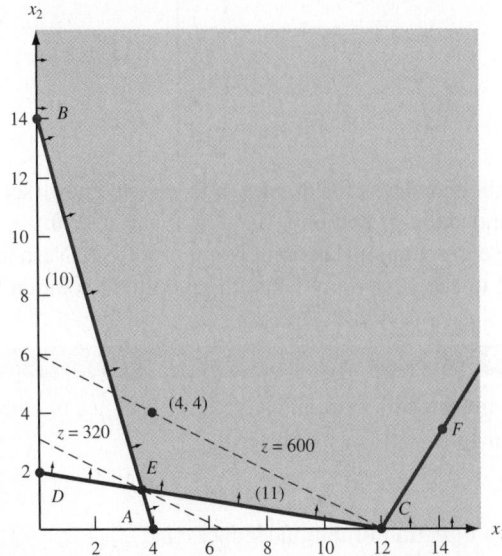

FIGURE 4
Expressing $F = (14, 4)$
Using Theorem 2

From Figure 4, we see that to move from bfs C to point F we need to move up and to the right along a line having slope $\frac{4-0}{14-12} = 2$. This line corresponds to the direction of unboundedness

$$\mathbf{d} = \begin{bmatrix} 2 \\ 4 \\ 22 \\ 52 \end{bmatrix}$$

Letting

$$\mathbf{b}_1 = \begin{bmatrix} 12 \\ 0 \\ 56 \\ 0 \end{bmatrix} \quad \text{and} \quad \mathbf{x} = \begin{bmatrix} 14 \\ 4 \\ 78 \\ 52 \end{bmatrix}$$

we may write $\mathbf{x} = \mathbf{d} + \mathbf{b}_1$, which is the desired representation.

4.4 Why Does an LP Have an Optimal bfs?

Consider an LP with objective function max $\mathbf{cx}$ and constraints $A\mathbf{x} = \mathbf{b}$. Suppose this LP has an optimal solution. We now sketch a proof of the fact that the LP has an optimal bfs.

THEOREM 3

If an LP has an optimal solution, then it has an optimal bfs.

Proof   Let $\mathbf{x}$ be an optimal solution to our LP. Because $\mathbf{x}$ is feasible, Theorem 2 tells us that we may write $\mathbf{x} = \mathbf{d} + \sum_{i=1}^{i=k} \sigma_i \mathbf{b}_i$, where $\mathbf{d}$ is $\mathbf{0}$ or a direction of unboundedness and $\mathbf{b}_1, \mathbf{b}_2, \ldots, \mathbf{b}_k$ are the LP's bfs. Also, $\sum_{i=1}^{i=k} \sigma_i = 1$ and $\sigma_i \geq 0$. If $\mathbf{cd} > \mathbf{0}$, then for any $k > 0$, $k\mathbf{d} + \sum_{i=1}^{i=k} \sigma_i \mathbf{b}_i$ is feasible, and as k grows larger and larger, the objective function value approaches infinity. This contradicts the fact that the LP has an optimal solution. If $\mathbf{cd} < \mathbf{0}$, then the feasible point $\sum_{i=1}^{i=k} \sigma_i \mathbf{b}_i$ has a larger objective function value than $\mathbf{x}$. This contradicts the optimality of $\mathbf{x}$. In short, we have shown that if $\mathbf{x}$ is optimal, then $\mathbf{cd} = 0$. Now the objective function value for $\mathbf{x}$ is given by

$$\mathbf{cx} = \mathbf{cd} + \Sigma_{i=1}^{i=k} \sigma_i \mathbf{cb}_i = \Sigma_{i=1}^{i=k} \sigma_i \mathbf{cb}_i$$

Suppose that $\mathbf{b}_1$ is the bfs with the largest objective function value. Because $\Sigma_{i=1}^{i=k} \sigma_i = 1$ and $\sigma_i \geq 0$,

$$\mathbf{cb}_1 \geq \mathbf{cx}$$

Because $\mathbf{x}$ is optimal, this shows that $\mathbf{b}_1$ is also optimal, and the LP does indeed have an optimal bfs.

Adjacent Basic Feasible Solutions

Before describing the simplex algorithm in general terms, we need to define the concept of an adjacent basic feasible solution.

DEFINITION ■ For any **LP** with m constraints, two basic feasible solutions are said to be **adjacent** if their sets of basic variables have $m - 1$ basic variables in common. ■

For example, in Figure 3, two basic feasible solutions will be adjacent if they have $2 - 1 = 1$ basic variable in common. Thus, the bfs corresponding to point E in Figure 3 is adjacent to the bfs corresponding to point C. Point E is not, however, adjacent to bfs F. Intuitively, two basic feasible solutions are adjacent if they both lie on the same edge of the boundary of the feasible region.

We now give a general description of how the simplex algorithm solves LPs in a max problem.

Step 1 Find a bfs to the LP. We call this bfs the initial basic feasible solution. In general, the most recent bfs will be called the current bfs, so at the beginning of the problem the initial bfs is the current bfs.

Step 2 Determine if the current bfs is an optimal solution to the LP. If it is not, then find an adjacent bfs that has a larger z-value.

Step 3 Return to step 2, using the new bfs as the current bfs.

If an LP in standard form has m constraints and n variables, then there may be a basic solution for each choice of nonbasic variables. From n variables, a set of $n - m$ nonbasic variables (or equivalently, m basic variables) can be chosen in

$$\binom{n}{m} = \frac{n!}{(n - m)!m!}$$

different ways. Thus, an LP can have at most

$$\binom{n}{m}$$

basic solutions. Because some basic solutions may not be feasible, an LP can have at most

$$\binom{n}{m}$$

basic feasible solutions. If we were to proceed from the current bfs to a better bfs (without ever repeating a bfs), then we would surely find the optimal bfs after examining at most

$$\binom{n}{m}$$

basic feasible solutions. This means (assuming that no bfs is repeated) that the simplex algorithm will find the optimal bfs after a finite number of calculations. We return to this discussion in Section 4.11.

In principle, we could enumerate all basic feasible solutions to an LP and find the bfs with the largest z-value. The problem with this approach is that even small LPs have a very large number of basic feasible solutions. For example, an LP in standard form that has 20 variables and 10 constraints might have (if each basic solution were feasible) up to

$$\binom{20}{10} = 184,756$$

basic feasible solutions. Fortunately, vast experience with the simplex algorithm indicates that when this algorithm is applied to an n-variable, m-constraint LP in standard form, an optimal solution is usually found after examining fewer than $3m$ basic feasible solutions. Thus, for a 20-variable, 10-constraint LP in standard form, the simplex will usually find the optimal solution after examining fewer than $3(10) = 30$ basic feasible solutions. Compared with the alternative of examining 184,756 basic solutions, the simplex is quite efficient![†]

Geometry of Three-Dimensional LPs

Consider the following LP:

$$
\begin{aligned}
\max z = \ & x_1 + 2x_2 + 2x_3 \\
\text{s.t.} \quad & 2x_1 + x_2 \quad\quad\ \leq 8 \\
& \quad\quad\quad\quad\ x_3 \leq 10 \\
& x_1, x_2, x_3 \geq 0
\end{aligned}
$$

The set of points satisfying a linear inequality in three (or any number of) dimensions is a **half-space.** For example, the set of points in three dimensions satisfying $2x_1 + x_2 \leq 8$ is a half-space. Thus, the feasible region for our LP is the intersection of the following five half-spaces: $2x_1 + x_2 \leq 8$, $x_3 \leq 10$, $x_1 \geq 0$, $x_2 \geq 0$, and $x_3 \geq 0$. The intersection of half-spaces is called a **polyhedron.** The feasible region for our LP is the prism pictured in Figure 5.

On each face (or facet) of the feasible region, one constraint (or sign restriction) is binding for all points on that face. For example, the constraint $2x_1 + x_2 \leq 8$ is binding for all points on the face $ABCD$; $x_3 \geq 0$ is binding on face ABF; $x_3 \leq 10$ is binding on face DEC; $x_2 \geq 0$ is binding on face $ADEF$; $x_1 \geq 0$ is binding on face $CBFE$.

Clearly, the corner (or extreme) points of the LP's feasible region are A, B, C, D, E, and F. In this case, the correspondence between the bfs and corner points is as shown in Table 2.

To illustrate the concept of adjacent basic feasible solutions, note that corner points A, E, and B are adjacent to corner point F. Thus, if the simplex algorithm begins at F, then we can be sure that the next bfs to be considered will be A, E, or B.

[†]In solving many LPs with 50 variables and $m \leq 50$ constraints, Chvàtal (1983) found that the simplex algorithm examined an average of $2m$ basic feasible solutions before finding an LP's optimal solution.

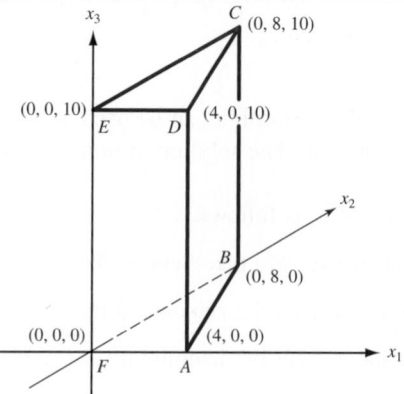

FIGURE 5
Feasible Region in
Three Dimensions

Points in figure: C $(0, 8, 10)$; $(0, 0, 10)$; $(4, 0, 10)$; E; D; x_2; B $(0, 8, 0)$; $(0, 0, 0)$; $(4, 0, 0)$; x_1; F; A; x_3

TABLE 2

Correspondence between bfs and Corner Points

Basic Variables	Basic Feasible Solution	Corresponds to Corner Point
x_1, x_3	$x_1 = 4, x_3 = 10, x_2 = s_1 = s_2 = 0$	D
s_1, s_2	$s_1 = 8, s_2 = 10, x_1 = x_2 = x_3 = 0$	F
s_1, x_3	$s_1 = 8, x_3 = 10, x_1 = x_2 = s_2 = 0$	E
x_2, x_3	$x_2 = 8, x_3 = 10, x_1 = s_1 = s_2 = 0$	C
x_2, s_2	$x_2 = 8, s_2 = 10, x_1 = x_3 = s_1 = 0$	B
x_1, s_2	$x_1 = 4, s_2 = 10, x_2 = x_3 = s_1 = 0$	A

PROBLEMS

Group A

1 For the Giapetto problem (Example 1 in Chapter 3), show how the basic feasible solutions to the LP in standard form correspond to the extreme points of the feasible region.

2 For the Dorian problem (Example 2 in Chapter 3), show how the basic feasible solutions to the LP in standard form correspond to the extreme points of the feasible region.

3 Widgetco produces two products: 1 and 2. Each requires the amounts of raw material and labor, and sells for the price given in Table 3.

Up to 350 units of raw material can be purchased at $2 per unit, while up to 400 hours of labor can be purchased at $1.50 per hour. To maximize profit, Widgetco must solve the following LP:

$$\max z = 2x_1 + 2.5x_2$$
$$\text{s.t.} \quad x_1 + 2x_2 \le 350 \quad \text{(Raw material)}$$
$$2x_1 + x_2 \le 400 \quad \text{(Labor)}$$
$$x_1, x_2 \ge 0$$

Here, x_i = number of units of product i produced. Demonstrate the correspondence between corner points and basic feasible solutions.

TABLE 3

	Product 1	Product 2
Raw material	1 unit	2 units
Labor	2 hours	1 hour
Sales price	$7	$8

4 For the Leather Limited problem, represent the point $(10, 20)$ in the form $\mathbf{cd} + \Sigma_{i=1}^{i=k} \sigma_i \mathbf{b}_i$.

5 For the Dorian problem, represent the point $(10, 40)$ in the form $\mathbf{cd} + \Sigma_{i=1}^{i=k} \sigma_i \mathbf{b}_i$.

Group B

6 For an LP in standard form with constraints $A\mathbf{x} = \mathbf{b}$ and $\mathbf{x} \ge \mathbf{0}$, show that $\mathbf{d}$ is a direction of unboundedness if and only if $A\mathbf{d} = \mathbf{0}$ and $\mathbf{d} \ge \mathbf{0}$.

7 Recall that Example 5 of Chapter 3 is an unbounded LP. Find a direction of unboundedness along which we can move for which the objective function becomes arbitrarily large.

4.5 The Simplex Algorithm

We now describe how the simplex algorithm can be used to solve LPs in which the goal is to maximize the objective function. The solution of minimization problems is discussed in Section 4.4.

The simplex algorithm proceeds as follows:

Step 1 Convert the LP to standard form (see Section 4.1).

Step 2 Obtain a bfs (if possible) from the standard form.

Step 3 Determine whether the current bfs is optimal.

Step 4 If the current bfs is not optimal, then determine which nonbasic variable should become a basic variable and which basic variable should become a nonbasic variable to find a new bfs with a better objective function value.

Step 5 Use EROs to find the new bfs with the better objective function value. Go back to step 3.

In performing the simplex algorithm, write the objective function

$$z = c_1x_1 + c_2x_2 + \cdots + c_nx_n$$

in the form

$$z - c_1x_1 - c_2x_2 - \cdots - c_nx_n = 0$$

We call this format the **row 0 version** of the objective function (row 0 for short).

EXAMPLE 2   Dakota Furniture Company

The Dakota Furniture Company manufactures desks, tables, and chairs. The manufacture of each type of furniture requires lumber and two types of skilled labor: finishing and carpentry. The amount of each resource needed to make each type of furniture is given in Table 4.

Currently, 48 board feet of lumber, 20 finishing hours, and 8 carpentry hours are available. A desk sells for $60, a table for $30, and a chair for $20. Dakota believes that demand for desks and chairs is unlimited, but at most five tables can be sold. Because the available resources have already been purchased, Dakota wants to maximize total revenue. Defining the decision variables as

$$x_1 = \text{number of desks produced}$$
$$x_2 = \text{number of tables produced}$$
$$x_3 = \text{number of chairs produced}$$

TABLE 4
Resource Requirements for Dakota Furniture

Resource	Desk	Table	Chair
Lumber (board ft)	8	6	1
Finishing hours	4	2	1.5
Carpentry hours	2	1.5	0.5

it is easy to see that Dakota should solve the following LP:

$$\max z = 60x_1 + 30x_2 + 20x_3$$

$$\begin{array}{llll}
\text{s.t.} & 8x_1 + 6x_2 + x_3 \le 48 & \text{(Lumber constraint)} \\
& 4x_1 + 2x_2 + 1.5x_3 \le 20 & \text{(Finishing constraint)} \\
& 2x_1 + 1.5x_2 + 0.5x_3 \le 8 & \text{(Carpentry constraint)} \\
& x_2 \le 5 & \text{(Limitation on table demand)} \\
& x_1, x_2, x_3 \ge 0
\end{array}$$

Convert the LP to Standard Form

We begin the simplex algorithm by converting the constraints of the LP to the standard form discussed in Section 4.1. Then we convert the LP's objective function to the row 0 format. To put the constraints in standard form, we simply add slack variables s_1, s_2, s_3, and s_4, respectively, to the four constraints. We label the constraints row 1, row 2, row 3, and row 4, and add the sign restrictions $s_i \ge 0$ ($i = 1, 2, 3, 4$). Note that the row 0 format for our objective function is

$$z - 60x_1 - 30x_2 - 20x_3 = 0$$

Putting rows 1–4 together with row 0 and the sign restrictions yields the equations and basic variables shown in Table 5. A system of linear equations (such as canonical form 0, shown in Table 5) in which each equation has a variable with a coefficient of 1 in that equation (and a zero coefficient in all other equations) is said to be in *canonical form*. We will soon see that if the right-hand side of each constraint in a canonical form is nonnegative, a basic feasible solution can be obtained by inspection.[†]

From Section 4.2, we know that the simplex algorithm begins with an initial basic feasible solution and attempts to find better ones. After obtaining a canonical form, we therefore search for the initial bfs. By inspection, we see that if we set $x_1 = x_2 = x_3 = 0$, we can solve for the values of s_1, s_2, s_3, and s_4 by setting s_i equal to the right-hand side of row i.

$$\text{BV} = \{s_1, s_2, s_3, s_4\} \quad \text{and} \quad \text{NBV} = \{x_1, x_2, x_3\}$$

TABLE 5
Canonical Form 0

Row			Basic Variable
0	$z - 60x_1 - 30x_2 - 20x_3$	$= 0$	$z = 0$
1	$8x_1 + 6x_2 + x_3 + s_1$	$= 48$	$s_1 = 48$
2	$4x_1 + 2x_2 + 1.5x_3 + s_2$	$= 20$	$s_2 = 20$
3	$2x_1 + 1.5x_2 + 0.5x_3 + s_3$	$= 8$	$s_3 = 8$
4	$x_2 + s_4$	$= 5$	$s_4 = 5$

[†]If a canonical form with nonnegative right-hand sides is not readily available, however, then the techniques described in Sections 4.12 and 4.13 can be used to find a canonical form and a basic feasible solution.

The basic feasible solution for this set of basic variables is $s_1 = 48$, $s_2 = 20$, $s_3 = 8$, $s_4 = 5$, $x_1 = x_2 = x_3 = 0$. Observe that each basic variable may be associated with the row of the canonical form in which the basic variable has a coefficient of 1. Thus, for canonical form 0, s_1 may be thought of as the basic variable for row 1, as may s_2 for row 2, s_3 for row 3, and s_4 for row 4.

To perform the simplex algorithm, we also need a basic (although not necessarily non-negative) variable for row 0. Because z appears in row 0 with a coefficient of 1, and z does not appear in any other row, we use z as its basic variable. With this convention, the basic feasible solution for our initial canonical form has

$$BV = \{z, s_1, s_2, s_3, s_4\} \quad \text{and} \quad NBV = \{x_1, x_2, x_3\}$$

For this basic feasible solution, $z = 0$, $s_1 = 48$, $s_2 = 20$, $s_3 = 8$, $s_4 = 5$, $x_1 = x_2 = x_3 = 0$.

As this example indicates, a slack variable can be used as a basic variable for an equation if the right-hand side of the constraint is nonnegative.

Is the Current Basic Feasible Solution Optimal?

Once we have obtained a basic feasible solution, we need to determine whether it is optimal; if the bfs is not optimal, then we try to find a bfs adjacent to the initial bfs with a larger z-value. To do this, we try to determine whether there is any way that z can be increased by increasing some nonbasic variable from its current value of zero while holding all other nonbasic variables at their current values of zero. If we solve for z by rearranging row 0, then we obtain

$$z = 60x_1 + 30x_2 + 20x_3 \tag{9}$$

For each nonbasic variable, we can use (9) to determine whether increasing a nonbasic variable (and holding all other nonbasic variables at zero) will increase z. For example, suppose we increase x_1 by 1 (holding the other nonbasic variables x_2 and x_3 at zero). Then (9) tells us that z will increase by 60. Similarly, if we choose to increase x_2 by 1 (holding x_1 and x_3 at zero), then (9) tells us that z will increase by 30. Finally, if we choose to increase x_3 by 1 (holding x_1 and x_2 at zero), then (9) tells us that z will increase by 20. Thus, increasing any of the nonbasic variables will increase z. Because a unit increase in x_1 causes the largest rate of increase in z, we choose to increase x_1 from its current value of zero. If x_1 is to increase from its current value of zero, then it will have to become a basic variable. For this reason, we call x_1 the **entering variable.** Observe that x_1 has the most negative coefficient in row 0.

Determine the Entering Variable

We choose the entering variable (in a max problem) to be the nonbasic variable with the most negative coefficient in row 0 (ties may be broken in an arbitrary fashion). Because each one-unit increase of x_1 increases z by 60, we would like to make x_1 as large as possible. What limits how large we can make x_1? Note that as x_1 increases, the values of the current basic variables (s_1, s_2, s_3, and s_4) will change. This means that increasing x_1 may cause a basic variable to become negative. With this in mind, we look at how increasing x_1 (while holding $x_2 = x_3 = 0$) changes the values of the current set of basic variables. From row 1, we see that $s_1 = 48 - 8x_1$ (remember that $x_2 = x_3 = 0$). Because the sign restriction $s_1 \geq 0$ must be satisfied, we can only increase x_1 as long as $s_1 \geq 0$, or $48 - 8x_1 \geq 0$, or $x_1 \leq \frac{48}{8} = 6$. From row 2, $s_2 = 20 - 4x_1$. We can only increase x_1 as long as

$s_2 \geq 0$, so x_1 must satisfy $20 - 4x_1 \geq 0$ or $x_1 \leq \frac{20}{4} = 5$. From row 3, $s_3 = 8 - 2x_1$ so $x_1 \leq \frac{8}{2} = 4$. Similarly, we see from row 4 that $s_4 = 5$. Thus, whatever the value of x_1, s_4 will be nonnegative. Summarizing,

$$s_1 \geq 0 \quad \text{for} \quad x_1 \leq \frac{48}{8} = 6$$

$$s_2 \geq 0 \quad \text{for} \quad x_1 \leq \frac{20}{4} = 5$$

$$s_3 \geq 0 \quad \text{for} \quad x_1 \leq \frac{8}{2} = 4$$

$$s_4 \geq 0 \quad \text{for all values of } x_1$$

This means that to keep all the basic variables nonnegative, the largest that we can make x_1 is min $\{\frac{48}{8}, \frac{20}{4}, \frac{8}{2}\} = 4$. If we make $x_1 > 4$, then s_3 will become negative, and we will no longer have a basic feasible solution. Notice that each row in which the entering variable had a positive coefficient restricted how large the entering variable could become. Also, for any row in which the entering variable had a positive coefficient, the row's basic variable became negative when the entering variable exceeded

$$\frac{\text{Right-hand side of row}}{\text{Coefficient of entering variable in row}} \tag{10}$$

If the entering variable has a nonpositive coefficient in a row (such as x_1 in row 4), the row's basic variable will remain positive for all values of the entering variable. Using (10), we can quickly compute how large x_1 can become before a basic variable becomes negative.

$$\text{Row 1 limit on } x_1 = \frac{48}{8} = 6$$

$$\text{Row 2 limit on } x_1 = \frac{20}{4} = 5$$

$$\text{Row 3 limit on } x_1 = \frac{8}{2} = 4$$

Row 4 limit on x_1 = no limit   (Because coefficient of x_1 in row 4 is nonpositive)

We can state the following rule for determining how large we can make an entering variable.

The Ratio Test

When entering a variable into the basis, compute the ratio in (10) for every constraint in which the entering variable has a positive coefficient. The constraint with the smallest ratio is called the **winner of the ratio test.** The smallest ratio is the largest value of the entering variable that will keep all the current basic variables nonnegative. In our example, row 3 was the winner of the ratio test for entering x_1 into the basis.

Find a New Basic Feasible Solution: Pivot in the Entering Variable

Returning to our example, we know that the largest we can make x_1 is 4. For x_1 to equal 4, it must become a basic variable. Looking at rows 1–4, we see that if we make x_1 a basic variable in row 1, then x_1 will equal $\frac{48}{8} = 6$; in row 2, x_1 will equal $\frac{20}{4} = 5$; in row 3, x_1 will equal $\frac{8}{2} = 4$. Also, because x_1 does not appear in row 4, x_1 cannot be made a basic variable in row 4. Thus, if we want to make $x_1 = 4$, then we have to make it a basic variable in row 3. The fact that row 3 was the winner of the ratio test illustrates the following rule.

In Which Row Does the Entering Variable Become Basic?

Always make the entering variable a basic variable in a row that wins the ratio test (ties may be broken arbitrarily).

To make x_1 a basic variable in row 3, we use elementary row operations to make x_1 have a coefficient of 1 in row 3 and a coefficient of 0 in all other rows. This procedure is called **pivoting** on row 3; and row 3 is the **pivot row.** The final result is that x_1 replaces s_3 as the basic variable for row 3. The term in the pivot row that involves the entering basic variable is called the **pivot term.** Proceeding as we did when we studied the Gauss–Jordan method in Chapter 2, we make x_1 a basic variable in row 3 by performing the following EROs.

ERO 1 Create a coefficient of 1 for x_1 in row 3 by multiplying row 3 by $\frac{1}{2}$. The resulting row (marked with a prime to show it is the first iteration) is

$$x_1 + 0.75x_2 + 0.25x_3 + 0.5s_3 = 4 \qquad \textbf{(row 3')}$$

ERO 2 To create a zero coefficient for x_1 in row 0, replace row 0 with 60(row 3') + row 0.

$$z + 15x_2 - 5x_3 + 30s_3 = 240 \qquad \textbf{(row 0')}$$

ERO 3 To create a zero coefficient for x_1 in row 1, replace row 1 with -8(row 3') + row 1.

$$-x_3 + s_1 - 4s_3 = 16 \qquad \textbf{(row 1')}$$

ERO 4 To create a zero coefficient for x_1 in row 2, replace row 2 with -4(row 3') + row 2.

$$-x_2 + 0.5x_3 + s_2 - 2s_3 = 4 \qquad \textbf{(row 2')}$$

Because x_1 does not appear in row 4, we don't need to perform an ero to eliminate x_1 from row 4. Thus, we may write the "new" row 4 (call it row 4' to be consistent with other notation) as

$$x_2 + s_4 = 5 \qquad \textbf{(row 4')}$$

Putting rows 0'–4' together, we obtain the canonical form shown in Table 6.

Looking for a basic variable in each row of the current canonical form, we find that

$$\text{BV} = \{z, s_1, s_2, x_1, s_4\} \qquad \text{and} \qquad \text{NBV} = \{s_3, x_2, x_3\}$$

Thus, canonical form 1 yields the basic feasible solution $z = 240$, $s_1 = 16$, $s_2 = 4$, $x_1 = 4$, $s_4 = 5$, $x_2 = x_3 = s_3 = 0$. We could have predicted that the value of z in canonical form 1 would be 240 from the fact that each unit by which x_1 is increased increases z by 60. Because x_1 was increased by 4 units (from $x_1 = 0$ to $x_1 = 4$), we would expect that

$$\text{Canonical form 1 } z\text{-value} = \text{initial } z\text{-value} + 4(60)$$
$$= 0 + 240 = 240$$

TABLE 6
Canonical Form 1

Row							Basic Variable
Row 0'	z	$+\ 15x_2 -$	$5x_3$		$+\ 30s_3$	$=240$	$z = 240$
Row 1'		$-$	$x_3 + s_1$		$-\ 4s_3$	$=16$	$s_1 = 16$
Row 2'		$-\ x_2 +$	$0.5x_3$	$+\ s_2 -$	$2s_3$	$=4$	$s_2 = 4$
Row 3'	$x_1 +$	$0.75x_2 +$	$0.25x_3$		$+\ 0.5s_3$	$=4$	$x_1 = 4$
Row 4'		x_2			$+\ s_4$	$=5$	$s_4 = 5$

In obtaining canonical form 1 from the initial canonical form, we have gone from one bfs to a better (larger z-value) bfs. Note that the initial bfs and the improved bfs are adjacent. This follows because the two basic feasible solutions have $4 - 1 = 3$ basic variables (s_1, s_2, and s_4) in common (excluding z, which is a basic variable in every canonical form). Thus, we see that in going from one canonical form to the next, we have proceeded from one bfs to a better adjacent bfs. The procedure used to go from one bfs to a better adjacent bfs is called an **iteration** (or sometimes, a *pivot*) of the simplex algorithm.

We now try to find a bfs that has a still larger z-value. We begin by examining canonical form 1 (Table 6) to see if we can increase z by increasing the value of some nonbasic variable (while holding all other nonbasic variables equal to zero). Rearranging row $0'$ to solve for z yields

$$z = 240 - 15x_2 + 5x_3 - 30s_3 \tag{11}$$

From (11), we see that increasing the nonbasic variable x_2 by 1 (while holding $x_3 = s_3 = 0$) will decrease z by 15. We don't want to do that! Increasing the nonbasic variable s_3 by 1 (holding $x_2 = x_3 = 0$) will decrease z by 30. Again, we don't want to do that. On the other hand, increasing x_3 by 1 (holding $x_2 = s_3 = 0$) will increase z by 5. Thus, we choose to enter x_3 into the basis. Recall that our rule for determining the entering variable is to choose the variable with the most negative coefficient in the current row 0. Because x_3 is the only variable with a negative coefficient in row $0'$, it should be entered into the basis.

Increasing x_3 by 1 will increase z by 5, so it is to our advantage to make x_3 as large as possible. We can increase x_3 as long as the current basic variables (s_1, s_2, x_1, and s_4) remain nonnegative. To determine how large x_3 can be, we must solve for the values of the current basic variables in terms of x_3 (holding $x_2 = s_3 = 0$). We obtain

From row $1'$: $s_1 = 16 + x_3$

From row $2'$: $s_2 = 4 - 0.5x_3$

From row $3'$: $x_1 = 4 - 0.25x_3$

From row $4'$: $s_4 = 5$

These equations tell us that $s_1 \geq 0$ and $s_4 \geq 0$ will hold for all values of x_3. From row $2'$, we see that $s_2 \geq 0$ will hold if $4 - 0.5x_3 \geq 0$, or $x_3 \leq \frac{4}{0.5} = 8$. From row $3'$, $x_1 \geq 0$ will hold if $4 - 0.25x_3 \geq 0$, or $x_3 \leq \frac{4}{0.25} = 16$. This shows that the largest we can make x_3 is min $\{\frac{4}{0.5}, \frac{4}{0.25}\} = 8$. This fact could also have been discovered by using (10) and the ratio test, as follows:

Row $1'$: no ratio $\qquad$ (x_3 has negative coefficient in row 1)

Row $2'$: $\dfrac{4}{0.5} = 8$

Row $3'$: $\dfrac{4}{0.25} = 16$

Row $4'$: no ratio $\qquad$ (x_3 has a nonpositive coefficient in row 4)

Thus, the smallest ratio occurs in row $2'$, and row $2'$ wins the ratio test. This means that we should use EROs to make x_3 a basic variable in row $2'$.

ERO 1 Create a coefficient of 1 for x_3 in row $2'$ by replacing row $2'$ with 2(row $2'$):

$$-2x_2 + x_3 + 2s_2 - 4s_3 = 8 \tag{row 2''}$$

ERO 2 Create a coefficient of 0 for x_3 in row $0'$ by replacing row $0'$ with 5(row 2)$''$ + row $0'$:

$$z + 5x_2 + 10s_2 + 10s_3 = 280 \tag{row 0''}$$

TABLE 7
Canonical Form 2

Row							Basic Variable
0″	z	$+ \quad 5x_2$		$+ \quad 10s_2 + 10s_3$		$= 280$	$z = 280$
1″		$- \quad 2x_2$	$+ s_1 +$	$2s_2 - \quad 8s_3$		$= 24$	$s_1 = 24$
2″		$- \quad 2x_2 + x_3$		$+ \quad 2s_2 - \quad 4s_3$		$= 8$	$x_3 = 8$
3″	$x_1 + 1.25x_2$			$- \quad 0.5s_2 + 1.5s_3$		$= 2$	$x_1 = 2$
4″		x_2			$+ s_4$	$= 5$	$s_4 = 5$

ERO 3 Create a coefficient of 0 for x_3 in row 1′ by replacing row 1′ with row 2″ + row 1′:

$$-2x_2 + s_1 + 2s_2 - 8s_3 = 24 \qquad \text{(row 1″)}$$

ERO 4 Create a coefficient of 0 for x_3 in row 3′, by replacing row 3′ with $-\frac{1}{4}(\text{row 2″}) + 3'$:

$$x_1 + 1.25x_2 - 0.5s_2 + 1.5s_3 = 2 \qquad \text{(row 3″)}$$

Because x_3 already has a zero coefficient in row 4′, we may write

$$x_2 + s_4 = 5 \qquad \text{(row 4″)}$$

Combining rows 0″–4″ gives the canonical form shown in Table 7.

Looking for a basic variable in each row of canonical form 2, we find

$$\text{BV} = \{z, s_1, x_3, x_1, s_4\} \qquad \text{and} \qquad \text{NBV} = \{s_2, s_3, x_2\}$$

Canonical form 2 yields the following bfs: $z = 280$, $s_1 = 24$, $x_3 = 8$, $x_1 = 2$, $s_4 = 5$, $s_2 = s_3 = x_2 = 0$. We could have predicted that canonical form 2 would have $z = 280$ from the fact that each unit of the entering variable x_3 increased z by 5, and we have increased x_3 by 8 units. Thus,

$$\text{Canonical form 2 } z\text{-value} = \text{canonical form 1 } z\text{-value} + 8(5)$$
$$= 240 + 40 = 280$$

Because the bfs's for canonical forms 1 and 2 have (excluding z) $4 - 1 = 3$ basic variables in common (s_1, s_4, x_1), they are adjacent basic feasible solutions.

Now that the second iteration (or pivot) of the simplex algorithm has been completed, we examine canonical form 2 to see if we can find a better bfs. If we rearrange row 0″ and solve for z, we obtain

$$z = 280 - 5x_2 - 10s_2 - 10s_3 \qquad \text{(12)}$$

From (12), we see that increasing x_2 by 1 (while holding $s_2 = s_3 = 0$) will decrease z by 5; increasing s_2 by 1 (holding $s_3 = x_2 = 0$) will decrease z by 10; increasing s_3 by 1 (holding $x_2 = s_2 = 0$) will decrease z by 10. Thus, increasing any nonbasic variable will cause z to decrease. This might lead us to believe that our current bfs from canonical form 2 is an optimal solution. This is indeed correct! To see why, look at (12). We know that any feasible solution to the Dakota Furniture problem must have $x_2 \geq 0$, $s_2 \geq 0$, and $s_3 \geq 0$, and $-5x_2 \leq 0$, $-10s_2 \leq 0$, and $-10s_3 \leq 0$. Combining these inequalities with (12), it is clear that any feasible solution must have $z = 280 +$ terms that are ≤ 0, and $z \leq 280$. Our current bfs from canonical form 2 has $z = 280$, so it must be optimal.

The argument that we just used to show that canonical form 2 is optimal revolved around the fact that each of its nonbasic variables had a nonnegative coefficient in row 0″.

This means that we can determine whether a canonical form's bfs is optimal by applying the following simple rule.

Is a Canonical Form Optimal (Max Problem)?

A canonical form is optimal (for a max problem) if each nonbasic variable has a non-negative coefficient in the canonical form's row 0.

REMARKS **1** The coefficient of a decision variable in row 0 is often referred to as the variable's **reduced cost.** Thus, in our optimal canonical form, the reduced costs for x_1 and x_3 are 0, and the reduced cost for x_2 is 5. The reduced cost of a nonbasic variable is the amount by which the value of z will decrease if we increase the value of the nonbasic variable by 1 (while all the other nonbasic variables remain equal to 0). For example, the reduced cost for the variable "tables" (x_2) in canonical form 2 is 5. From (12), we see that increasing x_2 by 1 will reduce z by 5. Note that because all basic variables (except z, of course) must have zero coefficients in row 0, the reduced cost for a basic variable will always be 0. In Chapters 5 and 6, we discuss the concept of reduced costs in much greater detail.

These comments are correct only if the values of all the basic variables remain nonnegative after the nonbasic variable is increased by 1. Increasing x_2 to 1 leaves x_1, x_3, and s_1 all nonnegative, so our comments are valid.

2 From canonical form 2, we see that the optimal solution to the Dakota Furniture problem is to manufacture 2 desks ($x_1 = 2$) and 8 chairs ($x_3 = 8$). Because $x_2 = 0$, no tables should be made. Also, $s_1 = 24$ is reasonable because only $8 + 8(2) = 24$ board feet of lumber are being used. Thus, $48 - 24 = 24$ board feet of lumber are not being used. Similarly, $s_4 = 5$ makes sense because, although up to 5 tables could have been produced, 0 tables are actually being produced. Thus, the slack in constraint 4 is $5 - 0 = 5$. Because $s_2 = s_3 = 0$, all available finishing and carpentry hours are being utilized, so the finishing and carpentry constraints are binding.

3 We have chosen the entering variable to be the one with the most negative coefficient in row 0, but this may not always lead us quickly to the optimal bfs (see Review Problem 11). Actually, even if we choose the variable with the smallest (in absolute value) negative coefficient, the simplex algorithm will eventually find the LP's optimal solution.

4 Although any variable with a negative row 0 coefficient may be chosen to enter the basis, the pivot row *must* be chosen by the ratio test. To show this formally, suppose that we have chosen to enter x_i into the basis, and in the current tableau x_i is a basic variable in row k. Then row k may be written as

$$\bar{a}_{ki}x_i + \cdots = \bar{b}_k$$

Consider any other constraint (say, row j) in the canonical form. Row j in the current canonical form may be written as

$$\bar{a}_{ji}x_i + \cdots = \bar{b}_j$$

If we pivot on row k, row k becomes

$$x_i + \cdots \quad = \frac{\bar{b}_k}{\bar{a}_{ki}}$$

The new row j after the pivot will be obtained by adding $-\bar{a}_{ji}$ times the last equation to row j of the current canonical form. This yields a new row j of

$$0x_i + \cdots \quad = \bar{b}_j - \frac{\bar{b}_k\bar{a}_{ji}}{\bar{a}_{ki}}$$

We know that after the pivot, each constraint must have a nonnegative right-hand side. Thus, $\bar{a}_{ki} > 0$ must hold to ensure that row k has a nonnegative right-hand side after the pivot. Suppose $\bar{a}_{ji} > 0$. Then, to ensure that row j will have a nonnegative right-hand side after the pivot, we must have

$$\frac{\bar{b}_j - \bar{b}_k\bar{a}_{ji}}{\bar{a}_{ki}} \geq 0$$

or (because $\bar{a}_{ji} > 0$)

$$\frac{\bar{b}_j}{\bar{a}_{ji}} \geq \frac{\bar{b}_k}{\bar{a}_{ki}}$$

Thus, row k must be a "winner" of the ratio test to ensure that row j will have a nonnegative right-hand side after the pivot is completed.

If $\bar{a}_{ji} \leq 0$, then the right-hand side of row j will surely be nonnegative after the pivot. This follows because

$$-\frac{\bar{b}_k \bar{a}_{ji}}{\bar{a}_{ki}} \geq 0$$

will now hold.

As promised earlier, we have outlined an algorithm that proceeds from one bfs to a better bfs. The algorithm stops when an optimal solution has been found. The convergence of the simplex algorithm is discussed further in Section 4.11.

Summary of the Simplex Algorithm for a Max Problem

Step 1   Convert the LP to standard form.

Step 2   Find a basic feasible solution. This is easy if all the constraints are $\leq$ with non-negative right-hand sides. Then the slack variable s_i may be used as the basic variable for row i. If no bfs is readily apparent, then use the techniques discussed in Sections 4.12 and 4.13 to find a bfs.

Step 3   If all nonbasic variables have nonnegative coefficients in row 0, then the current bfs is optimal. If any variables in row 0 have negative coefficients, then choose the variable with the most negative coefficient in row 0 to enter the basis. We call this variable the *entering variable.*

Step 4   Use EROs to make the entering variable the basic variable in any row that wins the ratio test (ties may be broken arbitrarily). After the EROs have been used to create a new canonical form, return to step 3, using the current canonical form.

When using the simplex algorithm to solve problems, there should never be a constraint with a negative right-hand side (it is okay for row 0 to have a negative right-hand side; see Section 4.6). A constraint with a negative right-hand side is usually the result of an error in the ratio test or in performing one or more EROs. If one (or more) of the constraints has a negative right-hand side, then there is no longer a bfs, and the rules of the simplex algorithm may not lead to a better bfs.

Representing Simplex Tableaus

Rather than writing each variable in every constraint, we often used a shorthand display called a **simplex tableau.** For example, the canonical form

$$
\begin{aligned}
z + 3x_1 + x_2 \quad\quad\quad &= 6 \\
x_1 \quad\quad + s_1 \quad &= 4 \\
2x_1 + x_2 \quad\quad + s_2 &= 3
\end{aligned}
$$

would be written in abbreviated form as shown in Table 8 (rhs = right-hand side). This format makes it very easy to spot basic variables: Just look for columns having a single entry of 1 and all other entries equal to 0 (s_1 and s_2). In our use of simplex tableaus, we will encircle the pivot term and denote the winner of the ratio test by *.

TABLE 8
A Simplex Tableau

z	x_1	x_2	s_1	s_2	rhs	Basic Variable
1	3	1	0	0	6	$z = 6$
0	1	0	1	0	4	$s_1 = 4$
0	2	1	0	1	3	$s_2 = 3$

PROBLEMS

Group A

1 Use the simplex algorithm to solve the Giapetto problem (Example 1 in Chapter 3).

2 Use the simplex algorithm to solve the following LP:

$$\max z = 2x_1 + 3x_2$$
$$\text{s.t.} \quad x_1 + 2x_2 \le 6$$
$$2x_1 + x_2 \le 8$$
$$x_1, x_2 \ge 0$$

3 Use the simplex algorithm to solve the following problem:

$$\max z = 2x_1 - x_2 + x_3$$
$$\text{s.t.} \quad 3x_1 + x_2 + x_3 \le 60$$
$$x_1 - x_2 + 2x_3 \le 10$$
$$x_1 + x_2 - x_3 \le 20$$
$$x_1, x_2, x_3 \ge 0$$

4 Suppose you want to solve the Dorian problem (Example 2 in Chapter 3) by the simplex algorithm. What difficulty would occur?

5 Use the simplex algorithm to solve the following LP:

$$\max z = x_1 + x_2$$
$$\text{s.t.} \quad 4x_1 + x_2 \le 100$$
$$x_1 + x_2 \le 80$$
$$x_1 \le 40$$
$$x_1, x_2 \ge 0$$

6 Use the simplex algorithm to solve the following LP:

$$\max z = x_1 + x_2 + x_3$$
$$\text{s.t.} \quad x_1 + 2x_2 + 2x_3 \le 20$$
$$2x_1 + x_2 + 2x_3 \le 20$$
$$2x_1 + 2x_2 + x_3 \le 20$$
$$x_1, x_2, x_3 \ge 0$$

Group B

7 It has been suggested that at each iteration of the simplex algorithm, the entering variable should be (in a maximization problem) the variable that would bring about the greatest increase in the objective function. Although this usually results in fewer pivots than the rule of entering the most negative row 0 entry, the greatest increase rule is hardly ever used. Why not?

4.6 Using the Simplex Algorithm to Solve Minimization Problems

There are two different ways that the simplex algorithm can be used to solve minimization problems. We illustrate these methods by solving the following LP:

$$\min z = 2x_1 - 3x_2$$
$$\text{s.t.} \quad x_1 + x_2 \le 4$$
$$x_1 - x_2 \le 6 \qquad \text{(LP 2)}$$
$$x_1, x_2 \ge 0$$

Method 1

The optimal solution to LP 2 is the point (x_1, x_2) in the feasible region for LP 2 that makes $z = 2x_1 - 3x_2$ the smallest. Equivalently, we may say that the optimal solution to LP 2 is the point in the feasible region that makes $-z = -2x_1 + 3x_2$ the largest. This means that we can find the optimal solution to LP 2 by solving LP 2':

$$\max -z = -2x_1 + 3x_2$$
$$\text{s.t.} \quad x_1 + x_2 \le 4$$
$$x_1 - x_2 \le 6 \qquad \textbf{(LP 2')}$$
$$x_1, x_2 \ge 0$$

In solving LP 2', we will use $-z$ as the basic variable for row 0. After adding slack variables s_1 and s_2 to the two constraints, we obtain the initial tableau in Table 9. Because x_2 is the only variable with a negative coefficient in row 0, we enter x_2 into the basis. The ratio test indicates that x_2 should enter the basis in the first constraint, row 1. The resulting tableau is shown in Table 10. Because each variable in row 0 has a nonnegative coefficient, this is an optimal tableau. Thus, the optimal solution to LP 2' is $-z = 12$, $x_2 = 4$, $s_2 = 10$, $x_1 = s_1 = 0$. Then the optimal solution to LP 2 is $z = -12$, $x_2 = 4$, $s_2 = 10$, $x_1 = s_1 = 0$. Substituting the values of x_1 and x_2 into LP 2's objective function, we obtain

$$z = 2x_1 - 3x_2 = 2(0) - 3(4) = -12$$

In summary, multiply the objective function for the min problem by -1 and solve the problem as a maximization problem with objective function $-z$. The optimal solution to the max problem will give you the optimal solution to the min problem. Remember that (optimal z-value for min problem) $= -$(optimal objective function value z for max problem).

Method 2

A simple modification of the simplex algorithm can be used to solve min problems directly. Modify Step 3 of the simplex as follows: If all nonbasic variables in row 0 have nonpositive coefficients, then the current bfs is optimal. If any nonbasic variable in row

TABLE 9
Initial Tableau for LP 2—Method 1

$-z$	x_1	x_2	s_1	s_2	rhs	Basic Variable	Ratio
1	2	-3	0	0	0	$-z = 0$	
0	1	①	1	0	4	$s_1 = 4$	$\frac{4}{1} = 4^*$
0	1	-1	0	1	6	$s_2 = 6$	None

TABLE 10
Optimal Tableau for LP 2—Method 1

$-z$	x_1	x_2	s_1	s_2	rhs	Basic Variable
1	5	0	3	0	12	$-z = 12$
0	1	1	1	0	4	$x_2 = 4$
0	2	0	1	1	10	$s_2 = 10$

TABLE 11
Initial Tableau for LP 2—Method 2

z	x_1	x_2	s_1	s_2	rhs	Basic Variable	Ratio
1	-2	3	0	0	0	$z = 0$	
0	1	①	1	0	4	$s_1 = 4$	$\frac{4}{1} = 4*$
0	1	-1	0	1	6	$s_2 = 6$	None

TABLE 12
Optimal Tableau for LP 2—Method 2

z	x_1	x_2	s_1	s_2	rhs	Basic Variable
1	-5	0	-3	0	-12	$z = -12$
0	1	1	1	0	4	$x_2 = 4$
0	2	0	1	1	10	$s_2 = 10$

0 has a positive coefficient, choose the variable with the "most positive" coefficient in row 0 to enter the basis.

This modification of the simplex algorithm works because increasing a nonbasic variable with a positive coefficient in row 0 will *decrease z*. If we use this method to solve LP 2, then our initial tableau will be as shown in Table 11. Because x_2 has the most positive coefficient in row 0, we enter x_2 into the basis. The ratio test says that x_2 should enter the basis in row 1, resulting in Table 12. Because each variable in row 0 has a nonpositive coefficient, this is an optimal tableau.[†] Thus, the optimal solution to LP 2 is (as we have already seen) $z = -12$, $x_2 = 4$, $s_2 = 10$, $x_1 = s_1 = 0$.

PROBLEMS

Group A

1 Use the simplex algorithm to find the optimal solution to the following LP:

$$\min z = 4x_1 - x_2$$
$$\text{s.t.} \quad 2x_1 + x_2 \le 8$$
$$x_2 \le 5$$
$$x_1 - x_2 \le 4$$
$$x_1, x_2 \ge 0$$

2 Use the simplex algorithm to find the optimal solution to the following LP:

$$\min z = -x_1 - x_2$$
$$\text{s.t.} \quad x_1 - x_2 \le 1$$
$$x_1 + x_2 \le 2$$
$$x_1, x_2 \ge 0$$

3 Use the simplex algorithm to find the optimal solution to the following LP:

$$\min z = 2x_1 - 5x_2$$
$$\text{s.t.} \quad 3x_1 + 8x_2 \le 12$$
$$2x_1 + 3x_2 \le 6$$
$$x_1, x_2 \ge 0$$

4 Use the simplex algorithm to find the optimal solution to the following LP:

$$\min z = -3x_1 + 8x_2$$
$$\text{s.t.} \quad 4x_1 + 2x_2 \le 12$$
$$2x_1 + 3x_2 \le 6$$
$$x_1, x_2 \ge 0$$

[†]To see that this tableau is optimal, note that from row 0, $z = -12 + 5x_1 + 3s_1$. Because $x_1 \ge 0$ and $s_1 \ge 0$, this shows that $z \ge -12$. Thus, the current bfs (which has $z = -12$) must be optimal.

4.7 Alternative Optimal Solutions

Recall from Example 3 of Section 3.3 that for some LPs, more than one extreme point is optimal. If an LP has more than one optimal solution, then we say that it has multiple or **alternative optimal solutions.** We show now how the simplex algorithm can be used to determine whether an LP has alternative optimal solutions.

Reconsider the Dakota Furniture example of Section 4.3, with the modification that tables sell for \$35 instead of \$30 (see Table 13). Because x_1 has the most negative coefficient in row 0, we enter x_1 into the basis. The ratio test indicates that x_1 should be entered in row 3. Now only x_3 has a negative coefficient in row 0, so we enter x_3 into the basis (see Table 14). The ratio test indicates that x_3 should enter the basis in row 2. The resulting, optimal, tableau is given in Table 15. As in Section 4.3, this tableau indicates that the optimal solution to the Dakota Furniture problem is $s_1 = 24$, $x_3 = 8$, $x_1 = 2$, $s_4 = 5$, and $x_2 = s_2 = s_3 = 0$.

TABLE 13

Initial Tableau for Dakota Furniture (\$35/Table)

z	x_1	x_2	x_3	s_1	s_2	s_3	s_4	rhs	Basic Variable	Ratio
1	−60	−35	−20	0	0	0	0	0	$z = 0$	
0	8	6	1	1	0	0	0	48	$s_1 = 48$	$\frac{48}{8} = 6$
0	4	2	1.5	0	1	0	0	20	$s_2 = 20$	$\frac{20}{4} = 5$
0	②	1.5	0.5	0	0	1	0	8	$s_3 = 8$	$\frac{8}{2} = 4^*$
0	0	1	0	0	0	0	1	5	$s_4 = 5$	None

TABLE 14

First Tableau for Dakota Furniture (\$35/Table)

z	x_1	x_2	x_3	s_1	s_2	s_3	s_4	rhs	Basic Variable	Ratio
1	0	10	−5	0	0	30	0	240	$z = 240$	
0	0	0	−1	1	0	−4	0	16	$s_1 = 16$	None
0	0	−1	(0.5)	0	1	−2	0	4	$s_2 = 4$	$\frac{4}{0.5} = 8^*$
0	1	0.75	0.25	0	0	0.5	0	4	$x_1 = 4$	$\frac{4}{0.25} = 16$
0	0	1	0	0	0	0	1	5	$s_4 = 5$	None

TABLE 15

Second (and Optimal) Tableau for Dakota Furniture (\$35/Table)

z	x_1	x_2	x_3	s_1	s_2	s_3	s_4	rhs	Basic Variable
1	0	0	0	0	10	10	0	280	$z = 280$
0	0	−2	0	1	2	−8	0	24	$s_1 = 24$
0	0	−2	1	0	2	−4	0	8	$x_3 = 8$
0	1	(1.25)	0	0	−0.5	1.5	0	2	$x_1 = 2^*$
0	0	1	0	0	0	0	1	5	$s_4 = 5$

TABLE **16**
Another Optimal Tableau for Dakota Furniture ($35/Table)

z	x_1	x_2	x_3	s_1	s_2	s_3	s_4	rhs	Basic Variable
1	0	0	0	0	10	10	0	280	$z = 280$
0	1.6	0	0	1	1.2	−5.6	0	27.2	$s_1 = 27.2$
0	1.6	0	1	0	1.2	−1.6	0	11.2	$x_3 = 11.2$
0	0.8	1	0	0	−0.4	1.2	0	1.6	$x_2 = 1.6$
0	−0.8	0	0	0	0.4	−1.2	1	3.4	$s_4 = 3.4$

Recall that all basic variables must have a zero coefficient in row 0 (or else they wouldn't be basic variables). However, in our optimal tableau, there is a nonbasic variable, x_2, which also has a zero coefficient in row 0. Let us see what happens if we enter x_2 into the basis. The ratio test indicates that x_2 should enter the basis in row 3 (check this). The resulting tableau is given in Table 16. The important thing to notice is that *because x_2 has a zero coefficient in the optimal tableau's row 0, the pivot that enters x_2 into the basis does not change row 0.* This means that all variables in our new row 0 will still have nonnegative coefficients. Thus, our new tableau is also optimal. Because the pivot has not changed the value of z, an alternative optimal solution for the Dakota example is $z = 280$, $s_1 = 27.2$, $x_3 = 11.2$, $x_2 = 1.6$, $s_4 = 3.4$, and $x_1 = s_3 = s_2 = 0$.

In summary, if tables sell for $35, Dakota can obtain $280 in sales revenue by manufacturing 2 desks and 8 chairs or by manufacturing 1.6 tables and 11.2 chairs. Thus, Dakota has multiple (or alternative) optimal extreme points.

As stated in Chapter 3, it can be shown that any point on the line segment joining two optimal extreme points will also be optimal. To illustrate this idea, let's write our two optimal extreme points:

$$\text{Optimal extreme point 1} = \begin{bmatrix} x_1 \\ x_2 \\ x_3 \end{bmatrix} = \begin{bmatrix} 2 \\ 0 \\ 8 \end{bmatrix}$$

$$\text{Optimal extreme point 2} = \begin{bmatrix} x_1 \\ x_2 \\ x_3 \end{bmatrix} = \begin{bmatrix} 0 \\ 1.6 \\ 11.2 \end{bmatrix}$$

Thus, for $0 \le c \le 1$,

$$\begin{bmatrix} x_1 \\ x_2 \\ x_3 \end{bmatrix} = c \begin{bmatrix} 2 \\ 0 \\ 8 \end{bmatrix} + (1 - c) \begin{bmatrix} 0 \\ 1.6 \\ 11.2 \end{bmatrix} = \begin{bmatrix} 2c \\ 1.6 - 1.6c \\ 11.2 - 3.2c \end{bmatrix}$$

will be optimal. This shows that although the Dakota Furniture example has only two optimal extreme points, there are an infinite number of optimal solutions to the Dakota problem. For example, by choosing $c = 0.5$, we obtain the optimal solution $x_1 = 1$, $x_2 = 0.8$, $x_3 = 9.6$.

If there is no nonbasic variable with a zero coefficient in row 0 of the optimal tableau, then the LP has a unique optimal solution (see Problem 3). Even if there is a nonbasic variable with a zero coefficient in row 0 of the optimal tableau, it is possible that the LP may not have alternative optimal solutions (see Review Problem 28).

PROBLEMS

Group A

1 Show that if a toy soldier sold for $28, then the Giapetto problem would have alternative optimal solutions.

2 Show that the following LP has alternative optimal solutions; find three of them.

$$\max z = -3x_1 + 6x_2$$
$$\text{s.t.} \quad 5x_1 + 7x_2 \le 35$$
$$-x_1 + 2x_2 \le 2$$
$$x_1, x_2 \ge 0$$

3 Find alternative optimal solutions to the following LP:

$$\max z = x_1 + x_2$$
$$\text{s.t.} \quad x_1 + x_2 + x_3 \le 1$$
$$x_1 \qquad + 2x_3 \le 1$$
$$\text{All } x_i \ge 0$$

4 Find all optimal solutions to the following LP:

$$\max z = 3x_1 + 3x_2$$
$$\text{s.t.} \quad x_1 + x_2 \le 1$$
$$\text{All } x_i \ge 0$$

5 How many optimal basic feasible solutions does the following LP have?

$$\max z = 2x_1 + 2x_2$$
$$\text{s.t.} \quad x_1 + x_2 \le 6$$
$$2x_1 + x_2 \le 13$$
$$\text{All } x_i \ge 0$$

Group B

6 Suppose you have found this optimal tableau (Table 17) for a maximization problem. Use the fact that each nonbasic variable has a strictly positive coefficient in row 0 to show that $x_1 = 4$, $x_2 = 3$, $s_1 = s_2 = 0$ is the unique optimal solution to this LP. (*Hint:* Can any extreme point having $s_1 > 0$ or $s_2 > 0$ have $z = 10$?)

7 Explain why the set of optimal solutions to an LP is a convex set.

8 Consider an LP with the optimal tableau shown in Table 18.

a Does this LP have more than one bfs that is optimal?

b How many optimal solutions does this LP have? (*Hint:* If the value of x_3 is increased, then how does this change the values of the basic variables and the z-value?)

9 Characterize all optimal solutions to the following LP:

$$\max z = -8x_5$$
$$\text{s.t.} \quad x_1 + \qquad x_3 + 3x_4 + 2x_5 = 2$$
$$x_2 + 2x_3 + 4x_4 + 5x_5 = 5$$
$$\text{All } x_i \ge 0$$

TABLE 17

z	x_1	x_2	s_1	s_2	rhs
1	0	0	2	3	10
0	1	0	3	2	4
0	0	1	1	1	3

TABLE 18

z	x_1	x_2	x_3	x_4	rhs
1	0	0	0	2	2
0	1	0	-1	1	2
0	0	1	-2	3	3

4.8 Unbounded LPs

Recall from Section 3.3 that for some LPs, there exist points in the feasible region for which z assumes arbitrarily large (in max problems) or arbitrarily small (in min problems) values. When this situation occurs, we say that LP is unbounded. In this section, we show how the simplex algorithm can be used to determine whether an LP is unbounded.

EXAMPLE 3   **Breadco Bakeries: An Unbounded LP**

Breadco Bakeries bakes two kinds of bread: French and sourdough. Each loaf of French bread can be sold for 36¢, and each loaf of sourdough bread for 30¢. A loaf of French bread requires 1 yeast packet and 6 oz of flour; sourdough requires 1 yeast packet and 5 oz of flour. At present, Breadco has 5 yeast packets and 10 oz of flour. Additional yeast packets can be purchased at 3¢ each, and additional flour at 4¢/oz. Formulate and solve an LP that can be used to maximize Breadco's profits (= revenues − costs).

Solution   Define

$$x_1 = \text{number of loaves of French bread baked}$$
$$x_2 = \text{number of loaves of sourdough bread baked}$$
$$x_3 = \text{number of yeast packets purchased}$$
$$x_4 = \text{number of ounces of flour purchased}$$

Then Breadco's objective is to maximize z = revenues − costs, where

$$\text{Revenues} = 36x_1 + 30x_2 \quad \text{and} \quad \text{Costs} = 3x_3 + 4x_4$$

Thus, Breadco's objective function is

$$\max z = 36x_1 + 30x_2 - 3x_3 - 4x_4$$

Breadco faces the following two constraints:

Constraint 1   Number of yeast packages used to bake bread cannot exceed available yeast plus purchased yeast.

Constraint 2   Ounces of flour used to bake breads cannot exceed available flour plus purchased flour.

Because

$$\text{Available yeast} + \text{purchased yeast} = 5 + x_3$$
$$\text{Available flour} + \text{purchased flour} = 10 + x_4$$

Constraint 1 may be written as

$$x_1 + x_2 \le 5 + x_3 \quad \text{or} \quad x_1 + x_2 - x_3 \le 5$$

and Constraint 2 may be written as

$$6x_1 + 5x_2 \le 10 + x_4 \quad \text{or} \quad 6x_1 + 5x_2 - x_4 \le 10$$

Adding the sign restrictions $x_i \ge 0$ ($i = 1, 2, 3, 4$) yields the following LP:

$$\max z = 36x_1 + 30x_2 - 3x_3 - 4x_4$$
$$\text{s.t.} \quad x_1 + x_2 - x_3 \qquad \le 5 \qquad \text{(Yeast constraint)}$$
$$6x_1 + 5x_2 \qquad - x_4 \le 10 \qquad \text{(Flour constraint)}$$
$$x_1, x_2, x_3, x_4 \ge 0$$

Adding slack variables s_1 and s_2 to the two constraints, we obtain the tableau in Table 19.

TABLE 19
Initial Tableau for Breadco

z	x_1	x_2	x_3	x_4	s_1	s_2	rhs	Basic Variable	Ratio
1	−36	−30	3	4	0	0	0	$z = 0$	
0	1	1	−1	0	1	0	5	$s_1 = 5$	$\frac{5}{1} = 5$
0	⑥	5	0	−1	0	1	10	$s_2 = 10$	$\frac{10}{6} = \frac{5}{3}*$

TABLE 20

First Tableau for Breadco

z	x_1	x_2	x_3	x_4	s_1	s_2	rhs	Basic Variable	Ratio
1	0	0	3	-2	0	6	60	$z = 60$	
0	0	$\frac{1}{6}$	-1	$\boxed{\frac{1}{6}}$	1	$-\frac{1}{6}$	$\frac{10}{3}$	$s_1 = \frac{10}{3}$	$(\frac{10}{3})/(\frac{1}{6}) = 20^*$
0	1	$\frac{5}{6}$	0	$-\frac{1}{6}$	0	$\frac{1}{6}$	$\frac{5}{3}$	$s_2 = \frac{5}{3}$	None

TABLE 21

Second Tableau for Breadco

z	x_1	x_2	x_3	x_4	s_1	s_2	rhs	Basic Variable	Ratio
1	0	2	-9	0	12	4	100	$z = 100$	
0	0	1	-6	1	6	-1	20	$x_4 = 20$	None
0	1	1	-1	0	1	0	5	$x_1 = 5$	None

Because $-36 < -30$, we enter x_1 into the basis. The ratio test indicates that x_1 should enter the basis in row 2. Entering x_1 into the basis in row 2 yields the tableau in Table 20. Because x_4 has the only negative coefficient in row 0, we enter x_4 into the basis. The ratio test indicates that x_4 should enter the basis in row 1, with the resulting tableau in Table 21. Because x_3 has the most negative coefficient in row 0, we would like to enter x_3 into the basis. The ratio test, however, fails to indicate the row in which x_3 should enter the basis. What is happening? Going back to the basic ideas that led us to the ratio test, we see that as x_3 is increased (holding the other nonbasic variables at zero), the current basic variables, x_4 and x_1, change as follows:

$$x_4 = 20 + 6x_3 \tag{13}$$

$$x_1 = 5 + x_3 \tag{14}$$

As x_3 is increased, both x_4 and x_1 increase. This means that no matter how large we make x_3, the inequalities $x_4 \geq 0$ and $x_1 \geq 0$ will still be true. Because each unit by which we increase x_3 will increase z by 9, we can find points in the feasible region for which z assumes an arbitrarily large value. For example, can we find a feasible point with $z \geq 1,000$? To do this, we need to increase z by $1,000 - 100 = 900$. Each unit by which x_3 is increased will increase z by 9, so increasing x_3 by $\frac{900}{9} = 100$ should give us $z = 1,000$. If we set $x_3 = 100$ (and hold the other nonbasic variables at zero), then (13) and (14) show that x_4 and x_1 must now equal

$$x_4 = 20 + 6(100) = 620$$
$$x_1 = 5 + (100) = 105$$

Thus, $x_1 = 105$, $x_3 = 100$, $x_4 = 620$, $x_2 = 0$ is a point in the feasible region with $z = 1,000$. In a similar fashion, we can find points in the feasible region having arbitrarily large z-values. This means the Breadco problem is an unbounded LP.

From the Breadco example, we see that an unbounded LP occurs in a max problem if there is a nonbasic variable with a negative coefficient in row 0 and there is no constraint that limits how large we can make the nonbasic variable. This situation will occur if a nonbasic variable (such as x_3) has a negative coefficient in row 0 and nonpositive coefficients in each constraint. To summarize, *an unbounded LP for a max problem occurs when a variable with a negative coefficient in row 0 has a nonpositive coefficient in each constraint.*

If an LP is unbounded, one will eventually come to a tableau where one wants to enter a variable (such as x_3) into the basis, but the ratio test will fail. This is probably the easiest way to spot an unbounded LP.

As we noted in Chapter 3, an unbounded LP is usually caused by an incorrect formulation. In the Breadco example, we obtained an unbounded LP because we allowed Breadco to pay $3 + 6(4) = 27¢$ for the ingredients in a loaf of French bread and then sell the loaf for 36¢. Thus, each loaf of French bread earns a profit of 9¢. Because unlimited purchases of yeast and flour are allowed, it is clear that our model allows Breadco to manufacture as much French bread as it desires, thereby earning arbitrarily large profits. This is the cause of the unbounded LP.

Of course, our formulation of the Breadco example ignored several aspects of reality. First, we assumed that demand for Breadco's products is unlimited. Second, we ignored the fact that certain resources to make bread (such as ovens and labor) are in limited supply. Finally, we made the unrealistic assumption that unlimited quantities of yeast and flour could be purchased.

Unbounded LPs and Directions of Unboundedness

Consider an LP with an objective function $c_1x_1 + c_2x_2 + \cdots + c_nx_n$. Let $\mathbf{c} = [c_1 \quad c_2 \ldots c_n]$. If the LP is a maximization problem, then the LP will be unbounded if and only if it has a direction of unboundedness $\mathbf{d}$ satisfying $\mathbf{cd} > 0$. If the LP is a minimization problem, then the LP will be unbounded if and only if it has a direction of unboundedness $\mathbf{d}$ satisfying $\mathbf{cd} < 0$. I[n Example 3, the last tableau shows us that if we start at the point

$$\begin{bmatrix} 5 \\ 0 \\ 0 \\ 20 \\ 0 \\ 0 \end{bmatrix}$$

(the variables are listed in the same order they are listed in the tableau), we can find a direction of unboundedness as follows. Every unit by which x_3 is increased will maintain feasibility if we increase x_1 by one unit and x_4 by six units and leave x_2, s_1, and s_2 unchanged. Because we can increase x_3 without limit, this indicates that

$$\mathbf{d} = \begin{bmatrix} 1 \\ 0 \\ 1 \\ 6 \\ 0 \\ 0 \end{bmatrix}$$

is a direction of unboundedness. Because

$$\mathbf{cd} = [36 \quad 30 \quad -3 \quad -4 \quad 0 \quad 0] \begin{bmatrix} 1 \\ 0 \\ 1 \\ 6 \\ 0 \\ 0 \end{bmatrix} = 9$$

we know that LP is unbounded. This follows because each time we move in the direction **d** an amount that increases x_3 by one unit, we increase z by 9, and we can move as far as we want in the direction **d**.

PROBLEMS

Group A

1 Show that the following LP is unbounded:

$$\max z = 2x_2$$
$$\text{s.t.} \quad x_1 - x_2 \leq 4$$
$$-x_1 + x_2 \leq 1$$
$$x_1, x_2 \geq 0$$

Find a point in the feasible region with $z \geq 10,000$.

2 State a rule that can be used to determine if a min problem has an unbounded optimal solution (that is, z can be made arbitrarily small). Use the rule to show that

$$\min z = -2x_1 - 3x_2$$
$$\text{s.t.} \quad x_1 - x_2 \leq 1$$
$$x_1 - 2x_2 \leq 2$$
$$x_1, x_2 \geq 0$$

is an unbounded LP.

3 Suppose that in solving an LP, we obtain the tableau in Table 22. Although x_1 can enter the basis, this LP is unbounded. Why?

4 Use the simplex method to solve Problem 10 of Section 3.3.

TABLE 22

z	x_1	x_2	x_3	x_4	rhs
1	-3	-2	0	0	0
0	1	-1	1	0	3
0	2	0	0	1	4

5 Show that the following LP is unbounded:

$$\max z = x_1 + 2x_2$$
$$\text{s.t.} \quad -x_1 + x_2 \leq 2$$
$$-2x_1 + x_2 \leq 1$$
$$x_1, x_2 \geq 0$$

6 Show that the following LP is unbounded:

$$\min z = -x_1 - 3x_2$$
$$\text{s.t.} \quad x_1 - 2x_2 \leq 4$$
$$-x_1 + x_2 \leq 3$$
$$x_1, x_2 \geq 0$$

4.9 The LINDO Computer Package

LINDO (Linear Interactive and Discrete Optimizer) was developed by Linus Schrage (1986). It is a user-friendly computer package that can be used to solve linear, integer, and quadratic programming problems.[†] Appendix A to this chapter gives a brief explanation of how LINDO can be used to solve LPs. In this section, we explain how the information on a LINDO printout is related to our discussion of the simplex algorithm.

We begin by discussing the LINDO ouput for the Dakota Furniture example (see Figure 6). LINDO allows the user to name the variables, so we define

$$\text{DESKS} = \text{number of desks produced}$$
$$\text{TABLES} = \text{number of tables produced}$$
$$\text{CHAIRS} = \text{number of chairs produced}$$

Then the Dakota formulation in the first block of Figure 6 is

[†]See Chapter 9 for a discussion of integer programming and Chapter 11 for a discussion of quadratic programming.

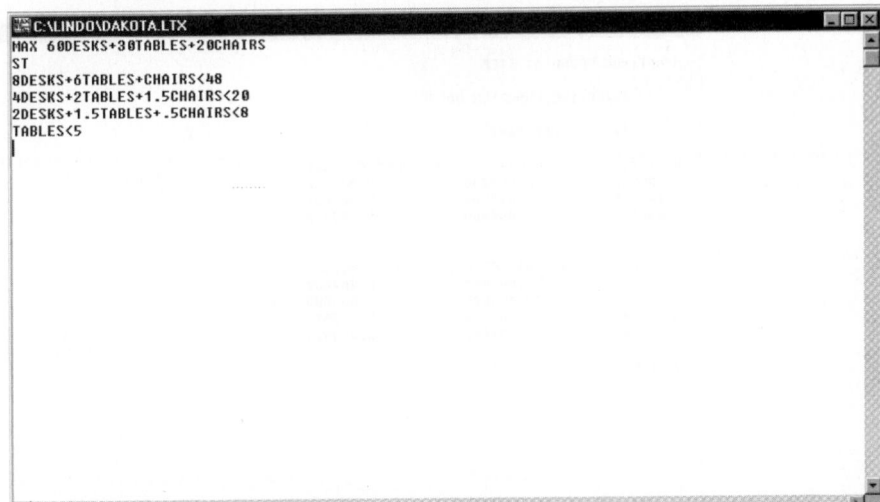

FIGURE 6
LINDO Output for
Dakota Furniture

$$\max 60 \text{ DESKS} + 30 \text{ TABLES} + 20 \text{ CHAIRS} \quad \text{(Row 1)}$$

s.t.
$$8 \text{ DESKS} + 6 \text{ TABLES} + \text{ CHAIRS} \leq 48 \text{ (Row 2)} \quad \text{(Lumber constraint)}$$
$$4 \text{ DESKS} + 2 \text{ TABLES} + 1.5 \text{ CHAIRS} \leq 20 \text{ (Row 3)} \quad \text{(Finishing constraint)}$$
$$2 \text{ DESKS} + 1.5 \text{ TABLES} + 0.5 \text{ CHAIRS} \leq 8 \text{ (Row 4)} \quad \text{(Carpentry constraint)}$$
$$\text{TABLES} \leq 5 \text{ (Row 5)}$$
$$\text{DESKS, TABLES, CHAIRS} \geq 0$$

(LINDO assumes that all variables are nonnegative, so the nonnegativity constraints need not be input to the computer.) To be consistent with LINDO, we have labeled the objective function row 1 and the constraint rows 2–5.

To enter this problem in LINDO, make sure the screen contains a blank window, or work area, with "Untitled" at the top of the work area. If necessary, a new window can be opened by selecting New from the File menu or by clicking on the New File button.

The first statement in a LINDO model is always the objective. Enter the objective much like you would write it in equation form:

$$\text{MAX } 60 \text{ DESKS} + 30 \text{ TABLES} + 20 \text{ CHAIRS}$$

This tells LINDO to maximize the objective function. Proceed by entering the constraints as follows:

$$\text{SUBJECT TO (OR s.t.)}$$
$$8 \text{ DESKS} + 6 \text{ TABLES} + \text{ CHAIRS} < 48$$
$$4 \text{ DESKS} + 2 \text{ TABLES} + 1.5 \text{ CHAIRS} < 20$$
$$2 \text{ DESKS} + 1.5 \text{ TABLES} + .5 \text{ CHAIRS} < 8$$
$$\text{TABLES} < 5$$

Your screen will now look like the one in Figure 6. Note that LINDO automatically assumes that all decision variables are nonnegative.

To save the file for later use, select Save from the File menu and when asked for a file name replace the * symbol with a name of your choice (we chose *Dakota*). Do not type over the characters .LTX. You may now use the File Open command to retrieve the problem.

Dakota

To solve the model, proceed as follows:

1 From the Solve menu, select the Solve command or click the button with a bull's-eye.

```
LP OPTIMUM FOUND AT STEP      2

        OBJECTIVE FUNCTION VALUE

    1)      280.0000

VARIABLE        VALUE           REDUCED COST
   DESKS        2.000000          0.000000
  TABLES        0.000000          5.000000
  CHAIRS        8.000000          0.000000


    ROW    SLACK OR SURPLUS    DUAL PRICES
    2)        24.000000         0.000000
    3)         0.000000        10.000000
    4)         0.000000        10.000000
    5)         5.000000         0.000000

NO. ITERATIONS=      2
```

FIGURE 7

2 When asked if you want to do a range (sensitivity analysis) choose No. We will explain how to interpret a range or sensitivity analysis in Chapter 6.

3 When the solution is completed, a display showing the status of the Solve command will be present. After reviewing the displayed information, select Close.

4 You should now see your input data overlaying a display labeled "Reports Window." Click anywhere in the Reports window, and your input data will be removed from the foreground. Move to the top of the screen using the single arrow at the right of the screen, and your screen should now look like that in Figure 7.

Looking now at the LINDO output in Figure 7, we see

LP OPTIMUM FOUND AT STEP 2

indicating that LINDO found the optimal solution after two iterations (or pivots) of the simplex algorithm.

OBJECTIVE FUNCTION VALUE 280.000000

indicates that the optimal z-value is 280.

VALUE

gives the value of the variable in the optimal LP solution. Thus, the optimal solution calls for Dakota to produce 2 desks, 0 tables, and 8 chairs.

SLACK OR SURPLUS

gives the value of the slack or excess ("surplus variable" is another name for excess variable) in the optimal solution. Thus,

$$s_1 = \text{slack for row 2 on LINDO output} = 24$$
$$s_2 = \text{slack for row 3 on LINDO output} = 0$$
$$s_3 = \text{slack for row 4 on LINDO output} = 0$$
$$s_4 = \text{slack for row 5 on LINDO output} = 5$$

REDUCED COST

gives the coefficient of the variable in row 0 of the optimal tableau (in a max problem). As discussed in Section 4.3, the reduced cost for each basic variable must be 0. For a non-

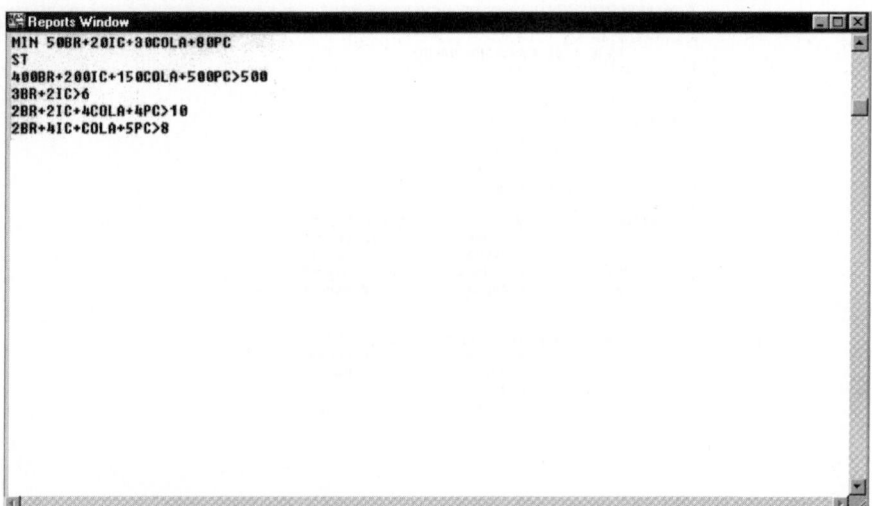

FIGURE 8

```
Reports Window
MIN 50BR+20IC+30COLA+80PC
ST
400BR+200IC+150COLA+500PC>500
3BR+2IC>6
2BR+2IC+4COLA+4PC>10
2BR+4IC+COLA+5PC>8
```

basic variable x_j, the reduced cost is the amount by which the optimal z-value is decreased if x_j is increased by 1 unit (and all other nonbasic variables remain equal to 0). In the LINDO output for the Dakota problem, the reduced cost is 0 for each of the basic variables (DESKS and CHAIRS). Also, the reduced cost for TABLES is 5. This means that if Dakota were forced to produce a table, revenue would decrease by $5.

For a minimization problem, the LP Optimum, Objective Function Value, and Slack and Surplus columns are interpreted as described. But the reduced cost for a variable is $-$(coefficient of variable in optimal row 0). Thus, in a min problem, the reduced cost for a basic variable will again be zero, but the reduced cost for a nonbasic variable x_j will be the amount by which the optimal z-value increases if x_j is increased by 1 unit (and all other nonbasic variables remain equal to 0).

To illustrate the interpretation of the LINDO output for a minimization problem, let's look at the LINDO output for the diet problem of Section 3.4 (see Figure 9). If we let

$$BR = \text{brownies eaten daily}$$
$$IC = \text{scoops of chocolate ice cream eaten daily}$$
$$COLA = \text{number of bottles of soda drunk daily}$$
$$PC = \text{pieces of pineapple cheesecake eaten daily}$$

then the diet problem may be formulated as

$$
\begin{array}{llll}
\min & 50\ BR + 20\ IC + 30\ COLA + 80\ PC & & \\
\text{s.t.} & 400\ BR + 200\ IC + 150\ COLA + 500\ PC \geq 500 & & \text{(Calorie constraint)} \\
& 3\ BR + 2\ IC \geq 6 & & \text{(Chocolate constraint)} \\
& 2\ BR + 2\ IC + 4\ COLA + 4\ PC \geq 10 & & \text{(Sugar constraint)} \\
& 2\ BR + 4\ IC + COLA + 5\ PC \geq 8 & & \text{(Fat constraint)} \\
& BR, IC, COLA, PC \geq 0 & &
\end{array}
$$

The Value column shows that the optimal solution is to eat three scoops of chocolate ice cream daily and drink one bottle of soda daily. The Objective Function Value on the LINDO output indicates that the cost of this diet is 90¢. The Slack or Surplus column shows that the first constraint (calories) has an excess of 250 calories and that the fourth

```
Reports Window                                              _ □ ×

        OBJECTIVE FUNCTION VALUE


LP OPTIMUM FOUND AT STEP      2

        OBJECTIVE FUNCTION VALUE

    1)      90.00000

VARIABLE         VALUE        REDUCED COST
     BR        0.000000         27.500000
     IC        3.000000          0.000000
   COLA        1.000000          0.000000
     PC        0.000000         50.000000


    ROW    SLACK OR SURPLUS    DUAL PRICES
    2)       250.000000         0.000000
    3)         0.000000        -2.500000
    4)         0.000000        -7.500000
    5)         5.000000         0.000000

NO. ITERATIONS=        2
```

FIGURE 9

constraint (fat) has an excess of 5 oz. Thus, the calorie and fat constraints are nonbinding. The chocolate and sugar constraints have no excess and are therefore binding constraints.

From the Reduced Cost column, we see that if we were forced to eat a brownie (while keeping $PC = 0$), the minimum cost of the daily diet would increase by 27.5¢, and if we were forced to eat a piece of pineapple cheesecake (while holding $BR = 0$), the minimum cost of the daily diet would increase by 50¢.

The Tableau Command

If, after obtaining the optimal solution to the Dakota furniture problem, you close the Reports window and select the Tableau command (under the Reports menu), LINDO will display the optimal tableau (see Figure 10). Remembering that the first constraint is row 2 in LINDO, we see that $BV = \{s_1, CHAIRS, DESKS, s_4\}$. Thus, for example, SLK5 on the LINDO output corresponds to s_4. The artificial variable (ART) listed as basic in row 1 is z; thus, row 0 of the optimal tableau is $z + 5TABLES + 10s_2 + 10s_3 = 280$.

When you have installed LINDO on your hard drive, the LINDO formulation for the Dakota and Diet problems will be in the directory C:\WINSTON\LINDO\SAMPLES.

See Appendix A of Chapter 4 for further discussion of LINDO.

```
THE TABLEAU
   ROW   (BASIS)     DESKS      TABLES     CHAIRS     SLK   2     SLK   3
    1    ART          .000       5.000       .000       .000      10.000
    2    SLK   2      .000      -2.000       .000      1.000       2.000
    3     CHAIRS      .000      -2.000      1.000       .000       2.000
    4      DESKS     1.000       1.250       .000       .000       -.500
    5    SLK   5      .000       1.000       .000       .000        .000

   ROW   SLK   4     SLK   5
    1    10.000       .000     280.000
    2    -8.000       .000      24.000
    3    -4.000       .000       8.000
    4     1.500       .000       2.000
    5      .000      1.000       5.000
```

FIGURE 10
Example of TABLEAU
Command

4.10 Matrix Generators, LINGO, and Scaling of LPs

Many LPs solved in practice contain thousands of constraints and decision variables. Few users of linear programming would want to input the constraints and objective function each time such an LP is to be solved. For this reason, most actual applications of LP use a **matrix generator** to simplify the inputting of the LP. A matrix generator allows the user to input the relevant parameters that determine the LP's objective function and constraints; it then generates the LP formulation from that information. For example, let's consider the Sailco example from Section 3.10. If we were dealing with a planning horizon of 200 periods, then this problem would involve 400 constraints and 600 decision variables—clearly too many for convenient input. A matrix generator for this problem would require the user to input only the following information for each period: cost of producing a sailboat with regular-time labor, cost with overtime labor, demand, and holding costs. From this information, the matrix generator would generate the LP's objective function and constraints, call up an LP software package (such as LINDO) and solve the problem. Finally, an output analyzer would be written to display the output in a user-friendly format.

The LINGO Package

The package LINGO is an example of a sophisticated matrix generator (and much more!). LINGO is an optimization modeling language that enables the user to create many (perhaps thousands) of constraints or objective function terms by typing one line. To illustrate how LINGO works, we will solve the Sailco problem (Example 12 of Chapter 3).

Solving the Sailco Problem

Sail.lng

The LINGO model follows (it is the file Sail.lng on your disk).

```
MODEL:
 1]  SETS:
 2]  QUARTERS/Q1,Q2,Q3,Q4/:TIME,DEM,RP,OP,INV;
 3]  ENDSETS
 4]  MIN=@SUM(QUARTERS:400*RP+450*OP+20*INV);
 5]  @FOR(QUARTERS(I):RP(I)<40);
 6]  @FOR(QUARTERS(I)|TIME(I)#GT#1:
 7]  INV(I)=INV(I-1)+RP(I)+OP(I)-DEM(I););
 8]  INV(1)=10+RP(1)+OP(1)-DEM(1);
 9]  DATA:
10]  DEM=40,60,75,25;
11]  TIME=1,2,3,4;
12]  ENDDATA
END
```

To begin setting up a model with LINGO, think of the objects or sets that define the problem. For Sailco, the four quarters (Q1, Q2, Q3, and Q4) help define the problem. For each quarter we determine the objects that must be known to find an optimal production schedule—demand (DEM), regular-time production (RP), overtime production (OP), and end-of-quarter inventory (INV). The first three lines of the Sailco program define these objects. **SETS:** begins the definition of the sets needed to model the problem and **ENDSETS** ends it. The effect of line 2 is to define four quarters: Q1, Q2, Q3, and Q4. For each quarter, line 2 creates time (indicating if the quarter is the first, second, third, or fourth quarter); the demand for sailboats; the regular-time and overtime production levels; and the ending inventory. Now that these sets and objects have been defined, we can use them to build a model (containing an objective function and constraints). LINGO will solve for the RP, OP, and INV once we input (in the DATA section of the program) the demands and numbers of the quarters.

Line 4 creates the objective function; **MIN** = indicates that we are minimizing. @**SUM**(QUARTERS: followed by 400*RP + 450*OP + 20*INV means sum 400*RP + 450*OP + 20*INV over all quarters. Thus for each quarter we compute 400*(regular-time production) + 450*(overtime production) + 20*(ending inventory). Notice that line 4 creates the proper objective function whether there are 4, 40, 400, or 4,000 quarters!

Line 5 says that for each quarter, RP cannot exceed 40. Again, if there were 400 quarters in the planning horizon, this statement would generate 400 constraints.

Together, lines 6 and 7 create constraints for all quarters (except the first) that ensure that

Ending Inventory for Quarter i = (Ending Inventory for Quarter $i - 1$)

$$+ \text{(Quarter } i \text{ Production)} - \text{(Quarter } i \text{ Demand)}$$

Notice that unlike LINDO, variables are allowed on the right side of a constraint (and numbers are allowed on the left side).

Line 8 creates the constraint ensuring that

(Ending Quarter 1 Inventory) = (Beginning Quarter 1 Inventory)

$$+ \text{(Quarter 1 Production)} - \text{(Quarter 1 Demand)}$$

Lines 9–12 input the needed data (the number of the quarter and the demand for each quarter). The DATA section must begin with a **DATA:** statement and end with an **END-DATA** statement. As with LINDO, a LINGO program ends with an **END** statement.

Notice that once we have created the LINGO model to solve the Sailco example, we can easily edit the model to solve any *n*-period production-scheduling model. If we were solving a 12-quarter problem, we would simply edit (see Remark 3 later) line 2 to QUAR-TERS/1..12/:TIME,DEM,RP,OP,INV;. Then enter the 12 quarterly demands in line 10 and change Line 11 to TIME=1,2,3,4,5,6,7,8,9,10,11,12;. To find the optimal solution to the problem either select the Solve command from the LINGO menu or click the button with a bull's-eye.

In this example, we will also look at how to use some of the editing capabilities of LINGO. Type the first four lines of this model just as you normally would. This will define the sets section and the objective function, and should appear as follows:

```
SETS:
  QUARTERS/Q1,Q2,Q3,Q4/:TIME,DEM,RP,OP,INV;
ENDSETS
MIN = @SUM(QUARTERS:400*RP+450*OP+20*INV);
```

The next line required is the @FOR statement that restricts regular-time production (RP) to values less than 40. Instead of typing in this entire statement, use LINGO's Paste Function command as follows:

1 From the Edit menu, select Paste Function. Notice that you do not have to click on this, but only highlight it, and a submenu appears.

2 From the submenu, select Set, and another submenu appears listing various @ functions.

3 Select the @FOR function, and a general form of the @FOR statement will appear in your input window.

4 Replace the general terms of the function with your specific parameters. This statement should then appear as follows:

```
@FOR(QUARTERS(I):RP(I)<40);
```

Because another @FOR statement is needed to further define constraints on all quarters, you could type this in or use the Paste Function command again. Using additional Edit commands, however, will allow you to copy and paste a portion of the previous @FOR statement instead of retyping it. Do this as follows:

1 Place your cursor at the beginning of the @FOR statement previously typed.

2 Hold down the left mouse button and drag the mouse to highlight the portion of the statement that can be reused, as shown below:

```
@FOR(QUARTERS(I):RP(I)<40);
```

3 From the Edit menu, select Copy (or use the shortcut Ctrl+C) to copy the highlighted text.

4 Place the cursor at the beginning of the next blank line and press Ctrl+V to paste the copied text.

You can now type in the remainder of this line, and the following lines as shown below.

```
@FOR(QUARTERS(I)|TIME(I) #GT#1:
INV(I)=INV(I-1)+RP(I)+OP(I)-DEM(I););
INV(1)=10+RP(1)+OP(1)-DEM(1);
DATA:
  DEM=40,60,75,25;
  TIME=1,2,3,4;
ENDDATA
END
```

While the Copy command only saved a few keystrokes in this example, it can save significantly more steps when you have repetition within a model. In a similar manner, the Cut command can remove highlighted portions for placement elsewhere within a model. The input for this example is saved in the file Sail.lng.

After solving the model, the first portion of the Reports window should indicate an objective value of $78,450, as shown in the output screen in Figure 11.

LINGO and the Post Office Problem

We now show how to use LINGO to solve the Post Office Scheduling example (Example 7) from Chapter 3. The following LINGO model (file Post.lng) can be used to solve this problem.

```
MODEL:
  1] SETS:
  2] DAYS/1..7/:RQMT,START;
  3] ENDSETS
  4] MIN=@SUM(DAYS:START);
  5] @FOR(DAYS(I):@SUM(DAYS(J)|
  6] (J#GT#I+2)#OR#(J#LE#I#AND#J#GT#I-5):
  7] START(J))>RQMT(I););
  8] DATA:
  9] RQMT=17,13,15,19,14,16,11;
 10] ENDDATA
END
```

Line 1 defines the sets needed to solve the problem. Line 2 defines the days of the week (Monday, Tuesday, . . . , Sunday) and associates each with two quantities: the number of workers needed (RQMT) and the number of workers that will begin work on that day of the week (START). Line 3 ends the definitions of the sets.

In Line 4, we create an objective function by summing the number of workers starting work on each day of the week. Lines 5–7 create for each day of the week the constraint that ensures the number of employees working that day is at least as large as the day's requirement. For DAY(I), lines 5 and 6 sum the number of employees starting work over the values of J satisfying $J > I + 2$ or $J \leq I$ and $J > I - 5$. For instance for $I = 1$, this

```
MODEL:
SETS:
QUARTERS/Q1,Q2,Q3,Q4/:TIME,DEM,RP,OP,INV;
ENDSETS
MIN=@SUM(QUARTERS:400*RP+450*OP+20*INV);
@FOR(QUARTERS(I):RP(I)<40);
@FOR(QUARTERS(I)|TIME(I)#GT#1:
INV(I)=INV(I-1)+RP(I)+OP(I)-DEM(I););
INV(1)=10+RP(1)+OP(1)-DEM(1);
DATA:
DEM=40,60,75,25;
TIME=1,2,3,4;
ENDDATA
 END


MIN    400 RP( Q1) + 450 OP( Q1) + 20 INV( Q1) + 400 RP( Q2)
       + 450 OP( Q2) + 20 INV( Q2) + 400 RP( Q3) + 450 OP( Q3)
       + 20 INV( Q3) + 400 RP( Q4) + 450 OP( Q4) + 20 INV( Q4)
   SUBJECT TO
2]    RP( Q1) <=    40
3]    RP( Q2) <=    40
4]    RP( Q3) <=    40
5]    RP( Q4) <=    40
6]-  INV( Q1) - RP( Q2) - OP( Q2) + INV( Q2) =  - 60
7]-  INV( Q2) - RP( Q3) - OP( Q3) + INV( Q3) =  - 75
8]-  INV( Q3) - RP( Q4) - OP( Q4) + INV( Q4) =  - 25
9]-  RP( Q1) - OP( Q1) + INV( Q1) =  - 30
   END

Global optimal solution found at step:           7
Objective value:                          78450.00
```

Variable	Value	Reduced Cost
TIME(Q1)	1.000000	0.0000000
TIME(Q2)	2.000000	0.0000000
TIME(Q3)	3.000000	0.0000000
TIME(Q4)	4.000000	0.0000000
DEM(Q1)	40.00000	0.0000000
DEM(Q2)	60.00000	0.0000000
DEM(Q3)	75.00000	0.0000000
DEM(Q4)	25.00000	0.0000000
RP(Q1)	40.00000	0.0000000
RP(Q2)	40.00000	0.0000000
RP(Q3)	40.00000	0.0000000
RP(Q4)	25.00000	0.0000000
OP(Q1)	0.0000000	20.00000
OP(Q2)	10.00000	0.0000000
OP(Q3)	35.00000	0.0000000
OP(Q4)	0.0000000	50.00000
INV(Q1)	10.00000	0.0000000
INV(Q2)	0.0000000	20.00000
INV(Q3)	0.0000000	70.00000
INV(Q4)	0.0000000	420.0000

Row	Slack or Surplus	Dual Price
1	78450.00	1.000000
2	0.0000000	30.00000
3	0.0000000	50.00000
4	0.0000000	50.00000
5	15.00000	0.0000000
6	0.0000000	450.0000
7	0.0000000	450.0000
8	0.0000000	400.0000
9	0.0000000	430.0000

FIGURE 11

generates the sum START(1) + START(4) + START(5) + START(6) + START(7), which is indeed the number of workers working on Day 1 (Monday). Line 7 (in concert with lines 5 and 6) then ensures that the number of employees working on Day I is at least as large as the number needed on Day I [RQMT(I)]. Line 8 begins the DATA section of the program. In Line 9, we input the requirements for each day of the week.

See Appendix B of Chapter 4 for further discussion of LINGO. Chapters 7, 8, 9, 11, and 14 contain many more examples of problems solved with LINGO.

Scaling of LPs

We close our discussion of computer packages by noting that an LP package may have trouble solving LPs in which there are nonzero coefficients that are either very small or very large in absolute value. If such coefficients are present, then LINDO will respond with a message that the LP is poorly scaled. The LINDO manual recommends that the user define the units of the objective function, right-hand sides, and decision variables so that no nonzero coefficients have absolute values of more than 100,000 or less than 0.0001.

PROBLEMS

Group A

1 A company produces three products. The per-unit profit, labor usage, and pollution produced per unit are given in Table 23. At most, 3 million labor hours can be used to produce the three products, and government regulations require that the company produce at most 2 lb of pollution. If we let x_i = units produced of product i, then the appropriate LP is

max $z = 6x_1 + 4x_2 + 3x_3$

s.t. $\quad 4x_1 + \quad\quad 3x_2 + \quad\quad\quad 2x_3 \leq 3,000,000$
$\quad 0.000003x_1 + 0.000002x_2 + 0.000001x_3 \leq 2$
$$x_1, x_2, x_3 \geq 0$$

a Explain why this LP is poorly scaled.

b Eliminate the scaling problem by redefining the units of the objective function, decision variables, and right-hand sides.

2 Use LINGO to solve Problem 1 in Section 3.5.

3 Use LINGO to solve Example 14 of Chapter 3.

4 The **product mix** problem occurs when we manufacture N products. Each unit produced of a given product uses a given amount of M resources. Each unit produced of product j earns a profit p_j. A quantity r_i of resource i is available. Formulate a LINGO model that could be used to maximize

profit in this situation. Then use it to solve the product mix problem defined by the data in Tables 24 and 25. Assume that a fractional number of vehicles are allowed.

5 The **media mix** problem occurs when a company has N media in which the company can place an ad. There are K groups of people the company wishes to reach, and the company wishes its ads to be seen at least e_i times by members of group i. An ad on media j costs c_j dollars and reaches a_{ij} members of group i. The goal is to minimize the cost of ensuring that the desired number of people in each group see the ads. Set up a LINGO model that can be used to solve any media mix problem. Then solve the media mix problem defined by the data in Tables 26 and 27. Assume that a fractional number of ads is feasible.

TABLE 25

Resource	Quantity Available
Steel	50 tons
Rubber	10 tons
Labor	150 hours

TABLE 23

Product	Profit ($)	Labor Usage (Hrs)	Pollution (Lb)
1	6	4	0.000003 lb
2	4	3	0.000002 lb
3	3	2	0.000001 lb

TABLE 24

	Cars	Trucks	Trains
Steel used (tons)	2	3	5
Rubber used (tons)	.3	.7	.2
Labor used (hrs)	10	12	20
Unit profit ($)	800	1,500	2,500

TABLE 26

Group	Needed Exposures (in Millions)
Children	15
Men	40
Women	50

TABLE 27

| No. Watching (million) | Program | | |
	Sponge Bob	Friends	Dawson's Creek
Children	3	1	0
Men	1	15	4
Women	2	20	9
Unit cost ($)	30,000	360,000	80,000

TABLE 28

	District									
	1	2	3	4	5	6	7	8	9	10
Whites	400	200	150	300	400	100	200	300	250	150
Blacks	200	150	100	120	80	90	140	160	100	60

	Distance (Miles)									
	1	2	3	4	5	6	7	8	9	10
High School 1	1	2	3	2	3	4	2	3	1	2
High School 2	2	1	3	3	4	2	1	2	2	3
High School 3	3	3	2	1	2	3	2	2	3	1

6 Consider the following **school redistricting problem.** There are I districts in a city and J high schools in the city. The distance between District i and High School j is d_{ij} miles. District i has w_i white and b_i black residents. Each high school must have between L and U students. In the interests of racial harmony, the percentages of blacks at each high school must be between 80% and 120% of the percentage of black students in the entire city.

a Set up a LINGO model that can be used to minimize the total distance that students will have to travel in order to meet the racial balance requirements.

b Use your model to solve the problem defined by the data in Table 28.

c What might be some alternative objective functions for this situation?

d Do you see any other problems with our model?

4.11 Degeneracy and the Convergence of the Simplex Algorithm

Theoretically, the simplex algorithm (as we have described it) can fail to find the optimal solution to an LP. However, LPs arising from actual applications seldom exhibit this unpleasant behavior. For the sake of completeness, however, we now discuss the type of situation in which the simplex can fail. Our discussion depends crucially on the following relationship (for a max problem) between the z-values for the current bfs and the new bfs (that is, the bfs after the next pivot):

z-value for new bfs $= z$-value of current bfs

$-$ (value of entering variable in new bfs)(coefficient   **(15)**

of entering variable in row 0 of current bfs)

Equation (15) follows, because each unit by which the entering variable is increased will increase z by $-$ (coefficient of entering variable in row 0 of current bfs). Recall that (coefficient of entering variable in row 0) < 0 and (value of entering variable in new bfs) ≥ 0. Combining these facts with (15), we can deduce the following facts:

1 If (value of entering variable in new bfs) > 0, then (z-value for new bfs) $>$ (z-value for current bfs).

2 If (value of entering variable in new bfs) $= 0$, then (z-value for new bfs) $=$ (z-value for current bfs).

For the moment, assume that the LP we are solving has the following property: In each of the LP's basic feasible solutions, all of the basic variables are positive (positive means > 0). An LP with this property is a **nondegenerate LP.**

If we are using the simplex to solve a nondegenerate LP, fact 1 in the foregoing list tells us that each iteration of the simplex will *increase* z. This implies that when the simplex is used to solve a nondegenerate LP, it is impossible to encounter the same bfs twice.

To see this, suppose that we are at a basic feasible solution (call it bfs 1) that has $z = 20$. Fact 1 shows that our next pivot will take us to a bfs (call it bfs 2) and has $z > 20$. Because no future pivot can decrease z, we can never return to a bfs having $z = 20$. Thus, we can never return to bfs 1. Now recall that every LP has only a finite number of basic feasible solutions. Because we can never repeat a bfs, this argument shows that when we use the simplex algorithm to solve a nondegenerate LP, we are guaranteed to find the optimal solution in a finite number of iterations. For example, suppose we are solving a nondegenerate LP with 10 variables and 5 constraints. Such an LP has at most

$$\binom{10}{5} = 252$$

basic feasible solutions. We will never repeat a bfs, so we know that for this problem, the simplex is guaranteed to find an optimal solution after at most 252 pivots.

However, the simplex may fail for a degenerate LP.

DEFINITION ■ An LP is **degenerate** if it has at least one bfs in which a basic variable is equal to zero. ■

The following LP is degenerate:

$$\max z = 5x_1 + 2x_2$$
$$\text{s.t.} \quad x_1 + x_2 \le 6$$
$$x_1 - x_2 \le 0 \tag{16}$$
$$x_1, x_2 \ge 0$$

What happens when we use the simplex algorithm to solve (16)? After adding slack variables s_1 and s_2 to the two constraints, we obtain the initial tableau in Table 29. In this bfs, the basic variable $s_2 = 0$. Thus, (16) is a degenerate LP. Any bfs that has at least one basic variable equal to zero (or, equivalently, at least one constraint with a zero right-hand side) is a **degenerate bfs.** Because $-5 < -2$, we enter x_1 into the basis. The winning ratio is 0. This means that after x_1 enters the basis, x_1 will equal zero in the new bfs. After doing the pivot, we obtain the tableau in Table 30. Our new bfs has the same z-value as

TABLE **29**
A Degenerate LP

z	x_1	x_2	s_1	s_2	rhs	Basic Variable	Ratio
1	-5	-2	0	0	0	$z = 0$	
0	1	1	1	0	6	$s_1 = 6$	6
0	①	-1	0	1	0	$s_2 = 0$	0*

TABLE **30**
First Tableau for (16)

z	x_1	x_2	s_1	s_2	rhs	Basic Variable	Ratio
1	0	-7	0	5	0	$z = 0$	
0	0	②	1	-1	6	$s_1 = 6$	$\frac{6}{2} = 3$*
0	1	-1	0	1	0	$x_1 = 0$	None

TABLE 31
Optimal Tableau for (16)

z	x_1	x_2	s_1	s_2	rhs	Basic Variable
1	0	0	3.5	1.5	21	$z = 21$
0	0	1	0.5	−0.5	3	$x_2 = 3$
0	1	0	0.5	0.5	3	$x_1 = 3$

the old bfs. This is consistent with fact 2. In the new bfs, all variables have exactly the same values as they had before the pivot! Thus, our new bfs is also degenerate. Continuing with the simplex, we enter x_2 in row 1. The resulting tableau is shown in Table 31. This is an optimal tableau, so the optimal solution to (16) is $z = 21$, $x_2 = 3$, $x_1 = 3$, $s_1 = s_2 = 0$.

We can now explain why the simplex may have problems in solving a degenerate LP. Suppose we are solving a degenerate LP for which the optimal z-value is $z = 30$. If we begin with a bfs that has, say, $z = 20$, we know (look at the LP we just solved) that it is possible for a pivot to leave the value of z unchanged. This means that it is possible for a sequence of pivots like the following to occur:

$$\text{Initial bfs (bfs 1): } z = 20$$
$$\text{After first pivot (bfs 2): } z = 20$$
$$\text{After second pivot (bfs 3): } z = 20$$
$$\text{After third pivot (bfs 4): } z = 20$$
$$\text{After fourth pivot (bfs 1): } z = 20$$

In this situation, we encounter the same bfs twice. This occurrence is called **cycling.** If cycling occurs, then we will loop, or cycle, forever among a set of basic feasible solutions and never get to the optimal solution ($z = 30$, in our example). Cycling can indeed occur (see Problem 3 at the end of this section). Fortunately, the simplex algorithm can be modified to ensure that cycling will never occur [see Bland (1977) or Dantzig (1963) for details].[†] For a practical example of cycling, see Kotiah and Slater (1973).

If an LP has many degenerate basic feasible solutions (or a bfs with many basic variables equal to zero), then the simplex algorithm is often very inefficient. To see why, look at the feasible region for (16) in Figure 12, the shaded triangle BCD. The extreme points of the feasible region are B, C, and D. Following the procedure outlined in Section 4.2, let's look at the correspondence between the basic feasible solutions to (16) and the extreme points of its feasible region (see Table 32). Three sets of basic variables correspond to extreme point C. It can be shown that for an LP with n decision variables to be degenerate, $n + 1$ or more of the LP's constraints (including the sign restrictions $x_i \geq 0$ as constraints) must be binding at an extreme point.

In (16), the constraints $x_1 - x_2 \leq 0$, $x_1 \geq 0$, and $x_2 \geq 0$ are all binding at point C. Each extreme point at which three or more constraints are binding will correspond to more than one set of basic variables. For example, at point C, s_1 must be one of the basic variables, but the other basic variable may be x_2, x_1, or s_2.

[†]Bland showed that cycling can be avoided by applying the following rules (assume that slack and excess variables are numbered $x_{n+1}, x_{n+2}, \ldots$):

1 Choose as the entering variable (in a max problem) the variable with a negative coefficient in row 0 that has the smallest subscript.

2 If there is a tie in the ratio test, then break the tie by choosing the winner of the ratio test so that the variable leaving the basis has the smallest subscript.

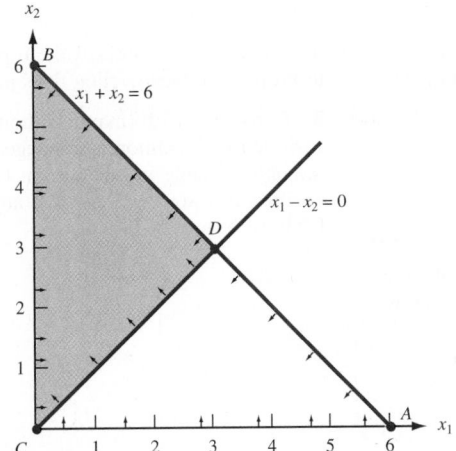

FIGURE 12
Feasible Region for the
LP (16)

TABLE 32
Three Sets of Basic Variables Correspond to Corner Point C

Basic Variables	Basic Feasible Solution	Corresponds to Extreme Point
x_1, x_2	$x_1 = x_2 = 3, s_1 = s_2 = 0$	D
x_1, s_1	$x_1 = 0, s_1 = 6, x_2 = s_2 = 0$	C
x_1, s_2	$x_1 = 6, s_2 = -6, x_2 = s_1 = 0$	Infeasible
x_2, s_1	$x_2 = 0, s_1 = 6, x_1 = s_2 = 0$	C
x_2, s_2	$x_2 = 6, s_2 = 6, s_1 = x_1 = 0$	B
s_1, s_2	$s_1 = 6, s_2 = 0, x_1 = x_2 = 0$	C

We can now discuss why the simplex algorithm often is an inefficient method for solving degenerate LPs. Suppose an LP is degenerate. Then there may be many sets (maybe hundreds) of basic variables that correspond to some nonoptimal extreme point. The simplex algorithm might encounter all these sets of basic variables before it finds that it was at a nonoptimal extreme point. This problem was illustrated (on a small scale) in solving (16): The simplex took two pivots before it found that point C was suboptimal. Fortunately, some degenerate LPs have a special structure that enables us to solve them by methods other than the simplex (see, for example, the discussion of the assignment problem in Chapter 7).

PROBLEMS

Group A

1 Even if an LP's initial tableau is nondegenerate, later tableaus may exhibit degeneracy. Degenerate tableaus often occur in the tableau following a tie in the ratio test. To illustrate this, solve the following LP:

$$\max z = 5x_1 + 3x_2$$
$$\text{s.t.} \quad 4x_1 + 2x_2 \leq 12$$
$$4x_1 + x_2 \leq 10$$
$$x_1 + x_2 \leq 4$$
$$x_1, x_2 \geq 0$$

Also graph the feasible region and show which extreme points correspond to more than one set of basic variables.

2 Find the optimal solution to the following LP:

$$\min z = -x_1 - x_2$$
$$\text{s.t.} \quad x_1 + x_2 \leq 1$$
$$-x_1 + x_2 \leq 0$$
$$x_1, x_2 \geq 0$$

3 Show that if ties in the ratio test are broken by favoring row 1 over row 2, then cycling occurs when the following LP is solved by the simplex:

$$\max z = 2x_1 + 3x_2 - x_3 - 12x_4$$
$$\text{s.t} \quad -2x_1 - 9x_2 + x_3 + 9x_4 \le 0$$
$$\frac{x_1}{3} + x_2 - \frac{x_3}{3} - 2x_4 \le 0$$
$$x_i \ge 0 \quad (i = 1, 2, 3, 4)$$

4 Show that if ties are broken in favor of lower-numbered rows, then cycling occurs when the simplex method is used to solve the following LP:

$$\max z = -3x_1 + x_2 - 6x_3$$
$$9x_1 + x_2 - 9x_3 - 2x_4 \le 0$$
$$x_1 + \frac{x_2}{3} - 2x_3 - \frac{x_4}{3} \le 0$$
$$-9x_1 - x_2 + 9x_3 + 2x_4 \le 1$$
$$x_i \ge 0 \quad (i = 1, 2, 3, 4)$$

5 Show that if Bland's Rule to prevent cycling is applied to Problem 4, then cycling does not occur.

6 Consider an LP (maximization problem) in which each basic feasible solution is nondegenerate. Suppose that x_i is the only variable in our current tableau having a negative coefficient in row 0. Show that any optimal solution to the LP must have $x_i > 0$.

4.12 The Big M Method

Recall that the simplex algorithm requires a starting bfs. In all the problems we have solved so far, we found a starting bfs by using the slack variables as our basic variables. If an LP has any $\ge$ or equality constraints, however, a starting bfs may not be readily apparent. Example 4 will illustrate that a bfs may be hard to find. When a bfs is not readily apparent, the Big M method (or the two-phase simplex method of Section 4.13) may be used to solve the problem. In this section, we discuss the **Big M method,** a version of the simplex algorithm that first finds a bfs by adding "artificial" variables to the problem. The objective function of the original LP must, of course, be modified to ensure that the artificial variables are all equal to 0 at the conclusion of the simplex algorithm. The following example illustrates the Big M method.

EXAMPLE 4   **Bevco**

Bevco manufactures an orange-flavored soft drink called Oranj by combining orange soda and orange juice. Each ounce of orange soda contains 0.5 oz of sugar and 1 mg of vitamin C. Each ounce of orange juice contains 0.25 oz of sugar and 3 mg of vitamin C. It costs Bevco 2¢ to produce an ounce of orange soda and 3¢ to produce an ounce of orange juice. Bevco's marketing department has decided that each 10-oz bottle of Oranj must contain at least 20 mg of vitamin C and at most 4 oz of sugar. Use linear programming to determine how Bevco can meet the marketing department's requirements at minimum cost.

Solution   Let

$$x_1 = \text{number of ounces of orange soda in a bottle of Oranj}$$
$$x_2 = \text{number of ounces of orange juice in a bottle of Oranj}$$

Then the appropriate LP is

$$\min z = 2x_1 + 3x_2$$
$$\text{s.t.} \quad \tfrac{1}{2}x_1 + \tfrac{1}{4}x_2 \le 4 \qquad \text{(Sugar constraint)} \tag{17}$$
$$x_1 + 3x_2 \ge 20 \qquad \text{(Vitamin C constraint)}$$

$$x_1 + x_2 = 10 \qquad \text{(10 oz in bottle of Oranj)}$$
$$x_1, x_2 \geq 0$$

(The solution will be continued later in this section.)

To put (17) into standard form, we add a slack variable s_1 to the sugar constraint and subtract an excess variable e_2 from the vitamin C constraint. After writing the objective function as $z - 2x_1 - 3x_2 = 0$, we obtain the following standard form:

$$
\begin{array}{lll}
\text{Row 0:} & z - 2x_1 - 3x_2 & = 0 \\
\text{Row 1:} & \tfrac{1}{2}x_1 + \tfrac{1}{4}x_2 + s_1 & = 4 \\
\text{Row 2:} & x_1 + 3x_2 \quad - e_2 & = 20 \\
\text{Row 3:} & x_1 + x_2 & = 10
\end{array}
\tag{18}
$$

All variables nonnegative

In searching for a bfs, we see that $s_1 = 4$ could be used as a basic (and feasible) variable for row 1. If we multiply row 2 by -1, we see that $e_2 = -20$ could be used as a basic variable for row 2. Unfortunately, $e_2 = -20$ violates the sign restriction $e_2 \geq 0$. Finally, in row 3 there is no readily apparent basic variable. Thus, in order to use the simplex to solve (17), rows 2 and 3 each need a basic (and feasible) variable. To remedy this problem, we simply "invent" a basic feasible variable for each constraint that needs one. Because these variables are created by us and are not real variables, we call them **artificial variables.** If an artificial variable is added to row i, we label it a_i. In the current problem, we need to add an artificial variable a_2 to row 2 and an artificial variable a_3 to row 3. The resulting set of equations is

$$
\begin{array}{ll}
z - 2x_1 - 3x_2 & = 0 \\
\tfrac{1}{2}x_1 + \tfrac{1}{4}x_2 + s_1 & = 4 \\
x_1 + 3x_2 \quad - e_2 + a_2 & = 20 \\
x_1 + x_2 \qquad\qquad + a_3 & = 10
\end{array}
\tag{18}
$$

We now have a bfs: $z = 0$, $s_1 = 4$, $a_2 = 20$, $a_3 = 10$. Unfortunately, there is no guarantee that the optimal solution to (18) will be the same as the optimal solution to (17). In solving (18), we might obtain an optimal solution in which one or more artificial variables are positive. Such a solution may not be feasible in the original problem (17). For example, in solving (18), the optimal solution may easily be shown to be $z = 0$, $s_1 = 4$, $a_2 = 20$, $a_3 = 10$, $x_1 = x_2 = 0$. This "solution" contains no vitamin C and puts 0 ounces of soda in a bottle, so it cannot possibly solve our original problem! If the optimal solution to (18) is to solve (17), then we must make sure that the optimal solution to (18) sets all artificial variables equal to zero. In a min problem, we can ensure that all the artificial variables will be zero by adding a term Ma_i to the objective function for each artificial variable a_i. (In a max problem, add a term $-Ma_i$ to the objective function.) Here M represents a "very large" positive number. Thus, in (18), we would change our objective function to

$$\min z = 2x_1 + 3x_2 + Ma_2 + Ma_3$$

Then row 0 will change to

$$z - 2x_1 - 3x_2 - Ma_2 - Ma_3 = 0$$

Modifying the objective function in this way makes it extremely costly for an artificial variable to be positive. With this modified objective function, it seems reasonable that the optimal solution to (18) will have $a_2 = a_3 = 0$. In this case, the optimal solution to (18) will solve the original problem (17). It sometimes happens, however, that in solving the analog of (18), some of the artificial variables may assume positive values in the optimal solution. If this occurs, the original problem has no feasible solution.

For obvious reasons, the method we have just outlined is often called the Big M method. We now give a formal description of the Big M method.

Description of Big M Method

Step 1 Modify the constraints so that the right-hand side of each constraint is non-negative. This requires that each constraint with a negative right-hand side be multiplied through by -1. Remember that if you multiply an inequality by any negative number, the direction of the inequality is reversed. For example, our method would transform the inequality $x_1 + x_2 \geq -1$ into $-x_1 - x_2 \leq 1$. It would also transform $x_1 - x_2 \leq -2$ into $-x_1 + x_2 \geq 2$.

Step 1′ Identify each constraint that is now (after step 1) an = or $\geq$ constraint. In step 3, we will add an artificial variable to each of these constraints.

Step 2 Convert each inequality constraint to standard form. This means that if constraint i is a $\leq$ constraint, we add a slack variable s_i, and if constraint i is a $\geq$ constraint, we subtract an excess variable e_i.

Step 3 If (after step 1 has been completed) constraint i is a $\geq$ or = constraint, add an artificial variable a_i. Also add the sign restriction $a_i \geq 0$.

Step 4 Let M denote a very large positive number. If the LP is a min problem, add (for each artificial variable) Ma_i to the objective function. If the LP is a max problem, add (for each artificial variable) $-Ma_i$ to the objective function.

Step 5 Because each artificial variable will be in the starting basis, all artificial variables must be eliminated from row 0 before beginning the simplex. This ensures that we begin with a canonical form. In choosing the entering variable, remember that M is a very large positive number. For example, $4M - 2$ is more positive than $3M + 900$, and $-6M - 5$ is more negative than $-5M - 40$. Now solve the transformed problem by the simplex. If all artificial variables are equal to zero in the optimal solution, then we have found the optimal solution to the original problem. If any artificial variables are positive in the optimal solution, then the original problem is infeasible.[†]

When an artificial variable leaves the basis, its column may be dropped from future tableaus because the purpose of an artificial variable is only to get a starting basic feasible solution. Once an artificial variable leaves the basis, we no longer need it. Despite this fact, we often maintain the artificial variables in all tableaus. The reason for this will become apparent in Section 6.7.

Solution **Example 4 (Continued)**

Step 1 Because none of the constraints has a negative right-hand side, we don't have to multiply any constraint through by -1.

[†]We have ignored the possibility that when the LP (with the artificial variables) is solved, the final tableau may indicate that the LP is unbounded. If the final tableau indicates the LP is unbounded and all artificial variables in this tableau equal zero, then the original LP is unbounded. If the final tableau indicates that the LP is unbounded and at least one artificial variable is positive, then the original LP is infeasible. See Bazaraa and Jarvis (1990) for details.

Step 1' Constraints 2 and 3 will require artificial variables.

Step 2 Add a slack variable s_1 to row 1 and subtract an excess variable e_2 from row 2. The result is

$$\min z = 2x_1 + 3x_2$$
$$\text{Row 1:} \quad \tfrac{1}{2}x_1 + \tfrac{1}{4}x_2 + s_1 \qquad\qquad = 4$$
$$\text{Row 2:} \quad x_1 + 3x_2 \qquad - e_2 \qquad = 20$$
$$\text{Row 3:} \quad x_1 + x_2 \qquad\qquad = 10$$

Step 3 Add an artificial variable a_2 to row 2 and an artificial variable a_3 to row 3. The result is

$$\min z = 2x_1 + 3x_2$$
$$\text{Row 1:} \quad \tfrac{1}{2}x_1 + \tfrac{1}{4}x_2 + s_1 \qquad\qquad\qquad = 4$$
$$\text{Row 2:} \quad x_1 + 3x_2 \qquad - e_2 + a_2 \qquad = 20$$
$$\text{Row 3:} \quad x_1 + x_2 \qquad\qquad + a_3 = 10$$

From this tableau, we see that our initial bfs will be $s_1 = 4$, $a_2 = 20$, and $a_3 = 10$.

Step 4 Because we are solving a min problem, we add $Ma_2 + Ma_3$ to the objective function (if we were solving a max problem, we would add $-Ma_2 - Ma_3$). This makes a_2 and a_3 very unattractive, and the act of minimizing z will cause a_2 and a_3 to be zero. The objective function is now

$$\min z = 2x_1 + 3x_2 + Ma_2 + Ma_3$$

Step 5 Row 0 is now

$$z - 2x_1 - 3x_2 - Ma_2 - Ma_3 = 0$$

Because a_2 and a_3 are in our starting bfs (that's why we introduced them), they must be eliminated from row 0. To eliminate a_2 and a_3 from row 0, simply replace row 0 by row $0 + M(\text{row 2}) + M(\text{row 3})$. This yields

$$\text{Row 0:} \quad z - \qquad 2x_1 - \qquad 3x_2 \qquad - Ma_2 - Ma_3 = 0$$
$$M(\text{row 2}): \qquad\qquad Mx_1 + \qquad 3Mx_2 - Me_2 + Ma_2 \qquad = 20M$$
$$M(\text{row 3}): \qquad\qquad Mx_1 + \qquad Mx_2 \qquad\qquad + Ma_3 = 10M$$
$$\text{New row 0:} \quad z + (2M - 2)x_1 + (4M - 3)x_2 - Me_2 \qquad\qquad = 30M$$

Combining the new row 0 with rows 1–3 yields the initial tableau shown in Table 33.

We are solving a min problem, so the variable with the most positive coefficient in row 0 should enter the basis. Because $4M - 3 > 2M - 2$, variable x_2 should enter the basis. The ratio test indicates that x_2 should enter the basis in row 2, which means the artificial variable a_2 will leave the basis. The most difficult part of doing the pivot is eliminating

TABLE 33
Initial Tableau for Bevco

z	x_1	x_2	s_1	e_2	a_2	a_3	rhs	Basic Variable	Ratio
1	$2M - 2$	$4M - 3$	0	$-M$	0	0	$30M$	$z = 30M$	
0	$\tfrac{1}{2}$	$\tfrac{1}{4}$	1	0	0	0	4	$s_1 = 4$	16
0	1	③	0	-1	1	0	20	$a_2 = 20$	$\tfrac{20}{3}$*
0	1	1	0	0	0	1	10	$a_3 = 10$	10

TABLE 34

First Tableau for Bevco

z	x_1	x_2	s_1	e_2	a_2	a_3	rhs	Basic Variable	Ratio
1	$\frac{2M-3}{3}$	0	0	$\frac{M-3}{3}$	$\frac{3-4M}{3}$	0	$\frac{60+10M}{3}$	$z = \frac{60+10M}{3}$	
0	$\frac{5}{12}$	0	1	$\frac{1}{12}$	$-\frac{1}{12}$	0	$\frac{7}{3}$	$s_1 = \frac{7}{3}$	$\frac{28}{5}$
0	$\frac{1}{3}$	1	0	$-\frac{1}{3}$	$\frac{1}{3}$	0	$\frac{20}{3}$	$x_2 = \frac{20}{3}$	20
0	$\left(\frac{2}{3}\right)$	0	0	$\frac{1}{3}$	$-\frac{1}{3}$	1	$\frac{10}{3}$	$a_3 = \frac{10}{3}$	5*

x_2 from row 0. First, replace row 2 by $\frac{1}{3}$(row 2). Thus, the new row 2 is

$$\tfrac{1}{3}x_1 + x_2 - \tfrac{1}{3}e_2 + \tfrac{1}{3}a_2 = \tfrac{20}{3}$$

We can now eliminate x_2 from row 0 by adding $-(4M - 3)$(new row 2) to row 0 or $(3 - 4M)$(new row 2) + row 0. Now

$$(3 - 4M)(\text{new row 2}) =$$

$$\frac{(3 - 4M)x_1}{3} + (3 - 4M)x_2 - \frac{(3 - 4M)e_2}{3} + \frac{(3 - 4M)a_2}{3} = \frac{20(3 - 4M)}{3}$$

Row 0: $z + (2M - 2)x_1 + (4M - 3)x_2 - Me_2 = 30M$

New row 0: $z + \dfrac{(2M - 3)x_1}{3} + \dfrac{(M - 3)e_2}{3} + \dfrac{(3 - 4M)a_2}{3} = \dfrac{60 + 10M}{3}$

After using EROs to eliminate x_2 from row 1 and row 3, we obtain the tableau in Table 34. Because $\frac{2M-3}{3} > \frac{M-3}{3}$, we next enter x_1 into the basis. The ratio test indicates that x_1 should enter the basis in the third row of the current tableau. Then a_3 will leave the basis, and our next tableau will have $a_2 = a_3 = 0$. To enter x_1 into the basis in row 3, we first replace row 3 by $\frac{3}{2}$(row 3). Thus, new row 3 will be

$$x_1 + \frac{e_2}{2} - \frac{a_2}{2} + \frac{3a_3}{2} = 5$$

To eliminate x_1 from row 0, we replace row 0 by row 0 + $(3 - 2M)$(new row 3)/3.

Row 0: $z + \dfrac{(2M - 3)x_1}{3} + \dfrac{(M - 3)e_2}{3} + \dfrac{(3 - 4M)a_2}{3} = \dfrac{60 + 10M}{3}$

$\dfrac{(3 - 2M)(\text{new row 3})}{3} : \dfrac{(3 - 2M)x_1}{3} + \dfrac{(3 - 2M)e_2}{6} + \dfrac{(2M - 3)a_2}{6}$

$$+ \frac{(3 - 2M)a_3}{2} = \frac{15 - 10M}{3}$$

New row 0: $z - \dfrac{e_2}{2} + \dfrac{(1 - 2M)a_2}{2} + \dfrac{(3 - 2M)a_3}{2} = 25$

New row 1 and new row 2 are computed as usual, yielding the tableau in Table 35. Because all variables in row 0 have nonpositive coefficients, this is an optimal tableau; all artificial variables are equal to zero in this tableau, so we have found the optimal solution to the Bevco problem: $z = 25$, $x_1 = x_2 = 5$, $s_1 = \frac{1}{4}$, $e_2 = 0$. This means that Bevco can hold the cost of producing a 10-oz bottle of Oranj to 25¢ by mixing 5 oz of orange soda and 5 oz of orange juice. Note that the a_2 column could have been dropped after a_2 left the basis (at the conclusion of the first pivot), and the a_3 column could have been dropped after a_3 left the basis (at the conclusion of the second pivot).

TABLE 35
Optimal Tableau for Bevco

z	x_1	x_2	s_1	e_2	a_2	a_3	rhs	Basic Variable
1	0	0	0	$-\frac{1}{2}$	$\frac{1-2M}{2}$	$\frac{3-2M}{2}$	25	$z = 25$
0	0	0	1	$-\frac{1}{8}$	$\frac{1}{8}$	$-\frac{5}{8}$	$\frac{1}{4}$	$s_1 = \frac{1}{4}$
0	0	1	0	$-\frac{1}{2}$	$\frac{1}{2}$	$-\frac{1}{2}$	5	$x_2 = 5$
0	1	0	0	$\frac{1}{2}$	$-\frac{1}{2}$	$\frac{3}{2}$	5	$x_1 = 5$

How to Spot an Infeasible LP

We now modify the Bevco problem by requiring that a 10-oz bottle of Oranj contain at least 36 mg of vitamin C. Even 10 oz of orange juice contain only $3(10) = 30$ mg of vitamin C, so we know that Bevco cannot possibly meet the new vitamin C requirement. This means that Bevco's LP should now have no feasible solution. Let's see how the Big M method reveals the LP's infeasibility. We have changed Bevco's LP to

$$\min z = 2x_1 + 3x_2$$
$$\text{s.t.} \quad \tfrac{1}{2}x_1 + \tfrac{1}{4}x_2 \leq 4 \qquad \text{(Sugar constraint)}$$
$$x_1 + 3x_2 \geq 36 \qquad \text{(Vitamin C constraint)} \qquad \textbf{(19)}$$
$$x_1 + x_2 = 10 \qquad \text{(10 oz constraint)}$$
$$x_1, x_2 \geq 0$$

After going through Steps 1–5 of the Big M method, we obtain the initial tableau in Table 36. Because $4M - 3 > 2M - 2$, we enter x_2 into the basis. The ratio test indicates that x_2 should be entered in row 3, causing a_3 to leave the basis. After entering x_2 into the basis, we obtain the tableau in Table 37. Because each variable has a nonpositive coefficient in row 0, this is an optimal tableau. The optimal solution indicated by this tableau is $z = 30 + 6M$, $s_1 = \frac{3}{2}$, $a_2 = 6$, $x_2 = 10$, $a_3 = e_2 = x_1 = 0$. An artificial variable (a_2) is positive in the optimal tableau, so Step 5 shows that the original LP has no feasible solution.[†] In summary, *if any artificial variable is positive in the optimal Big M tableau, then the original LP has no feasible solution.*

TABLE 36
Initial Tableau for Bevco (Infeasible)

z	x_1	x_2	s_1	e_2	a_2	a_3	rhs	Basic Variable	Ratio
1	$2M - 2$	$4M - 3$	0	$-M$	0	0	$46M$	$z = 46M$	
0	$\frac{1}{2}$	$\frac{1}{4}$	1	0	0	0	4	$s_1 = 4$	16
0	1	3	0	-1	1	0	36	$a_2 = 36$	12
0	1	①	0	0	0	1	10	$a_3 = 10$	10*

[†]To explain why (19) can have no feasible solution, suppose that it does $(\bar{x}_1, \bar{x}_2)$. Clearly, if we set $a_3 = a_2 = 0$, $(\bar{x}_1, \bar{x}_2)$ will be feasible for our modified LP (the LP with artificial variables). If we substitute $(\bar{x}_1, \bar{x}_2)$ into the modified objective function ($z = 2\bar{x}_1 + 3\bar{x}_2 + Ma_2 + Ma_3$), we obtain $z = 2\bar{x}_1 + 3\bar{x}_2$ (this follows because $a_3 = a_2 = 0$). Because M is large, this z-value is certainly less than $6M + 30$. This contradicts the fact that the best z-value for our modified objective function is $6M + 30$. This means that our original LP (19) must have no feasible solution.

TABLE 37
Tableau Indicating Infeasibility for Bevco (Infeasible)

z	x_1	s_2	s_1	e_2	a_2	a_3	rhs	Basic Variable
1	$1 - 2M$	0	0	$-M$	0	$3 - 4M$	$30 + 6M$	$z = 6M + 30$
0	$\frac{1}{4}$	0	1	0	0	$-\frac{1}{4}$	$\frac{3}{2}$	$s_1 = \frac{3}{2}$
0	-2	0	0	-1	1	-3	6	$a_2 = 6$
0	1	1	0	0	0	1	10	$x_2 = 10$

Note that when the Big M method is used, it is difficult to determine how large M should be. Generally, M is chosen to be at least 100 times larger than the largest coefficient in the original objective function. The introduction of such large numbers into the problem can cause roundoff errors and other computational difficulties. For this reason, most computer codes solve LPs by using the two-phase simplex method (described in Section 4.13).

PROBLEMS

Group A

Use the Big M method to solve the following LPs:

1 min $z = 4x_1 + 4x_2 + x_3$
s.t. $\quad x_1 + x_2 + x_3 \leq 2$
$\quad 2x_1 + x_2 \quad\quad \leq 3$
$\quad 2x_1 + x_2 + 3x_3 \geq 3$
$\quad\quad x_1, x_2, x_3 \geq 0$

2 min $z = 2x_1 + 3x_2$
s.t. $\quad 2x_1 + x_2 \geq 4$
$\quad x_1 - x_2 \geq -1$
$\quad\quad x_1, x_2 \geq 0$

3 max $z = 3x_1 + x_2$
s.t. $\quad x_1 + x_2 \geq 3$
$\quad 2x_1 + x_2 \leq 4$
$\quad x_1 + x_2 = 3$
$\quad\quad x_1, x_2 \geq 0$

4 min $z = 3x_1$
s.t. $\quad 2x_1 + x_2 \geq 6$
$\quad 3x_1 + 2x_2 = 4$
$\quad\quad x_1, x_2 \geq 0$

5 min $z = x_1 + x_2$
s.t. $\quad 2x_1 + x_2 + x_3 = 4$
$\quad x_1 + x_2 + 2x_3 = 2$
$\quad\quad x_1, x_2, x_3 \geq 0$

6 min $z = x_1 + x_2$
s.t. $\quad x_1 + x_2 = 2$
$\quad 2x_1 + 2x_2 = 4$
$\quad\quad x_1, x_2 \geq 0$

4.13 The Two-Phase Simplex Method[†]

When a basic feasible solution is not readily available, the two-phase simplex method may be used as an alternative to the Big M method. In the two-phase simplex method, we add artificial variables to the same constraints as we did in the Big M method. Then we find a bfs to the original LP by solving the Phase I LP. In the Phase I LP, the objective function is to minimize the sum of all artificial variables. At the completion of Phase I, we reintroduce the original LP's objective function and determine the optimal solution to the original LP.

The following steps describe the two-phase simplex method. Note that steps 1–3 for the two-phase simplex are identical to steps 1–3 for the Big M method.

[†]This section covers topics that may be omitted with no loss of continuity.

Step 1 Modify the constraints so that the right-hand side of each constraint is nonnegative. This requires that each constraint with a negative right-hand side be multiplied through by -1.

Step 1′ Identify each constraint that is now (after step 1) an $=$ or $\geq$ constraint. In step 3, we will add an artificial variable to each constraint.

Step 2 Convert each inequality constraint to the standard form. If constraint i is a $\leq$ constraint, then add a slack variable s_i. If constraint i is a $\geq$ constraint, subtract an excess variable e_i.

Step 3 If (after step 1′) constraint i is a $\geq$ or $=$ constraint, add an artificial variable a_i. Also add the sign restriction $a_i \geq 0$.

Step 4 For now, ignore the original LP's objective function. Instead solve an LP whose objective function is min w' = (sum of all the artificial variables). This is called the **Phase I LP.** The act of solving the Phase I LP will force the artificial variables to be zero.

Because each $a_i \geq 0$, solving the Phase I LP will result in one of the following three cases:

Case 1 The optimal value of w' is greater than zero. In this case, the original LP has no feasible solution.

Case 2 The optimal value of w' is equal to zero, and no artificial variables are in the optimal Phase I basis. In this case, we drop all columns in the optimal Phase I tableau that correspond to the artificial variables. We now combine the original objective function with the constraints from the optimal Phase I tableau. This yields the **Phase II LP.** The optimal solution to the Phase II LP is the optimal solution to the original LP.

Case 3 The optimal value of w' is equal to zero and at least one artificial variable is in the optimal Phase I basis. In this case, we can find the optimal solution to the original LP if at the end of Phase I we drop from the optimal Phase I tableau all nonbasic artificial variables and any variable from the original problem that has a negative coefficient in row 0 of the optimal Phase I tableau.

Before solving examples illustrating Cases 1–3, we briefly discuss why $w' > 0$ corresponds to the original LP having no feasible solution and $w' = 0$ corresponds to the original LP having at least one feasible solution.

Phases I and II Feasible Solutions

Suppose the original LP is infeasible. Then the only way to obtain a feasible solution to the Phase I LP is to let at least one artificial variable be positive. In this situation, $w' > 0$ (Case 1) will result. On the other hand, if the original LP has a feasible solution, then this feasible solution (with all $a_i = 0$) is feasible in the Phase I LP and yields $w' = 0$. This means that if the original LP has a feasible solution, the optimal Phase I solution will have $w' = 0$. We now work through examples of Cases 1 and 2 of the two-phase simplex method.

EXAMPLE 5   **Two-Phase Simplex: Case 2**

First we use the two-phase simplex to solve the Bevco problem of Section 4.12. Recall that the Bevco problem was

$$\min z = 2x_1 + 3x_2$$
$$\text{s.t.} \quad \tfrac{1}{2}x_1 + \tfrac{1}{4}x_2 \leq 4$$
$$x_1 + 3x_2 \geq 20$$
$$x_1 + x_2 = 10$$
$$x_1, x_2 \geq 0$$

Solution As in the Big M method, steps 1–3 transform the constraints into

$$\tfrac{1}{2}x_1 + \tfrac{1}{4}x_2 + s_1 \qquad\qquad = 4$$
$$x_1 + 3x_2 \qquad - e_2 + a_2 \qquad = 20$$
$$x_1 + x_2 \qquad\qquad + a_3 = 10$$

Step 4 yields the following Phase I LP:

$$\min w' = a_2 + a_3$$
$$\text{s.t.} \quad \tfrac{1}{2}x_1 + \tfrac{1}{4}x_2 + s_1 \qquad\qquad = 4$$
$$x_1 + 3x_2 \qquad - e_2 + a_2 \qquad = 20$$
$$x_1 + x_2 \qquad\qquad + a_3 = 10$$

This set of equations yields a starting bfs for Phase I ($s_1 = 4$, $a_2 = 20$, $a_3 = 10$).

Note, however, that the row 0 for this tableau ($w' - a_2 - a_3 = 0$) contains the basic variables a_2 and a_3. As in the Big M method, a_2 and a_3 must be eliminated from row 0 before we can solve Phase I. To eliminate a_2 and a_3 from row 0, simply add row 2 and row 3 to row 0:

$$\begin{array}{lll}
\text{Row 0:} & w' & - a_2 - a_3 = 0 \\
+ \text{ Row 2:} & x_1 + 3x_2 - e_2 + a_2 & = 20 \\
+ \text{ Row 3:} & x_1 + x_2 & + a_3 = 10 \\
= \text{ New row 0:} & w' + 2x_1 + 4x_2 - e_2 & = 30
\end{array}$$

Combining the new row 0 with the Phase I constraints yields the initial Phase I tableau in Table 38. Because the Phase I problem is *always* a min problem (even if the original LP is a max problem), we enter x_2 into the basis. The ratio test indicates that x_2 will enter the basis in row 2, with a_2 exiting the basis. After performing the necessary EROs, we obtain the tableau in Table 39. Because $5 < 20$ and $5 < \tfrac{28}{5}$, x_1 enters the basis in row 3. Thus, a_3 will leave the basis. Because a_2 and a_3 will be nonbasic after the current pivot is completed, we already know that the next tableau will be optimal for Phase I. A glance at the tableau in Table 40 confirms this fact.

Because $w' = 0$, Phase I has been concluded. The basic feasible solution $s_1 = \tfrac{1}{4}$, $x_2 = 5$, $x_1 = 5$ has been found. No artificial variables are in the optimal Phase I basis, so the problem is an example of Case 2. We now drop the columns for the artificial variables a_2 and a_3 (we no longer need them) and reintroduce the original objective function.

$$\min z = 2x_1 + 3x_2 \qquad \text{or} \qquad z - 2x_1 - 3x_2 = 0$$

Because x_1 and x_2 are both in the optimal Phase I basis, they must be eliminated from the Phase II row 0. We add 3(row 2) + 2(row 3) of the optimal Phase I tableau to row 0.

$$\begin{array}{lll}
\text{Phase II row 0:} & z - 2x_1 - 3x_2 & = 0 \\
+ 3(\text{row 2}): & 3x_2 - \tfrac{3}{2}e_2 & = 15 \\
+ 2(\text{row 3}): & 2x_1 \qquad + e_2 & = 10 \\
= \text{ New Phase II row 0:} & z \qquad\qquad - \tfrac{1}{2}e_2 & = 25
\end{array}$$

We now begin Phase II with the following set of equations:

$$\min z - \tfrac{1}{2}e_2 = 25$$
$$s_1 - \tfrac{1}{8}e_2 = \tfrac{1}{4}$$
$$x_2 - \tfrac{1}{2}e_2 = 5$$
$$x_1 + \tfrac{1}{2}e_2 = 5$$

TABLE 38
Initial Phase I Tableau for Bevco

w'	x_1	x_2	s_1	e_2	a_2	a_3	rhs	Basic Variable	Ratio
1	2	4	0	-1	0	0	30	$w' = 30$	
0	$\frac{1}{2}$	$\frac{1}{4}$	1	0	0	0	4	$s_1 = 4$	16
0	1	③	0	-1	1	0	20	$a_2 = 20$	$\frac{20}{3}$*
0	1	1	0	0	0	1	10	$a_3 = 10$	10

TABLE 39
Phase I Tableau for Bevco after One Iteration

w'	x_1	x_2	s_1	e_2	a_2	a_3	rhs	Basic Variable	Ratio
1	$\frac{2}{3}$	0	0	$\frac{1}{3}$	$-\frac{4}{3}$	0	$\frac{10}{3}$	$w' = \frac{10}{3}$	
0	$\frac{5}{12}$	0	1	$\frac{1}{12}$	$-\frac{1}{12}$	0	$\frac{7}{3}$	$s_1 = \frac{7}{3}$	$\frac{28}{5}$
0	$\frac{1}{3}$	1	0	$-\frac{1}{3}$	$\frac{1}{3}$	0	$\frac{20}{3}$	$x_2 = \frac{20}{3}$	20
0	⟨$\frac{2}{3}$⟩	0	0	$\frac{1}{3}$	$-\frac{1}{3}$	1	$\frac{10}{3}$	$a_3 = \frac{10}{3}$	5*

TABLE 40
Optimal Phase I Tableau for Bevco

w'	x_1	x_2	s_1	e_2	a_2	a_3	rhs	Basic Variable
1	0	0	0	0	-1	-1	0	$w' = 0$
0	0	0	1	$-\frac{1}{8}$	$\frac{1}{8}$	$-\frac{5}{8}$	$\frac{1}{4}$	$s_1 = \frac{1}{4}$
0	0	1	0	$-\frac{1}{2}$	$\frac{1}{2}$	$-\frac{1}{2}$	5	$x_2 = 5$
0	1	0	0	$\frac{1}{2}$	$-\frac{1}{2}$	$\frac{3}{2}$	5	$x_1 = 5$

This is optimal. Thus, in this problem, Phase II requires no pivots to find an optimal solution. If the Phase II row 0 does not indicate an optimal tableau, then simply continue with the simplex until an optimal row 0 is obtained. In summary, our optimal Phase II tableau shows that the optimal solution to the Bevco problem is $z = 25$, $x_1 = 5$, $x_2 = 5$, $s_1 = \frac{1}{4}$, and $e_2 = 0$. This agrees, of course, with the optimal solution found by the Big M method in Section 4.12.

EXAMPLE 6   **Two-Phase Simplex: Case I**

To illustrate Case 1, we now modify Bevco's problem so that 36 mg of vitamin C are required. From Section 4.12, we know that this problem is infeasible. This means that the optimal Phase I solution should have $w' > 0$ (Case 1). To show that this is true, we begin with the original problem:

$$\min z = 2x_1 + 3x_2$$
$$\text{s.t.} \quad \tfrac{1}{2}x_1 + \tfrac{1}{4}x_2 \leq 4$$
$$x_1 + 3x_2 \geq 36$$
$$x_1 + x_2 = 10$$
$$x_1, x_2 \geq 0$$

TABLE 41

Initial Phase I Tableau for Bevco (Infeasible)

w'	x_1	x_2	s_1	e_2	a_2	a_3	rhs	Basic Variable	Ratio
1	2	4	0	-1	0	0	46	$w' = 46$	
0	$\frac{1}{2}$	$\frac{1}{4}$	1	0	0	0	4	$s_1 = 4$	16
0	1	3	0	-1	1	0	36	$a_2 = 36$	12
0	1	①	0	0	0	1	10	$a_3 = 0$	10*

TABLE 42

Tableau Indicating Infeasibility for Bevco (Infeasible)

w'	x_1	x_2	s_1	e_2	a_2	a_3	rhs	Basic Variable
1	-2	0	0	-1	0	-4	6	$w' = 6$
0	$\frac{1}{4}$	0	1	0	0	$-\frac{1}{4}$	$\frac{3}{2}$	$s_1 = \frac{3}{2}$
0	-2	0	0	-1	1	-3	6	$a_2 = 6$
0	1	1	0	0	0	1	10	$x_2 = 10$

Solution After completing steps 1–4 of the two-phase simplex, we obtain the following Phase I problem:

$$\min w' = a_2 + a_3$$
$$\text{s.t.} \quad \tfrac{1}{2}x_1 + \tfrac{1}{4}x_2 + s_1 \qquad\qquad\qquad = 4$$
$$x_1 + 3x_2 \qquad - e_2 + a_2 \qquad = 36$$
$$x_1 + x_2 \qquad\qquad\qquad + a_3 = 10$$

From this set of equations, we see that the initial Phase I bfs is $s_1 = 4$, $a_2 = 36$, and $a_3 = 10$. Because the basic variables a_2 and a_3 occur in the Phase I objective function, they must be eliminated from the Phase I row 0. To do this, we add rows 2 and 3 to row 0:

Row 0:	w'	$- a_2 - a_3 = 0$	
+ Row 2:	$x_1 + 3x_2 - e_2 + a_2$	$= 36$	
+ Row 3:	$x_1 + x_2$	$+ a_3 = 10$	
= New row 0:	$w' + 2x_1 + 4x_2 - e_2$	$= 46$	

With the new row 0, the initial Phase I tableau is as shown in Table 41. Because $4 > 2$, we should enter x_2 into the basis. The ratio test indicates that x_2 should enter the basis in row 3, forcing a_3 to leave the basis. The resulting tableau is shown in Table 42. No variable in row 0 has a positive coefficient, so this is an optimal Phase I tableau, and since the optimal value of w' is $6 > 0$, the original LP must have no feasible solution. This is reasonable, because if the original LP had a feasible solution, it would have been feasible in the Phase I LP (after setting $a_2 = a_3 = 0$). This feasible solution would have yielded $w' = 0$. Because the simplex could not find a Phase I solution with $w' = 0$, the original LP must have no feasible solution.

REMARKS 1 As with the Big M method, the column for any artificial variable may be dropped from future tableaus as soon as the artificial variable leaves the basis. Thus, when we solved the Bevco problem, a_2's column could have been dropped after the first Phase I pivot, and a_3's column could have been dropped after the second Phase I pivot.

2 It can be shown that (barring ties for the entering variable and in the ratio test) the Big M method and Phase I of the two-phase method make the same sequence of pivots. Despite this equivalence, most computer codes utilize the two-phase method to find a bfs. This is because M, being a large positive number, may cause roundoff errors and other computational difficulties. The two-phase method does not introduce any large numbers into the objective function, so it avoids this problem.

EXAMPLE 7   Two-Phase Simplex: Case 3

Use the two-phase simplex method to solve the following LP:

$$\max z = 40x_1 + 10x_2 + 7x_5 + 14x_6$$

$$\begin{aligned}
\text{s.t.} \quad & x_1 - x_2 + 2x_5 = 0 \\
& -2x_1 + x_2 - 2x_5 = 0 \\
& x_1 + x_3 + x_5 - x_6 = 3 \\
& 2x_2 + x_3 + x_4 + 2x_5 + x_6 = 4 \\
& \text{All } x_i \geq 0
\end{aligned}$$

Solution   We may use x_4 as a basic variable for the fourth constraint and use artificial variables a_1, a_2, and a_3 as basic variables for the first three constraints. Our Phase I objective is to minimize $w = a_1 + a_2 + a_3$. After adding the first three constraints to $w - a_1 - a_2 - a_3 = 0$, we obtain the initial Phase I tableau shown in Table 43.

Even though x_5 has the most positive coefficient in row 0, we choose to enter x_3 into the basis (as a basic variable in row 3). We see that this will immediately yield $w = 0$. Our final Phase I tableau is shown in Table 44.

Because $w = 0$, we now have an optimal Phase I tableau. Two artificial variables remain in the basis (a_1 and a_2) at a zero level. We may now drop the artificial variable a_3 from our first Phase II tableau. The only original variable with a negative coefficient in the optimal Phase I tableau is x_1, so we may drop x_1 from all future tableaus. This is because from the optimal Phase I tableau we find $w = x_1$. This implies that x_1 can never become positive during Phase II, so we may drop x_1 from all future tableaus. Because $z - 40x_1 - 10x_2 - 7x_5 - 14x_6 = 0$ contains no basic variables, our initial tableau for Phase II is as in Table 45.

TABLE 43

W	x_1	x_2	x_3	x_4	x_5	x_6	a_1	a_2	a_3	rhs	Basic Variable
1	0	0	1	0	1	−1	0	0	0	3	$w = 3$
0	1	−1	0	0	2	0	1	0	0	0	$a_1 = 0$
0	−2	1	0	0	−2	0	0	1	0	0	$a_2 = 0$
0	1	0	①	0	1	−1	0	0	1	3	$a_3 = 3$
0	0	2	1	1	2	1	0	0	0	4	$x_4 = 4$

TABLE 44

W	x_1	x_2	x_3	x_4	x_5	x_6	a_1	a_2	a_3	rhs	Basic Variable
1	−1	0	0	0	0	0	0	0	−1	0	$w = 0$
0	1	−1	0	0	2	0	1	0	0	0	$a_1 = 0$
0	−2	1	0	0	−2	0	0	1	0	0	$a_2 = 0$
0	1	0	1	0	1	−1	0	0	1	3	$x_3 = 3$
0	−1	2	0	1	1	2	0	0	−1	1	$x_4 = 1$

TABLE 45

z	x_2	x_3	x_4	x_5	x_6	a_1	a_2	rhs	Basic Variables
1	-10	0	0	-7	-14	0	0	0	$z = 0$
0	-1	0	0	2	0	1	0	0	$a_1 = 0$
0	1	0	0	-2	0	0	1	0	$a_2 = 0$
0	0	1	0	1	-1	0	0	3	$x_3 = 3$
0	2	0	1	1	②	0	0	1	$x_4 = 1$

TABLE 46

z	x_2	x_3	x_4	x_5	x_6	a_1	a_2	rhs	Basic Variables
1	4	0	7	0	0	0	0	7	$z = 7$
0	0	0	0	2	0	1	0	0	$a_1 = 0$
0	1	0	0	0	0	0	1	0	$a_2 = 0$
0	1	1	$\frac{1}{2}$	$\frac{3}{2}$	0	0	0	$\frac{7}{2}$	$x_3 = \frac{7}{2}$
0	0	0	$\frac{1}{2}$	$\frac{1}{2}$	1	0	0	$\frac{1}{2}$	$x_4 = \frac{1}{2}$

We now enter x_6 into the basis in row 4 and obtain the optimal tableau shown in Table 46.

The optimal solution to our original LP is $z = 7$, $x_3 = 7/2$, $x_4 = \frac{1}{2}$, $x_2 = x_5 = x_6 = x_3 = 0$.

PROBLEMS

Group A

1 Use the two-phase simplex method to solve the Section 4.12 problems.

2 Explain why the Phase I LP will usually have alternative optimal solutions.

4.14 Unrestricted-in-Sign Variables

In solving LPs with the simplex algorithm, we used the ratio test to determine the row in which the entering variable became a basic variable. Recall that the ratio test depended on the fact that any feasible point required all variables to be nonnegative. Thus, if some variables are allowed to be unrestricted in sign (urs), the ratio test and therefore the simplex algorithm are no longer valid. In this section, we show how an LP with unrestricted-in-sign variables can be transformed into an LP in which all variables are required to be nonnegative.

For each urs variable x_i, we begin by defining two new variables x_i' and x_i''. Then substitute $x_i' - x_i''$ for x_i in each constraint and in the objective function. Also add the sign restrictions $x_i' \geq 0$ and $x_i'' \geq 0$. The effect of this substitution is to express x_i as the difference of the two nonnegative variables x_i' and x_i''. Because all variables are now required to be nonnegative, we can proceed with the simplex. As we will soon see, no basic feasible solution can have both $x_i' > 0$ and $x_i'' > 0$. This means that for any basic feasible solution, each urs variable x_i must fall into one of the following three cases:

Case 1 $x_i' > 0$ and $x_i'' = 0$. This case occurs if a bfs has $x_i > 0$. In this case, $x_i = x_i' - x_i'' = x_i'$. Thus, $x_i = x_i'$. For example, if $x_i = 3$ in a bfs, this will be indicated by $x_i' = 3$ and $x_i'' = 0$.

Case 2 $x_i' = 0$ and $x_i'' > 0$. This case occurs if $x_i < 0$. Because $x_i = x_i' - x_i''$, we obtain $x_i = -x_i''$. For example, if $x_i = -5$ in a bfs, we will have $x_i' = 0$ and $x_i'' = 5$. Then $x_i = 0 - 5 = -5$.

Case 3 $x_i' = x_i'' = 0$. In this case, $x_i = 0 - 0 = 0$.

In solving the following example, we will learn why no bfs can ever have both $x_i' > 0$ and $x_i'' > 0$.

EXAMPLE 8   Using urs Variables

A baker has 30 oz of flour and 5 packages of yeast. Baking a loaf of bread requires 5 oz of flour and 1 package of yeast. Each loaf of bread can be sold for 30¢. The baker may purchase additional flour at 4¢/oz or sell leftover flour at the same price. Formulate and solve an LP to help the baker maximize profits (revenues − costs).

Solution   Define

$$x_1 = \text{number of loaves of bread baked}$$
$$x_2 = \text{number of ounces by which flour supply is increased by cash transactions}$$

Therefore, $x_2 > 0$ means that x_2 oz of flour were purchased, and $x_2 < 0$ means that $-x_2$ ounces of flour were sold ($x_2 = 0$ means no flour was bought or sold). After noting that $x_1 \geq 0$ and x_2 is urs, the appropriate LP is

$$\max z = 30x_1 - 4x_2$$
$$\text{s.t.} \quad 5x_1 \leq 30 + x_2 \qquad \text{(Flour constraint)}$$
$$x_1 \leq 5 \qquad \text{(Yeast constraint)}$$
$$x_1 \geq 0,\ x_2 \text{ urs}$$

Because x_2 is urs, we substitute $x_2' - x_2''$ for x_2 in the objective function and constraints. This yields

$$\max z = 30x_1 - 4x_2' + 4x_2''$$
$$\text{s.t.} \quad 5x_1 \leq 30 + x_2' - x_2''$$
$$x_1 \leq 5$$
$$x_1, x_2', x_2'' \geq 0$$

After transforming the objective function to row 0 form and adding slack variables s_1 and s_2 to the two constraints, we obtain the initial tableau in Table 47. Notice that the x_2' column is simply the negative of the x_2'' column. We will see that *no matter how many pivots we make, the x_2' column will always be the negative of the x_2'' column.* (See Problem 6 for a proof of this assertion.)

Because x_1 has the most negative coefficient in row 0, x_1 enters the basis—in row 2. The resulting tableau is shown in Table 48. Again note that the x_2' column is the negative of the x_2'' column.

Because x_2'' now has the most negative coefficient in row 0, we enter x_2'' into the basis in row 1. The resulting tableau is shown in Table 49. Observe that the x_2' column is still the negative of the x_2'' column. This is an optimal tableau, so the optimal solution to the baker's problem is $z = 170$, $x_1 = 5$, $x_2'' = 5$, $x_2' = 0$, $s_1 = s_2 = 0$. Thus, the baker can earn a profit of 170¢ by baking 5 loaves of bread. Because $x_2 = x_2' - x_2'' = 0 - 5 = -5$,

TABLE 47
Initial Tableau for urs LP

z	x_1	x_2'	x_2''	s_1	s_2	rhs	Basic Variable	Ratio
1	−30	4	−4	0	0	0	$z = 0$	
0	5	−1	1	1	0	30	$s_1 = 30$	6
0	①	0	0	0	1	5	$s_2 = 5$	5*

TABLE 48
First Tableau for urs LP

z	x_1	x_2'	x_2''	s_1	s_2	rhs	Basic Variable	Ratio
1	0	4	−4	0	30	150	$z = 150$	
0	0	−1	①	1	−5	5	$s_1 = 5$	5*
0	1	0	0	0	1	5	$x_1 = 5$	None

TABLE 49
Optimal Tableau for urs LP

z	x_1	x_2'	x_2''	s_1	s_2	rhs	Basic Variable
1	0	0	0	4	10	170	$z = 170$
0	0	−1	1	1	−5	5	$x_2'' = 5$
0	1	0	0	0	1	5	$x_1 = 5$

the baker should sell 5 oz of flour. It is optimal for the baker to sell flour, because having 5 packages of yeast limits the baker to manufacturing at most 5 loaves of bread. These 5 loaves of bread use 5(5) = 25 oz of flour, so 30 − 25 = 5 oz of flour are left to sell.

The variables x_2' and x_2'' will never both be basic variables in the same tableau. To see why, suppose that x_2'' is basic (as it is in the optimal tableau). Then the x_2'' column will contain a single 1 and have every other entry equal to 0. The x_2' column is always the negative of the x_2'' column, so the x_2' column will contain a single −1 and have all other entries equal to 0. Such a tableau cannot also have x_2' as a basic feasible variable. The same reasoning shows that if x_i is urs, then x_i' and x_i'' cannot both be basic variables in the same tableau. This means that in any tableau, x_i', x_i'', or both must equal 0 and that one of Cases 1–3 must always occur.

The following example shows how urs variables can be used to model the production-smoothing costs discussed in the Sailco example of Section 3.10.

EXAMPLE 9   Modeling Production-Smoothing Costs

Mondo Motorcycles is determining its production schedule for the next four quarters. Demand for motorcycles will be as follows: quarter 1—40; quarter 2—70; quarter 3—50; quarter 4—20. Mondo incurs four types of costs.

1 It costs Mondo $400 to manufacture each motorcycle.

2 At the end of each quarter, a holding cost of $100 per motorcycle is incurred.

3 Increasing production from one quarter to the next incurs costs for training employees. It is estimated that a cost of $700 per motorcycle is incurred if production is increased from one quarter to the next.

4 Decreasing production from one quarter to the next incurs costs for severance pay, decreasing morale, and so forth. It is estimated that a cost of $600 per motorcycle is incurred if production is decreased from one quarter to the next.

All demands must be met on time, and a quarter's production may be used to meet demand for the current quarter. During the quarter immediately preceding quarter 1, 50 Mondos were produced. Assume that at the beginning of quarter 1, no Mondos are in inventory. Formulate an LP that minimizes Mondo's total cost during the next four quarters.

Solution To express inventory and production costs, we define for $t = 1, 2, 3, 4$,

$$p_t = \text{number of motorcycles produced during quarter } t$$
$$i_t = \text{inventory at end of quarter } t$$

To determine smoothing costs (costs 3 and 4), we define

$$x_t = \text{amount by which quarter } t \text{ production exceeds quarter } t - 1 \text{ production}$$

Because x_t is unrestricted in sign, we may write $x_t = x_t' - x_t''$, where $x_t' \geq 0$ and $x_t'' \geq 0$. We know that if $x_t \geq 0$, then $x_t = x_t'$ and $x_t'' = 0$. Also, if $x_t \leq 0$, then $x_t = -x_t''$ and $x_t' = 0$. This means that

$x_t' = $ increase in quarter t production over quarter $t - 1$ production

    ($x_t' = 0$ if period t production is less than period $t - 1$ production)

$x_t'' = $ decrease in quarter t production from quarter $t - 1$ production

    ($x_t' = 0$ if period t production is more than period $t - 1$ production)

For example, if $p_1 = 30$ and $p_2 = 50$, we have $x_2 = 50 - 30 = 20$, $x_2' = 20$, $x_2'' = 0$. Similarly, if $p_1 = 30$ and $p_2 = 15$, we have $x_2 = 15 - 30 = -15$, $x_2' = 0$, and $x_2'' = 15$. The variables x_t' and x_t'' can now be used to express the smoothing costs for quarter t. We may now express Mondo's total cost as

Total cost = production cost + inventory cost

    + smoothing cost due to increasing production

    + smoothing cost due to decreasing production

$$= 400(p_1 + p_2 + p_3 + p_4) + 100(i_1 + i_2 + i_3 + i_4)$$
$$+ 700(x_1' + x_2' + x_3' + x_4') + 600(x_1'' + x_2'' + x_3'' + x_4'')$$

To complete the formulation, we add two types of constraints. First we need inventory constraints (as in the Sailco problem of Section 3.10) that relate the inventory from the current quarter to the past quarter's inventory and the current quarter's production. For quarter t, the inventory constraint takes the form

Quarter t inventory = (quarter $t - 1$ inventory) + (quarter t production)

    − (quarter t demand)

For $t = 1, 2, 3, 4$, respectively, this yields the following four constraints:

$$i_1 = 0 + p_1 - 40 \qquad i_2 = i_1 + p_2 - 70$$
$$i_3 = i_2 + p_3 - 50 \qquad i_4 = i_3 + p_4 - 20$$

The sign restrictions $i_t \geq 0$ ($t = 1, 2, 3, 4$) ensure that each quarter's demands will be met on time.

The second type of constraint reflects the fact that p_t, p_{t-1}, x'_t, and x''_t are related. This relationship is captured by

$$(\text{quarter } t \text{ production}) - (\text{quarter } t - 1 \text{ production}) = x_t = x'_t - x''_t$$

For $t = 1, 2, 3, 4$, this relation yields the following four constraints:

$$p_1 - 50 = x'_1 - x''_1 \qquad p_2 - p_1 = x'_2 - x''_2$$
$$p_3 - p_2 = x'_3 - x''_3 \qquad p_4 - p_3 = x'_4 - x''_4$$

Combining the objective function, the four inventory constraints, the last four constraints, and the sign restrictions ($i_t, p_t, x'_t, x''_t \geq 0$ for $t = 1, 2, 3, 4$), we obtain the following LP:

$$\min z = 400p_1 + 400p_2 + 400p_3 + 400p_4 + 100i_1 + 100i_2 + 100i_3 + 100i_4$$
$$+ 700x'_1 + 700x'_2 + 700x'_3 + 700x'_4 + 600x''_1 + 600x''_2 + 600x''_3 + 600x''_4$$

$$\text{s.t.} \quad i_1 = 0 + p_1 - 40$$
$$i_2 = i_1 + p_2 - 70$$
$$i_3 = i_2 + p_3 - 50$$
$$i_4 = i_3 + p_4 - 20$$
$$p_1 - 50 = x'_1 - x''_1$$
$$p_2 - p_1 = x'_2 - x''_2$$
$$p_3 - p_2 = x'_3 - x''_3$$
$$p_4 - p_3 = x'_4 - x''_4$$
$$i_t, p_t, x'_t, x''_t \geq 0 \quad (t = 1, 2, 3, 4)$$

As in Example 7, the column for x'_t in the constraints is the negative of the x''_t column. Thus, as in Example 7, no bfs to Mondo's LP can have both $x'_t > 0$ and $x''_t > 0$. This means that x'_t actually is the increase in production during quarter t, and x''_t actually is the amount by which production decreases during quarter t.

There is another way to show that the optimal solution will not have both $x'_t > 0$ and $x''_t > 0$. Suppose, for example, that $p_2 = 70$ and $p_1 = 60$. Then the constraint

$$p_2 - p_1 = 70 - 60 = x'_2 - x''_2 \tag{20}$$

can be satisfied by many combinations of x'_2 and x''_2. For example, $x'_2 = 10$ and $x''_2 = 0$ will satisfy (20), as will $x'_2 = 20$, and $x''_2 = 10$; $x'_2 = 40$ and $x''_2 = 30$; and so on. If $p_2 - p_1 = 10$, the optimal LP solution will always choose $x'_2 = 10$ and $x''_2 = 0$ over any other possibility. To see why, look at Mondo's objective function. If $x'_2 = 10$ and $x''_2 = 0$, then x'_2 and x''_2 contribute $10(700) = \$7,000$ in smoothing costs. On the other hand, any other choice of x'_2 and x''_2 satisfying (20) will contribute more than $\$7,000$ in smoothing costs. For example, $x'_2 = 20$ and $x''_2 = 10$ contributes $20(700) + 10(600) = \$20,000$ in smoothing costs. We are minimizing total cost, so the simplex will never choose a solution where $x'_t > 0$ and $x''_t > 0$ both hold.

The optimal solution to Mondo's problem is $p_1 = 55, p_2 = 55, p_3 = 50, p_4 = 50$. This solution incurs a total cost of $\$95,000$. The optimal production schedule produces a total of 210 Mondos. Because total demand for the four quarters is only 180 Mondos, there will be an ending inventory of $210 - 180 = 30$ Mondos. Note that this is in contrast to the Sailco inventory model of Section 3.10, in which ending inventory was always 0. The optimal solution to the Mondo problem has a nonzero inventory in quarter 4, because for the quarter 4 inventory to be 0, quarter 4 production must be lower than quarter 3 production. Rather than incur the excessive smoothing costs associated with this strategy, the optimal solution opts for holding 30 Mondos in inventory at the end of quarter 4.

PROBLEMS

Group A

1 Suppose that Mondo no longer must meet demands on time. For each quarter that demand for a motorcycle is unmet, a penalty or shortage cost of $110 per motorcycle short is assessed. Thus, demand can now be backlogged. All demands must be met, however, by the end of quarter 4. Modify the formulation of the Mondo problem to allow for backlogged demand. (*Hint:* Unmet demand corresponds to $i_t \le 0$. Thus, i_t is now urs, and we must substitute $i_t = i_t' - i_t''$. Now i_t'' will be the amount of demand that is unmet at the end of quarter t.)

2 Use the simplex algorithm to solve the following LP:

$$\max z = 2x_1 + x_2$$
$$\text{s.t.} \quad 3x_1 + x_2 \le 6$$
$$x_1 + x_2 \le 4$$
$$x_1 \ge 0, \, x_2 \text{ urs}$$

Group B

3 During the next three months, Steelco faces the following demands for steel: 100 tons (month 1); 200 tons (month 2); 50 tons (month 3). During any month, a worker can produce up to 15 tons of steel. Each worker is paid $5,000 per month. Workers can be hired or fired at a cost of $3,000 per worker fired and $4,000 per worker hired (it takes 0 time to hire a worker). The cost of holding a ton of steel in inventory for one month is $100. Demand may be backlogged at a cost of $70 per ton month. That is, if 1 ton of month 1 demand is met during month 3, then a backlogging cost of $140 is incurred. At the beginning of month 1, Steelco has 8 workers. During any month, at most 2 workers can be hired. All demand must be met by the end of month 3. The raw material used to produce a ton of steel costs $300. Formulate an LP to minimize Steelco's costs.

4 Show how you could use linear programming to solve the following problem:

$$\max z = |2x_1 - 3x_2|$$
$$\text{s.t.} \quad 4x_1 + x_2 \le 4$$
$$2x_1 - x_2 \le 0.5$$
$$x_1, x_2 \ge 0$$

FIGURE 13

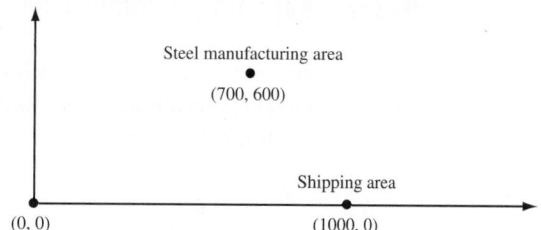

Steel manufacturing area
● (700, 600)

Shipping area

(0, 0)   (1000, 0)

5[†] Steelco's main plant currently has a steel manufacturing area and shipping area located as shown in Figure 13 (distances are in feet). The company must determine where to locate a casting facility and an assembly and storage facility to minimize the daily cost of moving material through the plant. The number of trips made each day are as shown in Table 50.

Assuming that all travel is in only an east–west or north–south direction, formulate an LP that can be used to determine where the casting and assembly and storage plants should be located in order to minimize daily transportation costs. (*Hint:* If the casting facility has coordinates (c1, c2), how should the constraint $c1 - 700 = e_1 - w_1$ be interpreted?)

6 Show that after any number of pivots the coefficient of x_i' in each row of the simplex tableau will equal the negative of the coefficient of x_i'' in the same row.

7 Clothco manufactures pants. During each of the next six months they can sell *up to* the numbers of pants given in Table 51.

Demand that is not met during a month is lost. Thus, for example, Clothco can sell up to 500 pants during month 1. A pair of pants sells for $40, requires 2 hours of labor, and uses $10 of raw material. At the beginning of month 1, Clothco has 4 workers. A worker can work at making pants up to 200 hours per month, and is paid $2,000 per month (irrespective of how many hours worked). At the beginning of each month, workers can be hired and fired. It costs

TABLE 50

From	To	Daily Number of Trips	Cost (c) Per 100 Feet Traveled
Casting	Assembly and storage	40	10
Steel manufacturing	Casting	8	10
Steel manufacturing	Assembly and storage	8	10
Shipping	Assembly and storage	2	20

TABLE 51

Month	Maximum Demand
1	500
2	600
3	300
4	400
5	300
6	800

[†]Based on Love and Yerex (1976).

$1,500 to hire and $1,000 to fire a worker. A holding cost of $5 per pair of pants is assessed against each month's ending inventory.

Determine how Clothco can maximize its profit for the next six months. Ignore the fact that during each month the number of hired and fired workers must be an integer.

4.15 Karmarkar's Method for Solving LPs

We now give a brief description of Karmarkar's method for solving LPs. For a more detailed explanation, see Section 10.6. Karmarkar's method requires that the LP be placed in the following form:

$$\min z = \mathbf{cx}$$
$$\text{s.t.} \quad K\mathbf{x} = 0$$
$$x_1 + x_2 + \cdots x_n = 1$$
$$x_i \geq 0$$

and that

1 The point $\mathbf{x}^0 = [\frac{1}{n} \quad \frac{1}{n} \quad \cdots \quad \frac{1}{n}]$ must be feasible for this LP.

2 The optimal z-value for the LP equals 0.

Surprisingly, any LP can be put in this form. Karmarkar's method uses a transformation from projective geometry to create a set of transformed variables $y_1, y_2, \ldots, y_n$. This transformation (call it f) will always transform the current point into the "center" of the feasible region in the space defined by the transformed variables. If the transformation takes the point $\mathbf{x}$ into the point $\mathbf{y}$, we write $f(\mathbf{x}) = \mathbf{y}$. The algorithm begins in the transformed space by moving from $f(\mathbf{x}^0)$ in the transformed space in a "good" direction (a direction that tends to improve z and maintains feasibility). This yields a point $\mathbf{y}^1$ in the transformed space, which is close to the boundary of the feasible region. Our new point is $\mathbf{x}^1$, satisfying $f(\mathbf{x}^1) = \mathbf{y}^1$. The procedure is repeated (this time $\mathbf{x}^1$ replaces $\mathbf{x}^0$) until the z-value for $\mathbf{x}^k$ is sufficiently close to 0.

If our current point is $\mathbf{x}^k$, then the transformation will have the property that $f(\mathbf{x}^k) = [\frac{1}{n} \quad \frac{1}{n} \quad \cdots \quad \frac{1}{n}]$. Thus, in transformed space, we are always moving away from the "center" of the feasible region.

Karmarkar's method has been shown to be a **polynomial time algorithm.** This implies that if an LP of size n is solved by Karmarkar's method, then there exist positive numbers a and b such that for any n, an LP of size n can be solved in a time of at most an^b.[†]

In contrast to Karmarkar's method, the simplex algorithm is an **exponential time algorithm** for solving LPs. If an LP of size n is solved by the simplex, then there exists a positive number c such that for any n, the simplex algorithm will find the optimal solution in a time of at most $c2^n$. For large enough n (for positive a, b, and c), $c2^n > an^b$. This means that, in theory, a polynomial time algorithm is superior to an exponential time algorithm. Preliminary testing of Karmarkar's method (by Karmarkar) has shown that for large LPs arising in actual application, this method may be up to 50 times as fast as the simplex algorithm. Hopefully, Karmarkar's method will enable researchers to solve many large LPs that currently require a prohibitively large amount of computer time when solved by the simplex. If Karmarkar's method lives up to its early promise, the ability to formulate LP models will be even more important in the near future than it is today.

Karmarkar's method has recently been utilized by the Military Airlift Command to determine how often to fly various routes, and which aircraft to use. The resulting LP con-

---

[†]The size of an LP may be defined as the number of symbols needed to represent the LP in binary notation.

tained 150,000 variables and 12,000 constraints and was solved in one hour of computer time using Karmarkar's method. Using the simplex method, an LP with similar structure containing 36,000 variables and 10,000 constraints required four hours of computer time. Delta Airlines has used Karmarkar's method to develop monthly schedules for 7,000 pilots and more than 400 aircraft. When the project is completed, Delta expects to have saved millions of dollars.

4.16 Multiattribute Decision Making in the Absence of Uncertainty: Goal Programming

In some situations, a decision maker may face multiple objectives, and there may be no point in an LP's feasible region satisfying all objectives. In such a case, how can the decision maker choose a satisfactory decision? **Goal programming** is one technique that can be used in such situations. The following example illustrates the main ideas of goal programming.

EXAMPLE 10   **Burnit Goal Programming**

The Leon Burnit Advertising Agency is trying to determine a TV advertising schedule for Priceler Auto Company. Priceler has three goals:

Goal 1   Its ads should be seen by at least 40 million high-income men (HIM).

Goal 2   Its ads should be seen by at least 60 million low-income people (LIP).

Goal 3   Its ads should be seen by at least 35 million high-income women (HIW).

Leon Burnit can purchase two types of ads: those shown during football games and those shown during soap operas. At most, $600,000 can be spent on ads. The advertising costs and potential audiences of a one-minute ad of each type are shown in Table 52. Leon Burnit must determine how many football ads and soap opera ads to purchase for Priceler.

Solution   Let

$$x_1 = \text{number of minutes of ads shown during football games}$$
$$x_2 = \text{number of minutes of ads shown during soap operas}$$

Then any feasible solution to the following LP would meet Priceler's goals:

$$
\begin{aligned}
\text{min (or max) } z = 0x_1 + 0x_2 \quad &\text{(or any other objective function)}\\
\text{s.t.} \quad 7x_1 + 3x_2 \geq 40 \quad &\text{(HIM constraint)}\\
10x_1 + 5x_2 \geq 60 \quad &\text{(LIP constraint)}\\
5x_1 + 4x_2 \geq 35 \quad &\text{(HIW constraint)}\\
100x_1 + 60x_2 \leq 600 \quad &\text{(Budget constraint)}\\
x_1, x_2 \geq 0
\end{aligned}
$$

(21)

From Figure 14, we find that no point that satisfies the budget constraint meets all three of Priceler's goals. Thus, (21) has no feasible solution. It is impossible to meet all of Priceler's goals, so Burnit might ask Priceler to identify, for each goal, a cost (per-unit short of meeting each goal) that is incurred for failing to meet the goal. Suppose Priceler determines that

TABLE 52
Cost and Number of Viewers of Ads for Priceler

Ad	Millions of Viewers			Cost ($)
	HIM	LIP	HIW	
Football	7	10	5	100,000
Soap opera	3	5	4	60,000

Each million exposures by which Priceler falls short of the HIM goal costs Priceler a $200,000 penalty because of lost sales.

Each million exposures by which Priceler falls short of the LIP goal costs Priceler a $100,000 penalty because of lost sales.

Each million exposures by which Priceler falls short of the HIW goal costs Priceler a $50,000 penalty because of lost sales.

Burnit can now formulate an LP that minimizes the cost incurred in deviating from Priceler's three goals. The trick is to transform each inequality constraint in (21) that represents one of Priceler's goals into an equality constraint. Because we don't know whether the cost-minimizing solution will undersatisfy or oversatisfy a given goal, we need to define the following variables:

$$s_i^+ = \text{amount by which we numerically exceed the } i\text{th goal}$$
$$s_i^- = \text{amount by which we are numerically under the } i\text{th goal}$$

The s_i^+ and s_i^- are referred to as **deviational variables.** For the Priceler problem, we assume that each s_i^+ and s_i^- is measured in millions of exposures. Using the deviational variables, we can rewrite the first three constraints in (21) as

$$7x_1 + 3x_2 + s_1^- - s_1^+ = 40 \qquad \text{(HIM constraint)}$$
$$10x_1 + 5x_2 + s_2^- - s_2^+ = 60 \qquad \text{(LIP constraint)}$$
$$5x_1 + 4x_2 + s_3^- - s_3^+ = 35 \qquad \text{(HIW constraint)}$$

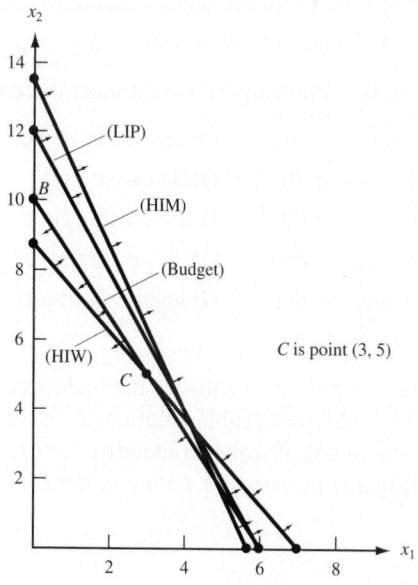

FIGURE 14
Constraints for Priceler

C is point (3, 5)

192

CHAPTER **4** The Simplex Algorithm and Goal Programming

For example, suppose that $x_1 = 5$ and $x_2 = 2$. This advertising schedule yields $7(5) + 3(2) = 41$ million HIM exposures. This exceeds the HIM goal by $41 - 40 = 1$ million exposures, so $s_1^- = 0$ and $s_1^+ = 1$. Also, this schedule yields $10(5) + 5(2) = 60$ million LIP exposures. This exactly meets the LIP requirement, and $s_2^- = s_2^+ = 0$. Finally, this schedule yields $5(5) + 4(2) = 33$ million HIW exposures. We are numerically under the HIW goal by $35 - 33 = 2$ million exposures, so $s_3^- = 2$ and $s_3^+ = 0$.

Suppose Priceler wants to minimize the total penalty from the lost sales. In terms of the deviational variables, the total penalty from lost sales (in thousands of dollars) caused by deviation from the three goals is $200s_1^- + 100s_2^- + 50s_3^-$. The objective function coefficient for the variable associated with goal i is called the **weight** for goal i. The most important goal has the largest weight, and so on. Thus, in the Priceler example, goal 1 (HIM) is most important, goal 2 (LIP) is second most important, and goal 3 (HIW) is least important.

Burnit can minimize the penalty from Priceler's lost sales by solving the following LP:

$$\min z = 200s_1^- + 100s_2^- + 50s_3^-$$

$$
\begin{aligned}
\text{s.t.} \quad & 7x_1 + 3x_2 + s_1^- - s_1^+ = 40 && \text{(HIM constraint)} \\
& 10x_1 + 5x_2 + s_2^- - s_2^+ = 60 && \text{(LIP constraint)} \\
& 5x_1 + 4x_2 + s_3^- - s_3^+ = 35 && \text{(HIW constraint)} \\
& 100x_1 + 60x_2 \le 600 && \text{(Budget constraint)}
\end{aligned}
$$

$$\text{All variables nonnegative}$$

(22)

The optimal solution to this LP is $z = 250$, $x_1 = 6$, $x_2 = 0$, $s_1^+ = 2$, $s_2^+ = 0$, $s_3^+ = 0$, $s_1^- = 0$, $s_2^- = 0$, $s_3^- = 5$. This meets goal 1 and goal 2 (the goals with the highest costs, or weights, for each unit of deviation from the goal) but fails to meet the least important goal (goal 3).

REMARKS If failure to meet goal i occurs when the attained value of an attribute is numerically smaller than the desired value of goal i, then a term involving s_i^- will appear in the objective function. If failure to meet goal i occurs when the attained value of an attribute is numerically larger than the desired value of goal i, then a term involving s_i^+ will appear in the objective function. Also, if we want to meet a goal exactly and a penalty is assessed for going both over and under a goal, then terms involving both s_i^- and s_i^+ will occur in the objective function.

Suppose we modify the Priceler example by deciding that the budget restriction of $600,000 is a goal. If we decide that a $1 penalty is assessed for each dollar by which this goal is unmet, then the appropriate goal programming formulation would be

$$\min z = 200s_1^- + 100s_2^- + 50s_3^- + s_4^+$$

$$
\begin{aligned}
\text{s.t.} \quad & 7x_1 + 3x_2 + s_1^- - s_1^+ = 40 && \text{(HIM constraint)} \\
& 10x_1 + 5x_2 + s_2^- - s_2^+ = 60 && \text{(LIP constraint)} \\
& 5x_1 + 4x_2 + s_3^- - s_3^+ = 35 && \text{(HIW constraint)} \\
& 100x_1 + 60x_2 + s_4^- - s_4^+ = 600 && \text{(Budget constraint)}
\end{aligned}
$$

$$\text{All variables nonnegative}$$

In contrast to our previous optimal solution, the optimal solution to this LP is $z = 33\frac{1}{3}$, $x_1 = 4\frac{1}{3}$, $x_2 = 3\frac{1}{3}$, $s_1^+ = \frac{1}{3}$, $s_2^+ = 0$, $s_3^+ = 0$, $s_4^+ = 33\frac{1}{3}$, $s_1^- = 0$, $s_2^- = 0$, $s_3^- = 0$, $s_4^- = 0$. Thus, when we define the budget restriction to be a goal, the optimal solution is to meet all three advertising goals by going $33\frac{1}{3}$ thousand over budget.

Preemptive Goal Programming

In our LP formulation of the Burnit example, we assumed that Priceler could exactly determine the relative importance of the three goals. For instance, Priceler determined that the HIM goal was $\frac{200}{100} = 2$ times as important as the LIP goal, and the LIP goal was $\frac{100}{50} = 2$ times as important as the HIW goal. In many situations, however, a decision maker may not be able to determine precisely the relative importance of the goals. When this is the case, *preemptive goal programming* may prove to be a useful tool. To apply preemptive goal programming, the decision maker must rank his or her goals from the most important (goal 1) to least important (goal n). The objective function coefficient for the variable representing goal i will be P_i. We assume that

$$P_1 >>> P_2 >>> P_3 >>> \cdots >>> P_n$$

Thus, the weight for goal 1 is much larger than the weight for goal 2, the weight for goal 2 is much larger than the weight for goal 3, and so on. This definition of the $P_1, P_2, \ldots, P_n$ ensures that the decision maker first tries to satisfy the most important (goal 1) goal. Then, among all points that satisfy goal 1, the decision maker tries to come as close as possible to satisfying goal 2, and so forth. We continue in this fashion until the only way we can come closer to satisfying a goal is to increase the deviation from a higher-priority goal.

For the Priceler problem, the preemptive goal programming formulation is obtained from (22) by replacing (22)'s objective function by $P_1 s_1^- + P_2 s_2^- + P_3 s_3^-$. Thus, the preemptive goal programming formulation of the Priceler problem is

$$\min z = P_1 s_1^- + P_2 s_2^- + P_3 s_3^-$$

$$\begin{aligned}
\text{s.t.} \quad & 7x_1 + 3x_2 + s_1^- - s_1^+ = 40 && \text{(HIM constraint)} \\
& 10x_1 + 5x_2 + s_2^- - s_2^+ = 60 && \text{(LIP constraint)} \\
& 5x_1 + 4x_2 + s_3^- - s_3^+ = 35 && \text{(HIW constraint)} \\
& 100x_1 + 60x_2 \leq 600 && \text{(Budget constraint)}
\end{aligned}$$

(23)

All variables nonnegative

Assume the decision maker has n goals. To apply preemptive goal programming, we must separate the objective function into n components, where component i consists of the objective function term involving goal i. We define

$$z_i = \text{objective function term involving goal } i$$

For the Priceler example, $z_1 = P_1 s_1^-$, $z_2 = P_2 s_2^-$, and $z_3 = P_3 s_3^-$. Preemptive goal programming problems can be solved by an extension of the simplex known as the **goal programming simplex.** To prepare a problem for solution by the goal programming simplex, we must compute n row 0's, with the ith row 0 corresponding to goal i. Thus, for the Priceler problem, we have

$$\text{Row 0 (goal 1): } z_1 - P_1 s_1^- = 0$$
$$\text{Row 0 (goal 2): } z_2 - P_2 s_2^- = 0$$
$$\text{Row 0 (goal 3): } z_3 - P_3 s_3^- = 0$$

From (23), we find that $BV = \{s_1^-, s_2^-, s_3^-, s_4\}$ ($s_4 = $ slack variable for fourth constraint) is a starting basic feasible solution that could be used to solve (23) via the simplex algorithm (or goal programming simplex algorithm). As with the regular simplex, we must first eliminate all variables in the starting basis from each row 0. Adding P_1 (HIM constraint) to row 0 (goal 1) yields

$$\text{Row 0 (goal 1): } z_1 + 7P_1 x_1 + 3P_1 x_2 - P_1 s_1^+ = 40P_1 \qquad \text{(HIM)}$$

Adding P_2 (LIP constraint) to row 0 (goal 2) yields

$$\text{Row 0 (goal 2): } z_2 + 10P_2x_1 + 5P_2x_2 - P_2s_2^+ = 60P_2 \qquad \text{(LIP)}$$

Adding P_3 (HIW constraint) to row 0 (goal 3) yields

$$\text{Row 0 (goal 3): } z_3 + 5P_3x_1 + 4P_3x_2 - P_3s_3^+ = 35P_3 \qquad \text{(HIW)}$$

The Priceler problem can now be solved by the goal programming simplex.

The differences between the goal programming simplex and the ordinary simplex are as follows:

1 The ordinary simplex has a single row 0, whereas the goal programming simplex requires n row 0's (one for each goal).

2 In the goal programming simplex, the following method is used to determine the entering variable: Find the highest-priority goal (goal i') that has not been met (or find the highest-priority goal i' having $z_{i'} > 0$). Find the variable with the most positive coefficient in row 0 (goal i') and enter this variable (subject to the following restriction) into the basis. This will reduce $z_{i'}$ and ensure that we come closer to meeting goal i'. *If, however, a variable has a negative coefficient in row 0 associated with a goal having a higher priority than i', then the variable cannot enter the basis.* Entering such a variable in the basis would increase the deviation from some higher-priority goal. If the variable with the most positive coefficient in row 0 (goal i') cannot be entered into the basis, then try to find another variable with a positive coefficient in row 0 (goal i'). If no variable for row 0 (goal i') can enter the basis, then there is no way to come closer to meeting goal i' without increasing the deviation from some higher-priority goal. In this case, move on to row 0 (goal $i' + 1$) in an attempt to come closer to meeting goal $i' + 1$.

3 When a pivot is performed, row 0 for each goal must be updated.

4 A tableau will yield the optimal solution if all goals are satisfied (that is, $z_1 = z_2 = \cdots = z_n = 0$), or if each variable that can enter the basis and reduce the value of $z_{i'}'$ for an unsatisfied goal i' will increase the deviation from some goal i having a higher priority than goal i'.

We now use the goal programming simplex to solve the Priceler example. In each tableau, the row 0's are listed in order of the goal's priorities (from highest priority to lowest priority). The initial tableau is Table 53. The current bfs is $s_1^- = 40$, $s_2^- = 60$, $s_3^- = 35$, $s_4 = 600$. Because $z_1 = 40P_1$, goal 1 is not satisfied. To reduce the penalty associated with not meeting goal 1, we enter the variable with the most positive coefficient (x_1) in row 0 (HIM). The ratio test indicates that x_1 should enter the basis in the HIM constraint.

After entering x_1 into the basis, we obtain Table 54. The current basic solution is $x_1 = \frac{40}{7}$, $s_2^- = \frac{20}{7}$, $s_3^- = \frac{45}{7}$, $s_4 = \frac{200}{7}$. Because $s_1^- = 0$ and $z_1 = 0$, goal 1 is now satisfied. We now try to satisfy goal 2 (while ensuring that the higher-priority goal 1 is still satisfied). The variable with the most positive coefficient in row 0 (LIP) is s_1^+. Observe that entering s_1^+ into the basis will not increase z_1 [because the coefficient of s_1^+ in row 0 (HIM) is 0]. Thus, after entering s_1^+ into the basis, goal 1 will still be satisfied. The ratio test indicates that s_1^+ could enter the basis in either the LIP or the budget constraint. We arbitrarily choose to enter s_1^+ into the basis in the budget constraint.

After pivoting s_1^+ into the basis, we obtain Table 55. Because $z_1 = z_2 = 0$, goals 1 and 2 are met. Because $z_3 = 5P_3$, however, goal 3 is unmet. The current bfs is $x_1 = 6$, $s_2^- = 0$, $s_3^- = 5$, $s_1^+ = 2$. We now try to come closer to meeting goal 3 (without violating either goal 1 or goal 2). Because x_2 is the only variable with a positive coefficient in row 0 (HIW), the only way to come closer to meeting goal 3 (HIW) is to enter x_2 into the basis. Observe, however, that x_2 has a negative coefficient in row 0 for goal 2 (LIP). Thus,

TABLE 53

Initial Tableau for Preemptive Goal Programming for Priceler

	x_1	x_2	s_1^+	s_2^+	s_3^+	s_1^-	s_2^-	s_3^-	s_4	rhs
Row 0 (HIM)	$7P_1$	$3P_1$	$-P_1$	0	0	0	0	0	0	$z_1 = 40P_1$
Row 0 (LIP)	$10P_2$	$5P_2$	0	$-P_2$	0	0	0	0	0	$z_2 = 60P_2$
Row 0 (HIW)	$5P_3$	$4P_3$	0	0	$-P_3$	0	0	0	0	$z_3 = 35P_3$
HIM	⑦	3	−1	0	0	1	0	0	0	40
LIP	10	5	0	−1	0	0	1	0	0	60
HIW	5	4	0	0	−1	0	0	1	0	35
Budget	100	60	0	0	0	0	0	0	1	600

TABLE 54

First Tableau for Preemptive Goal Programming for Priceler

	x_1	x_2	s_1^+	s_2^+	s_3^+	s_1^-	s_2^-	s_3^-	s_4	rhs
Row 0 (HIM)	0	0	0	0	0	$-P_1$	0	0	0	$z_1 = 0$
Row 0 (LIP)	0	$\frac{5P_2}{7}$	$\frac{10P_2}{7}$	$-P_2$	0	$-\frac{10P_2}{7}$	0	0	0	$z_2 = \frac{20P_2}{7}$
Row 0 (HIW)	0	$\frac{13P_3}{7}$	$\frac{5P_3}{7}$	0	$-P_3$	$-\frac{5P_3}{7}$	0	0	0	$z_3 = \frac{45P_3}{7}$
HIM	1	$\frac{3}{7}$	$-\frac{1}{7}$	0	0	$\frac{1}{7}$	0	0	0	$\frac{40}{7}$
LIP	0	$\frac{5}{7}$	$\frac{10}{7}$	−1	0	$-\frac{10}{7}$	1	0	0	$\frac{20}{7}$
HIW	0	$\frac{13}{7}$	$\frac{5}{7}$	0	−1	$-\frac{5}{7}$	0	1	0	$\frac{45}{7}$
Budget	0	$\frac{120}{7}$	⑩⓪⁄⑦ ($\frac{100}{7}$)	0	0	$-\frac{100}{7}$	0	0	1	$\frac{200}{7}$

TABLE 55

Optimal Tableau for Preemptive Goal Programming for Priceler

	x_1	x_2	s_1^+	s_2^+	s_3^+	s_1^-	s_2^-	s_3^-	s_4	rhs
Row 0 (HIM)	0	0	0	0	0	$-P_1$	0	0	0	$z_1 = 0$
Row 0 (LIP)	0	$-P_2$	0	$-P_2$	0	0	0	0	$-\frac{P_2}{10}$	$z_2 = 0$
Row 0 (HIW)	0	P_3	0	0	$-P_3$	0	0	0	$-\frac{P_3}{20}$	$z_3 = 5P_3$
HIM	1	$\frac{3}{5}$	0	0	0	0	0	0	$\frac{1}{100}$	6
LIP	0	−1	0	−1	0	0	1	0	$-\frac{1}{10}$	0
HIW	0	1	0	0	−1	0	0	1	$-\frac{1}{20}$	5
Budget	0	$\frac{6}{5}$	1	0	0	−1	0	0	$\frac{7}{100}$	2

the only way we can come closer to meeting goal 3 (HIW) is to violate a higher-priority goal, goal 2 (LIP). This is therefore an optimal tableau. The preemptive goal programming solution is to purchase 6 minutes of football ads and no soap opera ads. Goals 1 and 2 (HIM and LIP) are met, and Priceler falls 5 million exposures short of meeting goal 3 (HIW).

If the analyst has access to a computerized goal programming code, then by reordering the priorities assigned to the goals, many solutions can be generated. From among these solutions, the decision maker can choose a solution that she feels best fits her preferences. Table 56 lists the solutions found by the preemptive goal programming method for each possible set of priorities. Thus, we see that different ordering of priorities can lead to different advertising strategies.

TABLE 56
Optimal Solutions for Priceler Found by Preemptive Goal Programming

Priorities			Optimal				
			x_1 Value	x_2 Value	Deviations from		
Highest	Second Highest	Lowest			HIM	LIP	HIW
HIM	LIP	HIW	6	0	0	0	5
HIM	HIW	LIP	5	$\frac{5}{3}$	0	$\frac{5}{3}$	$\frac{10}{3}$
LIP	HIM	HIW	6	0	0	0	5
LIP	HIW	HIM	6	0	0	0	5
HIW	HIM	LIP	3	5	4	5	0
HIW	LIP	HIM	3	5	4	5	0

When a preemptive goal programming problem involves only two decision variables, the optimal solution can be found graphically. For example, suppose HIW is the highest-priority goal, LIP is the second-highest, and HIM is the lowest. From Figure 14, we find that the set of points satisfying the highest-priority goal (HIW) and the budget constraint is bounded by the triangle *ABC*. Among these points, we now try to come as close as we can to satisfying the second-highest-priority goal (LIP). Unfortunately, no point in triangle *ABC* satisfies the LIP goal. We see from the figure, however, that among all points satisfying the highest-priority goal, point *C* (*C* is where the HIW goal is exactly met and the budget constraint is binding) is the unique point that comes the closest to satisfying the LIP goal. Simultaneously solving the equations

$$5x_1 + 4x_2 = 35 \qquad \text{(HIW goal exactly met)}$$
$$100x_1 + 60x_2 = 600 \qquad \text{(Budget constraint binding)}$$

we find that point $C = (3, 5)$. Thus, for this set of priorities, the preemptive goal programming solution is to purchase 3 football game ads and 5 soap opera ads.

Goal programming is not the only approach used to analyze multiple objective decision-making problems under certainty. See Steuer (1985) and Zionts and Wallenius (1976) for other approaches to multiple objective decision making under certainty.

Using LINDO or LINGO to Solve Preemptive Goal Programming Problems

Readers who do not have access to a computer program that will solve preemptive goal programming problems may still use LINDO (or any other LP package) to solve them. To illustrate how LINDO can be used to solve a preemptive goal programming problem, let's look at the Priceler example with our original set of priorities (HIM followed by LIP followed by HIW).

We begin by asking LINDO to minimize the deviation from the highest-priority (HIM) goal by solving the following LP:

$$\min z = s_1^-$$

$$\begin{array}{lll}
\text{s.t.} & 7x_1 + 3x_2 + s_1^- - s_1^+ = 40 & \text{(HIM constraint)} \\
& 10x_1 + 5x_2 + s_2^- - s_2^+ = 60 & \text{(LIP constraint)} \\
& 5x_1 + 4x_2 + s_3^- - s_3^+ = 35 & \text{(HIW constraint)} \\
& 100x_1 + 60x_2 \le 600 & \text{(Budget constraint)}
\end{array}$$

All variables nonnegative

Goal 1 (HIM) can be met, so LINDO reports an optimal z-value of 0. We now want to come as close as possible to meeting goal 2 while ensuring that the deviation from goal 1 remains at its current level (0). Using an objective function of s_2^- (to minimize goal 2) we add the constraint $s_1^- = 0$ (to ensure that goal 1 is still met) and ask LINDO to solve

$$\min z = s_2^-$$

$$
\begin{array}{llll}
\text{s.t.} & 7x_1 + 3x_2 + s_1^- - s_1^+ = 40 & \text{(HIM constraint)} \\
& 10x_1 + 5x_2 + s_2^- - s_2^+ = 60 & \text{(LIP constraint)} \\
& 5x_1 + 4x_2 + s_3^- - s_3^+ = 35 & \text{(HIW constraint)} \\
& 100x_1 + 60x_2 \le 600 & \text{(Budget constraint)} \\
& s_1^- = 0
\end{array}
$$

All variables nonnegative

Because goals 1 and 2 can be simultaneously met, this LP will also yield an optimal z-value of 0. We now come as close as possible to meeting goal 3 (HIW) while keeping the deviations from goals 1 and 2 at their current levels. This requires LINDO to solve the following LP:

$$\min z = s_3^-$$

$$
\begin{array}{llll}
\text{s.t.} & 7x_1 + 3x_2 + s_1^- - s_1^+ = 40 & \text{(HIM constraint)} \\
& 10x_1 + 5x_2 + s_2^- - s_2^+ = 60 & \text{(LIP constraint)} \\
& 5x_1 + 4x_2 + s_3^- - s_3^+ = 35 & \text{(HIW constraint)} \\
& 100x_1 + 60x_2 + s_3^- - s_3^+ \le 600 & \text{(Budget constraint)} \\
& s_1^- = 0 \\
& s_2^- = 0
\end{array}
$$

All variables nonnegative

Of course, the LINDO (or LINGO) full-screen editor makes it easy to go from one step of the goal programming problem to the next. To go from step i to step $i + 1$, simply modify your objective function to minimize the deviation from the $i + 1$ highest-priority goal and add a constraint that ensures that the deviation from the ith highest-priority goal remains at its current level.

REMARKS **1** The optimal solution to this LP is $z = 5$, $x_1 = 6$, $x_2 = 0$, $s_1^- = 0$, $s_2^- = 0$, $s_3^- = 5$, $s_1^+ = 2$, $s_2^+ = 0$, $s_3^+ = 0$, which agrees with the solution obtained by the preemptive goal programming method. The z-value of 5 indicates that if goals 1 and 2 are met, then the best that Priceler can do is to come within 5 million exposures of meeting goal 3.
2 By the way, suppose we could only have come within two units of meeting goal 1. When solving our second LP, we would have added the constraint $s_1^- = 2$ (instead of $s_1^- = 0$).
3 The goal programming methodology of this section can be applied without any changes when some or all of the decision variables are restricted to be integer or 0–1 variables (see Problems 11, 12, and 14).
4 Using LINGO, the goal programming methodology of this section can be applied without any changes even if the objective function or some of the constraints are nonlinear.

PROBLEMS

Group A

1 Graphically determine the preemptive goal programming solution to the Priceler example for the following priorities:

    a LIP is highest-priority goal, followed by HIW and then HIM.

    b HIM is highest-priority goal, followed by LIP and then HIW.

    c HIM is highest-priority goal, followed by HIW and then LIP.

d HIW is highest-priority goal, followed by HIM and then LIP.

2 Fruit Computer Company is ready to make its annual purchase of computer chips. Fruit can purchase chips (in lots of 100) from three suppliers. Each chip is rated as being of excellent, good, or mediocre quality. During the coming year, Fruit will need 5,000 excellent chips, 3,000 good chips, and 1,000 mediocre chips. The characteristics of the chips purchased from each supplier are shown in Table 57. Each year, Fruit has budgeted $28,000 to spend on chips. If Fruit does not obtain enough chips of a given quality, then the company may special-order additional chips at $10 per excellent chip, $6 per good chip, and $4 per mediocre chip. Fruit assesses a penalty of $1 for each dollar by which the amount paid to suppliers 1–3 exceeds the annual budget. Formulate and solve an LP to help Fruit minimize the penalty associated with meeting the annual chip requirements. Also use preemptive goal programming to determine a purchasing strategy. Let the budget constraint have the highest priority, followed in order by the restrictions on excellent, good, and mediocre chips.

3 Highland Appliance must determine how many color TVs and VCRs should be stocked. It costs Highland $300 to purchase a color TV and $200 to purchase a VCR. A color TV requires 3 sq yd of storage space, and a VCR requires 1 sq yd of storage space. The sale of a color TV earns Highland a profit of $150, and the sale of a VCR earns Highland a profit of $100. Highland has set the following goals (listed in order of importance):

Goal 1 A maximum of $20,000 can be spent on purchasing color TVs and VCRs.
Goal 2 Highland should earn at least $11,000 in profits from the sale of color TVs and VCRs.
Goal 3 Color TVs and VCRs should use no more than 200 sq yd of storage space.

Formulate a preemptive goal programming model that Highland could use to determine how many color TVs and VCRs to order. How would the preemptive goal formulation be modified if Highland's goal were to have a profit of exactly $11,000?

4 A company produces two products. Relevant information for each product is shown in Table 58. The company has a goal of $48 in profits and incurs a $1 penalty for each dollar it falls short of this goal. A total of 32 hours of labor are available. A $2 penalty is incurred for each hour of overtime (labor over 32 hours) used, and a $1 penalty is incurred for each hour of available labor that is unused. Marketing considerations require that at least 10 units of product 2 be produced. For each unit (of either product) by which production falls short of demand, a penalty of $5 is assessed.

   a Formulate an LP that can be used to minimize the penalty incurred by the company.

   b Suppose the company sets (in order of importance) the following goals:

Goal 1 Avoid underutilization of labor.
Goal 2 Meet demand for product 1.
Goal 3 Meet demand for product 2.
Goal 4 Do not use any overtime.

Formulate and solve a preemptive goal programming model for this situation.

TABLE 57

Supplier	Characteristics of a Lot of 100 Chips			Price Per 100 Chips ($)
	Excellent	Good	Mediocre	
1	60	20	20	400
2	50	35	15	300
3	40	20	40	250

TABLE 58

	Product 1	Product 2
Labor required	4 hours	2 hours
Contribution to profit	$4	$2

5[†] Deancorp produces sausage by blending together beef head, pork chuck, mutton, and water. The cost per pound, fat per pound, and protein per pound for these ingredients is given in Table 59. Deancorp needs to produce 100 lb of sausage and has set the following goals, listed in order of priority:

Goal 1 Sausage should consist of at least 15% protein.
Goal 2 Sausage should consist of at most 8% fat.
Goal 3 Cost per pound of sausage should not exceed 8¢.

Formulate a preemptive goal programming model for Deancorp.

6[‡] The Touche Young accounting firm must complete three jobs during the next month. Job 1 will require 500 hours of work, job 2 will require 300 hours of work, and job 3 will require 100 hours of work. Currently, the firm consists of 5 partners, 5 senior employees, and 5 junior employees, each of whom can work up to 40 hours per month. The dollar amount (per hour) that the company can bill depends on the type of accountant who is assigned to each job, as shown in Table 60. (The X indicates that a junior employee does not have enough experience to work on job 1.) All jobs must be completed. Touche Young has also set the following goals, listed in order of priority:

Goal 1 Monthly billings should exceed $68,000.
Goal 2 At most, 1 partner should be hired.
Goal 3 At most, 3 senior employees should be hired.
Goal 4 At most, 5 junior employees should be hired.

TABLE 59

	Head	Chuck	Mutton	Moisture
Fat (per lb)	.05	.24	.11	0
Protein (per lb)	.20	.26	.08	0
Cost (in ¢)	.12	9	8	0

[†]Based on Steuer (1984).
[‡]Based on Welling (1977).

TABLE 60

	Job 1	Job 2	Job 3
Partner	160	120	110
Senior employee	120	90	70
Junior employee	X	50	40

Formulate a preemptive goal programming model for this situation.

7 There are four teachers in the Faber College Business School. Each semester, 200 students take each of the following courses: marketing, finance, production, and statistics. The "effectiveness" of each teacher in teaching each class is given in Table 61. Each teacher can teach a total of 200 students during the semester. The dean has set a goal of obtaining an average teaching effectiveness level of about 6 in each course. Deviations from this goal in any course are considered equally important. Formulate a goal programming model that can be used to determine the semester's teaching assignments.

Group B

8[†] Faber College is admitting students for the class of 2008. It has set four goals for this class, listed in order of priority:

Goal 1 Entering class should be at least 5,000 students.
Goal 2 Entering class should have an average SAT score of at least 640.
Goal 3 Entering class should consist of at least 25 percent out-of-state students.
Goal 4 At least 2,000 members of the entering class should not be nerds.

The applicants received by Faber are categorized in Table 62. Formulate a preemptive goal programming model that could determine how many applicants of each type should be admitted. Assume that all applicants who are admitted will decide to attend Faber.

9[‡] During the next four quarters, Wivco faces the following demands for globots: quarter 1—13 globots; quarter 2—14 globots; quarter 3—12 globots; quarter 4—15 globots. Globots may be produced by regular-time labor or by

TABLE 61

Teacher	Marketing	Finance	Production	Statistics
1	7	5	8	2
2	7	8	9	4
3	3	5	7	9
4	5	5	6	7

[†]Based on Lee and Moore, "University Admissions Planning" (1974).
[‡]Based on Lee and Moore, "Production Scheduling" (1974).

TABLE 62

Home State	SAT Score	No. of Nerds	No. of Non-Nerds
In-state	700	1500	400
In-state	600	1300	700
In-state	500	500	500
Out-of-state	700	350	50
Out-of-state	600	400	400
Out-of-state	500	400	600

overtime labor. Production capacity (number of globots) and production costs during the next four quarters are shown in Table 63. Wivco has set the following goals in order of importance:

Goal 1 Meet each quarter's demand on time.
Goal 2 Inventory at the end of each quarter cannot exceed 3 units.
Goal 3 Total production cost should be held below $250.

Formulate a preemptive goal programming model that could be used to determine Wivco's production schedule during the next four quarters. Assume that at the beginning of the first quarter 1 globot is in inventory.

10 Ricky's Record Store now employs five full-time employees and three part-time employees. The normal workload is 40 hours per week for full-time and 20 hours per week for part-time employees. Each full-time employee is paid $6 per hour for work up to 40 hours per week and can sell 5 records per hour. A full-time employee who works overtime is paid $10 per hour. Each part-time employee is paid $3 per hour and can sell 3 records per hour. It costs Ricky $6 to buy a record, and each record sells for $9. Ricky has weekly fixed expenses of $500. He has established the following weekly goals, listed in order of priority:

Goal 1 Sell at least 1,600 records per week.
Goal 2 Earn a profit of at least $2,200 per week.
Goal 3 Full-time employees should work at most 100 hours of overtime.
Goal 4 To increase their sense of job security, the number of hours by which each full-time employee fails to work 40 hours should be minimized.

Formulate a preemptive goal programming model that could be used to determine how many hours per week each employee should work.

TABLE 63

	Regular-Time		Overtime	
Quarter	Capacity	Cost/Unit	Capacity	Cost/Unit
1	9	$4	5	$6
2	10	$4	5	$7
3	11	$5	5	$8
4	12	$6	5	$9

11 Gotham City is trying to determine the type and location of recreational facilities to be built during the next decade. Four types of facilities are under consideration: golf courses, swimming pools, gymnasiams, and tennis courts. Six sites are under consideration. If a golf course is built, it must be built at either site 1 or site 6. Other facilities may be built at sites 2–5. The available land (in thousands of square feet) at each site is given in Table 64.

The cost of building each facility (in thousands of dollars), the annual maintenance cost (in thousands of dollars) for each facility, and the land (in thousands of square feet) required for each facility are given in Table 65.

The number of user days (in thousands) for each type of facility depends on where it is built. The dependence is given in Table 66.

a Consider the following set of priorities:

Priority 1 Limit land use at each site to the land available.
Priority 2 Construction costs should not exceed $1.2 million.
Priority 3 User days should exceed 200,000.
Priority 4 Annual maintenance costs should not exceed $200,000.

For this set of priorities, use preemptive goal programming to determine the type and location of recreation facilities in Gotham City.

b Consider the following set of priorities:

Priority 1 Limit land use at each site to the land available.
Priority 2 User days should exceed 200,000.
Priority 3 Construction costs should not exceed $1.2 million.
Priority 4 Annual maintenance costs should not exceed $200,000.

For this set of priorities, use preemptive goal programming to determine the type and location of recreation facilities in Gotham City.[†]

TABLE 64

	Site			
	2	3	4	5
Land	70	80	95	120

TABLE 65

Site	Construction Cost	Maintenance Cost	Land Required
Golf	340	80	Not relevant
Swimming	300	36	29
Gymnasium	840	50	38
Tennis courts	85	17	45

[†]Based on Taylor and Keown (1984).

TABLE 66

Site	1	2	3	4	5	6
Golf	31	X	X	X	X	27
Swimming	X	25	21	32	32	X
Gymnasium	X	37	29	28	38	X
Tennis courts	X	20	23	22	20	X

12 A small aerospace company is considering eight projects:

Project 1 Develop an automated test facility.
Project 2 Barcode all company inventory and machinery.
Project 3 Introduce a CAD/CAM system.
Project 4 Buy a new lathe and deburring system.
Project 5 Institute FMS (flexible manufacturing system).
Project 6 Install a LAN (local area network).
Project 7 Develop AIS (artificial intelligence simulation).
Project 8 Set up a TQM (total quality management) initiative.

Each project has been rated on five attributes: return on investment (ROI), cost, productivity improvement, worker requirements, and degree of technological risk. These ratings are given in Table 67.

The company has set the following five goals (listed in order of priority):

Goal 1 Achieve a return on investment of at least $3,250.
Goal 2 Limit cost to $1,300.
Goal 3 Achieve a productivity improvement of at least 6.
Goal 4 Limit manpower use to 108.
Goal 5 Limit technological risk to a total of 4.

Use preemptive goal programming to determine which projects should be undertaken.

13 The new president has just been elected and has set the following economic goals (listed from highest to lowest priority):

Goal 1 Balance the budget (this means revenues are at least as large as costs).
Goal 2 Cut spending by at most $150 billion.
Goal 3 Raise at most $550 billion in taxes from the rich.
Goal 4 Raise at most $350 billion in taxes from the poor.

Currently, the government spends $1 trillion (a trillion = 1,000 billion) per year. Revenue can be raised in two ways: through a gas tax and an income tax. You must determine

G = per gallon tax rate (in cents)
LTR = % tax rate charged on first $30,000 of income
HTR = % tax rate charged on any income earned more than $30,000
C = cut in spending (in billions)

If the government chooses G, LTR, and HTR, then the revenue given in Table 68 (in billions) is raised. Of course, the tax rate on income more than $30,000 must be at least as large as the tax rate on the first $30,000 of income. Formulate a preemptive goal programming model to help the president meet his goals.

TABLE **67**

	Project							
	1	**2**	**3**	**4**	**5**	**6**	**7**	**8**
ROI ($)	2,070	456	670	350	495	380	1,500	480
Cost ($)	900	240	335	700	410	190	500	160
Productivity improvement	3	2	2	0	1	0	3	2
Manpower needed	18	18	27	36	42	6	48	24
Degree of risk	3	2	4	1	1	0	2	3

TABLE **68**

	Low Income	High Income
Gas tax	G	.5G
Tax on income up to $30,000	20LTR	5LTR
Tax on income above $30,000	0	15HTR

TABLE **69**

Project	NPV (in millions)	Annual Growth Rate	Probability of Success	Cost (in millions)
1	40	20	.75	220
2	30	16	.70	140
3	60	12	.75	280
4	45	8	.90	240
5	55	18	.65	300
6	40	18	.60	200
7	90	19	.65	440

14 HAL computer must determine which of seven research and development (R&D) projects to undertake. For each project four quantities are of interest:

a the net present value (NPV in millions of dollars) of the project

b the annual growth rate in sales generated by the project

c the probability that the project will succeed

d the cost (in millions of dollars) of the project

The relevant information is given in Table 69. HAL has set the following four goals:

Goal 1 The total NPV of all chosen projects should be at least $200 million.

Goal 2 The average probability of success for all projects chosen should be at least .75.

Goal 3 The average growth rate of all projects chosen should be at least 15%.

Goal 4 The total cost of all chosen projects should be at most $1 billion.

For the following sets of priorities, use preemptive (integer) goal programming to determine which projects should be selected.

Priority Set 1 2>>>4>>>1>>>3
Priority Set 2 1>>>3>>>4>>>2

4.17 Using the Excel Solver to Solve LPs

Excel has the capability to solve linear (and often nonlinear) programming problems. In this section, we show how to use the Excel Solver[†] to find the optimal solution to the diet problem of Section 3.4 and the inventory example of Section 3.10.

The key to solving an LP on a spreadsheet is to set up a spreadsheet that tracks everything of interest (costs or profits, resource usage, etc.). Next, identify the cells of interest that can be varied. These are called **changing cells.** After defining the changing cells, identify the cell that contains your objective function as the **target cell.** Next, we identify our constraints and tell the Solver to solve the problem. At this point, the optimal solution to our problem will be placed in the spreadsheet.

[†]To activate the Excel Solver for the first time, select Tools and then select Add-Ins. Checking the Solver Add-in box will cause Excel to open Solver whenever you check Tools and then Solver.

Using the Excel Solver to Solve the Diet Problem

Diet1.xls

In file Diet1.xls, we set up a spreadsheet model of the diet problem (Example 6 of Chapter 3). To begin (see Figure 15) we enter headings for each type of food in B3:E3. In the range B4:E4, we input trial values for the amount of each food eaten. For example, Figure 15 indicates that we are considering eating three brownies, four scoops of chocolate ice cream, five bottles of cola, and six pieces of pineapple cheesecake. To see if the diet in Figure 15 is an "optimal" diet, we must determine its cost as well as the calories, chocolate, sugar, and fat it provides. In the range B5:E5, we input the per-unit cost for each available food. Then we compute the cost of the diet in cell F5.

We could compute the cost of the diet in cell F5 with the formula

$$=B4 \cdot B5 + C4 \cdot C5 + D4 \cdot D5 + E4 \cdot E5$$

but it is easier to enter the formula

$$=SUMPRODUCT(B\$4:E\$4, B5:E5)$$

The =SUMPRODUCT function requires two ranges as inputs. The first cell in the first range is multiplied by the first cell in the second range; then the second cell in the first range is multiplied by the second cell in the second range; and so on. All of these products are then added. Essentially, the =SUMPRODUCT function duplicates the notion of scalar products of vectors discussed in Section 2.1. Thus, in cell F5 the =SUMPRODUCT function computes total cost as $(3)(50) + 4(20) + 5(30) + 6(80) = 860$ cents.

In the range B6:E6, we enter the calories in each food; in B7:E7, the chocolate content; in B8:E8, the sugar content; and in B9:E9, the fat content. Copying the formula in F5 to the cell range F6:F9 now computes the calories, chocolate, sugar, and fat contained in the diet defined by the values in B4:E4. Note that the =SUMPRODUCT function makes it easy to create many constraints by entering one formula and using the copy command.

In the cell range H6:H9, we have listed the minimum daily requirement for each nutrient. From Figure 15, we see that our current diet is feasible (meets daily requirements for each nutrient) and costs $8.70. We now describe how to use Solver to find the optimal solution to the diet problem.

Step 1 From the Tools menu, select Solver. The dialog box in Figure 16 will appear.

Step 2 Move the mouse to the Set Target Cell portion of the dialog box and click (or type in the cell address) on your target cell (total cost in cell F5) and select Min. This tells Solver to minimize total cost.

Step 3 Move the mouse to the By Changing Cells portion of the dialog box and click on the changing cells (B4:E4). This tells Solver it can change the amount eaten of each food.

Step 4 Click on the Add button to add constraints. The screen in Figure 17 will appear. Move to the Cell Reference part of the Add Constraint dialog box and select F6:F9. Then move to the dropdown box and select >=. Finally, click on the constraint portion of the dialog box and select H6:H9. Choose OK because there are no more constraints. If you

	A	B	C	D	E	F	G	H
1			Feasible					
2			solution to Diet Problem					
3			Brownie	Choc IC	Cola	Pine Cheese	Totals		Required
4	Eaten	3	4	5	6			
5	Cost	50	20	30	80	860		
6	Calories	400	200	150	500	5750	>=	500
7	Chocolate	3	2	0	0	17	>=	6
8	Sugar	2	2	4	4	58	>=	10
9	Fat	2	4	1	5	57	>=	8

FIGURE 15

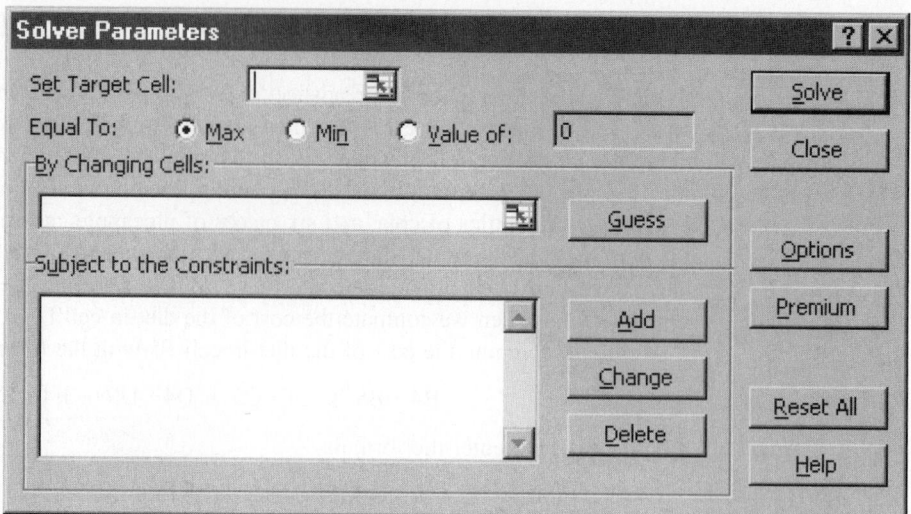

FIGURE 16

need to add more constraints, choose Add. From the main Solver box you may change a constraint by selecting Change or delete a constraint by selecting Delete.

We have now created four constraints. Solver will ensure that the changing cells are chosen so F6>=H6, F7>=H7, F8>=H8, and F9>=H9. In short, the diet will be chosen to ensure that enough calories, chocolate, sugar, and fat are eaten.

Our Solver window should now look like Figure 18.

Step 5   Before solving the problem, we need to tell Solver that all changing cells must be nonnegative. We must also tell Solver that we have a linear model. If we do not tell Solver the model is linear, then Solver will not know it should use the simplex method to solve the problem, and Solver may get an incorrect answer. We may accomplish both of these goals by selecting options. The screen in Figure 19 will appear. Checking Assume Non-Negative ensures that all changing cells will be nonnegative. Checking the Assume Linear Model box ensures that Solver will use the simplex method to solve our LP. Sometimes in a poorly scaled LP (one with both large and small numbers present in the objective function, right-hand sides, or constraints), the Solver will not recognize an LP as a linear model. Checking the Use Automatic Scaling box minimizes the chances that a poorly scaled LP will be interpreted as a nonlinear model. By the way, Max Time is the maximum time the Solver will run before prompting the user about whether to terminate the solution procedure. Iterations is the maximum number of simplex pivots the Solver will make before asking the user whether the solution procedure will continue. The Precision setting describes how much "error" is tolerated before deciding a constraint is not satisfied. For example, with a precision of .001, a changing cell with a value of $-.0009$ would be deemed to satisfy a nonnegativity constraint. The Tolerance and Convergence settings will be discussed in Chapter 8.

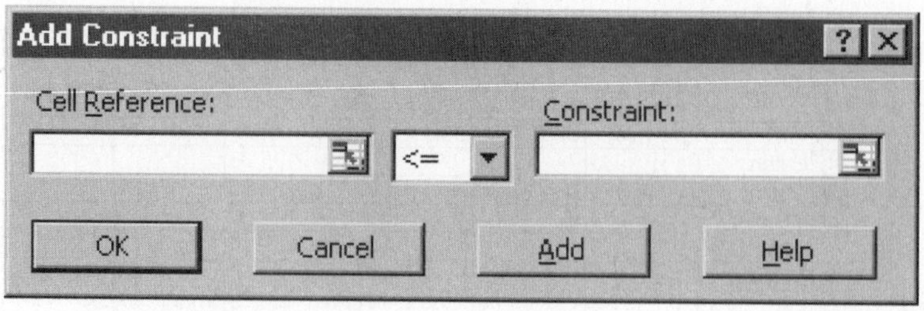

FIGURE 17

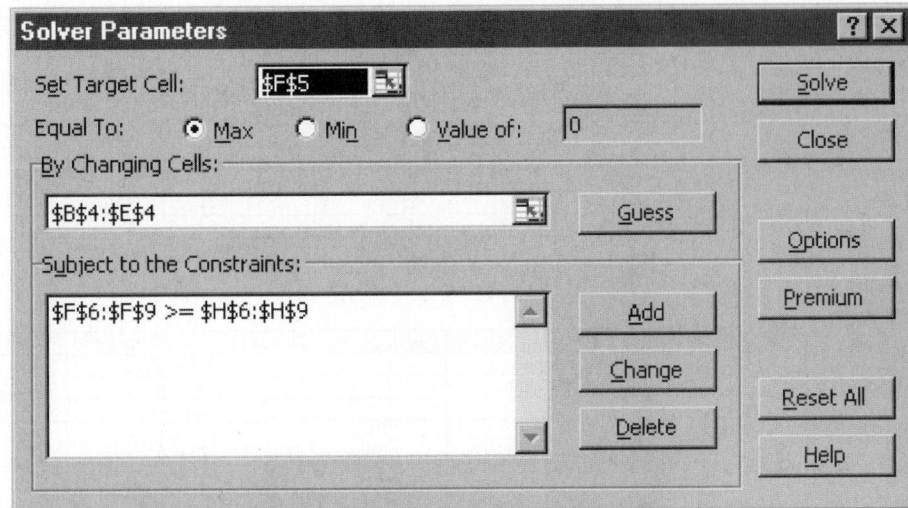

FIGURE 18

Step 6 After choosing OK from the Solver Options box, we then select Solve. Solver yields the optimal solution shown in Figure 20.

Just like LINDO, the Solver says the minimum cost is 90 cents. The minimum cost is obtained by eating no brownies, 3 oz of chocolate ice cream, 1 bottle of cola, and no pineapple cheesecake.

Using the Solver to Solve the Sailco Example

Sailco.xls

We now set up a spreadsheet (Sailco.xls) to solve the Sailco example (Example 12 of Chapter 3). See Figure 21. For each month, we need to keep track of our beginning inventory, ending inventory, and cost. Note that for each month

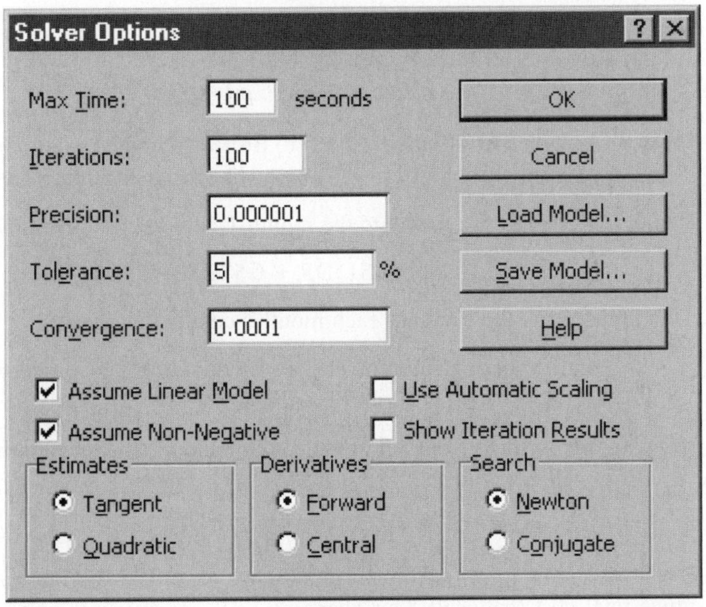

FIGURE 19

FIGURE 20

	A	B	C	D	E	F	G	H
1			Optimal Solution					
2			to the Diet Problem					
3		Brownie	Choc IC	Cola	Pine Cheese	Totals		Required
4	Eaten	0	3	1	0			
5	Cost	50	20	30	80	90		
6	Calories	400	200	150	500	750	>=	500
7	Chocolate	3	2	0	0	6	>=	6
8	Sugar	2	2	4	4	10	>=	10
9	Fat	2	4	1	5	13	>=	8

FIGURE 21

	A	B	C	D	E	F	G	H	I	J	K
1			Optimal solution					RT unit cost	$ 400.00		
2			to Sailco problem					OT unit cost	$ 450.00		
3								Unit Holding cost	$ 20.00		
4	Month	Beg Inventory	OT Production	RT Production		RT Capacity	Demand	Ending Inventory			Monthly Cost
5	1	10	0	40	<=	40	40	10	>=	0	$ 16,200.00
6	2	10	10	40	<=	40	60	0	>=	0	$ 20,500.00
7	3	0	35	40	<=	40	75	0	>=	0	$ 31,750.00
8	4	0	0	25	<=	40	25	0	>=	0	$ 10,000.00
9										Total Cost	$ 78,450.00

Monthly cost = 400(regular-time production) + 450(overtime production) + 20(unit holding cost)

Ending inventory = beginning inventory + monthly production − monthly demand

Step 1 Enter unit costs in I1:I3, regular-time monthly capacities in F5:F8, demands in G5:G8, and beginning month 1 inventory in B5.

Step 2 Enter trial values of each month's overtime and regular-time production in C5:D8.

Step 3 Determine month 1 ending inventory in H5 with the formula

$$=B5 + C5 + D5 − G5$$

This implements the following relationship:

Ending inventory = beginning inventory + monthly production − monthly demand

Step 4 Set month 2 beginning inventory to month 1 ending inventory by entering in cell B6 the formula

$$=H5$$

Step 5 Copying the formula from B5 to B6:B8 computes beginning inventory for months 2–4. Copying the formula from H5 to H6:H8 computes ending inventory for months 2–4.

Step 6 In cell K5, we compute the month 1 cost with the formula

$$=\$I\$1*D5 + C5*\$I\$2 + \$I\$3*H5$$

This implements the fact that each month's cost is given by

Monthly cost = 400(regular-time production) + 450(overtime production) + 20(unit holding cost)

Copying this formula from K5 to K6:K8 computes costs for months 2–4. We compute total cost in cell K9 with the formula

$$=SUM(K5:K8)$$

Step 7 We now fill in our Solver dialog box as shown in Figure 22. Our goal is to minimize total cost (cell K9). Our changing cells are overtime and regular-time production

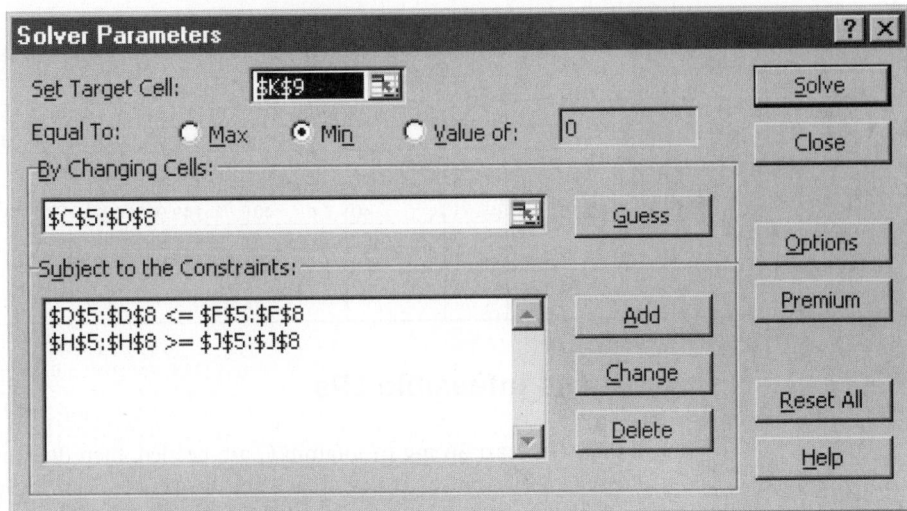

FIGURE 22

(C5:D8). We must ensure that each month's regular-time production is at most 40 (D5:D8 <=F5:F8). Finally, constraining each month's ending inventory to be non-negative (H5:H8 >= J5:J8) ensures that each month's demand is met on time. In Options we check Assume Linear Model, Assume Non-Negative, and Use Automatic Scaling. After choosing Solve, we find the optimal solution shown in Figure 21. A minimum cost of $78,450 is achieved by producing 40 units with regular-time production during months 1–3, 25 units of regular-time production during month 4, 10 units of overtime production during month 2, and 35 units of overtime production during month 3.

Using the Value of Option

Sailco.xls

Recall that in the Sailco problem the minimum cost was $78,450. Suppose that we wanted to find a solution that yielded a cost of exactly $90,000. Then we may use the Solver's Value of option. Simply fill in the Solver dialog box as shown in Figure 23 (see the sheet titled Cost of $90,000 in file Sailco.xls).

Solver yields the solution in Figure 24. Note that Solver found a feasible solution having a total cost of exactly $90,000.

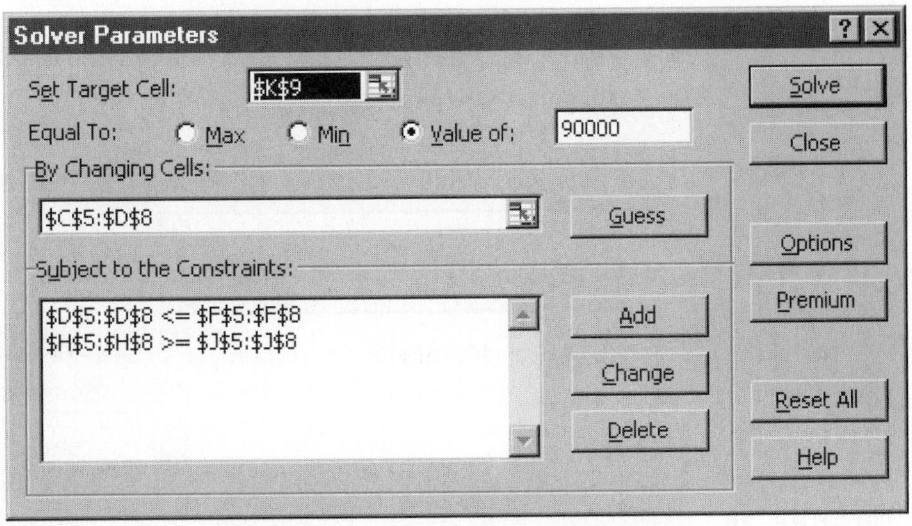

FIGURE 23

FIGURE 24

	A	B	C	D	E	F	G	H	I	J	K
1			Optimal solution					RT unit cost	$ 400.00		
2			to Sailco problem					OT unit cost	$ 450.00		
3								Unit Holding cost	$ 20.00		
4	Month	Beg Inventory	OT Production	RT Production		RT Capacity	Demand	Ending Inventory			Monthly Cost
5	1	10	179.090909	0	<=	40	40	149.0909091	>=	0	$ 83,572.73
6	2	149.0909	0	0	<=	40	60	89.09090909	>=	0	$ 1,781.82
7	3	89.09091	0	0	<=	40	75	14.09090909	>=	0	$ 281.82
8	4	14.09091	0	10.909091	<=	40	25	0	>=	0	$ 4,363.64
9										Total Cost	$ 90,000.00

Solver and Infeasible LPs

Bevco.xls

Recall that if at least 36 mg of vitamin C are needed, then the Bevco problem (Example 4 of this chapter) is infeasible. We have set this problem up in Solver in file Bevco.xls. Figure 25 shows the spreadsheet, and Figure 26 shows the Solver window.

When we choose Solve, we obtain the message shown in Figure 27. This indicates that the LP has no feasible solution.

Solver and Unbounded LPs

Breadco.xls

Recall that Example 3 of this chapter was an unbounded LP. The file Breadco.xls (see Figure 28) contains a Solver formulation of this LP. Figure 29 contains the Solver window for the Breadco example. When we choose Solve, we obtain the message in Figure 30.

	A	B	C	D	E	F
1						
2	Infeasible LP					
3						
4					Total Cost	
5		Soda	Juice		3 0	
6	Amount	0	10			
7	Unit cost	2	3			
8				Available		Needed
9	Sugar	0.5	0.25	2.5	>=	4
10	Vitamin C	1	3	30	>=	36
11	Total oz.	1	1	10	=	10

FIGURE 25

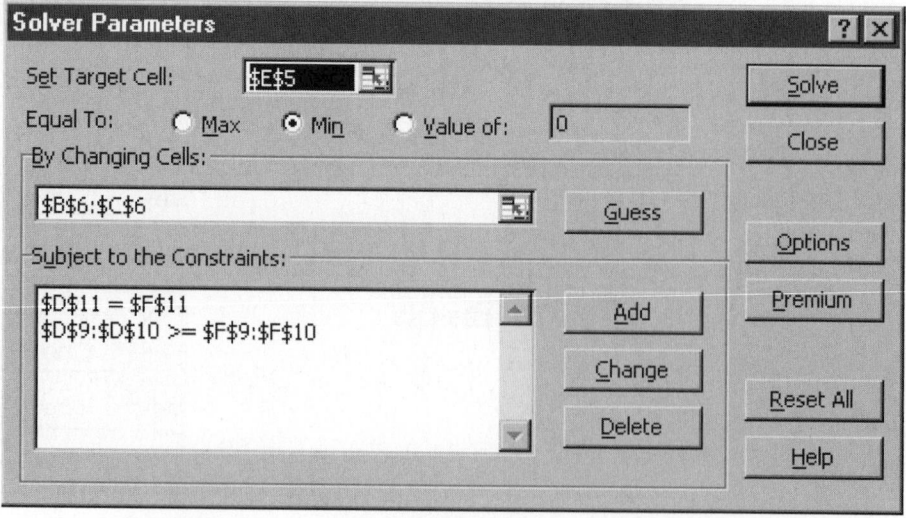

FIGURE 26

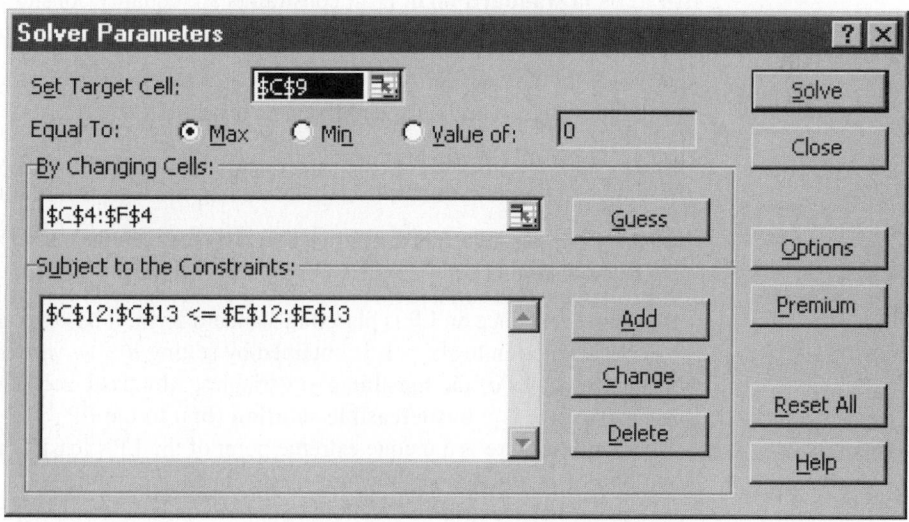

Solver Results

Solver could not find a feasible solution.

- ⦿ Keep Solver Solution
- ○ Restore Original Values

Reports:
Feasibility
Feasibility-Bounds

[OK] [Cancel] [Save Scenario...] [Help]

FIGURE 27

	A	B	C	D	E	F	G	H
1	Unbounded LP							
2								
3			FB Baked	SD Baked	Yeast bought	Flour bought		Originally we have
4			5	0	0	20		
5		Price or cost	36	30	3	4		
6		Yeast needed	1	1				5
7		Flour needed	6	5				10
8								
9		Profit	100					
10								
11			Used		Available			
12		Yeast	5	<=	5			
13		Flour	30	<=	30			

FIGURE 28

Solver Parameters

Set Target Cell: $C$9

Equal To: ⦿ Max   ○ Min   ○ Value of: 0

By Changing Cells:
$C$4:$F$4   [Guess]

Subject to the Constraints:
$C$12:$C$13 <= $E$12:$E$13

[Add] [Change] [Delete]

[Solve] [Close] [Options] [Premium] [Reset All] [Help]

FIGURE 29

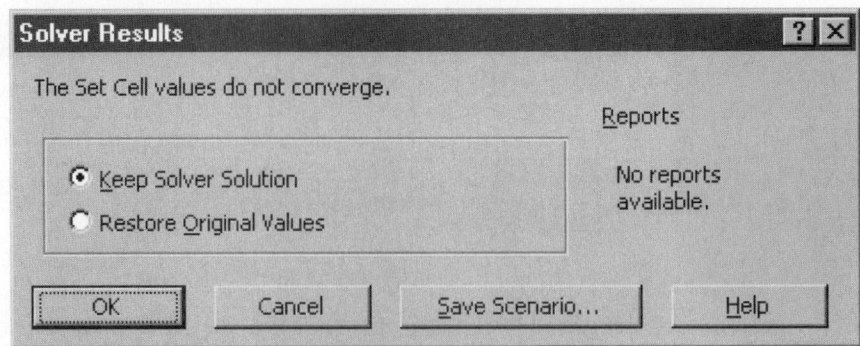

FIGURE 30

The message "Set Cell values do not converge" indicates an unbounded LP; that is, there are values of the changing cells that satisfy all constraints and yield arbitrarily large profit.

PROBLEMS

Group A

Use Excel Solver to find the optimal solution to the following problems:

1 Problem 2 of Section 3.4

2 Example 7 of Chapter 3

3 Example 11 of Chapter 3

4 Problem 3 of Section 3.10

5 Example 14 of Section 3.12

Group B

6 Problem 4 of Section 3.11

7 Problem 5 of Section 3.11

8 Problem 3 of Section 3.12

9 Problem 5 of Section 3.12

SUMMARY   Preparing an LP for Solution by the Simplex

An LP is in **standard form** if all constraints are equality constraints and all variables are nonnegative. To place an LP in standard form, we do the following:

Step 1   If the ith constraint is a $\leq$ constraint, then we convert it to an equality constraint by adding a slack variable s_i and the sign restriction $s_i \geq 0$.

Step 2   If the ith constraint is a $\geq$ constraint, then we convert it to an equality constraint by subtracting an excess variable e_i and adding the sign restriction $e_i \geq 0$.

Step 3   If the variable x_i is unrestricted in sign (urs), replace x_i in both the objective function and constraints by $x_i' - x_i''$, where $x_i' \geq 0$ and $x_i'' \geq 0$.

Suppose that once an LP is placed in standard form, it has m constraints and n variables.

A basic solution to $A\mathbf{x} = \mathbf{b}$ is obtained by setting $n - m$ variables equal to 0 and solving for the values of the remaining m variables. Any basic solution in which all variables are nonnegative is a **basic feasible solution** (bfs) to the LP.

For any LP, there is a unique extreme point of the LP's feasible region corresponding to each bfs. Also, at least one bfs corresponds to each extreme point of the feasible region.

If an LP has an optimal solution, then there is an extreme point that is optimal. Thus, in searching for an optimal solution to an LP, we may restrict our search to the LP's basic feasible solutions.

The Simplex Algorithm

If the LP is in standard form and a bfs is readily apparent, then the simplex algorithm (for a max problem) proceeds as follows:

Step 1 If all nonbasic variables have nonnegative coefficients in row 0, then the current bfs is optimal. If any variables in row 0 have negative coefficients, then choose the variable with the most negative coefficient in row 0 to enter the basis.

Step 2 For each constraint in which the entering variable has a positive coefficient, compute the following ratio:

$$\frac{\text{Right-hand side of constraint}}{\text{Coefficient of entering variable in constraint}}$$

Any constraint attaining the smallest value of this ratio is the winner of the **ratio test.** Use EROs to make the entering variable a basic variable in any constraint that wins the ratio test. Return to step 1.

If the LP (a max problem) is **unbounded,** then we eventually reach a tableau in which a nonbasic variable has a negative coefficient in row 0 and a nonpositive coefficient in each constraint. Otherwise (barring the extremely rare occurrence of *cycling*), the simplex algorithm will find an optimal solution to an LP.

If a bfs is not readily apparent, then the Big M method or the two-phase simplex method must be used to obtain a bfs.

The Big M Method

Step 1 Modify the constraints so that the right-hand side of each constraint is nonnegative.

Step 1' Identify each constraint that is now (after step 1) an $=$ or $\geq$ constraint. In step 3, we will add an artificial variable to each of these constraints.

Step 2 Convert each inequality constraint to standard form.

Step 3 If (after step 1 has been completed) constraint i is a $\geq$ or $=$ constraint, then add an artificial variable a_i and the sign restriction $a_i \geq 0$.

Step 4 Let M denote a very large positive number. If the LP is a min problem, then add (for each artificial variable) Ma_i to the objective function. For a max problem, add $-Ma_i$.

Step 5 Because each artificial variable will be in the starting basis, each must be eliminated from row 0 before beginning the simplex. If all artificial variables are equal to 0 in the optimal solution, then we have found the optimal solution to the original problem. If any artificial variables are positive in the optimal solution, then the original problem is infeasible.

The Two-Phase Method

Step 1 Modify the constraints so that the right-hand side of each constraint is nonnegative.

Step 1' Identify each constraint that is now (after step 1) an $=$ or $\geq$ constraint. In step 3, we will add an artificial variable to each of these constraints.

Step 2 Convert each inequality constraint to the standard form.

Step 3 If (after step 1') constraint i is a $\geq$ or $=$ constraint, then add an artificial variable a_i and the sign restriction $a_i \geq 0$.

Step 4 For now, ignore the original LP's objective function. Instead, solve an LP whose objective function is min $w' =$ (sum of all the artificial variables). This is called the **Phase I LP.**

Because each $a_i \geq 0$, solving the Phase I LP will result in one of the following three cases:

Case 1 The optimal value of w' is greater than zero. In this case, the original LP has no feasible solution.

Case 2 The optimal value of w' is equal to zero, and no artificial variables are in the optimal Phase I basis. In this case, drop all columns in the optimal Phase I tableau that correspond to the artificial variables and combine the original objective function with the constraints from the optimal Phase I tableau. This yields the **Phase II LP.** The optimal solution to the Phase II LP and the original LP are the same.

Case 3 The optimal value of w' is equal to zero, and at least one artificial variable is in the optimal Phase I basis. In this case, we can find the optimal solution to the original LP if, at the end of Phase I, we drop from the optimal Phase I tableau all nonbasic artificial variables and any variable from the original problem that has a negative coefficient in row 0 of the optimal Phase I tableau.

Solving Minimization Problems

To solve a minimization problem by the simplex, choose as the entering variable the nonbasic variable in row 0 with the most positive coefficient. A tableau or canonical form is optimal if each variable in row 0 has a nonpositive coefficient.

Alternative Optimal Solutions

If a nonbasic variable has a zero coefficient in row 0 of an optimal tableau and the nonbasic variable can be pivoted into the basis, the LP may have **alternative optimal solutions.** If two basic feasible solutions are optimal, then any point on the line segment joining the two optimal basic feasible solutions is also an optimal solution to the LP.

Unrestricted-in-Sign Variables

If we replace a urs variable x_i with $x_i' - x_i''$, the LP's optimal solution will have x_i', x_i'' or both x_i' and x_i'' equal to zero.

REVIEW PROBLEMS

Group A

1 Use the simplex algorithm to find *two* optimal solutions to the following LP:

$$\max z = 5x_1 + 3x_2 + x_3$$
$$\text{s.t.} \quad x_1 + x_2 + 3x_3 \leq 6$$
$$5x_1 + 3x_2 + 6x_3 \leq 15$$
$$x_3, x_1, x_2 \geq 0$$

2 Use the simplex algorithm to find the optimal solution to the following LP:

$$\min z = -4x_1 + x_2$$
$$\text{s.t.} \quad 3x_1 + x_2 \leq 6$$
$$\text{s.t.} \quad -x_1 + 2x_2 \leq 0$$
$$x_1, x_2 \geq 0$$

3 Use the Big M method and the two-phase method to find the optimal solution to the following LP:

$$\max z = 5x_1 - x_2$$
$$\text{s.t.} \quad 2x_1 + x_2 = 6$$
$$x_1 + x_2 \le 4$$
$$x_1 + 2x_2 \le 5$$
$$x_1, x_2 \ge 0$$

4 Use the simplex algorithm to find the optimal solution to the following LP:

$$\max z = 5x_1 - x_2$$
$$\text{s.t.} \quad x_1 - 3x_2 \le 1$$
$$x_1 - 4x_2 \le 3$$
$$x_1, x_2 \ge 0$$

5 Use the simplex algorithm to find the optimal solution to the following LP:

$$\min z = -x_1 - 2x_2$$
$$\text{s.t.} \quad 2x_1 + x_2 \le 5$$
$$x_1 + x_2 \le 3$$
$$x_1, x_2 \ge 0$$

6 Use the Big M method and the two-phase method to find the optimal solution to the following LP:

$$\max z = x_1 + x_2$$
$$\text{s.t.} \quad 2x_1 + x_2 \ge 3$$
$$3x_1 + x_2 \le 3.5$$
$$x_1 + x_2 \le 1$$
$$x_1, x_2 \ge 0$$

7 Use the simplex algorithm to find *two* optimal solutions to the following LP. How many optimal solutions does this LP have? Find a third optimal solution.

$$\max z = 4x_1 + x_2$$
$$\text{s.t.} \quad 2x_1 + 3x_2 \le 4$$
$$x_1 + x_2 \le 1$$
$$4x_1 + x_2 \le 2$$
$$x_1, x_2 \ge 0$$

8 Use the simplex method to find the optimal solution to the following LP:

$$\max z = 5x_1 + x_2$$
$$\text{s.t.} \quad 2x_1 + x_2 \le 6$$
$$x_1 - x_2 \le 0$$
$$x_1, x_2 \ge 0$$

9 Use the Big M method and the two-phase method to find the optimal solution to the following LP:

$$\min z = -3x_1 + x_2$$
$$\text{s.t.} \quad x_1 - 2x_2 \ge 2$$
$$-x_1 + x_2 \ge 3$$
$$x_1, x_2 \ge 0$$

10 Suppose that in the Dakota Furniture problem, 10 types of furniture could be manufactured. To obtain an optimal solution, how many types of furniture (at the most) would have to be manufactured?

11 Consider the following LP:

$$\max z = 10x_1 + x_2$$
$$\text{s.t.} \quad x_1 \le 1$$
$$20x_1 + x_2 \le 100$$
$$x_1, x_2 \ge 0$$

a Find all the basic feasible solutions for this LP.

b Show that when the simplex is used to solve this LP, every basic feasible solution must be examined before the optimal solution is found.

By generalizing this example, Klee and Minty (1972) constructed (for $n = 2, 3, \ldots$) an LP with n decision variables and n constraints for which the simplex algorithm examines $2^n - 1$ basic feasible solutions before the optimal solution is found. Thus, there exists an LP with 10 variables and 10 constraints for which the simplex requires $2^{10} - 1 = 1{,}023$ pivots to find the optimal solution. Fortunately, such "pathological" LPs rarely occur in practical applications.

12 Productco produces three products. Each product requires labor, lumber, and paint. The resource requirements, unit price, and variable cost (exclusive of raw materials) for each product are given in Table 70. Currently, 900 labor hours, 1,550 gallons of paint, and 1,600 board feet of lumber are available. Additional labor can be purchased at $6 per hour, additional paint at $2 per gallon, and additional lumber at $3 per board foot. For the following two sets of priorities, use preemptive goal programming to determine an optimal production schedule. For set 1:

Priority 1 Obtain profit of at least $10,500.
Priority 2 Purchase no additional labor.
Priority 3 Purchase no additional paint.
Priority 4 Purchase no additional lumber.

For set 2:

Priority 1 Purchase no additional labor.
Priority 2 Obtain profit of at least $10,500.
Priority 3 Purchase no additional paint.
Priority 4 Purchase no additional lumber.

13 Jobs at Indiana University are rated on three factors:

Factor 1 Complexity of duties
Factor 2 Education required
Factor 3 Mental and or visual demands

For each job at IU, the requirement for each factor has been rated on a scale of 1–4, with a 4 in factor 1 representing high complexity of duty, a 4 in factor 2 representing high educational requirement, and a 4 in factor 3 representing high mental and/or visual demands.

TABLE 70

Product	Labor	Lumber	Paint	Price ($)	Variable Cost ($)
1	1.5	2	3	26	10
2	3	3	2	28	6
3	2	4	2	31	7

IU wants to determine a formula for grading each job. To do this, it will assign a point value to the score for each factor that a job requires. For example, suppose level 2 of factor 1 yields a point total of 10, level 3 of factor 2 yields a point total of 20, and level 3 of factor 3 yields a point value of 30. Then a job with these requirements would have a point total of $10 + 20 + 30$. A job's hourly salary equals half its point total.

IU has two goals (listed in order of priority) in setting up the points given to each level of each job factor.

Goal 1 When increasing the level of a factor by one, the points should increase by at least 10. For example, level 2 of factor 1 should earn at least 10 more points than level 1 of factor 1. Goal 1 is to minimize the sum of deviations from this requirement.

Goal 2 For the benchmark jobs in Table 71, the actual point total for each job should come as close as possible to the point total listed in the table. Goal 2 is to minimize the sum of the absolute deviations of the point totals from the desired scores.

Use preemptive goal programming to come up with appropriate point totals. What salary should a job with skill levels of 3 for each factor be paid?

14 A hospital outpatient clinic performs four types of operations. The profit per operation, as well as the minutes of X-ray time and laboratory time used are given in Table 72. The clinic has 500 private rooms and 500 intensive care rooms. Type 1 and Type 2 operations require a patient to stay in an intensive care room for one day while Type 3 and Type 4 operations require a patient to stay in a private room for one day. Each day the hospital is required to perform at least 100 operations of each type. The hospital has set the following goals:

Goal 1 Earn a daily profit of at least $100,000.
Goal 2 Use at most 50 hours daily of X-ray time.
Goal 3 Use at most 40 hours daily of laboratory time.

The cost per unit deviation from each goal is as follows:

Goal 1 Cost of $1 for each dollar by which profit goal is unmet

Goal 2 Cost of $10 for each hour by which X-ray goal is unmet

Goal 3 Cost of $8 for each hour by which laboratory goal is unmet

Formulate a goal programming model to minimize the daily cost incurred due to failing to meet the hospital's goals.

Group B

15 Consider a maximization problem with the optimal tableau in Table 73. The optimal solution to this LP is $z = 10, x_3 = 3, x_4 = 5, x_1 = x_2 = 0$. Determine the second-best bfs to this LP. (*Hint:* Show that the second-best solution must be a bfs that is one pivot away from the optimal solution.)

16 A camper is considering taking two types of items on a camping trip. Item 1 weighs a_1 lb, and item 2 weighs a_2 lb. Each type 1 item earns the camper a benefit of c_1 units, and each type 2 item earns the camper c_2 units. The knapsack can hold items weighing at most b lb.

a Assuming that the camper can carry a fractional number of items along on the trip, formulate an LP to maximize benefit.

b Show that if

$$\frac{c_2}{a_2} \geq \frac{c_1}{a_1}$$

then the camper can maximize benefit by filling a knapsack with $\frac{b}{a_2}$ type 2 items.

c Which of the linear programming assumptions are violated by this formulation of the camper's problem?

17 You are given the tableau shown in Table 74 for a maximization problem. Give conditions on the unknowns a_1, a_2, a_3, b, and c that make the following statements true:

a The current solution is optimal.

b The current solution is optimal, and there are alternative optimal solutions.

c The LP is unbounded (in this part, assume that $b \geq 0$).

18 Suppose we have obtained the tableau in Table 75 for a maximization problem. State conditions on a_1, a_2, a_3, b, c_1, and c_2 that are required to make the following statements true:

a The current solution is optimal, and there are alternative optimal solutions.

b The current basic solution is not a basic feasible solution.

TABLE 71

Job	Factor Level			Desired Score
	1	2	3	
1	4	4	4	105
2	3	3	2	93
3	2	2	2	75
4	1	1	2	68

TABLE 72

	Type of Operation			
	1	2	3	4
Profit ($)	200	150	100	80
X-ray time (minutes)	6	5	4	3
Laboratory time (minutes)	5	4	3	2

TABLE 73

z	x_1	x_2	x_3	x_4	rhs
1	2	1	0	0	10
0	3	2	1	0	3
0	4	3	0	1	5

TABLE 74

z	x_1	x_2	x_3	x_4	x_5	rhs
1	$-c$	2	0	0	0	10
0	-1	a_1	1	0	0	4
0	a_2	-4	0	1	0	1
0	a_3	3	0	0	1	b

TABLE 75

z	x_1	x_2	x_3	x_4	x_5	x_6	rhs
1	c_1	c_2	0	0	0	0	10
0	4	a_1	1	0	a_2	0	b
0	-1	-5	0	1	-1	0	2
0	a_3	-3	0	0	-4	1	3

c The current basic solution is a degenerate bfs.

d The current basic solution is feasible, but the LP is unbounded.

e The current basic solution is feasible, but the objective function value can be improved by replacing x_6 as a basic variable with x_1.

19 Suppose we are solving a maximization problem and the variable x_r is about to leave the basis.

a What is the coefficient of x_r in the current row 0?

b Show that after the current pivot is performed, the coefficient of x_r in row 0 cannot be less than zero.

c Explain why a variable that has left the basis on a given pivot cannot re-enter the basis on the next pivot.

20 A bus company believes that it will need the following number of bus drivers during each of the next five years: year 1—60 drivers; year 2—70 drivers; year 3—50 drivers; year 4—65 drivers; year 5—75 drivers. At the beginning of each year, the bus company must decide how many drivers should be hired or fired. It costs $4,000 to hire a driver and $2,000 to fire a driver. A driver's salary is $10,000 per year. At the beginning of year 1, the company has 50 drivers. A driver hired at the beginning of a year may be used to meet the current year's requirements and is paid full salary for the current year. Formulate an LP to minimize the bus company's salary, hiring, and firing costs over the next five years.

21 Shoemakers of America forecasts the following demand for each of the next six months: month 1—5,000 pairs; month 2—6,000 pairs; month 3—5,000 pairs; month 4—9,000 pairs; month 5—6,000 pairs; month 6—5,000 pairs. It takes a shoemaker 15 minutes to produce a pair of shoes. Each shoemaker works 150 hours per month plus up to 40 hours per month of overtime. A shoemaker is paid a regular salary of $2,000 per month plus $50 per hour for overtime. At the beginning of each month, Shoemakers can either hire or fire workers. It costs the company $1,500 to hire a worker and $1,900 to fire a worker. The monthly holding cost per pair of shoes is 3% of the cost of producing a pair of shoes with regular-time labor. (The raw materials in a pair of shoes cost $10.) Formulate an LP that minimizes the cost of

meeting (on time) the demands of the next six months. At the beginning of month 1, Shoemakers has 13 workers.

22 Monroe County is trying to determine where to place the county fire station. The locations of the county's four major towns are given in Figure 31. Town 1 is at (10, 20); town 2 is at (60, 20); town 3 is at (40, 30); town 4 is at (80, 60). Town 1 averages 20 fires per year; town 2, 30 fires; town 3, 40 fires; and town 4, 25 fires. The county wants to build the fire station in a location that minimizes the average distance that a fire engine must travel to respond to a fire. Since most roads run in either an east–west or a north–south direction, we assume that the fire engine can only do the same. Thus, if the fire station were located at (30, 40) and a fire occurred at town 4, the fire engine would have to travel $(80 - 30) + (60 - 40) = 70$ miles to the fire. Use linear programming to determine where the fire station should be located. (*Hint:* If the fire station is to be located at the point (x, y) and there is a town at the point (a, b), define variables e, w, n, s (east, west, north, south) that satisfy the equations $x - a = w - e$ and $y - b = n - s$. It should now be easy to obtain the correct LP formulation.)

23[†] During the 1972 football season, the games shown in Table 76 were played by the Miami Dolphins, the Buffalo Bills, and the New York Jets. Suppose that on the basis of these games, we want to rate these three teams. Let $M =$ Miami rating, $J =$ Jets rating, and $B =$ Bills rating. Given values of M, J, and B, you would predict that when, for example, the Bills play Miami, Miami is expected to win by $M - B$ points. Thus, for the first Miami–Bills game, your prediction would have been in error by $|M - B - 1|$ points. Show how linear programming can be used to determine ratings for each team that minimize the sum of the prediction errors for all games.

FIGURE 31

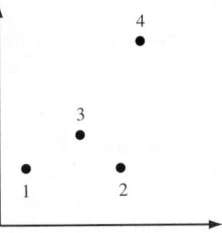

TABLE 76

Miami	Bills	Jets
27	—	17
28	—	24
24	23	—
30	16	—
—	24	41
—	3	41

[†]Based on Wagner (1954).

At the conclusion of the season, this method has been used to determine ratings for college football and college basketball. What problems could be foreseen if this method were used to rate teams early in the season?

24 During the next four quarters, Dorian Auto must meet (on time) the following demands for cars: quarter 1—4,000; quarter 2—2,000; quarter 3—5,000; quarter 4—1,000. At the beginning of quarter 1, there are 300 autos in stock, and the company has the capacity to produce at most 3,000 cars per quarter. At the beginning of each quarter, the company can change production capacity by one car. It costs $100 to increase quarterly production capacity. It costs $50 per quarter to maintain one car of production capacity (even if it is unused during the current quarter). The variable cost of producing a car is $2,000. A holding cost of $150 per car is assessed against each quarter's ending inventory. It is required that at the end of quarter 4, plant capacity must be at least 4,000 cars. Formulate an LP to minimize the total cost incurred during the next four quarters.

25 Ghostbusters, Inc., exorcises (gets rid of) ghosts. During each of the next three months, the company will receive the following number of calls from people who want their ghosts exorcised: January, 100 calls; February, 300 calls; March, 200 calls. Ghostbusters is paid $800 for each ghost exorcised during the month in which the customer calls. Calls need not be responded to during the month they are made, but if a call is responded to one month after it is made, then Ghostbusters loses $100 in future goodwill, and if a call is responded to two months after it is made, Ghostbusters loses $200 in goodwill. Each employee of Ghostbusters can exorcise 10 ghosts during a month. Each employee is paid a salary of $4,000 per month. At the beginning of January, the company has 8 workers. Workers can be hired and trained (in 0 time) at a cost of $5,000 per worker. Workers can be fired at a cost of $4,000 per worker. Formulate an LP to maximize Ghostbusters' profit (revenue less costs) over the next three months. Assume that all calls must be handled by the end of March.

26 Carco uses robots to manufacture cars. The following demands for cars must be met (not necessarily on time, but all demands must be met by end of quarter 4): quarter 1—600; quarter 2—800; quarter 3—500; quarter 4—400. At the beginning of the quarter, Carco has two robots. Robots can be purchased at the beginning of each quarter, but a maximum of two per quarter can be purchased. Each robot can build as many as 200 cars per quarter. It costs $5,000 to purchase a robot. Each quarter, a robot incurs $500 in maintenance costs (even if it is not used to build any cars). Robots can also be sold at the beginning of each quarter for $3,000. At the end of each quarter, a holding cost of $200 per car is incurred. If any demand is backlogged, then a cost of $300 per car is incurred for each quarter the demand is backlogged.

At the end of quarter 4, Carco must have at least two robots. Formulate an LP to minimize the total cost incurred in meeting the next four quarters' demands for cars.

27 Suppose we have found an optimal tableau for an LP, and the bfs for that tableau is nondegenerate. Also suppose that there is a nonbasic variable in row 0 with a zero coefficient. Prove that the LP has more than one optimal solution.

28 Suppose the bfs for an optimal tableau is degenerate, and a nonbasic variable in row 0 has a zero coefficient. Show by example that either of the following cases may hold:

Case 1 The LP has more than one optimal solution.
Case 2 The LP has a unique optimal solution.

29 You are the mayor of Gotham City, and you must determine a tax policy for the city. Five types of taxes are used to raise money:

a Property taxes. Let p = property tax percentage rate.

b A sales tax on all items except food, drugs, and durable goods. Let s = sales tax percentage rate.

c A sales tax on durable goods. Let d = durable goods sales tax percentage rate.

d A gasoline sales tax. Let g = gasoline tax sales percentage rate.

e A sales tax on food and drugs. Let f = sales tax on food and drugs.

The city consists of three groups of people: low-income (LI), middle-income (MI), and high-income (HI). The amount of revenue (in millions of dollars) raised from each group by setting a particular tax at a 1% level is given in Table 77.

For example, a 3% tax on durable good sales will raise $360 million from low-income people. Your tax policy must satisfy the following:

Restriction 1 The tax burden on MI people cannot exceed $2.8 billion.
Restriction 2 The tax burden on HI people cannot exceed $2.4 billion.
Restriction 3 The total revenue raised must exceed the current level of $6.5 billion.
Restriction 4 s must be between 1% and 3%.

Given these restrictions, the mayor has set the following three goals:

Goal P Keep the property tax rate less than 3%.
Goal LI Limit the tax burden on LI people to $2 billion.
Goal Suburbs If their tax burden becomes too high, 20% of the LI people, 20% of the MI people, and 40% of the HI people may consider moving to the suburbs. Suppose that this will happen if their total tax burden exceeds $1.5 billion. To discourage this exodus, the suburb goal is to keep the total tax burden on these people below $1.5 billion.

Use goal programming to determine an optimal tax policy if the mayor's goals follow the following set of priorities:

$$LI >>> P >>> Suburbs^{\dagger}$$

TABLE 77

	p	s	d	g	f
LI	900	300	120	30	90
MI	1,200	400	100	20	60
HI	1,000	250	60	10	40

†Based on Chrisman, Fry, Reeves, Lewis, and Weinstein (1989).

APPENDIX A   LINDO Menu Commands and Statements

Menu Commands

LINDO's commands can be accessed from a convenient menu similar to those of other Windows programs. The main menu includes six submenus along the top of the screen that list the various commands. When you click on one of the submenus—File, Edit, Solve, Reports, Window, or Help—a pull-down menu appears with the various commands. You can select commands just like you would in most Window programs—by either clicking on the command with your mouse or pressing the underlined letter in the command name when the appropriate submenu is highlighted. Many commands also have shortcut keys assigned to them (F2, Ctrl+Z, etc.). As an added convenience, some of the most often used commands also may be accessed with icons located in a tool bar at the top of the screen. The following sections briefly describe the various menu commands and list the applicable shortcuts and icons.

File Menu

The File menu commands allow you to manipulate your LINDO data files in various ways. You can use this menu to open, close, save, and print files, as well as perform various tasks unique to LINDO. A description of the File commands follows.

COMMAND		DESCRIPTION
New	F2	Creates a new window for entering input data.
Open	F3	Opens an existing file. Dialog boxes allow you to select from various file types and locations.
View	F4	Opens an existing file for viewing only. No changes can be made to the file.
Save	F5	Saves the window. You can save input data (a model), a Reports window, or a command window. Data can be saved in the following formats: *.LTX, a text format that can be edited with word processing software; *.LPK, for saving compiled models in a "packed" format, but without any special formatting or comments; and *.MPS, the machine-independent industry standard format for transferring LP problems between LINDO and other LP software.
Save As...	F6	Saves the active window with a specified file name. This is useful for renaming a revised file, while keeping the original file intact.
Close	F7	Closes the active window. If the window contains new input data, then you will be asked if you want to save the changes.
Print	F8	Sends the active window to your printer.
Printer Setup...	F9	Selects the printer and various options for print format.
Log Output...	F10	Sends all subsequent screen activity that would normally be sent to the Reports window to a text file. When you have specified a log file location, a check will appear in the File menu by the Log Output line. To disable Log Output, simply select the command again.

COMMAND	DESCRIPTION
Take Commands... F11	"Takes" a LINGO batch file with commands and text for automated operation. A model could be put in memory, solved, and the solution placed in the Reports window and saved to a file. If you use the Batch command before the beginning of the model text, the model and the commands contained in the file, as well as the solution, would be visible in the Reports window.
Basis Read F12	Retrieves a solution to a model that was saved using the Basis Save command.
Basis Save Shift+F2	Saves the solution for the active model to disk with a specified file name.
Title Shift+F3	Displays the title of the active model, if one has been included with the optional Title statement in the model.
Date Shift+F4	Opens a Reports window and displays the current date and time based on your computer's clock.
Elapsed Time Shift+F5	Opens a Reports window and displays the total time elapsed in your current LINDO session.
Exit Shift+F6	Quits LINDO.

Edit Menu

The Edit menu commands allow you to perform basic editing tasks common to most Windows applications, as well as perform various tasks unique to LINDO. A description of the Edit commands follows.

COMMAND	DESCRIPTION
Undo Ctrl+Z	Reverses the last action.
Cut Ctrl+X	Removes any selected text and places it on the clipboard for pasting.
Copy Ctrl+C	Copies selected text to the clipboard for pasting.
Paste Ctrl+V	Inserts or pastes clipboard contents at the insertion point.
Clear Delete	Deletes selected text without placing it on the clipboard.
Find/Replace... Ctrl+F	Searches the active window to find selected text and replaces it with text entered in the "Replace with" box.
Options Alt+O	Allows viewing and changing of various parameters used in LINDO sessions.
Go To Line... Ctrl+T	Allows you to move the cursor to any specified line in the active window.
Paste Symbol... Ctrl+P	Allows you to paste variable names and reserved symbols into the active window.
Select All Ctrl+A	Selects all of the active window for cutting and copying.
Clear All	Deletes the entire contents of the active window.
Choose New Font	Selects a new font for the text in the active window.

Solve Menu

The Solve menu commands are used after you have entered data and are ready to obtain a solution. A description of the Solve command follows.

COMMAND		DESCRIPTION
Solve Ctrl+S		Sends the model in the active window to the LINDO solver to obtain the solution.
Compile Model Ctrl+E		Translates the model into the arithmetic format required by the LINDO solver. Models are also automatically compiled when you use the Solve command.
Debug Ctrl+D		Helps determine problems with infeasible and unbounded models. Sufficient and necessary sets (rows) can be identified, as can crucial constraints—those that make an infeasible model feasible if dropped from the model.
Pivot Ctrl+N		Causes LINDO to perform the next step in the solution process, allowing linear programming problems to be solved step-by-step.
Preemptive Goal Ctrl+G		Performs Lexico optimization (a form of goal programming) on a model.

Reports Menu

The Reports menu commands allow you to specify how LINDO reports are generated. Descriptions of the Reports commands follow.

COMMAND		DESCRIPTION
Solution Alt+0 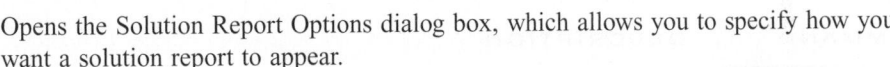		Opens the Solution Report Options dialog box, which allows you to specify how you want a solution report to appear.
Range Alt+1		Creates a range report, or sensitivity analysis, for the active model window.
Parametrics Alt+2		Performs a parametric analysis on the right-hand side of a constraint.
Statistics Alt+3		Displays key statistics for the model in the active window.
Peruse Alt+4		Used to view reports on selected portions of the current model's solution or structure.
Picture Alt+5		Creates a display of the current model in matrix form. The nonzero coefficients of the matrix may be displayed as either text or graphic.
Basis Picture Alt+6		Displays a text-format report with a "picture" of the current basis, ordering the rows and columns according to the last inversion or triangularization performed by the solver. The Basis Picture report is sent to the Reports window.
Tableau Alt+7		Displays the simplex tableau for the active model. This permits observation of the simplex algorithm at each step.
Formulation Alt+8		Displays all, or selected segments, of your model in the Reports window.
Show Column Alt+9		Displays a selected column without the rest of the model.
Positive Definite		Checks for a guarantee of global optimality in a quadratic model.

Window Menu

The Window menu commands allow you to adjust active command and status windows, as well as organize the display of multiple windows. Descriptions of the Window commands follow.

COMMAND	DESCRIPTION
Open Command Window Alt+C	Provides access to LINDO's command-line interface, where you may enter commands at the colon prompt.
Open Status Window	Opens LINDO's Solver Status window, which displays information about the optimizer status, such as number of iterations and elapsed run time. This window also appears when you select Solve from the Solve menu.
Send to Back Ctrl+B	Sends the frontmost window to the back.
Cascade Alt+A	Arranges all open windows in a cascade fashion from upper left to lower right, with the active window on top.
Tile Alt+T	Arranges all open windows so they each occupy equivalent space within the program window.
Close All Alt+X	Closes all active windows.
Arrange Icons Alt+I	Moves icons representing minimized windows so that they are arranged across the bottom of the screen.
List of Windows	At the bottom of the Window menu, a list of the open windows is displayed. The active window is checked.

Help Menu

The Help menu commands provide access to LINDO's online help. Descriptions of the Help commands follow.

COMMAND	DESCRIPTION
Contents  F1	Displays the contents of the help section. The second icon (with the arrow and question mark) enables context-sensitive help, where the cursor indicator will change to a question mark, and help will be provided specifically for a command selected.
Search for Help on... Alt+F1	Searches the help section for a word or topic.
How to Use Help Ctrl+F1	Provides assistance in learning to use the online help system.
About LINDO...	Displays the initial startup screen with general information about LINDO.

Optional Modeling Statements

Besides the basic elements of a model, LINDO recognizes several optional statements that may appear after the END statement. These statements provide additional modeling capabilities, such as placing additional limits on variables. Descriptions of these statements follow.

STATEMENT	DESCRIPTION
FREE <Variable>	Removes all bounds on a variable, allowing it to take on any real value—positive or negative.
GIN <Variable>	Restricts a variable to be a general integer (i.e., in the set of non-negative integers).
INT <Variable>	Restricts a variable to be a binary integer (i.e., either 0 or 1).
SLB <Variable> <Value>	Sets a simple lower bound for a variable (i.e., SLB X 10 would required that X be greater than or equal to 10).

SUB \<Variable\> \<Value\>	Sets a simple upper bound for a variable (i.e., SUB X 10 would require that X be less than or equal to 10).
QCP \<Constraint\>	Indicates the first "real" constraints in a quadratic programming model.
TITLE \<Title\>	Allows you to attach a title to your model. The title can then be displayed using the Title command in the File menu.

APPENDIX B   Getting Started with LINGO

Welcome to the LINGO portion of this text. This appendix will give you brief background information on LINGO and help you install the software. Subsequent chapters will describe features of the software and how to apply the software on sample problems.

What Is LINGO?

LINGO is an interactive computer-software package that can be used to solve linear, integer, and nonlinear programming problems. It can be applied in similar situations to those of LINDO, but it offers more flexibility in terms of how models are expressed. Unlike LINDO, LINGO allows parentheses and variables on the right-hand side of an equation. Constraints can therefore be written in original form and do not have to be rewritten with constants on the right-hand side. LINGO is also capable of generating large models with relatively few lines of input. The program also provides a vast library of mathematical, statistical, and probability functions and greater ability to read data from external files and worksheets.

LINGO Fundamentals

Much like LINDO, LINGO can be used to solve problems interactively from the keyboard or solve problems using files created elsewhere—either self-contained or as part of an integrated program containing customized code and LINGO optimization libraries. This appendix will primarily focus on the first method, that of solving problems interactively. More information on the other methods is available from LINDO Systems, Inc.

Entering a model in the Windows version of LINGO is similar to typing in a Windows word-processing format: You simply type in model data much as you would write it if solving a problem manually. The inner window initially labeled "untitled" is provided to accept input data. LINGO also contains basic editing commands for cutting, copying, and pasting text. These tools, and other features, are found in the window commands discussed in Appendix C.

The required elements of LINGO are similar to those of LINDO: LINGO also requires an objective, one or more variables, and one or more constraints. Unlike LINDO, however, LINGO constraints are *not* preceded by any special terms such as SUBJECT TO or SUCH THAT.

LINGO follows a syntax similar to that of LINDO, with the following differences:

- LINGO statements end with semicolons.
- LINGO includes additional mathematical operators, as discussed in Appendix C. An asterisk is required to denote multiplication.
- Parentheses may be included to define the order of mathematical operations if you wish.
- Variable names can be up to 32 characters long.

APPENDIX C LINGO Menu Commands and Functions

Menu Commands

LINGO's commands can be accessed from a convenient menu similar to those of other Windows programs. The main menu includes five submenus along the top of the screen that list the various commands. When you click on one of the submenus—File, Edit, LINGO, Window, or Help—a pull-down menu appears with the various commands. You can select commands just like you would in most Window programs—by either clicking on the command with your mouse or pressing the underlined letter in the command name when the appropriate submenu is highlighted. Many commands also have shortcut keys assigned to them (F2, Ctrl1+Z, etc.). As an additional convenience, some of the most often used commands may also be accessed with icons located in a tool bar at the top of the screen.

File Menu

The File menu commands allow you to manipulate your LINGO data files in various ways. You can use this menu to open, close, save, and print files, as well as perform various tasks unique to LINGO. Descriptions of the File commands follow.

COMMAND		DESCRIPTION
New	F2	Creates a new window for entering input data.
Open	F3	Opens an existing file. Dialog boxes allow you to select from various file types and locations.
Save	F4	Saves the active window. You can save input data (a model), a reports window, or a command window.
Save As...	F5	Saves the active window with a specified file name. This is useful for renaming a revised file while keeping the original file intact.
Close	F6	Closes the active window. If the window contains new input data, then you will be asked if you want to save the changes.
Print	F7	Sends the active window to your printer.
Printer Setup...	F8	Selects the printer and various options for print format.
Log Output...	F9	Sends all subsequent screen activity that would normally be sent to the Reports window to a text file. When you have specified a log file location, a check will appear in the File menu by the Log Output line. To disable Log Output, simply select the command again.
Take Commands	F11	"Takes" a LINGO batch file with commands and model text for automated operation. A model could be put in memory, solved, and the solution placed in the Reports window saved to a file. If you use the BATCH command before the beginning of the model text, then the model and the commands contained in the file would be visible in the Reports window, as well as the solution.
Import LINDO file	F12	Opens a file that contains a LINDO model in LINDO TAKE format, translating the model into a format acceptable to LINGO.
Exit	F10	Quits LINGO.

Edit Menu

The Edit menu commands allow you to perform basic editing tasks common to most Windows applications, as well as perform various tasks unique to LINGO. Descriptions of the Edit commands follow.

COMMAND	DESCRIPTION
Undo Ctrl+Z	Reverses the last action.
Cut Ctrl+X	Removes any selected text and places it on the clipboard for pasting.
Copy Ctrl+C	Copies selected text to the clipboard for pasting.
Paste Ctrl+V	Inserts clipboard contents at the insertion point.
Clear Delete	Deletes selected text without placing it on the clipboard.
Find/Replace... Ctrl+F	Searches the active window to find selected text and replace it with text entered in the "Replace with" box.
Go To Line... Ctrl+T	Allows you to move the cursor to any specified line in the active window.
Match Parenthesis Ctrl+P	Finds the close parenthesis that corresponds to the selected open parenthesis.
Paste Function	Pastes any of LINGO's built-in functions at the current insertion point. After selecting this command, another submenu appears with the various function categories.
Select All Ctrl+A	Selects all of the active window for cutting and copying.
Choose New Font	Selects a new font for the text in the active window.

LINGO Menu

The LINGO menu commands are used after you have entered data and are ready to obtain a solution. Descriptions of the LINGO commands follow.

COMMAND	DESCRIPTION
Solve Ctrl+S	Sends the model in the active window to the LINGO solver.
Solution... Ctrl+O	Opens the Solution Report Options dialog box, which allows you to specify how you want a solution report to appear.
Range Ctrl+R	Displays a range report, which shows over what ranges you can change coefficients without changing optimal values.
Look... Ctrl+L	Displays all or selected lines of a model.
Generate... Ctrl+S	Creates another version of the current model in algebraic, LINDO, or MPS format. Can be used to number rows and display the model in a more readable format. The GEN command provides a similar capability from the command window.
Export to Spreadsheet Ctrl+E	Exports selected variable values to named ranges in a spreadsheet. A spreadsheet must first be created with ranges sized to accommodate the exported values. The ranges *must* contain numbers. Selecting this command will produce a dialog box that requests the

template and output worksheets (spreadsheet file names), variables to export, and the range to which the values are to be exported. The variables and range are entered in pairs and added to the list of variable and range pairs by clicking the add button.

Options Alt+O   Allows viewing and changing of various parameters used in LINGO sessions.

Workspace Limit Allocates memory to LINGO. If you enter "None," LINGO will use all available memory.
Ctrl+S

Window Menu

The Window menu commands allow you to adjust any open command and status windows, as well as organize the display of multiple windows. Descriptions of the Window commands follow.

COMMAND	DESCRIPTION
Open Command Window	Provides access to LINGO's command-line interface, where you may enter commands at the colon prompt.
Open Status Window	Opens LINGO's Solver Status window, which displays information about the optimizer status, such as number of iterations and elapsed run time. This window also appears when you select Solve from the LINGO menu.

Send to Back   Sends the frontmost window to the back.
Alt+B

Close All Alt+X Closes all active windows.

Cascade Alt+A Arranges all open windows in a cascade fashion from upper left to lower right, with the active window on top.

Tile Alt+T Arranges all open windows so they each occupy equivalent space within the program window.

Arrange Icons Arranges icons representing minimized windows across the bottom of the screen.
Alt+I

List of Windows At the bottom of the Window menu, a list of the open windows is displayed. The active window is checked.

Help Menu

The Help menu commands provide access to LINGO's on-line help. Descriptions of the Help commands follow.

COMMAND   **DESCRIPTION**

Contents Displays the contents of the help section. The second icon (the arrow with the question mark) enables context-sensitive help; the cursor indicator will change to a question mark, and help will be provided specifically for a command selected.
F1

Search for Help on... Searches the help section for a word or topic.
Alt+F1

How to Use Help Provides assistance in learning to use the online help system.
Ctrl+F1

About LINGO... Displays the initial startup screen with general information about LINGO.

Functions

LINGO has seven main functions—standard operators, file import, financial, mathematical, set-looping, variable-domain, and probability, and an assortment of other functions. Most of these functions are available through the menu commands. The LINGO software includes a detailed description of its functions in online help screens; therefore, only a brief description of LINGO functions is offered here.

Standard Operators

Standard operators include arithmetic operators (i.e., ^, *, /, +, and −), logical operators (#EQ#, #NE#, #GT#, #GE#, #LT#, and #L3#) for determining set membership, and equality–inequality operators (<, =, >, <=, and >=) for specifying whether the left-hand side of an expression should be less than, equal to, or greater than the right-hand side. These operators constitute some of the most basic functions available in LINGO. Note that the "greater than" and "less than" symbols (> and >) are interpreted as "loose" inequalities [i.e., greater than or equal to ($\geq$) and less than or equal to ($\leq$), respectively]. You typically type these operators in at the keyboard rather than access them from a window command.

File Import Functions

File import functions allow you to import text and data from external sources. The @FILE function lets you import text or data from an ASCII file, and the @IMPORT function lets you import data only from a worksheet.

Financial Functions

Financial functions include the @FPA(I,N) function, which returns the present value of an annuity; and the @FPL(I,N) function, which returns the present value of a lump sum of $1 N periods from now if the interest rate is I per period. I is not a percentage but rather a non-negative number representing the interest rate.

Mathematical Functions

Mathematical functions include the following general and trigonometric functions: @ABS(X), @COS(X), @EXP(X), @LGM(X), @LOG(X), @SIGN(X), @SIN(X), @SMAX(list), @SMIN(list), @TAN(X). Combinations of the three basic trigonometric functions (sine, cosine, and tangent) may be used to obtain other trigonometric functions.

Set-Looping Functions

Set-looping functions include @FOR (set_name : constraint_expressions), @MAX (set_name : expression), @MIN (set_name : expression), and @SUM (set_name : expression). These functions operate over an entire set, producing a single result in all cases, except the @FOR function, which generates constraints independently for each element of the set.

Variable Domain Functions

The variable domain functions place additional restrictions on variables and attributes. They include the following: @BND(L, X, U), @BIN(X), @FREE(X), and @GIN(X).

Probability Functions

LINGO provides common statistical capabilities through its probability functions: @PSN(X), @PSL(X), @PPS(A,X), @PPL(A,X), @PBN(P,N,X), @PHG(POP,G,N,X), @PEL(A,X), @PEB(A,X), @PFS(A,X,C), @PFD(N,D,X), @PFD(N,D,X), @PCX(N,X), @PTD(N,X), and @RAND(X).

Other functions provided by LINGO include @IN (set_name, set_element), @SIZE (set_name), @WARN('text', condition), @WRAP(I,N), and @USER. These functions provide a variety of capabilities in addition to those of the categories above.

REFERENCES

There are many fine linear programming texts, including the following books:

Bazaraa, M., and J. Jarvis. *Linear Programming and Network Flows.* New York: Wiley, 1990.

Bersitmas, D., and J. Tsitsiklis. *Introduction to Linear Optimization.* Belmont, Mass.: Athena Publishing, 1997.

Bradley, S., A. Hax, and T. Magnanti. *Applied Mathematical Programming.* Reading, Mass.: Addison-Wesley, 1977.

Chvàtal, V. *Linear Programming.* San Francisco: Freeman, 1983.

Dantzig, G. *Linear Programming and Extensions.* Princeton, N.J.: Princeton University Press, 1963.

Gass, S. *Linear Programming: Methods and Applications,* 5th ed. New York: McGraw-Hill, 1985.

Luenberger, D. *Linear and Nonlinear Programming,* 2d ed. Reading, Mass.: Addison-Wesley, 1984.

Murty, K. *Linear Programming.* New York: Wiley, 1983.

Nash, S., and A. Sofer. *Linear and Nonlinear Programming.* New York: McGraw-Hill, 1995.

Nering, E., and A. Tucker. *Linear Programs and Related Problems.* New York: Academic Press, 1993.

Simmons, D. *Linear Programming for Operations Research.* Englewood Cliffs, N.J.: Prentice Hall, 1972.

Simonnard, M. *Linear Programming.* Englewood Cliffs, N.J.: Prentice Hall, 1966.

Wu, N., and R. Coppins. *Linear Programming and Extensions.* New York: McGraw-Hill, 1981.

Bland, R. "New Finite Pivoting Rules for the Simplex Method," *Mathematics of Operations Research* 2(1977):103–107. Describes simple, elegant approach to prevent cycling.

Dantzig, G., and N. Thapa. *Linear Programming.* New York: Springer-Verlag, 1997.

Karmarkar, N. "A New Polynomial Time Algorithm for Linear Programming," *Combinatorica* 4(1984):373–395. Karmarkar's method for solving LPs.

Klee, V., and G. Minty. "How Good Is the Simplex Algorithm?" In *Inequalities—III.* New York: Academic Press, 1972. Describes LPs for which the simplex method examines every basic feasible solution before finding the optimal solution.

Kotiah, T., and N. Slater. "On Two-Server Poisson Queues with Two Types of Customers," *Operations Research* 21(1973):597–603. Describes an actual application that led to an LP in which cycling occurred.

Love, R., and L. Yerex. "An Application of a Facilities Location Model in the Prestressed Concrete Industry," *Interfaces* 6(no.4, 1976):45–49.

Papadimitriou, C., and K. Steiglitz. *Combinatorial Optimization: Algorithms and Complexity.* Englewood Cliffs, N.J.: Prentice Hall, 1982. More discussion of polynomial time and exponential time algorithms.

Schrage, L. *User's Manual for LINDO.* Palo Alto, Calif.: Scientific Press, 1990. Gives complete details of LINDO.

Schrage, L. *User's Manual for LINGO.* Chicago, Ill.: LINDO Systems Inc., 1991. Gives complete details of LINGO.

Schrage, L. *User's Manual for What's Best.* Chicago, Ill.: LINDO Systems Inc., 1993. Gives complete details of What's Best.

Wagner, H. "Linear Programming Techniques for Regression Analysis," *Journal of the American Statistical Association* 54(1954):206–212.

5

Sensitivity Analysis: An Applied Approach

In this chapter, we discuss how changes in an LP's parameters affect the optimal solution. This is called *sensitivity analysis*. We also explain how to use the LINDO output to answer questions of managerial interest such as "What is the most money a company would be willing to pay for an extra hour of labor?" We begin with a graphical explanation of sensitivity analysis.

5.1 A Graphical Introduction to Sensitivity Analysis

Sensitivity analysis is concerned with how changes in an LP's parameters affect the optimal solution.

Reconsider the Giapetto problem of Section 3.1:

$$\max z = 3x_1 + 2x_2$$
$$\begin{aligned} \text{s.t.} \quad & 2x_1 + x_2 \leq 100 && \text{(Finishing constraint)} \\ & x_1 + x_2 \leq 80 && \text{(Carpentry constraint)} \\ & x_1 \leq 40 && \text{(Demand constraint)} \\ & x_1, x_2 \geq 0 \end{aligned}$$

where

$$x_1 = \text{number of soldiers produced per week}$$
$$x_2 = \text{number of trains \ produced per week}$$

The optimal solution to this problem is $z = 180$, $x_1 = 20$, $x_2 = 60$ (point B in Figure 1), and it has x_1, x_2, and s_3 (the slack variable for the demand constraint) as basic variables. How would changes in the problem's objective function coefficients or right-hand sides change this optimal solution?

Graphical Analysis of the Effect of a Change in an Objective Function Coefficient

If the contribution to profit of a soldier were to increase sufficiently, then it seems reasonable that it would be optimal for Giapetto to produce more soldiers (that is, s_3 would become nonbasic). Similarly, if the contribution to profit of a soldier were to decrease sufficiently, then it would become optimal for Giapetto to produce only trains (x_1 would now be nonbasic). We now show how to determine the values of the contribution to profit for soldiers for which the current optimal basis will remain optimal.

Let c_1 be the contribution to profit by each soldier. For what values of c_1 does the current basis remain optimal?

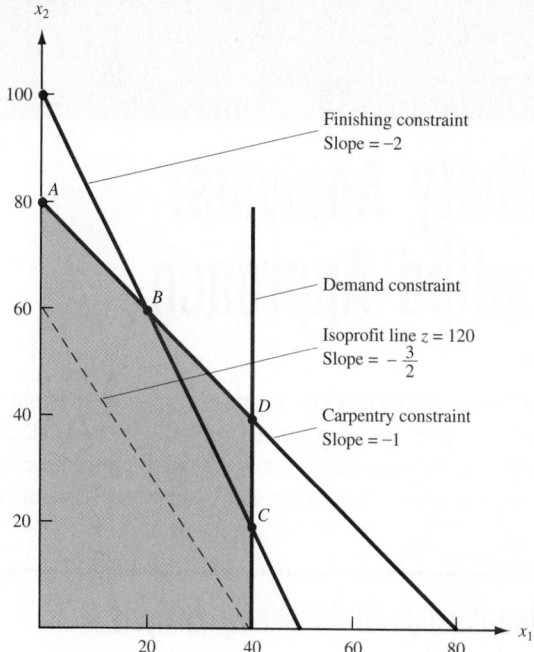

FIGURE 1
Analysis of Range of Values for Which c_1 Remains Optimal in Giapetto Problem

Currently, $c_1 = 3$, and each isoprofit line has the form $3x_1 + 2x_2 =$ constant, or

$$x_2 = -\frac{3x_1}{2} + \frac{\text{constant}}{2}$$

and each isoprofit line has a slope of $-\frac{3}{2}$. From Figure 1, we see that if a change in c_1 causes the isoprofit lines to be flatter than the carpentry constraint, then the optimal solution will change from the current optimal solution (point B) to a new optimal solution (point A). If the profit for each soldier is c_1, the slope of each isoprofit line will be $-\frac{c_1}{2}$. Because the slope of the carpentry constraint is -1, the isoprofit lines will be flatter than the carpentry constraint if $-\frac{c_1}{2} > -1$, or $c_1 < 2$, and the current basis will no longer be optimal. The new optimal solution will be $(0, 80)$, point A in Figure 1.

If the isoprofit lines are steeper than the finishing constraint, then the optimal solution will change from point B to point C. The slope of the finishing constraint is -2. If $-\frac{c_1}{2} < -2$, or $c_1 > 4$, then the current basis is no longer optimal and point C, $(40, 20)$, will be optimal. In summary, we have shown that (if all other parameters remain unchanged) the current basis remains optimal for $2 \leq c_1 \leq 4$, and Giapetto should still manufacture 20 soldiers and 60 trains. Of course, even if $2 \leq c_1 \leq 4$, Giapetto's profit will change. For instance, if $c_1 = 4$, then Giapetto's profit will now be $4(20) + 2(60) = \$200$ instead of $\$180$.

Graphical Analysis of the Effect of a Change in a Right-Hand Side on the LP's Optimal Solution

A graphical analysis can also be used to determine whether a change in the right-hand side of a constraint will make the current basis no longer optimal. Let b_1 be the number of available finishing hours. Currently, $b_1 = 100$. For what values of b_1 does the current basis remain optimal? From Figure 2, we see that a change in b_1 shifts the finishing constraint parallel to its current position. The current optimal solution (point B in Figure 2)

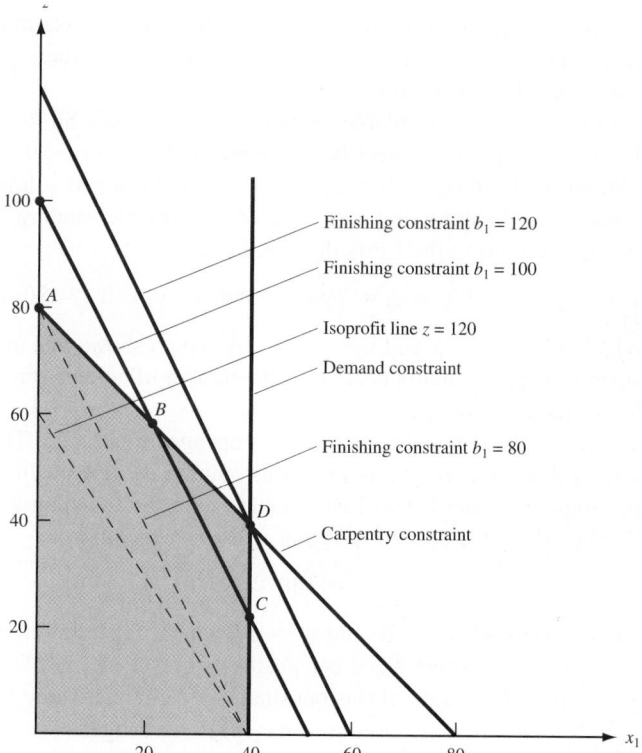

FIGURE 2
Range of Values on Finishing Hours for Which Current Basis Remains Optimal in Giapetto Problem

is where the carpentry and finishing constraints are binding. If we change the value of b_1, then *as long as the point where the finishing and carpentry constraints are binding remains feasible, the optimal solution will still occur where the finishing and carpentry constraints intersect.* From Figure 2, we see that if $b_1 > 120$, then the point where the finishing and carpentry constraints are both binding will lie on the portion of the carpentry constraint below point D. Note that at point D, $2(40) + 40 = 120$ finishing hours are used. In this region, $x_1 > 40$, and the demand constraint for soldiers is not satisfied. Thus, for $b_1 > 120$, the current basis will no longer be optimal. Similarly, if $b_1 < 80$, then the carpentry and finishing constraints will be binding at an infeasible point having $x_1 < 0$, and the current basis will no longer be optimal. Note that at point A, $0 + 80 = 80$ finishing hours are used. Thus (if all other parameters remain unchanged), the current basis remains optimal if $80 \leq b_1 \leq 120$.

Note that although for $80 \leq b_1 \leq 120$, the current basis remains optimal, *the values of the decision variables and the objective function value change.* For example, if $80 \leq b_1 \leq 100$, then the optimal solution will change from point B to some other point on the line segment AB. Similarly, if $100 \leq b_1 \leq 120$, then the optimal solution will change from point B to some other point on the line BD.

As long as the current basis remains optimal, it is a routine matter to determine how a change in the right-hand side of a constraint changes the values of the decision variables. To illustrate the idea, let $b_1 =$ number of available finishing hours. If we change b_1 to $100 + \Delta$, then we know that the current basis remains optimal for $-20 \leq \Delta \leq 20$. Note that as b_1 changes (as long as $-20 \leq \Delta \leq 20$), the optimal solution to the LP is still the point where the finishing-hour and carpentry-hour constraints are binding. Thus, if $b_1 = 100 + \Delta$, we can find the new values of the decision variables by solving

$$2x_1 + x_2 = 100 + \Delta \quad \text{and} \quad x_1 + x_2 = 80$$

This yields $x_1 = 20 + \Delta$ and $x_2 = 60 - \Delta$. Thus, an increase in the number of available finishing hours results in an increase in the number of soldiers produced and a decrease in the number of trains produced.

If b_2 (the number of available carpentry hours) equals $80 + \Delta$, then it can be shown (see Problem 2) that the current basis remains optimal for $-20 \leq \Delta \leq 20$. If we change the value of b_2 (keeping $-20 \leq \Delta \leq 20$), then the optimal solution to the LP is still the point where the finishing and carpentry constraints are binding. Thus, if $b_2 = 80 + \Delta$, the optimal solution to the LP is the solution to

$$2x_1 + x_2 = 100 \qquad \text{and} \qquad x_1 + x_2 = 80 + \Delta$$

This yields $x_1 = 20 - \Delta$ and $x_2 = 60 + 2\Delta$, which shows that an increase in the amount of available carpentry hours decreases the number of soldiers produced and increases the number of trains produced.

Suppose b_3, the demand for soldiers, is changed to $40 + \Delta$. Then it can be shown (see Problem 3) that the current basis remains optimal for $\Delta \geq -20$. For Δ in this range, the optimal solution to the LP will still occur where the finishing and carpentry constraints are binding. Thus, the optimal solution will be the solution to

$$2x_1 + x_2 = 100 \qquad \text{and} \qquad x_1 + x_2 = 80$$

Of course, this yields $x_1 = 20$ and $x_2 = 60$, which illustrates an important fact. Consider a constraint with positive slack (or positive excess) in an LP's optimal solution; if we change the right-hand side of this constraint in the range where the current basis remains optimal, then the optimal solution to the LP is unchanged.

Shadow Prices

As we will see in Sections 5.2 and 5.3, it is often important for managers to determine how a change in a constraint's right-hand side changes the LP's optimal z-value. With this in mind, we define the **shadow price** for the ith constraint of an LP to be the amount by which the optimal z-value is improved—increased in a max problem and decreased in a min problem—if the right-hand side of the ith constraint is increased by 1. This definition applies only if the change in the right-hand side of Constraint i leaves the current basis optimal.

For any two-variable LP, it is a simple matter to determine each constraint's shadow price. For example, we know that if $100 + \Delta$ finishing hours are available (assuming the current basis remains optimal), then the LP's optimal solution is $x_1 = 20 + \Delta$ and $x_2 = 60 - \Delta$. Then the optimal z-value will equal $3x_1 + 2x_2 = 3(20 + \Delta) + 2(60 - \Delta) = 180 + \Delta$. Thus, as long as the current basis remains optimal, a one-unit increase in the number of available finishing hours will increase the optimal z-value by \$1. So the shadow price of the first (finishing hours) constraint is \$1.

For the second (carpentry hours) constraint, we know that if $80 + \Delta$ carpentry hours are available (and the current basis remains optimal), then the optimal solution to the LP is $x_1 = 20 - \Delta$ and $x_2 = 60 + 2\Delta$. Then the new optimal z-value is $3x_1 + 2x_2 = 3(20 - \Delta) + 2(60 + 2\Delta) = 180 + \Delta$. So a one-unit increase in the number of finishing hours will increase the optimal z-value by \$1 (as long as the current basis remains optimal). Thus, the shadow price of the second (carpentry hour) constraint is \$1.

We now find the shadow price of the third (demand) constraint. If the right-hand side is $40 + \Delta$, then (as long as the current basis remains optimal) the optimal values of the decision variables remain unchanged. Then the optimal z-value will also remain unchanged, which shows that the shadow price of the third (demand) constraint is \$0. It turns out that whenever the slack or excess variable for a constraint is positive in an LP's optimal solution, the constraint will have a zero shadow price.

Suppose that the current basis remains optimal as we increase the right-hand side of the ith constraint of an LP by Δb_i. ($\Delta b_i < 0$ means that we are decreasing the right-hand side of the ith constraint.) Then each unit by which Constraint i's right-hand side is increased will increase the optimal z-value (for a max problem) by the shadow price. Thus, the new optimal z-value is given by

(New optimal z-value) = (old optimal z-value) + (Constraint i's shadow price) Δb_i **(1)**

For a minimization problem,

(New optimal z-value) = (old optimal z-value) − (Constraint i's shadow price) Δb_i **(2)**

For example, if 95 carpentry hours are available, then $\Delta b_2 = 15$, and the new z-value is given by

$$\text{New optimal } z\text{-value} = 180 + 15(1) = \$195$$

We will continue our discussion of shadow prices in Sections 5.2 and 5.3.

Importance of Sensitivity Analysis

Sensitivity analysis is important for several reasons. In many applications, the values of an LP's parameters may change. For example, the prices at which soldiers and trains are sold or the availability of carpentry and finishing hours may change. If a parameter changes, then sensitivity analysis often makes it unnecessary to solve the problem again. For example, if the profit contribution of a soldier increased to $3.50, we would not have to solve the Giapetto problem again, because the current solution remains optimal. Of course, solving the Giapetto problem again would not be much work, but solving an LP with thousands of variables and constraints again would be a chore. A knowledge of sensitivity analysis often enables the analyst to determine from the original solution how changes in an LP's parameters change its optimal solution.

Recall that we may be uncertain about the values of parameters in an LP. For example, we might be uncertain about the weekly demand for soldiers. With the graphical method, it can be shown that if the weekly demand for soldiers is at least 20, then the optimal solution to the Giapetto problem is still (20, 60) (see Problem 3 at the end of this section). Thus, even if Giapetto is uncertain about the demand for soldiers, the company can be fairly confident that it is still optimal to produce 20 soldiers and 60 trains.

PROBLEMS

Group A

1 Show that if the contribution to profit for trains is between $1.50 and $3, the current basis remains optimal. If the contribution to profit for trains is $2.50, then what would be the new optimal solution?

2 Show that if available carpentry hours remain between 60 and 100, the current basis remains optimal. If between 60 and 100 carpentry hours are available, would Giapetto still produce 20 soldiers and 60 trains?

3 Show that if the weekly demand for soldiers is at least 20, then the current basis remains optimal, and Giapetto should still produce 20 soldiers and 60 trains.

4 For the Dorian Auto problem (Example 2 in Chapter 3),

a Find the range of values on the cost of a comedy ad for which the current basis remains optimal.

b Find the range of values on the cost of a football ad for which the current basis remains optimal.

c Find the range of values for required HIW exposures for which the current basis remains optimal. Determine the new optimal solution if $28 + \Delta$ million HIW exposures are required.

d Find the range of values for required HIM exposures for which the current basis remains optimal. Determine the new optimal solution if $24 + \Delta$ million HIM exposures are required.

e Find the shadow price of each constraint.

f If 26 million HIW exposures are required, determine the new optimal z-value.

5 Radioco manufactures two types of radios. The only scarce resource that is needed to produce radios is labor. At present, the company has two laborers. Laborer 1 is willing to work up to 40 hours per week and is paid \$5 per hour. Laborer 2 will work up to 50 hours per week for \$6 per hour. The price as well as the resources required to build each type of radio are given in Table 1.

Letting x_i be the number of Type i radios produced each week, Radioco should solve the following LP:

$$\max z = 3x_1 + 2x_2$$
$$\text{s.t.} \quad x_1 + 2x_2 \le 40$$
$$2x_1 + x_2 \le 50$$
$$x_1, x_2 \ge 0$$

a For what values of the price of a Type 1 radio would the current basis remain optimal?

b For what values of the price of a Type 2 radio would the current basis remain optimal?

TABLE 1

	Radio 1		Radio 2	
Price (\$)	Resource Required	Price (\$)	Resource Required	
25	Laborer 1: 1 hour	22	Laborer 1: 2 hours	
	Laborer 2: 2 hours		Laborer 2: 2 hours	
	Raw material cost: \$5		Raw material cost: \$4	

c If laborer 1 were willing to work only 30 hours per week, then would the current basis remain optimal? Find the new optimal solution to the LP.

d If laborer 2 were willing to work up to 60 hours per week, then would the current basis remain optimal? Find the new optimal solution to the LP.

e Find the shadow price of each constraint.

5.2 The Computer and Sensitivity Analysis

If an LP has more than two decision variables, the range of values for a right-hand side (or objective function coefficient) for which the current basis remains optimal cannot be determined graphically. These ranges can be computed by hand calculations (see Section 6.3), but this is often tedious, so they are usually determined by packaged computer programs. In this section, we discuss the interpretation of the sensitivity analysis information on the LINDO output.

To obtain a sensitivity report in LINDO, select Yes when asked (after solving LP) whether you want a Range analysis. To obtain sensitivity report in LINGO, go to Options and select Range (after solving LP). If this does not work, then go to Options and choose the General Solver tab. Then go to Dual Computations and select the Ranges and Values option.

EXAMPLE 1   **Winco Products 1**

Winco sells four types of products. The resources needed to produce one unit of each and the sales prices are given in Table 2. Currently, 4,600 units of raw material and 5,000 labor hours are available. To meet customer demands, exactly 950 total units must be produced. Customers also demand that at least 400 units of product 4 be produced. Formulate an LP that can be used to maximize Winco's sales revenue.

Solution   Let x_i = number of units of product i produced by Winco.

$$\max z = 4x_1 + 6x_2 + 7x_3 + 8x_4$$
$$\text{s.t.} \quad x_1 + x_2 + x_3 + x_4 = 950$$
$$x_4 \ge 400$$
$$2x_1 + 3x_2 + 4x_3 + 7x_4 \le 4,600$$
$$3x_1 + 4x_2 + 5x_3 + 6x_4 \le 5,000$$
$$x_1, x_2, x_3, x_4 \ge 0$$

TABLE 2
Costs and Resource Requirements for Winco

Resource	Product 1	Product 2	Product 3	Product 4
Raw material	2	3	4	7
Hours of labor	3	4	5	6
Sales price ($)	4	6	7	8

The LINDO output for this LP is given in Figure 3.

When we discuss the interpretation of the LINDO output for minimization problems, we will refer to the following example.

EXAMPLE 2   Tucker Inc.

Tucker Inc. must produce 1,000 Tucker automobiles. The company has four production plants. The cost of producing a Tucker at each plant, along with the raw material and labor needed, is shown in Table 3.

```
MAX      4  X1 + 6  X2 + 7  X3  + 8  X4
SUBJECT TO
      2)   X1  +  X2  +  X3  +  X4  =        950
      3)   X4   >=      400
      4)   2  X1  +  3  X2  +  4  X3 + 7  X4  <=     4600
      5)   3  X1  +  4  X2  +  5  X3 + 6  X4  <=     5000
END

LP OPTIMUM FOUND AT STEP          4
                 OBJECTIVE FUNCTION VALUE
             1)   6650.00000

VARIABLE          VALUE          REDUCED COST
    X1           .000000           1.000000
    X2          400.000000          .000000
    X3          150.000000          .000000
    X4          400.000000          .000000


    ROW        SLACK OR SURPLUS     DUAL PRICES
    2)            .000000           3.000000
    3)            .000000          -2.000000
    4)            .000000           1.000000
    5)          250.000000          .000000

NO. ITERATIONS=          4

RANGES IN WHICH THE BASIS IS UNCHANGED:

                    OBJ COEFFICIENT RANGES
VARIABLE    CURRENT       ALLOWABLE        ALLOWABLE
             COEF         INCREASE         DECREASE
    X1     4.000000       1.000000         INFINITY
    X2     6.000000        .666667          .500000
    X3     7.000000       1.000000          .500000
    X4     8.000000       2.000000         INFINITY

                    RIGHTHAND SIDE RANGES
    ROW    CURRENT       ALLOWABLE        ALLOWABLE
            RHS          INCREASE         DECREASE
     2     950.000000    50.000000        100.000000
     3     400.000000    37.000000        125.000000
     4    4600.000000    250.000000       150.000000
     5    5000.000000    INFINITY         250.000000
```

FIGURE 3
LINDO Output
for Winco

TABLE 3
Cost and Requirements for Producing a Tucker

Plant	Cost (in Thousands of Dollars)	Labor	Raw Material
1	15	2	3
2	10	3	4
3	9	4	5
4	7	5	6

The autoworkers' labor union requires that at least 400 cars be produced at plant 3; 3,300 hours of labor and 4,000 units of raw material are available for allocation to the four plants. Formulate an LP whose solution will enable Tucker Inc. to minimize the cost of producing 1,000 cars.

Solution Let x_i = number of cars produced at plant i. Then, expressing the objective function in thousands of dollars, the appropriate LP is

$$\min z = 15x_1 + 10x_2 + 9x_3 + 7x_4$$
$$\text{s.t.} \quad x_1 + x_2 + x_3 + x_4 = 1000$$
$$x_3 \geq 400$$
$$2x_1 + 3x_2 + 4x_3 + 5x_4 \leq 3300$$
$$3x_1 + 4x_2 + 5x_3 + 6x_4 \leq 4000$$
$$x_1, x_2, x_3, x_4 \geq 0$$

The LINDO output for this LP is given in Figure 4.

Objective Function Coefficient Ranges

Recall from Section 5.1 that (at least in a two-variable problem) we can determine the range of values for an objective function coefficient for which the current basis remains optimal. For each objective function coefficient, this range is given in the OBJECTIVE COEFFICIENT RANGES portion of the LINDO output. The ALLOWABLE INCREASE (AI) section indicates the amount by which an objective function coefficient can be increased with the current basis remaining optimal. Similarly, the ALLOWABLE DE-CREASE (AD) section indicates the amount by which an objective function coefficient can be decreased with the current basis remaining optimal. To illustrate these ideas, let c_i be the objective function coefficient for x_i in Example 1. If c_1 is changed, then the current basis remains optimal if

$$-\infty = 4 - \infty \leq c_1 \leq 4 + 1 = 5$$

If c_2 is changed, then the current basis remains optimal if

$$5.5 = 6 - 0.5 \leq c_2 \leq 6 + 0.666667 = 6.666667$$

We will refer to the range of variables of c_i for which the current basis remains optimal as the **allowable range** for c_i. As discussed in Section 5.1, if c_i remains in its allowable range then the values of the decision variables remain unchanged, although the optimal z-value may change. The following examples illustrate these ideas.

```
MIN       15 X1 + 10  X2  + 9  X3  + 7  X4
SUBJECT TO
     2)    X1  +  X2  +  X3  +  X4  =      1000
     3)    X3   >=       400
     4)    2 X1  +  3  X2  +  4  X3 + 5  X4  <=    3300
     5)    3 X1  +  4  X2  +  5  X3 + 6  X4  <=    4000
END

LP OPTIMUM FOUND AT STEP          3
              OBJECTIVE FUNCTION VALUE
              1)    11600.0000

VARIABLE             VALUE         REDUCED COST
     X1         400.000000            .000000
     X2         200.000000            .000000
     X3         400.000000            .000000
     X4            .000000           7.000000


     ROW      SLACK OR SURPLUS      DUAL PRICES
     2)            .000000         -30.000000
     3)            .000000          -4.000000
     4)         300.000000            .000000
     5)            .000000           5.000000


NO. ITERATIONS=          3

RANGES IN WHICH THE BASIS IS UNCHANGED:

                      OBJ COEFFICIENT RANGES
VARIABLE    CURRENT      ALLOWABLE        ALLOWABLE
             COEF        INCREASE         DECREASE
    X1    15.000000      INFINITY         3.500000
    X2    10.000000      2.000000         INFINITY
    X3     9.000000      INFINITY         4.000000
    X4     7.000000      INFINITY         7.000000

                      RIGHTHAND SIDE RANGES
    ROW    CURRENT      ALLOWABLE        ALLOWABLE
            RHS         INCREASE         DECREASE
     2   1000.000000    66.666660        100.000000
     3    400.000000   100.000000        400.000000
     4   3300.000000     INFINITY        300.000000
     5   4000.000000   300.000000        200.000000
```

FIGURE 4
LINDO Output
for Tucker

a Suppose Winco raises the price of product 2 by 50¢ per unit. What is the new optimal solution to the LP?

b Suppose the sales price of product 1 is increased by 60¢ per unit. What is the new optimal solution to the LP?

c Suppose the sales price of product 3 is decreased by 60¢. What is the new optimal solution to the LP?

Solution **a** Because the AI for c_2 is \$0.666667, and we are increasing c_2 by only \$0.5, the current basis remains optimal. The optimal values of the decision variables remain unchanged ($x_1 = 0$, $x_2 = 400$, $x_3 = 150$, and $x_4 = 400$ is still optimal). The new optimal z-value may be determined in two ways. First, we may simply substitute the optimal values of the decision variables into the new objective function, yielding

$$\text{New optimal } z\text{-value} = 4(0) + 6.5(400) + 7(150) + 8(400) = \$6,850$$

Another way to see that the new optimal z-value is \$6,850 is to observe the only difference in sales revenue: Each unit of product 2 brings in 50¢ more in revenue. Thus, total revenue should increase by 400(.50) = \$200, so

$$\text{New } z\text{-value} = \text{original } z\text{-value} + 200 = \$6,850$$

b The AI for c_1 is 1, so the current basis remains optimal, and the optimal values of the decision variables remain unchanged. Because the value of x_1 in the optimal solution is 0, the change in the sales price for product 1 will not change the optimal z-value—it will remain $6,650.

c For c_3, AD = .50, so the current basis is no longer optimal. Without resolving the problem by hand or on the computer, we cannot determine the new optimal solution.

Reduced Costs and Sensitivity Analysis

The REDUCED COST portion of the LINDO output gives us information about how changing the objective function coefficient for a nonbasic variable will change the LP's optimal solution. For simplicity, let's assume that the current optimal bfs is nondegenerate (that is, if the LP has m constraints, then the current optimal solution has m variables assuming positive values). For any nonbasic variable x_k, the reduced cost is the amount by which the objective function coefficient of x_k must be improved before the LP will have an optimal solution in which x_k is a basic variable. If the objective function coefficient of a nonbasic variable x_k is improved by its reduced cost, then the LP will have alternative optimal solutions—at least one in which x_k is a basic variable, and at least one in which x_k is not a basic variable. If the objective function coefficient of a nonbasic variable x_k is improved by more than its reduced cost, then (barring degeneracy) any optimal solution to the LP will have x_k as a basic variable and $x_k > 0$. To illustrate these ideas, note that in Example 1 the basic variables associated with the optimal solution are x_2, x_3, x_4, and s_4 (the slack for the labor constraint). The nonbasic variable x_1 has a reduced cost of $1. This implies that if we increase x_1's objective function coefficient (in this case, the sales price per unit of x_1) by exactly $1, then there will be alternative optimal solutions, at least one of which will have x_1 as a basic variable. If we increase x_1's objective function coefficient by more than $1, then (because the current optimal bfs is nondegenerate) any optimal solution to the LP will have x_1 as a basic variable (with $x_1 > 0$). Thus, the reduced cost for x_1 is the amount by which x_1 "misses the optimal basis." We must keep a close watch on x_1's sales price, because a slight increase will change the LP's optimal solution.

Let's now consider Example 2, a minimization problem. Here the basic variables associated with the optimal solution are x_1, x_2, x_3, and s_3 (the slack variable for the labor constraint). Again, the optimal bfs is nondegenerate. The nonbasic variable x_4 has a reduced cost of 7 ($7,000), so we know that if the cost of producing x_4 is decreased by 7, then there will be alternative optimal solutions. In at least one of these optimal solutions, x_4 will be a basic variable. If the cost of producing x_4 is lowered by more than 7, then (because the current optimal solution is nondegenerate) any optimal solution to the LP will have x_4 as a basic variable (with $x_4 > 0$).

Right-Hand Side Ranges

Recall from Section 5.1 that we can determine (at least for a two-variable problem) the range of values for a right-hand side within which the current basis remains optimal. This information is given in the RIGHTHAND SIDE RANGES section of the LINDO output. To illustrate, consider the first constraint in Example 1. Currently, the right-hand side of this constraint (call it b_1) is 950. The current basis remains optimal if b_1 is decreased by up to 100 (the allowable decrease, or AD, for b_1) or increased by up to 50 (the allowable increase, or AI, for b_1). Thus, the current basis remains optimal if

$$850 = 950 - 100 \leq b_1 \leq 950 + 50 = 1{,}000$$

We call this the allowable range for b_1. Even if a change in the right-hand side of a constraint leaves the current basis optimal, the LINDO output does not provide sufficient information to determine the new values of the decision variables. However, the LINDO output does allow us to determine the LP's new optimal z-value.

Shadow Prices and Dual Prices

In Section 5.1, we defined the shadow price of an LP's ith constraint to be the amount by which the optimal z-value of the LP is improved if the right-hand side is increased by one unit (assuming this change leaves the current basis optimal). If, after a change in a constraint's right-hand side, the current basis is no longer optimal, then the shadow prices of *all* constraints may change. We will discuss this further in Section 5.4. The shadow price for each constraint is found in the DUAL PRICES section of the LINDO output. If we increase the right-hand side of the ith constraint by an amount Δb_i—a decrease in b_i implies that $\Delta b_i < 0$—and the new right-hand side value for Constraint i remains within the allowable range for the right-hand side given in the RIGHTHAND SIDE RANGES section of the output, then formulas (1) and (2) may be used to determine the optimal z-value after a right-hand side is changed. The following example illustrates how shadow prices may be used to determine how a change in a right-hand side affects the optimal z-value.

EXAMPLE 4   Interpretation of RHS Sensitivity Analysis

a   In Example 1, suppose that a total of 980 units must be produced. Determine the new optimal z-value.

b   In Example 1, suppose that 4,500 units of raw material are available. What is the new optimal z-value? What if only 4,400 units of raw material are available?

c   In Example 2, suppose that 4,100 units of raw material are available. Find the new optimal z-value.

d   In Example 2, suppose that exactly 950 cars must be produced. What will be the new optimal z-value?

Solution   **a** $\Delta b_1 = 30$. Because the allowable increase is 50, the current basis remains optimal, and the shadow price of $3 remains applicable. Then (1) yields

$$\text{New optimal } z\text{-value} = 6{,}650 + 30(3) = \$6{,}740$$

Here we see that (as long as the current basis remains optimal) each additional unit of demand increases revenues by $3.

b $\Delta b_3 = -100$. Because the allowable decrease is 150, the shadow price of $1 remains valid. Then (1) yields

$$\text{New optimal } z\text{-value} = 6{,}650 - 100(1) = \$6{,}550$$

Thus (as long as the current basis remains optimal), a decrease in available raw material of one unit decreases revenue by $1. If only 4,400 units of raw material are available, then $\Delta b_3 = -200$. Because the allowable decrease is 150, we cannot determine the new optimal z-value.

c $\Delta b_4 = 100$. The dual (or shadow) price is 5 (thousand). The current basis remains optimal, so (2) yields

$$\text{New optimal } z\text{-value} = 11{,}600 - 100(5) = 11{,}100 \ (\$11{,}100{,}000)$$

Thus, as long as the current basis remains optimal, each additional unit of raw material decreases costs by $5,000.

d $\Delta b_1 = -50$. The allowable decrease is 100, so the shadow price of -30 (thousand) and (2) yield

New optimal z-value $= 11{,}600 - (-50)(-30) = 10{,}100 = \$10{,}100{,}000$

Thus, each unit by which demand is reduced (as long as the current basis remains optimal) decreases costs by $30,000.

Let's give an interpretation to the shadow price for each constraint in Examples 1 and 2. Again, all discussions are assuming that we are within the allowable range where the current basis remains optimal. The shadow price of $3 for Constraint 1 in Example 1 implies that each one-unit increase in total demand will increase sales revenues by $3. The shadow price of $-\$2$ for Constraint 2 implies that each unit increase in the requirement for product 4 will decrease revenue by $2. The shadow price of $1 for Constraint 3 implies that an additional unit of raw material given to Winco (for no cost) increases total revenue by $1. Finally, the shadow price of $0 for Constraint 4 implies that an additional unit of labor given to Winco (at no cost) will not increase total revenue. This is reasonable; at present, 250 of the available 5,000 labor hours are not being used, so why should we expect additional labor to raise revenues?

The shadow price of $-\$30$ (thousand) for Constraint 1 of Example 2 means that each extra car that must be produced will decrease costs by $-\$30{,}000$ (or increase costs by $30,000). The shadow price of $-\$4$ (thousand) for Constraint 2 means that an extra car that the firm is forced to produce at plant 3 will decrease costs by $-\$4{,}000$ (or increase costs by $4,000). The shadow price of $0 for the third constraint means that an extra hour of labor given to Tucker will decrease costs by $0. Thus, if Tucker is given an additional hour of labor then costs are unchanged. This is reasonable; now 300 hours of available labor are unused. The shadow price for Constraint 4 is $5 (thousand), which means that if Tucker were given an additional unit of raw material, then costs would decrease by $5,000.

Signs of Shadow Prices

A $\geq$ constraint will always have a nonpositive shadow price; a $\leq$ constraint will always have a nonnegative shadow price; and an equality constraint may have a positive, negative, or zero shadow price. To see why this is true, observe that adding points to an LP's feasible region can only improve the optimal z-value or leave it the same. Eliminating points from an LP's feasible region can only make the optimal z-value worse or leave it the same. For example, let's look at the shadow price of the raw-material constraint (a $\leq$ constraint) in Example 1. Why must this shadow price be nonnegative? The shadow price of the raw-material constraint represents the improvement in the optimal z-value if 4,601 units (instead of 4,600) of raw material are available. Having an additional unit of raw material available adds points to the feasible region—points for which Winco uses $>$ 4,600 but $\leq$ 4,601 units of raw material—so we know that the optimal z-value must increase or stay the same. Thus, the shadow price of this $\leq$ constraint must be nonnegative.

Similarly, let's consider the shadow price of the $x_4 \geq 400$ constraint in Example 1. Increasing the right-hand side of this constraint to 401 eliminates points from the feasible region (points for which Winco produces ≥ 400 but < 401 units of product 4). Thus, the optimal z-value must decrease or stay the same, implying that the shadow price of this constraint must be nonpositive. Similar reasoning shows that for a minimization problem, a $\geq$ constraint will have a nonpositive shadow price, and a $\leq$ constraint will have a nonnegative shadow price.

An equality constraint's shadow price may be positive, negative, or zero. To see why, consider the following two LPs:

$$\max z = x_1 + x_2$$
$$\text{s.t.} \quad x_1 + x_2 = 1 \qquad \qquad \textbf{(LP 1)}$$
$$x_1, x_2 \geq 0$$

$$\max z = x_1 + x_2$$
$$\text{s.t.} \quad -x_1 - x_2 = -1 \qquad \qquad \textbf{(LP 2)}$$
$$x_1, x_2 \geq 0$$

Both LPs have the same feasible region and set of optimal solutions (the portion of the line segment $x_1 + x_2 = 1$ in the first quadrant). However, LP 1's constraint has a shadow price of $+1$, whereas LP 2's constraint has a shadow price of -1. Thus, the sign of the shadow price for an equality constraint may either be positive, negative, or zero.

Sensitivity Analysis and Slack and Excess Variables

It can be shown (see Section 6.10) that for any inequality constraint, the product of the values of the constraint's slack or excess variable and the constraint's shadow price must equal 0. This implies that any constraint whose slack or excess variable is > 0 will have a zero shadow price. It also implies that any constraint with a nonzero shadow price must be binding (have slack or excess equal to 0). To illustrate these ideas, consider the labor constraint in Example 1. This constraint has positive slack, so its shadow price must be 0. This is reasonable, because slack $= 250$ for this constraint indicates that 250 hours of currently available labor are unused at present. Thus, an extra hour of labor would not increase revenues. Now consider the raw material constraint of Example 1. Because this constraint has a nonzero shadow price, it must have slack $= 0$. This is reasonable; the nonzero shadow price means that additional raw material will increase revenue. This can be the case only if all currently available raw material is now being used.

For constraints with nonzero slack or excess, the value of the slack or excess variable is related to the ALLOWABLE INCREASE and ALLOWABLE DECREASE sections of the RIGHTHAND SIDE RANGES portion of the LINDO output. This relationship is detailed in Table 4.

For any constraint having positive slack or excess, the optimal z-value and values of the decision variables remain unchanged within the right-hand side's allowable range. To illustrate these ideas, consider the labor constraint in Example 1. Because slack $= 250$, we see from Table 4 that $\text{AI} = \infty$ and $\text{AD} = 250$. Thus, the current basis remains optimal

TABLE 4
Allowable Increases and Decreases for Constraints
with Nonzero Slack or Excess

Type of Constraint	AI for rhs	AD for rhs
$\leq$	∞	$=$ Value for slack
$\geq$	$=$ Value of excess	$= \infty$

for $4{,}750 \leq$ available labor $\leq \infty$. Within this range, both the optimal z-value and values of the decision variables remain unchanged.

Degeneracy and Sensitivity Analysis

When the optimal solution to an LP is degenerate, caution must be used when interpreting the LINDO output. Recall from Section 4.11 that a bfs is degenerate if at least one basic variable in the optimal solution equals 0. For an LP with m constraints, if the LINDO output indicates that less than m variables are positive, then the optimal solution is a degenerate bfs. To illustrate, consider the following LP:

$$\max z = 6x_1 + 4x_2 + 3x_3 + 2x_4$$
$$\text{s.t.} \quad 2x_1 + 3x_2 + x_3 + 2x_4 \leq 400$$
$$x_1 + x_2 + 2x_3 + x_4 \leq 150$$
$$2x_1 + x_2 + x_3 + .5x_4 \leq 200$$
$$3x_1 + x_2 \qquad\quad x_4 \leq 250$$
$$x_1, x_2, x_3, x_4 \geq 0$$

The LINDO output for this LP is in Figure 5. The LP has four constraints and in the optimal solution only two variables are positive, so the optimal solution is a degenerate bfs. By the way, using the **TABLEAU** command indicates that the optimal basis is BV = $\{x_2, x_3, s_3, x_1\}$.

We now discuss three "oddities" that may occur when the optimal solution found by LINDO is degenerate.

Oddity 1 In the RANGES IN WHICH THE BASIS IS UNCHANGED, at least one constraint will have a 0 AI or AD. This means that for at least one constraint, the DUAL PRICE can tell us about the new z-value for either an increase or decrease in the right-hand side, but not both.

To understand Oddity 1, consider the first constraint. Its AI is 0. This means that the first constraint's DUAL PRICE of .50 cannot be used to determine a new z-value resulting from any increase in the first constraint's right-hand side.

Oddity 2 For a nonbasic variable to become positive, its objective function coefficient may have to be improved by more than its REDUCED COST.

To understand Oddity 2, consider the nonbasic variable x_4; its REDUCED COST is 1.5. If we increase its objective function coefficient by 2, however, we still find that the new optimal solution has $x_4 = 0$. This oddity occurs because the increase changes the set of basic variables but not the LP's optimal solution.

Oddity 3 Increasing a variable's objective function coefficient by more than its AI or decreasing it by more than its AD may leave the optimal solution to the LP the same.

Oddity 3 is similar to Oddity 2. To understand it, consider the nonbasic variable x_4. Its AI is 1.5. If we increase its objective function coefficient by 2, however, we still find that the new optimal solution is unchanged. This oddity occurs because the increase changes the set of basic variables but not the LP's optimal solution.

We close this section by noting that our discussions apply only if one objective function coefficient or one right-hand side is changed. If more than one objective function coefficient or the right-hand side is changed, it is sometimes still possible to use the LINDO output to determine whether the current basis remains optimal. See Section 6.4 for details.

```
MAX        6  X1  +  4  X2  +  3  X3  +  2  X4
SUBJECT TO
       2)     2  X1  +  3 X2  +  X3  +  2  X4  <= 400
       3)        X1  +  X2  +  2  X3  +  X4  <=      150
       4)     2  X1  +  X2  +  X3  +  0.5  X4  <=      200
       5)     3  X1  +  X2  +  X4  <=      250
END

LP OPTIMUM FOUND AT STEP           3
              OBJECTIVE FUNCTION VALUE
          1)   700.00000

   VARIABLE          VALUE          REDUCED COST
       X1         50.000000            .000000
       X2        100.000000            .000000
       X3          .000000             .000000
       X4          .000000            1.500000


      ROW     SLACK OR SURPLUS       DUAL PRICES
      2)          .000000              .500000
      3)          .000000             1.250000
      4)          .000000              .000000
      5)          .000000             1.250000

NO. ITERATIONS=           3

RANGES IN WHICH THE BASIS IS UNCHANGED:

                  OBJ COEFFICIENT RANGES
   VARIABLE   CURRENT        ALLOWABLE        ALLOWABLE
              COEF           INCREASE         DECREASE
       X1    6.000000        3.000000        3.000000
       X2    4.000000        5.000000        1.000000
       X3    3.000000        3.000000        2.142857
       X4    2.000000        1.500000        INFINITY

                  RIGHTHAND SIDE RANGES
     ROW    CURRENT        ALLOWABLE        ALLOWABLE
            RHS            INCREASE         DECREASE
      2   400.000000        .000000        200.000000
      3   150.000000        .000000          .000000
      4   200.000000       INFINITY          .000000
      5   250.000000        .000000        120.000000

THE TABLEAU
    ROW    (BASIS)    X1       X2       X3       X4      SLK   2
      1    ART       .000     .000     .000    1.500     .500
      2     X2       .000    1.000     .000     .500     .500
      3     X3       .000     .000    1.000     .167    -.167
      4    SLK  4    .000     .000     .000    -.500     .000
      5     X1      1.000     .000     .000     .167    -.167

    ROW    SLK   3    SLK   4    SLK   5
      1    1.250      .000     1.250     700.000
      2    -.250      .000     -.250     100.000
      3     .583      .000     -.083       .000
      4    -.500     1.000     -.500       .000
      5     .083      .000      .417      50.000
```

FIGURE 5

PROBLEMS

Group A

1 Farmer Leary grows wheat and corn on his 45-acre farm. He can sell at most 140 bushels of wheat and 120 bushels of corn. Each acre planted with wheat yields 5 bushels, and each acre planted with corn yields 4 bushels. Wheat sells for $30 per bushel, and corn sells for $50 per bushel. To harvest an acre of wheat requires 6 hours of labor; 10 hours are needed to harvest an acre of corn. Up to 350 hours of labor can be purchased at $10 per hour. Let $A1$ = acres planted with wheat; $A2$ = acres planted with corn; and L = hours of labor that are purchased. To maximize profits, Leary should solve the following LP:

$$\max z = 150A1 + 200A2 - 10L$$
$$\text{s.t.} \quad A1 + A2 \leq 45$$
$$6A1 + 10A2 - L \leq 0$$
$$L \leq 350$$
$$5A1 \leq 140$$
$$4A2 \leq 120$$
$$A1, A2, L \geq 0$$

Use the LINDO output in Figure 6 to answer the following questions:

a If only 40 acres of land were available, what would Leary's profit be?

b If the price of wheat dropped to $26, what would be the new optimal solution to Leary's problem?

c Use the SLACK portion of the output to determine the allowable increase and allowable decrease for the amount of wheat that can be sold. If only 130 bushels of wheat could be sold, then would the answer to the problem change?

2 Carco manufactures cars and trucks. Each car contributes $300 to profit, and each truck contributes $400. The resources required to manufacture a car and a truck are shown in Table 5. Each day, Carco can rent up to 98 Type 1 machines at a cost of $50 per machine. The company has 73 Type 2 machines and 260 tons of steel available. Marketing considerations dictate that at least 88 cars and at

TABLE 5

Vehicle	Days on Type 1 Machine	Days on Type 2 Machine	Tons of Steel
Car	0.8	0.6	2
Truck	1	0.7	3

least 26 trucks be produced. Let x_1 = number of cars produced daily; x_2 = number of trucks produced daily; and m_1 = Type 1 machines rented daily.

To maximize profit, Carco should solve the LP in Figure 7. Use the LINDO output to answer the following questions:

a If each car contributed $310 to profit, what would be the new optimal solution to the problem?

FIGURE 6
LINDO Output for Wheat and Corn

```
MAX      150  A1 + 200  A2  -  10 L
SUBJECT TO
    2)     A1 + A2 <=    45
    3)     6 A1 + 10 A2 - L <=   0
    4)     L <=    350
    5)     5 A1 <=   140
    6)     4 A2 <=   120
END

LP OPTIMUM FOUND AT STEP          4

              OBJECTIVE FUNCTION VALUE

            1)   4250.00000

VARIABLE         VALUE          REDUCED COST
    A1         25.000000           .000000
    A2         20.000000           .000000
    L         350.000000           .000000

    ROW     SLACK OR SURPLUS     DUAL PRICES
    2)           .000000         75.000000
    3)           .000000         12.500000
    4)           .000000          2.500000
    5)         15.000000           .000000
    6)         40.000000           .000000

NO. ITERATIONS=       4

RANGES IN WHICH THE BASIS IS UNCHANGED:

                OBJ COEFFICIENT RANGES
VARIABLE  CURRENT      ALLOWABLE        ALLOWABLE
          COEF         INCREASE         DECREASE
    A1  150.000000   10.000000        30.000000
    A2  200.000000   50.000000        10.000000
    L   -10.000000   INFINITY          2.500000

                RIGHTHAND SIDE RANGES
    ROW   CURRENT      ALLOWABLE        ALLOWABLE
          RHS          INCREASE         DECREASE
    2    45.000000    1.200000         6.666667
    3     .000000    40.000000        12.000000
    4   350.000000   40.000000        12.000000
    5   140.000000   INFINITY         15.000000
    6   120.000000   INFINITY         40.000000
```

FIGURE 7
LINDO Output for Carco

```
MAX      300  X1 + 400   X2  -   50  M1
SUBJECT TO
    2)     0.8 X1  +   X2  -  M1  <=    0
    3)     M1 <=       98
    4)     0.6  X1 +  0.7  X2 <=    73
    5)     2 X1  + 3  X2  <=    260
    6)     X1 >=   88
    7)     X2 >=   26
END

LP OPTIMUM FOUND AT STEP           4

              OBJECTIVE FUNCTION VALUE

            1)   32540.0000

VARIABLE         VALUE          REDUCED COST
    X1         88.000000           .000000
    X2         27.600000           .000000
    M1         98.000000           .000000

    ROW     SLACK OR SURPLUS     DUAL PRICES
    2)           .000000        400.000000
    3)           .000000        350.000000
    4)           .879999           .000000
    5)          1.200003           .000000
    6)           .000000        -20.000000
    7)          1.599999           .000000

NO. ITERATIONS=       4

RANGES IN WHICH THE BASIS IS UNCHANGED:

                OBJ COEFFICIENT RANGES
VARIABLE  CURRENT      ALLOWABLE        ALLOWABLE
          COEF         INCREASE         DECREASE
    X1  300.000000   20.000000        INFINITY
    X2  400.000000   INFINITY         25.000000
    M1  -50.000000   INFINITY        350.000000

                RIGHTHAND SIDE RANGES
    ROW   CURRENT      ALLOWABLE        ALLOWABLE
          RHS          INCREASE         DECREASE
    2     .000000      .400001         1.599999
    3    98.000000     .400001         1.599999
    4    73.000000    INFINITY          .879999
    5   260.000000    INFINITY         1.200003
    6    88.000000    1.999999         3.000008
    7    26.000000    1.599999        INFINITY
```

b If Carco were required to produce at least 86 cars, what would Carco's profit become?

3 Consider the diet problem discussed in Section 3.4. Use the LINDO output in Figure 8 to answer the following questions.

a If a Brownie costs 30¢, then what would be the new optimal solution to the problem?

b If a bottle of cola cost 35¢, then what would be the new optimal solution to the problem?

c If at least 8 oz of chocolate were required, then what would be the cost of the optimal diet?

d If at least 600 calories were required, then what would be the cost of the optimal diet?

e If at least 9 oz of sugar were required, then what would be the cost of the optimal diet?

FIGURE 8
LINDO Output for Diet Problem

```
MAX      50 BR + 20 IC + 30 COLA + 80 PC
SUBJECT TO
    2)    400 BR + 200 IC + 150 COLA
                           + 500 PC >=   500
    3)     3  BR  +  2  IC   >=   6
    4)     2  BR  +  2  IC  +  4 COLA
                           +  4   PC  >=  10
    5)     2  BR  +  4  IC  +    COLA
                           +  5   PC   >=  8

END

LP OPTIMUM FOUND AT STEP              2

            OBJECTIVE FUNCTION VALUE

        1)   90.0000000

VARIABLE        VALUE         REDUCED COST
    BR         .000000         27.500000
    IC        3.000000          .000000
    COLA      1.000000          .000000
    PC         .000000         50.000000


    ROW    SLACK OR SURPLUS    DUAL PRICES
    2)       250.000000          .000000
    3)         .000000         -2.500000
    4)         .000000         -7.500000
    5)        5.000000          .000000

NO. ITERATIONS=        2

RANGES IN WHICH THE BASIS IS UNCHANGED:

               OBJ COEFFICIENT RANGES
VARIABLE   CURRENT     ALLOWABLE      ALLOWABLE
            COEF       INCREASE       DECREASE
    BR   50.000000     INFINITY      27.500000
    IC   20.000000    18.333330       5.000000
    COLA 30.000000    10.000000      30.000000
    PC   80.000000     INFINITY      50.000000

               RIGHTHAND SIDE RANGES
    ROW    CURRENT     ALLOWABLE      ALLOWABLE
            RHS        INCREASE       DECREASE
    2   500.000000    250.000000     INFINITY
    3     6.000000      4.000000      2.857143
    4    10.000000     INFINITY       4.000000
    5     8.000000      5.000000     INFINITY
```

f What would the price of pineapple cheesecake have to be before it would be optimal to eat cheesecake?

g What would the price of a brownie have to be before it would be optimal to eat a brownie?

h Use the SLACK or SURPLUS portion of the LINDO output to determine the allowable increase and allowable decrease for the fat constraint. If 10 oz of fat were required, then would the optimal solution to the problem change?

4 Gepbab Corporation produces three products at two different plants. The cost of producing a unit at each plant is shown in Table 6. Each plant can produce a total of 10,000 units. At least 6,000 units of product 1, at least 8,000 units of product 2, and at least 5,000 units of product 3 must be produced. To minimize the cost of meeting these demands, the following LP should be solved:

$$\min z = 5x_{11} + 6x_{12} + 8x_{13} + 8x_{21} + 7x_{22} + 10x_{23}$$

$$
\begin{aligned}
\text{s.t.} \quad & x_{11} + x_{12} + x_{13} && \leq && 10{,}000 \\
& x_{21} + x_{22} + x_{23} && \leq && 10{,}000 \\
& x_{11} + x_{21} && \geq && 6{,}000 \\
& x_{12} + x_{22} && \geq && 8{,}000 \\
& x_{13} + x_{23} && \geq && 5{,}000 \\
& \text{All variables} && \geq && 0
\end{aligned}
$$

Here, x_{ij} = number of units of product j produced at plant i. Use the LINDO output in Figure 9 to answer the following questions:

a What would the cost of producing product 2 at plant 1 have to be for the firm to make this choice?

b What would total cost be if plant 1 had 9,000 units of capacity?

c If it cost $9 to produce a unit of product 3 at plant 1, then what would be the new optimal solution?

5 Mondo produces motorcycles at three plants. At each plant, the labor, raw material, and production costs (excluding labor cost) required to build a motorcycle are as shown in Table 7. Each plant has sufficient machine capacity to produce up to 750 motorcycles per week. Each of Mondo's workers can work up to 40 hours per week and is paid $12.50 per hour worked. Mondo has a total of 525 workers and now owns 9,400 units of raw material. Each week, at least 1,400 Mondos must be produced. Let x_1 = motorcycles produced at plant 1; x_2 = motorcycles produced at plant 2; and x_3 = motorcycles produced at plant 3.

The LINDO output in Figure 10 enables Mondo to minimize the variable cost (labor + production) of meeting demand. Use the output to answer the following questions:

a What would be the new optimal solution to the problem if the production cost at plant 1 were only $40?

TABLE 6

Plant	Product ($)		
	1	2	3
1	5	6	8
2	8	7	10

FIGURE 9
LINDO Output for Gepbab

```
MAX     5  X11 + 6 X12 + 8 X13 + 8 X21
                              + 7 X22 + 10 X23
SUBJECT TO
      2)    X11  +  X12  +  X13 <=   10000
      3)    X21  +  X22  +  X23 <=   10000
      4)    X11  +  X21       >=    6000
      5)    X12  +  X22       >=    8000
      6)    X13  +  X23       >=    5000
END

LP OPTIMUM FOUND AT STEP           5

              OBJECTIVE FUNCTION VALUE

           1)   128000.000

VARIABLE        VALUE          REDUCED COST
   X11       6000.0000000        .000000
   X12         .000000          1.000000
   X13       4000.0000000        .000000
   X21         .000000          1.000000
   X22       8000.0000000        .000000
   X23       1000.0000000        .000000


    ROW    SLACK OR SURPLUS    DUAL PRICES
     2)        .000000         2.000000
     3)      1000.000000        .000000
     4)        .000000        -7.000000
     5)        .000000        -7.000000
     6)        .000000       -10.000000

NO. ITERATIONS=        5

RANGES IN WHICH THE BASIS IS UNCHANGED:

              OBJ COEFFICIENT RANGES
VARIABLE    CURRENT      ALLOWABLE     ALLOWABLE
            COEF         INCREASE      DECREASE
   X11    5.000000      1.000000      7.000000
   X12    6.000000      INFINITY      1.000000
   X13    8.000000      1.000000      1.000000
   X21    8.000000      INFINITY      1.000000
   X22    7.000000      1.000000      7.000000
   X23   10.000000      1.000000      1.000000

              RIGHTHAND SIDE RANGES
   ROW    CURRENT        ALLOWABLE      ALLOWABLE
          RHS            INCREASE       DECREASE
    2   10000.000000   1000.000000   1000.000000
    3   10000.000000    INFINITY     1000.000000
    4    6000.000000   1000.000000   1000.000000
    5    8000.000000   1000.000000   8000.000000
    6    5000.000000   1000.000000   1000.000000
```

TABLE 7

Plant	Labor Needed (Hours)	Raw Material Needed (Units)	Production Cost ($)
1	20	5	50
2	16	8	80
3	10	7	100

FIGURE 10
LINDO Output for Mondo

```
MAX     300  X1 + 280  X2  +  225  X3
SUBJECT TO
      2)   20 X1  + 16 X2  + 10  X3 <=   21000
      3)    5  X1  + 8  X2  + 7  X3 <=    9400
      4)    X1  <=      750
      5)    X2  <=      750
      6)    X3  <=      750
      7)    X1  +  X2  +  X3  >=    1400
END

LP OPTIMUM FOUND AT STEP            3

              OBJECTIVE FUNCTION VALUE

           1)   357750.000

VARIABLE        VALUE          REDUCED COST
   X1        350.000000         .000000
   X2        300.000000         .000000
   X3        750.000000         .000000


    ROW    SLACK OR SURPLUS    DUAL PRICES
     2)      1700.000000        .000000
     3)        .000000         6.666668
     4)       400.000000        .000000
     5)       450.000000        .000000
     6)        .000000        61.666660
     7)        .000000      -333.333300

NO. ITERATIONS=        3

RANGES IN WHICH THE BASIS IS UNCHANGED:

              OBJ COEFFICIENT RANGES
VARIABLE    CURRENT      ALLOWABLE     ALLOWABLE
            COEF         INCREASE      DECREASE
   X1     300.000000     INFINITY     20.000000
   X2     280.000000    20.000010     92.499990
   X3     225.000000    61.666660     INFINITY

              RIGHTHAND SIDE RANGES
   ROW    CURRENT        ALLOWABLE       ALLOWABLE
          RHS            INCREASE        DECREASE
    2   21000.000000     INFINITY     1700.000000
    3    9400.000000   1050.000000    900.000000
    4     750.000000     INFINITY     400.000000
    5     750.000000     INFINITY     450.000000
    6     750.000000    450.000000    231.818200
    7    1400.000000    63.750000     131.250000
```

b How much money would Mondo save if the capacity of plant 3 were increased by 100 motorcycles?

c By how much would Mondo's cost increase if it had to produce one more motorcycle?

6 Steelco uses coal, iron, and labor to produce three types of steel. The inputs (and sales price) for one ton of each type of steel are shown in Table 8. Up to 200 tons of coal can be purchased at a price of $10 per ton. Up to 60 tons of iron can be purchased at $8 per ton, and up to 100 labor hours can be purchased at $5 per hour. Let x_1 = tons of steel 1 produced; x_2 = tons of steel 2 produced; and x_3 = tons of steel 3 produced.

The LINDO output that yields a maximum profit for the company is given in Figure 11. Use the output to answer the following questions.

a What would profit be if only 40 tons of iron could be purchased?

TABLE 8

Steel	Coal Required (Tons)	Iron Required (Tons)	Labor Required (Hours)	Sales Price ($)
1	3	1	1	51
2	2	0	1	30
3	1	1	1	25

FIGURE 11
LINDO Output for Steelco

```
MAX        8 X1 + 5  X2  + 2  X3
SUBJECT TO
    2)      3 X1  +  2  X2  +  X3 <=  200
    3)       X1  +  X3     <=     60
    4)       X1  +  X2  +  X3 <=  100
END

LP OPTIMUM FOUND AT STEP              2

        OBJECTIVE FUNCTION VALUE

    1)    530.000000

VARIABLE         VALUE          REDUCED COST
    X1         60.000000          .000000
    X2         10.000000          .000000
    X3          .000000          1.000000


    ROW    SLACK OR SURPLUS    DUAL PRICES
    2)          .000000         2.500000
    3)          .000000          .500000
    4)        30.000000          .000000

NO. ITERATIONS=           2

RANGES IN WHICH THE BASIS IS UNCHANGED:

              OBJ COEFFICIENT RANGES
VARIABLE   CURRENT     ALLOWABLE      ALLOWABLE
            COEF       INCREASE       DECREASE
    X1    8.000000     INFINITY        .500000
    X2    5.000000      .333333       5.000000
    X3    2.000000     1.000000       INFINITY

            RIGHTHAND SIDE RANGES
  ROW    CURRENT     ALLOWABLE      ALLOWABLE
          RHS        INCREASE       DECREASE
   2   200.000000   60.000000     20.000000
   3    60.000000    6.666667     60.000000
   4   100.000000    INFINITY     30.000000
```

b What is the smallest price per ton for steel 3 that would make it desirable to produce it?

c Find the new optimal solution if steel 1 sold for $55 per ton.

Group B

7 Shoeco must meet (on time) the following demands for pairs of shoes: month 1—300; month 2—500; month 3—100; and month 4—100. At the beginning of month 1, 50 pairs of shoes are on hand, and Shoeco has three workers. A worker is paid $1,500 per month. Each worker can work up to 160 hours per month before receiving overtime. During any month, each worker may be forced to work up to 20 hours of overtime; workers are paid $25 per hour for overtime labor. It takes 4 hours of labor and $5 of raw material to produce each pair of shoes. At the beginning of each month, workers can be hired or fired. Each hired worker costs $1,600, and each fired worker costs $2,000. At the end of each month, a holding cost of $30 per pair of shoes is assessed. Formulate an LP that can be used to minimize the total cost of meeting the next four months' demands. Then use LINDO to solve the LP. Finally, use the LINDO printout to answer the questions that follow these hints (which may help in the formulation.) Let

x_t = Pairs of shoes produced during month t with nonovertime labor

o_t = Pairs of shoes produced during month t with overtime labor

i_t = Inventory of pairs of shoes at end of month t

h_t = Workers hired at beginning of month t

f_t = Workers fired at beginning of month t

w_t = Workers available for month t (after month t hiring and firing)

Four types of constraints will be needed:

Type 1 Inventory equations. For example, during month 1, $i_1 = 50 + x_1 + o_1 - 300$.

Type 2 Relate available workers to hiring and firing. For month 1, for example, the following constraint is needed: $w_1 = 3 + h_1 - f_1$.

Type 3 For each month, the amount of shoes made with nonovertime labor is limited by the number of workers. For example, for month 1, the following constraint is needed: $4x_1 \leq 160w_1$.

Type 4 For each month, the number of overtime labor hours used is limited by the number of workers. For example, for month 1, the following constraint is needed: $4(o_1) \leq 20w_1$.

For the objective function, the following costs must be considered:

1 Workers' salaries

2 Hiring costs

3 Firing costs

4 Holding costs

5 Overtime costs

6 Raw-material costs

a Describe the company's optimal production plan, hiring policy, and firing policy. Assume that it is acceptable to have a fractional number of workers, hirings, or firings.

b If overtime labor during month 1 costs $16 per hour, should any overtime labor be used?

c If the cost of firing workers during month 3 were $1,800, what would be the new optimal solution to the problem?

d If the cost of hiring workers during month 1 were $1,700, what would be the new optimal solution to the problem?

e By how much would total costs be reduced if demand in month 1 were 100 pairs of shoes?

f What would the total cost become if the company had 5 workers at the beginning of month 1 (before month 1's hiring or firing takes place)?

g By how much would costs increase if demand in month 2 were increased by 100 pairs of shoes?

8 Consider the LP:

$$\max \quad 9x_1 + 8x_2 + 5x_3 + 4x_4$$

$$\text{s.t.} \quad x_1 \qquad\qquad + x_4 \le 200$$
$$x_2 + x_3 \qquad \le 150$$
$$x_1 + x_2 + x_3 \qquad \le 350$$
$$2x_1 + x_2 + x_3 + x_4 \le 550$$
$$x_1, x_2, x_3, x_4 \ge 0$$

a Solve this LP with LINDO and use your output to show that the optimal solution is degenerate.

b Use your LINDO output to find an example of Oddities 1–3.

5.3 Managerial Use of Shadow Prices

In this section, we will discuss the managerial significance of shadow prices. In particular, we will learn how shadow prices can often be used to answer the following question: What is the maximum amount that a manager should be willing to pay for an additional unit of a resource? To answer this question, we usually focus our attention on the shadow price of the constraint that describes the availability of the resource. We now discuss four examples of the interpretation of shadow prices.

EXAMPLE 5   **Winco Products 2**

In Example 1, what is the most that Winco should be willing to pay for an additional unit of raw material? How about an extra hour of labor?

Solution   Because the shadow price of the raw-material-availability constraint is 1, an extra unit would increase total revenue by $1. Thus, Winco could pay up to $1 for an extra unit of raw material and be as well off as it is now. This means that Winco should be willing to pay up to $1 for an extra unit of raw material. The labor-availability constraint has a shadow price of 0. This means that an extra hour of labor will not increase revenues, so Winco should not be willing to pay anything for an extra hour of labor. (Note that this discussion is valid because the AIs for the labor and raw-material constraints both exceed 1.)

EXAMPLE 6   **Winco Products 3**

Let's reconsider Example 1 with the following changes. Suppose as many as 4,600 units of raw material are available, but they must be purchased at a cost of $4 per unit. Also, as many as 5,000 hours of labor are available, but they must be purchased at a cost of $6 per hour. The per-unit sales price of each product is as follows: product 1—$30; product 2—$42; product 3—$53; product 4—$72. A total of 950 units must be produced, of which at least 400 must be product 4. Determine the maximum amount that the firm should be willing to pay for an extra unit of raw material and an extra hour of labor.

Solution   The contribution to profit from one unit of each product may be computed as follows:

$$\text{Product 1:} \quad 30 - 4(2) - 6(3) = \$4$$
$$\text{Product 2:} \quad 42 - 4(3) - 6(4) = \$6$$
$$\text{Product 3:} \quad 53 - 4(4) - 6(5) = \$7$$
$$\text{Product 4:} \quad 72 - 4(7) - 6(6) = \$8$$

Thus, Winco's profit is $4x_1 + 6x_2 + 7x_3 + 8x_4$. To maximize profit, Winco should solve the same LP as in Example 1, and the relevant LINDO output is again Figure 3. To determine the most Winco should be willing to pay for an extra unit of raw material, note that the shadow price of the raw material constraint may be interpreted as follows: If Winco has the right to buy one more unit of raw material (at $4 per unit), then profits increase by $1. Thus, paying $4 + $1 = $5 for an extra unit of raw material will increase profits by $1 − $1 = $0. So Winco could pay up to $5 for an extra unit of raw material and still be better off. For the raw-material constraint, the shadow price of $1 represents a *premium* above and beyond the current price Winco is willing to pay for an extra unit of raw material.

The shadow price of the labor-availability constraint is $0, which means that the right to buy an extra hour of labor at $4 an hour will not increase profits. Unfortunately, all this tells us is that at the current price of $4 per hour, Winco should buy no more labor.

EXAMPLE 7	Farmer Leary's Shadow Price

Consider the Farmer Leary problem (Problem 1 in Section 5.2).

a What is the most that Leary should pay for an additional hour of labor?

b What is the most that Leary should pay for an additional acre of land?

Solution **a** From the $L \le 350$ constraint's shadow price of 2.5, we see that if 351 hours of labor are available, then (after paying $10 for another hour of labor) profits increase by $2.50. So if Leary pays $10 + $2.50 = $12.50 for an extra hour of labor, profits would increase by $2.50 − $2.50 = $0. This implies that Leary should be willing to pay up to $12.50 for another hour of labor.

To look at it another way, the shadow price of the $6A1 + 10A2 − L \le 0$ constraint is 12.5. This means that if the constraint $6A1 + 10A2 \le L$ were replaced by the constraint $6A1 + 10A2 \le L + 1$, profits would increase by $12.50. So if one extra hour of labor were "given" to Leary (at zero cost), profits would increase by $12.50. Thus, Leary should be willing to pay up to $12.50 for an extra hour of labor.

b If 46 acres of land were available, profits would increase by $75 (the shadow price of the $A1 + A2 \le 45$ constraint). This includes the cost ($0) of purchasing an additional acre of land. Thus, Leary should be willing to pay up to $75 for an extra acre of land.

We now illustrate some of the managerial insights that can be gained by analyzing the shadow prices for a minimization problem.

EXAMPLE 8	Tucker Inc.'s Shadow Price

The following questions refer to Example 2.

a What is the most that Tucker should pay for an extra hour of labor?

b What is the most that Tucker should pay for an extra unit of raw material?

c A new customer is willing to purchase 20 cars at a price of $25,000 per vehicle. Should Tucker fill her order?

Solution **a** Because the shadow price of the labor-availability constraint (row 4) is 0, an extra hour of labor reduces costs by $0. Thus, Tucker should not pay anything for an extra hour of labor.

b Because the shadow price of the raw-material-availability constraint (row 5) is 5 (thousand dollars), an additional unit of raw material reduces costs by $5,000. Thus, Tucker should be willing to pay up to $5,000 for an extra unit of raw material.

c The allowable increase for the constraint $x_1 + x_2 + x_3 + x_4 = 1,000$ is 66.666660. Because the shadow price of this constraint is -30 (thousand dollars), we know that if Tucker fills the order, its costs will increase by $-20(-30,000) = \$600,000$. So Tucker should not fill the order.

In Example 8, the astute reader may notice that each car costs at most \$15,000 to produce. How is it then possible that a unit increase in the number of cars that must be produced increases costs by \$30,000? To see why this is the case, we re-solved Tucker's LP after increasing the number of cars that had to be produced to 1,001. The new optimal solution has $z = 11,630$, $x_1 = 404$, $x_2 = 197$, $x_3 = 400$, $x_4 = 0$. We now see why increasing demand by one car raises costs by \$30,000. To produce one more car, Tucker must produce four more Type 1 cars and three fewer Type 2 cars. This ensures that Tucker still uses only 4,400 units of raw material, but it increases total cost by $4(15,000) - 3(10,000) = \$30,000$!

PROBLEMS

Group A

1 In Problem 2 of Section 5.2, what is the most that Carco should be willing to pay for an extra ton of steel?

2 In Problem 2 of Section 5.2, what is the most that Carco should be willing to pay to rent an additional Type 1 machine for one day?

3 In Problem 3 of Section 5.2, what is the most that one should be willing to pay for an additional ounce of chocolate?

4 In Problem 4 of Section 5.2, how much should Gepbab be willing to pay for another unit of capacity at plant 1?

5 In Problem 5 of Section 5.2, suppose that Mondo could purchase an additional unit of raw material at a cost of \$6. Should the company do it? Explain.

6 In Problem 6 of Section 5.2, what is the most that Steelco should be willing to pay for an extra ton of coal?

7 In Problem 6 of Section 5.2, what is the most that Steelco should be willing to pay for an extra ton of iron?

8 In Problem 6 of Section 5.2, what is the most that Steelco should be willing to pay for an extra hour of labor?

9 In Problem 7 of Section 5.2, suppose that a new customer wishes to buy a pair of shoes during month 1 for \$70. Should Shoeco oblige him?

10 In Problem 7 of Section 5.2, what is the most the company would be willing to pay for having one more worker at the beginning of month 1?

11 In solving part (c) of Example 8, a manager reasons as follows: The average cost of producing a car is \$11,600 up to 1,000 cars. Therefore, if a customer is willing to pay me \$25,000 for a car, I should certainly fill his order. What is wrong with this reasoning?

5.4 What Happens to the Optimal z-Value If the Current Basis Is No Longer Optimal?

In Section 5.2, we used shadow prices to determine the new optimal z-value if the right-hand side of a constraint were changed but remained in the range where the current basis remains optimal. Suppose we change the right-hand side of a constraint to a value where the current basis is no longer optimal. In this situation, the LINDO Parametrics feature can be used to determine how the shadow price of a constraint and the optimal z-value change.

We illustrate the use of the Parametrics feature by varying the amount of raw material available in Example 1. Suppose we want to determine how the optimal z-value and shadow price change as the amount of available raw material varies between 0 and 10,000 units. We first realize that with little raw material available, the LP will be infeasible. To begin, we change the amount of raw material available to 0. We then obtain from the Range and Sensitivity Analysis results that row 4 has an Allowable Decrease of −3,900. This indicates that if at least 3,900 units of raw material are available, the problem will be feasible. We therefore change the rhs of the raw material constraint to 3,900 and solve the LP. After finding the optimal solution, select Reports Parametrics. From the dialog box, choose row 4 and set the value to 10,000. We will choose Text output. We obtain the output shown in Figure 12.

From Figure 12 we find that if the amount of available raw material is 3,900, then the shadow price (or dual price) for raw material is now $2, and the optimal z-value is 5,400. The current basis remains optimal until rm = 4,450; between rm = 3,900 and rm = 4,450, each unit increase in rm will increase the optimal z-value by the shadow price of $2. Thus, when rm = 4,450, the optimal z-value will be

$$5,400 + 2(4,450 - 3,900) = \$6,500$$

From Figure 12, we see that when rm = 4,450, x_3 enters the basis and x_1 exits. The shadow price of rm is now $1, and each additional unit of rm (up to the next change of basis) will increase the optimal z-value by $1. The next basis change occurs when rm = 4,850. At this point, the new optimal z-value may be computed as (optimal z-value for rm = 6,500) + (4,850 − 4,450)($1) = $6,900. When rm = 4,850, we pivot in SLACK3 (the slack variable for row 3 or constraint 2), and SLACK5 exits. The new shadow price for rm is $0. Thus when rm > 4,850, we see that an additional unit of rm will not increase the optimal z-value. This discussion is summarized in Figure 13, which shows the optimal z-value as a function of the amount of available raw material.

For any LP, a graph of the optimal objective function value as a function of a right-hand side will consist of several straight-line segments of possibly differing slopes. (Such a function is called a *piecewise linear function*.) The slope of each straight-line segment is just the constraint's shadow price. At points where the optimal basis changes (points B, C, and D in Figure 13), the slope of the graph may change. For a ≤ constraint in a maximization problem, the slope of each line segment must be nonnegative—more of a re-

```
RIGHTHANDSIDE PARAMETRICS REPORT FOR ROW: 4

      VAR          VAR    PIVOT     RHS       DUAL PRICE      OBJ
      OUT ........  IN     ROW       VAL       BEFORE PIVOT    VAL
      ............
                                    3900.00    2.00000         5400.00
      X1 ........   X3      2        4450.00    2.00000         6500.00
SLK    5   SLK      3       5        4850.00    1.00000         6900.00
      X3   SLK      4       2        5250.00   -0.333067E-15    6900.00
                                    10000.0    0.555112E-16    6900.00
```

FIGURE **12**

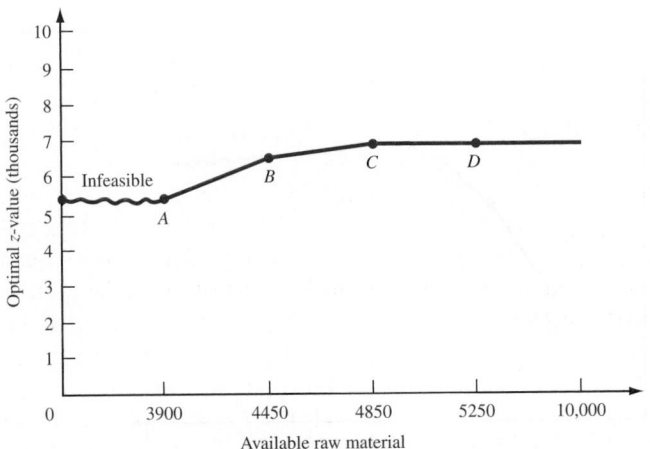

FIGURE **13**
Optimal z-Value versus
Raw Material

source can't hurt. In a maximization problem, the slopes of successive line segments for a ≤ constraint will be nonincreasing. This is simply a consequence of diminishing returns; as we obtain more of a resource (and availability of other resources is held constant), the value of an additional unit of the resource cannot increase.

For a ≥ constraint in a maximization problem, the graph of the optimal z-value as a function of the right-hand side will again be a piecewise linear function. The slope of each line segment will be nonpositive (corresponding to the fact that a ≥ constraint has a non-positive shadow price). The slopes of successive line segments will be nonincreasing. For the $x_4 \geq 400$ constraint in Example 1, plotting the optimal z-value as a function of the constraint's right-hand side yields the graph in Figure 14.

For an equality constraint in a maximization problem, the graph of the optimal z-value as a function of right-hand side will again be piecewise linear. The slopes of each line segment may be positive or negative, but the slopes of successive line segments will again be nonincreasing. For the constraint $x_1 + x_2 + x_3 + x_4 = 950$ in Example 1, we obtain the graph in Figure 15.

For a minimization problem, the plot of the optimal z-value against a constraint's right-hand side is again a piecewise linear function. For all minimization problems, the slopes of successive line segments will be nondecreasing. For a ≤ constraint, the slope of each line segment is nonpositive; for a ≥ constraint, the slope is nonnegative; and for an equality constraint, the slope may be positive or negative.

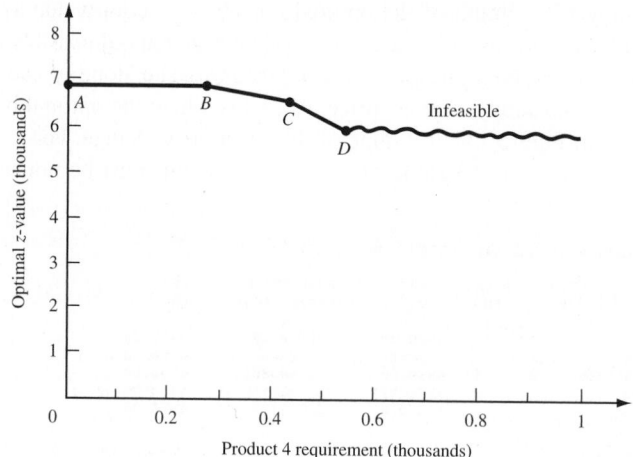

FIGURE 14
Optimal z-Value versus Product 4 Requirement

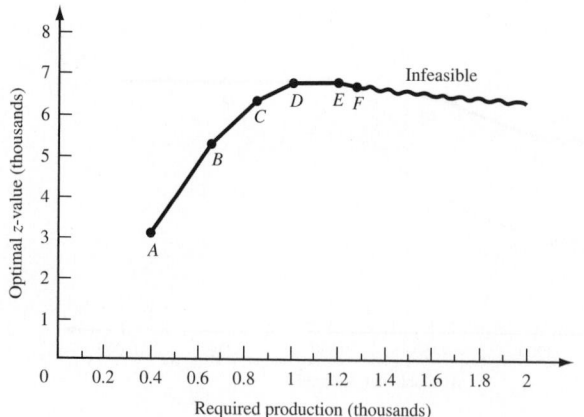

FIGURE 15
Optimal z-Value versus Production Requirement

Effect of Change in Objective Function Coefficient on Optimal z-Value

We now discuss how to find the graph of the optimal objective function value as a function of a variable's objective function coefficient. To see how this works, let's reconsider the Giapetto problem.

$$\max z = 3x_1 + 2x_2$$
$$\text{s.t.} \quad 2x_1 + x_2 \le 100$$
$$x_1 + x_2 \le 80$$
$$x_1 \quad\quad \le 40$$
$$x_1, x_2 \ge 0$$

Let c_1 = objective function coefficient for x_1. Currently, we have $c_1 = 3$. We want to determine how the optimal z-value depends on c_1. To determine this relationship, we must find, for each value of c_1, the optimal values of the decision variables. Recall from Figure 1 (p. 228) that point $A = (0, 80)$ is optimal if the isoprofit line is flatter than the carpentry constraint. Also note that point $B = (20, 60)$ is optimal if the slope of the isoprofit line is steeper than the carpentry constraint and flatter than the finishing-hour constraint. Finally, point $C = (40, 20)$ is optimal if the slope of the isoprofit line is steeper than the slope of the finishing-hour constraint. A typical isoprofit line is $c_1 x_1 + 2x_2 = k$, so we know that the slope of a typical isoprofit line is $-\frac{c_1}{2}$. This implies that point A is optimal if $-\frac{c_1}{2} \ge -1$ (or $c_1 \le 2$). We also find that point B is optimal if $-2 \le -\frac{c_1}{2} \le -1$ (or $2 \le c_1 \le 4$). Finally, point C is optimal if $-\frac{c_1}{2} \le -2$ (or $c_1 \ge 4$). By substituting the optimal values of the decision variables into the objective function ($c_1 x_1 + 2x_2$), we obtain the following information:

Value of c_1	Optimal z-value
$0 \le c_1 \le 2$	$c_1(0) + 2(80) = \$160$
$2 \le c_1 \le 4$	$c_1(20) + 2(60) = 120 + 20c_1$
$c_1 \ge 4$	$c_1(40) + 2(20) = 40 + 40c_1$

The relationship between c_1 and the optimal z-value is portrayed graphically in Figure 16. As seen in the figure, the graph of the optimal z-value as a function of c_1 is a piecewise linear function. The slope of each line segment in the graph is equal to the value of x_1 in the optimal solution. In a maximization problem, it can be shown (see Problem 5) that as the value of an objective function coefficient increases, the value of the variable in the LP's optimal solution cannot decrease. Thus, the slope of the graph of the optimal z-value as a function of an objective function coefficient will be nondecreasing.

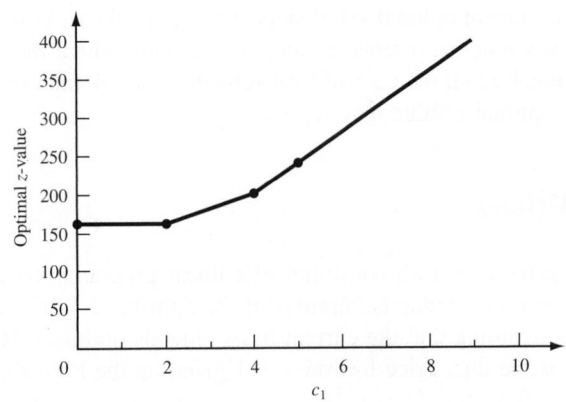

FIGURE 16
Optimal z-Value
versus c_1

Similarly, in a minimizing problem, the graph of the optimal z-value as a function of a variable x_i's objective function coefficient c_i is a piecewise linear function. Again, the slope of each line segment is equal to the optimal value of x_i in the bfs corresponding to the line segment. It can be shown (see Problem 6) that the optimal x_i-value is a nonincreasing function of c_i. Thus, in a minimization problem, the graph of the optimal z-value as a function of c_i will be a piecewise linear function having a nonincreasing slope.

PROBLEMS

Group A

In what follows, b_i represents the right-hand side of an LP's ith constraint.

1 Use the LINDO **PARA** command to graph the optimal z-value for Example 1 as a function of b_4.

2 Use the **PARA** command to graph the optimal z-value for Example 2 as a function of b_1. Then answer the same questions for b_2, b_3, and b_4, respectively.

3 For the Giapetto example of Section 3.1, graph the optimal z-value as a function of x_2's objective function coefficient. Also graph the optimal z-value as a function of b_1, b_2, and b_3.

4 For the Dorian Auto example (Example 2 in Chapter 3), let c_1 be the objective function coefficient of x_1. Determine the optimal z-value as a function of c_1.

Group B

5 For Example 1, suppose that we increase the sales price of a product. Show that in the new optimal solution, the amount produced of that product cannot decrease.

6 For Example 2, suppose that we increase the cost of producing a type of car. Show that in the new optimal solution to the LP, the number of cars produced of that type cannot increase.

7 Consider the Sailco problem (Example 12 in Chapter 3). Suppose we want to consider how profit will be affected if we change the number of sailboats that can be produced each month with regular-time labor. How can we use the **PARA** command to answer this question? (*Hint:* Let c = change in number of sailboats that can be produced each month with regular-time labor. Change the right-hand side of some constraints to $40 + c$ and add another constraint to the problem.)

SUMMARY   Graphical Sensitivity Analysis

To determine whether the current basis remains optimal after changing an objective function coefficient, note that changing the objective function coefficient of a variable changes the slope of the isoprofit line. The current basis remains optimal as long as the current optimal solution is the last point in the feasible region to make contact with isoprofit lines as we move in the direction of increasing z (for a max problem). If the current basis remains optimal, the values of the decision variables remain unchanged, but the optimal z-value may change.

To determine if the current basis remains optimal after changing the right-hand side of a constraint, begin by finding the constraints (possibly including sign restrictions) that are binding for the current optimal solution. As we change the right-hand side of a constraint, the current basis remains optimal as long as the point where the constraints are binding remains feasible. Even if the current basis remains optimal, the values of the decision variables and the optimal z-value may change.

Shadow Prices

The **shadow price** of the ith constraint of a linear programming problem is the amount by which the optimal z-value is improved if the right-hand side of the ith constraint is increased by 1 (assuming that the current basis remains optimal). The shadow price of the ith constraint is the dual price for row $i + 1$ given on the LINDO output.

If the right-hand side of the ith constraint is increased by Δb_i, then (assuming the current basis remains optimal) the new optimal z-value for a maximization problem may be found as follows:

(New optimal z-value) = (old optimal z-value) + (Constraint i's shadow price) Δb_i (1)

For a minimization problem, the new optimal z-value may be found from

(New optimal z-value) = (old optimal z-value) − (Constraint i's shadow price)Δb_i (2)

Objective Function Coefficient Range

The OBJ COEFFICIENT RANGE portion of the LINDO output gives the range of values for an objective function coefficient for which the current basis remains optimal. Within this range, the values of the decision variables remain unchanged, but the optimal z-value may or may not change.

Reduced Cost

For any nonbasic variable, the reduced cost for the variable is the amount by which the nonbasic variable's objective function coefficient must be improved before that variable will become a basic variable in some optimal solution to the LP.

Right-Hand Side Range

If the right-hand side of a constraint remains within the RIGHTHAND SIDE RANGES value given on the LINDO printout, then the current basis remains optimal, and the dual price may be used to determine how a change in the right-hand side changes the optimal z-value. Even if the right-hand side of a constraint remains within the RIGHTHAND SIDE RANGES value on the LINDO output, then the values of the decision variables will probably change.

Signs of Shadow Prices

A $\geq$ constraint will have a nonpositive shadow price; a $\leq$ constraint will have a nonnegative shadow price; and an equality constraint may have a positive, negative, or zero shadow price.

Optimal z-Value as a Function of a Constraint's Right-Hand Side

In all cases, the optimal z-value will be a piecewise linear function of a constraint's right-hand side. The exact form of the function is as shown in Table 9.

Optimal z-Value as a Function of an Objective Function Coefficient

In a maximization problem, the optimal z-value will be a nondecreasing, piecewise linear function of an objective function coefficient. The slope will be a nondecreasing function of the objective function coefficient.

TABLE 9

Type of LP	Type of Constraint	Slopes of Each Piecewise Linear Segment Are
Maximization	≤	Nonnegative and nonincreasing
Maximization	≥	Nonpositive and nonincreasing
Maximization	=	Unrestricted in sign and nonincreasing
Minimization	≤	Nonpositive and nondecreasing
Minimization	≥	Nonnegative and nondecreasing
Minimization	=	Unrestricted in sign and nondecreasing

In a minimization problem, the optimal z-value will be a nondecreasing, piecewise linear function of an objective function coefficient. The slope will be a nonincreasing function of the objective function coefficient.

REVIEW PROBLEMS

Group A

1 HAL produces two types of computers: PCs and VAXes. The computers are produced in two locations: New York and Los Angeles. New York can produce up to 800 computers and Los Angeles up to 1,000 computers. HAL can sell up to 900 PCs and 900 VAXes. The profit associated with each production site and computer sale is as follows: New York—PC, $600; VAX, $800; Los Angeles—PC, $1,000; VAX, $1,300. The skilled labor required to build each computer at each location is as follows: New York— PC, 2 hours; VAX, 2 hours; Los Angeles—PC, 3 hours; VAX, 4 hours. A total of 4,000 hours of labor are available. Labor is purchased at a cost of $20 per hour. Let

$$XNP = \text{PCs produced in New York}$$
$$XLP = \text{PCs produced in Los Angeles}$$
$$XNV = \text{VAXes produced in New York}$$
$$XLV = \text{VAXes produced in Los Angeles}$$

Use the LINDO printout in Figure 17 to answer the following questions:

a If 3,000 hours of skilled labor were available, what would be HAL's profit?

b Suppose an outside contractor offers to increase the capacity of New York to 850 computers at a cost of $5,000. Should HAL hire the contractor?

c By how much would the profit for a VAX produced in Los Angeles have to increase before HAL would want to produce VAXes in Los Angeles?

d What is the most HAL should pay for an extra hour of labor?

2 Vivian's Gem Company produces two types of gems: Types 1 and 2. Each Type 1 gem contains 2 rubies and 4 diamonds. A Type 1 gem sells for $10 and costs $5 to produce. Each Type 2 gem contains 1 ruby and 1 diamond. A Type 2 gem sells for $6 and costs $4 to produce. A total of 30 rubies and 50 diamonds are available. All gems that are produced can be sold, but marketing considerations

dictate that at least 11 Type 1 gems be produced. Let $x_1 =$ number of Type 1 gems produced and $x_2 =$ number of Type 2 gems produced. Assume that Vivian wants to maximize profit. Use the LINDO printout in Figure 18 to answer the following questions:

a What would Vivian's profit be if 46 diamonds were available?

b If Type 2 gems sold for only $5.50, what would be the new optimal solution to the problem?

c What would Vivian's profit be if at least 12 Type 1 gems had to be produced?

3 Wivco produces product 1 and product 2 by processing raw material. Up to 90 lb of raw material may be purchased at a cost of $10/lb. One pound of raw material can be used to produce either 1 lb of product 1 or 0.33 lb of product 2. Using a pound of raw material to produce a pound of product 1 requires 2 hours of labor or 3 hours to produce 0.33 lb of product 2. A total of 200 hours of labor are available, and at most 40 pounds of product 2 can be sold. Product 1 sells for $13/lb and product 2, $40/lb. Let

$$RM = \text{pounds of raw material processed}$$
$$P1 = \text{pounds of raw material used to produce product 1}$$
$$P2 = \text{pounds of raw material used to produce product 2}$$

To maximize profit, Wivco should solve the following LP:

$$\max z = 13P1 + 40(0.33)P2 - 10RM$$
$$\begin{aligned} \text{s.t.} \quad RM &\geq P1 + P2 \\ 2P1 + 3P2 &\leq 200 \\ RM &\leq 90 \\ 0.33P2 &\leq 40 \\ P1, P2, RM &\geq 0 \end{aligned}$$

Use the LINDO output in Figure 19 to answer the following questions:

a If only 87 lb of raw material could be purchased, what would be Wivco's profits?

FIGURE **17**
LINDO Output for HAL

```
MAX        600  XNP + 1000 XLP + 800 XNV
                            + 1300 XLV - 20 L

SUBJECT TO
      2)        2 XNP + 3 XLP + 2 XNV
                            + 4 XLV - L <=     0
      3)        XNP  +  XNV   <=   800
      4)        XLP  +  XLV   <=  1000
      5)        XNP  +  XLP   <=   900
      6)        XNV  +  XLV   <=   900
      7)        L    <=    4000

END

LP OPTIMUM FOUND AT STEP            3

                OBJECTIVE FUNCTION VALUE

           1)   1360000.00

VARIABLE         VALUE         REDUCED COST
   XNP          .000000         200.000000
   XLP        800.000000          .000000
   XNV        800.000000          .000000
   XLV          .000000          33.333370
   L         4000.000000          .000000


 ROW     SLACK OR SURPLUS     DUAL PRICES
  2)         .000000         333.333300
  3)         .000000         133.333300
  4)       200.000000          .000000
  5)       100.000000          .000000
  6)       100.000000          .000000
  7)         .000000         313.333300


NO. ITERATIONS=            3

RANGES IN WHICH THE BASIS IS UNCHANGED:

              OBJ COEFFICIENT RANGES
VARIABLE      CURRENT      ALLOWABLE     ALLOWABLE
               COEF        INCREASE      DECREASE
  XNP     600.000000    200.000000      INFINITY
  XLP    1000.000000    200.000000     25.000030
  XNV     800.000000      INFINITY    133.333300
  XLV    1300.000000     33.333370      INFINITY
  L       -20.000000      INFINITY    313.333300

              RIGHTHAND SIDE RANGES
 ROW    CURRENT      ALLOWABLE      ALLOWABLE
         RHS         INCREASE       DECREASE
  2       .000000   300.000000    2400.000000
  3    800.000000   100.000000     150.000000
  4   1000.000000     INFINITY     200.000000
  5    900.000000     INFINITY     100.000000
  6    900.000000     INFINITY     100.000000
  7   4000.000000   300.000000    2400.000000
```

b If product 2 sold for $39.50/lb, what would be the new optimal solution to Wivco's problem?

c What is the most that Wivco should pay for another pound of raw material?

d What is the most that Wivco should pay for another hour of labor?

4 Zales Jewelers uses rubies and sapphires to produce two types of rings. A Type 1 ring requires 2 rubies, 3 sapphires, and 1 hour of jeweler's labor. A Type 2 ring requires 3 rubies, 2 sapphires, and 2 hours of jeweler's labor. Each Type 1 ring sells for $400; type 2 sells for $500. All rings

FIGURE **18**
LINDO Output for Vivian's Gem

```
MAX        5 X1 + 2  X2
SUBJECT TO
      2)        2 X1  +  X2 <=   30
      3)        4 X1  +  X2 <=   50
      4)        X1  >=     11

END

LP OPTIMUM FOUND AT STEP            2

                OBJECTIVE FUNCTION VALUE

           1)   67.0000000

VARIABLE         VALUE         REDUCED COST
   X1          11.000000         .000000
   X2           6.000000         .000000


 ROW     SLACK OR SURPLUS     DUAL PRICES
  2)        2.000000         0.000000
  3)         .000000         2.000000
  4)         .000000        -3.000000

NO. ITERATIONS=            2

RANGES IN WHICH THE BASIS IS UNCHANGED:

              OBJ COEFFICIENT RANGES
VARIABLE      CURRENT      ALLOWABLE     ALLOWABLE
               COEF        INCREASE      DECREASE
  X1     5.000000      3.000000        INFINITY
  X2     2.000000      INFINITY        .750000

              RIGHTHAND SIDE RANGES
 ROW    CURRENT      ALLOWABLE      ALLOWABLE
         RHS         INCREASE       DECREASE
  2   30.000000      INFINITY      2.000000
  3   50.000000      2.000000      6.000000
  4   11.000000      1.500000      1.000000
```

produced by Zales can be sold. At present, Zales has 100 rubies, 120 sapphires, and 70 hours of jeweler's labor. Extra rubies can be purchased at a cost of $100 per ruby. Market demand requires that the company produce at least 20 Type 1 rings and at least 25 Type 2. To maximize profit, Zales should solve the following LP:

$X1$ = Type 1 rings produced

$X2$ = Type 2 rings produced

R = number of rubies purchased

max $z = 400X1 + 500X2 - 100R$

s.t.
$$2X1 + 3X2 - R \leq 100$$
$$3X1 + 2X2 \leq 120$$
$$X1 + 2X2 \leq 70$$
$$X1 \geq 20$$
$$X2 \geq 25$$
$$X1, X2 \geq 0$$

Use the LINDO output in Figure 20 to answer the following questions:

a Suppose that instead of $100, each ruby costs $190. Would Zales still purchase rubies? What would be the new optimal solution to the problem?

b Suppose that Zales were only required to produce at least 23 Type 2 rings. What would Zales' profit now be?

FIGURE **19**
LINDO Output for Wivco

```
MAX        13 P1 + 13.2 P2 - 10 RM
SUBJECT TO
     2) - P1 - P2 + RM >=   0
     3)   2 P1 + 3 P2  <= 200
     4)      RM     <= 90
     5)   0.33 P2   <=  40
END
```

LP OPTIMUM FOUND AT STEP   3

OBJECTIVE FUNCTION VALUE

1)   274.000000

VARIABLE	VALUE	REDUCED COST
P1	70.000000	0.000000
P2	20.000000	0.000000
RM	90.000000	0.000000

ROW	SLACK OR SURPLUS	DUAL PRICES
2)	0.000000	-12.600000
3)	0.000000	0.200000
4)	0.000000	2.600000
5)	33.400002	0.000000

NO. ITERATIONS=   3

RANGES IN WHICH THE BASIS IS UNCHANGED:

OBJ COEFFICIENT RANGES

VARIABLE	CURRENT COEF	ALLOWABLE INCREASE	ALLOWABLE DECREASE
P1	13.000000	0.200000	0.866667
P2	13.200000	1.300000	0.200000
RM	-10.000000	INFINITY	2.600000

RIGHTHAND SIDE RANGES

ROW	CURRENT RHS	ALLOWABLE INCREASE	ALLOWABLE DECREASE
2	0.000000	23.333334	10.000000
3	200.000000	70.000000	20.000000
4	90.000000	10.000000	23.333334
5	40.000000	INFINITY	33.400002

FIGURE **20**
LINDO Output for Zales

```
MAX      400 X1 +  500 X2  -  100 R
SUBJECT TO
     2)    2 X1 +  3 X2  -  R <=    100
     3)    3 X1 +  2 X2 <= 120
     4)      X1 +  2 X2 <= 70
     5)      X1  >=   20
     6)      X2  >=   25
END
```

LP OPTIMUM FOUND AT STEP   2

OBJECTIVE FUNCTION VALUE

1)   19000.0000

VARIABLE	VALUE	REDUCED COST
X1	20.000000	0.000000
X2	25.000000	0.000000
R	15.000000	0.000000

ROW	SLACK OR SURPLUS	DUAL PRICES
2)	0.000000	100.000000
3)	10.000000	0.000000
4)	0.000000	200.000000
5)	0.000000	0.000000
6)	0.000000	-200.000000

NO. ITERATIONS=   2

RANGES IN WHICH THE BASIS IS UNCHANGED:

OBJ COEFFICIENT RANGES

VARIABLE	CURRENT COEF	ALLOWABLE INCREASE	ALLOWABLE DECREASE
X1	400.000000	INFINITY	100.000000
X2	500.000000	200.000000	INFINITY
R	-100.000000	100.000000	100.000000

RIGHTHAND SIDE RANGES

ROW	CURRENT RHS	ALLOWABLE INCREASE	ALLOWABLE DECREASE
2	100.000000	15.000000	INFINITY
3	120.000000	INFINITY	10.000000
4	70.000000	3.333333	0.000000
5	20.000000	0.000000	INFINITY
6	25.000000	0.000000	2.500000

c What is the most that Zales would be willing to pay for another hour of jeweler's labor?

d What is the most that Zales would be willing to pay for another sapphire?

5 Beerco manufactures ale and beer from corn, hops, and malt. Currently, 40 lb of corn, 30 lb of hops, and 40 lb of malt are available. A barrel of ale sells for $40 and requires 1 lb of corn, 1 lb of hops, and 2 lb of malt. A barrel of beer sells for $50 and requires 2 lb of corn, 1 lb of hops, and 1 lb of malt. Beerco can sell all ale and beer that is produced. Assume that Beerco's goal is to maximize total sales revenue and solve the following LP:

$$\max z = 40\text{ALE} + 50\text{BEER}$$

s.t.	ALE + 2BEER ≤ 40	(Corn constraint)
	ALE + BEER ≤ 30	(Hops constraint)
	2ALE + BEER ≤ 40	(Malt constraint)
	ALE, BEER ≥ 0	

ALE = barrels of ale produced, and BEER = barrels of beer produced.

a Graphically find the range of values for the price of ale for which the current basis remains optimal.

b Graphically find the range of values for the price of beer for which the current basis remains optimal.

c Graphically find the range of values for the amount of available corn for which the current basis remains optimal. What is the shadow price of the corn constraint?

d Graphically find the range of values for the amount of available hops for which the current basis remains optimal. What is the shadow price of the hops constraint?

e Graphically find the range of values for the amount of available malt for which the current basis remains optimal. What is the shadow price of the malt constraint?

f Find the shadow price of each constraint if the constraints were expressed in ounces instead of pounds.

g Draw a graph of the optimal z-value as a function of the price of ale.

h Draw a graph of the optimal z-value as a function of the amount of available corn.

i Draw a graph of the optimal z-value as a function of the amount of available hops.

j Draw a graph of the optimal z-value as a function of the amount of available malt.

6 Gepbab Production Company uses labor and raw material to produce three products. The resource requirements and sales price for the three products are as shown in Table 10. Currently, 60 units of raw material are available. Up to 90 hours of labor can be purchased at $1 per hour. To maximize Gepbab profits, solve the following LP:

$$\max z = 6X1 + 8X2 + 13X3 - L$$
$$\text{s.t.} \quad 3X1 + 4X2 + 6X3 - L \leq 0$$
$$2X1 + 2X2 + 5X3 \leq 60$$
$$L \leq 90$$
$$X1, X2, X3, L \geq 0$$

Here, X_i = units of product i produced, and L = number of labor hours purchased. Use the LINDO output in Figure 21 to answer the following questions:

a What is the most the company should pay for another unit of raw material?

b What is the most the company should pay for another hour of labor?

c What would product 1 have to sell for to make it desirable for the company to produce it?

d If 100 hours of labor could be purchased, what would the company's profit be?

e Find the new optimal solution if product 3 sold for $15.

7 Giapetto, Inc., sells wooden soldiers and wooden trains. The resources used to produce a soldier and train are shown in Table 11. A total of 145,000 board feet of lumber and 90,000 hours of labor are available. As many as 50,000 soldiers and 50,000 trains can be sold, with trains selling for $55 and soldiers for $32. In addition to producing trains and soldiers itself, Giapetto can buy (from an outside supplier) extra soldiers at $27 each and extra trains at $50 each. Let

SM = thousands of soldiers manufactured
SB = thousands of soldiers bought at $27
TM = thousands of trains manufactured
TB = thousands of trains bought at $50

Then Giapetto can maximize profit by solving the LP in the LINDO printout in Figure 22. Use this printout to answer the following questions. (*Hint:* Think about the units of the constraints and objective function.)

a If Giapetto could purchase trains for $48 per train, then what would be the new optimal solution to the LP? Explain.

TABLE 10

Resource	Product		
	1	2	3
Labor (hours)	3	4	6
Raw material (units)	2	2	5
Sales price ($)	6	8	13

FIGURE 21
LINDO Output for Gepbab

```
MAX       6 X1 + 8 X2 + 13 X3 - L
SUBJECT TO
    2)    3 X1 + 4 X2 + 6 X3 - L  <=    0
    3)    2 X1 + 2 X2 + 5 X3 <=    60
    4)    L     <=    90
END

LP OPTIMUM FOUND AT STEP          3

              OBJECTIVE FUNCTION VALUE

        1)   97.5000000

VARIABLE          VALUE          REDUCED COST
   X1           .000000            .250000
   X2         11.250000            .000000
   X3          7.500000            .000000
   L          90.000000            .000000

   ROW     SLACK OR SURPLUS     DUAL PRICES
   2)          .000000          1.750000
   3)          .000000           .500000
   4)          .000000           .750000

NO. ITERATIONS=         3

RANGES IN WHICH THE BASIS IS UNCHANGED:

                 OBJ COEFFICIENT RANGES
VARIABLE    CURRENT     ALLOWABLE      ALLOWABLE
             COEF       INCREASE       DECREASE
  X1      6.000000      .250000        INFINITY
  X2      8.000000      .666667         .666667
  X3     13.000000     3.000000        1.000000
  L      -1.000000     INFINITY         .750000

                 RIGHTHAND SIDE RANGES
  ROW     CURRENT     ALLOWABLE      ALLOWABLE
           RHS        INCREASE       DECREASE
   2       .000000    30.000000     18.000000
   3     60.000000    15.000000     15.000000
   4     90.000000    30.000000     18.000000
```

TABLE 11

	Soldier	Train
Lumber (board ft)	3	5
Labor (hours)	2	4

b What is the most Giapetto would be willing to pay for another 100 board feet of lumber? For another 100 hours of labor?

c If 60,000 labor hours are available, what would Giapetto's profit be?

d If only 40,000 trains could be sold, what would Giapetto's profit be?

8 Wivco produces two products: 1 and 2. The relevant data are shown in Table 12. Each week, up to 400 units of raw material can be purchased at a cost of $1.50 per unit. The company employs four workers, who work 40 hours per week. (Their salaries are considered a fixed cost.) Workers are paid $6 per hour to work overtime. Each week, 320 hours of machine time are available.

FIGURE 22
LINDO Output for Giapetto

```
MAX        32 SM + 55 TM + 5 SB + 5  TB
SUBJECT TO
      2)    3 SM + 5 TM    <=   145
      3)    2 SM + 4 TM    <=   90
      4)    SM + SB        <=   50
      5)    TM + TB        <=   50
END

LP OPTIMUM FOUND AT STEP        4

              OBJECTIVE FUNCTION VALUE

          1)    1715.00000

VARIABLE        VALUE           REDUCED COST
    SM       45.000000             .000000
    TM         .000000            4.000000
    SB        5.000000             .000000
    TB       50.000000             .000000

 ROW    SLACK OR SURPLUS    DUAL PRICES
  2)       10.000000           .000000
  3)         .000000         13.500000
  4)         .000000          5.000000
  5)         .000000          5.000000

NO. ITERATIONS=       4

RANGES IN WHICH THE BASIS IS UNCHANGED:

                  OBJ COEFFICIENT RANGES
VARIABLE    CURRENT      ALLOWABLE      ALLOWABLE
             COEF        INCREASE       DECREASE
   SM      32.000000     INFINITY       2.000000
   TM      55.000000     4.000000       INFINITY
   SB       5.000000     2.000000       5.000000
   TB       5.000000     INFINITY       4.000000

                  RIGHTHAND SIDE RANGES
ROW     CURRENT      ALLOWABLE      ALLOWABLE
         RHS         INCREASE       DECREASE
 2     145.000000    INFINITY       10.000000
 3      90.000000    6.666667       90.000000
 4      50.000000    INFINITY       5.000000
 5      50.000000    INFINITY       50.000000
```

TABLE 12

	Product 1	Product 2
Selling price ($)	15	8
Labor required (hours)	0.75	0.50
Machine time required (hours)	1.5	0.80
Raw material required (units)	2	1

In the absence of advertising, 50 units of product 1 and 60 units of product 2 will be demanded each week. Advertising can be used to stimulate demand for each product. Each dollar spent on advertising product 1 increases its demand by 10 units, and each dollar spent for product 2 increases its demand by 15 units. At most $100 can be spent on advertising. Define

$P1$ = number of units of product 1 produced each week

$P2$ = number of units of product 2 produced each week

OT = number of hours of overtime labor used each week

RM = number of units of raw materials purchased each week

$A1$ = dollars spent each week on advertising product 1

$A2$ = dollars spent each week on advertising product 2

Then Wivco should solve the following LP:

$$\max z = 15PI + 8P2 - 6(OT) - 1.5RM - A1 - A2$$

$$\text{s.t.} \quad P1 - 10A1 \le 50 \tag{1}$$
$$P2 - 15A2 \le 60 \tag{2}$$
$$0.75P1 + 0.5P2 \le 160 + (OT) \tag{3}$$
$$2P1 + P2 \le RM \tag{4}$$
$$RM \le 400 \tag{5}$$
$$A1 + A2 \le 100 \tag{6}$$
$$1.5P1 + 0.8P2 \le 320 \tag{7}$$

All variables non-negative

Use LINDO to solve this LP. Then use the computer output to answer the following questions:

a If overtime cost only $4 per hour, would Wivco use it?

b If each unit of product 1 sold for $15.50, would the current basis remain optimal? What would be the new optimal solution?

c What is the most that Wivco should be willing to pay for another unit of raw material?

d How much would Wivco be willing to pay for another hour of machine time?

e If each worker were required (as part of the regular workweek) to work 45 hours per week, what would the company's profits be?

f Explain why the shadow price of row (1) is 0.10. (*Hint:* If the right-hand side of (1) were increased from 50 to 51, then in the absence of advertising for product 1, 51 units of product 1 could now be sold each week.)

9 In this problem, we discuss how shadow prices can be interpreted for blending problems (see Section 3.8). To illustrate the ideas, we discuss Problem 2 of Section 3.8. If we define

$$x_{6J} = \text{pounds of grade 6 oranges in juice}$$
$$x_{9J} = \text{pounds of grade 9 oranges in juice}$$
$$x_{6B} = \text{pounds of grade 6 oranges in bags}$$
$$x_{9B} = \text{pounds of grade 9 oranges in bags}$$

then the appropriate formulation is

$$\max z = 0.45(x_{6J} + x_{9J}) + 0.30(x_{6B} + x_{9B})$$

$$\text{s.t.} \quad x_{6J} + x_{6B} \le 120{,}000 \quad \text{(Grade 6 constraint)}$$

$$x_{9J} + x_{9B} \le 100{,}000 \quad \text{(Grade 9 constraint)}$$

$$\frac{6x_{6J} + 9x_{9J}}{x_{6J} + x_{9J}} \ge 8 \quad \text{(Orange juice constraint)} \tag{1}$$

$$\frac{6x_{6B} + 9x_{9B}}{x_{6B} + x_{9B}} \ge 7 \quad \text{(Bags constraint)} \tag{2}$$

$$x_{6J}, x_{9J}, x_{6B}, x_{9B} \ge 0$$

Constraints (1) and (2) are examples of blending constraints because they specify the proportion of grade 6 and grade 9 oranges that must be blended to manufacture orange juice and bags of oranges. It would be useful to determine how a slight change in the standards for orange juice and bags of oranges would affect profit. At the end of this problem, we explain how to use the shadow prices of Constraints (1) and (2) to answer the following questions:

a Suppose that the average grade for orange juice is increased to 8.1. Assuming the current basis remains optimal, by how much would profits change?

b Suppose the average grade requirements for bags of oranges is decreased to 6.9. Assuming the current basis remains optimal, by how much would profits change?

The shadow price for both (1) and (2) is -0.15. The optimal solution to the O.J. problem is $x_{6J} = 26,666.67$, $x_{9J} = 53,333.33$, $x_{6B} = 93,333.33$, $x_{9B} = 46,666.67$. To interpret the shadow prices of blending constraints (1) and (2), *we assume that a slight change in the quality standard for a product will not significantly change the quantity of the product that is produced.*

Now note that (1) may be written as

$$6x_{6J} + 9x_{9J} \geq 8(x_{6J} + x_{9J}) \quad \text{or} \quad -2x_{6J} + x_{9J} \geq 0$$

If the quality standard for orange juice is changed to $8 + \Delta$, then (1) can be written as

$$6x_{6J} + 9x_{9J} \geq (8 + \Delta)(x_{6J} + x_{9J})$$

or

$$-2x_{6J} + x_{9J} \geq \Delta(x_{6J} + x_{9J})$$

Because we are assuming that changing orange juice quality from 8 to $8 + \Delta$ does not change the amount of orange juice produced, $x_{6J} + x_{9J}$ will remain equal to 80,000, and (1) will become

$$-2x_{6J} + x_{9J} \geq 80,000\Delta$$

Using the definition of shadow price, answer parts (a) and (b).

10 Use LINDO to solve the Sailco problem of Section 3.10, then use the output to answer the following questions:

a If month 1 demand decreased to 35 sailboats, what would be the total cost of satisfying the demands during the next four months?

b If the cost of producing a sailboat with regular-time labor during month 1 were $420, what would be the new optimal solution to the Sailco problem?

c Suppose a new customer is willing to pay $425 for a sailboat. If his demand must be met during month 1, should Sailco fill the order? How about if his demand must be met during month 4?

11 Autoco has three assembly plants located in various parts of the country. The first plant (built in 1937 and located in Norwood, Ohio) requires 2 hours of labor and 1 hour of machine time to assemble one automobile. The second plant (built in 1958 and located in Bakersfield, California) requires 1.5 hours of labor and 1.5 hours of machine time to assemble one automobile. The third plant (built in 1981 and located in Kingsport, Tennessee) requires 1.1 hours of labor and 2.5 hours of machine time to assemble one automobile.

The firm pays $30 per hour of labor and $10 per hour of machine time at each of its plants. The first plant has a capacity of 1,000 hours of machine time per day; the second, 900 hours; and the third, 2,000 hours. The manufacturer's production target is 1,800 automobiles per day.

The production department sets each plant's schedule by solving a linear programming problem designed to identify the cost-minimizing pattern of assembly across the three plants.

a Use LINDO to determine the cost-minimizing method of meeting Autoco's daily production target.

b The UWA local in Norwood, Ohio, has proposed wage concessions at that plant to raise employment. What is the *smallest* decrease in the wage rate at that plant that would increase employment there?

c What is the cost of assembling an extra automobile given the current output level of 1,800 automobiles? Would your answer be different if the production target were only 1,000 automobiles? Why or why not?

d A team of production specialists has indicated that the auto manufacturer can achieve efficiencies at its Bakersfield plant by reconfiguring the assembly line. The reconfiguration has the effect of increasing the productivity of the labor at this plant from 1.5 hours to 1 hour per automobile. By how much will the firm's costs fall as a result of this change, assuming that it continues to produce 1,800 automobiles?

e If 2,000 autos must be produced, by how much would costs increase?

f If labor costs $32 per hour in Bakersfield, California, what would be the new solution to the problem?

12 Machinco produces four products, requiring time on two machines and two types (skilled and unskilled) of labor. The amount of machine time and labor (in hours) used by each product and the sales prices are given in Table 13. Each month, 700 hours are available on machine 1 and 500 hours on machine 2. Each month, Machinco can purchase up to 600 hours of skilled labor at $8 per hour and up to 650 hours of unskilled labor at $6 per hour. Formulate an LP that will enable Machinco to maximize its monthly profit. Solve this LP and use the output to answer the following questions:

a By how much does the price of product 3 have to increase before it becomes optimal to produce it?

b If product 1 sold for $290, then what would be the new optimal solution to the problem?

c What is the most Machinco would be willing to pay for an extra hour of time on each machine?

d What is the most Machinco would be willing to pay for an extra hour of each type of labor?

e If up to 700 hours of skilled labor could be purchased each month, then what would be Machinco's monthly profits?

13 A company produces tools at two plants and sells them to three customers. The cost of producing 1,000 tools at a

TABLE 13

Product	Machine 1	Machine 2	Skilled	Unskilled	Sales ($)
1	11	4	8	7	300
2	7	6	5	8	260
3	6	5	4	7	220
4	5	4	6	4	180

plant and shipping them to a customer is given in Table 14. Customers 1 and 3 pay $200 per thousand tools; customer 2 pays $150 per thousand tools. To produce 1,000 tools at plant 1, 200 hours of labor are needed, while 300 hours are needed at plant 2. A total of 5,500 hours of labor are available for use at the two plants. Additional labor hours can be purchased at $20 per labor hour. Plant 1 can produce up to 10,000 tools and plant 2, up to 12,000 tools. Demand by each customer is assumed unlimited. If we let X_{ij} = tools (in thousands) produced at plant i and shipped to customer j then the company should solve the LP on the LINDO printout in Figure 23. Use this printout to answer the following questions:

a If it costs $70 to produce 1,000 tools at plant 1 and ship them to customer 1, what would be the new solution to the problem?

b If the price of an additional hour of labor were reduced to $4, would the company purchase any additional labor?

c A consultant offers to increase plant 1's production capacity by 5,000 tools for a cost of $400. Should the company take her offer?

d If the company were given 5 extra hours of labor, what would its profit become?

14 Solve Review Problem 24 of Chapter 3 on LINDO and answer the following questions:

a For which type of DRGs should the hospital seek to increase demand?

b What resources are in excess supply? Which resources should the hospital expand?

c What is the most the hospital should be willing to pay additional nurses?

15 Old Macdonald's 200-acre farm sells wheat, alfalfa, and beef. Wheat sells for $30 per bushel, alfalfa sells for $200 per bushel, and beef sells for $300 per ton. Up to 1,000 bushels of wheat and up to 1,000 bushels of alfalfa can be sold, but demand for beef is unlimited. If an acre of land is devoted to raising wheat, alfalfa, or beef, the yield and the required labor are given in Table 15. As many as 2,000 hours of labor can be purchased at $15 per hour. Each acre devoted to beef requires 5 bushels of alfalfa. The LINDO output in Figure 24 shows how to maximize profit, use it to answer the following questions. The variables are as follows:

W = acres devoted to wheat
AS = bushels of alfalfa sold
A = acres devoted to alfalfa
B = acres devoted to beef

TABLE 14

Plant	Customer ($)		
	1	2	3
1	60	30	160
2	130	70	170

FIGURE 23
LINDO Output for Problem 13

```
MAX    140 X11 + 120 X12 + 40 X13
              + 70 X21 + 80 X22 + 30 X23 - 20 L
SUBJECT TO
       2) X11 + X12 + X13      <=      10
       3) X21 + X22 + X23      <=      12
       4) 200 X11 + 200 X12 + 200 X13 + 300 X21
              + 300 X22 + 300 X23 - L <= 5500
END

LP OPTIMUM FOUND AT STEP          2

            OBJECTIVE FUNCTION VALUE

        1)    2333.3330

VARIABLE          VALUE          REDUCED COST
  X11          10.000000            .000000
  X12            .000000          20.000000
  X13            .000000         100.000000
  X21            .000000          10.000000
  X22          11.666670            .000000
  X23            .000000          50.000000
  L              .000000          19.733330

  ROW      SLACK OR SURPLUS      DUAL PRICES
  2)            .000000          86.666660
  3)            .333333            .000000
  4)            .000000            .266667

NO. ITERATIONS=          2

RANGES IN WHICH THE BASIS IS UNCHANGED:

                  OBJ COEFFICIENT RANGES
VARIABLE    CURRENT      ALLOWABLE      ALLOWABLE
              COEF       INCREASE       DECREASE
  X11     140.000000     INFINITY      20.000000
  X12     120.000000    20.000000      INFINITY
  X13      40.000000   100.000000      INFINITY
  X21      70.000000    10.000000      INFINITY
  X22      80.000000   130.000000      10.000000
  X23      30.000000    50.000000      INFINITY
  L       -20.000000    19.733330      INFINITY

                  RIGHTHAND SIDE RANGES
  ROW     CURRENT       ALLOWABLE      ALLOWABLE
            RHS         INCREASE       DECREASE
   2     10.000000     17.500000       .500000
   3     12.000000     INFINITY        .333333
   4   5500.000000    100.000000    3500.000000
```

TABLE 15

Crop	Yield/Acre	Labor/Acre (Hours)
Wheat	50 bushels	30
Alfalfa	100 bushels	20
Beef	10 tons	50

AB = bushels of alfalfa devoted to beef
L = hours of labor purchased

a How much must the price of a bushel of wheat increase before it becomes profitable to grow wheat?

b What is the most Old Macdonald should pay for another hour of labor?

FIGURE **24**
LINDO Output for Old Macdonald

```
MAX        1500 W + 200 AS + 3000 B - 15  L
SUBJECT TO
    2)     50  W    <=    1000
    3)     AS   <=    1000
    4)     AS + AB - 100  A  =     0
    5)     -  5  B +  AB  =     0
    6)     W  +  B  +  A <=  200
    7)     L  <=  2000
    8)     30 W + 50 B - L + 20  A  <=   0
END

LP OPTIMUM FOUND AT STEP           1

              OBJECTIVE FUNCTION VALUE

         1)    275882.300

VARIABLE        VALUE         REDUCED COST
    W           .000000        264.705800
    AS        1000.000000        .000000
    B          35.294120         .000000
    L        2000.000000         .000000
    AB        176.470600         .000000
    A          11.764710         .000000


ROW     SLACK OR SURPLUS      DUAL PRICES
 2)       1000.000000           .000000
 3)          .000000         188.235300
 4)          .000000          11.764710
 5)          .000000         -11.764710
 6)        152.941200           .000000
 7)          .000000          43.823530
 8)          .000000          58.823530

NO. ITERATIONS=          1

RANGES IN WHICH THE BASIS IS UNCHANGED:

                OBJ COEFFICIENT RANGES
VARIABLE    CURRENT      ALLOWABLE      ALLOWABLE
            COEF         INCREASE       DECREASE
    W    1500.000000    264.705800      INFINITY
    AS    200.000000     INFINITY      188.235300
    B    3000.000000  48000.000000     449.999800
    L     -15.000000     INFINITY       43.823530
    AB       .000000   9599.999000      89.999980
    A        .000000     INFINITY     8999.998000

                RIGHTHAND SIDE RANGES
ROW     CURRENT      ALLOWABLE       ALLOWABLE
        RHS          INCREASE        DECREASE
 2   1000.000000     INFINITY      1000.000000
 3   1000.000000   8999.999000     1000.000000
 4       .000000   1200.000000     8999.999000
 5       .000000   8999.999000      180.000000
 6    200.000000     INFINITY       152.941200
 7   2000.000000   7428.571000     1800.000000
 8       .000000   7428.571000     1800.000000
```

c What is the most Old Macdonald should pay for another bushel of alfalfa?

d What would be the new optimal solution if alfalfa sold for $20 for bushel?

16 Cornco produces two products: PS and QT. The sales price for each product and the maximum quantity of each that can be sold during each of the next three months are given in Table 16.

Each product must be processed through two assembly lines: 1 and 2. The number of hours required by each product on each assembly line are given in Table 17.

The number of hours available on each assembly line during each month are given in Table 18.

Each unit of PS requires 4 pounds of raw material; each unit of QT requires 3 pounds. As many as 710 units of raw material can be purchased at $3 per pound. At the beginning of month 1, 10 units of PS and 5 units of QT are available. It costs $10 to hold a unit of either product in inventory for a month. Solve this LP on LINDO and use your output to answer the following questions:

a Find the new optimal solution if it costs $11 to hold a unit of PS in inventory at the end of month 1.

b Find the company's new optimal solution if 210 hours on line 1 are available during month 1.

c Find the company's new profit level if 109 hours are available on line 2 during month 3.

d What is the most Cornco should be willing to pay for an extra hour of line 1 time during month 2?

e What is the most Cornco should be willing to pay for an extra pound of raw material?

f What is the most Cornco should be willing to pay for an extra hour of line 1 time during month 3?

g Find the new optimal solution if PS sells for $50 during month 2.

h Find the new optimal solution if QT sells for $50 during month 3.

i Suppose spending $20 on advertising would increase demand for QT in month 2 by 5 units. Should the advertising be done?

TABLE 16

Product	Month 1 Price ($)	Month 1 Demand	Month 2 Price ($)	Month 2 Demand	Month 3 Price ($)	Month 3 Demand
PS	40	50	60	45	55	50
QT	35	43	40	50	44	40

TABLE 17

Product	Hours Line 1	Hours Line 2
PS	3	2
QT	2	2

TABLE 18

Line	Month 1	Month 2	Month 3
1	1,200	160	190
2	2,140	150	110

6

Sensitivity Analysis and Duality

Two of the most important topics in linear programming are sensitivity analysis and duality. After studying these important topics, the reader will have an appreciation of the beauty and logic of linear programming and be ready to study advanced linear programming topics such as those discussed in Chapter 10.

In Section 6.1, we illustrate the concept of sensitivity analysis through a graphical example. In Section 6.2, we use our knowledge of matrices to develop some important formulas, which are used in Sections 6.3 and 6.4 to develop the mechanics of sensitivity analysis. The remainder of the chapter presents the important concept of duality. Duality provides many insights into the nature of linear programming, gives us the useful concept of shadow prices, and helps us understand sensitivity analysis. It is a necessary basis for students planning to take advanced topics in linear and nonlinear programming.

6.1 A Graphical Introduction to Sensitivity Analysis

Sensitivity analysis is concerned with how changes in an LP's parameters affect the LP's optimal solution.

Reconsider the Giapetto problem of Section 3.1:

$$\max z = 3x_1 + 2x_2$$
$$\text{s.t.} \quad 2x_1 + x_2 \le 100 \quad \text{(Finishing constraint)}$$
$$x_1 + x_2 \le 80 \quad \text{(Carpentry constraint)}$$
$$x_1 \quad \le 40 \quad \text{(Demand constraint)}$$
$$x_1, x_2 \ge 0$$

where

$$x_1 = \text{number of soldiers produced per week}$$
$$x_2 = \text{number of trains produced per week}$$

The optimal solution to this problem is $z = 180$, $x_1 = 20$, $x_2 = 60$ (point B in Figure 1), and it has x_1, x_2, and s_3 (the slack variable for the demand constraint) as basic variables. How would changes in the problem's objective function coefficients or right-hand sides change this optimal solution?

Graphical Analysis of the Effect of a Change in an Objective Function Coefficient

If the contribution to profit of a soldier were to increase sufficiently, then it would be optimal for Giapetto to produce more soldiers (s_3 would become nonbasic). Similarly, if the

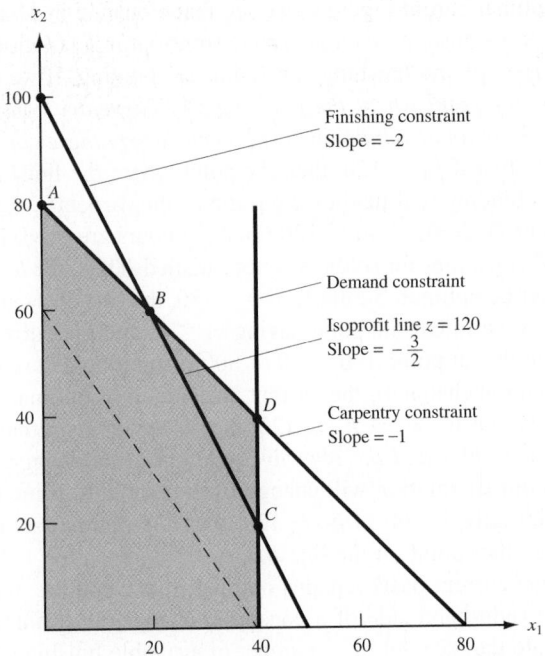

FIGURE 1
Analysis of Range of Values for Which c_1 Remains Optimal in Giapetto Problem

contribution to profit of a soldier were to decrease sufficiently, it would be optimal for Giapetto to produce only trains (x_1 would now be nonbasic). We now show how to determine the values of the contribution to profit for soldiers for which the current optimal basis will remain optimal.

Let c_1 be the contribution to profit by each soldier. For what values of c_1 does the current basis remain optimal?

At present, $c_1 = 3$, and each isoprofit line has the form $3x_1 + 2x_2 = $ constant, or $x_2 = -\frac{3x}{2} + \frac{\text{constant}}{2}$, and each isoprofit line has a slope of $-\frac{3}{2}$. From Figure 1, we see that if a change in c_1 causes the isoprofit lines to be flatter than the carpentry constraint, then the optimal solution will change from the current optimal solution (point B) to a new optimal solution (point A). If the profit for each soldier is c_1, then the slope of each isoprofit line will be $-\frac{c_1}{2}$. Because the slope of the carpentry constraint is -1, the isoprofit lines will be flatter than the carpentry constraint if $-\frac{c_1}{2} > -1$, or $c_1 < 2$, and the current basis will no longer be optimal. The new optimal solution will be (0, 80), point A in Figure 1.

If the isoprofit lines are steeper than the finishing constraint, then the optimal solution will change from point B to point C. The slope of the finishing constraint is -2. If $-\frac{c_1}{2} < -2$, or $c_1 > 4$, then the current basis is no longer optimal, and point C (40, 20) will be optimal. In summary, we have shown that (if all other parameters remain unchanged) the current basis remains optimal for $2 \leq c_1 \leq 4$, and Giapetto should still manufacture 20 soldiers and 60 trains. Of course, even if $2 \leq c_1 \leq 4$, Giapetto's profit will change. For instance, if $c_1 = 4$, Giapetto's profit will now be $4(20) + 2(60) = \$200$ instead of $180.

Graphical Analysis of the Effect of a Change in a Right-Hand Side on the LP's Optimal Solution

A graphical analysis can also be used to determine whether a change in the right-hand side of a constraint will make the current basis no longer optimal. Let b_1 be the number of available finishing hours. Currently, $b_1 = 100$. For what values of b_1 does the current basis remain optimal? From Figure 2, we see that a change in b_1 shifts the finishing constraint parallel to its current position. The current optimal solution (point B in Figure 2) is where the carpentry and finishing constraints are binding. If we change the value of b_1, then *as long as the point where the finishing and carpentry constraints are binding remains feasible, the optimal solution will still occur where these constraints intersect.* From Figure 2, we see that if $b_1 > 120$, then the point where the finishing and carpentry constraints are both binding will lie on the portion of the carpentry constraint below point D. Note that at point D, $2(40) + 40 = 120$ finishing hours are used. In this region, $x_1 > 40$, and the demand constraint for soldiers is not satisfied. Thus, for $b_1 > 120$, the current basis will no longer be optimal. Similarly, if $b_1 < 80$, the carpentry and finishing constraints will be binding at an infeasible point having $x_1 < 0$, and the current basis will no longer be optimal. Note that at point A, $0 + 80 = 80$ finishing hours are used. Thus (if all other parameters remain unchanged), the current basis remains optimal if $80 \le b_1 \le 120$.

Note that although for $80 \le b_1 \le 120$, the current basis remains optimal, *the values of the decision variables and the objective function value change.* For example, if $80 \le b_1 \le 100$, the optimal solution will change from point B to some other point on the line segment AB. Similarly, if $100 \le b_1 \le 120$, then the optimal solution will change from point B to some other point on the line BD.

As long as the current basis remains optimal, it is a routine matter to determine how a change in the right-hand side of a constraint changes the values of the decision variables. To illustrate the idea, let b_1 = number of available finishing hours. If we change b_1 to $100 + \Delta$, we know that the current basis remains optimal for $-20 \le \Delta \le 20$. Note that as b_1 changes (as long as $-20 \le \Delta \le 20$), the optimal solution to the LP is still the point where the finishing-hour and carpentry-hour constraints are binding. Thus, if $b_1 = 100 + \Delta$, we can find the new values of the decision variables by solving

$$2x_1 + x_2 = 100 + \Delta \qquad \text{and} \qquad x_1 + x_2 = 80$$

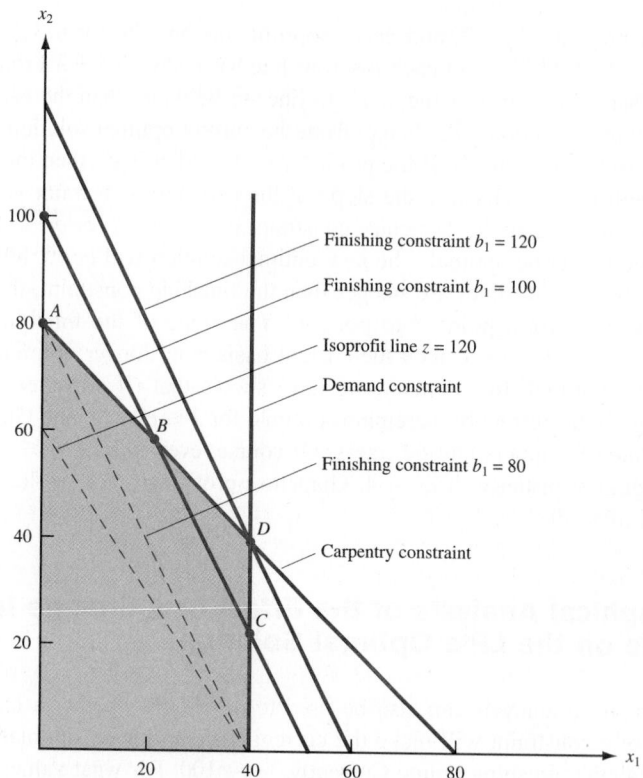

FIGURE 2
Range of Values of
Finishing Hours for
Which Current Basis
Remains Optimal in
Giapetto Problem

This yields $x_1 = 20 + \Delta$ and $x_2 = 60 - \Delta$. Thus, an increase in the number of available finishing hours results in an increase in the number of soldiers produced and a decrease in the number of trains produced.

If b_2 (the number of available carpentry hours) equals $80 + \Delta$, it can be shown (see Problem 2) that the current basis remains optimal for $-20 \leq \Delta \leq 20$. If we change the value of b_2 (keeping $-20 \leq \Delta \leq 20$), then the optimal solution to the LP is still the point where the finishing and carpentry constraints are binding. Thus, if $b_2 = 80 + \Delta$, the optimal solution to the LP is the solution to

$$2x_1 + x_2 = 100 \qquad \text{and} \qquad x_1 + x_2 = 80 + \Delta$$

This yields $x_1 = 20 - \Delta$ and $x_2 = 60 + 2\Delta$, which shows that an increase in the amount of available carpentry hours decreases the number of soldiers produced and increases the number of trains produced.

Suppose b_3, the demand for soldiers, is changed to $40 + \Delta$. Then it can be shown (see Problem 3) that the current basis remains optimal for $\Delta \geq -20$. For Δ in this range, the optimal solution to the LP will still occur where the finishing and carpentry constraints are binding. Thus, the optimal solution will be the solution to

$$2x_1 + x_2 = 100 \qquad \text{and} \qquad x_1 + x_2 = 80$$

Of course, this yields $x_1 = 20$ and $x_2 = 60$, which illustrates an important point. In a constraint with positive slack (or positive excess) in an LP's optimal solution, if we change the right-hand side of the constraint to a value in the range where the current basis remains optimal, the optimal solution to the LP is unchanged.

Shadow Prices

As we will see in Section 6.8, it is often important for managers to determine how a change in a constraint's right-hand side changes the LP's optimal z-value. With this in mind, we define the **shadow price** for the ith constraint of an LP to be the amount by which the optimal z-value is improved (improvement means increase in a max problem and decrease in a min problem) if the right-hand side of the ith constraint is increased by 1. This definition applies only if the change in the right-hand side of Constraint i leaves the current basis optimal.

For any two-variable LP, it is a simple matter to determine each constraint's shadow price. For example, we know that if $100 + \Delta$ finishing hours are available (assuming that the current basis remains optimal), then the LP's optimal solution is $x_1 = 20 + \Delta$ and $x_2 = 60 - \Delta$. Then the optimal z-value will equal $3x_1 + 2x_2 = 3(20 + \Delta) + 2(60 - \Delta) = 180 + \Delta$. Thus, as long as the current basis remains optimal, a unit increase in the number of available finishing hours will increase the optimal z-value by $1. So the shadow price of the first (finishing hour) constraint is $1.

For the second (carpentry hour) constraint, we know that if $80 + \Delta$ carpentry hours are available (and the current basis remains optimal), then the optimal solution to the LP is $x_1 = 20 - \Delta$ and $x_2 = 60 + 2\Delta$. Then the new optimal z-value is $3x_1 + 2x_2 = 3(20 - \Delta) + 2(60 + 2\Delta) = 180 + \Delta$. Thus, a unit increase in the number of carpentry hours will increase the optimal z-value by $1 (as long as the current basis remains optimal). So the shadow price of the second (carpentry hour) constraint is $1.

We now find the shadow price of the third (demand) constraint. If the right-hand side is $40 + \Delta$, then the optimal values of the decision variables remain unchanged, as long as the current basis remains optimal. Then the optimal z-value will also remain unchanged, which shows that the shadow price of the third (demand) constraint is $0. It turns out that whenever the slack variable or excess variable for a constraint is positive in an LP's optimal solution, the constraint will have a zero shadow price.

Suppose we increase the right-hand side of the ith constraint of an LP by $\Delta b_i(\Delta b_i < 0$ means that we are decreasing the right-hand side) and the current basis remains optimal. Then each unit by which Constraint i's right-hand side is increased will increase the optimal z-value (for a max problem) by the shadow price. Thus, the new optimal z-value is given by

(New optimal z-value) = (old optimal z-value) + (Constraint i's shadow price) Δb_i

For a minimization problem,

(New optimal z-value) = (old optimal z-value) − (Constraint i's shadow price) Δb_i

For example, if 95 carpentry hours are available, then $\Delta b_2 = 15$, and the new z-value is given by

$$\text{New optimal } z\text{-value} = 180 + 15(1) = \$195$$

We will continue our discussion of shadow prices in Section 6.8.

Importance of Sensitivity Analysis

Sensitivity analysis is important for several reasons. In many applications, the values of an LP's parameters may change. For example, the prices at which soldiers and trains are sold may change, as may the availability of carpentry and finishing hours. If a parameter changes, sensitivity analysis often makes it unnecessary to solve the problem again. For example, if the profit contribution of a soldier increased to $3.50, we would not have to solve the Giapetto problem again because the current solution remains optimal. Of course, solving the Giapetto problem again would not be much work, but solving an LP with thousands of variables and constraints again would be a chore. A knowledge of sensitivity analysis often enables the analyst to determine from the original solution how changes in an LP's parameters change the optimal solution.

Recall that we may be uncertain about the values of parameters in an LP, for example, the weekly demand for soldiers. With the graphical method, it can be shown that if the weekly demand for soldiers is at least 20, then the optimal solution to the Giapetto problem is still (20, 60) (see Problem 3 at the end of this section). Thus, even if Giapetto is uncertain about the demand for soldiers, the company can still be fairly confident that it is optimal to produce 20 soldiers and 60 trains.

Of course, the graphical approach is not useful for sensitivity analysis on an LP with more than two variables. Before learning how to perform sensitivity analysis on an arbitrary LP, we need to use our knowledge of matrices to express simplex tableaus in matrix form. This is the subject of Section 6.2.

PROBLEMS

Group A

1 Show that if the contribution to profit for trains is between $1.50 and $3, the current basis remains optimal. If the contribution to profit for trains is $2.50, what would be the new optimal solution?

2 Show that if available carpentry hours remain between 60 and 100, the current basis remains optimal. If between 60 and 100 carpentry hours are available, then would Giapetto still produce 20 soldiers and 60 trains?

3 Show that if the weekly demand for soldiers is at least 20, the current basis remains optimal, and Giapetto should still produce 20 soldiers and 60 trains.

4 For the Dorian Auto problem (Example 2 in Chapter 3),

a Find the range of values of the cost of a comedy ad for which the current basis remains optimal.

b Find the range of values of the cost of a football ad for which the current basis remains optimal.

c Find the range of values of required HIW exposures for which the current basis remains optimal. Determine the new optimal solution if $28 + \Delta$ million HIW exposures are required.

d Find the range of values of required HIM exposures for which the current basis remains optimal. Determine the new optimal solution if $24 + \Delta$ million HIM exposures are required.

e Find the shadow price of each constraint.

f If 26 million HIW exposures are required, determine the new optimal z-value.

5 Radioco manufactures two types of radios. The only scarce resource needed to produce radios is labor. The company now has two laborers. Laborer 1 is willing to work as many as 40 hours per week and is paid $5 per hour. Laborer 2 is willing to work up to 50 hours per week and is paid $6 per hour. The price as well as the resources required to build each type of radio are given in Table 1.

Letting x_i be the number of type i radios produced each week, show that Radioco should solve the following LP:

$$\max z = 3x_1 + 2x_2$$
$$\text{s.t.} \quad x_1 + 2x_2 \leq 40$$
$$2x_1 + x_2 \leq 50$$
$$x_1, x_2 \geq 0$$

TABLE 1

	Radio 1		Radio 2	
Price ($)		Resource Required	Price ($)	Resource Required
25		Laborer 1: 1 hour	22	Laborer 1: 2 hours
		Laborer 2: 2 hours		Laborer 2: 2 hours
		Raw material cost: $5		Raw material cost: $4

a For what values of the price of a Type 1 radio would the current basis remain optimal?

b For what values of the price of a Type 2 radio would the current basis remain optimal?

c If laborer 1 were willing to work only 30 hours per week, would the current basis remain optimal? Find the new optimal solution to the LP.

d If laborer 2 were willing to work up to 60 hours per week, would the current basis remain optimal? Find the new optimal solution to the LP.

e Find the shadow price of each constraint.

6.2 Some Important Formulas

In this section, we use our knowledge of matrices to show how an LP's optimal tableau can be expressed in terms of the LP's parameters. The formulas developed in this section are used in our study of sensitivity analysis, duality, and advanced LP topics.

Assume that we are solving a max problem that has been prepared for solution by the Big M method and that at this point, the LP has m constraints and n variables. Although some of these variables may be slack, excess, or artificial, we choose to label them $x_1, x_2, \ldots, x_n$. Then the LP may be written as

$$\max z = c_1 x_1 + c_2 x_2 + \cdots + c_n x_n$$
$$\text{s.t.} \quad a_{11}x_1 + a_{12}x_2 + \cdots + a_{1n}x_n = b_1$$
$$a_{21}x_1 + a_{22}x_2 + \cdots + a_{2n}x_n = b_2$$
$$\vdots \qquad \vdots \qquad\qquad \vdots \qquad \vdots \tag{1}$$
$$a_{m1}x_1 + a_{m2}x_2 + \cdots + a_{mn}x_n = b_m$$
$$x_i \geq 0 \quad (i = 1, 2, \ldots, n)$$

Throughout this chapter, we use the Dakota Furniture problem of Section 4.5 (without the $x_2 \leq 5$ constraint) as an example. For the Dakota problem, the analog of LP (1) is

$$\max z = 60x_1 + 30x_2 + 20x_3 + 0s_1 + 0s_2 + 0s_3$$
$$\text{s.t.} \quad 8x_1 + 6x_2 + x_3 + s_1 = 48$$
$$4x_1 + 2x_2 + 1.5x_3 + s_2 = 20 \tag{1'}$$
$$2x_1 + 1.5x_2 + 0.5x_3 + s_3 = 8$$
$$x_1, x_2, x_3, s_1, s_2, s_3 \geq 0$$

Suppose we have found the optimal solution to (1). Let BV_i be the basic variable for row i of the optimal tableau. Also define $BV = \{BV_1, BV_2, \ldots, BV_m\}$ to be the set of basic variables in the optimal tableau, and define the $m \times 1$ vector

$$\mathbf{x}_{BV} = \begin{bmatrix} x_{BV_1} \\ x_{BV_2} \\ \vdots \\ x_{BV_m} \end{bmatrix}$$

We also define

NBV = the set of nonbasic variables in the optimal tableau

$\mathbf{x}_{NBV} = (n - m) \times 1$ vector listing the nonbasic variables (in any desired order)

To illustrate these definitions, we recall that the optimal tableau for the Dakota problem is

$$\begin{aligned}
z &+ 5x_2 + &+ 10s_2 + 10s_3 &= 280 \\
&- 2x_2 + &+ s_1 + 2s_2 - 8s_3 &= 24 \\
&- 2x_2 + x_3 &+ 2s_2 - 4s_3 &= 8 \\
&x_1 + 1.25x_2 &- 0.5s_2 + 1.5s_3 &= 2
\end{aligned} \tag{2}$$

For this optimal tableau, $BV_1 = s_1$, $BV_2 = x_3$, and $BV_3 = x_1$. Then

$$\mathbf{x}_{BV} = \begin{bmatrix} s_1 \\ x_3 \\ x_1 \end{bmatrix}$$

We may choose NBV $= \{x_2, s_2, s_3\}$. Then

$$\mathbf{x}_{NBV} = \begin{bmatrix} x_2 \\ s_2 \\ s_3 \end{bmatrix}$$

Using our knowledge of matrix algebra, we can express the optimal tableau in terms of BV and the original LP (1). Recall that $c_1, c_2, \ldots, c_n$ are the objective function coefficients for the variables $x_1, x_2, \ldots, x_n$ (some of these may be slack, excess, or artificial variables).

DEFINITION ■ $\mathbf{c}_{BV}$ is the $1 \times m$ row vector $[c_{BV_1} \quad c_{BV_2} \quad \cdots \quad c_{BV_m}]$. ■

Thus, the elements of $\mathbf{c}_{BV}$ are the objective function coefficients for the optimal tableau's basic variables. For the Dakota problem, BV $= \{s_1, x_3, x_1\}$. Then from (1′) we find that $\mathbf{c}_{BV} = [0 \quad 20 \quad 60]$.

DEFINITION ■ $\mathbf{c}_{NBV}$ is the $1 \times (n - m)$ row vector whose elements are the coefficients of the nonbasic variables (in the order of NBV). ■

If we choose to list the nonbasic variables for the Dakota problem in the order NBV $= \{x_2, s_2, s_3\}$, then $\mathbf{c}_{NBV} = [30 \quad 0 \quad 0]$.

DEFINITION ■ The $m \times m$ matrix B is the matrix whose jth column is the column for BV_j in (1). ■

For the Dakota problem, the first column of B is the s_1 column in (1′); the second, the x_3 column; and the third, the x_1 column. Thus,

$$B = \begin{bmatrix} 1 & 1 & 8 \\ 0 & 1.5 & 4 \\ 0 & 0.5 & 2 \end{bmatrix}$$

DEFINITION ■ $\mathbf{a}_j$ is the column (in the constraints) for the variable x_j in (1). ■

For example, in the Dakota problem,

$$\mathbf{a}_2 = \begin{bmatrix} 6 \\ 2 \\ 1.5 \end{bmatrix} \quad \text{and} \quad \mathbf{a} \text{ (for } s_1) = \begin{bmatrix} 1 \\ 0 \\ 0 \end{bmatrix}$$

DEFINITION ■ N is the $m \times (n - m)$ matrix whose columns are the columns for the nonbasic variables (in the NBV order) in (1). ■

If for the Dakota problem, we write NBV $= \{x_2, s_2, s_3\}$, then

$$N = \begin{bmatrix} 6 & 0 & 0 \\ 2 & 1 & 0 \\ 1.5 & 0 & 1 \end{bmatrix}$$

DEFINITION ■ The $m \times 1$ column vector $\mathbf{b}$ is the right-hand side of the constraints in (1). ■

For the Dakota problem,

$$\mathbf{b} = \begin{bmatrix} 48 \\ 20 \\ 8 \end{bmatrix}$$

We write b_i for the right-hand side of the ith constraint in the original Dakota problem: $b_2 = 20$.

We can now use matrix algebra to determine how an LP's optimal tableau (with set of basic variables BV) is related to the original LP in the form (1).

Expressing the Constraints in Any Tableau in Terms of B^{-1} and the Original LP

We begin by observing that (1) may be written as

$$z = \mathbf{c}_{BV}\mathbf{x}_{BV} + \mathbf{c}_{NBV}\mathbf{x}_{NBV}$$
$$\text{s.t.} \quad B\mathbf{x}_{BV} + N\mathbf{x}_{NBV} = \mathbf{b} \tag{3}$$
$$\mathbf{x}_{BV}, \, \mathbf{x}_{NBV} \geq 0$$

Using the format of (3), the Dakota problem can be written as

$$\max z = \begin{bmatrix} 0 & 20 & 60 \end{bmatrix} \begin{bmatrix} s_1 \\ x_3 \\ x_1 \end{bmatrix} + \begin{bmatrix} 30 & 0 & 0 \end{bmatrix} \begin{bmatrix} x_2 \\ s_2 \\ s_3 \end{bmatrix}$$

$$\text{s.t.} \quad \begin{bmatrix} 1 & 1 & 8 \\ 0 & 1.5 & 4 \\ 0 & 0.5 & 2 \end{bmatrix} \begin{bmatrix} s_1 \\ x_3 \\ x_1 \end{bmatrix} + \begin{bmatrix} 6 & 0 & 0 \\ 2 & 1 & 0 \\ 1.5 & 0 & 1 \end{bmatrix} \begin{bmatrix} x_2 \\ s_2 \\ s_3 \end{bmatrix} = \begin{bmatrix} 48 \\ 20 \\ 8 \end{bmatrix}$$

$$\begin{bmatrix} s_1 \\ x_3 \\ x_1 \end{bmatrix} \geq \begin{bmatrix} 0 \\ 0 \\ 0 \end{bmatrix}, \quad \begin{bmatrix} x_2 \\ s_2 \\ s_3 \end{bmatrix} \geq \begin{bmatrix} 0 \\ 0 \\ 0 \end{bmatrix}$$

Multiplying the constraints in (3) through by B^{-1}, we obtain

$$B^{-1}B\mathbf{x}_{BV} + B^{-1}N\mathbf{x}_{NBV} = B^{-1}\mathbf{b} \quad \text{or} \quad \mathbf{x}_{BV} + B^{-1}N\mathbf{x}_{NBV} = B^{-1}\mathbf{b} \qquad (4)$$

In (4), BV_i occurs with a coefficient of 1 in the ith constraint and a zero coefficient in each other constraint. Thus, BV is the set of the basic variables for (4), and (4) yields the constraints for the optimal tableau.

For the Dakota problem, the Gauss–Jordan method can be used to show that

$$B^{-1} = \begin{bmatrix} 1 & 2 & -8 \\ 0 & 2 & -4 \\ 0 & -0.5 & 1.5 \end{bmatrix}$$

Then (4) yields

$$\begin{bmatrix} s_1 \\ x_3 \\ x_1 \end{bmatrix} + \begin{bmatrix} 1 & 2 & -8 \\ 0 & 2 & -4 \\ 0 & -0.5 & 1.5 \end{bmatrix}\begin{bmatrix} 6 & 0 & 0 \\ 2 & 1 & 0 \\ 1.5 & 0 & 1 \end{bmatrix}\begin{bmatrix} x_2 \\ s_2 \\ s_3 \end{bmatrix} = \begin{bmatrix} 1 & 2 & -8 \\ 0 & 2 & -4 \\ 0 & -0.5 & 1.5 \end{bmatrix}\begin{bmatrix} 48 \\ 20 \\ 8 \end{bmatrix}$$

or

$$\begin{bmatrix} s_1 \\ x_3 \\ x_1 \end{bmatrix} + \begin{bmatrix} -2 & 2 & -8 \\ -2 & 2 & -4 \\ 1.25 & -0.5 & 1.5 \end{bmatrix}\begin{bmatrix} x_2 \\ s_2 \\ s_3 \end{bmatrix} = \begin{bmatrix} 24 \\ 8 \\ 2 \end{bmatrix} \qquad (4')$$

Of course, these are the constraints for the Dakota optimal tableau, (2).

From (4), we see that the column of a nonbasic variable x_j in the constraints of the optimal tableau is given by B^{-1} [column for x_j in (1)] $= B^{-1}\mathbf{a}_j$. For example, the x_2 column is B^{-1} (first column of N) $= B^{-1}\mathbf{a}_2$. From (4), we also find that the right-hand side of the constraints is the vector $B^{-1}\mathbf{b}$. The following two equations summarize the preceding discussion:

$$\text{Column for } x_j \text{ in optimal tableau's constraints} = B^{-1}\mathbf{a}_j \qquad (5)$$

$$\text{Right-hand side of optimal tableau's constraints} = B^{-1}\mathbf{b} \qquad (6)$$

To illustrate (5), we find:

$$\begin{array}{l} \text{Column for } x_2 \\ \text{in Dakota optimal tableau} \end{array} = B^{-1}\mathbf{a}_2$$

$$= \begin{bmatrix} 1 & 2 & -8 \\ 0 & 2 & -4 \\ 0 & -0.5 & 1.5 \end{bmatrix}\begin{bmatrix} 6 \\ 2 \\ 1.5 \end{bmatrix} = \begin{bmatrix} -2 \\ -2 \\ 1.25 \end{bmatrix}$$

To illustrate (6), we compute:

$$\begin{array}{l} \text{Right-hand side of constraints} \\ \text{in Dakota optimal tableau} \end{array} = B^{-1}\mathbf{b}$$

$$= \begin{bmatrix} 1 & 2 & -8 \\ 0 & 2 & -4 \\ 0 & -0.5 & 1.5 \end{bmatrix}\begin{bmatrix} 48 \\ 20 \\ 8 \end{bmatrix} = \begin{bmatrix} 24 \\ 8 \\ 2 \end{bmatrix}$$

Determining the Optimal Tableau's Row 0 in Terms of the Initial LP

We now show how to express row 0 of the optimal tableau in terms of BV and the original LP (1). To begin, we multiply the constraints (expressed in the form $B\mathbf{x}_{BV} + N\mathbf{x}_{NBV} = \mathbf{b}$) through by the vector $\mathbf{c}_{BV}B^{-1}$:

$$\mathbf{c}_{BV}\mathbf{x}_{BV} + \mathbf{c}_{BV}B^{-1}N\mathbf{x}_{NBV} = \mathbf{c}_{BV}B^{-1}\mathbf{b} \tag{7}$$

and rewrite the original objective function, $z = c_{BV}\mathbf{x}_{BV} + c_{NBV}\mathbf{x}_{NBV}$, as

$$z - c_{BV}\mathbf{x}_{BV} - c_{NBV}\mathbf{x}_{NBV} = 0 \tag{8}$$

By adding (7) to (8), we can eliminate the optimal tableau's basic variables and obtain its row 0:

$$z + (\mathbf{c}_{BV}B^{-1}N - \mathbf{c}_{NBV})\mathbf{x}_{NBV} = \mathbf{c}_{BV}B^{-1}\mathbf{b} \tag{9}$$

From (9), the coefficient of x_j in row 0 is

$$\mathbf{c}_{BV}B^{-1} \text{ (column of } N \text{ for } x_j) - \text{(coefficient for } x_j \text{ in } \mathbf{c}_{NBV}) = \mathbf{c}_{BV}B^{-1}\mathbf{a}_j - c_j$$

and the right-hand side of row 0 is $\mathbf{c}_{BV}B^{-1}\mathbf{b}$.

To help summarize the preceding discussion, we let $\bar{c}_j$ be the coefficient of x_j in the optimal tableau's row 0. Then we have shown that

$$\bar{c}_j = \mathbf{c}_{BV}B^{-1}\mathbf{a}_j - c_j \tag{10}$$

and

$$\text{Right-hand side of optimal tableau's row 0} = \mathbf{c}_{BV}B^{-1}\mathbf{b} \tag{11}$$

To illustrate the use of (10) and (11), we determine row 0 of the Dakota problem's optimal tableau. Recall that

$$\mathbf{c}_{BV} = \begin{bmatrix} 0 & 20 & 60 \end{bmatrix} \qquad \text{and} \qquad B^{-1} = \begin{bmatrix} 1 & 2 & -8 \\ 0 & 2 & -4 \\ 0 & -0.5 & 1.5 \end{bmatrix}$$

Then $\mathbf{c}_{BV}B^{-1} = \begin{bmatrix} 0 & 10 & 10 \end{bmatrix}$, and from (10) we find that the coefficients of the nonbasic variables in row 0 of the optimal tableau are

$$\bar{c}_2 = \mathbf{c}_{BV}B^{-1}\mathbf{a}_2 - c_2 = \begin{bmatrix} 0 & 10 & 10 \end{bmatrix}\begin{bmatrix} 6 \\ 2 \\ 1.5 \end{bmatrix} - 30 = 20 + 15 - 30 = 5$$

and

$$\text{Coefficient of } s_2 \text{ in optimal row 0} = \mathbf{c}_{BV}B^{-1}\begin{bmatrix} 0 \\ 1 \\ 0 \end{bmatrix} - 0 = 10$$

$$\text{Coefficient of } s_3 \text{ in optimal row 0} = \mathbf{c}_{BV}B^{-1}\begin{bmatrix} 0 \\ 0 \\ 1 \end{bmatrix} - 0 = 10$$

Of course, the optimal tableau's basic variables (x_1, x_3, and s_1) will have zero coefficients in row 0.

From (11), the right-hand side of row 0 is

$$\mathbf{c}_{BV}B^{-1}\mathbf{b} = \begin{bmatrix} 0 & 10 & 10 \end{bmatrix} \begin{bmatrix} 48 \\ 20 \\ 8 \end{bmatrix} = 280$$

Putting it all together, we see that row 0 is

$$z + 5x_2 + 10s_2 + 10s_3 = 280$$

Of course, this result agrees with (2).

Simplifying Formula (10) for Slack, Excess, and Artificial Variables

Formula (10) can be greatly simplified if x_j is a slack, excess, or artificial variable. For example, if x_j is the slack variable s_i, the coefficient of s_i in the objective function is 0, and the column for s_i in the original tableau has 1 in row i and 0 in all other rows. Then (10) yields

$$\text{Coefficient of } s_i \text{ in optimal row } 0 = i\text{th element of } \mathbf{c}_{BV}B^{-1} - 0 \qquad (10')$$
$$= i\text{th element of } \mathbf{c}_{BV}B^{-1}$$

Similarly, if x_j is the excess variable e_i, then the coefficient of e_i in the objective function is 0 and the column for e_i in the original tableau has -1 in row i and 0 in all other rows. Then (10) reduces to

$$\text{Coefficient of } e_i \text{ in optimal row } 0 = -(i\text{th element of } \mathbf{c}_{BV}B^{-1}) - 0 \qquad (10'')$$
$$= -(i\text{th element of } \mathbf{c}_{BV}B^{-1})$$

Finally, if x_j is an artificial variable a_i, then the objective function coefficient of a_i (for a max problem) is $-M$ and the original column for a_i has 1 in row i and 0 in all other rows. Then (10) reduces to

$$\text{Coefficient of } a_i \text{ in optimal row } 0 = (i\text{th element of } \mathbf{c}_{BV}B^{-1}) - (-M) \qquad (10''')$$
$$= (i\text{th element of } \mathbf{c}_{BV}B^{-1}) + (M)$$

The derivations of this section have not been easy. Fortunately, use of (5), (6), (10), and (11) does not require a complete understanding of the derivations. A summary of the formulas derived in this section for computing an optimal tableau from the initial LP follows.

Summary of Formulas for Computing the Optimal Tableau from the Initial LP

$$x_j \text{ column in optimal tableau's constraints} = B^{-1}\mathbf{a}_j \qquad (5)$$

$$\text{Right-hand side of optimal tableau's constraints} = B^{-1}\mathbf{b} \qquad (6)$$

$$\bar{c}_j = \mathbf{c}_{BV}B^{-1}\mathbf{a}_j - c_j \qquad (10)$$

Coefficient of slack variable s_i in optimal row 0

$$= i\text{th element of } \mathbf{c}_{BV}B^{-1} \qquad (10')$$

Coefficient of excess variable e_i in optimal row 0

$$= -(i\text{th element of } \mathbf{c}_{BV}B^{-1}) \qquad (10'')$$

Coefficient of artificial variable a_i in optimal row 0

$$= (i\text{th element of } \mathbf{c}_{BV}B^{-1}) + M \quad \text{(max problem)} \qquad (10''')$$

$$\text{Right-hand side of optimal row } 0 = \mathbf{c}_{BV}B^{-1}\mathbf{b} \qquad (11)$$

We must first find B^{-1} because it is necessary in order to compute all parts of the optimal tableau. Similarly, we must find $\mathbf{c}_{BV}B^{-1}$ to compute the optimal tableau's row 0.

The following example is another illustration of the use of the preceding formulas.

EXAMPLE 1   **Computing Optimal Tableau**

For the following LP, the optimal basis is BV $= \{x_2, s_2\}$. Compute the optimal tableau.

$$\max z = x_1 + 4x_2$$
$$\text{s.t.} \quad x_1 + 2x_2 \leq 6$$
$$2x_1 + x_2 \leq 8$$
$$x_1, x_2 \geq 0$$

Solution   After adding slack variables s_1 and s_2, we obtain the analog of (1):

$$\max z = x_1 + 4x_2$$
$$\text{s.t.} \quad x_1 + 2x_2 + s_1 \qquad = 6$$
$$2x_1 + x_2 \qquad + s_2 = 8$$

First we compute B^{-1}. Because

$$B = \begin{bmatrix} 2 & 0 \\ 1 & 1 \end{bmatrix}$$

we find B^{-1} by applying the Gauss–Jordan method to the following matrix:

$$B|I_2 = \begin{bmatrix} 2 & 0 & | & 1 & 0 \\ 1 & 1 & | & 0 & 1 \end{bmatrix}$$

The reader should verify that

$$B^{-1} = \begin{bmatrix} \frac{1}{2} & 0 \\ -\frac{1}{2} & 1 \end{bmatrix}$$

Use (5) and (6) to determine the optimal tableau's constraints. Because

$$\mathbf{a}_1 = \begin{bmatrix} 1 \\ 2 \end{bmatrix}$$

the column for x_1 in the optimal tableau is

$$B^{-1}\mathbf{a}_1 = \begin{bmatrix} \frac{1}{2} & 0 \\ -\frac{1}{2} & 1 \end{bmatrix} \begin{bmatrix} 1 \\ 2 \end{bmatrix} = \begin{bmatrix} \frac{1}{2} \\ \frac{3}{2} \end{bmatrix}$$

The other nonbasic variable is s_1. The column for s_1 in the original problem is

$$\begin{bmatrix} 1 \\ 0 \end{bmatrix}$$

so (5) yields

$$\text{Column for } s_1 \text{ in optimal tableau} = \begin{bmatrix} \frac{1}{2} & 0 \\ -\frac{1}{2} & 1 \end{bmatrix} \begin{bmatrix} 1 \\ 0 \end{bmatrix} = \begin{bmatrix} \frac{1}{2} \\ -\frac{1}{2} \end{bmatrix}$$

Because

$$\mathbf{b} = \begin{bmatrix} 6 \\ 8 \end{bmatrix}$$

(6) yields

$$\text{Right-hand side of optimal tableau} = \begin{bmatrix} \frac{1}{2} & 0 \\ -\frac{1}{2} & 1 \end{bmatrix} \begin{bmatrix} 6 \\ 8 \end{bmatrix} = \begin{bmatrix} 3 \\ 5 \end{bmatrix}$$

Because BV is listed as $\{x_2, s_2\}$, x_2 is the basic variable for row 1, and s_2 is the basic variable for row 2. Thus, the constraints of the optimal tableau are

$$\frac{1}{2}x_1 + x_2 + \frac{1}{2}s_1 \qquad = 3$$
$$\frac{3}{2}x_1 \qquad - \frac{1}{2}s_1 + s_2 = 5$$

Because $\mathbf{c}_{BV} = [4 \quad 0]$,

$$\mathbf{c}_{BV}B^{-1} = [4 \quad 0] \begin{bmatrix} \frac{1}{2} & 0 \\ -\frac{1}{2} & 1 \end{bmatrix} = [2 \quad 0]$$

Then (10) yields

$$\text{Coefficient of } x_1 \text{ in row 0 of optimal tableau} = \mathbf{c}_{BV}B^{-1}\mathbf{a}_1 - c_1$$

$$= [2 \quad 0] \begin{bmatrix} 1 \\ 2 \end{bmatrix} - 1 = 1$$

From (10′)

$$\text{Coefficient of } s_1 \text{ in optimal tableau} = \text{First element of } \mathbf{c}_{BV}B^{-1} = 2$$

Because

$$\mathbf{b} = \begin{bmatrix} 6 \\ 8 \end{bmatrix}$$

(11) shows that the right-hand side of the optimal tableau's row 0 is

$$\mathbf{c}_{BV}B^{-1}\mathbf{b} = [2 \quad 0] \begin{bmatrix} 6 \\ 8 \end{bmatrix} = 12$$

Of course, the basic variables x_2 and s_2 will have zero coefficients in row 0. Thus, the optimal tableau's row 0 is $z + x_1 + 2s_1 = 12$, and the complete optimal tableau is

$$z + x_1 \qquad + 2s_1 \qquad = 12$$
$$\frac{1}{2}x_1 + x_2 + \frac{1}{2}s_1 \qquad = 3$$
$$\frac{3}{2}x_1 \qquad - \frac{1}{2}s_1 + s_2 = 5$$

We have used the formulas of this section to create an LP's optimal tableau, but they can also be used to create the tableau for *any* set of basic variables. This observation will be important when we study the revised simplex method in Section 10.1.

PROBLEMS

Group A

1 For the following LP, x_1 and x_2 are basic variables in the optimal tableau. Use the formulas of this section to determine the optimal tableau.

$$\max z = 3x_1 + x_2$$
$$\text{s.t.} \qquad 2x_1 - x_2 \le 2$$
$$-x_1 + x_2 \le 4$$
$$x_1, x_2 \ge 0$$

2 For the following LP, x_2 and s_1 are basic variables in the optimal tableau. Use the formulas of this section to determine the optimal tableau.

$$\max z = -x_1 + x_2$$
$$\text{s.t.} \quad 2x_1 + x_2 \le 4$$
$$x_1 + x_2 \le 2$$
$$x_1, x_2 \ge 0$$

6.3 Sensitivity Analysis

We now explore how changes in an LP's parameters (objective function coefficients, right-hand sides, and technological coefficients) change the optimal solution. As described in Section 6.1, the study of how an LP's optimal solution depends on its parameters is called *sensitivity analysis.* Our discussion focuses on maximization problems and relies heavily on the formulas of Section 6.2. (The modifications for min problems are straightforward; see Problem 8 at the end of this section.)

As in Section 6.2, we let BV be the set of basic variables in the optimal tableau. Given a change (or changes) in an LP, we want to determine whether BV remains optimal. The mechanics of sensitivity analysis hinge on the following important observation. *From Chapter 4, we know that a simplex tableau (for a max problem) for a set of basic variables BV is optimal if and only if each constraint has a nonnegative right-hand side and each variable has a nonnegative coefficient in row 0.* This follows, because if each constraint has a nonnegative right-hand side, then BV's basic solution is feasible, and if each variable in row 0 has a nonnegative coefficient, then there can be no basic feasible solution with a higher z-value than BV. Our observation implies that whether a tableau is feasible and optimal depends only on the right-hand sides of the constraints and on the coefficients of each variable in row 0. For example, if an LP has variables $x_1, x_2, \ldots, x_6$, the following partial tableau would be optimal:

$$z + 2x_2 + x_4 + x_6 = 6$$
$$= 1$$
$$= 2$$
$$= 3$$

This tableau's optimality is not affected by the parts of the tableau that are omitted.

Suppose we have solved an LP and have found that BV is an optimal basis. We can use the following procedure to determine if any change in the LP will cause BV to be no longer optimal.

Step 1 Using the formulas of Section 6.2, determine how changes in the LP's parameters change the right-hand side and row 0 of the optimal tableau (the tableau having BV as the set of basic variables).

Step 2 If each variable in row 0 has a non-negative coefficient and each constraint has a nonnegative right-hand side, then BV is still optimal. Otherwise, BV is no longer optimal.

If BV is no longer optimal, then you can find the new optimal solution by using the Section 6.2 formulas to recreate the entire tableau for BV and then continuing the simplex algorithm with the BV tableau as your starting tableau.

There can be two reasons why a change in an LP's parameters causes BV to be no longer optimal. First, a variable (or variables) in row 0 may have a negative coefficient. In this case, a better (larger z-value) bfs can be obtained by pivoting in a nonbasic variable with a negative coefficient in row 0. If this occurs, we say that BV is now a **suboptimal basis.** Second, a constraint (or constraints) may now have a negative right-hand side. In this case, at least one member of BV will now be negative and BV will no longer yield a bfs. If this occurs, we say that BV is now an **infeasible basis.**

We illustrate the mechanics of sensitivity analysis in the Dakota Furniture example. Recall that

$$x_1 = \text{number of desks manufactured}$$
$$x_2 = \text{number of tables manufactured}$$
$$x_3 = \text{number of chairs manufactured}$$

The objective function for the Dakota problem was

$$\max z = 60x_1 + 30x_2 + 20x_3$$

and the initial tableau was

$$
\begin{aligned}
z - 60x_1 - 30x_2 - 20x_3 && = 0 & \\
8x_1 + 6x_2 + x_3 + s_1 && = 48 & \quad \text{(Lumber constraint)} \\
4x_1 + 2x_2 + 1.5x_3 + s_2 && = 20 & \quad \text{(Finishing constraint)} \\
2x_1 + 1.5x_2 + 0.5x_3 + s_3 && = 8 & \quad \text{(Carpentry constraint)}
\end{aligned}
\tag{12}
$$

The optimal tableau was

$$
\begin{aligned}
z + 5x_2 + 10s_2 + 10s_3 &= 280 \\
- 2x_2 + s_1 + 2s_2 - 8s_3 &= 24 \\
- 2x_2 + x_3 + 2s_2 - 4s_3 &= 8 \\
x_1 + 1.25x_2 - 0.5s_2 + 1.5s_3 &= 2
\end{aligned}
\tag{13}
$$

Note that BV = $\{s_1, x_3, x_1\}$ and NBV = $\{x_2, s_2, s_3\}$. The optimal bfs is $z = 280$, $s_1 = 24$, $x_3 = 8$, $x_1 = 2$, $x_2 = 0$, $s_2 = 0$, $s_3 = 0$.

We now discuss how six types of changes in an LP's parameters change the optimal solution:

Change 1   Changing the objective function coefficient of a nonbasic variable

Change 2   Changing the objective function coefficient of a basic variable

Change 3   Changing the right-hand side of a constraint

Change 4   Changing the column of a nonbasic variable

Change 5   Adding a new variable or activity

Change 6   Adding a new constraint (see Section 6.11)

Changing the Objective Function Coefficient of a Nonbasic Variable

In the Dakota problem, the only nonbasic decision variable is x_2 (tables). Currently, the objective function coefficient of x_2 is $c_2 = 30$. How would a change in c_2 affect the optimal solution to the Dakota problem? More specifically, for what values of c_2 would BV = $\{s_1, x_3, x_1\}$ remain optimal?

Suppose we change the objective function coefficient of x_2 from 30 to $30 + \Delta$. Then Δ represents the amount by which we have changed c_2 from its current value. For what values of Δ will the current set of basic variables (the current basis) remain optimal? We begin by determining how changing c_2 from 30 to $30 + \Delta$ will change the BV tableau. Note that B^{-1} and **b** are unchanged, and therefore, from (6), the right-hand side of BV's tableau ($B^{-1}\mathbf{b}$) has not changed, so BV is still feasible. Because x_2 is a nonbasic variable, $\mathbf{c}_{BV}$ has not changed. From (10), we can see that the only variable whose row 0 coefficient will be changed by a change in c_2 is x_2. Thus, BV will remain optimal if $\bar{c}_2 \geq 0$, and BV will be suboptimal if $\bar{c}_2 < 0$. In this case, z could be improved by entering x_2 into the basis.

We have

$$\mathbf{a}_2 = \begin{bmatrix} 6 \\ 2 \\ 1.5 \end{bmatrix}$$

and $c_2 = 30 + \Delta$. Also, from Section 6.2, we know that $\mathbf{c}_{BV}B^{-1} = \begin{bmatrix} 0 & 10 & 10 \end{bmatrix}$. Now (10) shows that

$$\bar{c}_2 = \begin{bmatrix} 0 & 10 & 10 \end{bmatrix} \begin{bmatrix} 6 \\ 2 \\ 1.5 \end{bmatrix} - (30 + \Delta) = 35 - 30 - \Delta = 5 - \Delta$$

Thus, $\bar{c}_2 \geq 0$ holds, and BV will remain optimal, if $5 - \Delta \geq 0$, or $\Delta \leq 5$. Similarly, $\bar{c}_2 < 0$ holds if $\Delta > 5$, but then BV is no longer optimal. This means that if the price of tables is decreased or increased by \$5 or less, BV remains optimal. Thus, for $c_2 \leq 30 + 5 = 35$, BV remains optimal.

If BV remains optimal after a change in a nonbasic variable's objective function coefficient, the values of the decision variables and the optimal z-value remain unchanged. This is because a change in the objective function coefficient for a nonbasic variable leaves the right-hand side of row 0 and the constraints unchanged. For example, if the price of tables increases to \$33 ($c_2 = 33$), the optimal solution to the Dakota problem remains unchanged (Dakota should still make 2 desks and 8 chairs, and $z = 280$). On the other hand, if $c_2 > 35$, BV will no longer be optimal, because $\bar{c}_2 < 0$. In this case, we find the new optimal solution by recreating the BV tableau and then using the simplex algorithm. For example, if $c_2 = 40$, we know that the only part of the BV tableau that will change is the coefficient of x_2 in row 0. If $c_2 = 40$, then

$$\bar{c}_2 = \begin{bmatrix} 0 & 10 & 10 \end{bmatrix} \begin{bmatrix} 6 \\ 2 \\ 1.5 \end{bmatrix} - 40 = -5$$

Now the BV "final" tableau is as shown in Table 2. This is not an optimal tableau (it is suboptimal), and we can increase z by making x_2 a basic variable in row 3. The resulting tableau is given in Table 3. This is an optimal tableau. Thus, if $c_2 = 40$, the optimal solution to the Dakota problem changes to $z = 288$, $s_1 = 27.2$, $x_3 = 11.2$, $x_2 = 1.6$, $x_1 = 0$, $s_2 = 0$, $s_3 = 0$. In this case, the increase in the price of tables has made tables sufficiently more attractive to induce Dakota to manufacture them. Note that after changing a nonbasic variable's objective function coefficient, it may, in general, take more than one pivot to find the new optimal solution.

There is a more insightful way to show that the current basis in the Dakota problem remains optimal as long as the price of tables is decreased or increased by \$5 or less. From the optimal row 0 in (13), we see that if $c_2 = 30$, then

$$z = 280 - 10s_2 - 10s_3 - 5x_2$$

This tells us that each table that Dakota manufactures will decrease revenue by \$5 (in other words, the reduced cost for tables is 5). If we increase the price of tables by more than \$5, each table would now increase Dakota's revenue. For example, if $c_2 = 36$, each table would increase revenues by $6 - 5 = \$1$ and Dakota should manufacture tables. Thus, as before, we see that for $\Delta > 5$, the current basis is no longer optimal. This analysis yields another interpretation of the reduced cost of a nonbasic variable: *The reduced cost for a nonbasic variable (in a max problem) is the maximum amount by which the*

TABLE 2
"Final" (Suboptimal) Dakota Tableau ($40/Table)

										Basic Variable	Ratio
z	$-$	$5x_2$			$+$	$10s_2$	$+$	$10s_3$	$= 280$	$z = 280$	
	$-$	$2x_2$	$+ s_1$	$+$		$2s_2$	$-$	$8s_3$	$= 24$	$s_1 = 24$	None
	$-$	$2x_2$	$+ x_3$	$+$		$2s_2$	$-$	$4s_3$	$= 8$	$x_3 = 8$	None
x_1	$+$	$\boxed{1.25x_2}$				$- 0.5s_2$	$+$	$1.5s_3$	$= 2$	$x_1 = 2$	1.6*

TABLE 3
Optimal Dakota Tableau ($40/Table)

							Basic Variable
$z +$	$4x_1$		$+$	$8s_2$	$+$	$16s_3 = 288$	$z = 288$
	$1.6x_1$	$+ s_1$	$+$	$1.2s_2$	$-$	$5.6s_3 = 27.2$	$s_1 = 27.2$
	$1.6x_1$	$+ x_3$	$+$	$1.2s_2$	$-$	$1.6s_3 = 11.2$	$x_3 = 11.2$
	$0.8x_1 + x_2$			$- 0.4s_2$	$+$	$1.2s_3 = 1.6$	$x_2 = 1.6$

variable's objective function coefficient can be increased before the current basis becomes suboptimal, and it becomes optimal for the nonbasic variable to enter the basis.

In summary, if the objective function coefficient for a nonbasic variable x_j is changed, the current basis remains optimal if $\bar{c}_j \geq 0$. If $\bar{c}_j < 0$, then the current basis is no longer optimal, and x_j will be a basic variable in the new optimal solution.

Changing the Objective Function Coefficient of a Basic Variable

In the Dakota problem, the decision variables x_1 (desks) and x_3 (chairs) are basic variables. We now explain how a change in the objective function coefficient of a basic variable will affect an LP's optimal solution. We begin by analyzing how this change affects the BV tableau. Because we are not changing B (or therefore B^{-1}) or $\mathbf{b}$, (6) shows that the right-hand side of each constraint will remain unchanged, and BV will remain feasible. Because we are changing $\mathbf{c}_{BV}$, however, so $\mathbf{c}_{BV}B^{-1}$ will change. From (10), we see that a change in $\mathbf{c}_{BV}B^{-1}$ may change more than one coefficient in row 0. To determine whether BV remains optimal, we must use (10) to recompute row 0 for the BV tableau. If each variable in row 0 still has a nonnegative coefficient, BV remains optimal. Otherwise, BV is now suboptimal. To illustrate the preceding ideas, we analyze how a change in the objective function coefficient for x_1 (desks) from its current value of $c_1 = 60$ affects the optimal solution to the Dakota problem.

Suppose that c_1 is changed to $60 + \Delta$, changing $\mathbf{c}_{BV}$ to $\mathbf{c}_{BV} = [0 \quad 20 \quad 60 + \Delta]$. To compute the new row 0, we need to know B^{-1}. We could (as in Section 6.2) use the Gauss–Jordan method to compute B^{-1}. Recall that this method begins by writing down the 3×6 matrix $B|I_3$:

$$B|I_3 = \begin{bmatrix} 1 & 1 & 8 & 1 & 0 & 0 \\ 0 & 1.5 & 4 & 0 & 1 & 0 \\ 0 & 0.5 & 2 & 0 & 0 & 1 \end{bmatrix}$$

Then we use EROs to transform the first three columns of $B|I_3$ to I_3. At this point, the last three columns of the resulting matrix will be B^{-1}.

It turns out that when we solved the Dakota problem by the simplex algorithm, without realizing it, we found B^{-1}. To see why this is the case, note that in going from the initial Dakota tableau (12) to the optimal Dakota tableau (13) we performed a series of EROs on the constraints. These EROs transformed the constraint columns corresponding to the initial basis (s_1, s_2, s_3)

$$
\text{from} \quad
\begin{array}{ccc} s_1 & s_2 & s_3 \end{array}
\begin{bmatrix} 1 & 0 & 0 \\ 0 & 1 & 0 \\ 0 & 0 & 1 \end{bmatrix}
\quad \text{to} \quad
\begin{array}{ccc} s_1 & s_2 & s_3 \end{array}
\begin{bmatrix} 1 & 2 & -8 \\ 0 & 2 & -4 \\ 0 & -0.5 & 1.5 \end{bmatrix}
$$

These same EROs have transformed the columns corresponding to $BV = \{s_1, x_3, x_1\}$

$$
\text{from} \quad B =
\begin{array}{ccc} s_1 & x_3 & x_1 \end{array}
\begin{bmatrix} 1 & 1 & 8 \\ 0 & 1.5 & 4 \\ 0 & 0.5 & 2 \end{bmatrix}
\quad \text{to} \quad
\begin{array}{ccc} s_1 & x_3 & x_1 \end{array}
\begin{bmatrix} 1 & 0 & 0 \\ 0 & 1 & 0 \\ 0 & 0 & 1 \end{bmatrix}
$$

This means that in solving the Dakota problem by the simplex algorithm, we have used EROs to transform B to I_3. These same EROs transformed I_3 into

$$
\begin{bmatrix} 1 & 2 & -8 \\ 0 & 2 & -4 \\ 0 & -0.5 & 1.5 \end{bmatrix} = B^{-1}
$$

We have discovered an extremely important fact: *For any simplex tableau, B^{-1} is the $m \times m$ matrix consisting of the columns in the current tableau that correspond to the initial tableau's set of basic variables (taken in the same order).* This means that if the starting basis for an LP consists entirely of slack variables, then B^{-1} for the optimal tableau is simply the columns for the slack variables in the constraints of the optimal tableau. In general, if the starting basic variable for the ith constraint is the artificial variable a_i, then the ith column of B^{-1} will be the column for a_i in the optimal tableau's constraints. Thus, we need not use the Gauss–Jordan method to find the optimal tableau's B^{-1}. We have already found B^{-1} by performing the simplex algorithm.

We can now compute what $\mathbf{c}_{BV}B^{-1}$ will be if $c_1 = 60 + \Delta$:

$$
\mathbf{c}_{BV}B^{-1} = \begin{bmatrix} 0 & 20 & 60 + \Delta \end{bmatrix} \begin{bmatrix} 1 & 2 & -8 \\ 0 & 2 & -4 \\ 0 & -0.5 & 1.5 \end{bmatrix} \tag{14}
$$

$$
= \begin{bmatrix} 0 & 10 - 0.5\Delta & 10 + 1.5\Delta \end{bmatrix}
$$

Observe that for $\Delta = 0$, (14) yields the original $\mathbf{c}_{BV}B^{-1}$. We can now compute the new row 0 corresponding to $c_1 = 60 + \Delta$. After noting that

$$
\mathbf{a}_1 = \begin{bmatrix} 8 \\ 4 \\ 2 \end{bmatrix}, \mathbf{a}_2 = \begin{bmatrix} 6 \\ 2 \\ 1.5 \end{bmatrix}, \mathbf{a}_3 = \begin{bmatrix} 1 \\ 1.5 \\ 0.5 \end{bmatrix}, c_1 = 60 + \Delta, c_2 = 30, c_3 = 20
$$

we can use (10) to compute the new row 0. Because s_1, x_3, and x_1 are basic variables, their coefficients in row 0 must still be 0. The coefficient of each nonbasic variable in the new row 0 is as follows:

$$
\bar{c}_2 = \mathbf{c}_{BV}B^{-1}\mathbf{a}_2 - c_2 = \begin{bmatrix} 0 & 10 - 0.5\Delta & 10 + 1.5\Delta \end{bmatrix} \begin{bmatrix} 6 \\ 2 \\ 1.5 \end{bmatrix} - 30 = 5 + 1.25\Delta
$$

Coefficient of s_2 in row 0 = second element of $\mathbf{c}_{BV}B^{-1} = 10 - 0.5\Delta$

Coefficient of s_3 in row 0 = third element of $\mathbf{c}_{BV}B^{-1} = 10 + 1.5\Delta$

Thus, row 0 of the optimal tableau is now

$$z + (5 + 1.25\Delta)x_2 + (10 - 0.5\Delta)s_2 + (10 + 1.5\Delta)s_3 = ?$$

From the new row 0, we see that BV will remain optimal if and only if the following hold:

$$5 + 1.25\Delta \geq 0 \quad (\text{true iff}^\dagger \ \Delta \geq -4)$$
$$10 - 0.5\Delta \geq 0 \quad (\text{true iff} \ \Delta \geq 20)$$
$$10 + 1.5\Delta \geq 0 \quad (\text{true iff} \ \Delta \geq -(20/3))$$

This means that the current basis remains optimal as long as $\Delta \geq -4$, $\Delta \leq 20$, and $\Delta \geq -\frac{20}{3}$. From Figure 3, we see that the current basis will remain optimal if and only if $-4 \leq \Delta \leq 20$: If c_1 is decreased by \$4 or less or increased by up to \$20, the current basis remains optimal. Thus, as long as $56 = 60 - 4 \leq c_1 \leq 60 + 20 = 80$, the current basis remains optimal. If $c_1 < 56$ or $c_1 > 80$, the current basis is no longer optimal.

If the current basis remains optimal, then the values of the decision variables don't change because $B^{-1}\mathbf{b}$ remains unchanged. The optimal z-value does change, however. To illustrate this, suppose $c_1 = 70$. Because $56 \leq 70 \leq 80$, we know that the current basis remains optimal. Thus, Dakota should still manufacture 2 desks ($x_1 = 2$) and 8 chairs ($x_3 = 8$). However, changing c_1 to 70 changes z to $z = 70x_1 + 30x_2 + 20x_3$. This changes z to $70(2) + 20(8) = \$300$. Another way to see that z is now \$300 is to note that we have increased the revenue from each desk by $70 - 60 = \$10$. Dakota is making 2 desks, so revenue should increase by $2(10) = \$20$, and new revenue $= 280 + 20 = \$300$.

When the Current Basis Is No Longer Optimal

Recall that if $c_1 < 56$ or $c_1 > 80$, then the current basis is no longer optimal. Intuitively, if the price of desks is decreased sufficiently (with all other prices held constant), desks will no longer be worth making. Our analysis shows that this occurs if the price of desks is decreased by more than \$4. The reader should verify (see Problem 2 at the end of this section) that if $c_1 < 56$, x_1 is no longer a basic variable in the new optimal solution. On the other hand, if $c_1 > 80$, desks have become profitable enough to make the current basis suboptimal; desks are now so attractive that we want to make more of them. To do this, we must force another variable out of the basis. Suppose $c_1 = 100$. Because $100 > 80$, we know that the current basis is no longer optimal. How can we determine the new optimal solution? Simply create the optimal tableau for $c_1 = 100$ and proceed with the simplex. If $c_1 = 100$, then $\Delta = 100 - 60 = 40$, and the new row 0 will have

$$\bar{c}_1 = 0, \quad \bar{c}_2 = 5 + 1.25\Delta = 55, \quad \bar{c}_3 = 0,$$

s_1 coefficient in row 0 = 0

s_2 coefficient in row 0 $= 10 - 0.5\Delta = -10$

s_3 coefficient in row 0 $= 10 + 1.5\Delta = 70$

$$\text{Right-hand side of row 0} = \mathbf{c}_{BV}B^{-1}\mathbf{b} = \begin{bmatrix} 0 & -10 & 70 \end{bmatrix} \begin{bmatrix} 48 \\ 20 \\ 8 \end{bmatrix} = 360$$

From (6), changing c_1 does not change the constraints in the BV tableau. This means that if $c_1 = 100$, then the BV tableau is as given in Table 4. BV $= \{s_1, x_3, x_1\}$ is now subopti-

---

†"If and only if"

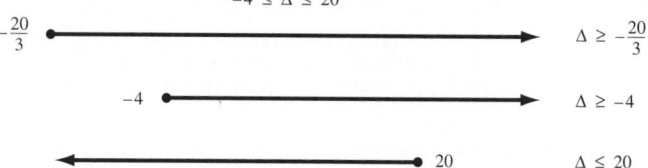

TABLE 4
"Final" (Suboptimal) Tableau If $c_1 = 100$

								Basic Variable	Ratio
z	$+$	$55x_2$		$-$	$10s_2 +$	$70s_3 = 360$		$z = 360$	
	$-$	$2x_2$	$+ s_1 +$		$2s_2 -$	$8s_3 = 24$		$s_1 = 24$	12
	$-$	$2x_2 + x_3$	$+$		$(2s_2) -$	$4s_3 = 8$		$x_3 = 8$	4*
	$x_1 +$	$1.25x_2$		$-$	$0.5s_2 +$	$1.5s_3 = 2$		$x_1 = 2$	None

TABLE 5
Optimal Dakota Tableau If $c_1 = 100$

							Basic Variable
z	$+$	$45x_2 +$	$5x_3$	$+$	$50s_3 = 400$		$z = 400$
			$- x_3 + s_1$	$-$	$4s_3 = 16$		$s_1 = 16$
	$-$	$x_2 +$	$0.5x_3$	$+ s_2 -$	$2s_3 = 4$		$s_2 = 4$
	$x_1 +$	$0.75x_2 +$	$0.25x_3$	$+$	$0.5s_3 = 4$		$x_1 = 4$

mal. To find the new Dakota optimal solution, we enter s_2 into the basis in row 2 (Table 5). This is an optimal tableau. If $c_1 = 100$, then the new optimal solution to the Dakota problem is $z = 400$, $s_1 = 16$, $s_2 = 4$, $x_1 = 4$, $x_2 = 0$, $x_3 = 0$. Notice that increasing the profitability of desks has caused Dakota to stop making chairs. The resources that were previously used to make the chairs are now used to make $4 - 2 = 2$ extra desks.

In summary, if the objective function coefficient of a basic variable x_j is changed, then the current basis remains optimal if the coefficient of every variable in row 0 of the BV tableau remains nonnegative. If any variable in row 0 has a negative coefficient, then the current basis is no longer optimal.

Interpretation of the Objective Coefficient Ranges Block of the LINDO Output

To obtain a sensitivity report in LINDO, select Yes when asked (after solving LP) whether you want a Range analysis. To obtain a sensitivity report in LINGO, go to Options and select Range (after solving LP). If this does not work, go to Options, choose the General Solver tab, and then go to Dual Computations and select the Ranges and Values option.

In the OBJ COEFFICIENT RANGES block of the LINDO (or LINGO) computer output, we see the amount by which each variable's objective function coefficient may be changed before the current basis becomes suboptimal (assuming all other LP parameters are held constant). Look at the LINDO output for the Dakota problem (Figure 4). For each variable, the CURRENT COEF column gives the current value of the variable's objective function coefficient. For example, the objective function coefficient for DESKS is 60. The ALLOWABLE INCREASE column gives the maximum amount by which the objective

```
MAX          60 DESKS + 30 TABLES + 20 CHAIRS
SUBJECT TO
        2)   8 DESKS + 6 TABLES + CHAIRS <=      48
        3)   4 DESKS + 2 TABLES + 1.5 CHAIRS <=  20
        4)   2 DESKS + 1.5 TABLES + 0.5 CHAIRS <=   8
END

    LP OPTIMUM FOUND  AT STEP        2

        OBJECTIVE FUNCTION VALUE

    1)        280.000000

VARIABLE         VALUE       REDUCED COST
    DESKS         2.000000       0.000000
    TABLES        0.000000       5.000000
    CHAIRS        8.000000       0.000000

    ROW     SLACK OR SURPLUS     DUAL PRICES
    2)          24.000000          0.000000
    3)           0.000000         10.000000
    4)           0.000000         10.000000

NO. ITERATIONS=      2

    RANGES IN WHICH THE BASIS IS UNCHANGED

                        OBJ COEFFICIENT RANGES
VARIABLE      CURRENT        ALLOWABLE       ALLOWABLE
              COEF           INCREASE        DECREASE
    DESKS     60.000000      20.000000        4.000000
    TABLES    30.000000       5.000000       INFINITY
    CHAIRS    20.000000       2.500000        5.000000

                        RIGHTHAND SIDE RANGES
    ROW       CURRENT        ALLOWABLE       ALLOWABLE
              RHS            INCREASE        DECREASE
    2         48.000000      INFINITY        24.000000
    3         20.000000       4.000000        4.000000
    4          8.000000       2.000000        1.333333
```

FIGURE 4
LINDO Output for Dakota Furniture

function coefficient of a variable can be increased with the current basis remaining optimal (assuming all other LP parameters stay constant). For example, if the objective function coefficient for DESKS is increased above $60 + 20 = 80$, then the current basis is no longer optimal. Similarly, the ALLOWABLE DECREASE column gives the maximum amount by which the objective function coefficient of a variable can be decreased with the current basis remaining optimal (assuming all other LP parameters constant). If the objective function coefficient for DESKS drops below $60 - 4 = 56$, the current basis is no longer optimal. In summary, we see from the LINDO output that if the objective function coefficient for DESKS is changed, the current basis remains optimal if

$$56 = 60 - 4 \leq \text{objective coefficient for DESKS} \leq 60 + 20 = 80$$

Of course, this agrees with our earlier computations.

Changing the Right-Hand Side of a Constraint

Effect on the Current Basis

In this section, we examine how the optimal solution to an LP changes if the right-hand side of a constraint is changed. Because **b** does not appear in (10), changing the right-hand side of a constraint will leave row 0 of the optimal tableau unchanged; changing a right-hand side cannot cause the current basis to become suboptimal. From (5) and (6), however, we see that a change in the right-hand side of a constraint will affect the right-hand side of the constraints in the optimal tableau. *As long as the right-hand side of each* *constraint in the optimal tableau remains nonnegative, the current basis remains feasible and optimal. If at least one right-hand side in the optimal tableau becomes negative, then the current basis is no longer feasible and therefore no longer optimal.*

Suppose we want to determine how changing the amount of finishing hours (b_2) affects the optimal solution to the Dakota problem. Currently, $b_2 = 20$. If we change b_2 to $20 + \Delta$, then from (6), the right-hand side of the constraints in the optimal tableau will become

$$B^{-1}\begin{bmatrix} 48 \\ 20 + \Delta \\ 8 \end{bmatrix} = \begin{bmatrix} 1 & 2 & -8 \\ 0 & 2 & -4 \\ 0 & -0.5 & 1.5 \end{bmatrix}\begin{bmatrix} 48 \\ 20 + \Delta \\ 8 \end{bmatrix}$$

$$= \begin{bmatrix} 24 + 2\Delta \\ 8 + 2\Delta \\ 2 - 0.5\Delta \end{bmatrix}$$

Of course, for $\Delta = 0$, the right-hand side reduces to the right-hand side of the original optimal tableau. If this does not happen, then an error has been made.

It can be shown (see Problem 9) that if the right-hand side of the ith constraint is increased by Δ, then the right-hand side of the optimal tableau is given by (original right-hand side of the optimal tableau) $+\Delta$(column i of B^{-1}). Because the second column of B^{-1} is

$$\begin{bmatrix} 2 \\ 2 \\ -0.5 \end{bmatrix} \quad \text{and the original right-hand side is} \quad \begin{bmatrix} 24 \\ 8 \\ 2 \end{bmatrix}$$

we again find that the right-hand side of the constraints in the optimal tableau is

$$\begin{bmatrix} 24 + 2\Delta \\ 8 + 2\Delta \\ 2 - 0.5\Delta \end{bmatrix}$$

For the current basis to remain optimal, we require that the right-hand side of each constraint in the optimal tableau remain nonnegative. This means that the current basis will remain optimal if and only if the following hold:

$$24 + 2\Delta \ge 0 \quad \text{(true iff } \Delta \ge -12)$$
$$8 + 2\Delta \ge 0 \quad \text{(true iff } \Delta \ge -4)$$
$$2 - 0.5\Delta \ge 0 \quad \text{(true iff } \Delta \le 4)$$

As long as $\Delta \ge -12$, $\Delta \ge -4$, and $\Delta \le 4$, the current basis remains feasible and therefore optimal. From Figure 5, we see that for $-4 \le \Delta \le 4$, the current basis remains feasible and therefore optimal. This means that for $20 - 4 \le b_2 \le 20 + 4$, or $16 \le b_2 \le 24$, the current basis remains optimal: If between 16 and 24 finishing hours are available, BV = $\{s_1, x_3, x_1\}$ remains optimal, and Dakota should still manufacture desks and chairs. If $b_2 > 24$ or if $b_2 < 16$, however, the current basis becomes infeasible and is no longer optimal.

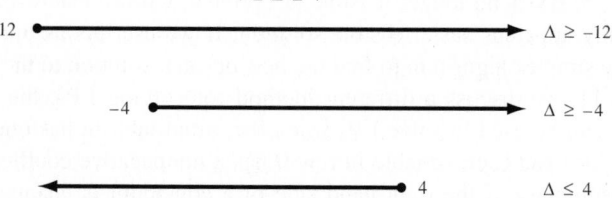

FIGURE 5
Determination of Range of Values on b_2 for Which Current Basis Remains Optimal

Effect on Decision Variables and z

Even if the current basis remains optimal ($16 \leq b_2 \leq 24$), *the values of the decision variables and z change.* This was illustrated in our graphical discussion of sensitivity analysis in Section 6.1. To see how the values of the objective function and decision variables change, recall that the values of the basic variables in the optimal solution are given by $B^{-1}\mathbf{b}$ and the optimal z-value is given by $\mathbf{c}_{BV}B^{-1}\mathbf{b}$. Changing $\mathbf{b}$ will change the values of the basic variables and the optimal z-value. To illustrate this, suppose that 22 finishing hours are available. Because $16 \leq 22 \leq 24$, the current basis remains optimal and, from (6), the new values of the basic variables are as follows (the same basis remains optimal, so the nonbasic variables remain equal to 0):

$$\begin{bmatrix} s_1 \\ x_3 \\ x_1 \end{bmatrix} = B^{-1}\mathbf{b} = \begin{bmatrix} 1 & 2 & -8 \\ 0 & 2 & -4 \\ 0 & -0.5 & 1.5 \end{bmatrix} \begin{bmatrix} 48 \\ 22 \\ 8 \end{bmatrix} = \begin{bmatrix} 28 \\ 12 \\ 1 \end{bmatrix}$$

If 22 finishing hours were available, then Dakota should manufacture 12 chairs and only 1 desk.

To determine how a change in a right-hand side affects the optimal z-value, we may use formula (11). If 22 finishing hours are available, we find that

$$\text{New } z\text{-value} = \mathbf{c}_{BV}B^{-1}(\text{new } \mathbf{b}) = \begin{bmatrix} 0 & 10 & 10 \end{bmatrix} \begin{bmatrix} 48 \\ 22 \\ 8 \end{bmatrix} = 300$$

In Section 6.8, we explain how the important concept of shadow price can be used to determine how changes in a right-hand side change the optimal z-value.

When the Current Basis Is No Longer Optimal

If we change a right-hand side enough that the current basis is no longer optimal, how can we determine the new optimal basis? Suppose we change b_2 to 30. Because $b_2 > 24$, we know that the current basis is no longer optimal. If we re-create the optimal tableau, we see from the formulas of Section 6.2 that the only part of the optimal tableau that will change is the right-hand side of row 0 and the constraints. From (6), the right-hand side of the constraints in the tableau for BV = $\{s_1, x_3, x_1\}$ is

$$B^{-1}\mathbf{b} = \begin{bmatrix} 1 & 2 & -8 \\ 0 & 2 & -4 \\ 0 & -0.5 & 1.5 \end{bmatrix} \begin{bmatrix} 48 \\ 30 \\ 8 \end{bmatrix} = \begin{bmatrix} 44 \\ 28 \\ -3 \end{bmatrix}$$

From (11), the right-hand side of row 0 is now

$$\mathbf{c}_{BV}B^{-1}\mathbf{b} = \begin{bmatrix} 0 & 10 & 10 \end{bmatrix} \begin{bmatrix} 48 \\ 30 \\ 8 \end{bmatrix} = 380$$

The tableau for the optimal basis, BV = $\{s_1, x_3, x_1\}$, is now as shown in Table 6. Because $x_1 = -3$, BV is no longer feasible or optional. Unfortunately, this tableau does not yield a readily apparent basic feasible solution. If we use this as our initial tableau, we can't use the simplex algorithm to find the new optimal solution to the Dakota problem. In Section 6.11, we discuss a different method for solving LPs, the dual simplex algorithm, which can be used to solve LPs when the initial tableau has one or more negative right-hand sides and each variable in row 0 has a nonnegative coefficient.

In summary, if the right-hand side of a constraint is changed, then the current basis remains optimal if the right-hand side of each constraint in the tableau remains non-

TABLE 6
Final (Infeasible) Dakota Tableau If $b_2 = 30$

						Basic Variable
z	$+ 5x_2$		$+ 10s_2 +$	$10s_3 = 380$		$z = 480$
	$- 2x_2$	$+ s_1 +$	$2s_2 -$	$8s_3 = 44$		$s_1 = 44$
	$- 2x_2 + x_3$		$+ 2s_2 -$	$4s_3 = 28$		$x_3 = 28$
x_1	$+ 1.25x_2$		$- 0.5s_2 +$	$1.5s_3 = -3$		$x_1 = -3$

negative. If the right-hand side of any constraint is negative, then the current basis is infeasible, and a new optimal solution must be found.

Interpretation of the Right-Hand Side Ranges Block of the LINDO Output

The block of the LINDO (or LINGO) output labeled RIGHTHAND SIDE RANGES (see Figure 4) gives information concerning the amount by which a right-hand side can be changed before the current basis becomes infeasible (all other LP parameters constant). The CURRENT RHS column gives the current right-hand side of each constraint. Thus, for row 3 (the second constraint), the current right-hand side is 20. The ALLOWABLE INCREASE column is the maximum amount by which the right-hand side of the constraint can be increased with the current basis remaining optimal (all other LP parameters constant). For example, if the amount of available finishing hours (second constraint) is increased by up to 4 hours, then the current basis remains optimal. Similarly, the ALLOWABLE DECREASE column gives the maximum amount by which the right-hand side of a constraint can be decreased with the current basis remaining optimal (all other LP parameters constant). If the amount of available finishing hours is decreased by more than 4 hours, then the current basis is no longer optimal. In summary, if the number of finishing hours is changed (all other LP parameters constant), the current basis remains optimal if

$$16 = 20 - 4 \leq \text{available finishing hours} \leq 20 + 4 = 24$$

Changing the Column of a Nonbasic Variable

Currently, 6 board feet of lumber, 2 finishing hours, and 1.5 carpentry hours are required to make a table that can be sold for $30. Also x_2 (the variable for tables) is a nonbasic variable in the optimal solution. This means that Dakota should not manufacture any tables now. Suppose, however, that the price of tables increased to $43 and, because of changes in production technology, a table required 5 board feet of lumber, 2 finishing hours, and 2 carpentry hours. Would this change the optimal solution to the Dakota problem? Here we are changing elements of the column for x_2 in the original problem (including the objective function). Changing the column for a nonbasic variable such as tables leaves B (and B^{-1}) and $\mathbf{b}$ unchanged. Thus, the right-hand side of the optimal tableau remains unchanged. A glance at (10) also shows that the only part of row 0 that is changed is $\bar{c}_2$; the current basis will remain optimal if and only if $\bar{c}_2 \geq 0$ holds. We now use (10) to compute the new coefficient of x_2 in row 0. This process is called **pricing out** x_2. From (10),

$$\bar{c}_2 = \mathbf{c}_{BV}B^{-1}\mathbf{a}_2 - c_2$$

Note that $c_{BV}B^{-1}$ still equals [0   10   10], but a_2 and c_2 have changed to

$$c_2 = 43 \qquad \text{and} \qquad a_2 = \begin{bmatrix} 5 \\ 2 \\ 2 \end{bmatrix}$$

Now

$$\bar{c}_2 = [0 \quad 10 \quad 10] \begin{bmatrix} 5 \\ 2 \\ 2 \end{bmatrix} - 43 = -3 < 0$$

Because $\bar{c}_2 < 0$, the current basis is no longer optimal. The fact that $\bar{c}_2 = -3$ means that each table that Dakota manufactures now increases revenues by \$3. It is clearly to Dakota's advantage to enter x_2 into the basis. To find the new optimal solution to the Dakota problem, we recreate the tableau for BV $= \{s_1, x_3, x_1\}$ and then apply the simplex algorithm. From (5), the column for x_2 in the constraint portion of the BV tableau is now

$$B^{-1}a_2 = \begin{bmatrix} 1 & 2 & -8 \\ 0 & 2 & -4 \\ 0 & -0.5 & 1.5 \end{bmatrix} \begin{bmatrix} 5 \\ 2 \\ 2 \end{bmatrix} = \begin{bmatrix} -7 \\ -4 \\ 2 \end{bmatrix}$$

The tableau for BV $= \{s_1, x_3, x_1\}$ is now as shown in Table 7. To find the new optimal solution, we enter x_2 into the basis in row 3. This yields the optimal tableau in Table 8. Thus, the new optimal solution to the Dakota problem is $z = 283$, $s_1 = 31$, $x_3 = 12$, $x_2 = 1$, $x_1 = 0$, $s_2 = 0$, $s_3 = 0$. After the column for the nonbasic variable x_2 (tables) has been changed, Dakota should manufacture 12 chairs and 1 table. In summary, if the column of a nonbasic variable x_j is changed, then the current basis remains optimal if $\bar{c}_j \geq 0$. If $\bar{c}_j < 0$, then the current basis is no longer optimal and x_j will be a basic variable in the new optimal solution.

If the column of a basic variable is changed, then it is usually difficult to determine whether the current basis remains optimal. This is because the change may affect both B (and hence B^{-1}) and c_{BV} and thus the entire row 0 and the entire right-hand side of the optimal

TABLE 7
"Final" (Suboptimal) Dakota Tableau for New Method of Making Tables

		Basic Variable
$z \quad - 3x_2 \qquad\qquad + 10s_2 + 10s_3 = 280$		$z = 280$
$- 7x_2 \quad + s_1 + \ 2s_2 - \ 8s_3 = 24$		$s_1 = 24$
$- 4x_2 + x_3 \qquad + \ 2s_2 - \ 4s_3 = 8$		$x_3 = 8$
$x_1 + \textcircled{2x_2} \qquad\qquad - 0.5s_2 + 1.5s_3 = 2$		$x_1 = 2^*$

TABLE 8
Optimal Dakota Tableau for New Method of Making Tables

		Basic Variable
$z + 1.5x_1 \qquad\qquad + 9.25s_2 + 12.25s_3 = 283$		$z = 283$
$3.5x_1 \qquad + s_1 + 0.25s_2 - \ 2.75s_3 = 31$		$s_1 = 31$
$2x_1 \qquad + x_3 \qquad + \quad s_2 - \qquad s_3 = 12$		$x_3 = 12$
$0.5x_1 + x_2 \qquad\qquad - 0.25s_2 + 0.75s_3 = 1$		$x_1 = 1$

tableau. As always, the current basis would remain optimal if and only if each variable has a nonnegative coefficient in row 0 and each constraint has a nonnegative right-hand side.

Adding a New Activity

In many situations, opportunities arise to undertake new activities. For example, in the Dakota problem, the company may be presented with the opportunity to manufacture additional types of furniture, such as footstools. If a new activity is available, we can evaluate it by applying the method utilized to determine whether the current basis remains optimal after a change in the column of a nonbasic variable. The following example illustrates the approach.

Suppose that Dakota is considering making footstools. A stool sells for $15 and requires 1 board foot of lumber, 1 finishing hour, and 1 carpentry hour. Should the company manufacture any stools?

To answer this question, define x_4 to be the number of footstools manufactured by Dakota. The initial tableau is now changed by the introduction of the x_4 column. Our new initial tableau is

$$
\begin{aligned}
z - 60x_1 - 30x_2 - 20x_3 - 15x_4 & & = 0 \\
8x_1 + 6x_2 + x_3 + x_4 + s_1 & & = 48 \\
4x_1 + 2x_2 + 1.5x_3 + x_4 & + s_2 & = 20 \\
2x_1 + 1.5x_2 + 0.5x_3 + x_4 & + s_3 & = 8
\end{aligned}
\tag{15}
$$

We call the addition of the x_4 column to the problem **adding a new activity.** How will the addition of the new activity change the optimal BV $= \{s_1, x_3, x_1\}$ tableau? From (6), we see that the right-hand sides of all constraints in the optimal tableau will remain unchanged. From (10), we see that the coefficient of each of the old variables in row 0 will remain unchanged. We must, of course, compute $\bar{c}_4$, the coefficient of the new activity in row 0 of the optimal tableau. The right-hand side of each constraint in the optimal tableau is unchanged and the only variable in row 0 that can have a negative coefficient is x_4, so the current basis will remain optimal if $\bar{c}_4 \geq 0$ or become nonoptimal if $\bar{c}_4 < 0$.

To determine whether a new activity causes the current basis to be no longer optimal, price out the new activity. Because

$$
c_4 = 15 \quad \text{and} \quad \mathbf{a}_4 = \begin{bmatrix} 1 \\ 1 \\ 1 \end{bmatrix}
$$

we may use (10) to price out x_4. The result is

$$
\bar{c}_4 = [0 \quad 10 \quad 10] \begin{bmatrix} 1 \\ 1 \\ 1 \end{bmatrix} - 15 = 5
$$

Because $\bar{c}_4 \geq 0$, the current basis is still optimal. Equivalently, the reduced cost of footstools is $5. This means that each stool manufactured will decrease revenues by $5. For this reason, we choose not to manufacture any stools.

In summary, if a new column (corresponding to a variable x_j) is added to an LP, then the current basis remains optimal if $\bar{c}_j \geq 0$. If $\bar{c}_j < 0$, then the current basis is no longer optimal and x_j will be a basic variable in the new optimal solution. Table 9 presents a summary of sensitivity analyses for a maximization problem. When applying the techniques of this section to a minimization problem, just remember that a tableau is optimal if and only if each variable has a *nonpositive* coefficient in row 0 and the right-hand side of each constraint is nonnegative.

TABLE 9
Summary of Sensitivity Analysis (Max Problem)

Change in Initial Problem	Effect on Optimal Tableau	Current Basis Is Still Optimal If:
Changing nonbasic objective function coefficient c_j	Coefficient of x_j in optimal row 0 is changed	Coefficient of x_j in row 0 for current basis is still nonnegative
Changing basic objective function coefficient c_j	Entire row 0 may change	Each variable still has a nonnegative coefficient in row 0
Changing right-hand side of a constraint	Right-hand side of constraints and row 0 are changed	Right-hand side of each constraint is still nonnegative
Changing the column of a nonbasic variable x_j or adding a new variable x_j	Changes the coefficient for x_j in row 0 and x_j's constraint column in optimal tableau	The coefficient of x_j in row 0 is still nonnegative

PROBLEMS

Group A

1 In the Dakota problem, show that the current basis remains optimal if c_3, the price of chairs, satisfies $15 \le c_3 \le 22.5$. If $c_3 = 21$, find the new optimal solution. Also, if $c_3 = 25$, find the new optimal solution.

2 If $c_1 = 55$ in the Dakota problem, show that the new optimal solution does not produce any desks.

3 In the Dakota problem, show that if the amount of lumber (board ft) available (b_1) satisfies $b_1 \ge 24$, the current basis remains optimal. If $b_1 = 30$, find the new optimal solution.

4 Show that if tables sell for $50 and use 1 board ft of lumber, 3 finishing hours, and 1.5 carpentry hours, the current basis for the Dakota problem will no longer be optimal. Find the new optimal solution.

5 Dakota Furniture is considering manufacturing home computer tables. A home computer table sells for $36 and uses 6 board ft of lumber, 2 finishing hours, and 2 carpentry hours. Should the company manufacture any home computer tables?

6 Sugarco can manufacture three types of candy bar. Each candy bar consists totally of sugar and chocolate. The compositions of each type of candy bar and the profit earned from each candy bar are shown in Table 10. Fifty oz of sugar and 100 oz of chocolate are available. After defining x_i to be the number of Type i candy bars manufactured, Sugarco should solve the following LP:

$$\max z = 3x_1 + 7x_2 + 5x_3$$
$$\text{s.t.} \quad x_1 + x_2 + x_3 \le 50 \quad \text{(Sugar constraint)}$$
$$2x_1 + 3x_2 + x_3 \le 100 \quad \text{(Chocolate constraint)}$$
$$x_1, x_2, x_3 \ge 0$$

After adding slack variables s_1 and s_2, the optimal tableau is as shown in Table 11. Using this optimal tableau, answer the following questions:

a For what values of Type 1 candy bar profit does the current basis remain optimal? If the profit for a Type 1 candy bar were 7¢, what would be the new optimal solution to Sugarco's problem?

b For what values of Type 2 candy bar profit would the current basis remain optimal? If the profit for a Type 2 candy bar were 13¢, then what would be the new optimal solution to Sugarco's problem?

c For what amount of available sugar would the current basis remain optimal?

d If 60 oz of sugar were available, what would be Sugarco's profit? How many of each candy bar should the company make? Could these questions be answered if only 30 oz of sugar were available?

e Suppose a Type 1 candy bar used only 0.5 oz of sugar and 0.5 oz of chocolate. Should Sugarco now make Type 1 candy bars?

TABLE 10

Bar	Amount of Sugar (Ounces)	Amount of Chocolate (Ounces)	Profit (Cents)
1	1	2	3
2	1	3	7
3	1	1	5

TABLE 11

z	x_1	x_2	x_3	s_1	s_2	rhs	Basic Variable
1	3	0	0	4	1	300	$z = 300$
0	$\frac{1}{2}$	0	1	$\frac{3}{2}$	$-\frac{1}{2}$	25	$x_3 = 25$
0	$\frac{1}{2}$	1	0	$-\frac{1}{2}$	$\frac{1}{2}$	25	$x_2 = 25$

f Sugarco is considering making Type 4 candy bars. A Type 4 candy bar earns 17¢ profit and requires 3 oz of sugar and 4 oz of chocolate. Should Sugarco manufacture any Type 4 candy bars?

7 The following questions refer to the Giapetto problem (Section 3.1). Giapetto's LP was

$$\max z = 3x_1 + 2x_2$$

$$\begin{aligned}
\text{s.t.} \quad & 2x_1 + x_2 \le 100 && \text{(Finishing constraint)} \\
& x_1 + x_2 \le 80 && \text{(Carpentry constraint)} \\
& x_1 \le 40 && \text{(Limited demand for soldiers)}
\end{aligned}$$

(x_1 = soldiers and x_2 = trains). After adding slack variables s_1, s_2, and s_3, the optimal tableau is as shown in Table 12. Use this optimal tableau to answer the following questions:

a Show that as long as soldiers (x_1) contribute between $2 and $4 to profit, the current basis remains optimal. If soldiers contribute $3.50 to profit, find the new optimal solution to the Giapetto problem.

b Show that as long as trains (x_2) contribute between $1.50 and $3.00 to profit, the current basis remains optimal.

c Show that if between 80 and 120 finishing hours are available, the current basis remains optimal. Find the new optimal solution to the Giapetto problem if 90 finishing hours are available.

d Show that as long as the demand for soldiers is at least 20, the current basis remains optimal.

e Giapetto is considering manufacturing toy boats. A toy boat uses 2 carpentry hours and 1 finishing hour. Demand for toy boats is unlimited. If a toy boat contributes $3.50 to profit, should Giapetto manufacture any toy boats?

Group B

8 Consider the Dorian Auto problem (Example 2 of Chapter 3):

$$\min z = 50x_1 + 100x_2$$

$$\begin{aligned}
\text{s.t.} \quad & 7x_1 + 2x_2 \ge 28 && \text{(HIW)} \\
& 2x_1 + 12x_2 \ge 24 && \text{(HIM)} \\
& x_1, x_2 \ge 0
\end{aligned}$$

(x_1 = number of comedy ads, and x_2 = number of football ads). The optimal tableau is given in Table 13. Remember that for a min problem, a tableau is optimal if and only if each variable has a nonpositive coefficient in row 0 and the right-hand side of each constraint is nonnegative.

a Find the range of values of the cost of a comedy ad (currently $50,000) for which the current basis remains optimal.

b Find the range of values of the number of required HIW exposures (currently 28 million) for which the current basis remains optimal. If 40 million HIW exposures were required, what would be the new optimal solution?

c Suppose an ad on a news program costs $110,000 and reaches 12 million HIW and 7 million HIM. Should Dorian advertise on the news program?

9 Show that if the right-hand side of the ith constraint is increased by Δ, then the right-hand side of the optimal tableau is given by (original right-hand side of the optimal tableau) + Δ(column i of B^{-1}).

TABLE 12

z	x_1	x_2	s_1	s_2	s_3	rhs	Basic Variable
1	0	0	1	1	0	180	$z = 180$
0	1	0	1	−1	0	20	$x_1 = 20$
0	0	1	−1	2	0	60	$x_2 = 60$
0	0	0	−1	1	1	20	$s_3 = 20$

TABLE 13

z	x_1	x_2	e_1	e_2	a_1	a_2	rhs
1	0	0	−5	−7.5	$5 - M$	$7.5 - M$	320
0	1	0	$-\frac{3}{20}$	$\frac{1}{40}$	$\frac{3}{20}$	$-\frac{1}{40}$	3.6
0	0	1	$\frac{1}{40}$	$-\frac{7}{80}$	$-\frac{1}{40}$	$\frac{7}{80}$	1.4

6.4 Sensitivity Analysis When More Than One Parameter Is Changed: The 100% Rule[†]

In this section, we show how to use the LINDO output to determine whether the current basis remains optimal when more than one objective function coefficient or right-hand side is changed.

The 100% Rule for Changing Objective Function Coefficients

Depending on whether the objective function coefficient of any variable with a zero reduced cost in the optimal tableau is changed, there are two cases to consider:

[†]This section covers topics that may be omitted with no loss of continuity.

Case 1 All variables whose objective function coefficients are changed have nonzero reduced costs in the optimal row 0.

Case 2 At least one variable whose objective function coefficient is changed has a reduced cost of zero.

In Case 1, the current basis remains optimal if and only if the objective function coefficient for each variable remains within the allowable range[†] given on the LINDO printout (see Problem 10 at the end of this section). If the current basis remains optimal, then both the values of the decision variables and objective function remain unchanged. If the objective function coefficient for any variable is outside its allowable range, then the current basis is no longer optimal.

The following two examples of Case 1 refer to the diet problem of Section 3.4. The LINDO printout for this problem is given in Figure 6.

EXAMPLE 2   100% Rule for Objective Functional Coefficients 1

Suppose the price of a brownie increases to 60¢ and a piece of pineapple cheesecake decreases to 50¢. Does the current basis remain optimal? What would be the new optimal solution?

Solution Both brownies and pineapple cheesecake have nonzero reduced costs, so we are in Case 1. From Figure 6 and the Case 1 discussion, we see that the current basis remains optimal if and only if

$$22.5 = 50 - 27.5 \leq \text{cost of a brownie} \leq 50 + \infty = \infty$$
$$30 = 80 - 50 \leq \text{cost of a piece of cheesecake} \leq 80 + \infty = \infty$$

Because the new prices satisfy both of these conditions, the current basis remains optimal. Also the optimal z-value and optimal value of the decision variables remain unchanged.

EXAMPLE 3   100% Rule for Objective Functional Coefficients 2

If prices drop to 40¢ for a brownie and 25¢ for a piece of pineapple cheesecake, is the current basis still optimal?

Solution From Figure 6, we see that Case 1 again applies. The cost of a brownie remains in its allowable range, but the price of pineapple cheesecake does not. Thus, the current basis is no longer optimal, and the problem must be solved again.

In Case 2, we can often show that the current basis remains optimal by using the **100% Rule.** Let

c_j = original objective function coefficient for x_j

Δc_j = change in c_j

I_j = maximum allowable increase in c_j for which current basis remains optimal (from LINDO output)

D_j = maximum allowable decrease in c_j for which current basis remains optimal (from LINDO output)

[†]The allowable range for c_j is the range of values for which the current basis remains optimal (assuming that only c_j is changed).

```
MIN      50 BR + 20 IC + 30 COLA + 80 PC
SUBJECT TO
    2)    400 BR + 200 IC + 150 COLA + 500 PC >=    500
    3)    3 BR + 2 IC >=    6
    4)    2 BR + 2 IC + 4 COLA + 4 PC >=    10
    5)    2 BR + 4 IC +  COLA + 5 PC >=    8
END

    LP OPTIMUM FOUND   AT STEP      5

        OBJECTIVE FUNCTION VALUE

  1)        90.000000

VARIABLE       VALUE        REDUCED COST
      BR       0.000000       27.500000
      IC       3.000000        0.000000
    COLA       1.000000        0.000000
      PC       0.000000       50.000000

 ROW     SLACK OR SURPLUS    DUAL PRICES
    2)      250.000000        0.000000
    3)        0.000000       -2.500000
    4)        0.000000       -7.500000
    5)        5.000000        0.000000

NO. ITERATIONS=      5

    RANGES IN WHICH THE BASIS IS UNCHANGED

                      OBJ COEFFICIENT RANGES
VARIABLE       CURRENT      ALLOWABLE      ALLOWABLE
                COEF        INCREASE       DECREASE
      BR      50.000000     INFINITY      27.500000
      IC      20.000000    18.333334       5.000000
    COLA      30.000000    10.000000      30.000000
      PC      80.000000     INFINITY      50.000000

                      RIGHTHAND SIDE RANGES
 ROW           CURRENT      ALLOWABLE      ALLOWABLE
                RHS         INCREASE       DECREASE
    2        500.000000    250.000000      INFINITY
    3          6.000000      4.000000       2.857143
    4         10.000000      INFINITY       4.000000
    5          8.000000      5.000000       INFINITY
```

FIGURE 6
LINDO Output for
Diet Problem

For each variable x_j, we define the ratio r_j:

$$\text{If} \quad \Delta c_j \geq 0, \qquad r_j = \frac{\Delta c_j}{I_j}$$

$$\text{If} \quad \Delta c_j \leq 0, \qquad r_j = \frac{-\Delta c_j}{D_j}$$

If c_j is unchanged, then $r_j = 0$. Thus, r_j measures the ratio of the actual change in c_j to the maximum allowable change in c_j that would keep the current basis optimal. If only one objective function coefficient were being changed, then the current basis would remain optimal if $r_j \leq 1$ (or equivalently, if r_j, expressed as a percentage, were less than or equal to 100%). The 100% Rule for objective function coefficients is a generalization of this idea. It states that if $\Sigma r_j \leq 1$, then we can be sure that the current basis remains optimal. If $\Sigma r_j > 1$, then the current basis may or may not be optimal; we can't be sure. If the current basis does remain optimal, then the values of the decision variables remain unchanged, but the optimal z-value may change. The reader is referred to Bradley, Hax, and Magnanti (1977) for a proof of the 100% Rule. We sketch the proof in Problem 11 at the end of this section.

The following two examples of Case 2 refer to the Dakota Furniture problem and illustrate the use of the 100% Rule.

EXAMPLE 4   **Basis No Longer Optimal**

Suppose the desk price increases to $70 and chairs decrease to $18. Does the current basis remain optimal? What is the new optimal z-value?

Solution   Because both desks and chairs have zero reduced costs (they are basic variables), we must apply the 100% Rule to determine whether the current basis remains optimal. Returning to the notation that x_1 = desks, x_2 = tables, and x_3 = chairs, we may write

$$\Delta c_1 = 70 - 60 = 10, \quad I_1 = 20, \quad \text{so} \quad r_1 = \tfrac{10}{20} = 0.5$$

$$\Delta c_3 = 18 - 20 = -2, \quad D_3 = 5, \quad \text{so} \quad r_3 = \tfrac{2}{5} = 0.4$$

$$\Delta c_2 = 0, \quad \text{so} \quad r_2 = 0$$

Because $r_1 + r_2 + r_3 = 0.9 \le 1$, the current basis remains optimal. Another way of looking at it: We changed c_1 50% of the amount it was "allowed" to change and c_3 40% of the amount it was "allowed" to change. Because 50% + 40% = 90% $\le$ 100%, the current basis remains optimal.

The current basis remains optimal, so the values of the decision variables do not change. Note that the revenue from each desk has increased by $10 and the revenue from each chair has decreased by $2. Dakota is still producing 2 desks and 8 chairs, so revenue increases by $2(10) - 8(2) = \$4$ and is now $280 + 4 = \$284$.

EXAMPLE 5   **100% Rule and Optimal Basis 1**

Show that if the price of tables increases to $33 and desk prices decrease to $58, the 100% Rule does not tell us whether the current basis is still optimal.

Solution   For this situation,

$$\Delta c_1 = 58 - 60 = -2, \quad D_1 = 4, \quad \text{so} \quad r_1 = \tfrac{2}{4} = 0.5$$

$$\Delta c_2 = 33 - 30 = 3, \quad I_2 = 5, \quad \text{so} \quad r_2 = \tfrac{3}{5} = 0.6$$

$$\Delta c_3 = 0, \quad \text{so} \quad r_3 = 0$$

Because $r_1 + r_2 + r_3 = 0.5 + 0.6 + 0 = 1.1 > 1$, the 100% Rule yields no information about whether the current basis is optimal.

The 100% Rule for Changing Right-Hand Sides

Depending on whether any of the constraints whose right-hand sides are being modified are binding constraints, there are two cases to consider:

Case 1   All constraints whose right-hand sides are being modified are nonbinding constraints.

Case 2   At least one of the constraints whose right-hand side is being modified is a binding constraint (that is, has zero slack or zero excess).

In Case 1, the current basis remains optimal if and only if each right-hand side remains within its allowable range.[†] Then the values of the decision variables and optimal objective function remain unchanged. If the right-hand side for any constraint is outside its allowable range, then the current basis is no longer optimal (see Problem 12 at the end of this section). The following examples for the diet problem illustrate the application of Case 1.

[†]The allowable range for a right-hand side b_i is the range of values for which the current basis remains optimal (assuming the other LP parameters remain unchanged).

EXAMPLE 6   **New Optimal Solution**

Suppose the calorie requirement is decreased to 400 calories and the fat requirement is increased to 10 oz. Does the current basis remain optimal? What is the new optimal solution?

Solution Both the calorie and fat constraints are nonbinding, so Case 1 applies. From Figure 6, we see that the allowable ranges for the calorie and fat constraints are

$$-\infty = 500 - \infty \leq \text{calorie requirement} \leq 500 + 250 = 750$$
$$-\infty = 8 - \infty \leq \text{fat requirement} \leq 8 + 5 = 13$$

The new calorie and fat requirements both remain within their allowable ranges, so the current basis remains optimal. The optimal z-value and the values of the decision variables remain unchanged.

EXAMPLE 7   **Basis No Longer Optimal**

Suppose the calorie requirement is decreased to 400 calories and the fat requirement is increased to 15 oz. Is the current basis still optimal?

Solution The fat requirement is no longer in its allowable range, so the current basis is no longer optimal.

In Case 2, we can often show that the current basis remains optimal via another version of the 100% Rule. Let

b_j = current right-hand side of the jth constraint (from row $j + 1$ on LINDO output)

Δb_j = change in b_j

I_j = maximum allowable increase in b_j for which the current basis remains optimal (from LINDO output)

D_j = maximum allowable decrease in b_j for which the current basis remains optimal (from LINDO output)

For each constraint, compute the ratio r_j:

$$\text{If} \quad \Delta b_j \geq 0, \quad r_j = \frac{\Delta b_j}{I_j}$$

$$\text{If} \quad \Delta b_j \leq 0, \quad r_j = \frac{-\Delta b_j}{I_j}$$

If only the jth right-hand side is changed, then the current basis remains optimal if $r_j \leq 1$. Also note that r_j is the fraction of the maximum allowable change (in the sense that the current basis remains optimal) in b_j that has occurred. The 100% Rule states that if $\Sigma r_j \leq 1$, then the current basis remains optimal. If $\Sigma r_j > 1$, then the current basis may or may not be optimal; we can't be sure (see Problem 13 at the end of this section for a sketch of the proof of this result). The following examples illustrate the use of the 100% Rule for right-hand sides.

EXAMPLE 8   **Basis Remains Optimal**

In the Dakota problem, suppose 22 finishing hours and 9 carpentry hours are available. Does the current basis remain optimal?

Solution The finishing and carpentry constraints are binding, so we are in Case 2 and need to use the 100% Rule.

$$\Delta b_1 = 0, \qquad \text{so} \qquad r_1 = 0$$
$$\Delta b_2 = 22 - 20 = 2, \qquad I_2 = 4, \qquad \text{so} \qquad r_2 = \frac{2}{4} = 0.5$$
$$\Delta b_3 = 9 - 8 = 1, \qquad I_3 = 2, \qquad \text{so} \qquad r_3 = \frac{1}{2} = 0.5$$

Because $r_1 + r_2 + r_3 = 1$, the current basis remains optimal.

EXAMPLE 9   **100% Rule and Optimal Basis 2**

In the diet problem, suppose the chocolate requirement is increased to 8 oz and the sugar requirement is reduced to 7 oz. Does the current basis remain optimal?

Solution   The chocolate and sugar constraints are binding, so we are in Case 2 and need to use the 100% Rule.

$$\Delta b_2 = 8 - 6 = 2, \qquad I_2 = 4, \qquad \text{so} \qquad r_2 = \frac{2}{4} = 0.5$$
$$\Delta b_3 = 7 - 10 = -3, \qquad D_3 = 4, \qquad \text{so} \qquad r_3 = \frac{3}{4} = 0.75$$
$$\Delta b_1 = \Delta b_4 = 0, \quad \text{so} \qquad r_1 = r_4 = 0$$

Because $r_1 + r_2 + r_3 + r_4 = 1.25 > 1$, the 100% Rule yields no information about whether the current basis remains optimal.

PROBLEMS

Group A

The following questions refer to the diet problem:

1 If the cost of a brownie is 70¢ and a piece of cheesecake costs 60¢, does the current basis remain optimal?

2 If the cost of a brownie is 20¢ and a piece of cheesecake is $1, does the current basis remain optimal?

3 If the fat requirement is reduced to 3 oz and the calorie requirement is increased to 800 calories, does the current basis remain optimal?

4 If the fat requirement is 6 oz and the calorie requirement is 600 calories, does the current basis remain optimal?

5 If the price of a bottle of soda is 15¢ and a piece of cheesecake is 60¢, show that the current basis remains optimal. What will be the new optimal solution to the diet problem?

6 If 8 oz of chocolate and 60 calories are required, show that the current basis remains optimal.

The following questions refer to the Dakota problem.

7 Suppose that the price of a desk is $65, a table is $25, and a chair is $18. Show that the current basis remains optimal. What is the new optimal z-value?

8 Suppose that 60 board ft of lumber and 23 finishing hours are available. Show that the current basis remains optimal.

9 Suppose 40 board ft of lumber, 21 finishing hours, and 8.5 carpentry hours are available. Show that the current basis remains optimal.

Group B

10 Prove the Case 1 result for the objective function coefficients.

11 To illustrate the validity of the 100% Rule for objective function coefficients, consider an LP with four decision variables (x_1, x_2, x_3, and x_4) and two constraints in which x_1 and x_2 are basic variables in the optimal basis. Suppose (if only a single objective function coefficient is changed) the current basis is known to be optimal for $L_1 \leq c_1 \leq U_1$ and $L_2 \leq c_2 \leq U_2$. Suppose we change c_1 to $c_1' = c_1 + \Delta c_1$ and c_2 to $c_2' = c_2 + \Delta c_2$, where $\Delta c_1 > 0$ and $\Delta c_2 < 0$. Let

$$\frac{\Delta c_1}{U_1 - c_1} = r_1 \quad \text{and} \quad \frac{-\Delta c_2}{c_2 - L_2} = r_2$$

Show that if $r_1 + r_2 \leq 1$, the current basis remains optimal. *Hint: Any variable x_j prices out to $\mathbf{c}_{BV}B^{-1}\mathbf{a}_j - c_j$. To show that for the new values of c_1 and c_2, all variables still price out nonnegative, use the fact that

$$[c_1', c_2'] = r_1[U_1, c_2] + r_2[c_1, L_2] + (1 - r_1 - r_2)[c_1, c_2]$$

12 Prove the Case 1 result for right-hand sides. Use the fact that if a constraint is nonbinding in the optimal solution, then its slack or excess variable is in the optimal basis, and the corresponding column of B^{-1} will have a single 1 and all other elements equal to 0.

13 In this problem, we sketch a proof of the 100% Rule for right-hand sides. Consider an LP with two constraints and right-hand sides b_1 and b_2. Suppose that if only one right-hand side is changed, the current basis remains optimal for $L_1 \leq b_1 \leq U_1$ and $L_2 \leq b_2 \leq U_2$. Suppose we change the right-hand sides to $b_1' = b_1 + \Delta b_1$ and $b_2' = b_2 + \Delta b_2$, where $\Delta b_1 > 0$ and $\Delta b_2 < 0$. Let

$$r_1 = \frac{\Delta b_1}{U_1 - b_1} \quad \text{and} \quad r_2 = \frac{-\Delta b_2}{b_2 - L_2}$$

Show that if $r_1 + r_2 \leq 1$, the current basis remains optimal. (*Hint:* You must show that

$$B^{-1}\begin{bmatrix} b_1' \\ b_2' \end{bmatrix} \geq \begin{bmatrix} 0 \\ 0 \end{bmatrix}$$

Use the fact that

$$[b_1', b_2'] = r_1[U_1, b_2] + r_2[b_1, L_2] + (1 - r_1 - r_2)[b_1, b_2]$$

to show this.)

6.5 Finding the Dual of an LP

Associated with any LP is another LP, called the **dual.** Knowing the relation between an LP and its dual is vital to understanding advanced topics in linear and nonlinear programming. This relation is important because it gives us interesting economic insights. Knowledge of duality will also provide additional insights into sensitivity analysis.

In this section, we explain how to find the dual of any LP; in Section 6.6, we discuss the economic interpretation of the dual; and in Sections 6.7–6.10, we discuss the relation that exists between an LP and its dual.

When taking the dual of a given LP, we refer to the given LP as the **primal.** If the primal is a max problem, then the dual will be a min problem, and vice versa. For convenience, we define the variables for the max problem to be $z, x_1, x_2, \ldots, x_n$ and the variables for the min problem to be $w, y_1, y_2, \ldots, y_m$. We begin by explaining how to find the dual of a max problem in which all variables are required to be nonnegative and all constraints are $\leq$ constraints (called a **normal max problem**). A normal max problem may be written as

$$\max z = c_1 x_1 + c_2 x_2 + \cdots + c_n x_n$$

$$\begin{aligned}
\text{s.t.} \quad & a_{11}x_1 + a_{12}x_2 + \cdots + a_{1n}x_n \leq b_1 \\
& a_{21}x_1 + a_{22}x_2 + \cdots + a_{2n}x_n \leq b_2 \\
& \quad \vdots \qquad \quad \vdots \qquad \qquad \quad \vdots \\
& a_{m1}x_1 + a_{m2}x_2 + \cdots + a_{mn}x_n \leq b_m \\
& x_j \geq 0 \quad (j = 1, 2, \ldots, n)
\end{aligned}$$

(16)

The dual of a normal max problem such as (16) is defined to be

$$\min w = b_1 y_1 + b_2 y_2 + \cdots + b_m x_m$$

$$\begin{aligned}
\text{s.t.} \quad & a_{11}y_1 + a_{21}y_2 + \cdots + a_{m1}y_m \geq c_1 \\
& a_{12}y_1 + a_{22}y_2 + \cdots + a_{m2}y_m \geq c_2 \\
& \quad \vdots \qquad \quad \vdots \qquad \qquad \quad \vdots \\
& a_{1n}y_1 + a_{2n}y_2 + \cdots + a_{mn}y_m \geq c_n \\
& y_i \geq 0 \quad (i = 1, 2, \ldots, m)
\end{aligned}$$

(17)

A min problem such as (17) that has all $\geq$ constraints and all variables nonnegative is called a **normal min problem.** If the primal is a normal min problem such as (17), then we define the dual of (17) to be (16).

Finding the Dual of a Normal Max or Min Problem

A tabular approach makes it easy to find the dual of an LP. If the primal is a normal max problem, then it can be read across (Table 14); the dual is found by reading down. Similarly, if the primal is a normal min problem, we find it by reading down; the dual is found

min w		max z				
		$(x_1 \geq 0)$	$(x_2 \geq 0)$	$\cdots$	$(x_n \geq 0)$	
		x_1	x_2		x_n	
$(y_1 \geq 0)$	y_1	a_{11}	a_{12}	$\cdots$	a_{1n}	$\leq b_1$
$(y_2 \geq 0)$	y_2	a_{21}	a_{22}	$\cdots$	a_{2n}	$\leq b_2$
$\vdots$	$\vdots$	$\vdots$	$\vdots$		$\vdots$	$\vdots$
$(y_m \geq 0)$	y_m	a_{m1}	a_{m2}	$\cdots$	a_{mn}	$\leq b_m$
		$\geq c_1$	$\geq c_2$		$\geq c_n$	

by reading across in the table. We illustrate the use of the table by finding the dual of the Dakota problem and the dual of the diet problems. The Dakota problem is

$$\max z = 60x_1 + 30x_2 + 20x_3$$
$$\text{s.t.} \quad 8x_1 + 6x_2 + x_3 \leq 48 \quad \text{(Lumber constraint)}$$
$$4x_1 + 2x_2 + 1.5x_3 \leq 20 \quad \text{(Finishing constraint)}$$
$$2x_1 + 1.5x_2 + 0.5x_3 \leq 8 \quad \text{(Carpentry constraint)}$$
$$x_1, x_2, x_3 \geq 0$$

where

$$x_1 = \text{number of desks manufactured}$$
$$x_2 = \text{number of tables manufactured}$$
$$x_3 = \text{number of chairs manufactured}$$

Using the format of Table 14, we read the Dakota problem across in Table 15. Then, reading down, we find the Dakota dual to be

$$\min w = 48y_1 + 20y_2 + 8y_3$$
$$\text{s.t.} \quad 8y_1 + 4y_2 + 2y_3 \geq 60$$
$$6y_1 + 2y_2 + 1.5y_3 \geq 30$$
$$y_1 + 1.5y_2 + 0.5y_3 \geq 20$$
$$y_1, y_2, y_3 \geq 0$$

The tabular method of finding the dual makes it clear that the ith dual *constraint* corresponds to the ith primal *variable* x_i. For example, the first dual constraint corresponds to x_1 (desks), because each number comes from the x_1 (desk) column of the primal. Simi-

min w		max z			
		$(x_1 \geq 0)$	$(x_2 \geq 0)$	$(x_3 \geq 0)$	
		x_1	x_2	x_3	
$(y_1 \geq 0)$	y_1	8	6	1	≤ 48
$(y_2 \geq 0)$	y_2	4	2	1.5	≤ 20
$(y_3 \geq 0)$	y_3	2	1.5	0.5	≤ 8
		≥ 60	≥ 30	≥ 20	

larly, the second dual constraint corresponds to x_2 (tables), and the third dual constraint corresponds to x_3 (chairs). In a similar fashion, dual *variable* y_i is associated with the ith primal constraint. For example, y_1 is associated with the first primal constraint (lumber constraint), because each coefficient of y_1 in the dual comes from the lumber constraint, or the availability of lumber. The importance of these correspondences between the primal and the dual will become clear in Section 6.6.

We now find the dual of the diet problem. Because the diet problem is a min problem, we follow the convention of using w to denote the objective function and y_1, y_2, y_3, and y_4 for the variables. Then the diet problem may be written as

$$\min w = 50y_1 + 20y_2 + 30y_3 + 80y_4$$

$$
\begin{array}{llll}
\text{s.t.} & 400y_1 + 200y_2 + 150y_3 + 500y_4 \geq 500 & \text{(Calorie constraint)} \\
& 3y_1 + 2y_2 \geq 6 & \text{(Chocolate constraint)} \\
& 2y_1 + 2y_2 + 4y_3 + 4y_4 \geq 10 & \text{(Sugar constraint)} \\
& 2y_1 + 4y_2 + y_3 + 5y_4 \geq 8 & \text{(Fat constraint)} \\
& y_1, y_2, y_3, y_4 \geq 0 &
\end{array}
$$

where

y_1 = number of brownies eaten daily

y_2 = number of scoops of chocolate ice cream eaten daily

y_3 = bottles of soda drunk daily

y_4 = pieces of pineapple cheesecake eaten daily

The primal is a normal min problem, so we can read it down, and read its dual across, in Table 16. We find that the dual of the diet problem is

$$\max z = 500x_1 + 6x_2 + 10x_3 + 8x_4$$

$$
\begin{array}{ll}
\text{s.t.} & 400x_1 + 3x_2 + 2x_3 + 2x_4 \leq 50 \\
& 200x_1 + 2x_2 + 2x_3 + 4x_4 \leq 20 \\
& 150x_1 + 4x_3 + x_4 \leq 30 \\
& 500x_1 + 4x_3 + 5x_4 \leq 80 \\
& x_1, x_2, x_3, x_4 \geq 0
\end{array}
$$

As in the Dakota problem, we see that the ith dual constraint corresponds to the ith primal variable. For example, the third dual constraint may be thought of as the soda constraint. Also, the ith dual variable corresponds to the ith primal constraint. For example, x_3 (the third dual variable) may be thought of as the dual sugar variable.

TABLE 16
Finding the Dual of the Diet Problem

min w		max z			
		$(x_1 \geq 0)$	$(x_2 \geq 0)$	$(x_3 \geq 0)$	$(x_4 \geq 0)$
		x_1	x_2	x_3	x_4	
$(y_1 \geq 0)$	y_1	400	3	2	2	≤ 50
$(y_2 \geq 0)$	y_2	200	2	2	4	≤ 20
$(y_3 \geq 0)$	y_3	150	0	4	1	≤ 30
$(y_4 \geq 0)$	y_4	500	0	4	5	≤ 80
		≥ 500	≥ 6	≥ 10	≥ 8	

Finding the Dual of a Nonnormal LP

Unfortunately, many LPs are not normal max or min problems. For example,

$$\max z = 2x_1 + x_2$$
$$\text{s.t.} \quad x_1 + x_2 = 2$$
$$2x_1 - x_2 \geq 3 \qquad (18)$$
$$x_1 - x_2 \leq 1$$
$$x_1 \geq 0, x_2 \text{ urs}$$

is not a normal max problem because it has a $\geq$ constraint, an equality constraint, and an unrestricted-in-sign variable. As another example of a nonnormal LP, consider

$$\min w = 2y_1 + 4y_2 + 6y_3$$
$$\text{s.t.} \quad y_1 + 2y_2 + y_3 \geq 2$$
$$y_1 \qquad - y_3 \geq 1$$
$$y_2 + y_3 = 1 \qquad (19)$$
$$2y_1 + \quad y_2 \qquad \leq 3$$
$$y_1 \text{ urs}, y_2, y_3 \geq 0$$

This LP is not a normal min problem because it contains an equality constraint, a $\leq$ constraint, and an unrestricted-in-sign variable.

Fortunately, an LP can be transformed into normal form (either (16) or (17)). To place a max problem into normal form, we proceed as follows:

Step 1 Multiply each $\geq$ constraint by -1, converting it into a $\leq$ constraint. For example, in (18), $2x_1 - x_2 \geq 3$ would be transformed into $-2x_1 + x_2 \leq -3$.

Step 2 Replace each equality constraint by two inequality constraints (a $\leq$ constraint and a $\geq$ constraint). Then convert the $\geq$ constraint to a $\leq$ constraint. For example, in (18), we would replace $x_1 + x_2 = 2$ by the two inequalities $x_1 + x_2 \geq 2$ and $x_1 + x_2 \leq 2$. Then we would convert $x_1 + x_2 \geq 2$ to $-x_1 - x_2 \leq -2$. The net result is that $x_1 + x_2 = 2$ is replaced by the two inequalities $x_1 + x_2 \leq 2$ and $-x_1 - x_2 \leq -2$.

Step 3 As in Section 4.14, replace each urs variable x_i by $x_i = x' - x''_i$, where $x'_i \geq 0$ and $x''_i \geq 0$. In (18), we would replace x_2 by $x'_2 - x''_2$.

After these transformations are complete, (18) has been transformed into the following (equivalent) LP:

$$\max z = 2x_1 + x'_2 - x''_2$$
$$\text{s.t.} \quad x_1 + x'_2 - x''_2 \leq 2$$
$$-x_1 - x'_2 + x''_2 \leq -2$$
$$-2x_1 + x'_2 - x''_2 \leq -3 \qquad (18')$$
$$x_1 - x'_2 + x''_2 \leq 1$$
$$x_1, x'_2, x''_2 \geq 0$$

Because (18′) is a normal max problem, we could use (16) and (17) to find the dual of (18′).

If the primal is not a normal min problem, then we can transform it into a normal min problem as follows:

Step 1 Convert each $\leq$ constraint into a $\geq$ constraint by multiplying through by -1. For example, in (19), $2y_1 + y_2 \leq 3$ is transformed into $-2y_1 - y_2 \geq -3$.

Step 2 Replace each equality constraint by a $\leq$ constraint and a $\geq$ constraint. Then transform the $\leq$ constraint into a $\geq$ constraint. For example, in (19), the constraint $y_2 + y_3 = 1$ is equivalent to $y_2 + y_3 \leq 1$ and $y_2 + y_3 \geq 1$. Transforming $y_2 + y_3 \leq 1$ into $-y_2 - y_3 \geq -1$, we see that we can replace the constraint $y_2 + y_3 = 1$ by the two constraints $y_2 + y_3 \geq 1$ and $-y_2 - y_3 \geq -1$.

Step 3 Replace any urs variable y_i by $y_i = y_i' - y_i''$, where $y_i' \geq 0$ and $y_i'' \geq 0$. Applying these steps to (19) yields the following standard min problem:

$$
\begin{aligned}
\min w = 2y_1' - 2y_1'' &+ 4y_2 + 6y_3 \\
\text{s.t.} \quad y_1' - y_1'' + 2y_2 &+ y_3 \geq 2 \\
y_1' - y_1'' \quad &- y_3 \geq 1 \\
y_2 &+ y_3 \geq 1 \\
- y_2 &- y_3 \geq -1 \\
-2y_1' + 2y_1'' - y_2 \quad & \geq -3 \\
y_1', y_1'', y_2, y_3 & \geq 0
\end{aligned}
\tag{19$'$}
$$

Because (19$'$) is a normal min problem in standard form, we may use (16) and (17) to find its dual.

We can find the dual of a nonnormal LP without going through the transformations that we have described by using the following rules.[†]

Finding the Dual of a Nonnormal Max Problem

Step 1 Fill in Table 14 so that the primal can be read across.

Step 2 After making the following changes, the dual can be read down in the usual fashion: (a) If the ith primal constraint is a $\geq$ constraint, then the corresponding dual variable y_i must satisfy $y_i \leq 0$. (b) If the ith primal constraint is an equality constraint, then the dual variable y_i is now unrestricted in sign. (c) If the ith primal variable is urs, then the ith dual constraint will be an equality constraint.

When this method is applied to (18), the Table 14 format yields Table 17. We note with an asterisk (*) the places where the rules must be used to determine part of the dual. For example, x_2 urs causes the second dual constraint to be an equality constraint. Also, the first primal constraint being an equality constraint makes y_1 urs, and the second primal constraint being a $\geq$ constraint makes $y_2 \leq 0$. Filling in the missing information across from the appropriate asterisk yields Table 18. Reading the dual down, we obtain

$$
\begin{aligned}
\min w = 2y_1 &+ 3y_2 + y_3 \\
\text{s.t.} \quad y_1 + 2y_2 &+ y_3 \geq 2 \\
y_1 - y_2 &- y_3 = 1 \\
y_1 \text{ urs}, \; y_2 &\leq 0, \; y_3 \geq 0
\end{aligned}
$$

In Section 6.8, we give an intuitive explanation of why an equality constraint yields an unrestricted-in-sign dual variable and why a $\geq$ constraint yields a negative dual variable.

We can use the following rules to take the dual of a nonnormal min problem.

[†]In Problems 5 and 6 at the end of this section, we show that these rules are consistent with taking the dual of the transformed LP via (16) and (17).

TABLE 17

Finding the Dual of LP (18)

min w		max z ($x_1 \geq 0$)	(x_2 urs)*	
		x_1	x_2	
	y_1	1	1	$=2$*
	y_2	2	-1	≥ 3*
($y_3 \geq 0$)	y_3	1	-1	≤ 1
		≥ 2	$=1$	

TABLE 18

Finding the Dual of LP (18) (Continued)

min w		max z ($x_1 \geq 0$)	(x_2 urs)	
		x_1	x_2	
(y_1 urs)	y_1	1	1	$=2$
($y_2 \leq 0$)	y_2	2	-1	≥ 3
($y_3 \geq 0$)	y_3	1	-1	≤ 1
		≥ 2	$=1$	

Finding the Dual of a Nonnormal Min Problem

Step 1 Write out the primal so it can be read down in Table 14.

Step 2 Except for the following changes, the dual can be read across the table: (a) If the ith primal constraint is a $\leq$ constraint, then the corresponding dual variable x_i must satisfy $x_i \leq 0$. (b) If the ith primal constraint is an equality constraint, then the corresponding dual variable x_i will be urs. (c) If the ith primal variable y_i is urs, then the ith dual constraint is an equality constraint.

When this method is applied to (19), we get Table 19. Asterisks (*) show where the new rules must be used to determine parts of the dual. Because y_1 is urs, the first dual constraint is an equality. The third primal constraint is an equality, so dual variable x_3 is urs. Finally, because the fourth primal constraint is a $\leq$ constraint, the fourth dual variable x_4 must satisfy $x_4 \leq 0$. We can now complete the table (see Table 20). Reading the dual across, we obtain

TABLE 19

Finding the Dual of LP (19)

min w		max z ($x_1 \geq 0$)	($x_2 \geq 0$)			
		x_1	x_2	x_3	x_4	
(y_1 urs)*	y_1	1	1	0	2	2
($y_2 \geq 0$)	y_2	2	0	1	1	≤ 4
($y_3 \geq 0$)	y_3	1	-1	1	0	≤ 6
		≥ 2	≥ 1	$=1$*	≤ 3*	

TABLE **20**

Finding the Dual of LP (19) (Continued)

min w		max z				
		$(x_1 \geq 0)$	$(x_2 \geq 0)$	$(x_3$ urs$)$	$(x_4 \leq 0)$	
		x_1	x_2	x_3	x_4	
$(y_1$ urs$)$	y_1	1	1	0	2	$=2$
$(y_2 \geq 0)$	y_2	2	0	1	1	≤ 4
$(y_3 \geq 0)$	y_3	1	-1	1	0	≤ 6
		≥ 2	≥ 1	$=1$	≤ 3	

$$\max z = 2x_1 + x_2 + x_3 + 3x_4$$
$$\text{s.t.} \quad x_1 + x_2 \qquad + 2x_4 = 2$$
$$2x_1 \qquad + x_3 + x_4 \leq 4$$
$$x_1 - x_2 + x_3 \qquad \leq 6$$
$$x_1, x_2 \geq 0, x_3 \text{ urs}, x_4 \leq 0$$

The reader may verify that with these rules, the dual of the dual is always the primal. This is easily seen from the Table 14 format, because when you take the dual of the dual you are changing the LP back to its original position.

PROBLEMS

Group A

Find the duals of the following LPs:

1 $\max z = 2x_1 + x_2$
s.t. $-x_1 + x_2 \leq 1$
$x_1 + x_2 \leq 3$
$x_1 - 2x_2 \leq 4$
$x_1, x_2 \geq 0$

2 $\min w = y_1 - y_2$
s.t. $2y_1 + y_2 \geq 4$
$y_1 + y_2 \geq 1$
$y_1 + 2y_2 \geq 3$
$y_1, y_2 \geq 0$

3 $\max z = 4x_1 - x_2 + 2x_3$
s.t. $x_1 + x_2 \leq 5$
$2x_1 + x_2 \leq 7$
$2x_2 + x_3 \geq 6$
$x_1 + x_3 = 4$
$x_1 \geq 0, x_2, x_3 \text{ urs}$

4 $\min w = 4y_1 + 2y_2 - y_3$
s.t. $y_1 + 2y_2 \leq 6$
$y_1 - y_2 + 2y_3 = 8$
$y_1, y_2 \geq 0, y_3 \text{ urs}$

Group B

5 This problem shows why the dual variable for an equality constraint should be urs.

a Use the rules given in the text to find the dual of

$$\max z = x_1 + 2x_2$$
$$\text{s.t.} \quad 3x_1 + x_2 \leq 6$$
$$2x_1 + x_2 = 5$$
$$x_1, x_2 \geq 0$$

b Now transform the LP in part (a) to the normal form. Using (16) and (17), take the dual of the transformed LP. Use y_2' and y_2'' as the dual variables for the two primal constraints derived from $2x_1 + x_2 = 5$.

c Make the substitution $y_2 = y_2' - y_2''$ in the part (b) answer. Now show that the two duals obtained in parts (a) and (b) are equivalent.

6 This problem shows why a dual variable y_i corresponding to a $\geq$ constraint in a max problem must satisfy $y_i \leq 0$.

a Using the rules given in the text, find the dual of

$$\max z = 3x_1 + x_2$$
$$\text{s.t.} \quad x_1 + x_2 \leq 1$$
$$-x_1 + x_2 \geq 2$$
$$x_1, x_2 \geq 0$$

b Transform the LP of part (a) into a normal max problem. Now use (16) and (17) to find the dual of the transformed LP. Let $\bar{y}_2$ be the dual variable corresponding to the second primal constraint.

c Show that, defining $\bar{y}_2 = -y_2$, the dual in part (a) is equivalent to the dual in part (b).

6.6 Economic Interpretation of the Dual Problem

Interpreting the Dual of a Max Problem

The dual of the Dakota problem is

$$\min w = 48y_1 + 20y_2 + 8y_3$$
$$\text{s.t.} \quad 8y_1 + 4y_2 + 2y_3 \geq 60 \quad \text{(Desk constraint)}$$
$$6y_1 + 2y_2 + 1.5y_3 \geq 30 \quad \text{(Table constraint)} \qquad \text{(20)}$$
$$y_1 + 1.5y_2 + 0.5y_3 \geq 20 \quad \text{(Chair constraint)}$$
$$y_1, y_2, y_3 \geq 0$$

The first dual constraint is associated with desks, the second with tables, and the third with chairs. Also, y_1 is associated with lumber, y_2 with finishing hours, and y_3 with carpentry hours. The relevant information about the Dakota problem is shown in Table 21.

We are now ready to interpret the Dakota dual (20). Suppose an entrepreneur wants to purchase all of Dakota's resources. Then the entrepreneur must determine the price he or she is willing to pay for a unit of each of Dakota's resources. With this in mind, we define

$$y_1 = \text{price paid for 1 board ft of lumber}$$
$$y_2 = \text{price paid for 1 finishing hour}$$
$$y_3 = \text{price paid for 1 carpentry hour}$$

The resource prices y_1, y_2, and y_3 should be determined by solving the Dakota dual (20). The total price that should be paid for these resources is $48y_1 + 20y_2 + 8y_3$. Because the cost of purchasing the resources is to be minimized,

$$\min w = 48y_1 + 20\,y_2 + 8y_3$$

is the objective function for the Dakota dual.

In setting resource prices, what constraints does the entrepreneur face? Resource prices must be set high enough to induce Dakota to sell. For example, the entrepreneur must offer Dakota at least $60 for a combination of resources that includes 8 board feet of lumber, 4 finishing hours, and 2 carpentry hours, because Dakota could, if it desires, use these resources to produce a desk that can be sold for $60. The entrepreneur is offering $8y_1 + 4y_2 + 2y_3$ for the resources used to produce a desk, so he or she must choose y_1, y_2, and y_3 to satisfy

$$8y_1 + 4y_2 + 2y_3 \geq 60$$

But this is just the first (or desk) constraint of the Dakota dual. Similar reasoning shows that at least $30 must be paid for the resources used to produce a table (6 board feet of lumber, 2 finishing hours, and 1.5 carpentry hours). This means that y_1, y_2, and y_3 must satisfy

$$6y_1 + 2y_2 + 1.5y_3 \geq 30$$

TABLE 21
Relevant Information for Dakota Problem

Resource	Resource/Product			Amount of Resource Available
	Desk	Table	Chair	
Lumber (board ft)	8	6	1	48
Finishing (hours)	4	2	1.5	20
Carpentry (hours)	2	1.5	0.5	8
Selling price ($)	60	30	20	

This is the second (or table) constraint of the Dakota dual.

Similarly, the third (or chair) dual constraint,

$$y_1 + 1.5y_2 + 0.5y_3 \geq 20$$

states that at least $20 (the price of a chair) must be paid for the resources needed to produce a chair (1 board foot of lumber, 1.5 finishing hours, and 0.5 carpentry hour). The sign restrictions $y_1 \geq 0$, $y_2 \geq 0$, and $y_3 \geq 0$ must also hold. Putting everything together, we see that the solution to the dual of the Dakota problem does yield prices for lumber, finishing hours, and carpentry hours. The preceding discussion also shows that the ith dual variable does indeed correspond in a natural way to the ith primal constraint.

In summary, when the primal is a normal max problem, the dual variables are related to the value of the resources available to the decision maker. For this reason, the dual variables are often referred to as **resource shadow prices.** A more thorough discussion of shadow prices is given in Section 6.8.

Interpreting the Dual of a Min Problem

To interpret the dual of a min problem, we consider the dual of the diet problem of Section 3.4. In Section 6.5, we found that the diet problem dual was

$$\max z = 500x_1 + 6x_2 + 10x_3 + 8x_4$$

$$
\begin{array}{lll}
\text{s.t.} & 400x_1 + 3x_2 + 2x_3 + 2x_4 \leq 50 & \text{(Brownie constraint)} \\
& 200x_1 + 2x_2 + 2x_3 + 4x_4 \leq 20 & \text{(Ice cream constraint)} \\
& 150x_1 \qquad\quad + 4x_3 + x_4 \leq 30 & \text{(Soda constraint)} \\
& 500x_1 \qquad\quad + 4x_3 + 5x_4 \leq 80 & \text{(Cheesecake constraint)} \\
& x_1, x_2, x_3, x_4 \geq 0
\end{array}
$$

(21)

The data for the diet problem are shown in Table 22. To interpret (21), suppose Candice is a "nutrient" salesperson who sells calories, chocolate, sugar, and fat. She wants to ensure that a dieter will meet all of his or her daily requirements by purchasing calories, sugar, fat, and chocolate. Then Candice must determine

$$x_1 = \text{price per calorie to charge dieter}$$
$$x_2 = \text{price per ounce of chocolate to charge dieter}$$
$$x_3 = \text{price per ounce of sugar to charge dieter}$$
$$x_4 = \text{price per ounce of fat to charge dieter}$$

Candice wants to maximize her revenue from selling the dieter the daily ration of required nutrients. Because she will receive $500x_1 + 6x_2 + 10x_3 + 8x_4$ cents in revenue from the dieter, her objective is to

$$\max z = 500x_1 + 6x_2 + 10x_3 + 8x_4$$

TABLE 22
Relevant Information for Diet Problem

	Calories	Chocolate (Ounces)	Sugar (Ounces)	Fat (Ounces)	Price (Cents)
Brownie	400	3	2	2	50
Ice cream	200	2	2	4	20
Soda	150	0	4	1	30
Cheesecake	500	0	4	5	80
Requirements	500	6	10	8	

This is the objective function for the dual of the diet problem. But in setting nutrient prices, Candice must set prices low enough so that it will be in the dieter's economic interest to purchase all nutrients from her. For example, by purchasing a brownie for 50¢, the dieter can obtain 400 calories, 3 oz of chocolate, 2 oz of sugar, and 2 oz of fat. So Candice cannot charge more than 50¢ for this combination of nutrients. This leads to the following (brownie) constraint:

$$400x_1 + 3x_2 + 2x_3 + 2x_4 \leq 50$$

the first constraint in the diet problem dual. Similar reasoning yields the second dual (ice cream) constraint, the third (soda constraint), and the fourth (cheesecake constraint). Again, the sign restrictions $x_1 \geq 0$, $x_2 \geq 0$, $x_3 \geq 0$, and $x_4 \geq 0$ must be satisfied.

Our discussion shows that the optimal value of x_i may be interpreted as a price for 1 unit of the nutrient associated with the ith dual constraint. Thus, x_1 would be the price for 1 calorie, x_2 would be the price for 1 oz of chocolate, and so on. Again, we see that it is reasonable to associate the ith dual variable (x_i) and the ith primal constraint.

In summary, we have shown that when the primal is a normal max problem or a normal min problem, the dual problem has an intuitive economic interpretation. In Section 6.8, we explain more about the proper interpretation of the dual variables.

PROBLEM

Group A

1 Find the dual of Example 3 in Chapter 3 (an auto company) and give an economic interpretation of the dual problem.

2 Find the dual of Example 2 in Chapter 3 (Dorian Auto) and give an economic interpretation of the dual problem.

6.7 The Dual Theorem and Its Consequences

In this section, we discuss one of the most important results in linear programming: the Dual Theorem. In essence, the Dual Theorem states that the primal and dual have equal optimal objective function values (if the problems have optimal solutions). This result is interesting in its own right, but we will see that in proving the Dual Theorem, we gain many important insights into linear programming.

To simplify the exposition, we assume that the primal is a normal max problem with m constraints and n variables. Then the dual problem will be a normal min problem with m variables and n constraints. In this case, the primal and the dual may be written as follows:

$$\begin{aligned} \max z = c_1x_1 + c_2x_2 + \cdots + c_nx_n \\ \text{s.t.} \quad a_{11}x_1 + a_{12}x_2 + \cdots + a_{1n}x_n \leq b_1 \\ a_{21}x_1 + a_{22}x_2 + \cdots + a_{2n}x_n \leq b_2 \\ \vdots \qquad \vdots \qquad \qquad \vdots \qquad \vdots \\ a_{i1}x_1 + a_{i2}x_2 + \cdots + a_{in}x_n \leq b_i \\ \vdots \qquad \vdots \qquad \qquad \vdots \qquad \vdots \\ a_{m1}x_1 + a_{m2}x_2 + \cdots + a_{mn}x_n \leq b_m \\ x_j \geq 0 \quad (j = 1, 2, \ldots, n) \end{aligned}$$

Primal Problem

(22)

$$\min w = b_1 y_1 + b_2 y_2 + \cdots + b_m y_m$$

$$\text{s.t.} \quad a_{11} y_1 + a_{21} y_2 + \cdots + a_{m1} y_m \geq c_1$$

$$a_{12} y_1 + a_{22} y_2 + \cdots + a_{m2} y_m \geq c_2$$

$$\vdots \qquad \vdots \qquad \qquad \vdots \qquad \vdots$$

Dual Problem (23)

$$a_{1j} y_1 + a_{2j} y_2 + \cdots + a_{mj} y_m \geq c_j$$

$$\vdots \qquad \vdots \qquad \qquad \vdots \qquad \vdots$$

$$a_{1n} y_1 + a_{2n} y_2 + \cdots + a_{mn} y_m \geq c_n$$

$$y_i \geq 0 \quad (i = 1, 2, \ldots, m)$$

Weak Duality

If we choose any feasible solution to the primal and any feasible solution to the dual, the w-value for the feasible dual solution will be at least as large as the z-value for the feasible primal solution. This result is formally stated in Lemma 1.

LEMMA 1

(Weak Duality). Let

$$\mathbf{x} = \begin{bmatrix} x_1 \\ x_2 \\ \vdots \\ x_n \end{bmatrix}$$

be any feasible solution to the primal and $\mathbf{y} = [y_1 \quad y_2 \quad \cdots \quad y_m]$ be any feasible solution to the dual. Then (z-value for $\mathbf{x}$) $\leq$ (w-value for $\mathbf{y}$).

Proof Because $y_i \geq 0$, multiplying the ith primal constraint in (22) by y_i will yield the following valid inequality:

$$y_i a_{i1} x_1 + y_i a_{i2} x_2 + \cdots + y_i a_{in} x_n \leq b_i y_i \quad (i = 1, 2, \ldots, m) \tag{24}$$

Adding the m inequalities in (24), we find that

$$\sum_{i=1}^{i=m} \sum_{j=1}^{j=n} y_i a_{ij} x_j \leq \sum_{i=1}^{i=m} b_i y_i \tag{25}$$

Because $x_j \geq 0$, multiplying the jth dual constraint in (23) by x_j yields the following valid inequality:

$$x_j a_{1j} y_1 + x_j a_{2j} y_2 + \cdots + x_j a_{mj} y_m \geq c_j x_j \quad (j = 1, 2, \ldots, n) \tag{26}$$

Adding the n inequalities in (26) yields

$$\sum_{i=1}^{i=m} \sum_{j=1}^{j=n} y_i a_{ij} x_j \geq \sum_{j=1}^{j=n} c_j x_j \tag{27}$$

Combining (25) and (27), we obtain

$$\sum_{j=1}^{j=n} c_j x_j \leq \sum_{i=1}^{i=m} \sum_{j=1}^{j=n} y_i a_{ij} x_j \leq \sum_{i=1}^{i=m} b_i y_i$$

which is the desired result.

If a feasible solution to either the primal or the dual is readily available, weak duality can be used to obtain a bound on the optimal objective function value for the other problem. For example, in looking at the Dakota problem, it is easy to see that $x_1 = x_2 = x_3 = 1$ is primal feasible. This solution has a z-value of $60 + 30 + 20 = 110$. Weak duality now implies that any dual feasible solution (y_1, y_2, y_3) must satisfy

$$48y_1 + 20y_2 + 8y_3 \geq 110$$

Because the dual is a min problem, and any dual feasible solution must have $w \geq 110$, this means that the optimal w-value for the dual ≥ 110 (see Figure 7). This shows that weak duality enables us to use any primal feasible solution to bound the optimal value of the dual objective function.

Analogously, we can use any feasible solution to the dual to develop a bound on the optimal value of the primal objective function. For example, looking at the Dakota dual, it can readily be verified that $y_1 = 10$, $y_2 = 10$, $y_3 = 0$ is dual feasible. This dual solution has a dual objective function value of $48(10) + 20(10) + 8(0) = 680$. From weak duality, we see that any primal feasible solution

$$\begin{bmatrix} x_1 \\ x_2 \\ x_3 \end{bmatrix}$$

must satisfy

$$60x_1 + 30x_2 + 20x_3 \leq 680$$

Because the primal is a max problem and every primal feasible solution has $z \leq 680$, we may conclude that the optimal primal objective function value ≤ 680 (see Figure 8).

If we define

$$\mathbf{b} = \begin{bmatrix} b_1 \\ b_2 \\ \vdots \\ b_m \end{bmatrix} \quad \text{and} \quad \mathbf{c} = [c_1 \quad c_2 \quad \cdots \quad c_n]$$

then for a point

$$\mathbf{x} = \begin{bmatrix} x_1 \\ x_2 \\ \vdots \\ x_n \end{bmatrix}$$

the primal objective function value may be written as $\mathbf{cx}$, and for a point $\mathbf{y} = [y_1 \quad y_2 \quad \cdots \quad y_m]$ the dual objective function value may be written as $\mathbf{yb}$. We now use weak duality to prove the following important result.

FIGURE 7
Illustration of Weak Duality

$z = 110$

No dual feasible point has $w < 110$ | $w \geq 110$ must hold for all dual feasible points

w

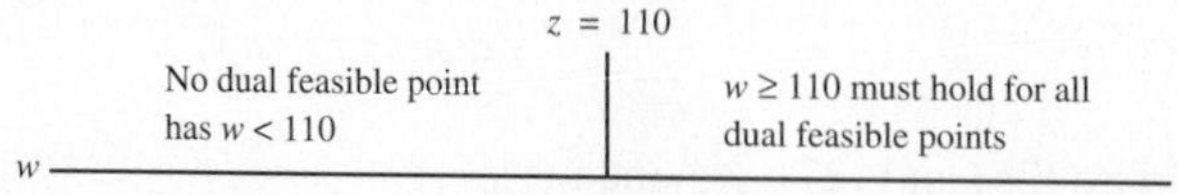

FIGURE 8
Illustration of Weak Duality

$w = 680$

$z \leq 680$ must hold for all primal feasible points | No primal feasible point has $z > 680$

z

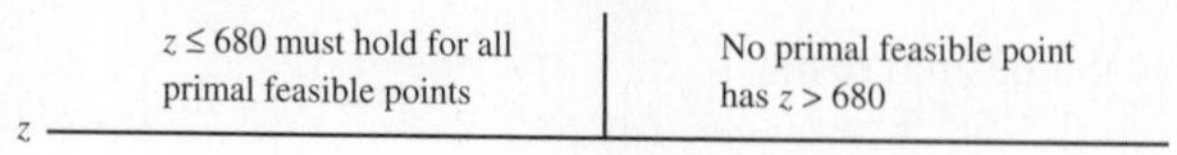

Let

$$\bar{\mathbf{x}} = \begin{bmatrix} \bar{x}_1 \\ \bar{x}_2 \\ \vdots \\ \bar{x}_n \end{bmatrix}$$

be a feasible solution to the primal and $\bar{\mathbf{y}} = [\bar{y}_1 \quad \bar{y}_2 \quad \cdots \quad \bar{y}_m]$ be a feasible solution to the dual. If $\mathbf{c}\bar{\mathbf{x}} = \bar{\mathbf{y}}\mathbf{b}$, then $\bar{\mathbf{x}}$ is optimal for the primal and $\bar{\mathbf{y}}$ is optimal for the dual.

Proof   From weak duality we know that for any primal feasible point $\mathbf{x}$,

$$\mathbf{c}\mathbf{x} \leq \bar{\mathbf{y}}\mathbf{b}$$

Thus, any primal feasible point must yield a z-value that does not exceed $\bar{\mathbf{y}}\mathbf{b}$. Because $\bar{\mathbf{x}}$ is primal feasible and has a primal objective function value of $\mathbf{c}\bar{\mathbf{x}} = \bar{\mathbf{y}}\mathbf{b}$, $\bar{\mathbf{x}}$ must be primal optimal. Similarly, because $\bar{\mathbf{x}}$ is primal feasible, weak duality implies that for any dual feasible point $\mathbf{y}$,

$$\mathbf{c}\bar{\mathbf{x}} \leq \mathbf{y}\mathbf{b}$$

Thus, any dual feasible point must yield an objective function value exceeding $\mathbf{c}\bar{\mathbf{x}}$. Because $\bar{\mathbf{y}}$ is dual feasible and has a dual objective function value $\bar{\mathbf{y}}\mathbf{b} = \mathbf{c}\bar{\mathbf{x}}$, $\bar{\mathbf{y}}$ must be an optimal solution for the dual.

We use the Dakota problem to illustrate the use of Lemma 2. The reader may verify that

$$\bar{\mathbf{x}} = \begin{bmatrix} 2 \\ 0 \\ 8 \end{bmatrix}$$

is primal feasible and that $\bar{\mathbf{y}} = [0 \quad 10 \quad 10]$ is dual feasible. Because $\mathbf{c}\bar{\mathbf{x}} = \bar{\mathbf{y}}\mathbf{b} = 280$, Lemma 2 implies that $\bar{\mathbf{x}}$ is optimal for the Dakota primal, and $\bar{\mathbf{y}}$ is optimal for the Dakota dual. Lemma 2 plays an important role in our proof of the Dual Theorem.

The Dual Theorem

Before proceeding with our proof of the Dual Theorem, we note that weak duality can be used to prove the following results.

If the primal is unbounded, then the dual problem is infeasible.

Proof   See Problem 7 at the end of this section.

If the dual is unbounded, then the primal is infeasible.

Proof   See Problem 8 at the end of this section.

Lemmas 3 and 4 describe the relation between the primal and dual in two relatively unimportant cases.[†]

These cases are of limited interest. We are primarily interested in the relation between the primal and dual when the primal has an optimal solution. In what follows, we let $\bar{z}$ = optimal primal objective function value and $\bar{w}$ = optimal dual objective function value. If the primal has an optimal solution, then the following important result (the Dual Theorem) describes the relation between the primal and the dual.

THEOREM 1

The Dual Theorem

Suppose BV is an optimal basis for the primal. Then $\mathbf{c}_{BV}B^{-1}$ is an optimal solution to the dual. Also, $\bar{z} = \bar{w}$.

Proof The argument used to prove the Dual Theorem includes the following steps:

1 Use the fact that BV is an optimal basis for the primal to show that $\mathbf{c}_{BV}B^{-1}$ is dual feasible.

2 Show that the optimal primal objective function value = the dual objective function for $\mathbf{c}_{BV}B^{-1}$.

3 We have found a primal feasible solution (from BV) and a dual feasible solution ($\mathbf{c}_{BV}B^{-1}$) that have equal objective function values. From Lemma 2, we can now conclude that $\mathbf{c}_{BV}B^{-1}$ is optimal for the dual and $\bar{z} = \bar{w}$.

We now verify step 1 for the case where the primal is a normal maximization problem with n variables and m constraints.[‡] After adding slack variables $s_1, s_2, \ldots, s_m$ to the primal, we write the primal and dual problems as follows:

Primal Problem

$$\max z = c_1 x_1 + c_2 x_2 + \cdots + c_n x_n$$

$$\begin{aligned}
\text{s.t.} \quad a_{11}x_1 + a_{12}x_2 + \cdots + a_{1n}x_n + s_1 \quad\quad\quad &= b_1 \\
a_{21}x_1 + a_{22}x_2 + \cdots + a_{2n}x_n \quad\quad + s_2 \quad\quad &= b_2 \\
\vdots \quad\quad \vdots \quad\quad\quad \vdots \quad\quad\quad\quad &\quad\quad \vdots \\
a_{m1}x_1 + a_{m2}x_2 + \cdots + a_{mn}x_n \quad\quad\quad\quad + s_m &= b_m
\end{aligned}$$

$$x_j \geq 0 \quad (j = 1, 2, \ldots, n); \quad s_i \geq 0 \quad (i = 1, 2, \ldots, m)$$

(28)

Dual Problem

$$\min w = b_1 y_1 + b_2 y_2 + \cdots + b_m y_m$$

$$\begin{aligned}
\text{s.t.} \quad a_{11}y_1 + a_{21}y_2 + \cdots + a_{m1}y_m &\geq c_1 \\
a_{12}y_1 + a_{22}y_2 + \cdots + a_{m2}y_m &\geq c_2 \\
\vdots \quad\quad \vdots \quad\quad\quad \vdots \quad\quad\quad\quad &\quad\quad \vdots \\
a_{1n}y_1 + a_{2n}y_2 + \cdots + a_{mn}y_m &\geq c_n
\end{aligned}$$

$$y_i \geq 0 \quad (i = 1, 2, \ldots, m)$$

(29)

Let BV be an optimal basis for the primal, and define $\mathbf{c}_{BV}B^{-1} = [y_1, y_2, \ldots, y_m]$. Thus, for the optimal basis BV, y_i is the ith element of $\mathbf{c}_{BV}B^{-1}$. BV is optimal for

---

[†]It can happen that both the primal and the dual can be infeasible, as in the following example:

Primal
$$\begin{aligned}
\max z &= x_2 \\
\text{s.t.} \quad x_1 \quad\quad &\leq -1 \\
-x_2 &\leq 1 \\
x_1, x_2 &\geq 0
\end{aligned}$$

Dual
$$\begin{aligned}
\min w &= -y_1 + y_2 \\
\text{s.t.} \quad y_1 \quad &\geq 0 \\
-y_2 &\geq 1 \\
y_1, y_2 &\geq 0
\end{aligned}$$

[‡]Our proof can easily be modified to handle the situation where the primal is not a normal max problem.

the primal, so the coefficient of each variable in row 0 of BV's primal tableau must be nonnegative. From (10), the coefficient of x_j in row 0 of the BV tableau ($\bar{c}_j$) is given by

$$\bar{c}_j = \mathbf{c}_{BV}B^{-1}\mathbf{a}_j - c_j$$

$$= [y_1 \quad y_2 \quad \cdots \quad y_m] \begin{bmatrix} a_{1j} \\ a_{2j} \\ \vdots \\ a_{mj} \end{bmatrix} - c_j$$

$$= y_1 a_{1j} + y_2 a_{2j} + \cdots + y_m a_{mj} - c_j$$

But we know that $\bar{c}_j \geq 0$, so for $j = 1, 2, \ldots, n$,

$$y_1 a_{1j} + y_2 a_{2j} + \cdots + y_m a_{mj} - c_j \geq 0$$

Thus, $\mathbf{c}_{BV}B^{-1}$ satisfies each of the n dual constraints.

Because BV is an optimal basis for the primal, we also know that each slack variable has a nonnegative coefficient in the BV primal tableau. From (10'), we find that the coefficient of s_i in BV's row 0 is y_i, the ith element of $\mathbf{c}_{BV}B^{-1}$. Thus, for $i = 1, 2, \ldots, m$, $y_i \geq 0$. We have shown that $\mathbf{c}_{BV}B^{-1}$ satisfies all n constraints in (29) and that all the elements of $\mathbf{c}_{BV}B^{-1}$ are nonnegative. Thus, $\mathbf{c}_{BV}B^{-1}$ is indeed dual feasible.

Step 2 of the Dual Theorem proof requires that we show

Dual objective function value for $\mathbf{c}_{BV}B^{-1}$

$$= \text{primal objective function value for BV} \qquad \textbf{(30)}$$

From (11), we know that the primal objective function value for BV is $\mathbf{c}_{BV}B^{-1}\mathbf{b}$. But the dual objective function value for the dual feasible solution $\mathbf{c}_{BV}B^{-1}$ is

$$b_1 y_1 + b_2 y_2 + \cdots + b_m y_m = [y_1 \quad y_2 \quad \cdots \quad y_m] \begin{bmatrix} b_1 \\ b_2 \\ \vdots \\ b_m \end{bmatrix} = \mathbf{c}_{BV}B^{-1}\mathbf{b}$$

Thus, (30) is valid.

We have shown that steps 1 and 2 of the Dual Theorem proof are valid. Step 3 now completes our proof of the Dual Theorem.

REMARKS 1 In step 1 of the Dual Theorem proof, we showed that a basis BV that is feasible for the primal is optimal if and only if $\mathbf{c}_{BV}B^{-1}$ is dual feasible. In Section 6.9, we use this result to gain useful insights into sensitivity analysis.

2 When we find the optimal solution to the primal by using the simplex algorithm, we have also found the optimal solution to the dual.

To see why Remark 2 is true, suppose that the primal is a normal max problem with m constraints. To use the simplex to solve this problem, we must add a slack variable s_i to the ith primal constraint. Suppose BV is an optimal basis for the primal. Then the Dual Theorem tells us that $\mathbf{c}_{BV}B^{-1} = [y_1 \quad y_2 \quad \cdots \quad y_m]$ is the optimal solution to the dual. Recall from (10'), however, that y_i is the coefficient of s_i in row 0 of the optimal (BV) primal tableau. Thus, we have shown that *if the primal is a normal max problem, then the optimal value of the ith dual variable is the coefficient of s_i in row 0 of the optimal primal tableau.*

TABLE **23**
Optimal Solution to the Dakota Problem

					Basic Variable
z	$+ \quad 5x_2$	$+ \, 10s_2 + 10s_3 = 280$			$z_1 = 280$
	$- \quad 2x_2$	$+ \, s_1 + \quad 2s_2 - \quad 8s_3 = 24$			$s_1 = 24$
	$- \quad 2x_2 + x_3$	$+ \quad 2s_2 - \quad 4s_3 = 8$			$x_3 = 8$
x_1	$+ \, 1.25x_2$	$- \, 0.5s_2 + 1.5s_3 = 2$			$x_1 = 2$

We use the Dakota problem to illustrate Remark 2. The optimal tableau for the Dakota problem is shown in Table 23. The optimal primal solution is $z = 280$, $s_1 = 24$, $x_3 = 8$, $x_1 = 2$, $x_2 = 0$, $s_2 = 0$, $s_3 = 0$. From the preceding discussion, the optimal dual solution is $y_1 = 0$, $y_2 = 10$, $y_3 = 10$, $w = 48(0) + 20(10) + 8(10) = 280$. Observe that the optimal primal and dual objective function values are equal, as required by the Dual Theorem.

Of course, we may always compute the optimal dual solution directly by solving

$$\mathbf{c}_{BV}B^{-1} = [0 \quad 20 \quad 60] \begin{bmatrix} 1 & 2 & -8 \\ 0 & 2 & -4 \\ 0 & -0.5 & 1.5 \end{bmatrix} = [0 \quad 10 \quad 10]$$

Of course, the two methods of obtaining the dual solution agree.

If the primal has $\geq$ or equality constraints, then we can still find the optimal dual solution from the optimal primal tableau. To see how this is done, recall that the Dual Theorem tells us that the optimal value of the ith dual variable (y_i) is the ith element of $\mathbf{c}_{BV}B^{-1}$. From (10″), we see that if the ith constraint of the primal is a $\geq$ constraint, then

Optimal value of ith dual variable $= y_i = -$(coefficient of e_i in the optimal row 0)

The coefficient of e_i in the optimal row 0 must be nonnegative, so this shows that if the ith constraint in the primal is a $\geq$ constraint, then $y_i \leq 0$. This agrees with our previous convention (see Section 6.5) that a $\geq$ constraint must have a nonpositive dual variable. From (10‴), we see that if the ith primal constraint is an equality constraint, then

$$y_i = \text{(coefficient of } a_i \text{ in optimal row 0)} - M$$

Although the coefficient of a_i in the optimal row 0 must be nonnegative, the fact that M is a large positive number means that $y_i \geq 0$ or $y_i \leq 0$ is possible. This agrees with our previous convention, which stated that the dual variable for an equality constraint is urs.

How to Read the Optimal Dual Solution from Row 0 of the Optimal Tableau If the Primal Is a Max Problem

$\dfrac{\text{Optimal value of dual variable } y_i}{\text{if Constraint } i \text{ is a} \leq \text{constraint}} = \text{coefficient of } s_i \text{ in optimal row 0}$ (31)

$\dfrac{\text{Optimal value of dual variable } y_i}{\text{if Constraint } i \text{ is a} \geq \text{constraint}} = -\text{(coefficient of } e_i \text{ in optimal row 0)}$ (31′)

$\dfrac{\text{Optimal value of dual variable } y_i}{\text{if Constraint } i \text{ is an equality constraint}} = \text{(coefficient of } a_i \text{ in optimal row 0)} - M$ (31″)

The following example illustrates how to find the optimal dual solution to a problem with $\leq$, $\geq$, and equality constraints.

EXAMPLE 10   **Finding the Dual Solution to a Nonnormal Max Problem**

To solve the following LP,

$$\max z = 3x_1 + 2x_2 + 5x_3$$
$$\text{s.t.} \quad x_1 + 3x_2 + 2x_3 \leq 15$$
$$2x_2 - x_3 \geq 5 \tag{32}$$
$$2x_1 + x_2 - 5x_3 = 10$$
$$x_1, x_2, x_3 \geq 0$$

we add a slack variable s_1, subtract an excess variable e_2, and add two artificial variables a_2 and a_3. The optimal tableau for (32) is given in Table 24. From this tableau, the optimal solution is $z = \frac{565}{23}$, $x_3 = \frac{15}{23}$, $x_2 = \frac{65}{23}$, $x_1 = \frac{120}{23}$, $s_1 = e_2 = a_2 = a_3 = 0$. Use this information to find the optimal solution to the dual of (32).

Solution   Following the steps in Section 6.5, we find the dual of (32) from the tableau in Table 25:

$$\min w = 15y_1 + 5y_2 + 10y_3$$
$$\text{s.t.} \quad y_1 \quad\quad + 2y_3 \geq 3$$
$$3y_1 + 2y_2 + y_3 \geq 2 \tag{33}$$
$$2y_1 - y_2 - 5y_3 \geq 5$$
$$y_1 \geq 0, y_2 \leq 0, y_3 \text{ urs}$$

From (31) and the optimal primal tableau, we can find the optimal solution to (33) as follows:
Because the first primal constraint is a $\leq$ constraint, we see from (31) that $y_1 = $ coefficient of s_1 in optimal row $0 = \frac{51}{23}$. The second primal constraint is a $\geq$ constraint, so we see from (31') that $y_2 = -(\text{coefficient of } e_2 \text{ in optimal row 0}) = -\frac{58}{23}$. Because the third constraint is an equality constraint, we see from (31'') that $y_3 = (\text{coefficient of } a_3 \text{ in the optimal row 0}) - M = \frac{9}{23}$.

TABLE 24
Optimal Tableau for LP (32)

z	x_1	x_2	x_3	s_1	e_2	a_2	a_3	rhs	Basic Variable
1	0	0	0	$\frac{51}{23}$	$\frac{58}{23}$	$M - \frac{58}{23}$	$M + \frac{9}{23}$	$\frac{565}{23}$	$z = \frac{565}{23}$
0	0	0	1	$\frac{4}{23}$	$\frac{5}{23}$	$-\frac{5}{23}$	$-\frac{2}{23}$	$\frac{15}{23}$	$x_3 = \frac{15}{23}$
0	0	1	0	$\frac{2}{23}$	$-\frac{9}{23}$	$\frac{9}{23}$	$-\frac{1}{23}$	$\frac{65}{23}$	$x_2 = \frac{65}{23}$
0	1	0	0	$\frac{9}{23}$	$\frac{17}{23}$	$-\frac{17}{23}$	$\frac{7}{23}$	$\frac{120}{23}$	$x_1 = \frac{120}{23}$

TABLE 25
Finding the Dual of LP (32)

min w		max z		
		$(x_1 \geq 0)$	$(x_2 \geq 0)$	$(x_3 \geq 0)$
		x_1	x_2	x_3	
$(y_1 \geq 0)$	y_1	1	3	2	≤ 15
$(y_2 \leq 0)$	y_2	0	2	-1	$\geq 5*$
$(y_3 \text{ urs})$	y_3	2	1	-5	$= 10*$
		≥ 3	≥ 2	≥ 5	

By the Dual Theorem, the optimal dual objective function value w must equal $\frac{565}{23}$. In summary, the optimal dual solution is

$$\overline{w} = \tfrac{565}{23}, \; y_1 = \tfrac{51}{23}, \; y_2 = -\tfrac{58}{23}, \; y_3 = \tfrac{9}{23}$$

The reader should check that this solution is indeed feasible (all dual constraints are satisfied with equality) and that

$$\overline{w} = 15\left(\tfrac{51}{23}\right) + 5\left(-\tfrac{58}{23}\right) + 10\left(\tfrac{9}{23}\right) = \tfrac{565}{23}$$

Even if the primal is a min problem, we may read the optimal dual solution from the optimal primal tableau.

How to Read the Optimal Dual Solution from Row 0 of the Optimal Tableau If the Primal Is a Min Problem

$$\begin{array}{ll} \text{Optimal value of dual variable } x_i \\ \text{if Constraint } i \text{ is a} \leq \text{constraint} \end{array} = \text{coefficient of } s_i \text{ in optimal row 0}$$

$$\begin{array}{ll} \text{Optimal value of dual variable } x_i \\ \text{if Constraint } i \text{ is a} \geq \text{constraint} \end{array} = -(\text{coefficient of } e_i \text{ in optimal row 0})$$

$$\begin{array}{ll} \text{Optimal value of dual variable } x_i \\ \text{if Constraint } i \text{ is an equality} \\ \text{constraint} \end{array} = (\text{coefficient of } a_i \text{ in optimal row 0}) + M$$

To illustrate how the optimal solution to the dual of a min problem may be read from the optimal primal tableau, consider

$$\min w = 3y_1 + 2y_2 + y_3$$
$$\text{s.t.} \quad y_1 + y_2 + \; y_3 \geq 4$$
$$\quad\quad\quad\quad y_2 - \; y_3 \leq 2$$
$$y_1 + y_2 + 2y_3 = 6$$
$$y_1, y_2, y_3 \geq 0$$

The optimal tableau for this problem is given in Table 26. Thus, the optimal primal solution is $w = 6$, $y_2 = y_3 = 2$, $y_1 = 0$. The dual of the preceding LP is

$$\max z = 4x_1 + 2x_2 + 6x_3$$
$$\text{s.t.} \quad x_1 \quad\quad + \; x_3 \leq 3$$
$$\quad\quad x_1 + x_2 + \; x_3 \leq 2$$
$$\quad\quad x_1 - x_2 + 2x_3 \leq 1$$
$$x_1 \geq 0, \, x_2 \leq 0, \, x_3 \text{ urs}$$

TABLE 26
Finding the Optimal Solution to the Dual When Primal Is a Min Problem

w	y_1	y_2	y_3	e_1	s_2	a_1	a_3	rhs
1	−1	0	0	−3	0	$3 - M$	$-1 - M$	6
0	1	1	0	−2	0	2	−1	2
0	−1	0	0	3	1	−3	2	2
0	0	0	1	1	0	−1	1	2

From the optimal primal tableau, we find that the optimal dual solution is $z = 6$, $x_1 = 3$, $x_2 = 0$, $x_3 = -1$.

PROBLEMS

Group A

1 The following questions refer to the Giapetto problem (see Problem 7 of Section 6.3).

a Find the dual of the Giapetto problem.

b Use the optimal tableau of the Giapetto problem to determine the optimal dual solution.

c Verify that the Dual Theorem holds in this instance.

2 Consider the following LP:

$$\max z = -2x_1 - x_2 + x_3$$
$$\text{s.t.} \quad x_1 + x_2 + x_3 \le 3$$
$$x_2 + x_3 \ge 2$$
$$x_1 \qquad + x_3 = 1$$
$$x_1, x_2, x_3 \ge 0$$

a Find the dual of this LP.

b After adding a slack variable s_1, subtracting an excess variable e_2, and adding artificial variables a_2 and a_3, row 0 of the LP's optimal tableau is found to be

$$z + 4x_1 + e_2 + (M - 1)a_2 + (M + 2)a_3 = 0$$

Find the optimal solution to the dual of this LP.

3 For the following LP,

$$\max z = -x_1 + 5x_2$$
$$\text{s.t.} \quad x_1 + 2x_2 \le 0.5$$
$$-x_1 + 3x_2 \le 0.5$$
$$x_1, x_2 \ge 0$$

row 0 of the optimal tableau is $z + 0.4s_1 + 1.4s_2 = ?$ Determine the optimal z-value for the given LP.

4 The following questions refer to the Bevco problem of Section 4.10.

a Find the dual of the Bevco problem.

b Use the optimal tableau for the Bevco problem that is given in Section 4.10 to find the optimal solution to the dual. Verify that the Dual Theorem holds in this instance.

5 Consider the following linear programming problem:

$$\max z = 4x_1 + x_2$$
$$\text{s.t.} \quad 3x_1 + 2x_2 \le 6$$
$$6x_1 + 3x_2 \le 10$$
$$x_1, x_2 \ge 0$$

Suppose that in solving this problem, row 0 of the optimal tableau is found to be $z + 2x_2 + s_2 = \frac{20}{3}$. Use the Dual Theorem to prove that the computations must be incorrect.

6 Show that (for a max problem) if the ith primal constraint is a $\ge$ constraint, then the optimal value of the ith dual variable may be written as (coefficient of a_i in optimal row 0) $- M$.

Group B

7 In this problem, we use weak duality to prove Lemma 3.

a Show that Lemma 3 is equivalent to the following: If the dual is feasible, then the primal is bounded. (*Hint:* Do you remember, from plane geometry, what the contrapositive is?)

b Use weak duality to show the validity of the form of Lemma 3 given in part (a). (*Hint:* If the dual is feasible, then there must be a dual feasible point having a w-value of, say, w_o. Now use weak duality to show that the primal is bounded.)

8 Following along the lines of Problem 7, use weak duality to prove Lemma 4.

9 Use the information given in Problem 8 of Section 6.3 to determine the dual of the Dorian Auto problem and its optimal solution.

6.8 Shadow Prices

We now return to the concept of shadow price that was discussed in Section 6.1. A more formal definition follows.

DEFINITION ■ The **shadow price** of the ith constraint is the amount by which the optimal z-value is improved (increased in a max problem and decreased in a min problem) if we increase b_i by 1 (from b_i to $b_i + 1$).[†]

[†]This assumes that after the right-hand side of Constraint i has been changed to $b_i + 1$, the current basis remains optimal.

By using the Dual Theorem, we can easily determine the shadow price of the ith constraint. To illustrate, we find the shadow price of the second constraint (finishing hours) of the Dakota problem. Let $\mathbf{c}_{BV}B^{-1} = [y_1 \ \ y_2 \ \ y_3] = [0 \ \ 10 \ \ 10]$ be the optimal solution to the dual of the max problem. From the Dual Theorem, we know that

Optimal z-value when rhs of constraints are ($b_1 = 48, b_2 = 20, b_3 = 8$)

$$= 48y_1 + 20y_2 + 8y_3 \quad \textbf{(34)}$$

What happens to the optimal z-value for the Dakota problem if b_2 (currently 20 finishing hours) is increased by 1 unit (to 21 hours)? We know that changing a right-hand side may cause the current basis to no longer be optimal (see Section 6.3). For the moment, however, we assume that the current basis remains optimal when we increase b_2 by 1. Then $\mathbf{c}_{BV}$ and B^{-1} remain unchanged, so the optimal solution to the dual of the Dakota problem remains unchanged.

We next find

Optimal z-value when rhs of finishing constraint is $21 = 48y_1 + 21y_2 + 8y_3$ \quad \textbf{(35)}

Subtracting (34) from (35) yields

Change in optimal z-value if finishing hours are increased by 1

= shadow price for finishing constraint 2 \qquad\qquad\qquad\qquad\qquad **(36)**

$= y_2 = 10$

This example shows that *the shadow price of the ith constraint of a max problem is the optimal value of the ith dual variable.* The shadow prices are the dual variables, so we know that the shadow price for a $\leq$ constraint will be nonnegative; for a $\geq$ constraint, nonpositive; and for an equality constraint, unrestricted in sign. The examples discussed later in this section give intuitive justifications for these sign conventions.

Similar reasoning can be used to show that if (in a maximization problem) the right-hand side of the ith constraint is increased by an amount Δb_i, then (assuming the current basis remains optimal) the new optimal z-value may be found from

New optimal z-value = old optimal z-value + Δb_i(Constraint i shadow price) \quad **(37)**

For a minimization problem, the shadow price of the ith constraint is the amount by which a unit increase in the right-hand side improves, or decreases, the optimal z-value (assuming that the current basis remains optimal). It can be shown that the shadow price of the ith constraint of a min problem = −(optimal value of the ith dual variable). If the right-hand side is increased by an amount Δb_i, then (assuming the current basis remains optimal) the new optimal z-value may be found from

New optimal z-value = old optimal z-value − Δb_i(Constraint i shadow price) \quad **(37′)**

The following three examples should clarify the shadow price concept.

EXAMPLE 11　　**Shadow Prices for Normal Max Problem**

For the Dakota problem:

1　Find and interpret the shadow prices

2　If 18 finishing hours were available, what would be Dakota's revenue? (It can be shown by the methods of Section 6.3 that if $16 \leq$ finishing hours ≤ 24, the current basis remains optimal.)

3　If 9 carpentry hours were available, what would be Dakota's revenue? (For $\frac{20}{3} \leq$ carpentry hours ≤ 10, the current basis remains optimal.)

4 If 30 board feet of lumber were available, what would be Dakota's revenue? (For $24 \leq$ lumber $\leq \infty$, the current basis remains optimal.)

5 If 30 carpentry hours were available, why couldn't the shadow price for the carpentry constraint be used to determine the new z-value?

Solution
1 In Section 6.7, we found the optimal solution to the Dakota dual to be $y_1 = 0$, $y_2 = 10$, $y_3 = 10$. Thus, the shadow price for the lumber constraint is 0; for the finishing constraint, 10; and for the carpentry constraint, 10. The fact that the lumber constraint has a shadow price of 0 means that increasing the amount of available lumber by 1 board foot (or *any* amount) will not increase revenue. This is reasonable because we are currently using only 24 of the available 48 board feet of lumber, so adding any more will not do Dakota any good. Dakota's revenue would increase by $10 if 1 more finishing hour were available. Similarly, 1 more carpentry hour would increase Dakota's revenue by $10. In this problem, the shadow price of the ith constraint may be thought of as the maximum amount that the company would pay for an extra unit of the resource associated with the ith constraint. For example, an extra carpentry hour would raise revenue by $y_3 = \$10$ (see Example 12 for a max problem in which this interpretation is invalid). Thus, Dakota could pay up to $10 for an extra carpentry hour and still be better off. Similarly, the company would be willing to pay nothing ($0) for an extra board foot of lumber and up to $10 for an extra finishing hour. To answer questions 2–4, we apply (37), using the fact that the old z-value $= 280$.

2 $y_2 = 10$, $\Delta b_2 = 18 - 20 = -2$. The current basis is still optimal because $16 \leq 18 \leq 24$. Then (37) yields (new revenue) $= 280 + 10(-2) = \$260$.

3 $y_3 = 10$, $\Delta b_3 = 9 - 8 = 1$. Because $\frac{20}{3} \leq 9 \leq 10$, the current basis remains optimal. Then (37) yields (new revenue) $= 280 + 10(1) = \$290$.

4 $y_1 = 0$, $\Delta b_1 = 30 - 48 = -18$. Because $24 \leq 30 \leq \infty$, the current basis is still optimal. Then (37) yields (new revenue) $= 280 + 0(-18) = \$280$.

5 If $b_3 = 30$, the current basis is no longer optimal, because $30 > 10$. This means that BV (and therefore $\mathbf{c}_{BV}B^{-1}$) changes, and we cannot use the current set of shadow prices to determine the new revenue level.

Intuitive Explanation of the Sign of Shadow Prices

We can now give an intuitive explanation of why (in a max problem) the shadow price of a $\leq$ constraint will always be nonnegative. Consider the following situation: We are given two LP max problems (LP 1 and LP 2) that have the same objective functions. Suppose that every point that is feasible for LP 1 is also feasible for LP 2. This means that LP 2's feasible region contains all the points in LP 1's feasible region and possibly some other points. Then the optimal z-value for LP 2 must be at least as large as the optimal z-value for LP 1. To see this, suppose that point x' (with z-value z') is optimal for LP 1. Because x' is also feasible for LP 2 (which has the same objective function as LP 1), LP 2 can attain a z-value of z' (by using the feasible point x'). It is also possible that by using one of the points feasible for only LP 2 (and not for LP 1), LP 2 might do better than z'. In short, *adding points to the feasible region of a max problem cannot decrease the optimal z-value.*

We can use this observation to show why a $\leq$ constraint must have a nonnegative shadow price. For the Dakota problem, if we increase the right-hand side of the carpentry constraint by 1 (from 8 to 9), we see that all points that were originally feasible remain feasible, and some new points (which use > 8 and ≤ 9 carpentry hours) may be feasible. Thus, the optimal z-value cannot decrease, and the shadow price for the carpentry constraint must be nonnegative.

The purpose of the following example is to show that (contrary to what many books say) the shadow price of a $\leq$ constraint is not always the maximum price you would be willing to pay for an additional unit of a resource.

EXAMPLE 12   Shadow Price as a Premium

Leatherco manufactures belts and shoes. A belt requires 2 square yards of leather and 1 hour of skilled labor. A pair of shoes requires 3 sq yd of leather and 2 hours of skilled labor. As many as 25 sq yd of leather and 15 hours of skilled labor can be purchased at a price of $5/sq yd of leather and $10/hour of skilled labor. A belt sells for $23, and a pair of shoes sells for $40. Leatherco wants to maximize profits (revenues $-$ costs). Formulate an LP that can be used to maximize Leatherco's profits. Then find and interpret the shadow prices for this LP.

Solution   Define

$$x_1 = \text{number of belts produced}$$
$$x_2 = \text{number of pairs of shoes produced}$$

After noting that

$$\text{Cost/belt} = 2(5) + 1(10) = \$20$$
$$\text{Cost/pair of shoes} = 3(5) + 2(10) = \$35$$

we find that Leatherco's objective function is

$$\text{max } z = (23 - 20)x_1 + (40 - 35)x_2 = 3x_1 + 5x_2$$

Leatherco faces the following two constraints:

Constraint 1   Leatherco can use at most 25 sq yd of leather.

Constraint 2   Leatherco can use at most 15 hours of skilled labor.

Constraint 1 is expressed by

$$2x_1 + 3x_2 \leq 25 \qquad \text{(Leather constraint)}$$

while Constraint 2 is expressed by

$$x_1 + 2x_2 \leq 15 \qquad \text{(Skilled-labor constraint)}$$

After adding the sign restrictions $x_1 \geq 0$ and $x_2 \geq 0$, we obtain the following LP:

$$\text{max } z = 3x_1 + 5x_2$$
$$\text{s.t.} \quad 2x_1 + 3x_2 \leq 25 \qquad \text{(Leather constraint)}$$
$$\quad x_1 + 2x_2 \leq 15 \qquad \text{(Skilled-labor constraint)}$$
$$\quad x_1, x_2 \geq 0$$

After adding slack variables s_1 and s_2 to the leather and skilled-labor constraints, respectively, we obtain the optimal tableau shown in Table 27. Thus, the optimal solution to Leatherco's problem is $z = 40$, $x_1 = 5$, $x_2 = 5$. The shadow prices are

$$y_1 = \text{leather shadow price} = \text{coefficient of } s_1 \text{ in optimal row } 0 = 1$$
$$y_2 = \text{skilled-labor shadow price} = \text{coefficient of } s_2 \text{ in optimal row } 0 = 1$$

TABLE 27
Optimal Tableau for Leatherco

		Basic Variable
z	$+\ s_1 +\ s_2 = 40$	$z\ =\ 40$
x_1	$+\ 2s_1 -\ 3s_2 = 5$	$x_1\ =\ 5$
	$x_2 -\ s_1 + 2s_2 = 5$	$x_2\ =\ 5$

The meaning of the leather shadow price is that if one more square yard of leather were available, then Leatherco's objective function (profits) would increase by \$1. Let's look further at what happens if an additional square yard of leather is available. Because s_1 is nonbasic, the extra square yard of leather will be purchased. Also, because s_2 is nonbasic, we will still use all available labor. This means that the \$1 increase in profits includes the cost of purchasing an extra square yard of leather. If the availability of an extra square yard of leather increases profits by \$1, then it must be increasing revenue by $1 + 5 = \$6$. Thus, the maximum amount Leatherco should pay for an extra square yard of leather is \$6 (not \$1).

Another way to see this is as follows: If we purchase another square yard of leather at the current price of \$5, profits increase by $y_1 = \$1$. If we purchase another square yard of leather at a price of $\$6 = \$5 + \$1$, then profits increase by $\$1 - \$1 = \$0$. Thus, the most Leatherco would be willing to pay for an extra square yard of leather is \$6.

Similarly, the most Leatherco would be willing to pay for an extra hour of labor is $y_2 +$ (cost of an extra hour of skilled labor) $= 1 + 10 = \$11$. In this problem, we see that the shadow price for a resource represents the *premium* over and above the cost of the resource that Leatherco would be willing to pay for an extra unit of resource.

The two preceding examples show that we must be careful when interpreting the shadow price of a $\leq$ constraint. Remember that the shadow price for a constraint in a max problem is the amount by which the objective function increases if the right-hand side is increased by 1.

The following example illustrates the interpretation of the shadow prices of $\geq$ and equality constraints.

EXAMPLE 13   Shadow Prices for $\geq$ and $=$ Constraints

Steelco has received an order for 100 tons of steel. The order must contain at least 3.5 tons of nickel, at most 3 tons of carbon, and exactly 4 tons of manganese. Steelco receives \$20/ton for the order. To fill the order, Steelco can combine four alloys, whose chemical composition is given in Table 28. Steelco wants to maximize the profit (revenues − costs) obtained from filling the order. Formulate the appropriate LP. Also find and interpret the shadow prices for each constraint.

Solution   After we define x_i = number of tons of alloy i used to fill the order, Steelco's LP is seen to be

$$\max z = (20 - 12)x_1 + (20 - 10)x_2 + (20 - 8)x_3 + (20 - 6)x_4$$

$$\begin{aligned}
\text{s.t.} \quad & 0.06x_1 + 0.03x_2 + 0.02x_3 + 0.01x_4 \geq 3.5 && \text{(Nickel constraint)} \\
& 0.03x_1 + 0.02x_2 + 0.05x_3 + 0.06x_4 \leq 3 && \text{(Carbon constraint)} \\
& 0.08x_1 + 0.03x_2 + 0.02x_3 + 0.01x_4 = 4 && \text{(Manganese constraint)} \\
& x_1 + \quad x_2 + \quad x_3 + \quad x_4 = 100 && \text{(Order size = 100 tons)} \\
& x_1, x_2, x_3, x_4 \geq 0
\end{aligned}$$

TABLE 28

Relevant Information for Steelco

Cement	Alloy (%)			
	1	2	3	4
Nickel	6	3	2	1
Carbon	3	2	5	6
Manganese	8	3	2	1
Cost/ton ($)	12	10	8	6

After adding a slack variable s_2, subtracting an excess variable e_1, and adding artificial variables a_1, a_3, and a_4, the following optimal solution is obtained: $z = 1,000$, $s_2 = 0.25$, $x_1 = 25$, $x_2 = 62.5$, $x_4 = 12.5$, $e_1 = 0$, $x_3 = 0$. The optimal row 0 is

$$z + 400e_1 + (M - 400)a_1 + (M + 200)a_3 + (M + 16)a_4 = 1,000$$

Using (31), (31′), and (31″), we obtain

Shadow price of nickel constraint = −(coefficient of e_1 in optimal row 0)

= −400

Shadow price of carbon constraint = coefficient of s_2 in optimal row 0

= 0

Shadow price of manganese constraint = (coefficient of a_3 in optimal row 0) − M

= 200

Shadow price of order size constraint = (coefficient of a_4 in optimal row 0) − M

= 16

By the sensitivity analysis procedures of Section 6.3, it can be shown that the current basis remains optimal if $3.46 \leq b_1 \leq 3.6$. As long as the nickel requirement is in this range, increasing the nickel requirement by an amount Δb_1 will increase Steelco's profits by $-400 \, \Delta b_1$. For example, increasing the nickel requirement to 3.55 tons ($\Delta b_1 = 0.05$) would "increase" (actually decrease) profits by $-400(0.05) = \$20$. The nickel constraint has a negative shadow price because increasing the right-hand side of the nickel constraint makes it harder to satisfy the nickel constraint. In fact, an increase in the nickel requirement forces Steelco to use more of the expensive type 1 alloy. This raises costs and lowers profits. As we have already seen, the shadow price of a $\geq$ constraint (in a max problem) will always be nonpositive, because increasing the right-hand side of a $\geq$ constraint eliminates points from the feasible region. Thus, the optimal z-value must decrease or remain unchanged.

By the Section 6.3 sensitivity analysis procedures, for $2.75 \leq b_2 \leq \infty$, the current basis remains optimal. As stated before, the carbon constraint has a zero shadow price. This means that if we increase Steelco's carbon requirement, Steelco's profit will not change. Intuitively, this is because our present optimal solution contains only $2.75 < 3$ tons of carbon. Thus, relaxing the carbon requirement won't enable Steelco to reduce costs, so Steelco's profit will remain unchanged.

By the sensitivity analysis procedures, the current basis remains optimal if $3.83 \leq b_3 \leq 4.07$. The shadow price of the third (manganese) constraint is 200, so we know that as long as the manganese requirement remains in the given range, increasing it by an amount of Δb_3 will increase profit by $200\Delta b_3$. For example, if the manganese requirement were 4.05 tons ($\Delta b_3 = 0.05$), then profits would increase by $(0.05)200 = \$10$.

By the sensitivity analysis procedures, the current basis remains optimal if $91.67 \leq b_4 \leq 103.12$. Because the shadow price of the fourth (order size) constraint is 16, increasing the order size by Δb_4 tons (with nickel, carbon, and manganese requirements unchanged) would increase profits by $16\Delta b_4$. For example, the profit from a 103-ton order that required ≥ 3.5 tons of nickel, ≤ 3 tons of carbon, and exactly 4 tons of manganese would be $1,000 + 3(16) = \$1,048$.

In this problem, both equality constraints had positive shadow prices. In general, we know that it is possible for an equality constraint's dual variable (and shadow price) to be negative. If this occurs, then the equality constraint will have a negative shadow price. To illustrate this possibility, suppose that Steelco's customer required exactly 4.5 tons of manganese in the order. Because $4.5 > 4.07$, the current basis is no longer optimal. If we solve Steelco's LP again, it can be shown that the shadow price for the manganese constraint has changed to -54.55. This means that an increase in the manganese requirement will decrease Steelco's profits.

Interpretation of the Dual Prices Column of the LINDO Output

For a max problem, LINDO gives the values of the shadow prices in the DUAL PRICES column of the output. The dual price for row $i + 1$ on the LINDO output is the shadow price for the ith constraint and the optimal value for the ith dual variable. Thus, in Figure 4, we see that for the Dakota problem,

$$y_1 = \text{shadow price for lumber constraint} = \text{row 2 dual price} = 0$$
$$y_2 = \text{shadow price for finishing constraint} = \text{row 3 dual price} = 10$$
$$y_3 = \text{shadow price for carpentry constraint} = \text{row 4 dual price} = 10$$

For a maximization problem, the vector $c_{BV}B^{-1}$ (needed for pricing out new activities) is the same as the vector of dual prices given in the LINDO output. For the Dakota problem, we would price out new activities using $c_{BV}B^{-1} = \begin{bmatrix} 0 & 10 & 10 \end{bmatrix}$.

For a minimization problem, the entry in the DUAL PRICE column for any constraint is the shadow price. Thus, from the LINDO printout in Figure 6, we find that the shadow prices for the constraints in the diet problem are as follows: calorie $= 0$; chocolate $= -2.5¢$; sugar $= -7.5¢$; and fat $= 0$. This implies that

1 Increasing the calorie requirement by 1 will leave the cost of the optimal diet unchanged.

2 Increasing the chocolate requirement by 1 oz will decrease the cost of the optimal diet by $-2.5¢$ (that is, increase the cost of the optimal diet by $2.5¢$).

3 Increasing the sugar requirement by 1 oz will decrease the cost of the optimal diet by $-7.5¢$ (that is, increase the cost of the optimal diet by $7.5¢$).

4 Increasing the fat requirement by 1 oz will leave the cost of the optimal diet unchanged.

The entry in the DUAL PRICE column for any constraint is, however, the negative of the constraint's dual variable. Thus, for the diet problem, we see from Figure 6 that the optimal dual solution to the diet problem is given by $c_{BV}B^{-1} = \begin{bmatrix} 0 & 2.5 & 7.5 & 0 \end{bmatrix}$. When pricing out a new activity for a minimization problem, use the negative of each dual price as the corresponding element of $\mathbf{c}_{BV}B^{-1}$.

Remember that for any LP, the dual prices remain valid only as long as the current basis remains optimal. As stated in Section 6.3, the range of right-hand side values for which the current basis remains optimal may be obtained from the RIGHTHAND SIDE RANGES block of the LINDO output.

Degeneracy and Sensitivity Analysis

When the optimal solution to an LP is degenerate, caution must be used when interpreting the LINDO output. Recall from Section 4.11 that a bfs is degenerate if at least one basic variable in the optimal solution equals 0. For an LP with m constraints, if the LINDO output indicates that less than m variables are positive, then the optimal solution is a degenerate bfs. To illustrate, consider the following LP:

$$\max z = 6X_1 + 4X_2 + 3X_3 + 2X_4$$

$$
\begin{aligned}
\text{s.t.} \quad 2X_1 + 3X_2 + \;\; X_3 + 2X_4 &\leq 400 \\
X_1 + \;\; X_2 + 2X_3 + \;\; X_4 &\leq 150 \\
2X_1 + \;\; X_2 + \;\; X_3 + .5X_4 &\leq 200 \\
3X_1 + \;\; X_2 + \qquad\quad\; X_4 &\leq 250 \\
X_1, X_2, X_3, X_4 &\geq 0
\end{aligned}
$$

The LINDO output for this LP is in Figure 9. The LP has four constraints and only two positive variables in the optimal solution, so the bfs is degenerate. By the way, results from using the **TABLEAU** command indicate that the optimal basis is BV = $\{X_2, X_3, S_3, X_1\}$.

We now discuss three "oddities" that may occur when the optimal solution found by LINDO is degenerate.

Oddity 1 At least one constraint's RANGE IN WHICH THE BASIS IS UNCHANGED will have a 0 ALLOWABLE INCREASE or ALLOWABLE DECREASE. This means that for at least one constraint the DUAL PRICE can tell us about the new z-value for either an increase or a decrease in the constraint's right-hand side, but not both.

To understand Oddity 1, consider the first constraint. Its AI is 0. Thus, its DUAL PRICE of .50 cannot be used to determine a new z-value resulting from any increase in the first constraint's right-hand side.

Oddity 2 For a nonbasic variable to become positive, its objective function coefficient may have to be improved by more than its REDUCED COST.

To understand Oddity 2 consider the nonbasic variable X_4; its REDUCED COST is 1.5. If we increase X_4's objective function coefficient by 2, however, we still find that the new optimal solution has $X_4 = 0$ because the change affects the set of basic variables, but not the LP's optimal solution. If we increase X_4's objective function coefficient by 4.5 or more, then we find that X_4 is positive.

Oddity 3 If you increase a variable's objective function coefficient by more than its AI or decrease it by more than its AD, then the optimal solution to the LP may remain the same.

Oddity 3 is similar to Oddity 2. To understand Oddity 3, consider the nonbasic variable X_4; its AI is 1.5. If we increase X_4's objective function coefficient by 2, however, we still find that the new optimal solution is unchanged. This oddity occurs because the change affects the set of basic variables but not the LP's optimal solution.

```
MAX        6 X1 + 4 X2 + 3 X3 + 2 X4
SUBJECT TO
      2)    2 X1 + 3 X2 + X3 + 2 X4 <=      400
      3)    X1 + X2 + 2 X3 + X4 <=   150
      4)    2 X1 + X2 + X3 + 0.5 X4 <=   200
      5)    3 X1 + X2 + X4 <=   250
END

LP OPTIMUM FOUND AT STEP  3

           OBJECTIVE FUNCTION VALUE

      1)      700.00000

   VARIABLE        VALUE        REDUCED COST
     X1         50.000000          .000000
     X2        100.000000          .000000
     X3           .000000          .000000
     X4           .000000         1.500000


   ROW     SLACK  OR SURPLUS      DUAL PRICES
     2)          .000000           .500000
     3)          .000000          1.250000
     4)          .000000           .000000
     5)          .000000          1.250000

NO. ITERATIONS=         3

RANGES IN WHICH THE BASIS IS UNCHANGED:

                        OBJ COEFFICIENT RANGES
   VARIABLE       CURRENT       ALLOWABLE       ALLOWABLE
                  COEF          INCREASE        DECREASE
     X1         6.000000       3.000000        3.000000
     X2         4.000000       5.000000        1.000000
     X3         3.000000       3.000000        2.142857
     X4         2.000000       1.500000        INFINITY

                        RIGHTHAND SIDE RANGES
   ROW           CURRENT       ALLOWABLE       ALLOWABLE
                 RHS           INCREASE        DECREASE
     2         400.000000        .000000      200.000000
     3         150.000000        .000000         .000000
     4         200.000000      INFINITY          .000000
     5         250.000000        .000000      120.000000


THE TABLEAU
   ROW  (BASIS)        X1      X2      X3      X4     SLK    2
     1 ART            .000    .000    .000   1.500      .500
     2      X2        .000   1.000    .000    .500      .500
     3      X3        .000    .000   1.000    .167     -.167
     4 SLK   4        .000    .000    .000   -.500      .000
     5      X1       1.000    .000    .000    .167     -.167


   ROW   SLK   3  SLK   4  SLK   5
     1   1.250     .000   1.250  700.000
     2   -.250     .000   -.250  100.000
     3    .583     .000   -.083     .000
     4   -.500    1.000   -.500     .000
     5    .083     .000    .417   50.000
```

FIGURE 9

PROBLEMS

Group A

1   Use the Dual Theorem to prove (37).

2   The following questions refer to the Sugarco problem (Problem 6 of Section 6.3):

   a   Find the shadow prices for the Sugarco problem.

b   If 60 oz of sugar were available, what would be Sugarco's profit?

c   How about 40 oz of sugar?

d   How about 30 oz of sugar?

TABLE 29

Resource	Product 1	Product 2
Skilled labor (hours)	3	4
Unskilled labor (hours)	2	3
Raw material (units)	1	2

3 Suppose we are working with a min problem and increase the right-hand side of a $\geq$ constraint. What can happen to the optimal z-value?

4 Suppose we are working with a min problem and increase the right-hand side of a $\leq$ constraint. What can happen to the optimal z-value?

5 A company manufactures two products (1 and 2). Each unit of product 1 can be sold for $15, and each unit of product 2 for $25. Each product requires raw material and two types of labor (skilled and unskilled) (see Table 29). Currently, the company has available 100 hours of skilled labor, 70 hours of unskilled labor, and 30 units of raw material. Because of marketing considerations, at least 3 units of product 2 must be produced.

a Explain why the company's goal is to maximize revenue.

b The relevant LP is

$$\max z = 15x_1 + 25x_2$$
$$\text{s.t.} \quad 3x_1 + 4x_2 \leq 100 \quad \text{(Skilled labor constraint)}$$
$$2x_1 + 3x_2 \leq 70 \quad \text{(Unskilled labor constraint)}$$
$$x_1 + 2x_2 \leq 30 \quad \text{(Raw material constraint)}$$
$$x_2 \geq 3 \quad \text{(Product 2 constraint)}$$
$$x_1, x_2 \geq 0$$

The optimal tableau for this problem has the following row 0:

$$z + 15s_3 + 5e_4 + (M - 5)a_4 = 435$$

The optimal solution to the LP is $z = 435$, $x_1 = 24$, $x_2 = 3$. Find and interpret the shadow price of each constraint. How much would the company be willing to pay for an additional unit of each type of labor? How much would it be willing to pay for an extra unit of raw material?

c Assuming the current basis remains optimal (it does), what would the company's revenue be if 35 units of raw material were available?

d With the current basis optimal, what would the company's revenue be if 80 hours of skilled labor were available?

e With the current basis optimal, what would the company's new revenue be if at least 5 units of product 2 were required? How about if at least 2 units of product 2 were required?

6 Suppose that the company in Problem 5 owns no labor and raw material but can purchase them at the following prices: as many as 100 hours of skilled labor at $3/hour, 70 hours of unskilled labor at $2/hour, and 30 units of raw material at $1 per unit of raw material. If the company's goal is to maximize profit, show that the appropriate LP is

$$\max z = x_1 + 5x_2$$
$$\text{s.t.} \quad 3x_1 + 4x_2 \leq 100$$
$$2x_1 + 3x_2 \leq 70$$
$$x_1 + 2x_2 \leq 30$$
$$x_2 \geq 3$$
$$x_1, x_2 \geq 0$$

The optimal row 0 for this LP is

$$z + 1.5x_1 + 2.5s_3 + Ma_4 = 75$$

and the optimal solution is $z = 75$, $x_1 = 0$, $x_2 = 15$. In answering parts (a) and (b), assume that the current basis remains optimal.

a How much should the company pay for an extra unit of raw material?

b How much should the company pay for an extra hour of skilled labor? Unskilled labor? (Be careful here!)

7 For the Dorian problem (see Problem 8 of Section 6.3), answer the following questions:

a What would Dorian's cost be if 40 million HIW exposures were required?

b What would Dorian's cost be if only 20 million HIM exposures were required?

8 If it seems difficult to believe that the shadow price of an equality constraint should be urs, try this problem. Consider the following two LPs:

$$\max z = x_2$$
$$\textbf{(LP 1)} \quad \text{s.t.} \quad x_1 + x_2 = 2$$
$$x_1, x_2 \geq 0$$

$$\max z = x_2$$
$$\textbf{(LP 2)} \quad \text{s.t.} \quad -x_1 - x_2 = -2$$
$$x_1, x_2 \geq 0$$

In which LP will the constraint have a positive shadow price? Which will have a negative shadow price?

Group B

9 For the Dakota problem, suppose that 22 finishing hours and 9 carpentry hours are available. What would be the new optimal z-value? [*Hint:* Use the 100% Rule to show that the current basis remains optimal, and mimic (34)–(36).]

10 For the diet problem, suppose at least 8 oz of chocolate and at least 9 oz of sugar are required (with other requirements remaining the same). What is the new optimal z-value?

11 Consider the LP:

$$\max z = 9x_1 + 8x_2 + 5x_3 + 4x_4$$
$$\text{s.t.} \quad x_1 \qquad\qquad + x_4 \leq 200$$
$$x_2 + x_3 \qquad \leq 150$$
$$x_1 + x_2 + x_3 \qquad \leq 350$$
$$2x_1 + x_2 + x_3 + x_4 \leq 550$$
$$x_1, x_2, x_3, x_4 \geq 0$$

a Solve this LP with LINDO and use your output to show that the optimal solution is degenerate.

b Use your LINDO output to find an example of Oddities 1–3.

6.9 Duality and Sensitivity Analysis

Our proof of the Dual Theorem demonstrated the following result: *Assuming that a set of basic variables BV is feasible, then BV is optimal (that is, each variable in row 0 has a nonnegative coefficient) if and only if the associated dual solution* ($\mathbf{c}_{BV}B^{-1}$) *is dual feasible.*

This result can be used for an alternative way of doing the following types of sensitivity analysis (see list of changes at the beginning of Section 6.3).

Change 1   Changing the objective function coefficient of a nonbasic variable

Change 4   Changing the column of a nonbasic variable

Change 5   Adding a new activity

In each case, the change leaves BV feasible. BV will remain optimal if the BV row 0 remains nonnegative. Primal optimality and dual feasibility are equivalent, so we see that *the above changes will leave the current basis optimal if and only if the current dual solution* $\mathbf{c}_{BV}B^{-1}$ *remains dual feasible.* If the current dual solution is no longer dual feasible, then BV will be suboptimal, and a new optimal solution must be found.

We illustrate the duality-based approach to sensitivity analysis by reworking some of the Section 6.3 illustrations. Recall that these illustrations dealt with the Dakota problem:

$$\max z = 60x_1 + 30x_2 + 20x_3$$

$$\begin{array}{llll}
\text{s.t.} & 8x_1 + 6x_2 + x_3 \leq 48 & \text{(Lumber constraint)} \\
& 4x_1 + 2x_2 + 1.5x_3 \leq 20 & \text{(Finishing constraint)} \\
& 2x_1 + 1.5x_2 + 0.5x_3 \leq 8 & \text{(Carpentry constraint)} \\
& x_1, x_2, x_3 \geq 0
\end{array}$$

The optimal solution was $z = 280$, $s_1 = 24$, $x_3 = 8$, $x_1 = 2$, $x_2 = 0$, $s_2 = 0$, $s_3 = 0$. The only nonbasic decision variable in the optimal solution is x_2 (tables). The dual of the Dakota problem is

$$\min w = 48y_1 + 20y_2 + 8y_3$$

$$\begin{array}{llll}
\text{s.t.} & 8y_1 + 4y_2 + 2y_3 \geq 60 & \text{(Desk constraint)} \\
& 6y_1 + 2y_2 + 1.5y_3 \geq 30 & \text{(Table constraint)} \\
& y_1 + 1.5y_2 + 0.5y_3 \geq 20 & \text{(Chair constraint)} \\
& y_1, y_2, y_3 \geq 0
\end{array}$$

Recall that the optimal dual solution—and therefore the constraint shadow prices—are $y_1 = 0$, $y_2 = 10$, $y_3 = 10$. We now show how knowledge of duality can be applied to sensitivity analysis.

EXAMPLE 14   **Changing Objective Function Coefficient of Nonbasic Variable**

We want to change the objective function coefficient of a nonbasic variable. Let c_2 be the coefficient of x_2 (tables) in the Dakota objective function. In other words, c_2 is the price at which a table is sold. For what values of c_2 will the current basis remain optimal?

Solution   If $y_1 = 0$, $y_2 = 10$, $y_3 = 10$ remains dual feasible, then the current basis—and the values of all the variables—are unchanged. Note that if the objective function coefficient for x_2 is changed, then the first and third dual constraints remain unchanged, but the second (table) dual constraint is changed to

$$6y_1 + 2y_2 + 1.5y_3 \geq c_2$$

If $y_1 = 0$, $y_2 = 10$, $y_3 = 10$ satisfies this inequality, then dual feasibility (and therefore primal optimality) is maintained. Thus, the current basis remains optimal if c_2 satisfies $6(0) + 2(10) + 1.5(10) \geq c_2$, or $c_2 \leq 35$. This shows that for $c_2 \leq 35$, the current basis remains optimal. Conversely, if $c_2 > 35$, the current basis is no longer optimal. This agrees with the result obtained in Section 6.3.

Using shadow prices, we may give an alternative interpretation of this result. We can use shadow prices to compute the implied value of the resources needed to construct a table (see Table 30). A table uses $35 worth of resources, so the only way producing tables can increase Dakota's revenues is if a table sells for more than $35. Thus, the current basis fails to be optimal if $c_2 > 35$, and the current basis remains optimal if $c_2 \leq 35$.

TABLE **30**
Why a Table Is Profitable at > $35/Table

Resource in a Table	Shadow Price of Resource ($)	Amount of Resource Used	Value of Resource Used
Lumber	0	6 board ft	0(6) = $0
Finishing	10	2 hours	10(2) = $20
Carpentry	10	1.5 hours	10(1.5) = $15
			Total: = $35

EXAMPLE **15**   **Changing a Nonbasic Variable**

We want to change the column for a nonbasic activity. Suppose a table sells for $43 and uses 5 board feet of lumber, 2 finishing hours, and 2 carpentry hours. Does the current basis remain optimal?

Solution   Changing the column for the nonbasic variable "tables" leaves the first and third dual constraints unchanged but changes the second to

$$5y_1 + 2y_2 + 2y_3 \geq 43$$

Because $y_1 = 0$, $y_2 = 10$, $y_3 = 10$ does not satisfy the new second dual constraint, dual feasibility is not maintained, and the current basis is no longer optimal. In terms of shadow prices, this result is reasonable (see Table 31). Each table uses $40 worth of resources and sells for $43, so Dakota can increase its revenue by $43 - 40 = \$3$ for each table that is produced. Thus, the current basis is no longer optimal, and x_2 (tables) will be basic in the new optimal solution.

TABLE **31**
Shadow Price Interpretation of Table Production Decision ($40/Table)

Resource in a Table	Shadow Price of Resource ($)	Amount of Resource Used	Value of Resource Used ($)
Lumber	0	5 board ft	0(5) = $0
Finishing	10	2 hours	10(2) = $20
Carpentry	10	2 hours	10(2) = $20
			Total: = $40

EXAMPLE 16   Adding a New Activity

We want to add a new activity. Suppose Dakota is considering manufacturing footstools (x_4). A footstool sells for $15 and uses 1 board foot of lumber, 1 finishing hour, and 1 carpentry hour. Does the current basis remain optimal?

Solution   Introducing the new activity (footstools) leaves the three dual constraints unchanged, but the new variable x_4 adds a new dual constraint (corresponding to footstools). The new dual constraint will be

$$y_1 + y_2 + y_3 \geq 15$$

The current basis remains optimal if $y_1 = 0$, $y_2 = 10$, $y_3 = 10$ satisfies the new dual constraint. Because $0 + 10 + 10 \geq 15$, the current basis remains optimal. In terms of shadow prices, a stool utilizes $1(0) = \$0$ worth of lumber, $1(10) = \$10$ worth of finishing hours, and $1(10) = \$10$ worth of carpentry time. A stool uses $0 + 10 + 10 = \$20$ worth of resources and sells for only $15, so Dakota should not make footstools, and the current basis remains optimal.

PROBLEMS

Group A

1 For the Dakota problem, suppose computer tables sell for $35 and use 6 board feet of lumber, 2 hours of finishing time, and 1 hour of carpentry time. Is the current basis still optimal? Interpret this result in terms of shadow prices.

2 The following questions refer to the Sugarco problem (Problem 6 of Section 6.3):

a For what values of profit on a Type 1 candy bar does the current basis remain optimal?

b If a Type 1 candy bar used 0.5 oz of sugar and 0.75 oz of chocolate, would the current basis remain optimal?

c A Type 4 candy bar is under consideration. A Type 4 candy bar yields a 10¢ profit and uses 2 oz of sugar and 1 oz of chocolate. Does the current basis remain optimal?

3 Suppose, in the Dakota problem, a desk still sells for $60 but now uses 8 board ft of lumber, 4 finishing hours, and 15 carpentry hours. Determine whether the current basis remains optimal. What is wrong with the following reasoning?

The change in the column for desks leaves the second and third dual constraints unchanged and changes the first to

$$8y_1 + 4y_2 + 15y_3 \geq 60$$

Because $y_1 = 0$, $y_2 = 10$, $y_3 = 10$ satisfies the new dual constraint, the current basis remains optimal.

6.10 Complementary Slackness

The Theorem of Complementary Slackness is an important result that relates the optimal primal and dual solutions. To state this theorem, we assume that the primal is a normal max problem with variables $x_1, x_2, \ldots, x_n$ and $m \leq$ constraints. Let $s_1, s_2, \ldots, s_m$ be the slack variables for the primal. Then the dual is a normal min problem with variables $y_1, y_2, \ldots, y_m$ and $n \geq$ constraints. Let $e_1, e_2, \ldots, e_n$ be the excess variables for the dual. A statement of the Theorem of Complementary Slackness follows.

Let

$$
\mathbf{x} = \begin{bmatrix} x_1 \\ x_2 \\ \vdots \\ x_n \end{bmatrix}
$$

be a feasible primal solution and $\mathbf{y} = [y_1 \ \ y_2 \ \ \cdots \ \ y_m]$ be a feasible dual solution. Then $\mathbf{x}$ is primal optimal and $\mathbf{y}$ is dual optimal if and only if

$$s_i y_i = 0 \qquad (i = 1, 2, \ldots, m) \tag{38}$$

$$e_j x_j = 0 \qquad (j = 1, 2, \ldots, n) \tag{39}$$

In Problem 4 at the end of this section, we sketch the proof of the Theorem of Complementary Slackness, but first we discuss the intuitive meaning of this theorem.

From (38), it follows that the optimal primal and dual solutions must satisfy

$$\text{ith primal slack} > 0 \text{ implies ith dual variable} = 0 \tag{40}$$

$$\text{ith dual variable} > 0 \text{ implies ith primal slack} = 0 \tag{41}$$

From (39), it follows that the optimal primal and dual solutions must satisfy

$$\text{jth dual excess} > 0 \text{ implies jth primal variable} = 0 \tag{42}$$

$$\text{jth primal variable} > 0 \text{ implies jth dual excess} = 0 \tag{43}$$

From (40) and (42), we see that if *a constraint in either the primal or dual is nonbinding (has either $s_i > 0$ or $e_j > 0$), then the corresponding variable in the other (or complementary) problem must equal* 0. Hence the name **complementary slackness.**

To illustrate the interpretation of the Theorem of Complementary Slackness, we return to the Dakota problem. Recall that the primal is

$$
\begin{aligned}
\max z = {}& 60x_1 + 30x_2 + 20x_3 \\
\text{s.t.} \quad & 8x_1 + 6x_2 + x_3 \le 48 \qquad \text{(Lumber constraint)} \\
& 4x_1 + 2x_2 + 1.5x_3 \le 20 \qquad \text{(Finishing constraint)} \\
& 2x_1 + 1.5x_2 + 0.5x_3 \le 8 \qquad \text{(Carpentry constraint)} \\
& x_1, x_2, x_3 \ge 0
\end{aligned}
$$

and the dual is

$$
\begin{aligned}
\min w = {}& 48y_1 + 20y_2 + 8y_3 \\
\text{s.t.} \quad & 8y_1 + 4y_2 + 2y_3 \ge 60 \qquad \text{(Desk constraint)} \\
& 6y_1 + 2y_2 + 1.5y_3 \ge 30 \qquad \text{(Table constraint)} \\
& y_1 + 1.5y_2 + 0.5y_3 \ge 20 \qquad \text{(Chair constraint)} \\
& y_1, y_2, y_3 \ge 0
\end{aligned}
$$

The optimal primal solution is

$$z = 280, \qquad x_1 = 2, \qquad x_2 = 0, \qquad x_3 = 8$$
$$s_1 = 48 - (8(2) + 6(0) + 1(8)) = 24$$
$$s_2 = 20 - (4(2) + 2(0) + 1.5(8)) = 0$$
$$s_3 = 8 - (2(2) + 1.5(0) + 0.5(8)) = 0$$

The optimal dual solution is

$$w = 280, \quad y_1 = 0, \quad y_2 = 10, \quad y_3 = 10$$
$$e_1 = (8(0) + 4(10) + 2(10)) - 60 = 0$$
$$e_2 = (6(0) + 2(10) + 1.5(10)) - 30 = 5$$
$$e_3 = (1(0) + 1.5(10) + 0.5(10)) - 20 = 0$$

For the Dakota problem, (38) reduces to

$$s_1 y_1 = s_2 y_2 = s_3 y_3 = 0$$

which is indeed satisfied by the optimal primal and dual solutions. Also, (39) reduces to

$$e_1 x_1 = e_2 x_2 = e_3 x_3 = 0$$

which is also satisfied by the optimal primal and dual solutions.

We now illustrate the interpretation of (40)–(43). Note that (40) tells us that because the optimal primal solution has $s_1 > 0$, the optimal dual solution must have $y_1 = 0$. In the context of the Dakota problem, this means that positive slack in the lumber constraint implies that lumber must have a zero shadow price. Slack in the lumber constraint means that extra lumber would not be used, so an extra board foot of lumber should indeed be worthless.

Equation (41) tells us that because $y_2 > 0$ in the optimal dual solution, $s_2 = 0$ must hold in the optimal primal solution. This is reasonable because $y_2 > 0$ means that an extra finishing hour has some value. This can only occur if we are at present using all available finishing hours (or equivalently, if $s_2 = 0$).

Observe that (42) tells us that because $e_2 > 0$ in the optimal dual solution, $x_2 = 0$ must hold in the optimal primal solution. This is reasonable because $e_2 = 6y_1 + 2y_2 + 1.5y_3 - 30$. Because y_1, y_2, and y_3 are resource shadow prices, e_2 may be written as

$$e_2 = (\text{value of resources used by table}) - (\text{sales price of a table})$$

Thus, if $e_2 > 0$, tables are selling for a price that is less than the value of the resources used to make 1 unit of x_2 (tables). This means that no tables should be made (or equivalently, that $x_2 = 0$). This shows that $e_2 > 0$ in the optimal dual solution implies that $x_2 = 0$ must hold in the optimal primal solution.

Note that for the Dakota problem, (43) tells us that $x_1 > 0$ for the optimal primal solution implies that $e_1 = 0$. This result simply reflects the following important fact. *For any variable x_j in the optimal primal basis, the marginal revenue obtained from producing a unit of x_j must equal the marginal cost of the resources used to produce a unit of x_j.* This is a consequence of the fact that each basic variable must have a zero coefficient in row 0 of the optimal primal tableau. In short, (43) is simply the LP version of the well-known economic maxim that an optimal production strategy must have marginal revenue equal marginal cost.

To be more specific, observe that $x_1 > 0$ means that desks are in the optimal basis. Then

$$\text{Marginal revenue obtained by manufacturing desk} = \$60$$

To compute the marginal cost of manufacturing a desk (in terms of shadow prices), note that

$$\text{Cost of lumber in desk} = 8(0) = \$0$$
$$\text{Cost of finishing hours used to make a desk} = 4(10) = \$40$$
$$\text{Cost of carpentry hours used to make a desk} = 2(10) = \$20$$
$$\text{Marginal cost of producing a desk} = 0 + 40 + 20 = \$60$$

Thus, for desks, marginal revenue is equal to marginal cost.

Using Complementary Slackness to Solve LPs

If the optimal solution to the primal or dual is known, complementary slackness can sometimes be used to determine the optimal solution to the complementary problem. For example, suppose we were told that the optimal solution to the Dakota problem was $z = 280$, $x_1 = 2$, $x_2 = 0$, $x_3 = 8$, $s_1 = 24$, $s_2 = 0$, $s_3 = 0$. Can we use Theorem 2 to help us find the optimal solution to the Dakota dual? Because $s_1 > 0$, (40) tells us that the optimal dual solution must have $y_1 = 0$. Because $x_1 > 0$ and $x_3 > 0$, (43) implies that the optimal dual solution must have $e_1 = 0$, and $e_3 = 0$. This means that for the optimal dual solution, the first and third constraints must be binding. We know that $y_1 = 0$, so we know that the optimal values of y_2 and y_3 may be found by solving the first and third dual constraints as equalities (with $y_1 = 0$). Thus, the optimal values of y_2 and y_3 must satisfy

$$4y_2 + 2y_3 = 60 \quad \text{and} \quad 1.5y_2 + 0.5y_3 = 20$$

Solving these equations simultaneously shows that the optimal dual solution must have $y_2 = 10$ and $y_3 = 10$. Thus, complementary slackness has helped us find the optimal dual solution $y_1 = 0$, $y_2 = 10$, $y_3 = 10$. (From the Dual Theorem, we know, of course, that the optimal dual solution must have $\overline{w} = 280$.)

PROBLEMS

Group A

1 Glassco manufactures glasses: wine, beer, champagne, and whiskey. Each type of glass requires time in the molding shop, time in the packaging shop, and a certain amount of glass. The resources required to make each type of glass are given in Table 32. Currently, 600 minutes of molding time, 400 minutes of packaging time, and 500 oz of glass are available. Assuming that Glassco wants to maximize revenue, the following LP should be solved:

$$\max z = 6x_1 + 10x_2 + 9x_3 + 20x_4$$
s.t. $4x_1 + 9x_2 + 7x_3 + 10x_4 \leq 600$ (Molding constraint)

     $x_1 + x_2 + 3x_3 + 40x_4 \leq 400$ (Packaging constraint)

     $3x_1 + 4x_2 + 2x_3 + x_4 \leq 500$ (Glass constraint)

$$x_1, x_2, x_3, x_4 \geq 0$$

It can be shown that the optimal solution to this LP is $z = \frac{2800}{3}$, $x_1 = \frac{400}{3}$, $x_4 = \frac{20}{3}$, $x_2 = 0$, $x_3 = 0$, $s_1 = 0$, $s_2 = 0$, $s_3 = \frac{280}{3}$.

a Find the dual of the Glassco problem.

b Using the given optimal primal solution and the Theorem of Complementary Slackness, find the optimal solution to the dual of the Glassco problem.

c Find an example of each of the complementary slackness conditions, (40)–(43). As in the text, interpret each example in terms of shadow prices.

2 Use the Theorem of Complementary Slackness to show that in the LINDO output, the SLACK or SURPLUS and DUAL PRICE entries for any row cannot both be positive.

3 Consider the following LP:

$$\max z = 5x_1 + 3x_2 + x_3$$
s.t. $2x_1 + x_2 + x_3 \leq 6$

     $x_1 + 2x_2 + x_3 \leq 7$

$$x_1, x_2, x_3 \geq 0$$

Graphically solve the dual of this LP. Then use complementary slackness to solve the max problem.

TABLE 32

| | Glass | | | |
	X_1 Wine	X_2 Beer	X_3 Champagne	X_4 Whiskey
Molding time	4 minutes	9 minutes	7 minutes	10 minutes
Packaging time	1 minute	1 minute	3 minutes	40 minutes
Glass	3 oz	4 oz	2 oz	1 oz
Selling price	$6	$10	$9	$20

4 Let $\mathbf{x} = [x_1 \quad x_2 \quad x_3 \quad s_1 \quad s_2 \quad s_3]$ be a primal feasible point for the Dakota problem and $\mathbf{y} = [y_1 \quad y_2 \quad y_3 \quad e_1 \quad e_2 \quad e_3]$ be a dual feasible point.

a Multiply the ith constraint (in standard form) of the primal by y_i and sum the resulting constraints.

b Multiply the jth dual constraint (in standard form) by x_j and sum them.

c Compute: part (a) answer minus part (b) answer.

d Use the part (c) answer and the Dual Theorem to show that if $\mathbf{x}$ is primal optimal and $\mathbf{y}$ is dual optimal, then (38) and (39) hold.

e Use the part (c) answer to show that if (38) and (39) both hold, then $\mathbf{x}$ is primal optimal and $\mathbf{y}$ is dual optimal. (*Hint:* Look at Lemma 2.)

6.11 The Dual Simplex Method

When we use the simplex method to solve a max problem (we will refer to the max problem as a primal), we begin with a primal feasible solution (because each constraint in the initial tableau has a nonnegative right-hand side). At least one variable in row 0 of the initial tableau has a negative coefficient, so our initial primal solution is not dual feasible. Through a sequence of simplex pivots, we maintain primal feasibility and obtain an optimal solution when dual feasibility (a non-negative row 0) is attained. In many situations, however, it is easier to solve an LP by beginning with a tableau in which each variable in row 0 has a nonnegative coefficient (so the tableau is dual feasible) and at least one constraint has a negative right-hand side (so the tableau is primal infeasible). The dual simplex method maintains a nonnegative row 0 (dual feasibility) and eventually obtains a tableau in which each right-hand side is nonnegative (primal feasibility). At this point, an optimal tableau has been obtained. Because this technique maintains dual feasibility, it is called the **dual simplex method.**

Dual Simplex Method for a Max Problem

Step 1 Is the right-hand side of each constraint nonnegative? If so, an optimal solution has been found; if not, at least one constraint has a negative right-hand side, and we go to step 2.

Step 2 Choose the most negative basic variable as the variable to leave the basis. The row in which the variable is basic will be the pivot row. To select the variable that enters the basis, we compute the following ratio for each variable x_j that has a *negative* coefficient in the pivot row:

$$\frac{\text{Coefficient of } x_j \text{ in row 0}}{\text{Coefficient of } x_j \text{ in pivot row}}$$

Choose the variable with the smallest ratio (absolute value) as the entering variable. This form of the ratio test maintains a dual feasible tableau (all variables in row 0 have nonnegative coefficients). Now use EROs to make the entering variable a basic variable in the pivot row.

Step 3 If there is any constraint in which the right-hand side is negative and each variable has a nonnegative coefficient, then the LP has no feasible solution. If no constraint indicating infeasibility is found, return to step 1.

To illustrate the case of an infeasible LP, suppose the dual simplex method yielded a constraint such as $x_1 + 2x_2 + x_3 = -5$. Because $x_1 \geq 0$, $2x_2 \geq 0$, and $x_3 \geq 0$, $x_1 + 2x_2 + x_3 \geq 0$, and the constraint $x_1 + 2x_2 + x_3 = -5$ cannot be satisfied. In this case, the original LP must be infeasible.

Three uses of the dual simplex follow:

1 Finding the new optimal solution after a constraint is added to an LP

2 Finding the new optimal solution after changing a right-hand side of an LP

3 Solving a normal min problem

Finding the New Optimal Solution After a Constraint Is Added to an LP

The dual simplex method is often used to find the new optimal solution to an LP after a constraint is added. When a constraint is added, one of the following three cases will occur:

Case 1 The current optimal solution satisfies the new constraint.

Case 2 The current optimal solution does not satisfy the new constraint, but the LP still has a feasible solution.

Case 3 The additional constraint causes the LP to have no feasible solutions.

If Case 1 occurs, then the current optimal solution satisfies the new constraint, and the current solution remains optimal. To illustrate why this is true, suppose we have added the constraint $x_1 + x_2 + x_3 \leq 11$ to the Dakota problem. The current optimal solution ($z = 280$, $x_1 = 2$, $x_2 = 0$, $x_3 = 8$) satisfies this constraint. To see why this solution remains optimal after the constraint $x_1 + x_2 + x_3 \leq 11$ is added, recall that adding a constraint to an LP either leaves the feasible region unchanged or eliminates points from the feasible region. In this case, the Section 6.8 discussion tells us that adding a constraint (to a max problem) either reduces the optimal z-value or leaves it unchanged. This means that if we add the constraint $x_1 + x_2 + x_3 \leq 11$ to the Dakota problem, the new optimal z-value can be at most 280. The current solution is still feasible and has $z = 280$, so it must still be optimal.

If Case 2 occurs, the current solution is no longer feasible, so it can no longer be optimal. The dual simplex method can be used to determine the new optimal solution. Suppose that in the Dakota problem, marketing considerations dictate that at least 1 table be manufactured. This adds the constraint $x_2 \geq 1$. Because the current optimal solution has $x_2 = 0$, it is no longer feasible and cannot be optimal. To find the new optimal solution, we subtract an excess variable e_4 from the constraint $x_2 \geq 1$. This yields the constraint $x_2 - e_4 = 1$. If we multiply this constraint through by -1, we obtain $-x_2 + e_4 = -1$, and we can use e_4 as a basic variable for this constraint. Appending this constraint to the optimal Dakota tableau yields Table 33.

Because we are using the row 0 from an optimal tableau, each variable has a nonnegative coefficient in row 0, and we may proceed with the dual simplex method. The variable $e_4 = -1$ is the most negative basic variable, so e_4 will exit from the basis, and row 4 will be the pivot row. Because x_2 is the only variable with a negative coefficient in row 4, x_2 must enter into the basis (see Table 34).

This is an optimal tableau. Thus, if the constraint $x_2 \geq 1$ is added to the Dakota problem, the optimal solution becomes $z = 275$, $s_1 = 26$, $x_3 = 10$, $x_1 = \frac{3}{4}$, $x_2 = 1$, which has reduced Dakota's objective function (revenue) by \$5 (the reduced cost for tables).

If we had wanted to, we could simply have added the constraint $x_2 \geq 1$ to the original Dakota initial tableau and used the regular simplex method to solve the problem. This would have entailed adding an artificial variable to the $x_2 \geq 1$ constraint and would probably have required many pivots. When we use the dual simplex to solve a problem again after a constraint has been added, we are taking advantage of the fact that we have already

TABLE 33

"Old" Optimal Dakota Tableau If $x_2 \geq 1$ Is Required

						Basic Variable
z	$+ 5x_2$		$+ 10s_2 + 10s_3$		$= 280$	$z = 280$
	$- 2x_2$	$+ s_1 +$	$2s_2 - 8s_3$		$= 24$	$s_1 = 24$
	$- 2x_2 + x_3$	$+$	$2s_2 - 4s_3$		$= 8$	$x_3 = 8$
	$x_1 + \frac{5}{4}x_2$		$- \frac{1}{2}s_2 + \frac{3}{2}s_3$		$= 2$	$x_1 = 2$
	$- \boxed{x_2}$			$+ e_4$	$= -1$	$e_4 = -1$

TABLE 34

"New" Optimal Dakota Tableau If $x_2 \geq 1$ Is Required

				Basic Variable
z		$+ 10s_2 + 10s_3 + 5e_4 = 275$		$z = 275$
	$s_1 +$	$2s_2 - 8s_3 - 2e_4 = 26$		$s_1 = 26$
	$x_3 +$	$2s_2 - 4s_3 - 2e_4 = 10$		$x_3 = 10$
	x_1	$- \frac{1}{2}s_2 + \frac{3}{2}s_3 + \frac{5}{4}e_4 = \frac{3}{4}$		$x_1 = \frac{3}{4}$
	x_2	$- e_4 = 1$		$x_2 = 1$

obtained a nonnegative row 0 and that most of our right-hand sides have nonnegative coefficients. This is why the dual simplex usually requires relatively few pivots to find a new optimal solution when a constraint is added to an LP.

If Case 3 occurs, step 3 of the dual simplex method allows us to show that the LP is now infeasible. To illustrate the idea, suppose we add the constraint $x_1 + x_2 \geq 12$ to the Dakota problem. After subtracting an excess variable e_4 from this constraint, we obtain

$$x_1 + x_2 - e_4 = 12 \qquad \text{or} \qquad -x_1 - x_2 + e_4 = -12$$

Appending this constraint to the optimal Dakota tableau yields Table 35.

Because x_1 appears in the new constraint, it seems that x_1 can no longer be used as a basic variable for row 3. To remedy this problem, we eliminate x_1 (and in general all basic variables) from the new constraint by replacing row 4 by row 3 + row 4 (see Table 36). Because $e_4 = -10$ is the most negative basic variable, e_4 will leave the basis and row 4 will be the pivot row. The variable s_2 is the only one with a negative coefficient in row 4, so s_2 enters the basis and becomes a basic variable in row 4 (see Table 37). Now x_3 must leave the basis, and row 2 will be the pivot row. Because x_2 is the only variable in row 2 with a negative coefficient, x_2 now enters the basis (see Table 38). Because $x_1 \geq 0$, $x_3 \geq 0$, $2s_3 \geq 0$, and $3e_4 \geq 0$, the left side of row 3 must be nonnegative and cannot equal -20. Hence, the Dakota problem with the additional constraint $x_1 + x_2 \geq 12$ has no feasible solution.

TABLE 35

"Old" Optimal Dakota Tableau If $x_1 + x_2 \geq 12$ Is Required

						Basic Variable
z	$+ 5x_2$		$+ 10s_2 + 10s_3$		$= 280$	$z = 280$
	$- 2x_2$	$+ s_1 +$	$2s_2 - 8s_3$		$= 24$	$s_1 = 24$
	$- 5x_2 + x_3$	$+$	$2s_2 - 4s_3$		$= 8$	$x_3 = 8$
	$x_1 + 1.25x_2$		$- 0.5s_2 + 1.5s_3$		$= 2$	$x_1 = 2$
	$- x_1 - x_2$			$+ e_4$	$= -12$	$e_4 = -12$

TABLE 36
e_4 Is Now a Basic Variable in Row 4

							Basic Variable
z	$+$	$5x_2$	$+ 10s_2 + 10s_3$	$= 280$			$z = 280$
	$-$	$2x_2$	$+ s_1 + 2s_2 - 8s_3$	$= 24$			$s_1 = 24$
	$-$	$2x_2 + x_3$	$+ 2s_2 - 4s_3$	$= 8$			$x_3 = 8$
x_1	$+$	$1.25x_2$	$- 0.5s_2 + 1.5s_3$	$= 2$			$x_1 = 2$
		$0.25x_2$	$\boxed{- 0.5s_2} + 1.5s_3 + e_4$	$= -10$			$e_4 = -10$

TABLE 37
s_2 Enters the Basis in Row 4

						Basic Variable
z	$+ 10x_2$		$+ 40s_3 + 20e_4 = 80$			$z = 80$
	$- x_2$	$+ s_1$	$- 2s_3 + 4e_4 = -16$			$s_1 = -16$
	$- \boxed{x_2} + x_3$		$+ 2s_3 + 4e_4 = -32$			$x_3 = -32$
x_1	$+ x_2$		$- e_4 = 12$			$x_2 = 12$
	$- 0.5x_2$		$+ s_2 - 3s_3 - 2e_4 = 20$			$s_2 = 20$

TABLE 38
Tableau Indicating Infeasibility of Dakota Example When $x_1 + x_2 \geq 12$ Is Required

						Basic Variable
z	$+ 10x_3$		$+ 60s_3 + 60e_4 = -240$			$z = -240$
	$- x_3 + s_1$		$- 4s_3 = 16$			$s_1 = 16$
	$x_2 - x_3$		$- 2s_3 - 4e_4 = 32$			$x_2 = 32$
x_1	$+ x_3$		$+ 2s_3 + 3e_4 = -20$			$x_1 = -20$
	$- 0.5x_3$		$+ s_2 - 4s_3 - 4e_4 = 36$			$s_2 = 36$

Finding the New Optimal Solution After Changing a Right-Hand Side

If the right-hand side of a constraint is changed and the current basis becomes infeasible, the dual simplex can be used to find the new optimal solution. To illustrate, suppose that 30 finishing hours are now available. In Section 6.3, we showed that this changed the current optimal tableau to that shown in Table 39.

Because each variable in row 0 has a non-negative coefficient, the dual simplex method may be used to find the new optimal solution. The variable x_1 is the most negative one, so x_1 must leave the basis, and row 3 will be the pivot row. Because s_2 has the only negative coefficient in row 3, s_2 will enter the basis (see Table 40).

This is an optimal tableau. If 30 finishing hours are available, the new optimal solution to the Dakota problem is to manufacture 16 chairs, 0 tables, and 0 desks. Of course, if we change the right-hand side of a constraint, it is possible that the LP will be infeasible. Step 3 of the dual simplex algorithm will indicate whether this is the case.

TABLE 39
"Old" Optimal Dakota Tableau If 30 Finishing Hours Are Available

							Basic Variable
z	$+$	$5x_2$		$+$	$10s_2 +$	$10s_3 = 380$	$z = 380$
	$-$	$2x_2$	$+ s_1 +$		$2s_2 -$	$8s_3 = 44$	$s_1 = 44$
	$-$	$2x_2 + x_3$		$+$	$2s_2 -$	$4s_3 = 28$	$x_3 = 28$
$x_1 +$		$1.25x_2$		$-$	$(0.5s_2) +$	$1.5s_3 = -3$	$x_1 = -3$

TABLE 40
"New" Optimal Dakota Tableau If 30 Finishing Hours Are Available

					Basic Variable
$z + 20x_1 + 30x_2$		$+ 40s_3 = 320$			$z = 320$
$4x_1 + 3x_2$	$+ s_1$	$- 2s_3 = 32$			$s_1 = 32$
$4x_1 + 3x_2 + x_3$		$+ 2s_3 = 16$			$x_3 = 16$
$- 2x_1 - 2.5x_2$		$+ s_2 - 3s_3 = 6$			$x_1 = 6$

Solving a Normal Min Problem

To illustrate how the dual simplex can be used to solve a normal min problem, we solve the following LP:

$$\min z = x_1 + 2x_2$$
$$\text{s.t.} \quad x_1 - 2x_2 + x_3 \geq 4$$
$$2x_1 + x_2 - x_3 \geq 6$$
$$x_1, x_2, x_3 \geq 0$$

We begin by multiplying z by -1 to convert the LP to a max problem with objective function $z' = -x_1 - 2x_2$. After subtracting excess variables e_1 and e_2 from the two constraints, we obtain the initial tableau in Table 41. Each variable has a nonnegative coefficient in row 0, so the dual simplex method can be applied. Before proceeding, we need to find the basic variables for the constraints. If we multiply each constraint through by -1, we can use e_1 and e_2 as basic variables. This yields the tableau in Table 42. At least one constraint has a negative right-hand side, so this is not an optimal tableau, and we proceed to step 2.

We choose the most negative basic variable (e_2) to exit from the basis. Because e_2 is basic in row 2, row 2 will be the pivot row. To determine the entering variable, we find the following ratios:

$$x_1 \text{ ratio} = 1/-2 = -\tfrac{1}{2}$$
$$x_2 \text{ ratio} = 2/-1 = -2$$

The smaller ratio (in absolute value) is the x_1 ratio, so we use EROs to enter x_1 into the basis in row 2 (see Table 43).[†]

There is no constraint indicating infeasibility (step 3), so we return to step 1. The first constraint has a negative right-hand side, so the tableau is not optimal, and we go to step 2. Because $e_1 = -1$ is the most negative basic variable, e_1 will exit from the basis, and

---

[†]The interested reader may verify that if we had made an error in performing the ratio test and had chosen x_2 to enter the basis, then a negative coefficient in row 0 would have resulted, and dual feasibility would have been destroyed.

TABLE 41

Initial Tableau for Solving Normal Min Problem

$$z' + x_1 + 2x_2 \qquad\qquad = 0$$
$$x_1 - 2x_2 + x_3 - e_1 \qquad = 4$$
$$2x_1 + x_2 - x_3 \qquad - e_2 = 6$$

TABLE 42

Initial Tableau in Canonical Form

	Basic Variable
$z' + x_1 + 2x_1 \qquad\qquad = 0$	$z' = 0$
$- x_1 + 2x_2 - x_3 + e_1 \qquad = -4$	$e_1 = -4$
$-\boxed{2x_1} - x_2 + x_3 \qquad + e_2 = -6$	$e_2 = -6$

TABLE 43

First Dual Simplex Tableau

	Basic Variable
$z' \quad + \frac{3}{2}x_2 + \frac{1}{2}x_3 \qquad + \frac{1}{2}e_2 = -3$	$z' = -3$
$\frac{5}{2}x_2 - \boxed{\frac{3}{2}x_3} + e_1 - \frac{1}{2}e_2 = -1$	$e_1 = -1$
$x_1 + \frac{1}{2}x_2 - \frac{1}{2}x_3 \qquad - \frac{1}{2}e_2 = 3$	$x_1 = 3$

row 1 will be the pivot row. The possible entering variables are x_3 and e_2. The relevant ratios are

$$x_3 \text{ ratio} = \frac{\frac{1}{2}}{-\frac{3}{2}} = -\frac{1}{3}$$

$$e_2 \text{ ratio} = \frac{\frac{1}{2}}{-\frac{1}{2}} = -1$$

The smallest ratio (in absolute value) is $-\frac{1}{3}$, so x_3 will enter the basis in row 1. After pivoting in x_3, the new tableau is as shown in Table 44.[†] Each right-hand side is non-negative, so this is an optimal tableau. The original problem was a min problem, so the optimal solution to the original min problem is $z = \frac{10}{3}$, $x_1 = \frac{10}{3}$, $x_3 = \frac{2}{3}$, and $x_2 = 0$.

Observe that each dual simplex tableau (except the optimal dual simplex tableau) has a z'-value exceeding the optimal z'-value. For this reason, we say that the dual simplex tableaus are superoptimal. As the dual simplex proceeds, each pivot brings us closer to a primal feasible solution. Each pivot (barring degeneracy) decreases z', and we are "less superoptimal." Once primal feasibility is obtained, our solution is optimal.

TABLE 44

Optimal Tableau for Dual Simplex Example

	Basic Variable
$z' \quad + \frac{7}{3}x_2 \qquad + \frac{1}{3}e_1 + \frac{1}{3}e_2 = -\frac{10}{3}$	$z' = -\frac{10}{3}$
$-\frac{5}{3}x_2 + x_3 - \frac{2}{3}e_1 + \frac{1}{3}e_2 = \frac{2}{3}$	$x_3 = \frac{2}{3}$
$x_1 - \frac{1}{3}x_2 \qquad - \frac{1}{3}e_1 - \frac{1}{3}e_2 = \frac{10}{3}$	$x_1 = \frac{10}{3}$

[†]If we had chosen to enter into the basis any variable with a positive coefficient in the pivot row, then we would have ended up with some negative entries in row 0. This is why any variable that is entered into the basis must have a negative coefficient in the pivot row.

PROBLEMS

Group A

1 Use the dual simplex method to solve the following LP:

$$\max z = -2x_1 - x_3$$
$$\text{s.t.} \quad x_1 + x_2 - x_3 \geq 5$$
$$x_1 - 2x_2 + 4x_3 \geq 8$$
$$x_1, x_2, x_3 \geq 0$$

2 In solving the following LP, we obtain the optimal tableau shown in Table 45.

$$\max z = 6x_1 + x_2$$
$$\text{s.t.} \quad x_1 + x_2 \leq 5$$
$$2x_1 + x_2 \leq 6$$
$$x_1, x_2 \geq 0$$

a Find the optimal solution to this LP if we add the constraint $3x_1 + x_2 \leq 10$.

b Find the optimal solution if we add the constraint $x_1 - x_2 \geq 6$.

TABLE 45

			Basic Variable
$z + 2x_2$	$+ 3s_2 = 18$		$z_1 = 18$
$0.5x_2 + s_1 - 0.5s_2 = 2$			$s_1 = 2$
$x_1 + 0.5x_2$	$+ 0.5s_2 = 3$		$x_1 = 3$

c Find the optimal solution if we add the constraint $8x_1 + x_2 \leq 12$.

3 Find the new optimal solution to the Dakota problem if only 20 board ft of lumber are available.

4 Find the new optimal solution to the Dakota problem if 15 carpentry hours are available.

6.12 Data Envelopment Analysis[†]

Often we wonder if a university, hospital, restaurant, or other business is operating efficiently. The **Data Envelopment Analysis (DEA) method** can be used to answer this question. Our presentation is based on Callen (1991). To illustrate how DEA works, let's consider a group of three hospitals. To simplify matters, we assume that each hospital "converts" two inputs into three different outputs. The two inputs used by each hospital are

Input 1 = capital (measured by the number of hospital beds)

Input 2 = labor (measured in thousands of labor hours used during a month)

The outputs produced by each hospital are

Output 1 = hundreds of patient-days during month for patients under age 14

Output 2 = hundreds of patient-days during month for patients between 14 and 65

Output 3 = hundreds of patient-days during month for patients over 65

Suppose that the inputs and outputs for the three hospitals are as given in Table 46.
To determine whether a hospital is efficient, let's define t_r = price or value of one unit of output r and w_s = cost of one unit of input s. The *efficiency* of hospital i is defined to be

$$\frac{\text{value of hospital } i\text{'s outputs}}{\text{cost of hospital } i\text{'s inputs}}$$

For the data in Table 46, we find the efficiency of each hospital to be as follows:

$$\text{Hospital 1 efficiency} = \frac{9t_1 + 4t_2 + 16t_3}{5w_1 + 14w_2}$$

$$\text{Hospital 2 efficiency} = \frac{5t_1 + 7t_2 + 10t_3}{8w_1 + 15w_2}$$

$$\text{Hospital 3 efficiency} = \frac{4t_1 + 9t_2 + 13t_3}{7w_1 + 12w_2}$$

[†]This section may be omitted without loss of continuity.

TABLE 46

Inputs and Outputs for Hospitals

Hospital	Inputs		Outputs		
	1	2	1	2	3
1	5	14	9	4	16
2	8	15	5	7	10
3	7	12	4	9	13

The DEA approach uses the following four ideas to determine if a hospital is efficient.

1 No hospital can be more than 100% efficient. Thus, the efficiency of each hospital must be less than or equal to 1. For hospital 1, we find that $(9t_1 + 4t_2 + 16t_3)/(5w_1 + 14w_2) \leq 1$. Multiplying both sides of this inequality by $(5w_1 + 14w_2)$ (this is the trick we used to simplify blending constraints in Section 3.8!) yields the LP constraint $5w_1 + 14w_2 - 9t_1 - 4t_2 - 16t_3 \geq 0$.

2 Suppose we are interested in evaluating the efficiency of hospital i. We attempt to choose output prices (t_1, t_2, and t_3) and input costs (w_1 and w_2) that maximize efficiency. If the efficiency of hospital i equals 1, then it is efficient; if the efficiency is less than 1, then it is inefficient.

3 To simplify computations, we may scale the output prices so that the cost of hospital i's inputs equals 1. Thus, for hospital 2 we add the constraint $8w_1 + 15w_2 = 1$.

4 We must ensure that each input cost and output price is strictly positive. If, for example, $t_i = 0$, then DEA could not detect an inefficiency involving output i; if $w_j = 0$, then DEA could not detect an inefficiency involving input j.

Points (1)–(4) lead to the following LPs for testing the efficiency of each hospital.

Hospital 1 LP

$$\max z = 9t_1 + 4t_2 + 16t_3 \tag{1}$$

$$\text{s.t.} \quad -9t_1 - 4t_2 - 16t_3 + 5w_1 + 14w_2 \geq 0 \tag{2}$$

$$-5t_1 - 7t_2 - 10t_3 + 8w_1 + 15w_2 \geq 0 \tag{3}$$

$$-4t_1 - 9t_2 - 13t_3 + 7w_1 + 12w_2 \geq 0 \tag{4}$$

$$5w_1 + 14w_2 = 1 \tag{5}$$

$$t_1 \geq .0001 \tag{6}$$

$$t_2 \geq .0001 \tag{7}$$

$$t_3 \geq .0001 \tag{8}$$

$$w_1 \geq .0001 \tag{9}$$

$$w_2 \geq .0001 \tag{10}$$

Hospital 2 LP

$$\max z = 5t_1 + 7t_2 + 10t_3 \tag{1}$$

$$\text{s.t.} \quad -9t_1 - 4t_2 - 16t_3 + 5w_1 + 14w_2 \geq 0 \tag{2}$$

$$-5t_1 - 7t_2 - 10t_3 + 8w_1 + 15w_2 \geq 0 \tag{3}$$

$$-4t_1 - 9t_2 - 13t_3 + 7w_1 + 12w_2 \geq 0 \tag{4}$$

$$8w_1 + 15w_2 = 1 \tag{5}$$

$$t_1 \geq .0001 \tag{6}$$

$$t_2 \geq .0001 \quad (7)$$
$$t_3 \geq .0001 \quad (8)$$
$$w_1 \geq .0001 \quad (9)$$
$$w_2 \geq .0001 (10)$$

Hospital 3 LP $\quad \max z = 4t_1 + 9t_2 + 13t_3 \qquad\qquad (1)$

$$\text{s.t.} \quad -9t_1 - 4t_2 - 16t_3 + 5w_1 + 14w_2 \geq 0 \qquad (2)$$
$$-5t_1 - 7t_2 - 10t_3 + 8w_1 + 15w_2 \geq 0 \qquad (3)$$
$$-4t_1 - 9t_2 - 13t_3 + 7w_1 + 12w_2 \geq 0 \qquad (4)$$
$$7w_1 + 12w_2 = 1 \qquad (5)$$
$$t_1 \geq .0001 \quad (6)$$
$$t_2 \geq .0001 \quad (7)$$
$$t_3 \geq .0001 \quad (8)$$
$$w_1 \geq .0001 \quad (9)$$
$$w_2 \geq .0001 (10)$$

Let's see how the hospital 1 LP incorporates points (1)–(4). Point (1) maximizes the efficiency of hospital 1. This is because Constraint (5) implies that the total cost of hospital 1's inputs equal 1. Constraints (2)–(4) ensure that no hospital is more than 100% efficient. Constraints (6)–(10) ensure that each input cost and output price is strictly positive (the .0001 right-hand side is arbitrary; any small positive number may be used).

The LINDO output for these LPs is given in Figures 10(a)–(c). From the optimal ob-

```
MAX  9 T1 + 4 T2 + 16 T3
SUBJECT TO
      2)   - 9 T1 - 4 T2 - 16 T3 + 5 W1 + 14 W2 >=    0
      3)   - 5 T1 - 7 T2 - 10 T3 + 8 W1 + 15 W2 >=    0
      4)   - 4 T1 - 9 T2 - 13 T3 + 7 W1 + 12 W2 >=    0
      5)     W1 >=     0.0001
      6)     W2 >=     0.0001
      7)     T1 >=     0.0001
      8)     T2 >=     0.0001
      9)     T3 >=     0.0001
     10)     5 W1 + 14 W2 =     1
END

LP OPTIMUM FOUND AT STEP        6

          OBJECTIVE FUNCTION VALUE

       1)     1.00000000

VARIABLE          VALUE          REDUCED COST
      T1          .110889           .000000
      T2          .000100           .000000
      T3          .000100           .000000
      W1          .000100           .000000
      W2          .071393           .000000


      ROW    SLACK OR SURPLUS     DUAL PRICES
       2)          .000000         -1.000000
       3)          .515548           .000000
       4)          .411659           .000000
       5)          .000000           .000000
       6)          .071293           .000000
       7)          .110789           .000000
       8)          .000000           .000000
       9)          .000000           .000000
      10)          .000000          1.000000

NO. ITERATIONS=        6
```

FIGURE 10(a)
Hospital 1 LP

```
            MAX    5 T1 + 7 T2 + 10 T3
            SUBJECT TO
                   2)   - 9 T1 - 4 T2 - 16 T3 + 5 W1 + 14 W2  >=     0
                   3)   - 5 T1 - 7 T2 - 10 T3 + 8 W1 + 15 W2  >=     0
                   4)   - 4 T1 - 9 T2 - 13 T3 + 7 W1 + 12 W2  >=     0
                   5)     8 W1 + 15 W2 =    1
                   6)     W1   >=   0.0001
                   7)     W2   >=   0.0001
                   8)     T1   >=   0.0001
                   9)     T2   >=   0.0001
                  10)     T3   >=   0.0001
            END

            LP OPTIMUM FOUND AT STEP          0

                   OBJECTIVE FUNCTION VALUE

                 1)      .773030000

            VARIABLE           VALUE          REDUCED COST
                  T1          .079821            .000000
                  T2          .053275            .000000
                  T3          .000100            .000000
                  W1          .000100            .000000
                  W2          .066613            .000000


                 ROW     SLACK OR SURPLUS     DUAL PRICES
                  2)          .000000          -.261538
                  3)          .226970           .000000
                  4)          .000000          -.661538
                  5)          .000000           .773333
                  6)          .000000          -.248206
                  7)          .066513           .000000
                  8)          .079721           .000000
                  9)          .053175           .000000
                 10)          .000000         -2.784615

            NO. ITERATIONS=          0
```

FIGURE 10(b)
Hospital 2 LP

jective function value to each LP we find that

$$\text{Hospital 1 efficiency} = 1$$
$$\text{Hospital 2 efficiency} = .773$$
$$\text{Hospital 3 efficiency} = 1$$

Thus we find that hospital 2 is inefficient and hospitals 1 and 3 are efficient.

REMARK   **1**   An easy way to create the hospital 2 LP is to use LINDO to modify the objective function of the hospital 1 LP and the constraint $5w_1 + 14w_2 = 1$. Then it is easy to modify the hospital 2 LP to create the hospital 3 LP.

Using LINGO to Run a DEA

DEA.lng

The following LINGO program (see file DEA.lng) will solve our hospital DEA problem. When faced with another DEA problem, we begin by changing the numbers of inputs, outputs, and units. Next we change the resource usage and outputs for each unit. Finally, by changing number to (say) 1, we can evaluate the efficiency of unit 1. If the optimal objective function value for unit 1 is less than 1, then unit 1 is inefficient. Otherwise, unit 1 is efficient.

```
SETS:
INPUTS/1..2/:COSTS;
OUTPUTS/1..3/:PRICES;
```

```
           MAX  4 TI + 9 + T2 + 13 T3
           SUBJECT TO
                 2) - 9 T1 - 4 T2 - 16 T3 + 5 W1 + 14 W2 >= 0
                 3) - 5 T1 - 7 T2 - 10 T3 + 8 W1 + 15 W2 >= 0
                 4) - 4 T1 - 9 T2 - 13 T3 + 7 W1 + 12 W2 >= 0
                 5)   W1 >=   0.0001
                 6)   W2 >=   0.0001
                 7)   T1 >=   0.0001
                 8)   T2 >=   0.0001
                 9)   T3 >=   0.0001
                10)  7 W1 + 12 W2 = 1
           END

           LP OPTIMUM FOUND AT STEP 7

               OBJECTIVE FUNCTION VALUE

               1)    1.00000000

           VARIABLE        VALUE        REDUCED COST
              T1          .099815          .000000
              T2          .066605          .000000
              T3          .000100          .000000
              W1          .000100          .000000
              W2          .083275          .000000


           ROW     SLACK OR SURPLUS     DUAL PRICES
            2)         .000000            .000000
            3)         .283620            .000000
            4)         .000000          -1.000000
            5)         .000000            .000000
            6)         .083175            .000000
            7)         .099715            .000000
            8)         .066505            .000000
            9)         .000000            .000000
           10)         .000000           1.000000
```

FIGURE 10(c)

Hospital 3 LP   NO. ITERATIONS=   7

```
UNITS/1..3/;
UNIN(UNITS,INPUTS):USED;
UNOUT(UNITS,OUTPUTS):PRODUCED;
ENDSETS
NUMBER=2;
@FOR(UNITS(J)|j#EQ#NUMBER:MAX=@SUM(OUTPUTS(I):PRICES(I)*PRODUCED(J,I)));
@FOR(UNITS(J)|J#EQ#NUMBER:@SUM(INPUTS(I):COSTS(I)*USED(J,I))=1);
@FOR(INPUTS(I):COSTS(I)>=.0001);
@FOR(OUTPUTS(I):PRICES(I)>=.0001);
@FOR(UNITS(I):@SUM(INPUTS(J):COSTS(J)*USED(I,J))>=@SUM(OUTPUTS(J):PRICES(J)*PRODUCED(I,J))
);
DATA:
USED=5,14,
     8,15,
     7,12;
PRODUCED=9,4,16,
         5,7,10,
         4,9,13;
ENDDATA
END
```

Dual Prices and DEA

The DUAL PRICES section of the LINDO output gives us great insight into Hospital 2's (or any organization's found inefficient by DEA) inefficiency. Consider all hospitals whose efficiency constraints have nonzero dual prices in the hospital 2 LP (Figure 10b). (In our example, hospitals 1 and 3 have nonzero dual prices.) If we average the output vectors and input vectors for these hospitals (using the absolute value of the dual price for each hospital as the weight) we obtain the following:

Averaged Output Vector

$$.261538 \begin{bmatrix} 9 \\ 4 \\ 16 \end{bmatrix} + .661538 \begin{bmatrix} 4 \\ 9 \\ 13 \end{bmatrix} = \begin{bmatrix} 5 \\ 7 \\ 12.785 \end{bmatrix}$$

Averaged Input Vector

$$.261538 \begin{bmatrix} 5 \\ 14 \end{bmatrix} + .661538 \begin{bmatrix} 7 \\ 12 \end{bmatrix} = \begin{bmatrix} 5.938 \\ 11.6 \end{bmatrix}$$

Suppose we create a composite hospital by combining .261538 of hospital 1 with .661538 of hospital 3. The averaged output vector tells us that the composite hospital produces the same amount of outputs 1 and 2 as hospital 2, but the composite hospital produces $12.785 - 10 = 2.785$ more of output 3 (patient days for more than 65 patients). From the averaged input vector for the composite hospital, we find that the composite hospital uses less of each input than does hospital 2. We now see exactly where hospital 2 is inefficient!

By the way, the objective function value of .7730 for the hospital 2 LP implies that the more efficient composite hospital produces its superior outputs by using at most 77.30% as much of each input. Note that

Input 1 used by composite hospital $< .7730$ * (Input 1 used by hospital 2) $= 6.2186$

and

Input 2 used by composite hospital $= .7730$ * (Input 2 used by hospital 2) $= 11.6$

An explanation of why the dual prices are needed to find a composite hospital that is superior to an inefficient hospital is given in Problems 5–7.

PROBLEMS

Group A

1 The Salem Board of Education wants to evaluate the efficiency of the town's four elementary schools. The three outputs of the schools are defined to be

Output 1 = average reading score

Output 2 = average mathematics score

Output 3 = average self-esteem score

The three inputs to the schools are defined to be

Input 1 = average educational level of mothers (defined by highest grade completed—12 = high school graduate; 16 = college graduate, and so on).

Input 2 = number of parent visits to school (per child)

Input 3 = teacher to student ratio

The relevant information for the four schools is given in Table 47. Determine which (if any) schools are inefficient. For any inefficient school, determine the nature of the inefficiency.

2 Pine Valley Bank has three branches. You have been assigned to evaluate the efficiency of each. The following inputs and outputs are to be used for the study.

TABLE 47

School	Inputs 1	2	3	Outputs 1	2	3
1	13	4	.05	9	7	6
2	14	5	.05	10	8	7
3	11	6	.06	11	7	8
4	15	8	.08	9	9	9

Input 1 = labor hours used (hundreds per month)

Input 2 = space used (in hundreds of square feet)

Input 3 = supplies used per month (in dollars)

Output 1 = loan applications per month

Output 2 = deposits processed per month (in thousands)

Output 3 = checks processed per month (in thousands)

The relevant information is given in Table 48. Use this data to determine if any bank branches are inefficient. If any

TABLE 48

	Inputs			Outputs		
Bank	1	2	3	1	2	3
1	15	20	50	200	15	35
2	14	23	51	220	18	45
3	16	19	51	210	17	20

TABLE 49

	Inputs		Outputs	
Precinct	1	2	1	2
1	200	60	6	8
2	300	90	8	9.5
3	400	120	10	11

bank branches are inefficient, determine the nature of the inefficiency.

3 You have been assigned to evaluate the efficiency of the Port Charles Police Department. Three precincts are to be evaluated. The inputs and outputs for each precinct are as follows:

Input 1 = number of police officers

Input 2 = number of vehicles used

Output 1 = number of patrol units responding to service requests (thousands per year)

Output 2 = number of convictions obtained each year (in hundreds)

You are given the data in Table 49. Use this information to determine which precincts, if any, are inefficient. For any inefficient precincts, determine the nature of the inefficiency.

4 You have been assigned by Indiana University to evaluate the relative efficiency of four degree-granting units: Business; Education; Arts and Sciences; and Health, Physical Education, and Recreation (HPER). You are given the

information in Table 50. Use DEA to find all inefficient units. Comment on the nature of the inefficiencies you found.

Group B

5 Explain why the amount of each output produced by the composite hospital obtained by averaging hospitals 1 and 3 (with the absolute value of the dual prices as weights) is at least as large as the amount of the corresponding output produced by hospital 2. (*Hint:* Price out variables t_1, t_2, and t_3, and use the fact that the coefficient of these variables in row 0 of the optimal tableau must equal 0.)

6 Explain why the dual price for the $8w_1 + 15w_2 = 1$ constraint must equal the optimal z-value for the hospital 2 LP.

7 a Explain why the amount of each input used by the composite hospital is at most (efficiency of hospital 2) * (the amount of the corresponding input used by hospital 2). (*Hint:* Price out w_1 and w_2 and use Problem 6.)

b Explain why the amount of each input used by the composite hospital is no larger than the amount of the corresponding input used by hospital 2.

TABLE 50

	Faculty	Support Staff	Supply Budget (in Millions)	Credit Hours (in Thousands)	Research Publications
Business	150	70	5	15	225
Education	60	20	3	5.4	70
Arts and Sciences	800	140	20	56	1,300
HPER	30	15	1	2.1	40

SUMMARY Graphical Sensitivity Analysis

To determine whether the current basis remains optimal after changing an objective function coefficient, note that the change affects the slope of the isoprofit line. The current basis remains optimal as long as the current optimal solution is the last point in the feasible region to make contact with isoprofit lines as we move in the direction of increasing z (for a max problem). If the current basis remains optimal, then the values of the decision variables remain unchanged, but the optimal z-value may change.

To determine whether the current basis remains optimal after changing the right-hand side of a constraint, find the constraints (possibly including sign restrictions) that are binding for the current optimal solution. As we change the right-hand side of a constraint, the current basis remains optimal as long as the point where the constraints are binding remains feasible. Even if the current basis remains optimal, the values of the decision variables and the optimal z-value may change.

Shadow Prices

The **shadow price** of the ith constraint of a linear programming problem is the amount by which the optimal z-value is improved if the right-hand side is increased by 1. The shadow price of the ith constraint is the DUAL PRICE for row $i + 1$ in the LINDO output.

Notation

BV_i = basic variable for ith constraint in the optimal tableau

$\mathbf{c}_{\text{BV}}$ = row vector whose ith element is the objective function coefficient for BV_i in the LP

$\mathbf{a}_j$ = column for variable x_j in constraints of original LP

$\mathbf{b}$ = right-hand side vector for original LP

$\bar{c}_j$ = coefficient of x_j in row 0 of the optimal tableau

How to Compute Optimal Tableau from Initial LP

$$\text{Column for } x_j \text{ in optimal tableau's constraints} = B^{-1}\mathbf{a}_j \tag{5}$$

$$\text{Right-hand side of optimal tableau's constraints} = B^{-1}\mathbf{b} \tag{6}$$

$$\bar{c}_j = \mathbf{c}_{\text{BV}}B^{-1}\mathbf{a}_j - c_j \tag{10}$$

$$\text{Coefficient of slack variable } s_i \text{ in optimal row 0} = \\ i\text{th element of } \mathbf{c}_{\text{BV}}B^{-1} \tag{10'}$$

$$\text{Coefficient of excess variable } e_i \text{ in optimal row 0} = \\ -(i\text{th element of } \mathbf{c}_{\text{BV}}B^{-1}) \tag{10''}$$

$$\text{Coefficient of artificial variable } a_i \text{ in optimal row 0} = \\ (i\text{th element of } \mathbf{c}_{\text{BV}}B^{-1}) + M \tag{10'''}$$

$$\text{Right-hand side of optimal row 0} = \mathbf{c}_{\text{BV}}B^{-1}\mathbf{b} \tag{11}$$

Sensitivity Analysis

For a max problem, a tableau is optimal if and only if each variable has a nonnegative coefficient in row 0 and each constraint has a nonnegative right-hand side. For a min problem, a tableau is optimal if and only if each variable has a nonpositive coefficient in row 0 and each constraint has a nonnegative right-hand side.

If the current basis remains optimal after changing the objective function coefficient of a nonbasic variable, the values of the decision variables and the optimal z-value remain unchanged. With a basic variable, the values of the decision variables remain unchanged, but the optimal z-value may change. Both the values of the decision variables and the optimal z-value may change after changing a right-hand side. The new values of the decision variables may be found by computing B^{-1} (new right-hand side vector). The new optimal z-value may be determined by using shadow prices or Equation (11).

Objective Function Coefficient Range

The OBJ COEFFICIENT RANGES section of the LINDO output gives the range of values for an objective function coefficient for which the current basis remains optimal. Within this range, the values of the decision variables remain unchanged, but the optimal z-value may or may not change.

Reduced Cost

For any nonbasic variable, the reduced cost for the variable is the amount by which its objective function coefficient must be improved before that variable will be a basic variable in some optimal solution to the LP.

Right-Hand Side Range

If the right-hand side of a constraint remains within the RIGHTHAND SIDE RANGE of the LINDO printout, the current basis remains optimal, and the LINDO listing for the constraint's dual price may be used to determine how the change affects the optimal z-value. Even if the right-hand side of a constraint remains within the range, the values of the decision variables will probably change.

Finding the Dual of an LP

For a normal (all $\leq$ constraints and all variables nonnegative) max problem or a normal min (all $\geq$ constraints and all variables nonnegative) problem, we find the dual as follows:

If we read the primal across in Table 14, we read the dual down. If we read the primal down in Table 14, we read the dual across. We use x_i's and z as variables for a maximization problem and y_j's and w as variables for a minimization problem.

To find the dual of a nonnormal max problem:

Step 1 Fill in Table 14 so that the primal can be read across.

Step 2 After making the following changes, the dual can be read down in the usual fashion: (a) If the ith primal constraint is a $\geq$ constraint, the corresponding dual variable y_i must satisfy $y_i \leq 0$. (b) If the ith primal constraint is an equality, then the dual variable y_i is now urs. (c) If the ith primal variable is urs, then the ith dual constraint will be an equality.

To find the dual of a nonnormal min problem:

Step 1 Write out the primal so it can be read down in Table 14.

Step 2 Except for the following changes, the dual can be read across the table: (a) If the ith primal constraint is a $\leq$ constraint, then the corresponding dual variable x_i must satisfy $x_i \leq 0$. (b) If the ith primal constraint is an equality, then the corresponding dual variable x_i will be urs. (c) If the ith primal variable y_i is urs, then the ith dual constraint is an equality.

The Dual Theorem

Suppose BV is an optimal basis for the primal. Then $\mathbf{c}_{\text{BV}} B^{-1}$ is an optimal solution to the dual. Also, $\bar{z} = \bar{w}$.

Finding the Optimal Solution to the Dual of an LP

If the primal is a max problem, then the optimal dual solution may be read from row 0 of the optimal tableau by using the following rules:

$$\begin{array}{l} \text{Optimal value of dual variable } y_i \\ \text{if Constraint } i \text{ is a } \leq \text{ constraint} \end{array} = \text{coefficient of } s_i \text{ in optimal row 0} \qquad \textbf{(31)}$$

$$\begin{array}{l} \text{Optimal value of dual variable } y_i \\ \text{if Constraint } i \text{ is a } \geq \text{ constraint} \end{array} = -(\text{coefficient of } e_i \text{ in optimal row 0}) \qquad \textbf{(31}')$$

$$\begin{array}{l} \text{Optimal value of dual variable } y_i \\ \text{if Constraint } i \text{ is an equality} \\ \text{constraint} \end{array} = (\text{coefficient of } a_i \text{ in optimal row 0}) - M \quad \textbf{(31}'')$$

If the primal is a min problem, then the optimal dual solution may be read from row 0 of the optimal tableau by using the following rules:

$$\begin{array}{l} \text{Optimal value of dual variable } x_i \\ \text{if Constraint } i \text{ is a } \leq \text{ constraint} \end{array} = \text{coefficient of } s_i \text{ in optimal row 0}$$

$$\begin{array}{l} \text{Optimal value of dual variable } x_i \\ \text{if Constraint } i \text{ is a } \geq \text{ constraint} \end{array} = -(\text{coefficient of } e_i \text{ in optimal row 0})$$

$$\begin{array}{l} \text{Optimal value of dual variable } x_i \\ \text{if Constraint } i \text{ is an equality} \\ \text{constraint} \end{array} = (\text{coefficient of } a_i \text{ in optimal row 0}) + M$$

Shadow Prices (Again)

For a maximization LP, the shadow price of the ith constraint is the value of the ith dual variable in the optimal dual solution. For a minimization LP, the shadow price of the ith constraint $= -(i$th dual variable in the optimal dual solution). The shadow price of the ith constraint is found in row $i + 1$ of the DUAL PRICES portion of the LINDO printout.

$$\begin{array}{l} \text{New optimal } z\text{-value} = (\text{old optimal } z\text{-value}) \\ \qquad + (\text{Constraint } i \text{ shadow price}) \, \Delta b_i \quad \text{(max problem)} \end{array} \qquad \textbf{(37)}$$

$$\begin{array}{l} \text{New optimal } z\text{-value} = (\text{old optimal } z\text{-value}) \\ \qquad - (\text{Constraint } i \text{ shadow price}) \, \Delta b_i \quad \text{(min problem)} \end{array} \qquad \textbf{(37}')$$

A $\geq$ constraint will have a nonpositive shadow price; a $\leq$ constraint will have a nonnegative shadow price; and an equality constraint may have a positive, negative, or zero shadow price.

Duality and Sensitivity Analysis

Our proof of the Dual Theorem showed that if a set of basic variables BV is feasible, then BV is optimal (that is, each variable in row 0 has a nonnegative coefficient) if and only if the associated dual solution, $\mathbf{c}_{\text{BV}}B^{-1}$, is dual feasible.

This result can be used to yield an alternative way of doing the following types of sensitivity analysis:

Change 1   Changing the objective function coefficient of a nonbasic variable

Change 4   Changing the column of a nonbasic variable

Change 5   Adding a new activity

In each case, simply determine whether a change in the original LP maintains dual feasibility. If dual feasibility is maintained, then the current basis remains optimal. If dual feasibility is not maintained, then the current basis is no longer optimal.

Complementary Slackness

THEOREM 3

Let

$$\mathbf{x} = \begin{bmatrix} x_1 \\ x_2 \\ \vdots \\ x_n \end{bmatrix}$$

be a feasible primal solution and $\mathbf{y} = \begin{bmatrix} y_1 & y_2 & \cdots & y_m \end{bmatrix}$ be a feasible dual solution. Then $\mathbf{x}$ is primal optimal and $\mathbf{y}$ is dual optimal if and only if

$$s_i y_i = 0 \quad (i = 1, 2, \ldots, m) \tag{38}$$
$$e_j x_j = 0 \quad (j = 1, 2, \ldots, n) \tag{39}$$

The Dual Simplex Method

The dual simplex method can be applied (to a max problem) whenever there is a basic solution in which each variable has a nonnegative coefficient in row 0. If we have found such a basic solution, then the dual simplex method proceeds as follows:

Step 1 If the right-hand side of each constraint is nonnegative, then an optimal solution has been found; if not, then at least one constraint has a negative right-hand side, and we go to step 2.

Step 2 Choose the most negative basic variable as the variable to leave the basis. The row in which this variable is basic will be the pivot row. To select the variable that enters the basis, compute the following ratio for each variable x_j that has a *negative* coefficient in the pivot row:

$$\frac{\text{Coefficient of } x_j \text{ in row } 0}{\text{Coefficient of } x_j \text{ in the pivot row}}$$

Choose the variable that has the smallest ratio (absolute value) as the entering variable. Use EROs to make the entering variable a basic variation in the pivot row.

Step 3 If there is any constraint in which the right-hand side is negative and each variable has a nonnegative coefficient, then the LP has no feasible solution. Infeasibility would be indicated by the presence (after possibly several pivots) of a constraint such as $x_1 + 2x_2 + x_3 = -5$. If no constraint indicating infeasibility is found, return to step 1.

The dual simplex method is often used in the following situations:

1 Finding the new optimal solution after a constraint is added to an LP

2 Finding the new optimal solution after changing an LP's right-hand side

3 Solving a normal min problem

REVIEW PROBLEMS

All problems from Sections 5.2 and 5.3 are relevant, along with Chapter 5 Review Problems 1, 2, 6, and 7.

Group A

1 Consider the following LP and its optimal tableau (Table 51):

$$\max z = 4x_1 + x_2$$
$$\text{s.t.} \quad x_1 + 2x_2 = 6$$
$$x_1 - x_2 \geq 3$$
$$2x_1 + x_2 \leq 10$$
$$x_1, x_2 \geq 0$$

a Find the dual of this LP and its optimal solution.

b Find the range of values of b_3 for which the current basis remains optimal. If $b_3 = 11$, what would be the new optimal solution?

2 For the LP in Problem 1, graphically determine the range of values on c_1 for which the current basis remains optimal. (*Hint:* The feasible region is a line segment.)

3 Consider the following LP and its optimal tableau (Table 52):

$$\max z = 5x_1 + x_2 + 2x_3$$
$$\text{s.t.} \quad x_1 + x_2 + x_3 \leq 6$$
$$6x_1 + x_3 \leq 8$$
$$x_2 + x_3 \leq 2$$
$$x_1, x_2, x_3 \geq 0$$

a Find the dual to this LP and its optimal solution.

b Find the range of values of c_1 for which the current basis remains optimal.

c Find the range of values of c_2 for which the current basis remains optimal.

4 Carco manufactures cars and trucks. Each car contributes $300 to profit and each truck, $400. The

TABLE 51

z	x_1	x_2	e_2	s_3	a_1	a_2	rhs
1	0	0	0	$\frac{7}{3}$	$M - \frac{2}{3}$	M	$\frac{58}{3}$
0	0	1	0	$-\frac{1}{3}$	$\frac{2}{3}$	0	$\frac{2}{3}$
0	1	0	0	$\frac{2}{3}$	$-\frac{1}{3}$	0	$\frac{14}{3}$
0	0	0	1	1	-1	-1	1

TABLE 52

z	x_1	x_2	x_3	s_1	s_2	s_3	rhs
1	0	$\frac{1}{6}$	0	0	$\frac{5}{6}$	$\frac{7}{6}$	9
0	0	$\frac{1}{6}$	0	1	$-\frac{1}{6}$	$-\frac{5}{6}$	3
0	1	$-\frac{1}{6}$	0	0	$\frac{1}{6}$	$-\frac{1}{6}$	1
0	0	1	1	0	0	1	2

TABLE 53

Vehicle	Days on Type 1 Machine	Days on Type 2 Machine	Tons of Steel
Car	0.8	0.6	2
Truck	1	0.7	3

resources required to manufacture a car and a truck are shown in Table 53. Each day, Carco can rent up to 98 Type 1 machines at a cost of $50 per machine. The company now has 73 Type 2 machines and 260 tons of steel available. Marketing considerations dictate that at least 88 cars and at least 26 trucks be produced. Let

$$X1 = \text{number of cars produced daily}$$
$$X2 = \text{number of trucks produced daily}$$
$$M1 = \text{type 1 machines rented daily}$$

To maximize profit, Carco should solve the LP given in Figure 11. Use the LINDO output to answer the following questions:

a If cars contributed $310 to profit, what would be the new optimal solution to the problem?

b What is the most that Carco should be willing to pay to rent an additional Type 1 machine for 1 day?

c What is the most that Carco should be willing to pay for an extra ton of steel?

d If Carco were required to produce at least 86 cars, what would Carco's profit become?

e Carco is considering producing jeeps. A jeep contributes $600 to profit and requires 1.2 days on machine 1, 2 days on machine 2, and 4 tons of steel. Should Carco produce any jeeps?

5 The following LP has the optimal tableau shown in Table 54.

$$\max z = 4x_1 + x_2$$
$$\text{s.t.} \quad 3x_1 + x_2 \geq 6$$
$$2x_1 + x_2 \geq 4$$
$$x_1 + x_2 = 3$$
$$x_1, x_2 \geq 0$$

a Find the dual of this LP and its optimal solution.

b Find the range of values of the objective function coefficient of x_2 for which the current basis remains optimal.

c Find the range of values of the objective function coefficient of x_1 for which the current basis remains optimal.

6 Consider the following LP and its optimal tableau (Table 55):

$$\max z = 3x_1 + x_2 - x_3$$
$$\text{s.t.} \quad 2x_1 + x_2 + x_3 \leq 8$$
$$4x_1 + x_2 - x_3 \leq 10$$
$$x_1, x_2, x_3 \geq 0$$

FIGURE 11
LINDO Output for Carco (Problem 4)

```
MAX        300 X1 + 400 X2 - 50 M1
SUBJECT TO
     2)    0.8 X1 + X2 - M1 <=       0
     3)      M1 <=   98
     4)    0.6 X1 + 0.7 X2 <=    73
     5)    2 X1 + 3 X2 <=    260
     6)      X1 >=    88
     7)      X2 >=    26
END

   LP OPTIMUM FOUND  AT STEP      1

        OBJECTIVE FUNCTION VALUE

   1)        32540.0000

VARIABLE          VALUE      REDUCED COST
     X1        88.000000        0.000000
     X2        27.599998        0.000000
     M1        98.000000        0.000000


ROW       SLACK OR SURPLUS    DUAL PRICES
     2)         0.000000      400.000000
     3)         0.000000      350.000000
     4)         0.879999        0.000000
     5)         1.200003        0.000000
     6)         0.000000      -20.000000
     7)         1.599999        0.000000


NO. ITERATIONS=            1

   RANGES IN WHICH THE BASIS IS UNCHANGED

                      OBJ COEFFICIENT RANGES
VARIABLE       CURRENT        ALLOWABLE       ALLOWABLE
               COEF           INCREASE        DECREASE
     X1      300.000000      20.000000       INFINITY
     X2      400.000000      INFINITY        25.000000
     M1      -50.000000      INFINITY        350.000000

                      RIGHTHAND SIDE RANGES
ROW            CURRENT        ALLOWABLE       ALLOWABLE
               RHS            INCREASE        DECREASE
     2        0.000000       0.400001        1.599999
     3       98.000000       0.400001        1.599999
     4       73.000000       INFINITY        0.879999
     5      260.000000       INFINITY        1.200003
     6       88.000000       1.999999        3.000008
     7       26.000000       1.599999        INFINITY
```

TABLE 54

z	x_1	x_2	e_1	e_2	a_1	a_2	a_3	rhs
1	0	3	0	0	M	M	$M + 4$	12
0	1	1	0	0	0	0	1	3
0	0	2	1	0	-1	0	3	3
0	0	1	0	1	0	-1	2	2

TABLE 55

z	x_1	x_2	x_3	s_1	s_2	rhs
1	0	0	1	$\frac{1}{2}$	$\frac{1}{2}$	9
0	0	1	3	2	-1	6
0	1	0	-1	$-\frac{1}{2}$	$\frac{1}{2}$	1

a Find the dual of this LP and its optimal solution.

b Find the range of values of b_2 for which the current basis remains optimal. If $b_2 = 12$, what is the new optimal solution?

7 Consider the following LP:
$$\max z = 3x_1 + 4x_2$$
$$\text{s.t.} \quad 2x_1 + x_2 \leq 8$$
$$4x_1 + x_2 \leq 10$$
$$x_1, x_2 \geq 0$$

The optimal solution to this LP is $z = 32$, $x_1 = 0$, $x_2 = 8$, $s_1 = 0$, $s_2 = 2$. Graphically find the range of values of c_1 for which the current basis remains optimal.

8 Wivco produces product 1 and product 2 by processing raw material. As much as 90 lb of raw material may be purchased at a cost of $10/lb. One pound of raw material can be used to produce either 1 lb of product 1 or 0.33 lb of product 2. Using a pound of raw material to produce a pound of product 1 requires 2 hours of labor or 3 hours to produce 0.33 lb of product 2. A total of 200 hours of labor are available, and at most 40 pounds of product 2 can be sold. Product 1 sells for $13/lb, and product 2 sells for $40/lb. Let

RM = pounds of raw material processed

P1 = pounds of raw material used to produce product 1

P2 = pounds of raw material used to produce product 2

To maximize profit, Wivco should solve the following LP:

$$\max z = 13P1 + 40(0.33)P2 - 10RM$$

$$\text{s.t.} \quad RM \geq P1 + P2$$
$$2P1 + 3P2 \leq 200$$
$$RM \leq 90$$
$$0.33P2 \leq 40$$
$$P1, P2, RM \geq 0$$

Use the LINDO output in Figure 12 to answer the following questions:

a If only 87 lb of raw material could be purchased, what would be Wivco's profits?

b If product 2 sold for $39.50/lb, what would be the new optimal solution?

c What is the most that Wivco should pay for another pound of raw material?

d What is the most that Wivco should pay for another hour of labor?

e Suppose that 1 lb of raw material could also be used to produce 0.8 lb of product 3, which sells for $24/lb. Processing 1 lb of raw material into 0.8 lb of product 3 requires 7 hours of labor. Should Wivco produce any of product 3?

9 Consider the following LP and its optimal tableau (Table 56):

$$\max z = 3x_1 + 4x_2 + x_3$$

$$\text{s.t.} \quad x_1 + x_2 + x_3 \leq 50$$
$$2x_1 - x_2 + x_3 \geq 15$$
$$x_1 + x_2 = 10$$
$$x_1, x_2, x_3 \geq 0$$

a Find the dual of this LP and its optimal solution.

b Find the range of values of the objective function coefficient of x_1 for which the current basis remains optimal.

c Find the range of values of the objective function coefficient for x_2 for which the current basis remains optimal.

FIGURE 12
LINDO Output for Wivco (Problem 8)

```
MAX     13 P1 + 13.2 P2 - 10 RM
SUBJECT TO
     2)  - P1 - P2 + RM >= 0
     3)   2 P1 + 3 P2 <=   200
     4)    RM <=   90
     5)   0.33 P2 <=  40
END

    LP OPTIMUM FOUND  AT STEP     3

         OBJECTIVE FUNCTION VALUE

  1)       274.000000

VARIABLE        VALUE        REDUCED COST
     P1      70.000000          0.000000
     P2      20.000000          0.000000
     RM      90.000000          0.000000

  ROW    SLACK OR SURPLUS     DUAL PRICES
    2)       0.000000         -12.600000
    3)       0.000000          0.200000
    4)       0.000000          2.600000
    5)      33.400002          0.000000

NO. ITERATIONS=     3

    RANGES IN WHICH THE BASIS IS UNCHANGED

                     OBJ COEFFICIENT RANGES
VARIABLE      CURRENT      ALLOWABLE      ALLOWABLE
               COEF        INCREASE       DECREASE
     P1      13.000000      0.200000       0.866667
     P2      13.200000      1.300000       0.200000
     RM     -10.000000      INFINITY       2.600000

                     RIGHTHAND SIDE RANGES
  ROW         CURRENT      ALLOWABLE      ALLOWABLE
               RHS         INCREASE       DECREASE
    2         0.000000     23.333334      10.000000
    3       200.000000     70.000000      20.000000
    4        90.000000     10.000000      23.333334
    5        40.000000     INFINITY       33.400002
```

TABLE 56

TABLE 56

z	x_1	x_2	x_3	s_1	e_2	a_2	a_3	rhs
1	1	0	0	1	0	M	$M+3$	80
0	-3	0	0	1	1	-1	-2	15
0	0	0	1	1	0	0	-2	40
0	1	1	0	0	0	0	1	10

TABLE 57

z	x_1	x_2	s_1	s_2	rhs
1	0	0	0	1	10
0	0	$\frac{1}{3}$	1	$-\frac{2}{3}$	$\frac{4}{3}$
0	1	$\frac{7}{3}$	0	$\frac{1}{3}$	$\frac{10}{3}$

10 Consider the following LP and its optimal tableau (Table 57):

$$\max z = 3x_1 + 2x_2$$
$$\text{s.t.} \quad 2x_1 + 5x_2 \le 8$$
$$3x_1 + 7x_2 \le 10$$
$$x_1, x_2 \ge 0$$

a Find the dual of this LP and its optimal solution.

b Find the range of values of b_2 for which the current basis remains optimal. Also find the new optimal solution if $b_2 = 5$.

11 Consider the following LP:

$$\max z = 3x_1 + x_2$$
$$\text{s.t.} \quad 2x_1 + x_2 \le 8$$
$$4x_1 + x_2 \le 10$$
$$x_1, x_2 \ge 0$$

The optimal solution to this LP is $z = 9$, $x_1 = 1$, $x_2 = 6$. Graphically find the range of values of b_2 for which the current basis remains optimal.

12 Farmer Leary grows wheat and corn on his 45-acre farm. He can sell at most 140 bushels of wheat and 120 bushels of corn. Each planted acre yields either 5 bushels of wheat or 4 bushels of corn. Wheat sells for $30 per bushel, and corn sells for $50 per bushel. Six hours of labor are needed to harvest an acre of wheat, and 10 hours are needed to harvest an acre of corn. As many as 350 hours of labor can be purchased at $10 per hour. Let

A1 = acres planted with wheat

A2 = acres planted with corn

L = hours of labor that are purchased

To maximize profits, farmer Leary should solve the following LP:

$$\max z = 150A1 + 200A2 - 10L$$
$$\text{s.t.} \quad A1 + A2 \le 45$$
$$6A1 + 10A2 - L \le 0$$
$$L \le 350$$
$$5A1 \le 140$$
$$4A2 \le 120$$
$$A1, A2, L \ge 0$$

Use the LINDO output in Figure 13 to answer the following questions:

a What is the most that Leary should pay for an additional hour of labor?

b What is the most that Leary should pay for an additional acre of land?

c If only 40 acres of land were available, what would be Leary's profit?

d If the price of wheat dropped to $26, what would be the new optimal solution?

e Farmer Leary is considering growing barley. Demand for barley is unlimited. An acre yields 4 bushels of barley and requires 3 hours of labor. If barley sells for $30 per bushel, should Leary produce any barley?

13 Consider the following LP and its optimal tableau (Table 58):

$$\max z = 4x_1 + x_2 + 2x_3$$
$$\text{s.t.} \quad 8x_1 + 3x_2 + x_3 \le 2$$
$$6x_1 + x_2 + x_3 \le 8$$
$$x_1, x_2, x_3 \ge 0$$

a Find the dual to this LP and its optimal solution.

b Find the range of values of the objective function coefficient of x_3 for which the current basis remains optimal.

c Find the range of values of the objective function coefficient of x_1 for which the current basis remains optimal.

14 Consider the following LP and its optimal tableau (Table 59):

$$\max z = 3x_1 + x_2$$
$$\text{s.t.} \quad 2x_1 + x_2 \le 4$$
$$3x_1 + 2x_2 \ge 6$$
$$4x_1 + 2x_2 = 7$$
$$x_1 \ge 0, x_2 \ge 0$$

a Find the dual to this LP and its optimal solution.

b Find the range of values of the right-hand side of the third constraint for which the current basis remains optimal. Also find the new optimal solution if the right-hand side of the third constraint were $\frac{15}{2}$.

15 Consider the following LP:

$$\max z = 3x_1 + x_2$$
$$\text{s.t.} \quad 4x_1 + x_2 \le 7$$
$$5x_1 + 2x_2 \le 10$$
$$x_1, x_2 \ge 0$$

The optimal solution to this LP is $z = \frac{17}{3}$, $x_1 = \frac{4}{3}$, $x_2 = \frac{5}{3}$. Use the graphical approach to determine the range of values for the right-hand side of the second constraint for which the current basis remains optimal.

16 Zales Jewelers uses rubies and sapphires to produce two types of rings. A Type 1 ring requires 2 rubies, 3 sapphires, and 1 hour of jeweler's labor. A Type 2 ring requires 3 rubies, 2 sapphires, and 2 hours of jeweler's labor. Each Type 1 ring sells for $400, and each Type 2 ring sells for $500. All rings produced by Zales can be sold. Zales now has 100 rubies, 120 sapphires, and 70 hours of jeweler's

FIGURE 13

LINDO Output for Wheat/Corn (Problem 12)

```
MAX 150A1+200A2-10L
ST
A1+A2<45
6A1+10A2-L<0
L<350
5A1<140
4A2<120
END
```

LP OPTIMUM FOUND AT STEP   4

    OBJECTIVE FUNCTION VALUE

   1)   4250.000

VARIABLE	VALUE	REDUCED COST
A1	25.000000	0.000000
A2	20.000000	0.000000
L	350.000000	0.000000

ROW	SLACK OR SURPLUS	DUAL PRICES
2)	0.000000	75.000000
3)	0.000000	12.500000
4)	0.000000	2.500000
5)	15.000000	0.000000
6)	40.000000	0.000000

NO. ITERATIONS=   4

RANGES IN WHICH THE BASIS IS UNCHANGED:

OBJ COEFFICIENT RANGES

VARIABLE	CURRENT COEF	ALLOWABLE INCREASE	ALLOWABLE DECREASE
A1	150.000000	10.000000	30.000000
A2	200.000000	50.000000	10.000000
L	-10.000000	INFINITY	2.500000

RIGHTHAND SIDE RANGES

ROW	CURRENT RHS	ALLOWABLE INCREASE	ALLOWABLE DECREASE
2	45.000000	1.200000	6.666667
3	0.000000	40.000000	12.000000
4	350.000000	40.000000	12.000000
5	140.000000	INFINITY	15.000000
6	120.000000	INFINITY	40.000000

TABLE 58

z	x_1	x_2	x_3	s_1	s_2	rhs
1	8	1	0	0	2	16
0	2	2	0	1	−1	4
0	6	1	1	0	1	8

TABLE 59

z	x_1	x_2	s_1	e_2	a_2	a_3	rhs
1	0	0	0	1	$M-1$	$M+\frac{3}{2}$	$\frac{9}{2}$
0	0	0	1	0	0	$-\frac{1}{2}$	$\frac{1}{2}$
0	0	1	0	−2	2	$-\frac{3}{2}$	$\frac{3}{2}$
0	1	0	0	1	−1	1	1

labor. Extra rubies can be purchased at a cost of $100 per ruby. Market demand requires that the company produce at least 20 Type 1 rings and at least 25 Type 2 rings. To maximize profit, Zales should solve the following LP:

$$X1 = \text{Type 1 rings produced}$$
$$X2 = \text{Type 2 rings produced}$$
$$R = \text{number of rubies purchased}$$

$$\max z = 400X1 + 500X2 - 100R$$

$$
\begin{aligned}
\text{s.t.} \quad & 2X1 + 3X2 - R \le 100 \\
& 3X1 + 2X2 \le 120 \\
& X1 + 2X2 \le 70 \\
& X1 \ge 20 \\
& X2 \ge 25 \\
& X1, X2 \ge 0
\end{aligned}
$$

Use the LINDO output in Figure 14 to answer the following questions:

a Suppose that instead of $100, each ruby costs $190. Would Zales still purchase rubies? What would be the new optimal solution to the problem?

FIGURE **14**
LINDO Output for Jewelry (Problem 16)

```
MAX     400 X1 + 500 X2 - 100 R
SUBJECT TO
       2)    2 X1 + 3 X2 - R <= 100
       3)    3 X1 + 2 X2 <=  120
       4)     X1 + 2 X2 <=   70
       5)     X1 >=    20
       6)     X2 >=    25
END

   LP OPTIMUM FOUND    AT STEP  2

          OBJECTIVE FUNCTION VALUE

 1)       19000.0000

 VARIABLE          VALUE         REDUCED COST
       X1        20.000000          0.000000
       X2        25.000000          0.000000
        R        15.000000          0.000000

    ROW     SLACK OR SURPLUS      DUAL PRICES
      2)         0.000000         100.000000
      3)        10.000000           0.000000
      4)         0.000000         200.000000
      5)         0.000000           0.000000
      6)         0.000000        -200.000000

 NO. ITERATIONS=     2

  RANGES IN WHICH THE BASIS IS UNCHANGED

                       OBJ COEFFICIENT RANGES
 VARIABLE        CURRENT        ALLOWABLE       ALLOWABLE
                  COEF          INCREASE        DECREASE
      X1        400.000000       INFINITY      100.000000
      X2        500.000000     200.000000        INFINITY
       R       -100.000000     100.000000      100.000000

                       RIGHTHAND SIDE RANGES
    ROW         CURRENT        ALLOWABLE       ALLOWABLE
                  RHS          INCREASE        DECREASE
      2        100.000000      15.000000        INFINITY
      3        120.000000       INFINITY       10.000000
      4         70.000000       3.333333        0.000000
      5         20.000000       0.000000        INFINITY
      6         25.000000       0.000000        2.500000
```

b Suppose that Zales were only required to produce at least 23 Type 2 rings. What would Zales' profit now be?

c What is the most that Zales would be willing to pay for another hour of jeweler's labor?

d What is the most that Zales would be willing to pay for another sapphire?

e Zales is considering producing Type 3 rings. Each Type 3 ring can be sold for $550 and requires 4 rubies, 2 sapphires, and 1 hour of jeweler's labor. Should Zales produce any Type 3 rings?

17 Use the dual simplex method to solve the following LP:

$$\max z = -2x_1 - x_2$$
$$\text{s.t.} \quad x_1 + x_2 \geq 5$$
$$x_1 - 2x_2 \geq 8$$
$$x_1, x_2 \geq 0$$

18 Consider the following LP:

$$\max z = -4x_1 - x_2$$
$$\text{s.t.} \quad 4x_1 + 3x_2 \geq 6$$
$$x_1 + 2x_2 \leq 3$$
$$3x_1 + x_2 = 3$$
$$x_1, x_2 \geq 0$$

After subtracting an excess variable e_1 from the first constraint, adding a slack variable s_2 to the second constraint, and adding artificial variables a_1 and a_3 to the first and third constraints, the optimal tableau for this LP is as shown in Table 60.

a Find the dual to this LP and its optimal solution.

b If we changed this LP to

$$\max z = -4x_1 - x_2 - x_3$$
$$\text{s.t.} \quad 4x_1 + 3x_2 + x_3 \geq 6$$
$$x_1 + 2x_2 + x_3 \leq 3$$
$$3x_1 + x_2 + x_3 = 3$$
$$x_1, x_2, x_3 \geq 0$$

would the current optimal solution remain optimal?

TABLE 60

z	x_1	x_2	e_1	s_2	a_1	a_3	rhs
1	0	0	0	$\frac{1}{5}$	M	$M - \frac{7}{5}$	$-\frac{18}{5}$
0	0	1	0	$\frac{3}{5}$	0	$-\frac{1}{5}$	$\frac{6}{5}$
0	1	0	0	$-\frac{1}{5}$	0	$\frac{2}{5}$	$\frac{3}{5}$
0	0	0	1	1	-1	1	0

TABLE 61

Radio 1		Radio 2	
Price ($)	Resource Required	Price ($)	Resource Required
25	Laborer 1: 1 hour	22	Laborer 1: 2 hours
	Laborer 2: 2 hours		Laborer 2: 2 hours
	Raw material cost: $5		Raw material cost: $4

TABLE 62

z	x_1	x_2	s_1	s_2	rhs
1	0	0	$\frac{1}{3}$	$\frac{4}{3}$	80
0	1	0	$-\frac{1}{3}$	$\frac{2}{3}$	20
0	0	1	$\frac{2}{3}$	$-\frac{1}{3}$	10

19 Consider the following LP:

$$\max z = -2x_1 + 6x_2$$
$$\text{s.t.} \quad x_1 + x_2 \geq 2$$
$$-x_1 + x_2 \leq 1$$
$$x_1, x_2 \geq 0$$

This LP is unbounded. Use this fact to show that the following LP has no feasible solution:

$$\min 2y_1 + y_2$$
$$\text{s.t.} \quad y_1 - y_2 \geq -2$$
$$y_1 + y_2 \geq 6$$
$$y_1 \leq 0, y_2 \geq 0$$

20 Use the Theorem of Complementary Slackness to find the optimal solution to the following LP and its dual:

$$\max z = 3x_1 + 4x_2 + x_3 + 5x_4$$
$$\text{s.t.} \quad x_1 + 2x_2 + x_3 + 2x_4 \leq 5$$
$$2x_1 + 3x_2 + x_3 + 3x_4 \leq 8$$
$$x_1, x_2, x_3, x_4 \geq 0$$

21 $z = 8$, $x_1 = 2$, $x_2 = 0$ is the optimal solution to the following LP:

$$\max z = 4x_1 + x_2$$
$$\text{s.t.} \quad 3x_1 + x_2 \leq 6$$
$$5x_1 + 3x_2 \leq 15$$
$$x_1, x_2 \geq 0$$

Use the graphical approach to answer the following questions:

a Determine the range of values of c_1 for which the current basis remains optimal.

b Determine the range of values of c_2 for which the current basis remains optimal.

c Determine the range of values of b_1 for which the current basis remains optimal.

d Determine the range of values of b_2 for which the current basis remains optimal.

22 Radioco manufactures two types of radios. The only scarce resource that is needed to produce radios is labor. The company now has two laborers. Laborer 1 is willing to work up to 40 hours per week and is paid $5 per hour. Laborer 2 is willing to work up to 50 hours per week and is paid $6 per hour. The price as well as the resources required to build each type of radio are given in Table 61.

a Letting x_i be the number of type i radios produced each week, show that Radioco should solve the following LP (its optimal tableau is given in Table 62):

$$\max z = 3x_1 + 2x_2$$
$$\text{s.t.} \quad x_1 + 2x_2 \leq 40$$
$$2x_1 + x_2 \leq 50$$
$$x_1, x_2 \geq 0$$

b For what values of the price of a Type 1 radio would the current basis remain optimal?

c For what values of the price of a Type 2 radio would the current basis remain optimal?

d If laborer 1 were willing to work only 30 hours per week, would the current basis remain optimal?

e If laborer 2 were willing to work as many as 60 hours per week, would the current basis remain optimal?

f If laborer 1 were willing to work an additional hour, what is the most that Radioco should pay?

g If laborer 2 were willing to work only 48 hours, what would Radioco's profits be? Verify your answer by determining the number of radios of each type that would be produced.

h A Type 3 radio is under consideration for production. The specifications of a Type 3 radio are as follows: price, $30; 2 hours from laborer 1; 2 hours from laborer 2; cost of raw materials, $3. Should Radioco manufacture any Type 3 radios?

23 Beerco manufactures ale and beer from corn, hops, and malt. Currently, 40 lb of corn, 30 lb of hops, and 40 lb of malt are available. A barrel of ale sells for $40 and requires 1 lb of corn, 1 lb of hops, and 2 lb of malt. A barrel of beer sells for $50 and requires 2 lb of corn, 1 lb of hops, and 1 lb of malt. Beerco can sell all ale and beer that is produced. To maximize total sales revenue, Beerco should solve the following LP:

$$\max z = 40\text{ALE} + 50\text{BEER}$$
$$\text{s.t.} \quad \text{ALE} + 2\text{BEER} \leq 40 \quad \text{(Corn constraint)}$$
$$\text{ALE} + \text{BEER} \leq 30 \quad \text{(Hops constraint)}$$

TABLE 63

z	Ale	Beer	s_1	s_2	s_3	rhs
1	0	0	20	0	10	1,200
0	0	1	$\frac{2}{3}$	0	$-\frac{1}{3}$	$\frac{40}{3}$
0	0	0	$-\frac{1}{3}$	1	$-\frac{1}{3}$	$\frac{10}{3}$
0	1	0	$-\frac{1}{3}$	0	$\frac{2}{3}$	$\frac{40}{3}$

$$2\text{ALE} + \text{BEER} \le 40 \qquad \text{(Malt constraint)}$$
$$\text{ALE, BEER} \ge 0$$

ALE = barrels of ale produced, and BEER = barrels of beer produced. An optimal tableau for this LP is shown in Table 63.

a Write down the dual to Beerco's LP and find its optimal solution.

b Find the range of values of the price of ale for which the current basis remains optimal.

c Find the range of values of the price of beer for which the current basis remains optimal.

d Find the range of values of the amount of available corn for which the current basis remains optimal.

e Find the range of values of the amount of available hops for which the current basis remains optimal.

f Find the range of values of the amount of available malt for which the current basis remains optimal.

g Suppose Beerco is considering manufacturing malt liquor. A barrel of malt liquor requires 0.5 lb of corn, 3 lb of hops, and 3 lb of malt and sells for $50. Should Beerco manufacture any malt liquor?

h Suppose we express the Beerco constraints in ounces. Write down the new LP and its dual.

i What is the optimal solution to the dual of the new LP? (*Hint:* Think about what happens to $c_{BV}B^{-1}$. Use the idea of shadow prices to explain why the dual to the original LP (pounds) and the dual to the new LP (ounces) should have different optimal solutions.)

Group B

24 Consider the following LP:

$$\max z = -3x_1 + x_2 + 2x_3$$
$$\text{s.t.} \qquad x_2 + 2x_3 \le 3$$
$$-x_1 + 3x_3 \le -1$$
$$-2x_1 - 3x_2 \le -2$$
$$x_1, x_2, x_3 \ge 0$$

a Find the dual to this LP and show that it has the same feasible region as the original LP.

b Use weak duality to show that the optimal objective function value for the LP (and its dual) must be 0.

25 Consider the following LP:

$$\max z = 2x_1 + x_2 + x_3$$
$$\text{s.t.} \quad x_1 \qquad + x_3 \le 1$$
$$x_2 + x_3 \le 2$$
$$x_1 + x_2 \qquad \le 3$$
$$x_1, x_2, x_3 \ge 0$$

TABLE 64

	Product 1	Product 2
Selling price	$15	$8
Labor required	0.75 hour	0.50 hour
Machine time required	1.5 hours	0.80 hour
Raw material required	2 units	1 unit

It is given that

$$\begin{bmatrix} 1 & 0 & 1 \\ 0 & 1 & 1 \\ 1 & 1 & 0 \end{bmatrix}^{-1} = \begin{bmatrix} \frac{1}{2} & -\frac{1}{2} & \frac{1}{2} \\ -\frac{1}{2} & \frac{1}{2} & \frac{1}{2} \\ \frac{1}{2} & \frac{1}{2} & -\frac{1}{2} \end{bmatrix}$$

a Show that the basic solution with basic variables x_1, x_2, and x_3 is optimal. Find the optimal solution.

b Write down the dual to this LP and find its optimal solution.

c Show that if we multiply the right-hand side of each constraint by a non-negative constant k, then the new optimal solution is obtained simply by multiplying the value of each variable in the original optimal solution by k.

26 Wivco produces two products: 1 and 2. The relevant data are shown in Table 64. Each week, as many as 400 units of raw material can be purchased at a cost of $1.50 per unit. The company employs four workers, who work 40 hours per week (their salaries are considered a fixed cost). Workers can be asked to work overtime and are paid $6 per hour for overtime work. Each week, 320 hours of machine time are available.

In the absence of advertising, 50 units of product 1 and 60 units of product 2 will be demanded each week. Advertising can be used to stimulate demand for each product. Each dollar spent on advertising product 1 increases its demand by 10 units; each dollar spent for product 2 increases its demand by 15 units. At most $100 can be spent on advertising. Define

P1 = number of units of product 1 produced each week
P2 = number of units of product 2 produced each week
OT = number of hours of overtime labor used each week
RM = number of units of raw material purchased each week
A1 = dollars spent each week on advertising product 1
A2 = dollars spent each week on advertising product 2

Then Wivco should solve the following LP:

$$\max z = 15\text{P1} + 8\text{P2} - 6(\text{OT}) - 1.5\text{RM} - \text{A1} - \text{A2}$$

s.t.	$\text{P1} - 10\text{A1} \le 50$	(1)
	$\text{P2} - 15\text{A2} \le 60$	(2)
	$0.75\text{P1} + 0.5\text{P2} \le 160 + (\text{OT})$	(3)
	$2\text{P1} + \text{P2} \le \text{RM}$	(4)
	$\text{RM} \le 400$	(5)
	$\text{A1} + \text{A2} \le 100$	(6)
	$1.5\text{P1} + 0.8\text{P2} \le 320$	(7)
	All variables non-negative	

Use LINDO to solve this LP. Then use the computer output to answer the following questions:

a If overtime were only $4 per hour, would Wivco use it?

b If each unit of product 1 sold for $15.50, would the current basis remain optimal? What would be the new optimal solution?

c What is the most that Wivco should be willing to pay for another unit of raw material?

d How much would Wivco be willing to pay for another hour of machine time?

e If each worker were required (as part of the regular workweek) to work 45 hours per week, what would the company's profits be?

f Explain why the shadow price of row (1) is 0.10. (*Hint:* If the right-hand side of (1) were increased from 50 to 51, then in the absence of advertising for product 1, 51 units could now be sold each week.)

g Wivco is considering producing a new product (product 3). Each unit sells for $17 and requires 2 hours of labor, 1 unit of raw material, and 2 hours of machine time. Should Wivco produce any of product 3?

h If each unit of product 2 sold for $10, would the current basis remain optimal?

27 The following question concerns the Rylon example discussed in Section 3.9. After defining

RB = ounces of Regular Brute produced annually

LB = ounces of Luxury Brute produced annually

RC = ounces of Regular Chanelle produced annually

LC = ounces of Luxury Chanelle produced annually

RM = pounds of raw material purchased annually

the LINDO output in Figure 15 was obtained for this problem. Use this output to answer the following questions:

a Interpret the shadow price of each constraint.

FIGURE 15
LINDO Output for Brute/Chanelle (Problem 27)

```
MAX      7 RB + 14 LB + 6 RC + 10 LC - 3 RM
SUBJECT TO
      2)     RM <=   4000
      3)    3 LB + 2 LC +  RM <=   6000
      4)    RM + LB - 3 RM =    0
      5)    RC + LC - 4 RM =    0
END

   LP OPTIMUM FOUND   AT STEP     6

        OBJECTIVE FUNCTIONS VALUE

  1)         172666.672

VARIABLE       VALUE         REDUCED COST
     RB     11333.333008       0.000000
     LB       666.666687       0.000000
     RC     16000.000000       0.000000
     LC         0.000000       0.666667
     RM      4000.000000       0.000000

   ROW    SLACK OR SURPLUS     DUAL PRICES
     2)        0.000000         39.666668
     3)        0.000000          2.333333
     4)        0.000000          7.000000
     5)        0.000000          6.000000

NO.  ITERATIONS=    6

    RANGES IN WHICH THE BASIS IS UNCHANGED

                     OBJ COEFFICIENT RANGES
VARIABLE        CURRENT      ALLOWABLE       ALLOWABLE
                COEF         INCREASE        DECREASE
     RB        7.000000       1.000000       11.900001
     LB       14.000000     119.000000        1.000000
     RC        6.000000      INFINITY         0.666667
     LC       10.000000       0.666667       INFINITY
     RM       -3.000000      INFINITY        39.666668

                     RIGHTHAND SIDE RANGES
   ROW          CURRENT      ALLOWABLE       ALLOWABLE
                RHS          INCREASE        DECREASE
     2       4000.000000    2000.000000     3400.000000
     3       6000.000000   33999.996094     2000.000000
     4          0.000000     INFINITY      11333.333008
     5          0.000000     INFINITY      16000.000000
```

b If the price of RB were to increase by 50¢, what would be the new optimal solution to the Rylon problem?

c If 8,000 laboratory hours were available each year, but only 2,000 lb of raw material were available each year, would Rylon's profits increase or decrease? [*Hint:* Use the 100% Rule to show that the current basis remains optimal. Then use reasoning analogous to (34)–(37) to determine the new objective function value.]

d Rylon is considering expanding its laboratory capacity. Two options are under consideration:

Option 1 For a cost of $10,000 (incurred now), annual laboratory capacity can be increased by 1,000 hours.

Option 2 For a cost of $200,000 (incurred now), annual laboratory capacity can be increased by 10,000 hours.

Suppose that all other aspects of the problem remain unchanged and that future profits are discounted, with the interest rate being $11\frac{1}{9}\%$ per year. Which option, if any, should Rylon choose?

e Rylon is considering purchasing a new type of raw material. Unlimited quantities can be purchased at $8/lb. It requires 3 laboratory hours to process a pound of the new raw material. Each processed pound yields 2 oz of RB and 1 oz of RC. Should Rylon purchase any of the new material?

28 Consider the following two LPs:

$$\max z = c_1x_1 + c_2x_2$$
$$\text{s.t.} \quad a_{11}x_1 + a_{12}x_2 \le b_1 \qquad \textbf{(LP 1)}$$
$$a_{21}x_1 + a_{22}x_2 \le b_2$$
$$x_1, x_2 \ge 0$$

$$\max z = 100c_1x_1 + 100c_2x_2$$
$$\text{s.t.} \quad 100a_{11}x_1 + 100a_{12}x_2 \le b_1 \qquad \textbf{(LP 2)}$$
$$100a_{21}x_1 + 100a_{22}x_2 \le b_2$$
$$x_1, x_2 \ge 0$$

Suppose that BV $= \{x_1, x_2\}$ is an optimal basis for both LPs, and the optimal solution to LP 1 is $x_1 = 50$, $x_2 = 500$, $z = 550$. Also suppose that for LP 1, the shadow price of both Constraint 1 and Constraint 2 $= \frac{100}{3}$. Find the optimal solution to LP 2 and the optimal solution to the dual of LP 2. (*Hint:* If we multiply each number in a matrix by 100, what happens to B^{-1}?)

29 The following questions pertain to the Star Oil capital budgeting example of Section 3.6. The LINDO output for this problem is shown in Figure 16.

a Find and interpret the shadow price for each constraint.

b If the NPV of investment 1 were $5 million, would the optimal solution to the problem change?

c If the NPV of investment 2 and investment 4 were each decreased by 25%, would the optimal solution to the problem change? (This part requires knowledge of the 100% Rule.)

d Suppose that Star Oil's investment budget were changed to $50 million at time 0 and $15 million at time 1. Would Star be better off? (This part requires knowledge of the 100% Rule.)

e Suppose a new investment (investment 6) is available. Investment 6 yields an NPV of $10 million and re-

quires a cash outflow of $5 million at time 0 and $10 million at time 1. Should Star Oil invest any money in investment 6?

30 The following questions pertain to the Finco investment example of Section 3.11. The LINDO output for this problem is shown in Figure 17.

a If Finco has $2,000 more on hand at time 0, by how much would their time 3 cash increase?

b Observe that if Finco were given a dollar at time 1, the cash available for investment at time 1 would now be $0.5A + 1.2C + 1.08S_0 + 1$. Use this fact and the shadow price of Constraint 2 to determine by how much Finco's time 3 cash position would increase if an extra dollar were available at time 1.

c By how much would Finco's time 3 cash on hand change if Finco were given an extra dollar at time 2?

d If investment D yielded $1.80 at time 3, would the current basis remain optimal?

e Suppose that a super money market fund yielded 25% for the period between time 0 and time 1. Should Finco invest in this fund at time 0?

f Show that if the investment limitations of $75,000 on investments A, B, C, and D were all eliminated, the current basis would remain optimal. (Knowledge of the 100% Rule is required for this part.) What would be the new optimal z-value?

g A new investment (investment F) is under consideration. One dollar invested in investment F generates the following cash flows: time 0, $-\$1.00$; time 1, $+\$1.10$; time 2, $+\$0.20$; time 3, $+\$0.10$. Should Finco invest in investment F?

31 In this problem, we discuss how shadow prices can be interpreted for blending problems (see Section 3.8). To illustrate the ideas, we discuss Problem 2 of Section 3.8. If we define

x_{6J} = pounds of grade 6 oranges in juice
x_{9J} = pounds of grade 9 oranges in juice
x_{6B} = pounds of grade 6 oranges in bags
x_{9B} = pounds of grade 9 oranges in bags

then the appropriate formulation is

$$\max z = 0.45(x_{6J} + x_{9J}) + 0.30(x_{6B} + x_{9B})$$

$$\text{s.t.} \quad x_{6J} + x_{6B} \le 120,000 \quad \text{(Grade 6 constraint)}$$

$$x_{9J} + x_{9B} \le 100,000 \quad \text{(Grade 9 constraint)}$$

$$(1) \quad \frac{6x_{6J} + 9x_{9J}}{x_{6J} + x_{9J}} \ge 8 \quad \text{(Orange Juice constraint)}$$

$$(2) \quad \frac{6x_{6B} + 9x_{9B}}{x_{6B} + x_{9B}} \ge 7 \quad \text{(Bags constraint)}$$

$$x_{6J}, x_{9J}, x_{6B}, x_{9B} \ge 0$$

Constraints (1) and (2) are examples of blending constraints, because they specify the proportion of grade 6 and grade 9 oranges that must be blended to manufacture orange juice and bags of oranges. It would be useful to determine how a slight change in the standards for orange juice and bags of

FIGURE **16**
LINDO Output for Star Oil (Problem 29)

```
MAX     13 X1 + 16 X2 + 16 X3 + 14 X4 + 39 X5
SUBJECT TO
        2)   11 X1 + 53 X2 + 5 X3 + 5 X4 + 29 X5 <= 40
        3)    3 X1 + 6 X2 + 5 X3 + X4 + 34 X5 <= 20
        4)    X1 <=   1
        5)    X2 <=   1
        6)    X3 <=   1
        7)    X4 <=   1
        8)    X5 <=   1
END

    LP OPTIMUM FOUND    AT STEP  5

        OBJECTIVE FUNCTION VALUE

1)        57.4490166

VARIABLE          VALUE         REDUCED COST
      X1         1.000000         0.000000
      X2         0.200860         0.000000
      X3         1.000000         0.000000
      X4         1.000000         0.000000
      X5         0.288084         0.000000

    ROW     SLACK OR SURPLUS     DUAL PRICES
      2)         0.000000         0.190418
      3)         0.000000         0.984644
      4)         0.000000         7.951474
      5)         0.799140         0.000000
      6)         0.000000        10.124693
      7)         0.000000        12.063268
      8)         0.711916         0.000000

NO. ITERATIONS=     5

    RANGES IN WHICH THE BASIS IS UNCHANGED

                    OBJ COEFFICIENT RANGES
VARIABLE        CURRENT       ALLOWABLE       ALLOWABLE
                 COEF         INCREASE        DECREASE
      X1       13.000000      INFINITY        7.951474
      X2       16.000000      45.104530       9.117648
      X3       16.000000      INFINITY       10.124693
      X4       14.000000      INFINITY       12.063268
      X5       39.000000      51.666668      30.245283

                    RIGHTHAND SIDE RANGES
    ROW         CURRENT       ALLOWABLE       ALLOWABLE
                 RHS          INCREASE        DECREASE
      2        40.000000      38.264709       9.617647
      3        20.000000      11.275863       8.849057
      4         1.000000       1.139373       1.000000
      5         1.000000      INFINITY        0.799140
      6         1.000000       1.995745       1.000000
      7         1.000000       2.319149       1.000000
      8         1.000000      INFINITY        0.711916
```

oranges would affect profit. At the end of this problem, we explain how to use the shadow prices of Constraints (1) and (2) to answer the following questions:

a Suppose that the average grade for orange juice is increased to 8.1. Assuming the current basis remains optimal, by how much would profits change?

b Suppose the average grade requirement for bags of oranges is decreased to 6.9. Assuming the current basis remains optimal, by how much would profits change?

The shadow price for both (1) and (2) is -0.15. The optimal solution is $x_{6J} = 26,666.67$, $x_{9J} = 53,333.33$, $x_{6B} =$ 93,333.33, $x_{9B} = 46,666.67$. To interpret the shadow prices of blending Constraints (1) and (2), *we assume that a slight change in the quality standard for a product will not significantly change the quantity of the product that is produced.*

Now note that (1) may be written as

$$6x_{6J} + 9x_{9J} \geq 8(x_{6J} + x_{9J}), \qquad \text{or} \qquad -2x_{6J} + x_{9J} \geq 0$$

If the quality standard for orange juice is changed to $8 + \Delta$, then (1) can be written as

$$6x_{6J} + 9x_{9J} \geq (8 + \Delta)(x_{6J} + x_{9J})$$

or

FIGURE 17
LINDO Output for Finco (Problem 30)

```
MAX     B + 1.9 D + 1.5 E + 1.08  S2
SUBJECT TO
      2)   D + A + C  +  SO =        100000
      3) - B + 0.5 A + 1.2 C + 1.08 SO -  S1 =   0
      4)  0.5 B - E- S2 + A + 1.08 S1 =   0
      5)   A <=  75000
      6)   B <=  75000
      7)   C <=  75000
      8)   D <=  75000
      9)   E <=  75000

END

   LP OPTIMUM FOUND  AT STEP     8

         OBJECTIVE FUNCTION VALUE

1)        218500.000

VARIABLE        VALUE         REDUCED COST
       B      30000.000000       0.000000
       D      40000.000000       0.000000
       E      75000.000000       0.000000
      S2          0.000000       0.040000
       A      60000.000000       0.000000
       C          0.000000       0.028000
      SO          0.000000       0.215200
      S1          0.000000       0.350400

    ROW   SLACK OR SURPLUS     DUAL PRICES
      2)        0.000000        1.900000
      3)        0.000000       -1.560000
      4)        0.000000       -1.120000
      5)    15000.000000        0.000000
      6)    45000.000000        0.000000
      7)    75000.000000        0.000000
      8)    35000.000000        0.000000
      9)        0.000000        0.380000

NO. ITERATIONS=      8

   RANGES IN WHICH THE BASIS IS UNCHANGED

                       OBJ COEFFICIENT RANGES
VARIABLE      CURRENT       ALLOWABLE      ALLOWABLE
              COEF          INCREASE       DECREASE
      B       1.000000      0.029167       0.284416
      D       1.900000      0.475000       0.050000
      E       1.500000      INFINITY       0.380000
     S2       1.080000      0.040000       INFINITY
      A       0.000000      0.050000       0.058333
      C       0.000000      0.028000       INFINITY
     SO       0.000000      0.215200       INFINITY
     S1       0.000000      0.350400       INFINITY

                       RIGHTHAND SIDE RANGES
 ROW          CURRENT       ALLOWABLE      ALLOWABLE
              RHS           INCREASE       DECREASE
  2       100000.000000     35000.000000   40000.000000
  3            0.000000     37500.000000   56250.000000
  4            0.000000     18750.000000   43750.000000
  5        75000.000000     INFINITY       15000.000000
  6        75000.000000     INFINITY       45000.000000
  7        75000.000000     INFINITY       75000.000000
  8        75000.000000     INFINITY       35000.000000
  9        75000.000000     18750.000000   43750.000000
```

$$-2x_{6J} + x_{9J} \geq \Delta(x_{6J} + x_{9J})$$

Because we are assuming that changing orange juice quality from 8 to $8 + \Delta$ does not change the amount produced, $x_{6J} + x_{9J}$ will remain equal to 80,000, and (1) will become

$$-2x_{6J} + x_{9J} \geq 80,000\Delta$$

Using the definition of shadow price, now answer parts (a) and (b).

32 Ballco manufactures large softballs, regular softballs, and hardballs. Each type of ball requires time in three departments: cutting, sewing, and packaging, as shown in

TABLE 65

Balls	Cutting Time	Sewing Time	Packaging Time
Regular softballs	15	15	3
Large softballs	10	15	4
Hardballs	8	4	2

Table 65 (in minutes). Because of marketing considerations, at least 1,000 regular softballs must be produced. Each regular softball can be sold for $3, each large softball, for $5; and each hardball, for $4. A total of 18,000 minutes of cutting time, 18,000 minutes of sewing time, and 9,000 minutes of packaging time are available. Ballco wants to maximize sales revenue. If we define

RS = number of regular softballs produced
LS = number of large softballs produced
HB = number of hardballs produced

then the appropriate LP is

max $z = 3RS + 5LS + 4HB$

s.t. $15RS + 10LS + 8HB \leq 18,000$ (Cutting constraint)

$15RS + 15LS + 4HB \leq 18,000$ (Sewing constraint)

$3RS + 4LS + 2HB \leq 9,000$ (Packaging constraint)

$RS \geq 1,000$ (Demand constraint)

RS, LS, HB ≥ 0

The optimal tableau for this LP is shown in Table 66.

a Find the dual of the Ballco problem and its optimal solution.

b Show that the Ballco problem has an alternative optimal solution. Find it. How many minutes of sewing time are used by the alternative optimal solution?

c By how much would an increase of 1 minute in the amount of available sewing time increase Ballco's revenue? How can this answer be reconciled with the fact that the sewing constraint is binding? (*Hint:* Look at the answer to part (b).)

d Assuming the current basis remains optimal, how would an increase of 100 in the regular softball requirement affect Ballco's revenue?

33 Consider the following LP:

max $z = c_1 x_1 + c_2 x_2$

s.t. $3x_1 + 4x_2 \leq 6$

$2x_1 + 3x_2 \leq 4$

$x_1, x_2 \geq 0$

The optimal tableau for this LP is

$z + s_1 + 2s_2 = 14$

$x_1 + 3s_1 - 4s_2 = 2$

$x_2 - 2s_1 + 3s_2 = 0$

Without doing any pivots, determine c_1 and c_2.

34 Consider the following LP and its partial optimal tableau (Table 67):

max $z = 20x_1 + 10x_2$

s.t. $x_1 + x_2 = 150$

$x_1 \leq 40$

$x_2 \geq 20$

$x_1, x_2 \geq 0$

a Complete the optimal tableau.

b Find the dual to this LP and its optimal solution.

35 Consider the following LP and its optimal tableau (Table 68):

max $z = c_1 x_1 + c_2 x_2$

s.t. $a_{11} x_1 + a_{12} x_2 \leq b_1$

$a_{21} x_1 + a_{22} x_2 \leq b_2$

$x_1, x_2 \geq 0$

Determine c_1, c_2, b_1, b_2, a_{11}, a_{12}, a_{21}, and a_{22}.

36 Consider an LP with three $\leq$ constraints. The right-hand sides are 10, 15, and 20, respectively. In the optimal tableau, s_2 is a basic variable in the second constraint, which has a right-hand side of 12. Determine the range of values of b_2 for which the current basis remains optimal. (*Hint:* If rhs of Constraint 2 is $15 + \Delta$, this should help in finding the rhs of the optimal tableau.)

37 Use LINDO to solve the Sailco problem of Section 3.10. Then use the output to answer the following questions:

a If month 1 demand decreased to 35 sailboats, what would be the total cost of satisfying the demands during the next four months?

b If the cost of producing a sailboat with regular-time labor during month 1 were $420, what would be the new optimal solution?

c Suppose a new customer is willing to pay $425 for a sailboat. If his demand must be met during month 1, should Sailco fill the order? How about if his demand must be met during month 4?

TABLE 66

z	RS	LS	HB	s_1	s_2	s_3	e_4	a_4	rhs
1	0	0	0	0.5	0	0	4.5	$M - 4.5$	4,500
0	0	0	1	0.19	-0.125	0	0.94	-0.94	187.5
0	0	1	0	-0.05	0.10	0	0.75	-0.75	150
0	0	0	0	-0.17	-0.15	1	-1.88	1.88	5,025
0	1	0	0	0	0	0	-1	1	1,000

TABLE 67

z	x_1	x_2	s_2	e_3	a_1	a_3	rhs
1	0	0		0			1,900
0	0	0	−1	1	1	−1	90
0	1	0	1	0	0	0	40
0	0	1	−1	0	1	0	110

TABLE 68

z	x_1	x_2	s_1	s_2	b
1	0	0	2	3	$\frac{5}{2}$
0	1	0	3	2	$\frac{5}{2}$
0	1	1	1	1	1

REFERENCES

The following texts contain extensive discussions of sensitivity analysis and duality:

Bazaraa, M., and J. Jarvis. *Linear Programming and Network Flows*. New York: Wiley, 1990.

Bersitmas, D., and Tsitsiklis, J. *Introduction to Linear Optimization*. Belmont, Mass.: Athena, 1997.

Bradley, S., A. Hax, and T. Magnanti. *Applied Mathematical Programming*. Reading, Mass.: Addison-Wesley, 1977.

Dantzig, G. *Linear Programming and Extensions*. Princeton, N.J.: Princeton University Press, 1963.

Dantzig, G., and Thapa, N. *Linear Programming*. New York: Springer-Verlag, 1997.

Gass, S. *Linear Programming: Methods and Applications,* 5th ed. New York: McGraw-Hill, 1985.

Luenberger, D. *Linear and Nonlinear Programming,* 2d ed. Reading, Mass.: Addison-Wesley, 1984.

Murty, K. *Linear Programming*. New York: Wiley, 1983.

Nash, S., and Sofer, A. *Linear and Nonlinear Programming*. New York: McGraw-Hill, 1995.

Nering, E., and Tucker, A. *Linear Programs and Related Problems*. New York: Academic Press, 1993.

Simmons, D. *Linear Programming for Operations Research*. Englewood Cliffs, N.J.: Prentice Hall, 1972.

Simonnard, M. *Linear Programming*. Englewood Cliffs, N.J.: Prentice Hall, 1966.

Wu, N., and R. Coppins. *Linear Programming and Extensions*. New York: McGraw-Hill, 1981.

The following contains a lucid discussion of DEA:

Callen, J. "Data Envelopment Analysis: Practical Survey and Managerial Accounting Applications," *Journal of Management Accounting Research* 3(1991):35–57.

Transportation, Assignment, and Transshipment Problems

In this chapter, we discuss three special types of linear programming problems: transportation, assignment, and transshipment. Each of these can be solved by the simplex algorithm, but specialized algorithms for each type of problem are much more efficient.

7.1 Formulating Transportation Problems

We begin our discussion of transportation problems by formulating a linear programming model of the following situation.

EXAMPLE 1 **Powerco Formulation**

Powerco has three electric power plants that supply the needs of four cities.[†] Each power plant can supply the following numbers of kilowatt-hours (kwh) of electricity: plant 1—35 million; plant 2—50 million; plant 3—40 million (see Table 1). The peak power demands in these cities, which occur at the same time (2 P.M.), are as follows (in kwh): city 1—45 million; city 2—20 million; city 3—30 million; city 4—30 million. The costs of sending 1 million kwh of electricity from plant to city depend on the distance the electricity must travel. Formulate an LP to minimize the cost of meeting each city's peak power demand.

Solution To formulate Powerco's problem as an LP, we begin by defining a variable for each decision that Powerco must make. Because Powerco must determine how much power is sent from each plant to each city, we define (for $i = 1, 2, 3$ and $j = 1, 2, 3, 4$)

x_{ij} = number of (million) kwh produced at plant i and sent to city j

In terms of these variables, the total cost of supplying the peak power demands to cities 1–4 may be written as

$$
\begin{aligned}
& 8x_{11} + 6x_{12} + 10x_{13} + 9x_{14} \quad &\text{(Cost of shipping power from plant 1)}\\
+\ & 9x_{21} + 12x_{22} + 13x_{23} + 7x_{24} \quad &\text{(Cost of shipping power from plant 2)}\\
+\ & 14x_{31} + 9x_{32} + 16x_{33} + 5x_{34} \quad &\text{(Cost of shipping power from plant 3)}
\end{aligned}
$$

Powerco faces two types of constraints. First, the total power supplied by each plant cannot exceed the plant's capacity. For example, the total amount of power sent from plant

---

[†]This example is based on Aarvik and Randolph (1975).

TABLE 1

Shipping Costs, Supply, and Demand for Powerco

From	To				Supply (million kwh)
	City 1	City 2	City 3	City 4	
Plant 1	$8	$6	$10	$9	35
Plant 2	$9	$12	$13	$7	50
Plant 3	$14	$9	$16	$5	40
Demand (million kwh)	45	20	30	30	

1 to the four cities cannot exceed 35 million kwh. Each variable with first subscript 1 represents a shipment of power from plant 1, so we may express this restriction by the LP constraint

$$x_{11} + x_{12} + x_{13} + x_{14} \leq 35$$

In a similar fashion, we can find constraints that reflect plant 2's and plant 3's capacities. Because power is supplied by the power plants, each is a **supply point.** Analogously, a constraint that ensures that the total quantity shipped from a plant does not exceed plant capacity is a **supply constraint.** The LP formulation of Powerco's problem contains the following three supply constraints:

$$x_{11} + x_{12} + x_{13} + x_{14} \leq 35 \quad \text{(Plant 1 supply constraint)}$$
$$x_{21} + x_{22} + x_{23} + x_{24} \leq 50 \quad \text{(Plant 2 supply constraint)}$$
$$x_{31} + x_{32} + x_{33} + x_{34} \leq 40 \quad \text{(Plant 3 supply constraint)}$$

Second, we need constraints that ensure that each city will receive sufficient power to meet its peak demand. Each city demands power, so each is a **demand point.** For example, city 1 must receive at least 45 million kwh. Each variable with second subscript 1 represents a shipment of power to city 1, so we obtain the following constraint:

$$x_{11} + x_{21} + x_{31} \geq 45$$

Similarly, we obtain a constraint for each of cities 2, 3, and 4. A constraint that ensures that a location receives its demand is a **demand constraint.** Powerco must satisfy the following four demand constraints:

$$x_{11} + x_{21} + x_{31} \geq 45 \quad \text{(City 1 demand constraint)}$$
$$x_{12} + x_{22} + x_{32} \geq 20 \quad \text{(City 2 demand constraint)}$$
$$x_{13} + x_{23} + x_{33} \geq 30 \quad \text{(City 3 demand constraint)}$$
$$x_{14} + x_{24} + x_{34} \geq 30 \quad \text{(City 4 demand constraint)}$$

Because all the x_{ij}'s must be nonnegative, we add the sign restrictions $x_{ij} \geq 0$ ($i = 1, 2, 3; j = 1, 2, 3, 4$).

Combining the objective function, supply constraints, demand constraints, and sign restrictions yields the following LP formulation of Powerco's problem:

$$\min z = 8x_{11} + 6x_{12} + 10x_{13} + 9x_{14} + 9x_{21} + 12x_{22} + 13x_{23} + 7x_{24}$$
$$+ 14x_{31} + 9x_{32} + 16x_{33} + 5x_{34}$$
$$\text{s.t.} \quad x_{11} + x_{12} + x_{13} + x_{14} \leq 35 \quad \text{(Supply constraints)}$$
$$x_{21} + x_{22} + x_{23} + x_{24} \leq 50$$
$$x_{31} + x_{32} + x_{33} + x_{34} \leq 40$$

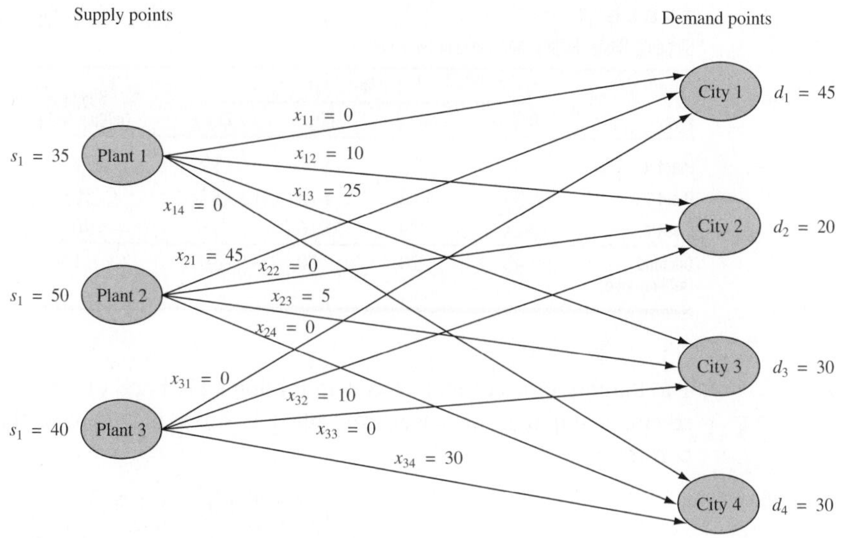

FIGURE 1
Graphical Representation of Powerco Problem and Its Optimal Solution

$$
\begin{aligned}
x_{11} + x_{21} + x_{31} &\geq 45 \qquad \text{(Demand constraints)} \\
x_{12} + x_{22} + x_{32} &\geq 20 \\
x_{13} + x_{23} + x_{33} &\geq 30 \\
x_{14} + x_{24} + x_{34} &\geq 30 \\
x_{ij} \geq 0 \quad (i = 1, 2, 3; \; &j = 1, 2, 3, 4)
\end{aligned}
$$

In Section 7.3, we will find that the optimal solution to this LP is $z = 1020$, $x_{12} = 10$, $x_{13} = 25$, $x_{21} = 45$, $x_{23} = 5$, $x_{32} = 10$, $x_{34} = 30$. Figure 1 is a graphical representation of the Powerco problem and its optimal solution. The variable x_{ij} is represented by a line, or arc, joining the ith supply point (plant i) and the jth demand point (city j).

General Description of a Transportation Problem

In general, a transportation problem is specified by the following information:

1 A set of m *supply points* from which a good is shipped. Supply point i can supply at most s_i units. In the Powerco example, $m = 3$, $s_1 = 35$, $s_2 = 50$, and $s_3 = 40$.

2 A set of n *demand points* to which the good is shipped. Demand point j must receive at least d_j units of the shipped good. In the Powerco example, $n = 4$, $d_1 = 45$, $d_2 = 20$, $d_3 = 30$, and $d_4 = 30$.

3 Each unit produced at supply point i and shipped to demand point j incurs a *variable cost* of c_{ij}. In the Powerco example, $c_{12} = 6$.

Let

$$x_{ij} = \text{number of units shipped from supply point } i \text{ to demand point } j$$

then the general formulation of a transportation problem is

$$\min \sum_{i=1}^{i=m} \sum_{j=1}^{j=n} c_{ij} x_{ij}$$

$$\text{s.t.} \quad \sum_{j=1}^{j=n} x_{ij} \leq s_i \quad (i = 1, 2, \ldots, m) \qquad \text{(Supply constraints)}$$

(1)

$$\sum_{i=1}^{i=m} x_{ij} \geq d_j \quad (j = 1, 2, \ldots, n) \qquad \text{(Demand constraints)}$$

$$x_{ij} \geq 0 \quad (i = 1, 2, \ldots, m; j = 1, 2, \ldots, n)$$

If a problem has the constraints given in (1) and is a *maximization* problem, then it is still a transportation problem (see Problem 7 at the end of this section). If

$$\sum_{i=1}^{i=m} s_i = \sum_{j=1}^{j=n} d_j$$

then total supply equals total demand, and the problem is said to be a **balanced transportation problem.**

For the Powerco problem, total supply and total demand both equal 125, so this is a balanced transportation problem. In a balanced transportation problem, all the constraints must be binding. For example, in the Powerco problem, if any supply constraint were non-binding, then the remaining available power would not be sufficient to meet the needs of all four cities. For a balanced transportation problem, (1) may be written as

$$\min \sum_{i=1}^{i=m} \sum_{j=1}^{j=n} c_{ij} x_{ij}$$

$$\text{s.t.} \quad \sum_{j=1}^{j=n} x_{ij} = s_i \quad (i = 1, 2, \ldots, m) \qquad \text{(Supply constraints)}$$

(2)

$$\sum_{i=1}^{i=m} x_{ij} = d_j \quad (j = 1, 2, \ldots, n) \qquad \text{(Demand constraints)}$$

$$x_{ij} \geq 0 \quad (i = 1, 2, \ldots, m; j = 1, 2, \ldots, n)$$

Later in this chapter, we will see that it is relatively simple to find a basic feasible solution for a balanced transportation problem. Also, simplex pivots for these problems do not involve multiplication and reduce to additions and subtractions. For these reasons, it is desirable to formulate a transportation problem as a balanced transportation problem.

Balancing a Transportation Problem If Total Supply Exceeds Total Demand

If total supply exceeds total demand, we can balance a transportation problem by creating a **dummy demand point** that has a demand equal to the amount of excess supply. Because shipments to the dummy demand point are not real shipments, they are assigned a cost of zero. Shipments to the dummy demand point indicate unused supply capacity. To understand the use of a dummy demand point, suppose that in the Powerco problem, the demand for city 1 were reduced to 40 million kwh. To balance the Powerco problem, we would add a dummy demand point (point 5) with a demand of $125 - 120 = 5$ million kwh. From each plant, the cost of shipping 1 million kwh to the dummy is 0. The optimal solution to this balanced transportation problem is $z = 975$, $x_{13} = 20$, $x_{12} = 15$, $x_{21} = 40$, $x_{23} = 10$, $x_{32} = 5$, $x_{34} = 30$, and $x_{35} = 5$. Because $x_{35} = 5$, 5 million kwh of plant 3 capacity will be unused (see Figure 2).

A transportation problem is specified by the supply, the demand, and the shipping costs, so the relevant data can be summarized in a **transportation tableau** (see Table 2). The square, or **cell,** in row i and column j of a transportation tableau corresponds to the

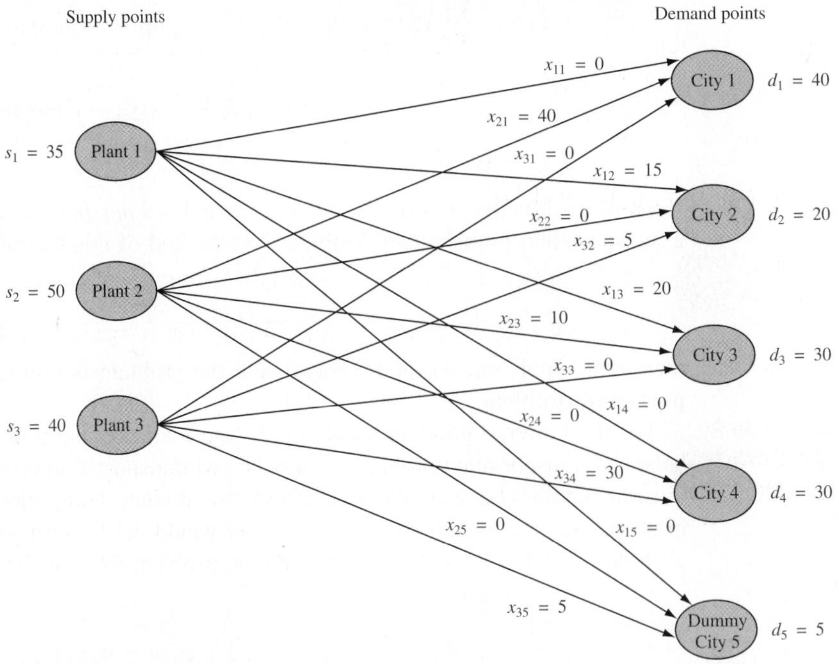

FIGURE 2
Graphical
Representation of
Unbalanced Powerco
Problem and Its
Optimal Solution (with
Dummy Demand Point)

$s_1 = 35$ Plant 1

$s_2 = 50$ Plant 2

$s_3 = 40$ Plant 3

$x_{11} = 0$ City 1 $d_1 = 40$
$x_{21} = 40$
$x_{31} = 0$
$x_{12} = 15$
$x_{22} = 0$ City 2 $d_2 = 20$
$x_{32} = 5$
$x_{13} = 20$
$x_{23} = 10$
$x_{33} = 0$ City 3 $d_3 = 30$
$x_{14} = 0$
$x_{24} = 0$
$x_{34} = 30$ City 4 $d_4 = 30$
$x_{25} = 0$
$x_{15} = 0$
$x_{35} = 5$ Dummy City 5 $d_5 = 5$

Supply points

Demand points

TABLE 2
A Transportation Tableau

				Supply
c_{11}	c_{12}	$\cdots$	c_{1n}	s_1
c_{21}	c_{22}	$\cdots$	c_{2n}	s_2
$\vdots$	$\vdots$	$\vdots$	$\vdots$	$\vdots$
c_{m1}	c_{m2}	$\cdots$	c_{mn}	s_m
d_1	d_2	$\cdots$	d_n	

Demand

TABLE 3
Transportation Tableau
for Powerco

	City 1	City 2	City 3	City 4	Supply
Plant 1	8	6 — 10	10 — 25	9	35
Plant 2	9 — 45	12	13 — 5	7	50
Plant 3	14	9 — 10	16	5 — 30	40
Demand	45	20	30	30	

variable x_{ij}. If x_{ij} is a basic variable, its value is placed in the lower left-hand corner of the ijth cell of the tableau. For example, the balanced Powerco problem and its optimal solution could be displayed as shown in Table 3. The tableau format implicitly expresses the supply and demand constraints through the fact that the sum of the variables in row i must equal s_i and the sum of the variables in column j must equal d_j.

Balancing a Transportation Problem If Total Supply Is Less Than Total Demand

If a transportation problem has a total supply that is strictly less than total demand, then the problem has no feasible solution. For example, if plant 1 had only 30 million kwh of capacity, then a total of only 120 million kwh would be available. This amount of power would be insufficient to meet the total demand of 125 million kwh, and the Powerco problem would no longer have a feasible solution.

When total supply is less than total demand, it is sometimes desirable to allow the possibility of leaving some demand unmet. In such a situation, a penalty is often associated with unmet demand. Example 2 illustrates how such a situation can yield a balanced transportation problem.

EXAMPLE 2	Handling Shortages

Two reservoirs are available to supply the water needs of three cities. Each reservoir can supply up to 50 million gallons of water per day. Each city would like to receive 40 million gallons per day. For each million gallons per day of unmet demand, there is a penalty. At city 1, the penalty is $20; at city 2, the penalty is $22; and at city 3, the penalty is $23. The cost of transporting 1 million gallons of water from each reservoir to each city is shown in Table 4. Formulate a balanced transportation problem that can be used to minimize the sum of shortage and transport costs.

Solution   In this problem,

$$\text{Daily supply} = 50 + 50 = 100 \text{ million gallons per day}$$
$$\text{Daily demand} = 40 + 40 + 40 = 120 \text{ million gallons per day}$$

To balance the problem, we add a dummy (or shortage) *supply point* having a supply of $120 - 100 = 20$ million gallons per day. The cost of shipping 1 million gallons from the dummy supply point to a city is just the shortage cost per million gallons for that city. Table 5 shows the balanced transportation problem and its optimal solution. Reservoir 1 should send 20 million gallons per day to city 1 and 30 million gallons per day to city 2, whereas reservoir 2 should send 10 million gallons per day to city 2 and 40 million gallons per day to city 3. Twenty million gallons per day of city 1's demand will be unsatisfied.

TABLE 4
Shipping Costs for Reservoir

From	To		
	City 1	City 2	City 3
Reservoir 1	$7	$8	$10
Reservoir 2	$9	$7	$8

TABLE 5
Transportation Tableau
for Reservoir

	City 1	City 2	City 3	Supply
Reservoir 1	7 20	8 30	10	50
Reservoir 2	9	7 10	8 40	50
Dummy (shortage)	20 20	22	23	20
Demand	40	40	40	

Modeling Inventory Problems as Transportation Problems

Many inventory planning problems can be modeled as balanced transportation problems. To illustrate, we formulate a balanced transportation model of the Sailco problem of Section 3.10.

EXAMPLE 3   **Setting Up an Inventory Problem as a Transportation Problem**

Sailco Corporation must determine how many sailboats should be produced during each of the next four quarters (one quarter is three months). Demand is as follows: first quarter, 40 sailboats; second quarter, 60 sailboats; third quarter, 75 sailboats; fourth quarter, 25 sailboats. Sailco must meet demand on time. At the beginning of the first quarter, Sailco has an inventory of 10 sailboats. At the beginning of each quarter, Sailco must decide how many sailboats should be produced during the current quarter. For simplicity, we assume that sailboats manufactured during a quarter can be used to meet demand for the current quarter. During each quarter, Sailco can produce up to 40 sailboats at a cost of $400 per sailboat. By having employees work overtime during a quarter, Sailco can produce additional sailboats at a cost of $450 per sailboat. At the end of each quarter (after production has occurred and the current quarter's demand has been satisfied), a carrying or holding cost of $20 per sailboat is incurred. Formulate a balanced transportation problem to minimize the sum of production and inventory costs during the next four quarters.

Solution   We define supply and demand points as follows:

$$\text{Point } 1 = \text{initial inventory} \quad (s_1 = 10)$$
$$\text{Point } 2 = \text{quarter 1 regular-time (RT) production} \quad (s_2 = 40)$$
$$\text{Point } 3 = \text{quarter 1 overtime (OT) production} \quad (s_3 = 150)$$
$$\text{Point } 4 = \text{quarter 2 RT production} \quad (s_4 = 40)$$
$$\textbf{Supply Points} \quad \text{Point } 5 = \text{quarter 2 OT production} \quad (s_5 = 150)$$
$$\text{Point } 6 = \text{quarter 3 RT production} \quad (s_6 = 40)$$
$$\text{Point } 7 = \text{quarter 3 OT production} \quad (s_7 = 150)$$
$$\text{Point } 8 = \text{quarter 4 RT production} \quad (s_8 = 40)$$
$$\text{Point } 9 = \text{quarter 4 OT production} \quad (s_9 = 150)$$

There is a supply point corresponding to each source from which demand for sailboats can be met:

$$\text{Point 1} = \text{quarter 1 demand} \quad (d_1 = 40)$$
$$\text{Point 2} = \text{quarter 2 demand} \quad (d_2 = 60)$$
Demand Points $\quad \text{Point 3} = \text{quarter 3 demand} \quad (d_3 = 75)$
$$\text{Point 4} = \text{quarter 4 demand} \quad (d_4 = 25)$$
$$\text{Point 5} = \text{dummy demand point} \quad (d_5 = 770 - 200 = 570)$$

A shipment from, say, quarter 1 RT to quarter 3 demand means producing 1 unit on regular time during quarter 1 that is used to meet 1 unit of quarter 3's demand. To determine, say, c_{13}, observe that producing 1 unit during quarter 1 RT and using that unit to meet quarter 3 demand incurs a cost equal to the cost of producing 1 unit on quarter 1 RT plus the cost of holding a unit in inventory for $3 - 1 = 2$ quarters. Thus, $c_{13} = 400 + 2(20) = 440$.

Because there is no limit on the overtime production during any quarter, it is not clear what value should be chosen for the supply at each overtime production point. Total demand $= 200$, so at most $200 - 10 = 190$ (-10 is for initial inventory) units will be produced during any quarter. Because 40 units must be produced on regular time before any units are produced on overtime, overtime production during any quarter will never exceed $190 - 40 = 150$ units. Any unused overtime capacity will be "shipped" to the dummy demand point. To ensure that no sailboats are used to meet demand during a quarter prior to their production, a cost of M (M is a large positive number) is assigned to any cell that corresponds to using production to meet demand for an earlier quarter.

TABLE 6
Transportation Tableau for Sailco

	1	2	3	4	Dummy	Supply
Initial	0 / 10	20	40	60	0	10
Qtr 1 RT	400 / 30	420 / 10	440	460	0	40
Qtr 1 OT	450	470	490	510	0 / 150	150
Qtr 2 RT	M	400 / 40	420	440	0	40
Qtr 2 OT	M	450 / 10	470	490	0 / 140	150
Qtr 3 RT	M	M	400 / 40	420	0	40
Qtr 3 OT	M	M	450 / 35	470	0 / 115	150
Qtr 4 RT	M	M	M	400 / 25	0 / 15	40
Qtr 4 OT	M	M	M	450	0 / 150	150
Demand	40	60	75	25	570	

Total supply = 770 and total demand = 200, so we must add a dummy demand point with a demand of 770 − 200 = 570 to balance the problem. The cost of shipping a unit from any supply point to the dummy demand point is 0.

Combining these observations yields the balanced transportation problem and its optimal solution shown in Table 6. Thus, Sailco should meet quarter 1 demand with 10 units of initial inventory and 30 units of quarter 1 RT production; quarter 2 demand with 10 units of quarter 1 RT, 40 units of quarter 2 RT, and 10 units of quarter 2 OT production; quarter 3 demand with 40 units of quarter 3 RT and 35 units of quarter 3 OT production; and finally, quarter 4 demand with 25 units of quarter 4 RT production.

In Problem 12 at the end of this section, we show how this formulation can be modified to incorporate other aspects of inventory problems (backlogged demand, perishable inventory, and so on).

Solving Transportation Problems on the Computer

To solve a transportation problem with LINDO, type in the objective function, supply constraints, and demand constraints. Other menu-driven programs are available that accept the shipping costs, supply values, and demand values. From these values, the program can generate the objective function and constraints.

LINGO can be used to easily solve any transportation problem. The following LINGO model can be used to solve the Powerco example (file Trans.lng).

Trans.lng

```
MODEL:
  1]SETS:
  2]PLANTS/P1,P2,P3/:CAP;
  3]CITIES/C1,C2,C3,C4/:DEM;
  4]LINKS(PLANTS,CITIES):COST,SHIP;
  5]ENDSETS
  6]MIN=@SUM(LINKS:COST*SHIP);
  7]@FOR(CITIES(J):
  8]@SUM(PLANTS(I):SHIP(I,J))>DEM(J));
  9]@FOR(PLANTS(I):
 10]@SUM(CITIES(J):SHIP(I,J))<CAP(I));
 11]DATA:
 12]CAP=35,50,40;
 13]DEM=45,20,30,30;
 14]COST=8,6,10,9,
 15]9,12,13,7,
 16]14,9,16,5;
 17]ENDDATA
END
```

Lines 1–5 define the **SETS** needed to generate the objective function and constraints. In line 2, we create the three power plants (the supply points) and specify that each has a capacity (given in the **DATA** section). In line 3, we create the four cities (the demand points) and specify that each has a demand (given in the **DATA** section). The **LINK** statement in line 4 creates a LINK(I,J) as I runs over all PLANTS and J runs over all CITIES. Thus, objects LINK(1,1), LINK (1,2), LINK(1,3), LINK(1,4), LINK(2,1), LINK (2,2), LINK(2,3), LINK(2,4), LINK(3,1), LINK (3,2), LINK(3,3), LINK(3,4) are created and stored in this order. Attributes with multiple subscripts are stored so that the rightmost subscripts advance most rapidly. Each LINK has two attributes: a per-unit shipping cost [(COST), given in the **DATA** section] and the amount shipped (SHIP), for which LINGO will solve.

Line 6 creates the objective function. We sum over all links the product of the unit shipping cost and the amount shipped. Using the **@FOR** and **@SUM** operators, lines 7–8 generate all demand constraints. They ensure that for each city, the sum of the amount shipped into the city will be at least as large as the city's demand. Note that the extra parenthesis after SHIP(I,J) in line 8 is to close the **@SUM** operator, and the extra parenthesis after DEM(J) is to close the **@FOR** operator. Using the **@FOR** and **@SUM** operators, lines 9–10 generate all supply constraints. They ensure that for each plant, the total shipped out of the plant will not exceed the plant's capacity.

Lines 11–17 contain the data needed for the problem. Line 12 defines each plant's capacity, and line 13 defines each city's demand. Lines 14–16 contain the unit shipping cost from each plant to each city. These costs correspond to the ordering of the links described previously. **ENDDATA** ends the data section, and **END** ends the program. Typing **GO** will solve the problem.

This program can be used to solve any transportation problem. If, for example, we wanted to solve a problem with 15 supply points and 10 demand points, we would change line 2 to create 15 supply points and line 3 to create 10 demand points. Moving to line 12, we would type in the 15 plant capacities. In line 13, we would type in the demands for the 10 demand points. Then in line 14, we would type in the 150 shipping costs. Observe that the part of the program (lines 6–10) that generates the objective function and constraints remains unchanged! Notice also that our LINGO formulation does not require that the transportation problem be balanced.

Obtaining LINGO Data from an Excel Spreadsheet

Often it is easier to obtain data for a LINGO model from a spreadsheet. For example, shipping costs for a transportation problem may be the end result of many computations. As an example, suppose we have created the capacities, demands, and shipping costs for the Powerco model in the file Powerco.xls (see Figure 3). We have created capacities in the cell range F9:F11 and named the range Cap. As you probably know, you can name a range of cells in Excel by selecting the range and clicking in the name box in the upper left-hand corner of your spreadsheet. Then type the range name and hit the Enter key. In a similar fashion, name the city demands (in cells B12:E12) with the name Demand and the unit shipping costs (in cells B4:E6) with the name Costs.

Powerco.xls

FIGURE 3

	A	B	C	D	E	F	G	H
1		OPTIMAL SOLUTION	FOR	POWERCO		COSTS		
2	COSTS		CITY			1020		
3	PLANT		1	2	3	4		
4		1	8	6	10	9		
5		2	9	12	13	7		
6		3	14	9	16	5		
7	SHIPMENTS		CITY			SHIPPED		SUPPLIES
8	PLANT		1	2	3	4		
9		1	0	10	25	0	35	<=	35
10		2	45	0	5	0	50	<=	50
11		3	0	10	0	30	40	<=	40
12	RECEIVED		45	20	30	30		
13			>=	>=	>=	>=		
14	DEMANDS		45	20	30	30		

Using an **@OLE** statement, LINGO can read from a spreadsheet the values of data that are defined in the Sets portion of a program. The LINGO program (see file Transpspread.lng) needed to read our input data from the Powerco.xls file is shown below.

```
MODEL:
SETS:
PLANTS/P1,P2,P3/:CAP;
CITIES/C1,C2,C3,C4/:DEM;
LINKS(PLANTS,CITIES):COST,SHIP;
ENDSETS
MIN=@SUM(LINKS:COST*SHIP);
@FOR(CITIES(J):
@SUM(PLANTS(I):SHIP(I,J))>DEM(J));
@FOR(PLANTS(I);
@SUM(CITIES(J):SHIP(I,J))<CAP(I));
DATA:
CAP, DEM, COST=@OLE('C:\MPROG\POWERCO.XLS','Cap','Demand','Costs');
ENDDATA
  END

The key statement is

CAP, DEM, COST=@OLE('C:\MPROG\POWERCO.XLS','Cap','Demand','Costs');.
```

This statement reads the defined data sets CAP, DEM, and COSTS from the Powerco.xls spreadsheet. Note that the full path location of our Excel file (enclosed in single quotes) must be given first followed by the spreadsheet range names that contain the needed data. The range names are paired with the data sets in the order listed. Therefore, CAP values are found in range Cap and so on. The **@OLE** statement is very powerful, because a spreadsheet will usually greatly simplify the creation of data for a LINGO program.

Spreadsheet Solution of Transportation Problems

In the file Powerco.xls, we show how easy it is to use the Excel Solver to find the optimal solution to a transportation problem. After entering the plant capacities, city demands, and unit shipping costs as shown, we enter trial values of the units shipped from each plant to each city in the range B9:E11. Then we proceed as follows:

Step 1 Compute the total amount shipped out of each city by copying from F9 to F10:F11 the formula

$$=SUM(B9:E9)$$

Step 2 Compute the total received by each city by copying from B12 to C12:E12 the formula

$$=SUM(B9:B11)$$

Step 3 Compute the total shipping cost in cell F2 with the formula

$$=SUMPRODUCT(B9:E11,Costs)$$

Note that the =SUMPRODUCT function works on rectangles as well as rows or columns of numbers. Also, we have named the range of unit shipping costs (B4:E6) as COSTS.

Step 4 We now fill in the Solver window shown in Figure 4. We minimize total shipping costs (F2) by changing units shipped from each plant to each city (B9:E11). We constrain amount received by each city (B12:E12) to be at least each city's demand (range name Demand). We constrain the amount shipped out of each plant (F9:F11) to be at most each plant's capacity (range name Cap). After checking the Assume Nonnegative option and Assume Linear Model option, we obtain the optimal solution shown in Figure 3. Note, of course, that the objective function of the optimal solution found by Excel equals the ob-

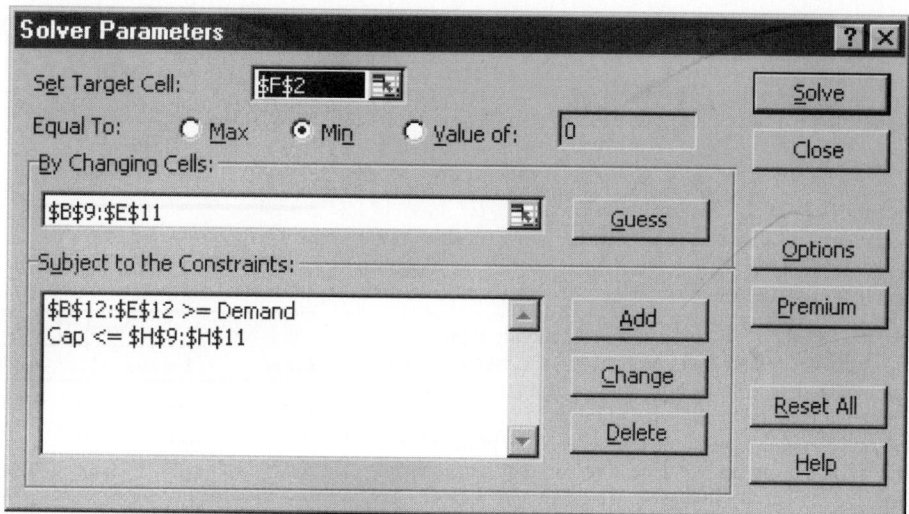

FIGURE 4

jective function value found by LINGO and our hand solution. If the problem had multiple optimal solutions, then it is possible that the values of the shipments found by LINGO, Excel, and our hand solution might be different.

PROBLEMS

Group A

1 A company supplies goods to three customers, who each require 30 units. The company has two warehouses. Warehouse 1 has 40 units available, and warehouse 2 has 30 units available. The costs of shipping 1 unit from warehouse to customer are shown in Table 7. There is a penalty for each unmet customer unit of demand: With customer 1, a penalty cost of $90 is incurred; with customer 2, $80; and with customer 3, $110. Formulate a balanced transportation problem to minimize the sum of shortage and shipping costs.

2 Referring to Problem 1, suppose that extra units could be purchased and shipped to either warehouse for a total cost of $100 per unit and that all customer demand must be met. Formulate a balanced transportation problem to minimize the sum of purchasing and shipping costs.

3 A shoe company forecasts the following demands during the next six months: month 1—200; month 2—260; month 3—240; month 4—340; month 5—190; month 6—150. It costs $7 to produce a pair of shoes with regular-time labor (RT) and $11 with overtime labor (OT). During each month, regular production is limited to 200 pairs of shoes, and

overtime production is limited to 100 pairs. It costs $1 per month to hold a pair of shoes in inventory. Formulate a balanced transportation problem to minimize the total cost of meeting the next six months of demand on time.

4 Steelco manufactures three types of steel at different plants. The time required to manufacture 1 ton of steel (regardless of type) and the costs at each plant are shown in Table 8. Each week, 100 tons of each type of steel (1, 2, and 3) must be produced. Each plant is open 40 hours per week.

   a Formulate a balanced transportation problem to minimize the cost of meeting Steelco's weekly requirements.

   b Suppose the time required to produce 1 ton of steel depends on the type of steel as well as on the plant at which it is produced (see Table 9, page 372). Could a transportation problem still be formulated?

5 A hospital needs to purchase 3 gallons of a perishable medicine for use during the current month and 4 gallons for use during the next month. Because the medicine is

TABLE 7

From	To		
	Customer 1	Customer 2	Customer 3
Warehouse 1	$15	$35	$25
Warehouse 2	$10	$50	$40

TABLE 8

Plant	Cost ($)			Time (minutes)
	Steel 1	Steel 2	Steel 3	
1	60	40	28	20
2	50	30	30	16
3	43	20	20	15

TABLE 9

Plant	Time (minutes)		
	Steel 1	Steel 2	Steel 3
1	15	12	15
2	15	15	20
3	10	10	15

perishable, it can only be used during the month of purchase. Two companies (Daisy and Laroach) sell the medicine. The medicine is in short supply. Thus, during the next two months, the hospital is limited to buying at most 5 gallons from each company. The companies charge the prices shown in Table 10. Formulate a balanced transportation model to minimize the cost of purchasing the needed medicine.

6 A bank has two sites at which checks are processed. Site 1 can process 10,000 checks per day, and site 2 can process 6,000 checks per day. The bank processes three types of checks: vendor, salary, and personal. The processing cost per check depends on the site (see Table 11). Each day, 5,000 checks of each type must be processed. Formulate a balanced transportation problem to minimize the daily cost of processing checks.

7[†] The U.S. government is auctioning off oil leases at two sites: 1 and 2. At each site, 100,000 acres of land are to be auctioned. Cliff Ewing, Blake Barnes, and Alexis Pickens are bidding for the oil. Government rules state that no bidder can receive more than 40% of the land being auctioned. Cliff has bid $1,000/acre for site 1 land and $2,000/acre for site 2 land. Blake has bid $900/acre for site 1 land and $2,200/acre for site 2 land. Alexis has bid $1,100/acre for site 1 land and $1,900/acre for site 2 land. Formulate a balanced transportation model to maximize the government's revenue.

TABLE 10

Company	Current Month's Price per Gallon ($)	Next Month's Price per Gallon ($)
Daisy	800	720
Laroach	710	750

TABLE 11

Checks	Site (c)	
	1	2
Vendor	5	3
Salary	4	4
Personal	2	5

[†]This problem is based on Jackson (1980).

TABLE 12

From ($)	To ($)	
	England	Japan
Field 1	1	2
Field 2	2	1

TABLE 13

Auditor	Project ($)		
	1	2	3
1	120	150	190
2	140	130	120
3	160	140	150

8 The Ayatola Oil Company controls two oil fields. Field 1 can produce up to 40 million barrels of oil per day, and field 2 can produce up to 50 million barrels of oil per day. At field 1, it costs $3 to extract and refine a barrel of oil; at field 2, the cost is $2. Ayatola sells oil to two countries: England and Japan. The shipping cost per barrel is shown in Table 12. Each day, England is willing to buy up to 40 million barrels (at $6 per barrel), and Japan is willing to buy up to 30 million barrels (at $6.50 per barrel). Formulate a balanced transportation problem to maximize Ayatola's profits.

9 For the examples and problems of this section, discuss whether it is reasonable to assume that the proportionality assumption holds for the objective function.

10 Touche Young has three auditors. Each can work as many as 160 hours during the next month, during which time three projects must be completed. Project 1 will take 130 hours; project 2, 140 hours; and project 3, 160 hours. The amount per hour that can be billed for assigning each auditor to each project is given in Table 13. Formulate a balanced transportation problem to maximize total billings during the next month.

Group B

11[‡] Paperco recycles newsprint, uncoated paper, and coated paper into recycled newsprint, recycled uncoated paper, and recycled coated paper. Recycled newsprint can be produced by processing newsprint or uncoated paper. Recycled coated paper can be produced by recycling any type of paper. Recycled uncoated paper can be produced by processing uncoated paper or coated paper. The process used to produce recycled newsprint removes 20% of the input's pulp, leaving 80% of the input's pulp for recycled paper. The process used to produce recycled coated paper removes 10% of the input's pulp. The process used to produce recycled uncoated paper removes 15% of the input's pulp. The purchasing costs, processing costs, and availability of each type of paper are shown in Table 14. To meet demand,

[‡]This problem is based on Glassey and Gupta (1974).

TABLE **14**

	Purchase Cost per Ton of Pulp ($)	Processing Cost per Ton of Input ($)	Availability
Newsprint	10		500
Coated paper	9		300
Uncoated paper	8		200
NP used for RNP		3	
NP used for RCP		4	
UCP used for RNP		4	
UCP used for RUP		1	
UCP used for RCP		6	
CP used for RUP		5	
CP used for RCP		3	

Paperco must produce at least 250 tons of recycled newsprint pulp, at least 300 tons of recycled uncoated paper pulp, and at least 150 tons of recycled coated paper pulp. Formulate a balanced transportation problem that can be used to minimize the cost of meeting Paperco's demands.

12 Explain how each of the following would modify the formulation of the Sailco problem as a balanced transportation problem:

a Suppose demand could be backlogged at a cost of $30/sailboat/month. (*Hint:* Now it is permissible to ship from, say, month 2 production to month 1 demand.)

b If demand for a sailboat is not met on time, the sale is lost and an opportunity cost of $450 is incurred.

c Sailboats can be held in inventory for a maximum of two months.

d At a cost of $440/sailboat, Sailco can purchase up to 10 sailboats/month from a subcontractor.

7.2 Finding Basic Feasible Solutions for Transportation Problems

Consider a balanced transportation problem with m supply points and n demand points. From (2), we see that such a problem contains $m + n$ equality constraints. From our experience with the Big M method and the two-phase simplex method, we know it is difficult to find a bfs if all of an LP's constraints are equalities. Fortunately, the special structure of a balanced transportation problem makes it easy for us to find a bfs.

Before describing three methods commonly used to find a bfs to a balanced transportation problem, we need to make the following important observation. *If a set of values for the x_{ij}'s satisfies all but one of the constraints of a balanced transportation problem, then the values for the x_{ij}'s will automatically satisfy the other constraint.* For example, in the Powerco problem, suppose a set of values for the x_{ij}'s is known to satisfy all the constraints with the exception of the first supply constraint. Then this set of x_{ij}'s must supply $d_1 + d_2 + d_3 + d_4 = 125$ million kwh to cities 1–4 and supply $s_2 + s_3 = 125 - s_1 = 90$ million kwh from plants 2 and 3. Thus, plant 1 must supply $125 - (125 - s_1) = 35$ million kwh, so the x_{ij}'s must also satisfy the first supply constraint.

The preceding discussion shows that when we solve a balanced transportation problem, we may omit from consideration any one of the problem's constraints and solve an LP having $m + n - 1$ constraints. We (arbitrarily) assume that the first supply constraint is omitted from consideration.

In trying to find a bfs to the remaining $m + n - 1$ constraints, you might think that any collection of $m + n - 1$ variables would yield a basic solution. Unfortunately, this is not the case. For example, consider (3), a balanced transportation problem. (We omit the costs because they are not needed to find a bfs.)

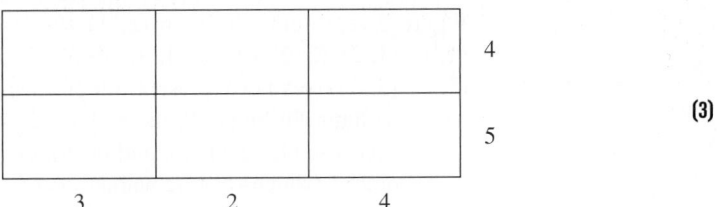

(3)

In matrix form, the constraints for this balanced transportation problem may be written as

$$\begin{bmatrix} 1 & 1 & 1 & 0 & 0 & 0 \\ 0 & 0 & 0 & 1 & 1 & 1 \\ 1 & 0 & 0 & 1 & 0 & 0 \\ 0 & 1 & 0 & 0 & 1 & 0 \\ 0 & 0 & 1 & 0 & 0 & 1 \end{bmatrix} \begin{bmatrix} x_{11} \\ x_{12} \\ x_{13} \\ x_{21} \\ x_{22} \\ x_{23} \end{bmatrix} = \begin{bmatrix} 4 \\ 5 \\ 3 \\ 2 \\ 4 \end{bmatrix} \tag{3'}$$

After dropping the first supply constraint, we obtain the following linear system:

$$\begin{bmatrix} 0 & 0 & 0 & 1 & 1 & 1 \\ 1 & 0 & 0 & 1 & 0 & 0 \\ 0 & 1 & 0 & 0 & 1 & 0 \\ 0 & 0 & 1 & 0 & 0 & 1 \end{bmatrix} \begin{bmatrix} x_{11} \\ x_{12} \\ x_{13} \\ x_{21} \\ x_{22} \\ x_{23} \end{bmatrix} = \begin{bmatrix} 5 \\ 3 \\ 2 \\ 4 \end{bmatrix} \tag{3''}$$

A basic solution to (3″) must have four basic variables. Suppose we try BV = $\{x_{11}, x_{12}, x_{21}, x_{22}\}$. Then

$$B = \begin{bmatrix} 0 & 0 & 1 & 1 \\ 1 & 0 & 1 & 0 \\ 0 & 1 & 0 & 1 \\ 0 & 0 & 0 & 0 \end{bmatrix}$$

For $\{x_{11}, x_{12}, x_{21}, x_{22}\}$ to yield a basic solution, it must be possible to use EROs to transform B to I_4. Because rank $B = 3$ and EROs do not change the rank of a matrix, there is no way that EROs can be used to transform B into I_4. Thus, BV = $\{x_{11}, x_{12}, x_{21}, x_{22}\}$ cannot yield a basic solution to (3″). Fortunately, the simple concept of a loop may be used to determine whether an arbitrary set of $m + n - 1$ variables yields a basic solution to a balanced transportation problem.

DEFINITION ■ An ordered sequence of at least four different cells is called a **loop** if

1 Any two consecutive cells lie in either the same row or same column

2 No three consecutive cells lie in the same row or column

3 The last cell in the sequence has a row or column in common with the first cell in the sequence ■

In the definition of a loop, the first cell is considered to follow the last cell, so the loop may be thought of as a closed path. Here are some examples of the preceding definition: Figure 5 represents the loop (2, 1)–(2, 4)–(4, 4)–(4, 1). Figure 6 represents the loop (1, 1)–(1, 2)–(2, 2)–(2, 3)–(4, 3)–(4, 5)–(3, 5)–(3, 1). In Figure 7, the path (1, 1)–(1, 2)–(2, 3)–(2, 1) does not represent a loop, because (1, 2) and (2, 3) do not lie in the same row or column. In Figure 8, the path (1, 2)–(1, 3)–(1, 4)–(2, 4)–(2, 2) does not represent a loop, because (1, 2), (1, 3), and (1, 4) all lie in the same row.

Theorem 1 (which we state without proof) shows why the concept of a loop is important.

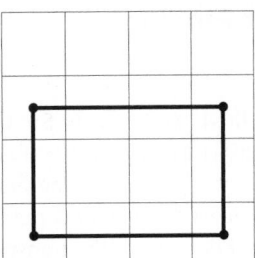

FIGURE 5

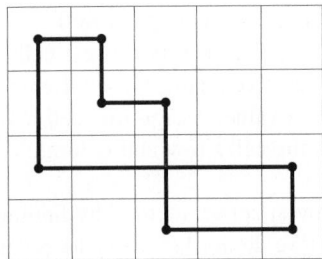

FIGURE 6

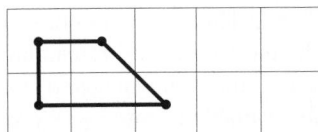

FIGURE 7

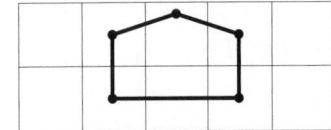

FIGURE 8

Theorem 1 follows from the fact that a set of $m + n - 1$ cells contains no loop if and only if the $m + n - 1$ columns corresponding to these cells are linearly independent. Because $(1, 1)$–$(1, 2)$–$(2, 2)$–$(2, 1)$ is a loop, Theorem 1 tells us that $\{x_{11}, x_{12}, x_{22}, x_{21}\}$ cannot yield a basic solution for $(3'')$. On the other hand, no loop can be formed with the cells $(1, 1)$–$(1, 2)$–$(1, 3)$–$(2, 1)$, so $\{x_{11}, x_{12}, x_{13}, x_{21}\}$ will yield a basic solution to $(3'')$.

We are now ready to discuss three methods that can be used to find a basic feasible solution for a balanced transportation problem:

1 northwest corner method

2 minimum-cost method

3 Vogel's method

Northwest Corner Method for Finding a Basic Feasible Solution

To find a bfs by the northwest corner method, we begin in the upper left (or northwest) corner of the transportation tableau and set x_{11} as large as possible. Clearly, x_{11} can be no larger than the smaller of s_1 and d_1. If $x_{11} = s_1$, cross out the first row of the transportation tableau; this indicates that no more basic variables will come from row 1. Also change d_1 to $d_1 - s_1$. If $x_{11} = d_1$, cross out the first column of the transportation tableau; this indicates that no more basic variables will come from column 1. Also change s_1 to $s_1 - d_1$. If $x_{11} = s_1 = d_1$, cross out either row 1 or column 1 (but not both). If you cross out row 1, change d_1 to 0; if you cross out column 1, change s_1 to 0.

Continue applying this procedure to the most northwest cell in the tableau that does not lie in a crossed-out row or column. Eventually, you will come to a point where there is only one cell that can be assigned a value. Assign this cell a value equal to its row or column demand, and cross out both the cell's row and column. A basic feasible solution has now been obtained.

We illustrate the use of the northwest corner method by finding a bfs for the balanced transportation problem in Table 15. (We do not list the costs because they are not needed to apply the algorithm.) We indicate the crossing out of a row or column by placing an $\times$ by the row's supply or column's demand.

To begin, we set $x_{11} = \min\{5, 2\} = 2$. Then we cross out column 1 and change s_1 to $5 - 2 = 3$. This yields Table 16. The most northwest remaining variable is x_{12}. We set $x_{12} = \min\{3, 4\} = 3$. Then we cross out row 1 and change d_2 to $4 - 3 = 1$. This yields Table 17. The most northwest available variable is now x_{22}. We set $x_{22} = \min\{1, 1\} = 1$. Because both the supply and demand corresponding to the cell are equal, we may cross out either row 2 or column 2 (but not both). For no particular reason, we choose to cross out row 2. Then d_2 must be changed to $1 - 1 = 0$. The resulting tableau is Table 18. At the next step, this will lead to a *degenerate* bfs.

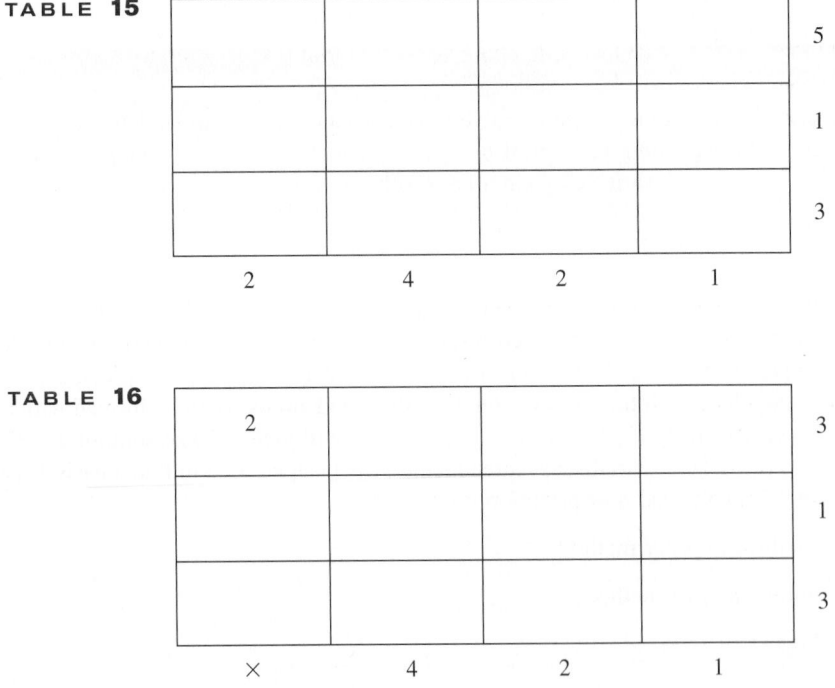

TABLE 15

TABLE 16

TABLE **17**

2	3			×
				1
				3
×	1	2	1	

TABLE **18**

2	3			×
	1			×
				3
×	0	2	1	

The most northwest available cell is now x_{32}, so we set $x_{32} = \min\{3, 0\} = 0$. Then we cross out column 2 and change s_3 to $3 - 0 = 3$. The resulting tableau is Table 19. We now set $x_{33} = \min\{3, 2\} = 2$. Then we cross out column 3 and reduce s_3 to $3 - 2 = 1$. The resulting tableau is Table 20. The only available cell is x_{34}. We set $x_{34} = \min\{1, 1\} = 1$. Then we cross out row 3 and column 4. No cells are available, so we are finished. We have obtained the bfs $x_{11} = 2$, $x_{12} = 3$, $x_{22} = 1$, $x_{32} = 0$, $x_{33} = 2$, $x_{34} = 1$.

Why does the northwest corner method yield a bfs? The method ensures that no basic variable will be assigned a negative value (because no right-hand side ever becomes nega-

TABLE **19**

2	3			×
	1			×
	0			3
×	×	2	1	

TABLE **20**

2	3			×
	1			×
	0	2		1
×	×	×	1	

tive) and also that each supply and demand constraint is satisfied (because every row and column is eventually crossed out). Thus, the northwest corner method yields a feasible solution.

To complete the northwest corner method, $m + n$ rows and columns must be crossed out. The last variable assigned a value results in a row and column being crossed out, so the northwest corner method will assign values to $m + n - 1$ variables. The variables chosen by the northwest corner method cannot form a loop, so Theorem 1 implies that the northwest corner method must yield a bfs.

Minimum-Cost Method for Finding a Basic Feasible Solution

The northwest corner method does not utilize shipping costs, so it can yield an initial bfs that has a very high shipping cost. Then determining an optimal solution may require several pivots. The minimum-cost method uses the shipping costs in an effort to produce a bfs that has a lower total cost. Hopefully, fewer pivots will then be required to find the problem's optimal solution.

To begin the minimum-cost method, find the variable with the smallest shipping cost (call it x_{ij}). Then assign x_{ij} its largest possible value, $\min\{s_i, d_j\}$. As in the northwest corner method, cross out row i or column j and reduce the supply or demand of the noncrossed-out row or column by the value of x_{ij}. Then choose from the cells that do not lie in a crossed-out row or column the cell with the minimum shipping cost and repeat the procedure. Continue until there is only one cell that can be chosen. In this case, cross out both the cell's row and column. Remember that (with the exception of the last variable) if a variable satisfies both a supply and demand constraint, only cross out a row or column, not both.

To illustrate the minimum cost method, we find a bfs for the balanced transportation problem in Table 21. The variable with the minimum shipping cost is x_{22}. We set $x_{22} = \min\{10, 8\} = 8$. Then we cross out column 2 and reduce s_2 to $10 - 8 = 2$ (Table 22). We could now choose either x_{11} or x_{21} (both having shipping costs of 2). We arbitrarily choose x_{21} and set $x_{21} = \min\{2, 12\} = 2$. Then we cross out row 2 and change d_1 to $12 - 2 = 10$ (Table 23). Now we set $x_{11} = \min\{5, 10\} = 5$, cross out row 1, and change d_1 to $10 - 5 = 5$ (Table 24). The minimum cost that does not lie in a crossed-out row or column is x_{31}. We set $x_{31} = \min\{15, 5\} = 5$, cross out column 1, and reduce s_3 to $15 - 5 = 10$ (Table 25). Now we set $x_{33} = \min\{10, 4\} = 4$, cross out column 3, and reduce s_3 to $10 - 4 = 6$ (Table 26). The only cell that we can choose is x_{34}. We set $x_{34} = \min\{6, 6\}$ and cross out both row 3 and column 4. We have now obtained the bfs: $x_{11} = 5$, $x_{21} = 2$, $x_{22} = 8$, $x_{31} = 5$, $x_{33} = 4$, and $x_{34} = 6$.

Because the minimum-cost method chooses variables with small shipping costs to be basic variables, you might think that this method would always yield a bfs with a relatively low total shipping cost. The following problem shows how the minimum-cost method can be fooled into choosing a relatively high-cost bfs.

TABLE 21

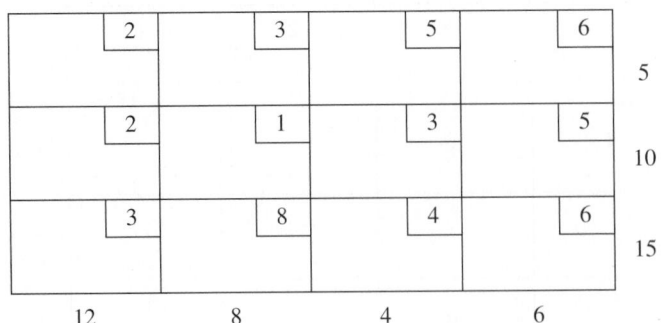

TABLE 22

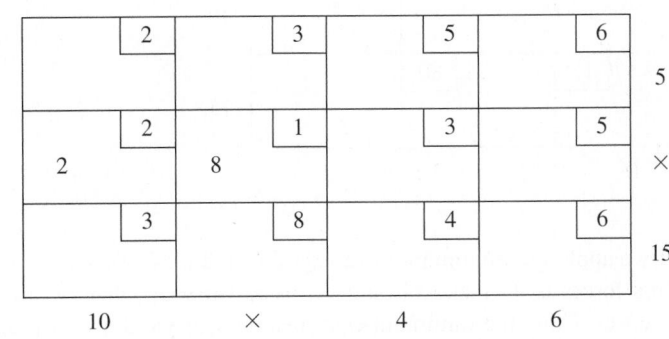

[2]	[3]	[5]	[6]	5
[2]	8 [1]	[3]	[5]	2
[3]	[8]	[4]	[6]	15
12	×	4	6	

TABLE 23

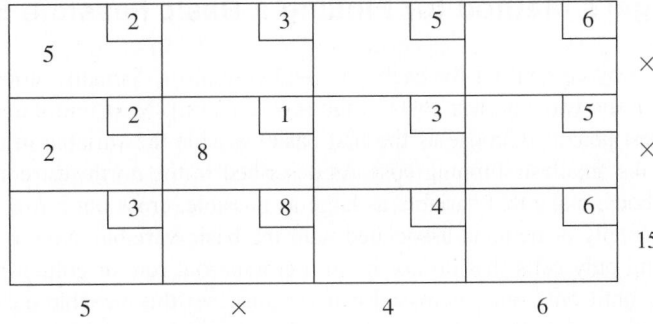

[2]	[3]	[5]	[6]	5
2 [2]	8 [1]	[3]	[5]	×
[3]	[8]	[4]	[6]	15
10	×	4	6	

TABLE 24

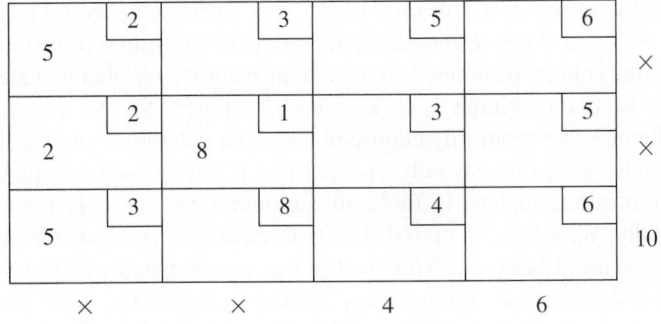

5 [2]	[3]	[5]	[6]	×
2 [2]	8 [1]	[3]	[5]	×
[3]	[8]	[4]	[6]	15
5	×	4	6	

TABLE 25

5 [2]	[3]	[5]	[6]	×
2 [2]	8 [1]	[3]	[5]	×
5 [3]	[8]	[4]	[6]	10
×	×	4	6	

TABLE 26

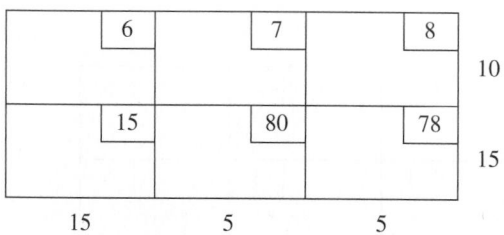

	2	3	5	6	
5					×
	2	1	3	5	
2	8				×
	3	8	4	6	
5		4			6
	×	×	×	6	

TABLE 27

	6	7	8	
				10
	15	80	78	
				15
	15	5	5	

If we apply the minimum-cost method to Table 27, we set $x_{11} = 10$ and cross out row 1. This forces us to make x_{22} and x_{23} basic variables, thereby incurring their high shipping costs. Thus, the minimum-cost method will yield a costly bfs. Vogel's method for finding a bfs usually avoids extremely high shipping costs.

Vogel's Method for Finding a Basic Feasible Solution

Begin by computing for each row (and column) a "penalty" equal to the difference between the two smallest costs in the row (column). Next find the row or column with the largest penalty. Choose as the first basic variable the variable in this row or column that has the smallest shipping cost. As described in the northwest corner and minimum-cost methods, make this variable as large as possible, cross out a row or column, and change the supply or demand associated with the basic variable. Now recompute new penalties (using only cells that do not lie in a crossed-out row or column), and repeat the procedure until only one uncrossed cell remains. Set this variable equal to the supply or demand associated with the variable, and cross out the variable's row and column. A bfs has now been obtained.

We illustrate Vogel's method by finding a bfs to Table 28. Column 2 has the largest penalty, so we set $x_{12} = \min\{10, 5\} = 5$. Then we cross out column 2 and reduce s_1 to $10 - 5 = 5$. After recomputing the new penalties (observe that after a column is crossed out, the column penalties will remain unchanged), we obtain Table 29. The largest penalty now occurs in column 3, so we set $x_{13} = \min\{5, 5\}$. We may cross out either row 1 or column 3. We arbitrarily choose to cross out column 3, and we reduce s_1 to $5 - 5 = 0$. Because each row has only one cell that is not crossed out, there are no row penalties. The resulting tableau is Table 30. Column 1 has the only (and, of course, the largest) penalty. We set $x_{11} = \min\{0, 15\} = 0$, cross out row 1, and change d_1 to $15 - 0 = 15$. The result is Table 31. No penalties can be computed, and the only cell that is not in a crossed-out row or column is x_{21}. Therefore, we set $x_{21} = 15$ and cross out both column 1 and row 2. Our application of Vogel's method is complete, and we have obtained the bfs: $x_{11} = 0$, $x_{12} = 5$, $x_{13} = 5$, and $x_{21} = 15$ (see Table 32).

TABLE 28

			Supply	Row Penalty
6	7	8	10	$7 - 6 = 1$
15	80	78	15	$78 - 15 = 63$
Demand 15	5	5		
Column Penalty $15 - 6 = 9$	$80 - 7 = 73$	$78 - 8 = 70$		

TABLE 29

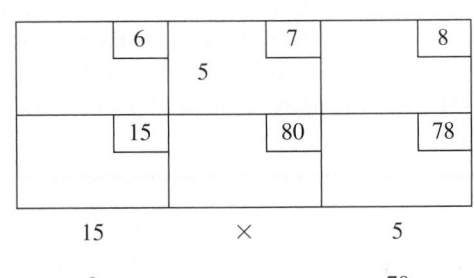

			Supply	Row Penalty
6 5	7	8	5	$8 - 6 = 2$
15	80	78	15	$78 - 15 = 63$
Demand 15	×	5		
Column Penalty 9	—	70		

TABLE 30

			Supply	Row Penalty
6 5	7 5	8	0	—
15	80	78	15	—
Demand 15	×	×		
Column Penalty 9	—	—		

TABLE 31

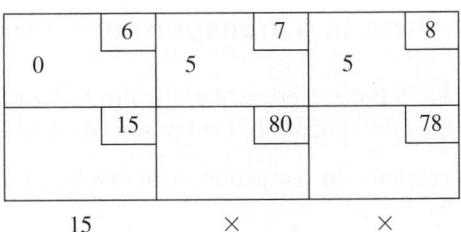

			Supply	Row Penalty
6 0	7 5	8 5	×	—
15	80	78	15	—
Demand 15	×	×		
Column Penalty —	—	—		

TABLE 32

	6	7	8	
0		5	5	10
	15	80	78	
15				15
	15	5	5	

Observe that Vogel's method avoids the costly shipments associated with x_{22} and x_{23}. This is because the high shipping costs resulted in large penalties that caused Vogel's method to choose other variables to satisfy the second and third demand constraints.

Of the three methods we have discussed for finding a bfs, the northwest corner method requires the least effort, and Vogel's method requires the most effort. Extensive research [Glover et al. (1974)] has shown, however, that when Vogel's method is used to find an initial bfs, it usually takes substantially fewer pivots than if the other two methods had been used. For this reason, the northwest corner and minimum-cost methods are rarely used to find a basic feasible solution to a large transportation problem.

PROBLEMS

Group A

1 Use the northwest corner method to find a bfs for Problems 1, 2, and 3 of Section 7.1.

2 Use the minimum-cost method to find a bfs for Problems 4, 7, and 8 of Section 7.1. (*Hint:* For a maximization problem, call the minimum-cost method the maximum-profit method or the maximum-revenue method.)

3 Use Vogel's method to find a bfs for Problems 5 and 6 of Section 7.1.

4 How should Vogel's method be modified to solve a maximization problem?

7.3 The Transportation Simplex Method

In this section, we show how the simplex algorithm simplifies when a transportation problem is solved. We begin by discussing the pivoting procedure for a transportation problem.

Recall that when the pivot row was used to eliminate the entering basic variable from other constraints and row 0, many multiplications were usually required. In solving a transportation problem, however, *pivots require only additions and subtractions.*

How to Pivot in a Transportation Problem

By using the following procedure, the pivots for a transportation problem may be performed within the confines of the transportation tableau:

Step 1 Determine (by a criterion to be developed shortly) the variable that should enter the basis.

Step 2 Find the loop (it can be shown that there is only one loop) involving the entering variable and some of the basic variables.

Step 3 Counting *only cells in the loop,* label those found in step 2 that are an even number (0, 2, 4, and so on) of cells away from the entering variable as *even* cells. Also label those that are an odd number of cells away from the entering variable as *odd* cells.

Step 4   Find the odd cell whose variable assumes the smallest value. Call this value θ. The variable corresponding to this odd cell will leave the basis. To perform the pivot, decrease the value of each odd cell by θ and increase the value of each even cell by θ. The values of variables not in the loop remain unchanged. The pivot is now complete. If $\theta = 0$, then the entering variable will equal 0, and an odd variable that has a current value of 0 will leave the basis. In this case, a degenerate bfs existed before and will result after the pivot. If more than one odd cell in the loop equals θ, you may arbitrarily choose one of these odd cells to leave the basis; again, a degenerate bfs will result.

We illustrate the pivoting procedure on the Powerco example. When the northwest corner method is applied to the Powerco example, the bfs in Table 33 is found. For this bfs, the basic variables are $x_{11} = 35$, $x_{21} = 10$, $x_{22} = 20$, $x_{23} = 20$, $x_{33} = 10$, and $x_{34} = 30$.

Suppose we want to find the bfs that would result if x_{14} were entered into the basis. The loop involving x_{14} and some of the basic variables is

$$\begin{array}{cccccc} E & O & E & O & E & O \\ (1, 4) & (3, 4) & (3, 3) & (2, 3) & (2, 1) & (1, 1) \end{array}$$

In this loop, (1, 4), (3, 3), and (2, 1) are the even cells, and (1, 1), (3, 4), and (2, 3) are the odd cells. The odd cell with the smallest value is $x_{23} = 20$. Thus, after the pivot, x_{23} will have left the basis. We now add 20 to each of the even cells and subtract 20 from each of the odd cells. The bfs in Table 34 results. Because each row and column has as many $+20$s as -20s, the new solution will satisfy each supply and demand constraint. By choosing the smallest odd variable (x_{23}) to leave the basis, we have ensured that all variables will remain nonnegative. Thus, the new solution is feasible. There is no loop involving the cells (1, 1), (1, 4), (2, 1), (2, 2), (3, 3), and (3, 4), so the new solution is a bfs. After the pivot, the new bfs is $x_{11} = 15$, $x_{14} = 20$, $x_{21} = 30$, $x_{22} = 20$, $x_{33} = 30$, and $x_{34} = 10$, and all other variables equal 0.

TABLE 33
Northwest Corner Basic Feasible Solution for Powerco

35				35
10	20	20		50
		10	30	40
45	20	30	30	

TABLE 34
New Basic Feasible Solution After x_{14} Is Pivoted into Basis

35 − 20			0 + 20	35
10 + 20	20	20 − 20 (nonbasic)		50
		10 + 20	30 − 20	40
45	20	30	30	

The preceding illustration of the pivoting procedure makes it clear that each pivot in a transportation problem involves only additions and subtractions. Using this fact, we can show that *if all the supplies and demands for a transportation problem are integers, then the transportation problem will have an optimal solution in which all the variables are integers.* Begin by observing that, by the northwest corner method, we can find a bfs in which each variable is an integer. Each pivot involves only additions and subtractions, so each bfs obtained by performing the simplex algorithm (including the optimal solution) will assign all variables integer values. The fact that a transportation problem with integer supplies and demands has an optimal integer solution is useful, because it ensures that we need not worry about whether the Divisibility Assumption is justified.

Pricing Out Nonbasic Variables (Based on Chapter 6)

To complete our discussion of the transportation simplex, we now show how to compute row 0 for any bfs. From Section 6.2, we know that for a bfs in which the set of basic variables is BV, the coefficient of the variable x_{ij} (call it $\bar{c}_{ij}$) in the tableau's row 0 is given by

$$\bar{c}_{ij} = \mathbf{c}_{BV}B^{-1}\mathbf{a}_{ij} - c_{ij}$$

where c_{ij} is the objective function coefficient for x_{ij} and $\mathbf{a}_{ij}$ is the column for x_{ij} in the original LP (we are assuming that the first supply constraint has been dropped).

Because we are solving a minimization problem, the current bfs will be optimal if all the $\bar{c}_{ij}$'s are nonpositive; otherwise, we enter into the basis the variable with the most positive $\bar{c}_{ij}$.

After determining $\mathbf{c}_{BV}B^{-1}$, we can easily determine $\bar{c}_{ij}$. Because the first constraint has been dropped, $\mathbf{c}_{BV}B^{-1}$ will have $m + n - 1$ elements. We write

$$\mathbf{c}_{BV}B^{-1} = [u_2 \quad u_3 \quad \cdots \quad u_m \quad v_1 \quad v_2 \quad \cdots \quad v_n]$$

where $u_2, u_3, \ldots, u_m$ are the elements of $\mathbf{c}_{BV}B^{-1}$ corresponding to the $m - 1$ supply constraints, and $v_1, v_2, \ldots, v_n$ are the elements of $\mathbf{c}_{BV}B^{-1}$ corresponding to the n demand constraints.

To determine $\mathbf{c}_{BV}B^{-1}$, we use the fact that in any tableau, each basic variable x_{ij} must have $\bar{c}_{ij} = 0$. Thus, for each of the $m + n - 1$ variables in BV,

$$\mathbf{c}_{BV}B^{-1}\mathbf{a}_{ij} - c_{ij} = 0 \tag{4}$$

For a transportation problem, the equations in (4) are very easy to solve. To illustrate the solution of (4), we find $\mathbf{c}_{BV}B^{-1}$ for (5), by applying the northwest corner method bfs to the Powerco problem.

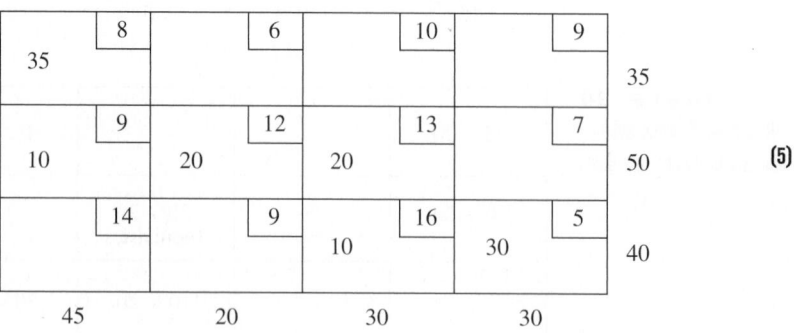

(5)

For this bfs, BV $= \{x_{11}, x_{21}, x_{22}, x_{23}, x_{33}, x_{34}\}$. Applying (4) we obtain

$$\bar{c}_{11} = [u_2 \quad u_3 \quad v_1 \quad v_2 \quad v_3 \quad v_4] \begin{bmatrix} 0 \\ 0 \\ 1 \\ 0 \\ 0 \\ 0 \end{bmatrix} - 8 = v_1 - 8 = 0$$

$$\bar{c}_{21} = [u_2 \quad u_3 \quad v_1 \quad v_2 \quad v_3 \quad v_4] \begin{bmatrix} 1 \\ 0 \\ 1 \\ 0 \\ 0 \\ 0 \end{bmatrix} - 9 = u_2 + v_1 - 9 = 0$$

$$\bar{c}_{22} = [u_2 \quad u_3 \quad v_1 \quad v_2 \quad v_3 \quad v_4] \begin{bmatrix} 1 \\ 0 \\ 0 \\ 1 \\ 0 \\ 0 \end{bmatrix} - 12 = u_2 + v_2 - 12 = 0$$

$$\bar{c}_{23} = [u_2 \quad u_3 \quad v_1 \quad v_2 \quad v_3 \quad v_4] \begin{bmatrix} 1 \\ 0 \\ 0 \\ 0 \\ 1 \\ 0 \end{bmatrix} - 13 = u_2 + v_3 - 13 = 0$$

$$\bar{c}_{33} = [u_2 \quad u_3 \quad v_1 \quad v_2 \quad v_3 \quad v_4] \begin{bmatrix} 0 \\ 1 \\ 0 \\ 0 \\ 1 \\ 0 \end{bmatrix} - 16 = u_3 + v_3 - 16 = 0$$

$$\bar{c}_{34} = [u_2 \quad u_3 \quad v_1 \quad v_2 \quad v_3 \quad v_4] \begin{bmatrix} 0 \\ 1 \\ 0 \\ 0 \\ 0 \\ 1 \end{bmatrix} - 5 = u_3 + v_4 - 5 = 0$$

For each basic variable x_{ij} (except those having $i = 1$), we see that (4) reduces to $u_i + v_j = c_{ij}$. If we define $u_1 = 0$, we see that (4) reduces to $u_i + v_j = c_{ij}$ for all basic variables. Thus, to solve for $\mathbf{c}_{\mathrm{BV}}B^{-1}$, we must solve the following system of $m + n$ equations: $u_1 = 0$, $u_i + v_j = c_{ij}$ for all basic variables.

For (5), we find $\mathbf{c}_{BV}B^{-1}$ by solving

$$u_1 = 0 \tag{6}$$
$$u_1 + v_1 = 8 \tag{7}$$
$$u_2 + v_1 = 9 \tag{8}$$
$$u_2 + v_2 = 12 \tag{9}$$
$$u_2 + v_3 = 13 \tag{10}$$
$$u_3 + v_3 = 16 \tag{11}$$
$$u_3 + v_4 = 5 \tag{12}$$

From (7), $v_1 = 8$. From (8), $u_2 = 1$. Then (9) yields $v_2 = 11$, and (10) yields $v_3 = 12$. From (11), $u_3 = 4$. Finally, (12) yields $v_4 = 1$. For each nonbasic variable, we now compute $\bar{c}_{ij} = u_i + v_j - c_{ij}$. We obtain

$$\bar{c}_{12} = 0 + 11 - 6 = 5 \qquad \bar{c}_{13} = 0 + 12 - 10 = 2$$
$$\bar{c}_{14} = 0 + 1 - 9 = -8 \qquad \bar{c}_{24} = 1 + 1 - 7 = -5$$
$$\bar{c}_{31} = 4 + 8 - 14 = -2 \qquad \bar{c}_{32} = 4 + 11 - 9 = 6$$

Because $\bar{c}_{32}$ is the most positive $\bar{c}_{ij}$, we would next enter x_{32} into the basis. Each unit of x_{32} that is entered into the basis will decrease Powerco's cost by \$6.

How to Determine the Entering Nonbasic Variable (Based on Chapter 5)

For readers who have not covered Chapter 6, we now discuss how to determine whether a bfs is optimal, and, if it is not, how to determine which nonbasic variable should enter the basis. Let $-u_i$ ($i = 1, 2, \ldots, m$) be the shadow price of the ith supply constraint, and let $-v_j$ ($j = 1, 2, \ldots, n$) be the shadow price of the jth demand constraint. We assume that the first supply constraint has been dropped, so we may set $-u_1 = 0$. From the definition of shadow price, if we were to increase the right-hand side of the ith supply and jth demand constraint by 1, the optimal z-value would decrease by $-u_i - v_j$. Equivalently, if we were to decrease the right-hand side of the ith supply and jth demand constraint by 1, the optimal z-value would increase by $-u_i - v_j$. Now suppose x_{ij} is a nonbasic variable. Should we enter x_{ij} into the basis? Observe that if we increase x_{ij} by 1, costs directly increase by c_{ij}. Also, increasing x_{ij} by 1 means that one less unit will be shipped from supply point i and one less unit will be shipped to demand point j. This is equivalent to reducing the right-hand sides of the ith supply constraint and jth demand constraint by 1. This will increase z by $-u_i - v_j$. Thus, increasing x_{ij} by 1 will increase z by a total of $c_{ij} - u_i - v_j$. So if $c_{ij} - u_i - v_j \geq 0$ (or $u_i + v_j - c_{ij} \leq 0$) for all nonbasic variables, the current bfs will be optimal. If, however, a nonbasic variable x_{ij} has $c_{ij} - u_i - v_j < 0$ (or $u_i + v_j - c_{ij} > 0$), then z can be decreased by $u_i + v_j - c_{ij}$ per unit of x_{ij} by entering x_{ij} into the basis. Thus, we may conclude that if $u_i + v_j - c_{ij} \leq 0$ for all nonbasic variables, then the current bfs is optimal. Otherwise, the nonbasic variable with the most positive value of $u_i + v_j - c_{ij}$ should enter the basis. How do we find the u_i's and v_j's? The coefficient of a nonbasic variable x_{ij} in row 0 of any tableau is the amount by which a unit increase in x_{ij} will decrease z, so we can conclude that the coefficient of any nonbasic variable (and, it turns out, any basic variable) in row 0 is $u_i + v_j - c_{ij}$. So we may solve for the u_i's and v_j's by solving the following system of equations: $u_1 = 0$ and $u_i + v_j - c_{ij} = 0$ for all basic variables.

To illustrate the previous discussion, consider the bfs for the Powerco problem shown in (5).

	8	6	10	9	
	35				35
	9	12	13	7	
	10	20	20		50
	14	9	16	5	
		10	30		40
	45	20	30	30	

(5)

We find the u_i's and v_j's by solving

$$u_1 = 0 \tag{6}$$
$$u_1 + v_1 = 8 \tag{7}$$
$$u_2 + v_1 = 9 \tag{8}$$
$$u_2 + v_2 = 12 \tag{9}$$
$$u_2 + v_3 = 13 \tag{10}$$
$$u_3 + v_3 = 16 \tag{11}$$
$$u_3 + v_4 = 5 \tag{12}$$

From (7), $v_1 = 8$. From (8), $u_2 = 1$. Then (9) yields $v_2 = 11$, and (10) yields $v_3 = 12$. From (11), $u_3 = 4$. Finally, (12) yields $v_4 = 1$. For each nonbasic variable, we now compute $\bar{c}_{ij} = u_i + v_j - c_{ij}$. We obtain

$$\bar{c}_{12} = 0 + 11 - 6 = 5 \qquad \bar{c}_{13} = 0 + 12 - 10 = 2$$
$$\bar{c}_{14} = 0 + 1 - 9 = -8 \qquad \bar{c}_{24} = 1 + 1 - 7 = -5$$
$$\bar{c}_{31} = 4 + 8 - 14 = -2 \qquad \bar{c}_{32} = 4 + 11 - 9 = 6$$

Because $\bar{c}_{32}$ is the most positive $\bar{c}_{ij}$, we would next enter x_{32} into the basis. Each unit of x_{32} that is entered into the basis will decrease Powerco's cost by $6.

We can now summarize the procedure for using the transportation simplex to solve a transportation (min) problem.

Summary and Illustration of the Transportation Simplex Method

Step 1 If the problem is unbalanced, balance it.

Step 2 Use one of the methods described in Section 7.2 to find a bfs.

Step 3 Use the fact that $u_1 = 0$ and $u_i + v_j = c_{ij}$ for all basic variables to find the $[u_1 \quad u_2 \quad \ldots \quad u_m \quad v_1 \quad v_2 \quad \ldots \quad v_n]$ for the current bfs.

Step 4 If $u_i + v_j - c_{ij} \leq 0$ for all nonbasic variables, then the current bfs is optimal. If this is not the case, then we enter the variable with the most positive $u_i + v_j - c_{ij}$ into the basis using the pivoting procedure. This yields a new bfs.

Step 5 Using the new bfs, return to steps 3 and 4.

For a maximization problem, proceed as stated, but replace step 4 by step 4'.

Step 4' If $u_i + v_j - c_{ij} \geq 0$ for all nonbasic variables, then the current bfs is optimal. Otherwise, enter the variable with the most negative $u_i + v_j - c_{ij}$ into the basis using the pivoting procedure described earlier.

We illustrate the procedure for solving a transportation problem by solving the Powerco problem. We begin with the bfs (5). We have already determined that x_{32} should enter the basis. As shown in Table 35, the loop involving x_{32} and some of the basic variables is (3, 2)–(3, 3)–(2, 3)–(2, 2). The odd cells in this loop are (3, 3) and (2, 2). Because $x_{33} = 10$ and $x_{22} = 20$, the pivot will decrease the value of x_{33} and x_{22} by 10 and increase the value of x_{32} and x_{23} by 10. The resulting bfs is shown in Table 36. The u_i's and v_j's for the new bfs were obtained by solving

$$u_1 = 0 \qquad u_2 + v_3 = 13$$
$$u_2 + v_2 = 12 \qquad u_2 + v_1 = 9$$
$$u_3 + v_4 = 5 \qquad u_3 + v_2 = 9$$
$$u_1 + v_1 = 8$$

In computing $\bar{c}_{ij} = u_i + v_j - c_{ij}$ for each nonbasic variable, we find that $\bar{c}_{12} = 5$, $\bar{c}_{24} = 1$, and $\bar{c}_{13} = 2$ are the only positive $\bar{c}_{ij}$'s. Thus, we next enter x_{12} into the basis. The loop involving x_{12} and some of the basic variables is (1, 2)–(2, 2)–(2, 1)–(1, 1). The odd cells are (2, 2) and (1, 1). Because $x_{22} = 10$ is the smallest entry in an odd cell, we decrease x_{22} and x_{11} by 10 and increase x_{12} and x_{21} by 10. The resulting bfs is shown in Table 37. For this bfs, the u_i's and v_j's were determined by solving

$$u_1 = 0 \qquad u_1 + v_2 = 6$$
$$u_2 + v_1 = 9 \qquad u_3 + v_2 = 9$$
$$u_1 + v_1 = 8 \qquad u_3 + v_4 = 5$$
$$u_2 + v_3 = 13$$

In computing $\bar{c}_{ij}$ for each nonbasic variable, we find that the only positive $\bar{c}_{ij}$ is $\bar{c}_{13} = 2$. Thus, x_{13} enters the basis. The loop involving x_{13} and some of the basic variables is

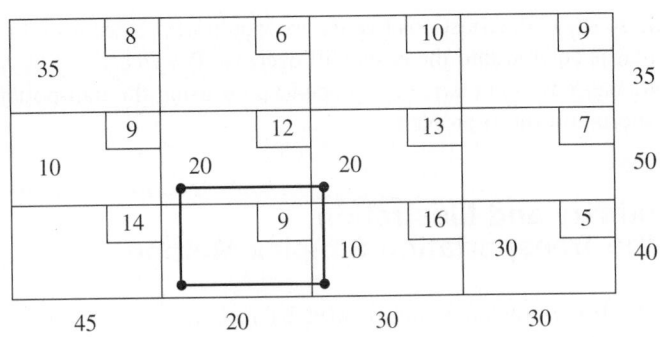

TABLE 35
Loop Involving
Entering Variable x_{32}

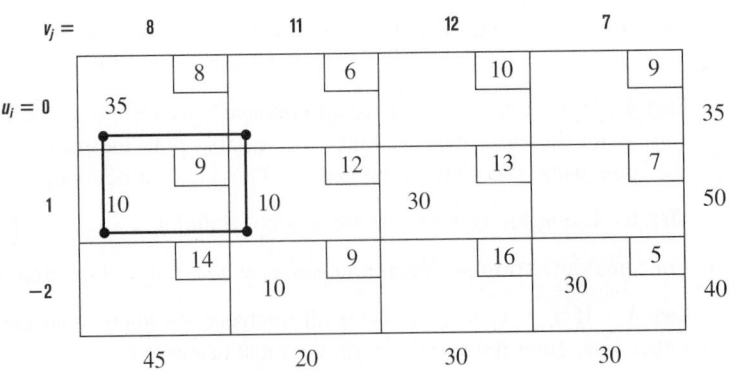

TABLE 36
x_{32} Has Entered the Basis,
and x_{12} Enters Next

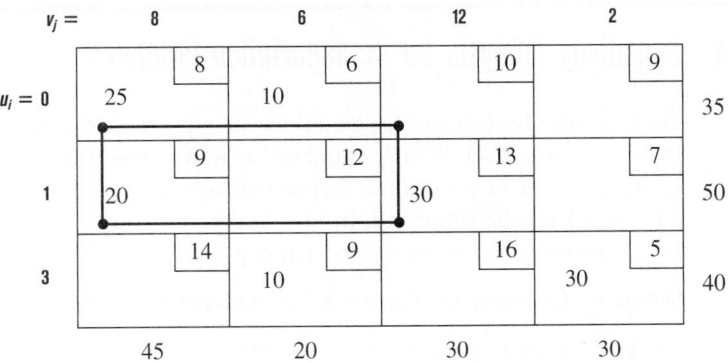

TABLE 37
x_{12} Has Entered the Basis, and x_{13} Enters Next

$v_j =$	8	6	12	2	
$u_i = 0$	8 25	6 10	10	9	35
1	9 20	12	13 30	7	50
3	14	9 10	16	5 30	40
	45	20	30	30	

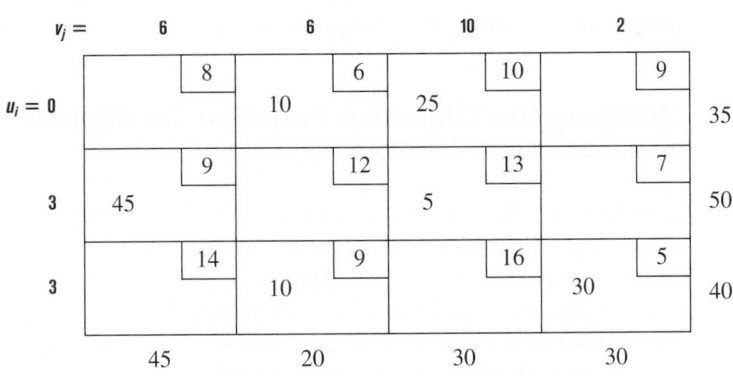

TABLE 38
Optimal Tableau for Powerco

$v_j =$	6	6	10	2	
$u_i = 0$	8	6 10	10 25	9	35
3	9 45	12 5	13	7	50
3	14	9 10	16	5 30	40
	45	20	30	30	

$(1, 3)–(2, 3)–(2, 1)–(1, 1)$. The odd cells are x_{23} and x_{11}. Because $x_{11} = 25$ is the smallest entry in an odd cell, we decrease x_{23} and x_{11} by 25 and increase x_{13} and x_{21} by 25. The resulting bfs is shown in Table 38. For this bfs, the u_i's and v_j's were obtained by solving

$$u_1 = 0 \qquad u_2 + v_3 = 13$$
$$u_2 + v_1 = 9 \qquad u_1 + v_3 = 10$$
$$u_3 + v_4 = 5 \qquad u_3 + v_2 = 9$$
$$u_1 + v_2 = 6$$

The reader should check that for this bfs, all $\bar{c}_{ij} \leq 0$, so an optimal solution has been obtained. Thus, the optimal solution to the Powerco problem is $x_{12} = 10$, $x_{13} = 25$, $x_{21} = 45$, $x_{23} = 5$, $x_{32} = 10$, $x_{34} = 30$, and

$$z = 6(10) + 10(25) + 9(45) + 13(5) + 9(10) + 5(30) = \$1,020$$

PROBLEMS

Group A

Use the transportation simplex to solve Problems 1–8 in Section 7.1. Begin with the bfs found in Section 7.2.

7.4 Sensitivity Analysis for Transportation Problems[†]

We have already seen that for a transportation problem, the determination of a bfs and of row 0 for a given set of basic variables, as well as the pivoting procedure, all simplify. It should therefore be no surprise that certain aspects of the sensitivity analysis discussed in Section 6.3 can be simplified. In this section, we discuss the following three aspects of sensitivity analysis for the transportation problem:

Change 1 Changing the objective function coefficient of a nonbasic variable.

Change 2 Changing the objective function coefficient of a basic variable.

Change 3 Increasing a single supply by Δ and a single demand by Δ.

We illustrate three changes using the Powerco problem. Recall from Section 7.3 that the optimal solution for the Powerco problem was $z = \$1,020$; the optimal tableau is Table 39.

Changing the Objective Function Coefficient of a Nonbasic Variable

As in Section 6.3, changing the objective function coefficient of a nonbasic variable x_{ij} will leave the right-hand side of the optimal tableau unchanged. Thus, the current basis will still be feasible. We are not changing $\mathbf{c}_{BV}B^{-1}$, so the u_i's and v_j's remain unchanged. In row 0, only the coefficient of x_{ij} will change. Thus, as long as the coefficient of x_{ij} in the optimal row 0 is nonpositive, the current basis remains optimal.

To illustrate the method, we answer the following question: For what range of values of the cost of shipping 1 million kwh of electricity from plant 1 to city 1 will the current basis remain optimal? Suppose we change c_{11} from 8 to $8 + \Delta$. For what values of Δ will the current basis remain optimal? Now $\bar{c}_{11} = u_1 + v_1 - c_{11} = 0 + 6 - (8 + \Delta) = -2 - \Delta$. Thus, the current basis remains optimal for $-2 - \Delta \leq 0$, or $\Delta \geq -2$, and $c_{11} \geq 8 - 2 = 6$.

Changing the Objective Function Coefficient of a Basic Variable

Because we are changing $\mathbf{c}_{BV}B^{-1}$, the coefficient of each nonbasic variable in row 0 may change, and to determine whether the current basis remains optimal, we must find the new u_i's and v_j's and use these values to price out all nonbasic variables. The current basis remains optimal as long as all nonbasic variables price out nonpositive. To illustrate the idea, we determine for the Powerco problem the range of values of the cost of shipping 1 million kwh from plant 1 to city 3 for which the current basis remains optimal.

Suppose we change c_{13} from 10 to $10 + \Delta$. Then the equation $\bar{c}_{13} = 0$ changes from $u_1 + v_3 = 10$ to $u_1 + v_3 = 10 + \Delta$. Thus, to find the u_i's and v_j's, we must solve the following equations:

$$u_1 = 0 \qquad u_3 + v_2 = 9$$
$$u_2 + v_1 = 9 \qquad u_1 + v_3 = 10 + \Delta$$
$$u_1 + v_2 = 6 \qquad u_3 + v_4 = 5$$
$$u_2 + v_3 = 13$$

[†]This section covers topics that may be omitted with no loss of continuity.

		City 1	City 2	City 3	City 4	Supply
	$v_j =$	6	6	10	2	
Plant 1	$u_i = 0$	8	6 — 10	10 — 25	9	35
Plant 2	3	9 — 45	12	13 — 5	7	50
Plant 3	3	14	9 — 10	16	5 — 30	40
Demand		45	20	30	30	

TABLE 39
Optimal Tableau for Powerco

Solving these equations, we obtain $u_1 = 0$, $v_2 = 6$, $v_3 = 10 + \Delta$, $v_1 = 6 + \Delta$, $u_2 = 3 - \Delta$, $u_3 = 3$, and $v_4 = 2$.

We now price out each nonbasic variable. The current basis will remain optimal as long as each nonbasic variable has a nonpositive coefficient in row 0.

$$\bar{c}_{11} = u_1 + v_1 - 8 = \Delta - 2 \le 0 \qquad \text{for } \Delta \le 2$$
$$\bar{c}_{14} = u_1 + v_4 - 9 = -7$$
$$\bar{c}_{22} = u_2 + v_2 - 12 = -3 - \Delta \le 0 \qquad \text{for } \Delta \ge -3$$
$$\bar{c}_{24} = u_2 + v_4 - 7 = -2 - \Delta \le 0 \qquad \text{for } \Delta \ge -2$$
$$\bar{c}_{31} = u_3 + v_1 - 14 = -5 + \Delta \le 0 \qquad \text{for } \Delta \le 5$$
$$\bar{c}_{33} = u_3 + v_3 - 16 = \Delta - 3 \le 0 \qquad \text{for } \Delta \le 3$$

Thus, the current basis remains optimal for $-2 \le \Delta \le 2$, or $8 = 10 - 2 \le c_{13} \le 10 + 2 = 12$.

Increasing Both Supply s_i and Demand d_j by Δ

Observe that this change maintains a balanced transportation problem. Because the u_i's and v_j's may be thought of as the negative of each constraint's shadow prices, we know from (37′) of Chapter 6 that if the current basis remains optimal,

$$\text{New } z\text{-value} = \text{old } z\text{-value} + \Delta u_i + \Delta v_j$$

For example, if we increase plant 1's supply and city 2's demand by 1 unit, then (new cost) $= 1,020 + 1(0) + 1(6) = \$1,026$.

We may also find the new values of the decision variables as follows:

1 If x_{ij} is a basic variable in the optimal solution, then increase x_{ij} by Δ.

2 If x_{ij} is a nonbasic variable in the optimal solution, then find the loop involving x_{ij} and some of the basic variables. Find an odd cell in the loop that is in row i. Increase the value of this odd cell by Δ and go around the loop, alternately increasing and then decreasing current basic variables in the loop by Δ.

To illustrate the first situation, suppose we increase s_1 and d_2 by 2. Because x_{12} is a basic variable in the optimal solution, the new optimal solution will be the one shown in Table 40. The new optimal z-value is $1,020 + 2u_1 + 2v_2 = \$1,032$. To illustrate the second situation, suppose we increase both s_1 and d_1 by 1. Because x_{11} is a nonbasic variable in the current optimal solution, we must find the loop involving $x_{\cdot 11}$ and some of the

TABLE 40
Optimal Tableau for Powerco If
$s_1 = 35 + 2 = 37$ and
$d_2 = 20 + 2 = 22$

	$v_j =$	City 1 — 6	City 2 — 6	City 3 — 10	City 4 — 2	Supply
Plant 1	$u_i = 0$	8	6 — 12	10 — 25	9	37
Plant 2	3	9 — 45	12	13 — 5	7	50
Plant 3	3	14	9 — 10	16	5 — 30	40
Demand		45	22	30	30	

TABLE 41
Optimal Tableau for Powerco If
$s_1 = 35 + 1 = 36$ and
$d_1 = 45 + 1 = 46$

	$v_j =$	City 1 — 6	City 2 — 6	City 3 — 10	City 4 — 2	Supply
Plant 1	$u_i = 0$	8	6 — 10	10 — 26	9	36
Plant 2	3	9 — 46	12	13 — 4	7	50
Plant 3	3	14	9 — 10	16	5 — 30	40
Demand		46	20	30	30	

basic variables. The loop is $(1, 1)$–$(1, 3)$–$(2, 3)$–$(2, 1)$. The odd cell in the loop and row 1 is x_{13}. Thus, the new optimal solution will be obtained by increasing both x_{13} and x_{21} by 1 and decreasing x_{23} by 1. This yields the optimal solution shown in Table 41. The new optimal z-value is found from (new z-value) $= 1,020 + v_1 + v_1 = \$1,026$. Observe that if both s_1 and d_1 were increased by 6, the current basis would be infeasible. (Why?)

PROBLEMS

Group A

The following problems refer to the Powerco example.

1 Determine the range of values of c_{14} for which the current basis remains optimal.

2 Determine the range of values of c_{34} for which the current basis remains optimal.

3 If s_2 and d_3 are both increased by 3, what is the new optimal solution?

4 If s_3 and d_3 are both decreased by 2, what is the new optimal solution?

5 Two plants supply three customers with medical supplies. The unit costs of shipping from the plants to the customers, along with the supplies and demands, are given in Table 42.

a The company's goal is to minimize the cost of meeting customers' demands. Find two optimal bfs for this transportation problem.

b Suppose that customer 2's demand increased by one unit. By how much would costs increase?

TABLE 42

From	To			Supply
	Customer 1	Customer 2	Customer 3	
Plant 1	$55	$65	$80	35
Plant 2	$10	$15	$25	50
Demand	10	10	10	

7.5 Assignment Problems

Although the transportation simplex appears to be very efficient, there is a certain class of transportation problems, called assignment problems, for which the transportation simplex is often very inefficient. In this section, we define assignment problems and discuss an efficient method that can be used to solve them.

EXAMPLE 4   **Machine Assignment Problem**

Machineco has four machines and four jobs to be completed. Each machine must be assigned to complete one job. The time required to set up each machine for completing each job is shown in Table 43. Machineco wants to minimize the total setup time needed to complete the four jobs. Use linear programming to solve this problem.

Solution   Machineco must determine which machine should be assigned to each job. We define (for $i, j = 1, 2, 3, 4$)

$$x_{ij} = 1 \text{ if machine } i \text{ is assigned to meet the demands of job } j$$
$$x_{ij} = 0 \text{ if machine } i \text{ is not assigned to meet the demands of job } j$$

Then Machineco's problem may be formulated as

$$\min z = 14x_{11} + 5x_{12} + 8x_{13} + 7x_{14} + 2x_{21} + 12x_{22} + 6x_{23} + 5x_{24}$$
$$+ 7x_{31} + 8x_{32} + 3x_{33} + 9x_{34} + 2x_{41} + 4x_{42} + 6x_{43} + 10x_{44}$$

$$
\begin{aligned}
\text{s.t.} \quad & x_{11} + x_{12} + x_{13} + x_{14} = 1 && \text{(Machine constraints)} \\
& x_{21} + x_{22} + x_{23} + x_{24} = 1 \\
& x_{31} + x_{32} + x_{33} + x_{34} = 1 \\
& x_{41} + x_{42} + x_{43} + x_{44} = 1 \\
& x_{11} + x_{21} + x_{31} + x_{41} = 1 && \text{(Job constraints)} \\
& x_{12} + x_{22} + x_{32} + x_{42} = 1 \\
& x_{13} + x_{23} + x_{33} + x_{43} = 1 \\
& x_{14} + x_{24} + x_{34} + x_{44} = 1 \\
& x_{ij} = 0 \quad \text{ or } \quad x_{ij} = 1
\end{aligned}
$$

(13)

The first four constraints in (13) ensure that each machine is assigned to a job, and the last four ensure that each job is completed. If $x_{ij} = 1$, then the objective function will pick up the time required to set up machine i for job j; if $x_{ij} = 0$, then the objective function will not pick up the time required.

Ignoring for the moment the $x_{ij} = 0$ or $x_{ij} = 1$ restrictions, we see that Machineco faces a balanced transportation problem in which each supply point has a supply of 1 and each

TABLE 43
Setup Times for Machineco

Machine	Time (Hours)			
	Job 1	Job 2	Job 3	Job 4
1	14	5	8	7
2	2	12	6	5
3	7	8	3	9
4	2	4	6	10

demand point has a demand of 1. In general, an **assignment problem** is a balanced transportation problem in which all supplies and demands are equal to 1. Thus, an assignment problem is characterized by knowledge of the cost of assigning each supply point to each demand point. The assignment problem's matrix of costs is its **cost matrix.**

All the supplies and demands for the Machineco problem (and for any assignment problem) are integers, so our discussion in Section 7.3 implies that all variables in Machineco's optimal solution must be integers. Because the right-hand side of each constraint is equal to 1, each x_{ij} must be a nonnegative integer that is no larger than 1, so each x_{ij} must equal 0 or 1. This means that we can ignore the restrictions that $x_{ij} = 0$ or 1 and solve (13) as a balanced transportation problem. By the minimum cost method, we obtain the bfs in Table 44. The current bfs is highly degenerate. (In any bfs to an $m \times m$ assignment problem, there will always be m basic variables that equal 1 and $m - 1$ basic variables that equal 0.)

We find that $\bar{c}_{43} = 1$ is the only positive $\bar{c}_{ij}$. We therefore enter x_{43} into the basis. The loop involving x_{43} and some of the basic variables is $(4, 3)$–$(1, 3)$–$(1, 2)$–$(4, 2)$. The odd variables in the loop are x_{13} and x_{42}. Because $x_{13} = x_{42} = 0$, either x_{13} or x_{42} will leave

TABLE 44
Basic Feasible Solution for Machineco

	$v_j =$	Job 1 3	Job 2 4	Job 3 8	Job 4 7	
Machine 1	$u_i = 0$	14	5 1	8 0	7 0	1
Machine 2	-2	2	12	6	5 1	1
Machine 3	-5	7	8	3 1	9	1
Machine 4	-1	2 1	4 0	6	10	1
		1	1	1	1	

TABLE 45
x_{43} Has Entered the Basis

	$v_j =$	Job 1 3	Job 2 5	Job 3 7	Job 4 7	
Machine 1	$u_i = 0$	14	5 1	8	7 0	1
Machine 2	-2	2	12	6	5 1	1
Machine 3	-4	7	8	3 1	9	1
Machine 4	-1	2 1	4 0	6 0	10	1
		1	1	1	1	

the basis. We arbitrarily choose x_{13} to leave the basis. After performing the pivot, we obtain the bfs in Table 45. All $\bar{c}_{ij}$'s are now nonpositive, so we have obtained an optimal assignment: $x_{12} = 1$, $x_{24} = 1$, $x_{33} = 1$, and $x_{41} = 1$. Thus, machine 1 is assigned to job 2, machine 2 is assigned to job 4, machine 3 is assigned to job 3, and machine 4 is assigned to job 1. A total setup time of $5 + 5 + 3 + 2 = 15$ hours is required.

The Hungarian Method

Looking back at our initial bfs, we see that it was an optimal solution. We did not know that it was optimal, however, until performing one iteration of the transportation simplex. This suggests that the high degree of degeneracy in an assignment problem may cause the transportation simplex to be an inefficient way of solving assignment problems. For this reason (and the fact that the algorithm is even simpler than the transportation simplex), the Hungarian method is usually used to solve assignment (min) problems:

Step 1 Find the minimum element in each row of the $m \times m$ cost matrix. Construct a new matrix by subtracting from each cost the minimum cost in its row. For this new matrix, find the minimum cost in each column. Construct a new matrix (called the reduced cost matrix) by subtracting from each cost the minimum cost in its column.

Step 2 Draw the minimum number of lines (horizontal, vertical, or both) that are needed to cover all the zeros in the reduced cost matrix. If m lines are required, then an optimal solution is available among the covered zeros in the matrix. If fewer than m lines are needed, then proceed to step 3.

Step 3 Find the smallest nonzero element (call its value k) in the reduced cost matrix that is uncovered by the lines drawn in step 2. Now subtract k from each uncovered element of the reduced cost matrix and add k to each element that is covered by two lines. Return to step 2.

REMARKS 1 To solve an assignment problem in which the goal is to maximize the objective function, multiply the profits matrix through by -1 and solve the problem as a minimization problem.
2 If the number of rows and columns in the cost matrix are unequal, then the assignment problem is unbalanced. The Hungarian method may yield an incorrect solution if the problem is unbalanced. Thus, any assignment problem should be balanced (by the addition of one or more dummy points) before it is solved by the Hungarian method.
3 In a large problem, it may not be easy to find the minimum number of lines needed to cover all zeros in the current cost matrix. For a discussion of how to find the minimum number of lines needed, see Gillett (1976). It can be shown that if j lines are required, then only j "jobs" can be assigned to zero costs in the current matrix. This explains why the algorithm terminates when m lines are required.

Solution of Machineco Example by the Hungarian Method

We illustrate the Hungarian method by solving the Machineco problem (see Table 46).

Step 1 For each row, we subtract the row minimum from each element in the row, obtaining Table 47. We now subtract 2 from each cost in column 4, obtaining Table 48.

Step 2 As shown, lines through row 1, row 3, and column 1 cover all the zeros in the reduced cost matrix. From remark 3, it follows that only three jobs can be assigned to zero costs in the current cost matrix. Fewer than four lines are required to cover all the zeros, so we proceed to step 3.

TABLE 46
Cost Matrix for Machineco

				Row Minimum
14	5	8	7	5
2	12	6	5	2
7	8	3	9	3
2	4	6	10	2

TABLE 47
Cost Matrix After Row Minimums Are Subtracted

9	0	3	2
0	10	4	3
4	5	0	6
0	2	4	8
Column Minimum 0	0	0	2

TABLE 48
Cost Matrix After Column Minimums Are Subtracted

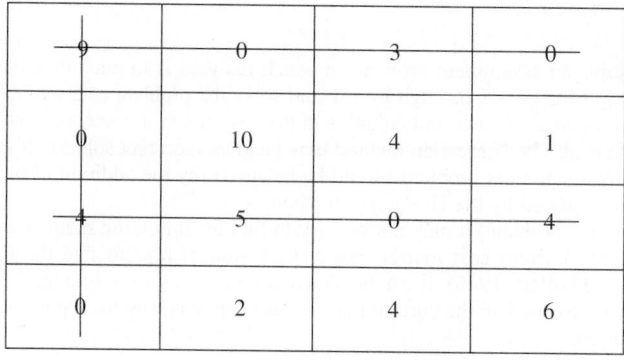

9	0	3	0
0	10	4	1
4	5	0	4
0	2	4	6

Step 3 The smallest uncovered element equals 1, so we now subtract 1 from each uncovered element in the reduced cost matrix and add 1 to each twice-covered element. The resulting matrix is Table 49. Four lines are now required to cover all the zeros. Thus, an optimal solution is available. To find an optimal assignment, observe that the only covered 0 in column 3 is x_{33}, so we must have $x_{33} = 1$. Also, the only available covered zero in column 2 is x_{12}, so we set $x_{12} = 1$ and observe that neither row 1 nor column 2 can be used again. Now the only available covered zero in column 4 is x_{24}. Thus, we choose $x_{24} = 1$ (which now excludes both row 2 and column 4 from further use). Finally, we choose $x_{41} = 1$.

TABLE 49
Four Lines Required; Optimal Solution Is Available

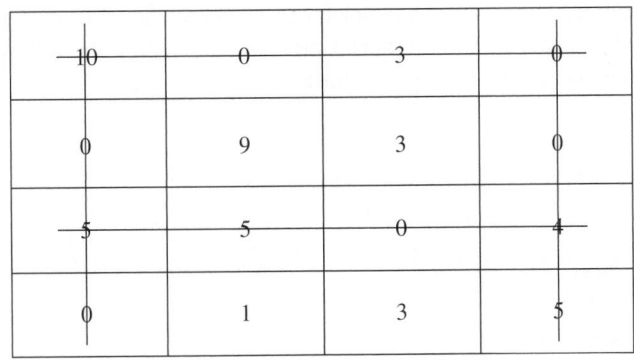

10	0	3	0
0	9	3	0
5	5	0	4
0	1	3	5

Thus, we have found the optimal assignment $x_{12} = 1$, $x_{24} = 1$, $x_{33} = 1$, and $x_{41} = 1$. Of course, this agrees with the result obtained by the transportation simplex.

Intuitive Justification of the Hungarian Method

To give an intuitive explanation of why the Hungarian algorithm works, we need to discuss the following result: *If a constant is added to each cost in a row (or column) of a balanced transportation problem, then the optimal solution to the problem is unchanged.* To show why the result is true, suppose we add k to each cost in the first row of the Machineco problem. Then

New objective function = old objective function + $k(x_{11} + x_{12} + x_{13} + x_{14})$

Because any feasible solution to the Machineco problem must have $x_{11} + x_{12} + x_{13} + x_{14} = 1$,

New objective function = old objective function + k

Thus, the optimal solution to the Machineco problem remains unchanged if a constant k is added to each cost in the first row. A similar argument applies to any other row or column.

Step 1 of the Hungarian method consists (for each row and column) of subtracting a constant from each element in the row or column. Thus, step 1 creates a new cost matrix having the same optimal solution as the original problem. Step 3 of the Hungarian method is equivalent (see Problem 7 at the end of this section) to adding k to each cost that lies in a covered row and subtracting k from each cost that lies in an uncovered column (or vice versa). Thus, step 3 creates a new cost matrix with the same optimal solution as the initial assignment problem. Each time step 3 is performed, at least one new zero is created in the cost matrix.

Steps 1 and 3 also ensure that all costs remain nonnegative. Thus, the net effect of steps 1 and 3 of the Hungarian method is to create a sequence of assignment problems (with nonnegative costs) that all have the same optimal solution as the original assignment problem. Now consider an assignment problem in which all costs are nonnegative. Any feasible assignment in which all the x_{ij}'s that equal 1 have zero costs must be optimal for such an assignment problem. Thus, when step 2 indicates that m lines are required to cover all the zeros in the cost matrix, an optimal solution to the original problem has been found.

Computer Solution of Assignment Problems

To solve assignment problems in LINDO, type in the objective function and constraints. Also, many menu-driven programs require the user to input only a list of supply and demand points (such as jobs and machines, respectively) and a cost matrix. LINGO can also be used to easily solve assignment problems, including the following model to solve the Machineco example (file Assign.lng).

Assign.lng

```
MODEL:
  1]SETS:
  2]MACHINES/1..4/;
  3]JOBS/1..4/;
  4]LINKS(MACHINES,JOBS):COST,ASSIGN;
  5]ENDSETS
  6]MIN=@SUM(LINKS:COST*ASSIGN);
  7]@FOR(MACHINES(I):
  8]@SUM(JOBS(J):ASSIGN(I,J))<1);
  9]@FOR(JOBS(J):
 10]@SUM(MACHINES(I):ASSIGN(I,J))>1);
 11]DATA:
 12]COST = 14,5,8,7,
 13]2,12,6,5,
 14]7,8,3,9,
 15]2,4,6,10;
 16]ENDDATA
END
```

Line 2 defines the four supply points (machines), and line 3 defines the four demand points (jobs). In line 4, we define each possible combination of jobs and machines (16 in all) and associate with each combination an assignment cost [for example COST(1, 2) = 5] and a variable ASSIGN(I,J). ASSIGN(I,J) equals 1 if machine i is used to perform job j; it equals 0 otherwise. Line 5 ends the definition of sets.

Line 6 expresses the objective function by summing over all possible (I,J) combinations the product of the assignment cost and ASSIGN(I,J). Lines 7–8 limit each MACHINE to performing at most one job by forcing (for each machine) the sum of ASSIGN(I,J) over all JOBS to be at most 1. Lines 9–10 require that each JOB be completed by forcing (for each job) the sum of ASSIGN(I,J) over all MACHINES to be at least 1.

Lines 12–16 input the cost matrix.

Observe that this LINGO program can (with simple editing) be used to solve any assignment problem (even if it is not balanced!). For example, if you had 10 machines available to perform 8 jobs, you would edit line 2 to indicate that there are 10 machines (replace 1..4 with 1..10). Then edit line 3 to indicate that there are 8 jobs. Finally, in line 12, you would type the 80 entries of your cost matrix, following "COST=" and you would be ready to roll!

REMARK 1 From our discussion of the Machineco example, it is unnecessary to force the ASSIGN(I,J) to equal 0 or 1; this will happen automatically!

PROBLEMS

Group A

1 Five employees are available to perform four jobs. The time it takes each person to perform each job is given in Table 50. Determine the assignment of employees to jobs that minimizes the total time required to perform the four jobs.

2† Doc Councillman is putting together a relay team for the 400-meter relay. Each swimmer must swim 100 meters of breaststroke, backstroke, butterfly, or freestyle. Doc believes that each swimmer will attain the times given in

†This problem is based on Machol (1970).

TABLE 50

Person	Job 1	Job 2	Job 3	Job 4
	\multicolumn{4}{c}{Time (hours)}			
1	22	18	30	18
2	18	—	27	22
3	26	20	28	28
4	16	22	—	14
5	21	—	25	28

Note: Dashes indicate person cannot do that particular job.

Table 51. To minimize the team's time for the race, which swimmer should swim which stroke?

3 Tom Cruise, Freddy Prinze Jr., Harrison Ford, and Matt LeBlanc are marooned on a desert island with Jennifer Aniston, Courteney Cox, Gwyneth Paltrow, and Julia Roberts. The "compatibility measures" in Table 52 indicate how much happiness each couple would experience if they spent all their time together. The happiness earned by a couple is proportional to the fraction of time they spend together. For example, if Freddie and Gwyneth spend half their time together, they earn happiness of $\frac{1}{2}(9) = 4.5$.

a Let x_{ij} be the fraction of time that the ith man spends with the jth woman. The goal of the eight people is to maximize the total happiness of the people on the island. Formulate an LP whose optimal solution will yield the optimal values of the x_{ij}'s.

b Explain why the optimal solution in part (a) will have four $x_{ij} = 1$ and twelve $x_{ij} = 0$. The optimal solution requires that each person spend all his or her time with one person of the opposite sex, so this result is often referred to as the Marriage Theorem.

c Determine the marriage partner for each person.

d Do you think the Proportionality Assumption of linear programming is valid in this situation?

4 A company is taking bids on four construction jobs. Three people have placed bids on the jobs. Their bids (in thousands of dollars) are given in Table 53 (a * indicates that the person did not bid on the given job). Person 1 can do only one job, but persons 2 and 3 can each do as many as two jobs. Determine the minimum cost assignment of persons to jobs.

5 Greydog Bus Company operates buses between Boston and Washington, D.C. A bus trip between these two cities takes 6 hours. Federal law requires that a driver rest for four or more hours between trips. A driver's workday consists of two trips: one from Boston to Washington and one from Washington to Boston. Table 54 gives the departure times for the buses. Greydog's goal is to minimize the total downtime for all drivers. How should Greydog assign crews to trips? *Note:* It is permissible for a driver's "day" to overlap midnight. For example, a Washington-based driver can be assigned to the Washington–Boston 3 P.M. trip and the Boston–Washington 6 A.M. trip.

6 Five male characters (Billie, John, Fish, Glen, and Larry) and five female characters (Ally, Georgia, Jane, Rene, and Nell) from *Ally McBeal* are marooned on a desert island. The problem is to determine what percentage of time each woman on the island should spend with each man. For example, Ally could spend 100% of her time with John or she could "play the field" by spending 20% of her time with each man. Table 55 shows a "happiness index" for each potential pairing of a man and woman. For example, if Larry and Rene spend all their time together, they earn 8 units of happiness for the island.

a Play matchmaker and determine an allocation of each man and woman's time that earns the maximum total happiness for the island. Assume that happiness earned by a couple is proportional to the amount of time they spend together.

b Explain why the optimal solution to this problem will, for any matrix of "happiness indices," always involve each woman spending all her time with one man.

TABLE 51

Swimmer	Time (seconds)			
	Free	Breast	Fly	Back
Gary Hall	54	54	51	53
Mark Spitz	51	57	52	52
Jim Montgomery	50	53	54	56
Chet Jastremski	56	54	55	53

TABLE 52

	JA	CC	GP	JR
TC	7	5	8	2
FP	7	8	9	4
HF	3	5	7	9
ML	5	5	6	7

TABLE 53

Person	Job			
	1	2	3	4
1	50	46	42	40
2	51	48	44	*
3	*	47	45	45

TABLE 54

Trip	Departure Time	Trip	Departure Time
Boston 1	6 A.M.	Washington 1	5:30 A.M.
Boston 2	7:30 A.M.	Washington 2	9 A.M.
Boston 3	11.30 A.M.	Washington 3	3 P.M.
Boston 4	7 P.M.	Washington 4	6:30 P.M.
Boston 5	12:30 A.M.	Washington 5	12 midnight

TABLE 55

	Ally	Georgia	Jane	Rene	Nell
Billie	8	6	4	7	5
John	5	7	6	4	9
Fish	10	6	5	2	10
Glen	1	0	0	0	0
Larry	5	7	9	8	6

c What assumption made in the problem is needed for the Marriage Theorem to hold?

Group B

7 Any transportation problem can be formulated as an assignment problem. To illustrate the idea, determine an assignment problem that could be used to find the optimal solution to the transportation problem in Table 56. (*Hint:* You will need five supply and five demand points).

8 The Chicago board of education is taking bids on the city's four school bus routes. Four companies have made the bids in Table 57.

    a Suppose each bidder can be assigned only one route. Use the assignment method to minimize Chicago's cost of running the four bus routes.

TABLE 56

TABLE 57

Company	Bids			
	Route 1	Route 2	Route 3	Route 4
1	$4,000	$5,000	—	—
2	—	$4,000	—	$4,000
3	$3,000	—	$2,000	—
4	—	—	$4,000	$5,000

    b Suppose that each company can be assigned two routes. Use the assignment method to minimize Chicago's cost of running the four bus routes. (*Hint:* Two supply points will be needed for each company.)

9 Show that step 3 of the Hungarian method is equivalent to performing the following operations: (1) Add k to each cost that lies in a covered row. (2) Subtract k from each cost that lies in an uncovered column.

10 Suppose c_{ij} is the smallest cost in row i and column j of an assignment problem. Must $x_{ij} = 1$ in any optimal assignment?

7.6 Transshipment Problems

A transportation problem allows only shipments that go directly from a supply point to a demand point. In many situations, shipments are allowed between supply points or between demand points. Sometimes there may also be points (called *transshipment points*) through which goods can be transshipped on their journey from a supply point to a demand point. Shipping problems with any or all of these characteristics are transshipment problems. Fortunately, the optimal solution to a transshipment problem can be found by solving a transportation problem.

    In what follows, we define a **supply point** to be a point that can send goods to another point but cannot receive goods from any other point. Similarly, a **demand point** is a point that can receive goods from other points but cannot send goods to any other point. A **transshipment point** is a point that can both receive goods from other points and send goods to other points. The following example illustrates these definitions ("—" indicates that a shipment is impossible).

EXAMPLE 5	Transshipment

Widgetco manufactures widgets at two factories, one in Memphis and one in Denver. The Memphis factory can produce as many as 150 widgets per day, and the Denver factory can produce as many as 200 widgets per day. Widgets are shipped by air to customers in Los Angeles and Boston. The customers in each city require 130 widgets per day. Because of the deregulation of airfares, Widgetco believes that it may be cheaper to first fly some widgets to New York or Chicago and then fly them to their final destinations. The costs of flying a widget are shown in Table 58. Widgetco wants to minimize the total cost of shipping the required widgets to its customers.

TABLE **58**

Shipping Costs for Transshipments

From	To ($)					
	Memphis	**Denver**	**N.Y.**	**Chicago**	**L.A.**	**Boston**
Memphis	0	—	8	13	25	28
Denver	—	0	15	12	26	25
N.Y.	—	—	0	6	16	17
Chicago	—	—	6	0	14	16
L.A.	—	—	—	—	0	—
Boston	—	—	—	—	—	0

In this problem, Memphis and Denver are supply points, with supplies of 150 and 200 widgets per day, respectively. New York and Chicago are transshipment points. Los Angeles and Boston are demand points, each with a demand of 130 widgets per day. A graphical representation of possible shipments is given in Figure 9.

We now describe how the optimal solution to a transshipment problem can be found by solving a transportation problem. Given a transshipment problem, we create a balanced transportation problem by the following procedure (assume that total supply exceeds total demand):

Step 1 If necessary, add a dummy demand point (with a supply of 0 and a demand equal to the problem's excess supply) to balance the problem. Shipments to the dummy and from a point to itself will, of course, have a zero shipping cost. Let s = total available supply.

Step 2 Construct a transportation tableau as follows: A row in the tableau will be needed for each supply point and transshipment point, and a column will be needed for each demand point and transshipment point. Each supply point will have a supply equal to its original supply, and each demand point will have a demand equal to its original demand. Let s = total available supply. Then each transshipment point will have a supply equal to (point's original supply) + s and a demand equal to (point's original demand) + s. This ensures that any transshipment point that is a net supplier will have a net outflow equal to the point's original supply, and, similarly, a net demander will have a net inflow equal to the point's original demand. Although we don't know how much will be shipped through each transshipment point, we can be sure that the total amount will not exceed s. This explains why we add s to the supply and demand at each transshipment point. By adding the same amounts to the supply and demand, we ensure that the net outflow at each transshipment point will be correct, and we also maintain a balanced transportation tableau.

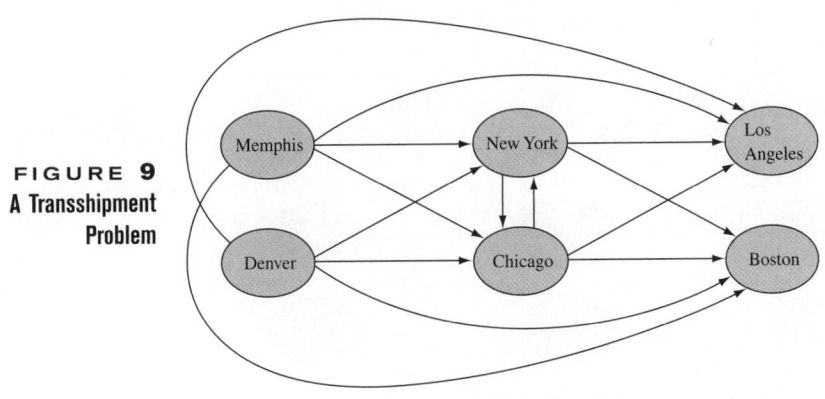

FIGURE 9
A Transshipment Problem

For the Widgetco example, this procedure yields the transportation tableau and its optimal solution given in Table 59. Because s = (total supply) = 150 + 200 = 350 and (total demand) = 130 + 130 = 260, the dummy demand point has a demand of 350 − 260 = 90. The other supplies and demands in the transportation tableau are obtained by adding s = 350 to each transshipment point's supply and demand.

In interpreting the solution to the transportation problem created from a transshipment problem, we simply ignore the shipments to the dummy and from a point to itself. From Table 59, we find that Widgetco should produce 130 widgets at Memphis, ship them to New York, and transship them from New York to Los Angeles. The 130 widgets produced at Denver should be shipped directly to Boston. The net outflow from each city is

Memphis:	130 + 20	= 150
Denver:	130 + 70	= 200
N.Y.:	220 + 130 − 130 − 220	= 0
Chicago:	350 − 350	= 0
L.A.:	−130	
Boston:	−130	
Dummy:	−20 − 70	= −90

A negative net outflow represents an inflow. Observe that each transshipment point (New York and Chicago) has a net outflow of 0; whatever flows into the transshipment point must leave the transshipment point. A graphical representation of the optimal solution to the Widgetco example is given in Figure 10.

Suppose that we modify the Widgetco example and allow shipments between Memphis and Denver. This would make Memphis and Denver transshipment points and would add columns for Memphis and Denver to the Table 59 tableau. The Memphis row in the tableau would now have a supply of 150 + 350 = 500, and the Denver row would have

TABLE 59
Representation of Transshipment Problem as Balanced Transportation Problem

	N.Y.		Chicago		L.A.		Boston		Dummy		Supply
Memphis		8		13		25		28		0	150
	130								20		
Denver		15		12		26		25		0	200
							130		70		
N.Y.		0		6		16		17		0	350
	220				130						
Chicago		6		0		14		16		0	350
			350								
Demand	350		350		130		130		90		

FIGURE 10
Optimal Solution to Widgetco

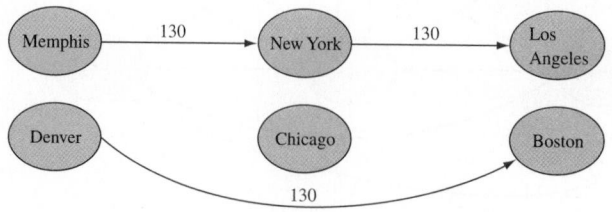

a supply of 200 + 350 = 550. The new Memphis column would have a demand of 0 + 350 = 350, and the new Denver column would have a demand of 0 + 350 = 350. Finally, suppose that shipments between demand points L.A. and Boston were allowed. This would make L.A. and Boston transshipment points and add rows for L.A. and Boston. The supply for both the L.A. and Boston rows would be 0 + 350 = 350. The demand for both the L.A. and Boston columns would now be 130 + 350 = 480.

PROBLEMS

Group A

1 General Ford produces cars at L.A. and Detroit and has a warehouse in Atlanta; the company supplies cars to customers in Houston and Tampa. The cost of shipping a car between points is given in Table 60 ("—" means that a shipment is not allowed). L.A. can produce as many as 1,100 cars, and Detroit can produce as many as 2,900 cars. Houston must receive 2,400 cars, and Tampa must receive 1,500 cars.

    a Formulate a balanced transportation problem that can be used to minimize the shipping costs incurred in meeting demands at Houston and Tampa.

    b Modify the answer to part (a) if shipments between L.A. and Detroit are not allowed.

    c Modify the answer to part (a) if shipments between Houston and Tampa are allowed at a cost of $5.

2 Sunco Oil produces oil at two wells. Well 1 can produce as many as 150,000 barrels per day, and well 2 can produce as many as 200,000 barrels per day. It is possible to ship oil directly from the wells to Sunco's customers in Los Angeles and New York. Alternatively, Sunco could transport oil to the ports of Mobile and Galveston and then ship it by tanker to New York or Los Angeles. Los Angeles requires 160,000 barrels per day, and New York requires 140,000 barrels per day. The costs of shipping 1,000 barrels between two points are shown in Table 61. Formulate a transshipment model (and equivalent transportation model) that could be used to minimize the transport costs in meeting the oil demands of Los Angeles and New York.

3 In Problem 2, assume that before being shipped to Los Angeles or New York, all oil produced at the wells must be refined at either Galveston or Mobile. To refine 1,000 barrels of oil costs $12 at Mobile and $10 at Galveston. Assuming that both Mobile and Galveston have infinite refinery capacity,

TABLE 60

From	To ($)				
	L.A.	**Detroit**	**Atlanta**	**Houston**	**Tampa**
L.A.	0	140	100	90	225
Detroit	145	0	111	110	119
Atlanta	105	115	0	113	78
Houston	89	109	121	0	—
Tampa	210	117	82	—	0

TABLE 61

From	To ($)					
	Well 1	**Well 2**	**Mobile**	**Galveston**	**N.Y.**	**L.A.**
Well 1	0	—	10	13	25	28
Well 2	—	0	15	12	26	25
Mobile	—	—	0	6	16	17
Galveston	—	—	6	0	14	16
N.Y.	—	—	—	—	0	15
L.A.	—	—	—	—	15	0

Note: Dashes indicate shipments that are not allowed.

formulate a transshipment and balanced transportation model to minimize the daily cost of transporting and refining the oil requirements of Los Angeles and New York.

4 Rework Problem 3 under the assumption that Galveston has a refinery capacity of 150,000 barrels per day and Mobile has one of 180,000 barrels per day. (*Hint:* Modify the method used to determine the supply and demand at each transshipment point to incorporate the refinery capacity restrictions, but make sure to keep the problem balanced.)

5 General Ford has two plants, two warehouses, and three customers. The locations of these are as follows:

    Plants: Detroit and Atlanta
    Warehouses: Denver and New York
    Customers: Los Angeles, Chicago, and Philadelphia

Cars are produced at plants, then shipped to warehouses, and finally shipped to customers. Detroit can produce 150 cars per week, and Atlanta can produce 100 cars per week. Los Angeles requires 80 cars per week; Chicago, 70; and Philadelphia, 60. It costs $10,000 to produce a car at each plant, and the cost of shipping a car between two cities is given in Table 62. Determine how to meet General Ford's weekly demands at minimum cost.

Group B

6[†] A company must meet the following demands for cash at the beginning of each of the next six months: month 1,

†Based on Srinivasan (1974).

TABLE 62

	To ($)	
From	Denver	New York
Detroit	1,253	637
Atlanta	1,398	841

	To ($)		
From	Los Angeles	Chicago	Philadelphia
Denver	1,059	996	1,691
New York	2,786	802	100

TABLE 63

	Month of Sale					
Bond	1	2	3	4	5	6
1	$0.21	$0.19	$0.17	$0.13	$0.09	$0.05
2	$0.50	$0.50	$0.50	$0.33	$0	$0
3	$1.00	$1.00	$1.00	$1.00	$1.00	$0

$200; month 2, $100; month 3, $50; month 4, $80; month 5, $160; month 6, $140. At the beginning of month 1, the company has $150 in cash and $200 worth of bond 1, $100 worth of bond 2, and $400 worth of bond 3. The company will have to sell some bonds to meet demands, but a penalty will be charged for any bonds sold before the end of month 6. The penalties for selling $1 worth of each bond are as shown in Table 63.

a Assuming that all bills must be paid on time, formulate a balanced transportation problem that can be used to minimize the cost of meeting the cash demands for the next six months.

b Assume that payment of bills can be made after they are due, but a penalty of 5¢ per month is assessed for each dollar of cash demands that is postponed for one month. Assuming all bills must be paid by the end of month 6, develop a transshipment model that can be used to minimize the cost of paying the next six months' bills. (*Hint:* Transshipment points are needed, in the form Ct = cash available at beginning of month t after bonds for month t have been sold, but before month t demand is met. Shipments into Ct occur from bond sales and $Ct - 1$. Shipments out of Ct occur to $Ct + 1$ and demands for months $1, 2, \ldots . t$.)

SUMMARY Notation

$$m = \text{number of supply points}$$

$$n = \text{number of demand points}$$

$$x_{ij} = \text{number of units shipped from supply point } i \text{ to demand point } j$$

$$c_{ij} = \text{cost of shipping 1 unit from supply point } i \text{ to demand point } j$$

$$s_i = \text{supply at supply point } i$$

$$d_j = \text{demand at demand point } j$$

$$\bar{c}_{ij} = \text{coefficient of } x_{ij} \text{ in row 0 of a given tableau}$$

$$\mathbf{a}_{ij} = \text{column for } x_{ij} \text{ in transportation constraints}$$

A transportation problem is **balanced** if total supply equals total demand. To use the methods of this chapter to solve a transportation problem, the problem must first be balanced by use of a dummy supply or a dummy demand point. A balanced transportation problem may be written as

$$\min \sum_{i=1}^{i=m} \sum_{j=1}^{j=n} c_{ij}x_{ij}$$

$$\text{s.t.} \quad \sum_{j=1}^{j=n} x_{ij} = s_i \quad (i = 1, 2, \ldots, m) \quad \text{(Supply constraints)}$$

$$\sum_{i=1}^{i=m} x_{ij} = d_j \quad (j = 1, 2, \ldots, n) \quad \text{(Demand constraints)}$$

$$x_{ij} \geq 0 \quad (i = 1, 2, \ldots, m; j = 1, 2, \ldots, n)$$

Finding Basic Feasible Solutions
for Balanced Transportation Problems

We can find a bfs for a balanced transportation problem by the northwest corner method, the minimum-cost method, or Vogel's method. To find a bfs by the northwest corner method, begin in the upper left-hand (or northwest) corner of the transportation tableau and set x_{11} as large as possible. Clearly, x_{11} can be no larger than the smaller of s_1 and d_1. If $x_{11} = s_1$, then cross out the first row of the transportation tableau; this indicates that no more basic variables will come from row 1 of the tableau. Also change d_1 to $d_1 - s_1$. If $x_{11} = d_1$, then cross out the first column of the transportation tableau and change s_1 to $s_1 - d_1$. If $x_{11} = s_1 = d_1$, cross out either row 1 or column 1 (but not both) of the transportation tableau. If you cross out row 1, change d_1 to 0; if you cross out column 1, change s_1 to 0. Continue applying this procedure to the most northwest cell in the tableau that does not lie in a crossed-out row or column. Eventually, you will come to a point where there is only one cell that can be assigned a value. Assign this cell a value equal to its row or column demand, and cross out both the cell's row and its column. A basic feasible solution has now been obtained.

Finding the Optimal Solution
for a Transportation Problem

Step 1 If the problem is unbalanced, balance it.

Step 2 Use one of the methods described in Section 7.2 to find a bfs.

Step 3 Use the fact that $u_1 = 0$ and $u_i + v_j = c_{ij}$ for all basic variables to find the $[u_1 \quad u_2 \ldots u_m \quad v_1 \quad v_2 \ldots v_n]$ for the current bfs.

Step 4 If $u_i + v_j - c_{ij} \le 0$ for all nonbasic variables, then the current bfs is optimal. If this is not the case, then we enter the variable with the most positive $u_i + v_j - c_{ij}$ into the basis. To do this, find the loop. Then, *counting only cells in the loop,* label the even cells. Also label the odd cells. Now find the odd cell whose variable assumes the smallest value, θ. The variable corresponding to this odd cell will leave the basis. To perform the pivot, decrease the value of each odd cell by θ and increase the value of each even cell by θ. The values of variables not in the loop remain unchanged. The pivot is now complete. If $\theta = 0$, then the entering variable will equal 0, and an odd variable that has a current value of 0 will leave the basis. In this case, a degenerate bfs will result. If more than one odd cell in the loop equals θ, you may arbitrarily choose one of these odd cells to leave the basis; again, a degenerate bfs will result. The pivoting yields a new bfs.

Step 5 Using the new bfs, return to steps 3 and 4.

For a maximization problem, proceed as stated, but replace step 4 by step 4'.

Step 4' If $u_i + v_j - c_{ij} \ge 0$ for all nonbasic variables, the current bfs is optimal. Otherwise, enter the variable with the most negative $u_i + v_j - c_{ij}$ into the basis using the pivoting procedure.

Assignment Problems

An **assignment problem** is a balanced transportation problem in which all supplies and demands equal 1. An $m \times m$ assignment problem may be efficiently solved by the Hungarian method:

Step 1 Find the minimum element in each row of the cost matrix. Construct a new matrix by subtracting from each cost the minimum cost in its row. For this new matrix, find the minimum cost in each column. Construct a new matrix (reduced cost matrix) by subtracting from each cost the minimum cost in its column.

Step 2 Cover all the zeros in the reduced cost matrix using the minimum number of lines needed. If m lines are required, then an optimal solution is available among the covered zeros in the matrix. If fewer than m lines are needed, then proceed to step 3.

Step 3 Find the smallest nonzero element (k) in the reduced cost matrix that is uncovered by the lines drawn in step 2. Now subtract k from each uncovered element and add k to each element that is covered by two lines. Return to step 2.

REMARKS **1** To solve an assignment problem in which the goal is to maximize the objective function, multiply the profits matrix through by -1 and solve it as a minimization problem.
2 If the number of rows and columns in the cost matrix are unequal, then the problem is unbalanced. The Hungarian method may yield an incorrect solution if the problem is unbalanced. Thus, any assignment problem should be balanced (by the addition of one or more dummy points) before it is solved by the Hungarian method.

Transshipment Problems

A transshipment problem allows shipment between supply points and between demand points, and it may also contain transshipment points through which goods may be shipped on their way from a supply point to a demand point. Using the following method, a transshipment problem may be transformed into a balanced transportation problem.

Step 1 If necessary, add a dummy demand point (with a supply of 0 and a demand equal to the problem's excess supply) to balance the problem. Shipments to the dummy and from a point to itself will, of course, have a zero shipping cost. Let s = total available supply.

Step 2 Construct a transportation tableau creating a row for each supply point and transshipment point, and a column for each demand point and transshipment point. Each supply point will have a supply equal to its original supply, and each demand point will have a demand equal to its original demand. Let s = total available supply. Then each transshipment point will have a supply equal to (point's original supply) + s and a demand equal to (point's original demand) + s.

Sensitivity Analysis for Transportation Problems

Following the discussion of sensitivity analysis in Chapter 6, we can analyze how a change in a transportation problem affects the problem's optimal solution.

Change 1 Changing the objective function coefficient of a nonbasic variable. As long as the coefficient of x_{ij} in the optimal row 0 is nonpositive, the current basis remains optimal.

Change 2 Changing the objective function coefficient of a basic variable. To see whether the current basis remains optimal, find the new u_i's and v_j's and use these values to price out all nonbasic variables. The current basis remains optimal as long as all nonbasic variables have a nonpositive coefficient in row 0.

Change 3 Increasing both supply s_i and demand d_j by Δ.

$$\text{New } z\text{-value} = \text{old } z\text{-value} + \Delta u_i + \Delta v_j$$

We may find the new values of the decision variables as follows:

1 If x_{ij} is a basic variable in the optimal solution, then increase x_{ij} by Δ.

2 If x_{ij} is a nonbasic variable in the optimal solution, find the loop involving x_{ij} and some of the basic variables. Find an odd cell in the loop that is in row i. Increase the value of this odd cell by Δ and go around the loop, alternately increasing and then decreasing current basic variables in the loop by Δ.

REVIEW PROBLEMS

Group A

1 Televco produces TV picture tubes at three plants. Plant 1 can produce 50 tubes per week; plant 2, 100 tubes per week; and plant 3, 50 tubes per week. Tubes are shipped to three customers. The profit earned per tube depends on the site where the tube was produced and on the customer who purchases the tube (see Table 64). Customer 1 is willing to purchase as many as 80 tubes per week; customer 2, as many as 90; and customer 3, as many as 100. Televco wants to find a shipping and production plan that will maximize profits.

a Formulate a balanced transportation problem that can be used to maximize Televco's profits.

b Use the northwest corner method to find a bfs to the problem.

c Use the transportation simplex to find an optimal solution to the problem.

2 Five workers are available to perform four jobs. The time it takes each worker to perform each job is given in Table 65. The goal is to assign workers to jobs so as to minimize the total time required to perform the four jobs. Use the Hungarian method to solve the problem.

3 A company must meet the following demands for a product: January, 30 units; February, 30 units; March, 20 units. Demand may be backlogged at a cost of $5/unit/month. All demand must be met by the end of March. Thus, if 1 unit of January demand is met during March, a backlogging cost of $5(2) = \$10$ is incurred. Monthly production capacity and unit production cost during each month are given in Table 66. A holding cost of $20/unit is assessed on the inventory at the end of each month.

a Formulate a balanced transportation problem that could be used to determine how to minimize the total cost (including backlogging, holding, and production costs) of meeting demand.

b Use Vogel's method to find a basic feasible solution.

c Use the transportation simplex to determine how to meet each month's demand. Make sure to give an interpretation of your optimal solution (for example, 20 units of month 2 demand is met from month 1 production).

4 Appletree Cleaning has five maids. To complete cleaning my house, they must vacuum, clean the kitchen, clean the bathroom, and do general straightening up. The time it takes each maid to do each job is shown in Table 67. Each maid

TABLE 64

From	To ($)		
	Customer 1	Customer 2	Customer 3
Plant 1	75	60	69
Plant 2	79	73	68
Plant 3	85	76	70

TABLE 66

Month	Production Capacity	Unit Production Cost
January	35	$400
February	30	$420
March	35	$410

TABLE 65

Worker	Time (Hours)			
	Job 1	Job 2	Job 3	Job 4
1	10	15	10	15
2	12	8	20	16
3	12	9	12	18
4	6	12	15	18
5	16	12	8	12

TABLE 67

Maid	Time (Hours)			
	Vacuum	Clean Kitchen	Clean Bathroom	Straighten Up
1	6	5	2	1
2	9	8	7	3
3	8	5	9	4
4	7	7	8	3
5	5	5	6	4

is assigned one job. Use the Hungarian method to determine assignments that minimize the total number of maid-hours needed to clean my house.

5[†]   Currently, State University can store 200 files on hard disk, 100 files in computer memory, and 300 files on tape. Users want to store 300 word-processing files, 100 packaged-program files, and 100 data files. Each month a typical word-processing file is accessed eight times; a typical packaged-program file, four times; and a typical data file, two times. When a file is accessed, the time it takes for the file to be retrieved depends on the type of file and on the storage medium (see Table 68).

   a If the goal is to minimize the total time per month that users spend accessing their files, formulate a balanced transportation problem that can be used to determine where files should be stored.

   b Use the minimum cost method to find a bfs.

   c Use the transportation simplex to find an optimal solution.

6   The Gotham City police have just received three calls for police. Five cars are available. The distance (in city blocks) of each car from each call is given in Table 69. Gotham City wants to minimize the total distance cars must travel to respond to the three police calls. Use the Hungarian method to determine which car should respond to which call.

7   There are three school districts in the town of Busville. The number of black and white students in each district are shown in Table 70. The Supreme Court requires the schools in Busville to be racially balanced. Thus, each school must have exactly 300 students, and each school must have the same number of black students. The distances between districts are shown in Table 70.

TABLE 68

Storage Medium	Time (Minutes)		
	Word Processing	Packaged Program	Data
Hard disk	5	4	4
Memory	2	1	1
Tape	10	8	6

TABLE 69

Car	Distance (Blocks)		
	Call 1	Call 2	Call 3
1	10	11	18
2	6	7	7
3	7	8	5
4	5	6	4
5	9	4	7

[†]This problem is based on Evans (1984).

TABLE 70

District	No. of Students		Distance to (Miles)	
	Whites	Blacks	District 2	District 3
1	210	120	3	5
2	210	30	—	4
3	180	150	—	—

Formulate a balanced transportation problem that can be used to determine the minimum total distance that students must be bused while still satisfying the Supreme Court's requirements. Assume that a student who remains in his or her own district will not be bused.

8   Using the northwest corner method to find a bfs, find (via the transportation simplex) an optimal solution to the transportation (minimization) problem shown in Table 71.

9   Solve the following LP:
$$\min z = 2x_1 + 3x_2 + 4x_3 + 3x_4$$
$$\text{s.t.} \quad x_1 + x_2 \qquad\qquad \leq 4$$
$$x_3 + x_4 \leq 5$$
$$x_1 \qquad + x_3 \qquad \geq 3$$
$$x_2 \qquad + x_4 \geq 6$$
$$\min x_j \geq 0 \quad (j = 1, 2, 3, 4)$$

10   Find the optimal solution to the balanced transportation problem in Table 72 (minimization).

11   In Problem 10, suppose we increase s_i to 16 and d_3 to 11. The problem is still balanced, and because 31 units (instead of 30 units) must be shipped, one would think that the total shipping costs would be increased. Show that the total shipping cost has actually decreased by $2, however. This is called the "more for less" paradox. Explain why increasing both the supply and the demand has decreased cost. Using the theory of shadow prices, explain how one could have predicted that increasing s_1 and d_3 by 1 would decrease total cost by $2.

12   Use the northwest corner method, the minimum-cost method, and Vogel's method to find basic feasible solutions to the transportation problem in Table 73.

13   Find the optimal solution to Problem 12.

TABLE 71

	12	14	16	60
	14	13	19	50
	17	15	18	40
	40	70	10	

TABLE 72

4	2	4	15
12	8	4	15
10	10	10	

TABLE 73

20	11	3	6	5
5	9	10	2	10
18	7	4	1	15
3	3	12	12	

TABLE 74

From	To ($)			
	Dallas	Houston	N.Y.	Chicago
L.A.	300	110	—	—
San Diego	420	100	—	—
Dallas	—	—	450	550
Houston	—	—	470	530

TABLE 75

Plant	Cars Available
Atlanta	5,000
Boston	6,000
Chicago	4,000
L.A.	3,000

TABLE 76

Warehouse	Cars Required
Memphis	6,000
Milwaukee	4,000
N.Y.	4,000
Denver	2,000
San Francisco	2,000

TABLE 77

	Memphis	Milwaukee	N.Y.	Denver	S.F.
Atlanta	371	761	841	1,398	2,496
Boston	1,296	1,050	206	1,949	3,095
Chicago	530	87	802	996	2,142
L.A.	1,817	2,012	2,786	1,059	379

14 Oilco has oil fields in San Diego and Los Angeles. The San Diego field can produce 500,000 barrels per day, and the Los Angeles field can produce 400,000 barrels per day. Oil is sent from the fields to a refinery, either in Dallas or in Houston (assume that each refinery has unlimited capacity). It costs $700 to refine 100,000 barrels of oil at Dallas and $900 at Houston. Refined oil is shipped to customers in Chicago and New York. Chicago customers require 400,000 barrels per day of refined oil; New York customers require 300,000 barrels per day. The costs of shipping 100,000 barrels of oil (refined or unrefined) between cities are given in Table 74. Formulate a balanced transportation model of this situation.

15 For the Powerco problem, find the range of values of c_{24} for which the current basis remains optimal.

16 For the Powerco problem, find the range of values of c_{23} for which the current basis remains optimal.

17 A company produces cars in Atlanta, Boston, Chicago, and Los Angeles. The cars are then shipped to warehouses in Memphis, Milwaukee, New York City, Denver, and San Francisco. The number of cars available at each plant is given in Table 75.

Each warehouse needs to have available the number of cars given in Table 76.

The distance (in miles) between the cities is given in Table 77.

a Assuming that the cost (in dollars) of shipping a car equals the distance between two cities, determine an optimal shipping schedule.

b Assuming that the cost (in dollars) of shipping a car equals the square root of the distance between two cities, determine an optimal shipping schedule.

18 During the next three quarters, Airco faces the following demands for air conditioner compressors: quarter 1—200; quarter 2—300; quarter 3—100. As many as 240 air compressors can be produced during each quarter. Production costs/compressor during each quarter are given in Table 78. The cost of holding an air compressor in inventory is $100/quarter. Demand may be backlogged (as long as it is met by the end of quarter 3) at a cost of $60/compressor/quarter. Formulate the tableau for a balanced transportation problem whose solution tells Airco how to minimize the total cost of meeting the demands for quarters 1–3.

19 A company is considering hiring people for four types of jobs. It would like to hire the number of people in Table 79 for each type of job.

Four types of people can be hired by the company. Each type is qualified to perform two types of jobs according to

TABLE 78

Quarter 1	Quarter 2	Quarter 3
$200	$180	$240

TABLE 79

	Job			
	1	2	3	4
Number of people	30	30	40	20

TABLE 80

	Type of Person			
	1	2	3	4
Jobs qualified for	1 and 3	2 and 3	3 and 4	1 and 4

Table 80. A total of 20 Type 1, 30 Type 2, 40 Type 3, and 20 Type 4 people have applied for jobs. Formulate a balanced transportation problem whose solution will tell the company how to maximize the number of employees assigned to suitable jobs. (*Note:* Each person can be assigned to at most one job.)

20 During each of the next two months you can produce as many as 50 units/month of a product at a cost of $12/unit during month 1 and $15/unit during month 2. The customer is willing to buy as many as 60 units/month during each of the next two months. The customer will pay $20/unit during month 1, and $16/unit during month 2. It costs $1/unit to hold a unit in inventory for a month. Formulate a balanced transportation problem whose solution will tell you how to maximize profit.

Group B

21[†] The Carter Caterer Company must have the following number of clean napkins available at the beginning of each of the next four days: day 1—15; day 2—12; day 3—18; day 4—6. After being used, a napkin can be cleaned by one of two methods: fast service or slow service. Fast service costs 10¢ per napkin, and a napkin cleaned via fast service is available for use the day after it is last used. Slow service costs 6¢ per napkin, and these napkins can be reused two days after they are last used. New napkins can be purchased for a cost of 20¢ per napkin. Formulate a balanced transportation problem to minimize the cost of meeting the demand for napkins during the next four days.

22 Braneast Airlines must staff the daily flights between New York and Chicago shown in Table 81. Each of Braneast's crews lives in either New York or Chicago. Each day a crew must fly one New York–Chicago and one Chicago–New

[†]This problem is based on Jacobs (1954).

TABLE 81

Flight	Leave Chicago	Arrive New York	Flight	Leave New York	Arrive Chicago
1	6 A.M.	10 A.M.	1	7 A.M.	9 A.M.
2	9 A.M.	1 P.M.	2	8 A.M.	10 A.M.
3	12 noon	4 P.M.	3	10 A.M.	12 noon
4	3 P.M.	7 P.M.	4	12 noon	2 P.M.
5	5 P.M.	9 P.M.	5	2 P.M.	4 P.M.
6	7 P.M.	11 P.M.	6	4 P.M.	6 P.M.
7	8 P.M.	12 midnight	7	6 P.M.	8 P.M.

York flight with at least 1 hour of downtime between flights. Braneast wants to schedule the crews to minimize the total downtime. Set up an assignment problem that can be used to accomplish this goal. (*Hint:* Let $x_{ij} = 1$ if the crew that flies flight i also flies flight j, and $x_{ij} = 0$ otherwise. If $x_{ij} = 1$, then a cost c_{ij} is incurred, corresponding to the downtime associated with a crew flying flight i and flight j.) Of course, some assignments are not possible. Find the flight assignments that minimize the total downtime. How many crews should be based in each city? Assume that at the end of the day, each crew must be in its home city.

23 A firm producing a single product has three plants and four customers. The three plants will produce 3,000, 5,000, and 5,000 units, respectively, during the next time period. The firm has made a commitment to sell 4,000 units to customer 1, 3,000 units to customer 2, and at least 3,000 units to customer 3. Both customers 3 and 4 also want to buy as many of the remaining units as possible. The profit associated with shipping a unit from plant i to customer j is given in Table 82. Formulate a balanced transportation problem that can be used to maximize the company's profit.

24 A company can produce as many as 35 units/month. The demands of its primary customers must be met on time each month; if it wishes, the company may also sell units to secondary customers each month. A $1/unit holding cost is assessed against each month's ending inventory. The relevant data are shown in Table 83. Formulate a balanced transportation problem that can be used to maximize profits earned during the next three months.

25 My home has four valuable paintings that are up for sale. Four customers are bidding for the paintings. Customer 1 is willing to buy two paintings, but each other customer is willing to purchase at most one painting. The prices that each customer is willing to pay are given in Table 84. Use

TABLE 82

From	To Customer ($)			
	1	2	3	4
Plant 1	65	63	62	64
Plant 2	68	67	65	62
Plant 3	63	60	59	60

TABLE 83

Month	Production Cost/Unit ($)	Primary Demand	Available for Secondary Demand	Sales Price/Unit ($)
1	13	20	15	15
2	12	15	20	14
3	13	25	15	16

TABLE 84

Customer	Bid for ($)			
	Painting 1	Painting 2	Painting 3	Painting 4
1	8	11	—	—
2	9	13	12	7
3	9	—	11	—
4	—	—	12	9

the Hungarian method to determine how to maximize the total revenue received from the sale of the paintings.

26 Powerhouse produces capacitors at three locations: Los Angeles, Chicago, and New York. Capacitors are shipped from these locations to public utilities in five regions of the country: northeast (NE), northwest (NW), midwest (MW), southeast (SE), and southwest (SW). The cost of producing and shipping a capacitor from each plant to each region of the country is given in Table 85. Each plant has an annual production capacity of 100,000 capacitors. Each year, each region of the country must receive the following number of capacitors: NE, 55,000; NW, 50,000; MW, 60,000; SE, 60,000; SW, 45,000. Powerhouse feels shipping costs are too high, and the company is therefore considering building one or two more production plants. Possible sites are Atlanta and Houston. The costs of producing a capacitor and shipping it to each region of the country are given in Table 86. It costs $3 million (in current dollars) to build a new plant, and operating each plant incurs a fixed cost (in addition to variable shipping and production costs) of $50,000 per year. A plant at Atlanta or Houston will have the capacity to produce 100,000 capacitors per year.

Assume that future demand patterns and production costs will remain unchanged. If costs are discounted at a rate of $11\frac{1}{9}\%$ per year, how can Powerhouse minimize the present value of all costs associated with meeting current and future demands?

TABLE 85

From	To ($)				
	NE	NW	MW	SE	SW
L.A.	27.86	4.00	20.54	21.52	13.87
Chicago	8.02	20.54	2.00	6.74	10.67
N.Y.	2.00	27.86	8.02	8.41	15.20

TABLE 86

From	To ($)				
	NE	NW	MW	SE	SW
Atlanta	8.41	21.52	6.74	3.00	7.89
Houston	15.20	13.87	10.67	7.89	3.00

27[†] During the month of July, Pittsburgh resident B. Fly must make four round-trip flights between Pittsburgh and Chicago. The dates of the trips are as shown in Table 87. B. Fly must purchase four round-trip tickets. Without a discounted fare, a round-trip ticket between Pittsburgh and Chicago costs $500. If Fly's stay in a city includes a weekend, then he gets a 20% discount on the round-trip fare. If his stay in a city is at least 21 days, then he receives a 35% discount; and if his stay is more than 10 days, then he receives a 30% discount. Of course, only one discount can be applied toward the purchase of any ticket. Formulate and solve an assignment problem that minimizes the total cost of purchasing the four round-trip tickets. (*Hint:* Let $x_{ij} = 1$ if a round-trip ticket is purchased for use on the ith flight out of Pittsburgh and the jth flight out of Chicago. Also think about where Fly should buy a ticket if, for example, $x_{21} = 1$.)

28 Three professors must be assigned to teach six sections of finance. Each professor must teach two sections of finance, and each has ranked the six time periods during which finance is taught, as shown in Table 88. A ranking of 10 means that the professor wants to teach that time, and a ranking of 1 means that he or she does not want to teach at that time. Determine an assignment of professors to sections that will maximize the total satisfaction of the professors.

29[‡] Three fires have just broken out in New York. Fires 1 and 2 each require two fire engines, and fire 3 requires three fire engines. The "cost" of responding to each fire depends on the time at which the fire engines arrive. Let t_{ij} be the time (in minutes) when the jth engine arrives at fire i. Then the cost of responding to each fire is as follows:

$$\text{Fire 1:} \quad 6t_{11} + 4t_{12}$$
$$\text{Fire 2:} \quad 7t_{21} + 3t_{22}$$
$$\text{Fire 3:} \quad 9t_{31} + 8t_{32} + 5t_{33}$$

Three fire companies can respond to the three fires. Company 1 has three engines available, and companies 2

TABLE 87

Leave Pittsburgh	Leave Chicago
Monday, July 1	Friday, July 5
Tuesday, July 9	Thursday, July 11
Monday, July 15	Friday, July 19
Wednesday, July 24	Thursday, July 25

[†]Based on Hansen and Wendell (1982).
[‡]Based on Denardo, Rothblum, and Swersey (1988).

TABLE 88

Professor	9 A.M.	10 A.M.	11 A.M.	1 P.M.	2 P.M.	3 P.M.
1	8	7	6	5	7	6
2	9	9	8	8	4	4
3	7	6	9	6	9	9

TABLE 89

Company	Fire 1	Fire 2	Fire 3
1	6	7	9
2	5	8	11
3	6	9	10

and 3 each have two engines available. The time (in minutes) it takes an engine to travel from each company to each fire is shown in Table 89.

a Formulate and solve a transportation problem that can be used to minimize the cost associated with as-

signing the fire engines. (*Hint:* Seven demand points will be needed.)

b Would the formulation in part (a) still be valid if the cost of fire 1 were $4t_{11} + 6t_{12}$?

REFERENCES

The following six texts discuss transportation, assignment, and transshipment problems:

Bazaraa, M., and J. Jarvis. *Linear Programming and Network Flows.* New York: Wiley, 1990.

Bradley, S., A. Hax, and T. Magnanti. *Applied Mathematical Programming.* Reading, Mass.: Addison-Wesley, 1977.

Dantzig, G. *Linear Programming and Extensions.* Princeton, N.J.: Princeton University Press, 1963.

Gass, S. *Linear Programming: Methods and Applications,* 5th ed. New York: McGraw-Hill, 1985.

Murty, K. *Linear Programming.* New York: Wiley, 1983.

Wu, N., and R. Coppins. *Linear Programming and Extensions.* New York: McGraw-Hill, 1981.

Aarvik, O., and P. Randolph. "The Application of Linear Programming to the Determination of Transmission Line Fees in an Electrical Power Network," *Interfaces* 6(1975):17–31.

Denardo, E., U. Rothblum, and A. Swersey. "Transportation Problem in Which Costs Depend on Order of Arrival," *Management Science* 34(1988):774–784.

Evans, J. "The Factored Transportation Problem," *Management Science* 30(1984):1021–1024.

Gillett, B. *Introduction to Operations Research: A Computer-Oriented Algorithmic Approach.* New York: McGraw-Hill, 1976.

Glassey, R., and V. Gupta. "A Linear Programming Analysis of Paper Recycling," *Management Science* 21(1974): 392–408.

Glover, F., et al. "A Computational Study on Starting Procedures, Basis Change Criteria and Solution Algorithms for Transportation Problems," *Management Science* 20(1974):793–813. This article discusses the computational efficiency of various methods used to find basic feasible solutions for transportation problems.

Hansen, P., and R. Wendell. "A Note on Airline Commuting," *Interfaces* 11(no. 12, 1982):85–87.

Jackson, B. "Using LP for Crude Oil Sales at Elk Hills: A Case Study," *Interfaces* 10(1980):65–70.

Jacobs, W. "The Caterer Problem," *Naval Logistics Research Quarterly* 1(1954):154–165.

Machol, R. "An Application of the Assignment Problem," *Operations Research* 18(1970):745–746.

Srinivasan, P. "A Transshipment Model for Cash Management Decisions," *Management Science* 20(1974): 1350–1363.

Wagner, H., and D. Rubin. "Shadow Prices: Tips and Traps for Managers and Instructors," *Interfaces* 20(no. 4, 1990):150–157.

8

Network Models

Many important optimization problems can best be analyzed by means of a graphical or network representation. In this chapter, we consider four specific network models–shortest-path problems, maximum-flow problems, CPM–PERT project-scheduling models, and minimum-spanning tree problems–for which efficient solution procedures exist. We also discuss minimum-cost network flow problems (MCNFPs), of which transportation, assignment, transshipment, shortest-path, and maximum-flow problems and the CPM project-scheduling models are all special cases. Finally, we discuss a generalization of the transportation simplex, the network simplex, which can be used to solve MCNFPs. We begin the chapter with some basic terms used to describe graphs and networks.

8.1 Basic Definitions

A **graph,** or **network,** is defined by two sets of symbols: nodes and arcs. First, we define a set (call it V) of points, or **vertices.** The vertices of a graph or network are also called **nodes.**

We also define a set of arcs A.

DEFINITION ■ An **arc** consists of an ordered pair of vertices and represents a possible direction of motion that may occur between vertices. ■

For our purposes, if a network contains an arc (j, k), then motion is possible from node j to node k. Suppose nodes 1, 2, 3, and 4 of Figure 1 represent cities, and each arc represents a (one-way) road linking two cities. For this network, $V = \{1, 2, 3, 4\}$ and $A = \{(1, 2), (2, 3), (3, 4), (4, 3), (4, 1)\}$. For the arc (j, k), node j is the **initial node,** and node k is the **terminal node.** The arc (j, k) is said to go from node j to node k. Thus, the arc $(2, 3)$ has initial node 2 and terminal node 3, and it goes from node 2 to node 3. The arc $(2, 3)$ may be thought of as a (one-way) road on which we may travel from city 2 to city 3. In Figure 1, the arcs show that travel is allowed from city 3 to city 4, and from city 4 to city 3, but that travel between the other cities may be one way only.

Later, we often discuss a group or collection of arcs. The following definitions are convenient ways to describe certain groups or collections of arcs.

DEFINITION ■ A sequence of arcs such that every arc has exactly one vertex in common with the previous arc is called a **chain.** ■

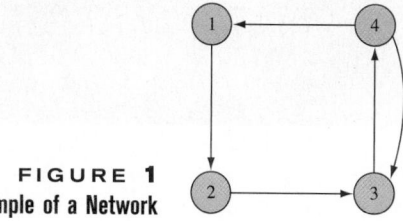

FIGURE 1
Example of a Network

DEFINITION ■   A **path** is a chain in which the terminal node of each arc is identical to the initial node of the next arc. ■

For example, in Figure 1, (1, 2)–(2, 3)–(4, 3) is a chain but not a path; (1, 2)–(2, 3)–(3, 4) is a chain *and* a path. The path (1, 2)–(2, 3)–(3, 4) represents a way to travel from node 1 to node 4.

8.2 Shortest-Path Problems

In this section, we assume that each arc in the network has a length associated with it. Suppose we start at a particular node (say, node 1). The problem of finding the shortest path (path of minimum length) from node 1 to any other node in the network is called a **shortest-path problem.** Examples 1 and 2 are shortest-path problems.

EXAMPLE 1   **Shortest Path**

Let us consider the Powerco example (Figure 2). Suppose that when power is sent from plant 1 (node 1) to city 1 (node 6), it must pass through relay substations (nodes 2–5). For any pair of nodes between which power can be transported, Figure 2 gives the distance (in miles) between the nodes. Thus, substations 2 and 4 are 3 miles apart, and power cannot be sent between substations 4 and 5. Powerco wants the power sent from plant 1 to city 1 to travel the minimum possible distance, so it must find the shortest path in Figure 2 that joins node 1 to node 6.

If the cost of shipping power were proportional to the distance the power travels, then knowing the shortest path between plant 1 and city 1 in Figure 2 (and the shortest path between plant *i* and city *j* in similar diagrams) would be necessary to determine the shipping costs for the transportation version of the Powerco problem discussed in Chapter 7.

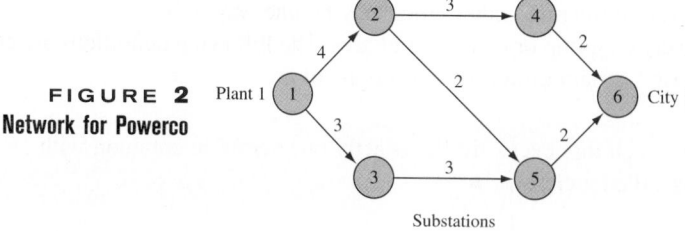

FIGURE 2
Network for Powerco

EXAMPLE 2   **Equipment Replacement**

I have just purchased (at time 0) a new car for $12,000. The cost of maintaining a car during a year depends on its age at the beginning of the year, as given in Table 1. To avoid the high maintenance costs associated with an older car, I may trade in my car and purchase a new car. The price I receive on a trade-in depends on the age of the car at the time of trade-in (see Table 2). To simplify the computations, we assume that at any time, it costs $12,000 to purchase a new car. My goal is to minimize the net cost (purchasing costs + maintenance costs − money received in trade-ins) incurred during the next five years. Formulate this problem as a shortest-path problem.

Solution   Our network will have six nodes (1, 2, 3, 4, 5, and 6). Node i is the beginning of year i. For $i < j$, an arc (i, j) corresponds to purchasing a new car at the beginning of year i and keeping it until the beginning of year j. The length of arc (i, j) (call it c_{ij}) is the total net cost incurred in owning and operating a car from the beginning of year i to the beginning of year j if a new car is purchased at the beginning of year i and this car is traded in for a new car at the beginning of year j. Thus,

$$c_{ij} = \text{maintenance cost incurred during years } i, i + 1, \ldots, j - 1$$
$$+ \text{ cost of purchasing car at beginning of year } i$$
$$- \text{ trade-in value received at beginning of year } j$$

Applying this formula to the information in the problem yields (all costs are in thousands)

$c_{12} = 2 + 12 - 7 = 7$ $c_{16} = 2 + 4 + 5 + 9 + 12 + 12 - 0 = 44$

$c_{13} = 2 + 4 + 12 - 6 = 12$ $c_{23} = 2 + 12 - 7 = 7$

$c_{14} = 2 + 4 + 5 + 12 - 2 = 21$ $c_{24} = 2 + 4 + 12 - 6 = 12$

$c_{15} = 2 + 4 + 5 + 9 + 12 - 1 = 31$ $c_{25} = 2 + 4 + 5 + 12 - 2 = 21$

TABLE **1**

Car Maintenance Costs

Age of Car (Years)	Annual Maintenance Cost ($)
0	2,000
1	4,000
2	5,000
3	9,000
4	12,000

TABLE **2**

Car Trade-in Prices

Age of Car (Years)	Trade-in Price
1	7,000
2	6,000
3	2,000
4	1,000
5	0

FIGURE **3**
Network for Minimizing
Car Costs

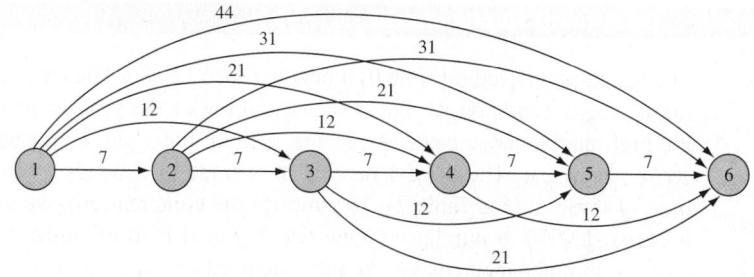

$$c_{26} = 2 + 4 + 5 + 9 + 12 - 1 = 31 \qquad c_{45} = 2 + 12 - 7 = 7$$

$$c_{34} = 2 + 12 - 7 = 7 \qquad\qquad\quad c_{46} = 2 + 4 + 12 - 6 = 12$$

$$c_{35} = 2 + 4 + 12 - 6 = 12 \qquad\quad c_{56} = 2 + 12 - 7 = 7$$

$$c_{36} = 2 + 4 + 5 + 12 - 2 = 21$$

We now see that the length of any path from node 1 to node 6 is the net cost incurred during the next five years corresponding to a particular trade-in strategy. For example, suppose I trade in the car at the beginning of year 3 and next trade in the car at the end of year 5 (the beginning of year 6). This strategy corresponds to the path 1–3–6 in Figure 3. The length of this path ($c_{13} + c_{36}$) is the total net cost incurred during the next five years if I trade in the car at the beginning of year 3 and at the beginning of year 6. Thus, the length of the shortest path from node 1 to node 6 in Figure 3 is the minimum net cost that can be incurred in operating a car during the next five years.

Dijkstra's Algorithm

Assuming that all arc lengths are nonnegative, the following method, known as **Dijkstra's algorithm,** can be used to find the shortest path from a node (say, node 1) to all other nodes. To begin, we label node 1 with a permanent label of 0. Then we label each node i that is connected to node 1 by a single arc with a "temporary" label equal to the length of the arc joining node 1 to node i. Each other node (except, of course, for node 1) will have a temporary label of ∞. Choose the node with the smallest temporary label and make this label permanent.

Now suppose that node i has just become the $(k + 1)$th node to be given a permanent label. Then node i is the kth closest node to node 1. At this point, the temporary label of any node (say, node i') is the length of the shortest path from node 1 to node i' that passes only through nodes contained in the $k - 1$ closest nodes to node 1. For each node j that now has a temporary label and is connected to node i by an arc, we replace node j's temporary label with

$$\min \begin{cases} \text{node } j\text{'s current temporary label} \\ \text{node } i\text{'s permanent label } + \text{ length of arc } (i, j) \end{cases}$$

(Here, $\min\{a, b\}$ is the smaller of a and b.) The new temporary label for node j is the length of the shortest path from node 1 to node j that passes only through nodes contained in the k closest nodes to node 1. We now make the smallest temporary label a permanent label. The node with this new permanent label is the $(k + 1)$th closest node to node 1. Continue this process until all nodes have a permanent label. To find the shortest path from node 1 to node j, work backward from node j by finding nodes having labels differing by exactly the length of the connecting arc. Of course, if we want the shortest path from node 1 to node j, we can stop the labeling process as soon as node j receives a permanent label.

To illustrate Dijkstra's algorithm, we find the shortest path from node 1 to node 6 in Figure 2. We begin with the following labels (a * represents a permanent label, and the ith number is the label of the node i): [0* 4 3 ∞ ∞ ∞]. Node 3 now has the smallest temporary label. We therefore make node 3's label permanent and obtain the following labels:

$$[0* \quad 4 \quad 3* \quad \infty \quad \infty \quad \infty]$$

We now know that node 3 is the closest node to node 1. We compute new temporary labels for all nodes that are connected to node 3 by a single arc. In Figure 2 that is node 5.

$$\text{New node 5 temporary label} = \min\{\infty, 3 + 3\} = 6$$

Node 2 now has the smallest temporary label; we now make node 2's label permanent. We now know that node 2 is the second closest node to node 1. Our new set of labels is

$$[0* \quad 4* \quad 3* \quad \infty \quad 6 \quad \infty]$$

Because nodes 4 and 5 are connected to the newly permanently labeled node 2, we must change the temporary labels of nodes 4 and 5. Node 4's new temporary label is min $\{\infty, 4 + 3\} = 7$ and node 5's new temporary label is min $\{6, 4 + 2\} = 6$. Node 5 now has the smallest temporary label, so we make node 5's label permanent. We now know that node 5 is the third closest node to node 1. Our new labels are

$$[0* \quad 4* \quad 3* \quad 7 \quad 6* \quad \infty]$$

Only node 6 is connected to node 5, so node 6's temporary label will change to min $\{\infty, 6 + 2\} = 8$. Node 4 now has the smallest temporary label, so we make node 4's label permanent. We now know that node 4 is the fourth closest node to node 1. Our new labels are

$$[0* \quad 4* \quad 3* \quad 7* \quad 6* \quad 8]$$

Because node 6 is connected to the newly permanently labeled node 4, we must change node 6's temporary label to min $\{8, 7 + 2\} = 8$. We can now make node 6's label permanent. Our final set of labels is [0* 4* 3* 7* 6* 8*]. We can now work backward and find the shortest path from node 1 to node 6. The difference between node 6's and node 5's permanent labels is $2 = $ length of arc $(5, 6)$, so we go back to node 5. The difference between node 5's and node 2's permanent labels is $2 = $ length of arc $(2, 5)$, so we may go back to node 2. Then, of course, we must go back to node 1. Thus, 1–2–5–6 is a shortest path (of length 8) from node 1 to node 6. Observe that when we were at node 5, we could also have worked backward to node 3 and obtained the shortest path 1–3–5–6.

The Shortest-Path Problem as a Transshipment Problem

Finding the shortest path between node i and node j in a network may be viewed as a transshipment problem. Simply try to minimize the cost of sending one unit from node i to node j (with all other nodes in the network being transshipment points), where the cost of sending one unit from node k to node k' is the length of arc (k, k') if such an arc exists and is M (a large positive number) if such an arc does not exist. As in Section 7.6, the cost of shipping one unit from a node to itself is zero. Following the method described in Section 7.6, this transshipment problem may be transformed into a balanced transportation problem.

TABLE 3
Transshipment Representation of Shortest-Path Problem and Optimal Solution (1)

Node	2	3	4	5	6	Supply
1	4 (1)	3	M	M	M	1
2	0	M	3	2 (1)	M	1
3	M	0 (1)	M	3	M	1
4	M	M	0 (1)	M	2	1
5	M	M	M	0	2 (1)	1
Demand	1	1	1	1	1	

To illustrate the preceding ideas, we formulate the balanced transportation problem associated with finding the shortest path from node 1 to node 6 in Figure 2. We want to send one unit from node 1 to node 6. Node 1 is a supply point, node 6 is a demand point, and nodes 2, 3, 4, and 5 will be transshipment points. Using $s = 1$, we obtain the balanced transportation problem shown in Table 3. This transportation problem has two optimal solutions:

1 $z = 4 + 2 + 2 = 8$, $x_{12} = x_{25} = x_{56} = x_{33} = x_{44} = 1$ (all other variables equal 0). This solution corresponds to the path 1–2–5–6.

2 $z = 3 + 3 + 2 = 8$, $x_{13} = x_{35} = x_{56} = x_{22} = x_{44} = 1$ (all other variables equal 0). This solution corresponds to the path 1–3–5–6.

REMARK After formulating a shortest-path problem as a transshipment problem, the problem may be solved easily by using LINGO or a spreadsheet optimizer. See Section 7.1 for details.

PROBLEMS

Group A

1 Find the shortest path from node 1 to node 6 in Figure 3.

2 Find the shortest path from node 1 to node 5 in Figure 4.

3 Formulate Problem 2 as a transshipment problem.

4 Use Dijkstra's algorithm to find the shortest path from node 1 to node 4 in Figure 5. Why does Dijkstra's algorithm fail to obtain the correct answer?

FIGURE 4
Network for Problem 2

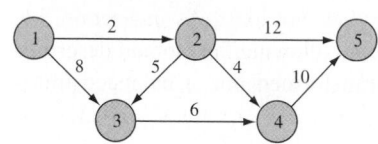

FIGURE 5
Network for Problem 4

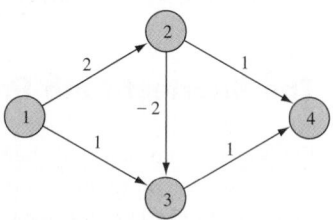

5 Suppose it costs $10,000 to purchase a new car. The annual operating cost and resale value of a used car are shown in Table 4. Assuming that one now has a new car, determine a replacement policy that minimizes the net costs of owning and operating a car for the next six years.

TABLE **4**

Age of Car (Years)	Resale Value ($)	Operating Cost ($)
1	7,000	300 (year 1)
2	6,000	500 (year 2)
3	4,000	800 (year 3)
4	3,000	1,200 (year 4)
5	2,000	1,600 (year 5)
6	1,000	2,200 (year 6)

TABLE **6**

Year	Purchase Cost ($)
1	170,000
2	190,000
3	210,000
4	250,000
5	300,000

6 It costs $40 to buy a telephone from the department store. Assume that I can keep a telephone for at most five years and that the estimated maintenance cost each year of operation is as follows: year 1, $20; year 2, $30; year 3, $40; year 4, $60; year 5, $70. I have just purchased a new telephone. Assuming that a telephone has no salvage value, determine how to minimize the total cost of purchasing and operating a telephone for the next six years.

7 At the beginning of year 1, a new machine must be purchased. The cost of maintaining a machine i years old is given in Table 5.

The cost of purchasing a machine at the beginning of each year is given in Table 6.

There is no trade-in value when a machine is replaced. Your goal is to minimize the total cost (purchase plus maintenance) of having a machine for five years. Determine the years in which a new machine should be purchased.

Group B

8[†] A library must build shelving to shelve 200 4-inch high books, 100 8-inch high books, and 80 12-inch high books.

TABLE **5**

Age at Beginning of Year	Maintenance Cost for Next Year ($)
0	38,000
1	50,000
2	97,000
3	182,000
4	304,000

[†]Based on Ravindran (1971).

Each book is 0.5 inch thick. The library has several ways to store the books. For example, an 8-inch high shelf may be built to store all books of height less than or equal to 8 inches, and a 12-inch high shelf may be built for the 12-inch books. Alternatively, a 12-inch high shelf might be built to store all books. The library believes it costs $2,300 to build a shelf and that a cost of $5 per square inch is incurred for book storage. (Assume that the area required to store a book is given by height of storage area times book's thickness.)

Formulate and solve a shortest-path problem that could be used to help the library determine how to shelve the books at minimum cost. (*Hint:* Have nodes 0, 4, 8, and 12, with c_{ij} being the total cost of shelving all books of height $> i$ and $\leq j$ on a single shelf.)

9 A company sells seven types of boxes, ranging in volume from 17 to 33 cubic feet. The demand and size of each box is given in Table 7. The variable cost (in dollars) of producing each box is equal to the box's volume. A fixed cost of $1,000 is incurred to produce any of a particular box. If the company desires, demand for a box may be satisfied by a box of larger size. Formulate and solve a shortest-path problem whose solution will minimize the cost of meeting the demand for boxes.

10 Explain how by solving a single transshipment problem you can find the shortest path from node 1 in a network to *each other node* in the network.

TABLE **7**

	Box						
	1	2	3	4	5	6	7
Size	33	30	26	24	19	18	17
Demand	400	300	500	700	200	400	200

8.3 Maximum-Flow Problems

Many situations can be modeled by a network in which the arcs may be thought of as having a capacity that limits the quantity of a product that may be shipped through the arc. In these situations, it is often desired to transport the maximum amount of flow from a starting point (called the **source**) to a terminal point (called the **sink**). Such problems are called **maximum-flow problems.** Several specialized algorithms exist to solve maximum-flow problems. In this section, we begin by showing how linear programming can be used to solve a maximum-flow problem. Then we discuss the Ford–Fulkerson (1962) method for solving maximum-flow problems.

LP Solution of Maximum-Flow Problems

EXAMPLE 3   Maximum Flow

Sunco Oil wants to ship the maximum possible amount of oil (per hour) via pipeline from node *so* to node *si* in Figure 6. On its way from node *so* to node *si*, oil must pass through some or all of stations 1, 2, and 3. The various arcs represent pipelines of different diameters. The maximum number of barrels of oil (millions of barrels per hour) that can be pumped through each arc is shown in Table 8. Each number is called an **arc capacity.** Formulate an LP that can be used to determine the maximum number of barrels of oil per hour that can be sent from *so* to *si*.

Solution Node *so* is called the *source* node because oil flows out of it but no oil flows into it. Analogously, node *si* is called the *sink* node because oil flows into it and no oil flows out of it. For reasons that will soon become clear, we have added an artificial arc a_0 from the sink to the source. The flow through a_0 is not actually oil, hence the term **artificial arc.**

To formulate an LP that will yield the maximum flow from node *so* to *si*, we observe that Sunco must determine how much oil (per hour) should be sent through arc (i, j). Thus, we define

x_{ij} = millions of barrels of oil per hour that will pass through arc (i,j) of pipeline

As an example of a possible flow (termed a *feasible flow*), consider the flow indentified by the numbers in parentheses in Figure 6.

$$x_{so,1} = 2, \quad x_{13} = 0, \quad x_{12} = 2, \quad x_{3,si} = 0, \quad x_{2,si} = 2, \quad x_{si,so} = 2, \quad x_{so,2} = 0$$

FIGURE 6
Network for Sunco Oil

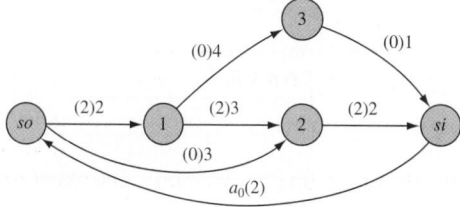

TABLE 8
Arc Capacities for
Sunco Oil

Arc	Capacity
$(so, 1)$	2
$(so, 2)$	3
$(1, 2)$	3
$(1, 3)$	4
$(3, si)$	1
$(2, si)$	2

For a flow to be feasible, it must have two characteristics:

$$0 \leq \text{flow through each arc} \leq \text{arc capacity} \tag{1}$$

and

$$\text{Flow into node } i = \text{flow out of node } i \tag{2}$$

We assume that no oil gets lost while being pumped through the network, so at each node, a feasible flow must satify (2), the *conservation-of-flow* constraint. The introduction of the artificial arc a_0 allows us to write the conservation-of-flow constraint for the source and sink.

If we let x_0 be the flow through the artificial arc, then conservation of flow implies that x_0 = total amount of oil entering the sink. Thus, Sunco's goal is to maximize x_0 subject to (1) and (2):

$$\max z = x_0$$

$$
\begin{aligned}
\text{s.t.} \quad & x_{so,1} \leq 2 && \text{(Arc capacity constraints)} \\
& x_{so,2} \leq 3 \\
& x_{12} \leq 3 \\
& x_{2,si} \leq 2 \\
& x_{13} \leq 4 \\
& x_{3,si} \leq 1 \\
& x_0 = x_{so,1} + x_{so,2} && \text{(Node } so \text{ flow constraint)} \\
& x_{so,1} = x_{12} + x_{13} && \text{(Node 1 flow constraint)} \\
& x_{so,2} + x_{12} = x_{2,si} && \text{(Node 2 flow constraint)} \\
& x_{13} = x_{3,si} && \text{(Node 3 flow constraint)} \\
& x_{3,si} + x_{2,si} = x_0 && \text{(Node } si \text{ flow constraint)} \\
& x_{ij} \geq 0
\end{aligned}
$$

One optimal solution to this LP is $z = 3$, $x_{so,1} = 2$, $x_{13} = 1$, $x_{12} = 1$, $x_{so,2} = 1$, $x_{3,si} = 1$, $x_{2,si} = 2$, $x_0 = 3$. Thus, the maximum possible flow of oil from node so to si is 3 million barrels per hour, with 1 million barrels each sent via the following paths: so–1–2–si, so–1–3–si, and so–2–si.

The linear programming formulation of maximum-flow problems is a special case of the minimum-cost network flow problem (MCNFP) discussed in Section 8.5. A generalization of the transportation simplex (known as the *network simplex*) can be used to solve MCNFPs.

Before discussing the Ford–Fulkerson method for solving maximum-flow problems, we give two examples for situations in which a maximum-flow problem might arise.

EXAMPLE 4   Airline Maximum-Flow

Fly-by-Night Airlines must determine how many connecting flights daily can be arranged between Juneau, Alaska, and Dallas, Texas. Connecting flights must stop in Seattle and then stop in Los Angeles or Denver. Because of limited landing space, Fly-by-Night is limited to making the number of daily flights between pairs of cities shown in Table 9. Set up a maximum-flow problem whose solution will tell the airline how to maximize the number of connecting flights daily from Juneau to Dallas.

TABLE 9
Arc Capacities for Fly-by-Night Airlines

Cities	Maximum Number of Daily Flights
Juneau–Seattle (J, S)	3
Seattle–L.A. (S, L)	2
Seattle–Denver (S, De)	3
L.A.–Dallas (L, D)	1
Denver–Dallas (De, D)	2

FIGURE 7
Network for Fly-by-Night Airlines

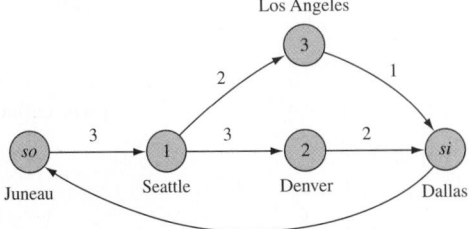

Solution The appropriate network is given in Figure 7. Here the capacity of arc (i, j) is the maximum number of daily flights between city i and city j. The optimal solution to this maximum flow problem is $z = x_0 = 3$, $x_{J,S} = 3$, $x_{S,L} = 1$, $x_{S,De} = 2$, $x_{L,D} = 1$, $x_{De,D} = 2$. Thus, Fly-by-Night can send three flights daily connecting Juneau and Dallas. One flight connects via Juneau–Seattle–L.A.–Dallas, and two flights connect via Juneau–Seattle–Denver–Dallas.

EXAMPLE 5   Matchmaking

Five male and five female entertainers are at a dance. The goal of the matchmaker is to match each woman with a man in a way that maximizes the number of people who are matched with compatible mates. Table 10 describes the compatibility of the entertainers. Draw a network that makes it possible to represent the problem of maximizing the number of compatible pairings as a maximum-flow problem.

Solution Figure 8 is the appropriate network. In Figure 8, there is an arc with capacity 1 joining the source to each man, an arc with capacity 1 joining each pair of compatible mates, and an arc with capacity 1 joining each woman to the sink. The maximum flow in this network is the number of compatible couples that can be created by the matchmaker. For ex-

TABLE 10
Compatibilities for Matching

	Loni Anderson	Meryl Streep	Katharine Hepburn	Linda Evans	Victoria Principal
Kevin Costner	—	C	—	—	—
Burt Reynolds	C	—	—	—	—
Tom Selleck	C	C	—	—	—
Michael Jackson	C	C	—	—	C
Tom Cruise	—	—	C	C	C

Note: C indicates compatibility.

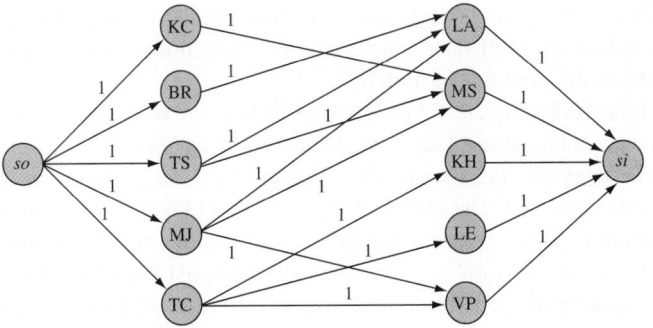

FIGURE 8
Network for
Matchmaker

ample, if the matchmaker pairs KC and MS, BR and LA, MJ and VP, and TC and KH, a flow of 4 from source to sink would be obtained. (This turns out to be a maximum flow for the network.)

To see why our network representation correctly models the matchmaker's problem, note that because the arc joining each woman to the sink has a capacity of 1, conservation of flow ensures that each woman will be matched with at most one man. Similarly, because each arc from the source to a man has a capacity of 1, each man can be paired with at most one woman. Because arcs do not exist between noncompatible mates, we can be sure that a flow of k units from source to sink represents an assignment of men to women in which k compatible couples are created.

Solving Maximum-Flow Problems with LINGO

The maximum flow in a network can be found using LINDO, but LINGO greatly lessens the effort needed to communicate the necessary information to the computer. The following LINGO program (in the file Maxflow.lng) can be used to find the maximum flow from source to sink in Figure 6.

Maxflow.lng

```
MODEL:
  1]SETS:
  2]NODES/1..5/;
  3]ARCS(NODES,NODES)/1,2  1,3  2,3  2,4  3,5  4,5  5,1/
  4]:CAP,FLOW;
  5]ENDSETS
  6]MAX=FLOW (5,1);
  7]@FOR(ARCS(I,J):FLOW(I,J)<CAP(I,J));
  8]@FOR(NODES(I):@SUM(ARCS(J,I):FLOW(J,I))
  9]=@SUM(ARCS(I,J):FLOW(I,J)));
 10]DATA:
 11]CAP=2,3,3,4,2,1,1000;
 12]ENDDATA
END
```

If some nodes are identified by numbers, then LINGO will not allow you to identify other nodes with names involving letters. Thus, we have identified node 1 in line 2 with node *so* in Figure 6 and node 5 in line 2 with node *si*. Also nodes 1, 2, and 3 in Figure 6 correspond to nodes 2, 3, and 4, respectively, in line 2 of our LINGO program. Thus, line 2 defines the nodes of the flow network. In line 3, we define the arcs of the network by listing them (separated by spaces). For example, 1, 2 represents the arc from the source to node 1 in Figure 6 and 5,1 is the artificial arc. In line 4, we indicate that an arc capacity and a flow are associated with each arc. Line 5 ends the definition of the relevant sets.

In line 6, we indicate that our objective is to maximize the flow through the artificial arc (this equals the flow into the sink). Line 7 specifies the arc capacity constraints; for each arc, the flow through the arc cannot exceed the arc's capacity. Lines 8 and 9 create the conservation of flow constraints. For each node I, they ensure that the flow into node I equals the flow out of node I.

Line 10 begins the DATA section. In line 11, we input the arc capacities. Note that we have given the artificial arc a large capacity of 1,000. Line 12 ends the DATA section and the **END** statement ends the program. Typing GO yields the solution, a maximum flow of 3 previously described. The values of the variable FLOW(I,J) give the flow through each arc.

Note that this program can be used to find the maximum flow in any network. Begin by listing the network's nodes in line 2. Then list the network's arcs in line 3. Finally, list the capacity of each arc in the network in line 11, and you are ready to find the maximum flow in the network!

The Ford–Fulkerson Method for Solving Maximum-Flow Problems

We assume that a feasible flow has been found (letting the flow in each arc equal zero gives a feasible flow), and we turn our attention to the following important questions:

Question 1 Given a feasible flow, how can we tell if it is an optimal flow (that is, maximizes x_0)?

Question 2 If a feasible flow is nonoptimal, how can we modify the flow to obtain a new feasible flow that has a larger flow from the source to the sink?

First, we answer question 2. We determine which of the following properties is possessed by each arc in the network:

Property 1 The flow through arc (i, j) is below the capacity of arc (i, j). In this case, the flow through arc (i, j) can be increased. For this reason, we let I represent the set of arcs with this property.

Property 2 The flow in arc (i, j) is positive. In this case, the flow through arc (i, j) can be reduced. For this reason, we let R be the set of arcs with this property.

As an illustration of the definitions of I and R, consider the network in Figure 9. The arcs in this figure may be classified as follows: $(so, 1)$ is in I and R; $(so, 2)$ is in I; $(1, si)$ is in R; $(2, si)$ is in I; and $(2, 1)$ is in I.

We can now describe the Ford–Fulkerson labeling procedure used to modify a feasible flow in an effort to increase the flow from the source to the sink.

Step 1 Label the source.

Step 2 Label nodes and arcs (except for arc a_0) according to the following rules: (1) If node x is labeled, then node y is unlabeled and arc (x, y) is a member of I; then label node y and arc (x, y). In this case, arc (x, y) is called a **forward arc.** (2) If node y is unlabeled, node x is labeled and arc (y, x) is a member of R; label node y and arc (y, x). In this case, (y, x) is called a **backward arc.**

FIGURE 9
Illustration of I and R arcs

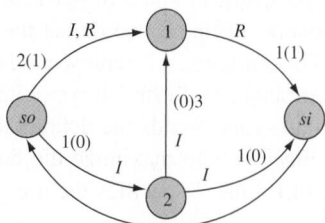

Step 3 Continue this labeling process until the sink has been labeled or until no more vertices can be labeled.

If the labeling process results in the sink being labeled, then there will be a chain of labeled arcs (call it C) leading from the source to the sink. By adjusting the flow of the arcs in C, we can maintain a feasible flow and increase the total flow from source to sink. To see this, observe that C must consist of one of the following:

Case 1 C consists entirely of forward arcs.

Case 2 C contains both forward and backward arcs.[†]

In each case, we can obtain a new feasible flow that has a larger flow from source to sink than the current feasible flow. In Case 1, the chain C consists entirely of forward arcs. For each forward arc in C, let $i(x, y)$ be the amount by which the flow in arc (x, y) can be increased without violating the capacity constraint for arc (x, y) . Let

$$k = \min_{(x, y)\in C} i(x, y)$$

Then $k > 0$. To create a new flow, increase the flow through each arc in C by k units. No capacity constraints are violated, and conservation of flow is still maintained. Thus, the new flow is feasible, and the new feasible flow will transport k more units from source to sink than does the current feasible flow.

We use Figure 10 to illustrate Case 1. Currently, 2 units are being transported from source to sink. The labeling procedure results in the sink being labeled by the chain $C = (so, 1) - (1, 2) - (2, si)$. Each arc is in I, and $i(so, 1) = 5 - 2 = 3; i(1, 2) = 3 - 2 = 1$; and $i(2, si) = 4 - 2 = 2$. Hence, $k = \min(3, 1, 2) = 1$. Thus, an improved feasible flow can be obtained by increasing the flow on each arc in C by 1 unit. The resulting flow transports 3 units from source to sink (see Figure 11).

In Case 2, the chain C leading from the source to the sink contains both backward and forward arcs. For each backward arc in C, let $r(x, y)$ be the amount by which the flow through arc (x, y) can be reduced. Also define

$$k_1 = \min_{x, y\in C\cap R} r(x, y) \quad \text{and} \quad k_2 = \min_{x, y\in C\cap I} i(x, y)$$

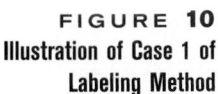

FIGURE 10
Illustration of Case 1 of Labeling Method

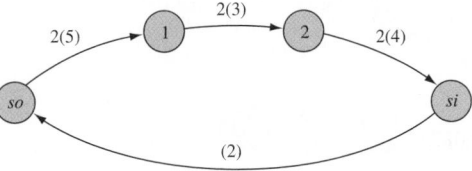

Flow from source to sink = 2
Chain is $(so, 1) - (1, 2) - (2, si)$

FIGURE 11
Improved Flow from Source to Sink: Case 1

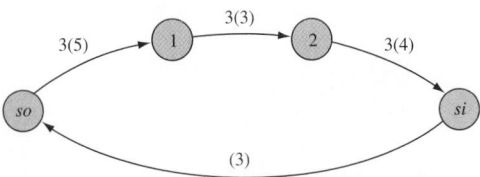

Flow from source to sink = 3

[†]Because we exclude arc a_0 from the labeling procedure, no chain made entirely of backward arcs can lead from source to sink.

Of course, both k_1 and k_2 and min (k_1, k_2) are > 0. To increase the flow from source to sink (while maintaining a feasible flow), decrease the flow in all of C's backward arcs by min (k_1, k_2) and increase the flow in all of C's forward arcs by min(k_1, k_2). This will maintain conservation of flow and ensure that no arc capacity constraints are violated. Because the last arc in C is a forward arc leading into the sink, we have found a new feasible flow and have increased the total flow into the sink by min(k_1, k_2). We now adjust the flow in the arc a_0 to maintain conservation of flow. To illustrate Case 2, suppose we have found the feasible flow in Figure 12. For this flow, $(so, 1) \in R$; $(so, 2) \in I$; $(1, 3) \in I$; $(1, 2) \in I$ and R; $(2, si) \in R$; and $(3, si) \in I$.

We begin by labeling arc $(so, 2)$ and node 2 (thus $(so, 2)$ is a forward arc). Then we label arc $(1, 2)$ and node 1. Arc $(1, 2)$ is a backward arc, because node 1 was unlabeled before we labeled arc $(1, 2)$, and arc $(1, 2)$ is in R. Nodes so, 1, and 2 are labeled, so we can label arc $(1, 3)$ and node 3. [Arc $(1, 3)$ is a forward arc, because node 3 has not yet been labeled.] Finally we label arc $(3, si)$ and node si. Arc $(3, si)$ is a forward arc, because node si has not yet been labeled. We have now labeled the sink via the chain $C = (so, 2) - (1, 2) - (1, 3) - (3, si)$. With the exception of arc $(1, 2)$, all arcs in the chain are forward arcs. Because $i(so, 2) = 3$; $i(1, 3) = 4$; $i(3, si) = 1$; and $r(1, 2) = 2$, we have

$$\min_{(x, y) \in C \cap R} r(x, y) = 2 \quad \text{and} \quad \min_{(x, y) \in C \cap I} i(x, y) = 1$$

Thus, we can increase the flow on all forward arcs in C by 1 and decrease the flow in all backward arcs by 1. The new result, pictured in Figure 13, has increased the flow from source to sink by 1 unit (from 2 to 3). We accomplish this by diverting 1 unit that was transported through the arc $(1, 2)$ to the path $1-3-si$. This enabled us to transport an extra unit from source to sink via the path $so-2-si$. Observe that the concept of a backward arc was needed to find this improved flow.

If the sink cannot be labeled, then the current flow is optimal. The proof of this fact relies on the concept of a *cut* for a network.

DEFINITION ■ Choose any set of nodes V' that contains the sink but does not contain the source. Then the set of arcs (i, j) with i not in V' and j a member of V' is a **cut** for the network. ■

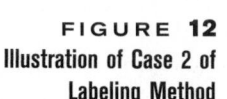

FIGURE 12
Illustration of Case 2 of Labeling Method

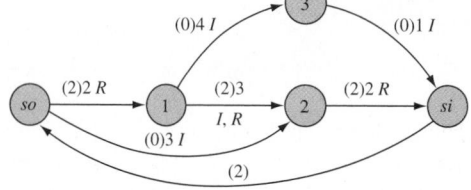

Flow from source to sink = 2
Chain is $(so, 2) - (1, 2) - (1, 3) - (3, si)$

FIGURE 13
Improved Flow from Source to Sink: Case 2

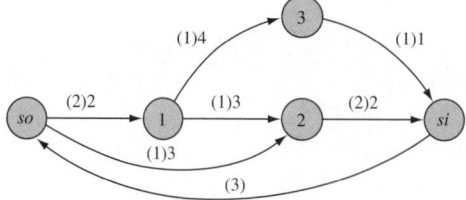

Flow from source to sink = 3

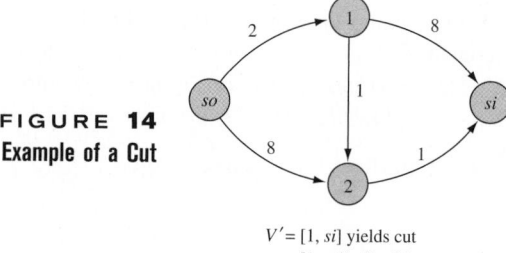

FIGURE 14
Example of a Cut

$V' = [1, si]$ yields cut
$[(so, 1), (2, si)]$

The **capacity** of a cut is the sum of the capacities of the arcs in the cut.   ■

In short, a cut is a set of arcs whose removal from the network makes it impossible to travel from the source to the sink. A network may have many cuts. For example, in the network in Figure 14, $V' = \{1, si\}$ yields the cut containing the arcs $(so, 1)$ and $(2, si)$, which has capacity $2 + 1 = 3$. The set $V' = \{1, 2, si\}$ yields the cut containing the arcs $(so, 1)$ and $(so, 2)$, which has capacity $2 + 8 = 10$.

Lemma 1 and Lemma 2 indicate the connection between cuts and maximum flows.

LEMMA 1

The flow from source to sink for any feasible flow is less than or equal to the capacity of *any* cut.

Proof   Consider an arbitrary cut specified by a set of nodes V' that contains the sink but does not contain the source. Let V be all other nodes in the network. Also let x_{ij} be the flow in arc (i, j) for any feasible flow and f be the flow from source to sink for this feasible flow. Summing the flow balance equations (flow out of node i − flow into node $i = 0$) over all nodes i in V, we find that the terms involving arcs (i, j) having i and j both members of V will cancel, and we obtain

$$\sum_{\substack{i \in V; \\ j \in V'}} x_{ij} - \sum_{\substack{i \in V'; \\ j \in V}} x_{ij} = f \tag{3}$$

Now the first sum in (3) equals the capacity of the cut. Each x_{ij} is nonnegative, so we see that $f \leq$ capacity of the cut, which is the desired result.

Lemma 1 is analogous to the weak duality result discussed in Chapter 6. From Lemma 1, we see that the capacity of any cut is an upper bound for the maximum flow from source to sink. Thus, if we can find a feasible flow and a cut for which the flow from source to sink equals the capacity of the cut, then we have found the maximum flow from source to sink.

Suppose that we find a feasible flow and cannot label the sink. Let CUT be the cut corresponding to the set of unlabeled nodes.

LEMMA 2

If the sink cannot be labeled, then

Capacity of CUT = current flow from source to sink

Proof   Let V' be the set of unlabeled nodes and V be the set of labeled nodes. Consider an arc (i, j) such that i is in V and j is in V'. Then we know that $x_{ij} =$ capacity of arc (i, j) must hold; otherwise, we could label node j (via a forward arc) and node j would not be in V'. Now consider an arc (i, j) such that i is in V' and j is in V. Then $x_{ij} = 0$ must hold; otherwise, we could label node i (via a backward arc) and node i would not be in V'. Now (3) shows that the current flow must satisfy

Capacity of CUT = current flow from source to sink

which is the desired result.

From the remarks following Lemma 1, when the sink cannot be labeled, the maximum flow from source to sink has been obtained.

Summary and Illustration of the Ford–Fulkerson Method

Step 1   Find a feasible flow (setting each arc's flow to zero will do).

Step 2   Using the labeling procedure, try to label the sink. If the sink cannot be labeled, then the current feasible flow is a maximum flow; if the sink is labeled, then go on to step 3.

Step 3   Using the method previously described, adjust the feasible flow and increase the flow from the source to the sink. Return to step 2.

To illustrate the Ford–Fulkerson method, we find the maximum flow from source to sink for Sunco Oil, Example 3 (see Figure 6). We begin by letting the flow in each arc equal zero. We then try to label the sink—label the source, and then arc $(so, 1)$ and node 1; then label arc $(1, 2)$ and node 2; finally, label arc $(2, si)$ and node si. Thus, $C =$ $(so, 1)–(1, 2)–(2, si)$. Each arc in C is a forward arc, so we can increase the flow through each arc in C by min $(2, 3, 2) = 2$ units. The resulting flow is pictured in Figure 15.

As we saw previously (Figure 12), we can label the sink by using the chain $C =$ $(so, 2)–(1, 2)–(1, 3)–(3, si)$. We can increase the flow through the forward arcs $(so, 2)$, $(1, 3)$, and $(3, si)$ by 1 unit and decrease the flow through the backward arc $(1, 2)$ by 1 unit. The resulting flow is pictured in Figure 16. It is now impossible to label the sink. Any attempt to label the sink must begin by labeling arc $(so, 2)$ and node 2; then we could label arc $(1, 2)$ and arc $(1, 3)$. But there is no way to label the sink.

We can verify that the current flow is maximal by finding the capacity of the cut corresponding to the set of unlabeled vertices (in this case, si). The cut corresponding to si is the set of arcs $(2, si)$ and $(3, si)$, with capacity $2 + 1 = 3$. Thus, Lemma 1 implies that any feasible flow can transport at most 3 units from source to sink. Our current flow transports 3 units from source to sink, so it must be an optimal flow.

Another example of the Ford–Fulkerson method is given in Figure 17. Note that without the concept of a backward arc, we could not have obtained the maximum flow of 7

FIGURE **15**
Network for Sunco Oil
(Increased Flow)

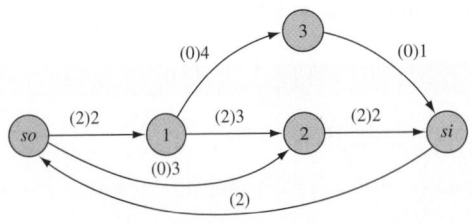

Flow from source to sink = 2
Label sink by $(so, 2) – (1, 2) – (1, 3) – (3, si)$

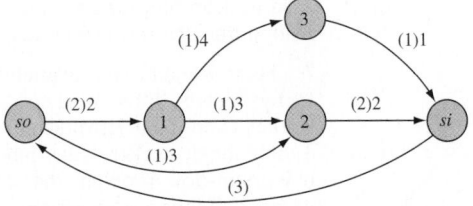

FIGURE **16**
Network for Sunco Oil
(Optimal Flow)

Flow from source to sink = 3
Since sink cannot be labeled, this is an optimal flow

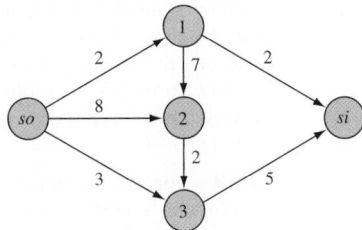

a Original network

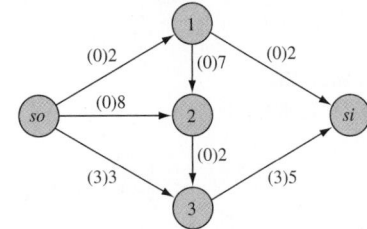

b Label sink by $so - 3 - si$ (adds 3 units
of flow using only forward arcs)

FIGURE **17**
Example of
Ford–Fulkerson Method

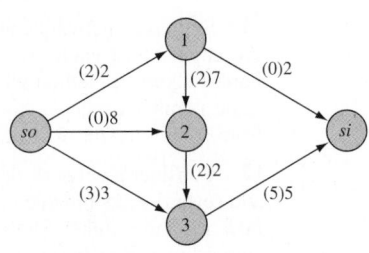

c Label sink by $so - 1 - 2 - 3 - si$ (adds 2 units
of flow using only forward arcs)

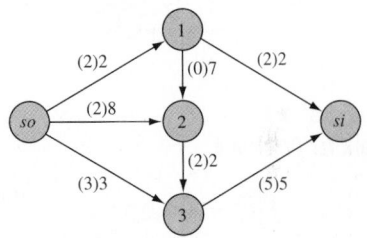

d Label sink by $so - 2 - 1 - si$ (adds 2 units
of flow using backward arc $(1, 2)$;
maximum flow of 7 has been obtained)

units from source to sink. The minimum cut (with capacity 7, of course) corresponds to
nodes 1, 3, and si and consists of arcs $(so, 1)$, $(so, 3)$ and $(2, 3)$.

PROBLEMS

Group A

1–3 Figures 18–20 show the networks for Problems 1–3.
Find the maximum flow from source to sink in each network.
Find a cut in the network whose capacity equals the
maximum flow in the network. Also, set up an LP that could
be used to determine the maximum flow in the network.

FIGURE **18**
Network for Problem 1

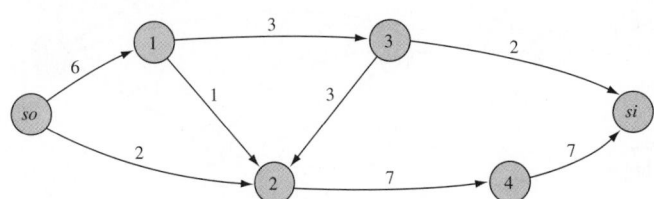

FIGURE **19**
Network for Problem 2

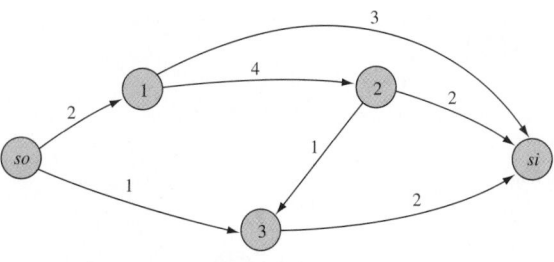

FIGURE **20**
Nework for Problem 3

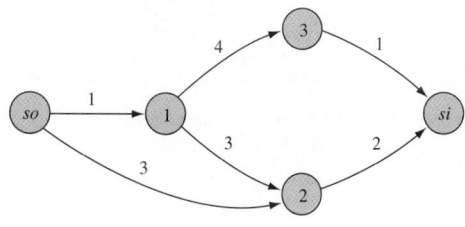

FIGURE **21**
Network for Problem 4

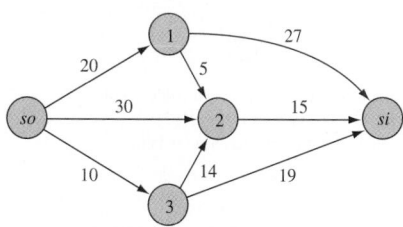

FIGURE **22**
Network for Problem 5

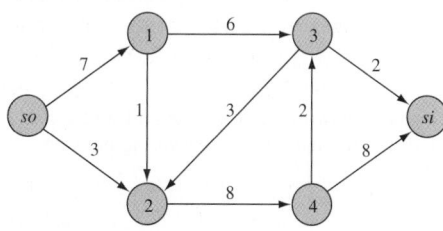

4–5 For the networks in Figures 21 and 22, find the maximum flow from source to sink. Also find a cut whose capacity equals the maximum flow in the network.

6 Seven types of packages are to be delivered by five trucks. There are three packages of each type, and the capacities of the five trucks are 6, 4, 5, 4, and 3 packages, respectively. Set up a maximum-flow problem that can be used to determine whether the packages can be loaded so that no truck carries two packages of the same type.

7 Four workers are available to perform jobs 1–4. Unfortunately, three workers can do only certain jobs: worker 1, only job 1; worker 2, only jobs 1 and 2; worker 3, only job 2; worker 4, any job. Draw the network for the maximum-flow problem that can be used to determine whether all jobs can be assigned to a suitable worker.

8 The Hatfields, Montagues, McCoys, and Capulets are going on their annual family picnic. Four cars are available to transport the families to the picnic. The cars can carry the following number of people: car 1, four; car 2, three; car 3, three; and car 4, four. There are four people in each family, and no car can carry more than two people from any one family. Formulate the problem of transporting the maximum possible number of people to the picnic as a maximum-flow problem.

9–10 For the networks in Figures 23 and 24, find the maximum flow from source to sink. Also find a cut whose capacity equals the maximum flow in the network.

Group B

11 Suppose a network contains a finite number of arcs and the capacity of each arc is an integer. Explain why the Ford–Fulkerson method will find the maximum flow in the finite number of steps. Also show that the maximum flow from source to sink will be an integer.

12 Consider a network flow problem with several sources and several sinks in which the goal is to maximize the total flow into the sinks. Show how such a problem can be converted into a maximum-flow problem having only a single source and a single sink.

FIGURE **23**

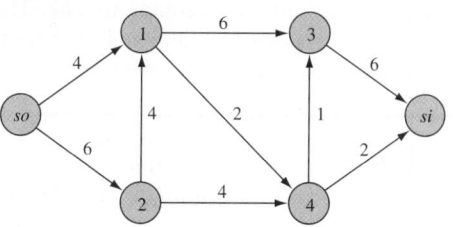

FIGURE **24**

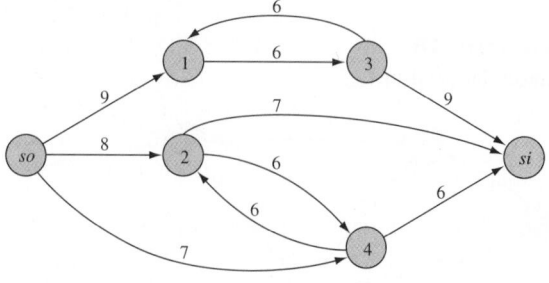

13 Suppose the total flow into a node of a network is restricted to 10 units or less. How can we represent this restriction via an arc capacity constraint? (This still allows us to use the Ford–Fulkerson method to find the maximum flow.)

14 Suppose as many as 300 cars per hour can travel between any two of the cities 1, 2, 3, and 4. Set up a maximum-flow problem that can be used to determine how many cars can be sent in the next two hours from city 1 to city 4. (*Hint:* Have portions of the network represent $t = 0$, $t = 1$, and $t = 2$.)

15 Fly-by-Night Airlines is considering flying three flights. The revenue from each flight and the airports used by each flight are shown in Table 11. When Fly-by-Night uses an airport, the company must pay the following landing fees (independent of the number of flights using the airport): airport 1, $300; airport 2, $700; airport 3, $500. Thus, if flights 1 and 3 are flown, a profit of $900 + 800 - 300 - 700 - 500 = \$200$ will be earned. Show that for the network in Figure 25 (maximum profit) = (total revenue from all flights) − (capacity of minimal cut). Explain how this result can be used to help Fly-by-Night maximize profit (even if it has hundreds of possible flights). (*Hint:* Consider any set of flights F (say, flights 1 and 3). Consider the cut corresponding

TABLE 11

Flight	Revenue ($)	Airport Used
1	900	1 and 2
2	600	2
3	800	2 and 3

to the sink, the nodes associated with the flights not in F, and the nodes associated with the airports not used by F. Show that (capacity of this cut) = (revenue from flights not in F) + (costs associated with airports used by F).)

16 During the next four months, a construction firm must complete three projects. Project 1 must be completed within three months and requires 8 months of labor. Project 2 must be completed within four months and requires 10 months of labor. Project 3 must be completed at the end of two months and requires 12 months of labor. Each month, 8 workers are available. During a given month, no more than 6 workers can work on a single job. Formulate a maximum-flow problem that could be used to determine whether all three projects can be completed on time. (*Hint:* If the maximum flow in the network is 30, then all projects can be completed on time.)

FIGURE 25
Network for Problem 15

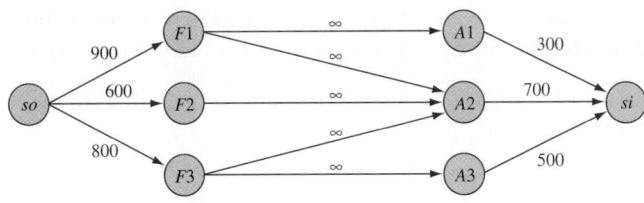

8.4 CPM and PERT

Network models can be used as an aid in scheduling large complex projects that consist of many activities. If the duration of each activity is known with certainty, then the **critical path method (CPM)** can be used to determine the length of time required to complete a project. CPM also can be used to determine how long each activity in the project can be delayed without delaying the completion of the project. CPM was developed in the late 1950s by researchers at DuPont and Sperry Rand.

If the duration of the activities is not known with certainty, the Program Evaluation and Review Technique (PERT) can be used to estimate the probability that the project will be completed by a given deadline. PERT was developed in the late 1950s by consultants working on the development of the Polaris missile. CPM and PERT were given a major share of the credit for the fact that the Polaris missile was operational two years ahead of schedule.

CPM and PERT have been successfully used in many applications, including:

1 Scheduling construction projects such as office buildings, highways, and swimming pools

2 Scheduling the movement of a 400-bed hospital from Portland, Oregon, to a suburban location

3 Developing a countdown and "hold" procedure for the launching of space flights

4 Installing a new computer system

5 Designing and marketing a new product

6 Completing a corporate merger

7 Building a ship

To apply CPM and PERT, we need a list of the activities that make up the project. The project is considered to be completed when all the activities have been completed. For each activity, there is a set of activities (called the **predecessors** of the activity) that must be completed before the activity begins. A project network is used to represent the precedence relationships between activities. In our discussion, activities will be represented by directed arcs, and nodes will be used to represent the completion of a set of activities. (For this reason, we often refer to the nodes in our project network as **events.**) This type of project network is called an **AOA (activity on arc)** network.[†]

To understand how an AOA network represents precedence relationships, suppose that activity A is a predecessor of activity B. Each node in an AOA network represents the completion of one or more activities. Thus, node 2 in Figure 26 represents the completion of activity A and the beginning of activity B. Suppose activities A and B must be completed before activity C can begin. In Figure 27, node 3 represents the event that activities A and B are completed. Figure 28 shows activity A as a predecessor of both activities B and C.

Given a list of activities and predecessors, an AOA representation of a project (called a **project network** or **project diagram**) can be constructed by using the following rules:

1 Node 1 represents the start of the project. An arc should lead from node 1 to represent each activity that has no predecessors.

FIGURE 26
Activity A Must Be Completed Before Activity B Can Begin

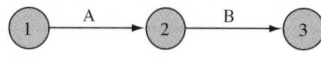

FIGURE 27
Activities A and B Must Be Completed Before Activity C Can Begin

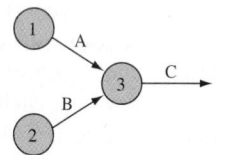

FIGURE 28
Activity A Must Be Completed Before Activities B and C Can Begin

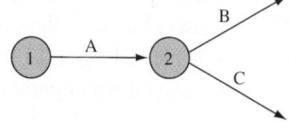

[†]In an AON (activity on node) project network, the nodes of the network are used to represent activities. See Wiest and Levy (1977) for details.

FIGURE **29**
Violation of Rule 5

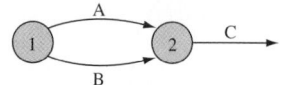

FIGURE **30**
Use of Dummy Activity

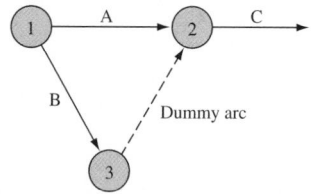

2 A node (called the **finish node**) representing the completion of the project should be included in the network.

3 Number the nodes in the network so that the node representing the completion of an activity always has a larger number than the node representing the beginning of an activity (there may be more than one numbering scheme that satisfies rule 3).

4 An activity should not be represented by more than one arc in the network.

5 Two nodes can be connected by at most one arc.

To avoid violating rules 4 and 5, it is sometimes necessary to utilize a **dummy activity** that takes zero time. For example, suppose activities A and B are both predecessors of activity C and can begin at the same time. In the absence of rule 5, we could represent this by Figure 29. However, because nodes 1 and 2 are connected by more than one arc, Figure 29 violates rule 5. By using a dummy activity (indicated by a dotted arc), as in Figure 30, we may represent the fact that A and B are both predecessors of C. Figure 30 ensures that activity C cannot begin until both A and B are completed, but it does not violate rule 5. Problem 10 at the end of this section illustrates how dummy activities may be needed to avoid violating rule 4.

Example 6 illustrates a project network.

EXAMPLE 6	**Drawing a Project Network**

Widgetco is about to introduce a new product (product 3). One unit of product 3 is produced by assembling 1 unit of product 1 and 1 unit of product 2. Before production begins on either product 1 or 2, raw materials must be purchased and workers must be trained. Before products 1 and 2 can be assembled into product 3, the finished product 2 must be inspected. A list of activities and their predecessors and of the duration of each activity is given in Table 12. Draw a project diagram for this project.

TABLE 12

Duration of Activities and Predecessor Relationships for Widgetco

Activity	Predecessors	Duration (Days)
A = train workers	—	6
B = purchase raw materials	—	9
C = produce product 1	A, B	8
D = produce product 2	A, B	7
E = test product 2	D	10
F = assemble products 1 and 2	C, E	12

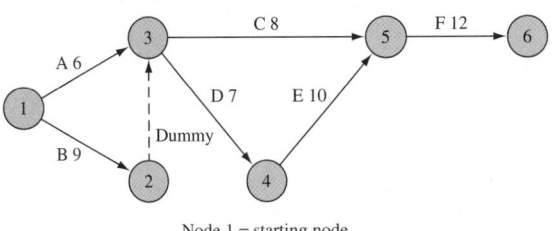

FIGURE 31
Project Diagram for
Widgetco

Node 1 = starting node
Node 6 = finish node

Solution Observe that although we list only C and E as predecessors of F, it is actually true that activities A, B, and D must also be completed before F begins. C cannot begin until A and B are completed, and E cannot begin until D is completed, however, so it is redundant to state that A, B, and D are predecessors of F. Thus, in drawing the project network, we need only be concerned with the immediate predecessors of each activity.

The AOA network for this project is given in Figure 31 (the number above each arc represents activity duration in days). Node 1 is the beginning of the project, and node 6 is the finish node representing completion of the project. The dummy arc (2, 3) is needed to ensure that rule 5 is not violated.

The two key building blocks in CPM are the concepts of early event time (ET) and late event time (LT) for an event.

DEFINITION ■ The **early event time** for node i, represented by $ET(i)$, is the earliest time at which the event corresponding to node i can occur. ■

The **late event time** for node i, represented by $LT(i)$, is the latest time at which the event corresponding to node i can occur without delaying the completion of the project. ■

Computation of Early Event Time

To find the early event time for each node in the project network, we begin by noting that because node 1 represents the start of the project, $ET(1) = 0$. We then compute $ET(2)$, $ET(3)$, and so on, stopping when ET(finish node) has been calculated. To illustrate how $ET(i)$ is calculated, suppose that for the segment of a project network in Figure 32, we have already determined that $ET(3) = 6$, $ET(4) = 8$, and $ET(5) = 10$. To determine $ET(6)$, observe that the earliest time that node 6 can occur is when the activities corresponding to arc (3, 6), (4, 6), and (5, 6) have *all* been completed.

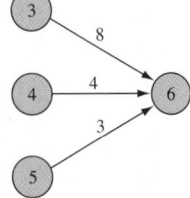

FIGURE 32
Determination of $ET(6)$

$$ET(6) = \max \begin{cases} ET(3) + 8 = 14 \\ ET(4) + 4 = 12 \\ ET(5) + 3 = 13 \end{cases}$$

Thus, the earliest time that node 6 can occur is 14, and $ET(6) = 14$.

From this example, it is clear that computation of $ET(i)$ requires (for $j < i$) knowledge of one or more of the $ET(j)$'s. This explains why we begin by computing the predecessor ETs. In general, if $ET(1)$, $ET(2)$, ... , $ET(i - 1)$ have been determined, then we compute $ET(i)$ as follows:

Step 1 Find each prior event to node i that is connected by an arc to node i. These events are the **immediate predecessors** of node i.

Step 2 To the ET for each immediate predecessor of the node i add the duration of the activity connecting the immediate predecessor to node i.

Step 3 $ET(i)$ equals the maximum of the sums computed in step 2.

We now compute the $ET(i)$'s for Example 6. We begin by observing that $ET(1) = 0$. Node 1 is the only immediate predecessor of node 2, so $ET(2) = ET(1) + 9 = 9$. The immediate predecessors of node 3 are nodes 1 and 2. Thus,

$$ET(3) = \max \begin{cases} ET(1) + 6 = 6 \\ ET(2) + 0 = 9 \end{cases} = 9$$

Node 4's only immediate predecessor is node 3. Thus, $ET(4) = ET(3) + 7 = 16$. Node 5's immediate predecessors are nodes 3 and 4. Thus,

$$ET(5) = \max \begin{cases} ET(3) + 8 = 17 \\ ET(4) + 10 = 26 \end{cases} = 26$$

Finally, node 5 is the only immediate predecessor of node 6. Thus, $ET(6) = ET(5) + 12 = 38$. Because node 6 represents the completion of the project, we see that the earliest time that product 3 can be assembled is 38 days from now.

It can be shown that $ET(i)$ is the length of the longest path in the project network from node 1 to node i.

Computation of Late Event Time

To compute the $LT(i)$'s, we begin with the finish node and work backward (in descending numerical order) until we determine $LT(1)$. The project in Example 6 can be completed in 38 days, so we know that $LT(6) = 38$. To illustrate how $LT(i)$ is computed for nodes other than the finish node, suppose we are working with a network (Figure 33) for which we have already determined that $LT(5) = 24$, $LT(6) = 26$, and $LT(7) = 28$. In this situation, how can we compute $LT(4)$? If the event corresponding to node 4 occurs after $LT(5) - 3$, node 5 will occur after $LT(5)$, and the completion of the project will be delayed.

FIGURE 33
Computation of $LT(4)$

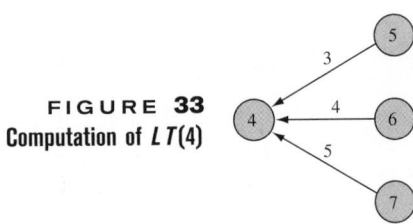

Similarly, if node 4 occurs after $LT(6) - 4$ or if node 4 occurs after $LT(7) - 5$, the completion of the project will be delayed. Thus,

$$LT(4) = \min \begin{cases} LT(5) - 3 = 21 \\ LT(6) - 4 = 22 = 21 \\ LT(7) - 5 = 23 \end{cases}$$

In general, if $LT(j)$ is known for $j > i$, we can find $LT(i)$ as follows:

Step 1   Find each node that occurs after node i and is connected to node i by an arc. These events are the **immediate successors** of node i.

Step 2   From the LT for each immediate successor to node i, subtract the duration of the activity joining the successor the node i.

Step 3 $LT(i)$ is the smallest of the differences determined in step 2.

We now compute the $LT(i)$'s for Example 6. Recall that $LT(6) = 38$. Because node 6 is the only immediate successor of node 5, $LT(5) = LT(6) - 12 = 26$. Node 4's only immediate successor is node 5. Thus, $LT(4) = LT(5) - 10 = 16$. Nodes 4 and 5 are immediate successors of node 3. Thus,

$$LT(3) = \min \begin{cases} LT(4) - 7 = 9 \\ LT(5) - 8 = 18 \end{cases}$$

Node 3 is the only immediate successor of node 2. Thus, $LT(2) = LT(3) - 0 = 9$. Finally, node 1 has nodes 2 and 3 as immediate successors. Thus,

$$LT(1) = \min \begin{cases} LT(3) - 6 = 3 \\ LT(2) - 9 = 0 \end{cases}$$

Table 13 summarizes our computations for Example 6. If $LT(i) = ET(i)$, any delay in the occurrence of node i will delay the completion of the project. For example, because $LT(4) = ET(4)$, any delay in the occurrence of node 4 will delay the completion of the project.

Total Float

Before the project is begun, the duration of an activity is unknown, and the duration of each activity used to construct the project network is just an estimate of the activity's actual completion time. The concept of total float of an activity can be used as a measure of how important it is to keep each activity's duration from greatly exceeding our estimate of its completion time.

TABLE 13
ET and *LT* for Widgetco

Node	ET(i)	LT(i)
1	0	0
2	9	9
3	9	9
4	16	16
5	26	26
6	38	38

For an arbitrary arc representing activity (i, j), the **total float,** represented by $TF(i, j)$, of the activity represented by (i, j) is the amount by which the starting time of activity (i, j) could be delayed beyond its earliest possible starting time without delaying the completion of the project (assuming no other activities are delayed). ■

Equivalently, the total float of an activity is the amount by which the duration of the activity can be increased without delaying the completion of the project.

If we define t_{ij} to be the duration of activity (i, j), then $TF(i, j)$ can easily be expressed in terms of $LT(j)$ and $ET(i)$. Activity (i, j) begins at node i. If the occurrence of node i, or the duration of activity (i, j), is delayed by k time units, then activity (i, j) will be completed at time $ET(i) + k + t_{ij}$. Thus, the completion of the project will not be delayed if

$$ET(i) + k + t_{ij} \leq LT(j) \qquad \text{or} \qquad k \leq LT(j) - ET(i) - t_{ij}$$

Therefore,

$$TF(i, j) = LT(j) - ET(i) - t_{ij}$$

For Example 6, the $TF(i, j)$ are as follows:

Activity B: $TF(1, 2) = LT(2) - ET(1) - 9 = 0$

Activity A: $TF(1, 3) = LT(3) - ET(1) - 6 = 3$

Activity D: $TF(3, 4) = LT(4) - ET(3) - 7 = 0$

Activity C: $TF(3, 5) = LT(5) - ET(3) - 8 = 9$

Activity E: $TF(4, 5) = LT(5) - ET(4) - 10 = 0$

Activity F: $TF(5, 6) = LT(6) - ET(5) - 12 = 0$

Dummy activity: $TF(2, 3) = LT(3) - ET(2) - 0 = 0$

Finding a Critical Path

If an activity has a total float of zero, then any delay in the start of the activity (or the duration of the activity) will delay the completion of the project. In fact, increasing the duration of an activity by Δ days will increase the length of the project by Δ days. Such an activity is critical to the completion of the project on time.

Any activity with a total float of zero is a **critical activity.** ■

A path from node 1 to the finish node that consists entirely of critical activities is called a **critical path.** ■

In Figure 31, activities B, D, E, F, and the dummy activity are critical activities and the path 1–2–3–4–5–6 is the critical path (it is possible for a network to have more than one critical path). A critical path in any project network is the longest path from the start node to the finish node (see Problem 2 in Section 8.5).

Any delay in the duration of a critical activity will delay the completion of the project, so it is advisable to monitor closely the completion of critical activities.

Free Float

As we have seen, the total float of an activity can be used as a measure of the flexibility in the duration of an activity. For example, activity A can take up to 3 days longer than its scheduled duration of 6 days without delaying the completion of the project. Another measure of the flexibility available in the duration of an activity is free float.

The **free float** of the activity corresponding to arc (i, j), denoted by $FF(i, j)$, is the amount by which the starting time of the activity corresponding to arc (i, j) (or the duration of the activity) can be delayed without delaying the start of any later activity beyond its earliest possible starting time. ■

Suppose the occurrence of node i, or the duration of activity (i, j), is delayed by k units. Then the earliest that node j can occur is $ET(i) + t_{ij} + k$. Thus, if $ET(i) + t_{ij} + k \leq ET(j)$, or $k \leq ET(j) - ET(i) - t_{ij}$, then node j will not be delayed. If node j is not delayed, then no other activities will be delayed beyond their earliest possible starting times. Therefore,

$$FF(i, j) = ET(j) - ET(i) - t_{ij}$$

For Example 6, the $FF(i, j)$ are as follows:

$$\begin{aligned}
\text{Activity B:} \quad & FF(1, 2) = 9 - 0 - 9 = 0 \\
\text{Activity A:} \quad & FF(1, 3) = 9 - 0 - 6 = 3 \\
\text{Activity D:} \quad & FF(3, 4) = 16 - 9 - 7 = 0 \\
\text{Activity C:} \quad & FF(3, 5) = 26 - 9 - 8 = 9 \\
\text{Activity E:} \quad & FF(4, 5) = 26 - 16 - 10 = 0 \\
\text{Activity F:} \quad & FF(5, 6) = 38 - 26 - 12 = 0
\end{aligned}$$

For example, because the free float for activity C is 9 days, a delay in the start of activity C (or in the occurrence of node 3) or a delay in the duration of activity C of more than 9 days will delay the start of some later activity (in this case, activity F).

Using Linear Programming to Find a Critical Path

Although the previously described method for finding a critical path in a project network is easily programmed on a computer, linear programming can also be used to determine the length of the critical path. Define

$$x_j = \text{the time that the event corresponding to node } j \text{ occurs}$$

For each activity (i, j), we know that before node j occurs, node i must occur and activity (i, j) must be completed. This implies that for each arc (i, j) in the project network, $x_j \geq x_i + t_{ij}$. Let F be the node that represents completion of the project. Our goal is to minimize the time required to complete the project, so we use an objective function of $z = x_F - x_1$.

To illustrate how linear programming can be used to find the length of the critical path, we apply the preceding approach to Example 6. The appropriate LP is

$$\begin{aligned}
\min z = x_6 - x_1 \\
\text{s.t.} \quad x_3 &\geq x_1 + 6 \quad &&\text{(Arc (1, 3) constraint)} \\
x_2 &\geq x_1 + 9 \quad &&\text{(Arc (1, 2) constraint)} \\
x_5 &\geq x_3 + 8 \quad &&\text{(Arc (3, 5) constraint)} \\
x_4 &\geq x_3 + 7 \quad &&\text{(Arc (3, 4) constraint)}
\end{aligned}$$

$$x_5 \geq x_4 + 10 \quad \text{(Arc (4, 5) constraint)}$$
$$x_6 \geq x_5 + 12 \quad \text{(Arc (5, 6) constraint)}$$
$$x_3 \geq x_2 \quad \text{(Arc (2, 3) constraint)}$$

All variables urs

An optimal solution to this LP is $z = 38$, $x_1 = 0$, $x_2 = 9$, $x_3 = 9$, $x_4 = 16$, $x_5 = 26$, and $x_6 = 38$. This indicates that the project can be completed in 38 days.

This LP has many alternative optimal solutions. In general, the value of x_i in any optimal solution may assume any value between $ET(i)$ and $LT(i)$. All optimal solutions to this LP, however, will indicate that the length of any critical path is 38 days.

A critical path for this project network consists of a path from the start of the project to the finish in which each arc in the path corresponds to a constraint having a dual price of -1. From the LINDO output in Figure 34, we find, as before, that 1–2–3–4–5–6 is a critical path. For each constraint with a dual price of -1, increasing the duration of the activity corresponding to that constraint by Δ days will increase the duration of the project by Δ days. For example, an increase of Δ days in the duration of activity B will increase the duration of the project by Δ days. This assumes that the current basis remains optimal.

Crashing the Project

In many situations, the project manager must complete the project in a time that is less than the length of the critical path. For instance, suppose Widgetco believes that to have any chance of being a success, product 3 must be available for sale before the competitor's product hits the market. Widgetco knows that the competitor's product is scheduled to hit the market 26 days from now, so Widgetco must introduce product 3 within 25 days. Because the critical path in Example 6 has a length of 38 days, Widgetco will have to expend additional resources to meet the 25-day project deadline. In such a situation, linear programming can often be used to determine the allocation of resources that minimizes the cost of meeting the project deadline.

Suppose that by allocating additional resources to an activity, Widgetco can reduce the duration of any activity by as many as 5 days. The cost per day of reducing the duration of an activity is shown in Table 14. To find the minimum cost of completing the project by the 25-day deadline, define variables A, B, C, D, E, and F as follows:

A = number of days by which duration of activity A is reduced

$\vdots \qquad\qquad\qquad\qquad \vdots$

F = number of days by which duration of activity F is reduced

x_j = time that the event corresponding to node j occurs

Then Widgetco should solve the following LP:

$$\min z = 10A + 20B + 3C + 30D + 40E + 50F$$
$$\text{s.t.} \quad A \leq 5$$
$$B \leq 5$$
$$C \leq 5$$
$$D \leq 5$$
$$E \leq 5$$
$$F \leq 5$$

```
           MIN    X6 - X1
           SUBJECT TO
                2)  - X1 + X3 >=   6
                3)  - X1 + X2 >=   9
                4)  - X3 + X5 >=   8
                5)  - X3 + X4 >=   7
                6)    X5 - X4 >=  10
                7)    X6 - X5 >=  12
                8)    X3 - X2 >=   0
           END


               LP OPTIMUM FOUND  AT STEP      7

                  OBJECTIVE FUNCTION VALUE

        1)          38.0000000

           VARIABLE          VALUE          REDUCED COST
                 X6        38.000000            0.000000
                 X1         0.000000            0.000000
                 X3         9.000000            0.000000
                 X2         9.000000            0.000000
                 X5        26.000000            0.000000
                 X4        16.000000            0.000000


            ROW       SLACK OR SURPLUS      DUAL PRICES
                2)         3.000000            0.000000
                3)         0.000000           -1.000000
                4)         9.000000            0.000000
                5)         0.000000           -1.000000
                6)         0.000000           -1.000000
                7)         0.000000           -1.000000
                8)         0.000000           -1.000000

           NO. ITERATIONS=         7


              RANGES IN WHICH THE BASIS IS UNCHANGED

                             OBJ COEFFICIENT RANGES
           VARIABLE       CURRENT       ALLOWABLE        ALLOWABLE
                          COEF          INCREASE         DECREASE
                 X6     1.000000        INFINITY         0.000000
                 X1    -1.000000        INFINITY         0.000000
                 X3     1.000000        INFINITY         0.000000
                 X2     1.000000        INFINITY         0.000000
                 X5     1.000000        INFINITY         0.000000
                 X4     1.000000        INFINITY         0.000000

                              RIGHTHAND SIDE RANGES
            ROW           CURRENT       ALLOWABLE        ALLOWABLE
                          RHS           INCREASE         DECREASE
                 2       6.000000       3.000000         INFINITY
                 3       9.000000       INFINITY         3.000000
                 4       8.000000       9.000000         INFINITY
                 5       7.000000       INFINITY         9.000000
                 6      10.000000       INFINITY         9.000000
                 7      12.000000       INFINITY        38.000000
                 8       0.000000       INFINITY         3.000000
```

FIGURE 34
LINDO Output
for Widgetco

TABLE 14

A	B	C	D	E	F
$10	$20	$3	$30	$40	$50

$$x_2 \geq x_1 + 9 - B \qquad \text{(Arc (1, 2) constraint)}$$
$$x_3 \geq x_1 + 6 - A \qquad \text{(Arc (1, 3) constraint)}$$
$$x_5 \geq x_3 + 8 - C \qquad \text{(Arc (3, 5) constraint)}$$
$$x_4 \geq x_3 + 7 - D \qquad \text{(Arc (3, 4) constraint)}$$
$$x_5 \geq x_4 + 10 - E \qquad \text{(Arc (4, 5) constraint)}$$
$$x_6 \geq x_5 + 12 - F \qquad \text{(Arc (5, 6) constraint)}$$
$$x_3 \geq x_2 + 0 \qquad \text{(Arc (2, 3) constraint)}$$
$$x_6 - x_1 \leq 25$$
$$A, B, C, D, E, F \geq 0, \; x_j \text{urs}$$

The first six constraints stipulate that the duration of each activity can be reduced by at most 5 days. As before, the next seven constraints ensure that event j cannot occur until after node i occurs and activity (i, j) is completed. For example, activity B (arc $(1, 2)$) now has a duration of $9 - B$. Thus, we need the constraint $x_2 \geq x_1 + (9 - B)$. The constraint $x_6 - x_1 \leq 25$ ensures that the project is completed within the 25-day deadline. The objective function is the total cost incurred in reducing the duration of the activities. An optimal solution to this LP is $z = \$390$, $x_1 = 0$, $x_2 = 4$, $x_3 = 4$, $x_4 = 6$, $x_5 = 13$, $x_6 = 25$, $A = 2$, $B = 5$, $C = 0$, $D = 5$, $E = 3$, $F = 0$. After reducing the durations of projects B, A, D, and E by the given amounts, we obtain the project network pictured in Figure 35. The reader should verify that A, B, D, E, and F are critical activities and that 1–2–3–4–5–6 and 1–3–4–5–6 are both critical paths (each having length 25). Thus, the project deadline of 25 days can be met for a cost of \$390.

Using LINGO to Determine the Critical Path

Many computer packages (such as Microsoft Project) enable the user to determine (among other things!) the critical path(s) and critical activities in a project network. You can always find a critical path and critical activities using LINDO, but LINGO makes it very easy to communicate the necessary information to the computer. The following LINGO program (file Widget1.lng) generates the objective function and constraints needed to find the critical path for the project network of Example 6 via linear programming.

Widget1.lng

```
MODEL:
 1]SETS:
 2]NODES/1..6/:TIME;
 3]ARCS(NODES,NODES)/
 4]1,2   1,3   2,3   3,4   3,5   4,5    5,6/:DUR;
 5]ENDSETS
 6]MIN=TIME(6)-TIME(1);
 7]@FOR(ARCS(I,J):TIME(J)>TIME(I)+DUR(I,J));
 8]DATA:
 9]DUR=9,6,0,7,8,10,12;
10]ENDDATA
END
```

Line 1 begins the SETS portion of the program. In line 2, we define the six nodes of the project network and associate with each node a time that the events corresponding to

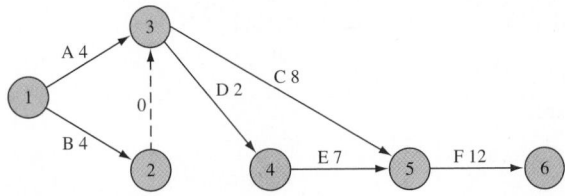

FIGURE 35
Duration of Activities after Crashing

```
MIN    -ET(1 + ET(6
SUBJECT TO
2)-  ET(1 + ET(2 >=  9
3)-  ET(1 + ET(3 >=  6
4)-  ET(2 + ET(3 >=  0
5)-  ET(3 + ET(4 >=  7
6)-  ET(3 + ET(5 >=  8
7)-  ET(4 + ET(5 >=  10
8)-  ET(5 + ET(6 >=  12
END

LP OPTIMUM FOUND AT STEP        6
OBJECTIVE VALUE =    38.0000000

                   VARIABLE          VALUE         REDUCED COST
                    ET( 1)       0.0000000E+00    0.0000000E+00
                    ET( 2)       9.000000         0.0000000E+00
                    ET( 3)       9.000000         0.0000000E+00
                    ET( 4)       16.00000         0.0000000E+00
                    ET( 5)       26.00000         0.0000000E+00
                    ET( 6)       38.00000         0.0000000E+00
                 DUR( 1, 2)      9.000000         0.0000000E+00
                 DUR( 1, 3)      6.000000         0.0000000E+00
                 DUR( 2, 3)      0.0000000E+00    0.0000000E+00
                 DUR( 3, 4)      7.000000         0.0000000E+00
                 DUR( 3, 5)      8.000000         0.0000000E+00
                 DUR( 4, 5)      10.00000         0.0000000E+00
                 DUR( 5, 6)      12.00000         0.0000000E+00

                      ROW    SLACK OR SURPLUS      DUAL PRICE
                       1       38.00000          1.000000
                       2       0.0000000E+00    -1.000000
                       3       3.000000          0.0000000E+00
                       4       0.0000000E+00    -1.000000
                       5       0.0000000E+00    -1.000000
                       6       9.000000          0.0000000E+00
                       7       0.0000000E+00    -1.000000
                       8       0.0000000E+00    -1.000000
```

FIGURE 36

the node occurs. For example, TIME(3) represents the time when activities A and B have just been completed. In line 3, we generate the arcs in the project network by listing them (separated by spaces). For example, arc (3, 4) represents activity D. In line 4, we associate a duration (DUR) of each activity with each arc. Line 5 ends the SETS section of the program.

Line 6 specifies the objective, to minimize the time it takes to complete the project. For each arc defined in line 3, line 7 creates a constraint analagous to $x_j \geq x_i + t_{ij}$.

Line 8 begins the DATA section of the program. In line 9, we list the duration of each activity. Line 10 concludes the data entry and the **END** statement concludes the program. The output from this LINGO model is given in Figure 36, where by following the arcs corresponding to constraints having dual prices of -1, we find the critical path to be 1–2–3–4–5–6.

To find the critical path in any network we would begin by listing the nodes, arcs, and activity durations in our program. Then we would modify the objective function created by line 6 to reflect the number of nodes in the network. For example, if there were 10 nodes in the project network, we would change line 6 to **MIN**=TIME(10)–TIME(1); and we would be ready to go!

Widget2.lng

The following LINGO program (file Widget2.lng) enables the user to determine the critical path and total float at each node for Example 6 without using linear programming.

```
MODEL:
  1]MODEL:
  2]SETS:
  3]NODES/1..6/:ET,LT;
  4]ARCS(NODES,NODES)/1,2   1,3   2,3   3,4   3,5   4,5   5,6/:DUR,TFLOAT;
  5]ENDSETS
  6]DATA:
  7]DUR = 9,6,0,7,8,10,12;
```

```
 8]ENDDATA
 9]ET(1)=0;
10]@FOR(NODES(J) | J#GT#1:
11]ET(J) = @MAX(ARCS(I,J): ET(I)+DUR(I,J)););
12]LNODE=@SIZE(NODES);
13]LT(LNODE) = ET(LNODE);
14]@FOR(NODES(I) | I#LT#LNODE:
15]LT(I) = @MIN(ARCS(I,J): LT(J) - DUR(I,J)););
16]@FOR(ARCS(I,J):TFLOAT(I,J)=LT(J)-ET(I)-DUR(I,J));
END
```

In line 3, we define the nodes of the project network and associate an early event time (ET) and late event time (LT) with each node. We define the arcs of the project network by listing them in line 4. With each arc we associate the duration of the arc's activity and the total float of the activity. In line 7, we input the duration of each activity.

To begin the computation of the ET(J)'s for each node, we set ET(1) = 0 in line 9. In lines 10–11, we compute ET(J) for all other nodes. For J > 1 ET(J) is the maximum value of ET(I) + DUR(I, J) for all (I, J) such that (I, J) is an arc in the network. By using the **@SIZE** function, which returns the number of elements in a set, we identify the finish node in the network in line 12. Thus, line 12 defines node 6 as the last node. In line 13, we set LT(6) = ET(6). Lines 14–15 work backward from node 6 toward node 1 to compute the LT(I)'s. For every node I other than the last node (6), LT(I) is the minimum of LT(J) − DUR (I, J), where the minimum is taken over all (I, J) such that (I, J) is an arc in the project network.

Finally, line 16 computes the total float for each activity (I, J) from total float for activity (I, J) = LT(Node J) − ET(Node I) − Duration (I, J). All activities whose total float equals 0 are critical activities.

After inputting a list of nodes, arcs, and activity durations we can use this program to analyze any project network (without changing any of lines 9–16). It is also easy to write a LINGO program that can be used to crash the network (see Problem 14).

PERT: Program Evaluation and Review Technique

CPM assumes that the duration of each activity is known with certainty. For many projects, this is clearly not applicable. PERT is an attempt to correct this shortcoming of CPM by modeling the duration of each activity as a random variable. For each activity, PERT requires that the project manager estimate the following three quantities:

$$a = \text{estimate of the activity's duration under the most favorable conditions}$$

$$b = \text{estimate of the activity's duration under the least favorable conditions}$$

$$m = \text{most likely value for the activity's duration}$$

Let $\mathbf{T}_{ij}$ (random variables are printed in boldface) be the duration of activity (i, j). PERT requires the assumption that $\mathbf{T}_{ij}$ follows a beta distribution. The specific definition of a beta distribution need not concern us, but it is important to realize that it can approximate a wide range of random variables, including many positively skewed, negatively skewed, and symmetric random variables. If $\mathbf{T}_{ij}$ follows a beta distribution, then it can be shown that the mean and variance of $\mathbf{T}_{ij}$ may be approximated by

$$E(\mathbf{T}_{ij}) = \frac{a + 4m + b}{6} \tag{4}$$

$$\text{var}\mathbf{T}_{ij} = \frac{(b - a)^2}{36} \tag{5}$$

PERT requires the assumption that the durations of all activities are independent. Then for any path in the project network, the mean and variance of the time required to complete the activities on the path are given by

$$\sum_{(i,\,j)\in\text{path}} E(\mathbf{T}_{ij}) = \text{expected duration of activities on any path} \qquad (6)$$

$$\sum_{(i,\,j)\in\text{path}} \text{var}\mathbf{T}_{ij} = \text{variance of duration of activities on any path} \qquad (7)$$

Let **CP** be the random variable denoting the total duration of the activities on a critical path found by CPM. PERT assumes that the critical path found by CPM contains enough activities to allow us to invoke the Central Limit Theorem and conclude that

$$\mathbf{CP} = \sum_{(i,\,j)\in\text{critical path}} \mathbf{T}_{ij}$$

is normally distributed. With this assumption, (4)–(7) can be used to answer questions concerning the probability that the project will be completed by a given date. For example, suppose that for Example 6, a, b, and m for each activity are shown in Table 15. Now (4) and (5) yield

$$E(\mathbf{T}_{12}) = \frac{\{5 + 13 + 36\}}{6} = 9 \qquad \text{var}\mathbf{T}_{12} = \frac{(13 - 5)^2}{36} = 1.78$$

$$E(\mathbf{T}_{13}) = \frac{\{2 + 10 + 24\}}{6} = 6 \qquad \text{var}\mathbf{T}_{13} = \frac{(10 - 2)^2}{36} = 1.78$$

$$E(\mathbf{T}_{35}) = \frac{\{3 + 13 + 32\}}{6} = 8 \qquad \text{var}\mathbf{T}_{35} = \frac{(13 - 3)^2}{36} = 2.78$$

$$E(\mathbf{T}_{34}) = \frac{\{1 + 13 + 28\}}{6} = 7 \qquad \text{var}\mathbf{T}_{34} = \frac{(13 - 1)^2}{36} = 4$$

$$E(\mathbf{T}_{45}) = \frac{\{8 + 12 + 40\}}{6} = 10 \qquad \text{var}\mathbf{T}_{45} = \frac{(12 - 8)^2}{36} = 0.44$$

$$E(\mathbf{T}_{56}) = \frac{\{9 + 15 + 48\}}{6} = 12 \qquad \text{var}\mathbf{T}_{56} = \frac{(15 - 9)^2}{36} = 1$$

Of course, the fact that arc (2, 3) is a dummy arc yields

$$E(\mathbf{T}_{23}) = \text{var } \mathbf{T}_{23} = 0$$

Recall that the critical path for Example 6 was 1–2–3–4–5–6. From Equations (6) and (7),

$$E(\mathbf{CP}) = 9 + 0 + 7 + 10 + 12 = 38$$
$$\text{var } \mathbf{CP} = 1.78 + 0 + 4 + 0.44 + 1 = 7.22$$

Then the standard deviation for **CP** is $(7.22)^{1/2} = 2.69$.

TABLE **15**
a, b, and **m** for Activities in Widgetco

Activity	a	b	m
(1, 2)	5	13	9
(1, 3)	2	10	6
(3, 5)	3	13	8
(3, 4)	1	13	7
(4, 5)	8	12	10
(5, 6)	9	15	12

Applying the assumption that **CP** is normally distributed, we can answer questions such as the following: What is the probability that the project will be completed within 35 days? To answer this question, we must also make the following assumption: *No matter what the durations of the project's activities turn out to be, 1–2–3–4–5–6 will be a critical path.* This assumption implies that the probability that the project will be completed within 35 days is just $P(\textbf{CP} \leq 35)$. Standardizing and applying the assumption that **CP** is normally distributed, we find that **Z** is a standardized normal random variable with mean 0 and variance 1. The cumulative distribution function for a normal random variable is tabulated in Table 16. For example, $P(\textbf{Z} \leq -1) = 0.1587$ and $P(\textbf{Z} \leq 2) = 0.9772$. Thus,

$$P(\textbf{CP} \leq 35) = P\left(\frac{\textbf{CP} - 38}{2.69} \leq \frac{35 - 38}{2.69}\right) = P(\textbf{Z} \leq -1.12) = .13$$

where $F(-1.12) = .13$ may be obtained using the NORMSDIST function in Excel. Entering the formula =NORMSDIST(x) returns the probability that a standard normal random variable with mean 0 and standard deviation 1 is less than or equal to x. For example =NORMDIST(-1.12) yields .1313.

Difficulties with PERT

There are several difficulties with PERT:

1 The assumption that the activity durations are independent is difficult to justify.

2 Activity durations may not follow a beta distribution.

3 The assumption that the critical path found by CPM will always be the critical path for the project may not be justified.

The last difficulty is the most serious. For example, in our analysis of Example 6, we assumed that 1–2–3–4–5–6 would always be the critical path. If, however, activity A were significantly delayed and activity B were completed ahead of schedule, then the critical path might be 1–3–4–5–6.

Here is a more concrete example of the fact that (because of the uncertain duration of activities) the critical path found by CPM may not actually be the path that determines the completion date of the project. Consider the simple project network in Figure 37. As-

TABLE 16
a, b, and m for Figure 37

Activity	a	b	m
A	1	9	5
B	6	14	10
C	5	7	6
D	7	9	8

FIGURE 37
Project Network to Illustrate Difficulties with PERT

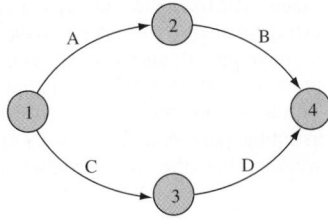

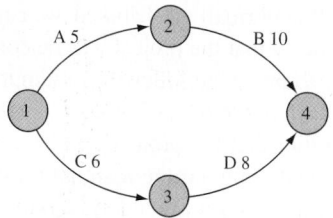

FIGURE 38
Network to Determine
Critical Path If Each
Activity's Duration
Equals *m*

TABLE 17
Probability That Each Arc
Is on a Critical Path

Activity	Probability
A	$\frac{17}{27}$
B	$\frac{17}{27}$
C	$\frac{12}{27}$
D	$\frac{12}{27}$

sume that for each activity in Table 16, *a*, *b*, and *m* each occur with probability $\frac{1}{3}$. If CPM were applied (using the expected duration of each activity as the duration of the activity), then we would obtain the network in Figure 38. For this network, the critical path is 1–2–4. In actuality, however, the critical path could be 1–3–4. For example, if the optimistic duration of B (6 days) occurred and all other activities had a duration *m*, then 1–3–4 would be the critical path in the network. If we assume that the durations of the four activities are independent random variables, then using elementary probability (see Problem 11 at the end of this section), it can be shown that there is a $\frac{10}{27}$ probability that 1–3–4 is the critical path, a $\frac{15}{27}$ chance that 1–2–4 is the critical path, and a $\frac{2}{27}$ chance that 1–2–4 and 1–3–4 will both be critical paths. This example shows that one must be cautious in designating an activity as critical. In this situation, the probability that each activity is actually a critical activity is shown in Table 17.

When the duration of activities is uncertain, the best way to analyze a project is to use a Monte Carlo simulation add-in for Excel. In Chapter 23, we will show how to use the Excel add-in @Risk to perform Monte Carlo simulations. With @Risk, we can easily determine the probability that a project is completed on time and determine the probability that each activity is critical.

PROBLEMS

Group A

1 What problem would arise if the network in Figure 39 were a portion of a project network?

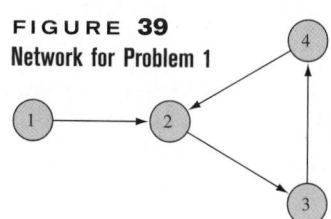

FIGURE 39
Network for Problem 1

2 A company is planning to manufacture a product that consists of three parts (A, B, and C). The company anticipates that it will take 5 weeks to design the three parts and to determine the way in which these parts must be assembled to make the final product. Then the company estimates that it will take 4 weeks to make part A, 5 weeks to make part B, and 3 weeks to make part C. The company must test part A after it is completed (this takes 2 weeks). The assembly line process will then proceed as follows: assemble parts A and B (2 weeks) and then attach part C (1 week). Then the final product must undergo 1 week of testing. Draw the project network and find the critical path, total float, and free float for each activity. Also set up the LP that could be used to find the critical path.

When determining the critical path in Problems 3 and 4, assume that m = activity duration.

3 Consider the project network in Figure 40. For each activity, you are given the estimates of a, b, and m in Table 18. Determine the critical path for this network, the total float for each activity, the free float for each activity, and the probability that the project is completed within 40 days. Also set up the LP that could be used to find the critical path.

4 The promoter of a rock concert in Indianapolis must perform the tasks shown in Table 19 before the concert can be held (all durations are in days).

a Draw the project network.

b Determine the critical path.

c If the advance promoter wants to have a 99% chance of completing all preparations by June 30, when should work begin on finding a concert site?

d Set up the LP that could be used to find the project's critical path.

5 Consider the (simplified) list of activities and predecessors that are involved in building a house (Table 20).

FIGURE 40
Network for Problem 3

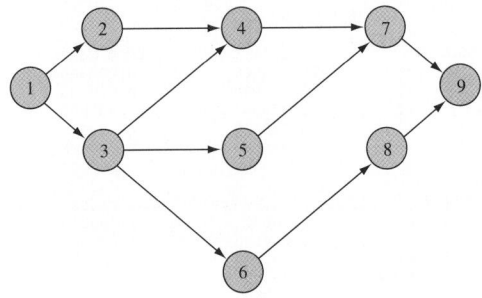

TABLE 18

Activity	a	b	m
(1, 2)	4	8	6
(1, 3)	2	8	4
(2, 4)	1	7	3
(3, 4)	6	12	9
(3, 5)	5	15	10
(3, 6)	7	18	12
(4, 7)	5	12	9
(5, 7)	1	3	2
(6, 8)	2	6	3
(7, 9)	10	20	15
(8, 9)	6	11	9

TABLE 19

Activity	Description	Immediate Predecessors	a	b	m
A	Find site	—	2	4	3
B	Find engineers	A	1	3	2
C	Hire opening act	A	2	10	6
D	Set radio and TV ads	C	1	3	2
E	Set up ticket agents	A	1	5	3
F	Prepare electronics	B	2	4	3
G	Print advertising	C	3	7	5
H	Set up transportation	C	0.5	1.5	1
I	Rehearsals	F, H	1	2	1.5
J	Last-minute details	I	1	3	2

TABLE 20

Activity	Description	Immediate Predecessors	Duration (Days)
A	Build foundation	—	5
B	Build walls and ceilings	A	8
C	Build roof	B	10
D	Do electrical wiring	B	5
E	Put in windows	B	4
F	Put on siding	E	6
G	Paint house	C, F	3

a Draw a project network, determine the critical path, find the total float for each activity, and find the free float for each activity.

b Suppose that by hiring additional workers, the duration of each activity can be reduced. The costs per day of reducing the duration of the activities are given in Table 21. Write down the LP to be solved to minimize the total cost of completing the project within 20 days.

6 Horizon Cable is about to expand its cable TV offerings in Smalltown by adding MTV and other exciting stations. The activities in Table 22 must be completed before the service expansion is completed.

a Draw the project network and determine the critical path for the network, the total float for each activity, and the free float for each activity.

b Set up the LP that can be used to find the project's critical path.

7 When an accounting firm audits a corporation, the first phase of the audit involves obtaining "knowledge of the business." This phase of the audit requires the activities in Table 23.

a Draw the project network and determine the critical path for the network, the total float for each activity, and the free float for each activity. Also set up the LP that can be used to find the project's critical path.

TABLE 21

Activity	Cost per Day of Reducing Duration of Activity ($)	Maximum Possible Reduction in Duration of Activity (Days)
Foundation	30	2
Walls and ceiling	15	3
Roof	20	1
Electrical wiring	40	2
Windows	20	2
Siding	30	3
Paint	40	1

TABLE 22

Activity	Description	Immediate Predecessors	Duration (Weeks)
A	Choose stations	—	2
B	Get town council to approve expansion	A	4
C	Order converters needed to expand service	B	3
D	Install new dish to receive new stations	B	2
E	Install converters	C, D	10
F	Change billing system	B	4

TABLE 23

Activity	Description	Immediate Predecessors	Duration (Days)
A	Determining terms of engagement	—	3
B	Appraisal of auditability risk and materiality	A	6
C	Identification of types of transactions and possible errors	A	14
D	Systems description	C	8
E	Verification of systems description	D	4
F	Evaluation of internal controls	B, E	8
G	Design of audit approach	F	9

TABLE 24

Activity	Cost per Day of Reducing Duration of Activity ($)	Maximum Possible Reduction in Duration of Activity (Days)
A	100	3
B	80	4
C	60	5
D	70	2
E	30	4
F	20	4
G	50	4

FIGURE 41
LINDO Output for Problem 8

```
MIN      X6 - X1
   SUBJECT TO
      2) - X1 + X2 >=   5
      3) - X2 + X3 >=   8
      4) - X3 + X4 >=   4
      5) - X3 + X5 >=  10
      6) - X4 + X5 >=   6
      7)   X6 - X3 >=   5
      8)   X6 - X5 >=   3
   END


      LP OPTIMUM FOUND  AT STEP      6

         OBJECTIVE FUNCTION VALUE

      1)      26.0000000

   VARIABLE        VALUE         REDUCED COST
        X6      26.000000          0.000000
        X1       0.000000          0.000000
        X2       5.000000          0.000000
        X3      13.000000          0.000000
        X4      17.000000          0.000000
        X5      23.000000          0.000000

      ROW    SLACK OR SURPLUS     DUAL PRICES
       2)       0.000000         -1.000000
       3)       0.000000         -1.000000
       4)       0.000000         -1.000000
       5)       0.000000          0.000000
       6)       0.000000         -1.000000
       7)       8.000000          0.000000
       8)       0.000000         -1.000000

   NO. ITERATIONS=        6


RANGES IN WHICH THE BASIS IS UNCHANGED

                       OBJ COEFFICIENT RANGES
   VARIABLE     CURRENT       ALLOWABLE      ALLOWABLE
                COEF          INCREASE       DECREASE
        X6      1.000000      INFINITY       0.000000
        X1     -1.000000      INFINITY       0.000000
        X2      0.000000      INFINITY       0.000000
        X3      0.000000      INFINITY       0.000000
        X4      0.000000      INFINITY       0.000000
        X5      0.000000      INFINITY       0.000000

                       RIGHTHAND SIDE RANGES
   ROW         CURRENT       ALLOWABLE      ALLOWABLE
               RHS           INCREASE       DECREASE
       2       5.000000      INFINITY       5.000000
       3       8.000000      INFINITY      13.000000
       4       4.000000      0.000000       8.000000
       5      10.000000      INFINITY       0.000000
       6       6.000000      0.000000       8.000000
       7       5.000000      8.000000      INFINITY
       8       3.000000      INFINITY       8.000000
```

b Assume that the project must be completed in 30 days. The duration of each activity can be reduced by incurring the costs shown in Table 24. Formulate an LP that can be used to minimize the cost of meeting the project deadline.

8 The LINDO output in Figure 41 can be used to determine the critical path for Problem 5. Use this output to do the following:

a Draw the project diagram.

b Determine the length of the critical path and the critical activities for this project.

9 Explain why an activity's free float can never exceed the activity's total float.

10 A project is complete when activities A–E are completed. The predecessors of each activity are shown in Table 25. Draw the appropriate project diagram. (*Hint:* Don't violate rule 4.)

11 Determine the probabilities that 1–2–4 and 1–3–4 are critical paths for Figure 37.

12 Given the information in Table 26, **(a)** draw the appropriate project network, and **(b)** find the critical path.

13 The government is going to build a high-speed computer in Austin, Texas. Once the computer is designed (D), we can select the exact site (S), the building contractor (C), and the operating personnel (P). Once the site is selected, we can begin erecting the building (B). We can start manufacturing the computer (COM) and preparing the operations manual (M) only after contractor is selected. We can begin training the computer operators (T) when the operating manual and personnel selection are completed. When the computer and the building are both finished, the computer may be installed (I). Then the computer is considered operational. Draw a project network that could be used to determine when the project is operational.

14 Write a LINGO program that can be used to crash the project network of Example 6 with the crashing costs given in Table 14.

15 Consider the project diagram in Figure 42. This project must be completed in 90 days. The time required to complete each activity can be reduced by up to five days at the costs given in Table 27.

Formulate an LP whose solution will enable us to minimize the cost of completing the project in 90 days.

16–17 Find the critical path, total float, and free float for each activity in the project networks of Figures 43 and 44.

TABLE 25

Activity	Predecessors
A	—
B	A
C	A
D	B
E	B, C

TABLE 26

Activity	Immediate Predecessors	Duration (Days)
A	—	3
B	—	3
C	—	1
D	A, B	3
E	A, B	3
F	B, C	2
G	D, E	4
H	E	3

FIGURE 42

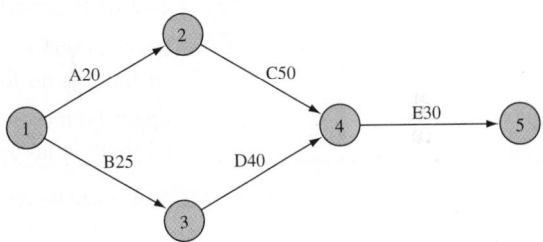

TABLE 27

Activity	Cost of Reducing Activities Duration by 1 Day ($)
A	300
B	200
C	350
D	260
E	320

FIGURE 43

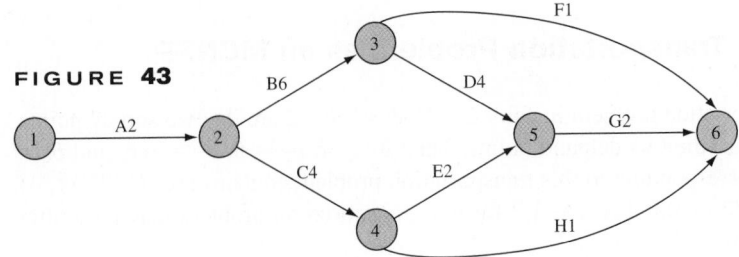

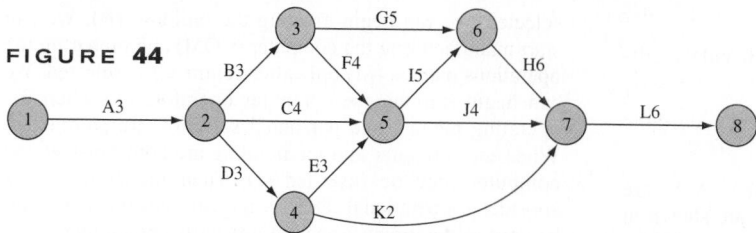

FIGURE 44

8.5 Minimum-Cost Network Flow Problems

The transportation, assignment, transshipment, shortest-path, maximum flow, and CPM problems are all special cases of the minimum-cost network flow problem (MCNFP). Any MCNFP can be solved by a generalization of the transportation simplex called the **network simplex.**

To define an MCNFP, let

x_{ij} = number of units of flow sent from node i to node j through arc (i, j)

b_i = net supply (outflow − inflow) at node i

c_{ij} = cost of transporting 1 unit of flow from node i to node j via arc (i, j)

L_{ij} = lower bound on flow through arc (i, j)
(if there is no lower bound, let $L_{ij} = 0$)

U_{ij} = upper bound on flow through arc (i, j)
(if there is no upper bound, let $U_{ij} = \infty$)

Then the MCNFP may be written as

$$\min \sum_{\text{all arcs}} c_{ij}x_{ij}$$

$$\text{s.t.} \quad \sum_{j} x_{ij} - \sum_{k} x_{ki} = b_i \quad \text{(for each node } i \text{ in the network)} \tag{8}$$

$$L_{ij} \le x_{ij} \le U_{ij} \quad \text{(for each arc in the network)} \tag{9}$$

Constraints (8) stipulate that the net flow out of node i must equal b_i. Constraints (8) are referred to as the **flow balance equations** for the network. Constraints (9) ensure that the flow through each arc satisfies the arc capacity restrictions. In all our previous examples, we have set $L_{ij} = 0$.

Let us show that transportation and maximum-flow problems are special cases of the minimum-cost network flow problem.

Formulating a Transportation Problem as an MCNFP

Consider the transportation problem in Table 28. Nodes 1 and 2 are the two supply points, and nodes 3 and 4 are the two demand points. Then $b_1 = 4$, $b_2 = 5$, $b_3 = -6$, and $b_4 = -3$. The network corresponding to this transportation problem contains arcs $(1, 3)$, $(1, 4)$, $(2, 3)$, and $(2, 4)$ (see Figure 45). The LP for this transportation problem may be written as shown in Table 29.

The first two constraints are the supply constraints, and the last two constraints are (after being multiplied by -1) the demand constraints. Because this transportation problem

TABLE **28**

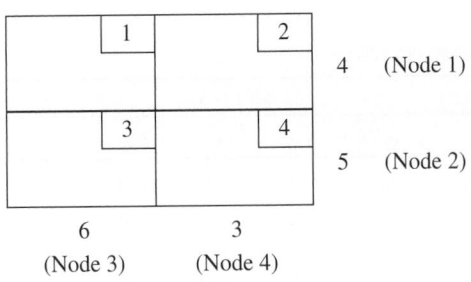

1	2	4 (Node 1)
3	4	5 (Node 2)
6 (Node 3)	3 (Node 4)	

FIGURE **45**
Representation of
Transportation Problem
as an MCNFP

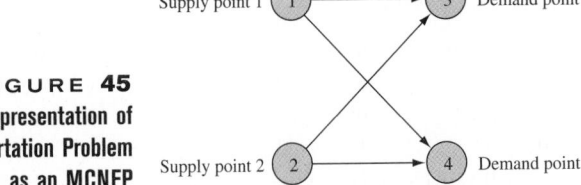

TABLE **29**
MCNFP Representation of Transportation Problem

			min $z = x_{13} + 2x_{14} + 3x_{23} + 4x_{24}$			
x_{13}	x_{14}	x_{23}	x_{24}		rhs	Constraint
1	1	0	0	=	4	Node 1
0	0	1	1	=	5	Node 2
−1	0	−1	0	=	−6	Node 3
0	−1	0	−1	=	−3	Node 4
			All variables non-negative			

has no arc capacity restrictions, the flow balance equations are the only constraints. We note that if the problem had not been balanced, we could not have formulated the problem as an MCNFP. This is because if total supply exceeded total demand, we would not know with certainty the net outflow at each supply point. Thus, to formulate a transportation (or a transshipment) problem as an MCNFP, it may be necessary to add a dummy point.

Formulating a Maximum-Flow Problem as an MCNFP

To see how a maximum-flow problem fits into the minimum-cost network flow context, consider the problem of finding the maximum flow from source to sink in the network of Figure 6. After creating an arc a_0 joining the sink to the source, we have $b_{so} = b_1 = b_2 = b_3 = b_{si} = 0$. Then the LP constraints for finding the maximum flow in Figure 6 may be written as shown in Table 30.

The first five constraints are the flow balance equations for the nodes of the network, and the last six constraints are the arc capacity constraints. Because there is no upper limit on the flow through the artificial arc, there is no arc capacity constraint for a_0.

The flow balance equations in any MCNFP have the following important property: *Each variable x_{ij} has a coefficient of $+1$ in the node i flow balance equation, a coefficient of -1 in the node j flow balance equation, and a coefficient of 0 in all other flow balance equations.* For example, in a transportation problem, the variable x_{ij} will have a coeffi-

TABLE 30
MCNFP Representation of Maximum-Flow Problem

				min $z = x_0$					
$x_{so,1}$	$x_{so,2}$	x_{13}	x_{12}	$x_{3,si}$	$x_{2,si}$	x_0		rhs	Constraint
1	1	0	0	0	0	-1	$=$	0	Node so
-1	0	1	1	0	0	0	$=$	0	Node 1
0	-1	0	-1	0	1	0	$=$	0	Node 2
0	0	-1	0	1	0	0	$=$	0	Node 3
0	0	0	0	-1	-1	1	$=$	0	Node si
1	0	0	0	0	0	0	$\leq$	2	Arc $(so, 1)$
0	1	0	0	0	0	0	$\leq$	3	Arc $(so, 2)$
0	0	1	0	0	0	0	$\leq$	4	Arc $(1, 3)$
0	0	0	1	0	0	0	$\leq$	3	Arc $(1, 2)$
0	0	0	0	1	0	0	$\leq$	1	Arc $(3, si)$
0	0	0	0	0	1	0	$\leq$	2	Arc $(2, si)$
				All variables nonnegative					

cient of $+1$ in the flow balance equation for supply point i, a coefficient of -1 in the flow balance equation for demand point j, and a coefficient of 0 in all other flow balance equations. Even if the constraints of an LP do not appear to contain the flow balance equations of a network, clever transformation of an LP's constraints can often show that an LP is equivalent to an MCNFP (see Problem 6 at the end of this section).

An MCNFP can be solved by a generalization of the transportation simplex known as the *network simplex algorithm* (see Section 8.7). As with the transportation simplex, the pivots in the network simplex involve only additions and subtractions. This fact can be used to prove that if all the b_i's and arc capacities are integers, then in the optimal solution to an MCNFP, all the variables will be integers. Computer codes that use the network simplex can quickly solve even extremely large network problems. For example, MCNFPs with 5,000 nodes and 600,000 arcs have been solved in under 10 minutes. To use a network simplex computer code, the user need only input a list of the network's nodes and arcs, the c_{ij}'s and arc capacity for each arc, and the b_i's for each node. The network simplex is efficient and easy to use, so it is extremely important to formulate an LP, if at all possible, as an MCNFP.

To close this section, we formulate a simple traffic assignment problem as an MCNFP.

EXAMPLE 7 **Traffic MCNFP**

Each hour, an average of 900 cars enter the network in Figure 46 at node 1 and seek to travel to node 6. The time it takes a car to traverse each arc is shown in Table 31. In Figure 46, the number above each arc is the maximum number of cars that can pass by any point on the arc during a one-hour period. Formulate an MCNFP that minimizes the total time required for all cars to travel from node 1 to node 6.

Solution Let

$$x_{ij} = \text{number of cars per hour that traverse the arc from node } i \text{ to node } j$$

Then we want to minimize

$$z = 10x_{12} + 50x_{13} + 70x_{25} + 30x_{24} + 30x_{56} + 30x_{45} + 60x_{46} + 60x_{35} + 10x_{34}$$

We are given that $b_1 = 900$, $b_2 = b_3 = b_4 = b_5 = 0$, and $b_6 = -900$ (we will not introduce the artificial arc connecting node 6 to node 1). The constraints for this MCNFP are shown in Table 32.

FIGURE **46**
Representation of
Traffic Example as
MCNFP

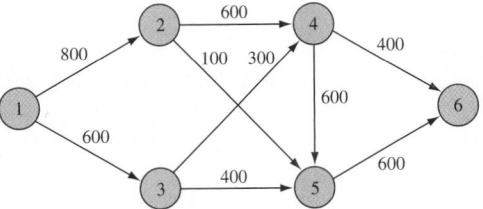

TABLE **31**
Travel Times for Traffic
Example

Arc	Time (Minutes)
(1, 2)	10
(1, 3)	50
(2, 5)	70
(2, 4)	30
(5, 6)	30
(4, 5)	30
(4, 6)	60
(3, 5)	60
(3, 4)	10

TABLE **32**
MCNFP Representation of
Traffic Example

X_{12}	X_{13}	X_{24}	X_{25}	X_{34}	X_{35}	X_{45}	X_{46}	X_{56}		rhs	Constraint
1	1	0	0	0	0	0	0	0	=	900	Node 1
−1	0	1	1	0	0	0	0	0	=	0	Node 2
0	−1	0	0	1	1	0	0	0	=	0	Node 3
0	0	−1	0	−1	0	1	1	0	=	0	Node 4
0	0	0	−1	0	−1	−1	0	1	=	0	Node 5
0	0	0	0	0	0	0	−1	−1	=	−900	Node 6
1	0	0	0	0	0	0	0	0	≤	800	Arc (1, 2)
0	1	0	0	0	0	0	0	0	≤	600	Arc (1, 3)
0	0	1	0	0	0	0	0	0	≤	600	Arc (2, 4)
0	0	0	1	0	0	0	0	0	≤	100	Arc (2, 5)
0	0	0	0	1	0	0	0	0	≤	300	Arc (3, 4)
0	0	0	0	0	1	0	0	0	≤	400	Arc (3, 5)
0	0	0	0	0	0	1	0	0	≤	600	Arc (4, 5)
0	0	0	0	0	0	0	1	0	≤	400	Arc (4, 6)
0	0	0	0	0	0	0	0	1	≤	600	Arc (5, 6)

All variables non-negative

Solving an MCNFP with LINGO

Traffic.lng

The following LINGO program (file Traffic.lng) can be used to find the optimal solution
to Example 7 (or any MCNFP).

```
MODEL:
  1]  SETS:
  2]  NODES/1..6/:SUPP;
  3]  ARCS(NODES,NODES)/1,2  1,3  2,4  2,5  3,4  3,5  4,5  4,6  5,6/
  4]  :CAP,FLOW,COST;
  5]  ENDSETS
  6]  MIN=@SUM(ARCS:COST*FLOW);
  7]  @FOR(ARCS(I,J):FLOW(I,J)<CAP(I,J));
  8]  @FOR(NODES(I):-@SUM(ARCS)(J,I):FLOW(J,I))
  9]  +@SUM(ARCS(I,J):FLOW(I,J))=SUPP(I));
 10]  DATA:
 11]  COST=10,50,30,70,10,60,30,60,30;
 12]  SUPP=900,0,0,0,0,-900;
 13]  CAP=800,600,600,100,300,400,600,400,600;
 14]  ENDDATA
END
```

In line 2, we define the network's nodes and associate a net supply (flow out−flow in) with each node. The supplies data are entered in line 12. In line 3, we define, by listing, the arcs in the network and in line 4 associate a capacity (CAP), a flow (FLOW), and a cost-per-unit-shipped (COST) with each arc. The unit shipping costs data are entered in line 11. Line 6 generates the objective function by summing over all arcs (unit cost for arc)*(flow through arc). Line 7 generates each arc's capacity constraint (arc capacities data are entered in line 13). For each node, lines 8–9 generate the conservation-of-flow constraint. They imply that for each node I, −(flow into node I) + (flow out of node I) = (supply of node I). When solved on LINGO, we find that the solution to Example 7 is $z = 95,000$ minutes, $x_{12} = 700$, $x_{13} = 200$, $x_{24} = 600$, $x_{25} = 100$, $x_{34} = 200$, $x_{45} = 400$, $x_{46} = 400$, $x_{56} = 500$.

Our LINGO program can be used to solve any MCNFP. Just input the set of nodes, supplies, arcs, and unit shipping cost; hit **GO** and you are done!

PROBLEMS

Note: To formulate a problem as an MCNFP, you should draw the appropriate network and determine the c_{ij}'s, the b_i's, and the arc capacities.

Group A

1 Formulate the problem of finding the shortest path from node 1 to node 6 in Figure 2 as an MCNFP. (*Hint:* Think of finding the shortest path as the problem of minimizing the total cost of sending 1 unit of flow from node 1 to node 6.)

2 a Find the dual of the LP that was used to find the length of the critical path for Example 6 of Section 8.4.

b Show that the answer in part (a) is an MCNFP.

c Explain why the optimal objective function value for the LP found in part (a) is the longest path in the project network from node 1 to node 6. Why does this justify our earlier claim that the critical path in a project network is the longest path from the start node to the finish node?

3 Fordco produces cars in Detroit and Dallas. The Detroit plant can produce as many as 6,500 cars, and the Dallas plant can produce as many as 6,000 cars. Producing a car costs $2,000 in Detroit and $1,800 in Dallas. Cars must be shipped to three cities. City 1 must receive 5,000 cars, city 2 must receive 4,000 cars, and city 3 must receive 3,000 cars. The cost of shipping a car from each plant to each city is given in Table 33. At most, 2,200 cars may be sent from a given plant to a given city. Formulate an MCNFP that can be used to minimize the cost of meeting demand.

4 Each year, Data Corporal produces as many as 400 computers in Boston and 300 computers in Raleigh. Los Angeles customers must receive 400 computers, and 300 computers must be supplied to Austin customers. Producing a computer costs $800 in Boston and $900 in Raleigh. Computers are transported by plane and may be sent through Chicago. The costs of sending a computer between pairs of cities are shown in Table 34.

a Formulate an MCNFP that can be used to minimize the total (production + distribution) cost of meeting Data Corporal's annual demand.

TABLE 33

From	To ($)		
	City 1	City 2	City 3
Detroit	800	600	300
Dallas	500	200	200

TABLE 34

From	To ($)		
	Chicago	Austin	Los Angeles
Boston	80	220	280
Raleigh	100	140	170
Chicago	—	40	50

b How would you modify the part (a) formulation if at most 200 units could be shipped through Chicago? [*Hint:* Add an additional node and arc to this part (a) network.]

5 Oilco has oil fields in San Diego and Los Angeles. The San Diego field can produce 500,000 barrels per day, and the Los Angeles field can produce 400,000 barrels per day. Oil is sent from the fields to a refinery, in either Dallas or Houston (assume each refinery has unlimited capacity). To refine 100,000 barrels costs $700 at Dallas and $900 at Houston. Refined oil is shipped to customers in Chicago and New York. Chicago customers require 400,000 barrels per day, and New York customers require 300,000 barrels per day. The costs of shipping 100,000 barrels of oil (refined or unrefined) between cities are shown in Table 35.

a Formulate an MCNFP that can be used to determine how to minimize the total cost of meeting all demands.

b If each refinery had a capacity of 500,000 barrels per day, how would the part (a) answer be modified?

Group B

6 Workco must have the following number of workers available during the next three months: month 1, 20; month 2, 16; month 3, 25. At the beginning of month 1, Workco has no workers. It costs Workco $100 to hire a worker and $50 to fire a worker. Each worker is paid a salary of $140/month. We will show that the problem of determining a hiring and firing strategy that minimizes the total cost incurred during the next three (or in general, the next n) months can be formulated as an MCNFP.

a Let

x_{ij} = number of workers hired at beginning of month i and fired after working till end of month $j - 1$

(if $j = 4$, the worker is never fired). Explain why the following LP will yield a minimum-cost hiring and firing strategy:

TABLE 35

From	To ($)			
	Dallas	Houston	New York	Chicago
Los Angeles	300	110	—	—
San Diego	420	100	—	—
Dallas	—	—	450	550
Houston	—	—	470	530

$$\min z = 50(x_{12} + x_{13} + x_{23})$$
$$+ 100(x_{12} + x_{13} + x_{14} + x_{23} + x_{24} + x_{34})$$
$$+ 140(x_{12} + x_{23} + x_{34})$$
$$+ 280(x_{13} + x_{24}) + 420x_{14}$$

s.t.   (1) $x_{12} + x_{13} + x_{14} \qquad - e_1 = 20$
                      (Month 1 constraint)

        (2) $x_{13} + x_{14} + x_{23} + x_{24} - e_2 = 16$
                      (Month 2 constraint)

        (3) $x_{14} + x_{24} + x_{34} \qquad - e_3 = 25$
                      (Month 3 constraint)

$$x_{ij} \geq 0$$

b To obtain an MCNFP, replace the constraints in part (a) by

   i   Constraint (1);

  ii   Constraint (2) − Constraint (1);

 iii   Constraint (3) − Constraint (2);

 iv   − (Constraint (3)).

Explain why an LP with Constraints (i)–(iv) is an MCNFP.

c Draw the network corresponding to the MCNFP obtained in answering part (b).

7[†] Braneast Airlines must determine how many airplanes should serve the Boston–New York–Washington air corridor and which flights to fly. Braneast may fly any of the daily flights shown in Table 36. The fixed cost of operating an airplane is $800/day. Formulate an MCNFP that can be used to maximize Braneast's daily profits. (*Hint:* Each node in the network represents a city and a time. In addition to arcs representing flights, we must allow for the possibility that an airplane will stay put for an hour or more. We must ensure that the model includes the fixed cost of operating a plane. To include this cost, the following three arcs might be included in the network: from Boston 7 P.M. to Boston 9 A.M.; from New York 7 P.M. to New York 9 A.M.; and from Washington 7 P.M. to Washington 9 A.M.)

8 Daisymay Van Line moves people between New York, Philadelphia, and Washington, D.C. It takes a van one day to travel between any two of these cities. The company incurs costs of $1,000 per day for a van that is fully loaded and traveling, $800 per day for an empty van that travels, $700 per day for a fully loaded van that stays in a city, and $400 per day for an empty van that remains in a city. Each day of the week, the loads described in Table 37 must be shipped. On Monday, for example, two trucks must be sent from Philadelphia to New York (arriving on Tuesday). Also, two trucks must be sent from Philadelphia to Washington on Friday (assume that Friday shipments must arrive on Monday). Formulate an MCNFP that can be used to minimize the cost of meeting weekly requirements. To simplify the formulation, assume that the requirements repeat each week. Then it seems plausible to assume that any of the company's trucks will begin each week in the same city in which it began the previous week.

[†]This problem is based on Glover et al. (1982).

TABLE 36

Leaves		Arrives		Flight Revenue	Variable Cost of Flight (S)
City	Time	City	Time		
N.Y.	9 A.M.	Wash.	10 A.M.	$900	400
N.Y.	2 P.M.	Wash.	3 P.M.	$600	350
N.Y.	10 A.M.	Bos.	11 A.M.	$800	400
N.Y.	4 P.M.	Bos.	5 P.M.	$1,200	450
Wash.	9 A.M.	N.Y.	10 A.M.	$1,100	400
Wash.	3 P.M.	N.Y.	4 P.M.	$900	350
Wash.	10 A.M.	Bos.	12 noon	$1,500	700
Wash.	5 P.M.	Bos.	7 P.M.	$1,800	900
Bos.	10 A.M.	N.Y.	11 A.M.	$900	500
Bos.	2 P.M.	N.Y.	3 P.M.	$800	450
Bos.	11 A.M.	Wash.	1 P.M.	$1,100	600
Bos.	3 P.M.	Wash.	5 P.M.	$1,200	650

TABLE 37

Trip	Monday	Tuesday	Wednesday	Thursday	Friday
Phil.–N.Y.	2	—	—	—	—
Phil.–Wash.	—	2	—	—	2
N.Y.–Phil.	3	2	—	—	—
N.Y.–Wash.	—	—	2	2	—
N.Y.–Phil.	1	—	—	—	—
Wash.–N.Y.	—	—	1	—	1

8.6 Minimum Spanning Tree Problems

Suppose that each arc (i, j) in a network has a length associated with it and that arc (i, j) represents a way of connecting node i to node j. For example, if each node in a network represents a computer at State University, then arc (i, j) might represent an underground cable that connects computer i with computer j. In many applications, we want to determine the set of arcs in a network that connect all nodes such that the sum of the length of the arcs is minimized. Clearly, such a group of arcs should contain no loop. (A loop is often called a *closed path* or *cycle*.) For example, in Figure 47, the sequence of arcs $(1, 2)$–$(2, 3)$–$(3, 1)$ is a loop.

DEFINITION ■ For a network with n nodes, a **spanning tree** is a group of $n - 1$ arcs that connects all nodes of the network and contains no loops. ■

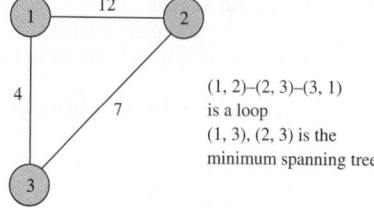

FIGURE 47
Illustration of Loop and Minimum Spanning Tree

(1, 2)–(2, 3)–(3, 1) is a loop
(1, 3), (2, 3) is the minimum spanning tree

In Figure 47, there are three spanning trees:

1 Arcs (1, 2) and (2, 3)

2 Arcs (1, 2) and (1, 3)

3 Arcs (1, 3) and (2, 3)

A spanning tree of minimum length in a network is a **minimum spanning tree (MST).** In Figure 47, the spanning tree consisting of arcs (1, 3) and (2, 3) is the unique minimum spanning tree.

The following method (MST algorithm) may be used to find a minimum spanning tree.

Step 1 Begin at any node i, and join node i to the node in the network (call it node j) that is closest to node i. The two nodes i and j now form a connected set of nodes $C = \{i, j\}$, and arc (i, j) will be in the minimum spanning tree. The remaining nodes in the network (call them C') are referred to as the *unconnected* set of nodes.

Step 2 Now choose a member of C' (call it n) that is closest to some node in C. Let m represent the node in C that is closest to n. Then the arc (m, n) will be in the minimum spanning tree. Now update C and C'. Because n is now connected to $\{i, j\}$, C now equals $\{i, j, n\}$ and we must eliminate node n from C'.

Step 3 Repeat this process until a minimum spanning tree is found. Ties for closest node and arc to be included in the minimum spanning tree may be broken arbitrarily.

At each step the algorithm chooses the shortest arc that can be used to expand C, so the algorithm is often referred to as a "greedy" algorithm. It is remarkable that the act of being "greedy" at each step of the algorithm can never force us later to follow a "bad arc." In Example 1 of Chapter 9 we will see that for some types of problems, a greedy algorithm may not yield an optimal solution! A justification of the MST algorithm is given in Problem 3 at the end of this section. Example 8 illustrates the algorithm.

EXAMPLE 8   **MST Algorithm**

The State University campus has five minicomputers. The distance between each pair of computers (in city blocks) is given in Figure 48. The computers must be interconnected by underground cable. What is the minimum length of cable required? Note that if no arc is drawn connecting a pair of nodes, this means that (because of underground rock formations) no cable can be laid between these two computers.

Solution We want to find the minimum spanning tree for Figure 48.

Iteration 1 Following the MST algorithm, we arbitrarily choose to begin at node 1. The closest node to node 1 is node 2. Now $C = \{1, 2\}$, $C' = \{3, 4, 5\}$, and arc (1, 2) will be in the minimum spanning tree (see Figure 49a).

Iteration 2 Node 5 is closest (two blocks distant) to C. Because node 5 is two blocks from node 1 and from node 2, we may include either arc (2, 5) or arc (1, 5) in the minimum spanning tree. We arbitrarily choose to include arc (2, 5). Then $C = \{1, 2, 5\}$ and $C' = \{3, 4\}$ (see Figure 49b).

Iteration 3 Node 3 is two blocks from node 5, so we may include arc (5, 3) in the minimum spanning tree. Now $C = \{1, 2, 3, 5\}$ and $C' = 4$ (see Figure 49c).

Iteration 4 Node 5 is the closest node to node 4, so we add arc (5, 4) to the minimum spanning tree (see Figure 49d).

We have now obtained the minimum spanning tree consisting of arcs (1, 2), (2, 5), (5, 3), and (5, 4). The length of the minimum spanning tree is $1 + 2 + 2 + 4 = 9$ blocks.

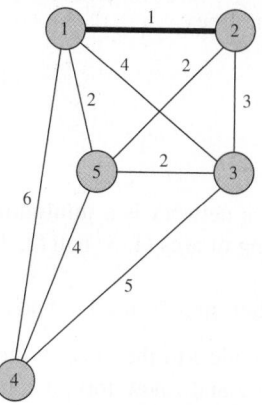

FIGURE **48**
Distances between
State University
Computers

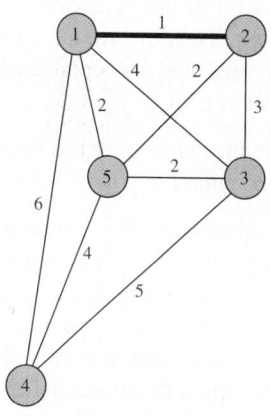

$C = [1, 2]$
$C' = [3, 4, 5]$

a Iteration 1

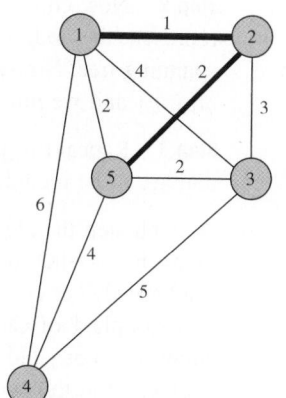

$C = [1, 2, 5]$
$C' = [3, 4]$

b Iteration 2

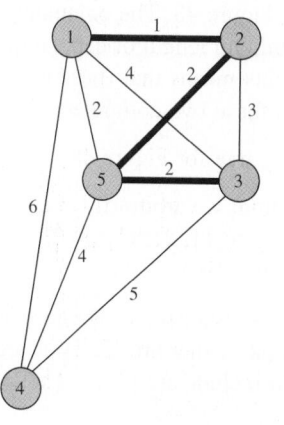

$C = [1, 2, 3, 5]$
$C' = [4]$

FIGURE **49**
MST Algorithm for
Computer Example

c Iteration 3

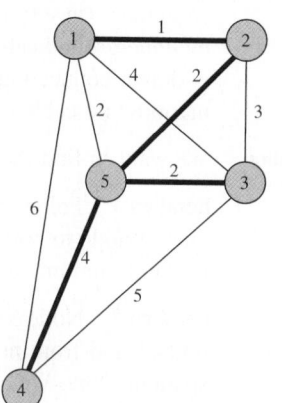

Arcs (1, 2), (2, 5), (5, 3),
and (5, 4) are the MST

d Iteration 4: MST has been found

PROBLEMS

Group A

1 The distances (in miles) between the Indiana cities of Gary, Fort Wayne, Evansville, Terre Haute, and South Bend are shown in Table 38. It is necessary to build a state road system that connects all these cities. Assume that for political reasons no road can be built connecting Gary and Fort Wayne, and no road can be built connecting South Bend and Evansville. What is the minimum length of road required?

2 The city of Smalltown consists of five subdivisions. Mayor John Lion wants to build telephone lines to ensure that all the subdivisions can communicate with each other. The distances between the subdivisions are given in Figure 50. What is the minimum length of telephone line required? Assume that no telephone line can be built between subdivisions 1 and 4.

Group B

3 In this problem, we explain why the MST algorithm works. Define

S = minimum spanning tree

C_t = nodes connected after iteration t of MST algorithm has been completed

C_t' = nodes not connected after iteration t of MST algorithm has been completed

A_t = set of arcs in minimum spanning tree after t iterations of MST algorithm have been completed

FIGURE 50
Network for Problem 2

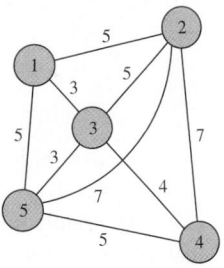

Suppose the MST algorithm does not yield a minimum spanning tree. Then, for some t, it must be the case that all arcs in A_{t-1} are in S, but the arc chosen at iteration t (call it a_t) of the MST algorithm is not in S. Then S must contain some arc a_t' that leads from a node in C_{t-1} to a node in C_{t-1}'. Show that by replacing arc a_t' with arc a_t, we can obtain a shorter spanning tree than S. This contradiction proves that all arcs chosen by the MST algorithm must be in S. Thus, the MST algorithm does indeed find a minimum spanning tree.

4 a Three cities are at the vertices of an equilateral triangle of unit length. Flying Lion Airlines needs to supply connecting service between these three cities. What is the minimum length of the two routes needed to supply the connecting service?

b Now suppose Flying Lion Airlines adds a hub at the "center" of the equilateral triangle. Show that the length of the routes needed to connect the three cities has decreased by 13%. *(Note:* It has been shown that no matter how many "hubs" you add and no matter how many points must be connected, you can never save more than 13% of the total distance needed to "span" all the original points by adding hubs.)[†]

TABLE 38

	Gary	Fort Wayne	Evansville	Terre Haute	South Bend
Gary	—	132	217	164	58
Fort Wayne	132	—	290	201	79
Evansville	217	290	—	113	303
Terre Haute	164	201	113	—	196
South Bend	58	79	303	196	—

8.7 The Network Simplex Method[‡]

In this section, we describe how the simplex algorithm simplifies for MCNFPs. To simplify our presentation, we assume that for each arc, $L_{ij} = 0$. Then the information needed to describe an MCNFP of the form (8)–(9) may be summarized graphically as in Figure 51. We will denote the c_{ij} for each arc by the symbol $, and the other number on each arc will represent the arc's upper bound (U_{ij}). The b_i for any node with nonzero outflow will be listed in parentheses. Thus, Figure 51 represents an MCNFP with $c_{12} = 5$, $c_{25} = 2$, $c_{13} = 4$, $c_{35} = 8$,

[†]Based on Peterson (1990).
[‡]This section covers topics that may be omitted with no loss of continuity.

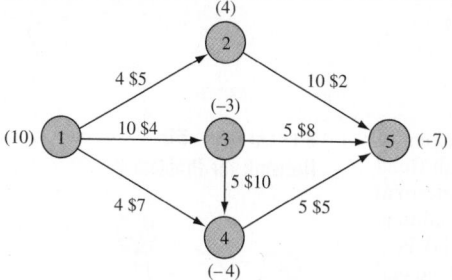

FIGURE 51
Graphical
Representation of
an MCNFP

$c_{14} = 7$, $c_{34} = 10$, $c_{45} = 5$, $b_1 = 10$, $b_2 = 4$, $b_3 = -3$, $b_4 = -4$, $b_5 = -7$, $U_{12} = 4$, $U_{25} = 10$, $U_{13} = 10$, $U_{35} = 5$, $U_{14} = 4$, $U_{34} = 5$, $U_{45} = 5$. For the network simplex to be used, we must have $\Sigma b_i = 0$; usually this can be ensured by adding a dummy node.

Recall that when we used the simplex method to solve a transportation problem, the following aspects of the simplex algorithm simplified: finding a basic feasible solution, computing the coefficient of a nonbasic variable in row 0, and pivoting. We now describe how these aspects of the simplex algorithm simplify when we are solving an MCNFP.

Basic Feasible Solutions for MCNFPs

How can we determine whether a feasible solution to an MCNFP is a bfs? Begin by observing that any bfs to an MCNFP will contain three types of variables:

1 Basic variables: In the absence of degeneracy, each basic variable x_{ij} will satisfy $L_{ij} < x_{ij} < U_{ij}$; with degeneracy, it is possible for a basic variable x_{ij} to equal arc (i, j)'s upper or lower bound.

2 Nonbasic variables x_{ij}: These equal arc (i, j)'s upper bound U_{ij}.

3 Nonbasic variables x_{ij}: These equal arc (i, j)'s lower bound L_{ij}.

Suppose we are solving an MCNFP with n nodes. In solving an MCNFP, we consider the n conservation-of-flow constraints and ignore the upper- and lower-bound constraints (for reasons that will soon become apparent). As in the transportation problem, any solution satisfying $n - 1$ of the conservation-of-flow constraints will automatically satisfy the last conservation-of-flow constraint, so we may drop one such constraint. This means that a bfs to an n-node MCNFP will have $n - 1$ basic variables. Suppose we choose a set of $n - 1$ variables (or arcs). How can we determine whether this set of $n - 1$ variables yields a basic feasible solution? A set of $n - 1$ variables will yield a bfs if and only if the arcs corresponding to the basic variables form a spanning tree for the network. For example, consider the MCNFP in Figure 52. In Figure 53, we give a bfs for this MCNFP. The basic variables are x_{13}, x_{35}, x_{25}, and x_{45}. The variables $x_{12} = 5$ and $x_{14} = 4$ are nonbasic vari-

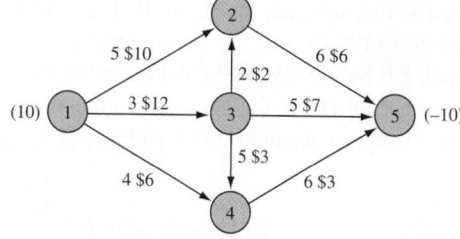

FIGURE 52
Example of an MCNFP

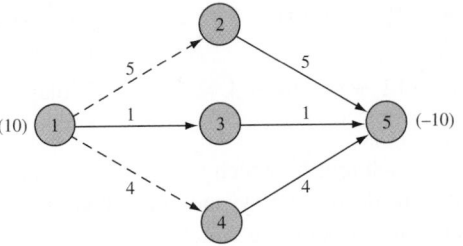

FIGURE 53
Example of a bfs
for an MCNFP

ables at their upper bound. (Such variables will be indicated by dashed arcs.) Because the arcs (1, 3), (3, 5), (2, 5), and (4, 5) form a spanning tree (they connect all nodes of the graph and do not contain any cycles), we know that this is a bfs. As will soon become clear, a bfs for small problems can often be obtained by trial and error.

Computing Row 0 for Any bfs

For any given bfs, how do we determine the objective function coefficient for a nonbasic variable? Suppose we arbitrarily choose to drop the conservation-of-flow constraint for node 1. For a given bfs, let $c_{BV}B^{-1} = [y_2 \quad y_3 \quad \cdots \quad y_n]$. Each variable x_{ij} will have a $+1$ coefficient in the node i flow constraint and a -1 coefficient in the node j constraint. If we define $y_1 = 0$, then the coefficient of x_{ij} in row 0 of a given tableau may be written as $\bar{c}_{ij} = y_i - y_j - c_{ij}$. Each basic variable must have $\bar{c}_{ij} = 0$, so we can find $y_1, y_2, \ldots, y_n$ by solving the following system of linear equations:

$$y_1 = 0, \quad y_i - y_j = c_{ij} \quad \text{for each basic variable}$$

The $y_1, y_2, \ldots, y_n$ corresponding to a bfs are often called the **simplex multipliers** for the bfs.

How can we determine whether a bfs is optimal? For a bfs to be optimal, it must be possible to improve (decrease) the value of z by changing the value of a nonbasic variable. Note that $\bar{c}_{ij} \le 0$ if and only if increasing x_{ij} cannot decrease z. Also note that $\bar{c}_{ij} \ge 0$ if and only if decreasing x_{ij} cannot decrease z. These observations can be used to show that a bfs is optimal if and only if the following conditions are met:

1 If a variable $x_{ij} = L_{ij}$, then an increase in x_{ij} cannot result in a decrease in z. Thus, if $x_{ij} = L_{ij}$ and the bfs is optimal, then $\bar{c}_{ij} \le 0$ must hold.

2 If a variable $x_{ij} = U_{ij}$, then a decrease in x_{ij} cannot result in a decrease in z. Thus, if $x_{ij} = U_{ij}$ and the bfs is optimal, then $\bar{c}_{ij} \ge 0$ must hold.

If conditions 1 and 2 are not met, then z can be improved (barring degeneracy) by pivoting into the basis any nonbasic variable violating either condition. To illustrate, let's determine the objective function coefficient for each nonbasic variable in the simplex tableau corresponding to the bfs in Figure 53. To find $y_1, y_2, y_3, y_4,$ and y_5, we solve the following set of equations:

$$y_1 = 0, \quad y_1 - y_3 = 12, \quad y_2 - y_5 = 6, \quad y_3 - y_5 = 7, \quad y_4 - y_5 = 3$$

The solutions to these equations are $y_1 = 0, y_2 = -13, y_3 = -12, y_4 = -16,$ and $y_5 = -19$. We now "price out" each nonbasic variable and obtain

$\bar{c}_{12} = y_1 - y_2 - c_{12} = 0 - (-13) - 10 = 3$ (Satisfies optimality condition for nonbasic variable at upper bound)

$\bar{c}_{14} = y_1 - y_4 - c_{14} = 0 - (-16) - 6 = 10$ (Satisfies optimality condition for nonbasic variable at upper bound)

$$\bar{c}_{32} = y_3 - y_2 - c_{32} = -12 - (-13) - 2 = -1 \qquad \text{(Satisfies optimality condition for nonbasic variable at lower bound)}$$

$$\bar{c}_{34} = y_3 - y_4 - c_{34} = -12 - (-16) - 3 = 1 \qquad \text{(Violates optimality condition for nonbasic variable at lower bound)}$$

Because $\bar{c}_{34} = 1 > 0$, each unit by which we increase x_{34} (x_{34} is at its lower bound, so it's okay to increase it) will decrease z by one unit. Thus, we can improve z by entering x_{34} into the basis. Note that if a nonbasic variable x_{ij} at its upper bound had $\bar{c}_{ij} < 0$, then we could decrease z by entering x_{ij} into the basis and decreasing x_{ij}. We now show that when solving an MCNFP, the pivot step may be performed almost by inspection.

Pivoting in the Network Simplex

As we have just shown, for the bfs in Figure 53, we want to enter x_{34} into the basis. To do this, note that if we add the arc (3, 4) to the set of arcs corresponding to the current set of basic variables, a cycle (or loop) will be formed. To enter x_{34} into the basis, note that $x_{34} = 0$ is at its lower bound, we want to increase x_{34}. Suppose we try to increase x_{34} by θ. The values of all variables after x_{34} is entered into the basis may be found by invoking the conservation-of-flow constraints. In Figure 54, we find that arc (3, 4), (4, 5), and (3, 5) form a cycle. After the pivot, all variables corresponding to arcs not in the cycle will remain unchanged, but when we set $x_{34} = \theta$, the values of the variables corresponding to arcs in the cycle will change. Setting $x_{34} = \theta$ increases the flow into node 4 by θ, so the flow out of node 4 must increase by θ. This requires $x_{45} = 4 + \theta$. Because the flow into node 5 has now increased by θ, conservation of flow requires that $x_{35} = 1 - \theta$. The pivot leaves all other variables unchanged. To find the new values of the variables, observe that we want to increase x_{34} by as much as possible. We can increase x_{34} to the point where a basic variable first attains its upper or lower bound. Thus, arc (3, 4) implies that $\theta \leq 5$; arc (3, 5) requires $1 - \theta \geq 0$ or $\theta \leq 1$; arc (4, 5) requires $4 + \theta \leq 6$ or $\theta \leq 2$. So the best we can do is set $\theta = 1$. The basic variable that first hits its upper or lower bound as θ is increased is chosen to exit the basis (in case of a tie, we can choose the exiting variable arbitrarily). Now x_{35} exits the basis, and the new bfs is shown in Figure 55. The spanning tree corresponding to the current set of basic variables is (1, 3), (3, 4), (4, 5), and (2, 5). We now compute the coefficient of each nonbasic variable in row 0. To begin, we solve the following set of equations:

$$y_1 = 0, \qquad y_1 - y_3 = 12, \qquad y_3 - y_4 = 3, \qquad y_2 - y_5 = 6, \qquad y_4 - y_5 = 3$$

This yields $y_1 = 0$, $y_2 = -12$, $y_3 = -12$, $y_4 = -15$, and $y_5 = -18$.

The nonbasic variables that currently equal their upper bounds will have row 0 coefficients of

$$\bar{c}_{12} = 0 - (-12) - 10 = 2 \qquad \text{and} \qquad \bar{c}_{14} = 0 - (-15) - 6 = 9$$

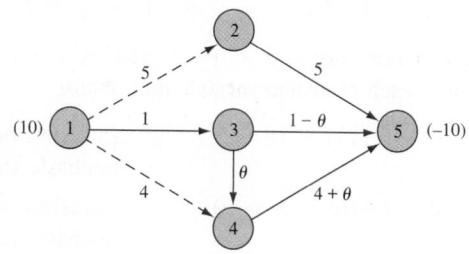

FIGURE 54
Cycle (3, 4), (4, 5), (3, 5) Helps Us Pivot in x_{34}

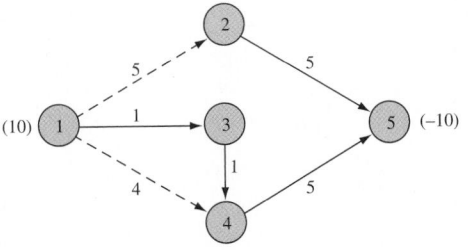

FIGURE 55
New bfs ($\theta = 1$) After
x_{34} Enters and x_{35} Exits

The nonbasic variables that currently equal their lower bounds will have row 0 coefficients of

$$\bar{c}_{32} = -12 - (-12) - 2 = -2 \quad \text{and} \quad \bar{c}_{35} = -12 - (-18) - 7 = -1$$

Because each nonbasic variable at its upper bound has $\bar{c}_{ij} \geq 0$, and each nonbasic variable at its lower bound has $\bar{c}_{ij} \leq 0$, the current bfs is optimal. Thus, the optimal solution to the MCNFP in Figure 52 is

$$\text{Upper bounded variables:} \quad x_{12} = 5, \quad x_{14} = 4$$
$$\text{Lower bounded variables:} \quad x_{32} = x_{35} = 0$$
$$\text{Basic variables:} \quad x_{13} = 1, x_{34} = 1, x_{25} = 5, x_{45} = 5$$

Summary of the Network Simplex Method

Step 1 Determine a starting bfs. The $n - 1$ basic variables will correspond to a spanning tree. Indicate nonbasic variables at their upper bound by dashed arcs.

Step 2 Compute $y_1, y_2, \ldots y_n$ (often called the simplex multipliers) by solving $y_1 = 0$, $y_i - y_j = c_{ij}$ for all basic variables x_{ij}. For all nonbasic variables, determine the row 0 coefficient $\bar{c}_{ij}$ from $\bar{c}_{ij} = y_i - y_j - c_{ij}$. The current bfs is optimal if $\bar{c}_{ij} \leq 0$ for all $x_{ij} = L_{ij}$ and $\bar{c}_{ij} \geq 0$ for all $x_{ij} = U_{ij}$. If the bfs is not optimal, choose the nonbasic variable that most violates the optimality conditions as the entering basic variable.

Step 3 Identify the cycle (there will be exactly one!) created by adding the arc corresponding to the entering variable to the current spanning tree of the current bfs. Use conservation of flow to determine the new values of the variables in the cycle. The variable that exits the basis will be the variable that first hits its upper or lower bound as the value of the entering basic variable is changed.

Step 4 Find the new bfs by changing the flows of the arcs in the cycle found in step 3. Now go to step 2.

Example 9 illustrates the network simplex.

EXAMPLE 9 | **Network Simplex Solution to MCNFP**

Use the network simplex to solve the MCNFP in Figure 56.

Solution A bfs requires that we find a spanning tree (three arcs that connect nodes 1, 2, 3, and 4 and do not form a cycle). Any arcs not in the spanning tree may be set equal to their upper or lower bound. By trial and error, we find the bfs in Figure 57 involving the spanning tree (1, 2), (1, 3), and (2, 4).
To find $y_1, y_2, y_3,$ and y_4 we solve

$$y_1 = 0, \quad y_1 - y_2 = 4, \quad y_2 - y_4 = 3, \quad y_1 - y_3 = 3$$

FIGURE **56**
Example of
Network Simplex

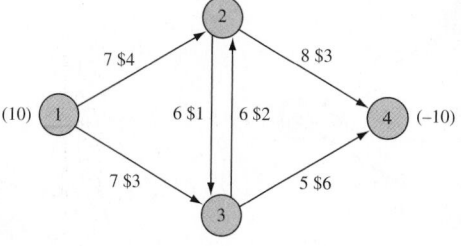

FIGURE **57**
bfs for Example 9

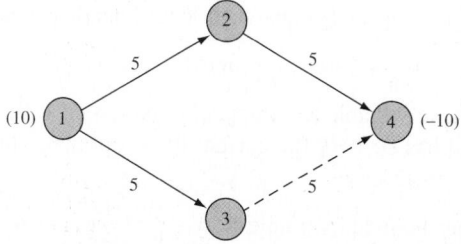

This yields $y_1 = 0$, $y_2 = -4$, $y_3 = -3$, and $y_4 = -7$. The row 0 coefficients for each nonbasic variable are

$$\bar{c}_{34} = -3 - (-7) - 6 = -2 \quad \text{(Violates optimality condition)}$$
$$\bar{c}_{23} = -4 - (-3) - 1 = -2 \quad \text{(Satisfies optimality condition)}$$
$$\bar{c}_{32} = -3 - (-4) - 2 = -1 \quad \text{(Satisfies optimality condition)}$$

Thus, x_{34} enters the basis. We set $x_{34} = 5 - \theta$ and obtain the cycle in Figure 58. From arc $(1, 2)$, we find $5 + \theta \le 7$ or $\theta \le 2$. From arc $(1, 3)$, we find $5 - \theta \ge 0$ or $\theta \le 5$. From arc $(2, 4)$, we find $5 + \theta \le 8$ or $\theta \le 3$. From arc $(3, 4)$, we find $5 - \theta \ge 0$ or $\theta \le 5$. Thus, we can set $\theta = 2$. Now x_{12} exits the basis at its upper bound, and x_{34} enters, yielding the bfs in Figure 59.

The new bfs is associated with the spanning tree $(1, 3)$, $(2, 4)$, and $(3, 4)$. Solving for the new values of the simplex multipliers, we obtain

$$y_1 = 0, \quad y_1 - y_3 = 3, \quad y_3 - y_4 = 6, \quad y_2 - y_4 = 3$$

This yields $y_1 = 0$, $y_2 = -6$, $y_3 = -3$, $y_4 = -9$. The coefficient of each nonbasic variable in row 0 is given by

$$\bar{c}_{12} = 0 - (-6) - 4 = 2 \qquad \text{(Satisfies optimality condition)}$$
$$\bar{c}_{23} = -6 - (-3) - 1 = -4 \qquad \text{(Satisfies optimality condition)}$$
$$\bar{c}_{32} = -3 - (-6) - 2 = 1 \qquad \text{(Violates optimality condition)}$$

Now x_{32} enters the basis, yielding the cycle in Figure 60. From arc $(2, 4)$, we find $7 + \theta \le 8$ or $\theta \le 1$); from arc $(3, 4)$, we find $3 - \theta \ge 0$ or $\theta \le 3$. From arc $(3, 2)$, we find $\theta \le 6$. So we now set $\theta = 1$ and have x_{24} exit from the basis at its upper bound. The new bfs is given in Figure 61.

The current set of basic values corresponds to the spanning tree $(1, 3)$, $(3, 2)$, and $(3, 4)$. The new values of the simplex multipliers are found by solving

$$y_1 = 0, \quad y_1 - y_3 = 3, \quad y_3 - y_2 = 2, \quad y_3 - y_4 = 6$$

which yields $y_1 = 0$, $y_2 = -5$, $y_3 = -3$, $y_4 = -9$. The coefficient of each nonbasic variable in row 0 is now

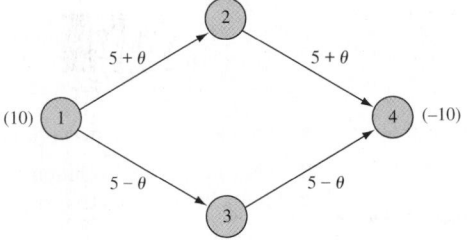

FIGURE 58
Cycle Created When
x_{34} **Enters the Basis**

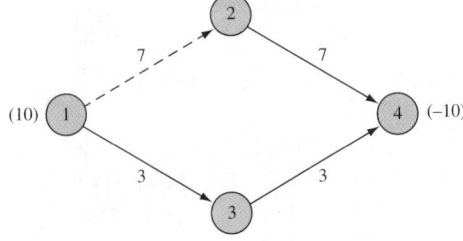

FIGURE 59
bfs After x_{12} Exits
and x_{34} Enters

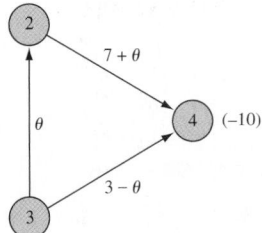

FIGURE 60
Cycle Created When
x_{32} **Enters Basis**

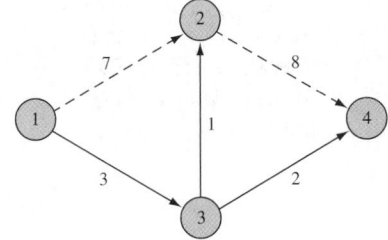

FIGURE 61
New bfs When x_{32}
Enters and x_{24} Exits

$$\bar{c}_{23} = -5 - (-3) - 1 = -3 \qquad \text{(Satisfies optimality condition)}$$
$$\bar{c}_{12} = 0 - (-5) - 4 = 1 \qquad \text{(Satisfies optimality condition)}$$
$$\bar{c}_{24} = -5 - (-9) - 3 = 1 \qquad \text{(Satisfies optimality condition)}$$

Thus, the current bfs is optimal. The optimal solution to the MCNFP is

Basic variables: $\quad x_{13} = 3, \qquad x_{32} = 1, \qquad x_{34} = 2$

Nonbasic variables at their upper bound: $\qquad x_{12} = 7, \qquad x_{24} = 8$

Nonbasic variable at lower bound: $\quad x_{23} = 0$

The optimal z-value is obtained from

$$z = 7(4) + 3(3) + 1(2) + 8(3) + 2(6) = \$75$$

PROBLEMS

Group A

1 Consider the problem of finding the shortest path from node 1 to node 6 in Figure 2.

   a Formulate this problem as an MCNFP.

   b Find a bfs in which x_{12}, x_{24}, and x_{46} are positive. (*Hint:* A degenerate bfs will be obtained.)

   c Use the network simplex to find the shortest path from node 1 to node 6.

2 For the MCNFP in Figure 62, find a bfs.

3 Find the optimal solution to the MCNFP in Figure 63 using the bfs in Figure 64 as a starting basis.

4 Find a bfs for the network in Figure 65.

5 Find the optimal solution to the MCNFP in Figure 66 using the bfs in Figure 67 as a starting basis.

FIGURE **65**

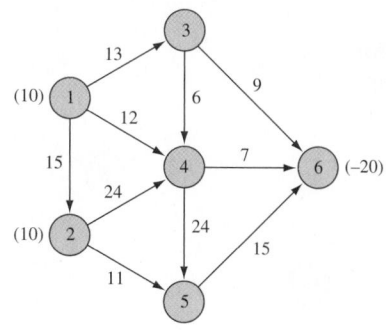

FIGURE **62**

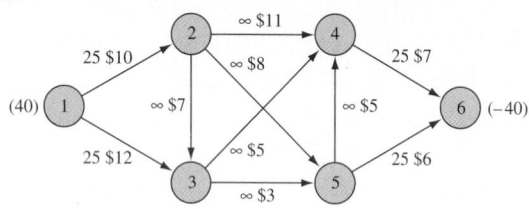

FIGURE **66**

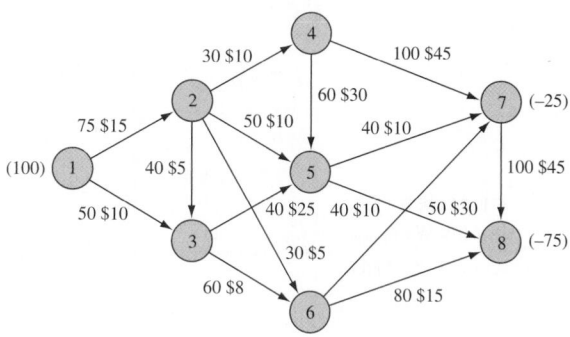

FIGURE **63**

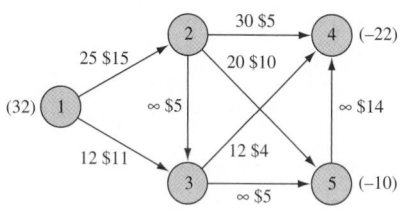

FIGURE **64**

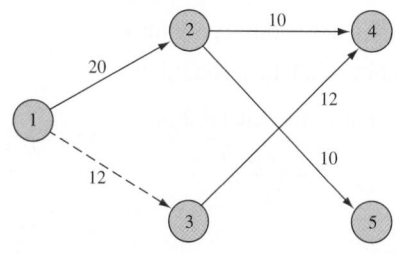

FIGURE **67**

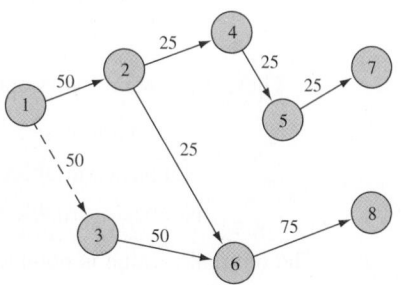

S U M M A R Y   Shortest-Path Problems

Suppose we want to find the shortest path from node 1 to node j in a network in which all arcs have nonnegative lengths.

Dijkstra's Algorithm

1 Label node 1 with a permanent label of 0. Then label each arc connected to node 1 by a single arc with a "temporary" label equal to the length of the arc joining node 1 and node i. Remaining nodes will have a temporary label of ∞. Choose the node with the smallest temporary label and make this label permanent.

2 Suppose that node i is the $(k + 1)$th node to be given a permanent label. For each node j that now has a temporary label and is connected to node i by an arc, replace node j's temporary label with min {node j's current temporary label, (node i's permanent label) + length of arc (i, j)}. Make the smallest temporary label a permanent label. Continue this process until all nodes have permanent labels. To find the shortest path from node 1 to node j, work backward from node j by finding nodes having labels differing by exactly the length of the connecting arc. If the shortest path from node 1 to node j is desired, stop the labeling process as soon as node j receives a permanent label.

The Shortest-Path Problem as a Transshipment Problem

To find the shortest path from node 1 to node j, try to minimize the cost of sending one unit from node 1 to node j (with all other nodes in the network being transshipment points), where the cost of sending one unit from node k to node k' is the length of arc (k, k') if such an arc exists and is M (a large positive number) if such an arc does not exist. As in Section 7.6, the cost of shipping one unit from a node to itself is zero.

Maximum-Flow Problems

We can find the maximum flow from source to sink in a network by linear programming or by the Ford–Fulkerson method.

Finding Maximum Flow by Linear Programming

Let

$$x_0 = \text{flow through artificial arc going from sink to source}$$

Then to find the maximum flow from source to sink, maximize x_0 subject to the following two sets of constraints:

1 The flow through each arc must be nonnegative and cannot exceed the arc capacity.

2 Flow into node i = flow out of node i (Conservation of flow)

Finding Maximum Flow by the Ford–Fulkerson Method

Let

$$I = \text{set of arcs in which flow may be increased}$$
$$R = \text{set of arcs in which flow may be reduced}$$

Step 1   Find a feasible flow (setting each arc's flow to zero will do).

Step 2   Using the following procedure, try to find a chain of labeled arcs and nodes that can be used to label the sink. Label the source. Then label vertices and arcs (except for arc a_0) according to the following rules: (1) If vertex x is labeled, then vertex y is unlabeled and arc (x, y) is a member of I; then label vertex y and arc (x, y). Arc (x, y) is called a **forward arc.** (2) If vertex y is unlabeled, then vertex x is labeled and arc (y, x) is a member of R; then label vertex y and arc (y, x). Arc (y, x) is called a **backward arc.**

If the sink cannot be labeled, the current feasible flow is a maximum flow; if the sink is labeled, go on to step 3.

Step 3   If the chain used to label the sink consists entirely of forward arcs, the flow through each of the forward arcs in the chain may be increased, thereby increasing the flow from source to sink. If the chain used to label the sink consists of both forward and backward arcs, increase the flow in each forward arc in the chain and decrease the flow in each backward arc in the chain. Again, this will increase the flow from source to sink. Return to step 2.

Critical Path Method

Assuming the duration of each activity is known, the critical path method (CPM) may be used to find the duration of a project.

Rules for Constructing an AOA Project Diagram

1   Node 1 represents the start of the project. An arc should lead from node 1 to represent each activity that has no predecessors.

2   A node (called the finish node) representing the completion of the project should be included in the network.

3   Number the nodes in the network so that the node representing the completion of an activity always has a larger number than the node representing the beginning of an activity (there may be more than one numbering scheme that satisfies rule 3).

4   An activity should not be represented by more than one arc in the network.

5   Two nodes can be connected by at most one arc.

To avoid violating rules 4 and 5, it is sometimes necessary to utilize a **dummy activity** that takes zero time.

Computation of Early Event Time

The early event time for node i, denoted $ET(i)$, is the earliest time at which the event corresponding to node i can occur. We compute $ET(i)$ as follows:

Step 1   Find each prior event to node i that is connected by an arc to node i. These events are the **immediate predecessors** of node i.

Step 2   To the ET for each immediate predecessor of node i, add the duration of the activity connecting the immediate predecessor to node i.

Step 3 $ET(i)$ equals the maximum of the sums computed in step 2.

Computation of Late Event Time

The late event time for node i, denoted $LT(i)$, is the latest time at which the event corresponding to node i can occur without delaying the completion of the project. We compute $LT(i)$ as follows:

Step 1 Find each node that occurs after node i and is connected to node i by an arc. These events are the **immediate successors** of node i.

Step 2 From the LT for each immediate successor to node i, subtract the duration of the activity joining the successor to node i.

Step 3 $LT(i)$ is the smallest of the differences determined in step 2.

Total Float

For an arbitrary arc representing activity (i, j), the total float (denoted $TF(i, j)$ of the activity represented by (i, j) is the amount by which the starting time of activity (i, j) could be delayed beyond its earliest possible starting time without delaying the completion of the project (assuming no other activities are delayed):

$$TF(i, j) = LT(j) - ET(i) - t_{ij} \quad [t_{ij} = \text{duration of activity represented by arc } (i, j)]$$

Any activity with a total float of zero is a **critical activity.** A path from node 1 to the finish node that consists entirely of critical activities is called a **critical path.** Any critical path (there may be more than one in a project network) is the longest path in the network from the start node (node 1) to the finish node. If the start of a critical activity is delayed, or if the duration of a critical activity is longer than expected, then the completion of the project will be delayed.

Free Float

The free float of the activity corresponding to arc (i, j), denoted by $FF(i, j)$, is the amount by which the starting time of the activity corresponding to arc (i, j) (or the duration of the activity) can be delayed without delaying the start of any later activity beyond its earliest possible starting time:

$$FF(i, j) = ET(j) - ET(i) - t_{ij}$$

Linear programming can be used to find a critical path and the duration of the project. Let

$$x_j = \text{time at which node } j \text{ in project network occurs}$$
$$F = \text{node representing finish or completion of the project}$$

To find a critical path, minimize $z = x_F - x_1$ subject to

$$x_j \geq x_i + t_{ij} \quad \text{or} \quad x_j - x_i \geq t_{ij} \quad \text{for each arc}$$
$$x_j \text{ urs}$$

The optimal objective function value is the length of any critical path (or time to project completion). To find a critical path, simply find a path from node 1 to node F for which each arc in the path is represented by an arc (i, j) whose constraint $(x_j - x_i \geq t_{ij})$ has a dual price of -1.

Linear programming can also be used to determine the minimum-cost method of reducing the duration of activities (crashing) to meet a project completion deadline.

PERT

If the durations of the project's activities are not known with certainty, then PERT may be used to estimate the probability that the project will be completed in a specified amount of time. PERT requires that for each activity the following three numbers be specified:

a = estimate of the activity's duration under the most favorable conditions

b = estimate of the activity's duration under the least favorable conditions

m = most likely value for the activity's duration

If the estimates a, b, and m refer to the activity represented by arc (i, j), then $\mathbf{T}_{ij}$ is the random variable representing the duration of the activity represented by arc (i, j). $\mathbf{T}_{ij}$ has (approximately) the following properties:

$$E(\mathbf{T}_{ij}) = \frac{a + 4m + b}{6}$$

$$\text{var}\mathbf{T}_{ij} = \frac{(b - a)^2}{36}$$

Then

$$\sum_{(i, j) \in \text{path}} E(\mathbf{T}_{ij}) = \text{expected duration of activities on any path}$$

$$\sum_{(i, j) \in \text{path}} \text{var}\mathbf{T}_{ij} = \text{variance of duration of activities on any path}$$

Assuming (sometimes incorrectly) that the critical path found by CPM is the critical path, and assuming that the duration of the critical path is normally distributed, the preceding equations may be used to estimate the probability that the project will be completed within any specified length of time.

Minimum-Cost Network Flow Problems

The transportation, assignment, transshipment, shortest-path, maximum-flow, and critical path problems are all special cases of the minimum-cost network flow problem (MCNFP).

x_{ij} = number of units of flow sent from node i to node j through arc (i, j)

b_i = net supply (outflow $-$ inflow) at node i

c_{ij} = cost of transporting one unit of flow from node i to node j via arc (i, j)

L_{ij} = lower bound on flow through arc (i, j) (if there is no lower bound, let $L_{ij} = 0$)

U_{ij} = upper bound on flow through arc (i, j) (if there is no upper bound, let $U_{ij} = \infty$)

Then an MCNFP may be written as

$$\min \sum_{\text{all arcs}} c_{ij}x_{ij}$$

$$\text{s.t.} \quad \sum_{j} x_{ij} - \sum_{k} x_{ki} = b_i \qquad \text{(for each node } i \text{ in the network)}$$

$$L_{ij} \leq x_{ij} \leq U_{ij} \qquad \text{(for each arc in the network)}$$

The first set of constraints are the **flow balance equations,** and the second set of constraints express limitations on arc capacities.

Any MCNFP may be solved by a computer code using the **network simplex;** the user need only input the nodes and arcs in the network, the c_{ij}'s and arc capacity for each arc, and the b_i's for each node. Formulation of a problem as an MCNFP may require adding a dummy point to the problem.

Minimum Spanning Tree Problems

The following method (MST algorithm) may be used to find a minimum spanning tree for a network:

Step 1 Begin at any node i, and join node i to the node in the network (node j) that is closest to node i. The two nodes i and j now form a connected set of nodes $C = \{i, j\}$ and arc (i, j) will be in the minimum spanning tree. The remaining nodes in the network (C') are the unconnected set of nodes.

Step 2 Choose a member of $C'(n)$ that is closest to some node in C. Let m represent the node in C that is closest to n. Then the arc (m, n) will be in the minimum spanning tree. Update C and C'. Because n is now connected to $\{i, j\}$, C now equals $\{i, j, n\}$, and we must eliminate node n from C'.

Step 3 Repeat this process until a minimum spanning tree is found. Ties for closest node and arc may be broken arbitrarily.

Network Simplex Method

Step 1 Determine a starting bfs. The $n - 1$ basic variables will correspond to a spanning tree. Indicate nonbasic variables at their upper bound by dashed arcs.

Step 2 Compute $y_1, y_2, \ldots y_n$ (often called the *simplex multipliers*) by solving $y_1 = 0$, $y_i - y_j = c_{ij}$ for all basic variables x_{ij}. For all nonbasic variables, determine the row 0 coefficient $\bar{c}_{ij}$ from $\bar{c}_{ij} = y_i - y_j - c_{ij}$. The current bfs is optimal if $\bar{c}_{ij} \leq 0$ for all $x_{ij} = L_{ij}$ and $\bar{c}_{ij} \geq 0$ for all $x_{ij} = U_{ij}$. If the bfs is not optimal, then choose the nonbasic variable that most violates the optimality conditions as the entering basic variable.

Step 3 Identify the cycle (there will be exactly one!) created by adding the arc corresponding to the entering variable to the current spanning tree of the current bfs. Use conservation of flow to determine the new values of the variables in the cycle. The variable that first hits its upper or lower bound as the value of the entering basic variable is changed exits the basis.

Step 4 Find the new bfs by changing the flows of the arcs in the cycle found in step 3. Go to step 2.

R E V I E W   P R O B L E M S

Group A

1 A truck must travel from New York to Los Angeles. As shown in Figure 68, a variety of routes are available. The number associated with each arc is the number of gallons of fuel required by the truck to traverse the arc.

a Use Dijkstra's algorithm to find the route from New York to Los Angeles that uses the minimum amount of gas.

b Formulate a balanced transportation problem that could be used to find the route from New York to Los Angeles that uses the minimum amount of gas.

c Formulate as an MCNFP the problem of finding the New York to Los Angeles route that uses the minimum amount of gas.

FIGURE **68**
Network for Problem 1

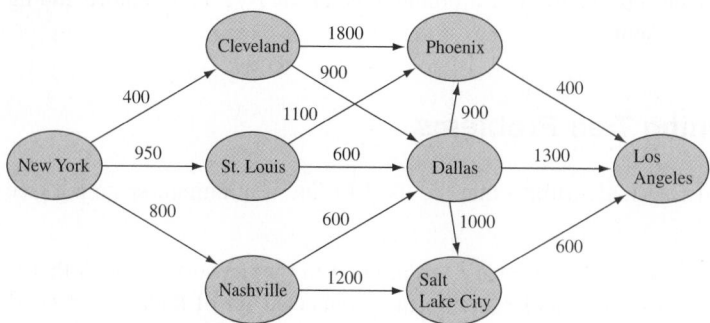

2 Telephone calls from New York to Los Angeles are transported as follows: The call is sent first to either Chicago or Memphis, then routed through either Denver or Dallas, and finally sent to Los Angeles. The number of phone lines joining each pair of cities is shown in Table 39.

a Formulate an LP that can be used to determine the maximum number of calls that can be sent from New York to Los Angeles at any given time.

b Use the Ford–Fulkerson method to determine the maximum number of calls that can be sent from New York to Los Angeles at any given time.

TABLE 39

Cities	No. of Telephone Lines
N.Y.–Chicago	500
N.Y.–Memphis	400
Chicago–Denver	300
Chicago–Dallas	250
Memphis–Denver	200
Memphis–Dallas	150
Denver–L.A.	400
Dallas–L.A.	350

3 Before a new product can be introduced, the activities in Table 40 must be completed (all times are in weeks).

a Draw the project diagram.

b Determine all critical paths and critical activities.

c Determine the total float and free float for each activity.

d Set up an LP that can be used to determine the critical path.

e Formulate an MCNFP that can be used to find the critical path.

f It is now 12 weeks before Christmas. What is the probability that the product will be in the stores before Christmas?

g The duration of each activity can be reduced by up to 2 weeks at the following cost per week: A, $80; B, $60; C, $30; D, $60; E, $40; F, $30; G, $20. Assuming that the duration of each activity is known with certainty, formulate an LP that will minimize the cost of getting the product into the stores by Christmas.

4 During the next three months, Shoemakers, Inc. must meet (on time) the following demands for shoes: month 1, 1,000 pairs; month 2, 1,500 pairs; month 3, 1,800 pairs. It takes 1 hour of labor to produce a pair of shoes. During each of the next three months, the following number of regular-time labor hours are available: month 1, 1,000 hours; month 2, 1,200 hours; month 3, 1,200 hours. Each month, the company can require workers to put in up to 400 hours of overtime. Workers

TABLE 40

Activity	Description	Predecessors	Duration	a	b	m
A	Design the product	—	6	2	10	6
B	Survey the market	—	5	4	6	5
C	Place orders for raw materials	A	3	2	4	3
D	Receive raw materials	C	2	1	3	2
E	Build prototype of product	A, D	3	1	5	3
F	Develop ad campaign	B	2	3	5	4
G	Set up plan for mass production	E	4	2	6	4
H	Deliver product to stores	G, F	2	0	4	2

are paid only for the hours they work, and a worker receives $4 per hour for regular-time work and $6 per hour for overtime work. At the end of each month, a holding cost of $1.50 per pair of shoes is incurred. Formulate an MCNFP that can be used to minimize the total cost incurred in meeting the demands of the next three months. A formulation requires drawing the appropriate network and determining the c_{ij}'s, b_i's, and arc capacities. How would you modify your answer if demand could be backlogged (all demand must still be met by the end of month 3) at a cost of $20/pair/month?

5 Find a minimum spanning tree for the network in Figure 68.

6 A company produces a product at two plants, 1 and 2. The unit production cost and production capacity during each period are given in Table 41. The product is instantaneously shipped to the company's only customer according to the unit shipping costs given in Table 42. If a unit is produced and shipped during period 1, it can still be used to meet a period 2 demand, but a holding cost of $13 per unit in inventory is assessed. At the end of period 1, at most six units may be held in inventory. Demands are as follows: period 1, 9; period 2, 11. Formulate an MCNFP that can be used to minimize the cost of meeting all demands on time. Draw the network and determine the net outflow at each node, the arc capacities, and shipping costs.

7 A project is considered completed when activities A–F have all been completed. The duration and predecessors of each activity are given in Table 43. The LINDO output in Figure 69 can be used to determine the critical path for this project.

a Use the LINDO output to draw the project network. Indicate the activity represented by each arc.

b Determine a critical path in the network. What is the earliest the project can be completed?

8[†] State University has three professors who each teach four courses per year. Each year, four sections of marketing, finance, and production must be offered. At least one section of each class must be offered during each semester (fall and spring). Each professor's time preference and preference for teaching various courses are given in Table 44.

TABLE 41

	Unit Production Cost ($)	Capacity
Plant 1 (period 1)	33	7
Plant 1 (period 2)	43	4
Plant 2 (period 1)	30	9
Plant 2 (period 2)	41	9

TABLE 42

	Period 1	Period 2
Plant 1 to customer	$51	$60
Plant 2 to customer	$42	$71

†Based on Mulvey (1979).

FIGURE 69

```
MIN     X6 - X1
SUBJECT TO
      2)  - X1 + X3 >=    3
      3)    X4 - X2 >=    1
      4)  - X3 + X4 >=    0
      5)  - X4 + X5 >=    7
      6)  - X3 + X5 >=    5
      7)    X6 - X5 >=    5
      8)    X3 - X2 >=    0
      9)  - X1 + X2 >=    2
END

    LP OPTIMUM FOUND AT STEP    3

        OBJECTIVE FUNCTION VALUE

   1)        15.0000000

VARIABLE         VALUE        REDUCED COST
      X6      15.000000         0.000000
      X1       0.000000         0.000000
      X3       3.000000         0.000000
      X4       3.000000         0.000000
      X2       2.000000         0.000000
      X5      10.000000         0.000000

ROW        SLACK OR SURPLUS     DUAL PRICES
   2)        0.000000          -1.000000
   3)        0.000000           0.000000
   4)        0.000000          -1.000000
   5)        0.000000          -1.000000
   6)        2.000000           0.000000
   7)        0.000000          -1.000000
   8)        1.000000           0.000000
   9)        0.000000           0.000000

NO. ITERATIONS=       3
```

TABLE 43

Activity	Duration	Immediate Predecessors
A	2	—
B	3	—
C	1	A
D	5	A, B
E	7	B, C
F	5	D, E

The total satisfaction a professor earns teaching a class is the sum of the semester satisfaction and the course satisfaction. Thus, professor 1 derives a satisfaction of $3 + 6 = 9$ from teaching marketing during the fall semester. Formulate an MCNFP that can be used to assign professors to courses so as to maximize the total satisfaction of the three professors.

Group B

9[†] During the next two months, Machineco must meet (on time) the demands for three types of products shown in Table 45. Two machines are available to produce these

†This problem is based on Brown, Geoffrion, and Bradley (1981).

TABLE 44

	Professor 1	Professor 2	Professor 3
Fall Preference	3	5	4
Spring Preference	4	3	4
Marketing	6	4	5
Finance	5	6	4
Production	4	5	6

TABLE 45

Month	Product 1	Product 2	Product 3
1	50 units	70 units	80 units
2	60 units	90 units	120 units

products. Machine 1 can only produce products 1 and 2, and machine 2 can only produce products 2 and 3. Each machine can be used for up to 40 hours per month. Table 46 shows the time required to produce one unit of each product (independent of the type of machine); the cost of producing one unit of each product on each type of machine; and the cost of holding one unit of each product in inventory for one month. Formulate an MCNFP that could be used to minimize the total cost of meeting all demands on time.

TABLE 46

Product	Production Time (minutes)	Production Cost ($) Machine 1	Production Cost ($) Machine 2	Holding Cost ($)
1	30	40	—	15
2	20	45	60	10
3	15	—	55	5

REFERENCES

Brown, G., A. Geoffrion, and G. Bradley. "Production and Sales Planning with Limited Shared Tooling at the Key Operation," *Management Science* 27(1981):247–259.

Glover, F., et al. "The Passenger-Mix Problem in the Scheduled Airlines," *Interfaces* 12(1982):73–80.

Mulvey, M. "Strategies in Modeling: A Personnel Example," *Interfaces* 9(no. 3, 1979):66–75.

Peterson, I. "Proven Path for Limiting Shortest Shortcut," *Science News* December 22, 1990: 389.

Ravidran, A. "On Compact Book Storage in Libraries," *Opsearch* 8(1971).

The following three texts contain an overview of networks at an elementary level:

Chachra, V., P. Ghare, and J. Moore. *Applications of Graph Theory Algorithms.* New York: North-Holland, 1979.

Mandl, C. *Applied Network Optimization.* Orlando, Fla.: Academic Press, 1979.

Phillips, D., and A. Diaz. *Fundamentals of Network Analysis.* Englewood Cliffs, N.J.: Prentice Hall, 1981.

The two best comprehensive references on network models are:

Ahuja, R., Magnanti, T., and Orlin, J. *Network Flows: Theory Algorithms and Applications.* Englewood-Cliffs, N.J.: Prentice-Hall, 1993.

Bersetkas, D. *Linear Network Optimization: Algorithms and Codes.* Cambridge, Mass.: MIT Press, 1991.

Detailed discussion of methods for solving shortest path problems can be found in the following three texts:

Denardo, E. *Dynamic Programming: Theory and Applications.* Englewood Cliffs, N.J.: Prentice Hall, 1982.

Evans, T., and E. Minieka. *Optimization Algorithms for Networks and Graphs.* New York: Dekker, 1992. Also discusses minimum spanning tree algorithms.

Hu, T. *Combinatorial Algorithms.* Reading, Mass.: Addison-Wesley, 1982. Also discusses minimum spanning tree algorithms.

Evans and Minieka (1992) and Hu (1982) discuss the maximum-flow problem in detail, as do the following three texts:

Ford, L., and D. Fulkerson. *Flows in Networks.* Princeton, N.J.: Princeton University Press, 1962.

Jensen, P., and W. Barnes. *Network Flow Programming.* New York: Wiley, 1980.

Lawler, E. *Combinatorial Optimization: Networks and Matroids.* Chicago: Holt, Rinehart & Winston, 1976.

Excellent discussions of CPM and PERT are contained in:

Hax, A., and D. Candea. *Production and Inventory Management.* Englewood Cliffs, N.J.: Prentice Hall, 1984.

Wiest, J., and F. Levy. *A Management Guide to PERT/CPM,* 2d ed. Englewood Cliffs, N.J.: Prentice Hall, 1977.

Jensen and Barnes (1980) and the following references each contain a detailed discussion of the network simplex method used to solve an MCNFP.

Chvàtal, V. *Linear Programming.* San Francisco: Freeman, 1983.

Shapiro, J. *Mathematical Programming: Structures and Algorithms.* New York: Wiley, 1979.

Wu, N., and R. Coppins. *Linear Programming and Extensions.* New York: McGraw-Hill, 1981.

An excellent discussion of applications of MCNFPs is contained in the following:

Glover, F., D. Klingman, and N. Phillips. *Network Models and Their Applications in Practice.* New York: Wiley, 1992.

Integer Programming

Recall that we defined integer programming problems in our discussion of the Divisibility Assumption in Section 3.1. Simply stated, an *integer programming problem* (IP) is an LP in which some or all of the variables are required to be non-negative integers.[†]

In this chapter (as for LPs in Chapter 3), we find that many real-life situations may be formulated as IPs. Unfortunately, we will also see that IPs are usually much harder to solve than LPs.

In Section 9.1, we begin with necessary definitions and some introductory comments about IPs. In Section 9.2, we explain how to formulate integer programming models. We also discuss how to solve IPs on the computer with LINDO, LINGO, and Excel Solver. In Sections 9.3–9.8, we discuss other methods used to solve IPs.

9.1 Introduction to Integer Programming

An IP in which all variables are required to be integers is called a **pure integer programming problem.** For example,

$$\max z = 3x_1 + 2x_2$$
$$\text{s.t.} \quad x_1 + x_2 \leq 6 \tag{1}$$
$$x_1, x_2 \geq 0, \; x_1, x_2 \text{ integer}$$

is a pure integer programming problem.

An IP in which only some of the variables are required to be integers is called a **mixed integer programming problem.** For example,

$$\max z = 3x_1 + 2x_2$$
$$\text{s.t.} \quad x_1 + x_2 \leq 6$$
$$x_1, x_2 \geq 0, \; x_1 \text{ integer}$$

is a mixed integer programming problem (x_2 is not required to be an integer).

An integer programming problem in which all the variables must equal 0 or 1 is called a 0–1 IP. In Section 9.2, we see that 0–1 IPs occur in surprisingly many situations.[‡] The following is an example of a 0–1 IP:

$$\max z = x_1 - x_2$$
$$\text{s.t.} \quad x_1 + 2x_2 \leq 2$$
$$2x_1 - x_2 \leq 1 \tag{2}$$
$$x_1, x_2 = 0 \text{ or } 1$$

Solution procedures especially designed for 0–1 IPs are discussed in Section 9.7.

[†]A nonlinear integer programming problem is an optimization problem in which either the objective function or the left-hand side of some of the constraints are nonlinear functions and some or all of the variables must be integers. Such problems may be solved with LINGO or Excel Solver.

[‡]Actually, any pure IP can be reformulated as an equivalent 0–1 IP (Section 9.7).

The concept of LP relaxation of an integer programming problem plays a key role in the solution of IPs.

DEFINITION ■ The LP obtained by omitting all integer or 0–1 constraints on variables is called the **LP relaxation** of the IP. ■

For example, the LP relaxation of (1) is

$$\max z = 3x_1 + 2x_2$$
$$\text{s.t.} \quad x_1 + x_2 \leq 6 \qquad \qquad (1')$$
$$x_1, x_2 \geq 0$$

and the LP relaxation of (2) is

$$\max z = x_1 - x_2$$
$$\text{s.t.} \quad x_1 + 2x_2 \leq 2$$
$$2x_1 - x_2 \leq 1 \qquad \qquad (2')$$
$$x_1, x_2 \geq 0$$

Any IP may be viewed as the LP relaxation plus additional constraints (the constraints that state which variables must be integers or be 0 or 1). Hence, the LP relaxation is a less constrained, or more relaxed, version of the IP. This means that *the feasible region for any IP must be contained in the feasible region for the corresponding LP relaxation.* For any IP that is a max problem, this implies that

$$\text{Optimal } z\text{-value for LP relaxation} \geq \text{optimal } z\text{-value for IP} \qquad (3)$$

This result plays a key role when we discuss the solution of IPs.

To shed more light on the properties of integer programming problems, we consider the following simple IP:

$$\max z = 21x_1 + 11x_2$$
$$\text{s.t.} \quad 7x_1 + 4x_2 \leq 13 \qquad \qquad (4)$$
$$x_1, x_2 \geq 0; x_1, x_2 \text{ integer}$$

From Figure 1, we see that the feasible region for this problem consists of the following set of points: $S = \{(0, 0), (0, 1), (0, 2), (0, 3), (1, 0), (1, 1)\}$. Unlike the feasible region for any LP, the one for (4) is not a convex set. By simply computing and comparing the z-values for each of the six points in the feasible region, we find the optimal solution to (4) is $z = 33$, $x_1 = 0$, $x_2 = 3$.

If the feasible region for a pure IP's LP relaxation is bounded, as in (4), then the feasible region for the IP will consist of a finite number of points. In theory, such an IP could be solved (as described in the previous paragraph) by enumerating the z-values for each feasible point and determining the feasible point having the largest z-value. The problem with this approach is that most actual IPs have feasible regions consisting of billions of feasible points. In such cases, a complete enumeration of all feasible points would require a large amount of computer time. As we explain in Section 9.3, IPs often are solved by cleverly enumerating all the points in the IP's feasible region.

Further study of (4) sheds light on other interesting properties of IPs. Suppose that a naive analyst suggests the following approach for solving an IP: First solve the LP relaxation; then round off (to the nearest integer) each variable that is required to be an integer and that assumes a fractional value in the optimal solution to the LP relaxation.

Applying this approach to (4), we first find the optimal solution to the LP relaxation: $x_1 = \frac{13}{7}$, $x_2 = 0$. Rounding this solution yields the solution $x_1 = 2$, $x_2 = 0$ as a possible

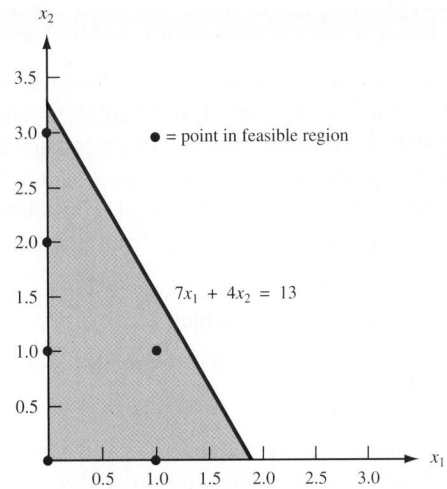

FIGURE 1
Feasible Region for
Simple IP (4)

optimal solution to (4). But $x_1 = 2$, $x_2 = 0$ is infeasible for (4), so it cannot possibly be the optimal solution to (4). Even if we round x_1 downward (yielding the candidate solution $x_1 = 1$, $x_2 = 0$), we do not obtain the optimal solution ($x_1 = 0$, $x_2 = 3$ is the optimal solution).

For some IPs, it can even turn out that every roundoff of the optimal solution to the LP relaxation is infeasible. To see this, consider the following IP:

$$\max z = 4x_1 + x_2$$
$$\text{s.t.} \quad 2x_1 + x_2 \le 5$$
$$2x_1 + 3x_2 = 5$$
$$x_1, x_2 \ge 0; \; x_1, x_2 \text{ integer}$$

The optimal solution to the LP relaxation for this IP is $z = 10$, $x_1 = \frac{5}{2}$, $x_2 = 0$. Rounding off this solution, we obtain either the candidate $x_1 = 2$, $x_2 = 0$ or the candidate $x_1 = 3$, $x_2 = 0$. Neither candidate is a feasible solution to the IP.

Recall from Chapter 4 that the simplex algorithm allowed us to solve LPs by going from one basic feasible solution to a better one. Also recall that in most cases, the simplex algorithm examines only a small fraction of all basic feasible solutions before the optimal solution is obtained. This property of the simplex algorithm enables us to solve relatively large LPs by expending a surprisingly small amount of computational effort. Analogously, one would hope that an IP could be solved via an algorithm that proceeded from one feasible integer solution to a better feasible integer solution. Unfortunately, no such algorithm is known.

In summary, even though the feasible region for an IP is a subset of the feasible region for the IP's LP relaxation, the IP is usually much more difficult to solve than the IP's LP relaxation.

9.2 Formulating Integer Programming Problems

In this section, we show how practical solutions can be formulated as IPs. After completing this section, the reader should have a good grasp of the art of developing integer programming formulations. We begin with some simple problems and gradually build to more complicated formulations. Our first example is a capital budgeting problem reminiscent of the Star Oil problem of Section 3.6.

EXAMPLE 1　　**Capital Budgeting IP**

Stockco is considering four investments. Investment 1 will yield a net present value (NPV) of $16,000; investment 2, an NPV of $22,000; investment 3, an NPV of $12,000; and investment 4, an NPV of $8,000. Each investment requires a certain cash outflow at the present time: investment 1, $5,000; investment 2, $7,000; investment 3, $4,000; and investment 4, $3,000. Currently, $14,000 is available for investment. Formulate an IP whose solution will tell Stockco how to maximize the NPV obtained from investments 1–4.

Solution　As in LP formulations, we begin by defining a variable for each decision that Stockco must make. This leads us to define a 0–1 variable:

$$x_j(j=1, 2, 3, 4) = \begin{cases} 1 & \text{if investment } j \text{ is made} \\ 0 & \text{otherwise} \end{cases}$$

For example, $x_2 = 1$ if investment 2 is made, and $x_2 = 0$ if investment 2 is not made.

The NPV obtained by Stockco (in thousands of dollars) is

$$\text{Total NPV obtained by Stockco} = 16x_1 + 22x_2 + 12x_3 + 8x_4 \qquad \text{(5)}$$

To see this, note that if $x_j = 1$, then (5) includes the NPV of investment j, and if $x_j = 0$, (5) does not include the NPV of investment j. This means that whatever combination of investments is undertaken, (5) gives the NPV of that combination of projects. For example, if Stockco invests in investments 1 and 4, then an NPV of $16,000 + 8,000 = $24,000 is obtained. This combination of investments corresponds to $x_1 = x_4 = 1$, $x_2 = x_3 = 0$, so (5) indicates that the NPV for this investment combination is $16(1) + 22(0) + 12(0) + 8(1) = $24 (thousand). This reasoning implies that Stockco's objective function is

$$\max z = 16x_1 + 22x_2 + 12x_3 + 8x_4 \qquad \text{(6)}$$

Stockco faces the constraint that at most $14,000 can be invested. By the same reasoning used to develop (5), we can show that

$$\text{Total amount invested (in thousands of dollars)} = 5x_1 + 7x_2 + 4x_3 + 3x_4 \qquad \text{(7)}$$

For example, if $x_1 = 0$, $x_2 = x_3 = x_4 = 1$, then Stockco makes investments 2, 3, and 4. In this case, Stockco must invest $7 + 4 + 3 = $14 (thousand). Equation (7) yields a total amount invested of $5(0) + 7(1) + 4(1) + 3(1) = $14 (thousand). Because at most $14,000 can be invested, x_1, x_2, x_3, and x_4 must satisfy

$$5x_1 + 7x_2 + 4x_3 + 3x_4 \le 14 \qquad \text{(8)}$$

Combining (6) and (8) with the constraints $x_j = 0$ or 1 ($j = 1, 2, 3, 4$) yields the following 0–1 IP:

$$\max z = 16x_1 + 22x_2 + 12x_3 + 8x_4$$
$$\text{s.t.} \quad 5x_1 + 7x_2 + 4x_3 + 3x_4 \le 14 \qquad \text{(9)}$$
$$x_j = 0 \text{ or } 1 \quad (j = 1, 2, 3, 4)$$

REMARKS　**1** In Section 9.5, we show that the optimal solution to (9) is $x_1 = 0$, $x_2 = x_3 = x_4 = 1$, $z = $42,000. Hence, Stockco should make investments 2, 3, and 4, but not 1. Investment 1 yields a higher NPV per dollar invested than any of the others (investment 1 yields $3.20 per dollar invested, investment 2, $3.14; investment 3, $3; and investment 4, $2.67), so it may seem surprising that investment 1 is not undertaken. To see why the optimal solution to (9) does not involve making the "best" investment, note that any investment combination that includes investment 1 cannot use more than $12,000. This means that using investment 1 forces Stockco to forgo investing $2,000. On the other hand, the optimal investment combination uses all $14,000 of the investment budget. This en-

TABLE **1**
Weights and Benefits for
Items in Josie's Knapsack

Item	Weight (Pounds)	Benefit
1	5	16
2	7	22
3	4	12
4	3	8

ables the optimal combination to obtain a higher NPV than any combination that includes investment 1. If, as in Chapter 3, fractional investments were allowed, the optimal solution to (9) would be $x_1 = x_2 = 1$, $x_3 = 0.50$, $x_4 = 0$, $z = \$44,000$, and investment 1 would be used. This simple example shows that the choice of modeling a capital budgeting problem as a linear programming or as an integer programming problem can significantly affect the optimal solution to the problem.

2 Any IP, such as (9), that has only one constraint is referred to as a **knapsack problem.** Suppose that Josie Camper is going on an overnight hike. There are four items Josie is considering taking along on the trip. The weight of each item and the benefit Josie feels she would obtain from each item are listed in Table 1.

Suppose Josie's knapsack can hold up to 14 lb of items. For $j = 1, 2, 3, 4$, define

$$x_j = \begin{cases} 1 \text{ if Josie takes item } j \text{ on the hike} \\ 0 \text{ otherwise} \end{cases}$$

Then Josie can maximize the total benefit by solving (9).

In the following example, we show how the Stockco formulation can be modified to handle additional constraints.

EXAMPLE 2 Capital Budgeting (Continued)

Modify the Stockco formulation to account for each of the following requirements:

1 Stockco can invest in at most two investments.

2 If Stockco invests in investment 2, they must also invest in investment 1.

3 If Stockco invests in investment 2, they cannot invest in investment 4.

Solution **1** Simply add the constraint

$$x_1 + x_2 + x_3 + x_4 \leq 2 \tag{10}$$

to (9). Because any choice of three or four investments will have $x_1 + x_2 + x_3 + x_4 \geq 3$, (10) excludes from consideration all investment combinations involving three or more investments. Thus, (10) eliminates from consideration exactly those combinations of investments that do not satisfy the first requirement.

2 In terms of x_1 and x_2, this requirement states that if $x_2 = 1$, then x_1 must also equal 1. If we add the constraint

$$x_2 \leq x_1 \quad \text{or} \quad x_2 - x_1 \leq 0 \tag{11}$$

to (9), then we will have taken care of the second requirement. To show that (11) is equivalent to requirement 2, we consider two possibilities: either $x_2 = 1$ or $x_2 = 0$.

Case 1 $x_2 = 1$. If $x_2 = 1$, then the (11) implies that $x_1 \geq 1$. Because x_1 must equal 0 or 1, this implies that $x_1 = 1$, as required by 2.

Case 2 $x_2 = 0$. In this case, (11) reduces to $x_1 \geq 0$, which allows $x_1 = 0$ or $x_1 = 1$. In short, if $x_2 = 0$, (11) does not restrict the value of x_1. This is also consistent with requirement 2.

In summary, for any value of x_2, (11) is equivalent to requirement 2.

3 Simply add the constraint

$$x_2 + x_4 \leq 1 \qquad\qquad (12)$$

to (9). We now show that for the two cases $x_2 = 1$ and $x_2 = 0$, (12) is equivalent to the third requirement.

Case 1 $x_2 = 1$. In this case, we are investing in investment 2, and requirement 3 implies that Stockco cannot invest in investment 4 (that is, x_4 must equal 0). Note that if $x_2 = 1$, then (12) does imply $1 + x_4 \leq 1$, or $x_4 \leq 0$. Thus, if $x_2 = 1$, then (12) is consistent with requirement 3.

Case 2 $x_2 = 0$. In this case, requirement 3 does not restrict the value of x_4. Note that if $x_2 = 0$, then (12) reduces to $x_4 \leq 1$, which also leaves x_4 free to equal 0 or 1.

Fixed-Charge Problems

Example 3 illustrates an important trick that can be used to formulate many location and production problems as IPs.

EXAMPLE 3 **Fixed-Charge IP**

Gandhi Cloth Company is capable of manufacturing three types of clothing: shirts, shorts, and pants. The manufacture of each type of clothing requires that Gandhi have the appropriate type of machinery available. The machinery needed to manufacture each type of clothing must be rented at the following rates: shirt machinery, $200 per week; shorts machinery, $150 per week; pants machinery, $100 per week. The manufacture of each type of clothing also requires the amounts of cloth and labor shown in Table 2. Each week, 150 hours of labor and 160 sq yd of cloth are available. The variable unit cost and selling price for each type of clothing are shown in Table 3. Formulate an IP whose solution will maximize Gandhi's weekly profits.

Solution As in LP formulations, we define a decision variable for each decision that Gandhi must make. Clearly, Gandhi must decide how many of each type of clothing should be manufactured each week, so we define

$$x_1 = \text{number of shirts produced each week}$$
$$x_2 = \text{number of shorts produced each week}$$
$$x_3 = \text{number of pants produced each week}$$

TABLE 2
Resource Requirements for Gandhi

Clothing Type	Labor (Hours)	Cloth (Square Yards)
Shirt	3	4
Shorts	2	3
Pants	6	4

TABLE 3
Revenue and Cost Information for Gandhi

Clothing Type	Sales Price ($)	Variable Cost ($)
Shirt	12	6
Shorts	8	4
Pants	15	8

Note that the cost of renting machinery depends only on the types of clothing produced, not on the amount of each type of clothing. This enables us to express the cost of renting machinery by using the following variables:

$$y_1 = \begin{cases} 1 & \text{if any shirts are manufactured} \\ 0 & \text{otherwise} \end{cases}$$

$$y_2 = \begin{cases} 1 & \text{if any shorts are manufactured} \\ 0 & \text{otherwise} \end{cases}$$

$$y_3 = \begin{cases} 1 & \text{if any pants are manufactured} \\ 0 & \text{otherwise} \end{cases}$$

In short, if $x_j > 0$, then $y_j = 1$, and if $x_j = 0$, then $y_j = 0$. Thus, Gandhi's weekly profits = (weekly sales revenue) − (weekly variable costs) − (weekly costs of renting machinery). Also,

$$\text{Weekly cost of renting machinery} = 200y_1 + 150y_2 + 100y_3 \qquad \text{(13)}$$

To justify (13), note that it picks up the rental costs only for the machines needed to manufacture those products that Gandhi is actually manufacturing. For example, suppose that shirts and pants are manufactured. Then $y_1 = y_3 = 1$ and $y_2 = 0$, and the total weekly rental cost will be $200 + 100 = \$300$.

Because the cost of renting, say, shirt machinery does not depend on the number of shirts produced, the cost of renting each type of machinery is called a **fixed charge.** A fixed charge for an activity is a cost that is assessed whenever the activity is undertaken at a nonzero level. The presence of fixed charges will make the formulation of the Gandhi problem much more difficult.

We can now express Gandhi's weekly profits as

$$\text{Weekly profit} = (12x_1 + 8x_2 + 15x_3) - (6x_1 + 4x_2 + 8x_3)$$
$$- (200y_1 + 150y_2 + 100y_3)$$
$$= 6x_1 + 4x_2 + 7x_3 - 200y_1 - 150y_2 - 100y_3$$

Thus, Gandhi wants to maximize

$$z = 6x_1 + 4x_2 + 7x_3 - 200y_1 - 150y_2 - 100y_3$$

Because its supply of labor and cloth is limited, Gandhi faces the following two constraints:

Constraint 1   At most, 150 hours of labor can be used each week.

Constraint 2   At most, 160 sq yd of cloth can be used each week.

Constraint 1 is expressed by

$$3x_1 + 2x_2 + 6x_3 \le 150 \qquad \text{(Labor constraint)} \qquad \text{(14)}$$

Constraint 2 is expressed by

$$4x_1 + 3x_2 + 4x_3 \leq 160 \qquad \text{(Cloth constraint)} \qquad \text{(15)}$$

Observe that $x_j > 0$ and x_j integer ($j = 1, 2, 3$) must hold along with $y_j = 0$ or 1 ($j = 1, 2, 3$). Combining (14) and (15) with these restrictions and the objective function yields the following IP:

$$\max z = 6x_1 + 4x_2 + 7x_3 - 200y_1 - 150y_2 - 100y_3$$
$$\text{s.t.} \quad 3x_1 + 2x_2 + 6x_3 \leq 150$$
$$4x_1 + 3x_2 + 4x_3 \leq 160 \qquad \text{(IP 1)}$$
$$x_1, x_2, x_3 \geq 0; \; x_1, x_2, x_3 \text{ integer}$$
$$y_1, y_2, y_3 = 0 \text{ or } 1$$

The optimal solution to this problem is found to be $x_1 = 30$, $x_3 = 10$, $x_2 = y_1 = y_2 = y_3 = 0$. This cannot be the optimal solution to Gandhi's problem because it indicates that Gandhi can manufacture shirts and pants without incurring the cost of renting the needed machinery. The current formulation is incorrect because the variables y_1, y_2, and y_3 are not present in the constraints. This means that there is nothing to stop us from setting $y_1 = y_2 = y_3 = 0$. Setting $y_i = 0$ is certainly less costly than setting $y_i = 1$, so a minimum-cost solution to (IP 1) will always set $y_i = 0$. Somehow we must modify (IP 1) so that whenever $x_i > 0$, $y_i = 1$ must hold. The following trick will accomplish this goal. Let M_1, M_2, and M_3 be three large positive numbers, and add the following constraints to (IP 1):

$$x_1 \leq M_1 y_1 \qquad \text{(16)}$$
$$x_2 \leq M_2 y_2 \qquad \text{(17)}$$
$$x_3 \leq M_3 y_3 \qquad \text{(18)}$$

Adding (16)–(18) to IP 1 will ensure that if $x_i > 0$, then $y_i = 1$. To illustrate, let us show that (16) ensures that if $x_1 > 0$, then $y_1 = 1$. If $x_1 > 0$, then y_1 cannot be 0. For if $y_1 = 0$, then (16) would imply $x_1 \leq 0$ or $x_1 = 0$. Thus, if $x_1 > 0$, $y_1 = 1$ must hold. If any shirts are produced ($x_1 > 0$), (16) ensures that $y_1 = 1$, and the objective function will include the cost of the machinery needed to manufacture shirts. Note that if $y_1 = 1$, then (16) becomes $x_1 \leq M_1$, which does not unnecessarily restrict the value of x_1. If M_1 were not chosen large, however (say, $M_1 = 10$), then (16) would unnecessarily restrict the value of x_1. In general, M_i should be set equal to the maximum value that x_i can attain. In the current problem, at most 40 shirts can be produced (if Gandhi produced more than 40 shirts, the company would run out of cloth), so we can safely choose $M_1 = 40$. The reader should verify that we can choose $M_2 = 53$ and $M_3 = 25$.

If $x_1 = 0$, (16) becomes $0 \leq M_1 y_1$. This allows either $y_1 = 0$ or $y_1 = 1$. Because $y_1 = 0$ is less costly than $y_1 = 1$, the optimal solution will choose $y_1 = 0$ if $x_1 = 0$. In summary, we have shown that if (16)–(18) are added to (IP 1), then $x_i > 0$ will imply $y_i = 1$, and $x_i = 0$ will imply $y_i = 0$.

The optimal solution to the Gandhi problem is $z = \$75$, $x_3 = 25$, $y_3 = 1$. Thus, Gandhi should produce 25 pants each week.

The Gandhi problem is an example of a **fixed-charge problem.** In a fixed-charge problem, there is a cost associated with performing an activity at a nonzero level that does not depend on the level of the activity. Thus, in the Gandhi problem, if we make any shirts at all (no matter how many we make), we must pay the fixed charge of $200 to rent a shirt machine. Problems in which a decision maker must choose where to locate facilities are often fixed-charge problems. The decision maker must choose where to locate various facilities (such as plants, warehouses, or business offices), and a fixed charge is often associated with building or operating a facility. Example 4 is a typical location problem involving the idea of a fixed charge.

EXAMPLE 4   **The Lockbox Problem**

J. C. Nickles receives credit card payments from four regions of the country (West, Midwest, East, and South). The average daily value of payments mailed by customers from each region is as follows: the West, $70,000; the Midwest, $50,000; the East, $60,000; the South, $40,000. Nickles must decide where customers should mail their payments. Because Nickles can earn 20% annual interest by investing these revenues, it would like to receive payments as quickly as possible. Nickles is considering setting up operations to process payments (often referred to as lockboxes) in four different cities: Los Angeles, Chicago, New York, and Atlanta. The average number of days (from time payment is sent) until a check clears and Nickles can deposit the money depends on the city to which the payment is mailed, as shown in Table 4. For example, if a check is mailed from the West to Atlanta, it would take an average of 8 days before Nickles could earn interest on the check. The annual cost of running a lockbox in any city is $50,000. Formulate an IP that Nickles can use to minimize the sum of costs due to lost interest and lockbox operations. Assume that each region must send all its money to a single city and that there is no limit on the amount of money that each lockbox can handle.

Solution   Nickles must make two types of decisions. First, Nickles must decide where to operate lockboxes. We define, for $j = 1, 2, 3, 4,$

$$y_j = \begin{cases} 1 & \text{if a lockbox is operated in city } j \\ 0 & \text{otherwise} \end{cases}$$

Thus, $y_2 = 1$ if a lockbox is operated in Chicago, and $y_3 = 0$ if no lockbox is operated in New York. Second, Nickles must determine where each region of the country should send payments. We define (for $i, j = 1, 2, 3, 4$)

$$x_{ij} = \begin{cases} 1 & \text{if region } i \text{ sends payments to city } j \\ 0 & \text{otherwise} \end{cases}$$

For example, $x_{12} = 1$ if the West sends payments to Chicago, and $x_{23} = 0$ if the Midwest does not send payments to New York.

Nickles wants to minimize (total annual cost) = (annual cost of operating lockboxes) + (annual lost interest cost). To determine how much interest Nickles loses annually, we must determine how much revenue would be lost if payments from region i were sent to region j. For example, how much in annual interest would Nickles lose if customers from the West region sent payments to New York? On any given day, 8 days' worth, or $8(70,000) = \$560,000$ of West payments will be in the mail and will not be earning in-

TABLE 4
Average Number of Days from Mailing of Payment Until Payment Clears

| From | To | | | |
	City 1 (Los Angeles)	City 2 (Chicago)	City 3 (New York)	City 4 (Atlanta)
Region 1 West	2	6	8	8
Region 2 Midwest	6	2	5	5
Region 3 East	8	5	2	5
Region 4 South	8	5	5	2

terest. Because Nickles can earn 20% annually, each year West funds will result in $0.20(560,000) = \$112,000$ in lost interest. Similar calculations for the annual cost of lost interest for each possible assignment of a region to a city yield the results shown in Table 5. The lost interest cost from sending region i's payments to city j is only incurred if $x_{ij} = 1$, so Nickles's annual lost interest costs (in thousands) are

$$\begin{aligned}
\text{Annual lost interest costs} = {} & 28x_{11} + 84x_{12} + 112x_{13} + 112x_{14} \\
& + 60x_{21} + 20x_{22} + 50x_{23} + 50x_{24} \\
& + 96x_{31} + 60x_{32} + 24x_{33} + 60x_{34} \\
& + 64x_{41} + 40x_{42} + 40x_{43} + 16x_{44}
\end{aligned}$$

The cost of operating a lockbox in city i is incurred if and only if $y_i = 1$, so the annual lockbox operating costs (in thousands) are given by

$$\text{Total annual lockbox operating cost} = 50y_1 + 50y_2 + 50y_3 + 50y_4$$

Thus, Nickles's objective function may be written as

$$\begin{aligned}
\min z = {} & 28x_{11} + 84x_{12} + 112x_{13} + 112x_{14} \\
& + 60x_{21} + 20x_{22} + 50x_{23} + 50x_{24} \\
& + 96x_{31} + 60x_{32} + 24x_{33} + 60x_{34} \\
& + 64x_{41} + 40x_{42} + 40x_{43} + 16x_{44} \\
& + 50y_1 + 50y_2 + 50y_3 + 50y_4
\end{aligned} \tag{19}$$

Nickles faces two types of constraints.

Type 1 Constraint   Each region must send its payments to a single city.

Type 2 Constraint   If a region is assigned to send its payments to a city, that city must have a lockbox.

TABLE **5**
Calculation of Annual Lost Interest

Assignment	Annual Lost Interest Cost (\$)
West to L.A.	$0.20(70,000)2 = 28,000$
West to Chicago	$0.20(70,000)6 = 84,000$
West to N.Y.	$0.20(70,000)8 = 112,000$
West to Atlanta	$0.20(70,000)8 = 112,000$
Midwest to L.A.	$0.20(50,000)6 = 60,000$
Midwest to Chicago	$0.20(50,000)2 = 20,000$
Midwest to N.Y.	$0.20(50,000)5 = 50,000$
Midwest to Atlanta	$0.20(50,000)5 = 50,000$
East to L.A.	$0.20(60,000)8 = 96,000$
East to Chicago	$0.20(60,000)5 = 60,000$
East to N.Y.	$0.20(60,000)2 = 24,000$
East to Atlanta	$0.20(60,000)5 = 60,000$
South to L.A.	$0.20(40,000)8 = 64,000$
South to Chicago	$0.20(40,000)5 = 40,000$
South to N.Y	$0.20(40,000)5 = 40,000$
South to Atlanta	$0.20(40,000)2 = 16,000$

The type 1 constraints state that for region i ($i = 1, 2, 3, 4$) exactly one of x_{i1}, x_{i2}, x_{i3}, and x_{i4} must equal 1 and the others must equal 0. This can be accomplished by including the following four constraints:

$$x_{11} + x_{12} + x_{13} + x_{14} = 1 \quad \text{(West region constraint)} \tag{20}$$
$$x_{21} + x_{22} + x_{23} + x_{24} = 1 \quad \text{(Midwest region constraint)} \tag{21}$$
$$x_{31} + x_{32} + x_{33} + x_{34} = 1 \quad \text{(East region constraint)} \tag{22}$$
$$x_{41} + x_{42} + x_{43} + x_{44} = 1 \quad \text{(South region constraint)} \tag{23}$$

The type 2 constraints state that if

$$x_{ij} = 1 \quad \text{(that is, customers in region } i \text{ send payments to city } j) \tag{24}$$

then y_j must equal 1. For example, suppose $x_{12} = 1$. Then there must be a lockbox at city 2, so $y_2 = 1$ must hold. This can be ensured by adding 16 constraints of the form

$$x_{ij} \leq y_j \quad (i = 1, 2, 3, 4; j = 1, 2, 3, 4) \tag{25}$$

If $x_{ij} = 1$, then (25) ensures that $y_j = 1$, as desired. Also, if $x_{1j} = x_{2j} = x_{3j} = x_{4j} = 0$, then (25) allows $y_j = 0$ or $y_j = 1$. As in the fixed-charge example, the act of minimizing costs will result in $y_j = 0$. In summary, the constraints in (25) ensure that Nickles pays for a lockbox at city i if it uses a lockbox at city i.

Combining (19)–(23) with the 4(4) = 16 constraints in (25) and the 0–1 restrictions on the variables yields the following formulation:

$$\min z = 28x_{11} + 84x_{12} + 112x_{13} + 112x_{14} + 60x_{21} + 20x_{22} + 50x_{23} + 50x_{24}$$
$$+ 96x_{31} + 60x_{32} + 24x_{33} + 60x_{34} + 64x_{41} + 40x_{42} + 40x_{43} + 16x_{44}$$
$$+ 50y_1 + 50y_2 + 50y_3 + 50y_4$$

s.t. $\quad x_{11} + x_{12} + x_{13} + x_{14} = 1 \quad$ (West region constraint)

$\qquad x_{21} + x_{22} + x_{23} + x_{24} = 1 \quad$ (Midwest region constraint)

$\qquad x_{31} + x_{32} + x_{33} + x_{34} = 1 \quad$ (East region constraint)

$\qquad x_{41} + x_{42} + x_{43} + x_{44} = 1 \quad$ (South region constraint)

$\qquad x_{11} \leq y_1, x_{21} \leq y_1, x_{31} \leq y_1, x_{41} \leq y_1, x_{12} \leq y_2, x_{22} \leq y_2, x_{32} \leq y_2, x_{42} \leq y_2,$

$\qquad x_{13} \leq y_3, x_{23} \leq y_3, x_{33} \leq y_3, x_{43} \leq y_3, x_{14} \leq y_4, x_{24} \leq y_4, x_{34} \leq y_4, x_{44} \leq y_4$

$\qquad$ All x_{ij} and y_j = 0 or 1

The optimal solution is $z = 242$, $y_1 = 1$, $y_3 = 1$, $x_{11} = 1$, $x_{23} = 1$, $x_{33} = 1$, $x_{43} = 1$. Thus, Nickles should have a lockbox operation in Los Angeles and New York. West customers should send payments to Los Angeles, and all other customers should send payments to New York.

There is an alternative way of modeling the Type 2 constraints. Instead of the 16 constraints of the form $x_{ij} \leq y_j$, we may include the following four constraints:

$$x_{11} + x_{21} + x_{31} + x_{41} \leq 4y_1 \quad \text{(Los Angeles constraint)}$$
$$x_{12} + x_{22} + x_{32} + x_{42} \leq 4y_2 \quad \text{(Chicago constraint)}$$
$$x_{13} + x_{23} + x_{33} + x_{43} \leq 4y_3 \quad \text{(New York constraint)}$$
$$x_{14} + x_{24} + x_{34} + x_{44} \leq 4y_4 \quad \text{(Atlanta constraint)}$$

For the given city, each constraint ensures that if the lockbox is used, then Nickles must pay for it. For example, consider $x_{14} + x_{24} + x_{34} + x_{44} \leq 4y_4$. The lockbox in Atlanta is used if $x_{14} = 1$, $x_{24} = 1$, $x_{34} = 1$, or $x_{44} = 1$. If any of these variables equals 1, then the Atlanta constraint ensures that $y_4 = 1$, and Nickles must pay for the lockbox. If all these variables are 0, then the act of minimizing costs will cause $y_4 = 0$, and the cost of the Atlanta lockbox will not be incurred. Why does the right-hand side of each constraint equal 4? This ensures that for each city, it is possible to send money from all four regions to the city. In Section 9.3, we discuss which of the two alternative formulations of the lockbox problem is easier for a computer to solve. The answer may surprise you!

Set-Covering Problems

The following example is typical of an important class of IPs known as set-covering problems.

EXAMPLE 5   Facility-Location Set-Covering Problem

There are six cities (cities 1–6) in Kilroy County. The county must determine where to build fire stations. The county wants to build the minimum number of fire stations needed to ensure that at least one fire station is within 15 minutes (driving time) of each city. The times (in minutes) required to drive between the cities in Kilroy County are shown in Table 6. Formulate an IP that will tell Kilroy how many fire stations should be built and where they should be located.

Solution   For each city, Kilroy must determine whether to build a fire station there. We define the 0–1 variables x_1, x_2, x_3, x_4, x_5, and x_6 by

$$x_i = \begin{cases} 1 & \text{if a fire station is built in city } i \\ 0 & \text{otherwise} \end{cases}$$

Then the total number of fire stations that are built is given by $x_1 + x_2 + x_3 + x_4 + x_5 + x_6$, and Kilroy's objective function is to minimize

$$z = x_1 + x_2 + x_3 + x_4 + x_5 + x_6$$

What are Kilroy's constraints? Kilroy must ensure that there is a fire station within 15 minutes of each city. Table 7 indicates which locations can reach the city in 15 minutes or less. To ensure that at least one fire station is within 15 minutes of city 1, we add the constraint

$$x_1 + x_2 \geq 1 \quad \text{(City 1 constraint)}$$

This constraint ensures that $x_1 = x_2 = 0$ is impossible, so at least one fire station will be built within 15 minutes of city 1. Similarly the constraint

$$x_1 + x_2 + x_6 \geq 1 \quad \text{(City 2 constraint)}$$

ensures that at least one fire station will be located within 15 minutes of city 2. In a similar fashion, we obtain constraints for cities 3–6. Combining these six constraints with the

TABLE 6
Time Required to Travel between Cities in Kilroy County

From	To					
	City 1	City 2	City 3	City 4	City 5	City 6
City 1	0	10	20	30	30	20
City 2	10	0	25	35	20	10
City 3	20	25	0	15	30	20
City 4	30	35	15	0	15	25
City 5	30	20	30	15	0	14
City 6	20	10	20	25	14	0

TABLE 7
Cities within 15 Minutes of Given City

City	Within 15 Minutes
1	1, 2
2	1, 2, 6
3	3, 4
4	3, 4, 5
5	4, 5, 6
6	2, 5, 6

objective function (and with the fact that each variable must equal 0 or 1), we obtain the following 0–1 IP:

$$\min z = x_1 + x_2 + x_3 + x_4 + x_5 + x_6$$

$$
\begin{aligned}
\text{s.t.} \quad x_1 + x_2 & \geq 1 \quad \text{(City 1 constraint)} \\
x_1 + x_2 + x_6 & \geq 1 \quad \text{(City 2 constraint)} \\
x_3 + x_4 & \geq 1 \quad \text{(City 3 constraint)} \\
x_3 + x_4 + x_5 & \geq 1 \quad \text{(City 4 constraint)} \\
x_4 + x_5 + x_6 & \geq 1 \quad \text{(City 5 constraint)} \\
x_2 + x_5 + x_6 & \geq 1 \quad \text{(City 6 constraint)}
\end{aligned}
$$

$$x_i = 0 \text{ or } 1 \quad (i = 1, 2, 3, 4, 5, 6)$$

One optimal solution to this IP is $z = 2$, $x_2 = x_4 = 1$, $x_1 = x_3 = x_5 = x_6 = 0$. Thus, Kilroy County can build two fire stations: one in city 2 and one in city 4.

As noted, Example 5 represents a class of IPs known as **set-covering problems.** In a set-covering problem, each member of a given set (call it set 1) must be "covered" by an acceptable member of some set (call it set 2). The objective in a set-covering problem is to minimize the number of elements in set 2 that are required to cover all the elements in set 1. In Example 5, set 1 is the cities in Kilroy County, and set 2 is the set of fire stations. The station in city 2 covers cities 1, 2, and 6, and the station in city 4 covers cities 3, 4, and 5. Set-covering problems have many applications in areas such as airline crew scheduling, political districting, airline scheduling, and truck routing.

Either–Or Constraints

The following situation commonly occurs in mathematical programming problems. We are given two constraints of the form

$$f(x_1, x_2, \ldots, x_n) \leq 0 \tag{26}$$

$$g(x_1, x_2, \ldots, x_n) \leq 0 \tag{27}$$

We want to ensure that at least one of (26) and (27) is satisfied, often called **either–or constraints.** Adding the two constraints (26′) and (27′) to the formulation will ensure that at least one of (26) and (27) is satisfied:

$$f(x_1, x_2, \ldots, x_n) \leq My \tag{26′}$$

$$g(x_1, x_2, \ldots, x_n) \leq M(1 - y) \tag{27′}$$

In (26′) and (27′), y is a 0–1 variable, and M is a number chosen large enough to ensure that $f(x_1, x_2, \ldots, x_n) \leq M$ and $g(x_1, x_2, \ldots, x_n) \leq M$ are satisfied for all values of $x_1, x_2, \ldots, x_n$ that satisfy the other constraints in the problem.

Let us show that the inclusion of constraints (26′) and (27′) is equivalent to at least one of (26) and (27) being satisfied. Either $y = 0$ or $y = 1$. If $y = 0$, then (26′) and (27′) become $f \leq 0$ and $g \leq M$. Thus, if $y = 0$, then (26) (and possibly (27)) must be satisfied. Similarly, if $y = 1$, then (26′) and (27′) become $f \leq M$ and $g \leq 0$. Thus, if $y = 1$, then (27) (and possibly (26)) must be satisfied. Therefore, whether $y = 0$ or $y = 1$, (26′) and (27′) ensure that at least one of (26) and (27) is satisfied.

The following example illustrates the use of either–or constraints.

EXAMPLE 6   Either–Or Constraint

Dorian Auto is considering manufacturing three types of autos: compact, midsize, and large. The resources required for, and the profits yielded by, each type of car are shown in Table 8. Currently, 6,000 tons of steel and 60,000 hours of labor are available. For production of a type of car to be economically feasible, at least 1,000 cars of that type must be produced. Formulate an IP to maximize Dorian's profit.

Solution   Because Dorian must determine how many cars of each type should be built, we define

$$x_1 = \text{number of compact cars produced}$$
$$x_2 = \text{number of midsize cars produced}$$
$$x_3 = \text{number of large cars produced}$$

Then contribution to profit (in thousands of dollars) is $2x_1 + 3x_2 + 4x_3$, and Dorian's objective function is

$$\max z = 2x_1 + 3x_2 + 4x_3$$

We know that if any cars of a given type are produced, then at least 1,000 cars of that type must be produced. Thus, for $i = 1, 2, 3$, we must have $x_i \leq 0$ or $x_i \geq 1,000$. Steel and labor are limited, so Dorian must satisfy the following five constraints:

Constraint 1 $x_1 \leq 0$ or $x_1 \geq 1,000$.

Constraint 2 $x_2 \leq 0$ or $x_2 \geq 1,000$.

Constraint 3 $x_3 \leq 0$ or $x_3 \geq 1,000$.

Constraint 4   The cars produced can use at most 6,000 tons of steel.

Constraint 5   The cars produced can use at most 60,000 hours of labor.

TABLE 8
Resources and Profits for Three Types of Cars

Resource	Car Type		
	Compact	Midsize	Large
Steel required	1.5 tons	3 tons	5 tons
Labor required	30 hours	25 hours	40 hours
Profit yielded ($)	2,000	3,000	4,000

From our previous discussion, we see that if we define $f(x_1, x_2, x_3) = x_1$ and $g(x_1, x_2, x_3) = 1{,}000 - x_1$, we can replace Constraint 1 by the following pair of constraints:

$$x_1 \leq M_1 y_1$$
$$1{,}000 - x_1 \leq M_1(1 - y_1)$$
$$y_1 = 0 \text{ or } 1$$

To ensure that both x_1 and $1{,}000 - x_1$ will never exceed M_1, it suffices to choose M_1 large enough so that M_1 exceeds 1,000 and x_1 is always less than M_1. Building $\frac{60{,}000}{30} = 2{,}000$ compacts would use all available labor (and still leave some steel), so at most 2,000 compacts can be built. Thus, we may choose $M_1 = 2{,}000$. Similarly, Constraint 2 may be replaced by the following pair of constraints:

$$x_2 \leq M_2 y_2$$
$$1{,}000 - x_2 \leq M_2(1 - y_2)$$
$$y_2 = 0 \text{ or } 1$$

You should verify that $M_2 = 2{,}000$ is satisfactory. Similarly, Constraint 3 may be replaced by

$$x_3 \leq M_3 y_3$$
$$1{,}000 - x_3 \leq M_3(1 - y_3)$$
$$y_3 = 0 \text{ or } 1$$

Again, you should verify that $M_3 = 1{,}200$ is satisfactory. Constraint 4 is a straightforward resource constraint that reduces to

$$1.5x_1 + 3x_2 + 5x_3 \leq 6{,}000 \qquad \text{(Steel constraint)}$$

Constraint 5 is a straightforward resource usage constraint that reduces to

$$30x_1 + 25x_2 + 40x_3 \leq 60{,}000 \qquad \text{(Labor constraint)}$$

After noting that $x_i \geq 0$ and that x_i must be an integer, we obtain the following IP:

$$\max z = 2x_1 + 3x_2 + 4x_3$$
$$\text{s.t.} \qquad x_1 \leq 2{,}000y_1$$
$$1{,}000 - x_1 \leq 2{,}000(1 - y_1)$$
$$x_2 \leq 2{,}000y_2$$
$$1{,}000 - x_2 \leq 2{,}000(1 - y_2)$$
$$x_3 \leq 1{,}200y_3$$
$$1{,}000 - x_3 \leq 1{,}200(1 - y_3)$$
$$1.5x_1 + 3x_2 + 5x_3 \leq 6{,}000 \qquad \text{(Steel constraint)}$$
$$30x_1 + 25x_2 + 40x_3 \leq 60{,}000 \qquad \text{(Labor constraint)}$$
$$x_1, x_2, x_3 \geq 0; x_1, x_2, x_3 \text{ integer}$$
$$y_1, y_2, y_3 = 0 \text{ or } 1$$

The optimal solution to the IP is $z = 6{,}000$, $x_2 = 2{,}000$, $y_2 = 1$, $y_1 = y_3 = x_1 = x_3 = 0$. Thus, Dorian should produce 2,000 midsize cars. If Dorian had not been required to manufacture at least 1,000 cars of each type, then the optimal solution would have been to produce 570 compacts and 1,715 midsize cars.

If–Then Constraints

In many applications, the following situation occurs: We want to ensure that if a constraint $f(x_1, x_2, \ldots, x_n) > 0$ is satisfied, then the constraint $g(x_1, x_2, \ldots, x_n) \geq 0$ must be satisfied, while if $f(x_1, x_2, \ldots, x_n) > 0$ is not satisfied, then $g(x_1, x_2, \ldots, x_n) \geq 0$ may or may not be satisfied. In short, we want to ensure that $f(x_1, x_2, \ldots, x_n) > 0$ implies $g(x_1, x_2, \ldots, x_n) \geq 0$.

To ensure this, we include the following constraints in the formulation:

$$-g(x_1, x_2, \ldots, x_n) \leq My \tag{28}$$

$$f(x_1, x_2, \ldots, x_n) \leq M(1 - y) \tag{29}$$

$$y = 0 \text{ or } 1$$

As usual, M is a large positive number. (M must be chosen large enough so that $f \leq M$ and $-g \leq M$ hold for all values of $x_1, x_2, \ldots, x_n$ that satisfy the other constraints in the problem.) Observe that if $f > 0$, then (29) can be satisfied only if $y = 0$. Then (28) implies $-g \leq 0$, or $g \geq 0$, which is the desired result. Thus, if $f > 0$, then (28) and (29) ensure that $g \geq 0$. Also, if $f > 0$ is not satisfied, then (29) allows $y = 0$ or $y = 1$. By choosing $y = 1$, (28) is automatically satisfied. Thus, if $f > 0$ is not satisfied, then the values of $x_1, x_2, \ldots, x_n$ are unrestricted and $g < 0$ or $g \geq 0$ are both possible.

To illustrate the use of this idea, suppose we add the following constraint to the Nickles lockbox problem: If customers in region 1 send their payments to city 1, then no other customers may send their payments to city 1. Mathematically, this restriction may be expressed by

$$\text{If } x_{11} = 1, \quad \text{then} \quad x_{21} = x_{31} = x_{41} = 0 \tag{30}$$

Because all x_{ij} must equal 0 or 1, (30) may be written as

$$\text{If } x_{11} > 0, \quad \text{then} \quad x_{21} + x_{31} + x_{41} \leq 0, \quad \text{or} \quad -x_{21} - x_{31} - x_{41} \geq 0 \tag{30'}$$

If we define $f = x_{11}$ and $g = -x_{21} - x_{31} - x_{41}$, we can use (28) and (29) to express (30') [and therefore (30)] by the following two constraints:

$$x_{21} + x_{31} + x_{41} \leq My$$

$$x_{11} \leq M(1 - y)$$

$$y = 0 \text{ or } 1$$

Because $-g$ and f can never exceed 3, we can choose $M = 3$ and add the following constraints to the original lockbox formulation:

$$x_{21} + x_{31} + x_{41} \leq 3y$$

$$x_{11} \leq 3(1 - y)$$

$$y = 0 \text{ or } 1$$

Integer Programming and Piecewise Linear Functions†

The next example shows how 0–1 variables can be used to model optimization problems involving piecewise linear functions. A **piecewise linear function** consists of several straight-line segments. The piecewise linear function in Figure 2 is made of four straight-line segments. The points where the slope of the piecewise linear function changes (or the range of definition of the function ends) are called the **break points** of the function. Thus, 0, 10, 30, 40, and 50 are the break points of the function pictured in Figure 2.

†This section covers topics that may be omitted with no loss of continuity.

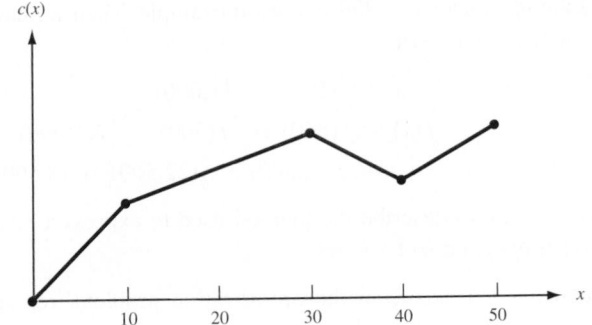

FIGURE **2**
A Piecewise
Linear Function

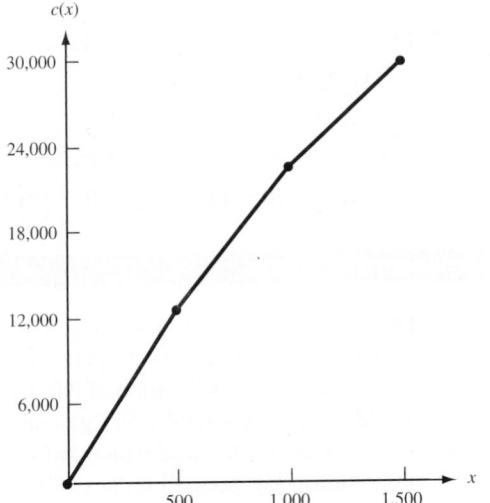

FIGURE **3**
Cost of Purchasing Oil

To illustrate why piecewise linear functions can occur in applications, suppose we manufacture gasoline from oil. In purchasing oil from our supplier, we receive a quantity discount. The first 500 gallons of oil purchased cost 25¢ per gallon; the next 500 gallons cost 20¢ per gallon; and the next 500 gallons cost 15¢ per gallon. At most, 1,500 gallons of oil can be purchased. Let x be the number of gallons of oil purchased and $c(x)$ be the cost (in cents) of purchasing x gallons of oil. For $x \leq 0$, $c(x) = 0$. Then for $0 \leq x \leq 500$, $c(x) = 25x$. For $500 \leq x \leq 1,000$, $c(x) =$ (cost of purchasing first 500 gallons at 25¢ per gallon) + (cost of purchasing next $x - 500$ gallons at 20¢ per gallon) $= 25(500) + 20(x - 500) = 20x + 2,500$. For $1,000 \leq x \leq 1,500$, $c(x) =$ (cost of purchasing first 1,000 gallons) + (cost of purchasing next $x - 1,000$ gallons at 15¢ per gallon) $= c(1,000) + 15(x - 1,000) = 7,500 + 15x$. Thus, $c(x)$ has break points 0, 500, 1,000, and 1,500 and is graphed in Figure 3.

A piecewise linear function is not a linear function, so one might think that linear programming could not be used to solve optimization problems involving these functions. By using 0–1 variables, however, piecewise linear functions can be represented in linear form. Suppose that a piecewise linear function $f(x)$ has break points $b_1, b_2, \ldots, b_n$. For some k $(k = 1, 2, \ldots, n - 1)$, $b_k \leq x \leq b_{k+1}$. Then, for some number z_k ($0 \leq z_k \leq 1$), x may be written as

$$x = z_k b_k + (1 - z_k) b_{k+1}$$

Because $f(x)$ is linear for $b_k \leq x \leq b_{k+1}$, we may write

$$f(x) = z_k f(b_k) + (1 - z_k) f(b_{k+1})$$

To illustrate the idea, take $x = 800$ in our oil example. Then we have $b_2 = 500 \leq 800 \leq 1{,}000 = b_3$, and we may write

$$x = \tfrac{2}{5}(500) + \tfrac{3}{5}(1{,}000)$$
$$f(x) = f(800) = \tfrac{2}{5}f(500) + \tfrac{3}{5}f(1{,}000)$$
$$= \tfrac{2}{5}(12{,}500) + \tfrac{3}{5}(22{,}500) = 18{,}500$$

We are now ready to describe the method used to express a piecewise linear function via linear constraints and 0–1 variables:

Step 1 Wherever $f(x)$ occurs in the optimization problem, replace $f(x)$ by $z_1 f(b_1) + z_2 f(b_2) + \cdots + z_n f(b_n)$.

Step 2 Add the following constraints to the problem:

$$z_1 \leq y_1, z_2 \leq y_1 + y_2, z_3 \leq y_2 + y_3, \ldots, z_{n-1} \leq y_{n-2} + y_{n-1}, z_n \leq y_{n-1}$$
$$y_1 + y_2 + \cdots + y_{n-1} = 1$$
$$z_1 + z_2 + \cdots + z_n = 1$$
$$x = z_1 b_1 + z_2 b_2 + \cdots + z_n b_n$$
$$y_i = 0 \text{ or } 1 \quad (i = 1, 2, \ldots, n-1); \qquad z_i \geq 0 \quad (i = 1, 2, \ldots, n)$$

EXAMPLE 7   IP with Piecewise Linear Functions

Euing Gas produces two types of gasoline (gas 1 and gas 2) from two types of oil (oil 1 and oil 2). Each gallon of gas 1 must contain at least 50 percent oil 1, and each gallon of gas 2 must contain at least 60 percent oil 1. Each gallon of gas 1 can be sold for 12¢, and each gallon of gas 2 can be sold for 14¢. Currently, 500 gallons of oil 1 and 1,000 gallons of oil 2 are available. As many as 1,500 more gallons of oil 1 can be purchased at the following prices: first 500 gallons, 25¢ per gallon; next 500 gallons, 20¢ per gallon; next 500 gallons, 15¢ per gallon. Formulate an IP that will maximize Euing's profits (revenues − purchasing costs).

Solution Except for the fact that the cost of purchasing additional oil 1 is a piecewise linear function, this is a straightforward blending problem. With this in mind, we define

$$x = \text{amount of oil 1 purchased}$$
$$x_{ij} = \text{amount of oil } i \text{ used to produce gas } j \quad (i, j = 1, 2)$$

Then (in cents)

$$\text{Total revenue} - \text{cost of purchasing oil 1} = 12(x_{11} + x_{21}) + 14(x_{12} + x_{22}) - c(x)$$

As we have seen previously,

$$c(x) = \begin{cases} 25x & (0 \leq x \leq 500) \\ 20x + 2{,}500 & (500 \leq x \leq 1{,}000) \\ 15x + 7{,}500 & (1{,}000 \leq x \leq 1{,}500) \end{cases}$$

Thus, Euing's objective function is to maximize

$$z = 12x_{11} + 12x_{21} + 14x_{12} + 14x_{22} - c(x)$$

Euing faces the following constraints:

Constraint 1 Euing can use at most $x + 500$ gallons of oil 1.

Constraint 2 Euing can use at most 1,000 gallons of oil 2.

Constraint 3 The oil mixed to make gas 1 must be at least 50% oil 1.

Constraint 4 The oil mixed to make gas 2 must be at least 60% oil 1.

Constraint 1 yields

$$x_{11} + x_{12} \leq x + 500$$

Constraint 2 yields

$$x_{21} + x_{22} \leq 1,000$$

Constraint 3 yields

$$\frac{x_{11}}{x_{11} + x_{21}} \geq 0.5 \quad \text{or} \quad 0.5x_{11} - 0.5x_{21} \geq 0$$

Constraint 4 yields

$$\frac{x_{12}}{x_{12} + x_{22}} \geq 0.6 \quad \text{or} \quad 0.4x_{12} - 0.6x_{22} \geq 0$$

Also all variables must be nonnegative. Thus, Euing Gas must solve the following optimization problem:

$$\max z = 12x_{11} + 12x_{21} + 14x_{12} + 14x_{22} - c(x)$$

$$\text{s.t.} \quad x_{11} \quad + \quad x_{12} \quad \leq x + 500$$
$$x_{21} \quad + \quad x_{22} \leq 1,000$$
$$0.5x_{11} - 0.5x_{21} \quad \geq 0$$
$$0.4x_{12} - 0.6x_{22} \geq 0$$
$$x_{ij} \geq 0, 0 \leq x \leq 1,500$$

Because $c(x)$ is a piecewise linear function, the objective function is not a linear function of x, and this optimization is not an LP. By using the method described earlier, however, we can transform this problem into an IP. After recalling that the break points for $c(x)$ are 0, 500, 1,000, and 1,500, we proceed as follows:

Step 1 Replace $c(x)$ by $c(x) = z_1 c(0) + z_2 c(500) + z_3 c(1,000) + z_4 c(1,500)$.

Step 2 Add the following constraints:

$$x = 0z_1 + 500z_2 + 1,000z_3 + 1,500z_4$$
$$z_1 \leq y_1, z_2 \leq y_1 + y_2, z_3 \leq y_2 + y_3, z_4 \leq y_3$$
$$z_1 + z_2 + z_3 + z_4 = 1, \quad y_1 + y_2 + y_3 = 1$$
$$y_i = 0 \text{ or } 1 \ (i = 1, 2, 3); z_i \geq 0 \ (i = 1, 2, 3, 4)$$

Our new formulation is the following IP:

$$\max z = 12x_{11} + 12x_{21} + 14x_{12} + 14x_{22} - z_1 c(0) - z_2 c(500)$$
$$- z_3 c(1,000) - z_4 c(1,500)$$

$$\text{s.t.} \quad x_{11} \quad + \quad x_{12} \quad \leq x + 500$$
$$x_{21} \quad + \quad x_{22} \leq 1,000$$
$$0.5x_{11} - 0.5x_{21} \quad \geq 0$$
$$0.4x_{12} - 0.6x_{22} \geq 0$$

$$x = 0z_1 + 500z_2 + 1,000z_3 + 1,500z_4 \tag{31}$$
$$z_1 \leq y_1 \tag{32}$$
$$z_2 \leq y_1 + y_2 \tag{33}$$
$$z_3 \leq y_2 + y_3 \tag{34}$$

$$z_4 \leq y_3 \tag{35}$$

$$y_1 + y_2 + y_3 = 1 \tag{36}$$

$$z_1 + z_2 + z_3 + z_4 = 1 \tag{37}$$

$$y_i = 0 \text{ or } 1 \quad (i = 1, 2, 3); z_i \geq 0 \quad (i = 1, 2, 3, 4)$$

$$x_{ij} \geq 0$$

To see why this formulation works, observe that because $y_1 + y_2 + y_3 = 1$ and $y_i = 0$ or 1, exactly one of the y_i's will equal 1, and the others will equal 0. Now, (32)–(37) imply that if $y_i = 1$, then z_i and z_{i+1} may be positive, but all the other z_i's must equal 0. For instance, if $y_2 = 1$, then $y_1 = y_3 = 0$. Then (32)–(35) become $z_1 \leq 0$, $z_2 \leq 1$, $z_3 \leq 1$, and $z_4 \leq 0$. These constraints force $z_1 = z_4 = 0$ and allow z_2 and z_3 to be any nonnegative number less than or equal to 1. We can now show that (31)–(37) correctly represent the piecewise linear function $c(x)$. Choose any value of x, say $x = 800$. Note that $b_2 = 500 \leq 800 \leq 1,000 = b_3$. For $x = 800$, what values do our constraints assign to y_1, y_2, and y_3? The value $y_1 = 1$ is impossible, because if $y_1 = 1$, then $y_2 = y_3 = 0$. Then (34)–(35) force $z_3 = z_4 = 0$. Then (31) reduces to $800 = x = 500z_2$, which cannot be satisfied by $z_2 \leq 1$. Similarly, $y_3 = 1$ is impossible. If we try $y_2 = 1$ (32) and (35) force $z_1 = z_4 = 0$. Then (33) and (34) imply $z_2 \leq 1$ and $z_3 \leq 1$. Now (31) becomes $800 = x = 500z_2 + 1,000z_3$. Because $z_2 + z_3 = 1$, we obtain $z_2 = \frac{2}{5}$ and $z_3 = \frac{3}{5}$. Now the objective function reduces to

$$12x_{11} + 12x_{21} + 14x_{21} + 14x_{22} - \frac{2c(500)}{5} - \frac{3c(1,000)}{5}$$

Because

$$c(800) = \frac{2c(500)}{5} + \frac{3c(1,000)}{5}$$

our objective function yields the correct value of Euing's profits!

The optimal solution to Euing's problem is $z = 12,500$, $x = 1,000$, $x_{12} = 1,500$, $x_{22} = 1,000$, $y_3 = z_3 = 1$. Thus, Euing should purchase 1,000 gallons of oil 1 and produce 2,500 gallons of gas 2.

In general, constraints of the form (31)–(37) ensure that if $b_i \leq x \leq b_{i+1}$, then $y_i = 1$ and only z_i and z_{i+1} can be positive. Because $c(x)$ is linear for $b_i \leq x \leq b_{i+1}$, the objective function will assign the correct value to $c(x)$.

If a piecewise linear function $f(x)$ involved in a formulation has the property that the slope of $f(x)$ becomes less favorable to the decision maker as x increases, then the tedious IP formulation we have just described is unnecessary.

EXAMPLE 8   **Media Selection with Piecewise Linear Functions**

Dorian Auto has a $20,000 advertising budget. Dorian can purchase full-page ads in two magazines: *Inside Jocks* (IJ) and *Family Square* (FS). An exposure occurs when a person reads a Dorian Auto ad for the first time. The number of exposures generated by each ad in IJ is as follows: ads 1–6, 10,000 exposures; ads 7–10, 3,000 exposures; ads 11–15, 2,500 exposures; ads 16+, 0 exposures. For example, 8 ads in IJ would generate $6(10,000) + 2(3,000) = 66,000$ exposures. The number of exposures generated by each ad in FS is as follows: ads 1–4, 8,000 exposures; ads 5–12, 6,000 exposures; ads 13–15, 2,000 exposures; ads 16+, 0 exposures. Thus, 13 ads in FS would generate $4(8,000) +$

$8(6,000) + 1(2,000) = 82,000$ exposures. Each full-page ad in either magazine costs $1,000. Assume there is no overlap in the readership of the two magazines. Formulate an IP to maximize the number of exposures that Dorian can obtain with limited advertising funds.

Solution   If we define

$$x_1 = \text{number of IJ ads yielding 10,000 exposures}$$
$$x_2 = \text{number of IJ ads yielding 3,000 exposures}$$
$$x_3 = \text{number of IJ ads yielding 2,500 exposures}$$
$$y_1 = \text{number of FS ads yielding 8,000 exposures}$$
$$y_2 = \text{number of FS ads yielding 6,000 exposures}$$
$$y_3 = \text{number of FS ads yielding 2,000 exposures}$$

then the total number of exposures (in thousands) is given by

$$10x_1 + 3x_2 + 2.5x_3 + 8y_1 + 6y_2 + 2y_3$$

Thus, Dorian wants to maximize

$$z = 10x_1 + 3x_2 + 2.5x_3 + 8y_1 + 6y_2 + 2y_3$$

Because the total amount spent (in thousands) is just the toal number of ads placed in both magazines, Dorian's budget constraint may be written as

$$x_1 + x_2 + x_3 + y_1 + y_2 + y_3 \leq 20$$

The statement of the problem implies that $x_1 \leq 6$, $x_2 \leq 4$, $x_3 \leq 5$, $y_1 \leq 4$, $y_2 \leq 8$, and $y_3 \leq 3$ all must hold. Adding the sign restrictions on each variable and noting that each variable must be an integer, we obtain the following IP:

$$\max z = 10x_1 + 3x_2 + 2.5x_3 + 8y_1 + 6y_2 + 2y_3$$
$$\text{s.t.} \quad x_1 + x_2 + x_3 + y_1 + y_2 + y_3 \leq 20$$
$$x_1 \leq 6$$
$$x_2 \leq 4$$
$$x_3 \leq 5$$
$$y_1 \leq 4$$
$$y_2 \leq 8$$
$$y_3 \leq 3$$
$$x_i, y_i \text{ integer} \quad (i = 1, 2, 3)$$
$$x_i, y_i \geq 0 \quad (i = 1, 2, 3)$$

Observe that the statement of the problem implies that x_2 cannot be positive unless x_1 assumes its maximum value of 6. Similarly, x_3 cannot be positive unless x_2 assumes its maximum value of 4. Because x_1 ads generate more exposures than x_2 ads, however, the act of maximizing ensures that x_2 will be positive only if x_1 has been made as large as possible. Similarly, because x_3 ads generate fewer exposures than x_2 ads, x_3 will be positive only if x_2 assumes its maximum value. (Also, y_2 will be positive only if $y_1 = 4$, and y_3 will be positive only if $y_2 = 8$.)

The optimal solution to Dorian's IP is $z = 146,000$, $x_1 = 6$, $x_2 = 2$, $y_1 = 4$, $y_2 = 8$, $x_3 = 0$, $y_3 = 0$. Thus, Dorian will place $x_1 + x_2 = 8$ ads in IJ and $y_1 + y_2 = 12$ ads in FS.

In Example 8, additional advertising in a magazine yielded diminishing returns. This ensured that x_i (y_i) would be positive only if x_{i-1} (y_{i-1}) assumed its maximum value. If additional advertising generated increasing returns, then this formulation would not yield the correct solution. For example, suppose that the number of exposures generated by each IJ ad was as follows: ads 1–6, 2,500 exposures; ads 7–10, 3,000 exposures; ads 11–15, 10,000 exposures. Suppose also that the number of exposures generated by each FS is as follows: ads 1–4, 2,000 exposures; ads 5–12, 6,000 exposures; ads 13–15, 8,000 exposures.

If we define

$$x_1 = \text{number of IJ ads generating 2,500 exposures}$$
$$x_2 = \text{number of IJ ads generating 3,000 exposures}$$
$$x_3 = \text{number of IJ ads generating 10,000 exposures}$$
$$y_1 = \text{number of FS ads generating 2,000 exposures}$$
$$y_2 = \text{number of FS ads generating 6,000 exposures}$$
$$y_3 = \text{number of FS ads generating 8,000 exposures}$$

the reasoning used in the previous example would lead to the following formulation:

$$\max z = 2.5x_1 + 3x_2 + 10x_3 + 2y_1 + 6y_2 + 8y_3$$
$$\text{s.t.} \quad x_1 + x_2 + x_3 + y_1 + y_2 + y_3 \leq 20$$
$$x_1 \leq 6$$
$$x_2 \leq 4$$
$$x_3 \leq 5$$
$$y_1 \leq 4$$
$$y_2 \leq 8$$
$$y_3 \leq 3$$
$$x_i, y_i \text{ integer} \quad (i = 1, 2, 3)$$
$$x_i, y_i \leq 0 \quad (i = 1, 2, 3)$$

The optimal solution to this IP is $x_3 = 5$, $y_3 = 3$, $y_2 = 8$, $x_2 = 4$, $x_1 = 0$, $y_1 = 0$, which cannot be correct. According to this solution, $x_1 + x_2 + x_3 = 9$ ads should be placed in IJ. If 9 ads were placed in IJ, however, then it must be that $x_1 = 6$ and $x_2 = 3$. Therefore, we see that the type of formulation used in the Dorian Auto example is correct only if the piecewise linear objective function has a less favorable slope for larger values of x. In our second example, the effectiveness of an ad increased as the number of ads in a magazine increased, and the act of maximizing will not ensure that x_i can be positive only if x_{i-1} assumes its maximum value. In this case, the approach used in the Euing Gas example would yield a correct formulation (see Problem 8).

Solving IPs with LINDO

LINDO can be used to solve pure or mixed IPs. In addition to the optimal solution, the LINDO output for an IP gives shadow prices and reduced costs. Unfortunately, the shadow prices and reduced costs refer to subproblems generated during the branch-and-bound solution—*not* to the IP. Unlike linear programming, there is no well-developed theory of sensitivity analysis for integer programming. The reader interested in a discussion of sensitivity analysis for IPs should consult Williams (1985).

To use LINDO to solve an IP, begin by entering the problem as if it were an LP. After typing in the **END** statement (to designate the end of the LP constraints), type for each 0–1 variable x the following statement:

INTE x

Thus, for an IP in which x and y are 0–1 variables, the following statements would be typed after the **END** statement:

INTE x

INTE y

A variable (say, w) that can assume any non-negative integer value is indicated by the **GIN** statement. Thus, if w may assume the values 0, 1, 2, . . . , we would type the following statement after the **END** statement:

GIN w

To tell LINDO that the first n variables appearing in the formulation must be 0–1 variables, use the command **INT** n.

To tell LINDO that the first n variables appearing in the formulation may assume any non-negative integer value, use the command **GIN** n.

To illustrate how to use LINDO to solve IPs, we show how to solve Example 3 with LINDO. We typed the following input (file Gandhi):

```
MAX       6 X1 + 4 X2 + 7 X3 - 200 Y1 - 150 Y2 - 100 Y3
SUBJECT TO
      2)    3 X1 + 2 X2 + 6 X3 <= 150
      3)    4 X1 + 3 X2 + 4 X3 <= 160
      4)    X1 - 40 Y1 <= 0
      5)    X2 - 53 Y2 <= 0
      6)    X3 - 25 Y3 <= 0
END
GIN       X1
GIN       X2
GIN       X3
INTE      Y1
INTE      Y2
INTE      Y3
```

Thus we see that X1, X2, and X3 can be any nonnegative integer, while Y1, Y2, and Y3 must equal 0 or 1. By the way, we could have typed GIN 3 to ensure that X1, X2, and X3 must be nonnegative integers. The optimal solution found by LINDO is given in Figure 4.

Solving IPs with LINGO

LINGO can also be used to solve IPs. To indicate that a variable must equal 0 or 1 use the **@BIN** operator (see the following example). To indicate that a variable must equal a non-negative integer, use the **@GIN** operator. We illustrate how LINGO is used to solve IPs with Example 4 (the Lockbox Problem). The following LINGO program (file Lock.lng) can be used to solve Example 4 (or any reasonably sized lockbox program).

```
MODEL:
  1]SETS:
  2]REGIONS/W,MW,E,S/:DEMAND;
  3]CITIES/LA,CHIC,NY,ATL/:Y;
  4]LINKS(REGIONS,CITIES):DAYS,COST,ASSIGN;
  5]ENDSETS
  6]MIN=@SUM(CITIES:50000*Y)+@SUM(LINKS:COST*ASSIGN);
  7]@FOR(LINKS(I,J):ASSIGN(I,J) < Y(J));
  8]@FOR(REGIONS(I):
  9]@SUM(CITIES(J):ASSIGN(I,J))=1);
 10]@FOR(CITIES(I):@BIN(Y(I));););
```

```
MAX      6 X1 + 4 X2 + 7 X3 - 200 Y1 - 150 Y2 - 100 Y3
SUBJECT TO
        2)    3 X1 + 2 X2 + 6 X3 <=    150
        3)    4 X1 + 3 X2 + 4 X3 <=  · 160
        4)    X1 - 40 Y1 <=    0
        5)    X2 - 53 Y2 <=    0
        6)    X3 - 25 Y3 <=    0
END
GIN       X1
GIN       X2
GIN       X3
INTE      Y1
INTE      Y2
INTE      Y3


        OBJECTIVE FUNCTION VALUE

    1)      75.000000

VARIABLE        VALUE        REDUCED COST
      X1       .000000       -6.000000
      X2       .000000       -4.000000
      X3      25.000000      -7.000000
      Y1       .000000      200.000000
      Y2       .000000      150.000000
      Y3      1.000000      100.000000

    ROW    SLACK OR SURPLUS     DUAL PRICES
    2)          .000000          .000000
    3)        60.000000          .000000
    4)          .000000          .000000
    5)          .000000          .000000
    6)          .000000          .000000

NO. ITERATIONS=        11
BRANCHES=      1 DETERM.=  1.000E     0
```

FIGURE 4

```
11]@FOR(LINKS(I,J):@BIN(ASSIGN(I,J)););
12]@FOR(LINKS(I,J):COST(I,J)=.20*DEMAND(I)*DAYS(I,J));
13]DATA:
14]DAYS=2,6,8,8,
15]6,2,5,5,
16]8,5,2,5,
17]8,5,5,2;
18]DEMAND=70000,50000,60000,40000;
19]ENDDATA
END
```

In line 2, we define the four regions of the country and associate a daily demand for cash payments from each region. Line 3 specifies the four cities where a lockbox may be built. With each city I, we associate a 0–1 variable (Y(I)) that equals 1 if a lockbox is built in the city or 0 otherwise. In line 4, we create a "link" (LINK(I,J)) between each region of the country and each potential lockbox site. Associated with each link are the following quantities:

1 The average number of days (DAYS) it takes a check to clear when mailed from region I to city J. This information is given in the DATA section.

2 The annual lost interest cost for funds sent from region i (COST) incurred if region I sends its money to city J.

3 A 0–1 variable ASSIGN(I,J) which equals 1 if region I sends its money to city J and 0 otherwise.

In line 6, we compute the total cost by summing 50000*Y(I) over all cities. This computes the total annual cost of running lockboxes. Then we sum COST*ASSIGN over all links. This picks up the total annual lost interest cost. The line 7 constraints ensure that (for all combinations of I and J) if region I sends its money to city J, then $Y(J) = 1$. This forces us to pay for lockboxes we use. Lines 8–9 ensure that each region of the country sends its money to some city. Line 10 ensures that each $Y(I)$ equals 0 or 1. Line 11 ensures that each ASSIGN(I,J) equals 0 or 1 (actually we do not need this statement; see Problem 44). We compute the lost annual interest cost if region I sends its money to city J in line 12. This duplicates the calculations in Table 5. Note that an * is needed to ensure that multiplications are performed.

In lines 14–17, we input the average number of days required for a check to clear when it is sent from region I to city J. In line 18, we input the daily demand for each region.

Note that to obtain the objective function and constraints we selected the Model window and then chose LINDO, Generate, Display Model. See Figure 8.

Using the Excel Solver to Solve IP Problems

Gandhi.xls

It is easy to use the Excel Solver to solve integer programming problems. The file Gandhi.xls contains a spreadsheet solution to Example 3. See Figure 7 for the optimal solution. In our spreadsheet, the changing cells J4:J6 (the number of each product produced) must be integers. To tell the Solver that these changing cells must be integers, just select Add Constraint and point to the cells J4:J6. Then select int from the drop-down arrow in the middle.

The changing cells K4:K6 are the binary fixed charge variables. To tell the Solver that these changing cells must be binary, select Add Constraint and point to cells K4:K6. Then select bin from the drop-down arrow. See Figure 6.

From Figure 7, we find that the optimal solution (as found with LINDO) is to make 25 pairs of pants.

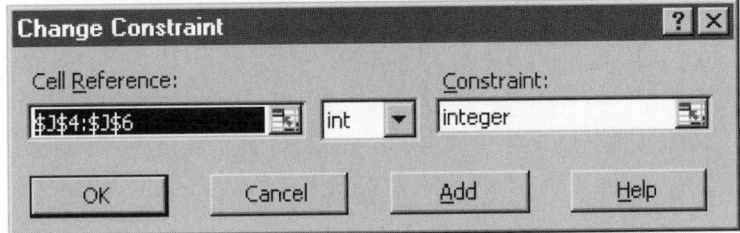

FIGURE 5

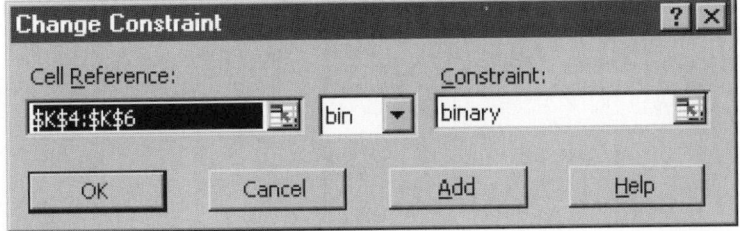

FIGURE 6

FIGURE 7

	A	B	C	D	E	F	G	H	I	J	K
1	Gandhi										
2											
3				Labor hours used	Cloth yards used	Unit price	Unit cost	Unit profit	Fixed Cost	Number Made	Binary variable
4			Shirt	3	4	$ 12.00	$ 6.00	$ 6.00	$ 200.00	0	0
5			Shorts	2	3	$ 8.00	$ 4.00	$ 4.00	$ 150.00	0	0
6			Pants	6	4	$ 15.00	$ 8.00	$ 7.00	$ 100.00	25	1
7		Resource Constraints									
8			Used		Available				Fixed charge	$ 100.00	
9		Labor	150	<=	150				Variable cost	$ 200.00	
10		Cloth	100	<=	160				Revenue	$ 375.00	
11									Profit	$ 75.00	
12		Fixed Charge Constraints	Number Made		Logical Upper Bound	Max possible to make					
13		Shirts	0	<=	0	40					
14		Shorts	0	<=	0	53.33333					
15		Pants	25	<=	25	25					

```
MIN     50000 Y(ATL + 50000 Y(NY + 50000 Y(CHIC + 50000 Y(LA + 16000 ASSIGNSA
      + 40000 ASSIGNSN + 40000 ASSIGNSC + 64000 ASSIGNSL + 60000 ASSIGNEA
      + 24000 ASSIGNEN + 60000 ASSIGNEC + 96000 ASSIGNEL + 50000 ASSIGNMW
      + 50000 ASSIGNMW + 20000 ASSIGNMW + 60000 ASSIGNMW + 112000 ASSIGNWA
      + 112000 ASSIGNWN + 84000 ASSIGNWC + 28000 ASSIGNWL
SUBJECT TO
 2) - Y(LA + ASSIGNWL <=    0
 3) - Y(CHIC + ASSIGNWC <=    0
 4) - Y(NY + ASSIGNWN <=    0
 5) - Y(ATL + ASSIGNWA <=    0
 6) - Y(LA + ASSIGNMW <=    0
 7) - Y(CHIC + ASSIGNMW <=    0
 8) - Y(NY + ASSIGNMW <=    0
 9) - Y(ATL + ASSIGNMW <=    0
10) - Y(LA + ASSIGNEL <=    0
11) - Y(CHIC + ASSIGNEC <=    0
12) - Y(NY + ASSIGNEN <=    0
13) - Y(ATL + ASSIGNEA <=    0
14) - Y(LA + ASSIGNSL <=    0
15) - Y(CHIC + ASSIGNSC <=    0
16) - Y(NY + ASSIGNSN <=    0
17) - Y(ATL + ASSIGNSA <=    0
18)   ASSIGNWA + ASSIGNWN + ASSIGNWC + ASSIGNWL =    1
19)   ASSIGNMW + ASSIGNMW + ASSIGNMW + ASSIGNMW =    1
20)   ASSIGNEA + ASSIGNEN + ASSIGNEC + ASSIGNEL =    1
21)   ASSIGNSA + ASSIGNSN + ASSIGNSC + ASSIGNSL =    1
END
INTE    20


[ERROR CODE: 96]
WARNING: SEVERAL LINGO NAMES MAY HAVE BEEN TRANSFORMED INTO A
SINGLE LINDO NAME.

LP OPTIMUM FOUND AT STEP     14
OBJECTIVE VALUE =   242000.000
ENUMERATION COMPLETE. BRANCHES=      0 PIVOTS=    14
```

FIGURE 8

LAST INTEGER SOLUTION IS THE BEST FOUND
RE-INSTALLING BEST SOLUTION...

VARIABLE	VALUE	REDUCED COST
DEMAND(W)	70000.00	0.0000000E+00
DEMAND(MW)	50000.00	0.0000000E+00
DEMAND(E)	60000.00	0.0000000E+00
DEMAND(S)	40000.00	0.0000000E+00
Y(LA)	1.000000	50000.00
Y(CHIC)	0.0000000E+00	50000.00
Y(NY)	1.000000	50000.00
Y(ATL)	0.0000000E+00	50000.00
DAYS(W, LA)	2.000000	0.0000000E+00
DAYS(W, CHIC)	6.000000	0.0000000E+00
DAYS(W, NY)	8.000000	0.0000000E+00
DAYS(W, ATL)	8.000000	0.0000000E+00
DAYS(MW, LA)	6.000000	0.0000000E+00
DAYS(MW, CHIC)	2.000000	0.0000000E+00
DAYS(MW, NY)	5.000000	0.0000000E+00
DAYS(MW, ATL)	5.000000	0.0000000E+00
DAYS(E, LA)	8.000000	0.0000000E+00
DAYS(E, CHIC)	5.000000	0.0000000E+00
DAYS(E, NY)	2.000000	0.0000000E+00
DAYS(E, ATL)	5.000000	0.0000000E+00
DAYS(S, LA)	8.000000	0.0000000E+00
DAYS(S, CHIC)	5.000000	0.0000000E+00
DAYS(S, NY)	5.000000	0.0000000E+00
DAYS(S, ATL)	2.000000	0.0000000E+00
COST(W, LA)	28000.00	0.0000000E+00
COST(W, CHIC)	84000.00	0.0000000E+00
COST(W, NY)	112000.0	0.0000000E+00
COST(W, ATL)	112000.0	0.0000000E+00
COST(MW, LA)	60000.00	0.0000000E+00
COST(MW, CHIC)	20000.00	0.0000000E+00
COST(MW, NY)	50000.00	0.0000000E+00
COST(MW, ATL)	50000.00	0.0000000E+00
COST(E, LA)	96000.00	0.0000000E+00
COST(E, CHIC)	60000.00	0.0000000E+00
COST(E, NY)	24000.00	0.0000000E+00
COST(E, ATL)	60000.00	0.0000000E+00
COST(S, LA)	64000.00	0.0000000E+00
COST(S, CHIC)	40000.00	0.0000000E+00
COST(S, NY)	40000.00	0.0000000E+00
COST(S, ATL)	16000.00	0.0000000E+00
ASSIGN(W, LA)	1.000000	28000.00
ASSIGN(W, CHIC)	0.0000000E+00	84000.00
ASSIGN(W, NY)	0.0000000E+00	112000.0
ASSIGN(W, ATL)	0.0000000E+00	112000.0
ASSIGN(MW, LA)	0.0000000E+00	60000.00
ASSIGN(MW, CHIC)	0.0000000E+00	20000.00
ASSIGN(MW, NY)	1.000000	50000.00
ASSIGN(MW, ATL)	0.0000000E+00	50000.00
ASSIGN(E, LA)	0.0000000E+00	96000.00
ASSIGN(E, CHIC)	0.0000000E+00	60000.00
ASSIGN(E, NY)	1.000000	24000.00
ASSIGN(E, ATL)	0.0000000E+00	60000.00
ASSIGN(S, LA)	0.0000000E+00	64000.00
ASSIGN(S, CHIC)	0.0000000E+00	40000.00
ASSIGN(S, NY)	1.000000	40000.00
ASSIGN(S, ATL)	0.0000000E+00	16000.00

FIGURE 8
(Continued)

```
          ROW    SLACK OR SURPLUS      DUAL PRICE
           1       242000.0          -1.000000
           2       0.0000000E+00      0.0000000E+00
           3       0.0000000E+00      0.0000000E+00
           4       1.000000           0.0000000E+00
           5       0.0000000E+00      0.0000000E+00
           6       1.000000           0.0000000E+00
           7       0.0000000E+00      0.0000000E+00
           8       0.0000000E+00      0.0000000E+00
           9       0.0000000E+00      0.0000000E+00
          10       1.000000           0.0000000E+00
          11       0.0000000E+00      0.0000000E+00
          12       0.0000000E+00      0.0000000E+00
          13       0.0000000E+00      0.0000000E+00
          14       1.000000           0.0000000E+00
          15       0.0000000E+00      0.0000000E+00
          16       0.0000000E+00      0.0000000E+00
          17       0.0000000E+00      0.0000000E+00
          18       0.0000000E+00      0.0000000E+00
          19       0.0000000E+00      0.0000000E+00
          20       0.0000000E+00      0.0000000E+00
          21       0.0000000E+00      0.0000000E+00
          22       0.0000000E+00     -1.000000
          23       0.0000000E+00      0.0000000E+00
          24       0.0000000E+00      0.0000000E+00
          25       0.0000000E+00      0.0000000E+00
          26       0.0000000E+00      0.0000000E+00
          27       0.0000000E+00      0.0000000E+00
          28       0.0000000E+00     -1.000000
          29       0.0000000E+00      0.0000000E+00
          30       0.0000000E+00      0.0000000E+00
          31       0.0000000E+00      0.0000000E+00
          32       0.0000000E+00     -1.000000
          33       0.0000000E+00      0.0000000E+00
          34       0.0000000E+00      0.0000000E+00
          35       0.0000000E+00      0.0000000E+00
          36       0.0000000E+00     -1.000000
          37       0.0000000E+00      0.0000000E+00
```

FIGURE 8
(Continued)

PROBLEMS

Group A

1 Coach Night is trying to choose the starting lineup for the basketball team. The team consists of seven players who have been rated (on a scale of 1 = poor to 3 = excellent) according to their ball-handling, shooting, rebounding, and defensive abilities. The positions that each player is allowed to play and the player's abilities are listed in Table 9.

The five-player starting lineup must satisfy the following restrictions:

1 At least 4 members must be able to play guard, at least 2 members must be able to play forward, and at least 1 member must be able to play center.

2 The average ball-handling, shooting, and rebounding level of the starting lineup must be at least 2.

3 If player 3 starts, then player 6 cannot start.

4 If player 1 starts, then players 4 and 5 must both start.

5 Either player 2 or player 3 must start.

Given these constraints, Coach Night wants to maximize the total defensive ability of the starting team. Formulate an IP that will help him choose his starting team.

2 Because of excessive pollution on the Momiss River, the state of Momiss is going to build pollution control stations. Three sites (1, 2, and 3) are under consideration. Momiss is

TABLE 9

Player	Position	Ball-Handling	Shooting	Rebounding	Defense
1	G	3	3	1	3
2	C	2	1	3	2
3	G-F	2	3	2	2
4	F-C	1	3	3	1
5	G-F	3	3	3	3
6	F-C	3	1	2	3
7	G-F	3	2	2	1

interested in controlling the pollution levels of two pollutants (1 and 2). The state legislature requires that at least 80,000 tons of pollutant 1 and at least 50,000 tons of pollutant 2 be removed from the river. The relevant data for this problem are shown in Table 10. Formulate an IP to minimize the cost of meeting the state legislature's goals.

3 A manufacturer can sell product 1 at a profit of $2/unit and product 2 at a profit of $5/unit. Three units of raw material are needed to manufacture 1 unit of product 1, and

TABLE 10

| Site | Cost of Building Station ($) | Cost of Treating 1 Ton Water ($) | Amount Removed per Ton of Water | |
			Pollutant 1	Pollutant 2
1	100,000	20	0.40	0.30
2	60,000	30	0.25	0.20
3	40,000	40	0.20	0.25

TABLE 11

| From | To ($) | | |
	Region 1	Region 2	Region 3
New York	20	40	50
Los Angeles	48	15	26
Chicago	26	35	18
Atlanta	24	50	35

6 units of raw material are needed to manufacture 1 unit of product 2. A total of 120 units of raw material are available. If any of product 1 is produced, a setup cost of $10 is incurred, and if any of product 2 is produced, a setup cost of $20 is incurred. Formulate an IP to maximize profits.

4 Suppose we add the following restriction to Example 1 (Stockco): If investments 2 and 3 are chosen, then investment 4 must be chosen. What constraints would be added to the formulation given in the text?

5 How would the following restrictions modify the formulation of Example 6 (Dorian car sizes)? (Do each part separately.)

  a If midsize cars are produced, then compacts must also be produced.

  b Either compacts or large cars must be manufactured.

6 To graduate from Basketweavers University with a major in operations research, a student must complete at least two math courses, at least two OR courses, and at least two computer courses. Some courses can be used to fulfill more than one requirement: Calculus can fulfill the math requirement; operations research, math and OR requirements; data structures, computer and math requirements; business statistics, math and OR requirements; computer simulation, OR and computer requirements; introduction to computer programming, computer requirement; and forecasting, OR and math requirements.

  Some courses are prerequisites for others: Calculus is a prerequisite for business statistics; introduction to computer programming is a prerequisite for computer simulation and for data structures; and business statistics is a prerequisite for forecasting. Formulate an IP that minimizes the number of courses needed to satisfy the major requirements.

7 In Example 7 (Euing Gas), suppose that $x = 300$. What would be the values of y_1, y_2, y_3, z_1, z_2, z_3, and z_4? How about if $x = 1,200$?

8 Formulate an IP to solve the Dorian Auto problem for the advertising data that exhibit increasing returns as more ads are placed in a magazine (pages 495–496).

9 How can integer programming be used to ensure that the variable x can assume only the values 1, 2, 3, and 4?

10 If x and y are integers, how could you ensure that $x + y \le 3$, $2x + 5y \le 12$, or both are satisfied by x and y?

11 If x and y are both integers, how would you ensure that whenever $x \le 2$, then $y \le 3$?

12 A company is considering opening warehouses in four cities: New York, Los Angeles, Chicago, and Atlanta. Each warehouse can ship 100 units per week. The weekly fixed cost of keeping each warehouse open is $400 for New York, $500 for Los Angeles, $300 for Chicago, and $150 for Atlanta. Region 1 of the country requires 80 units per week, region 2 requires 70 units per week, and region 3 requires 40 units per week. The costs (including production and shipping costs) of sending one unit from a plant to a region are shown in Table 11. We want to meet weekly demands at minimum cost, subject to the preceding information and the following restrictions:

  1 If the New York warehouse is opened, then the Los Angeles warehouse must be opened.

  2 At most two warehouses can be opened.

  3 Either the Atlanta or the Los Angeles warehouse must be opened.

Formulate an IP that can be used to minimize the weekly costs of meeting demand.

13 Glueco produces three types of glue on two different production lines. Each line can be utilized by up to seven workers at a time. Workers are paid $500 per week on production line 1, and $900 per week on production line 2. A week of production costs $1,000 to set up production line 1 and $2,000 to set up production line 2. During a week on a production line, each worker produces the number of units of glue shown in Table 12. Each week, at least 120 units of glue 1, at least 150 units of glue 2, and at least 200 units of glue 3 must be produced. Formulate an IP to minimize the total cost of meeting weekly demands.

14[†] The manager of State University's DED computer wants to be able to access five different files. These files are scattered on 10 disks as shown in Table 13. The amount of storage required by each disk is as follows: disk 1, 3K; disk 2, 5K; disk 3, 1K; disk 4, 2K; disk 5, 1K; disk 6, 4K; disk 7, 3K; disk 8, 1K; disk 9, 2K; disk 10, 2K.

  a Formulate an IP that determines a set of disks requiring the minimum amount of storage such that each

TABLE 12

| Production Line | Glue | | |
	1	2	3
1	20	30	40
2	50	35	45

†Based on Day (1965).

TABLE 13

File	Disk 1	2	3	4	5	6	7	8	9	10
1	x	x		x	x			x	x	
2	x		x							
3		x			x		x			x
4			x			x		x		
5	x	x		x			x	x		x

file is on at least one of the disks. For a given disk, we must either store the entire disk or store none of the disk; we cannot store part of a disk.

b Modify your formulation so that if disk 3 or disk 5 is used, then disk 2 must also be used.

15 Fruit Computer produces two types of computers: Pear computers and Apricot computers. Relevant data are given in Table 14. A total of 3,000 chips and 1,200 hours of labor are available. Formulate an IP to help Fruit maximize profits.

16 The Lotus Point Condo Project will contain both homes and apartments. The site can accommodate up to 10,000 dwelling units. The project must contain a recreation project: either a swimming–tennis complex or a sailboat marina, but not both. If a marina is built, then the number of homes in the project must be at least triple the number of apartments in the project. A marina will cost $1.2 million, and a swimming–tennis complex will cost $2.8 million. The developers believe that each apartment will yield revenues with an NPV of $48,000, and each home will yield revenues with an NPV of $46,000. Each home (or apartment) costs $40,000 to build. Formulate an IP to help Lotus Point maximize profits.

17 A product can be produced on four different machines. Each machine has a fixed setup cost, variable production costs per-unit-processed, and a production capacity given in Table 15. A total of 2,000 units of the product must be produced. Formulate an IP whose solution will tell us how to minimize total costs.

TABLE 14

Computer	Labor	Chips	Equipment Costs ($)	Selling Price ($)
Pear	1 hour	2	5,000	400
Apricot	2 hours	5	7,000	900

TABLE 15

Machine	Fixed Cost ($)	Variable Cost per Unit ($)	Capacity
1	1,000	20	900
2	920	24	1,000
3	800	16	1,200
4	700	28	1,600

TABLE 16

	Book 1	2	3	4	5
Maximum Demand	5,000	4,000	3,000	4,000	3,000
Variable Cost ($)	25	20	15	18	22
Sales Price ($)	50	40	38	32	40
Fixed Cost ($ Thousands)	80	50	60	30	40

18 Use LINDO, LINGO, or Excel Solver to find the optimal solution to the following IP:

Bookco Publishers is considering publishing five textbooks. The maximum number of copies of each textbook that can be sold, the variable cost of producing each textbook, the sales price of each textbook, and the fixed cost of a production run for each book are given in Table 16. Thus, for example, producing 2,000 copies of book 1 brings in a revenue of 2,000(50) = $100,000 but costs 80,000 + 25(2,000) = $130,000. Bookco can produce at most 10,000 books if it wants to maximize profit.

19 Comquat owns four production plants at which personal computers are produced. Comquat can sell up to 20,000 computers per year at a price of $3,500 per computer. For each plant the production capacity, the production cost per computer, and the fixed cost of operating a plant for a year are given in Table 17. Determine how Comquat can maximize its yearly profit from computer production.

20 WSP Publishing sells textbooks to college students. WSP has two sales reps available to assign to the A–G state area. The number of college students (in thousands) in each state is given in Figure 9. Each sales rep must be assigned to two adjacent states. For example, a sales rep could be assigned to A and B, but not A and D. WSP's goal is to

TABLE 17

Plant	Production Capacity	Plant Fixed Cost ($ Million)	Cost per Computer ($)
1	10,000	9	1,000
2	8,000	5	1,700
3	9,000	3	2,300
4	6,000	1	2,900

FIGURE 9

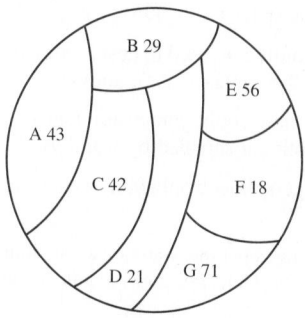

maximize the number of total students in the states assigned to the sales reps. Formulate an IP whose solution will tell you where to assign the sales reps. Then use LINDO to solve your IP.

21 Eastinghouse sells air conditioners. The annual demand for air conditioners in each region of the country is as follows: East, 100,000; South, 150,000; Midwest, 110,000; West, 90,000. Eastinghouse is considering building the air conditioners in four different cities: New York, Atlanta, Chicago, and Los Angeles. The cost of producing an air conditioner in a city and shipping it to a region of the country is given in Table 18. Any factory can produce as many as 150,000 air conditioners per year. The annual fixed cost of operating a factory in each city is given in Table 19. At least 50,000 units of the Midwest demand for air conditioners must come from New York, or at least 50,000 units of the Midwest demand must come from Atlanta. Formulate an IP whose solution will tell Eastinghouse how to minimize the annual cost of meeting demand for air conditioners.

22 Consider the following puzzle. You are to pick out 4 three-letter "words" from the following list:

DBA DEG ADI FFD GHI BCD FDF BAI

For each word, you earn a score equal to the position that the word's third letter appears in the alphabet. For example, DBA earns a score of 1, DEG earns a score of 7, and so on. Your goal is to choose the four words that maximize your total score, subject to the following constraint: The sum of the positions in the alphabet for the first letter of each word chosen must be at least as large as the sum of the positions in the alphabet for the second letter of each word chosen. Formulate an IP to solve this problem.

23 At a machine tool plant, five jobs must be completed each day. The time it takes to do each job depends on the machine used to do the job. If a machine is used at all, there is a setup time required. The relevant times are given in Table 20. The company's goal is to minimize the sum of the setup and machine operation times needed to complete all

TABLE 18

City	Price by Region ($)			
	East	South	Midwest	West
New York	206	225	230	290
Atlanta	225	206	221	270
Chicago	230	221	208	262
Los Angeles	290	270	262	215

TABLE 19

City	Annual Fixed Cost ($ Million)
New York	6
Atlanta	5.5
Chicago	5.8
Los Angeles	6.2

TABLE 20

Machine	Job					Machine Setup Time (Minutes)
	1	2	3	4	5	
1	42	70	93	X	X	30
2	X	85	45	X	X	40
3	58	X	X	37	X	50
4	58	X	55	X	38	60
5	X	60	X	54	X	20

jobs. Formulate and solve (with LINDO, LINGO, or Excel Solver) an IP whose solution will do this.

Group B

24[†] Breadco Bakeries is a new bakery chain that sells bread to customers throughout the state of Indiana. Breadco is considering building bakeries in three locations: Evansville, Indianapolis, and South Bend. Each bakery can bake as many as 900,000 loaves of bread each year. The cost of building a bakery at each site is $5 million in Evansville, $4 million in Indianapolis, and $4.5 million in South Bend. To simplify the problem, we assume that Breadco has only three customers, whose demands each year are 700,000 loaves (customer 1); 400,000 loaves (customer 2); and 300,000 loaves (customer 3). The total cost of baking and shipping a loaf of bread to a customer is given in Table 21.

Assume that future shipping and production costs are discounted at a rate of $11\frac{1}{9}\%$ per year. Assume that once built, a bakery lasts forever. Formulate an IP to minimize Breadco's total cost of meeting demand (present and future). (*Hint:* You will need the fact that for $x < 1$, $a + ax + ax^2 + ax^3 + \cdots = a/(1 - x)$.) How would you modify the formulation if either Evansville or South Bend must produce at least 800,000 loaves per year?

25[‡] Speaker's Clearinghouse must disburse sweepstakes checks to winners in four different regions of the country: Southeast (SE), Northeast (NE), Far West (FW), and Midwest (MW). The average daily amount of the checks written to winners in each region of the country is as follows: SE, $40,000; NE, $60,000; FW, $30,000; MW, $50,000. Speaker's must issue the checks the day they find out a customer has won. They can delay winners from quickly cashing their checks by giving a winner a check drawn on an out-of-the-way bank (this will cause the check to clear

TABLE 21

From	To		
	Customer 1	Customer 2	Customer 3
Evansville	16¢	34¢	26¢
Indianapolis	40¢	30¢	35¢
South Bend	45¢	45¢	23¢

[†]Based on Efroymson and Ray (1966).
[‡]Based on Shanker and Zoltners (1972).

slowly). Four bank sites are under consideration: Frosbite Falls, Montana (FF), Redville, South Carolina (R), Painted Forest, Arizona (PF), and Beanville, Maine (B). The annual cost of maintaining an account at each bank is as follows: FF, $50,000; R, $40,000; PF, $30,000; B, $20,000. Each bank has a requirement that the average daily amount of checks written cannot exceed $90,000. The average number of days it takes a check to clear is given in Table 22. Assuming that money invested by Speaker's earns 15% per year, where should the company have bank accounts, and from which bank should a given customer's check be written?

26[†] Governor Blue of the state of Berry is attempting to get the state legislature to gerrymander Berry's congressional districts. The state consists of 10 cities, and the numbers of registered Republicans and Democrats (in thousands) in each city are shown in Table 23. Berry has five congressional representatives. To form congressional districts, cities must be grouped according to the following restrictions:

1 All voters in a city must be in the same district.

2 Each district must contain between 150,000 and 250,000 voters (there are no independent voters).

Governor Blue is a Democrat. Assume that each voter always votes a straight party ticket. Formulate an IP to help Governor Blue maximize the number of Democrats who will win congressional seats.

27[‡] The Father Domino Company sells copying machines. A major factor in making a sale is Domino's quick service. Domino sells copiers in six cities: Boston, New York, Philadelphia, Washington, Providence, and Atlantic City. The annual sales of copiers projected depend on whether a service representative is within 150 miles of a city (see Table 24).

Each copier costs $500 to produce and sells for $1,000. The annual cost per service representative is $80,000. Domino must determine in which of its markets to base a service representative. Only Boston, New York, Philadelphia, and Washington are under consideration as bases for service representative. The distance (in miles) between the cities is shown in Table 25. Formulate an IP that will help Domino maximize annual profits.

28[§] Thailand inducts naval draftees at three drafting centers. Then the draftees must each be sent to one of three naval bases for training. The cost of transporting a draftee from a drafting center to a base is given in Table 26. Each year, 1,000 men are inducted at center 1; 600 at center 2; and 700 at center 3. Base 1 can train 1,000 men a year, base 2, 800 men; and base 3, 700 men. After the inductees are trained, they are sent to Thailand's main naval base (B). They may be transported on either a small ship or a large ship. It costs $5,000 plus $2 per mile to use a small ship. A small ship can transport up to 200 men to the main base and may visit up to two bases on its way to the main base. Seven small and five large ships are available. It costs $10,000 plus $3 per mile to use a large ship. A large ship may visit up to

TABLE 22

Region	FF	R	PF	B
SE	7	2	6	5
NE	8	4	5	3
FW	4	8	2	11
MW	5	4	7	5

TABLE 23

City	Republicans	Democrats
1	80	34
2	60	44
3	40	44
4	20	24
5	40	114
6	40	64
7	70	14
8	50	44
9	70	54
10	70	64

TABLE 24

Representative Within 150 Miles?	Sales					
	Boston	N.Y.	Phila.	Wash.	Prov.	Atl. City
Yes	700	1,000	900	800	400	450
No	500	750	700	450	200	300

TABLE 25

	Boston	N.Y.	Phila.	Wash.
Boston	0	222	310	441
New York	222	0	89	241
Philadelphia	310	89	0	146
Washington	441	241	146	0
Providence	47	186	255	376
Atlantic City	350	123	82	178

TABLE 26

From	To ($)		
	Base 1	Base 2	Base 3
Center 1	200	200	300
Center 2	300	400	220
Center 3	300	400	250

[†]Based on Garfinkel and Nemhauser (1970).
[‡]Based on Gelb and Khumawala (1984).

[§]Based on Choypeng, Puakpong, and Rosenthal (1986).

three bases on its way to the main base and may transport up to 500 men. The possible "tours" for each type of ship are given in Table 27.

Assume that the assignment of draftees to training bases is done using the transportation method. Then formulate an IP that will minimize the total cost incurred in sending the men from the training bases to the main base. (*Hint:* Let y_{ij} = number of men sent by tour i from base j to main base (B) on a small ship, x_{ij} = number of men sent by tour i from base j to B on a large ship, S_i = number of times tour i is used by a small ship, and L_i = number of times tour i is used by a large ship.)

29 You have been assigned to arrange the songs on the cassette version of Madonna's latest album. A cassette tape has two sides (1 and 2). The songs on each side of the cassette must total between 14 and 16 minutes in length. The length and type of each song are given in Table 28. The assignment of songs to the tape must satisfy the following conditions:

1 Each side must have exactly two ballads.

2 Side 1 must have at least three hit songs.

3 Either song 5 or song 6 must be on side 1.

4 If songs 2 and 4 are on side 1, then song 5 must be on side 2.

Explain how you could use an integer programming formulation to determine whether there is an arrangement of songs satisfying these restrictions.

30 Cousin Bruzie of radio station WABC schedules radio commercials in 60-second blocks. This hour, the station has sold commercial time for commercials of 15, 16, 20, 25, 30, 35, 40, and 50 seconds. Formulate an integer programming model that can be used to determine the minimum number of 60-second blocks of commercials that must be scheduled to fit in all the current hour's commercials. (*Hint:* Certainly no more than eight blocks of time are needed. Let $y_i = 1$ if block i is used and $y_i = 0$ otherwise).

31[†] A Sunco oil delivery truck contains five compartments, holding up to 2,700, 2,800, 1,100, 1,800, and 3,400 gallons of fuel, respectively. The company must deliver three types of fuel (super, regular, and unleaded) to a customer. The demands, penalty per gallon short, and the maximum allowed shortage are given in Table 29. Each compartment of the truck can carry only one type of gasoline. Formulate an IP whose solution will tell Sunco how to load the truck in a way that minimizes shortage costs.

32[‡] Simon's Mall has 10,000 sq ft of space to rent and wants to determine the types of stores that should occupy the mall. The minimum number and maximum number of each type of store (along with the square footage of each type) is given in Table 30. The annual profit made by each type of store will, of course, depend on how many stores of that type are in the mall. This dependence is given in Table 31 (all profits are in units of $10,000). Thus, if there are two department stores in the mall, each department store earns $210,000 profit per year. Each store pays 5% of its annual profit as rent to Simon's. Formulate an IP whose solution will tell Simon's how to maximize rental income from the mall.

33[§] Boris Milkem's financial firm owns six assets. The expected sales price (in millions of dollars) for each asset is given in Table 32. If asset 1 is sold in year 2, the firm receives $20 million. To maintain a regular cash flow, Milkem must sell at least $20 million of assets during year 1, at least $30 million worth during year 2, and at least $35 million worth during year 3. Set up an IP that Milkem can

TABLE 27

Tour Number	Locations Visited	Miles Traveled
1	B–1–B	370
2	B–1–2–B	515
3	B–2–3–B	665
4	B–2–B	460
5	B–3–B	600
6	B–1–3–B	640
7	B–1–2–3–B	720

TABLE 28

Song	Type	Length (in minutes)
1	Ballad	4
2	Hit	5
3	Ballad	3
4	Hit	2
5	Ballad	4
6	Hit	3
7		5
8	Ballad and hit	4

TABLE 29

Type of Gasoline	Demand	Cost per Gallon Short ($)	Maximum Allowed Shortage
Super	2,900	10	500
Regular	4,000	8	500
Unleaded	4,900	6	500

TABLE 30

Store Type	Square Footage	Minimum	Maximum
Jewelry	500	1	3
Shoe	600	1	3
Department	1,500	1	3
Book	700	0	3
Clothing	900	1	3

[†]Based on Brown (1987).
[‡]Based on Bean et al. (1988).
[§]Based on Bean, Noon, and Salton (1987).

TABLE 31

Type of Store	Number of Stores		
	1	2	3
Jewelry	9	8	7
Shoe	10	9	5
Department	27	21	20
Book	16	9	7
Clothing	17	13	10

TABLE 32

Asset	Sold In		
	Year 1	Year 2	Year 3
1	15	20	24
2	16	18	21
3	22	30	36
4	10	20	30
5	17	19	22
6	19	25	29

TABLE 33

Alarm Box	Two Closest Ladder Companies
1	2, 3
2	3, 4
3	1, 5
4	2, 6
5	3, 6
6	4, 7
7	5, 7

TABLE 34

Boiler Number	Minimum Steam	Maximum Steam	Cost/Ton ($)
1	500	1,000	10
2	300	900	8
3	400	800	6

TABLE 35

Turbine Number	Minimum	Maximum	Kwh per Ton of Steam	Processing Cost per Ton ($)
1	300	600	4	2
2	500	800	5	3
3	600	900	6	4

use to determine how to maximize total revenue from assets sold during the next three years. In implementing this model, how could the idea of a rolling planning horizon be used?

34[†] The Smalltown Fire Department currently has seven conventional ladder companies and seven alarm boxes. The two closest ladder companies to each alarm box are given in Table 33. The city fathers want to maximize the number of conventional ladder companies that can be replaced with tower ladder companies. Unfortunately, political considerations dictate that a conventional company can be replaced only if, after replacement, at least one of the two closest companies to each alarm box is still a conventional company.

    a Formulate an IP that can be used to maximize the number of conventional companies that can be replaced by tower companies.

    b Suppose $y_k = 1$ if conventional company k is replaced. Show that if we let $z_k = 1 - y_k$, the answer in part (a) is equivalent to a set-covering problem.

35[‡] A power plant has three boilers. If a given boiler is operated, it can be used to produce a quantity of steam (in tons) between the minimum and maximum given in Table 34. The cost of producing a ton of steam on each boiler is also given. Steam from the boilers is used to produce power on three turbines. If operated, each turbine can process an amount of steam (in tons) between the minimum and maximum given in Table 35. The cost of processing a ton of steam and the power produced by each turbine is also given. Formulate an IP that can be used to minimize the cost of producing 8,000 kwh of power.

36[§] An Ohio company, Clevcinn, consists of three subsidiaries. Each has the respective average payroll, unemployment reserve fund, and estimated payroll given in Table 36. (All figures are in millions of dollars.) Any employer in the state of Ohio whose reserve/average payroll ratio is less than 1 must pay 20% of its estimated payroll in unemployment insurance premiums or 10% if the ratio is at least 1. Clevcinn can aggregate its subsidiaries and label them as separate employers. For instance, if subsidiaries 2 and 3 are aggregated, they must pay 20% of their combined payroll in unemployment insurance premiums. Formulate an IP that can be used to determine which subsidiaries should be aggregated.

37 The Indiana University Business School has two rooms that each seat 50 students, one room that seats 100 students, and one room that seats 150 students. Classes are held five hours a day. The four types of requests for rooms are listed in Table 37. The business school must decide how many requests of each type should be assigned to each type of room. Penalties for each type of assignment are given in Table 38. An X means that a request must be satisfied by a room of adequate size. Formulate an IP whose solution will tell the business school how to assign classes to rooms in a way that minimizes total penalties.

[†]Based on Walker (1974).
[‡]Based on Cavalieri, Roversi, and Ruggeri (1971).

[§]Based on Salkin (1979).

TABLE 36

Subsidiary	Average Payroll	Reserve	Estimated Payroll
1	300	400	350
2	600	510	400
3	800	600	500

TABLE 37

Type	Size Room Requested (Seats)	Hours Requested	Number of Requests
1	50	2, 3, 4	3
2	150	1, 2, 3	1
3	100	5	1
4	50	1, 2	2

TABLE 38

Size Requested	Sizes Used to Satisfy Request			Penalty
	50	100	150	
50	0	2	4	100* (Hours requested)
100	X	0	1	100* (Hours requested)
150	X	X	0	100* (Hours requested)

38 A company sells seven types of boxes, ranging in volume from 17 to 33 cubic feet. The demand and size of each box are given in Table 39. The variable cost (in dollars) of producing each box is equal to the box's volume. A fixed cost of $1,000 is incurred to produce any of a particular box. If the company desires, demand for a box may be satisfied by a box of larger size. Formulate and solve (with LINDO, LINGO, or Excel Solver) an IP whose solution will minimize the cost of meeting the demand for boxes.

39 Huntco produces tomato sauce at five different plants. The capacity (in tons) of each plant is given in Table 40. The tomato sauce is stored at one of three warehouses. The per-ton cost (in hundreds of dollars) of producing tomato sauce at each plant and shipping it to each warehouse is given in Table 41. Huntco has four customers. The cost of shipping a ton of sauce from each warehouse to each customer is as given in Table 42. Each customer must be delivered the amount (in tons) of sauce given in Table 43.

TABLE 39

	Box						
	1	2	3	4	5	6	7
Size	33	30	26	24	19	18	17
Demand	400	300	500	700	200	400	200

TABLE 40

	Plant				
	1	2	3	4	5
Tons	300	200	300	200	400

TABLE 41

From	To		
	Warehouse 1	Warehouse 2	Warehouse 3
Plant 1	8	10	12
Plant 2	7	5	7
Plant 3	8	6	5
Plant 4	5	6	7
Plant 5	7	6	5

TABLE 42

From	To			
	Customer 1	Customer 2	Customer 3	Customer 4
Warehouse 1	40	80	90	50
Warehouse 2	70	70	60	80
Warehouse 3	80	30	50	60

TABLE 43

	Customer			
	1	2	3	4
Demand	200	300	150	250

a Formulate a balanced transportation problem whose solution will tell us how to minimize the cost of meeting the customer demands.

b Modify this problem if these are annual demands and there is a fixed annual cost of operating each plant and warehouse. These costs (in thousands) are given in Table 44.

40 To satisfy telecommunication needs for the next 20 years, Telstar Corporation estimates that the number of circuits required between the United States and Germany, France, Switzerland, and the United Kingdom will be as given in Table 45.

Two types of circuits may be created: cable and satellite. Two types of cable circuits (TA7 and TA8) are available. The fixed cost of building each type of cable and the circuit capacity of each type are as given in Table 46.

TA7 and TA8 cable go underseas from the United States to the English Channel. Thus, it costs an additional amount to extend these circuits to other European countries. The annual variable cost per circuit is given in Table 47.

TABLE 44[†]

Facility	Fixed Annual Cost (in Thousands) $
Plant 1	35
Plant 2	45
Plant 3	40
Plant 4	42
Plant 5	40
Warehouse 1	30
Warehouse 2	40
Warehouse 3	30

[†]Based on Geoffrion and Graves (1974).

TABLE 45

Country	Required Circuits
France	20,000
Germany	60,000
Switzerland	16,000
United Kingdom	60,000

TABLE 46

Cable Type	Fixed Operating Cost ($ Billion)	Capacity
TA7	1.6	8,500
TA8	2.3	37,800

TABLE 47

Country	Variable Cost per Circuit ($)
France	0
Germany	310
Switzerland	290
United Kingdom	0

To create and use a satellite circuit, Telstar must launch a satellite, and each country using the satellite must have an earth station(s) to receive the signal. It costs $3 billion to launch a satellite. Each launched satellite can handle up to 140,000 circuits. All earth stations have a maximum capacity of 190 circuits and cost $6,000 per year to operate. Formulate an integer programming model to help determine how to supply the needed circuits and minimize total cost incurred during the next 20 years.

Then use LINDO (or LINGO) to find a near optimal solution. LINDO after 300 pivots did not think it had an optimal solution! By the way, do not require that the number of cable or satellite circuits in a country be integers, or your

model will never get solved! For some variables, however, the integer requirement is vital![†]

41 A large drug company must determine how many sales representatives to assign to each of four sales districts. The cost of having n representatives in a district is ($88,000 + $80,000n$) per year. If a rep is based in a given district, the time it takes to complete a call on a doctor is given in Table 48 (times are in hours).

Each sales rep can work up to 160 hours per month. Each month the number of calls given in Table 49 must be made in each district. A fractional number of representatives in a district is not permissible. Determine how many representatives should be assigned to each district.

42[‡] In this assignment, we will use integer programming and the concept of bond duration to show how Wall Street firms can select an optimal bond portfolio. The *duration* of a bond (or any stream of payments) is defined as follows: Let $C(t)$ be the payment of the bond at time t ($t = 1, 2, \ldots, n$). Let r = market interest rate. If the time-weighted average of the bond's payments is given by:

$$\sum_{t=1}^{t=n} tC(t)/(1 + r)^t$$

and the market price P of the bond is given by:

$$\sum_{t=1}^{t=n} C(t)/(1 + r)^t$$

then the duration of the bond D is given by:

$$D = (1/P) \sum_{t=1}^{n} \frac{tC(t)}{(1 + r)^t}$$

Thus, the duration of a bond measures the "average" time (in years) at which a randomly chosen $1 of NPV is received. Suppose an insurance company needs to make payments of $20,000 every six months for the next 10 years. If the market

TABLE 48

Rep's Base District	Actual Sales Call District			
	1	2	3	4
1	1	4	5	7
2	4	1	3	5
3	5	3	1	2
4	7	5	2	1

TABLE 49

District	Number of Calls
1	50
2	80
3	100
4	60

[†]Based on Calloway, Cummins, and Freeland (1990).
[‡]Based on Strong (1989).

rate of interest is 10% per year, then this stream of payments has an NPV of \$251,780 and a duration of 4.47 years. If we want to minimize the sensitivity of our bond portfolio to interest risk and still meet our payment obligations, then it has been shown that we should invest \$251,780 at the beginning of year 1 in a bond portfolio having a duration equal to the duration of the payment stream.

Suppose the only cost of owning a bond portfolio is the transaction cost associated with the cost of purchasing the bonds. Let's suppose six bonds are available. The payment streams for these six bonds are given in Table 50. The transaction cost of purchasing any units of bond i equals \$500 + \$5 per bond purchased. Thus, purchasing one unit of bond 1 costs \$505 and purchasing 10 units of bond 1 costs \$550. Assume that a fractional number of bond i unit purchases is permissible, but in the interests of diversification at most 100 units of any bond can be purchased. Treasury bonds may also be purchased (with no transaction cost). A treasury bond costs \$980 and has a duration of .25 year (90 days).

After computing the price and duration for each bond, use integer programming to determine the immunized bond portfolio that incurs the smallest transaction costs. You may assume the duration of your portfolio is a weighted average of the durations of the bonds included in the portfolio, where the weight associated with each bond is equal to the money invested in that bond.

43 Ford has four automobile plants. Each is capable of producing the Taurus, Lincoln, or Escort, but it can only produce one of these cars. The fixed cost of operating each plant for a year and the variable cost of producing a car of each type at each plant are given in Table 51.

Ford faces the following restrictions:

a Each plant can produce only one type of car.

b The total production of each type of car must be at a single plant; that is, for example, if any Tauruses are made at plant 1, then all Tauruses must be made there.

c If plants 3 and 4 are used, then plant 1 must also be used.

TABLE 50

	Available Bonds					
Year	Bond 1	Bond 2	Bond 3	Bond 4	Bond 5	Bond 6
1	50	100	130	20	100	120
2	60	90	130	20	100	100
3	70	80	130	20	100	80
4	80	70	130	20	100	140
5	90	60	130	20	100	100
6	100	50	130	80	100	90
7	110	40	130	40	100	110
8	120	30	130	150	100	130
9	130	20	130	200	100	180
10	1,010	1,040	1,130	1,200	1,100	950

TABLE 51

Plant	Fixed Cost (\$)	Variable Cost (\$)		
		Taurus	Lincoln	Escort
1	7 billion	12,000	16,000	9,000
2	6 billion	15,000	18,000	11,000
3	4 billion	17,000	19,000	12,000
4	2 billion	19,000	22,000	14,000

Each year, Ford must produce 500,000 of each type of car. Formulate an IP whose solution will tell Ford how to minimize the annual cost of producing cars.

44 Venture capital firm JD is trying to determine in which of 10 projects it should invest. It knows how much money is available for investment each of the next N years, the NPV of each project, and the cash required by each project during each of the next N years (see Table 52).

a Write a LINGO program to determine the projects in which JD should invest.

b Use your LINGO program to determine which of the 10 projects should be selected. Each project requires cash investment during the next three years. During year 1, \$80 million is available for investment. During year 2, \$60 million is available for investment. During year 3, \$70 million is available for investment. (All figures are in millions of dollars.)

45 Write a LINGO program that can solve a fixed-charge problem of the type described in Example 3. Assume there is a limited demand for each product. Then use your program to solve a four-product, three-resource fixed-charge problem with the parameters shown in Tables 53, 54, and 55.

TABLE 52

Investment (\$ Million)	Project									
	1	2	3	4	5	6	7	8	9	10
Year 1	6	9	12	15	18	21	24	27	30	35
Year 2	3	5	7	9	11	13	15	17	19	21
Year 3	5	7	9	12	12	14	16	11	20	24
NPV	20	30	40	50	60	70	80	90	100	130

TABLE 53

Resource	Resource Availability
1	40
2	60
3	80

TABLE **54**

Product	Demand	Unit Profit Contribution ($)	Fixed Charge ($)
1	40	2	30
2	60	5	40
3	65	6	50
4	70	7	60

TABLE **55**

Resource Usage	Product			
	1	2	3	4
1	1	2	3.5	4
2	5	6	7	9
3	3	4	5	6

9.3 The Branch-and-Bound Method for Solving Pure Integer Programming Problems

In practice, most IPs are solved by using the technique of branch-and-bound. Branch-and-bound methods find the optimal solution to an IP by efficiently enumerating the points in a subproblem's feasible region. Before explaining how branch-and-bound works, we need to make the following elementary but important observation: *If you solve the LP relaxation of a pure IP and obtain a solution in which all variables are integers, then the optimal solution to the LP relaxation is also the optimal solution to the IP.*

To see why this observation is true, consider the following IP:

$$\max z = 3x_1 + 2x_2$$
$$\text{s.t.} \quad 2x_1 + x_2 \leq 6$$
$$x_1, x_2 \geq 0; \ x_1, x_2 \text{ integer}$$

The optimal solution to the LP relaxation of this pure IP is $x_1 = 0$, $x_2 = 6$, $z = 12$. Because this solution gives integer values to all variables, the preceding observation implies that $x_1 = 0$, $x_2 = 6$, $z = 12$ is also the optimal solution to the IP. Observe that the feasible region for the IP is a subset of the points in the LP relaxation's feasible region (see Figure 10). Thus, the optimal z-value for the IP cannot be larger than the optimal z-value for the LP relaxation. This means that the optimal z-value for the IP must be ≤ 12. But the point $x_1 = 0$, $x_2 = 6$, $z = 12$ is feasible for the IP and has $z = 12$. Thus, $x_1 = 0$, $x_2 = 6$, $z = 12$ must be optimal for the IP.

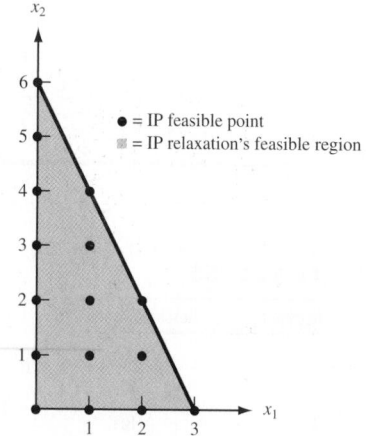

FIGURE 10
Feasible Region for an IP and Its LP Relaxation

EXAMPLE 9   Branch-and-Bound Method

The Telfa Corporation manufactures tables and chairs. A table requires 1 hour of labor and 9 square board feet of wood, and a chair requires 1 hour of labor and 5 square board feet of wood. Currently, 6 hours of labor and 45 square board feet of wood are available. Each table contributes $8 to profit, and each chair contributes $5 to profit. Formulate and solve an IP to maximize Telfa's profit.

Solution   Let

$$x_1 = \text{number of tables manufactured}$$
$$x_2 = \text{number of chairs manufactured}$$

Because x_1 and x_2 must be integers, Telfa wants to solve the following IP:

$$\max z = 8x_1 + 5x_2$$
$$\text{s.t.} \quad x_1 + x_2 \leq 6 \quad \text{(Labor constraint)}$$
$$9x_1 + 5x_2 \leq 45 \quad \text{(Wood constraint)}$$
$$x_1, x_2 \geq 0; x_1, x_2 \text{ integer}$$

The branch-and-bound method begins by solving the LP relaxation of the IP. If all the decision variables assume integer values in the optimal solution to the LP relaxation, then the optimal solution to the LP relaxation will be the optimal solution to the IP. We call the LP relaxation subproblem 1. Unfortunately, the optimal solution to the LP relaxation is $z = \frac{165}{4}$, $x_1 = \frac{15}{4}$, $x_2 = \frac{9}{4}$ (see Figure 11). From Section 9.1, we know that (optimal z-value for IP) $\leq$ (optimal z-value for LP relaxation). This implies that the optimal z-value for the IP cannot exceed $\frac{165}{4}$. Thus, the optimal z-value for the LP relaxation is an **upper bound** for Telfa's profit.

Our next step is to partition the feasible region for the LP relaxation in an attempt to find out more about the location of the IP's optimal solution. We arbitrarily choose a variable that is fractional in the optimal solution to the LP relaxation—say, x_1. Now observe that every point in the feasible region for the IP must have either $x_1 \leq 3$ or $x_1 \geq 4$. (Why can't a feasible solution to the IP have $3 < x_1 < 4$?) With this in mind, we "branch" on the variable x_1 and create the following two additional subproblems:

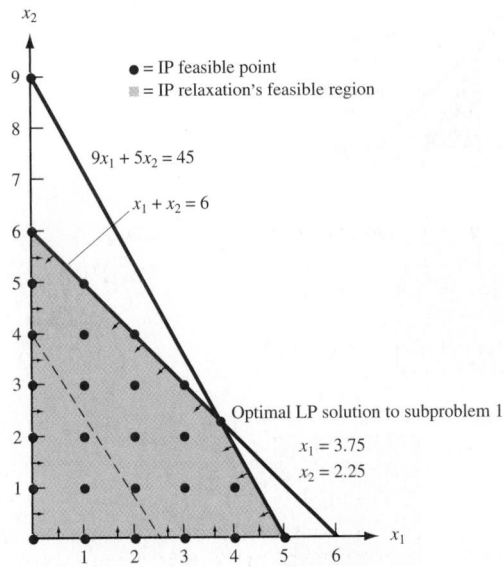

FIGURE 11
Feasible Region for Telfa Problem

Subproblem 2 Subproblem 1 + Constraint $x_1 \geq 4$.

Subproblem 3 Subproblem 1 + Constraint $x_1 \leq 3$.

Observe that neither subproblem 2 nor subproblem 3 includes any points with $x_1 = \frac{15}{4}$. This means that the optimal solution to the LP relaxation cannot recur when we solve subproblem 2 or subproblem 3.

From Figure 12, we see that every point in the feasible region for the Telfa IP is included in the feasible region for subproblem 2 or subproblem 3. Also, the feasible regions for subproblems 2 and 3 have no points in common. Because subproblems 2 and 3 were created by adding constraints involving x_1, we say that subproblems 2 and 3 were created by **branching** on x_1.

We now choose any subproblem that has not yet been solved as an LP. We arbitrarily choose to solve subproblem 2. From Figure 12, we see that the optimal solution to subproblem 2 is $z = 41$, $x_1 = 4$, $x_2 = \frac{9}{5}$ (point C). Our accomplishments to date are summarized in Figure 13.

A display of all subproblems that have been created is called a **tree**. Each subproblem is referred to as a **node** of the tree, and each line connecting two nodes of the tree is called an **arc**. The constraints associated with any node of the tree are the constraints for the LP relaxation plus the constraints associated with the arcs leading from subproblem 1 to the node. The label t indicates the chronological order in which the subproblems are solved.

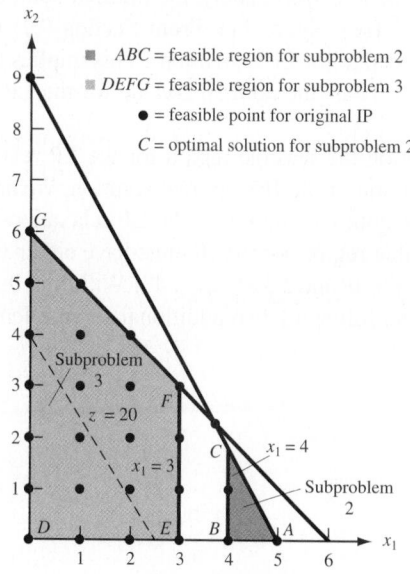

FIGURE 12
Feasible Region for Subproblems 2 and 3 of Telfa Problem

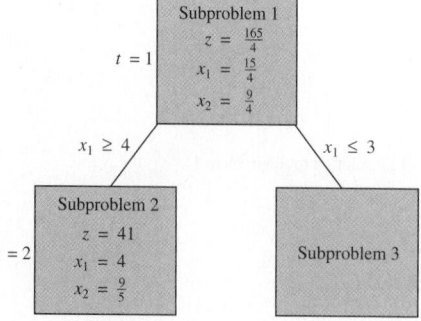

FIGURE 13
Telfa Subproblems 1 and 2 Solved

The optimal solution to subproblem 2 did not yield an all-integer solution, so we choose to use subproblem 2 to create two new subproblems. We choose a fractional-valued variable in the optimal solution to subproblem 2 and then branch on that variable. Because x_2 is the only fractional variable in the optimal solution to subproblem 2, we branch on x_2. We partition the feasible region for subproblem 2 into those points having $x_2 \geq 2$ and $x_2 \leq 1$. This creates the following two subproblems:

Subproblem 4   Subproblem 1 + Constraints $x_1 \geq 4$ and $x_2 \geq 2$ = subproblem 2 + Constraint $x_2 \geq 2$.

Subproblem 5   Subproblem 1 + Constraints $x_1 \geq 4$ and $x_2 \leq 1$ = subproblem 2 + Constraint $x_2 \leq 1$.

The feasible regions for subproblems 4 and 5 are displayed in Figure 14. The set of unsolved subproblems consists of subproblems 3, 4, and 5. We now choose a subproblem to solve. For reasons that are discussed later, we choose to solve the most recently created subproblem. (This is called the LIFO, or last-in-first-out, rule.) The LIFO rule implies that we should next solve subproblem 4 or subproblem 5. We arbitrarily choose to solve subproblem 4. From Figure 14 we see that subproblem 4 is infeasible. Thus, subproblem 4 cannot yield the optimal solution to the IP. To indicate this fact, we place an $\times$ by subproblem 4 (see Figure 15). Because any branches emanating from subproblem 4 will yield no useful information, it is fruitless to create them. When further branching on a subproblem cannot yield any useful information, we say that the subproblem (or node) is **fathomed.** Our results to date are displayed in Figure 15.

Now the only unsolved subproblems are subproblems 3 and 5. The LIFO rule implies that subproblem 5 should be solved next. From Figure 14, we see that the optimal solution to subproblem 5 is point I in Figure 14: $z = \frac{365}{9}$, $x_1 = \frac{40}{9}$, $x_2 = 1$. This solution does not yield any immediately useful information, so we choose to partition subproblem 5's feasible region by branching on the fractional-valued variable x_1. This yields two new subproblems (see Figure 16).

Subproblem 6   Subproblem 5 + Constraint $x_1 \geq 5$.

Subproblem 7   Subproblem 5 + Constraint $x_1 \leq 4$.

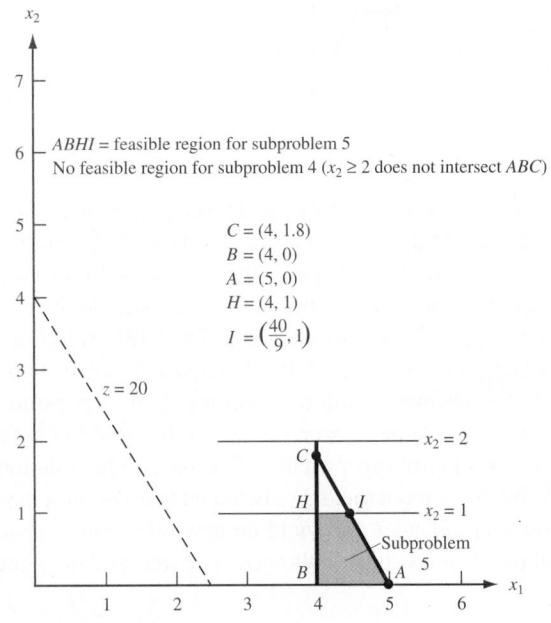

FIGURE 14
Feasible Regions for Subproblems 4 and 5 of Telfa Problem

$ABHI$ = feasible region for subproblem 5
No feasible region for subproblem 4 ($x_2 \geq 2$ does not intersect ABC)

$C = (4, 1.8)$
$B = (4, 0)$
$A = (5, 0)$
$H = (4, 1)$
$I = \left(\frac{40}{9}, 1\right)$

$z = 20$

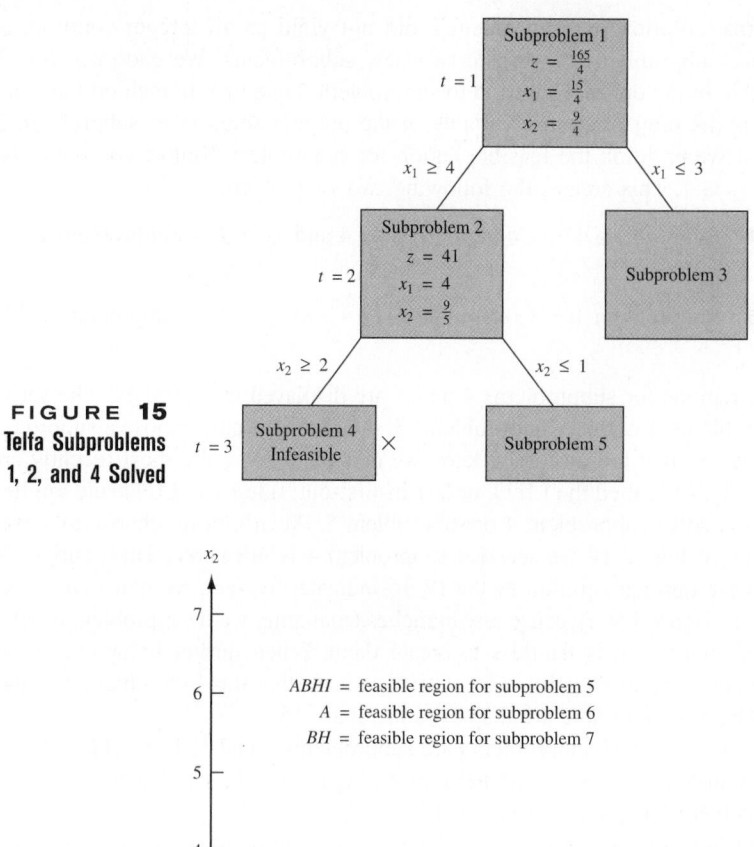

FIGURE **15**
Telfa Subproblems
1, 2, and 4 Solved

FIGURE **16**
Feasible Regions for
Subproblems 6 and 7
of Telfa Problem

Together, subproblems 6 and 7 include all integer points that were included in the feasible region for subproblem 5. Also, no point having $x_1 = \frac{40}{9}$ can be in the feasible region for subproblem 6 or subproblem 7. Thus, the optimal solution to subproblem 5 will not recur when we solve subproblems 6 and 7. Our tree now looks as shown in Figure 17.

Subproblems 3, 6, and 7 are now unsolved. The LIFO rule implies that we next solve subproblem 6 or subproblem 7. We arbitrarily choose to solve subproblem 7. From Figure 16, we see that the optimal solution to subproblem 7 is point H: $z = 37$, $x_1 = 4$, $x_2 = 1$. Both x_1 and x_2 assume integer values, so this solution is feasible for the original IP. We now know that subproblem 7 yields a feasible integer solution with $z = 37$. We also know that subproblem 7 cannot yield a feasible integer solution having $z > 37$. Thus, further branching on subproblem 7 will yield no new information about the optimal solution to the IP, and subproblem has been fathomed. The tree to date is pictured in Figure 18.

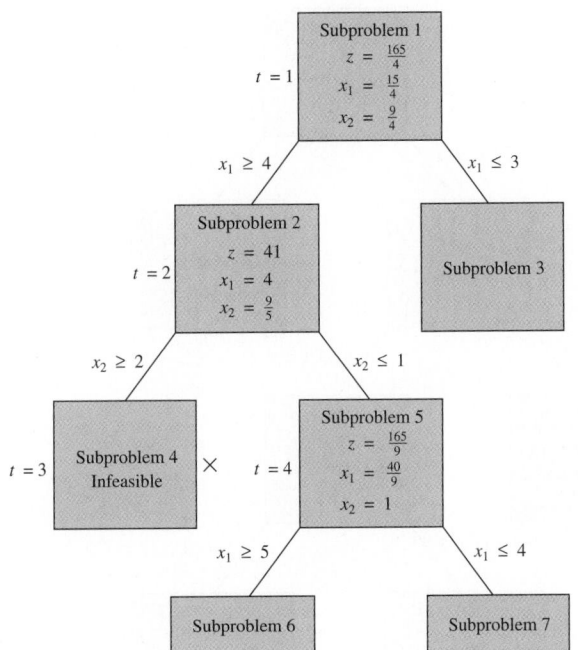

FIGURE 17
Telfa Subproblems
1, 2, 4, and 5 Solved

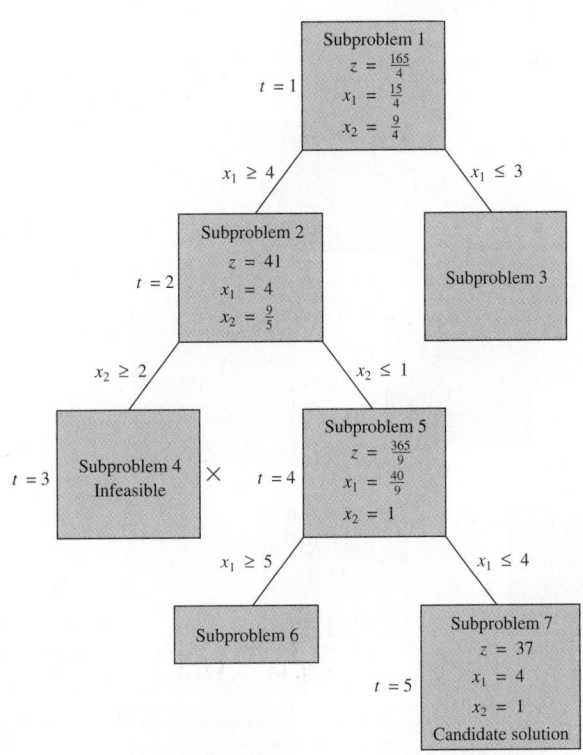

FIGURE 18
Branch-and-Bound Tree
After Five Subproblems
Have Been Solved

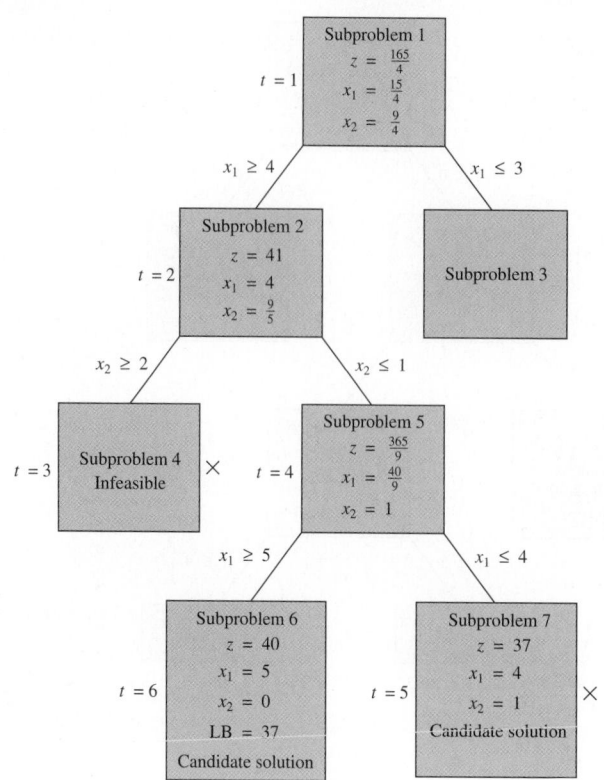

FIGURE 19
Branch-and-Bound Tree
After Six Subproblems
Have Been Solved

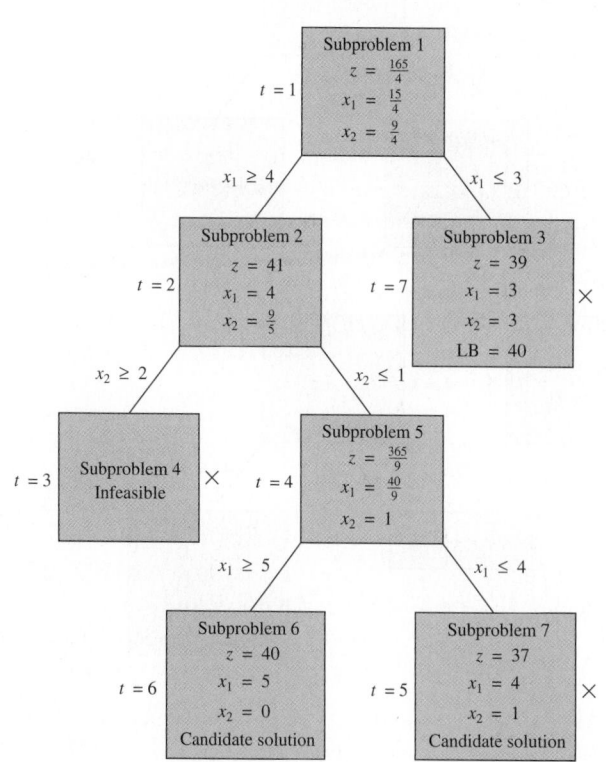

FIGURE 20
Final Branch-and-Bound
Tree for Telfa Problem

A solution obtained by solving a subproblem in which all variables have integer values is a **candidate solution.** Because the candidate solution may be optimal, we must keep a candidate solution until a better feasible solution to the IP (if any exists) is found. We have a feasible solution to the original IP with $z = 37$, so we may conclude that the optimal z-value for the IP ≥ 37. Thus, the z-value for the candidate solution is a **lower bound** on the optimal z-value for the original IP. We note this by placing the notation $LB = 37$ in the box corresponding to the *next* solved subproblem (see Figure 19).

The only remaining unsolved subproblems are 6 and 3. Following the LIFO rule, we next solve subproblem 6. From Figure 16, we find that the optimal solution to subproblem 6 is point A: $z = 40$, $x_1 = 5$, $x_2 = 0$. All decision variables have integer values, so this is a candidate solution. Its z-value of 40 is larger than the z-value of the best previous candidate (candidate 7 with $z = 37$). Thus, subproblem 7 cannot yield the optimal solution of the IP (we denote this fact by placing an $\times$ by subproblem 7). We also update our LB to 40. Our progress to date is summarized in Figure 20.

Subproblem 3 is the only remaining unsolved problem. From Figure 12, we find that the optimal solution to subproblem 3 is point F: $z = 39$, $x_1 = x_2 = 3$. Subproblem 3 cannot yield a z-value exceeding the current lower bound of 40, so it cannot yield the optimal solution to the original IP. Therefore, we place an $\times$ by it in Figure 20. From Figure 20, we see that there are no remaining unsolved subproblems, and that only subproblem 6 can yield the optimal solution to the IP. Thus, the optimal solution to the IP is for Telfa to manufacture 5 tables and 0 chairs. This solution will contribute $40 to profits.

In using the branch-and-bound method to solve the Telfa problem, we have implicitly enumerated all points in the IP's feasible region. Eventually, all such points (except for the optimal solution) are eliminated from consideration, and the branch-and-bound procedure is complete. To show that the branch-and-bound procedure actually does consider all points in the IP's feasible region, we examine several possible solutions to the Telfa problem and show how the procedure found these points to be nonoptimal. For example, how do we know that $x_1 = 2$, $x_2 = 3$ is not optimal? This point is in the feasible region for subproblem 3, and we know that all points in the feasible region for subproblem 3 have $z \leq 39$. Thus, our analysis of subproblem 3 shows that $x_1 = 2$, $x_2 = 3$ cannot beat $z = 40$ and cannot be optimal. As another example, why isn't $x_1 = 4$, $x_2 = 2$ optimal? Following the branches of the tree, we find that $x_1 = 4$, $x_2 = 2$ is associated with subproblem 4. Because no point associated with subproblem 4 is feasible, $x_1 = 4$, $x_2 = 2$ must fail to satisfy the constraints for the original IP and thus cannot be optimal for the Telfa problem. In a similar fashion, the branch-and-bound analysis has eliminated all points x_1, x_2 (except for the optimal solution) from consideration.

For the simple Telfa problem, the use of the branch-and-bound method may seem like using a cannon to kill a fly, but for an IP in which the feasible region contains a large number of integer points, the procedure can be very efficient for eliminating nonoptimal points from consideration. For example, suppose we are applying the branch-and-bound method and our current $LB = 42$. Suppose we solve a subproblem that contains 1 million feasible points for the IP. If the optimal solution to this subproblem has $z < 42$, then we have eliminated 1 million nonoptimal points by solving a single LP!

The key aspects of the branch-and-bound method for solving pure IPs (mixed IPs are considered in the next section) may be summarized as follows:

Step 1 If it is unnecessary to branch on a subproblem, then it is fathomed. The following three situations result in a subproblem being fathomed: (1) The subproblem is infeasible; (2) the subproblem yields an optimal solution in which all variables have integer values; and (3) the optimal z-value for the subproblem does not exceed (in a max problem) the current LB.

Step 2 A subproblem may be eliminated from consideration in the following situations: (1) The subproblem is infeasible (in the Telfa problem, subproblem 4 was eliminated for this reason); (2) the LB (representing the z-value of the best candidate to date) is at least as large as the z-value for the subproblem (in the Telfa problem, subproblems 3 and 7 were eliminated for this reason).

Recall that in solving the Telfa problem by the branch-and-bound procedure, many seemingly arbitrary choices were made. For example, when x_1 and x_2 were both fractional in the optimal solution to subproblem 1, how did we determine the branching variable? Or how did we determine which subproblem should next be solved? The manner in which these questions are answered can result in trees that differ greatly in size and in the computer time required to find an optimal solution. Through experience and ingenuity, practitioners of the procedure have developed guidelines on how to make the necessary decisions.

Two general approaches are commonly used to determine which subproblems should be solved next. The most widely used is the LIFO rule, which chooses to solve the most recently created subproblem.[†] LIFO leads us down one side of the branch-and-bound tree (as in the Telfa problem) and quickly finds a candidate solution. Then we backtrack our way up to the top of the other side of the tree. For this reason, the LIFO approach is often called **backtracking.**

The second commonly used method is **jumptracking.** When branching on a node, the jumptracking approach solves all the problems created by the branching. Then it branches again on the node with the best z-value. Jumptracking often jumps from one side of the tree to the other. It usually creates more subproblems and requires more computer storage than backtracking. The idea behind jumptracking is that moving toward the subproblems with good z-values should lead us more quickly to the best z-value.

If two or more variables are fractional in a subproblem's optimal solution, then on which variable should we branch? Branching on the fractional-valued variable that has the greatest economic importance is often the best strategy. In the Nickles example, suppose the optimal solution to a subproblem had y_1 and x_{12} fractional. Our rule would say to branch on y_1 because y_1 represents the decision to operate (or not operate) a lockbox in city 1, and this is presumably a more important decision than whether region 1 payments should be sent to city 2. When more than one variable is fractional in a subproblem solution, many computer codes will branch on the lowest-numbered fractional variable. Thus, if an integer programming computer code requires that variables be numbered, they should be numbered in order of their economic importance (1 = most important).

REMARKS   **1** For some IP's, the optimal solution to the LP relaxation will also be the optimal solution to the IP. Suppose the constraints of the IP are written as $A\mathbf{x} = \mathbf{b}$. If the determinant[‡] of every square submatrix of A is $+1$, -1, or 0, we say that the matrix A is **unimodular.** If A is unimodular and each element of $\mathbf{b}$ is an integer, then the optimal solution to the LP relaxation will assign all variables integer values [see Shapiro (1979) for a proof] and will therefore be the optimal solution to the IP. It can be shown that the constraint matrix of any MCNFP is unimodular. Thus, as was discussed in Chapter 8, any MCNFP in which each node's net outflow and each arc's capacity are integers will have an integer-valued solution.

2 As a general rule, the more an IP looks like an MCNFP, the easier the problem is to solve by branch-and-bound methods. Thus, in formulating an IP, it is good to choose a formulation in which as many variables as possible have coefficients of $+1$, -1, and 0. To illustrate this idea, recall that the formulation of the Nickles (lockbox) problem given in Section 9.2 contained 16 constraints of the following form:

Formulation 1 $\qquad\qquad x_{ij} \leq y_j \ (i = 1, 2, 3, 4; j = 1, 2, 3, 4)$ $\qquad\qquad$ **(25)**

[†]For two subproblems created at the same time, many sophisticated methods have been developed to determine which one should be solved first. See Taha (1975) for details.

[‡]The determinant of a matrix is defined in Section 2.6.

As we have already seen in Section 9.2, if the 16 constraints in (25) are replaced by the following 4 constraints, then an equivalent formulation results:

Formulation 2

$$x_{11} + x_{21} + x_{31} + x_{41} \leq 4y_1$$
$$x_{12} + x_{22} + x_{32} + x_{42} \leq 4y_2$$
$$x_{13} + x_{23} + x_{33} + x_{43} \leq 4y_3$$
$$x_{14} + x_{24} + x_{34} + x_{44} \leq 4y_4$$

Because formulation 2 has $16 - 4 = 12$ fewer constraints than formulation 1, one might think that formulation 2 would require less computer time to find the optimal solution. This turns out to be untrue. To see why, recall that the branch-and-bound method begins by solving the LP relaxation of the IP. The feasible region of the LP relaxation of formulation 2 contains many more noninteger points than the feasible region of formulation 1. For example, the point $y_1 = y_2 = y_3 = y_4$ $= \frac{1}{4}$, $x_{11} = x_{22} = x_{33} = x_{44} = 1$ (all other x_{ij}'s equal 0) is in the feasible region for the LP relaxation of formulation 2, but not for formulation 1. The branch-and-bound method must eliminate all noninteger points before obtaining the optimal solution to the IP, so it seems reasonable that formulation 2 will require more computer time than formulation 1. Indeed, when the LINDO package was used to find the optimal solution to formulation 1, the LP relaxation yielded the optimal solution. But 17 subproblems were solved before the optimal solution was found for formulation 2. Note that formulation 2 contains the terms $4y_1$, $4y_2$, $4y_3$, and $4y_4$. These terms "disturb" the network-like structure of the lockbox problem and cause the branch-and-bound method to be less efficient.

3 When solving an IP in the real world, we are usually happy with a near-optimal solution. For example, suppose that we are solving a lockbox problem and the LP relaxation yields a cost of $200,000. This means that the optimal solution to the lockbox IP will certainly have a cost of at least $200,000. If we find a candidate solution during the course of the branch-and-bound procedure that has a cost of, say, $205,000, why bother to continue with the branch-and-bound procedure? Even if we found the optimal solution to the IP, it could not save more than $5,000 in costs over the candidate solution with $z = 205,000$. It might even cost more than $5,000 in computer time to find the optimal lockbox solution. For this reason, the branch-and-bound procedure is often terminated when a candidate solution is found with a z-value close to the z-value of the LP relaxation.

4 Subproblems for branch-and-bound problems are often solved using some variant of the dual simplex algorithm. To illustrate this, we return to the Telfa example. The optimal tableau for the LP relaxation of the Telfa problem is

$$z \quad + 1.25s_1 + 0.75s_2 = 41.25$$
$$x_2 + 2.25s_1 - 0.25s_2 = 2.25$$
$$x_1 \quad - 1.25s_1 + 0.25s_2 = 3.75$$

After solving the LP relaxation, we solved subproblem 2, which is just subproblem 1 plus the constraint $x_1 \geq 4$. Recall that the dual simplex is an efficient method for finding the new optimal solution to an LP when we know the optimal tableau and a new constraint is added to the LP. We have added the constraint $x_1 \geq 4$ (which may be written as $x_1 - e_3 = 4$). To utilize the dual simplex, we must eliminate the basic variable x_1 from this constraint and use e_3 as a basic variable for $x_1 - e_3 = 4$. Adding $-$(second row of optimal tableau) to the constraint $x_1 - e_3 = 4$, we obtain the constraint $1.25s_1 - 0.25s_2 - e_3 = 0.25$. Multiplying this constraint through by -1, we obtain $-1.25s_1 + 0.25s_2 + e_3 = -0.25$. After adding this constraint to subproblem 1's optimal tableau, we obtain the tableau in Table 56. The dual simplex method states that we should enter a variable from row 3 into the basis. Because s_1 is the only variable with a negative coefficient in row 3, s_1 will enter the basis in row 3. After the pivot, we obtain the (optimal) tableau in Table 57. Thus, the optimal solution to subproblem 2 is $z = 41$, $x_2 = 1.8$, $x_1 = 4$, $s_1 = 0.20$.

TABLE 56
Initial Tableau for Solving Subproblem 2 by Dual Simplex

	Basic Variable
$z \quad + 1.25s_1 + 0.75s_2 = 41.25$	$z = 41.25$
$x_2 + 2.25s_1 - 0.25s_2 = 2.25$	$x_2 = 2.25$
$x_1 \quad - 1.25s_1 + 0.25s_2 = 3.75$	$x_1 = 3.75$
$ \quad - 1.25s_1 + 0.25s_2 + e_3 = -0.25$	$e_3 = -0.25$

TABLE 57
Optimal Tableau for Solving Subproblem 2 by Dual Simplex

				Basic Variable
z	$+$	$s_2 +$	$e_3 = 41$	$z = 41$
	x_2	$+ 0.20s_2 +$	$1.8e_3 = 1.8$	$x_2 = 1.8$
	x_1	$-$	$e_3 = 4$	$x_1 = 4$
		$s_1 - 0.20s_2 -$	$0.80e_3 = 0.20$	$s_1 = 0.20$

5 In Problem 8, we show that if we create two subproblems by adding the constraints $x_k \leq i$ and $x_k \geq i + 1$, then the optimal solution to the first subproblem will have $x_k = i$ and the optimal solution to the second subproblem will have $x_k = i + 1$. This observation is very helpful when we graphically solve subproblems. For example, we know the optimal solution to subproblem 5 of Example 9 will have $x_2 = 1$. Then we can find the value of x_1 that solves subproblem 5 by choosing x_1 to be the largest integer satisfying all constraints when $x_2 = 1$.

Solver Tolerance Option for Solving IPs

When solving integer programming problems with the Excel Solver, you may go to Options and set a tolerance. A tolerance value of, say, .20, causes the Excel Solver to stop when a feasible solution is found that has an objective function value within 20% of the optimal z-value for the problem's LP relaxation. For instance, in Example 9, the optimal z-value for the LP relaxation was 41.25. With a tolerance of .20, the Solver would stop whenever a feasible integer solution is found with a z-value exceeding $(1 - .2)(41.25) = 33$. Thus, if we solved Example 9 with the Excel Solver and found a feasible integer solution having $z = 35$, then the Solver would stop because this solution would be within 20% of the LP relaxation bound.

Why set a nonzero tolerance? For many large IP problems, it might take a long time (weeks or months!) to find an optimal solution. It might take much less time to find a near-optimal solution (say, within 5% of the optimal LP relaxation). In this case, we would be much better off with a near-optimal solution, and use of the tolerance option might be appropriate.

PROBLEMS

Group A

Use branch-and-bound to solve the following IPs:

1 $\max z = 5x_1 + 2x_2$
s.t. $3x_1 + x_2 \leq 12$
  $x_1 + x_2 \leq 5$
  $x_1, x_2 \geq 0; x_1, x_2$ integer

2 The Dorian Auto example of Section 3.2.

3 $\max z = 2x_1 + 3x_2$
s.t. $x_1 + 2x_2 \leq 10$
  $3x_1 + 4x_2 \leq 25$
  $x_1, x_2 \geq 0; x_1, x_2$ integer

4 $\max z = 4x_1 + 3x_2$
s.t. $4x_1 + 9x_2 \leq 26$
  $8x_1 + 5x_2 \leq 17$
  $x_1, x_2 \geq 0; x_1, x_2$ integer

5 $\max z = 4x_1 + 5x_2$
s.t. $x_1 + 4x_2 \geq 5$
  $3x_1 + 2x_2 \geq 7$
  $x_1, x_2 \geq 0; x_1, x_2$ integer

6 $\max z = 4x_1 + 5x_2$
s.t. $3x_1 + 2x_2 \leq 10$
  $x_1 + 4x_2 \leq 11$
  $3x_1 + 3x_2 \leq 13$
  $x_1, x_2 \geq 0; x_1, x_2$ integer

7 Use the branch-and-bound method to find the optimal solution to the following IP:

$\max z = 7x_1 + 3x_2$
s.t. $2x_1 + x_2 \leq 9$
  $3x_1 + 2x_2 \leq 13$
  $x_1, x_2 \geq 0; x_1, x_2$ integer

8 Suppose we have branched on a subproblem (call it subproblem 0, having optimal solution SOL0) and have obtained the following two subproblems:

Subproblem 1 Subproblem 0 + Constraint $x_1 \leq i$.
Subproblem 2 Subproblem 0 + Constraint $x_1 \geq i + 1$ (i is some integer).

Prove that there will exist at least one optimal solution to subproblem 1 having $x_1 = i$ and at least one optimal solution to subproblem 2 having $x_1 = i + 1$. [*Hint:* Suppose an optimal solution to subproblem 1 (call it SOL1) has $x_1 = \bar{x}_1$, where $\bar{x}_1 < i$. For some number c ($0 < c < 1$), $c(\text{SOL0}) + (1 - c)\text{SOL1}$ will have the following three properties:

a The value of x_1 in $c(\text{SOL0}) + (1 - c)\text{SOL1}$ will equal i.

b $c(\text{SOL0}) + (1 - c)\text{SOL1}$ will be feasible in subproblem 1.

c The z-value for $c(\text{SOL0}) + (1 - c)\text{SOL1}$ will be at least as good as the z-value for SOL1.

Explain how this result can help when we graphically solve branch-and-bound problems.]

9 During the next five periods, the demands in Table 58 must be met on time. At the beginning of period 1, the

TABLE 58

	Period				
	1	**2**	**3**	**4**	**5**
Demand	220	280	360	140	270

inventory level is 0. Each period that production occurs a setup cost of \$250 and a per-unit production cost of \$2 are incurred. At the end of each period a per-unit holding cost of \$1 is incurred.

a Solve for the cost-minimizing production schedule using the following decision variables: x_t = units produced during month t and $y_t = 1$ if any units are produced during period t, $y_t = 0$ otherwise.

b Solve for the cost-minimizing production schedule using the following variables: y_t's defined in part (a) and x_{it} = number of units produced during period i to satisfy period t demand.

c Which formulation took LINDO or LINGO less time to solve?

d Give an intuitive explanation of why the part (b) formulation is solved faster than the part (a) formulation.

9.4 The Branch-and-Bound Method for Solving Mixed Integer Programming Problems

Recall that, in a mixed IP, some variables are required to be integers and others are allowed to be either integers or nonintegers. To solve a mixed IP by the branch-and-bound method, modify the method described in Section 9.3 by branching only on variables that are required to be integers. Also, for a solution to a subproblem to be a candidate solution, it need only assign integer values to those variables that are required to be integers. To illustrate, let us solve the following mixed IP:

$$\max z = 2x_1 + x_2$$
$$\text{s.t.} \quad 5x_1 + 2x_2 \leq 8$$
$$x_1 + x_2 \leq 3$$
$$x_1, x_2 \geq 0; \ x_1 \text{ integer}$$

As before, we begin by solving the LP relaxation of the IP. The optimal solution of the LP relaxation is $z = \frac{11}{3}, x_1 = \frac{2}{3}, x_2 = \frac{7}{3}$. Because x_2 is allowed to be fractional, we do not branch on x_2; if we did so, we would be excluding points having x_2 values between 2 and 3, and we don't want to do that. Thus, we must branch on x_1. This yields subproblems 2 and 3 in Figure 21.

We next choose to solve subproblem 2. The optimal solution to subproblem 2 is the candidate solution $z = 3, x_1 = 0, x_2 = 3$. We now solve subproblem 3 and obtain the candidate solution $z = \frac{7}{2}, x_1 = 1, x_2 = \frac{3}{2}$. The z-value from the subproblem 3 candidate exceeds the z-value for the subproblem 2 candidate, so subproblem 2 can be eliminated from consideration, and the subproblem 3 candidate ($z = \frac{7}{2}, x_1 = 1, x_2 = \frac{3}{2}$) is the optimal solution to the mixed IP.

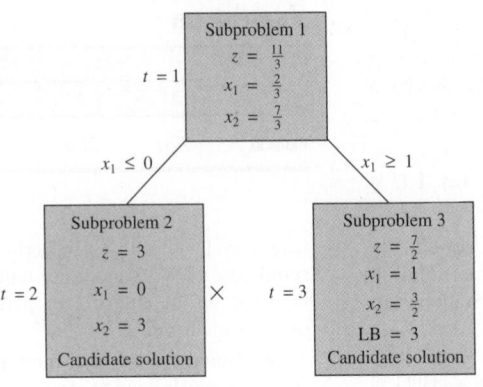

FIGURE 21
Branch-and-Bound
Tree for Mixed IP

PROBLEMS

Group A

Use the branch-and-bound method to solve the following IPs:

1
$$\max z = 3x_1 + x_2$$
$$\text{s.t.} \quad 5x_1 + 2x_2 \le 10$$
$$4x_1 + x_2 \le 7$$
$$x_1, x_2 \ge 0; \, x_2 \text{ integer}$$

2
$$\min z = 3x_1 + x_2$$
$$\text{s.t.} \quad x_1 + 5x_2 \ge 8$$
$$x_1 + 2x_2 \ge 4$$
$$x_1, x_2 \ge 0; \, x_1 \text{ integer}$$

3
$$\max z = 4x_1 + 3x_2 + x_3$$
$$\text{s.t.} \quad 3x_1 + 2x_2 + x_3 \le 7$$
$$2x_1 + x_2 + 2x_3 \le 11$$
$$x_2, x_3 \text{ integer}, \, x_1, x_2, x_3 \ge 0$$

9.5 Solving Knapsack Problems by the Branch-and-Bound Method

In Section 9.2, we learned that a knapsack problem is an IP with a single constraint. In this section, we discuss knapsack problems in which each variable must equal 0 or 1 (see Problem 1 at the end of this section for an explanation of how any knapsack problem can be reformulated so that each variable must equal 0 or 1). A knapsack problem in which each variable must equal 0 or 1 may be written as

$$\max z = c_1x_1 + c_2x_2 + \cdots + c_nx_n$$
$$\text{s.t.} \quad a_1x_1 + a_2x_2 + \cdots + a_nx_n \le b \tag{38}$$
$$x_i = 0 \text{ or } 1 \quad (i = 1, 2, \ldots, n)$$

Recall that c_i is the benefit obtained if item i is chosen, b is the amount of an available resource, and a_i is the amount of the available resource used by item i.

When knapsack problems are solved by the branch-and-bound method, two aspects of the method greatly simplify. Because each variable must equal 0 or 1, branching on x_i will yield an $x_i = 0$ and an $x_i = 1$ branch. Also, the LP relaxation (and other subproblems) may be solved by inspection. To see this, observe that $\frac{c_i}{a_i}$ may be interpreted as the benefit item i earns for each unit of the resource used by item i. Thus, the best items have the largest values of $\frac{c_i}{a_i}$, and the worst items have the smallest values of $\frac{c_i}{a_i}$. To solve any subproblem resulting from a knapsack problem, compute all the ratios $\frac{c_i}{a_i}$. Then put the best item in the knapsack. Then put the second-best item in the knapsack. Continue in this fashion until the best remaining item will overfill the knapsack. Then fill the knapsack with as much of this item as possible.

To illustrate, we solve the LP relaxation of

$$\max z = 40x_1 + 80x_2 + 10x_3 + 10x_4 + 4x_5 + 20x_6 + 60x_7$$

$$\text{s.t.} \quad 40x_1 + 50x_2 + 30x_3 + 10x_4 + 10x_5 + 40x_6 + 30x_7 \leq 100 \qquad \textbf{(39)}$$

$$x_i = 0 \text{ or } 1 \quad (i = 1, 2, \ldots, 7)$$

We begin by computing the $\frac{c_i}{a_i}$ ratios and ordering the variables from best to worst (see Table 59). To solve the LP relaxation of (39), we first choose item 7 ($x_7 = 1$). Then $100 - 30 = 70$ units of the resource remain. Now we include the second-best item (item 2) in the knapsack by setting $x_2 = 1$. Now $70 - 50 = 20$ units of the resource remain. Item 4 and item 1 have the same $\frac{c_i}{a_i}$ ratio, so we can next choose either of these items. We arbitrarily choose to set $x_4 = 1$. Then $20 - 10 = 10$ units of the resource remain. The best remaining item is item 1. We now fill the knapsack with as much of item 1 as we can. Because only 10 units of the resource remain, we set $x_1 = \frac{10}{40} = \frac{1}{4}$. Thus an optimal solution to the LP relaxation of (39) is $z = 80 + 60 + 10 + (\frac{1}{4})(40) = 160$, $x_2 = x_7 = x_4 = 1$, $x_1 = \frac{1}{4}$, $x_3 = x_5 = x_6 = 0$.

To show how the branch-and-bound method can be used to solve a knapsack problem, let us find the optimal solution to the Stockco capital budgeting problem (Example 1). Recall that this problem was

$$\max z = 16x_1 + 22x_2 + 12x_3 + 8x_4$$

$$\text{s.t.} \quad 5x_1 + 7x_2 + 4x_3 + 3x_4 \leq 14$$

$$x_j = 0 \text{ or } 1$$

The branch-and-bound tree for this problem is shown in Figure 22. From the tree, we find that the optimal solution to Example 1 is $z = 42$, $x_1 = 0$, $x_2 = x_3 = x_4 = 1$. Thus, we should invest in investments 2, 3, and 4 and earn an NPV of $42,000. As discussed in Section 9.2, the "best" investment is not used.

REMARKS The method we used in traversing the tree of Figure 22 is as follows:

1 We used the LIFO approach to determine which subproblem should be solved.

2 We arbitrarily chose to solve subproblem 3 before subproblem 2. To solve subproblem 3, we first set $x_3 = 1$ and then solved the resulting knapsack problem. After setting $x_3 = 1$, $14 - 4 = \$10$ million was still available for investment. Applying the technique used to solve the LP relaxation of a knapsack problem yielded the following optimal solution to subproblem 3: $x_3 = 1$, $x_1 = 1$, $x_2 = \frac{5}{7}$, $x_4 = 0$, $z = 16 + (\frac{5}{7})(22) + 12 = \frac{306}{7}$. Other subproblems were solved similarly; of course, if a subproblem specified $x_i = 0$, the optimal solution to that subproblem could not use investment i.

TABLE 59
Ordering Items from Best to Worst in a Knapsack Problem

Item	$\dfrac{c_i}{a_i}$	Ranking (1 = best, 7 = worst)
1	1	3.5 (tie for third or fourth)
2	$\frac{8}{5}$	2
3	$\frac{1}{3}$	7
4	1	3.5
5	$\frac{4}{10}$	6
6	$\frac{1}{2}$	5
7	2	1

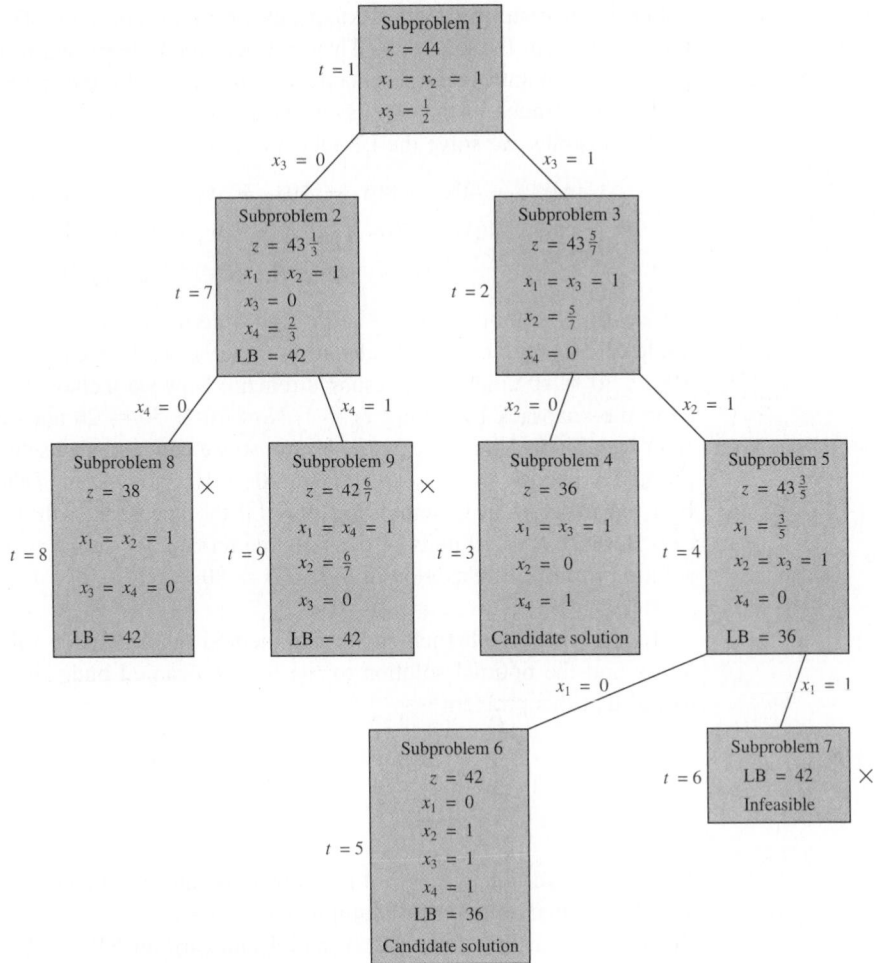

FIGURE **22**
Branch-and-Bound
Tree for Stockco
Knapsack Problem

3 Subproblem 4 yielded the candidate solution $x_1 = x_3 = x_4 = 1$, $z = 36$. We then set LB $= 36$.
4 Subproblem 6 yielded a candidate solution with $z = 42$. Thus, subproblem 4 was eliminated from consideration, and the LB was updated to 42.
5 Subproblem 7 was infeasible because it required $x_1 = x_2 = x_3 = 1$, and such a solution requires at least \$16 million.
6 Subproblem 8 was eliminated because its z-value ($z = 38$) did not exceed the current LB of 42.
7 Subproblem 9 had a z-value of $42\frac{6}{7}$. Because the z-value for any all-integer solution must also be an integer, this meant that branching on subproblem 9 could never yield a z-value larger than 42. Thus, further branching on subproblem 9 could not beat the current LB of 42, and subproblem 9 was eliminated from consideration.

In Chapter 13, we show how dynamic programming can be used to solve knapsack problems.

PROBLEMS

Group A

1 Show how the following problem can be expressed as a knapsack problem in which all variables must equal 0 or 1. NASA is determining how many of three types of objects should be brought on board the space shuttle. The weight

and benefit of each of the items are given in Table 60. If the space shuttle can carry a maximum of 26 lb of items 1–3, which items should be taken on the space shuttle?

TABLE 60

Item	Benefit	Weight (Pounds)
1	10	3
2	15	4
3	17	5

TABLE 61

Item	Value ($)	Volume (Cubic Feet)
Bedroom set	60	800
Dining room set	48	600
Stereo	14	300
Sofa	31	400
TV set	10	200

TABLE 62

Project	Cash Outflow at Time 0 ($)	NPV ($)
1	3	5
2	5	8
3	2	3
4	4	7

2 I am moving from New Jersey to Indiana and have rented a truck that can haul up to 1,100 cu ft of furniture. The volume and value of each item I am considering moving on the truck are given in Table 61. Which items should I bring to Indiana? To solve this problem as a knapsack problem, what unrealistic assumptions must we make?

3 Four projects are available for investment. The projects require the cash flows and yield the net present values (NPV) (in millions) shown in Table 62. If $6 million is available for investment at time 0, find the investment plan that maximizes NPV.

9.6 Solving Combinatorial Optimization Problems by the Branch-and-Bound Method

Loosely speaking, a **combinatorial optimization problem** is any optimization problem that has a finite number of feasible solutions. A branch-and-bound approach is often the most efficient way to solve them. Three examples of combinatorial optimization problems follow:

1 Ten jobs must be processed on a single machine. You know the time it takes to complete each job and the time at which each job must be completed (the job's due date). What ordering of the jobs minimizes the total delay of the 10 jobs?

2 A salesperson must visit each of 10 cities once before returning to his home. What ordering of the cities minimizes the total distance the salesperson must travel before returning home? Not surprisingly, this problem is called the *traveling salesperson problem* (TSP).

3 Determine how to place eight queens on a chessboard so that no queen can capture any other queen (see Problem 7 at the end of this section).

In each of these problems, many possible solutions must be considered. For instance, in Problem 1, the first job to be processed can be one of 10 jobs, the next job can be one of 9 jobs, and so on. Thus, even for this relatively small problem there are $10(9)(8) \cdots (1) = 10! = 3,628,000$ possible ways to schedule the jobs. A combinatorial optimization problem may have many feasible solutions, so it can require a great deal of computer time to enumerate all possible solutions explicitly. For this reason, branch-and-bound methods are often used for *implicit* enumeration of all possible solutions to a combinatorial optimization problem. As we will see, the branch-and-bound method should take advantage of the structure of the particular problem that is being solved.

To illustrate how branch-and-bound methods are used to solve combinatorial optimization problems, we show how the approach can be used to solve Problems 1 and 2 of the preceding list.

Branch-and-Bound Approach
for Machine-Scheduling Problem

Example 10 illustrates how a branch-and-bound approach may be used to schedule jobs on a single machine. See Baker (1974) and Hax and Candea (1984) for a discussion of other branch-and-bound approaches to machine-scheduling problems.

EXAMPLE 10	Branch-and-Bound Machine Scheduling

Four jobs must be processed on a single machine. The time required to process each job and the date the job is due are shown in Table 63. The delay of a job is the number of days after the due date that a job is completed (if a job is completed on time or early, the job's delay is zero). In what order should the jobs be processed to minimize the total delay of the four jobs?

Solution   Suppose the jobs are processed in the following order: job 1, job 2, job 3, and job 4. Then the delays shown in Table 64 would occur. For this sequence, total delay = $0 + 6 + 3 + 7 = 16$ days. We now describe a branch-and-bound approach for solving this type of machine-scheduling problem.

Because a possible solution to the problem must specify the order in which the jobs are processed, we define

$$x_{ij} = \begin{cases} 1 & \text{if job } i \text{ is the } j\text{th job to be processed} \\ 0 & \text{otherwise} \end{cases}$$

The branch-and-bound approach begins by partitioning all solutions according to the job that is *last* processed. Any sequence of jobs must process some job last, so each sequence of jobs must have $x_{14} = 1$, $x_{24} = 1$, $x_{34} = 1$, or $x_{44} = 1$. This yields four branches with nodes 1–4 in Figure 23. After we create a node by branching, we obtain a lower bound on the total delay (D) associated with the node. For example, if $x_{44} = 1$, we know that job 4 is the last job to be processed. In this case, job 4 will be completed at the end of day $6 + 4 + 5 + 8 = 23$ and will be $23 - 16 = 7$ days late. Thus, any schedule having

TABLE 63
Durations and Due Date of Jobs

Job	Days Required to Complete Job	Due Date
1	6	End of day 8
2	4	End of day 4
3	5	End of day 12
4	8	End of day 16

TABLE 64
Delays Incurred If Jobs Are Processed in the Order 1-2-3-4

Job	Completion Time of Job	Delay of Job
1	6	0
2	$6 + 4 = 10$	$10 - 4 = 6$
3	$6 + 4 + 5 = 15$	$15 - 12 = 3$
4	$6 + 4 + 5 + 8 = 23$	$23 - 16 = 7$

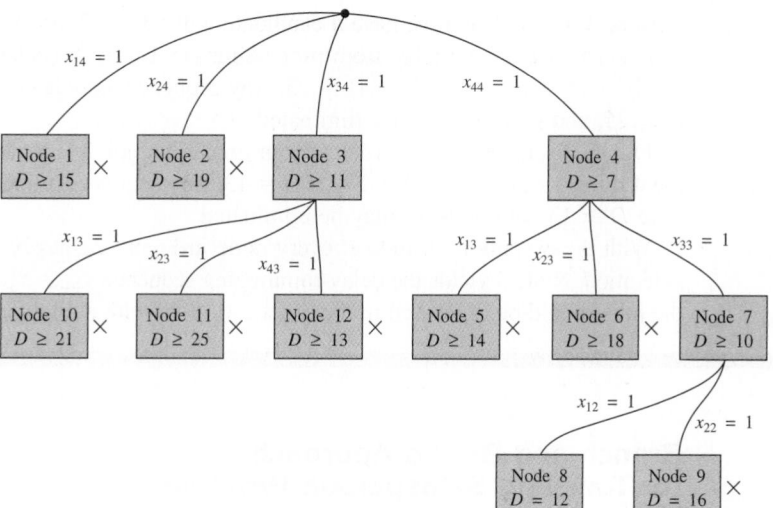

FIGURE 23
Branch-and-Bound Tree for Machine-Scheduling Problem

$x_{44} = 1$ must have $D \geq 7$. Thus, we write $D \geq 7$ inside node 4 of Figure 23. Similar reasoning shows that any sequence of jobs having $x_{34} = 1$ will have $D \geq 11$, $x_{24} = 1$ will have $D \geq 19$, and $x_{14} = 1$ will have $D \geq 15$. We have no reason to exclude any of nodes 1–4 from consideration as part of the optimal job sequence, so we choose to branch on a node. We use the jumptracking approach and branch on the node that has the smallest bound on D: node 4. Any job sequence associated with node 4 must have $x_{13} = 1$, $x_{23} = 1$, or $x_{33} = 1$. Branching on node 4 yields nodes 5–7 in Figure 23. For each new node, we need a lower bound for the total delay. For example, at node 7, we know from our analysis of node 1 that job 4 will be processed last and will be delayed by 7 days. For node 7, we know that job 3 will be the third job processed. Thus, job 3 will be completed after $6 + 4 + 5 = 15$ days and will be $15 - 12 = 3$ days late. Any sequence associated with node 7 must have $D \geq 7 + 3 = 10$ days. Similar reasoning shows that node 5 must have $D \geq 14$, and node 6 must have $D \geq 18$. We still do not have any reason to eliminate any of nodes 1–7 from consideration, so we again branch on a node. The jumptracking approach directs us to branch on node 7. Any job sequence associated with node 7 must have either job 1 or job 2 as the second job processed. Thus, any job sequence associated with node 7 must have $x_{12} = 1$ or $x_{22} = 1$. Branching on node 7 yields nodes 8 and 9 in Figure 23.

Node 9 corresponds to processing the jobs in the order 1–2–3–4. This sequence yields a total delay of 7 (for job 4) + 3 (for job 3) + $(6 + 4 - 4)$ (for job 2) + 0 (for job 1) = 16 days. Node 9 is a feasible sequence and may be considered a candidate solution having $D = 16$. We now know that any node that cannot have a total delay of less than 16 days can be eliminated.

Node 8 corresponds to the sequence 2–1–3–4. This sequence has a total delay of 7 (for job 4) + 3 (for job 3) + $(4 + 6 - 8)$ (for job 1) + 0 (for job 2) = 12 days. Node 8 is a feasible sequence and may be viewed as a candidate solution with $D = 12$. Because node 8 is better than node 9, node 9 may be eliminated from consideration.

Similarly, node 5 (having $D \geq 14$), node 6 (having $D \geq 18$), node 1 (having $D \geq 15$), and node 2 (having $D \geq 19$) can be eliminated. Node 3 cannot yet be eliminated, because it is still possible for node 3 to yield a sequence having $D = 11$. Thus, we now branch on node 3. Any job sequence associated with node 3 must have $x_{13} = 1$, $x_{23} = 1$, or $x_{43} = 1$, so we obtain nodes 10–12.

For node 10, $D \geq$ (delay from processing job 3 last) + (delay from processing job 1 third) = $11 + (6 + 4 + 8 - 8) = 21$. Because any sequence associated with node 10 must have $D \geq 21$ and we have a candidate with $D = 12$, node 10 may be eliminated.

For node 11, $D \geq$ (delay from processing job 3 last) + (delay from processing job 2 third) = $11 + (6 + 4 + 8 - 4) = 25$. Any sequence associated with node 11 must have $D \geq 25$, and node 11 may be eliminated.

Finally, for node 12, $D \geq$ (delay from processing job 3 last) + (delay from processing job 4 third) = $11 + (6 + 4 + 8 - 16) = 13$. Any sequence associated with node 12 must have $D \geq 13$, and node 12 may be eliminated.

With the exception of node 8, every node in Figure 23 has been eliminated from consideration. Node 8 yields the delay-minimizing sequence $x_{44} = x_{33} = x_{12} = x_{21} = 1$. Thus, the jobs should be processed in the order 2–1–3–4, with a total delay of 12 days resulting.

Branch-and-Bound Approach for Traveling Salesperson Problem

EXAMPLE 11 **Traveling Salesperson Problem**

Joe State lives in Gary, Indiana. He owns insurance agencies in Gary, Fort Wayne, Evansville, Terre Haute, and South Bend. Each December, he visits each of his insurance agencies. The distance between each agency (in miles) is shown in Table 65. What order of visiting his agencies will minimize the total distance traveled?

Solution Joe must determine the order of visiting the five cities that minimizes the total distance traveled. For example, Joe could choose to visit the cities in the order 1–3–4–5–2–1. Then he would travel a total of $217 + 113 + 196 + 79 + 132 = 737$ miles.

To tackle the traveling salesperson problem, define

$$x_{ij} = \begin{cases} 1 & \text{if Joe leaves city } i \text{ and travels next to city } j \\ 0 & \text{otherwise} \end{cases}$$

Also, for $i \neq j$,

$$c_{ij} = \text{distance between cities } i \text{ and } j$$
$$c_{ii} = M, \text{ where } M \text{ is a large positive number}$$

It seems reasonable that we might be able to find the answer to Joe's problem by solving an assignment problem having a cost matrix whose ijth element is c_{ij}. For instance, suppose we solved this assignment problem and obtained the solution $x_{12} = x_{24} = x_{45} = x_{53} = x_{31} = 1$. Then Joe should go from Gary to Fort Wayne, from Fort Wayne to Terre Haute, from Terre Haute to South Bend, from South Bend to Evansville, and from Evansville to Gary. This solution can be written as 1–2–4–5–3–1. An itinerary that begins and ends at the same city and visits each city once is called a **tour.**

TABLE 65
Distance between Cities in Traveling Salesperson Problem

Day	Gary	Fort Wayne	Evansville	Terre Haute	South Bend
City 1 Gary	0	132	217	164	58
City 2 Fort Wayne	132	0	290	201	79
City 3 Evansville	217	290	0	113	303
City 4 Terre Haute	164	201	113	0	196
City 5 South Bend	58	79	303	196	0

If the solution to the preceding assignment problem yields a tour, then it is the optimal solution to the traveling salesperson problem. (Why?) Unfortunately, the optimal solution to the assignment problem need not be a tour. For example, the optimal solution to the assignment problem might be $x_{15} = x_{21} = x_{34} = x_{43} = x_{52} = 1$. This solution suggests going from Gary to South Bend, then to Fort Wayne, and then back to Gary. This solution also suggests that if Joe is in Evansville he should go to Terre Haute and then to Evansville (see Figure 24). Of course, if Joe begins in Gary, this solution will never get him to Evansville or Terre Haute. This is because the optimal solution to the assignment problem contains two **subtours.** A subtour is a round trip that does not pass through all cities. The current assignment contains the two subtours 1–5–2–1 and 3–4–3. If we could exclude all feasible solutions that contain subtours and then solve the assignment problem, we would obtain the optimal solution to the traveling salesperson problem. This is not easy to do, however. In most cases, a branch-and-bound approach is the most efficient approach for solving a TSP.

Several branch-and-bound approaches have been developed for solving TSPs [see Wagner (1975)]. We describe an approach here in which the subproblems reduce to assignment problems. To begin, we solve the preceding assignment problem, in which, for $i \neq j$, the cost c_{ij} is the distance between cities i and j and $c_{ii} = M$ (this prevents a person in a city from being assigned to visit that city itself). Because this assignment problem contains no provisions to prevent subtours, it is a relaxation (or less constrained problem) of the original traveling salesperson problem. Thus, if the optimal solution to the assignment problem is feasible for the traveling salesperson problem (that is, if the assignment solution contains no subtours), then it is also optimal for the traveling salesperson problem. The results of the branch-and-bound procedure are given in Figure 25.

We first solve the assignment problem in Table 66 (referred to as subproblem 1). The optimal solution is $x_{15} = x_{21} = x_{34} = x_{43} = x_{52} = 1$, $z = 495$. This solution contains two subtours (1–5–2–1 and 3–4–3) and cannot be the optimal solution to Joe's problem.

We now branch on subproblem 1 in a way that will prevent one of subproblem 1's subtours from recurring in solutions to subsequent subproblems. We choose to exclude the subtour 3–4–3. Observe that the optimal solution to Joe's problem must have either $x_{34} = 0$ or $x_{43} = 0$ (if $x_{34} = x_{43} = 1$, the optimal solution would have the subtour 3–4–3). Thus, we can branch on subproblem 1 by adding the following two subproblems:

Subproblem 2 Subproblem 1 + ($x_{34} = 0$, or $c_{34} = M$).

Subproblem 3 Subproblem 1 + ($x_{43} = 0$, or $c_{43} = M$).

We now arbitrarily choose subproblem 2 to solve, applying the Hungarian method to the cost matrix as shown in Table 67. The optimal solution to subproblem 2 is $z = 652$, $x_{14} = x_{25} = x_{31} = x_{43} = x_{52} = 1$. This solution includes the subtours 1–4–3–1 and 2–5–2, so this cannot be the optimal solution to Joe's problem.

We now branch on subproblem 2 in an effort to exclude the subtour 2–5–2. We must ensure that either x_{25} or x_{52} equals zero. Thus, we add the following two subproblems:

Subproblem 4 Subproblem 2 + ($x_{25} = 0$, or $c_{25} = M$).

Subproblem 5 Subproblem 2 + ($x_{52} = 0$, or $c_{52} = M$).

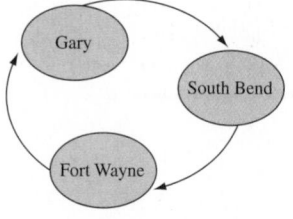
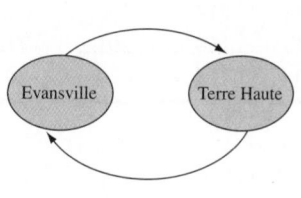

FIGURE 24
Example of Subtours
in Traveling
Salesperson Problem

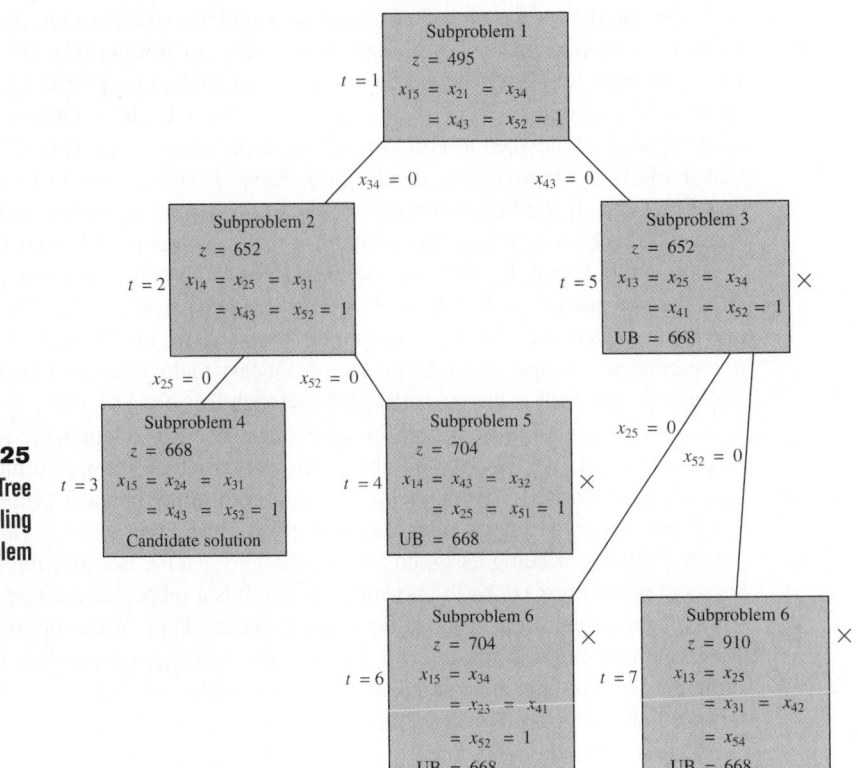

FIGURE 25
Branch-and-Bound Tree for Traveling Salesperson Problem

TABLE 66
Cost Matrix for Subproblem 1

	City 1	City 2	City 3	City 4	City 5
City 1	M	132	217	164	58
City 2	132	M	290	201	79
City 3	217	290	M	113	303
City 4	164	201	113	M	196
City 5	58	79	303	196	M

TABLE 67
Cost Matrix for Subproblem 2

	City 1	City 2	City 3	City 4	City 5
City 1	M	132	217	164	58
City 2	132	M	290	201	79
City 3	217	290	M	M	303
City 4	164	201	113	M	196
City 5	58	79	303	196	M

Following the LIFO approach, we should next solve subproblem 4 or subproblem 5. We arbitrarily choose to solve subproblem 4. Applying the Hungarian method to the cost matrix shown in Table 68, we obtain the optimal solution $z = 668$, $x_{15} = x_{24} = x_{31} = x_{43} = x_{52} = 1$. This solution contains no subtours and yields the tour 1–5–2–4–3–1. Thus, subproblem 4 yields a candidate solution with $z = 668$. Any node that cannot yield a z-value < 668 may be eliminated from consideration.

Following the LIFO rule, we next solve subproblem 5, applying the Hungarian method to the matrix in Table 69. The optimal solution to subproblem 5 is $z = 704$, $x_{14} = x_{43} = x_{32} = x_{25} = x_{51} = 1$. This solution is a tour, but $z = 704$ is not as good as the subproblem 4 candidate's $z = 668$. Thus, subproblem 5 may be eliminated from consideration.

Only subproblem 3 remains. We find the optimal solution to the assignment problem in Table 70, $x_{13} = x_{25} = x_{34} = x_{41} = x_{52} = 1$, $z = 652$. This solution contains the subtours 1–3–4–1 and 2–5–2. Because $652 < 668$, however, it is still possible for subproblem 3 to yield a solution with no subtours that beats $z = 668$. Thus, we now branch on subproblem 3 in an effort to exclude the subtours. Any feasible solution to the traveling salesperson problem that emanates from subproblem 3 must have either $x_{25} = 0$ or $x_{52} = 0$ (why?), so we create subproblems 6 and 7.

Subproblem 6   Subproblem 3 + ($x_{25} = 0$, or $c_{25} = M$).

Subproblem 7   Subproblem 3 + ($x_{52} = 0$, or $c_{52} = M$).

TABLE **68**
Cost Matrix for Subproblem 4

	City 1	City 2	City 3	City 4	City 5
City 1	M	132	217	164	58
City 2	132	M	290	201	M
City 3	217	290	M	M	303
City 4	164	201	113	M	196
City 5	58	79	303	196	M

TABLE **69**
Cost Matrix for Subproblem 5

	City 1	City 2	City 3	City 4	City 5
City 1	M	132	217	164	58
City 2	132	M	290	201	79
City 3	217	290	M	M	303
City 4	164	201	113	M	196
City 5	58	M	303	196	M

TABLE **70**
Cost Matrix for Subproblem 3

	City 1	City 2	City 3	City 4	City 5
City 1	M	132	217	164	58
City 2	132	M	290	201	79
City 3	217	290	M	113	303
City 4	164	201	M	M	196
City 5	58	79	303	196	M

We next choose to solve subproblem 6. The optimal solution to subproblem 6 is $x_{15} = x_{34} = x_{23} = x_{41} = x_{52} = 1, z = 704$. This solution contains no subtours, but its z-value of 704 is inferior to the candidate solution from subproblem 4, so subproblem 6 cannot yield the optimal solution to the problem.

The only remaining subproblem is subproblem 7. The optimal solution to subproblem 7 is $x_{13} = x_{25} = x_{31} = x_{42} = x_{54} = 1, z = 910$. Again, $z = 910$ is inferior to $z = 668$, so subproblem 7 cannot yield the optimal solution.

Subproblem 4 thus yields the optimal solution: Joe should travel from Gary to South Bend, from South Bend to Fort Wayne, from Fort Wayne to Terre Haute, from Terre Haute to Evansville, and from Evansville to Gary. Joe will travel a total distance of 668 miles.

Heuristics for TSPs

When using branch-and-bound methods to solve TSPs with many cities, large amounts of computer time may be required. For this reason, **heuristic methods,** or **heuristics,** which quickly lead to a good (but not necessarily optimal) solution to a TSP, are often used. A heuristic is a method used to solve a problem by trial and error when an algorithmic approach is impractical. Heuristics often have an intuitive justification. We now discuss two heuristics for the TSP: the nearest-neighbor and the cheapest-insertion heuristics.

To apply the nearest-neighbor heuristic (NNH), we begin at any city and then "visit" the nearest city. Then we go to the unvisited city closest to the city we have most recently visited. Continue in this fashion until a tour is obtained. We now apply the NNH to Example 11. We arbitrarily choose to begin at city 1. City 5 is the closest city to city 1, so we have now generated the arc 1–5. Of cities 2, 3, and 4, city 2 is closest to city 5, so we have now generated the arcs 1–5–2. Of cities 3 and 4, city 4 is closest to city 2. We now have generated the arcs 1–5–2–4. Of course, we must next visit city 3 and then return to city 1; this yields the tour 1–5–2–4–3–1. In this case, the NNH yields an optimal tour. If we had begun at city 3, however, the reader should verify that the tour 3–4–1–5–2–3 would be obtained. This tour has length $113 + 164 + 58 + 79 + 290 = 704$ miles and is not optimal. Thus, the NNH need not yield an optimal tour. A popular heuristic is to apply the NNH beginning at each city and then take the best tour obtained.

In the cheapest-insertion heuristic (CIH), we begin at any city and find its closest neighbor. Then we create a subtour joining those two cities. Next, we replace an arc in the subtour [say, arc (i, j)] by the combination of two arcs—(i, k) and (k, j), where k is not in the current subtour—that will increase the length of the subtour by the smallest (or cheapest) amount. Let c_{ij} be the length of arc (i, j). Note that if arc (i, j) is replaced by arcs (i, k) and (k, j), then a length $c_{ik} + c_{kj} - c_{ij}$ is added to the subtour. Then we continue with this procedure until a tour is obtained. Suppose we begin the CIH at city 1. City 5 is closest to city 1, so we begin with the subtour $(1, 5)$–$(5, 1)$. Then we could replace $(1, 5)$ by $(1, 2)$–$(2, 5)$, $(1, 3)$–$(3, 5)$, or $(1, 4)$–$(4, 5)$. We could also replace arc $(5, 1)$ by $(5, 2)$–$(2, 1)$, $(5, 3)$–$(3, 1)$, or $(5, 4)$–$(4, 1)$. The calculations used to determine which arc of $(1,5)$–$(5,1)$ should be replaced are given in Table 71 (* indicates the correct replacement). As seen in the table, we may replace either $(1, 5)$ or $(5, 1)$. We arbitrarily choose to replace arc $(1, 5)$ by arcs $(1, 2)$ and $(2, 5)$. We currently have the subtour $(1, 2)$–$(2, 5)$–$(5, 1)$. We must now replace an arc (i, j) of this subtour by the arcs (i, k) and (k, j), where $k = 3$ or 4. The relevant computations are shown in Table 72.

We now replace $(1, 2)$ by arcs $(1, 4)$ and $(4, 2)$. This yields the subtour $(1, 4)$–$(4, 2)$–$(2, 5)$–$(5, 1)$. An arc (i, j) in this subtour must now be replaced by arcs $(i, 3)$ and $(3, j)$. The relevant computations are shown in Table 73. We now replace arc $(1, 4)$ by arcs $(1, 3)$ and $(3, 4)$. This yields the tour $(1, 3)$–$(3, 4)$–$(4, 2)$–$(2, 5)$–$(5, 1)$. In this example, the CIH yields an optimal tour—but, in general, the CIH does not necessarily do so.

TABLE 71
Determining Which Arc of (1, 5)-(5, 1) Is Replaced

Arc Replaced	Arcs Added to Subtour	Added Length
(1, 5)*	(1, 2)–(2, 5)	$c_{12} + c_{25} - c_{15} = 153$
(1, 5)	(1, 3)–(3, 5)	$c_{13} + c_{35} - c_{15} = 462$
(1, 5)	(1, 4)–(4, 5)	$c_{14} + c_{45} - c_{15} = 302$
(5, 1)*	(5, 2)–(2, 1)	$c_{52} + c_{21} - c_{51} = 153$
(5, 1)	(5, 3)–(3, 1)	$c_{53} + c_{31} - c_{51} = 462$
(5, 1)	(5, 4)–(4, 1)	$c_{54} + c_{41} - c_{51} = 302$

TABLE 72
Determining Which Arc of (1, 2)-(2, 5)-(5, 1) Is Replaced

Arc Replaced	Arcs Added	Added Length
(1, 2)	(1, 3)–(3, 2)	$c_{13} + c_{32} - c_{12} = 375$
(1, 2)*	(1, 4)–(4, 2)	$c_{14} + c_{42} - c_{12} = 233$
(2, 5)	(2, 3)–(3, 5)	$c_{23} + c_{35} - c_{25} = 514$
(2, 5)	(2, 4)–(4, 5)	$c_{24} + c_{45} - c_{25} = 318$
(5, 1)	(5, 3)–(3, 1)	$c_{53} + c_{31} - c_{51} = 462$
(5, 1)	(5, 4)–(4, 1)	$c_{54} + c_{41} - c_{51} = 302$

TABLE 73
Determining Which Arc of (1, 4)-(4, 2)-(2, 5)-(5, 1) Is Replaced

Arc Replaced	Arcs Added	Added Length
(1, 4)*	(1, 3)–(3, 4)	$c_{13} + c_{34} - c_{14} = 166$
(4, 2)	(4, 3)–(3, 2)	$c_{43} + c_{32} - c_{42} = 202$
(2, 5)	(2, 3)–(3, 5)	$c_{23} + c_{35} - c_{25} = 514$
(5, 1)	(5, 3)–(3, 1)	$c_{53} + c_{31} - c_{51} = 462$

Evaluation of Heuristics

The following three methods have been suggested for evaluating heuristics:

1 Performance guarantees

2 Probabilistic analysis

3 Empirical analysis

A performance guarantee for a heuristic gives a worst-case bound on how far away from optimality a tour constructed by the heuristic can be. For the NNH, it can be shown that for any number r, a TSP can be constructed such that the NNH yields a tour that is r times as long as the optimal tour. Thus, in a worst-case scenario, the NNH fares poorly. For a symmetric TSP satisfying the triangle inequality (that is, for which $c_{ij} = c_{ji}$ and $c_{ik} \leq c_{ij} + c_{jk}$ for all i, j, and k), it has been shown that the length of the tour obtained by the CIH cannot exceed twice the length of the optimal tour.

In probabilistic analysis, a heuristic is evaluated by assuming that the location of cities follows some known probability distribution. For example, we might assume that the cities are independent random variables that are uniformly distributed on a cube of unit length, width, and height. Then, for each heuristic, we would compute the following ratio:

$$\frac{\text{Expected length of the path found by the heuristic}}{\text{Expected length of an optimal tour}}$$

The closer the ratio is to 1, the better the heuristic.

For empirical analysis, heuristics are compared to the optimal solution for a number of problems for which the optimal tour is known. As an illustration, for five 100-city TSPs, Golden, Bodin, Doyle, and Stewart (1980) found that the NNH—taking the best of all solutions found when the NNH was applied beginning at each city—produced tours that averaged 15% longer than the optimal tour. For the same set of problems, it was found that the CIH (again applying the best solution obtained by applying CIH to all cities) produced tours that also averaged 15% longer than the optimal tour.

REMARKS **1** Golden, Bodin, Doyle, and Stewart (1980) describe a heuristic that regularly comes within 2–3% of the optimal tour.
2 It is also important to compare heuristics with regard to computer running time and ease of implementation.
3 For an excellent discussion of heuristics, see Chapters 5–7 of Lawler (1985).

An Integer Programming Formulation of the TSP

We now discuss how to formulate an IP whose solution will solve a TSP. We note, however, that the formulation of this section becomes unwieldy and inefficient for large TSPs. Suppose the TSP consists of cities 1, 2, 3, . . . , N. For $i \neq j$ let c_{ij} = distance from city i to city j and let $c_{ii} = M$, where M is a very large number (relative to the actual distances in the problem). Setting $c_{ii} = M$ ensures that we will not go to city i immediately after leaving city i. Also define

$$x_{ij} = \begin{cases} 1 & \text{if the solution to TSP goes from city } i \text{ to city } j \\ 0 & \text{otherwise} \end{cases}$$

Then the solution to a TSP can be found by solving

$$\min z = \sum_i \sum_j c_{ij} x_{ij} \tag{40}$$

$$\text{s.t.} \quad \sum_{i=1}^{i=N} x_{ij} = 1 \quad (\text{for } j = 1, 2, \ldots, N) \tag{41}$$

$$\sum_{j=1}^{j=N} x_{ij} = 1 \quad (\text{for } i = 1, 2, \ldots, N) \tag{42}$$

$$u_i - u_j + N x_{ij} \leq N - 1 \quad (\text{for } i \neq j; i = 2, 3, \ldots, N; j = 2, 3, \ldots, N) \tag{43}$$

$$\text{All } x_{ij} = 0 \text{ or } 1, \text{ All } u_j \geq 0$$

The objective function (40) gives the total length of the arcs included in a tour. The constraints in (41) ensure that we arrive once at each city. The constraints in (42) ensure that we leave each city once. The constraints in (43) are the key to the formulation. They ensure the following:

1 Any set of x_{ij}'s containing a subtour will be infeasible [that is, they violate (43)].

2 Any set of x_{ij}'s that forms a tour will be feasible [there will exist a set of u_j's that satisfy (43)].

To illustrate that any set of x_{ij}'s containing a subtour will violate (43), consider the subtour illustration given in Figure 24. Here $x_{15} = x_{21} = x_{43} = x_{43} = x_{52} = 1$. This assignment contains the two subtours 1–5–2–1 and 3–4–3. Choose the subtour that does *not* contain city 1 (3–4–3) and write down the constraints in (43) corresponding to the arcs in this subtour. We obtain $u_3 - u_4 + 5x_{34} \le 4$ and $u_4 - u_3 + 5x_{43} \le 4$. Adding these constraints yields $5(x_{34} + x_{43}) \le 8$. Clearly, this rules out the possibility that $x_{43} = x_{34} = 1$, so the subtour 3–4–3 (and any other subtour!) is ruled out by the constraints in (43).

We now show that for any set of x_{ij}'s that does not contain a subtour, there exist values of the u_j's that will satisfy all constraints in (43). Assume that city 1 is the first city visited (we visit all cities eventually, so this is okay). Let t_i = the position in the tour where city i is visited. Then setting $u_i = t_i$ will satisfy all constraints in (43). To illustrate, consider the tour 1–3–4–5–2–1. Then we choose $u_1 = 1$, $u_2 = 5$, $u_3 = 2$, $u_4 = 3$, $u_5 = 4$. We now show that with this choice of the u_i's all constraints in (43) are satisfied. First, consider any constraint corresponding to an arc having $x_{ij} = 1$. For example, the constraint corresponding to x_{52} is $u_5 - u_2 + 5x_{52} \le 4$. Because city 2 immediately follows city 5, $u_5 - u_2 = -1$. Then the constraint for x_{52} in (43) reduces to $-1 + 5 \le 4$, which is true. Now consider a constraint corresponding to an x_{ij} (say, x_{32}) satisfying $x_{ij} = 0$. For x_{32}, we obtain the constraint $u_3 - u_2 + 5x_{32} \le 4$. This reduces to $u_3 - u_2 \le 4$. Because $u_3 \le 5$ and $u_2 > 1$, $u_3 - u_2$ cannot exceed $5 - 2$.

This shows that the formulation defined by (40)–(43) eliminates from consideration all sequences of N cities that begin in city 1 and include a subtour. We have also shown that this formulation does not eliminate from consideration any sequence of N cities beginning in city 1 that does not include a subtour. Thus, (40)–(43) will (if solved) yield the optimal solution to the TSP.

Using LINGO to Solve TSPs

TSP.lng

The IP described in (40)–(43) can easily be implemented with the following LINGO program (file TSP.lng).

```
MODEL:
 1]SETS:
 2]CITY/1..5/:U;
 3]LINK(CITY,CITY):DIST,X;
 4]ENDSETS
 5]DATA:
 6]DIST= 50000 132 217 164 58
 7]132 50000 290 201 79
 8]217 290 50000 113 303
 9]164 201 113 50000 196
10]58 79 303 196 5000;
11]ENDDATA
12]N=@SIZE(CITY);
13]MIN=@SUM(LINK:DIST*X);
14]@FOR(CITY(K):@SUM(CITY(I):X(I,K))=1;);
15]@FOR(CITY(K):@SUM(CITY(J):X(K,J))=1;);
16]@FOR(CITY(K):@FOR(CITY(J)|J#GT#1#AND#K#GT#1:
17]U(J)-U(K)+N*X(J,K)<N-1;));
18]@FOR(LINK:@BIN(X););
END
```

In line 2, we define our five cities and associate a U(J) with city J. In line 3, we create the arcs joining each combination of cities. With the arc from city I to city J, we associate the distance between city I and J and a 0–1 variable X(I,J), which equals 1 if city J immediately follows city I in a tour.

In lines 6–10, we input the distance between the cities given in Example 11. Note that the distance between city I and itself is assigned a large number, to ensure that city I does not follow itself.

In line 12, we use **@SIZE** to compute the number of cities (we use this in line 17). In line 13, we create the objective function by summing over each link (I,J) the product of the distance between cities I and J and X(I,J). Line 14 ensures that for each city we enter the city exactly once. Line 15 ensures that for each city we leave the city exactly once. Lines 16–17 create the constraints in (43). Note that we only create these constraints for combinations J,K where J > 1 and K > 1. This agrees with (43). Note that when J = K line 17 generates constraints of the form N*X(J,J) ≤ N − 1, which imply that all X(J,J) = 0. In line 18, we ensure that each X(I,J) = 0 or 1. We need not constrain the U(J)'s, because LINGO assumes they are nonnegative. *Note:* Even for small TSPs, this formulation will exceed the capacity of student LINGO.

PROBLEMS

Group A

1 Four jobs must be processed on a single machine. The time required to perform each job and the due date for each job are shown in Table 74. Use the branch-and-bound method to determine the order of performing the jobs that minimizes the total time the jobs are delayed.

2 Each day, Sunco manufactures four types of gasoline: lead-free premium (LFP), lead-free regular (LFR), leaded premium (LP), and leaded regular (LR). Because of cleaning and resetting of machinery, the time required to produce a batch of gasoline depends on the type of gasoline last produced. For example, it takes longer to switch between a lead-free gasoline and a leaded gasoline than it does to switch between two lead-free gasolines. The time (in minutes) required to manufacture each day's gasoline requirements are shown in Table 75. Use a branch-and-bound approach to determine the order in which the gasolines should be produced each day.

3 A Hamiltonian path in a network is a closed path that passes exactly once through each node in the network before

returning to its starting point. Taking a four-city TSP as an example, explain why solving a TSP is equivalent to finding the shortest Hamiltonian path in a network.

4 There are four pins on a printed circuit. The distance between each pair of pins (in inches) is given in Table 76.

a Suppose we want to place three wires between the pins in a way that connects all the wires and uses the minimum amount of wire. Solve this problem by using one of the techniques discussed in Chapter 8.

b Now suppose that we again want to place three wires between the pins in a way that connects all the wires and uses the minimum amount of wire. Also suppose that if more than two wires touch a pin, a short circuit will occur. Now set up a traveling salesperson problem that can be used to solve this problem. (*Hint:* Add a pin 0 such that the distance between pin 0 and any other pin is 0.)

5 a Use the NNH to find a solution to the TSP in Problem 2. Begin with LFR.

b Use the CIH to find a solution to the TSP in Problem 2. Begin with the subtour LFR–LFP–LFR.

6 LL Pea stores clothes at five different locations. Several times a day it sends an "order picker" out to each location to pick up orders. Then the order picker must return to the packaging area. Describe a TSP that could be used to minimize the time needed to pick up orders and return to the packaging area.

Group B

7 Use branch-and-bound to determine a way (if any exists to place four queens on a 4 × 4 chessboard so that no queen can capture another queen. (*Hint:* Let $x_{ij} = 1$ if a queen is placed in row i and column j of the chessboard and $x_{ij} = 0$ otherwise. Then branch as in the machine-delay problem.

TABLE 74

Job	Time to Perform Job (Minutes)	Due Date of Job
1	7	End of minute 14
2	5	End of minute 13
3	9	End of minute 18
4	11	End of minute 15

TABLE 75

Last-Produced Gasoline	Gas to Be Next Produced			
	LFR	LFP	LR	LP
LFR	—	50	120	140
LFP	60	—	140	110
LR	90	130	—	60
LP	130	120	80	—

Note: Assume that the last gas produced yesterday precedes the first gas produced today.

TABLE 76

	1	2	3	4
1	0	1	2	2
2	1	0	3	2.9
3	2	3	0	3
4	2	2.9	3	0

Many nodes may be eliminated from consideration because they are infeasible. For example, the node associated with the arcs $x_{11} = x_{22} = 1$ is infeasible, because the two queens can capture each other.)

8 Although the Hungarian method is an efficient method for solving an assignment problem, the branch-and-bound method can also be used to solve an assignment problem. Suppose a company has five factories and five warehouses. Each factory's requirements must be met by a single warehouse, and each warehouse can be assigned to only one factory. The costs of assigning a warehouse to meet a factory's demand (in thousands) are shown in Table 77.

Let $x_{ij} = 1$ if warehouse i is assigned to factory j and 0 otherwise. Begin by branching on the warehouse assigned to factory 1. This creates the following five branches: $x_{11} = 1$, $x_{21} = 1$, $x_{31} = 1$, $x_{41} = 1$, and $x_{51} = 1$. How can we obtain a lower bound on the total cost associated with a branch? Examine the branch $x_{21} = 1$. If $x_{21} = 1$, no further assignments can come from row 2 or column 1 of the cost matrix. In determining the factory to which each of the unassigned warehouses (1, 3, 4, and 5) is assigned, we cannot do better than assign each to the smallest cost in the warehouse's row (excluding the factory 1 column). Thus, the minimum-cost assignment having $x_{21} = 1$ must have a total cost of at least $10 + 10 + 9 + 5 + 5 = 39$.

Similarly, in determining the warehouse to which each of the unassigned factories (2, 3, 4, and 5) is assigned, we cannot do better than to assign each to the smallest cost in the factory's column (excluding the warehouse 2 row). Thus, the minimum-cost assignment having $x_{21} = 1$ must have a total cost of at least $10 + 9 + 5 + 5 + 7 = 36$. Thus, the total cost of any assignment having $x_{21} = 1$ must be at least $\max(36, 39) = 39$. So, if branching ever leads to a candidate solution having a total cost of 39 or less, the $x_{21} = 1$ branch may be eliminated from consideration. Use this idea to solve the problem by branch-and-bound.

9[†] Consider a long roll of wallpaper that repeats its pattern every yard. Four sheets of wallpaper must be cut from the roll. With reference to the beginning (point 0) of the wallpaper, the beginning and end of each sheet are located as shown in Table 78. Thus, sheet 1 begins 0.3 yd from the beginning of the roll (and 1.3 yd from the beginning of the roll) and sheet 1 ends 0.7 yd from the beginning of the roll (and 1.7 yd from the beginning of the roll). Assume we are

TABLE 78

Sheet	Beginning (Yards)	End (Yards)
1	0.3	0.7
2	0.4	0.8
3	0.2	0.5
4	0.7	0.9

at the beginning of the roll. In what order should the sheets be cut to minimize the total amount of wasted paper? Assume that a final cut is made to bring the roll back to the beginning of the pattern.

10[‡] A manufacturer of printed circuit boards uses programmable drill machines to drill six holes in each board. The x and y coordinates of each hole are given in Table 79. The time (in seconds) it takes the drill machine to move from one hole to the next is equal to the distance between the points. What drilling order minimizes the total time that the drill machine spends moving between holes?

11 Four jobs must be processed on a single machine. The time required to perform each job, the due date, and the penalty (in dollars) per day the job is late are given in Table 80.

Use branch-and-bound to determine the order of performing the jobs that will minimize the total penalty costs due to delayed jobs.

TABLE 79

x	y	Hole
1	2	1
3	1	2
5	3	3
7	2	4
8	3	5

TABLE 77

Warehouse	Factory ($)				
	1	2	3	4	5
1	5	15	20	25	10
2	10	12	5	15	19
3	5	17	18	9	11
4	8	9	10	5	12
5	9	10	5	11	7

TABLE 80

Job	Time (Days)	Due Date	Penalty
1	4	Day 4	4
2	5	Day 2	5
3	2	Day 13	7
4	3	Day 8	2

[†]Based on Garfinkle (1977).

[‡]Based on Magirou (1986).

9.7 Implicit Enumeration

The method of implicit enumeration is often used to solve 0–1 IPs. Implicit enumeration uses the fact that each variable must equal 0 or 1 to simplify both the branching and bounding components of the branch-and-bound process and to determine efficiently when a node is infeasible.

Before discussing implicit enumeration, we show how any pure IP may be expressed as a 0–1 IP: Simply express each variable in the original IP as the sum of powers of 2. For example, suppose the variable x_i is required to be an integer. Let n be the smallest integer such that we can be sure that $x_i < 2^{n+1}$. Then x_i may be (uniquely) expressed as the sum of $2^0, 2^1, \ldots, 2^{n-1}, 2^n$, and

$$x_i = u_n 2^n + u_{n-1} 2^{n-1} + \cdots + u_2 2^2 + 2u_1 + u_0 \tag{44}$$

where $u_i = 0$ or 1 $(i = 0, 1, \ldots, n)$.

To convert the original IP to a 0–1 IP, replace each occurrence of x_i by the right side of (44). For example, suppose we know that $x_i \leq 100$. Then $x_i < 2^{6+1} = 128$. Then (44) yields

$$x_i = 64u_6 + 32u_5 + 16u_4 + 8u_3 + 4u_2 + 2u_1 + u_0 \tag{45}$$

where $u_i = 0$ or 1 $(i = 0, 1, \ldots, 6)$. Then replace each occurrence of x_i by the right side of (45). How can we find the values of the u's corresponding to a given value of x_i? Suppose $x_i = 93$. Then u_6 will be the largest multiple of $2^6 = 64$ that is contained in 93. This yields $u_6 = 1$; then the rest of the right side of (45) must equal $93 - 64 = 29$. Then u_5 will be the largest multiple of $2^5 = 32$ contained in 29. This yields $u_5 = 0$. Then u_4 will be the largest multiple of $2^4 = 16$ contained in 29. This yields $u_4 = 1$. Continuing in this fashion, we obtain $u_3 = 1$, $u_2 = 1$, $u_1 = 0$, and $u_0 = 1$. Thus $93 = 2^6 + 2^4 + 2^3 + 2^2 + 2^0$.

We will soon discover that 0–1 IP's are generally easier to solve than other pure IP's. Why, then, don't we transform every pure IP into a 0–1 IP? Simply because transforming a pure IP into a 0–1 IP greatly increases the number of variables. However, many situations (such as lockbox and knapsack problems) naturally yield 0–1 problems. Thus, it is certainly worthwhile to learn how to solve 0–1 IPs.

The tree used in the implicit enumeration method is similar to those used to solve 0–1 knapsack problems in Section 9.5. Each branch of the tree will specify, for some variable x_i, that $x_i = 0$ or $x_i = 1$. At each node, the values of some of the variables are specified. For instance, suppose a 0–1 problem has variables $x_1, x_2, x_3, x_4, x_5, x_6$, and part of the tree looks like Figure 26. At node 4, the values of x_3, x_4, and x_2 are specified. These variables are referred to as **fixed variables**. All variables whose values are unspecified at a node are called **free variables**. Thus, at node 4, x_1, x_5, and x_6 are free variables. For any node, a

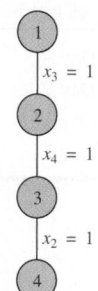

FIGURE 26
Illustration of Free and Fixed Variables

specification of the values of all the free variables is called a **completion** of the node. Thus $x_1 = 1$, $x_5 = 1$, $x_6 = 0$ is a completion of node 4.

We are now ready to outline the three main ideas used in implicit enumeration.

1 Suppose we are at any node. Given the values of the fixed variables at that node, is there an easy way to find a good completion of that node that is feasible in the original 0–1 IP? To answer this question, we complete the node by setting each free variable equal to the value (0 or 1) that makes the objective function largest (in a max problem) or smallest (in a min problem). If this completion of the node is feasible, then it is certainly the best feasible completion of the node, and further branching of the node is unnecessary. Suppose we are solving

$$\max z = 4x_1 + 2x_2 - x_3 + 2x_4$$
$$\text{s.t.} \quad x_1 + 3x_2 - x_3 - 2x_4 \geq 1$$
$$x_i = 0 \text{ or } 1 \quad (i = 1, 2, 3, 4)$$

If we are at a node (call it node 4) where $x_1 = 0$ and $x_2 = 1$ are fixed, then the best we can do is set $x_3 = 0$ and $x_4 = 1$. Because $x_1 = 0$, $x_2 = 1$, $x_3 = 0$, and $x_4 = 1$ is feasible in the original problem, we have found the best feasible completion of node 4. Thus, node 4 is fathomed and $x_1 = 0$, $x_2 = 1$, $x_3 = 0$, $x_4 = 1$ (along with its z-value of 4) may be used as a candidate solution.

2 Even if the best completion of a node is not feasible, the best completion gives us a bound on the best objective function value that can be obtained via a feasible completion of the node. This bound can often be used to eliminate a node from consideration. For example, suppose we have previously found a candidate solution with $z = 6$, and our objective is to maximize

$$z = 4x_1 + 2x_2 + x_3 - x_4 + 2x_5$$

Also suppose that we are at a node where the fixed variables are $x_1 = 0$, $x_2 = 1$, and $x_3 = 1$. Then the best completion of this node is $x_4 = 0$ and $x_5 = 1$. This yields a z-value of $2 + 1 + 2 = 5$. Because $z = 5$ cannot beat the candidate with $z = 6$, we can immediately eliminate this node from consideration (whether or not the completion is feasible is irrelevant).

3 At any node, is there an easy way to determine if all completions of the node are infeasible? Suppose we are at node 4 of Figure 26 and one of the constraints is

$$-2x_1 + 3x_2 + 2x_3 - 3x_4 - x_5 + 2x_6 \leq -5 \qquad \text{(46)}$$

Is there any completion of node 4 that can satisfy this constraint? We assign values to the free variables that make the left side of (46) as small as possible. If this completion of node 4 won't satisfy (46), then certainly no completion of node 4 can. Thus, we set $x_1 = 1$, $x_5 = 1$, and $x_6 = 0$. Substituting these values and the values of the fixed variables, we obtain $-2 + 3 + 2 - 3 - 1 \leq -5$. This inequality does not hold, so no completion of node 4 can satisfy (46). No completion of node 4 can be feasible for the original problem, and node 4 may be eliminated from consideration.

In general, we check whether a node has a feasible completion by looking at each constraint and assigning each free variable the best value (as described in Table 81) for satisfying the constraint.[†] If even one constraint is not satisfied by its most feasible completion, then we know that the node has no feasible completion. In this case, the node cannot yield the optimal solution to the original IP.

[†]Each equality constraint should be replaced by a $\leq$ and a $\geq$ constraint.

TABLE 81

How to Determine Whether a Node Has a Completion
Satisfying a Given Constraint

Type of Constraint	Sign of Free Variable's Coefficient in Constraint	Value Assigned to Free Variable in Feasibility Check
$\leq$	$+$	0
$\leq$	$-$	1
$\geq$	$+$	1
$\geq$	$-$	0

We note, however, that even if a node has no feasible completion, our crude infeasibility check may not reveal that the node has no feasible completion until we have moved further down the tree to a node where there are more fixed variables. If we have failed to obtain any information about a node, we now branch on a free variable x_i and add two new nodes: one with $x_i = 1$ and another with $x_i = 0$.

EXAMPLE 12 | **Implicit Enumeration**

Use implicit enumeration to solve the following 0–1 IP:

$$\max z = -7x_1 - 3x_2 - 2x_3 - x_4 - 2x_5$$
$$\text{s.t.} \quad -4x_1 - 2x_2 + x_3 - 2x_4 - x_5 \leq -3 \tag{47}$$
$$-4x_1 - 2x_2 - 4x_3 + x_4 + 2x_5 \leq -7 \tag{48}$$
$$x_i = 0 \text{ or } 1 \quad (i = 1, 2, 3, 4, 5)$$

Solution At the beginning (node 1), all variables are free. We first check whether the best completion of node 1 is feasible. The best completion of node 1 is $x_1 = 0$, $x_2 = 0$, $x_3 = 0$, $x_4 = 0$, $x_5 = 0$, which is not feasible (it violates both constraints). We now check to see whether node 1 has no feasible completion. Checking (47) for feasibility, we set $x_1 = 1$, $x_2 = 1$, $x_3 = 0$, $x_4 = 1$, $x_5 = 1$. This satisfies (47) (it yields $-9 \leq -3$). We now check (48) for feasibility by setting $x_1 = 1$, $x_2 = 1$, $x_3 = 1$, $x_4 = 0$, $x_5 = 0$. This completion of node 1 satisfies (48) (it yields $-10 \leq -7$). Thus, node 1 has a feasible completion satisfying (48). Therefore, our infeasibility check does not allow us to classify node 1 as having no feasible completion. We now choose to branch on a free variable: arbitrarily, x_1. This yields two new nodes: node 2 with the constraint $x_1 = 1$ and node 3 with the constraint $x_1 = 0$ (see Figure 27).

We now choose to analyze node 2. The best completion of node 2 is $x_1 = 1$, $x_2 = 0$, $x_3 = 0$, $x_4 = 0$, and $x_5 = 0$. Unfortunately, this completion is not feasible. We now try to determine whether node 2 has a feasible completion. We check whether $x_1 = 1$, $x_2 = 1$, $x_3 = 0$, $x_4 = 1$, $x_5 = 1$ satisfies (47) (this yields $-9 \leq -3$). Then we check whether $x_1 = 1$, $x_2 = 1$, $x_3 = 1$, $x_4 = 0$, $x_5 = 0$ satisfies (48) (this yields $-10 \leq -7$). Thus, our infeasibility check has yielded no information about whether node 2 has a feasible completion.

We now choose to branch on node 2, arbitrarily, on the free variable x_2. This yields nodes 4 and 5 in Figure 28. Using the LIFO rule, we choose to next analyze node 5. The best completion of node 5 is $x_1 = 1$, $x_2 = 0$, $x_3 = 0$, $x_4 = 0$, $x_5 = 0$. Again, this completion is infeasible. We now perform a feasibility check on node 5. We determine whether $x_1 = 1$, $x_2 = 0$, $x_3 = 0$, $x_4 = 1$, $x_5 = 1$ satisfies (47) (this yields $-7 \leq -3$). Then we check whether $x_1 = 1$, $x_2 = 0$, $x_3 = 1$, $x_4 = 0$, $x_5 = 0$ satisfies (48) (this yields $-8 \leq -7$). Again our feasibility check has yielded no information. Thus, we branch on node 5, arbitrarily choosing the free variable x_3. This adds nodes 6 and 7 in Figure 29.

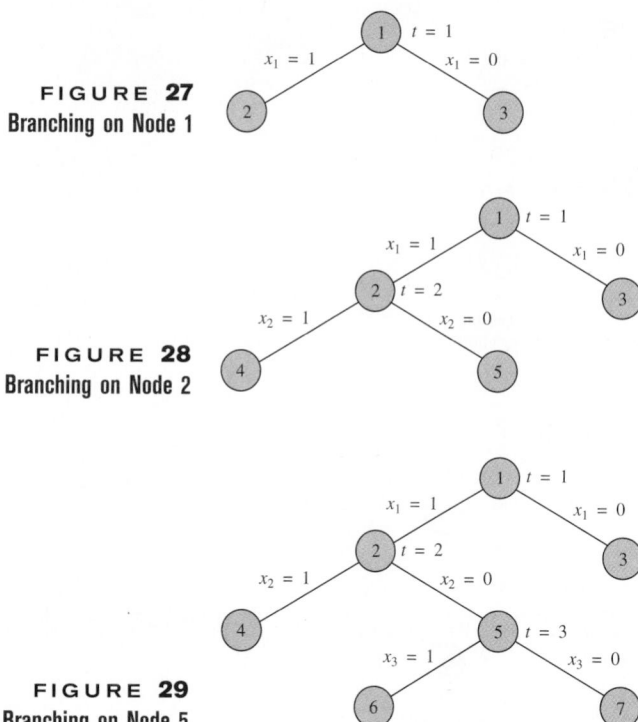

FIGURE 27
Branching on Node 1

FIGURE 28
Branching on Node 2

FIGURE 29
Branching on Node 5

Applying the LIFO rule, we next choose to analyze node 6. The best completion of node 6 is $x_1 = 1$, $x_2 = 0$, $x_3 = 1$, $x_4 = 0$, $x_5 = 0$, $z = -9$. This point is feasible, so we have found a candidate solution with $z = -9$. Using the LIFO rule, we next analyze node 7. The best completion of node 7 is $x_1 = 1$, $x_2 = 0$, $x_3 = 0$, $x_4 = 0$, $x_5 = 0$, $z = -7$. Because $z = -7$ is better than $z = -9$, it is possible for node 7 to beat the current candidate. Thus, we must check node 7 to see whether it has any feasible completion. We see whether $x_1 = 1$, $x_2 = 0$, $x_3 = 0$, $x_4 = 1$, $x_5 = 1$ satisfies (47) (this yields $-7 \le -3$). Then we see whether $x_1 = 1$, $x_2 = 0$, $x_3 = 0$, $x_4 = 0$, $x_5 = 0$ satisfies (48) (this yields $-4 \le -7$). This means that no completion of node 7 can satisfy (48). Thus, node 7 has no feasible completion, and it may be eliminated from consideration (indicated by an $\times$ in Figure 30).

The LIFO rule now indicates that we should analyze node 4. The best completion of node 4 is $x_1 = 1$, $x_2 = 1$, $x_3 = 0$, $x_4 = 0$, $x_5 = 0$. This solution has $z = -10$. Thus, node 4 cannot beat the previous candidate solution from node 6 (having $z = -9$), and node 4 may be eliminated from consideration.

We are now facing the tree in Figure 31, where only node 3 remains to be analyzed. The best completion of node 3 is $x_1 = 0$, $x_2 = 0$, $x_3 = 0$, $x_4 = 0$, $x_5 = 0$. This point is infeasible. This point has $z = 0$, however, so it is possible that node 3 can yield a feasible solution that is better than our current candidate (with $z = -9$). We now check whether node 3 has any feasible completion: Does $x_1 = 0$, $x_2 = 1$, $x_3 = 1$, $x_4 = 1$, $x_5 = 1$ satisfy (47)? This yields $-5 \le -3$, so node 3 does have a completion satisfying (47). Then we see whether node 3 has any completion satisfying (48): Does $x_1 = 0$, $x_2 = 1$, $x_3 = 1$, $x_4 = 0$, $x_5 = 0$ satisfy (48)? This yields $-6 \le -7$, which is untrue. Thus, node 3 has no completion satisfying (48), and node 3 may be eliminated from consideration. We now have the tree in Figure 32.

Because there are no nodes left to analyze, the node 6 candidate with $z = -9$ must be optimal. Thus, $x_1 = 1$, $x_2 = 0$, $x_3 = 1$, $x_4 = 0$, $x_5 = 0$, $z = -9$ is the optimal solution to

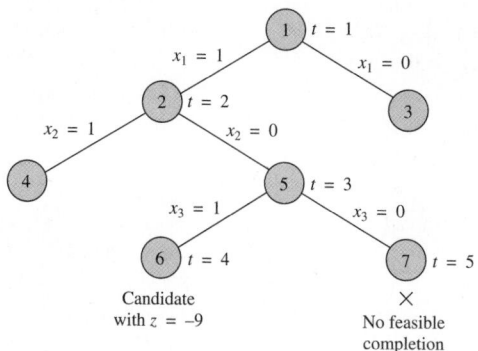

FIGURE 30
Node 6 Yields a Candidate Solution, and Node 7 Has No Feasible Completion

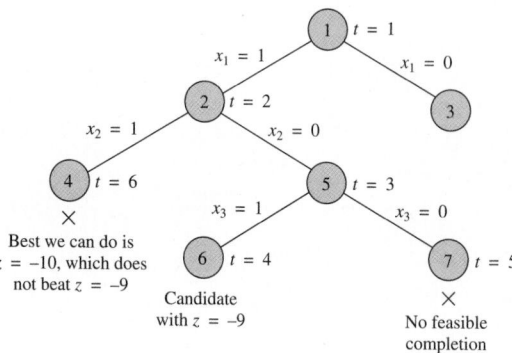

FIGURE 31
Node 4 Cannot Beat Node 6 Candidate

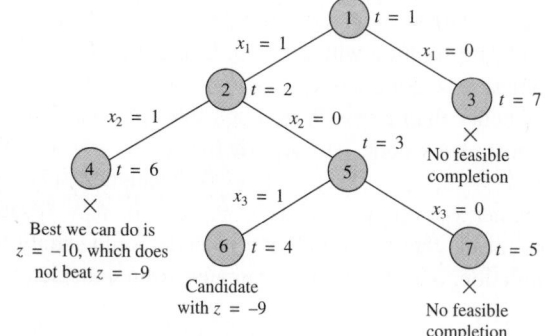

FIGURE 32
Node 3 Has No Feasible Completion

the 0–1 IP. Note that every possible point $(x_1, x_2, x_3, x_4, x_5)$ where $x_i = 0$ or 1 has been implicitly considered, and all but the optimal solution have been eliminated. For example, for the point $x_1 = 1$, $x_2 = 1$, $x_3 = 1$, $x_4 = 1$, $x_5 = 0$, the analysis of node 4 shows that this point cannot be optimal because it cannot have a z-value of better than -9. As another example, the point $x_1 = 0$, $x_2 = 1$, $x_3 = 1$, $x_4 = 1$, $x_5 = 1$ cannot be optimal, because our analysis of node 3 shows that no completion can be feasible.

The use of subtler infeasibility tests (called **surrogate constraints**) can often reduce the number of nodes that must be examined before an optimal solution is found. For example, consider a 0–1 IP with the following constraints:

$$x_1 + x_2 + x_3 + x_4 + x_5 \leq 2 \tag{49}$$

$$x_1 - x_2 + x_3 - x_4 - x_5 \geq 1 \tag{50}$$

Suppose we are at a node where $x_1 = x_2 = 1$. To check whether this node has a feasible completion, we would first see whether $x_1 = 1$, $x_2 = 1$, $x_3 = 0$, $x_4 = 0$, $x_5 = 0$ satisfies (49) (it does). Then we would see whether $x_1 = 1$, $x_2 = 1$, $x_3 = 1$, $x_4 = 0$, $x_5 = 0$ (50) (it does). In this situation, our crude infeasibility tests do not yet indicate that this node is infeasible. Observe, however, that because $x_1 = x_2 = 1$, the only way to satisfy (49) is by choosing $x_3 = x_4 = x_5 = 0$, but this completion of the $x_1 = x_2 = 1$ node fails to satisfy (50). Thus, the node with $x_1 = x_2 = 1$ will have no feasible completion. Eventually, our crude infeasibility test would have indicated this fact, but we might have been forced to examine several more nodes before we found that the node with $x_1 = x_2 = 1$ had no feasible completion. In a more complex problem, a subtler infeasibility test that combined information from both constraints might have enabled us to examine fewer nodes. Of course, a subtler infeasibility test would require more computation, so it might not be worth the effort. For a discussion of surrogate constraints, see Salkin (1975), Taha (1975), and Nemhauser and Wolsey (1988).

As with any branch-and-bound algorithm, many arbitrary choices determine the efficiency of the implicit enumeration algorithm. See Salkin, Taha, and Nemhauser and Wolsey for further discussion of implicit enumeration techniques.

PROBLEMS

Group A

Use implicit enumeration to solve the following 0–1 IPs:

1   max $z = 3x_1 + x_2 + 2x_3 - x_4 + x_5$
   s.t. $2x_1 + x_2 \qquad - 3x_4 \qquad \leq 1$
            $x_1 + 2x_2 - 3x_3 - x_4 + 2x_5 \geq 2$
                    $x_i = 0$ or 1

2   max $z = 2x_1 - x_2 + x_3$
   s.t. $x_1 + 2x_2 - x_3 \leq 1$
         $x_1 + x_2 + x_3 \leq 2$
             $x_i = 0$ or 1

3   Finco is considering investing in five projects. Each requires a cash outflow at time 0 and yields an NPV as described in Table 82 (all dollars in millions). At time 0, $10 million is available for investment. Projects 1 and 2 are mutually exclusive (that is, Finco cannot undertake both). Similarly, projects 3 and 4 are mutually exclusive. Also, project 2 cannot be undertaken unless project 5 is undertaken. Use implicit enumeration to determine which projects should be undertaken to maximize NPV.

TABLE 82

Project	Time 0 Cash Outflow ($)	NPV ($)
1	4	5
2	6	9
3	5	6
4	4	3
5	3	2

4   Use implicit enumeration to find the optimal solution to Example 5 (the set-covering problem).

5   Use implicit enumeration to solve Problem 1 of Section 9.2.

Group B

6   Why are the values of $u_0, u_1, \ldots, u_n$ in (44) unique?

9.8 The Cutting Plane Algorithm[†]

In previous portions of this chapter, we have described in some detail how branch-and-bound methods can be used to solve IPs. In this section, we discuss an alternative method, **the cutting plane algorithm.** We illustrate the cutting plane algorithm by solving the Telfa Corporation problem (Example 9). Recall from Section 9.3 that this problem was

[†]This section covers topics that may be omitted with no loss of continuity.

TABLE 83

z	x_1	x_2	s_1	s_2	rhs
1	0	0	1.25	0.75	41.25
0	0	1	2.25	−0.25	2.25
0	1	0	−1.25	0.25	3.75

$$\max z = 8x_1 + 5x_2$$
$$\text{s.t.} \quad x_1 + x_2 \leq 6$$
$$9x_1 + 5x_2 \leq 45$$
$$x_1, x_2 \geq 0; \; x_1, x_2 \text{ integer}$$

(51)

After adding slack variables s_1 and s_2, we found the optimal tableau for the LP relaxation of the Telfa example to be as shown in Table 83.

To apply the cutting plane method, we begin by choosing any constraint in the LP relaxation's optimal tableau in which a basic variable is fractional. We arbitrarily choose the second constraint, which is

$$x_1 - 1.25s_1 + 0.25s_2 = 3.75 \tag{52}$$

We now define $[x]$ to be the largest integer less than or equal to x. For example, $[3.75] = 3$ and $[-1.25] = -2$. Any number x can be written in the form $[x] + f$, where $0 \leq f < 1$. We call f the fractional part of x. For example, $3.75 = 3 + 0.75$, and $-1.25 = -2 + 0.75$. In (51)'s optimal tableau, we now write each variable's coefficient and the constraint's right-hand side in the form $[x] + f$, where $0 \leq f < 1$. Now (52) may be written as

$$x_1 - 2s_1 + 0.75s_1 + 0s_2 + 0.25s_2 = 3 + 0.75 \tag{53}$$

Putting all terms with integer coefficients on the left side and all terms with fractional coefficients on the right side yields

$$x_1 - 2s_1 + 0s_2 - 3 = 0.75 - 0.75s_1 - 0.25s_2 \tag{54}$$

The cutting plane algorithm now suggests adding the following constraint to the LP relaxation's optimal tableau:

Right-hand side of (54) ≤ 0

or

$$0.75 - 0.75s_1 - 0.25s_2 \leq 0 \tag{55}$$

This constraint is called (for reasons that will soon become apparent) a **cut.** We now show that a cut generated by this method has two properties:

1 Any feasible point for the IP will satisfy the cut.

2 The current optimal solution to the LP relaxation will not satisfy the cut.

Thus, a cut "cuts off" the current optimal solution to the LP relaxation, but not any feasible solutions to the IP. When the cut to the LP relaxation is added, we hope we will obtain a solution where all variables are integer-valued. If so, we have found the optimal solution to the original IP. If our new optimal solution (to the LP relaxation plus the cut) has some fractional-valued variables, then we generate another cut and continue the process. Gomory (1958) showed that this process will yield an optimal solution to the IP after a finite number of cuts. Before finding the optimal solution to the IP (51), we show why the cut (55) satisfies properties 1 and 2.

We now show that any feasible solution to the IP (51) will satisfy the cut (55). Consider any point that is feasible for the IP. For such a point, x_1 and x_2 take on integer values, and the point must be feasible in the LP relaxation of (51). Because (54) is just a rearrangement of the optimal tableau's second constraint, any feasible point for the IP must satisfy (54). Any feasible solution to the IP must have $s_1 \geq 0$ and $s_2 \geq 0$. Because $0.75 < 1$, any feasible solution to the IP will make the right-hand side of (54) less than 1. Also note that for any point that is feasible for the IP, the left-hand side of (54) will be an integer. Thus, for any feasible point to the IP, the right-hand side must be an integer that is less than 1. This implies that any point that is feasible for the IP satisfies (55), so our cut does not eliminate any feasible integer points from consideration!

We now show that the current optimal solution to the LP relaxation cannot satisfy the cut (55). The current optimal solution to the LP relaxation has $s_1 = s_2 = 0$. Thus, it cannot satisfy (55). This argument works because 0.75 (the fractional part of the right-hand side of the second constraint) is greater than 0. Thus, if we choose any constraint whose right-hand side in the optimal tableau is fractional, we can cut off the LP relaxation's optimal solution.

The effect of the cut (55) can be seen in Figure 33; all points feasible for the IP (51) satisfy the cut (55), but the current optimal solution to the LP relaxation ($x_1 = 3.75$ and $x_2 = 2.25$) does not. To obtain the graph of the cut, we replaced s_1 by $6 - x_1 - x_2$ and s_2 by $45 - 9x_1 - 5x_2$. This enabled us to rewrite the cut as $3x_1 + 2x_2 \leq 15$.

We now add (55) to the LP relaxation's optimal tableau and use the dual simplex to solve the resulting LP. Cut (55) may be written as $-0.75s_1 - 0.25s_2 \leq -0.75$. After adding a slack variable s_3 to this constraint, we obtain the tableau shown in Table 84.

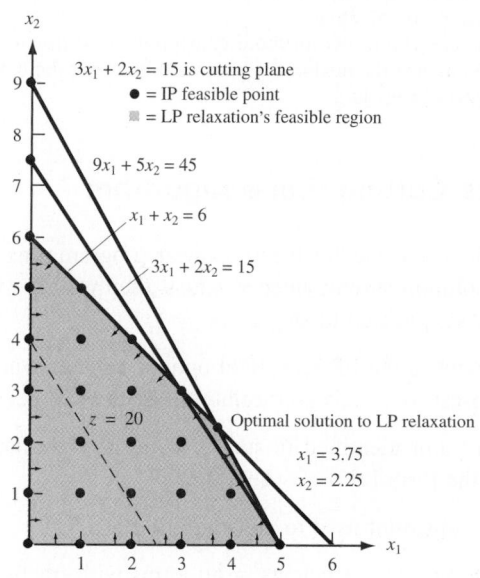

FIGURE 33
Example of Cutting Plane

$3x_1 + 2x_2 = 15$ is cutting plane
● = IP feasible point
▪ = LP relaxation's feasible region
$9x_1 + 5x_2 = 45$
$x_1 + x_2 = 6$
$3x_1 + 2x_2 = 15$
Optimal solution to LP relaxation
$x_1 = 3.75$
$x_2 = 2.25$
$z = 20$

TABLE 84
Cutting Plane Tableau After Adding Cut (55)

z	x_1	x_2	s_1	s_2	s_3	rhs
1	0	0	1.25	0.75	0	41.25
0	0	1	2.25	−0.25	0	2.25
0	1	0	−1.25	0.25	0	3.75
0	0	0	−0.75	−0.25	1	−0.75

TABLE **85**
Optimal Tableau for Cutting Plane

z	x_1	x_2	s_1	s_2	s_3	rhs
1	0	0	0	0.33	1.67	40
0	0	1	0	-1	3	0
0	1	0	0	0.67	-1.67	5
0	0	0	1	0.33	-1.33	1

The dual simplex ratio test indicates that s_1 should enter the basis in the third constraint. The resulting tableau is given in Table 85, which yields the optimal solution $z = 40$, $x_1 = 5$, $x_2 = 0$.

Recall that a cut does not eliminate any points that are feasible for the IP. This means that whenever we solve the LP relaxation to an IP with several cuts as additional constraints and find an optimal solution in which all variables are integers, we have solved our original IP. Because x_1 and x_2 are integers in our current optimal solution, this point must be optimal for (51). Of course, if the first cut had not yielded the optimal solution to the IP, we would have kept adding cuts until we obtained an optimal tableau in which all variables were integers.

REMARKS **1** The algorithm requires that all coefficients of variables in the constraints and all right-hand sides of constraints be integers. This is to ensure that if the original decision variables are integers, then the slack and excess variables will also be integers. Thus, a constraint such as $x_1 + 0.5x_2 \leq 3.6$ must be replaced by $10x_1 + 5x_2 \leq 36$.

2 If at any stage of the algorithm, two or more constraints have fractional right-hand sides, then best results are often obtained if the next cut is generated by using the constraint whose right-hand side has the fractional part closest to $\frac{1}{2}$.

Summary of the Cutting Plane Algorithm

Step 1 Find the optimal tableau for the IP's linear programming relaxation. If all variables in the optimal solution assume integer values, then we have found an optimal solution to the IP; otherwise, proceed to step 2.

Step 2 Pick a constraint in the LP relaxation optimal tableau whose right-hand side has the fractional part closest to $\frac{1}{2}$. This constraint will be used to generate a cut.

Step 2a For the constraint identified in step 2, write its right-hand side and each variables's coefficient in the form $[x] + f$, where $0 \leq f < 1$.

Step 2b Rewrite the constraint used to generate the cut as

All terms with integer coefficients = all terms with fractional coefficients

Then the cut is

All terms with fractional coefficients ≤ 0

Step 3 Use the dual simplex to find the optimal solution to the LP relaxation, with the cut as an additional constraint. If all variables assume integer values in the optimal solution, we have found an optimal solution to the IP. Otherwise, pick the constraint with the most fractional right-hand side and use it to generate another cut, which is added to the tableau. We continue this process until we obtain a solution in which all variables are integers. This will be an optimal solution to the IP.

PROBLEMS

Group A

1 Consider the following IP:

$$\max z = 14x_1 + 18x_2$$
$$\text{s.t.} \quad -x_1 + 3x_2 \leq 6$$
$$7x_1 + x_2 \leq 35$$
$$x_1, x_2 \geq 0; \ x_1, x_2 \text{ integer}$$

The optimal tableau for this IP's linear programming relaxation is given in Table 86. Use the cutting plane algorithm to solve this IP.

2 Consider the following IP:

$$\min z = 6x_1 + 8x_2$$
$$\text{s.t.} \quad 3x_1 + x_2 \geq 4$$
$$x_1 + 2x_2 \geq 4$$
$$x_1, x_2 \geq 0; \ x_1, x_2 \text{ integer}$$

The optimal tableau for this IP's linear programming relaxation is given in Table 87. Use the cutting plane algorithm to find the optimal solution.

3 Consider the following IP:

$$\max z = 2x_1 - 4x_2$$
$$\text{s.t.} \quad 2x_1 + x_2 \leq 5$$
$$-4x_1 + 4x_2 \leq 5$$
$$x_1, x_2 \geq 0; \ x_1, x_2 \text{ integer}$$

The optimal tableau for this IP's linear programming relaxation is given in Table 88. Use the cutting plane algorithm to find the optimal solution.

TABLE 87

z	x_1	x_2	e_1	e_2	rhs
1	0	0	$-\frac{4}{5}$	$-\frac{18}{5}$	$\frac{88}{5}$
0	1	0	$-\frac{2}{5}$	$\frac{1}{5}$	$\frac{4}{5}$
0	0	1	$\frac{1}{5}$	$-\frac{3}{5}$	$\frac{8}{5}$

TABLE 86

z	x_1	x_2	s_1	s_2	rhs
1	0	0	$\frac{56}{11}$	$\frac{30}{11}$	126
0	0	1	$\frac{7}{22}$	$\frac{1}{22}$	$\frac{7}{2}$
0	1	0	$-\frac{1}{22}$	$\frac{3}{22}$	$\frac{9}{2}$

TABLE 88

z	x_1	x_2	s_1	s_2	rhs
1	0	0	$-\frac{2}{3}$	$-\frac{5}{6}$	$-\frac{15}{2}$
0	1	0	$\frac{1}{3}$	$-\frac{1}{12}$	$\frac{5}{4}$
0	0	1	$\frac{1}{3}$	$\frac{1}{6}$	$\frac{5}{2}$

SUMMARY

Integer programming problems (IP's) are usually much harder to solve than linear programming problems.

Integer Programming Formulations

Most integer programming formulations involve **0–1 variables.**

Fixed-Charge Problems

Suppose activity i incurs a fixed charge if undertaken at any positive level. Let

$$x_i = \text{level of activity } i$$

$$y_i = \begin{cases} 1 & \text{if activity } i \text{ is undertaken at positive level } (x_i > 0) \\ 0 & \text{if } x_i = 0 \end{cases}$$

Then a constraint of the form $x_i \leq M_i y_i$ must be added to the formulation. Here, M_i must be large enough to ensure that x_i will be less than or equal to M_i.

Either–Or Constraints

Suppose we want to ensure that at least one of the following two constraints (and possibly both) are satisfied:

$$f(x_1, x_2, \ldots, x_n) \le 0 \qquad (26)$$

$$g(x_1, x_2, \ldots, x_n) \le 0 \qquad (27)$$

Adding the following two constraints to the formulation will ensure that at least one of (26) and (27) is satisfied:

$$f(x_1, x_2, \ldots, x_n) \le My \qquad (26')$$

$$g(x_1, x_2, \ldots, x_n) \le M(1 - y) \qquad (27')$$

In (26′) and (27′), y is a 0–1 variable, and M is a number chosen large enough to ensure that $f(x_1, x_2, \ldots, x_n) \le M$ and $g(x_1, x_2, \ldots, x_n) \le M$ are satisfied for all values of $x_1, x_2, \ldots, x_n$ that satisfy the other constraints in the problem.

If–Then Constraints

Suppose we want to ensure that $f(x_1, x_2, \ldots, x_n) > 0$ implies $g(x_1, x_2, \ldots, x_n) \ge 0$. Then we include the following constraints in the formulation:

$$-g(x_1, x_2, \ldots, x_n) \le My \qquad (28')$$

$$f(x_1, x_2, \ldots, x_n) \le M(1 - y) \qquad (29)$$

$$y = 0 \text{ or } 1$$

Here, M is a large positive number, chosen large enough so that $f \le M$ and $-g \le M$ hold for all values of $x_1, x_2, \ldots, x_n$ that satisfy the other constraints in the problem.

How to Model a Piecewise Linear Function $f(x)$ with 0–1 Variables

Suppose the piecewise linear function $f(x)$ has break points $b_1, b_2, \ldots, b_n$.

Step 1 Wherever $f(x)$ occurs in the optimization problem, replace $f(x)$ by $z_1 f(b_1) + z_2 f(b_2) + \cdots + z_n f(b_n)$.

Step 2 Add the following constraints to the problem:

$$z_1 \le y_1, z_2 \le y_1 + y_2, z_3 \le y_2 + y_3, \ldots, z_{n-1} \le y_{n-2} + y_{n-1}, z_n \le y_{n-1}$$

$$y_1 + y_2 + \cdots + y_{n-1} = 1$$

$$z_1 + z_2 + \cdots + z_n = 1$$

$$x = z_1 b_1 + z_2 b_2 + \cdots + z_n b_n$$

$$y_i = 0 \text{ or } 1 \ (i = 1, 2, \ldots, n - 1); z_i \ge 0 \ (i = 1, 2, \ldots, n)$$

Branch-and-Bound Method

Usually, IPs are solved by some version of the **branch-and-bound** procedure. Branch-and-bound methods implicitly enumerate all possible solutions to an IP. By solving a single **subproblem,** many possible solutions may be eliminated from consideration.

Branch-and-Bound for Pure IP's

Subproblems are generated by branching on an appropriately chosen fractional-valued variable x_i. Suppose that in a given subproblem (call it old subproblem), x_i assumes a fractional value between the integers i and $i + 1$. Then the two newly generated subproblems are

New Subproblem 1   Old subproblem + Constraint $x_i \leq i$.

New Subproblem 2   Old subproblem + Constraint $x_i \geq i + 1$.

If it is unnecessary to branch on a subproblem, then we say it is **fathomed.** The following three situations (for a max problem) result in a subproblem being fathomed: (1) The subproblem is infeasible, thus it cannot yield the optimal solution to the IP. (2) The subproblem yields an optimal solution in which all variables have integer values. If this optimal solution has a better z-value than any previously obtained solution that is feasible in the IP, then it becomes a **candidate solution,** and its z-value becomes the current lower bound (LB) on the optimal z-value for the IP. In this case, the current subproblem may yield the optimal solution to the IP. (3) The optimal z-value for the subproblem does not exceed (in a max problem) the current LB, so it may be eliminated from consideration.

Branch-and-Bound for Mixed IPs

When branching on a fractional variable, only branch on those required to be integers.

Branch-and-Bound for Knapsack Problems

Subproblems may easily be solved by first putting the best (in terms of benefit per-unit weight) item in the knapsack, then the next best, and so on, until a fraction of an item is used to completely fill the knapsack.

Branch-and-Bound to Minimize Delay on a Single Machine

Begin the branching by determining which job should be processed last. Suppose there are n jobs. At a node where the jth job to be processed, $(j + 1)$th job to be processed, . . . , nth job to be processed are fixed, a lower bound on the total delay is given by (delay of jth job to be processed) + (delay of $(j + 1)$th job to be processed) + $\cdots$ + (delay of nth job to be processed).

Branch-and-Bound for Traveling Salesperson Problem

Subproblems are assignment problems. If the optimal solution to a subproblem contains no subtours, then it is a feasible solution to the traveling salesperson problem. Create new subproblems by branching to exclude a subtour. Eliminate a subproblem if its optimal z-value is inferior to the best previously found feasible solution.

Heuristics for the TSP

To apply the nearest-neighbor heuristic (NNH), we begin at any city and then "visit" the nearest city. Then we go to the unvisited city closest to the city we have most recently visited. We continue in this fashion until a tour is obtained. After applying this procedure beginning at each city, we take the best tour found.

In the cheapest-insertion heuristic (CIH), we begin at any city and find its closest neighbor. Then we create a subtour joining those two cities. Next, we replace an arc in the subtour [say, arc (i, j)] by the combination of two arcs—(i, k) and (k, j), where k is not in the current subtour—that will increase the length of the subtour by the smallest (or cheapest) amount. We continue with this procedure until a tour is obtained. After applying this procedure beginning with each city, we take the best tour found.

Implicit Enumeration

In a 0–1 IP, implicit enumeration may be used to find an optimal solution. When branching at a node, create two new subproblems by (for some free variable x_i) adding constraints $x_i = 0$ and $x_i = 1$. If the best completion of a node is feasible, then we need not branch on the node. If the best completion is feasible and better than the current candidate solution, then the current node yields a new LB (in a max problem) and may be optimal. If the best completion is feasible and is not better than the current candidate solution, then the current node may be eliminated from consideration. If at a given node, there is at least one constraint that is not satisfied by any completion of the node, then the node cannot yield a feasible solution nor an optimal solution to the IP.

Cutting Plane Algorithm

Step 1 Find the optimal tableau for the IP's linear programming relaxation. If all variables in the optimal solution assume integer values, we have found an optimal solution to the IP; otherwise, proceed to step 2.

Step 2 Pick a constraint in the LP relaxation optimal tableau whose right-hand side has the fractional part closest to $\frac{1}{2}$. This constraint will be used to generate a cut.

Step 2a For the constraint identified in step 2, write its right-hand side and each variable's coefficient in the form $[x] + f$, where $0 \leq f < 1$.

Step 2b Rewrite the constraint used to generate the cut as

All terms with integer coefficients = all terms with fractional coefficients

Then the cut is

All terms with fractional coefficients ≤ 0

Step 3 Use the dual simplex to find the optimal solution to the LP relaxation, with the cut as an additional constraint. If all variables assume integer values in the optimal solution, then we have found an optimal solution to the IP. Otherwise, pick the constraint with the most fractional right-hand side and use it to generate another cut, which is added to the tableau. We continue this process until we obtain a solution in which all variables are integers. This will be an optimal solution to the IP.

REVIEW PROBLEMS

Group A

1 In the Sailco problem of Section 3.10, suppose that a fixed cost of $200 is incurred during each quarter that production takes place. Formulate an IP to minimize Sailco's total cost of meeting the demands for the four quarters.

2 Explain how you would use integer programming and piecewise linear functions to solve the following optimization problem. (*Hint:* Approximate x^2 and y^2 by piecewise linear functions.)

$$\max z = 3x^2 + y^2$$
$$\text{s.t.} \quad x + y \leq 1$$
$$x, y \geq 0$$

3[†] The Transylvania Olympic Gymnastics Team consists of six people. Transylvania must choose three people to enter both the balance beam and floor exercises. They must also enter a total of four people in each event. The score that each individual gymnast can attain in each event is shown in Table 89. Formulate an IP to maximize the total score attained by the Transylvania gymnasts.

4[‡] A court decision has stated that the enrollment of each high school in Metropolis must be at least 20 percent black. The numbers of black and white high school students in each of the city's five school districts are shown in Table 90. The distance (in miles) that a student in each district must travel to each high school is shown in Table 91. School board policy requires that all the students in a given district attend the same school. Assuming that each school must have an enrollment of at least 150 students, formulate an IP that will minimize the total distance that Metropolis students must travel to high school.

5 The Cubs are trying to determine which of the following free agent pitchers should be signed: Rick Sutcliffe (RS), Bruce Sutter (BS), Dennis Eckersley (DE), Steve Trout (ST), Tim Stoddard (TS). The cost of signing each pitcher and the number of victories each pitcher will add to the Cubs are shown in Table 92. Subject to the following restrictions, the Cubs want to sign the pitchers who will add the most victories to the team.

TABLE 89

Gymnast	Balance Beam	Floor Exercise
1	8.8	7.9
2	9.4	8.3
3	9.2	8.5
4	7.5	8.7
5	8.7	8.1
6	9.1	8.6

TABLE 90

District	Whites	Blacks
1	80	30
2	70	5
3	90	10
4	50	40
5	60	30

[†]Based on Ellis and Corn (1984).
[‡]Based on Liggett (1973).

TABLE 91

District	High School 1	High School 2
1	1	2
2	0.5	1.7
3	0.8	0.8
4	1.3	0.4
5	1.5	0.6

TABLE 92

Pitcher	Cost of Signing Pitcher ($) Millions	Victories Added to Cubs
RS	6	6 (righty)
BS	4	5 (righty)
DE	3	3 (righty)
ST	2	3 (lefty)
TS	2	2 (righty)

a At most, $12 million can be spent.
b If DE and ST are signed, then BS cannot be signed.
c At most two right-handed pitchers can be signed.
d The Cubs cannot sign both BS and RS.

Formulate an IP to help the Cubs determine who they should sign.

6 State University must purchase 1,100 computers from three vendors. Vendor 1 charges $500 per computer plus a delivery charge of $5,000. Vendor 2 charges $350 per computer plus a delivery charge of $4,000. Vendor 3 charges $250 per computer plus a delivery charge of $6,000. Vendor 1 will sell the university at most 500 computers; vendor 2, at most 900; and vendor 3, at most 400. Formulate an IP to minimize the cost of purchasing the needed computers.

7 Use the branch-and-bound method to solve the following IP:

$$\max z = 3x_1 + x_2$$
$$\text{s.t.} \quad 5x_1 + x_2 \leq 12$$
$$2x_1 + x_2 \leq 8$$
$$x_1, x_2 \geq 0; x_1, x_2 \text{ integer}$$

8 Use the branch-and-bound method to solve the following IP:

$$\min z = 3x_1 + x_2$$
$$\text{s.t.} \quad 2x_1 - x_2 \leq 6$$
$$x_1 + x_2 \leq 4$$
$$x_1, x_2 \geq 0; x_1 \text{ integer}$$

9 Use the branch-and-bound method to solve the following IP:

$$\max z = x_1 + 2x_2$$
$$\text{s.t.} \quad x_1 + x_2 \leq 10$$
$$2x_1 + 5x_2 \leq 30$$
$$x_1, x_2 \geq 0; x_1, x_2 \text{ integer}$$

10 Consider a country where there are 1¢, 5¢, 10¢, 20¢, 25¢, and 50¢ pieces. You work at the Two-Twelve Convenience Store and must give a customer 91¢ in change. Formulate an IP that can be used to minimize the number of coins needed to give the correct change. Use what you know about knapsack problems to solve the IP by the branch-and-bound method. (*Hint:* We need only solve a 90¢ problem.)

11 Use the branch-and-bound approach to find the optimal solution to the traveling salesperson problem shown in Table 93.

12 Use the implicit enumeration method to find the optimal solution to Problem 5.

13 Use the implicit enumeration method to find the optimal solution to the following 0–1 IP:

$$\max z = 5x_1 - 7x_2 + 10x_3 + 3x_4 - x_5$$

$$\text{s.t.} \quad -x_1 - 3x_2 + 3x_3 - x_4 - 2x_5 \le 0$$

$$2x_1 - 5x_2 + 3x_3 - 2x_4 - 2x_5 \le 3$$

$$- x_2 + x_3 + x_4 - x_5 \ge 2$$

$$\text{All variables 0 or 1}$$

14 A soda delivery truck starts at location 1 and must deliver soda to locations 2, 3, 4, and 5 before returning to location 1. The distance between these locations is given in Table 94. The soda truck wants to minimize the total distance traveled. In what order should the delivery truck make its deliveries?

15 At Blair General Hospital, six types of surgical operations are performed. The types of operations each surgeon is qualified to perform (indicated by an X) are given in Table 95. Suppose that surgeon 1 and surgeon 2 dislike each other and cannot be on duty at the same time. Formulate

TABLE 95

Surgeon	Operation					
	1	2	3	4	5	6
1	x	x		x		
2			x		x	x
3			x		x	
4	x					x
5		x				
6					x	x

an IP whose solution will determine the minimum number of surgeons required so that the hospital can perform all types of surgery.

16 Eastinghouse ships 12,000 capacitors per month to their customers. The capacitors may be produced at three different plants. The production capacity, fixed monthly cost of operation, and variable cost of producing a capacitor at each plant are given in Table 96. The fixed cost for a plant is incurred only if the plant is used to make any capacitors. Formulate an integer programming model whose solution will tell Eastinghouse how to minimize their monthly costs of meeting their customers' demands.

17[†] Newcor's steel mill has received an order for 25 tons of steel. The steel must be 5% carbon and 5% molybdenum by weight. The steel is manufactured by combining three types of metal: steel ingots, scrap steel, and alloys. Four steel ingots are available for purchase. The weight (in tons), cost per ton, carbon and molybdenum content of each ingot are given in Table 97.

Three types of alloys can be purchased. The cost per ton and chemical makeup of each alloy are given in Table 98.

TABLE 93

City	City				
	1	2	3	4	5
1	—	3	1	7	2
2	3	—	4	4	2
3	1	4	—	4	2
4	7	4	4	—	7
5	2	2	2	7	—

TABLE 94

Location	Location				
	1	2	3	4	5
1	0	20	4	10	25
2	20	0	5	30	10
3	4	5	0	6	6
4	10	25	6	0	20
5	35	10	6	20	0

TABLE 96

Plant	Fixed Cost (in $ Thousands)	Variable Cost ($)	Production Capacity
1	80	20	6,000
2	40	25	7,000
3	30	30	6,000

TABLE 97

Ingot	Weight	Cost per Ton ($)	Carbon %	Molybdenum %
1	5	350	5	3
2	3	330	4	3
3	4	310	5	4
4	6	280	3	4

[†]Based on Westerberg, Bjorklund, and Hultman (1977).

TABLE 98

Alloy	Cost per Ton ($)	Carbon %	Molybdenum %
1	500	8	6
2	450	7	7
3	400	6	

Steel scrap may be purchased at a cost of $100 per ton. Steel scrap contains 3% carbon and 9% molybdenum. Formulate a mixed integer programming problem whose solution will tell Newcor how to minimize the cost of filling their order.

18[†] Monsanto annually produces 359 million lb of the chemical maleic anhydride. A total of four reactors are available to produce maleic anhydride. Each reactor can be run on one of three settings. The cost (in thousands of dollars) and pounds produced (in millions) annually for each reactor and each setting are given in Table 99. A reactor can only be run on one setting for the entire year. Set up an IP whose solution will tell Monsanto the minimum-cost method to meet its annual demand for maleic anhydride.

19[‡] Hallco runs a day shift and a night shift. No matter how many units are produced, the only production cost during a shift is a setup cost. It costs $8,000 to run the day shift and $4,500 to run the night shift. Demand for the next two days is as follows: day 1, 2,000; night 1, 3,000; day 2, 2,000; night 2, 3,000. It costs $1 per unit to hold a unit in inventory for a shift. Determine a production schedule that minimizes the sum of setup and inventory costs. All demand must be met on time.

20[‡] After listening to a seminar on the virtues of the Japanese theory of production, Hallco has cut its day shift setup cost to $1,000 per shift and its night shift setup cost to $3,500 per shift. Determine a production schedule that minimizes the sum of setup and inventory costs. All demand must be met on time. Show that the decrease in setup costs has actually *raised* the average inventory level!

Group B

21[§] Gotham City has been divided into eight districts. The time (in minutes) it takes an ambulance to travel from one district to another is shown in Table 100. The population of each district (in thousands) is as follows: district 1, 40; district 2, 30; district 3, 35; district 4, 20; district 5, 15; district 6, 50; district 7, 45; district 8, 60. The city has only two ambulances and wants to locate them to maximize the number of people who live within 2 minutes of an ambulance. Formulate an IP to accomplish this goal.

22 A company must complete three jobs. The amounts of processing time (in minutes) required are shown in Table 101. A job cannot be processed on machine j unless for all $i < j$ the job has completed its processing on machine i. Once a job begins its processing on machine j, the job cannot be preempted on machine j. The flow time for a job is the difference between its completion time and the time at which the job begins its first stage of processing. Formulate an IP whose solution can be used to minimize the average flow time of the three jobs. (*Hint:* Two types of constraints will be needed: Constraint type 1 ensures that a job cannot begin to be processed on a machine until all earlier portions of the job are completed. You will need five constraints of this type. Constraint type 2 ensures that only one job will occupy a machine at any given time. For example, on machine 1, either job 1 is completed before job 2 begins, or job 2 is completed before job 1 begins.)

TABLE 99

Reactor	Setting	Cost ($ Thousands)	Pounds
1	1	50	80
1	2	80	140
1	3	100	170
2	1	65	100
2	2	90	140
2	3	120	215
3	1	70	112
3	2	90	153
3	3	110	195
4	1	40	65
4	2	60	105
4	3	70	130

TABLE 100

District	1	2	3	4	5	6	7	8
1	10	3	4	6	8	9	8	10
2	3	0	5	4	8	6	12	9
3	4	5	0	2	2	3	5	7
4	6	4	2	0	3	2	5	4
5	8	8	2	3	0	2	2	4
6	9	6	3	2	2	0	3	2
7	8	12	5	5	2	3	0	2
8	10	9	7	4	4	2	2	0

TABLE 101

Job	Machine 1	2	3	4
1	20	—	25	30
2	15	20	—	18
3	—	35	28	—

[†]Based on Boykin (1985).
[‡]Based on Zangwill (1992).

[§]Based on Eaton et al. (1985).

23 Arthur Ross, Inc., must complete many corporate tax returns during the period February 15–April 15. This year the company must begin and complete the five jobs shown in Table 102 during this eight-week period. Arthur Ross employs four full-time accountants who normally work 40 hours per week. If necessary, however, they will work up to 20 hours of overtime per week for which they are paid $100 per hour. Use integer programming to determine how Arthur Ross can minimize the overtime cost incurred in completing all jobs by April 15.

24[†] PSI believes it will need the amounts of generating capacity shown in Table 103 during the next five years. The company has a choice of building (and then operating) power plants with the specifications shown in Table 104. Formulate an IP to minimize the total costs of meeting the generating capacity requirements of the next five years.

25[†] Reconsider Problem 24. Suppose that at the beginning of year 1, power plants 1–4 have been constructed and are in operation. At the beginning of each year, PSI may shut down a plant that is operating or reopen a shut-down plant.

TABLE 102

Job	Duration (Weeks)	Accountant Hours Needed per Week
1	3	120
2	4	160
3	3	80
4	2	80
5	4	100

TABLE 103

Year	Generating Capacity (Million kwh)
1	80
2	100
3	120
4	140
5	160

TABLE 104

Plant	Generating Capacity (Million kwh)	Construction Cost ($ Millions)	Annual Operating Cost ($ Millions)
1	70	20	1.5
2	50	16	0.8
3	60	18	1.3
4	40	14	0.6

†Based on Muckstadt and Wilson (1968).

The costs associated with reopening or shutting down a plant are shown in Table 105. Formulate an IP to minimize the total cost of meeting the demands of the next five years. (*Hint:* Let

$X_{it} = 1$ if plant i is operated during year t

$Y_{it} = 1$ if plant i is shut down at end of year t

$Z_{it} = 1$ if plant i is reopened at beginning of year t

You must ensure that if $X_{it} = 1$ and $X_{i,t+1} = 0$, then $Y_{it} = 1$. You must also ensure that if $X_{i,t-1} = 0$ and $X_{it} = 1$, then $Z_{it} = 1$.)

26[‡] Houseco Developers is considering erecting three office buildings. The time required to complete each and the number of workers required to be on the job at all times are shown in Table 106. Once a building is completed, it brings in the following amount of rent per year: building 1, $50,000; building 2, $30,000; building 3, $40,000. Houseco faces the following constraints:

a During each year, 60 workers are available.

b At most, one building can be started during any year.

c Building 2 must be completed by the end of year 4.

Formulate an IP that will maximize the total rent earned by Houseco through the end of year 4.

27 Four trucks are available to deliver milk to five groceries. The capacity and daily operating cost of each truck are shown in Table 107. The demand of each grocery store can be supplied by only one truck, but a truck may deliver to more than one grocery. The daily demands of each grocery are as follows: grocery 1, 100 gallons; grocery 2, 200 gallons; grocery 3, 300 gallons; grocery 4, 500 gallons; grocery 5, 800 gallons. Formulate an IP that can be used to minimize the daily cost of meeting the demands of the four groceries.

TABLE 105

Plant	Reopening Cost ($ Million)	Shutdown Cost ($ Millions)
1	1.9	1.7
2	1.5	1.2
3	1.6	1.3
4	1.1	0.8

TABLE 106

Building	Duration of Project (Years)	Number of Workers Required
1	2	30
2	2	20
3	3	20

‡Based on Peiser and Andrus (1983).

TABLE 107

Truck	Capacity (Gallons)	Daily Operating Cost ($)
1	400	45
2	500	50
3	600	55
4	1,100	60

TABLE 108

	Auditor Cost ($)			
	Northeast	Midwest	West	South
New York	1,100	1,400	1,900	1,400
Chicago	1,200	1,000	1,500	1,200
Los Angeles	1,900	1,700	1,100	1,400
Atlanta	1,300	1,400	1,500	1,050

TABLE 109

Project	Required Workers	Revenue ($)
1	1,4,5,8	10,000
2	2,3,7,10	15,000
3	1,6,8,9	6,000
4	2,3,5,10	8,000
5	1,6,7,9	12,000
6	2,4,8,10	9,000

28[†] The State of Texas frequently does tax audits of companies doing business in Texas. These companies often have headquarters located outside the state, so auditors must be sent to out-of-state locations. Each year, auditors must make 500 trips to cities in the Northeast, 400 trips to cities in the Midwest, 300 trips to cities in the West, and 400 trips to cities in the South. Texas is considering basing auditors in Chicago, New York, Atlanta, and Los Angeles. The annual cost of basing auditors in any city is $100,000. The cost of

TABLE 110

	Worker									
	1	2	3	4	5	6	7	8	9	10
Retainer ($)	800	500	600	700	800	600	400	500	400	500

TABLE 111

	Project					
	1	2	3	4	5	6
Fee ($)	250	300	250	300	175	180

TABLE 112

District	Coordinates		Tons	Cost ($ Millions)	
	x	y		Fixed	Variable
1	4	3	49	2	310
2	2	5	874	1	40
3	10	8	555	1	51
4	2	8	352	1	341
5	5	3	381	3	131
6	4	5	428	2	182
7	10	5	985	1	20
8	5	1	105	2	40
9	5	8	258	4	177
10	1	7	210	2	75

sending an auditor from any of these cities to a given region of the country is given in Table 108. Formulate an IP whose solution will minimize the annual cost of conducting out-of-state audits.

29 A consulting company has 10 employees, each of whom can work on at most two team projects. Six projects are under consideration. Each project requires 4 of our 10 workers. The required workers and the revenue earned from each project are shown in Table 109.

Each worker who is used on *any project* must be paid the retainer shown in Table 110.

Finally, each worker on a project is paid the project fee shown in Table 111.

How can we maximize our profit?

30 New York City has 10 trash districts and is trying to determine which of the districts should be a site for dumping trash. It costs $1,000 to haul one ton of trash one mile. The location of each district, the number of tons of trash produced per year by the district, the annual fixed cost (in millions of dollars) of running a dumping site, and the variable cost (per ton) of processing a ton of trash at a site are shown in Table 112.

[†]Based on Fitzsimmons and Allen (1983).

TABLE **113**

City	Calls Required
San Antonio	2
Phoenix	3
Los Angeles	6
Seattle	3
Detroit	4
Minneapolis	2
Chicago	7
Atlanta	5
New York	9
Boston	5
Philadelphia	4

For example, district 3 is located at coordinates (10,8). District 3 produces 555 tons of trash a year, and it costs $1 million per year in fixed costs to operate a dump site in district 3. Each ton of trash processed at site 3 incurs $51 in variable costs. Each dump site can handle at most 1,500 tons of trash. Each district must send all its trash to a single site. Determine how to locate the dump sites in order to minimize total cost per year.

31 You are the sales manager for Eli Lilly. You want to have sales headquarters located in four of the cities in Table 113. The number of sales calls (in thousands) that must be made in each city are given in Table 113. For example, San Antonio requires 2,000 calls and is 602 miles from Phoenix. The distance between each pair of cities is given in Table 114 and in file Test1.xls. Where should the headquarters be located to minimize the total distance that must be traveled to make the needed calls?

32 Alcoa produces 100-, 200-, and 300-foot-long aluminum ingots for customers. This week's demand for ingots is shown in Table 115.

Alcoa has 4 furnaces in which ingots can be produced. During a week, each furnace can be operated for 50 hours. Because ingots are produced by cutting long strips of aluminum, longer ingots take less time to produce than shorter ingots. If a furnace is devoted completely to producing one type of ingot; the number it can produce in a week is shown in Table 116.

For example, furnace 1 could produce 350 300-foot ingots per week. The material in an ingot costs $10 per foot. If a customer wants a 100- or 200-foot ingot, then she will accept an ingot of that length or longer. How can Alcoa minimize the material costs incurred in meeting required weekly demands?

33[‡] In treating a brain tumor with radiation, physicians want the maximum amount of radiation possible to bombard

the tissue containing the tumors. The constraint is, however, that there is a maximum amount of radiation that normal tissue can handle without suffering tissue damage. Physicians must therefore decide how to aim the radiation so as to maximize the radiation that hits the tumor tissue subject to the constraint of not damaging the normal tissue. As a simple example of this situation, suppose six types of radiation beams (beams differ in where they are aimed and their intensity) can be aimed at a tumor. The region containing the tumor has been divided into six regions: three regions contain tumors and three contain normal tissue. The amount of radiation delivered to each region by each type of beam is shown in Table 117.

If each region of normal tissue can handle at most 40 units of radiation, then which beams should be used to maximize the total amount of radiation received by the tumors?

34 It is currently the beginning of 2003. Gotham City is trying to sell municipal bonds to support improvements in recreational facilities and highways. The face value and due date at which principal comes due of the bonds are in Table 118.

Gold and Silver (GS) wants to underwrite Gotham City's bonds. A proposal to Gotham for underwriting this issue consists of the following:

- An interest rate (3%, 4%, 5%, 6%, or 7%) for each bond. Coupons are paid annually
- An up-front premium paid by GS to Gotham City

GS has determined the fair prices (in thousands) for possible bonds as shown in Table 119.

For example, if GS underwrites the bond maturing in 2006 at 5%, then it would charge Gotham City $444,000 for that bond. GS is constrained to use at most three different interest rates. GS wants to make a profit of at least $46,000. GS profit is given by

(Sales price of bonds) − (Face value of bonds)

− (Premium)

To maximize the chances that GS will get Gotham City's business, GS wants to minimize the total cost of the bond issue to Gotham City. The total cost of the bond issue to Gotham City is given by

(Total interests on bonds) − (Premium)

For example, if the year 2005 bond is issued at a 4% rate, then Gotham City must pay 2 years of coupon interest or $2*(.04)*(\$700,000) = \$56,000$ of interest.

What assignment of interest rates to each bond and up-front premium ensures that GS makes the desired profit (if it gets the contract) and maximizes the chances of GS getting Gotham City's business?

35 When you lease 800-phone numbers from AT&T for telemarketing, AT&T uses a Solver model to tell you where you should locate calling centers to minimize your operating costs over a 10-year horizon. To illustrate the model, suppose you are considering 7 calling center locations: Boston, New York, Charlotte, Dallas, Chicago, L.A., and Omaha. We know the average cost (in dollars) incurred if a telemarketing call is made from any of these cities to any region of the country. We also know the hourly wage that we must pay workers in each city (see Table 120).

We assume that an average call requires 4 minutes. We make calls 250 days per year, and the average number of

[‡]Based on "Radiotherapy Design Using Mathematical Programming Models," by D. Sonderman and P. Abrahamson, *Operations Research*, Vol. 33, No. 4 (1985):705–725.

TABLE 114

	San Antonio	Phoenix	Los Angeles	Seattle	Detroit	Minneapolis	Chicago	Atlanta	New York	Boston	Philadelphia
San Antonio	—	602	1,376	1,780	1,262	1,140	1,060	935	1,848	2,000	1,668
Phoenix	602	—	851	1,193	1,321	1,026	1,127	1,290	2,065	2,201	1,891
Los Angeles	1,376	851	—	971	2,088	1,727	1,914	2,140	2,870	2,995	2,702
Seattle	1,780	1,193	971	—	1,834	1,432	1,734	2,178	2,620	2,707	2,486
Detroit	1,262	1,321	2,088	1,834	—	403	205	655	801	912	654
Minneapolis	1,140	1,026	1,727	1,432	403	—	328	876	1,200	1,304	1,057
Chicago	1,060	1,127	1,914	1,734	205	328	—	564	957	1,082	794
Atlanta	935	1,290	2,140	2,178	655	876	564	—	940	1,096	765
New York	1,848	2,065	2,870	2,620	801	1,200	957	940	—	156	180
Boston	2,000	2,201	2,995	2,707	912	1,304	1,082	1,096	156	—	333
Philadelphia	1,668	1,891	2,702	2,486	654	1,057	794	765	180	333	—

TABLE 115

Ingot (ft)	Demand
100	700
200	300
300	150

TABLE 116

Furnace	Ingot Length 100'	200'	300'
1	230	340	350
2	230	260	280
3	240	300	310
4	200	280	300

TABLE 117

Normal 1	2	3	Tumor 1	2	3	Beam
16	12	8	20	12	6	1
12	10	6	18	15	8	2
9	8	13	13	10	17	3
4	12	12	6	18	16	4
9	4	11	13	5	14	5
8	7	7	10	10	10	6

TABLE 118

Due Date	Principal ($ Thousands)
2005	700
2006	450
2007	250
2008	600
2009	300

TABLE 119

Interest Rate (%)	Amount at Maturity ($ Thousands) 2005	2006	2007	2008	2009
3	695	427	233	504	248
4	701	433	235	522	256
5	715	444	247	548	268
6	731	460	255	575	288
7	750	478	269	605	307

calls made per day to each region of the country is shown in Table 121.

The cost of building a calling center in each possible location is in Table 122.

Each calling center can make as many as 5,000 calls per day. Given this information, how can we minimize the discounted cost (at 10% per year) of running the telemarketing operation for 10 years? Assume all wage and calling costs are paid at the end of each year.

36 Cook County needs to build two hospitals. There are nine cities where the hospitals can be built. The number of hospital visits made annually by the inhabitants of each city and the x and y coordinates of each city are as shown in Table 123.

To minimize the total distance patients must travel to hospitals, where should the hospitals be located? (*Hint:* Use Lookup functions to generate the distances between each pair of cities.)

TABLE 120

Cost Call	New England	Middle Atlantic	Southeast	Southwest	Great Lakes	Plains	Rocky Mountains	Pacific	Hourly Wage ($)
Boston	1.2	1.4	1.1	2.6	2	2.2	2.8	2.2	14
New York	1.3	1	1.3	2.2	1.8	1.9	2.5	2.8	16
Charlotte	1.5	1.4	0.9	1.9	2.1	2.3	2.6	3.3	11
Dallas	2	1.8	1.2	1	1.7	2.2	1.8	2.7	12
Chicago	2.1	1.9	2.3	1.5	0.9	1.3	1.2	2.2	13
LA	2.5	2.1	1.9	1.2	1.7	1.5	1.4	1	18
Omaha	2.2	2.1	2	1.3	1.4	0.6	0.9	1.5	10

TABLE 121

Region	Daily Calls
New England	1,000
Middle Atlantic	2,000
Southeast	2,000
Southwest	2,000
Great Lakes	3,000
Plains	1,000
Rocky Mountain	2,000
Pacific	4,000

TABLE 123

City	x	y	Visits
1	0	0	3,000
2	10	3	4,000
3	12	15	5,000
4	14	13	6,000
5	16	9	4,000
6	18	6	3,000
7	8	12	2,000
8	6	10	4,000
9	4	8	1,200

TABLE 122[†]

City	Building Cost ($ Millions)
Boston	2.7
New York	3
Charlotte	2.1
Dallas	2.1
Chicago	2.4
LA	3.6
Omaha	2.1

[†]Based on Spencer, T., Brigandi, A., Dargon D., and Sheehan, M., "AT&T's Telemarketing Site Selection System Offers Customer Support," *Interfaces,* Vol. 20, no. 1, 1990.

REFERENCES

The following eight texts offer a more advanced discussion of integer programming:

Garfinkel, R., and G. Nemhauser. *Integer Programming.* New York: Wiley, 1972.

Nemhauser, G., and L. Wolsey. *Integer and Combinatorial Optimization.* New York: Wiley, 1999.

Parker, G., and R. Rardin. *Discrete Optimization.* San Diego: Academic Press, 1988.

Salkin, H. *Integer Programming.* Reading, Mass.: Addison-Wesley, 1975.

Schrijver, A. *Theory of Linear and Integer Programming.* New York: Wiley, 1998.

Shapiro, J. *Mathematical Programming: Structures and Algorithms.* New York: Wiley, 1979.

Taha, H. *Integer Programming: Theory, Applications, and Computations.* Orlando, Fla.: Academic Press, 1975. Also details branch-and-bound methods for traveling salesperson problem.

Wolsey, L. *Integer Programming.* New York: Wiley, 1998.

The following three texts contain extensive discussion of the art of formulating integer programming problems:

Plane, D., and C. McMillan. *Discrete Optimization: Integer Programming and Network Analysis for Management Decisions.* Englewood Cliffs, N.J.: Prentice Hall, 1971.

Wagner, H. *Principles of Operations Research,* 2d ed. Englewood Cliffs, N.J.: Prentice Hall, 1975. Also details branch-and-bound methods for traveling salesperson problem.

Williams, H. *Model Building in Mathematical Programming,* 4th ed. New York: Wiley, 1999.

Recently, the techniques of Lagrangian Relaxation and Benders' Decomposition have been used to solve many large integer programming problems. Discussion of these techniques is beyond the scope of the text. The reader interested in Lagrangian Relaxation should read Shapiro (1979), Nemhauser and Wolsey (1988), or

Fisher, M. "An Applications-Oriented Guide to Lagrangian Relaxation," *Interfaces* 15(no. 2, 1985):10–21.

Geoffrion, A. "Lagrangian Relaxation for Integer Programming," in *Mathematical Programming Study 2: Approaches to Integer Programming,* ed. M. Balinski. New York: North-Holland, 1974, pp. 82–114.

The reader interested in Benders' Decomposition should read Shapiro (1979), Taha (1975), Nemhauser and Wolsey (1988), or the following reference:

Geoffrion, A., and G. Graves. "Multicommodity Distribution System Design by Benders' Decomposition," *Management Science* 20(1974):822–844.

Baker, K. *Introduction to Sequencing and Scheduling.* New York: Wiley, 1974. Discusses branch-and-bound methods for traveling salesperson and machine-scheduling problems.

Bean, J., C. Noon, and J. Salton. "Asset Divestiture at Homart Development Company," *Interfaces* 17(no. 1, 1987):48–65.

Bean, J., et al. "Selecting Tenants in a Shopping Mall," *Interfaces* 18(no. 2, 1988):1–10.

Boykin, R. "Optimizing Chemical Production at Monsanto," *Interfaces* 15(no. 1, 1985):88–95.

Brown, G., et al. "Real-Time Wide Area Dispatch of Mobil Tank Trucks," *Interfaces* 17(no. 1, 1987):107–120.

Calloway, R., M. Cummins, and J. Freeland, "Solving Spreadsheet-Based Integer Programming Models: An Example from International Telecommunications," *Decision Sciences* 21(1990):808–824.

Cavalieri, F., A. Roversi, and R. Ruggeri. "Use of Mixed Integer Programming to Investigate Optimal Planning Policy for a Thermal Power Station and Extension to Capacity," *Operational Research Quarterly* 22(1971): 221–236.

Choypeng, P., P. Puakpong, and R. Rosenthal. "Optimal Ship Routing and Personnel Assignment for Naval Recruitment in Thailand," *Interfaces* 16(no. 4, 1986):47–52.

Day, R. "On Optimal Extracting from a Multiple File Data Storage System: An Application of Integer Programming," *Operations Research* 13(1965):482–494.

Eaton, D., et al. "Determining Emergency Medical Service Vehicle Deployment in Austin, Texas," *Interfaces* 15(1985):96–108.

Efroymson, M., and T. Ray. "A Branch-Bound Algorithm for Plant Location," *Operations Research* 14(1966):361–368.

Ellis, P., and R. Corn, "Using Bivalent Integer Programming to Select Teams for Intercollegiate Women's Gymnastics Competition," *Interfaces* 14(1984):41–46.

Fitzsimmons, J., and L. Allen. "A Warehouse Location Model Helps Texas Comptroller Select Out-of-State Audit Offices," *Interfaces* 13 (no. 5, 1983):40–46.

Garfinkel, R. "Minimizing Wallpaper Waste I: A Class of Traveling Salesperson Problems," *Operations Research* 25(1977):741–751.

Garfinkel, R., and G. Nemhauser. "Optimal Political Districting by Implicit Enumeration Techniques," *Management Science* 16(1970):B495–B508.

Gelb, B., and B. Khumawala. "Reconfiguration of an Insurance Company's Sales Regions," *Interfaces* 14(1984):87–94.

Golden, B., L. Bodin, T. Doyle, and W. Stewart. "Approximate Traveling Salesmen Algorithms," *Operations Research* 28(1980):694–712. Contains an excellent discussion of heuristics for the TSP.

Gomory, R. "Outline of an Algorithm for Integer Solutions to Linear Programs," *Bulletin of the American Mathematical Society* 64(1958):275–278.

Hax, A., and D. Candea. *Production and Inventory Management.* Englewood Cliffs, N.J.: Prentice Hall, 1984. Branch-and-bound methods for machine-scheduling problems.

Lawler, L., et al. *The Traveling Salesman Problem.* New York: Wiley, 1985. Everything you ever wanted to know about this problem.

Liggett, R. "The Application of an Implicit Enumeration Algorithm to the School Desegregation Problem," *Management Science* 20(1973):159–168.

Magirou, V.F. "The Efficient Drilling of Printed Circuit Boards," *Interfaces* 16(no. 4, 1984):13–23.

Muckstadt, J., and R. Wilson. "An Application of Mixed Integer Programming Duality to Scheduling Thermal Generating Systems," *IEEE Transactions on Power Apparatus and Systems* (1968):1968–1978.

Peiser, R., and S. Andrus. "Phasing of Income-Producing Real Estate," *Interfaces* 13(1983):1–11.

Salkin, H., and C. Lin. "Aggregation of Subsidiary Firms for Minimal Unemployment Compensation Payments via Integer Programming," *Management Science* 25(1979):405–408.

Shanker, R., and A. Zoltners. "The Corporate Payments Problem," *Journal of Bank Research* (1972):47–53.

Strong, R. "LP Solves Problem: Eases Duration Matching Process," *Pension and Investment Age* 17(no. 26, 1989):21.

Walker, W. "Using the Set Covering Problem to Assign Fire Companies to Firehouses," *Operations Research* 22(1974):275–277.

Westerberg, C., B. Bjorklund, and E. Hultman. "An Application of Mixed Integer Programming in a Swedish Steel Mill," *Interfaces* 7(no. 2, 1977):39–43.

Zangwill, W. "The Limits of Japanese Production Theory," *Interfaces* 22(no. 5, 1992):14–25.

10

Advanced Topics in Linear Programming[†]

In this chapter, we discuss six advanced linear programming topics: the revised simplex method, the product form of the inverse, column generation, the Dantzig–Wolfe decomposition algorithm, the simplex method for upper-bounded variables, and Karmarkar's method for solving LPs. The techniques discussed are often utilized to solve large linear programming problems. The results of Section 6.2 play a key role throughout this chapter.

10.1 The Revised Simplex Algorithm

In Section 6.2, we demonstrated how to create an optimal tableau from an initial tableau, given an optimal set of basic variables. Actually, the results of Section 6.2 can be used to create a tableau corresponding to *any set of basic variables*. To show how to create a tableau for any set of basic variables BV, we first describe the following notation (assume the LP has m constraints):

BV = any set of basic variables (the first element of BV is the basic variable in the first constraint, the second variable in BV is the basic variable in the second constraint, and so on; thus, BV_j is the basic variable for constraint j in the desired tableau)

$\mathbf{b}$ = right-hand-side vector of the original tableau's constraints

$\mathbf{a}_j$ = column for x_j in the constraints of the original problem

B = $m \times m$ matrix whose jth column is the column for BV_j in the original constraints

c_j = coefficients of x_j in the objective function

$\mathbf{c}_{BV}$ = $1 \times m$ row vector whose jth element is the objective function coefficient for BV_j

$\mathbf{u}_i$ = $m \times 1$ column vector with ith element 1 and all other elements equal to zero

Summarizing the formulas of Section 6.2, we write:

$$B^{-1}\mathbf{a}_j = \text{column for } x_j \text{ in BV tableau} \tag{1}$$

$$\mathbf{c}_{BV}B^{-1}\mathbf{a}_j - c_j = \text{coefficient of } x_j \text{ in row 0} \tag{2}$$

$$B^{-1}\mathbf{b} = \text{right-hand side of constraints in BV tableau} \tag{3}$$

$$\mathbf{c}_{BV}B^{-1}\mathbf{u}_i = \text{coefficient of slack variable } s_i \text{ in BV in row 0} \tag{4}$$

[†]This chapter covers topics that may be omitted with no loss of continuity.

$$\mathbf{c}_{BV}B^{-1}(-\mathbf{u}_i) = \text{coefficient of excess variable } e_i \text{ in BV row 0} \tag{5}$$

$$M + \mathbf{c}_{BV}B^{-1}\mathbf{u}_i = \text{coefficient of artificial variable } a_i \text{ in BV row 0} \tag{6}$$
$$\text{(in a max problem)}$$

$$\mathbf{c}_{BV}B^{-1}\mathbf{b} = \text{right-hand side of BV row 0} \tag{7}$$

If we know BV, B^{-1}, and the original tableau, formulas (1)–(7) enable us to compute any part of the simplex tableau for any set of basic variables BV. This means that if a computer is programmed to perform the simplex algorithm, then all the computer needs to store on any pivot is the current set of basic variables, B^{-1}, and the initial tableau. Then (1)–(7) can be used to generate any portion of the simplex tableau. This idea is the basis of the revised simplex algorithm.

We illustrate the revised simplex algorithm by using it to solve the Dakota problem of Chapter 6. Recall that after adding slack variables s_1, s_2, and s_3, the initial tableau (tableau 0) for the Dakota problem is

$$\max z = 60x_1 + 30x_2 + 20x_3$$
$$\text{s.t.} \quad 8x_1 + 6x_2 + x_3 + s_1 \qquad\qquad = 48$$
$$4x_1 + 2x_2 + 1.5x_3 \qquad + s_2 \qquad = 20$$
$$2x_1 + 1.5x_2 + 0.5x_3 \qquad\qquad + s_3 = 8$$

No matter how many pivots have been completed, B^{-1} for the current tableau will simply be the 3×3 matrix whose jth column is the column for s_j in the current tableau. Thus, for the original tableau BV(0), the set of basic variables is given by

$$BV(0) = \{s_1, s_2, s_3\}$$
$$NBV(0) = \{x_1, x_2, x_3\}$$

We let B_i be the columns in the original LP that correspond to the basic variables for tableau i. Then

$$B_0^{-1} = B_0 = \begin{bmatrix} 1 & 0 & 0 \\ 0 & 1 & 0 \\ 0 & 0 & 1 \end{bmatrix}$$

We can now determine which nonbasic variable should enter the basis by computing the coefficient of each nonbasic variable in the current row 0. This procedure is often referred to as **pricing out** the nonbasic variable. From (2)–(5), we see that we can't price out the nonbasic variables until we have determined $\mathbf{c}_{BV}B_0^{-1}$. Because $\mathbf{c}_{BV} = [0 \ \ 0 \ \ 0]$, we have

$$\mathbf{c}_{BV}B_0^{-1} = [0 \ \ 0 \ \ 0]\begin{bmatrix} 1 & 0 & 0 \\ 0 & 1 & 0 \\ 0 & 0 & 1 \end{bmatrix} = [0 \ \ 0 \ \ 0]$$

We now use (2) to price out each nonbasic variable:

$$\bar{c}_1 = [0 \ \ 0 \ \ 0]\begin{bmatrix} 8 \\ 4 \\ 2 \end{bmatrix} - 60 = -60$$

$$\bar{c}_2 = [0 \quad 0 \quad 0] \begin{bmatrix} 6 \\ 2 \\ 1.5 \end{bmatrix} - 30 = -30$$

$$\bar{c}_3 = [0 \quad 0 \quad 0] \begin{bmatrix} 1 \\ 1.5 \\ 0.5 \end{bmatrix} - 20 = -20$$

Because x_1 has the most negative coefficient in the current row 0, x_1 should enter the basis. To continue the simplex, all we need to know about the new tableau is the new set of basic variables, BV(1), and the corresponding B_1^{-1}. To determine BV(1), we find the row in which x_1 enters the basis. We compute the column for x_1 in the current tableau and the right-hand side of the current tableau.

From(1),

$$\text{Column for } x_1 \text{ in current tableau} = \begin{bmatrix} 1 & 0 & 0 \\ 0 & 1 & 0 \\ 0 & 0 & 1 \end{bmatrix} \begin{bmatrix} 8 \\ 4 \\ 2 \end{bmatrix} = \begin{bmatrix} 8 \\ 4 \\ 2 \end{bmatrix}$$

From (3),

$$\text{Right-hand side of current tableau} = \begin{bmatrix} 1 & 0 & 0 \\ 0 & 1 & 0 \\ 0 & 0 & 1 \end{bmatrix} \begin{bmatrix} 48 \\ 20 \\ 8 \end{bmatrix} = \begin{bmatrix} 48 \\ 20 \\ 8 \end{bmatrix}$$

We now use the ratio test to determine the row in which x_1 should enter the basis. The appropriate ratios are row 1, $\frac{48}{8} = 6$; row 2, $\frac{20}{4} = 5$; and row 3, $\frac{8}{2} = 4$. Thus, x_1 should enter the basis in row 3. This means that our new tableau (tableau 1) will have BV(1) = $\{s_1, s_2, x_1\}$ and NBV(1) = $\{s_3, x_2, x_3\}$.

The new B^{-1} will be the columns of s_1, s_2, and s_3 in the new tableau. To determine the new B^{-1}, look at the column in tableau 0 for the entering variable x_1. From this column, we see that in going from tableau 0 to tableau 1, we must perform the following EROs:

1 Multiply row 3 of tableau 0 by $\frac{1}{2}$.

2 Replace row 1 of tableau 0 by -4(row 3 of tableau 0) + row 1 of tableau 0.

3 Replace row 2 of tableau 0 by -2(row 3 of tableau 0) + row 2 of tableau 0.

Applying these EROs to B_0^{-1} yields

$$B_1^{-1} = \begin{bmatrix} 1 & 0 & -4 \\ 0 & 1 & -2 \\ 0 & 0 & \frac{1}{2} \end{bmatrix}$$

We can now price out all the nonbasic variables for the new tableau. First we compute

$$\mathbf{c}_{BV}B_1^{-1} = [0 \quad 0 \quad 60] \begin{bmatrix} 1 & 0 & -4 \\ 0 & 1 & -2 \\ 0 & 0 & \frac{1}{2} \end{bmatrix} = [0 \quad 0 \quad 30]$$

Then use (2) and (4) to price out tableau 1's nonbasic variables:

$$\bar{c}_2 = [0 \quad 0 \quad 30] \begin{bmatrix} 6 \\ 2 \\ 1.5 \end{bmatrix} - 30 = 15$$

$$\bar{c}_3 = [0 \quad 0 \quad 30] \begin{bmatrix} 1 \\ 1.5 \\ 0.5 \end{bmatrix} - 20 = -5$$

$$\text{Coefficient of } s_3 \text{ in row } 0 = [0 \quad 0 \quad 30] \begin{bmatrix} 0 \\ 0 \\ 1 \end{bmatrix} - 0 = 30$$

Because x_3 is the only variable with a negative coefficient in row 0 of tableau 1, we enter x_3 into the basis. To determine the new set of basic variables, BV(2), and the corresponding B_2^{-1}, we find the row in which x_3 enters the basis and compute

$$x_3 \text{ column in tableau } 1 = B_1^{-1}\mathbf{a}_3 = \begin{bmatrix} 1 & 0 & -4 \\ 0 & 1 & -2 \\ 0 & 0 & 0.5 \end{bmatrix} \begin{bmatrix} 1 \\ 1.5 \\ 0.5 \end{bmatrix} = \begin{bmatrix} -1 \\ 0.5 \\ 0.25 \end{bmatrix}$$

$$\text{Right-hand side of tableau } 1 = B_1^{-1}\mathbf{b} = \begin{bmatrix} 1 & 0 & -4 \\ 0 & 1 & -2 \\ 0 & 0 & 0.5 \end{bmatrix} \begin{bmatrix} 48 \\ 20 \\ 8 \end{bmatrix} = \begin{bmatrix} 16 \\ 4 \\ 4 \end{bmatrix}$$

The appropriate ratios for determining where x_3 should enter the basis are row 1, none; row 2, $\frac{4}{0.5} = 8$; and row 3, $\frac{4}{0.25} = 16$. Hence, x_3 should enter the basis in row 2. Then tableau 2 will have BV(2) = $\{s_1, x_3, x_1\}$ and NBV(2) = $\{s_2, s_3, x_2\}$.

To compute B_2^{-1}, note that to make x_3 a basic variable in row 2, we must perform the following EROs on tableau 1:

1 Replace row 2 of tableau 1 by 2(row 2 of tableau 1).

2 Replace row 1 of tableau 1 by 2(row 2 of tableau 1) + row 1 of tableau 1.

3 Replace row 3 of tableau 1 by $-\frac{1}{2}$(row 2 of tableau 1) + row 3 of tableau 1.

Applying these EROs to B_1^{-1}, we obtain

$$B_2^{-1} = \begin{bmatrix} 1 & 2 & -8 \\ 0 & 2 & -4 \\ 0 & -0.5 & 1.5 \end{bmatrix}$$

We now price out the nonbasic variables in tableau 2. First we compute

$$\mathbf{c}_{BV}B_2^{-1} = [0 \quad 20 \quad 60] \begin{bmatrix} 1 & 2 & -8 \\ 0 & 2 & -4 \\ 0 & -0.5 & 1.5 \end{bmatrix} = [0 \quad 10 \quad 10]$$

Then we price out the nonbasic variables x_2, s_2, and s_3:

$$\bar{c}_2 = [0 \quad 10 \quad 10] \begin{bmatrix} 6 \\ 2 \\ 1.5 \end{bmatrix} - 30 = 5$$

$$\text{Coefficient of } s_2 \text{ in row } 0 = [0 \quad 10 \quad 10] \begin{bmatrix} 0 \\ 1 \\ 0 \end{bmatrix} - 0 = 10$$

$$\text{Coefficient of } s_3 \text{ in row } 0 = [0 \quad 10 \quad 10]\begin{bmatrix} 0 \\ 0 \\ 1 \end{bmatrix} - 0 = 10$$

Each nonbasic variable has a nonnegative coefficient in row 0, so tableau 2 is an optimal tableau. To find the optimal solution, we find the right-hand side of tableau 2. From (3), we obtain

$$\text{Right-hand side of tableau } 2 = \begin{bmatrix} 1 & 2 & -8 \\ 0 & 2 & -4 \\ 0 & -0.5 & 1.5 \end{bmatrix} \begin{bmatrix} 48 \\ 20 \\ 8 \end{bmatrix} = \begin{bmatrix} 24 \\ 8 \\ 2 \end{bmatrix}$$

Because $BV(2) = \{s_1, x_3, x_1\}$, the optimal solution to the Dakota problem is

$$\begin{bmatrix} s_1 \\ x_3 \\ x_1 \end{bmatrix} = \begin{bmatrix} 24 \\ 8 \\ 2 \end{bmatrix}$$

or $s_1 = 24$, $x_3 = 8$, $x_1 = 2$, $x_2 = s_2 = s_3 = 0$. The optimal z-value may be found from (7):

$$\mathbf{c}_{BV}B_2^{-1}\mathbf{b} = [0 \quad 10 \quad 10]\begin{bmatrix} 48 \\ 20 \\ 8 \end{bmatrix} = 280$$

A summary of the revised simplex method (for a max problem) follows:

Step 0 Note the columns from which the current B^{-1} will be read. Initially, $B^{-1} = I$.

Step 1 For the current tableau, compute $\mathbf{c}_{BV}B^{-1}$.

Step 2 Price out all nonbasic variables in the current tableau. If each nonbasic variable prices out to be nonnegative, then the current basis is optimal. If the current basis is not optimal, then enter into the basis the nonbasic variable with the most negative coefficient in row 0. Call this variable x_k.

Step 3 To determine the row in which x_k enters the basis, compute x_k's column in the current tableau ($B^{-1}\mathbf{a}_k$) and compute the right-hand side of the current tableau ($B^{-1}\mathbf{b}$). Then use the ratio test to determine the row in which x_k should enter the basis. We now know the set of basic variables (BV) for the new tableau.

Step 4 Use the column for x_k in the current tableau to determine the EROs needed to enter x_k into the basis. Perform these EROs on the current B^{-1}. This will yield the new B^{-1}. Return to step 1.

Most linear programming computer codes use some version of the revised simplex to solve LPs. Knowing the current tableau's B^{-1} and the initial tableau is all that is needed to obtain the next tableau, so the computational effort required to solve an LP by the revised simplex depends primarily on the size of B^{-1}. Suppose the LP being solved has m constraints and n variables. Then each B^{-1} will be an $m \times m$ matrix, and the effort required to solve an LP will depend primarily on the number of constraints (not the number of variables). This fact has important computational implications. For example, if we are solving an LP that has 500 constraints and 10 variables, the LP's dual will have 10 constraints and 500 variables. Then all the B^{-1}'s for the dual will be 10×10 matrices, and all the B^{-1}'s for the primal will be 500×500. Thus, it will be much easier to solve the dual than to solve the primal. In this situation, computation can be greatly reduced by solving the dual and reading the optimal primal solution from the SHADOW PRICE or DUAL VARIABLE section of a computer printout.

PROBLEMS

Group A

Use the revised simplex method to solve the following LPs:

1

$$\max z = 3x_1 + x_2 + x_3$$
$$\text{s.t.} \quad x_1 + x_2 + x_3 \le 6$$
$$2x_1 \quad\quad - x_3 \le 4$$
$$x_2 + x_3 \le 2$$
$$x_1, x_2, x_3 \ge 0$$

2

$$\max z = 4x_1 + x_2$$
$$\text{s.t.} \quad x_1 + x_2 \le 4$$
$$2x_1 + x_2 \ge 6$$
$$3x_2 \ge 6$$
$$x_1, x_2, x_3 \ge 0$$

Remember that B^{-1} is always found under the columns corresponding to the starting basis.)

3

$$\min z = 3x_1 + x_2 - 3x_3$$
$$\text{s.t.} \quad x_1 - x_2 + x_3 \le 4$$
$$x_1 \quad\quad + x_3 \le 6$$
$$2x_2 - x_3 \le 5$$
$$x_1, x_2, x_3 \ge 0$$

10.2 The Product Form of the Inverse

Much of the computation in the revised simplex algorithm is concerned with updating B^{-1} from one tableau to the next. In this section, we develop an efficient method to update B^{-1}.

Suppose we are solving an LP with m constraints. Assume that we have found that x_k should enter the basis, in row r. Let the column for x_k in the current tableau be

$$\begin{bmatrix} \bar{a}_{1k} \\ \bar{a}_{2k} \\ \vdots \\ \bar{a}_{mk} \end{bmatrix}$$

Define the $m \times m$ matrix E:

$$\text{(column } r\text{)}$$

$$E = \begin{bmatrix} 1 & 0 & \cdots & -\dfrac{\bar{a}_{1k}}{\bar{a}_{rk}} & \cdots & 0 & 0 \\ 0 & 1 & \cdots & -\dfrac{\bar{a}_{2k}}{\bar{a}_{rk}} & \cdots & 0 & 0 \\ \vdots & \vdots & & \vdots & & \vdots & \vdots \\ 0 & 0 & \cdots & \dfrac{1}{\bar{a}_{rk}} & \cdots & 0 & 0 \\ \vdots & \vdots & & \vdots & & \vdots & \vdots \\ 0 & 0 & \cdots & -\dfrac{\bar{a}_{m-1,k}}{\bar{a}_{rk}} & \cdots & 1 & 0 \\ 0 & 0 & \cdots & -\dfrac{\bar{a}_{mk}}{\bar{a}_{rk}} & \cdots & 0 & 1 \end{bmatrix} \text{(row } r\text{)}$$

In short, E is simply I_m with column r replaced by the column vector

$$\begin{bmatrix} -\dfrac{\bar{a}_{1k}}{\bar{a}_{rk}} \\[2mm] -\dfrac{\bar{a}_{2k}}{\bar{a}_{rk}} \\[2mm] \vdots \\[2mm] \dfrac{1}{\bar{a}_{rk}} \\[2mm] \vdots \\[2mm] -\dfrac{\bar{a}_{m-1,k}}{\bar{a}_{rk}} \\[2mm] -\dfrac{\bar{a}_{mk}}{\bar{a}_{rk}} \end{bmatrix}$$

DEFINITION ■ A matrix (such as E) that differs from the identity matrix in only one column is called an **elementary matrix.** ■

We now show that

$$B^{-1} \text{ for new tableau} = E(B^{-1} \text{ for current tableau}) \qquad (8)$$

To see why this is true, note that the EROs used to go from the current tableau to the new tableau boil down to

$$\text{Row } r \text{ of new } B^{-1} = \left(\frac{1}{\bar{a}_{rk}}\right)(\text{row } r \text{ of current } B^{-1}) \qquad (9)$$

and for $i \neq r$,

$$\begin{aligned} &\text{Row } i \text{ of new } B^{-1} \\ &= (\text{row } i \text{ of current } B^{-1}) - \left(\frac{\bar{a}_{ik}}{\bar{a}_{rk}}\right)(\text{row } r \text{ of current } B^{-1}) \end{aligned} \qquad (10)$$

Recall from Section 2.1 that

$$\text{Row } i \text{ of } E(\text{current } B^{-1}) = (\text{row } i \text{ of } E)(\text{current } B^{-1}) \qquad (11)$$

Combining (11) with the definition of E, we find that

$$\text{Row } r \text{ of } E(\text{current } B^{-1}) = \left(\frac{1}{\bar{a}_{rk}}\right)(\text{row } r \text{ of current } B^{-1})$$

and for $i \neq r$,

$$\begin{aligned} &\text{Row } i \text{ of } E(\text{current } B^{-1}) \\ &= (\text{row } i \text{ of current } B^{-1}) - \left(\frac{\bar{a}_{ik}}{\bar{a}_{rk}}\right)(\text{row } r \text{ of current } B^{-1}) \end{aligned}$$

Hence, (8) does agree with (9) and (10). Thus, we can use (8) to find the new B^{-1} from the current B^{-1}.

Define the initial tableau to be tableau 0, and let E_i be the elementary matrix E associated with the ith simplex tableau. Recall that $B_0^{-1} = I_m$. We now write

$$B_1^{-1} = E_0 B_0^{-1} = E_0$$

Then

$$B_2^{-1} = E_1 B_1^{-1} = E_1 E_0$$

and, in general,

$$B_k^{-1} = E_{k-1} E_{k-2} \cdots E_1 E_0 \tag{12}$$

Equation (12) is called the **product form of the inverse.** Most linear programming computer codes utilize the revised simplex method and compute successive B^{-1}'s by using the product form of the inverse.

EXAMPLE 1   Product Form of the Inverse

Use the product form of the inverse to compute B_1^{-1} and B_2^{-1} for the Dakota problem that was solved by the revised simplex in Section 10.1.

Solution   Recall that in tableau 0, x_1 entered the basis in row 3. Hence, for tableau 0, $r = 3$ and $k = 1$. For tableau 0,

$$\begin{bmatrix} \bar{a}_{11} \\ \bar{a}_{21} \\ \bar{a}_{31} \end{bmatrix} = \begin{bmatrix} 8 \\ 4 \\ 2 \end{bmatrix}$$

Then

$$E_0 = \begin{bmatrix} 1 & 0 & -\frac{8}{2} \\ 0 & 1 & -\frac{4}{2} \\ 0 & 0 & \frac{1}{2} \end{bmatrix} = \begin{bmatrix} 1 & 0 & -4 \\ 0 & 1 & -2 \\ 0 & 0 & \frac{1}{2} \end{bmatrix}$$

$$B_1^{-1} = \begin{bmatrix} 1 & 0 & -4 \\ 0 & 1 & -2 \\ 0 & 0 & \frac{1}{2} \end{bmatrix} \begin{bmatrix} 1 & 0 & 0 \\ 0 & 1 & 0 \\ 0 & 0 & 1 \end{bmatrix} = \begin{bmatrix} 1 & 0 & -4 \\ 0 & 1 & -2 \\ 0 & 0 & \frac{1}{2} \end{bmatrix}$$

As we proceeded from tableau 1 to tableau 2, x_3 entered the basis in row 2. Hence, in computing E_1, we set $r = 2$ and $k = 3$. To compute E_1, we need to find the column for the entering variable (x_3) in tableau 1:

$$\begin{bmatrix} \bar{a}_{13} \\ \bar{a}_{23} \\ \bar{a}_{33} \end{bmatrix} = B_1^{-1} \mathbf{a}_3 = \begin{bmatrix} 1 & 0 & -4 \\ 0 & 1 & -2 \\ 0 & 0 & 0.5 \end{bmatrix} \begin{bmatrix} 1 \\ 1.5 \\ 0.5 \end{bmatrix} = \begin{bmatrix} -1 \\ 0.5 \\ 0.25 \end{bmatrix}$$

As before, x_3 enters the basis in row 2. Then

$$E_1 = \begin{bmatrix} 1 & -(-\frac{1}{0.5}) & 0 \\ 0 & \frac{1}{0.5} & 0 \\ 0 & -\frac{0.25}{0.50} & 1 \end{bmatrix} = \begin{bmatrix} 1 & 2 & 0 \\ 0 & 2 & 0 \\ 0 & -0.5 & 1 \end{bmatrix}$$

and (as before)

$$B_2^{-1} = E_1 B_1^{-1} = \begin{bmatrix} 1 & 2 & 0 \\ 0 & 2 & 0 \\ 0 & -0.5 & 1 \end{bmatrix} \begin{bmatrix} 1 & 0 & -4 \\ 0 & 1 & -2 \\ 0 & 0 & 0.5 \end{bmatrix} = \begin{bmatrix} 1 & 2 & -8 \\ 0 & 2 & -4 \\ 0 & -0.5 & 1.5 \end{bmatrix}$$

In the next two sections, we use the product form of the inverse in our study of column generation and of the Dantzig–Wolfe decomposition algorithm.

PROBLEM

Group A

For the problems of Section 10.1, use the product form of the inverse to perform the revised simplex method.

10.3 Using Column Generation to Solve Large-Scale LPs

We have already seen that the revised simplex algorithm requires less computation than the simplex algorithm of Chapter 4. In this section, we discuss the method of column generation, devised by Gilmore and Gomory (1961). For LPs that have many variables, column generation can be used to increase the efficiency of the revised simplex algorithm. Column generation is also a very important component of the Dantzig–Wolfe decomposition algorithm, which is discussed in Section 10.4. To explain the idea of column generation, we solve a simple version of the classic *cutting stock problem*.

EXAMPLE 2   Odds and Evens

Woodco sells 3-ft, 5-ft, and 9-ft pieces of lumber. Woodco's customers demand 25 3-ft boards, 20 5-ft boards, and 15 9-ft boards. Woodco, who must meet its demands by cutting up 17-ft boards, wants to minimize the waste incurred. Formulate an LP to help Woodco accomplish its goal, and solve the LP by column generation.

Solution   Woodco must decide how each 17-ft board should be cut. Hence, each decision corresponds to a way in which a 17-ft board can be cut. For example, one decision variable would correspond to a board being cut into three 5-ft boards, which would incur waste of $17 - 15 = 2$ ft. Many possible ways of cutting a board need not be considered. For example, it would be foolish to cut a board into one 9-ft and one 5-ft piece; we could just as easily cut the board into a 9-ft piece, a 5-ft piece, *and* a 3-ft piece. In general, any cutting pattern that leaves 3 ft or more of waste need not be considered because we could use the waste to obtain one or more 3-ft boards. Table 1 lists the sensible ways to cut a 17-ft board.

TABLE 1
Ways to Cut a Board in the Cutting Stock Problem

| | Number of | | | |
Combination	3-ft Boards	5-ft Boards	9-ft Boards	Waste (Feet)
1	5	0	0	2
2	4	1	0	0
3	2	2	0	1
4	2	0	1	2
5	1	1	1	0
6	0	3	0	2

We now define

$$x_i = \text{number of 17-ft boards cut according to combination } i$$

and formulate Woodco's LP:

$$\text{Woodco's waste} + \text{total customer demand} = \text{total length of board cut}$$

Because

$$\text{Total customer demand} = 25(3) + 20(5) + 15(9) = 310 \text{ ft}$$
$$\text{Total length of boards cut} = 17(x_1 + x_2 + x_3 + x_4 + x_5 + x_6)$$

we write

$$\text{Woodco's waste (in feet)} = 17x_1 + 17x_2 + 17x_3 + 17x_4 + 17x_5 + 17x_6 - 310$$

Then Woodco's objective function is to minimize

$$\min z = 17x_1 + 17x_2 + 17x_3 + 17x_4 + 17x_5 + 17x_6 - 310$$

This is equivalent to minimizing

$$17(x_1 + x_2 + x_3 + x_4 + x_5 + x_6)$$

which is equivalent to minimizing

$$x_1 + x_2 + x_3 + x_4 + x_5 + x_6$$

Hence, Woodco's objective function is

$$\min z = x_1 + x_2 + x_3 + x_4 + x_5 + x_6 \tag{13}$$

This means that Woodco can minimize its total waste by minimizing the number of 17-ft boards that are cut.

Woodco faces the following three constraints:

Constraint 1   At least 25 3-ft boards must be cut.

Constraint 2   A t least 20 5-ft boards must be cut.

Constraint 3   At least 15 9-ft boards must be cut.

Because the total number of 3-ft boards that are cut is given by $5x_1 + 4x_2 + 2x_3 + 2x_4 + x_5$, Constraint 1 becomes

$$5x_1 + 4x_2 + 2x_3 + 2x_4 + x_5 \geq 25 \tag{14}$$

Similarly, Constraint 2 becomes

$$x_2 + 2x_3 + x_5 + 3x_6 \geq 20 \tag{15}$$

and Constraint 3 becomes

$$x_4 + x_5 \geq 15 \tag{16}$$

Note that the coefficient of x_i in the constraint for k-ft boards is just the number of k-ft boards yielded if a board is cut according to combination i.

It is clear that the x_i should be required to assume integer values. Despite this fact, in problems with large demands, a near-optimal solution can be obtained by solving the cutting stock problem as an LP and then rounding all fractional variables upward. This procedure may not yield the best possible integer solution, but it usually yields a near-optimal integer solution. For this reason, we concentrate on the LP version of the cutting stock problem. Combining the sign restrictions with (13)–(16), we obtain the following LP:

$$\min z = x_1 + x_2 + x_3 + x_4 + x_5 + x_6$$

$$\begin{array}{rll}
\text{s.t.} \quad 5x_1 + 4x_2 + 2x_3 + 2x_4 + x_5 & \geq 25 & \text{(3-ft constraint)} \\
x_2 + 2x_3 \quad + x_5 + 3x_6 & \geq 20 & \text{(5-ft constraint)} \\
x_4 + x_5 & \geq 15 & \text{(9-ft constraint)} \\
x_1, x_2, x_3, x_4, x_5, x_6 & \geq 0
\end{array}$$

(17)

Note that x_1 only occurs in the 3-ft constraint (because combination 1 yields only 3-ft boards), and x_6 occurs in the 5-ft constraint (because combination 6 yields only 5-ft boards). This means that x_1 and x_6 can be used as starting basic variables for the 3-ft and 5-ft constraints. Unfortunately, none of combinations 1–6 yields only 9-ft boards, so the 9-ft constraint has no obvious basic variable. To avoid having to add an artificial variable to the 9-ft constraint, we define combination 7 to be the cutting combination that yields only one 9-ft board. Also, define x_7 to be the number of boards cut according to combination 7. Clearly, x_7 will be equal to zero in the optimal solution, but inserting x_7 in the starting basis allows us to avoid using the Big M or the two-phase simplex method. Note that the column for x_7 in the LP constraints will be

$$\begin{bmatrix} 0 \\ 0 \\ 1 \end{bmatrix}$$

and a term x_7 will be added to the objective function. We can now use BV $= \{x_1, x_6, x_7\}$ as a starting basis for LP (17). If we let the tableau for this basis be tableau 0, then we have

$$B_0 = \begin{bmatrix} 5 & 0 & 0 \\ 0 & 3 & 0 \\ 0 & 0 & 1 \end{bmatrix}$$

$$B_0^{-1} = \begin{bmatrix} \frac{1}{5} & 0 & 0 \\ 0 & \frac{1}{3} & 0 \\ 0 & 0 & 1 \end{bmatrix}$$

Then

$$\mathbf{c}_{BV}B_0^{-1} = \begin{bmatrix} 1 & 1 & 1 \end{bmatrix} \begin{bmatrix} \frac{1}{5} & 0 & 0 \\ 0 & \frac{1}{3} & 0 \\ 0 & 0 & 1 \end{bmatrix} = \begin{bmatrix} \frac{1}{5} & \frac{1}{3} & 1 \end{bmatrix}$$

If we now priced out each nonbasic variable it would tell us which variable should enter the basis. However, in a large-scale cutting stock problem, there may be thousands of variables, so pricing out each nonbasic variable would be an extremely tedious chore. This is the type of situation in which column generation comes into play. Because we are solving a minimization problem, we want to find a column that will price out positive (have a positive coefficient in row 0). In the cutting stock problem, each column, or variable, represents a combination for cutting up a board: A variable is specified by three numbers: a_3, a_5, and a_9, where a_i is the number of i-ft boards yielded by cutting one 17-ft board according to the given combination. For example, the variable x_2 is specified by $a_3 = 4$, $a_5 = 1$, and $a_9 = 0$. The idea of column generation is to search efficiently for a column that will price out favorably (positive in a min problem and negative in a max problem). For our current basis, a combination specified by a_3, a_5, and a_9 will price out as

$$\mathbf{c}_{BV}B_0^{-1} \begin{bmatrix} a_3 \\ a_5 \\ a_9 \end{bmatrix} - 1 = \frac{1}{5}a_3 + \frac{1}{3}a_5 + a_9 - 1$$

Note that a_3, a_5, and a_9 must be chosen so they don't use more than 17 ft of wood. We also know that a_3, a_5, and a_9 must be nonnegative integers. In short, for any combination, a_3, a_5, and a_9 must satisfy

$$3a_3 + 5a_5 + 9a_9 \leq 17 \quad (a_3 \geq 0, a_5 \geq 0, a_9 \geq 0; a_3, a_5, a_9 \text{ integer}) \tag{18}$$

We can now find the combination that prices out most favorably by solving the following knapsack problem:

$$\max z = \tfrac{1}{5}a_3 + \tfrac{1}{3}a_5 + a_9 - 1$$
$$\text{s.t.} \quad 3a_3 + 5a_5 + 9a_9 \leq 17 \tag{19}$$
$$a_3, a_5, a_9 \geq 0; a_3, a_5, a_9 \text{ integer}$$

Because (19) is a knapsack problem (without 0–1 restrictions on the variables), it can easily be solved by using the branch-and-bound procedure outlined in Section 9.5.

The resulting branch-and-bound tree is given in Figure 1. For example, to solve Problem 6 in Figure 1, we first set $a_5 = 1$ (because $a_5 \geq 1$ is necessary). Then we have 12 ft left in the knapsack, and we choose to make a_9 (the best item) as large as possible. Because $a_9 \geq 1$, we set $a_9 = 1$. This leaves 3 ft, so we set $a_3 = 1$ to fill the knapsack. From Figure 1, we find that the optimal solution to LP (19) is $z = \frac{8}{15}$, $a_3 = a_5 = a_9 = 1$. This corresponds to combination 5 and variable x_5. Hence, x_5 prices out $\frac{8}{15}$, and entering x_5 into the basis will decrease Woodco's waste. To enter x_5 into the basis, we create the right-hand side of the current tableau and the x_5 column of the current tableau.

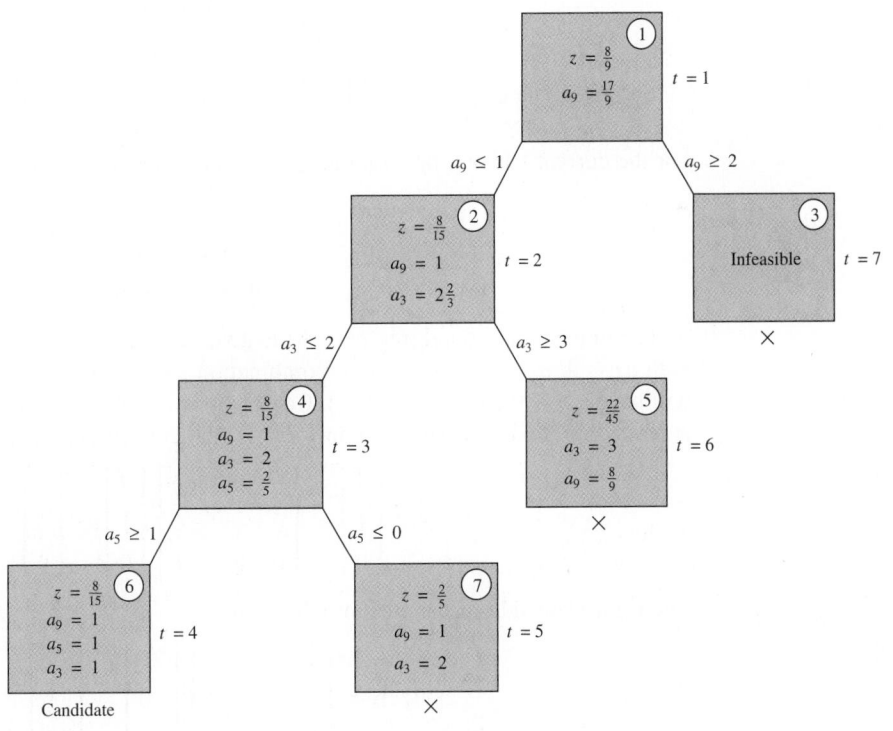

FIGURE 1
Branch-and-Bound Tree for IP (19)

$$x_5 \text{ column in current tableau} = B_0^{-1}\begin{bmatrix} 1 \\ 1 \\ 1 \end{bmatrix} = \begin{bmatrix} \frac{1}{5} & 0 & 0 \\ 0 & \frac{1}{3} & 0 \\ 0 & 0 & 1 \end{bmatrix}\begin{bmatrix} 1 \\ 1 \\ 1 \end{bmatrix} = \begin{bmatrix} \frac{1}{5} \\ \frac{1}{3} \\ 1 \end{bmatrix}$$

$$\text{Right-hand side of current tableau} = B_0^{-1}\mathbf{b} = \begin{bmatrix} \frac{1}{5} & 0 & 0 \\ 0 & \frac{1}{3} & 0 \\ 0 & 0 & 1 \end{bmatrix}\begin{bmatrix} 25 \\ 20 \\ 15 \end{bmatrix} = \begin{bmatrix} 5 \\ \frac{20}{3} \\ 15 \end{bmatrix}$$

The ratio test indicates that x_5 should enter the basis in row 3. This yields BV(1) = $\{x_1, x_6, x_5\}$. Using the product form of the inverse, we obtain

$$B_1^{-1} = E_0 B_0^{-1} = \begin{bmatrix} 1 & 0 & -\frac{1}{5} \\ 0 & 1 & -\frac{1}{3} \\ 0 & 0 & 1 \end{bmatrix}\begin{bmatrix} \frac{1}{5} & 0 & 0 \\ 0 & \frac{1}{3} & 0 \\ 0 & 0 & 1 \end{bmatrix}$$

$$= \begin{bmatrix} \frac{1}{5} & 0 & -\frac{1}{5} \\ 0 & \frac{1}{3} & -\frac{1}{3} \\ 0 & 0 & 1 \end{bmatrix}$$

Now

$$\mathbf{c}_{BV}B_1^{-1} = \begin{bmatrix} 1 & 1 & 1 \end{bmatrix}\begin{bmatrix} \frac{1}{5} & 0 & -\frac{1}{5} \\ 0 & \frac{1}{3} & -\frac{1}{3} \\ 0 & 0 & 1 \end{bmatrix} = \begin{bmatrix} \frac{1}{5} & \frac{1}{3} & \frac{7}{15} \end{bmatrix}$$

With our new set of shadow prices ($\mathbf{c}_{BV}B_1^{-1}$), we can again use column generation to determine whether there is any combination that should be entered into the basis. For the current set of shadow prices, a combination specified by a_3, a_5, and a_9 prices out to

$$\begin{bmatrix} \frac{1}{5} & \frac{1}{3} & \frac{7}{15} \end{bmatrix}\begin{bmatrix} a_3 \\ a_5 \\ a_9 \end{bmatrix} - 1 = \frac{1}{5}a_3 + \frac{1}{3}a_5 + \frac{7}{15}a_9 - 1$$

For the current tableau, the column generation procedure yields the following problem:

$$\begin{aligned} \max z &= \tfrac{1}{5}a_3 + \tfrac{1}{3}a_5 + \tfrac{7}{15}a_9 - 1 \\ \text{s.t.} \quad & 3a_3 + 5a_5 + 9a_9 \le 17 \\ & a_3,\, a_5,\, a_9 \ge 0;\ a_3,\, a_5,\, a_9 \text{ integer} \end{aligned} \qquad \text{(20)}$$

The branch-and-bound tree for (20) is given in Figure 2. We see that the combination with $a_3 = 4$, $a_5 = 1$, and $a_9 = 0$ (combination 2) will price out better than any other (it will have a row 0 coefficient of $\frac{2}{15}$). Combination 2 prices out most favorably, so we now enter x_2 into the basis. The column for x_2 in the current tableau is

$$B_1^{-1}\begin{bmatrix} 4 \\ 1 \\ 0 \end{bmatrix} = \begin{bmatrix} \frac{1}{5} & 0 & -\frac{1}{5} \\ 0 & \frac{1}{3} & -\frac{1}{3} \\ 0 & 0 & 1 \end{bmatrix}\begin{bmatrix} 4 \\ 1 \\ 0 \end{bmatrix} = \begin{bmatrix} \frac{4}{5} \\ \frac{1}{3} \\ 0 \end{bmatrix}$$

The right-hand side of the current tableau is

$$B_1^{-1}\mathbf{b} = \begin{bmatrix} \frac{1}{5} & 0 & -\frac{1}{5} \\ 0 & \frac{1}{3} & -\frac{1}{3} \\ 0 & 0 & 1 \end{bmatrix}\begin{bmatrix} 25 \\ 20 \\ 15 \end{bmatrix} = \begin{bmatrix} 2 \\ \frac{5}{3} \\ 15 \end{bmatrix}$$

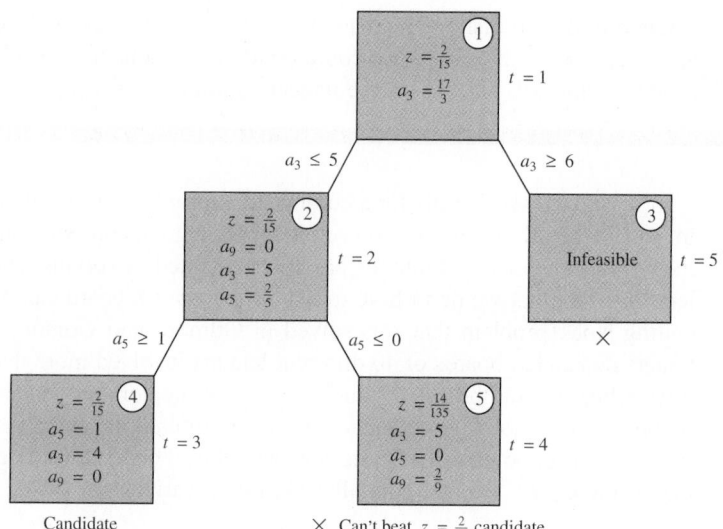

FIGURE 2
Branch-and-Bound
Tree for IP (20)

Node 1: $z = \frac{2}{15}$, $t = 1$, $a_3 = \frac{17}{3}$

$a_3 \le 5$ (left), $a_3 \ge 6$ (right)

Node 2: $z = \frac{2}{15}$, $a_9 = 0$, $a_3 = 5$, $a_5 = \frac{2}{5}$, $t = 2$

Node 3: Infeasible, $t = 5$

$a_5 \ge 1$ (left), $a_5 \le 0$ (right)

Node 4: $z = \frac{2}{15}$, $a_5 = 1$, $a_3 = 4$, $a_9 = 0$, $t = 3$ — Candidate

Node 5: $z = \frac{14}{135}$, $a_3 = 5$, $a_5 = 0$, $a_9 = \frac{2}{9}$, $t = 4$

$\times$ Can't beat $z = \frac{2}{15}$ candidate

The ratio test indicates the x_2 should enter the basis in row 1. Hence, $BV(2) = \{x_2, x_6, x_5\}$. Using the product form of the inverse, we find that

$$E_1 = \begin{bmatrix} \frac{5}{4} & 0 & 0 \\ -\frac{5}{12} & 1 & 0 \\ 0 & 0 & 1 \end{bmatrix}$$

Then

$$B_2^{-1} = E_1 B_1^{-1} = \begin{bmatrix} \frac{5}{4} & 0 & 0 \\ -\frac{5}{12} & 1 & 0 \\ 0 & 0 & 1 \end{bmatrix} \begin{bmatrix} \frac{1}{5} & 0 & -\frac{1}{5} \\ 0 & \frac{1}{3} & -\frac{1}{3} \\ 0 & 0 & 1 \end{bmatrix} = \begin{bmatrix} \frac{1}{4} & 0 & -\frac{1}{4} \\ -\frac{1}{12} & \frac{1}{3} & -\frac{1}{4} \\ 0 & 0 & 1 \end{bmatrix}$$

The new set of shadow prices is given by

$$\mathbf{c}_{BV}B_2^{-1} = \begin{bmatrix} 1 & 1 & 1 \end{bmatrix} \begin{bmatrix} \frac{1}{4} & 0 & -\frac{1}{4} \\ -\frac{1}{12} & \frac{1}{3} & -\frac{1}{4} \\ 0 & 0 & 1 \end{bmatrix} = \begin{bmatrix} \frac{1}{6} & \frac{1}{3} & \frac{1}{2} \end{bmatrix}$$

For this set of shadow prices, a combination specified by a_3, a_5, and a_9 will price out to $\frac{1}{6}a_3 + \frac{1}{3}a_5 + \frac{1}{2}a_9 - 1$. Thus, the column-generation procedure requires us to solve the following problem:

$$\max z = \tfrac{1}{6}a_3 + \tfrac{1}{3}a_5 + \tfrac{1}{2}a_9 - 1$$
$$\text{s.t.} \quad 3a_3 + 5a_5 + 9a_9 \le 17 \tag{21}$$
$$a_3, a_5, a_9 \ge 0; \ a_3, a_5, a_9 \text{ integer}$$

The branch-and-bound tree for IP (21) is left as an exercise (see Problem 1 at the end of this section). The optimal z-value for (21) is found to be $z = 0$. This means that no combination can price out favorably. Hence, our current basic solution must be an optimal solution. To find the values of the basic variables in the optimal solution, we find the right-hand side of the current tableau:

$$B_2^{-1}\mathbf{b} = \begin{bmatrix} \frac{1}{4} & 0 & -\frac{1}{4} \\ -\frac{1}{12} & \frac{1}{3} & -\frac{1}{4} \\ 0 & 0 & 1 \end{bmatrix} \begin{bmatrix} 25 \\ 20 \\ 15 \end{bmatrix} = \begin{bmatrix} \frac{5}{2} \\ \frac{5}{6} \\ 15 \end{bmatrix}$$

Therefore, the optimal solution to Woodco's cutting stock problem is given by $x_2 = \frac{5}{2}$, $x_6 = \frac{5}{6}$, $x_5 = 15$. If desired, we could obtain a "reasonable" integer solution by rounding x_2 and x_6 upward. This yields the integer solution $x_2 = 3$, $x_6 = 1$, $x_5 = 15$.

If we have a starting bfs for a cutting stock problem, we need not list all possible ways in which a board may be cut. At each iteration, a good combination (one that will improve the z-value when entered into the basis) is generated by solving a branch-and-bound problem. The fact that we don't have to list all the ways a board can be cut is very helpful; a cutting stock problem that was solved in Gilmore and Gomory (1961) for which customers demanded boards of 40 different lengths involved more than 100 million possible ways a board could be cut. At the last stage of the column-generation procedure for this problem, solving a single branch-and-bound problem indicated that none of the 100 million (nonbasic) ways would price out favorably. This method is certainly more pleasant than using $\mathbf{c}_{BV}B^{-1}$ to price out all 100 million variables!

PROBLEMS

Group A

1 Show that the optimal solution to IP (21) has $z = 0$.

2 Use column generation to solve a cutting stock problem in which 15-ft boards are cut to satisfy the following requirements: 10 3-ft boards, 20 5-ft boards, and 15 8-ft boards.

3 Use column generation to solve a cutting stock problem in which 15-ft boards are cut to meet the folowing requirements: 80 4-ft boards, 50 6-ft boards, and 100 7-ft boards.

10.4 The Dantzig–Wolfe Decomposition Algorithm

In many LPs, the constraints and variables may be decomposed in the following manner:

Constraints in set 1 only involve variables in Variable set 1.
Constraints in set 2 only involve variables in Variable set 2.

$$\vdots$$

Constraints in set k only involve variables in Variable set k.

Constraints in set $k + 1$ may involve any variable. The constraints in set $k + 1$ are referred to as the **central constraints.** LPs that can be decomposed in this fashion can often be solved efficiently by the Dantzig–Wolfe decomposition algorithm.

EXAMPLE 3 **Decomposition**

Steelco manufactures two types of steel (steel 1 and steel 2) at two locations (plants 1 and 2). Three resources are needed to manufacture a ton of steel: iron, coal, and blast furnace time. The two plants have different types of furnaces, so the resources needed to manufacture a ton of steel depend on the location (see Table 2). Each plant has its own coal mine. Each day, 12 tons of coal are available at plant 1 and 15 tons at plant 2. Coal cannot be shipped between plants. Each day, plant 1 has 10 hours of blast furnace time available, and plant 2 has 4 hours available. Iron ore is mined in a mine located midway between the two plants; 80 tons of iron are available each day. Each ton of steel 1 can be sold for $170/ton, and each ton of steel 2 can be sold for $160/ton. All steel that is sold is shipped to a single customer. It costs $80 to ship a ton of steel from plant 1, and $100

TABLE **2**
Resource Requirements for Steelco

Product (1 Ton)	Iron Required (Tons)	Coal Required (Tons)	Blast Furnace Time Requested (Hours)
Steel 1 at plant 1	8	3	2
Steel 2 at plant 1	6	1	1
Steel 1 at plant 2	7	3	1
Steel 2 at plant 2	5	2	1

a ton from plant 2. Assuming that the only variable cost is the shipping cost, formulate and solve an LP to maximize Steelco's revenues less shipping costs.

Solution   Define

$$x_1 = \text{tons of steel 1 produced daily at plant 1}$$
$$x_2 = \text{tons of steel 2 produced daily at plant 1}$$
$$x_3 = \text{tons of steel 1 produced daily at plant 2}$$
$$x_4 = \text{tons of steel 2 produced daily at plant 2}$$

Steelco's revenue is given by $170(x_1 + x_3) + 160(x_2 + x_4)$, and Steelco's shipping cost is $80(x_1 + x_2) + 100(x_3 + x_4)$. Therefore, Steelco wants to maximize

$$z = (170 - 80)x_1 + (160 - 80)x_2 + (170 - 100)x_3 + (160 - 100)x_4$$
$$= 90x_1 + 80x_2 + 70x_3 + 60x_4$$

Steelco faces the following five constraints:

Constraint 1   At plant 1, no more than 12 tons of coal can be used daily.

Constraint 2   At plant 1, no more than 10 hours of blast furnace time can be used daily.

Constraint 3   At plant 2, no more than 15 tons of coal can be used daily.

Constraint 4   At plant 2, no more than 4 hours of blast furnace time can be used daily.

Constraint 5   At most, 80 tons of iron ore can be used daily.

Constraints 1–5 lead to the following five LP constraints:

$3x_1 + x_2 \leq 12$	(Plant 1 coal constraint)	**(23)**
$2x_1 + x_2 \leq 10$	(Plant 1 furnace constraint)	**(24)**
$3x_3 + 2x_4 \leq 15$	(Plant 2 coal constraint)	**(25)**
$x_3 + x_4 \leq 4$	(Plant 2 furnace constraint)	**(26)**
$8x_1 + 6x_2 + 7x_3 + 5x_4 \leq 80$	(Iron ore constraint)	**(27)**

We also need the sign restrictions $x_i \geq 0$. Putting it all together, we write Steelco's LP as

max $z = 90x_1 + 80x_2 + 70x_3 + 60x_4$

s.t.			
$3x_1 + x_2$	≤ 12	(Plant 1 coal constraint)	**(22)**
$2x_1 + x_2$	≤ 10	(Plant 1 furnace constraint)	**(23)**
$3x_3 + 2x_4 \leq 15$		(Plant 2 coal constraint)	**(24)**
$x_3 + x_4 \leq 4$		(Plant 2 furnace constraint)	**(25)**
$8x_1 + 6x_2 + 7x_3 + 5x_4 \leq 80$		(Iron ore constraint)	**(26)**

$$x_1, x_2, x_3, x_4 \geq 0$$

Using our definition of decomposition, we may decompose the Steelco LP in the following manner:

Variable set 1 x_1 and x_2 (plant 1 variables).

Variable set 2 x_3 and x_4 (plant 2 variables).

Constraint 1   (22) and (23) (plant 1 constraints).

Constraint 2   (24) and (25) (plant 2 constraints).

Constraint 3   (26).

Constraint set 1 and Variable set 1 involve activities at plant 1 and do not involve x_3 and x_4 (which represent plant 2 activities). Constraint set 2 and Variable set 2 involve activities at plant 2 and do not involve x_1 and x_2 (plant 1 activities). Constraint set 3 may be thought of as a centralized constraint that interrelates the two sets of variables. (Solution to be continued.)

Problems in which several plants manufacture several products can easily be decomposed along the lines of Example 3.

To efficiently solve LPs that decompose along the lines of Example 3, Dantzig and Wolfe developed the Dantzig–Wolfe decomposition algorithm. To simplify our discussion of this algorithm, we assume we are solving an LP in which each subproblem has a bounded feasible region.[†] The decomposition algorithm depends on the results in Theorem 1.

THEOREM 1

Suppose the feasible region for an LP is bounded and the extreme points (or basic feasible solutions) of the LP's feasible region are $P_1, P_2, \ldots, P_k$. Then any point $\mathbf{x}$ in the LP's feasible region may be written as a linear combination of $P_1, P_2, \ldots, P_k$. In other words, there exist weights $\mu_1, \mu_2, \ldots, \mu_k$ satisfying

$$\mathbf{x} = \mu_1 P_1 + \mu_2 P_2 + \cdots + \mu_k P_k \tag{27}$$

Moreover, the weights $\mu_1, \mu_2, \ldots, \mu_k$ in (27) may be chosen such that

$$\mu_1 + \mu_2 + \cdots + \mu_k = 1 \quad \text{and} \quad \mu_i \geq 0 \quad \text{for } i = 1, 2, \ldots, k \tag{28}$$

Any linear combination of vectors for which the weights satisfy (28) is called a **convex combination.** Thus, Theorem 1 states that if an LP's feasible region is bounded, then any point within may be written as a convex combination of the extreme points of the LP's feasible region.

We illustrate Theorem 1 by showing how it applies to the LPs defined by Constraint set 1 and Constraint set 2 of Example 3. To begin, we look at the feasible region defined by the sign restrictions $x_1 \geq 0$ and $x_2 \geq 0$ and Constraint set 1 (consisting of (22) and (23)). This feasible region is the interior and the boundary of the shaded quadrilateral $P_1P_2P_3P_4$ in Figure 3. The extreme points are $P_1 = [0 \quad 0]$, $P_2 = [4 \quad 0]$, $P_3 = [2 \quad 6]$, and $P_4 = [0 \quad 10]$. For this feasible region, Theorem 1 states that any point

$$\begin{bmatrix} x_1 \\ x_2 \end{bmatrix}$$

[†]See Bradley, Hax, and Magnanti (1977) for a discussion of decomposition that includes the case where at least one subproblem has an unbounded feasible region.

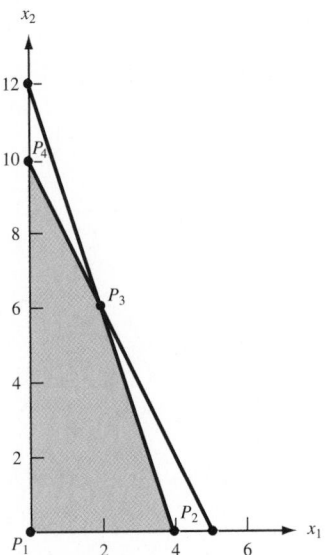

FIGURE 3
Feasible Region for
Constraint Set 1

in the feasible region for Constraint set 1 may be written as

$$\begin{bmatrix} x_1 \\ x_2 \end{bmatrix} = \mu_1 \begin{bmatrix} 0 \\ 0 \end{bmatrix} + \mu_2 \begin{bmatrix} 4 \\ 0 \end{bmatrix} + \mu_3 \begin{bmatrix} 2 \\ 6 \end{bmatrix} + \mu_4 \begin{bmatrix} 0 \\ 10 \end{bmatrix} = \begin{bmatrix} 4\mu_2 + 2\mu_3 \\ 6\mu_3 + 10\mu_4 \end{bmatrix}$$

where $\mu_i \geq 0 (i = 1, 2, 3, 4)$ and $\mu_1 + \mu_2 + \mu_3 + \mu_4 = 1$. For example, the point

$$\begin{bmatrix} 2 \\ 2 \end{bmatrix}$$

is in the feasible region $P_1 P_2 P_3 P_4$. A glance at Figure 3 shows that

$$\begin{bmatrix} 2 \\ 2 \end{bmatrix}$$

may be written as a linear combination of $\mathbf{P}_1, \mathbf{P}_2$, and $\mathbf{P}_3$. A little algebra shows that

$$\begin{bmatrix} 2 \\ 2 \end{bmatrix} = \frac{1}{3} \begin{bmatrix} 0 \\ 0 \end{bmatrix} + \frac{1}{3} \begin{bmatrix} 4 \\ 0 \end{bmatrix} + \frac{1}{3} \begin{bmatrix} 2 \\ 6 \end{bmatrix}$$

As another illustration of Theorem 1, consider the feasible region defined by the sign restrictions $x_3 \geq 0$ and $x_4 \geq 0$ and Constraint set 2 [(24) and (25)]. The feasible region for this LP is the shaded area $Q_1 Q_2 Q_3$ in Figure 4. The extreme points are $\mathbf{Q}_1 = (0, 0)$, $\mathbf{Q}_2 = (4, 0)$, and $\mathbf{Q}_3 = (0, 4)$. Theorem 1 tells us that any point

$$\begin{bmatrix} x_3 \\ x_4 \end{bmatrix}$$

that is in the feasible region for Constraint set 2 may be written as

$$\begin{bmatrix} x_3 \\ x_4 \end{bmatrix} = \mu_1 \begin{bmatrix} 0 \\ 0 \end{bmatrix} + \mu_2 \begin{bmatrix} 4 \\ 0 \end{bmatrix} + \mu_3 \begin{bmatrix} 0 \\ 4 \end{bmatrix}$$

where $\mu_i \geq 0$ and $\mu_1 + \mu_2 + \mu_3 = 1$. For example, the feasible point

$$\begin{bmatrix} 2 \\ 1 \end{bmatrix}$$

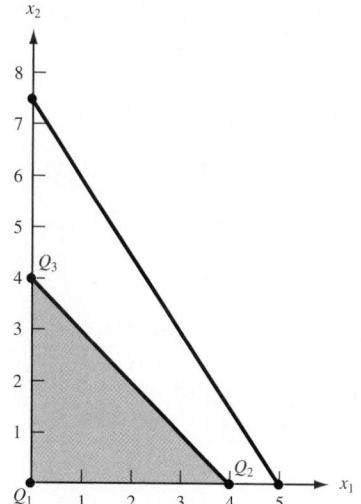

FIGURE 4
Feasible Region for
Constraint Set 2

may be written as

$$\begin{bmatrix} 2 \\ 1 \end{bmatrix} = \frac{1}{4}\begin{bmatrix} 0 \\ 0 \end{bmatrix} + \frac{1}{2}\begin{bmatrix} 4 \\ 0 \end{bmatrix} + \frac{1}{4}\begin{bmatrix} 0 \\ 4 \end{bmatrix}$$

For our purposes, it is not important to know how to determine the set of weights corresponding to a particular feasible point. The decomposition algorithm does not require us to be able to find the weights for an arbitrary point.

To explain the basic ideas of the decomposition algorithm, we assume that the set of variables has been decomposed into set 1 and set 2. The reader should have no trouble generalizing to a situation where the set of variables is decomposed into more than two sets of variables.

The Dantzig–Wolfe decomposition algorithm proceeds as follows:

Step 1 Let the variables in Variable set 1 be $x_1, x_2, \ldots, x_{n_1}$. Express the variables as a convex combination (see Theorem 1) of the extreme points of the feasible region for Constraint set 1 (the constraints that only involve the variables in Variable set 1). If we let $\mathbf{P}_1, \mathbf{P}_2, \ldots, \mathbf{P}_k$ be the extreme points of this feasible region, then any point

$$\begin{bmatrix} x_1 \\ x_2 \\ \vdots \\ x_{n_1} \end{bmatrix}$$

in the feasible region for Constraint set 1 may be written in the form

$$\begin{bmatrix} x_1 \\ x_2 \\ \vdots \\ x_{n_1} \end{bmatrix} = \mu_1\mathbf{P}_1 + \mu_2\mathbf{P}_2 + \cdots + \mu_k\mathbf{P}_k \tag{29}$$

where $\mu_1 + \mu_2 + \cdots + \mu_k = 1$ and $\mu_i \geq 0$ ($i = 1, 2, \ldots, k$).

Step 2 Express the variables in Variable set 2, $x_{n_1+1}, x_{n_1+2}, \ldots, x_n$, as a convex combination of the extreme points of Constraint set 2's feasible region. If we let the extreme points of the feasible region be $\mathbf{Q}_1, \mathbf{Q}_2, \ldots, \mathbf{Q}_m$, then any point in Constraint set 2's feasible region may be written as

$$\begin{bmatrix} x_{n_1+1} \\ x_{n_1+2} \\ \vdots \\ x_n \end{bmatrix} = \lambda_1 \mathbf{Q}_1 + \lambda_2 \mathbf{Q}_2 + \cdots + \lambda_m \mathbf{Q}_m \tag{30}$$

where $\lambda_i \geq 0$ $(i = 1, 2, \ldots, m)$ and $\lambda_1 + \lambda_2 + \cdots + \lambda_m = 1$.

Step 3 Using (29) and (30), express the LP's objective function and centralized constraints in terms of the μ_i's and the λ_i's. After adding the constraints (called convexity constraints) $\mu_1 + \mu_2 + \cdots + \mu_k = 1$ and $\lambda_1 + \lambda_2 + \cdots + \lambda_m = 1$ and the sign restrictions $\mu_i \geq 0$ $(i = 1, 2, \ldots, k)$ and $\lambda_i \geq 0$ $(i = 1, 2, \ldots, m)$, we obtain the following LP, which is referred to as the **restricted master:**

$$\max \text{ (or min) [objective function in terms of } \mu_i\text{'s and } \lambda_i\text{'s]}$$

$$\text{s.t.} \quad \text{[central constraints in terms of } \mu_i\text{'s and } \lambda_i\text{'s]}$$

$$\mu_1 + \mu_2 + \cdots + \mu_k = 1 \quad \text{(Convexity constraints)}$$

$$\lambda_1 + \lambda_2 + \cdots + \lambda_m = 1$$

$$\mu_i \geq 0 \quad (i = 1, 2, \ldots, k) \quad \text{(Sign restrictions)}$$

$$\lambda_i \geq 0 \quad (i = 1, 2, \ldots, m)$$

In many large-scale LPs, the restricted master may have millions of variables (corresponding to the many basic feasible solutions of extreme points for each constraint set). Fortunately, however, we rarely have to write down the entire restricted master; all we need is to generate the column in the restricted master that corresponds to a specific μ_i or λ_i.

Step 4 Assume that a basic feasible solution for the restricted master is readily available.[†] Then use the column generation method of Section 10.3 to solve the restricted master.

Step 5 Substitute the optimal values of the μ_i's and λ_i's found in step 4 into (29) and (30). This will yield the optimal values of $x_1, x_2, \ldots, x_n$.

Solution **Example 3 (Continued)** For Example 3, we have already seen that

$$\text{Variable set } 1 = \{x_1, x_2\} \tag{22}$$

$$\text{Constraint set } 1 = \begin{cases} 3x_1 + x_2 \leq 12 \\ 2x_1 + x_2 \leq 10 \end{cases} \tag{23}$$

We have also seen that the feasible region for Constraint set 1 has four extreme points, and any feasible point

$$\begin{bmatrix} x_1 \\ x_2 \end{bmatrix}$$

for Constraint set 1 may be written as

$$\begin{bmatrix} x_1 \\ x_2 \end{bmatrix} = \mu_1 \begin{bmatrix} 0 \\ 0 \end{bmatrix} + \mu_2 \begin{bmatrix} 4 \\ 0 \end{bmatrix} + \mu_3 \begin{bmatrix} 2 \\ 6 \end{bmatrix} + \mu_4 \begin{bmatrix} 0 \\ 10 \end{bmatrix} = \begin{bmatrix} 4\mu_2 + 2\mu_3 \\ 6\mu_3 + 10\mu_4 \end{bmatrix} \tag{29'}$$

[†]If this is not the case, then the two-phase simplex method must be used. See Bradley, Hax, and Magnanti (1977) for details.

where $\mu_1 + \mu_2 + \mu_3 + \mu_4 = 1$ and $\mu_i \geq 0$.

$$\text{Variable set 2} = x_3 \text{ and } x_4 \tag{24}$$

$$\text{Constraint set 2} = \begin{cases} 3x_3 + 2x_4 \leq 15 \\ x_3 + x_4 \leq 4 \end{cases} \tag{25}$$

Any point

$$\begin{bmatrix} x_3 \\ x_4 \end{bmatrix}$$

in the feasible region for Constraint set 2 may be written as

$$\begin{bmatrix} x_3 \\ x_4 \end{bmatrix} = \lambda_1 \begin{bmatrix} 0 \\ 0 \end{bmatrix} + \lambda_2 \begin{bmatrix} 4 \\ 0 \end{bmatrix} + \lambda_3 \begin{bmatrix} 0 \\ 4 \end{bmatrix} = \begin{bmatrix} 4\lambda_2 \\ 4\lambda_3 \end{bmatrix} \tag{30'}$$

where $\lambda_1 + \lambda_2 + \lambda_3 = 1$ and $\lambda_i \geq 0$ $(i = 1, 2, 3)$.

We now obtain the restricted master by substituting (29') and (30') into the objective function and the centralized constraint. The objective function for (21) becomes

$$90x_1 + 80x_2 + 70x_3 + 60x_4 = 90(4\mu_2 + 2\mu_3) + 80(6\mu_3 + 10\mu_4) + 70(4\lambda_2) + 60(4\lambda_3)$$
$$= 360\mu_2 + 660\mu_3 + 800\mu_4 + 280\lambda_2 + 240\lambda_3$$

The centralized constraint becomes

$$8(4\mu_2 + 2\mu_3) + 6(6\mu_3 + 10\mu_4) + 7(4\lambda_2) + 5(4\lambda_3) \leq 80$$

or

$$32\mu_2 + 52\mu_3 + 60\mu_4 + 28\lambda_2 + 20\lambda_3 \leq 80$$

After adding a slack variable s_1 to this constraint and writing down the convexity constraints and the sign restrictions, we obtain the following restricted master program:

$$\max z = 360\mu_2 + 660\mu_3 + 800\mu_4 + 280\lambda_2 + 240\lambda_3$$
$$\text{s.t.} \quad 32\mu_2 + 52\mu_3 + 60\mu_4 + 28\lambda_2 + 20\lambda_3 + s_1 = 80$$
$$\mu_1 + \mu_2 + \mu_3 + \mu_4 = 1$$
$$\lambda_1 + \lambda_2 + \lambda_3 = 1$$
$$\mu_i, \lambda_i \geq 0$$

There is a more insightful way to obtain the column for a variable in the restricted master. Recall that each variable in the restricted master corresponds to an extreme point for the feasible region of Constraint set 1 or Constraint set 2. As an example, let's focus on how to find the column in the restricted master for a variable μ_i, which corresponds to an extreme point

$$\begin{bmatrix} x_1 \\ x_2 \end{bmatrix}$$

for Constraint set 1. Because x_1 and x_2 correspond to activity at plant 1, we may consider any specification of x_1 and x_2 as a "proposal" from plant 1. For example, the point

$$\begin{bmatrix} 2 \\ 6 \end{bmatrix}$$

corresponds to plant 1 proposing to produce 2 tons of type 1 steel and 6 tons of type 2 steel. Then the weight μ_i may be thought of as a fraction of the proposal corresponding to extreme point $\mathbf{P}_i$ that is included in the actual production schedule. For example, because

$$\begin{bmatrix} 2 \\ 2 \end{bmatrix} = \frac{1}{3}\mathbf{P}_1 + \frac{1}{3}\mathbf{P}_2 + \frac{1}{3}\mathbf{P}_3$$

we may think of

$$\begin{bmatrix} 2 \\ 2 \end{bmatrix}$$

as consisting of one-third of plant 1 proposal $\mathbf{P}_1$, one-third of plant 1 proposal $\mathbf{P}_2$, and one-third of plant 1 proposal $\mathbf{P}_3$.

We can now describe an easy method to determine the column for any variable in the restricted master. Suppose we want to determine the column for the extreme point

$$\begin{bmatrix} x_1 \\ x_2 \end{bmatrix}$$

corresponding to the weight μ_i. If we include a fraction μ_i of the extreme point

$$\begin{bmatrix} x_1 \\ x_2 \end{bmatrix}$$

what will this contribute to the objective function? If $\mu_i = 1$, then

$$\mu_i \begin{bmatrix} x_1 \\ x_2 \end{bmatrix}$$

will contribute $90x_1 + 80x_2$ to the objective function. By the Proportionality Assumption, if we use a fraction μ_i of the extreme point

$$\begin{bmatrix} x_1 \\ x_2 \end{bmatrix}$$

then it will contribute $\mu_i(90x_1 + 80x_2)$ to the objective function. Similarly, if $\mu_i = 1$, then

$$\mu_i \begin{bmatrix} x_1 \\ x_2 \end{bmatrix}$$

will contribute $8x_1 + 6x_2$ of iron usage. Thus, for an arbitrary value of μ_i,

$$\mu_i \begin{bmatrix} x_1 \\ x_2 \end{bmatrix}$$

will contribute an amount $\mu_i(8x_1 + 6x_2)$ to the left-hand side of the iron ore usage constraint.

To be more specific, let's use the reasoning we have just described to determine the column in the restricted master for the weight μ_3 corresponding to the extreme point

$$\begin{bmatrix} 2 \\ 6 \end{bmatrix}$$

Our logic shows that the left-hand side of the objective function involving μ_3 is μ_3 $[90(2) + 80(6)] = 660\mu_3$. Similarly, the term involving μ_3 on the left-hand side of the iron ore constraint will be $\mu_3[8(2) + 6(6)] = 52\mu_3$. Also, μ_3 will have a coefficient of 1 in the first convexity constraint and a zero coefficient in the other convexity constraint. (If the reader understood how we obtained the μ_3 column, there should be little trouble with what follows; readers who are confused should reread the last two pages before continuing.)

We now solve the restricted master by using the revised simplex method and column generation. We refer to our initial tableau as tableau 0. Then BV(0) = $\{s_1, \mu_1, \lambda_1\}$. Also,

$$B_0 = \begin{bmatrix} 1 & 0 & 0 \\ 0 & 1 & 0 \\ 0 & 0 & 1 \end{bmatrix}, \quad \text{so} \quad B_0^{-1} = \begin{bmatrix} 1 & 0 & 0 \\ 0 & 1 & 0 \\ 0 & 0 & 1 \end{bmatrix}$$

Because s_1, μ_1, and λ_1 don't appear in the objective function of the restricted master, we have $\mathbf{c}_{BV} = [0 \quad 0 \quad 0]$, and the tableau 0 shadow prices are given by

$$\mathbf{c}_{BV} B_0^{-1} = [0 \quad 0 \quad 0] \begin{bmatrix} 1 & 0 & 0 \\ 0 & 1 & 0 \\ 0 & 0 & 1 \end{bmatrix} = [0 \quad 0 \quad 0]$$

We now apply the idea of column generation in two stages. First, we determine whether there is any weight μ_i associated with Constraint set 1 that prices out favorably (because we are solving a max problem, a negative coefficient in row 0 is favorable). A weight μ_i associated with an extreme point

$$\begin{bmatrix} x_1 \\ x_2 \end{bmatrix}$$

of Constraint set 1 will have the following column in the restricted master:

$$\text{Objective function coefficient for } \mu_i = 90x_1 + 80x_2$$

$$\text{Column in constraints for } \mu_i = \begin{bmatrix} 8x_1 + 6x_2 \\ 1 \\ 0 \end{bmatrix}$$

From this information, we see that in tableau 0, the column for the weight μ_i corresponding to

$$\begin{bmatrix} x_1 \\ x_2 \end{bmatrix}$$

will price out to

$$\mathbf{c}_{BV} B_0^{-1} \begin{bmatrix} 8x_1 + 6x_2 \\ 1 \\ 0 \end{bmatrix} - (90x_1 + 80x_2) = -90x_1 - 80x_2$$

Since

$$\begin{bmatrix} x_1 \\ x_2 \end{bmatrix}$$

must satisfy Constraint set 1 (or the plant 1 constraints), the weight μ_i that prices out most negatively will be the weight associated with the extreme point that is the optimal solution to the following LP:

Tableau 0
Plant 1 Subproblem

$$\min z = -90x_1 - 80x_2$$
$$\text{s.t.} \quad 3x_1 + x_2 \le 12$$
$$2x_1 + x_2 \le 10$$
$$x_1, x_2 \ge 0$$

Solving the plant 1 subproblem graphically, we obtain the solution $z = -800$, $x_1 = 0$, $x_2 = 10$. This means that the weight μ_i associated with the extreme point

$$\begin{bmatrix} 0 \\ 10 \end{bmatrix}$$

will price out most negatively. Recall that

$$\mathbf{P}_4 = \begin{bmatrix} 0 \\ 10 \end{bmatrix}$$

This means that μ_4 will price out with a coefficient of -800 in the restricted master.

We now look at the weights associated with Constraint set 2 and try to determine the weight λ_i that will price out most negatively. The λ_i corresponding to an extreme point

$$\begin{bmatrix} x_3 \\ x_4 \end{bmatrix}$$

of Constraint set 2 will have the following column in the restricted master:

$$\text{Objective function coefficient for } \lambda_i = 70x_3 + 60x_4$$

$$\text{Column in constraints for } \lambda_i = \begin{bmatrix} 7x_3 + 5x_4 \\ 0 \\ 1 \end{bmatrix}$$

This means that the λ_i corresponding to the extreme point

$$\begin{bmatrix} x_3 \\ x_4 \end{bmatrix}$$

will price out to

$$\mathbf{c}_{BV}B_0^{-1} \begin{bmatrix} 7x_3 + 5x_4 \\ 0 \\ 1 \end{bmatrix} - (70x_3 + 60x_4) = -70x_3 - 60x_4$$

Note that

$$\begin{bmatrix} x_3 \\ x_4 \end{bmatrix}$$

must satisfy Constraint set 2. Thus, the extreme point whose weight λ_i prices out most favorably will be the solution to the following LP:

Tableau 0
Plant 2 Subproblem

$$\min z = -70x_3 - 60x_4$$
$$\text{s.t.} \quad 3x_3 + 2x_4 \leq 15$$
$$x_3 + x_4 \leq 4$$
$$x_3, x_4 \geq 0$$

The optimal solution to this LP is $z = -280$, $x_3 = 4$, $x_4 = 0$. Because

$$\begin{bmatrix} 4 \\ 0 \end{bmatrix} = \mathbf{Q}_2$$

λ_2 prices out the most negatively of all the λ_i's. But μ_4 prices out more negatively than λ_2, so we enter μ_4 into the basis (by using the revised simplex procedure). To do this, we need to find the column for μ_4 in tableau 0 and also find the right-hand side of tableau 0. The column for μ_4 in tableau 0 is

$$B_0^{-1} \begin{bmatrix} 8(0) + 6(10) \\ 1 \\ 0 \end{bmatrix} = \begin{bmatrix} 60 \\ 1 \\ 0 \end{bmatrix}$$

and the right-hand side of tableau 0 is

$$B_0^{-1}\mathbf{b} = \begin{bmatrix} 1 & 0 & 0 \\ 0 & 1 & 0 \\ 0 & 0 & 1 \end{bmatrix} \begin{bmatrix} 80 \\ 1 \\ 1 \end{bmatrix} = \begin{bmatrix} 80 \\ 1 \\ 1 \end{bmatrix}$$

The ratio test now indicates that μ_4 should enter the basis in the second constraint. Then $BV(1) = \{s_1, \mu_4, \lambda_1\}$. Because

$$E_0 = \begin{bmatrix} 1 & -60 & 0 \\ 0 & 1 & 0 \\ 0 & 0 & 1 \end{bmatrix}$$

$$B_1^{-1} = E_0 B_0^{-1} = \begin{bmatrix} 1 & -60 & 0 \\ 0 & 1 & 0 \\ 0 & 0 & 1 \end{bmatrix}$$

The objective function coefficient for μ_4 is $90(0) + 80(10) = 800$, so the new set of shadow prices may be found from

$$\mathbf{c}_{BV}B_1^{-1} = \begin{bmatrix} 0 & 800 & 0 \end{bmatrix} \begin{bmatrix} 1 & -60 & 0 \\ 0 & 1 & 0 \\ 0 & 0 & 1 \end{bmatrix} = \begin{bmatrix} 0 & 800 & 0 \end{bmatrix}$$

We now try to find the weight that prices out most negatively in the current tableau. As before, we solve the current tableau's plant 1 and plant 2 subproblems. Also, as before, a weight μ_i that corresponds to a Constraint 1 extreme point

$$\begin{bmatrix} x_1 \\ x_2 \end{bmatrix}$$

will price out to

$$\mathbf{c}_{BV}B_1^{-1} \begin{bmatrix} 8x_1 + 6x_2 \\ 1 \\ 0 \end{bmatrix} - (90x_1 + 80x_2)$$

$$= \begin{bmatrix} 0 & 800 & 0 \end{bmatrix} \begin{bmatrix} 8x_1 + 6x_2 \\ 1 \\ 0 \end{bmatrix} - (90x_1 + 80x_2) = 800 - 90x_1 - 80x_2$$

Because

$$\begin{bmatrix} x_1 \\ x_2 \end{bmatrix}$$

must satisfy Constraint set 1, the μ_i that prices out most favorably will correspond to the point

$$\begin{bmatrix} x_1 \\ x_2 \end{bmatrix}$$

that solves the following LP:

Tableau 1
Plant 1 Subproblem

$$\min z = 800 - 90x_1 - 80x_2$$
$$\text{s.t.} \quad 3x_1 + x_2 \le 12$$
$$2x_1 + x_2 \le 10$$
$$x_1, x_2 \ge 0$$

The optimal solution to this LP is $z = 0$, $x_1 = 0$, $x_2 = 10$. This means that no μ_i can price out favorably. We now solve the plant 2 subproblem in an effort to find a λ_i that prices out favorably. A λ_i corresponding to an extreme point

$$\begin{bmatrix} x_3 \\ x_4 \end{bmatrix}$$

of Constraint set 2 will price out to

$$\mathbf{c}_{BV}B_1^{-1} \begin{bmatrix} 7x_3 + 5x_4 \\ 0 \\ 1 \end{bmatrix} - (70x_3 + 60x_4) = -70x_3 - 60x_4$$

Because

$$\begin{bmatrix} x_3 \\ x_4 \end{bmatrix}$$

must satisfy the plant 2 constraints, the λ_i that will price out most negatively will correspond to the extreme point

$$\begin{bmatrix} x_3 \\ x_4 \end{bmatrix}$$

that solves the plant 2 subproblem for tableau 1:

Tableau 1
Plant 2 Subproblem

$$\min z = -70x_3 - 60x_4$$
$$\text{s.t.} \quad 3x_3 + 2x_4 \leq 15$$
$$x_3 + x_4 \leq 4$$
$$x_3, x_4 \geq 0$$

The optimal solution to this LP is $x_3 = 4$, $x_4 = 0$, $z = -280$. This means that the λ_i corresponding to

$$\begin{bmatrix} 4 \\ 0 \end{bmatrix}$$

prices out to -280. Because

$$\begin{bmatrix} 4 \\ 0 \end{bmatrix} = \mathbf{Q}_2$$

λ_2 prices out to -280. No μ_i has priced out negatively, so the best we can do is to enter λ_2 into the basis. To enter λ_2 into the basis, we need the column for λ_2 in tableau 1 and the right-hand side for tableau 1. The column for λ_2 in tableau 1 is given by

$$B_1^{-1} \begin{bmatrix} 7(4) + 5(0) \\ 0 \\ 1 \end{bmatrix} = \begin{bmatrix} 1 & -60 & 0 \\ 0 & 1 & 0 \\ 0 & 0 & 1 \end{bmatrix} \begin{bmatrix} 28 \\ 0 \\ 1 \end{bmatrix} = \begin{bmatrix} 28 \\ 0 \\ 1 \end{bmatrix}$$

and the right-hand side of tableau 1 is

$$B_1^{-1}\mathbf{b} = \begin{bmatrix} 1 & -60 & 0 \\ 0 & 1 & 0 \\ 0 & 0 & 1 \end{bmatrix} \begin{bmatrix} 80 \\ 1 \\ 1 \end{bmatrix} = \begin{bmatrix} 20 \\ 1 \\ 1 \end{bmatrix}$$

The ratio test indicates that λ_2 should enter the basis in row 1. Thus, BV(2) = $\{\lambda_2, \mu_4, \lambda_1\}$. Because

$$E_1 = \begin{bmatrix} \frac{1}{28} & 0 & 0 \\ 0 & 1 & 0 \\ -\frac{1}{28} & 0 & 1 \end{bmatrix}$$

$$B_2^{-1} = E_1 B_1^{-1} = \begin{bmatrix} \frac{1}{28} & 0 & 0 \\ 0 & 1 & 0 \\ -\frac{1}{28} & 0 & 1 \end{bmatrix} \begin{bmatrix} 1 & -60 & 0 \\ 0 & 1 & 0 \\ 0 & 0 & 1 \end{bmatrix} = \begin{bmatrix} \frac{1}{28} & -\frac{60}{28} & 0 \\ 0 & 1 & 0 \\ -\frac{1}{28} & \frac{60}{28} & 1 \end{bmatrix}$$

To compute $\mathbf{c}_{BV} B_2^{-1}$, note that λ_2 has a coefficient of $70x_3 + 60x_4 = 70(4) + 60(0) = 280$ in the objective function of the restricted master. Recall that μ_4 has an objective coefficient of 800 in the restricted master objective function, and λ_1 has an objective function coefficient of 0 in the restricted master. Then the new set of shadow prices is

$$\mathbf{c}_{BV} B_2^{-1} = [280 \quad 800 \quad 0] \begin{bmatrix} \frac{1}{28} & -\frac{60}{28} & 0 \\ 0 & 1 & 0 \\ -\frac{1}{28} & \frac{60}{28} & 1 \end{bmatrix} = [10 \quad 200 \quad 0]$$

By solving the plant 1 subproblem for tableau 2, we can determine whether any μ_i prices out favorably. The μ_i corresponding to

$$\begin{bmatrix} x_1 \\ x_2 \end{bmatrix}$$

prices out to

$$\mathbf{c}_{BV} B_2^{-1} \begin{bmatrix} 8x_1 + 6x_2 \\ 1 \\ 0 \end{bmatrix} - (90x_1 + 80x_2)$$

$$= [10 \quad 200 \quad 0] \begin{bmatrix} 8x_1 + 6x_2 \\ 1 \\ 0 \end{bmatrix} - (90x_1 + 80x_2) = 200 - 10x_1 - 20x_2$$

Thus, we have

Tableau 2
Plant 1 Subproblem

$$\min z = 200 - 10x_1 - 20x_2$$
$$\text{s.t.} \quad 3x_1 + x_2 \le 12$$
$$2x_1 + x_2 \le 10$$
$$x_1, x_2 \ge 0$$

The optimal solution to this LP is $z = 0$, $x_1 = 0$, $x_2 = 10$. As before, this means that no μ_i can price out favorably.

To determine whether the λ_i corresponding to the extreme point

$$\begin{bmatrix} x_3 \\ x_4 \end{bmatrix}$$

should be entered into the basis, observe that it prices out to

$$[10 \quad 200 \quad 0] \begin{bmatrix} 7x_3 + 5x_4 \\ 0 \\ 1 \end{bmatrix} - (70x_3 + 60x_4) = -10x_4$$

Because

$$\begin{bmatrix} x_3 \\ x_4 \end{bmatrix}$$

must satisfy Constraint set 2, the λ_i that prices out most favorably will be the λ_i associated with the point

$$\begin{bmatrix} x_3 \\ x_4 \end{bmatrix}$$

that solves the following LP:

Tableau 2
Plant 2 Subproblem

$$\min z = -10x_4$$
$$\text{s.t.} \quad 3x_3 + 2x_4 \le 15$$
$$x_3 + x_4 \le 4$$
$$x_3, x_4 \ge 0$$

This LP has the solution $z = -40$, $x_3 = 0$, $x_4 = 4$. Thus, the λ_i corresponding to

$$\begin{bmatrix} 0 \\ 4 \end{bmatrix} = \mathbf{Q}_3$$

should enter the basis, and λ_3 should be entered into the basis. The λ_3 column in tableau 2 is

$$B_2^{-1} \begin{bmatrix} 7(0) + 5(4) \\ 0 \\ 1 \end{bmatrix} = \begin{bmatrix} \frac{1}{28} & -\frac{60}{28} & 0 \\ 0 & 1 & 0 \\ -\frac{1}{28} & \frac{60}{28} & 1 \end{bmatrix} \begin{bmatrix} 20 \\ 0 \\ 1 \end{bmatrix} = \begin{bmatrix} \frac{20}{28} \\ 0 \\ \frac{8}{28} \end{bmatrix}$$

Tableau 2's right-hand side is

$$B_2^{-1}\mathbf{b} = \begin{bmatrix} \frac{1}{28} & -\frac{60}{28} & 0 \\ 0 & 1 & 0 \\ -\frac{1}{28} & \frac{60}{28} & 1 \end{bmatrix} \begin{bmatrix} 80 \\ 1 \\ 1 \end{bmatrix} = \begin{bmatrix} \frac{20}{28} \\ 1 \\ \frac{8}{28} \end{bmatrix}$$

The ratio test indicates that λ_3 should enter the basis in Constraint 1 or Constraint 3; we arbitrarily choose Constraint 1. Thus, BV(3) = $\{\lambda_3, \mu_4, \lambda_1\}$. Because

$$E_2 = \begin{bmatrix} \frac{28}{20} & 0 & 0 \\ 0 & 1 & 0 \\ -\frac{2}{5} & 0 & 1 \end{bmatrix}$$

$$B_3^{-1} = E_2 B_2^{-1} = \begin{bmatrix} \frac{28}{20} & 0 & 0 \\ 0 & 1 & 0 \\ -\frac{2}{5} & 0 & 1 \end{bmatrix} \begin{bmatrix} \frac{1}{28} & -\frac{60}{28} & 0 \\ 0 & 1 & 0 \\ -\frac{1}{28} & \frac{60}{28} & 1 \end{bmatrix} = \begin{bmatrix} \frac{1}{20} & -3 & 0 \\ 0 & 1 & 0 \\ -\frac{1}{20} & 3 & 1 \end{bmatrix}$$

λ_3 corresponds to

$$\begin{bmatrix} 0 \\ 4 \end{bmatrix}$$

so the coefficient of λ_3 in the objective function of the restricted master is $70x_3 + 60x_4 = 70(0) + 60(4) = 240$. The μ_4 and λ_1 coefficients in the objective function have already been found to be 800 and 0, respectively, so we have $\mathbf{c}_{BV} = [240 \quad 800 \quad 0]$, and the new set of shadow prices is given by

$$\mathbf{c}_{BV}B_3^{-1} = [240 \quad 800 \quad 0] \begin{bmatrix} \frac{1}{20} & -3 & 0 \\ 0 & 1 & 0 \\ -\frac{1}{20} & 3 & 1 \end{bmatrix} = [12 \quad 80 \quad 0]$$

With these shadow prices, the μ_i corresponding to the extreme point

$$\begin{bmatrix} x_1 \\ x_2 \end{bmatrix}$$

will price out to

$$[12 \quad 80 \quad 0] \begin{bmatrix} 8x_1 + 6x_2 \\ 1 \\ 0 \end{bmatrix} - (90x_1 + 80x_2) = 80 + 6x_1 - 8x_2$$

Then we have

Tableau 3
Plant 1 Subproblem

$\min z = 80 + 6x_1 - 8x_2$

s.t. $\quad 3x_1 + x_2 \leq 12$

$\quad\quad 2x_1 + x_2 \leq 10$

$\quad\quad x_1, x_2 \geq 0$

The optimal solution to this LP is $z = 0$, $x_1 = 0$, $x_2 = 10$. Again, this means that no μ_i prices out favorably.

Using the new shadow prices, we now determine whether any λ_i will price out favorably. If no λ_i prices out favorably, then we will have found an optimal tableau. The λ_i corresponding to

$$\begin{bmatrix} x_3 \\ x_4 \end{bmatrix}$$

will price out to

$$[12 \quad 80 \quad 0] \begin{bmatrix} 7x_3 + 5x_4 \\ 0 \\ 1 \end{bmatrix} - (70x_3 + 60x_4) = 14x_3$$

Then we have

Tableau 3
Plant 2 Subproblem

$\min z = 14x_3$

s.t. $\quad 3x_3 + 2x_4 \leq 15$

$\quad\quad x_3 + x_4 \leq 4$

$\quad\quad x_3, x_4 \geq 0$

The optimal solution to this LP is $z = 0$, $x_3 = x_4 = 0$. This means that no λ_i can price out favorably. Because no μ_i or λ_i prices out favorably for tableau 3, tableau 3 must be an optimal tableau for the restricted master. Recall that $BV(3) = \{\lambda_3, \mu_4, \lambda_1\}$. Thus,

$$\begin{bmatrix} \lambda_3 \\ \mu_4 \\ \lambda_1 \end{bmatrix} = B_3^{-1} = \begin{bmatrix} \frac{1}{20} & -3 & 0 \\ 0 & 1 & 0 \\ -\frac{1}{20} & 3 & 1 \end{bmatrix} \begin{bmatrix} 80 \\ 1 \\ 1 \end{bmatrix} = \begin{bmatrix} 1 \\ 1 \\ 0 \end{bmatrix}$$

Thus the optimal solution to the restricted master is $\lambda_3 = 1$, $\mu_4 = 1$, $\lambda_1 = 0$, and all other weights equal 0.

We can now use the representation of the Constraint set 1 feasible region as a convex combination of its extreme points to determine that the optimal value of

$$\begin{bmatrix} x_1 \\ x_2 \end{bmatrix}$$

is given by

$$\begin{bmatrix} x_1 \\ x_2 \end{bmatrix} = 0\mathbf{P}_1 + 0\mathbf{P}_2 + 0\mathbf{P}_3 + \mathbf{P}_4 = \begin{bmatrix} 0 \\ 10 \end{bmatrix}$$

Similarly, we can use the representation of the Constraint set 2 feasible region as a convex combination of its extreme points to determine that the optimal value of

$$\begin{bmatrix} x_3 \\ x_4 \end{bmatrix}$$

is given by

$$\begin{bmatrix} x_3 \\ x_4 \end{bmatrix} = 0\mathbf{Q}_1 + 0\mathbf{Q}_2 + \mathbf{Q}_3 = \begin{bmatrix} 0 \\ 4 \end{bmatrix}$$

Then the optimal solution to Steelco's problem is $x_2 = 10$, $x_4 = 4$, $x_1 = x_3 = 0$, $z = 1040$. Thus, Steelco can maximize its net profit by manufacturing 10 tons of steel 2 at plant 1 and 4 tons of steel 2 at plant 2.

REMARKS 1 If there are k sets of variables, then the restricted master will contain the central constraints and k convexity constraints (one convexity constraint for each set of variables). For each tableau, there will also be k subproblems that must be solved (one for the weights associated with the extreme points of the constraint set corresponding to each set of variables). After solving these subproblems, use the revised simplex algorithm to enter into the basis the weight that prices out most favorably.
2 A major virtue of decomposition is that solving several relatively small LPs is often much easier than solving one large LP. For example, consider an analog of Example 3 in which there are five plants and each plant has 50 constraints. Also suppose that there are 40 central constraints. Then the master problem will involve a 45×45 B^{-1}, and each subproblem will involve a 50×50 B^{-1}. Solving the original LP would involve a 290×290 B^{-1}. Clearly, storing a 290×290 matrix requires more computer memory than storing five 50×50 matrices and a 45×45 matrix. This illustrates how decomposition greatly reduces storage requirements.
3 Decomposition has an interesting economic interpretation. What is the meaning of the shadow prices for the restricted master of Example 3? For each tableau, the shadow price for the central constraint (reflecting the limited amount of iron ore) is the amount by which an extra unit of iron would increase profits. It can be shown that for any tableau, the shadow price for the plant i ($i = 1, 2$) convexity constraint is the profit obtained from the current mix of extreme points being used at plant i less the value of the centralized resource (calculated via the centralized shadow price) required by the current mix of extreme points that is being used at plant i. For example, in tableau 3, the shadow price for the plant 1 convexity constraint is 80. Currently, plant 1 is utilizing the mix $x_1 = 0$ and $x_2 = 10$. This mix yields a profit of $80(10) = \$800$, and it uses $6(10) = 60$ tons of iron worth $60(12) = \$720$. Thus, the plant 1 convexity constraint has a shadow price of $800 - 720 = \$80$. This means that if Δ of the plant 1 weight were taken away, profits would be reduced by 80Δ.

We can now give an economic interpretation of the pricing-out procedure that we use to generate our subproblems. If we are at tableau 3, what are the benefits and costs if we try to introduce the μ_i associated with the extreme point

$$\begin{bmatrix} x_1 \\ x_2 \end{bmatrix}$$

into the basis? Recall that for tableau 3, the iron shadow price is 12 and the plant 1 convexity constraint has a shadow price of 80. In determining whether μ_i should enter the basis, we must balance

$$\text{Increased profits for } \mu_i = \text{profits earned by } \mu_i \begin{bmatrix} x_1 \\ x_2 \end{bmatrix}$$

$$= 90(\mu_i x_1) + 80(\mu_i x_2)$$

against the costs incurred if μ_i is entered into the basis.

If we enter μ_i into the basis, we incur two costs: first, \$12 for each ton of iron used. This amounts to a cost of $12[8(\mu_i x_1) + 6(\mu_i x_2)]$. By entering μ_i into the basis, we are also diverting a fraction μ_i of the available plant 1 weights away from the current mix. This incurs an opportunity cost of $80\mu_i$. Hence,

$$\text{Increase in cost from entering } \mu_i \text{ into basis} = 96\mu_i x_1 + 72\mu_i x_2 + 80\mu_i$$

This means that entering μ_i into the basis can increase profits if and only if

$$90\mu_i x_1 + 80\mu_i x_2 > 96\mu_i x_1 + 72\mu_i x_2 + 80\mu_i$$

Canceling the μ_i's from both sides, we see that μ_i will price out favorably if

$$90x_1 + 80x_2 > 96x_1 + 72x_2 + 80 \qquad \text{or} \qquad 0 > 80 + 6x_1 - 8x_2$$

Thus, the best μ_i will be the μ_i associated with the extreme point

$$\begin{bmatrix} x_1 \\ x_2 \end{bmatrix}$$

that minimizes $80 + 6x_1 - 8x_2$. This is indeed the objective function for the plant 1 tableau 3 subproblem.

This discussion shows that the Dantzig–Wolfe decomposition algorithm combines centralized information (from the shadow prices of the centralized constraints) with local information (the shadow price of each plant's convexity constraint) in an effort to determine which weights should be entered into the basis (or equivalently, which extreme points from each plant should be used).

PROBLEMS

Group A

Use the Dantzig–Wolfe decomposition algorithm to solve the following problems:

1
$$\max z = 7x_1 + 5x_2 + 3x_3$$
$$\text{s.t.} \quad x_1 + 2x_2 + x_3 \leq 10$$
$$x_2 + x_3 \leq 5$$
$$x_1 \qquad \leq 3$$
$$2x_2 + x_3 \leq 8$$
$$x_1, x_2, x_3 \geq 0$$

2
$$\max z = 4x_1 + 2x_2 + 3x_3 + 4x_4 + 2x_5$$
$$\text{s.t.} \quad x_1 + 2x_2 + x_3 \qquad \leq 8$$
$$x_1 + 2x_2 + 2x_3 \qquad \leq 8$$
$$x_4 + x_5 \leq 3$$
$$x_1, x_2, x_3, x_4, x_5 \geq 0$$

3
$$\max z = 3x_1 + 6x_2 + 5x_3$$
$$\text{s.t.} \quad x_1 + 2x_2 + x_3 \leq 4$$
$$2x_1 + 3x_2 + 2x_3 \leq 6$$
$$x_1 + x_2 \qquad \leq 2$$
$$2x_1 + x_2 \qquad \leq 3$$
$$x_1, x_2, x_3 \geq 0$$

(*Hint:* There is no law against having only one set of variables and one subproblem.)

4 Give an economic interpretation to explain why λ_3 priced out favorably in the plant 2 tableau 2 subproblem.

5 Give an example to show why Theorem 1 does not hold for an LP with an unbounded feasible region.

10.5 The Simplex Method for Upper-Bounded Variables

Often, LPs contain many constraints of the form $x_i \leq u_i$ (where u_i is a constant). For example, in a production-scheduling problem, there may be many constraints of the type $x_i \leq u_i$, where

$$x_i = \text{period } i \text{ production}$$
$$u_i = \text{period } i \text{ production capacity}$$

Because a constraint of the form $x_i \leq u_i$ provides an upper bound on x_i, it is called an **upper-bound constraint.** Because $x_i \leq u_i$ is a legal LP constraint, we can clearly use the ordinary simplex method to solve an LP that has upper-bound constraints. However, if an LP contains several upper-bound constraints, then the procedure described in this section (called the simplex method for upper-bounded variables) is much more efficient than the ordinary simplex algorithm.

To efficiently solve an LP with upper-bound constraints, we allow the variable x_i to be nonbasic if $x_i = 0$ (the usual criterion for a nonbasic variable) or if $x_i = u_i$. To accomplish this, we use the following gimmick: For each variable x_i that has an upper-bound constraint $x_i \leq u_i$, we define a new variable x_i' by the relationship $x_i + x_i' = u_i$, or $x_i = u_i - x_i'$. Note that if $x_i = 0$, then $x_i' = u_i$, whereas if $x_i = u_i$, then $x_i' = 0$. Whenever we want x_i to equal its upper bound of u_i, we simply replace x_i by $u_i - x_i'$. This is called an **upper-bound substitution.**

We are now ready to describe the simplex method for upper-bounded variables. We assume that a basic solution is available and that we are solving a max problem. As usual, at each iteration, we choose to increase the variable x_i that has the most negative coefficient in row 0. Three possible occurrences, or bottlenecks, can restrict the amount by which we increase x_i:

Bottleneck 1 x_i cannot exceed its upper bound of u_i.

Bottleneck 2 x_i increases to a point where it causes one of the current basic variables to become negative. The smallest value of x_i that will cause one of the current basic variables to become negative may be found by expressing each basic variable in terms of x_i (recall that we used this idea in Chapter 4, in discussing the simplex algorithm).

Bottleneck 3 x_i increases to a point where it causes one of the current basic variables to exceed its upper bound. As in bottleneck 2, the smallest value of x_i for which this bottleneck occurs can be found by expressing each basic variable in terms of x_i.

Let BN_k ($k = 1, 2, 3$) be the value of x_i where bottleneck k occurs. Then x_i can be increased only to a value of $\min\{BN_1, BN_2, BN_3\}$. The smallest of BN_1, BN_2, and BN_3 is called the winning bottleneck. If the winning bottleneck is BN_1, then we make an upper-bound substitution on x_i by replacing x_i by $u_i - x_i'$. If the winning bottleneck is BN_2, then we enter x_i into the basis in the row corresponding to the basic variable that caused BN_2 to occur. If the winning bottleneck is BN_3, then we make an upper-bound substitution of the variable x_j (by replacing x_j by $u_j - x_j'$) that reaches its upper bound when $x_i = BN_3$. Then we enter x_i into the basis in the row for which x_j was a basic variable.

After following this procedure, we examine the new row 0. If each variable has a nonnegative coefficient in row 0, then we have obtained an optimal tableau. Otherwise, we try to increase the variable with the most negative coefficient in row 0. Our procedure ensures (through BN_1 and BN_3) that no upper-bound constraint is ever violated and (through BN_2) that all of the nonnegativity constraints are satisfied.

EXAMPLE 4   Simplex with Upper Bounds 1

Solve the following LP:

$$\max z = 4x_1 + 2x_2 + 3x_3$$
$$\text{s.t.} \quad 2x_1 + x_2 + x_3 \le 10$$
$$x_1 + \tfrac{1}{2}x_2 + \tfrac{1}{2}x_3 \le 6$$
$$2x_1 + 2x_2 + 4x_3 \le 20$$
$$x_1 \qquad\qquad \le 4$$
$$x_2 \qquad \le 3$$
$$x_3 \le 1$$
$$x_1, x_2, x_3 \ge 0$$

Solution   The initial tableau for this problem is given in Table 3. Because x_1 has the most negative coefficient in row 0, we try to increase x_1 as much as we can. The three bottlenecks for x_1 are computed as follows: x_1 cannot exceed its upper bound of 4, so $BN_1 = 4$. To compute BN_2, we solve for the current set of basic variables in terms of x_1:

$$s_1 = 10 - 2x_1 \qquad (s_1 \ge 0 \text{ iff } x_1 \le 5)$$
$$s_2 = 6 - x_1 \qquad (s_2 \ge 0 \text{ iff } x_1 \le 6)$$
$$s_3 = 20 - 2x_1 \qquad (s_3 \ge 0 \text{ iff } x_1 \le 10)$$

Hence, $BN_2 = \min\{5, 6, 10\} = 5$. The current basic variables ($\{s_1, s_2, s_3\}$) have no upper bounds, so there is no value of BN_3. Then the winning bottleneck is $\min\{4, 5\} = 4 = BN_1$. Thus, we must make an upper-bound substitution on x_1 by replacing x_1 by $4 - x_1'$. The resulting tableau is Table 4.

Because x_3 has the most negative coefficient in row 0, we try to increase x_3 as much as possible. The x_3 bottlenecks are computed as follows: x_3 cannot exceed its upper bound of 1, so $BN_1 = 1$. For BN_2, we solve for the current set of basic variables in terms of x_3:

$$s_1 = 2 - x_3 \qquad (s_1 \ge 0 \text{ iff } x_3 \le 2)$$
$$s_2 = 2 - \tfrac{1}{2}x_3 \qquad (s_2 \ge 0 \text{ iff } x_3 \le 4)$$
$$s_3 = 12 - 4x_3 \qquad (s_3 \ge 0 \text{ iff } x_3 \le 3)$$

Thus, $BN_2 = \min\{2, 4, 3\} = 2$. Because s_1, s_2, and s_3 do not have an upper bound, there is no BN_3. The winning bottleneck is $\min\{1, 2\} = BN_1 = 1$, so we make an upper-bound substitution on x_3 by replacing x_3 by $1 - x_3'$. The resulting tableau is Table 5.

Because x_2 now has the most negative coefficient in row 0, we try to increase x_2. The computation of the bottlenecks follows: For BN_1, x_2 cannot exceed its upper bound of 3, so $BN_1 = 3$. For BN_2,

TABLE 3
Initial Tableau for Example 5

		Basic Variable
$z - 4x_1 - 2x_2 - 3x_3$	$= 0$	$z = 0$
$2x_1 + x_2 + x_3 + s_1$	$= 10$	$s_1 = 10$
$x_1 + \tfrac{1}{2}x_2 + \tfrac{1}{2}x_3 \qquad + s_2$	$= 6$	$s_2 = 6$
$2x_1 + 2x_2 + 4x_3 \qquad\qquad + s_3$	$= 20$	$s_3 = 20$

TABLE 4
Replacing x_1 by $4 - x_1'$

		Basic Variable
$z + 4x_1' - 2x_2 - 3x_3$	$= 16$	$z = 16$
$-2x_1' + x_2 + x_3 + s_1$	$= 2$	$s_1 = 2$
$-x_1' + \frac{1}{2}x_2 + \frac{1}{2}x_3 + s_2$	$= 2$	$s_2 = 2$
$-2x_1' + 2x_2 + 4x_3 + s_3$	$= 12$	$s_3 = 12$

TABLE 5
Replacing x_3 by $1 - x_3'$

		Basic Variable
$z + 4x_1' - 2x_2 + 3x_3'$	$= 19$	$z = 19$
$-2x_1' + x_2 - x_3' + s_1$	$= 1$	$s_1 = 1$
$-x_1' + \frac{1}{2}x_2 - \frac{1}{2}x_3' + s_2$	$= \frac{3}{2}$	$s_2 = \frac{3}{2}$
$-2x_1' + 2x_2 - 4x_3' + s_3$	$= 8$	$s_3 = 8$

TABLE 6
Optimal Tableau for Example 4

		Basic Variable
$z + x_3' + 2s_1$	$= 21$	$z = 21$
$-2x_1' + x_2 - x_3' + s_1$	$= 1$	$x_2 = 1$
$-\frac{1}{2}s_1 + s_2$	$= 1$	$s_2 = 1$
$2x_1' - 2x_3' - 2s_1 + s_3$	$= 6$	$s_3 = 6$

$$s_1 = 1 - x_2 \qquad (s_1 \geq 0 \text{ iff } x_2 \leq 1)$$
$$s_2 = \frac{3}{2} - \frac{1}{2}x_2 \qquad (s_2 \geq 0 \text{ iff } x_2 \leq 3)$$
$$s_3 = 8 - 2x_2 \qquad (s_3 \geq 0 \text{ iff } x_2 \leq 4)$$

Thus, $BN_2 = \min\{1, 3, 4\} = 1$. Note that BN_2 occurs because s_1 is forced to zero. None of the basic variables in the current set has an upper-bound constraint, so there is no BN_3. The winning bottleneck is $\min\{3, 1\} = 1 = BN_2$, so x_2 will enter the basis in the row in which s_1 was a basic variable (row 1). After the pivot is performed, the new tableau is Table 6. Because each variable has a nonnegative coefficient in row 0, this is an optimal tableau. Thus, the optimal solution to the LP is $z = 21$, $s_2 = 1$, $x_2 = 1$, $s_3 = 6$, $x_1' = 0$, $s_1 = 0$, $x_3' = 0$. Because $x_1' = 4 - x_1$ and $x_3' = 1 - x_3$, we also have $x_1 = 4$ and $x_3 = 1$.

EXAMPLE 5   **Simplex with Upper Bounds 2**

Solve the following LP:

$$\max z = 6x_3$$
$$\text{s.t.} \quad x_1 - x_3 = 6$$
$$x_2 + 2x_3 = 8$$
$$x_1 \leq 8, x_2 \leq 10, x_3 \leq 5; x_1, x_2, x_3 \geq 0$$

TABLE 7
Initial Tableau for Example 6

		Basic Variable
z $\qquad -6x_3 = 0$		$z = 0$
$x_1 \quad - \;\; x_3 = 6$		$x_1 = 6$
$x_2 + 2x_3 = 8$		$x_2 = 8$

TABLE 8
Replacing x_1 by $8 - x_1'$

		Basic Variable
$z \qquad -6x_3 = 0$		$z = 0$
$x_1' \quad + \;\;\boxed{x_3} = 2$		$x_1' = 6$
$x_2 + 2x_3 = 8$		$x_2 = 8$

TABLE 9
Optimal Tableau for Example 5

		Basic Variable
$z + 6x_1' \qquad\qquad = 12$		$z = 12$
$x_1' \quad + x_3 = 2$		$x_3 = 2$
$-2x_1' + x_2 \qquad = 4$		$x_2 = 4$

Solution After putting the objective function in our standard row 0 format, we obtain the tableau in Table 7. Fortunately, the basic feasible solution $z = 0$, $x_1 = 6$, $x_2 = 8$, $x_3 = 0$ is readily apparent. We can now proceed with the simplex method for upper-bounded variables. Because x_3 has the most negative coefficient in row 0, we try to increase x_3. Because x_3 cannot exceed its upper bound of 5, $BN_1 = 5$. To compute BN_2,

$$x_1 = 6 + x_3 \qquad (x_1 \geq 0 \text{ iff } x_3 \geq -6)$$
$$x_2 = 8 - 2x_3 \qquad (x_2 \geq 0 \text{ iff } x_3 \leq 4)$$

Thus, all the current basic variables will remain nonnegative as long as $x_3 \leq 4$. Hence, $BN_2 = 4$. For BN_3, note that $x_1 \leq 8$ will hold iff $6 + x_3 \leq 8$, or $x_3 \leq 2$. Also, $x_2 \leq 10$ will hold iff $8 - 2x_3 \leq 10$, or $x_3 \geq -1$. Thus, for $x_3 \leq 2$, each basic variable remains less than or equal to its upper bound, so $BN_3 = 2$. Note that BN_3 occurs when the basic variable x_1 attains its upper bound. The winning bottleneck is $\min\{5, 4, 2\} = 2 = BN_3$, so the largest that we can make x_3 is 2, and the bottleneck occurs because x_1 attains its upper bound of 8. Thus, we make an upper-bound substitution on x_1 by replacing x_1 by $8 - x_1'$. The resulting tableau is

$$z \qquad\qquad\quad -6x_3 = 0$$
$$-x_1' \qquad -\;\; x_3 = -2$$
$$x_2 + 2x_3 = 8$$

After rewriting $-x_1' - x_3 = -2$ as $x_1' + x_3 = 2$, we obtain the tableau in Table 8.

Because x_1, the variable that caused BN_3, was basic in row 1, we now make x_3 a basic variable in row 1. After the pivot, we obtain the tableau in Table 9, which is optimal. Thus, the optimal solution to the LP is $z = 12$, $x_3 = 2$, $x_2 = 4$, $x_1' = 0$. Because $x_1' = 0$, $x_1 = 8 - x_1' = 8$.

To illustrate the efficiencies obtained by using the simplex algorithm with upper bounds, suppose we are solving an LP (call it LP 1) with 100 variables, each having an upper-bound constraint, with five other constraints. If we were to solve LP 1 by the revised simplex method, the B^{-1} for each tableau would be a 105×105 matrix. If we were to use the simplex method for upper-bounded variables, however, the B^{-1} for each tableau would be only a 5×5 matrix. Although the computation of the winning bottleneck in each iteration is more complicated than the ordinary ratio test, solving LP 1 by the simplex method for upper-bounded variables would still be much more efficient than by the ordinary revised simplex.

PROBLEMS

Use the upper-bounded simplex algorithm to solve the following LPs:

Group A

1

$$\max z = 4x_1 + 3x_2 + 5x_3$$

s.t.
$$2x_1 + 2x_2 + x_3 + x_4 \quad\quad \leq 9$$
$$4x_1 - x_2 - x_3 \quad\quad + x_5 \leq 6$$
$$2x_2 + x_3 \quad\quad\quad\quad \leq 5$$
$$x_1 \quad\quad\quad\quad\quad\quad \leq 2$$
$$x_2 \quad\quad\quad\quad\quad \leq 3$$
$$x_3 \quad\quad\quad \leq 4$$
$$x_4 \quad \leq 5$$
$$x_5 \leq 7$$
$$x_1, x_2, x_3, x_4, x_5 \geq 0$$

2

$$\min z = -4x_1 - 9x_2$$

s.t.
$$3x_1 + 5x_2 \leq 6$$
$$5x_1 + 6x_2 \leq 10$$
$$2x_1 - 3x_2 \leq 4$$
$$x_1 \quad\quad \leq 2$$
$$x_2 \leq 1$$
$$x_1, x_2 \geq 0$$

3

$$\max z = 4x_1 + 3x_2$$

s.t.
$$2x_1 - x_2 \leq 1$$
$$x_1 + 6x_2 \leq 6$$
$$x_2 \leq 5$$
$$x_1, x_2 \geq 0$$

4 Suppose an LP contained lower-bound constraints of the following form: $x_j \geq L_j$. Suggest an algorithm that could be used to solve such a problem efficiently.

10.6 Karmarkar's Method for Solving LPs

As discussed in Section 4.13, Karmarkar's method for solving LPs is a polynomial time algorithm. This is in contrast to the simplex algorithm, an exponential time algorithm. Unlike the ellipsoid method (another polynomial time algorithm), Karmarkar's method appears to solve many LPs faster than does the simplex algorithm. In this section, we give a description of the basic concepts underlying Karmarkar's method. Note that several versions of Karmarkar's method are computationally more efficient than the version we describe; our goal is simply to introduce the reader to the exciting ideas used in Karmarkar's method. For a more detailed description of Karmarkar's method, see Hooker (1986), Parker and Rardin (1988), and Murty (1989).

Karmarkar's method is applied to an LP in the following form:

$$\min z = \mathbf{cx}$$
$$\text{s.t.} \quad A\mathbf{x} = \mathbf{0}$$
$$x_1 + x_2 + \cdots + x_n = 1 \tag{31}$$
$$\mathbf{x} \geq \mathbf{0}$$

In (31), $\mathbf{x} = [x_1 \quad x_2 \quad \cdots \quad x_n]^T$, A is an $m \times n$ matrix, $\mathbf{c} = [c_1 \quad c_2 \quad \cdots \quad c_n]$ and $\mathbf{0}$ is an n-dimensional column vector of zeros. The LP must also satisfy

$$[\tfrac{1}{n} \quad \tfrac{1}{n} \quad \cdots \quad \tfrac{1}{n}]^T \qquad \text{is feasible} \tag{32}$$

$$\text{Optimal } z\text{-value} = 0 \tag{33}$$

Although it may seem unlikely that an LP would have the form (31) and satisfy (32)–(33), it is easy to show that any LP may be put in a form such that (31)–(33) are satisfied. We will demonstrate this at the end of this section.

The following three concepts play a key role in Karmarkar's method:

1 Projection of a vector onto the set of $\mathbf{x}$ satisfying $A\mathbf{x} = \mathbf{0}$

2 Karmarkar's centering transformation

3 Karmarkar's potential function

We now discuss the first two concepts, leaving a discussion of Karmarkar's potential function to the end of the section. Before discussing the ideas just listed, we need a definition.

DEFINITION ■ The ***n*-dimensional unit simplex** S is the set of points $[x_1 \quad x_2 \quad \cdots \quad x_n]^T$ satisfying $x_1 + x_2 + \cdots + x_n = 1$ and $x_j \geq 0, j = 1, 2, \ldots, n.$ ■

Projection

Suppose we are given a point $\mathbf{x}^0$ that is feasible for (31), and we want to move from $\mathbf{x}^0$ to another feasible point (call it $\mathbf{x}^1$) that, for some fixed vector $\mathbf{v}$, will have a larger value of $\mathbf{vx}$. Suppose that we find $\mathbf{x}^1$ by moving away from $\mathbf{x}^0$ in a direction $\mathbf{d} = [d_1 \quad d_2 \quad \cdots \quad d_n]$. For $\mathbf{x}^1$ to be feasible, $\mathbf{d}$ must satisfy $A\mathbf{d} = \mathbf{0}$ and $d_1 + d_2 + \cdots + d_n = 0$. If we choose the direction $\mathbf{d}$ that solves the optimization problem

$$\max \mathbf{vd}$$
$$\text{s.t.} \qquad A\mathbf{d} = \mathbf{0}$$
$$d_1 + d_2 + \cdots + d_n = 0$$
$$\|\mathbf{d}\| = 1$$

then we will be moving in the "feasible" direction that maximizes the increase in $\mathbf{vx}$ per unit of length moved. The direction $\mathbf{d}$ that solves this optimization problem is given by the **projection** of $\mathbf{v}$ onto the set of $\mathbf{x} = [x_1 \quad x_2 \quad \cdots \quad x_n]^T$ satisfying $A\mathbf{x} = \mathbf{0}$ and $x_1 + x_2 + \cdots + x_n = 0$. The projection of $\mathbf{v}$ onto the set of $\mathbf{x}$ satisfying $A\mathbf{x} = \mathbf{0}$ and $x_1 + x_2 + \cdots + x_n = 0$ is given by $[I - B^T(BB^T)^{-1}B]\mathbf{v}$, where B is the $(m + 1) \times n$ matrix whose first m rows are A and whose last row is a vector of 1's.

Geometrically, what does it mean to project a vector $\mathbf{v}$ onto the set of $\mathbf{x}$ satisfying $A\mathbf{x} = \mathbf{0}$? It can be shown that any vector $\mathbf{v}$ may be written (uniquely) in the form $\mathbf{v} = \mathbf{p} + \mathbf{w}$, where $\mathbf{p}$ satisfies $A\mathbf{p} = \mathbf{0}$ and $\mathbf{w}$ is perpendicular to all vectors $\mathbf{x}$ satisfying $A\mathbf{x} = \mathbf{0}$. Then $\mathbf{p}$ is the projection of $\mathbf{v}$ onto the set of $\mathbf{x}$ satisfying $A\mathbf{x} = \mathbf{0}$. An example of this idea is given in Figure 5, where $\mathbf{v} = [-2 \quad -1 \quad 7]$ is projected onto the set of three-dimensional vectors satisfying $x_3 = 0$ (the x_1–x_2-plane). In this case, we decompose $\mathbf{v}$ as $\mathbf{v} = [-2 \quad -1 \quad 0] + [0 \quad 0 \quad 7]$. Thus, $\mathbf{p} = [-2 \quad -1 \quad 0]$. It is easy to show that $\mathbf{p}$ is the vector in the set of $\mathbf{x}$ satisfying $A\mathbf{x} = \mathbf{0}$ that is "closest" to $\mathbf{v}$. This is apparent from Figure 5.

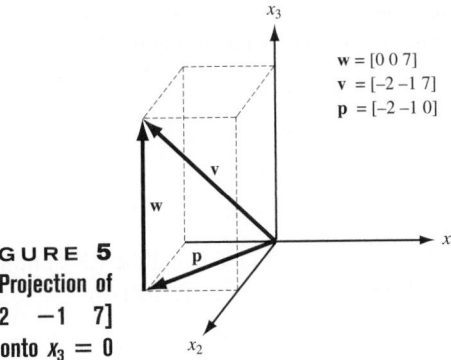

FIGURE 5
Projection of
$[-2 \quad -1 \quad 7]$
onto $x_3 = 0$

$\mathbf{w} = [0\ 0\ 7]$
$\mathbf{v} = [-2\ -1\ 7]$
$\mathbf{p} = [-2\ -1\ 0]$

Karmarkar's Centering Transformation

Given a feasible point (in (31)) $\mathbf{x}^k = [x_1^k \quad x_2^k \quad \cdots \quad x_n^k]$ in S having $x_j^k > 0, j = 1, 2, \ldots, n$, we write the **centering transformation** associated with the point $\mathbf{x}^k$ as $f([x_1 \quad x_2 \quad \cdots \quad x_n] \| \mathbf{x}^k)$. If $\mathbf{x}^k$ is a point in S, then $f([x_1 \quad x_2 \quad \cdots \quad x_n] \| \mathbf{x}^k)$ transforms a point $[x_1 \quad x_2 \quad \cdots \quad x_n]^T$ in S into a point $[y_1 \quad y_2 \quad \cdots \quad y_n]^T$ in S, where

$$y_j = \frac{\dfrac{x_j}{x_j^k}}{\displaystyle\sum_{r=1}^{r=n} \dfrac{x_r}{x_r^k}} \tag{34}$$

Let $\text{Diag}(\mathbf{x}^k)$ be the $n \times n$ matrix with all off-diagonal entries equal to 0 and $\text{Diag}(\mathbf{x}^k)_{ii} = x_i^k$. The centering transformation specified by (34) can be shown to have the properties listed in Lemma 1.

LEMMA 1

Karmarkar's centering transformation has the following properties:

$$f(\mathbf{x}^k \mid \mathbf{x}^k) = [\tfrac{1}{n} \quad \tfrac{1}{n} \quad \cdots \quad \tfrac{1}{n}]^T \tag{35}$$

For $\mathbf{x} \neq \mathbf{x}'$, $\quad f(\mathbf{x} \mid \mathbf{x}^k) \neq f(\mathbf{x}' \mid \mathbf{x}^k) \tag{36}$

$f(\mathbf{x} \mid \mathbf{x}^k) \in S \tag{37}$

For any point $[y_1 \quad y_2 \quad \cdots \quad y_n]^T$ in S, there is a unique point $\tag{38}$
$[x_1 \quad x_2 \quad \cdots \quad x_n]^T$ in S satisfying
$f([x_1 \quad x_2 \quad \cdots \quad x_n]^T \mid \mathbf{x}^k) = [y_1 \quad y_2 \quad \cdots \quad y_n]^T \tag{38'}$

The point $[x_1 \quad x_2 \quad \cdots \quad x_n]^T$ is given by

$$x_j = \frac{x_j^k y_j}{\displaystyle\sum_{r=1}^{r=n} x_r^k y_r}$$

If $[x_1 \quad x_2 \quad \cdots \quad x_n]^T$ and $[y_1 \quad y_2 \quad \cdots \quad y_n]^T$ satisfy (38'), we write $f^{-1}([y_1 \quad y_2 \quad \cdots \quad y_n]^T \mid \mathbf{x}^k) = [x_1 \quad x_2 \quad \cdots \quad x_n]^T$.

A point $\mathbf{x}$ in S will satisfy $A\mathbf{x} = \mathbf{0}$ $\quad$ if $\quad$ $A[\text{Diag}(\mathbf{x}^k)]f(\mathbf{x} \mid \mathbf{x}^k) = \mathbf{0}$ $\quad$ (39)

(See Problem 5 for a proof of Lemma 1.)

To illustrate the centering transformation, consider the following LP:

$$\min z = x_1 + 3x_2 - 3x_3$$

$$\text{s.t.} \quad x_2 - x_3 = 0$$

$$x_1 + x_2 + x_3 = 1 \tag{40}$$

$$x_i \geq 0$$

This LP is of the form (31); the point $\left[\frac{1}{3} \quad \frac{1}{3} \quad \frac{1}{3}\right]^T$ is feasible, and the LP's optimal z-value is 0. The feasible point $\left[\frac{1}{4} \quad \frac{3}{8} \quad \frac{3}{8}\right]$ yields the following transformation:

$$f([x_1 \quad x_2 \quad x_3] \mid [\tfrac{1}{4} \quad \tfrac{3}{8} \quad \tfrac{3}{8}])$$

$$= \left[\frac{4x_1}{4x_1 + \dfrac{8x_2}{3} + \dfrac{8x_3}{3}} \quad \frac{\dfrac{8x_2}{3}}{4x_1 + \dfrac{8x_2}{3} + \dfrac{8x_3}{3}} \quad \frac{\dfrac{8x_3}{3}}{4x_1 + \dfrac{8x_2}{3} + \dfrac{8x_3}{3}} \right]$$

For example,

$$f([\tfrac{1}{3} \quad \tfrac{1}{3} \quad \tfrac{1}{3}] \mid [\tfrac{1}{4} \quad \tfrac{3}{8} \quad \tfrac{3}{8}]) = [\tfrac{12}{28} \quad \tfrac{8}{28} \quad \tfrac{8}{28}]$$

We now refer to the variables $x_1, x_2, \ldots, x_n$ as being the *original* space and the variables $y_1, y_2, \ldots, y_n$ as being the *transformed* space. The unit simplex involving variables $y_1, y_2, \ldots, y_n$ will be called the transformed unit simplex. We now discuss the intuitive meaning of (35)–(39). Equation (35) implies that $f(\cdot \mid x^k)$ maps x^k into the "center" of the transformed unit simplex. Equations (36)–(37) imply that any point in S is transformed into a point in the transformed unit simplex, and no two points in S can yield the same point in the transformed unit simplex (that is, f is a one-to-one mapping). Equation (38) implies that for any point y in the transformed unit simplex, there is a point x in S that is transformed into y. The formula for the x that is transformed into y is also given. Thus, (36)–(38) imply that f is a one-to-one and an onto mapping from S to S. Finally, (39) states that feasible points in the original problem correspond to points y in the transformed unit simplex that satisfy $A[\text{Diag}(x^k)]y = 0$.

Description and Example of Karmarkar's Method

We assume that we will be satisfied with a feasible point having an optimal z-value $< \epsilon$ (for some small ϵ). Karmarkar's method proceeds as follows:

Step 1 Begin at the feasible point $x^0 = \left[\frac{1}{n} \quad \frac{1}{n} \quad \cdots \quad \frac{1}{n}\right]^T$ and set $k = 0$.

Step 2 Stop if $cx^k < \epsilon$. If not, go to step 3.

Step 3 Find the new point $y^{k+1} = [y_1^{k+1} \quad y_2^{k+1} \quad \cdots \quad y_n^{k+1}]^T$ in the transformed unit simplex given by

$$y^{k+1} = \left[\tfrac{1}{n} \quad \tfrac{1}{n} \quad \cdots \quad \tfrac{1}{n}\right]^T - \frac{\theta(I - P^T(PP^T)^{-1}P)[\text{Diag}(x^k)]c^T}{\|c_p\|\sqrt{n(n-1)}}$$

Here, $\|c_p\| = $ the length of $(I - P^T(PP^T)^{-1}P)[\text{Diag}(x^k)]c^T$, P is the $(m+1) \times n$ matrix whose first m rows are $A[\text{Diag}(x^k)]$ and whose last row is a vector of 1's, and $0 < \theta < 1$ is chosen to ensure convergence of the algorithm. $\theta = \frac{1}{4}$ is known to ensure convergence.

Now obtain a new point x^{k+1} in the original space by using the centering transformation to determine the point corresponding to y^{k+1}. That is, $x^{k+1} = f^{-1}(y^{k+1} \mid x^k)$. Increase k by 1 and return to step 2.

REMARKS **1** In step 3, we move from the "center" of the transformed unit simplex in a direction opposite to the projection of $\text{Diag}(\mathbf{x}^k)\mathbf{c}^T$ onto the transformation of the feasible region (the set of $\mathbf{y}$ satisfying $A[\text{Diag}(\mathbf{x}^k)]\mathbf{y} = \mathbf{0}$). From our discussion of the projection, this ensures that we maintain feasibility (in the transformed space) and move in a direction that maximizes the rate of decrease of $[\text{Diag}(\mathbf{x}^k)]\mathbf{c}^T$.
2 By moving a distance

$$\frac{\theta}{\sqrt{n(n-1)}}$$

from the center of the transformed unit simplex, we ensure that $\mathbf{y}^{k+1}$ will remain in the interior of the transformed unit simplex.
3 When we use the inverse of Karmarkar's centering transformation to transform $\mathbf{y}^{k+1}$ back into $\mathbf{x}^{k+1}$, the definition of projection and (39) imply that $\mathbf{x}^{k+1}$ will be feasible for the original LP (see Problem 6).
4 Why do we project $[\text{Diag}(\mathbf{x}^k)]\mathbf{c}^T$ rather than $\mathbf{c}^T$ onto the transformed feasible region? The answer to this question must await our discussion of Karmarkar's potential function. Problem 7 provides another explanation of why we project $[\text{Diag}(\mathbf{x}^k)]\mathbf{c}^T$ rather than $\mathbf{c}^T$.

We now work out the first iteration of Karmarkar's method when applied to (40), choosing $\epsilon = 0.10$.

First Iteration of Karmarkar's Method

Step 1 $\mathbf{x}^0 = [\frac{1}{3} \quad \frac{1}{3} \quad \frac{1}{3}]^T$ and $k = 0$.

Step 2 $\mathbf{x}^0$ yields $z = \frac{1}{3} > 0.10$, so we must proceed to step 3.

Step 3

$$A = [0 \quad 1 \quad -1], \qquad \text{Diag}(\mathbf{x}^k) = \begin{bmatrix} \frac{1}{3} & 0 & 0 \\ 0 & \frac{1}{3} & 0 \\ 0 & 0 & \frac{1}{3} \end{bmatrix}$$

$$A[\text{Diag}(\mathbf{x}^k)] = [0 \quad \frac{1}{3} \quad -\frac{1}{3}], \qquad P = \begin{bmatrix} 0 & \frac{1}{3} & -\frac{1}{3} \\ 1 & 1 & 1 \end{bmatrix}$$

$$PP^T = \begin{bmatrix} \frac{2}{9} & 0 \\ 0 & 3 \end{bmatrix}, \qquad (PP^T)^{-1} = \begin{bmatrix} \frac{9}{2} & 0 \\ 0 & \frac{1}{3} \end{bmatrix}$$

$$(I - P^T(PP^T)^{-1}P) = \begin{bmatrix} \frac{2}{3} & -\frac{1}{3} & -\frac{1}{3} \\ -\frac{1}{3} & \frac{1}{6} & \frac{1}{6} \\ -\frac{1}{3} & \frac{1}{6} & \frac{1}{6} \end{bmatrix}, \qquad \mathbf{c} = [1 \quad 3 \quad -3]$$

$$[\text{Diag } \mathbf{x}^k]\mathbf{c}^T = \begin{bmatrix} \frac{1}{3} \\ 1 \\ -1 \end{bmatrix}$$

$$(I - P^T(PP^T)^{-1}P)[\text{Diag } \mathbf{x}^k]\mathbf{c}^T = [\frac{2}{9} \quad -\frac{1}{9} \quad -\frac{1}{9}]$$

Now, (using $\theta = 0.25$), we obtain

$$\mathbf{y}^1 = [\frac{1}{3} \quad \frac{1}{3} \quad \frac{1}{3}]^T - \frac{0.25[\frac{2}{9} \quad -\frac{1}{9} \quad -\frac{1}{9}]^T}{\sqrt{3(2)}\|[\frac{2}{9} \quad -\frac{1}{9} \quad -\frac{1}{9}]\|}$$

Because

$$\|[\frac{2}{9} \quad -\frac{1}{9} \quad -\frac{1}{9}]\|^T = \sqrt{(\frac{2}{9})^2 + (-\frac{1}{9})^2 + (-\frac{1}{9})^2}$$
$$= \frac{\sqrt{6}}{9}$$

we obtain

$$\mathbf{y}^1 = [\tfrac{1}{3} \quad \tfrac{1}{3} \quad \tfrac{1}{3}]^T - [\tfrac{6}{72} \quad -\tfrac{3}{72} \quad -\tfrac{3}{72}]^T = [\tfrac{1}{4} \quad \tfrac{3}{8} \quad \tfrac{3}{8}]^T$$

Using (38′), we now obtain $\mathbf{x}^1 = [x_1^1 \quad x_2^1 \quad x_3^1]^T$ from

$$x_1^1 = \frac{\tfrac{1}{3}(\tfrac{1}{4})}{\tfrac{1}{3}(\tfrac{1}{4}) + \tfrac{1}{3}(\tfrac{3}{8}) + \tfrac{1}{3}(\tfrac{3}{8})} = \tfrac{1}{4}$$

$$x_2^1 = \frac{\tfrac{1}{3}(\tfrac{3}{8})}{\tfrac{1}{3}(\tfrac{1}{4}) + \tfrac{1}{3}(\tfrac{3}{8}) + \tfrac{1}{3}(\tfrac{3}{8})} = \tfrac{3}{8}$$

$$x_3^1 = \frac{\tfrac{1}{3}(\tfrac{3}{8})}{\tfrac{1}{3}(\tfrac{1}{4}) + \tfrac{1}{3}(\tfrac{3}{8}) + \tfrac{1}{3}(\tfrac{3}{8})} = \tfrac{3}{8}$$

Thus, $\mathbf{x}^1 = [\tfrac{1}{4} \quad \tfrac{3}{8} \quad \tfrac{3}{8}]^T$. It will always be the case (see Problem 3) that $\mathbf{x}^1 = \mathbf{y}^1$, but for $k > 1$, $\mathbf{x}^k$ need not equal $\mathbf{y}^k$. Note that for $\mathbf{x}^1$, we have $z = \tfrac{1}{4} + 3(\tfrac{3}{8}) - 3(\tfrac{3}{8}) = \tfrac{1}{4} < \tfrac{1}{3}$ (the z-value for $\mathbf{x}^0$).

Potential Function

Because we are projecting $[\mathrm{Diag}(\mathbf{x}^k)]\mathbf{c}^T$ rather than $\mathbf{c}^T$, we cannot be sure that each iteration of Karmarkar's method will decrease z. In fact, it is possible for $\mathbf{c}\mathbf{x}^{k+1} > \mathbf{c}\mathbf{x}^k$ to occur. To explain why Karmarkar projects $[\mathrm{Diag}(\mathbf{x}^k)]\mathbf{c}^T$, we need to discuss Karmarkar's potential function. For $\mathbf{x} = [x_1 \quad x_2 \quad \cdots \quad x_n]^T$, we define the potential function $f(\mathbf{x})$ by

$$f(\mathbf{x}) = \sum_{j=1}^{j=n} \ln\left(\frac{\mathbf{c}\mathbf{x}^T}{x_j}\right)$$

Karmarkar showed that if we project $[\mathrm{Diag}(\mathbf{x}^k)]\mathbf{c}^T$ (not $\mathbf{c}^T$) onto the feasible region in the transformed space, then for some $\delta > 0$, it will be true that for $k = 0, 1, 2, \ldots,$

$$f(\mathbf{x}^k) - f(\mathbf{x}^{k+1}) \geq \delta \tag{41}$$

Inequality (41) states that each iteration of Karmarkar's method decreases the potential function by an amount bounded away from 0. Karmarkar shows that if the potential function evaluated at $\mathbf{x}^k$ is small enough, then $\mathbf{z} = \mathbf{c}\mathbf{x}^k$ will be near 0. Because $f(\mathbf{x}^k)$ is decreased by at least δ per iteration, it follows that by choosing k sufficiently large, we can ensure that the z-value for $\mathbf{x}^k$ is less than ϵ.

Putting an LP in Standard Form for Karmarkar's Method

We now show how to convert any LP to the form defined by (31)–(33). To illustrate, we show how to transform the following LP

$$
\begin{aligned}
\max z = {}& 3x_1 + x_2 \\
\text{s.t.} \quad & 2x_1 - x_2 \leq 2 \\
& x_1 + 2x_2 \leq 5 \\
& x_1, x_2 \geq 0
\end{aligned}
\tag{42}
$$

into the form defined by (31)–(33).

We begin by finding the dual of (42).

$$\min w = 2y_1 + 5y_2$$

$$\text{s.t.} \quad 2y_1 + y_2 \geq 3$$

$$-y_1 + 2y_2 \geq 1 \tag{42'}$$

$$y_1, y_2 \geq 0$$

From the Dual Theorem (Theorem 1 of Chapter 6), we know that if (x_1, x_2) is feasible in (42), (y_1, y_2) is feasible in (42′), and the z-value for (x_1, x_2) in (42) equals the w-value for (y_1, y_2) in (42′), then (x_1, x_2) is optimal for (42). This means that any feasible solution to the following set of constraints will yield the optimal solution to (42):

$$3x_1 + x_2 - 2y_1 - 5y_2 = 0$$

$$2x_1 - x_2 \qquad\qquad \leq 2$$

$$x_1 + 2x_2 \qquad\qquad \leq 5$$

$$2y_1 + y_2 \geq 3 \tag{43}$$

$$-y_1 + 2y_2 \geq 1$$

$$\text{All variables} \geq 0$$

Inserting slack and excess variables into (43) yields

$$3x_1 + x_2 - 2y_1 - 5y_2 \qquad\qquad = 0$$

$$2x_1 - x_2 \qquad\qquad + s_1 = 2$$

$$x_1 + 2x_2 \qquad\qquad + s_2 = 5$$

$$2y_1 + y_2 - e_1 = 3 \tag{44}$$

$$-y_1 + 2y_2 - e_2 = 1$$

$$\text{All variables} \geq 0$$

We now find a number M such that any feasible solution to (44) will satisfy

$$\text{sum of all variables in (44)} \leq M \tag{45}$$

and add constraint (45) to (44). Being conservative, we can see that any values of the variables that yield an optimal primal solution to (42) and an optimal dual solution to (42′) will have no variable exceeding 10. This would yield $M = 10(8) = 80$. We then add a slack variable (dummy variable d_1) to (45). Our new goal is then to find a feasible solution to

$$3x_1 + x_2 - 2y_1 - 5y_2 \qquad\qquad = 0$$

$$2x_1 - x_2 \qquad\qquad + s_1 = 2$$

$$x_1 + 2x_2 \qquad\qquad + s_2 = 5$$

$$2y_1 + y_2 - e_1 = 3 \tag{46}$$

$$-y_1 + 2y_2 - e_2 = 1$$

$$x_1 + x_2 + y_1 + y_2 + s_1 + s_2 + e_1 + e_2 + d_1 = 80$$

$$\text{All variables} \geq 0$$

We now define a new dummy variable d_2; $d_2 = 1$. We can use this new variable to "homogenize" the constraints in (46), which have nonzero right-hand sides. To do this, we add the appropriate multiple of the constraint $d_2 = 1$ to each constraint in (46) (except the last constraint) having a nonzero right-hand side. For example we add $-2(d_2 = 1)$ to the constraint $2x_1 - x_2 + s_1 = 2$. We also replace the last constraint in (46) by the following two constraints:

(a) Add $d_2 = 1$ to the last constraint

(b) Subtract M times $(d_2 = 1)$ from (46)

Together (a) and (b) are equivalent to $d_2 = 1$ and the last constraint in (46).
We now seek a feasible solution to

$$
\begin{aligned}
3x_1 + x_2 - 2y_1 - 5y_2 &= 0 \\
2x_1 - x_2 \quad\quad\quad\quad + s_1 - 2d_2 &= 0 \\
x_1 + 2x_2 \quad\quad\quad\quad + s_2 - 5d_2 &= 0 \\
2y_1 + y_2 - e_1 - 3d_2 &= 0 \\
- y_1 + 2y_2 - e_2 - d_2 &= 0 \\
x_1 + x_2 + y_1 + y_2 + s_1 + s_2 + e_1 + e_2 + d_1 - 80d_2 &= 0 \\
x_1 + x_2 + y_1 + y_2 + s_1 + s_2 + e_1 + e_2 + d_1 + d_2 &= 81 \\
\text{All variables} &\geq 0
\end{aligned}
\tag{47}
$$

Now we make the following change of variables in (47):

$$x_j = (M + 1)x_j', \; y_j = (M + 1)y_j', \; s_j = (M + 1)s_j', \; e_j = (M + 1)e_j',$$
$$d_j = (M + 1)d_j' \; (j = 1, 2)$$

This yields

$$
\begin{aligned}
3x_1' + x_2' - 2y_1' - 5y_2' &= 0 \\
2x_1' - x_2' \quad\quad\quad\quad + s_1' - 2d_2' &= 0 \\
x_1' + 2x_2' \quad\quad\quad\quad + s_2' - 5d_2' &= 0 \\
2y_1' + y_2' - e_1' - 3d_2' &= 0 \\
- y_1' + 2y_2' - e_2' - d_2' &= 0 \\
x_1' + x_2' + y_1' + y_2' + s_1' + s_2' + e_1' + e_2' + d_1' - 80d_2' &= 0 \\
x_1' + x_2' + y_1' + y_2' + s_1' + s_2' + e_1' + e_2' + d_1' + d_2' &= 1 \\
\text{All variables} &\geq 0
\end{aligned}
\tag{48}
$$

We now ensure that a point that sets all variables equal is feasible in (48). (Recall that this is requirement (33) for Karmarkar's method.) To accomplish this, we first add a dummy variable d_3' to the last constraint in (48) and then add a multiple of d_3' to each of the other constraints. This multiple is chosen so that the sum of the coefficients of all variables in each constraint (except the last) will equal 0. This yields LP (49).

$$
\begin{aligned}
\min z = d_3' \\
\text{s.t.} \quad 3x_1' + x_2' - 2y_1' - 5y_2' + \quad\quad\quad\quad\quad\quad 3d_3' &= 0 \\
2x_1' - x_2' \quad\quad\quad\quad + s_1' - 2d_2' + d_3' &= 0 \\
x_1' + 2x_2' \quad\quad\quad\quad + s_2' - 5d_2' + d_3' &= 0 \\
2y_1' + y_2' - e_1' - 3d_2' + d_3' &= 0 \\
- y_1' + 2y_2' - e_2' - d_2' + d_3' &= 0 \\
x_1' + x_2' + y_1' + y_2' + s_1' + s_2' + e_1' + e_2' + d_1' - 80d_2' + 71d_3' &= 0 \\
x_1' + x_2' + y_1' + y_2' + s_1' + s_2' + e_1' + e_2' + d_1' + d_2' + d_3' &= 1 \\
\text{All variables} &\geq 0
\end{aligned}
\tag{49}
$$

In (49) the point $x_1' = x_2' = y_1' = y_2' = s_1' = s_2' = e_1' = e_2' = d_1' = d_2' = d_3' = 1/11$ is feasible. Because d_3' should equal 0 in a feasible solution to (48), we need to have (49) minimize d_3'. If (48) is feasible, then the minimum value of d_3', in (49) will equal 0, and the values of the remaining variables in an optimal solution to (49) will yield a feasible solution to (48). The values of x_1 and x_2 in the optimal solution to (49) will yield an optimal solution to our original LP (42). The LP in (49) satisfies (31)–(33) and is ready for solution by Karmarkar's method.

PROBLEMS

Group A

1 Perform one iteration of Karmarkar's method for the following LP:

$$\min z = x_1 + 2x_2 - x_3$$
$$\text{s.t.} \quad x_1 \qquad - x_3 = 0$$
$$x_1 + x_2 + x_3 = 1$$
$$x_1, x_2, x_3 \geq 0$$

2 Perform one iteration of Karmarkar's method for the following LP:

$$\min z = x_1 - x_2 + 6x_3$$
$$\text{s.t.} \quad x_1 - x_2 \qquad = 0$$
$$x_1 + x_2 + x_3 = 1$$
$$x_1, x_2, x_3 \geq 0$$

3 Prove that in Karmarkar's method, $x^1 = y^1$.

4 Perform two iterations of Karmarkar's method for the following LP:

$$\min z = 2x_2$$
$$\text{s.t.} \quad x_1 + x_2 - 2x_3 = 0$$
$$x_1 + x_2 + x_3 = 1$$
$$x_1, x_2, x_3 \geq 0$$

Group B

5 Prove Lemma 1.

6 Show that the point x^k in Karmarkar's method is feasible for the original LP.

7 Given a point y^k in Karmarkar's method, express the LP's original objective function as a function of y^k. Use the answer to this question to give a reason why $[\text{Diag}(x^k)]c^T$ is projected, rather than c^T.

SUMMARY The Revised Simplex Method and the Product Form of the Inverse

Step 0 Note the columns from which the current B^{-1} will be read. Initially $B^{-1} = I$.

Step 1 For the current tableau, compute $c_{BV}B^{-1}$.

Step 2 Price out all nonbasic variables in the current tableau. If (for a max problem) each nonbasic variable prices out nonnegative, the current basis is optimal. If the current basis is not optimal, enter into the basis the nonbasic variable with the most negative coefficient in row 0. Call this variable x_k.

Step 3 To determine the row in which x_k enters the basis, compute x_k's column in the current tableau ($B^{-1}a_k$) and compute the right-hand side of the current tableau ($B^{-1}b$). Then use the ratio test to determine the row in which x_k should enter the basis. We now know the set of basic variables (BV) for the new tableau.

Step 4 Use the column for x_k in the current tableau to determine the EROs needed to enter x_k into the basis. Perform these EROs on the current B^{-1} to yield the new B^{-1}. Return to step 1.

Alternatively, we may use the product form of the inverse to update B^{-1}. Suppose we have found that x_k should enter the basis in row r. Let the column for x_k in the current tableau be

$$\begin{bmatrix} \bar{a}_{1k} \\ \bar{a}_{2k} \\ \vdots \\ \bar{a}_{mk} \end{bmatrix}$$

Define the $m \times m$ matrix E by

(column r)

$$
E = \begin{bmatrix}
1 & 0 & \cdots & -\dfrac{\bar{a}_{1k}}{\bar{a}_{rk}} & \cdots & 0 & 0 \\
0 & 1 & \cdots & -\dfrac{\bar{a}_{2k}}{\bar{a}_{rk}} & \cdots & 0 & 0 \\
\vdots & \vdots & & \vdots & & \vdots & \vdots \\
0 & 0 & \cdots & \dfrac{1}{\bar{a}_{rk}} & \cdots & 0 & 0 \\
\vdots & \vdots & & \vdots & & \vdots & \vdots \\
0 & 0 & \cdots & -\dfrac{\bar{a}_{m-1,k}}{\bar{a}_{rk}} & \cdots & 1 & 0 \\
0 & 0 & \cdots & -\dfrac{\bar{a}_{mk}}{\bar{a}_{rk}} & \cdots & 0 & 1
\end{bmatrix} \quad \text{(row } r\text{)}
$$

Then

$$B^{-1} \text{ for new tableau} = E(B^{-1} \text{ for current tableau})$$

Return to step 1.

Column Generation

When an LP has many variables, it is very time-consuming to price out each nonbasic variable individually. The column generation approach lets us determine the nonbasic variable that prices out most favorably by solving a subproblem (such as the branch-and-bound problems in the cutting stock problem).

Dantzig–Wolfe Decomposition Method

In many LPs, the constraints and variables may be decomposed in the following manner:

Constraints in set 1 only involve variables in Variable set 1.

Constraints in set 2 only involve variables in Variable set 2.

$$\vdots$$

Constraints in set k only involve variables in Variable set k.

Constraints in set $k + 1$ may involve any variable. The constraints in set $k + 1$ are referred to as the **central constraints.**

LPs that can be decomposed in this fashion can often be efficiently solved by the Dantzig–Wolfe decomposition algorithm. The following explanation assumes that $k = 2$.

Step 1 Let the variables in Variable set 1 be $x_1, x_2, \ldots, x_{n_1}$. Express the variables in Variable set 1 as a convex combination of the extreme points of the feasible region for Constraint set 1 (the constraints that involve only the variables in Variable set 1). If we let $\mathbf{P}_1, \mathbf{P}_2, \ldots, \mathbf{P}_k$ be the extreme points of this feasible region, then any point

$$
\begin{bmatrix} x_1 \\ x_2 \\ \vdots \\ x_{n_1} \end{bmatrix}
$$

in the feasible region for Constraint set 1 may be written in the form

$$
\begin{bmatrix} x_1 \\ x_2 \\ \vdots \\ x_{n_1} \end{bmatrix} = \mu_1 \mathbf{P}_1 + \mu_2 \mathbf{P}_2 + \cdots + \mu_k \mathbf{P}_k \tag{29}
$$

where $\mu_1 + \mu_2 + \cdots + \mu_k = 1$ and $\mu_i \geq 0$ $(i = 1, 2, \ldots, k)$.

Step 2 Express the variables in Variable set 2, $x_{n_1+1}, x_{n_1+2}, \ldots, x_n$, as a convex combination of the extreme points of Constraint set 2's feasible region. If we let the extreme points of the feasible region be $\mathbf{Q}_1, \mathbf{Q}_2, \ldots, \mathbf{Q}_m$, then any point in Constraint set 2's feasible region may be written as

$$
\begin{bmatrix} x_{n_1+1} \\ x_{n_1+2} \\ \vdots \\ x_n \end{bmatrix} = \lambda_1 \mathbf{Q}_1 + \lambda_2 \mathbf{Q}_2 + \cdots + \lambda_m \mathbf{Q}_m \tag{30}
$$

where $\lambda_i \geq 0$ $(i = 1, 2, \ldots, m)$ and $\lambda_1 + \lambda_2 + \cdots + \lambda_m = 1$.

Step 3 Using (29) and (30), express the LP's objective function and centralized constraints in terms of the μ_i's and the λ_i's. After adding the constraints (called convexity constraints), $\mu_1 + \mu_2 + \cdots + \mu_k = 1$ and $\lambda_1 + \lambda_2 + \cdots + \lambda_m = 1$ and the sign restrictions $\mu_i \geq 0$ $(i = 1, 2, \ldots, k)$ and $\lambda_i \geq 0$ $(i = 1, 2, \ldots, m)$, we obtain the following LP, which is referred to as the **restricted master:**

$$
\max \text{ (or min) [objective function in terms of } \mu_i\text{'s and } \lambda_i\text{'s]}
$$

s.t. [central constraints in terms of μ_i's and λ_i's]

$$
\mu_1 + \mu_2 + \cdots + \mu_k = 1 \quad \text{(Convexity constraints)}
$$
$$
\lambda_1 + \lambda_2 + \cdots + \lambda_m = 1
$$
$$
\mu_i \geq 0 \quad (i = 1, 2, \ldots, k) \quad \text{(Sign restrictions)}
$$
$$
\lambda_i \geq 0 \quad (i = 1, 2, \ldots, m)
$$

Step 4 Assume that a basic feasible solution for the restricted master is readily available. Then use the column generation method of Section 10.3 to determine whether there is any μ_i or λ_i that can improve the z-value for the restricted master. If so, use the revised simplex method to enter that variable into the basis. Otherwise, the current tableau is optimal for the restricted master. If the current tableau is not optimal, continue with column generation until an optimal solution is found.

Step 5 Substitute the optimal values of the μ_i's and λ_i's found in step 4 into (29) and (30). This will yield the optimal values of $x_1, x_2, \ldots, x_n$.

The Simplex Method for Upper-Bounded Variables

For each variable x_i that has an upper-bound constraint $x_i \leq u_i$, we define a new variable x_i' by the relationship $x_i + x_i' = u_i$, or $x_i = u_i - x_i'$.

At each iteration, we choose (for a max problem) to increase the variable x_i with the most negative coefficient in row 0. Three possible occurrences, or bottlenecks, can restrict the amount by which we increase x_i:

Bottleneck 1 x_i cannot exceed its upper bound of u_i.

Bottleneck 2 x_i increases to a point where it causes one of the current basic variables to become negative.

Bottleneck 3 x_i increases to a point where it causes one of the current basic variables to exceed its upper bound.

Let BN_k ($k = 1, 2, 3$) be the value of x_i where bottleneck k occurs. Then x_i can only be increased to a value of $\min\{BN_1, BN_2, BN_3\}$, the winning bottleneck. If the winning bottleneck is BN_1, then we make an upper-bound substitution on x_i by replacing x_i by $u_i - x_i'$. If the winning bottleneck is BN_2, then we enter x_i into the basis in the row corresponding to the basic variable that caused BN_2 to occur. If the winning bottleneck is BN_3, we make an upper-bound substitution on the variable x_j (by replacing x_j by $u_j - x_j'$) that reaches its upper bound when $x_i = BN_3$. Then enter x_i into the basis in the row for which x_j was a basic variable.

After following this procedure, examine the new row 0. If each variable has a non-negative coefficient, we have obtained an optimal tableau. Otherwise, we try to increase the variable with the most negative coefficient in row 0.

Karmarkar's Method

Step 1 Begin at the feasible point $\mathbf{x}^0 = [\frac{1}{n} \quad \frac{1}{n} \quad \cdots \quad \frac{1}{n}]^T$ and set $k = 0$.

Step 2 Stop if $\mathbf{c}\mathbf{x}^k < \epsilon$. If not, go to step 3.

Step 3 Find the new point $\mathbf{y}^{k+1} = [y_1^{k+1} \quad y_2^{k+1} \quad \cdots \quad y_n^{k+1}]^T$ in the transformed unit simplex given by

$$\mathbf{y}^{k+1} = [\tfrac{1}{n} \quad \tfrac{1}{n} \quad \cdots \quad \tfrac{1}{n}]^T - \frac{\theta(I - P^T(PP^T)^{-1}P)[\mathrm{Diag}(\mathbf{x}^k)]\mathbf{c}^T}{\|\mathbf{c}_p\|\sqrt{n(n-1)}}$$

Here, $\|\mathbf{c}_p\|$ = the length of $(I - P^T(PP^T)^{-1}P)[\mathrm{Diag}(\mathbf{x}^k)]\mathbf{c}^T$, P is the $(m + 1) \times n$ matrix whose first m rows are $A[\mathrm{Diag}(\mathbf{x}^k)]$ and whose last row is a vector of 1's, and $0 < \theta < 1$ is chosen to ensure convergence of the algorithm. $\theta = \frac{1}{4}$ is known to ensure convergence.

Now obtain a new point $\mathbf{x}^{k+1}$ in the original space by using the centering transformation to determine the point corresponding to $\mathbf{y}^{k+1}$. That is, $\mathbf{x}^{k+1} = f^{-1}(\mathbf{y}^{k+1}|\mathbf{x}^k)$. Increase k by 1 and return to step 2.

REVIEW PROBLEMS

Group A

1 Use the revised simplex with the product form of the inverse to solve the following LP:

$$\max z = 4x_1 + 3x_2 + x_3$$
$$\text{s.t.} \quad 3x_1 + 2x_2 + x_3 \le 6$$
$$x_2 + x_3 \le 3$$
$$x_1 \quad + x_3 \le 2$$
$$x_1, x_2, x_3 \ge 0$$

2 Use the column generation technique to solve a cutting stock problem in which a customer demands 20 3-ft boards, 25 4-ft boards, and 30 5-ft boards, and demand is met by cutting up 14-ft boards.

3 Use the Dantzig–Wolfe decompostion method to solve the following LP:

$$\min z = 2x_1 - x_2 + x_3 - x_4$$
$$\text{s.t.} \quad x_1 + 2x_2 \quad \le 4$$

$$x_1 - x_2 \leq 1$$
$$x_3 - 3x_4 \leq 7$$
$$2x_3 + x_4 \leq 10$$
$$x_1 + 3x_2 - x_3 - 2x_4 \leq 10$$
$$x_i \geq 0 \ (i = 1, 2, 3, 4)$$

4 Consider the following situation:

a Two types of cars are produced at three production plants and are demanded by three customers.

b You are given the cost of producing each type of car at each plant and the cost of shipping each type of car from each plant to each customer.

c You are given the production capacity of each plant (for each type of car).

d You are also told that at most one half of the total number of cars demanded by customer 1 can be met from plant 1 production.

Explain how you would use decomposition to minimize the cost of meeting the customers' demands.

5 A company manufactures products 1 and 2. A total of 100 hours of production time is available at each plant. The times required to produce a unit of each product at each plant are shown in Table 10, and the profits earned for a unit

of each product produced at each plant are shown in Table 11. At most, 35 units of each product can be sold. Use decomposition to determine how the company can maximize profits.

TABLE 10

| Plant | Hours | |
	Product 1	Product 2
1	2	3
2	3	4

TABLE 11

| Plant | Profit per Product ($) | |
	Product 1	Product 2
1	8	6
2	10	8

REFERENCES

The following three references are classic works that detail methods used to solve large LPs:

Beale, E. *Mathematical Programming in Practice.* Pittman, 1968.

Lasdon, L. *Optimization Theory for Large Systems.* New York: Macmillan, 1970.

Orchard-Hays, W. *Advanced LP Computing Techniques.* New York: McGraw-Hill, 1968.

The following three references contain excellent discussions of Dantzig–Wolfe decomposition:

Bradley, S., A. Hax, and T. Magnanti. *Applied Mathematical Programming.* Reading, Mass.: Addison-Wesley, 1977.

Chvàtal, V. *Linear Programming.* San Francisco: Freeman, 1983.

Shapiro, J. *Mathematical Programming: Structures and Algorithms.* New York: Wiley, 1979.

The following two references discuss column generation and the cutting stock problem:

Gilmore, P., and R. Gomory. "A Linear Programming Approach to the Cutting Stock Problem," *Operations Research* 9(1961):849–859.

———. "A Linear Programming Approach to the Cutting Stock Problem: Part II," *Operations Research* 11(1963):863–888.

The following three references contain lucid discussions of Karmarkar's method:

Hooker, J. N. "Karmarkar's Linear Programming Algorithm," *Interfaces* 16(no. 4, 1986)75–90.

Murty, K. G. *Linear Complementarity, Linear and Nonlinear Programming.* Berlin, Germany: Heldermann Verlag, 1989.

Parker, G., and R. Rardin. *Discrete Optimization.* San Diego: Academic Press, 1988.

11

Nonlinear Programming

In previous chapters, we have studied linear programming problems. For an LP, our goal was to maximize or minimize a linear function subject to linear constraints. But in many interesting maximization and minimization problems, the objective function may not be a linear function, or some of the constraints may not be linear constraints. Such an optimization problem is called a *nonlinear programming* problem (NLP). In this chapter, we discuss techniques used to solve NLPs.

We begin with a review of material from differential calculus, which will be needed for our study of nonlinear programming.

11.1 Review of Differential Calculus

Limits

The idea of a limit is one of the most basic ideas in calculus.

DEFINITION ■ The equation

$$\lim_{x \to a} f(x) = c$$

means that as x gets closer to a (but not equal to a), the value of $f(x)$ gets arbitrarily close to c. ■

It is also possible that $\lim_{x \to a} f(x)$ may not exist.

EXAMPLE 1 | Limits

1 Show that $\lim_{x \to 2} x^2 - 2x = 2^2 - 2(2) = 0$.

2 Show that $\lim_{x \to 0} \frac{1}{x}$ does not exist.

Solution 1 To verify this result, evaluate $x^2 - 2x$ for values of x close to, but not equal to 2.

2 To verify this result, observe that as x gets near 0, $\frac{1}{x}$ becomes either a very large positive number or a very large negative number. Thus, as x approaches 0, $\frac{1}{x}$ will not approach any single number.

Continuity

DEFINITION ■ A function $f(x)$ is **continuous** at a point a if

$$\lim_{x \to a} f(x) = f(a)$$

If $f(x)$ is not continuous at $x = a$, we say that $f(x)$ is **discontinuous** (or has a discontinuity) at a. ■

EXAMPLE 2 | **Continuous Functions**

Bakeco orders sugar from Sugarco. The per-pound purchase price of the sugar depends on the size of the order (see Table 1). Let

$$x = \text{number of pounds of sugar purchased by Bakeco}$$
$$f(x) = \text{cost of ordering } x \text{ pounds of sugar}$$

Then

$$f(x) = 25x \text{ for } 0 \le x < 100$$
$$f(x) = 20x \text{ for } 100 \le x \le 200$$
$$f(x) = 15x \text{ for } x > 200$$

For all values of x, determine if x is continuous or discontinuous.

Solution From Figure 1, it is clear that

$$\lim_{x \to 100} f(x) \quad \text{and} \quad \lim_{x \to 200} f(x)$$

do not exist. Thus, $f(x)$ is discontinuous at $x = 100$ and $x = 200$ and is continuous for all other values of x satisfying $x \ge 0$.

TABLE 1
Price of Sugar Paid by Bakeco

Size of Order	Price per Pound (c)
$0 \le x < 100$	25
$100 \le x \le 200$	20
$x > 200$	15

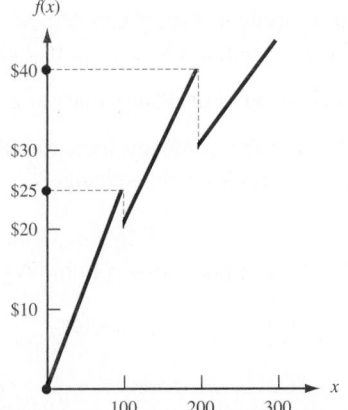

FIGURE 1
Cost of Purchasing Sugar for Bakeco

TABLE 2
Rules for Finding the Derivative of a Function

Function	Derivative of Function
a	0
x	1
$af(x)$	$af'(x)$
$f(x) + g(x)$	$f'(x) + g'(x)$
x^n	nx^{n-1}
e^x	e^x
a^x	$a^x \ln a$
$\ln x$	$\frac{1}{x}$
$[f(x)]^n$	$n[f(x)]^{n-1}f'(x)$
$e^{f(x)}$	$e^{f(x)}f'(x)$
$a^{f(x)}$	$a^{f(x)}f'(x)\ln a$
$\ln f(x)$	$\dfrac{f'(x)}{fx}$
$f(x)g(x)$	$f(x)g'(x) + f'(x)g(x)$
$\dfrac{f(x)}{g(x)}$	$\dfrac{g(x)f'(x) - f(x)g'(x)}{g(x)^2}$

Differentiation

DEFINITION ■ The **derivative** of a function $f(x)$ at $x = a$ [written $f'(a)$] is defined to be

$$\lim_{\Delta x \to 0} \frac{f(a + \Delta x) - f(a)}{\Delta x} \quad ■$$

If this limit does not exist, then $f(x)$ has no derivative at $x = a$.

We may think of $f'(a)$ as the slope of $f(x)$ at $x = a$. Thus, if we begin at $x = a$ and increase x by a small amount Δ (Δ may be positive or negative), then $f(x)$ will increase by an amount approximately equal to $\Delta f'(a)$. If $f'(a) > 0$, then $f(x)$ is increasing at $x = a$, whereas if $f'(a) < 0$, then $f(x)$ is decreasing at $x = a$. The derivatives of many functions can be found via application of the rules in Table 2 (a represents an arbitrary constant). Example 3 illustrates the use and interpretation of the derivative.

EXAMPLE 3   Product Profitability

If a company charges a price p for a product, then it can sell $3e^{-p}$ thousand units of the product. Then, $f(p) = 3,000pe^{-p}$ is the company's revenue if it charges a price p.

1 For what values of p is $f(p)$ decreasing? For what values of p is $f(p)$ increasing?

2 Suppose the current price is \$4 and the company increases the price by 5¢. By approximately how much would the company's revenue change?

Solution We have

$$f'(p) = -3,000pe^{-p} + 3,000e^{-p} = 3,000e^{-p}(1 - p)$$

1 For $p < 1$, $f'(p) > 0$ and $f(p)$ is increasing, whereas for $p > 1$, $f'(p) < 0$ and $f(p)$ is decreasing.

2 Using the interpretation of $f'(4)$ as the slope of $f(p)$ at $p = 4$ (with $\Delta p = 0.05$), we see that the company's revenue would increase by approximately

$$0.05(3{,}000e^{-4})(1-4) = -8.24$$

In actuality, of course, the company's revenue would increase by

$$f(4.05) - f(4) = 3{,}000(4.05)e^{-4.05} - 3{,}000(4)e^{-4}$$
$$= 211.68 - 219.79 = -8.11$$

Higher Derivatives

We define $f^{(2)}(a) = f''(a)$ to be the derivative of the function $f'(x)$ at $x = a$. Similarly, we can define (if it exists) $f^{(n)}(a)$ to be the derivative of $f^{(n-1)}(x)$ at $x = a$. Thus, for Example 3,

$$f''(p) = 3{,}000e^{-p}(-1) - 3{,}000e^{-p}(1-p)$$

Taylor Series Expansion

In the Taylor series expansion of a function $f(x)$, given that $f^{(n+1)}(x)$ exists for every point on the interval $[a, b]$, we can write for any h satisfying $0 \le h \le b - a$,

$$f(a + h) = f(a) + \sum_{i=1}^{i=n} \frac{f^{(i)}(a)}{i!} h^i + \frac{f^{(n+1)}(p)}{(n+1)!} h^{n+1} \tag{1}$$

where (1) will hold for some number p between a and $a + h$. Equation (1) is the **nth-order Taylor series expansion** of $f(x)$ about a.

EXAMPLE 4	Taylor Series Expansion

Find the first-order Taylor series expansion of e^{-x} about $x = 0$.

Solution Because $f'(x) = -e^{-x}$ and $f''(x) = e^{-x}$, we know that (1) will hold on any interval $[0, b]$. Also, $f(0) = 1, f'(0) = -1$, and $f''(x) = e^{-x}$. Then (1) yields the following first-order Taylor series expansion for e^{-x} about $x = 0$:

$$e^{-h} = f(h) = 1 - h + \frac{h^2 e^{-p}}{2}$$

This equation holds for some p between 0 and h.

Partial Derivatives

We now consider a function f of $n > 1$ variables $(x_1, x_2, \ldots, x_n)$, using the notation $f(x_1, x_2, \ldots, x_n)$ to denote such a function.

The **partial derivative** of $f(x_1, x_2, \ldots, x_n)$ with respect to the variable x_i is written $\dfrac{\partial f}{\partial x_i}$, where

$$\frac{\partial f}{\partial x_i} = \lim_{\Delta x_i \to 0} \frac{f(x_1, \ldots, x_i + \Delta x_i, \ldots, x_n) - f(x_1, \ldots, x_i, \ldots, x_n)}{\Delta x_i}$$ ■

Intuitively, if x_i is increased by Δ (and all other variables are held constant), then for small values of Δ, the value of $f(x_1, x_2, \ldots, x_n)$ will increase by approximately $\Delta \dfrac{\partial f}{\partial x_i}$. We find $\dfrac{\partial f}{\partial x_i}$ by treating all variables other than x_i as constants and finding the derivatives of $f(x_1, x_2, \ldots, x_n)$. More generally, suppose that for each i, we increase x_i by a small amount Δx_i. Then the value of f will increase by approximately

$$\sum_{i=1}^{i=n} \frac{\partial f}{\partial x_i} \Delta x_i$$

EXAMPLE 5   When Is a Function Increasing?

The demand $f(p, a) = 30{,}000 p^{-2} a^{1/6}$ for a product depends on $p =$ product price (in dollars) and $a =$ dollars spent advertising the product. Is demand an increasing or decreasing function of price? Is demand an increasing or decreasing function of advertising expenditure? If $p = 10$ and $a = 1{,}000{,}000$, then by how much (approximately) will a \$1 cut in price increase demand?

Solution

$$\frac{\partial f}{\partial p} = 30{,}000(-2p^{-3})a^{1/6} = -60{,}000 p^{-3} a^{1/6} < 0$$

$$\frac{\partial f}{\partial a} = 30{,}000 p^{-2} \left(\frac{a^{-5/6}}{6} \right) = 5{,}000 p^{-2} a^{-5/6} > 0$$

Thus, an increase in price (with advertising held constant) will decrease demand, while an increase in advertising (with price held constant) will increase demand. Because

$$\frac{\partial f}{\partial p} (10, 1{,}000{,}000) = -60{,}000 \left(\frac{1}{1{,}000} \right) (1{,}000{,}000)^{1/6} = -600$$

a \$1 price cut will increase demand by approximately $(-1)(-600)$, or 600 units.

We will also use *second-order partial derivatives* extensively. We use the notation $\dfrac{\partial^2}{\partial x_i \partial x_j}$ to denote a second-order partial derivative. To find $\dfrac{\partial^2}{\partial x_i \partial x_j}$, we first find $\dfrac{\partial f}{\partial x_i}$ and then take its partial derivative with respect to x_j. If the second-order partials exist and are everywhere continuous, then

$$\frac{\partial^2 f}{\partial x_i \partial x_j} = \frac{\partial^2 f}{\partial x_j \partial x_i}$$

EXAMPLE 6   Second-Order Partial Derivatives

For $f(p, a) = 30{,}000 p^{-2} a^{1/6}$, find all second-order partial derivatives.

Solution

$$\frac{\partial^2 f}{\partial p^2} = -60,000(-3p^{-4})a^{1/6} = \frac{180,000a^{1/6}}{p^4}$$

$$\frac{\partial^2 f}{\partial a^2} = 5,000p^{-2}\left(\frac{-5a^{-11/6}}{6}\right) = -\frac{25,000p^{-2}a^{-11/6}}{6}$$

$$\frac{\partial^2 f}{\partial a \partial p} = 5,000(-2p^{-3})a^{-5/6} = -10,000p^{-3}a^{-5/6}$$

$$\frac{\partial^2 f}{\partial p \partial a} = -60,000p^{-3}\left(\frac{a^{-5/6}}{6}\right) = -10,000p^{-3}a^{-5/6}$$

Observe that for $p \neq 0$ and $a \neq 0$,

$$\frac{\partial^2 f}{\partial a \partial p} = \frac{\partial^2 f}{\partial p \partial a}$$

PROBLEMS

Group A

1 Find $\lim_{h\to 0} \dfrac{3h + h^2}{h}$.

2 It costs Sugarco 25¢/lb to purchase the first 100 lb of sugar, 20¢/lb to purchase the next 100 lb, and 15¢ to buy each additional pound. Let $f(x)$ be the cost of purchasing x pounds of sugar. Is $f(x)$ continuous at all points? Are there any points where $f(x)$ has no derivative?

3 Find $f'(x)$ for each of the following functions:

a xe^{-x}

b $\dfrac{x^2}{x^2 + 1}$

c e^{3x}

d $(3x + 2)^{-2}$

e $\ln x^3$

4 Find all first- and second-order partial derivatives for $f(x_1, x_2) = x_1^2 e^{x_2}$.

5 Find the second-order Taylor series expansion of $\ln x$ about $x = 1$.

Group B

6 Let $q = f(p)$ be the demand for a product when the price is p. For a given price p, the price elasticity E of the product is defined by

$$E = \frac{\text{percentage change in demand}}{\text{percentage change in price}}$$

If the change in price (Δp) is small, this formula reduces to

$$E = \frac{\frac{\Delta q}{q}}{\frac{\Delta p}{p}} = \left(\frac{p}{q}\right)\left(\frac{dq}{dp}\right)$$

a Would you expect $f(p)$ to be positive or negative?

b Show that if $E < -1$, a small decrease in price will increase the firm's total revenue (in this case, we say that demand is *elastic*).

c Show that if $-1 < E < 0$, a small price decrease will decrease total revenue (in this case, we say demand is *inelastic*).

7 Suppose that if x dollars are spent on advertising during a given year, $k(1 - e^{-cx})$ customers will purchase a product ($c > 0$).

a As x grows large, the number of customers purchasing the product approaches a limit. Find this limit.

b Can you give an interpretation for k?

c Show that the sales response from a dollar of advertising is proportional to the number of potential customers who are not purchasing the product at present.

8 Let the total cost of producing x units, $c(x)$, be given by $c(x) = kx^{1-b}$ ($0 < b < 1$). This cost curve is called the *learning* or *experience cost curve*.

a Show that the cost of producing a unit is a decreasing function of the number of units that have been produced.

b Suppose that each time the number of units produced is doubled, the per-unit product cost drops to $r\%$ of its previous value (because workers learn how to perform their jobs better). Show that $r = 100(2^{-b})$.

9 If a company has m hours of machine time and w hours of labor, it can produce $3m^{1/3}w^{2/3}$ units of a product. Currently, the company has 216 hours of machine time and 1,000 hours of labor. An extra hour of machine time costs $100, and an extra hour of labor costs $50. If the company has $100 to invest in purchasing additional labor and machine time, would it be better off buying 1 hour of machine time or 2 hours of labor?

11.2 Introductory Concepts

DEFINITION ■ A general **nonlinear programming problem** (NLP) can be expressed as follows: Find the values of decision variables $x_1, x_2, \ldots, x_n$ that

$$\max \quad (\text{or min}) \; z = f(x_1, x_2, \ldots, x_n)$$

$$\text{s.t.} \quad g_1(x_1, x_2, \ldots, x_n) \quad (\leq, =, \text{or} \geq) \; b_1$$

$$\text{s.t.} \quad g_2(x_1, x_2, \ldots, x_n) \quad (\leq, =, \text{or} \geq) \; b_2 \qquad \qquad \textbf{(2)}$$

$$\vdots$$

$$g_m(x_1, x_2, \ldots, x_n) \quad (\leq, =, \text{or} \geq) \; b_m \quad ■$$

As in linear programming, $f(x_1, x_2, \ldots, x_n)$ is the NLP's **objective function,** and $g_1(x_1, x_2, \ldots, x_n) (\leq, =, \text{or} \geq) b_1, \ldots, g_m(x_1, x_2, \ldots, x_n) (\leq, =, \text{or} \geq) b_m$ are the NLP's **constraints.** An NLP with no constraints is an **unconstrained NLP.**

The set of all points $(x_1, x_2, \ldots, x_n)$ such that x_i is a real number is R^n. Thus, R^1 is the set of all real numbers. The following subsets of R^1 (called intervals) will be of particular interest:

$$[a, b] = \text{all } x \text{ satisfying } a \leq x \leq b$$
$$[a, b) = \text{all } x \text{ satisfying } a \leq x < b$$
$$(a, b] = \text{all } x \text{ satisfying } a < x \leq b$$
$$(a, b) = \text{all } x \text{ satisfying } a < x < b$$
$$[a, \infty) = \text{all } x \text{ satisfying } x \geq a$$
$$(-\infty, b] = \text{all } x \text{ satisfying } x \leq b$$

The following definitions are analogous to the corresponding definitions for LPs given in Section 3.1.

DEFINITION ■ The **feasible region** for NLP (2) is the set of points $(x_1, x_2, \ldots, x_n)$ that satisfy the m constraints in (2). A point in the feasible region is a *feasible point,* and a point that is not in the feasible region is an *infeasible point.* ■

Suppose (2) is a maximization problem.

DEFINITION ■ Any point $\bar{x}$ in the feasible region for which $f(\bar{x}) \geq f(x)$ holds for all points x in the feasible region is an **optimal solution** to the NLP. [For a minimization problem, $\bar{x}$ is the optimal solution if $f(\bar{x}) \leq f(x)$ for all feasible x.] ■

Of course, if $f, g_1, g_2, \ldots, g_m$ are all linear functions, then (2) is a linear programming problem and may be solved by the simplex algorithm.

Examples of NLPs

EXAMPLE 7   **Profit Maximization**

It costs a company c dollars per unit to manufacture a product. If the company charges p dollars per unit for the product, customers demand $D(p)$ units. To maximize profits, what price should the firm charge?

Solution   The firm's decision variable is p. Since the firm's profit is $(p - c)D(p)$, the firm wants to solve the following unconstrained maximization problem: $\max(p - c)D(p)$.

<hr/>

EXAMPLE 8   **Production Maximization**

If K units of capital and L units of labor are used, a company can produce KL units of a manufactured good. Capital can be purchased at \$4/unit and labor can be purchased at \$1/unit. A total of \$8 is available to purchase capital and labor. How can the firm maximize the quantity of the good that can be manufactured?

Solution   Let K = units of capital purchased and L = units of labor purchased. Then K and L must satisfy $4K + L \leq 8$, $K \geq 0$, and $L \geq 0$. Thus, the firm wants to solve the following constrained maximization problem:

$$\max z = KL$$
$$\text{s.t.} \quad 4K + L \leq 8$$
$$K, L \geq 0$$

Solving NLPs with LINGO

Cap.lng

LINGO may be used to solve NLPs on a PC. Figure 2 (file Cap.lng) contains the LINGO formulation and output for Example 8. From the Value column, we see that LINGO has found the solution $K = 1$ and $L = 4$, which has an objective function value of 4. As we shall soon see, this is indeed the optimal solution to Example 8. However, in general, there is no guarantee that the solution found by LINGO is an optimal solution. Throughout this chapter, we will detail the circumstances in which you can be sure that LINGO will find the optimal solution to an NLP.

Note that the $\wedge$ symbol is used to indicate raising to a power and * indicates multiplication. LINGO has several built-in functions including

- ABS(X) = absolute value of X
- EXP(X) = e^x
- LOG(X) = natural logarithm of X

In Sections 11.9 and 11.10, we will discuss the Price column of the LINGO output. We will not discuss the Reduced Cost column.

Differences Between NLPs and LPs

Recall from Chapter 3 that the feasible region for any LP is a convex set (that is, if A and B are feasible for an LP, then the entire line segment joining A and B is also feasible). Also recall that if an LP has an optimal solution, then there is an extreme point of the feasible region that is optimal. We will soon see, however, that even if the feasible region for an NLP is a convex set, the optimal solution (unlike the optimal solution for an LP) need not be an extreme point of the NLP's feasible region. The previous example illustrates this idea. Figure 3 shows graphically the feasible region (bounded by triangle ABC) for the example and the isoprofit curves $KL = 1$, $KL = 2$, and $KL = 4$. We see that the optimal solution to the example occurs where an isoprofit curve is tangent to the boundary of the feasible region. Thus, the optimal solution to the example is $z = 4$, $K = 1$, $L = 4$ (point D). Of course, point D is not an extreme point of the NLP's feasible region. For this ex-

FIGURE **2**

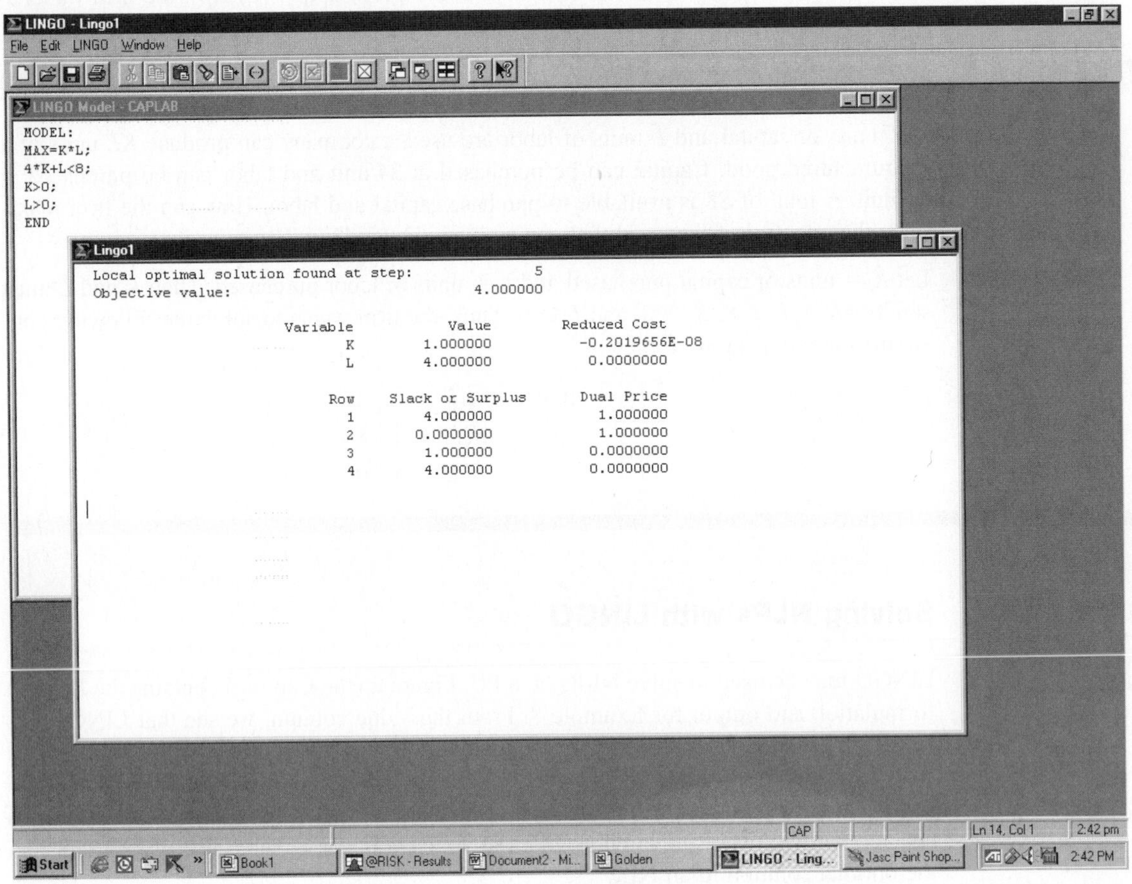

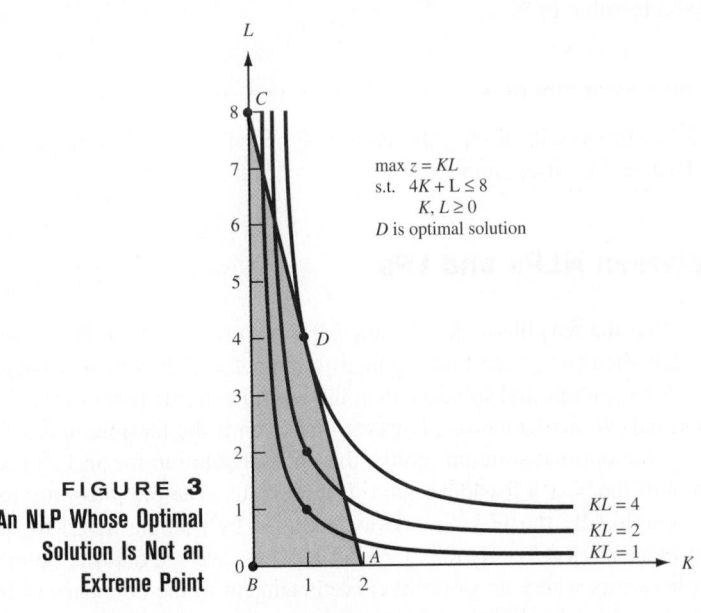

FIGURE **3**
An NLP Whose Optimal
Solution Is Not an
Extreme Point

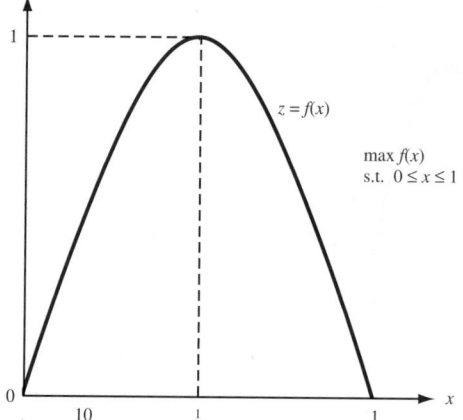

FIGURE 4
An NLP Whose Optimal
Solution Is Not on
Boundary of Feasible
Region

ample (and many other NLPs with linear constraints), the optimal solution fails to be an extreme point of the feasible region because the isoprofit curves are not straight lines. In fact, the optimal solution for an NLP may not be on the boundary of the feasible region. For example, consider the following NLP:

$$\max z = f(x)$$
$$\text{s.t.} \quad 0 \leq x \leq 1$$

where $f(x)$ is pictured in Figure 4. The optimal solution for this NLP is $z = 1$, $x = \frac{1}{2}$. Of course, $x = \frac{1}{2}$ is not on the boundary of the feasible region.

Local Extremum

DEFINITION ■ For any NLP (maximization), a feasible point $x = (x_1, x_2, \ldots, x_n)$ is a **local maximum** if for sufficiently small ϵ, any feasible point $x' = (x'_1, x'_2, \ldots, x'_n)$ having $|x_i - x'_i| < \epsilon$ $(i = 1, 2, \ldots, n)$ satisfies $f(x) \geq f(x')$. ■

In short, a point x is a local maximum if $f(x) \geq f(x')$ for all feasible x' that are close to x. Analogously, for a minimization problem, a point x is a local minimum if $f(x) \leq f(x')$ holds for all feasible x' that are close to x. A point that is a local maximum or a local minimum is called a **local,** or **relative, extremum.**

For an LP (max problem), any local maximum is an optimal solution to the LP. (Why?) For a general NLP, however, this may not be true. For example, consider the following NLP:

$$\max z = f(x)$$
$$\text{s.t.} \quad 0 \leq x \leq 10$$

where $f(x)$ is given in Figure 5. Points A, B, and C are all local maxima, but point C is the unique optimal solution to the NLP.

Unlike an LP, an NLP may not satisfy the Proportionality and Additivity assumptions. For instance, in Example 8, increasing L by 1 will increase z by K. Thus, the effect on z of increasing L by 1 depends on K. This means that the example does not satisfy the Additivity Assumption.

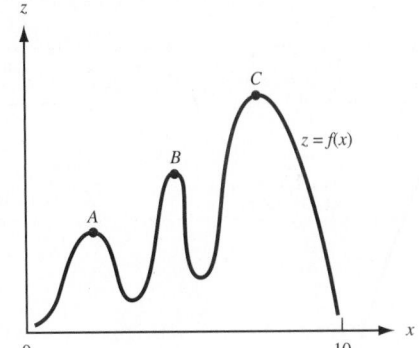

FIGURE 5
A Local Maximum May
Not Be the Optimal
Solution to an NLP

The NLP

$$\max z = x^{1/3} + y^{1/3}$$
$$\text{s.t.} \quad x + y = 1$$
$$x, y \geq 0$$

does not satisfy the Proportionality Assumption, because doubling the value of x does not double the contribution of x to the objective function.

More Examples of NLP Formulations

We now give three more examples of nonlinear programming formulations.

EXAMPLE 9   **Oilco NLP**

Oilco produces three types of gasoline: regular, unleaded, and premium. All three are produced by combining lead and crude oil brought in from Alaska and Texas. The required sulphur content, octane levels, minimum daily demand (in gallons), and sales price per gallon of each type of gasoline are given in Table 3. The crude brought in from Alaska is made by blending two types of crude: Alaska1 and Alaska2. The Alaska crude is blended in Alaska and shipped via pipeline to Oilco's Texas refinery. At most, 10,000 gallons of crude per day can be shipped from Alaska. The sulphur content, octane level, daily maximum amount available (in gallons) and purchase cost (per gallon) for each type of Alaska crude, Texas crude, and lead are given in Table 4. Of course, unleaded gasoline can contain no lead. Formulate an NLP to help Oilco maximize the daily profit obtained from selling gasoline.[†]

Solution   After defining the following decision variables:

R = gallons of regular gasoline produced daily

U = gallons of unleaded gasoline produced daily

P = gallons of premium gasoline produced daily

$A1$ = gallons of Alaska1 crude purchased daily

$A2$ = gallons of Alaska2 crude purchased daily

T = gallons of Texas crude purchased daily

L = gallons of lead purchased daily

[†]Based on Haverly (1978).

TABLE 3

Type of Gasoline	Sulphur Content (%)	Octane Level	Minimum Daily Demand (Gallons)	Sales Price ($)
Regular	≤ 3	≥ 90	5,000	.86
Unleaded	≤ 3	≥ 88	5,000	.93
Premium	≤ 2.8	≥ 94	5,000	1.06

TABLE 4

Type of Input	Sulphur Content (%)	Octane Level	Maximum Availability (Gallons)	Cost (per Gallon) ($)
Alaska 1	4	91	0	.78
Alaska 2	1	97	0	.88
Texas	2	83	11,000	.75
Lead	0	800	6,000	1.30

SA = sulphur content of crude purchased from Alaska

OA = octane level of crude purchased from Alaska

A = total gallons of crude purchased from Alaska

LP = gallons of lead used daily to make premium gasoline

TP = gallons of Texas crude used daily to make premium gasoline

AP = gallons of Alaska crude used daily to make premium gasoline

TU = Texas crude used daily to make unleaded gasoline

AU = Alaska crude used daily to make unleaded gasoline

AR = Alaska crude used daily to make regular gasoline

TR = Texas crude used daily to make regular gasoline

LR = gallons of lead used daily to make regular gasoline

Alas.lng

we find the appropriate formulation in the LINGO printout given in Figure 6 (file Alas.lng).

The objective function maximizes daily revenues (86 * R + 93 * U + 106 * P) less the daily costs of purchasing crude (78 * A1 + 88 * A2 + 75 * T + 130 * L). Rows 2–4 specify that the amount of each input cannot exceed its daily availability. Rows 5–7 ensure that the minimum demand requirements for each gasoline are met.

The percentage sulphur content (as a decimal) of Alaska crude in terms of the amount of each type of Alaska crude purchased is defined in row 8. Similarly, the octane level of Alaska crude in terms of the amount of each type of Alaska crude purchased is defined in row 9. Row 10 defines the total amount of Alaska crude purchased as the sum of the amount of Alaska1 and Alaska2 purchased. Similarly, row 11 expresses the amount of premium produced as the sum of its lead, Texas crude, and Alaska crude inputs, and row 12 expresses the amount of unleaded gasoline produced as the sum of its inputs. Rows 13–15 tell us that inputs are fully consumed in production—all lead is used to produce premium or regular gasoline; all Alaskan crude is used to make premium, unleaded, or regular gasoline; and all Texas crude is used to make premium, unleaded, or regular gasoline.

Row 16 requires that the average octane level of the inputs used to produce regular gasoline is at least 90. Notice that this is not a linear constraint because of the presence of the term AR * OA. Similarly, row 17 (again, not a linear constraint) ensures that the

```
MODEL:
   1) MAX= 86 * R + 93 * U + 106 * P - 78 * A1 - 88 * A2 - 75 * T - 130 *
      L ;
   2) A < 10000 ;
   3) T < 11000 ;
   4) L < 6000 ;
   5) R > 5000 ;
   6) U > 5000 ;
   7) P > 5000 ;
   8) SA = ( .04 * A1 + .01 * A2 ) / A ;
   9) OA = ( 91 * A1 + 97 * A2 ) / A ;
  10) A = A1 + A2 ;
  11) P = LP + TP + AP ;
  12) U = TU + AU ;
  13) L = LP + LR ;
  14) A = AP + AU + AR ;
  15) T = TP + TU + TR ;
  16) ( AR * OA + 83 * TR + 800 * LR ) / R > 90 ;
  17) ( AP * OA + 83 * TP + 800 * LP ) / P > 94 ;
  18) ( AU * OA + TU * 83 ) / U > 88 ;
  19) ( SA * AR + .02 * TR ) / R < .03 ;
  20) ( SA * AP + .02 * TP ) / P < .028 ;
  21) ( SA * AU + .02 * TU ) / U < .03 ;
  22) LP > 0 ;
  23) TP > 0 ;
  24) AP > 0 ;
  25) TU > 0 ;
  26) AU > 0 ;
  27) LR > 0 ;
  28) TR > 0 ;
  29) AR > 0 ;
  30) R = TR + AR + LR ;
END


SOLUTION STATUS:  OPTIMAL TO TOLERANCES.  DUAL CONDITIONS:  SATISFIED.

                OBJECTIVE FUNCTION VALUE

        1)     443237.052541

   VARIABLE         VALUE          REDUCED COST
        R       5000.000000          .000000
        U       5000.000000          .000000
        P      11134.965633          .000000
       A1       9047.622772          .000000
       A2        952.377228          .000000
        T      11000.000000          .000000
        L        134.965633          .000000
        A      10000.000000          .000000
       SA           .037143          .000000
       OA         91.571426          .000000
       LP        121.210474          .000000
       TP       6863.136139          .000000
       AP       4150.619020          .000000
       TU       2083.333333          .000000
       AU       2916.666667          .000000
       LR         13.755159          .000000
       AR       2932.714313          .000000
       TR       2053.530528          .000000


   ROW    SLACK OR SURPLUS          PRICE
    2)          .000000          26.965066
    3)          .000000          30.626062
    4)      5865.034367            .000000
    5)          .000000         -19.864023
    6)          .000000         -12.796034
    7)      6134.965633            .000000
    8)          .000000       -2332388.904850
    9)          .000000        5004.725331
   10)          .000000          41.786554
   11)          .000000         102.804532
   12)          .000000         -13.948311
   13)          .000000        -130.000000
```

FIGURE 6
Oilco Problem and Solution

(continued)

14)	.000000	-105.917442
15)	.000000	-105.626062
16)	-.000001	-169.971719
17)	.000001	-378.525763
18)	-.000001	-8166.740734
19)	.000000	.000000
20)	.001828	.000000
21)	.000000	3998380.979742
22)	121.210474	.000000
23)	6863.136139	.000000
24)	4150.619020	.000000
25)	2083.333333	.000000
26)	2916.666667	.000000
27)	13.755159	.000000
28)	2053.530528	.000000
29)	2932.714313	.000000
30)	.000000	102.804532

FIGURE 6
(Continued)

average octane level of the inputs used to produce premium gasoline is at least 94, and row 18 (again, not a linear constraint) that the octane level of the unleaded gasoline inputs is at least 88.

Row 19 (again a nonlinear constraint due to the presence of the term SA * AR) ensures that regular gasoline contains at most 3% sulphur; row 20, that premium gasoline contains at most 2.8% sulphur; and row 21, that unleaded gasoline contains at most 3% sulphur.

The amount of each input used to produce each output must be non-negative, required by rows 22–29. Row 30 specifies that the amount of regular gasoline sold must equal the sum of the inputs used to produce regular gasoline.

When solved on LINGO, we obtain a solution with a profit of $4,432.37 (remember the objective function is in cents) earned by producing 5,000 gallons of regular gasoline (with 13.76 gallons of lead, 2,932.71 gallons of Alaska crude, and 2,053.53 gallons of Texas crude); 5,000 gallons of unleaded gasoline (with 2,916.67 gallons of Alaska crude and 2,083.33 gallons of Texas crude); and 11,134.97 gallons of premium gasoline (using 121.21 gallons of lead, 6,863.14 gallons of Texas crude, and 4,150.62 gallons of Alaska crude). The mix of 10,000 gallons of Alaska crude was 90.48% Alaska1 and 9.52% Alaska2.

In Section 11.10, we will discuss how we can be sure that the solution found by LINGO is optimal.

REMARK By using a nonlinear blending model to optimize production of its gasoline products Texaco saves at least $30 million per year. See Dewitt et al. (1989) for details.

EXAMPLE 10 **Warehouse Location**

Truckco is trying to determine where it should locate a single warehouse. The positions in the x–y plane (in miles) of four customers and the number of shipments made annually to each customer are given in Table 5. Truckco wants to locate the warehouse to minimize the total distance trucks must travel annually from the warehouse to the four customers.

Solution Define

$$X = x\text{-coordinate of warehouse}$$

$$Y = y\text{-coordinate of warehouse}$$

$$Di = \text{Distance from customer } i \text{ to warehouse}$$

Ware.lng

The appropriate NLP is given in the LINGO printout in Figure 7 (file Ware.lng). The objective function minimizes the total distance trucks travel each year from the warehouse to the four customers. Rows 2–5 define the distance from each customer to the warehouse in

TABLE 5

Customer	Coordinate x	Coordinate y	Number of Shipments
1	5	10	200
2	10	5	150
3	0	12	200
4	12	0	300

```
MODEL:
    1) MIN= 200 * D1 + 150 * D2 + 200 * D3 + 300 * D4 ;
    2) D1 = ( ( X - 5 ) ^ 2 + ( Y - 10 ) ^ 2 ) ^ .5 ;
    3) D2 = ( ( X - 10 ) ^ 2 + ( Y - 5 ) ^ 2 ) ^ .5 ;
    4) D3 = ( X ^ 2 + ( Y - 12 ) ^ 2 ) ^ .5 ;
    5) D4 = ( ( X - 12 ) ^ 2 + Y ^ 2 ) ^ .5 ;
END


SOLUTION STATUS:  OPTIMAL TO TOLERANCES.  DUAL CONDITIONS:  UNSATISFIED.

              OBJECTIVE FUNCTION VALUE

        1)      5456.539688

        VARIABLE        VALUE        REDUCED COST
           D1         6.582238         .000000
           D2          .686433         .000000
           D3        11.634119         .000000
           D4         5.701011         .000000
            X         9.314167         .000176
            Y         5.028701         .000167

        ROW    SLACK OR SURPLUS         PRICE
        2)          .000000       -200.000000
        3)          .000000       -150.000000
        4)          .000000       -200.000000
        5)          .000000       -300.000000
```

FIGURE 7
Truckco Problem and Solution

terms of the warehouse location. LINGO located the warehouse at $X = 9.31$ and $Y = 5.03$. Each year the truck will travel a total of 5,456.54 miles from the warehouse to the customers.

EXAMPLE 11   Tire Production

Firerock produces rubber used for tires by combining three ingredients: rubber, oil, and carbon black. The cost in cents per pound of each ingredient is given in Table 6.

The rubber used in automobile tires must have a hardness of between 25 and 35, an elasticity of at least 16, and a tensile strength of at least 12. To manufacture a set of four automobile tires, 100 pounds of product is needed. The rubber used to make a set of four tires must contain between 25 and 60 pounds of rubber and at least 50 pounds of carbon black. If we define

$$R = \text{pounds of rubber in mixture used to produce four tires}$$
$$O = \text{pounds of oil in mixture used to produce four tires}$$
$$C = \text{pounds of carbon black used to produce four tires}$$

then statistical analysis has shown that the hardness, elasticity, and tensile strength of a 100-pound mixture of rubber, oil, and carbon black is as follows:

Tensile strength $= 12.5 - .10(O) - .001\,(O)^2$

Elasticity $= 17 + .35R - .04(O) - .002(R)^2$

Hardness $= 34 + .10R + .06(O) - .3(C) + .001(R)(O) + .005(O)^2 + .001C^2$

TABLE 6

Product	Cost (Cents/Pound)
Rubber	4
Oil	1
Carbon black	7

Formulate an NLP whose solution will tell Firerock how to minimize the cost of producing the rubber product needed to manufacture a set of automobile tires.[†]

Solution After defining

$$TS = \text{Tensile strength of mixture}$$
$$E = \text{Elasticity of mixture}$$
$$H = \text{Hardness of mixture}$$

Rubber.lng the LINGO program in Figure 8 (file Rubber.lng) gives the correct formulation. Row 1 minimizes the cost of producing the needed rubber product. Rows 2–4 express the tensile strength, elasticity, and hardness, respectively, of the mixture in terms of its component ingredients. Observe that tensile strength, elasticity, and hardness are each nonlinear functions of R, O, and C. Row 5 requires that we combine 100 pounds of inputs to produce

FIGURE 8

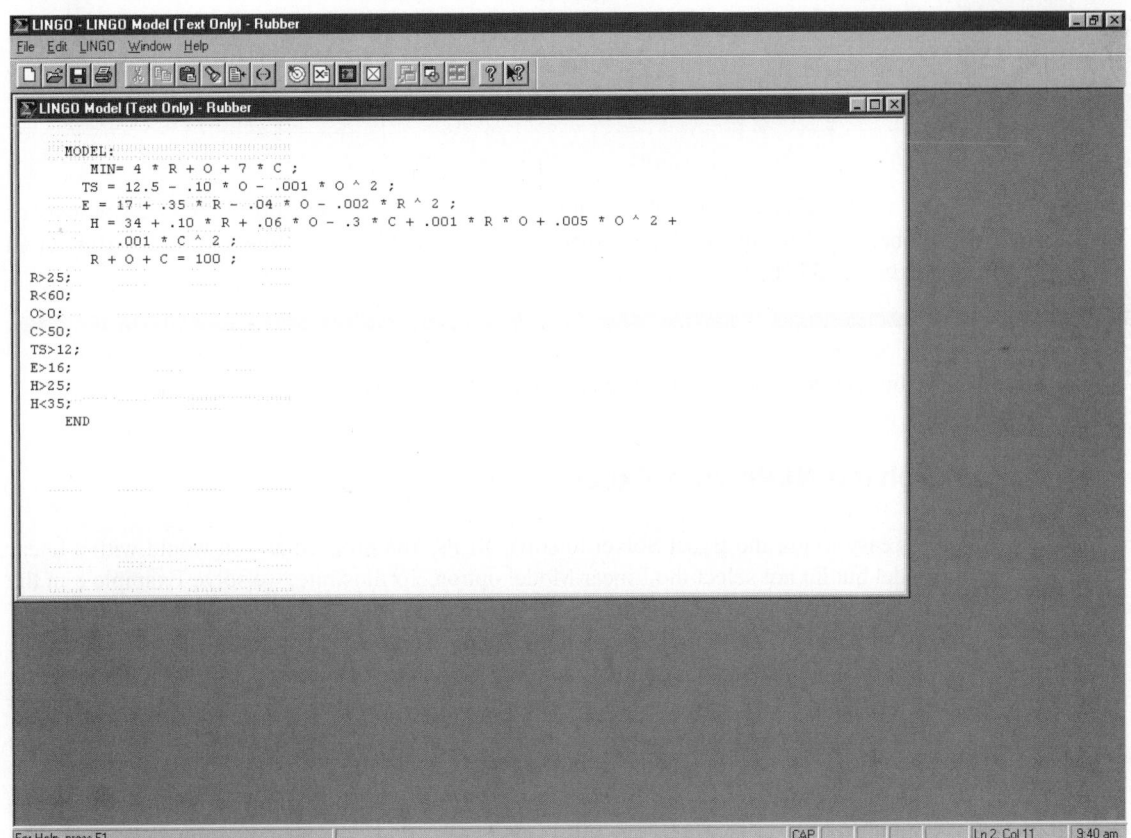

```
MODEL:
    MIN= 4 * R + O + 7 * C ;
    TS = 12.5 - .10 * O - .001 * O ^ 2 ;
    E = 17 + .35 * R - .04 * O - .002 * R ^ 2 ;
    H = 34 + .10 * R + .06 * O - .3 * C + .001 * R * O + .005 * O ^ 2 +
        .001 * C ^ 2 ;
    R + O + C = 100 ;
R>25;
R<60;
O>0;
C>50;
TS>12;
E>16;
H>25;
H<35;
    END
```

[†]Based on Nicholson (1971).

FIGURE **9**

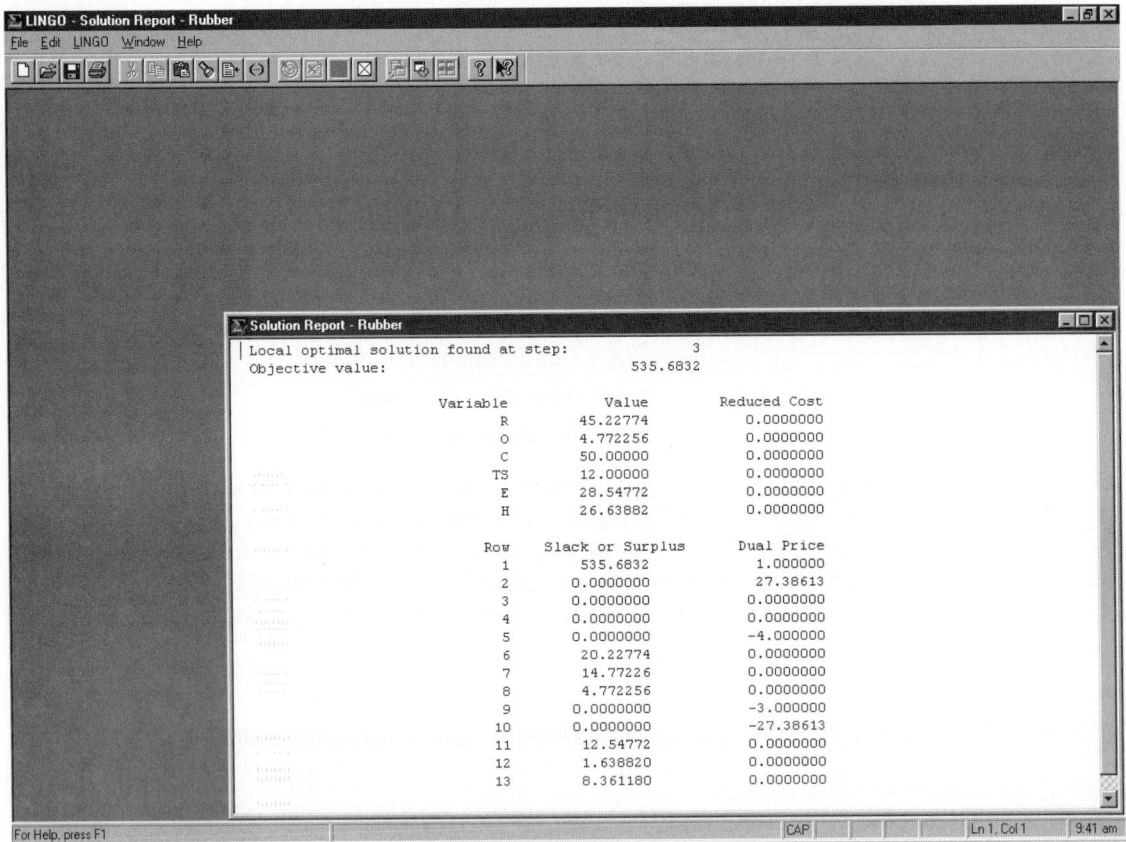

our final rubber product. The solution found by LINGO (Figure 9) is 45.23 pounds of rubber, 4.77 pounds of oil, and 50 pounds of carbon black. The total cost of the 100-pound mixture is \$5.36.

In Section 11.9, we will discuss whether or not this solution is optimal.

Solving NLPs with Excel

Caplabor.xls

It is easy to use the Excel Solver to solve NLPs. You proceed as you would with a linear model but do not select the Linear Model option. To illustrate, we solve Example 8 in the file Caplabor.xls (see Figure 10).

Our changing cells are the capital and labor purchased (cells C5 and D5, respectively). Our target cell is the total number of units produced (computed in cell C8). Our constraint is that the total spent (in cell B11) is less than or equal to \$8. Of course, the quantity of capital and labor purchased must be nonnegative. Our Solver window is shown in Figure 11. We find that the optimal solution is $K = 1$, $L = 4$, and $z = 8$.

For NLPs having multiple local optimal solutions, the Excel Solver may fail to find the optimal solution, because it may pick a local extremum that is not a global extremum. To illustrate, consider the following NLP:

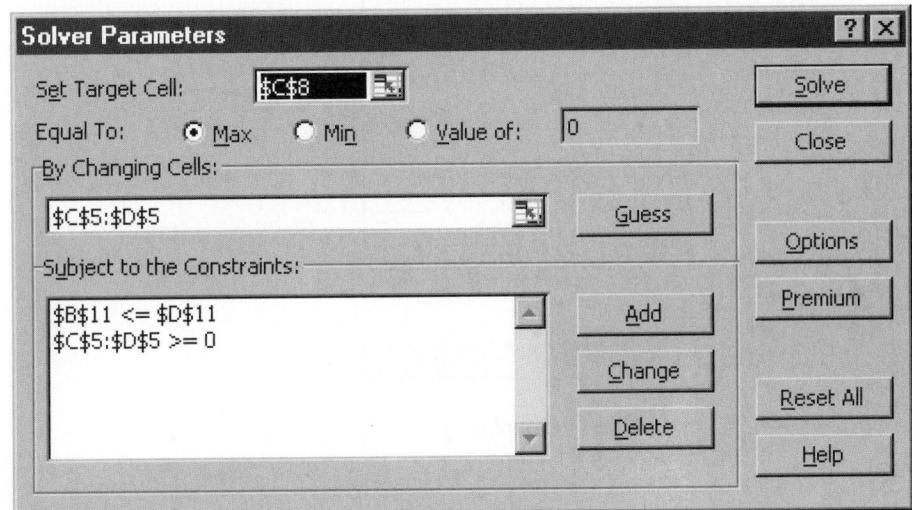

	A	B	C	D
3				
4			Capital	Labor
5		Purchased	1	4
6		Cost	$ 4.00	$ 1.00
7				
8		Units produced	4	
9				
10		Total spent		Available
11		$ 8.00	<=	$ 8.00

FIGURE 10

Solver Parameters

Set Target Cell: $C$8

Equal To: ● Max ○ Min ○ Value of: 0

By Changing Cells:

$C$5:$D$5

Subject to the Constraints:

$B$11 <= $D$11
$C$5:$D$5 >= 0

[Solve] [Close] [Guess] [Options] [Premium] [Add] [Change] [Reset All] [Delete] [Help]

FIGURE 11

$$\max z = (x - 1)(x - 2)(x - 3)(x - 4)(x - 5)$$
$$\text{s.t.} \quad x \geq 1$$
$$x \leq 5$$

Multiple.xls

The graph of this function is shown in Figure 12. Note there are two local maxima for this problem. In file Multiple.xls, we solve this problem twice. The first time, we begin with $x = 2$ and we find the optimal solution, which is $x = 1.36$ and $z = 3.63$ (see Figure 13).

The second time, we begin with $x = 3.5$ and we find the other local maximum, $x = 3.54$ and $z = 1.42$ (see Figure 14). The reason for this is that when we start with $x = 3.5$, the Solver soon hits $x = 3.54$ and finds that the objective function cannot be improved by small moves in either direction. Both LINGO and Solver use calculus-based methods (to be described later in this chapter) to solve NLPs. Any calculus-based approach to solving NLPs runs the risk of finding a local extremum that is not a global extremum. Evolutionary algorithms do not have this drawback; see Chapters 14 and 15 of *Mathematical Programming: Applications and Algorithms* for a discussion of evolutionary algorithms.

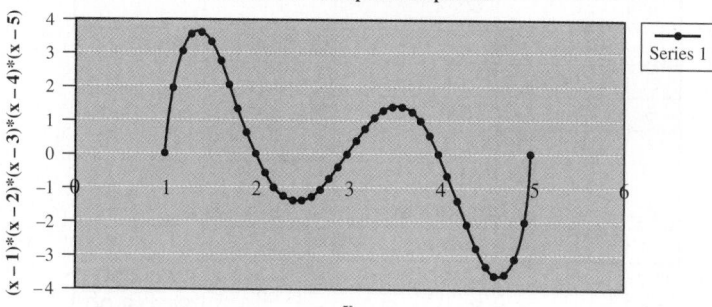

FIGURE 12

	C	D
1		
2		
3		
4	**Start with** **x =2**	right answer!
5		
6		
7	x	f(x)
8	1.355567	3.631432208

FIGURE 13

	C	D
4	**Start with** **x =3.5**	Wrong answer!
5		
6		
7	x	f(x)
8	3.543912	1.418696626

FIGURE 14

PROBLEMS

Group A

1 Q & H Company advertises on soap operas and football games. Each soap opera ad costs $50,000, and each football game ad costs $100,000. Giving all figures in millions of viewers, if S soap opera ads are bought, they will be seen by $5\sqrt{S}$ men and $20\sqrt{S}$ women. If F football ads are bought, they will be seen by $17\sqrt{F}$ men and $7\sqrt{F}$ women. Q & H wants at least 40 million men and at least 60 million women to see its ads.

    a Formulate an NLP that will minimize Q & H's cost of reaching sufficient viewers.

    b Does the NLP violate the Proportionality and Additivity Assumptions?

    c Suppose that the number of women reached by F football ads and S soap opera ads is $7\sqrt{F} + 20\sqrt{S} - 0.2\sqrt{FS}$. Why might this be a more realistic representation of the number of women viewers seeing Q & H's ads?

2 The area of a triangle with sides of length a, b, and c is $\sqrt{s(s - a)(s - b)(s - c)}$, where s is half the perimeter of the triangle. We have 60 ft of fence and want to fence a triangular-shaped area. Formulate an NLP that will enable us to maximize the fenced area.

3 The energy used in compressing a gas (in three stages) from an initial pressure I to a final pressure F is given by

$$K\left\{\sqrt{\frac{p_1}{I}} + \sqrt{\frac{p_2}{p_1}} + \sqrt{\frac{F}{p_2}} - 3\right\}$$

Formulate an NLP whose solution describes how to minimize the energy used in compressing the gas.

4 Use LINGO to solve Problem 1.

5 Use LINGO to solve Problem 2.

6 Use LINGO to solve Problem 3. Use $I = 64$ and $F = 1,000$.

7 For Example 6 of Chapter 8, let A = number of days duration of A is reduced, B = number of days duration of B is reduced, and so on. Suppose that the cost of crashing each activity is as follows:

A, $5A^2$; B, $20B^2$; C, $2C^2$; D, $20D^2$; E, $10E^2$; F, $15F^2$

and that each activity may be "crashed" to a duration of 0 days, if desired. Formulate an NLP that will minimize the cost of finishing the project in 25 days or less.

8 Beerco has $100,000 to spend on advertising in four markets. The sales revenue (in thousands of dollars) that can be created in each market by spending x_i thousand dollars in market i is given in Table 7. To maximize sales revenue, how much money should be spent in each market?

9 Widgetco produces widgets at plant 1 and plant 2. It costs $20x^{1/2}$ to produce x units at plant 1 and $40x^{1/3}$ to produce x units at plant 2. Each plant can produce as many as 70 units. Each unit produced can be sold for $10. At most, 120 widgets can be sold. Formulate an NLP whose solution will tell Widgetco how to maximize profit.

10 Three cities are located at the vertices of an equilateral triangle. An airport is to be built at a location that minimizes the total distance from the airport to the three cities. Formulate an NLP whose solution will tell us where to build the airport. Then solve your NLP on LINGO.

11 The yield of a chemical process depends on the length of time T (in minutes) that the process is run and the temperature TEMP (in degrees centigrade) at which the process is operated. This dependence is described by the equation

YIELD = $87 - 1.4T' + .4$TEMP$' - 2.2T'^2 -$
$$3.2\text{TEMP}'^2 - 4.9(T')(\text{TEMP}')$$

where $T' = (T - 90)/10$ and TEMP$' = $ (TEMP $-150)/5$. T must be between 60 and 120 minutes, while TEMP must be between 100 and 200 degrees. Set up an NLP that could be used to maximize the yield of the process. Use LINGO to solve your NLP.

TABLE 7

Market	Sales Revenue
1	$10x_1^{.4}$
2	$8x_2^{.5}$
3	$12x_3^{.3}$
4	$16x_4^{.6}$

Group B

12 Consider Problem 5 of Section 3.8 with the following modification: Suppose that we can add a chemical called Superquality (SQ) to improve the quality level of gasoline and heating oil. If we add an amount x of SQ to each barrel of gasoline we improve its quality level by $x^{.5}$ over what it would have been. If we add an amount x of SQ to each barrel of heating oil we improve its quality level by $.6x^{.6}$ over what it would have been. The amount of SQ added to heating oil cannot exceed (by weight) 5% of the oils used to make heating oil. Similarly, the amount of SQ added to gasoline cannot exceed (by weight) 5% of the oils used to make gasoline. SQ may be purchased at a cost of $20 per pound. Formulate (and solve with LINGO) an NLP that will help CEO Adam Chandler maximize his profits.

13 A salesperson for Fuller Brush has three options: quit, put forth a low-effort level, or put forth a high-effort level. Suppose for simplicity that each salesperson will either sell $0, $5,000, or $50,000 worth of brushes. The probability of each sales amount depends on the effort level in the manner described in Table 8.

If the salesperson is paid w, then he or she earns a benefit $w^{1/2}$. Low effort costs the salesperson 0 benefit units, while high effort costs 50 benefit units. If the salesperson were to quit Fuller and work elsewhere, then he or she could earn a benefit of 20. Fuller wants all salespeople to put forth a high-effort level. The question is how to minimize the cost of doing it. The company cannot observe the level of effort put forth by a salesperson, but they can observe the size of his or her sale. Thus, the wage is completely determined by the size of the sale. Fuller must then determine w_0 = wage paid for $0 in sale, w_{5000} = wage paid for $5,000 in sales, and $w_{50,000}$ = wage paid for $50,000 in sales. These wages must be set so that the salespeople value the expected benefit from high effort more than quitting and more than low effort. Formulate (and solve on LINGO) an NLP that can be used to ensure that all salespeople put forth high effort. This problem is an example of *agency theory.*[†]

TABLE 8

Size of Sale ($)	Effort Level	
	Low	High
0	.6	.3
5,000	.3	.2
50,000	.1	.5

[†]Based on Grossman and Hart (1983).

11.3 Convex and Concave Functions

Convex and concave functions play an extremely important role in the study of nonlinear programming problems.

Let $f(x_1, x_2, \ldots, x_n)$ be a function that is defined for all points $(x_1, x_2, \ldots, x_n)$ in a convex set S.[†]

DEFINITION ■ A function $f(x_1, x_2, \ldots, x_n)$ is a **convex function** on a convex set S if for any $x' \in S$ and $x'' \in S$

$$f[cx' + (1 - c)x''] \leq cf(x') + (1 - c)f(x'') \tag{3}$$

holds for $0 \leq c \leq 1$. ■

DEFINITION ■ A function $f(x_1, x_2, \ldots, x_n)$ is a **concave function** on a convex set S if for any $x' \in S$ and $x'' \in S$

$$f[cx' + (1 - c)x''] \geq cf(x') + (1 - c)f(x'') \tag{4}$$

holds for $0 \leq c \leq 1$. ■

From (3) and (4), we see that $f(x_1, x_2, \ldots, x_n)$ is a convex function if and only if $-f(x_1, x_2, \ldots, x_n)$ is a concave function, and conversely.

To gain some insights into these definitions, let $f(x)$ be a function of a single variable. From Figure 15 and inequality (3), we find that $f(x)$ is convex if and only if the line segment joining any two points on the curve $y = f(x)$ is never below the curve $y = f(x)$. Similarly, Figure 16 and inequality (4) show that $f(x)$ is a concave function if and only if the straight line joining any two points on the curve $y = f(x)$ is never above the curve $y = f(x)$.

FIGURE 15
A Convex Function

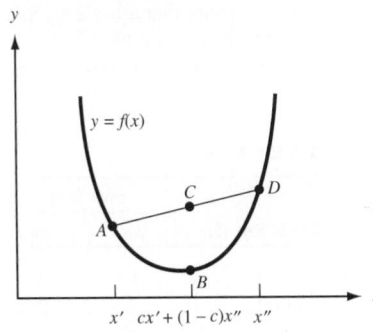

Point $A = (x', f(x'))$
Point $D = (x'', f(x''))$
Point $C = (cx' + (1-c)x'', cf(x') + (1-c)f(x''))$
Point $B = (cx' + (1-c)x'', f(cx' + (1-c)x''))$
From figure: $f(cx' + (1-c)x'') \leq cf(x') + (1-c)f(x'')$

FIGURE 16
A Concave Function

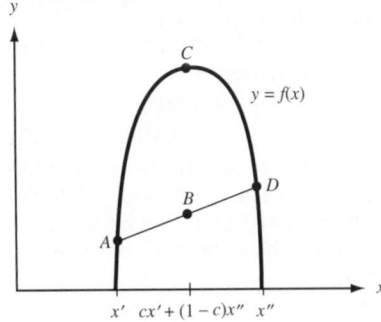

Point $A = (x', f(x'))$
Point $D = (x'', f(x''))$
Point $C = (cx' + (1-c)x'', f(cx' + (1-c)x''))$
Point $B = (cx' + (1-c)x'', cf(x') + (1-c)f(x''))$
From figure: $f(cx' + (1-c)x'') \geq cf(x') + (1-c)f(x'')$

[†]Recall from Chapter 3 that a set S is convex if $x' \in S$ and $x'' \in S$ imply that all points on the line segment joining x' and x'' are members of S. This ensures that $cx' + (1 - c)x''$ will be a member of S.

EXAMPLE 12 | **Convex and Concave Functions**

For $x \geq 0$, $f(x) = x^2$ and $f(x) = e^x$ are convex functions and $f(x) = x^{1/2}$ is a concave function. These facts are evident from Figure 17.

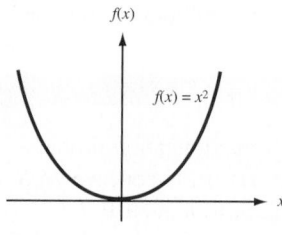

a Convex

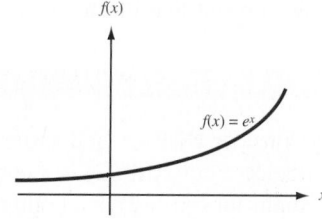

b Convex

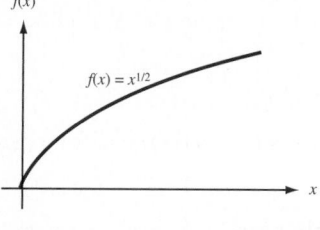

c Concave

FIGURE 17
Examples of Convex and Concave Functions

EXAMPLE 13 | **Sum of Convex Functions**

It can be shown (see Problem 12 at the end of this section) that the sum of two convex functions is convex and the sum of two concave functions is concave. Thus, $f(x) = x^2 + e^x$ is a convex function.

EXAMPLE 14 | **Neither Convex nor Concave Function**

Because the line segment AB lies below $y = f(x)$ and the line segment BC lies above $y = f(x)$, $f(x)$ as pictured in Figure 18 is not a convex or a concave function.

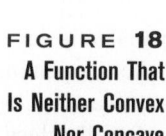

FIGURE 18
A Function That Is Neither Convex Nor Concave

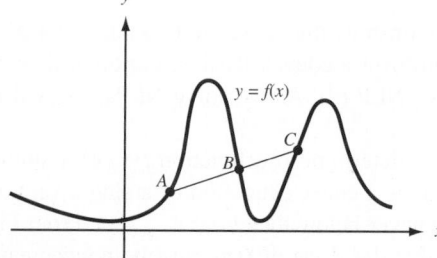

EXAMPLE 15 | **Both Convex and Concave Linear Function**

A linear function of the form $f(x) = ax + b$ is both a convex and a concave function. This follows from

$$f[cx' + (1 - c)x''] = a[cx' + (1 - c)x''] + b$$
$$= c(ax' + b) + (1 - c)(ax'' + b)$$
$$= cf(x') + (1 - c)f(x'')$$

Both (3) and (4) hold with equality, so $f(x) = ax + b$ is both a convex and a concave function.

Before discussing how to determine whether a given function is convex or concave, we prove a result that illustrates the importance of convex and concave functions.

THEOREM 1

Consider NLP (2) and assume it is a maximization problem. Suppose the feasible region S for NLP (2) is a convex set. If $f(x)$ is concave on S, then any local maximum for NLP (2) is an optimal solution to this NLP.

Proof If Theorem 1 is false, then there must be a local maximum $\bar{x}$ that is not an optimal solution to NLP (2). Let S be the feasible region for NLP (2) (we have assumed that S is a convex set). Then, for some $x \in S, f(x) > f(\bar{x})$. The inequality (4) implies that for any c satisfying $0 < c < 1$,

$$f[c\bar{x} + (1 - c)x] \geq cf(\bar{x}) + (1 - c)f(x)$$
$$> cf(\bar{x}) + (1 - c)f(\bar{x}) \qquad [\text{from } f(x) > f(\bar{x})]$$
$$= f(\bar{x})$$

Now observe that for c arbitrarily near 1, $c\bar{x} + (1 - c)x$ is feasible (because S is convex) and is near $\bar{x}$. Thus, $\bar{x}$ cannot be a local maximum. This contradiction proves Theorem 1.

Similar reasoning can be used to prove Theorem 1$'$ (see Problem 11 at the end of this section).

THEOREM 1$'$

Consider NLP (2) and assume it is a minimization problem. Suppose the feasible region S for NLP (2) is a convex set. If $f(x)$ is convex on S, then any local minimum for NLP (2) is an optimal solution to this NLP.

Theorems 1 and 1$'$ demonstrate that if we are maximizing a concave function (or minimizing a convex function) over a convex feasible region S, then any local maximum (or local minimum) will solve NLP (2). As we solve NLPs, we will repeatedly apply Theorems 1 and 1$'$.

We now explain how to determine if a function $f(x)$ of a single variable is convex or concave. Recall that if $f(x)$ is a convex function of a single variable, the line joining any two points on $y = f(x)$ is never below the curve $y = f(x)$. From Figures 9 and 10, we see that $f(x)$ convex implies that the slope of $f(x)$ must be nondecreasing for all values of x.

THEOREM 2

Suppose $f''(x)$ exists for all x in a convex set S. Then $f(x)$ is a convex function on S if and only if $f''(x) \geq 0$ for all x in S.

Because $f(x)$ is convex if and only if $-f(x)$ is concave, Theorem 2′ must also be true.

THEOREM 2′

Suppose $f''(x)$ exists for all x in a convex set S. Then $f(x)$ is a concave function on S if and only if $f''(x) \leq 0$ for all x in S.

EXAMPLE 16   Determining If a Function Is Convex or Concave

1. Show that $f(x) = x^2$ is a convex function on $S = R^1$.

2. Show that $f(x) = e^x$ is a convex function on $S = R^1$.

3. Show that $f(x) = x^{1/2}$ is a concave function on $S = (0, \infty)$.

4. Show that $f(x) = ax + b$ is both a convex and a concave function on $S = R^1$.

Solution
1. $f''(x) = 2 \geq 0$, so $f(x)$ is convex on $S = R^1$.

2. $f''(x) = e^x \geq 0$, so $f(x)$ is convex on $S = R^1$.

3. $f''(x) = -x^{-3/2}/4 \leq 0$, so $f(x)$ is a concave function on $S(0, \infty)$.

4. $f''(x) = 0$, so $f(x)$ is both convex and concave on $S = R^1$.

How can we determine whether a function $f(x_1, x_2, \ldots, x_n)$ of n variables is convex or concave on a set $S \subset R^n$? We assume that $f(x_1, x_2, \ldots, x_n)$ has continous second-order partial derivatives. Before stating the criterion used to determine whether $f(x_1, x_2, \ldots, x_n)$ is convex or concave, we require three definitions.

DEFINITION ■   The **Hessian** of $f(x_1, x_2, \ldots, x_n)$ is the $n \times n$ matrix whose ijth entry is

$$\frac{\partial^2 f}{\partial x_i \partial x_j} \quad ■$$

We let $H(x_1, x_2, \ldots, x_n)$ denote the value of the Hessian at $(x_1, x_2, \ldots, x_n)$. For example, if $f(x_1, x_2) = x_1^3 + 2x_1 x_2 + x_2^2$, then

$$H(x_1, x_2) = \begin{bmatrix} 6x_1 & 2 \\ 2 & 2 \end{bmatrix}$$

DEFINITION ■   An *ith principal minor* of an $n \times n$ matrix is the determinant of any $i \times i$ matrix obtained by deleting $n - i$ rows and the corresponding $n - i$ columns of the matrix.   ■

Thus, for the matrix

$$\begin{bmatrix} -2 & -1 \\ -1 & -4 \end{bmatrix}$$

the first principal minors are -2 and -4, and the second principal minor is $-2(-4) - (-1)(-1) = 7$. For any matrix, the first principal minors are just the diagonal entries of the matrix.

of an $n \times n$ matrix is the determinant of the $k \times k$ matrix obtained by deleting the last $n - k$ rows and columns of the matrix. ■

We let $H_k(x_1, x_2, \ldots, x_n)$ be the *k*th leading principal minor of the Hessian matrix evaluated at the point $(x_1, x_2, \ldots, x_n)$. Thus, if $f(x_1, x_2) = x_1^3 + 2x_1x_2 + x_2^2$, then $H_1(x_1, x_2) = 6x_1$, and $H_2(x_1, x_2) = 6x_1(2) - 2(2) = 12x_1 - 4$.

By applying Theorems 3 and 3′ (stated below, without proof), the Hessian matrix can be used to determine whether $f(x_1, x_2, \ldots, x_n)$ is a convex or a concave (or neither) function on a convex set $S \subset R^n$. [See Bazaraa and Shetty pages 91–93 (1993) for proof of Theorems 3 and 3′.]

THEOREM 3

Suppose $f(x_1, x_2, \ldots, x_n)$ has continuous second-order partial derivatives for each point $x = (x_1, x_2, \ldots, x_n) \in S$. Then $f(x_1, x_2, \ldots, x_n)$ is a convex function on S if and only if for each $x \in S$, all principal minors of H are nonnegative.

EXAMPLE 17   Using the Hessian to Ascertain Convexity or Concavity 1

Show that $f(x_1, x_2) = x_1^2 + 2x_1x_2 + x_2^2$ is a convex function on $S = R^2$.

Solution   We find that

$$H(x_1, x_2) = \begin{bmatrix} 2 & 2 \\ 2 & 2 \end{bmatrix}$$

The first principal minors of the Hessian are the diagonal entries (both equal $2 \geq 0$). The second principal minor is $2(2) - 2(2) = 0 \geq 0$. For any point, all principal minors of H are nonnegative, so Theorem 3 shows that $f(x_1, x_2)$ is a convex function on R^2.

THEOREM 3′

Suppose $f(x_1, x_2, \ldots, x_n)$ has continuous second-order partial derivatives for each point $x = (x_1, x_2, \ldots, x_n) \in S$. Then $f(x_1, x_2, \ldots, x_n)$ is a concave function on S if and only if for each $x \in S$ and $k = 1, 2, \ldots, n$, all nonzero principal minors have the same sign as $(-1)^k$.

EXAMPLE 18   Using the Hessian to Ascertain Convexity or Concavity 2

Show that $f(x_1, x_2) = -x_1^2 - x_1x_2 - 2x_2^2$ is a concave function on R^2.

Solution   We find that

$$H(x_1, x_2) = \begin{bmatrix} -2 & -1 \\ -1 & -4 \end{bmatrix}$$

The first principal minors are the diagonal entries of the Hessian (-2 and -4). These are both nonpositive. The second principal minor is the determinant of $H(x_1, x_2)$ and equals $-2(-4) - (-1)(-1) = 7 > 0$. Thus, $f(x_1, x_2)$ is a concave function on R^2.

EXAMPLE 19   Using the Hessian to Ascertain Convexity or Concavity 3

Show that for $S = R^2$, $f(x_1, x_2) = x_1^2 - 3x_1x_2 + 2x_2^2$ is not a convex or a concave function.

Solution   We have

$$H(x_1, x_2) = \begin{bmatrix} 2 & -3 \\ -3 & 4 \end{bmatrix}$$

The first principal minors of the Hessian are 2 and 4. Because both the first principal minors are positive, $f(x_1, x_2)$ cannot be concave. The second principal minor is $2(4) - (-3)(-3) = -1 < 0$. Thus, $f(x_1, x_2)$ cannot be convex. Together, these facts show that $f(x_1, x_2)$ cannot be a convex or a concave function.

EXAMPLE 20   Using the Hessian to Ascertain Convexity or Concavity 4

Show that for $S = R^3$, $f(x_1, x_2, x_3) = x_1^2 + x_2^2 + 2x_3^2 - x_1x_2 - x_2x_3 - x_1x_3$ is a convex function.

Solution   The Hessian is given by

$$H(x_1, x_2, x_3) = \begin{bmatrix} 2 & -1 & -1 \\ -1 & 2 & -1 \\ -1 & -1 & 4 \end{bmatrix}$$

By deleting rows (and columns) 1 and 2 of Hessian, we obtain the first-order principal minor $4 > 0$. By deleting rows (and columns) 1 and 3 of Hessian, we obtain the first-order principal minor $2 > 0$. By deleting rows (and columns) 2 and 3 of Hessian, we obtain the first-order principal minor $2 > 0$.

By deleting row 1 and column 1 of Hessian, we find the second-order principal minor

$$\det \begin{bmatrix} 2 & -1 \\ -1 & 4 \end{bmatrix} = 7 > 0.$$

By deleting row 2 and column 2 of Hessian, we find the second-order principal minor

$$\det \begin{bmatrix} 2 & -1 \\ -1 & 4 \end{bmatrix} = 7 > 0$$

By deleting row 3 and column 3 of Hessian, we find the second-order principal minor

$$\det \begin{bmatrix} 2 & -1 \\ -1 & 2 \end{bmatrix} = 3 > 0.$$

The third-order principal minor is simply the determinant of the Hessian itself. Expanding by row 1 cofactors we find the third-order principal minor

$$2[(2)(4) - (-1)(-1)] - (-1)[(-1)(4) - (-1)(-1)]$$
$$+(-1)[(-1)(-1) - (-1)(2)] = 14 - 5 - 3 = 6 > 0.$$

Because for all (x_1, x_2, x_3) all principal minors of the Hessian are nonnegative, we have shown that $f(x_1, x_2, x_3)$ is a convex function on R^3.

PROBLEMS

Group A

On the given set S, determine whether each function is convex, concave, or neither.

1 $f(x) = x^3$; $S = [0, \infty)$

2 $f(x) = x^3$; $S = R^1$

3 $f(x) = \frac{1}{x}$; $S = (0, \infty)$

4 $f(x) = x^a$ $(0 \le a \le 1)$; $S = (0, \infty)$

5 $f(x) = \ln x$; $S = (0, \infty)$

6 $f(x_1, x_2) = x_1^3 + 3x_1x_2 + x_2^2$; $S = R^2$

7 $f(x_1, x_2) = x_1^2 + x_2^2$; $S = R^2$

8 $f(x_1, x_2) = -x_1^2 - x_1x_2 - 2x_2^2$; $S = R^2$

9 $f(x_1, x_2, x_3) = -x_1^2 - x_2^2 - 2x_3^2 + .5x_1x_2$; $S = R^3$

10 For what values of a, b, and c will $ax_1^2 + bx_1x_2 + cx_2^2$ be a convex function on R^2? A concave function on R^2?

Group B

11 Prove Theorem 1'.

12 Show that if $f(x_1, x_2, \ldots, x_n)$ and $g(x_1, x_2, \ldots, x_n)$ are convex functions on a convex set S, then $h(x_1, x_2, \ldots, x_n) = f(x_1, x_2, \ldots, x_n) = g(x_1, x_2, \ldots, x_n)$ is a convex function on S.

13 If $f(x_1, x_2, \ldots, x_n)$ is a convex function on a convex set S, show that for $c \ge 0$, $g(x, x_2, \ldots, x_n) = cf(x_1, x_2, \ldots, x_n)$ is a convex function on S, and for $c \le 0$, $g(x_1, x_2, \ldots, x_n) = cf(x_1, x_2, \ldots, x_n)$ is a concave function on S.

14 Show that if $y = f(x)$ is a concave function on R^1, then $z = \frac{1}{f(x)}$ is a convex function [assume that $f(x) > 0$].

15 A function $f(x_1, x_2, \ldots, x_n)$ is *quasi-concave* on a convex set $S \subset R^n$ if $x' \in S, x'' \in S$, and $0 \le c \le 1$ implies

$$f[cx' + (1 - c)x''] \ge \min[f(x'), f(x'')]$$

Show that if f is concave on R^1, then f is quasi-concave. Which of the functions in Figure 19 is quasi-concave? Is a quasi-concave function necessarily a concave function?

16 From Problem 12, it follows that the sum of concave functions is concave. Is the sum of quasi-concave functions necessarily quasi-concave?

17 Suppose a function's Hessian has both positive and negative entries on its diagonal. Show that the function is neither concave nor convex.

18 Show that if $f(x)$ is a non-negative, increasing concave function, then $\ln[f(x)]$ is also a concave function.

19 Show that if a function $f(x_1, x_2, \ldots, x_n)$ is quasi-concave on a convex set S, then for any number a the set $S_a =$ all points satisfying $f(x_1, x_2, \ldots, x_n) \ge a$ is a convex set.

20 Show that Theorem 1 is untrue if f is a quasi-concave function.

21 Suppose the constraints of an NLP are of the form $g_i(x_1, x_2, \ldots, x_n) \le b_i (i = 1, 2, \ldots m)$. Show that if each of the g_i is a convex function, then the NLP's feasible region is convex.

Group C

22 If $f(x_1, x_2)$ is a concave function on R^2, show that for any number a, the set of (x_1, x_2) satisfying $f(x_1, x_2) \ge a$ is a convex set.

23 Let Z be a $N(0, 1)$ random variable, and let $F(x)$ be the cumulative distribution function for Z. Show that on $S = (-\infty, 0]$, $F(x)$ is an increasing convex function, and on $S = [0, \infty)$, $F(x)$ is an increasing concave function.

24 Recall the Dakota LP discussed in Chapter 6. Let $v(L, FH, CH)$ be the maximum revenue that can be earned when L sq board ft of lumber, FH finishing hours, and CH carpentry hours are available.

  a Show that $v(L, FH, CH)$ is a concave function.

  b Explain why this result shows that the value of each additional available unit of a resource must be a nonincreasing function of the amount of the resource that is available.

FIGURE 19

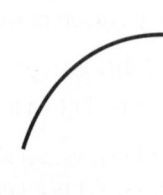

a   b   c

11.4 Solving NLPs with One Variable

In this section, we explain how to solve the NLP

$$\text{max (or min) } f(x)$$
$$\text{s.t.} \quad x \in [a, b] \tag{5}$$

[If $b = \infty$, then the feasible region for NLP (5) is $x \geq a$, and if $a = -\infty$, then the feasible region for (5) is $x \leq b$.]

To find the optimal solution to (5), we find all local maxima (or minima). A point that is a local maximum or a local minimum for (5) is called a local extremum. Then the optimal solution to (5) is the local maximum (or minimum) having the largest (or smallest) value of $f(x)$. Of course, if $a = -\infty$ or $b = \infty$, then (5) may have no optimal solution (see Figure 20).

There are three types of points for which (5) can have a local maximum or minimum (these points are often called *extremum candidates*):

Case 1 Points where $a < x < b$, and $f'(x) = 0$ [called a stationary point of $f(x)$].

Case 2 Points where $f'(x)$ does not exist.

Case 3 Endpoints a and b of the interval $[a, b]$.

Case 1. Points Where $a < x < b$ and $f'(x) = 0$

Suppose $a < x < b$, and $f'(x_0)$ exists. If x_0 is a local maximum or a local minimum, then $f'(x_0) = 0$. To see this, look at Figures 21a and 21b. From Figure 21a, we see that if $f'(x_0) > 0$, then there are points x_1 and x_2 near x_0 where $f(x_1) < f(x_0)$ and $f(x_2) > f(x_0)$. Thus, if $f'(x_0) > 0$, x_0 cannot be a local maximum or a local minimum. Similarly, Figure 21b shows that if $f'(x_0) < 0$, then x_0 cannot be a local maximum or a local minimum. From Figures 21c and 21d, however, we see $f'(x_0) = 0$, then x_0 may be a local maximum or a local minimum. Unfortunately, Figure 21e shows that $f'(x_0)$ can equal zero without x_0 being a local maximum or a local minimum. From Figure 21c, we see that if $f'(x)$ changes from positive to negative as we pass through x_0, then x_0 is a local maximum. Thus, if $f''(x_0) < 0$, x_0 is a local maximum. Similarly, from Figure 21d, we see that if $f'(x)$ changes from negative to positive as we pass through x_0, x_0 is a local minimum. Thus, if $f''(x_0) > 0$, x_0 is a local minimum.

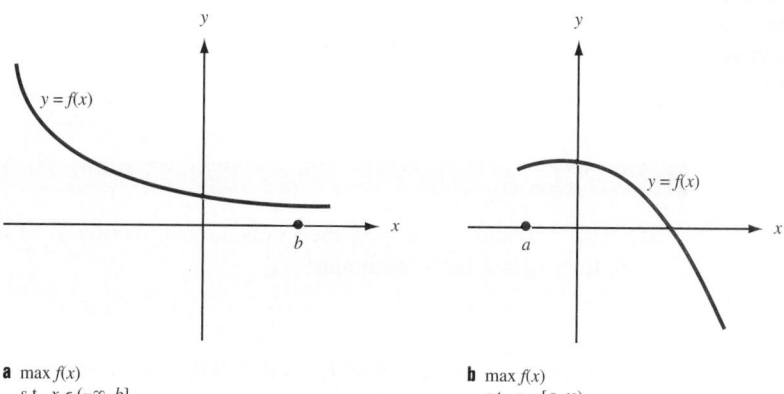

FIGURE 20
NLPs with No Solution

a max $f(x)$
s.t. $x \in (-\infty, b]$

b max $f(x)$
s.t. $x \in [a, \infty)$

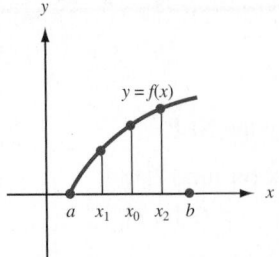

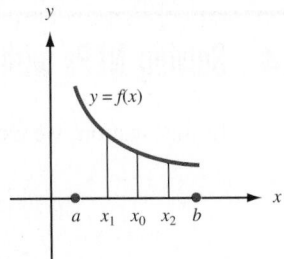

a $f'(x_0) > 0$
$f(x_1) < f(x_0)$
$f(x_2) > f(x_0)$
x_0 not a local extremum

b $f'(x_0) < 0$
$f(x_1) > f(x_0)$
$f(x_2) < f(x_0)$
x_0 not a local extremum

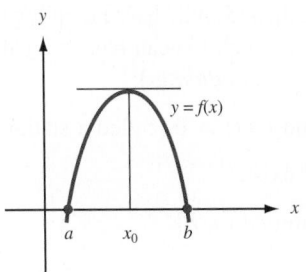

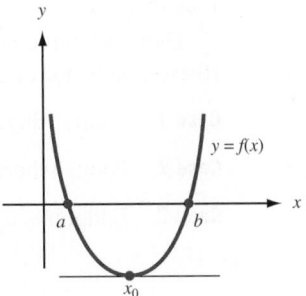

c $f'(x_0) = 0$
For $x < x_0, f'(x) > 0$
For $x > x_0, f'(x) < 0$
x_0 is a local maximum

d $f'(x_0) = 0$
For $x < x_0, f'(x) < 0$
For $x > x_0, f'(x) > 0$
x_0 is a local maximum

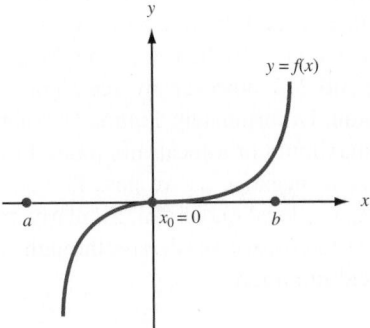

FIGURE 21
How to Determine
Whether x_0 Is a
Local Maximum or a
Local Minimum When
$f'(x_0)$ Exists

e $x_0 = 0$ not a local maximum
or a local minimum
but $f'(x_0) = 0$

THEOREM 4

If $f'(x_0) = 0$ and $f''(x_0) < 0$, then x_0 is a local maximum. If $f'(x_0) = 0$ and $f''(x_0) > 0$, then x_0 is a local minimum.

What happens if $f'(x_0) = 0$ and $f''(x_0) = 0$ (this is the case in Figure 21e)? In this case, we determine whether x_0 is a local maximum or a local minimum by applying Theorem 5.

We omit the proofs of Theorems 4 and 5. [They follow in a straightforward fashion by applying the definition of a local maximum and a local minimum to the Taylor series expansion of $f(x)$ about x_0.] Theorem 4 is a special case of Theorem 5. We ask you to prove Theorems 4 and 5 in Problems 16 and 17.

Case 2. Points Where $f'(x)$ Does Not Exist

If $f(x)$ does not have a derivative at x_0, x_0 may be a local maximum, a local minimum, or neither (see Figure 22). In this case, we determine whether x_0 is a local maximum or a local minimum by checking values of $f(x)$ at points $x_1 < x_0$ and $x_2 > x_0$ near x_0. The four possible cases that can occur are summarized in Table 9.

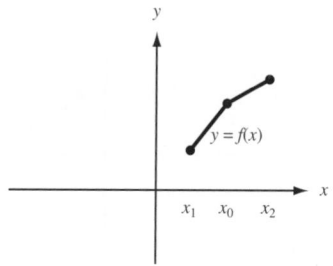

a x_0 not a local extremum

b x_0 not a local extremum

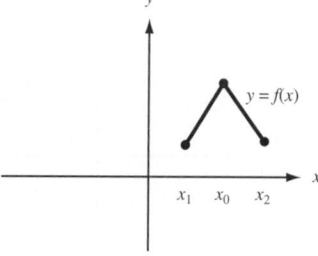

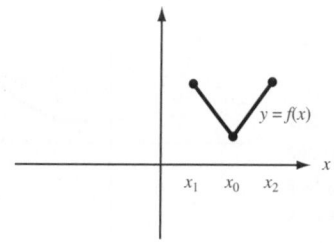

FIGURE 22
How to Determine Whether x_0 Is a Local Maximum or a Local Minimum When $f'(x_0)$ Does Not Exist

c x_0 is a local maximum

d x_0 is a local minimum

TABLE **9**
How to Determine Whether a Point Where $f'(x)$ Does Not Exist Is a Local Maximum
or a Local Minimum

Relationship Between $f(x_0)$, $f(x_1)$, and $f(x_2)$	x_0	Figure
$f(x_0) > f(x_1)$; $f(x_0) < f(x_2)$	Not local extremum	16a
$f(x_0) < f(x_1)$; $f(x_0) > f(x_2)$	Not local extremum	16b
$f(x_0) \geq f(x_1)$; $f(x_0) \geq f(x_2)$	Local maximum	16c
$f(x_0) \leq f(x_1)$; $f(x_0) \leq f(x_2)$	Local minimum	16d

Case 3. Endpoints *a* and *b* of [*a*, *b*]

From Figure 23, we see that

If $f'(a) > 0$, then *a* is a local minimum.

If $f'(a) < 0$, then *a* is a local maximum.

If $f'(b) > 0$, then *b* is a local maximum.

If $f'(b) < 0$, then *b* is a local minimum.

If $f'(a) = 0$ or $f'(b) = 0$, draw a sketch like Figure 22 to determine whether *a* or *b* is a local extremum.

The following examples illustrate how these ideas can be applied to solve NLPs of the form (5).

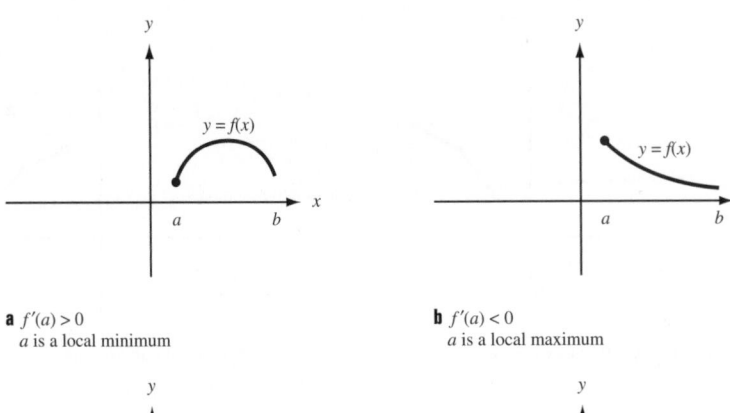

a $f'(a) > 0$
   a is a local minimum

b $f'(a) < 0$
   a is a local maximum

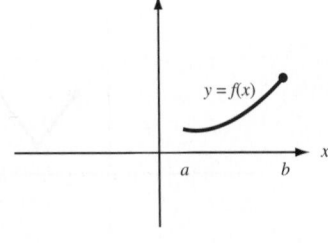

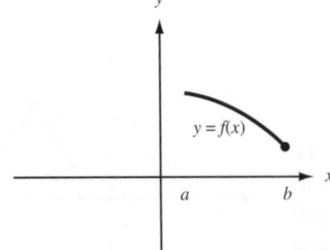

FIGURE 23
How to Determine Whether x_0 Is a Local Maximum or a Local Minimum If x_0 Is an Endpoint

c $f'(b) > 0$
   b is a local maximum

d $f'(b) < 0$
   b is a local minimum

EXAMPLE 21 **Profit Maximization by Monopolist**

It costs a monopolist $5/unit to produce a product. If he produces x units of the product, then each can be sold for $10 - x$ dollars ($0 \le x \le 10$). To maximize profit, how much should the monopolist produce?

Solution Let $P(x)$ be the monopolist's profit if he produces x units. Then

$$P(x) = x(10 - x) - 5x = 5x - x^2 \quad (0 \le x \le 10)$$

Thus, the monopolist wants to solve the following NLP:

$$\max P(x)$$
$$\text{s.t.} \quad 0 \le x \le 10$$

We now classify all extremum candidates:

Case 1 $P'(x) = 5 - 2x$, so $P'(2.5) = 0$. Because $P''(x) = -2$, $x = 2.5$ is a local maximum yielding a profit of $P(2.5) = 6.25$.

Case 2 $P'(x)$ exists for all points in [0, 10], so there are no Case 2 candidates.

Case 3 $a = 0$ has $P'(0) = 5 > 0$, so $a = 0$ is a local minimum; $b = 10$ has $P'(10) = -15 < 0$, so $b = 10$ is a local minimum.

Thus, $x = 2.5$ is the only local maximum. This means that the monopolist's profits are maximized by choosing $x = 2.5$.

Observe that $P''(x) = -2$ for all values of x. This shows that $P(x)$ is a concave function. Any local maximum for $P(x)$ must be the optimal solution to the NLP. Thus, Theorem 1 implies that once we have determined that $x = 2.5$ is a local maximum, we know that it is the optimal solution to the NLP.

EXAMPLE 22 **Finding Global Maximum When Endpoint Is a Maximum**

Let

$$f(x) = 2 - (x - 1)^2 \quad \text{for} \quad 0 \le x < 3$$
$$f(x) = -3 + (x - 4)^2 \quad \text{for} \quad 3 \le x \le 6$$

Find

$$\max f(x)$$
$$\text{s.t.} \quad 0 \le x \le 6$$

Solution **Case 1** For $0 \le x < 3$, $f'(x) = -2(x - 1)$ and $f''(x) = -2$. For $3 < x \le 6$, $f'(x) = 2(x - 4)$ and $f''(x) = 2$. Thus, $f'(1) = f'(4) = 0$. Because $f''(1) < 0$, $x = 1$ is a local maximum. Because $f''(4) > 0$, $x = 4$ is a local minimum.

Case 2 From Figure 24, we see that $f(x)$ has no derivative at $x = 3$ (for x slightly less than 3, $f'(x)$ is near -4, and for x slightly bigger than 3, $f'(x)$ is near -2). Because $f(2.9) = -1.61$, $f(3) = -2$, and $f(3.1) = -2.19$, $x = 3$ is not a local extremum.

Case 3 Because $f'(0) = 2 > 0$, $x = 0$ is a local minimum. Because $f'(6) = 4 > 0$, $x = 6$ is a local maximum.

Thus, on [0, 6], $f(x)$ has a local maximum for $x = 1$ and $x = 6$. Because $f(1) = 2$ and $f(6) = 1$, we find that the optimal solution to the NLP occurs for $x = 1$.

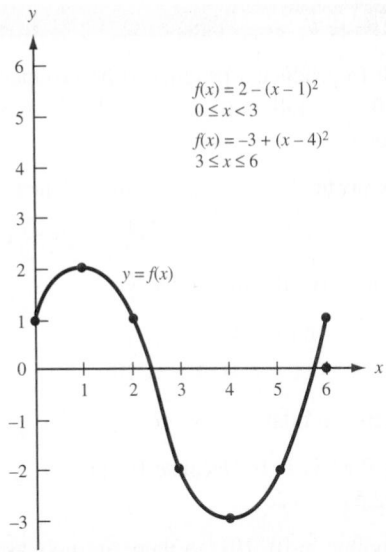

FIGURE 24
Graph for Example 22

$f(x) = 2 - (x - 1)^2$
$0 \le x < 3$

$f(x) = -3 + (x - 4)^2$
$3 \le x \le 6$

$y = f(x)$

Pricing and Nonlinear Optimization

An important business decision is the determination of the profit-maximizing price that should be charged for a product. Demand for a product is often modeled as a linear function of price

$$\text{Demand} = a - b(\text{price})$$

where a and b are constants. If the linear demand function is relevant, then profit from a product with a unit cost of c is given by

$$(\text{Price} - c)*[a - b(\text{price})].$$

This implies that profit is a concave function of price, and Solver should find the profit-maximizing price. In this section, we give two Solver models (based on Dolan and Simon, 1997) that can be used to determine optimal prices.

Our first model tackles the following problem: As exchange rates fluctuate, how should a U.S. company change the overseas price of its product? To be more specific, suppose Eli Daisy is selling a drug in Germany. Its goal is to maximize its profit in dollars, but when the drug is sold in Germany, it receives marks. To maximize Daisy's dollar profit, how should the price in marks vary with the exchange rate? To illustrate the ideas involved, consider the following example.

EXAMPLE 23 **Pricing When Exchange Rates Change**

The drug taxoprol costs $60 to produce. Currently, the exchange rate is .667 $/mark, and we are charging 150 marks for taxoprol. Current demand for taxoprol is 100 units, and it is estimated that the elasticity for taxoprol is 2.5. Assuming a linear demand curve, determine how the price (in marks) for taxoprol should vary with the exchange rate.

Solution We begin by determining the linear demand curve that relates demand to the price in marks. Currently, demand is 100, and the price is 150 marks. Recall from economics that the price elasticity of a product is the percentage decrease in sales that will result from a

1% increase in price. Price elasticity is 2.5, so a 1% increase in price (to 151.5) will result in a 2.5% decrease in demand (to $100 - 2.5 = 97.5$). In the file Intprice.xls (sheet Linear Demand), we entered these two points in B12:C13. We now find the slope and intercept of the demand curve.

$$\text{Slope} = \frac{97.5 - 100}{151.5 - 150} = -1.6667$$

(because demand is larger than 1 in absolute value, demand is elastic).

The slope is computed in D13 with the formula

$$=(B13 - B12)/(A13 - A12)$$
$$\text{Intercept} = 100 + (-150)(-1.6667) = 350$$

The intercept is computed in D14 with the formula

$$=B12 + (-A12)*(D13)$$

Thus, demand $= 350 - 1.6667$(price in marks). (See Figure 25.)

We can now compute (for a trial set of prices) our profit for exchange rates ranging from .4 \$/mark to 1 \$/mark. Then we use Solver to find the set of prices maximizing the *sum* of these profits. This will ensure that we will have found a profit-maximizing price for a variety of exchange rates.

Step 1 Enter trial values for the exchange rate (\$/mark) in the cell range B4:J4.

Step 2 In B5:J5, enter the unit cost in dollars (\$60).

Step 3 In B6:J6, enter trial prices (in marks) for taxoprol.

Step 4 Observe that demand for each exchange rate is given by $350 - 1.66667*$(price in marks)

In cells B7:J7, we determine the demand for each exchange rate. In B7, we find the demand for the exchange rate of .6667 \$/mark with the formula

$$=\$D\$14 + \$D\$13*B6$$

Copying this formula to the range C7:J7 computes the demand for all other exchange rates.

Step 5 Observe that profit in dollars is given by

$$[(\$/\text{mark})*\text{price in marks} - \text{cost in dollars}]*(\text{demand})$$

FIGURE **25**

	A	B	C	D	E	F	G	H	I	J	K
1	Price dependence										
2	on exchange rate										
3											
4	Current $/DM	0.666667	0.4	0.5	0.6	0.666667	0.7	0.8	0.9	1	
5	Unit Cost US $	60	60	60	60	60	60	60	60	60	
6	Current price DM	149.9999	179.9999	164.9999	154.9999	149.9999	147.8571	142.4999	138.3333	134.9999	
7	Current demand	100.0002	50.00015	75.00014	91.6668	100.0002	103.5716	112.5001	119.4446	125.0001	
8	Current profit US$	4000.005	600	1687.5	3025	4000.005	4505.357	6075	7704.167	9375	Total Profit
9	Elasticity	2.5	2.5	2.5	2.5	2.5	2.5	2.5	2.5	2.5	36972.03
10											
11	Price DM	demand									
12	150	100									
13	151.5	97.5	slope	-1.666667							
14	Demand = 350-(5/3)*price		intercept	350							

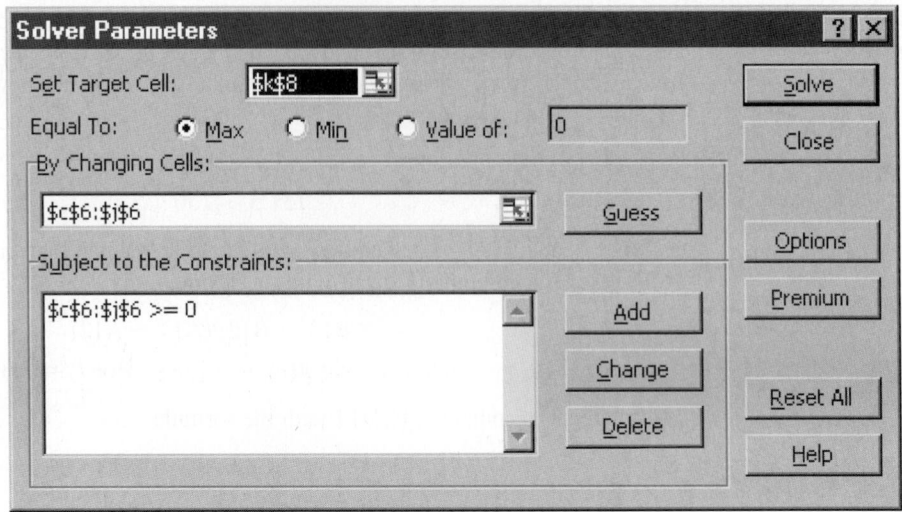

FIGURE **26**

In cells B8:J8, we compute the dollar profit (for the trial prices) for each exchange rate. In B8, we find the profit for our current exchange rate (.66667 \$/mark) and current price (150 marks) with the formula

$$=(B4*B6 - B5)*B7$$

Copying this formula to the cell range C8:J8 computes profits for all other exchange rates.

Step 6 In cell K8, we add the profit for all exchange rates with the formula

$$= SUM(C8:J8)$$

Step 7 We now use the Solver to determine the profit-maximizing price for each exchange rate. By changing the nonnegative prices for each exchange rate (C6:J6), we can maximize the sum of the profits (K8). Because each price only affects the profit for the exchange rate in its own column, this ensures that we find the profit-maximizing price for each exchange rate. For example, for .66667 \$/mark, the optimal price is 150 marks. Note that if the mark drops in value by 25% to .5 \$/mark, we only raise the cost in marks by 10% ($\frac{165 - 150}{150} = .10$). Because of the elastic demand, profit maximization does not call for making the German customers absorb all of the loss in dollars due to depreciation of the mark. Our Solver window is as shown in Figure 26.

How can we use Solver to determine a profit-maximizing price? One way is to derive a demand curve by breaking the market into segments and identifying a low price, a medium price, and a high price. For each of these prices and market segments, ask company experts to estimate product demand. Then we can use Excel's trend-curve-fitting capabilities to fit a quadratic function that can be used to estimate each segment's demand for different prices. Finally, we can add the segment demand curves to derive an aggregate demand curve and use the Solver to determine the profit-maximizing price. The procedure is illustrated in Example 24 (based on Dolan and Simon (1996)).

EXAMPLE 24 | Pricing a Candy Bar

A candy bar costs 55 cents to produce. We are considering charging a price of between $1.10 and $1.50 for this candy bar. For a price of $1.10, $1.30, and $1.50, the marketing department estimates the demand for the candy bar in the three regions where the candy bar will be sold (see Table 10). What price will maximize profit?

Solution

Expdemand.xls

Step 1 We begin by fitting a quadratic curve to the three demands specified in Table 10 for each region. See file Expdemand.xls. For example, for region 1 we use the X-Y Chart Wizard option to plot D4:E6. Click the points on the graph until they turn yellow and choose Insert Trendline Polynomial (2) and check the equation option to make sure the quadratic equation that exactly fits the three points is listed.

Thus we estimate region 1 demand (see Figure 27):

$$= -87.5*(price)^2 + 195*(price) - 73.625$$

Similarly, in regions 2 and 3 we find the following demand equations (see Figures 28 and 29):

$$\text{Region 2 demand} = -75*(price)^2 + 155*(price) - 47.75$$
$$\text{Region 3 demand} = -12.5*(price)^2 - 5*(price) + 44.625$$

Step 2 We now enter a trial price in cell H4 and determine in cells I4:K4 the demand (in thousands of units) for that price in each region:

$$\text{Region 1 demand (cell I4)} = -87.5*H4\char`^2 + 195*H4 - 73.625$$
$$\text{Region 2 demand (cell J4)} = -75*(H4)\char`^2 + 155*H4 - 47.75$$
$$\text{Region 3 demand (cell K4)} = -12.5*(H4)\char`^2 - 5*H4 + 44.625$$

Step 3 In cell L4, compute the total demand (in thousands of units) with the formula

$$= SUM(I4:K4)$$

Step 4 In cell I6, compute our profit (in thousands of dollars):

$$= (H4 - I2)*L4$$

Step 5 We are now ready to invoke the Solver to find the profit-maximizing price. We simply maximize profit (cell I6), with price (H4) being a changing cell. Because

TABLE 10

Price ($) (Unit cost: 0.55)	Demand (in Thousands)		
	Region 1	Region 2	Region 3
Low (1.10)	35	32	24
Medium (1.30)	32	27	17
High (1.50)	22	16	9

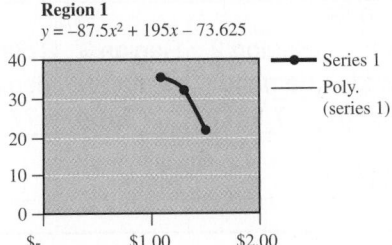

FIGURE 27

our demand curves are, in theory, only valid for prices between \$1.10 and \$1.50, we add the constraints H4≥1.10 and H4≤1.50 (see Figure 30 for the Solver window).

Why is the model nonlinear? As Figure 31 shows, we find the profit-maximizing price to be \$1.29.

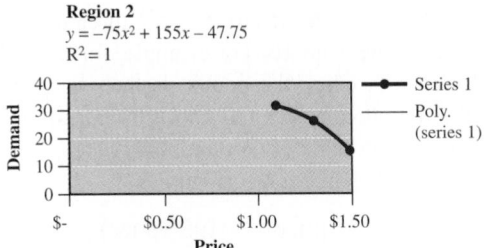

Region 2
$y = -75x^2 + 155x - 47.75$
$R^2 = 1$

FIGURE **28**

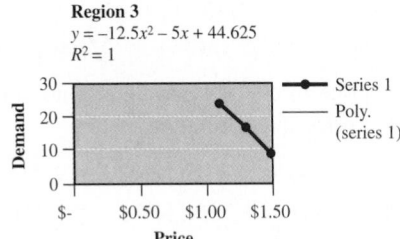

Region 3
$y = -12.5x^2 - 5x + 44.625$
$R^2 = 1$

FIGURE **29**

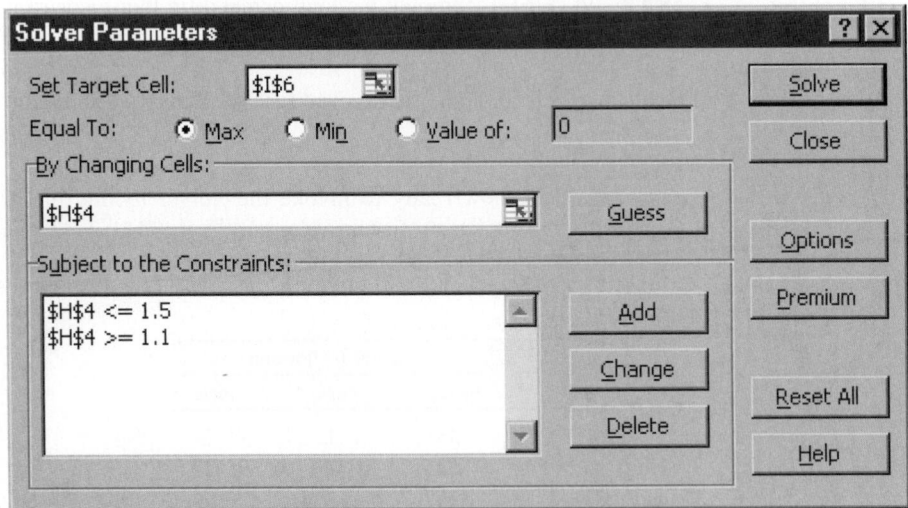

FIGURE **30**

	H	I	J	K	L
2	Variable cost	0.55			
3	Price	Region 1 demand	Region 2 demand	Region 3 demand	Total demand
4	1.286325018	32.42807	27.53297	17.51047	77.47152
5					
6	Profit	57.04422			
7	(000's)				

FIGURE **31**

If we are trying to maximize a function $f(x)$ that is a product of several functions, it is often easier to maximize $\ln[f(x)]$. Because ln is an increasing function, we know that any x solving max $z' = \ln[f(x)]$ subject to $x \in S$ will also solve max $z = f(x)$ over $x \in S$. See Problem 4 for application of this idea.

Solving One-Variable NLPs with LINGO

If you are maximizing a concave objective function $f(x)$ (or even if the logarithm of the objective function in a maximization problem is a concave function), you can be certain that LINGO will find the optimal solution to the NLP

$$\max z = f(x)$$
$$\text{s.t.} \quad a \leq x \leq b$$

Thus, if we solved Example 21 on LINGO, we would be confident that it had found the correct answer. In Example 22, however, we could not be sure that LINGO would find the maximum value of $f(x)$ on the interval $[0, 6]$.

Similarly, if you are minimizing a convex objective function, then you know that LINGO will find the optimal solution to the NLP

$$\min z = f(x)$$
$$\text{s.t.} \quad a \leq x \leq b$$

If you are trying to minimize a nonconvex function or maximize a nonconcave function of a one-variable NLP subject to the constraint $a \leq x \leq b$, then LINGO may find a local extremum that does not solve the NLP. In such situations, the user can influence the solution found by LINGO by inputting a starting value for x with the INIT command. For example, if we direct LINGO to solve

$$\min z = x \sin(\pi x)$$
$$\text{s.t.} \quad 0 \leq x \leq 6$$

LINGO may find the local minimum $x = 1.564$. This is because as a default, LINGO first guesses that $x = 0$, and at $x = 1.564$ the conditions for a local minimum [$f'(x) = 0$ and $f''(x) > 0$] are satisfied. A sketch of the function $x \sin(\pi x)$ reveals that another local minimum occurs for x between 5 and 6. By using the INIT command, we may direct LINGO to start near $x = 5$. Then LINGO does indeed find the optimal solution ($x = 5.52$) to the NLP. For instance, to have LINGO start with $x1=2$ and $x2=3$, we would add the following section to our LINGO program:

INIT:

$x1=2$;

$x2=3$;

ENDINIT

PROBLEMS

Group A

1 It costs a company $100 in variable costs to produce an air conditioner, plus a fixed cost of $5,000 if any air conditioners are produced. If the company spends x dollars on advertising, then it can sell $x^{1/2}$ air conditioners at $300 each. How can the company maximize its profit? If the fixed cost of producing any air conditioners were $20,000, what should the company do?

2 If a monopolist produces q units, she can charge $100 - 4q$ dollars/unit. The fixed cost of production is \$50, and the variable per-unit cost is \$2. How can the monopolist maximize profits? If a sales tax of \$2/unit must be paid by the monopolist, then would she increase or decrease production?

3 Show that for all x, $e^x \geq x + 1$. [Hint: Let $f(x) = e^x - x - 1$. Show that

$$\min f(x)$$
$$\text{s.t.} \quad x \in R$$

occurs for $x = 0$.]

4 Suppose that in n "at bats," a baseball player gets x hits. Suppose we want to estimate the player's probability (p) of getting a hit on each "at bat." The method of maximum likelihood estimates p by $\hat{p}$, where $\hat{p}$ maximizes the probability of observing x hits in n "at bats." Show that the method of maximum likelihood would choose $\hat{p} = \frac{x}{n}$.

5 Find the optimal solution to

$$\max x^3$$
$$\text{s.t.} \quad -1 \leq x \leq 1$$

6 Find the optimal solution to

$$\min x^3 - 3x^2 + 2x - 1$$
$$\text{s.t.} \quad -2 \leq x \leq 4$$

7 During the Reagan administration, economist Arthur Laffer became famous for his Laffer curve, which implied that an increase in the tax rate might decrease tax revenues, while a decrease in the tax rate might increase tax revenues. This problem illustrates the idea behind the Laffer curve. Suppose that if an individual puts in a degree of effort e, he or she earns a revenue of $10e^{1/2}$. Also suppose that an individual associates a cost of e with an effort level of e. Suppose further that the tax rate is T. This means that each individual gets to keep a fraction $1 - T$ of before-tax revenue. Show that $T = .5$ maximizes the government's tax revenues. Thus, if the tax rate were 60%, then a cut in the tax rate would increase revenues.

8 The cost per day of running a hospital is $200{,}000 + .002x^2$ dollars, where $x =$ patients served per day. What size hospital minimizes the per-patient cost of running the hospital?

9 Each morning during rush hour, 10,000 people want to travel from New Jersey to New York City. If a person takes the subway, the trip lasts 40 minutes. If x thousand people per morning drive to New York, it takes $20 + 5x$ minutes to make the trip. This problem illustrates a basic fact of life: If people are left to their own devices, they will cause more congestion than need actually occur!

   a Show that if people are left to their own devices, an average of 4,000 people will travel by road from New Jersey to New York. Here you should assume that people will divide up between the subways and roads in a way that makes the average travel time by road = average travel time by subway. When this "equilibrium" occurs, nobody has an incentive to switch from road to subway or subway to road.

   b Show that the average travel time per person is minimized if 2,000 people travel by road.

10 Currently, the exchange rate is 100 yen per dollar. In Japan, we sell a product that costs \$5 to produce for 700 yen. The product has an elasticity of 3. For exchange rates varying from 70 to 130 yen per dollar, determine the optimal product price in Japan and the profit in dollars. Assume a linear demand curve. Current demand is assumed to equal 100.

11 It costs \$250 to produce an X-Box. We are trying to determine the selling price for the X-Box. Prices between \$200 and \$400 are under consideration, with demand for prices of \$200, \$250, \$350, and \$400 given below. Suppose MSFT earns \$10 in profit for each game that an X-Box owner purchases. Determine the optimal price and associated profit for the case in which an average X-Box owner buys 10 games.

Console Price (\$)	Demand
200	2.00E+06
250	1.20E+06
350	6.00E+05
400	2.00E+05
Unit cost	\$250

12 You are the publisher of a new magazine. The variable cost of printing and distributing each weekly copy of the magazine is \$0.25. You are thinking of charging between \$0.50 and \$1.30 per week for the magazine. The estimated numbers of subscribers (in millions) for weekly prices of \$0.50, \$0.80, and \$1.30 are as follows:

Price	Demand (Millions)
0.5	2.00
0.8	1.20
1.3	0.30

What price will maximize weekly profit from the magazine?

Group B

13 It costs a company $c(x)$ dollars to produce x units. The curve $y = c'(x)$ is called the firm's marginal cost curve. (Why?) The firm's average cost curve is given by $z = \frac{c(x)}{x}$. Let x^* be the production level that minimizes the company's average cost. Give conditions under which the marginal cost curve intersects the average cost curve at x^*.

14 When a machine is t years old, it earns revenue at a rate of e^{-t} dollars per year. After t years of use, the machine can be sold for $\frac{1}{t+1}$ dollars.

   a When should the machine be sold to maximize total revenue?

b If revenue is discounted continuously (so that $1 of revenue received t years from now is equivalent to e^{-rt} dollars of revenue received now, how would the answer in part (a) change?

15[†] Suppose a company must service customers lying in an area of A sq mi with n warehouses. Kolesar and Blum have shown that the average distance between a warehouse and a customer is

$$\sqrt{\frac{A}{n}}$$

Assume that it costs the company $60,000 per year to maintain a warehouse and $400,000 to build a warehouse. (Assume that a $400,000 cost is equivalent to forever incurring a cost of $40,000 per year.) The company fills 160,000 orders per year, and the shipping cost per order is $1 per mile. If the company serves an area of 100 sq mi, then how many warehouses should it have?

16 Prove Theorem 4.

17 Prove Theorem 5.

11.5 Golden Section Search

Consider a function $f(x)$. [For some x, $f'(x)$ may not exist.] Suppose we want to solve the following NLP:

$$\max f(x)$$
$$\text{s.t.} \quad a \leq x \leq b \tag{6}$$

It may be that $f'(x)$ does not exist, or it may be difficult to solve the equation $f'(x) = 0$. In either case, it may be difficult to use the methods of the previous section to solve this NLP. In this section, we discuss how (6) can be solved if $f(x)$ is a special type of function (a unimodal function).

DEFINITION ■   A function $f(x)$ is **unimodal** on $[a, b]$ if for some point $\bar{x}$ on $[a, b]$, $f(x)$ is strictly increasing on $[a, \bar{x}]$ and strictly decreasing on $[\bar{x}, b]$. ■

If $f(x)$ is unimodal on $[a, b]$, then $f(x)$ will have only one local maximum ($\bar{x}$) on $[a, b]$ and that local maximum will solve (6). (See Figure 32.) Let $\bar{x}$ denote the optimal solution to (6).

Without any further information, all we can say is that the optimal solution to (6) is some point on the interval $[a, b]$. By evaluating $f(x)$ at two points x_1 and x_2 (assume $x_1 < x_2$) on $[a, b]$, we may reduce the size of the interval in which the solution to (6) must

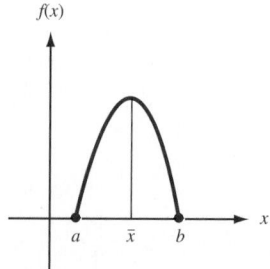
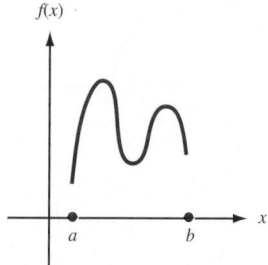

FIGURE 32
Definition of a
Unimodal Function

a A unimodal function on $[a, b]$
$\bar{x}$ = local maximum and solution to
$\max f(x)$
s.t. $a \leq x \leq b$

b A function that is not unimodal on $[a, b]$

[†]Based on Kolesar and Blum (1973).

lie. After evaluating $f(x_1)$ and $f(x_2)$, one of three cases must occur. In each case, we can show that the optimal solution to (6) will lie in a subset of $[a, b]$.

Case 1 $f(x_1) < f(x_2)$. Because $f(x)$ is increasing for at least part of the interval $[x_1, x_2]$, the fact that $f(x)$ is unimodal shows that the optimal solution to (6) cannot occur on $[a, x_1]$. Thus, in Case 1, $\bar{x} \in (x_1, b]$. (See Figure 33.)

Case 2 $f(x_1) = f(x_2)$. For some part of the interval $[x_1, x_2]$, $f(x)$ must be decreasing, and the optimal solution to (6) must occur for some $\bar{x} < x_2$. Thus, in Case 2, $\bar{x} \in [a, x_2]$. (See Figure 34.)

Case 3 $f(x_1) > f(x_2)$. In this case, $f(x)$ begins decreasing before x reaches x_2. Thus, $\bar{x} \in [a, x_2)$. (See Figure 35.)

The interval in which $\bar{x}$ must lie—either $[a, x_2)$ or $(x_1, b]$—is called the **interval of uncertainty.**

Many search algorithms use these ideas to reduce the interval of uncertainty [see Bazaraa and Shetty (1993, Section 8.1)]. Most of these algorithms proceed as follows:

Step 1 Begin with the region of uncertainty for x being $[a, b]$. Evaluate $f(x)$ at two judiciously chosen points x_1 and x_2.

Step 2 Determine which of Cases 1–3 holds, and find a reduced interval of uncertainty.

Step 3 Evaluate $f(x)$ at two new points (the algorithm specifies how the two new points are chosen). Return to step 2 unless the length of the interval of uncertainty is sufficiently small.

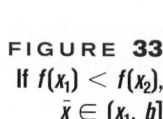
FIGURE 33
If $f(x_1) < f(x_2)$, $\bar{x} \in (x_1, b]$

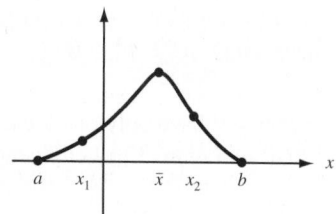

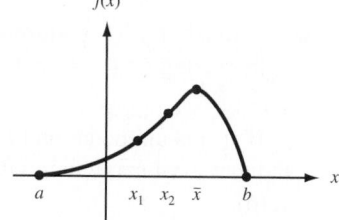

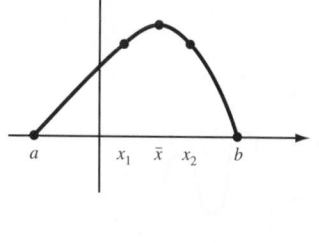
FIGURE 34
If $f(x_1) = f(x_2)$, $\bar{x} \in [a, x_2)$

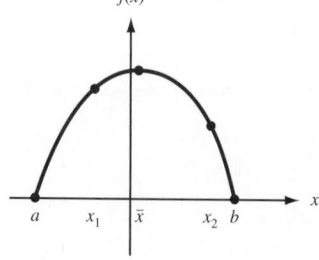
FIGURE 35
If $f(x_1) > f(x_2)$, $\bar{x} \in [a, x_2)$

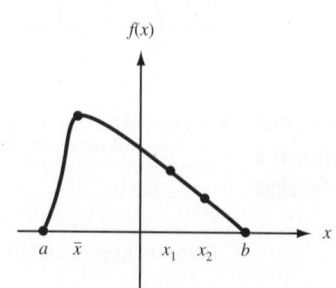

We discuss in detail one such search algorithm: Golden Section Search. In using this algorithm to solve (6) for a unimodal function $f(x)$, we will see that when we choose two new points at step 3, one will always coincide with a point at which we have previously evaluated $f(x)$.

Let r be the unique positive root of the quadratic equation $r^2 + r = 1$. Then the quadratic formula yields that

$$r = \frac{5^{1/2} - 1}{2} = 0.618$$

(See Problem 3 at the end of this section for an explanation of why r is referred to as the Golden Section.) Golden Section Search begins by evaluating $f(x)$ at points x_1 and x_2, where $x_1 = b - r(b - a)$, and $x_2 = a + r(b - a)$ (see Figure 36). From this figure, we see that to find x_1, we move a fraction r of the interval from the right endpoint of the interval; to find x_2, we move a fraction r of the interval from the left endpoint. Then Golden Section Search generates two new points, at which $f(x)$ should again be evaluated with the following moves:

New Left-Hand Point Move a distance equal to a fraction r of the current interval of uncertainty from the right endpoint of the interval of uncertainty.

New Right-Hand Point Move a distance equal to a fraction r of the current interval of uncertainty from the left endpoint of the interval.

From our discussion of Cases 1–3, we know that if $f(x_1) < f(x_2)$, then $\bar{x} \in (x_1, b]$, whereas if $f(x_1) \geq f(x_2)$, then $\bar{x} \in [a, x_2)$. If $f(x_1) < f(x_2)$, then the reduced interval of uncertainty has length $b - x_1 = r(b - a)$, and if $f(x_1) \geq f(x_2)$, then the reduced interval of uncertainty has a length $x_2 - a = r(b - a)$. Thus, after evaluating $f(x_1)$ and $f(x_2)$, we have reduced the interval of uncertainty to a length $r(b - a)$.

Each time $f(x)$ is evaluated at two points and the interval of uncertainty is reduced, we say that an iteration of Golden Section Search has been completed. Define

$$L_k = \text{length of the interval of uncertainty}$$
$$\text{after } k \text{ iterations of the algorithm have been completed}$$

$$I_k = \text{interval of uncertainty}$$
$$\text{after } k \text{ iterations have been completed}$$

Then we see that $L_1 = r(b - a)$, and $I_1 = [a, x_2)$ or $I_1 = (x_1, b]$.

Following this procedure, we generate two new points, x_3 and x_4, at which $f(x)$ must be evaluated.

Case 1 $f(x_1) < f(x_2)$. The new interval of uncertainty, $(x_1, b]$, has length $b - x_1 = r(b - a)$. Then (see Figure 37a)

$$x_3 = \text{new left-hand point} = b - r(b - x_1) = b - r^2(b - a)$$
$$x_4 = \text{new right-hand point} = x_1 + r(b - x_1)$$

The new left-hand point, x_3, will equal the old right-hand point, x_2. To see this, use the fact that $r^2 = 1 - r$ to conclude that $x_3 = b - r^2(b - a) = b - (1 - r)(b - a) = a + r(b - a) = x_2$.

FIGURE **36**
Location of x_1
and x_2 for Golden
Section Search

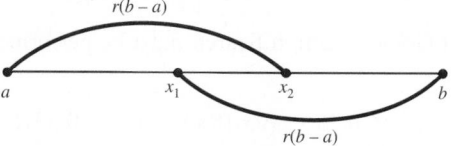

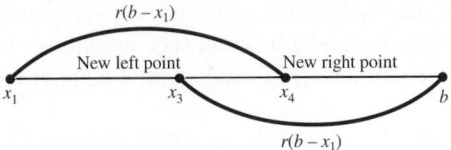

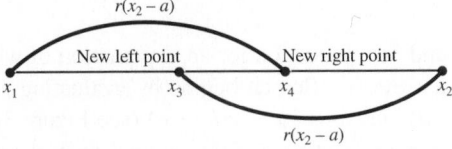

a If $f(x_1) < f(x_2)$, new interval of uncertainty is $(x_1, b]$

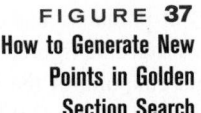

FIGURE **37**
How to Generate New
Points in Golden
Section Search

b If $f(x_1) \geq f(x_2)$, new interval of uncertainty is $[a, x_2)$

Case 2 $f(x_1) \geq f(x_2)$. The new interval of uncertainty, $[a, x_2)$, has length $x_2 - a = r(b - a)$. Then (see Figure 37b)

$$x_3 = \text{new left-hand point} = x_2 - r(x_2 - a)$$
$$x_4 = \text{new right-hand point} = a + r(x_2 - a) = a + r^2(b - a)$$

The new right-hand point, x_4, will equal the old left-hand point, x_1. To see this, use the fact that $r^2 = 1 - r$ to conclude that $x_4 = a + r^2(b - a) = a + (1 - r)(b - a) = b - r(b - a) = x_1$.

Now the values of $f(x_3)$ and $f(x_4)$ can be used to further reduce the length of the interval of uncertainty. At this point, two iterations of Golden Section Search have been completed.

We have shown that at each iteration of Golden Section Search, $f(x)$ must be evaluated at only one of the new points. It is easy to see that $L_2 = rL_1 = r^2(b - a)$ and, in general, $L_k = rL_{k-1}$ yields that $L_k = r^k(b - a)$. Thus, if we want our final interval of uncertainty to have a length $< \epsilon$, we must perform k iterations of Golden Section Search, where $r^k(b - a) < \epsilon$.

EXAMPLE 25   **Golden Section Search**

Use Golden Section Search to find

$$\max -x^2 - 1$$
$$\text{s.t.} \quad -1 \leq x \leq 0.75$$

with the final interval of uncertainty having a length less than $\frac{1}{4}$.

Solution Here $a = -1$, $b = 0.75$, and $b - a = 1.75$. To determine the number k of iterations of Golden Section Search that must be performed, we solve for k using $1.75(0.618^k) < 0.25$, or $0.618^k < \frac{1}{7}$. Taking logarithms to base e of both sides, we obtain

$$k \ln 0.618 < \ln \frac{1}{7}$$
$$k(-0.48) < -1.95$$
$$k > \frac{1.95}{0.48} = 4.06$$

Thus, five iterations of Golden Section Search must be performed. We first determine x_1 and x_2:

$$x_1 = 0.75 - (0.618)(1.75) = -0.3315$$
$$x_2 = -1 + (0.618)(1.75) = 0.0815$$

Then $f(x_1) = -1.1099$ and $f(x_2) = -1.0066$. Because $f(x_1) < f(x_2)$, the new interval of uncertainty is $I_1 = (x_1, b] = (-0.3315, 0.75]$, and we have that $x_3 = x_2$. Of course, $L_1 = 0.75 + 0.3315 = 1.0815$. We now determine the two new points x_3 and x_4:

$$x_3 = x_2 = 0.0815$$
$$x_4 = -0.3315 + 0.618(1.0815) = 0.3369$$

Now $f(x_3) = f(x_2) = -1.0066$ and $f(x_4) = -1.1135$. Because $f(x_3) > f(x_4)$, the new interval of uncertainty is $I_2 = [-0.3315, x_4) = [-0.3315, 0.3369)$, and x_6 will equal x_3. Also, $L_2 = 0.3369 + 0.3315 = 0.6684$. Then

$$x_5 = 0.3369 - 0.618(0.6684) = -0.0762$$
$$x_6 = x_3 = 0.0815$$

Note that $f(x_5) = -1.0058$ and $f(x_6) = f(x_3) = -1.0066$. Because $f(x_5) > f(x_6)$, the new interval of uncertainty is $I_3 = [-0.3315, x_6) = [-0.3315, 0.0815)$ and $L_3 = 0.0815 + 0.3315 = 0.4130$. Because $f(x_6) < f(x_5)$, we have that $x_5 = x_8$ and $f(x_8) = -1.0058$. Now

$$x_7 = 0.0815 - 0.618(0.413) = -0.1737$$
$$x_8 = x_5 = -0.0762$$

and $f(x_7) = -1.0302$. Because $f(x_8) > f(x_7)$, the new interval of uncertainty is $I_4 = (x_7, 0.0815] = (-0.1737, 0.0815]$, and $L_4 = 0.0815 + 0.1737 = 0.2552$. Also, $x_9 = x_8$ will hold. Finally,

$$x_9 = x_8 = -0.0762$$
$$x_{10} = -0.1737 + 0.618(0.2552) = -0.016$$

Now $f(x_9) = f(x_8) = -1.0058$ and $f(x_{10}) = -1.0003$. Because $f(x_{10}) > f(x_9)$, the new interval of uncertainty is $I_5 = (x_9, 0.0815] = (-0.0762, 0.0815]$ and $L_5 = 0.0815 + 0.0762 = 0.1577 < 0.25$ (as desired).

Thus, we have determined that

$$\max \quad -x^2 - 1$$
$$\text{s.t.} \quad -1 \le x \le 0.75$$

must lie within the interval $(-0.0762, 0.0815]$. (Of course, the actual maximum occurs for $\bar{x} = 0$.)

Golden Section Search can be applied to a minimization problem by multiplying the objective function by -1. This assumes that the modified objective function is unimodal.

Using Spreadsheets to Conduct Golden Section Search

Golden.xls

Figure 38 (file Golden.xls) displays an implementation of Golden Section Search on Lotus 1-2-3. We begin by entering the left-hand and right-hand endpoints ($a = -1, b = .75$) of the interval of uncertainty for Example 25 in cells A2 and B2. We compute r by entering the formula (5^.5–1)/2 into G2. Then, we name the cell G2 as the range R (with the **INSERT NAME CREATE** sequence of commands). In all subsequent formulas, R refers to the range R and assumes the value of r computed in G2. We compute the initial left-hand point x_1 by entering the formula =B2–R*(B2–A2) in C2 and the initial right-hand point x_2 by entering the formula =A2+R*(B2–A2) in D2. In effect, the formulas in C2 and D2 implement Figure 36. We evaluate $f(x_1)$ by entering $-(C2^.2-1)$ in E2 and $f(x_2)$ by entering $-(D2)^2-1$ in F2.

FIGURE **38**
Golden Section Search
for Example 25

	A	B	C	D	E	F	G
1	LEFTPTUNC	RIGTPTUNC	LEFTPT	RIGHTPT	F(LEFTPT)	F(RIGHTPT)	R
2	.1	0.75	-0.33156	0.081559	-1.10993169	-1.00665195	0.618034
3	-0.33155948	0.75	0.081559	0.336881	-1.00665195	-1.11348883	
4	-0.33155948	0.336881039	-0.07624	0.081559	-1.00581222	-1.00665195	FIGURE
5	-0.33155948	0.08155948	-0.17376	-0.07624	-1.03019326	-1.00581222	25
6	-0.17376208	0.08155948	-0.07624	-0.01596	-1.00581222	-1.00025487	GOLDEN
7	-0.07623792	0.08155948	-0.01596	0.021286	-1.00025487	-1.0004531	SECTION
8							SEARCH

In A3, we determine the new left point of the interval of uncertainty by entering the formula $=\textbf{IF}(E2<F2, C2, A2)$. This ensures that if $f(x_1) < f(x_2)$, then the new left point of the interval of uncertainty equals the last left-hand point where the function is evaluated (x_1); while if $f(x_1) \geq f(x_2)$, then the new left-hand point of uncertainty equals the old left-hand endpoint (a). Similarly, in B3 we determine the new right-hand endpoint of the interval of uncertainty. In C3, we compute the new left-hand point (x_3) where the function is evaluated by entering the formula $=\textbf{IF}(E2<F2,D2,D2-R*(D2-A2))$. If $f(x_1) < f(x_2)$, then this formula ensures that the new left-hand point (x_3) will equal the old right-hand point (x_2); if $f(x_1) \geq f(x_2)$, then the new left-hand point (x_3) will equal $x_2 - r(x_2 - a)$ [this equals D2–R*(D2–A2)]. In D3, we compute the new right-hand point (x_4) by entering the formula $=\textbf{IF}(E2<F2,C2+R*(B2-C2), C2)$. If $f(x_1) < f(x_2)$, then the new right-hand point (x_4) will equal $x_1 + r(b - x_1)$ [this equals C2+R*(B2–C2)]; if $f(x_1) \geq f(x_2)$, then the new right-hand point will equal the old left-hand point (x_1) (which equals C2). In E3, we evaluate the function at the new left-hand point by entering $-(C4)^2-1$, and in F3 we evaluate the function at the new right-hand endpoint by entering $-(D4)^2-1$.

Now copying the formulas from the range A3:F3 to the range A3:F7 will generate four more iterations of Golden Section Search.

PROBLEMS

Group A

1 Use Golden Section Search to determine (within an interval of 0.8) the optimal solution to

$$\max x^2 + 2x$$
$$\text{s.t.} \quad -3 \leq x \leq 5$$

2 Use Golden Section Search to determine (within an interval of 0.6) the optimal solution to

$$\max x - e^x$$
$$\text{s.t.} \quad -1 \leq x \leq 3$$

3 Consider a line segment [0, 1] that is divided into two parts (Figure 39). The line segment is said to be divided into the Golden Section if

$$\frac{\text{Length of whole line}}{\text{Length of larger part of line}}$$
$$= \frac{\text{length of larger part of line}}{\text{length of smaller part of line}}$$

Show that for the line segment to be divided into the Golden Section,

$$r = \frac{5^{1/2} - 1}{2}$$

4 Hughesco is interested in determining how cutting fluid jet pressure (p) affects the useful life of a machine tool (t), using the data in Table 11. Pressure p is constrained to be between 0 and 600 pounds per square inch (psi). Use Golden Section Search to estimate (within 50 units) the value of p that maximizes useful tool life. Assume that t is a unimodal function of p.

FIGURE 39

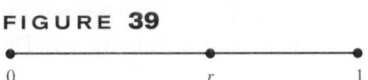

0   r 1

TABLE 11

p (Pounds per Square Inch)	t (Minutes)
229	39
371	81
458	82
513	79
425	84
404	85
392	84

11.6 Unconstrained Maximization and Minimization with Several Variables

We now discuss how to find an optimal solution (if it exists) or a local extremum for the following unconstrained NLP:

$$\max \text{ (or min) } f(x_1, x_2, \ldots, x_n)$$
$$\text{s.t.} \quad (x_1, x_2, \ldots, x_n) \in R^n \tag{7}$$

We assume that the first and second partial derivatives of $f(x_1, x_2, \ldots, x_n)$ exist and are continuous at all points. Let

$$\frac{\partial f(\bar{x})}{\partial x_i}$$

be the partial derivative of $f(x_1, x_2, \ldots, x_n)$ with respect to x_i, evaluated at $\bar{x}$. A necessary condition for $\bar{x} = (\bar{x}_1, \bar{x}_2, \ldots, \bar{x}_n)$ to be a local extremum for NLP (7) is given in Theorem 6.

THEOREM 6

If $\bar{x}$ is a local extremum for (6), then $\dfrac{\partial f(\bar{x})}{\partial x_i} = 0$.

To see why Theorem 6 holds, suppose $\bar{x}$ is a local extremum for (7)—say, a local maximum. If $\dfrac{\partial f(\bar{x})}{\partial x_i} > 0$ holds for any i, then by slightly increasing x_i (and holding all other variables constant), we can find a point x' near $\bar{x}$ with $f(x') > f(\bar{x})$. This would contradict the fact that $\bar{x}$ is a local maximum. Similarly, if $\bar{x}$ is a local maximum for (7) and $\dfrac{\partial f(\bar{x})}{\partial x_i} < 0$, then by slightly decreasing x_i (and holding all other variables constant), we can find a point x'' near $\bar{x}$ with $f(x'') > f(\bar{x})$. Thus, if $\bar{x}$ is a local maximum for (7), then $\dfrac{\partial f(\bar{x})}{\partial x_i} = 0$ must hold for $i = 1, 2, \ldots, n$. A similar argument shows that if $\bar{x}$ is a local minimum, then $\dfrac{\partial f(\bar{x})}{\partial x_i} = 0$ must hold for $i = 1, 2, \ldots, n$.

DEFINITION ■ A point $\bar{x}$ having $\dfrac{\partial f(\bar{x})}{\partial x_i} = 0$ for $i = 1, 2, \ldots, n$ is called a **stationary point** of f. ■

The following three theorems give conditions (involving the Hessian of f) under which a stationary point is a local minimum, a local maximum, or not a local extremum.

THEOREM 7

If $H_k(\bar{x}) > 0$, $k = 1, 2, \ldots, n$, then a stationary point $\bar{x}$ is a local minimum for NLP (7).

THEOREM 7'

If, for $k = 1, 2, \ldots, n$, $H_k(\bar{x})$ is nonzero and has the same sign as $(-1)^k$, then a stationary point $\bar{x}$ is a local maximum for NLP (7).

If $H_n(\bar{x}) \neq 0$ and the conditions of Theorems 7 and 7' do not hold, then a stationary point $\bar{x}$ is not a local extremum.

If a stationary point $\bar{x}$ is not a local extremum, then it is called a **saddle point.** If $H_n(\bar{x}) = 0$ for a stationary point $\bar{x}$, then $\bar{x}$ may be a local minimum, a local maximum, or a saddle point, and the preceding tests are inconclusive.

From Theorems 1 and 7', we know that if $f(x_1, x_2, \ldots, x_n)$ is a concave function (and NLP (7) is a max problem), then any stationary point for (7) is an optimal solution to (7). From Theorems 1' and 7, we know that if $f(x_1, x_2, \ldots, x_n)$ is a convex function [and NLP (7) is a min problem], then any stationary point for (7) is an optimal solution to (7).

EXAMPLE 26   Monopolistic Pricing with Multiple Customer Types

A monopolist producing a single product has two types of customers. If q_1 units are produced for customer 1, then customer 1 is willing to pay a price of $70 - 4q_1$ dollars. If q_2 units are produced for customer 2, then customer 2 is willing to pay a price of $150 - 15q_2$ dollars. For $q > 0$, the cost of manufacturing q units is $100 + 15q$ dollars. To maximize profit, how much should the monopolist sell to each customer?

Solution   Let $f(q_1, q_2)$ be the monopolist's profit if she produces q_i units for customer i. Then (assuming some production takes place)

$$f(q_1, q_2) = q_1(70 - 4q_1) + q_2(150 - 15q_2) - 100 - 15q_1 - 15q_2$$

To find the stationary point(s) for $f(q_1, q_2)$, we set

$$\frac{\partial f}{\partial q_1} = 70 - 8q_1 - 15 = 0 \qquad (\text{for } q_1 = \tfrac{55}{8})$$

$$\frac{\partial f}{0q_2} = 150 - 30q_2 - 15 = 0 \qquad (\text{for } q_2 = \tfrac{9}{2})$$

Thus, the only stationary point of $f(q_1, q_2)$ is $(\tfrac{55}{8}, \tfrac{9}{2})$. Next we find the Hessian for $f(q_1, q_2)$.

$$H(q_1, q_2) = \begin{bmatrix} -8 & 0 \\ 0 & -30 \end{bmatrix}$$

Since the first leading principal minor of H is $-8 < 0$, and the second leading principal minor of H is $(-8)(-30) = 240 > 0$, Theorem 7' shows that $(\tfrac{55}{8}, \tfrac{9}{2})$ is a local maximum. Also, Theorem 3' implies that $f(q_1, q_2)$ is a concave function [on the set of points S of (q_1, q_2) satisfying $q_1 \geq 0$, $q_2 \geq 0$, and $q_1 + q_2 > 0$]. Thus, Theorem 1 implies that $(\tfrac{55}{8}, \tfrac{9}{2})$ maximizes profit among all production possibilities (with the possible exception of no production). Then $(\tfrac{55}{8}, \tfrac{9}{2})$ yields a profit of

$$f(q_1, q_2) = \tfrac{55}{8}(70 - \tfrac{220}{8}) + \tfrac{9}{2}[150 - 15(\tfrac{9}{2})] - 100 - 15(\tfrac{55}{8} + \tfrac{9}{2}) = \$392.81$$

The profit from producing $(\tfrac{55}{8}, \tfrac{9}{2})$ exceeds the profit of \$0 that is obtained by producing nothing, so $(\tfrac{55}{8}, \tfrac{9}{2})$ solves the NLP; the monopolist should sell $\tfrac{55}{8}$ units to customer 1 and $\tfrac{9}{2}$ units to customer 2.

EXAMPLE 27 | **Least Squares Estimation**

Suppose the grade-point average (GPA) for a student can be accurately predicted from the student's score on the GMAT (Graduate Management Admissions Test). More specifically, suppose that the ith student observed has a GPA of y_i and a GMAT score of x_i. How can we use the **least squares method** to estimate a hypothesized relation of the form $y_i = a + bx_i$?

Solution   Let $\hat{a}$ be our estimate of a and $\hat{b}$ our estimate of b. Given that for students $i = 1, 2, \ldots, n$ we have observed $(x_1, y_1), (x_2, y_2), \ldots, (x_n, y_n)$, $\hat{e}_i = y_i - (\hat{a} + \hat{b}x_i)$ is our error in estimating the GPA of student i. The least squares method chooses $\hat{a}$ and $\hat{b}$ to minimize

$$f(a, b) = \sum_{i=1}^{i=n} \hat{e}_i^2 = \sum_{i=1}^{i=n} (y_i - a - bx_i)^2$$

Since

$$\frac{\partial f}{\partial a} = -2 \sum_{i=1}^{i=n} (y_i - a - bx_i) \quad \text{and} \quad \frac{\partial f}{\partial b} = -2 \sum_{i=1}^{i=n} (y_i - a - bx_i)x_i$$

$\dfrac{\partial f}{\partial a} = \dfrac{\partial f}{\partial b} = 0$ will hold for the point $(\hat{a}, \hat{b})$ satisfying

$$\sum_{i=1}^{i=n} (y_i - a - bx_i) = 0 \quad \text{or} \quad \sum_{i=1}^{i=n} y_i = na + b \sum_{i=1}^{i=n} x_i$$

and

$$\sum_{i=1}^{i=n} x_i(y_i - a - bx_i) = 0 \quad \text{or} \quad \sum_{i=1}^{i=n} x_iy_i = a \sum_{i=1}^{i=n} x_i + b \sum_{i=1}^{i=n} x_i^2$$

These are the well-known **normal equations.** Does the solution $(\hat{a}, \hat{b})$ to the normal equations minimize $f(a, b)$? To answer this question, we must compute the Hessian for $f(a, b)$:

$$\frac{\partial^2 f}{\partial a^2} = 2n, \quad \frac{\partial^2 f}{\partial b^2} = 2 \sum_{i=1}^{i=n} x_i^2, \quad \frac{\partial^2 f}{\partial a \partial b} = \frac{\partial^2 f}{\partial b \partial a} = 2 \sum_{i=1}^{i=n} x_i$$

Thus,

$$H = \begin{bmatrix} 2n & 2\sum_{i=1}^{i=n} x_i \\ 2\sum_{i=1}^{i=n} x_i & 2\sum_{i=1}^{i=n} x_i^2 \end{bmatrix}$$

Since $H_1(\hat{a}, \hat{b}) = 2n > 0$, $(\hat{a}, \hat{b})$ will be a local minimum if

$$H_2(\hat{a}, \hat{b}) = 4n \sum_{i=1}^{i=n} x_i^2 - 4\left(\sum_{i=1}^{i=n} x_i\right)^2 > 0$$

In Example 31 of Section 11.8, we show that

$$n \sum_{i=1}^{i=n} x_i^2 \geq \left(\sum_{i=1}^{i=n} x_i\right)^2$$

with equality holding if and only if $x_1 = x_2 = \cdots = x_n$. Thus, if at least two of the x_i's are different, Theorem 7' implies that $(\hat{a}, \hat{b})$ will be a local minimum. $H(a, b)$ does not depend on the values of a and b, so this reasoning (and Theorem 3) shows that if at least two of the x_i's are different, then $f(a, b)$ is a convex function. If at least two of the x_i's are different, then Theorem 1' shows that $(\hat{a}, \hat{b})$ minimizes $f(a, b)$.

EXAMPLE 28 **Finding Maxima, Minima, and Saddle Points**

Find all local maxima, local minima, and saddle points for $f(x_1, x_2) = x_1^2 x_2 + x_2^3 x_1 - x_1 x_2$.

Solution We have

$$\frac{\partial f}{\partial x_1} = 2x_1 x_2 + x_2^3 - x_2, \qquad \frac{\partial f}{\partial x_2} = x_1^2 + 3x_2^2 x_1 - x_1$$

Thus, $\dfrac{\partial f}{\partial x_1} = \dfrac{\partial f}{\partial x_2} = 0$ requires

$$2x_1 x_2 + x_2^3 - x_2 = 0 \qquad \text{or} \qquad x_2(2x_1 + x_2^2 - 1) = 0 \tag{8}$$
$$x_1^2 + 3x_2^2 x_1 - x_1 = 0 \qquad \text{or} \qquad x_1(x_1 + 3x_2^2 - 1) = 0 \tag{9}$$

For (8) to hold, either (i) $x_2 = 0$ or (ii) $2x_1 + x_2^2 - 1 = 0$ must hold. For (9) to hold, either (iii) $x_1 = 0$ or (iv) $x_1 + 3x_2^2 - 1 = 0$ must hold.

Thus, for (x_1, x_2) to be a stationary point, we must have:

(i) and (iii) hold. This is only true at $(0, 0)$.

(i) and (iv) hold. This is only true at $(1, 0)$.

(ii) and (iii) hold. This is only true at $(0, 1)$ and $(0, -1)$.

(ii) and (iv) hold. This requires that $x_2^2 = 1 - 2x_1$ and $x_1 + 3(1 - 2x_1) - 1 = 0$ hold.

Then

$$x_1 = \frac{2}{5} \qquad \text{and} \qquad x_2 = \frac{5^{1/2}}{5} \qquad \text{or} \qquad -\frac{5^{1/2}}{5}$$

Thus, $f(x_1, x_2)$ has the following stationary points:

$$(0, 0), (1, 0), (0, 1), (0, -1), \left(\frac{2}{5}, \frac{5^{1/2}}{5}\right) \qquad \text{and} \qquad \left(\frac{2}{5}, -\frac{5^{1/2}}{5}\right)$$

Also,

$$H(x_1, x_2) = \begin{bmatrix} 2x_2 & 2x_1 + 3(x_2)^2 - 1 \\ 2x_1 + 3(x_2)^2 - 1 & 6x_1 x_2 \end{bmatrix}$$

$$H(0, 0) = \begin{bmatrix} 0 & -1 \\ -1 & 0 \end{bmatrix}$$

Because $H_1(0, 0) = 0$, the conditions of Theorems 7 and 7' cannot be satisfied. Because $H_2(0, 0) = -1 \neq 0$, Theorem 7'' now implies that $(0, 0)$ is a saddle point.

$$H(1, 0) = \begin{bmatrix} 0 & 1 \\ 1 & 0 \end{bmatrix}$$

Then $H_1(1, 0) = 0$ and $H_2(1, 0) = -1$, so by Theorem 7'' $(1, 0)$ is also a saddle point. Since

$$H(0, 1) = \begin{bmatrix} 2 & 2 \\ 2 & 0 \end{bmatrix}$$

we have $H_1(0, 1) = 2 > 0$ (so the hypotheses of Theorem 7' cannot be satisfied) and $H_2(0, 1) = -4$ (so the hypothesis of Theorem 7 cannot be satisfied). Because $H_2(0, 1) \neq 0$, $(0, 1)$ is a saddle point.

For $\left(\dfrac{2}{5}, -\dfrac{5^{1/2}}{5}\right)$, we have

$$H\left(\frac{2}{5}, -\frac{5^{1/2}}{5}\right) = \begin{bmatrix} -\dfrac{2}{5^{1/2}} & \dfrac{2}{5} \\[2mm] \dfrac{2}{5} & -\dfrac{12}{5(5)^{1/2}} \end{bmatrix}$$

Thus,

$$H_1\left(\frac{2}{5}, -\frac{5^{1/2}}{5}\right) = -\frac{2}{5^{1/2}} < 0 \qquad \text{and} \qquad H_2\left(\frac{2}{5}, -\frac{5^{1/2}}{5}\right) = \frac{20}{25} > 0$$

Thus, Theorem 7' shows that $\left(\dfrac{2}{5}, -\dfrac{5^{1/2}}{5}\right)$ is a local maximum. Finally,

$$H\left(\frac{2}{5}, \frac{5^{1/2}}{5}\right) = \begin{bmatrix} \dfrac{2}{5^{1/2}} & \dfrac{2}{5} \\[2mm] \dfrac{2}{5} & \dfrac{12}{5(5)^{1/2}} \end{bmatrix}$$

Since $H_1\left(\dfrac{2}{5}, \dfrac{5^{1/2}}{5}\right) = \dfrac{2}{5^{1/2}} > 0$ and $H_2\left(\dfrac{2}{5}, \dfrac{5^{1/2}}{5}\right) = \dfrac{20}{25} > 0$, Theorem 7 shows that $\left(\dfrac{2}{5}, \dfrac{5^{1/2}}{5}\right)$ is a local minimum.

When Does LINGO Find the Optimal Solution to an Unconstrained NLP?

If you are maximizing a concave function (with no constraints) or minimizing a convex function (with no constraints), you can be sure that any solution found by LINGO is the optimal solution to your problem. In Example 27, for instance, our work shows that $f(a, b)$ is a convex function, so we know that LINGO would correctly find the least squares line fitting a set of points.

PROBLEMS

Group A

1 A company has n factories. Factory i is located at point (x_i, y_i), in the x–y plane. The company wants to locate a warehouse at a point (x, y) that minimizes

$$\sum_{i=1}^{i=n} (\text{distance from factory } i \text{ to warehouse})^2$$

Where should the warehouse be located?

2 A company can sell all it produces of a given output for \$2/unit. The output is produced by combining two inputs. If q_1 units of input 1 and q_2 units of input 2 are used, then the

company can produce $q_1^{1/3} + q_2^{2/3}$ units of the output. If it costs \$1 to purchase a unit of input 1 and \$1.50 to purchase a unit of input 2, then how can the company maximize its profit?

3 (Collusive Duopoly Model) There are two firms producing widgets. It costs the first firm q_1 dollars to produce q_1 widgets and the second firm $0.5q_2^2$ dollars to produce q_2 widgets. If a total of q widgets are produced, consumers will pay \$200 $- q$ for each widget. If the two manufacturers want to collude in an attempt to maximize the sum of their profits, how many widgets should each company produce?

4 It costs a company \$6/unit to produce a product. If it charges a price p and spends a dollars on advertising, it can sell $10,000p^{-2}a^{1/6}$ units of the product. Find the price and advertising level that will maximize the company's profits.

5 A company manufactures two products. If it charges a price p_i for product i, it can sell q_i units of product i, where $q_1 = 60 - 3p_1 + p_2$ and $q_2 = 80 - 2p_2 + p_1$. It costs \$25 to produce a unit of product 1 and \$72 to produce a unit of product 2. How many units of each product should be produced to maximize profits?

6 Find all local maxima, local minima, and saddle points for $f(x_1, x_2) = x_1^3 - 3x_1x_2^2 + x_2^4$.

7 Find all local maxima, local minima, and saddle points for $f(x_1, x_2) = x_1x_2 + x_2x_3 + x_1x_3$.

Group B

8[†] (Cournot Duopoly Model) Let's reconsider Problem 3. The Cournot solution to this situation is obtained as follows: Firm i will produce $\bar{q}_i$, where if firm 1 changes its production

level from $\bar{q}_1$ (and firm 2 still produces $\bar{q}_2$), then firm 1's profit will decrease. Also, if firm 2 changes its production level from $\bar{q}_2$ (and firm 1 still produces $\bar{q}_1$), then firm 2's profit will decrease. If firm i produces $\bar{q}_i$, this solution is stable, because if either firm changes its production level, it will do worse. Find $\bar{q}_1$ and $\bar{q}_2$.

9 In the Bloomington Girls Club basketball league, the following games have been played: team A beat team B by 7 points, team C beat team A by 8 points, team B beat team C by 6 points, and team B beat team C by 9 points. Let A, B, and C represent "ratings" for each team in the sense that if, say, team A plays team B, then we predict that team A will defeat team B by $A - B$ points. Determine values of A, B, and C that best fit (in the least squares sense) these results. To obtain a unique set of ratings, it may be helpful to add the constraint $A + B + C = 0$. This ensures that an "average" team will have a rating of 0.

11.7 The Method of Steepest Ascent

Suppose we want to solve the following unconstrained NLP:

$$\max z = f(x_1, x_2, \ldots, x_n)$$
$$\text{s.t.} \quad (x_1, x_2, \ldots, x_n) \in R^n \tag{10}$$

Our discussion in Section 11.6 shows that if $f(x_1, x_2, \ldots, x_n)$ is a concave function, then the optimal solution to (10) (if there is one) will occur at a stationary point $\bar{x}$ having

$$\frac{\partial f(\bar{x})}{\partial x_1} = \frac{\partial f(\bar{x})}{\partial x_2} = \cdots = \frac{\partial f(\bar{x})}{\partial x_n} = 0$$

In Examples 26 and 28, it was easy to find a stationary point, but in many problems, it may be difficult. In this section, we discuss the *method of steepest ascent*, which can be used to approximate a function's stationary point.

DEFINITION ■ Given a vector $\mathbf{x} = (x_1, x_2, \ldots, x_n) \in R^n$, the **length** of $\mathbf{x}$ (written $\| \mathbf{x} \|$) is

$$\| \mathbf{x} \| = (x_1^2 + x_2^2 + \cdots + x_n^2)^{1/2} \quad ■$$

Recall from Section 2.1 that any n-dimensional vector represents a direction in R^n. Unfortunately, for any direction, there are an infinite number of vectors representing that direction. For example, the vectors $(1, 1)$, $(2, 2)$, and $(3, 3)$ all represent the same direction (moving at a positive 45° angle) in R^2. For any vector $\mathbf{x}$, the vector $\mathbf{x}/\| \mathbf{x} \|$ will have a length of 1 and will define the same direction as $\mathbf{x}$ (see Problem 1 at the end of this section). Thus, with any direction in R^n, we may associate a vector of length 1 (called a unit vector). For example, because $\mathbf{x} = (1, 1)$ has $\| \mathbf{x} \| = 2^{1/2}$, the direction defined by $\mathbf{x} = (1, 1)$ is associated with the unit vector $(1/2^{1/2}, 1/2^{1/2})$. For any vector $\mathbf{x}$, the unit vector

---

[†]Based on Cournot (1897).

$\mathbf{x}/\|\mathbf{x}\|$ is called the **normalized** version of $\mathbf{x}$. Henceforth, any direction in R^n will be described by the normalized vector defining that direction. Thus, the direction in R^2 defined by $(1, 1), (2, 2), (3, 3), \ldots$ will be described by the normalized vector

$$\left(\frac{1}{2^{1/2}}, \frac{1}{2^{1/2}}\right)$$

Consider a function $f(x_1, x_2, \ldots, x_n)$, all of whose partial derivatives exist at every point.

DEFINITION ■ The **gradient vector** for $f(x_1, x_2, \ldots, x_n)$, written $\nabla f(\mathbf{x})$, is given by

$$\nabla f(\mathbf{x}) = \left[\frac{\partial f(\mathbf{x})}{\partial x_1}, \frac{\partial f(\mathbf{x})}{\partial x_2}, \ldots, \frac{\partial f(\mathbf{x})}{\partial x_n}\right] \quad ■$$

$\nabla f(\mathbf{x})$ defines the direction

$$\frac{\nabla f(\mathbf{x})}{\|\nabla f(\mathbf{x})\|}$$

For example, if $f(x_1, x_2) = x_1^2 + x_2^2$, then $\nabla f(\mathbf{x}_1, \mathbf{x}_2) = (2x_1, 2x_2)$. Thus, $\nabla f(3, 4) = (6, 8)$. Because $\|\nabla f(3, 4)\| = 10$, $\nabla f(3, 4)$ defines the direction $(\frac{6}{10}, \frac{8}{10}) = (0.6, 0.8)$.

At any point $\bar{\mathbf{x}}$ that lies on the curve $f(x_1, x_2, \ldots, x_n) = f(\bar{\mathbf{x}})$, the vector

$$\frac{\nabla f(\bar{\mathbf{x}})}{\|\nabla f(\bar{\mathbf{x}})\|}$$

will be perpendicular to the curve $f(x_1, x_2, \ldots, x_n) = f(\bar{x})$ (see Problem 5 at the end of this section). For example, let $f(x_1, x_2) = x_1^2 + x_2^2$. Then at $(3, 4)$,

$$\frac{\nabla f(3, 4)}{\|\nabla f(3, 4)\|} = (0.6, 0.8)$$

is perpendicular to $x_1^2 + x_2^2 = 25$ (see Figure 40).

From the definition of $\dfrac{\partial f(\mathbf{x})}{\partial x_i}$, it follows that if the value of x_i is increased by a small amount δ, the value of $f(\mathbf{x})$ will increase by approximately $\delta\dfrac{\partial f(\mathbf{x})}{\partial x_i}$. Suppose we move from a point $\mathbf{x}$ a small length δ in a direction defined by a normalized column vector $\mathbf{d}$. By how much does $f(\mathbf{x})$ increase? The answer is that $f(\mathbf{x})$ increases by δ times the scalar product of $\dfrac{\nabla f(\mathbf{x})}{\|\nabla f(\mathbf{x})\|}$ and $\mathbf{d}$ $\left(\text{written } \dfrac{\delta\,\nabla f(\mathbf{x})\cdot\mathbf{d}}{\|\nabla f(\mathbf{x})\|}\right)$. Thus, if $\dfrac{\nabla f(\mathbf{x})\cdot\mathbf{d}}{\|\nabla f(\mathbf{x})\|} > 0$, moving

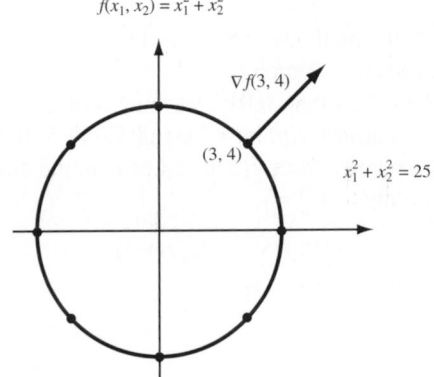

FIGURE **40**
$\nabla f(3, 4)$ Is
Perpendicular to
$f(x_1, x_2)$ at $(3, 4)$

in a direction $\mathbf{d}$ away from $\mathbf{x}$ will increase the value of $f(\mathbf{x})$, and if $\dfrac{\nabla f(\mathbf{x}) \cdot \mathbf{d}}{\|\nabla f(\mathbf{x})\|} < 0$, moving in a direction $\mathbf{d}$ away from $\mathbf{x}$ will decrease $f(\mathbf{x})$. For example, suppose $f(x_1, x_2) = x_1^2 + x_2^2$ and we move a length δ in a $45°$ direction away from the point $(3, 4)$. By how much will the value of $f(x_1, x_2)$ change? A $45°$ direction is represented by the vector $\left(\dfrac{1}{2^{1/2}}, \dfrac{1}{2^{1/2}}\right)$ and $\dfrac{\nabla f(3, 4)}{\|\nabla f(3, 4)\|} = (0.6, 0.8)$, so the value of $f(x_1, x_2)$ will increase by approximately

$$\delta[0.6 \quad 0.8] \begin{bmatrix} \dfrac{1}{2^{1/2}} \\[2mm] \dfrac{1}{2^{1/2}} \end{bmatrix} = 0.99\delta$$

Recall from Section 11.6 that the optimal solution $\bar{\mathbf{v}}$ to (10) must satisfy $\nabla f(\bar{\mathbf{v}}) = 0$. Now suppose that we are at a point $\mathbf{v}_0$ and want to find a point $\bar{\mathbf{v}}$ that solves (10). In an attempt to find $\bar{\mathbf{v}}$, it seems reasonable to move away from $\mathbf{v}_0$ in a direction that maximizes the rate (at least locally) at which $f(x_1, x_2, \ldots, x_n)$ increases. Lemma 1 proves useful here (see Review Problem 22).

LEMMA 1

Suppose we are at a point $\mathbf{v}$ and we move from $\mathbf{v}$ a small distance δ in a direction $\mathbf{d}$. Then for a given δ, the maximal increase in the value of $f(x_1, x_2, \ldots, x_n)$ will occur if we choose

$$\mathbf{d} = \frac{\nabla f(\mathbf{x})}{\|\nabla f(\mathbf{x})\|}$$

In short, if we move a small distance away from $\mathbf{v}$ and we want $f(x_1, x_2, \ldots, x_n)$ to increase as quickly as possible, then we should move in the direction of $\nabla f(\mathbf{v})$.

We are now ready to describe the method of steepest ascent. Begin at any point $\mathbf{v}_0$. Moving in the direction of $\nabla f(\mathbf{v}_0)$ will result in a maximum rate of increase for f, so we begin by moving away from $\mathbf{v}_0$ in the direction of $\nabla f(\mathbf{v}_0)$. For some nonnegative value of t, we move to a point $\mathbf{v}_1 = \mathbf{v}_0 + t\nabla f(\mathbf{v}_0)$. The maximum possible improvement in the value of f (for a max problem) that can be attained by moving away from $\mathbf{v}_0$ in the direction of $\nabla f(\mathbf{v}_0)$ results from moving to $\mathbf{v}_1 = \mathbf{v}_0 + t_0\nabla f(\mathbf{v}_0)$, where t_0 solves the following one-dimensional optimization problem:

$$\max f(\mathbf{v}_0 + t_0\nabla f(\mathbf{v}_0)) \tag{11}$$
$$\text{s.t.} \quad t_0 \geq 0$$

NLP (11) may be solved by the methods of Section 11.4 or, if necessary, by a search procedure such as the Golden Section Search.

If $\|\nabla f(\mathbf{v}_1)\|$ is small (say, less than 0.01), we may terminate the algorithm with the knowledge that $\mathbf{v}_1$ is near a stationary point $\bar{\mathbf{v}}$ having $\nabla f(\bar{\mathbf{v}}) = 0$. If $\|\nabla f(\mathbf{v}_1)\|$ is not sufficiently small, then we move away from $\mathbf{v}_1$ a distance t_1 in the direction of $\|\nabla f(\mathbf{v}_1)\|$. As before, we choose t_1 by solving

$$\max f(\mathbf{v}_1 + t_1\nabla f(\mathbf{v}_1))$$
$$\text{s.t.} \quad t_1 \geq 0$$

We are now at the point $\mathbf{v}_2 = \mathbf{v}_1 + t_1\nabla f(\mathbf{v}_1)$. If $\|\nabla f(\mathbf{v}_2)\|$ is sufficiently small, then we terminate the algorithm and choose $\mathbf{v}_2$ as our approximation to a stationary point of $f(x_1, x_2, \ldots, x_n)$. Otherwise, we continue in this fashion until we reach a point $\mathbf{v}_n$ having $\|\nabla f(\mathbf{v}_n)\|$ sufficiently small. Then we choose $\mathbf{v}_n$ as our approximation to a stationary point of $f(x_1, x_2, \ldots, x_n)$.

This algorithm is called the **method of steepest ascent** because to generate points, we always move in the direction that maximizes the rate at which f increases (at least locally).

EXAMPLE 29   **Steepest Ascent Example**

Use the method of steepest ascent to approximate the solution to

$$\max z = -(x_1 - 3)^2 - (x_2 - 2)^2 = f(x_1, x_2)$$
$$\text{s.t.} \quad (x_1, x_2) \in R^2$$

Solution   We arbitrarily choose to begin at the point $\mathbf{v}_0 = (1, 1)$. Because $\nabla f(x_1, x_2) = (-2(x_1 - 3), -2(x_2 - 2))$, we have $\nabla f(1, 1) = (4, 2)$. Thus, we must choose t_0 to maximize

$$f(t_0) = f[(1, 1) + t_0(4, 2)] = f(1 + 4t_0, 1 + 2t_0) = -(-2 + 4t_0)^2 - (-1 + 2t_0)^2$$

Setting $f'(t_0) = 0$, we obtain

$$-8(-2 + 4t_0) - 4(-1 + 2t_0) = 0$$
$$20 - 40t_0 = 0$$
$$t_0 = 0.5$$

Our new point is $\mathbf{v}_1 = (1, 1) + 0.5(4, 2) = (3, 2)$. Now $\nabla f(3, 2) = (0, 0)$, and we terminate the algorithm. Because $f(x_1, x_2)$ is a concave function, we have found the optimal solution to the NLP.

PROBLEMS

Group A

1 For any vector $\mathbf{x}$, show that the vector $\mathbf{x}/\|\mathbf{x}\|$ has unit length.

2 Use the method of steepest ascent to approximate the optimal solution to the following problem: $\max z = -(x_1 - 2)^2 - x_1 - x_2^2$. Begin at the point $(2.5, 1.5)$.

3 Use steepest ascent to approximate the optimal solution to the following problem: $\max z = 2x_1x_2 + 2x_2 - x_1^2 - 2x_2^2$. Begin at the point $(0.5, 0.5)$. Note that at later iterations, successive points are very close together. Variations of steepest ascent have been developed to deal with this problem [see Bazaraa and Shetty (1993, Section 8.6)].

Group B

4 How would you modify the method of steepest ascent if each variable x_1 were constrained to lie in an interval $[a_i, b_i]$?

Group C

5 Show that at any point $\bar{\mathbf{x}} = (\bar{x}_1, \bar{x}_2)$, $\nabla f(\bar{\mathbf{x}})$ is perpendicular to the curve $f(x_1, x_2) = f(\bar{x}_1, \bar{x}_2)$. (*Hint:* Two vectors are perpendicular if their scalar product equals zero.)

11.8 Lagrange Multipliers

Lagrange multipliers can be used to solve NLPs in which all the constraints are equality constraints. We consider NLPs of the following type:

$$\max \text{ (or min) } z = f(x_1, x_2, \ldots, x_n)$$
$$\text{s.t.} \quad g_1(x_1, x_2, \ldots, x_n) = b_1$$
$$g_2(x_1, x_2, \ldots, x_n) = b_2 \tag{12}$$
$$\vdots$$
$$g_m(x_1, x_2, \ldots, x_n) = b_m$$

To solve (12), we associate a **multiplier** λ_i with the ith constraint in (12) and form the **Lagrangian**

$$L(x_1, x_2, \ldots, x_n, \lambda_1, \lambda_2, \ldots, \lambda_m) = f(x_1, x_2, \ldots, x_n)$$
$$+ \sum_{i=1}^{i=m} \lambda_i[b_i - g_i(x_1, x_2, \ldots, x_n)] \tag{13}$$

Then we attempt to find a point $(\bar{x}_1, \bar{x}_2, \ldots, \bar{x}_n, \bar{\lambda}_1, \bar{\lambda}_2, \ldots, \bar{\lambda}_m)$ that maximizes (or minimizes) $L(x_1, x_2, \ldots, x_n, \lambda_1, \lambda_2, \ldots, \lambda_m)$. In many situations, $(\bar{x}_1, \bar{x}_2, \ldots, \bar{x}_n)$ will solve (12). Suppose that (12) is a maximization problem. If $(\bar{x}_1, \bar{x}_2, \ldots, \bar{x}_n, \bar{\lambda}_1, \bar{\lambda}_2, \ldots, \bar{\lambda}_m)$ maximizes L, then at $(\bar{x}_1, \bar{x}_2, \ldots, \bar{x}_n, \bar{\lambda}_1, \bar{\lambda}_2, \ldots, \bar{\lambda}_n)$

$$\frac{\partial L}{\partial \lambda_i} = b_i - g_i(x_1, x_2, \ldots, x_n) = 0$$

Here $\dfrac{\partial L}{\partial \lambda_i}$ is the partial derivative of L with respect to λ_i. This shows that $(\bar{x}_1, \bar{x}_2, \ldots, \bar{x}_n)$ will satisfy the constraints in (12). To show that $(\bar{x}_1, \bar{x}_2, \ldots, \bar{x}_n)$ solves (12), let $(x'_1, x'_2, \ldots, x'_n)$ be any point that is in (12)'s feasible region. Since $(\bar{x}_1, \bar{x}_2, \ldots, \bar{x}_n, \bar{\lambda}_1, \bar{\lambda}_2, \ldots, \bar{\lambda}_m)$ maximizes L, for any numbers $\lambda'_1, \lambda'_2, \ldots, \lambda'_m$ we have

$$L(\bar{x}_1, \bar{x}_2, \ldots, \bar{x}_n, \bar{\lambda}_1, \bar{\lambda}_2, \ldots, \bar{\lambda}_m) \geq L(x'_1, x'_2, \ldots, x'_n, \lambda'_1, \lambda'_2, \ldots \lambda'_m) \tag{14}$$

Since $(\bar{x}_1, \bar{x}_2, \ldots, \bar{x}_n)$ and $(x'_1, x'_2, \ldots, x'_n)$ are both feasible in (12), the terms in (13) involving the λ's are all zero, and (14) becomes $f(\bar{x}_1, \bar{x}_2, \ldots, \bar{x}_n) \geq f(x'_1, x'_2, \ldots, x'_n)$. Thus, $(\bar{x}_1, \bar{x}_2, \ldots, \bar{x}_n)$ does solve (12). In short, if $(\bar{x}_1, \bar{x}_2, \ldots, \bar{x}_n, \bar{\lambda}_1, \bar{\lambda}_2, \ldots, \bar{\lambda}_m)$ solves the unconstrained maximization problem

$$\max L(x_1, x_2, \ldots, x_n, \lambda_1, \lambda_2, \ldots, \lambda_m) \tag{15}$$

then $(\bar{x}_1, \bar{x}_2, \ldots, \bar{x}_n)$ solves (12).

From Section 11.6, we know that for $(\bar{x}_1, \bar{x}_2, \ldots, \bar{x}_n, \bar{\lambda}_1, \bar{\lambda}_2, \ldots, \bar{\lambda}_m)$ to solve (15), it is necessary that at $(\bar{x}_1, \bar{x}_2, \ldots, \bar{x}_n, \bar{\lambda}_1, \bar{\lambda}_2, \ldots, \bar{\lambda}_m)$,

$$\frac{\partial L}{\partial x_1} = \frac{\partial L}{\partial x_2} = \cdots = \frac{\partial L}{\partial x_n} = \frac{\partial L}{\partial \lambda_1} = \frac{\partial L}{\partial \lambda_2} = \cdots = \frac{\partial L}{\partial \lambda_m} = 0 \tag{16}$$

Theorem 8 gives conditions implying that any point $(\bar{x}_1, \bar{x}_2, \ldots, \bar{x}_n, \bar{\lambda}_1, \bar{\lambda}_2, \ldots, \bar{\lambda}_m)$ that satisfies (16) will yield an optimal solution $(\bar{x}_1, \bar{x}_2, \ldots, \bar{x}_n)$ to (12).

THEOREM 8

Suppose (12) is a maximization problem. If $f(x_1, x_2, \ldots, x_n)$ is a concave function and each $g_i(x_1, x_2, \ldots, x_n)$ is a linear function, then any point $(\bar{x}_1, \bar{x}_2, \ldots, \bar{x}_n, \bar{\lambda}_1, \bar{\lambda}_2, \ldots, \bar{\lambda}_m)$ satisfying (16) will yield an optimal solution $(\bar{x}_1, \bar{x}_2, \ldots, \bar{x}_n)$ to (12).

Suppose (12) is a minimization problem. If $f(x_1, x_2, \ldots, x_n)$ is a convex function and each $g_i(x_1, x_2, \ldots, x_n)$ is a linear function, then any point $(\bar{x}_1, \bar{x}_2, \ldots, \bar{x}_n, \bar{\lambda}_1, \bar{\lambda}_2, \ldots, \bar{\lambda}_m)$ satisfying (16) will yield an optimal solution $(\bar{x}_1, \bar{x}_2, \ldots, \bar{x}_n)$ to (12).

Even if the hypotheses of these theorems fail to hold, it is possible that any point satisfying (16) will solve (12). See the appendix of Henderson and Quandt (1980) for details.

Geometrical Interpretation of Lagrange Multipliers

From (16) we know that for the point $\bar{x} = (\bar{x}_1, \bar{x}_2, \ldots, \bar{x}_n)$ to solve (12) it is necessary that at $\bar{x}$

$$\frac{\partial L}{\partial x_j} = 0 \quad \text{for } j = 1, 2, \ldots, n$$

This is equivalent to saying that there exist numbers $\lambda_1, \lambda_2, \ldots \lambda_m$ such that at the point $\bar{x}$

$$\nabla f = \sum_{i=1}^{i=m} \lambda_i \, \nabla g_i \tag{17}$$

To see why this is so, note that the jth component of the left-hand side of (17) is

$$\frac{\partial f}{\partial x_j}$$

and the jth component of the right-hand side is

$$\sum_{i=1}^{i=m} \lambda_i \, \frac{\partial g_i}{\partial x_j}$$

Thus, (17) implies that for $j = 1, 2, \ldots, n$

$$\frac{\partial f}{\partial x_j} - \sum_{i=1}^{i=m} \lambda_i \, \frac{\partial g_i}{\partial x_j} = 0 \quad \text{or} \quad \frac{\partial L}{\partial x_j} = 0$$

Another way to look at (17) is as follows: For $\bar{x}$ to solve (12), it is necessary that at $\bar{x}$, ∇f is a linear combination of the constraint gradients.

For an optimization problem with one constraint it is easy to see why (17) must hold at a solution to (12). If (12) has one constraint, then (17) is equivalent to the statement that the gradient of the objective function and the constraint are parallel. The necessity of this condition is illustrated in Figure 41. Here $z = 3$ is the optimal z-value when we try to maximize $f(x_1, x_2)$, subject to $g(x_1, x_2) = 0$. At the optimal point in Figure 41, $\nabla f = \lambda \nabla g$, where $\lambda < 0$.

To see why (17) must hold for an optimal solution to (12), let's consider the following NLP:

$$\max z = f(x_1, x_2, x_3)$$
$$\text{s.t.} \quad g_1(x_1, x_2, x_3) = 0 \tag{18}$$
$$g_2(x_1, x_2, x_3) = 0$$

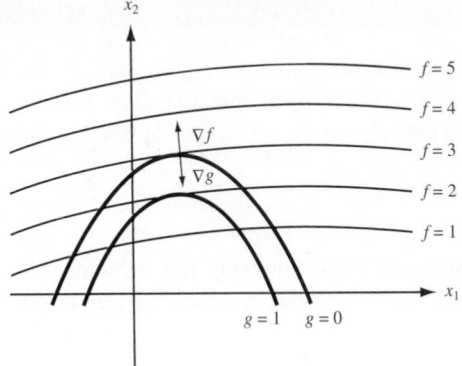

FIGURE 41
One-Constraint
Example of (17)

Suppose $\bar{x} = (\bar{x}_1, \bar{x}_2, \bar{x}_3)$ is an optimal solution to (18). We claim that for any $c \neq 0$, the following system of equations can have no solution (all gradients are evaluated at $\bar{x}$).

$$\begin{bmatrix} \nabla g_1 \\ \nabla g_2 \\ \nabla f \end{bmatrix} \begin{bmatrix} d_1 \\ d_2 \\ d_3 \end{bmatrix} = \begin{bmatrix} 0 \\ 0 \\ c \end{bmatrix} \tag{19}$$

To see why (19) can have no solution, suppose that it has a solution for some $c > 0$. [If (19) has a solution with $c < 0$, then a similar argument holds.] This solution defines a direction $\mathbf{d}$ in three dimensions. If we move in the direction $\mathbf{d}$ a small distance ϵ away from $\bar{x}$ we can find a feasible point $\bar{x} + \epsilon\mathbf{d}$ for (18) that has a larger z-value than $\bar{x}$. This would contradict the optimality of $\bar{x}$. To see that $\bar{x} + \epsilon\mathbf{d}$ is feasible in (18), note that for $i = 1, 2$ (19) implies that $g_i(\bar{x} + \epsilon\mathbf{d})$ is approximately equal to

$$g_i(\bar{x}) + \sum_{j=1}^{j=3} \frac{\partial g_i(\bar{x})}{\partial x_j} \left(\frac{\epsilon d_j}{\| \mathbf{d} \|} \right) = g_i(\bar{x}) = 0$$

Also $f(\bar{x} + \epsilon\mathbf{d})$ is approximately equal to

$$f(\bar{x}) + \sum_{j=1}^{j=3} \frac{\partial f}{\partial x_j} \left(\frac{\epsilon d_j}{\| \mathbf{d} \|} \right) = f(\bar{x}) + c\epsilon / \| \mathbf{d} \| > f(\bar{x})$$

This means that if $\bar{x}$ solves (18), then (19) can have no solution for $c \neq 0$. From Section 2.4, we know that (19) can have no solution if and only if the rank of the matrix on the left side of (19) is less than or equal to 2. This means that $\nabla f, \nabla g_1, \nabla g_2$ at $\bar{x}$ are linearly dependent vectors. Thus, a nontrivial linear combination of $\nabla f, \nabla g_1$, and ∇g_2 must add up to the zero vector. If we assume that ∇g_1 and ∇g_2 are linearly independent (the usual case), then (17) must hold.

Lagrange Multipliers and Sensitivity Analysis

The Lagrange multipliers λ_i can be used in sensitivity analysis. If the right-hand side of the ith constraint is increased by a small amount Δb_i (in either a maximization or minimization problem), then the optimal z-value for (12) will increase by approximately $\sum_{i=1}^{i=m} (\Delta b_i)\lambda_i$. This result is proven in Problem 9 of this section. In particular, if we increase the right-hand side of only constraint i by Δb_i, then the optimal z-value of (12) will increase by $(\Delta b_i)\lambda_i$.

The two examples that follow illustrate the use of Lagrange multipliers. In most cases, the easiest way to find a point $(\bar{x}_1, \bar{x}_2, \ldots, \bar{x}_n, \bar{\lambda}_1, \bar{\lambda}_2, \ldots, \bar{\lambda}_m)$ satisfying (16) is to first solve for $\bar{x}_1, \bar{x}_2, \ldots, \bar{x}_n$ in terms of $\bar{\lambda}_1, \bar{\lambda}_2, \ldots, \bar{\lambda}_m$. Then determine the values of the $\bar{\lambda}_i$'s by substituting these relations into the constraints of (12). Finally, use the values of the $\bar{\lambda}_i$'s to determine $\bar{x}_1, \bar{x}_2, \ldots, \bar{x}_n$.

EXAMPLE 30   **Lagrange Multiplier in Advertising**

A company is planning to spend \$10,000 on advertising. It costs \$3,000 per minute to advertise on television and \$1,000 per minute to advertise on radio. If the firm buys x minutes of television advertising and y minutes of radio advertising, then its revenue in thousands of dollars is given by $f(x, y) = -2x^2 - y^2 + xy + 8x + 3y$. How can the firm maximize its revenue?

Solution   We want to solve the following NLP:

$$\max z = -2x^2 - y^2 + xy + 8x + 3y$$
$$\text{s.t.} \quad 3x + y = 10$$

Then $L(x, y, \lambda) = -2x^2 - y^2 + xy + 8x + 3y + \lambda(10 - 3x - y)$. We set

$$\frac{\partial L}{\partial x} = \frac{\partial L}{\partial y} = \frac{\partial L}{\partial \lambda} = 0$$

This yields

$$\frac{\partial L}{\partial x} = -4x + y + 8 - 3\lambda = 0 \tag{20}$$

$$\frac{\partial L}{\partial y} = -2y + x + 3 - \lambda = 0 \tag{21}$$

$$\frac{\partial L}{\partial \lambda} = 10 - 3x - y = 0 \tag{22}$$

Observe that $10 - 3x - y = 0$ reduces to the constraint $3x + y = 10$. Equation (20) yields $y = 3\lambda - 8 + 4x$, and (21) yields $x = \lambda - 3 + 2y$. Thus, $y = 3\lambda - 8 + 4(\lambda - 3 + 2y) = 7\lambda - 20 + 8y$, or

$$y = \tfrac{20}{7} - \lambda \tag{23}$$
$$x = \lambda - 3 + 2(\tfrac{20}{7} - \lambda) = \tfrac{19}{7} - \lambda \tag{24}$$

Substituting (23) and (24) into (22) yields $10 - 3(\tfrac{19}{7} - \lambda) - (\tfrac{20}{7} - \lambda) = 0$, or $4\lambda - 1 = 0$, or $\lambda = \tfrac{1}{4}$. Then (23) and (24) yield

$$\bar{y} = \tfrac{20}{7} - \tfrac{1}{4} = \tfrac{73}{28}$$
$$\bar{x} = \tfrac{19}{7} - \tfrac{1}{4} = \tfrac{69}{28}$$

The Hessian for $f(x, y)$ is

$$H(x, y) = \begin{bmatrix} -4 & 1 \\ 1 & -2 \end{bmatrix}$$

Since each first-order principal minor is negative, and $H_2(x, y) = 7 > 0$, $f(x, y)$ is concave. The constraint is linear, so Theorem 8 shows that the Lagrange multiplier method does yield the optimal solution to the NLP.

Thus, the firm should purchase $\tfrac{69}{28}$ minutes of television time and $\tfrac{73}{28}$ minutes of radio time. Since $\lambda = \tfrac{1}{4}$, spending an extra Δ (thousands) (for small Δ) would increase the firm's revenues by approximately \$0.25Δ (thousands).

In general, if the firm had a dollars to spend on advertising, then it could be shown that $\lambda = \tfrac{11-a}{4}$ (see Problem 1 at the end of this section). We see that as more money is spent on advertising, the increase to revenue for each additional advertising dollar becomes smaller.

EXAMPLE 31   Lagrange Multiplier and Optimal Solution

Given numbers $x_1, x_2, \ldots, x_n$, show that

$$n \sum_{i=1}^{i=n} x_i^2 \geq \left(\sum_{i=1}^{i=n} x_i \right)^2$$

with equality holding only if $x_1 = x_2 = \cdots = x_n$.

Solution Suppose that $x_1 + x_2 + \cdots + x_n = c$. Consider the NLP

$$\min z = \sum_{i=1}^{i=n} x_i^2$$

$$\text{s.t.} \quad \sum_{i=1}^{i=n} x_i = c \tag{25}$$

To solve (25), we form

$$L(x_1, x_2, \ldots, x_n, \lambda) = x_1^2 + x_2^2 + \cdots + x_n^2 + \lambda(c - x_1 - x_2 - \cdots - x_n)$$

Then to solve (25) we need to find $(x_1, x_2, \ldots, x_n, \lambda)$ that satisfy

$$\frac{\partial L}{\partial x_i} = 2x_i - \lambda = 0 \qquad (i = 1, 2, \ldots, n) \quad \text{and}$$

$$\frac{\partial L}{\partial \lambda} = c - x_1 - x_2 - \cdots - x_n = 0$$

From $\dfrac{\partial L}{\partial x_i} = 0$, we obtain $2\bar{x}_1 = 2\bar{x}_2 = \cdots = 2\bar{x}_n = \bar{\lambda}$, or $x_i = \dfrac{\bar{\lambda}}{2}$. From $\dfrac{\partial L}{\partial \lambda} = 0$, we obtain $c - \dfrac{n\bar{\lambda}}{2} = 0$, or $\bar{\lambda} = \dfrac{2c}{n}$. The objective function is convex (it is the sum of n convex functions), and the constraint is linear. Thus, Theorem 8′ shows that the Lagrange multiplier method does yield an optimal solution to (25); it has

$$\bar{x}_i = \frac{\left(\dfrac{2c}{n} \right)}{2} = \frac{c}{n} \quad \text{and} \quad z = n \left(\frac{c^2}{n^2} \right) = \frac{c^2}{n}$$

Thus, if

$$\sum_{i=1}^{i=n} x_i = c$$

then

$$n \sum_{i=1}^{i=n} x_i^2 \geq n \left(\frac{c^2}{n} \right) = \left(\sum_{i=1}^{i=n} x_i \right)^2$$

with equality holding if and only if $x_1 = x_2 = \cdots = x_n$.

If we are trying to maximize a function $f(x_1, x_2, \ldots, x_n)$ that is a product of several functions, then it is often easier to maximize $\ln [f(x_1, x_2, \ldots, x_n)]$. Since $\ln$ is an increasing function, we know that any x^* maximizing $\ln [f(x_1, x_2, \ldots, x_n)]$ over any set of possible values for $(x_1, x_2, \ldots, x_n)$ will also maximize $f(x_1, x_2, \ldots, x_n)$ over the same set of possible values for $(x_1, x_2, \ldots, x_n)$. See Problem 2 for an application of this idea.

Solving NLP with Equality Constraints on LINGO

Adv.lng

If the hypotheses of Theorem 8 or Theorem 8′ hold for a problem, LINGO will find the optimal solution to the NLP. You will receive the messages OPTIMAL TO TOLERANCES and DUAL CONDITIONS: SATISFIED. "Optimal to Tolerances" means that LINGO is sure that it has found a local extremum. "Dual Conditions: Satisfied" means that LINGO is sure that the point it has found satisfies (16). Figure 42 (file Adv.lng) contains the LINGO printout for Example 28.

Interpretation of the LINGO Price Column

For a maximization problem, the LINGO PRICE column yields the Lagrange multiplier for each constraint. Thus, if the right-hand side of Constraint i in a maximization problem is increased by a small amount Δ, then the optimal z-value is increased by approximately Δ (PRICE for Constraint i). The PRICE column in Figure 42 implies that in Example 30 spending an extra Δ thousand dollars on advertising will increase revenues by approximately 0.25Δ (thousands).

For a minimization problem, the LINGO PRICE column yields the negative of the Lagrange multiplier for each constraint. Thus, if the right-hand side of Constraint i in a minimization problem is increased by a small amount Δ, then the optimal z-value will increase by approximately $\Delta(-\text{PRICE for Constraint } i)$.

```
MODEL:
  1) MAX= - 2 * X ^ 2 - Y ^ 2 + X * Y + 8 * X + 3 * Y ;
  2) 3 * X + Y = 10 ;
  3) X > 0 ;
  4) Y > 0 ;
END


SOLUTION STATUS:  OPTIMAL TO TOLERANCES.  DUAL CONDITIONS:  SATISFIED.

           OBJECTIVE FUNCTION VALUE

      1)        15.017855

   VARIABLE        VALUE        REDUCED COST
        X         2.464283         .000000
        Y         2.607140         .000003

   ROW    SLACK OR SURPLUS          PRICE
    2)       -.000010            .249996
    3)       2.464283            .000000
    4)       2.607140            .000000
```

FIGURE 42
Optimal Solution for Example 28

PROBLEMS

Group A

1 For Example 30, show that if a dollars are available for advertising, then an extra dollar spent on advertising will increase revenues by approximately $\frac{11-a}{4}$.

2 It costs me $2 to purchase an hour of labor and $1 to purchase a unit of capital. If L hours of labor and K units of capital are available, then $L^{2/3}K^{1/3}$ machines can be produced. If I have $10 to purchase labor and capital, what is the maximum number of machines that can be produced?

3 In Problem 2, what is the minimum cost method of producing 6 machines?

4 A beer company has divided Bloomington into two territories. If x_1 dollars are spent on promotion in territory 1, then $6x_1^{1/2}$ cases of beer can be sold there; and if x_2 dollars are spent on promotion in territory 2, then $4x_2^{1/2}$ cases of beer can be sold there. Each case of beer sold in territory 1 sells for $10 and incurs $5 in shipping and production costs.

Each case of beer sold in territory 2 sells for $9 and incurs $4 in shipping and production costs. A total of $100 is available for promotion. How can the beer company maximize profits? If an extra dollar could be spent on promotion, by approximately how much would profits increase? By how much would revenues increase?

Group B

5 We must invest all our money in two stocks: x and y. The variance of the annual return on one share of stock x is var x, and the variance of the annual return on one share of stock y is var y. Assume that the covariance between the annual return for one share of x and one share of y is cov(x, y). If we invest $a\%$ of our money in stock x and $b\%$ in stock y, then the variance of our return is given by a^2var x + b^2var y + 2ab cov(x, y). We want to minimize the variance of the return on our invested money. What percentage of the money should be invested in each stock?

6 As in Problem 5, assume that we must determine the percentage of our money that is invested in stocks x and y. A choice of a and b is called a *portfolio*. A portfolio is efficient if there exists no other portfolio whose return has a higher mean return and lower variance, or a higher mean return and the same variance, or a lower variance with the same mean return. Let $\bar{x}$ be the mean return on stock x and $\bar{y}$ be the mean return on stock y. Consider the following NLP:

$$\max z = c[a\bar{x} + b\bar{y}]$$
$$- (1 - c)[a^2\text{var } x + b^2\text{var } y$$
$$+ 2ab\text{cov}(x, y)]$$
$$\text{s.t.} \quad a + b = 1$$
$$a, b \geq 0$$

Suppose that $1 > c > 0$. Show that any solution to this NLP is an efficient portfolio.

7 Suppose product i (i = 1, 2) costs c_i per unit. If $x_i(i = 1, 2)$ units of products 1 and 2 are purchased, then a utility $x_1^a x_2^{1-a}(0 < a < 1)$ is received.

a If d are available to purchase products 1 and 2, how many of each type should be purchased?

b Show that an increase in the cost of product i decreases the number of units of product i that should be purchased.

c Show that an increase in the cost of product i does not change the number of units of the other product that should be purchased.

8 Suppose that a cylindrical soda can must have a volume of 26 cu in. If the soda company wants to minimize the surface area of the soda can, what should be the ratio of the height of the can to the radius of the can? (*Hint:* The volume of a right circular cylinder is $\pi r^2 h$, and the surface area of a right circular cylinder is $2\pi r^2 + 2\pi rh$, where r = the radius of the cylinder and h = the height of the cylinder.)

9 Show that if the right-hand side of the ith constraint is increased by a small amount Δb_i (in either a maximization or minimization problem), then the optimal z-value for (11) will increase by approximately $\sum_{i=1}^{i=m} (\Delta b_i)\lambda_i$.

11.9 The Kuhn–Tucker Conditions

In this section, we discuss necessary and sufficient conditions for $\bar{x} = (\bar{x}_1, \bar{x}_2, \ldots, \bar{x}_n)$ to be an optimal solution for the following NLP:

$$\max \text{ (or min) } f(x_1, x_2, \ldots, x_n)$$
$$\text{s.t.} \quad g_1(x_1, x_2, \ldots, x_n) \leq b_1$$
$$g_2(x_1, x_2, \ldots, x_n) \leq b_2 \quad \text{(26)}$$
$$\vdots$$
$$g_m(x_1, x_2, \ldots, x_n) \leq b_m$$

To apply the results of this section, all the NLP's constraints must be $\leq$ constraints. A constraint of the form $h(x_1, x_2, \ldots, x_n) \geq b$ must be rewritten as $-h(x_1, x_2, \ldots, x_n) \leq -b$. For example, the constraint $2x_1 + x_2 \geq 2$ should be rewritten as $-2x_1 - x_2 \leq -2$. A constraint of the form $h(x_1, x_2, \ldots, x_n) = b$ must be replaced by $h(x_1, x_2, \ldots, x_n) \leq b$ and $-h(x_1, x_2, \ldots, x_n) \leq -b$. For example, $2x_1 + x_2 = 2$ would be replaced by $2x_1 + x_2 \leq 2$ and $-2x_1 - x_2 \leq -2$.

Theorems 9 and 9′ give conditions (the **Kuhn–Tucker,** or **KT, conditions**) that are necessary for a point $\bar{x} = (\bar{x}_1, \bar{x}_2, \ldots, \bar{x}_n)$ to solve (26). The partial derivative of a function f with respect to a variable x_j evaluated at $\bar{x}$ is written

$$\frac{\partial f(\bar{x})}{\partial x_j}$$

For the theorems of this section to hold, the functions $g_1, g_2, \ldots, g_m$ must satisfy certain regularity conditions (usually called **constraint qualifications**). We will briefly discuss one constraint qualification at the end of the section. [For a detailed discussion of constraint qualifications we refer the reader to Chapter 5 of Bazaraa and Shetty (1993).]

When the constraints are linear, these regularity assumptions are always satisfied. In other situations (particularly when some of the constraints are equality constraints), the regularity conditions may not be satisfied. We assume that all problems we consider satisfy these regularity conditions.

THEOREM 9

Suppose (26) is a maximization problem. If $\bar{x} = (\bar{x}_1, \bar{x}_2, \ldots, \bar{x}_n)$ is an optimal solution to (26), then $\bar{x} = (\bar{x}_1, \bar{x}_2, \ldots, \bar{x}_n)$ must satisfy the m constraints in (26), and there must exist multipliers $\lambda_1, \lambda_2, \ldots, \lambda_m$ satisfying

$$\frac{\partial f(\bar{x})}{\partial x_j} - \sum_{i=1}^{i=m} \bar{\lambda}_i \frac{\partial g_i(\bar{x})}{\partial x_j} = 0 \qquad (j = 1, 2, \ldots, n) \tag{27}$$

$$\bar{\lambda}_i[b_i - g_i(\bar{x})] = 0 \qquad (i = 1, 2, \ldots, m) \tag{28}$$

$$\bar{\lambda}_i \geq 0 \qquad (i = 1, 2, \ldots, m) \tag{29}$$

THEOREM 9'

Suppose (26) is a minimization problem. If $\bar{x} = (\bar{x}_1, \bar{x}_2, \ldots, \bar{x}_n)$ is an optimal solution to (26), then $\bar{x} = (\bar{x}_1, \bar{x}_2, \ldots, \bar{x}_n)$ must satisfy the m constraints in (26), and there must exist multipliers $\lambda_1, \lambda_2, \ldots, \lambda_m$ satisfying

$$\frac{\partial f(\bar{x})}{\partial x_j} + \sum_{i=1}^{i=m} \bar{\lambda}_i \frac{\partial g_i(\bar{x})}{\partial x_j} = 0 \qquad (j = 1, 2, \ldots, n)$$

$$\bar{\lambda}_i[b_i - g_i(\bar{x})] = 0 \qquad (i = 1, 2, \ldots, m)$$

$$\bar{\lambda}_i \geq 0 \qquad (i = 1, 2, \ldots, m)$$

Like the Lagrange multipliers of the preceding section, the multiplier $\bar{\lambda}_i$ associated with the K–T conditions may be thought of as the shadow price for the ith constraint in (26). Suppose (26) is a maximization problem. If the right-hand side of the ith constraint is increased from b_i to $b_i + \Delta$ (for Δ small), the optimal objective function value will increase by approximately $\Delta\lambda_i$. Suppose (26) is a minimization problem. If the right-hand side of the ith constraint is increased from b_i to $b_i + \Delta$ (for Δ small), then the optimal objective function value is decreased by $\Delta\lambda_i$.

Bearing in mind this interpretation of the multipliers as shadow prices, we may interpret (27)–(29) for a max problem. Suppose we consider each constraint in (26) to be a resource-usage constraint. That is, at $\bar{x} = (\bar{x}_1, \bar{x}_2, \ldots, \bar{x}_n)$ we use $g_i(\bar{x}_1, \bar{x}_2, \ldots, \bar{x}_n)$ units of resource i, and b_i units of resource i are available. If we increase the value of x_j by a small amount Δ, then the value of the objective function increases by

$$\frac{\partial f(\bar{x})}{\partial x_j} \Delta$$

Changing the value of x_j to $\bar{x}_j + \Delta$ also changes the ith constraint to

$$g_i(\bar{x}) + \frac{\partial g_i(\bar{x})}{\partial x_j} \Delta \leq b_i \quad \text{or} \quad g_i(\bar{x}) \leq b_i - \frac{\partial g_i(\bar{x})}{\partial x_j} \Delta$$

Thus, increasing x_j by Δ has the effect of increasing the right-hand side of the ith constraint by

$$-\frac{\partial g_i(\bar{x})}{\partial x_j} \Delta$$

These changes in the right-hand sides of the constraints will increase the value of z by approximately

$$-\Delta \sum_{i=1}^{i=m} \bar{\lambda}_i \frac{\partial g_i(\bar{x})}{\partial x_j}$$

In total, the approximate change in z due to increasing x_j by Δ is

$$\Delta \left[\frac{\partial f(\bar{x})}{\partial x_j} - \sum_{i=1}^{i=m} \bar{\lambda}_i \frac{\partial g_i(\bar{x})}{\partial x_j} \right]$$

If the term in brackets is larger than zero, we can increase f by choosing $\Delta > 0$. On the other hand, if this term is smaller than zero, we can increase f by choosing $\Delta < 0$. Thus, for $\bar{x}$ to be optimal, (27) must hold.

Condition (28) is a generalization of the complementary slackness conditions for LPs discussed in Section 6.10. Condition (28) implies that

$$\text{If } \bar{\lambda}_i > 0, \quad \text{then} \quad g_i(\bar{x}) = b_i \quad (\text{ith constraint binding}) \tag{28$'$}$$
$$\text{If } g_i(\bar{x}) < b_i, \quad \text{then} \quad \bar{\lambda}_i = 0 \tag{28$''$}$$

Suppose the constraint $g_i(x_1, x_2, \ldots, x_n) \leq b_i$ is a resource-usage constraint representing the fact that at most b_i units of the ith resource can be used. Then (28$'$) states that if an additional unit of the resource associated with the ith constraint is to have any value, then the current optimal solution must use all b_i units of the ith resource currently available. On the other hand, (28$''$) states that if some of the ith resource currently available is unused, then additional amounts of the ith resource have no value.

If for $\Delta > 0$, we increase the right-hand side of the ith constraint from b_i to $b_i + \Delta$, then the optimal objective function value must increase or stay the same, because the increase adds points to the problem's feasible region. Increasing the right-hand side of the ith constraint by Δ increases the optimal objective function value by $\Delta\bar{\lambda}_i$, so it must be that $\bar{\lambda}_i \geq 0$. This is why (29) is included in the K–T conditions.

In many situations, the K–T conditions are applied to NLPs in which the variables must be non-negative. For example, we may want to use the K–T conditions to find the optimal solution to

$$\max \text{ (or min) } z = f(x_1, x_2, \ldots, x_n)$$
$$\text{s.t.} \quad g_1(x_1, x_2, \ldots, x_n) \leq b_1$$
$$g_2(x_1, x_2, \ldots, x_n) \leq b_2$$
$$\vdots$$
$$g_m(x_1, x_2, \ldots, x_n) \leq b_m \tag{30}$$
$$-x_1 \leq 0$$
$$-x_2 \leq 0$$
$$\vdots$$
$$-x_n \leq 0$$

If we associate multipliers $\mu_1, \mu_2, \ldots, \mu_n$ with the non-negativity constraints in (30), Theorems 9 and 9' reduce to Theorems 10 and 10'.

THEOREM 10

Suppose (30) is a maximization problem. If $\bar{x} = (\bar{x}_1, \bar{x}_2, \ldots, \bar{x}_n)$ is an optimal solution to (30), then $\bar{x} = (\bar{x}_1, \bar{x}_2, \ldots, \bar{x}_n)$ must satisfy the constraints in (30) and there must exist multipliers $\lambda_1, \lambda_2, \ldots, \lambda_m, \bar{\mu}_1, \bar{\mu}_2, \ldots, \bar{\mu}_n$ satisfying

$$\frac{\partial f(\bar{x})}{\partial x_j} - \sum_{i=1}^{i=m} \bar{\lambda}_i \frac{\partial g_i(\bar{x})}{\partial x_j} + \mu_j = 0 \qquad (j = 1, 2, \ldots, n) \tag{31}$$

$$\bar{\lambda}_i[b_i - g_i(\bar{x})] = 0 \qquad (i = 1, 2, \ldots, m) \tag{32}$$

$$\left[\frac{\partial f(\bar{x})}{\partial x_j} - \sum_{i=1}^{i=m} \bar{\lambda}_i \frac{\partial g_i(\bar{x})}{\partial x_j}\right] \bar{x}_j = 0 \qquad (j = 1, 2, \ldots, n) \tag{33}$$

$$\bar{\lambda}_i \geq 0 \qquad (i = 1, 2, \ldots, m) \tag{34}$$

$$\bar{\mu}_j \geq 0 \qquad (j = 1, 2, \ldots, n) \tag{35}$$

Because $\bar{\mu}_j \geq 0$, (31) is equivalent to

$$\frac{\partial f(\bar{x})}{\partial x_j} - \sum_{i=1}^{i=m} \bar{\lambda}_i \frac{\partial g_i(\bar{x})}{\partial x_j} \leq 0 \qquad (j = 1, 2, \ldots, n) \tag{31'}$$

Then (31)–(34), the K–T conditions for a maximization problem with nonnegativity constraints, may be rewritten as

$$\frac{\partial f(\bar{x})}{\partial x_j} - \sum_{i=1}^{i=m} \bar{\lambda}_i \frac{\partial g_i(\bar{x})}{\partial x_j} \leq 0 \qquad (j = 1, 2, \ldots, n) \tag{31'}$$

$$\bar{\lambda}_i[b_i - g_i(\bar{x})] = 0 \qquad (i = 1, 2, \ldots, m) \tag{32'}$$

$$\left[\frac{\partial f(\bar{x})}{\partial x_j} - \sum_{i=1}^{i=m} \bar{\lambda}_i \frac{\partial g_i(\bar{x})}{\partial x_j}\right] \bar{x}_j = 0 \qquad (j = 1, 2, \ldots, n) \tag{33'}$$

$$\bar{\lambda}_i \geq 0 \qquad (i = 1, 2, \ldots, m) \tag{34'}$$

THEOREM 10'

Suppose (30) is a minimization problem. If $\bar{x} = (\bar{x}_1, \bar{x}_2, \ldots, \bar{x}_n)$ is an optimal solution to (30), then $\bar{x} = (\bar{x}_1, \bar{x}_2, \ldots, \bar{x}_n)$ must satisfy the constraints in (30), and there must exist multipliers $\lambda_1, \lambda_2, \ldots, \lambda_m, \bar{\mu}_1, \bar{\mu}_2, \ldots, \bar{\mu}_n$ satisfying

$$\frac{\partial f(\bar{x})}{\partial x_j} + \sum_{i=1}^{i=m} \bar{\lambda}_i \frac{\partial g_i(\bar{x})}{\partial x_j} - \mu_j = 0 \qquad (j = 1, 2, \ldots, n) \tag{36}$$

$$\bar{\lambda}_i[b_i - g_i(\bar{x})] = 0 \qquad (i = 1, 2, \ldots, m) \tag{37}$$

$$\left[\frac{\partial f(\bar{x})}{\partial x_j} + \sum_{i=1}^{i=m} \bar{\lambda}_i \frac{\partial g_i(\bar{x})}{\partial x_j}\right] \bar{x}_j = 0 \qquad (j = 1, 2, \ldots, n) \tag{38}$$

$$\bar{\lambda}_i \geq 0 \qquad (i = 1, 2, \ldots, m) \tag{39}$$

$$\bar{\mu}_j \geq 0 \qquad (j = 1, 2, \ldots, n) \tag{40}$$

Because $\bar{\mu}_j \geq 0$, (36) may be written as

$$\frac{\partial f(\bar{x})}{\partial x_j} + \sum_{i=1}^{i=m} \bar{\lambda}_i \frac{\partial g_i(\bar{x})}{\partial x_j} \geq 0 \tag{36'}$$

Then (36)–(39), the K–T conditions for a minimization problem with nonnegativity constraints, may be rewritten as

$$\frac{\partial f(\bar{x})}{\partial x_j} + \sum_{i=1}^{i=m} \bar{\lambda}_i \frac{\partial g_i(\bar{x})}{\partial x_j} \geq 0 \qquad (j = 1, 2, \ldots, n) \tag{36'}$$

$$\bar{\lambda}_i[b_i - g_i(\bar{x})] = 0 \qquad (i = 1, 2, \ldots, m) \tag{37'}$$

$$\left[\frac{\partial f(\bar{x})}{\partial x_j} + \sum_{i=1}^{i=m} \bar{\lambda}_i \frac{\partial g_i(\bar{x})}{\partial x_j}\right] \bar{x}_j = 0 \qquad (j = 1, 2, \ldots, n) \tag{38'}$$

$$\bar{\lambda}_i \geq 0 \qquad (i = 1, 2, \ldots, m) \tag{39'}$$

Theorems 9, 9', 10, and 10' give conditions that are *necessary* for a point $\bar{x} = (\bar{x}_1, \bar{x}_2, \ldots, \bar{x}_n)$ to be an optimal solution to (26) or (30). The following two theorems give conditions that are *sufficient* for $\bar{x} = (\bar{x}_1, \bar{x}_2, \ldots, \bar{x}_n)$ to be an optimal solution to (26) or (30) (see Bazaraa and Shetty (1993)).

THEOREM 11

Suppose (26) is a maximization problem. If $f(x_1, x_2, \ldots, x_n)$ is a concave function and $g_1(x_1, x_2, \ldots, x_n), \ldots, g_m(x_1, x_2, \ldots, x_n)$ are convex functions, then any point $\bar{x} = (\bar{x}_1, \bar{x}_2, \ldots, \bar{x}_n)$ satisfying the hypotheses of Theorem 9 is an optimal solution to (26). Also, if (30) is a maximization problem, $f(x_1, x_2, \ldots, x_n)$ is a concave function, and $g_1(x_1, x_2, \ldots, x_n), \ldots, g_m(x_1, x_2, \ldots, x_n)$ are convex functions, then any point $\bar{x} = (\bar{x}_1, \bar{x}_2, \ldots, \bar{x}_n)$ satisfying the hypotheses of Theorem 10 is an optimal solution to (30).

THEOREM 11'

Suppose (26) is a minimization problem. If $f(x_1, x_2, \ldots, x_n)$ is a convex function and $g_1(x_1, x_2, \ldots, x_n), \ldots, g_m(x_1, x_2, \ldots, x_n)$ are convex functions, then any point $\bar{x} = (\bar{x}_1, \bar{x}_2, \ldots, \bar{x}_n)$ satisfying the hypotheses of Theorem 9' is an optimal solution to (26). Also, if (30) is a minimization problem, $f(x_1, x_2, \ldots, x_n)$ is a convex function, and $g_1(x_1, x_2, \ldots, x_n), \ldots, g_m(x_1, x_2, \ldots, x_n)$ are convex functions, then any point $\bar{x} = (\bar{x}_1, \bar{x}_2, \ldots, \bar{x}_n)$ satisfying the hypotheses of Theorem 10' is an optimal solution to (30).

REMARK The reason that the hypotheses of Theorems 11 and 11' require that each $g_i(x_1, x_2, \ldots, x_n)$ be convex is that this ensures the feasible region for (26) or (30) is a convex set (see Problem 21 of Section 11.3).

Geometrical Interpretation of Kuhn–Tucker Conditions

It is easy to show that conditions (27)–(29) of Theorem 9 will hold at a point $\bar{x}$ if and only if ∇f is a non-negative linear combination of $\nabla g_1, \nabla g_2, \ldots, \nabla g_m$, and the weight multiplying ∇g_i in this linear combination equals 0 if the ith constraint in (26) is nonbinding.

In short, (27)–(29) are equivalent to the existence of $\lambda_i \geq 0$ such that

$$\nabla f(\bar{x}) = \sum_{i=1}^{i=m} \lambda_i \nabla g_i(\bar{x}) \tag{41}$$

and each constraint that is nonbinding at $\bar{x}$ has $\lambda_i = 0$.

Figures 43 and 44 illustrate (41). In Figure 43, we are trying to solve (the feasible region is shaded)

$$\min z = f(x_1, x_2)$$
$$\text{s.t.} \quad g_1(x_1, x_2) \leq 0$$
$$g_2(x_1, x_2) \leq 0$$

At $\bar{x}$, (41) holds with both constraints binding and we have $\lambda_1 > 0$ and $\lambda_2 > 0$. In Figure 44, we are again trying to solve (feasible region is again shaded)

$$\min z = f(x_1, x_2)$$
$$\text{s.t.} \quad g_1(x_1, x_2) \leq 0$$
$$g_2(x_1, x_2) \leq 0$$

Here, the second constraint is nonbinding so (41) must hold with $\lambda_2 = 0$.

The following two examples illustrate the use of the K–T conditions.

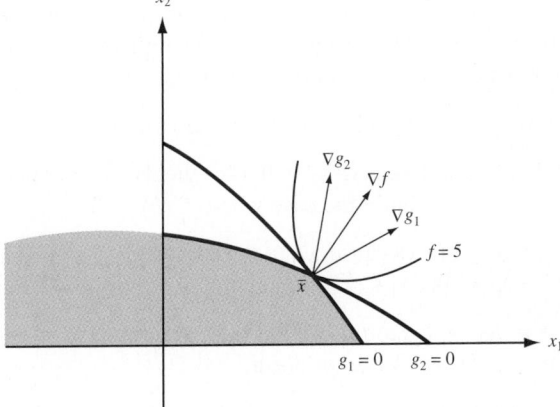

FIGURE 43
Example of Kuhn–Tucker Conditions: Both Constraints Binding

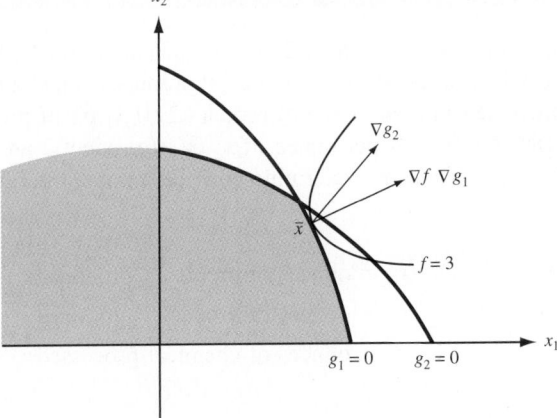

FIGURE 44
Example of Kuhn–Tucker Conditions: One Constraint Binding and One Constraint Nonbinding

EXAMPLE 32 **Interpretation of Kuhn–Tucker Conditions**

Describe the optimal solution to

$$\max f(x)$$

$$\text{s.t.} \quad a \leq x \leq b \tag{42}$$

Solution From Section 11.4, we know {assuming that $f'(x)$ exists for all x on the interval $[a, b]$} that the optimal solution to this problem must occur at a [with $f'(a) \leq 0$], at b [with $f'(b) \geq 0$], or at a point having $f'(x) = 0$. How do the K–T conditions yield these three cases?

We write (42) as

$$\max f(x)$$

$$\text{s.t.} \quad -x \leq -a$$

$$x \leq b$$

Then (27)–(29) yield

$$f'(x) + \lambda_1 - \lambda_2 = 0 \tag{43}$$

$$\lambda_1(-a + x) = 0 \tag{44}$$

$$\lambda_2(b - x) = 0 \tag{45}$$

$$\lambda_1 \geq 0 \tag{46}$$

$$\lambda_2 \geq 0 \tag{47}$$

In using the K–T conditions to solve NLPs, it is useful to note that each multiplier λ_i must satisfy $\lambda_i = 0$ or $\lambda_i > 0$. Thus, in attempting to find values of x, λ_1, and λ_2 that satisfy (43)–(47), we must consider the following four cases:

Case 1 $\lambda_1 = \lambda_2 = 0$. From (43), we obtain the case $f'(\bar{x}) = 0$.

Case 2 $\lambda_1 = 0$, $\lambda_2 > 0$. Because $\lambda_2 > 0$, (45) yields $\bar{x} = b$. Then (43) yields $f'(b) = \lambda_2$, and because $\lambda_2 > 0$, we obtain the case where $f'(b) > 0$.

Case 3 $\lambda_1 > 0$, $\lambda_2 = 0$. Because $\lambda_1 > 0$, (44) yields $\bar{x} = a$. Then (43) yields the case where $f'(a) = -\lambda_1 < 0$.

Case 4 $\lambda_1 > 0$, $\lambda_2 > 0$. From (44) and (45), we obtain $\bar{x} = a$ and $\bar{x} = b$. This contradiction indicates that Case 4 cannot occur.

The constraints are linear, so Theorem 11 shows that if $f(x)$ is concave, then (43)–(47) yield the optimal solution to (42).

EXAMPLE 33 **Production Process**

A monopolist can purchase up to 17.25 oz of a chemical for \$10/oz. At a cost of \$3/oz, the chemical can be processed into an ounce of product 1; or, at a cost of \$5/oz, the chemical can be processed into an ounce of product 2. If x_1 oz of product 1 are produced, it sells for a price of \$30 $- x_1$ per ounce. If x_2 oz of product 2 are produced, it sells for a price of \$50 $- 2x_2$ per ounce. Determine how the monopolist can maximize profits.

Solution Let

$$x_1 = \text{ounces of product 1 produced}$$

$$x_2 = \text{ounces of product 2 produced}$$

$$x_3 = \text{ounces of chemical processed}$$

Then we want to solve the following NLP:

$$\max z = x_1(30 - x_1) + x_2(50 - 2x_2) - 3x_1 - 5x_2 - 10x_3$$

$$\text{s.t.} \quad x_1 + x_2 \le x_3 \quad \text{or} \quad x_1 + x_2 - x_3 \le 0 \qquad (48)$$

$$x_3 \le 17.25$$

Of course, we should add the constraints $x_1, x_2, x_3 \ge 0$. However, because the optimal solution to (48) satisfies the nonnegativity constraints, it also will be optimal for an NLP consisting of (48) with the nonnegativity constraints.

Observe that the objective function in (48) is the sum of concave functions (and is therefore concave), and the constraints are convex (because they are linear). Thus, Theorem 11 shows that the K–T conditions are necessary and sufficient for (x_1, x_2, x_3) to be an optimal solution to (48). From Theorem 9, the K–T conditions become

$$30 - 2x_1 - 3 - \lambda_1 = 0 \qquad (49)$$

$$50 - 4x_2 - 5 - \lambda_1 = 0 \qquad (50)$$

$$-10 + \lambda_1 - \lambda_2 = 0 \qquad (51)$$

$$\lambda_1(-x_1 - x_2 + x_3) = 0 \qquad (52)$$

$$\lambda_2(17.25 - x_3) = 0 \qquad (53)$$

$$\lambda_1 \ge 0 \qquad (54)$$

$$\lambda_2 \ge 0 \qquad (55)$$

As in the previous example, there are four cases to consider:

Case 1 $\lambda_1 = \lambda_2 = 0$. This case cannot occur, because (51) would be violated.

Case 2 $\lambda_1 = 0, \lambda_2 > 0$. If $\lambda_1 = 0$, then (51) implies $\lambda_2 = -10$. This would violate (55).

Case 3 $\lambda_1 > 0, \lambda_2 = 0$. From (51), we obtain $\lambda_1 = 10$. Now (49) yields $x_1 = 8.5$, and (50) yields $x_2 = 8.75$. From (52), we obtain $x_1 + x_2 = x_3$, so $x_3 = 17.25$. Thus, $\bar{x}_1 = 8.5$, $\bar{x}_2 = 8.75$, $\bar{x}_3 = 17.25$, $\bar{\lambda}_1 = 10$, $\bar{\lambda}_2 = 0$ satisfies the K–T conditions.

Case 4 $\lambda_1 > 0, \lambda_2 > 0$. Case 3 yields an optimal solution, so we need not consider Case 4.

Thus, the optimal solution to (48) is to buy 17.25 oz of the chemical and produce 8.5 oz of product 1 and 8.75 oz of product 2. For Δ small, $\bar{\lambda}_1 = 10$ indicates that if an extra Δ oz of the chemical were obtained at no cost, then profits would increase by 10Δ. (Can you see why?) From (51), we find that $\bar{\lambda}_2 = 0$. This implies that the right to purchase an extra Δ oz of the chemical would not increase profits. (Can you see why?)

Constraint Qualifications

Unless a constraint qualification or regularity condition is satisfied at an optimal point $\bar{x}$, the Kuhn–Tucker conditions may fail to hold at $\bar{x}$. There are many constraint qualifications, but we choose to discuss the Linear Independence Constraint Qualification: Let $\bar{x}$ be an optimal solution to NLP (26) or (30). If all g_i are continuous, and the gradients of all binding constraints (including any binding nonnegativity constraints on $x_1, x_2, \ldots, x_n$) at $\bar{x}$ form a set of linearly independent vectors, then the Kuhn–Tucker conditions must hold at $\bar{x}$.

The following example shows that if the Linear Independence Constraint Qualification fails to hold, then the Kuhn–Tucker conditions may fail to hold at the optimal solution to an NLP.

EXAMPLE 34   **Necessity of Constraint Qualification**

Show that the Kuhn–Tucker conditions fail to hold at the optimal solution to the following NLP:

$$\max z = x_1$$
$$\text{s.t.} \quad x_2 - (1 - x_1)^3 \leq 0 \tag{56}$$
$$x_1 \geq 0, x_2 \geq 0$$

Solution   If $x_1 > 1$, then the first constraint in (56) implies that $x_2 < 0$. Thus, the optimal z-value for (56) cannot exceed 1. Because $x_1 = 1$ and $x_2 = 0$ is feasible and yields $z = 1$, $(1, 0)$ must be the optimal solution to NLP (56).

From Theorem 10, the following are two of the Kuhn–Tucker conditions for (56).

$$1 + 3\lambda_1(1 - x_1)^2 = -\mu_1 \tag{57}$$
$$\mu_1 \geq 0 \tag{58}$$

At the optimal solution $(1, 0)$, (57) implies $\mu_1 = -1$, which contradicts (58). Thus, the Kuhn–Tucker conditions are not satisfied at $(1, 0)$. We now show that at the point $(1, 0)$ the Linear Independence Constraint Qualification is violated. At $(1, 0)$ the constraints $x_2 - (1 - x_1)^3 \leq 0$ and $x_2 \geq 0$ are binding. Then

$$\nabla(x_2 - (1 - x_1)^3) = [0, 1]$$
$$\nabla(-x_2) = [0, -1]$$

Because $[0, 1] + [0, -1] = [0, 0]$, these gradients are linearly dependent. Thus, at $(1,0)$ the gradients of the binding constraints are linearly dependent, and the constraint qualification is not satisfied.

Solving NLPs with Inequality (and Possibly Equality) Constraints on LINGO

LINGO does not require that all constraints be put in the form (26) or (30). Constraints may be input as less than or equal, equal, or greater than or equal to constraints. If your problem satisfies the hypotheses of Theorem 11 or Theorem 11′, then you can know that LINGO will find the optimal solution to your problem. You will know that LINGO has found a point satisfying the Kuhn–Tucker conditions if you see the message DUAL CONDITIONS: SATISFIED. For instance, we can be sure that LINGO would find the optimal solution to Example 33.

For the LINGO printouts given in Section 11.2 for Examples 9–11, we cannot be sure that LINGO has found the optimal solution to any of these problems. Example 9 (Figure 6) fails to satisfy the hypotheses of Theorem 11 because the left-hand side of rows 16–18 are not concave functions and the left-hand side of rows 19–21 are not convex functions. To see if LINGO has actually found the optimal solution to the NLP, we used the INIT command to input a wide variety of starting solutions (focusing on values of R, U, and P). We could not find any solution that was better than the solution in Figure 6, so we are fairly confident that LINGO has found the optimal solution to Example 9. Similarly, Examples 10 and 11 do not satisfy the hypotheses of Theorem 11′, so we cannot be sure that LINGO has found an optimal solution to these problems (even though LINGO has found a point satisfying Kuhn–Tucker conditions!). Again, however, extensive use of the INIT command failed to turn up any better solutions, so we are fairly confident that LINGO has found the optimal solution to Examples 10 and 11.

Interpretation of Price Column on LINGO Output

If the right-hand side of Constraint i (the type of constraint does not matter) in an NLP is increased by a small amount Δ, then the optimal z-value is *improved* by approximately Δ(PRICE for Constraint i). Thus, in a maximization problem increasing the right-hand side of the ith constraint by a small amount Δb_i will result in the optimal z-value increasing by approximately Δb_i(price of Constraint i); in a minimization problem increasing the right-hand side of the ith constraint by a small amount Δb_i will result in the optimal z-value decreasing by approximately Δb_i(Price of Constraint i).

PROBLEMS

Group A

1[†] A power company faces demands during both peak and off-peak times. If a price of p_1 dollars per kilowatt-hour is charged during the peak time, customers will demand $60 - 0.5\,p_1$ kwh of power. If a price of p_2 dollars is charged during the off-peak time, then customers will demand $40 - p_2$ kwh. The power company must have sufficient capacity to meet demand during both the peak and off-peak times. It costs \$10 per day to maintain each kilowatt-hour of capacity. Determine how the power company can maximize daily revenues less operating costs.

2 Use the K–T conditions to find the optimal solution to the following NLP:

$$\max z = x_1 - x_2$$
$$\text{s.t.} \quad x_1^2 + x_2^2 \le 1$$

3 Consider the Giapetto problem of Section 3.1:

$$\max z = 3x_1 + 2x_2$$
$$\text{s.t.} \quad 2x_1 + x_2 \le 100$$
$$x_1 + x_2 \le 80$$
$$x_1 \quad \le 40$$
$$x_1 \quad \ge 0$$
$$x_2 \ge 0$$

Find the K–T conditions for this problem and discuss their relation to the dual of the Giapetto LP and the complementary slackness conditions for the LP.

4 If the feasible region for (26) is bounded and contains its boundary points, then it can be shown that (26) has an optimal solution. Suppose that the regularity conditions are valid but that the hypotheses of Theorems 11 and 11′ are not valid. If we can prove that only one point satisfies the K–T conditions, then why must that point be the optimal solution to the NLP?

5 A total of 160 hours of labor are available each week at \$15/hour. Additional labor can be purchased at \$25/hour. Capital can be purchased in unlimited quantities at a cost of \$5/unit of capital. If K units of capital and L units of labor are available during a week, then $L^{1/2}K^{1/3}$ machines can be produced. Each machine sells for \$270. How can the firm maximize its weekly profits?

6 Use the K–T conditions to find the optimal solution to the following NLP:

$$\min z = (x_1 - 1)^2 + (x_2 - 2)^2$$
$$\text{s.t.} \quad -x_1 + x_2 = 1$$
$$x_1 + x_2 \le 2$$
$$x_1, x_2 \ge 0$$

7 For Example 31, explain why $\bar{\lambda}_1 = 10$ and $\bar{\lambda}_2 = 0$. (*Hint:* Think about the economic principle that for each product produced, marginal revenue must equal marginal cost.)

8 Use the K–T conditions to find the optimal solution to the following NLP:

$$\max z = -x_1^2 - x_2^2 + 4x_1 + 6x_2$$
$$\text{s.t.} \quad x_1 + x_2 \le 6$$
$$x_1 \quad \le 3$$
$$x_2 \le 4$$
$$x_1, x_2 \ge 0$$

9 Use the K–T conditions to find the optimal solution to the following NLP:

$$\min z = e^{-x_1} + e^{-2x_2}$$
$$\text{s.t.} \quad x_1 + x_2 \le 1$$
$$x_1, x_2 \ge 0$$

10 Use the K–T conditions to find the optimal solution to the following NLP:

$$\min z = (x_1 - 3)^2 + (x_2 - 5)^2$$
$$\text{s.t.} \quad x_1 + x_2 \le 7$$
$$x_1, x_2 \ge 0$$

For Problems 11–15, use LINGO to solve the problem. Then explain whether you are sure the program has found the optimal solution.

11 Solve Problem 7 of Section 11.2.

12 Solve Problem 8 of Section 11.2.

13 Solve Problem 11 of Section 11.2.

14 Solve Problem 15 of Section 11.2.

15 Solve Problem 16 of Section 11.2.

[†]Based on Littlechild, "Peak Loads" (1970).

16 We must determine the percentage of our money to be invested in stocks x and y. Let a = percentage of money invested in x and $b = 1 - a$ = percentage of money invested in y. A choice of a and b is called a *portfolio*. A portfolio is efficient if there exists no other portfolio whose return has a higher mean return and lower variance, or a higher mean return and the same variance, or a lower variance with the same mean return. Let $\bar{x}$ be the mean return on stock x and $\bar{y}$ be the mean return on stock y. The variance of the annual return on one share of stock x is var x, and the variance of the annual return on one share of stock y is var y. Assume that the covariance between the annual return for one share of x and one share of y is cov(x, y). If we invest $a\%$ of our money in stock x and $b\%$ in stock y, the variance of the return is given by

$$a^2 \text{ var } x + b^2 \text{ var } y + 2ab \text{ cov}(x, y)$$

Consider the following NLP:

$$\max z = a\bar{x} + b\bar{y}$$
$$\text{s.t.} \quad a^2\text{var } x + b^2\text{var } y + 2ab \text{ cov}(x, y) \le v^*,$$
$$a + b = 1$$

where v^* is a given non-negative number.

a Show that any solution to this NLP is an efficient portfolio.

b Show that as v^* ranges over all non-negative numbers, all efficient portfolios are obtained.

11.10 Quadratic Programming

Consider an NLP whose objective function is the sum of terms of the form $x_1^{k_1} x_2^{k_2} \ldots x_n^{k_n}$. The degree of the term $x_1^{k_1} x_2^{k_1} \ldots x_n^{k_n}$ is $k_1 + k_2 + \cdots k_n$. Thus, the degree of the term $x_1^2 x_2$ is 3, and the degree of the term $x_1 x_2$ is 2. An NLP whose constraints are linear and whose objective is the sum of terms of the form $x_1^{k_1} x_2^{k_2} \ldots x_n^{k_n}$ (with each term having a degree of 2, 1, or 0) is a **quadratic programming problem** (QPP).

Several algorithms can be used to solve QPPs [see Bazaraa and Shetty (1993, Chapter 11)]. We discuss here the application of quadratic programming to portfolio selection and show how LINGO can be used to solve QPPs. We also describe Wolfe's method for solving QPPs.

Quadratic Programming and Portfolio Selection

Consider an investor who has a fixed amount of money that can be invested in several investments. It is often assumed that an investor wants to maximize the expected return from his investments (portfolio) while simultaneously ensuring that the risk of his portfolio is small (as measured by the variance of the return earned by the portfolio). Unfortunately, the return on stocks that yield a large expected return is usually highly variable. Thus, one often approaches the problem of selecting a portfolio by choosing an acceptable minimum expected return and finding the portfolio with the minimum variance that attains an acceptable expected return. For example, an investor may seek the minimum variance portfolio that yields a 12% expected return. By varying the minimum acceptable expected return, the investor may obtain and compare several desirable portfolios.

These ideas reduce the portfolio selection problem to a quadratic programming problem. To see this, we need to observe that given random variables $\mathbf{X}_1, \mathbf{X}_2, \cdots, \mathbf{X}_n$ and constants a, b, and k,

$$E(\mathbf{X}_1 + \mathbf{X}_2 + \cdots + \mathbf{X}_n) = E(\mathbf{X}_1) + E(\mathbf{X}_2) + \cdots + E(\mathbf{X}_n) \tag{59}$$

$$\text{var }(\mathbf{X}_1 + \mathbf{X}_2 + \cdots + \mathbf{X}_n) = \text{var } \mathbf{X}_1 + \text{var } \mathbf{X}_2 + \cdots + \text{var } \mathbf{X}_n + \sum_{i \ne j} \text{cov}(\mathbf{X}_i, \mathbf{X}_j) \tag{60}$$

$$E(k\mathbf{X}_i) = kE(\mathbf{X}_i) \tag{61}$$

$$\text{var }(k\mathbf{X}_i) = k^2\text{var } \mathbf{X}_i \tag{62}$$

$$\text{cov}(a\mathbf{X}_i, b\mathbf{X}_j) = ab \text{ cov }(\mathbf{X}_i, \mathbf{X}_j) \tag{63}$$

Here, cov$(\mathbf{X}, \mathbf{Y})$ is the covariance between random variables $\mathbf{X}$ and $\mathbf{Y}$. In the following example, we show how the portfolio selection problem reduces to a quadratic programming problem.

EXAMPLE 35 **Portfolio Optimization**

I have $1,000 to invest in three stocks. Let S_i be the random variable representing the annual return on $1 invested in stock i. Thus, if $S_i = 0.12$, $1 invested in stock i at the beginning of a year was worth $1.12 at the end of the year. We are given the following information: $E(S_1) = 0.14$, $E(S_2) = 0.11$, $E(S_3) = 0.10$, var $S_1 = 0.20$, var $S_2 = 0.08$, var $S_3 = 0.18$, cov $(S_1, S_2) = 0.05$, cov $(S_1, S_3) = 0.02$, cov $(S_2, S_3) = 0.03$. Formulate a QPP that can be used to find the portfolio that attains an expected annual return of at least 12% and minimizes the variance of the annual dollar return on the portfolio.

Solution Let x_j = number of dollars invested in stock $j(j = 1, 2, 3)$. Then the annual return on the portfolio is $(x_1S_1 + x_2S_2 + x_3S_3)/1,000$ and the expected annual return on the portfolio is [by (59) and (61)]:

$$\frac{x_1E(S_1) + x_2E(S_2) + x_3E(S_3)}{1,000}$$

To ensure that the portfolio has an expected return of at least 12%, we must include the following constraint in the formulation:

$$\frac{0.14x_1 + 0.11x_2 + 0.10x_3}{1,000} \geq 0.12 = 0.14x_1 + 0.11x_2 + 0.10x_3 \geq 0.12 \, (1,000) = 120$$

Of course, we must also include the constraint $x_1 + x_2 + x_3 = 1,000$. We assume that the amount invested in a stock must be nonnegative (that is, no short sales of stock are allowed) and add the constraints $x_1, x_2, x_3 \geq 0$. Our objective is simply to minimize the variance of the portfolio's final value. From (60), the variance of the final value is given by

$$
\begin{aligned}
\text{var } (x_1S_1 + x_2S_2 + x_3S_3) = {} & \text{var } (x_1S_1) + \text{var } (x_2S_2) + \text{var } (x_3S_3) \\
& + 2 \text{ cov}(x_1S_1, x_2S_2) + 2 \text{ cov}(x_1S_1, x_3S_3) \\
& + 2 \text{ cov}(x_2S_2, x_3S_3) \\
= {} & x_1^2 \text{ var } S_1 + x_2^2 \text{ var } S_2 + x_3^2 \text{ var } S_3 + 2x_1x_2\text{cov}(S_1, S_2) \\
& + 2x_1x_3\text{cov}(S_1, S_3) + 2x_2x_3 \text{ cov}(S_2, S_3) \\
& \text{[from Equations (62) and (63)]} \\
= {} & 0.20x_1^2 + 0.08x_2^2 + 0.18x_3^2 + 0.10x_1x_2 \\
& + 0.04x_1x_3 + 0.06x_2x_3
\end{aligned}
$$

Observe that each term in the last expression for the portfolio's variance is of degree 2. Thus, we have an NLP with linear constraints and an objective function consisting of terms of degree 2. To obtain the minimum variance portfolio yielding an expected return of at least 12%, we must solve the following QPP:

$$
\begin{aligned}
\min z = {} & 0.20x_1^2 + 0.08x_2^2 + 0.18x_3^2 + 0.10x_1x_2 + 0.04x_1x_3 + 0.06x_2x_3 \\
\text{s.t.} \quad & 0.14x_1 + 0.11x_2 + 0.10x_3 \geq 120 \\
& x_1 + x_2 + x_3 = 1,000 \\
& x_1, x_2, x_3 \geq 0
\end{aligned}
$$

(64)

REMARKS 1 The idea of using quadratic programming to determine optimal portfolios comes from Markowitz (1959) and is part of the work that won him the Nobel Prize in economics.
2 In Problem 9, we will discuss how to use actual data to estimate the mean and variance of the return on an investment, as well as the covariance of the returns on pairs of investments.
3 In Problem 10, we explore Sharpe's (1963) single-factor model, which greatly simplifies portfolio optimization.

4 In reality, transaction costs are incurred when investments are bought and sold. In Problem 11, we explore how transaction costs change portfolio optimization models.

Solving NLPs with LINGO

Port.lng

When LINGO solves nonlinear programming problems, it assumes all variables are non-negative. The following LINGO model (file Port.lng) can be used to solve the portfolio selection problem, Example 33.

```
MODEL:
 1]SETS:
 2]STOCKS/1..3/:MEAN,AMT;
 3]PAIRS(STOCKS,STOCKS):COV;
 4]ENDSETS
 5]MIN=@SUM(PAIRS(I,J):AMT(I)*AMT(J)*COV(I,J));
 6]@SUM(STOCKS:AMT)=1000;
 7]@SUM(STOCKS:AMT*MEAN)>RQRT;
 8]DATA:
 9]MEAN= .14,.11,.10;
10]RQRT=120;
11]COV= .2,.05,.02,
12].05,.08,.03,
13].02,.03,.18;
14]ENDDATA
END
```

Line 2 defines the set of available investments, and associates with each the mean return per dollar invested (MEAN) and the amount placed in each investment (AMT). Line 3 associates with stocks I and J the quantity $COV(I, J) = COV(\mathbf{X}_i, \mathbf{X}_j)$. Note that $COV(I, I) = VAR\ \mathbf{X}_i$. Line 5 minimizes the variance of the portfolio. We compute the variance of the portfolio (in dollars^2) by summing over all pairs (I, J) of investments $AMT(I)$ * $AMT(J)$ * $COV(I, J)$. Lines 6 and 7 ensure that the total amount invested will equal $1,000 and that the expected return on the portfolio will exceed our required rate of return (RQRT), respectively. (Note that RQRT is input in line 10 of the DATA section.)

The expected annual return on $1 placed in each investment is defined in line 9. Lines 11–13 construct the covariance matrix to complete the model. After selecting the solution, we obtain the optimal solution: z-value = 75,238 dollars^2, AMT(1) = $380.95, AMT(2) = $476.19, and AMT(3) = $142.86. See Figure 45.

```
MODEL:
    1)   MIN= .20 * X1 ^ 2 + .08 * X2 ^ 2 + .18 * X3 ^ 2 + .10 * X1 * X2 +
         .04 * X1 * X3 + .06 * X2 * X3 ;
    2)   .14 * X1 + .11 * X2 + .10 * X3 > 120 ;
    3)   X1 + X2 + X3 = 1000 ;
    4)   X1 > 0 ;
    5)   X2 > 0 ;
    6)   X3 > 0 ;
END


SOLUTION STATUS:  OPTIMAL TO TOLERANCES.  DUAL CONDITIONS:  SATISFIED.

            OBJECTIVE FUNCTION VALUE

        1)      75238.095110

    VARIABLE        VALUE         REDUCED COST
        X1       380.952379          .000000
        X2       476.190470         -.000001
        X3       142.857151          .000000

        ROW    SLACK OR SURPLUS          PRICE
        2)             .000000     -2761.906304
        3)             .000000       180.952513
        4)          380.952379          .000000
        5)          476.190470          .000000
        6)          142.857151          .000000
```

FIGURE 45

By modifying the data of our LINGO model, we could easily solve for a variance-minimizing portfolio that attains a desired expected return when many stocks are available.

Spreadsheet Solution of NLP

Port.xls

We now illustrate how to use the Excel Solver to solve Example 35. Figure 46 (file Port.xls) shows the solution to Example 35 obtained with Solver, using the following procedure.

Solving Portfolio Optimization Problems with Excel Solver

We now show how to use the Excel Solver to solve a portfolio optimization problem. The key is to note that formulas (60) and (62) imply that for random variables $X_1, X_2, \ldots, X_n$:

$$\text{var}(c_1X_1 + c_2X_{2+} + \cdots + c_nX_n) = [c_1, c_2, \ldots, c_n](\text{covariance matrix})[c_1, c_2, \ldots, c_n]^T$$

Here is how we proceed.

Step 1 In A3:C3, enter trial values for the amount invested in each stock.

Step 2 In cell D3, compute the total invested with the formula

$$=\text{SUM(A3:C3)}$$

Step 3 In cell D5, compute the expected dollar return on the portfolio with the formula

$$=\text{SUMPRODUCT(A5:C5,A3:C3)}$$

Step 4 In cell D8, compute the variance of the portfolio's final value with the following array formula

$$=\text{MMULT(A3:C3,MMULT(A8:C10,TRANSPOSE(A3:C3)))}$$

This formula multiplies the vector of amounts invested in each stock times the covariance matrix times the transpose of the vector of amounts invested in each stock. (*Note:* You must hit Control Shift Enter for this formula to work.)

Step 5 Now complete the Solver dialog box as shown in Figure 46. We minimize variance of dollar profit (cell D8). We invest exactly \$1,000 (D3 = F3) and ensure that we

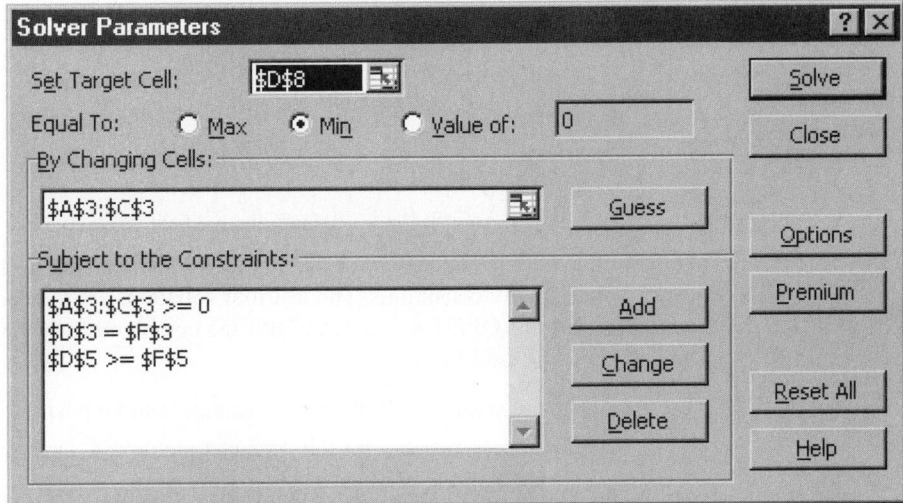

FIGURE 46

FIGURE 47

	A	B	C	D	E	F
1		**PORTFOLIO**	**EXAMPLE**			
2	X1	X2	X3	TOTALINV		
3	380.9523849	476.190461	142.8571541	1000	=	1000
4	E(X1)	E(X2)	E(X3)	MEANRET		
5	0.14	0.11	0.1	120	>=	120
6	COVARIANCE					
7	MATRIX			PORTVAR		
8	0.2	0.05	0.02	75238.09525		
9	0.05	0.08	0.03			
10	0.02	0.03	0.18			

earn an expected return of at least \$120 (D5>=F5). Constraining the amount placed in each investment to be nonnegative rules out short sales. From Figure 47, we find the same optimal solution as we found with LINGO.

Wolfe's Method for Solving Quadratic Programming Problems

Wolfe's method may be used to solve QPPs in which all variables must be nonnegative. We illustrate the method by solving the following QPP:

$$\min z = -x_1 - x_2 + (\tfrac{1}{2})x_1^2 + x_2^2 - x_1 x_2$$
$$\text{s.t.} \quad x_1 + x_2 \le 3$$
$$-2x_1 - 3x_2 \le -6$$
$$x_1, x_2 \ge 0$$

The objective function may be shown to be convex, so any point satisfying the Kuhn–Tucker conditions (36′)–(39′) will solve this QPP. After employing excess variables e_1 for the x_1 constraint and e_2 for the x_2 constraint in (36′), e_2' for the constraint $-2x_1 - 3x_2 \le -6$, and a slack variable s_1' for the constraint $x_1 + x_2 \le 3$, the K–T conditions may be written as

$$x_1 - 1 - x_2 + \lambda_1 - 2\lambda_2 - e_1 = 0 \qquad \text{[x_1 constraint in (36′)]}$$
$$2x_2 - 1 - x_1 + \lambda_1 - 3\lambda_2 - e_2 = 0 \qquad \text{[x_2 constraint in (36′)]}$$
$$x_1 + x_2 + s_1' = 3$$
$$2x_1 + 3x_2 - e_2' = 6$$

All variables nonnegative

$$\lambda_2 e_2' = 0, \qquad \lambda_1 s_1' = 0, \qquad e_1 x_1 = 0, \qquad e_2 x_2 = 0$$

Observe that with the exception of the last four equations, the K–T conditions are all linear or nonnegativity constraints. The last four equations are the complementary slackness conditions for this QPP. For a general QPP, the complementary slackness conditions may be verbally expressed by

e_i from x_i constraint in (36′) and x_i cannot both be positive

Slack or excess variable for the ith constraint and λ_i cannot both be positive

(65)

To find a point satisfying the K–T conditions (except for the complementary slackness conditions), Wolfe's method simply applies a modified version of Phase I of the two-phase simplex method. We first add an artificial variable to each constraint in the K–T conditions that does not have an obvious basic variable, and then we attempt to minimize the sum of the artificial variables. To ensure that the final solution (with all artificial variables equal to zero) satisfies the complementary slackness conditions (65), Wolfe's method modifies the simplex's choice of the entering variable as follows:

1 Never perform a pivot that would make the e_i from the ith constraint in (36′) and x_i both basic variables.

2 Never perform a pivot that would make the slack (or excess) variable for the ith constraint and λ_i both basic variables.

To apply Wolfe's method to our example, we must solve the following LP:

$$\min w = a_1 + a_2 + a_2'$$

$$\text{s.t.} \quad x_1 - x_2 + \lambda_1 - 2\lambda_2 - e_1 + a_1 = 1$$

$$-x_1 + 2x_2 + \lambda_1 - 3\lambda_2 - e_2 + a_2 = 1$$

$$x_1 + x_2 + s_1' = 3$$

$$2x_1 + 3x_2 - e_2' + a_2' = 6$$

$$\text{All variables nonnegative}$$

After eliminating the artificial variables from row 0, we obtain the tableau in Table 12. The current basic feasible solution is $w = 8$, $a_1 = 1$, $a_2 = 1$, $s_1' = 3$, $a_2' = 6$. Since x_2 has the most positive coefficient in row 0, we choose to enter x_2 into the basis. The resulting tableau is Table 13. The current basic feasible solution is $w = 6$, $a_1 = \frac{3}{2}$, $x_2 = \frac{1}{2}$, $s_1' = \frac{5}{2}$, $a_2' = \frac{9}{2}$. Since x_1 has the most positive coefficient in row 0, we now enter x_1 into the basis. The resulting tableau is Table 14.

The current basic feasible solution is $w = \frac{6}{7}$, $a_1 = \frac{6}{7}$, $x_2 = \frac{8}{7}$, $s_1' = \frac{4}{7}$, $x_1 = \frac{9}{7}$. The simplex method recommends that λ_1 should enter the basis. However, Wolfe's modification of the simplex method for selecting the entering variable does not allow λ_1 and s_1' to both

TABLE 12
Initial Tableau for Wolfe's Method

w	x_1	x_2	λ_1	λ_2	e_1	e_2	s_1'	e_2'	a_1	a_2	a_2'	rhs
1	2	4	2	-5	-1	-1	0	-1	0	0	0	8
0	1	-1	1	-2	-1	0	0	0	1	0	0	1
0	-1	②	1	-3	0	-1	0	0	0	1	0	1
0	1	1	0	0	0	0	1	0	0	0	0	3
0	2	3	0	0	0	0	0	-1	0	0	1	6

TABLE 13
First Tableau for Wolfe's Method

w	x_1	x_2	λ_1	λ_2	e_1	e_2	s_1'	e_2'	a_1	a_2	a_2'	rhs
1	4	0	0	1	-1	1	0	-1	0	-2	0	6
0	$\frac{1}{2}$	0	$\frac{3}{2}$	$-\frac{7}{2}$	-1	$-\frac{1}{2}$	0	0	1	$\frac{1}{2}$	0	$\frac{3}{2}$
0	$-\frac{1}{2}$	1	$\frac{1}{2}$	$-\frac{3}{2}$	0	$-\frac{1}{2}$	0	0	0	$\frac{1}{2}$	0	$\frac{1}{2}$
0	$\frac{3}{2}$	0	$-\frac{1}{2}$	$\frac{3}{2}$	0	$\frac{1}{2}$	1	0	0	$-\frac{1}{2}$	0	$\frac{5}{2}$
0	⑦⁄₂	0	$-\frac{3}{2}$	$\frac{9}{2}$	0	$\frac{3}{2}$	0	-1	0	$-\frac{3}{2}$	1	$\frac{9}{2}$

TABLE 14
Second Tableau for Wolfe's Method

W	x_1	x_2	λ_1	λ_2	e_1	e_2	s'_1	e'_2	a_1	a_2	a'_2	rhs
1	0	0	$\frac{12}{7}$	$-\frac{29}{7}$	-1	$-\frac{5}{7}$	0	$\frac{1}{7}$	0	$-\frac{2}{7}$	$-\frac{8}{7}$	$\frac{6}{7}$
0	0	0	$\frac{12}{7}$	$-\frac{29}{7}$	-1	$-\frac{5}{7}$	0	$\frac{1}{7}$	1	$\frac{5}{7}$	$-\frac{1}{7}$	$\frac{6}{7}$
0	0	1	$\frac{2}{7}$	$-\frac{6}{7}$	0	$-\frac{2}{7}$	0	$-\frac{1}{7}$	0	$\frac{2}{7}$	$\frac{1}{7}$	$\frac{8}{7}$
0	0	0	$\frac{1}{7}$	$-\frac{3}{7}$	0	$-\frac{1}{7}$	1	$\left(\frac{3}{7}\right)$	0	$\frac{1}{7}$	$-\frac{3}{7}$	$\frac{4}{7}$
0	1	0	$-\frac{3}{7}$	$\frac{9}{7}$	0	$\frac{3}{7}$	0	$-\frac{2}{7}$	0	$\frac{3}{7}$	$\frac{2}{7}$	$\frac{9}{7}$

TABLE 15
Third Tableau for Wolfe's Method

W	x_1	x_2	λ_1	λ_2	e_1	e_2	s'_1	e'_2	a_1	a_2	a'_2	rhs
1	0	0	$\frac{5}{3}$	-4	-1	$-\frac{2}{3}$	$-\frac{1}{3}$	0	0	$-\frac{1}{3}$	-1	$\frac{2}{3}$
0	0	0	$\left(\frac{5}{3}\right)$	-4	-1	$-\frac{2}{3}$	$-\frac{1}{3}$	0	1	$\frac{2}{3}$	0	$\frac{2}{3}$
0	0	1	$\frac{1}{3}$	-1	0	$-\frac{1}{3}$	$\frac{1}{3}$	0	0	$\frac{1}{3}$	0	$\frac{4}{3}$
0	0	0	$\frac{1}{3}$	-1	0	$-\frac{1}{3}$	$\frac{7}{3}$	1	0	$\frac{1}{3}$	-1	$\frac{4}{3}$
0	1	0	$-\frac{1}{3}$	1	0	$\frac{1}{3}$	$\frac{2}{3}$	0	0	$-\frac{1}{3}$	0	$\frac{5}{3}$

TABLE 16
Optimal Tableau for Wolfe's Method

W	x_1	x_2	λ_1	λ_2	e_1	e_2	s'_1	e'_2	a_1	a_2	a'_2	rhs
1	0	0	0	0	0	0	0	0	-1	-1	-1	0
0	0	0	1	$-\frac{12}{5}$	$-\frac{3}{5}$	$-\frac{2}{5}$	$-\frac{1}{5}$	0	$\frac{3}{5}$	$\frac{2}{5}$	0	$\frac{2}{5}$
0	0	1	0	$-\frac{1}{5}$	$\frac{1}{5}$	$-\frac{1}{5}$	$\frac{2}{5}$	0	$-\frac{1}{5}$	$\frac{1}{5}$	0	$\frac{6}{5}$
0	0	0	0	$-\frac{1}{5}$	$\frac{1}{5}$	$-\frac{1}{5}$	$\frac{12}{5}$	1	$-\frac{1}{5}$	$\frac{1}{5}$	-1	$\frac{6}{5}$
0	1	0	0	$\frac{1}{5}$	$-\frac{1}{5}$	$\frac{1}{5}$	$\frac{3}{5}$	0	$\frac{1}{5}$	$-\frac{1}{5}$	0	$\frac{9}{5}$

be basic variables. Thus, λ_1 cannot enter the basis. Because e'_2 is the only other variable with a positive coefficient in row 0, we now enter e'_2 into the basis. The resulting tableau is Table 15. The current basic feasible solution is $w = \frac{2}{3}$, $a_1 = \frac{2}{3}$, $x_2 = \frac{4}{3}$, $e'_2 = \frac{4}{3}$, and $x_1 = \frac{5}{3}$. Because s'_1 is now a nonbasic variable, we can enter λ_1 into the basis. The resulting tableau is Table 16. This is (finally!) an optimal tableau. Because $w = 0$, we have found a solution that satisfies the Kuhn–Tucker conditions and is optimal for the QPP. Thus, the optimal solution to the QPP is $x_1 = \frac{9}{5}$, $x_2 = \frac{6}{5}$. From the optimal tableau, we also find that $\lambda_1 = \frac{2}{5}$ and $\lambda_2 = 0$ (because $e'_2 = \frac{6}{5} > 0$, we know that $\lambda_2 = 0$ must hold).

Wolfe's method is guaranteed to obtain the optimal solution to a QPP if all leading principal minors of the objective function's Hessian are positive. Otherwise, Wolfe's method may not converge in a finite number of pivots. In practice, the method of **complementary pivoting** is most often used to solve QPPs. Unfortunately, space limitations preclude a discussion of complementary pivoting. The interested reader is referred to Shapiro (1979).

PROBLEMS

Group A

1 We are considering investing in three stocks. The random variable S_i represents the value one year from now of \$1 invested in stock i. We are given that $E(S_1) = 1.15$, $E(S_2) = 1.21$, $E(S_3) = 1.09$; var $S_1 = 0.09$, var $S_2 = 0.04$, var $S_3 = 0.01$; $cov(S_1, S_2) = 0.006$, $cov(S_1, S_3) = -0.004$, and $cov(S_2, S_3) = 0.005$. We have \$100 to invest and want to have an expected return of at least 15% during the next year. Formulate a QPP to find the portfolio of minimum variance that attains an expected return of at least 15%.

2 Show that the objective function for Example 35 is convex [it can be shown that the variance of any portfolio is a convex function of $(x_1, x_2, \dots, x_n)$].

3 In Figure 45, interpret the entries in the PRICE column for rows 2 and 3.

4 Fruit Computer Company produces Pear and Apricot computers. If the company charges a price p_1 for Pear computers and p_2 for Apricot computers, it can sell q_1 Pear and q_2 Apricot computers, where $q_1 = 4,000 - 10p_1 + p_2$, and $q_2 = 2,000 - 9p_2 + 0.8p_1$. Manufacturing a Pear computer requires 2 hours of labor and 3 computer chips. An Apricot computer uses 3 hours of labor and 1 computer chip. Currently, 5,000 hours of labor and 4,500 chips are available. Formulate a QPP to maximize Fruit's revenue. Use the K–T conditions (or LINGO) to find Fruit's optimal pricing policy. What is the most that Fruit should pay for another hour of labor? What is the most that Fruit should pay for another computer chip?

5 Use Wolfe's method to solve the following QPP:

$$\min z = 2x_1^2 - x_2$$
$$\text{s.t.} \quad 2x_1 - x_2 \le 1$$
$$x_1 + x_2 \le 1$$
$$x_1, x_2 \ge 0$$

6 Use Wolfe's method to solve the following QPP:

$$\min x_1 + 2x_2^2$$
$$\text{s.t.} \quad x_1 + x_2 \le 2$$
$$2x_1 + x_2 \le 3$$
$$x_1, x_2 \ge 0$$

7 In an electrical network, the power loss incurred when a current of I amperes flows through a resistance of R ohms is I^2R watts. In Figure 48, 710 amperes of current must be sent from node 1 to node 4. The current flowing through each node must satisfy conservation of flow. For example, for node 1, $710 =$ flow through 1-ohm resistor + flow through 4-ohm resistor. Remarkably, nature determines the current flow through each resistor by minimizing the total power loss in the network.

a Formulate a QPP whose solution will yield the current flowing through each resistor.

b Use LINGO to determine the current flowing through each resistor.

FIGURE **48**

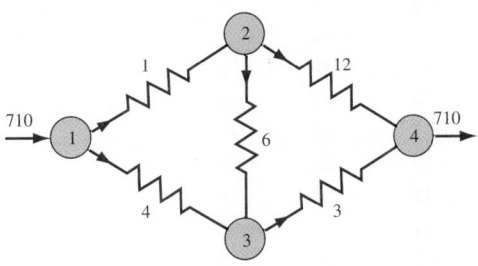

8 Use Wolfe's method to find the optimal solution to the following QPP:

$$\min z = x_1^2 + x_2^2 - 2x_1 - 3x_2 + x_1x_2$$
$$\text{s.t.} \quad x_1 + 2x_2 \le 2$$
$$x_1, x_2 \ge 0$$

Group B

9 (This problem requires some knowledge of regression.) In Table 17, you are given the annual returns on three different types of assets (T-bills, stocks, and gold) (file Invest68.xls) during the years 1968–1988. For example, \$1 invested in T-bills at the beginning of 1978 grew to \$1.07 by the end of 1978. You have \$1,000 to invest in these three investments. Your goal is to minimize the variance of the annual dollar return of your portfolio subject to the constraint that the expected return on the portfolio for a one-year period be at least 10%. Determine how much money should be invested in each investment. Use a spreadsheet to compute the mean, standard deviation, and variance of the return on each asset. To compute the covariance between each pair of assets, remember that an estimate of the covariance of T-bills and gold is given by $cov(T, G) = s_T s_G r_{TG}$ (where $s_T =$ standard deviation of return on T-bills; $s_G =$ standard deviation of return on gold). Note that $r_{TG} = \pm(R^2)^{1/2}$, where the sign of r is the same as the slope of the least squares line.

In addition to determining the amount to be invested in each asset, answer the following two questions.

a I am 95% sure that the increase in the value of my assets during the next year will be between _____ and _____ .

b I am 95% sure that the percentage annual return on my portfolio will be between _____ and _____ .

10 (Refer to Problem 9 data.) Suppose that the return on the ith asset may be estimated as $\mu_i + \beta_i M + \epsilon_i$, where M is the return on the market. Assume that the ϵ_i are independent and that the standard deviation of ϵ_i may be estimated by the standard error of the estimate from the

TABLE 17
Annual Returns on Assets

Year	Stocks	Gold	T-Bills
1968	11	11	5
1969	-9	8	7
1970	4	-14	7
1971	14	14	4
1972	19	44	4
1973	-15	66	7
1974	-27	64	8
1975	37	0	6
1976	24	-22	5
1977	-7	18	5
1978	7	31	7
1979	19	59	10
1980	33	99	11
1981	-5	-25	15
1982	22	4	11
1983	23	-11	9
1984	6	-15	10
1985	32	-12	8
1986	19	16	6
1987	5	22	5
1988	17	-2	6

regression, with the return on the market as independent variable and the return on the ith asset as the dependent variable. Now you can express the variance of the portfolio without calculating the covariance between each pair of investments. (*Hint:* The variance of the market will enter into your equation). Use the estimated regression equation to estimate the mean return on the ith asset as a function of the return on the market.

For the data in Problem 9, formulate an NLP that can be used to find the minimum variance portfolio yielding an expected return of at least 10%. Why is this method useful when many potential investments are available?

11 (Refer to Problem 9 data.) Suppose that you now hold 30% of your investment in stocks, 50% in T-bills, and 20% in gold. Assume that transactions incur costs. Every $100 of stocks traded costs you $1, every $100 of your gold portfolio traded costs you $2, and every $1 of your T-bill portfolio traded costs you 5¢. Find the minimum variance portfolio that yields, after transaction costs, an expected return of at least 10%. (*Hint:* Define variables for the dollars bought or sold in each investment.)

11.11 Separable Programming†

Many NLPs are of the following form:

$$\max \ (\text{or min}) \ z = \sum_{j=1}^{j=n} f_j(x_j)$$

$$\text{s.t.} \quad \sum_{j=1}^{j=n} g_{ij}(x_j) \le b_i \qquad (i = 1, 2, \ldots, m)$$

Because the decision variables appear in separate terms of the objective function and the constraints, NLPs of this form are called **separable programming problems.** Separable programming problems are often solved by approximating each $f_j(x_j)$ and $g_{ij}(x_j)$ by a piecewise linear function (see Section 9.2). Before describing the separable programming technique, we give an example of a separable programming problem.

EXAMPLE 36   **Separable Programming**

Oilco must determine how many barrels of oil to extract during each of the next two years. If Oilco extracts x_1 million barrels during year 1, each barrel can be sold for $30 - x_1$. If Oilco extracts x_2 million barrels during year 2, each barrel can be sold for $35 - x_2$. The cost of extracting x_1 million barrels during year 1 is x_1^2 million dollars, and the cost of extracting x_2 million barrels during year 2 is $2x_2^2$ million dollars. A total of 20 million

†This section covers topics that may be omitted with no loss of continuity.

barrels of oil are available, and at most \$250 million can be spent on extraction. Formulate an NLP to help Oilco maximize profits (revenues less costs) for the next two years.

Solution Define

$$x_1 = \text{millions of barrels of oil extracted during year 1}$$
$$x_2 = \text{millions of barrels of oil extracted during year 2}$$

Then the appropriate NLP is

$$\max z = x_1(30 - x_1) + x_2(35 - x_2) - x_1^2 - 2x_2^2$$
$$= 30x_1 + 35x_2 - 2x_1^2 - 3x_2^2$$

$$\text{s.t.} \quad x_1^2 + 2x_2^2 \le 250 \tag{66}$$
$$x_1 + x_2 \le 20$$
$$x_1, x_2 \ge 0$$

This is a separable programming problem with $f_1(x_1) = 30x_1 - 2x_1^2$, $f_2(x_2) = 35x_2 - 3x_2^2$, $g_{11}(x_1) = x_1^2$, $g_{12}(x_2) = 2x_2^2$, $g_{21}(x_1) = x_1$, and $g_{22}(x_2) = x_2$.

Before approximating the functions f_j and g_{ij} by piecewise linear functions, we must determine (for $j = 1, 2, \ldots, n$) numbers a_j and b_j such that we are sure that the value of x_j in the optimal solution will satisfy $a_j \le x_j \le b_j$. For the previous example, $a_1 = a_2 = 0$ and $b_1 = b_2 = 20$ will suffice. Next, for each variable x_j we choose grid points $p_{j1}, p_{j2}, \ldots, p_{jk}$ with $a_j = p_{j1} \le p_{j2} \le \cdots \le p_{jk} = b_j$ (to simplify notation, we assume that each variable has the same number of grid points). For the previous example, we use five grid points for each variable: $p_{11} = p_{21} = 0$, $p_{12} = p_{22} = 5$, $p_{13} = p_{23} = 10$, $p_{14} = p_{24} = 15$, $p_{15} = p_{25} = 20$. The essence of the separable programming method is to approximate each function f_j and g_{ij} as if it were a linear function on each interval $[p_{j,r-1}, p_{j,r}]$.

More formally, suppose $p_{j,r} \le x_j \le p_{j,r+1}$. Then for some δ ($0 \le \delta \le 1$), $x_j = \delta p_{j,r} + (1 - \delta)p_{j,r+1}$. We approximate $f_j(x_j)$ and $g_{ij}(x_j)$ (see Figure 49) by

$$\hat{f}_j(x_j) = \delta f_j(p_{j,r}) + (1 - \delta)f_j(p_{j,r+1})$$
$$\hat{g}_{ij}(x_j) = \delta g_{ij}(p_{j,r}) + (1 - \delta)g_{ij}(p_{j,r+1})$$

For example, how would we approximate $f_1(12)$? Because $f_1(10) = 30(10) - 2(10)^2 = 100$, $f_1(15) = 30(15) - 2(15)^2 = 0$, and $12 = 0.6(10) + 0.4(15)$, we approximate $f_1(12)$ by $\hat{f}_1(12) = 0.6(100) + 0.4(0) = 60$ (see Figure 50).

More formally, to approximate a separable programming problem, we add constraints of the form

$$\delta_{j1} + \delta_{j2} + \cdots + \delta_{j,k} = 1 \quad (j = 1, 2, \ldots, n) \tag{67}$$
$$x_j = \delta_{j1}p_{j1} + \delta_{j2}p_{j2} + \cdots + \delta_{j,k}p_{j,k} \quad (j = 1, 2, \ldots, n) \tag{68}$$
$$\delta_{j,r} \ge 0 \quad (j = 1, 2, \ldots, n; r = 1, 2, \ldots, k) \tag{69}$$

FIGURE 49
The Separable Programming Approximation of $f_j(x_j)$

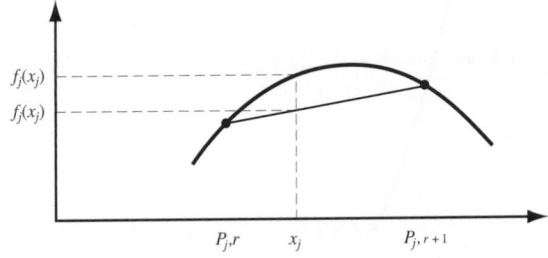

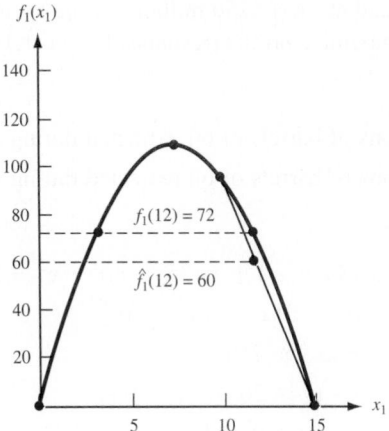

FIGURE 50
Approximation of $f_1(12)$

Then we replace $f_j(x_j)$ by

$$\hat{f}_j(x_j) = \delta_{j1} f_j(p_{j1}) + \delta_{j2} f_j(p_{j2}) + \cdots + \delta_{j,k} f_j(p_{j,k}) \tag{70}$$

and replace $g_{ij}(x_j)$ by

$$\hat{g}_{ij}(x_j) = \delta_{j1} g_{ij}(p_{j1}) + \delta_{j2} g_{ij}(p_{j2}) + \cdots + \delta_{j,k} g_{ij}(p_{j,k}) \tag{71}$$

To ensure accuracy of the approximations in (70) and (71), we must be sure that for each $j(j = 1, 2, \ldots, n)$ at most two of the $\delta_{j,k}$'s are positive. Also, for a given j, suppose that two $\delta_{j,k}$'s are positive. If $\delta_{j,k'}$ is positive, then the other positive $\delta_{j,k}$ must be either $\delta_{j,k'-1}$ or $\delta_{j,k'+1}$ (we say that $\delta_{j,k'}$ is adjacent to $\delta_{j,k'-1}$ and $\delta_{j,k'+1}$). To see the reason for these restrictions, suppose we want $x_1 = 12$. Then our approximations will be most accurate if $\delta_{13} = 0.6$ and $\delta_{14} = 0.4$. In this case, we approximate $f_1(12)$ by $0.6f_1(10) + 0.4f_1(15)$. We certainly don't want to have $\delta_{11} = 0.4$ and $\delta_{15} = 0.6$. This would yield $x_1 = 0.4(0) + 0.6(20) = 12$, but it would approximate $f_1(12)$ by $f_1(12) = 0.4f_1(0) + 0.6f_1(20)$, and in most cases this would be a poor approximation of $f_1(12)$ (see Figure 51). For the approximating problem to yield a good approximation to the functions f_i and $g_{j,k}$, we must add the following **adjacency assumption:** For $j = 1, 2, \ldots, n$, at most two $\delta_{j,k}$'s can be positive. If for a given j, two $\delta_{j,k}$'s are positive, then they must be adjacent.

Thus, the approximating problem consists of an objective function obtained from (70) and constraints obtained from (67), (68), (69), and (71) and the adjacency assumption.

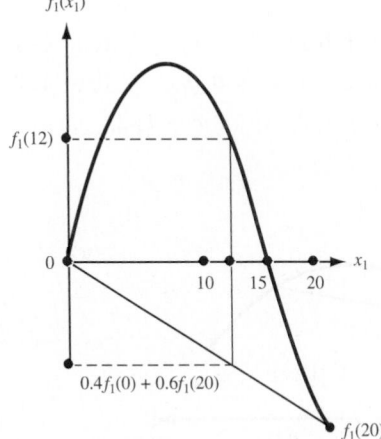

FIGURE 51
Violating the Adjacency
Assumption Results in
a Poor Approximation
of $f_1(12)$

Actually, the constraints (68) are only used to transform the values of the $\delta_{j,k}$'s into values for the original decision variables (the x_j's) and are not needed to determine the optimal values of the $\delta_{j,k}$'s. The constraints (68) need not be part of the approximating problem, and the **approximating problem** for a separable programming problem may be written as follows:

$$\max \text{ (or min) } \hat{z} = \sum_{j=1}^{j=n} [\delta_{j1}f_j(p_{j1}) + \delta_{j2}f_j(p_{j2}) + \cdots + \delta_{j,k}f_j(p_{j,k})]$$

$$\text{s.t.} \quad \sum_{j=1}^{j=n} [\delta_{j1}g_{ij}(p_{j1}) + \delta_{j2}g_{ij}(p_{j2}) + \cdots + \delta_{j,k}g_{ij}(p_{jk})] \leq b_i \quad (i = 1, 2, \ldots, m)$$

$$\delta_{j1} + \delta_{j2} + \cdots + \delta_{j,k} = 1 \quad (j = 1, 2, \ldots, n)$$

$$\delta_{j,r} \geq 0 \quad (j = 1, 2, \ldots, n; r = 1, 2, \ldots, k)$$

Adjacency assumption

For the previous example, we have

$$f_1(0) = 0, \quad f_1(5) = 100, \quad f_1(10) = 100, \quad f_1(15) = 0, \quad f_1(20) = -200$$
$$f_2(0) = 0, \quad f_2(5) = 100, \quad f_2(10) = 50, \quad f_2(15) = -150, \quad f_2(20) = -500$$
$$g_{11}(0) = 0, \quad g_{11}(5) = 25, \quad g_{11}(10) = 100, \quad g_{11}(15) = 225, \quad g_{11}(20) = 400$$
$$g_{12}(0) = 0, \quad g_{12}(5) = 50, \quad g_{12}(10) = 200, \quad g_{12}(15) = 450, \quad g_{12}(20) = 800$$
$$g_{21}(0) = 0, \quad g_{21}(5) = 5, \quad g_{21}(10) = 10, \quad g_{21}(15) = 15, \quad g_{21}(20) = 20$$
$$g_{22}(0) = 0, \quad g_{22}(5) = 5, \quad g_{22}(10) = 10, \quad g_{22}(15) = 15, \quad g_{22}(20) = 20$$

Applying (70) to the objective function of (66) yields an approximating objective function of

$$\max \hat{z} = 100\delta_{12} + 100\delta_{13} - 200\delta_{15} + 100\delta_{22} + 50\delta_{23} - 150\delta_{24} - 500\delta_{25}$$

Constraint (67) yields the following two constraints:

$$\delta_{11} + \delta_{12} + \delta_{13} + \delta_{14} + \delta_{15} = 1$$
$$\delta_{21} + \delta_{22} + \delta_{23} + \delta_{24} + \delta_{25} = 1$$

Constraint (68) yields the following two constraints:

$$x_1 = 5\delta_{12} + 10\delta_{13} + 15\delta_{14} + 20\delta_{15}$$
$$x_2 = 5\delta_{22} + 10\delta_{23} + 15\delta_{24} + 20\delta_{25}$$

Applying (71) transforms the two constraints in (66) to

$$25\delta_{12} + 100\delta_{13} + 225\delta_{14} + 400\delta_{15} + 50\delta_{22} + 200\delta_{23} + 450\delta_{24} + 800\delta_{25} \leq 250$$
$$5\delta_{12} + 10\delta_{13} + 15\delta_{14} + 20\delta_{15} + 5\delta_{22} + 10\delta_{23} + 15\delta_{24} + 20\delta_{25} \leq 20$$

After adding the sign restrictions, (68), and the adjacency assumption, we obtain

$$\max \hat{z} = 100\delta_{12} + 100\delta_{13} - 200\delta_{15} + 100\delta_{22} + 50\delta_{23} - 150\delta_{24} - 500\delta_{25}$$
$$\text{s.t.} \quad \delta_{11} + \delta_{12} + \delta_{13} + \delta_{14} + \delta_{15} = 1$$
$$\delta_{21} + \delta_{22} + \delta_{23} + \delta_{24} + \delta_{25} = 1$$
$$25\delta_{12} + 100\delta_{13} + 225\delta_{14} + 400\delta_{15} + 50\delta_{22} + 200\delta_{23} + 450\delta_{24} + 800\delta_{25} \leq 250$$
$$5\delta_{12} + 10\delta_{13} + 15\delta_{14} + 20\delta_{15} + 5\delta_{22} + 10\delta_{23} + 15\delta_{24} + 20\delta_{25} \leq 20$$
$$\delta_{j,k} \geq 0(j = 1, 2; k = 1, 2, 3, 4, 5)$$

Adjacency assumption

At first glance, the approximating problem may appear to be a linear programming problem. If we attempt to solve the approximating problem by the simplex, however, we may violate the adjacency assumption. To avoid this difficulty, we solve approximating problems via the simplex algorithm with the following restricted entry rule: If, for a given j all $\delta_{j,k} = 0$, then any $\delta_{j,k}$ may enter the basis. If, for a given j, a single $\delta_{j,k}$ (say, $\delta_{j,k'}$) is positive, then $\delta_{j,k'-1}$ or $\delta_{j,k'+1}$ may enter the basis, but no other $\delta_{j,k}$ may enter the basis. If, for a given j, two $\delta_{j,k}$'s are positive, then no other $\delta_{j,k}$ can enter the basis.

There are two situations in which solving the approximating problem via the ordinary simplex will yield a solution that automatically satisfies the adjacency assumption. If the separable programming problem is a maximization problem, then each $f_j(x_j)$ is concave, and each $g_{ij}(x_j)$ is convex, then any solution to the approximating problem obtained via the ordinary simplex will automatically satisfy the adjacency assumption. Also, if the separable programming problem is a minimization problem, each $f_j(x_j)$ is convex, and each $g_{ij}(x_j)$ is convex, then any solution to the approximating problem obtained via the ordinary simplex will automatically satisfy the adjacency assumption. Problem 3 at the end of this section indicates why this is the case.

In these two special cases, it can also be shown that as the maximum value of the distance between two adjacent grid points approaches zero, the optimal solution to the approximating problem approaches the optimal solution to the separable programming problem [see Bazaraa and Shetty (1993, p. 450)].

For the previous example, each $f_j(x_j)$ is concave and each $g_{ij}(x_j)$ is convex, so to find the optimal solution to the approximating problem, we may use the simplex and ignore the restricted entry rule. The optimal solution to the approximating problem for the previous example is $\delta_{12} = \delta_{22} = 1$. This yields $x_1 = 1(5) = 5$, $x_2 = 1(5) = 5$, $\hat{z} = 200$. Compare this with the actual optimal solution to the previous example, which is $x_1 = 7.5$, $x_2 = 5.83$, $z = 214.58$.

PROBLEMS

Group A

Set up an approximating problem for the following separable programming problems:

1
$$\min z = x_1^2 + x_2^2$$
$$\text{s.t.} \quad x_1^2 + 2x_2^2 \leq 4$$
$$x_1^2 + x_2^2 \leq 6$$
$$x_1, x_2 \geq 0$$

2
$$\max z = x_1^2 - 5x_1 + x_2^2 - 5x_2 - x_3$$
$$\text{s.t.} \quad x_1 + x_2 + x_3 \leq 4$$
$$x_1^2 - x_2 \leq 3$$
$$x_1, x_2, x_3 \geq 0$$

Group B

3 This problem will give you an idea why the restricted entry rule is unnecessary when (for a maximization problem) each $f_j(x_j)$ is concave and each $g_{ij}(x_j)$ is convex. Consider the Oilco example. When we solve the approximating problem by the simplex, show that a solution that violates the adjacency assumption cannot be obtained. For example, why can the simplex not yield a solution (x^*) of $\delta_{11} = 0.4$ and $\delta_{15} = 0.6$? To show that this cannot occur, find a feasible solution to the approximating problem that has a larger

$\hat{z}$-value than x^*. [*Hint:* Show that the solution that is identical to x^* with the exception that $\delta_{11} = 0$, $\delta_{15} = 0$, $\delta_{13} = 0.6$, and $\delta_{14} = 0.4$ is feasible for the approximating problem [use the convexity of $g_{ij}(x_j)$ for this part] and has a larger $\hat{z}$-value than x^* [use concavity of $f_j(x_j)$ for this part].]

4 Suppose an NLP appears to be separable except for the fact that a term of the form $x_i x_j$ appears in the objective function or constraints. Show that an NLP of this type can be made into a separable programming problem by defining two new variables y_i and y_j by $x_i = \frac{1}{2}(y_i + y_j)$ and $x_j = \frac{1}{2}(y_i - y_j)$. Use this technique to transform the following NLP into a separable programming problem:

$$\max z = x_1^2 + 3x_1x_2 - x_2^2$$
$$\text{s.t.} \quad x_1x_2 \leq 4$$
$$x_1^2 + x_2 \leq 6$$
$$x_1, x_2 \geq 0$$

11.12 The Method of Feasible Directions[†]

In Section 11.7, we used the method of steepest ascent to solve an unconstrained NLP. We now describe a modification of that method—the **feasible directions method,** which can be used to solve NLPs with linear constraints. Suppose we want to solve

$$\max z = f(\mathbf{x})$$
$$\text{s.t.} \quad A\mathbf{x} \le \mathbf{b} \tag{72}$$
$$\mathbf{x} \ge \mathbf{0}$$

where $\mathbf{x} = [x_1, x_2, \ldots, x_n]^T$, A is an $m \times n$ matrix, $\mathbf{0}$ is an n-dimensional column vector consisting entirely of zeros, $\mathbf{b}$ is an $m \times 1$ vector, and $f(\mathbf{x})$ is a concave function.

To begin, we must find (perhaps by using the Big M method or the two-phase simplex algorithm) a feasible solution $\mathbf{x}^0$ satisfying the constraints $A\mathbf{x} \le \mathbf{b}$. We now try to find a direction in which we can move away from $\mathbf{x}^0$. This direction should have two properties:

1 When we move away from $\mathbf{x}^0$, we remain feasible.

2 When we move away from $\mathbf{x}^0$, we increase the value of z.

From Section 11.7, we know that if $\nabla f(\mathbf{x}^0) \cdot \mathbf{d} > 0$ and we move a small distance away from $\mathbf{x}^0$ in a direction $\mathbf{d}$, then $f(\mathbf{x})$ will increase. We choose to move away from $\mathbf{x}^0$ in a direction $\mathbf{d}^0 - \mathbf{x}^0$, where $\mathbf{d}^0$ is an optimal solution to the following LP:

$$\max z = \nabla f(\mathbf{x}^0) \cdot \mathbf{d}$$
$$\text{s.t.} \quad A\mathbf{d} \le \mathbf{b} \tag{73}$$
$$\mathbf{d} \ge \mathbf{0}$$

Here $\mathbf{d} = [d_1 \quad d_2 \ldots d_n]^T$. Observe that if $\mathbf{d}^0$ solves (73) (and $\mathbf{x}^0$ does not), then $\nabla f(\mathbf{x}^0) \cdot \mathbf{d}^0 > \nabla f(\mathbf{x}^0) \cdot \mathbf{x}^0$, or $\nabla f(\mathbf{x}^0) \cdot (\mathbf{d}^0 - \mathbf{x}^0) > 0$. This means that moving a small distance from $\mathbf{x}^0$ in a direction $\mathbf{d}^0 - \mathbf{x}^0$ will increase z.

We now choose our new point $\mathbf{x}^1$ to be $\mathbf{x}^1 = \mathbf{x}^0 + t_0(\mathbf{d}^0 - \mathbf{x}^0)$, where t_0 solves

$$\max f[\mathbf{x}^0 + t_0(\mathbf{d}^0 - \mathbf{x}^0)]$$
$$0 \le t_0 \le 1$$

It can be shown that $f(\mathbf{x}^1) \ge f(\mathbf{x}^0)$ will hold, and that if $f(\mathbf{x}^1) = f(\mathbf{x}^0)$, then $\mathbf{x}^0$ is the optimal solution to (72). Thus, unless $\mathbf{x}^0$ is optimal, $\mathbf{x}^1$ will have a z-value larger than $\mathbf{x}^0$. It is easy to show that $\mathbf{x}^1$ is a feasible point. Observe that

$$A\mathbf{x}^1 = A[\mathbf{x}^0 + t_0(\mathbf{d}^0 - \mathbf{x}^0)] = (1 - t_0)A\mathbf{x}^0 + t_0 A\mathbf{d}^0 \le (1 - t_0)\mathbf{b} + t_0\mathbf{b} = \mathbf{b}$$

where the last inequality follows from the fact that both $\mathbf{x}^0$ and $\mathbf{d}^0$ satisfy the NLP's constraints and $0 \le t_0 \le 1$. $\mathbf{x}^1 \ge \mathbf{0}$ follows easily from $\mathbf{x}^0 \ge \mathbf{0}$, $\mathbf{d}^0 \ge \mathbf{0}$, and $0 \le t_0 \le 1$.

We now choose to move away from $\mathbf{x}^1$ in any direction $\mathbf{d}^1 - \mathbf{x}^1$, where $\mathbf{d}^1$ is an optimal solution to the following LP:

$$\max z = \nabla f(\mathbf{x}^1) \cdot \mathbf{d}$$
$$\text{s.t.} \quad A\mathbf{d} \le \mathbf{b}$$
$$\mathbf{d} \ge \mathbf{0}$$

Then we choose a new point $\mathbf{x}^2$ to be given by $\mathbf{x}^2 = \mathbf{x}^1 + t_1(\mathbf{d}^1 - \mathbf{x}^1)$, where t_1 solves

$$\max f[\mathbf{x}^1 + t_1(\mathbf{d}^1 - \mathbf{x}^1)]$$
$$0 \le t_1 \le 1$$

[†]This section covers topics that may be omitted with no loss of continuity.

Again, $\mathbf{x}^2$ will be feasible, and $f(\mathbf{x}^2) \geq f(\mathbf{x}^1)$ will hold. Also, if $f(\mathbf{x}^2) = f(\mathbf{x}^1)$, then $\mathbf{x}^1$ is the optimal solution to NLP (72).

We continue in this fashion and generate directions of movement $\mathbf{d}^2, \mathbf{d}^3, \ldots, \mathbf{d}^{n-1}$ and new points $\mathbf{x}^3, \mathbf{x}^4, \ldots, \mathbf{x}^n$. We terminate the algorithm if $\mathbf{x}^k = \mathbf{x}^{k-1}$. This means that $\mathbf{x}^{k-1}$ is an optimal solution to NLP (72). If the values of f are strictly increasing at each iteration of the method, then (as with the method of steepest ascent) we terminate the method whenever two successive points are very close together.

After the point $\mathbf{x}^k$ has been determined, an upper bound on the optimal z-value for (72) is available. It can be shown that if $f(x_1, x_x, \ldots, x_n)$ is concave, then

$$[\text{Optimal } z\text{-value for (71)}] \leq f(\mathbf{x}^k) + \nabla(\mathbf{x}^k) \cdot [\mathbf{d}^k - \mathbf{x}^k]^T \qquad \textbf{(74)}$$

Thus, if $f(\mathbf{x}^k)$ is near the upper bound on the optimal z-value obtained from (74), then we may terminate the algorithm.

The version of the feasible directions method we have discussed was developed by Frank and Wolfe. For a discussion of other feasible direction methods, we refer the reader to Chapter 11 of Bazaraa and Shetty (1993).

The following example illustrates the method of feasible directions.

EXAMPLE 37   Method of Feasible Directions

Perform two iterations of the feasible directions method on the following NLP:

$$\max z = f(x, y) = 2xy + 4x + 6y - 2x^2 - 2y^2$$
$$\text{s.t.} \quad x + y \leq 2$$
$$x, y \geq 0$$

Begin at the point (0,0).

Solution $\nabla f(x, y) = [2y - 4x + 4 \quad 6 + 2x - 4y]$, so $\nabla f(0, 0) = [4 \quad 6]$. We find a direction to move away from $[0 \quad 0]$ by solving the following LP:

$$\max z = 4d_1 + 6d_2$$
$$\text{s.t.} \quad d_1 + d_2 \leq 2$$
$$d_1, d_2 \geq 0$$

The optimal solution to this LP is $d_1 = 0$ and $d_2 = 2$. Thus, $\mathbf{d}^0 = [0 \quad 2]^T$. Since $\mathbf{d}^0 - \mathbf{x}^0 = [0 \quad 2]^T$, we now choose $\mathbf{x}^1 = [0 \quad 0]^T + t_0[0 \quad 2]^T = [0 \quad 2t_0]^T$, where t_0 solves

$$\max f(0, 2t) = 12t - 8t^2$$
$$0 \leq t \leq 1$$

Letting $g(t) = 12t - 8t^2$, we find $g'(t) = 12 - 16t = 0$ for $t = 0.75$. Since $g''(t) < 0$, we know that $t_0 = 0.75$. Thus $\mathbf{x}^1 = [0, 1.5]^T$. At this point, $z = f(0, 1.5) = 4.5$. We now have [via (74) with $k = 0$] the following upper bound on the NLP's optimal z-value:

$$(\text{Optimal } z\text{-value}) \leq f(0, 0) + [4 \quad 6] \cdot [0 \quad 2]^T = 12$$

Now $\nabla(\mathbf{x}^1) = f(0, 1.5) = [7 \quad 0]$. We now find the direction $\mathbf{d}^2$ to move away from $\mathbf{x}^1$ by solving

$$\max z = 7d_1$$
$$\text{s.t.} \quad d_1 + d_2 \leq 2$$
$$d_1, d_2 \geq 0$$

The optimal solution to this LP is $\mathbf{d}^1 = [2 \quad 0]^T$. Now we find $\mathbf{x}^2 = [0 \quad 1.5]^T + t_1\{[2 \quad 0]^T - [0 \quad 1.5]^T\} = [2t_1 \quad 1.5 - 1.5t_1]^T$, where t_1 is the optimal solution to

$$\max f(2t, 1.5 - 1.5t)$$
$$0 \le t \le 1$$

Now $f(2t, 1.5 - 1.5t) = 4.5 - 18.5t^2 + 14t$. Letting $g(t) = 4.5 - 18.5t^2 + 14t$, we find $g'(t) = 14 - 37t = 0$ for $t = \frac{14}{37}$. Since $g''(t) = -37 < 0$, we find that $t_1 = \frac{14}{37}$. Thus, $\mathbf{x}^2 = [\frac{28}{37} \quad \frac{69}{74}]^T = [0.76 \quad 0.93]^T$. Now we have $z = f(0.76, 0.93) = 7.15$. From (74) (with $k = 1$), we find

$$(\text{Optimal } z\text{-value}) \le 4.5 + [7 \quad 0] \cdot \{[2 \quad 0]^T - [0 \quad 1.5]^T\} = 18.5$$

Since our first upper bound on the optimal z-value (12) is a better bound than 18.5, we ignore this bound.

Actually, the NLP's optimal solution is $z = 8.17$, $x = .83$, and $y = 1.17$.

PROBLEMS

Group A

Perform two iterations of the method of feasible directions for each of the following NLPs.

1 $\max z = 4x + 6y - 2x^2 - 2xy - 2y^2$

s.t. $x + 2y \le 2$

$x, y \ge 0$

Begin at the point $(\frac{1}{2}, \frac{1}{2})$.

2 $\max z = 3xy - x^2 - y^2$

s.t. $3x + y \le 4$

$x, y \ge 0$

Begin at the point $(1, 0)$.

11.13 Pareto Optimality and Trade-Off Curves[†]

In a multiattribute decision-making situation in the absence of uncertainty, we often search for *Pareto optimal* solutions. We will assume that our decision maker has two objectives, and that the set of feasible points under consideration must satisfy a given set of constraints.

DEFINITION ■ A solution (call it A) to a multiple-objective problem is **Pareto optimal** if no other feasible solution is at least as good as A with respect to every objective and strictly better than A with respect to at least one objective. ■

If we define the concept of *dominated solution* as follows, we can rephrase our definition of Pareto optimality.

DEFINITION ■ A feasible solution B **dominates** a feasible solution A to a multiple-objective problem if B is at least as good as A with repect to every objective and is strictly better than A with respect to at least one objective. ■

[†]This section covers topics that may be omitted with no loss of continuity.

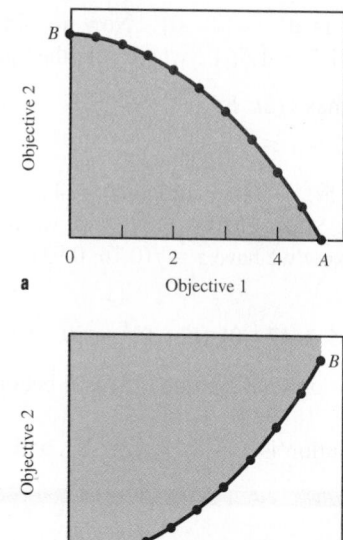

FIGURE 52 **b**

Thus, the Pareto optimal solutions are the set of all undominated feasible solutions.

If we graph the "score" of all Pareto optimal solutions in the x–y plane with the x-axis score being the score on objective 1 and the y-axis score being the score on objective 2, the graph is often called an **efficient frontier** or a **trade-off curve.**

To illustrate, suppose that the set of feasible solutions to a multiple-objective problem is the shaded region bounded by the curve AB and the first quadrant in Figure 52a. If we desire to maximize both objectives 1 and 2, then the curve AB is the set of Pareto optimal points.

For another illustration, suppose the set of feasible solutions to a multiple-objective problem is all shaded points in the first quadrant bounded from below by the curve AB in Figure 52b. If our goal is to maximize objective 1 and minimize objective 2, then the curve AB is the set of Pareto optimal points.

We will illustrate the concept of Pareto optimality (and how to determine Pareto optimal solutions) with the following example.

EXAMPLE 38 | **Profit Pollution Trade-Off Curve**

Chemco is considering producing three products. The per-unit contribution to profit, labor requirements, raw material used per unit produced, and pollution produced per unit of product are given in Table 18. Currently, 1,300 labor hours and 1,000 units of raw material are available. Chemco's two objectives are to maximize profit and minimize pollution produced. Graph the trade-off curve for this problem.

Solution If we define x_i = number of units of product i produced, then Chemco's two objectives may be written as follows:

Objective 1 Profit = $10x_1 + 9x_2 + 8x_3$

Objective 2 Pollution = $10x_1 + 6x_2 + 3x_3$

TABLE 18
Data for Chemco

	Product		
	1	2	3
Profit ($)	10	9	8
Labor (hours)	4	3	2
Raw material (units)	3	2	2
Pollution (units)	10	6	3

We will graph pollution on the x-axis and profit on the y-axis. The values of the decision variables must satisfy the following constraints:

$$4x_1 + 3x_2 + 2x_3 \leq 1,300 \qquad \text{(Labor constraint)} \qquad (75)$$

$$3x_1 + 2x_2 + 2x_3 \leq 1,000 \qquad \text{(Raw material constraint)} \qquad (76)$$

$$x_i \geq 0 \quad (i = 1, 2, 3) \qquad (77)$$

We can find a Pareto optimal solution by choosing to optimize either of our objectives, subject to the constraints (75)–(77). We begin by maximizing profit. To do this we must solve the following LP:

$$\max z = 10x_1 + 9x_2 + 8x_3$$

$$\text{s.t.} \quad 4x_1 + 3x_2 + 2x_3 \leq 1,300 \quad \text{(Labor constraint)}$$

$$3x_1 + 2x_2 + 2x_3 \leq 1,000 \quad \text{(Raw material constraint)} \qquad (78)$$

$$x_i \geq 0 \quad (i = 1, 2, 3)$$

When this LP is solved on LINDO, we find its unique optimal solution to be (call it A) $z = 4,300$, $x_1 = 0$, $x_2 = 300$, and $x_3 = 200$. This solution yields a pollution level of $6(300) + 3(200) = 2,400$ units. We claim this solution is Pareto optimal. To see this, note that for this solution not to be Pareto optimal, there would have to be a solution satisfying (75)–(77) that yielded $z \geq 4,300$ and pollution $\leq 2,400$, with at least one of these inequalities holding strictly. Since $x_1 = 0, x_2 = 300, x_3 = 200$ is the unique solution to (78), there is no feasible solution besides A satisfying (75)–(77) that can have $z \geq 4,300$. Thus, A cannot be dominated.

To find other Pareto optimal solutions, we choose any level of pollution (call it POLL) and solve the following LP:

$$\max z = 10x_1 + 9x_2 + 8x_3$$

$$\text{s.t.} \quad 4x_1 + 3x_2 + 2x_3 \leq 1,300 \quad \text{(Labor constraint)}$$

$$3x_1 + 2x_2 + 2x_3 \leq 1,000 \quad \text{(Raw material constraint)} \qquad (79)$$

$$10x_1 + 6x_2 + 3x_3 \leq \text{POLL}$$

$$x_i \geq 0 \quad (i = 1, 2, 3)$$

Let PROF be the (unique) optimal z-value when this LP is solved. For each value of POLL, the point (POLL, PROF) will be on the trade-off curve. To see this, note that any point (POLL′, PROF′) dominating (POLL, PROF) must have PROF′ $\geq$ PROF. The fact that (POLL, PROF) is the unique solution to (79) implies that all feasible points (with the exception of [POLL, PROF]) having PROF′ $\geq$ PROF must have POLL′ $>$ POLL.

This means that (POLL, PROF) cannot be dominated, so it is on the trade-off curve. Choosing any value of POLL $>$ 2,400 yields no new points on the trade-off curve. (Why?) Thus, as our next step we choose POLL = 2,300. Then LINDO yields an optimal z-value of 4,266.67 and $10x_1 + 6x_2 + 3x_3 = 2,300$. Thus, the point (2,300, 4,266.67) is on the

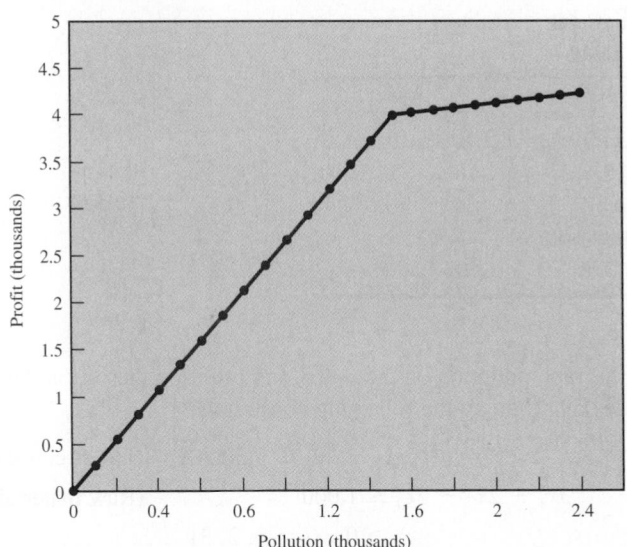

FIGURE 53
Example of
Trade-Off Curve

trade-off curve. Next, we change POLL to 2,200 and obtain the point (2,200, 4,233.33) on the trade-off curve. Continuing in this fashion, setting POLL = 2,100, 2,000, 1,900, . . . 0 we obtain the trade-off curve between profit and pollution given in Figure 53.

In a multiple-objective problem in which both the constraints and objectives are linear functions, the trade-off curve will be a piecewise linear curve (that is, the graph will consist of a number of line segments of different slopes). We now give an example of a trade-off curve for a problem in which the objectives are nonlinear functions.

EXAMPLE 39   Nonlinear Trade-Off Curve

Proctor and Ramble places ads on football games and soap operas. If F one-minute ads are placed on football games and S one-minute ads are placed on soap operas, then the number of men and women reached (in millions) and the cost (in thousands) of the ads are given in Table 19. P & R has a $1 million advertising budget and its two objectives are to maximize the number of men and the number of women who see its ads. Construct a trade-off curve for this situation.

Solution   To find a first point on the trade-off curve, let us ignore the goal of maximizing the number of women who see our ads and just maximize the number of men who see our ads. This requires that we solve the following NLP:

$$\max z = 20\sqrt{F} + 4\sqrt{S}$$
$$\text{s.t.} \quad 100F + 60S \leq 1,000 \tag{80}$$
$$F \geq 0, S \geq 0$$

TABLE 19
Data for Advertising

Type of Ad	Men Reached	Women Reached	Cost per Ad ($ Thousands)
Football	$20\sqrt{F}$	$4\sqrt{F}$	100
Soap opera	$4\sqrt{S}$	$15\sqrt{S}$	60

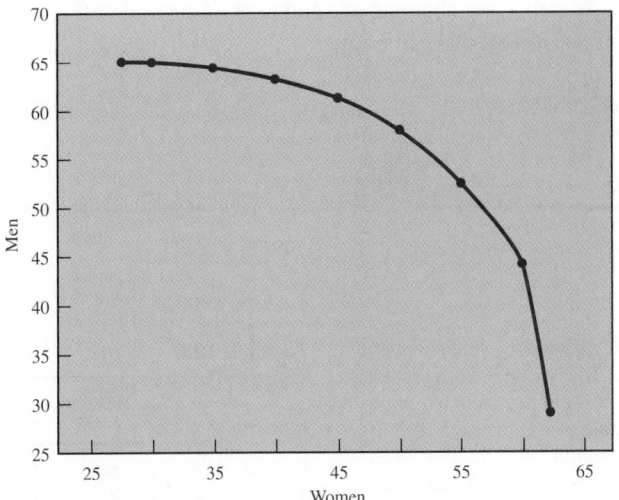

FIGURE 54
Trade-Off Curve for
Advertising Example

LINGO yields the optimal solution $z = 65.32$, $F = 9.38$, $S = 1.04$. This solution reaches $4\sqrt{9.38} + 15\sqrt{1.04} = 27.55$ million women. If we choose to place the women objective on the x-axis and the men objective on the y-axis, this yields the point $(27.55, 65.32)$ on the trade-off curve. To obtain other points on the trade-off curve, choose any value $W \geq 0$ and add the constraint $4\sqrt{F} + 15\sqrt{S} \geq W$ to (80).

This yields NLP (81):

$$\max z = 20\sqrt{F} + 4\sqrt{S}$$
$$\text{s.t.} \quad 100F + 60S \leq 1,000$$
$$4\sqrt{F} + 15\sqrt{S} \geq W \tag{81}$$
$$F \geq 0, S \geq 0$$

Suppose the optimal solution to (81) is unique and yields a z-value of M. Then the point of (W, M) is on the trade-off curve. To see this, note that any point (W', M') dominating (W, M) must have $W' \geq W$. The fact that (W, M) is the unique solution to (81) implies that all such feasible points [with the exception of (W, M)] will have $M' < M$. This means that (W, M) cannot be dominated, so it is on the trade-off curve. Using LINGO to solve (81) with $W = 30, 35, 40, 45, 50, 55, 60,$ and 62.5 yields the trade-off curve drawn in Figure 54. By the way, we cut the curve off at $W = 62.5$, because the budget constraint limits the maximum number of women watching ads to 62.5.

Summary of Trade-Off Curve Procedure

The procedure we have used to construct trade-off curves between two objectives may be summarized as follows:

Step 1 Choose an objective (say, objective 1) and determine the best value of this objective that can be attained (call it v_1). For the solution attaining v_1, find the value of objective 2 (call it v_2). Then (v_1, v_2) is a point on the trade-off curve.

Step 2 For values v of objective 2 that are better than v_2, solve the optimization problem in step 1 with the additional constraint: The value of objective 2 is at least as good as v. Varying v (over values of v preferred to v_2) will give you other points on the trade-off curve.

Step 3 In step 1, we obtained one endpoint of the trade-off curve. If we determine the best value of objective 2 that can be attained, we obtain the other endpoint of the trade-off curve.

REMARK In situations when there are more than two objectives, it is often helpful to examine trade-off curves between different pairs of objectives.

PROBLEMS

Group A

1 Widgetco produces two types of widgets. Each widget is made of steel and aluminum and is assembled with skilled labor. The resources used and the per-unit profit contribution (ignoring cost of overtime labor purchased) for each type of widget are given in Table 20. Currently, 200 units of steel and 300 units of aluminum and 300 hours of labor are available. Extra overtime labor can be purchased for $10 per hour. Construct an exchange curve between the objectives of maximizing profit and minimizing overtime labor.

2 Plantco produces three products. Three workers work for Plantco, and the company must determine which product(s) each worker should produce. The number of units each worker would produce if he or she spent the whole day producing each type of product are given in Table 21.

The company is also interested in maximizing the happiness of its workers. The amount of happiness "earned" by a worker who spends the entire day producing a given product is given in Table 22.

Construct a trade-off curve between the objectives of maximizing total units produced daily and total worker happiness.

3 If a company spends a on advertising and charges a price of p per unit, then it sells $1,000 - 100p + 20a^{1/2}$ units of the product. The per-unit cost of producing the product is $6. Construct a trade-off curve between the objectives of profit and units sold.

TABLE 22

Worker	Product 1	2	3
1	6	8	10
2	6	5	9
3	9	10	8

4 GMCO produces three types of cars: compacts, mid-size, and large. The variable cost per car (in thousands of dollars) and production capacity for each type of car are given in Table 23.

The annual demand for each type of car depends on the prices of the three types of cars, given in Table 24. Here PC = price charged for compact car (in thousands of dollars), and so on.

Suppose that each compact gets 30 mpg, each medium car gets 25 mpg, and each large car gets 18 mpg. GMCO wants to keep the planet pollution-free, so in addition to maximizing profit, it wants to maximize the average miles per gallon attained by the cars it sells. Use LINGO to construct a trade-off curve between these objectives.

5 Consider the discussion of crashing the length of the Widgetco project given in Section 8.4. For this example, construct a trade-off curve between cost of crashing the project and duration of the project.

6 For Example 35 of Section 11.10, construct a trade-off curve between the chosen portfolio's expected return and variance. This is often called the *efficient frontier.*

TABLE 20

Resource	Type of Widget 1	2
Steel (lbs)	6	12
Aluminum (lbs)	8	20
Skilled labor (hours)	11	24
Profit contribution ($)	500	1,100

TABLE 21

Worker	Product 1	2	3
1	20	12	10
2	12	15	9
3	6	5	10

TABLE 23

Type of Car	Variable Cost ($ Thousands)	Production Capacity (per Year)
Compact	10	2,000
Medium	14	1,500
Large	18	1,000

TABLE 24

Type of Car	Demand for Car
Compact	$2,500 - 100(PC) + 3(PM)$
Medium	$1,800 - 30 (PM) + 2(PC) + PL$
Large	$1,300 - 20 (PL) + PM$

Convex and Concave Functions

A function $f(x_1, x_2, \ldots, x_n)$ is a **convex function** on a convex set S if for any $x' \in S$ and $x'' \in S$

$$f[cx' + (1 - c)x''] \leq cf(x') + (1 - c)f(x'') \tag{3}$$

holds for $0 \leq c \leq 1$.

A function $f(x_1, x_2, \ldots, x_n)$ is a **concave function** on a convex set S if for any $x' \in S$ and $x'' \in S$

$$f[cx' + (1 - c)x''] \geq cf(x') + (1 - c)f(x'') \tag{4}$$

holds for $0 \leq c \leq 1$.

Consider a general NLP. Suppose the feasible region S for an NLP is a convex set. If $f(x)$ is a concave (convex) function of S, then any local maximum (minimum) for the NLP is an optimal solution to the NLP.

Suppose $f''(x)$ exists for all x in a convex set S. Then $f(x)$ is a convex (concave) function of S if and only if $f''(x) \geq 0[f''(x) \leq 0]$ for all x in S.

Suppose $f(x_1, x_2, \ldots, x_n)$ has continuous second-order partial derivatives for each point $x = (x_1, x_2, \ldots, x_n) \in S$. Then $f(x_1, x_2, \ldots, x_n)$ is a convex function on S if and only if for each $x \in S$, all principal minors of H are nonnegative.

Suppose $f(x_1, x_2, \ldots, x_n)$ has continuous second-order partial derivatives for each point $x = (x_1, x_2, \ldots, x_n) \in S$. Then $f(x_1, x_2, \ldots, x_n)$ is a concave function on S if and only if for each $x \in S$ and $k = 1, 2, \ldots, n$, all nonzero principal minors have the same sign as $(-1)^k$.

Solving NLPs with One Variable

To find an optimal solution to

$$\max \text{ (or min) } f(x)$$
$$\text{s.t.} \quad x \in [a, b]$$

we must consider the following three types of points:

Case 1 Points where $f'(x) = 0$ [a stationary point of $f(x)$].

Case 3 Points where $f'(x)$ does not exist.

Case 3 Endpoints a and b of the interval $[a, b]$.

If $f'(x_0) = 0$, $f''(x_0) < 0$, and $a < x_0 < b$, then x_0 is a local maximum. If $f'(x_0) = 0$, $f''(x_0) > 0$, and $a < x_0 < b$, then x_0 is a local minimum.

Golden Section Search

To determine (within ϵ) the optimal solution to

$$\max f(x)$$
$$\text{s.t.} \quad a \leq x \leq b$$

we can perform k iterations [where $r^k(b - a) < \epsilon$] of Golden Section Search. New points are generated as follows:

New Left-Hand Point Move a distance equal to a fraction r of the current interval of uncertainty from the right endpoint of the interval of uncertainty.

New Right-Hand Point Move a distance equal to a fraction r of the current interval of uncertainty from the left endpoint of the interval.

At each iteration, one of the new points will equal an old point.

Unconstrained Maximization and Minimization Problems with Several Variables

A local extremum $\bar{x}$ for

$$\text{max (or min) } f(x_1, x_2, \ldots, x_n)$$
$$\text{s.t.} \quad (x_1, x_2, \ldots, x_n) \in R^n \tag{7}$$

must satisfy $\dfrac{\partial f(\bar{x})}{\partial x_i} = 0$ for $i = 1, 2, \ldots, n$.

If $H_k(\bar{x}) > 0$ $(k = 1, 2, \ldots, n)$, then a stationary point $\bar{x}$ is a local minimum for (7).
If, for $0 \ k = 1, 2, \ldots, n$, $H_k(\bar{x})$ has the same sign as $(-1)^k$, then a stationary point $\bar{x}$ is a local maximum for (7).
If $H_n(\bar{x}) \neq 0$ and the conditions of Theorems 7 and 7' do not hold, then a stationary point $\bar{x}$ is not a local extremum.

The Method of Steepest Ascent

The method of steepest ascent can be used to solve problems of the following type:

$$\text{max } z = f(x_1, x_2, \ldots, x_n)$$
$$\text{s.t.} \quad (x_1, x_2, \ldots, x_n) \in R^n$$

To find a new point with a larger z-value, we move away from the current point ($\mathbf{v}$) in the direction of $\nabla f(\mathbf{v})$. The distance we move away from $\mathbf{v}$ is chosen to maximize the value of the function at the new point. We stop when $\|\nabla f(\mathbf{v})\|$ is sufficiently close to zero.

Lagrange Multipliers

Lagrange multipliers are used to solve NLPs of the following type:

$$\text{max (or min) } z = f(x_1, x_2, \ldots, x_n)$$
$$\text{s.t.} \quad g_1(x_1, x_2, \ldots, x_n) = b_1$$
$$g_2(x_1, x_2, \ldots, x_n) = b_2 \tag{12}$$
$$\vdots$$
$$g_m(x_1, x_2, \ldots, x_n) = b_m$$

To solve (12), form the Lagrangian

$$L(x_1, x_2, \ldots, x_n, \lambda_1, \lambda_2, \ldots, \lambda_m) = f(x_1, x_2, \ldots, x_n) + \sum_{i=1}^{i=m} \lambda_i[b_i - g_i(x_1, x_2, \ldots, x_n)]$$

and look for points $(\bar{x}_1, \bar{x}_2, \ldots, \bar{x}_n, \bar{\lambda}_1, \bar{\lambda}_2, \ldots, \bar{\lambda}_m)$ for which

$$\frac{\partial L}{\partial x_1} = \frac{\partial L}{\partial x_2} = \cdots = \frac{\partial L}{\partial x_n} = \frac{\partial L}{\partial \lambda_1} = \frac{\partial L}{\partial \lambda_2} = \cdots = \frac{\partial L}{\partial \lambda_m} = 0$$

The Kuhn–Tucker Conditions

The Kuhn–Tucker conditions are used to solve NLPs of the following type:

$$\max \text{ (or min) } f(x_1, x_2, \ldots, x_n)$$
$$\text{s.t.} \quad g_1(x_1, x_2, \ldots, x_n) \leq b_1$$
$$g_2(x_1, x_2, \ldots, x_n) \leq b_2 \qquad \text{(26)}$$
$$\vdots$$
$$g_m(x_1, x_2, \ldots, x_n) \leq b_m$$

Suppose (26) is a maximization problem. If $\bar{x} = (\bar{x}_1, \bar{x}_2, \ldots, \bar{x}_n)$ is an optimal solution to (26), then $\bar{x} = (\bar{x}_1, \bar{x}_2, \ldots, \bar{x}_n)$ must satisfy the m constraints in (26), and there must exist multipliers $\lambda_1, \lambda_2, \ldots, \lambda_m$ satisfying

$$\frac{\partial f(\bar{x})}{\partial x_j} - \sum_{i=1}^{i=m} \bar{\lambda}_i \frac{\partial g_i(\bar{x})}{\partial x_j} = 0 \qquad (j = 1, 2, \ldots, n)$$
$$\bar{\lambda}_i[b_i - g_i(\bar{x})] = 0 \qquad (i = 1, 2, \ldots, m)$$
$$\bar{\lambda}_i \geq 0 \qquad (i = 1, 2, \ldots, m)$$

Suppose (26) is a minimization problem. If $\bar{x} = (\bar{x}_1, \bar{x}_2, \ldots, \bar{x}_n)$ is an optimal solution to (26), then $\bar{x} = (\bar{x}_1, \bar{x}_2, \ldots, \bar{x}_n)$ must satisfy the m constraints in (26), and there must exist multipliers $\lambda_1, \lambda_2, \ldots, \lambda_m$ satisfying

$$\frac{\partial f(\bar{x})}{\partial x_j} + \sum_{i=1}^{i=m} \bar{\lambda}_i \frac{\partial g_i(\bar{x})}{\partial x_j} = 0 \qquad (j = 1, 2, \ldots, n)$$
$$\bar{\lambda}_i[b_i - g_i(\bar{x})] = 0 \qquad (i = 1, 2, \ldots, m)$$
$$\bar{\lambda}_i \geq 0 \qquad (i = 1, 2, \ldots, m)$$

The Kuhn–Tucker conditions are **necessary** conditions for a point to solve (26). If the $g_i(x_1, x_2, \ldots, x_n)$ are convex functions and the objective function $f(x_1, x_2, \ldots, x_n)$ is concave (convex), then for a maximization (minimization) problem, any point satisfying the Kuhn–Tucker conditions will yield an optimal solution to (26).

Quadratic Programming

A quadratic programming problem (QPP) is an NLP in which each term in the objective function is of degree 2, 1, or 0 and all constraints are linear. Wolfe's method (a modified version of the two-phase simplex) may also be used to solve QPPs.

Separable Programming

If an NLP can be written in the following form:

$$\max \text{ (or min) } z = \sum_{j=1}^{j=n} f_j(x_j)$$
$$\text{s.t.} \quad \sum_{j=1}^{j=n} g_{ij}(x_j) \leq b_i \qquad (i = 1, 2, \ldots, m)$$

it is a **separable programming problem.** To approximate the optimal solution to a separable programming problem, we solve the following **approximating problem:**

$$\max \text{ (or min) } \hat{z} = \sum_{j=1}^{j=n} [\delta_{j1} f_j(p_{j1}) + \delta_{j2} f_j(p_{j2}) + \cdots + \delta_{j,k} f_j(p_{j,k})]$$

$$\text{s.t.} \quad \sum_{j=1}^{j=n} [\delta_{j1} g_{ij}(p_{j1}) + \delta_{j2} g_{ij}(p_{j2}) + \cdots + \delta_{j,k} g_{ij}(p_{j,k})] \leq b_i \quad (i = 1, 2, \ldots, m)$$

$$\delta_{j1} + \delta_{j2} + \cdots + \delta_{j,k} = 1 \quad (j = 1, 2, \ldots, n)$$

$$\delta_{j,r} \geq 0 \quad (j = 1, 2, \ldots, n; r = 1, 2, \ldots, k)$$

(For $j = 1, 2, \ldots, n$, at most two $\delta_{j,k}$'s can be positive. If for a given j, two $\delta_{j,k}$'s are positive, they must be adjacent.)

The Method of Feasible Directions

To solve

$$\max z = f(\mathbf{x})$$
$$\text{s.t.} \quad A\mathbf{x} \leq \mathbf{b}$$
$$\mathbf{x} \geq \mathbf{0}$$

we begin with a feasible solution $\mathbf{x}^0$. Let $\mathbf{d}^0$ be a solution to

$$\max z = \nabla f(\mathbf{x}^0) \cdot \mathbf{d}$$
$$\text{s.t.} \quad A\mathbf{d} \leq \mathbf{b}$$
$$\mathbf{d} \geq \mathbf{0}$$

Choose our new point $\mathbf{x}^1$ to be $\mathbf{x}^1 = \mathbf{x}^0 + t_0(\mathbf{d}^0 - \mathbf{x}^0)$, where t_0 solves

$$\max f[\mathbf{x}^0 + t_0(\mathbf{d}^0 - \mathbf{x}^0)]$$
$$0 \leq t_0 \leq 1$$

Let $\mathbf{d}^1$ be a solution to

$$\max z = \nabla f(\mathbf{x}^1) \cdot \mathbf{d}$$
$$\text{s.t.} \quad A\mathbf{d} \leq \mathbf{b}$$
$$\mathbf{d} \geq \mathbf{0}$$

Choose our new point $\mathbf{x}^2$ to be $\mathbf{x}^2 = \mathbf{x}^1 + t_1(\mathbf{d}^1 - \mathbf{x}^1)$, where t_1 solves

$$\max f[\mathbf{x}^1 + t_1(\mathbf{d}^1 - \mathbf{x}^1)]$$
$$0 \leq t_1 \leq 1$$

Continue generating points $\mathbf{x}^3, \ldots, \mathbf{x}^k$ in this fashion until $\mathbf{x}^k = \mathbf{x}^{k-1}$ or successive points are sufficiently close together.

Summary of Trade-Off Curve Procedure

The procedure we have used to construct trade-off curves between two objectives may be summarized as follows:

Step 1 Choose an objective—say, objective 1—and determine its best attainable value v_1. For the solution attaining v_1, find the value of objective 2, v_2. Then (v_1, v_2) is a point on the trade-off curve.

Step 2 For values v of objective 2 that are better than v_2, solve the optimization problem in step 1 with the additional constraint that the value of objective 2 is at least as good as v. Varying v (over values of v preferred to v_2) will give you other points on the trade-off curve.

Step 3 In step 1 we obtained one endpoint of the trade-off curve. If we determine the best value of objective 2 that can be attained, we obtain the other endpoint of the trade-off curve.

REVIEW PROBLEMS

Group A

1 Show that $f(x) = e^{-x}$ is a convex function on R^1.

2 Five of a store's major customers are located as in Figure 55. Determine where the store should be located to minimize the sum of the squares of the distances that each customer would have to travel to the store. Can you generalize this result to the case of n customers located at points $x_1, x_2, \ldots, x_n$?

3 A company uses a raw material to produce two types of products. When processed, each unit of raw material yields 2 units of product 1 and 1 unit of product 2. If x_1 units of product 1 are produced, then each unit can be sold for $49 - x_1$, if x_2 units of product 2 are produced, then each unit can be sold for $30 - 2x_2$. It costs $5 to purchase and process each unit of raw material.

   a Use the Kuhn–Tucker conditions to determine how the company can maximize profits.

   b Use LINGO or Wolfe's method to determine how the company can maximize profits.

   c What is the most that the company would be willing to pay for an extra unit of raw material?

4 Show that $f(x) = |x|$ is a convex function on R^1.

5 Use Golden Section Search to locate, within 0.5, the optimal solution to

$$\max 3x - x^2$$
$$\text{s.t.} \quad 0 \le x \le 5$$

6 Perform two iterations of the method of steepest ascent in an attempt to maximize

$$f(x_1, x_2) = (x_1 + x_2)e^{-(x_1 + x_2)} - x_1$$

Begin at the point (0,1).

7 The cost of producing x units of a product during a month is x^2 dollars. Find the minimum cost method of producing 60 units during the next three months. Can you generalize this result to the case where the cost of producing x units during a month is an increasing convex function?

8 Solve the following NLP:

$$\max z = xyw$$
$$\text{s.t.} \quad 2x + 3y + 4w = 36$$

9 Solve the following NLP:

$$\min z = \frac{50}{x} + \frac{20}{y} + xy$$
$$\text{s.t.} \quad x \ge 1, y \ge 1$$

10 If a company charges a price p for a product and spends $a on advertising, it can sell $10,000 + 5\sqrt{a} - 100p$ units of the product. If the product costs $10 per unit to produce, then how can the company maximize profits?

11 With L labor hours and M machine hours, a company can produce $L^{1/3}M^{2/3}$ computer disk drives. Each disk drive sells for $150. If labor can be purchased at $50 per hour and machine hours can be purchased at $100 per hour, determine how the company can maximize profits.

Group B

12 In time t, a tree can grow to a size $F(t)$, where $F'(t) \ge 0$ and $F''(t) < 0$. Assume that for large t, $F'(t)$ is near 0. If the tree is cut at time t, then a revenue $F(t)$ is received. Assume that revenues are discounted continuously at a rate r, so $1 received at time t is equivalent to e^{-rt} received at time 0. The goal is to cut the tree at the time t^* that maximizes discounted revenue. Show that the tree should be cut at the time t^* satisfying the equation

$$r = \frac{F'(t^*)}{F(t^*)}$$

In the answer, explain why (if $\frac{F'(0)}{F(0)} > r$) this equation has a unique solution. Also show that the answer is a maximum, not a minimum. [*Hint:* Why is it sufficient to choose t^* to maximize $\ln(e^{-rt}F(t)$?]

13 Suppose we are hiring a weather forecaster to predict the probability that next summer will be rainy or sunny. The following suggests a method that can be used to ensure that the forecaster is accurate. Suppose that the actual probability of rain next summer is q. For simplicity, we assume that the summer can only be rainy or sunny. If the forecaster announces a probability p that the summer will be rainy, then she receives a payment of $1 - (1 - p)^2$ if the summer is rainy and a payment of $1 - p^2$ if the summer is sunny. Show that the forecaster will maximize expected profits by announcing that the probability of a rainy summer is q.

14 Show that if $b > a \ge e$, then $a^b > b^a$. Use this result to show that $e^\pi > \pi^e$. [*Hint:* Show that $\max(\frac{\ln x}{x})$ over $x \ge a$ occurs for $x = a$.]

FIGURE 55

15 Consider the points $(0, 0)$, $(1, 1)$, and $(2, 3)$. Formulate an NLP whose solution will yield the circle of smallest radius enclosing these three points. Use LINGO to solve the NLP.

16 The cost of producing x units of a product during a month is $x^{1/2}$ dollars. Show that the minimum cost method of producing 40 units during the next two months is to produce all 40 units during a single month. Is it possible to generalize this result to the case where the cost of producing x units during a month is an increasing concave function?

17 Consider the problem

$$\max z = f(x)$$
$$\text{s.t.} \quad a \leq x \leq b$$

a Suppose $f(x)$ is a convex function that has derivatives for all values of x. Show that $x = a$ or $x = b$ must be optimal for the NLP. (Draw a picture.)

b Suppose $f(x)$ is a convex function for which $f'(x)$ may not exist. Show that $x = a$ or $x = b$ must be optimal for the NLP. (Use the definition of a convex function.)

18 Reconsider Problem 2. Suppose that the store should now be located to minimize the total distance that customers must walk to the store. Where should the store be located? (*Hint:* Use Problem 4 and the fact that for any convex function a local minimum will solve the NLP; then show that locating the store where one of the customers lives yields a local minimum.) Can the result be generalized?

19[†] A company uses raw material to produce two products. For c dollars, a unit of raw material can be purchased and processed into k_1 units of product 1 and k_2 units of product 2. If x_1 units of product 1 are produced, they can be sold at $p_1(x_1)$ dollars per unit. If x_2 units of product 2 are produced, they can be sold at $p_2(x_2)$ dollars per unit. Let z be the number of units of raw material that are purchased and processed. To maximize profits (ignoring non-negativity constraints), the following NLP should be solved:

$$\max w = x_1 p_1(x_1) + x_2 p_2(x_2) - cz$$
$$\text{s.t.} \quad x_1 \leq k_1 z$$
$$\quad x_2 \leq k_2 z$$

a Write down the Kuhn–Tucker conditions for this problem. Let $\bar{x}_1, \bar{x}_2, \bar{\lambda}_1, \bar{\lambda}_2$ represent the optimal solution to this problem.

b Consider a modified version of the problem. The company can now purchase each unit of product 1 for $\bar{\lambda}_1$ dollars and each unit of product 2 for $\bar{\lambda}_2$ dollars. Show that if the company tries to maximize profits in this situation, it will, as in part (*a*), produce $\bar{x}_1$ units of product 1 and $\bar{x}_2$ units of product 2. Also, show that profit and production costs will remain unchanged.

c Give an interpretation of $\bar{\lambda}_1$ and $\bar{\lambda}_2$ that might be useful to the company's accountant.

20 The area of a triangle with sides of length a, b, and c is $\sqrt{s(s - a)(s - b)(s - c)}$, where s is half the perimeter of the triangle. We have 60 ft of fence and want to fence a triangular-shaped area. Determine how to maximize the fenced area.

21 The energy used in compressing a gas (in three stages) from an initial pressure I to a final pressure F is given by

$$K \left\{ \sqrt{\frac{p_1}{I}} + \sqrt{\frac{p_2}{p_1}} + \sqrt{\frac{F}{p_2}} - 3 \right\}$$

Determine how to minimize the energy used in compressing the gas.

22 Prove Lemma 1 (use Lagrange multipliers).

REFERENCES

The following books emphasize the theoretical aspects of nonlinear programming:

Bazaraa, M., H. Sherali, and C. Shetty. *Nonlinear Programming: Theory and Algorithms.* New York: John Wiley, 1993.

Bertsetkas, D. *Nonlinear Programming.* Cambridge, Mass.: Athena Publishing, 1995.

Luenberger, D. *Linear and Nonlinear Programming.* Reading, Mass.: Addison-Wesley, 1984.

Mangasarian, O. *Nonlinear Programming.* New York: McGraw-Hill, 1969.

McCormick, G. *Nonlinear Programming: Theory, Algorithms, and Applications.* New York: Wiley, 1983.

Shapiro, J. *Mathematical Programming: Structures and Algorithms.* New York: Wiley, 1979.

Zangwill, W. *Nonlinear Programming.* Englewood Cliffs, N.J.: Prentice Hall, 1969.

The following book emphasizes various nonlinear programming algorithms:

Rao, S. *Optimization Theory and Applications.* New Delhi: Wiley Eastern Ltd., 1979.

[†]Based on Littlechild, "Marginal Pricing" (1970).

12

Review of Calculus and Probability

We review in this chapter some basic topics in calculus and probability, which will be useful in later chapters.

12.1 Review of Integral Calculus

In our study of random variables, we often require a knowledge of the basics of integral calculus, which will be briefly reviewed in this section.

Consider two functions: $f(x)$ and $F(x)$. If $F'(x) = f(x)$, we say that $F(x)$ is the **indefinite integral** of $f(x)$. The fact that $F(x)$ is the indefinite integral of $f(x)$ is written

$$F(x) = \int f(x)\, dx$$

The following rules may be used to find the indefinite integrals of many functions (C is an arbitrary constant):

$$\int (1)\, dx = x + C$$

$$\int af(x)\, dx = a \int f(x)\, dx \qquad\qquad (a \text{ is any constant})$$

$$\int [f(x) + g(x)]\, dx = \int f(x)\, dx + \int g(x)\, dx$$

$$\int x^n\, dx = \frac{x^{n+1}}{n+1} + C \qquad\qquad (n \neq -1)$$

$$\int x^{-1}\, dx = \ln x + C$$

$$\int e^x\, dx = e^x + C$$

$$\int a^x\, dx = \frac{a^x}{\ln a} + C \qquad\qquad (a > 0, a \neq 1)$$

$$\int [f(x)]^n f'(x)\, dx = \frac{[f(x)]^{n+1}}{n+1} + C \qquad\qquad (n \neq -1)$$

$$\int f(x)^{-1} f'(x)\, dx = \ln f(x) + C$$

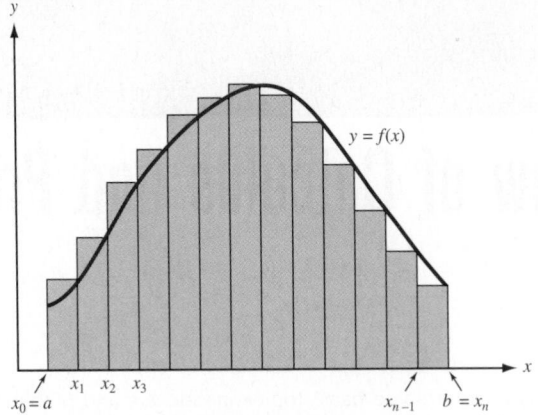

FIGURE **1**
Relation of Area and
Definite Integral

For two functions $u(x)$ and $v(x)$,

$$\int u(x)v'(x)\, dx = u(x)v(x) - \int v(x)u'(x)\, dx \qquad \text{(Integration by parts)}$$

$$\int e^{f(x)}f'(x)\, dx = e^{f(x)} + C$$

$$\int a^{f(x)}f'(x)\, dx = \frac{a^{f(x)}}{\ln a} + C \qquad\qquad (a > 0, a \neq 1)$$

The concept of an integral is important for the following reasons. Consider a function $f(x)$ that is continuous for all points satisfying $a \leq x \leq b$. Let $x_0 = a$, $x_1 = x_0 + \Delta$, $x_2 = x_1 + \Delta, \ldots, x_i = x_{i-1} + \Delta, x_n = x_{n-1} + \Delta = b$, where $\Delta = \frac{b-a}{n}$. From Figure 1, we see that as Δ approaches zero (or equivalently, as n grows large),

$$\sum_{i=1}^{i=n} f(x_i)\, \Delta$$

will closely approximate the area under the curve $y = f(x)$ between $x = a$ and $x = b$. If $f(x)$ is continuous for all x satisfying $a \leq x \leq b$, it can be shown that the area under the curve $y = f(x)$ between $x = a$ and $x = b$ is given by

$$\lim_{\Delta \to 0} \sum_{i=1}^{i=n} f(x_i)\Delta$$

which is written as

$$\int_a^b f(x)\, dx$$

or the **definite integral** of $f(x)$ from $x = a$ to $x = b$. The **Fundamental Theorem of Calculus** states that if $f(x)$ is continuous for all x satisfying $a \leq x \leq b$, then

$$\int_a^b f(x)\, dx = F(b) - F(a)$$

where $F(x)$ is any indefinite integral of $f(x)$. $F(b) - F(a)$ is often written as $[F(x)]_a^b$. Example 1 illustrates the use of the definite integral.

EXAMPLE 1   Customer Arrivals at a Bank

Suppose that at time t (measured in hours, and the present $t = 0$), the rate $a(t)$ at which customers enter a bank is $a(t) = 100t$. During the next 2 hours, how many customers will enter the bank?

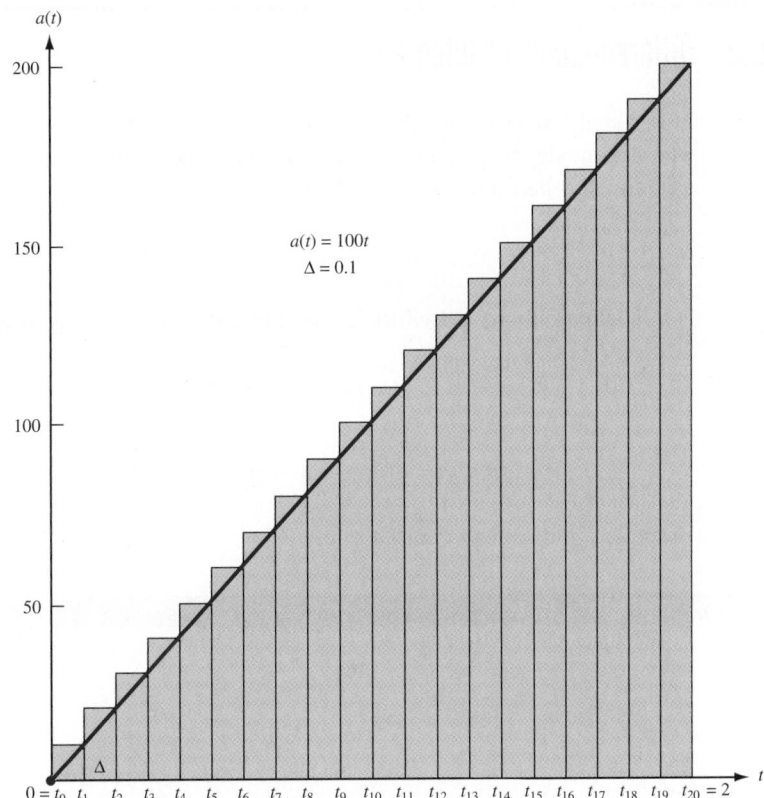

FIGURE 2
Relation of Total
Arrivals in Next 2 Hours
to Area under $a(t)$
Curve

Solution   Let $t_0 = 0$, $t_1 = t_0 + \Delta$, $t_2 = t_1 + \Delta$, ..., $t_n = t_{n-1} + \Delta = 2$ (of course, $\Delta = \frac{2}{n}$). Between time t_{i-1} and time t_i, approximately $100t_i\Delta$ customers will arrive. Therefore, the total number of customers to arrive during the next 2 hours will equal

$$\lim_{\Delta \to 0} \sum_{i=1}^{i=n} 100t_i\Delta$$

(see Figure 2). From the Fundamental Theorem of Calculus,

$$\lim_{\Delta \to 0} \sum_{i=1}^{i=n} 100t_i\Delta = \int_0^2 (100t)\, dt = [50t^2]_0^2 = 200 - 0 = 200$$

Thus, 200 customers will arrive during the next 2 hours.

PROBLEMS

Group A

1   The present is $t = 0$. At a time t years from now, I earn income at a rate e^{2t}. How much money do I earn during the next 5 years?

2   If money is continuously discounted at a rate of $r\%$ per year, then \$1 earned t years in the future is equivalent to e^{-rt} dollars earned at the present time. Use this fact to determine the present value of the income earned in Problem 1.

3   At time 0, a company has I units of inventory in stock. Customers demand the product at a constant rate of d units per year (assume that $I \geq d$). The cost of holding 1 unit of stock in inventory for a time Δ is \$$h\Delta$. Determine the total holding cost incurred during the next year.

12.2 Differentiation of Integrals

In our study of inventory theory in Chapter 16, we will have to differentiate a function whose value depends on an integral. Let $f(x, y)$ be a function of variables x and y, and let $g(y)$ and $h(y)$ be functions of y. Then

$$F(y) = \int_{g(y)}^{h(y)} f(x, y)\, dx$$

is a function only of y. **Leibniz's rule for differentiating an integral** states that

If $\quad F(y) = \int_{g(y)}^{h(y)} f(x, y)\, dx, \quad$ then

$$F'(y) = h'(y)f(h(y), y) - g'(y)f(g(y), y) + \int_{g(y)}^{h(y)} \frac{\partial f(x, y)}{\partial y}\, dx$$

Example 2 illustrates Leibniz's rule.

EXAMPLE 2   **Leibniz's Rule**

For

$$F(y) = \int_1^{y^2} \frac{y\, dx}{x}$$

find $F'(y)$.

Solution   We have that $f(x, y) = \dfrac{y}{x}$, $h(y) = y^2$, $h'(y) = 2y$, $\dfrac{\partial f}{\partial y} = \dfrac{1}{x}$, $g(y) = 1$, $g'(y) = 0$. Then

$$F'(y) = 2y\left(\frac{y}{y^2}\right) - 0\left(\frac{y}{1}\right) + \int_1^{y^2} \frac{dx}{x}$$
$$= 2 + [\ln x]_1^{y^2} = 2 + \ln y^2 - 0 = 2 + 2\ln y$$

PROBLEMS

Group A

For each of the following functions, use Leibniz's rule to find $F'(y)$:

1 $F(y) = \int_y^{y^2} (2y + x)\, dx$

2 $F(y) = \int_0^y yx^2\, dx$

3 $F(y) = \int_0^y 6(5 - x)f(x)\, dx + \int_y^\infty 4(x - 5)f(x)\, dx$

12.3 Basic Rules of Probability

In this section, we review some basic rules and definitions that you may have encountered during your previous study of probability.

DEFINITION ■   Any situation where the outcome is uncertain is called an **experiment**. ■

For example, drawing a card from a deck of cards would be an experiment.

DEFINITION ■ For any experiment, the **sample space** S of the experiment consists of all possible outcomes for the experiment. ■

For example, if we toss a die and are interested in the number of dots showing, then $S = \{1, 2, 3, 4, 5, 6\}$.

DEFINITION ■ An **event** E consists of any collection of points (set of outcomes) in the sample space. ■

A collection of events $E_1, E_2, \ldots, E_n$ is said to be a **mutually exclusive** collection of events if for $i \neq j$ ($i = 1, 2, \ldots, n$ and $j = 1, 2, \ldots, n$), E_i and E_j have no points in common. ■

With each event E, we associate an event $\overline{E}$. $\overline{E}$ consists of the points in the sample space that are not in E. With each event E, we also associate a number $P(E)$, which is the probability that event E will occur when we perform the experiment. The probabilities of events must satisfy the following rules of probability:

Rule 1 For any event E, $P(E) \geq 0$.

Rule 2 If $E = S$ (that is, if E contains all points in the sample space), then $P(E) = 1$.

Rule 3 If $E_1, E_2, \ldots, E_n$ is a mutually exclusive collection of events, then

$$P(E_1 \cup E_2 \cup \cdots \cup E_n) = \sum_{k=1}^{k=n} P(E_k)$$

Rule 4 $P(\overline{E}) = 1 - P(E)$.

DEFINITION ■ For two events E_1 and E_2, $P(E_2|E_1)$ (the **conditional probability** of E_2 given E_1) is the probability that the event E_2 will occur given that event E_1 has occurred. Then

$$P(E_2|E_1) = \frac{P(E_1 \cap E_2)}{P(E_1)} \qquad ■ \qquad (1)$$

Suppose events E_1 and E_2 both occur with positive probability. Events E_1 and E_2 are **independent** if and only if $P(E_2|E_1) = P(E_2)$ (or equivalently, $P(E_1|E_2) = P(E_1)$). ■

Thus, events E_1 and E_2 are independent if and only if knowledge that E_1 has occurred does not change the probability that E_2 has occurred, and vice versa. From (1), E_1 and E_2 are independent if and only if

$$\frac{P(E_1 \cap E_2)}{P(E_1)} = P(E_2) \qquad \text{or} \qquad P(E_1 \cap E_2) = P(E_1)\,P(E_2) \qquad (2)$$

EXAMPLE 3 **Drawing a Card**

Suppose we draw a single card from a deck of 52 cards.

1 What is the probability that a heart or spade is drawn?

2 What is the probability that the drawn card is not a 2?

3 Given that a red card has been drawn, what is the probability that it is a diamond? Are the events

$$E_1 = \text{red card is drawn}$$
$$E_2 = \text{diamond is drawn}$$

independent events?

4 Show that the events

$$E_1 = \text{spade is drawn}$$
$$E_2 = 2 \text{ is drawn}$$

are independent events.

Solution **1** Define the events

$$E_1 = \text{heart is drawn}$$
$$E_2 = \text{spade is drawn}$$

E_1 and E_2 are mutually exclusive events with $P(E_1) = P(E_2) = \frac{1}{4}$. We seek $P(E_1 \cup E_2)$. From probability rule 3,

$$P(E_1 \cup E_2) = P(E_1) + P(E_2) = (\tfrac{1}{4}) + (\tfrac{1}{4}) = \tfrac{1}{2}$$

2 Define event $E = $ a 2 is drawn. Then $P(E) = \frac{4}{52} = \frac{1}{13}$. We seek $P(\bar{E})$. From probability rule 4, $P(\bar{E}) = 1 - \frac{1}{13} = \frac{12}{13}$.

3 From (1),

$$P(E_2|E_1) = \frac{P(E_1 \cap E_2)}{P(E_1)}$$

$$P(E_1 \cap E_2) = P(E_2) = \frac{13}{52} = \frac{1}{4}$$

$$P(E_1) = \frac{26}{52} = \frac{1}{2}$$

Thus,

$$P(E_2|E_1) = \frac{\frac{1}{4}}{\frac{1}{2}} = \frac{1}{2}$$

Since $P(E_2) = \frac{1}{4}$, we see that $P(E_2|E_1) \neq P(E_2)$. Thus, E_1 and E_2 are not independent events. (This is because knowing that a red card was drawn increases the probability that a diamond was drawn.)

4 $P(E_1) = \frac{13}{52} = \frac{1}{4}$, $P(E_2) = \frac{4}{52} = \frac{1}{13}$, and $P(E_1 \cap E_2) = \frac{1}{52}$. Since $P(E_1)\,P(E_2) = P(E_1 \cap E_2)$, E_1 and E_2 are independent events. Intuitively, since $\frac{1}{4}$ of all cards in the deck are spades and $\frac{1}{4}$ of all 2's in the deck are spades, knowing that a 2 has been drawn does not change the probability that the card drawn was a spade.

PROBLEMS

Group A

1 Suppose two dice are tossed (for each die, it is equally likely that 1, 2, 3, 4, 5, or 6 dots will show).

a What is the probability that the total of the two dice will add up to 7 or 11?

b What is the probability that the total of the two dice will add up to a number other than 2 or 12?

c Are the events

E_1 = first die shows a 3

E_2 = total of the two dice is 6

independent events?

d Are the events

E_1 = first die shows a 3

E_2 = total of the two dice is 7

independent events?

e Given that the total of the two dice is 5, what is the probability that the first die showed 2 dots?

f Given that the first die shows 5, what is the probability that the total of the two dice is even?

12.4 Bayes' Rule

An important decision often depends on the "state of the world." For example, we may want to know whether a person has tuberculosis. Then we would be concerned with the probability of the following states of the world:

S_1 = person has tuberculosis

S_2 = person does not have tuberculosis

More generally, n mutually exclusive states of the world $(S_1, S_2, \ldots, S_n)$ may occur. The states of the world are **collectively exhaustive:** $S_1, S_2, \ldots, S_n$ include all possibilities. Suppose a decision maker assigns a probability $P(S_i)$ to S_i. $P(S_i)$ is the **prior probability** of S_i. To obtain more information about the state of the world, the decision maker may observe the outcome of an experiment. Suppose that for each possible outcome O_j and each possible state of the world S_i, the decision maker knows $P(O_j|S_i)$, the **likelihood** of the outcome O_j given state of the world S_i. Bayes' rule combines prior probabilities and likelihoods with the experimental outcomes to determine a post-experimental probability, or **posterior probability,** for each state of the world. To derive Bayes' rule, observe that (1) implies that

$$P(S_i|O_j) = \frac{P(S_i \cap O_j)}{P(O_j)} \tag{3}$$

From (1), it also follows that

$$P(S_i \cap O_j) = P(O_j|S_i)P(S_i) \tag{4}$$

The states of the world $S_1, S_2, \ldots, S_n$ are collectively exhaustive, so the experimental outcome O_j (if it occurs) must occur with one of the S_i (see Figure 3). Since $S_1 \cap O_j$, $S_2 \cap O_j, \ldots, S_n \cap O_j$ are mutually exclusive events, probability rule 3 implies that

$$P(O_j) = P(S_1 \cap O_j) + P(S_2 \cap O_j) + \cdots + P(S_n \cap O_j) \tag{5}$$

The probabilities of the form $P(S_i \cap O_j)$ are often referred to as **joint probabilities,** and the probabilities $P(O_j)$ are called **marginal probabilities.** Substituting (4) into (5), we obtain

$$P(O_j) = \sum_{k=1}^{k=n} P(O_j|S_k)P(S_k) \tag{6}$$

FIGURE 3
Illustration of
Equation (5)

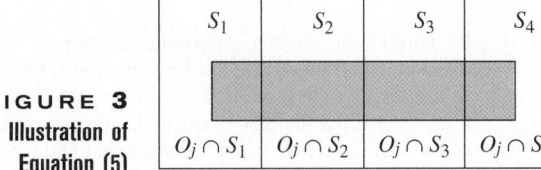

$$P(O_j) = P(O_j \cap S_1) + P(O_j \cap S_2)$$
$$+ P(O_j \cap S_3) + P(O_j \cap S_4)$$

Shaded area = outcome O_j

Substituting (4) and (6) into (3) yields **Bayes' rule:**

$$P(S_i|O_j) = \frac{P(O_j|S_i)P(S_i)}{\displaystyle\sum_{k=1}^{k=n} P(O_j|S_k)P(S_k)} \tag{7}$$

The following example illustrates the use of Bayes' rule.

EXAMPLE 4   Bayes' Rule

Suppose that 1% of all children have tuberculosis (TB). When a child who has TB is given the Mantoux test, a positive test result occurs 95% of the time. When a child who does not have TB is given the Mantoux test, a positive test result occurs 1% of the time. Given that a child is tested and a positive test result occurs, what is the probability that the child has TB?

Solution   The states of the world are

$$S_1 = \text{child has TB}$$
$$S_2 = \text{child does not have TB}$$

The possible experimental outcomes are

$$O_1 = \text{positive test result}$$
$$O_2 = \text{nonpositive test result}$$

We are given the prior probabilities $P(S_1) = .01$ and $P(S_2) = .99$ and the likelihoods $P(O_1|S_1) = .95$, $P(O_1|S_2) = .01$, $P(O_2|S_1) = .05$, and $P(O_2|S_2) = .99$. We seek $P(S_1|O_1)$. From (7),

$$P(S_1|O_1) = \frac{P(O_1|S_1)P(S_1)}{P(O_1|S_1)P(S_1) + P(O_1|S_2)P(S_2)}$$

$$= \frac{.95(.01)}{.95(.01) + .01(.99)} = \frac{95}{194} = .49$$

The reason a positive test result implies only a 49% chance that the child has TB is that many of the 99% of all children who do not have TB will test positive. For example, in a typical group of 10,000 children, 9,900 will not have TB and $.01(9,900) = 99$ children will yield a positive test result. In the same group of 10,000 children, $.01(10,000) = 100$ children will have TB and $.95(100) = 95$ children will yield a positive test result. Thus, the probability that a positive test result indicates TB is $\frac{95}{95+99} = \frac{95}{194}$.

PROBLEMS

Group A

1 A desk contains three drawers. Drawer 1 contains two gold coins. Drawer 2 contains one gold coin and one silver coin. Drawer 3 contains two silver coins. I randomly choose a drawer and then randomly choose a coin. If a silver coin is chosen, what is the probability that I chose drawer 3?

2 Cliff Colby wants to determine whether his South Japan oil field will yield oil. He has hired geologist Digger Barnes to run tests on the field. If there is oil in the field, there is a 95% chance that Digger's tests will indicate oil. If the field contains no oil, there is a 5% chance that Digger's tests will indicate oil. If Digger's tests indicate that there is no oil in the field, what is the probability that the field contains oil? Before Digger conducts the test, Cliff believes that there is a 10% chance that the field will yield oil.

3 A customer has approached a bank for a loan. Without further information, the bank believes there is a 4% chance that the customer will default on the loan. The bank can run a credit check on the customer. The check will yield either a favorable or an unfavorable report. From past experience, the bank believes that P(favorable report being received)| customer will default) $= \frac{1}{40}$, and P(favorable report| customer will not default) $= \frac{99}{100}$. If a favorable report is received, what is the probability that the customer will default on the loan?

4 Of all 40-year-old women, 1% have breast cancer. If a woman has breast cancer, a mammogram will give a positive indication for cancer 90% of the time. If a woman does not have breast cancer, a mammogram will give a positive indication for cancer 9% of the time. If a 40-year-old woman's mammogram gives a positive indication for cancer, what is the probability that she has cancer?

5 Three out of every 1,000 low-risk 50-year-old males have colon cancer. If a man has colon cancer, a test for

hidden blood in the stool will indicate hidden blood half the time. If he does not have colon cancer, a test for hidden blood in the stool will indicate hidden blood 3% of the time. If the hidden-blood test turns out positive for a low-risk 50-year-old male, what is the chance that he has colon cancer?

Group B

6 You have made it to the final round of "Let's Make a Deal." You know there is $1 million behind either door 1, door 2, or door 3. It is equally likely that the prize is behind any of the three. The two doors without a prize have nothing behind them. You randomly choose door 2, but before door 2 is opened Monte reveals that there is no prize behind door 3. You now have the opportunity to switch and choose door 1. Should you switch? Assume that Monte plays as follows: Monte knows where the prize is and will open an empty door, but he cannot open door 2. If the prize is really behind door 2, Monte is equally likely to open door 1 or door 3. If the prize is really behind door 1, Monte must open door 3. If the prize is really behind door 3, Monte must open door 1. What is your decision?

12.5 Random Variables, Mean, Variance, and Covariance

The concepts of random variables, mean, variance, and covariance are employed in several later chapters.

DEFINITION ■ A **random variable** is a function that associates a number with each point in an experiment's sample space. We denote random variables by boldface capital letters (usually **X**, **Y**, or **Z**). ■

Discrete Random Variables

DEFINITION ■ A random variable is **discrete** if it can assume only discrete values $x_1, x_2, \ldots$. A discrete random variable **X** is characterized by the fact that we know the probability that $\mathbf{X} = x_i$ (written $P(\mathbf{X} = x_1)$). ■

$P(\mathbf{X} = x_i)$ is the **probability mass function** (pmf) for the random variable **X**.

DEFINITION ■ The **cumulative distribution function** $F(x)$ for any random variable **X** is defined by $F(x) = P(\mathbf{X} \le x)$. For a discrete random variable **X**,

$$F(x) = \sum_{\substack{\text{all } x \\ \text{having } x_k \le x}} P(\mathbf{X} = x_k) \quad ■$$

An example of a discrete random variable follows.

EXAMPLE 5   Tossing a Die

Let **X** be the number of dots that show when a die is tossed. Then for $i = 1, 2, 3, 4, 5, 6$, $P(\mathbf{X} = i) = \frac{1}{6}$. The cumulative distribution function (cdf) for **X** is shown in Figure 4.

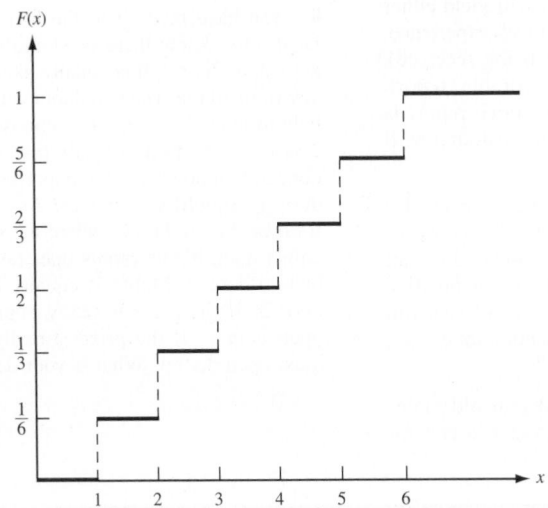

FIGURE 4
Cumulative Distribution
Function for Example 5

Continuous Random Variables

If, for some interval, the random variable **X** can assume all values on the interval, then **X** is a **continuous** random variable. Probability statements about a continuous random variable **X** require knowing **X**'s **probability density function** (pdf). The probability density function $f(x)$ for a random variable **X** may be interpreted as follows: For Δ small,

$$P(x \leq \mathbf{X} \leq x + \Delta) \cong \Delta f(x)$$

From Figure 5, we see that for a random variable **X** having density function $f(x)$,

$$\text{Area 1} = P(a \leq \mathbf{X} \leq a + \Delta) \cong \Delta f(a)$$

and

$$\text{Area 2} = P(b \leq \mathbf{X} \leq b + \Delta) \cong \Delta f(b)$$

Thus, for a random variable **X** with density function $f(x)$ as given in Figure 5, values of **X** near a are much more likely to occur than values of **X** near b.

From our previous discussion of the Fundamental Theorem of Calculus, it follows that

$$P(a \leq \mathbf{X} \leq b) = \int_a^b f(x)\, dx$$

Thus, for a continuous random variable, any area under the random variable's pdf corresponds to a probability. Using the concept of area as probability, we see that the cdf for a continuous random variable **X** with density $f(x)$ is given by

$$F(a) = P(\mathbf{X} \leq a) = \int_{-\infty}^a f(x)\, dx$$

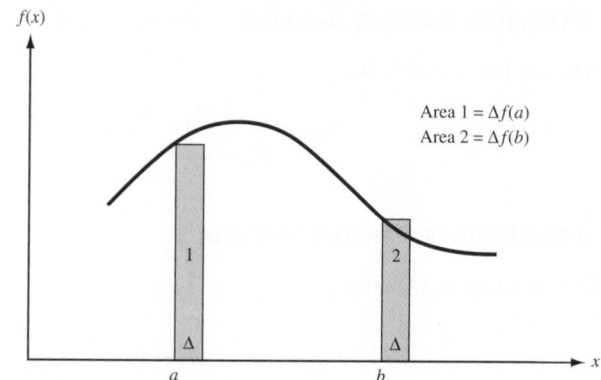

FIGURE 5
Illustration of
Probability Density
Function

EXAMPLE 6 **Cumulative Distribution Function**

Consider a continuous random variable **X** having a density function $f(x)$ given by

$$f(x) = \begin{cases} 2x & \text{if } 0 \le x \le 1 \\ 0 & \text{otherwise} \end{cases}$$

Find the cdf for **X**. Also find $P(\frac{1}{4} \le \mathbf{X} \le \frac{3}{4})$.

Solution For $a \le 0$, $F(a) = 0$. For $0 \le a \le 1$,

$$F(a) = \int_0^a 2x \, dx = a^2$$

For $a \ge 1$, $F(a) = 1$. $F(a)$ is graphed in Figure 6.

$$P(\tfrac{1}{4} \le \mathbf{X} \le \tfrac{3}{4}) = \int_{1/4}^{3/4} 2x \, dx = [x^2]_{1/4}^{3/4} = (\tfrac{9}{16}) - (\tfrac{1}{16}) = \tfrac{1}{2}$$

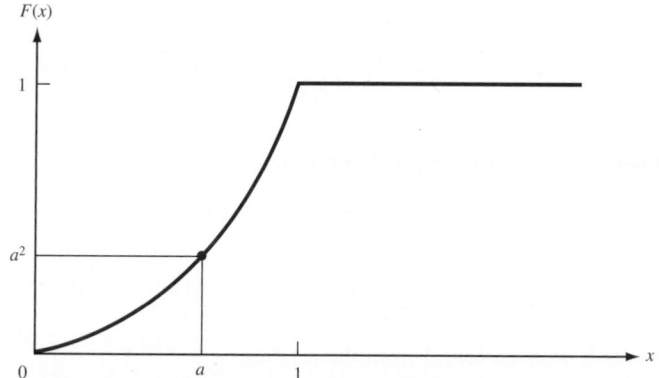

FIGURE 6
Cumulative Distribution
Function for Example 6

Mean and Variance of a Random Variable

The **mean** (or expected value) and **variance** are two important measures that are often used to summarize information contained in a random variable's probability distribution. The mean of a random variable **X** (written $E(\mathbf{X})$) is a measure of central location for the random variable.

Mean of a Discrete Random Variable

For a discrete random variable $\mathbf{X}$,

$$E(\mathbf{X}) = \sum_{\text{all } k} x_k P(\mathbf{X} = x_k) \tag{8}$$

Mean of a Continuous Random Variable

For a continuous random variable,

$$E(\mathbf{X}) = \int_{-\infty}^{\infty} x f(x)\, dx \tag{9}$$

Observe that in computing $E(\mathbf{X})$, each possible value of a random variable is weighted by its probability of occurring. Thus, the mean of a random variable is essentially the random variable's center of mass.

For a function $h(\mathbf{X})$ of a random variable $\mathbf{X}$ (such as $\mathbf{X}^2$ and $e^{\mathbf{X}}$), $E[h(\mathbf{X})]$ may be computed as follows. If $\mathbf{X}$ is a discrete random variable,

$$E[h(\mathbf{X})] = \sum_{\text{all } k} h(x_k) P(\mathbf{X} = x_k) \tag{8'}$$

If $\mathbf{X}$ is a continuous random variable,

$$E[h(\mathbf{X})] = \int_{-\infty}^{\infty} h(x) f(x)\, dx \tag{9'}$$

The variance of a random variable $\mathbf{X}$ (written as var $\mathbf{X}$) measures the dispersion or spread of $\mathbf{X}$ about $E(\mathbf{X})$. Then var $\mathbf{X}$ is defined to be $E[\mathbf{X} - E(\mathbf{X})]^2$.

Variance of a Discrete Random Variable

For a discrete random variable $\mathbf{X}$, (8′) yields

$$\text{var } \mathbf{X} = \sum_{\text{all } k} [x_k - E(\mathbf{X})]^2 P(\mathbf{X} = x_k) \tag{10}$$

Variance of a Continuous Random Variable

For a continuous random variable $\mathbf{X}$, (9′) yields

$$\text{var } \mathbf{X} = \int_{-\infty}^{\infty} [x - E(\mathbf{X})]^2 f(x)\, dx \tag{11}$$

Also, var $\mathbf{X}$ may be found from the relation

$$\text{var } \mathbf{X} = E(\mathbf{X}^2) - E(\mathbf{X})^2 \tag{12}$$

For any random variable $\mathbf{X}$, $(\text{var } \mathbf{X})^{1/2}$ is the **standard deviation** of $\mathbf{X}$ (written σ_x).

Examples 7 and 8 illustrate the computation of mean and variance for a discrete and a continuous random variable.

EXAMPLE 7 | **Discrete Random Variable**

Consider the discrete random variable $\mathbf{X}$ having $P(\mathbf{X} = i) = \frac{1}{6}$ for $i = 1, 2, 3, 4, 5, 6$. Find $E(\mathbf{X})$ and var $\mathbf{X}$.

Solution

$$E(\mathbf{X}) = (\tfrac{1}{6})(1 + 2 + 3 + 4 + 5 + 6) = \tfrac{21}{6} = \tfrac{7}{2}$$
$$\text{var } \mathbf{X} = (\tfrac{1}{6})[(1 - 3.5)^2 + (2 - 3.5)^2 + (3 - 3.5)^2$$
$$+ (4 - 3.5)^2 + (5 - 3.5)^2 + (6 - 3.5)^2] = \tfrac{35}{12}$$

EXAMPLE 8   Continuous Random Variable

Find the mean and variance for the continuous random variable $\mathbf{X}$ having the following density function:

$$f(x) = \begin{cases} 2x & \text{if } 0 \leq x \leq 1 \\ 0 & \text{otherwise} \end{cases}$$

Solution

$$E(\mathbf{X}) = \int_0^1 x(2x)\, dx = \left[\frac{2x^3}{3}\right]_0^1 = \frac{2}{3}$$

$$\text{var } \mathbf{X} = \int_0^1 \left(x - \frac{2}{3}\right)^2 2x\, dx = \int_0^1 \left(x^2 - \frac{4x}{3} + \frac{4}{9}\right) 2x\, dx$$

$$= \left[\frac{2x^4}{4} - \frac{8x^3}{9} + \frac{8x^2}{18}\right]_0^1 = \frac{1}{18}$$

Independent Random Variables

DEFINITION ■ Two random variables $\mathbf{X}$ and $\mathbf{Y}$ are **independent** if and only if for any two sets A and B,

$$P(\mathbf{X} \in A \text{ and } \mathbf{Y} \in B) = P(\mathbf{X} \in A)P(\mathbf{Y} \in B) \quad ■$$

From this definition, it can be shown that $\mathbf{X}$ and $\mathbf{Y}$ are independent random variables if and only if knowledge about the value of $\mathbf{Y}$ does not change the probability of any event involving $\mathbf{X}$. For example, suppose $\mathbf{X}$ and $\mathbf{Y}$ are independent random variables. This implies that where $\mathbf{Y} = 8$, $\mathbf{Y} = 10$, $\mathbf{Y} = 0$, or $\mathbf{Y} = $ anything else, $P(\mathbf{X} \geq 10)$ will be the same. If $\mathbf{X}$ and $\mathbf{Y}$ are independent, then $E(\mathbf{XY}) = E(\mathbf{X})E(\mathbf{Y})$. (The random variable $\mathbf{XY}$ has an expected value equal to the product of the expected value of $\mathbf{X}$ and the expected value of $\mathbf{Y}$.)

The definition of independence generalizes to situations where more than two random variables are of interest. Loosely speaking, a group of n random variables is independent if knowledge of the values of any subset of the random variables does not change our view of the distribution of any of the other random variables. (See Problem 5 at the end of this section.)

Covariance of Two Random Variables

An important concept in the study of financial models is covariance. For two random variables $\mathbf{X}$ and $\mathbf{Y}$, the **covariance** of $\mathbf{X}$ and $\mathbf{Y}$ (written $\text{cov}(\mathbf{X}, \mathbf{Y})$) is defined by

$$\text{cov}(\mathbf{X}, \mathbf{Y}) = E\{[\mathbf{X} - E(\mathbf{X})][\mathbf{Y} - E(\mathbf{Y})]\} \tag{13}$$

If $\mathbf{X} > E(\mathbf{X})$ tends to occur when $\mathbf{Y} > E(\mathbf{Y})$, and $\mathbf{X} < E(\mathbf{X})$ tends to occur when $\mathbf{Y} < E(\mathbf{Y})$, then $\text{cov}(\mathbf{X}, \mathbf{Y})$ will be positive. On the other hand, if $\mathbf{X} > E(\mathbf{X})$ tends to occur when $\mathbf{Y} < E(\mathbf{Y})$, and $\mathbf{X} < E(\mathbf{X})$ tends to occur when $\mathbf{Y} > E(\mathbf{Y})$, then $\text{cov}(\mathbf{X}, \mathbf{Y})$ will be negative. The value of cov($\mathbf{X}$, $\mathbf{Y}$) measures the association (actually, linear association) between random variables $\mathbf{X}$ and $\mathbf{Y}$. It can be shown that if $\mathbf{X}$ and $\mathbf{Y}$ are independent random variables, then cov($\mathbf{X}$, $\mathbf{Y}$) = 0. (However, cov($\mathbf{X}$, $\mathbf{Y}$) = 0 can hold even if $\mathbf{X}$ and $\mathbf{Y}$ are not independent random variables. See Problem 6 at the end of this section for an example.)

EXAMPLE 9   Gotham City Summers

Each summer in Gotham City is classified as being either a rainy summer or a sunny summer. The profits earned by Gotham City's two leading industries (the Gotham City Hotel and the Gotham City Umbrella Store) depend on the summer's weather, as shown in Table 1. Of all summers, 20% are rainy, and 80% are sunny. Let $\mathbf{H}$ and $\mathbf{U}$ be the following random variables:

$\mathbf{H}$ = profit earned by Gotham City Hotel during a summer

$\mathbf{U}$ = profit earned by Gotham City Umbrella Store during a summer

Find cov($\mathbf{H}$,$\mathbf{U}$).

Solution   We find that

$$E(\mathbf{H}) = .2(-1{,}000) + .8(2{,}000) = \$1{,}400$$
$$E(\mathbf{U}) = .2(4{,}500) + .8(-500) = \$500$$

With probability .20, Gotham City has a rainy summer. Then

$$[\mathbf{H} - E(\mathbf{H})][\mathbf{U} - E(\mathbf{U})] = (-1{,}000 - 1{,}400)(4{,}500 - 500) = -9{,}600{,}000(\text{dollars})^2$$

With probability .80, Gotham City has a sunny summer. Then

$$[\mathbf{H} - E(\mathbf{H})][\mathbf{U} - E(\mathbf{U})] = (2{,}000 - 1{,}400)(-500 - 500) = -600{,}000(\text{dollars})^2$$

Thus,

$$\text{cov}(\mathbf{H},\mathbf{U}) = E\{[\mathbf{H} - E(\mathbf{H})][\mathbf{U} - E(\mathbf{U})]\} = .20(-9{,}600{,}000) + .80(-600{,}000)$$
$$= -2{,}400{,}000(\text{dollars})^2$$

The fact that cov($\mathbf{H}$,$\mathbf{U}$) is negative indicates that when one industry does well, the other industry tends to do poorly.

TABLE 1
Profits for Gotham City Covariance

Type of Summer	Hotel Profit	Umbrella Profit
Rainy	−$1,000	$4,500
Sunny	$2,000	−$500

Mean, Variance, and Covariance for Sums of Random Variables

From given random variables $\mathbf{X}_1$ and $\mathbf{X}_2$, we often create new random variables (c is a constant): $c\mathbf{X}_1$, $\mathbf{X}_1 + c$, $\mathbf{X}_1 + \mathbf{X}_2$. The following rules can be used to express the mean, variance, and covariance of these random variables in terms of $E(\mathbf{X}_1)$, $E(\mathbf{X}_2)$, var $\mathbf{X}_1$, var $\mathbf{X}_2$, and cov($\mathbf{X}_1$, $\mathbf{X}_2$). Examples 10 and 11 illustrate the use of these rules.

$$E(c\mathbf{X}_1) = cE(\mathbf{X}_1) \tag{14}$$

$$E(\mathbf{X}_1 + c) = E(\mathbf{X}_1) + c \tag{15}$$

$$E(\mathbf{X}_1 + \mathbf{X}_2) = E(\mathbf{X}_1) + E(\mathbf{X}_2) \tag{16}$$

$$\text{var } c\mathbf{X}_1 = c^2 \text{var } \mathbf{X}_1 \tag{17}$$

$$\text{var}(\mathbf{X}_1 + c) = \text{var } \mathbf{X}_1 \tag{18}$$

If $\mathbf{X}_1$ and $\mathbf{X}_2$ are independent random variables,

$$\text{var}(\mathbf{X}_1 + \mathbf{X}_2) = \text{var } \mathbf{X}_1 + \text{var } \mathbf{X}_2 \tag{19}$$

In general,

$$\text{var}(\mathbf{X}_1 + \mathbf{X}_2) = \text{var } \mathbf{X}_1 + \text{var } \mathbf{X}_2 + 2\text{cov}(\mathbf{X}_1, \mathbf{X}_2) \tag{20}$$

For random variables $\mathbf{X}_1, \mathbf{X}_2, \ldots, \mathbf{X}_n$,

$$\text{var}(\mathbf{X}_1 + \mathbf{X}_2 + \cdots + \mathbf{X}_n) = \text{var } \mathbf{X}_1 + \text{var } \mathbf{X}_2 + \cdots + \text{var } \mathbf{X}_n + \sum_{i \neq j} \text{cov}(\mathbf{X}_i, \mathbf{X}_j) \tag{21}$$

Finally, for constants a and b,

$$\text{cov}(a\mathbf{X}_1, b\mathbf{X}_2) = ab\,\text{cov}(\mathbf{X}_1, \mathbf{X}_2) \tag{22}$$

EXAMPLE 10   Tossing a Die: Mean and Variance

I pay \$1 to play the following game: I toss a die and receive \$3 for each dot that shows. Determine the mean and variance of my profit.

Solution   Let $\mathbf{X}$ be the random variable representing the number of dots that show when the die is tossed. Then my profit is given by the value of the random variable $3\mathbf{X} - 1$. From Example 7, we know that $E(\mathbf{X}) = \frac{7}{2}$ and $\text{var } \mathbf{X} = \frac{35}{12}$. In turn, Equations (15) and (14) yield

$$E(3\mathbf{X} - 1) = E(3\mathbf{X}) - 1 = 3E(\mathbf{X}) - 1 = 3(\tfrac{7}{2}) - 1 = \tfrac{19}{2}$$

From Equations (18) and (17), respectively,

$$\text{var}(3\mathbf{X} - 1) = \text{var}(3\mathbf{X}) = 9(\text{var } \mathbf{X}) = 9(\tfrac{35}{12}) = \tfrac{315}{12}$$

EXAMPLE 11   Gotham City Profit: Mean and Variance

In Example 9, suppose I owned both the hotel and the umbrella store. Find the mean and the variance of the total profit I would earn during a summer.

Solution   My total profits are given by the random variable $\mathbf{H} + \mathbf{U}$. From Equation (16) and Example 9,

$$E(\mathbf{H} + \mathbf{U}) = E(\mathbf{H}) + E(\mathbf{U}) = 1{,}400 + 500 = \$1{,}900$$

Now

$$\text{var } \mathbf{H} = .2(-1{,}000 - 1{,}400)^2 + .8(2{,}000 - 1{,}400)^2 = 1{,}440{,}000(\text{dollars})^2$$
$$\text{var } \mathbf{U} = .2(4{,}500 - 500)^2 + .8(-500 - 500)^2 = 4{,}000{,}000(\text{dollars})^2$$

From Example 9, $\text{cov}(\mathbf{H}, \mathbf{U}) = -2{,}400{,}000 \text{ (dollars)}^2$. Then Equation (20) yields

$$\begin{aligned}
\text{var}(\mathbf{H} + \mathbf{U}) &= \text{var } \mathbf{H} + \text{var } \mathbf{U} + 2\text{cov}(\mathbf{H}, \mathbf{U}) \\
&= 1{,}440{,}000(\text{dollars})^2 + 4{,}000{,}000(\text{dollars})^2 - 2(2{,}400{,}000)(\text{dollars})^2 \\
&= 640{,}000(\text{dollars})^2
\end{aligned}$$

Thus, **H** + **U** has a smaller variance than either **H** or **U**. This is because by owning both the hotel and umbrella store, we will always have, regardless of the weather, one industry that does well and one that does poorly. This reduces the spread, or variability, of our profits.

PROBLEMS

Group A

1 I have 100 items of a product in stock. The probability mass function for the product's demand **D** is $P(\mathbf{D} = 90) = P(\mathbf{D} = 100) = P(\mathbf{D} = 110) = \frac{1}{3}$.

a Find the mass function, mean, and variance of the number of items sold.

b Find the mass function, mean, and variance of the amount of demand that will be unfilled because of lack of stock.

2 I draw 5 cards from a deck (replacing each card immediately after it is drawn). I receive \$4 for each heart that is drawn. Find the mean and variance of my total payoff.

3 Consider a continuous random variable **X** with the density function (called the *exponential density*)

$$f(x) = \begin{cases} e^{-x} & \text{if } x \geq 0 \\ 0 & \text{otherwise} \end{cases}$$

a Find and sketch the cdf for **X**.

b Find the mean and variance of **X**. (*Hint:* Use integration by parts.)

c Find $P(1 \leq \mathbf{X} \leq 2)$.

4 I have 100 units of a product in stock. The demand **D** for the item is a continuous random variable with the following density function:

$$f(d) = \begin{cases} \frac{1}{40} & \text{if } 80 \leq d \leq 120 \\ 0 & \text{otherwise} \end{cases}$$

a Find the probability that supply is insufficient to meet demand.

b What is the expected number of items sold? What is the variance of the number of items sold?

5 An urn contains 10 red balls and 30 blue balls.

a Suppose you draw 4 balls from the urn. Let $\mathbf{X}_i$ be the number of red balls drawn on the ith ball ($\mathbf{X}_i = 0$ or 1). After each ball is drawn, it is put back into the urn. Are the random variables $\mathbf{X}_1$, $\mathbf{X}_2$, $\mathbf{X}_3$, and $\mathbf{X}_4$ independent random variables?

b Repeat part (a) for the case in which the balls are not put back in the urn after being drawn.

Group B

6 Let **X** be the following discrete random variable: $P(\mathbf{X} = -1) = P(\mathbf{X} = 0) = P(\mathbf{X} = 1) = \frac{1}{3}$. Let $\mathbf{Y} = \mathbf{X}^2$. Show that $\text{cov}(\mathbf{X}, \mathbf{Y}) = 0$, but **X** and **Y** are not independent random variables.

12.6 The Normal Distribution

The most commonly used probability distribution in this book is the normal distribution. In this section, we discuss some useful properties of the normal distribution.

DEFINITION ■ A continuous random variable **X** has a normal distribution if for some μ and $\sigma > 0$, the random variable has the following density function:

$$f(x) = \frac{1}{\sigma(2\pi)^{1/2}} \exp\left[-\frac{(x - \mu)^2}{2\sigma^2}\right] \quad ■$$

If a random variable **X** is normally distributed with a mean μ and variance σ^2, we write that **X** is $N(\mu, \sigma^2)$. It can be shown that for a normal random variable, $E(\mathbf{X}) = \mu$ and var $\mathbf{X} = \sigma^2$ (the standard deviation of **X** is σ). The normal density functions for several values of σ and a single value of μ are shown in Figure 7.

For any normal distribution, the normal density is symmetric about μ (that is, $f(\mu + a) = f(\mu - a)$). Also, as σ increases, the probability that the random variable assumes a value within c of μ (for any $c > 0$) decreases. Thus, as σ increases, the normal distribution becomes more spread out. The properties are illustrated in Figure 7.

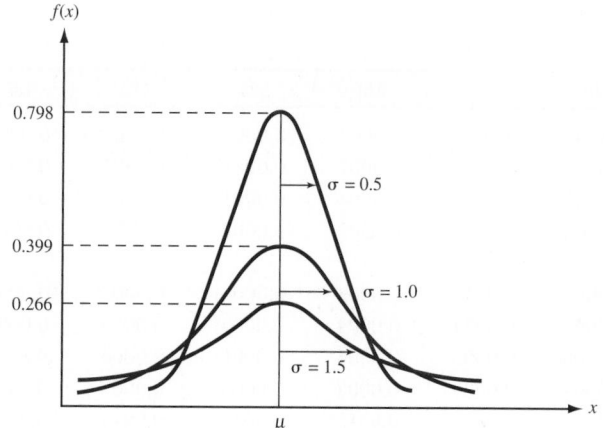

FIGURE 7
Some Examples of
Normal Distributions

Useful Properties of Normal Distributions

Property 1 If $\mathbf{X}$ is $N(\mu, \sigma^2)$, then $c\mathbf{X}$ is $N(c\mu, c^2\sigma^2)$.

Property 2 If $\mathbf{X}$ is $N(\mu, \sigma^2)$, then $\mathbf{X} + c$ (for any constant c) is $N(\mu + c, \sigma^2)$.

Property 3 If $\mathbf{X}_1$ is $N(\mu_1, \sigma_1^2)$, $\mathbf{X}_2$ is $N(\mu_2, \sigma_2^2)$, and $\mathbf{X}_1$ and $\mathbf{X}_2$ are independent, then $\mathbf{X}_1 + \mathbf{X}_2$ is $N(\mu_1 + \mu_2, \sigma_1^2 + \sigma_2^2)$.

Finding Normal Probabilities via Standardization

If $\mathbf{Z}$ is a random variable that is $N(0, 1)$, then $\mathbf{Z}$ is said to be a standardized normal random variable. In Table 2, $F(z) = P(\mathbf{Z} \leq z)$ is tabulated. For example,

$$P(\mathbf{Z} \leq -1) = F(-1) = .1587$$

and

$$P(\mathbf{Z} \geq 2) = 1 - P(\mathbf{Z} \leq 2) = 1 - F(2) = 1 - .9772 = .0228.$$

If $\mathbf{X}$ is $N(\mu, \sigma^2)$, then $(\mathbf{X} - \mu)/\sigma$ is $N(0, 1)$. This follows, because by property 2 of the normal distribution, $\mathbf{X} - \mu$ is $N(\mu - \mu, \sigma^2) = N(0, \sigma^2)$. Then by property 1, $\frac{x-\mu}{\sigma}$ is $N(\frac{0}{\sigma}, \frac{\sigma^2}{\sigma^2}) = N(0, 1)$. The last equality enables us to use Table 2 to find probabilities for any normal random variable, not just an $N(0, 1)$ random variable. Suppose $\mathbf{X}$ is $N(\mu, \sigma^2)$ and we want to find $P(a \leq \mathbf{X} \leq b)$. To find this probability from Table 2, we use the following relations (this procedure is called **standardization**):

$$P(a \leq \mathbf{X} \leq b) = P\left(\frac{a - \mu}{\sigma} \leq \frac{\mathbf{X} - \mu}{\sigma} \leq \frac{b - \mu}{\sigma}\right)$$

$$= P\left(\frac{a - \mu}{\sigma} \leq \mathbf{Z} \leq \frac{b - \mu}{\sigma}\right)$$

$$= F\left(\frac{b - \mu}{\sigma}\right) - F\left(\frac{a - \mu}{\sigma}\right)$$

The Central Limit Theorem

If $\mathbf{X}_1, \mathbf{X}_2, \ldots, \mathbf{X}_n$ are independent random variables, then for n sufficiently large (usually $n \geq 30$ will do, but the actual size of n depends on the distributions of $\mathbf{X}_1, \mathbf{X}_2, \ldots, \mathbf{X}_n$),

TABLE 2
Standard Normal Cumulative Probabilities[†]

z	0.00	0.01	0.02	0.03	0.04	0.05	0.06	0.07	0.08	0.09
−3.8	0.0001	0.0001	0.0001	0.0001	0.0001	0.0001	0.0001	0.0001	0.0001	0.0001
−3.7	0.0001	0.0001	0.0001	0.0001	0.0001	0.0001	0.0001	0.0001	0.0001	0.0001
−3.6	0.0002	0.0002	0.0001	0.0001	0.0001	0.0001	0.0001	0.0001	0.0001	0.0001
−3.5	0.0002	0.0002	0.0002	0.0002	0.0002	0.0002	0.0002	0.0002	0.0002	0.0002
−3.4	0.0003	0.0003	0.0003	0.0003	0.0003	0.0003	0.0003	0.0003	0.0003	0.0002
−3.3	0.0005	0.0005	0.0005	0.0004	0.0004	0.0004	0.0004	0.0004	0.0004	0.0003
−3.2	0.0007	0.0007	0.0006	0.0006	0.0006	0.0006	0.0006	0.0005	0.0005	0.0005
−3.1	0.0010	0.0009	0.0009	0.0009	0.0008	0.0008	0.0008	0.0008	0.0007	0.0007
−3.0	0.0014	0.0013	0.0013	0.0012	0.0012	0.0011	0.0011	0.0011	0.0010	0.0010
−2.9	0.0019	0.0018	0.0018	0.0017	0.0016	0.0016	0.0015	0.0015	0.0014	0.0014
−2.8	0.0026	0.0025	0.0024	0.0023	0.0023	0.0022	0.0021	0.0021	0.0020	0.0019
−2.7	0.0035	0.0034	0.0033	0.0032	0.0031	0.0030	0.0029	0.0028	0.0027	0.0026
−2.6	0.0047	0.0045	0.0044	0.0043	0.0041	0.0040	0.0039	0.0038	0.0037	0.0036
−2.5	0.0062	0.0060	0.0059	0.0057	0.0055	0.0054	0.0052	0.0051	0.0049	0.0048
−2.4	0.0082	0.0080	0.0078	0.0076	0.0073	0.0071	0.0069	0.0068	0.0066	0.0064
−2.3	0.0107	0.0104	0.0102	0.0099	0.0096	0.0094	0.0091	0.0089	0.0087	0.0084
−2.2	0.0139	0.0136	0.0132	0.0129	0.0125	0.0122	0.0119	0.0116	0.0113	0.0110
−2.1	0.0179	0.0174	0.0170	0.0166	0.0162	0.0158	0.0154	0.0150	0.0146	0.0143
−2.0	0.0228	0.0222	0.0217	0.0212	0.0207	0.0202	0.0197	0.0192	0.0188	0.0183
−1.9	0.0287	0.0281	0.0274	0.0268	0.0262	0.0256	0.0250	0.0244	0.0239	0.0233
−1.8	0.0359	0.0351	0.0344	0.0336	0.0329	0.0322	0.0314	0.0307	0.0301	0.0294
−1.7	0.0446	0.0436	0.0427	0.0418	0.0409	0.0401	0.0392	0.0384	0.0375	0.0367
−1.6	0.0548	0.0537	0.0526	0.0516	0.0505	0.0495	0.0485	0.0475	0.0465	0.0455
−1.5	0.0668	0.0655	0.0643	0.0630	0.0618	0.0606	0.0594	0.0582	0.0571	0.0559
−1.4	0.0808	0.0793	0.0778	0.0764	0.0749	0.0735	0.0721	0.0708	0.0694	0.0681
−1.3	0.0968	0.0951	0.0934	0.0918	0.0901	0.0885	0.0869	0.0853	0.0838	0.0823
−1.2	0.1151	0.1131	0.1112	0.1093	0.1075	0.1057	0.1038	0.1020	0.1003	0.0985
−1.1	0.1357	0.1335	0.1314	0.1292	0.1271	0.1251	0.1230	0.1210	0.1190	0.1170
−1.0	0.1587	0.1562	0.1539	0.1515	0.1492	0.1469	0.1446	0.1423	0.1401	0.1379
−0.9	0.1841	0.1814	0.1788	0.1762	0.1736	0.1711	0.1685	0.1660	0.1635	0.1611
−0.8	0.2119	0.2090	0.2061	0.2033	0.2005	0.1977	0.1949	0.1922	0.1894	0.1867
−0.7	0.2420	0.2389	0.2358	0.2327	0.2297	0.2266	0.2236	0.2206	0.2177	0.2148
−0.6	0.2743	0.2709	0.2676	0.2643	0.2611	0.2578	0.2546	0.2514	0.2483	0.2451
−0.5	0.3085	0.3050	0.3015	0.2981	0.2946	0.2912	0.2877	0.2843	0.2810	0.2776
−0.4	0.3446	0.3409	0.3372	0.3336	0.3300	0.3264	0.3228	0.3192	0.3156	0.3121
−0.3	0.3821	0.3783	0.3745	0.3707	0.3669	0.3632	0.3594	0.3557	0.3520	0.3483
−0.2	0.4207	0.4168	0.4129	0.4090	0.4052	0.4013	0.3974	0.3936	0.3897	0.3859
−0.1	0.4602	0.4562	0.4522	0.4483	0.4443	0.4404	0.4364	0.4325	0.4286	0.4247
−0.0	0.5000	0.4960	0.4920	0.4880	0.4840	0.4801	0.4761	0.4721	0.4681	0.4641

Source: Reprinted by permission from David E. Kleinbaum, Lawrence L. Kupper, and Keith E. Muller, *Applied Regression Analysis and Other Multivariable Methods,* 2nd edition. Copyright © 1988 PWS-KENT Publishing Company.

[†]*Note:* Table entry is the area under the standard normal curve to the left of the indicated z-value, thus giving $P(Z \leq z)$.

TABLE 2

Standard Normal Cumulative Probabilities (Continued)

z	0.00	0.01	0.02	0.03	0.04	0.05	0.06	0.07	0.08	0.09
0.0	0.5000	0.5040	0.5080	0.5120	0.5160	0.5199	0.5239	0.5279	0.5319	0.5359
0.1	0.5398	0.5438	0.5478	0.5517	0.5557	0.5596	0.5636	0.5675	0.5714	0.5753
0.2	0.5793	0.5832	0.5871	0.5910	0.5948	0.5987	0.6026	0.6064	0.6103	0.6141
0.3	0.6179	0.6217	0.6255	0.6293	0.6331	0.6368	0.6406	0.6443	0.6480	0.6517
0.4	0.6554	0.6591	0.6628	0.6664	0.6700	0.6736	0.6772	0.6808	0.6844	0.6879
0.5	0.6915	0.6950	0.6985	0.7019	0.7054	0.7088	0.7123	0.7157	0.7190	0.7224
0.6	0.7257	0.7291	0.7324	0.7357	0.7389	0.7422	0.7454	0.7486	0.7517	0.7549
0.7	0.7580	0.7611	0.7642	0.7673	0.7703	0.7734	0.7764	0.7794	0.7823	0.7852
0.8	0.7881	0.7910	0.7939	0.7967	0.7995	0.8023	0.8051	0.8078	0.8106	0.8133
0.9	0.8159	0.8186	0.8212	0.8238	0.8264	0.8289	0.8315	0.8340	0.8365	0.8389
1.0	0.8413	0.8438	0.8461	0.8485	0.8508	0.8531	0.8554	0.8577	0.8599	0.8621
1.1	0.8643	0.8665	0.8686	0.8708	0.8729	0.8749	0.8770	0.8790	0.8810	0.8830
1.2	0.8849	0.8869	0.8888	0.8907	0.8925	0.8943	0.8962	0.8980	0.8997	0.9015
1.3	0.9032	0.9049	0.9066	0.9082	0.9099	0.9115	0.9131	0.9147	0.9162	0.9177
1.4	0.9192	0.9207	0.9222	0.9236	0.9251	0.9265	0.9279	0.9292	0.9306	0.9319
1.5	0.9332	0.9345	0.9357	0.9370	0.9382	0.9394	0.9406	0.9418	0.9429	0.9441
1.6	0.9452	0.9463	0.9474	0.9484	0.9495	0.9505	0.9515	0.9525	0.9535	0.9545
1.7	0.9554	0.9564	0.9673	0.9582	0.9591	0.9599	0.9608	0.9616	0.9625	0.9633
1.8	0.9641	0.9649	0.9656	0.9664	0.9671	0.9678	0.9686	0.9683	0.9699	0.9706
1.9	0.9713	0.9719	0.9726	0.9732	0.9738	0.9744	0.9750	0.9756	0.9762	0.9767
2.0	0.9772	0.9778	0.9783	0.9788	0.9793	0.9798	0.9803	0.9808	0.9812	0.9817
2.1	0.9821	0.9826	0.9830	0.9834	0.9838	0.9842	0.9846	0.9850	0.9854	0.9857
2.2	0.9861	0.9864	0.9868	0.9871	0.9875	0.9878	0.9881	0.9884	0.9887	0.9890
2.3	0.9893	0.9896	0.9898	0.9901	0.9904	0.9906	0.9909	0.9911	0.9913	0.9916
2.4	0.9918	0.9920	0.9922	0.9924	0.9927	0.9929	0.9931	0.9932	0.9934	0.9936
2.5	0.9938	0.9940	0.9941	0.9943	0.9945	0.9946	0.9948	0.9949	0.9951	0.9952
2.6	0.9953	0.9955	0.9956	0.9957	0.9959	0.9960	0.9961	0.9962	0.9963	0.9964
2.7	0.9965	0.9966	0.9967	0.9968	0.9969	0.9970	0.9971	0.9972	0.9973	0.9974
2.8	0.9974	0.9975	0.9976	0.9977	0.9977	0.9978	0.9979	0.9979	0.9980	0.9981
2.9	0.9981	0.9982	0.9982	0.9983	0.9984	0.9984	0.9985	0.9985	0.9986	0.9986
3.0	0.9986	0.9987	0.9987	0.9988	0.9988	0.9989	0.9989	0.9989	0.9990	0.9990
3.1	0.9990	0.9991	0.9991	0.9991	0.9992	0.9992	0.9992	0.9992	0.9993	0.9993
3.2	0.9993	0.9993	0.9994	0.9994	0.9994	0.9994	0.9994	0.9995	0.9995	0.9995
3.3	0.9995	0.9995	0.9995	0.9996	0.9996	0.9996	0.9996	0.9996	0.9996	0.9997
3.4	0.9997	0.9997	0.9997	0.9997	0.9997	0.9997	0.9997	0.9997	0.9997	0.9998
3.5	0.9998	0.9998	0.9998	0.9998	0.9998	0.9998	0.9998	0.9998	0.9998	0.9998
3.6	0.9998	0.9998	0.9999	0.9999	0.9999	0.9999	0.9999	0.9999	0.9999	0.9999
3.7	0.9999	0.9999	0.9999	0.9999	0.9999	0.9999	0.9999	0.9999	0.9999	0.9999
3.8	0.9999	0.9999	0.9999	0.9999	0.9999	0.9999	0.9999	0.9999	0.9999	0.9999
3.9	1.0000									

the random variable $\mathbf{X} = \mathbf{X}_1 + \mathbf{X}_2 + \cdots + \mathbf{X}_n$ may be closely approximated by a normal random variable $\mathbf{X}'$ that has $E(\mathbf{X}') = E(\mathbf{X}_1) + E(\mathbf{X}_2) + \cdots + E(\mathbf{X}_n)$ and var $\mathbf{X}' =$ var $\mathbf{X}_1 +$ var $\mathbf{X}_2 + \cdots +$ var $\mathbf{X}_n$. This result is known as the Central Limit Theorem. When we say that $\mathbf{X}'$ closely approximates $\mathbf{X}$, we mean that $P(a \leq \mathbf{X} \leq b)$ is close to $P(a \leq \mathbf{X}' \leq b)$.

Finding Normal Probabilities with Excel

Probabilities involving a standard normal variable can be determined with Excel, using the =NORMSDIST function. The S in NORMSDIST stands for *standardized normal.* For example, $P(\mathbf{Z} \leq -1)$ can be found by entering the formula

$$=\text{NORMSDIST}(-1)$$

Excel returns the value .1587. See Figure 8 and file Normal.xls.

The =NORMDIST function can be used to determine a normal probability for any normal (not just a standard normal) random variable. If $\mathbf{X}$ is $N(\mu, \sigma^2)$, then entering the formula

$$=\text{NORMSDIST}(a,\mu,\sigma,1)$$

will return $P(\mathbf{X} \leq a)$. The "1" ensures that Excel returns the cumulative normal probability. Changing the last argument to "0" causes Excel to return the height of the normal density function for $\mathbf{X} = a$. As an example, we know that IQs follow $N(100, 225)$. The fraction of people with IQs of 90 or less is computed with the formula

$$=\text{NORMDIST}(90,100,15,1)$$

Excel yields .2525. See Figure 8 and file Normal.xls.

The height of the density for $N(100, 225)$ for $\mathbf{X} = 100$ is computed with the formula

$$=\text{NORMDIST}(100,100,15,0)$$

Excel yields .026596.

By varying the first argument in the =NORMDIST function, we may graph a normal density. See Figure 9 and sheet density of file Normal.xls.

Consider a given normal random variable $\mathbf{X}$, with mean μ and standard deviation σ. In many situations, we want to answer questions such as the following. (1) Eli Lilly believes that the year's demand for Prozac will be normally distributed, with $\mu = 60$ million d.o.t. (days of therapy) and $\sigma = 5$ million d.o.t. How many units should be produced this year if Lilly wants to have only a 1% chance of running out of Prozac? (2) Family income in Bloomington is normally distributed, with $\mu = \$30,000$ and $\sigma = \$8,000$. The poorest 10% of all families in Bloomington are eligible for federal aid. What should the aid cutoff be?

In the first example, we want the 99th percentile of Prozac demand. That is, we seek the number $\mathbf{X}$ such that there is only a 1% chance that demand will exceed $\mathbf{X}$ and a 99% chance

	E	F	G	H
7				
8				
9	P(Z<=-1)	0.158655	normsdist(-1)	
10	P(IQ<90)	0.252492	normdist(90,100,15,1)	
11	density for IQ=100	0.026596	normdist(100,100,15,0)	

FIGURE 8

FIGURE **9**

	D	E	F	G	H	I	J	K	L	M	N	O	P
1													
2		IQ	Density										
3		45	3.2018E-05										
4		50	0.00010282										
5		55	0.00029546										
6		60	0.00075973										
7		65	0.00174813										
8		70	0.0035994										
9		75	0.00663181										
10		80	0.010934										
11		85	0.01613138										
12		90	0.02129653										
13		95	0.02515888										
14		100	0.02659615										
15		105	0.02515888										
16		110	0.02129653										
17		115	0.01613138										
18		120	0.010934										
19		125	0.00663181										
20		130	0.0035994										
21		135	0.00174813										
22		140	0.00075973										
23		145	0.00029546										
24		150	0.00010282										
25		155	3.2018E-05										

Normal Density for IQ's

that it will be less than **X**. In the second example, we want the 10th percentile of family income in Bloomington. That is, we seek the number **X** such that there is only a 10% chance that family income will be less than **X** and a 90% chance that it will exceed **X**.

Suppose we want to find the pth percentile (expressed as a decimal) of a normal random variable **X** with mean μ and standard deviation σ. Simply enter the following formula into Excel:

$$=\text{NORMINV}(p,\mu,\sigma)$$

This will return the number x having the property that $P(\mathbf{X} \leq x) = p$, as desired. We now can solve the two examples described above.

EXAMPLE 12   **Prozac Demand**

Eli Lilly believes that the year's demand for Prozac will be normally distributed, with $\mu = 60$ million d.o.t. (days of therapy) and $\sigma = 5$ million d.o.t. How many units should be produced this year if Lilly wants to have only a 1% chance of running out of Prozac?

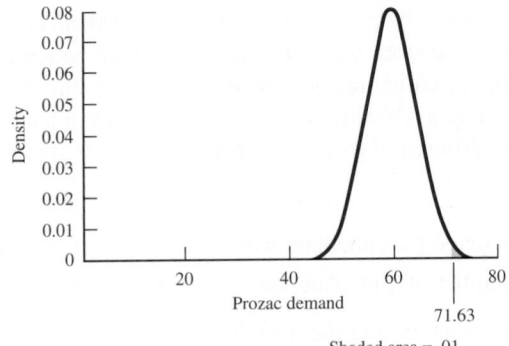

FIGURE 10
99th Percentile of
Prozac Demand

Solution Letting $\mathbf{X}$ = annual demand for Prozac, we seek a value x such that $P(\mathbf{X} \geq x) = .01$ or $P(\mathbf{X} \leq x) = .99$. Thus, we seek the 99th percentile of Prozac demand, which we find (in millions) with the formula

$$=\text{NORMINV}(.99,60,5)$$

Excel returns 71.63, so Lilly must produce 71,630,000 d.o.t. This assumes, of course, that Lilly begins the year with no Prozac on hand. If the company had a beginning inventory of 10 million d.o.t., it would need to produce 61,630,000 d.o.t. during the current year. Figure 10 displays the 99th percentile of Prozac demand.

EXAMPLE 13   Family Income

Family income in Bloomington is normally distributed, with μ = \$30,000 and σ = \$8,000. The poorest 10% of all families in Bloomington are eligible for federal aid. What should the aid cutoff be?

Solution If $\mathbf{X}$ = income of a Bloomington family, we seek an x such that $P(\mathbf{X} \leq x) = .10$. Thus, we seek the 10th percentile of Bloomington family income, which we find with the statement

$$=\text{NORMINV}(.10,30000,8000)$$

Excel returns \$19,747.59. Thus, aid should be given to all families with incomes smaller than \$19,749.59. Figure 11 displays the 10th percentile of family income.

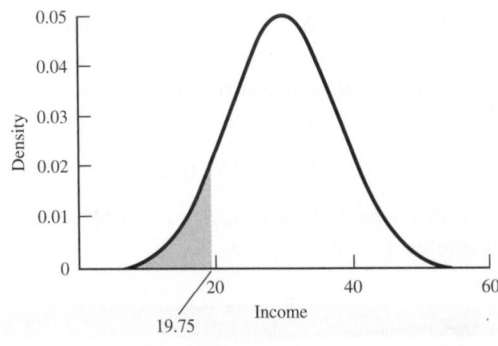

FIGURE 11
10th Percentile of
Family Income

EXAMPLE 14   Stocking Chocolate Bars

Daily demand for chocolate bars at the Gillis Grocery has a mean of 100 and a variance of 3,000 (chocolate bars)^2. At present, the store has 3,500 chocolate bars in stock. What is the probability that the store will run out of chocolate bars during the next 30 days? Also, how many should Gillis have on hand at the beginning of a 30-day period if the store wants to have only a 1% chance of running out during the 30-day period? Assume that the demands on different days are independent random variables.

Solution Let

$$\mathbf{X}_i = \text{demand for chocolate bars on day } i \quad (i = 1, 2, \ldots, 30)$$
$$\mathbf{X} = \text{number of chocolate bars demanded in next 30 days}$$

Gillis will run out of stock during the next 30 days if $\mathbf{X} \geq 3,500$. The Central Limit Theorem implies that $\mathbf{X} = \mathbf{X}_1 + \mathbf{X}_2 + \cdots + \mathbf{X}_{30}$ can be closely approximated by a normal distribution $\mathbf{X}'$ with $E(\mathbf{X}') = 30(100) = 3,000$ and var $\mathbf{X}' = 30(3,000) = 90,000$ and

$\sigma_{x'} = (90,000)^{1/2} = 300$. Then we approximate the probability that Gillis will run out of stock during the next 30 days by

$$P(\mathbf{X}' \geq 3,500) = P\left(\frac{\mathbf{X}' - 3000}{300} \geq \frac{3,500 - 3,000}{300}\right)$$
$$= P(\mathbf{Z} \geq 1.67) = 1 - P(\mathbf{Z} \leq 1.67)$$
$$= 1 - F(1.67) = 1 - .9525 = .0475$$

Let c = number of chocolate bars that should be stocked to have only a 1% chance of running out of chocolate bars within the next 30 days. We seek c satisfying $P(\mathbf{X}' \geq c) = .01$, or

$$P\left(\frac{\mathbf{X}' - 3,000}{300} \geq \frac{c - 3,000}{300}\right) = .01$$

This is equivalent to

$$P\left(\mathbf{Z} \geq \frac{c - 3,000}{300}\right) = .01$$

Since $F(2.33) = P(\mathbf{Z} \leq 2.33) = .99$,

$$\frac{c - 3,000}{300} = 2.33 \quad \text{or} \quad c = 3,699$$

Thus, if Gillis has 3,699 chocolate bars in stock, there is a 1% probability that the store will run out during the next 30 days. (We have defined running out of chocolate bars as having no chocolate bars left at the end of 30 days.)

Alternatively, we could find the probability that the demand is at least 3,500 with the Excel formula

$$=1 - \text{NORMDIST}(3500,3000,300,1)$$

This formula returns .0475.

We could also have used Excel to determine the level that must be stocked to have a 1% chance of running out as the 99th percentile of the demand distribution. Simply use the formula

$$=\text{NORMINV}(.99,3000,300)$$

This formula returns the value 3,699.

PROBLEMS

Group A

1 The daily demand for milk (in gallons) at Gillis Grocery is $N(1,000, 100)$. How many gallons must be in stock at the beginning of the day if Gillis is to have only a 5% chance of running out of milk by the end of the day?

2 Before burning out, a light bulb gives $\mathbf{X}$ hours of light, where $\mathbf{X}$ is $N(500, 400)$. If we have 3 bulbs, what is the probability that they will give a total of at least 1,460 hours of light?

Group B

3 The number of traffic accidents occurring in Bloomington in a single day has a mean and a variance of 3. What is the probability that during a given year (365-day period), there will be at least 1,000 traffic accidents in Bloomington?

4 Suppose that the number of ounces of soda put into a Pepsi can is normally distributed, with $\mu = 12.05$ oz and $\sigma = .03$ oz.

    a Legally, a can must contain at least 12 oz of soda. What fraction of cans will contain at least 12 oz of soda?

b What fraction of cans will contain under 11.9 oz of soda?

c What fraction of cans will contain between 12 and 12.08 oz of soda?

d 1% of all cans will contain more than ——— oz.

e 10% of all cans will contain less than ——— oz.

f Pepsi controls the mean content in a can by setting a timer. For what mean should the timer be set so that only 1 in 1,000 cans will be underfilled?

g Every day, Pepsi produces 10,000 cans. The government inspects 10 randomly chosen cans per day. If at least two are underfilled, Pepsi is fined $10,000. Given that $\mu = 12.05$ oz and $\sigma = .03$ oz, what is the chance that Pepsi will be fined on a given day?

5 Suppose the annual return on Disney stock follows a normal distribution, with mean .12 and standard deviation .30.

a What is the probability that Disney's value will decrease during a year?

b What is the probability that the return on Disney during a year will be at least 20%?

c What is the probability that the return on Disney during a year will be between −6% and 9%?

d There is a 5% chance that the return on Disney during a year will be greater than or equal to ——— .

e There is a 1% chance that the return on Disney during a year will be less than ——— .

f There is a 95% chance that the return on Disney during a year will be between ——— and ——— .

6 The daily demand for six-packs of Coke at Mr. D's follows a normal distribution, with a mean of 120 and a standard deviation of 30. Every Monday, the delivery driver delivers Coke to Mr. D's. If the store wants to have only a 1% chance of running out of Coke by the end of the week, how many six-packs should be ordered for the week? (Assume that orders can be placed Sunday at midnight.)

7 The Coke factory fills bottles of soda by setting a timer on a filling machine. It has been observed that the number of ounces the machine puts in a bottle has a standard deviation of .05 oz. If 99.9% of all bottles are to have at least 16 oz of soda, to what amount should the average amount be set? (*Hint:* Use the Excel Goal Seek feature.)

8 We assemble a large part by joining two smaller parts together. In the past, the smaller parts we have produced have had a mean length of 1″ and a standard deviation of .01″. Assume that the lengths of the smaller parts are normally distributed and are independent.

a What fraction of the larger parts are more than 2.05″ in diameter?

b What fraction of the larger parts are between 1.96″ and 2.02″ in diameter?

9 Weekly Ford sales follow a normal distribution, with a mean of 50,000 cars and a standard deviation of 14,000 cars.

a There is a 1% chance that Ford will sell more than ——— cars during the next year.

b The chance that Ford will sell between 2.4 and 2.7 million cars during the next year is——— .

10 Warren Dinner has invested in nine different investments. The profits earned on the different investments are independent. The return on each investment follows a normal distribution, with a mean of $500 and a standard deviation of $100.

a There is a 1% chance that the total return on the nine investment is less than——— .

b The probability that Warren's total return is between $4,000 and $5,200 is——— .

12.7 z-Transforms

Consider a discrete random variable **X** whose only possible values are nonnegative integers. For $n = 0, 1, 2, \ldots$, let $P(\mathbf{X} = n) = a_n$. We define (for $|z| \leq 1$) the **z-transform of X** (call it $p_{\mathbf{X}}^T(z)$) to be

$$E(z^{\mathbf{X}}) = \sum_{n=0}^{n=\infty} a_n z^n$$

To see why z-transforms are useful, note that

$$\left[\frac{dp_{\mathbf{X}}^T(z)}{dz}\right]_{z=1} = \left[\sum_{n=1}^{n=\infty} nz^{n-1}a_n\right]_{z=1} = E(\mathbf{X})$$

Also note that

$$\left[\frac{d^2 p_{\mathbf{X}}^T(z)}{dz^2}\right]_{z=1} = \left[\sum_{n=1}^{n=\infty} n(n-1)z^{n-2}a_n\right]_{z=1} = E(\mathbf{X}^2) - E(\mathbf{X})$$

This implies that we can find the mean, the second moment ($E(\mathbf{X}^2)$), and variance of $\mathbf{X}$ from the following relationships:

$$E(\mathbf{X}) = \left[\frac{dp_{\mathbf{X}}^T(z)}{dz} \right]_{z=1} \tag{23}$$

$$E(\mathbf{X}^2) = \left[\frac{d^2 p_{\mathbf{X}}^T(z)}{dz^2} \right]_{z=1} + \left[\frac{dp_{\mathbf{X}}^T(z)}{dz} \right]_{z=1} \tag{24}$$

$$\text{var } \mathbf{X} = \left[\frac{d^2 p_{\mathbf{X}}^T(z)}{dz^2} \right]_{z=1} + \left[\frac{dp_{\mathbf{X}}^T(z)}{dz} \right]_{z=1} - \left(\left[\frac{dp_{\mathbf{X}}^T(z)}{dz} \right]_{z=1} \right)^2 \tag{25}$$

The following examples illustrate the power of z-transforms.

EXAMPLE 15   *z*-Transform for the Binomial Random Variable

Suppose we toss a coin n times, and the probability of obtaining heads each time is p. Let $q = 1 - p$. If successive coin tosses are independent events, then the mass function describing the random variable $\mathbf{X} =$ number of heads is the well-known **binomial random variable** defined by

$$P(\mathbf{X} = j) = \frac{n!}{j!(n-j)!} \, p^j (q)^{n-j}, j = 0, 1, 2, \ldots, n$$

The z-transform for the random variable $\mathbf{X}$ is given by

$$p_{\mathbf{X}}^T(z) = \sum_{j=0}^{j=n} \frac{n!}{j!(n-j)!} \, p^j (q)^{n-j} z^j = \sum_{j=0}^{j=n} \frac{n!}{j!(n-j)!} \, (pz)^j (q)^{n-j} = (pz + q)^n$$

We can now use the z-transform to determine the mean and variance of the binomial random variable. Note that

$$\frac{dp_{\mathbf{X}}^T(z)}{dz} = np(pz + q)^{n-1} \quad \text{and} \quad \frac{d^2 p_{\mathbf{X}}^T(z)}{dz^2} = n(n-1)p^2(pz + q)^{n-2}$$

For $z = 1$, we find

$$\frac{dp_{\mathbf{X}}^T(z)}{dz} = np \quad \text{and} \quad \frac{d^2 p_{\mathbf{X}}^T(z)}{dz^2} = n(n-1)p^2$$

Then from (23), we find $E(\mathbf{X}) = np$, and from (25), we find that var $\mathbf{X} = n(n-1)p^2 + np - (np)^2 = npq$.

EXAMPLE 16   *z*-Transform for a Geometric Random Variable

Let the random variable $\mathbf{X}$ be defined as the number of coin tosses needed to obtain the first heads, given that successive tosses are independent, the probability that each toss is heads is given by p, and the probability that each coin is tails is given by $q = 1 - p$. Then $\mathbf{X}$ follows a **geometric random variable**, where $P(\mathbf{X} = j) = pq^{j-1}$ ($j = 1, 2, \ldots, n$).

Then $p_{\mathbf{X}}^T(z) = \sum_{j=1}^{j=\infty} pq^{j-1}z^j$. For $x < 1$, we know that $a + ax + ax^2 + \cdots = \dfrac{a}{1 - x}$. Therefore, $p_{\mathbf{X}}^T(z) = \dfrac{pz}{1 - qz}$. We find that

$$\frac{dp_{\mathbf{X}}^T(z)}{dz} = \frac{p}{(1 - qz)^2} \quad \text{and} \quad \frac{d^2 p_{\mathbf{X}}^T(z)}{dz^2} = \frac{2 pq}{(1 - qz)^3}$$

Letting $z = 1$ (Equation (23)) tells us that

$$E(\mathbf{X}) = \frac{p}{p^2} = \frac{1}{p} \quad \text{and} \quad \text{var } \mathbf{X} = \frac{2pq}{p^3} + \frac{1}{p} - \frac{1}{p^2} = \frac{q}{p^2}$$

Suppose $\mathbf{X}_1, \mathbf{X}_2, \ldots, \mathbf{X}_n$ are independent random variables. Let $\mathbf{S} = \mathbf{X}_1 + \mathbf{X}_2 + \cdots + \mathbf{X}_n$. Then it is easy to prove (see Problem 2) that

$$p_{\mathbf{S}}^T(z) = p_{\mathbf{X}_1}^T(z) \cdots p_{\mathbf{X}_n}^T(z) \tag{26}$$

To see the usefulness of this result, reconsider Example 15. Let $\mathbf{X}_i =$ number of heads on the ith toss of a coin. Then number of heads on n tosses of a coin is given by $\mathbf{X}_1 + \mathbf{X}_2 + \cdots + \mathbf{X}_n$. For each $\mathbf{X}_i$, we have that $p_{\mathbf{X}_i}^T(z) = pz + q$. Then from (26), we find that $p_{\mathbf{X}}^T(z) = (pz + q)^n$. Of course, this agrees with the z-transform we obtained in Example 15.

PROBLEMS

Group A

1 For a given μ, the **Poisson random variable** has the mass function $P(\mathbf{X} = k) = e^{-\mu} \dfrac{\mu^n}{n!}$ $(k = 0, 1, 2, \ldots)$. Find the mean and variance of a Poisson random variable.

Group B

2 Prove Equation (26).

3 Suppose we toss a coin. Successive coin tosses are independent and yield heads with probability p. The negative binomial random variable with parameter k assumes a value n if it takes n failures until the kth success occurs. Use z-transforms to determine the probability mass function for the negative binomial random variable.

Hint: The number of ways of making k choices from the numbers $0, 1, \ldots, n$ add up to n is given by $\dfrac{(n + k - 1)!}{n!(k - 1)!}$.

SUMMARY   Formulas for Determining Indefinite Integrals

$$\int (1) \, dx = x + C$$

$$\int af(x) \, dx = a \int f(x) \, dx \qquad (a \text{ is any constant})$$

$$\int [f(x) + g(x)] \, dx = \int f(x) \, dx + \int g(x) \, dx$$

$$\int x^n \, dx = \frac{x^{n+1}}{n + 1} + C \qquad (n \neq -1)$$

$$\int x^{-1} \, dx = \ln x + C$$

$$\int e^x \, dx = e^x + C$$

$$\int a^x \, dx = \frac{a^x}{\ln a} + C \qquad (a > 0, a \neq 1)$$

$$\int [f(x)]^n f'(x)\, dx = \frac{[f(x)]^{n+1}}{n+1} + C \qquad (n \neq -1)$$

$$\int f(x)^{-1} f'(x)\, dx = \ln f(x) + C$$

For two functions $u(x)$ and $v(x)$,

$$\int u(x)v'(x)\, dx = u(x)v(x) - \int v(x)u'(x)\, dx \qquad \text{(Integration by parts)}$$

$$\int e^{f(x)} f'(x)\, dx = e^{f(x)} + C$$

$$\int a^{f(x)} f'(x)\, dx = \frac{a^{f(x)}}{\ln a} + C \qquad (a > 0,\ a \neq 1)$$

Leibniz's Rule for Differentiating an Integral

If $\qquad F(y) = \displaystyle\int_{g(y)}^{h(y)} f(x, y)\, dx,$ $\qquad$ then

$$F'(y) = h'(y)f(h(y),\ y) - g'(y)f(g(y),\ y) + \int_{g(y)}^{h(y)} \frac{\partial f(x, y)}{\partial y}\, dx$$

Probability

Basic Rules

Rule 1 $\quad$ For any event E, $P(E) \geq 0$.

Rule 2 $\quad$ If $E = S$ (that is, if E contains all points in the sample space), then $P(E) = 1$.

Rule 3 $\quad$ If $E_1, E_2, \ldots, E_n$ is a mutually exclusive collection of events, then

$$P(E_1 \cup E_2 \cup \cdots \cup E_n) = \sum_{k=1}^{k=n} P(E_k)$$

Rule 4 $\quad P(\bar{E}) = 1 - P(E)$.

Formula for Conditional Probability

$$P(E_2|E_1) = \frac{P(E_1 \cap E_2)}{P(E_1)} \tag{1}$$

Bayes' Rule

$$P(S_i|O_j) = \frac{P(O_j|S_i)P(S_i)}{\displaystyle\sum_{k=1}^{k=n} P(O_j|S_k)P(S_k)} \tag{7}$$

Random Variables, Mean, Variance, and Covariance

Mean of a Discrete Random Variable

$$E(\mathbf{X}) = \sum_{\text{all } k} x_k P(\mathbf{X} = x_k) \tag{8}$$

Mean of a Continuous Random Variable

$$E(\mathbf{X}) = \int_{-\infty}^{\infty} xf(x) \, dx \tag{9}$$

Variance of a Discrete Random Variable

$$\text{var } \mathbf{X} = \sum_{\text{all } k} [x_k - E(\mathbf{X})]^2 \, P(\mathbf{X} = x_k) \tag{10}$$

Variance of a Continuous Random Variable

$$\text{var } \mathbf{X} = \int_{-\infty}^{\infty} [x - E(\mathbf{X})]^2 f(x) \, dx \tag{11}$$

Covariance of Two Random Variables

$$\text{cov}(\mathbf{X}, \mathbf{Y}) = E\{[\mathbf{X} - E(\mathbf{X})][\mathbf{Y} - E(\mathbf{Y})]\} \tag{13}$$

Mean, Variance, and Covariance for Sums of Random Variables

$$E(c\mathbf{X}_1) = cE(\mathbf{X}_1) \tag{14}$$
$$E(\mathbf{X}_1 + c) = E(\mathbf{X}_1) + c \tag{15}$$
$$E(\mathbf{X}_1 + \mathbf{X}_2) = E(\mathbf{X}_1) + E(\mathbf{X}_2) \tag{16}$$
$$\text{var } c\mathbf{X}_1 = c^2 \text{var } \mathbf{X}_1 \tag{17}$$
$$\text{var}(\mathbf{X}_1 + c) = \text{var } \mathbf{X}_1 \tag{18}$$

If $\mathbf{X}_1$ and $\mathbf{X}_2$ are independent random variables,

$$\text{var}(\mathbf{X}_1 + \mathbf{X}_2) = \text{var } \mathbf{X}_1 + \text{var } \mathbf{X}_2 \tag{19}$$

In general,

$$\text{var}(\mathbf{X}_1 + \mathbf{X}_2) = \text{var } \mathbf{X}_1 + \text{var } \mathbf{X}_2 + 2\text{cov}(\mathbf{X}_1, \mathbf{X}_2) \tag{20}$$

For random variables $\mathbf{X}_1, \mathbf{X}_2, \ldots, \mathbf{X}_n$,

$$\text{var}(\mathbf{X}_1 + \mathbf{X}_2 + \cdots + \mathbf{X}_n) = \text{var } \mathbf{X}_1 + \text{var } \mathbf{X}_2 + \cdots$$
$$+ \text{var } \mathbf{X}_n + \sum_{i \neq j} \text{cov}(\mathbf{X}_i, \mathbf{X}_j) \tag{21}$$

$$\text{cov}(a\mathbf{X}_1, b\mathbf{X}_2) = ab \, \text{cov}(\mathbf{X}_1, \mathbf{X}_2) \tag{22}$$

Useful Properties of the Normal Distribution

Property 1 If $\mathbf{X}$ is $N(\mu, \sigma^2)$, then $c\mathbf{X}$ is $N(c\mu, c^2\sigma^2)$.

Property 2 If $\mathbf{X}$ is $N(\mu, \sigma^2)$, then $\mathbf{X} + c$ (for any constant c) is $N(\mu + c, \sigma^2)$.

Property 3 If X_1 is $N(\mu_1, \sigma_1^2)$, X_2 is $N(\mu_2, \sigma_2^2)$, and X_1 and X_2 are independent, then $X_1 + X_2$ is $N(\mu_1 + \mu_2, \sigma_1^2 + \sigma_2^2)$.

If X is $N(\mu, \sigma^2)$, then

$$P(a \le X \le b) = F\left(\frac{b - \mu}{\sigma}\right) - F\left(\frac{a - \mu}{\sigma}\right)$$

where $F(x) = P(Z \le x)$ and Z is $N(0, 1)$.

z-Transforms

We define (for $|z| \le 1$) the **z-transform of X** (call it $p_X^T(z)$) to be

$$E(z^X) = \sum_{n=0}^{n=\infty} a_n z^n$$

We can find the mean, the second moment ($E(X^2)$), and variance of X from the following relationships:

$$E(X) = \left[\frac{dp_X^T(z)}{dz}\right]_{z=1} \tag{23}$$

$$E(X^2) = \left[\frac{d^2 p_X^T(z)}{dz^2}\right]_{z=1} + \left[\frac{dp_X^T(z)}{dz}\right]_{z=1} \tag{24}$$

$$\text{var } X = \left[\frac{d^2 p_X^T(z)}{dz^2}\right]_{z=1} + \left[\frac{dp_X^T(z)}{dz}\right]_{z=1} - \left(\left[\frac{dp_X^T(z)}{dz}\right]_{z=1}\right)^2 \tag{25}$$

REVIEW PROBLEMS

Group A

1 Let $f(x) = xe^{-x}$.
  a Find $f'(x)$ and $f''(x)$.
  b For what values of x is $f(x)$ increasing? Decreasing?
  c Find the first-order Taylor series expansion for $f(x)$ about $x = 1$.

2 Let $f(x_1, x_2) = x_1 \ln(x_2 - x_1)$. Determine all first-order and second-order partial derivatives.

3 Some t years from now, air conditioners are sold at a rate of t per year. How many air conditioners will be sold during the next five years?

4 Let X be a continuous random variable with density function

$$f(x) = \begin{cases} \frac{4-x}{k} & \text{if } 0 \le x \le 4 \\ 0 & \text{otherwise} \end{cases}$$

  a What is k?
  b Find the cdf for X.
  c Find $E(X)$ and var X.
  d Find $P(2 \le X \le 5)$.

5 Let X_i be the price (in dollars) of stock i one year from now. X_1 is $N(15, 100)$ and X_2 is $N(20, 2025)$. Today I buy three shares of stock 1 for \$12/share and two shares of stock 2 for \$17/share. Assume that X_1 and X_2 are independent random variables.

  a Find the mean and variance of the value of my stocks one year from now.
  b What is the probability that one year from now I will have earned at least a 30% return on my investment?
  c If X_1 and X_2 were not independent, why would it be difficult to answer parts (a) and (b)?

Group B

6 An airplane has four engines. On a flight from New York to Paris, each engine has a 0.001 chance of failing. The plane will crash if at any time two or fewer engines are working properly. Assume that the failures of different engines are independent.

  a What is the probability that the plane will crash?
  b Given that engine 1 will not fail during the flight, what is the probability that the plane will crash?

c Given that engine 1 will fail during the flight, what is the probability that the plane will not crash?

7 Suppose that each engine can be inspected before the flight. After inspection, each engine is labeled as being in either good or bad condition. You are given that

P(inspection says engine is in good condition | engine will fail) = .001

P(inspection says engine is in bad condition | engine will fail) = .999

P(inspection says engine is in good condition | engine will not fail) = .995

P(inspection says engine is in bad condition | engine will not fail) = .005

a If the inspection indicates the engine is in bad condition, what is the probability that the engine will fail on the flight?

b If an inspector randomly inspects an engine (that is, with probability .001 she chooses an engine that is about to fail, and with probability .999 she chooses an engine that is not about to fail), what is the probability that she will make an error in her evaluation of the engine?

REFERENCES

Allen, R. *Mathematical Analysis for Economists*. New York: St. Martin's Press, 1938. An advanced calculus review.

Byrkit, D., and S. Shamma. *Calculus for Business and Economics*. New York: Van Nostrand, 1981. A review of calculus at a beginning level.

Chiang, A. *Fundamental Methods of Mathematical Economics*. New York: McGraw-Hill, 1978. An advanced calculus review.

Harnett, D. *Statistical Methods*. Reading, Mass.: Addison-Wesley, 1982. A review of probability.

Winkler, R., and W. Hays. *Statistics: Probability, Inference, and Decision*. Chicago: Holt, Rinehart & Winston, 1971. A review of probability.

13

Decision Making under Uncertainty

We have all had to make important decisions where we were uncertain about factors that were relevant to the decisions. In this chapter, we study situations in which decisions are made in an uncertain environment.

The following model encompasses several aspects of making a decision in the absence of certainty. The decision maker first chooses an action a_i from a set $A = \{a_1, a_2, \ldots, a_k\}$ of available actions. Then the state of the world is observed; with probability p_j, the state of the world is observed to be $s_j \in S = \{s_1, s_2, \ldots, s_n\}$. If action a_i is chosen and the state of the world is s_j, the decision maker receives a reward r_{ij}. We refer to this model as the *state-of-the-world decision-making model*.

This chapter presents the basic theory of decision making under uncertainty: the widely used Von Neumann–Morgenstern utility model, and the use of decision trees for making decisions at different points in time. We close by looking at decision making with multiple objectives.

13.1 Decision Criteria

In this section, we consider four decision criteria that can be used to make decisions under uncertainty.

EXAMPLE 1 | **Newspaper Vendor**

News vendor Phyllis Pauley sells newspapers at the corner of Kirkwood Avenue and Indiana Street, and each day she must determine how many newspapers to order. Phyllis pays the company 20¢ for each paper and sells the papers for 25¢ each. Newspapers that are unsold at the end of the day are worthless. Phyllis knows that each day she can sell between 6 and 10 papers, with each possibility being equally likely. Show how this problem fits into the state-of-the-world model.

Solution In this example, the members of $S = \{6, 7, 8, 9, 10\}$ are the possible values of the daily demand for newspapers. We are given that $p_6 = p_7 = p_8 = p_9 = p_{10} = \frac{1}{5}$. Phyllis must choose an action (the number of papers to order each day) from $A = \{6, 7, 8, 9, 10\}$.

If Phyllis purchases i papers and j papers are demanded, then i papers are purchased at a cost of $20i$¢, and $\min(i, j)$ papers are sold for 25¢ each.[†] Thus, if Phyllis purchases i papers and j papers are demanded, she earns a net profit of r_{ij}, where

$$r_{ij} = 25i - 20i = 5i \qquad (i \leq j)$$
$$r_{ij} = 25j - 20i \qquad (i \geq j)$$

The values of r_{ij} are tabulated in Table 1.

[†] $\min(i, j)$ is the smaller of i and j.

TABLE 1
Rewards for News Vendor

Papers Ordered	Papers Demanded				
	6	7	8	9	10
6	30¢	30¢	30¢	30¢	30¢
7	10¢	35¢	35¢	35¢	35¢
8	−10¢	15¢	40¢	40¢	40¢
9	−30¢	−5¢	20¢	45¢	45¢
10	−50¢	−25¢	0¢	25¢	50¢

Dominated Actions

Why did we not consider the possibility that Phyllis would order 1, 2, 3, 4, 5, or more than 10 papers? Answering this question involves the idea of a dominated action.

DEFINITION ■ An action a_i is **dominated** by an action $a_{i'}$ if for all $s_j \in S$, $r_{ij} \leq r_{i'j}$, and for some state $s_{j'}$, $r_{ij'} < r_{i'j'}$. ■

If action a_i is dominated, then in no state of the world is a_i better than $a_{i'}$, and in at least one state of the world a_i is inferior to $a_{i'}$. Thus, if action a_i is dominated, there is no reason to choose a_i ($a_{i'}$ would be a better choice).

If Phyllis orders i papers ($i = 1, 2, 3, 4, 5$), she will earn (for all states of the world) a profit of $5i¢$. From the table of rewards, we see that, for $i = 1, 2, 3, 4, 5$, ordering 6 papers dominates ordering i papers ($j' = 6, 7, 8, 9,$ or 10 will do). Similarly, the reader should check that ordering i papers ($i > 11$) is dominated by ordering 10 papers (see Problem 3 at the end of this section). A quick check shows that none of the actions in $A = \{6, 7, 8, 9, 10\}$ are dominated. Thus, Phyllis should indeed choose her action from $A = \{6, 7, 8, 9, 10\}$.

We now discuss four criteria that can be used to choose an action.

The Maximin Criterion

For each action, determine the worst outcome (smallest reward). The maximin criterion chooses the action with the "best" worst outcome.

DEFINITION ■ The **maximin criterion** chooses the action a_i with the largest value of $\min_{j \in S} r_{ij}$. ■

For Example 1, we obtain the results in Table 2. Thus, the maximin criterion recommends ordering 6 papers. This ensures that Phyllis will, no matter what the state of the world, earn a profit of at least 30¢. The maximin criterion is concerned with making the worst possible outcome as pleasant as possible. Unfortunately, choosing a decision to mitigate the worst case may prevent the decision maker from taking advantage of good fortune. For example, if Phyllis follows the maximin criterion, she will never make less than 30¢, but she will never make more than 30¢.

TABLE 2
Computation of Maximin Decision for News Vendor

Papers Ordered	Worst State of the World	Reward in Worst State of the World
6	6, 7, 8, 9, 10	30¢
7	6	10¢
8	6	−10¢
9	6	−30¢
10	6	−50¢

The Maximax Criterion

For each action, determine the best outcome (largest reward). The maximax criterion chooses the action with the "best" best outcome.

DEFINITION ■ The **maximax criterion** chooses the action a_i with the largest value of $\max_{j \in S} r_{ij}$. ■

For Example 1, we obtain the results in Table 3. Thus, the maximax criterion would recommend ordering 10 papers. In the best state (when 10 papers are demanded), this yields a profit of 50¢. Of course, making a decision according to the maximax criterion leaves Phyllis open to the disastrous possibility that only 6 papers will be demanded, in which case she loses 50¢.

Minimax Regret

The minimax regret criterion (developed by L. J. Savage) uses the concept of opportunity cost to arrive at a decision. For each possible state of the world s_j, find an action $i^*(j)$ that maximizes r_{ij}. That is, $i^*(j)$ is the best possible action to choose if the state of the world is actually s_j. Then for any action a_i and state s_j, the opportunity loss or regret for a_i in s_j is $r_{i^*(j),j} - r_{ij}$. For example, if $j = 7$ papers are demanded, the best decision is to order $i^*(7) = 7$ papers, yielding a profit of $r_{77} = 7(25) - 7(20) = 35$¢. Suppose we chose to order $i = 6$ papers. Since $r_{67} = 6(25) - 6(20) = 30$¢, the opportunity loss or regret for $i = 6$ and $j = 7$ is $35 - 30 = 5$¢. Thus, if we order 6 papers and 7 papers are demanded, in hindsight we realize that by making the optimal choice (ordering 7 papers) for the actual state of the world (7 papers demanded), we would have done 5¢ better than we did by ordering 6 papers. Table 4 shows the opportunity cost or regret matrix for Example 1.

TABLE 3
Computation of Maximax Decision for News Vendor

Papers Ordered	State Yielding Best Outcome	Best Outcome
6	6, 7, 8, 9, 10	30¢
7	7, 8, 9, 10	35¢
8	8, 9, 10	40¢
9	9, 10	45¢
10	10	50¢

The minimax regret criterion chooses an action by applying the minimax criterion to the regret matrix. In other words, the minimax regret criterion attempts to avoid disappointment over what might have been. From the regret matrix in Table 4, we obtain the minimax regret decision in Table 5. Thus, the minimax regret criterion recommends ordering 6 or 7 papers.

The Expected Value Criterion

The expected value criterion chooses the action that yields the largest expected reward. For Example 1, the expected value criterion would recommend ordering 6 or 7 papers (see Table 6).

The decision-making criteria discussed in this section may seem reasonable, but many people make decisions without using any of them. A more comprehensive model of individual decision making, the Von Neumann–Morgenstern utility model, is discussed in Section 13.2.

TABLE 4

Regret Matrix for News Vendor

Papers Ordered	Papers Demanded				
	6	7	8	9	10
6	$30 - 30 = 0¢$	$35 - 30 = 5¢$	$40 - 30 = 10¢$	$45 - 30 = 15¢$	$50 - 30 = 20¢$
7	$30 - 10 = 20¢$	$35 - 35 = 0¢$	$40 - 35 = 5¢$	$45 - 35 = 10¢$	$50 - 35 = 15¢$
8	$30 + 10 = 40¢$	$35 - 15 = 20¢$	$40 - 40 = 0¢$	$45 - 40 = 5¢$	$50 - 40 = 10¢$
9	$30 + 30 = 60¢$	$35 + 5 = 40¢$	$40 - 20 = 20¢$	$45 - 45 = 0¢$	$50 - 45 = 5¢$
10	$30 + 50 = 80¢$	$35 + 25 = 60¢$	$40 - 0 = 40¢$	$45 - 25 = 20¢$	$50 - 50 = 0¢$

TABLE 5

Computation of Minimax Regret Decision for News Vendor

Papers Ordered	Maximum Regret
6	20¢
7	20¢
8	40¢
9	60¢
10	80¢

TABLE 6

Computation of Expected Value Decision for News Vendor

Papers Ordered	Expected Reward
6	$\frac{1}{5}(30 + 30 + 30 + 30 + 30) = 30¢$
7	$\frac{1}{5}(10 + 35 + 35 + 35 + 35) = 30¢$
8	$\frac{1}{5}(-10 + 15 + 40 + 40 + 40) = 25¢$
9	$\frac{1}{5}(-30 - 5 + 20 + 45 + 45) = 15¢$
10	$\frac{1}{5}(-50 - 25 + 0 + 25 + 50) = 0¢$

PROBLEMS

Group A

1 Pizza King and Noble Greek are two competing restaurants. Each must determine simultaneously whether to undertake small, medium, or large advertising campaigns. Pizza King believes that it is equally likely that Noble Greek will undertake a small, a medium, or a large advertising campaign. Given the actions chosen by each restaurant, Pizza King's profits are as shown in Table 7. For the maximin, maximax, and minimax regret criteria, determine Pizza King's choice of advertising campaign.

TABLE 7

Pizza King Chooses	Noble Greek Chooses		
	Small	Medium	Large
Small	$6,000	$5,000	$2,000
Medium	$5,000	$6,000	$1,000
Large	$9,000	$6,000	$0

2 Sodaco is considering producing a new product: Chocovan soda. Sodaco estimates that the annual demand for Chocovan, D (in thousands of cases), has the following mass function: $P(D = 30) = .30$, $P(D = 50) = .40$, $P(D = 80) = .30$. Each case of Chocovan sells for $5 and incurs a variable cost of $3. It costs $800,000 to build a plant to produce Chocovan. Assume that if $1 is received every year (forever), this is equivalent to receiving $10 at the present time. Considering the reward for each action and state of the world to be in terms of net present value, use each decision criterion of this section to determine whether Sodaco should build the plant.

3 For Example 1, show that ordering 11 or more papers is dominated by ordering 10 papers.

Group B

4 Suppose that Pizza King and Noble Greek stop advertising but must determine the price they will charge for each pizza sold. Pizza King believes that Noble Greek's price is a random variable D having the following mass function: $P(D = \$6) = .25$, $P(D = \$8) = .50$, $P(D = \$10) = .25$. If Pizza King charges a price p_1 and Noble Greek charges a price p_2, Pizza King will sell $100 + 25(p_2 - p_1)$ pizzas. It costs Pizza King $4 to make a pizza. Pizza King is considering charging $5, $6, $7, $8, or $9 for a pizza. Use each decision criterion of this section to determine the price that Pizza King should charge.

5 Alden Construction is bidding against Forbes Construction for a project. Alden believes that Forbes's bid is a random variable B with the following mass function: $P(B = \$6,000) = .40$, $P(B = \$8,000) = .30$, $P(B = \$11,000) = .30$. It will cost Alden $6,000 to complete the project. Use each of the decision criteria of this section to determine Alden's bid. Assume that in case of a tie, Alden wins the bidding. (*Hint:* Let p = Alden's bid. For $p \leq 6,000$, $6,000 < p \leq 8,000$, $8,000 < p \leq 11,000$, and $p > 11,000$, determine Alden's profit in terms of Alden's bid and Forbes's bid.)

13.2 Utility Theory

We now show how the Von Neumann–Morgenstern concept of a utility function can be used as an aid to decision making under uncertainty.

Consider a situation in which a person will receive, for $i = 1, 2, \ldots, n$, a reward r_i with probability p_i. This is denoted as the **lottery** $(p_1, r_1; p_2, r_2; \ldots; p_n, r_n)$. A lottery is often represented by a tree in which each branch stands for a possible outcome of the lottery, and the number on each branch represents the probability that the outcome will occur. Thus, the lottery $(\frac{1}{4}, \$500; \frac{3}{4}, \$0)$ could be denoted by

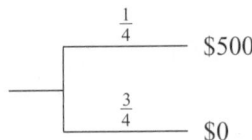

Suppose we are asked to choose between two lotteries (L_1 and L_2). With certainty, lottery L_1 yields $10,000:

$$L_1 \underline{\quad\quad 1 \quad\quad} \$10,000$$

Lottery L_2 consists of tossing a coin. If heads comes up, we receive $30,000, and if tails comes up, we receive $0:

$$L_2 \quad \begin{array}{c} \overset{\frac{1}{2}}{\rule{3cm}{0.4pt}} \$30{,}000 \\ \underset{\frac{1}{2}}{\rule{3cm}{0.4pt}} \$0 \end{array}$$

L_1 yields an expected reward of $10,000, and L_2 yields an expected reward of $(\frac{1}{2})(30{,}000) + (\frac{1}{2})(0) = \$15{,}000$. Although L_2 has a larger expected value than L_1, most people prefer L_1 to L_2 because L_1 offers the certainty of a relatively large payoff, whereas L_2 yields a substantial $(\frac{1}{2})$ chance of earning a reward of $0. In short, most people prefer L_1 to L_2 because L_1 involves less risk (or uncertainty) than L_2.

Our goal is to determine a method that a person can use to choose between lotteries. Suppose he or she must choose to play L_1 or L_2 but not both. We write L_1pL_2 if the person prefers L_1. We write L_1iL_2 if he or she is indifferent between choosing L_1 and L_2. If L_1iL_2, we say that L_1 and L_2 are **equivalent lotteries.** Finally, we write L_2pL_1 if the decision maker prefers L_2.

Suppose we ask a decision maker to rank the following lotteries:

$$L_1 \overset{1}{\rule{2cm}{0.4pt}} \$10{,}000 \qquad L_2 \begin{array}{c} \overset{.50}{\rule{2cm}{0.4pt}} \$30{,}000 \\ \underset{.50}{\rule{2cm}{0.4pt}} \$0 \end{array}$$

$$L_3 \overset{1}{\rule{2cm}{0.4pt}} \$0 \qquad L_4 \begin{array}{c} \overset{.02}{\rule{2cm}{0.4pt}} -\$10{,}000 \\ \underset{.98}{\rule{2cm}{0.4pt}} \$500 \end{array}$$

The Von Neumann–Morgenstern approach to ranking these lotteries is as follows. Begin by identifying the most favorable ($30,000) and the least favorable (−$10,000) outcomes that can occur. For all other possible outcomes ($r_1 = \$10{,}000$, $r_2 = \$500$, and $r_3 = \$0$), the decision maker is asked to determine a probability p_i such that he or she is indifferent between two lotteries:

$$\overset{1}{\rule{2cm}{0.4pt}} r_i \qquad \text{and} \qquad \begin{array}{c} \overset{p_i}{\rule{2cm}{0.4pt}} \$30{,}000 \\ \underset{1-p_i}{\rule{2cm}{0.4pt}} -\$10{,}000 \end{array}$$

Suppose that for $r_1 = \$10{,}000$, the decision maker is indifferent between

$$\overset{1}{\rule{2cm}{0.4pt}} \$10{,}000 \qquad \text{and} \qquad \begin{array}{c} \overset{.90}{\rule{2cm}{0.4pt}} \$30{,}000 \\ \underset{.10}{\rule{2cm}{0.4pt}} -\$10{,}000 \end{array} \qquad (1)$$

and for $r_2 = \$500$, indifferent between

$$\overset{1}{\rule{2cm}{0.4pt}} \$500 \qquad \text{and} \qquad \begin{array}{c} \overset{.62}{\rule{2cm}{0.4pt}} \$30{,}000 \\ \underset{.38}{\rule{2cm}{0.4pt}} -\$10{,}000 \end{array} \qquad (2)$$

and for $r_3 = \$0$, indifferent between

$$\underline{\hspace{2cm}}^1\hspace{0.3cm}\$0 \qquad \text{and}$$

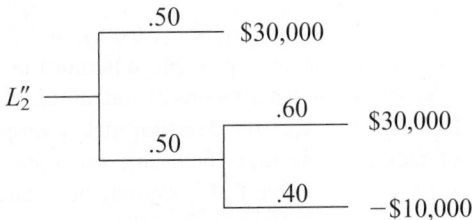
$$.60 \quad \$30,000$$
$$.40 \quad -\$10,000$$

(3)

Using (1)–(3), the decision maker can construct lotteries L_1', L_2', L_3', and L_4' such that $L_i' i L_i$ and each L_i' involves only the best ($\$30,000$) and the worst ($-\$10,000$) possible outcomes. Thus, from (1), we find that $L_1 i L_1'$, where

$$L_1' \begin{cases} .90 \quad \$30,000 \\ .10 \quad -\$10,000 \end{cases}$$

From (3), we find that $L_2 i L_2''$, where

$$L_2'' \begin{cases} .50 \quad \$30,000 \\ .50 \begin{cases} .60 \quad \$30,000 \\ .40 \quad -\$10,000 \end{cases} \end{cases}$$

L_2'' is a compound lottery in which with probability .50 we receive $\$30,000$ and with probability .50 we play a lottery yielding a .60 chance at $\$30,000$ and a .40 chance at $-\$10,000$. More formally, a lottery L is a **compound lottery** if for some i, there is a probability p_i that the decision maker's reward is to play another lottery L'. The following is an example of a compound lottery:

$$L \begin{cases} .50 \begin{cases} .60 \quad \$6 \\ .40 \quad -\$4 \end{cases} \\ .50 \quad -\$4 \end{cases} \qquad (L')$$

Thus, with probability .50, L yields a reward of $-\$4$, and with probability .50, L causes us to play L'. If a lottery is not a compound lottery, it is a **simple lottery.**

Returning to our discussion of L_2'', we observe that L_2'' is a lottery that yields a .50 + .50(.60) = .80 chance at $\$30,000$ and a .40(.50) = .20 chance at $-\$10,000$. Thus, $L_2 i L_2'' i L_2'$, where

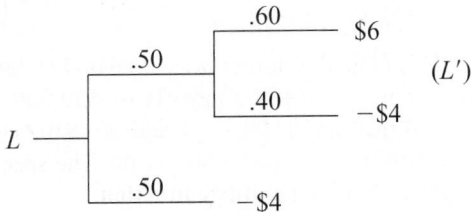
$$L_2' \begin{cases} .80 \quad \$30,000 \\ .20 \quad -\$10,000 \end{cases}$$

Similarly, using (3), we find that $L_3 i L_3'$, where

$$L_3' \begin{cases} .60 \quad \$30,000 \\ .40 \quad -\$10,000 \end{cases}$$

Using (2), we find that the decision maker is indifferent between L_4 and L_4'', where

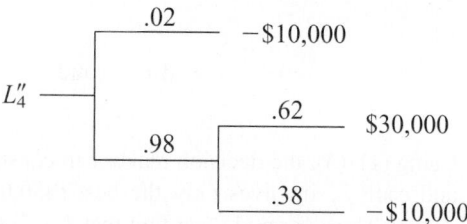

In actuality, however, L_4'' yields a $.98(.62) = .6076$ chance at \$30,000 and a $.02 + .38(.98) = .3924$ chance at $-\$10,000$. Thus, $L_4 i L_4'' i L_4'$, where

```
        .6076
       ┌─────── $30,000
  L_4' ─┤
       └─────── −$10,000
        .3924
```

Since $L_i i L_i'$, we may rank L_1, L_2, L_3, and L_4 by ranking L_1', L_2', L_3', and L_4'. Consider two lotteries whose only possible outcomes are \$30,000 (the most favorable outcome) and $-\$10,000$ (the least favorable outcome). If he or she is given a choice between two lotteries of this type, the decision maker simply chooses the lottery with the larger chance of receiving the most favorable outcome. Applying this idea to L_1' through L_4' yields $L_1' p L_2' p L_4' p L_3'$. Since $L_i i L_i'$, we may conclude that $L_1 p L_2 p L_4 p L_3$.

We now give a more formal description of the process that we have used to rank L_1, L_2, L_3, and L_4. The **utility** of the reward r_i, written $u(r_i)$, is the number q_i such that the decision maker is indifferent between the following two lotteries:

$$\frac{\quad 1 \quad}{} r_i \quad \text{and} \quad \begin{array}{l} \rule{1cm}{0.4pt}\; q_i \rule{1cm}{0.4pt}\;\text{Most favorable outcome} \\[4pt] \rule{1cm}{0.4pt}\; 1-q_i \rule{1cm}{0.4pt}\;\text{Least favorable outcome} \end{array}$$

This definition forces $u(\text{least favorable outcome}) = 0$ and $u(\text{most favorable outcome}) = 1$. For our possible payoffs of \$30,000, $-\$10,000$, \$0, \$500, and \$10,000, we first find that $u(\$30,000) = 1$ and $u(-\$10,000) = 0$. Then (1)–(3) yield $u(\$10,000) = .90$, $u(\$500) = .62$, and $u(\$0) = .60$. The specification of $u(r_i)$ for all rewards r_i is called the decision maker's **utility function.**

For a given lottery $L = (p_1, r_1; p_2, r_2; \ldots; p_n, r_n)$, define the expected utility of the lottery L, written $E(U \text{ for } L)$, by

$$E(U \text{ for } L) = \sum_{i=1}^{i=n} p_i u(r_i)$$

Thus, in our example

$$E(U \text{ for } L_1) = 1(.90) = .90$$
$$E(U \text{ for } L_2) = .50(1) + .50(.60) = .80$$
$$E(U \text{ for } L_3) = 1(.60) = .60$$
$$E(U \text{ for } L_4) = .02(0) + .98(.62) = .6076$$

Recall that we found that $L_i i L_i'$, where L_i' yielded an $E(U \text{ for } L_i)$ chance at \$30,000 and a $1 - E(U \text{ for } L_i)$ chance at $-\$10,000$. Thus, in choosing between lotteries L_1', L_2', L_3', and L_4' (or equivalently, L_1, L_2, L_3, and L_4), we simply chose the lottery with the largest expected utility. Given two lotteries L_1 and L_2, we may choose between them via the expected utility criteria:

$$L_1pL_2 \quad \text{if and only if} \quad E(U \text{ for } L_1) > E(U \text{ for } L_2)$$
$$L_2pL_1 \quad \text{if and only if} \quad E(U \text{ for } L_2) > E(U \text{ for } L_1)$$
$$L_1iL_2 \quad \text{if and only if} \quad E(U \text{ for } L_2) = E(U \text{ for } L_1)$$

Von Neumann–Morgenstern Axioms

Von Neumann and Morgenstern proved that if a person's preferences satisfy the following axioms, then he or she should choose between lotteries by using the expected utility criterion.

Axiom 1: Complete Ordering Axiom

For any two rewards r_1 and r_2, one of the following must be true: The decision maker (1) prefers r_1 to r_2, (2) prefers r_2 to r_1, or (3) is indifferent between r_1 and r_2. Also, if the person prefers r_1 to r_2 and r_2 to r_3, then he or she must prefer r_1 to r_3 (transitivity of preferences).

In our discussion, we used the Complete Ordering Axiom to determine the most and least favorable outcomes.

Axiom 2: Continuity Axiom

If the decision maker prefers r_1 to r_2 and r_2 to r_3, then for some $c(0 < c < 1)$, L_1iL_2, where

In our informal discussion, we used the Continuity Axiom when we found, for example, that L_3iL_3', where

Axiom 3: Independence Axiom

Suppose the decision maker is indifferent between rewards r_1 and r_2. Let r_3 be any other reward. Then for any c $(0 < c < 1)$, L_1iL_2, where

L_1 and L_2 differ only in that L_1 has a probability c of yielding a reward r_1, whereas L_2 has a probability c of yielding a reward r_2. Thus, the Independence Axiom implies that the decision maker views a chance c at r_1 and a chance c at r_2 to be of identical value, and this view holds for all values of c and r_3. We applied the Independence Axiom when we used (3) to claim that $L_2 i L_2''$, where

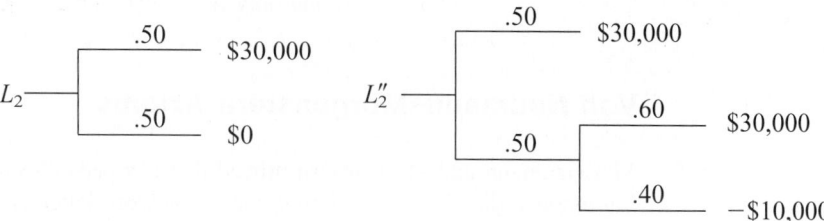

Axiom 4: Unequal Probability Axiom

Suppose the decision maker prefers reward r_1 to reward r_2. If two lotteries have only r_1 and r_2 as their possible outcomes, he or she will prefer the lottery with the higher probability of obtaining r_1.

We used the Unequal Probability Axiom when we concluded, for example, that L_1' was preferred to L_2' (because L_1' had a .90 chance at $30,000 and L_2' had only a .80 chance at $30,000).

Axiom 5: Compound Lottery Axiom

Suppose that when all possible outcomes are considered, a compound lottery L yields (for $i = 1, 2, \ldots, n$) a probability p_i of receiving a reward r_i. Then $L i L'$, where L' is the simple lottery $(p_1, r_1; p_2, r_2; \ldots; p_n, r_n)$.

For example, consider the following compound lottery:

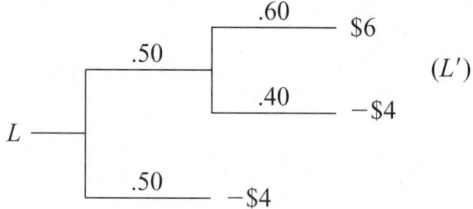

L yields a $.50 + .50(.40) = .70$ chance at $-$4 and a $.50(.60) = .30$ chance at $6. Thus, $L i L''$, where

$$
\begin{array}{ccc}
 & \underline{.70} & -\$4 \\
L'' & & \\
 & \underline{.30} & \$6
\end{array}
$$

In our informal discussion, we used the Compound Lottery Axiom when, for example, we stated that the compound equivalent of L_2 (L_2'')

$$
\begin{array}{ccccc}
 & \underline{.50} & & \$30,000 & \\
L_2'' & & & & \\
 & & & \underline{.60} & \$30,000 \\
 & \underline{.50} & & & \\
 & & & \underline{.40} & -\$10,000
\end{array}
$$

was equivalent to the following simple lottery:

$$.50 + .50(.60) = .80 \quad \text{---} \quad \$30,000$$

$$.50(.40) = .20 \quad \text{---} \quad -\$10,000$$

Why We May Assume u(Worst Outcome) = 0 and u(Best Outcome) = 1

Up to now, we have assumed that u(least favorable outcome) = 0 and u(most favorable outcome) = 1. Even if a decision maker's utility function does not have these values, we can transform his or her utility function (without changing the preferences among lotteries) into a utility function having u(least favorable outcome) = 0 and u(most favorable outcome) = 1.

LEMMA 1

Given a utility function $u(x)$, define for any $a > 0$ and any b the function $v(x) = au(x) + b$. Given any two lotteries L_1 and L_2, it will be the case that

1 A decision maker using $u(x)$ as his or her utility function will have $L_1 p L_2$ if and only if a decision maker using $v(x)$ as his or her utility function will have $L_1 p L_2$.

2 A decision maker using $u(x)$ as his or her utility function will have $L_1 i L_2$ if and only if a decision maker using $v(x)$ as his or her utility function will have $L_1 i L_2$.

Proof Let

$$L_1 = (p_1, r_1; p_2, r_2; \ldots; p_n, r_n)$$
$$L_2 = (p_1', r_1'; p_2', r_2'; \ldots; p_m', r_m')$$

Suppose the decision maker using $u(x)$ prefers L_1 to L_2. Then by the expected utility criterion, we know that

$$\sum_{i=1}^{i=n} p_i u(r_i) > \sum_{i=1}^{i=m} p_i' u(r_i') \tag{4}$$

Now the $v(x)$ decision maker will have $L_1 p L_2$ if

$$\sum_{i=1}^{i=n} p_i[au(r_i) + b] > \sum_{i=1}^{i=m} p_i'[au(r_i') + b] \tag{5}$$

Since

$$\sum_{i=1}^{i=n} p_i = \sum_{i=1}^{i=m} p_i' = 1$$

(5) simplifies to

$$a \sum_{i=1}^{i=n} p_i u(r_i) + b > a \sum_{i=1}^{i=m} p_i' u(r_i') + b \tag{6}$$

Since $a > 0$, (6) follows from (4). Thus, if the $u(x)$ decision maker has $L_1 p L_2$, the $v(x)$ decision maker has $L_1 p L_2$. Similarly, if (6) holds, then (4) will hold. Thus, if the $v(x)$ decision maker has $L_1 p L_2$, the $u(x)$ decision maker will also have $L_1 p L_2$. A similar argument can be used to prove part (2) of Lemma 1.

Using Lemma 1, we can show that without changing how an individual ranks lotteries, we can transform the decision maker's utility function into one having u(least favorable outcome) $= 0$ and u(most favorable outcome) $= 1$. To illustrate, let's reconsider ranking lotteries L_1–L_4. Suppose our decision maker's utility function had $u(-\$10{,}000) = -5$ and $u(\$30{,}000) = 10$. Define $v(x) = au(x) + b$. Choose a and b so that $v(\$30{,}000) = 10a + b = 1$ and $v(-\$10{,}000) = -5a + b = 0$. Then $a = \frac{1}{15}$ and $b = \frac{1}{3}$. Then by Lemma 1, the utility function $v(x) = \frac{u(x)}{15} + \frac{1}{3}$ will yield the same ranking of lotteries as does $u(x)$, and we will have constructed $v(x)$ so that $v(\$30{,}000) = 1$ and $v(-\$10{,}000) = 0$. Thus, we see that without loss of generality, we may assume that u(least favorable outcome) $= 0$ and u(most favorable outcome) $= 1$.

Estimating an Individual's Utility Function

How might we estimate an individual's (call her Jill) utility function? We begin by assuming that the least favorable outcome (say, $-\$10{,}000$) has a utility of 0 and that the most favorable outcome (say, $\$30{,}000$) has a utility of 1. Next we define a number $x_{1/2}$ having $u(x_{1/2}) = \frac{1}{2}$. To determine $x_{1/2}$, ask Jill for the number (call it $x_{1/2}$) that makes her indifferent between

$$
\underbrace{\hspace{2cm}}_{x_{1/2}}\qquad \text{and} \qquad
\begin{array}{l}
\frac{1}{2} \quad \$30{,}000 \quad \text{(Most favorable outcome)}\\[1em]
\frac{1}{2} \quad -\$10{,}000 \quad \text{(Least favorable outcome)}
\end{array}
$$

Since Jill is indifferent between the two lotteries, they must have the same expected utility. Thus, $u(x_{1/2}) = (\frac{1}{2})(1) + (\frac{1}{2})(0) = \frac{1}{2}$.

This procedure yields a point $x_{1/2}$ having $u(x_{1/2}) = \frac{1}{2}$. Suppose Jill states that $x_{1/2} = -\$3{,}400$. Using $x_{1/2}$ and the least favorable outcome ($-\$10{,}000$) as possible outcomes, we can construct a lottery that can be used to determine the point $x_{1/4}$ having a utility of $\frac{1}{4}$ (that is, $u(x_{1/4}) = \frac{1}{4}$). Point $x_{1/4}$ must be such that Jill is indifferent between

$$
\underbrace{\hspace{2cm}}_{x_{1/4}}\qquad \text{and} \qquad
\begin{array}{l}
\frac{1}{2} \quad x_{1/2} = -\$3{,}400\\[1em]
\frac{1}{2} \quad -\$10{,}000 \quad \text{(Least favorable outcome)}
\end{array}
$$

Then $u(x_{1/4}) = (\frac{1}{2})(\frac{1}{2}) + (\frac{1}{2})(0) = \frac{1}{4}$. Thus, $x_{1/4}$ will satisfy $u(x_{1/4}) = \frac{1}{4}$. Suppose Jill states that $x_{1/4} = -\$8{,}000$. This gives us another point on Jill's utility function.

Jill can now use the $x_{1/2}$ and $\$30{,}000$ outcomes to construct a lottery that will yield a value $x_{3/4}$ satisfying $u(x_{3/4}) = \frac{3}{4}$. (How?) Suppose that $x_{3/4} = \$8{,}000$. Similarly, outcomes of $x_{1/4}$ and $-\$10{,}000$ can be used to construct a lottery that will yield a value $x_{1/8}$ satisfying $u(x_{1/8}) = \frac{1}{8}$. Now Jill's utility function can be approximated by drawing a curve (smooth, we hope) joining the points

$$(-\$10{,}000, 0), (x_{1/8}, 1/8), (x_{1/4}, 1/4), \ldots, (\$30{,}000, 1)$$

The result is shown in Figure 1. Unfortunately, if a decision maker's preferences violate any of the preceding axioms (such as transitivity), this procedure may not yield a smooth curve. If it does not yield a relatively smooth curve, more sophisticated procedures for assessing utility functions must be used (see Keeney and Raiffa (1976)).

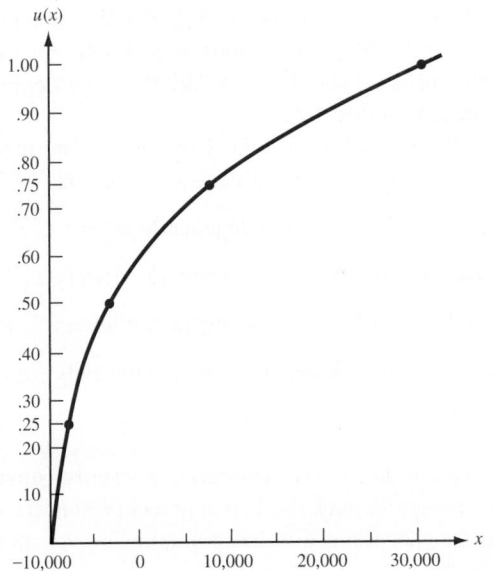

FIGURE 1
Jill's Utility Function

Relation between an Individual's Utility Function and His or Her Attitude toward Risk

A decision maker's utility function contains information about his or her attitude toward risk. To discuss this information, we need to define the concepts of a lottery's certainty equivalent and risk premium.

DEFINITION ■ The **certainty equivalent** of a lottery L, written $CE(L)$, is the number $CE(L)$ such that the decision maker is indifferent between the lottery L and receiving a certain payoff of $CE(L)$. ■

For example, we saw earlier that Jill was indifferent between

$$\underline{\hspace{2cm}} -\$3,400 \qquad \text{and} \qquad L \underset{\frac{1}{2}}{\overset{\frac{1}{2}}{<}} \begin{matrix} \$30,000 \\ -\$10,000 \end{matrix}$$

Thus, $CE(L) = -\$3,400$.

DEFINITION ■ The **risk premium** of a lottery L, written $RP(L)$, is given by $RP(L) = EV(L) - CE(L)$, where $EV(L)$ is the expected value of the lottery's outcomes. ■

For example, if

$$L \underset{\frac{1}{2}}{\overset{\frac{1}{2}}{<}} \begin{matrix} \$30,000 \\ -\$10,000 \end{matrix}$$

then $EV(L) = (\frac{1}{2})(\$30,000) + (\frac{1}{2})(-\$10,000) = \$10,000$. We have already seen that $CE(L) = -\$3,400$. Thus, $RP(L) = 10,000 - (-3,400) = \$13,400$; Jill values L at \$13,400 less than its expected value, because she does not like the large degree of uncertainty that is associated with the reward yielded by L.

Let a **nondegenerate lottery** be any lottery in which more than one outcome can occur. With respect to attitude toward risk, a decision maker is

1 Risk-averse if and only if for any nondegenerate lottery L, $RP(L) > 0$

2 Risk-neutral if and only if for any nondegenerate lottery L, $RP(L) = 0$

3 Risk-seeking if and only if for any nondegenerate lottery L, $RP(L) < 0$

An individual's attitude toward risk depends on the concavity (or convexity) of his or her utility function.

DEFINITION ■   A function $u(x)$ is said to be **strictly concave** (or **strictly convex**) if for any two points on the curve $y = u(x)$, the line segment joining those two points lies entirely (with the exception of its endpoints) below (or above) the curve $y = u(x)$.   ■

If $u(x)$ is differentiable, then $u(x)$ will be strictly concave if and only if $u''(x) < 0$ for all x and $u(x)$ will be strictly convex if and only if $u''(x) > 0$ for all x. It can easily be shown that a decision maker with a utility function $u(x)$ is

1 Risk-averse if and only if $u(x)$ is strictly concave

2 Risk-neutral if and only if $u(x)$ is a linear function (if $u(x)$ is both convex and concave)

3 Risk-seeking if and only if $u(x)$ is strictly convex

To illustrate these definitions, we show that a decision maker with a concave utility function $u(x)$ exhibits risk-averse behavior (has $RP(L) > 0$). Consider a binary lottery L (a lottery with only two possible outcomes):

$$L \quad \begin{array}{c} \overset{p}{\rule{3cm}{0.4pt}} x_1 \\ \underset{1-p}{\rule{3cm}{0.4pt}} x_2 \end{array} \qquad \text{(Assume } x_1 < x_2)$$

Suppose $u(x)$ is strictly concave. Then, from Figure 2, we see that

$$E(U \text{ for } L) = p\,u(x_1) + (1 - p)u(x_2) = y\text{-coordinate of point 1}$$

Since $CE(L)$ is the value x^* having $u(x^*) = E(U \text{ for } L)$, Figure 2 shows that $CE(L) < EV(L)$, so $RP(L) > 0$. This follows because the strict concavity of $u(x)$ implies that the line segment joining the points $(x_1, u(x_1))$ and $(x_2, u(x_2))$ lies below the curve $u(x)$.

We can also give an algebraic proof that $u(x)$ strictly concave implies that $RP(L) = EV(L) - CE(L) > 0$. Recall that for

$$L \quad \begin{array}{c} \overset{p}{\rule{3cm}{0.4pt}} x_1 \\ \underset{1-p}{\rule{3cm}{0.4pt}} x_2 \end{array}$$

$EV(L) = px_1 + (1 - p)x_2$. Now the strict concavity of $u(x)$ implies that $u[px_1 + (1 - p)x_2] > pu(x_1) + (1 - p)u(x_2) = E(U \text{ for } L)$. Thus, the decision maker prefers $px_1 + (1 - p)x_2 = EV(L)$ with certainty to the prospect of playing L. The certainty equivalent

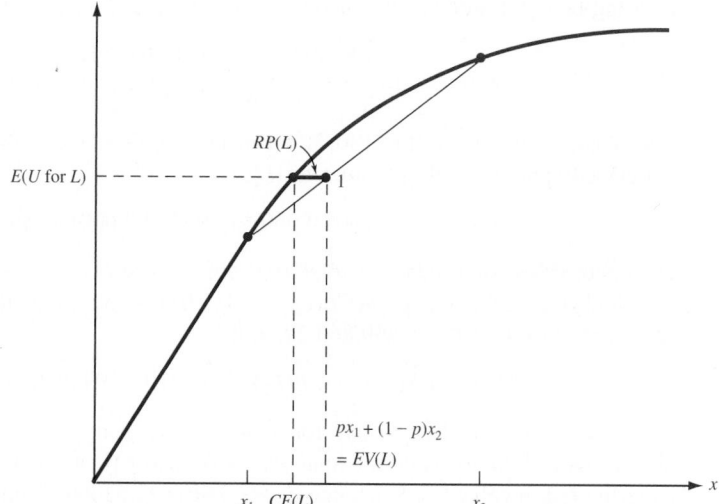

FIGURE 2
Why a Concave Utility Function Implies Risk-Averse Behavior

of L must be less than $px_1 + (1 - p)x_2 = EV(L)$. This implies that $RP(L) = EV(L) - CE(L) > 0$, and the decision maker exhibits risk-averse behavior. In Problem 4 at the end of this section, the reader will be asked to show that if $u(x)$ is strictly convex, the decision maker exhibits risk-seeking behavior.

If the decision maker is risk-neutral (that is, $u(x) = ax + b$), he or she chooses among lotteries via the expected reward criterion of Section 13.1 (see Problem 5 at the end of this section). Thus, when ranking lotteries, a risk-neutral decision maker considers only the expected value (and not the risk) of the lotteries.

Example 2 illustrates the concepts of risk premium, certainty equivalent, and risk aversion.

EXAMPLE 2   **Joan's Assets**

Joan's utility function for her asset position x is given by $u(x) = x^{1/2}$. Currently, Joan's assets consist of $10,000 in cash and a $90,000 home. During a given year, there is a .001 chance that Joan's home will be destroyed by fire or other causes. How much would Joan be willing to pay for an insurance policy that would replace her home if it were destroyed?

Solution   Let x = annual insurance premium. Then Joan must choose between the following lotteries:

Asset Position

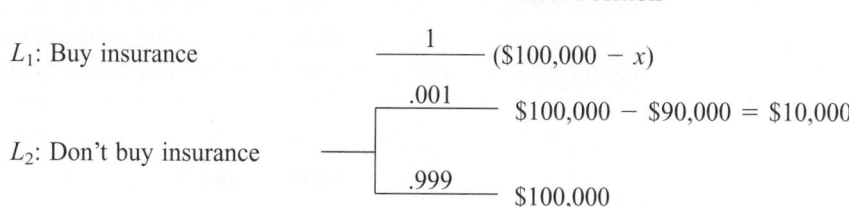

L_1: Buy insurance

L_2: Don't buy insurance

Joan will prefer L_1 to L_2 if L_1's expected utility exceeds L_2's expected utility. Thus, $L_1 p L_2$ if and only if

$$(100{,}000 - x)^{1/2} > .001(10{,}000)^{1/2} + .999(100{,}000)^{1/2}$$
$$> .10 + 315.91154$$
$$> 316.01154$$

Squaring both sides of the last inequality we find that $L_1 p L_2$ if and only if

$$100{,}000 - x > (316.01154)^2$$
$$x < \$136.71$$

Thus, Joan would pay up to $136.71 for insurance. Of course, if $p = \$136.71$, $L_1 i L_2$. Let's compute the risk premium for L_2:

$$EV(L_2) = .001(10{,}000) + .999(100{,}000) = \$99{,}910$$

(an expected loss of $100{,}000 - 99{,}910 = \$90$). Since $E(U$ for $L_2) = 316.01154$, we can find $CE(L_2)$ from the relation $u(CE(L_2)) = 316.01154$, or $[CE(L_2)]^{1/2} = 316.01154$. Thus, $CE(L_2) = (316.01154)^2 = \$99{,}863.29$, and

$$RP(L_2) = EV(L_2) - CE(L_2) = 99{,}910 - 99{,}863.29 = \$46.71$$

Therefore, Joan is willing to pay for annual home insurance $46.71 more than the expected loss of $90. (Recall that Joan was willing to pay up to $90 + 46.71 = \$136.71$ to avoid the risk involved in her home being destroyed.) Joan exhibits risk-averse behavior ($RP(L_2) > 0$). Since

$$u''(x) = \frac{-x^{-3/2}}{4} < 0$$

$u(x)$ is strictly concave, and $RP(L) > 0$ would hold for any nondegenerate lottery.

In reality, many people exhibit both risk-seeking behavior (they purchase lottery tickets, go to Las Vegas) and risk-averse behavior (they buy home insurance). A person whose utility function contains both convex and concave segments may exhibit both risk-averse and risk-seeking behavior. Consider a decision maker whose utility function $u(x)$ for change in current asset position is given in Figure 3. If forced to choose between

$$L_1 \xrightarrow{\ 1\ } 0 \quad \text{and} \quad L_2 \begin{cases} .10 & \$2{,}500 \\ .90 & -\$300 \end{cases}$$

what would this person do?

From Figure 3, we find that $u(0) = .20$, $u(2{,}500) = .50$, and $u(-300) = .18$. Thus, $E(U$ for $L_1) = .20$ and $E(U$ for $L_2) = .10(.50) + .90(.18) = .212$. Thus, $L_2 p L_1$. This means that L_2 has a certainty equivalent of at least $0. Since $EV(L_2) = -\$20$, this implies that $RP(L_2) = EV(L_2) - CE(L_2) < 0$. The decision maker exhibits risk-seeking behavior in this situation, because for changes in asset position between $0 and $2,500, $u(x)$ is a convex function.

Now suppose the decision maker can, for $200, insure himself against a loss of $2,000, which occurs with probability .08. Then he must choose between

$$L_3 \xrightarrow{\ 1\ } -\$200 \quad \text{and} \quad L_4 \begin{cases} .08 & -\$2{,}000 \\ .92 & \$0 \end{cases}$$

From Figure 3, $u(-200) = .19$, $u(0) = .20$, and $u(-2{,}000) = 0$. Thus, $E(U$ for $L_3) = .19$ and $E(U$ for $L_4) = .80(0) + .92(.20) = .184$, and $L_3 p L_4$. This shows that $CE(L_4) < -\$200$. Since $EV(L_4) = .08(-2{,}000) + .92(0) = -\$160$, $RP(L_4) = EV(L_4) - CE(L_4) > 0$, and the decision maker is exhibiting risk-averse behavior, because $u(x)$ is concave for $-2{,}000 < x < 0$. Thus, if his utility function has both convex and concave segments, a person can exhibit both risk-seeking and risk-averse behavior.

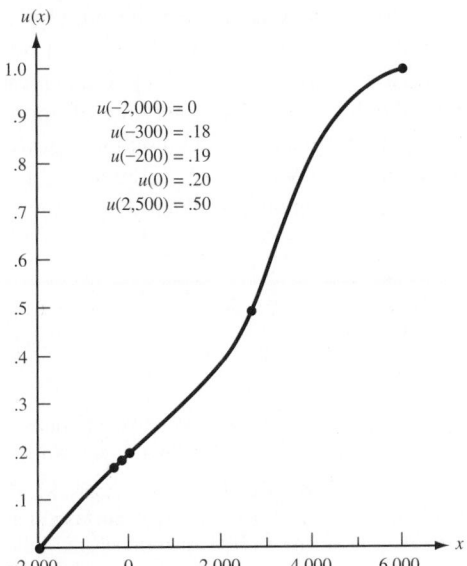

$u(-2{,}000) = 0$
$u(-300) = .18$
$u(-200) = .19$
$u(0) = .20$
$u(2{,}500) = .50$

FIGURE 3
A Utility Function That Exhibits Both Risk-Seeking and Risk-Averse Behavior

Exponential Utility

Classes of "ready-made" utility functions have been developed. One important class is called **exponential utility** and has been used in many financial investment analyses. An exponential utility function has only one adjustable numerical parameter, and there are straightforward ways to discover the most appropriate value of this parameter for a particular individual or company. So the advantage of using an exponential utility function is that it is relatively easy to assess. The drawback is that exponential utility functions do not capture all types of attitudes toward risk. Nevertheless, their ease of use has made them popular.

An exponential utility function has the following form:

$$U(x) = 1 - e^{-x/R}$$

Here, x is a monetary value (a payoff if positive, a cost if negative), $U(x)$ is the utility of this value, and $R > 0$ is an adjustable parameter called the **risk tolerance.** Basically, the risk tolerance measures how much risk the decision maker will tolerate. The larger the value of R, the less risk averse the decision maker is. That is, a person with a large value of R is more willing to take risks than a person with a small value of R.

To assess a person's (or company's) exponential utility function, we need only assess the value of R. There are a couple of tips for doing this. First, it has been shown that the risk tolerance is approximately equal to that dollar amount R such that the decision maker is indifferent between the following two options:

- Option 1: Obtain no payoff at all
- Option 2: Obtain a payoff of R dollars or a loss of $R/2$ dollars, depending on the flip of a fair coin

For example, if I am indifferent between a bet where I win \$1,000 or lose \$500, with probability 0.5 each, and not betting at all, then my R is approximately \$1,000. From this criterion it certainly makes intuitive sense that a wealthier person (or company) ought to have a larger value of R. This has been found in practice.

A second tip for finding R is based on empirical evidence found by Ronald Howard, a prominent decision analyst. Through his consulting experience with several large companies, he discovered tentative relationships between risk tolerance and several financial variables—net sales, net income, and equity. (See Howard (1992).) Specifically, he found that R was approximately 6.4% of net sales, 124% of net income, and 15.7% of equity for the companies he studied. For example, according to this prescription, a company with net sales of $30 million should have a risk tolerance of approximately $1.92 million. Howard admits that these percentages are only guidelines. However, they do indicate that larger and more profitable companies tend to have larger values of R, which means that they are more willing to take risks involving given dollar amounts.

PROBLEMS

Group A

1 Suppose my utility function for asset position x is given by $u(x) = \ln x$.

   a Am I risk-averse, risk-neutral, or risk-seeking?

   b I now have $20,000 and am considering the following two lotteries:

      L_1: With probability 1, I lose $1,000.

      L_2: With probability .9, I gain $0.

          With probability .1, I lose $10,000.

Determine which lottery I prefer and the risk premium of L_2.

2 Answer Problem 1 for a utility function $u(x) = x^2$.

3 Answer Problem 1 for a utility function $u(x) = 2x + 1$.

4 Show that a decision maker who has a strictly convex utility function will exhibit risk-seeking behavior.

5 Show that a decision maker who has a linear utility function will rank two lotteries according to their expected value.

6 A decision maker has a utility function for monetary gains x given by $u(x) = (x + 10,000)^{1/2}$.

   a Show that the person is indifferent between the status quo and

     L: With probability $\frac{1}{3}$, he or she gains $80,000

        With probability $\frac{2}{3}$, he or she loses $10,000

   b If there is a 10% chance that a painting valued at $10,000 will be stolen during the next year, what is the most (per year) that the decision maker would be willing to pay for insurance covering the loss of the painting?

7 Patty is trying to determine which of two courses to take. If she takes the operations research course, she believes that she has a 10% chance of receiving an A, a 40% chance for a B, and a 50% chance for a C. If Patty takes a statistics course, she has a 70% chance for a B, a 25% chance for a C, and a 5% chance for a D. Patty is indifferent between

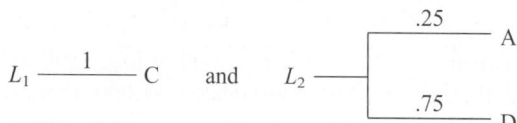

She is also indifferent between

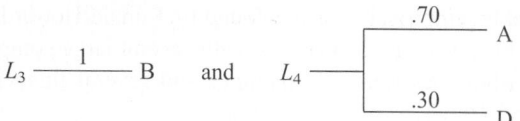

If Patty wants to take the course that maximizes the expected utility of her final grade, which course should she take?

8 We are going to invest $1,000 for a period of 6 months. Two potential investments are available: T-bills and gold. If the $1,000 is invested in T-bills, we are certain to end the 6-month period with $1,296. If we invest in gold, there is a $\frac{3}{4}$ chance that we will end the 6-month period with $400 and a $\frac{1}{4}$ chance that we will end the 6-month period with $10,000. If we end up with x dollars, our utility function is given by $u(x) = x^{1/2}$. Should we invest in gold or T-bills?

9 We now have $5,000 in assets and are given a choice between investment 1 and investment 2. With investment 1, 80% of the time we increase our asset position by $295,000, and 20% of the time we increase our asset position by $95,000. With investment 2, 50% of the time we increase our asset position by $595,000, and 50% of the time we increase our asset position by $5,000. Our utility function for final asset position x is $u(x)$. We are given the following values for $u(x)$: $u(0) = 0$, $u(640,000) = .80$, $u(810,000) = .90$, $u(0) = 0$, $u(90,000) = .30$, $u(1,000,000) = 1$, $u(490,000) = .7$.

   a Are we risk-averse, risk-seeking, or risk-neutral? Explain.

   b Will we prefer investment 1 or investment 2?

10 My current income is $40,000. I believe that I owe $8,000 in taxes. For $500, I can hire a CPA to review my tax return; there is a 20% chance that she will save me $4,000 in taxes. My utility function for (disposable income) = (current income) − (taxes) − (payment to accountant) is given by $\sqrt{x}$ where x is disposable income. Should I hire the CPA?

Group B

11[†] (The Allais Paradox) Suppose we are offered a choice between the following two lotteries:

     L_1: With probability 1, we receive $1 million.

     L_2: With probability .10, we receive $5 million.

         With probability .89, we receive $1 million.

         With probability .01, we receive $0.

Which lottery do we prefer? Now consider the following two lotteries:

[†]Based on Allais (1953).

L_3: With probability .11, we receive \$1 million.

With probability .89, we receive \$0.

L_4: With probability .10, we receive \$5 million.

With probability .90, we receive \$0.

Which lottery do we prefer? Suppose (like most people), we prefer L_1 to L_2. Show that L_3 must have a larger expected utility than L_4.

12 (The St. Petersburg Paradox) Let L represent the following lottery. I toss a coin until it comes up heads. If the first heads is obtained on the nth toss of the coin, I receive a payoff of $\$2^n$.

a If I were a risk-neutral decision maker, what would be the certainty equivalent of L? Is this reasonable?

b If a decision maker's utility function for increasing wealth by x dollars is given by $u(x) = \log_2(x)$, what would be the certainty equivalent of L?

13 Joe is a risk-averse decision maker. Which of the following lotteries will he prefer?

L_1: With probability .10, Joe loses \$100.

With probability .90, Joe receives \$0.

L_2: With probability .10, Joe loses \$190.

With probability .90, Joe receives \$10.

14[†] (The Ellsberg Paradox) An urn contains 90 balls. It is known that 30 are red and that each of the other 60 is either yellow or black. One ball will be drawn at random from the urn. Consider the following four options:

Option 1 We receive \$1,000 if a red ball is drawn.

Option 2 We receive \$1,000 if a yellow ball is drawn.

Option 3 We receive \$1,000 if a yellow or black ball is drawn.

Option 4 We receive \$1,000 if a red or black ball is drawn.

a Explain why most people prefer option 1 over option 2 and also prefer option 3 over option 4.

b If we prefer option 1 to option 2, explain why we should also prefer option 4 over option 3.

15 Although the Von Neumann–Morgenstern axioms seem plausible, there are many reasonable situations in which people appear to violate these axioms. For example, suppose

[†]Based on Ellsberg (1961).

TABLE 8

	Starting Salary	Location	Opportunity for Advancement
Job 1	E	S	G
Job 2	G	E	S
Job 3	S	G	E

a recent college graduate must choose between three job offers on the basis of starting salary, location of job, and opportunity for advancement. Given two job offers that are satisfactory with regard to all three attributes, the graduate will decide between two job offers by choosing the one that is superior on at least two of the three attributes. Suppose he or she has three job offers and has rated each one as shown in Table 8 (E = excellent, G = good, and S = satisfactory). Show that the graduate's preferences among these jobs violate the Complete Ordering Axiom.

Group C

16 Suppose my utility function for my asset position is $u(x) = x^{1/2}$. I have \$10,000 at present. Consider the following lottery:

L: With probability $\frac{1}{2}$, L yields a payoff of \$1,025.

With probability $\frac{1}{2}$, L yields a payoff of $-\$199$.

a If I don't have the right to play L, find an equation that when solved would yield the amount I would be willing to pay for the right to play L. This is called the **buying price** of lottery L.

b If I have the right to play L, what is the least I would accept from somebody who wanted to buy the right to play L? (After someone else buys L, I can't play L.) This is called the **selling price** of lottery L.

c Answer part (b) for the case that I have \$1,000.

d Suppose that my utility function for my asset position is $u(x) = 1 - e^{-x}$. Show that for all possible asset positions, the buying price of L and the selling price of L will remain the same. Show that for all asset positions, the buying price of L will equal the selling price of L.

13.3 Flaws in Expected Maximization of Utility: Prospect Theory and Framing Effects

The axioms underlying expected maximization of utility (EMU) seem reasonable, but in practice people's decisions often deviate from the predictions of EMU. Psychologists Tversky and Kahneman[‡] (1981) developed **prospect theory** and **framing effects for values** to try and explain why people deviate from the predictions of EMU.

[‡]In 2002, Kahneman received the Nobel Prize for Economics, in large part honoring his work with Tversky. Tversky was not awarded the prize because he died in 1996 (Nobel Prizes are not given posthumously).

Prospect Theory

Here is one example of a decision that cannot be explained by EMU. Ask a person to choose between lottery 1 and lottery 2:

> Lottery 1: $30 for certain
>
> Lottery 2:   80% chance at $45 and 20% chance at $0

Most people prefer lottery 1 to lottery 2. Next ask the same person to choose between lottery 3 and lottery 4:

> Lottery 3:   20% chance at $45 and 80% chance at $0
>
> Lottery 4:   25% at $30 and 75% chance at $0

Most people choose lottery 3 over lottery 4. Now let $u(0) = 0$ and $u(45) = 1$. A decision maker following EMU will choose lottery 1 over lottery 2 if and only if $u(30) > .8$. A decision maker following EMU will choose lottery 3 over lottery 4 if and only if $.2 > .25u(30)$ or $u(30) < .8$. This implies that a believer in EMU cannot choose lottery 1 over lottery 2 and lottery 3 over lottery 4. Thus, for this situation, the choices of most people contradict EMU. Tversky and Kahneman developed prospect theory to explain the decision-making paradox we have just described. Prospect theory assumes that we do not treat probabilities as they are given in a decision-making problem. Instead, the decision maker treats a probability p for an event as a "distorted" probability $\Pi(p)$. A $\Pi(p)$ function that seems to explain many paradoxes is shown in Figure 4.

The shape of the $\Pi(p)$ function in the figure implies that individuals are more sensitive to changes in probability when the probability of an event is small (near 0) or large (near 1). The equation we used to construct our $\Pi(p)$ curve is $\Pi(p) = 1.89799p - 3.55995p^2 + 2.662549p^3$. How does prospect theory explain our paradox? From the values of $\Pi(p)$ given in Figure 5, we can compare the expected "prospects" of lottery 1 versus lottery 2 and lottery 3 versus lottery 4.

> Prospect for lottery 1: $u(30)$
>
> Prospect for lottery 2:   .602
>
> Prospect for lottery 3:   .258
>
> Prospect for lottery 4:   .293$u(30)$.

Thus, lottery 1 is preferred to lottery 2 if $u(30) > .602$, while lottery 3 is preferred to lottery 4 if $.258 > .293u(30)$ or $u(30) < .258/.293 = .88$. Our paradox evaporates, because for many people, $u(30)$ will be between .602 and .88!

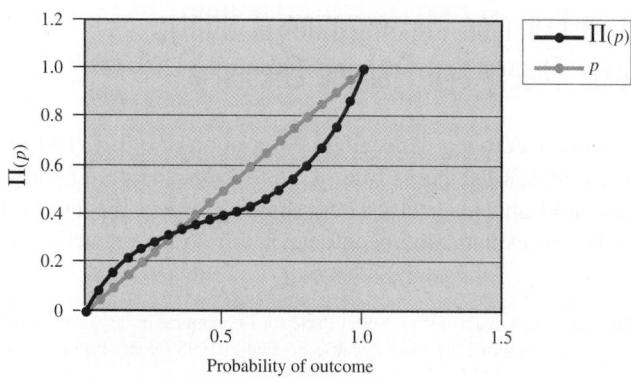

FIGURE 4
Weight Applied to Utility

	C	D
15	0.1	0.156803
16	0.15	0.213497
17	0.2	0.258382
18	0.25	0.293455
19	0.3	0.320713
20	0.35	0.342153
21	0.4	0.359771
22	0.45	0.375565
23	0.5	0.391531
24	0.55	0.409667
25	0.6	0.431969
26	0.65	0.460434
27	0.7	0.497059
28	0.75	0.543841
29	0.8	0.602778
30	0.85	0.675865
31	0.9	0.7651
32	0.95	0.872479
33	1	1

FIGURE 5
Lottery Prospects

Framing

The idea of framing is based on the fact that people often set their utility function from the standpoint of a frame or status quo from which they view the current situation. Most people's utility functions treat a loss of a given value as being more important than a gain of an identical value. This is reflected in the utility function shown in Figure 6, which is convex for losses and concave for gains.

To see how framing can explain the failure of EMU, consider the following problem that Tversky and Kahneman gave to a group of students. The US is preparing for the outbreak of a disease that is expected to kill 600 people. Two alternative programs have been proposed:

Program I: 200 people are saved.

Program II: With probability $\frac{1}{3}$, 600 people are saved.

Most students preferred program I, probably because with program II there is a large risk of saving nobody. Since the programs are phrased in terms of lives saved, most people take the frame or reference point for this problem to be no lives saved or 600 people dead. Since the effect of each program is expressed in gains, and the utility function is concave for gains, we find that $u(200) = u((\frac{2}{3}) 0 + (\frac{1}{3})600)) > (\frac{1}{3})u(600) + (\frac{2}{3})u(0) = (\frac{1}{3})u(600)$. This implies, of course, that the person chooses program I over program II.

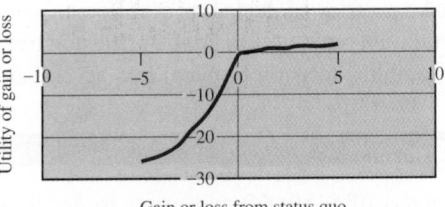

FIGURE 6
Utility Function for Framing

Gain or loss from status quo

Next, Tversky and Kahneman rephrased the problem as follows:

Program I: 400 people die.

Program II: With probability $\frac{2}{3}$, 600 people die.

Now most people choose program II. Note that both program I's are identical, as are both program II's. Why do most people choose program II for the second phrasing of the alternatives? The second phrasing shifts most people's reference points from "No lives saved" (in first phrasing) to "Nobody dies." The outcomes are expressed as losses (deaths), so the convexity of the utility curve for losses implies that

$$(\tfrac{2}{3})u(-600) = (\tfrac{2}{3})u(-600) + (\tfrac{1}{3})u(0) > u((\tfrac{2}{3})(-600) + \tfrac{1}{3}(0)) = u(-400)$$

This implies, of course, that the person chooses program II over program I.

PROBLEMS

Group A

1 Explain how prospect theory and/or framing explains the Allais Paradox. (See Problem 11 of Section 13.2.)

2 Suppose a decision maker has a utility function $u(x) = x^{1/3}$. We flip a fair coin and receive $10 for heads and $0 for tails.

a Using expected utility theory, determine the certainty equivalent of this lottery.

b Using $\Pi(p) = 1.89799p - 3.55995p^2 + 2.662549p^3$, use prospect theory to determine the certainty equivalent of the lottery.

c Intuively explain why your answer in part (b) is smaller than your answer in part (a).

d What implications does this problem have for the method used in Section 13.2 to estimate a person's utility function?

3 You are given a choice between lottery 1 and lottery 2. You are also given a choice between lottery 3 and lottery 4.

Lottery 1: A sure gain of $240

Lottery 2: 25% chance to gain $1,000 and 75% chance to gain nothing

Lottery 3: A sure loss of $750

Lottery 4: A 75% chance to lose $1,000 and a 25% chance of losing nothing

84% of all people prefer lottery 1 over lottery 2, and 87% choose lottery 4 over lottery 3.

a Explain why the choice of lottery 1 over lottery 2 and lottery 4 over lottery 3 contradicts expected utility maximization. (*Hint:* Compare lottery 1 + lottery 4 to lottery 2 + lottery 3.)

b Can you explain this anomalous behavior?

4 Tversky and Kahneman asked 72 respondents to choose between lottery 1 and lottery 2 and lottery 3 and lottery 4.

Lottery 1: A .001 chance at winning $5,000 and a .999 chance of winning $0

Lottery 2: A sure gain of $5

Lottery 3: A .001 chance of losing $5,000 and a .999 chance of losing $0

Lottery 4: A sure loss of $5

More than 75% of all participants preferred lottery 1 to lottery 2 and lottery 4 to lottery 3.

a Which choices would be made by a risk-averse decision maker?

b Which choices would be made by a risk-seeking decision maker?

c How does the observed behavior of the participants contradict expected utility maximization?

d How does prospect theory resolve the contradiction?

13.4 Decision Trees

Often, people must make a series of decisions at different points in time. Then decision trees can be used to determine optimal decisions. A decision tree enables a decision maker to decompose a large complex decision problem into several smaller problems.

EXAMPLE 3   **Colaco Marketing**

Colaco currently has assets of $150,000 and wants to decide whether to market a new chocolate-flavored soda, Chocola. Colaco has three alternatives:

Alternative 1 Test market Chocola locally, then utilize the results of the market study to determine whether or not to market Chocola nationally.

Alternative 2 Immediately (without test marketing) market Chocola nationally.

Alternative 3 Immediately (without test marketing) decide not to market Chocola nationally.

In the absence of a market study, Colaco believes that Chocola has a 55% chance of being a national success and a 45% chance of being a national failure. If Chocola is a national success, Colaco's asset position will increase by $300,000, and if Chocola is a national failure, Colaco's asset position will decrease by $100,000.

   If Colaco performs a market study (at a cost of $30,000), there is a 60% chance that the study will yield favorable results (referred to as a *local success*) and a 40% chance that the study will yield unfavorable results (referred to as a *local failure*). If a local success is observed, there is an 85% chance that Chocola will be a national success. If a local failure is observed, there is only a 10% chance that Chocola will be a national success. If Colaco is risk-neutral (wants to maximize its expected final asset position), what strategy should the company follow?

Solution To draw a decision tree that represents Colaco's problem, we begin at the present and proceed toward future events and decisions. The decision tree in Figure 7 is constructed with two kinds of forks: decision forks (denoted by □) and event forks (denoted by ○).

   A **decision fork** represents a point in time when Colaco has to make a decision. Each branch emanating from a decision fork represents a possible decision. An example of a decision fork occurs when Colaco must determine whether or not to test market Chocola.

   An **event fork** is drawn when outside forces determine which of several random events will occur. Each branch of an event fork represents a possible outcome, and the number on each branch represents the probability that the event will occur. For example, if Colaco decides to test market Chocola, the company faces the following event fork when observing the results of the test market study:

   A branch of a decision tree is a **terminal branch** if no forks emanate from the branch. Thus, the branches indicating National success and National failure are terminal branches of Colaco's decision tree. Since we are maximizing expected final asset position at each terminal branch, we must enter the final asset position that will result if the path leading to the given terminal branch occurs. For example, the terminal branch National failure that follows Local failure leads to a final asset position of $150,000 - 30,000 - 100,000 = \$20,000$. If we were maximizing expected revenues, we would enter revenues on each terminal branch.

   To determine the decisions that will maximize Colaco's expected final asset position, we work backward (sometimes called "folding back the tree") from right to left.[†] At each

---

[†]See Chapters 17 and 18 for an explanation of working backward (often called *dynamic programming*).

FIGURE 7
Colaco's Decision Tree (Risk-Neutral)

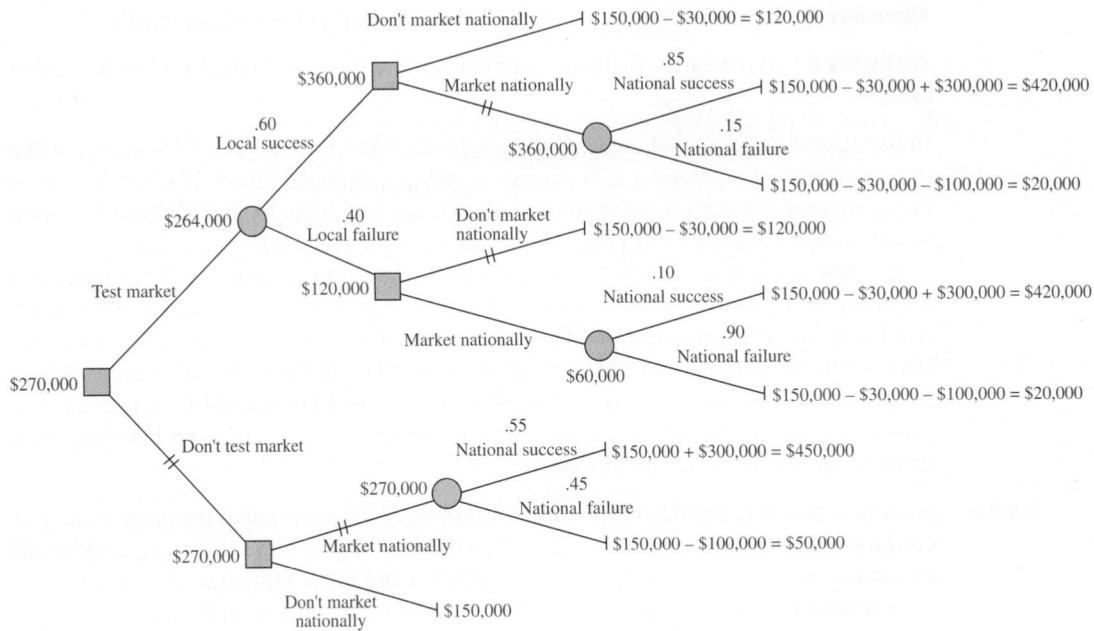

event fork, we calculate the expected final asset position and enter it in ○. At each decision fork, we denote by ‖ the decision that maximizes the expected final asset position and enter the expected final asset position associated with that decision in □. We continue working backward in this fashion until we reach the beginning of the tree. Then the optimal sequence of decisions can be obtained by following the ‖.

We begin by determining the expected final asset positions for the following three event forks:

1   Market nationally after Local success. Here we have an expected final asset position of .85(420,000) + .15(20,000) = $360,000.

2   Market nationally after Local failure. Here we have an expected final asset position of .10(420,000) + .90(20,000) = $60,000.

3   Market nationally after Don't test market. Here we have an expected final asset position of .55(450,000) + .45(50,000) = $270,000.

We may now evaluate three decision forks:

1   Decision after Local success. Market nationally yields a larger expected final asset position than Don't market nationally, so we ‖ Market nationally and enter an expected final asset position of $360,000.

2   Decision after Local failure. Don't market nationally yields a larger expected final asset position than Market nationally, so we ‖ Don't market nationally and enter an expected final asset position of $120,000.

3   Decision for Don't test market. Market nationally yields a larger expected final asset position than Don't market nationally, so we ‖ Market nationally and enter an expected final asset position of $270,000.

We now must evaluate the event fork emanating from the Test market decision. This event fork yields an expected final asset position of .60(360,000) + .40(120,000) = $264,000, which is entered in ○.

All that remains is to determine the correct decision at the decision fork Test market versus Don't test market. We have found that Test market yields an expected final asset position of $264,000, and Don't test market yields an expected final asset position of $270,000. Thus, we ‖ Don't test market and enter $270,000 in □.

We have now reached the beginning of the tree and have found that Colaco's optimal decision is Don't test market and then Market nationally. This strategy will yield an expected final asset position of $270,000. Observe that the decision tree also tells us that if we had test marketed and then acted optimally (Market nationally after Local success and Don't market nationally after Local failure), we would have obtained an expected final asset position of $264,000.

Incorporating Risk Aversion into Decision Tree Analysis

Note that Colaco's optimal strategy yields a .45 chance that the company will end up with a relatively small final asset position of $50,000. On the other hand, the strategy of test marketing and acting optimally on the results of the test market study yields only a (.60)(.15) = .09 chance that Colaco's asset position will be below $100,000. (Why?) Thus, if Colaco is a risk-averse decision maker, the strategy of immediately marketing nationally may not reflect the company's preference.

To illustrate how risk aversion may be incorporated into decision tree analysis, suppose that Colaco has the risk-averse utility function $u(x)$ in Figure 8 (x = final asset position). (How do we know that this utility function exhibits risk aversion?) *To determine Colaco's optimal decisions (that is, the decisions that maximize expected utility), simply replace each final asset position x_0 with its utility $u(x_0)$. Then at each event fork, compute the expected utility of Colaco's final asset position, and at each decision fork, choose the branch having the largest expected utility.*

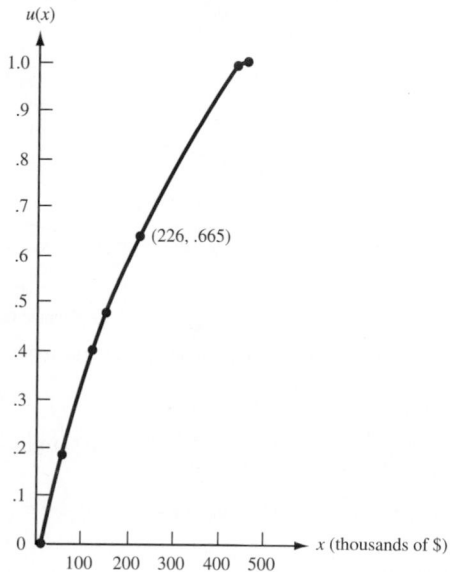

FIGURE 8
Colaco's Utility Function

We find from Figure 8 that $u(\$450,000) = 1$, $u(\$420,000) = .99$, $u(\$150,000) = .48$, $u(\$120,000) = .40$, $u(\$50,000) = .19$, and $u(\$20,000) = 0$. Substituting these values into the decision tree of Figure 7 yields the decision tree in Figure 9. We compute the expected utility at the following three event forks:

1 Market nationally after Local success. Here we have an expected utility of $.85(.99) + .15(0) = .8415$.

2 Market nationally after Local failure. Here we have an expected utility of $.10(.99) + .90(0) = .099$.

3 Market nationally after Don't test market. Here we have an expected utility of $.55(1) + .45(.19) = .6355$.

We may now evaluate three decision forks:

1 Decision after Local success. Market nationally yields a larger expected utility than Don't market nationally, so for this fork we ‖ Market nationally and enter an expected utility of .8415.

2 Decision after Local failure. Don't market nationally yields a larger expected utility than Market nationally, so for this fork we ‖ Don't market nationally and enter an expected utility of .40.

3 Decision for Don't test market. Market nationally yields a larger expected utility than Don't market nationally, so for this fork we ‖ Market nationally and enter an expected utility of .6355.

We now must evaluate the event fork emanating from the Test market decision. This event fork yields an expected utility of $.60(.8415) + .40(.40) = .6649$, which is entered in ○. All that remains is to determine the correct decision at the decision fork Test market versus Don't test market. We know that Test market yields an expected utility of .6649,

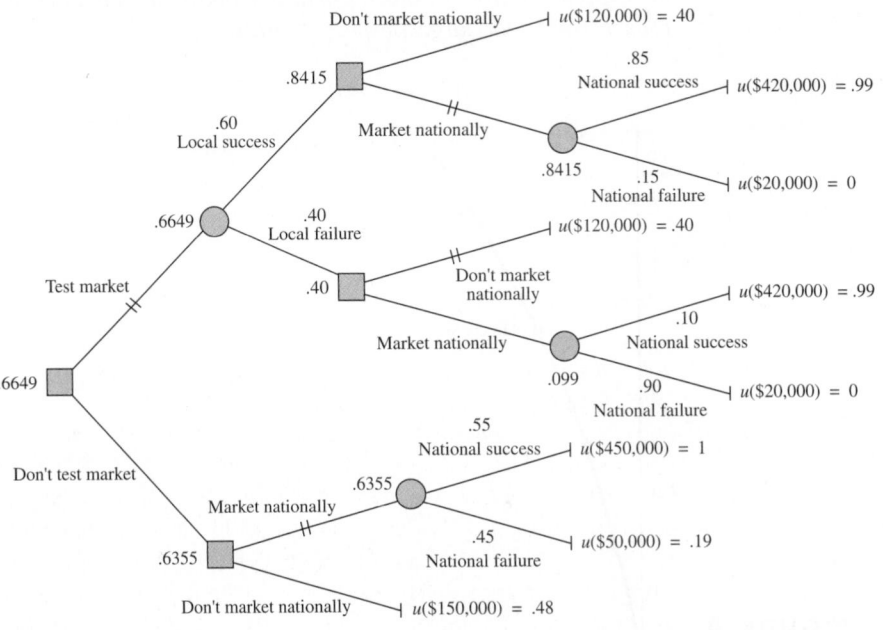

FIGURE 9
Colaco's Decision Tree
(Risk-Averse)

$u(\$226,000) = .6649$, so this situation is equivalent to a certain asset position of $226,000.

and Don't test market yields an expected utility of .6355, so we ‖ Test market and enter an expected utility of .6649 in □.

We have now reached the beginning of the tree and have found that Colaco's optimal decision is to begin by test marketing. If a local success is observed, then Colaco should market Chocola nationally; if a local failure is observed, then Colaco should not market Chocola nationally. This optimal strategy yields only a $.60(.15) = .09$ chance that Colaco will have a final asset position of less than \$100,000. This reflects the risk-averse nature of the utility function in Figure 8. Also, we see from Figure 8 that $u(\$226,000) = .665$. Since Colaco views the current situation as having an expected utility of .6649, this means that the company considers the current situation equivalent to a certain asset position of \$226,000. Thus, if somebody offered to pay more than $226,000 - 150,000 = \$76,000$ to buy the rights to Chocola, Colaco should take the offer. This is because receiving more than \$76,000 for the rights to Chocola would bring Colaco's asset position to more than $150,000 + 76,000 = \$226,000$, and this situation has a higher expected utility than .665.

Expected Value of Sample Information

Decision trees can be used to measure the value of sample or test market information. To illustrate how this is done, we again assume that Colaco is risk-neutral. What is the value of the information that would be obtained by test marketing Chocola?

We begin by determining Colaco's expected final asset position if the company acts optimally and the test market study is costless. We call this expected final asset position Colaco's **expected value with sample information** (EVWSI). From Figure 7, we see that if we Test market and then act optimally, we will now have an expected final asset position of $264,000 + 30,000 = \$294,000$. Since \$294,000 is larger than the expected asset position of the Don't test market branch (\$270,000), we find that EVWSI = \$294,000.

We next determine the largest expected final asset position that Colaco would obtain if the test market study were not available. We call this the **expected value with original information** (EVWOI). From the Don't test market branch of Figure 7, we find EVWOI = \$270,000. Now the expected value of the test market information, referred to as **expected value of sample information** (EVSI), is defined to be EVSI = EVWSI − EVWOI.

In the Colaco example, EVSI is the most that Colaco can pay for the test market information and still be at least as well off as without the test market information. Thus, for the Colaco example, EVSI = $294,000 - 270,000 = \$24,000$. Since the cost of the test market study (\$30,000) exceeds EVSI, Colaco should not (as we already know) conduct the test market study.

Expected Value of Perfect Information

We can modify the analysis used to determine EVSI to find the value of perfect information. By **perfect information** we mean that all uncertain events that can affect Colaco's final asset position still occur with the given probabilities (so there is still a .55 chance of Chocola being a national success and a .45 chance that Chocola will be a national failure), but Colaco finds out whether Chocola is a national success or a national failure *before* making the decision to market Chocola nationally or not. This information can then be used to determine Colaco's optimal marketing strategy. Thus, **expected value with perfect information** (EVWPI) is found by drawing a decision tree in which the decision maker has perfect information about which state has occurred before making a decision. Then the **expected value of perfect information** (EVPI) is given by EVPI = EVWPI − EVWOI.

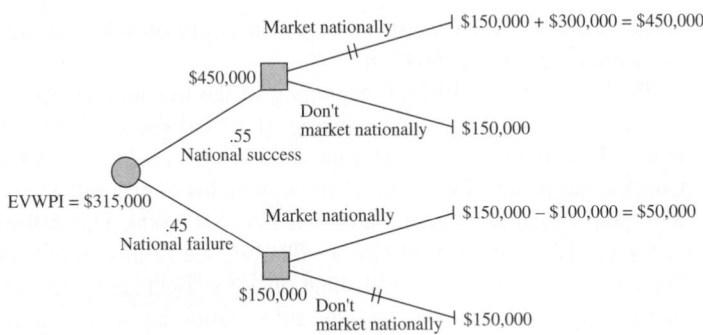

FIGURE 10
Expected Value with Perfect Information (EVWPI) for Colaco

For the Colaco example, we find from Figure 10 that EVWPI = $315,000. Then EVPI = 315,000 − 270,000 = $45,000. Thus, a perfect (one that was always correct) test marketing study would be worth $45,000. EVPI is a useful upper bound on the value of sample or test market information; that is, no sample or test market information (no matter how good) can be worth more than $45,000.

EXAMPLE 4 **Art Dealer**

An art dealer's client is willing to buy the painting *Sunplant* at $50,000. The dealer can buy the painting today for $40,000 or can wait a day and buy the painting tomorrow (if it has not been sold) for $30,000. The dealer may also wait another day and buy the painting (if it is still available) for $26,000. At the end of the third day, the painting will no longer be available for sale. Each day, there is a .60 probability that the painting will be sold. What strategy maximizes the dealer's expected profit?

Solution The decision tree for this example is given in Figure 11. The key to drawing this decision tree is that each day, the dealer must choose between buying the painting and waiting another day. Of course, waiting might mean that the dealer may never be able to buy the painting. As we see from the decision tree, the dealer should buy the painting on the first day.

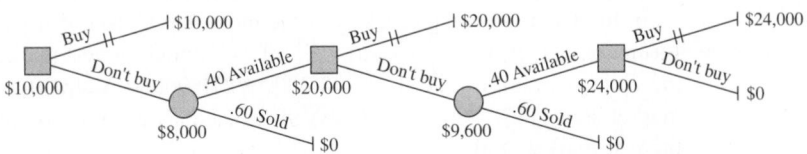

FIGURE 11
Decision Tree for Example 4

PROBLEMS

Group A

1 Oilco must determine whether or not to drill for oil in the South China Sea. It costs $100,000, and if oil is found, the value is estimated to be $600,000. At present, Oilco believes there is a 45% chance that the field contains oil. Before drilling, Oilco can hire (for $10,000) a geologist to obtain more information about the likelihood that the field will contain oil. There is a 50% chance that the geologist will issue a favorable report and a 50% chance of an unfavorable report. Given a favorable report, there is an

80% chance that the field contains oil. Given an unfavorable report, there is a 10% chance that the field contains oil. Determine Oilco's optimal course of action. Also determine EVSI and EVPI.

2 The decision sciences department is trying to determine which of two copying machines to purchase. Both machines will satisfy the department's needs for the next ten years. Machine 1 costs $2,000 and has a maintenance agreement, which, for an annual fee of $150, covers all repairs. Machine 2 costs $3,000, and its annual maintenance cost is a random variable. At present, the decision sciences department believes there is a 40% chance that the annual maintenance cost for machine 2 will be $0, a 40% chance it will be $100, and a 20% chance it will be $200.

Before the purchase decision is made, the department can have a trained repairer evaluate the quality of machine 2. If the repairer believes that machine 2 is satisfactory, there is a 60% chance that its annual maintenance cost will be $0 and a 40% chance it will be $100. If the repairer believes that machine 2 is unsatisfactory, there is a 20% chance that the annual maintenance cost will be $0, a 40% chance it will be $100, and a 40% chance it will be $200. If there is a 50% chance that the repairer will give a satisfactory report, what is EVSI? If the repairer charges $40, what should the decision sciences department do? What is EVPI?

3 I am managing the Chicago Cubs. Suppose there is a runner on first base with nobody out and we want to determine whether we should bunt. Assume that a bunt will yield one of two results: (1) With probability .80, the bunt will be successful, in which case the batter is out and the runner on first base advances to second base. (2) With probability .20 the bunt is unsuccessful and the runner on first base is out trying to advance to second base and the batter is safe at first base.

The expected number of runs that the Cubs will score in an inning in various situations is given in Table 9.

a If our goal is to maximize the expected number of runs scored in an inning, should we bunt? Despite this answer, why do you think teams bunt?

b If we are considering stealing second base with nobody out, what chance of success is needed for stealing second to be an optimal decision?

4 The Nitro Fertilizer Company is developing a new fertilizer. If Nitro markets the product and it is successful, the company will earn a $50,000 profit; if it is unsuccessful, the company will lose $35,000. In the past, similar products have been successful 60% of the time. At a cost of $5,000, the effectiveness of the new fertilizer can be tested. If the test result is favorable, there is an 80% chance that the fertilizer will be successful. If the test result is unfavorable, there is only a 30% chance that the fertilizer will be successful. There is a 60% chance of a favorable test result and a 40% chance of an unfavorable test result. Determine Nitro's optimal strategy. Also find EVSI and EVPI.

5 During the summer, Olympic swimmer Adam Johnson swims every day. On sunny summer days, he goes to an outdoor pool, where he may swim for no charge. On rainy days, he must go to a domed pool. At the beginning of the summer, he has the option of purchasing a $15 season pass to the domed pool, which allows him use for the entire summer. If he doesn't buy the season pass, he must pay $1 each time he goes there. Past meteorological records indicate that there is a 60% chance that the summer will be sunny (in which case there is an average of 6 rainy days during the summer) and a 40% chance the summer will be rainy (an average of 30 rainy days during the summer).

Before the summer begins, Adam has the option of purchasing a long-range weather forecast for $1. The forecast predicts a sunny summer 80% of the time and a rainy summer 20% of the time. If the forecast predicts a sunny summer, there is a 70% chance that the summer will actually be sunny. If the forecast predicts a rainy summer, there is an 80% chance that the summer will actually be rainy. Assuming that Adam's goal is to minimize his total expected cost for the summer, what should he do? Also find EVSI and EVPI.

6 Pete is considering placing a bet on the NCAA playoff game between Indiana and Purdue. Without any further information, he believes that each team has an equal chance to win. If he wins the bet, he will win $10,000; if he loses, he will lose $11,000. Before betting, he may pay Bobby $1,000 for his inside prediction on the game; 60% of the time, Bobby will predict that Indiana will win and 40% of the time, Bobby will predict that Purdue will win. When Bobby says that IU will win, IU has a 70% chance of winning, and when Bobby says that Purdue will win, IU has only a 20% chance of winning. Determine how Pete can maximize his total expected profit. What is EVSI? What is EVPI?

7 Erica is going to fly to London on August 5 and return home on August 20. It is now July 1. On July 1, she may buy a one-way ticket (for $350) or a round-trip ticket (for $660). She may also wait until August 1 to buy a ticket. On August 1, a one-way ticket will cost $370, and a round-trip ticket will cost $730. It is possible that between July 1 and August 1, her sister (who works for the airline) will be able to obtain a free one-way ticket for Erica. The probability that her sister will obtain the free ticket is .30. If Erica has bought a round-trip ticket on July 1 and her sister has obtained a free ticket, she may return "half" of her round-trip to the airline. In this case, her total cost will be $330 plus a $50 penalty. Use a decision tree approach to determine how to minimize Erica's expected cost of obtaining round-trip transportation to London.

8 I am a contestant on the TV show *Remote Jeopardy*, which works as follows. I am first asked a question about Stupid Videos. If I answer correctly, I earn $100. I believe that I have an 80% chance of answering such a question correctly. If I answer incorrectly, the game is over, and I win nothing. If I answer correctly, I may leave with $100 or go on and answer a question about Stupid TV Shows. If I answer this question correctly, I earn another $300, but if I answer incorrectly, I lose all previous earnings and am sent home. My chance of answering this question correctly is .60. If I answer the Stupid TV Shows question correctly, I

TABLE 9

On-Base Situation	Number of Outs	Expected Number of Runs
Runner on first	0	0.813
Runner on first	1	0.498
Runner on second	1	0.671
Runner on second	0	1.194
No base runners	1	0.243

may leave with my "earnings" or go on and answer a question about Statistics. If I answer this question correctly, I earn another $500, but if I answer it incorrectly, I lose all previous earnings and am sent home. My chance of answering this question correctly is .40. Draw a decision tree that can be used to maximize my expected earnings. What are my expected earnings?

Group B

In many decision tree problems, the decision maker's goal is to maximize the probability of a favorable event occurring. To incorporate this goal into a decision tree, simply give a reward of 1 to any terminal branch that results in the favorable event occurring and a reward of 0 to any terminal branch that results in the favorable event not occurring. Then maximizing expected reward is the same as maximizing the probability that the favorable event will occur. Use this idea to solve the next two problems.

9 The American chess master Jonathan Meller is playing the Soviet expert Yuri Gasparov in a two-game exhibition match. Each win earns a player one point, and each draw earns a half point. The player who has the most points after two games wins the match. If the players are tied after two games, they play until one wins a game; then the first player to win a game wins the match. During each game, Meller has two possible approaches: to play a daring strategy or to play a conservative strategy. His probabilities of winning, losing, and drawing when he follows each strategy are shown in Table 10. To maximize his probability of winning the match, what should the American do?

10 Yvonne Delaney is playing Chris Becker a single point for the women's world tennis championship. She has won the coin toss and elected to serve. If she tries a hard serve, her probability of getting the serve into play is .60. Given that the hard serve is in play, she has a .60 chance of winning the point. If she tries a soft serve, her probability of getting the serve in play is .90, but if the soft serve is in play, her probability of winning the point is only .50. To maximize her probability of winning the point, what should Yvonne do?

11[†] The Indiana Hoosiers trail the Purdue Boilermakers by a 14–0 score late in the fourth quarter of a football game. Indiana's guardian angel has informed Indiana that before the game ends they will have the ball two more times, and they will score a touchdown each time. The Indiana coach is indifferent between a tie and the following lottery: a 40% chance at beating Purdue and a 60% chance at losing to Purdue. Indiana's kicker has never missed an extra point, and Indiana has been successful on 35% of all two-point conversion attempts. After each touchdown (worth six points), Indiana must decide whether or not to attempt a one-point or a two-point conversion. Help the Indiana coach maximize his expected utility.

12 Edwina, a commodities broker, has acquired an option to buy 1,000 oz of gold at $50/oz. If she takes the option and if Congress relaxes import quotas, she can sell the gold for $80/oz. If she takes the option and Congress does not relax the import quotas, however, the company will lose $10/oz.

TABLE 10

Strategy	Win	Loss	Draw
Daring	.45	.55	0
Conservative	0	.10	.90

Edwina believes that there is a 50% chance that the government will relax the quota. She also has the option of waiting until Congress decides whether to relax the import quota. If she adopts this strategy, however, there is a 70% chance that some other broker will have already taken the option.

a If Edwina is risk-neutral, what should she do?

b If Edwina's utility function for a change x in her asset position is given by $u(x) = (10,000 + x)^{1/2}$, what should she do?

13 We are going to see the movie *Fatal Repulsion.* There are three parking lots we may park in. One is one block east of the theater (call this lot −1); one lot is directly behind the theater (lot 0); and one lot is one block west of the theater (lot 1). We are approaching the theater from the east. There is an 80% chance that lot −1 will have a vacant space, a 60% chance that lot 0 will, and an 80% chance that lot 1 will. Once we pass a lot, we can't go back to it. Assume that when we are at a given parking lot, we can determine whether it has any vacant spaces, but we can't see any of the other lots. Our dates for the evening will assess us a penalty equal to the distance (in blocks) that we park from the theater. If we find no space, they will assess a penalty of 10 (and never go out with us again). What strategy minimizes our expected penalty? Answer the same question if there is a 70% chance that lot 0 has a vacant space.

14[‡] A patient enters the hospital with severe abdominal pains. Based on past experience, Doctor Craig believes there is a 28% chance that the patient has appendicitis and a 72% chance that the patient has nonspecific abdominal pains. Dr. Craig may operate on the patient now or wait 12 hours to gain a more accurate diagnosis. In 12 hours, Dr. Craig will surely know whether the patient has appendicitis. The problem is that in the meantime, the patient's appendix may perforate (if he has appendicitis), thereby making the operation much more dangerous. Again based on past experience, Dr. Craig believes that if he waits 12 hours, there is a 6% chance that the patient will end up with a perforated appendix, a 22% chance the patient will end up with "normal" appendicitis, and a 72% chance that the patient will end up with nonspecific abdominal pain. From past experience, Dr. Craig assesses the probabilities shown in Table 11 of the patient dying. Assume that Dr. Craig's goal is to maximize the probability that the patient will survive. Use a decision tree to help Dr. Craig make the right decision.

15 a Suppose you are given a choice between the following options:

A_1: Win $30 for sure

A_2: 80% chance of winning $45 and 20% chance of winning nothing

---

[†]Based on Porter (1967).

[‡]Based on Clarke (1981).

TABLE **11**

Situation	Probability That Patient Will Die
Operation on patient with appendicitis	.0009
Operation on patient with nonspecific abdominal pain	.0004
Operation on perforated appendix	.0064
No operation on patient with nonspecific abdominal pain	0

B_1: 25% chance of winning $30

B_2: 20% chance of winning $45

Most people prefer A_1 to A_2 and B_2 to B_1. Explain why this behavior violates the assumption that decision makers maximize expected utility.

b Now suppose you play the following game: You have a 75% chance of winning nothing and a 25% chance of playing the second stage of the game. If you reach the second stage, you have a choice of two options (C_1 and C_2), but your choice must be made now, before you reach the second stage.

C_1: Win $30 for sure

C_2: 80% chance of winning $45

Most people choose C_1 over C_2 and B_2 to B_1 (from part (a)). Explain why this again violates the assumption of expected utility maximization. Tversky and Kahneman (1981) speculate that most people are attracted to the sure $30 in the second stage, even though the second stage may never be reached! Note that B_1 and C_1 both give $30 with the same probability, and B_2 and C_2 both yield $45 with the same probability. It appears that people do not act very rationally![†]

16 You have just been chosen to appear on Hoosier Millionaire! The rules are as follows: There are four hidden cards. One says "STOP" and the other three have dollar amounts of $150,000, $200,000, and $1,000,000. You get to choose a card. If the card says "STOP," you win no money. At any time you may quit and keep the largest amount of money that has appeared on any card you have chosen, or continue. If you continue and choose the stop card, however, you win no money. As an example, you may first choose the $150,000 card, then the $200,000 card, and then you may choose to quit and receive $200,000!

a If you goal is to maximize your expected payoff, what strategy should you follow?

b My utility function for an increase in cash satisfies $u(0) = 0$, $u(\$40,000) = .25$, $u(\$120,000) = .50$, $u(\$400,000) = .75$, and $u(\$1,000,000) = 1$. After drawing a curve through these points, determine a strategy that maximizes my expected utility. You might want to use your own utility function.

[†]Based on Tversky and Kahneman (1981).

13.5 Bayes' Rule and Decision Trees

The Colaco example and many other decision tree problems share several common features.

There are several states of the world. Different states of the world result in different payoffs to the decision maker. In the Colaco example, the two states of the world were that Chocola is a national success (*NS*) or a national failure (*NF*). We are also given (before the test marketing, if any, is done) estimates of the probabilities of each state of the world. These are called **prior probabilities.** In the Colaco example, the prior probabilities are $p(NS) = .55$ and $p(NF) = .45$.

In different states of the world, different decisions may be optimal. In the Colaco example, the company should market nationally if the state of the world is *NS* and not market nationally if the state of the world is *NF*.

It may be desirable to purchase information that gives the decision maker more fore-knowledge about the state of the world. This may enable the decision maker to make better decisions. For instance, in the Colaco example, the information obtained from test marketing might help Colaco decide whether or not Chocola should be marketed nationally.

The decision maker receives information by observing the outcomes of an experiment. Let $s_1, s_2, \ldots, s_n$ denote the possible states of the world, and let $o_1, o_2, \ldots, o_m$ be the possible outcomes of the experiment. Often, the decision maker is given the conditional probabilities $p(s_i|o_j)(i = 1, 2, \ldots, n; j = 1, 2, \ldots, m)$. Given knowledge of the outcome of the experiment, these probabilities give new values for the probability of each state of the world. The probabilities $p(s_i|o_j)$ are called **posterior probabilities.**

In the Colaco example, the experiment was the test-marketing procedure, and the two possible outcomes were LF = local failure and LS = local success. The posterior probabilities were given to be

$$p(NS|LS) = .85, \qquad p(NS|LF) = .10,$$
$$p(NF|LS) = .15, \qquad p(NF|LF) = .90$$

Thus, the knowledge of a local test market success would greatly increase Colaco's estimate of the probability of national success, and the knowledge of a local test market failure would greatly decrease Colaco's estimate of the probability of a national success. The posterior probabilities just listed were used to define the event forks in the decision tree that followed the action Test market.

In many situations, however, we may be given the prior probabilities $p(s_i)$ for each state of the world, and instead of being given the posterior probabilities $p(s_i|o_j)$, we might be given the **likelihoods** $p(o_j|s_i)$. For each state of the world, the likelihoods give the probability of observing each experimental outcome. Thus, in the Colaco example, we might be given the prior probabilities $p(NS) = .55$ and $p(NF) = .45$ and the likelihoods $p(LS|NS) = \frac{51}{55}, p(LF|NS) = \frac{4}{55}, p(LS|NF) = \frac{9}{45}$, and $p(LF|NF) = \frac{36}{45}$.

To clarify the meaning of likelihoods, suppose that 55 products that have been national successes had previously been test marketed; of these 55 products, 51 were local successes and 4 were local failures. This would have led us to estimate $p(LS|NS)$ as $\frac{51}{55}$ and $p(LF|NS)$ as $\frac{4}{55}$.

To complete the decision tree in Figure 7, we still need to know the posterior probabilities $p(NS|LS)$, $p(NF|LS)$, $p(NS|LF)$, and $p(NF|LF)$. With the help of Bayes' rule (see Section 12.4), we can use the prior probabilities and likelihoods to determine the needed posterior probabilities. To begin the computation of the posterior probabilities, we need to determine the joint probabilities of each state of the world and experimental outcome (that is, we must determine $p(NS \cap LS)$, $p(NS \cap LF)$, $p(NF \cap LS)$, and $p(NF \cap LF)$). We obtain these joint probabilities by using the definition of conditional probability:

$$p(NS \cap LS) = p(NS)p(LS|NS) = .55(\tfrac{51}{55}) = .51$$
$$p(NS \cap LF) = p(NS)p(LF|NS) = .55(\tfrac{4}{55}) = .04$$
$$p(NF \cap LS) = p(NF)p(LS|NF) = .45(\tfrac{9}{45}) = .09$$
$$p(NF \cap LF) = p(NF)p(LF|NF) = .45(\tfrac{36}{45}) = .36$$

Next we compute the probability of each possible experimental outcome (often called a *marginal probability*) $p(LS)$ and $p(LF)$:

$$p(LS) = p(NS \cap LS) + p(NF \cap LS) = .51 + .09 = .60$$
$$p(LF) = p(NS \cap LF) + p(NF \cap LF) = .04 + .36 = .40$$

Now Bayes' rule can be applied to obtain the desired posterior probabilities:

$$p(NS|LS) = \frac{p(NS \cap LS)}{p(LS)} = \frac{.51}{.60} = .85$$

$$p(NF|LS) = \frac{p(NF \cap LS)}{p(LS)} = \frac{.09}{.60} = .15$$

$$p(NS|LF) = \frac{p(NS \cap LF)}{p(LF)} = \frac{.04}{.40} = .10$$

$$p(NF|LF) = \frac{p(NF \cap LF)}{p(LF)} = \frac{.36}{.40} = .90$$

These posterior probabilities can be used to complete the decision tree in Figure 7.

In summary, to find posterior probabilities, we go through the following three-step process:

Step 1 Determine the joint probabilities of the form $p(s_i \cap o_j)$ by multiplying the prior probability ($p(s_i)$) times the likelihood ($p(o_j|s_i)$).

Step 2 Determine the probabilities of each experimental outcome $p(o_j)$ by summing up all joint probabilities of the form $p(s_k \cap o_j)$.

Step 3 Determine each posterior probability ($p(s_i|o_j)$) by dividing the joint probability ($p(s_i \cap o_j)$) by the probability of the experimental outcome o_j ($p(o_j)$).

We now give a complete example of a decision tree analysis that requires use of Bayes' rule.

EXAMPLE 5   **Fruit Computer Company**

Fruit Computer Company manufactures memory chips in lots of ten chips. From past experience, Fruit knows that 80% of all lots contain 10% (1 out of 10) defective chips, and 20% of all lots contain 50% (5 out of 10) defective chips. If a good (that is, 10% defective) batch of chips is sent on to the next stage of production, processing costs of $1,000 are incurred, and if a bad batch (50% defective) is sent on to the next stage of production, processing costs of $4,000 are incurred. Fruit also has the alternative of reworking a batch at a cost of $1,000. A reworked batch is sure to be a good batch. Alternatively, for a cost of $100, Fruit can test one chip from each batch in an attempt to determine whether the batch is defective. Determine how Fruit can minimize the expected total cost per batch. Also compute EVSI and EVPI.

Solution   We will multiply costs by -1 and work with maximizing $-$(total cost). This enables us to use the EVSI and EVPI formulas of Section 13.4. There are two states of the world:

$$G = \text{batch is good}$$
$$B = \text{batch is bad}$$

We are given the following prior probabilities:

$$p(G) = .80 \quad \text{and} \quad p(B) = .20$$

Fruit has the option of performing an experiment: inspecting one chip per batch. The possible outcomes of the experiment are

$$D = \text{defective chip is observed}$$
$$ND = \text{nondefective chip is observed}$$

We are given the following likelihoods:

$$p(D|G) = .10, \quad p(ND|G) = .90, \quad p(D|B) = .50, \quad P(ND|B) = .50$$

To complete the decision tree in Figure 12, we need to determine the posterior probabilities $p(B|D)$, $p(G|D)$, $p(B|ND)$, and $p(G|ND)$. We begin by computing joint probabilities:

$$p(D \cap G) = p(G)p(D|G) = .80(.10) = .08$$
$$p(D \cap B) = p(B)p(D|B) = .20(.50) = .10$$
$$p(ND \cap G) = p(G)p(ND|G) = .80(.90) = .72$$
$$p(ND \cap B) = p(B)p(ND|B) = .20(.50) = .10$$

We then compute the probability of each experimental outcome:

$$p(D) = p(D \cap G) + p(D \cap B) = .08 + .10 = .18$$
$$p(ND) = p(ND \cap G) + p(ND \cap B) = .72 + .10 = .82$$

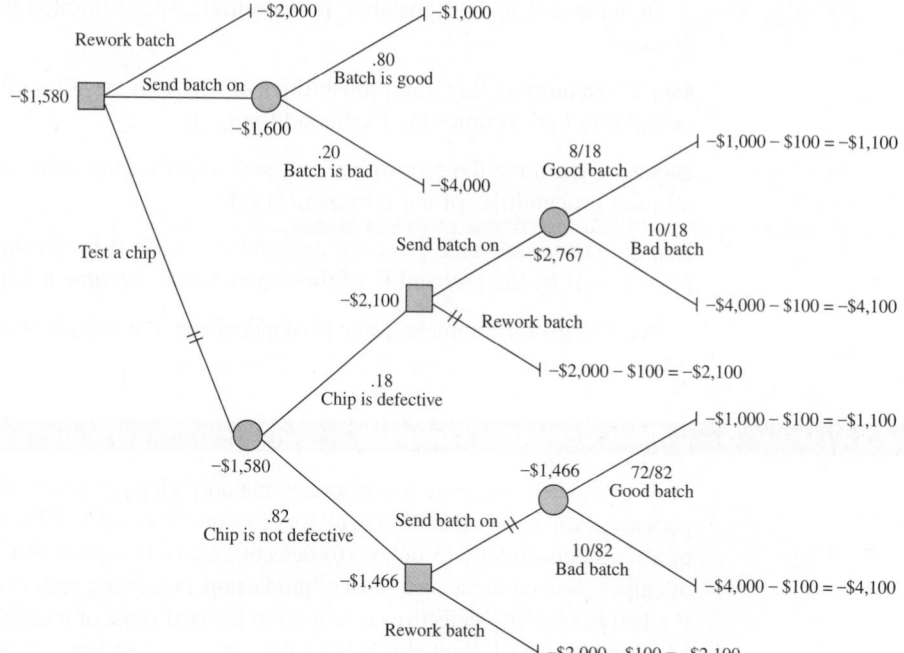

FIGURE 12
Illustration of Use of Bayes' Rule in Decision Tree for Fruit Computer Co.

Then we use Bayes' rule to determine the required posterior probabilities:

$$p(B|D) = \frac{p(D \cap B)}{p(D)} = \frac{.10}{.18} = \frac{5}{9}$$

$$p(G|D) = \frac{p(D \cap G)}{p(D)} = \frac{.08}{.18} = \frac{4}{9}$$

$$p(B|ND) = \frac{p(ND \cap B)}{p(ND)} = \frac{.10}{.82} = \frac{10}{82}$$

$$p(G|ND) = \frac{p(ND \cap G)}{p(ND)} = \frac{.72}{.82} = \frac{72}{82}$$

These posterior probabilities are used to complete the tree in Figure 12. Straightforward computations show that the optimal strategy is to test a chip. If the chip is defective, rework the batch. If the chip is not defective, send the batch on. An expected cost of $1,580 is incurred.

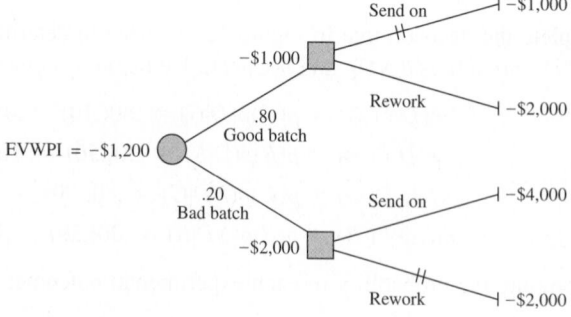

FIGURE 13
Expected Value with Perfect Information (EVWPI) for Fruit Computer

To find EVSI, suppose that testing one chip in a batch were costless. Then the Test chip branch of the tree would have its expected value increased by $100 (to −$1,480). Then we would have EVWSI = −$1,480 and EVWOI = −$1,600. Then EVSI = EVWSI − EVWOI = −$1,480 − (−$1,600) = $120.

To find EVPI, we use the tree in Figure 13. We find EVWPI = −$1,200. Then EVPI = EVWPI − EVWOI = −$1,200 − (−$1,600) = $400.

Using LINGO to Compute Posterior Probabilities

The following LINGO program can be used to compute the posterior probabilities for Example 5 (or any other situation).

```
MODEL:
   1]SETS:
   2]ST/G,B/:PR;
   3]OUT/D,ND/:MARG;
   4]SXO(ST,OUT):POST,JOINT,LIKE;
   5]ENDSETS
   6]DATA:
   7]PR=.8,.2;
   8]LIKE=.1,.9,.5,.5;
   9]ENDDATA
  10]@FOR(SXO(I,J):JOINT(I,J)=PR(I)*LIKE(I,J););
  11]@FOR(OUT(J):MARG(J)=@SUM(ST(I):JOINT(I,J)););
  12]@FOR(SXO(I,J):POST(I,J)=JOINT(I,J)/MARG(J););
  13]END
```

Line 2 defines the states (G and B) and associates a prior probability with each state. The prior probabilities are input by the user in the DATA section of the program. Line 3 defines the set of possible experimental outcomes and associates a marginal probability (MARG) with each outcome. The values of MARG are computed in line 11. In line 4, we create the set SXO, consisting of (G, D), (G, ND), (B, D), and (B, ND), and associate with each member of this set the following:

1 A posterior probability (POST); for example, POST$(G, D) = p(G|D)$. The values of POST are computed in line 12.

2 A joint probability (JOINT); for example, JOINT$(G, ND) = p(G \cap ND)$. The values of JOINT are computed in line 10.

3 A likelihood (LIKE); for example, LIKE$(B, D) = p(D|B)$. The likelihoods are input in the DATA section.

Lines 6–9 input the relevant data. Recall that attributes with multiple subscripts (such as SXO) are stored so that the rightmost subscripts advance most rapidly. This helps us determine the order in which to input the values of LIKE.

In line 10, we compute all joint probabilities JOINT(I, J) by multiplying the prior probability PR(I) by the likelihood LIKE(I, J). In line 11 we compute each marginal probability MARG(J) by summing over I all joint probabilities involving J. In line 12, we compute each posterior probability POST(I, J) (really this is $p(I|J)$) by dividing $p(I \cap J)$ by the marginal probability of outcome $J(p(J))$.

PROBLEMS

Group A

1 A customer has approached a bank for a $50,000 one-year loan at 12% interest. If the bank does not approve the loan, the $50,000 will be invested in bonds that earn a 6% annual return. Without further information, the bank feels that there is a 4% chance that the customer will totally default on the loan. If the customer totally defaults, the bank loses $50,000. At a cost of $500, the bank can thoroughly investigate the customer's credit record and supply a favorable or unfavorable recommendation. Past experience indicates that

$$p(\text{favorable recommendation}|$$
$$\text{customer does not default}) = \tfrac{77}{96}$$
$$p(\text{favorable recommendation}|$$
$$\text{customer defaults}) = \tfrac{1}{4}$$

How can the bank maximize its expected profits? Also find EVSI and EVPI.

2 A nuclear power company is deciding whether or not to build a nuclear power plant at Diablo Canyon or at Roy Rogers City. The cost of building the power plant is $10 million at Diablo and $20 million at Roy Rogers City. If the company builds at Diablo, however, and an earthquake occurs at Diablo during the next five years, construction will be terminated and the company will lose $10 million (and will still have to build a power plant at Roy Rogers City). A priori, the company believes there is a 20% chance that an earthquake will occur at Diablo during the next five years. For $1 million, a geologist can be hired to analyze the fault structure at Diablo Canyon. He will either predict that an earthquake will occur or that an earthquake will not occur. The geologist's past record indicates that he will predict an earthquake on 95% of the occasions for which an earthquake will occur and no earthquake on 90% of the occasions for which an earthquake will not occur. Should the power company hire the geologist? Also find EVSI and EVPI.

3 Farmer Jones must determine whether to plant corn or wheat. If he plants corn and the weather is warm, he earns $8,000; if he plants corn and the weather is cold, he earns $5,000. If he plants wheat and the weather is warm, he earns $7,000; if he plants wheat and the weather is cold, he earns $6,500. In the past, 40% of all years have been cold and 60% have been warm. Before planting, Jones can pay $600 for an expert weather forecast. If the year is actually cold, there is a 90% chance that the forecaster will predict a cold year. If the year is actually warm, there is an 80% chance that the forecaster will predict a warm year. How can Jones maximize his expected profits? Also find EVSI and EVPI.

4 The NBS television network earns an average of $400,000 from a hit show and loses an average of $100,000 on a flop. Of all shows reviewed by the network, 25% turn out to be hits and 75% turn out to be flops. For $40,000, a market research firm will have an audience view a pilot of a prospective show and give its view about whether the show will be a hit or a flop. If a show is actually going to be a hit, there is a 90% chance that the market research firm will predict the show to be a hit. If the show is actually going to be a flop, there is an 80% chance that the market research firm will predict the show to be a flop. Determine how the network can maximize its expected profits. Also find EVSI and EVPI.

5 We are thinking of filming the Don Harnett story. We know that if the film is a flop, we will lose $4 million, and if the film is a success, we will earn $15 million. Beforehand, we believe that there is a 10% chance that the Don Harnett story will be a hit. Before filming, we have the option of paying the noted movie critic Roger Alert $1 million for his view of the film. In the past, Alert has predicted 60% of all actual hits to be hits and 90% of all actual flops to be flops. We want to maximize our expected profits. Use a decision tree to determine our best strategy. What is EVSI? What is EVPI?

Group B

6 Abdul has one die in his left hand and one in his right hand. One die has six dots painted on each face, and the other has one dot painted on two of the faces and six dots painted on each of the other four faces. Greta is to pick one die (either "left" or "right") and will receive $10 for each dot painted on the die that is picked. Before choosing, Greta may pay Abdul $15, and he will toss the die in his left hand and tell her how many dots are painted on the face that comes up. Use a decision tree to determine how to maximize Greta's profit. Also determine EVSI and EVPI.

7 Pat Sajork has two drawers. One drawer contains three gold coins, and the other contains one gold coin and two silver coins. We are allowed to choose one drawer, and we will be paid $500 for each gold coin and $100 for each silver coin in that drawer. Before choosing, we may pay Pat $200, and he will draw a randomly selected coin (each of the six coins has an equal chance of being chosen) and tell us whether it is gold or silver. For instance, Pat may say that he drew a gold coin from drawer 1. Should we pay Pat $200? What is EVSI? What is EVPI?

8 Joe owns a coin that is either a fair coin or a two-headed coin. Imelda believes that there is a $\tfrac{1}{2}$ chance that the coin is two-headed. She must guess what kind of coin Joe has. If she guesses correctly, she pays Joe nothing, but if she guesses incorrectly, she must pay Joe $2. Before guessing, she may pay 30¢ to see the result of a single coin toss (heads or tails). Determine how Imelda should minimize her expected loss. Also determine EVSI and EVPI.

9 The government is attempting to determine whether immigrants should be tested for a contagious disease. Let's assume that the decision will be made on a financial basis. Assume that each immigrant who is allowed into the country and has the disease costs the United States $100,000, and each immigrant who enters and does not have the disease will contribute $10,000 to the national economy. Assume that 10% of all potential immigrants have the disease. The government may admit all immigrants, admit no immigrants, or test immigrants for the disease before determining whether they should be admitted. It costs $100 to test a person for the disease; the test result is either positive or negative. If the test result is positive, the person *definitely* has the disease. However, 20% of all people who *do* have the disease test negative. A person who does not have the disease always tests negative. The government's goal is to maximize (per potential immigrant) expected benefits minus expected costs. Use a decision tree to aid in this undertaking. Also determine EVSI and EVPI.

10[†] Many colleges face the problem of whether athletes should be tested for drug use. Define

[†]Based on Feinstein (1990).

$c_1 =$ Cost if athlete is falsely accused of drug use

$c_2 =$ Cost if a drug user is not identified

$c_3 =$ Cost due to invasion of privacy if a nonuser is tested

Suppose that 5% of all athletes are drug users, and that the test used is 90% reliable. This means that if an athlete uses drugs, there is a 90% chance that the test will detect it, and if the athlete does not use drugs, there is a 90% chance that the test will show no drug use.

a If $c_1 = 10$, $c_2 = 5$, and $c_3 = 1$, should the college test athletes for drugs?

b Prove that if $c_1 > c_2 > c_3$, then the college should not test for drugs.

13.6 Decision Making with Multiple Objectives

In previously considered decision problems, the decision maker made a choice based on how each possible action affected a single variable (or attribute). For example, in the news vendor problem, the number of papers ordered was determined by how this affected Phyllis's profits. Similarly, in the Colaco example, Colaco's decision depended on how each of its strategies affected its final asset position.

In many situations, however, the action chosen depends on how each possible action affects more than one attribute or variable. Four examples follow. (1) Suppose that Joe Bunker wants to buy a new car. In choosing which car to buy, Joe may consider the following attributes of each car:

Attribute 1 Size of car

Attribute 2 Fuel economy of car (miles per gallon)

Attribute 3 Style of car

Attribute 4 Price of car

(2) Suppose Joe Bunker has just graduated from the nation's top business school, Business School (B.S.) University, and has received five job offers. In choosing which to accept, Joe will consider the following attributes of each job:

Attribute 1 Starting salary of job

Attribute 2 Location of job

Attribute 3 Degree of interest Joe has in doing the work involved in the particular job

Attribute 4 Long-term opportunities associated with job

(3) Gotham City must determine where to locate a new jetport. In determining the site three factors (or attributes) must be considered:

Attribute 1 Accessibility of jetport for residents of Gotham City

Attribute 2 Degree of noise pollution caused by the jetport (if the jetport is placed in a densely populated area, noise pollution will be more serious than if the jetport is placed in a sparsely populated area)

Attribute 3 Size of the jetport (determined in part by the amount of land available at the jetport site)

(4) Wivco Toy Corporation is introducing a new product (a globot). Wivco must determine the price to charge for each globot. Two factors (market share and profits) will affect the pricing decision.

In these four examples, the decision maker chooses an action by determining how each possible action affects the relevant attributes. Such problems are called **multiattribute decision problems.**

Multiattribute Decision Making in the Absence of Uncertainty: Goal Programming

Suppose a woman believes that there are n attributes that will determine her decision. Let $x_i(a)$ be the value of the ith attribute associated with an alternative a. She associates a value $v(x_1(a), x_2(a), \ldots, x_n(a))$ with the alternative a. The function $v(x_1, x_2, \ldots, x_n)$ is the decision maker's **value function.** If A represents the decision maker's set of possible decisions, then she should choose the alternative a^* (with level x_i^* of attribute i) satisfying

$$\max_{a \in A} v(x_1(a), x_2(a), \ldots, x_n(a)) = v(x_1^*, x_2^*, \ldots, x_n^*)$$

Alternatively, the decision maker can associate a cost $c(x_1(a), x_2(a), \ldots, x_n(a))$ with the alternative a. The function $c(x_1, x_2, \ldots, x_n)$ is her **cost function.** If A represents the decision maker's set of possible decisions, then she should choose the alternative a^* (with level x_i^* of attribute i) satisfying

$$\min_{a \in A} c(x_1(a), x_2(a), \ldots, x_n(a)) = c(x_1^*, x_2^*, \ldots, x_n^*)$$

A particular form of the value or cost function is of special interest.

DEFINITION ■ A value function $v(x_1, x_2, \ldots, x_n)$ is an **additive value function** if there exist n functions $v_1(x_1), v_2(x_2), \ldots, v_n(x_n)$ satisfying

$$v(x_1, x_2, \ldots, x_n) = \sum_{i=1}^{i=n} v_i(x_i) \quad ■ \tag{7}$$

A cost function $c(x_1, x_2, \ldots, x_n)$ is an **additive cost function** if there exist n functions $c_1(x_1), c_2(x_2), \ldots, c_n(x_n)$ satisfying

$$c(x_1, x_2, \ldots, x_n) = \sum_{i=1}^{i=n} c_i(x_i) \quad ■ \tag{8}$$

Under what conditions will a decision maker have an additive value (or cost) function? Before answering this question, we need some more definitions.

DEFINITION ■ An attribute (call it attribute 1) is **preferentially independent** (pi) of another attribute (attribute 2) if preferences for values of attribute 1 do not depend on the value of attribute 2. ■

To illustrate the concept of preferential independence, we consider Joe's search for a job following graduation. In this situation, attribute 1 would be preferentially independent of attribute 2 if, for any possible job location, a higher starting salary is preferred to a lower salary.

As another illustration of preferential independence, suppose that the Griswold family is trying to determine how to spend Sunday afternoon. Let the two relevant attributes be

Attribute 1 Choice of activity (either picnic or go to see movie *Antarctic Vacation*)
Attribute 2 Sunday afternoon's weather (either sunny or rainy)

Suppose that on a sunny day, the picnic is preferred to the movie, but on a rainy day, the movie is preferred to the picnic. Then attribute 1 is not preferentially independent of attribute 2.

DEFINITION ■   If attribute 1 is pi of attribute 2, and attribute 2 is pi of attribute 1, then attribute 1 is **mutually preferentially independent** (mpi) of attribute 2.   ■

Again refer to Joe's search for a job. Suppose Joe's five job offers are located in Los Angeles, Chicago, Dallas, New York, and Indianapolis. If, for any given salary level, Joe prefers to work in Los Angeles, then attribute 2 is pi of attribute 1. If attribute 1 were also pi of attribute 2, then attributes 1 and 2 would be mpi.

The concept of mutual preferential independence can be generalized to sets of attributes.

DEFINITION ■   A set of attributes S is **mutually preferentially independent** (mpi) of a set of attributes S' if (1) the values of the attributes in S' do not affect preferences for the values of attributes in S, and (2) the values of attributes in S do not affect preferences for the values of attributes in S'.   ■

In the example of Joe's purchase of a new car, let S = attributes 1 and 2, and S' = attributes 3 and 4. Then for S to be mpi of S', it must be the case that (1) Joe's preferences for size and fuel economy are unaffected by a car's style and price, and (2) Joe's preferences for car style and price are unaffected by the car's size and fuel economy. Thus, if S and S' were mpi, we could conclude that if for a given style and price level, Joe preferred A_1 (a large car getting 15 mpg) to A_2 (a small car getting 25 mpg), then for any style and price level, Joe would prefer A_1 to A_2.

DEFINITION ■   A set of attributes $1, 2, \ldots, n$ is **mutually preferentially independent** (mpi) if for all subsets S of $\{1, 2, \ldots, n\}$, S is mpi of $\bar{S}$. ($\bar{S}$ is all members of $\{1, 2, \ldots, n\}$ that are not included in S.)   ■

It is easy to see that if there are only two attributes (1 and 2), the attributes are mpi if and only if attribute 1 is mpi of attribute 2.

The following result gives a condition ensuring that the decision maker will have an additive value (or cost) function.

THEOREM 1

If the set of attributes $1, 2, \ldots, n$ is mpi, the decision maker's preferences can be represented by an additive value (or cost) function.

This is not an obvious result. (For a proof, see Keeney and Raiffa (1976, Chapter 3).) To illustrate the result, suppose that the decision maker's value function for two attributes is given by

$$v(x_1, x_2) = x_1 + x_1 x_2 + x_2 \tag{9}$$

A decision maker with value function (2) would, for example, prefer (6, 6) to (4, 8) (because $v(6, 6) = 48$ and $v(4, 8) = 44$). The reader should verify that for (2), attribute 1 is pi of attribute 2, and attribute 2 is pi of attribute 1 (see Problem 3 at the end of this section). Thus, attributes 1 and 2 are mpi, and Theorem 1 implies that the decision maker's preferences can be represented by an additive function. To demonstrate this, define new attributes $1'$ and $2'$ as

$$\text{Value of attribute } 1' = x_1' = x_1 + x_2$$
$$\text{Value of attribute } 2' = x_2' = x_1 - x_2$$

Consider the additive value function

$$v'(x_1', x_2') = x_1' + \frac{(x_1')^2}{4} - \frac{(x_2')^2}{4}$$

It is easy to show that $v'(x_1', x_2') = v(x_1, x_2)$. Thus, $v'(x_1', x_2')$ represents the decision maker's preferences. For example, of the following two alternatives

$$\text{Alternative 1: } x_1 = 6, x_2 = 6$$
$$\text{Alternative 2: } x_1 = 4, x_2 = 8$$

we already know that the decision maker prefers alternative 1. In terms of the new attributes $1'$ and $2'$, we have

$$\text{Alternative 1: } x_1' = 12, x_2' = 0$$
$$\text{Alternative 2: } x_1' = 12, x_2' = -4$$

Then

$$v'(12, 0) = \text{value of alternative } 1 = 12 + \frac{12^2}{4} = 48$$

$$v'(12, -4) = \text{value of alternative } 2 = 12 + \frac{12^2}{4} - \frac{(-4)^2}{4} = 44$$

Therefore, in this example, the additive value function $v'(x_1', x_2')$ replicates the decision maker's preferences.

Multiattribute Utility Functions

In Section 13.2, we described how the Von Neumann–Morgenstern utility theory could be used to make decisions under uncertainty when only one attribute affected the decision maker's preference and attitude toward risk. In this section, we discuss the extension of utility theory to situations in which more than one attribute affects the decision maker's preferences and attitude toward risk. Even in this case, a decision maker who subscribes to the Von Neumann–Morgenstern axioms will still choose the lottery or the alternative that maximizes his or her expected utility. When more than one attribute affects a decision maker's preferences, the person's utility function is called a **multiattribute utility function.** We restrict ourselves here to explaining how to assess and use multiattribute utility functions when only two attributes are operative. The reader seeking a more detailed discussion of multiattribute utility functions is referred to Bunn (1984) and (at a more advanced level) the classic work by Keeney and Raiffa (1976).

Suppose a decision maker's preferences and attitude toward risk depend on two attributes. Let

$$x_i = \text{level of attribute } i$$

$$u(x_1, x_2) = \text{utility associated with level } x_1 \text{ of attribute 1 and level } x_2 \text{ of attribute 2}$$

How can we find a utility function $u(x_1, x_2)$ such that choosing a lottery or alternative that maximizes the expected value of $u(x_1, x_2)$ will yield a decision consistent with the decision maker's preferences and attitude toward risk?

In general, determination of $u(x_1, x_2)$ (or, in the case of n attributes, determination of $u(x_1, x_2, \ldots, x_n)$) is a difficult matter. Under certain conditions, however, the assessment of a utility function $u(x_1, x_2)$ is greatly simplified.

Properties of Multiattribute Utility Functions

DEFINITION ■ Attribute 1 is **utility independent** (ui) of attribute 2 if preferences for lotteries involving different levels of attribute 1 do not depend on the level of attribute 2. ■

Let's reconsider the problem of Wivco Toy Corporation. Wivco is introducing a new product (a globot) and must determine what price to charge for each globot. Two factors (market share and profits) will affect Wivco's pricing decision. Let

$$x_1 = \text{Wivco's market share}$$
$$x_2 = \text{Wivco's profits (millions of dollars)}$$

Suppose that Wivco is indifferent between

$$L_1 \quad \overset{\frac{1}{2}}{\underset{\frac{1}{2}}{\rule{0pt}{0pt}}} \quad \begin{array}{l} 10\%, \$5 \\ 30\%, \$5 \end{array} \qquad \text{and} \qquad L_1' \overset{1}{\rule{2em}{0.4pt}} 16\%, \$5$$

If attribute 1 (market share) is ui of attribute 2 (profit), Wivco would also be indifferent between

$$L_2 \quad \overset{\frac{1}{2}}{\underset{\frac{1}{2}}{\rule{0pt}{0pt}}} \quad \begin{array}{l} 10\%, \$20 \\ 30\%, \$20 \end{array} \qquad \text{and} \qquad L_2' \overset{1}{\rule{2em}{0.4pt}} 16\%, \$20$$

In short, if market share is ui of profit, then for any level of profits, a $\frac{1}{2}$ chance at a 10% market share and a $\frac{1}{2}$ chance at a 30% market share has a certainty equivalent of a 16% market share.

DEFINITION ■ If attribute 1 is ui of attribute 2, and attribute 2 is ui of attribute 1, then attributes 1 and 2 are **mutually utility independent** (mui). ■

If attributes 1 and 2 are mui, it can be shown that the decision maker's utility function $u(x_1, x_2)$ must be of the following form:

$$u(x_1, x_2) = k_1 u_1(x_1) + k_2 u_2(x_2) + k_3 u_1(x_1)u_2(x_2) \qquad \text{(10)}$$

In (10), k_1, k_2, and k_3 are constants, and $u_1(x_1)$ and $u_2(x_2)$ are functions of x_1 and x_2, respectively. Equation (10) is often called a **multilinear utility function.**

To show that a multilinear utility function exhibits mui, we assume that Wivco's utility function is of the form (10) and that Wivco is indifferent between L_1 and L_1'. If Wivco exhibits mui, then Wivco should also be indifferent between L_2 and L_2'. Using (10), we can now show that $L_1 i L_1'$ implies $L_2 i L_2'$. First, (10) and $L_1 i L_1'$ imply

$$\frac{1}{2}[k_1 u_1(10) + k_2 u_2(5) + k_3 u_1(10)u_2(5)]$$
$$+ \frac{1}{2}[k_1 u_1(30) + k_2 u_2(5) + k_3 u_1(30)u_2(5)]$$
$$= k_1 u_1(16) + k_2 u_2(5) + k_3 u_1(16)u_2(5)$$

Simplifying this equation yields (if $k_1 \neq 0$)

$$\tfrac{1}{2}[u_1(10) + u_1(30)] = u_1(16) \tag{11}$$

Using (11), we find

$$
\begin{aligned}
E(U \text{ for } L_2) &= \tfrac{1}{2}[k_1 u_1(10) + k_2 u_2(20) + k_3 u_1(10)u_2(20)] \\
&\quad + \tfrac{1}{2}[k_1 u_1(30) + k_2 u_2(20) + k_3 u_1(30)u_2(20)] \\
&= k_1 u_1(16) + k_2 u_2(20) + k_3 u_1(16)u_2(20) \\
&= E(U \text{ for } L_2')
\end{aligned}
$$

Thus, we see that a multilinear utility function of the form (10) implies that attribute 1 is ui of attribute 2. Similarly, it can be shown that (10) implies that attribute 2 is ui of attribute 1. Thus, (10) implies that attributes 1 and 2 are mui. It can also be shown that if x_1 and x_2 are mui, then $u(x_1, x_2)$ must be of the form (10) (see Keeney and Raiffa (1976)).

THEOREM 2

Attributes 1 and 2 are mui if and only if the decision maker's utility function $u(x_1, x_2)$ is a multilinear function of the form

$$u(x_1, x_2) = k_1 u_1(x_1) + k_2 u_2(x_2) + k_3 u_1(x_1)u_2(x_2) \tag{10}$$

The determination of a decision maker's utility function $u(x_1, x_2)$ can be further simplified if it exhibits **additive independence.** Before defining additive independence, we must define x_1 (best) or x_2 (best) to be the most favorable level of attribute 1 or 2 that can occur; also, x_1 (worst) or x_2 (worst) is the least favorable level of attribute 1 or 2 that can occur.

DEFINITION ■ A decision maker's utility function exhibits **additive independence** if the decision maker is indifferent between

Essentially, additive independence of attributes 1 and 2 implies that preferences over lotteries involving only attribute 1 (or only attribute 2) depend only on the marginal distribution for possible values of attribute 1 (or of attribute 2) and do not depend on the joint distribution of the possible values of attributes 1 and 2.

If attributes 1 and 2 are mui and the decision maker's utility function exhibits additive independence, it is easy to show that in (10), $k_3 = 0$ must hold. As in Section 2.2, simply scale $u_1(x_1)$ and $u_2(x_2)$ such that

$$
\begin{aligned}
u_1(x_1(\text{best})) &= 1 \qquad u_1(x_1(\text{worst})) = 0 \\
u_2(x_2(\text{best})) &= 1 \qquad u_2(x_2(\text{worst})) = 0
\end{aligned}
$$

Now (10) implies

$$
\begin{aligned}
u(x_1(\text{best}), x_2(\text{best})) &= k_1 + k_2 + k_3 \qquad u(x_1(\text{worst}), x_2(\text{worst})) = 0 \\
u(x_1(\text{best}), x_2(\text{worst})) &= k_1 \qquad\qquad\quad u(x_1(\text{worst}), x_2(\text{best})) = k_2
\end{aligned}
$$

Then additive independence implies that

$$\tfrac{1}{2}(k_1 + k_2 + k_3) + \tfrac{1}{2}(0) = \tfrac{1}{2}(k_1) + \tfrac{1}{2}(k_2)$$
$$k_3 = 0$$

Thus, if attributes 1 and 2 are mui and the decision maker's utility function exhibits additive independence, his or her utility function is of the following additive form:

$$u(x_1, x_2) = k_1 u_1(x_1) + k_2 u_2(x_2) \tag{12}$$

Assessment of Multiattribute Utility Functions

If attributes 1 and 2 are mui, how can we determine $u_1(x_1)$, $u_2(x_2)$, k_1, k_2, and k_3? To determine $u_1(x_1)$ and $u_2(x_2)$, we apply the technique that was used to assess utility functions in Section 13.2. We illustrate by determining $u_1(x_1)$. Let $u_1(x_1(\text{best})) = 1$ and $u_1(x_1(\text{worst})) = 0$. Next determine a value of attribute 1 (call it $x_1(\tfrac{1}{2})$) having $u_1(x_1(\tfrac{1}{2})) = \tfrac{1}{2}$. By the definition of $u_1(x_1(\tfrac{1}{2}))$ and mui, the decision maker is (for any value of x_2) indifferent between

Thus, $x_1(\tfrac{1}{2})$ may be determined from the fact that the certainty equivalent of L is $(x_1(\tfrac{1}{2})$, $x_2)$. In a similar fashion, we can determine values $x_1(\tfrac{1}{4})$ and $x_1(\tfrac{3}{4})$ of the first attribute satisfying $u_1(x_1(\tfrac{1}{4})) = \tfrac{1}{4}$ and $u_1(x_1(\tfrac{3}{4})) = \tfrac{3}{4}$. Continuing in this fashion, we may approximate $u_1(x_1)$ and $u_2(x_2)$.

To find k_1, k_2, and k_3, we begin by rescaling $u_1(x_1)$, $u_2(x_2)$, and $u(x_1, x_2)$ such that

$$u(x_1(\text{best}), x_2(\text{best})) = 1, \qquad u(x_1(\text{worst}), x_2(\text{worst})) = 0,$$

$$u_1(x_1(\text{best})) = 1, \qquad u_1(x_1(\text{worst})) = 0, \qquad u_2(x_2(\text{best})) = 1, \qquad u_2(x_2(\text{worst})) = 0$$

Now (10) yields

$$u(x_1(\text{best}), x_2(\text{worst})) = k_1(1) + k_2(0) + k_3(0) = k_1$$

Thus, k_1 can be determined from the fact that the decision maker is indifferent between

Similarly (see Problem 3 at the end of this section), $u(x_1(\text{worst}), x_2(\text{best})) = k_2$ and k_2 can be determined from the fact that the decision maker is indifferent between

To determine k_3, observe that from (10) and

$$u(x_1(\text{best}), x_2(\text{best})) = u_1(x_1(\text{best})) = u_2(x_2(\text{best})) = 1$$

we find that

$$1 = u(x_1(\text{best}), x_2(\text{best})) = k_1(1) + k_2(1) + k_3(1) = k_1 + k_2 + k_3$$

Thus, $k_1 + k_2 + k_3 = 1$, or $k_3 = 1 - k_1 - k_2$. Of course, if the decision maker's utility function exhibits additive independence, then $k_3 = 0$.

The procedure to be used in assessing a multiattribute utility function (when there are two attributes) may be summarized as follows:

Step 1 Check whether attributes 1 and 2 are mui. If they are, go on to step 2. If the attributes are not mui, the assessment of the multiattribute utility function is beyond the scope of our discussion. (See Keeney and Raiffa (1976, Section 5.7).)

Step 2 Check for additive independence.

Step 3 Assess $u_1(x_1)$ and $u_2(x_2)$.

Step 4 Determine k_1, k_2, and (if there is no additive independence) k_3.

Step 5 Check to see whether the assessed utility function is really consistent with the decision maker's preferences. To do this, set up several lotteries and use the expected utility of each to rank the lotteries from most to least favorable. Then ask the decision maker to rank the lotteries from most to least favorable. If the assessed utility function is consistent with the decision maker's preferences, the ranking of the lotteries obtained from the assessed utility function should closely resemble the decision maker's ranking of the lotteries.

Example 6 illustrates the assessment and use of a multiattribute utility function.

EXAMPLE 6   **Fruit Computer Co.**

Fruit Computer Company is certain that its market share during 2005 will be between 10% and 50% of the microcomputer market. Fruit is also sure that its profits during 2005 will be between $5 million and $30 million. Assess Fruit's multiattribute utility function where $u(x_1, x_2)$, where

$$x_1 = \text{Fruit's market share during 2005}$$
$$x_2 = \text{Fruit's profit during 2005 (in millions of dollars)}$$

Solution   We begin by checking for mui. It is helpful to draw a diagram (see Figure 14) that displays various levels of each attribute. First, we check whether attribute 1 (market share) is ui of attribute 2 (profit). We ask Fruit for the certainty equivalent of a $\frac{1}{2}$ chance at the

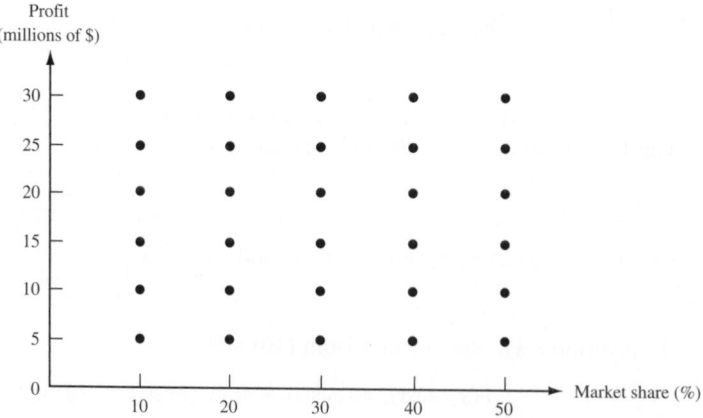

FIGURE 14
Possible Levels of Each
Attribute for Fruit
Computer Company

worst market share (10%) and a $\frac{1}{2}$ chance at the best market share (50%), with x_2 fixed at some level (say $x_2 = 15 million). Suppose the certainty equivalent of

is (30%, \$15). To determine whether attribute 1 is ui of attribute 2, we fix attribute 2 at some other level (say, $x_2 = 20 million) and find Fruit's certainty equivalent for the following lottery:

If attribute 1 is ui of attribute 2, the certainty equivalent of this lottery should be close to (30%, \$20). For other values of x_2 (say, $x_2 = $5, $10, $25, and $30), we check if the certainty equivalent of the lottery

is close to (30%, x_2). Suppose this is the case. Then we repeat this procedure with other values of market share replacing 10% and 50%. If similar results ensue, then attribute 1 is ui of attribute 2. In analogous fashion, we can determine whether attribute 2 is ui of attribute 1. If attribute 1 is ui of attribute 2, and attribute 2 is ui of attribute 1, the two attributes are mui. Let's assume that attributes 1 and 2 are (at least approximately) mui and proceed to step 2 (checking for additive independence).

To check for additive independence, we must determine whether Fruit is indifferent between

Suppose that Fruit is not indifferent between these lotteries. Then Fruit's utility function will not exhibit additive independence. We now know that $u(x_1, x_2)$ may be written as

$$u(x_1, x_2) = k_1 u_1(x_1) + k_2 u_2(x_2) + k_3 u_1(x_1) u_2(x_2)$$

We now proceed to step 3 (assessing $u_1(x_1)$ and $u_2(x_2)$). Suppose we obtain the results shown in Figure 15. To complete the assessment of Fruit's multiattribute utility function, we must determine k_1, k_2, and k_3 (step 4). To find k_1, we ask Fruit to determine the number k_1 that makes Fruit indifferent between

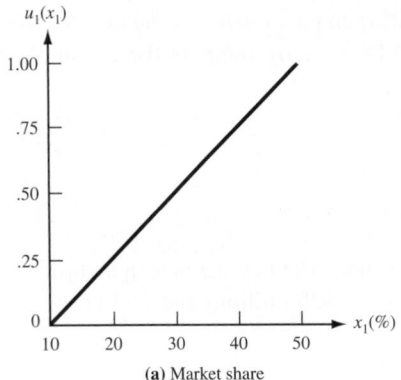

(a) Market share

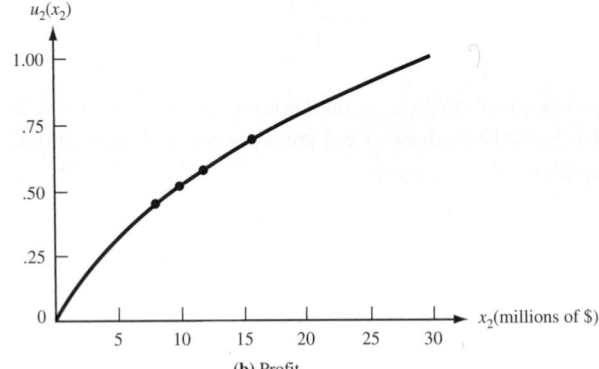

FIGURE 15
$u_1(x_1)$ and $u_2(x_2)$ for
Fruit Computer

(b) Profit

Suppose that for $k_1 = 0.6$, Fruit is indifferent between these two lotteries. Similarly, k_2 is the number that makes Fruit indifferent between

$$\underline{}^{1} 10\%, \$30 \qquad \text{and}$$

$$
\begin{array}{l}
k_2 50\%, \$30 \\
\hline
1 - k_2 10\%, \$5
\end{array}
$$

Suppose that for $k_2 = 0.5$, Fruit is indifferent between these two lotteries. Now $k_3 = 1 - k_1 - k_2 = -0.1$, and Fruit's multiattribute utility function is

$$u(x_1, x_2) = 0.6u_1(x_1) + 0.5u_2(x_2) - 0.1u_1(x_1)u_2(x_2) \qquad \text{(13)}$$

where $u_1(x_1)$ and $u_2(x_2)$ are sketched in Figure 15. Note that

$$\frac{\partial u(x_1, x_2)}{\partial x_1} = 0.6u_1'(x_1) - 0.1u_1'(x_1)u_2(x_2)$$

Thus, as Fruit's profit increases, we see that the utility gained from an additional point of market share decreases. Similarly, if $k_3 > 0$, then as profit increases, the benefit gained from an additional point of market share would increase. As outlined in step 5, we should now check whether this multiattribute utility function is consistent with Fruit's preferences.

Use of Multiattribute Utility Functions

To illustrate how a multiattribute utility function might be used, suppose that Fruit must determine whether to mount a small or a large advertising campaign during the coming year. Fruit believes there is a $\frac{1}{2}$ probability that its main rival, CSL Computers, will mount a small TV ad campaign and a $\frac{1}{2}$ probability that CSL will mount a large TV ad campaign. At the end of the current year, Fruit's market share and profits (in millions of dollars) will be as shown in Table 12. Fruit must determine which of the following lotteries has a larger expected utility: From Figure 15, we find that $u_1(15) = 0.125$, $u_1(25) = 0.375$, $u_1(35) = 0.625$, $u_2(8) = 0.45$, $u_2(10) = 0.53$, $u_2(12) = 0.58$, and $u_2(16) = 0.70$. Then

$$u(25\%, \$16) = 0.6(.375) + 0.5(.7) - 0.1(.375)(.7) = .549$$
$$u(15\%, \$12) = 0.6(.125) + 0.5(.58) - 0.1(.125)(.58) = .358$$
$$u(35\%, \$8) = 0.6(.625) + 0.5(.45) - 0.1(.625)(.45) = .572$$
$$u(25\%, \$10) = 0.6(.375) + 0.5(.53) - 0.1(.375)(.53) = .470$$

Then

$$E(U \text{ for small ad campaign}) = (\tfrac{1}{2})(.549) + (\tfrac{1}{2})(.358) = .454$$
$$E(U \text{ for large ad campaign}) = (\tfrac{1}{2})(.572) + (\tfrac{1}{2})(.470) = .521$$

Thus, during the current year, Fruit should mount a large ad campaign.

TABLE 12
Effect of Advertising on Market Share and Profit

	CSL Chooses	
Fruit Chooses	Small Ad Campaign	Large Ad Campaign
Small ad campaign	25%, $16	15%, $12
Large ad campaign	35%, $8	25%, $10

PROBLEMS

Group A

1 National Express Carriers is interested in two attributes:

Attribute 1 The average cost of delivering a letter (known to be between $1 and $5)

Attribute 2 Percentage of all letters reaching their destination on time (known to be between 70% and 100%)

  a Would National's multiattribute utility function exhibit mui?

  b Would National's utility function be additive?

  c Assume that attributes 1 and 2 exhibit mui. Suppose National is indifferent between ($1, 70%) for certain and the following lottery:

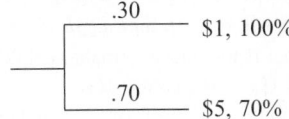

Also assume that National is indifferent between ($5, 100%) for certain and

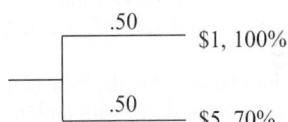

Find National's multiattribute utility function. Express National's multiattribute utility function in terms of $u_1(x_1)$ and $u_2(x_2)$.

2 Keeney and Raiffa (1976) discuss the assessment of a blood bank's multiattribute utility function. For simplicity, we assume that the blood bank must determine at the beginning of each week how many pints of blood should be ordered. Any blood left over at the end of the week spoils (it is outdated). For the blood bank, two attributes of interest are as follows:

Attribute 1 Number of pints of blood by which ordered blood falls short of the week's demand (the weekly shortage). The weekly shortage is known to be always between 0 and 10 pints.

Attribute 2 Number of pints of blood that are outdated (known to be always between 0 and 10 pints)

Assume that attributes 1 and 2 exhibit mui.

a Suppose the blood bank is indifferent between

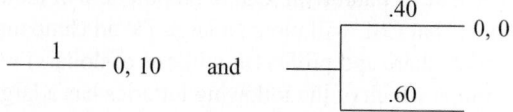

and between

Let x_1 = value of attribute 1, and x_2 = value of attribute 2. Also suppose that

$$u_1(x_1) = .58 \exp\left(1 - \frac{x_1}{10}\right) - .58$$

and

$$u_2(x_2) = 1 - \frac{x_2^2}{100}$$

Determine the blood bank's multiattribute utility function.

b Suppose that each week there is a $\frac{1}{2}$ chance that the demand for blood will be 25 pints and a $\frac{1}{2}$ chance it will be 35 pints. Would the blood bank be better off ordering 28 pints, 30 pints, or 32 pints?

3 Show that the method for determining k_2 described in the text is valid.

4 Gotham City is trying to determine how many ambulances it should have and how to staff them. Each ambulance may be staffed with paramedics or emergency medical technicians. Paramedics are considered to provide better service and are paid higher salaries. Budgetary limitations have forced the city to choose between the following two alternatives:

Alternative 1 Four ambulances, two staffed with emergency medical technicians and two staffed with paramedics

Alternative 2 Three ambulances, all staffed with paramedics

The city authorities believe that the following two attributes determine the city's satisfaction with ambulance service:

Attribute 1 Time until an ambulance reaches a patient

Attribute 2 Percentage of ambulance calls handled by paramedics. Assume that Gotham City's multiattribute utility function $u(x_1, x_2)$ exhibits mui and that

$$u_1(x_1) = 1 - \frac{x_1^2}{900} \quad \text{and} \quad u_2(x_2) = \frac{x_2^2}{10,000}$$

The time for an ambulance to reach a patient is always between 0 and 30 minutes. The city authorities are indifferent between

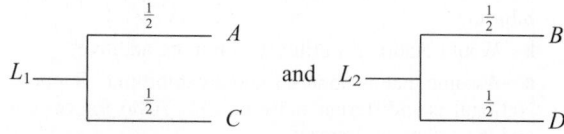

The city authorities are also indifferent between

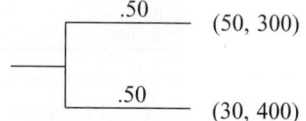

Assume that if an ambulance is available when a call comes in, then the ambulance will arrive in 5 minutes; if an ambulance is not available when a call comes in, it will arrive in 20 minutes. With three ambulances, one will be immediately available 60% of the time, and with four ambulances, one will be immediately available 80% of the time.

a Determine the city authorities' multiattribute utility function.

b Which alternative should they choose?

5 Public service Indiana (PSI) is considering two sites for a nuclear power plant. The following two attributes will influence its determination about where to build the plant:

Attribute 1 Cost of the plant (in millions of dollars)

Attribute 2 Acres of land damaged by building the plant

Assume that PSI's multiattribute utility function is given by $u_1(x_1, x_2) = .70u_1(x_1) + .20u_2(x_2) + .10u_1(x_1)u_2(x_2)$, where $u_1(x_1) = .1 + \exp(-.1x_1)$ and $u_2(x_2) = 2.5 - 2.5 \exp(.0006x_2 - .48)$.

Two locations for the power plant are under consideration. Location 1 is equivalent to the following lottery:

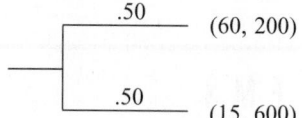

and location 2 is equivalent to the following lottery:

Which location should be chosen?

Group B

6 Consider the four points A, B, C, and D in Figure 16. Assume that more of each attribute is desirable and that a decision maker's utility function exhibits mui. Consider the following two lotteries:

a Show that if $k_3 > 0$, then $L_1 p L_2$.

b Show that if $k_3 < 0$, then $L_2 p L_1$.

c Show that if the decision maker exhibits additive independence ($k_3 = 0$), then $L_1 i L_2$.

d Let attribute 1 = performance of Germany on the eastern front near the end of World War II, and attribute 2 = performance of Germany on the western front. A high level of an attribute means that Germany did well,

FIGURE **16**

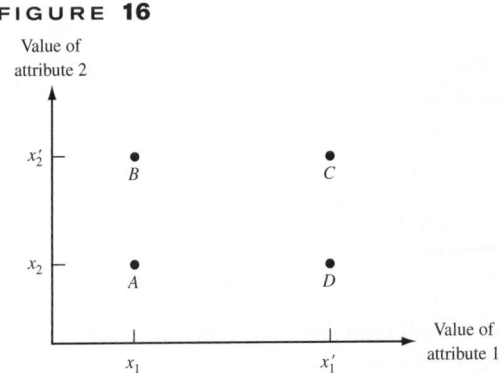

Value of
attribute 2

and a low level of an attribute means that Germany did poorly. Suppose that Germany will suffer defeat if it performs poorly on either front. If these attributes exhibit mui, what would be the sign of k_3?

e General Motors has domestic and international divisions. Let attribute 1 = profits in the domestic division and attribute 2 = profits in the international division. Suppose General Motors is reasonably happy if at least one division has a good year but is very unhappy if both divisions have a bad year. If these attributes exhibit mui, what would be the sign of k_3?

13.7 The Analytic Hierarchy Process

In Section 13.6, we discussed situations in which a decision maker chooses between alternatives on the basis of how well the alternatives meet various objectives. For example, in determining which job offer to accept, a job seeker (call her Jane) might choose between the offers by determining how well each one meets the following four objectives:

Objective 1 High starting salary (SAL)

Objective 2 Quality of life in city where job is located (QL)

Objective 3 Interest in work (IW)

Objective 4 Job location near family and relatives (NF)

When multiple objectives are important to a decision maker, it may be difficult to choose between alternatives. For example, one job offer may offer the highest starting salary, but it may score poorly on the other three objectives. Another job offer may meet objectives 2–4 but have a low starting salary. In such a case, it may be difficult for Jane to choose between job offers. Thomas Saaty's **analytic hierarchy process** (AHP) provides a powerful tool that can be used to make decisions in situations involving multiple objectives.

To illustrate how the AHP works, let's suppose that Jane has three job offers and must determine which offer to accept. For the ith objective (in this example, $i = 1, 2, 3, 4$), the AHP generates (by a method to be described shortly) a weight w_i ($i = 1, 2, 3, 4$) for the ith objective. For convenience, the chosen weights always sum to 1. Suppose that for this example, we have found Jane's weights to be

$$w_1 = .5115, \qquad w_2 = .0986, \qquad w_3 = .2433, \qquad w_4 = .1466$$

(These weights fail to add up to 1 due to rounding.) The weights indicate that a high starting salary is the most important objective, followed by interest in work, nearness to family, and quality of life in the city where the job is located.

Next suppose (again by a method that is soon to be described) that Jane can determine how well each job "scores" on each objective. For example, suppose Jane determines that each job scores on each objective as shown in Table 13. For example, job 1 best meets the objective of a high starting salary but "scores" worst on all other objectives.

Given Jane's weights and the score of each job on each objective, how can she determine which job offer to accept? For the jth job offer ($j = 1, 2, 3$), compute job offer j's overall score as follows:

$$\sum_{i=1}^{i=4} w_i \text{ (job offer } j\text{'s score on objective } i)$$

TABLE 13
Jane's "Score" for Each Job and Objective

Objective	Job 1	Job 2	Job 3
Salary	.571	.286	.143
Quality of life	.159	.252	.589
Interest in work	.088	.669	.243
Proximity to family	.069	.426	.506

Now choose the job offer with the highest overall score. Note that the overall score gives more weight to a job offer's score on the more important objectives. Computing each job's overall score, we obtain

$$\text{Job 1 overall score} = .5115(.571) + .0986(.159) + .2433(.088)$$
$$+ .1466(.069) = .339$$
$$\text{Job 2 overall score} = .5115(.286) + .0986(.252) + .2433(.669)$$
$$+ .1466(.426) = .396$$
$$\text{Job 3 overall score} = .5115(.143) + .0986(.589) + .2433(.243)$$
$$+ .1466(.506) = .265$$

Thus, the AHP would indicate that Jane should accept job 2.

Obtaining Weights for Each Objective

Suppose there are n objectives. We begin by writing down an $n \times n$ matrix (known as the **pairwise comparison matrix**) A. The entry in row i and column j of A (call it a_{ij}) indicates how much more important objective i is than objective j. "Importance" is to be measured on an integer-valued 1–9 scale, with each number having the interpretation shown in Table 14. For all i, it is necessary that $a_{ii} = 1$. If, for example, $a_{13} = 3$, objective 1 is weakly more important than objective 3. If $a_{ij} = k$, then for consistency, it is necessary that $a_{ji} = \frac{1}{k}$. Thus, if $a_{13} = 3$, then $a_{31} = \frac{1}{3}$ must hold.

Suppose that Jane has identified the following pairwise comparison matrix for her four objectives (SAL = high salary; QL = high quality of life; IW = interest in work; NF = nearness to family):

$$
\begin{array}{c c}
 & \begin{array}{cccc} \text{SAL} & \text{QL} & \text{IW} & \text{NF} \end{array} \\
\begin{array}{c} \text{SAL} \\ \text{QL} \\ \text{IW} \\ \text{NF} \end{array} &
\begin{bmatrix}
1 & 5 & 2 & 4 \\
\frac{1}{5} & 1 & \frac{1}{2} & \frac{1}{2} \\
\frac{1}{2} & 2 & 1 & 2 \\
\frac{1}{4} & 2 & \frac{1}{2} & 1
\end{bmatrix}
\end{array}
$$

Unfortunately, some of Jane's pairwise comparisons are inconsistent. To illustrate the meaning of consistency, note that since $a_{13} = 2$, she feels SAL is twice as important as IW. Since $a_{32} = 2$, she also believes that IW is twice as important as QL. Consistency of preferences would imply that Jane should feel that SAL is $2(2) = 4$ times as important as QL. Since $a_{12} = 5$, however, Jane believes that SAL is 5 times as important as QL. This shows that Jane's pairwise comparisons exhibit a slight inconsistency. Slight inconsistencies are common and do not cause serious difficulties. An index that can be used to measure the consistency of Jane's preferences will be discussed later in this section.

TABLE **14**
Interpretation of Entries in a Pairwise Comparison Matrix

Value of a_{ij}	Interpretation
1	Objective i and j are of equal importance.
3	Objective i is weakly more important than objective j.
5	Experience and judgment indicate that objective i is strongly more important than objective j.
7	Objective i is very strongly or demonstrably more important than objective j.
9	Objective i is absolutely more important than objective j.
2, 4, 6, 8	Intermediate values—for example, a value of 8 means that objective i is midway between strongly and absolutely more important than objective j.

Suppose there are n objectives. Let w_i = the weight given to objective i. To describe how the AHP determines the w_i's, let's suppose the decision maker is perfectly consistent. Then her pairwise comparison matrix should be of the following form:

$$A = \begin{bmatrix} \dfrac{w_1}{w_1} & \dfrac{w_1}{w_2} & \cdots & \dfrac{w_1}{w_n} \\[2mm] \dfrac{w_2}{w_1} & \dfrac{w_2}{w_2} & \cdots & \dfrac{w_2}{w_n} \\[2mm] \vdots & \vdots & & \vdots \\[2mm] \dfrac{w_n}{w_1} & \dfrac{w_n}{w_2} & \cdots & \dfrac{w_n}{w_n} \end{bmatrix} \tag{13}$$

For example, suppose that $w_1 = \frac{1}{2}$ and $w_2 = \frac{1}{6}$. Then objective 1 is three times as important as objective 2, so

$$a_{12} = \frac{w_1}{w_2} = 3$$

Now suppose that a consistent decision maker has a pairwise comparison matrix A of the form (13). How can we recover the vector $\mathbf{w} = [w_1 \quad w_2 \quad \cdots \quad w_n]$ from A? Consider the system of n equations

$$A\mathbf{w}^T = \Delta \mathbf{w}^T \tag{14}$$

where Δ is an unknown number and $\mathbf{w}^T$ is an unknown n-dimensional column vector. For any number Δ, (14) always has the trivial solution $\mathbf{w} = [0 \quad 0 \quad \cdots \quad 0]$. It can be shown that if A is the pairwise comparison matrix of a perfectly consistent decision maker (that is, if A is of the form (13)) and we do not allow $\Delta = 0$, then the only nontrivial solution to (14) is $\Delta = n$ and $\mathbf{w} = [w_1 \quad w_2 \quad \cdots \quad w_n]$. This shows that for a consistent decision maker, the weights w_i can be obtained from the only nontrivial solution to (14). Now suppose that the decision maker is not perfectly consistent. Let $\Delta_{\max}$ be the largest number for which (14) has a nontrivial solution (call this solution $\mathbf{w}_{\max}$). If the decision maker's comparisons do not deviate very much from perfect consistency, we would expect $\Delta_{\max}$ to be close to n and $\mathbf{w}_{\max}$ to be close to $\mathbf{w}$. Saaty verified that this intuition is indeed correct and suggested approximating $\mathbf{w}$ by $\mathbf{w}_{\max}$. Saaty also proposed measuring the decision maker's consistency by looking how close $\Delta_{\max}$ is to n. The software package Expert Choice gives (among other outputs) exact values of $\Delta_{\max}$ and $\mathbf{w}_{\max}$ and a measure of the decision maker's consistency.

In what follows, we outline a simple method (easily implemented on any spreadsheet) that can be used to approximate Δ_{max} and $\mathbf{w}_{max}$ and an index of consistency.

To approximate $\mathbf{w}_{max}$, we use the following two-step procedure:

Step 1   For each of A's columns, do the following. Divide each entry in column i of A by the sum of the entries in column i. This yields a new matrix (call it A_{norm}, for normalized) in which the sum of the entries in each column is 1. For Jane's pairwise comparison matrix, step 1 yields

$$A_{norm} = \begin{bmatrix} .5128 & .5000 & .5000 & .5333 \\ .1026 & .1000 & .1250 & .0667 \\ .2564 & .2000 & .2500 & .2667 \\ .1282 & .2000 & .1250 & .1333 \end{bmatrix}$$

Step 2   To find an approximation to $\mathbf{w}_{max}$ (to be used as our estimate of $\mathbf{w}$), proceed as follows. Estimate w_i as the average of the entries in row i of A_{norm}. This yields (as previously stated)

$$w_1 = \frac{.5128 + .5000 + .5000 + .5333}{4} = .5115$$

$$w_2 = \frac{.1026 + .1000 + .1250 + .0667}{4} = .0986$$

$$w_3 = \frac{.2564 + .2000 + .2500 + .2667}{4} = .2433$$

$$w_4 = \frac{.1282 + .2000 + .1250 + .1333}{4} = .1466$$

Intuitively, why does w_1 approximate the weight that objective 1 (salary) should be given? The percentage of the weight that SAL is given in pairwise comparisons of each objective to SAL is .5128. Similarly, .50 represents the percentage of total weight that SAL is given in pairwise comparisons of each objective to QL. Thus, we see that the four numbers averaged to obtain w_1 each represents in some way a measure of the total weight attached to SAL. Thus, averaging these numbers should give a good estimate of the percentage of the total weight that should be given to SAL.

Checking for Consistency

We can now use the following four-step procedure to check for the consistency of the decision maker's comparisons. (From now on, $\mathbf{w}$ denotes our estimate of the decision maker's weights.)

Step 1   Compute $A\mathbf{w}^T$. For our example, we obtain

$$A\mathbf{w}^T = \begin{bmatrix} 1 & 5 & 2 & 4 \\ \frac{1}{5} & 1 & \frac{1}{2} & \frac{1}{2} \\ \frac{1}{2} & 2 & 1 & 2 \\ \frac{1}{4} & 2 & \frac{1}{2} & 1 \end{bmatrix} \begin{bmatrix} .5115 \\ .0986 \\ .2433 \\ .1466 \end{bmatrix} = \begin{bmatrix} 2.0775 \\ 0.3959 \\ 0.9894 \\ 0.5933 \end{bmatrix}$$

Step 2   Compute

$$\frac{1}{n} \sum_{i=1}^{i=n} \frac{i\text{th entry in } A\mathbf{w}^T}{i\text{th entry in } \mathbf{w}^T}$$

$$= \left(\frac{1}{4}\right) \left\{ \frac{2.0775}{.5115} + \frac{.3959}{.0986} + \frac{.9894}{.2433} + \frac{.5933}{.1466} \right\}$$

$$= 4.05$$

TABLE **15**

Values of the
Random Index (RI)

n	RI
2	0
3	.58
4	.90
5	1.12
6	1.24
7	1.32
8	1.41
9	1.45
10	1.51

Step 3 Compute the **consistency index** (CI) as follows:

$$CI = \frac{(\text{Step 2 result}) - n}{n - 1} = \frac{4.05 - 4}{3} = .017$$

Step 4 Compare CI to the random index (RI) for the appropriate value of n, shown in Table 15.

For a perfectly consistent decision maker (see Problem 5), the ith entry in $A\mathbf{w}^T = n$ (ith entry of $\mathbf{w}^T$). This implies that a perfectly consistent decision maker has CI = 0. The values of RI in Table 15 give the average value of CI if the entries in A were chosen at random, subject to the constraint that all diagonal entries must equal 1 and

$$a_{ij} = \frac{1}{a_{ji}}$$

If CI is sufficiently small, the decision maker's comparisons are probably consistent enough to give useful estimates of the weights for his or her objective function. If $\frac{CI}{RI} < .10$, the degree of consistency is satisfactory, but if $\frac{CI}{RI} > .10$, serious inconsistencies may exist, and the AHP may not yield meaningful results. In our example, $\frac{CI}{RI} = \frac{.017}{.90} = .019 < .10$; thus, Jane's pairwise comparison matrix does not exhibit any serious inconsistencies.

Finding the Score of an Alternative for an Objective

We have now described how to determine the objective function weights that we earlier used to help Jane determine which job offer to accept. We now determine how well each job "satisfies" or "scores" on each objective. To determine these scores, we construct for each objective a pairwise comparison matrix in which the rows and columns are Jane's possible decisions (in this case, job offers). For SAL, suppose we obtain the following pairwise comparison matrix:

$$
\begin{array}{c}
\\
\text{Job 1} \\
\text{Job 2} \\
\text{Job 3}
\end{array}
\begin{array}{c}
\begin{array}{ccc}
\text{Job 1} & \text{Job 2} & \text{Job 3}
\end{array} \\
\left[
\begin{array}{ccc}
1 & 2 & 4 \\
\frac{1}{2} & 1 & 2 \\
\frac{1}{4} & \frac{1}{2} & 1
\end{array}
\right]
\end{array}
$$

Thus, for example, with respect to salary, job 1 is better (between weakly and strongly) than job 3. We can now apply our procedure for generating weights to the SAL pairwise comparison matrix. We obtain

$$A_{\text{norm}} = \begin{bmatrix} .571 & .571 & .571 \\ .286 & .286 & .286 \\ .143 & .143 & .143 \end{bmatrix}$$

This yields $\mathbf{w} = [.571\ .286\ .143]$. These weights indicate how well each job "scores" with respect to the SAL objective. As previously stated in Table 13, we obtain

$$\text{Job 1 salary score} = .571$$
$$\text{Job 2 salary score} = .286$$
$$\text{Job 3 salary score} = .143$$

Since all three columns of the pairwise comparison matrix for salary are identical, Jane's pairwise comparisons for salary exhibit perfect consistency.

Suppose Jane's pairwise comparison matrix for quality of life (QL) is as follows:

$$\begin{array}{c} \\ \text{Job 1} \\ \text{Job 2} \\ \text{Job 3} \end{array} \begin{array}{ccc} \text{Job 1} & \text{Job 2} & \text{Job 3} \\ \begin{bmatrix} 1 & \frac{1}{2} & \frac{1}{3} \\ 2 & 1 & \frac{1}{3} \\ 3 & 3 & 1 \end{bmatrix} \end{array}$$

Then

$$A_{\text{norm}} = \begin{bmatrix} \frac{1}{6} & \frac{1}{9} & \frac{1}{5} \\ \frac{1}{3} & \frac{2}{9} & \frac{1}{5} \\ \frac{1}{2} & \frac{6}{9} & \frac{3}{5} \end{bmatrix}$$

and we obtain

$$\text{Job 1 quality of life score} = \frac{\frac{1}{6} + \frac{1}{9} + \frac{1}{5}}{3} = .159$$

$$\text{Job 2 quality of life score} = \frac{\frac{1}{3} + \frac{2}{9} + \frac{1}{5}}{3} = .252$$

$$\text{Job 3 quality of life score} = \frac{\frac{1}{2} + \frac{6}{9} + \frac{3}{5}}{3} = .589$$

For interest in work, suppose the pairwise comparison matrix is as follows:

$$\begin{array}{c} \\ \text{Job 1} \\ \text{Job 2} \\ \text{Job 3} \end{array} \begin{array}{ccc} \text{Job 1} & \text{Job 2} & \text{Job 3} \\ \begin{bmatrix} 1 & \frac{1}{7} & \frac{1}{3} \\ 7 & 1 & 3 \\ 3 & \frac{1}{3} & 1 \end{bmatrix} \end{array}$$

It can easily be shown that

$$\text{Job 1 interest in work score} = .088$$
$$\text{Job 2 interest in work score} = .669$$
$$\text{Job 3 interest in work score} = .243$$

Finally, for nearness to family, suppose the pairwise comparison matrix is as follows:

$$
\begin{array}{c c c c}
 & \text{Job 1} & \text{Job 2} & \text{Job 3} \\
\begin{array}{c} \text{Job 1} \\ \text{Job 2} \\ \text{Job 3} \end{array} &
\left[\begin{array}{ccc}
1 & \frac{1}{4} & \frac{1}{7} \\
4 & 1 & 2 \\
7 & 2 & 1
\end{array} \right]
\end{array}
$$

Routine calculations yield

Job 1 score for nearness to family $= .069$

Job 2 score for nearness to family $= .426$

Job 3 score for nearness to family $= .506$

As described earlier, we can now "synthesize" the objective weights with the scores of each job on each objective to obtain an overall score for each alternative (in this case, each job offer). As before, we find that job offer 2 is most preferred, followed by job offer 1, with job offer 3 the least preferred.

We close by noting that AHP has been applied by decision makers in countless areas, including accounting, finance, marketing, energy resource planning, microcomputer selection, sociology, architecture, and political science. See Zahedi (1986) and Saaty (1988) for a discussion of applications of AHP.

Implementing AHP on a Spreadsheet

AHP.xls

Figure 17 illustrates how easy it is to implement AHP on a spreadsheet (file AHP.xls). Enter in the pairwise comparison matrix for objectives in B7:E10. In B12 enter the formula =B7/**SUM**(B$7:B$10) and copy this to the range B12:E15, yielding A_{norm} for objectives. Compute the weight for salary in F12 with the command **AVERAGE**(B12:E12). Copy this to F12:F15 to compute the weights of the remaining objectives. In a similar fashion, the normalized matrices and weights for each objective are obtained.

To determine the score for job 1, enter into F17 the formula

$$=\text{F\$12} * \text{F21} + \text{F\$13} * \text{F29} + \text{F\$14} * \text{F37} + \text{F\$15} * \text{F45}$$

Copying this formula to F17:F19 computes the score for jobs 2 and 3. Again, we see that job 2 receives the highest score (indicated by ****).

To compute the consistency index for the pairwise comparison matrix for objectives, the Excel matrix multiplication function MMULT is used, computing $A\mathbf{w}^T$ in the range C2:C5. In the range D2:D5 compute (*i*th entry in $A\mathbf{w}^T$)/(*i*th entry in $\mathbf{w}^T$). Finally, in E2 compute the CI, using the formula (**AVERAGE**(D2:D5) − 4)/3.

Mmult.xls

Using the Excel MMULT function, it is easy to multiply matrices. To illustrate, we will use Excel to find the matrix product *AB* (see Figure 18 and file Mmult.xls). We proceed as follows:

Step 1 Enter *A* and *B* in D2:F3 and D5:E7, respectively.

Step 2 Select the range (D9:E10) in which the product *AB* will be computed.

Step 3 In the upper left-hand corner (D9) of the selected range, type the formula

$$= \text{MMULT}(\text{D2:F3},\text{D5:E7})$$

Then hit **CONTROL SHIFT ENTER** (not just ENTER), and the desired matrix product will be computed. Note that MMULT is an array function, not an ordinary spreadsheet function. This explains why we must preselect the range for AB and use CONTROL SHIFT ENTER.

FIGURE **17**
AHP Spreadsheet

	A	B	C	D	E	F	G
1		CONSISTENCY	INDEX	AwT/wT	CI		
2	IMPLEMENTING		2.0774038	4.0610902	0.0158569		
3	AHP	AwT=	0.3958173	4.0160976			
4	ON		0.9894231	4.0671937			
5	A SPREADSHEET		0.5932692	4.0459016			
6	OBJECTIVES MATRIX	SAL	QL	IW	NF		
7	SAL	1	5	2	4		
8	QL	0.2	1	0.5	0.5		
9	IW	0.5	2	1	2		
10	NF	0.25	2	0.5	1		
11	ANORM(OBJECTIVES)	SAL	QL	NF	IW	WEIGHTS	
12	SAL	0.512820513	0.5	0.5	0.5333333	0.5115385	SAL
13	QL	0.102564103	0.1	0.125	0.0666667	0.0985577	QL
14	NF	0.256410256	0.2	0.25	0.2666667	0.2432692	IW
15	IW	0.128205128	0.2	0.125	0.1333333	0.1466346	NF
16	SALARY MATRIX	JOB1	JOB2	JOB3			
17	JOB1	1	2	4	JOB1SC=	0.3395156	
18	JOB2	0.5	1	2	JOB2SC=	0.3960857	****
19	JOB3	0.25	0.5	1	JOB3SC=	0.2643988	
20	ANORM(SALARY)	JOB1	JOB2	JOB3		WEIGHTS	
21	JOB1	0.571428571	0.5714286	0.5714286		0.5714286	JOB1
22	JOB2	0.285714286	0.2857143	0.2857143		0.2857143	JOB2
23	JOB3	0.142857143	0.1428571	0.1428571		0.1428571	JOB3
24	QL MATRIX	JOB1	JOB2	JOB3			
25	JOB1	1	0.5	0.3333333			
26	JOB2	2	1	0.3333333			
27	JOB3	3	3	1			
28	ANORM(QL)	JOB1	JOB2	JOB3		WEIGHTS	
29	JOB1	0.166666667	0.1111111	0.2		0.1592593	JOB1
30	JOB2	0.333333333	0.2222222	0.2		0.2518519	JOB2
31	JOB3	0.5	0.6666667	0.6		0.5888889	JOB3
32	IW MATRIX	JOB1	JOB2	JOB3			
33	JOB1	1	0.1428571	0.3333333			
34	JOB2	7	1	3			
35	JOB3	3	0.3333333	1			
36	ANORM(IW)	JOB1	JOB2	JOB3		WEIGHTS	
37	JOB1	0.090909091	0.0967742	0.0769231		0.0882021	JOB1
38	JOB2	0.636363636	0.6774194	0.6923077		0.6686969	JOB2
39	JOB3	0.272727273	0.2258065	0.2307692		0.243101	JOB3
40	NF MATRIX	JOB1	JOB2	JOB3			
41	JOB1	1	0.25	0.1428571			
42	JOB2	4	1	2			
43	JOB3	7	2	1			
44	ANORM(NF)	JOB1	JOB2	JOB3		WEIGHTS	
45	JOB1	0.083333333	0.0769231	0.0454545		0.0685703	JOB1
46	JOB2	0.333333333	0.3076923	0.6363636		0.4257964	JOB2
47	JOB3	0.583333333	0.6153846	0.3181818		0.5056333	JOB3

	A	B	C	D	E	F
1	MatrixMultiplication					
2				1	1	2
3			A	2	1	3
4						
5			B	1	1	
6				2	3	
7				1	2	
8						
9				5	8	
10			C	7	11	
11						

FIGURE 18

PROBLEMS

Group A

1 Each professor's annual salary increase is determined by performance in three areas: teaching, research, and service to the university. The administration has come up with the following pairwise comparison matrix for these objectives:

	Teaching	Research	Service
Teaching	1	$\frac{1}{3}$	5
Research	3	1	7
Service	$\frac{1}{5}$	$\frac{1}{7}$	1

The administration has compared two professors with regard to their teaching, research, and service over the past year. The pairwise comparison matrices are as follows. For teaching:

	Professor 1	Professor 2
Professor 1	1	4
Professor 2	$\frac{1}{4}$	1

For research:

	Professor 1	Professor 2
Professor 1	1	$\frac{1}{3}$
Professor 2	3	1

For service:

	Professor 1	Professor 2
Professor 1	1	6
Professor 2	$\frac{1}{6}$	1

a Which professor should receive a bigger raise?

b Does the AHP indicate how large a raise each professor should be given?

c Check the pairwise comparison matrix for consistency.

2 A business is about to purchase a new personal computer. Three objectives are important in determining which computer should be purchased: cost, user-friendliness, and software availability. The pairwise comparison matrix for these objectives is as follows:

	Cost	User-friendliness	Software availability
Cost	1	$\frac{1}{4}$	$\frac{1}{5}$
User-friendliness	4	1	$\frac{1}{2}$
Software availability	5	2	1

Three computers are being considered for purchase. The performance of each computer with regard to each objective is indicated by the following pairwise comparison matrices. For cost (low cost is good, high cost is bad!):

	Computer 1	Computer 2	Computer 3
Computer 1	1	3	5
Computer 2	$\frac{1}{3}$	1	2
Computer 3	$\frac{1}{5}$	$\frac{1}{2}$	1

For user-friendliness:

	Computer 1	Computer 2	Computer 3
Computer 1	1	$\frac{1}{3}$	$\frac{1}{2}$
Computer 2	3	1	5
Computer 3	2	$\frac{1}{5}$	1

For software availability:

	Computer 1	Computer 2	Computer 3
Computer 1	1	$\frac{1}{3}$	$\frac{1}{7}$
Computer 2	3	1	$\frac{1}{5}$
Computer 3	7	5	1

a Which computer should be purchased?

b Check the pairwise comparison matrices for consistency.

3 Woody is ready to select his mate for life and has determined that beauty, intelligence, and personality are the key factors in selecting a satisfactory mate. His pairwise comparison matrix for these objectives is as follows:

	Beauty	Intelligence	Personality
Beauty	1	3	5
Intelligence	$\frac{1}{3}$	1	3
Personality	$\frac{1}{5}$	$\frac{1}{3}$	1

Three women (Jennifer Lopez, Britney Spears, and Mandy Moore) are begging to be Woody's mate. His views of these women's beauty, intelligence and personality are given in the following pairwise comparison matrices.

Beauty:

	Jennifer	Britney	Mandy
Jennifer	1	5	3
Britney	$\frac{1}{5}$	1	$\frac{1}{2}$
Mandy	$\frac{1}{3}$	2	1

Intelligence:

	Jennifer	Britney	Mandy
Jennifer	1	$\frac{1}{6}$	$\frac{1}{4}$
Britney	6	1	2
Mandy	4	$\frac{1}{2}$	1

Personality:

	Jennifer	Britney	Mandy
Jennifer	1	4	$\frac{1}{4}$
Britney	$\frac{1}{4}$	1	$\frac{1}{9}$
Mandy	4	9	1

a Whom should Woody choose as his lifetime mate?

b Evaluate all pairwise comparison matrices for consistency.

4 In determining where to invest my money, two objectives—expected rate of return and degree of risk—are considered equally important. Two investments (1 and 2) have the following pairwise comparison matrices: Expected return:

	Investment 1	Investment 2
Investment 1	1	$\frac{1}{2}$
Investment 2	2	1

Degree of risk:

	Investment 1	Investment 2
Investment 1	1	3
Investment 2	$\frac{1}{3}$	1

a How should I rank these investments?

b Now suppose another investment (investment 3) is available. Suppose the pairwise comparison matrices for these investments are as follows. Expected return:

	Investment 1	Investment 2	Investment 3
Investment 1	1	$\frac{1}{2}$	4
Investment 2	2	1	8
Investment 3	$\frac{1}{4}$	$\frac{1}{8}$	1

Degree of risk:

	Investment 1	Investment 2	Investment 3
Investment 1	1	3	$\frac{1}{2}$
Investment 2	$\frac{1}{3}$	1	$\frac{1}{6}$
Investment 3	2	6	1

c Observe that the entries in the comparison matrices for investments 1 and 2 have not changed. How should I now rank the investments? Contrast my ranking of investments 1 and 2 with the answer from part (a).

5 Show that for a perfectly consistent decision maker, the ith entry in $A\mathbf{w}^T = n$ (ith entry of $\mathbf{w}^T$).

6 A consumer is trying to determine which type of frozen dinner to eat. He considers three attributes to be important: taste, nutritional value, and price. Nutritional value is considered to be determined by cholesterol and sodium levels. Three types of dinners are under consideration. The pairwise comparison matrix for the three attributes is as follows:

	Taste	Nutrition	Price
Taste	1	3	$\frac{1}{2}$
Nutrition	$\frac{1}{3}$	1	$\frac{1}{5}$
Price	2	5	1

Between the three frozen dinners the pairwise comparison matrix for each attribute is as follows. For taste:

	Dinner 1	Dinner 2	Dinner 3
Taste 1	1	5	3
Taste 2	$\frac{1}{5}$	1	$\frac{1}{2}$
Taste 3	$\frac{1}{3}$	2	1

For sodium:

	Dinner 1	Dinner 2	Dinner 3
Sodium 1	1	$\frac{1}{7}$	$\frac{1}{3}$
Sodium 2	7	1	2
Sodium 3	3	$\frac{1}{2}$	1

For cholesterol:

	Dinner 1	Dinner 2	Dinner 3
Cholesterol 1	1	$\frac{1}{8}$	$\frac{1}{4}$
Cholesterol 2	8	1	2
Cholesterol 3	4	$\frac{1}{2}$	1

For price:

	Dinner 1	Dinner 2	Dinner 3
Price 1	1	4	$\frac{1}{2}$
Price 2	$\frac{1}{4}$	1	$\frac{1}{6}$
Price 3	2	6	1

To determine how each dinner rates on nutrition you will need the following pairwise comparison matrix for cholesterol and sodium:

	Cholesterol	Sodium
Cholesterol	1	5
Sodium	$\frac{1}{5}$	1

Which frozen dinner would he prefer? (*Hint:* Nutrition score for a dinner = (score of dinner on sodium) * (weight for sodium) + (score for dinner on cholesterol) * (weight for cholesterol).)

7 You are trying to determine which MBA program to attend. You have been accepted at two programs: Indiana and Northwestern. You have chosen three attributes to use in helping you make your decision:

Attribute 1 Cost
Attribute 1 Starting salary
Attribute 1 Ambience of school (can we party there?!!)

Your pairwise comparison matrix for these attributes is as follows:

	Cost	Starting salary	Ambience
Cost	1	$\frac{1}{4}$	2
Starting salary	4	1	7
Ambience	$\frac{1}{2}$	$\frac{1}{7}$	1

For each attribute the pairwise comparison matrix for Indiana and Northwestern is as follows. For cost:

	Indiana	Northwestern
Indiana	1	6
Northwestern	$\frac{1}{6}$	1

For starting salary:

	Indiana	Northwestern
Indiana	1	$\frac{1}{3}$
Northwestern	3	1

For ambience:

	Indiana	Northwestern
Indiana	1	4
Northwestern	$\frac{1}{4}$	1

Which MBA program should you attend?

8 You have been hired by Arthur Ross to determine which of the following accounts receivable procedures should be used in an audit of the Keating Five and Dime Store:

a Analytic review
b Confirmations
c Test of subsequent collections (receipts)

The three criteria used to distinguish between the procedures are as follows:

a Reliability
b Cost
c Validity

The pairwise comparison matrix for the three criteria is as follows:

Reliability	1	5	7
Cost	$\frac{1}{5}$	1	2
Validity	$\frac{1}{7}$	$\frac{1}{2}$	1

For the reliability criterion the pairwise comparison matrix of the three procedures is as follows:

Analytical review	1	$\frac{1}{6}$	$\frac{1}{2}$
Confirmations	6	1	4
Test of subsequent collections	2	$\frac{1}{4}$	1

For the cost criterion the pairwise comparison matrix of the three procedures is as follows:

Analytical review	1	5	3
Confirmations	$\frac{1}{5}$	1	$\frac{1}{2}$
Test of subsequent collections	$\frac{1}{3}$	2	1

For the validity criterion the pairwise comparison matrix of the three procedures is as follows:

Analytical review	1	3	2
Confirmations	$\frac{1}{3}$	1	$\frac{1}{2}$
Test of subsequent collections	$\frac{1}{2}$	2	1

Use the AHP to determine which auditing procedure should be used. Also check the first pairwise comparison matrix for consistency.[†]

9 You are trying to determine which of two secretarial candidates (Jack and Jill) to hire. The three objectives that are important to your decision are personality, typing ability, and intelligence. You have assessed the following pairwise comparison matrix:

	Personality	Typing ability	Intelligence
Personality	1	$\frac{1}{4}$	$\frac{1}{3}$
Typing ability	4	1	$\frac{1}{2}$
Intelligence	3	2	1

The "score" of each employee on each objective is as follows:

	Personality	Typing ability	Intelligence
Jack	.4	.6	.2
Jill	.6	.4	.8

If you follow the AHP method which employee should be hired?

[†]Based on Lin, Mock, and Wright (1984).

S U M M A R Y Decision Criteria

In the **state-of-the-world model,** the decision maker first chooses an action a_i from a set $A = \{a_1, a_2, \ldots, a_k\}$ of available actions. With probability p_j the state of the world is observed to be $s_j \in S = \{s_1, s_2, \ldots, s_n\}$. If action a_i is chosen and the state of the world is s_j, the decision maker receives a reward r_{ij}.

The **maximin criterion** chooses the action a_i with the largest value of $\min_{j \in S} r_{ij}$. The **maximax criterion** chooses the action a_i with the largest value of $\max_{j \in S} r_{ij}$. In each state, the **minimax regret criterion** chooses an action by applying the minimax criterion to the regret matrix. The **expected value criterion** chooses the decision that yields the largest expected reward.

Utility Theory

A decision maker who subscribes to the Von Neumann–Morgenstern axioms, when facing a choice between several lotteries, should choose the lottery with the largest expected utility.

The **certainty equivalent** of a lottery L, written $CE(L)$, is the number $CE(L)$ such that the decision maker is indifferent between the lottery L and receiving a certain payoff of $CE(L)$. For a given lottery L, the **risk premium**, written $RP(L)$, is given by $RP(L) = EV(L) - CE(L)$.

A decision maker is **risk-averse** if and only if for any nondegenerate lottery L, $RP(L) > 0$. A risk-averse decision maker has a strictly concave utility function. A decision maker is **risk-neutral** if and only if for any nondegenerate lottery L, $RP(L) = 0$. A risk-neutral decision maker has a linear utility function. A decision maker is **risk-seeking** if and only if for any nondegenerate lottery L, $RP(L) < 0$. A risk-seeking decision maker has a strictly convex utility function.

Prospect Theory and Framing

Tversky and Kahneman resolved several flaws in EMU by developing prospect theory and framing. Prospect theory assumes that we do not treat probabilities as they are given in a decision-making problem. Instead, the decision maker treats a probability p for an event as a "distorted" probability $\Pi(p)$. The idea of framing is based on the fact that people often set their utility function from the standpoint of a frame or status quo from which they view the current situation.

Decision Trees

To determine the optimal decisions in a decision tree, we work backward (folding back the tree) from right to left. First assume that the decision maker is risk-neutral and wants to maximize final asset position. At each event fork, we calculate the expected final asset position and enter it in ○. At each decision fork, we denote by ‖ the decision that maximizes the expected final asset position and enter the expected final asset position associated with that decision in □. We continue working backward in this fashion until we reach the beginning of the tree. Then the optimal sequence of decisions can be obtained by following the ‖.

To incorporate a decision maker's utility function into a decision tree analysis, simply replace each final asset position x_0 by its utility $u(x_0)$. Then at each event fork, compute expected utility, and at each decision fork, choose the branch having the largest expected utility.

The **expected value of sample information** (EVSI) measures the value associated with test or sample information: EVSI = EVWSI − EVWOI. **Expected value with perfect information** (EVWPI) is found by drawing a decision tree in which the decision maker has perfect information about which state has occurred before the decision must be made. Then the **expected value of perfect information** (EVPI) is given by EVPI = EVWPI − EVWOI.

Bayes' Rule and Decision Trees

We use Bayes' rule in decision tree analysis when we are given prior probabilities and (for each state of the world) the likelihood that an experimental outcome will occur. Bayes' rule is then used to compute the probability that each experimental outcome will occur and (for each experimental outcome) the posterior probability of each state of the world. Then the decision tree analysis proceeds as already described.

Decision Making with Multiple Objectives

Attribute 1 is **preferentially independent** (pi) of attribute 2 if preferences for values of attribute 1 do not depend on the value of attribute 2.

A set of attributes S is **mutually preferentially independent** (mpi) of a set of attributes S' if (1) the values of the attributes in S' do not affect preferences for the values of attributes in S; (2) the values of attributes in S do not affect preferences for the values of attributes in S'.

A set of attributes $1, 2, \ldots, n$ is mutually preferentially independent (mpi) if for all subsets S of $\{1, 2, \ldots, n\}$, S is mpi of $\bar{S}$. ($\bar{S}$ is all members of $\{1, 2, \ldots, n\}$ that are not included in S.)

THEOREM 1

If the set of attributes $1, 2, \ldots, n$ is mpi, the decision maker's preferences can be represented by an additive value (or cost) function.

Multiattribute Utility Functions

Attribute 1 is **utility independent** (ui) of attribute 2 if preferences for lotteries involving different levels of attribute 1 do not depend on the level of attribute 2.

If attribute 1 is ui of attribute 2, and attribute 2 is ui of attribute 1, then attributes 1 and 2 are **mutually utility independent** (mui).

THEOREM 2

Attributes 1 and 2 are mui if and only if the decision maker's utility function $u(x_1, x_2)$ is a multilinear function of the form

$$u(x_1, x_2) = k_1 u_1(x_1) + k_2 u_2(x_2) + k_3 u_1(x_1)u_2(x_2) \qquad (10)$$

A decision maker's utility function exhibits **additive independence** if the decision maker is indifferent between

and

If attributes 1 and 2 are mpi and the decision maker's utility function exhibits additive independence, the decision maker's utility function is of the following additive form:

$$u(x_1, x_2) = k_1 u_1(x_1) + k_2 u_2(x_2)$$

The following procedure is used to assess multiattribute utility functions:

Step 1 Check whether attributes 1 and 2 are mui. If they are, go on to step 2. If the attributes are not mui, the assessment of the multiattribute utility function is beyond the scope of our discussion.

Step 2 Check for additive independence.

Step 3 Assess $u_1(x_1)$ and $u_2(x_2)$.

Step 4 Determine k_1, k_2, and (if there is no additive independence) k_3.

Step 5 Check to see whether the assessed utility function is really consistent with the decision maker's preferences. To do this, set up several lotteries and use the expected utility of each lottery to rank the lotteries from most to least favorable. Then ask the decision maker to rank the lotteries from most to least favorable. If the assessed utility function is consistent with the decision maker's preferences, the ranking of lotteries obtained from the assessed utility function should closely resemble the decision maker's ranking of the lotteries.

Analytic Hierarchy Process (AHP)

The AHP is often used to make decisions in situations when there are multiple objectives. Given a pairwise comparison matrix A, we can approximate the weights for each attribute as follows:

Step 1 For each of A's columns, do the following. Divide each entry in column i of A by the sum of the entries in column i. This yields a new matrix(A_{norm}), in which the sum of the entries in each column is 1.

Step 2 To find an approximation to $\mathbf{w}_{\text{max}}$, which will be used as our estimate of $\mathbf{w}$, proceed as follows. Estimate w_i as the average of the entries in row i of A_{norm}. To find the best decision, determine an overall score for a decision as follows:

$$\text{Decision score} = \sum_i w_i(\text{decision score on objective } i)$$

Now choose the decision with the largest score.

To check for consistency in pairwise comparision matrices, we use the following four-step process. ($\mathbf{w}$ denotes our estimate of the decision maker's weights.)

Step 1 Compute $A\mathbf{w}^T$.

Step 2 Compute

$$\frac{1}{n}\sum_{i=1}^{i=n}\frac{i\text{th entry in } A\mathbf{w}^T}{i\text{th entry in } \mathbf{w}^T}$$

Step 3 Compute the **consistency index** (CI) as follows:

$$CI = \frac{(\text{Step 2 result}) - n}{n - 1}$$

Step 4 Compare CI to the random index (RI) for the appropriate value of n. If $\frac{CI}{RI} < .10$, the degree of consistency is satisfactory, but if $\frac{CI}{RI} > .10$, serious inconsistencies may exist, and the AHP may not yield meaningful results.

REVIEW PROBLEMS

Group A

1 We have $1,000 to invest. All the money must be placed in one of three investments: gold, stock, or money market certificates. If $1,000 is placed in an investment, the value of the investment one year from now depends on the state of the economy (see Table 16). Assume that each state of the economy is equally likely. For each of the following decision criteria, determine the optimal decision:

a maximin

b maximax

TABLE 16

Value of $1,000	State 1	State 2	State 3
Money market certificate	$1,100	$1,100	$1,100
Stock	$1,000	$1,100	$1,200
Gold	$1,600	$300	$1,400

TABLE 17

	Economy Has Good Year	Economy Has Bad Year
Yield on stocks	22%	10%
Yield on bonds	16%	14%

c minimax regret

d expected value

2 In Problem 1, suppose that the utility function for the value of the investment (x) one year from now is given by $u(x) = \ln x$. Determine which investment we should choose. Could we have predicted this answer without a table of logarithms?

3 Consider the following four lotteries:

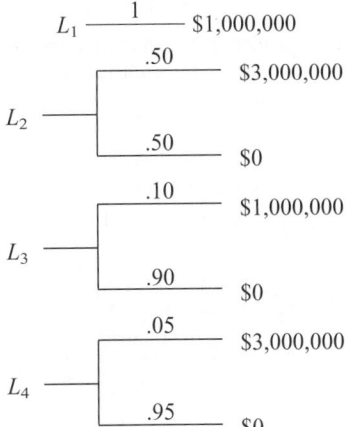

a Most people prefer L_1 to L_2 and L_4 to L_3. Explain why.

b Suppose a decision maker subscribes to the Von Neumann–Morgenstern axioms and prefers L_1 to L_2. Show that he or she must also prefer L_3 to L_4.

4 Jay Boyville Corporation is being sued by Lark Dent. Lark can settle out of court and win $40,000, or go to court. If Lark goes to court, there is a 30% chance that she will win the case. If she wins, a small and a large settlement are equally likely (a small settlement nets $50,000, and a large settlement nets $300,000).

a If Lark is risk-neutral, what should she do? What should Lark do if her utility function for an increase x in her cash position is given by $u(x) = x^{1/2}$?

b For $10,000, Lark can hire a consultant who will predict who will win the trial. The consultant is correct 90% of the time. Should she hire the consultant? (Assume Lark is risk-neutral.) What is EVSI?

c If Lark is risk-neutral, what is EVPI?

5 Rollo Megabux has $1 million to invest in stocks or bonds. The percentage yield on each investment during the coming year depends on whether the economy has a good or a bad year (see Table 17). It is equally likely that the economy will have a good or a bad year.

a If Rollo is risk-neutral, how should he invest his money?

b For $10,000, Rollo can hire a consulting firm to forecast the state of the economy. The consulting firm's forecasts have the following properties:

$P(\text{good forecast}|\text{economy good}) = .80$

$P(\text{good forecast}|\text{economy bad}) = .20$

Should Rollo hire the consulting firm? What are EVSI and EVPI?

6 Willy Mutton has three potential bank robberies lined up. His chance of success and the size of the take are given in Table 18: These robberies must be attempted in order; if you "pass" on a robbery you may not go on to the next robbery. If Willie is caught, he loses all his money. What strategy maximizes his expected "take"?

7 Let

x_1 = undergraduate grade point average (GPA) of a student applying to State U's MBA program

x_2 = GMAT score of the same student

Suppose that preference between applicants is based on the following value function:

$$v(x_1, x_2) = 200x_1 + x_2 - 0.1x_2(x_1)^2$$

a Would the MBA program prefer a student with a 3.8 GPA and a 500 GMAT score to a student with a 3.0 GPA and a 710 GMAT score?

b Does this value function exhibit mutual preferential independence?

8 The Pine Valley Board of Education is trying to determine its multiattribute utility function with respect to the following attributes:

Attribute 1 Average score of students on an English achievement test

Attribute 2 Average score of students on a mathematics achievement test

TABLE 18

Robbery	Chance of Success	Size of Take (in millions of dollars)
1	.60	7
2	.80	6
3	.70	5

The board believes that both attributes range between 70% and 90% correct answers. The board is indifferent between

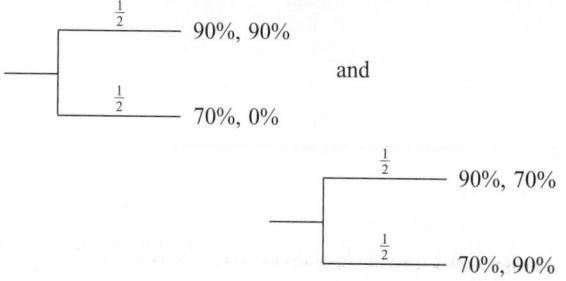

and

For any level x_2 of attribute 2, the board is also indifferent between

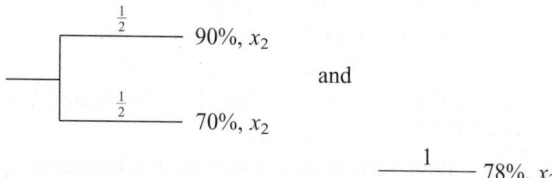

and

For any level x_1 of attribute 1, the board is indifferent between

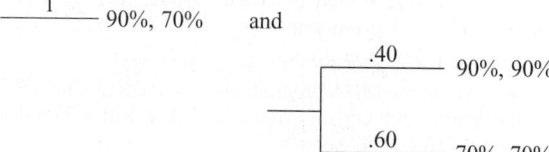

The board is also indifferent between

——— 90%, 70%   and

    .40 ——— 90%, 90%

    .60 ——— 70%, 70%

Finally, the board is indifferent between

——— 70%, 90%   and

    .60 ——— 90%, 90%

    .40 ——— 70%, 70%

The board must decide which of two instructional techniques should be utilized in the Pine Valley schools. Technique 1 is equivalent to the following lottery:

    $\frac{1}{2}$ ——— 78%, 76%

    $\frac{1}{2}$ ——— 70%, 90%

Technique 2 is equivalent to the following lottery:

    .60 ——— 90%, 90%

    .40 ——— 70%, 70%

Would the board prefer technique 1 or technique 2?

9 BeatTrop Foods is trying to choose one of three companies to merge with. In making this decision seven factors are important:

Factor 1 Contribution to profitability
Factor 2 Growth potential
Factor 3 Labor environment
Factor 4 R&D ability of company
Factor 5 Organizational fit
Factor 6 Relative size
Factor 7 Industry commonality

The pairwise comparison for these factors is as follows:

$$
\begin{array}{c c c c c c c c}
 & 1 & 2 & 3 & 4 & 5 & 6 & 7 \\
1 & 1 & 3 & 7 & 5 & 1 & 7 & 1 \\
2 & \frac{1}{3} & 1 & 9 & 1 & 1 & 5 & 1 \\
3 & \frac{1}{7} & \frac{1}{9} & 1 & \frac{1}{7} & \frac{1}{5} & \frac{1}{2} & \frac{1}{4} \\
4 & \frac{1}{5} & 1 & 7 & 1 & \frac{1}{4} & 7 & \frac{1}{3} \\
5 & 1 & 1 & 5 & 4 & 1 & 5 & 3 \\
6 & \frac{1}{7} & \frac{1}{5} & 2 & \frac{1}{7} & \frac{1}{5} & 1 & \frac{1}{6} \\
7 & 1 & 1 & 4 & 3 & \frac{1}{3} & 6 & 1
\end{array}
$$

The three contenders for merger have the following pairwise comparison matrices for each factor:

Factor 1
$$
\begin{array}{c c c c}
 & 1 & 2 & 3 \\
1 & 1 & 9 & 3 \\
2 & \frac{1}{9} & 1 & \frac{1}{5} \\
3 & \frac{1}{3} & 5 & 1
\end{array}
$$

Factor 2
$$
\begin{array}{c c c c}
 & 1 & 2 & 3 \\
1 & 1 & 7 & 4 \\
2 & \frac{1}{7} & 1 & \frac{1}{3} \\
3 & \frac{1}{4} & 3 & 1
\end{array}
$$

Factor 3
$$
\begin{array}{c c c c}
 & 1 & 2 & 3 \\
1 & 1 & \frac{1}{5} & \frac{1}{3} \\
2 & 5 & 6 & 2 \\
3 & 3 & \frac{1}{2} & 1
\end{array}
$$

Factor 4
$$
\begin{array}{c c c c}
 & 1 & 2 & 3 \\
1 & 1 & 6 & 3 \\
2 & \frac{1}{6} & 1 & \frac{1}{2} \\
3 & \frac{1}{3} & 2 & 1
\end{array}
$$

Factor 5
$$
\begin{array}{c c c c}
 & 1 & 2 & 3 \\
1 & 1 & \frac{1}{9} & \frac{1}{5} \\
2 & 9 & \frac{1}{2} & 4 \\
3 & 5 & \frac{1}{4} & 1
\end{array}
$$

Factor 6
$$
\begin{array}{c c c c}
 & 1 & 2 & 3 \\
1 & 1 & \frac{1}{7} & \frac{1}{4} \\
2 & 7 & 1 & 3 \\
3 & 3 & \frac{1}{3} & 1
\end{array}
$$

Factor 7
$$
\begin{array}{c c c c}
 & 1 & 2 & 3 \\
1 & 1 & \frac{1}{7} & \frac{1}{3} \\
2 & 7 & 1 & 3 \\
3 & 3 & \frac{1}{3} & 1
\end{array}
$$

Use the AHP to determine the company with which BeatTrop should prefer to merge.

10 You are trying to determine which city to live in. New York and Chicago are under consideration. Four objectives will determine your decision: affordability of housing,

TABLE 19

Affordability of housing	.50
Cultural opportunities	.10
Quality of schools and universities	.20
Crime level	.20

TABLE 20

	New York	Chicago
Affordability of housing	.30	.70
Cultural opportunities	.70	.30
Crime level	.40	.60

cultural opportunities, quality of schools and universities, and crime level. The weight for each objective is in Table 19. For each objective (except for quality of schools and universities) New York and Chicago scores are as given in Table 20. Suppose that the score for each city on the quality of schools and universities depends on two things: a score on public school quality and a score on university quality. The pairwise comparison matrix for public school and university quality is as follows:

$$\begin{array}{c} & \begin{array}{cc} \text{Public school} & \text{University} \\ \text{quality} & \text{quality} \end{array} \\ \begin{array}{c} \text{Public school quality} \\ \text{University quality} \end{array} & \left[\begin{array}{cc} 1 & 4 \\ \frac{1}{4} & 1 \end{array} \right] \end{array}$$

To see how each city scores on public school quality and university quality use the following pairwise comparison matrices. For public school quality:

$$\begin{array}{c} & \begin{array}{cc} \text{New York} & \text{Chicago} \end{array} \\ \begin{array}{c} \text{New York} \\ \text{Chicago} \end{array} & \left[\begin{array}{cc} 1 & 4 \\ \frac{1}{4} & 1 \end{array} \right] \end{array}$$

For university quality:

$$\begin{array}{c} & \begin{array}{cc} \text{New York} & \text{Chicago} \end{array} \\ \begin{array}{c} \text{New York} \\ \text{Chicago} \end{array} & \left[\begin{array}{cc} 1 & \frac{1}{3} \\ 3 & 1 \end{array} \right] \end{array}$$

You should now be able to come up with a score for each city on the quality of schools and universities objective. Now determine where you should live.

Group B

11 In Problem 5, suppose Rollo cannot hire the consulting firm, and his utility function for ending cash position is $u(x) = \ln x$. How much money should he invest in stocks and bonds?

12 At present, littering is punished by a $50 fine, and there is a 10% chance that a litterer will be brought to justice. To cut down on littering, Gotham City is considering two alternatives:

Alternative 1 Raise the littering fine by 20% (to $60).
Alternative 2 Hire more police and increase by 20% the probability that a litterer will be brought to justice (to a 12% probability that a litterer will be caught).

Assuming that all Gotham City residents are risk-averse, which alternative will lead to a larger reduction in littering?

Group C

13[†] In Section 13.2, we discussed the concept of the risk premium of a lottery and a risk-averse decision maker. In many situations, we would like to measure the degree of risk aversion associated with a utility function, and how a decision maker's risk aversion depends on his or her wealth. In this problem, we develop **Pratt's measure of absolute risk aversion.** Consider Ivana, who has initial wealth W and utility function $u(w)$ for final wealth position w. She has placed money in a small investment. The investment will increase her wealth by a random amount $\mathbf{X}$, with $E(\mathbf{X}) = 0$. We want to investigate how the risk premium of $\mathbf{X}$ depends on W. Let $RP(W, \mathbf{X})$ be the risk premium associated with investment $\mathbf{X}$ if the decision maker's wealth is W.

a Explain why $RP(W, \mathbf{X})$ satisfies the following equation:

$E(\text{Utility for wealth level of } W + \mathbf{X})$
$\quad = \text{utility of wealth level } [W - RP(W, \mathbf{X})]$

b Perform a second-order Taylor series expansion on E(utility for wealth level of $W + \mathbf{X}$) about W.

c Perform a first-order Taylor series expansion on utility of wealth level $[W - RP(W, \mathbf{X})]$ about W.

d Equating the answers in (b) and (c) (disregard the remainder terms), show that

$$RP(W, \mathbf{X}) = \frac{-\text{var}(\mathbf{X})u''(W)}{2u'(W)}$$

e Pratt's measure of absolute risk aversion at wealth level W, called $ARA(W)$, is defined to be twice the amount of risk premium per unit of variance when a decision maker is faced with a small lottery that has a zero expected value. Use your answer in part (d) to explain why

$$ARA(W) = \frac{-u''(W)}{u'(W)}$$

f If $ARA(W)$ is an increasing function of W, then $u(w)$ is said to exhibit increasing risk aversion, and if $ARA(W)$ is a decreasing function of W, then $u(w)$ exhibits decreasing risk aversion. Is increasing or decreasing risk aversion more consistent with most people's behavior?
Determine whether the following utility functions exhibit increasing or decreasing risk aversion:

g $u(w) = \ln w$

h $u(w) = w^{1/2}$

i $u(w) = aw - bw^2$, where $w < \frac{a}{2b}$. Explain how the answer indicates that a quadratic utility function probably is not an accurate representation of most people's preferences.

[†]Based on Pratt (1964).

REFERENCES

The following books discuss decision making under uncertainty at an intermediate level:

Bunn, D. *Applied Decision Analysis.* New York: McGraw-Hill, 1984.

Vatter, P., et al. *Quantitative Methods in Management: Text and Cases.* Homewood, Ill.: Irwin, 1978.

Winkler, R. *Introduction to Bayesian Inference and Decision.* New York: Holt, Rinehart & Winston, 1972.

For discussion of decision making under uncertainty at a more advanced level, readers should consult the next five books:

French, S. *Decision Theory.* New York: Wiley, 1986.

Keeney, R., and H. Raiffa. *Decision Making with Multiple Objectives.* New York: Wiley, 1976.

Raiffa, H. *Decision Analysis.* Reading, Mass.: Addison-Wesley, 1968.

Watson, S., and D. Buede. *Decision Synthesis.* Cambridge, England: Cambridge Press, 1987.

Winterfeldt, D., and W. Edwards. *Decision Analysis and Behavioral Research.* Cambridge, England: Cambridge Press, 1986.

Allais, M. "Le comportement de l'homme rationnel devant le risque: Critique des postulats et axioms de l'école Americaine," *Econometrica* 21(1953):503–546.

Clarke, J. "Applications of Decision Analysis to Clinical Medicine," *Interfaces* 17(no. 2, 1981):27–34.

Ellsberg, D. "Risk, Ambiguity, and the Savage Axioms," *Quarterly Journal of Economics* 75(1961):643–669.

Feinstein, C. "Deciding Whether to Test Student Athletes for Drug Use," *Interfaces* 20(no. 3, 1990):80–87.

Howard, R. "Heathens, Heretics, and Cults: The Religious Spectrum of Decision Aiding," *Interfaces* 22(no. 6, 1992):15–27.

Porter, R. "Extra-Point Strategy in Football," *American Statistician* 21(1967):14–15.

Pratt, J. "Risk Aversion in the Small and the Large," *Econometrica* 32(1964):122–136.

Tversky, A., and D. Kahneman. "The Framing of Decisions and the Psychology of Choice," *Science* 211(1981): 453–458.

Alternative approaches to multiattribute decision making under certainty are discussed in the following:

Steuer, R. *Multiple Criteria Optimization.* New York: Wiley, 1985.

Zeleny, M. *Multiple Criteria Decision Making.* New York: McGraw-Hill, 1982.

Zionts, S., and J. Wallenius. "An Interactive Programming Method for Solving the Multiple Criteria Problem," *Management Science* 22(1976): 652–663.

The following are recommended for elementary discussions of multiattribute utility theory:

Bunn, D. *Applied Decision Analysis.* New York: McGraw-Hill, 1984.

Keeney, R. "An Illustrated Procedure for Accessing Multiattributed Utility Functions," *Sloan Management Review* 14(1972):37–50.

This classic gives a comprehensive discussion of multiattribute utility theory.

Keeney, R., and H. Raiffa. *Decision Making with Multiple Objectives.* New York: Wiley, 1976.

The following references discuss the AHP:

Golden, B., E. Wasil, and P. Harkey. *The Analytic Hierarchy Process.* Heidelberg, Germany: Springer-Verlag, 1989.

Lin, W., T. Mock, and A. Wright. "The Use of AHP as an Aid in Planning the Nature and Extent of Audit Procedures," *Auditing: A Journal of Practice and Theory* 4(no. 1, 1984):89–99.

Saaty, T. *The Analytic Hierarchy Process.* Pittsburgh, Pa.: 1988.

Zahedi, F. "The Analytic Hierarchy Process—a Survey of the Method and Its Applications," *Interfaces* 16(no. 4, 1986):96–108.

14

Game Theory

In previous chapters, we have encountered many situations in which a *single* decision maker chooses an optimal decision without reference to the effect that the decision has on other decision makers (and without reference to the effect that the decisions of others have on him or her). In many business situations, however, two or more decision makers simultaneously choose an action, and the action chosen by each player affects the rewards earned by the other players. For example, each fast-food company must determine an advertising and pricing policy for its product, and each company's decision will affect the revenues and profits of other fast-food companies.

Game theory is useful for making decisions in cases where two or more decision makers have conflicting interests. Most of our study of game theory deals with situations where there are only two decision makers (or players), but we briefly study *n*-person (where $n > 2$) game theory also. We begin our study of game theory with a discussion of two-player games in which the players have no common interest.

14.1 Two-Person Zero-Sum and Constant-Sum Games: Saddle Points

Characteristics of Two-Person Zero-Sum Games

1 There are two players (called the *row* player and the *column* player).

2 The row player must choose 1 of m strategies. Simultaneously, the column player must choose 1 of n strategies.

3 If the row player chooses his ith strategy and the column player chooses his jth strategy, then the row player receives a reward of a_{ij} and the column player loses an amount a_{ij}. Thus, we may think of the row player's reward of a_{ij} as coming from the column player.

Such a game is called a **two-person zero-sum game,** which is represented by the matrix in Table 1 (the game's **reward matrix**). As previously stated, a_{ij} is the row player's

TABLE 1
Example of Two-Person Zero-Sum Game

Row Player's Strategy	Column Player's Strategy			
	Column 1	Column 2	$\cdots$	Column n
Row 1	a_{11}	a_{12}	$\cdots$	a_{1n}
Row 2	a_{21}	a_{22}	$\cdots$	a_{2n}
$\vdots$	$\vdots$	$\vdots$		$\vdots$
Row m	a_{m1}	a_{m2}	$\cdots$	a_{mn}

reward (and the column player's loss) if the row player chooses his ith strategy and the column player chooses his jth column strategy.

For example, in the two-person zero-sum game in Table 2, the row player would receive two units (and the column player would lose two units) if the row player chose his second strategy and the column player chose his first strategy.

A two-person zero-sum game has the property that for any choice of strategies, the sum of the rewards to the players is zero. In a zero-sum game, every dollar that one player wins comes out of the other player's pocket, so the two players have totally conflicting interests. Thus, cooperation between the two players would not occur.

John von Neumann and Oskar Morgenstern developed a theory of how two-person zero-sum games should be played, based on the following assumption.

Basic Assumption of Two-Person Zero-Sum Game Theory

Each player chooses a strategy that enables him to do the best he can, given that his opponent *knows the strategy he is following.* Let's use this assumption to determine how the row and column players should play the two-person zero-sum game in Table 3.

How should the row player play this game? If he chooses row 1, then the assumption implies that the column player will choose column 1 or column 2 and hold the row player to a reward of four units (the smallest number in row 1 of the game matrix). Similarly, if the row player chooses row 2, then the assumption implies that the column player will choose column 3 and hold the row player's reward to one unit (the smallest or minimum number in the second row of the game matrix). If the row player chooses row 3, then he will be held to the smallest number in the third row (5). Thus, the assumption implies that the row player should choose the row having the largest minimum. Because max (4, 1, 5) = 5, the row player should choose row 3. By choosing row 3, the row player can ensure that he will win at least max (row minimum) = five units.

From the column player's viewpoint, if he chooses column 1, then the row player will choose the strategy that makes the column player's losses as large as possible (and the row player's winnings as large as possible). Thus, if the column player chooses column 1, then the row player will choose row 3 (because the largest number in the first column is the 6 in the third row). Similarly, if the column player chooses column 2, then the row player will again choose row 3, because 5 = max (4, 3, 5). Finally, if the column player chooses column 3, the row player will choose row 1, causing the column player to lose 10 = max

TABLE 2

$$\begin{bmatrix} 1 & 2 & 3 & -1 \\ 2 & 1 & -2 & 0 \end{bmatrix}$$

TABLE 3
A Game with a Saddle Point

Row Player's Strategy	Column Player's Strategy			Row Minimum
	Column 1	Column 2	Column 3	
Row 1	4	4	10	4
Row 2	2	3	1	1
Row 3	6	5	7	5
Column Maximum	6	5	10	

(10, 1, 7) units. Thus, the column player can hold his losses to min (column maximum) = min (6, 5, 10) = 5 by choosing column 2.

We have shown that the row player can ensure that he will win at least five units and the column player can hold the row player's winnings to at most five units. Thus, the only rational outcome of this game is for the row player to win exactly five units; the row player cannot expect to win more than five units, because the column player (by choosing column 2) can hold the row player's winnings to five units.

The game matrix we have just analyzed has the property of satisfying the **saddle point condition:**

$$\max_{\substack{\text{all} \\ \text{rows}}} (\text{row minimum}) = \min_{\substack{\text{all} \\ \text{columns}}} (\text{column maximum}) \qquad (1)$$

Any two-person zero-sum game satisfying (1) is said to have a **saddle point.** If a two-person zero-sum game has a saddle point, then the row player should choose any strategy (row) attaining the maximum on the left side of (1). The column player should choose any strategy (column) attaining the minimum on the right side of (1). Thus, for the game we have just analyzed, a saddle point occurred where the row player chose row 3 and the column player chose column 2. The row player could make sure of receiving a reward of at least five units (by choosing the optimal strategy of row 3), and the column player could ensure that the row player would receive a reward of at most five units (by choosing the optimal strategy of column 2). If a game has a saddle point, then we call the common value of both sides of (1) the **value** (v) of the game to the row player. Thus, this game has a value of 5.

An easy way to spot a saddle point is to observe that the reward for a saddle point must be the smallest number in its row and the largest number in its column (see Problem 4 at the end of this section). Thus, like the center point of a horse's saddle, a saddle point for a two-person zero-sum game is a local minimum in one direction (looking across the row) and a local maximum in another direction (looking up and down the column).

A saddle point can also be thought of as an **equilibrium point** in that neither player can benefit from a unilateral change in strategy. For example, if the row player were to change from the optimal strategy of row 3 (to either row 1 or row 2), his reward would decrease, while if the column player changed from his optimal strategy of column 2 (to either column 1 or column 3), the row player's reward (and the column player's losses) would increase. Thus, a saddle point is stable in that neither player has an incentive to move away from it.

Many two-person zero-sum games do not have saddle points. For example, the game in Table 4 does not have a saddle point, because

$$\max (\text{row minimum}) = -1 < \min (\text{column maximum}) = +1$$

In Sections 14.2 and 14.3, we explain how to find the value and the optimal strategies for two-person zero-sum games that do not have saddle points.

TABLE 4
A Game with No Saddle Point

Row Player's Strategy	Column Player's Strategy		Row Minimum
	Column 1	Column 2	
Row 1	-1	$+1$	-1
Row 2	$+1$	-1	-1
Column Maximum	$+1$	$+1$	

Two-Person Constant-Sum Games

Even if a two-person game is not zero-sum, two players can still be in total conflict. To illustrate this, we now consider two-person constant-sum games.

DEFINITION ■   A **two-person constant-sum game** is a two-player game in which, for any choice of both player's strategies, the row player's reward and the column player's reward add up to a constant value c. ■

Of course, a two-person zero-sum game is just a two-person constant-sum game with $c = 0$. A two-person constant-sum game maintains the feature that the row and column players are in total conflict, because a unit increase in the row player's reward will always result in a unit decrease in the column player's reward. In general, the optimal strategies and value for a two-person constant-sum game may be found by the same methods used to find the optimal strategies and value for a two-person zero-sum game.

EXAMPLE 1   **Constant Sum TV Game**

During the 8 to 9 P.M. time slot, two networks are vying for an audience of 100 million viewers. The networks must simultaneously announce the type of show they will air in that time slot. The possible choices for each network and the number of network 1 viewers (in millions) for each choice are shown in Table 5. For example, if both networks choose a western, the matrix indicates that 35 million people will watch network 1 and $100 - 35 = 65$ million people will watch network 2. Thus, we have a two-person constant-sum game with $c = 100$ (million). Does this game have a saddle point? What is the value of the game to network 1?

Solution   Looking at the row minima, we find that by choosing a soap opera, network 1 can be sure of at least max (15, 45, 14) = 45 million viewers. Looking at the column maxima, we find that by choosing a western, network 2 can hold network 1 to at most min (45, 58, 70) = 45 million viewers. Because

$$\text{max (row minimum)} = \text{min (column maximum)} = 45$$

we find that Equation (1) is satisfied. Thus, network 1's choosing a soap opera and network 2's choosing a western yield a saddle point; neither side will do better if it unilaterally changes strategy (check this). Thus, the value of the game to network 1 is 45 million viewers, and the value of the game to network 2 is $100 - 45 = 55$ million viewers. The optimal strategy for network 1 is to choose a soap opera, and the optimal strategy for network 2 is to choose a western.

TABLE 5
A Constant-Sum Game

Network 1	Network 2 Western	Soap Opera	Comedy	Row Minimum
Western	35	15	60	15
Soap Opera	45	58	50	45
Comedy	38	14	70	14
Column Maximum	45	58	70	

PROBLEMS

Group A

1 Find the value and optimal strategy for the game in Table 6.

2 Find the value and the optimal strategies for the two-person zero-sum game in Table 7.

Group B

3 Mad Max wants to travel from New York to Dallas by the shortest possible route. He may travel over the routes shown in Table 8. Unfortunately, the Wicked Witch can block one road leading out of Atlanta and one road leading out of Nashville. Mad Max will not know which roads have been blocked until he arrives at Atlanta or Nashville. Should Mad Max start toward Atlanta or Nashville? Which routes should the Wicked Witch block?

Group C

4 Explain why the reward for a saddle point must be the smallest number in its row and the largest number in its column. Suppose a reward is the smallest in its row and the largest in its column. Must that reward yield a saddle point? (*Hint:* Think about the idea of weak duality discussed in Chapter 6.)

TABLE 8

Route	Length of Route (Miles)
New York–Atlanta	800
New York–Nashville	900
Nashville–St. Louis	400
Nashville–New Orleans	200
Atlanta–St. Louis	300
Atlanta–New Orleans	600
St. Louis–Dallas	500
New Orleans–Dallas	300

TABLE 6

2	2
1	3

TABLE 7

4	5	5	8
6	7	6	9
5	7	5	4
6	6	5	5

14.2 Two-Person Zero-Sum Games: Randomized Strategies, Domination, and Graphical Solution

In the previous section, we found that not all two-person zero-sum games have saddle points. We now discuss how to find the value and optimal strategies for a two-person zero-sum game that does not have a saddle point. We begin with the simple game of Odds and Evens.

EXAMPLE 2   Odds and Evens

Two players (called Odd and Even) simultaneously choose the number of fingers (1 or 2) to put out. If the sum of the fingers put out by both players is odd, then Odd wins $1 from Even. If the sum of the fingers is even, then Even wins $1 from Odd. We consider the row player to be Odd and the column player to be Even. Determine whether this game has a saddle point.

Solution   This is a zero-sum game, with the reward matrix shown in Table 9. Because max (row minimum) = −1 and min (column maximum) = +1, Equation (1) is not satisfied, and this game has no saddle point. All we know is that Odd can be sure of a reward of at least

TABLE 9
Reward Matrix for Odds and Evens

Row Player (Odd)	Column Player (Even)		Row Minimum
	1 Finger	2 Fingers	
1 Finger	-1	$+1$	-1
2 Fingers	$+1$	-1	-1
Column Maximum	$+1$	$+1$	

-1, and Even can hold Odd to a reward of at most $+1$. Thus, it is unclear how to determine the value of the game and the optimal strategies. Observe that for any choice of strategies by both players, there is a player who can benefit by unilaterally changing her strategy. For example, if both players put out one finger, then Odd could have increased her reward from -1 to $+1$ by putting out two fingers. Thus, no choice of strategies by the player is stable. We now determine optimal strategies and the value for this game.

Randomized or Mixed Strategies

To progress further with the analysis of Example 2 (and other games without saddle points), we must expand the set of allowable strategies for each player to include **randomized strategies.** Until now, we have assumed that each time a player plays a game, the player will choose the same strategy. Why not allow each player to select a probability of playing each strategy? For Example 2, we might define

$$x_1 = \text{probability that Odd puts out one finger}$$
$$x_2 = \text{probability that Odd puts out two fingers}$$
$$y_1 = \text{probability that Even puts out one finger}$$
$$y_2 = \text{probability that Even puts out two fingers}$$

If $x_1 \geq 0$, $x_2 \geq 0$, and $x_1 + x_2 = 1$, then (x_1, x_2) is a randomized, or mixed, strategy for Odd. For example, the mixed strategy $(\frac{1}{2}, \frac{1}{2})$ could be realized by Odd if she tossed a coin before each play of the game and put out one finger for heads and two fingers for tails. Similarly, if $y_1 \geq 0$, $y_2 \geq 0$, and $y_1 + y_2 = 1$, then (y_1, y_2) is a mixed strategy for Even.

Any mixed strategy $(x_1, x_2, \ldots, x_m)$ for the row player is a **pure strategy** if any of the x_i equals 1. Similarly, any mixed strategy $(y_1, y_2, \ldots, y_n)$ for the column player is a pure strategy if any of the y_i equals 1. A pure strategy is a special case of a mixed strategy in which a player always chooses the same action. Recall from Section 14.1 that the game in Table 10 had a value of 5 (corresponding to a saddle point), so the row player's optimal strategy could be represented as the pure strategy $(0, 0, 1)$, and the column player's optimal strategy could be represented as the pure strategy $(0, 1, 0)$.

We continue to assume that both players will play two-person zero-sum games in accordance with the basic assumption of Section 14.1. In the context of randomized strate-

TABLE 10

4	4	10
2	3	1
6	5	7

gies, the assumption (from the standpoint of Odd) may be stated as follows: Odd should choose x_1 and x_2 to maximize her expected reward under the assumption that Even knows the value of x_1 and x_2.

It is important to realize that even though we assume that Even knows the values of x_1 and x_2, on a particular play of the game, she is not assumed to know Odd's actual strategy choice until the instant the game is played.

Graphical Solution of Odds and Evens

Finding Odd's Optimal Strategy

With this version of the basic assumption, we can determine the optimal strategy for Odd. Because $x_1 + x_2 = 1$, we know that $x_2 = 1 - x_1$. Thus, any mixed strategy may be written as $(x_1, 1 - x_1)$, and it suffices to determine the value of x_1. Suppose Odd chooses a particular mixed strategy $(x_1, 1 - x_1)$. What is Odd's expected reward against each of Even's strategies? If Even puts out one finger, then Odd will receive a reward of -1 with probability x_1 and a reward of $+1$ with probability $x_2 = 1 - x_1$. Thus, if Even puts out one finger and Odd chooses the mixed strategy $(x_1, 1 - x_1)$, then Odd's expected reward is

$$(-1)x_1 + (+1)(1 - x_1) = 1 - 2x_1$$

As a function of x_1, this expected reward is drawn as line segment AC in Figure 1. Similarly, if Even puts out two fingers and Odd chooses the mixed strategy $(x_1, 1 - x_1)$, Odd's expected reward is

$$(+1)(x_1) + (-1)(1 - x_1) = 2x_1 - 1$$

which is line segment DE in Figure 1.

Suppose Odd chooses the mixed strategy $(x_1, 1 - x_1)$. Because Even is assumed to know the value of x_1, for any value of x_1 Even will choose the strategy (putting out one or two fingers) that yields a smaller expected reward for Odd. From Figure 1, we see that, as a function of x_1, Odd's expected reward will be given by the y-coordinate in DBC. Odd wants to maximize her expected reward, so she should choose the value of x_1 corresponding to point B. Point B occurs where the line segments AC and DE intersect, or

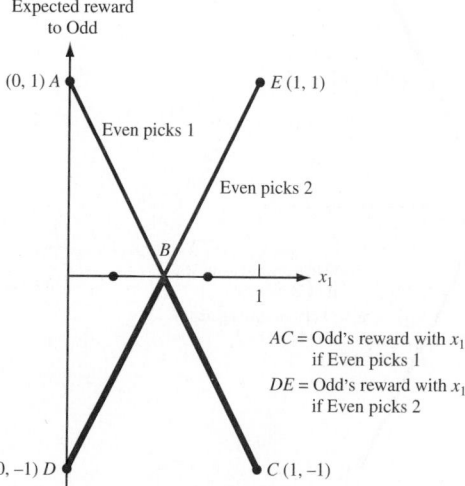

FIGURE 1
Choosing Odd's Strategy

where $1 - 2x_1 = 2x_1 - 1$. Solving this equation, we obtain $x_1 = \frac{1}{2}$. Thus, Odd should choose the mixed strategy $(\frac{1}{2}, \frac{1}{2})$. The reader should verify that against each of Even's strategies, $(\frac{1}{2}, \frac{1}{2})$ yields an expected reward of zero. Thus, zero is a **floor** on Odd's expected reward, because by choosing the mixed strategy $(\frac{1}{2}, \frac{1}{2})$, Odd can be sure that (for any choice of Even's strategy) her expected reward will always be at least zero.

Finding Even's Optimal Strategy

We now consider how Even should choose a mixed strategy (y_1, y_2). Again, because $y_2 = 1 - y_1$, we may ask how Even should choose a mixed strategy $(y_1, 1 - y_1)$. The basic assumption implies that Even should choose y_1 to minimize her expected losses (or, equivalently, minimize Odd's expected reward) under the assumption that Odd knows the value of y_1. Suppose Even chooses the mixed strategy $(y_1, 1 - y_1)$. What will Odd do? If Odd puts out one finger, then her expected reward is

$$(-1)y_1 + (+1)(1 - y_1) = 1 - 2y_1$$

which is line segment AC in Figure 2. If Odd puts out two fingers, then her expected reward is

$$(+1)(y_1) + (-1)(1 - y_1) = 2y_1 - 1$$

which is line segment DE in Figure 2. Because Odd is assumed to know the value of y_1, she will put out the number of fingers corresponding to max $(1 - 2y_1, 2y_1 - 1)$. Thus, for a given value of y_1, Odd's expected reward (and Even's expected loss) will be given by the y-coordinate on the piecewise linear curve ABE.

Now Even chooses the mixed strategy $(y_1, 1 - y_1)$ that will make Odd's expected reward as small as possible. Thus, Even should choose the value of y_1 corresponding to the lowest point on ABE (point B). Point B is where the line segments AC and DE intersect, or where $1 - 2y_1 = 2y_1 - 1$, or $y_1 = \frac{1}{2}$. The basic assumption implies that Even should choose the mixed strategy $(\frac{1}{2}, \frac{1}{2})$. For this mixed strategy, Even's expected loss (and Odd's expected reward) is zero. We say that zero is a **ceiling** on Even's expected loss

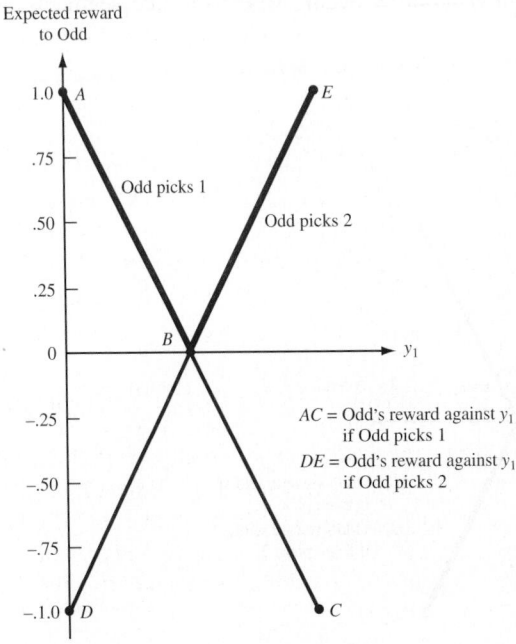

FIGURE 2
Choosing Even's Strategy

TABLE 11

How to Make a Nonoptimal Strategy Pay the Price

Odd's Mixed Strategy	Even Can Choose	Odd's Expected Reward (Even's expected losses)
$x_1 < \frac{1}{2}$	2 fingers	< 0 (on BD in Figure 1)
$x_1 > \frac{1}{2}$	1 finger	< 0 (on BC in Figure 1)
Even's Mixed Strategy	**Odd Can Choose**	**Odd's Expected Reward (Even's expected losses)**
$y_1 < \frac{1}{2}$	1 finger	> 0 (on AB in Figure 2)
$y_1 > \frac{1}{2}$	2 fingers	> 0 (on BE in Figure 2)

(or Odd's expected reward), because by choosing the mixed strategy $(\frac{1}{2}, \frac{1}{2})$, Even can ensure that her expected loss (for any choice of strategies by Odd) will not exceed zero.

More on the Idea of Value and Optimal Strategies

For the game of Odds and Evens, the row player's *floor* and the column player's *ceiling* are equal. This is not a coincidence. When each player is allowed to choose mixed strategies, the row player's floor will always equal the column player's ceiling. In Section 14.3, we use the Dual Theorem of Chapter 6 to prove this interesting result. We call the common value of the floor and ceiling the **value** of the game to the row player. Any mixed strategy for the row player that guarantees that the row player gets an expected reward at least equal to the value of the game is an **optimal strategy** for the row player. Similarly, any mixed strategy for the column player that guarantees that the column player's expected loss is no more than the value of the game is an optimal strategy for the column player. Thus, for Example 2, we have shown that the value of the game is zero, the row player's optimal strategy is $(\frac{1}{2}, \frac{1}{2})$, and the column player's optimal strategy is $(\frac{1}{2}, \frac{1}{2})$.

Example 2 illustrates that by allowing mixed strategies, we have enabled each player to find an optimal strategy in that *if the row player departs from her optimal strategy, the column player may have a strategy that reduces the row player's expected reward below the value of the game, and if the column player departs from her optimal strategy, the row player may have a strategy that increases her expected reward above the value of the game.* Table 11 illustrates this idea for the game of Odds and Evens.

For example, suppose that Odd chooses a nonoptimal mixed strategy with $x_1 < \frac{1}{2}$. Then, by choosing two fingers, Even ensures that Odd's expected reward can be read from BD in Figure 1. This means that if Odd chooses a mixed strategy having $x_1 < \frac{1}{2}$, then her expected reward can be negative (less than the value of the game).

To close this section, we find the value and optimal strategies for a more complicated game.

EXAMPLE 3   **Coin Toss Game with Bluffing**

A fair coin is tossed, and the result is shown to player 1. Player 1 must then decide whether to pass or bet. If player 1 passes, then he must pay player 2 $1. If player 1 bets, then player 2 (who does not know the result of the coin toss) may either fold or call the bet. If player 2 folds, then she pays player 1 $1. If player 2 calls and the coin comes up heads, then she pays player 1 $2; if player 2 calls and the coin comes up tails, then player 1 must pay her

$2. Formulate this as a two-person zero-sum game. Then graphically determine the value of the game and each player's optimal strategy.

Solution Player 1's strategies may be represented as follows: **PP,** pass on heads and pass on tails; **PB,** pass on heads and bet on tails; **BP,** bet on heads and pass on tails; and **BB,** bet on heads and bet on tails. Player 2 simply has the two strategies call and fold. For each choice of strategies, player 1's expected reward is as shown in Table 12.

To illustrate these computations, suppose player 1 chooses **BP** and player 2 calls. Then with probability $\frac{1}{2}$, heads is tossed. Then player 1 bets, is called, and wins $2 from player 2. With probability $\frac{1}{2}$, tails is tossed. In this case, player 1 passes and pays player 2 $1. Thus, if player 1 chooses **BP** and player 2 calls, then player 1's expected reward is $(\frac{1}{2})(2) + (\frac{1}{2})(-1) = \$0.50$. For each line in Table 12, the first term in the expectation corresponds to heads being tossed, and the second term corresponds to tails being tossed.

Example 3 may be described as the two-person zero-sum game represented by the reward matrix in Table 13. Because max (row minimum) = 0 < min (column maximum) = $\frac{1}{2}$, this game does not have a saddle point. Observe that player 1 would be unwise ever to choose the strategy **PP,** because (for each of player 2's strategies) player 1 could do better than **PP** by choosing either **BP** or **BB.** In general, a strategy i for a given player is **dominated** by a strategy i' if, for each of the other player's possible strategies, the given player does at least as well with strategy i' as he or she does with strategy i, and if for at least one of the other player's strategies, strategy i' is superior to strategy i. A player may eliminate all dominated strategies from consideration. We have just shown that for player 1, **BP**

TABLE 12
Computation of Reward Matrix for Example 3

	Player 1's Expected Reward
PP vs. call	$(\frac{1}{2})(-1) + (\frac{1}{2})(-1) = -\$1$
PP vs. fold	$(\frac{1}{2})(-1) + (\frac{1}{2})(-1) = -\$1$
PB vs. call	$(\frac{1}{2})(-1) + (\frac{1}{2})(-2) = -\$1.50$
PB vs. fold	$(\frac{1}{2})(-1) + (\frac{1}{2})(1) = \$0$
BP vs. call	$(\frac{1}{2})(2) + (\frac{1}{2})(-1) = \$0.50$
BP vs. fold	$(\frac{1}{2})(1) + (\frac{1}{2})(-1) = \$0$
BB vs. call	$(\frac{1}{2})(2) + (\frac{1}{2})(-2) = \$0$
BB vs. fold	$(\frac{1}{2})(1) + (\frac{1}{2})(1) = \$1$

TABLE 13
Reward Matrix for Example 3

Player 1	Player 2 Call	Fold	Row Minimum
PP	-1	-1	-1
PB	$-\frac{3}{2}$	0	$-\frac{3}{2}$
BP	$\frac{1}{2}$	0	0
BB	0	1	0
Column Maximum	$\frac{1}{2}$	1	

or **BB** dominates **PP.** Similarly, the reader should be able to show that player 1's **PB** strategy is dominated by **BP** or **BB.** After eliminating the dominated strategies **PP** and **PB,** we are left with the game matrix shown in Table 14.

As with Odds and Evens, this game has no saddle point, and we proceed with a graphical solution. Let

$$x_1 = \text{probability that player 1 chooses } \textbf{BP}$$
$$x_2 = 1 - x_1 = \text{probability that player 1 chooses } \textbf{BB}$$
$$y_1 = \text{probability that player 2 chooses call}$$
$$y_2 = 1 - y_1 = \text{probability that player 2 chooses fold}$$

To determine the optimal strategy for player 1, observe that for any value of x_1, her expected reward against calling is

$$(\tfrac{1}{2})(x_1) + 0(1 - x_1) = \frac{x_1}{2}$$

which is line segment AB in Figure 3. Against folding, player 1's expected reward is

$$0(x_1) + 1(1 - x_1) = 1 - x_1$$

which is line segment CD in Figure 3. Player 2 is assumed to know the value of x_1, so player 1's expected reward (as a function of x_1) is given by the piecewise linear curve AED

TABLE 14
Reward Matrix for Example 3 After Dominated
Strategies Have Been Eliminated

Player 1	Player 2		Row Minimum
	Call	Fold	
BP	$\frac{1}{2}$	0	0
BB	0	1	0
Column Maximum	$\frac{1}{2}$	1	

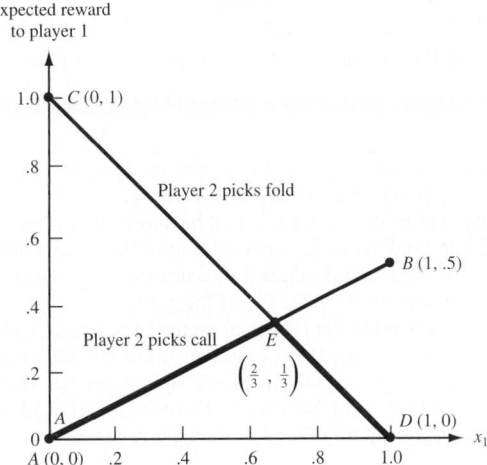

FIGURE 3
How Player 1 Chooses Optimal Strategy in Example 3

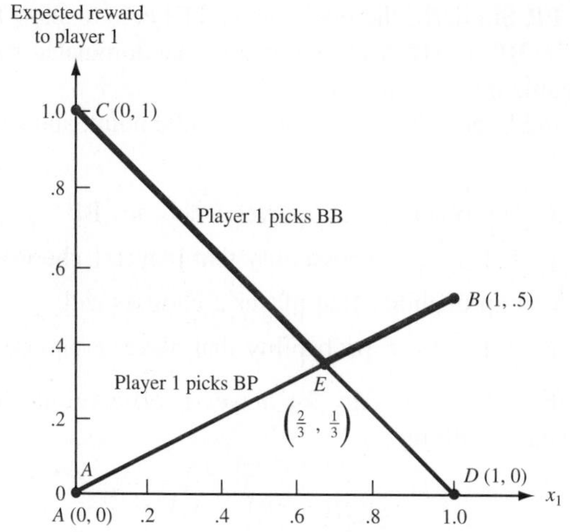

FIGURE 4
How Player 2 Chooses Optimal Strategy in Example 3

in Figure 3. Thus, to maximize her expected reward, player 1 should choose the value of x_1 corresponding to point E, which solves $x_1/2 = 1 - x_1$, or $x_1 = \frac{2}{3}$. Then $x_2 = 1 - \frac{2}{3} = \frac{1}{3}$, and player 1's expected reward against either of player 2's strategies is $\frac{x_1}{2}$ (or $1 - x_1$) $= \frac{1}{3}$.

How should player 2 choose y_1? (Remember, $y_2 = 1 - y_1$.) For a given value of y_1, suppose player 1 chooses **BP.** Then her expected reward is

$$(\tfrac{1}{2})(y_1) + 0(1 - y_1) = \frac{y_1}{2}$$

which is line segment AB in Figure 4. For a given value of y_1, suppose player 1 chooses **BB.** Then her expected reward is

$$0(y_1) + 1(1 - y_1) = 1 - y_1$$

which is line segment CD in Figure 4. Thus, for a given value of y_1, player 1 will choose a strategy that causes his expected reward to be given by the piecewise linear curve CEB in Figure 4. Knowing this, player 2 should choose the value of y_1 corresponding to point E in Figure 4. The value of y_1 at point E is the solution to $\frac{y_1}{2} = 1 - y_1$, or $y_1 = \frac{2}{3}$ (and $y_2 = \frac{1}{3}$). You should check that no matter what player 1 does, player 2's mixed strategy $(\frac{2}{3}, \frac{1}{3})$ ensures that player 1 earns an expected reward of $\frac{1}{3}$.

In summary, the value of the game is $\frac{1}{3}$ to player 1; the optimal mixed strategy for player 1 is $(\frac{2}{3}, \frac{1}{3})$; and the optimal strategy for player 2 is also $(\frac{2}{3}, \frac{1}{3})$.

REMARKS

1 Observe that player 1 should bet $\frac{1}{3}$ of the time that she has a losing coin. Thus, our simple model indicates that player 1's optimal strategy includes bluffing.

2 In Problem 4 at the end of this section, it will be shown that if player 1 deviates from her optimal strategy, player 2 can hold her to an expected reward that is less than the value ($\frac{1}{3}$) of the game. Similarly, Problem 5 will show that if player 2 deviates from her optimal strategy, player 1 can earn an expected reward in excess of the value ($\frac{1}{3}$) of the game.

3 Although we have only applied the graphical method to games in which each player (after dominated strategies have been eliminated) has only two strategies, the graphical approach can be used to solve two-person zero-sum games in which only one player has two strategies (games in which the reward matrix is $2 \times n$ or $m \times 2$). We choose, however, to solve all non-2×2 two-person games by the linear programming method outlined in the next section.

PROBLEMS

Group A

1 Find the value and the optimal strategies for the two-person zero-sum game in Table 15.

2 Player 1 writes an integer between 1 and 20 on a slip of paper. Without showing this slip of paper to player 2, player 1 tells player 2 what he has written. Player 1 may lie or tell the truth. Player 2 must then guess whether or not player 1 has told the truth. If caught in a lie, player 1 must pay player 2 $10; if falsely accused of lying, player 1 collects $5 from player 2. If player 1 tells the truth and player 2 guesses that player 1 has told the truth, then player 1 must pay $1 to player 2. If player 1 lies and player 2 does not guess that player 1 has lied, player 1 wins $5 from player 2. Determine the value of this game and each player's optimal strategy.

3 Find the value and optimal strategies for the two-person zero-sum game in Table 16.

4 For Example 3, show that if player 1 deviates from her optimal strategy, then player 2 can ensure that player 1 earns an expected reward that is less than the value ($\frac{1}{3}$) of the game.

5 For Example 3, show that if player 2 deviates from her optimal strategy, then player 1 can ensure that she earns an expected reward that is more than the value ($\frac{1}{3}$) of the game.

6 Two competing firms must simultaneously determine how much of a product to produce. The total profit earned by the two firms is always $1,000. If firm 1's production level is low and firm 2's is also low, then firm 1 earns a profit of $500; if firm 1's level is low and 2's is high, then firm 1's profit is $400. If firm 1's production level is high and so is firm 2's, then firm 1's profit is $600; but if firm 1's level is high while firm 2's level is low, then firm 1's profit is only $300. Find the value and optimal strategies for this constant-sum game.

7 Mo and Bo each have a quarter and a penny. Simultaneously, they each display a coin. If the coins match, then Mo wins both coins; if they don't match, then Bo wins both coins. Determine optimal strategies for this game.

Group B

8 State University is about to play Ivy College for the state tennis championship. The State team has two players (A and B), and the Ivy team has three players (X, Y, and Z). The following facts are known about the players' relative abilities:

X will always beat B; Y will always beat A; A will always beat Z. In any other match, each player has a $\frac{1}{2}$ chance of winning. Before State plays Ivy, the State coach must determine who will play first singles and who will play second singles. The Ivy coach (after choosing which two players will play singles) must also determine who will play first singles and second singles. Assume that each coach wants to maximize the expected number of singles matches won by the team. Use game theory to determine optimal strategies for each coach and the value of the game to each team.

9 Consider a two-person zero-sum game with the reward matrix in Table 17. Suppose this game does not have a saddle point. Show that the optimal strategy for the row player is to play the first row a fraction $(d - c)/(a + d - b - c)$ of the time and the optimal strategy for the column player is to play the first column a fraction $(d - b)/(a + d - b - c)$ of the time.

10 Consider the following simplified version of football. On each play the offense chooses to run or pass. At the same time, the defense chooses to play a run defense or pass defense. The number of yards gained on each play is determined by the reward matrix in Table 18. The offense's goal is to maximize the average yards gained per play.

   a Use Problem 9 to show that the offense should run 10/17 of the time.

   b Suppose that the effectiveness of a pass against the run defense improves. Use the results of Problem 9 to show that the offense should pass less! Can you give an explanation for this strange phenomenon?

11 Use the idea of dominated strategies to determine optimal strategies for the reward matrix in Table 19.

TABLE 17

a	b
c	d

TABLE 18

Offense	Defense	
	Run	Pass
Run	1	8
Pass	10	0

TABLE 15

1	2	3
2	0	3

TABLE 16

2	1	3
4	3	2

TABLE 19

−5	−10	−1	−10	2	−1
−1	2	−10	7	−5	20
2	7	−5	−10	−10	7
7	20	−1	−1	−1	2
20	7	−10	7	−1	−10

14.3 Linear Programming and Zero-Sum Games

Linear programming can be used to find the value and optimal strategies (for the row and column players) for any two-person zero-sum game. To illustrate the main ideas, we consider the well-known game Stone, Paper, Scissors.

EXAMPLE 4   Stone, Paper, Scissors

Two players simultaneously utter one of the three words *stone, paper,* or *scissors* and show corresponding hand signs. If both players utter the same word, then the game is a draw. Otherwise, one player wins $1 from the other player according to the following: Scissors defeats (cuts) paper, paper defeats (covers) stone, and stone defeats (breaks) scissors. Find the value and optimal strategies for this two-person zero-sum game. The solution is given later in this section.

The reward matrix is shown in Table 20. Observe that no strategies are dominated and that the game does not have a saddle point. To determine optimal mixed strategies for the row and the column player, define

$$x_1 = \text{probability that row player chooses stone}$$
$$x_2 = \text{probability that row player chooses paper}$$
$$x_3 = \text{probability that row player chooses scissors}$$
$$y_1 = \text{probability that column player chooses stone}$$
$$y_2 = \text{probability that column player chooses paper}$$
$$y_3 = \text{probability that column player chooses scissors}$$

The Row Player's LP

If the row player chooses the mixed strategy (x_1, x_2, x_3), then her expected reward against each of the column player's strategies is as shown in Table 21. Suppose the row player chooses the mixed strategy (x_1, x_2, x_3). By the basic assumption, the column player will choose a strategy that makes the row player's expected reward equal to min $(x_2 - x_3, -x_1 + x_3, x_1 - x_2)$. Then the row player should choose (x_1, x_2, x_3) to make min $(x_2 - x_3, -x_1 + x_3, x_1 - x_2)$ as *large* as possible. To obtain an LP formulation (called the row player's LP) that will yield the row player's optimal strategy, observe that for any values of x_1, x_2, and x_3, min $(x_2 - x_3, -x_1 + x_3, x_1 - x_2)$ is just the largest number (call it v) that is simulta-

TABLE 20
Reward Matrix for Stone, Paper, Scissors

Row Player	Column Player			
	Stone	Paper	Scissors	Row Minimum
Stone	0	−1	+1	−1
Paper	+1	0	−1	−1
Scissors	−1	+1	0	−1
Column Maximum	+1	+1	+1	

TABLE 21

Expected Reward to Row Player in Stone, Paper, Scissors

Column Player Chooses	Row Player's Expected Reward If Row Player Chooses (x_1, x_2, x_3)
Stone	$x_2 - x_3$
Paper	$-x_1 + x_3$
Scissors	$x_1 - x_2$

neously less than or equal to $x_2 - x_3$, $-x_1 + x_3$, and $x_1 - x_2$. After noting that x_1, x_2, and x_3 must satisfy $x_1 \geq 0$, $x_2 \geq 0$, $x_3 \geq 0$, and $x_1 + x_2 + x_3 = 1$, we see that the row player's optimal strategy can be found by solving the following LP:

$$\max z = v$$

$$
\begin{aligned}
\text{s.t.} \quad & v \leq x_2 - x_3 && \text{(Stone constraint)} \\
& v \leq -x_1 + x_3 && \text{(Paper constraint)} \\
& v \leq x_1 - x_2 && \text{(Scissors constraint)} \\
& x_1 + x_2 + x_3 = 1 \\
& x_1, x_2, x_3 \geq 0; \ v \text{ urs}
\end{aligned}
$$

(2)

Note that there is a constraint in (2) for each of the column player's strategies. The value of v in the optimal solution to (2) is the row player's *floor*, because no matter what strategy (pure or mixed) is chosen by the column player, the row player is sure to receive an expected reward of at least v.

The Column Player's LP

How should the column player choose an optimal mixed strategy (y_1, y_2, y_3)? Suppose the column player has chosen the mixed strategy (y_1, y_2, y_3). For each of the row player's strategies, we may compute the row player's expected reward if the column player chooses (y_1, y_2, y_3) (see Table 22). The row player is assumed to know (y_1, y_2, y_3), the row player will choose a strategy to ensure that she obtains an expected reward of max $(-y_2 + y_3, y_1 - y_3, -y_1 + y_2)$. Thus, the column player should choose (y_1, y_2, y_3) to make max $(-y_2 + y_3, y_1 - y_3, -y_1 + y_2)$ as *small* as possible. To obtain an LP formulation that will yield the column player's optimal strategies, observe that for any choice of (y_1, y_2, y_3), max $(-y_2 + y_3, y_1 - y_3, -y_1 + y_2)$ will equal the smallest number that is simultaneously greater than or equal to $-y_2 + y_3$, $y_1 - y_3$, and $-y_1 + y_2$ (call this number w). Also note that for (y_1, y_2, y_3) to be a mixed strategy, (y_1, y_2, y_3) must satisfy $y_1 + y_2 + y_3 = 1$, $y_1 \geq 0$, $y_2 \geq 0$, and $y_3 \geq 0$. Thus, the column player may find his optimal strategy by solving the following LP:

$$\min z = w$$

$$
\begin{aligned}
\text{s.t.} \quad & w \geq -y_2 + y_3 && \text{(Stone constraint)} \\
& w \geq y_1 - y_3 && \text{(Paper constraint)} \\
& w \geq -y_1 + y_2 && \text{(Scissors constraint)} \\
& y_1 + y_2 + y_3 = 1 \\
& y_1, y_2, y_3 \geq 0; \ w \text{ urs}
\end{aligned}
$$

(3)

Observe that (3) contains a constraint corresponding to each of the row player's strategies. Also, the optimal objective function w for (3) is a *ceiling* on the column player's expected

TABLE 22
Expected Reward to Row Player in
Stone, Paper, Scissors

Row Player Chooses	Row Player's Expected Reward If Column Player Chooses (y_1, y_2, y_3)
Stone	$-y_2 + y_3$
Paper	$y_1 - y_3$
Scissors	$-y_1 + y_2$

losses (or the row player's expected reward), because by choosing a mixed strategy (y_1, y_2, y_3) that solves (3), the column player can ensure that his expected losses will be (against any of the row player's strategies) at most w.

Relation Between the Row and the Column Player's LPs

It is easy to show that the column player's LP is the dual of the row player's LP. Begin by rewriting the row player's LP (2) as

$$
\begin{aligned}
\max z = v \\
\text{s.t.} \quad -x_2 + x_3 + v &\le 0 \\
x_1 \quad\quad - x_3 + v &\le 0 \\
-x_1 + x_2 \quad\quad + v &\le 0 \\
x_1 + x_2 + x_3 \quad\quad &= 1 \\
x_1, x_2, x_3 \ge 0; \; v \text{ urs} &
\end{aligned}
\tag{4}
$$

Let the dual variables for the constraints in (4) be y_1, y_2, y_3, and w, respectively. We can now show that the dual of the row player's LP is the column player's LP. As in Section 6.5, we read the row player's LP across in Table 23 and find the dual of the row player's LP by reading down. Recall that the dual constraint corresponding to the variable v will be an equality constraint (because v is urs), and the dual variable w corresponding to the primal constraint $x_1 + x_2 + x_3 = 1$ will be urs (because $x_1 + x_2 + x_3 = 1$ is an equality constraint). Reading down in Table 23, we find the dual of the row player's LP (4) to be

$$
\begin{aligned}
\min z = w \\
\text{s.t.} \quad y_2 - y_3 + w &\ge 0 \\
-y_1 \quad\quad + y_3 + w &\ge 0 \\
y_1 - y_2 \quad\quad + w &\ge 0 \\
y_1 + y_2 + y_3 \quad\quad &= 1 \\
y_1, y_2, y_3 \ge 0; \; w \text{ urs} &
\end{aligned}
$$

After transposing all terms involving y_1, y_2, and y_3 in the first three constraints to the right-hand side, we see that the last LP is the same as the column player's LP (3). Thus, the dual of the row player's LP is the column player's LP. (Of course, the dual of the column player's LP would be the row player's LP.)

It is easy to show that both the row player's LP (2) and the column player's LP (3) have an optimal solution (that is, neither LP can be infeasible or unbounded). Then the Dual Theorem of Section 6.7 implies that v, the optimal objective function value for the row player's LP, and w, the optimal objective function value for the column player's LP, are equal. Thus, the row player's floor equals the column player's ceiling. This result is often

TABLE **23**
Dual or Row Player's LP

Min		Max				
		x_1	x_2	x_3	v	
y_1 $(y_1 \geq 0)$		0	-1	1	1	≤ 0
y_2 $(y_2 \geq 0)$		1	0	-1	1	≤ 0
y_3 $(y_3 \geq 0)$		-1	1	0	1	≤ 0
w (urs)		1	1	1	0	$= 1$
		≥ 0	≥ 0	≥ 0	$= 1$	

known as the **Minimax Theorem.** We call the common value of v and w the **value** of the game to the row player. As in Sections 14.1 and 14.2, the row player can (by playing an optimal strategy) guarantee that her expected reward will at least equal the value of the game. Similarly, the column player can (by playing an optimal strategy) guarantee that his expected losses will not exceed the value of the game. It can also be shown (see Problem 6 at the end of this section) that the optimal strategies obtained via linear programming represent a stable equilibrium, because neither player can improve his or her situation by a unilateral change in strategy.

For the Stone, Paper, Scissors game, the optimal solution to the row player's LP (2) is $w = 0$, $x_1 = \frac{1}{3}$, $x_2 = \frac{1}{3}$, $x_3 = \frac{1}{3}$, and the optimal solution to the column player's LP (3) is $v = 0$, $y_1 = \frac{1}{3}$, $y_2 = \frac{1}{3}$, $y_3 = \frac{1}{3}$. Note that the first solution is feasible in (2), and the second solution is feasible in (3). Each solution yields an objective function value of zero, so Lemma 2 in Chapter 6 shows that $x_1 = \frac{1}{3}$, $x_2 = \frac{1}{3}$, $x_3 = \frac{1}{3}$ is optimal for the row player's LP, and $y_1 = \frac{1}{3}$, $y_2 = \frac{1}{3}$, $y_3 = \frac{1}{3}$ is optimal for the column player's LP.

The complementary slackness theory of linear programming (discussed in Section 6.10) could have been used to find the optimal strategies and value for Stone, Paper, Scissors (as well as other games). Before showing how, we state the row and the column player's LPs for a general two-person zero-sum game.

Consider a two-person zero-sum game with the reward matrix shown in Table 24. The reasoning used to derive (2) and (3) yields the following LPs:

$$\max z = v$$

Row Player's LP

$$\text{s.t.} \quad v \leq a_{11}x_1 + a_{21}x_2 + \cdots + a_{m1}x_m \quad \text{(Column 1 constraint)}$$
$$v \leq a_{12}x_1 + a_{22}x_2 + \cdots + a_{m2}x_m \quad \text{(Column 2 constraint)}$$
$$\vdots$$
$$v \leq a_{1n}x_1 + a_{2n}x_2 + \cdots + a_{mn}x_m \quad \text{(Column n constraint)}$$
$$x_1 + x_2 + \cdots + x_m = 1$$
$$x_i \geq 0 \quad (i = 1, 2, \ldots, m); \ v \text{ urs}$$

(5)

$$\min z = w$$

Column Player's LP

$$\text{s.t.} \quad w \geq a_{11}y_1 + a_{12}y_2 + \cdots + a_{1n}y_n \quad \text{(Row 1 constraint)}$$
$$w \geq a_{21}y_1 + a_{22}y_2 + \cdots + a_{2n}y_n \quad \text{(Row 2 constraint)}$$
$$\vdots$$
$$w \geq a_{m1}y_1 + a_{m2}y_2 + \cdots + a_{mn}y_n \quad \text{(Row m constraint)}$$
$$y_1 + y_2 + \cdots + y_n = 1$$
$$y_j \geq 0 \quad (j = 1, 2, \ldots, n); \ w \text{ urs}$$

(6)

TABLE **24**
A General Two-Person Zero-Sum Game

Row Player	Column Player			
	Strategy 1	Strategy 2	$\cdots$	Strategy n
Strategy 1	a_{11}	a_{12}	$\cdots$	a_{1n}
Strategy 2	a_{21}	a_{22}	$\cdots$	a_{2n}
$\vdots$	$\vdots$	$\vdots$	$\vdots$	$\vdots$
Strategy m	a_{m1}	a_{m2}	$\cdots$	a_{mn}

In (5), x_i = probability that the row player chooses row i, and in (6), y_j = probability that the column player chooses column j. The jth constraint ($j = 1, 2, \ldots, n$) in the row player's LP implies that her expected reward against column j must at least equal v; otherwise, the column player could hold the row player's expected reward below v by choosing column j. Similarly, the ith ($i = 1, 2, \ldots, m$) constraint in the column player's LP implies that if the row player chooses row i, then the column player's expected losses cannot exceed w; if this were not the case, the row player could obtain an expected reward that exceeded w by choosing row i.

How to Solve the Row and the Column Players' LPs

It is easy to show (see Problem 9 at the end of this section) that if we add a constant c to each entry in a game's reward matrix, the optimal strategies for each player remain unchanged, but the optimal values of w and v (and thus the value of the game) are both increased by c. Let A be the original reward matrix. Suppose we add $c = |$most negative entry in reward matrix$|$ to each element of A. Call the new reward matrix A'. A' is a two-person constant-sum game. (Why?) Let $\bar{v}$ and $\bar{w}$ be the optimal objective function values for the row and the column players' LP's for A, and let $\bar{v}'$ and $\bar{w}'$ denote the same quantities for the game A'. Because A' will have no negative rewards, $\bar{v}' \geq 0$ and $\bar{w}' \geq 0$ must hold. Thus, when solving the row and the column players' LP's for A', we may assume that $v' \geq 0$ and $w' \geq 0$ and ignore v urs and w urs. Then the optimal strategies for A' will be identical to the optimal strategies for A, and the value of A' = (value of A) + c, or value of A = (value of A') − c.

In solving small games by hand, it is often helpful to use the constraint $x_1 + x_2 + \cdots + x_m = 1$ to eliminate one of the x_i's from the row player's LP and to use the constraint $y_1 + y_2 + \cdots + y_n = 1$ to eliminate one of the y_i's from the column player's LP. Then (as illustrated by Examples 5 and 6, which follow), the complementary slackness results of Section 6.10 can often be used to solve the row and the column player's LP's simultaneously.

EXAMPLE 4   **Stone, Paper, Scissors (Continued)**

Solution   The most negative element in the Stone, Paper, Scissors reward matrix is -1. Therefore, we add $|-1| = 1$ to each element of the reward matrix. This yields the constant-sum game shown in Table 25. The row player's LP is as follows:

$$\max v'$$
$$\text{s.t.} \quad v' \leq x_1 + 2x_2$$
$$v' \leq x_2 + 2x_3$$
$$v' \leq 2x_1 + x_3$$
$$x_1 + x_2 + x_3 = 1$$
$$x_1, x_2, x_3, v' \geq 0$$

TABLE 25
Modified Reward Matrix for Stone, Paper, Scissors

	Column Player		
Row Player	Stone	Paper	Scissors
Stone	1	0	2
Paper	2	1	0
Scissors	0	2	1

Substituting $x_3 = 1 - x_1 - x_2$ transforms the row player's LP into the following LP:

$$\max v'$$

$$\text{s.t.} \quad \text{(a)} \quad v' - x_1 - 2x_2 \leq 0 \quad (y_1, \text{ or column 1, constraint})$$
$$\text{(b)} \quad v' + 2x_1 + x_2 \leq 2 \quad (y_2, \text{ or column 2, constraint})$$
$$\text{(c)} \quad v' - x_1 + x_2 \leq 1 \quad (y_3, \text{ or column 3, constraint})$$
$$x_1, x_2, v' \geq 0$$

(7)

The column player's LP is as follows:

$$\min w'$$

$$\text{s.t.} \quad w' \geq y_1 + 2y_3 \quad (x_1, \text{ or row 1, constraint})$$
$$w' \geq 2y_1 + y_2 \quad (x_2, \text{ or row 2, constraint})$$
$$w' \geq 2y_2 + y_3 \quad (x_3, \text{ or row 3, constraint})$$
$$y_1 + y_2 + y_3 = 1$$
$$y_1, y_2, y_3, w' \geq 0$$

Substituting $y_3 = 1 - y_1 - y_2$ transforms the column player's LP into the following LP:

$$\min w'$$

$$\text{s.t.} \quad \text{(a)} \quad w' + y_1 + 2y_2 \geq 2 \quad (x_1, \text{ or row 1, constraint})$$
$$\text{(b)} \quad w' - 2y_1 - y_2 \geq 0 \quad (x_2, \text{ or row 2, constraint})$$
$$\text{(c)} \quad w' + y_1 - y_2 \geq 1 \quad (x_3, \text{ or row 3, constraint})$$
$$y_1, y_2, w' \geq 0$$

(8)

Stone, Paper, Scissors appears to be a fair game, so we might conjecture that $v = w = 0$. This would make $v' = w' = 0 + 1 = 1$. Let's try this and conjecture that constraints (7a) and (7b) are binding in the optimal solution to (7). If this is the case, then solving (7a) and (7b) simultaneously (with $v' = 1$) yields $x_1 = x_2 = \frac{1}{3}$. Because $x_1 = \frac{1}{3}, x_2 = \frac{1}{3}, w' = 1$ satisfies (7c) with equality, we have obtained a feasible solution to the row player's LP. Suppose this solution is optimal for the row player's LP. Then by complementary slackness (see Section 6.10), $x_1 > 0$ and $x_2 > 0$ would imply that the first two dual constraints in (8) must be binding in the optimal solution to (8). Solving (8a) and (8b) simultaneously (using $w' = 1$) yields $y_1 = y_2 = \frac{1}{3}, w' = 1$. This solution is dual feasible. Thus, we have found a primal feasible and a dual feasible solution, both of which have the same objective function value and are optimal. Thus:

1 The value of Stone, Paper, Scissors is $v' - 1 = 0$.

2 The optimal strategy for the row player is $(\frac{1}{3}, \frac{1}{3}, \frac{1}{3})$.

3 The optimal strategy for the column player is $(\frac{1}{3}, \frac{1}{3}, \frac{1}{3})$.

Suppose we had not been able to conjecture that $v' = w' = 1$. Then the row player's LP (7) would have had three unknowns (x_1, x_2, and v'), and we might have hoped that the optimal solution to (7) occurred where all three constraints (7a)–(7c) were binding. Solving (7a)–(7c) simultaneously yields $v' = 1$, $x_1 = x_2 = \frac{1}{3}, x_3 = 1 - \frac{2}{3} = \frac{1}{3}$. If this is the optimal solution to the row player's LP, then complementary slackness implies that constraints (8a)–(8c) must all be binding. Simultaneously solving (8a)–(8c) yields $w' = 1, y_1 = y_2 = \frac{1}{3}, y_3 = 1 - \frac{2}{3} = \frac{1}{3}$. Again we have obtained a primal feasible point and a dual feasible point having the same objective function value, and both solutions must be optimal.

EXAMPLE 5   Using Complementary Slackness to Solve a Two-Person Zero-Sum Game

Find the value and optimal strategies for the two-person zero-sum game in Table 26.

Solution   The game has no saddle point and no dominated strategies, so we set up the row and the column players' LP's. All entries in the reward matrix are nonnegative, so we are sure that the value of the game is nonnegative. The row and the column players' LP's for this game are as follows:

$$\max v$$
$$\text{s.t.} \quad v \leq 30x_1 + 60x_2$$
$$v \leq 40x_1 + 10x_2$$
$$v \leq 36x_1 + 36x_2 \tag{9}$$
$$x_1 + x_2 = 1$$
$$x_1, x_2, v \geq 0$$

Substituting $x_2 = 1 - x_1$ into the row player's LP yields

$$\max v$$
$$\text{s.t.} \quad \text{(a) } v + 30x_1 \leq 60 \quad (y_1, \text{ or column 1, constraint})$$
$$\text{(b) } v - 30x_1 \leq 10 \quad (y_2, \text{ or column 2, constraint}) \tag{9'}$$
$$\text{(c) } v \qquad \leq 36 \quad (y_3, \text{ or column 3, constraint})$$
$$x_1, v \geq 0$$

Similarly, we find

$$\min w$$
$$\text{s.t.} \quad \text{(a) } w \geq 30y_1 + 40y_2 + 36y_3$$
$$\text{(b) } w \geq 60y_1 + 10y_2 + 36y_3 \tag{10}$$
$$y_1 + y_2 + y_3 = 1$$
$$y_1, y_2, y_3, w \geq 0$$

and substituting $y_3 = 1 - y_1 - y_2$ into the column player's LP yields

$$\min w$$
$$\text{s.t.} \quad \text{(a) } w + 6y_1 - 4y_2 \geq 36 \quad (x_1, \text{ or row 1, constraint})$$
$$\text{(b) } w - 24y_1 + 26y_2 \geq 36 \quad (x_2, \text{ or row 2, constraint}) \tag{10'}$$
$$y_1, y_2, w \geq 0$$

TABLE 26
Reward Matrix for Example 5

	Column Player			Row Minimum
Row Player	30	40	36	30
	60	10	36	10
Column Maximum	60	40	36	

When using complementary slackness to solve an LP and its dual, it is usually easier to first examine the LP with the smaller number of variables. Thus, we first examine (9′). We assume that (9′a) and (9′b) are both binding in the optimal solution to the row player's LP. Then $v = 35$, $x_1 = \frac{5}{6}$, $x_2 = \frac{1}{6}$ would be the optimal solution to the row player's LP. This solution is feasible in the row player's LP and makes constraint (9′c) nonbinding. If this solution is optimal for the row player's LP, then complementary slackness implies that (10′a) and (10′b) must both be binding and $y_3 = 0$ must hold. This implies that $y_1 + y_2 = 1$, or $y_2 = 1 - y_1$. Trying $w = 35$ and substituting $y_2 = 1 - y_1$ in (10′a) and (10′b) yields $y_1 = y_2 = \frac{1}{2}$. Thus, we have found a feasible solution ($v = 35$, $x_1 = \frac{5}{6}$, $x_2 = \frac{1}{6}$) to the row player's LP and a feasible solution ($w = 35$, $y_1 = \frac{1}{2}$, $y_2 = \frac{1}{2}$, $y_3 = 0$) to the column player's LP, both of which have the same objective function value. We have therefore found the value of the game and the optimal strategy for each player.

In closing, we note that while the column player's third strategy is not dominated by column 1 or column 2, he should still never choose column 3. Why?

In the following two-person zero-sum game, the complementary slackness method does not yield optimal strategies.

EXAMPLE 6   Two-Finger Morra

Two players in the game of Two-Finger Morra simultaneously put out either one or two fingers. Each player must also announce the number of fingers that he believes his opponent has put out. If neither or both players correctly guess the number of fingers put out by the opponent, the game is a draw. Otherwise, the player who guesses correctly wins (from the other player) the sum (in dollars) of the fingers put out by the two players. If we let (i, j) represent the strategy of putting out i fingers and guessing the opponent has put out j fingers, the appropriate reward matrix is as shown in Table 27.

Solution   Again, this game has no saddle point and no dominated strategies. To ensure that the value of the game is nonnegative, we add 4 to each entry in the reward matrix. This yields the reward matrix in Table 28. For this game, the row player's and the column player's LP's are as follows (recall that the value for the original Two-Finger Morra game $= v' - 4$):

max v'

s.t.

Row Player's LP

(a)	$v' \leq 4x_1 + 2x_2 + 7x_3 + 4(1 - x_1 - x_2 - x_3)$	(y_1 constraint)
(b)	$v' \leq 6x_1 + 4x_2 + 4x_3 + (1 - x_1 - x_2 - x_3)$	(y_2 constraint)
(c)	$v' \leq x_1 + 4x_2 + 4x_3 + 8(1 - x_1 - x_2 - x_3)$	(y_3 constraint)
(d)	$v' \leq 4x_1 + 7x_2 + 4(1 - x_1 - x_2 - x_3)$	(y_4 constraint)

$$x_1, x_2, x_3, v' \geq 0$$

max w'

s.t.

Column Player's LP

(a)	$w' \geq 4y_1 + 6y_2 + y_3 + 4(1 - y_1 - y_2 - y_3)$	(x_1 constraint)
(b)	$w' \geq 2y_1 + 4y_2 + 4y_3 + 7(1 - y_1 - y_2 - y_3)$	(x_2 constraint)
(c)	$w' \geq 7y_1 + 4y_2 + 4y_3$	(x_3 constraint)
(d)	$w' \geq 4y_1 + y_2 + 8y_3 + 4(1 - y_1 - y_2 - y_3)$	(x_4 constraint)

$$y_1, y_2, y_3, w' \geq 0$$

An attempt to use complementary slackness to solve the row and the column players' LP's fails, because the optimal strategies for both players are degenerate. (Try complementary

TABLE **27**
Reward Matrix for Two-Finger Morra

Row Player		Column Player			Row Minimum
	(1, 1)	(1, 2)	(2, 1)	(2, 2)	
(1, 1)	0	2	−3	0	−3
(1, 2)	−2	0	0	3	−2
(2, 1)	3	0	0	−4	−4
(2, 2)	0	−3	4	0	−3
Column Maximum	3	2	4	3	

TABLE **28**
Transformed Reward Matrix for Two-Finger Morra

Row Player		Column Player		
	(1, 1)	(1, 2)	(2, 1)	(2, 2)
(1, 1)	4	6	1	4
(1, 2)	2	4	4	7
(2, 1)	7	4	4	0
(2, 2)	4	1	8	4

slackness and see what happens.) Using LINDO (or the simplex) to solve the LP's yields the following solutions: For the row player's problem, $v' = 4$, $x_1 = 0$, $x_2 = \frac{3}{5}$, $x_3 = \frac{2}{5}$, $x_4 = 0$ or $v' = 4$, $x_1 = 0$, $x_2 = \frac{4}{7}$, $x_3 = \frac{3}{7}$, $x_4 = 0$; for the column player's problem, $w' = 4$, $y_1 = 0$, $y_2 = \frac{3}{5}$, $y_3 = \frac{2}{5}$, $y_4 = 0$ or $w' = 4$, $y_1 = 0$, $y_2 = \frac{4}{7}$, $y_3 = \frac{3}{7}$, $y_4 = 0$. Each player's LP has alternative optimal solutions, so each player actually has an infinite number of optimal strategies. For example, for any c satisfying $0 \le c \le 1$, $x_1 = 0$, $x_2 = \frac{3c}{5} + \frac{4(1-c)}{7}$, $x_3 = \frac{2c}{5} + \frac{3(1-c)}{7}$, $x_4 = 0$ would be an optimal strategy for the row player. Of course, the value (to the row player) of the original Two-Finger Morra game is $v' - 4 = 0$.

REMARKS **1** Observe that both players have the same optimal strategies. (This is no accident; see Problem 5 at the end of this section.)

2 Also note that if each player utilizes his optimal strategy, then neither player will ever lose or win any money. This illustrates the fact that if both players follow the basic assumption of two-person zero-sum game theory, then conservative play will generally result.

3 Finally, we see that if each player uses his optimal strategy, then a player never guesses the same number of fingers that he has actually put out. This fact is explained in Table 29.

Now suppose the row player chooses the optimal strategy $(0, \frac{3}{5}, \frac{2}{5}, 0)$. Then the column player only breaks even by playing $(1, 1)$ and loses an average of $\frac{1}{5}$ per play when he plays $(2, 2)$. Similarly, if the row player chooses the optimal strategy $(0, \frac{4}{7}, \frac{3}{7}, 0)$, the column player breaks even with $(2, 2)$

TABLE **29**
Expected Reward to Row Player

Row Plays Optimal Strategy	Expected Reward to Row	
	Column Plays	Column Plays
	(1, 1)	(2, 2)
$(0, \frac{3}{5}, \frac{2}{5}, 0)$	$-2(\frac{3}{5}) + 3(\frac{2}{5}) = 0$	$3(\frac{3}{5}) - 4(\frac{2}{5}) = \frac{1}{5}$
$(0, \frac{4}{7}, \frac{3}{7}, 0)$	$-2(\frac{4}{7}) + 3(\frac{3}{7}) = \frac{1}{7}$	$3(\frac{4}{7}) - 4(\frac{3}{7}) = 0$

TABLE 30

Expected Reward to Row Player If Column Player Plays $(\frac{1}{4}, \frac{1}{4}, \frac{1}{4}, \frac{1}{4})$

Row Chooses	Column Chooses	Reward to Row	Probability of Occurrence
(1, 2)	(1, 1)	−2	$(\frac{3}{5})(\frac{1}{4}) = \frac{3}{20}$
(1, 2)	(1, 2)	0	$(\frac{3}{5})(\frac{1}{4}) = \frac{3}{20}$
(1, 2)	(2, 1)	0	$(\frac{3}{5})(\frac{1}{4}) = \frac{3}{20}$
(1, 2)	(2, 2)	3	$(\frac{3}{5})(\frac{1}{4}) = \frac{3}{20}$
(2, 1)	(1, 1)	3	$(\frac{2}{5})(\frac{1}{4}) = \frac{2}{20}$
(2, 1)	(1, 2)	0	$(\frac{2}{5})(\frac{1}{4}) = \frac{2}{20}$
(2, 1)	(2, 1)	0	$(\frac{2}{5})(\frac{1}{4}) = \frac{2}{20}$
(2, 1)	(2, 2)	−4	$(\frac{2}{5})(\frac{1}{4}) = \frac{2}{20}$

and loses an average of $\frac{1}{7}$ per play when he plays (1, 1). Thus, putting out the same number of fingers as you guess cannot have a positive expected reward against the other player's optimal strategy.

The preceding discussion explains why the seemingly reasonable strategy $(\frac{1}{4}, \frac{1}{4}, \frac{1}{4}, \frac{1}{4})$ is not optimal for either player. For instance, if the column player chooses the strategy $(\frac{1}{4}, \frac{1}{4}, \frac{1}{4}, \frac{1}{4})$ and the row player plays the optimal strategy $(0, \frac{3}{5}, \frac{2}{5}, 0)$, the row player's expected reward may be computed as in Table 30. In this situation, the expected reward received by the row player is $-2(\frac{3}{20}) + 0(\frac{3}{20}) + 0(\frac{3}{20}) + 3(\frac{3}{20}) + 3(\frac{2}{20}) + 0(\frac{2}{20}) + 0(\frac{2}{20}) - 4(\frac{2}{20}) = \frac{1}{20}$. Another way to see this: Each time the column player chooses (1, 1), (1, 2), or (2, 1), the players break even, but on the plays for which the column player chooses (2, 2), the row player wins an average of $\frac{1}{5}$ unit. Thus, the row player's expected reward is $(\frac{1}{4})(\frac{1}{5}) = \frac{1}{20}$ unit.

4 In Odds and Evens and in Stone, Paper, Scissors, the optimal strategies may have been intuitively obvious, but the game of Two-Finger Morra shows that game theory can often yield subtle insights into how a two-person zero-sum game should be played.

Using LINDO or LINGO to Solve Two-Person Zero-Sum Games

To use LINDO to solve for the value and optimal strategies in a two-person zero-sum game, simply type in either the row or column player's problem. If, for example, you type in the row player's problem, your optimal z-value is the value of the game; your optimal values of the decision variables are the row player's optimal strategies; and the absolute value of the dual prices are the column player's optimal strategies. By the way, because v is unrestricted in sign, you should use the command **FREE** v after the **END** statement.

Game.lng

The following LINGO model (file Game.lng) can be used to solve for the value and optimal strategies for Two-Finger Morra (or any two-person zero-sum game).

```
MODEL:
 1]SETS:
 2]ROWS/1..4/:X;
 3]COLS/1..4/;
 4]MATRIX(ROWS,COLS):REW;
 5]ENDSETS
 6]@FOR(COLS(J):@SUM(ROWS(I):REW(I,J)*X(I))>V;);
 7]@SUM(ROWS(I):X(I))=1;
 8]MAX=V;
 9]@FREE(V);
10]DATA:
11]REW=0,2,-3,0,
12]-2,0,0,3,
13]3,0,0,-4,
14]0,-3,4,0;
15]ENDDATA
16]END
```

In line 2, we define the rows of our reward matrix, associating row i with X(I) = probability that the row player plays row i. In line 3, we define the columns of the reward matrix. In line 4, we create the reward matrix itself and define the reward REW(I,J) to the row player when row i and column j are played. For each column j, line 6 creates the constraint that $\sum_I \text{REW}(I,J)*X(I) \geq V$. In line 7, we ensure that the row player's probabilities sum to 1. Row 8 creates the objective function of max $z = v$. Row 9 uses the **@FREE** statement to allow v to be negative. In rows 11 through 14, we input the reward matrix.

To use this model to solve for optimal strategies in any two-person zero-sum game, change the number of rows and columns and change the entries in the reward matrix. Remember that the dual prices yield the column player's optimal strategies.

Summary of How to Solve a Two-Person Zero-Sum Game

To close our discussion of two-person zero-sum games, we summarize a procedure that can be used to find the value and optimal strategies for any two-person zero-sum (or constant-sum) game.

Step 1 Check for a saddle point. If the game has no saddle point, then go on to step 2.

Step 2 Eliminate any of the row player's dominated strategies. Looking at the reduced matrix (dominated rows crossed out), eliminate any of the column player's dominated strategies. Now eliminate any of the row player's dominated strategies. Continue in this fashion until no more dominated strategies can be found. Now proceed to step 3.

Step 3 If the game matrix is now 2×2, solve the game graphically. Otherwise, solve the game by using the linear programming methods of this section.

PROBLEMS

Group A

1 A soldier can hide in one of five foxholes (1, 2, 3, 4, or 5) (see Figure 5). A gunner has a single shot and may fire at any of the four spots A, B, C, or D. A shot will kill a soldier if the soldier is in a foxhole adjacent to the spot where the shot was fired. For example, a shot fired at spot B will kill the soldier if he is in foxhole 2 or 3, while a shot fired at spot D will kill the soldier if he is in foxhole 4 or 5. Suppose the gunner receives a reward of 1 if the soldier is killed and a reward of 0 if the soldier survives the shot.

   a Assuming this to be a zero-sum game, construct the reward matrix.

   b Find and eliminate all dominated strategies.

   c We are given that an optimal strategy for the soldier is to hide $\frac{1}{3}$ of the time in foxholes 1, 3, and 5. We are also told that for the gunner, an optimal strategy is to shoot $\frac{1}{3}$ of the time at A, $\frac{1}{3}$ of the time at D, and $\frac{1}{3}$ of the time at B or C. Determine the value of the game to the gunner.

   d Suppose the soldier chooses the following nonoptimal strategy: $\frac{1}{2}$ of the time, hide in foxhole 1; $\frac{1}{4}$ of the time, hide in foxhole 3; and $\frac{1}{4}$ of the time, hide in fox-

hole 5. Find a strategy for the gunner that ensures that his expected reward will exceed the value of the game.

   e Write down each player's LP and verify that the strategies given in part (c) are optimal strategies.

2 Find each player's optimal strategy and the value of the two-person zero-sum game in Table 31.

3 Find each player's optimal strategy and the value of the two-person zero-sum game in Table 32.

4 Two armies are advancing on two cities. The first army is commanded by General Custard and has four regiments;

TABLE 31

4	5	1	4
2	1	6	3
1	0	0	2

TABLE 32

2	4	6
3	1	5

FIGURE 5

① A ② B ③ C ④ D ⑤

the second army is commanded by General Peabody and has three regiments. At each city, the army that sends more regiments to the city captures both the city and the opposing army's regiments. If both armies send the same number of regiments to a city, then the battle at the city is a draw. Each army scores 1 point per city captured and 1 point per captured regiment. Assume that each army wants to maximize the difference between its reward and its opponent's reward. Formulate this situation as a two-person zero-sum game and solve for the value of the game and each player's optimal strategies.

Group B

5 A two-person zero-sum game with an $n \times n$ reward matrix A is a **symmetric** game if $A = -A^T$.

a Explain why a game having $A = -A^T$ is called a symmetric game.

b Show that a symmetric game must have a value of zero.

c Show that if $(\bar{x}_1, \bar{x}_2, \ldots, \bar{x}_n)$ is an optimal strategy for the row player, then $(\bar{x}_1, \bar{x}_2, \ldots, \bar{x}_n)$ is also an optimal strategy for the column player.

d What examples discussed in this chapter are symmetric games? How could the results of this problem make it easier to solve for the value and optimal strategies of a symmetric game?

6 For a two-person zero-sum game with an $m \times n$ reward matrix, let $\bar{x} = (\bar{x}_1, \bar{x}_2, \ldots, \bar{x}_m)$ be a solution to the row player's LP and $\bar{y} = (\bar{y}_1, \bar{y}_2, \ldots, \bar{y}_n)$ be a solution to the column player's LP. Show that if the row player departs from his optimal strategy, he cannot increase his expected reward against $\bar{y}$.

7 Interpret the complementary slackness conditions for the row and the column players' LP's.

8 Wivco has observed the daily production and the daily variable production costs of widgets at the New York City

plant. The data in Table 33 have been collected. Wivco believes that daily production and daily variable production costs are related as follows: For some numbers a and b,

Daily production cost = $a + b$(daily production)

Wivco wants to find estimates of a and b ($\hat{a}$ and $\hat{b}$) that minimize the maximum error (in absolute value) incurred in estimating daily production costs. For example, if Wivco chooses $\hat{a} = 3$ and $\hat{b} = 2$, then the predicted daily costs are shown in Table 34. In this case, the maximum error would be $3,000. Formulate an LP that can be used to find the optimal estimates $\hat{a}$ and $\hat{b}$.

9 Suppose we add a constant c to every element in a reward matrix A. Call the new game matrix A'. Show that A and A' have the same optimal strategies and that value of $A' = $ (value of A) $+ c$.

TABLE 33

Day	Production	Variable Production Cost ($)
1	4,000	9,000
2	6,000	12,000
3	7,000	14,000
4	1,000	5,000
5	3,000	8,000

TABLE 34

Day	Predicted Cost ($)	Absolute Error ($)
1	11,000	2,000
2	15,000	3,000
3	17,000	3,000
4	5,000	0
5	9,000	1,000

14.4 Two-Person Nonconstant-Sum Games

Most game-theoretic models of business situations are not constant-sum games, because it is unusual for business competitors to be in total conflict.

In this section, we briefly discuss the analysis of two-person nonconstant-sum games in which cooperation between the players is not allowed. We begin with a discussion of the famous Prisoner's Dilemma.

EXAMPLE 7 | **Prisoner's Dilemma**

Two prisoners who escaped and participated in a robbery have been recaptured and are awaiting trial for their new crime. Although they are both guilty, the Gotham City

TABLE 35

Reward Matrix for Prisoner's Dilemma

Prisoner 1	Prisoner 2	
	Confess	Don't Confess
Confess	$(-5, -5)$	$(0, -20)$
Don't confess	$(-20, 0)$	$(-1, -1)$

district attorney is not sure he has enough evidence to convict them. To entice them to testify against each other, the district attorney tells each prisoner the following: "If only one of you confesses and testifies against your partner, the person who confesses will go free while the person who does not confess will surely be convicted and given a 20-year jail sentence. If both of you confess, then you will both be convicted and sent to prison for 5 years. Finally, if neither of you confesses, I can convict you both of a misdemeanor and you will each get 1 year in prison." What should each prisoner do?

Solution   If we assume that the prisoners cannot communicate with each other, the strategies and rewards for each are as shown in Table 35. The first number in each cell of this matrix is the reward (negative, because years in prison is undesirable) to prisoner 1, and the second matrix in each cell is the reward to prisoner 2. Note that the sum of the rewards in each cell varies from a high of -2 $(-1 - 1)$ to a low of -20 $(-20 + 0)$. Thus, this is not a constant-sum two-player game.

Suppose each prisoner seeks to eliminate any dominated strategies from consideration. For each prisoner, the "confess" strategy dominates the "don't confess" strategy. If each prisoner follows his undominated ("confess") strategy, however, each prisoner will spend 5 years in jail. On the other hand, if each prisoner chooses the dominated "don't confess" strategy, then each prisoner will spend only 1 year in prison. Thus, if each prisoner chooses his dominated strategy, both are better off than if each prisoner chooses his undominated strategy.

DEFINITION ■   As in a two-person zero-sum game, a choice of strategy by each player (prisoner) is an **equilibrium point** if neither player can benefit from a unilateral change in strategy. ■

Thus, $(-5, -5)$ is an equilibrium point, because if either prisoner changes his strategy, then his reward decreases (from -5 to -20). Clearly, however, each prisoner is better off at the point $(-1, -1)$. To see that the outcome $(-1, -1)$ may not occur, observe that $(-1, -1)$ is not an equilibrium point, because if we are currently at the outcome $(-1, -1)$, either prisoner can increase his reward (from -1 to 0) by changing his strategy from "don't confess" to "confess" (that is, each prisoner can benefit from double-crossing his opponent). This illustrates an important aspect of the Prisoner's Dilemma type of game: If the players are cooperating (if each prisoner chooses "don't confess"), then each player can gain by double-crossing his opponent (assuming his opponent's strategy remains unchanged). If both players double-cross each other, however, then both will be worse off than if they had both chosen their cooperative strategy. This anomaly cannot occur in a two-person constant-sum game. (Why not?)

TABLE 36

TABLE 36
A General Prisoner's Dilemma
Reward Matrix

Player 1	Player 2	
	NC	C
NC	(P, P)	(T, S)
C	(S, T)	(R, R)

More formally, a Prisoner's Dilemma game may be described as in Table 36, where

NC = noncooperative action

C = cooperative action

P = punishment for not cooperating

S = payoff to person who is double-crossed

R = reward for cooperating if both players cooperate

T = temptation for double-crossing opponent

In a Prisoner's Dilemma game, (P, P) is an equilibrium point. This requires $P > S$. For (R, R) not to be an equilibrium point requires $T > R$. (This gives each player a temptation to double-cross his opponent.) The game is reasonable only if $R > P$. Thus, for Table 36 to represent a Prisoner's Dilemma game, we require that $T > R > P > S$. The Prisoner's Dilemma game is of interest because it explains why two adversaries often fail to cooperate with each other. This is illustrated by Examples 8 and 9.

EXAMPLE 8 **Advertising Prisoner's Dilemma Game**

Competing restaurants Hot Dog King and Hot Dog Chef are attempting to determine their advertising budgets for next year. The two restaurants will have combined sales of $240 million and can spend either $6 million or $10 million on advertising. If one restaurant spends more money than the other, then the restaurant that spends more money will have sales of $190 million. If both companies spend the same amount on advertising, then they will have equal sales. Each dollar of sales yields 10¢ of profit. Suppose each restaurant is interested in maximizing (contribution of sales to profit) − (advertising costs). Find an equilibrium point for this game.

Solution The appropriate reward matrix is shown in Table 37. If we identify spending $10 million on advertising as the noncooperative action and spending $6 million as the cooperative action, then $(2, 2)$ (corresponding to heavy advertising by both restaurants) is an equilibrium point. Although both restaurants are better off at $(6, 6)$ than at $(2, 2)$, $(6, 6)$ is unstable because either restaurant may gain by changing its strategy. Thus, to protect its market share, each restaurant must spend heavily on advertising.

TABLE 37
Reward Matrix for Advertising Game

Hot Dog King	Hot Dog Chef	
	Spend $10 Million	Spend $6 Million
Spend $10 million	(2, 2)	(9, −1)
Spend $6 million	(−1, 9)	(6, 6)

EXAMPLE 9 | Arms Race Prisoner's Dilemma

The Vulcans and the Klingons are engaged in an arms race in which each nation is assumed to have two possible strategies: develop a new missile or maintain the status quo. The reward matrix is assumed to be as shown in Table 38. This reward matrix is based on the assumption that if only one nation develops a new missile, the nation with the new missile will conquer the other nation. In this case, the conquering nation earns a reward of 20 units and the conquered nation loses 100 units. It is also assumed that the cost of developing a new missile is 10 units. Identify an equilibrium point for this game.

Solution Identifying "develop" as the noncooperative action and "maintain" as the cooperative action, we see that $(-10, -10)$ (both nations choosing their noncooperative action) is an equilibrium point. Although $(0, 0)$ leaves both nations better off than $(-10, -10)$, we see that in this situation, each nation can gain from a double-cross. Thus, $(0, 0)$ is not stable. This example shows how maintaining the balance of power may lead to an arms race.

TABLE 38
Reward Matrix for Arms Race Game

Vulcans	Klingons	
	Develop New Missile	Maintain Status Quo
Develop new missile	$(-10, -10)$	$(10, -100)$
Maintain status quo	$(-100, 10)$	$(0, 0)$

The following two-person nonconstant-sum game is not a Prisoner's Dilemma game.

EXAMPLE 10 | "Chicken" Game

Angry Max drives toward James Bound on a deserted road. Each person has two strategies: swerve or don't swerve. The reward matrix in Table 39 needs no explanation! Find the equilibrium point(s) for this game.

Solution For both $(5, -5)$ and $(-5, 5)$, neither player can gain by a unilateral change in strategy. Thus, $(5, -5)$ and $(-5, 5)$ are both equilibrium points.

TABLE 39
Reward Matrix for Swerve Game

Angry Max	James Bound	
	Swerve	Don't Swerve
Swerve	$(0, 0)$	$(-5, 5)$
Don't swerve	$(5, -5)$	$(-100, -100)$

Like constant-sum games, a nonconstant-sum game may fail to have an equilibrium point in pure strategies. It can be shown that if mixed strategies are allowed, then in any two-person nonconstant-sum game, each player has an equilibrium strategy (in that if one player plays her equilibrium strategy, the other player cannot benefit by deviating from her equilibrium strategy) [see Owen (1982, p. 127)]. For example, consider the two-

TABLE 40
A Game with No Equilibrium in Pure Strategies

Player 1	Player 2	
	Strategy 1	Strategy 2
Strategy 1	$(2, -1)$	$(-2, 1)$
Strategy 2	$(-2, 1)$	$(2, -1)$

person nonconstant-sum game in Table 40. For this game, the reader should verify that there is no equilibrium in pure strategies and also that each player's choice of the mixed strategy $(\frac{1}{2}, \frac{1}{2})$ is an equilibrium because neither player can benefit from a unilateral change in strategy (see Problem 4 at the end of this section). Owen (1999, Chapter 7) discusses two-person nonconstant-sum games in which the players are allowed to cooperate.

PROBLEMS

Group A

1 Find an equilibrium point (if one exists in pure strategies) for the two-person nonconstant-sum game in Table 41.

2 Find an equilibrium point in pure strategies (if any exists) for the two-person nonconstant-sum game in Table 42.

3 The New York City Council is ready to vote on two bills that authorize the construction of new roads in Manhattan and Brooklyn. If the two boroughs join forces, they can pass both bills, but neither borough by itself has enough power to pass a bill. If a bill is passed, then it will cost the taxpayers of each borough $1 million, but if roads are built in a borough, the benefits to the borough are estimated to be $10 million. The council votes on both bills simultaneously, and each councilperson must vote on the bills without knowing how anybody else will vote. Assuming that each borough supports its own bill, determine whether this game has any equilibrium points. Is this game analogous to the Prisoner's Dilemma? Explain why or why not.

Group B

4 Given that each player's goal is to maximize her expected reward, show that for the game in Table 43 each player's choice of the mixed strategy $(\frac{1}{2}, \frac{1}{2})$ is an equilibrium point.

TABLE 41

$(9, -1)$	$(-2, -3)$
$(8, 7)$	$(-9, 11)$

TABLE 42

$(9, 9)$	$(-10, 10)$
$(10, -10)$	$(-1, 1)$

TABLE 43

Player 1	Player 2	
	Strategy 1	Strategy 2
Strategy 1	$(2, -1)$	$(-2, 1)$
Strategy 2	$(-2, 1)$	$(2, -1)$

5[†] A Japanese electronics company and an American electronics company are both considering working on developing a superconductor. If both companies work on the superconductor, they will have to share the market, and each company will lose $10 billion. If only one company works on the superconductor, that company will earn $100 billion in profits. Of course, if neither company works on the superconductor, then each company earns profits of $0.

   a Formulate this situation as a two-person nonconstant-sum game. Does the game have any equilibrium points?

   b Now suppose the Japanese government offers the Japanese electronics company a $15 billion subsidy to work on the superconductor. Formulate the reward matrix for this game. Does this game have any equilibrium points?

   c Businesspeople have often said that a protectionist attitude toward trade can increase exports, but economists have usually argued that it will reduce exports. Whose viewpoint does this problem support?

[†]Based on "Protectionism Gets Clever" (1988).

14.5 Introduction to *n*-Person Game Theory

In many competitive situations, there are more than two competitors. With this in mind, we now turn our attention to games with three or more players. Let $N = \{1, 2, \ldots, n\}$ be the set of players. Any game with n players is an **n-person game.** For our purposes, an n-person game is specified by the game's characteristic function.

DEFINITION ■ For each subset S of N, the **characteristic function** v of a game gives the amount $v(S)$ that the members of S can be sure of receiving if they act together and form a coalition. ■

Thus, $v(S)$ can be determined by calculating the amount that members of S can get without any help from players who are not in S.

EXAMPLE 11 — The Drug Game

Joe Willie has invented a new drug. Joe cannot manufacture the drug himself, but he can sell the drug's formula to company 2 or company 3. The lucky company will split a $1 million profit with Joe Willie. Find the characteristic function for this game.

Solution Letting Joe Willie be player 1, company 2 be player 2, and company 3 be player 3, we find the characteristic function for this game to be:

$$v(\{\ \}) = v(\{1\}) = v(\{2\}) = v(\{3\}) = v(\{2, 3\}) = 0$$
$$v(\{1, 2\}) = v(\{1, 3\}) = v(\{1, 2, 3\}) = \$1,000,000$$

EXAMPLE 12 — The Garbage Game

Each of four property owners has one bag of garbage and must dump it on somebody's property. If b bags of garbage are dumped on the coalition of property owners, then the coalition receives a reward of $-b$. Find the characteristic function for this game.

Solution The best that the members of any coalition can do is to dump all of their garbage on the property of owners who are not in S. Thus, the characteristic function for the garbage game ($|S|$ is the number of players in S) is given by

$$v(\{S\}) = -(4 - |S|) \qquad (\text{if } |S| < 4) \tag{11}$$
$$v(\{1, 2, 3, 4\}) = -4 \qquad (\text{if } |S| = 4) \tag{11.1}$$

Equation (11.1) follows because if players are in S, they must dump their garbage on members of S.

EXAMPLE 13 — The Land Development Game

Player 1 owns a piece of land and values the land at $10,000. Player 2 is a subdivider who can develop the land and increase its worth to $20,000. Player 3 is a subdivider who can develop the land and increase its worth to $30,000. There are no other prospective buyers. Find the characteristic function for this game.

Solution Note that any coalition that does not contain player 1 has a worth or value of $0. Any other coalition has a value equal to the maximum value that a member of the coalition places on the piece of land. Thus, we obtain the following characteristic function:

$$v(\{1\}) = \$10{,}000, \quad v(\{\ \}) = v(\{2\}) = v(\{3\}) = \$0, \quad v(\{1, 2\}) = \$20{,}000,$$
$$v(\{1, 3\}) = \$30{,}000, \quad v(\{2, 3\}) = \$0, \quad v(\{1, 2, 3\}) = \$30{,}000$$

Consider any two subsets of sets A and B such that A and B have no players in common ($A \cap B = \emptyset$). Then for each of our examples (and any n-person game), the characteristic function must satisfy the following inequality:

$$v(A \cup B) \geq v(A) + v(B) \tag{12}$$

This property of the characteristic function is called **superadditivity.** Equation (12) is reasonable, because if the players in $A \cup B$ band together, one of their options (but not their only option) is to let the players in A fend for themselves and let the players in B fend for themselves. This would result in the coalition receiving an amount $v(A) + v(B)$. Thus, $v(A \cup B)$ must be at least as large as $v(A) + v(B)$.

There are many solution concepts for n-person games. A solution concept should indicate the reward that each player will receive. More formally, let $\mathbf{x} = \{x_1, x_2, \dots, x_n\}$ be a vector such that player i receives a reward x_i. We call such a vector a **reward vector.** A reward vector $\mathbf{x} = (x_1, x_2, \dots, x_n)$ is not a reasonable candidate for a solution unless $\mathbf{x}$ satisfies

$$v(N) = \sum_{i=1}^{i=n} x_i \qquad \text{(Group rationality)} \tag{13}$$

$$x_i \geq v(\{i\}) \quad \text{(for each } i \in N\text{)} \qquad \text{(Individual rationality)} \tag{14}$$

If $\mathbf{x}$ satisfies both (13) and (14), we say that $\mathbf{x}$ is an **imputation.** Equation (13) states that any reasonable reward vector must give all the players an amount that equals the amount that can be attained by the supercoalition consisting of all players. Equation (14) implies that player i must receive a reward at least as large as what he can get for himself ($v\{i\}$).

To illustrate the idea of an imputation, consider the payoff vectors for Example 13, shown in Table 44. Any solution concept for n-person games chooses some subset of the set of imputations (possibly empty) as the solution to the n-person game. In Sections 14.6 and 14.7, we discuss two solution concepts, the core and the Shapley value. See Owen (1999) for a discussion of other solution concepts for n-person games. The problems involving n-person game theory are at the end of Section 14.7.

TABLE 44
Examples of Imputation

x	Is x an Imputation?
($\$10{,}000$, $\$10{,}000$, $\$10{,}000$)	Yes
($\$5{,}000$, $\$2{,}000$, $\$5{,}000$)	No, $x_1 < v(\{1\})$, so (14) is violated
($\$12{,}000$, $\$19{,}000$, $-\$1000$)	No, (14) is violated
($\$11{,}000$, $\$11{,}000$, $\$11{,}000$)	No, (13) is violated

14.6 The Core of an *n*-Person Game

An important solution concept for an *n*-person game is the core. Before defining this, we must define the concept of **domination.** Given an imputation $\mathbf{x} = (x_1, x_2, \ldots, x_n)$, we say that the imputation $\mathbf{y} = (y_1, y_2, \ldots, y_n)$ *dominates* $\mathbf{x}$ through a coalition S (written $\mathbf{y} > {}^S\mathbf{x}$) if

$$\sum_{i \in S} y_i \leq v(S) \qquad \text{and for all } i \in S, \qquad y_i > x_i \tag{15}$$

If $\mathbf{y} > {}^S\mathbf{x}$, then both the following must be true:

1 Each member of S prefers $\mathbf{y}$ to $\mathbf{x}$.

2 Because $\sum_{i \in S} y_i \leq v(S)$, the members of S can attain the rewards given by $\mathbf{y}$.

Thus, if $\mathbf{y} > {}^S\mathbf{x}$, then $\mathbf{x}$ should not be considered a possible solution to the game, because the players in S can object to the rewards given by $\mathbf{x}$ and enforce their objection by banding together and thereby receiving the rewards given by $\mathbf{y}$ [because members of S can surely receive an amount equal to $v(S)$].

The founders of game theory, John von Neumann and Oskar Morgenstern, argued that a reasonable solution concept for an *n*-person game was the set of all undominated imputations.

DEFINITION ■ The **core** of an *n*-person game is the set of all undominated imputations. ■

Examples 14 and 15 illustrate the concept of domination.

EXAMPLE 14 **Dominance**

Consider a three-person game with the following characteristic function:

$$v(\{\ \}) = v(\{1\}) = v(\{2\}) = v(\{3\}) = 0$$
$$v(\{1, 2\}) = 0.1, \quad v(\{1, 3\}) = 0.2, \quad v(\{2, 3\}) = 0.2, \quad v(\{1, 2, 3\}) = 1$$

Let $\mathbf{x} = (0.05, 0.90, 0.05)$ and $\mathbf{y} = (0.10, 0.80, 0.10)$. Show that $\mathbf{y} > {}^{\{1,3\}}\mathbf{x}$.

Solution First, note that both $\mathbf{x}$ and $\mathbf{y}$ are imputations. Next, observe that with the imputation $\mathbf{y}$, players 1 and 3 both receive more than they receive with $\mathbf{x}$. Also, $\mathbf{y}$ gives the players in $\{1, 3\}$ a total of $0.10 + 0.10 = 0.20$. Because 0.20 does not exceed $v(\{1, 3\}) = 0.20$, it is reasonable to assume that players 1 and 3 can band together and receive a total reward of 0.20. Thus, players 1 and 3 will never allow the rewards given by $\mathbf{x}$ to occur.

EXAMPLE 15 **Dominance in Land Development Game**

For the land development game (Example 13), let $\mathbf{x} = (\$19{,}000, \$1{,}000, \$10{,}000)$ and $\mathbf{y} = (\$19{,}800, \$100, \$10{,}100)$. Show that $\mathbf{y} > {}^{\{1,3\}}\mathbf{x}$.

Solution We need only observe that players 1 and 3 both receive more from $\mathbf{y}$ than they receive from $\mathbf{x}$, and the total received by players 1 and 3 from $\mathbf{y}$ ($\$29{,}900$) does not exceed $v(\{1, 3\})$. If $\mathbf{x}$ were proposed as a solution to the land development game, player 1 would sell the land to player 3 and $\mathbf{y}$ (or some other imputation that dominates $\mathbf{x}$) would result. The important point is that $\mathbf{x}$ cannot occur, because players 1 and 3 will never allow $\mathbf{x}$ to occur.

We are now ready to show how to determine the core of an *n*-person game, for which Theorem 1 is often useful.

THEOREM 1

An imputation $\mathbf{x} = \{x_1, x_2, \ldots, x_n\}$ is in the core of an *n*-person game if and only if for each subset S of N,

$$\sum_{i \in S} x_i \geq v(S)$$

Theorem 1 states that an imputation $\mathbf{x}$ is in the core (that $\mathbf{x}$ is undominated) if and only if for every coalition S, the total of the rewards received by the players in S (according to $\mathbf{x}$) is at least as large as $v(S)$.

To illustrate the use of Theorem 1, we find the core of the three games discussed in Section 14.5.

EXAMPLE 11   The Drug Game (Continued)

Find the core of the drug game.

Solution   For this game, $\mathbf{x} = (x_1, x_2, x_3)$ will be an imputation if and only if

$$x_1 \geq 0 \tag{16}$$
$$x_2 \geq 0 \tag{17}$$
$$x_3 \geq 0 \tag{18}$$
$$x_1 + x_2 + x_3 = \$1{,}000{,}000 \tag{19}$$

Theorem 1 shows that $\mathbf{x} = (x_1, x_2, x_3)$ will be in the core if and only if x_1, x_2, and x_3 satisfy (16)–(19) and the following inequalities:

$$x_1 + x_2 \geq \$1{,}000{,}000 \tag{20}$$
$$x_1 + x_3 \geq \$1{,}000{,}000 \tag{21}$$
$$x_2 + x_3 \geq \$0 \tag{22}$$
$$x_1 + x_2 + x_3 \geq \$1{,}000{,}000 \tag{23}$$

To determine the core, note that if $\mathbf{x} = (x_1, x_2, x_3)$ is in the core, then x_1, x_2, and x_3 must satisfy the inequality generated by adding together inequalities (20)–(22). Adding (20)–(22) yields $2(x_1 + x_2 + x_3) \geq \$2{,}000{,}000$, or

$$x_1 + x_2 + x_3 \geq \$1{,}000{,}000 \tag{24}$$

By (19), $x_1 + x_2 + x_3 = \$1{,}000{,}000$. Thus, (20)–(22) must all be binding.[†] Simultaneously solving (20)–(22) as equalities yields $x_1 = \$1{,}000{,}000$, $x_2 = \$0$, $x_3 = \$0$. A quick check shows that ($\$1{,}000{,}000$, $\$0$, $\$0$) does satisfy (16)–(23). In summary, the core of this game is the imputation ($\$1{,}000{,}000$, $\$0$, $\$0$). Thus, the core emphasizes the importance of player 1.

REMARKS   1   In Section 14.7, we show that for this game, an alternative solution concept, the Shapley value, gives player 1 less than $\$1{,}000{,}000$ and gives both player 2 and player 3 some money.

[†]If (20), (21), or (22) were nonbinding, then for any point in the core, the sum of (20)–(22) would also be nonbinding. Because we know that (24) must be binding, this implies that for any point in the core, (20), (21), and (22) must all be binding.

2 For the drug game, if we choose an imputation that is not in the core, then we can show how it is dominated. Consider the imputation $\mathbf{x} = (\$900{,}000, \$50{,}000, \$50{,}000)$. If we let $\mathbf{y} = (\$925{,}000, \$75{,}000, \$0)$, then $\mathbf{y} >^{\{1,2\}} \mathbf{x}$.

EXAMPLE 12   The Garbage Game (Continued)

Determine the core of the garbage game.

Solution   Note that $\mathbf{x} = (x_1, x_2, x_3, x_4)$ will be an imputation if and only if x_1, x_2, x_3, and x_4 satisfy the following inequalities:

$$x_1 \geq -3 \tag{25}$$
$$x_2 \geq -3 \tag{26}$$
$$x_3 \geq -3 \tag{27}$$
$$x_4 \geq -3 \tag{28}$$
$$x_1 + x_2 + x_3 + x_4 = -4 \tag{29}$$

Applying Theorem 1 to all three-player coalitions, we find that for $\mathbf{x} = \{x_1, x_2, x_3, x_4\}$ to be in the core, it is necessary that x_1, x_2, x_3, and x_4 satisfy the following inequalities:

$$x_1 + x_2 + x_3 \geq -1 \tag{30}$$
$$x_1 + x_2 + x_4 \geq -1 \tag{31}$$
$$x_1 + x_3 + x_4 \geq -1 \tag{32}$$
$$x_2 + x_3 + x_4 \geq -1 \tag{33}$$

We now show that no imputation $\mathbf{x} = (x_1, x_2, x_3, x_4)$ can satisfy (30)–(33) and that the garbage game has an empty core. Consider an imputation $\mathbf{x} = (x_1, x_2, x_3, x_4)$. If $\mathbf{x}$ is to be in the core of the garbage game, $\mathbf{x}$ must satisfy the inequality generated by adding together (30)–(33):

$$3(x_1 + x_2 + x_3 + x_4) \geq -4 \tag{34}$$

Equation (29) implies that any imputation $\mathbf{x} = (x_1, x_2, x_3, x_4)$ must satisfy $x_1 + x_2 + x_3 + x_4 = -4$. Thus, (34) cannot hold. This means that no imputation $\mathbf{x} = (x_1, x_2, x_3, x_4)$ can satisfy (30)–(33) and the core of the garbage game is empty.

To understand why the garbage game has an empty core, consider the imputation $\mathbf{x} = (-2, -1, -1, 0)$, which treats players 1 and 2 unfairly. By joining, players 1 and 2 could ensure that the imputation $\mathbf{y} = (-1.5, -0.5, -1, -1)$ occurred. Thus, $\mathbf{y} >^{\{1,2\}} \mathbf{x}$. In a similar fashion, any imputation can be dominated by another imputation. We note that for a two-player version of the garbage game, the core consists of the imputation $(-1, -1)$, and for $n > 2$, the n-player garbage game has an empty core (see Problems 4 and 5 at the end of Section 14.7).

EXAMPLE 13   The Land Development Game (Continued)

Find the core of the land development game.

Solution   For the land development game, any imputation $\mathbf{x} = (x_1, x_2, x_3)$ must satisfy

$$x_1 \geq \$10{,}000 \tag{35}$$
$$x_2 \geq \$0 \tag{36}$$
$$x_3 \geq \$0 \tag{37}$$
$$x_1 + x_2 + x_3 = \$30{,}000 \tag{38}$$

An imputation $\mathbf{x} = (x_1, x_2, x_3)$ is in the core if and only if it satisfies the following inequalities:

$$x_1 + x_2 \geq \$20,000 \tag{39}$$

$$x_1 + x_3 \geq \$30,000 \tag{40}$$

$$x_2 + x_3 \geq \$0 \tag{41}$$

$$x_1 + x_2 + x_3 \geq \$30,000 \tag{42}$$

Adding (36) and (40), we find that if $\mathbf{x} = (x_1, x_2, x_3)$ is in the core, then x_1, x_2, and x_3 must satisfy $x_1 + x_2 + x_3 \geq \$30,000$. From (38), $x_1 + x_2 + x_3 = \$30,000$. Thus, (36) and (40) must be binding. This argument shows that for $\mathbf{x} = (x_1, x_2, x_3)$ to be in the core, x_1, x_2, and x_3 must satisfy

$$x_2 = \$0 \qquad \text{and} \qquad x_1 + x_3 = \$30,000 \tag{43}$$

Now (39) implies that

$$x_1 \geq \$20,000 \tag{44}$$

Thus, for $\mathbf{x} = (x_1, x_2, x_3)$ to be in the core, (43) and (44) must both be satisfied. Any vector in the core must also satisfy $x_3 \geq 0$ and $x_1 \leq \$30,000$, and any vector $\mathbf{x} = (x_1, x_2, x_3)$ satisfying (43), (44), $x_3 \geq \$0$, and $x_1 \leq \$30,000$ will be in the core of the land development game. Thus, if $\$20,000 \leq x_1 \leq \$30,000$, then any vector of the form $(x_1, \$0, \$30,000 - x_1)$ will be in the core of the land development game. The interpretation of the core is as follows: Player 3 outbids player 2 and purchases the land from player 1 for a price x_1 ($\$20,000 \leq x_1 \leq \$30,000$). Then player 1 receives a reward of x_1 dollars, and player 3 receives a reward of $\$30,000 - x_1$. Player 2 is shut out and receives nothing. In this example, the core contains an infinite number of points.

The problems involving n-person game theory are at the end of Section 14.7.

14.7 The Shapley Value[†]

In Section 14.6, we found that the core of the drug game gave all benefits or rewards to the game's most important player (the inventor of the drug). Now we discuss an alternative solution concept for n-person games, the **Shapley value,** which in general gives more equitable solutions than the core does.[‡]

For any characteristic function, Lloyd Shapley showed there is a unique reward vector $\mathbf{x} = (x_1, x_2, \ldots, x_n)$ satisfying the following axioms:

Axiom 1 Relabeling of players interchanges the players' rewards. Suppose the Shapley value for a three-person game is $\mathbf{x} = (10, 15, 20)$. If we interchange the roles of player 1 and player 3 [for example, if originally $v(\{1\}) = 10$ and $v(\{3\}) = 15$, we would make $v(\{1\}) = 15$ and $v(\{3\}) = 10$], then the Shapley value for the new game would be $\mathbf{x} = (20, 15, 10)$.

Axiom 2 $\sum_{i=1}^{i=n} x_i = v(N)$. This is simply group rationality.

Axiom 3 If $v(S - \{i\}) = v(S)$ holds for all coalitions S, then the Shapley value has $x_i = 0$. If player i adds no value to any coalition, then player i receives a reward of zero from the Shapley value.

Before stating Axiom 4, we define the sum of two n-person games. Let v and $\bar{v}$ be two characteristic functions for games with identical players. Define the game $(v + \bar{v})$ to be

---

[†]This section covers topics that can be omitted with no loss of continuity.
[‡]See Owen (1982) for an excellent discussion of the Shapley value. See also Shapley (1953).

the game with the characteristic function $(v + \bar{v})$ given by $(v + \bar{v})(S) = v(S) + \bar{v}(S)$. For example, if $v(\{1, 2\}) = 10$ and $\bar{v}(\{1, 2\}) = -3$, then in the game $(v + \bar{v})$ the coalition $\{1, 2\}$ would have $(v + \bar{v})(\{1, 2\}) = 10 - 3 = 7$.

Axiom 4 Let $\mathbf{x}$ be the Shapley value vector for game v, and let $\mathbf{y}$ be the Shapley value vector for game $\bar{v}$. Then the Shapley value vector for the game $(v + \bar{v})$ is the vector $\mathbf{x} + \mathbf{y}$.

The validity of this axiom has often been questioned, because adding rewards from two different games may be like adding apples and oranges. If Axioms 1–4 are assumed to be valid, however, Shapley proved the remarkable result in Theorem 2.

THEOREM 2

Given any n-person game with the characteristic function v, there is a unique reward vector $\mathbf{x} = (x_1, x_2, \ldots, x_n)$ satisfying Axioms 1–4. The reward of the ith player (x_i) is given by

$$x_i = \sum_{\substack{\text{all } S \text{ for which} \\ i \text{ is not in } S}} p_n(S)[v(S \cup \{i\}) - v(S)] \tag{45}$$

In (45),

$$p_n(S) = \frac{|S|!(n - |S| - 1)!}{n!} \tag{46}$$

where $|S|$ is the number of players in S, and for $n \geq 1$, $n! = n(n - 1) \cdots 2(1)$ $(0! = 1)$.

Although (45) seems complex, the equation has a simple interpretation. Suppose that players 1, 2, $\ldots, n$ arrive in a random order. That is, any of the $n!$ permutations of 1, 2, $\ldots, n$ has a $\frac{1}{n!}$ chance of being the order in which the players arrive. For example, if $n = 3$, then there is a $\frac{1}{3}! = \frac{1}{6}$ probability that the players arrive in any one of the following sequences:

$$1, 2, 3 \quad 2, 3, 1$$
$$1, 3, 2 \quad 3, 1, 2$$
$$2, 1, 3 \quad 3, 2, 1$$

Suppose that when player i arrives, he finds that the players in the set S have already arrived. If player i forms a coalition with the players who are present when he arrives, then player i adds $v(S \cup \{i\}) - v(S)$ to the coalition S. The probability that when player i arrives the players in the coalition S are present is $p_n(S)$. Then (45) implies that *player i's reward should be the expected amount that player i adds to the coalition made up of the players who are present when he or she arrives.*

We now show that $p_n(S)$ [as given by (46)] is the probability that when player i arrives, the players in the subset S will be present. Observe that the number of permutations of 1, 2, $\ldots, n$ that result in player i's arriving when the players in the coalition S are present is given by

$$\underbrace{|S|(|S| - 1)(|S| - 2) \cdots + (2)(1)}_{S \text{ arrives}} \underbrace{(1)}_{i \text{ arrives}} \underbrace{(n - |S| - 1)(n - |S| - 2) \cdots (2)(1)}_{\text{Players not in } S \cup \{i\} \text{ arrive}}$$

$$= |S|!(n - |S| - 1)!$$

Because there are a total of $n!$ permutations of $1, 2, \ldots, n$, the probability that player i will arrive and see the players in S is

$$\frac{|S|!(n - |S| - 1)!}{n!} = p_n(S)$$

We now compute the Shapley value for the drug game.

EXAMPLE 11 **The Drug Game (Continued)**

Find the Shapley value for the drug game.

Solution To compute x_1, the reward that player 1 should receive, we list all coalitions S for which player 1 is not a member. For each of these coalitions, we compute $v(S \cup \{i\}) - v(S)$ and $p_3(S)$ (see Table 45). Because player 1 adds (on the average)

$$(\tfrac{2}{6})(0) + (\tfrac{1}{6})(1{,}000{,}000) + (\tfrac{2}{6})(1{,}000{,}000) + (\tfrac{1}{6})(1{,}000{,}000) = \tfrac{\$4{,}000{,}000}{6}$$

the Shapley value concept recommends that player 1 receive a reward of $\frac{\$4{,}000{,}000}{6}$.

To compute the Shapley value for player 2, we require the information in Table 46. Thus, the Shapley value recommends a reward of

$$(\tfrac{1}{6})(1{,}000{,}000) = \tfrac{\$1{,}000{,}000}{6}$$

for player 2. The Shapley value must allocate a total of $v(\{1, 2, 3\}) = \$1{,}000{,}000$ to the players, so the Shapley value will recommend that player 3 receive $\$1{,}000{,}000 - x_1 - x_2 = \frac{\$1{,}000{,}000}{6}$.

TABLE 45
Computation of Shapley Value for Player 1
(Joe Willie)

S	$p_3(S)$	$v(S \cup \{1\}) - v(S)$
{ }	$\tfrac{2}{6}$	$0
{2}	$\tfrac{1}{6}$	$1,000,000
{2, 3}	$\tfrac{2}{6}$	$1,000,000
{3}	$\tfrac{1}{6}$	$1,000,000

TABLE 46
Computation of Shapley Value for Player 2

S	$p_3(S)$	$v(S \cup \{2\}) - v(S)$
{ }	$\tfrac{2}{6}$	$0
{1}	$\tfrac{1}{6}$	$1,000,000
{3}	$\tfrac{1}{6}$	$0
{1, 3}	$\tfrac{2}{6}$	$0

REMARKS 1 Recall that the core of this game assigned $1,000,000 to player 1 and no money to players 2 and 3. Thus, the Shapley value treats players 2 and 3 more fairly than the core. In general, the Shapley value provides more equitable solutions than the core.
2 For a game with few players, it may be easier to compute each player's Shapley value by using the fact that player i should receive the expected amount that she adds to the coalition present when she arrives. For Example 11, this method yields the computations in Table 47. Each of the six or-

TABLE 47

Alternative Method for Determining Shapley Value

Order of Arrival	Amount Added by Player's Arrival ($)		
	Player 1	Player 2	Player 3
1, 2, 3	0	1,000,000	0
1, 3, 2	0	0	1,000,000
2, 1, 3	1,000,000	0	0
2, 3, 1	1,000,000	0	0
3, 1, 2	1,000,000	0	0
3, 2, 1	1,000,000	0	0

derings of the arrivals of the players is equally likely, so we find that the Shapley value to each player is as follows:

$$x_1 = \frac{\$4,000,000}{6}, \qquad x_2 = \frac{\$1,000,000}{6}, \qquad x_3 = \frac{\$1,000,000}{6}$$

3 The Shapley value can be used as a measure of the power of individual members of a political or business organization. For example, the UN Security Council consists of five permanent members (who have veto power over any resolution) and ten nonpermanent members. For a resolution to pass the Security Council, it must receive at least nine votes, including the votes of all permanent members. Assigning a value of 1 to all coalitions that can pass a resolution and a value of 0 to all coalitions that cannot pass a resolution defines a characteristic function. For this characteristic function, it can be shown that the Shapley value of each permanent member is 0.1963 and of each nonpermanent member is 0.001865, giving 5(0.1963) + 10(0.001865) = 1. Thus, the Shapley value indicates that 5(0.1963) = 98.15% of the power in the Security Council resides with the permanent members.

As a final application of the Shapley value, we discuss how it can be used to determine a pricing schedule for landing fees at an airport.

EXAMPLE 16   **Airport Pricing**

Suppose three types of planes (Piper Cubs, DC-10s, and 707s) use an airport. A Piper Cub requires a 100-yd runway, a DC-10 requires a 150-yd runway, and a 707 requires a 400-yd runway. Suppose the cost (in dollars) of maintaining a runway for one year is equal to the length of the runway. Because 707s land at the airport, the airport will have a 400-yd runway. For simplicity, suppose that each year only one plane of each type lands at the airport. How much of the $400 annual maintenance cost should be charged to each plane?

Solution   Let player 1 = Piper Cub, player 2 = DC-10, and player 3 = 707. We can now define a three-player game in which the value to a coalition is the cost associated with the runway length needed to service the largest plane in the coalition. Thus, the characteristic function for this game (we list a cost as a negative revenue) would be

$$v(\{\ \}) = \$0, \quad v(\{1\}) = -\$100, \quad v(\{1, 2\}) = v(\{2\}) = -\$150,$$
$$v(\{3\}) = v(\{2, 3\}) = v(\{1, 3\}) = v(\{1, 2, 3\}) = -\$400$$

To find the Shapley value (cost) to each player, we assume that the three planes land in a random order, and we determine how much cost (on the average) each plane adds to the cost incurred by the planes that are already present (see Table 48). The Shapley cost for each player is as follows:

Player 1 cost = $(\frac{1}{6})(100 + 100) = \frac{\$200}{6}$

Player 2 cost = $(\frac{1}{6})(50 + 150 + 150) = \frac{\$350}{6}$

Player 3 cost = $(\frac{1}{6})(250 + 300 + 250 + 250 + 400 + 400) = \frac{\$1,850}{6}$

TABLE 48
Computation of Shapley Value for Airport Game

Order of Arrival	Probability of Order	Cost Added by Player's Arrival ($)		
		Player 1	Player 2	Player 3
1, 2, 3	$\frac{1}{6}$	100	50	250
1, 3, 2	$\frac{1}{6}$	100	0	300
2, 1, 3	$\frac{1}{6}$	0	150	250
2, 3, 1	$\frac{1}{6}$	0	150	250
3, 1, 2	$\frac{1}{6}$	0	0	400
3, 2, 1	$\frac{1}{6}$	0	0	400

Thus, the Shapley value concept suggests that the Piper Cub pay $33.33, the DC-10 pay $58.33, and the 707 pay $308.33.

In general, even if more than one plane of each type lands, it has been shown that the Shapley value for the airport problem allocates runway operating cost as follows: All planes that use a portion of the runway should divide equally the cost of that portion of the runway (see Littlechild and Owen (1973)). Thus, all planes should cover the cost of the first 100 yd of runway, the DC-10s and 707s should pay for the next $150 - 100 = 50$ yd of runway, and the 707s should pay for the last $400 - 150 = 250$ yd of runway. If there were ten Piper Cub landings, five DC-10 landings, and two 707 landings, the Shapley value concept would recommend that each Piper Cub pay $\frac{100}{10+5+2} = \$5.88$ in landing fees, each DC-10 pay $\$5.88 + \frac{150-100}{5+2} = \$13.03$, and each 707 pay $\$13.03 + \frac{400-150}{2} = \$138.03$.

PROBLEMS

Group A

1 Consider the four-player game with the following characteristic function:

$$v(\{1, 2, 3\}) = v(\{1, 2, 4\}) = v(\{1, 3, 4\})$$
$$= v(\{2, 3, 4\}) = 75$$
$$v(\{1, 2, 3, 4\}) = 100$$
$$v(\{3, 4\}) = 60$$
$$v(S) = 0 \text{ for all other coalitions}$$

Show that this game has an empty core.

2 Show that if $v(\{3, 4\})$ in Problem 1 were changed to 50, then the game's core would consist of a single point.

3 The game of Odd Man Out is a three-player coin toss game in which each player must choose heads or tails. If all the players make the same choice, the house pays each player $1; otherwise, the odd man out pays each of the other players $1.

    a Find the characteristic function for this game.

    b Find the core of this game.

    c Find the Shapley value for this game.

4 Show that for $n = 2$, the core of the garbage game is the imputation $(-1, -1)$.

5 Show that for $n > 2$, the n-player garbage game has an empty core.

6 For the four-player garbage game, find an imputation that dominates $(-1, -1, -1, -1)$.

7 The Gotham City airport runway is 5,000 ft long and costs $100,000 per year to maintain. Last year there were 2,000 landings at the airport. Four types of planes landed. The length of runway required by each type of plane and the number of landings of each type are shown in Table 49. Assuming that the cost of operating a length of runway is proportional to the length of the runway, how much per landing should be paid by each type of plane?

TABLE 49

Type of Plane	Number of Landings	Length of Runway (ft)
1	600	2,000
2	700	3,000
3	500	4,000
4	200	5,000

8 Consider the following three-person game:

$$v(\{\ \}) = 0, \qquad v(\{1\}) = 0.2,$$
$$v(\{2\}) = v(\{3\}) = 0, \quad v(\{1, 2\}) = 1.5,$$
$$v(\{1, 3\}) = 1.6, \qquad v(\{2, 3\}) = 1.8,$$
$$v(\{1, 2, 3\}) = 2$$

a Find the core of this game.

b Find the Shapley value for this game.

c Find an imputation dominating the imputation $(1, \frac{1}{2}, \frac{1}{2})$.

9 Howard Whose has left an estate of $200,000 to support his three ex-wives. Unfortunately, Howard's attorney has determined that each ex-wife needs the following amount of money to take care of Howard's children: wife 1—$100,000; wife 2—$200,000; wife 3—$300,000. Howard's attorney must determine how to divide the money among the three wives. He defines the value of a coalition S of ex-wives to be the maximum amount of money left for the ex-wives in S after all ex-wives not in S receive what they need. Using this definition, construct a characteristic function for this problem. Then determine the core and Shapley value for this game.

10 Indiana University leases WATS lines and is charged according to the following rules: $400 per month for each of the first five lines; $300 per month for each of the next five lines; $100 per month for each additional line. The College of Arts and Sciences makes 150 calls per hour, the School of Business makes 120 calls per hour, and the rest

of the university makes 30 calls per hour. Assume that each line can handle 30 calls per hour. Thus, the university will rent 10 WATS lines. The university wants to determine how much each part of the university should pay for long-distance phone service.

a Set up a characteristic function representation of the problem.

b Use the Shapley value to allocate the university's long-distance phone costs.

11 Three doctors have banded together to form a joint practice: the Port Charles Trio. The overhead for the practice is $40,000 per year. Each doctor brings in annual revenues and incurs annual variable costs as follows: doctor 1—$155,000 in revenue, $40,000 in variable cost; doctor 2—$160,000 in revenue, $35,000 in variable cost; and doctor 3—$140,000 in revenue, $38,000 in variable cost.

The Port Charles Trio wants to use game theory to determine how much each doctor should be paid. Determine the relevant characteristic function and show that the core of the game consists of an infinite number of points. Also determine the Shapley value of the game. Does the Shapley value give a reasonable division of the practice's profits?

Group B

12 Consider an n-person game in which the only winning coalitions are those containing player 1 and at least one other player. If a winning coalition receives a reward of $1, find the Shapley value to each player.

SUMMARY Two-Person Zero-Sum and Constant-Sum Games

John von Neumann and Oskar Morgenstern suggested that two-person zero-sum and constant-sum games be played according to the following basic assumption of two-person zero-sum game theory: Each player chooses a strategy that enables him to do the best he can, given that his opponent *knows the strategy he is following*.

A two-person zero-sum game has a saddle point if and only if

$$\max_{\substack{\text{all} \\ \text{rows}}} (\text{row minimum}) = \min_{\substack{\text{all} \\ \text{columns}}} (\text{column maximum}) \tag{1}$$

If a two-person zero-sum or constant-sum game has a saddle point, then the row player should choose any strategy (row) attaining the maximum on the left side of (1). The column player should choose any strategy (column) attaining the minimum on the right side of (1).

In general, we may use the following method to find the optimal strategies and the value of a two-person zero-sum or constant-sum game:

Step 1 Check for a saddle point. If the game has none, go on to step 2.

Step 2 Eliminate any of the row player's dominated strategies. Looking at the reduced matrix (dominated rows crossed out), eliminate any of the column player's dominated strategies and then those of the row player. Continue until no more dominated strategies can be found. Then proceed to step 3.

Step 3 If the game matrix is now 2×2, solve the game graphically. Otherwise, solve by using the linear programming method in Table 24.

The value of the game and the optimal strategies for the row and column players in the Table 24 reward matrix may be found by solving the row player's LP and the column player's LP, respectively.

The dual of the row (column) player's LP is the column (row) player's LP. The optimal objective function value for either the row or the column player's LP is the value of the game to the row player. If the row player departs from her optimal strategy, then she may receive an expected reward that is less than the value of the game. If the column player departs from his optimal strategy, then he may incur an expected loss that exceeds the value of the game. Complementary slackness may be used to simultaneously solve the row and the column players' LPs.

Two-Person Nonconstant-Sum Games

As in a two-person zero-sum game, a choice of strategy by each player is an **equilibrium point** if neither player can benefit from a unilateral change in strategy.

A two-person nonconstant-sum game of particular interest is Prisoner's Dilemma. If $T > R > P > S$, a reward matrix like the one in Table 36 will be a Prisoner's Dilemma game. For such a game, (NC, NC) (both players choosing a noncooperative action) is an equilibrium point.

n-Person Games

When more than two players are involved, the structure of a competitive situation may be summarized by the **characteristic function.** For each set of players S, the characteristic function v of a game gives the amount $v(S)$ that the members of S can be sure of receiving if they act together and form a coalition.

Let $\mathbf{x} = (x_1, x_2, \ldots, x_n)$ be a vector such that player i receives a reward x_i. We call such a vector a **reward vector.** A reward vector $\mathbf{x} = (x_1, x_2, \ldots, x_n)$ is an **imputation** if and only if

$$v(N) = \sum_{i=1}^{i=n} x_i \qquad \text{(Group rationality)} \qquad (13)$$

$$x_i \geq v(\{i\}) \quad \text{(for each } i \in N) \qquad \text{(Individual rationality)} \qquad (14)$$

The imputation $\mathbf{y} = (y_1, y_2, \ldots, y_n)$ **dominates** $\mathbf{x}$ through a coalition S (written $\mathbf{y} >^s \mathbf{x}$) if

$$\sum_{i \in S} y_i \leq v(S) \qquad \text{and for all } i \in S, \quad y_i > x_i \qquad (15)$$

The **core** and the **Shapley value** are two alternative solution concepts for n-person games. The *core* of an n-person game is the set of all undominated imputations. An imputation $\mathbf{x} = (x_1, x_2, \ldots, x_n)$ is in the core of an n-person game if and only if for each subset S of $N = \{1, 2, \ldots, n\}$

$$\sum_{i \in S} x_i \geq v(S)$$

The Shapley value gives a reward x_i to the ith player, where x_i is given by

$$x_i = \sum_{\substack{\text{all } S \text{ for which} \\ i \text{ is not in } S}} p_n(S)[v(S \cup \{i\}) - v(S)] \qquad (45)$$

In (45),

$$p_n(S) = \frac{|S|!(n - |S| - 1)!}{n!} \qquad (46)$$

Equation (45) implies that player i's reward should be the expected amount that player i adds to the coalition made up of the players who are present when player i arrives.

REVIEW PROBLEMS

Group A

1 Two competing firms are deciding whether to locate a new store at point A, B, or C. There are 52 prospective customers for the two stores. Twenty customers live in village A, 20 customers live in village B, and 12 customers live in village C (see Figure 6). Each customer will shop at the nearer store. If a customer is equidistant from both stores, then assume there is a $\frac{1}{2}$ chance that he or she will shop at either store. Each firm wants to maximize the expected number of customers that will shop at its store. Where should each firm locate its store? ($AB = BC = 10$ miles.)

2 A total of 90,000 customers frequent the Ruby and the Swamp supermarkets. To induce customers to enter, each store gives away a free item. Each week, the giveaway item is announced in the Monday newspaper. Of course, neither store knows which item the other store will choose to give away this week. Ruby's is considering giving away a carton of soda or a half gallon of milk. Swamp's is considering giving away a pound of butter or a half gallon of orange juice. For each possible choice of items, the number of customers who will stop at Ruby's during the current week is shown in Table 50. Each store wants to maximize its expected number of customers during the current week. Use game theory to determine an optimal strategy for each store and the value of the game. Interpret the value of the game.

3 Consider the two-person zero-sum game in Table 51.

   a Write down each player's LP.

   b We are told that player 1's optimal strategy has $x_1 > 0$, $x_2 > 0$, and $x_3 > 0$. Find the value of the game and each player's optimal strategies.

   c Suppose the column player plays the nonoptimal strategy $(\frac{1}{2}, \frac{1}{2}, 0)$. Show how the row player can earn an expected reward that exceeds the value of the game.

4 Find optimal strategies for each player and the value of the two-person zero-sum game in Table 52.

FIGURE 6

20 customers	20 customers	12 customers
● A	● B	● C

TABLE 50

Ruby Chooses	Swamp Chooses	
	Butter	**Orange Juice**
Soda	40,000	50,000
Milk	60,000	30,000

TABLE 51

$\frac{1}{2}$	-1	-1
-1	$\frac{1}{2}$	-1
-1	-1	1

TABLE 52

20	1	2
12	10	4
24	8	-2

5 Airway (a Midwestern department store chain) and Corvett (an Eastern department store chain) are determining whether to expand their geographical bases. The only viable manner by which expansion might be carried out is for a chain to open stores in the other's area. If neither chain expands, then Airway's profits will be $3 million and Corvett's will be $2 million. If Airway expands and Corvett does not, then Airway's profits will be $5 million, and Corvett will lose $2 million. If Airway does not expand and Corvett does, Airway will lose $1 million, and Corvett will earn $4 million. Finally, if both chains expand, Airway will earn $1 million and Corvett will earn $500,000 in profits. Determine the equilibrium points, if any, for this game.

6 The stock in Alden Corporation is held by three people. Person 1 owns 1%, person 2 owns 49%, and person 3 owns 50%. To pass a resolution at the annual stockholders' meeting, 51% of the stock is needed. A coalition receives a reward of 1 if it can pass a resolution and a reward of 0 if it cannot pass a resolution.

   a Find the characteristic function for this game.

   b Find the core of this game.

   c Find the Shapley value for this game.

   d Because $(\frac{1}{3}, \frac{1}{3}, \frac{1}{3})$ is not in the core, there must be an imputation dominating $(\frac{1}{3}, \frac{1}{3}, \frac{1}{3})$. Find one.

Group B

7 In addition to the core and the Shapley value, the stable set is an alternative solution concept for n-person games. A set I of imputations is called a **stable set** if each imputation in I is undominated and every imputation that is not in I is dominated by some member of I. Consider the three-person game in which all zero- and one-member coalitions have a characteristic function value of 0, and each two- and three-player coalition has a value of 1. Show that for this game $I = \{(\frac{1}{2}, \frac{1}{2}, 0), (0, \frac{1}{2}, \frac{1}{2}), (\frac{1}{2}, 0, \frac{1}{2})\}$ is a stable set.

REFERENCES

The following books take an elementary, applications-oriented approach to game theory:

Davis, M. *Game Theory: An Introduction.* New York: Basic Books, 1983.

Dixit, A., and B. Nalebuff, *Thinking Strategically.* New York: Norton, 1991.

McMillian, J. *Games, Strategies, and Managers.* New York: Oxford, 1992.

Poundstone, W. *The Prisoner's Dilemma.* New York: Doubleday, 1992.

Rapoport, A. *Two-Person Game Theory.* Ann Arbor, Mich.: University of Michigan Press, 1973.

The following classics are still worth reading:

Luce, R., and H. Raiffa. *Games and Decisions.* New York: Wiley, 1957.

Von Neumann, J., and O. Morgenstern. *Theory of Games and Economic Behavior.* Princeton, N.J.: Princeton University Press, 1944.

For the more mathematically inclined reader, the next eleven books are recommended:

Dutta, P. *Strategies and Games: Theory and Practice.* Cambridge, Mass.: MIT Press, 1999.

Friedman, J. *Game Theory with Applications to Economics.* New York: Oxford Press, 1990.

Fudenberg, D., and J. Tirole. *Game Theory.* Cambridge, Mass: MIT Press, 1991.

Gibbons, R. *Game Theory for Applied Economists.* Princeton, N.J.: Princeton University Press, 1992.

Gitnis, H. *Game Theory Evolving.* Princeton, N.J.: Princeton University Press, 2000.

Hargreaves, S., and Varoufakis, Y. *Game Theory: A Critical Introduction.* New York: Routledge, 1995.

Osborne, M., and Rubenstein, A. *A Course in Game Theory.* Cambridge, Mass.: MIT Press, 1994.

Owen, G. *Game Theory.* Orlando, Fla.: Academic Press, 1999.

Shubik, M. *Game Theory in the Social Sciences: Concepts and Solutions.* Cambridge, Mass.: MIT Press, 1982.

————. *A Game-Theoretic Approach to Political Economy.* Cambridge, Mass.: MIT Press, 1984.

Thomas, L. C. *Games, Theory and Applications.* Chichester, England: Ellis Horwood, 1986.

Vorobev, N. *Game Theory Lectures for Economists and Social Sciences.* New York: Springer-Verlag, 1977.

Littlechild, S., and G. Owen. "A Simple Expression for the Shapley Value in a Special Case," *Management Science* 20(1973): 370–372. Discusses applications of the Shapley value to airport landings.

"Protectionism Gets Clever," *The Economist* (November 21, 1988): 78.

Shapley, L. "Quota Solutions of *n*-Person Games." In *Contributions to the Theory of Games II,* ed. H. Kuhn and A. Tucker. Princeton, N.J.: Princeton University Press, 1953.

15

Deterministic EOQ Inventory Models

In this chapter, we begin our formal study of inventory modeling. In earlier chapters, we described how linear programming can be used to solve certain inventory problems. Our study of inventory will continue in Chapters 16, 18, and 19.

We begin by discussing some important concepts of inventory models. Then we develop versions of the famous economic order quantity (EOQ) model that can be used to make optimal inventory decisions when demand is deterministic (known in advance). In Chapters 16 and 19, we discuss models in which demand is allowed to be random.

15.1 Introduction to Basic Inventory Models

To meet demand on time, companies often keep on hand stock that is awaiting sale. The purpose of inventory theory is to determine rules that management can use to minimize the costs associated with maintaining inventory and meeting customer demand. Inventory models answer the following questions. (1) When should an order be placed for a product? (2) How large should each order be?

Costs Involved in Inventory Models

The inventory models considered in this book involve some or all of the following costs.

Ordering and Setup Cost

Many costs associated with placing an order or producing a good internally do not depend on the size of the order or on the production run. Costs of this type are referred to as the *ordering and setup cost*. For example, ordering cost would include the cost of paperwork and billing associated with an order. If the product is made internally rather than ordered from an external source, the cost of labor (and idle time) for setting up and shutting down a machine for a production run would be included in the ordering and setup cost.

Unit Purchasing Cost

This is simply the variable cost associated with purchasing a single unit. Typically, the unit purchasing cost includes the variable labor cost, variable overhead cost, and raw material cost associated with purchasing or producing a single unit. If goods are ordered from an external source, the unit purchase cost must include shipping cost.

Holding or Carrying Cost

This is the cost of carrying one unit of inventory for one time period. If the time period is a year, the carrying cost will be expressed in dollars per unit per year. The holding cost usually includes storage cost, insurance cost, taxes on inventory, and a cost due to the possibility of spoilage, theft, or obsolescence. Usually, however, the most significant component of holding cost is the opportunity cost incurred by tying up capital in inventory. For example, suppose that one unit of a product costs $100 and the company can earn 15% annually on its investments. Then holding one unit in inventory for one year is costing the company $0.15(100) = \$15$. When interest rates are high, most firms assume that their annual holding cost is 20%–40% of the unit purchase cost.

Stockout or Shortage Cost

When a customer demands a product and the demand is not met on time, a stockout, or shortage, is said to occur. If customers will accept delivery at a later date (no matter how late that date may be), we say that demands may be **back-ordered.** The case in which back-ordering is allowed is often referred to as the **backlogged demand** case. If no customer will accept late delivery, we are in the **lost sales** case. Of course, reality lies between these two extremes, but by determining optimal inventory policies for both the backlogged demand and the lost sales cases, we can get a ballpark estimate of what the optimal inventory policy should be.

Many costs are associated with stockouts. If back-ordering is allowed, placement of back orders usually results in an extra cost. Stockouts often cause customers to go elsewhere to meet current and future demands, resulting in lost sales and lost goodwill. Stockouts may also cause a company to fall behind in other aspects of its business and may force a plant to incur the higher cost of overtime production. Usually, the cost of a stockout is harder to measure than ordering, purchasing, or holding costs.

In this chapter, we study several versions of the classic economic order quantity (EOQ) model that was first developed in 1915 by F. W. Harris of Westinghouse Corporation. For the models in this chapter to be valid, certain assumptions must be satisfied.

Assumptions of EOQ Models

Repetitive Ordering

The ordering decision is repetitive, in the sense that it is repeated in a regular fashion. For example, a company that is ordering bearing assemblies will place an order, then see its inventory depleted, then place another order, and so on. This contrasts with one-time orders. For example, when a news vendor decides how many Sunday newspapers to order, only one order (per Sunday) will be placed. Problems where an order is placed just once are referred to as single-period inventory problems; these are discussed in Chapter 16.

Constant Demand

Demand is assumed to occur at a known, constant rate. This implies, for example, that if demand occurs at a rate of 1,000 units per year, the demand during any t-month period will be $\frac{1,000t}{12}$.

Constant Lead Time

The lead time for each order is a known constant, L. By the **lead time** we mean the length of time between the instant when an order is placed and the instant at which the order arrives. For example, if $L = 3$ months, then each order will arrive exactly 3 months after the order is placed.

Continuous Ordering

An order may be placed at any time. Inventory models that allow this are called **continuous review models.** If the amount of on-hand inventory is reviewed periodically and orders may be placed only periodically, we are dealing with a **periodic review model.** For example, if a firm reviews its on-hand inventory only at the end of each month and decides at this time whether an order should be placed, we are dealing with a periodic review model. Periodic review models are discussed in Chapters 16, 17, and 18.

Although the Constant Demand and Constant Lead Time assumptions may seem overly restrictive and unrealistic, there are many situations in which the models of this chapter provide good approximations to reality. Models in which demand is not deterministic are discussed in Chapters 16 and 19. Models in which demand is deterministic but occurs at a nonconstant rate have already been reviewed in our discussion of LP inventory models in Chapter 3 and are examined further in Chapter 18.

15.2 The Basic Economic Order Quantity Model

Assumptions of the Basic EOQ Model

For the basic EOQ model to hold, certain assumptions are required (for the sake of definiteness, we assume that the unit of time is one year):

1 Demand is deterministic and occurs at a constant rate.

2 If an order of any size (say, q units) is placed, an ordering and setup cost K is incurred.

3 The lead time for each order is zero.

4 No shortages are allowed.

5 The cost per unit-year of holding inventory is h.

We define D to be the number of units demanded per year. Then assumption 1 implies that during any time interval of length t years, an amount Dt is demanded.

The setup cost K of assumption 2 is in addition to a cost pq of purchasing or producing the q units ordered. Note that we are assuming that the unit purchasing cost p does not depend on the size of the order. This excludes many interesting situations, such as quantity discounts for larger orders. In Section 15.3, we discuss a model that allows quantity discounts.

Assumption 3 implies that each order arrives as soon as it is placed. We relax this assumption later in this section.

Assumption 4 implies that all demands must be met on time; a negative inventory position is not allowed. We relax this assumption in Section 15.5.

Assumption 5 implies that a carrying cost of h dollars will be incurred if 1 unit is held for one year, if 2 units are held for half a year, or if $\frac{1}{4}$ unit is held for four years. In short, if I units are held for T years, a holding cost of ITh is incurred.

Given these five assumptions, the EOQ model determines an ordering policy that minimizes the yearly sum of ordering cost, purchasing cost, and holding cost.

Derivation of Basic EOQ Model

We begin our derivation of the optimal ordering policy by making some simple observations. Since orders arrive instantaneously, we should never place an order when I, the inventory level, is greater than zero; if we place an order when $I > 0$, we are incurring an unnecessary holding cost. On the other hand, if $I = 0$, we must place an order to prevent a shortage from occurring. Together, these observations show that the policy that minimizes yearly costs must place an order whenever $I = 0$. At all instants when an order is placed, we are facing the same situation ($I = 0$). This means that each time we place an order, we should order the same quantity. We let q be the quantity that is ordered each time that $I = 0$.

We now determine the value of q that minimizes annual cost (call it q^*). Let $TC(q)$ be the total annual cost incurred if q units are ordered each time that $I = 0$. Note that

$$TC(q) = \text{annual cost of placing orders} + \text{annual purchasing cost}$$
$$+ \text{annual holding cost}$$

Since each order is for q units, $\frac{D}{q}$ orders per year will have to be placed so that the annual demand of D units is met. Hence

$$\frac{\text{Ordering cost}}{\text{Year}} = \left(\frac{\text{ordering cost}}{\text{order}}\right)\left(\frac{\text{orders}}{\text{year}}\right) = \frac{KD}{q}$$

For all values of q, the per-unit purchasing cost is p. Since we always purchase D units per year,

$$\frac{\text{Purchasing cost}}{\text{Year}} = \left(\frac{\text{purchasing cost}}{\text{unit}}\right)\left(\frac{\text{units purchased}}{\text{year}}\right) = pD$$

To compute the annual holding cost, note that if we hold I units for a period of one year, we incur a holding cost of $(I \text{ units})(1 \text{ year})(h \text{ dollars/unit/year}) = hI$ dollars.

Suppose the inventory level is not constant and varies over time. If the average inventory level during a length of time T is $\bar{I}$, the holding cost for the time period will be $hT\bar{I}$. This idea is illustrated in Figure 1. If we define $I(t)$ to be the inventory level at time t, then during the interval $[0, T]$ the total inventory cost is given by

$$h(\text{area from 0 to } T \text{ under the } I(t) \text{ curve}) = hT\bar{I}$$

The reader may verify that this result holds for the two cases graphed in Figure 1. More formally, $\bar{I}(T)$, the average inventory level from time 0 to time T, is given by

$$\bar{I}(t) = \frac{\int_0^T I(t)dt}{T}$$

and the total holding cost incurred between time 0 and time T is

$$\int_0^T hI(t)dt = hT\bar{I}(T)$$

To determine the annual holding cost, we need to examine the behavior of I over time. Assume that an order of size q has just arrived at time 0. Since demand occurs at a rate of D per year, it will take $\frac{q}{D}$ years for inventory to reach zero again. Since demand during any period of length t is Dt, the inventory level over any time interval will decline along a straight line of slope $-D$. When inventory reaches zero, an order of size q is placed and arrives instantaneously, raising the inventory level back to q. Given these observations, Figure 2 describes the behavior of I over time.

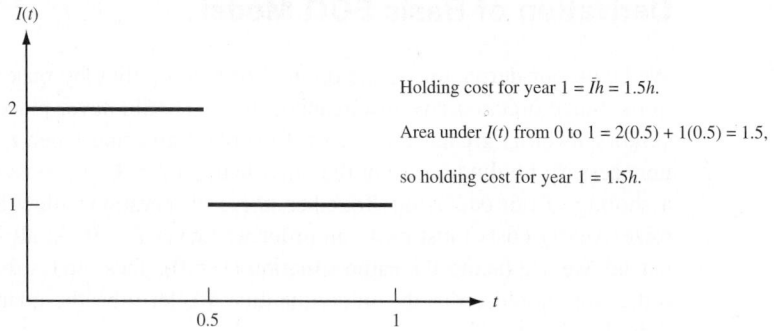

Holding cost for year 1 = $\bar{I}h$ = 1.5h.

Area under $I(t)$ from 0 to 1 = 2(0.5) + 1(0.5) = 1.5,

so holding cost for year 1 = 1.5h.

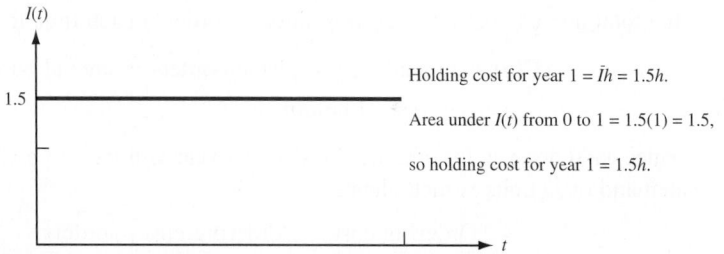

Holding cost for year 1 = $\bar{I}h$ = 1.5h.

Area under $I(t)$ from 0 to 1 = 1.5(1) = 1.5,

so holding cost for year 1 = 1.5h.

FIGURE 1
Holding Cost and
Average Inventory Level

A key concept in the study of EOQ models is the idea of a cycle.

DEFINITION ■ Any interval of time that begins with the arrival of an order and ends the instant before the next order is received is called a **cycle.** ■

Observe that Figure 2 simply consists of repeated cycles of length $\frac{q}{D}$. Hence, each year will contain

$$\frac{1}{\frac{q}{D}} = \frac{D}{q}$$

cycles. The average inventory during any cycle is simply half of the maximum inventory level attained during the cycle. This result will hold in any model for which demand occurs at a constant rate and no shortages are allowed. Thus, for our model, the average inventory level during a cycle will be $\frac{q}{2}$ units.

We are now ready to determine the annual holding cost. We write

$$\frac{\text{Holding cost}}{\text{Year}} = \left(\frac{\text{holding cost}}{\text{cycle}}\right)\left(\frac{\text{cycles}}{\text{year}}\right)$$

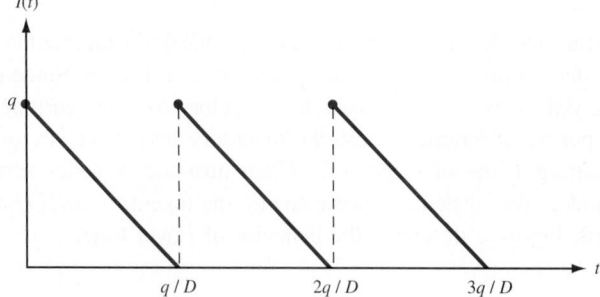

FIGURE 2
Behavior of $I(t)$ in
Basic EOQ Model

Since the average inventory level during each cycle is $\frac{q}{2}$ and each cycle is of length $\frac{q}{D}$,

$$\frac{\text{Holding cost}}{\text{Cycle}} = \frac{q}{2}\left(\frac{q}{D}\right)h = \frac{q^2 h}{2D}$$

Then

$$\frac{\text{Holding cost}}{\text{Year}} = \frac{q^2 h}{2D}\left(\frac{D}{q}\right) = \frac{hq}{2}$$

Combining ordering cost, purchasing cost, and holding cost, we obtain

$$TC(q) = \frac{KD}{q} + pD + \frac{hq}{2}$$

To find the value of q that minimizes $TC(q)$, we set $TC'(q)$ equal to zero. This yields

$$TC'(q) = -\frac{KD}{q^2} + \frac{h}{2} = 0 \tag{1}$$

Equation (1) is satisfied for $q = \pm(2\,KD/h)^{1/2}$. Since $q = -(2\,KD/h)^{1/2}$ makes no sense, let's hope that the **economic order quantity**, or EOQ,

$$q^* = \left(\frac{2KD}{h}\right)^{1/2} \tag{2}$$

minimizes $TC(q)$. Since $TC''(q) = 2KD/q^3 > 0$ for all $q > 0$, we know that $TC(q)$ is a convex function. Then Theorem 1′ of Chapter 11 implies that any point where $TC'(q) = 0$ will minimize $TC(q)$. Thus, q^* does indeed minimize total annual cost.

REMARKS   **1**   The EOQ does not depend on the unit purchasing price p, because the size of each order does not change the unit purchasing cost. Thus, the total annual purchasing cost is independent of q. In Section 15.3, we discuss models in which the size of the order changes the unit purchasing cost.
2   Since each order is for q^* units, a total of $\frac{D}{q^*}$ orders must be placed during each year.
3   To see whether the EOQ formula is reasonable, let's see how changes in certain parameters change q^*. For example, as K increases, we would expect the number of orders placed each year, $\frac{D}{q^*}$, to decrease. Equivalently, we would expect an increase in K to increase q^*. A glance at (2) shows that this is indeed the case. Analogously, an increase in h makes it more costly to hold inventory, so we would expect an increase in h to reduce the average inventory level, $\frac{q^*}{2}$. Equation (2) shows that an increase in h does reduce q^*; it also shows that the ratio of the ordering cost to the holding cost is the critical factor in determining q^*. For example, if both K and h are doubled, q^* remains unchanged. Also note that q^* is proportional to $D^{1/2}$. Thus, quadrupling demand will only double q^*.
4   It is not difficult to show that if the EOQ is ordered, then

$$\frac{\text{Holding cost}}{\text{Year}} = \frac{\text{ordering cost}}{\text{year}} \tag{3}$$

To prove this, note that

$$\frac{\text{Holding cost}}{\text{Year}} = \frac{hq^*}{2} = \frac{h}{2}\left(\frac{2KD}{h}\right)^{1/2} = \left(\frac{KDh}{2}\right)^{1/2}$$

$$\frac{\text{Ordering cost}}{\text{Year}} = \frac{KD}{q^*} = \frac{KD}{\left(\dfrac{2KD}{h}\right)^{1/2}} = \left(\frac{KDh}{2}\right)^{1/2}$$

Figure 3 illustrates the trade-off between holding cost and ordering cost. The figure confirms the fact that at q^*, the annual holding and ordering costs are the same.

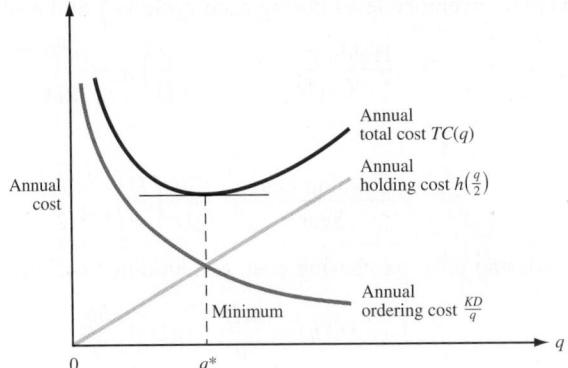

FIGURE 3
Trade-Off between Holding Cost and Ordering Cost

We illustrate the use of the EOQ formula with the following example.

EXAMPLE 1 **Braneast Airlines**

Braneast Airlines uses 500 taillights per year. Each time an order for taillights is placed, an ordering cost of $5 is incurred. Each light costs 40¢, and the holding cost is 8¢/light/year. Assume that demand occurs at a constant rate and shortages are not allowed. What is the EOQ? How many orders will be placed each year? How much time will elapse between the placement of orders?

Solution We are given that $K = \$5$, $h = \$0.08$/light/year, and $D = 500$ lights/year. The EOQ is

$$q^* = \left(\frac{2(5)(500)}{0.08}\right)^{1/2} = 250$$

Hence, the airline should place an order for 250 taillights each time that inventory reaches zero.

$$\frac{\text{Orders}}{\text{Year}} = \frac{D}{q^*} = \frac{500}{250} = \frac{2 \text{ orders}}{\text{year}}$$

The time between placement (or arrival) of orders is simply the length of a cycle. Since the length of each cycle is $\frac{q^*}{D}$, the time between orders will be

$$\frac{q^*}{D} = \frac{250}{500} = \frac{1}{2} \text{ year}$$

Sensitivity of Total Cost to Small Variations in the Order Quantity

In most situations, a slight deviation from the EOQ will result in only a slight increase in costs. For Example 1, let's see how deviations from the EOQ change the total annual cost. Since annual purchasing cost is unaffected by the order quantity, we focus our attention on how the annual holding and ordering costs are affected by changes in the order quantity. Let

$$HC(q) = \text{annual holding cost if the order quantity is } q$$
$$OC(q) = \text{annual ordering cost if the order quantity is } q$$

TABLE 1
Cost Calculations for Figure 4

q	HC(q)	OC(q)	HC(q) + OC(q)
50	2.0	50.00	52.00
100	4.0	25.00	29.00
150	6.0	16.67	22.67
200	8.0	12.50	20.50
220	8.8	11.36	20.16
240	9.6	10.42	20.02
250	10.0	10.00	20.00
260	10.4	9.62	20.02
280	11.2	8.93	20.13
300	12.0	8.33	20.33
350	14.0	7.14	21.14
400	16.0	6.25	22.25

We find that

$$HC(q) = \tfrac{1}{2}(0.08q) = 0.04q \qquad OC(q) = 5\left(\frac{500}{q}\right) = \frac{2,500}{q}$$

Using the information in Table 1, we obtain the sketch of $HC(q) + OC(q)$ given in Figure 4. The figure shows that $HC(q) + OC(q)$ is very flat near q^*. For example, ordering 20% more than the EOQ ($q = 300$) raises $HC(q) + OC(q)$ from 20 to 20.33 (an increase of under 2%).

The flatness of the $HC(q) + OC(q)$ curve is important, because it is often difficult to estimate h and K. Inaccurate estimation of h and K may result in a value of q that differs slightly from the actual EOQ. The flatness of the $HC(q) + OC(q)$ curve indicates that even a moderate error in the determination of the EOQ will only increase costs by a slight amount.

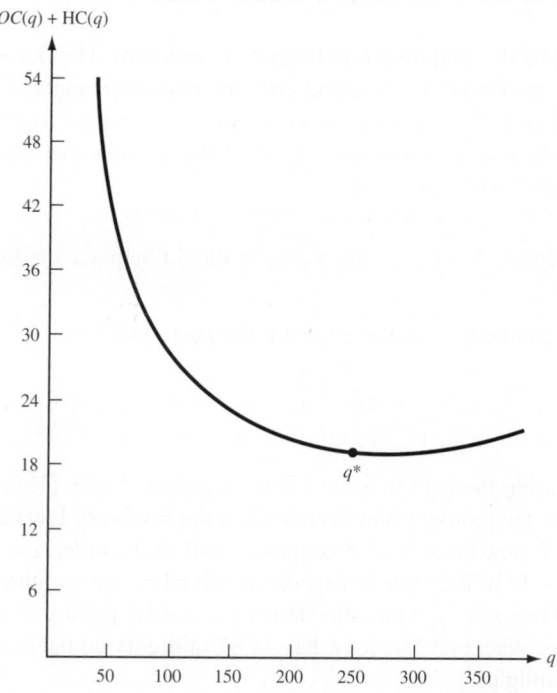

FIGURE 4
OC(q) + HC(q) for Braneast Example

Determination of EOQ When Holding Cost Is Expressed in Terms of Dollar Value of Inventory

Often, the annual holding cost is expressed in terms of the cost of holding one dollar's worth of inventory for one year. Suppose that h_d = cost of holding one dollar in inventory for one year. Then the cost of holding one unit of inventory for one year will be ph_d, and (2) may be written as

$$q^* = \left(\frac{2KD}{ph_d}\right)^{1/2} \tag{4}$$

EXAMPLE 2 Ordering Cameras

A department store sells 10,000 cameras per year. The store orders cameras from a regional warehouse. Each time an order is placed, an ordering cost of $5 is incurred. The store pays $100 for each camera, and the cost of holding $1 worth of inventory for a year is estimated to be the annual capital opportunity cost of 20¢. Determine the EOQ.

Solution We are given that $K = \$5$, $D = 10,000$ cameras per year, $h_d = 20$¢/dollar/year, and $p = \$100$ per camera. Then

$$q^* = \left(\frac{2(5)(10,000)}{(100)(0.20)}\right)^{1/2} = (5,000)^{1/2} = 70.71 \text{ cameras}$$

Hence, the EOQ recommends that the store order 70.71 cameras each time the inventory level reaches zero. Of course, the number of cameras ordered must be an integer. Since $TC(q)$ is a convex function of q, either $q = 70$ or $q = 71$ must minimize $TC(q)$. (If this seems difficult to believe, look at Figure 4.) Because of the flatness of the $HC(q) + OC(q)$ curve, it doesn't really matter whether the store chooses to order 70 or 71 cameras.

The Effect of a Nonzero Lead Time

We now allow the lead time L to be greater than zero. The introduction of a nonzero lead time leaves the annual holding and ordering costs unchanged. Hence, the EOQ still minimizes total costs. To prevent shortages from occurring and to minimize holding cost, each order must be placed at an inventory level that ensures that when each order arrives, the inventory level will equal zero.

DEFINITION ■ The inventory level at which an order should be placed is the **reorder point.** ■

To determine the reorder point for the basic EOQ model, two cases must be considered.

Case 1

Demand during the lead time does not exceed the EOQ. (This means that $LD \leq$ EOQ.) In this case, the reorder point occurs when the inventory level equals LD. Then the order will arrive L time units later, and upon arrival of the order, the inventory level will equal $LD - LD = 0$. In Example 1, suppose that it takes one month for a shipment of taillights to arrive. Then $L = \frac{1}{12}$ year, and Braneast's reorder point will be $(\frac{1}{12})(500) = 41.67$ taillights. Thus, whenever Braneast has 41.67 taillights on hand, an order should be placed for more taillights.

Case 2

Demand during the lead time exceeds the EOQ. (This means that $LD > $ EOQ.) In this case, the reorder point does not equal LD. Suppose that in Example 1, $L = 15$ months. Then $LD = (15/12)500 = 625$ taillights. Why can't we place an order each time the inventory level reaches 625 taillights? Since the EOQ = 250, our inventory level will never reach 625. To determine the correct reorder point, observe that orders are placed every six months. Suppose that an order has just arrived at time 0. Then an order must have been placed $L = 15$ months ago (at $T = -15$ months). Since orders arrive every six months, orders must be placed at $T = -9$ months, $T = -3$ months, $T = 3$ months, and so on. Since at $T = 0$ an order has just arrived, our inventory level at $T = 0$ is 250. Then at $T = 3$ (or any other point when an order is placed), the inventory level will equal $250 - (3/12)(500) = 125$. Thus, the reorder point is 125 taillights.

In general (see Problem 15), it can be shown that the reorder point equals the remainder when LD is divided by the EOQ. Thus, in our example, the reorder point is the remainder when 625 is divided by 250. This again yields a reorder point of 125 taillights.

The determination of the reorder point becomes extremely important when demand is random and stockouts can occur. In Sections 16.6 and 16.7, we discuss the problem of determining the reorder point when demand is random.

We close this section by giving an example of a noninventory problem that can be solved with the reasoning that we used to develop the EOQ.

EXAMPLE 3 **Bus Service**

Each hour, D students want to ride a bus from the student union to Fraternity Row. The administration places a value of h dollars on each hour that a student is forced to wait for a bus. It costs the university K dollars to send a bus from the student union to Fraternity Row. Assuming that demand occurs at a constant rate, how many buses should be sent each hour from the student union to Fraternity Row?

Solution Note that

$$\frac{\text{Total cost}}{\text{Hour}} = \frac{\text{cost of sending buses}}{\text{hour}} + \frac{\text{student waiting cost}}{\text{hour}}$$

Since demand occurs at a constant rate, buses should leave at regular intervals. This means that each bus that arrives at the student union will find the same number of students waiting. Let $q = $ number of students present when each bus arrives. Assuming that a bus has just arrived at time 0, "number of students waiting" displays the behavior shown in Figure 5. Then

$$\frac{\text{Cost of sending buses}}{\text{Hour}} = \left(\frac{K \text{ dollars}}{\text{bus}}\right)\left(\frac{\frac{D}{q}\text{buses}}{\text{hour}}\right) = \frac{\frac{KD}{q} \text{ dollars}}{\text{hour}}$$

From Figure 5, the average number of students waiting is $\frac{q}{2}$. Then

$$\frac{\text{Student waiting cost}}{\text{Hour}} = \left(\frac{q}{2} \text{ students}\right)\left(\frac{h \text{ dollars/student}}{\text{hour}}\right) = \frac{\frac{hq}{2} \text{ dollars}}{\text{hour}}$$

These computations show that

$$\frac{\text{Total cost}}{\text{Hour}} = \frac{hq}{2} + \frac{KD}{q}$$

This is identical to $HC(q) + OC(q)$ for the basic EOQ model. Hence, the optimal value of q for our busing problem is simply the EOQ. This means that the optimal value of q is

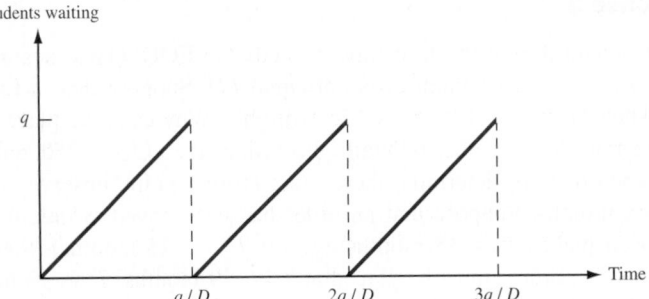

FIGURE 5
Evolution over Time of
Students Waiting
(Example 3)

$q^* = (\frac{2KD}{h})^{1/2}$. Since each bus picks up q^* students, $\frac{D}{q^*}$ buses should be sent each hour. From Figure 5, we see that the time between buses will be $\frac{q^*}{D}$ hours. For example, if $h =$ \$5/student/hour, $D = 100$ students/hour, and $K = $10/bus, we find that

$$q^* = \left(\frac{2(10)(100)}{5}\right)^{1/2} = 20$$

Then $\frac{100}{20} = 5$ buses/hour will leave the student union, and a bus will leave the student union every $\frac{1}{5}$ hour = 12 minutes.

Spreadsheet Template for the Basic EOQ Model

EOQ.xls

Figure 6 (file EOQ.xls) illustrates an Excel template for the basic EOQ model. The user inputs the values of K, h (say, per year), lead time (L), and D (again, per year). Cell A5 has been given the range name K; cell B5, H; cell C5, D; and cell A11, L. In A8, the EOQ is determined by the formula (2*K*D/H)^.5. In B8, we compute annual holding costs with the formula .5*A8*H. In D5, we compute orders per year for the EOQ with the formula =D/A8. In C8, we compute annual ordering costs for the EOQ with the formula =K*D5. In D8, we compute total annual cost for the EOQ with the formula =B8+C8. In B11, we compute the reorder point with the formula =**MOD**(L*D,A8). This yields the remainder obtained when L*D is divided by the EOQ. In Figure 6, we have input the data values for Example 1.

A	A	B	C	D
1	SIMPLE			
2	EOQ			
3	MODEL			
4	K	h	D	ORDERS/YR
5	5	0.08	500	2
6				
7	EOQ	HOLDING COSTS	ORDERING COSTS	TOTAL COST
8	250	10	10	20
9				
10	LEADTIME	REORDER POINT		
11	1.25	125		

FIGURE 6
Simple EOQ Model

Power-of-Two Ordering Policies

Suppose a company orders three products, and the EOQs for each product yield times between orders of 3.5 days, 5.6 days, and 9.2 days. It would rarely be the case that orders for different products would arrive on the same day. If we could somehow synchronize our reorder intervals so that orders for different products often arrived on the same day, we could greatly reduce our coordination costs. For example, we would need far fewer trucks to deliver our orders if we could synchronize their arrival. Roundy (1985) devised an elegant and simple method called **power-of-two ordering policies** to ensure that orders for multiple products are well synchronized. Let $q^* = $ EOQ. Then the optimal reorder interval for a product is $t^* = q^*/D$. We assume t^* is at least 1 day. Then for some $m \geq 0$, it must be true that $2^m \leq t^* \leq 2^{m+1}$. If $t^* \leq \sqrt{2} * 2^m$, we choose a reorder quantity corresponding to a reorder interval of 2^m. If $t^* \geq \sqrt{2} * 2^m$, we choose a reorder quantity corresponding to a reorder interval of 2^{m+1}. Roundy proved that using this method (called a *power-of-two* policy) to round the reorder interval to a neighboring power of 2 will increase the sum of fixed and holding costs at most 6%. The virtue of a power-of-two policy is that different products will frequently arrive at the same time. In many circumstances, this will greatly reduce coordination costs. For example, consider our three products with reorder intervals of 3.5 days, 5.6 days, and 9.2 days. Roundy's power-of-two policy would choose order quantities corresponding to reorder periods of 4, 4, and 8 days, respectively. Thus, products 1 and 2 always arrive together; half the time, product 3 arrives with product 2. In most circumstances, this policy will reduce coordination costs by more than the maximum possible 6% increase in total cost. We now give a proof of Roundy's result.

To begin, pick an arbitrary order quantity q' and define the total cost for this order quantity by

$$TC(q') = \frac{hq'}{2} + \frac{KD}{q'}$$

Then

$$\frac{TC(q')}{TC(q^*)} = \frac{\dfrac{hq'}{2} + \dfrac{KD}{q'}}{\sqrt{2KhD}} = \frac{q'}{2}\sqrt{\frac{h^2}{2KDh}} + \frac{1}{q'}\sqrt{\frac{K^2D^2}{2KDh}}$$

$$= \frac{q'}{2}\sqrt{\frac{h}{2KD}} + \frac{1}{2q'}\sqrt{\frac{2KD}{h}}$$

$$= \frac{q'}{2q^*} + \frac{q^*}{2q'}$$

$$= \frac{1}{2}\left(\frac{q'}{q^*} + \frac{q^*}{q'}\right)$$

Since $t^* = \dfrac{q^*}{D}$ and $t' = \dfrac{q'}{D}$, we find that

$$\frac{TC(t')}{TC(t^*)} = \frac{1}{2}\left(\frac{t'}{t^*} + \frac{t^*}{t'}\right) \tag{5}$$

We can now prove Roundy's result. We assume that t^* is at least 1 day. Then for some nonnegative integer m, $2^m \leq t^* \leq 2^{m+1}$.

THEOREM 1

If $t^* \leq 2^m(\sqrt{2})$, then the minimum-cost power-of-two ordering policy is to set $t = 2^m$. If $t^* \geq 2^m(\sqrt{2})$, then the minimum-cost power-of-two ordering policy is 2^{m+1}.

In either case, the total cost of the optimal power-of-two ordering policy will never be more than 6% higher than the total cost of the EOQ.

Proof Since $TC''(q) > 0$, we know that $TC(q)$ is a convex function of q. The convexity of $TC(q)$ implies that the optimal power-of-two reorder time interval is either 2^m or 2^{m+1}. From (5), 2^m will be the optimal power-of-two reorder time interval if and only if

$$\frac{1}{2}\left(\frac{2^m}{t^*} + \frac{t^*}{2^m}\right) \leq \frac{1}{2}\left(\frac{2^{m+1}}{t^*} + \frac{t^*}{2^{m+1}}\right) \tag{6}$$

Inequality (6) will hold if and only if

$$\frac{t^*}{2^{m+1}} \leq \frac{2^m}{t^*}$$

or $t^* \leq \sqrt{2}(2^m)$. We have now shown that if $t^* \leq 2^m(\sqrt{2})$, then the minimum-cost power-of-two ordering policy is to set $t = 2^m$. If $t^* \geq 2^m(\sqrt{2})$, then the minimum-cost power-of-two ordering policy is 2^{m+1}. This result shows that the optimal power-of-two ordering policy must choose a reorder time in the interval $\left[\frac{t^*}{\sqrt{2}}, \sqrt{2}t^*\right]$.

From (5), we now find that the maximum discrepancy between the total cost for the power-of-two ordering policy and the total cost for t^* will occur if the power-of-two reorder interval equals either $\sqrt{2}t^*$ or $\frac{t^*}{\sqrt{2}}$. In either case,

$$\frac{TC\left(\sqrt{2}t^* \text{ or } \frac{t^*}{\sqrt{2}}\right)}{TC(t^*)} = \frac{1}{2}\left(\frac{1}{\sqrt{2}} + \sqrt{2}\right) = 1.06$$

Thus, a power-of-two policy cannot cause an increase in total cost of more than 6%.

PROBLEMS

Group A

1 Each month, a gas station sells 4,000 gallons of gasoline. Each time the parent company refills the station's tanks, it charges the station $50 plus 70¢, per gallon. The annual cost of holding a gallon of gasoline is 30¢.

a How large should the station's orders be?

b How many orders per year will be placed?

c How long will it be between orders?

d Would the EOQ assumptions be satisfied in this situation? Why or why not?

e If the lead time is two weeks, what is the reorder point? If the lead time is ten weeks, what is the reorder point? Assume 1 week = $\frac{1}{52}$ year.

2[†] Money in my savings account earns interest at 10% annual rate. Each time I go to the bank, I waste 15 minutes in line. My time is worth $10 per hour. During each year, I need to withdraw $10,000 to pay my bills.

a How often should I go to the bank?

b Each time I go to the bank, how much money should I withdraw?

c If my need for cash increases, will I go to the bank more often or less often?

d If interest rates rise, will I go to the bank more often or less often?

e If the bank adds more tellers, will I go to the bank more often or less often?

3[‡] Father Dominic's Pizza Parlor receives 30 calls per hour for delivery of pizza. It costs Father Dominic's $10 to send out a truck to deliver pizzas. It is estimated that each minute a customer spends waiting for a pizza costs the pizza parlor 20¢ in lost future business.

a How often should Father Dominic's send out a truck?

b What would be the answer if a truck could only carry five pizzas?

[†]Based on Baumol (1952).

[‡]Based on Ignall and Kolesar (1972).

4 The efficiency of an inventory system is often measured by the **turnover ratio.** The turnover ratio (TR) is defined by

$$TR = \frac{\text{cost of goods sold during a year}}{\text{average value of on-hand inventory}}$$

a Does a high turnover ratio indicate an efficient inventory system?

b If the EOQ model is being used, determine TR in terms of K, D, h, and q.

c Suppose D is increased. Show that TR will also be increased.

5 Suppose we order three types of appliances for the appliance store Ohm City. The optimal reorder intervals are 9.2 days, 21.2 days, and 38.1 days. What would be the optimal power-of-two ordering policy?

6 Suppose we order three types of clothing for Ceiling Mart. The optimal reorder intervals are 92 days, 21 days, and 60 days. What would be the optimal power-of-two ordering policy?

Group B

7 Suppose we are ordering computer chips. Suppose that in each order, exactly 10% of all chips are defective. As soon as the order arrives, we find out which chips are defective and return them for a complete refund. What would be the optimal ordering policy in this situation?

8 Show that for $q \le q^*$, an order size of $q + q^*$ will have a lower cost than an order size of $q - q^*$. What is the managerial significance of this result?

9 Suppose that instead of ordering the EOQ q^*, we use the order quantity $0.8q^*$. Use Equation (3) to show that $HC(q) + OC(q)$ will have increased by 2.50%.

10 In terms of K, D, and h, what is the average length of time that an item spends in inventory before being used to meet demand? Explain how this result can be used to characterize a fast- or slow-moving item.

11 A drug store sells 30 bottles of antibiotics per week. Each time it orders antibiotics, there is a fixed ordering cost of $10 and a cost of $10/bottle. Assume that the annual holding cost is 20% of the cost of a bottle of antibiotics, and suppose antibiotics spoil and cannot be sold if they spend more than one week in inventory. When the drug store places an order, how many bottles of antibiotics should be ordered?

12 During each year, CSL Computer Company needs to train 27 service representatives. No matter how many students are trained, it costs $12,000 to run a training program. Since service reps earn a monthly salary of $1,500,

CSL does not want to train them before they are needed. Each training session takes one month.

a State the assumptions needed for the EOQ model to be applicable.

b How many service representatives should be in each training group?

c How many training programs should CSL undertake each year?

d How many trained service reps will be available when each training program begins?

13 A newspaper has 500,000 subscribers who pay $4 per month for the paper. It costs the company $200,000 to bill all its customers. Assume that the company can earn interest at a rate of 20% per year on all revenues. Determine how often the newspaper should bill its customers. (*Hint:* Look at unpaid subscriptions as the inventoried good.)

14 Consider a firm that knows that the price of the product it is ordering is going to increase permanently by $$X$. How much of the product should be ordered before the price increase goes into effect?

Here is one approach to this question: Suppose the firm orders Q units before the price increase goes into effect.

a What extra holding cost is incurred by ordering Q units now?

b How much in purchasing costs is saved by ordering Q units now?

c What value of Q maximizes purchasing cost savings less extra holding costs?

d Suppose that annual demand is 1,000 units, holding cost per unit-year is $7.50, and the price of the item is going to increase by $10. How large an order should be placed before the price increase goes into effect?

15 Show that the reorder point in the EOQ model equals the remainder when LD is divided by the EOQ.

16 The borough of Staten Island has two "sanitation districts." In district 1, street litter piles up at an average rate of 2,000 tons per week, and in district 2 at an average rate of 1,000 tons per week. Each district has 500 miles of streets. Staten Island has 10 sanitation crews and each crew can clean 50 miles per week of streets. To minimize the average level of the total amount of street litter in the two districts, how often should each district be cleaned? Assume that litter in a district grows at a constant rate until it is picked up (assume pickup is instantaneous). (*Hint:* Let p_i equal the average number of times that each district is cleaned per week. Then $p_1 + p_2 = 1$.)[†]

[†]Based on Riccio, Miller, and Little (1986).

15.3 Computing the Optimal Order Quantity When Quantity Discounts Are Allowed

Up to now, we have assumed that the annual purchasing cost does not depend on the order size. In Section 15.2, this assumption allowed us to ignore the annual purchasing cost when we computed the order quantity that minimizes total annual cost. In real life, however, suppliers often reduce the unit purchasing price for large orders. Such price reductions are referred to as *quantity discounts.* If a supplier gives quantity discounts, the annual purchasing cost will depend on the order size. If holding cost is expressed as a percentage of an item's purchasing cost, the annual holding cost will also depend on the order size. Since the annual purchasing cost now depends on the order size, we can no longer ignore purchasing cost while trading off holding cost against setup cost. Thus, the approach used in Section 15.2 to find the optimal quantity is no longer valid, and a new approach is needed.

If we let q be the quantity ordered each time an order is placed, the general quantity discount model analyzed in this section may be described as follows:

$$\text{If } q < b_1, \text{ each item costs } p_1 \text{ dollars.}$$
$$\text{If } b_1 \leq q < b_2, \text{ each item costs } p_2 \text{ dollars.}$$
$$\text{If } b_{k-2} \leq q < b_{k-1}, \text{ each item costs } p_{k-1} \text{ dollars.}$$
$$\text{If } b_{k-1} \leq q < b_k = \infty, \text{ each item costs } p_k \text{ dollars.}$$

Since $b_1, b_2, \ldots, b_{k-1}$ are points where a price change (or break) occurs, we refer to b_1, $b_2, \ldots, b_{k-1}$ as **price break points.** Since larger order quantities should be associated with lower prices, we have $p_k < p_{k-1} < \cdots < p_2 < p_1$. The following example illustrates the quantity discount model.

EXAMPLE 4   **Buying Disks**

A local accounting firm in Smalltown orders boxes of floppy disks (10 disks to a box) from a store in Megalopolis. The per-box price charged by the store depends on the number of boxes purchased (see Table 2). The accounting firm uses 10,000 disks per year. The cost of placing an order is assumed to be $100. The only holding cost is the opportunity cost of capital, which is assumed to be 20% per year. For this example, $b_1 = 100$, $b_2 = 300$, $p_1 = \$50.00$, $p_2 = \$49.00$, and $p_3 = \$48.50$.

The example is continued later in this section.

TABLE 2
Purchase Costs for Disks

No. of Boxes Ordered (q)	Price per Box
$0 \leq q < 100$	$50.00
$100 \leq q < 300$	$49.00
$q \geq 300$	$48.50

Before explaining how to find the order quantity minimizing total annual costs, we need the following definitions.

1 $TC_i(q)$ = total annual cost (including holding, purchasing, and ordering costs) if each order is for q units at a price p_i.

2 EOQ_i = quantity that minimizes total annual cost if, for any order quantity, the purchasing cost of the item is p_i.

3 EOQ_i is *admissible* if $b_{i-1} \leq EOQ_i < b_i$.

4 $TC(q)$ = actual annual cost if q items are ordered each time an order is placed. (We determine $TC(q)$ by using price p_i if $b_{i-1} \leq q < b_i$.)

Our goal is to find the value of q minimizing $TC(q)$. Figures 7a and 7b illustrate these definitions. Observe that in Figure 7a, EOQ_2 is admissible because $b_1 < EOQ_2 < b_2$, but EOQ_1 and EOQ_3 are not admissible. In each figure, $TC(q)$ is the solid portion of the curve. The dashed portion of each curve represents unattainable costs. For instance, in Figure 7b, $TC_2(q)$ is dotted for $q < b_1$, because the price is not p_2 for $q < b_1$. For $q < b_1$, total annual cost is given by the solid portion of $TC_1(q)$, because for $q < b_1$, the price is p_1, and for $q \geq b_1$, total annual cost is given by the solid portion of $TC_2(q)$.

In general, the value of q minimizing $TC(q)$ can be either a break point (see Figure 7b) or some EOQ_i (see Figure 7a).

The following observations are helpful in determining the point (break point or EOQ_i) that minimizes $TC(q)$.

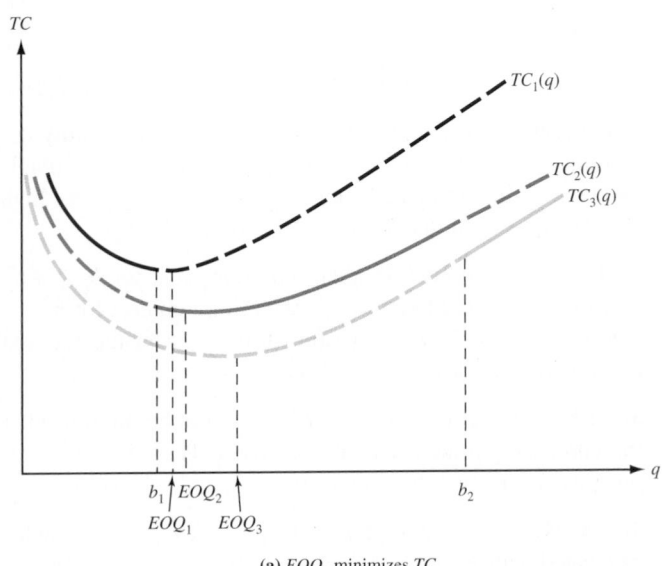

(a) EOQ_2 minimizes TC

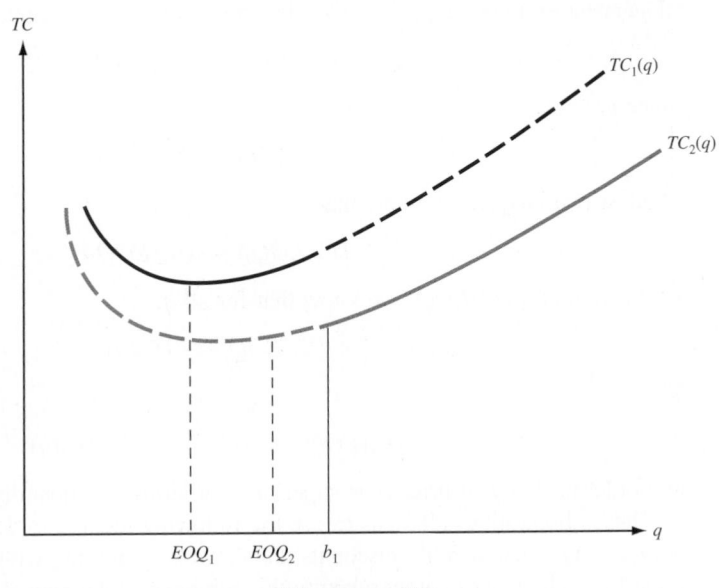

FIGURE 7
Illustrations of Definitions of $TC_i(q)$ and EOQ_i

(b) b_1 minimizes TC

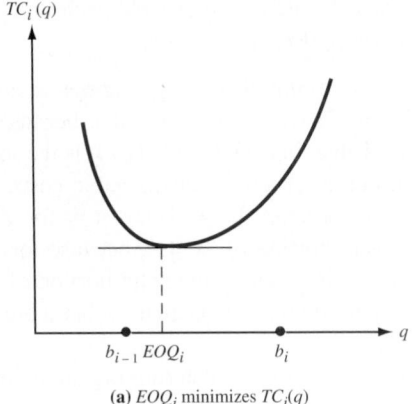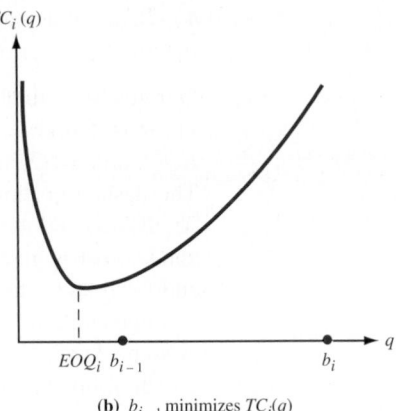

FIGURE 8
For $b_{i-1} \le q < b_i$,
What Value of q
Minimizes $TC_i(q)$?

(a) EOQ_i minimizes $TC_i(q)$ **(b)** b_{i-1} minimizes $TC_i(q)$

1 For any value of q,

$$TC_k(q) < TC_{k-1}(q) < \cdots < TC_2(q) < TC_1(q)$$

This observation is valid because for any order quantity q, $TC_k(q)$ will have the lowest holding and purchasing costs, since p_k is the lowest available price; $TC_1(q)$ will have the highest holding and purchasing costs, because p_1 is the highest available price. Thus, in Figure 7a, we find that $TC_3(q) < TC_2(q) < TC_1(q)$.

2 If EOQ_i is admissible, then minimum cost for $b_{i-1} \le q < b_i$ occurs for $q = EOQ_i$ (see Figure 8a). If $EOQ_i < b_{i-1}$, the minimum cost for $b_{i-1} \le q < b_i$ occurs for $q = b_{i-1}$ (see Figure 8b). This observation follows from the fact that $TC_i(q)$ decreases for $q < EOQ_i$ and increases for $q > EOQ_i$.

3 If EOQ_i is admissible, then $TC(q)$ cannot be minimized at an order quantity for which the purchasing price per item exceeds p_i. Thus, if EOQ_i is admissible, the optimal order quantity must occur for either price $p_i, p_{i+1}, \ldots,$ or p_k.

To see why observation 3 holds, suppose EOQ_i is admissible. Why can't an order quantity associated with a price $p_j > p_i$ have a lower cost than EOQ_i? Note that EOQ_i minimizes total annual cost if price is p_i and EOQ_j does not minimize total annual cost if price is p_i. Thus,

$$TC_i(EOQ_i) < TC_i(EOQ_j)$$

Since $p_j > p_i$,

$$TC_i(EOQ_j) < TC_j(EOQ_j)$$

The last two inequalities show that

$$TC_i(EOQ_i) < TC_j(EOQ_j)$$

By the definition of EOQ_j, we know that for all q,

$$TC_j(EOQ_j) \le TC_j(q)$$

Thus,

$$TC_i(EOQ_i) < TC_j(EOQ_j) \le TC_j(q)$$

and ordering EOQ_i at price p_i is superior to ordering any quantity at a higher price p_j.

These observations allow us to use the following method to determine the optimal order quantity when quantity discounts are allowed. Beginning with the lowest price, determine for each price the order quantity that minimizes total annual costs for $b_{i-1} \le q < b_i$

(call this order quantity q_i^*). Continue determining $q_k^*, q_{k-1}^*, \ldots$ until one of the q_i^*'s (call it q_i^*) is admissible; from observation 2, this will mean that $q_i^* = EOQ_i$. The optimal order quantity will be the member of $\{q_k^*, q_{k-1}^*, \ldots, q_i^*\}$ with the smallest value of $TC(q)$.

EXAMPLE 4   **Buying Disks (Continued)**

Each time an order is placed for disks, how many boxes of disks should be ordered? How many orders will be placed annually? What is the total annual cost of meeting the accounting firm's disk needs?

Solution Note that $K = \$100$ and $D = 1,000$ boxes per year. We first determine the best order quantity for $p_3 = \$48.50$ and $300 \leq q$. Then

$$EOQ_3 = \left(\frac{2(100)(1,000)}{0.2(48.50)} \right)^{1/2} = 143.59$$

Since $EOQ_3 < 300$, EOQ_3 is not admissible. Therefore, Figure 8b is relevant, and for $q \geq 300$, $TC_3(q)$ is minimized by $q_3^* = 300$.

We next consider $p_2 = \$49.00$ and $100 \leq q < 300$. Then

$$EOQ_2 = \left(\frac{2(100)(1,000)}{9.8} \right)^{1/2} = 142.86$$

Since $100 \leq EOQ_2 < 300$, EOQ_2 is admissible, and for a price $p_2 = \$49.00$, the best we can do is to choose $q_2^* = 142.86$; Figure 8a is relevant. Since q_2^* is admissible, $p_1 = \$50.00$ and $0 \leq q < 100$ cannot yield the order quantity minimizing $TC(q)$ (see observation 3). Thus, either $q_2^* = 142.86$ or $q_3^* = 300$ will minimize $TC(q)$. To determine which of these order quantities minimizes $TC(q)$, we must find the smaller of $TC_3(300)$ and $TC_2(142.86)$. For $q_3^* = 300$, the annual holding cost/item/year is $0.20(48.50) = \$9.70$. Thus, for q_3^*,

$$\text{Annual ordering cost} = 100(\tfrac{1,000}{300}) = \$333.33$$
$$\text{Annual purchasing cost} = 1,000(48.50) = \$48,500$$
$$\text{Annual holding cost} = (\tfrac{1}{2})(300)(9.7) = \$1,455$$
$$TC_3(300) = \$50,288.33$$

For $q_2^* = 142.86$, the annual holding cost/item/year is $0.20(49) = \$9.80$. Thus, for q_2^*,

$$\text{Annual ordering cost} = 100(\tfrac{1,000}{142.86}) = \$699.99$$
$$\text{Annual purchasing cost} = 1,000(49) = \$49,000$$
$$\text{Annual holding cost} = (\tfrac{1}{2})(142.86)(9.8) = \$700.01$$
$$TC_2(142.86) = \$50,400$$

Thus, $q_3^* = 300$ will minimize $TC(q)$.

Our analysis shows that each time an order is placed, 300 boxes of disks should be ordered. Then $\frac{1,000}{300} = 3.33$ orders are placed each year. As we have already seen, the minimum total annual cost is $\$50,288.33$.

A Spreadsheet Template for Quantity Discounts

Qd.xls

Figure 9 (file Qd.xls) illustrates how inventory problems with a quantity discount can be solved on a spreadsheet. In cell B2 (given the range name K), we enter K, the cost per order. In cell C2 (range name D) we enter D, the annual demand. In cell D2 (HD), we enter the annual cost of holding $1 of goods in inventory for one year.

FIGURE 9
Quantity Discount Calculations

A	A	B	C	D	E	F
1	QUANTITY	K	D	hperdollar		
2	DISCOUNT	100	1000	0.2		
3	CALCULATIONS					
4						
5	LEFTENDPOINT	RIGHT ENDPOINT	PRICE	EOQ	MINCOSTOQ	MINIMUM COST
6	0	100	$50.00	141.42135624	99.0000	$51,505.10
7	100	300	$49.00	142.85714286	142.8571	$50,400.00
8	300	10000	$48.50	143.59163172	300.0000	$50,288.33

In the cell range A6:C8, we enter (using the data from Example 4) the left-hand endpoint, right-hand endpoint, and price for each interval. Thus, for an order quantity ≥ 0 and < 100 the per-unit price is $50. Now observe that Figure 8 implies that for each interval the minimum cost in that interval is obtained as follows.

1 If the EOQ for the ith interval's price lies in the interval, then the EOQ for that interval obtains the minimum cost in the ith interval.

2 If the EOQ for the ith interval's price is smaller than the left-hand endpoint of the ith interval (b_{i-1}), then the minimum cost for that interval is attained by an order quantity of b_{i-1}. Here we set $b_0 = 0$.

3 If the EOQ for the ith interval's price is larger than the right-hand endpoint for the ith interval (b_i), then the minimum cost for that interval is attained by an order quantity of $b_i - 1$.

Our spreadsheet incorporates this logic as follows: In D6, we compute the EOQ for the interval $b_0 = 0 \leq$ order quantity $< 100 = b_1$ by entering the formula $(2*K*D/(HD*C6))^\wedge.5$. In E6, we enter the formula

$$=IF(AND(D6>=A6,D6<B6),D6,IF(D6<A6,A6,B6-1))$$

This statement computes the order quantity in the first interval that minimizes annual costs by implementing the logic described in (1)–(3) here. In F6, we compute the annual cost corresponding to the order quantity in E6. This is given by $(K*D/E6)+ D*C6+.5*H*D*C6*D6$.

In this formula, the first term is the annual cost of placing orders; the second term is the cost of purchasing one year's demand at the price for the first interval; and the third term is the annual holding cost (whose per-unit cost equals the price of item times the annual holding cost per dollar of inventory). Copying from the range D6:F6 to D6:F8 generates the minimum annual cost for the other two intervals. We see that the minimum annual cost is $50,288.33, and it is attained by an order quantity of 300.

PROBLEMS

Group A

1 A consulting firm is trying to determine how to minimize the annual costs associated with purchasing computer paper. Each time an order is placed an ordering cost of $20 is incurred. The price per box of computer paper depends on q, the number of boxes ordered (see Table 3). The annual holding cost is 20% of the dollar value of inventory. During each month, the consulting firm uses 80 boxes of computer paper. Determine the optimal order quantity and the number of orders placed each year.

2 Each year, Shopalot Stores sells 10,000 cases of soda. The company is trying to determine how many cases should be ordered each time. It costs $5 to process each order, and

TABLE 3

No. of Boxes Ordered	Price per Box
$q < 300$	$10.00
$300 \leq q < 500$	$9.80
$q \geq 500$	$9.70

TABLE 4

No. of Cases Ordered	Price per Case
$q < 200$	$4.40
$200 \leq q < 400$	$4.20
$q \geq 400$	$4.00

the cost of carrying a case of soda in inventory for one year is 20% of the purchase price. The soda supplier offers Shopalot the schedule of quantity discounts shown in Table 4 (q = number of cases ordered per order). Each time an order is placed, how many cases of soda should the company order?

3 A firm buys a product using the price schedule given in Table 5. The company estimates holding costs at 10% of purchase price per year and ordering costs at $40 per order. The firm's annual demand is 460 units.

a Determine how often the firm should order.

b Determine the size of each order.

c At what price should the firm order?

TABLE 5

Order Size	Price per Unit
0–99 units	$20.00
100–199	$19.50
200–499	$19.00
500 or more	$18.75

Group B

4 A hospital orders its thermometers from a hospital supply firm. The cost per thermometer depends on the order size q, as shown in Table 6. The annual holding cost is 25% of the purchasing cost. Let EOQ_{80} be the EOQ if the cost per thermometer is 80¢, and let EOQ_{79} be the EOQ if the cost per thermometer is 79¢.

a Explain why EOQ_{79} will be larger than EOQ_{80}.

b Explain why the optimal order quantity must be either EOQ_{79}, EOQ_{80}, or 100.

c If $EOQ_{80} > 100$, show that the optimal order quantity must be EOQ_{79}.

d If $EOQ_{80} < 100$ and $EOQ_{79} < 100$, show that the optimal order quantity must be either EOQ_{80} or 100.

e If $EOQ_{80} < 100$ and $EOQ_{79} > 100$, show that the optimal order quantity must be EOQ_{79}.

5 In Problem 4, suppose the cost per order is $1 and the monthly demand is 50 thermometers. What is the optimal order quantity? How small a discount could the supplier offer and still have the hospital accept the discount?

TABLE 6

Order Size	Price per Thermometer
$q < 100$	80¢
$q \geq 100$	79¢

15.4 The Continuous Rate EOQ Model

Many goods are produced internally rather than purchased from an outside supplier. In this situation, the EOQ assumption that each order arrives at the same instant seems unrealistic; it isn't possible to produce, say, 10,000 cars at the drop of a hard hat. If a company meets demand by making its own products, the continuous rate EOQ model will be more realistic than the traditional EOQ model. Again, we assume that demand is deterministic and occurs at a constant rate; we also assume that shortages are not allowed.

The continuous rate EOQ model assumes that a firm can produce a good at a rate of r units per time period (we again use one year as the time unit). This means that during any time period of length t, the firm can produce rt units. We define

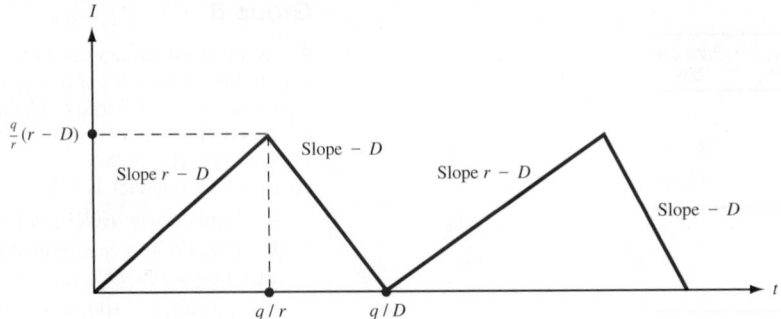

FIGURE 10
Variation of Inventory
for Continuous Rate
EOQ Model

q = number of units produced during each production run

K = cost of setting up a production run
(often due to idle time that occurs at the beginning or end of a production run)

h = cost of holding one unit in inventory for one year

D = annual demand for the product

Assuming that a production run begins at time 0, the variation of inventory over time is described by Figure 10. At the beginning of a production run, we are producing at a rate of r units per year, and demand is occurring at a rate of D units per year. Thus, until q units are produced, inventory increases at a rate of $r - D$ units per year. (Of course, $r \geq D$ must hold, or else demand could not be met.) At time $\frac{q}{r}$, q units will have been produced. At this time, the production run is complete, and inventory decreases at a rate of D units per year until a zero inventory position is reached. A zero inventory level will occur at time $\frac{q}{D}$. Then another production run begins.

Assuming that per-unit production costs are independent of run size, we must determine the value of q that minimizes

$$\frac{\text{Holding cost}}{\text{Year}} + \frac{\text{setup cost}}{\text{year}}$$

Since demand occurs at a constant rate, we know that (average inventory level) = $(\frac{1}{2})$(maximum inventory level). From Figure 10, we see that the maximum inventory level occurs at time $\frac{q}{r}$. Since between zero and $\frac{q}{r}$, the inventory level is increasing at a rate of $r - D$ units per year, the inventory level at time $\frac{q}{r}$ will be $(\frac{q}{r})(r - D)$. Then (average inventory level) = $(\frac{1}{2})(\frac{q}{r})(r - D)$, and

$$\frac{\text{Holding cost}}{\text{Year}} = h(\text{average inventory})(1 \text{ year}) = \frac{h(r - D)q}{2r}$$

Observe that the annual holding cost for the continuous rate EOQ model is the same as that for a conventional EOQ model in which the unit holding cost is $\frac{h(r - D)}{r}$. As usual,

$$\frac{\text{Ordering cost}}{\text{Year}} = \left(\frac{\text{ordering cost}}{\text{cycle}}\right)\left(\frac{\text{cycles}}{\text{year}}\right) = \frac{KD}{q}$$

The discussion shows that

$$\frac{\text{Holding cost}}{\text{Year}} + \frac{\text{ordering cost}}{\text{year}} = \frac{hq(r - D)}{2r} + \frac{KD}{q}$$

The last equation shows that the problem of minimizing the sum of annual holding and ordering costs for the continuous rate model is equivalent to solving an EOQ model with holding cost $\frac{h(r - D)}{r}$, ordering cost K, and annual demand D. Using this observation and the economic order quantity (or lot size) formula (2), we may immediately deduce that for the continuous rate EOQ model,

$$\text{Optimal run size} = \left(\frac{2KD}{\frac{h(r-D)}{r}} \right)^{1/2} = \left(\frac{2KDr}{h(r-D)} \right)^{1/2} \tag{7}$$

As usual, $\frac{D}{q}$ production runs must be made each year to meet the annual demand of D units. Using the fact that

$$\text{EOQ} = \left(\frac{2KD}{h} \right)^{1/2}$$

we may rewrite (7) as

$$\text{Optimal run size} = \text{EOQ} \left(\frac{r}{r-D} \right)^{1/2} \tag{8}$$

As r increases, production occurs at a more rapid rate. Hence, for large r, the rate model should approach the instantaneous delivery situation of the EOQ model. To see that this is the case, note that for r large, $\frac{r}{(r-D)}$ approaches 1. Then (8) shows that as r increases toward infinity, the optimal run size for the continuous rate model approaches the EOQ.

EXAMPLE 5   **Macho Auto Company**

Macho Auto Company needs to produce 10,000 car chassis per year. Each is valued at $2,000. The plant has the capacity to produce 25,000 chassis per year. It costs $200 to set up a production run, and the annual holding cost is 25¢, per dollar of inventory. Determine the optimal production run size. How many production runs should be made each year?

Solution   We are given that

$$r = 25,000 \text{ chassis per year}$$
$$D = 10,000 \text{ chassis per year}$$
$$h = 0.25(\$2,000)/\text{chassis/year} = \$500/\text{chassis/year}$$
$$K = \$200 \text{ per production run}$$

From (7),

$$\text{Optimal run size} = \left(\frac{2(200)(10,000)(25,000)}{500(25,000 - 10,000)} \right)^{1/2} = 115.47$$

Also, $\frac{10,000}{115.47} = 86.60$ production runs will be made each year.

Spreadsheet Template for the Continuous Rate EOQ Model

ConEOQ.xls

Figure 11 (file ConEOQ.xls) illustrates a template for the continuous rate EOQ model. In cell A6, the user inputs K; in B6, h; in C6, D; and in D6, the production rate r. In Figure 11 we have used the parameter values given in Example 5. In A8 (assigned the range name Q), the formula (2*K*D/H)^.5*(R/(R−D))^.5 (again we are using range names) computes the optimal run size. In B8, the formula D/Q computes the number of runs per year. In C8, we compute the annual cost (exclusive of purchasing costs) with the formula (H*Q*(R−D)/(2*R))+K*D/Q. The first term in this formula equals the annual cost of

		A	B	C	D
	1	CONTINUOUS			
	2	RATE			
	3	EOQ			
	4	MODEL			
	5	K	h	D	r
	6	200	500	10000	25000
	7	RUN SIZE	RUNS/YR	COST/YR	
	8	115.47005383793	86.60254	34641.016	

FIGURE 11
Continuous Rate
EOQ Model

holding inventory. This follows because, from Figure 10, the maximum level of inventory during a cycle is $q(r - D)/r$. The second term is the annual cost of placing orders.

PROBLEMS

Group A

1 Show that the optimal run size always exceeds the EOQ. Give an intuitive explanation for this result.

2 A company can produce 100 home computers per day. The setup cost for a production run is $1,000. The cost of holding a computer in inventory for one year is $300. Customers demand 2,000 home computers per month (assume that 1 month = 30 days and 360 days = 1 year). What is the optimal production run size? How many production runs must be made each year?

3 The production process at Father Dominic's Pizza can produce 400 pizza pies per day; the firm operates 250 days per year. Father Dominic's has a cost of $180 per production run and a holding cost of $5 per pizza-year. The pies are frozen immediately after they are produced and stored in a refrigerated warehouse with a current maximum capacity of 2,000 pies.

  a Annual demand is 37,500 pies per year. What production run size should be used?

  b What is the total annual cost incurred in meeting demand?

  c How many days per year will the company be producing pizza pies?

Group B

4 A company has the option of purchasing a good or manufacturing the item. If the item is purchased, the company will be charged $25 per unit plus a cost of $4 per order. If the company manufactures the item, it has a production capacity of 8,000 units per year. It costs $50 to set up a production run, and annual demand is 3,000 units per year. If the annual holding cost is 10% and the cost of manufacturing one unit is $23, determine whether the company should purchase or manufacture the item.

15.5 The EOQ Model with Back Orders Allowed

In many real-life situations, demand is not met on time, and shortages occur. When a shortage occurs, costs are incurred (because of lost business, the cost of placing special orders, loss of future goodwill, and so on). In this section, we modify the EOQ model of Section 15.2 to allow for the possibility of shortages. Let s be the cost of being short one unit for one year. The variables K, D, and h have their usual meanings. In most situations, s is very difficult to measure. We assume that all demand is backlogged and no sales are lost. To determine the order policy that minimizes annual costs, we define

$$q = \text{order quantity}$$
$$q - M = \text{maximum shortage that occurs under an ordering policy}$$

Equivalently (assuming a zero lead time), the firm will be $q - M$ units short each time an order is placed.

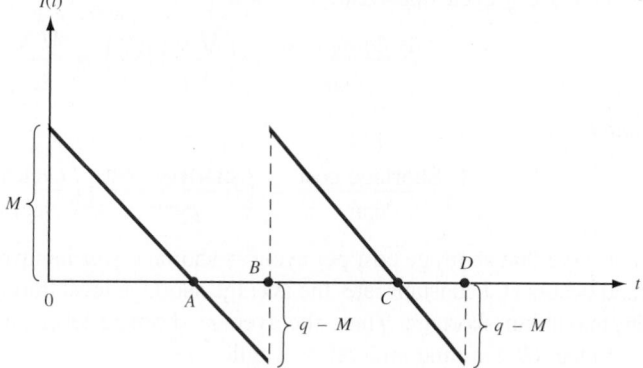

We assume that the lead time for each order is zero. Since an order is placed each time the firm is $q - M$ units short (or when the firm's inventory position is $M - q$), the firm's maximum inventory level will be $M - q + q = M$. For example, if $q = 500$ and $q - M = 100$, we know that an order for 500 units will be used to satisfy the backlogged demand for 100 units and will result in an inventory level of $500 - 100 = 400$ units.

Assuming that an order is placed at time 0, the evolution of the inventory level over time is described by Figure 12. Since purchasing costs do not depend on q and M, we can minimize annual costs by determining the values of q and M that minimize

$$\frac{\text{Holding cost}}{\text{Year}} + \frac{\text{shortage cost}}{\text{year}} + \frac{\text{order cost}}{\text{year}} \tag{9}$$

Notice that what happens between time 0 and time B is identical to what happens between time B and time D. For this reason, we call the time periods $0B$ and BD **cycles.** A cycle may also be thought of as the time interval between placement of orders. To determine holding cost per year and shortage cost per year, we begin by finding holding cost per cycle and shortage cost per cycle. This requires that we find the length of line segments $0A$ and AB in Figure 12. Since a zero inventory level occurs after M units have been demanded, we conclude that $0A = \frac{M}{D}$. Since a cycle ends when q units have been demanded, we conclude that $0B = \frac{q}{D}$. Then

$$\text{Length of } AB = (\text{length of } 0B) - (\text{length of } 0A) = \frac{q - M}{D}$$

Also note that since q units are ordered during each cycle, $\frac{D}{q}$ cycles (and orders) must be placed during each year. We can now express the costs in (9). Recall that

$$\frac{\text{Holding cost}}{\text{Year}} = \left(\frac{\text{holding cost}}{\text{cycle}}\right)\left(\frac{\text{cycles}}{\text{year}}\right)$$

and

$$\frac{\text{Holding cost}}{\text{Cycle}} = \text{holding cost from time 0 to time } A$$

From Figure 12, the average inventory level between time 0 and time A is simply $\frac{M}{2}$. Since $0A$ is of length $\frac{M}{D}$,

$$\frac{\text{Holding cost}}{\text{Cycle}} = \left(\frac{M}{2}\right)\left(\frac{M}{D}\right)h = \frac{M^2 h}{2D}$$

Since there are $\frac{D}{q}$ cycles per year,

$$\frac{\text{Holding cost}}{\text{Year}} = \left(\frac{M^2h}{2D}\right)\left(\frac{D}{q}\right) = \frac{M^2h}{2q}$$

Similarly,

$$\frac{\text{Shortage cost}}{\text{Year}} = \left(\frac{\text{shortage cost}}{\text{cycle}}\right)\left(\frac{\text{cycles}}{\text{year}}\right)$$

Also observe that shortage cost per cycle = shortage cost incurred during time AB. Since demand occurs at a constant rate, the average shortage level during time AB is simply half of the maximum shortage. Thus, the average shortage level on the time interval AB is $\frac{q-M}{2}$. Since AB is a time interval of length $\frac{q-M}{D}$,

$$\frac{\text{Shortage cost}}{\text{Cycle}} = \frac{1}{2}\left(q - M\right)\left(\frac{q-M}{D}\right)s = \frac{(q-M)^2s}{2D}$$

Since there are $\frac{D}{q}$ cycles per year,

$$\frac{\text{Shortage cost}}{\text{Year}} = \frac{(q-M)^2s}{2D}\left(\frac{D}{q}\right) = \frac{(q-M)^2s}{2q}$$

As usual, ordering cost per year $= \frac{KD}{q}$. Let $TC(q, M)$ be the total annual cost (excluding purchasing cost) if our order policy uses parameters q and M. From our discussion, we must choose q and M to minimize

$$TC(q, M) = \frac{M^2h}{2q} + \frac{(q-M)^2s}{2q} + \frac{KD}{q}$$

By using Theorem 3 of Chapter 11, we can show that $TC(q, M)$ is a convex function of q and M. From Theorem 1' and Theorem 7 of Chapter 11, the minimum value of $TC(q, M)$ will occur at the point where

$$\frac{\partial TC}{\partial q} = \frac{\partial TC}{\partial M} = 0$$

Some tedious algebra shows that $TC(q, M)$ is minimized for q^* and M^*:

$$q^* = \left[\frac{2KD(h+s)}{hs}\right]^{1/2} = \text{EOQ}\left(\frac{h+s}{s}\right)^{1/2}$$

$$M^* = \left[\frac{2KDs}{h(h+s)}\right]^{1/2} = \text{EOQ}\left(\frac{s}{h+s}\right)^{1/2}$$

$$\text{Maximum shortage} = q^* - M^*$$

As s approaches infinity, q^* and M^* both approach the EOQ, and the maximum shortage approaches zero. This is reasonable, because if s is large, the cost of a shortage is prohibitive, and we would expect the optimal ordering policy to incur very few, if any, shortages. In other words, if s is very large, we are facing (to all intents and purposes) the no-shortages-allowed situation of Section 15.2.

EXAMPLE 6   Smalltown Optometry Clinic

Each year, the Smalltown Optometry Clinic sells 10,000 frames for eyeglasses. The clinic orders frames from a regional supplier, which charges $15 per frame. Each order incurs an ordering cost of $50. Smalltown Optometry believes that the demand for frames can be backlogged and that the cost of being short one frame for one year is $15 (because of loss of future business). The annual holding cost for inventory is 30¢ per dollar value of inventory. What is the optimal order quantity? What is the maximum shortage that will occur? What is the maximum inventory level that will occur?

Solution We are given that

$$K = \$50$$
$$D = 10,000 \text{ frames per year}$$
$$h = 0.3(15) = \$4.50/\text{frame/year}$$
$$s = \$15/\text{frame/year}$$

Our formula for q^* and M^* now yield

$$q^* = \left(\frac{2(50)(10{,}000)(19.50)}{(4.50)(15)} \right)^{1/2} = 537.48$$

$$M^* = \left(\frac{2(50)(10{,}000)(15)}{(4.50)(19.50)} \right)^{1/2} = 413.45$$

Then the maximum shortage occurring will be $q^* - M^* = 124.03$ frames, and each order should be for 537 or 538 frames. A maximum inventory level of $M^* = 413.45$ frames will occur.

As in Section 15.4, suppose that production is not instantaneous and we can produce at a rate of r units per year. If shortages are allowed, it can be shown that

$$q^* = \left(\frac{2KDr(h+s)}{h(r-D)s} \right)^{1/2}$$

$$M^* = \frac{q^*(r-D)}{r} - \left(\frac{2KD(r-D)h}{sr(h+s)} \right)^{1/2}$$

The maximum shortage occurring in this case (call it S^*) is given by

$$S^* = \left(\frac{2KD(r-D)h}{sr(h+s)} \right)^{1/2}$$

Spreadsheet Template for the EOQ Model with Back Orders

BackEOQ.xls

Figure 13 (file BackEOQ.xls) illustrates a spreadsheet template for the EOQ model with back orders. In cells A6, B6, C6, and D6, we enter the values of K, D, h, and s, respectively, for Example 6. In A8 (given the range name Q), we compute the optimal order quantity with the formula (2*K*D*(H+S)/(H*S))^.5. In B8 (range name M), we com-

FIGURE 13
EOQ Model with Back Orders

A	A	B	C	D
1	EOQ			
2	MODEL			
3	WITH			
4	BACKORDERS			
5	K	D	h	s
6	50	10000	4.5	15
7	q*	M*	MAX SHORT	ANNUAL COST
8	537.48384989	413.44912	124.03473459	1860.52101884

pute the optimal value of M with the formula $(2*K*D*S/(H*(H+S)))^{.5}$. In C8, we compute the maximum shortage with the formula $Q-M$. In D8, we compute the annual total cost $TC(q, M)$ (exclusive of purchasing costs) with the formula $(M^2*H)/(2*Q))+((Q-M)^2*S/2*Q))+(K*D/Q)$.

PROBLEMS

Group A

1 Show that the optimal order quantity for the backlogged demand model is always at least as large as the EOQ but that the maximum inventory level for the backlogged demand model cannot exceed the EOQ.

2 A Mercedes dealer must pay $20,000 for each car purchased. The annual holding cost is estimated to be 25% of the dollar value of inventory. The dealer sells an average of 500 cars per year. He believes that demand is backlogged but estimates that if he is short one car for one year he will lose $20,000 worth of future profits. Each time the dealer places an order for cars, ordering cost amounts to $10,000. Determine the Mercedes dealer's optimal ordering policy. What is the maximum shortage that will occur?

Group B

3 Suppose that instead of measuring shortage in terms of cost per shortage year, a cost of S dollars is incurred for each unit the firm is short. This cost does not depend on the length of time before the backlogged demand is satisfied. Determine a new expression for $TC(q, M)$, and explain how to determine optimal values q^* and M^*.

4 For the model developed in this section, determine

a the average length of time it takes to meet demand for a unit.

b the fraction of all demanded units that are backordered.

15.6 When to Use EOQ Models

Demand is often irregular, or "lumpy." This may be caused by seasonality or other factors. If demand is irregular, the Constant Demand Assumption that was required for all the EOQ models will not be satisfied.

To determine whether the assumption of constant demand is reasonable, suppose that during n periods of time, demands $d_1, d_2, \ldots, d_n$ have been observed. Also, enough is known about future demands to make the assumption of deterministic demand a realistic one. To decide whether demand is sufficiently regular to justify use of EOQ models, Peterson and Silver (1998) recommend that the following computations be done:

1 Determine the estimate $\bar{d}$ of the average demand per period given by

$$\bar{d} = \frac{1}{n} \sum_{i=1}^{i=n} d_i$$

2 Determine an estimate of the variance of the per-period demand D from

$$\text{Est. var } \mathbf{D} = \frac{1}{n} \sum_{i=1}^{i=n} d_i^2 - \bar{d}^2$$

3 Determine an estimate of the relative variability of demand (called the **variability coefficient**). This quantity is labeled VC, where

$$VC = \frac{\text{est. var } \mathbf{D}}{\bar{d}^2}$$

Note that if all the d_i are equal, the estimate of the variance of $\mathbf{D}$ will equal zero. This will also make $VC = 0$. Hence, if VC is small, this indicates that the Constant Demand Assumption is reasonable. Research indicates that the EOQ should be used if $VC < 0.20$;

otherwise, demand is too irregular to justify the use of an EOQ model. (See Peterson and Silver (1998).)

If $VC > 0.20$, dynamic programming methods and the Silver–Meal heuristic, which are discussed in Chapter 18, may be used to determine optimal ordering policies.

As an example of the use of the VC formula, suppose that demands during the four quarters of the past year have been as follows: 80 units, 100 units, 130 units, and 90 units. Assuming that future demand is known to follow a similar pattern, should an EOQ model be used in this situation?

Since $\bar{d} = \frac{400}{4} = 100$ and est. var $\mathbf{D} = (\frac{1}{4})(80^2 + 100^2 + 130^2 + 90^2) - 100^2 = 350$, we have $VC = \frac{350}{(100)^2} = 0.035$. Since VC is smaller than 0.20, an EOQ model can be used in this situation.

In closing, we note that the EOQ models of this chapter require the implicit assumption that demands during different periods of time are independent. In other words, the EOQ models require that any knowledge about demand during one period of time gives no information about demand at any other point in time. If a firm's inventory needs are met through internal production, this is often an unrealistic assumption. For example, suppose a company needs to produce 5 units of product A by December 11 and that each unit of product A requires 2 units of product B and 3 units of product C. Once product B and product C are available, it takes ten days to assemble a unit of product A. Then the fact that there is a demand for 5 units of product A on December 11 *creates* a December 1 demand for 10 units of product B and 15 units of product C. Hence, the December 1 demand for products B and C *depends* on the December 11 product A demand. Our EOQ models do not take into account the dependence of demand that is present in many manufacturing situations. These can best be exploited by using material resource planning (MRP) systems.

REMARK Use the Excel commands =**AVERAGE,** to estimate the average demand for a given period, and =**VARP,** to estimate the variance in the demand for a given period.

PROBLEM

Group A

1 Observed demand for air conditioners during the last four quarters was as follows: fall, 100; winter, 50; spring, 150; summer, 300. Is it reasonable to use an EOQ model in this situation?

15.7 Multiple-Product EOQ Models

Suppose a company orders several products. Each time an order is received, shipments of some (but perhaps not all) of the products arrive. Each time an order arrives, there is a fixed cost associated with the order (for example, the cost of driving a truck to deliver the order), and there is another fixed cost associated with each product included in the order. How can we minimize the sum of annual holding and fixed costs? An example of this situation would be an appliance store that orders three different types of appliances from a supplier. For a low-demand product, it would be unreasonable to order the product each time a truck arrives. Chopra and Meindl (2001) devised a method to find a nearly optimal solution to this type of problem. To begin, we find the product that is most frequently ordered. Suppose that is product 1; we assume this product will be included in each order. We then set up a Solver model that determines the following changing cells:

- Number of orders received per year. Note that each order is assumed to contain a shipment of product 1.

- For all products other than product 1, the number of orders that need to be received before an order of the product is received. If, for example, product 2 should be contained in every third order, then the changing cell for product 2 will equal 3.

Given trial values of these quantities, we can easily determine the total fixed cost (sum of fixed cost for each product plus fixed cost for each order) and total holding cost for each product. The sum of these costs will be our target cell for Solver. Our model is highly nonlinear. It is necessary to use the Evolutionary Solver to find the optimal solution. Here is an example of the method.

EXAMPLE 7   **Ohm City Appliances**

Ohm City Appliances has three types of TVs delivered from Springfield TV. Figure 14 gives the annual demand, unit purchasing cost, annual holding cost (as a percentage of purchase cost), the fixed cost of ordering a product, and the fixed cost of placing an order. Determine an ordering policy that minimizes the sum of fixed and holding costs.

FIGURE 14

	A	B	C	D	E	F
4						
5						
6			Product 1	Product 2	Product 3	
7		annual demand	12000	1200	120	
8		unit cost	$ 500.00	$ 500.00	$ 500.00	
9		holding cost	0.2	0.2	0.2	
10		product order cost	$ 1,000.00	$ 1,000.00	$ 1,000.00	
11		eoq	489.8979486	154.9193338	48.989795	
12		orders per year	24.49489743	7.745966692	2.4494897	
13		overall order cost	$ 4,000.00			
14						
15		Orders per year P1	10.46135741			
16		Orders of P1 per P2	1			
17		Orders of P1 per P3	4			
18		Orders per Year P2	10.46135741			
19		Orders Per Year P3	2.615339354			
20						
21						
22						
23		Main annual order cost	$ 41,845.43			
24						
25			Prod 1	Prod 2	Prod 3	
26		Order quantity	1147.078675	114.7078675	45.883147	
27		Avg. Inventory	573.5393374	57.35393374	22.941573	
28		Annual Holding cost	$ 57,353.93	$ 5,735.39	$ 2,294.16	
29		Annual product ordering cost	$ 10,461.36	$ 10,461.36	$ 2,615.34	
30						
31						
32		Total Annual cost	$ 130,766.97			
33						

Our work is in file MultipleEOQ.xls. Also see Figure 14.

Step 1 In C11:E11, we compute the EOQ for each product by copying from C11 to D11:E11 the formula

$$=SQRT(2*C10*C7/(C9*C8))$$

Then, in C12:E12, we compute the number of times each product is ordered during a year by copying from C12 to D12:E12 the formula

$$=C7/C11$$

We find that product 1 is the most frequently ordered.

Step 2 In cell C15, we enter a trial value (not necessarily an integer) for the number of orders placed each year. In C16, we enter a trial value (which must be an integer) for the number of orders with product 1 that must be received before an order of product 2 is received. In C17, we enter a trial value (which must be an integer) for the number of orders with product 1 that must be received before an order of product 3 is obtained.

Step 3 In cell C23, we compute the total fixed cost associated with the orders as (number of orders per year)*(cost per order) with the formula

$$=C15*C13$$

Step 4 In cells C18 and C19, we compute the number of times product I (I = 2 or 3) are ordered each year by computing (orders per year)*(fraction of orders containing product I).

$$=\$C\$15/C16 \text{ (cell C18: orders of product 2 per year)}$$
$$=\$C\$15/C17 \text{ (cell C19: orders of product 3 per year)}$$

Step 5 In cell C26, we compute the size of each product 1 order as (annual product 1 demand)/(orders of product 1 received each year).

$$=C7/C15$$

In a similar fashion, we compute size of product 2 and product 3 orders in cells D26 and E26.

$$=D7/C18 \text{ (cell D26: product 2 order size)}$$
$$=E7/C19 \text{ (cell E26: product 3 order size)}$$

Step 6 In C27:E27, we compute the average inventory level for each product as half the order size. To do this, copy from C27 to D27:E27 the formula

$$=0.5*C26$$

Step 7 In C28:E28, we compute the annual holding cost for each product as (average inventory level for product)*(annual cost of holding one unit of product in inventory). To do this, copy from C28 to D28:E28 the formula

$$=C9*C8*C27$$

Step 8 In C29:E29, we compute the annual ordering cost for each product as (cost per order for product)*(times product is ordered per year). For example, in cell C29, we compute annual ordering cost for product 1 with the formula

$$=C10*C15.$$

Step 9 In cell C32, we compute total annual cost (exclusive of purchasing costs, which do not depend on ordering policy) with the formula

$$=SUM(C28:E29)+C23$$

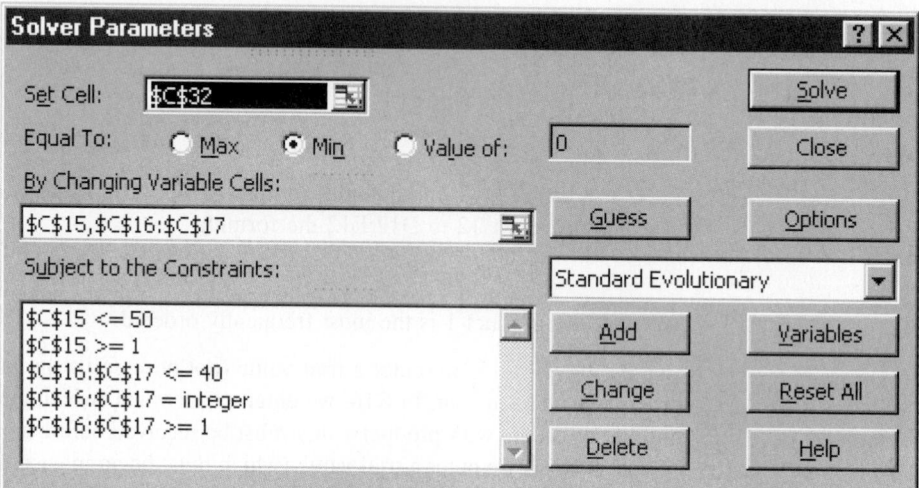

FIGURE 15
Solver Window for Ohm
City Appliances

Step 10   We now use Solver to find the cost-minimizing ordering policy. Figure 15 shows our Solver window.

We minimize total cost (C32) by changing the number of orders per year (C15) and the number of orders that must be placed before orders are placed for less frequently ordered products (C16 and C17). We require an integer for the number of orders before each less frequently ordered product is placed. Included are the lower and upper bounds that Evolutionary Solver requires for each changing cell.

We find that 10.46 truckloads of TVs should be received each year. Each truckload will contain 1,147 type 1 TVs and 114 type 2 TVs. 25% of all orders will include an order of 46 type 3 TVs. Note that the low-demand type 3 TVs are infrequently ordered.

PROBLEMS

Group A

1   Square City Appliance orders four types of washing machines. Table 7 gives the annual demand, purchasing cost, and annual holding cost (as a percentage of purchase cost), and the fixed cost of ordering a product. Determine an ordering policy that minimizes the sum of fixed and holding costs. Each time an order is delivered, a $10,000 cost is incurred. Determine an ordering policy to minimize annual cost of meeting demand.

2   In Problem 1, suppose that Square City manufactures the washing machines. The company can manufacture washing machines at a rate of 30,000 per year. What manufacturing policy will minimize the annual cost of meeting demand?

TABLE 7

	Product 1	Product 2	Product 3	Product 4
Annual demand	10,000	3,000	4,000	500
Unit purchasing cost	$400	$300	$200	$900
Holding cost percentage	.2	.2	.2	.2
Product order cost	$1,000	$1,000	$1,000	$1,000

SUMMARY   Notation

K = setup or ordering cost

h = cost of holding one unit in inventory for one unit of time

D = demand rate per unit time

r = rate at which firm can make product per unit time $(r > D)$

s = cost of being one unit short for one unit of time

Basic EOQ Model

$$\text{Order quantity} = q^* = \left(\frac{2KD}{h}\right)^{1/2}$$

$\frac{D}{q^*}$ orders are placed each unit of time.

Quantity Discount Model

If $q < b_1$, each item costs p_1 dollars.

If $b_1 \leq q < b_2$, each item costs p_2 dollars.

If $b_{k-2} \leq q < b_{k-1}$, each item costs p_{k-1} dollars.

If $b_{k-1} \leq q < b_k = \infty$, each item costs p_k dollars.

Beginning with the lowest price, determine for each price the order quantity (q_i^*) that minimizes total annual costs for $b_{i-1} \leq q < b_i$. Continue determining $q_k^*, q_{k-1}^*, \ldots$ until one of the q_i^*'s (call it q_i^*) is admissible; from observation 2, this will mean that $q_i^* = EOQ_{i'}$. The optimal order quantity will be the member of $\{q_k^*, q_{k-1}^*, \ldots, q_i^*\}$ with the smallest value of $TC(q)$.

If EOQ_i is admissible, then $q_i^* = EOQ_i$. If $EOQ_i < b_{i-1}$, then $q_i^* = b_{i-1}$.

Continuous Rate Model

$$\text{Optimal run size} = \left[\frac{2KDr}{h(r - D)}\right]^{1/2}$$

EOQ with Back Orders Allowed

q^* = optimal order quantity

M^* = maximum inventory level under optimal ordering policy

$q^* - M^*$ = maximum shortage occurring under optimal ordering policy

$$q^* = \left[\frac{2KD(h + s)}{hs}\right]^{1/2} = EOQ\left(\frac{h + s}{s}\right)^{1/2}$$

$$M^* = \left[\frac{2KDs}{h(h + s)}\right]^{1/2} = EOQ\left(\frac{s}{h + s}\right)^{1/2}$$

REVIEW PROBLEMS

Group A

1 Customers at Joe's Office Supply Store demand an average of 6,000 desks per year. Each time an order is placed, an ordering cost of $300 is incurred. The annual holding cost for a single desk is 25% of the $200 cost of a desk. One week elapses between the placement of an order and the arrival of the order. In parts (a)–(d), assume that no shortages are allowed.

  a Each time an order is placed, how many desks should be ordered?

  b How many orders should be placed each year?

  c Determine the total annual costs (excluding purchasing costs) of meeting the customers' demands for desks.

  d Determine the reorder point. If the lead time were five weeks, what would be the reorder point? (52 weeks = one year.)

  e How would the answers to parts (a) and (b) change if shortages were allowed and a cost of $80 is incurred if Joe's is short one desk for one year?

2 Suppose Joe's is considering manufacturing desks. It costs $250 to set up a production run, and Joe's has the capacity to manufacture up to 10,000 desks per year. What is the optimal production run size? How many production runs will be made each year?

3 A camera store sells an average of 100 cameras per month. The cost of holding a camera in inventory for a year is 30% of the price the camera shop pays for the camera. It costs $120 each time the camera store places an order with its supplier. The price charged per camera depends on the number of cameras ordered (see Table 8). Each time the camera store places an order, how many cameras should be ordered?

Group B

4 A company inventories two items. The relevant data for each item are shown in Table 9. Determine the optimal

inventory policy if no shortages are allowed and if the average investment in inventory is not allowed to exceed $700. If this constraint could be relaxed by $1, by how much would the company's annual costs decrease? (This problem requires knowledge of Section 11.8.)

5 A company produces three types of items. A single machine is used to produce the three items on a cyclical basis. The company has the policy that every item is produced once during each cycle, and it wants to determine the number of production cycles per year that will minimize the sum of holding and setup costs (no shortages are allowed). The following data are given:

P_i = number of units of product i that could be produced per year if the machine were entirely devoted to producing product i

D_i = annual demand for product i

K_i = cost of setting up production for product i

h_i = cost of holding one unit of product i in inventory for one year

  a Suppose there are N cycles per year. Assuming that during each cycle, a fraction $\frac{1}{N}$ of all demand for each product is met, determine the annual holding cost and the annual setup cost.

  b Let q_i^* be the number of units of product i produced during each cycle. Determine the optimal value of N (call it N^*) and q_i^*.

  c Let $EROQ_i$ be the optimal production run size for product i if the cyclical nature of the problem is ignored. Suppose q_i^* is much smaller than $EROQ_i$. What conclusion could be drawn?

  d Under certain circumstances, it might not be desirable to produce every item during each cycle. Which of the following factors would tend to make it undesirable to produce product i during each cycle: (1) Demand is relatively low. (2) The setup cost is relatively high. (3) The holding cost is relatively high.

TABLE 8

No. of Cameras Ordered	Price per Camera
1–10	$10.00
11–40	$9.00
41–100	$7.00
More than 100	$5.50

TABLE 9

	Item 1	Item 2
Annual demand	6,000	4,000
Per-unit cost	$4.00	$3.50
Annual holding cost	30% per year	25% per year
Price per order	$35	$20

REFERENCES

The following books emphasize applications over theory:

Brown, R. *Decision Rules for Inventory Management.* New York: Holt, Rinehart and Winston, 1967.

Chopra, S., and P. Meindl. *Supply Chain Management.* Englewood Cliffs, N.J.: Prentice Hall, 2001.

McLeavey, D., and S. Narasimhan. *Production Planning and Inventory Control.* Boston: Allyn and Bacon, 1985.

Peterson, R., and E. Silver. *Decision Systems for Inventory Management and Production Planning.* New York: Wiley, 1988.

Tersine, R. *Principles of Inventory and Materials Management.* New York: North-Holland, 1982.

Vollman, T., W. Berry, and C. Whybark. *Manufacturing Planning and Control Systems.* Homewood, Ill.: Irwin, 1998.

Zipkin, P. *Foundations of Inventory Management.* New York: Irwin-McGraw-Hill, 2000.

The following three books contain extensive discussions of inventory theory as well as applications:

Hadley, G., and T. Whitin. *Analysis of Inventory Systems.* Englewood Cliffs, N.J.: Prentice Hall, 1963.

Hax, A., and D. Candea. *Production and Inventory Management.* Englewood Cliffs, N.J.: Prentice Hall, 1984.

Johnson, L., and D. Montgomery. *Operations Research in Production, Planning, Scheduling, and Inventory Control.* New York: Wiley, 1974.

Baumol, W. "The Transactions Demand for Cash: An Inventory Theoretic Approach," *Quarterly Journal of Economics* 16(1952):545–556.

Ignall, E., and P. Kolesar. "Operating Characteristics of a Simple Shuttle under Local Dispatching Rules," *Operations Research* 20(1972):1077–1088.

Riccio, L., J. Miller, and A. Little."Polishing the Big Apple," *Interfaces* 16(no. 1, 1986):83–88.

Roundy, R. "98% Effective Integer Rate Lot-Sizing for One-Warehouse Multi-Retailer Systems," *Management Science* 31(1985):1416–1430.

16

Probabilistic Inventory Models

All the inventory models discussed in Chapter 15 require that demand during any period of time be known with certainty. In this chapter, we consider inventory models in which demand over a given time period is uncertain, or random; single-period inventory models, where a problem is ended once a single ordering decision has been made; single-period bidding models; versions of the EOQ model for uncertain demand that incorporate the important concepts of safety stock and service level; the periodic review (R, S) model; the ABC inventory classification system; and exchange curves.

16.1 Single-Period Decision Models

In many situations, a decision maker is faced with the problem of determining the value q for a variable (q may be the quantity ordered of an inventoried good, for example, or the bid on a contract). After q has been determined, the value d assumed by a random variable $\mathbf{D}$ is observed. Depending on the values of d and q, the decision maker incurs a cost $c(d, q)$. We assume that the person is risk-neutral and wants to choose q to minimize his or her expected cost. Since the decision is made only once, we call a model of this type a *single-period decision model.*

16.2 The Concept of Marginal Analysis

For the single-period model described in Section 16.1, we now assume that $\mathbf{D}$ is an integer-valued discrete random variable with $P(\mathbf{D} = d) = p(d)$. Let $E(q)$ be the decision maker's expected cost if q is chosen. Then

$$E(q) = \sum_d p(d)c(d, q)$$

In most practical applications, $E(q)$ is a convex function of q. Let q^* be the value of q that minimizes $E(q)$. If $E(q)$ is a convex function, the graph of $E(q)$ must look something like Figure 1. From the figure, we see that q^* is the smallest value of q for which

$$E(q^* + 1) - E(q^*) \geq 0 \tag{1}$$

Thus, if $E(q)$ is a convex function of q, we can find the value of q minimizing expected cost by finding the smallest value of q that satisfies Inequality (1). Note that $E(q + 1) - E(q)$ is the change in expected cost that occurs if we increase the decision variable q to $q + 1$.

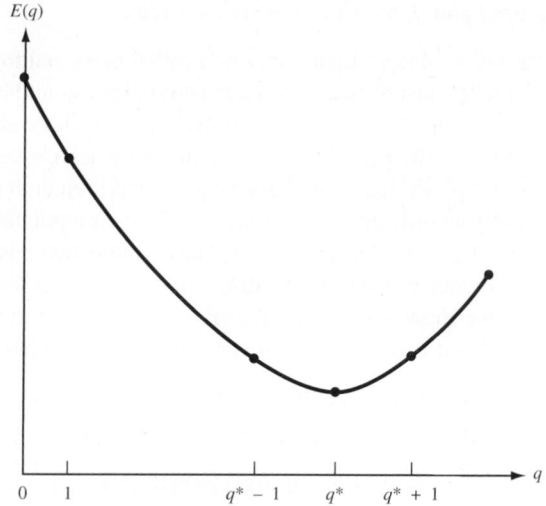

FIGURE 1
Determination of q^* by
Marginal Analysis

To determine q^*, we begin with $q = 0$. If $E(1) - E(0) \leq 0$, we can benefit by increasing q from 0 to 1. Now we check to see whether $E(2) - E(1) \leq 0$. If this is true, then increasing q from 1 to 2 will reduce expected cost. Continuing in this fashion, we see that increasing q by 1 will reduce expected costs up to the point where we try to increase q from q^* to $q^* + 1$. In this case, increasing q by 1 will increase expected cost. From Figure 1 (which is the appropriate picture if $E(q)$ is a convex function), we see that if $E(q^* + 1) - E(q^*) \geq 0$, then for $q \geq q^*$, $E(q + 1) - E(q) \geq 0$. Thus, q^* must be the value of q that minimizes $E(q)$. If $E(q)$ is not convex, this argument may not work. (See Problem 1 at the end of this section.)

Our approach determines q^* by repeatedly computing the effect of adding a marginal unit to the value of q. For this reason, it is often called **marginal analysis.** Marginal analysis is very useful if it is easy to determine a simple expression for $E(q + 1) - E(q)$. In the next section, we use marginal analysis to solve the classical news vendor problem.

PROBLEM

Group A

1 Suppose $E(q)$ is $E(0) = 8$, $E(1) = 6$, $E(2) = 5$, $E(3) = 7$, $E(4) = 6$, $E(5) = 5.5$, $E(6) = 4.5$, and $E(7) = 5$.

  a What value of q minimizes $E(q)$?

b If marginal analysis is used to determine the value of q that minimizes $E(q)$, what is the answer?

c Explain why marginal analysis fails to find the value of q that minimizes $E(q)$.

16.3 The News Vendor Problem: Discrete Demand

Organizations often face inventory problems where the following sequence of events occurs:

1 The organization decides how many units to order. We let q be the number of units ordered.

2 With probability $p(d)$, a demand of d units occurs. In this section, we assume that d must be a nonnegative integer. We let **D** be the random variable representing demand.

3 Depending on d and q, a cost $c(d, q)$ is incurred.

Problems that follow this sequence are often called **news vendor problems.** To see why, consider a vendor who must decide how many newspapers should be ordered each day from the newspaper plant. If the vendor orders too many papers, he or she will be left with many worthless newspapers at the end of the day. On the other hand, a vendor who orders too few newspapers will lose profit that could have been earned if enough newspapers to meet customer demand had been ordered, and customers will be disappointed. The news vendor must order the number of papers that properly balances these two costs. We have already encountered a news vendor problem in the discussion of decision theory in Section 13.1.

In this section, we show how marginal analysis can be used to solve news vendor problems when demand is a discrete random variable and $c(d, q)$ has the following form:

$$c(d, q) = c_o q + \text{(terms not involving } q) \qquad (d \leq q) \tag{2}$$

$$c(d, q) = -c_u q + \text{(terms not involving } q) \qquad (d \geq q + 1) \tag{2.1}$$

In (2), c_o is the per-unit cost of being overstocked. If $d \leq q$, we have ordered more than was demanded—that is, overstocked. If the size of the order is increased from q to $q + 1$, then (2) shows that the cost increases by c_o. Hence, c_o is the cost due to being overstocked by one extra unit. We refer to c_o as the **overstocking cost.** Similarly, if $d \geq q + 1$, we have understocked (ordered an amount less than demand). If $d \geq q + 1$ and we increase the size of the order by one unit, we are understocked by one less unit. Then (2.1) implies that the cost is reduced by c_u, so c_u is the per-unit cost of being understocked. We call c_u the **understocking cost.**

To derive the optimal order quantity via marginal analysis, let $E(q)$ be the expected cost if an order is placed for q units. We assume that the decision maker's goal is to find the value q^* that minimizes $E(q)$. If $c(d, q)$ can be described by (2) and (2.1), and $E(q)$ is a convex function of q, then marginal analysis can be used to determine q^*.

Following (1), we must determine the smallest value of q for which $E(q + 1) - E(q) \geq 0$. To calculate $E(q + 1) - E(q)$, we must consider two possibilities:

Case 1 $d \leq q$. In this case, ordering $q + 1$ units instead of q units causes us to be overstocked by one more unit. This increases cost by c_o. The probability that Case 1 will occur is simply $P(\mathbf{D} \leq q)$, where $\mathbf{D}$ is the random variable representing demand.

Case 2 $d \geq q + 1$. In this case, ordering $q + 1$ units instead of q units enables us to be short one less unit. This will decrease cost by c_u. The probability that Case 2 will occur is $P(\mathbf{D} \geq q + 1) = 1 - P(\mathbf{D} \leq q)$.

In summary, a fraction $P(\mathbf{D} \leq q)$ of the time, ordering $q + 1$ units will cost c_o more than ordering q units; and a fraction $1 - P(\mathbf{D} \leq q)$ of the time, ordering $q + 1$ units will cost c_u less than ordering q units. Thus, on the average, ordering $q + 1$ units will cost

$$c_o P(\mathbf{D} \leq q) - c_u[1 - P(\mathbf{D} \leq q)]$$

more than ordering q units.

More formally, we have shown that

$$E(q + 1) - E(q) = c_o P(\mathbf{D} \leq q) - c_u[1 - P(\mathbf{D} \leq q)]$$
$$= (c_o + c_u) P(\mathbf{D} \leq q) - c_u{}^\dagger$$

Then $E(q + 1) - E(q) \geq 0$ will hold if

$$(c_o + c_u) P(\mathbf{D} \leq q) - c_u \geq 0 \qquad \text{or} \qquad P(\mathbf{D} \leq q) \geq \frac{c_u}{c_o + c_u}$$

---

†Since $P(\mathbf{D} \leq q)$ increases as q increases, $E(q + 1) - E(q)$ will increase as q increases. Hence, if $c_o + c_u \geq 0$, $E(q)$ is a convex function of q, and our use of marginal analysis is justified.

Let $F(q) = P(\mathbf{D} \leq q)$ be the demand distribution function. Since marginal analysis is applicable, we have just shown that $E(q)$ will be minimized by the smallest value of q (call it q^*) satisfying

$$F(q^*) \geq \frac{c_u}{c_o + c_u} \tag{3}$$

The following example illustrates the use of (3).

EXAMPLE 1   Walton Bookstore Calendar Sales

In August, Walton Bookstore must decide how many of next year's nature calendars should be ordered. Each calendar costs the bookstore \$2 and is sold for \$4.50. After January 1, any unsold calendars are returned to the publisher for a refund of 75¢ per calendar. Walton believes that the number of calendars sold by January 1 follows the probability distribution shown in Table 1. Walton wants to maximize the expected net profit from calendar sales. How many calendars should the bookstore order in August?[†]

Solution   Let

$$q = \text{number of calendars ordered in August}$$

$$d = \text{number of calendars demanded by January 1}$$

If $d \leq q$, the costs shown in Table 2 are incurred (revenue is negative cost). From (2), $c_o = 1.25$.

If $d \geq q + 1$, the costs shown in Table 3 are incurred. From (2), $-c_u = -2.5$, or $c_u = 2.50$. Then

$$\frac{c_u}{c_o + c_u} = \frac{2.50}{3.75} = \frac{2}{3}$$

TABLE 1
Probability Mass Function for Calendar Sales

No. of Calendars Sold	Probability
100	.30
150	.20
200	.30
250	.15
300	.05

TABLE 2
Computation of Total Cost If $d \leq q$

	Cost
Buy q calendars at \$2/calendar	$2q$
Sell d calendars at \$4.50/calendar	$-4.50d$
Return $q - d$ calendars at 75¢/calendar	$-0.75(q - d)$
Total cost	$1.25q - 3.75d$

[†]Based on Barron (1985).

TABLE 3
Computation of Total Cost If $d \geq q + 1$

	Cost
Buy q calendars at $2/calendar	$2q$
Sell d calendars at $4.50/calendar	$-4.50q$
Total cost	$-2.50q$

From (3), Walton should order q^* calendars, where q^* is the smallest number for which $P(\mathbf{D} \leq q^*) \geq \frac{2}{3}$. As a function of q, $P(\mathbf{D} \leq q)$ increases only when $q = 100, 150, 200, 250,$ or 300. Also note that $P(\mathbf{D} \leq 100) = .30$, $P(\mathbf{D} \leq 150) = .50$, and $P(\mathbf{D} \leq 200) = .80$. Since $P(\mathbf{D} \leq 200)$ is greater than or equal to $\frac{2}{3}$, $q^* = 200$ calendars should be ordered.

REMARKS **1** In terms of marginal analysis, the probability of selling the 200th calendar that is ordered is $P(\mathbf{D} \geq 200) = .50$. This implies that the 200th calendar has a $1 - .50 = .50$ chance of being unsold. Thus, the 200th calendar will increase Walton's expected costs by $.50(-2.50) + .50(1.25) = -\$0.625$. Hence, the 200th calendar should be ordered. On the other hand, the probability that the 201st calendar will be sold is $P(\mathbf{D} \geq 201) = .20$, and the probability that the 201st calendar will not be sold is $1 - .20 = .80$. Therefore, the 201st calendar will increase expected costs by $.20(-2.50) + .80(1.25) = \$0.50$. Thus, the 201st calendar will increase expected costs and should not be ordered.

2 In Example 1, c_o and c_u could easily have been determined without recourse to (2) and (2.1). For example, being one more unit over actual demand increases Walton's costs by $2 - 0.75 = \$1.25$. Thus, $c_o = \$1.25$. Similarly, being one more unit under actual demand will cost Walton $4.50 - 2.00 = \$2.50$ in profit. Hence, $c_u = \$2.50$. If we are able to determine c_o and c_u without using Equations (2) and (2.1), we should do so. In more difficult problems, however, they can be very useful (see Examples 2 and 3).

PROBLEMS

Group A

1 In August 2003, a car dealer is trying to determine how many 2004 models should be ordered. Each car costs the dealer $10,000. The demand for the dealer's 2004 models has the probability distribution shown in Table 4. Each car is sold for $15,000. If the demand for 2004 cars exceeds the number of cars ordered in August, the dealer must reorder at a cost of $12,000 per car. If the demand for 2004 cars falls short, the dealer may dispose of excess cars in an end-of-model-year sale for $9,000 per car. How many 2004 models should be ordered in August?

2 Each day, a news vendor must determine how many *New York Herald Wonderfuls* to order. She pays 15¢ for each paper and sells each for 30¢. Any leftover papers are a total loss. From past experience, she believes that the number of papers she can sell each day is governed by the probability distribution shown in Table 5. How many papers should she order each day?

3 If c_u is fixed, will an increase in c_o increase or decrease the optimal order quantity?

TABLE 4

No. of Cars Demanded	Probability
20	.30
25	.15
30	.15
35	.20
40	.20

TABLE 5

No. of Papers Demanded	Probability
50	.30
70	.15
90	.25
110	.10
130	.20

TABLE 6

No. of Cells	Probability
50	.20
60	.15
70	.30
80	.10
90	.15
100	.10

TABLE 7

Number Needed	Probability
200	.03
275	.03
350	.03
400	.05
450	.40
500	.30
550	.06
600	.07
650	.03

TABLE 8

Copies Demanded	Probability
5,000	.30
6,000	.20
7,000	.40
8,000	.10

TABLE 9

Week of Birth	Probability
36	.05
37	.15
39	.20
40	.30
41	.15
42	.10
43	.05

4 If c_o is fixed, will an increase in c_u increase or decrease the optimal order quantity?

5 The power at Ice Station Lion is supplied via solar cells. Once a year, a plane flies in and sells solar cells to the ice station at a price of $20 per cell. Because of uncertainty about future power needs, the ice station can only guess the number of cells that will be required during the coming year (see probability distribution in Table 6). If the ice station runs out of solar cells, a special order must be placed at a cost of $30 per cell.

    a Assuming that the news vendor problem is relevant, how many cells should be ordered from the plane?

    b In part (a), what type of cost is being ignored?

6 The daily demand for substitute teachers in the Los Angeles teaching system follows the distribution given in Table 7. Los Angeles wants to know how many teachers to keep in the substitute teacher pool. Whether or not the substitute teacher is needed, it costs $30 per day to keep a substitute teacher in the pool. If not enough substitute teachers are available on a given day, regular teachers are used to cover classes at a cost of $54 per regular teacher. How many teachers should Los Angeles have in the substitute teacher pool?[†]

Group B

7 Every four years, Blockbuster Publishers revises its textbooks. It has been three years since the best-selling book, *The Joy of OR,* has been revised. At present, 2,000

[†]Based on Bruno (1970).

copies of the book are in stock, and Blockbuster must determine how many copies of the book should be printed for the next year. The sales department believes that sales during the next year are governed by the distribution in Table 8. Each copy of *Joy* sold during the next year brings the publisher $35 in revenues. Any copies left at the end of the next year cannot be sold at full price but can be sold for $5 to Bonds Ennoble and Gitano's bookstores. The cost of a printing of the book is $50,000 plus $15 per book printed. How many copies of *Joy* should be printed? Would the answer change if 4,000 copies were currently in stock?

8 Vivian and Wayne are planning on going to Lamaze natural childbirth classes. Lamaze classes meet once a week for five weeks. Each class gives 20% of the knowledge needed for "natural" childbirth. If Vivian and Wayne finish their classes before the birth of their child, they will forget during each week 5% of what they have learned in class. To maximize their expected knowledge at the time of childbirth, during which week of pregnancy should they begin classes? Assume that the number of weeks from conception to childbirth follows the probability distribution given in Table 9.

9[‡] Some universities allow an employee to put an amount q into an account at the beginning of each year, to be used for child-care expenses. The amount q is not subject to federal income tax. Assume that all other income is taxed by the federal government at a 40% rate. If child-care expenses for the year (call them d) are less than q, the employee in effect loses $q - d$ dollars in before-tax income. If child-care expenses exceed q, the employee must pay the excess out of his or her own pocket but may credit 25% of that as a savings on his or her state income tax.

Suppose Professor Muffy Rabbit believes that there is an equal chance that her child-care expenses for the coming year will be $3,000, $4,000, $5,000, $6,000, or $7,000. At the beginning of the year, how much money should she place in the child-care account?

[‡]Based on Rosenfeld (1986).

16.4 The News Vendor Problem: Continuous Demand

We now consider the news vendor scenario of Section 16.3 when demand $\mathbf{D}$ is a continuous random variable having density function $f(d)$. By modifying our marginal analysis argument of Section 16.3 (or by using Leibniz's rule for differentiating an integral—see Problem 7 at the end of this section), it can be shown that the decision maker's expected cost is minimized by ordering q^* units, where q^* is the smallest number satisfying

$$P(\mathbf{D} \leq q^*) \geq \frac{c_u}{c_o + c_u} \tag{4}$$

Since demand is a continuous random variable, we can find a number q^* for which (4) holds with equality. Hence, in this case, the optimal order quantity can be determined by finding the value of q^* satisfying

$$P(\mathbf{D} \leq q^*) = \frac{c_u}{c_o + c_u} \qquad \text{or} \qquad P(\mathbf{D} \geq q^*) = \frac{c_o}{c_o + c_u} \tag{5}$$

From (5), we see that it is optimal to order units up to the point where the last unit ordered has a chance

$$\frac{c_o}{c_o + c_u}$$

of being sold. Examples 2 and 3 illustrate the use of (5).

EXAMPLE 2   **ABA Room Reservations**

The American Bar Association (ABA) is holding its annual convention in Las Vegas. Six months before the convention begins, the ABA must decide how many rooms should be reserved in the convention hotel. At this time, the ABA can reserve rooms at a cost of $50 per room, but six months before the convention, the ABA does not know with certainty how many people will attend the convention. The ABA believes, however, that the number of rooms required is normally distributed, with a mean of 5,000 rooms and a standard deviation of 2,000 rooms. If the number of rooms required exceeds the number of rooms reserved at the convention hotel, extra rooms will have to be found at neighboring hotels at a cost of $80 per room. It is inconvenient for convention participants to stay at neighboring hotels. We measure this inconvenience by assessing an additional cost of $10 for each room obtained at a neighboring hotel. If the goal is to minimize the expected cost to the ABA and its members, how many rooms should the ABA reserve at the convention hotel?

Solution   Define

$$q = \text{number of rooms reserved}$$
$$d = \text{number of rooms actually required}$$

If $d \leq q$, then the only cost incurred is the cost of the rooms reserved in advance, so if $d \leq q$, the total cost is $50q$. Thus, $c_o = 50$. If $d \geq q + 1$, the following costs are incurred:

     Cost of reserving q rooms $= 50q$

     Cost of renting $d - q$ rooms in neighboring hotels $= 80(d - q)$

     Inconvenience cost to overflow participants $= 10(d - q)$

     Total cost $= 90d - 40q$ and $c_u = 40$

Since $\dfrac{c_u}{c_u + c_o} = \dfrac{40}{90} = \dfrac{4}{9}$, we see from (5) that the optimal number of rooms to reserve is the number q^* satisfying

$$P(\mathbf{D} \le q^*) = \tfrac{4}{9} \tag{6}$$

The Excel function NORMINV can be used to calculate q^*. Since

$$= \text{NORMINV}(4/9, 5000, 2000)$$

yields 4,720.58, the ABA should reserve 4,720 or 4,721 rooms.

EXAMPLE 3   **Airline Overbooking**

The ticket price for a New York–Indianapolis flight is \$200. Each plane can hold up to 100 passengers. Usually, some of the passengers who have purchased tickets for a flight fail to show up (no-shows). To protect against no-shows, the airline will try to sell more than 100 tickets for each flight. Federal law states that any ticketed customer who is unable to board the plane is entitled to compensation (say, \$100). Past data indicate that the number of no-shows for each New York–Indianapolis flight is normally distributed, with a mean of 20 and a standard deviation of 5. To maximize expected revenues less compensation costs, how many tickets should the airline sell for each flight? Assume that anybody who doesn't use a ticket receives a \$200 refund.

Solution   Let

$$q = \text{number of tickets sold by airline}$$
$$d = \text{number of no-shows}$$

Observe that $q - d$ will be the number of customers actually showing up for the flight. If $q - d \le 100$, then all customers who show up will board the flight, and the cost to the airline is $-200(q - d) = 200d - 200q$. If $q - d \ge 100$, then 100 passengers will board the plane (paying the airline $200(100) = \$20,000$), and $q - d - 100$ customers will be turned away. These $q - d - 100$ customers will receive compensation of $100(q - d - 100)$. Hence, if $q - d \ge 100$, the total cost to the airline is given by $100(q - d - 100) - 200(100) = 100(q - 100) - 100d - 20,000$. In summary, the net cost to the airline may be expressed as shown in Table 10.

If $q - 100$ is considered as a decision variable, we have a news vendor problem with $-c_u = -200$ (or $c_u = 200$) and $c_o = 100$. From (5), we should choose $q - 100$ to satisfy

$$P(\mathbf{D} \le q - 100) = \frac{c_u}{c_o + c_u} = \frac{2}{3} \tag{7}$$

The problem can be solved with the help of Excel. Since

$$= \text{NORMINV}(2/3, 120, 5)$$

TABLE 10
Computation of Total Cost

	Total Cost
$q - d \ge 100$ (or $d \le q - 100$)	$100(q - 100) - 100d - 20,000$
$q - d \le 100$ (or $d \ge q - 100$)	$200d - 200(q - 100) - 200(100)$

yields 122.15, we may conclude that the airline should attempt to sell 122 or 123 tickets. This means that once ticket sales have reached 122 (or 123), no more tickets should be sold for the flight. Of course, if fewer than 122 people want to purchase tickets for the flight, the airline should not refuse to sell anybody a ticket for the flight.

PROBLEMS

Group A

1 **a** In Example 3, why is it unrealistic to assume that the distribution of the number of no-shows is independent of q?

b If the number of no-shows were normally distributed with a mean of $.05q$ and a standard deviation of $.05q$, would we still have a news vendor problem?

2 Condo Construction Company is going to First National Bank for a loan. At the present time, the bank is willing to lend Condo up to $1 million, with interest costs of 10%. Condo believes that the amount of borrowed funds needed during the current year is normally distributed, with a mean of $700,000 and a standard deviation of $300,000. If Condo needs to borrow more money during the year, the company will have to go to Louie the Loan Shark. The cost per dollar borrowed from Louie is 25¢. To minimize expected interest costs for the year, how much money should Condo borrow from the bank?

3 Joe is selling Christmas trees to pay his college tuition. He purchases trees for $10 each and sells them for $25 each. The number of trees he can sell is normally distributed with a mean of 100 and standard deviation of 30. How many trees should Joe purchase?

4 A hot dog vendor at Wrigley Field sells hot dogs for $1.50 each. He buys them for $1.20 each. All the hot dogs he fails to sell at Wrigley Field during the afternoon can be sold that evening at Comiskey Park for $1 each. The daily demand for hot dogs at Wrigley Field is normally distributed with a mean of 40 and a standard deviation of 10.

a If the vendor buys hot dogs once a day, how many should he buy?

b If he buys 52 hot dogs, what is the probability that he will meet all of the day's demand for hot dogs at Wrigley?

Group B

5[†] Motorama TV estimates the annual demand for its TVs is (and will be in the future) normally distributed, with a mean of 6,000 and standard deviation of 2,000. Motorama

[†]Based on Virts and Garrett (1970).

must determine how much production capacity it should have. The cost of building enough production capacity to make 1,000 sets per year is $1,000,000 (equivalent in present value terms to a cost of $100,000 per year forever). Exclusive of the cost of building capacity, each set sold contributes $250 to profits. How much production capacity should Motorama have?

6 I. L. Pea is a well-known mail-order company. During the Christmas rush (from November 1 to December 15), the number of orders that I. L. Pea must fill each day (five days per week) is normally distributed, with a mean of 2,000 and a standard deviation of 500. I. L. Pea must determine how many employees should be working during the Christmas rush. Each employee works five days a week, eight hours a day, can process 50 orders per day, and is paid $10 per hour. If the full-time work force cannot handle the day's orders during regular hours, some employees will have to work overtime. Each employee is paid $15 per hour for overtime work. For example, if 300 orders are received in a day and there are four employees, then $300 - 4(50) = 100$ orders must be processed by employees who are working overtime. Since each employee can fill $\frac{50}{8} = 6.25$ orders per hour, I. L. Pea would need to pay workers $\frac{100}{6.25} = 16$ hours of overtime for that day. To minimize its expected labor costs, how many full-time employees should I. L. Pea employ during the Christmas rush?

7 Suppose demand is a continuous random variable having a probability density function $f(d)$, and $c(d, q)$ is given by Equation (2). Show that if q units are ordered, the expected cost $E(q)$ may be written as

$$E(q) = \int_0^q c_o q f(t)dt + \int_q^\infty (-c_u)q f(t)dt$$
$$+ \text{ (terms not involving } q \text{ in integrand)}$$

Now use Leibniz's rule to derive Equation (5).

16.5 Other One-Period Models

Many interesting single-period models in operations research cannot be easily handled by marginal analysis. In such situations, we express the decision maker's objective function (usually expected profit or expected cost) as a function $f(q)$ of the decision variable q.

Then we find a maximum or minimum of $f(q)$ by setting $f'(q) = 0$. In this section, we illustrate this idea by a brief discussion of a bidding model.

EXAMPLE 4 **Condo Construction Company**

Condo Construction Company is bidding on an important construction job. The job will cost $2 million to complete. One other company is bidding for the job. Condo believes that the opponent's bid is equally likely to be any amount between $2 million and $4 million. If Condo wants to maximize expected profit, what should its bid be?

Solution Let

$$\mathbf{B} = \text{random variable representing bid of Condo's opponent}$$
$$b = \text{actual bid of Condo's opponent}$$

Then $f(b)$, the density function for $\mathbf{B}$, is given by

$$f(b) = \begin{cases} \dfrac{1}{2,000,000} & (2,000,000 \leq b \leq 4,000,000) \\ 0 & \text{otherwise} \end{cases}$$

Let $q =$ Condo's bid. If $b > q$, Condo outbids the opponent and earns a profit of $q - 2,000,000$. On the other hand, if $b < q$, Condo is outbid by the opponent and earns nothing. The event $b = q$ has a zero probability of occurring and may be ignored. Let $E(q)$ be Condo's expected profit if it bids q. Then

$$E(q) = \int_{2,000,000}^{q} (0)f(b)db + \int_{q}^{4,000,000} (q - 2,000,000)f(b)db$$

Since $f(b) = \frac{1}{2,000,000}$ for $2,000,000 \leq b \leq 4,000,000$, we obtain

$$E(q) = \frac{(q - 2,000,000)(4,000,000 - q)}{2,000,000}$$

To find the value of q maximizing $E(q)$, we find

$$E'(q) = \frac{-(q - 2,000,000) + (4,000,000 - q)}{2,000,000} = \frac{6,000,000 - 2q}{2,000,000}$$

Hence, $E'(q) = 0$ for $q = 3,000,000$. Since $E''(q) = \frac{-2}{2,000,000} < 0$, we know that $E(q)$ is a concave function of q, and $q = 3,000,000$ does indeed maximize $E(q)$. Hence, Condo should bid $3 million. Condo's expected profit will be $E(3,000,000) = \$500,000$.

PROBLEMS

Group A

1 The City of Rulertown consists of the unit interval [0, 1] (see Figure 2). Rulertown needs to determine where to build the city's only fire station. It knows that for small Δx, the probability that a given fire occurs at a location between x and $x + \Delta x$ is $2x(\Delta x)$. Rulertown wants to minimize the average distance between the fire station and a fire. Where should the fire station be located?

FIGURE 2

| 0 | x | $x + \Delta x$ | 1 |

Group B

2 Assume that the Federal Reserve Board can control the growth rate of the U.S. money supply. Also assume that during a year in which the money supply grows by $x\%$, the Gross Domestic Product (GDP) grows by $\mathbf{Z}x\%$, where $\mathbf{Z}$ is a known random variable. The government has decided it wants the GDP to grow by $k\%$ each year. (Too high a growth rate causes excessive inflation, and too low a growth rate causes high unemployment.) To model the government's view, the government assesses a cost of $(d - k)^2$ during a year in which the GDP grows by $d\%$.

a Determine the growth rate of the money supply that should be set by the Federal Reserve Board if the goal is to minimize the expected cost to the government.

b Show that for a given value of $E(\mathbf{Z})$, an increase in var $\mathbf{Z}$ will decrease the optimal growth rate of the money supply found in part (a). (*Hint:* Use the fact that var $\mathbf{Z} = E(\mathbf{Z}^2) - E(\mathbf{Z})^2$.)

16.6 The EOQ with Uncertain Demand: The (r, q) and (s, S) Models

In this section, we discuss a modification of the EOQ that is used when lead time is nonzero and the demand during each lead time is random. We begin by assuming that all demand can be backlogged. As in Chapter 15, we assume a continuous review model, so that orders may be placed at any time, and we define

$$K = \text{ordering cost}$$
$$h = \text{holding cost/unit/year}$$
$$L = \text{lead time for each order (assumed to be known with certainty)}$$
$$q = \text{quantity ordered each time an order takes place}$$

We also require the following definitions:

$\mathbf{D} = $ random variable (assumed continuous) representing annual demand, with mean $E(\mathbf{D})$, variance var $\mathbf{D}$, and standard deviation $\sigma_{\mathbf{D}}$

$c_B = $ cost incurred for each unit short, which does not depend on how long it takes to make up stockout

$OHI(t) = $ on-hand inventory (amount of stock on hand) at time t

From Figure 3, we can see that $OHI(1) = 100$, $OHI(0) = 200$, and $OHI(6) = OHI(7) = 0$.

$B(t) = $ number of outstanding back orders at time t

$I(t) = $ net inventory level at time $t = OHI(t) - B(t)$

$r = $ inventory level at which order is placed (reorder point)

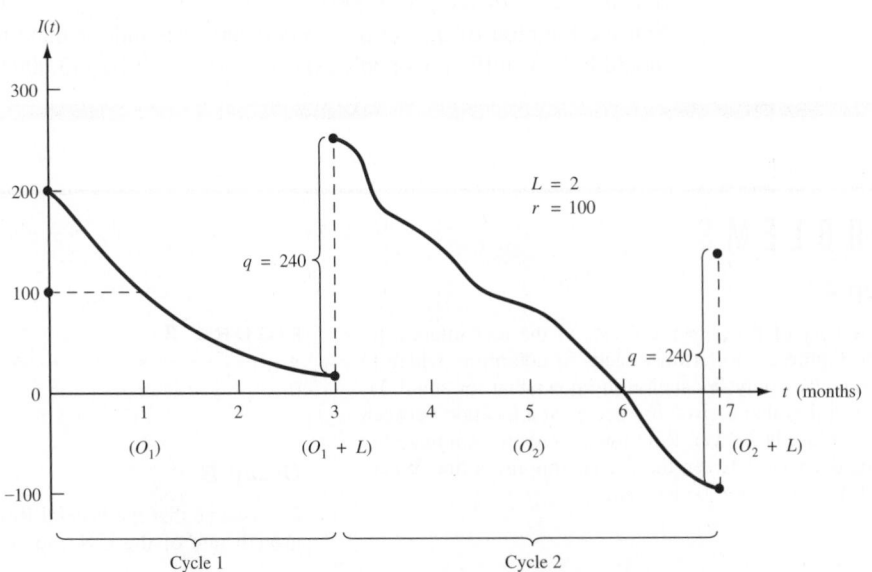

FIGURE 3
Evolution of Inventory over Time in Reorder Point Model

In Figure 3, $B(t) = 0$ for $0 \le t \le 6$ and $B(7) = 100$. $I(t)$ agrees with the inventory concept used in Chapter 15; $I(0) = 200 - 0 = 200$, $I(3) = 260 - 0 = 260$, and $I(7) = 0 - 100 = -100$. The reorder point $r = 100$; whenever the inventory level drops to r, an order is placed for q units.

$$\mathbf{X} = \text{random variable representing demand during lead time}$$

We assume that $\mathbf{X}$ is a continuous random variable having density function $f(x)$ and mean, variance, and standard deviation of $E(\mathbf{X})$, var $\mathbf{X}$, and $\sigma_{\mathbf{X}}$, respectively. If we assume that the demands at different points in time are independent, then it can be shown that the random lead time demand $\mathbf{X}$ satisfies

$$E(\mathbf{X}) = LE(\mathbf{D}), \qquad \text{var } \mathbf{X} = L(\text{var } \mathbf{D}), \qquad \sigma_{\mathbf{X}} = \sigma_{\mathbf{D}}\sqrt{L} \tag{8}$$

We assume that if $\mathbf{D}$ is normally distributed, then $\mathbf{X}$ will also be normally distributed.

Suppose we allow the lead time L to be a random variable (denoted by $\mathbf{L}$), with mean $E(\mathbf{L})$, variance var $\mathbf{L}$, and standard deviation $\sigma_{\mathbf{L}}$. If the length of the lead time is independent of the demand per unit time during the lead time, then

$$E(\mathbf{X}) = E(\mathbf{L})E(\mathbf{D}) \qquad \text{and} \qquad \text{var } \mathbf{X} = E(\mathbf{L})(\text{var } \mathbf{D}) + E(\mathbf{D})^2(\text{var } \mathbf{L}) \tag{8'}$$

We want to choose q and r to minimize the annual expected total cost (exclusive of purchasing cost). Before showing how optimal values of r and q can be found, we look at an illustration of how inventory evolves over time. Assume that an order of $q = 240$ units has just arrived at time 0. We also assume that $L = 2$. In Figure 3, orders of size q are placed at times $O_1 = 1$ and $O_2 = 5$. These orders are received at times $O_1 + L = 3$ and $O_2 + L = 7$, respectively. A **cycle** is defined to be the time interval between any two instants at which an order is received. Figure 3 contains two complete cycles: cycle 1, from arrival of order at time 0 to the instant before order arrives at time $O_1 + L = 3$; and cycle 2, from arrival of order at time $O_1 + L = 3$ to the instant before order arrives at time $O_2 + L = 7$.

During cycle 1, demand during lead time is less than r, so no shortage occurs. During cycle 2, however, demand during lead time exceeds r, so stockouts do occur between time 6 and time $O_2 + L = 7$. It should be clear that by increasing r, we can reduce the number of stockouts. Unfortunately, increasing r will force us to carry more inventory, thereby resulting in higher holding costs. Thus, an optimal value of r must represent some sort of trade-off between holding and stockout costs.

We now show how the optimal values of q and r may be determined.

Determination of Reorder Point: The Back-Ordered Case

The situation in which all demand must eventually be met and no sales are lost is called the **back-ordered case,** for which we show how to determine the reorder point and order quantity that minimize annual expected cost.

We assume each unit is purchased for the same price, so purchasing costs are fixed. Define $TC(q, r) =$ expected annual cost (excluding purchasing cost) incurred if each order is for q units and is placed when the reorder point is r. Then $TC(q, r) =$ (expected annual holding cost) + (expected annual ordering cost) + (expected annual cost due to shortages). To determine the optimal reorder point and order quantity, we assume that the average number of back orders is small relative to the average on-hand inventory level. In most cases, this assumption is reasonable, because shortages (if they occur at all) usually occur during only a small portion of a cycle. (See Problem 5 at the end of this section.) Then $I(t) = OHI(t) - B(t)$ yields

$$\text{Expected value of } I(t) \cong \text{expected value of } OHI(t) \tag{9}$$

We can now approximate the expected annual holding cost. We know that expected annual holding cost = h(expected value of on-hand inventory level). Then from (9), we can approximate expected annual holding cost by h(expected value of $I(t)$). As in Chapter 3, the expected value of $I(t)$ will equal the expected value of $I(t)$ during a cycle. Since the mean rate at which demand occurs is constant, we may write

$$\text{Expected value of } I(t) \text{ during a cycle}$$
$$= \tfrac{1}{2}[(\text{expected value of } I(t) \text{ at beginning of cycle}) \tag{10}$$
$$+ (\text{expected value of } I(t) \text{ at end of a cycle})]$$

At the end of a cycle (the instant before an order arrives), the inventory level will equal the inventory level at the reorder point (r) less the demand **X** during lead time. Thus, expected value of $I(t)$ at end of cycle = $r - E(\mathbf{X})$.

At the beginning of a cycle, the inventory level at the end of the cycle is augmented by the arrival of an order of size q. Thus, expected value of $I(t)$ at beginning of cycle = $r - E(\mathbf{X}) + q$. Now (10) yields

$$\text{Expected value of } I(t) \text{ during cycle} = \tfrac{1}{2}(r - E(\mathbf{X}) + r - E(\mathbf{X}) + q)$$
$$= \tfrac{q}{2} + r - E(\mathbf{X})$$

Thus, expected annual holding cost $\cong h(\tfrac{q}{2} + r - E(\mathbf{X}))$.

To determine the expected annual cost due to stockouts or back orders, we must define

$$\mathbf{B}_r = \text{random variable representing the number of stockouts}$$
$$\text{or back orders during a cycle if the reorder point is } r$$

Now

$$\text{Expected annual shortage cost} = \left(\frac{\text{expected shortage cost}}{\text{cycle}}\right)\left(\frac{\text{expected cycles}}{\text{year}}\right)$$

By the definition of $\mathbf{B}_r$,

$$\frac{\text{Expected shortage cost}}{\text{Cycle}} = c_B E(\mathbf{B}_r)$$

Since all demand will eventually be met, an average of $\frac{E(\mathbf{D})}{q}$ orders will be placed each year. Then

$$\frac{\text{Expected shortage cost}}{\text{Year}} = \frac{c_B E(\mathbf{B}_r) E(\mathbf{D})}{q}$$

Finally,

$$\text{Expected annual order cost} = K\left(\frac{\text{expected orders}}{\text{year}}\right) = \frac{K E(\mathbf{D})}{q}$$

Putting together the expected annual holding, shortage, and ordering costs, we obtain

$$TC(q, r) = h\left(\frac{q}{2} + r - E(\mathbf{X})\right) + \frac{c_B E(\mathbf{B}_r) E(\mathbf{D})}{q} + \frac{K E(\mathbf{D})}{q} \tag{11}$$

Using the method described in Section 11.5, we could find the values of q and r that minimize (11) by determining values q^* and r^* of q and r satisfying

$$\frac{\partial TC(q^*, r^*)}{\partial q} = \frac{\partial TC(q^*, r^*)}{\partial r} = 0 \tag{12}$$

In Review Problem 7 we show how LINGO can be used to determine values of q and r that exactly satisfy (12). In most cases, however, the value of q^* satisfying (12) is very close to the EOQ† of $(\frac{2KE(\mathbf{D})}{h})^{1/2}$. For this reason, we assume that the optimal order quantity q^* may be adequately approximated by the EOQ. Given a value q for the order quantity, we now show how marginal analysis can be used to determine a reorder point r^* that minimizes $TC(q, r)$.

If we assume a given value of q, the expected annual ordering cost is independent of r. Thus, in determining a value of r that minimizes $TC(q, r)$, we may concentrate on minimizing the sum of the expected annual holding and shortage costs. Following the marginal analysis approach of Sections 16.2–16.3, suppose we increase the reorder point (for Δ small) from r to $r + \Delta$ (with q fixed). Will this result in an increase or a decrease in $TC(q, r)$?

If we increase r to $r + \Delta$, the expected annual holding cost will increase by

$$h\left(\frac{q}{2} + r + \Delta - E(\mathbf{X})\right) - h\left(\frac{q}{2} + r - E(\mathbf{X})\right) = h\Delta$$

If we increase the reorder point from r to $r + \Delta$, expected annual stockout costs will be reduced, because of the fact that during any cycle in which lead time demand is at least r, the number of stockouts during the cycle will be reduced by Δ units. In other words, increasing the reorder point from r to $r + \Delta$ will reduce stockout costs by $c_B\Delta$ during a fraction $P(\mathbf{X} \geq r)$ of all cycles. Since there are an average of $\frac{E(\mathbf{D})}{q}$ cycles per year, increasing the reorder point from r to $r + \Delta$ will reduce expected annual stockout cost by

$$\frac{\Delta E(\mathbf{D})c_B P(\mathbf{X} \geq r)}{q}$$

Observe that as r increases, $P(\mathbf{X} \geq r)$ decreases, so as r increases, the expected reduction in expected annual shortage cost resulting from increasing the reorder point by Δ will decrease. This observation allows us to draw Figure 4.

Let r^* be the value of r for which marginal benefit equals marginal cost, or

$$\frac{\Delta E(\mathbf{D})c_B P(\mathbf{X} \geq r^*)}{q} = h\Delta$$

$$P(\mathbf{X} \geq r^*) = \frac{hq}{c_B E(\mathbf{D})}$$

Suppose that $r < r^*$. Then Figure 4 shows that if we increase the reorder point from r to r^*, we can save more in shortage cost than we lose in holding cost. Now suppose that $r > r^*$. Figure 4 shows that by reducing the reorder point from r to r^*, we can save more in holding cost than we lose in increased shortage cost. Thus, r^* does attain the optimal trade-off between shortage and holding costs. In summary, if we assume that the order quantity can be approximated by

$$\text{EOQ} = \left(\frac{2KE(\mathbf{D})}{h}\right)^{1/2}$$

then we have the reorder point r^* and the order quantity q^* for the back-ordered case:

$$q^* = \left(\frac{2KE(\mathbf{D})}{h}\right)^{1/2} \tag{13}$$

$$P(\mathbf{X} \geq r^*) = \frac{hq^*}{c_B E(\mathbf{D})}$$

If

$$\frac{hq^*}{c_B E(\mathbf{D})} > 1$$

†Brown (1967) has shown that for approximating the optimal value of q, the EOQ is usually acceptable unless EOQ $\leq \sigma_{\mathbf{X}}$.

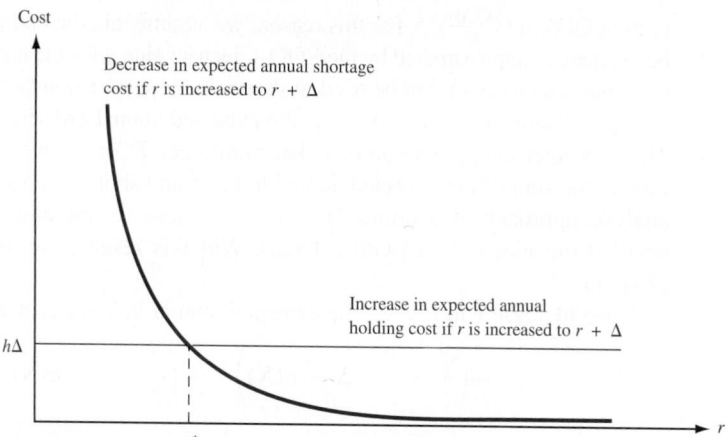

FIGURE 4
Trade-off between Holding Cost and Shortage Cost

Cost

Decrease in expected annual shortage cost if r is increased to $r + \Delta$

$h\Delta$

Increase in expected annual holding cost if r is increased to $r + \Delta$

r^*

r

then (13) will have no solution, and holding cost is prohibitively high relative to the stock-out cost. Management should set the reorder point at the smallest acceptable level. If (13) yields a negative value of r^*, management should also set the reorder point at the smallest acceptable level.

REMARKS **1** $P(X \geq r)$ is just the probability that a stockout will occur during a lead time. Also note that for h near zero, (13) yields a stockout probability near zero. For large c_B also, (13) yields a stockout probability near zero. Both of these results should be consistent with intuition.
2 After substituting the EOQ for q in (13), we may easily determine an approximately optimal value of r, the reorder point. Note that $r - E(X)$ is the amount in excess of expected lead time demand that is ordered to protect against the occurrence of stockouts during the lead time. For this reason, $r - E(X)$ is often referred to as **safety stock.**
3 From (11), we find that the expected annual cost of holding safety stock is $h(r - E(X)) = h$(safety stock level).

The following example illustrates the determination of the reorder point and safety stock level in the back-ordered demand case.

EXAMPLE 5 **Disk Stock**

Each year, a computer store sells an average of 1,000 boxes of disks. Annual demand for boxes of disks is normally distributed with a standard deviation of 40.8 boxes. The store orders disks from a regional distributor. Each order is filled in two weeks. The cost of placing each order is $50, and the annual cost of holding one box of disks in inventory is $10. The per-unit stockout cost (because of loss of goodwill and the cost of placing a special order) is assumed to be $20. The store is willing to assume that all demand is backlogged. Determine the proper order quantity, reorder point, and safety stock level for the computer store. Assume that annual demand is normally distributed. What is the probability that a stockout occurs during the lead time?

Solution We begin by determining the EOQ. Since $h = \$10$/box/year, $K = \$50$, and $E(\mathbf{D}) = 1,000$, we find that

$$\text{EOQ} = \left(\frac{2(50)(1,000)}{10}\right)^{1/2} = 100$$

We now substitute $q^* = 100$ in (13) and use (13) to determine the reorder point. To do this, we need to determine the probability distribution of $\mathbf{X}$, the lead time demand. Since $L = 2$ weeks, $\mathbf{X}$ will be normally distributed with

$$E(\mathbf{X}) = \frac{E(\mathbf{D})}{26} = \frac{1,000}{26} = 38.46 \qquad \text{and} \qquad \sigma_{\mathbf{X}} = \frac{\sigma_{\mathbf{D}}}{\sqrt{26}} = \frac{40.8}{\sqrt{26}} = 8$$

Since $c_B = \$20$, (13) now yields

$$P(\mathbf{X} \geq r) = \frac{10(100)}{20(1,000)} = .05 \qquad (14)$$

We use the Excel function NORMINV. Since

$$=\text{NORMINV}(0.95, 38.46, 8)$$

yields 51.62, we find that the safety stock level is $r - E(\mathbf{X}) = 51.62 - 38.46 = 13.16$.

To see how the reorder point and safety stock level would be affected by a variable lead time, suppose that the lead time has a mean of two weeks but also has a standard deviation of one week ($\frac{1}{52}$ year). Then (8') yields

$$\sigma_{\mathbf{X}}^2 = (\tfrac{1}{26})(40.8)^2 + (1,000)^2 \, (\tfrac{1}{52})^2 = 64.02 + 369.82 = 433.84$$
$$\sigma_{\mathbf{X}} = \sqrt{433.84} = 20.83$$

Assuming that the lead time demand is normally distributed, we would find that $r = 38.46 + 1.65(20.83) = 72.83$, and the safety stock held is $1.65(20.83) = 34.37$. Thus, the variability of the lead time has more than doubled the required safety stock level!

Determination of Reorder Point: The Lost Sales Case

We now assume that all stockouts result in lost sales and that a cost of c_{LS} dollars is incurred for each lost sale. (In addition to penalties for loss of future goodwill, c_{LS} should include profit lost because of a lost sale.)

As in the back-ordered case, we assume that the optimal order quantity can be adequately approximated by the EOQ and attempt to use marginal analysis to determine the optimal reorder point r^* (see Problem 6 at the end of this section). The optimal order quantity q^* and the reorder point r^* for the lost sales case are

$$q^* = \left(\frac{2KE(\mathbf{D})}{h} \right)^{1/2}$$
$$P(\mathbf{X} \geq r^*) = \frac{hq^*}{hq^* + c_{LS}E(\mathbf{D})} \qquad (15)$$

The key to the derivation of (15) is to realize that expected inventory in lost sales case = (expected inventory in back-ordered case) + (expected number of shortages per cycle). This equation follows because in the lost sales case, we find that during each cycle, an average of (expected shortages per cycle) fewer orders will be filled from inventory, thereby raising the average inventory level by an amount equal to expected shortages per cycle. Observe that the right-hand side of (15) is smaller than the right-hand side of (13). Thus, the lost sales assumption will yield a lower stockout probability (and a larger reorder point and safety stock level) than the back-ordered assumption.

To illustrate the use of (15), we continue our discussion of Example 5. Suppose that each box of disks sells for \$50 and costs the store \$30. Assuming that the stockout cost of $20 given in Example 5 represents lost goodwill, we obtain c_{LS} by adding the lost profit ($50 − $30) to the lost goodwill of $20. Thus, $c_{LS} = 20 + 20 = 40$. Recall from Example 5 that $E(\mathbf{D}) = 1,000$ boxes per year, $h = \$10/\text{box/year}$, EOQ = 100 boxes, and $K = \$50$. Now (15) yields

$$P(\mathbf{X} \geq r^*) = \frac{10(100)}{10(100) + 40(1,000)} = .024$$

Excel is used to compute r. Since

$$=\text{NORMINV}(.976, 38.46, 8)$$

yields 54.28, we find that $r = 54.28$. Thus, in the lost sales case, the safety stock level is $54.28 − 38.46 = 15.82$.

Continuous Review (r, q) Policies

A continuous review inventory policy, in which we order a quantity q whenever our inventory level reaches a reorder level r, is often called an **(r, q) policy.** An (r, q) policy is also called a **two-bin** policy, because it can easily be implemented by using two bins to store an item. For example, to implement a (30, 500) policy, we fill orders from bin 1 as long as bin 1 contains any items. As soon as bin 1 becomes empty, we know that the reorder point $r = 30$ has been reached, and we place an order for $q = 500$ units. When the order arrives, we bring the number of units in bin 2 up to 30, and place the remainder of the 500 units ordered in bin 1. Thus, whenever bin 1 has been emptied, we know that the reorder point has been reached.

Continuous Review (s, S) Policies

In our derivation of the best (r, q) policy, we assumed that an order could be placed exactly at the point when the inventory level reached the reorder point r. We used this assumption to compute the expected inventory level at the beginning and end of a cycle. Suppose that a demand for more than one unit can arrive at a particular time. Then an order may be triggered when the inventory level is less than r, and our computation of expected inventory level at the end and beginning of a cycle is then incorrect. For example, suppose $r = 30$ and our current inventory level is 35. If an order for 10 units arrives, an order will be placed when the inventory level is 25 (not $r = 30$), and this invalidates the computations that led to (11). From this discussion, we see that it is possible for the inventory level to "undershoot" the reorder point.

Note that this problem could not occur if all demands were for one unit, for then the inventory level would drop from (say) 32 to 31 and then to 30, and each order would be placed when the inventory level equaled the reorder point r. From this example, we see that if demands of size greater than one unit can occur at a point in time, then the (r, q) model may not yield a policy that minimizes expected annual cost.

In such situations, it has been shown that an **(s, S) policy** is optimal. To implement an (s, S) policy, we place an order whenever the inventory level is less than or equal to s. The size of the order is sufficient to raise the inventory level to S (assuming zero lead time). For example, if we were implementing a (5, 40) policy and the inventory level suddenly dropped from 7 to 3, we would immediately place an order for $40 − 3 = 37$ units. Exact computation of the optimal (s, S) policy is difficult. If we neglect the problem of the "undershoots," however, we may approximate the optimal (s, S) policy as follows. Set $S − s$ equal to the economic order quantity q. Then set s equal to the reorder point r obtained from (13) or (15). Finally, we obtain $S = r + q$. Thus, for Example 5 (with back orders allowed), we would set $s = 51.66$ and $S = 51.66 + 100 = 151.66$ and use (assuming that fractional demand is possible) a (51.66, 151.66) policy.

PROBLEMS

Group A

1 A hospital orders its blood from a regional blood bank. Each year, the hospital uses an average of 1,040 pints of Type O blood. Each order placed with the regional blood bank incurs a cost of $20. The lead time for each order is one week. It costs the hospital $20 to hold 1 pint of blood in inventory for a year. The per-pint stockout cost is estimated to be $50. Annual demand for Type O blood is normally distributed, with standard deviation of 43.26 pints. Determine the optimal order quantity, reorder point, and safety stock level. Assume that 52 weeks = 1 year and that all demand is backlogged. To use the techniques of this section, what unrealistic assumptions must be made? What (s, S) policy would be used in this situation?

2 Furnco sells secretarial chairs. Annual demand is normally distributed, with mean of 1,040 chairs and standard deviation of 50.99 chairs. Furnco orders its chairs from its flagship store. It costs $100 to place an order, and the lead time is two weeks. Furnco estimates that each stockout causes a loss of $50 in future goodwill. Furnco pays $60 for each chair and sells it for $100. The annual cost of holding a chair in inventory is 30% of its purchase cost.

   a Assuming that all demand is backlogged, what are the reorder point and the safety stock level?

   b Assuming that all stockouts result in lost sales, determine the optimal reorder point and the safety stock level.

3 We are given the following information for a product:

      Order cost = $50

      Annual demand = N(960, 3,072.49)

      Annual holding cost = $6/item/year

      Shortage cost = $80 per unit

      Lead time = one month

      Sales price = $40 per unit

      Product cost = $30 per unit

   a Determine the order quantity and the reorder point under the assumption that all demands are backordered.

   b Determine the order quantity and reorder point under the lost sales assumption.

4 The lead time demand for bathing suits is governed by the discrete random variable shown in Table 11. The company sells an average of 10,400 suits per year. The cost of placing an order for bathing suits is $30, and the cost of holding one bathing suit in inventory for a year is $3. The stockout cost is $3 per bathing suit. Use marginal analysis to determine the optimal order quantity and the reorder point.

TABLE 11

Lead Time Demand	Probability
180	.30
190	.30
200	.15
210	.10
220	.15

5 In Figure 3, assume that demand occurs at a constant rate during each cycle. Approximate the average level of on-hand inventory between $t = 0$ and $t = 7$. Also approximate the average number of shortages. Does the assumption that the average shortage level is small relative to the average level of on-hand inventory seem valid here?

Group B

6 In this problem, use marginal analysis to determine the optimal reorder point for the lost sales case.

   a Show that the average inventory level for the lost sales case may be written as

$$\tfrac{1}{2}[(r - E(\mathbf{X}) + E(\mathbf{B}_r)) + (r - E(\mathbf{X}) + E(\mathbf{B}_r) + q)]$$
$$= r - E(\mathbf{X}) + E(\mathbf{B}_r) + \tfrac{q}{2}$$

   b Although expected orders per year will no longer equal $\frac{E(\mathbf{D})}{q}$ (why?), we assume that the expected number of lost sales per year is relatively small. Thus, we may still assume that expected orders per year $= \frac{E(\mathbf{D})}{q}$. Now use marginal analysis to derive (15).

7 Suppose that a cost of S dollars (independent of the size of the stockout) is incurred whenever a stockout occurs during a cycle. Under the assumption of backlogged demand, use marginal analysis to determine the reorder point.

8 Explain the following statement: Faster-moving items require larger safety stocks than slower-moving items. (*Hint:* Does $\frac{q}{E(\mathbf{D})}$ large imply that an item is fast-moving or slow-moving?)

9 Suppose annual demand for a product is normally distributed, with a mean of 600 and a variance of 300. Suppose that the lead time for an order is always one month. Show (without using Equation (8)) that the lead time demand has mean 50, variance 25, and standard deviation 5. Assume that the demands during different one-month periods are independent, identically distributed random variables.

16.7 The EOQ with Uncertain Demand: The Service Level Approach to Determining Safety Stock Level

As we have previously stated, it is usually very difficult to determine accurately the cost of being one unit short. For this reason, managers often decide to control shortages by meeting a specified service level. In this section, we discuss two measures of service level:

Service Level Measure 1 SLM_1, the expected fraction (usually expressed as a percentage) of all demand that is met on time.

Service Level Measure 2 SLM_2, the expected number of cycles per year during which a shortage occurs.

Throughout this section, we assume that all shortages are backlogged. The following example illustrates the meaning of the two service level measures.

EXAMPLE 6   *SLM₁ and SLM₂*

Suppose that for a given inventory situation, average annual demand is 1,000 and the EOQ is 100. Demand during a lead time is random and is described by the probability distribution in Table 12. For a reorder point of 30 units, determine SLM_1 and SLM_2.

Solution   The expected demand during a lead time is $\frac{1}{5}(20) + \frac{1}{5}(30) + \frac{1}{5}(40) + \frac{1}{5}(50) + \frac{1}{5}(60) = 40$ units. With a reorder point of 30 units, we will reorder during each cycle at the instant when the inventory level hits 30 units. If the lead time demand during a cycle is 20 or 30 units, we will experience no shortage. During a cycle in which lead time demand is 40, a shortage of 10 units will occur; if lead time demand is 50, a shortage of 20 units will occur; if lead time demand is 60, a shortage of 30 units will occur. Hence, the expected number of units short per cycle is given by $\frac{1}{5}(0) + \frac{1}{5}(0) + \frac{1}{5}(10) + \frac{1}{5}(20) + \frac{1}{5}(30) = 12$.

Since the EOQ = 100 and all demand must eventually be met, the average number of orders placed each year will be $\frac{E(\mathbf{D})}{q} = \frac{1,000}{100} = 10$. Then the average number of shortages that occur during a year will equal $10(12) = 120$ units. Thus, each year, on the average, the demand for $1,000 - 120 = 880$ units is met on time. In this case, the $SLM_1 = \frac{880}{1,000} = 0.88$ or 88%. This shows that even if the reorder point is less than the mean lead time demand, a relatively high SLM_1 may result, because stockouts can only occur during the lead time, which is often a small portion of each cycle.

We now determine SLM_2 for a reorder point of 30. With a reorder point of 30, a stockout will occur during any cycle in which lead time demand exceeds 30 units. Thus, the probability of a stockout during a cycle $= P(\mathbf{X} = 40) + P(\mathbf{X} = 50) + P(\mathbf{X} = 60) = \frac{3}{5}$.

TABLE 12
Mass Function for Lead Time Demand

Lead Time Demand	Probability
20	$\frac{1}{5}$
30	$\frac{1}{5}$
40	$\frac{1}{5}$
50	$\frac{1}{5}$
60	$\frac{1}{5}$

Since there are an average of 10 cycles per year, the expected number of cycles per year that will result in shortages is $10 \left(\frac{3}{5} \right) = 6$. Thus, a reorder point of 30 yields $SLM_2 = 6$ stockouts per year.

Determination of Reorder Point and Safety Stock Level for SLM_1

Given a desired value of SLM_1, how do we determine a reorder point that provides the desired service level? Suppose we order the EOQ (q) and use a reorder point r. From Section 16.6,

$$\frac{\text{Expected shortages}}{\text{Cycle}} = E(\mathbf{B}_r)$$

$$\frac{\text{Expected shortages}}{\text{Year}} = \frac{E(\mathbf{B}_r)E(\mathbf{D})}{q}$$

Here, $E(\mathbf{D})$ is the average annual demand. Let SLM_1 be the percentage of all demand that is met on time. Then for given values of q (for the order quantity) and r (for the reorder point), we have

$$1 - SLM_1 = \frac{\text{expected shortages per year}}{\text{expected demand per year}} = \frac{E(\mathbf{B}_r)E(\mathbf{D})/q}{E(\mathbf{D})} = \frac{E(\mathbf{B}_r)}{q} \tag{16}$$

Equation (16) can be used to determine the reorder point that yields a desired service level. We now assume that the lead time demand is normally distributed, with mean $E(\mathbf{X})$ and standard deviation $\sigma_{\mathbf{X}}$. To use (16), we need to determine $E(\mathbf{B}_r)$. If $\mathbf{X}$ is normally distributed, the determination of $E(\mathbf{B}_r)$ requires a knowledge of the normal loss function.

DEFINITION ■ The **normal loss function**, $NL(y)$, is defined by the fact that $\sigma_{\mathbf{X}}NL(y)$ is the expected number of shortages that will occur during a lead time if (1) lead time demand is normally distributed with mean $E(\mathbf{X})$ and standard deviation $\sigma_{\mathbf{X}}$ and (2) the reorder point is $E(\mathbf{X}) + y\sigma_{\mathbf{X}}$. ■

In short, if we hold y standard deviations (in terms of lead time demand) of safety stock, then $NL(y)\sigma_{\mathbf{X}}$ is the expected number of shortages occurring during a lead time.

Since a larger reorder point leads to fewer shortages, we would expect $NL(y)$ to be a nonincreasing function of y. This is indeed the case. The function $NL(y)$ is tabulated in Table 13. For example, $NL(0) = 0.3989$ means that if the reorder point equals the expected lead time demand, and the standard deviation of lead time demand is $\sigma_{\mathbf{X}}$, then an average of $0.3989\sigma_{\mathbf{X}}$ shortages will occur during a lead time. Similarly, $NL(2) = 0.0085$ means that if the reorder point exceeds the mean lead time demand by $2\sigma_{\mathbf{X}}$, then an average of $0.0085\sigma_{\mathbf{X}}$ shortages will occur during a given lead time. $NL(y)$ is not tabulated for negative values of y. This is because it can be shown that for $y \leq 0$, $NL(y) = NL(-y) - y$. For example, $NL(-2) = NL(2) + 2 = 2.0085$. This means that if the reorder point is $2\sigma_{\mathbf{X}}$ less than the mean lead time demand, an average of $2.0085\sigma_{\mathbf{X}}$ shortages will occur during each cycle.

LINGO with the @**PSL** function may be used to compute values of the normal loss function. In LINGO, the program

```
MODEL:
  x = @PSL(2);
END
```

will yield $x = .0085$.

TABLE 13
The Normal Loss Function

x	NL(x)	x	NL(x)	x	NL(x)
0.00	0.3989	0.40	0.2304	0.80	0.1202
0.01	0.3940	0.41	0.2270	0.81	0.1181
0.02	0.3890	0.42	0.2236	0.82	0.1160
0.03	0.3841	0.43	0.2203	0.83	0.1140
0.04	0.3793	0.44	0.2169	0.84	0.1120
0.05	0.3744	0.45	0.2137	0.85	0.1100
0.06	0.3697	0.46	0.2104	0.86	0.1080
0.07	0.3649	0.47	0.2072	0.87	0.1061
0.08	0.3602	0.48	0.2040	0.88	0.1042
0.09	0.3556	0.49	0.2009	0.89	0.1023
0.10	0.3509	0.50	0.1978	0.90	0.1004
0.11	0.3464	0.51	0.1947	0.91	0.09860
0.12	0.3418	0.52	0.1917	0.92	0.09680
0.13	0.3373	0.53	0.1887	0.93	0.09503
0.14	0.3328	0.54	0.1857	0.94	0.09328
0.15	0.3284	0.55	0.1828	0.95	0.09156
0.16	0.3240	0.56	0.1799	0.96	0.08986
0.17	0.3197	0.57	0.1771	0.97	0.08819
0.18	0.3154	0.58	0.1742	0.98	0.08654
0.19	0.3111	0.59	0.1714	0.99	0.08491
0.20	0.3069	0.60	0.1687	1.00	0.08332
0.21	0.3027	0.61	0.1659	1.01	0.08174
0.22	0.2986	0.62	0.1633	1.02	0.08019
0.23	0.2944	0.63	0.1606	1.03	0.07866
0.24	0.2904	0.64	0.1580	1.04	0.07716
0.25	0.2863	0.65	0.1554	1.05	0.07568
0.26	0.2824	0.66	0.1528	1.06	0.07422
0.27	0.2784	0.67	0.1503	1.07	0.07279
0.28	0.2745	0.68	0.1478	1.08	0.07138
0.29	0.2706	0.69	0.1453	1.09	0.06999
0.30	0.2668	0.70	0.1429	1.10	0.06862
0.31	0.2630	0.71	0.1405	1.11	0.06727
0.32	0.2592	0.72	0.1381	1.12	0.06595
0.33	0.2555	0.73	0.1358	1.13	0.06465
0.34	0.2518	0.74	0.1334	1.14	0.02034
0.35	0.2481	0.75	0.1312	1.15	0.06210
0.36	0.2445	0.76	0.1289	1.16	0.06086
0.37	0.2409	0.77	0.1267	1.17	0.05964
0.38	0.2374	0.78	0.1245	1.18	0.05844
0.39	0.2339	0.79	0.1223	1.19	0.05726

(Continued)

TABLE **13**
(Continued)

x	NL(x)	x	NL(x)	x	NL(x)
1.20	0.05610	1.60	0.02324	2.00	0.008491
1.21	0.05496	1.61	0.02270	2.01	0.008266
1.22	0.05384	1.62	0.02217	2.02	0.008046
1.23	0.05274	1.63	0.02165	2.03	0.007832
1.24	0.05165	1.64	0.02114	2.04	0.007623
1.25	0.05059	1.65	0.02064	2.05	0.007418
1.26	0.04954	1.66	0.02015	2.06	0.007219
1.27	0.04851	1.67	0.01967	2.07	0.007024
1.28	0.04750	1.68	0.01920	2.08	0.006835
1.29	0.04650	1.69	0.01874	2.09	0.006649
1.30	0.04553	1.70	0.01829	2.10	0.006468
1.31	0.04457	1.71	0.01785	2.11	0.006292
1.32	0.04363	1.72	0.01742	2.12	0.006120
1.33	0.04270	1.73	0.01699	2.13	0.005952
1.34	0.04179	1.74	0.01658	2.14	0.005788
1.35	0.04090	1.75	0.01617	2.15	0.005628
1.36	0.04002	1.76	0.01578	2.16	0.005472
1.37	0.03916	1.77	0.01539	2.17	0.005320
1.38	0.03831	1.78	0.01501	2.18	0.005172
1.39	0.03748	1.79	0.01464	2.19	0.005028
1.40	0.03667	1.80	0.01428	2.20	0.004887
1.41	0.03587	1.81	0.01392	2.21	0.004750
1.42	0.03508	1.82	0.01357	2.22	0.004616
1.43	0.03431	1.83	0.01323	2.23	0.004486
1.44	0.03356	1.84	0.01290	2.24	0.004358
1.45	0.03281	1.85	0.01257	2.25	0.004235
1.46	0.03208	1.86	0.01226	2.26	0.004114
1.47	0.03137	1.87	0.01195	2.27	0.003996
1.48	0.03067	1.88	0.01164	2.28	0.003882
1.49	0.02998	1.89	0.01134	2.29	0.003770
1.50	0.02931	1.90	0.01105	2.30	0.003662
1.51	0.02865	1.91	0.01077	2.31	0.003556
1.52	0.02800	1.92	0.01049	2.32	0.003453
1.53	0.02736	1.93	0.01022	2.33	0.003352
1.54	0.02674	1.94	0.009957	2.34	0.003255
1.55	0.02612	1.95	0.009698	2.35	0.003159
1.56	0.02552	1.96	0.009445	2.36	0.003067
1.57	0.02494	1.97	0.009198	2.37	0.002977
1.58	0.02436	1.98	0.008957	2.38	0.002889
1.59	0.02380	1.99	0.008721	2.39	0.002804

(Continued)

TABLE **13**
(Continued)

x	NL(x)	x	NL(x)	x	NL(x)
2.40	0.002720	2.80	0.0007611	3.20	0.0001852
2.41	0.002640	2.81	0.0007359	3.21	0.0001785
2.42	0.002561	2.82	0.0007115	3.22	0.0001720
2.43	0.002484	2.83	0.0006879	3.23	0.0001657
2.44	0.002410	2.84	0.0006650	3.24	0.0001596
2.45	0.002337	2.85	0.0006428	3.25	0.0001537
2.46	0.002267	2.86	0.0006213	3.26	0.0001480
2.47	0.002199	2.87	0.0006004	3.27	0.0001426
2.48	0.002132	2.88	0.0005802	3.28	0.0001373
2.49	0.002067	2.89	0.0005606	3.29	0.0001322
2.50	0.002004	2.90	0.0005417	3.30	0.0001273
2.51	0.001943	2.91	0.0005233	3.31	0.0001225
2.52	0.001883	2.92	0.0005055	3.32	0.0001179
2.53	0.001826	2.93	0.0004883	3.33	0.0001135
2.54	0.001769	2.94	0.0004716	3.34	0.0001093
2.55	0.001715	2.95	0.0004555	3.35	0.0001051
2.56	0.001662	2.96	0.0004398	3.36	0.0001012
2.57	0.001610	2.97	0.0004247	3.37	0.00009734
2.58	0.001560	2.98	0.0004101	3.38	0.00009365
2.59	0.001511	2.99	0.0003959	3.39	0.00009009
2.60	0.001464	3.00	0.0003822	3.40	0.00008666
2.61	0.001418	3.01	0.0003689	3.41	0.00008335
2.62	0.001373	3.02	0.0003560	3.42	0.00008016
2.63	0.001330	3.03	0.0003436	3.43	0.00007709
2.64	0.001288	3.04	0.0003316	3.44	0.00007413
2.65	0.001247	3.05	0.0003199	3.45	0.00007127
2.66	0.001207	3.06	0.0003087	3.46	0.00006852
2.67	0.001169	3.07	0.0002978	3.47	0.00006587
2.68	0.001132	3.08	0.0002873	3.48	0.00006331
2.69	0.001095	3.09	0.0002771	3.49	0.00006085
2.70	0.001060	3.10	0.0002672	3.50	0.00005848
2.71	0.001026	3.11	0.0002577	3.51	0.00005620
2.72	0.0009928	3.12	0.0002485	3.52	0.00005400
2.73	0.0009607	3.13	0.0002396	3.53	0.00005188
2.74	0.0009295	3.14	0.0002311	3.54	0.00004984
2.75	0.0008992	3.15	0.0002227	3.55	0.00004788
2.76	0.0008699	3.16	0.0002147	3.56	0.00004599
2.77	0.0008414	3.17	0.0002070	3.57	0.00004417
2.78	0.0008138	3.18	0.0001995	3.58	0.00004242
2.79	0.0007870	3.19	0.0001922	3.59	0.00004073

(Continued)

TABLE **13**
(Continued)

x	NL(x)	x	NL(x)	x	NL(x)
3.60	0.00003911	3.75	0.00002103	3.90	0.00001108
3.61	0.00003755	3.76	0.00002016	3.91	0.00001061
3.62	0.00003605	3.77	0.00001933	3.92	0.00001016
3.63	0.00003460	3.78	0.00001853	3.93	0.00000972
3.64	0.00003321	3.79	0.00001776	3.94	0.000009307
3.65	0.00003188	3.80	0.00001702	3.95	0.000008908
3.66	0.00003059	3.81	0.00001632	3.96	0.000008525
3.67	0.00002935	3.82	0.00001563	3.97	0.000008158
3.68	0.00002816	3.83	0.00001498	3.98	0.000007806
3.69	0.00002702	3.84	0.00001435	3.99	0.000007469
3.70	0.00002592	3.85	0.00001375	4.00	0.000007145
3.71	0.00002486	3.86	0.00001317		
3.72	0.00002385	3.87	0.00001262		
3.73	0.00002287	3.88	0.00001208		
3.74	0.00002193	3.89	0.00001157		

Source: From R. Peterson and E. Silver, *Decision Systems for Inventory and Production Planning,*
© 1998 John Wiley & Sons, New York. Reprinted with permission.

Assuming normal lead time demand, we now determine the reorder point r that will yield a desired level of SLM_1 (expressed as a fraction). A reorder point of r corresponds to holding

$$y = \frac{r - E(\mathbf{X})}{\sigma_{\mathbf{X}}}$$

standard deviations of safety stock. Now the definition of the normal loss function implies that during a lead time, a reorder point of r will yield an expected number of shortages $E(\mathbf{B}_r)$ given by

$$E(\mathbf{B}_r) = \sigma_{\mathbf{X}} NL\left(\frac{r - E(\mathbf{X})}{\sigma_{\mathbf{X}}}\right) \tag{17}$$

Substituting (17) into (16), we obtain the reorder point for SLM_1 with normal lead time demand:

$$1 - SLM_1 = \frac{\sigma_{\mathbf{X}} NL\left(\dfrac{r - E(\mathbf{X})}{\sigma_{\mathbf{X}}}\right)}{q}$$

$$NL\left(\frac{r - E(\mathbf{X})}{\sigma_{\mathbf{X}}}\right) = \frac{q(1 - SLM_1)}{\sigma_{\mathbf{X}}} \tag{18}$$

With the exception of r, all quantities in (18) are known. Thus, (18) and Table 13 can be used to determine the reorder point corresponding to a given level of SLM_1.

EXAMPLE 7　Bads, Inc.

Bads, Inc., sells an average of 1,000 food processors each year. Each order for food processors placed by Bads costs $50. The lead time is one month. It costs $10 to hold a food processor in inventory for one year. Annual demand for food processors is normally distributed, with a standard deviation of 69.28. For each of the following values of SLM_1, determine the reorder point: 80%, 90%, 95%, 99%, 99.9%.

Solution Note that $E(\mathbf{D}) = 1,000$, $K = \$50$, and $h = \$10$, so

$$q = \left[\frac{2(50)(1,000)}{10} \right]^{1/2} = 100$$

Also,

$$E(\mathbf{X}) = (\tfrac{1}{12})(1,000) = 83.33 \qquad \text{and} \qquad \sigma_{\mathbf{X}} = \frac{69.28}{\sqrt{12}} = 20$$

From (18), the reorder point for an 80% value of SLM_1 must satisfy

$$NL \left(\frac{r - 83.33}{20} \right) = \frac{100(1 - 0.80)}{20} = 1$$

From Table 13, we find that 1 exceeds any of the tabulated values of the normal loss function. Thus, the value of r must make $\frac{r-83.33}{20}$ a negative number. A little trial and error reveals that $NL(-0.9) = NL(0.9) + 0.9 = 1.004$. Hence,

$$\frac{r - 83.33}{20} = -0.9$$

$$r = 83.33 - 20(0.9) = 65.33$$

For $SLM_1 = 0.90$, Equation (18) shows that the reorder point must satisfy

$$NL \left(\frac{r - 83.33}{20} \right) = \frac{(1 - 0.90)100}{20} = 0.5$$

Again, 0.5 exceeds all tabulated values of the normal loss function. Hence, $\frac{r-83.33}{20}$ must be a negative number. A little trial and error reveals that $N(-0.19) = N(0.19) + 0.19 = 0.5011$. Thus, the reorder point for a 90% service level must satisfy

$$\frac{r - 83.33}{20} = -0.19$$

$$r = 83.33 - 20(0.19) = 79.53$$

A 90% service level can be attained by a reorder point that is less than the expected lead time demand.

To attain a 95% service level, r must satisfy

$$NL \left(\frac{r - 83.33}{20} \right) = \frac{(1 - 0.95)100}{20} = 0.25$$

Since $NL(0.34) = 0.2518$,

$$\frac{r - 83.33}{20} = 0.34$$

$$r = 83.33 + 20(0.34) = 90.13$$

For a 99% service level,

$$NL \left(\frac{r - 83.33}{20} \right) = \frac{(1 - 0.99)100}{20} = 0.05$$

Since $NL(1.25) = 0.0506$, we see that

$$\frac{r - 83.33}{20} = 1.25$$

$$r = 83.33 + 20(1.25) = 108.33$$

TABLE 14
Reorder Points for Various Service Levels

SLM_1	Reorder Point
80%	65.33
90%	79.53
95%	90.13
99%	108.33
99.9%	127.13

Finally, for a 99.9% service level, r must satisfy

$$\frac{r - 83.33}{20} = \frac{(1 - 0.999)100}{20} = 0.005$$

Since $NL(2.19) = 0.005$,

$$\frac{r - 83.33}{20} = 2.19$$

$$r = 83.33 + 20(2.19) = 127.13$$

In summary, the reorder points corresponding to the various values of SLM_1 are given in Table 14. Notice that to go from an 80% to a 90% service level, we must increase the reorder point by 14.20, but to go from a 90% to a 99.9% service level, the reorder point must be increased by 47.60. For higher service levels, a much greater increase in the reorder point is required to cause a commensurate increase in the service level.

Using LINGO to Compute the Reorder Point Level for SLM_1

Using the **@PSL** function in LINGO, it is a simple matter to compute the reorder point level for SLM_1. For example, to compute the reorder point for Example 7 corresponding to $SLM_1 = .90$ in LINGO, we would use the program

```
MODEL:
1) @PSL((R - 83.33)/20) = 100*(1 - SLM1)/20;
2) SLM1 = .9;
```

This program yields $r = 79.57$. Note that by altering the right-hand side of line 2 we can quickly compute the reorder points for various values of SLM_1.

Using Excel to Compute the Normal Loss Function

It can be shown that

$$NL(y) = \text{(height of normal density at } y) - y*\text{(probability standard normal}$$
$$\text{is greater than or equal to } y)$$

Normalloss.xls

In the file Normalloss.xls, we therefore compute $NL(y)$ with the Excel formula

$$=\text{NORMDIST(D3,0,1,0)-D3*(1-NORMSDIST(D3))}$$

FIGURE 5

	B	C	D	E
1		Computing		
2		Normal Loss Function		
3		y	2	
4		NL(y)	0.008491	
5				
6				

FIGURE 5

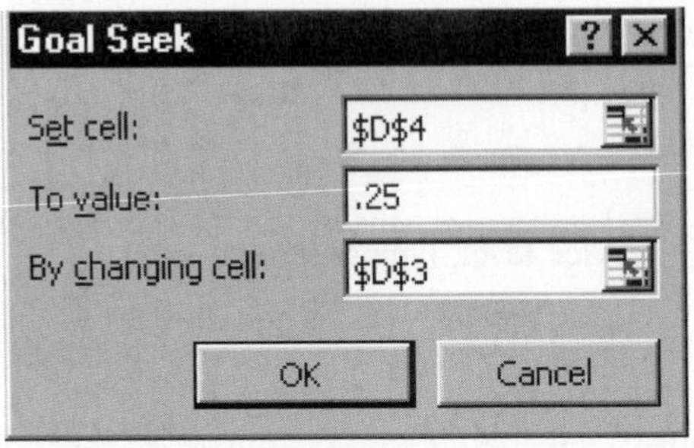

Goal Seek ? X

Set cell: $D$4

To value: .25

By changing cell: $D$3

OK   Cancel

FIGURE 6

	B	C	D	E
1		Computing		
2		Normal Loss Function		
3		y	0.344868	
4		NL(y)	0.25	
5				
6				

FIGURE 7

Recall that NORMDIST with last argument 0 computes the density function for a normal random variable, and NORMSDIST( ) computes the standardized normal cumulative probability. For example, we see from Figure 5 that (consistent with Table 14) $NL(2) = .008491$.

To illustrate the use of this spreadsheet, recall that in Example 7 we needed to find a value of y such that $NL(y) = .25$. To do this, we use Excel Goal Seek and fill in the Goal Seek dialog box as shown in Figure 6. This tells Excel to change cell D3 until cell D4 (the normal loss value) reaches .25. The result in Figure 7 shows us that $NL(.345) = .25$. Before doing Goal Seek, you should go to Tools Options Calculation Iteration and change the Maximum Change box to a very small number, such as .0000001. This makes Excel force the Set cell within .000001 of its desired value.

Determination of Reorder Point and Safety Stock Level for SLM_2

Suppose that a manager wants to hold sufficient safety stock to ensure that an average of s_0 cycles per year will result in a stockout. Given a reorder point of r, a fraction $P(\mathbf{X} > r)$ of all cycles will lead to a stockout. Since an average of $\dfrac{E(\mathbf{D})}{q}$ cycles per year will occur (remember we are assuming backlogging), an average of $\dfrac{P(\mathbf{X}>r)E(\mathbf{D})}{q}$ cycles per year will result in a stockout. Thus, given s_0, the reorder point is the smallest value of r satisfying

$$\frac{P(\mathbf{X} > r)\, E(\mathbf{D})}{q} \leq s_0 \qquad \text{or} \qquad P(\mathbf{X} > r) \leq \frac{s_0 q}{E(\mathbf{D})}$$

If $\mathbf{X}$ is a continuous random variable, then $P(\mathbf{X} > r) = P(\mathbf{X} \geq r)$. Thus, we obtain the reorder point r for SLM_2 for continuous lead time demand,

$$P(\mathbf{X} \geq r) = \frac{s_0 q}{E(\mathbf{D})} \tag{19}$$

and the reorder point for SLM_2 for discrete lead time demand, by choosing the smallest value of r satisfying

$$P(\mathbf{X} > r) \leq \frac{s_0 q}{E(\mathbf{D})} \tag{19'}$$

To illustrate the determination of the reorder point for SLM_2, we suppose that Bads, Inc., wants to ensure that stockouts occur during an average of two lead times per year. Recall from Example 7 that EOQ $= 100$, $E(\mathbf{D}) = 1,000$ units per year, and $\mathbf{X}$ is $N(83.33, 400)$. Now (19) yields $P(\mathbf{X} \geq r) = \frac{2(100)}{1,000} = .2$. The reorder point r is calculated using Excel. Since

$$=\text{NORMINV}(.8, 83.33, 20)$$

yields 100.16, we find that $r = 100.16$. The safety stock level yielding an average of two stockouts per year would be $100.16 - E(\mathbf{X}) = 16.83$.

PROBLEMS

Group A

1 For Problem 1 of Section 16.6, determine the reorder point that yields 80%, 90%, 95%, and 99% values of SLM_1. What reorder point would yield an average of 0.5 stockout per year?

2 For Problem 2 of Section 16.6, determine the reorder point that yields 80%, 90%, 95%, and 99% values of SLM_1. What reorder point would yield an average of two stockouts per year?

3 Suppose that the EOQ is 100, average annual demand is 1,000 units, and the lead time demand is a random variable having the distribution shown in Table 15.

   a What value of SLM_1 corresponds to a reorder point of 25?

   b If we wanted to attain a 95% value of SLM_1, what reorder point should we choose?

   c If we wanted an average of at most two stockouts per year, what reorder point should we choose?

TABLE 15

Lead Time Demand	Probability
10	$\frac{1}{6}$
15	$\frac{1}{4}$
20	$\frac{1}{4}$
25	$\frac{1}{12}$
30	$\frac{1}{4}$

4 A firm experiences demand with a mean of 100 units per day. Lead time demand is normally distributed, with a mean of 1,000 units and a standard deviation of 200 units. It costs $6 to hold one unit for one year. If the firm wants to meet 90% of all demand on time, what will be the annual cost of holding safety stock? (Assume that each order costs $50.)

16.8 (R, S) Periodic Review Policy[†]

In this section, we describe a widely used periodic review policy: the (R, S) policy. Before describing the operation of this policy, we need to define the concept of **on-order inventory level.** The on-order inventory level is simply the sum of on-hand inventory and inventory on order. Thus, if 30 units of a product are on hand, and we order 70 units (with a lead time of, say, one month), our on-order inventory level is 100.

[†]This section covers topics that may be omitted with no loss of continuity.

We can now describe the operation of the (R, S) inventory policy. Every R units of time (say, years), we review the on-hand inventory level and place an order to bring the on-order inventory level up to S. For example, if we were using a $(.25, 100)$ policy, we would review the inventory level at the end of each quarter. If $i < 100$ units were on hand, an order for $100 - i$ units would be placed. In general, an (R, S) policy will incur higher holding costs than a cost-minimizing (r, q) policy, but an (R, S) policy is usually easier to administer than a continuous review policy. With an (R, S) policy (unlike a continuous review policy), we can predict with certainty the times when an order will be placed. An (R, S) policy also allows a company to coordinate replenishments. For example, a company could use $R = 1$ month for all products ordered from the same supplier and then order all products from that supplier on the first day of each month.

We now assume that the review interval R has been determined and focus on the determination of a value for S that will minimize expected annual costs. Later in this section, we will discuss how to determine an appropriate value for R. We now assume that all shortages are backlogged and demand is a continuous random variable whose distribution remains unchanged over time. Finally, we assume that the per-unit purchase price is constant. This implies that annual purchasing costs do not depend on our choice of R and S. We define

$$R = \text{time (in years) between reviews}$$
$$\mathbf{D} = \text{demand (random) during a one-year period}$$
$$E(\mathbf{D}) = \text{mean demand during a one-year period}$$
$$K = \text{cost of placing an order}$$
$$J = \text{cost of reviewing inventory level}$$
$$h = \text{cost of holding one item in inventory for one year}$$
$$c_B = \text{cost per-unit short in the backlogged case (assumed to be}$$
$$\text{independent of the length of time until the order is filled)}$$
$$L = \text{lead time for each order (assumed constant)}$$
$$\mathbf{D}_{L+R} = \text{demand (random) during a time interval of length } L + R$$
$$E(\mathbf{D}_{L+R}) = \text{mean of } \mathbf{D}_{L+R}$$
$$\sigma_{\mathbf{D}_{L+R}} = \text{standard deviation of } \mathbf{D}_{L+R}$$

Given a value of R, we can now determine a value of S that minimizes expected annual costs. Our derivation mimics the derivation of (13). For a given choice of R and S, our expected costs are given by

$$\text{(Annual expected purchase costs)} + \text{(annual review costs)}$$
$$+ \text{(annual ordering costs)} + \text{(annual expected holding costs)}$$
$$+ \text{(annual expected shortage costs)}$$

Since $\frac{1}{R}$ reviews per year are placed, annual review costs are given by $\frac{J}{R}$. Also note that whenever an order is placed, the on-order inventory level will equal S. The only way that an order will not be placed at the next review point is if $\mathbf{D}_{L+R} = 0$. Since $\mathbf{D}_{L+R}$ is a continuous random variable, $\mathbf{D}_{L+R} = 0$ will occur with zero probability. Thus, an order is sure to be placed at the next review point (or any review point). This implies that annual ordering cost is given by $K(\frac{1}{R}) = \frac{K}{R}$. Observe that both the annual ordering cost and the review cost are independent of S. Thus, the value of S that minimizes annual expected costs will be the value of S that minimizes (annual expected holding costs) + (annual expected shortage costs).

To determine the annual expected holding cost for a given (R, S) policy, we first define a cycle to be the time interval between the arrival of orders. If we can determine the expected value of the average inventory level over a cycle, then expected annual holding cost is just h(expected value of on-hand inventory level over a cycle). As in our derivation of (11), we now assume that the average number of back orders is small relative to the average on-hand inventory level. Then, as in Section 16.6,

$$\text{Expected value of } I(t) \cong \text{expected value of } OHI(t)$$

Then expected value of $I(t)$ over a cycle may be approximated by 0.5(expected value of $I(t)$ right before an order arrives) + 0.5(expected value of $I(t)$ right after an order arrives).

Right before an order arrives, our maximum on-order inventory level (S) has been reduced by an average of $E(\mathbf{D}_{L+R})$. Thus, expected value of $I(t)$ right before an order arrives = $S - E(\mathbf{D}_{L+R})$.

Since $\frac{1}{R}$ orders are placed each year and an average of $E(\mathbf{D})$ units must be ordered each year, the average size of an order is $E(\mathbf{D})R$. Thus,

$$\text{Expected value of } I(t) \text{ right after an order arrives} = S - E(\mathbf{D}_{L+R}) + E(\mathbf{D})R$$

Then

$$\text{Expected value of } I(t) \text{ during a cycle} = S - E(\mathbf{D}_{L+R}) + \frac{E(\mathbf{D})R}{2}$$

Thus,

$$\text{Expected annual holding cost} = h\left[S - E(\mathbf{D}_{L+R}) + \frac{E(\mathbf{D})R}{2}\right]$$

From this expression, it follows that increasing S to $S + \Delta$ will increase expected annual holding costs by $h\Delta$.

We now focus on how an increase in S to $S + \Delta$ affects expected annual shortage costs. Then we can use marginal analysis to find the value of S that minimizes the sum of annual expected holding and shortage costs. Let's define the shortages "associated" with each order to be the shortages occurring in the time interval between the arrival of the order and the arrival of the next order. For example, an order placed at time 0 arrives at time L, and the next order will not arrive until time $R + L$. Thus, all shortages occurring between L and $R + L$ are associated with the time 0 order. Clearly, the sum of all shortages will equal the sum of the shortages associated with all orders. Let's again focus on the shortages associated with the time 0 order. Since the next order arrives at time $R + L$, and our time 0 order brought the on-order inventory level up to S, a shortage will be associated with the time 0 order if and only if the demand between time 0 and $R + L$ exceeds S. If a shortage occurs, the magnitude of the shortage will equal $\mathbf{D}_{L+R} - S$.

We can now use marginal analysis to determine (for a given R) the value of S that minimizes the sum of annual expected holding and shortage costs. If we increase S to $S + \Delta$, annual expected holding costs increase by $h\Delta$. Increasing S to $S + \Delta$ will decrease shortages associated with an order if $\mathbf{D}_{L+R} \geq S$. Thus, for a fraction $P(\mathbf{D}_{L+R} \geq S)$ of all orders, increasing S to $S + \Delta$ will save $c_B\Delta$ in shortage costs. Since $\frac{1}{R}$ orders are placed each year, increasing S to $S + \Delta$ will reduce expected annual shortage costs by $(\frac{1}{R})c_B\Delta P(\mathbf{D}_{L+R} \geq S)$. Marginal analysis then implies that the value of S minimizing the sum of annual expected holding and shortage costs will occur for the value of S satisfying

$$h\,\Delta = (\tfrac{1}{R})c_B\Delta P(\mathbf{D}_{L+R} \geq S)$$

or

$$P(\mathbf{D}_{L+R} \geq S) = \frac{Rh}{c_B} \tag{20}$$

Suppose that all shortages result in lost sales, and a cost of c_{LS} (including shortage cost plus lost profit) is incurred for each lost sale. Then the value of S minimizing the sum of annual expected holding and shortage costs is given by

$$P(\mathbf{D}_{L+R} \geq S) = \frac{Rh}{Rh + c_{LS}} \tag{21}$$

The following example illustrates the use of (20).

EXAMPLE 8 | **Lowland Appliance**

Lowland Appliance replenishes its stock of color TVs three times a year. Each order takes $\frac{1}{9}$ year to arrive. Annual demand for color TVs is $N(990, 1,600)$. The cost of holding one color TV in inventory for one year is $100. Assume that all shortages are backlogged, with a shortage cost of $150 per TV. When Lowland places an order, what should the on-order inventory be?

Solution   We are given that $R = \frac{1}{3}$ year, $L = \frac{1}{9}$ year, $R + L = \frac{4}{9}$ year, and $c_B = 150. $\mathbf{D}_{L+R}$ is normally distributed, with $E(\mathbf{D}_{L+R}) = \frac{4}{9}(990) = 440$ and $\sigma_{\mathbf{D}_{L+R}} = \sqrt{\frac{4}{9}}\sqrt{1,600} = 26.67$. From (20), S should be chosen to satisfy

$$P(\mathbf{D}_{L+R} \geq S) = \frac{(\frac{1}{3})\,100}{150} = .22$$

We use the Excel function NORMINV to compute s. Since

$$=\text{NORMINV}(0.78, 440, 26.67)$$

yields 460.59, when Lowland places an order for TVs, it should order enough to bring the on-order inventory level up to 460.59 (or 461) TVs. For example, if 160 TVs are in stock when a review takes place, $461 - 160 = 301$ TVs should be ordered.

Determination of R

Often, the review interval R is set equal to $\frac{EOQ}{E(\mathbf{D})}$. This makes the number of orders placed per year equal the number recommended if a simple EOQ model were used to determine the size of orders. Since each order is accompanied by a review, however, we must set the cost per order to $K + J$. This yields

$$\text{EOQ} = \sqrt{\frac{2(K + J)E(\mathbf{D})}{h}}$$

To illustrate the idea, suppose that it costs $500 to review the inventory level and $5,000 to place an order for TVs. Then

$$\text{EOQ} = \sqrt{\frac{2(5,500)(990)}{100}} = 330$$

This implies a review interval $R = \frac{330}{990} = \frac{1}{3}$ year.

Implementation of an (R, S) System

Retail stores (such as J. C. Penney's) often find an (R, S) policy easy to implement, because the quantity ordered equals the number of sales occurring during the period between reviews. For example, suppose a (1 month, 1,000) policy is being used, and orders are placed on the first day of each month. If 800 items were sold during January, then an order of 800 items must be placed at the beginning of February to bring the on-order inventory level back up to 1,000. By programming a computer to set monthly orders equal to monthly sales, such a policy can easily be implemented.

PROBLEMS

Group A

1 A hospital must order the drug Porapill from Daisy Drug Company. It costs $500 to place an order and $30 to review the hospital's inventory of the drug. Annual demand for the drug is $N(10,000, 640,000)$, and it costs $5 to hold one unit in inventory for one year. Orders arrive one month after being placed. Assume that all shortages are backlogged.

    a Estimate R and the number of orders per year that should be placed.

    b Using the answer in part (a), determine the optimal (R, S) inventory policy. Assume that the shortage cost per unit of the drug is $100.

2 Chicago's Treadway Tires Dealer must order tires from its national warehouse. It costs $10,000 to place an order and $400 to review the inventory level. Annual tire sales are $N(20,000, 4,000,000)$. It costs $10 per year to hold a tire in inventory, and each order arrives two weeks after being placed (52 weeks = 1 year). Assume that all shortages are backlogged.

    a Estimate R and the number of orders per year that should be placed.

    b Using the answer in part (a), determine the optimal (R, S) inventory policy. Assume that the shortage cost is $100 per tire.

3 Suppose we have found the optimal (R, S) policy for the back-ordered case and that $S = 50$. Is the following true or false?

    The optimal S for the lost sales case has $S > 50$.

16.9 The ABC Inventory Classification System

Many companies must develop inventory policies for thousands of items. In such a situation, a company cannot devote a great deal of attention to determining an "optimal" inventory policy for each item. The **ABC classification,** devised at General Electric during the 1950s, helps a company identify a small percentage of its items that account for a large percentage of the dollar value of annual sales. These items are called Type A items. Since most of the firm's inventory investment is in Type A items, concentrating effort on developing effective inventory control policies for these items should produce substantial savings.

Repeated studies have shown that in most companies, 5%–20% of all items stocked account for 55%–65% of sales; these are the Type A items. It has also been found that 20%–30% of all items account for 20%–40% of sales; these are called Type B items. Finally, it is often found that 50%–75% of all items account for only 5%–25% of sales; these are called Type C items. To illustrate how we determine which items are Type A, Type B, and Type C, consider a firm that stocks 100 items. We reorder the items as item 1, item 2, . . . , item 100, where item 1 generates the largest annual sales volume, item 2 generates the second largest annual sales volume, and so on. Then we plot the points (k, percentage of annual sales due to top k% of all items). For example, the point (20, 60) indicates that the top 20 items (from the standpoint of dollar sales) generate 60% of all sales. We then obtain a graph like Figure 8, where items 1–20 are Type A items, items 21–40 are Type B items, and items 41–100 are Type C items.

Since most of our inventory investment is in Type A items, high service levels will result in huge investments in safety stocks. Therefore, Hax and Candea (1984) recommend that SLM_1 be set at only 80%–85% for Type A items. Tight management control of ordering procedures is essential for Type A items; individual demand forecasts should be made for each Type A item. Also, every effort should be made to lower the lead time needed to receive orders or produce the item. If an (R, S) policy is used, R should be small—perhaps one week. This enables us to keep a close watch on inventory levels. Parameters such as estimates of annual mean demand, length of lead time, standard deviation of annual demand, and shortage costs should be reviewed fairly often.

For Type B items, Hax and Candea (1984) recommend that SLM_1 be set at 95%. Inventory policies for Type B items can generally be controlled by computer. Parameters for

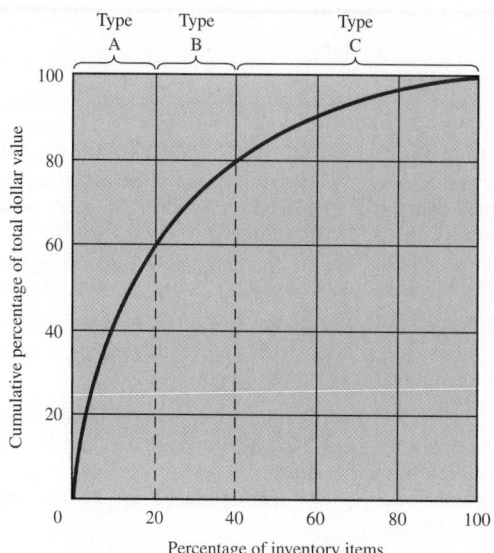

FIGURE 8
Example of ABC
Classification of
Inventory

Type B items should be reviewed less often than for Type A items.

For Type C items, a simple two-bin system is usually adequate. Parameters may be reviewed once or twice a year. Demand for Type C items may be forecast by simple extrapolation methods. A high value of SLM_1 (usually 98%–99%) is recommended. Little extra investment in safety stock will be required to maintain these high service levels.

DEVRO Incorporated, a producer of edible sausage casings, implemented an ABC analysis of its spare parts inventory and found that 2.5% of all items (the Type A items) accounted for 49% of all dollar usage, and 24.7% of all items (the Type B items) accounted for 38% of all dollar usage. By preparing requisition forms in advance for Type A and Type B items, DEVRO was able to substantially reduce the lead time needed to obtain those items. This helped DEVRO effect substantial savings in annual inventory costs. See Flowers and O'Neill (1978) for details.

PROBLEMS

Group A

1 Develop an ABC graph for the data in Table 16. Which items should be classified A, B, and C?

TABLE 16

Item	Annual Usage	Unit Cost (in dollars)
1	20,000	20
2	23,000	10
3	20,000	3
4	30,000	2
5	5,000	10
6	10,000	7
7	1,000	30
8	2,000	15
9	3,000	10
10	5,000	6

16.10 Exchange Curves

In many situations, it is difficult to estimate holding and shortage costs accurately. **Exchange curves** can be used in such situations to identify "reasonable" inventory policies. Consider a company that stocks two items (1 and 2). Many different ordering policies are possible. For example, the company may order item 1 five times a year and item 2 ten times a year (policy 1), or it may order each item once per year (policy 2). Clearly, policy 1 will result in higher ordering costs than policy 2, but policy 2 will result in higher holding costs and a higher average inventory level than policy 1. An exchange curve enables us to display graphically the trade-off between annual ordering costs and average inventory investment.

To illustrate the construction of an exchange curve, suppose a company stocks two items (item 1 and 2), and suppose that

c_i = cost of purchasing each unit of product i

h = cost of holding \$1 worth of either product in inventory for one year

K_i = order cost for product i

q_i = EOQ for product i

D_i = annual demand for product i

Then

$$q_i = \sqrt{\frac{2K_i D_i}{hc_i}}$$

Suppose the company wants to minimize the sum of annual ordering and holding costs. Then it should follow an EOQ policy for each product and order q_i of product i $\dfrac{D_i}{q_i}$ times per year. Two measures of effectiveness for this (or any other) ordering policy are

AII = average dollar value of inventory cost

AOC = annual ordering cost

If we follow the EOQ policy for each product, then

$$\begin{aligned}
\text{AII} &= \left(\frac{q_1}{2}\right)c_1 + \left(\frac{q_2}{2}\right)c_2 \\[2mm]
&= \left(\frac{1}{2}\right)\left\{ c_1\sqrt{\frac{2K_1 D_1}{c_1 h}} + c_2\sqrt{\frac{2K_2 D_2}{c_2 h}} \right\} \\[2mm]
&= \left(\frac{\sqrt{2}}{2\sqrt{h}}\right)\left\{ \sqrt{K_1 D_1 c_1} + \sqrt{K_2 D_2 c_2} \right\}
\end{aligned}$$

$$\begin{aligned}
\text{AOC} &= K_1\left(\frac{D_1}{q_1}\right) + K_2\left(\frac{D_2}{q_2}\right) \\[2mm]
&= K_1 D_1 \sqrt{\frac{c_1 h}{2K_1 D_1}} + K_2 D_2 \sqrt{\frac{c_2 h}{2K_2 D_2}} \\[2mm]
&= \left(\frac{\sqrt{2h}}{2}\right)\left\{ \sqrt{K_1 D_1 c_1} + \sqrt{K_2 D_2 c_2} \right\}
\end{aligned}$$

The expression for AII follows from the fact that the average inventory level of an item equals half the order quantity. The expression for AOC follows from the fact that $\dfrac{D_i}{q_i}$ orders per year are placed for item i.

Since h is often hard to estimate, let's suppose h is unknown and look at how a change in h affects AII and AOC. A plot of the points (AOC, AII) associated with each value of h is known as an **exchange curve.** For any point on the exchange curve, we see that

$$\text{AII(AOC)} = (\tfrac{1}{2})\{\sqrt{K_1 D_1 c_1} + \sqrt{K_2 D_2 c_2}\}^2 \tag{21}$$

This shows that the exchange curve is a hyperbola. Also, any point on the exchange curve satisfies $\frac{\text{AII}}{\text{AOC}} = \frac{1}{h}$ or $\frac{\text{AOC}}{\text{AII}} = h$. Thus, for any point on the exchange curve, the annual holding cost per dollar of inventory is the ratio of the x-coordinate to the y-coordinate. This shows how each point on the exchange curve can be identified with a value of h.

We now illustrate the computation of an exchange curve and show how the exchange curve can be used as an aid in decision making.

EXAMPLE 9 | **Exchange Curve**

A company stocks two products. Relevant information is given in Table 17.

1 Draw an exchange curve.

2 Currently, the company is ordering each product ten times per year. Use the exchange curve to demonstrate to management that this is an unsatisfactory ordering policy.

3 Suppose that management limits the company's average inventory investment to $10,000. Use the exchange curve to determine an appropriate ordering policy.

Solution **1** From (21), we find the equation of the exchange curve to be

$$(\text{AII})(\text{AOC}) = (\tfrac{1}{2})\{\sqrt{50(10,000)(200)} + \sqrt{80(20,000)(2.5)}\}^2$$
$$= 72,000,000$$

Some representative points on the exchange curve, along with the associated value of h, are given in Table 18. The exchange curve is graphed in Figure 9.

2 If the company orders each product ten times per year,

$$\text{AOC} = 10(\$50) + 10(\$80) = \$1,300$$
$$\text{AII} = \tfrac{1}{2}(1,000)(\$200) + \tfrac{1}{2}(2,000)(\$2.50) = \$102,500$$

TABLE 17
Relevant Information for Example 9

	K_i	D_i	c_i
Product 1	$50	10,000	$200
Product 2	$80	20,000	$2.50

TABLE 18
Points on Exchange Curve

AOC	AII	h
$2,000	$36,000	.06
$3,000	$24,000	.13
$4,000	$18,000	.22
$5,000	$14,400	.35
$6,000	$12,000	.50
$8,000	$9,000	.89

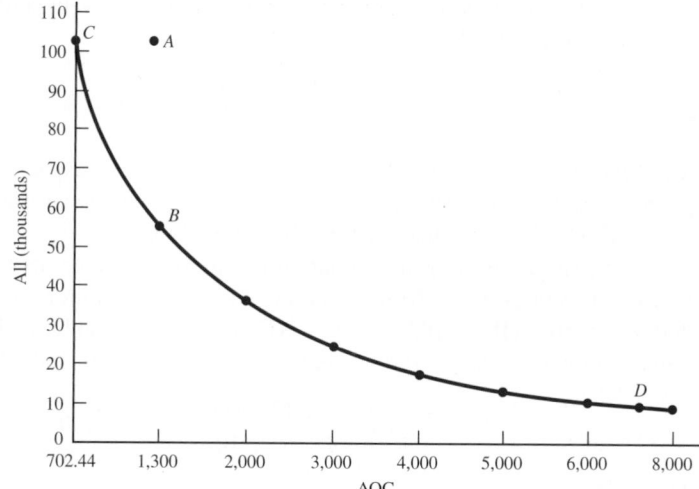

FIGURE 9
Example of an Exchange Curve

This is point A in Figure 9. Observe that point $B = (1,300, 55,385)$, corresponding to $h = .02$, yields the same AOC as the current policy, but a much lower AII. Also, point $C = (702.44, 102,500)$, corresponding to $h = .01$, yields the same AII as the current policy, but a much lower AOC. Thus, we can use the exchange curve to show the manager how to improve on the current ordering policy.

3 From the exchange curve, we find that $D = (7,200, 10,000)$ is on the exchange curve. Thus, for a \$10,000 AII, the best we can do is to hold ordering costs to \$7,200. Of course, the manager could opt for AII = \$9,000 and AOC = \$8,000 or one of many other possibilities. The point is that the exchange curve clarifies many of the options available to management.

Exchange Curves for Stockouts

Exchange curves can also be used to assess the trade-offs between average inventory investment (AII) and the expected number of lead times per year resulting in stockouts. To illustrate, consider a company stocking a single item for which

c = purchase cost per unit

K = setup cost

h = annual cost of holding one unit in inventory

c_B = cost of a stockout (we assume all items are back-ordered)

$E(\mathbf{D})$ = mean annual demand

q = economic order quantity

$\mathbf{X}$ = lead time demand

$E(\mathbf{X})$ = mean lead time demand

$\sigma_{\mathbf{X}}$ = standard deviation of lead time demand

r = reorder point (determined from Equation (13))

From (13), a fraction

$$\frac{qh}{c_B E(\mathbf{D})}$$

of all lead times will have a stockout. Since there are an average of $\frac{E(\mathbf{D})}{q}$ orders placed per year, an average of

$$\left(\frac{E(\mathbf{D})}{q}\right)\left(\frac{qh}{c_B E(\mathbf{D})}\right) = \frac{h}{c_B}$$

lead times per year will result in stockouts. We let SY = expected number of lead times per year resulting in stockouts. From (11), we know that the average inventory level is ($\frac{q}{2}$ + r − $E(\mathbf{X})$). Thus, we have AII = $c(\frac{q}{2} + r - E(\mathbf{X}))$.

An exchange curve for this situation is a graph of the points (AII, SY) corresponding to different values of c_B. To illustrate the construction of an exchange curve, let $E(\mathbf{X})$ = 200, $\sigma_X = 50$, $E(\mathbf{D}) = 100,000$, $K = \$12.50$, $h = \$10$, and $c = \$100$. We will find four points on the exchange curve by setting $c_B = \$1, \$5, \$10$, and $\$20$. First we find that

$$q = \sqrt{\frac{2(12.5)(100,000)}{10}} = 500$$

The stockout probabilities and SY are given in Table 19.

Using Table 2 in Chapter 12 or the Excel NORMSDIST( ) function, we can calculate the reorder point r for each value of c_B. Then we determine the average inventory level and AII = average inventory investment. These calculations are given in Table 20.

The exchange curve (based on the four points we have computed) is graphed in Figure 10. For example, the exchange curve shows us that if current AII is \$33,250, then for a \$3,400 increase in AII, we can reduce SY from 10 to 2, but an additional increase in AII of \$3,400 would decrease SY by less than 2.

Exchange Surfaces

Using more sophisticated techniques (see Gardner and Dannenbring (1979)), an **exchange surface** involving three or more quantities can be derived. The exchange surface in Figure 11 was derived from a sample of 500 items in a military distribution system. The

TABLE 19
Computation of SY

c_B	Stockout Probability = $\frac{qh}{c_B E(\mathbf{D})}$	$SY = \frac{h}{c_B}$
\$1	$\frac{500(10)}{1(100,000)} = .05$	$\frac{10}{1} = 10$
\$5	$\frac{500(10)}{5(100,000)} = .01$	$\frac{10}{5} = 2$
\$10	$\frac{500(10)}{10(100,000)} = .005$	$\frac{10}{10} = 1$
\$20	$\frac{500(10)}{20(100,000)} = .0025$	$\frac{10}{20} = 0.50$

TABLE 20
Calculation of AII

c_B	Reorder Point	Average Inventory Level	AII
\$1	200 + 50(1.65) = 282.5	250 + 282.5 − 200 = 332.5	\$33,250
\$5	200 + 50(2.33) = 316.5	250 + 316.5 − 200 = 366.5	\$36,650
\$10	200 + 50(2.58) = 329	250 + 329 − 200 = 379	\$37,900
\$20	200 + 50(2.81) = 340.5	250 + 340.5 − 200 = 390.5	\$39,050

FIGURE **10**
Exchange Curve for All
and *SY*

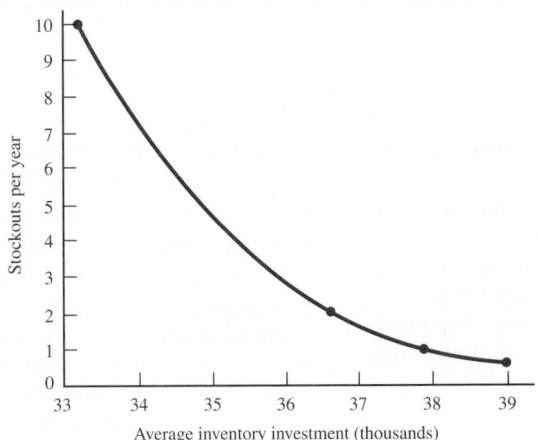

FIGURE **11**
Example of an
Exchange Surface[†]

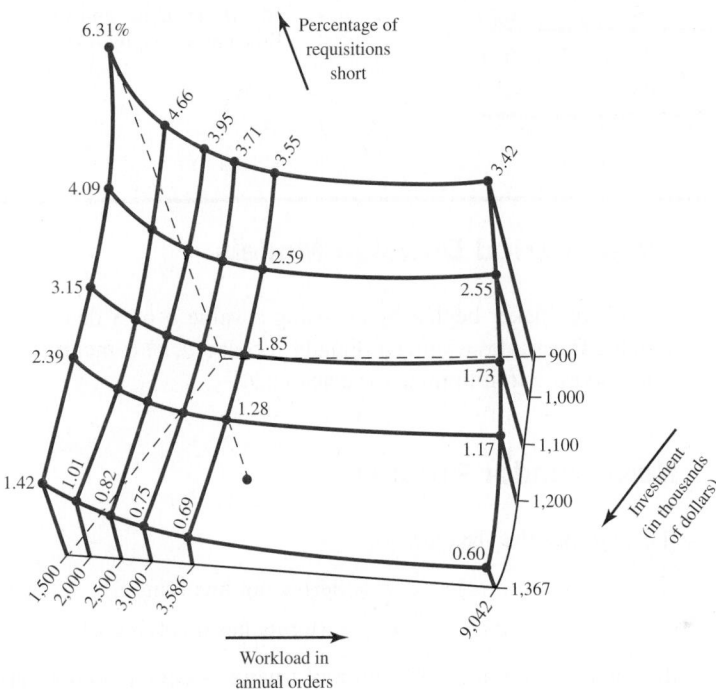

x-coordinate is the annual number of orders placed, the y-coordinate is the average inventory investment (in thousands of dollars), and the z-coordinate is the percentage of requests that yield shortages. For example, suppose the military has fixed a $900,000 average inventory investment. By varying the number of orders per year between 1,500 and 9,042, the military can vary the percentage of requests that yield shortages between 6.31% and 3.42%. Also, if annual orders are fixed at 3,000, then the percentage of requests yielding shortages can vary between 0.75% and 3.71%. An exchange surface makes it easy to identify the trade-offs involved between improving service, increased inventory investment, and increased work load (orders per year).

[†]Reprinted by permission of E. Gardner and D. Dannenbring, "Using Optimal Policy Surfaces to Analyze Aggregate Inventory Tradeoffs," *Management Science,* Vol. 25, No. 8, August 1979. Copyright 1979, the Institute of Management Sciences.

PROBLEMS

Group A

1 Consider a two-item inventory system with the attributes in Table 21.

a Draw an exchange curve for these products (use AOC and AII as the x- and y-coordinates).

b Currently, management is ordering each product twice a year. How can it improve on this strategy?

c The order costs correspond to machine setup times. Machine time is valued at $50 per hour. If management wants to limit machine setup time to 500 hours per year, what strategies are available?

2 Explain how to draw an exchange curve where the x-coordinate is AII and the y-coordinate is percentage of all requests for stock that result in shortages.

3 Consider the exchange surface in Figure 11. The current inventory policy has yielded 3,586 orders per year, an AII of $1,367,000, and 0.89% shortages.

a Without changing orders per year and AII, by how much can shortages be improved?

b If AII and shortages are maintained at current levels, by how much can orders per year be reduced?

c If shortages and orders per year are maintained at current levels, by how much can AII be reduced?

TABLE 21

	K_i	D_i	c_i
Product 1	$500	10,000	$2,000
Product 2	$800	20,000	$250

SUMMARY   Single-Period Decision Models

A decision maker begins by choosing a value q of a decision variable. Then a random variable **D** assumes a value d. Finally, a cost $c(d, q)$ is incurred. The decision maker's goal is to choose q to minimize expected cost.

News Vendor Problem

If $c(d, q)$ has the structure

$$c(d, q) = c_o q + \text{(terms not involving } q) \qquad (d \leq q) \tag{2}$$

$$c(d, q) = -c_u q + \text{(terms not involving } q) \qquad (d \geq q + 1) \tag{2.1}$$

the single-period decision model is a **news vendor problem.** Here

$$c_o = \text{per-unit overstocking cost}$$

$$c_u = \text{per-unit understocking cost}$$

If **D** is a discrete random variable, the optimal decision is given by the smallest value of q (q^*) satisfying

$$F(q^*) \geq \frac{c_u}{c_o + c_u} \tag{3}$$

If **D** is a continuous random variable, the optimal decision is the value of q (q^*) satisfying

$$P(\mathbf{D} \leq q^*) = \frac{c_u}{c_o + c_u} \tag{5}$$

Determination of Reorder Point and Order Quantity with Uncertain Demand: Minimizing Annual Expected Cost

Let

K = ordering cost

h = holding cost/unit/year

L = lead time for each order (assumed to be known with certainty)

q = order quantity

$\mathbf{D}$ = random variable representing annual demand, with mean $E(\mathbf{D})$, variance var $\mathbf{D}$, and standard deviation $\sigma_{\mathbf{D}}$

c_B = cost incurred for each unit short if shortages are backlogged

c_{LS} = cost (including lost profits, lost goodwill) incurred for each lost sale if each shortage results in a lost sale

$\mathbf{X}$ = random variable representing lead time demand

Then

$$E(\mathbf{X}) = LE(\mathbf{D}), \qquad \text{var } \mathbf{X} = L \text{ (var } \mathbf{D}\text{)}, \qquad \sigma_{\mathbf{X}} = \sqrt{L}\sigma_{\mathbf{D}}$$

and r is the reorder point, or inventory level at which an order should be placed. Safety stock, $r - E(\mathbf{X})$, is the amount of inventory held in excess of lead time demand to meet shortages that may occur before an order arrives.

Assume that the optimal order quantity can be reasonably approximated by the EOQ, $\mathbf{D}$ is a continuous random variable, and all shortages are backlogged. Then annual expected cost is minimized by q^* and r^* given by

$$q^* = \left(\frac{2KE(\mathbf{D})}{h}\right)^{1/2}$$

$$P(\mathbf{X} \geq r^*) = \frac{hq^*}{c_B E(\mathbf{D})}$$

(13)

Assume that the optimal order quantity can be reasonably approximated by the EOQ, $\mathbf{D}$ is a continuous random variable, and all shortages result in lost sales. Then annual expected cost is minimized by q^* and r^* satisfying

$$q^* = \left(\frac{2KE(\mathbf{D})}{h}\right)^{1/2}$$

$$P(\mathbf{X} \geq r^*) = \frac{hq^*}{hq^* + c_{LS}E(\mathbf{D})}$$

(15)

Determination of Reorder Point: The Service Level Approach

Since it may be difficult to determine the exact cost of a shortage or lost sale, it is often desirable to choose a reorder point that meets a desired service level. Two common measures of service level are

Service Level Measure 1 SLM_1, the expected fraction (usually expressed as a percentage) of all demand that is met on time.

Service Level Measure 2 SLM_2, the expected number of cycles per year during which a shortage occurs.

If lead time is normally distributed, then for a desired value SLM_1, the reorder point r is found from

$$NL\left(\frac{r - E(\mathbf{X})}{\sigma_{\mathbf{X}}}\right) = \frac{q(1 - SLM_1)}{\sigma_{\mathbf{X}}} \qquad (18)$$

where $NL(y)$ is the **normal loss function,** tabulated in Table 13, and q is the EOQ.

If lead time demand is a continuous random variable, and we desire $SLM_2 = s_0$ shortages per year, the reorder point r is given by

$$P(\mathbf{X} \geq r) = \frac{s_0 q}{E(\mathbf{D})} \qquad (19)$$

Again, q is the EOQ.

If lead time demand is a discrete random variable, and we desire $SLM_2 = s_0$ shortages per year, the reorder point is the smallest value of r satisfying

$$P(\mathbf{X} > r) \leq \frac{s_0 q}{E(\mathbf{D})} \qquad (19')$$

Again, q is the EOQ.

(R, S) Periodic Review Policy

Every R units of time, we review the inventory level and place an order to bring our on-hand inventory level up to S. Given a value of R, we determine the value of S from

$$P(\mathbf{D}_{L+R} \geq S) = \frac{Rh}{c_B}$$

ABC Classification

The 5%–20% of all items accounting for 55%–65% of sales are Type A items; the 20%–30% of all items accounting for 20%–40% of sales are Type B items; and the 50%–75% of all items that account for 5%–25% of all sales are Type C items. By concentrating effort on Type A (and possibly Type B) items, we can achieve substantial cost reductions.

Exchange Curves

Exchange curves (and exchange surfaces) are used to display trade-offs between various objectives. For example, an exchange curve may display the trade-off between annual ordering costs and average dollar level of inventory. An exchange curve can be used to compare how various ordering policies compare with respect to several objectives.

REVIEW PROBLEMS

Group A

1 The Chocochip Cookie Store bakes its cookies every morning before opening. It costs the store 15¢ to bake each cookie, and each cookie is sold for 35¢. At the end of the day, leftover cookies may be sold to a thrift bakery for 5¢ per cookie. The number of cookies sold each day is described by the discrete random variable in Table 22.

TABLE 22

Demand (dozens)	Probability
20	.30
30	.20
40	.20
50	.15
60	.15

TABLE 23

Cash Needs	Probability
$4,000	.30
$5,000	.20
$6,000	.10
$7,000	.30
$8,000	.10

a How many dozen cookies should be baked before the store opens?

b If the daily demand (in dozens) for cookies is $N(50, 400)$, how many dozen cookies should be baked? A description of the $N(\mu, \sigma^2)$ notation can be found in Section 1.7.

c If the daily demand (in dozens) for cookies has a density function

$$f(d) = \frac{e^{-d/50}}{50} \qquad (d \geq 0)$$

how many dozen cookies should be baked?

2 An optometrist orders eyeglass frames at a cost of $40 per frame and sells each frame for $70. Annual holding cost is 20% of the optometrist's cost of purchasing a frame. Each time frames are ordered, a cost of $200 is incurred. Because of lost goodwill, a cost of $50 is incurred each time a customer wants a frame that is not in stock. Frames are delivered one week after an order is placed. Annual demand for frames is $N(1,040, 15.73)$.

a Assuming all shortages are backlogged, determine the order quantity and reorder point.

b Assuming all shortages result in lost sales, determine the order quantity and reorder point.

c To meet 95% of all orders from stock, what should be the reorder point?

d To have shortages occur during an average of two lead times per year, what should be the reorder point?

3 We are given the following information about a product:

Cost of placing an order = $100
Cost per item = $5
Sale price per item = $8
Annual holding cost = 40% of cost of item
Annual demand = 5,000 units
Lead time demand = $N(20, 900)$

a If the reorder point that minimizes expected cost is 80, what is the shortage cost? (Assume backlogging.)

b If the reorder point that minimizes expected cost is 80, what is the shortage cost? (Assume lost sales.)

c What reorder point would meet 90% of all demand on time?

d What reorder point would result in a stockout occurring during an average of 0.5 lead time per year?

Group B

4 A business believes that its needs for cash during the next month are described by the random variable shown in Table 23. At the beginning of the month, the business has $10,000 available, and the business manager must determine how much of the money should be placed in an account bearing 24% annual interest. If any money must be withdrawn before the end of the month, all interest on the withdrawn money is forfeited, and a penalty equal to 2% of the withdrawn money must be paid. How much money should be placed in the 24% annual interest account?

5 A fur dealer buys fur coats for $100 each and sells them for $200 each. He believes that the demand for coats is $N(100, 100)$. Any coat not sold can be sold to a discount house for $100, but the fur dealer believes he must charge himself a cost of 10¢ per dollar invested in a fur coat that is sold at discount. How many coats should the dealer order? If the price at which the dealer sold his coats increased (assuming demand is unchanged), would he buy more or fewer coats?

6 A company currently has two warehouses. Each warehouse services half the company's demand, and the annual demand serviced by each warehouse is $N(10,000, 1,000,000)$. The lead time for meeting demand is $\frac{1}{10}$ year. The company wants to meet 95% of all demand on time. Assume that the EOQ at each warehouse is 2,000.

a How much safety stock must be held?

b Show that, if the company had only one warehouse, it would hold less safety stock than it does when it has two warehouses.

c A young MBA argues, "By having one central warehouse, I can reduce the total amount of safety stock needed to meet 95% of all customer demands on time. Therefore, we can save money by having only one central warehouse instead of several branch warehouses." How might this argument be rebutted?

7 Use LINGO to determine the values of q and r that minimize expected annual cost for Example 5. How close are your answers to those given in the text?

REFERENCES

The following references emphasize applications over theory:

Brown, R. *Decision Rules for Inventory Management.* New York: Holt, Rinehart and Winston, 1967.

Peterson, R., and E. Silver. *Decision Systems for Inventory Management and Production Planning.* New York: Wiley, 1998.

Tersine, R. *Principles of Inventory and Materials Management.* New York: North-Holland, 1982.

Vollman, T., W. Berry, and C. Whybark. *Manufacturing Planning and Control Systems.* Homewood, Ill.: Irwin, 1997.

The following references contain extensive theoretical discussions as well as applications:

Hadley, G., and T. Whitin. *Analysis of Inventory Systems.* Englewood Cliffs, N.J.: Prentice Hall, 1963.

Hax, A., and D. Candea. *Production and Inventory Management.* Englewood Cliffs, N.J.: Prentice Hall, 1984.

Johnson, L., and D. Montgomery. *Operations Research in Production, Scheduling, and Inventory Control.* New York: Wiley, 1974.

Barron, H. "Payoff Matrices Pay Off at Hallmark," *Interfaces* 15(no. 4, 1985):20–25.

Bruno, J. "The Use of Monte-Carlo Techniques for Determining the Size of Substitute Teacher Pools," *Socio-Economic Planning Science* 4(1970):415–428.

Flowers, D., and J. O'Neill. "An Application of Classical Inventory Analysis to a Spare Parts Inventory," *Interfaces* 8(no. 2, 1978):76–79.

Rosenfeld, D. "Optimal Management of Tax-Sheltered Employment Reimbursement Programs," *Interfaces* 16(no. 3, 1986):68–72.

Virts, J., and R. Garrett. "Weighting Risk in Capacity Expansion," *Harvard Business Review* 48(1970).

For a discussion of exchange curves and surfaces, see:

Gardner, E., and D. Dannenbring. "Using Optimal Policy Surfaces to Analyze Aggregate Inventory Tradeoffs," *Management Science* 25(1979):709–720.

17

Markov Chains

Sometimes we are interested in how a random variable changes over time. For example, we may want to know how the price of a share of stock or a firm's market share evolves. The study of how a random variable changes over time includes stochastic processes, which are explained in this chapter. In particular, we focus on a type of stochastic process known as a Markov chain. Markov chains have been applied in areas such as education, marketing, health services, finance, accounting, and production. We begin by defining the concept of a stochastic process. In the rest of the chapter, we will discuss the basic ideas needed for an understanding of Markov chains.

17.1 What Is a Stochastic Process?

Suppose we observe some characteristic of a system at discrete points in time (labeled 0, 1, 2, . . .). Let X_t be the value of the system characteristic at time t. In most situations, X_t is not known with certainty before time t and may be viewed as a random variable. A **discrete-time stochastic process** is simply a description of the relation between the random variables $X_0, X_1, X_2, \ldots$. Some examples of discrete-time stochastic processes follow.

EXAMPLE 1   **The Gambler's Ruin**

At time 0, I have \$2. At times 1, 2, . . . , I play a game in which I bet \$1. With probability p, I win the game, and with probability $1 - p$, I lose the game. My goal is to increase my capital to \$4, and as soon as I do, the game is over. The game is also over if my capital is reduced to \$0. If we define X_t to be my capital position after the time t game (if any) is played, then $X_0, X_1, \ldots, X_t$ may be viewed as a discrete-time stochastic process. Note that $X_0 = 2$ is a known constant, but X_1 and later X_t's are random. For example, with probability p, $X_1 = 3$, and with probability $1 - p$, $X_1 = 1$. Note that if $X_t = 4$, then X_{t+1} and all later X_t's will also equal 4. Similarly, if $X_t = 0$, then X_{t+1} and all later X_t's will also equal 0. For obvious reasons, this type of situation is called a *gambler's ruin* problem.

EXAMPLE 2   **Choosing Balls from an Urn**

An urn contains two unpainted balls at present. We choose a ball at random and flip a coin. If the chosen ball is unpainted and the coin comes up heads, we paint the chosen unpainted ball red; if the chosen ball is unpainted and the coin comes up tails, we paint the chosen unpainted ball black. If the ball has already been painted, then (whether heads or tails has been tossed) we change the color of the ball (from red to black or from black to red). To model this situation as a stochastic process, we define time t to be the time after the coin has been flipped for the tth time and the chosen ball has been painted. The state at any time may be described by the vector $[u \quad r \quad b]$, where u is the number of unpainted balls in the urn, r is the number of red balls in the urn, and b is the number of black balls in the urn. We are given that $\mathbf{X}_0 = [2 \quad 0 \quad 0]$. After the first coin toss, one ball will have been painted either red or black, and the state will be either $[1 \quad 1 \quad 0]$ or $[1 \quad 0 \quad 1]$. Hence, we can be sure that $\mathbf{X}_1 = [1 \quad 1 \quad 0]$ or $\mathbf{X}_1 = [1 \quad 0 \quad 1]$. Clearly, there must be some sort of relation between the $\mathbf{X}_t$'s. For example, if $\mathbf{X}_t = [0 \quad 2 \quad 0]$, we can be sure that $\mathbf{X}_{t+1}$ will be $[0 \quad 1 \quad 1]$.

EXAMPLE 3   CSL Computer Stock

Let $\mathbf{X}_0$ be the price of a share of CSL Computer stock at the beginning of the current trading day. Also, let $\mathbf{X}_t$ be the price of a share of CSL stock at the beginning of the tth trading day in the future. Clearly, knowing the values of $\mathbf{X}_0, \mathbf{X}_1, \ldots, \mathbf{X}_t$ tells us something about the probability distribution of $\mathbf{X}_{t+1}$; the question is, what does the past (stock prices up to time t) tell us about $\mathbf{X}_{t+1}$? The answer to this question is of critical importance in finance. (See Section 17.2 for more details.)

We close this section with a brief discussion of continuous-time stochastic processes. A **continuous-time stochastic process** is simply a stochastic process in which the state of the system can be viewed at any time, not just at discrete instants in time. For example, the number of people in a supermarket t minutes after the store opens for business may be viewed as a continuous-time stochastic process. (Models involving continuous-time stochastic processes are studied in Chapter 20.) Since the price of a share of stock can be observed at any time (not just the beginning of each trading day), it may be viewed as a continuous-time stochastic process. Viewing the price of a share of stock as a continuous-time stochastic process has led to many important results in the theory of finance, including the famous Black–Scholes option pricing formula.

17.2 What Is a Markov Chain?

One special type of discrete-time stochastic process is called a *Markov chain*. To simplify our exposition, we assume that at any time, the discrete-time stochastic process can be in one of a finite number of states labeled $1, 2, \ldots, s$.

DEFINITION ■ A discrete-time stochastic process is a **Markov chain** if, for $t = 0, 1, 2, \ldots$ and all states,

$$P(\mathbf{X}_{t+1} = i_{t+1} | \mathbf{X}_t = i_t, \mathbf{X}_{t-1} = i_{t-1}, \ldots, \mathbf{X}_1 = i_1, \mathbf{X}_0 = i_0)$$
$$= P(\mathbf{X}_{t+1} = i_{t+1} | \mathbf{X}_t = i_t) \quad ■ \tag{1}$$

Essentially, (1) says that the probability distribution of the state at time $t + 1$ depends on the state at time t (i_t) and does not depend on the states the chain passed through on the way to i_t at time t.

In our study of Markov chains, we make the further assumption that for all states i and j and all t, $P(\mathbf{X}_{t+1} = j | \mathbf{X}_t = i)$ is independent of t. This assumption allows us to write

$$P(\mathbf{X}_{t+1} = j | \mathbf{X}_t = i) = p_{ij} \tag{2}$$

where p_{ij} is the probability that given the system is in state i at time t, it will be in a state j at time $t + 1$. If the system moves from state i during one period to state j during the next period, we say that a **transition** from i to j has occurred. The p_{ij}'s are often referred to as the **transition probabilities** for the Markov chain.

Equation (2) implies that the probability law relating the next period's state to the current state does not change (or remains stationary) over time. For this reason, (2) is often called the **Stationarity Assumption.** Any Markov chain that satisfies (2) is called a **stationary Markov chain.**

Our study of Markov chains also requires us to define q_i to be the probability that the chain is in state i at time 0; in other words, $P(X_0 = i) = q_i$. We call the vector $\mathbf{q} = [q_1\ q_2\ \cdots\ q_s]$ the **initial probability distribution** for the Markov chain. In most applications, the transition probabilities are displayed as an $s \times s$ **transition probability matrix** P. The transition probability matrix P may be written as

$$P = \begin{bmatrix} p_{11} & p_{12} & \cdots & p_{1s} \\ p_{21} & p_{22} & \cdots & p_{2s} \\ \vdots & \vdots & & \vdots \\ p_{s1} & p_{s2} & \cdots & p_{ss} \end{bmatrix}$$

Given that the state at time t is i, the process must be somewhere at time $t + 1$. This means that for each i,

$$\sum_{j=1}^{j=s} P(X_{t+1} = j | P(X_t = i)) = 1$$

$$\sum_{j=1}^{j=s} p_{ij} = 1$$

We also know that each entry in the P matrix must be nonnegative. Hence, all entries in the transition probability matrix are nonnegative, and the entries in each row must sum to 1.

EXAMPLE 1   The Gambler's Ruin (Continued)

Find the transition matrix for Example 1.

Solution Since the amount of money I have after $t + 1$ plays of the game depends on the past history of the game only through the amount of money I have after t plays, we definitely have a Markov chain. Since the rules of the game don't change over time, we also have a stationary Markov chain. The transition matrix is as follows (state i means that we have i dollars):

$$
\begin{array}{c}
 & & \text{State} \\
 & \begin{array}{ccccc} \$0 & \$1 & \$2 & \$3 & \$4 \end{array} \\
P = \begin{array}{c} 0 \\ 1 \\ 2 \\ 3 \\ 4 \end{array} & \begin{bmatrix} 1 & 0 & 0 & 0 & 0 \\ 1-p & 0 & p & 0 & 0 \\ 0 & 1-p & 0 & p & 0 \\ 0 & 0 & 1-p & 0 & p \\ 0 & 0 & 0 & 0 & 1 \end{bmatrix}
\end{array}
$$

If the state is \$0 or \$4, I don't play the game anymore, so the state cannot change; hence, $p_{00} = p_{44} = 1$. For all other states, we know that with probability p, the next period's state will exceed the current state by 1, and with probability $1 - p$, the next period's state will be 1 less than the current state.

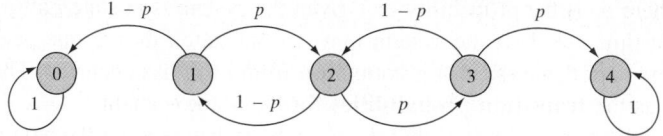

A transition matrix may be represented by a graph in which each node represents a state and arc (i, j) represents the transition probability p_{ij}. Figure 1 gives a graphical representation of Example 1's transition probability matrix.

EXAMPLE 2   Choosing Balls (Continued)

Find the transition matrix for Example 2.

Solution   Since the state of the urn after the next coin toss only depends on the past history of the process through the state of the urn after the current coin toss, we have a Markov chain. Since the rules don't change over time, we have a stationary Markov chain. The transition matrix for Example 2 is as follows:

$$
P = \begin{array}{c}
 & \text{State} \\
 & \begin{array}{cccccc}
[0\ \ 1\ \ 1] & [0\ \ 2\ \ 0] & [0\ \ 0\ \ 2] & [2\ \ 0\ \ 0] & [1\ \ 1\ \ 0] & [1\ \ 0\ \ 1]
\end{array} \\
\begin{array}{c}
[0\ \ 1\ \ 1] \\
[0\ \ 2\ \ 0] \\
[0\ \ 0\ \ 2] \\
[2\ \ 0\ \ 0] \\
[1\ \ 1\ \ 0] \\
[1\ \ 0\ \ 1]
\end{array}
\left[\begin{array}{cccccc}
0 & \frac{1}{2} & \frac{1}{2} & 0 & 0 & 0 \\
1 & 0 & 0 & 0 & 0 & 0 \\
1 & 0 & 0 & 0 & 0 & 0 \\
0 & 0 & 0 & 0 & \frac{1}{2} & \frac{1}{2} \\
\frac{1}{4} & \frac{1}{4} & 0 & 0 & 0 & \frac{1}{2} \\
\frac{1}{4} & 0 & \frac{1}{4} & 0 & \frac{1}{2} & 0
\end{array}\right]
\end{array}
$$

To illustrate the determination of the transition matrix, we determine the $[1\ \ 1\ \ 0]$ row of this transition matrix. If the current state is $[1\ \ 1\ \ 0]$, then one of the events shown in Table 1 must occur. Thus, the next state will be $[1\ \ 0\ \ 1]$ with probability $\frac{1}{2}$, $[0\ \ 2\ \ 0]$ with probability $\frac{1}{4}$, and $[0\ \ 1\ \ 1]$ with probability $\frac{1}{4}$. Figure 2 gives a graphical representation of this transition matrix.

TABLE 1
Computations of Transition Probabilities If Current State Is $[1\ \ 1\ \ 0]$

Event	Probability	New State
Flip heads and choose unpainted ball	$\frac{1}{4}$	$[0\ \ 2\ \ 0]$
Choose red ball	$\frac{1}{2}$	$[1\ \ 0\ \ 1]$
Flip tails and choose unpainted ball	$\frac{1}{4}$	$[0\ \ 1\ \ 1]$

FIGURE 2
Graphical
Representation of
Transition Matrix
for Urn

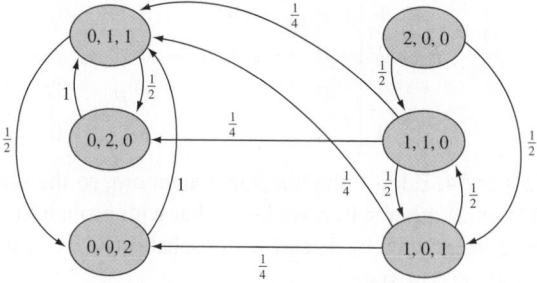

EXAMPLE 3   CSL Computer Stock (Continued)

In recent years, students of finance have devoted much effort to answering the question of whether the daily price of a stock share can be described by a Markov chain. Suppose the daily price of a stock share (such as CSL Computer stock) can be described by a Markov chain. What does that tell us? Simply that the probability distribution of tomorrow's price for one share of CSL stock depends *only* on today's price of CSL stock, *not* on the past prices of CSL stock. If the price of a stock share can be described by a Markov chain, the "chartists" who attempt to predict future stock prices on the basis of the patterns followed by past stock prices are barking up the wrong tree. For example, suppose the daily price of a share of CSL stock follows a Markov chain, and today's price for a share of CSL stock is $50. Then to predict tomorrow's price of a share of CSL stock, it does not matter whether the price has increased or decreased during each of the last 30 days. In either situation (or any other situation that might have led to today's $50 price), a prediction of tomorrow's stock price should be based only on the fact that today's price of CSL stock is $50. At this time, the consensus is that for most stocks the daily price of the stock can be described as a Markov chain. This idea is often referred to as the **efficient market hypothesis.**

PROBLEMS

Group A

1 In Smalltown, 90% of all sunny days are followed by sunny days, and 80% of all cloudy days are followed by cloudy days. Use this information to model Smalltown's weather as a Markov chain.

2 Consider an inventory system in which the sequence of events during each period is as follows. (1) We observe the inventory level (call it i) at the beginning of the period. (2) If $i \le 1$, $4 - i$ units are ordered. If $i \ge 2$, 0 units are ordered. Delivery of all ordered units is immediate. (3) With probability $\frac{1}{3}$, 0 units are demanded during the period; with probability $\frac{1}{3}$, 1 unit is demanded during the period; and with probability $\frac{1}{3}$, 2 units are demanded during the period. (4) We observe the inventory level at the beginning of the next period.

Define a period's state to be the period's beginning inventory level. Determine the transition matrix that could be used to model this inventory system as a Markov chain.

3 A company has two machines. During any day, each machine that is working at the beginning of the day has a $\frac{1}{3}$ chance of breaking down. If a machine breaks down during the day, it is sent to a repair facility and will be working two days after it breaks down. (Thus, if a machine breaks down during day 3, it will be working at the beginning of day 5.) Letting the state of the system be the number of machines working at the beginning of the day, formulate a transition probability matrix for this situation.

Group B

4 Referring to Problem 1, suppose that tomorrow's Smalltown weather depends on the last two days of Smalltown weather, as follows: (1) If the last two days have been sunny, then 95% of the time, tomorrow will be sunny. (2) If yesterday was cloudy and today is sunny, then 70% of the time, tomorrow will be sunny. (3) If yesterday was sunny and today is cloudy, then 60% of the time, tomorrow will be cloudy. (4) If the last two days have been cloudy, then 80% of the time, tomorrow will be cloudy.

Using this information, model Smalltown's weather as a Markov chain. If tomorrow's weather depended on the last three days of Smalltown weather, how many states will be needed to model Smalltown's weather as a Markov chain? (*Note:* The approach used in this problem can be used to model a discrete-time stochastic process as a Markov chain even if X_{t+1} depends on states prior to X_t, such as X_{t-1} in the current example.)

5 Let X_t be the location of your token on the Monopoly board after t dice rolls. Can X_t be modeled as a Markov chain? If not, how can we modify the definition of the state at time t so that $X_0, X_1, \ldots, X_t, \ldots$ would be a Markov chain? (*Hint:* How does a player go to Jail? In this problem, assume that players who are sent to Jail stay there until they roll doubles or until they have spent three turns there, whichever comes first.)

6 In Problem 3, suppose a machine that breaks down returns to service three days later (for instance, a machine that breaks down during day 3 would be back in working order at the beginning of day 6). Determine a transition probability matrix for this situation.

17.3 *n*-Step Transition Probabilities

Suppose we are studying a Markov chain with a known transition probability matrix P. (Since all chains that we will deal with are stationary, we will not bother to label our Markov chains as stationary.) A question of interest is: If a Markov chain is in state i at time m, what is the probability that n periods later the Markov chain will be in state j? Since we are dealing with a stationary Markov chain, this probability will be independent of m, so we may write

$$P(\mathbf{X}_{m+n} = j | \mathbf{X}_m = i) = P(\mathbf{X}_n = j | \mathbf{X}_0 = i) = P_{ij}(n)$$

where $P_{ij}(n)$ is called the **n-step probability** of a transition from state i to state j.

Clearly, $P_{ij}(1) = p_{ij}$. To determine $P_{ij}(2)$, note that if the system is now in state i, then for the system to end up in state j two periods from now, we must go from state i to some state k and then go from state k to state j (see Figure 3). This reasoning allows us to write

$$P_{ij}(2) = \sum_{k=1}^{k=s} (\text{probability of transition from } i \text{ to } k)$$
$$\times (\text{probability of transition from } k \text{ to } j)$$

Using the definition of P, the transition probability matrix, we rewrite the last equation as

$$P_{ij}(2) = \sum_{k=1}^{k=s} p_{ik}p_{kj} \tag{3}$$

The right-hand side of (3) is just the scalar product of row i of the P matrix with column j of the P matrix. Hence, $P_{ij}(2)$ is the ijth element of the matrix P^2. By extending this reasoning, it can be shown that for $n > 1$,

$$P_{ij}(n) = ij\text{th element of } P^n \tag{4}$$

Of course, for $n = 0$, $P_{ij}(0) = P(\mathbf{X}_0 = j | \mathbf{X}_0 = i)$, so we must write

$$P_{ij}(0) = \begin{cases} 1 & \text{if } j = i \\ 0 & \text{if } j \neq i \end{cases}$$

We illustrate the use of Equation (4) in Example 4.

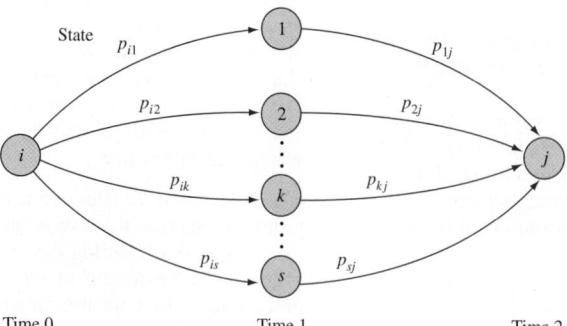

FIGURE 3
$P_{ij}(2) = p_{i1}p_{1j} +$
$p_{i2}p_{2j} + \cdots + p_{is}p_{sj}$

EXAMPLE 4   **The Cola Example**

Suppose the entire cola industry produces only two colas. Given that a person last purchased cola 1, there is a 90% chance that her next purchase will be cola 1. Given that a person last purchased cola 2, there is an 80% chance that her next purchase will be cola 2.

1 If a person is currently a cola 2 purchaser, what is the probability that she will purchase cola 1 two purchases from now?

2 If a person is currently a cola 1 purchaser, what is the probability that she will purchase cola 1 three purchases from now?

Solution We view each person's purchases as a Markov chain with the state at any given time being the type of cola the person last purchased. Hence, each person's cola purchases may be represented by a two-state Markov chain, where

$$\text{State } 1 = \text{person has last purchased cola 1}$$
$$\text{State } 2 = \text{person has last purchased cola 2}$$

If we define X_n to be the type of cola purchased by a person on her nth future cola purchase (present cola purchase $= X_0$), then $X_0, X_1, \ldots$ may be described as the Markov chain with the following transition matrix:

$$
P = \begin{array}{c} \text{Cola 1} \\ \text{Cola 2} \end{array} \begin{array}{cc} \text{Cola 1} & \text{Cola 2} \\ \left[\begin{array}{cc} .90 & .10 \\ .20 & .80 \end{array} \right] \end{array}
$$

We can now answer questions 1 and 2.

1 We seek $P(X_2 = 1 | X_0 = 2) = P_{21}(2) =$ element 21 of P^2:

$$
P^2 = \begin{bmatrix} .90 & .10 \\ .20 & .80 \end{bmatrix} \begin{bmatrix} .90 & .10 \\ .20 & .80 \end{bmatrix} = \begin{bmatrix} .83 & .17 \\ .34 & .66 \end{bmatrix}
$$

Hence, $P_{21}(2) = .34$. This means that the probability is .34 that two purchases in the future a cola 2 drinker will purchase cola 1. By using basic probability theory, we may obtain this answer in a different way (see Figure 4). Note that $P_{21}(2) = $ (probability that next purchase is cola 1 and second purchase is cola 1) + (probability that next purchase is cola 2 and second purchase is cola 1) $= p_{21}p_{11} + p_{22}p_{21} = (.20)(.90) + (.80)(.20) = .34$.

2 We seek $P_{11}(3) =$ element 11 of P^3:

$$
P^3 = P(P^2) = \begin{bmatrix} .90 & .10 \\ .20 & .80 \end{bmatrix} \begin{bmatrix} .83 & .17 \\ .34 & .66 \end{bmatrix} = \begin{bmatrix} .781 & .219 \\ .438 & .562 \end{bmatrix}
$$

Therefore, $P_{11}(3) = .781$.

FIGURE 4
Probability That Two Periods from Now, a Cola 2 Purchaser Will Purchase Cola 1 Is .20(.90) + .80(.20) = .34

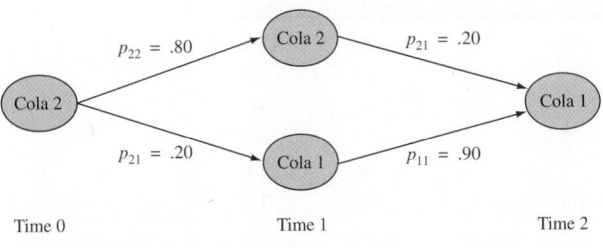

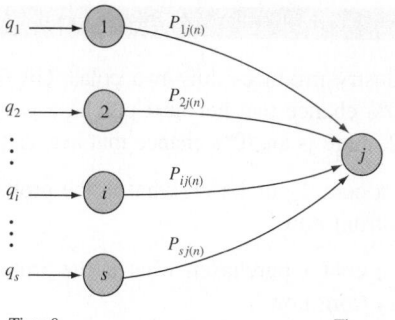

FIGURE 5
Determination of
Probability of Being in
State *j* at Time *n* When
Initial State Is Unknown

In many situations, we do not know the state of the Markov chain at time 0. As defined in Section 17.2, let q_i be the probability that the chain is in state i at time 0. Then we can determine the probability that the system is in state i at time n by using the following reasoning (see Figure 5).

Probability of being in state j at time n

$$= \sum_{i=1}^{i=s} (\text{probability that state is originally } i)$$

$$\times (\text{probability of going from } i \text{ to } j \text{ in } n \text{ transitions})$$

$$= \sum_{i=1}^{i=s} q_i P_{ij}(n)$$

$$= \mathbf{q}(\text{column } j \text{ of } P^n)$$

(5)

where $\mathbf{q} = [q_1 \quad q_2 \quad \cdots \quad q_s]$.

To illustrate the use of (5), we answer the following question: Suppose 60% of all people now drink cola 1, and 40% now drink cola 2. Three purchases from now, what fraction of all purchasers will be drinking cola 1? Since $\mathbf{q} = [.60 \quad .40]$ and $\mathbf{q}(\text{column 1 of } P^3) =$ probability that three purchases from now a person drinks cola 1, the desired probability is

$$[.60 \quad .40] \begin{bmatrix} .781 \\ .438 \end{bmatrix} = .6438$$

Hence, three purchases from now, 64% of all purchasers will be purchasing cola 1.

To illustrate the behavior of the n-step transition probabilities for large values of n, we have computed several of the n-step transition probabilities for the Cola example in Table 2.

TABLE 2
n-Step Transition Probabilities for Cola Drinkers

n	$P_{11}(n)$	$P_{12}(n)$	$P_{21}(n)$	$P_{22}(n)$
1	.90	.10	.20	.80
2	.83	.17	.34	.66
3	.78	.22	.44	.56
4	.75	.25	.51	.49
5	.72	.28	.56	.44
10	.68	.32	.65	.35
20	.67	.33	.67	.33
30	.67	.33	.67	.33
40	.67	.33	.67	.33

For large n, both $P_{11}(n)$ and $P_{21}(n)$ are nearly constant and approach .67. This means that for large n, no matter what the initial state, there is a .67 chance that a person will be a cola 1 purchaser. Similarly, we see that for large n, both $P_{12}(n)$ and $P_{22}(n)$ are nearly constant and approach .33. This means that for large n, no matter what the initial state, there is a .33 chance that a person will be a cola 2 purchaser. In Section 5.5, we make a thorough study of this settling down of the n-step transition probabilities.

REMARK   We can easily multiply matrices on a spreadsheet using the MMULT command, as discussed in Section 13.7.

PROBLEMS

Group A

1 Each American family is classified as living in an urban, rural, or suburban location. During a given year, 15% of all urban families move to a suburban location, and 5% move to a rural location; also, 6% of all suburban families move to an urban location, and 4% move to a rural location; finally, 4% of all rural families move to an urban location, and 6% move to a suburban location.

a If a family now lives in an urban location, what is the probability that it will live in an urban area two years from now? A suburban area? A rural area?

b Suppose that at present, 40% of all families live in an urban area, 35% live in a suburban area, and 25% live in a rural area. Two years from now, what percentage of American families will live in an urban area?

c What problems might occur if this model were used to predict the future population distribution of the United States?

2 The following questions refer to Example 1.

a After playing the game twice, what is the probability that I will have $3? How about $2?

b After playing the game three times, what is the probability that I will have $2?

3 In Example 2, determine the following n-step transition probabilities:

a After two balls are painted, what is the probability that the state is [0 2 0]?

b After three balls are painted, what is the probability that the state is [0 1 1]? (Draw a diagram like Figure 4.)

17.4 Classification of States in a Markov Chain

In Section 17.3, we mentioned the fact that after many transitions, the n-step transition probabilities tend to settle down. Before we can discuss this in more detail, we need to study how mathematicians classify the states of a Markov chain. We use the following transition matrix to illustrate most of the following definitions (see Figure 6).

$$P = \begin{bmatrix} .4 & .6 & 0 & 0 & 0 \\ .5 & .5 & 0 & 0 & 0 \\ 0 & 0 & .3 & .7 & 0 \\ 0 & 0 & .5 & .4 & .1 \\ 0 & 0 & 0 & .8 & .2 \end{bmatrix}$$

DEFINITION ■   Given two states i and j, a **path** from i to j is a sequence of transitions that begins in i and ends in j, such that each transition in the sequence has a positive probability of occurring.   ■

A state j is **reachable** from state i if there is a path leading from i to j.   ■

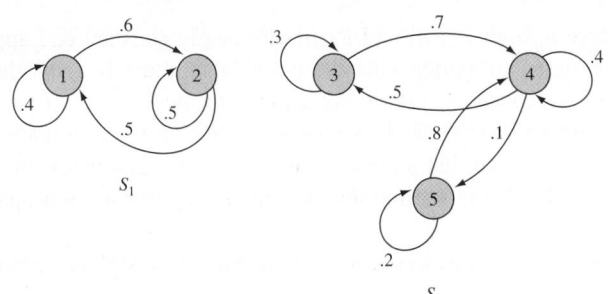

FIGURE 6
Graphical
Representation of
Transition Matrix

S_1

S_2

DEFINITION ■ Two states i and j are said to **communicate** if j is reachable from i, and i is reachable from j. ■

For the transition probability matrix P represented in Figure 6, state 5 is reachable from state 3 (via the path 3–4–5), but state 5 is not reachable from state 1 (there is no path from 1 to 5 in Figure 6). Also, states 1 and 2 communicate (we can go from 1 to 2 and from 2 to 1).

DEFINITION ■ A set of states S in a Markov chain is a **closed set** if no state outside of S is reachable from any state in S. ■

From the Markov chain with transition matrix P in Figure 6, $S_1 = \{1, 2\}$ and $S_2 = \{3, 4, 5\}$ are both closed sets. Observe that once we enter a closed set, we can never leave the closed set (in Figure 6, no arc begins in S_1 and ends in S_2 or begins in S_2 and ends in S_1).

DEFINITION ■ A state i is an **absorbing state** if $p_{ii} = 1$. ■

Whenever we enter an absorbing state, we never leave the state. In Example 1, the gambler's ruin, states 0 and 4 are absorbing states. Of course, an absorbing state is a closed set containing only one state.

DEFINITION ■ A state i is a **transient state** if there exists a state j that is reachable from i, but the state i is not reachable from state j. ■

In other words, a state i is transient if there is a way to leave state i that never returns to state i. In the gambler's ruin example, states 1, 2, and 3 are transient states. For example (see Figure 1), from state 2, it is possible to go along the path 2–3–4, but there is no way to return to state 2 from state 4. Similarly, in Example 2, [2 0 0], [1 1 0], and [1 0 1] are all transient states (in Figure 2, there is a path from [1 0 1] to [0 0 2], but once both balls are painted, there is no way to return to [1 0 1]).

After a large number of periods, the probability of being in any transient state i is zero. Each time we enter a transient state i, there is a positive probability that we will leave i forever and end up in the state j described in the definition of a transient state. Thus, eventually we are sure to enter state j (and then we will never return to state i). To illustrate, in Example 2, suppose we are in the transient state [1 0 1]. With probability 1, the unpainted ball will eventually be painted, and we will never reenter state [1 0 1] (see Figure 2).

DEFINITION ■ If a state is not transient, it is called a **recurrent state.** ■

In Example 1, states 0 and 4 are recurrent states (and also absorbing states), and in Example 2, [0 2 0], [0 0 2], and [0 1 1] are recurrent states. For the transition matrix P in Figure 6, all states are recurrent.

DEFINITION ■ A state i is **periodic** with period $k > 1$ if k is the smallest number such that all paths leading from state i back to state i have a length that is a multiple of k. If a recurrent state is not periodic, it is referred to as **aperiodic.** ■

For the Markov chain with transition matrix

$$Q = \begin{bmatrix} 0 & 1 & 0 \\ 0 & 0 & 1 \\ 1 & 0 & 0 \end{bmatrix}$$

each state has period 3. For example, if we begin in state 1, the only way to return to state 1 is to follow the path 1–2–3–1 for some number of times (say, m). (See Figure 7.) Hence, any return to state 1 will take $3m$ transitions, so state 1 has period 3. Wherever we are, we are sure to return three periods later.

DEFINITION ■ If all states in a chain are recurrent, aperiodic, and communicate with each other, the chain is said to be **ergodic.** ■

The gambler's ruin example is not an ergodic chain, because (for example) states 3 and 4 do not communicate. Example 2 is also not an ergodic chain, because (for example) [2 0 0] and [0 1 1] do not communicate. Example 4, the cola example, is an ergodic Markov chain. Of the following three Markov chains, P_1 and P_3 are ergodic, and P_2 is not ergodic.

$$P_1 = \begin{bmatrix} \frac{1}{3} & \frac{2}{3} & 0 \\ \frac{1}{2} & 0 & \frac{1}{2} \\ 0 & \frac{1}{4} & \frac{3}{4} \end{bmatrix} \quad \text{Ergodic}$$

$$P_2 = \begin{bmatrix} \frac{1}{2} & \frac{1}{2} & 0 & 0 \\ \frac{1}{2} & \frac{1}{2} & 0 & 0 \\ 0 & 0 & \frac{2}{3} & \frac{1}{3} \\ 0 & 0 & \frac{1}{4} & \frac{3}{4} \end{bmatrix} \quad \text{Nonergodic}$$

$$P_3 = \begin{bmatrix} \frac{1}{4} & \frac{1}{2} & \frac{1}{4} \\ \frac{2}{3} & \frac{1}{3} & 0 \\ 0 & \frac{2}{3} & \frac{1}{3} \end{bmatrix} \quad \text{Ergodic}$$

FIGURE 7
A Periodic Markov
Chain $k = 3$

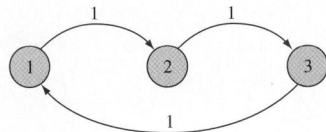

P_2 is not ergodic because there are two closed classes of states (class $1 = \{1, 2\}$ and class $2 = \{3, 4\}$), and the states in different classes do not communicate with each other.

After the next two sections, the importance of the concepts introduced in this section will become clear.

PROBLEMS

Group A

1 In Example 1, what is the period of states 1 and 3?

2 Is the Markov chain of Section 17.3, Problem 1, an ergodic Markov chain?

3 Consider the following transition matrix:

$$P = \begin{bmatrix} 0 & 0 & 1 & 0 & 0 & 0 \\ 0 & 0 & 0 & 0 & 0 & 1 \\ 0 & 0 & 0 & 0 & 1 & 0 \\ \frac{1}{4} & \frac{1}{4} & 0 & \frac{1}{2} & 0 & 0 \\ 1 & 0 & 0 & 0 & 0 & 0 \\ 0 & \frac{1}{3} & 0 & 0 & 0 & \frac{2}{3} \end{bmatrix}$$

a Which states are transient?

b Which states are recurrent?

c Identify all closed sets of states.

d Is this chain ergodic?

4 For each of the following chains, determine whether the Markov chain is ergodic. Also, for each chain, determine the recurrent, transient, and absorbing states.

$$P_1 = \begin{bmatrix} 0 & .8 & .2 \\ .3 & .7 & 0 \\ .4 & .5 & .1 \end{bmatrix} \qquad P_2 = \begin{bmatrix} .2 & .8 & 0 & 0 \\ 0 & 0 & .9 & .1 \\ .4 & .5 & .1 & 0 \\ 0 & 0 & 0 & 1 \end{bmatrix}$$

5 Fifty-four players (including Gabe Kaplan and James Garner) participated in the 1980 World Series of Poker. Each player began with $10,000. Play continued until one player had won everybody else's money. If the World Series of Poker were to be modeled as a Markov chain, how many absorbing states would the chain have?

6 Which of the following chains is ergodic?

$$P_1 = \begin{bmatrix} .4 & 0 & .6 \\ .3 & .3 & .4 \\ 0 & .5 & .5 \end{bmatrix} \qquad P_2 = \begin{bmatrix} .7 & 0 & 0 & .3 \\ .2 & .2 & .4 & .2 \\ .6 & .1 & .1 & .2 \\ .2 & 0 & 0 & .8 \end{bmatrix}$$

17.5 Steady-State Probabilities and Mean First Passage Times

In our discussion of the cola example (Example 4), we found that after a long time, the probability that a person's next cola purchase would be cola 1 approached .67 and .33 that it would be cola 2 (see Table 2). These probabilities *did not* depend on whether the person was initially a cola 1 or a cola 2 drinker. In this section, we discuss the important concept of steady-state probabilities, which can be used to describe the long-run behavior of a Markov chain.

The following result is vital to an understanding of steady-state probabilities and the long-run behavior of Markov chains.

THEOREM 1

Let P be the transition matrix for an s-state ergodic chain.[†] Then there exists a vector $\pi = [\pi_1 \quad \pi_2 \quad \cdots \quad \pi_s]$ such that

$$\lim_{n \to \infty} P^n = \begin{bmatrix} \pi_1 & \pi_2 & \cdots & \pi_s \\ \pi_1 & \pi_2 & \cdots & \pi_s \\ \vdots & \vdots & & \vdots \\ \pi_1 & \pi_2 & \cdots & \pi_s \end{bmatrix}$$

[†]To see why Theorem 1 fails to hold for a nonergodic chain, see Problems 11 and 12 at the end of this section. For a proof of this theorem, see Isaacson and Madsen (1976, Chapter 3).

Recall that the ijth element of P^n is $P_{ij}(n)$. Theorem 1 tells us that for any initial state i,

$$\lim_{n\to\infty} P_{ij}(n) = \pi_j$$

Observe that for large n, P^n approaches a matrix with identical rows. This means that after a long time, the Markov chain settles down, and (independent of the initial state i) there is a probability π_j that we are in state j.

The vector $\pi = [\pi_1 \quad \pi_2 \quad \cdots \quad \pi_s]$ is often called the **steady-state distribution,** or **equilibrium distribution,** for the Markov chain. For a given chain with transition matrix P, how can we find the steady-state probability distribution? From Theorem 1, observe that for large n and all i,

$$P_{ij}(n + 1) \cong P_{ij}(n) \cong \pi_j \tag{6}$$

Since $P_{ij}(n + 1) = (\text{row } i \text{ of } P^n)\,(\text{column } j \text{ of } P)$, we may write

$$P_{ij}(n + 1) = \sum_{k=1}^{k=s} P_{ik}(n)p_{kj} \tag{7}$$

If n is large, substituting (6) into (7) yields

$$\pi_j = \sum_{k=1}^{k=s} \pi_k p_{kj} \tag{8}$$

In matrix form, (8) may be written as

$$\pi = \pi P \tag{8'}$$

Unfortunately, the system of equations specified in (8) has an infinite number of solutions, because the rank of the P matrix always turns out to be $\leq s - 1$ (see Chapter 2, Review Problem 21). To obtain unique values of the steady-state probabilities, note that for any n and any i,

$$P_{i1}(n) + P_{i2}(n) + \cdots + P_{is}(n) = 1 \tag{9}$$

Letting n approach infinity in (9), we obtain

$$\pi_1 + \pi_2 + \cdots + \pi_s = 1 \tag{10}$$

Thus, after replacing any of the equations in (8) with (10), we may use (8) to solve for the steady-state probabilities.

To illustrate how to find the steady-state probabilities, we find the steady-state probabilities for Example 4, the cola example. Recall that the transition matrix for Example 4 was

$$P = \begin{bmatrix} .90 & .10 \\ .20 & .80 \end{bmatrix}$$

Then (8) or (8') yields

$$[\pi_1 \quad \pi_2] = [\pi_1 \quad \pi_2] \begin{bmatrix} .90 & .10 \\ .20 & .80 \end{bmatrix}$$

$$\pi_1 = .90\pi_1 + .20\pi_2$$

$$\pi_2 = .10\pi_1 + .80\pi_2$$

Replacing the second equation with the condition $\pi_1 + \pi_2 = 1$, we obtain the system

$$\pi_1 = .90\pi_1 + .20\pi_2$$

$$1 = \pi_1 + \pi_2$$

Solving for π_1 and π_2 we obtain $\pi_1 = \frac{2}{3}$ and $\pi_2 = \frac{1}{3}$. Hence, after a long time, there is a $\frac{2}{3}$ probability that a given person will purchase cola 1 and a $\frac{1}{3}$ probability that a given person will purchase cola 2.

Transient Analysis

A glance at Table 2 shows that for Example 4, the steady state is reached (to two decimal places) after only ten transitions. No general rule can be given about how quickly a Markov chain reaches the steady state, but if P contains very few entries that are near 0 or near 1, the steady state is usually reached very quickly. The behavior of a Markov chain before the steady state is reached is often called **transient** (or short-run) **behavior.** To study the transient behavior of a Markov chain, one simply uses the formulas for $P_{ij}(n)$ given in (4) and (5). It's nice to know, however, that for large n, the steady-state probabilities accurately describe the probability of being in any state.

Intuitive Interpretation of Steady-State Probabilities

An intuitive interpretation can be given to the steady-state probability equations (8). By subtracting $\pi_j p_{jj}$ from both sides of (8), we obtain

$$\pi_j(1 - p_{jj}) = \sum_{k \neq j} \pi_k p_{kj} \tag{11}$$

Equation (11) states that in the steady state,

$$\text{Probability that a particular transition leaves state } j$$
$$= \text{probability that a particular transition enters state } j \tag{12}$$

Recall that in the steady state, the probability that the system is in state j is π_j. From this observation, it follows that

$$\text{Probability that a particular transition leaves state } j$$
$$= (\text{probability that the current period begins in } j)$$
$$\times (\text{probability that the current transition leaves } j)$$
$$= \pi_j(1 - p_{jj})$$

and

$$\text{Probability that a particular transition enters state } j$$
$$= \sum_k (\text{probability that the current period begins in } k \neq j)$$
$$\times (\text{probability that the current transition enters } j)$$
$$= \sum_{k \neq j} \pi_k p_{kj}$$

Equation (11) is reasonable; if (11) were violated for any state, then for some state j, the right-hand side of (11) would exceed the left-hand side of (11). This would result in probability "piling up" at state j, and a steady-state distribution would not exist. Equation (11) may be viewed as saying that in the steady state, the "flow" of probability into each state must equal the flow of probability out of each state. This explains why steady-state probabilities are often called equilibrium probabilities.

Use of Steady-State Probabilities in Decision Making

EXAMPLE 5 **The Cola Example (Continued)**

In Example 4, suppose that each customer makes one purchase of cola during any week (52 weeks = 1 year). Suppose there are 100 million cola customers. One selling unit of cola costs the company $1 to produce and is sold for $2. For $500 million per year, an advertising firm guarantees to decrease from 10% to 5% the fraction of cola 1 customers who switch to cola 2 after a purchase. Should the company that makes cola 1 hire the advertising firm?

Solution At present, a fraction $\pi_1 = \frac{2}{3}$ of all purchases are cola 1 purchases. Each purchase of cola 1 earns the company a $1 profit. Since there are a total of 52(100,000,000), or 5.2 billion, cola purchases each year, the cola 1 company's current annual profit is

$$\tfrac{2}{3}(5,200,000,000) = \$3,466,666,667$$

The advertising firm is offering to change the P matrix to

$$P_1 = \begin{bmatrix} .95 & .05 \\ .20 & .80 \end{bmatrix}$$

For P_1, the steady-state equations become

$$\pi_1 = .95\pi_1 + .20\pi_2$$
$$\pi_2 = .05\pi_1 + .80\pi_2$$

Replacing the second equation by $\pi_1 + \pi_2 = 1$ and solving, we obtain $\pi_1 = .8$ and $\pi_2 = .2$. Now the cola 1 company's annual profit will be

$$(.80)(5,200,000,000) - 500,000,000 = \$3,660,000,000$$

Hence, the cola 1 company should hire the ad agency.

EXAMPLE 6 **Playing Monopoly**

With the assumption that each Monopoly player who goes to Jail stays until he or she rolls doubles or has spent three turns in Jail, the steady-state probability of a player landing on any Monopoly square has been determined by Ash and Bishop (1972) (see Table 3).[†] These steady-state probabilities can be used to measure the cost-effectiveness of various monopolies. For example, it costs $1,500 to build hotels on the Orange monopoly. Each time a player lands on a Tennessee Ave. or a St. James Place hotel, the owner of the monopoly receives $950, and each time a player lands on a New York Ave. hotel, the owner receives $1,000. From Table 3, we can compute the expected rent per dice roll earned by the Orange monopoly:

$$950(.0335) + 950(.0318) + 1,000(.0334) = \$95.44$$

Thus, per dollar invested, the Orange monopoly yields $\frac{95.44}{1,500} = \$0.064$ per dice roll.

Now let's consider the Green monopoly. To put hotels on the Green monopoly costs $3,000. If a player lands on a North Carolina Ave. or a Pacific Ave. hotel, the owner receives $1,275. If a player lands on a Pennsylvania Ave. hotel, the owner receives $1,400. From Table 3, the average revenue per dice roll earned from hotels on the Green monopoly is

$$1,275(.0294) + 1,275(.0300) + 1,400(.0279) = \$114.80$$

[†]This example is based on Ash and Bishop (1972).

TABLE 3
Steady-State Probabilities for Monopoly

n Position		Steady-State Probability
0	Go	.0346
1	Mediterranean Ave.	.0237
2	Community Chest 1	.0218
3	Baltic Ave.	.0241
4	Income tax	.0261
5	Reading RR	.0332
6	Oriental Ave.	.0253
7	Chance 1	.0096
8	Vermont Ave.	.0258
9	Connecticut Ave.	.0237
10	Visiting jail	.0254
11	St. Charles Place	.0304
12	Electric Co.	.0311
13	State Ave.	.0258
14	Virginia Ave.	.0288
15	Pennsylvania RR	.0313
16	St. James Place	.0318
17	Community Chest 2	.0272
18	Tennessee Ave.	.0335
19	New York Ave.	.0334
20	Free parking	.0336
21	Kentucky Ave.	.0310
22	Chance 2	.0125
23	Indiana Ave.	.0305
24	Illinois Ave.	.0355
25	B and O RR	.0344
26	Atlantic Ave.	.0301
27	Ventnor Ave.	.0299
28	Water works	.0315
29	Marvin Gardens	.0289
30	Jail	.1123
31	Pacific Ave.	.0300
32	North Carolina Ave.	.0294
33	Community Chest 3	.0263
34	Pennsylvania Ave.	.0279
35	Short Line RR	.0272
36	Chance 3	.0096
37	Park Place	.0245
38	Luxury tax	.0295
39	Boardwalk	.0295

Source: Reprinted by permission from R. Ash and R. Bishop, "Monopoly as a Markov Process," *Mathematics Magazine* 45(1972):26–29. Copyright © 1972 Mathematical Association of America.

Thus, per dollar invested, the Green monopoly yields only $\frac{114.80}{3,000} = \$0.038$ per dice roll.

This analysis shows that the Orange monopoly is superior to the Green monopoly. By the way, why does the Orange get landed on so often?

Mean First Passage Times

For an ergodic chain, let m_{ij} = expected number of transitions before we first reach state j, given that we are currently in state i; m_{ij} is called the **mean first passage time** from state i to state j. In Example 4, m_{12} would be the expected number of bottles of cola purchased by a person who just bought cola 1 before first buying a bottle of cola 2. Assume that we are currently in state i. Then with probability p_{ij}, it will take one transition to go from state i to state j. For $k \neq j$, we next go with probability p_{ik} to state k. In this case, it will take an average of $1 + m_{kj}$ transitions to go from i to j. This reasoning implies that

$$m_{ij} = p_{ij}(1) + \sum_{k \neq j} p_{ik}(1 + m_{kj})$$

Since

$$p_{ij} + \sum_{k \neq j} p_{ik} = 1$$

we may rewrite the last equation as

$$m_{ij} = 1 + \sum_{k \neq j} p_{ik} m_{kj} \qquad (13)$$

By solving the linear equations given in (13), we may find all the mean first passage times. It can be shown that

$$m_{ii} = \frac{1}{\pi_i}$$

This can simplify the use of (13).

To illustrate the use of (13), let's solve for the mean first passage times in Example 4. Recall that $\pi_1 = \frac{2}{3}$ and $\pi_2 = \frac{1}{3}$. Then

$$m_{11} = \frac{1}{\frac{2}{3}} = 1.5 \qquad \text{and} \qquad m_{22} = \frac{1}{\frac{1}{3}} = 3$$

Now (13) yields the following two equations:

$$m_{12} = 1 + p_{11}m_{12} = 1 + 0.9m_{12}, \qquad m_{21} = 1 + p_{22}m_{21} = 1 + 0.8m_{21}$$

Solving these two equations, we find that $m_{12} = 10$ and $m_{21} = 5$. This means, for example, that a person who last drank cola 1 will drink an average of ten bottles of soda before switching to cola 2.

Solving for Steady-State Probabilities and Mean First Passage Times on the Computer

Since we solve for steady-state probabilities and mean first passage times by solving a system of linear equations, we may use LINDO to determine them. Simply type in an objective function of 0, and type the equations you need to solve as your constraints.

Alternatively, you may use the following LINGO model (file Markov.lng) to determine steady-state probabilities and mean first passage times for an ergodic chain.

Markov.lng

```
MODEL:
1]
2]SETS:
3]STATE/1..2/:PI;
```

```
 4]SXS(STATE,STATE):TPROB,MFP;
 5]ENDSETS
 6]DATA:
 7]TPROB = .9,.1,
 8].2,.8;
 9]ENDDATA
10]@FOR(STATE(J)|J #LT# @SIZE(STATE):
11]PI(J) = @SUM(SXS(I,J): PI(I) * TPROB(I,J)););
12]@SUM(STATE:PI) = 1;
13]@FOR(SXS(I,J):MFP(I,J)=
14]1+@SUM(STATE(K)|K#NE#J:TPROB(I,K)*MFP(K,J)););
END
```

In line 3, we define the set of states and associate a steady-state probability ($PI(I)$) with each state I. In line 4, we create for each pairing of states (I, J) a transition probability ($TPROB(I, J)$) which equals p_{ij} and $MFP(I, J)$ which equals m_{ij}. The transition probabilities for the cola example are input in lines 7 and 8. In lines 10 and 11, we create (for each state except the highest-numbered state) the steady-state equation

$$PI(J) = \sum_I PI(I) * TPROB(I, J)$$

In line 12, we ensure that the steady-state probabilities sum to 1. In lines 13 and 14, we create the equations that must be solved to compute the mean first passage times. For each (I, J), lines 13–14 create the equation

$$MFP(I, J) = 1 + \sum_{K \neq J} TPROB(I, K) * MFP(K, J)$$

which is needed to compute the mean first passage times.

PROBLEMS

Group A

1 Find the steady-state probabilities for Problem 1 of Section 17.3.

2 For the gambler's ruin problem (Example 1), why is it unreasonable to talk about steady-state probabilities?

3 For each of the following Markov chains, determine the long-run fraction of the time that each state will be occupied.

a $\begin{bmatrix} \frac{2}{3} & \frac{1}{3} \\ \frac{1}{2} & \frac{1}{2} \end{bmatrix}$ **b** $\begin{bmatrix} .8 & .2 & 0 \\ 0 & .2 & .8 \\ .8 & .2 & 0 \end{bmatrix}$

c Find all mean first passage times for part (b).

4 At the beginning of each year, my car is in good, fair, or broken-down condition. A good car will be good at the beginning of next year with probability .85; fair with probability .10; or broken-down with probability .05. A fair car will be fair at the beginning of the next year with probability .70 or broken-down with probability .30. It costs $6,000 to purchase a good car; a fair car can be traded in for $2,000; and a broken-down car has no trade-in value and must immediately be replaced by a good car. It costs $1,000 per year to operate a good car and $1,500 to operate a fair car. Should I replace my car as soon as it becomes a fair car, or should I drive my car until it breaks down? Assume that the cost of operating a car during a year depends on the type of car on hand at the beginning of the year (after a new car, if any, arrives).

5 A square matrix is said to be doubly stochastic if its entries are all nonnegative and the entries in each row and each column sum to 1. For any ergodic, doubly stochastic matrix, show that all states have the same steady-state probability.

6 This problem will show why steady-state probabilities are sometimes referred to as stationary probabilities. Let $\pi_1, \pi_2, \ldots, \pi_s$ be the steady-state probabilities for an ergodic chain with transition matrix P. Also suppose that with probability π_i, the Markov chain begins in state i.

a What is the probability that after one transition, the system will be in state i? (*Hint:* Use Equation (8).)

b For any value of $n(n = 1, 2, \ldots)$, what is the probability that a Markov chain will be in state i after n transitions?

c Why are steady-state probabilities sometimes called stationary probabilities?

7 Consider two stocks. Stock 1 always sells for $10 or $20. If stock 1 is selling for $10 today, there is a .80 chance that it will sell for $10 tomorrow. If it is selling for $20 today, there is a .90 chance that it will sell for $20 tomorrow.

Stock 2 always sells for $10 or $25. If stock 2 sells today for $10, there is a .90 chance that it will sell tomorrow for $10. If it sells today for $25, there is a .85 chance that it will sell tomorrow for $25. On the average, which stock will sell for a higher price? Find and interpret all mean first passage times.

8 Three balls are divided between two containers. During each period a ball is randomly chosen and switched to the other container.

a Find (in the steady state) the fraction of the time that a container will contain 0, 1, 2, or 3 balls.

b If container 1 contains no balls, on the average how many periods will go by before it again contains no balls? (*Note:* This is a special case of the Ehrenfest Diffusion model, which is used in biology to model diffusion through a membrane.)

9 Two types of squirrels—gray and black—have been seen in Pine Valley. At the beginning of each year, we determine which of the following is true:

There are only gray squirrels in Pine Valley.

There are only black squirrels in Pine Valley.

There are both gray and black squirrels in Pine Valley.

There are no squirrels in Pine Valley.

Over the course of many years, the following transition matrix has been estimated.

	Gray	Black	Both	Neither
Gray	.7	.2	.05	.05
Black	.2	.6	.1	.1
Both	.1	.1	.8	0
Neither	.05	.05	.1	.8

a During what fraction of years will gray squirrels be living in Pine Valley?

b During what fraction of years will black squirrels be living in Pine Valley?

Group B

10 Payoff Insurance Company charges a customer according to his or her accident history. A customer who has had no accident during the last two years is charged a $100 annual premium. Any customer who has had an accident during each of the last two years is charged a $400 annual premium. A customer who has had an accident during only one of the last two years is charged an annual premium of $300. A customer who has had an accident during the last year has a 10% chance of having an accident during the current year. If a customer has not had an accident during the last year, there is only a 3% chance that he or she will have an accident during the current year. During a given year, what is the average premium paid by a Payoff customer? (*Hint:* In case of difficulty, try a four-state Markov chain.)

11 Consider the following nonergodic chain:

$$P = \begin{bmatrix} \frac{1}{2} & \frac{1}{2} & 0 & 0 \\ \frac{1}{2} & \frac{1}{2} & 0 & 0 \\ 0 & 0 & \frac{1}{3} & \frac{2}{3} \\ 0 & 0 & \frac{2}{3} & \frac{1}{3} \end{bmatrix}$$

a Why is the chain nonergodic?

b Explain why Theorem 1 fails for this chain. *Hint:* Find out if the following equation is true:

$$\lim_{n\to\infty} P_{12}(n) = \lim_{n\to\infty} P_{32}(n)$$

c Despite the fact that Theorem 1 fails, determine

$$\lim_{n\to\infty} P_{13}(n), \quad \lim_{n\to\infty} P_{21}(n),$$

$$\lim_{n\to\infty} P_{43}(n), \quad \lim_{n\to\infty} P_{41}(n)$$

12 Consider the following nonergodic chain:

$$P = \begin{bmatrix} 0 & 1 & 0 \\ 0 & 0 & 1 \\ 1 & 0 & 0 \end{bmatrix}$$

a Why is this chain nonergodic?

b Explain why Theorem 1 fails for this chain. (*Hint:* Show that $\lim_{n\to\infty} P_{11}(n)$ does not exist by listing the pattern that $P_{11}(n)$ follows as n increases.)

13 An important machine is known to never last more than four months. During its first month of operation, it fails 10% of the time. If the machine completes its first month, then it fails during its second month 20% of the time. If the machine completes its second month of operation, then it will fail during its third month 50% of the time. If the machine completes its third month, then it is sure to fail by the end of the fourth month. At the beginning of each month, we must decide whether or not to replace our machine with a new machine. It costs $500 to purchase a new machine, but if a machine fails during a month, we incur a cost of $1,000 (due to factory downtime) and must replace the machine (at the beginning of the next month) with a new machine. Three maintenance policies are under consideration:

Policy 1 Plan to replace a machine at the beginning of its fourth month of operation.

Policy 2 Plan to replace a machine at the beginning of its third month of operation.

Policy 3 Plan to replace a machine at the beginning of its second month of operation.

Which policy will give the lowest average monthly cost?

14 Each month, customers are equally likely to demand 1 or 2 computers from a Pearco dealer. All orders must be met from current stock. Two ordering policies are under consideration:

Policy 1 If ending inventory is 2 units or less, order enough to bring next month's beginning inventory to 4 units.

Policy 2 If ending inventory is 1 unit or less, order enough to bring next month's beginning inventory up to 3 units.

The following costs are incurred by Pearco:

It costs $4,000 to order a computer.

It costs $100 to hold a computer in inventory for a month.

It costs $500 to place an order for computers. This is in addition to the per-customer cost of $4,000.

Which ordering policy has a lower expected monthly cost?

15 The Gotham City Maternity Ward contains 2 beds. Admissions are made only at the beginning of the day. Each day, there is a .5 probability that a potential admission will arrive. A patient can be admitted only if there is an open bed at the beginning of the day. Half of all patients are discharged after one day, and all patients that have stayed one day are discharged at the end of their second day.

a What is the fraction of days where all beds are utilized?

b On the average, what percentage of the beds are utilized?

17.6 Absorbing Chains

Many interesting applications of Markov chains involve chains in which some of the states are absorbing and the rest are transient states. Such a chain is called an **absorbing chain.** Consider an absorbing Markov chain: If we begin in a transient state, then eventually we are sure to leave the transient state and end up in one of the absorbing states. To see why we are interested in absorbing chains, we consider the following two absorbing chains.

EXAMPLE 7   Accounts Receivable

The accounts receivable situation of a firm is often modeled as an absorbing Markov chain.[†] Suppose a firm assumes that an account is uncollectable if the account is more than three months overdue. Then at the beginning of each month, each account may be classified into one of the following states:

State 1   New account

State 2   Payment on account is one month overdue.

State 3   Payment on account is two months overdue.

State 4   Payment on account is three months overdue.

State 5   Account has been paid.

State 6   Account is written off as bad debt.

Suppose that past data indicate that the following Markov chain describes how the status of an account changes from one month to the next month:

	New	1 month	2 months	3 months	Paid	Bad debt
New	0	.6	0	0	.4	0
1 month	0	0	.5	0	.5	0
2 months	0	0	0	.4	.6	0
3 months	0	0	0	0	.7	.3
Paid	0	0	0	0	1	0
Bad debt	0	0	0	0	0	1

For example, if an account is two months overdue at the beginning of a month, there is a 40% chance that at the beginning of next month, the account will not be paid up (and therefore be three months overdue) and a 60% chance that the account will be paid up. To simplify our example, we assume that after three months, a debt is either collected or written off as a bad debt.

Once a debt is paid up or written off as a bad debt, the account is closed, and no further transitions occur. Hence, Paid and Bad Debt are absorbing states. Since every account

[†]This example is based on Cyert, Davidson, and Thompson (1963).

will eventually be paid up or written off as a bad debt, New, 1 Month, 2 Months, and 3 Months are transient states. For example, a two-month overdue account can follow the path 2 Months–Collected, but there is no return path from Collected to 2 Months.

A typical new account will be absorbed as either a collected debt or a bad debt. A question of major interest is: What is the probability that a new account will eventually be collected? The answer is worked out later in this section.

EXAMPLE 8   **Work-Force Planning**

The law firm of Mason and Burger employs three types of lawyers: junior lawyers, senior lawyers, and partners. During a given year, there is a .15 probability that a junior lawyer will be promoted to senior lawyer and a .05 probability that he or she will leave the firm. Also, there is a .20 probability that a senior lawyer will be promoted to partner and a .10 probability that he or she will leave the firm. There is a .05 probability that a partner will leave the firm. The firm never demotes a lawyer.

There are many interesting questions the law firm might want to answer. For example, what is the probability that a newly hired junior lawyer will leave the firm before becoming a partner? On the average, how long does a newly hired junior lawyer stay with the firm? The answers are worked out later in this section.

We model the career path of a lawyer through Mason and Burger as an absorbing Markov chain with the following transition probability matrix:

	Junior	Senior	Partner	Leave as NP	Leave as P
Junior	.80	.15	0	.05	0
Senior	0	.70	.20	.10	0
Partner	0	0	.95	0	.05
Leave as nonpartner	0	0	0	1	0
Leave as partner	0	0	0	0	1

The last two states are absorbing states, and all other states are transient. For example, Senior is a transient state, because there is a path from Senior to Leave as Nonpartner, but there is no path returning from Leave as Nonpartner to Senior (we assume that once a lawyer leaves the firm, he or she never returns).

For any absorbing chain, one might want to know certain things. (1) If the chain begins in a given transient state, and before we reach an absorbing state, what is the expected number of times that each state will be entered? How many periods do we expect to spend in a given transient state before absorption takes place? (2) If a chain begins in a given transient state, what is the probability that we end up in each absorbing state?

To answer these questions, we need to write the transition matrix with the states listed in the following order: transient states first, then absorbing states. For the sake of definiteness, let's assume that there are $s - m$ transient states $(t_1, t_2, \ldots, t_{s-m})$ and m absorbing states $(a_1, a_2, \ldots, a_m)$. Then the transition matrix for the absorbing chain may be written as follows:

$$P = \begin{array}{c} \\ s - m \text{ rows} \\ m \text{ rows} \end{array} \overset{\begin{array}{cc} s-m & m \\ \text{columns} & \text{columns} \end{array}}{\left[\begin{array}{c|c} Q & R \\ \hline 0 & I \end{array}\right]}$$

In this format, the rows and column of P correspond (in order) to the states $t_1, t_2, \ldots,$ $t_{s-m}, a_1, a_2, \ldots, a_m$. Here, I is an $m \times m$ identity matrix reflecting the fact that we can never leave an absorbing state: Q is an $(s - m) \times (s - m)$ matrix that represents transitions between transient states; R is an $(s - m) \times m$ matrix representing transitions from transient states to absorbing states; 0 is an $m \times (s - m)$ matrix consisting entirely of zeros. This reflects the fact that it is impossible to go from an absorbing state to a transient state.

Applying this notation to Example 7, we let

$$t_1 = \text{New}$$
$$t_2 = 1 \text{ Month}$$
$$t_3 = 2 \text{ Months}$$
$$t_4 = 3 \text{ Months}$$
$$a_1 = \text{Paid}$$
$$a_2 = \text{Bad Debt}$$

Then for Example 7, the transition probability matrix may be written as

	New	1 month	2 months	3 months	Paid	Bad debt
New	0	.6	0	0	.4	0
1 month	0	0	.5	0	.5	0
2 months	0	0	0	.4	.6	0
3 months	0	0	0	0	.7	.3
Paid	0	0	0	0	1	0
Bad debt	0	0	0	0	0	1

Then $s = 6$, $m = 2$, and

$$Q = \begin{bmatrix} 0 & .6 & 0 & 0 \\ 0 & 0 & .5 & 0 \\ 0 & 0 & 0 & .4 \\ 0 & 0 & 0 & 0 \end{bmatrix}_{4 \times 4} \qquad R = \begin{bmatrix} .4 & 0 \\ .5 & 0 \\ .6 & 0 \\ .7 & .3 \end{bmatrix}_{4 \times 2}$$

For Example 8, we let

$$t_1 = \text{Junior}$$
$$t_2 = \text{Senior}$$
$$t_3 = \text{Partner}$$
$$a_1 = \text{Leave as nonpartner}$$
$$a_2 = \text{Leave as partner}$$

and we may write the transition probability matrix as

	Junior	Senior	Partner	Leave as NP	Leave as P
Junior	.80	.15	0	.05	0
Senior	0	.70	.20	.10	0
Partner	0	0	.95	0	.05
Leave as nonpartner	0	0	0	1	0
Leave as partner	0	0	0	0	1

Then $s = 5$, $m = 2$, and

$$Q = \begin{bmatrix} .80 & .15 & 0 \\ 0 & .70 & .20 \\ 0 & 0 & .95 \end{bmatrix}_{3 \times 3} \qquad R = \begin{bmatrix} .05 & 0 \\ .10 & 0 \\ 0 & .05 \end{bmatrix}_{3 \times 2}$$

We can now find out some facts about absorbing chains (see Kemeny and Snell (1960). (1) If the chain begins in a given transient state, and before we reach an absorbing state, what is the expected number of times that each state will be entered? How many periods do we expect to spend in a given transient state before absorption takes place? *Answer:* If we are at present in transient state t_i, the expected number of periods that will be spent in transient state t_j before absorption is the ijth element of the matrix $(I - Q)^{-1}$. (See Problem 12 at the end of this section for a proof.) (2) If a chain begins in a given transient state, what is the probability that we end up in each absorbing state? *Answer:* If we are at present in transient state t_i, the probability that we will eventually be absorbed in absorbing state a_j is the ijth element of the matrix $(I - Q)^{-1} R$. (See Problem 13 at the end of this section for a proof.)

The matrix $(I - Q)^{-1}$ is often referred to as the **Markov chain's fundamental matrix.** The reader interested in further study of absorbing chains is referred to Kemeny and Snell (1960).

EXAMPLE 7   Accounts Receivable (Continued)

1 What is the probability that a new account will eventually be collected?

2 What is the probability that a one-month-overdue account will eventually become a bad debt?

3 If the firm's sales average $100,000 per month, how much money per year will go uncollected?

Solution From our previous discussion, recall that

$$Q = \begin{bmatrix} 0 & .6 & 0 & 0 \\ 0 & 0 & .5 & 0 \\ 0 & 0 & 0 & .4 \\ 0 & 0 & 0 & 0 \end{bmatrix} \qquad R = \begin{bmatrix} .4 & 0 \\ .5 & 0 \\ .6 & 0 \\ .7 & .3 \end{bmatrix}$$

Then

$$I - Q = \begin{bmatrix} 1 & -.6 & 0 & 0 \\ 0 & 1 & -.5 & 0 \\ 0 & 0 & 1 & -.4 \\ 0 & 0 & 0 & 1 \end{bmatrix}$$

By using the Gauss–Jordan method of Chapter 2, we find that

$$(I - Q)^{-1} = \begin{array}{c} \\ t_1 \\ t_2 \\ t_3 \\ t_4 \end{array} \begin{array}{c} \begin{matrix} t_1 & t_2 & t_3 & t_4 \end{matrix} \\ \begin{bmatrix} 1 & .60 & .30 & .12 \\ 0 & 1 & .50 & .20 \\ 0 & 0 & 1 & .40 \\ 0 & 0 & 0 & 1 \end{bmatrix} \end{array}$$

To answer questions 1–3, we need to compute

$$
(I - Q)^{-1}R = \begin{array}{c} t_1 \\ t_2 \\ t_3 \\ t_4 \end{array}
\begin{array}{c} \overset{a_1 \quad\; a_2}{\left[\begin{array}{cc} .964 & .036 \\ .940 & .060 \\ .880 & .120 \\ .700 & .300 \end{array}\right]} \end{array}
$$

Then

1 t_1 = New, a_1 = Paid. Thus, the probability that a new account is eventually collected is element 11 of $(I - Q)^{-1}R = .964$.

2 t_2 = 1 Month, a_2 = Bad Debt. Thus, the probability that a one-month overdue account turns into a bad debt is element 22 of $(I - Q)^{-1}R = .06$.

3 From answer 1, only 3.6% of all debts are uncollected. Since yearly accounts payable are \$1,200,000, on the average, $(.036)(1,200,000) = \$43,200$ per year will be uncollected.

EXAMPLE 8   Work-Force Planning (Continued)

1 What is the average length of time that a newly hired junior lawyer spends working for the firm?

2 What is the probability that a junior lawyer makes it to partner?

3 What is the average length of time that a partner spends with the firm (as a partner)?

Solution Recall that for Example 8,

$$
Q = \begin{bmatrix} .80 & .15 & 0 \\ 0 & .70 & .20 \\ 0 & 0 & .95 \end{bmatrix} \qquad R = \begin{bmatrix} .05 & 0 \\ .10 & 0 \\ 0 & .05 \end{bmatrix}
$$

Then

$$
I - Q = \begin{bmatrix} .20 & -.15 & 0 \\ 0 & .30 & -.20 \\ 0 & 0 & .05 \end{bmatrix}
$$

By using the Gauss–Jordan method of Chapter 2, we find that

$$
(I - Q)^{-1} = \begin{array}{c} t_1 \\ t_2 \\ t_3 \end{array}
\begin{array}{c} \overset{t_1 \quad\; t_2 \quad\;\; t_3}{\left[\begin{array}{ccc} 5 & 2.5 & 10 \\ 0 & \frac{10}{3} & \frac{40}{3} \\ 0 & 0 & 20 \end{array}\right]} \end{array}
$$

Then

$$
(I - Q)^{-1}R = \begin{array}{c} t_1 \\ t_2 \\ t_3 \end{array}
\begin{array}{c} \overset{a_1 \quad\; a_2}{\left[\begin{array}{cc} .50 & .50 \\ \frac{1}{3} & \frac{2}{3} \\ 0 & 1 \end{array}\right]} \end{array}
$$

Then

1 Expected time junior lawyer stays with firm = (expected time junior lawyer stays with firm as junior) + (expected time junior lawyer stays with firm as senior) + (expected time junior lawyer stays with firm as partner). Now

$$\text{Expected time as junior} = (I - Q)^{-1}_{11} = 5$$
$$\text{Expected time as senior} = (I - Q)^{-1}_{12} = 2.5$$
$$\text{Expected time as partner} = (I - Q)^{-1}_{13} = 10$$

Hence, the total expected time that a junior lawyer spends with the firm is $5 + 2.5 + 10 = 17.5$ years.

2 The probability that a new junior lawyer makes it to partner is just the probability that he or she leaves the firm as a partner. Since $t_1 = $ Junior Lawyer and $a_2 = $ Leave as Partner, the answer is element 12 of $(I - Q)^{-1}R = .50$.

3 Since $t_3 = $ Partner, we seek the expected number of years that are spent in t_3, given that we begin in t_3. This is just element 33 of $(I - Q)^{-1} = 20$ years. This is reasonable, because during each year, there is 1 chance in 20 that a partner will leave the firm, so it should take an average of 20 years before a partner leaves the firm.

REMARKS

IQinverse.xls

Computations with absorbing chains are greatly facilitated if we multiply matrices on a spreadsheet with the MMULT command and find the inverse of $(I - Q)$ with the MINVERSE function.

To use the Excel MINVERSE command to find $(I - Q)^{-1}$, we enter $(I - Q)$ into a spreadsheet (see cell range C4:E6 of file IQinverse.xls) and select the range (C8:E10) where we want to compute $(I - Q)^{-1}$. Next we type the formula

$$=\text{MINVERSE(C4:E6)}$$

in the upper left-hand corner (cell C8) of the output range C8:E10. Finally, we select **CONTROL SHIFT ENTER** (not just ENTER) to complete the computation of the desired inverse. The MINVERSE function must be entered with CONTROL SHIFT ENTER because it is an array function. We cannot edit or delete any part of a range computed by an array function. See Figure 8.

	B	C	D	E	F
2					
3					
4			0.2	-0.15	0	
5	I-Q	0	0.3	-0.2	
6		0	0	0.05	
7					
8			5	2.5	10	
9	(I-Q)⁻¹	0	3.333333	13.33333	
10		0	0	20	
11					

FIGURE 8

PROBLEMS

Group A

1[†] The State College admissions office has modeled the path of a student through State College as a Markov chain:

	F.	So.	J.	Sen.	Q.	G.
Freshman	.10	.80	0	0	.10	0
Sophmore	0	.10	.85	0	.05	0
Junior	0	0	.15	.80	.05	0
Senior	0	0	0	.10	.05	.85
Quits	0	0	0	0	1	0
Graduates	0	0	0	0	0	1

[†]Based on Bessent and Bessent (1980).

Each student's state is observed at the beginning of each fall semester. For example, if a student is a junior at the beginning of the current fall semester, there is an 80% chance that he will be a senior at the beginning of the next fall semester, a 15% chance that he will still be a junior, and a 5% chance that he will have quit. (We assume that once a student quits, he never reenrolls.)

a If a student enters State College as a freshman, how many years can he expect to spend as a student at State?

b What is the probability that a freshman graduates?

2[†] The *Herald Tribble* has obtained the following information about its subscribers: During the first year as subscribers, 20% of all subscribers cancel their subscriptions. Of those who have subscribed for one year, 10% cancel during the second year. Of those who have been subscribing for more than two years, 4% will cancel during any given year. On the average, how long does a subscriber subscribe to the *Herald Tribble*?

3 A forest consists of two types of trees: those that are 0–5 ft and those that are taller than 5 ft. Each year, 40% of all 0–5-ft tall trees die, 10% are sold for $20 each, 30% stay between 0 and 5 ft, and 20% grow to be more than 5 ft. Each year, 50% of all trees taller than 5 ft are sold for $50, 20% are sold for $30, and 30% remain in the forest.

  a What is the probability that a 0–5-ft tall tree will die before being sold?

  b If a tree (less than 5 ft) is planted, what is the expected revenue earned from that tree?

4[‡] Absorbing Markov chains are used in marketing to model the probability that a customer who is contacted by telephone will eventually buy a product. Consider a prospective customer who has never been called about purchasing a product. After one call, there is a 60% chance that the customer will express a low degree of interest in the product, a 30% chance of a high degree of interest, and a 10% chance the customer will be deleted from the company's list of prospective customers. Consider a customer who currently expresses a low degree of interest in the product. After another call, there is a 30% chance that the customer will purchase the product, a 20% chance the person will be deleted from the list, a 30% chance that the customer will still possess a low degree of interest, and a 20% chance that the customer will express a high degree of interest. Consider a customer who currently expresses a high degree of interest in the product. After another call, there is a 50% chance that the customer will have purchased the product, a 40% chance that the customer will still have a high degree of interest, and a 10% chance that the customer will have a low degree of interest.

  a What is the probability that a new prospective customer will eventually purchase the product?

  b What is the probability that a low-interest prospective customer will ever be deleted from the list?

  c On the average, how many times will a new prospective customer be called before either purchasing the product or being deleted from the list?

5 Each week, the number of acceptable-quality units of a drug that are processed by a machine is observed: >100, 50–100, 1–50, 0 (indicating that the machine was broken during the week). Given last week's observation, the probability distribution of next week's observation is as follows.

	>100	50–100	1–50	0
>100	.8	.1	.05	.05
50–100	.1	.6	.1	.2
1–50	.1	.1	.5	.3
0	0	0	0	1

For example, if we observe a week in which more than 100 units are produced, then there is a .10 chance that during the next week 50–100 units are produced.

  a Suppose last week the machine produced 200 units. On average, how many weeks will elapse before the machine breaks down?

  b Suppose last week the machine produced 50 units. On average, how many weeks will elapse before the machine breaks down?

6 I now have $2, and my goal is to have $6. I will repeatedly flip a coin that has a .4 chance of coming up heads. If the coin comes up heads, I win the amount I bet. If the coin comes up tails, I lose the amount of my bet. Let us suppose I follow the **bold strategy** of betting Min($6 − current asset position, current asset position). This strategy (see Section 19.3) maximizes my chance of reaching my goal. What is the probability that I reach my goal?

7 Suppose I toss a fair coin, and the first toss comes up heads. If I keep tossing the coin until I either see two consecutive heads or two consecutive tails, what is the probability that I will see two consecutive heads before I see two consecutive tails?

8 Suppose each box of Corn Snaps cereal contains one of five different Harry Potter trading cards. On the average, how many boxes of cereal will I have to buy to obtain a complete set of trading cards?

Group B

9 In the gambler's ruin problem (Example 1), assume $p = .60$.

  a What is the probability that I reach $4?

  b What is the probability that I am wiped out?

  c What is the expected duration of the game?

10[§] In caring for elderly patients at a mental hospital, a major goal of the hospital is successful placement of the patients in boarding homes or nursing homes. The movement of patients between the hospital, outside homes, and the absorbing state (death) may be described by the following Markov chain (the unit of time is one month):

	Hospital	Homes	Death
Hospital	.991	.003	.006
Homes	.025	.969	.006
Death	0	0	1

Each month that a patient spends in the hospital costs the state $655, and each month that a patient spends in a home costs the state $226. To improve the success rate of the placement of patients in homes, the state has recently begun a "geriatric resocialization program" (GRP) to prepare the patients for functioning in the homes. Some patients are placed in the GRP and then released to homes. These patients presumably are less likely to fail to adjust in the homes. Other patients continue to go directly from the hospital to homes without taking part in the GRP. The state pays $680 for each month that a patient spends in the GRP. The

[†]Based on Deming and Glasser (1968).
[‡]Based on Thompson and McNeal (1967).

[§]Based on Meredith (1973).

movement of the patients through various states is governed by the following Markov chain:

	GRP	Hos.	Homes (GRP)	Homes (Direct)	Dead
GRP	.854	.028	.112	0	.006
Hospital	.013	.978	0	.003	.006
Homes (GRP)	.025	0	.969	0	.006
Homes (Direct)	0	.025	0	.969	.006
Dead	0	0	0	0	1

a Does the GRP save the state money?

b Under the old system and under the GRP, compute the expected number of months that a patient spends in the hospital.

11 Freezco, Inc., sells refrigerators. The company has issued a warranty on all refrigerators that requires free replacement of any refrigerator that fails before it is three years old. We are given the following information: (1) 3% of all new refrigerators fail during their first year of operation; (2) 5% of all one-year-old refrigerators fail during their second year of operation; and (3) 7% of all two-year-old refrigerators fail during their third year of operation. A replacement refrigerator is not covered by the warranty.

a Use Markov chain theory to predict the fraction of all refrigerators that Freezco will have to replace.

b Suppose that it costs Freezco $500 to replace a refrigerator and that Freezco sells 10,000 refrigerators per year. If the company reduced the warranty period to two years, how much money in replacement costs would be saved?

12 For a Q matrix representing the transitions between transient states in an absorbing Markov chain, it can be shown that

$$(I - Q)^{-1} = I + Q + Q^2 + \cdots + Q^n + \cdots$$

a Explain why this expression for $(I - Q)^{-1}$ is plausible.

b Define m_{ij} = expected number of periods spent in transient state t_j before absorption, given that we begin in state t_i. (Assume that the initial period is spent in state t_i.) Explain why m_{ij} = (probability that we are in state t_j initially) + (probability that we are in state t_j after first transition) + (probability that we are in state t_j after second transition) + $\cdots$ + (probability that we are in state t_j after nth transition) + $\cdots$.

c Explain why the probability that we are in state t_j initially = ijth entry of the $(s - m) \times (s - m)$ identity matrix. Explain why the probability that we are in state t_j after nth transition = ijth entry of Q^n.

d Now explain why m_{ij} = ijth entry of $(I - Q)^{-1}$.

13 Define

b_{ij} = probability of ending up in absorbing state a_j given that we begin in transient state t_i

r_{ij} = ijth entry of R

q_{ik} = ikth entry of Q

$B = (s - m) \times m$ matrix whose ijth entry is b_{ij}

Suppose we begin in state t_i. On our first transition, three types of events may happen:

Event 1 We go to absorbing state a_j (with probability r_{ij}).

Event 2 We go to an absorbing state other than a_j (with probability $\sum_{k \neq j} r_{ik}$).

Event 3 We go to transient state t_k (with probability q_{ik}).

a Explain why

$$b_{ij} = r_{ij} + \sum_{k=1}^{k=s-m} q_{ik} b_{kj}$$

b Now show that $b_{ij} = ij$th entry of $(R + QB)$ and that $B = R + QB$.

c Show that $B = (I - Q)^{-1}R$ and that $b_{ij} = ij$th entry of $B = (I - Q)^{-1}R$.

14 Consider an LP with five basic feasible solutions and a unique optimal solution. Assume that the simplex method begins at the worst basic feasible solution, and on each pivot the simplex is equally likely to move to any better basic feasible solution. On the average, how many pivots will be required to find the optimal solution to the LP?

Group C

15 General Motors has three auto divisions (1, 2, and 3). It also has an accounting division and a management consulting division. The question is: What fraction of the cost of the accounting and management consulting divisions should be allocated to each auto division? We assume that the entire cost of the accounting and management consulting departments must be allocated to the three auto divisions. During a given year, the work of the accounting division and management consulting division is allocated as shown in Table 4.

For example, accounting spends 10% of its time on problems generated by the accounting department, 20% of its time on work generated by division 3, and so forth. Each year, it costs $63 million to run the accounting department and $210 million to run the management consulting department. What fraction of these costs should be allocated to each auto division? Think of $1 in costs incurred in accounting work. There is a .20 chance that this dollar should be allocated to each auto division, a .30 chance it should be allocated to consulting, and a .10 chance to accounting. If the dollar is allocated to an auto division, we know which division should be charged for that dollar. If the dollar is charged to consulting (for example), we repeat the process until the dollar is eventually charged to an auto division. Use knowledge of absorbing chains to figure out how to allocate the costs of running the accounting and management consulting departments among the three auto divisions.

16 A telephone sales force can model its contact with customers as a Markov chain. The six states of the chain are as follows:

State 1 Sale completed during most recent call

State 2 Sale lost during most recent call

State 3 New customer with no history

State 4 During most recent call, customer's interest level low

TABLE **4**

	Accounting	Management Consulting	Division 1	Division 2	Division 3
Accounting	10%	30%	20%	20%	20%
Management	30%	20%	30%	0%	20%

State 5 During most recent call, customer's interest level medium

State 6 During most recent call, customer's interest level high

Based on past phone calls, the following transition matrix has been estimated:

$$\begin{array}{c} \\ 1 \\ 2 \\ 3 \\ 4 \\ 5 \\ 6 \end{array} \begin{array}{cccccc} 1 & 2 & 3 & 4 & 5 & 6 \\ \left[\begin{array}{cccccc} 1 & 0 & 0 & 0 & 0 & 0 \\ 0 & 1 & 0 & 0 & 0 & 0 \\ .10 & .30 & 0 & .25 & .20 & .15 \\ .05 & .45 & 0 & .20 & .20 & .10 \\ .15 & .10 & 0 & .15 & .25 & .35 \\ .20 & .05 & 0 & .15 & .30 & .30 \end{array} \right] \end{array}$$

a For a new customer, determine the average number of calls made before the customer buys the product or the sale is lost.

b What fraction of new customers will buy the product?

c What fraction of customers currently having a low degree of interest will buy the product?

d Suppose a call costs $15 and a sale earns $190 in revenue. Determine the "value" of each type of customer.

17 Seas Beginning sells clothing by mail order. An important question is: When should the company strike a customer from its mailing list? At present, the company does so if a customer fails to order from six consecutive catalogs. Management wants to know if striking a customer after failure to order from four consecutive catalogs will result in a higher profit per customer.

The following data are available: Six percent of all customers who receive a catalog for the first time place an order. If a customer placed an order from the last-received catalog, then there is a 20% chance he or she will order from the next catalog. If a customer last placed an order one catalog ago, there is a 16% chance he or she will order from the next catalog received. If a customer last placed an order two catalogs ago, there is a 12% chance he or she will place an order from the next catalog received. If a customer last placed an order three catalogs ago, there is an 8% chance he or she will place an order from the next catalog received. If a customer last placed an order four catalogs ago, there is a 4% chance he or she will place an order from the next catalog received. If a customer last placed an order five catalogs ago, there is a 2% chance he or she will place an order from the next catalog received.

It costs $1 to send a catalog, and the average profit per order is $15. To maximize expected profit per customer, should Seas Beginning cancel customers after six nonorders or four nonorders?

Hint: Model each customer's evolution as a Markov chain with possible states New, 0, 1, 2, 3, 4, 5, Canceled. A customer's state represents the number of catalogs received since the customer last placed an order. "New" means the customer received a catalog for the first time. "Canceled" means that the customer has failed to order from six consecutive catalogs. For example, suppose a customer placed the following sequence of orders (O) and nonorders (NO):

NO NO O NO NO O O NO NO O NO NO NO NO NO
NO Canceled

Here we are assuming a customer is stricken from the mailing list after six consecutive nonorders. For this sequence of orders and nonorders, the states are (*i*th listed state occurs right before *i*th catalog is received)

New 1 2 0 1 2 0 0 1 2 0 1 2 3 4 5 Canceled

You should be able to figure (for each cancellation policy) the expected number of orders a customer will place before cancellation and the expected number of catalogs a customer will receive before cancellation. This will enable you to compute expected profit per customer.

17.7 Work-Force Planning Models†

Many organizations, like the Mason and Burger law firm of Example 8, employ several categories of workers. For long-term planning purposes, it is often useful to be able to predict the number of employees of each type who will (if present trends continue) be available in the steady state. Such predictions can be made via an analysis similar to the one in Section 17.5 of steady-state probabilities for Markov chains.

More formally, consider an organization whose members are classified at any point in time into one of *s* groups (labeled 1, 2, . . . , *s*). During every time period, a fraction p_{ij} of

†This section covers topics that may be omitted with no loss of continuity.

those who begin a time period in group i begin the next time period in group j. Also, during every time period, a fraction $p_{i,s+1}$ of all group i members leave the organization. Let P be the $s \times (s + 1)$ matrix whose ijth entry is p_{ij}. At the beginning of each time period, the organization hires H_i group i members. Let $N_i(t)$ be the number of group i members at the beginning of period t. A question of natural interest is whether $N_i(t)$ approaches a limit as t grows large (call the limit, if it exists, N_i). If each $N_i(t)$ does not approach a limit, we call $\mathbf{N} = (N_1, N_2, \ldots, N_s)$ the **steady-state census** of the organization.

If a steady-state census exists, we can find it by solving a system of s equations that is derived as follows: Simply note that for a steady-state census to exist, it must be true that in the steady state, for $i = 1, 2, \ldots, s$,

$$\text{Number of people entering group } i \text{ during each period} \tag{14}$$
$$= \text{number of people leaving group } i \text{ during each period}$$

After all, if (14) did not hold for all groups, then the number of people in at least one group would pile up as time progressed. We note that

$$\text{Number of people entering state } i \text{ during each period} = H_i + \sum_{k \neq i} N_k p_{ki}$$

$$\text{Number of people leaving state } i \text{ during each period} = N_i \sum_{k \neq i} p_{ik}$$

Then the equation used to compute the steady-state census is

$$H_i + \sum_{k \neq i} N_k p_{ki} = N_i \sum_{k \neq i} p_{ik} \ (i = 1, 2, \ldots, s) \tag{14'}$$

Note that $\sum_{k \neq i} p_{ik} = 1 - p_{ii}$. This can be used to simplify (14').

If a steady-state census does not exist, then (14') will have no solution. See Problem 6 for an example of this. Given the values of the p_{ij}'s and the H_i's, (14') can be used to solve for the steady-state census. Conversely, given the p_{ij}'s and a desired steady-state census, (14') can be used to determine a hiring policy (specified by values of $H_1, H_2, \ldots, H_s$) that attains the desired steady-state census. Some steady-state censuses may be impossible to maintain unless some of the H_i's are negative (corresponding to firing employees).

The following two examples illustrate the use of the steady-state census equation.

EXAMPLE 9   Steady-State Census

Suppose that each American can be classified into one of three groups: children, working adults, or retired people. During a one-year period, .959 of all children remain children, .04 of all children become working adults, and .001 of all children die. During any given year, .96 of all working adults remain working adults, .03 of all working adults retire, and .01 of all working adults die. Also, .95 of all retired people remain retired, and .05 of all retired people die. One thousand children are born each year.

1 Determine the steady-state census.

2 Each retired person receives a pension of $5,000 per year. The pension fund is funded by payments from working adults. How much money must each working adult contribute annually to the pension fund?

Solution   1 Let

$$\text{Group } 1 = \text{children}$$
$$\text{Group } 2 = \text{working adults}$$
$$\text{Group } 3 = \text{retired people}$$
$$\text{Group } 4 = \text{died}$$

We are given that $H_1 = 1{,}000$, $H_2 = H_3 = 0$, and

$$P = \begin{bmatrix} .959 & .040 & 0 & .001 \\ 0 & .960 & .030 & .010 \\ 0 & 0 & .950 & .050 \end{bmatrix}$$

Now (14) or (14') yields

Number entering group i each year = number leaving group i each year

$$1{,}000 = (.04 + .001)N_1 \qquad \text{(Children)}$$
$$.04N_1 = (.03 + .01)N_2 \qquad \text{(Working adults)}$$
$$.03N_2 = .05N_3 \qquad\qquad \text{(Retired people)}$$

Solving this system of equations, we find that $N_1 = 24{,}390$, $N_2 = 24{,}390.24$, and $N_3 = 14{,}634.14$.

2 Since in the steady state, there are 14,634.14 retired people, in the steady state they receive 14,634.14(5,000) dollars per year. Hence, each working adult must pay

$$\frac{14{,}634.14(5{,}000)}{24{,}390.24} = \$3{,}000 \text{ per year}$$

This result is reasonable, because in the steady state, there are $\frac{5}{3}$ as many working adults as there are retired people.

EXAMPLE 10 **The Mason and Burger Law Firm (continued)**

Let's return to the law firm of Mason and Burger (Example 8). Suppose the firm's long-term goal is to employ 50 junior lawyers, 30 senior lawyers, and 10 partners. To achieve this steady-state census, how many lawyers of each type should Mason and Burger hire each year?

Solution Let

Group 1 = junior lawyers

Group 2 = senior lawyers

Group 3 = partners

Group 4 = lawyers who have left firm

Mason and Burger want to obtain $N_1 = 50$, $N_2 = 30$, and $N_3 = 10$. Recall from Example 8 that

$$P = \begin{bmatrix} .80 & .15 & 0 & .05 \\ 0 & .70 & .20 & .10 \\ 0 & 0 & .95 & .05 \end{bmatrix}$$

Then (14) or (14') yields

Number entering group i = number leaving group i

$$H_1 = (.15 + .05)50 \qquad \text{(Junior lawyers)}$$
$$(.15)50 + H_2 = (.20 + .10)30 \qquad \text{(Senior lawyers)}$$
$$(.20)30 + H_3 = (.05)10 \qquad\qquad \text{(Partners)}$$

The unique solution to this system of equations is $H_1 = 10$, $H_2 = 1.5$, $H_3 = -5.5$. This means that to maintain the desired steady-state census, Mason and Burger would have to fire 5.5 partners each year. This is reasonable, because an average of .20(30) = 6 senior lawyers become partners every year, and once a senior lawyer becomes a partner, he or she stays a partner for an average of 20 years. This shows that to keep the number of partners down to 10, several partners must be released each year. An alternative solution might be to reduce (below its current value of .20) the fraction of senior lawyers who become partners during each year.

For more information on work-force planning models, the interested reader should consult the excellent book by Grinold and Marshall (1977).

Using LINGO to Solve for the Steady-State Census

Census.lng

The following LINGO model (file Census.lng) can be used to determine the steady-state census for a work-force planning problem:

```
MODEL:
 1]SETS:
 2]STATE/1..3/:N,H;
 3]SXS(STATE,STATE):TPROB;
 4]ENDSETS
 5]DATA:
 6]H=1000,0,0;
 7]TPROB=.959,.04,0,
 8]0,.96,.03,
 9]0,0,.95;
 10]ENDDATA
 11]@FOR(STATE(I):H(I)
 12]+@SUM(STATE(K)|K#NE#I:N(K)*TPROB(K,I))=
 13]N(I)*(1-TPROB(I,I)););
END
```

In line 2, we create the possible states and define for each state I the steady-state census level and number hired, $N(I)$ and $H(I)$, respectively. In line 3, we create for each pair (I, J) of states the probability TPROB (I, J) of going from state I in one period to state J during the next period. In line 6, we input the value of $H(I)$ for each state I. In lines 7 through 9, we input the TPROB (I, J) for Example 9. In lines 11 through 13, we create for each state I the equation $(14)'$. Note that we use the fact that

$$\sum_{K \neq I} \text{TPROB}(I, K) = 1 - \text{TPROB}(I, I)$$

Entering the **SOLVE** command will yield the steady-state census level $N(I)$ for state I. Note that by modifying the DATA portion of the program, we could also enter a desired steady-state census ($N(I)$) and have LINGO solve for a set of hiring levels ($H(I)$) which yield the desired steady-state census.

PROBLEMS

Group A

1 Refer to Problem 1 of Section 17.6. Suppose that each year, State College admits 7,000 freshmen, 500 sophomore transfers, and 500 junior transfers. In the long run, what will be the composition of the State College student body?

2 In Example 9, suppose that advances in medical science have reduced the annual death rate for retired people from 5% to 3%. By how much would this increase the annual pension contribution that a working adult would have to make to the pension fund?

3 New York City produces 1,000 tons of air pollution per day, Jersey City 100 tons, and Newark 50 tons. Each day, $\frac{1}{3}$ of New York's pollution is blown to Newark, $\frac{1}{3}$ dissipates, and $\frac{1}{3}$ remains in New York. Each day, $\frac{1}{3}$ of Jersey City's pollution is blown to New York, $\frac{1}{3}$ stays in Jersey City, and $\frac{1}{3}$ is blown to Newark. Each day, $\frac{1}{3}$ of Newark's pollution stays in Newark, and the rest is blown to Jersey City. On a typical day, which city will be the most polluted?

4 Money circulates among the Federation's three "capital" planets: Vulcan, Romulanville, and Klingonville. Ideally, the Federation would like to have $5 billion in circulation at each planet. Each month, $\frac{1}{3}$ of all the money at Vulcan leaves circulation, $\frac{1}{3}$ stays at Vulcan, and $\frac{1}{3}$ ends up in Klingonville. Each month, $\frac{1}{3}$ of the money at Romulanville remains in Romulanville, $\frac{1}{3}$ ends up in Klingonville, and $\frac{1}{3}$ ends up at Vulcan. Each month, $\frac{2}{3}$ of the money in Klingonville ends up in Romulanville, and $\frac{1}{3}$ stays in Klingonville. The Federation introduces money into the system at Vulcan. Is there any way to have a steady-state level of $5 billion in circulation at each planet?

Group B

5 All State University Business School faculty members are classified as tenured or untenured. Each year, 10% of the untenured faculty are granted tenure and 10% leave State University; 95% of the tenured faculty remain and 5% leave. The business school wants to maintain a faculty with 100 members, of which x% are untenured. Determine a hiring policy that will achieve this goal. For what values of x does this goal require firing tenured faculty members? Describe a hiring policy that maintains a faculty that is 10% untenured. Describe a hiring policy that maintains a faculty that is 40% untenured.

6 In the world of Never-Ever Land, one child is born at the beginning of each year. During each year, 90% of the children alive at the beginning of the year remain children, and 10% become adults. During each year, 90% of the adults alive at the beginning of the year remain adults and 10% of the adults become children.

   a Explain why no steady-state census exists.

   b Show that equation (14′) has no solution.

Group C

7[†] For simplicity, suppose that fresh blood obtained by a hospital will spoil if it is not transfused within five days. The hospital receives 100 pints of fresh blood daily from a local blood bank. Two policies are possible for determining the order in which blood is transfused (see Table 5). For example, under policy 1, blood has a 10% chance of being transfused during its first day at the hospital. Under policy 2, four-day-old blood has a 10% chance of being transfused.

   a A FIFO (first in, first out) blood-issuing policy issues "old" blood first, whereas a LIFO (last in, first out) policy issues "young" blood first. Which policy represents a LIFO policy, and which represents a FIFO policy?

TABLE 5

	Age of Blood (beginning of day)				
Chance of transfusion	0	1	2	3	4
Policy 1	.10	.20	.30	.40	.50
Policy 2	.50	.40	.30	.20	.10

   b For each policy, determine the probability that a new pint of blood will spoil.

   c For each policy, determine the average number of pints of blood in inventory.

   d For each policy, find the average age of transfused blood.

   e Comment on the relative merits of a FIFO policy and a LIFO policy.

8 Suppose that each week every American family buys a gallon of orange juice from company A or B or C. Let p_i = probability that a gallon produced by company i is of unsatisfactory quality. If the last gallon of juice purchased by a family is satisfactory, then the next week they will purchase a gallon of juice from the same company. If the last gallon of juice purchased by a family is not satisfactory, then the family will purchase a gallon from a competitor. Consider a week in which A families have purchased juice A, B families have purchased juice B, and C families have purchased juice C. Assume that families that switch brands during a period are allocated to the remaining brands in a manner proportionate to the current market shares of the other brands. Thus, if a family switches from brand A, there is a chance $B/(B + C)$ that they will switch to B and a chance $C/(B + C)$ that they will switch to C. Suppose that 1 million gallons of orange juice are purchased each week.

   a After a long time, what will be the market share for each firm? *Hint:* Show that for some k in the steady state, brand A will sell $k(p_B + p_C - p_A)$ gallons of juice each week, and conjecture the number of gallons of brands B and C that will be sold each week.

   b Suppose a 1% increase in market share is worth $10,000 per week to firm A. Also suppose that currently $p_A = .10$, $p_B = .15$, and $p_C = .20$. Firm A believes that for a cost of $1 million per year, it can cut the percentage of unsatisfactory juice cartons in half. Is this worthwhile?[‡]

9 The age-based probability that an American dies during a given year is shown in Table 6. For example, a fraction

TABLE 6

Age	Death Probability
0	0.007557
1–4	0.000383
5–9	0.000217
10–14	0.000896
15–24	0.001267
25–34	0.002213
35–44	0.004459
45–54	0.010941
55–64	0.025384
65–84	0.058031
85+	0.15327

[†]Based on Pegels and Jelmert (1970).

[‡]Based on Babich (1992).

.007557 of all babies die during their first year of life. Suppose 100 babies are born each year, and nobody lives to be older than 110.

a What is the average age of people in the United States?

b Suppose all people ages 21–65 work, and all people over age 65 are retired. If we want to pay each retiree $20,000 per year, how much money must each worker pay in to ensure that during each year, the retirement plan is self-financing?

SUMMARY

Let $\mathbf{X}_t$ be the value of a system's characteristic at time t. A **discrete-time stochastic process** is simply a description of the relation between the random variables $\mathbf{X}_0$, $\mathbf{X}_1$, $\mathbf{X}_2$, ... A discrete-time stochastic process is a **Markov chain** if, for $t = 0, 1, 2, \ldots$ and all states,

$$P(\mathbf{X}_{t+1} = i_{t+1}|\mathbf{X}_t = i_t, \mathbf{X}_{t-1} = i_{t-1}, \ldots, \mathbf{X}_1 = i_1, \mathbf{X}_0 = i_0)$$
$$= P(\mathbf{X}_{t+1} = i_{t+1}|\mathbf{X}_t = i_t)$$

For a stationary Markov chain, the **transition probability** p_{ij} is the probability that given the system is in state i at time t, the system will be in state j at time $t + 1$.

The vector $\mathbf{q} = [q_1 \quad q_2 \quad \cdots \quad q_s]$ is the **initial probability distribution** for the Markov chain. $P(\mathbf{X}_0 = i)$ is given by q_i.

n-Step Transition Probabilities

The **n-step transition probability,** $p_{ij}(n)$, is the probability that n periods from now, the state will be j, given that the current state is i. $P_{ij}(n) = ij$th element of P^n.

Given the intial probability vector $\mathbf{q}$, the probability of being in state j at time n is given by $\mathbf{q}$(column j of P^n).

Classification of States in a Markov Chain

Given two states i and j, a **path** from i to j is a sequence of transitions that begins in i and ends in j, such that each transition in the sequence has a positive probability of occurring. A state j is **reachable** from a state i if there is a path leading from i to j. Two states i and j are said to **communicate** if j is reachable from i, and i is reachable from j.

A set of states S in a Markov chain is a **closed set** if no state outside of S is reachable from any state in S.

A state i is an **absorbing state** if $p_{ii} = 1$. A state i is a **transient state** if there exists a state j that is reachable from i, but the state i is not reachable from state j.

If a state is not transient, it is a **recurrent state**. A state i is **periodic** with period $k >$ 1 if all paths leading from state i back to state i have a length that is a multiple of k. If a recurrent state is not periodic, it is **aperiodic**. If all states in a chain are recurrent, aperiodic, and communicate with each other, the chain is said to be **ergodic.**

Steady-State Probabilities

Let P be the transition probability matrix for an ergodic Markov chain with states 1, 2, . . . , s (with ijth element p_{ij}). After a large number of periods have elapsed, the probability (call it π_j) that the Markov chain is in state j is independent of the initial state. The long-run, or **steady-state,** probability π_j may be found by solving the following set of linear equations:

$$\pi_j = \sum_{k=1}^{k=s} \pi_k p_{kj} \qquad (j = 1, 2, \ldots, s; \text{ omit one of these equations})$$

$$\pi_1 + \pi_2 + \cdots + \pi_s = 1$$

Absorbing Chains

A Markov chain in which one or more states is an absorbing state is an **absorbing Markov chain.** To answer important questions about an absorbing Markov chain, we list the states in the following order: transient states first, then absorbing states. Assume there are $s - m$ transient states $(t_1, t_2, \ldots, t_{s-m})$ and m absorbing states $(a_1, a_2, \ldots, a_m)$. Write the transition probability matrix P as follows:

$$P = \begin{array}{c} \\ s - m \text{ rows} \\ m \text{ rows} \end{array} \overset{\begin{array}{cc} s - m & m \\ \text{columns} & \text{columns} \end{array}}{\left[\begin{array}{c|c} Q & R \\ \hline 0 & I \end{array} \right]}$$

The following questions may now be answered. (1) If the chain begins in a given transient state, and before we reach an absorbing state, what is the expected number of times that each state will be entered? How many periods do we expect to spend in a given transient state before absorption takes place? *Answer:* If we are at present in transient state t_i, the expected number of periods that will be spent in transient state t_j before absorption is the ijth element of the matrix $(I - Q)^{-1}$. (2) If a chain begins in a given transient state, what is the probability that we will end up in each absorbing state? *Answer:* If we are at present in transient state t_i, the probability that we will eventually be absorbed in absorbing state a_j is the ijth element of the matrix $(I - Q)^{-1}R$.

Work-Force Planning Models

For an organization in which each member is classified into one of s groups,

p_{ij} = fraction of members beginning a time period in group i who begin the next time period in group j

$p_{i,s+1}$ = fraction of all group i members who leave the organization during a period

$P = s \times (s + 1)$ matrix whose ijth entry is p_{ij}

H_i = number of group i members hired at the beginning of each period

N_i = limiting number (if it exists) of group i members

N_i may be found by equating the number of people per period who enter group i with the number of people per period who leave group i. Thus, $(N_1, N_2, \ldots, N_s)$ may be found by solving

$$H_i + \sum_{k \neq i} N_k p_{ki} = N_i \sum_{k \neq i} p_{ik} \qquad (i = 1, 2, \ldots, s)$$

REVIEW PROBLEMS

Group A

1 A machine is used to produce precision tools. If the machine is in good condition today, then 90% of the time, it will be in good condition tomorrow. If the machine is in bad condition today, then 80% of the time, it will be in bad condition tomorrow. If the machine is in good condition, it produces 100 tools per day. If the machine is in bad condition, it produces 60 tools per day. On the average, how many tools per day are produced?

2 Customers buy cars from three auto companies. Given the company from which a customer last bought a car, the probability that she will buy her next car from each company is as follows:

	Will Buy Next from		
Last Bought from	Co. 1	Co. 2	Co. 3
Co. 1	.80	.10	.10
Co. 2	.05	.85	.10
Co. 3	.10	.20	.70

a If someone currently owns a company 1 car, what is the probability that at least one of the next two cars she buys will be a company 1 car?

b At present, it costs company 1 an average of $5,000 to produce a car, and the average price a customer pays for one is $8,000. Company 1 is considering instituting a five-year warranty. It estimates that this will increase the cost per car by $300, but a market research survey indicates that the probabilities will change as follows:

	Will Buy Next from		
Last Bought from	Co. 1	Co. 2	Co. 3
Co. 1	.85	.10	.05
Co. 2	.10	.80	.10
Co. 3	.15	.10	.75

Should company 1 institute the five-year warranty?

3† A baseball team consists of 2 stars, 13 starters, and 10 substitutes. For tax purposes, the team owner must value the players. The value of each player is defined to be the total value of the salary he will earn until retirement. At the beginning of each season, the players are classified into one of four categories:

Category 1 Star (earns $1 million per year)
Category 2 Starter (earns $400,000 per year)
Category 3 Substitute (earns $100,000 per year)
Category 4 Retired (earns no more salary)

Given that a player is a star, starter, or substitute at the beginning of the current season, the probabilities that he will be a star, starter, substitute, or retired at the beginning of the next season are as follows:

	Next Season			
This Season	Star	Starter	Substitute	Retired
Star	.50	.30	.15	.05
Starter	.20	.50	.20	.10
Substitute	.05	.15	.50	.30
Retired	0	0	0	1

Determine the value of the team's players.

†Based on Flamholtz, Geis, and Perle (1984).

4 The best-selling college statistics text, *The Thrill of Statistics,* sells 5 million copies every fall. Some users keep the book, and some sell it back to the bookstore. Suppose that 90% of all students who buy a new book sell it back, 80% of all students who buy a once-used book sell it back, and 60% of all students who buy a twice-used book sell it back. If a book has been used four or more times, the cover falls off, and it cannot be sold back.

a In the steady state, how many new copies of the book will the publisher be able to sell each year?

b Suppose that a bookstore's profit on each type of book is as follows:

New book: $6
Once-used book: $3
Twice-used book: $2
Thrice-used book: $1

If the steady-state census is representative of the bookstore's sales, what will be its average profit per book?

5 Hearts Dog Food and Corporal Dog Food are battling tooth and nail for the nation's dog biscuit market. A dog owner buys one box of dog biscuits per month. If a dog owner's last purchase was a Hearts box of biscuits, there is a .8 chance that his next purchase will also be Hearts. If a dog owner's last purchase was a Corporal box of biscuits, there is a .9 chance that his next purchase will also be Corporal. It cost Hearts 80¢ to produce a box of biscuits, which sells for $1.

a If there are 40 million dog owners in the United States, what is Hearts' annual expected profit?

b If Hearts sells each box of biscuits for $100 − x$ cents ($0 \leq x \leq 20$), then a fraction $.8 + \frac{x}{100}$ of all dog owners whose last purchase was from Hearts will purchase their next box of biscuits from Hearts. How can Hearts maximize profit?

6 A small video store tracks the number of times per week a video is rented and estimates the following transition probabilities:

	5 times	4 times	3 times	2 times	1 time	0 time
5 times	.8	.1	.1	0	0	0
4 times	0	.7	.2	.1	0	0
3 times	0	0	.6	.3	.1	0
2 times	0	0	.5	.4	.1	0
1 time	0	0	0	0	.6	.4
0 time	0	0	0	0	0	1

For example, if a video was rented 5 times this week, then there is an 80% chance it will be rented 5 times next week, a 10% chance it will be rented 4 times, and a 10% chance it will be rented 3 times.

a Suppose a video was rented 5 times this week. On the average, how many times will it be rented during the next 2 weeks?

b Suppose a video was rented 5 times this week. On the average, how many more weeks will it be rented at least once?

c Suppose a video was rented 5 times this week. On the average, how many more times will it be rented?

7 Ross and Rachel have just tied the knot. The probability that they are happy each day depends on whether they were happy or sad during the last two days, in the following fashion:

Last two days	Happy	Sad
HH	.8	.2
HS	.5	.5
SH	.7	.3
SS	.4	.6

For example, if the newlyweds were sad two days ago and yesterday they were happy, then there is a 70% chance they will be happy tomorrow and a 30% chance they will be sad tomorrow. On what fraction of days will Ross and Rachel be happy?

8 Suppose that during a given year, 15% of all untenured processors leave a university (they are fired or find another job), and 15% are given tenure. Also assume that during each year, 5% of all tenured professors leave the university (via retirement or finding another job). If the university wants to have a faculty consisting of 200 untenured and 500 tenured professors, how many tenured and untenured professors should be hired each year?

Group B

9 At the beginning of a period, a company observes its inventory level. Then an order may be placed (and is instantaneously received). Finally, the period's demand is observed. We are given the following information: (1) A $2 cost is assessed against each unit of inventory on hand at the end of a period. (2) A $3 penalty is assessed against each unit of demand not met on time. Assume that all shortages result in lost sales. (3) Placing an order costs 50¢ per unit plus a $5 ordering cost. (4) During each period, demand is equally likely to equal 1, 2, or 3 units.

The company is considering the following ordering policy: At the end of any period, if the on-hand inventory is 1 unit or less, order sufficient units to bring the on-hand inventory level at the beginning of the next period up to 4 units.

a What fraction of the time will the on-hand inventory level at the end of each period be 0 unit? 1 unit? 2 units? 3 units? 4 units?

b Determine the average cost per period incurred by this ordering policy.

c Answer parts (a) and (b) if all shortages are backlogged. Assume that the cost for each unit backlogged is $3.

10[†] In problem 3, suppose that in evaluating a player's value, the owner must discount future salaries. Assume that $1 paid out in salary during the next season is equivalent to 90¢ paid out during the current season. Can you still determine the value of the team's players? (*Hint:* Modify

†Based on Flamholtz, Geis, and Perle (1984).

the probabilities in the transition probability matrix to account for the discounting of future salaries, or look at Problem 8 of Section 17.6.)

11[‡] During any month, Cashco has a .5 chance of receiving a $1,000 cash inflow and a .5 chance that there will be a $1,000 cash outflow. For every $1,000 in cash on hand at the end of a month, Cashco incurs a $15 cost (due to lost interest). At the beginning of each month, Cashco can adjust its on-hand cash balance upward or downward with the cost per transaction being $20. Cashco can never let the on-hand balance become negative. The company is considering the following two cash management policies:

Policy 1 At the beginning of a month in which the on-hand cash balance is $3,000, immediately reduce the cash balance to $1,000. At the beginning of a month in which the on-hand cash balance is $0, immediately bring the on-hand cash balance up to $1,000.

Policy 2 At the beginning of a month in which the on-hand cash balance is $4,000, immediately reduce the cash balance to $2,000. At the beginning of a month in which the on-hand cash balance is $0, immediately bring the on-hand cash balance up to $2,000.

Which policy will incur a smaller expected monthly cost (opportunity plus transaction)? The sequence of events during each month is as follows:

a Observe beginning cash balance

b Adjust (if desired) cash balance

c Cash balance changes

d Opportunity cost is assessed

12 In the game of craps, we roll a pair of six-sided dice. On the first throw, if we roll a 7 or an 11, we win right away. If we roll a 2, a 3, or a 12, we lose right away. If we first roll a total of 4, 5, 6, 8, 9, or 10, we keep rolling the dice until we get either a 7 or the total rolled on the first throw. If we get a 7, we lose. If we roll the same total as the first throw, we win. Use knowledge of Markov chains to determine our probability of winning at craps.

13 At the beginning of each day, a patient in a hospital is classified into one of three conditions: good, fair, or critical. At the beginning of the next day, the patient will either still be in the hospital and be in good, fair, or critical condition or will be discharged in one of three conditions: improved, unimproved, or dead. The transition probabilities for this situation are as follows:

	Good	Fair	Critical
Good	.65	.20	.05
Fair	.50	.30	.12
Critical	.51	.25	.20

	Improved	Unimproved	Dead
Good	.06	.03	.01
Fair	.03	.02	.03
Critical	.01	.01	.02

For example, a patient who begins the day in fair condition has a 12% chance of being in critical condition the next day

‡Based on Eppen and Fama (1970).

and a 3% chance of being discharged the next day in improved condition.

a Consider a patient who enters the hospital in good condition. On the average, how many days does this patient spend in the hospital?

b This morning there were 500 patients in good condition, 300 in fair condition, and 200 patients in critical condition in the hospital. Tomorrow morning the following admissions will be made: good condition, 50; fair condition, 40; critical condition, 30. Predict tomorrow morning's hospital census.

c The hospital's daily admissions are as follows: 20 patients in good condition, 10 patients in fair condition, and 10 patients in critical condition. On the average, how many patients of each type would you expect to see in the hospital?

d What fraction of patients who enter the hospital in good condition will leave the hospital in improved condition?

14 A major problem for a hospital is managing the database containing patient records. Blair General Hospital is considering two policies:

Policy 1 Dispose of a patient's records if he or she has not reentered the hospital in the last five years.
Policy 1 Dispose of a patient's records if he or she has not reentered the hospital in the last ten years.

The following information is available: If a patient has been hospitalized, there is a 30% chance he or she will reenter the hospital during the next year. If a patient has not been hospitalized during the last year, there is a 20% chance he or she will be hospitalized during the next year. If a patient has not been hospitalized during the last two years, there is a 10% chance he or she will be hospitalized during the next year. If a patient has not been hospitalized during

the last three years, there is a 5% chance he or she will be hospitalized during the next year. If a patient has not been hospitalized during the last four years, there is a 3% chance he or she will be hospitalized during the next year. If a patient has not been hospitalized during the last five years, there is a 2% chance he or she will be hospitalized during the next year. If a patient has not been hospitalized for at least six years, there is a 1% chance he or she will be hospitalized during the next year.

Assume that the hospital admits an average of 10,000 new patients each year. For each policy, estimate the number of patient records that will be in the system.[†]

15 Consider an n-state Markov chain in which each transition probability is positive and the transition matrix is symmetric; the entry in row I and column J of the transition matrix is identical to the entry in row J and column I.

a Why do we know that steady-state probabilities exist for this situation?

b What are the steady-state probabilities?

16[‡] The Euro was introduced on January 1, 2002 as the common currency for 15 European countries. Each Euro has a marking on the coin indicating the country of origin. For example, Euros minted in Portugal have a different marking than Euros minted in Spain. European politicians are interested in determining what fraction of Euros will eventually end up circulating in each country. For example, will 30% of all Euros circulate in France? How could Markov chains be used to answer this question? What parameters must be known before using Markov chain theory to solve this problem?

[†]Based on Liu, Wang, and Guh (1991).
[‡]Based on "Statisticians Count Euros and Find More Than Money," *New York Times,* July 2, 2002.

REFERENCES

Grinold, R., and K. Marshall. *Manpower Planning Models.* New York: North-Holland, 1977. Contains an extensive discussion of manpower planning models.

The following two references are the sources for two applications of Markov chains discussed in the chapter:

Meredith, J. "A Markovian Analysis of a Geriatric Ward," *Management Science* 19(1973):604–612.
Walker, J., and J. Lehmann. *100 Ways to Win at Monopoly.* New York: Dell, 1975.

The following books give both the theoretical aspects and many interesting applications of Markov chains:

Bhat, N. *Elements of Applied Stochastic Processes,* 2d ed. New York: Wiley, 1985.
Isaacson, D., and R. Madsen. *Markov Chains: Theory and Applications.* New York: Wiley, 1976.

Karlin, S., and H. Taylor. *A First Course in Stochastic Processes,* 2d ed. Orlando, Fla.: Academic Press, 1975.
Kemeny, J., and L. Snell. *Finite Markov Chains.* Princeton, N.J.: Van Nostrand, 1960.
Ash, R. and R. Bishop. "Monopoly as a Markov Process," *Mathematics Magazine* 45(1972):26–29.
Babich, P. "Customer Satisfaction: How Good Is Good Enough," *Quality Progress* (December 1992): 65–68.
Bessent, W., and A. Bessent. "Student Education Flow in a University Department: Results of a Markov Analysis," *Interfaces* 10(1980):52–59.
Cyert, R., M. Davidson, and G. Thompson. "Estimation of the Allowance for Doubtful Accounts by Markov Chains," *Management Science* 8(1963):287–303.
Deming, E., and G. Glasser. "A Markovian Analysis of the Life of Newspaper Subscriptions," *Management Science* 14(1968):B283–B294.
Eppen, G., and E. Fama. "Three Asset Cash Balance and Dynamic Portfolio Problems," *Management Science* 17(1970):311–319.

Flamholtz, E., G. Geis, and R. Perle. "A Markovian Model for the Valuation of Human Assets Acquired by an Organizational Purchase," *Interfaces* 14(1984):11–15.

Liu, C., K. Wang, and Y. Guh. "A Markov Chain Model for Medical Record Analysis," *Operations Research Quarterly* 42(no. 5, 1991):357–364.

Pegels, C., and A. Jelmert. "An Evaluation of Blood-Inventory Policies: A Markov Chain Application," *Operations Research* 18(1970):1097–1098.

Thompson, W., and J. McNeal. "Sales Planning and Control Using Absorbing Markov Chains," *Journal of Marketing Research* 4(1967):62–66.

18

Deterministic Dynamic Programming

Dynamic programming is a technique that can be used to solve many optimization problems. In most applications, dynamic programming obtains solutions by working backward from the end of a problem toward the beginning, thus breaking up a large, unwieldy problem into a series of smaller, more tractable problems.

We introduce the idea of working backward by solving two well-known puzzles and then show how dynamic programming can be used to solve network, inventory, and resource-allocation problems. We close the chapter by showing how to use spreadsheets to solve dynamic programming problems.

18.1 Two Puzzles[†]

In this section, we show how working backward can make a seemingly difficult problem almost trivial to solve.

EXAMPLE 1   Match Puzzle

Suppose there are 30 matches on a table. I begin by picking up 1, 2, or 3 matches. Then my opponent must pick up 1, 2, or 3 matches. We continue in this fashion until the last match is picked up. The player who picks up the last match is the loser. How can I (the first player) be sure of winning the game?

Solution   If I can ensure that it will be my opponent's turn when 1 match remains, I will certainly win. Working backward one step, if I can ensure that it will be my opponent's turn when 5 matches remain, I will win. The reason for this is that no matter what he does when 5 matches remain, I can make sure that when he has his next turn, only 1 match will remain. For example, suppose it is my opponent's turn when 5 matches remain. If my opponent picks up 2 matches, I will pick up 2 matches, leaving him with 1 match and sure defeat. Similarly, if I can force my opponent to play when 5, 9, 13, 17, 21, 25, or 29 matches remain, I am sure of victory. Thus, I cannot lose if I pick up $30 - 29 = 1$ match on my first turn. Then I simply make sure that my opponent will always be left with 29, 25, 21, 17, 13, 9, or 5 matches on his turn. Notice that we have solved this puzzle by working backward from the end of the problem toward the beginning. Try solving this problem without working backward!

EXAMPLE 2   Milk

I have a 9-oz cup and a 4-oz cup. My mother has ordered me to bring home exactly 6 oz of milk. How can I accomplish this goal?

[†]This section covers topics that may be omitted with no loss of continuity.

TABLE **1**
Moves in the Cup-and-Milk Problem

No. of Ounces in 9-oz Cup	No. of Ounces in 4-oz Cup
6	0
6	4
9	1
0	1
1	0
1	4
5	0
5	4
9	0
0	0

Solution   By starting near the end of the problem, I cleverly realize that the problem can easily be solved if I can somehow get 1 oz of milk into the 4-oz cup. Then I can fill the 9-oz cup and empty 3 oz from the 9-oz cup into the partially filled 4-oz cup. At this point, I will be left with 6 oz of milk. After I have this flash of insight, the solution to the problem may easily be described as in Table 1 (the initial situation is written last, and the final situation is written first).

PROBLEMS

Group A

1 Suppose there are 40 matches on a table. I begin by picking up 1, 2, 3, or 4 matches. Then my opponent must pick up 1, 2, 3, or 4 matches. We continue until the last match is picked up. The player who picks up the last match is the loser. Can I be sure of victory? If so, how?

2 Three players have played three rounds of a gambling game. Each round has one loser and two winners. The losing player must pay each winner the amount of money that the winning player had at the beginning of the round. At the end of the three rounds each player has $10. You are told that each player has won one round. By working backward, determine the original stakes of the three players. [*Note:* If the answer turns out to be (for example) 5, 15, 10, don't worry about which player had which stake; we can't really tell which player ends up with how much, but we can determine the numerical values of the original stakes.]

Group B

3 We have 21 coins and are told that one is heavier than any of the other coins. How many weighings on a balance will it take to find the heaviest coin? (*Hint:* If the heaviest coin is in a group of three coins, we can find it in one weighing. Then work backward to two weighings, and so on.)

4 Given a 7-oz cup and a 3-oz cup, explain how we can return from a well with 5 oz of water.

18.2 A Network Problem

Many applications of dynamic programming reduce to finding the shortest (or longest) path that joins two points in a given network. The following example illustrates how dynamic programming (working backward) can be used to find the shortest path in a network.

EXAMPLE 3   **Shortest Path**

Joe Cougar lives in New York City, but he plans to drive to Los Angeles to seek fame and fortune. Joe's funds are limited, so he has decided to spend each night on his trip at a friend's house. Joe has friends in Columbus, Nashville, Louisville, Kansas City, Omaha, Dallas, San Antonio, and Denver. Joe knows that after one day's drive he can reach Columbus, Nashville, or Louisville. After two days of driving, he can reach Kansas City, Omaha, or Dallas. After three days of driving, he can reach San Antonio or Denver. Finally, after four days of driving, he can reach Los Angeles. To minimize the number of miles traveled, where should Joe spend each night of the trip? The actual road mileages between cities are given in Figure 1.

Solution   Joe needs to know the shortest path between New York and Los Angeles in Figure 1. We will find it by working backward. We have classified all the cities that Joe can be in at the beginning of the nth day of his trip as stage n cities. For example, because Joe can only be in San Antonio or Denver at the beginning of the fourth day (day 1 begins when Joe leaves New York), we classify San Antonio and Denver as stage 4 cities. The reason for classifying cities according to stages will become apparent later.

The idea of working backward implies that we should begin by solving an easy problem that will eventually help us to solve a complex problem. Hence, we begin by finding the shortest path to Los Angeles from each city in which there is only one day of driving left (stage 4 cities). Then we use this information to find the shortest path to Los Angeles from each city for which only two days of driving remain (stage 3 cities). With this information in hand, we are able to find the shortest path to Los Angeles from each city that is three days distant (stage 2 cities). Finally, we find the shortest path to Los Angeles from each city (there is only one: New York) that is four days away.

To simplify the exposition, we use the numbers $1, 2, \ldots, 10$ given in Figure 1 to label the 10 cities. We also define c_{ij} to be the road mileage between city i and city j. For example, $c_{35} = 580$ is the road mileage between Nashville and Kansas City. We let $f_t(i)$ be the length of the shortest path from city i to Los Angeles, given that city i is a stage t city.[†]

Stage 4 Computations

We first determine the shortest path to Los Angeles from each stage 4 city. Since there is only one path from each stage 4 city to Los Angeles, we immediately see that $f_4(8) = 1{,}030$, the shortest path from Denver to Los Angeles simply being the *only* path from Denver to Los Angeles. Similarly, $f_4(9) = 1{,}390$, the shortest (and only) path from San Antonio to Los Angeles.

Stage 3 Computations

We now work backward one stage (to stage 3 cities) and find the shortest path to Los Angeles from each stage 3 city. For example, to determine $f_3(5)$, we note that the shortest path from city 5 to Los Angeles must be one of the following:

Path 1   Go from city 5 to city 8 and then take the shortest path from city 8 to city 10.

Path 2   Go from city 5 to city 9 and then take the shortest path from city 9 to city 10.

The length of path 1 may be written as $c_{58} + f_4(8)$, and the length of path 2 may be written as $c_{59} + f_4(9)$. Hence, the shortest distance from city 5 to city 10 may be written as

---

[†]In this example, keeping track of the stages is unnecessary; to be consistent with later examples, however, we do keep track.

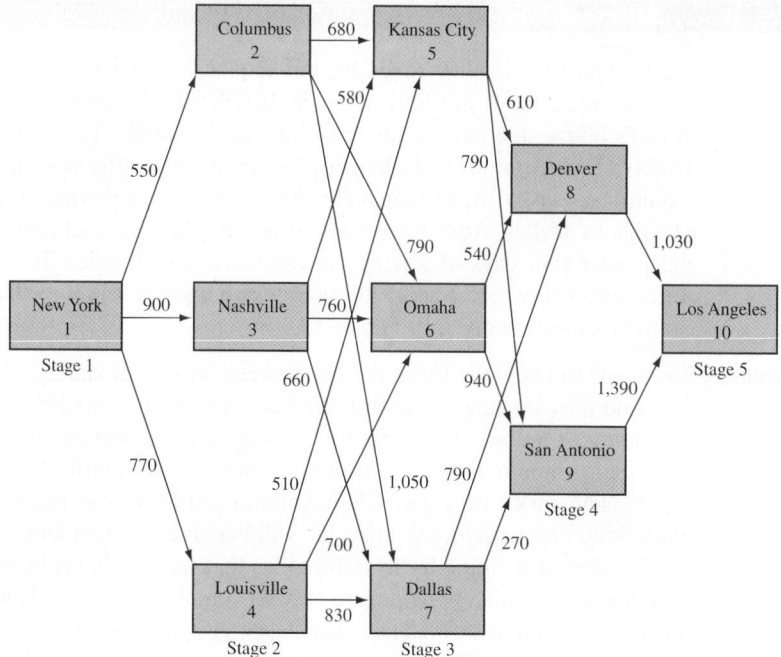

FIGURE 1
Joe's Trip Across the
United States

$$f_3(5) = \min \begin{cases} c_{58} + f_4(8) = 610 + 1{,}030 = 1{,}640^* \\ c_{59} + f_4(9) = 790 + 1{,}390 = 2{,}180 \end{cases}$$

[the * indicates the choice of arc that attains the $f_3(5)$]. Thus, we have shown that the shortest path from city 5 to city 10 is the path 5–8–10. Note that to obtain this result, we made use of our knowledge of $f_4(8)$ and $f_4(9)$.

Similarly, to find $f_3(6)$, we note that the shortest path to Los Angeles from city 6 must begin by going to city 8 or to city 9. This leads us to the following equation:

$$f_3(6) = \min \begin{cases} c_{68} + f_4(8) = 540 + 1{,}030 = 1{,}570^* \\ c_{69} + f_4(9) = 940 + 1{,}390 = 2{,}330 \end{cases}$$

Thus, $f_3(6) = 1{,}570$, and the shortest path from city 6 to city 10 is the path 6–8–10.

To find $f_3(7)$, we note that

$$f_3(7) = \min \begin{cases} c_{78} + f_4(8) = 790 + 1{,}030 = 1{,}820 \\ c_{79} + f_4(9) = 270 + 1{,}390 = 1{,}660^* \end{cases}$$

Therefore, $f_3(7) = 1{,}660$, and the shortest path from city 7 to city 10 is the path 7–9–10.

Stage 2 Computations

Given our knowledge of $f_3(5)$, $f_3(6)$, and $f_3(7)$, it is now easy to work backward one more stage and compute $f_2(2)$, $f_2(3)$, and $f_2(4)$ and thus the shortest paths to Los Angeles from city 2, city 3, and city 4. To illustrate how this is done, we find the shortest path (and its length) from city 2 to city 10. The shortest path from city 2 to city 10 must begin by going from city 2 to city 5, city 6, or city 7. Once this shortest path gets to city 5, city 6, or city 7, then it must follow a shortest path from that city to Los Angeles. This reasoning shows that the shortest path from city 2 to city 10 must be one of the following:

Path 1 Go from city 2 to city 5. Then follow a shortest path from city 5 to city 10. A path of this type has a total length of $c_{25} + f_3(5)$.

Path 2 Go from city 2 to city 6. Then follow a shortest path from city 6 to city 10. A path of this type has a total length of $c_{26} + f_3(6)$.

Path 3 Go from city 2 to city 7. Then follow a shortest path from city 7 to city 10. This path has a total length of $c_{27} + f_3(7)$. We may now conclude that

$$f_2(2) = \min \begin{cases} c_{25} + f_3(5) = 680 + 1{,}640 = 2{,}320^* \\ c_{26} + f_3(6) = 790 + 1{,}570 = 2{,}360 \\ c_{27} + f_3(7) = 1{,}050 + 1{,}660 = 2{,}710 \end{cases}$$

Thus, $f_2(2) = 2{,}320$, and the shortest path from city 2 to city 10 is to go from city 2 to city 5 and then follow the shortest path from city 5 to city 10 (5–8–10).

Similarly,

$$f_2(3) = \min \begin{cases} c_{35} + f_3(5) = 580 + 1{,}640 = 2{,}220^* \\ c_{36} + f_3(6) = 760 + 1{,}570 = 2{,}330 \\ c_{37} + f_3(7) = 660 + 1{,}660 = 2{,}320 \end{cases}$$

Thus, $f_2(3) = 2{,}220$, and the shortest path from city 3 to city 10 consists of arc 3–5 and the shortest path from city 5 to city 10 (5–8–10).

In similar fashion,

$$f_2(4) = \min \begin{cases} c_{45} + f_3(5) = 510 + 1{,}640 = 2{,}150^* \\ c_{46} + f_3(6) = 700 + 1{,}570 = 2{,}270 \\ c_{47} + f_3(7) = 830 + 1{,}660 = 2{,}490 \end{cases}$$

Thus, $f_2(4) = 2{,}150$, and the shortest path from city 4 to city 10 consists of arc 4–5 and the shortest path from city 5 to city 10 (5–8–10).

Stage 1 Computations

We can now use our knowledge of $f_2(2)$, $f_2(3)$, and $f_2(4)$ to work backward one more stage to find $f_1(1)$ and the shortest path from city 1 to city 10. Note that the shortest path from city 1 to city 10 must begin by going to city 2, city 3, or city 4. This means that the shortest path from city 1 to city 10 must be one of the following:

Path 1 Go from city 1 to city 2 and then follow a shortest path from city 2 to city 10. The length of such a path is $c_{12} + f_2(2)$.

Path 2 Go from city 1 to city 3 and then follow a shortest path from city 3 to city 10. The length of such a path is $c_{13} + f_2(3)$.

Path 3 Go from city 1 to city 4 and then follow a shortest path from city 4 to city 10. The length of such a path is $c_{14} + f_2(4)$. It now follows that

$$f_1(1) = \min \begin{cases} c_{12} + f_2(2) = 550 + 2{,}320 = 2{,}870^* \\ c_{13} + f_2(3) = 900 + 2{,}220 = 3{,}120 \\ c_{14} + f_2(4) = 770 + 2{,}150 = 2{,}920 \end{cases}$$

Determination of the Optimal Path

Thus, $f_1(1) = 2{,}870$, and the shortest path from city 1 to city 10 goes from city 1 to city 2 and then follows the shortest path from city 2 to city 10. Checking back to the $f_2(2)$ calculations, we see that the shortest path from city 2 to city 10 is 2–5–8–10. Translating the numerical labels into real cities, we see that the shortest path from New York to Los Angeles passes through New York, Columbus, Kansas City, Denver, and Los Angeles. This path has a length of $f_1(1) = 2{,}870$ miles.

Computational Efficiency of Dynamic Programming

For Example 3, it would have been an easy matter to determine the shortest path from New York to Los Angeles by enumerating all the possible paths [after all, there are only $3(3)(2) = 18$ paths]. Thus, in this problem, the use of dynamic programming did not really serve much purpose. For larger networks, however, dynamic programming is much more efficient for determining a shortest path than the explicit enumeration of all paths. To see this, consider the network in Figure 2. In this network, it is possible to travel from any node in stage k to any node in stage $k + 1$. Let the distance between node i and node j be c_{ij}. Suppose we want to determine the shortest path from node 1 to node 27. One way to solve this problem is explicit enumeration of all paths. There are 5^5 possible paths from node 1 to node 27. It takes five additions to determine the length of each path. Thus, explicitly enumerating the length of all paths requires $5^5(5) = 5^6 = 15{,}625$ additions.

Suppose we use dynamic programming to determine the shortest path from node 1 to node 27. Let $f_t(i)$ be the length of the shortest path from node i to node 27, given that node i is in stage t. To determine the shortest path from node 1 to node 27, we begin by finding $f_6(22)$, $f_6(23)$, $f_6(24)$, $f_6(25)$, and $f_6(26)$. This does not require any additions. Then we find $f_5(17)$, $f_5(18)$, $f_5(19)$, $f_5(20)$, $f_5(21)$. For example, to find $f_5(21)$ we use the following equation:

$$f_5(21) = \min_j \{c_{21,j} + f_6(j)\} \qquad (j = 22, 23, 24, 25, 26)$$

Determining $f_5(21)$ in this manner requires five additions. Thus, the calculation of all the $f_5(\cdot)$'s requires $5(5) = 25$ additions. Similarly, the calculation of all the $f_4(\cdot)$'s requires 25 additions, and the calculation of all the $f_3(\cdot)$'s requires 25 additions. The determination of all the $f_2(\cdot)$'s also requires 25 additions, and the determination of $f_1(1)$ requires 5 additions. Thus, in total, dynamic programming requires $4(25) + 5 = 105$ additions to find

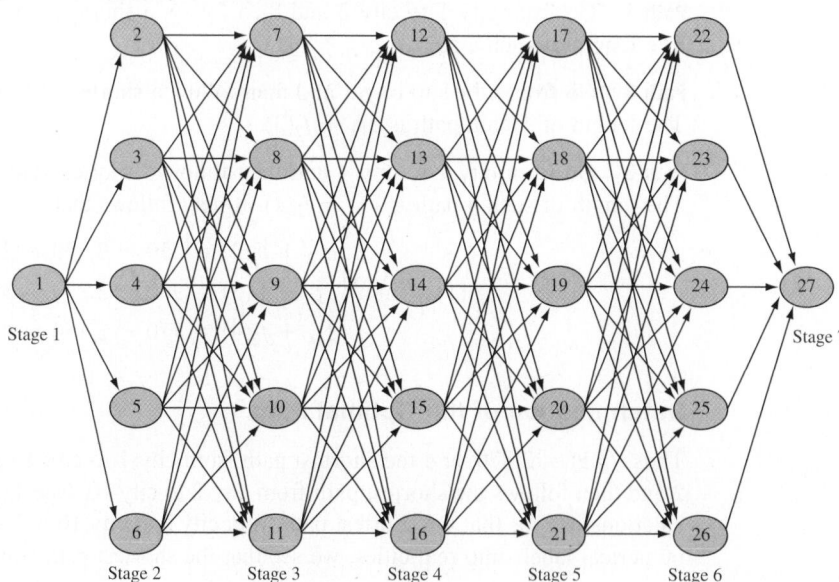

FIGURE 2
Illustration of Computational Efficiency of Dynamic Programming

the shortest path from node 1 to node 27. Because explicit enumeration requires 15,625 additions, we see that dynamic programming requires only 0.007 times as many additions as explicit enumeration. For larger networks, the computational savings effected by dynamic programming are even more dramatic.

Besides additions, determination of the shortest path in a network requires comparisons between the lengths of paths. If explicit enumeration is used, then $5^5 - 1 = 3,124$ comparisons must be made (that is, compare the length of the first two paths, then compare the length of the third path with the shortest of the first two paths, and so on). If dynamic programming is used, then for $t = 2, 3, 4, 5$, determination of each $f_t(i)$ requires $5 - 1 = 4$ comparisons. Then to compute $f_1(1)$, $5 - 1 = 4$ comparisons are required. Thus, to find the shortest path from node 1 to node 27, dynamic programming requires a total of $20(5 - 1) + 4 = 84$ comparisons. Again, dynamic programming comes out far superior to explicit enumeration.

Characteristics of Dynamic Programming Applications

We close this section with a discussion of the characteristics of Example 3 that are common to most applications of dynamic programming.

Characteristic 1

The problem can be divided into stages with a decision required at each stage. In Example 3, stage t consisted of those cities where Joe could be at the beginning of day t of his trip. As we will see, in many dynamic programming problems, the stage is the amount of time that has elapsed since the beginning of the problem. We note that in some situations, decisions are not required at every stage (see Section 18.5).

Characteristic 2

Each stage has a number of states associated with it. By a **state,** we mean the information that is needed at any stage to make an optimal decision. In Example 3, the state at stage t is simply the city where Joe is at the beginning of day t. For example, in stage 3, the possible states are Kansas City, Omaha, and Dallas. Note that to make the correct decision at any stage, Joe doesn't need to know how he got to his current location. For example, if Joe is in Kansas City, then his remaining decisions don't depend on how he goes to Kansas City; his future decisions just depend on the fact that he is now in Kansas City.

Characteristic 3

The decision chosen at any stage describes how the state at the current stage is transformed into the state at the next stage. In Example 3, Joe's decision at any stage is simply the next city to visit. This determines the state at the next stage in an obvious fashion. In many problems, however, a decision does not determine the next stage's state with certainty; instead, the current decision only determines the probability distribution of the state at the next stage.

Characteristic 4

Given the current state, the optimal decision for each of the remaining stages must not depend on previously reached states or previously chosen decisions. This idea is known as the **principle of optimality.** In the context of Example 3, the principle of optimality reduces to the following: Suppose the shortest path (call it R) from city 1 to city 10 is known to pass through city i. Then the portion of R that goes from city i to city 10 must be a shortest path from city i to city 10. If this were not the case, then we could create a path from city 1 to city 10 that was shorter than R by appending a shortest path from city i to city 10 to the portion of R leading from city 1 to city i. This would create a path from city 1 to city 10 that is shorter than R, thereby contradicting the fact that R is a shortest path from city 1 to city 10. For example, if the shortest path from city 1 to city 10 is known to pass through city 2, then the shortest path from city 1 to city 10 must include a shortest path from city 2 to city 10 (2–5–8–10). This follows because any path from city 1 to city 10 that passes through city 2 and does not contain a shortest path from city 2 to city 10 will have a length of c_{12} + [something bigger than $f_2(2)$]. Of course, such a path cannot be a shortest path from city 1 to city 10.

Characteristic 5

If the states for the problem have been classified into one of T stages, there must be a recursion that relates the cost or reward earned during stages $t, t + 1, \ldots, T$ to the cost or reward earned from stages $t + 1, t + 2, \ldots, T$. In essence, the recursion formalizes the working-backward procedure. In Example 3, our recursion could have been written as

$$f_t(i) = \min_j \{c_{ij} + f_{t+1}(j)\}$$

where j must be a stage $t + 1$ city and $f_5(10) = 0$.

We can now describe how to make optimal decisions. Let's assume that the initial state during stage 1 is i_1. To use the recursion, we begin by finding the optimal decision for each state associated with the last stage. Then we use the recursion described in characteristic 5 to determine $f_{T-1}(\cdot)$ (along with the optimal decision) for every stage $T - 1$ state. Then we use the recursion to determine $f_{T-2}(\cdot)$ (along with the optimal decision) for every stage $T - 2$ state. We continue in this fashion until we have computed $f_1(i_1)$ and the optimal decision when we are in stage 1 and state i_1. Then our optimal decision in stage 1 is chosen from the set of decisions attaining $f_1(i_1)$. Choosing this decision at stage 1 will lead us to some stage 2 state (call it state i_2) at stage 2. Then at stage 2, we choose any decision attaining $f_2(i_2)$. We continue in this fashion until a decision has been chosen for each stage.

In the rest of this chapter, we discuss many applications of dynamic programming. The presentation will seem easier if the reader attempts to determine how each problem fits into the network context introduced in Example 3. In the next section, we begin by studying how dynamic programming can be used to solve inventory problems.

PROBLEMS

Group A

1 Find the shortest path from node 1 to node 10 in the network shown in Figure 3. Also, find the shortest path from node 3 to node 10.

2 A sales representative lives in Bloomington and must be in Indianapolis next Thursday. On each of the days Monday, Tuesday, and Wednesday, he can sell his wares in Indianapolis, Bloomington, or Chicago. From past experience, he believes that he can earn $12 from spending a day in Indianapolis, $16 from spending a day in Bloomington, and $17 from spending a day in Chicago. Where should he spend the first three days

FIGURE 3

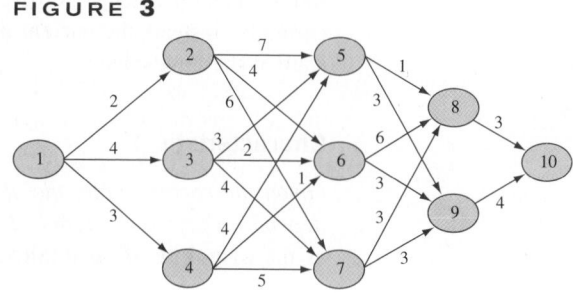

TABLE **2**

From	To		
	Indianapolis	Bloomington	Chicago
Indianapolis	—	5	2
Bloomington	5	—	7
Chicago	2	7	—

FIGURE **4**

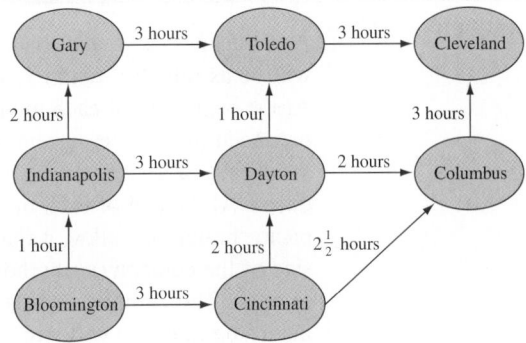

and nights of the week to maximize his sales income less travel costs? Travel costs are shown in Table 2.

Group B

3 I must drive from Bloomington to Cleveland. Several paths are available (see Figure 4). The number on each arc is the length of time it takes to drive between the two cities. For example, it takes 3 hours to drive from Bloomington to Cincinnati. By working backward, determine the shortest path (in terms of time) from Bloomington to Cleveland. [*Hint:* Work backward and don't worry about stages—only about states.]

18.3 An Inventory Problem

In this section, we illustrate how dynamic programming can be used to solve an inventory problem with the following characteristics:

1 Time is broken up into periods, the present period being period 1, the next period 2, and the final period *T*. At the beginning of period 1, the demand during each period is known.

2 At the beginning of each period, the firm must determine how many units should be produced. Production capacity during each period is limited.

3 Each period's demand must be met on time from inventory or current production. During any period in which production takes place, a fixed cost of production as well as a variable per-unit cost is incurred.

4 The firm has limited storage capacity. This is reflected by a limit on end-of-period inventory. A per-unit holding cost is incurred on each period's ending inventory.

5 The firm's goal is to minimize the total cost of meeting on time the demands for periods 1, 2, . . . , *T*.

In this model, the firm's inventory position is reviewed at the end of each period (say, at the end of each month), and then the production decision is made. Such a model is called a **periodic review model.** This model is in contrast to the continuous review models in which the firm knows its inventory position at all times and may place an order or begin production at any time.

If we exclude the setup cost for producing any units, the inventory problem just described is similar to the Sailco inventory problem that we solved by linear programming in Section 3.10. Here, we illustrate how dynamic programming can be used to determine a production schedule that minimizes the total cost incurred in an inventory problem that meets the preceding description.

EXAMPLE 4   Inventory

A company knows that the demand for its product during each of the next four months will be as follows: month 1, 1 unit; month 2, 3 units; month 3, 2 units; month 4, 4 units. At the beginning of each month, the company must determine how many units should be produced during the current month. During a month in which any units are produced, a setup cost of $3 is incurred. In addition, there is a variable cost of $1 for every unit produced. At the end of each month, a holding cost of 50¢ per unit on hand is incurred. Capacity limitations allow a maximum of 5 units to be produced during each month. The size of the company's warehouse restricts the ending inventory for each month to 4 units at most. The company wants to determine a production schedule that will meet all demands on time and will minimize the sum of production and holding costs during the four months. Assume that 0 units are on hand at the beginning of the first month.

Solution   Recall from Section 3.10 that we can ensure that all demands are met on time by restricting each month's ending inventory to be nonnegative. To use dynamic programming to solve this problem, we need to identify the appropriate state, stage, and decision. The stage should be defined so that when one stage remains, the problem will be trivial to solve. If we are at the beginning of month 4, then the firm would meet demand at minimum cost by simply producing just enough units to ensure that (month 4 production) + (month 3 ending inventory) = (month 4 demand). Thus, when one month remains, the firm's problem is easy to solve. Hence, we let time represent the stage. In most dynamic programming problems, the stage has something to do with time.

At each stage (or month), the company must decide how many units to produce. To make this decision, the company need only know the inventory level at the beginning of the current month (or the end of the previous month). Therefore, we let the state at any stage be the beginning inventory level.

Before writing a recursive relation that can be used to "build up" the optimal production schedule, we must define $f_t(i)$ to be the minimum cost of meeting demands for months $t, t + 1, \ldots, 4$ if i units are on hand at the beginning of month t. We define $c(x)$ to be the cost of producing x units during a period. Then $c(0) = 0$, and for $x > 0$, $c(x) = 3 + x$. Because of the limited storage capacity and the fact that all demand must be met on time, the possible states during each period are 0, 1, 2, 3, and 4. Thus, we begin by determining $f_4(0), f_4(1), f_4(2), f_4(3)$, and $f_4(4)$. Then we use this information to determine $f_3(0), f_3(1), f_3(2), f_3(3), f_3(4)$. Then we determine $f_2(0), f_2(1), f_2(2), f_2(3)$, and $f_2(4)$. Finally, we determine $f_1(0)$. Then we determine an optimal production level for each month. We define $x_t(i)$ to be a production level during month t that minimizes the total cost during months $t, t + 1, \ldots, 4$ if i units are on hand at the beginning of month t. We now begin to work backward.

Month 4 Computations

During month 4, the firm will produce just enough units to ensure that the month 4 demand of 4 units is met. This yields

$$f_4(0) = \text{cost of producing } 4 - 0 \text{ units} = c(4) = 3 + 4 = \$7 \text{ and } x_4(0) = 4 - 0 = 4$$
$$f_4(1) = \text{cost of producing } 4 - 1 \text{ units} = c(3) = 3 + 3 = \$6 \text{ and } x_4(1) = 4 - 1 = 3$$
$$f_4(2) = \text{cost of producing } 4 - 2 \text{ units} = c(2) = 3 + 2 = \$5 \text{ and } x_4(2) = 4 - 2 = 2$$
$$f_4(3) = \text{cost of producing } 4 - 3 \text{ units} = c(1) = 3 + 1 = \$4 \text{ and } x_4(3) = 4 - 3 = 1$$
$$f_4(4) = \text{cost of producing } 4 - 4 \text{ units} = c(0) = \$0 \quad \text{and} \quad x_4(4) = 4 - 4 = 0$$

Month 3 Computations

How can we now determine $f_3(i)$ for $i = 0, 1, 2, 3, 4$? The cost $f_3(i)$ is the minimum cost incurred during months 3 and 4 if the inventory at the beginning of month 3 is i. For each possible production level x during month 3, the total cost during months 3 and 4 is

$$(\tfrac{1}{2})(i + x - 2) + c(x) + f_4(i + x - 2) \tag{1}$$

This follows because if x units are produced during month 3, the ending inventory for month 3 will be $i + x - 2$. Then the month 3 holding cost will be $(\tfrac{1}{2})(i + x - 2)$, and the month 3 production cost will be $c(x)$. Then we enter month 4 with $i + x - 2$ units on hand. Since we proceed optimally from this point onward (remember the principle of optimality), the cost for month 4 will be $f_4(i + x - 2)$. We want to choose the month 3 production level to minimize (1), so we write

$$f_3(i) = \min_x \{(\tfrac{1}{2})(i + x - 2) + c(x) + f_4(i + x - 2)\} \tag{2}$$

In (2), x must be a member of $\{0, 1, 2, 3, 4, 5\}$, and x must satisfy $4 \geq i + x - 2 \geq 0$. This reflects the fact that the current month's demand must be met $(i + x - 2 \geq 0)$, and ending inventory cannot exceed the capacity of $4(i + x - 2 \leq 4)$. Recall that $x_3(i)$ is any value of x attaining $f_3(i)$. The computations for $f_3(0), f_3(1), f_3(2), f_3(3)$, and $f_3(4)$ are given in Table 3.

Month 2 Computations

We can now determine $f_2(i)$, the minimum cost incurred during months 2, 3, and 4 given that at the beginning of month 2, the on-hand inventory is i units. Suppose that month 2 production $= x$. Because month 2 demand is 3 units, a holding cost of $(\tfrac{1}{2})(i + x - 3)$ is

TABLE 3
Computations for $f_3(i)$

i	x	$(\tfrac{1}{2})(i + x - 2) + c(x)$	$f_4(i + x - 2)$	Total Cost Months 3, 4	$f_3(i)$ $x_3(i)$
0	2	$0 + 5 = 5$	7	$5 + 7 = 12^*$	$f_3(0) = 12$
0	3	$\tfrac{1}{2} + 6 = \tfrac{13}{2}$	6	$\tfrac{13}{2} + 6 = \tfrac{25}{2}$	$x_3(0) = 2$
0	4	$1 + 7 = 8$	5	$8 + 5 = 13$	
0	5	$\tfrac{3}{2} + 8 = \tfrac{19}{2}$	4	$\tfrac{19}{2} + 4 = \tfrac{27}{2}$	
1	1	$0 + 4 = 4$	7	$4 + 7 = 11$	$f_3(1) = 10$
1	2	$\tfrac{1}{2} + 5 = \tfrac{11}{2}$	6	$\tfrac{11}{2} + 6 = \tfrac{23}{2}$	$x_3(1) = 5$
1	3	$1 + 6 = 7$	5	$7 + 5 = 12$	
1	4	$\tfrac{3}{2} + 7 = \tfrac{17}{2}$	4	$\tfrac{17}{2} + 4 = \tfrac{25}{2}$	
1	5	$2 + 8 = 10$	0	$10 + 0 = 10^*$	
2	0	$0 + 0 = 0$	7	$0 + 7 = 7^*$	$f_3(2) = 7$
2	1	$\tfrac{1}{2} + 4 = \tfrac{9}{2}$	6	$\tfrac{9}{2} + 6 = \tfrac{21}{2}$	$x_3(2) = 0$
2	2	$1 + 5 = 6$	5	$6 + 5 = 11$	
2	3	$\tfrac{3}{2} + 6 = \tfrac{15}{2}$	4	$\tfrac{15}{2} + 4 = \tfrac{23}{2}$	
2	4	$2 + 7 = 9$	0	$9 + 0 = 9$	
3	0	$\tfrac{1}{2} + 0 = \tfrac{1}{2}$	6	$\tfrac{1}{2} + 6 = \tfrac{13}{2}^*$	$f_3(3) = \tfrac{13}{2}$
3	1	$1 + 4 = 5$	5	$5 + 5 = 10$	$x_3(3) = 0$
3	2	$\tfrac{3}{2} + 5 = \tfrac{13}{2}$	4	$\tfrac{13}{2} + 4 = \tfrac{21}{2}$	
3	3	$2 + 6 = 8$	0	$8 + 0 = 8$	
4	0	$1 + 0 = 1$	5	$1 + 5 = 6^*$	$f_3(4) = 6$
4	1	$\tfrac{3}{2} + 4 = \tfrac{11}{2}$	4	$\tfrac{11}{2} + 4 = \tfrac{19}{2}$	$x_3(4) = 0$
4	2	$2 + 5 = 7$	0	$7 + 0 = 7$	

incurred at the end of month 2. Thus, the total cost incurred during month 2 is $(\frac{1}{2})(i + x - 3) + c(x)$. During months 3 and 4, we follow an optimal policy. Since month 3 begins with an inventory of $i + x - 3$, the cost incurred during months 3 and 4 is $f_3(i + x - 3)$. In analogy to (2), we now write

$$f_2(i) = \min_x \{(\tfrac{1}{2})(i + x - 3) + c(x) + f_3(i + x - 3)\} \tag{3}$$

where x must be a member of $\{0, 1, 2, 3, 4, 5\}$ and x must also satisfy $0 \leq i + x - 3 \leq 4$. The computations for $f_2(0)$, $f_2(1)$, $f_2(2)$, $f_2(3)$, and $f_2(4)$ are given in Table 4.

Month 1 Computations

The reader should now be able to show that the $f_1(i)$'s can be determined via the following recursive relation:

$$f_1(i) = \min_x \{(\tfrac{1}{2})(i + x - 1) + c(x) + f_2(i + x - 1)\} \tag{4}$$

where x must be a member of $\{0, 1, 2, 3, 4, 5\}$ and x must satisfy $0 \leq i + x - 1 \leq 4$. Since the inventory at the beginning of month 1 is 0 units, we actually need only determine $f_1(0)$ and $x_1(0)$. To give the reader more practice, however, the computations for $f_1(1)$, $f_1(2)$, $f_1(3)$, and $f_1(4)$ are given in Table 5.

Determination of the Optimal Production Schedule

We can now determine a production schedule that minimizes the total cost of meeting the demand for all four months on time. Since our initial inventory is 0 units, the minimum cost for the four months will be $f_1(0) = \$20$. To attain $f_1(0)$, we must produce $x_1(0) = 1$

TABLE 4
Computations for $f_2(i)$

i	x	$(\tfrac{1}{2})(i + x - 3) + c(x)$	$f_3(i + x - 3)$	Total Cost Months 2–4	$f_2(i)$ $x_2(i)$
0	3	$0 + 6 = 6$	12	$6 + 12 = 18$	$f_2(0) = 16$
0	4	$\frac{1}{2} + 7 = \frac{15}{2}$	10	$\frac{15}{2} + 10 = \frac{35}{2}$	$x_2(0) = 5$
0	5	$1 + 8 = 9$	7	$9 + 7 = 16*$	
1	2	$0 + 5 = 5$	12	$5 + 12 = 17$	$f_2(1) = 15$
1	3	$\frac{1}{2} + 6 = \frac{13}{2}$	10	$\frac{13}{2} + 10 = \frac{33}{2}$	$x_2(1) = 4$
1	4	$1 + 7 = 8$	7	$8 + 7 = 15*$	
1	5	$\frac{3}{2} + 8 = \frac{19}{2}$	$\frac{13}{2}$	$\frac{19}{2} + \frac{13}{2} = 16$	
2	1	$0 + 4 = 4$	12	$4 + 12 = 16$	$f_2(2) = 14$
2	2	$\frac{1}{2} + 5 = \frac{11}{2}$	10	$\frac{11}{2} + 10 = \frac{31}{2}*$	$x_2(2) = 3$
2	3	$1 + 6 = 7$	7	$7 + 7 = 14*$	
2	4	$\frac{3}{2} + 7 = \frac{17}{2}$	$\frac{13}{2}$	$\frac{17}{2} + \frac{13}{2} = 15$	
2	5	$2 + 8 = 10$	6	$10 + 6 = 16$	
3	0	$0 + 0 = 0$	12	$0 + 12 = 12*$	$f_2(3) = 12$
3	1	$\frac{1}{2} + 4 = \frac{9}{2}$	10	$\frac{9}{2} + 10 = \frac{29}{2}$	$x_2(3) = 0$
3	2	$1 + 5 = 6$	7	$6 + 7 = 13$	
3	3	$\frac{3}{2} + 6 = \frac{15}{2}$	$\frac{13}{2}$	$\frac{15}{2} + \frac{13}{2} = 14$	
3	4	$2 + 7 = 9$	6	$9 + 6 = 15$	
4	0	$\frac{1}{2} + 0 = \frac{1}{2}$	10	$\frac{1}{2} + 10 = \frac{21}{2}*$	$f_2(4) = \frac{21}{2}$
4	1	$1 + 4 = 5$	7	$5 + 7 = 12$	$x_2(4) = 0$
4	2	$\frac{3}{2} + 5 = \frac{13}{2}$	$\frac{13}{2}$	$\frac{13}{2} + \frac{13}{2} = 13$	
4	3	$2 + 6 = 8$	6	$8 + 6 = 14$	

TABLE 5

Computations for $f_1(i)$

i	x	$(\frac{1}{2})(i + x - 1) + c(x)$	$f_2(i + x - 1)$	Total Cost	$f_1(i)$ $x_1(i)$
0	1	$0 + 4 = 4$	16	$4 + 16 = 20^*$	$f_1(0) = 20$
0	2	$\frac{1}{2} + 5 = \frac{11}{2}$	15	$\frac{11}{2} + 15 = \frac{41}{2}$	$x_1(0) = 1$
0	3	$1 + 6 = 7$	14	$7 + 14 = 21$	
0	4	$\frac{3}{2} + 7 = \frac{17}{2}$	12	$\frac{17}{2} + 12 = \frac{41}{2}$	
0	5	$2 + 8 = 10$	$\frac{21}{2}$	$10 + \frac{21}{2} = \frac{41}{2}$	
1	0	$0 + 0 = 0$	16	$0 + 16 = 16^*$	$f_1(1) = 16$
1	1	$\frac{1}{2} + 4 = \frac{9}{2}$	15	$\frac{9}{2} + 15 = \frac{39}{2}$	$x_1(1) = 0$
1	2	$1 + 5 = 6$	14	20	
1	3	$\frac{3}{2} + 6 = \frac{15}{2}$	12	$\frac{15}{2} + 12 = \frac{39}{2}$	
1	4	$2 + 7 = 9$	$\frac{21}{2}$	$9 + \frac{21}{2} = \frac{39}{2}$	
2	0	$\frac{1}{2} + 0 = \frac{1}{2}$	15	$\frac{1}{2} + 15 = \frac{31}{2}^*$	$f_1(2) = \frac{31}{2}$
2	1	$1 + 4 = 5$	14	$5 + 14 = 19$	$x_1(2) = 0$
2	2	$\frac{3}{2} + 5 = \frac{13}{2}$	12	$\frac{13}{2} + 12 = \frac{37}{2}$	
2	3	$2 + 6 = 8$	$\frac{21}{2}$	$8 + \frac{21}{2} = \frac{37}{2}$	
3	0	$1 + 0 = 1$	14	$1 + 14 = 15^*$	$f_1(3) = 15$
3	1	$\frac{3}{2} + 4 = \frac{11}{2}$	12	$\frac{11}{2} + 12 = \frac{35}{2}$	$x_1(3) = 0$
3	2	$2 + 5 = 7$	$\frac{21}{2}$	$7 + \frac{21}{2} = \frac{35}{2}$	
4	0	$\frac{3}{2} + 0 = \frac{3}{2}$	12	$\frac{3}{2} + 12 = \frac{27}{2}^*$	$f_1(4) = \frac{27}{2}$
4	1	$2 + 4 = 6$	$\frac{21}{2}$	$6 + \frac{21}{2} = \frac{33}{2}$	$x_1(4) = 0$

unit during month 1. Then the inventory at the beginning of month 2 will be $0 + 1 - 1 = 0$. Thus, in month 2, we should produce $x_2(0) = 5$ units. Then at the beginning of month 3, our beginning inventory will be $0 + 5 - 3 = 2$. Hence, during month 3, we need to produce $x_3(2) = 0$ units. Then month 4 will begin with $2 - 2 + 0 = 0$ units on hand. Thus, $x_4(0) = 4$ units should be produced during month 4. In summary, the optimal production schedule incurs a total cost of \$20 and produces 1 unit during month 1, 5 units during month 2, 0 units during month 3, and 4 units during month 4.

Note that finding the solution to Example 4 is equivalent to finding the shortest route joining the node $(1, 0)$ to the node $(5, 0)$ in Figure 5. Each node in Figure 5 corresponds to a state, and each column of nodes corresponds to all the possible states associated with a given stage. For example, if we are at node $(2, 3)$, then we are at the beginning of month 2, and the inventory at the beginning of month 2 is 3 units. Each arc in the network represents the way in which a decision (how much to produce during the current month) transforms the current state into next month's state. For example, the arc joining nodes $(1, 0)$ and $(2, 2)$ (call it arc 1) corresponds to producing 3 units during month 1. To see this, note that if 3 units are produced during month 1, then we begin month 2 with $0 + 3 - 1 = 2$ units. The length of each arc is simply the sum of production and inventory costs during the current period, given the current state and the decision associated with the chosen arc. For example, the cost associated with arc 1 would be $6 + (\frac{1}{2})2 = 7$. Note that some nodes in adjacent stages are not joined by an arc. For example, node $(2, 4)$ is not joined to node $(3, 0)$. The reason for this is that if we begin month 2 with 4 units, then at the beginning of month 3, we will have at least $4 - 3 = 1$ unit on hand. Also note that we have drawn arcs joining all month 4 states to the node $(5, 0)$, since having a positive inventory at the end of month 4 would clearly be suboptimal.

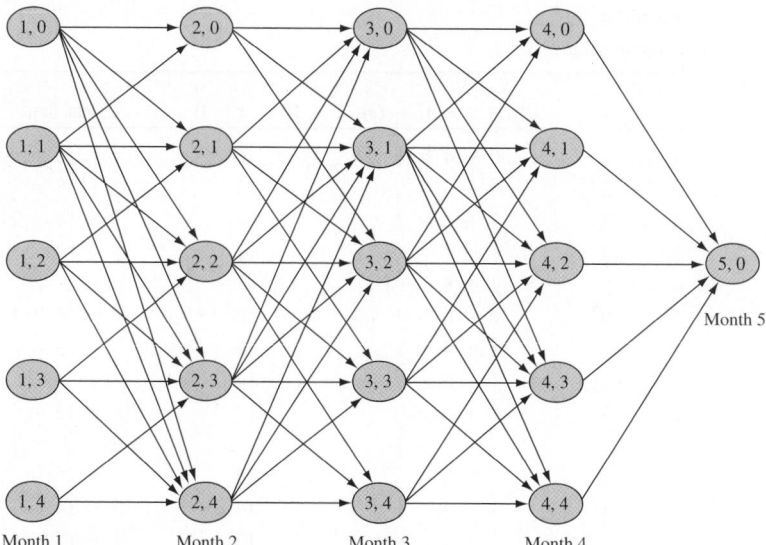

FIGURE 5
Network Representation of Inventory Example

Month 1   Month 2   Month 3   Month 4   Month 5

Returning to Example 4, the minimum-cost production schedule corresponds to the shortest path joining (1, 0) and (5, 0). As we have already seen, this would be the path corresponding to production levels of 1, 5, 0, and 4. In Figure 5, this would correspond to the path beginning at (1, 0), then going to (2, 0 + 1 − 1) = (2, 0), then to (3, 0 + 5 − 3) = (3, 2), then to (4, 2 + 0 − 2) = (4, 0), and finally to (5, 0 + 4 − 4) = (5, 0). Thus, our optimal production schedule corresponds to the path (1, 0)–(2, 0)–(3, 2)–(4, 0)–(5, 0) in Figure 5.

PROBLEMS

Group A

1 In Example 4, determine the optimal production schedule if the initial inventory is 3 units.

2 An electronics firm has a contract to deliver the following number of radios during the next three months; month 1, 200 radios; month 2, 300 radios; month 3, 300 radios. For each radio produced during months 1 and 2, a $10 variable cost is incurred; for each radio produced during month 3, a $12 variable cost is incurred. The inventory cost is $1.50 for each radio in stock at the end of a month. The cost of setting up for production during a month is $250.

Radios made during a month may be used to meet demand for that month or any future month. Assume that production during each month must be a multiple of 100. Given that the initial inventory level is 0 units, use dynamic programming to determine an optimal production schedule.

3 In Figure 5, determine the production level and cost associated with each of the following arcs:

  a (2, 3)–(3, 1)
  b (4, 2)–(5, 0)

18.4 Resource-Allocation Problems

Resource-allocation problems, in which limited resources must be allocated among several activities, are often solved by dynamic programming. Recall that we have solved such problems by linear programming (for instance, the Giapetto problem). To use linear programming to do resource allocation, three assumptions must be made:

Assumption 1 The amount of a resource assigned to an activity may be any nonnegative number.

Assumption 2 The benefit obtained from each activity is proportional to the amount of the resource assigned to the activity.

Assumption 3 The benefit obtained from more than one activity is the sum of the benefits obtained from the individual activities.

Even if assumptions 1 and 2 do not hold, dynamic programming can be used to solve resource-allocation problems efficiently when assumption 3 is valid and when the amount of the resource allocated to each activity is a member of a finite set.

EXAMPLE 5 **Resource Allocation**

Finco has $6,000 to invest, and three investments are available. If d_j dollars (in thousands) are invested in investment j, then a net present value (in thousands) of $r_j(d_j)$ is obtained, where the $r_j(d_j)$'s are as follows:

$$r_1(d_1) = 7d_1 + 2 \qquad\qquad (d_1 > 0)$$
$$r_2(d_2) = 3d_2 + 7 \qquad\qquad (d_2 > 0)$$
$$r_3(d_3) = 4d_3 + 5 \qquad\qquad (d_3 > 0)$$
$$r_1(0) = r_2(0) = r_3(0) = 0$$

The amount placed in each investment must be an exact multiple of $1,000. To maximize the net present value obtained from the investments, how should Finco allocate the $6,000?

Solution The return on each investment is not proportional to the amount invested in it [for example, $16 = r_1(2) \neq 2r_1(1) = 18$]. Thus, linear programming cannot be used to find an optimal solution to this problem.[†]

Mathematically, Finco's problem may be expressed as

$$\max\{r_1(d_1) + r_2(d_2) + r_3(d_3)\}$$
$$\text{s.t.} \quad d_1 + d_2 + d_3 = 6$$
$$d_j \text{ nonnegative integer} \quad (j = 1, 2, 3)$$

Of course, if the $r_j(d_j)$'s were linear, then we would have a knapsack problem like those we studied in Section 9.5.

To formulate Finco's problem as a dynamic programming problem, we begin by identifying the stage. As in the inventory and shortest-route examples, the stage should be chosen so that when one stage remains the problem is easy to solve. Then, given that the problem has been solved for the case where one stage remains, it should be easy to solve the problem where two stages remain, and so forth. Clearly, it would be easy to solve when only one investment was available, so we define stage t to represent a case where funds must be allocated to investments $t, t + 1, \ldots, 3$.

For a given stage, what must we know to determine the optimal investment amount? Simply how much money is available for investments $t, t + 1, \ldots, 3$. Thus, we define the state at any stage to be the amount of money (in thousands) available for investments t, $t + 1, \ldots, 3$. We can never have more than $6,000 available, so the possible states at any stage are 0, 1, 2, 3, 4, 5, and 6. We define $f_t(d_t)$ to be the maximum net present value (NPV) that can be obtained by investing d_t thousand dollars in investments $t, t + 1, \ldots,$ 3. Also define $x_t(d_t)$ to be the amount that should be invested in investment t to attain $f_t(d_t)$. We start to work backward by computing $f_3(0), f_3(1), \ldots, f_3(6)$ and then determine $f_2(0)$, $f_2(1), \ldots, f_2(6)$. Since $6,000 is available for investment in investments 1, 2, and 3, we

---

[†]The fixed-charge approach described in Section 9.2 could be used to solve this problem.

terminate our computations by computing $f_1(6)$. Then we retrace our steps and determine the amount that should be allocated to each investment (just as we retraced our steps to determine the optimal production level for each month in Example 4).

Stage 3 Computations

We first determine $f_3(0), f_3(1), \ldots, f_3(6)$. We see that $f_3(d_3)$ is attained by investing all available money (d_3) in investment 3. Thus,

$$
\begin{aligned}
f_3(0) &= 0 & x_3(0) &= 0 \\
f_3(1) &= 9 & x_3(1) &= 1 \\
f_3(2) &= 13 & x_3(2) &= 2 \\
f_3(3) &= 17 & x_3(3) &= 3 \\
f_3(4) &= 21 & x_3(4) &= 4 \\
f_3(5) &= 25 & x_3(5) &= 5 \\
f_3(6) &= 29 & x_3(6) &= 6
\end{aligned}
$$

TABLE 6
Computations for $f_2(0), f_2(1), \ldots, f_2(6)$

d_2	x_2	$r_2(x_2)$	$f_3(d_2 - x_2)$	NPV from Investments 2, 3	$f_2(d_2)$ $x_2(d_2)$
0	0	0	0	0*	$f_2(0) = 0$
					$x_2(0) = 0$
1	0	0	9	9	$f_2(1) = 10$
1	1	10	0	10*	$x_2(1) = 1$
2	0	0	13	13	$f_2(2) = 19$
2	1	10	9	19*	$x_2(2) = 1$
2	2	13	0	13	
3	0	0	17	17	$f_2(3) = 23$
3	1	10	13	23*	$x_2(3) = 1$
3	2	13	9	22	
3	3	16	0	16	
4	0	0	21	21	$f_2(4) = 27$
4	1	10	17	27*	$x_2(4) = 1$
4	2	13	13	26	
4	3	16	9	25	
4	4	19	0	19	
5	0	0	25	25	$f_2(5) = 31$
5	1	10	21	31*	$x_2(5) = 1$
5	2	13	17	30	
5	3	16	13	29	
5	4	19	9	28	
5	5	22	0	22	
6	0	0	29	29	$f_2(6) = 35$
6	1	10	25	35*	$x_2(6) = 1$
6	2	13	21	34	
6	3	16	17	33	
6	4	19	13	32	
6	5	22	9	31	
6	6	25	0	25	

TABLE 7
Computations for $f_1(6)$

d_1	x_1	$r_1(x_1)$	$f_2(6 - x_1)$	NPV from Investments 1–3	$f_1(6)$ $x_1(6)$
6	0	0	35	35	$f_1(6) = 49$
6	1	9	31	40	$x_1(6) = 4$
6	2	16	27	43	
6	3	23	23	46	
6	4	30	19	49*	
6	5	37	10	47	
6	6	44	0	44	

Stage 2 Computations

To determine $f_2(0), f_2(1), \ldots, f_2(6)$, we look at all possible amounts that can be placed in investment 2. To find $f_2(d_2)$, let x_2 be the amount invested in investment 2. Then an NPV of $r_2(x_2)$ will be obtained from investment 2, and an NPV of $f_3(d_2 - x_2)$ will be obtained from investment 3 (remember the principle of optimality). Since x_2 should be chosen to maximize the net present value earned from investments 2 and 3, we write

$$f_2(d_2) = \max_{x_2} \{r_2(x_2) + f_3(d_2 - x_2)\} \tag{5}$$

where x_2 must be a member of $\{0, 1, \ldots, d_2\}$. The computations for $f_2(0), f_2(1), \ldots, f_2(6)$ and $x_2(0), x_2(1), \ldots, x_2(6)$ are given in Table 6.

Stage 1 Computations

Following (5), we write

$$f_1(6) = \max_{x_1} \{r_1(x_1) + f_2(6 - x_1)\}$$

where x_1 must be a member of $\{0, 1, 2, 3, 4, 5, 6\}$. The computations for $f_1(6)$ are given in Table 7.

Determination of Optimal Resource Allocation

Since $x_1(6) = 4$, Finco invests $4,000 in investment 1. This leaves $6,000 - 4,000 = $2,000 for investments 2 and 3. Hence, Finco should invest $x_2(2) = $1,000 in investment 2. Then $1,000 is left for investment 3, so Finco chooses to invest $x_3(1) = $1,000 in investment 3. Therefore, Finco can attain a maximum net present value of $f_1(6) = $49,000 by investing $4,000 in investment 1, $1,000 in investment 2, and $1,000 in investment 3.

Network Representation of Resource Example

As with the inventory example of Section 18.3, Finco's problem has a network representation, equivalent to finding the *longest route* from (1, 6) to (4, 0) in Figure 6. In the figure, the node (t, d) represents the situation in which d thousand dollars is available for investments $t, t + 1, \ldots, 3$. The arc joining the nodes (t, d) and $(t + 1, d - x)$ has a length $r_t(x)$ corresponding to the net present value obtained by investing x thousand dollars in investment t. For example, the arc joining nodes (2, 4) and (3, 1) has a length $r_2(3) = $16,000, corresponding to the $16,000 net present value that can be obtained by invest-

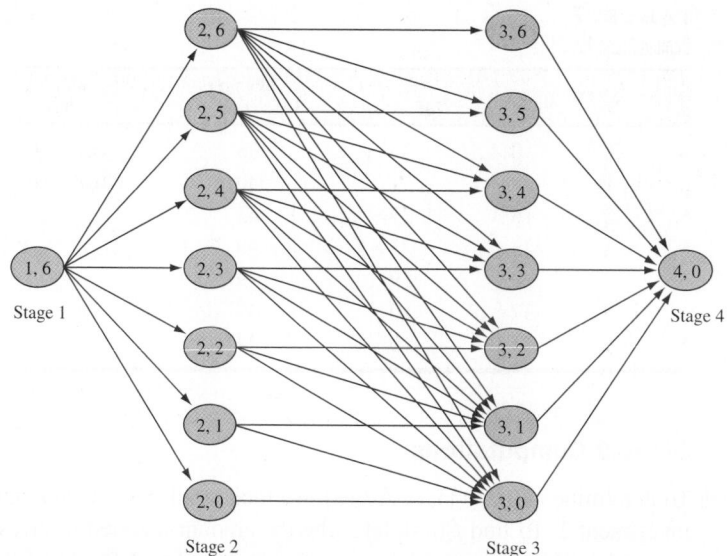

FIGURE 6
Network Representation
of Finco

ing $3,000 in investment 2. Note that not all pairs of nodes in adjacent stages are joined by arcs. For example, there is no arc joining the nodes (2, 4) and (3, 5); after all, if you have only $4,000 available for investments 2 and 3, how can you have $5,000 available for investment 3? From our computations, we see that the longest path from (1, 6) to (4, 0) is (1, 6)–(2, 2)–(3, 1)–(4, 0).

Generalized Resource Allocation Problem

We now consider a generalized version of Example 5. Suppose we have w units of a resource available and T activities to which the resource can be allocated. If activity t is implemented at a level x_t (we assume x_t must be a nonnegative integer), then $g_t(x_t)$ units of the resource are used by activity t, and a benefit $r_t(x_t)$ is obtained. The problem of determining the allocation of resources that maximizes total benefit subject to the limited resource availability may be written as

$$\max \sum_{t=1}^{t=T} r_t(x_t)$$

$$\text{s.t.} \quad \sum_{t=1}^{t=T} g_t(x_t) \leq w \tag{6}$$

where x_t must be a member of $\{0, 1, 2, \ldots\}$. Some possible interpretations of $r_t(x_t)$, $g_t(x_t)$, and w are given in Table 8.

To solve (6) by dynamic programming, define $f_t(d)$ to be the maximum benefit that can be obtained from activities $t, t + 1, \ldots, T$ if d units of the resource may be allocated to activities $t, t + 1, \ldots, T$. We may generalize the recursions of Example 5 to this situation by writing

$$f_{T+1}(d) = 0 \quad \text{for all } d$$
$$f_t(d) = \max_{x_t} \{r_t(x_t) + f_{t+1}[d - g_t(x_t)]\} \tag{7}$$

where x_t must be a nonnegative integer satisfying $g_t(x_t) \leq d$. Let $x_t(d)$ be any value of x_t that attains $f_t(d)$. To use (7) to determine an optimal allocation of resources to activities $1, 2, \ldots, T$, we begin by determining all $f_T(\cdot)$ and $x_T(\cdot)$. Then we use (7) to determine all $f_{T-1}(\cdot)$ and $x_{T-1}(\cdot)$, continuing to work backward in this fashion until all $f_2(\cdot)$ and $x_2(\cdot)$

TABLE 8
Examples of a Generalized Resource Allocation Problem

Interpretation of $r_t(x_t)$	Interpretation of $g_t(x_t)$	Interpretation of w
Benefit from placing x_t type t items in a knapsack	Weight of x_t type t items	Maximum weight that knapsack can hold
Grade obtained in course t if we study course t for x_t hours per week	Number of hours per week x_t spent studying course t	Total number of study hours available each week
Sales of a product in region t if x_t sales reps are assigned to region t	Cost of assigning x_t sales reps to region t	Total sales force budget
Number of fire alarms per week responded to within one minute if precinct t is assigned x_t engines	Cost per week of maintaining x_t fire engines in precinct t	Total weekly budget for maintaining fire engines

have been determined. To wind things up, we now calculate $f_1(w)$ and $x_1(w)$. Then we implement activity 1 at a level $x_1(w)$. At this point, we have $w - g_1[x_1(w)]$ units of the resource available for activities $2, 3, \ldots, T$. Then activity 2 should be implemented at a level of $x_2\{w - g_1[x_1(w)]\}$. We continue in this fashion until we have determined the level at which all activities should be implemented.

Solution of Knapsack Problems by Dynamic Programming

We illustrate the use of (7) by solving a simple knapsack problem (see Section 9.5). Then we develop an alternative recursion that can be used to solve knapsack problems.

EXAMPLE 6   Knapsack

Suppose a 10-lb knapsack is to be filled with the items listed in Table 9. To maximize total benefit, how should the knapsack be filled?

Solution   We have $r_1(x_1) = 11x_1$, $r_2(x_2) = 7x_2$, $r_3(x_3) = 12x_3$, $g_1(x_1) = 4x_1$, $g_2(x_2) = 3x_2$, and $g_3(x_3) = 5x_3$. Define $f_t(d)$ to be the maximum benefit that can be earned from a d-pound knapsack that is filled with items of Type $t, t + 1, \ldots, 3$.

Stage 3 Computations

Now (7) yields

$$f_3(d) = \max_{x_3}\{12x_3\}$$

TABLE 9
Weights and Benefits for Knapsack

Item	Weight (lb)	Benefit
1	4	11
2	3	7
3	5	12

where $5x_3 \leq d$ and x_3 is a nonnegative integer. This yields

$$f_3(10) = 24$$
$$f_3(5) = f_3(6) = f_3(7) = f_3(8) = f_3(9) = 12$$
$$f_3(0) = f_3(1) = f_3(2) = f_3(3) = f_3(4) = 0$$
$$x_3(10) = 2$$
$$x_3(9) = x_3(8) = x_3(7) = x_3(6) = x_3(5) = 1$$
$$x_3(0) = x_3(1) = x_3(2) = x_3(3) = x_3(4) = 0$$

Stage 2 Computations

Now (7) yields

$$f_2(d) = \max_{x_2} \{7x_2 + f_3(d - 3x_2)\}$$

where x_2 must be a nonnegative integer satisfying $3x_2 \leq d$. We now obtain

$$f_2(10) = \max \begin{cases} 7(0) + f_3(10) = 24^* & x_2 = 0 \\ 7(1) + f_3(7) = 19 & x_2 = 1 \\ 7(2) + f_3(4) = 14 & x_2 = 2 \\ 7(3) + f_3(1) = 21 & x_2 = 3 \end{cases}$$

Thus, $f_2(10) = 24$ and $x_2(10) = 0$.

$$f_2(9) = \max \begin{cases} 7(0) + f_3(9) = 12 & x_2 = 0 \\ 7(1) + f_3(6) = 19 & x_2 = 1 \\ 7(2) + f_3(3) = 14 & x_2 = 2 \\ 7(3) + f_3(0) = 21^* & x_2 = 3 \end{cases}$$

Thus, $f_2(9) = 21$ and $x_2(9) = 3$.

$$f_2(8) = \max \begin{cases} 7(0) + f_3(8) = 12 & x_2 = 0 \\ 7(1) + f_3(5) = 19^* & x_2 = 1 \\ 7(2) + f_3(2) = 14 & x_2 = 2 \end{cases}$$

Thus, $f_2(8) = 19$ and $x_2(8) = 1$.

$$f_2(7) = \max \begin{cases} 7(0) + f_3(7) = 12 & x_2 = 0 \\ 7(1) + f_3(4) = 7 & x_2 = 1 \\ 7(2) + f_3(1) = 14^* & x_2 = 2 \end{cases}$$

Thus, $f_2(7) = 14$ and $x_2(7) = 2$.

$$f_2(6) = \max \begin{cases} 7(0) + f_3(6) = 12 & x_2 = 0 \\ 7(1) + f_3(3) = 7 & x_2 = 1 \\ 7(2) + f_3(0) = 14^* & x_2 = 2 \end{cases}$$

Thus, $f_2(6) = 14$ and $x_2(6) = 2$.

$$f_2(5) = \max \begin{cases} 7(0) + f_3(5) = 12^* & x_2 = 0 \\ 7(1) + f_3(2) = 7 & x_2 = 1 \end{cases}$$

Thus, $f_2(5) = 12$ and $x_2(5) = 0$.

$$f_2(4) = \max \begin{cases} 7(0) + f_3(4) = 0 & x_2 = 0 \\ 7(1) + f_3(1) = 7^* & x_2 = 1 \end{cases}$$

Thus, $f_2(4) = 7$ and $x_2(4) = 1$.

$$f_2(3) = \max \begin{cases} 7(0) + f_3(3) = 0 & x_2 = 0 \\ 7(1) + f_3(0) = 7^* & x_2 = 1 \end{cases}$$

Thus, $f_2(3) = 7$ and $x_2(3) = 1$.

$$f_2(2) = 7(0) + f_3(2) = 0 \qquad x_2 = 0$$

Thus, $f_2(2) = 0$ and $x_2(2) = 0$.

$$f_2(1) = 7(0) + f_3(1) = 0 \qquad x_2 = 0$$

Thus, $f_2(1) = 0$ and $x_2(1) = 0$.

$$f_2(0) = 7(0) + f_3(0) = 0 \qquad x_2 = 0$$

Thus, $f_2(0) = 0$ and $x_2(0) = 0$.

Stage 1 Computations

Finally, we determine $f_1(10)$ from

$$f_1(10) = \max \begin{cases} 11(0) + f_2(10) = 24 & x_1 = 0 \\ 11(1) + f_2(6) \ = 25^* & x_1 = 1 \\ 11(2) + f_2(2) \ = 22 & x_1 = 2 \end{cases}$$

Determination of the Optimal Solution to Knapsack Problem

We have $f_1(10) = 25$ and $x_1(10) = 1$. Hence, we should include one Type 1 item in the knapsack. Then we have $10 - 4 = 6$ lb left for Type 2 and Type 3 items, so we should include $x_2(6) = 2$ Type 2 items. Finally, we have $6 - 2(3) = 0$ lb left for Type 3 items, and we include $x_3(0) = 0$ Type 3 items. In summary, the maximum benefit that can be gained from a 10-lb knapsack is $f_3(10) = 25$. To obtain a benefit of 25, one Type 1 and two Type 2 items should be included.

Network Representation of Knapsack Problem

Finding the optimal solution to Example 6 is equivalent to finding the longest path in Figure 7 from node (10, 1) to some stage 4 node. In Figure 7, for $t \leq 3$, the node (d, t) represents a situation in which d pounds of space may be allocated to items of Type $t, t + 1,$ $\ldots, 3$. The node $(d, 4)$ represents d pounds of unused space. Each arc from a stage t node to a stage $t + 1$ node represents a decision of how many Type t items are placed in the knapsack. For example, the arc from (10, 1) to (6, 2) represents placing one Type 1 item in the knapsack. This leaves $10 - 4 = 6$ lb for items of Types 2 and 3. This arc has a length of 11, representing the benefit obtained by placing one Type 1 item in the knapsack. Our solution to Example 6 shows that the longest path in Figure 7 from node (10, 1) to a stage 4 node is (10, 1)–(6, 2)–(0, 3)–(0, 4). We note that the optimal solution to a knapsack problem does not always use all the available weight. For example, the reader should verify that if a Type 1 item earned 16 units of benefit, the optimal solution would be to include two type 1 items, corresponding to the path (10, 1)–(2, 2)–(2, 3)–(2, 4). This solution leaves 2 lb of space unused.

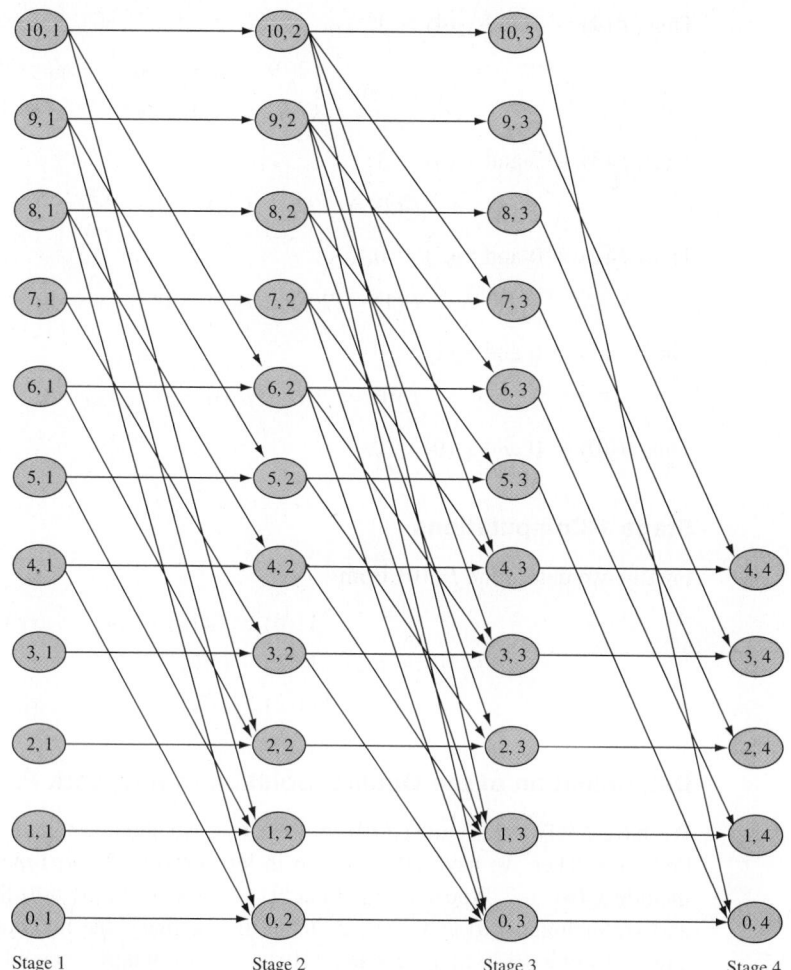

FIGURE 7
Network of
Representation of
Knapsack

Stage 1   Stage 2   Stage 3   Stage 4

An Alternative Recursion for Knapsack Problems

Other approaches can be used to solve knapsack problems by dynamic programming. The approach we now discuss builds up the optimal knapsack by first determining how to fill a small knapsack optimally and then, using this information, how to fill a larger knapsack optimally. We define $g(w)$ to be the maximum benefit that can be gained from a w-lb knapsack. In what follows, b_j is the benefit earned from a single Type j item, and w_j is the weight of a single Type j item. Clearly, $g(0) = 0$, and for $w > 0$,

$$g(w) = \max_j \{b_j + g(w - w_j)\} \tag{8}$$

where j must be a member of $\{1, 2, 3\}$, and j must satisfy $w_j \leq w$. The reasoning behind (8) is as follows: To fill a w-lb knapsack optimally, we must begin by putting some type of item into the knapsack. If we begin by putting a Type j item into a w-lb knapsack, the best we can do is earn $b_j + $ [best we can do from a $(w - w_j)$-lb knapsack]. After noting that a Type j item can be placed into a w-lb knapsack only if $w_j \leq w$, we obtain (8). We define $x(w)$ to be any type of item that attains the maximum in (8) and $x(w) = 0$ to mean that no item can fit into a w-lb knapsack.

To illustrate the use of (8), we re-solve Example 6. Because no item can fit in a 0-, 1-, or 2-lb knapsack, we have $g(0) = g(1) = g(2) = 0$ and $x(0) = x(1) = x(2) = 0$. Only a Type 2 item fits into a 3-lb knapsack, so we have that $g(3) = 7$ and $x(3) = 2$. Continuing, we find that

$$g(4) = \max \begin{cases} 11 + g(0) = 11^* & \text{(Type 1 item)} \\ 7 + g(1) = 7 & \text{(Type 2 item)} \end{cases}$$

Thus, $g(4) = 11$ and $x(4) = 1$.

$$g(5) = \max \begin{cases} 11 + g(1) = 11 & \text{(Type 1 item)} \\ 7 + g(2) = 7 & \text{(Type 2 item)} \\ 12 + g(0) = 12^* & \text{(Type 3 item)} \end{cases}$$

Thus, $g(5) = 12$ and $x(5) = 3$.

$$g(6) = \max \begin{cases} 11 + g(2) = 11 & \text{(Type 1 item)} \\ 7 + g(3) = 14^* & \text{(Type 2 item)} \\ 12 + g(1) = 12 & \text{(Type 3 item)} \end{cases}$$

Thus, $g(6) = 14$ and $x(6) = 2$.

$$g(7) = \max \begin{cases} 11 + g(3) = 18^* & \text{(Type 1 item)} \\ 7 + g(4) = 18^* & \text{(Type 2 item)} \\ 12 + g(2) = 12 & \text{(Type 3 item)} \end{cases}$$

Thus, $g(7) = 18$ and $x(7) = 1$ or $x(7) = 2$.

$$g(8) = \max \begin{cases} 11 + g(4) = 22^* & \text{(Type 1 item)} \\ 7 + g(5) = 19 & \text{(Type 2 item)} \\ 12 + g(3) = 19 & \text{(Type 3 item)} \end{cases}$$

Thus, $g(8) = 22$ and $x(8) = 1$.

$$g(9) = \max \begin{cases} 11 + g(5) = 23^* & \text{(Type 1 item)} \\ 7 + g(6) = 21 & \text{(Type 2 item)} \\ 12 + g(4) = 23^* & \text{(Type 3 item)} \end{cases}$$

Thus, $g(9) = 23$ and $x(9) = 1$ or $x(9) = 3$.

$$g(10) = \max \begin{cases} 11 + g(6) = 25^* & \text{(Type 1 item)} \\ 7 + g(7) = 25^* & \text{(Type 2 item)} \\ 12 + g(5) = 24 & \text{(Type 3 item)} \end{cases}$$

Thus, $g(10) = 25$ and $x(10) = 1$ or $x(10) = 2$. To fill the knapsack optimally, we begin by putting any $x(10)$ item in the knapsack. Let's arbitrarily choose a Type 1 item. This leaves us with $10 - 4 = 6$ lb to fill, so we now put an $x(10 - 4) = 2$ (Type 2) item in the knapsack. This leaves us with $6 - 3 = 3$ lb to fill, which we do with an $x(6 - 3) = 2$ (Type 2) item. Hence, we may attain the maximum benefit of $g(10) = 25$ by filling the knapsack with two Type 2 items and one Type 1 item.

A Turnpike Theorem

For a knapsack problem, let

$$c_j = \text{benefit obtained from each type } j \text{ item}$$
$$w_j = \text{weight of each type } j \text{ item}$$

In terms of benefit per unit weight, the best item is the item with the largest value of $\frac{c_j}{w_j}$. Assume there are n types of items that have been ordered, so that

$$\frac{c_1}{w_1} \geq \frac{c_2}{w_2} \geq \cdots \geq \frac{c_n}{w_n}$$

Thus, Type 1 items are the best, Type 2 items are the second best, and so on. Recall from Section 9.5 that it is possible for the optimal solution to a knapsack problem to use none of the best item. For example, the optimal solution to the knapsack problem

$$\max z = 16x_1 + 22x_2 + 12x_3 + 8x_4$$
$$\text{s.t.} \quad 5x_1 + 7x_2 + 5x_3 + 4x_4 \leq 14$$
$$x_i \text{ nonnegative integer}$$

is $z = 44$, $x_2 = 2$, $x_1 = x_3 = x_4 = 0$, and this solution does not use any of the best (Type 1) item. Assume that

$$\frac{c_1}{w_1} > \frac{c_2}{w_2}$$

Thus, there is a unique best item type. It can be shown that for some number w^*, it is optimal to use at least one Type 1 item if the knapsack is allowed to hold w pounds, where $w \geq w^*$. In Problem 6 at the end of this section, you will show that this result holds for

$$w^* = \frac{c_1 w_1}{c_1 - w_1 \left(\dfrac{c_2}{w_2} \right)}$$

Thus, for the knapsack problem

$$\max z = 16x_1 + 22x_2 + 12x_3 + 8x_4$$
$$\text{s.t.} \quad 5x_1 + 7x_2 + 5x_3 + 4x_4 \leq w$$
$$x_i \text{ nonnegative integer}$$

at least one Type 1 item will be used if

$$w \geq \frac{16(5)}{16 - 5(\frac{22}{7})} = 280$$

This result can greatly reduce the computation needed to solve a knapsack problem. For example, suppose that $w = 4,000$. We know that for $w \geq 280$, the optimal solution will use at least one Type 1 item, so we can conclude that the optimal way to fill a 4,000-lb knapsack will consist of one Type 1 item plus the optimal way to fill a knapsack of $4,000 - 5 = 3,995$ lb. Repeating this reasoning shows that the optimal way to fill a 4,000-lb knapsack will consist of $\frac{4,000-280}{5} = 744$ Type 1 items plus the optimal way to fill a knapsack of 280 lb. This reasoning substantially reduces the computation needed to determine how to fill a 4,000-lb knapsack. (Actually, the 280-lb knapsack will use at least one Type 1 item, so we know that to fill a 4,000-lb knapsack optimally, we can use 745 Type 1 items and then optimally fill a 275-lb knapsack.)

Why is this result referred to as a **turnpike theorem**? Think about taking an automobile trip in which our goal is to minimize the time needed to complete the trip. For a long enough trip, it may be advantageous to go slightly out of our way so that most of the trip will be spent on a turnpike, on which we can travel at the greatest speed. For a short trip, it may not be worth our while to go out of our way to get on the turnpike.

Similarly, in a long (large-weight) knapsack problem, it is always optimal to use some of the best items, but this may not be the case in a short knapsack problem. Turnpike results abound in the dynamic programming literature [see Morton (1979)].

PROBLEMS

Group A

1 J. R. Carrington has $4 million to invest in three oil well sites. The amount of revenue earned from site $i(i = 1, 2, 3)$ depends on the amount of money invested in site i (see Table 10). Assuming that the amount invested in a site must be an exact multiple of $1 million, use dynamic programming to determine an investment policy that will maximize the revenue J. R. will earn from his three oil wells.

2 Use either of the approaches outlined in this section to solve the following knapsack problem:

$$\max z = 5x_1 + 4x_2 + 2x_3$$
$$\text{s.t.} \quad 4x_1 + 3x_2 + 2x_3 \leq 8$$
$$x_1, x_2, x_3 \geq 0; x_1, x_2, x_3 \text{ integer}$$

3 The knapsack problem of Problem 2 can be viewed as finding the longest route in a particular network.

a Draw the network corresponding to the recursion derived from (7).

b Draw the network corresponding to the recursion derived from (8).

4 The number of crimes in each of a city's three police precincts depends on the number of patrol cars assigned to each precinct (see Table 11). Five patrol cars are available. Use dynamic programming to determine how many patrol cars should be assigned to each precinct.

TABLE 10

Amount Invested ($ Millions)	Revenue ($ Millions)		
	Site 1	Site 2	Site 3
0	4	3	3
1	7	6	7
2	8	10	8
3	9	12	13
4	11	14	15

TABLE 11

Precinct	No. of Patrol Cars Assigned to Precinct					
	0	1	2	3	4	5
1	14	10	7	4	1	0
2	25	19	16	14	12	11
3	20	14	11	8	6	5

5 Use dynamic programming to solve a knapsack problem in which the knapsack can hold up to 13 lb (see Table 12).

Group B

6 Consider a knapsack problem for which

$$\frac{c_1}{w_1} > \frac{c_2}{w_2}$$

Show that if the knapsack can hold w pounds, and $w \geq w^*$, where

$$w^* = \frac{c_1 w_1}{c_1 - w_1 \left(\dfrac{c_2}{w_2} \right)}$$

then the optimal solution to the knapsack problem must use at least one Type 1 item.

TABLE 12

Item	Weight (lb)	Benefit
1	3	12
2	5	25
3	7	50

18.5 Equipment-Replacement Problems

Many companies and customers face the problem of determining how long a machine should be utilized before it should be traded in for a new one. Problems of this type are called **equipment-replacement problems** and can often be solved by dynamic programming.

EXAMPLE 7 | **Equipment Replacement**

An auto repair shop always needs to have an engine analyzer available. A new engine analyzer costs $1,000. The cost m_i of maintaining an engine analyzer during its ith year of operation is as follows: $m_1 = \$60$, $m_2 = \$80$, $m_3 = \$120$. An analyzer may be kept for

FIGURE 8
Time Horizon for
Equipment
Replacement

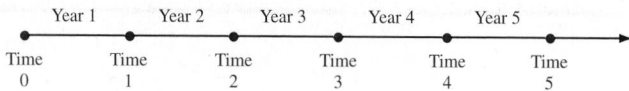

1, 2, or 3 years; after i years of use ($i = 1, 2, 3$), it may be traded in for a new one. If an i-year-old engine analyzer is traded in, a salvage value s_i is obtained, where $s_1 = \$800$, $s_2 = \$600$, and $s_3 = \$500$. Given that a new machine must be purchased now (time 0; see Figure 8), the shop wants to determine a replacement and trade-in policy that minimizes net costs = (maintenance costs) + (replacement costs) − (salvage value received) during the next 5 years.

Solution We note that after a new machine is purchased, the firm must decide when the newly purchased machine should be traded in for a new one. With this in mind, we define $g(t)$ to be the minimum net cost incurred from time t until time 5 (including the purchase cost and salvage value for the newly purchased machine) given that a new machine has been purchased at time t. We also define c_{tx} to be the net cost (including purchase cost and salvage value) of purchasing a machine at time t and operating it until time x. Then the appropriate recursion is

$$g(t) = \min_{x} \{c_{tx} + g(x)\} \qquad (t = 0, 1, 2, 3, 4) \tag{9}$$

where x must satisfy the inequalities $t + 1 \le x \le t + 3$ and $x \le 5$. Because the problem is over at time 5, no cost is incurred from time 5 onward, so we may write $g(5) = 0$.

To justify (9), note that after a new machine is purchased at time t, we must decide when to replace the machine. Let x be the time at which the replacement occurs. The replacement must be after time t but within 3 years of time t. This explains the restriction that $t + 1 \le x \le t + 3$. Since the problem ends at time 5, we must also have $x \le 5$. If we choose to replace the machine at time x, then what will be the cost from time t to time 5? Simply the sum of the cost incurred from the purchase of the machine to the sale of the machine at time x (which is by definition c_{tx}) and the total cost incurred from time x to time 5 (given that a new machine has just been purchased at time x). By the principle of optimality, the latter cost is, of course, $g(x)$. Hence, if we keep the machine that was purchased at time t until time x, then from time t to time 5, we incur a cost of $c_{tx} + g(x)$. Thus, x should be chosen to minimize this sum, and this is exactly what (9) does. We have assumed that maintenance costs, salvage value, and purchase price remain unchanged over time, so each c_{tx} will depend only on how long the machine is kept; that is, each c_{tx} depends only on $x - t$. More specifically,

$$c_{tx} = \$1,000 + m_1 + \cdots + m_{x-t} - s_{x-t}$$

This yields

$$c_{01} = c_{12} = c_{23} = c_{34} = c_{45} = 1,000 + 60 - 800 = \$260$$
$$c_{02} = c_{13} = c_{24} = c_{35} = 1,000 + 60 + 80 - 600 = \$540$$
$$c_{03} = c_{14} = c_{25} = 1,000 + 60 + 80 + 120 - 500 = \$760$$

We begin by computing $g(4)$ and work backward until we have computed $g(0)$. Then we use our knowledge of the values of x attaining $g(0)$, $g(1)$, $g(2)$, $g(3)$, and $g(4)$ to determine the optimal replacement strategy. The calculations follow.

At time 4, there is only one sensible decision (keep the machine until time 5 and sell it for its salvage value), so we find

$$g(4) = c_{45} + g(5) = 260 + 0 = \$260*$$

Thus, if a new machine is purchased at time 4, it should be traded in at time 5.

If a new machine is purchased at time 3, we keep it until time 4 or time 5. Hence,

$$g(3) = \min \begin{cases} c_{34} + g(4) = 260 + 260 = \$520^* & \text{(Trade at time 4)} \\ c_{35} + g(5) = 540 + 0 = \$540 & \text{(Trade at time 5)} \end{cases}$$

Thus, if a new machine is purchased at time 3, we should trade it in at time 4.

If a new machine is purchased at time 2, we trade it in at time 3, time 4, or time 5. This yields

$$g(2) = \min \begin{cases} c_{23} + g(3) = 260 + 520 = \$780 & \text{(Trade at time 3)} \\ c_{24} + g(4) = 540 + 260 = \$800 & \text{(Trade at time 4)} \\ c_{25} + g(5) = \$760^* & \text{(Trade at time 5)} \end{cases}$$

Thus, if we purchase a new machine at time 2, we should keep it until time 5 and then trade it in.

If a new machine is purchased at time 1, we trade it in at time 2, time 3, or time 4. Then

$$g(1) = \min \begin{cases} c_{12} + g(2) = 260 + 760 = \$1{,}020^* & \text{(Trade at time 2)} \\ c_{13} + g(3) = 540 + 520 = \$1{,}060 & \text{(Trade at time 3)} \\ c_{14} + g(4) = 760 + 260 = \$1{,}020^* & \text{(Trade at time 4)} \end{cases}$$

Thus, if a new machine is purchased at time 1, it should be traded in at time 2 or time 4.

The new machine that was purchased at time 0 may be traded in at time 1, time 2, or time 3. Thus,

$$g(0) = \min \begin{cases} c_{01} + g(1) = 260 + 1{,}020 = \$1{,}280^* & \text{(Trade at time 1)} \\ c_{02} + g(2) = 540 + 760 = \$1{,}300 & \text{(Trade at time 2)} \\ c_{03} + g(3) = 760 + 520 = \$1{,}280^* & \text{(Trade at time 3)} \end{cases}$$

Thus, the new machine purchased at time 0 should be replaced at time 1 or time 3. Let's arbitrarily choose to replace the time 0 machine at time 1. Then the new time 1 machine may be traded in at time 2 or time 4. Again we make an arbitrary choice and replace the time 1 machine at time 2. Then the time 2 machine should be kept until time 5, when it is sold for salvage value. With this replacement policy, we will incur a net cost of $g(0) = \$1{,}280$. The reader should verify that the following replacement policies are also optimal: (1) trading in at times 1, 4, and 5 and (2) trading in at times 3, 4, and 5.

We have assumed that all costs remain stationary over time. This assumption was made solely to simplify the computation of the c_{tx}'s. If we had relaxed the assumption of stationary costs, then the only complication would have been that the c_{tx}'s would have been messier to compute. We also note that if a short planning horizon is used, the optimal replacement policy may be extremely sensitive to the length of the planning horizon. Thus, more meaningful results can be obtained by using a longer planning horizon.

An equipment-replacement model was actually used by Phillips Petroleum to reduce costs associated with maintaining the company's stock of trucks (see Waddell (1983)).

Network Representation of Equipment-Replacement Problem

The reader should verify that our solution to Example 7 was equivalent to finding the shortest path from node 0 to node 5 in the network in Figure 9. The length of the arc joining nodes i and j is c_{ij}.

FIGURE 9
Network Representation
of Equipment
Replacement

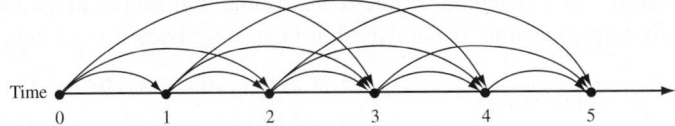

Time 0 1 2 3 4 5

An Alternative Recursion

There is another dynamic programming formulation of the equipment-replacement model. If we define the stage to be the time t and the state at any stage to be the age of the engine analyzer at time t, then an alternative dynamic programming recursion can be developed. Define $f_t(x)$ to be the minimum cost incurred from time t to time 5, given that at time t the shop has an x-year-old analyzer. The problem is over at time 5, so we sell the machine at time 5 and receive $-s_x$. Then $f_5(x) = -s_x$, and for $t = 0, 1, 2, 3, 4$,

$$f_t(3) = -500 + 1{,}000 + 60 + f_{t+1}(1) \qquad \text{(Trade)} \qquad \textbf{(10)}$$

$$f_t(2) = \min \begin{cases} -600 + 1{,}000 + 60 + f_{t+1}(1) & \text{(Trade)} \\ 120 + f_{t+1}(3) & \text{(Keep)} \end{cases} \qquad \textbf{(10.1)}$$

$$f_t(1) = \min \begin{cases} -800 + 1{,}000 + 60 + f_{t+1}(1) & \text{(Trade)} \\ 80 + f_{t+1}(2) & \text{(Keep)} \end{cases} \qquad \textbf{(10.2)}$$

$$f_0(0) = 1{,}000 + 60 + f_1(1) \qquad \text{(Keep)} \qquad \textbf{(10.3)}$$

The rationale behind Equations (10)–(10.3) is that if we have a 1- or 2-year-old analyzer, then we must decide between replacing the machine or keeping it another year. In (10.1) and (10.2), we compare the costs of these two options. For any option, the total cost from t until time 5 is the sum of the cost during the current year plus costs from time $t + 1$ to time 5. If we have a 3-year-old analyzer, then we must replace it, so there is no choice. The way we have defined the state means that it is only possible to be in state 0 at time 0. In this case, we must keep the analyzer for the first year (incurring a cost of $1,060). From this point on, a total cost of $f_1(1)$ is incurred. Thus, (10.3) follows. Since we know that $f_5(1) = -800$, $f_5(2) = -600$, and $f_5(3) = -500$, we can immediately compute all the $f_4(\cdot)$'s. Then we can compute the $f_3(\cdot)$'s. We continue in this fashion until $f_0(0)$ is determined (remember that we begin with a new machine). Then we follow our usual method for determining an optimal policy. That is, if $f_0(0)$ is attained by keeping the machine, then we keep the machine for a year and then, during year 1, we choose the action that attains $f_1(1)$. Continuing in this fashion, we can determine for each time whether or not the machine should be replaced. (See Problem 1 below.)

PROBLEMS

Group A

1 Use Equations (10)–(10.3) to determine an optimal replacement policy for the engine analyzer example.

2 Suppose that a new car costs $10,000 and that the annual operating cost and resale value of the car are as shown in Table 13. If I have a new car now, determine a replacement policy that minimizes the net cost of owning and operating a car for the next six years.

3 It costs $40 to buy a telephone from a department store. The estimated maintenance cost for each year of operation is shown in Table 14. (I can keep a telephone for at most five years.) I have just purchased a new telephone, and my old telephone has no salvage value. Determine how to minimize the total cost of purchasing and operating a telephone for the next six years.

TABLE 13

Age of Car (Years)	Resale Value ($)	Operating Cost ($)	
1	7,000	300	(year 1)
2	6,000	500	(year 2)
3	4,000	800	(year 3)
4	3,000	1,200	(year 4)
5	2,000	1,600	(year 5)
6	1,000	2,200	(year 6)

TABLE 14

Year	Maintenance Cost ($)
1	20
2	30
3	40
4	60
5	70

18.6 Formulating Dynamic Programming Recursions

In many dynamic programming problems (such as the inventory and shortest path examples), a given stage simply consists of all the possible states that the system can occupy at that stage. If this is the case, then the dynamic programming recursion (for a min problem) can often be written in the following form:

$$f_t(i) = \min\{(\text{cost during stage } t) + f_{t+1} \ (\text{new state at stage } t + 1)\} \qquad \text{(11)}$$

where the minimum in (11) is over all decisions that are allowable, or feasible, when the state at stage t is i. In (11), $f_t(i)$ is the minimum cost incurred from stage t to the end of the problem (say, the problem ends after stage T), given that at stage t the state is i.

Equation (11) reflects the fact that the minimum cost incurred from stage t to the end of the problem must be attained by choosing at stage t an allowable decision that minimizes the sum of the costs incurred during the current stage (stage t) plus the minimum cost that can be incurred from stage $t + 1$ to the end of the problem. Correct formulation of a recursion of the form (11) requires that we identify three important aspects of the problem:

Aspect 1 *The set of decisions that is allowable, or feasible, for the given state and stage.* Often, the set of feasible decisions depends on both t and i. For instance, in the inventory example of Section 18.3, let

$$d_t = \text{demand during month } t$$
$$i_t = \text{inventory at beginning of month } t$$

In this case, the set of allowable month t decisions (let x_t represent an allowable production level) consists of the members of $\{0, 1, 2, 3, 4, 5\}$ that satisfy $0 \le (i_t + x_t - d_t) \le 4$. Note how the set of allowable decisions at time t depends on the stage t and the state at time t, which is i_t.

Aspect 2 *We must specify how the cost during the current time period (stage t) depends on the value of t, the current state, and the decision chosen at stage t.* For instance, in the inventory example of Section 18.3, suppose a production level x_t is chosen during month t. Then the cost during month t is given by $c(x_t) + (\frac{1}{2})(i_t + x_t - d_t)$.

Aspect 3 *We must specify how the state at stage t + 1 depends on the value of t, the state at stage t, and the decision chosen at stage t.* Again referring to the inventory example, the month $t + 1$ state is $i_t + x_t - d_t$.

If you have properly identified the state, stage, and decision, then aspects 1–3 shouldn't be too hard to handle. A word of caution, however: Not all recursions are of the form (11). For instance, our first equipment-replacement recursion skipped over time $t + 1$.

This often occurs when the stage alone supplies sufficient information to make an optimal decision. We now work through several examples that illustrate the art of formulating dynamic programming recursions.

EXAMPLE 8 A Fishery

The owner of a lake must decide how many bass to catch and sell each year. If she sells x bass during year t, then a revenue $r(x)$ is earned. The cost of catching x bass during a year is a function $c(x, b)$ of the number of bass caught during the year and of b, the number of bass in the lake at the beginning of the year. Of course, bass do reproduce. To model this, we assume that the number of bass in the lake at the beginning of a year is 20% more than the number of bass left in the lake at the end of the previous year. Assume that there are 10,000 bass in the lake at the beginning of the first year. Develop a dynamic programming recursion that can be used to maximize the owner's net profits over a T-year horizon.

Solution
In problems where decisions must be made at several points in time, there is often a trade-off of current benefits against future benefits. For example, we could catch many bass early in the problem, but then the lake would be depleted in later years, and there would be very few bass to catch. On the other hand, if we catch very few bass now, we won't make much money early, but we can make a lot of money near the end of the horizon. In intertemporal optimization problems, dynamic programming is often used to analyze these complex trade-offs.

At the beginning of year T, the owner of the lake need not worry about the effect that the capture of bass will have on the future population of the lake. (At time T, there is no future!) So at the beginning of year T, the problem is relatively easy to solve. For this reason, we let time be the stage. At each stage, the owner of the lake must decide how many bass to catch. We define x_t to be the number of bass caught during year t. To determine an optimal value of x_t, the owner of the lake need only know the number of bass (call it b_t) in the lake at the beginning of year t. Therefore, the state at the beginning of year t is b_t.

We define $f_t(b_t)$ to be the maximum net profit that can be earned from bass caught during years $t, t + 1, \ldots, T$ given that b_t bass are in the lake at the beginning of year t. We may now dispose of aspects 1–3 of the recursion.

Aspect 1 What are the allowable decisions? During any year, we can't catch more bass than there are in the lake. Thus, in each state and for all t, $0 \leq x_t \leq b_t$ must hold.

Aspect 2 What is the net profit earned during year t? If x_t bass are caught during a year that begins with b_t bass in the lake, then the net profit is $r(x_t) - c(x_t, b_t)$.

Aspect 3 What will be the state during year $t + 1$? At the end of year t, there will be $b_t - x_t$ bass in the lake. By the beginning of year $t + 1$, these bass will have multiplied by 20%. This implies that at the beginning of year $t + 1$, $1.2(b_t - x_t)$ bass will be in the lake. Thus, the year $t + 1$ state will be $1.2(b_t - x_t)$.

We can now use (11) to develop the appropriate recursion. After year T, there are no future profits to consider, so

$$f_T(b_T) = \max_{x_T}\{r_T(x_T) - c(x_T, b_T)\}$$

where $0 \leq x_T \leq b_T$. Applying (11), we obtain

$$f_t(b_t) = \max\{r(x_t) - c(x_t, b_t) + f_{t+1}[1.2(b_t - x_t)]\} \tag{12}$$

where $0 \leq x_t \leq b_t$. To begin the computations, we first determine $f_T(b_T)$ for all values of b_T that might occur [b_T could be up to $10{,}000(1.2)^{T-1}$; why?]. Then we use (12) to work backward until $f_1(10,000)$ has been computed. Then, to determine an optimal fishing policy, we begin by choosing x_1 to be any value attaining the maximum in the (12) equation for $f_1(10,000)$. Then year 2 will begin with $1.2(10,000 - x_1)$ bass in the lake. This means that x_2 should be chosen to be any value attaining the maximum in the (12) equation for $f_2(1.2(10,000 - x_1))$. Continue in this fashion until the optimal values of $x_3, x_4, \ldots, x_T$ have been determined.

Incorporating the Time Value of Money into Dynamic Programming Formulations

A weakness of the current formulation is that profits received during later years are weighted the same as profits received during earlier years. As mentioned in the discussion of discounting (in Chapter 3), later profits should be weighted less than earlier profits. Suppose that for some $\beta < 1$, \$1 received at the beginning of year $t + 1$ is equivalent to β dollars received at the beginning of year t. We can incorporate this idea into the dynamic programming recursion by replacing (12) with

$$f_t(b_t) = \max_{x_t} \{r(x_t) - c(x_t, b_t) + \beta f_{t+1}[1.2(b_t - x_t)]\} \tag{12'}$$

where $0 \leq x_t \leq b_t$. Then we redefine $f_t(b_t)$ to be the maximum net profit *(in year t dollars)* that can be earned during years $t, t + 1, \ldots, T$. Since f_{t+1} is measured in year $t + 1$ dollars, multiplying it by β converts $f_{t+1}(\cdot)$ to year t dollars, which is just what we want. In Example 8, once we have worked backward and determined $f_1(10,000)$, an optimal fishing policy is found by using the same method that was previously described. This approach can be used to account for the time value of money in any dynamic programming formulation.

EXAMPLE 9 | **Power Plant**

An electric power utility forecasts that r_t kilowatt-hours (kwh) of generating capacity will be needed during year t (the current year is year 1). Each year, the utility must decide by how much generating capacity should be expanded. It costs $c_t(x)$ dollars to increase generating capacity by x kwh during year t. It may be desirable to reduce capacity, so x need not be nonnegative. During each year, 10% of the old generating capacity becomes obsolete and unusable (capacity does not become obsolete during its first year of operation). It costs the utility $m_t(i)$ dollars to maintain i units of capacity during year t. At the beginning of year 1, 100,000 kwh of generating capacity are available. Formulate a dynamic programming recursion that will enable the utility to minimize the total cost of meeting power requirements for the next T years.

Solution Again, we let time be the stage. At the beginning of year t, the utility must determine the amount of capacity (call it x_t) to add during year t. To choose x_t properly, all the utility needs to know is the amount of available capacity at the beginning of year t (call it i_t). Hence, we define the state at the beginning of year t to be the current capacity level. We may now dispose of aspects 1–3 of the formulation.

Aspect 1 What values of x_t are feasible? To meet year t's requirement of r_t, we must have $i_t + x_t \geq r_t$, or $x_t \geq r_t - i_t$. So the feasible x_t's are those values of x_t satisfying $x_t \geq r_t - i_t$.

Aspect 2 What cost is incurred during year t? If x_t kwh are added during a year that begins with i_t kwh of available capacity, then during year t, a cost $c_t(x_t) + m_t(i_t + x_t)$ is incurred.

Aspect 3 What will be the state at the beginning of year $t + 1$? At the beginning of year $t + 1$, the utility will have $0.9i_t$ kwh of old capacity plus the x_t kwh that have been added during year t. Thus, the state at the beginning of year $t + 1$ will be $0.9i_t + x_t$.

We can now use (11) to develop the appropriate recursion. Define $f_t(i_t)$ to be the minimum cost incurred by the utility during years $t, t + 1, \ldots, T$, given that i_t kwh of capacity are available at the beginning of year t. At the beginning of year T, there are no future costs to consider, so

$$f_T(i_T) = \min_{x_T} \{c_T(x_T) + m_T(i_T + x_T)\} \tag{13}$$

where x_T must satisfy $x_T \geq r_T - i_T$. For $t < T$,

$$f_t(i_t) = \min_{x_T} \{c_t(x_t) + m_t(i_t + x_t) + f_{t+1}(0.9i_t + x_t)\} \tag{14}$$

where x_t must satisfy $x_t \geq r_t - i_t$. If the utility does not start with any excess capacity, then we can safely assume that the capacity level would never exceed $r_{\text{MAX}} = \max_{t=1, 2, \ldots, T} \{r_t\}$. This means that we need consider only states $0, 1, 2, \ldots, r_{\text{MAX}}$. To begin computations, we use (13) to compute $f_T(0), f_T(1), \ldots, f_T(r_{\text{MAX}})$. Then we use (14) to work backward until $f_1(100,000)$ has been determined. To determine the optimal amount of capacity that should be added during each year, proceed as follows. During year 1, add an amount of capacity x_1 that attains the minimum in the (14) equation for $f_1(100,000)$. Then the utility will begin year 2 with $90,000 + x_1$ kwh of capacity. Then, during year 2, x_2 kwh of capacity should be added, where x_2 attains the minimum in the (14) equation for $f_2(90,000 + x_1)$. Continue in this fashion until the optimal value of x_T has been determined.

EXAMPLE 10   Wheat Sale

Farmer Jones now possesses $5,000 in cash and 1,000 bushels of wheat. During month t, the price of wheat is p_t. During each month, he must decide how many bushels of wheat to buy (or sell). There are three restrictions on each month's wheat transactions: (1) During any month, the amount of money spent on wheat cannot exceed the cash on hand at the beginning of the month; (2) during any month, he cannot sell more wheat than he has at the beginning of the month; and (3) because of limited warehouse capacity, the ending inventory of wheat for each month cannot exceed 1,000 bushels.

Show how dynamic programming can be used to maximize the amount of cash that farmer Jones has on hand at the end of six months.

Solution   Again, we let time be the stage. At the beginning of month t (the present is the beginning of month 1), farmer Jones must decide by how much to change the amount of wheat on hand. We define Δw_t to be the change in farmer Jones's wheat position during month t: $\Delta w_t \geq 0$ corresponds to a month t wheat purchase, and $\Delta w_t \leq 0$ corresponds to a month t sale of wheat. To determine an optimal value for Δw_t, we must know two things: the amount of wheat on hand at the beginning of month t (call it w_t) and the cash on hand at the beginning of month t, (call this c_t). We define $f_t(c_t, w_t)$ to be the maximum cash that farmer Jones can obtain at the end of month 6, given that farmer Jones has c_t dollars and w_t bushels of wheat at the beginning of month t. We now discuss aspects 1–3 of the formulation.

Aspect 1   What are the allowable decisions? If the state at time t is (c_t, w_t), then restrictions 1–3 limit Δw_t in the following manner:

$$p_t(\Delta w_t) \leq c_t \qquad \text{or} \qquad \Delta w_t \leq \frac{c_t}{p_t}$$

ensures that we won't run out of money at the end of month t. The inequality $\Delta w_t \geq -w_t$ ensures that during month t, we will not sell more wheat than we had at the beginning of month t; and $w_t + \Delta w_t \leq 1,000$, or $\Delta w_t \leq 1,000 - w_t$, ensures that we will end month t with at most 1,000 bushels of wheat. Putting these three restrictions together, we see that

$$-w_t \le \Delta w_t \le \min \left\{ \frac{c_t}{p_t}, 1{,}000 - w_t \right\}$$

will ensure that restrictions 1–3 are satisfied during month t.

Aspect 2 Since farmer Jones wants to maximize his cash on hand at the end of month 6, no benefit is earned during months 1 through 5. In effect, during months 1–5, we are doing bookkeeping to keep track of farmer Jones's position. Then, during month 6, we turn all of farmer Jones's assets into cash.

Aspect 3 If the current state is (c_t, w_t) and farmer Jones changes his month t wheat position by an amount Δw_t, what will be the new state at the beginning of month $t + 1$? Cash on hand will increase by $-(\Delta w_t)p_t$, and farmer Jones's wheat position will increase by Δw_t. Hence, the month $t + 1$ state will be $[c_t - (\Delta w_t)p_t, w_t + \Delta w_t]$.

We may now use (11) to develop the appropriate recursion. To maximize his cash position at the end of month 6, farmer Jones should convert his month 6 wheat into cash by selling all of it. This means that $\Delta w_6 = -w_6$. This leads to the following relation:

$$f_6(c_6, w_6) = c_6 + w_6 p_6 \tag{15}$$

Using (11), we obtain for $t < 6$

$$f_t(c_t, w_t) = \max_{\Delta w_t} \{0 + f_{t+1}[c_t - (\Delta w_t)p_t, w_t + \Delta w_t]\} \tag{16}$$

where Δw_t must satisfy

$$-w_t \le \Delta w_t \le \min \left\{ \frac{c_t}{p_t}, 1{,}000 - w_t \right\}$$

We begin our calculations by determining $f_6(c_6, w_6)$ for all states that can possibly occur during month 6. Then we use (16) to work backward until $f_1(5{,}000, 1{,}000)$ has been computed. Next, farmer Jones should choose Δw_1 to attain the maximum value in the (16) equation for $f_1(5{,}000, 1{,}000)$, and a month 2 state of $[5{,}000 - p_1(\Delta w_1), 1{,}000 + \Delta w_1]$ will ensue. Farmer Jones should next choose Δw_2 to attain the maximum value in the (16) equation for $f_2[5{,}000 - p_1(\Delta w_1), 1{,}000 + \Delta w_1]$. We continue in this manner until the optimal value of Δw_6 has been determined.

EXAMPLE 11   **Refinery Capacity**

Sunco Oil needs to build enough refinery capacity to refine 5,000 barrels of oil per day and 10,000 barrels of gasoline per day. Sunco can build refinery capacity at four locations. The cost of building a refinery at site t that has the capacity to refine x barrels of oil per day and y barrels of gasoline per day is $c_t(x, y)$. Use dynamic programming to determine how much capacity should be located at each site.

Solution If Sunco had only one possible refinery site, then the problem would be easy to solve. Sunco could solve a problem in which there were two possible refinery sites, and finally, a problem in which there were four refinery sites. For this reason, we let the stage represent the number of available oil sites. At any stage, Sunco must determine how much oil and gas capacity should be built at the given site. To do this, the company must know how much refinery capacity of each type must be built at the available sites. We now define $f_t(o_t, g_t)$ to be the minimum cost of building o_t barrels per day of oil refinery capacity and g_t barrels per day of gasoline refinery capacity at sites $t, t + 1, \ldots, 4$.

To determine $f_4(o_4, g_4)$, note that if only site 4 is available, Sunco must build a refinery at site 4 with o_4 barrels of oil capacity and g_4 barrels of gasoline capacity. This implies that $f_4(o_4, g_4) = c_4(o_4, g_4)$. For $t = 1, 2, 3$, we can determine $f_t(o_t, g_t)$ by noting that if we build a refinery at site t that can refine x_t barrels of oil per day and y_t barrels of gasoline per day, then we incur a cost of $c_t(x_t, y_t)$ at site t. Then we will need to build a total oil refinery capacity of $o_t - x_t$ and a gas refinery capacity of $g_t - y_t$ at sites $t + 1$, $t + 2, \ldots, 4$. By the principle of optimality, the cost of doing this will be $f_{t+1}(o_t - x_t, g_t - y)$. Since $0 \le x_t \le o_t$ and $0 \le y_t \le g_t$ must hold, we obtain the following recursion:

$$f_t(o_t, g_t) = \min \{c_t(o_t, g_t) + f_{t+1}(o_t - x_t, g_t - y_t)\} \tag{17}$$

where $0 \le x_t \le o_t$ and $0 \le y_t \le g_t$. As usual, we work backward until $f_1(5{,}000, 10{,}000)$ has been determined. Then Sunco chooses x_1 and y_1 to attain the minimum in the (17) equation for f_1 (5,000, 10,000). Then Sunco should choose x_2 and y_2 that attain the minimum in the (17) equation for $f_2(5{,}000 - x_1, 10{,}000 - y_1)$. Sunco continues in this fashion until optimal values of x_4 and y_4 are determined.

EXAMPLE 12   Traveling Salesperson

The traveling salesperson problem (see Section 9.6) can be solved by using dynamic programming. As an example, we solve the following traveling salesperson problem: It's the last weekend of the 2004 election campaign, and candidate Walter Glenn is in New York City. Before election day, Walter must visit Miami, Dallas, and Chicago and then return to his New York City headquarters. Walter wants to minimize the total distance he must travel. In what order should he visit the cities? The distances in miles between the four cities are given in Table 15.

Solution We know that Walter must visit each city exactly once, the last city he visits must be New York, and his tour originates in New York. When Walter has only one city left to visit, his problem is trivial: simply go from his current location to New York. Then we can work backward to a problem in which he is in some city and has only two cities left to visit, and finally we can find the shortest tour that originates in New York and has four cities left to visit. We therefore let the stage be indexed by the number of cities that Walter has already visited. At any stage, to determine which city should next be visited, we need to know two things: Walter's current location and the cities he has already visited. The state at any stage consists of the last city visited and the set of cities that have already been visited. We define $f_t(i, S)$ to be the minimum distance that must be traveled to complete a tour if the $t - 1$ cities in the set S have been visited and city i was the last city visited. We let c_{ij} be the distance between cities i and j.

Stage 4 Computations

We note that, at stage 4, it must be the case that $S = \{2, 3, 4\}$ (why?), and the only possible states are $(2, \{2, 3, 4\})$, $(3, \{2, 3, 4\})$, and $(4, \{2, 3, 4\})$. In stage 4, we must go from the current location to New York. This observation yields

TABLE 15
Distances for a Traveling Salesperson

	City			
	New York	Miami	Dallas	Chicago
1 New York	—	1,334	1,559	809
2 Miami	1,334	—	1,343	1,397
3 Dallas	1,559	1,343	—	921
4 Chicago	809	1,397	921	—

$$f_4(2, \{2, 3, 4\}) = c_{21} = 1,334^* \qquad \text{(Go from city 2 to city 1)}$$
$$f_4(3, \{2, 3, 4\}) = c_{31} = 1,559^* \qquad \text{(Go from city 3 to city 1)}$$
$$f_4(4, \{2, 3, 4\}) = c_{41} = 809^* \qquad \text{(Go from city 4 to city 1)}$$

Stage 3 Computations

Working backward to stage 3, we write

$$f_3(i, S) = \min_{\substack{j \notin S \\ \text{and } j \neq 1}} \{c_{ij} + f_4[j, S \cup \{j\}]\} \tag{18}$$

This result follows, because if Walter is now at city i and he travels to city j, he travels a distance c_{ij}. Then he is at stage 4, has last visited city j, and has visited the cities in $S \cup \{j\}$. Hence, the length of the rest of his tour must be $f_4(j, S \cup \{j\})$. To use (18), note that at stage 3, Walter must have visited $\{2, 3\}$, $\{2, 4\}$, or $\{3, 4\}$ and must next visit the nonmember of S that is not equal to 1. We can use (18) to determine $f_3(\cdot)$ for all possible states:

$$f_3(2, \{2, 3\}) = c_{24} + f_4(4, \{2, 3, 4\}) = 1,397 + 809 = 2,206^* \qquad \text{(Go from 2 to 4)}$$
$$f_3(3, \{2, 3\}) = c_{34} + f_4(4, \{2, 3, 4\}) = 921 + 809 = 1,730^* \qquad \text{(Go from 3 to 4)}$$
$$f_3(2, \{2, 4\}) = c_{23} + f_4(3, \{2, 3, 4\}) = 1,343 + 1,559 = 2,902^* \qquad \text{(Go from 2 to 3)}$$
$$f_3(4, \{2, 4\}) = c_{43} + f_4(3, \{2, 3, 4\}) = 921 + 1,559 = 2,480^* \qquad \text{(Go from 4 to 3)}$$
$$f_3(3, \{3, 4\}) = c_{32} + f_4(2, \{2, 3, 4\}) = 1,343 + 1,334 = 2,677^* \qquad \text{(Go from 3 to 2)}$$
$$f_3(4, \{3, 4\}) = c_{42} + f_4(2, \{2, 3, 4\}) = 1,397 + 1,334 = 2,731^* \qquad \text{(Go from 4 to 2)}$$

In general, we write, for $t = 1, 2, 3$,

$$f_t(i, S) = \min_{\substack{j \notin S \\ \text{and } j \neq 1}} \{c_{ij} + f_{t+1}[j, S \cup \{j\}]\} \tag{19}$$

This result follows, because if Walter is at present in city i and he next visits city j, then he travels a distance c_{ij}. The remainder of his tour will originate from city j, and he will have visited the cities in $S \cup \{j\}$. Hence, the length of the remainder of his tour must be $f_{t+1}(j, S \cup \{j\})$. Equation (19) now follows.

Stage 2 Computations

At stage 2, Walter has visited only one city, so the only possible states are $(2, \{2\})$, $(3, \{3\})$, and $(4, \{4\})$. Applying (19), we obtain

$$f_2(2, \{2\}) = \min \begin{cases} c_{23} + f_3(3, \{2, 3\}) = 1,343 + 1,730 = 3,073^* \\ \text{(Go from 2 to 3)} \\ c_{24} + f_3(4, \{2, 4\}) = 1,397 + 2,480 = 3,877 \\ \text{(Go from 2 to 4)} \end{cases}$$

$$f_2(3, \{3\}) = \min \begin{cases} c_{34} + f_3(4, \{3, 4\}) = 921 + 2,731 = 3,652 \\ \text{(Go from 3 to 4)} \\ c_{32} + f_3(2, \{2, 3\}) = 1,343 + 2,206 = 3,549^* \\ \text{(Go from 3 to 2)} \end{cases}$$

$$f_2(4, \{4\}) = \min \begin{cases} c_{42} + f_3(2, \{2, 4\}) = 1,397 + 2,902 = 4,299 \\ \text{(Go from 4 to 2)} \\ c_{43} + f_3(3, \{3, 4\}) = 921 + 2,677 = 3,598^* \\ \text{(Go from 4 to 3)} \end{cases}$$

Stage 1 Computations

Finally, we are back to stage 1 (where no cities have been visited). Since Walter is currently in New York and has visited no cities, the stage 1 state must be $f_1(1, \{\cdot\})$. Applying (19),

$$f_1(1, \{\cdot\}) = \min \begin{cases} c_{12} + f_2(2, \{2\}) = 1{,}334 + 3{,}073 = 4{,}407* \\ \text{(Go from 1 to 2)} \\ c_{13} + f_2(3, \{3\}) = 1{,}559 + 3{,}549 = 5{,}108 \\ \text{(Go from 1 to 3)} \\ c_{14} + f_2(4, \{4\}) = 809 + 3{,}598 = 4{,}407* \\ \text{(Go from 1 to 4)} \end{cases}$$

So from city 1 (New York), Walter may go to city 2 (Miami) or city 4 (Chicago). We arbitrarily have him choose to go to city 4. Then he must choose to visit the city that attains $f_2(4, \{4\})$, which requires that he next visit city 3 (Dallas). Then he must visit the city attaining $f_3(3, \{3, 4\})$, which requires that he next visit city 2 (Miami). Then Walter must visit the city attaining $f_4(2, \{2, 3, 4\})$, which means, of course, that he must next visit city 1 (New York). The optimal tour (1–4–3–2–1, or New York–Chicago–Dallas–Miami–New York) is now complete. The length of this tour is $f_1(1, \{\cdot\}) = 4{,}407$. As a check, note that

New York to Chicago distance = 809 miles

Chicago to Dallas distance = 921 miles

Dallas to Miami distance = 1,343 miles

Miami to New York distance = 1,334 miles

so the total distance that Walter travels is $809 + 921 + 1{,}343 + 1{,}334 = 4{,}407$ miles. Of course, if we had first sent him to city 2, we would have obtained another optimal tour (1–2–3–4–1) that would simply be a reversal of the original optimal tour.

Computational Difficulties in Using Dynamic Programming

For traveling salesperson problems that are large, the state space becomes very large, and the branch-and-bound approach outlined in Chapter 9 (along with other branch-and-bound approaches) is much more efficient than the dynamic programming approach outlined here. For example, for a 30-city problem, suppose we are at stage 16 (this means that 15 cities have been visited). Then it can be shown that there are more than 1 billion possible states. This brings up a problem that limits the practical application of dynamic programming. In many problems, *the state space becomes so large that excessive computational time is required to solve the problem by dynamic programming.* For instance, in Example 8, suppose that $T = 20$. It is possible that if no bass were caught during the first 20 years, then the lake might contain $10{,}000(1.2)^{20} = 383{,}376$ bass at the beginning of year 21. If we view this example as a network in which we need to find the longest route from the node (1, 10,000) (representing year 1 and 10,000 bass in the lake) to some stage 21 node, then stage 21 would have 383,377 nodes. Even a powerful computer would have difficulty solving this problem. Techniques to make problems with large state spaces computationally tractable are discussed in Bersetkas (1987) and Denardo (1982).

Nonadditive Recursions

The last two examples in this section differ from the previous ones in that the recursion does not represent $f_t(i)$ as the sum of the cost (or reward) incurred during the current period and future costs (or rewards) incurred during future periods.

EXAMPLE 13   **Minimax Shortest Route**

Joe Cougar needs to drive from city 1 to city 10. He is no longer interested in minimizing the length of his trip, but he is interested in minimizing the maximum altitude above sea level that he will encounter during his drive. To get from city 1 to city 10, he must follow a path in Figure 10. The length c_{ij} of the arc connecting city i and city j represents the maximum altitude (in thousands of feet above sea level) encountered when driving from city i to city j. Use dynamic programming to determine how Joe should proceed from city 1 to city 10.

Solution   To solve this problem by dynamic programming, note that for a trip that begins in city i and goes through stages $t, t + 1, \ldots, 5$, the maximum altitude that Joe encounters will be the maximum of the following two quantities: (1) the maximum altitude encountered on stages $t + 1, t + 2, \ldots, 5$ or (2) the altitude encountered when traversing the arc that begins in stage t. Of course, if we are in a stage 4 state, quantity 1 does not exist.

After defining $f_t(i)$ as the smallest maximum altitude that Joe can encounter in a trip from city i in stage t to city 10, this reasoning leads us to the following recursion:

$$f_4(i) = c_{i,10} \tag{20}$$
$$f_t(i) = \min_j \{\max[c_{ij}, f_{t+1}(j)]\} \qquad (t = 1, 2, 3)$$

where j may be any city such that there is an arc connecting city i and city j.

We first compute $f_4(7)$, $f_4(8)$, and $f_4(9)$ and then use (20) to work backward until $f_1(1)$ has been computed. We obtain the following results:

$$f_4(7) = 13* \qquad\qquad\qquad\qquad \text{(Go from 7 to 10)}$$
$$f_4(8) = 8* \qquad\qquad\qquad\qquad \text{(Go from 8 to 10)}$$
$$f_4(9) = 9* \qquad\qquad\qquad\qquad \text{(Go from 9 to 10)}$$

$$f_3(5) = \min \begin{cases} \max\,[c_{57}, f_4(7)] = 13 & \text{(Go from 5 to 7)} \\ \max\,[c_{58}, f_4(8)] = 8* & \text{(Go from 5 to 8)} \\ \max\,[c_{59}, f_4(9)] = 10 & \text{(Go from 5 to 9)} \end{cases}$$

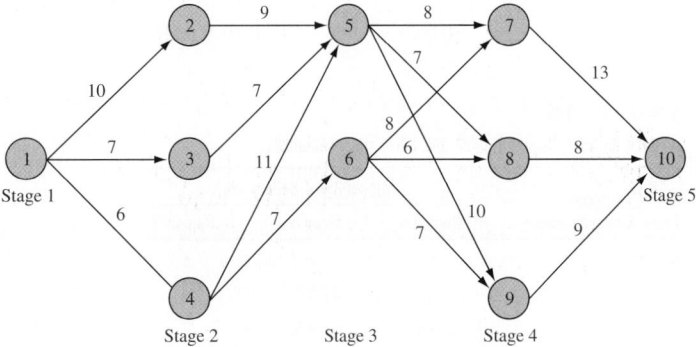

FIGURE 10
Joe's Trip
(Altitudes Given)

$$f_3(6) = \min \begin{cases} \max \ [c_{67}, f_4(7)] = 13 & \text{(Go from 6 to 7)} \\ \max \ [c_{68}, f_4(8)] = 8^* & \text{(Go from 6 to 8)} \\ \max \ [c_{69}, f_4(9)] = 9 & \text{(Go from 6 to 9)} \end{cases}$$

$$f_2(2) = \max \ [c_{25}, f_3(5)] = 9^* \qquad \text{(Go from 2 to 5)}$$

$$f_2(3) = \max \ [c_{35}, f_3(5)] = 8^* \qquad \text{(Go from 3 to 5)}$$

$$f_2(4) = \min \begin{cases} \max \ [c_{45}, f_3(5)] = 11 & \text{(Go from 4 to 5)} \\ \max \ [c_{46}, f_3(6)] = 8^* & \text{(Go from 4 to 6)} \end{cases}$$

$$f_1(1) = \min \begin{cases} \max \ [c_{12}, f_2(2)] = 10 & \text{(Go from 1 to 2)} \\ \max \ [c_{13}, f_2(3)] = 8^* & \text{(Go from 1 to 3)} \\ \max \ [c_{14}, f_2(4)] = 8^* & \text{(Go from 1 to 4)} \end{cases}$$

To determine the optimal strategy, note that Joe can begin by going from city 1 to city 3 or from city 1 to city 4. Suppose Joe begins by traveling to city 3. Then he should choose the arc attaining $f_2(3)$, which means he should next travel to city 5. Then Joe must choose the arc that attains $f_3(5)$, driving next to city 8. Then, of course, he must drive to city 10. Thus, the path 1–3–5–8–10 is optimal, and Joe will encounter a maximum altitude equal to $f_1(1) = 8,000$ ft. The reader should verify that the path 1–4–6–8–10 is also optimal.

EXAMPLE 14   Sales Allocation

Glueco is planning to introduce a new product in three different regions. Current estimates are that the product will sell well in each region with respective probabilities .6, .5, and .3. The firm has available two top sales representatives that it can send to any of the three regions. The estimated probabilities that the product will sell well in each region when 0, 1, or 2 additional sales reps are sent to a region are given in Table 16. If Glueco wants to maximize the probability that its new product will sell well in all three regions, then where should it assign sales representatives? You may assume that sales in the three regions are independent.

Solution   If Glueco had just one region to worry about and wanted to maximize the probability that the new product would sell in that region, then the proper strategy would be clear: Assign both sales reps to the region. We could then work backward and solve a problem in which Glueco's goal is to maximize the probability that the product will sell in two regions. Finally, we could work backward and solve a problem with three regions. We define $f_t(s)$ as the probability that the new product will sell in regions $t, t + 1, \ldots, 3$ if s sales reps are optimally assigned to these regions. Then

$$f_3(2) = .7 \qquad \text{(Assign 2 sales reps to region 3)}$$
$$f_3(1) = .55 \qquad \text{(Assign 1 sales rep to region 3)}$$
$$f_3(0) = .3 \qquad \text{(Assign 0 sales reps to region 3)}$$

TABLE 16
Relation between Regional Sales and Sales Representatives

No. of Additional Sales Representatives	Probability of Selling Well		
	Region 1	Region 2	Region 3
0	.6	.5	.3
1	.8	.7	.55
2	.85	.85	.7

Also, $f_1(2)$ will be the maximum probability that the product will sell well in all three regions. To develop a recursion for $f_2(\cdot)$ and $f_1(\cdot)$, we define p_{tx} to be the probability that the new product sells well in region t if x sales reps are assigned to region t. For example, $p_{21} = .7$. For $t = 1$ and $t = 2$, we then write

$$f_t(s) = \max_x \{p_{tx} f_{t+1}(s - x)\} \tag{21}$$

where x must be a member of $\{0, 1, \ldots, s\}$. To justify (21), observe that if s sales reps are available for regions $t, t + 1, \ldots, 3$ and x sales reps are assigned to region t, then

$$p_{tx} = \text{probability that product sells in region } t$$
$$f_{t+1}(s - x) = \text{probability that product sells well in regions } t + 1, \ldots, 3$$

Note that the sales in each region are independent. This implies that if x sales reps are assigned to region t, then the probability that the new product sells well in regions t, $t + 1, \ldots, 3$ is $p_{tx} f_{t+1}(s - x)$. We want to maximize this probability, so we obtain (21). Applying (21) yields the following results:

$$f_2(2) = \max \begin{cases} (.5)f_3(2 - 0) = .35 \\ (\text{Assign 0 sales reps to region 2}) \\ (.7)f_3(2 - 1) = .385^* \\ (\text{Assign 1 sales rep to region 2}) \\ (.85)f_3(2 - 2) = .255 \\ (\text{Assign 2 sales reps to region 2}) \end{cases}$$

Thus, $f_2(2) = .385$, and 1 sales rep should be assigned to region 2.

$$f_2(1) = \max \begin{cases} (.5)f_3(1 - 0) = .275^* \\ (\text{Assign 0 sales reps to region 2}) \\ (.7)f_3(1 - 1) = .21 \\ (\text{Assign 1 sales rep to region 2}) \end{cases}$$

Thus, $f_2(1) = .275$, and no sales reps should be assigned to region 2.

$$f_2(0) = (.5)f_3(0 - 0) = .15^*$$
$$(\text{Assign 0 sales reps to region 2})$$

Finally, we are back to the original problem, which is to find $f_1(2)$. Equation (21) yields

$$f_1(2) = \max \begin{cases} (.6)f_2(2 - 0) = .231^* \\ (\text{Assign 0 sales reps to region 1}) \\ (.8)f_2(2 - 1) = .220 \\ (\text{Assign 1 sales rep to region 1}) \\ (.85)f_2(2 - 2) = .1275 \\ (\text{Assign 2 sales reps to region 1}) \end{cases}$$

Thus, $f_1(2) = .231$, and no sales reps should be assigned to region 1. Then Glueco needs to attain $f_2(2 - 0)$, which requires that 1 sales rep be assigned to region 2. Glueco must next attain $f_3(2 - 1)$, which requires that 1 sales rep be assigned to region 3. In summary, Glueco can obtain a .231 probability of the new product selling well in all three regions by assigning 1 sales rep to region 2 and 1 sales rep to region 3.

PROBLEMS

Group A

1 At the beginning of year 1, Sunco Oil owns i_0 barrels of oil reserves. During year $t(t = 1, 2, \ldots, 10)$, the following events occur in the order listed: (1) Sunco extracts and refines x barrels of oil reserves and incurs a cost $c(x)$: (2) Sunco sells year t's extracted and refined oil at a price of p_t dollars per barrel; and (3) exploration for new reserves results in a discovery of b_t barrels of new reserves.

Sunco wants to maximize sales revenues less costs over the next 10 years. Formulate a dynamic programming recursion that will help Sunco accomplish its goal. If Sunco felt that cash flows in later years should be discounted, how should the formulation be modified?

2 At the beginning of year 1, Julie Ripe has D dollars (this includes year 1 income). During each year, Julie earns i dollars and must determine how much money she should consume and how much she should invest in Treasury bills. During a year in which Julie consumes d dollars, she earns a utility of $\ln d$. Each dollar invested in Treasury bills yields $1.10 in cash at the beginning of the next year. Julie's goal is to maximize the total utility she earns during the next 10 years.

a Why might $\ln d$ be a better indicator of Julie's utility than a function such as d^2?

b Formulate a dynamic programming recursion that will enable Julie to maximize the total utility she receives during the next 10 years. Assume that year t revenue is received at the beginning of year t.

3 Assume that during minute t (the current minute is minute 1), the following sequence of events occurs: (1) At the beginning of the minute, x_t customers arrive at the cash register; (2) the store manager decides how many cash registers should be operated during the current minute; (3) if s cash registers are operated and i customers are present (including the current minute's arrivals), $c(s, i)$ customers complete service; and (4) the next minute begins.

A cost of 10¢ is assessed for each minute a customer spends waiting to check out (this time includes checkout time). Assume that it costs $c(s)$ cents to operate s cash registers for 1 minute. Formulate a dynamic programming recursion that minimizes the sum of holding and service costs during the next 60 minutes. Assume that before the first minute's arrivals, no customers are present and that holding cost is assessed at the end of each minute.

4 Develop a dynamic programming formulation of the CSL Computer problem of Section 3.12.

5 To graduate from State University, Angie Warner needs to pass at least one of the three subjects she is taking this semester. She is now enrolled in French, German, and statistics. Angie's busy schedule of extracurricular activities allows her to spend only 4 hours per week on studying. Angie's probability of passing each course depends on the number of hours she spends studying for the course (see Table 17). Use dynamic programming to determine how many hours per week Angie should spend studying each subject. (*Hint:* Explain why maximizing the probability of

TABLE 17

Hours of Study per Week	Probability of Passing Course		
	French	German	Statistics
0	.20	.25	.10
1	.30	.30	.30
2	.35	.33	.40
3	.38	.35	.44
4	.40	.38	.50

TABLE 18

Component	No. of Actors Assigned to Component			
	0	1	2	3
Warp drive	.30	.55	.65	.95
Solar relay	.40	.50	.70	.90
Candy maker	.45	.55	.80	.98

passing at least one course is equivalent to minimizing the probability of failing all three courses.)

6 E.T. is about to fly home. For the trip to be successful, the ship's solar relay, warp drive, and candy maker must all function properly. E.T. has found three unemployed actors who are willing to help get the ship ready for takeoff. Table 18 gives, as a function of the number of actors assigned to repair each component, the probability that each component will function properly during the trip home. Use dynamic programming to help E.T. maximize the probability of having a successful trip home.

7 Farmer Jones is trying to raise a prize steer for the Bloomington 4-H show. The steer now weighs w_0 pounds. Each week, farmer Jones must determine how much food to feed the steer. If the steer weighs w pounds at the beginning of a week and is fed p pounds of food during a week, then at the beginning of the next week, the steer will weigh $g(w, p)$ pounds. It costs farmer Jones $c(p)$ dollars to feed the steer p pounds of food during a week. At the end of the 10th week (or equivalently, the beginning of the 11th week), the steer may be sold for $10/lb. Formulate a dynamic programming recursion that can be used to determine how farmer Jones can maximize profit from the steer.

Group B

8 MacBurger has just opened a fast-food restaurant in Bloomington. Currently, i_0 customers frequent MacBurger (we call these loyal customers), and $N - i_0$ customers frequent other fast-food establishments (we call these nonloyal customers). At the beginning of each month, MacBurger must decide how much money to spend on advertising. At the end of a month in which MacBurger spends d dollars on advertising, a fraction $p(d)$ of the loyal customers become nonloyal customers, and a fraction $q(d)$ of the nonloyal customers become loyal customers. During the next 12 months, MacBurger wants to spend D dollars on advertising. Develop a dynamic programming recursion that will enable MacBurger to maximize the number of loyal customers the company will have at the end of month 12. (Ignore the possibility of a fractional number of loyal customers.)

9 Public Service Indiana (PSI) is considering five possible locations to build power plants during the next 20 years. It will cost c_i dollars to build a plant at site i and h_i dollars to operate a site i plant for a year. A plant at site i can supply k_i kilowatt-hours (kwh) of generating capacity. During year t, d_t kwh of generating capacity are required. Suppose that at most one plant can be built during a year, and if it is decided to build a plant at site i during year t, then the site i plant can be used to meet the year t (and later) generating requirements. Initially, PSI has 500,000 kwh of generating capacity available. Formulate a recursion that PSI could use to minimize the sum of building and operating costs during the next 20 years.

10 During month t, a firm faces a demand for d_t units of a product. The firm's production cost during month t consists of two components. First, for each unit produced during month t, the firm incurs a variable production cost of c_t. Second, if the firm's production level during month $t - 1$ is x_{t-1} and the firm's production level during month t is x_t, then during month t, a smoothing cost of $5|x_t - x_{t-1}|$ will be incurred (see Section 16.12 for an explanation of smoothing costs). At the end of each month, a holding cost of h_t per unit is incurred. Formulate a recursion that will enable the firm to meet (on time) its demands over the next 12 months. Assume that at the beginning of the first month, 20 units are in inventory and that last month's production was 20 units. (*Hint:* The state during each month must consist of two quantities.)

11 The state of Transylvania consists of three cities with the following populations: city 1, 1.2 million people; city 2, 1.4 million people; city 3, 400,000 people. The Transylvania House of Representatives consists of three representatives. Given proportional representation, city 1 should have $d_1 = (\frac{1.2}{3}) = 1.2$ representatives; city 2 should have $d_2 = 1.4$ representatives; and city 3 should have $d_3 = 0.40$ representative. Each city must receive an integral number of representatives, so this is impossible. Transylvania has therefore decided to allocate x_i representatives to city i, where the allocation x_1, x_2, x_3 minimizes the maximum discrepancy between the desired and actual number of representatives received by a city. In short, Transylvania must determine x_1, x_2, and x_3 to minimize the largest of the following three numbers: $|x_1 - d_1|, |x_2 - d_2|, |x_3 - d_3|$. Use dynamic programming to solve Transylvania's problem.

12 A job shop has four jobs that must be processed on a single machine. The due date and processing time for each job are given in Table 19. Use dynamic programming to determine the order in which the jobs should be done so as to minimize the total lateness of the jobs. (The lateness of a job is simply how long after the job's due date the job is completed; for example, if the jobs are processed in the given order, then job 3 will be 2 days late, job 4 will be 4 days late, and jobs 1 and 2 will not be late.)

TABLE 19

Job	Processing Time (Days)	Due Date (Days from Now)
1	2	4
2	4	14
3	6	10
4	8	16

18.7 The Wagner–Whitin Algorithm and the Silver–Meal Heuristic[†]

The inventory example of Section 18.3 is a special case of the *dynamic lot-size model.*

Description of Dynamic Lot-Size Model

1 Demand d_t during period $t(t = 1, 2, \ldots, T)$ is known at the beginning of period 1.

2 Demand for period t must be met on time from inventory or from period t production. The cost $c(x)$ of producing x units during any period is given by $c(0) = 0$, and for $x > 0$, $c(x) = K + cx$, where K is a fixed cost for setting up production during a period, and c is the variable per-unit cost of production.

3 At the end of period t, the inventory level i_t is observed, and a holding cost hi_t is incurred. We let i_0 denote the inventory level before period 1 production occurs.

4 The goal is to determine a production level x_t for each period t that minimizes the total cost of meeting (on time) the demands for periods $1, 2, \ldots, T$.

[†]This section covers topics that may be omitted with no loss of continuity.

5 There is a limit c_t placed on period t's ending inventory.

6 There is a limit r_t placed on period t's production.

In this section, we consider these first four points. We let $x_t = $ period t production. Period t production can be used to meet period t demand.

EXAMPLE 15 **Dynamic Lot-Size Model**

We now determine an optimal production schedule for a five-period dynamic lot-size model with $K = \$250$, $c = \$2$, $h = \$1$, $d_1 = 220$, $d_2 = 280$, $d_3 = 360$, $d_4 = 140$, and $d_5 = 270$. We assume that the initial inventory level is zero. The solution to this example is given later in this section.

Discussion of the Wagner–Whitin Algorithm

If the dynamic programming approach outlined in Section 18.3 were used to find an optimal production policy for Example 15, we would have to consider the possibility of producing any amount between 0 and $d_1 + d_2 + d_3 + d_4 + d_5 = 1,270$ units during period 1. Thus, it would be possible for the period 2 state (period 2's entering inventory) to be $0, 1, \ldots, 1,270 - d_1 = 1,050$, and we would have to determine $f_2(0), f_2(1), \ldots, f_2(1,050)$. Using the dynamic programming approach of Section 18.3 to find an optimal production schedule for Example 15 would therefore require a great deal of computational effort. Fortunately, however, Wagner and Whitin (1958) have developed a method that greatly simplifies the computation of optimal production schedules for dynamic lot-size models. Lemmas 1 and 2 are necessary for the development of the Wagner–Whitin algorithm.

LEMMA 1

Suppose it is optimal to produce a positive quantity during a period t. Then for some $j = 0, 1, \ldots, T - t$, the amount produced during period t must be such that after period t's production, a quantity $d_t + d_{t+1} + \cdots + d_{t+j}$ will be in stock. In other words, if production occurs during period t, we must (for some j) produce an amount that exactly suffices to meet the demands for periods $t, t + 1, \ldots, t + j$.

Proof If the lemma is false, then for some t, some $j = 0, 1, \ldots, T - t - 1$, and some x satisfying $0 < x < d_{t+j+1}$, period t production must bring the stock level to $d_t + d_{t+1} + \cdots + d_{t+j} + x$, and at the beginning of period $t + j + 1$, our inventory level would be $x < d_{t+j+1}$. Thus, production must occur during period $t + j + 1$. By deferring production of x units from period t to period $t + j + 1$ (with all other production levels unchanged), we save $h(j + 1)x$ in holding costs while incurring no additional setup costs (because production is already occurring during period $t + j + 1$). Thus, it cannot have been optimal to bring our period t stock level to $d_t + d_{t+1} + \cdots + d_{t+j} + x$. This contradiction proves the lemma.

LEMMA 2

If it is optimal to produce anything during period t, then $i_{t-1} < d_t$. In other words, production cannot occur during period t unless there is insufficient stock to meet period t demand.

Proof If the lemma is false, there must be an optimal policy that (for some t) has $x_t > 0$ and $i_{t-1} \geq d_t$. If this is the case, then by deferring the period t production of x_t units to period $t + 1$, we save hx_t in holding costs and possibly K (if the optimal policy produces during period $t + 1$) in setup costs. Thus, any production schedule having $x_t > 0$ and $i_{t-1} \geq d_t$ cannot be optimal.

Lemma 2 shows that no production will occur until the first period t for which $i_{t-1} < d_t$, so production must occur during period t (or else period t's demand would not be met on time). Lemma 1 now implies that for some $j = 0, 1, \ldots, T - t$, period t production will be such that after period t's production, on-hand stock will equal $d_t + d_{t+1} + \cdots + d_{t+j}$. Then Lemma 2 implies that no production can occur until period $t + j + 1$. Since the entering inventory level for period $t + j + 1$ will equal zero, production must occur during period $t + j + 1$. During period $t + j + 1$, Lemma 1 implies that period $t + j + 1$ production will (for some k) equal $d_{t+j+1} + d_{t+j+2} + \cdots + d_{t+j+k}$ units. Then period $t + j + k + 1$ will begin with zero inventory, and production again occurs, and so on. *With the possible exception of the first period, production will occur only during periods in which beginning inventory is zero, and during each period in which beginning inventory is zero (and $d_t \neq 0$), production must occur.*

Using this insight, Wagner and Whitin developed a recursion that can be used to determine an optimal production policy. We assume that the initial inventory level is zero. (See Problem 1 at the end of this section if this is not the case.) Define f_t as the minimum cost incurred during periods $t, t + 1, \ldots, T$, given that at the beginning of period t, the inventory level is zero. Then $f_1, f_2, \ldots, f_T$ must satisfy

$$f_t = \min_{j = 0, 1, 2, \ldots, T-t} (c_{tj} + f_{t+j+1}) \tag{22}$$

where $f_{T+1} = 0$ and c_{tj} is the total cost incurred during periods $t, t + 1, \ldots, t + j$ if production during period t is exactly sufficient to meet demands for periods $t, t + 1, \ldots, t + j$. Thus,

$$c_{tj} = K + c(d_t + d_{t+1} + \cdots + d_{t+j}) + h[jd_{t+j} + (j - 1)d_{t+j-1} + \cdots + d_{t+1}]$$

where K is the setup cost incurred during period t, $c(d_t + d_{t+1} + \cdots + d_{t+j})$ is the variable production cost incurred during period t, and $h[jd_{t+j} + (j - 1)d_{t+j-1} + \cdots + d_{t+1}]$ is the holding cost incurred during periods $t, t + 1, \ldots, t + j$. For example, an amount d_{t+j} of period t production will be held in inventory for j periods (during periods $t, t + 1, \ldots, t + j - 1$), thereby incurring a holding cost of hjd_{t+j}.

To find an optimal production schedule by the Wagner–Whitin algorithm, begin by using (22) to find f_T. Then use (22) to compute $f_{T-1}, f_{T-2}, \ldots, f_1$. Once f_1 has been determined, an optimal production schedule may be easily obtained.

EXAMPLE 15   **Dynamic Lot-Size Model (continued)**

Solution To illustrate the Wagner–Whitin algorithm, we find an optimal production schedule for Example 15. The computations follow.

$$f_6 = 0$$
$$f_5 = 250 + 2(270) + f_6 = 790^* \qquad \text{(Produce for period 5)}$$

If we begin period 5 with zero inventory, we should produce enough during period 5 to meet period 5 demand.

$$f_4 = \min \begin{cases} 250 + 2(140) + f_5 = 1,320^* \\ \text{(Produce for period 4)} \\ 250 + 2(140 + 270) + 270 + f_6 = 1,340 \\ \text{(Produce for periods 4, 5)} \end{cases}$$

If we begin period 4 with zero inventory, we should produce enough during period 4 to meet the demand for period 4.

$$f_3 = \min \begin{cases} 250 + 2(360) + f_4 = 2,290 \\ \text{(Produce for period 3)} \\ 250 + 2(360 + 140) + 140 + f_5 = 2,180^* \\ \text{(Produce for periods 3, 4)} \\ 250 + 2(360 + 140 + 270) + 140 + 2(270) + f_6 = 2,470 \\ \text{(Produce for periods 3, 4, 5)} \end{cases}$$

If we begin period 3 with zero inventory, we should produce enough during period 3 to meet the demand for periods 3 and 4.

$$f_2 = \min \begin{cases} 250 + 2(280) + f_3 = 2,990^* \\ \text{(Produce for period 2)} \\ 250 + 2(280 + 360) + 360 + f_4 = 3,210 \\ \text{(Produce for periods 2, 3)} \\ 250 + 2(280 + 360 + 140) + 360 + 2(140) + f_5 = 3,240 \\ \text{(Produce for periods 2, 3, 4)} \\ 250 + 2(280 + 360 + 140 + 270) + 360 + 2(140) + 3(270) + f_6 = 3,800 \\ \text{(Produce for periods 2, 3, 4, 5)} \end{cases}$$

If we begin period 2 with zero inventory, we should produce enough during period 2 to meet the demand for period 2.

$$f_1 = \min \begin{cases} 250 + 2(220) + f_2 = 3,680^* \\ \text{(Produce for period 1)} \\ 250 + 2(220 + 280) + 280 + f_3 = 3,710 \\ \text{(Produce for periods 1, 2)} \\ 250 + 2(220 + 280 + 360) + 280 + 2(360) + f_4 = 4,290 \\ \text{(Produce for periods 1, 2, 3)} \\ 250 + 2(220 + 280 + 360 + 140) + 280 + 2(360) + 3(140) + f_5 = 4,460 \\ \text{(Produce for periods 1, 2, 3, 4)} \\ 250 + 2(220 + 280 + 360 + 140 + 270) + 280 \\ \qquad + 2(360) + 3(140) + 4(270) + f_6 = 5,290 \\ \text{(Produce for periods 1, 2, 3, 4, 5)} \end{cases}$$

If we begin period 1 with zero inventory, it is optimal to produce $d_1 = 220$ units during period 1; then we begin period 2 with zero inventory. Since f_2 is attained by producing period 2's demand, we should produce $d_2 = 280$ units during period 2; then we enter period 3 with zero inventory. Since f_3 is attained by meeting the demands for periods 3 and 4, we produce $d_3 + d_4 = 500$ units during period 3; then we enter period 5 with zero inventory and produce $d_5 = 270$ units during period 5. The optimal production schedule will incur at total cost of $f_1 = \$3,680$.

For Example 15, any optimal production schedule must produce exactly $d_1 + d_2 + d_3 + d_4 + d_5 = 1,270$ units, incurring variable production costs of $2(1,270) = \$2,540$. Thus, in computing the optimal production schedule, we may always ignore the variable production costs. This substantially simplifies the calculations.

The Silver–Meal Heuristic

The Silver–Meal (S–M) heuristic involves less work than the Wagner–Whitin algorithm and can be used to find a near-optimal production schedule. The S–M heuristic is based on the fact that our goal is to minimize average cost per period (for the reasons stated, variable production costs may be ignored). Suppose we are at the beginning of period 1 and are trying to determine how many periods of demand should be satisfied by period 1's production. During period 1, if we produce an amount sufficient to meet demand for the next t periods, then a cost of $TC(t) = K + HC(t)$ will be incurred (ignoring variable production costs). Here, $HC(t)$ is the holding cost incurred during the next t periods (including the current period) if production during the current period is sufficient to meet demand for the next t periods.

Let $AC(t) = \frac{TC(t)}{t}$ be the average per-period cost incurred during the next t periods. Since $\frac{1}{t}$ is a decreasing convex function of t, as t increases, $\frac{K}{t}$ decreases at a decreasing rate. In most cases, $\frac{HC(t)}{t}$ tends to be an increasing function of t (see Problem 4 at the end of this section). Thus, in most situations, an integer t^* can be found such that for $t < t^*$, $AC(t + 1) \le AC(t)$ and $AC(t^* + 1) \ge AC(t^*)$. The S–M heuristic recommends that period 1's production be sufficient to meet the demands for periods $1, 2, \ldots, t^*$ (if no t^* exists, period 1 production should satisfy the demand for periods $1, 2, \ldots, T$). Since t^* is a local (and perhaps a global) minimum for $AC(t)$, it seems reasonable that producing $d_1 + d_2 + \cdots + d_{t^*}$ units during period 1 will come close to minimizing the average per-period cost incurred during periods $1, 2, \ldots, t^*$. Next we apply the S–M heuristic while considering period $t^* + 1$ as the initial period. We find that during period $t^* + 1$, the demand for the next t_1^* periods should be produced. Continue in this fashion until the demand for period T has been produced.

To illustrate, we apply the S–M heuristic to Example 15. We have

$$TC(1) = 250 \qquad\qquad AC(1) = \frac{250}{1} = 250$$

$$TC(2) = 250 + 280 = 530 \qquad AC(2) = \frac{530}{2} = 265$$

Since $AC(2) \ge AC(1)$, $t^* = 1$, and the S–M heuristic dictates that we produce $d_1 = 220$ units during period 1. Then

$$TC(1) = 250 \qquad\qquad AC(1) = \frac{250}{1} = 250$$

$$TC(2) = 250 + 360 = 610 \qquad AC(2) = \frac{610}{2} = 305$$

Since $AC(2) \ge AC(1)$, the S–M heuristic recommends producing $d_2 = 280$ units during period 2. Then

$$TC(1) = 250 \qquad\qquad AC(1) = \frac{250}{1} = 250$$

$$TC(2) = 250 + 140 = 390 \qquad AC(2) = \frac{390}{2} = 195$$

$$TC(3) = 250 + 2(270) + 140 = 930 \qquad AC(3) = \frac{930}{3} = 310$$

Since $AC(3) \geq AC(2)$, period 3 production should meet the demand for the next two periods (periods 3 and 4). During period 3, we should produce $d_3 + d_4 = 500$ units. This brings us to period 5. Period 5 is the final period, so $d_5 = 270$ units should be produced during period 5.

For Example 15 (and many other dynamic lot-size problems), the S–M heuristic yields an optimal production schedule. In extensive testing, the S–M heuristic usually yielded a production schedule costing less than 1% above the optimal policy obtained by the Wagner–Whitin algorithm (see Peterson and Silver (1998)).

PROBLEMS

Group A

1 For Example 15, suppose we had an inventory of 200 units. What would be the optimal production schedule? What if the initial inventory were 400 units?

2 Use the Wagner–Whitin and Silver–Meal methods to find production schedules for the following dynamic lot-size problem: $K = \$50$, $h = \$0.40$, $d_1 = 10$, $d_2 = 60$, $d_3 = 20$, $d_4 = 140$, $d_5 = 90$.

3 Use the Wagner–Whitin and Silver–Meal methods to find production schedules for the following dynamic lot-size problem: $K = \$30$, $h = \$1$, $d_1 = 40$, $d_2 = 60$, $d_3 = 10$, $d_4 = 70$, $d_5 = 20$.

Group B

4 Explain why $HC(t)/t$ tends to be an increasing function of t.

18.8 Using Excel to Solve Dynamic Programming Problems[†]

In earlier chapters, we have seen that any LP problem can be solved with LINDO or LINGO, and any NLP can be solved with LINGO. Unfortunately, no similarly user-friendly package can be used to solve dynamic programming problems. LINGO can be used to solve DP problems, but student LINGO can only handle a very small problem. Fortunately, Excel can often be used to solve DP problems. Our three illustrations solve a knapsack problem (Example 6), a resource-allocation problem (Example 5), and an inventory problem (Example 4).

Solving Knapsack Problems on a Spreadsheet

Recall the knapsack problem of Example 6. The question is how to (using three types of items) fill a 10-lb knapsack and obtain the maximum possible benefit. Recall that $g(w) =$ maximum benefit that can be obtained from a w-lb knapsack. Recall that

$$g(w) = \max_{j}\{b_j + g(w - w_j)\} \tag{8}$$

where $b_j =$ benefit from a type j item and $w_j =$ weight of a type j item.

Dpknap.xls

In each row of the spreadsheet (see Figure 11 or file Dpknap.xls) we compute $g(w)$ for various values of w. We begin by entering $g(0) = g(1) = g(2) = 0$ and $g(3) = 7$; [$g(3) = 7$ follows because a 3-lb item is the only item that will fit in a 3-lb knapsack]. The

---

[†]This section covers topics that may be omitted with no loss of continuity.

A	A	B	C	D	E	F	G
1	KNAPSACK	ITEM1	ITEM2	ITEM3	g(SIZE)		FIGURE11
2	SIZE						KNAPSACK
3	0				0		PROBLEM
4	1				0		
5	2				0		
6	3				7		
7	4	11	7	-10000	11		
8	5	11	7	12	12		
9	6	11	14	12	14		
10	7	18	18	12	18		
11	8	22	19	19	22		
12	9	23	21	23	23		
13	10	25	25	24	25		
14							
15							
16							
17							
18							
19							
20							
21							
22							
23							
24							
25							
26							
27							
28							
29							
30							

FIGURE 11
Knapsack Problem

columns labeled ITEM1, ITEM2, and ITEM3 correspond to the terms $j = 1, 2, 3$, respectively, in (8). Thus, in the ITEM1 column, we should enter a formula to compute $b_1 + g(w - w_1)$; in the ITEM2 column, we should enter a formula to compute $b_2 + g(w - w_2)$; in the ITEM3 column, we should enter a formula to compute $b_3 + g(w - w_3)$. The only exception to this occurs when a w_j-lb item will not fit in a w-lb knapsack. In this situation, we enter a very negative number (such as 10,000) to ensure that a w_j-lb item will not be considered.

More specifically, in row 7, we want to compute $g(4)$. To do this, we enter the following formulas:

B7: 11 + E3   [This is $b_1 + g(4 - w_1)$]

C7: 7 + E4   [This is $b_2 + g(4 - w_2)$]

D7: $-10,000$ (This is because a 5-lb item will not fit in a 4-lb knapsack)

In E7, we compute $g(4)$ by entering the formula =MAX(B7:D7). In row 8, we compute $g(5)$ by entering the following formulas:

B8: 11 + E4

C8: 7 + E5

D8: 12 + E3

To compute $g(5)$, we enter =MAX(B8:D8) in E8. Now comes the fun part! Simply copy the formulas from the range B8:E8 to B8:E13. Then $g(10)$ will be computed in E13. We see that $g(10) = 25$. Because both item 1 and item 2 attain $g(10)$, we may begin filling a knapsack with a Type 1 or Type 2 item. We choose to begin with a Type 1 item. This leaves us with $10 - 4 = 6$ lb to fill. From row 9 we find that $g(6) = 14$ is attained by a Type 2 item. This leaves us with $6 - 3 = 3$ lb to fill. We also use a Type 2 item to attain $g(3) = 7$. This leaves us with 0 lb. Thus, we conclude that we can obtain 25 units of benefit by filling a 10-lb knapsack with two Type 2 items and one Type 1 item.

By the way, if we had been interested in filling a 100-lb knapsack, we would have copied the formulas from B8:E8 to B8:E103.

Solving a General Resource-Allocation Problem on a Spreadsheet

Solving a nonknapsack resource-allocation problem on a spreadsheet is more difficult. To illustrate, consider Example 5 in which we have $6,000 to allocate between three investments. Define $f_t(d)$ = maximum NPV obtained from investments $t, \ldots, 3$ given that d (in thousands) dollars are available for investments $t, \ldots, 3$. Then we may write

$$f_t(d) = \max_{0 \leq x \leq d} \{r_t(x) + f_{t+1}(d - x)\} \qquad (10)$$

where $f_4(d) = 0(d = 0, 1, 2, 3, 4, 5, 6)$, $r_t(x)$ = NPV obtained if x (in thousands) dollars are invested in investment t, and the maximization in (10) is only taken over integral values for d. Our subsequent discussion will be simplified if we define $J_t(d, x) = r_t(x) + f_{t+1}(d - x)$ and rewrite (10) as

$$f_t(d) = \max_{0 \leq x \leq d} \{J_t(d, x)\} \qquad (10')$$

Dpresour.xls

We begin the construction of the spreadsheet (Figure 12 and file Dpresour.xls) by entering the $r_t(x)$ in A1:H4. For example, $r_2(3) = 16$ is entered in E3. In rows 18–20, we have set up the computations to compute the $J_t(d, x)$. These computations require using the Excel =**HLOOKUP** command to look up the values of $r_t(x)$ (in rows 2–4) and $f_{t+1}(d - x)$ (in rows 11–14). For example, to compute $J_3(3, 1)$, we enter the following formula in I18:

=HLOOKUP(I$17,$B$1:$H$4,$A18+1)

+ HLOOKUP(I$16-I$17,$B$10:$H$14,$A18+1)

The portion =HLOOKUP(I$17,$B$1:$H$4,$A18+1) of the formula in cell I18 finds the column in B1:H4 whose first entry matches I17. Then we pick off the entry in row A18 + 1 of that column. This returns $r_3(1) = 9$. Note that H stands for horizontal lookup. The portion HLOOKUP(I$16-I$17,b10:h14,$A18+1) finds the column in B10:H14 whose first entry matches I16-I17. Then we pick off the entry in row A18 + 1 of that column. This yields $f_4(3 - 1) = 0$.

FIGURE 12
Resource Allocation

A	A	B	C	D	E	F	G	H	I	J	K	L	M
1	REWARD	0	1	2	3	4	5	6					
2	PERIOD3	0	9	13	17	21	25	29					
3	PERIOD2	0	10	13	16	19	22	25					
4	PERIOD1	0	9	16	23	30	37	44					
5													
6													
7	FIGURE 12												
8	RESOURCE	ALLOCATION											
9													
10	VALUE	0	1	2	3	4	5	6					
11	PERIOD4	0	0	0	0	0	0	0					
12	PERIOD3	0	9	13	17	21	25	29					
13	PERIOD2	0	10	19	23	27	31	35					
14	PERIOD1	0	10	19	28	35	42	49					
15													
16	d	0	1	1	2	2	2	3	3	3	3	4	4
17	x	0	0	1	0	1	2	0	1	2	3	0	1
18	1	0	0	9	0	9	13	0	9	13	17	0	9
19	2	0	9	10	13	19	13	17	23	22	16	21	27
20	3	0	10	9	19	19	16	23	28	26	23	27	32

FIGURE 12
(Continued)

A	N	O	P	Q	R	S	T	U	V	W	X	Y	Z
1													
2													
3													
4													
5													
6													
7													
8													
9													
10													
11													
12													
13													
14													
15													
16	4	4	4	5	5	5	5	5	5	6	6	6	6
17	2	3	4	0	1	2	3	4	5	0	1	2	3
18	13	17	21	0	9	13	17	21	25	0	9	13	17
19	26	25	19	25	31	30	29	28	22	29	35	34	33
20	35	33	30	31	36	39	42	40	37	35	40	43	46

A	AA	AB	AC	AD	AE	AF	AG	AH	AI	AJ	AK
1											
2											
3											
4											
5											
6											
7											
8											
9											
10											
11											
12											
13											
14											
15											
16	6	6	6	0	1	2	3	4	5	6	
17	4	5	6	ft(0)	ft(1)	ft(2)	ft(3)	ft(4)	ft(5)	ft(6)	t
18	21	25	29	0	9	13	17	21	25	29	3
19	32	31	25	0	10	19	23	27	31	35	2
20	49	47	44	0	10	19	28	35	42	49	1

We now copy any of the $J_t(d, x)$ formulas (such as the one in I18) to the range B18:AC20.

The $f_t(d)$ are computed in AD18:AJ20. We begin by manually entering in AD18:AJ18 the formulas used to compute $f_3(0), f_3(1), \ldots, f_3(6)$. These formulas are as follows:

AD18:	0	(Computes $f_3(0)$)
AE18:	=MAX(C18:D18)	(Computes $f_3(1)$)
AF18:	=MAX(E18:G18)	(Computes $f_3(2)$)
AG18:	=MAX(H18:K18)	(Computes $f_3(3)$)
AH18:	=MAX(L18:P18)	(Computes $f_3(4)$)
AI18:	=MAX(Q18:V18)	(Computes $f_3(5)$)
AJ18:	=MAX(W18:AC18)	(Computes $f_3(6)$)

We now copy these formulas from the range AD18:AJ18 to the range AD18:AJ20.

For our spreadsheet to work we must be able to compute the $J_t(d, x)$ by looking up the appropriate value of $f_t(d)$ in rows 11–14. Thus, in B11:H11, we enter a zero in each cell [because $f_4(d) = 0$ for all d]. In B12, we enter =AD18 [this is the cell in which $f_3(0)$ is computed]. We now copy this formula to the range B12:H14.

Note that rows 11–14 of our spreadsheet are defined in terms of rows 18–20, and rows 18–20 are defined in terms of rows 11–14. This creates **circularity** or **circular references** in our spreadsheet. To resolve the circular references in this (or any) spreadsheet, simply select Tools, Options, Calculations and select the Iteration box. This will cause Excel to resolve all circular references until the circularity is resolved.

To determine how $6,000 should be allocated to the three investments, note that $f_1(6) = 49$. Because $f_1(6) = J_1(6, 4)$, we allocate $4,000 to investment 1. Then we must find $f_2(6 - 4) = 19 = J_2(2, 1)$. We allocate $1,000 to investment 2. Finally, we find that $f_3(2 - 1) = J_3(1, 1)$ and allocate $1,000 to investment 3.

Solving an Inventory Problem on a Spreadsheet

We now show how to determine an optimal production policy for Example 4. An important aspect of this production problem is that each month's ending inventory must be between 0 and 4 units. We can ensure that this occurs by manually determining the allowable actions in each state. We will design our spreadsheet to ensure that the ending inventory for each month must be between 0 and 4 inclusive.

Our first step in setting up the spreadsheet (Figure 13, file Dpinv.xls) is to enter the production cost for each possible production level (0, 1, 2, 3, 4, 5) in B1:G2. Then we define $f_t(i)$ to be the minimum cost incurred in meeting demands for months $t, t + 1, \ldots, 4$ when i units are on hand at the beginning of month t. If d_t is month t's demand, then for $t = 1, 2, 3, 4$ we may write

$$f_t(i) = \min_{x|0 \le i+x-d_t \le 4} \{.5(i + x - d_t) + c(x) + f_{t+1}(i + x - d_t)\} \qquad (23)$$

where $c(x) = $ cost of producing x units during a month, and $f_5(i) = 0$ for $(i = 0, 1, 2, 3, 4)$.

If we define $J_t(i, x) = .5(i + x - d_t) + c(x) + f_{t+1}(i + x - d_t)$ we may write

$$f_t(i) = \min_{x|0 \le i+x-d_t \le 4} \{J_t(i, x)\}$$

FIGURE 13
Inventory Example

	A	B	C	D	E	F	G	H	I	J	K	L	M
1	PROD COST	0	1	2	3	4	5						
2		0	4	5	6	7	8						
3													
4	VALUE	-5	0	1	2	3	4	5					
5	M5	10000	0	0	0	0	0	10000					
6	M4	10000	7	6	5	4	0	10000					
7	M3	10000	12	10	7	6.5	6	10000					
8	M2	10000	16	15	14	12	10.5	10000					
9													
10		STATE	0	0	0	0	0	0	1	1	1	1	1
11		ACTION	0	1	2	3	4	5	0	1	2	3	4
12	DEMAND												
13	4		10000	10004	10005	10006	7	8.5	10000	10004	10005	6	7.5
14	2		10000	10004	12	12.5	13	13.5	10000	11	11.5	12	12.5
15	3		10000	10004	10005	18	17.5	16	10000	10004	17	16.5	15
16	1		10000	20	20.5	21	20.5	20.5	16	19.5	20	19.5	19.5
17													

	A	N	O	P	Q	R	S	T	U	V	W	X	Y	Z
1														
2														
3														
4														
5														
6														
7														
8														
9														
10		1	2	2	2	2	2	2	3	3	3	3	3	3
11		5	0	1	2	3	4	5	0	1	2	3	4	5
12														
13		9	10000	10004	5	6.5	8	9.5	10000	4	5.5	7	8.5	10
14		10	7	10.5	11	11.5	9	10010.5	6.5	10	10.5	8	10009.5	10011
15		16	10000	16	15.5	14	15	16	12	14.5	13	14	15	10010.5
16		10010.5	15.5	19	18.5	18.5	10009.5	10011	15	17.5	17.5	10008.5	10010	10011.5
17														

FIGURE **13**
(Continued)

A	AA	AB	AC	AD	AE	AF	AG	AH	AI	AJ	AK	AL
1												
2												
3												
4												
5												
6												
7												
8												
9												
10	4	4	4	4	4	4						
11	0	1	2	3	4	5	F(0)	F(1)	F(2)	F(3)	F(4)	
12												
13	0	4.5	6	7.5	9	10010.5	7	6	5	4	0	1
14	6	9.5	7	10008.5	10010	10011.5	12	10	7	6.5	6	2
15	10.5	12	13	14	10009.5	10011	16	15	14	12	10.5	3
16	13.5	16.5	10007.5	10009	10010.5	10012	20	16	15.5	15	13.5	4
17												

Next we compute $J_t(i, x)$ in A13:AF16. For example, to compute $J_4(0, 2)$, we enter the following formula in E13:

$$=\text{HLOOKUP}(E\$11,\$B\$1:\$G\$2,2)$$
$$+.5*+1\text{MAX}(E\$10+E\$11-\$A13,0)$$
$$+\text{HLOOKUP}(E\$10+E\$11-\$A13,\$B\$4:\$H\$8,1+\$AL13)$$

The first term in this sum yields $c(x)$ (this is because E\$11 is the production level). The second term gives the holding cost for the month (this is because E\$10+E\$11−\$A13 gives the month's ending inventory). The final term yields $f_{t+1}(i + x - d_t)$. This is because E\$10+E\$11−\$A13 is the beginning inventory for month $t + 1$. The reference to 1+\$AL13 in the final term ensures that we look up the value of $f_{t+1}(i + x - d_t)$ in the correct row [the values of the $f_{t+1}( )$ will be tabulated in C5:G8]. Copying the formula in E13 to the range C13:AF16 computes all the $J_t(i, x)$.

In AG13:AK16, we compute the $f_t(d)$. To begin, we enter the following formulas in cells AG13:AK13:

AG13:	=MIN(C13:H13)	[Computes $f_4(0)$]
AH13:	=MIN(I13:N13)	[Computes $f_4(1)$]
AI13:	=MIN(O13:T13)	[Computes $f_4(2)$]
AJ13:	=MIN(U13:Z13)	[Computes $f_4(3)$]
AK13:	=MIN(AA13:AF13)	[Computes $f_4(4)$]

To compute all the $f_t(i)$, we now copy from the range AG13:AK13 to the range AG13:AK16. For this to be successful, we need to have the correct values of the $f_t(i)$ in B5:H8. In columns B and H of rows 5–8, we enter 10,000 (or any large positive number). This ensures that it is very costly to end a month with an inventory that is negative or that exceeds 4. This will ensure that each month's ending inventory is between 0 and 4 inclusive. In the range C5:G5, we enter a 0 in each cell. This is because $f_5(i) = 0$ for $i = 0$, 1, 2, 3, 4. In cell C6, we enter +AG13; this enters the value of $f_1(0)$. By copying this formula to the range C6:G8, we have created a table of the $f_t(d)$, which can be used (in rows 13–16) to look up the $f_t(d)$.

As with the spreadsheet we used to solve Example 5, our current spreadsheet exhibits circular references. This is because rows 6–8 refer to rows 13–16, and rows 13–16 refer to rows 6–8. Pressing F9 several times, however, resolves the circular references. You also can resolve circular references by selecting Tools, Options, Calculations and checking the Iterations box.

For any initial inventory level, we can now compute the optimal production schedule. For example, suppose the inventory at the beginning of month 1 is 0. Then $f_1(0) = 20 = J_1(0, 1)$. Thus, it is optimal to produce 1 unit during month 1. Now we seek $f_2(0 + 1 - 1) = 16 = J_2(0, 5)$, so we produce 5 units during month 2. Then we seek $f_3(0 + 5 - 3) = 7 = J_3(2, 0)$, so we produce 0 units during month 3. Solving $f_4(2 + 0 - 2) = J_4(0, 4)$, we produce 4 units during month 4.

PROBLEMS

Group A

1 Use a spreadsheet to solve Problem 2 of Section 18.3.

2 Use a spreadsheet to solve Problem 4 of Section 18.4.

3 Use a spreadsheet to solve Problem 5 of Section 18.4.

SUMMARY

Dynamic programming solves a relatively complex problem by decomposing the problem into a series of simpler problems. First we solve a one-stage problem, then a two-stage problem, and finally a T-stage problem (T = total number of stages in the original problem).

In most applications, a decision is made at each stage (t = current stage), a reward is earned (or a cost is incurred) at each stage, and we go on to the stage $t + 1$ state.

Working Backward

In formulating dynamic programming recursions by working backward, it is helpful to remember that in most cases:

1 The **stage** is the mechanism by which we build up the problem.

2 The **state** at any stage gives the information needed to make the correct decision at the current stage.

3 In most cases, we must determine how the reward received (or cost incurred) during the current stage depends on the stage t decision, the stage t state, and the value of t.

4 We must also determine how the stage $t + 1$ state depends on the stage t decision, the stage t state, and the value of t.

5 If we define (for a minimization problem) $f_t(i)$ as the minimum cost incurred during stages $t, t + 1, \ldots, T$, given that the stage t state is i, then (in many cases) we may write $f_t(i) = \min \{(\text{cost during stage } t) + f_{t+1}(\text{new state at stage } t + 1)\}$, where the minimum is over all decisions allowable in state i during stage t.

6 We begin by determining all the $f_T(\cdot)$'s, then all the $f_{T-1}(\cdot)$'s, and finally f_1 (the initial state).

7 We then determine the optimal stage 1 decision. This leads us to a stage 2 state, at which we determine the optimal stage 2 decision. We continue in this fashion until the optimal stage T decision is found.

Wagner–Whitin Algorithm and Silver–Meal Heuristic for Dynamic Lot-Size Model

A periodic review inventory model in which each period's demand is known at the beginning of the problem is a **dynamic lot-size model.** A cost-minimizing production or ordering policy may be found via a backward recursion, a forward recursion, the Wagner–Whitin algorithm, or the Silver–Meal heuristic.

The Wagner–Whitin algorithm uses the fact that production occurs during a period if and only if the period's beginning inventory is zero. The decision during such a period is the number of consecutive periods of demand that production should meet.

During a period in which beginning inventory is zero, the Silver–Meal heuristic computes the average cost per period (setup plus holding) incurred in meeting the demand during the next k periods. If k^* minimizes this average cost, then the next k^* periods of demand should be met by the current period's production.

Computational Considerations

Dynamic programming is much more efficient than explicit enumeration of the total cost associated with each possible set of decisions that may be chosen during the T stages. Unfortunately, however, many practical applications of dynamic programming involve very large state spaces, and in these situations, considerable computational effort is required to determine optimal decisions.

REVIEW PROBLEMS

Group A

1 In the network in Figure 14, find the shortest path from node 1 to node 10 and the shortest path from node 2 to node 10.

2 A company must meet the following demands on time: month 1, 1 unit; month 2, 1 unit; month 3, 2 units; month 4, 2 units. It costs $4 to place an order, and a $2 per-unit holding cost is assessed against each month's ending inventory. At the beginning of month 1, 1 unit is available. Orders are delivered instantaneously.

   a Use a backward recursion to determine an optimal ordering policy.

   b Use the Wagner–Whitin method to determine an optimal ordering policy.

   c Use the Silver–Meal heuristic to determine an ordering policy.

3 Reconsider Problem 2, but now suppose that demands need not be met on time. Assume that all lost demand is backlogged and that a $1 per-unit shortage cost is assessed against the number of shortages incurred during each month. All demand must be met by the end of month 4. Use dynamic programming to determine an ordering policy that minimizes total cost.

4 Indianapolis Airlines has been told that it may schedule six flights per day departing from Indianapolis. The destination of each flight may be New York, Los Angeles,

FIGURE 14

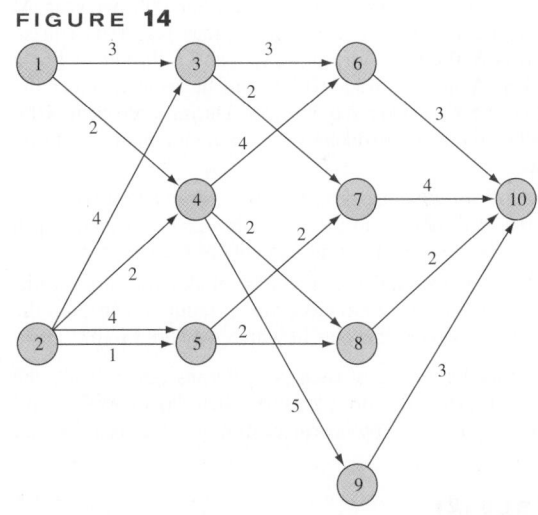

or Miami. Table 20 shows the contribution to the company's profit from any given number of daily flights from Indianapolis to each possible destination. Find the optimal number of flights that should depart Indianapolis for each destination. How would the answer change if the airline were restricted to only four daily flights?

TABLE 20

Destination	Profit per Flight ($)					
	Number of Planes					
	1	2	3	4	5	6
New York	80	150	210	250	270	280
Los Angeles	100	195	275	325	300	250
Miami	90	180	265	310	350	320

5 I am working as a cashier at the local convenience store. A customer's bill is $1.09, and he gives me $2.00. I want to give him change using the smallest possible number of coins. Use dynamic programming to determine how to give the customer his change. Does the answer suggest a general result about giving change? Resolve the problem if a 20¢ piece (in addition to other United States coins) were available.

6 A company needs to have a working machine during each of the next six years. Currently, it has a new machine. At the beginning of each year, the company may keep the machine or sell it and buy a new one. A machine cannot be kept for more than three years. A new machine costs $5,000. The revenues earned by a machine, the cost of maintaining it, and the salvage value that can be obtained by selling it at the end of a year depend on the age of the machine (see Table 21). Use dynamic programming to maximize the net profit earned during the next six years.

7 A company needs the following number of workers during each of the next five years: year 1, 15; year 2, 30; year 3, 10; year 4, 30; year 5, 20. At present, the company has 20 workers. Each worker is paid $30,000 per year. At the beginning of each year, workers may be hired or fired. It costs $10,000 to hire a worker and $20,000 to fire a worker. A newly hired worker can be used to meet the current year's worker requirement. During each year, 10% of all workers quit (workers who quit do not incur any firing cost).

a With dynamic programming, formulate a recursion that can be used to minimize the total cost incurred in meeting the worker requirements of the next five years.

b How would the recursion be modified if hired workers cannot be used to meet worker requirements until the year following the year in which they are hired?

8 At the beginning of each year, Barnes Carr Oil sets the world oil price. If a price p is set, then $D(p)$ barrels of oil will be demanded by world customers. We assume that during any year, each oil company sells the same number of barrels of oil. It costs Barnes Carr Oil c dollars to extract and refine each barrel of oil. Barnes Carr cannot set too high a price, however, because if a price p is set and there are currently N oil companies, then $g(p, N)$ oil companies will enter the oil business [$g(p, N)$ could be negative]. Setting too high a price will dilute future profits because of the entrance of new companies. Barnes Carr wants to maximize the discounted profit the company will earn over the next 20 years. Formulate a recursion that will aid Barnes Carr in meeting its goal. Initially, there are 10 oil companies.

9 For a computer to work properly, three subsystems of the computer must all function properly. To increase the reliability of the computer, spare units may be added to each system. It costs $100 to add a spare unit to system 1, $300 to system 2, and $200 to system 3. As a function of the number of added spares (a maximum of two spares may be added to each system), the probability that each system will work is given in Table 22. Use dynamic programming to maximize the probability that the computer will work properly, given that $600 is available for spare units.

Group B

10 During any year, I can consume any amount that does not exceed my current wealth. If I consume c dollars during a year, I earn c^a units of happiness. By the beginning of the next year, the previous year's ending wealth grows by a factor k.

a Formulate a recursion that can be used to maximize total utility earned during the next T years. Assume I originally have w_0 dollars.

b Let $f_t(w)$ be the maximum utility earned during years $t, t + 1, \ldots, T$, given that I have w dollars at the beginning of year t; and $c_t(w)$ be the amount that should be consumed during year t to attain $f_t(w)$. By working backward, show that for appropriately chosen constants a_t and b_t,

$$f_t(w) = b_t w^a \quad \text{and} \quad c_t(w) = a_t w$$

Interpret these results.

11 At the beginning of month t, farmer Smith has x_t bushels of wheat in his warehouse. He has the opportunity to sell wheat at a price s_t dollars per bushel and can buy wheat at p_t dollars per bushel. Farmer Smith's warehouse can hold at most C units at the end of each month.

a Formulate a recursion that can be used to maximize the total profit earned during the next T months.

b Let $f_t(x_t)$ be the maximum profit that can be earned during months $t, t + 1, \ldots, T$, given that x_t bushels of

TABLE 21

	Age of Machine at Beginning of Year		
	0 Year	1 Year	2 Years
Revenues ($)	4,500	3,000	1,500
Operating Costs ($)	500	700	1,100
Salvage Value at End of Year ($)	3,000	1,800	500

TABLE 22

Number of Spares	Probability That a System Works		
	System 1	System 2	System 3
0	.85	.60	.70
1	.90	.85	.90
2	.95	.95	.98

wheat are in the warehouse at the beginning of month t. By working backward, show that for appropriately chosen constants a_t and b_t,

$$f_t(x_t) = a_t + b_t x_t$$

c During any given month, show that the profit-maximizing policy has the following properties: (1) The amount sold during month t will equal either x_t or zero. (2) The amount purchased during a given month will be either zero or sufficient to bring the month's ending stock to C bushels.

REFERENCES

The following references are oriented toward applications and are written at an intermediate level:

Dreyfus, S., and A. Law. *The Art and Theory of Dynamic Programming.* Orlando, Fla.: Academic Press, 1977.

Nemhauser, G. *Introduction to Dynamic Programming.* New York: Wiley, 1966.

Wagner, H. *Principles of Operations Research,* 2d ed. Englewood Cliffs, N.J.: Prentice Hall, 1975.

The following five references are oriented toward theory and are written at a more advanced level:

Bellman, R. *Dynamic Programming.* Princeton, N.J.: Princeton University Press, 1957.

Bellman, R., and S. Dreyfus. *Applied Dynamic Programming.* Princeton, N.J.: Princeton University Press, 1962.

Bersetkas, D. *Dynamic Programming and Optimal Control,* vol. 1. Cambridge, Mass.: Athena Scientific, 2000.

Denardo, E. *Dynamic Programming: Theory and Applications.* Englewood Cliffs, N.J.: Prentice Hall, 1982.

Whittle, P. *Optimization Over Time: Dynamic Programming and Stochastic Control,* vol. 1. New York: Wiley, 1982.

Morton, T. "Planning Horizons for Dynamic Programs," *Operations Research* 27(1979):730–743. A discussion of turnpike theorems.

Peterson, R., and E. Silver. *Decision Systems for Inventory Management and Production Planning.* New York: Wiley, 1998. Discusses the Silver–Meal method.

Waddell, R. "A Model for Equipment Replacement Decisions and Policies," *Interfaces* 13(1983):1–8. An application of the equipment replacement model.

Wagner, H., and T. Whitin. "Dynamic Version of the Economic Lot Size Model," *Management Science* 5(1958):89–96. Discusses Wagner–Whitin method.

Probabilistic Dynamic Programming

Recall from our study of deterministic dynamic programming that many recursions were of the following form:

$$f_t \text{ (current state)} = \min_{\substack{\text{all feasible} \\ \text{decisions}}} \text{ (or max)} \{\text{costs during current stage} + f_{t+1} \text{ (new state)}\}$$

For all the examples in Chapter 18, a specification of the current state and current decision was enough to tell us *with certainty* the new state and the costs during the current stage. In many practical problems, these factors may not be known with certainty, even if the current state and decision are known. For example, in the inventory model of Section 18.3, we assumed that each period's demand was known at the beginning of the problem. In most situations, it would be more realistic to assume that period t's demand is a random variable whose value is not known until after period t's production decision is made. Even if we know the current period's state (beginning inventory level) and decision (production during the current period), the next period's state and the current period's cost will be random variables whose values are not known until the value of period t's demand is known. The Chapter 18 discussion simply does not apply to this problem.

In this chapter, we explain how to use dynamic programming to solve problems in which the current period's cost or the next period's state are random. We call these problems *probabilistic dynamic programming problems* (or PDPs). In a PDP, the decision maker's goal is usually to minimize expected (or expected discounted) cost incurred or to maximize expected (or expected discounted) reward earned over a given time horizon. Chapter 19 concludes with a brief study of *Markov decision processes*. A Markov decision process is just a probabilistic dynamic programming problem in which the decision maker faces an infinite horizon.

19.1 When Current Stage Costs Are Uncertain, but the Next Period's State Is Certain

For problems in this section, the next period's state is known with certainty, but the reward earned during the current stage is not known with certainty (given the current state and decision).

EXAMPLE 1   Milk Distribution

For a price of $1/gallon, the Safeco Supermarket chain has purchased 6 gallons of milk from a local dairy. Each gallon of milk is sold in the chain's three stores for $2/gallon. The dairy must buy back for 50¢/gallon any milk that is left at the end of the day. Unfortunately for Safeco, demand for each of the chain's three stores is uncertain. Past data indicate that the daily demand at each store is as shown in Table 1. Safeco wants to allo-

TABLE 1
Probability Distributions for Daily Milk Demand

	Daily Demand (gallons)	Probability
Store 1	1	.60
	2	0
	3	.40
Store 2	1	.50
	2	.10
	3	.40
Store 3	1	.40
	2	.30
	3	.30

cate the 6 gallons of milk to the three stores so as to maximize the expected net daily profit (revenues less costs) earned from milk. Use dynamic programming to determine how Safeco should allocate the 6 gallons of milk among the three stores.

Solution
With the exception of the fact that the demand (and therefore the revenue) is uncertain, this problem is very similar to the resource allocation problems studied in Section 18.4.

Observe that since Safeco's daily purchase costs are always $6, we may concentrate our attention on the problem of allocating the milk to maximize daily expected revenue earned from the 6 gallons.

Define

$$r_t(g_t) = \text{expected revenue earned from } g_t \text{ gallons assigned to store } t$$

$$f_t(x) = \text{maximum expected revenue earned from } x \text{ gallons assigned}$$
$$\text{to stores } t, t + 1, \ldots, 3$$

Since $f_3(x)$ must by definition be the expected revenue earned from assigning x gallons of milk to store 3, we see that $f_3(x) = r_3(x)$. For $t = 1, 2$, we may write

$$f_t(x) = \max_{g_t} \{r_t(g_t) + f_{t+1}(x - g_t)\} \tag{1}$$

where g_t must be a member of $\{0, 1, \ldots, x\}$. Equation (1) follows, because for any choice of g_t (the number of gallons assigned to store t), the expected revenue earned from store $t, t + 1, \ldots, 3$ will be the sum of the expected revenue earned from store t if g_t gallons are assigned to store t plus the maximum expected revenue that can be earned from the stores $t + 1, t + 2, \ldots, 3$ when $x - g_t$ gallons are assigned to these stores. To compute the optimal allocation of milk to the stores, we begin by computing $f_3(0), f_3(1), \ldots, f_3(6)$. Then we use Equation (1) to compute $f_2(0), f_2(1), \ldots, f_2(6)$. Finally we determine $f_1(6)$.

We begin by computing the $r_t(g_t)$'s. Note that it would be foolish to assign more than 3 gallons to any store. For this reason, we compute the $r_t(g_t)$'s only for $g_t = 0, 1, 2,$ or 3. As an example, we compute $r_3(2)$, the expected revenue earned if 2 gallons are assigned to store 3. If the demand at store 3 is for 2 or more gallons, both gallons assigned to store 3 will be sold, and $4 in revenue will be earned. If the demand at store 3 is 1 gallon, 1 gallon will be sold for $2, and 1 gallon will be returned for 50¢. Hence, if demand at store 3 is for 1 gallon, a revenue of $2.50 will be earned. Since there is a .60 chance that demand at store 3 will be for 2 or more gallons and a .40 chance that store 3 demand will be for 1 gallon, it follows that $r_3(2) = (.30 + .30)(4.00) + .40(2.50) = \$3.40$. Similar computations yield the following results:

$$r_3(0) = \$0 \qquad r_2(0) = \$0 \qquad r_1(0) = \$0$$
$$r_3(1) = \$2.00 \qquad r_2(1) = \$2.00 \qquad r_1(1) = \$2.00$$
$$r_3(2) = \$3.40 \qquad r_2(2) = \$3.25 \qquad r_1(2) = \$3.10$$
$$r_3(3) = \$4.35 \qquad r_2(3) = \$4.35 \qquad r_1(3) = \$4.20$$

We now use (1) to determine an optimal allocation of milk to stores. Let $g_t(x)$ be an allocation of milk to store t that attains $f_t(x)$. Then

$$f_3(0) = r_3(0) = 0 \qquad g_3(0) = 0$$
$$f_3(1) = r_3(1) = 2.00 \qquad g_3(1) = 1$$
$$f_3(2) = r_3(2) = 3.40 \qquad g_3(2) = 2$$
$$f_3(3) = r_3(3) = 4.35 \qquad g_3(3) = 3$$

We need not compute $f_3(4)$, $f_3(5)$, and $f_3(6)$, because an optimal allocation will never have more than 3 gallons to allocate to a single store (demand at any store is never more than 3 gallons).

Using (1) to work backward, we obtain

$$f_2(0) = r_2(0) + f_3(0 - 0) = 0 \qquad g_2(0) = 0$$

$$f_2(1) = \max \begin{cases} r_2(0) + f_3(1 - 0) = 2.00^* \\ r_2(1) + f_3(1 - 1) = 2.00^* \end{cases} \qquad g_2(1) = 0 \text{ or } 1$$

$$f_2(2) = \max \begin{cases} r_2(0) + f_3(2 - 0) = 0 + 3.40 = 3.40 \\ r_2(1) + f_3(2 - 1) = 2.00 + 2.00 = 4.00^* \\ r_2(2) + f_3(2 - 2) = 3.25 + 0 = 3.25 \end{cases} \qquad g_2(2) = 1$$

$$f_2(3) = \max \begin{cases} r_2(0) + f_3(3 - 0) = 0 + 4.35 = 4.35 \\ r_2(1) + f_3(3 - 1) = 2.00 + 3.40 = 5.40^* \\ r_2(2) + f_3(3 - 2) = 3.25 + 2.00 = 5.25 \\ r_2(3) + f_3(3 - 3) = 4.35 + 0 = 4.35 \end{cases} \qquad g_2(3) = 1$$

Note that in computing $f_2(4)$, $f_2(5)$, and $f_2(6)$, we need not consider any allocation for more than 3 gallons to store 2 or any that leaves more than 3 gallons for store 3.

$$f_2(4) = \max \begin{cases} r_2(1) + f_3(4 - 1) = 2.00 + 4.35 = 6.35 \\ r_2(2) + f_3(4 - 2) = 3.25 + 3.40 = 6.65^* \\ r_2(3) + f_3(4 - 3) = 4.35 + 2.00 = 6.35 \end{cases} \qquad g_2(4) = 2$$

$$f_2(5) = \max \begin{cases} r_2(2) + f_3(5 - 2) = 3.25 + 4.35 = 7.60 \\ r_2(3) + f_3(5 - 3) = 4.35 + 3.40 = 7.75^* \end{cases} \qquad g_2(5) = 3$$

$$f_2(6) = r_2(3) + f_3(6 - 3) = 4.35 + 4.35 = 8.70^* \qquad g_2(6) = 3$$

Finally,

$$f_1(6) = \max \begin{cases} r_1(0) + f_2(6 - 0) = 0 + 8.70 \\ r_1(1) + f_2(6 - 1) = 2.00 + 7.75 = 9.75^* \\ r_1(2) + f_2(6 - 2) = 3.10 + 6.65 = 9.75^* \\ r_1(3) + f_2(6 - 3) = 4.20 + 5.40 = 9.60 \end{cases} \qquad g_1(6) = 1 \text{ or } 2$$

Thus, we can either assign 1 or 2 gallons to store 1. Suppose we arbitrarily choose to assign 1 gallon to store 1. Then we have $6 - 1 = 5$ gallons for stores 2 and 3. Since $f_2(5)$ is attained by $g_2(5) = 3$, we assign 3 gallons to store 2. Then $5 - 3 = 2$ gallons are available for store 3. Since $g_3(2) = 2$, we assign 2 gallons to store 3. Note that although this policy obtains the maximum expected revenue, $f_1(6) = \$9.75$, the total revenue actually received on a given day may be more or less than \$9.75. For example, if demand at each store were 1 gallon, total revenue would be $3(2.00) + 3(0.50) = \$7.50$, whereas if demand at each store were 3 gallons, all the milk would be sold at \$2/gallon, and the total revenue would be $6(2.00) = \$12.00$.

PROBLEMS

Group A

1 In Example 1, find another allocation of milk that maximizes expected daily revenue.

2 Suppose that \$4 million is available for investment in three projects. The probability distribution of the net present value earned from each project depends on how much is invested in each project. Let I_t be the random variable

denoting the net present value earned by project t. The distribution of I_t depends on the amount of money invested in project t, as shown in Table 2 (a zero investment in a project always earns a zero NPV). Use dynamic programming to determine an investment allocation that maximizes the expected NPV obtained from the three investments.

TABLE 2
Investment Probability for Problem 2

	Investment (millions)	Probability		
Project 1	\$1	$P(I_1 = 2) = .6$	$P(I_1 = 4) = .3$	$P(I_1 = 5) = .1$
	\$2	$P(I_1 = 4) = .5$	$P(I_1 = 6) = .3$	$P(I_1 = 8) = .2$
	\$3	$P(I_1 = 6) = .4$	$P(I_1 = 7) = .5$	$P(I_1 = 10) = .1$
	\$4	$P(I_1 = 7) = .2$	$P(I_1 = 9) = .4$	$P(I_1 = 10) = .4$
Project 2	\$1	$P(I_2 = 1) = .5$	$P(I_2 = 2) = .4$	$P(I_2 = 4) = .1$
	\$2	$P(I_2 = 3) = .4$	$P(I_2 = 5) = .4$	$P(I_2 = 6) = .2$
	\$3	$P(I_2 = 4) = .3$	$P(I_2 = 6) = .3$	$P(I_2 = 8) = .4$
	\$4	$P(I_2 = 3) = .4$	$P(I_2 = 8) = .3$	$P(I_2 = 9) = .3$
Project 3	\$1	$P(I_3 = 0) = .2$	$P(I_3 = 4) = .6$	$P(I_3 = 5) = .2$
	\$2	$P(I_3 = 4) = .4$	$P(I_3 = 6) = .4$	$P(I_3 = 7) = .2$
	\$3	$P(I_3 = 5) = .3$	$P(I_3 = 7) = .4$	$P(I_3 = 8) = .3$
	\$4	$P(I_3 = 6) = .1$	$P(I_3 = 8) = .5$	$P(I_3 = 9) = .4$

19.2 A Probabilistic Inventory Model

In this section, we modify the inventory model of Section 18.3 to allow for uncertain demand. This will illustrate the difficulties involved in solving a PDP for which the state during the next period is uncertain (given the current state and current decision).

EXAMPLE 2 **Three-Period Production Policy**

Consider the following three-period inventory problem. At the beginning of each period, a firm must determine how many units should be produced during the current period. During a period in which x units are produced, a production cost $c(x)$ is incurred, where $c(0) =$

0, and for $x > 0$, $c(x) = 3 + 2x$. Production during each period is limited to at most 4 units. After production occurs, the period's random demand is observed. Each period's demand is equally likely to be 1 or 2 units. After meeting the current period's demand out of current production and inventory, the firm's end-of-period inventory is evaluated, and a holding cost of \$1 per unit is assessed. Because of limited capacity, the inventory at the end of each period cannot exceed 3 units. It is required that all demand be met on time. Any inventory on hand at the end of period 3 can be sold at \$2 per unit. At the beginning of period 1, the firm has 1 unit of inventory. Use dynamic programming to determine a production policy that minimizes the expected net cost incurred during the three periods.

Solution Define $f_t(i)$ to be the minimum expected net cost incurred during the periods $t, t + 1,$ $\ldots, 3$ when the inventory at the beginning of period t is i units. Then

$$f_3(i) = \min_x \{c(x) + (\tfrac{1}{2})(i + x - 1) + (\tfrac{1}{2})(i + x - 2) $$
$$- (\tfrac{1}{2})2(i + x - 1) - (\tfrac{1}{2})2(i + x - 2)\} \tag{2}$$

where x must be a member of $\{0, 1, 2, 3, 4\}$ and x must satisfy $(2 - i) \leq x \leq (4 - i)$.

Equation (2) follows, because if x units are produced during period 3, the net cost during period 3 is (expected production cost) + (expected holding cost) − (expected salvage value). If x units are produced, the expected production cost is $c(x)$, and there is a $\tfrac{1}{2}$ chance that the period 3 holding cost will be $i + x - 1$ and a $\tfrac{1}{2}$ chance that it will be $i + x - 2$. Hence, the period 3 expected holding cost will be $(\tfrac{1}{2})(i + x - 1) + (\tfrac{1}{2})(i + x - 2) = i + x - \tfrac{3}{2}$. Similar reasoning shows that the expected salvage value (a negative cost) at the end of period 3 will be $(\tfrac{1}{2})2(i + x - 1) + (\tfrac{1}{2})2(i + x - 2) = 2i + 2x - 3$. To ensure that period 3 demand is met, we must have $i + x \geq 2$, or $x \geq 2 - i$. Similarly, to ensure that ending period three inventory does not exceed 3 units, we must have $i + x - 1 \leq 3$, or $x \leq 4 - i$.

For $t = 1, 2$, we can derive the recursive relation for $f_t(i)$ by noting that for any month t production level x, the expected costs incurred during periods $t, t + 1, \ldots, 3$ are the sum of the expected costs incurred during period t and the expected costs incurred during periods $t + 1, t + 2, \ldots, 3$. As before, if x units are produced during month t, the expected cost during month t will be $c(x) + (\tfrac{1}{2})(i + x - 1) + (\tfrac{1}{2})(i + x - 2)$. (Note that during periods 1 and 2, no salvage value is received.) If x units are produced during month t, the expected cost during periods $t + 1, t + 2, \ldots, 3$ is computed as follows. Half of the time, the demand during period t will be 1 unit, and the inventory at the beginning of period $t + 1$ will be $i + x - 1$. In this situation, the expected costs incurred during periods $t + 1, t + 2, \ldots, 3$ (assuming we act optimally during these periods) is $f_{t+1}(i + x - 1)$. Similarly, there is a $\tfrac{1}{2}$ chance that the inventory at the beginning of period $t + 1$ will be $i + x - 2$. In this case, the expected cost incurred during periods $t + 1, t + 2, \ldots, 3$ will be $f_{t+1}(i + x - 2)$. In summary, the expected cost during periods $t + 1, t + 2, \ldots, 3$ will be $(\tfrac{1}{2})f_{t+1}(i + x - 1) + (\tfrac{1}{2})f_{t+1}(i + x - 2)$. With this in mind, we may write for $t = 1, 2$,

$$f_t(i) = \min_x [c(x) + (\tfrac{1}{2})(i + x - 1) + (\tfrac{1}{2})(i + x - 2) $$
$$+ (\tfrac{1}{2})f_{t+1}(i + x - 1) + (\tfrac{1}{2})f_{t+1}(i + x - 2)] \tag{3}$$

where x must be a member of $\{0, 1, 2, 3, 4\}$ and x must satisfy $(2 - i) \leq x \leq (4 - i)$.

Generalizing the reasoning that led to (3) yields the following important observation concerning the formulation of PDPs. Suppose the possible states during period $t + 1$ are $s_1, s_2, \ldots, s_n$ and the probability that the period $t + 1$ state will be s_i is p_i. Then the minimum expected cost incurred during periods $t + 1, t + 2, \ldots,$ end of the problem is

$$\sum_{i=1}^{i=n} p_i f_{t+1}(s_i)$$

where $f_{t+1}(s_i)$ is the minimum expected cost incurred from period $t + 1$ to the end of the problem, given that the state during period $t + 1$ is s_i.

We define $x_t(i)$ to be a period t production level attaining the minimum in (3) for $f_t(i)$. We now work backward until $f_1(1)$ is determined. The relevant computations are summarized in Tables 3, 4, and 5. Since each period's ending inventory must be nonnegative and cannot exceed 3 units, the state during each period must be 0, 1, 2, or 3.

As in Section 18.3, we begin by producing $x_1(1) = 3$ units during period 1. We cannot, however, determine period 2's production level until period 1's demand is observed. Also, period 3's production level cannot be determined until period 2's demand is observed. To illustrate the idea, we determine the optimal production schedule if period 1 and period

TABLE 3
Computations for $f_3(i)$

i	x	$c(x)$	Expected Holding Cost $(i + x - \frac{3}{2})$	Expected Salvage Value $(2i + 2x - 3)$	Total Expected Cost	$f_3(i)$ $x_3(i)$
3	0	0	$\frac{3}{2}$	3	$-\frac{3}{2}*$	$f_3(3) = -\frac{3}{2}$
3	1	5	$\frac{5}{2}$	5	$\frac{5}{2}$	$x_3(3) = 0$
2	0	0	$\frac{1}{2}$	1	$-\frac{1}{2}*$	$f_3(2) = -\frac{1}{2}$
2	1	5	$\frac{3}{2}$	3	$\frac{7}{2}$	$x_3(2) = 0$
2	2	7	$\frac{5}{2}$	5	$\frac{9}{2}$	
1	1	5	$\frac{1}{2}$	1	$\frac{9}{2}*$	$f_3(1) = \frac{9}{2}$
1	2	7	$\frac{3}{2}$	3	$\frac{11}{2}$	$x_3(1) = 1$
1	3	9	$\frac{5}{2}$	5	$\frac{13}{2}$	
0	2	7	$\frac{1}{2}$	1	$\frac{13}{2}*$	$f_3(0) = \frac{13}{2}$
0	3	9	$\frac{3}{2}$	3	$\frac{15}{2}$	$x_3(0) = 2$
0	4	11	$\frac{5}{2}$	5	$\frac{17}{2}$	

TABLE 4
Computations for $f_2(i)$

i	x	$c(x)$	Expected Holding Cost $(i + x - \frac{3}{2})$	Expected Future Cost $(\frac{1}{2})f_3(i + x - 1)$ $+(\frac{1}{2})f_3(i + x - 2))$	Total Expected Cost Periods 2,3	$f_2(i)$ $x_2(i)$
3	0	0	$\frac{3}{2}$	2	$\frac{7}{2}*$	$f_2(3) = \frac{7}{2}$
3	1	5	$\frac{5}{2}$	-1	$\frac{13}{2}$	$x_2(3) = 0$
2	0	0	$\frac{1}{2}$	$\frac{11}{2}$	$6*$	$f_2(2) = 6$
2	1	5	$\frac{3}{2}$	2	$\frac{17}{2}$	$x_2(2) = 0$
2	2	7	$\frac{5}{2}$	-1	$\frac{17}{2}$	
1	1	5	$\frac{1}{2}$	$\frac{11}{2}$	11	$f_2(1) = \frac{21}{2}$
1	2	7	$\frac{3}{2}$	2	$\frac{21}{2}*$	$x_2(1) = 2$ or 3
1	3	9	$\frac{5}{2}$	-1	$\frac{21}{2}*$	
0	2	7	$\frac{1}{2}$	$\frac{11}{2}$	13	$f_2(0) = \frac{25}{2}$
0	3	9	$\frac{3}{2}$	2	$\frac{25}{2}*$	$x_2(0) = 3$ or 4
0	4	11	$\frac{5}{2}$	-1	$\frac{25}{2}*$	

TABLE 5
Computations for $f_1(1)$

x	$c(x)$	Expected Holding Cost $(i + x - \frac{3}{2})$	Expected Future Cost $((\frac{1}{2})f_2(i + x - 1)$ $+ (\frac{1}{2})f_2(i + x - 2))$	Total Expected Cost Periods 1–3	$f_1(1)$ $x_1(1)$
1	5	$\frac{1}{2}$	$\frac{23}{2}$	17	$f_1(1) = \frac{65}{4}$
2	7	$\frac{3}{2}$	$\frac{33}{4}$	$\frac{67}{4}$	$x_1(1) = 3$
3	9	$\frac{5}{2}$	$\frac{19}{4}$	$\frac{65}{4}*$	

2 demands are both 2 units. Since $x_1(1) = 3$, 3 units will be produced during period 1. Then period 2 will begin with an inventory of $1 + 3 - 2 = 2$ units, so $x_2(2) = 0$ units should be produced. After period 2's demand of 2 units is met, period 3 will begin with $2 - 2 = 0$ units on hand. Thus, $x_3(0) = 2$ units will be produced during period 3.

In contrast, suppose that period 1 and period 2 demands are both 1 unit. As before, $x_1(1) = 3$ units will be produced during period 1. Then period 2 will begin with $1 + 3 - 1 = 3$ units, and $x_2(3) = 0$ units will be produced during period 2. Then period 3 will begin with $3 - 1 = 2$ units on hand, and $x_3(2) = 0$ units will be produced during period 3. Note that the optimal production policy has adapted to the low demand by reducing period 3 production. This example illustrates an important aspect of dynamic programming solutions for problems in which future states are not known with certainty at the beginning of the problem: *If a random factor (such as random demand) influences transitions from the period t state to the period t + 1 state, the optimal action for period t cannot be determined until period t's state is known.*

(s, S) Policies

Consider the following modification of the dynamic lot-size model of Section 18.7, for which there exists an optimal production policy called an (s, S) inventory policy:

1 The cost of producing $x > 0$ units during a period consists of a fixed cost K and a per-unit variable production cost c.

2 With a probability $p(x)$, the demand during a given period will be x.

3 A holding cost of h per unit is assessed on each period's ending inventory. If we are short, a per-unit shortage cost of d is incurred. (The case where no shortages are allowed may be obtained by letting d be very large.)

4 The goal is to minimize the total expected cost incurred during periods $1, 2, \ldots, T$.

5 All demands must be met by the end of period T.

For such an inventory problem, Scarf (1960) used dynamic programming to prove that there exists an optimal production policy of the following form: For each t ($t = 1, 2, \ldots, T$) there exists a pair of numbers (s_t, S_t) such that if i_{t-1}, the entering inventory for period t, is less than s_t, then an amount $S_t - i_{t-1}$ is produced; if $i_{t-1} \geq s_t$, then it is optimal not to produce during period t. Such a policy is called an **(s, S) policy.**

For Example 2, our calculations show that $s_2 = 2$, $S_2 = 3$ or 4, $s_3 = 2$, and $S_3 = 2$. Thus, if we enter period 2 with 1 or 0 units, we produce enough to bring our stock level (before meeting period 2 demand) up to 3 or 4 units. If we enter period 2 with more than 1 unit, then no production should take place during period 2.

PROBLEMS

Group A

1 For Example 2, suppose that the period 1 demand is 1 unit, and the period 2 demand is 2 units. What would be the optimal production schedule?

2 Re-solve Example 2 if the end-of-period holding cost is $2 per unit.

3 In Example 2, suppose that shortages are allowed, and each shortage results in a lost sale and a cost incurred of $3. Now re-solve Example 2.

Group B

4 Chip Bilton sells sweatshirts at State U football games. He is equally likely to sell 200 or 400 sweatshirts at each game. Each time Chip places an order, he pays $500 plus $5 for each sweatshirt he orders. Each sweatshirt sells for $8. A holding cost of $2 per shirt (because of the opportunity cost for capital tied up in sweatshirts as well as storage costs) is assessed against each shirt left at the end of a game. Chip can store at most 400 shirts after each game. Assuming that the number of shirts ordered by Chip must be a multiple of 100, determine an ordering policy that maximizes expected profits earned during the first three games of the season. Assume that any leftover sweatshirts have a value of $6.

19.3 How to Maximize the Probability of a Favorable Event Occurring[†]

There are many occasions on which the decision maker's goal is to maximize the probability of a favorable event occurring. For instance, a company may want to maximize its probability of reaching a specified level of annual profits. To solve such a problem, we assign a reward of 1 if the favorable event occurs and a reward of 0 if it does not occur. Then the maximization of expected reward will be equivalent to maximizing the probability that the favorable event will occur. Also, the maximum expected reward will equal the maximum probability of the favorable event occurring. The following two examples illustrate how this idea may be used to solve some fairly complex problems.

EXAMPLE 3   **Gambling Game**

A gambler has $2. She is allowed to play a game of chance four times, and her goal is to maximize her probability of ending up with a least $6. If the gambler bets b dollars on a play of the game, then with probability .40, she wins the game and increases her capital position by b dollars; with probability .60, she loses the game and decreases her capital by b dollars. On any play of the game, the gambler may not bet more money than she has available. Determine a betting strategy that will maximize the gambler's probability of attaining a wealth of at least $6 by the end of the fourth game. We assume that bets of zero dollars (that is, not betting) are permissible.

Solution   Define $f_t(d)$ to be the probability that by the end of game 4, the gambler will have at least $6, given that she acts optimally and has d dollars immediately before the game is played for the tth time. If we give the gambler a reward of 1 when her ending wealth is at least $6 and a reward of 0 if it is less, then $f_t(d)$ will equal the maximum expected reward that can

[†]This section covers topics that may be omitted with no loss of continuity.

be earned during games $t, t + 1, \ldots, 4$ if the gambler has d dollars immediately before the tth play of the game. As usual, we define $b_t(d)$ dollars to be a bet size that attains $f_t(d)$.

If the gambler is playing the game for the fourth and final time, her optimal strategy is clear: If she has $6 or more, don't bet anything, but if she has less than $6, bet enough money to ensure (if possible) that she will have $6 if she wins the last game. Note that if she begins game 4 with $0, $1, or $2, there is no way to win (no way to earn a reward of 1). This reasoning yields the following results:

$$
\begin{aligned}
f_4(0) &= 0 & b_4(0) &= \$0 \\
f_4(1) &= 0 & b_4(1) &= \$0 \text{ or } \$1 \\
f_4(2) &= 0 & b_4(2) &= \$0, \$1, \text{ or } \$2 \\
f_4(3) &= .40 & b_4(3) &= \$3 \\
f_4(4) &= .40 & b_4(4) &= \$2, \$3, \text{ or } \$4 \\
f_4(5) &= .40 & b_4(5) &= \$1, \$2, \$3, \$4, \text{ or } \$5
\end{aligned}
$$

For $d \geq 6$,

$$
f_4(d) = 1 \qquad b_4(d) = \$0, \$1, \ldots, \$(d - 6)
$$

For $t \leq 3$, we can find a recursion for $f_t(d)$ by noting that if the gambler has d dollars, is about to play the game for the tth time, and bets b dollars, then the following diagram summarizes what can occur:

With probability .40 win game t $f_{t+1}(d + b)$

                                         (Expected reward)

With probability .60 lose game t $f_{t+1}(d - b)$

Thus, if the gambler has d dollars at the beginning of game t and bets b dollars, the expected reward (or expected probability of reaching $6) will be $.4f_{t+1}(d + b) + .6f_{t+1}(d - b)$. This leads to the following recursion:

$$
f_t(d) = \max_b (.4f_{t+1}(d + b) + .6f_{t+1}(d - b)) \tag{4}
$$

where b must be a member of $\{0, 1, \ldots, d\}$. Then $b_t(d)$ is any bet size that attains the maximum in (4) for $f_t(d)$. Using (4), we work backward until $f_1(2)$ has been determined.

Stage 3 Computations

$$
f_3(0) = 0 \qquad b_3(0) = \$0
$$

$$
f_3(1) = \max \begin{cases} .4f_4(1) + .6f_4(1) = 0^* & \text{(Bet \$0)} \\ .4f_4(2) + .6f_4(0) = 0^* & \text{(Bet \$1)} \end{cases}
$$

Thus, $f_3(1) = 0$, and $b_3(1) = \$0$ or $\$1$.

$$
f_3(2) = \max \begin{cases} .4f_4(2) + .6f_4(2) = 0 & \text{(Bet \$0)} \\ .4f_4(3) + .6f_4(1) = .16^* & \text{(Bet \$1)} \\ .4f_4(4) + .6f_4(0) = .16^* & \text{(Bet \$2)} \end{cases}
$$

Thus, $f_3(2) = .16$, and $b_3(2) = \$1$ or $\$2$.

$$
f_3(3) = \max \begin{cases} .4f_4(3) + .6f_4(3) = .40^* & \text{(Bet \$0)} \\ .4f_4(4) + .6f_4(2) = .16 & \text{(Bet \$1)} \\ .4f_4(5) + .6f_4(1) = .16 & \text{(Bet \$2)} \\ .4f_4(6) + .6f_4(0) = .40^* & \text{(Bet \$3)} \end{cases}
$$

Thus, $f_3(3) = .40$, and $b_3(3) = \$0$ or $\$3$.

$$f_3(4) = \max \begin{cases} .4f_4(4) + .6f_4(4) = .40^* & \text{(Bet \$0)} \\ .4f_4(5) + .6f_4(3) = .40^* & \text{(Bet \$1)} \\ .4f_4(6) + .6f_4(2) = .40^* & \text{(Bet \$2)} \\ .4f_4(7) + .6f_4(1) = .40^* & \text{(Bet \$3)} \\ .4f_4(8) + .6f_4(0) = .40^* & \text{(Bet \$4)} \end{cases}$$

Thus, $f_3(4) = .40$, and $b_3(4) = \$0, \$1, \$2, \$3,$ or $\$4$.

$$f_3(5) = \max \begin{cases} .4f_4(5) + .6f_4(5) = .40 & \text{(Bet \$0)} \\ .4f_4(6) + .6f_4(4) = .64^* & \text{(Bet \$1)} \\ .4f_4(7) + .6f_4(3) = .64^* & \text{(Bet \$2)} \\ .4f_4(8) + .6f_4(2) = .40 & \text{(Bet \$3)} \\ .4f_4(9) + .6f_4(1) = .40 & \text{(Bet \$4)} \\ .4f_4(10) + .6f_4(0) = .40 & \text{(Bet \$5)} \end{cases}$$

Thus, $f_3(5) = .64$, and $b_3(5) = \$1$ or $\$2$. For $d \geq 6, f_3(d) = 1$, and $b_3(d) = \$0, \$1, \dots,$ $\$(d - 6)$.

Stage 2 Computations

$$f_2(0) = 0 \qquad b_2(0) = \$0$$

$$f_2(1) = \max \begin{cases} .4f_3(1) + .6f_3(1) = 0 & \text{(Bet \$0)} \\ .4f_3(2) + .6f_3(0) = .064^* & \text{(Bet \$1)} \end{cases}$$

Thus, $f_2(1) = .064$, and $b_2(1) = \$1$.

$$f_2(2) = \max \begin{cases} .4f_3(2) + .6f_3(2) = .16^* & \text{(Bet \$0)} \\ .4f_3(3) + .6f_3(1) = .16^* & \text{(Bet \$1)} \\ .4f_3(4) + .6f_3(0) = .16^* & \text{(Bet \$2)} \end{cases}$$

Thus, $f_2(2) = .16$, and $b_2(2) = \$0, \$1,$ or $\$2$.

$$f_2(3) = \max \begin{cases} .4f_3(3) + .6f_3(3) = .40^* & \text{(Bet \$0)} \\ .4f_3(4) + .6f_3(2) = .256 & \text{(Bet \$1)} \\ .4f_3(5) + .6f_3(1) = .256 & \text{(Bet \$2)} \\ .4f_3(6) + .6f_3(0) = .40^* & \text{(Bet \$3)} \end{cases}$$

Thus, $f_2(3) = .40$, and $b_2(3) = \$0$ or $\$3$.

$$f_2(4) = \max \begin{cases} .4f_3(4) + .6f_3(4) = .40 & \text{(Bet \$0)} \\ .4f_3(5) + .6f_3(3) = .496^* & \text{(Bet \$1)} \\ .4f_3(6) + .6f_3(2) = .496^* & \text{(Bet \$2)} \\ .4f_3(7) + .6f_3(1) = .40 & \text{(Bet \$3)} \\ .4f_3(8) + .6f_3(0) = .40 & \text{(Bet \$4)} \end{cases}$$

Thus, $f_2(4) = .496$, and $b_2(4) = \$1$ or $\$2$.

$$f_2(5) = \max \begin{cases} .4f_3(5) + .6f_3(5) = .64^* & \text{(Bet \$0)} \\ .4f_3(6) + .6f_3(4) = .64^* & \text{(Bet \$1)} \\ .4f_3(7) + .6f_3(3) = .64^* & \text{(Bet \$2)} \\ .4f_3(8) + .6f_3(2) = .496 & \text{(Bet \$3)} \\ .4f_3(9) + .6f_3(1) = .40 & \text{(Bet \$4)} \\ .4f_3(10) + .6f_3(0) = .40 & \text{(Bet \$5)} \end{cases}$$

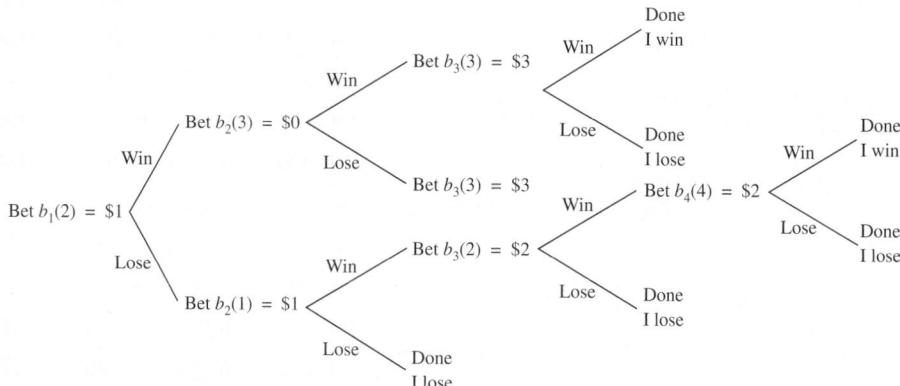

| Game 1 | Game 2 | Game 3 | Game 4 |

FIGURE 1
Ways Gambler Can
Reach $6

Thus, $f_2(5) = .64$, and $b_2(5) = \$0, \$1,$ or $\$2$. For $d \geq 6$, $f_2(d) = 1$ and $b_2(d) = \$0,$ $\$1, \ldots, \$(d - 6)$.

Stage 1 Computations

$$f_1(2) = \max \begin{cases} .4f_2(2) + .6f_2(2) = .16 & \text{(Bet \$0)} \\ .4f_2(3) + .6f_2(1) = .1984* & \text{(Bet \$1)} \\ .4f_2(4) + .6f_2(0) = .1984* & \text{(Bet \$2)} \end{cases}$$

Thus, $f_1(2) = .1984$, and $b_1(2) = \$1$ or $\$2$. Hence, the gambler has a .1984 chance of reaching $6. Suppose the gambler begins by betting $b_1(2) = \$1$. Then Figure 1 indicates the various possibilities that can occur. By following the strategy outlined in the figure, the gambler can reach her goal of $6 in two different ways. First, she can win game 1 and game 3. This will occur with probability $(.4)^2 = .16$. Second, the gambler can win if she loses the first game but wins the next three games. This will occur with probability $.6(.4)^3 = .0384$. Hence, the gambler's probability of reaching $6 is $.16 + .0384 = .1984 = f_1(2)$.

EXAMPLE 4   **Tennis Serves**

Martina McEnroe has two types of serves: a hard serve (H) and a soft serve (S).[†] The probability that Martina's hard serve will land in bounds is p_H, and the probability that her soft serve will land in bounds is p_S. If Martina's hard serve lands in bounds, there is a probability w_H that Martina will win the point. If Martina's soft serve lands in bounds, there is a probability w_S that Martina will win the point. We assume that $p_H < p_S$ and $w_H > w_S$. Martina's goal is to maximize the probability of winning a point on which she serves. Use dynamic programming to help Martina select an optimal serving strategy. Remember that if both serves are out of bounds, Martina loses the point.

Solution   To maximize Martina's probability of winning the point, we give her a reward of 1 if she wins the point and a reward of 0 if she loses the point. We also define $f_t(t = 1, 2)$ to be the probability that Martina wins a point if she plays optimally and is about to take her tth serve. To determine the optimal serving strategy, we work backward, beginning with f_2. If Martina serves hard on the second serve, she will win the point (and earn a reward

---

[†]Based on material by E. V. Denardo, personal communication.

of 1) with probability $p_H w_H$. Similarly, if she serves soft on the second serve, her expected reward is $p_S w_S$. Thus, we have

$$f_2 = \max \begin{cases} p_H w_H & \text{(Serve hard)} \\ p_S w_S & \text{(Serve soft)} \end{cases}$$

For the moment, let's assume that

$$p_S w_S > p_H w_H \tag{5}$$

If (5) holds, then Martina should serve soft on the second serve. In this situation, $f_2 = p_S w_S$.

To determine f_1, we need to look at what happens on the first serve. If Martina serves hard on the first serve, the events in Table 6 can occur, and Martina earns an expected reward of $p_H w_H + (1 - p_H)f_2$. If Martina serves soft on the first serve, then the events in Table 7 can occur, and Martina's expected reward is $p_S w_S + (1 - p_S)f_2$. We now write the following recursion for f_1:

$$f_1 = \max \begin{cases} p_H w_H + (1 - p_H)f_2 & \text{(Serve hard)} \\ p_S w_S + (1 - p_S)f_2 & \text{(Serve soft)} \end{cases}$$

From this equation, we see that Martina should serve hard on the first serve if

$$p_H w_H + (1 - p_H)f_2 \geq p_S w_S + (1 - p_S)f_2 \tag{6}$$

(If (6) is not satisfied, Martina should serve soft on the first serve.)

Continuing with the assumption that $p_S w_S > w_H p_H$ (which implies that $f_2 = p_S w_S$), we may substitute $f_2 = p_S w_S$ into (6) to obtain the result that Martina should serve hard on the first serve if

$$p_H w_H + (1 - p_H)p_S w_S \geq p_S w_S + (1 - p_S)p_S w_S$$

TABLE 6
Computation of Expected Reward If First Serve Is Hard

Event	Probability of Event	Expected Reward for Given Event
First serve in and Martina wins point	$p_H w_H$	1
First serve in and Martina loses point	$p_H (1 - w_H)$	0
First serve out of bounds	$1 - p_H$	f_2

TABLE 7
Computation of Expected Reward If First Serve Is Soft

Event	Probability of Event	Expected Reward for Given Event
First serve in and Martina wins point	$p_s w_s$	1
First serve in and Martina loses point	$p_s (1 - w_s)$	0
First serve out of bounds	$1 - p_s$	f_2

or

$$p_H w_H \geq p_S w_S (1 + p_H - p_S) \tag{7}$$

For example, if $p_H = .60$, $p_S = .90$, $w_H = .55$, and $w_S = .50$, then (5) and (7) are both satisfied, and Martina should serve hard on her first serve and soft on her second serve. On the other hand, if $p_H = .25$, $p_S = .80$, $w_H = .60$, and $w_S = .45$, then both serves should be soft. The reason for this is that in this case, the hard serve's advantage from the fact that w_H exceeds w_S is outweighed by the fact that a hard serve on the first serve greatly increases the chances of a double fault.

To complete our analysis, we must consider the situation where (5) does not hold. We now show that if

$$p_H w_H \geq p_S w_S \tag{8}$$

Martina should serve hard on both serves. Note that if (8) holds, then $f_2 = \max \{p_H w_H, p_S w_S\} = p_H w_H$, and Martina should serve hard on the second serve. Now (6) implies that Martina should serve hard on the first serve if

$$p_H w_H + (1 - p_H) p_H w_H \geq p_S w_S + (1 - p_S) p_H w_H$$

Upon rearrangement, the last inequality becomes

$$p_H w_H (1 + p_S - p_H) \geq p_S w_S$$

Dividing both sides of the last inequality by $p_S w_S$ shows that Martina should serve hard on the first serve if

$$\frac{p_H w_H}{p_S w_S} (1 + p_S - p_H) \geq 1$$

After noting that $p_H w_H \geq p_S w_S$ and $(1 + p_S - p_H) > 1$ (because $p_S > p_H$), we see that the last inequality holds. Thus, we have shown that if $p_H w_H \geq p_S w_S$, Martina should serve hard on both serves. This is reasonable, because if it is optimal to serve hard on the second (and this requires $p_H w_H \geq p_S w_S$), then it should be optimal to serve hard on the first serve, because the danger of double-faulting (which is the drawback to the hard serve) is less immediate on the first serve. Of course, Example 4 could have been solved using a decision tree; see Problem 10 of Section 13.4.

In our solution to Example 4, we have shown how Martina's optimal strategy depends on the values of the parameters defining the problem. This is a kind of sensitivity analysis like the one applied to linear programming problems in Chapters 5 and 6.

PROBLEMS

Group A

1 Vladimir Ulanowsky is playing Keith Smithson in a two-game chess match. Winning a game scores 1 match point, and drawing a game scores $\frac{1}{2}$ match point. After the two games are played, the player with more match points is declared the champion. If the two players are tied after two games, they continue playing until someone wins a game (the winner of that game will be the champion). During each game, Ulanowsky can play one of two ways: boldly or conservatively. If he plays boldly, he has a 45% chance of winning the game and a 55% chance of losing the game. If he plays conservatively, he has a 90% chance of drawing the game and a 10% chance of losing the game. Ulanowsky's goal is to maximize his probability of winning the match. Use dynamic programming to help him accomplish this goal. If this problem is solved correctly, even though Ulanowsky is the inferior player, his chance of winning the match is over $\frac{1}{2}$. Explain this anomalous result.

2 Dickie Hustler has $2 and is going to toss an unfair coin (probability .4 of heads) three times. Before each toss, he can bet any amount of money (up to what he now has). If heads comes up, Dickie wins the number of dollars he bets; if tails comes up, he loses the number of dollars he bets. Use dynamic programming to determine a strategy that maximizes Dickie's probability of having at least $5 after the third coin toss.

Group B

3 Supppose that Army trails by 14 points in the Army–Navy football game. Army's guardian angel has assured the Army coach that his team will have the ball two more times during the game and will score a touchdown (worth 6 points) each time it has the ball. The Army coach has also been assured that Navy will not score any more points. Suppose a win is assigned a value of 1, a tie is .3, and a loss is 0. Army's problem is to determine whether to go for 1 or 2 points after each touchdown. A 1-point conversion is always successful, and a 2-point conversion is successful only 40% of the time. The Army coach wants to maximize the expected reward earned from the outcome of the game. Use dynamic programming to determine an optimal strategy. Then prove the following result: *No matter what value is assigned to a tie, it is never optimal to use the following strategy: Go for a 1-point conversion after the first touchdown and go for a 2-point conversion after the second touchdown.* Note that this (suboptimal) strategy is the one most coaches follow!

19.4 Further Examples of Probabilistic Dynamic Programming Formulations

Many probabilistic dynamic programming problems can be solved using recursions of the following form (for max problems):

$$f_t(i) = \max_a \left\{ (\text{expected reward during stage } t|i, a) + \sum_j p(j|i, a, t)f_{t+1}(j) \right\} \quad (9)$$

In (9), $f_t(i)$ is the maximum expected reward that can be earned during stages $t, t + 1, \ldots$ end of the problem, given that the state at the beginning of stage t is i. The max in (9) is taken over all actions a that are feasible when the state at the beginning of stage t is i. In (9), $p(j|i, a, t)$ is the probability that the next period's state will be j, given that the current (stage t) state is i and action a is chosen. Hence, the summation in (9) represents the expected reward from stage $t + 1$ to the end of the problem. By choosing a to maximize the right-hand side of (9), we are choosing a to maximize the expected reward earned from stage t to the end of the problem, and this is what we want to do. The following are six examples of probabilistic dynamic programming formulations.

EXAMPLE 5   Sunco Oil Drilling

Sunco Oil has D dollars to allocate for drilling at sites $1, 2, \ldots, T$. If x dollars are allocated to site t, the probability is $q_t(x)$ that oil will be found on site t. Sunco estimates that if site t has any oil, it is worth r_t dollars. Formulate a recursion that could be used to enable Sunco to maximize the expected value of all oil found on sites $1, 2, \ldots, T$.

Solution   This is a typical resource allocation problem (see Example 1). Therefore, the stage should represent the number of sites, the decision for site t is how many dollars to allocate to site t, and the state is the number of dollars available to allocate to sites $t, t + 1, \ldots, T$. We therefore define $f_t(d)$ to be the maximum expected value of the oil that can be found on sites $t, t + 1, \ldots, T$ if d dollars are available to allocate to sites $t, t + 1, \ldots, T$.

We make the reasonable assumption that $q_T(x)$ is a nondecreasing function of x. If this is the case, then at stage T, all the money should be allocated to site T. This yields

$$f_T(d) = r_T q_T(d) + (1 - q_T(d))0 = r_T q_T(d)$$

For $t < T$,

$$f_t(d) = \max_x \{r_t q_t(x) + f_{t+1}(d - x)\}$$

where x must satisfy $0 \le x \le d$. The last recursion follows, because $r_t q_t(x)$ is the expected value of the reward for stage t, and since Sunco will have $d - x$ dollars available for sites

$t + 1, t + 2, \ldots, T, f_{t+1}(d - x)$ is the expected value of the oil that can be found by optimally drilling at sites $t + 1, t + 2, \ldots, T$. To solve the problem, we would work backward until $f_1(D)$ had been determined.

EXAMPLE 6   **Bass Fishing**

Each year, the owner of a lake must determine how many bass to capture and sell. During year t, a price p_t will be received for each bass that is caught. If the lake contains b bass at the beginning of year t, the cost of capturing x bass is $c_t(x|b)$. Between the time that year t's bass are caught and year $t + 1$ begins, the bass in the lake multiply by a random factor $\mathbf{D}$, where $P(\mathbf{D} = d) = q(d)$.

Formulate a dynamic programming recursion that can be used to determine a bass-catching strategy that will maximize the owner's net profit over the next ten years. At present, the lake contains 10,000 bass.

Solution   As in Example 8 of Chapter 18, the stage is the year, the state is the number of bass in the lake at the beginning of the year, and the decision is how many bass to catch during each year. We define $f_t(b)$ to be the maximum expected net profit that can be earned during the years $t, t + 1, \ldots, 10$ if the lake contains b bass at the beginning of year t. Then

$$f_{10}(b) = \max_x \{xp_{10} - c_{10}(x|b)\}$$

where $0 \le x \le b$, and for $t < 10$

$$f_t(b) = \max_x \left\{ xp_t - c_t(x|b) + \sum_d q(d)f_{t+1}(d(b - x)) \right\}$$

In this recursion, x must satisfy $0 \le x \le b$. To justify the recursion for $t < 10$, first note that the profits during year t are (with certainty) $xp_t - c_t(x|b)$. Then with probability $q(d)$, year $t + 1$'s state will be $d(b - x)$. It then follows that if x bass are caught during year t, the maximum expected net profit that can be earned during the years $t + 1, t + 2, \ldots, 10$ will be

$$\sum_d q(d)f_{t+1}(d(b - x))$$

Hence, the recursion chooses the number of bass during year t to maximize the sum of year t profits and future profits. To use this recursion, we work backward until $f_1(10,000)$ is computed. Then, after the number of bass in the lake at the beginning of year t is observed, we use the recursion to determine the number of bass that should be caught during year t.

EXAMPLE 7   **Waiting in Line**

When Sally Mutton arrives at the bank, 30 minutes remain on her lunch break. If Sally makes it to the head of the line and enters service before the end of her lunch break, she earns reward r. However, Sally does not enjoy waiting in lines, so to reflect her dislike for waiting in line, she incurs a cost c for each minute she waits. During a minute in which n people are ahead of Sally, there is a probability $p(x|n)$ that x people will complete their transactions. Suppose that when Sally arrives, 20 people are ahead of her in line. Use dynamic programming to determine a strategy for Sally that will maximize her expected net revenue (reward − waiting costs).

Solution   When Sally arrives at the bank, she must decide whether to join the line or to give up and leave. At any later time, she may also decide to leave if it is unlikely that she will be served by the end of her lunch break. If 1 minute remained, Sally's decision would be simple: She should stay in line if and only if her expected reward exceeds the cost of waiting for 1 minute (c). Then we can work backward to a problem with 2 minutes left, and so on. We define $f_t(n)$ to be the maximum expected net reward that Sally can receive from time t to the end of her lunch break if at time t, n people are ahead of her. We let $t = 0$ be the present and $t = 30$ be the end of the problem. Since $t = 29$ is the beginning of the last minute of the problem, we write

$$f_{29}(n) = \max \begin{cases} 0 & \text{(Leave)} \\ rp(n|n) - c & \text{(Stay)} \end{cases}$$

This follows because if Sally chooses to leave at time 29, she earns no reward and incurs no more costs. On the other hand, if she stays at time 29, she will incur a waiting cost of c (a revenue of $-c$) and with probability $p(n|n)$ will enter service and receive a reward r. Thus, if Sally stays, her expected net reward is $rp(n|n) - c$.

For $t < 29$, we write

$$f_t(n) = \max \begin{cases} 0 & \text{(Leave)} \\ rp(n|n) - c + \sum_{k<n} p(k|n)f_{t+1}(n - k) & \text{(Stay)} \end{cases}$$

The last recursion follows, because if Sally stays, she will earn an expected reward (as in the $t = 29$ case) of $rp(n|n) - c$ during the current minute, and with probability $p(k|n)$, there will be $n - k$ people ahead of her; in this case, her expected net reward from time $t + 1$ to time 30 will be $f_{t+1}(n - k)$. If Sally stays, her overall expected reward received from time $t + 1, t + 2, \ldots, 30$ will be

$$\sum_{k<n} p(k|n)f_{t+1}(n - k)$$

Of course, if n people complete their transactions during the current minute, the problem ends, and Sally's future net revenue will be zero.

To determine Sally's optimal waiting policy, we work backward until $f_0(20)$ is computed. If $f_0(20)$ is attained by "stay," Sally stays and sees how many people are ahead of her at time 1. She continues to stay until a situation arises for which the optimal action is "leave" or she begins to be served. In either case, the problem terminates.

Problems in which the decision maker can terminate the problem by choosing a particular action are known as **stopping rule problems;** they often have a special structure that simplifies the determination of optimal policies. See Ross (1983) for more information on stopping rule problems.

EXAMPLE 8 **Cash Management Policy**

E. J. Korvair Department Store is trying to determine an optimal cash management policy. During each day, the demand for cash may be described by a random variable $\mathbf{D}$, where $p(\mathbf{D} = d) = p(d)$. At the beginning of each day, the store sends an employee to the bank to deposit or withdraw funds. Each bank transaction costs K dollars. Then E. J.'s demand for cash is met by cash left from the previous day plus money withdrawn (or minus money deposited). At the end of the day, the store determines its cash balance at the store. If the cash balance is negative, a shortage cost of s dollars per dollar short is incurred. If the ending balance is positive, a cost of i dollars per dollar held is incurred (because of loss of interest that could have been earned by depositing cash in the bank). At the beginning of day 1, the store has $10,000 cash on hand and a bank balance of $100,000. Formulate a dynamic programming model that can be used to minimize the expected cost of filling the store's cash needs for the next 30 days.

Solution To determine how much money should be withdrawn or deposited, E. J. needs to know its cash on hand and bank balance at the beginning of the day. As usual, we let time be the stage. At the beginning of each stage (or day), E. J. must decide how much to withdraw from or deposit in the bank. We let $f_t(c, b)$ be the minimum expected cost incurred by the store during days $t, t + 1, \ldots, 30$, given that at the beginning of day t, the store has c dollars cash at the store and b dollars in the bank.

We observe that

$$f_{30}(c, b) = \min_x \left\{ K\delta(x) + \sum_{d \leq c+x} p(d)(c + x - d)i + \sum_{d \geq c+x} p(d)(d - c - x)s \right\} \quad (10)$$

Here, x is the amount of money transferred from the bank to the store (if $x < 0$ money is transferred from the store to the bank). Since the store cannot withdraw more than b dollars from the bank or deposit more than c dollars in the bank, x must satisfy $b \geq x \geq -c$. Also, in (10), $\delta(0) = 0$ and $\delta(x) = 1$ for $x \neq 0$. In short, $K\delta(x)$ picks up the transaction cost (if there is a transaction). If $d \leq c + x$, the store will end the day with $c + x - d$ dollars, so a cost of $i(c + x - d)$ is incurred (because of lost interest). Since this occurs with probability $p(d)$, the first sum in (10) represents the expected interest costs incurred during day 30. Also note that if $d \geq c + x$, the store will be $d - c - x$ dollars short, and a shortage cost of $s(d - c - x)$ will be incurred. Again, this cost is incurred with probability $p(d)$. Hence, the second sum in (10) is the expected shortage cost incurred during day 30.

For $t < 30$, we write

$$f_t(c, b) = \min_x \left\{ K\delta(x) + \sum_{d \leq c+x} p(d)(c + x - d)i \right. $$
$$\left. + \sum_{d \geq c+x} p(d)(d - c - x)s + \sum_d p(d)f_{t+1}(c + x - d, b - x) \right\} \quad (11)$$

As in (10), x must satisfy $b \geq x \geq -c$. Also, the term $K\delta(x)$ and the first two summations yield the expected cost incurred during day t. If day t demand is d, then at the beginning of day $t + 1$, the store will have $c + x - d$ dollars cash on hand and a bank balance of $b - x$. Thus, with probability $p(d)$, the store's expected cost during days $t + 1$, $t + 2, \ldots, 30$ will be $f_{t+1}(c + x - d, b - x)$. Weighting $f_{t+1}(c + x - d, b - x)$ by the probability that day t demand will be d, we see that the last sum in (11) is the expected cost incurred during days $t + 1, t + 2, \ldots, 30$. Hence, (11) is correct. To determine the optimal cash management policy, we would use (10) and (11) to work backward until $f_1(10,000, 100,000)$ has been computed.

EXAMPLE 9 | **Parking Spaces**

Robert Blue is trying to find a parking place near his favorite restaurant. He is approaching the restaurant from the west, and his goal is to park as nearby as possible. The available parking places are pictured in Figure 2. Robert is nearsighted and cannot see ahead; he can only see whether the space he is at now is empty. When Robert arrives at an empty space, he must decide whether to park there or to continue to look for a closer space. Once he passes a space, he cannot return to it. Robert estimates that the probability that space t is empty is p_t. If he does not end up with a parking space, he is embarrassed and incurs a cost M (M is a big positive number). If he does park in space t, he incurs a cost $|t|$. Show how Robert can use dynamic programming to develop a parking strategy that minimizes his expected cost.

Solution If Robert is at space T, his problem is easy to solve: park in space T if it is empty; otherwise, incur a cost of M. Then Robert can work backward until he determines what to do at space $-T$. For this reason, we let the space Robert is at represent the stage. In order to make a decision at any stage, all Robert must know is whether or not the space is empty (if a space is not empty, he must continue). Thus, the state at any stage is whether

FIGURE 2
Location of
Parking Places

$$\rightarrow \boxed{-T}\ \boxed{1-T}\ \boxed{2-T}\ \cdots\ \boxed{-2}\ \boxed{-1}\ \boxed{0}\ \boxed{1}\ \boxed{2}\ \cdots\ \boxed{T}$$

0 = Restaurant

or not the space is empty. Of course, if the space is empty, Robert's decision is whether to take the space or to continue.

We define

$f_t(o)$ = minimum expected cost if Robert is at space t and space t is occupied

$f_t(e)$ = minimum expected cost if Robert is at space t and space t is empty

If Robert is at space T, he will park in the space if it is empty (incurring a cost T) or incur a cost M if the space is occupied. Thus, we have $f_T(o) = M$ and $f_T(e) = T$.

For $t < T$, we write

$$f_t(o) = p_{t+1}f_{t+1}(e) + (1 - p_{t+1})f_{t+1}(o) \tag{12}$$

$$f_t(e) = \min \begin{cases} |t| & \text{(Take space } t) \\ p_{t+1}f_{t+1}(e) + (1 - p_{t+1})f_{t+1}(o) & \text{(Don't take space } t) \end{cases} \tag{13}$$

To justify (12), note that if space t is occupied, Robert must next look at space $t + 1$. With probability p_{t+1}, space $t + 1$ will be empty; in this case, Robert's expected cost will be $f_{t+1}(e)$. Similarly, with probability $(1 - p_{t+1})$, space $t + 1$ will be occupied, and Robert will incur an expected cost of $f_{t+1}(o)$. Thus, Robert's expected cost is

$$p_{t+1}f_{t+1}(e) + (1 - p_{t+1})f_{t+1}(o)$$

To justify (13), note that Robert can either take space t (incurring a cost of $|t|$) or continue. Thus, if Robert continues, his expected cost will be

$$p_{t+1}f_{t+1}(e) + (1 - p_{t+1})f_{t+1}(o)$$

Since Robert wants to minimize his expected cost, (13) follows. By using (12) and (13), Robert can work backward to compute $f_{-T}(e)$ and $f_{-T}(o)$. Then he will continue until he reaches an empty space at some location t for which the minimum in (13) is attained by taking space t. If no such empty space is reached, Robert will not find a space, and he will incur a cost M.

EXAMPLE 10 Safecracker

During month $t(t = 1, 2, \ldots, 60)$, expert safecracker Dirk Stack knows that he will be offered a role in a bank job that will pay him d_t dollars. There is, however, a probability p_t that month t's job will result in his capture. If Dirk is captured, all his money will be lost. Dirk's goal is to maximize his expected asset position at the end of month 60. Formulate a dynamic programming recursion that will help Dirk accomplish his goal. At the beginning of month 1, Dirk has $50,000.

Solution   At the beginning of month 60, Dirk has no future to consider and his problem is easy to solve, so we let time represent the stage. At the beginning of each month, Dirk must decide whether or not to take the current month's job offer. In order to make this decision, Dirk must know how much money he has at the beginning of the month. We define $f_t(d)$ to be Dirk's maximum expected asset position at the end of month 60, given that at the beginning of month t, Dirk has d dollars. Then

$$f_{60}(d) = \max \begin{cases} p_{60}(0) + (1 - p_{60})(d + d_{60}) & \text{(Accept month 60 job)} \\ d & \text{(Reject month 60 job)} \end{cases}$$

This result follows, because if Dirk takes the job during month 60, there is a probability p_{60} that he will be caught and end up with zero dollars and a probability $(1 - p_{60})$ that he will not be caught and end up with $d + d_{60}$ dollars. Of course, if Dirk does not take the month 60 job, he ends month 60 with d dollars.

Extending this reasoning yields, for $t < 60$,

$$f_t(d) = \max \begin{cases} p_t(0) + (1 - p_t)f_{t+1}(d + d_t) & \text{(Accept month } t \text{ job)} \\ f_{t+1}(d) & \text{(Reject month } t \text{ job)} \end{cases}$$

Note that if Dirk accepts month t's job, there is a probability p_t that he will be caught (and end up with zero) and a probability $(1 - p_t)$ that he will successfully complete month t's job and earn d_t dollars. In this case, Dirk will begin month $t + 1$ with $d + d_t$ dollars, and his expected final cash position will be $f_{t+1}(d + d_t)$. Of course, if Dirk rejects the month t job, he begins month $t + 1$ with d dollars, and his expected final cash position will be $f_{t+1}(d)$. Since Dirk wants to maximize his expected cash position at the end of month 60, the recursion follows. By using the recursion, Dirk can work backward to compute $f_1(50,000)$. Then he can decide whether to accept the month 1 job. Assuming he has not been caught, he can then determine whether to accept the month 2 job, and so on.

As described in Section 18.8, spreadsheets can be used to solve dynamic programming recursions. See Problems 14 and 15 for some examples of how spreadsheets can be used to solve PDPs.

PROBLEMS

Group A

1 The space shuttle is about to go up on another flight. With probability $p_t(z)$, it will use z type t fuel cells during the flight. The shuttle has room for at most W fuel cells. If at any time during the flight, all the type t fuel cells burn out, a cost c_t will be incurred. Assuming the goal is to minimize the expected cost due to fuel cell shortages, set up a dynamic programming model that could be used to determine how to stock the space shuttle with fuel cells. There are T different types of fuel cells.

2 At the beginning of each year, a firm observes its asset position (call it d) and may invest any amount x ($0 \le x \le d$) in a risky investment. During each year, the money invested doubles with probability p and is completely lost with probability $1 - p$. Independently of this investment, the firm's asset position increases by an amount y with probability q_y (y may be negative). If the firm's asset position is negative at the beginning of a year, it cannot invest any money during that year. The firm initially has \$10,000 in assets and wants to maximize its expected asset position ten years from now. Formulate a dynamic programming recursion that will help accomplish this goal.

3 Consider a machine that may be in any one of the states 0, 1, 2, ... At the beginning of each month, the state of the machine is observed, and it is decided whether to replace or keep the machine. If the machine is replaced, a new state 0 machine arrives instantaneously. It costs R dollars to replace

a machine. Each month that a state i machine is in operation, a maintenance cost of $c(i)$ is incurred. If a machine is in state i at the beginning of a month, then with probability p_{ij}, the machine will begin the next month in state j. At the beginning of the first month, we own a state i_0 machine. Assuming that the interest rate is 12% per year, formulate a dynamic programming recursion that could be used to minimize the expected discounted cost incurred during the next T months. Note that if we replace a machine at the beginning of a month, we incur a maintenance cost of $c(0)$ during the month, and with probability p_{0i}, we begin the next month with a state i machine.

4 In the time interval between t and $t - 1$ seconds before the departure of Braneast Airlines Flight 313, there is a probability p_t that the airline will receive a reservation for the flight and a probability $1 - p_t$ that the airline will receive no reservation. The flight can seat up to 100 passengers. At departure time, if r reservations have been accepted by the airline, there is a probability $q(y|r)$ that y passengers will show up for the flight. Each passenger who boards the flight adds \$500 to Braneast's revenues, but each passenger who shows up for the flight and cannot be seated receives \$200 in compensation. Formulate a dynamic programming recursion to enable the airline to maximize its expected revenue from Flight 313. Assume that no reservations are received more than 100,000 seconds before flight time.

5 At the beginning of each week, a machine is either running or broken down. If the machine runs throughout the week, it earns revenues of $100. If the machine breaks down during a week, it earns no revenue for that week. If the machine is running at the beginning of the week, we may perform maintenance on it to lessen the chance of a breakdown. If the maintenance is performed, a running machine has a .4 chance of breaking down during the week; if maintenance is not performed, a running machine has a .7 chance of breaking down during the week. Maintenance costs $20 per week. If the machine is broken down at the beginning of the week, it must be replaced or repaired. Both repair and replacement occur instantaneously. Repairing a machine costs $40, and there is a .4 chance that the repaired machine will break down during the week. Replacing a broken machine costs $90, but the new machine is guaranteed to run throughout the next week of operation. Use dynamic programming to determine a repair, replacement, and maintenance policy that maximizes the expected net profit earned over a four-week period. Assume that the machine is running at the beginning of the first week.

6 I own a single share of Wivco stock. I must sell my share at the beginning of one of the next 30 days. Each day, the price of the stock changes. With probability $q(x)$, the price tomorrow will increase by $x\%$ over today's stock price (x can be negative). For example, with probability $q(5)$, tomorrow's stock price will be 5% higher than today's. Show how dynamic programming can be used to determine a strategy that maximizes the expected revenue earned from selling the share of Wivco stock. Assume that at the beginning of the first day, the stock sells for $10 per share.

Group B

7 The National Cat Foundling Home encourages people to adopt its cats, but (because of limited funds) it allows each prospective owner to inspect only four cats before choosing one of them to take home. Ten-year-old Sara is eager to adopt a cat and agrees to abide by the following rules. A randomly selected cat is brought for Sara to see, and then Sara must either choose the cat or reject it. If the first cat is rejected, Sara sees another randomly selected cat and must accept or reject it. This procedure continues until Sara has selected her cat. Once Sara rejects a cat, she cannot go back later and choose it as her pet. Determine a strategy for Sara that will maximize her probability of ending up with the cat she actually prefers.

8 Consider the following probabilistic inventory model:

a At the beginning of each period, a firm observes its inventory position.

b Then the firm decides how many units to produce during the current period. It costs $c(x)$ dollars to produce x units during a period.

c With probability $q(d)$, d units are demanded during the period. From units on hand (including the current period's production), the firm satisfies as much of the demand as possible. The firm receives r dollars for each unit sold. For each unit of demand that is unsatisfied, a penalty cost p is incurred. All unsatisfied demand is assumed to be lost. For example, if the firm has 20 units available and current demand is 30, a revenue of $20r$

would be received, and a penalty of $10p$ would be incurred.

d If ending inventory is positive, a holding cost of $1 per unit is incurred.

e The next period now begins.

The firm's inital inventory is zero, and its goal is to minimize the expected cost over a 100-period horizon. Formulate a dynamic programming recursion that will help the firm accomplish its goal.

9 Martha and Ken Allen want to sell their house. At the beginning of each day, they receive an offer. We assume that from day to day, the sizes of the offers are independent random variables and that the probability that a given day's offer is for j dollars is p_j. An offer may be accepted during the day it is made or at any later date. For each day the house remains unsold, a maintenance cost of c dollars is incurred. The house must be sold within 30 days. Formulate a dynamic programming recursion that Martha and Ken can use to maximize their expected net profit (selling price − maintenance cost). Assume that the maintenance cost for a day is incurred before the current day's offer is received and that each offer is for an integer number of dollars.

10 An advertising firm has D dollars to spend on reaching customers in T separate markets. Market t consists of k_t people. If x dollars are spent on advertising in market t, the probability that a given person in market t will be reached is $p_t(x)$. Each person in market t who is reached will buy c_t units of the product. A person who is not reached will not buy any of the product. Formulate a dynamic programming recursion that could be used to maximize the expected number of units sold in T markets.

11 Georgia Stein is the new owner of the New York Yankees. Each season, Georgia must decide how much money to spend on the free agent draft. During each season, Georgia can spend any amount of money on free agents up to the team's capital position at the beginning of the season. If the Yankees finish in ith place during the season, their capital position increases by $R(i)$ dollars less the amount of money spent in the free agent draft. If the Yankees finished in ith place last season and spend d dollars on free agents during the off-season, the probability that the Yankees will finish in place j during the next season is $p_{ij}(d)(j = 1, 2, \ldots, 7)$. Last season, the Yankees finished in first place, and at the end of the season, they had a capital position of D dollars. Formulate a dynamic programming recursion that will enable the Yankees to maximize their expected cash position at the end of T seasons.

12 Bailey Bliss is the campaign manager for Walter Glenn's presidential campaign. He has D dollars to allocate to T winner-take-all primaries. If x_t dollars are allocated to primary t, then with probability $p_t(x_t)$, Glenn will win primary t and obtain v_t delegates. With probability $1 - p_t(x_t)$, Glenn loses primary t and obtains no delegates. Glenn needs K delegates to be nominated. Use dynamic programming to help Bliss maximize Glenn's probability of being nominated. What aspect of a real campaign does the present formulation ignore?

13 At 7 A.M., eight people leave their cars for repair at Harry's Auto Repair Shop. If person i's car is ready by time

t (7 A.M. = time 0, and so on), he will pay Harry $r_i(t)$ dollars. For example, if person 2's car must be ready by 2 P.M., we may have $r_2(8) = 0$. Harry estimates that with probability $p_i(t)$, it will take t hours to repair person i's car. Formulate a dynamic programming recursion that will enable Harry to maximize his expected revenue for the day. His workday ends at 5 P.M. = time 10.

14 In Example 10, suppose $p_t = t/60$ and $d_t = t$. Using a spreadsheet, solve for Dirk's optimal strategy. (*Hint:* The possible states are 50, 51, . . . , 1,880 (thousands).)

15 In Example 9, assume $T = 10$ and $p_t = |t|/10$. Using a spreadsheet, solve for Robert's optimal strategy.

19.5 Markov Decision Processes[†]

To use dynamic programming in a problem for which the stage is represented by time, one must determine the value of T, the number of time periods over which expected revenue or expected profit is maximized (or expected costs are minimized). T is referred to as the **horizon length.** For instance, in the equipment-replacement problem of Section 18.5, if our goal is to minimize costs over a 30-year period, then $T = 30$. Of course, it may be difficult for a decision maker to determine exactly the most suitable horizon length. In fact, when a decision maker is facing a long horizon and is not sure of the horizon length, it is more convenient to assume that the horizon length is infinite.

Suppose a decision maker's goal is to maximize the expected reward earned over an infinite horizon. In many situations, the expected reward earned over an infinite horizon may be unbounded. For example, if for any state and decision, the reward earned during a period is at least \$3, then the expected reward earned during an infinite number of periods will, no matter what decisions are chosen, be unbounded. In this situation, it is not clear how a decision maker should choose a decision. Two approaches are commonly used to resolve the problem of unbounded expected rewards over an infinite horizon.

1 We can discount rewards (or costs) by assuming that a \$1 reward received during the next period will have the same value as a reward of β dollars $(0 < \beta < 1)$ received during the current period. This is equivalent to assuming that the decision maker wants to maximize expected discounted reward. Let M be the maximum reward (over all possible states and choices of decisions) that can be received during a single period. Then the maximum expected discounted reward (measured in terms of current period dollars) that can be received over an infinite period horizon is

$$M + M\beta + M\beta^2 + \cdots = \frac{M}{1 - \beta} < \infty$$

Thus, discounting rewards (or costs) resolves the problem of an infinite expected reward.

2 The decision maker can choose to maximize the expected reward earned per period. Then he or she would choose a decision during each period in an attempt to maximize the average reward per period as given by

$$E\left(\lim_{n \to \infty} \frac{\text{reward earned during periods } 1, 2, \ldots, n}{n}\right)$$

Thus, if a \$3 reward were earned each period, the total reward earned during an infinite number of periods would be unbounded, but the average reward per period would equal \$3.

In our discussion of infinite horizon problems, we choose to resolve the problem of unbounded expected rewards by discounting rewards by a factor β per period. A brief discussion of the criterion of average reward per period is also included. Infinite horizon probabilistic dynamic programming problems are called **Markov decision processes** (or MDPs).

[†]This section covers topics that may be omitted with no loss of continuity.

Description of an MDP

An MDP is described by four types of information:

1 State space

2 Decision set

3 Transition probabilities

4 Expected rewards

State Space

At the beginning of each period, the MDP is in some state i, where i is a member of $S = \{1, 2, \ldots, N\}$. S is referred to as the MDP's **state space.**

Decision Set

For each state i, there is a finite set of allowable decisions, $D(i)$.

Transition Probabilities

Suppose a period begins in state i, and a decision $d \in D(i)$ is chosen. Then with probability $p(j|i, d)$, the next period's state will be j. The next period's state depends only on the current period's state and on the decision chosen during the current period (not on previous states and decisions). This is why we use the term *Markov* decision process.

Expected Rewards

During a period in which the state is i and a decision $d \in D(i)$ is chosen, an expected reward of r_{id} is received.

EXAMPLE 11   Machine Replacement

At the beginning of each week, a machine is in one of four conditions (states): excellent (E), good (G), average (A), or bad (B). The weekly revenue earned by a machine in each type of condition is as follows: excellent, $100; good, $80; average, $50; bad, $10. After observing the condition of a machine at the beginning of the week, we have the option of instantaneously replacing it with an excellent machine, which costs $200. The quality of a machine deteriorates over time, as shown in Table 8. For this situation, determine the state space, decision sets, transition probabilities, and expected rewards.

TABLE 8
Next Period's States of Machines

Present State of Machine	Probability That Machine Begins Next Week As			
	Excellent	Good	Average	Bad
Excellent	.7	.3	—	—
Good	—	.7	.3	—
Average	—	—	.6	.4
Bad	—	—	—	1.0 until replaced

Solution   The set of possible states is $S = \{E, G, A, B\}$. Let

$$R = \text{replace at beginning of current period}$$
$$NR = \text{do not replace during current period}$$

Since it is absurd to replace an excellent machine, we write

$$D(E) = \{NR\} \qquad D(G) = D(A) = D(B) = \{R, NR\}$$

We are given the following transition probabilities:

$$p(E|NR, E) = .7 \qquad p(G|NR, E) = .3 \qquad p(A|NR, E) = 0 \qquad p(B|NR, E) = 0$$
$$p(E|NR, G) = 0 \qquad p(G|NR, G) = .7 \qquad p(A|NR, G) = .3 \qquad p(B|NR, G) = 0$$
$$p(E|NR, A) = 0 \qquad p(G|NR, A) = 0 \qquad p(A|NR, A) = .6 \qquad p(B|NR, A) = .4$$
$$p(E|NR, B) = 0 \qquad p(G|NR, B) = 0 \qquad p(A|NR, B) = 0 \qquad p(B|NR, B) = 1$$

If we replace a machine with an excellent machine, the transition probabilities will be the same as if we had begun the week with an excellent machine. Thus,

$$p(E|G, R) = p(E|A, R) = p(E|B, R) = .7$$
$$p(G|G, R) = p(G|A, R) = p(G|B, R) = .3$$
$$p(A|G, R) = p(A|A, R) = p(A|B, R) = 0$$
$$p(B|G, R) = p(B|A, R) = p(B|B, R) = 0$$

If the machine is not replaced, then during the week, we receive the revenues given in the problem. Therefore, $r_{E,NR} = \$100$, $r_{G,NR} = \$80$, $r_{A,NR} = \$50$, and $r_{B,NR} = \$10$. If we replace a machine with an excellent machine, then no matter what type of machine we had at the beginning of the week, we receive $\$100$ and pay a cost of $\$200$. Thus, $r_{E,R} = r_{G,R} = r_{A,R} = r_{B,R} = -\$100$.

In an MDP, what criterion should be used to determine the correct decision? Answering this question requires that we discuss the idea of an **optimal policy** for an MDP.

DEFINITION ■   A **policy** is a rule that specifies how each period's decision is chosen.   ■

Period t's decision may depend on the prior history of the process. Thus, period t's decision can depend on the state during periods $1, 2, \ldots, t$ and the decisions chosen during periods $1, 2, \ldots, t - 1$.

DEFINITION ■   A policy δ is a **stationary policy** if whenever the state is i, the policy δ chooses (independently of the period) the same decision (call this decision $\delta(i)$).   ■

We let δ represent an arbitrary policy and Δ represent the set of all policies. Then

$\mathbf{X}_t$ = random variable for the state of MDP at the beginning of period t (for example, $\mathbf{X}_2, \mathbf{X}_3, \ldots, \mathbf{X}_n$)

X_1 = given state of the process at beginning of period 1 (initial state)

d_t = decision chosen during period t

$V_\delta(i)$ = expected discounted reward earned during an infinite number of periods, given that at beginning of period 1, state is i and stationary policy will be δ

Then

$$V_\delta(i) = E_\delta \left(\sum_{t=1}^{t=\infty} \beta^{t-1} r_{\mathbf{X}_t d_t} | X_1 = i \right)$$

where $E_\delta(\beta^{t-1} r_{\mathbf{X}_t d_t} | X_1 = i)$ is the expected discounted reward earned during period t, given that at the beginning of period 1, the state is i and stationary policy δ is followed.

In a maximization problem, we define

$$V(i) = \max_{\delta \in \Delta} V_\delta(i) \qquad \text{(14)}$$

In a minimization problem, we define

$$V(i) = \min_{\delta \in \Delta} V_\delta(i)$$

DEFINITION ■ If a policy δ^* has the property that for all $i \in S$

$$V(i) = V_{\delta^*}(i)$$

then δ^* is an **optimal policy. ■**

The existence of a single policy δ^* that simultaneously attains all N maxima in (14) is not obvious. If the r_{id}'s are bounded, Blackwell (1962) has shown that an optimal policy exists, and there is always a stationary policy that is optimal. (Even if the r_{id}'s are not bounded, an optimal policy may exist.)

We now consider three methods that can be used to determine an optimal stationary policy:

1 Policy iteration

2 Linear programming

3 Value iteration, or successive approximations

Policy Iteration

Value Determination Equations

Before we can explain the policy iteration method, we need to determine a system of linear equations that can be used to find $V_\delta(i)$ for $i \in S$ and any stationary policy δ. Let $\delta(i)$ be the decision chosen by the stationary policy δ whenever the process begins a period in state i. Then $V_\delta(i)$ can be found by solving the following system of N linear equations, the value determination equations:

$$V_\delta(i) = r_{i,\delta(i)} + \beta \sum_{j=1}^{j=N} p(j|i, \delta(i)) V_\delta(j) \qquad (i = 1, 2, \ldots, N) \qquad \text{(15)}$$

To justify (15), suppose we are in state i and we follow a stationary policy δ. The current period is period 1. Then the expected discounted reward earned during an infinite number of periods consists of $r_{i,\delta(i)}$ (the expected reward received during the current period) plus β (expected discounted reward, to beginning of period 2, earned from period 2 onward). But with probability $p(j|i,\delta(i))$, we will begin period 2 in state j and earn an expected discounted reward, back to period 2, of $V_\delta(j)$. Thus, the expected discounted reward, discounted back to the beginning of period 2 and earned from the beginning of period 2 onward, is given by

$$\sum_{j=1}^{j=N} p(j|i, \, \delta(i))V_\delta(j)$$

Equation (15) now follows.

To illustrate the use of the value determination equations, we consider the following stationary policy for the machine replacement example:

$$\delta(E) = \delta(G) = NR \qquad \delta(A) = \delta(B) = R$$

This policy replaces a bad or average machine and does not replace a good or excellent machine. For this policy, (15) yields the following four equations:

$$V_\delta(E) = 100 + .9(.7V_\delta(E) + .3V_\delta(G))$$
$$V_\delta(G) = 80 + .9(.7V_\delta(G) + .3V_\delta(A))$$
$$V_\delta(A) = -100 + .9(.7V_\delta(E) + .3V_\delta(G))$$
$$V_\delta(B) = -100 + .9(.7V_\delta(E) + .3V_\delta(G))$$

Solving these equations yields $V_\delta(E) = 687.81$, $V_\delta(G) = 572.19$, $V_\delta(A) = 487.81$, and $V_\delta(B) = 487.81$.

Howard's Policy Iteration Method

We now describe Howard's (1960) policy iteration method for finding an optimal stationary policy for an MDP (max problem).

Step 1 Policy evaluation—Choose a stationary policy δ and use the value determination equations to find $V_\delta(i)(i = 1, 2, \ldots, N)$.

Step 2 Policy improvement—For all states $i = 1, 2, \ldots, N$, compute

$$T_\delta(i) = \max_{d \in D(i)} \left(r_{id} + \beta \sum_{j=1}^{j=N} p(j|i, d)V_\delta(j) \right) \tag{16}$$

Since we can choose $d = \delta(i)$ for $i = 1, 2, \ldots, N$, $T_\delta(i) \geq V_\delta(i)$. If $T_\delta(i) = V_\delta(i)$ for $i = 1, 2, \ldots N$, then δ is an optimal policy. If $T_\delta(i) > V_\delta(i)$ for at least one state, then δ is not an optimal policy. In this case, modify δ so that the decision in each state i is the decision attaining the maximum in (16) for $T_\delta(i)$. This yields a new stationary policy δ' for which $V_{\delta'}(i) \geq V_\delta(i)$ for $i = 1, 2, \ldots N$, and for at least one state i', $V_{\delta'}(i') > V_\delta(i')$. Return to step 1, with policy δ' replacing policy δ.

In a minimization problem, we replace max in (16) with min. If $T_\delta(i) = V_\delta(i)$ for $i = 1, 2, \ldots, N$, then δ is an optimal policy. If $T_\delta(i) < V_\delta(i)$ for at least one state, then δ is not an optimal policy. In this case, modify δ so that the decision in each state i is the decision attaining the minimum in (16) for $T_\delta(i)$. This yields a new stationary policy δ' for which $V_{\delta'}(i) \leq V_\delta(i)$ for $i = 1, 2, \ldots N$, and for at least one state i', $V_{\delta'}(i') < V_\delta(i')$. Return to step 1, with policy δ' replacing policy δ.

The policy iteration method is guaranteed to find an optimal policy for the machine replacement example after evaluating a finite number of policies. We begin with the following stationary policy:

$$\delta(E) = \delta(G) = NR \qquad \delta(A) = \delta(B) = R$$

For this policy, we have already found that $V_\delta(E) = 687.81$, $V_\delta(G) = 572.19$, $V_\delta(A) = 487.81$, and $V_\delta(B) = 487.81$. We now compute $T_\delta(E)$, $T_\delta(G)$, $T_\delta(A)$, and $T_\delta(B)$. Since NR is the only possible decision in E,

$$T_\delta(E) = V_\delta(E) = 687.81$$

and $T_\delta(E)$ is attained by the decision NR.

$$T_\delta(G) = \max \begin{cases} -100 + .9(.7V_\delta(E) + .3V_\delta(G)) = 487.81 & (R) \\ 80 + .9(.7V_\delta(G) + .3V_\delta(A)) = V_\delta(G) = 572.19^* & (NR) \end{cases}$$

Thus, $T_\delta(G) = 572.19$ is attained by the decision NR.

$$T_\delta(A) = \max \begin{cases} -100 + .9(.7V_\delta(E) + .3V_\delta(G)) = 487.81 & (R) \\ 50 + .9(.6V_\delta(A) + .4V_\delta(B)) = 489.03^* & (NR) \end{cases}$$

Thus, $T_\delta(A) = 489.03$ is attained by the decision NR.

$$T_\delta(B) = \max \begin{cases} -100 + .9(.7V_\delta(E) + .3V_\delta(G)) = V_\delta(B) = 487.81^* & (R) \\ 10 + .9V_\delta(B) = 449.03 & (NR) \end{cases}$$

Thus, $T_\delta(B) = V_\delta(B) = 487.81$. We have found that $T_\delta(E) = V_\delta(E)$, $T_\delta(G) = V_\delta(G)$, $T_\delta(B) = V_\delta(B)$, and $T_\delta(A) > V_\delta(A)$. Thus, the policy δ is not optimal, and the policy δ' given by $\delta'(E) = \delta'(G) = \delta'(A) = NR$, $\delta'(B) = R$, is an improvement over δ. We now return to step 1 and solve the value determination equations for δ'. From (15), the value determination equations for δ' are

$$V_{\delta'}(E) = 100 + .9(.7V_{\delta'}(E) + .3V_{\delta'}(G))$$
$$V_{\delta'}(G) = 80 + .9(.7V_{\delta'}(G) + .3V_{\delta'}(A))$$
$$V_{\delta'}(A) = 50 + .9(.6V_{\delta'}(A) + .4V_{\delta'}(B))$$
$$V_{\delta'}(B) = -100 + .9(.7V_{\delta'}(E) + .3V_{\delta'}(G))$$

Solving these equations, we obtain $V_{\delta'}(E) = 690.23$, $V_{\delta'}(G) = 575.50$, $V_{\delta'}(A) = 492.35$, and $V_{\delta'}(B) = 490.23$. Observe that in each state i, $V_{\delta'}(i) > V_\delta(i)$. We now apply the policy iteration procedure to δ'. We compute

$$T_{\delta'}(E) = V_{\delta'}(E) = 690.23$$

$$T_{\delta'}(G) = \max \begin{cases} -100 + .9(.7V_{\delta'}(E) + .3V_{\delta'}(G)) = 490.23 & (R) \\ 80 + .9(.7V_{\delta'}(G) + .3V_{\delta'}(A)) = V_{\delta'}(G) = 575.50^* & (NR) \end{cases}$$

Thus, $T_{\delta'}(G) = V_{\delta'}(G) = 575.50$ is attained by NR.

$$T_{\delta'}(A) = \max \begin{cases} -100 + .9(.7V_{\delta'}(E) + .3V_{\delta'}(G)) = 490.23 & (R) \\ 50 + .9(.6V_{\delta'}(A) + .4V_{\delta'}(B)) = V_{\delta'}(A) = 492.35^* & (NR) \end{cases}$$

Thus, $T_{\delta'}(A) = V_{\delta'}(A) = 492.35$ is attained by NR.

$$T_{\delta'}(B) = \max \begin{cases} -100 + .9(.7V_{\delta'}(E) + .3V_{\delta'}(G)) = V_\delta(B) = 490.23^* & (R) \\ 10 + .9V_{\delta'}(B) = 451.21 & (NR) \end{cases}$$

Thus, $T_{\delta'}(B) = V_{\delta'}(B) = 490.23$ is attained by R.

For each state i, $T_{\delta'}(i) = V_{\delta'}(i)$. Thus, δ' is an optimal stationary policy. To maximize expected discounted rewards (profits), a bad machine should be replaced, but an excellent, good, or average machine should not be replaced. If we began period 1 with an excellent machine, an expected discounted reward of $690.23 could be earned.

Linear Programming

It can be shown (see Ross (1983)) that an optimal stationary policy for a maximization problem can be found by solving the following LP:

$$\min z = V_1 + V_2 + \cdots + V_N$$

$$\text{s.t.} \quad V_i - \beta \sum_{j=1}^{j=N} p(j|i, d)V_j \geq r_{id} \quad \text{(For each state } i \text{ and each } d \in d(i))$$

All variables urs

For a minimization problem, we solve the following LP:

$$\max z = V_1 + V_2 + \cdots + V_N$$

$$\text{s.t.} \quad V_i - \beta \sum_{j=1}^{j=N} p(j|i, d)V_j \leq r_{id} \quad \text{(For each state } i \text{ and each } d \in d(i))$$

All variables urs

The optimal solution to these LPs will have $V_i = V(i)$. Also, if a constraint for state i and decision d is binding (has no slack or excess), then decision d is optimal in state i.

REMARKS **1** In the objective function, the coefficient of each V_i may be any positive number.
2 If all the V_i's are nonnegative (this will surely be the case if all the r_{id}'s are nonnegative), we may assume that all variables are nonnegative. If it is possible for some state to have $V(i)$ negative, then we must replace each variable V_i by $V_i' - V_i''$, where both V_i' and V_i'' are nonnegative.
3 With LINDO, we may allow $V(i)$ to be negative with the statement **FREE** Vi. With LINGO, use the **@FREE** statement to allow a variable to assume a negative value.

Our machine replacement example yields the following LP:

$$\min z = V_E + V_G + V_A + V_B$$

$$\begin{aligned}
\text{s.t.} \quad & V_E \geq 100 + .9(.7V_E + .3V_G) && (NR \text{ in } E)\\
& V_G \geq 80 + .9(.7V_G + .3V_A) && (NR \text{ in } G)\\
& V_G \geq -100 + .9(.7V_E + .3V_G) && (R \text{ in } G)\\
& V_A \geq 50 + .9(.6V_A + .4V_B) && (NR \text{ in } A)\\
& V_A \geq -100 + .9(.7V_E + .3V_G) && (R \text{ in } A)\\
& V_B \geq 10 + .9V_B && (NR \text{ in } B)\\
& V_B \geq -100 + .9(.7V_E + .3V_G) && (R \text{ in } B)
\end{aligned}$$

All variables urs

The LINDO output for this LP yields $V_E = 690.23$, $V_G = 575.50$, $V_A = 492.35$, and $V_B = 490.23$. These values agree with those found via the policy iteration method. The LINDO output also indicates that the first, second, fourth, and seventh constraints have no slack. Thus, the optimal policy is to replace a bad machine and not to replace an excellent, good, or average machine.

Value Iteration

There are several versions of value iteration (see Denardo (1982)). We discuss for a maximization problem the simplest value iteration scheme, also known as *successive approx-*

imations. Let $V_t(i)$ be the maximum expected discounted reward that can be earned during t periods if the state at the beginning of the current period is i. Then

$$V_t(i) = \max_{d \in D(i)} \left\{ r_{id} + \beta \sum_{j=1}^{j=N} p(j|i, d)V_{t-1}(j) \right\} \qquad (t \geq 1)$$

$$V_0(i) = 0$$

This result follows, because during the current period, we earn an expected reward (in current dollars) of r_{id}, and during the next $t - 1$ periods, our expected discounted reward (in terms of period 2 dollars) is

$$\sum_{j=1}^{j=N} p(j|i, d)V_{t-1}(j)$$

Let $d_t(i)$ be the decision that must be chosen during period 1 in state i to attain $V_t(i)$. For an MDP with a finite state space and each $D(i)$ containing a finite number of elements, the most basic result in successive approximations states that for $i = 1, 2, \ldots, N$,

$$|V_t(i) - V(i)| \leq \frac{\beta^t}{1 - \beta} \max_{i,d} |r_{id}|$$

Recall that $V(i)$ is the maximum expected discounted reward earned during an infinite number of periods if the state is i at the beginning of the current period. Then

$$\lim_{t \to \infty} d_t(i) = \delta^*(i)$$

where $\delta^*(i)$ defines an optimal stationary policy. Since $\beta < 1$, for t sufficiently large, $V_t(i)$ will come arbitrarily close to $V(i)$. For instance, in the machine replacement example, $\beta = .9$ and $\max |r_{id}| = 100$. Thus, for all states, $V_{50}(i)$ would differ by at most $(.9)^{50}(\frac{100}{.10}) = \$5.15$ from $V(i)$. The equation

$$\lim_{t \to \infty} d_t(i) = \delta^*(i)$$

implies that for t sufficiently large, the decision that is optimal in state i for a t-period problem is also optimal in state i for an infinite horizon problem. This result is reminiscent of the turnpike theorem result for the knapsack problem that was discussed in Chapter 6.

Unfortunately, there is usually no easy way to determine a t^* such that for all i and $t \geq t^*$, $d_t(i) = \delta^*(i)$. (See Denardo (1982) for a partial result in this direction.) Despite this fact, value iteration methods usually obtain a satisfactory approximation to the $V(i)$ and $\delta^*(i)$ with less computational effort than is needed by the policy iteration method or by linear programming. Again, see Denardo (1982) for a discussion of this matter.

We illustrate the computation of V_1 and V_2 for the machine replacement example:

$$V_1(E) = 100 \qquad (NR)$$

$$V_1(G) = \max \begin{cases} 80^* & (NR) \\ -100 & (R) \end{cases} = 80$$

$$V_1(A) = \max \begin{cases} 50^* & (NR) \\ -100 & (R) \end{cases} = 50$$

$$V_1(B) = \max \begin{cases} 10^* & (NR) \\ -100 & (R) \end{cases} = 10$$

The * indicates the action attaining $V_1(i)$. Then

$$V_2(E) = 100 + .9(.7V_1(E) + .3V_1(G)) = 184.6 \qquad (NR)$$

$$V_2(G) = \max \begin{cases} 80 + .9(.7V_1(G) + .3V_1(A)) = 143.9^* & (NR) \\ -100 + .9(.7V_1(E) + .3V_1(G)) = -15.4 & (R) \end{cases}$$

$$V_2(A) = \max \begin{cases} 50 + .9(.6V_1(A) + .4V_1(B)) = 80.6^* & (NR) \\ -100 + .9(.7V_1(E) + .3V_1(G)) = -15.4 & (R) \end{cases}$$

$$V_2(B) = \max \begin{cases} 10 + .9V_1(B) = 19^* & (NR) \\ -100 + .9(.7V_1(E) + .3V_1(G)) = -15.4 & (R) \end{cases}$$

The * now indicates the decision $d_2(i)$ attaining $V_2(i)$. Observe that after two iterations of successive aproximations, we have not yet come close to the actual values of $V(i)$ and have not found it optimal to replace even a bad machine.

In general, if we want to ensure that all the $V_t(i)$'s are within ϵ of the corresponding $V(i)$, we would perform t^* iterations of successive approximations, where

$$\frac{\beta^{t^*}}{1 - \beta} \max_{i,d} |r_{id}| < \epsilon$$

There is no guarantee, however, that after t^* iterations of successive approximations, the optimal stationary policy will have been found.

Maximizing Average Reward per Period

We now briefly discuss how linear programming can be used to find a stationary policy that maximizes the expected per-period reward earned over an infinite horizon. Consider a decision rule or policy Q that chooses decision $d \in D(i)$ with probability $q_i(d)$ during a period in which the state is i. A policy Q will be a stationary policy if each $q_i(d)$ equals 0 or 1. To find a policy that maximizes expected reward per period over an infinite horizon, let π_{id} be the fraction of all periods in which the state is i and the decision $d \in D(i)$ is chosen. Then the expected reward per period may be written as

$$\sum_{i=1}^{i=N} \sum_{d \in D(i)} \pi_{id} r_{id} \tag{17}$$

What constraints must be satisfied by the π_{id}? First, all π_{id}'s must be nonnegative. Second,

$$\sum_{i=1}^{i=N} \sum_{d \in D(i)} \pi_{id} = 1$$

must hold. Finally, the fraction of all periods during which a transition occurs out of state j must equal the fraction of all periods during which a transition occurs into state j. This is identical to the restriction on steady-state probabilities for Markov chains discussed in Section 17.5. This yields (for $j = 1, 2, \ldots, n$),

$$\sum_{d \in D(j)} \pi_{jd}(1 - p(j|j, d)) = \sum_{d \in D(i)} \sum_{i \neq j} \pi_{id} p(j|i, d)$$

Rearranging the last equality yields (for $j = 1, 2, \ldots, N$)

$$\sum_{d \in D(j)} \pi_{jd} = \sum_{d \in D(i)} \sum_{i=1}^{i=N} \pi_{id} p(j|i, d)$$

Putting together our objective function (17) and all the constraints yields the following LP:

$$\max z = \sum_{i=1}^{i=N} \sum_{d \in D(i)} \pi_{id} r_{id}$$

$$\text{s.t.} \quad \sum_{i=1}^{i=N} \sum_{d \in D(i)} \pi_{id} = 1$$

(18)

$$\sum_{d \in D(j)} \pi_{jd} = \sum_{d \in D(i)} \sum_{i=1}^{i=N} \pi_{id} p(j|i, d)$$

$$(j = 1, 2, \ldots, N)$$

$$\text{All } \pi_{id's} \geq 0$$

It can be shown that this LP has an optimal solution in which for each i, at most one $\pi_{id} > 0$. This optimal solution implies that expected reward per period is maximized by a solution in which each $q_i(d)$ equals 0 or 1. Thus, the optimal solution to (18) will occur for a stationary policy. For states having $\pi_{id} = 0$, any decision may be chosen without affecting the expected reward per period.

We illustrate the use of (18) for Example 11 (machine replacement). For this example, (18) yields

$$\max z = 100\pi_{ENR} + 80\pi_{GNR} + 50\pi_{ANR} + 10\pi_{BNR} - 100(\pi_{GR} + \pi_{AR} + \pi_{BR})$$

$$\text{s.t.} \quad \pi_{ENR} + \pi_{GNR} + \pi_{ANR} + \pi_{BNR} + \pi_{GR} + \pi_{AR} + \pi_{BR} = 1$$

$$\pi_{ENR} = .7(\pi_{ENR} + \pi_{GR} + \pi_{AR} + \pi_{BR})$$

$$\pi_{GNR} + \pi_{GR} = .3(\pi_{GR} + \pi_{AR} + \pi_{BR} + \pi_{ENR}) + .7\pi_{GNR}$$

$$\pi_{AR} + \pi_{ANR} = .3\pi_{GNR} + .6\pi_{ANR}$$

$$\pi_{BR} + \pi_{BNR} = \pi_{BNR} + .4\pi_{ANR}$$

Using LINDO, we find the optimal objective function value for this LP to be $z = 60$. The only nonzero decision variables are $\pi_{ENR} = .35$, $\pi_{GNR} = .50$, $\pi_{AR} = .15$. Thus, an average of $60 profit per period can be earned by not replacing an excellent or good machine but replacing an average machine. Since we are replacing an average machine, the action chosen during a period in which a machine is in bad condition is of no importance.

PROBLEMS

Group A

1 A warehouse has an end-of-period capacity of 3 units. During a period in which production takes place, a setup cost of $4 is incurred. A $1 holding cost is assessed against each unit of a period's ending inventory. Also, a variable production cost of $1 per unit is incurred. During each period, demand is equally likely to be 1 or 2 units. All demand must be met on time, and $\beta = .8$. The goal is to minimize expected discounted costs over an infinite horizon.

a Use the policy iteration method to determine an optimal stationary policy.

b Use linear programming to determine an optimal stationary policy.

c Perform two iterations of value iteration.

2 Priceler Auto Corporation must determine whether or not to give consumers 8% or 11% financing on new cars. If Priceler gives 8% financing during the current month, the probability distribution of sales during the current month will be as shown in Table 9. If Priceler gives 11% financing during the current month, the probability distribution of sales during the current month will be as shown in Table 10. "Good" sales represents 400,000 sales per month, "bad" sales represents 300,000 sales per month. For example, if last month's sales were bad and Priceler gives 8% financing during the current month, there is a .40 chance that sales will be good during the current month. At 11% financing rates, Priceler earns $1,000 per car, and at 8% financing, Priceler earns $800 per car. Priceler's goal is to maximize

TABLE 9

Last Month's Sales	Current Month's Sales	
	Good	Bad
Good	.95	.05
Bad	.40	.60

TABLE 10

Last Month's Sales	Current Month's Sales	
	Good	Bad
Good	.80	.20
Bad	.20	.80

TABLE 11

Today's Price	Tomorrow's Price			
	$0	$1	$2	$3
$0	.5	.3	.1	.1
$1	.1	.5	.2	.2
$2	.2	.1	.5	.2
$3	.1	.1	.3	.5

expected discounted profit over an infinite horizon (use $\beta = .98$).

a Use the policy iteration method to determine an optimal stationary policy.

b Use linear programming to determine an optimal stationary policy.

c Perform two iterations of value iteration.

d Find a policy that maximizes average profit per month.

3 Suppose you are using the policy iteration method to determine an optimal policy for an MDP. How might you use LINDO to solve the value determination equations?

Group B

4 During any day, I may own either 0 or 1 share of a stock. The price of the stock is governed by the Markov chain shown in Table 11. At the beginning of a day in which I own a share of stock, I may either sell it at today's price or keep it. At the beginning of a day in which I don't own a share of stock, I may either buy a share of stock at today's price

or not buy a share. My goal is to maximize my expected discounted profit over an infinite horizon (use $\beta = .95$).

a Use the policy iteration method to determine an optimal stationary policy.

b Use linear programming to determine an optimal stationary policy.

c Perform two iterations of value iteration.

d Find a policy that maximizes average daily profit.

5 Ethan Sherwood owns two printing presses, on which he prints two types of jobs. At the beginning of each day, there is a .5 probability that a type 1 job will arrive, a .1 probability that a type 2 job will arrive, and a .4 probability that no job will arrive. Ethan receives $400 for completing a type 1 job and $200 for completing a type 2 job. (Payment for each job is received in advance.) Each type of job takes an average of three days to complete. To model this, we assume that each day a job is in press there is a $\frac{1}{3}$ probability that its printing will be completed at the end of the day. If both presses are busy at the beginning of the day, any arriving job is lost to the system. The crucial decision is when (if ever) Ethan should accept the less profitable type 2 job. Ethan's goal is to maximize expected discounted profit (use $\beta = .90$).

a Use the policy iteration method to determine an optimal stationary policy.

b Use linear programming to determine an optimal stationary policy.

c Perform two iterations of value iteration.

SUMMARY Key to Formulating Probabilistic Dynamic Programming Problems (PDPs)

Suppose the possible states during period $t + 1$ are $s_1, s_2, \ldots s_n$, and the probability that the period $t + 1$ state will be s_i is p_i. Then the minimum expected cost incurred during periods $t + 1, t + 2, \ldots,$ end of the problem is

$$\sum_{i=1}^{i=n} p_i f_{t+1}(s_i)$$

where $f_{t+1}(s_i)$ is the minimum expected cost incurred from period $t + 1$ to the end of the problem, given that the state during period $t + 1$ is s_i.

Maximizing the Probability of a Favorable Event Occurring

To maximize the probability that a favorable event will occur, assign a reward of 1 if the favorable event occurs and a reward of 0 if it does not occur.

Markov Decision Processes

A **Markov decision process** (MDP) is simply an infinite-horizon PDP. Let $V_\delta(i)$ be the expected discounted reward earned during an infinite number of periods, given that at the beginning of period 1, the state is i and the stationary policy δ is followed.

For a maximization problem, we define

$$V(i) = \max_{\delta \in \Delta} V_\delta(i)$$

For a minimization problem, we define

$$V(i) = \min_{\delta \in \Delta} V_\delta(i)$$

If a policy δ^* has the property that for all $i \in S$,

$$V(i) = V_{\delta^*}(i)$$

then δ^* is an **optimal policy.** We can use the **value determination equations** to determine $V_\delta(i)$:

$$V_\delta(i) = r_{i,\delta(i)} + \beta \sum_{j=1}^{j=N} p(j|i, \delta(i))V_\delta(j) \qquad (i = 1, 2, \ldots, N) \tag{15}$$

An optimal policy for an MDP may be determined by one of three methods:

1 Policy iteration

2 Linear programming

3 Value iteration, or successive approximations

Policy Iteration

A summary of Howard's policy iteration method for a maximization problem follows.

Step 1 Policy evaluation—Choose a stationary policy δ and use the value determination equations to find $V_\delta(i)(i = 1, 2, \ldots, N)$.

Step 2 Policy improvement—For all states $i = 1, 2, \ldots, N$, compute

$$T_\delta(i) = \max_{d \in D(i)} \left\{ r_{id} + \beta \sum_{j=1}^{j=N} p(j|i, d)V_\delta(j) \right\} \tag{16}$$

Since we can choose $d = \delta(i)$ for $i = 1, 2, \ldots, N$, $T_\delta(i) \geq V_\delta(i)$. If $T_\delta(i) = V_\delta(i)$ for $i = 1, 2, \ldots, N$, then δ is an optimal policy. If $T_\delta(i) > V_\delta(i)$ for at least one state, then δ is not an optimal policy. In this case, modify δ so that the decision in each state i is the decision attaining the maximum in (16) for $T_\delta(i)$. This yields a new stationary policy δ' for which $V_{\delta'}(i) \geq V_\delta(i)$ for $i = 1, 2, \ldots, N$, and for at least one state i', $V_{\delta'}(i') > V_\delta(i')$. Return to step 1, with policy δ' replacing policy δ.

Linear Programming

In a maximization problem, $V(i)$ for each state may be determined by solving the following LP:

$$\min z = V_1 + V_2 + \cdots + V_N$$

$$\text{s.t.} \quad V_i - \beta \sum_{j=1}^{j=N} p(j|i, d)V_j \geq r_{id} \qquad \text{(For each state } i \text{ and each } d \in D(i))$$

$$\text{All variables urs}$$

If the constraint for state i and decision d has no slack, then decision d is optimal in state i.

Value Iteration, or Successive Approximations

Let $V_t(i)$ be the maximum expected discounted reward that can be earned during t periods if the state at the beginning of the current period is i. Then

$$V_t(i) = \max_{d \in D(i)} \left\{ r_{id} + \beta \sum_{j=1}^{j=N} p(j|i, d)V_{t-1}(j) \right\} \qquad (t \geq 1)$$

$$V_0(i) = 0$$

As t grows large, $V_t(i)$ will approach $V(i)$. For t sufficiently large, the decision that is optimal in state i for a t-period problem is also optimal in state i for an infinite-horizon problem.

REVIEW PROBLEMS

Group A

1 A company has five sales representatives available for assignment to three sales districts. The sales in each district during the current year depend on the number of sales representatives assigned to the district and on whether the national economy has a bad or good year (see Table 12). In the Sales column for each district, the first number represents sales if the national economy had a bad year, and the second number represents sales if the economy had a good year. There is a .3 chance that the national economy will have a good year and a .7 chance that the national economy will have a bad year. Use dynamic programming to determine an assignment of sales representatives to districts that maximizes the company's expected sales.

TABLE 12

No. of Sales Reps Assigned to District	Sales (millions)		
	District 1	District 2	District 3
0	$1, $4	$2, $5	$3, $4
1	$2, $6	$4, $6	$5, $5
2	$3, $7	$5, $6	$6, $7
3	$4, $8	$6, $6	$7, $7

2 At the beginning of each period, a company must determine how many units to produce. A setup cost of $5 is incurred during each period in which production takes place. The production of each unit also incurs a $2 variable cost. All demand must be met on time, and there is a $1 per-unit holding cost on each period's ending inventory. During each period, it is equally likely that demand will equal 0 or 1 unit. Assume that each period's ending inventory cannot exceed 2 units.

a Use dynamic programming to minimize the expected costs incurred during three periods. Assume that the initial inventory is 0 units.

b Now suppose that each unit demanded can be sold for $4. If the demand is not met on time, the sale is lost. Use dynamic programming to maximize the expected profit earned during three periods. Assume that the initial inventory is 0 units.

c In parts (a) and (b), is an (s, S) policy optimal?

3 At Hot Dog Queen Restaurant, the following sequence of events occurs during each minute:

a With probability p, a customer arrives and waits in line.

b Hot Dog Queen determines the rate s at which customers are served. If any customers are in the restaurant, then with probability s, one of the customers completes service and leaves the restaurant. It costs $c(s)$ dollars per period to serve customers at a rate s. Each customer spends R dollars, and the customer's food costs Hot Dog Queen $R - 1$ dollars to prepare.

c For each customer in line at the end of the minute, a cost of h dollars is assessed (because of customer inconvenience).

d The next minute begins.

Formulate a recursion that could be used to maximize expected revenues less costs (including customer inconvenience costs) incurred during the next T minutes. Assume that initially there are no customers present.

4 At the beginning of 2004, the United States has B barrels of oil. If x barrels of oil are consumed during a year, then consumers earn a benefit (measured in dollars) of $u(x)$. The United States may spend money on oil exploration. If d dollars are spent during a year on oil exploration, then there is a probability $p(d)$ that an oil field (containing 500,000 barrels of oil) will be found. Formulate a recursion that can be used to maximize the expected discounted benefits less exploration expenditures earned from the beginning of 2004 to the end of the year 2539.

5 I am a contestant on the popular TV show "Tired of Fortune." During the bonus round, I will be asked up to four questions. For each question that is correctly answered, I win a certain amount of money. One incorrect answer, however, means that I lose all the money I have previously won, and the game is over. If I elect to pass, or not answer a question, the game is over, but I may keep what I have already won. The amount of money I win for each correct question and the probability that I will answer each question correctly are shown in Table 13.

a My goal is to maximize the expected amount of money won. Use dynamic programming to accomplish this goal.

b Suppose that I am allowed to pass, or not answer a question, and still go on to the next question. Now determine how to maximize the amount of money won.

6 A machine in excellent condition earns $100 profit per week, a machine in good condition earns $70 per week, and a machine in bad condition earns $20 per week. At the beginning of any week, a machine may be sent out for repairs at a cost of $90. A machine that is sent out for repairs returns in excellent condition at the beginning of the next week. If a machine is not repaired, the condition of the machine evolves in accordance with the Markov chain shown in Table 14. The company wants to maximize its expected discounted profit over an infinite horizon ($\beta = .9$).

TABLE 13

Question	Probability of Correct Answer	Money Won
1	.6	$10,000
2	.5	$20,000
3	.4	$30,000
4	.3	$40,000

TABLE 14

This Week	Next Week		
	Excellent	Good	Bad
Excellent	.7	.2	.1
Good	0	.7	.3
Bad	0	.1	.9

a Use policy iteration to determine an optimal stationary policy.

b Use linear programming to determine an optimal stationary policy.

c Perform two iterations of value iteration.

7 A country now has 10 units of capital. Each year, it may consume any amount of the available capital and invest the rest. Invested capital has a 50% chance of doubling and a 50% chance of losing half its value. For example, if the country invests 6 units of capital, there is a 50% chance that the 6 units will turn into 12 capital units and a 50% chance that the invested capital will turn into 3 units. What strategy should be used to maximize total expected consumption over a four-year period?

8 The Dallas Mavericks trail by two points and have the ball with 10 seconds remaining. They must decide whether to take a two- or a three-point shot. Assume that once the Mavericks take their shot, time expires. The probability that a two-point shot is successful is TWO, and the probability that a three-point shot is successful is THREE. If the game is tied, an overtime period will be played. Assume that there is a .5 chance the Mavericks will win in overtime. (*Note:* This problem is often used on Microsoft job interviews.)

a Give a rule based on the values of TWO and THREE that tells Dallas what to do.

b Typical values for an NBA team are TWO = .45 and THREE = .35. Based on this information, what strategy should most NBA teams follow?

9 At any time, the size of a tree is 0, 1, 2, or 3. We must decide when to harvest the tree. Each year, it costs $1 to maintain the tree. It costs $5 to harvest a tree. The sales price for a tree of each size is as follows:

Tree Size	Sales Price
0	$20
1	$30
2	$45
3	$49

The transition probability matrix for the size of the tree is as follows:

$$
\begin{array}{c} \\ 0 \\ 1 \\ 2 \\ 3 \end{array}
\begin{array}{cccc} 0 & 1 & 2 & 3 \end{array} \\
\begin{bmatrix} .8 & .2 & 0 & 0 \\ 0 & .9 & .1 & 0 \\ 0 & 0 & .7 & .3 \\ 0 & 0 & 0 & 1 \end{bmatrix}
$$

For example, 80% of all size 0 trees begin the next year as size 0 trees, and 20% of all size 0 trees begin the next year as size 1 trees. Assuming the discount factor for cash flows is .9 per year, determine an optimal harvesting strategy.

10 For $50, we can enter a raffle. We draw a certificate containing a number 100, 200, 300, . . . , 1,000. Each number is equally likely. At any time, we can redeem the highest-numbered certificate we have obtained so far for the face value of the certificate. We may enter the raffle as many times as we wish. Assuming no discounting, what strategy would maximize our expected profit? How does this model relate to the problem faced by an unemployed person who is searching for a job?

11 At the beginning of each year, an aircraft engine is in good, fair, or poor condition. It costs $500,000 to run a good engine for a year, $1 million to run a fair engine for a year, and $2 million to run a poor engine for a year. A fair engine can be overhauled for $2 million, and it immediately becomes a good engine. A poor engine can be replaced for $3 million, and it immediately becomes a good engine. The transition probability matrix for an engine is as follows:

	Good	Fair	Poor
Good	.7	.2	.1
Fair	0	.6	.4
Poor	0	0	1

The discount factor for costs is .9. What strategy minimizes expected discounted cost over an infinite horizon?

Group B

12 A syndicate of college students spends weekends gambling in Las Vegas. They begin week 1 with W dollars.

At the beginning of each week, they may wager any amount of their money at the gambling tables. If they wager d dollars, then with probability p, their wealth increases by d dollars, and with probability $1 - p$, their wealth decreases by d dollars. Their goal is to maximize their expected wealth at the end of T weeks.

a Show that if $p \geq \frac{1}{2}$, the students should bet all their money.

b Show that if $p < \frac{1}{2}$, the students should bet no money. (*Hint:* Define $f_t(w)$ as the maximum expected wealth at the end of week T, given that wealth is w dollars at the beginning of week t; by working backward, find an expression for $f_t(w)$.)

Group C

13 You have invented a new product: the HAL DVD player. Each of 1,000 potential customers places a different value on this product. A consumer's valuation is equally likely to be any number between $0 and $1,000. It costs $100 to produce the HAL player. During a year in which we set a price p for the product, all customers valuing the product at p or more will purchase the product. Each year, we set a price for the product. What pricing strategy will maximize our expected profit over three years? What commonly observed phenomenon does this problem illustrate?

REFERENCES

The following books contain elementary discussions of Markov decision processes and probabilistic dynamic programming:

Howard, R. *Dynamic Programming and Markov Processes.* Cambridge, Mass.: MIT Press, 1960.
Wagner, H. *Principles of Operations Research,* 2d ed. Englewood Cliffs, N.J.: Prentice Hall, 1975.

The following books treat Markov decision processes and probabilistic dynamic programming at a more advanced level:

Bersetkas, D. *Dynamic Programming and Optimal Control,* vols. 1 & 2. Cambridge, Mass.: Athena Publishing, 2000.
Heyman, D., and M. Sobel. *Stochastic Models in Operations Research,* vol. 2. New York: McGraw-Hill, 1984.
Kohlas, S. *Stochastic Methods of Operations Research.* Cambridge, U.K.: Cambridge University Press, 1982.
Puterman, M. *Markov Decision Processes: Discrete Stochastic Dynamic Programming.* New York: John Wiley, 1994.
Ross, S. *Introduction to Stochastic Dynamic Programming.* Orlando, Fla.: Academic Press, 1983.

Whittle, P. *Optimization Over Time: Dynamic Programming and Stochastic Control.* New York: Wiley, 1982.
White, D.J. *Markov Decision Processes.* New York: John Wiley, 1993.

Excellent one-chapter introductions to Markov decision processes are given in the following two books:

Denardo, E. *Dynamic Programming Theory and Applications.* Englewood Cliffs, N.J.: Prentice Hall, 1982.
Shapiro, J. *Mathematical Programming: Structures and Algorithms.* New York: Wiley, 1979.

Blackwell, D. "Discrete Dynamic Programming," *Annals of Mathematical Statistics* 33(1962):719–726. Indicates how one proves that a stationary policy is optimal for a Markov decision process.
Scarf, H. "The Optimality of (s, S) Policies for the Dynamic Inventory Problem," *Proceedings of the First Stanford Symposium on Mathematical Methods in the Social Sciences.* Stanford, Calif.: Stanford University Press, 1960. A proof of the optimality of (s, S) policies.

Queuing Theory

Each of us has spent a great deal of time waiting in lines. In this chapter, we develop mathematical models for waiting lines, or queues. In Section 20.1, we begin by discussing some terminology that is often used to describe queues. In Section 20.2, we look at some distributions (the exponential and the Erlang distributions) that are needed to describe queuing models. In Section 20.3, we introduce the idea of a birth–death process, which is basic to many queuing models involving the exponential distribution. The remainder of the chapter examines several models of queuing systems that can be used to answer questions like the following:

1 What fraction of the time is each server idle?

2 What is the expected number of customers present in the queue?

3 What is the expected time that a customer spends in the queue?

4 What is the probability distribution of the number of customers present in the queue?

5 What is the probability distribution of a customer's waiting time?

6 If a bank manager wants to ensure that only 1% of all customers will have to wait more than 5 minutes for a teller, how many tellers should be employed?

20.1 Some Queuing Terminology

To describe a queuing system, an input process and an output process must be specified. Some examples of input and output processes are given in Table 1.

The Input or Arrival Process

The input process is usually called the **arrival process.** Arrivals are called **customers.** In all models that we will discuss, we assume that no more than one arrival can occur at a given instant. For a case like a restaurant, this is a very unrealistic assumption. If more than one arrival can occur at a given instant, we say that **bulk arrivals** are allowed.

Usually, we assume that the arrival process is unaffected by the number of customers present in the system. In the context of a bank, this would imply that whether there are 500 or 5 people at the bank, the process governing arrivals remains unchanged.

There are two common situations in which the arrival process may depend on the number of customers present. The first occurs when arrivals are drawn from a small population. Suppose that there are only four ships in a naval shipyard. If all four ships are being repaired, then no ship can break down in the near future. On the other hand, if all four ships are at sea, a breakdown has a relatively high probability of occurring in the near

TABLE 1
Examples of Queuing Systems

Situation	Input Process	Output Process
Bank	Customers arrive at bank	Tellers serve the customers
Pizza parlor	Requests for pizza delivery are received	Pizza parlor sends out truck to deliver pizzas
Hospital blood bank	Pints of blood arrive	Patients use up pints of blood
Naval shipyard	Ships at sea break down and are sent to shipyard for repairs	Ships are repaired and return to sea

future. Models in which arrivals are drawn from a small population are called **finite source models.** Another situation in which the arrival process depends on the number of customers present occurs when the rate at which customers arrive at the facility decreases when the facility becomes too crowded. For example, if you see that the bank parking lot is full, you might pass by and come another day. If a customer arrives but fails to enter the system, we say that the customer has **balked.** The phenomenon of balking was described by Yogi Berra when he said, "Nobody goes to that restaurant anymore; it's too crowded."

If the arrival process is unaffected by the number of customers present, we usually describe it by specifying a probability distribution that governs the time between successive arrivals.

The Output or Service Process

To describe the output process (often called the service process) of a queuing system, we usually specify a probability distribution—the **service time distribution**—which governs a customer's service time. In most cases, we assume that the service time distribution is independent of the number of customers present. This implies, for example, that the server does not work faster when more customers are present.

In this chapter, we study two arrangements of servers: **servers in parallel** and **servers in series.** Servers are in parallel if all servers provide the same type of service and a customer need only pass through one server to complete service. For example, the tellers in a bank are usually arranged in parallel; any customer need only be serviced by one teller, and any teller can perform the desired service. Servers are in series if a customer must pass through several servers before completing service. An assembly line is an example of a series queuing system.

Queue Discipline

To describe a queuing system completely, we must also describe the queue discipline and the manner in which customers join lines.

The **queue discipline** describes the method used to determine the order in which customers are served. The most common queue discipline is the **FCFS discipline** (first come, first served), in which customers are served in the order of their arrival. Under the **LCFS discipline** (last come, first served), the most recent arrivals are the first to enter service. If we consider exiting from an elevator to be service, then a crowded elevator illustrates an LCFS discipline. Sometimes the order in which customers arrive has no effect on the order in which they are served. This would be the case if the next customer to enter service is randomly chosen from those customers waiting for service. Such a situation is referred to as the **SIRO discipline** (service in random order). When callers to an airline are put on hold, the luck of the draw often determines the next caller serviced by an operator.

Finally, we consider **priority queuing disciplines.** A priority discipline classifies each arrival into one of several categories. Each category is then given a priority level, and within each priority level, customers enter service on an FCFS basis. Priority disciplines are often used in emergency rooms to determine the order in which customers receive treatment, and in copying and computer time-sharing facilities, where priority is usually given to jobs with shorter processing times.

Method Used by Arrivals to Join Queue

Another factor that has an important effect on the behavior of a queuing system is the method that customers use to determine which line to join. For example, in some banks, customers must join a single line, but in other banks, customers may choose the line they want to join. When there are several lines, customers often join the shortest line. Unfortunately, in many situations (such as a supermarket), it is difficult to define the shortest line. If there are several lines at a queuing facility, it is important to know whether or not customers are allowed to switch, or jockey, between lines. In most queuing systems with multiple lines, jockeying is permitted, but jockeying at a toll booth plaza is not recommended.

20.2 Modeling Arrival and Service Processes

Modeling the Arrival Process

As previously mentioned, we assume that at most one arrival can occur at a given instant of time. We define t_i to be the time at which the ith customer arrives. To illustrate this, consider Figure 1. For $i \geq 1$, we define $T_i = t_{i+1} - t_i$ to be the ith interarrival time. Thus, in the figure, $T_1 = 8 - 3 = 5$, and $T_2 = 15 - 8 = 7$. In modeling the arrival process, we assume that the T_i's are independent, continuous random variables described by the random variable **A**. The independence assumption means, for example, that the value of T_2 has no effect on the value of T_3, T_4, or any later T_i. The assumption that each T_i is continuous is usually a good approximation of reality. After all, an interarrival time need not be exactly 1 minute or 2 minutes; it could just as easily be, say, 1.55892 minutes. The assumption that each interarrival time is governed by the same random variable implies that the distribution of arrivals is independent of the time of day or the day of the week. This is the assumption of stationary interarrival times. Because of phenomena such as rush hours, the assumption of stationary interarrival times is often unrealistic, but we may often approximate reality by breaking the time of day into segments. For example, if we were modeling traffic flow, we might break the day up into three segments: a morning rush hour segment, a midday segment, and an afternoon rush hour segment. During each of these segments, interarrival times may be stationary.

We assume that **A** has a density function $a(t)$. Recall from Section 12.5 that for small Δt, $P(t \leq \mathbf{A} \leq t + \Delta t)$ is approximately $\Delta t a(t)$. Of course, a negative interarrival time is impossible. This allows us to write

$$P(\mathbf{A} \leq c) = \int_0^c a(t)dt \quad \text{and} \quad P(\mathbf{A} > c) = \int_c^\infty a(t)dt \tag{1}$$

FIGURE 1
Definition of
Interarrival Times

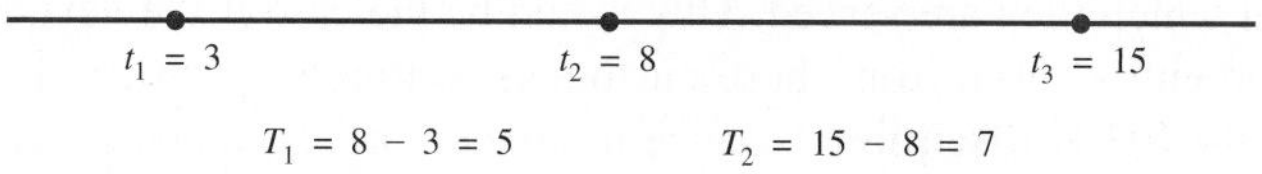

$t_1 = 3$ $t_2 = 8$ $t_3 = 15$

$T_1 = 8 - 3 = 5$ $T_2 = 15 - 8 = 7$

We define $\frac{1}{\lambda}$ to be the mean or average interarrival time. Without loss of generality, we assume that time is measured in units of hours. Then $\frac{1}{\lambda}$ will have units of hours per arrival. From Section 12.5, we may compute $\frac{1}{\lambda}$ from $a(t)$ by using the following equation:

$$\frac{1}{\lambda} = \int_0^\infty ta(t)dt \tag{2}$$

We define λ to be the **arrival rate,** which will have units of arrivals per hour.

In most applications of queuing, an important question is how to choose **A** to reflect reality and still be computationally tractable. The most common choice for **A** is the **exponential distribution.** An exponential distribution with parameter λ has a density $a(t) = \lambda e^{-\lambda t}$. Figure 2 shows the density function for an exponential distribution. We see that $a(t)$ decreases very rapidly for t small. This indicates that very long interarrival times are unlikely. Using Equation (2) and integration by parts, we can show that the average or mean interarrival time (call it $E(\mathbf{A})$) is given by

$$E(\mathbf{A}) = \frac{1}{\lambda} \tag{3}$$

Using the fact that var $\mathbf{A} = E(\mathbf{A}^2) - E(\mathbf{A})^2$, we can show that

$$\text{var } \mathbf{A} = \frac{1}{\lambda^2} \tag{4}$$

No-Memory Property of the Exponential Distribution

The reason the exponential distribution is often used to model interarrival times is embodied in the following lemma.

LEMMA 1

If **A** has an exponential distribution, then for all nonnegative values of t and h,

$$P(\mathbf{A} > t + h | \mathbf{A} \geq t) = P(\mathbf{A} > h) \tag{5}$$

Proof First note that from Equation (1), we have

$$P(\mathbf{A} > h) = \int_h^\infty \lambda e^{-\lambda t} = [-e^{-\lambda t}]_h^\infty = e^{-\lambda h} \tag{6}$$

Then

$$P(\mathbf{A} > t + h | \mathbf{A} \geq t) = \frac{P(\mathbf{A} > t + h \cap \mathbf{A} \geq t)}{P(\mathbf{A} \geq t)}$$

From (6),

$$P(\mathbf{A} > t + h \cap \mathbf{A} \geq t) = e^{-\lambda(t+h)} \quad \text{and} \quad P(\mathbf{A} \geq t) = e^{-\lambda t}$$

Thus,

$$P(\mathbf{A} > t + h | \mathbf{A} \geq t) = \frac{e^{-\lambda(t+h)}}{e^{-\lambda t}} = e^{-\lambda h} = P(\mathbf{A} > h)$$

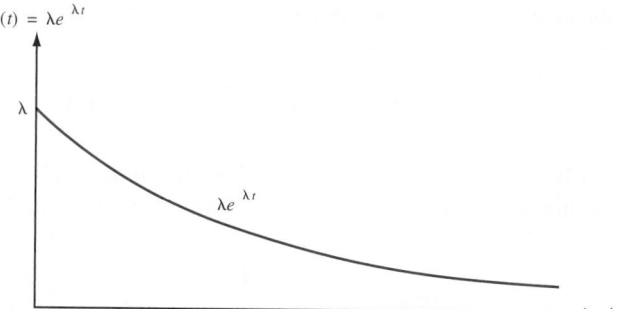

$a(t) = \lambda e^{\lambda t}$

λ

$\lambda e^{\lambda t}$

t

FIGURE 2
Density Function for
Exponential Distribution

It can be shown that no other density function can satisfy (5) (see Feller (1957)). For reasons that become apparent, a density that satisfies (5) is said to have the **no-memory property.** Suppose we are told that there has been no arrival for the last t hours (this is equivalent to being told that $\mathbf{A} \geq t$) and are asked what the probability is that there will be no arrival during the next h hours (that is, $\mathbf{A} > t + h$). Then (5) implies that this probability *does not depend on the value of* t, and for all values of t, this probability equals $P(\mathbf{A} > h)$. In short, if we know that at least t time units have elapsed since the last arrival occurred, then the distribution of the remaining time until the next arrival (h) does not depend on t. For example, if $h = 4$, then (5) yields, for $t = 5$, $t = 3$, $t = 2$, and $t = 0$,

$$P(\mathbf{A} > 9|\mathbf{A} \geq 5) = P(\mathbf{A} > 7|\mathbf{A} \geq 3) = P(\mathbf{A} > 6|\mathbf{A} \geq 2)$$
$$= P(\mathbf{A} > 4|\mathbf{A} \geq 0) = e^{-4\lambda}$$

The no-memory property of the exponential distribution is important, because it implies that if we want to know the probability distribution of the time until the next arrival, then *it does not matter how long it has been since the last arrival.* To put it in concrete terms, suppose interarrival times are exponentially distributed with $\lambda = 6$. Then the no-memory property implies that no matter how long it has been since the last arrival, the probability distribution governing the time until the next arrival has the density function $6e^{-6t}$. This means that to predict future arrival patterns, we need not keep track of how long it has been since the last arrival. This observation can appreciably simplify analysis of a queuing system.

To see that knowledge of the time since the last arrival does affect the distribution of time until the next arrival in most situations, suppose that $\mathbf{A}$ is discrete with $P(\mathbf{A} = 5) = P(\mathbf{A} = 100) = \frac{1}{2}$. If we are told that there has been no arrival during the last 6 time units, we *know with certainty* that it will be $100 - 6 = 94$ time units until the next arrival. On the other hand, if we are told that no arrival has occurred during the last time unit, then there is some chance that the time until the next arrival will be $5 - 1 = 4$ time units and some chance that it will be $100 - 1 = 99$ time units. Hence, in this situation, the distribution of the next interarrival time cannot easily be predicted with knowledge of the time that has elapsed since the last arrival.

Relation Between Poisson Distribution and Exponential Distribution

If interarrival times are exponential, the probability distribution of the number of arrivals occurring in any time interval of length t is given by the following important theorem.

THEOREM 1

Interarrival times are exponential with parameter λ if and only if the number of arrivals to occur in an interval of length t follows a Poisson distribution with parameter λt.

A discrete random variable $\mathbf{N}$ has a Poisson distribution with parameter λ if, for $n = 0, 1, 2, \ldots,$

$$P(\mathbf{N} = n) = \frac{e^{-\lambda}\lambda^n}{n!} \qquad (n = 0, 1, 2, \ldots) \tag{7}$$

If $\mathbf{N}$ is a Poisson random variable, it can be shown that $E(\mathbf{N}) = \text{var }\mathbf{N} = \lambda$. If we define $\mathbf{N}_t$ to be the number of arrivals to occur during any time interval of length t, Theorem 1 states that

$$P(\mathbf{N}_t = n) = \frac{e^{-\lambda t}(\lambda t)^n}{n!} \qquad (n = 0, 1, 2, \ldots)$$

Since $\mathbf{N}_t$ is Poisson with parameter λt, $E(\mathbf{N}_t) = \text{var }\mathbf{N}_t = \lambda t$. An average of λt arrivals occur during a time interval of length t, so λ may be thought of as the average number of arrivals per unit time, or the arrival rate.

What assumptions are required for interarrival times to be exponential? Theorem 2 provides a partial answer. Consider the following two assumptions:

1 Arrivals defined on nonoverlapping time intervals are independent (for example, the number of arrivals occurring between times 1 and 10 does not give us any information about the number of arrivals occurring between times 30 and 50).

2 For small Δt (and any value of t), the probability of one arrival occurring between times t and $t + \Delta t$ is $\lambda\Delta t + o(\Delta t)$, where $o(\Delta t)$ refers to any quantity satisfying

$$\lim_{\Delta t \to 0} \frac{o(\Delta t)}{\Delta t} = 0$$

Also, the probability of no arrival during the interval between t and $t + \Delta t$ is $1 - \lambda\Delta t + o(\Delta t)$, and the probability of more than one arrival occurring between t and $t + \Delta t$ is $o(\Delta t)$.

THEOREM 2

If assumptions 1 and 2 hold, then $\mathbf{N}_t$ follows a Poisson distribution with parameter λt, and interarrival times are exponential with parameter λ; that is, $a(t) = \lambda e^{-\lambda t}$.

In essence, Theorem 2 states that if the arrival rate is stationary, if bulk arrivals cannot occur, and if past arrivals do not affect future arrivals, then interarrival times will follow an exponential distribution with parameter λ, and the number of arrivals in any interval of length t is Poisson with parameter λt. The assumptions of Theorem 2 may appear to be very restrictive, but interarrival times are often exponential even if the assumptions of Theorem 2 are not satisfied (see Denardo (1982)). In Section 20.12, we discuss how to use data to test whether the hypothesis of exponential interarrival times is reasonable. In many applications, the assumption of exponential interarrival times turns out to be a fairly good approximation of reality.

Using Excel to Compute Poisson and Exponential Probabilities

Excel contains functions that facilitate the computation of probabilities concerning the Poisson and exponential random variables.

The syntax of the Excel POISSON function is as follows:

- =POISSON(x,MEAN,TRUE) gives the probability that a Poisson random variable with mean = Mean is less than or equal to x.

	A	B	C	D	E
3					
4	Poisson	Lambda			
5	P(X=40)	40	0.541918		
6					
7	P(X<=40)	40	0.062947		
8					
9	Exponential				
10		Lambda			
11	P(X<=10)	0.1	0.632121		
12	Density for X = 10	0.1	0.036788		
13					
14					
15					
16					

FIGURE 3

- =POISSON(x,MEAN,FALSE) gives probability that a Poisson random variable with mean = Mean is equal to x.

For example, if an average of 40 customers arrive per hour and arrivals follow a Poisson distribution then the function =POISSON(40,40,TRUE) yields the probability .542 that 40 or fewer customers arrive during an hour. The function =POISSON(40,40,FALSE) yields the probability .063 that *exactly* 40 customers arrive during an hour.

The syntax of the Excel EXPONDIST function is as follows:

- =EXPONDIST(x,LAMBDA,TRUE) gives the probability that an exponential random variable with parameter λ assumes a value less than or equal to x.

- =EXPONDIST(x,LAMBDA,FALSE) gives the value of the density function for an exponential random variable with parameter λ.

For example, suppose the average time between arrivals follows an exponential distribution with mean 10. Then $\lambda = .1$, and =EXPONDIST(10,0.1,TRUE) yields the probability .632 that the time between arrivals is 10 minutes or less.

The function =EXPONDIST(10,.1,FALSE) yields the height .037 of the density function for $x = 10$ and $\lambda = .1$. See file Poissexp.xls and Figure 3.

Poissexp.xls

Example 1 illustrates the relation between the exponential and Poisson distributions.

EXAMPLE 1 | **Beer Orders**

The number of glasses of beer ordered per hour at Dick's Pub follows a Poisson distribution, with an average of 30 beers per hour being ordered.

1 Find the probability that exactly 60 beers are ordered between 10 P.M. and 12 midnight.

2 Find the mean and standard deviation of the number of beers ordered between 9 P.M. and 1 A.M.

3 Find the probability that the time between two consecutive orders is between 1 and 3 minutes.

Solution **1** The number of beers ordered between 10 P.M. and 12 midnight will follow a Poisson distribution with parameter $2(30) = 60$. From Equation (7), the probability that 60 beers are ordered between 10 P.M. and 12 midnight is

$$\frac{e^{-60}60^{60}}{60!}$$

Alternatively, we can find the answer with the Excel function =POISSON(60,60,FALSE). This yields .051.

2 We have $\lambda = 30$ beers per hour; $t = 4$ hours. Thus, the mean number of beers ordered between 9 P.M. and 1 A.M. is $4(30) = 120$ beers. The standard deviation of the number of beers ordered between 10 P.M. and 1 A.M. is $(120)^{1/2} = 10.95$.

3 Let **X** be the time (in minutes) between successive beer orders. The mean number of orders per minute is exponential with parameter or rate $\frac{30}{60} = 0.5$ beer per minute. Thus, the probability density function of the time between beer orders is $0.5e^{-0.5t}$. Then

$$P(1 \le \mathbf{X} \le 3) = \int_1^3 (0.5e^{-0.5t})dt = e^{-0.5} - e^{-1.5} = .38$$

Alternatively, we can use Excel to find the answer with the formula

$$=\text{EXPONDIST}(3,.5,\text{TRUE}) - \text{EXPONDIST}(1,.5,\text{TRUE})$$

This yields a probability of .383.

The Erlang Distribution

If interarrival times do not appear to be exponential, they are often modeled by an Erlang distribution. An Erlang distribution is a continuous random variable (call it **T**) whose density function $f(t)$ is specified by two parameters: a rate parameter R and a shape parameter k (k must be a positive integer). Given values of R and k, the Erlang density has the following probability density function:

$$f(t) = \frac{R(Rt)^{k-1}e^{-Rt}}{(k-1)!} \qquad (t \ge 0) \tag{8}$$

Using integration by parts, we can show that if **T** is an Erlang distribution with rate parameter R and shape parameter k, then

$$E(\mathbf{T}) = \frac{k}{R} \qquad \text{and} \qquad \text{var } \mathbf{T} = \frac{k}{R^2} \tag{9}$$

To see how varying the shape parameter changes the shape of the Erlang distribution, we consider for a given value of λ, a family of Erlang distributions with rate parameter $k\lambda$ and shape parameter k. By (9), each of these Erlangs has a mean of $\frac{1}{\lambda}$. As k varies, the Erlang distribution takes on many shapes. For example, Figure 4 shows, for a given value of λ, the density functions for Erlang distributions having shape parameters 1, 2, 4, 6, and 20. For $k = 1$, the Erlang density looks similar to an exponential distribution; in fact, if we set $k = 1$ in (8), we find that for $k = 1$, the Erlang distribution is an exponential distribution with parameter R. As k increases, the Erlang distribution behaves more and more like a normal distribution. For extremely large values of k, the Erlang distribution approaches a random variable with zero variance (that is, a constant interarrival time). Thus, by varying k, we may approximate both skewed and symmetric distributions.

It can be shown that an Erlang distribution with shape parameter k and rate parameter $k\lambda$ has the same distribution as the random variable $\mathbf{A}_1 + \mathbf{A}_2 + \cdots + \mathbf{A}_k$, where each $\mathbf{A}_i$ is an exponential random variable with parameter $k\lambda$, and the $\mathbf{A}_i$'s are independent random variables.

If we model interarrival times as an Erlang distribution with shape parameter k, we are really saying that the interarrival process is equivalent to a customer going through k phases (each of which has the no-memory property) before arriving. For this reason, the shape parameter is often referred to as the *number of phases* of the Erlang distribution.

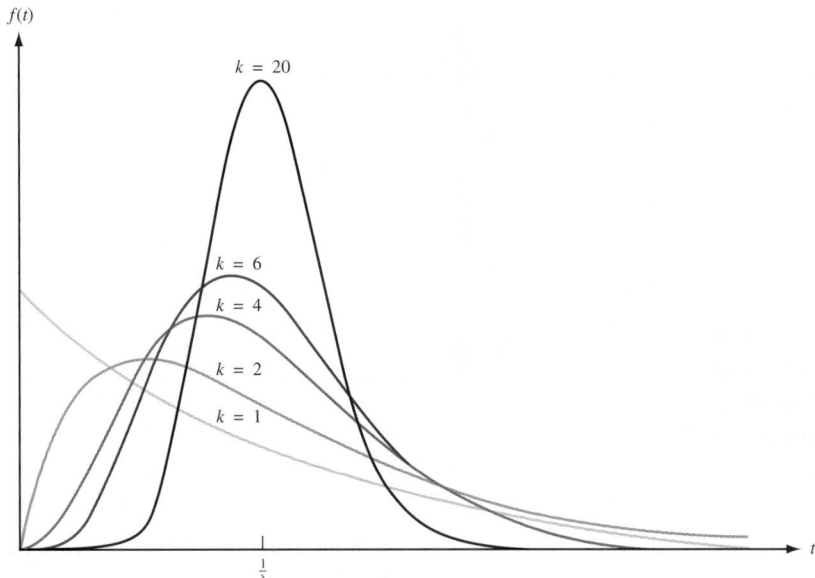

FIGURE 4
Density Functions for
Erlang Distributions

Modeling the Service Process

We now turn our attention to modeling the service process. We assume that the service times of different customers are independent random variables and that each customer's service time is governed by a random variable $\mathbf{S}$ having a density function $s(t)$. We let $\frac{1}{\mu}$ be the mean service time for a customer. Of course,

$$\frac{1}{\mu} = \int_0^\infty ts(t)dt$$

The variable $\frac{1}{\mu}$ will have units of hours per customer, so μ has units of customers per hour. For this reason, we call μ the service rate. For example, $\mu = 5$ means that if customers were always present, the server could serve an average of 5 customers per hour, and the average service time of each customer would be $\frac{1}{5}$ hour. As with interarrival times, we hope that service times can be accurately modeled as exponential random variables. If we can model a customer's service time as an exponential random variable, we can determine the distribution of a customer's remaining service time without having to keep track of how long the customer has been in service. Also note that if service times follow an exponential density $s(t) = \mu e^{-\mu t}$, then a customer's mean service time will be $\frac{1}{\mu}$.

As an example of how the assumption of exponential service times can simplify computations, consider a three-server system in which each customer's service time is governed by an exponential distribution $s(t) = \mu e^{-\mu t}$. Suppose all three servers are busy, and a customer is waiting (see Figure 5). What is the probability that the customer who is waiting will be the last of the four customers to complete service? From Figure 5, it is clear that the following will occur. One of customers 1–3 (say, customer 3) will be the first to complete service. Then customer 4 will enter service. By the no-memory property, customer 4's service time has the same distribution as the remaining service times of customers 1 and 2. Thus, by symmetry, customers 4, 1, and 2 will have the same chance of being the last customer to complete service. This implies that customer 4 has a $\frac{1}{3}$ chance of being the last customer to complete service. Without the no-memory property, this problem would be hard to solve, because it would be very difficult to determine the prob-

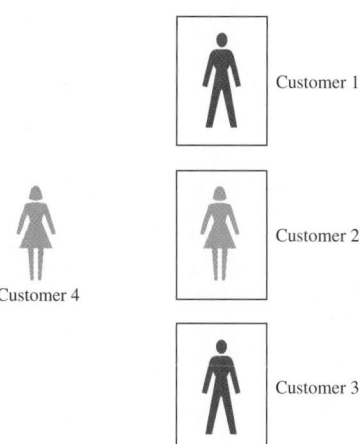

FIGURE 5
Example of Usefulness
of Exponential
Distribution

ability distribution of the remaining service time (after customer 3 completes service) of customers 1 and 2.

Unfortunately, actual service times may not be consistent with the no-memory property. For this reason, we often assume that $s(t)$ is an Erlang distribution with shape parameter k and rate parameter $k\mu$. From (9), this yields a mean service time of $\frac{1}{\mu}$. Modeling service times as an Erlang distribution with shape parameter k also implies that a customer's service time may be considered to consist of passage through k phases of service, in which the time to complete each phase has the no-memory property and a mean of $\frac{1}{k\mu}$ (see Figure 6). In many situations, an Erlang distribution can be closely fitted to observed service times.

In certain situations, interarrival or service times may be modeled as having zero variance; in this case, interarrival or service times are considered to be **deterministic.** For example, if interarrival times are deterministic, then each interarrival time will be exactly $\frac{1}{\lambda}$, and if service times are deterministic, each customer's service time will be exactly $\frac{1}{\mu}$.

The Kendall–Lee Notation for Queuing Systems

We have now developed enough terminology to describe the standard notation used to describe many queuing systems. The notation that we discuss in this section is used to describe a queuing system in which all arrivals wait in a single line until one of s identical parallel servers is free. Then the first customer in line enters service, and so on (see Figure 7). If, for example, the customer in server 3 is the next customer to complete service, then (assuming an FCFS discipline) the first customer in line would enter server 3. The next customer in line would enter service after the next service completion, and so on.

To describe such a queuing system, Kendall (1951) devised the following notation. Each queuing system is described by six characteristics:

$$1/2/3/4/5/6$$

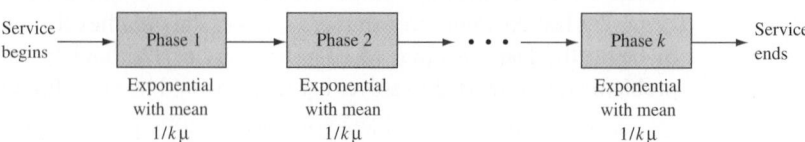

FIGURE 6
Representation of
Erlang Service Time

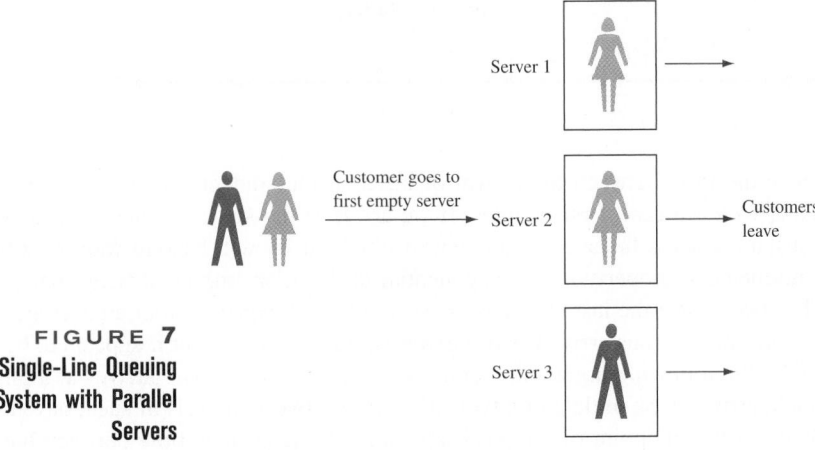

FIGURE 7
Single-Line Queuing System with Parallel Servers

Server 1

Customer goes to first empty server

Server 2

Customers leave

Server 3

The first characteristic specifies the nature of the arrival process. The following standard abbreviations are used:

M = Interarrival times are independent, identically distributed (iid) random variables having an exponential distribution.

D = Interarrival times are iid and deterministic.

E_k = Interarrival times are iid Erlangs with shape parameter k.

GI = Interarrival times are iid and governed by some general distribution.

The second characteristic specifies the nature of the service times:

M = Service times are iid and exponentially distributed.

D = Service times are iid and deterministic.

E_k = Service times are iid Erlangs with shape parameter k.

G = Service times are iid and follow some general distribution.

The third characteristic is the number of parallel servers. The fourth characteristic describes the queue discipline:

$$FCFS = \text{First come, first served}$$
$$LCFS = \text{Last come, first served}$$
$$SIRO = \text{Service in random order}$$
$$GD = \text{General queue discipline}$$

The fifth characteristic specifies the maximum allowable number of customers in the system (including customers who are waiting and customers who are in service). The sixth characteristic gives the size of the population from which customers are drawn. Unless the number of potential customers is of the same order of magnitude as the number of servers, the population size is considered to be infinite. In many important models 4/5/6 is $GD/\infty/\infty$. If this is the case, then 4/5/6 is often omitted.

As an illustration of this notation, $M/E_2/8/FCFS/10/\infty$ might represent a health clinic with 8 doctors, exponential interarrival times, two-phase Erlang service times, an FCFS queue discipline, and a total capacity of 10 patients.

The Waiting Time Paradox

We close this section with a brief discussion of an interesting paradox known as the waiting time paradox.

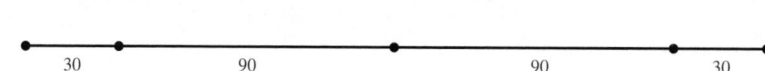

FIGURE 8
The Waiting
Time Paradox

Suppose the time between the arrival of buses at the student center is exponentially distributed, with a mean of 60 minutes. If we arrive at the student center at a randomly chosen instant, what is the average amount of time that we will have to wait for a bus?

The no-memory property of the exponential distribution implies that no matter how long it has been since the last bus arrived, we would still expect to wait an average of 60 minutes until the next bus arrived. This answer is indeed correct, but it appears to be contradicted by the following argument. On the average, somebody who arrives at a random time should arrive in the middle of a typical interval between arrivals of successive buses. If we arrive at the midpoint of a typical interval, and the average time between buses is 60 minutes, then we should have to wait, on the average, $(\frac{1}{2})60 = 30$ minutes for the next bus. Why is this argument incorrect? Simply because the typical interval between buses is *longer* than 60 minutes. The reason for this anomaly is that we are more likely to arrive during a longer interval than a shorter interval. Let's simplify the situation by assuming that half of all buses run 30 minutes apart and half of all buses run 90 minutes apart. One might think that since the average time between buses is 60 minutes, the average wait for a bus would be $(\frac{1}{2})60 = 30$ minutes, but this is incorrect. Look at a typical sequence of bus interarrival times (see Figure 8). Half of the interarrival times are 30 minutes, and half are 90 minutes. Clearly, there is a $\frac{90}{30+90} = \frac{3}{4}$ chance that one will arrive during a 90-minute interarrival time and a $\frac{30}{30+90} = \frac{1}{4}$ chance that one will arrive during a 30-minute interarrival time. Thus, the average-size interarrival time into which a customer arrives is $(\frac{3}{4})(90) + (\frac{1}{4})(30) = 75$ minutes. Since we do arrive, on the average, in the middle of an interarrival time, our average wait will be $(\frac{3}{4})(\frac{1}{2})90 + (\frac{1}{4})(\frac{1}{2})30 = 37.5$ minutes, which is longer than 30 minutes.

Returning to the case where interarrival times are exponential with mean 60 minutes, the average size of a typical interarrival time turns out to be 120 minutes. Thus, the average time that we will have to wait for a bus is $(\frac{1}{2})(120) = 60$ minutes. Note that if buses *always* arrived 60 minutes apart, then the average time a person would have to wait for a bus would be $(\frac{1}{2})(60) = 30$ minutes. In general, it can be shown that if **A** is the random variable for the time between buses, then the average time until the next bus (as seen by an arrival who is equally likely to come at any time) is given by

$$\frac{1}{2}\left(E(\mathbf{A}) + \frac{\text{var }\mathbf{A}}{E(\mathbf{A})}\right)$$

For our bus example, $\lambda = \frac{1}{60}$, so Equations (3) and (4) show that $E(\mathbf{A}) = 60$ minutes and var $\mathbf{A} = 3{,}600$ minutes^2. Substituting into this formula yields

$$\text{Expected waiting time} = \frac{1}{2}\left(60 + \frac{3{,}600}{60}\right) = 60 \text{ minutes}$$

PROBLEMS

Group A

1 Suppose I arrive at an $M/M/7/FCFS/8/\infty$ queuing system when all servers are busy. What is the probability that I will complete service before at least one of the seven customers in service?

2 The time between buses follows the mass function shown in Table 2. What is the average length of time one must wait for a bus?

TABLE 2

Time Between Buses	Probability
30 minutes	$\frac{1}{4}$
1 hour	$\frac{1}{4}$
2 hours	$\frac{1}{2}$

3 There are four sections of the third grade at Jefferson Elementary School. The number in each section is as follows: section 1, 20 students; section 2, 25 students; section 3, 35 students; section 4, 40 students. What is the average size of a third-grade section? Suppose the board of education randomly selects a Jefferson third-grader. On the average, how many students will be in her class?

4 The time between arrivals of buses follows an exponential distribution, with a mean of 60 minutes.

a What is the probability that exactly four buses will arrive during the next 2 hours?

b That at least two buses will arrive during the next 2 hours?

c That no buses will arrive during the next 2 hours?

d A bus has just arrived. What is the probability that it will be between 30 and 90 minutes before the next bus arrives?

5 During the year 2000, there was an average of .022 car accidents per person in the United States. Using your knowledge of the Poisson random variable, explain the truth in the statement, "Most drivers are better than average."

6 Suppose it is equally likely that a plane flight is 50%, 60%, 70%, 80%, or 90% full.

a What fraction of seats on a typical flight are full? This is known as the *flight load factor.*

b We are always complaining that there are never empty seats on our plane flights. Given the previous information, what is the average load factor on a plane trip I take?

7 An average of 12 jobs per hour arrive at our departmental printer.

a Use two different computations (one involving the Poisson and another the exponential random variable) to determine the probability that no job will arrive during the next 15 minutes.

b What is the probability that 5 or fewer jobs will arrive during the next 30 minutes?

20.3 Birth–Death Processes

In this section, we discuss the important idea of a birth–death process. We subsequently use birth–death processes to answer questions about several different types of queuing systems.

We define the number of people present in any queuing system at time t to be the **state** of the queuing system at time t. For $t = 0$, the state of the system will equal the number of people initially present in the system. Of great interest to us is the quantity $P_{ij}(t)$ which is defined as the probability that j people will be present in the queuing system at time t, given that at time 0, i people are present. Note that $P_{ij}(t)$ is analogous to the n-step transition probability $P_{ij}(n)$ (the probability that after n transitions, a Markov chain will be in state j, given that the chain began in state i), discussed in Chapter 17. Recall that for most Markov chains, the $P_{ij}(n)$ approached a limit π_j, which was independent of the initial state i. Similarly, it turns out that for many queuing systems, $P_{ij}(t)$ will, for large t, approach a limit π_j, which is independent of the initial state i. We call π_j the **steady state,** or equilibrium probability, of state j.

For the queuing systems that we will discuss, π_j may be thought of as the probability that at an instant in the distant future, j customers will be present. Alternatively, π_j may be thought of (for time in the distant future) as the fraction of the time that j customers are present. In most queuing systems, the value of $P_{ij}(t)$ for small t will critically depend on i, the number of customers initially present. For example, if t is small, then we would expect that $P_{50,1}(t)$ and $P_{1,1}(t)$ would differ substantially. However, if steady-state probabilities exist, then for large t, both $P_{50,1}(t)$ and $P_{1,1}(t)$ will be near π_1. The question of how large t must be before the steady state is approximately reached is difficult to answer. The behavior of $P_{ij}(t)$ before the steady state is reached is called the **transient behavior** of the queuing system. Analysis of the system's transient behavior will be discussed in Section 20.16. For now, when we analyze the behavior of a queuing system, we assume that the steady state has been reached. This allows us to work with the π_j's instead of the $P_{ij}(t)$'s.

We now discuss a certain class of continuous-time stochastic processes, called birth–death processes, which includes many interesting queuing systems. For a birth–death process, it is easy to determine the steady-state probabilities (if they exist).

A **birth–death process** is a continuous-time stochastic process for which the system's state at any time is a nonnegative integer (see Section 17.1 for a definition of a continuous-time stochastic process). If a birth–death process is in state j at time t, then the motion of the process is governed by the following laws.

Laws of Motion for Birth–Death Processes

Law 1 With probability $\lambda_j\Delta t + o(\Delta t)$, a birth occurs between time t and time $t + \Delta t$.[†] A birth increases the system state by 1, to $j + 1$. The variable λ_j is called the **birth rate** in state j. In most queuing systems, a birth is simply an arrival.

Law 2 With probability $\mu_j\Delta t + o(\Delta t)$, a death occurs between time t and time $t + \Delta t$. A death decreases the system state by 1, to $j - 1$. The variable μ_j is the **death rate** in state j. In most queuing systems, a death is a service completion. Note that $\mu_0 = 0$ must hold, or a negative state could occur.

Law 3 Births and deaths are independent of each other.

Laws 1–3 can be used to show that the probability that more than one event (birth or death) occurs between t and $t + \Delta t$ is $o(\Delta t)$. Note that any birth–death process is completely specified by knowledge of the birth rates λ_j and the death rates μ_j. Since a negative state cannot occur, any birth–death process must have $\mu_0 = 0$.

Relation of Exponential Distribution to Birth–Death Processes

Most queuing systems with exponential interarrival times and exponential service times may be modeled as birth–death processes. To illustrate why this is so, consider an $M/M/1/FCFS/\infty/\infty$ queuing system in which interarrival times are exponential with parameter λ and service times are exponentially distributed with parameter μ. If the state (number of people present) at time t is j, then the no-memory property of the exponential distribution implies that the probability of a birth during the time interval $[t, t + \Delta t]$ will not depend on how long the system has been in state j. This means that the probability of a birth occurring during $[t, t + \Delta t]$ will not depend on how long the system has been in state j and thus may be determined as if an arrival had just occurred at time t. Then the probability of a birth occurring during $[t, t + \Delta t]$ is

$$\int_0^{\Delta t} \lambda e^{-\lambda t}dt = 1 - e^{-\lambda\Delta t}$$

By the Taylor series expansion given in Section 11.1,

$$e^{-\lambda\Delta t} = 1 - \lambda\Delta t + o(\Delta t)$$

This means that the probability of a birth occurring during $[t, t + \Delta t]$ is $\lambda\Delta t + o(\Delta t)$. From this we may conclude that the birth rate in state j is simply the arrival rate λ.

To determine the death rate at time t, note that if the state is zero at time t, then nobody is in service, so no service completion can occur between t and $t + \Delta t$. Thus, $\mu_0 = 0$.

[†]Recall from Section 20.2 that $o(\Delta t)$ means that $\lim_{\Delta t \to 0} \dfrac{o(\Delta t)}{\Delta t} = 0$.

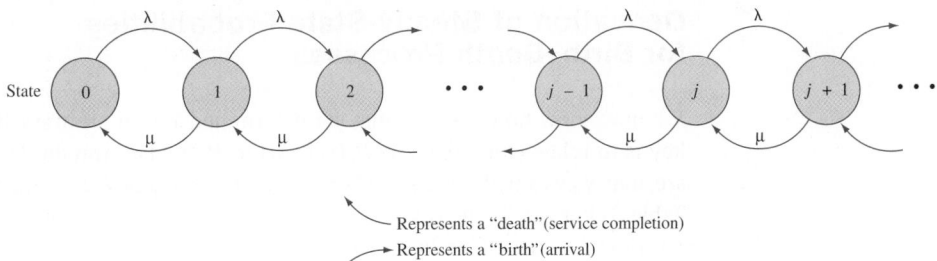

FIGURE 9
Rate Diagram for
***M/M*/1/FCFS/∞/∞**
Queuing System

⌣ Represents a "death"(service completion)
→ Represents a "birth"(arrival)

If the state at time t is $j \geq 1$, then we know (since there is only one server) that exactly one customer will be in service. The no-memory property of the exponential distribution then implies that the probability that a customer will complete service between t and $t + \Delta t$ is given by

$$\int_0^{\Delta t} \mu e^{-\mu t} dt = 1 - e^{-\mu \Delta t} = \mu \Delta t + o(\Delta t)$$

Thus, for $j \geq 1$, $\mu_j = \mu$. In summary, if we assume that service completions and arrivals occur independently, then an $M/M/1/FCFS/\infty/\infty$ queuing system is a birth–death process. The birth and death rates for the $M/M/1/FCFS/\infty/\infty$ queuing system may be represented in a rate diagram (see Figure 9).

More complicated queuing systems with exponential interarrival times and exponential service times may often be modeled as birth–death processes by adding the service rates for occupied servers and adding the arrival rates for different arrival streams. For example, consider an $M/M/3/FCFS/\infty/\infty$ queuing system in which interarrival times are exponential with $\lambda = 4$ and service times are exponential with $\mu = 5$. To model this system as a birth–death process, we would use the following parameters (see Figure 10):

$$\lambda_j = 4 \qquad\qquad\qquad\qquad\qquad\qquad\qquad (j = 0, 1, 2, \ldots)$$
$$\mu_0 = 0, \quad \mu_1 = 5, \quad \mu_2 = 5 + 5 = 10, \quad \mu_j = 5 + 5 + 5 = 15 \qquad (j = 3, 4, 5, \ldots)$$

If either interarrival times or service times are nonexponential, then the birth–death process model is not appropriate.[†] Suppose, for example, that service times are not exponential and we are considering an $M/G/1/FCFS/\infty/\infty$ queuing system. Since the service times for an $M/G/1/FCFS/\infty/\infty$ system may be nonexponential, the probability that a death (service completion) occurs between t and $t + \Delta t$ will depend on the time since the last service completion. This violates law 2, so we cannot model an $M/G/1/FCFS/\infty/\infty$ system as a birth–death process.

FIGURE 10
Rate Diagram for
***M/M*/3/FCFS/∞/∞**
Queuing System

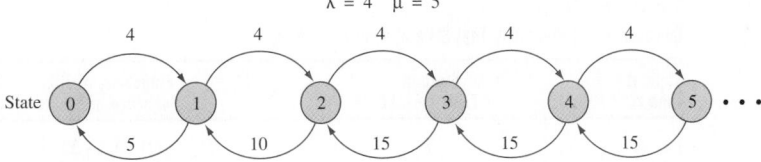

[†]A modified birth–death model can be developed if service times and interarrival times are Erlang distributions.

Derivation of Steady-State Probabilities for Birth–Death Processes

We now show how the π_j's may be determined for an arbitrary birth–death process. The key is to relate (for small Δt) $P_{ij}(t + \Delta t)$ to $P_{ij}(t)$. The way to do this is to note that there are four ways for the state at time $t + \Delta t$ to be j. For $j \geq 1$, the four ways are shown in Table 3. For $j \geq 1$, the probability that the state of the system will be $j - 1$ at time t and j at time $t + \Delta t$ is (see Figure 11)

$$P_{i,j-1}(t)(\lambda_{j-1}\Delta t + o(\Delta t))$$

Similar arguments yield (II) and (III). (IV) follows, because if the system is in a state other than $j, j - 1,$ or $j + 1$ at time t, then to end up in state j at time $t + \Delta t$, more than one event (birth or death) must occur between t and $t + \Delta t$. By law 3, this has probability $o(\Delta t)$. Thus,

$$P_{ij}(t + \Delta t) = (\text{I}) + (\text{II}) + (\text{III}) + (\text{IV})$$

After regrouping terms in this equation, we obtain

$$\begin{aligned} P_{ij}(t + \Delta t) = {}& P_{ij}(t) \\ & + \Delta t(\lambda_{j-1}P_{i,j-1}(t) + \mu_{j+1}P_{i,j+1}(t) - P_{ij}(t)\mu_j - P_{ij}(t)\lambda_j) \\ & + o(\Delta t)(P_{i,j-1}(t) + P_{i,j+1}(t) + 1 - 2P_{ij}(t)) \end{aligned} \tag{10}$$

Since the underlined term may be written as $o(\Delta t)$, we rewrite (10) as

$$P_{ij}(t + \Delta t) - P_{ij}(t) = \Delta t(\lambda_{j-1}P_{i,j-1}(t) + \mu_{j+1}P_{i,j+1}(t) - P_{ij}(t)\mu_j - P_{ij}(t)\lambda_j) + o(\Delta t)$$

Dividing both sides of this equation by Δt and letting Δt approach zero, we see that for all i and $j \geq 1$,

$$P'_{ij}(t) = \lambda_{j-1}P_{i,j-1}(t) + \mu_{j+1}P_{i,j+1}(t) - P_{ij}(t)\mu_j - P_{ij}(t)\lambda_j \tag{10'}$$

Since for $j = 0$, $P_{i,j-1}(t) = 0$ and $\mu_j = 0$, we obtain, for $j = 0$,

$$P'_{i,0}(t) = \mu_1 P_{i,1}(t) - \lambda_0 P_{i,0}(t)$$

This is an infinite system of differential equations. (A differential equation is simply an equation in which a derivative appears.) In theory, these equations may be solved for the $P_{ij}(t)$. In reality, however, this system of equations is usually extremely difficult to solve. All is not lost, however. We can use (10′) to obtain the steady-state probabilities π_j ($j = 0, 1, 2, \ldots$). As with Markov chains, we define the steady-state probability π_j to be

$$\lim_{t \to \infty} P_{ij}(t)$$

Then for large t and any initial state i, $P_{ij}(t)$ will not change very much and may be thought of as a constant. Thus, in the steady state (t large), $P'_{ij}(t) = 0$. In the steady state,

TABLE 3
Computations of Probability That State at Time $t + \Delta t$ Is j

State at Time t	State at Time $t + \Delta t$	Probability of This Sequence of Events
$j - 1$	j	$P_{i,j-1}(t)\,(\lambda_{j-1}\Delta t + o(\Delta t)) = (\text{I})$
$j + 1$	j	$P_{i,j+1}(t)\,(\mu_{j+1}\Delta t + o(\Delta t)) = (\text{II})$
j	j	$P_{i,j}(t)\,(1 - \mu_j\,\Delta t - \lambda_j\Delta t - 2o(\Delta t)) = (\text{III})$
Any other state	j	$o(\Delta t) = (\text{IV})$

FIGURE 11
Probability That State
Is $j - 1$ at Time t and
j at Time $t + \Delta t$ Is
$P_{i,j-1}(t)(\lambda_{j-1}(\Delta t) +$
$o(\Delta t))$

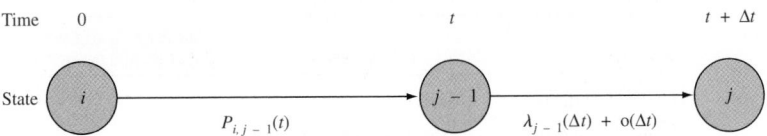

also, $P_{i,j-1}(t) = \pi_{j-1}$, $P_{i,j+1}(t) = \pi_{j+1}$, and $P_{ij}(t) = \pi_j$ will all hold. Substituting these relations into (10′), we obtain, for $j \geq 1$,

$$\lambda_{j-1}\pi_{j-1} + \mu_{j+1}\pi_{j+1} - \pi_j\mu_j - \pi_j\lambda_j = 0 \tag{10″}$$

$$\lambda_{j-1}\pi_{j-1} + \mu_{j+1}\pi_{j+1} = \pi_j(\lambda_j + \mu_j) \qquad (j = 1, 2, \ldots)$$

For $j = 0$, we obtain

$$\mu_1\pi_1 = \pi_0\lambda_0$$

Equations (10″) are an infinite system of *linear* equations that can be easily solved for the π_j's. Before discussing how to solve (10″), we give an intuitive derivation of (10″), based on the following observation: *At any time t that we observe a birth–death process, it must be true that for each state j, the number of times we have entered state j differs by at most 1 from the number of times we have left state j.*

Suppose that by time t, we have entered state 6 three times. Then one of the cases in Table 4 must have occurred. For example, if Case 2 occurs, we begin in state 6 and end up in some other state. Since we have observed three transitions into state 6 by time t, the following events (among others) must have occurred:

Start in state 6	Enter state 6 (second time)
Leave state 6 (first time)	Leave state 6 (third time)
Enter state 6 (first time)	Enter state 6 (third time)
Leave state 6 (second time)	Leave state 6 (fourth time)

Hence, if Case 2 occurs, then by time t, we must have left state 6 four times.

This observation suggests that for large t and for $j = 0, 1, 2, \ldots$ (and for any initial conditions), it will be true that

$$\frac{\text{Expected no. of departures from state } j}{\text{Unit time}}$$
$$= \frac{\text{Expected no. of entrances into state } j}{\text{Unit time}} \tag{11}$$

Assuming the system has settled down into the steady state, we know that the system spends a fraction π_j of its time in state j. We can now use (11) to determine the steady-state probabilities π_j. For $j \geq 1$, we can only leave state j by going to state $j + 1$ or state $j - 1$, so for $j \geq 1$, we obtain

$$\frac{\text{Expected no. of departures from state } j}{\text{Unit time}} = \pi_j(\lambda_j + \mu_j) \tag{12}$$

Since for $j \geq 1$ we can only enter state j from state $j - 1$ or state $j + 1$,

$$\frac{\text{Expected no. of entrances into state } j}{\text{Unit time}} = \pi_{j-1}\lambda_{j-1} + \pi_{j+1}\mu_{j+1} \tag{13}$$

Substituting (12) and (13) into (11) yields

$$\pi_{j-1}\lambda_{j-1} + \pi_{j+1}\mu_{j+1} = \pi_j(\lambda_j + \mu_j) \qquad (j = 1, 2, \ldots) \tag{14}$$

Initial State	State of Time *t*	Number of Transitions Out of State 6 by Time *t*
Case 1: state 6	State 6	3
Case 2: state 6	Any state except 6	4
Case 3: any state except state 6	State 6	2
Case 4: any state except state 6	Any state except 6	3

For $j = 0$, we know that $\mu_0 = \pi_{-1} = 0$, so we also have

$$\pi_1 \mu_1 = \pi_0 \lambda_0 \tag{14'}$$

Equations (14) and (14') are often called the **flow balance equations,** or **conservation of flow equations,** for a birth–death process. Note that (14) expresses the fact that in the steady state, the rate at which transitions occur into any state *i* must equal the rate at which transitions occur out of state *i*. If (14) did not hold for all states, then probability would "pile up" at some state, and a steady state would not exist.

Writing out the equations for (14) and (14'), we obtain the flow balance equations for a birth–death process:

$$
\begin{aligned}
(j = 0) \qquad & \pi_0 \lambda_0 = \pi_1 \mu_1 \\
(j = 1) \qquad & (\lambda_1 + \mu_1)\pi_1 = \lambda_0 \pi_0 + \mu_2 \pi_2 \\
(j = 2) \qquad & (\lambda_2 + \mu_2)\pi_2 = \lambda_1 \pi_1 + \mu_3 \pi_3 \\
& \qquad \vdots \\
(j\text{th equation}) \qquad & (\lambda_j + \mu_j)\pi_j = \lambda_{j-1}\pi_{j-1} + \mu_{j+1}\pi_{j+1}
\end{aligned}
\tag{15}
$$

Solution of Birth–Death Flow Balance Equations

To solve (15), we begin by expressing all the π_j's in terms of π_0. From the $(j = 0)$ equation, we obtain

$$\pi_1 = \frac{\pi_0 \lambda_0}{\mu_1}$$

Substituting this result into the $(j = 1)$ equation yields

$$\lambda_0 \pi_0 + \mu_2 \pi_2 = \frac{(\lambda_1 + \mu_1)\pi_0 \lambda_0}{\mu_1}$$

$$\mu_2 \pi_2 = \frac{\pi_0 (\lambda_0 \lambda_1)}{\mu_1}$$

Thus,

$$\pi_2 = \frac{\pi_0 (\lambda_0 \lambda_1)}{\mu_1 \mu_2}$$

We could now use the $(j = 3)$ equation to solve for π_3 in terms of π_0 and so on. If we define

$$c_j = \frac{\lambda_0 \lambda_1 \cdots \lambda_{j-1}}{\mu_1 \mu_2 \cdots \mu_j}$$

then it can be shown that

$$\pi_j = \pi_0 c_j \tag{16}$$

(See Problem 1 at the end of this section.) Since at any given time, we must be in some state, the steady-state probabilities must sum to 1:

$$\sum_{j=0}^{j=\infty} \pi_j = 1 \tag{17}$$

Substituting (16) into (17) yields

$$\pi_0 \left(1 + \sum_{j=1}^{j=\infty} c_j\right) = 1 \tag{18}$$

If $\sum_{j=1}^{j=\infty} c_j$ is finite, we can use (18) to solve for π_0:

$$\pi_0 = \frac{1}{1 + \sum_{j=1}^{j=\infty} c_j} \tag{19}$$

Then (16) can be used to determine $\pi_1, \pi_2, \ldots$. It can be shown that if $\sum_{j=1}^{j=\infty} c_j$ is infinite, then no steady-state distribution exists. The most common reason for a steady-state failing to exist is that the arrival rate is at least as large as the maximum rate at which customers can be served.

Using a Spreadsheet to Compute Steady-State Probabilities

The following example illustrates how a spreadsheet can be used to compute steady-state probabilities for a birth–death process.

EXAMPLE 2   **Indiana Bell**

Indiana Bell customer service representatives receive an average of 1,700 calls per hour. The time between calls follows an exponential distribution. A customer service representative can handle an average of 30 calls per hour. The time required to handle a call is also exponentially distributed. Indiana Bell can put up to 25 people on hold. If 25 people are on hold, a call is lost to the system. Indiana Bell has 75 service representatives.

1   What fraction of the time are all operators busy?

2   What fraction of all calls are lost to the system?

Bell.xls   **Solution**   In Figure 12 (file Bell.xls), we set up a spreadsheet to compute the steady-state probabilities for this birth–death process. We let the state i at any time equal the number of callers whose calls are being processed or are on hold. We have that for $i = 0, 1, 2, \ldots, 99$, $\lambda_i = 1{,}700$. The fact that any calls received when $75 + 25 = 100$ calls are in the system are lost to the system implies that $\lambda_{100} = 0$. Then no state $i > 100$ can occur (why?). We have $u_0 = 0$ and for $i = 1, 2, \ldots, 75$, $\mu_i = 30i$. For $i > 75$, $\mu_i = 30(75) = 2{,}250$.

To answer parts (1) and (2), we need to compute the steady-state probabilities $\pi_i = $ fraction of the time the state is i. In cells A4:A104, we enter the possible states of the system (0–100). To do this, enter 0 in cell A4 and 1 in A5. Then select the range A4:A5 and drag the cursor to A6:A104. In B4, type the arrival rate of 1,700 and just drag the cursor

	A	B	C	D	E	F
1						Prob(i>=75)
2	INDIANA	BELL	EXAMPLE		0	.012759326
3	STATE	LAMBDA	MU	CJ	PROB	
4	0	1700	0	1	2.451E-25	
5	1	1700	30	56.6666667	1.3889E-23	
6	2	1700	60	1605.55556	3.9352E-22	
7	3	1700	90	30327.1605	7.4332E-21	
8	4	1700	120	429634.774	1.053E-19	
9	5	1700	150	4869194.1	1.1934E-18	
10	6	1700	180	45986833.2	1.1271E-17	
11	7	1700	210	372274364	9.1244E-17	
12	8	1700	240	2636943411	6.4631E-16	
13	9	1700	270	1.6603E+10	4.0694E-15	
14	10	1700	300	9.4084E+10	2.306E-14	
15	11	1700	330	4.8467E+11	1.1879E-13	
16	12	1700	360	2.2887E+12	5.6097E-13	
17	13	1700	390	9.9765E+12	2.4452E-12	
18	14	1700	420	4.0381E+13	9.8974E-12	
19	15	1700	450	1.5255E+14	3.739E-11	
20	16	1700	480	5.4029E+14	1.3242E-10	
21	17	1700	510	1.801E+15	4.4141E-10	
22	18	1700	540	5.6697E+15	1.3896E-09	
23	19	1700	570	1.691E+16	4.1445E-09	
24	20	1700	600	4.791E+16	1.1743E-08	
25	21	1700	630	1.2928E+17	3.1687E-08	
26	22	1700	660	3.33E+17	8.1618E-08	
27	23	1700	690	8.2043E+17	2.0109E-07	
28	24	1700	720	1.9371E+18	4.7479E-07	
29	25	1700	750	4.3908E+18	1.0762E-06	
30	26	1700	780	9.5697E+18	2.3455E-06	
31	27	1700	810	2.0085E+19	4.9227E-06	
32	28	1700	840	4.0648E+19	9.9627E-06	
33	29	1700	870	7.9426E+19	1.9467E-05	
34	30	1700	900	1.5003E+20	3.6772E-05	
35	31	1700	930	2.7424E+20	6.7217E-05	
36	32	1700	960	4.8564E+20	0.00011903	
37	33	1700	990	8.3393E+20	0.00020439	
38	34	1700	1020	1.3899E+21	0.00034066	
39	35	1700	1050	2.2503E+21	0.00055154	
40	36	1700	1080	3.5421E+21	0.00086817	
41	37	1700	1110	5.4248E+21	0.00132962	
42	38	1700	1140	8.0897E+21	0.00198277	
43	39	1700	1170	1.1754E+22	0.00288095	
44	40	1700	1200	1.6652E+22	0.00408134	
45	41	1700	1230	2.3015E+22	0.00564088	
46	42	1700	1260	3.1052E+22	0.00761072	
47	43	1700	1290	4.0921E+22	0.01002963	
48	44	1700	1320	5.2701E+22	0.01291694	
49	45	1700	1350	6.6364E+22	0.01626578	
50	46	1700	1380	8.1753E+22	0.02003755	
51	47	1700	1410	9.8567E+22	0.02415875	
52	48	1700	1440	1.1636E+23	0.02852075	
53	49	1700	1470	1.3457E+23	0.03298318	
54	50	1700	1500	1.5251E+23	0.03738094	
55	51	1700	1530	1.6946E+23	0.04153437	
56	52	1700	1560	1.8467E+23	0.04526182	
57	53	1700	1590	1.9744E+23	0.04839314	
58	54	1700	1620	2.0719E+23	0.05078292	
59	55	1700	1650	2.1347E+23	0.0523218	
60	56	1700	1680	2.1601E+23	0.05294468	

FIGURE 12
Indiana Bell

	A	B	C	D	E	F
61	57	1700	1710	2.1E+23	0.0526351	
62	58	1700	1740	2.1E+23	0.0514251	
63	59	1700	1770	2.0E+23	0.0493913	
64	60	1700	1800	1.9E+23	0.0466473	
65	61	1700	1830	1.8E+23	0.0433336	
66	62	1700	1860	1.6E+23	0.039606	
67	63	1700	1890	1.5E+23	0.0356244	
68	64	1700	1920	1.3E+23	0.0315425	
69	65	1700	1950	1.1E+23	0.0274985	
70	66	1700	1980	9.6E+22	0.0236099	
71	67	1700	2010	8.1E+22	0.0199685	
72	68	1700	2040	6.8E+22	0.0166405	
73	69	1700	2070	5.6E+22	0.0136661	
74	70	1700	2100	4.5E+22	0.011063	
75	71	1700	2130	3.6E+22	0.0088296	
76	72	1700	2160	2.8E+22	0.0069492	
77	73	1700	2190	2.2E+22	0.0053944	
78	74	1700	2220	1.7E+22	0.0041308	
79	75	1700	2250	1.3E+22	0.0031211	
80	76	1700	2250	9.6E+21	0.0023581	
81	77	1700	2250	7.3E+21	0.0017817	
82	78	1700	2250	5.5E+21	0.0013462	
83	79	1700	2250	4.1E+21	0.0010171	
84	80	1700	2250	3.1E+21	0.0007685	
85	81	1700	2250	2.4E+21	0.0005806	
86	82	1700	2250	1.8E+21	0.0004387	
87	83	1700	2250	1.4E+21	0.0003315	
88	84	1700	2250	1.0E+21	0.0002504	
89	85	1700	2250	7.7E+20	0.0001892	
90	86	1700	2250	5.8E+20	0.000143	
91	87	1700	2250	4.4E+20	0.000108	
92	88	1700	2250	3.3E+20	0.0000816	
93	89	1700	2250	2.5E+20	0.0000617	
94	90	1700	2250	1.9E+20	0.0000466	
95	91	1700	2250	1.4E+20	0.0000352	
96	92	1700	2250	1.1E+20	0.0000266	
97	93	1700	2250	8.2E+19	0.0000201	
98	94	1700	2250	6.2E+19	0.0000152	
99	95	1700	2250	4.7E+19	0.0000115	
100	96	1700	2250	3.5E+19	0.0000087	
101	97	1700	2250	2.7E+19	0.0000065	
102	98	1700	2250	2.0E+19	0.0000049	
103	99	1700	2250	1.5E+19	0.0000037	
104	100	0	2250	1.2E+19	0.0000028	

FIGURE 12
(Continued)

down to B5:B104 to create the arrival rates for all states. To create the service rates, enter 0 in cell C4. Then enter 30 in C5 and 60 in cell C6. Then select the range C5:C6 and drag the cursor down to C79. This creates the service rates for states 0–75. In C80, enter 2,250 and drag that result down to C81:C104. This creates the service rate (2,250) for states 76–100. In the cell range D4:D104, we calculate the c_j's that are needed to compute the steady-state probabilities. To begin, we enter a 1 in D4. Since $c_1 = \lambda_0/\mu_1$, we enter =B4/C5 in cell D5. Since $c_2 = c_1\lambda_1/\mu_2$, we enter =D5*B5/C6 into D6. Copying from D6 to D7:D104 now generates the rest of the c_j's. In E4, we compute π_0 by entering =SUM(D$4:D$104). In E5, we compute π_1 by entering =D5*E$4. Copying from the range E5 to the range E5:E104 generates the rest of the steady-state probabilities. We can now answer questions (1) and (2).

1 We seek $\pi_{75} + \pi_{76} + \cdots + \pi_{100}$. To obtain this, we enter the command =SUM(E79:E104) in cell F2 and obtain .013.

2 An arriving call is turned away if the state equals 100. A fraction $\pi_{100} = .0000028$ of all arrivals will be turned away. Thus, the phone company is providing very good service!

In Sections 20.4–20.6 and 20.9–20.10, we apply the theory of birth–death processes to determine the steady-state probability distributions for a variety of queuing systems. Then we use the steady-state probability distributions to determine other quantities of interest (such as expected waiting time and expected number of customers in the system).

Birth–death models have been used to model phenomena other than queuing systems. For example, the number of firms in an industry can be modeled as a birth–death process: The state of the industry at any given time is the number of firms that are in business; a birth corresponds to a firm entering the industry; and a death corresponds to a firm going out of business.

PROBLEMS

Group A

1 Show that the values of the π_j's given in (16) do indeed satisfy the flow balance equations (14) and (14′).

2 My home uses two light bulbs. On average, a light bulb lasts for 22 days (exponentially distributed). When a light bulb burns out, it takes an average of 2 days (exponentially distributed) before I replace the bulb.

   a Formulate a three-state birth–death model of this situation.

   b Determine the fraction of the time that both light bulbs are working.

   c Determine the fraction of the time that no light bulbs are working.

Group B

3 You are doing an industry analysis of the Bloomington pizza industry. The rate (per year) at which pizza restaurants enter the industry is given by p, where p = price of a pizza in dollars. The price of a pizza is assumed to be max(0, $16 - .5F$), where F = number of pizza restaurants in Bloomington. During a given year, the probability that a pizza restaurant fails is $1/(10 + p)$. Create a birth–death model of this situation.

   a In the steady state, estimate the average number of pizza restaurants in Bloomington.

   b What fraction of the time will there be more than 20 pizza restaurants in Bloomington?

20.4 The $M/M/1/GD/\infty/\infty$ Queuing System and the Queuing Formula $L = \lambda W$

We now use the birth–death methodology explained in the previous section to analyze the properties of the $M/M/1/GD/\infty/\infty$ queuing system. Recall that the $M/M/1/GD/\infty/\infty$ queuing system has exponential interarrival times (we assume that the arrival rate per unit time is λ) and a single server with exponential service times (we assume that each customer's service time is exponential with rate μ). In Section 20.3, we showed that an $M/M/1/GD/\infty/\infty$ queuing system may be modeled as a birth–death process with the following parameters:

$$\lambda_j = \lambda \quad (j = 0, 1, 2, \ldots)$$
$$\mu_0 = 0 \tag{20}$$
$$\mu_j = \mu \quad (j = 1, 2, 3, \ldots)$$

Derivation of Steady-State Probabilities

We can use Equations (15)–(19) to solve for π_j, the steady-state probability that j customers will be present. Substituting (20) into (16) yields

$$\pi_1 = \frac{\lambda \pi_0}{\mu}, \qquad \pi_2 = \frac{\lambda^2 \pi_0}{\mu^2}, \qquad \ldots, \qquad \pi_j = \frac{\lambda^j \pi_0}{\mu^j} \tag{21}$$

We define $\rho = \frac{\lambda}{\mu}$. For reasons that will become apparent later, we call ρ the **traffic intensity** of the queuing system. Substituting (21) into (17) yields

$$\pi_0(1 + \rho + \rho^2 + \cdots) = 1 \tag{22}$$

We now assume that $0 \leq \rho < 1$. Then we evaluate the sum $S = 1 + \rho + \rho^2 + \cdots$ as follows: Multiplying S by ρ yields $\rho S = \rho + \rho^2 + \rho^3 + \cdots$. Then $S - \rho S = 1$, and

$$S = \frac{1}{1 - \rho} \tag{23}$$

Substituting (23) into (22) yields

$$\pi_0 = 1 - \rho \qquad (0 \leq \rho < 1) \tag{24}$$

Substituting (24) into (21) yields

$$\pi_j = \rho^j(1 - \rho) \qquad (0 \leq \rho < 1) \tag{25}$$

If $\rho \geq 1$, however, the infinite sum in (22) "blows up" (try $\rho = 1$, for example, and you get $1 + 1 + 1 + \cdots$). Thus, if $\rho \geq 1$, no steady-state distribution exists. Since $\rho = \frac{\lambda}{\mu}$, we see that if $\lambda \geq \mu$ (that is, the arrival rate is at least as large as the service rate), then no steady-state distribution exists.

If $\rho > 1$, it is easy to see why no steady-state distribution can exist. Suppose $\lambda = 6$ customers per hour and $\mu = 4$ customers per hour. Even if the server were working all the time, she could only serve an average of 4 people per hour. Thus, the average number of customers in the system would grow by at least $6 - 4 = 2$ customers per hour. This means that after a long time, the number of customers present would "blow up," and no steady-state distribution could exist. If $\rho = 1$, the nonexistence of a steady state is not quite so obvious, but our analysis does indicate that no steady state exists.

Derivation of L

Throughout the rest of this section, we assume that $\rho < 1$, ensuring that a steady-state probability distribution, as given in (25), does exist. We now use the steady-state probability distribution in (25) to determine several quantities of interest. For example, assuming that the steady state has been reached, the average number of customers present in the queuing system (call it L) is given by

$$L = \sum_{j=0}^{j=\infty} j\pi_j = \sum_{j=0}^{j=\infty} j\rho^j(1 - \rho)$$

$$= (1 - \rho)\sum_{j=0}^{j=\infty} j\rho^j$$

Defining

$$S' = \sum_{j=0}^{j=\infty} j\rho^j = \rho + 2\rho^2 + 3\rho^3 + \cdots$$

we see that $\rho S' = \rho^2 + 2\rho^3 + 3\rho^4 + \cdots$. Subtracting yields

$$S' - \rho S' = \rho + \rho^2 + \cdots = \frac{\rho}{1 - \rho}$$

Thus,

$$S' = \frac{\rho}{(1 - \rho)^2}$$

and

$$L = (1 - \rho)\frac{\rho}{(1 - \rho)^2} = \frac{\rho}{1 - \rho} = \frac{\lambda}{\mu - \lambda} \tag{26}$$

Derivation of L_q

In some circumstances, we are interested in the expected number of people waiting in line (or in the queue). We denote this number by L_q. Note that if 0 or 1 customer is present in the system, then nobody is waiting in line, but if j people are present ($j \geq 1$), there will be $j - 1$ people waiting in line. Thus, if we are in the steady state,

$$L_q = \sum_{j=1}^{j=\infty} (j - 1)\pi_j = \sum_{j=1}^{j=\infty} j\pi_j - \sum_{j=1}^{j=\infty} \pi_j$$

$$= L - (1 - \pi_0) = L - \rho$$

where the last equation follows from (24). Since $L = \frac{\rho}{1-\rho}$, we write

$$L_q = \frac{\rho}{1 - \rho} - \rho = \frac{\rho^2}{1 - \rho} = \frac{\lambda^2}{\mu(\mu - \lambda)} \tag{27}$$

Derivation of L_s

Also of interest is L_s, the expected number of customers in service. For an $M/M/1/GD/\infty/\infty$ queuing system,

$$L_s = 0\pi_0 + 1(\pi_1 + \pi_2 + \cdots) = 1 - \pi_0 = 1 - (1 - \rho) = \rho$$

Since every customer who is present is either in line or in service, it follows that for any queuing system (not just an $M/M/1/GD/\infty/\infty$ system), $L = L_s + L_q$. Thus, using our formulas for L and L_s, we could have determined L_q from

$$L_q = L - L_s = \frac{\rho}{1 - \rho} - \rho = \frac{\rho^2}{1 - \rho}$$

The Queuing Formula $L = \lambda W$

Often we are interested in the amount of time that a typical customer spends in a queuing system. We define W as the expected time a customer spends in the queuing system, including time in line plus time in service, and W_q as the expected time a customer spends waiting in line. Both W and W_q are computed under the assumption that the steady state has been reached. By using a powerful result known as **Little's queuing formula**, W and

W_q may be easily computed from L and L_q. We first define (for any queuing system or any subset of a queuing system) the following quantities:

λ = average number of arrivals *entering* the system per unit time

L = average number of customers present in the queuing system

L_q = average number of customers waiting in line

L_s = average number of customers in service

W = average time a customer spends in the system

W_q = average time a customer spends in line

W_s = average time a customer spends in service

In these definitions, all averages are steady-state averages. For most queuing systems, Little's queuing formula may be summarized as in Theorem 3.

THEOREM 3

For *any* queuing system in which a steady-state distribution exists, the following relations hold:

$$L = \lambda W \tag{28}$$
$$L_q = \lambda W_q \tag{29}$$
$$L_s = \lambda W_s \tag{30}$$

Before using these important results, we present an intuitive justification of (28). First note that both sides of (28) have the same units (we assume the unit of time is hours). This follows, because L is expressed in terms of number of customers, λ is expressed in terms of customers per hour, and W is expressed in hours. Thus, λW has the same units (customers) as L. For a rigorous proof of Little's theorem, see Ross (1970). We content ourselves with the following heuristic discussion.

Consider a queuing system in which customers are served on a first come, first served basis. An arbitrary arrival enters the system (assume that the steady state has been reached). This customer stays in the system until he completes service, and upon his departure, there will be (on the average) L customers present in the system. But when this customer leaves, who will be left in the system? Only those customers who arrive during the time the initial customer spends in the system. Since the initial customer spends an average of W hours in the system, an average of λW customers will arrive during his stay in the system. Hence, $L = \lambda W$. The "real" proof of $L = \lambda W$ is virtually independent of the number of servers, the interarrival time distribution, the service discipline, and the service time distribution. Thus, as long as a steady state exists, we may apply Equations (28)–(30) to any queuing system.

To illustrate the use of (28) and (29), we determine W and W_q for an $M/M/1/GD/\infty/\infty$ queuing system. From (26),

$$L = \frac{\rho}{1 - \rho}$$

Then (28) yields

$$W = \frac{L}{\lambda} = \frac{\rho}{\lambda(1 - \rho)} = \frac{1}{\mu - \lambda} \tag{31}$$

From (27), we obtain

$$L_q = \frac{\lambda^2}{\mu(\mu - \lambda)}$$

and (29) implies

$$W_q = \frac{L_q}{\lambda} = \frac{\lambda}{\mu(\mu - \lambda)} \tag{32}$$

Notice that (as expected) as ρ approaches 1, both W and W_q become very large. For ρ near zero, W_q approaches zero, but for small ρ, W approaches $\frac{1}{\mu}$, the mean service time.

The following three examples show applications of the formulas we have developed.

EXAMPLE 3	Drive-in Banking

An average of 10 cars per hour arrive at a single-server drive-in teller. Assume that the average service time for each customer is 4 minutes, and both interarrival times and service times are exponential. Answer the following questions:

1 What is the probability that the teller is idle?

2 What is the average number of cars waiting in line for the teller? (A car that is being served is not considered to be waiting in line.)

3 What is the average amount of time a drive-in customer spends in the bank parking lot (including time in service)?

4 On the average, how many customers per hour will be served by the teller?

Solution By assumption, we are dealing with an $M/M/1/GD/\infty/\infty$ queuing system for which $\lambda = 10$ cars per hour and $\mu = 15$ cars per hour. Thus, $\rho = \frac{10}{15} = \frac{2}{3}$.

1 From (24), $\pi_0 = 1 - \rho = 1 - \frac{2}{3} = \frac{1}{3}$. Thus, the teller will be idle an average of one-third of the time.

2 We seek L_q. From (27),

$$L_q = \frac{\rho^2}{1 - \rho} = \frac{(\frac{2}{3})^2}{1 - \frac{2}{3}} = \frac{4}{3} \quad \text{customers}$$

3 We seek W. From (28), $W = \frac{L}{\lambda}$. Then from (26).

$$L = \frac{\rho}{1 - \rho} = \frac{\frac{2}{3}}{1 - \frac{2}{3}} = 2 \quad \text{customers}$$

Thus, $W = \frac{2}{10} = \frac{1}{5}$ hour $= 12$ minutes (W will have the same units as λ).

4 If the teller were always busy, he would serve an average of $\mu = 15$ customers per hour. From part (1), we know that the teller is only busy two-thirds of the time. Thus, during each hour, the teller will serve an average of $(\frac{2}{3})(15) = 10$ customers. This must be the case, because in the steady state, 10 customers are arriving each hour, so each hour, 10 customers must leave the system.

EXAMPLE 4	Service Station

Suppose that all car owners fill up when their tanks are exactly half full.[†] At the present time, an average of 7.5 customers per hour arrive at a single-pump gas station. It takes an

[†]This example is based on Erickson (1973).

average of 4 minutes to service a car. Assume that interarrival times and service times are both exponential.

1 For the present situation, compute L and W.

2 Suppose that a gas shortage occurs and panic buying takes place. To model this phenomenon, suppose that all car owners now purchase gas when their tanks are exactly three-quarters full. Since each car owner is now putting less gas into the tank during each visit to the station, we assume that the average service time has been reduced to $3\frac{1}{3}$ minutes. How has panic buying affected L and W?

Solution **1** We have an $M/M/1/GD/\infty/\infty$ system with $\lambda = 7.5$ cars per hour and $\mu = 15$ cars per hour. Thus, $\rho = \frac{7.5}{15} = .50$. From (26), $L = \frac{.50}{1-.50} = 1$, and from (28), $W = \frac{L}{\lambda} = \frac{1}{7.5} = 0.13$ hour. Hence, in this situation, everything is under control, and long lines appear to be unlikely.

2 We now have an $M/M/1/GD/\infty/\infty$ system with $\lambda = 2(7.5) = 15$ cars per hour. (This follows because each car owner will fill up twice as often.) Now $\mu = \frac{60}{3.333} = 18$ cars per hour, and $\rho = \frac{15}{18} = \frac{5}{6}$. Then

$$L = \frac{\frac{5}{6}}{1 - \frac{5}{6}} = 5 \text{ cars} \quad \text{and} \quad W = \frac{L}{\lambda} = \frac{5}{15} = \frac{1}{3} \text{ hours} = 20 \text{ minutes}$$

Thus, panic buying has caused long lines.

Example 4 illustrates the fact that as ρ approaches 1, L and therefore W increase rapidly. Table 5 illustrates this fact.

A Queuing Optimization Model

Example 5 shows how queuing theory can be used as an aid in decision making.

TABLE 5
Relation between ρ and L for an $M/M/1/GD/\infty/\infty$ System

ρ	L for an $M/M/1/GD/\infty/\infty$ System
0.30	0.43
0.40	0.67
0.50	1.00
0.60	1.50
0.70	2.33
0.80	4.00
0.90	9.00
0.95	19.00
0.99	99.00

EXAMPLE 5   Tool Center

Machinists who work at a tool-and-die plant must check out tools from a tool center.[†] An average of ten machinists per hour arrive seeking tools. At present, the tool center is staffed by a clerk who is paid $6 per hour and who takes an average of 5 minutes to handle each request for tools. Since each machinist produces $10 worth of goods per hour, each hour that a machinist spends at the tool center costs the company $10. The company is deciding whether or not it is worthwhile to hire (at $4 per hour) a helper for the clerk. If the helper is hired, the clerk will take an average of only 4 minutes to process requests for tools. Assume that service and interarrival times are exponential. Should the helper be hired?

Solution   Problems in which a decision maker must choose between alternative queuing systems are called **queuing optimization problems.** In the current problem, the company's goal is to minimize the sum of the hourly service cost and the expected hourly cost due to the idle times of machinists. In queuing optimization problems, the component of cost due to customers waiting in line is referred to as the *delay cost.* Thus, the firm wants to minimize

$$\frac{\text{Expected cost}}{\text{Hour}} = \frac{\text{service cost}}{\text{hour}} + \frac{\text{expected delay cost}}{\text{hour}}$$

The computation of the hourly service cost is usually simple. The easiest way to compute the hourly delay cost is to note that

$$\frac{\text{Expected delay cost}}{\text{Hour}} = \left(\frac{\text{expected delay cost}}{\text{customer}}\right)\left(\frac{\text{expected customers}}{\text{hour}}\right)$$

In our problem,

$$\frac{\text{Expected delay cost}}{\text{Customer}} = \left(\frac{\$10}{\text{machinist-hour}}\right)\left(\begin{array}{l}\text{average hours machinist}\\\text{spends in system}\end{array}\right)$$

Thus,

$$\frac{\text{Expected delay cost}}{\text{Customer}} = 10W \quad \text{and} \quad \frac{\text{expected delay cost}}{\text{hour}} = 10W\lambda$$

We can now compare the expected cost per hour if the helper is not hired to the expected cost per hour if the helper is hired. If the helper is not hired, $\lambda = 10$ machinists per hour and $\mu = 12$ machinists per hour. From (31), $W = \frac{1}{12-10} = \frac{1}{2}$ hour. Since the clerk is paid $6 per hour, we have that

$$\frac{\text{Service cost}}{\text{Hour}} = \$6 \quad \text{and} \quad \frac{\text{expected delay cost}}{\text{hour}} = 10(\tfrac{1}{2})10 = \$50$$

[†]This example is based on Brigham (1955).

Thus, without the helper, the expected hourly cost is $6 + 50 = \$56$. With the helper, $\mu = 15$ customers per hour. Then $W = \frac{1}{15-10} = \frac{1}{5}$ hour and

$$\frac{\text{Expected delay cost}}{\text{Hour}} = 10(\tfrac{1}{5})(10) = \$20$$

Since the hourly service cost is now $6 + 4 = \$10$ per hour, the expected hourly cost with the helper is $20 + 10 = \$30$. Thus, the helper should be hired, because he saves $50 - 20 = \$30$ per hour in delay costs, which more than makes up for his \$4-per-hour salary.

The queuing formula $L = \lambda W$ is very general and can be applied to many situations that do not seem to be queuing problems. Think of any situation where a quantity (such as mortgage loan applications, potatoes at McDonald's, revenues from computer sales) flows through a system. If we let

$$L = \text{average amount of quantity present}$$
$$\lambda = \text{rate at which quantity arrives at system}$$
$$W = \text{average time a unit of quantity spends in system}$$

then $L = \lambda W$ or $W = L/\lambda$.

Here are some examples of $L = \lambda W$ in non-queuing situations.

EXAMPLE 6　Potatoes at McDonald's

Our local MacDonald's uses an average of 10,000 pounds of potatoes per week. The average number of pounds of potatoes on hand is 5,000. On the average, how long do potatoes stay in the restaurant before being used?

Solution　We are given that $L = 5,000$ pounds and $\lambda = 10,000$ pounds/week. Therefore, $W = 5,000$ pounds/(10,000 pounds/week) $= .5$ week.

EXAMPLE 7　Accounts Receivable

A local computer store sells \$300,000 worth of computers per year. On average accounts receivable are \$45,000. On average, how long does it take from the time a customer is billed until the store receives payment?

Solution　We are given that $L = \$45,000$ and $\lambda = \$300,000$/year. Therefore $W = \$45,000/(\$300,000/\text{year}) = .15$ year.

A Spreadsheet for the $M/M/1/GD/\infty/\infty$ Queuing System

MM1.xls

Figure 13 (file MM1.xls) gives a template that can be used to compute important quantities for the $M/M/1/GD/\infty/\infty$ queuing system. Simply input λ in cell A4 and μ in cell B4. L, L_q, L_s, W, W_q, and W_s are computed in rows 6 and 8. Column B prints out the steady-state probabilities (computed from (24) and (25)). We are assuming that λ and μ are such that the probability that more than 1,000 customers will be present is very small. In Figure 13, we have input the values of λ and μ for Example 3.

	A	B	C
1	M/M/1	QUEUE	
2			
3	LAMBDA?	MU?	RO
4	10	15	0.66666667
5	L	LQ	LS
6	2	1.33333333	0.66666667
7	W	WQ	WS
8	0.2	0.13333333	0.06666667
9	J	PI(J)	
10	0	0.33333333	
11	1	0.22222222	
12	2	0.14814815	
13	3	0.09876543	
14	4	0.06584362	
15	5	0.04389575	
16	6	0.02926383	
17	7	0.01950922	
18	8	0.01300615	
19	9	0.00867076	
20	10	0.00578051	
21	11	0.00385367	
22	12	0.00256912	
23	13	0.00171274	
24	14	0.00114183	
25	15	0.00076122	
26	16	0.00050748	
27	17	0.00033832	
28	18	0.00022555	
29	19	0.00015036	
30	20	0.00010024	
31	21	6.6829E-05	
32	22	4.4552E-05	
33	23	2.9702E-05	
34	24	1.9801E-05	
35	25	1.3201E-05	
36	26	8.8005E-06	
37	27	5.867E-06	
38	28	3.9113E-06	
39	29	2.6075E-06	
40	30	1.7384E-06	

FIGURE 13
*M/M/*1 Queue

A	A	B	C
4 1	31	0.0000012	
4 2	32	0.0000008	
4 3	33	0.0000005	
4 4	34	0.0000003	
4 5	35	0.0000002	
4 6	36	0.0000002	
4 7	37	0.0000001	
4 8	38	6.8E-08	
4 9	39	4.5E-08	
5 0	40	3.0E-08	
5 1	41	2.0E-08	
5 2	42	1.3E-08	
5 3	43	8.9E-09	
5 4	44	6.0E-09	
5 5	45	4.0E-09	
5 6	46	2.6E-09	
5 7	47	1.8E-09	
5 8	48	1.2E-09	
5 9	49	7.8E-10	
6 0	50	5.2E-10	
6 1	51	3.5E-10	
6 2	52	2.3E-10	
6 3	53	1.5E-10	
6 4	54	1.0E-10	
6 5	55	6.9E-11	
6 6	56	4.6E-11	
6 7	57	3.1E-11	
6 8	58	2.0E-11	
6 9	59	1.4E-11	
7 0	60	9.1E-12	
7 1	61	6.0E-12	
7 2	62	4.0E-12	
7 3	63	2.7E-12	
7 4	64	1.8E-12	
7 5	65	1.2E-12	
7 6	66	8.0E-13	
7 7	67	5.3E-13	
7 8	68	3.5E-13	
7 9	69	2.4E-13	
8 0	70	1.6E-13	

FIGURE 13
(Continued)

PROBLEMS

Group A

1[†] Each airline passenger and his or her luggage must be checked to determine whether he or she is carrying weapons onto the airplane. Suppose that at Gotham City Airport, an average of 10 passengers per minute arrive (interarrival times are exponential). To check passengers for weapons, the airport must have a checkpoint consisting of a metal detector and baggage X-ray machine. Whenever a check-point is in operation, two employees are required. A checkpoint can check an average of 12 passengers per minute (the time to check a passenger is exponential). Under the assumption that the airport has only one checkpoint, answer the following questions:

a What is the probability that a passenger will have to wait before being checked for weapons?

b On the average, how many passengers are waiting in line to enter the checkpoint?

[†]Based on Gilliam (1979).

c On the average, how long will a passenger spend at the checkpoint?

2 The Decision Sciences Department is trying to determine whether to rent a slow or a fast copier. The department believes that an employee's time is worth $15 per hour. The slow copier rents for $4 per hour and it takes an employee an average of 10 minutes to complete copying (exponentially distributed). The fast copier rents for $15 per hour and it takes an employee an average of 6 minutes to complete copying. An average of 4 employees per hour need to use the copying machine (interarrival times are exponential). Which machine should the department rent?

3 For an $M/M/1/GD/\infty/\infty$ queuing system, suppose that both λ and μ are doubled.
a How is L changed?
b How is W changed?
c How is the steady-state probability distribution changed?

4 A fast-food restaurant has one drive-through window. An average of 40 customers per hour arrive at the window. It takes an average of 1 minute to serve a customer. Assume that interarrival and service times are exponential.
a On the average, how many customers are waiting in line?
b On the average, how long does a customer spend at the restaurant (from time of arrival to time service is completed)?
c What fraction of the time are more than 3 cars waiting for service (this includes the car (if any) at the window)?

5 On a typical Saturday, Red Lobster serves 1,000 customers. The restaurant is open for 12 hours. On average, 150 customers are present. How long does an average customer spend in the restaurant?

6 Our local maternity ward delivers 1,500 babies per year. On the average, 5 beds in the maternity ward are filled. How long does the average mother stay in the maternity ward?

7 Assume that an average of 125 packets per second of information arrive to a router and that it takes an average of .002 second to process each packet. Assuming exponential interarrival and service times, answer the following questions.
a What is the average number of packets waiting for entry into the router?
b What is the probability that 10 or more packets are present?

Group B

8 Referring to Problem 1, suppose the airline wants to determine how many checkpoints to operate to minimize operating costs and delay costs over a ten-year period. Assume that the cost of delaying a passenger for 1 hour is $10 and that the airport is open every day for 16 hours per day. It costs $1 million to purchase, staff, and maintain a metal detector and baggage X-ray machine for a ten-year period. Finally, assume that each passenger is equally likely to enter a given checkpoint.

9[†] Each machine on Widgetco's assembly line gets out of whack an average of once a minute. Laborers are assigned to reset a machine that gets out of whack. The company pays each laborer c_s dollars per hour and estimates that each hour of idle machine time costs the company c_m dollars in lost production. Data indicate that the time between successive breakdowns of a machine and the time to reset a machine are exponential. Widgetco plans to assign each worker a certain number of machines to watch over and repair. Let M = total number of Widgetco machines, w = number of laborers hired by Widgetco, and $R = \frac{M}{w}$ = machines assigned to each laborer.
a Express Widgetco's hourly cost in terms of R and M.
b Show that the optimal value of R does not depend on the value of M.
c Use calculus to show that costs are minimized by choosing

$$R = \frac{\frac{\mu}{60}}{1 + \left(\frac{c_m}{c_s}\right)^{1/2}}$$

d Suppose c_m = 78¢ and c_s = $2.75. Widgetco has 200 machines, and a laborer can reset a machine in an average of 7.8 seconds. How can Widgetco minimize costs?
e In parts (a)–(d), we have tacitly assumed that at any point in time, the rate at which the machines assigned to a worker break down does not depend on the number of his or her assigned machines that are currently working properly. Does this assumption seem reasonable?

10 Consider an airport where taxis and customers arrive (exponential interarrival times) with respective rates of 1 and 2 per minute. No matter how many other taxis are present, a taxi will wait. If an arriving customer does not find a taxi, the customer immediately leaves.
a Model this system as a birth–death process (*Hint:* Determine what the state of the system is at any given time and draw a rate diagram.)
b Find the average number of taxis that are waiting for a customer.
c Suppose all customers who use a taxi pay a $2 fare. During a typical hour, how much revenue will the taxis receive?

11 A bank is trying to determine which of two machines should be rented to process checks. Machine 1 rents for $10,000 per year and processes 1,000 checks per hour. Machine 2 rents for $15,000 per year and processes 1,600 checks per hour. Assume that the machines work 8 hours a day, 5 days a week, 50 weeks a year. The bank must process an average of 800 checks per hour, and the average check processed is for $100. Assume an annual interest rate of 20%. Then determine the cost to the bank (in lost interest) for each hour that a check spends waiting for and undergoing processing. Assuming that interarrival times and service times are exponential, which machine should the bank rent?

12[‡] A tire plant must produce an average of 100 tires per day. The plant produces tires in a batch of size x. The plant

[†]Based on Vogel (1979).
[‡]Based on Karmarkar (1985).

manager must determine the batch size x that minimizes the time a batch spends in the plant. From the time a batch of tires arrives, it takes an average of $\frac{1}{20}$ of a day to set up the plant for production of tires. Once the plant is set up, it takes an average of $\frac{1}{150}$ day to produce each tire. Assume that the time to produce a batch of tires is exponentially distributed and that the time for a batch of tires to "arrive" is also exponentially distributed. Determine the batch size that minimizes the expected time a batch spends in the plant (from arrival of batch to time production of batch is completed).

13 A worker at the State Unemployment Office is responsible for processing a company's forms when it opens for business. The worker can process an average of 4 forms per week. In 2002, an average of 1.8 companies per week submitted forms for processing, and the worker had a backlog of .45 week. In 2003, an average of 3.9 companies per week submitted forms for processing, and the worker had a 5-week backlog. The poor worker was fired and sued to get his job back. The court said that since the amount of work submitted to the worker had approximately doubled, the worker's backlog should have also doubled. Since his backlog increased by more than a factor of 10, he must have been slacking off, so the state was justified in firing him. Use queuing theory to defend the worker (based on an actual case!).

14 For the $M/M/1/GD/\infty/\infty$ queuing model, show that the following results hold:

  a $W = (L + 1)W_s$.

  b $W_q = LW_s$.

  c Interpret the results in (a) and (b).

15 From the time a request for data is submitted until the request is fulfilled, a database takes an average of 3 seconds to respond to a request for data. We find that the database is idle around 20% of the time. Answer the following questions, assuming that the database can be modeled as an $M/M/1$ system.

  a What is the average service time per database query?

  b What is the average number of queries in the system?

  c What is the probability that 5 or more queries are present?

20.5 The $M/M/1/GD/c/\infty$ Queuing System

In this section, we analyze the $M/M/1/GD/c/\infty$ queuing system. Recall that this queuing system is an $M/M/1/GD/\infty/\infty$ system with a total capacity of c customers. The $M/M/1/GD/c/\infty$ system is identical to the $M/M/1/GD/\infty/\infty$ system except for the fact that when c customers are present, all arrivals are turned away and are forever lost to the system. As in Section 20.4, we assume that interarrival times are exponential with rate λ, and service times are exponential with rate μ. Then the $M/M/1/GD/c/\infty$ system may be modeled (see Figure 14) as a birth–death process with the following parameters:

$$\lambda_j = \lambda \quad (j = 0, 1, \ldots, c - 1)$$
$$\lambda_c = 0$$
$$\mu_0 = 0$$
$$\mu_j = \mu \quad (j = 1, 2, \ldots, c)$$

(33)

Since $\lambda_c = 0$, the system will never reach state $c + 1$ (or any higher-numbered state). As in Section 20.4, it is convenient to define $\rho = \frac{\lambda}{\mu}$. Then we can apply Equations (16)–(19) to find that if $\lambda \neq \mu$, the steady-state probabilities for the $M/M/1/GD/c/\infty$ model are given by

$$\pi_0 = \frac{1 - \rho}{1 - \rho^{c+1}}$$
$$\pi_j = \rho^j \pi_0 \quad (j = 1, 2, \ldots, c)$$
$$\pi_j = 0 \quad (j = c + 1, c + 2, \ldots)$$

(34)

FIGURE **14**
Rate Diagram for
$M/M/1/GD/c/\infty$
Queuing System

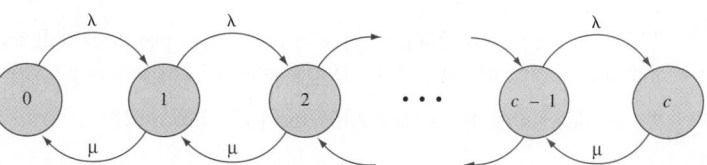

Combining (34) with the fact that $L = \sum_{j=0}^{j=c} j\pi_j$, we can show that when $\lambda \neq \mu$,

$$L = \frac{\rho[1 - (c + 1)\rho^c + c\rho^{c+1}]}{(1 - \rho^{c+1})(1 - \rho)} \qquad (35)$$

If $\lambda = \mu$, then all the c_j's in (16) equal 1, and all the π_j's must be equal. Hence, if $\lambda = \mu$, the steady-state probabilities for the $M/M/1/GD/c/\infty$ system are

$$\pi_j = \frac{1}{c + 1} \qquad (j = 0, 1, \ldots, c)$$

$$L = \frac{c}{2} \qquad (36)$$

As with the $M/M/1/GD/\infty/\infty$ system, $L_s = 0\pi_0 + 1(\pi_1 + \pi_2 + \cdots) = 1 - \pi_0$. As before, we may determine L_q from $L_q = L - L_s$.

Determination of W and W_q from (28) and (29) is a tricky matter. Recall that in (28) and (29), λ represents the average number of customers per unit time who *actually enter* the system. In our finite capacity model, an average of λ arrivals per unit time arrive, but $\lambda\pi_c$ of these arrivals find the system filled to capacity and leave. Thus, an average of $\lambda - \lambda\pi_c = \lambda(1 - \pi_c)$ arrivals per unit time will actually enter the system. Combining this fact with (28) and (29) yields

$$W = \frac{L}{\lambda(1 - \pi_c)} \qquad \text{and} \qquad W_q = \frac{L_q}{\lambda(1 - \pi_c)} \qquad (37)$$

For an $M/M/1/GD/c/\infty$ system, a steady state will exist even if $\lambda \geq \mu$. This is because, even if $\lambda \geq \mu$, the finite capacity of the system prevents the number of people in the system from "blowing up."

EXAMPLE 8   **Barber Shop**

A one-man barber shop has a total of 10 seats. Interarrival times are exponentially distributed, and an average of 20 prospective customers arrive each hour at the shop. Those customers who find the shop full do not enter. The barber takes an average of 12 minutes to cut each customer's hair. Haircut times are exponentially distributed.

1   On the average, how many haircuts per hour will the barber complete?

2   On the average, how much time will be spent in the shop by a customer who enters?

Solution   **1**   A fraction π_{10} of all arrivals will find the shop is full. Thus, an average of $\lambda(1 - \pi_{10})$ will enter the shop each hour. All entering customers will receive a haircut, so the barber will give an average of $\lambda(1 - \pi_{10})$ haircuts per hour. From our problem, $c = 10$, $\lambda = 20$ customers per hour, and $\mu = 5$ customers per hour. Then $\rho = \frac{20}{5} = 4$, and (34) yields

$$\pi_0 = \frac{1 - 4}{1 - 4^{11}}$$

and

$$\pi_{10} = 4^{10}\left(\frac{1 - 4}{1 - 4^{11}}\right) = \frac{-3(4^{10})}{1 - 4^{11}} = .75$$

Thus, an average of $20(1 - \frac{3}{4}) = 5$ customers per hour will receive haircuts. This means that an average of $20 - 5 = 15$ prospective customers per hour will not enter the shop.

2   To determine W, we use (35) and (37). From (35),

$$L = \frac{4[1 - 11(4^{10}) + 10(4^{11})]}{(1 - 4^{11})(1 - 4)} = 9.67 \text{ customers}$$

Then (37) yields

$$W = \frac{9.67}{20(1 - \frac{3}{4})} = 1.93 \text{ hours}$$

This barber shop is crowded, and the barber would be well advised to hire at least one more barber!

A Spreadsheet for the $M/M/1/GD/c/\infty$ Queuing System

MM1CAP.xls

Figure 15 (file MM1CAP.xls) gives a template that can be used to compute important quantities for the $M/M/1/GD/c/\infty$ queuing system. Input λ in cell B2, μ in cell C2, and c (we assume $c \le 1,000$) in cell D2. In cell F2, the steady-state probability that the state is c is given. This is the fraction of all arrivals who find the system full. In row 4, the quantities L, L_s, L_q, W, W_s, and W_q are computed. In column E, the steady-state probabilities are computed from equations (16)–(18). In Figure 15, we have input the data from Example 8.

PROBLEMS

Group A

1 A service facility consists of one server who can serve an average of 2 customers per hour (service times are exponential). An average of 3 customers per hour arrive at the facility (interarrival times are assumed exponential). The system capacity is 3 customers.

  a On the average, how many potential customers enter the system each hour?

  b What is the probability that the server will be busy?

2 An average of 40 cars per hour (interarrival times are exponentially distributed) are tempted to use the drive-in window at the Hot Dog King Restaurant. If a total of more than 4 cars are in line (including the car at the window) a car will not enter the line. It takes an average of 4 minutes (exponentially distributed) to serve a car.

  a What is the average number of cars waiting for the drive-in window (not including a car at the window)?

  b On the average, how many cars will be served per hour?

  c I have just joined the line at the drive-in window. On the average, how long will it be before I have received my food?

3 An average of 125 packets of information per minute arrive at an internet router. It takes an average of .002 second to process a packet of information. The router is designed to have a limited buffer to store waiting messages. Any message that arrives when the buffer is full is lost to the system. Assuming that interarrival and service times are

exponentially distributed, how big a buffer size is needed to ensure that at most 1 in a million messages is lost?

Group B

4 Show that if $\rho \ne 1$

$$1 + \rho + \rho^2 + \cdots + \rho^c = \frac{1 - \rho^{c+1}}{1 - \rho}$$

(*Hint:* Recall how we evaluated $1 + \rho + \rho^2 + \cdots$.)

5 Use the answer to Problem 3 to derive the steady-state probabilities for the $M/M/1/GD/c/\infty$ system given in Equation (34).

6 Two one-man barber shops sit side by side in Dunkirk Square. Each can hold a maximum of 4 people, and any potential customer who finds a shop full will not wait for a haircut. Barber 1 charges $11 per haircut and takes an average of 12 minutes to complete a haircut. Barber 2 charges $5 per haircut and takes an average of 6 minutes to complete a haircut. An average of 10 potential customers per hour arrive at each barber shop. Of course, a potential customer becomes an actual customer only if he finds that the shop is not full. Assuming that interarrival times and haircut times are exponential, which barber will earn more money?

7 A small mail order firm Seas Beginnings has one phone line. An average of 60 people per hour call in orders, and it takes an average of 1 minute to handle a call. Time between

FIGURE 15

	A	B	C	D	E	F	G
1	M/M/1/GD/c	LAMBDA?	MU?	c?	RO	PI(c)	TURNED AWAY
2		20	5	10	4	0.75000018	15.00000358
3		L	LS	LQ	W	WS	WQ
4		9.66666929	0.99999928	8.66667	1.93333524	0.2	1.733335241
5							
6							
7							
8							
9							
10							
11							
12	STATE	LAMBDA(J)	MU(J)	CJ	PROB	#IN QUEUE	COLA*COLE
13	0	20	0	1	7.1526E-07	0	0
14	1	20	5	4	2.861E-06	0	2.86102E-06
15	2	20	5	16	1.1444E-05	1	2.28882E-05
16	3	20	5	64	4.5776E-05	2	0.000137329
17	4	20	5	256	0.00018311	3	0.000732422
18	5	20	5	1024	0.00073242	4	0.00366211
19	6	20	5	4096	0.00292969	5	0.017578129
20	7	20	5	16384	0.01171875	6	0.08203127
21	8	20	5	65536	0.04687501	7	0.375000089
22	9	20	5	262144	0.18750004	8	1.687500402
23	10	0	5	1048576	0.75000018	9	7.500001788
24	11	0	5	0	0	10	0
25	12	0	5	0	0	11	0
26	13	0	5	0	0	12	0
27	14	0	5	0	0	13	0
28	15	0	5	0	0	14	0
29	16	0	5	0	0	15	0
30	17	0	5	0	0	16	0
31	18	0	5	0	0	17	0
32	19	0	5	0	0	18	0
33	20	0	5	0	0	19	0
34	21	0	5	0	0	20	0
35	22	0	5	0	0	21	0
36	23	0	5	0	0	22	0
37	24	0	5	0	0	23	0
38	25	0	5	0	0	24	0
39	26	0	5	0	0	25	0
40	27	0	5	0	0	26	0
41	28	0	5	0	0	27	0

calls and time to handle calls are exponentially distributed. If the phone line is busy, Seas Beginnings can put up to $c - 1$ people on hold. If $c - 1$ people are on hold, a caller gets a busy signal and calls a competitor (Air End). Seas Beginnings wants only 1% of all callers to get a busy signal. How many people should the company be able to put on hold?

20.6 The *M/M/s/GD/∞/∞* Queuing System

We now consider the *M/M/s/GD/∞/∞* system. We assume that interarrival times are exponential (with rate λ), service times are exponential (with rate μ), and there is a single line of customers waiting to be served at one of s parallel servers. If $j \leq s$ customers are present, then all j customers are in service; if $j > s$ customers are present, then all s servers are occupied, and $j - s$ customers are waiting in line. Any arrival who finds an idle server enters service immediately, but an arrival who does not find an idle server joins the queue of customers awaiting service. Banks and post office branches in which all customers wait in a single line for service can often be modeled as *M/M/s/GD/∞/∞* queuing systems.

To describe the *M/M/s/GD/∞/∞* system as a birth–death model, note that (as in the *M/M/1/GD/∞/∞* model) $\lambda_j = \lambda$ ($j = 0, 1, 2, \ldots$). If j servers are occupied, then service completions occur at a rate

$$\underbrace{\mu + \mu + \cdots = j\mu}_{j\mu\text{'s}}$$

Whenever j customers are present, min (j, s) servers will be occupied. Thus, $\mu_j = \min (j, s)\mu$. Summarizing, we find that the *M/M/s/GD/∞/∞* system can be modeled as a birth–death process (see Figure 16) with parameters

$$
\begin{aligned}
\lambda_j &= \lambda & (j = 0, 1, \ldots) \\
\mu_j &= j\mu & (j = 0, 1, \ldots, s) \\
\mu_j &= s\mu & (j = s + 1, s + 2, \ldots)
\end{aligned}
\tag{38}
$$

we define $\rho = \frac{\lambda}{s\mu}$. For $\rho < 1$, substituting (38) into (16)–(19) yields the following steady-state probabilities:

$$
\pi_0 = \cfrac{1}{\displaystyle\sum_{i=0}^{i=(s-1)} \frac{(s\rho)^i}{i!} + \frac{(s\rho)^s}{s!(1 - \rho)}}
\tag{39}
$$

$$
\pi_j = \frac{(s\rho)^j \pi_0}{j!} \qquad (j = 1, 2, \ldots, s)
\tag{39.1}
$$

$$
\pi_j = \frac{(s\rho)^j \pi_0}{s!s^{j-s}} \qquad (j = s, s + 1, s + 2, \ldots)
\tag{39.2}
$$

If $\rho \geq 1$, no steady state exists. In other words, if the arrival rate is at least as large as the maximum possible service rate ($\lambda \geq s\mu$), the system "blows up."

From (39.2) it can be shown that the steady-state probability that all servers are busy is given by

$$
P(j \geq s) = \frac{(s\rho)^s \pi_0}{s!(1 - \rho)}
\tag{40}
$$

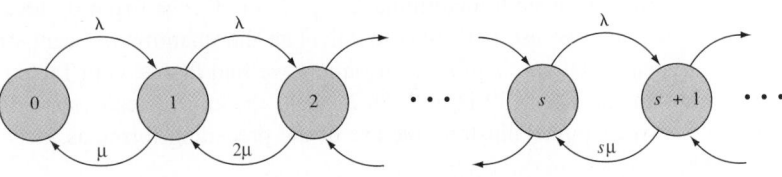

FIGURE 16
Rate Diagram for
M/M/s/GD/∞/∞
Queuing System

TABLE 6

$P(j \geq s)$ for the $M/M/s/GD/\infty/\infty$ Queuing System

ρ	$s = 2$	$s = 3$	$s = 4$	$s = 5$	$s = 6$	$s = 7$
.10	.02	.00	.00	.00	.00	.00
.20	.07	.02	.00	.00	.00	.00
.30	.14	.07	.04	.02	.01	.00
.40	.23	.14	.09	.06	.04	.03
.50	.33	.24	.17	.13	.10	.08
.55	.39	.29	.23	.18	.14	.11
.60	.45	.35	.29	.24	.20	.17
.65	.51	.42	.35	.30	.26	.21
.70	.57	.51	.43	.38	.34	.30
.75	.64	.57	.51	.46	.42	.39
.80	.71	.65	.60	.55	.52	.49
.85	.78	.73	.69	.65	.62	.60
.90	.85	.83	.79	.76	.74	.72
.95	.92	.91	.89	.88	.87	.85

Table 6 tabulates $P(j \geq s)$ for a variety of situations. It can also be shown that

$$L_q = \frac{P(j \geq s)\rho}{1 - \rho} \tag{41}$$

Then (28) yields

$$W_q = \frac{L_q}{\lambda} = \frac{P(j \geq s)}{s\mu - \lambda} \tag{42}$$

To determine L (and then W), we use the fact that $L = L_q + L_s$. Since $W_s = \frac{1}{\mu}$, Equation (30) shows that $L_s = \frac{\lambda}{\mu}$. Then

$$L = L_q + \frac{\lambda}{\mu} \tag{43}$$

Also,

$$W = \frac{L}{\lambda}$$
$$= \frac{L_q}{\lambda} + \frac{1}{\mu}$$
$$= W_q + \frac{1}{\mu} \tag{44}$$
$$= \frac{P(j \geq s)}{s\mu - \lambda} + \frac{1}{\mu}$$

When we need to determine L, L_q, W, or W_q, we begin by looking up $P(j \geq s)$ in Table 6. Then we use (41)–(44) to calculate the quantity we want. If we are interested in the steady-state probability distribution, we find $P(j \geq s)$ in Table 6 and then use (40) to obtain π_0. Then (39.1) and (39.2) yield the entire steady-state distribution. The following two examples illustrate the use of the preceding formulas.

EXAMPLE 9   **Bank Tellers**

Consider a bank with two tellers. An average of 80 customers per hour arrive at the bank and wait in a single line for an idle teller. The average time it takes to serve a customer is 1.2 minutes. Assume that interarrival times and service times are exponential. Determine

1   The expected number of customers present in the bank

2   The expected length of time a customer spends in the bank

3   The fraction of time that a particular teller is idle

Solution   1   We have an $M/M/2/GD/\infty/\infty$ system with $\lambda = 80$ customers per hour and $\mu = 50$ customers per hour. Thus $\rho = \frac{80}{2(50)} = 0.80 < 1$, so a steady state does exist. (For $\lambda \geq 100$, no steady state would exist.) From Table 6, $P(j \geq 2) = .71$. Then (41) yields

$$L_q = \frac{.80(.71)}{1 - .80} = 2.84 \text{ customers}$$

and from (43), $L = 2.84 + \frac{80}{50} = 4.44$ customers.

2   Since $W = \frac{L}{\lambda}$, $W = \frac{4.44}{80} = 0.055$ hour $= 3.3$ minutes.

3   To determine the fraction of time that a particular server is idle, note that he or she is idle during the entire time that $j = 0$ and half the time (by symmetry) that $j = 1$. The probability that a server is idle is given by $\pi_0 + 0.5\pi_1$. Using the fact that $P(j \geq 2) = .71$, we obtain π_0 from (40):

$$\pi_0 = \frac{s!P(j \geq s)(1 - \rho)}{(s\rho)^2} = \frac{2!(.71)(1 - .80)}{(1.6)^2} = .11$$

Now (39.1) yields

$$\pi_1 = \frac{(1.6)^1 \pi_0}{1!} = .176$$

Thus, the probability that a particular teller is idle is $\pi_0 + 0.5\pi_1 = .11 + 0.5(.176) = .198$. We could have determined π_0 directly from (39):

$$\pi_0 = \frac{1}{1 + \dfrac{[2(.80)]^1}{1!} + \dfrac{[2(.80)]^2}{2!(1 - .80)}} = \frac{1}{1 + 1.6 + 6.4} = \frac{1}{9}$$

This is consistent with our computation of $\pi_0 = .11$.

EXAMPLE 10   **Bank Staffing**

The manager of a bank must determine how many tellers should work on Fridays. For every minute a customer stands in line, the manager believes that a delay cost of 5¢ is incurred. An average of 2 customers per minute arrive at the bank. On the average, it takes a teller 2 minutes to complete a customer's transaction. It cost the bank $9 per hour to hire a teller. Interarrival times and service times are exponential. To minimize the sum of service costs and delay costs, how many tellers should the bank have working on Fridays?

Solution   Since $\lambda = 2$ customers per minute and $\mu = 0.5$ customer per minute, $\frac{\lambda}{s\mu} < 1$ requires that $\frac{4}{s} < 1$ or $s \geq 5$. Thus, there must be at least 5 tellers, or the number of customers present will "blow up." We now compute, for $s = 5, 6, \ldots,$

$$\frac{\text{Expected service cost}}{\text{Minute}} + \frac{\text{expected delay cost}}{\text{minute}}$$

Since each teller is paid $\frac{9}{60}$ = 15¢ per minute,

$$\frac{\text{Expected service cost}}{\text{Minute}} = 0.15s$$

As in Example 4,

$$\frac{\text{Expected delay cost}}{\text{Minute}} = \left(\frac{\text{expected customers}}{\text{minute}}\right)\left(\frac{\text{expected delay cost}}{\text{customer}}\right)$$

But

$$\frac{\text{Expected delay cost}}{\text{Customer}} = 0.05W_q$$

Since an average of 2 customers arrive per minute,

$$\frac{\text{Expected delay cost}}{\text{Minute}} = 2(0.05W_q) = 0.10W_q$$

For $s = 5$, $\rho = \frac{2}{.5(5)} = .80$ and $P(j \geq 5) = .55$. From (42),

$$W_q = \frac{.55}{5(.5) - 2} = 1.1 \text{ minutes}$$

Thus, for $s = 5$,

$$\frac{\text{Expected delay cost}}{\text{Minute}} = 0.10(1.1) = 11¢$$

and, for $s = 5$,

$$\frac{\text{Total expected cost}}{\text{Minute}} = 0.15(5) + 0.11 = 86¢$$

Since $s = 6$ has a service cost per minute of $6(0.15) = 90¢$, 6 tellers cannot have a lower total cost than 5 tellers. Hence, having 5 tellers serve is optimal. Putting it another way, adding an additional teller can save the bank at most 11¢ per minute in delay costs. Since an additional teller cost 15¢ per minute, it cannot be optimal to hire more than 5 tellers.

In addition to a customer's expected time in the system, the distribution of a customer's waiting time is of interest. For example, if all customers who have to wait more than 5 minutes at a supermarket checkout counter decide to switch to another store, the probability that a given customer will switch to another store equals $P(\mathbf{W} > 5)$. To determine this probability, we need to know the distribution of a customer's waiting time. For an $M/M/s/FCFS/\infty/\infty$ queuing system, it can be shown that

$$P(\mathbf{W} > t) = e^{-\mu t}\left\{1 + P(j \geq s)\frac{1 - \exp\left[-\mu t(s - 1 - s\rho)\right]}{s - 1 - s\rho}\right\}^\dagger \qquad (45)$$

$$P(\mathbf{W}_q > t) = P(j \geq s)\exp\left[-s\mu(1 - \rho)t\right] \qquad (46)$$

To illustrate the use of (45) and (46), suppose that in Example 7 (for $s = 5$), the bank manager wants to know the probability that a customer will have to wait in line for more than 10 minutes. For $s = 5$, $\rho = .80$, $P(j \geq 5) = .55$, and $\mu = 0.5$ customer per minute. (46) yields

$$P(\mathbf{W}_q > 10) = .55 \exp\left[-5(0.5)(1 - .80)(10)\right] = .55 \, e^{-5} = .004$$

†If $s - 1 = s\rho$, then $P(\mathbf{W} > t) = e^{-\mu t}(1 + P(j \geq s)\mu t)$.

Thus, the bank manager can be sure that the chance of a customer's having to wait more than 10 minutes is quite small.

A Spreadsheet for the *M/M/s/GD/∞/∞* Queuing System

Multiple.xls

Figure 17 (file Multiple.xls) gives a template that can be used to compute the important quantities for the *M/M/s/GD/∞/∞* queuing system. In cell B2 we input λ, in cell C2 we input μ, and in cell D2 we input s. In cell B6, we compute $P(j \geq s)$. In row 4, the quantities L, L_s, L_q, W, W_s, and W_q are computed. In A8, we compute $P(\mathbf{W}_q > t)$ for the value of t input in cell B8. In cell C8, we compute $P(\mathbf{W} > t)$ for the value of t input in cell B8. Steady-state probabilities are computed in column E (we are assuming that there is a small probability that more than 1,000 customers are present). In Figure 17, we have input data for Example 10 (with 5 servers).

Having a spreadsheet to compute quantities of interest for the *M/M/s* system enables us to use spreadsheet techniques such as Data Tables and Goal Seek to answer questions of interest. For example, reconsider Example 10. To determine the number of servers that minimizes expected cost per minute, we would like to vary the number of servers (starting with 5) and compute expected cost per minute for different numbers of servers. This is easily done with a **one-way data table.** (See Figure 18.)

Step 1   Enter the possible number of servers (5–8) in cells J5:J8.

Step 2   Enter the formula for expected cost per minute one column over to the right and one row above where the possible number of servers are listed. This is in cell K4.

$$=0.15*D2+B2*G4*0.05$$

Step 3   Highlight the *table range.* This includes the inputs, the calculated formula, and the range where values of the calculated formula are placed. In our example, the table range is J4:K8.

Step 4   Select Data Table and choose One-Way Table (because we are changing only one input, the number of servers).

Step 5   Fill in the dialog box as shown in Figure 19. This instructs Excel to repeatedly place the input values in the left-hand column of the table range in cell D2 (number of servers) and recalculate our formula (expected cost per minute, which is entered in cell K4). We then obtain the expected cost per minute for 5–8 servers. As before, we find that 5 servers yield the lowest expected cost per minute.

As another example of how we can use powerful spreadsheet tools to answer important queuing questions, suppose we want to know (for 5 servers) the 90th percentile of a customer's time in the system. That is, we wish to know the value of t that makes $P(\mathbf{W} > t)$ equal to .10. This may easily be determined with the Excel Goal Seek feature. Goal Seek enables us to find what value of one cell (the *changing* cell) causes a formula in another cell (the *set* cell) to assume a desired value (called the *to value*).

To use Goal Seek to find the 90th percentile of a customer's time in the system, we select Tools Goal Seek and fill in the dialog box as shown in Figure 20. This dialog box finds the value for t in B8 that makes $P(\mathbf{W} > t)$ (computed in C8) equal to .1. We find that with 5 servers, 10% of all customers will spend at least 6.7 minutes in the bank. See Figure 21.

We note that the precision of Goal Seek may be improved by selecting Tools Options Calculation and setting Maximum Change to a smaller number than the default value of .001. For example, a Maximum Change of .000001 ensures that upon completion of the Goal Seek operation, $P(\mathbf{W} > q)$ will be within .000001 of .10.

FIGURE 17

	A	B	C	D	E	F	G	H
1	M/M/s/GD	LAMBDA?	MU?	s?	RO			
2		2	0.5	5	0.8			
3		L	LS	LQ	W	WS	WQ	
4		6.21645022	4	2.21645022	3.10822511	1.999999999	1.108225109	
5	STATE	P(j>=s)						
6	1	0.55411255						
7	P(Wq>t)	t?	P(W>t)					
8	0.019390014	6.70521931	0.10000006					
9								
10								
11								
12	STATE	LAMBDA(J)	MU(J)	CJ	PROB	#IN QUEUE	COLA*COLE	COLE*COL
13	0	2	0	1	0.01298701	0	0	0
14	1	2	0.5	4	0.05194805	0	0.051948052	0
15	2	2	1	8	0.1038961	0	0.207792208	0
16	3	2	1.5	10.6666667	0.13852814	0	0.415584416	0
17	4	2	2	10.6666667	0.13852814	0	0.554112554	0
18	5	2	2.5	8.53333333	0.11082251	0	0.554112554	0
19	6	2	2.5	6.82666667	0.08865801	1	0.531948052	0.08865801
20	7	2	2.5	5.46133333	0.07092641	2	0.496484848	0.14185281
21	8	2	2.5	4.36906667	0.05674113	3	0.453929004	0.17022338
22	9	2	2.5	3.49525333	0.0453929	4	0.408536104	0.1815716
23	10	2	2.5	2.79620267	0.03631432	5	0.363143203	0.1815716
24	11	2	2.5	2.23696213	0.02905146	6	0.319566019	0.17430874
25	12	2	2.5	1.78956971	0.02324117	7	0.27889398	0.16268816
26	13	2	2.5	1.43165577	0.01859293	8	0.241708116	0.14874346
27	14	2	2.5	1.14532461	0.01487435	9	0.208240839	0.13386911
28	15	2	2.5	0.91625969	0.01189948	10	0.178492147	0.11899476
29	16	2	2.5	0.73300775	0.00951958	11	0.152313299	0.10471539
30	17	2	2.5	0.5864062	0.00761566	12	0.129466304	0.09138798
31	18	2	2.5	0.46912496	0.00609253	13	0.109665575	0.07920292
32	19	2	2.5	0.37529997	0.00487403	14	0.092606486	0.06823636
33	20	2	2.5	0.30023998	0.00389922	15	0.077984409	0.05848831
34	21	2	2.5	0.24019198	0.00311938	16	0.065506904	0.04991002
35	22	2	2.5	0.19215358	0.0024955	17	0.054901024	0.04242352
36	23	2	2.5	0.15372287	0.0019964	18	0.04591722	0.03593522
37	24	2	2.5	0.12297829	0.00159712	19	0.038330897	0.03034529
38	25	2	2.5	0.09838264	0.0012777	20	0.031942414	0.02555393
39	26	2	2.5	0.07870611	0.00102216	21	0.026576088	0.0214653
40	27	2	2.5	0.06296489	0.00081773	22	0.022078597	0.01798997
41	28	2	2.5	0.05037191	0.00065418	23	0.018317058	0.01504615
42	29	2	2.5	0.04029753	0.00052334	24	0.015176991	0.01256027
43	30	2	2.5	0.03223802	0.00041868	25	0.012560268	0.01046689
44	31	2	2.5	0.02579042	0.00033494	26	0.010383155	0.00870845
45	32	2	2.5	0.02063233	0.00026795	27	0.008574476	0.00723471
46	33	2	2.5	0.01650587	0.00021436	28	0.007073943	0.00600213
47	34	2	2.5	0.01320469	0.00017149	29	0.005830644	0.0049732
48	35	2	2.5	0.01056376	0.00013719	30	0.004801707	0.00411575
49	36	2	2.5	0.008451	0.00010975	31	0.003951119	0.00340235
50	37	2	2.5	0.0067608	8.7803E-05	32	0.003248698	0.00280968
51	38	2	2.5	0.00540864	7.0242E-05	33	0.0026692	0.00231799

	J	K
2		
3		
4	Servers	0.86082251
5	5	0.86082251
6	6	0.92847608
7	7	1.05900734
8	8	1.2029522

FIGURE **18**

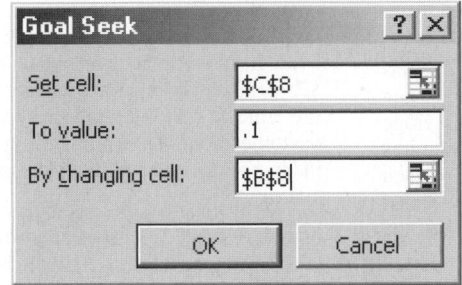

FIGURE **19**

FIGURE **20**

FIGURE **21**

	A	B	C
7	P(Wq>t)	t?	P(W>t)
8	0.019390014	6.70521931	0.10000006

Using LINGO for *M/M/s/GD/∞/∞* Computations

The LINGO function **@PEB( )** yields the probability that all servers are busy ($P(j \geq s)$) for an *M/M/s/GD/∞/∞* system. The **@PEB** function has two arguments: the first is the value of λ/μ and the second is the number of servers. Thus, for Example 9, **@PEB** $(80/50,2) = .711111$ yields $P(j \geq 2)$.

The **@PEB** function can be used to solve queuing optimization problems with LINGO. For instance, to determine the cost-minimizing number of servers in Example 10, we would input the following problem into LINGO:

```
MODEL:
1) MIN=.10*@PEB(4,S)/(.5*S-2) + .15*S;
2) S>5;
END
```

In line 1 .10*@PEB(4,S)/(.5*S−2) is the expected cost per minute due to customers waiting in line, while .15*S is the per-minute service cost. Line 2 follows, because we need at least 5 servers for a steady state to exist. LINGO outputs $S = 5$ with an objective function value of .860823 (this is expected cost per minute).

PROBLEMS

Group A

1 A supermarket is trying to decide how many cash registers to keep open. Suppose an average of 18 customers arrive each hour, and the average checkout time for a customer is 4 minutes. Interarrival times and service times are exponential, and the system may be modeled as an $M/M/s/GD/\infty/\infty$ queuing system. It costs $20 per hour to operate a cash register, and a cost of 25¢ is assessed for each minute the customer spends in the cash register area. How many registers should the store open?

2 A small bank is trying to determine how many tellers to employ. The total cost of employing a teller is $100 per day, and a teller can serve an average of 60 customers per day. An average of 50 customers per day arrive at the bank, and both service times and interarrival times are exponential. If the delay cost per customer-day is $100, how many tellers should the bank hire?

3 In this problem, all interarrival and service times are exponential.

a At present, the finance department and the marketing department each have one typist. Each typist can type 25 letters per day. Finance requires that an average of 20 letters per day be typed, and marketing requires that an average of 15 letters per day be typed. For each department, determine the average length of time elapsing between a request for a letter and completion of the letter.

b Suppose that the two typists were grouped into a typing pool; that is, each typist would be available to type letters for either department. For this arrangement, calculate the average length of time between a request for a letter and completion of the letter.

c Comment on the results of parts (a) and (b).

d Under the pooled arrangement, what is the probability that more than .200 day will elapse between a request for a letter and completion of the letter?

4 MacBurger's is attempting to determine how many servers (or lines) should be available during the breakfast shift. During each hour, an average of 100 customers arrive at the restaurant. Each line or server can handle an average of 50 customers per hour. A server costs $5 per hour, and the cost of a customer waiting in line for 1 hour is $20. Assuming that an $M/M/s/GD/\infty/\infty$ model is applicable, determine the number of lines that minimizes the sum of delay and service costs.

5 An average of 100 customers arrive each hour at the Gotham City Bank. The average service time for each customer is 1 minute. Service times and interarrival times are exponential. The manager wants to ensure that no more than 1% of all customers will have to wait in line for more than 5 minutes. If the bank follows the policy of having all customers join a single line, how many tellers must the bank hire?

6 An average of 90 patrons per hour arrive at a hotel lobby (interarrival times are exponential), waiting to check in. At present, there are 5 clerks, and patrons are waiting in a single line for the first available clerk. The average time for a clerk to service a patron is 3 minutes (exponentially distributed). Clerks earn $10 per hour, and the hotel assesses a waiting time cost of $20 for each hour that a patron waits in line.

a Compute the expected cost per hour of the current system.

b The hotel is considering replacing one clerk with an Automatic Clerk Machine (ACM). Management estimates that 20% of all patrons will use an ACM. An ACM takes an average of 1 minute to service a patron. It costs $48 per day (1 day = 8 hours) to operate an ACM. Should the hotel install the ACM? Assume that all customers who are willing to use the ACM wait in a single queue.

7 An average of 50 customers per hour arrive at a small post office. Interarrival times are exponentially distributed. Each window can serve an average of 25 customers per hour. Service times are exponentially distributed. It costs $25 per hour to open a window, and the post office values the time a customer spends waiting in line at $15 per customer-hour. To minimize expected hourly costs, how many postal windows should be opened?

8 An average of 300 customers per hour arrive at a huge branch of bank 2. It takes an average of 2 minutes to serve each customer. It costs $10 per hour to keep open a teller window, and the bank estimates that it will lose $50 in future profits for each hour that a customer waits in line. How many teller windows should bank 2 open?

9 An average of 40 students per hour arrive at the MBA computing lab. The average student uses a computer for 20 minutes. Assume exponential interarrival and service times.

a If we want the average time a student waits for a PC to be at most 10 minutes, how many computers should the lab have?

b If we want 95% of all students to spend 5 minutes or less waiting for a PC, how many PCs should the lab have?

10 A data storage system consists of 3 disk drives sharing a common queue. An average of 50 storage requests arrive per second. The average time required to service a request is .03 second. Assuming that interarrival times and service times are exponential, determine:

a The probability that a given disk drive is busy

b The probability that no disk drives are busy

c The probability that a job will have to wait

d The average number of jobs present in the storage system

11 A Northwest Airlines ticket counter forecasts that 200 people per hour will need to check in. It takes an average of two minutes to service a customer. Assume that interarrival times and service times are exponential and that all customers wait in a single line for the first available agent.

a If we want the average time a customer spends in line and in service to be 30 minutes or less, how many ticket agents should be on duty?

b If we want 95% of all customers to wait 45 minutes or less in line, how many ticket agents should be on duty?

Group B

12 An average of 100 customers per hour arrive at Gotham City Bank. It takes a teller an average of 2 minutes to serve a customer. Interarrival and service times are exponential. The bank currently has four tellers working. The bank manager wants to compare the following two systems with regard to average number of customers present in the bank and the probability that a customer will spend more than 8 minutes in the bank:

System 1 Each teller has her own line, and no jockeying between lines is permitted.

System 2 All customers wait in a single line for the first available teller.

If you were the bank manager, which system would you prefer?

13 A muffler shop has three mechanics. Each mechanic takes an average of 45 minutes to install a new muffler.

Suppose an average of 1 customer per hour arrives. What is the expected number of mechanics that are busy at any given time? Answer this question without assuming that service times and interarrival times are exponential.

14 Consider the following two queuing systems:

System 1 An $M/M/1$ system with arrival rate λ and service rate 3μ.

System 2 An $M/M/3$ system with arrival rate λ and each server working at rate μ.

Without doing extensive calculations, which system will have the smaller W and L? (*Hint:* Write down the birth–death parameters for each system. Then determine which system is more efficient.)

15 (Requires the use of a spreadsheet or LINGO) The Carco plant in Bedford produces windshield wipers for Fords. In a given day, each machine in the plant can produce 1,000 wipers. The plant operates 250 days per year, and Ford will need 3 million wipers per year. It costs $50,000 per year to operate a machine. For each day that a wiper is delayed, a cost of $100 (due to production downtime at other plants) is incurred. How many machines should the Ford plant have? Assume that interarrival times and service times are exponential.

20.7 The $M/G/\infty/GD/\infty/\infty$ and $GI/G/\infty/GD/\infty/\infty$ Models

There are many examples of systems in which a customer never has to wait for service to begin. In such a system, the customer's entire stay in the system may be thought of as his or her service time. Since a customer never has to wait for service, there is, in essence, a server available for each arrival, and we may think of such a system as an **infinite-server** (or self-service) system. Two examples of an infinite-server system are given in Table 7.

Using the Kendall–Lee notation, an infinite-server system in which interarrival and service times may follow arbitrary probability distributions may be written as $GI/G/\infty/GD/\infty/\infty$ queuing system. Such a system operates as follows:

1 Interarrival times are iid with common distribution **A**. Define $E(\mathbf{A}) = \frac{1}{\lambda}$. Thus, λ is the arrival rate.

2 When a customer arrives, he or she immediately enters service. Each customer's time in the system is governed by a distribution **S** having $E(\mathbf{S}) = \frac{1}{\mu}$.

TABLE 7
Examples of Infinite-Server Queuing Systems

Situation	Arrival	Service Time (time in system)	State of System
Industry	Firm enters industry	Time until firm leaves industry	Number of firms in industry
College program	Student enters program	Time student remains in program	Number of students in program

Let L be the expected number of customers in the system in the steady state, and W be the expected time that a customer spends in the system. By definition, $W = \frac{1}{\mu}$. Then Equation (30) implies that

$$L = \frac{\lambda}{\mu} \qquad (47)$$

Equation (47) does not require any assumptions of exponentiality. If interarrival times are exponential, it can be shown (even for an arbitrary service time distribution) that the steady-state probability that j customers are present (call it π_j) follows a Poisson distribution with mean $\frac{\lambda}{\mu}$. This implies that

$$\pi_j = \frac{(\frac{\lambda}{\mu})^j e^{-\lambda/\mu}}{j!}$$

The following example is a typical application of a $GI/G/\infty/GD/\infty/\infty$ system.

EXAMPLE 11 Smalltown Ice Cream Shops

During each year, an average of 3 ice cream shops open up in Smalltown. The average time that an ice cream shop stays in business is 10 years. On January 1, 2525, what is the average number of ice cream shops that you would find in Smalltown? If the time between the opening of ice cream shops is exponential, what is the probability that on January 1, 2525, there will be 25 ice cream shops in Smalltown?

Solution We are given that $\lambda = 3$ shops per year and $\frac{1}{\mu} = 10$ years per shop. Assuming that the steady state has been reached, there will be an average of $L = \lambda(\frac{1}{\mu}) = 3(10) = 30$ shops in Smalltown. If interarrivals of ice cream shops are exponential, then

$$\pi_{25} = \frac{(30)^{25} e^{-30}}{25!} = .05$$

Of course, we could also compute the probability that there are 25 ice cream shops with the Excel formula

$$=POISSON(30,25,0)$$

This yields .045.

PROBLEMS

Group A

1 Each week, the Columbus Record Club attracts 100 new members. Members remain members for an average of one year (1 year = 52 weeks). On the average, how many members will the record club have?

2 The State U doctoral program in business admits an average of 25 doctoral students each year. If a doctoral student spends an average of 4 years in residence at State U, how many doctoral students would one expect to find there?

3 There are at present 40 solar energy construction firms in the state of Indiana. An average of 20 solar energy construction firms open each year in the state. The average firm stays in business for 10 years. If present trends continue, what is the expected number of solar energy construction firms that will be found in Indiana? If the time between the entries of firms into the industry is exponentially distributed, what is the probability that (in the steady state) there will be more than 300 solar energy firms in business? (*Hint:* For large λ, the Poisson distribution can be approximated by a normal distribution.)

20.8 The $M/G/1/GD/\infty/\infty$ Queuing System

In this section, we consider a single-server queuing system in which interarrival times are exponential, but the service time distribution ($\mathbf{S}$) need not be exponential. Let λ be the arrival rate (assumed to be measured in arrivals per hour). Also define $\frac{1}{\mu} = E(\mathbf{S})$ and $\sigma^2 = \text{var } \mathbf{S}$.

In Kendall's notation, such a queuing system is described as an $M/G/1/GD/\infty/\infty$ queuing system. An $M/G/1/GD/\infty/\infty$ system is *not* a birth–death process, because the probability that a service completion occurs between t and $t + \Delta t$ when the state of the system at time t is j depends on the length of time since the last service completion (because service times no longer have the no-memory property). Thus, we cannot write the probability of a service completion between t and $t + \Delta t$ in the form $\mu \Delta t$, and a birth–death model is not appropriate.

Determination of the steady-state probabilities for an $M/G/1/GD/\infty/\infty$ queuing system is a difficult matter. Since the birth–death steady-state equations are no longer valid, a different approach must be taken. Markov chain theory is used to determine π_i', the probability that after the system has operated for a long time, i customers will be present at the instant immediately after a service completion occurs (see Problem 5 at the end of this section). It can be shown that $\pi_i' = \pi_i$, where π_i is the fraction of the time after the system has operated for a long time that i customers are present (see Kleinrock (1975)).

Fortunately, however, utilizing the results of Pollaczek and Khinchin, we may determine $L_q, L, L_s, W_q, W,$ and W_s. Pollaczek and Khinchin showed that for the $M/G/1/GD/\infty/\infty$ queuing system,

$$L_q = \frac{\lambda^2 \sigma^2 + \rho^2}{2(1 - \rho)} \tag{48}$$

where $\rho = \frac{\lambda}{\mu}$. Since $W_s = \frac{1}{\mu}$, (30) implies that $L_s = \lambda(\frac{1}{\mu}) = \rho$. Since $L = L_s + L_q$, we obtain

$$L = L_q + \rho \tag{49}$$

Then (29) and (28) imply that

$$W_q = \frac{L_q}{\lambda} \tag{50}$$

$$W = W_q + \frac{1}{\mu} \tag{51}$$

It can also be shown that π_0, the fraction of the time that the server is idle, is $1 - \rho$. (See Problem 2 at the end of this section.) This result is similar to the one for the $M/M/1/GD/\infty/\infty$ system.

To illustrate the use of (48)–(51), consider an $M/M/1/GD/\infty/\infty$ system with $\lambda = 5$ customers per hour and $\mu = 8$ customers per hour. From our study of the $M/M/1/GD/\infty/\infty$ model, we know that

$$L = \frac{\lambda}{\mu - \lambda} = \frac{5}{8 - 5} = \frac{5}{3} \text{ customers}$$

$$L_q = L - \rho = \frac{5}{3} - \frac{5}{8} = \frac{25}{24} \text{ customers}$$

$$W = \frac{L}{\lambda} = \frac{\frac{5}{3}}{5} = \frac{1}{3} \text{ hour}$$

$$W_q = \frac{L_q}{\lambda} = \frac{\frac{25}{24}}{5} = \frac{5}{24} \text{ hour}$$

From (3) and (4), we know that $E(\mathbf{S}) = \frac{1}{8}$ hour and var $\mathbf{S} = \frac{1}{64}$ hour^2. Then (48) yields

$$L_q = \frac{\frac{(5)^2}{64} + \left(\frac{5}{8}\right)^2}{2\left(1 - \frac{5}{8}\right)} = \frac{25}{24} \text{ customers}$$

$$L = L_q + \rho = \frac{25}{24} + \frac{5}{8} = \frac{40}{24} = \frac{5}{3} \text{ customers}$$

$$W_q = \frac{L_q}{\lambda} = \frac{\frac{25}{24}}{5} = \frac{5}{24} \text{ hour}$$

$$W = \frac{L}{\lambda} = \frac{\frac{5}{3}}{5} = \frac{1}{3} \text{ hour}$$

To demonstrate how the variance of the service time can significantly affect the efficiency of a queuing system, we consider an $M/D/1/GD/\infty/\infty$ queuing system having λ and μ identical to the $M/M/1/GD/\infty/\infty$ system that we have just analyzed. For this $M/D/1/GD/\infty/\infty$ model, $E(\mathbf{S}) = \frac{1}{8}$ hour and var $\mathbf{S} = 0$. Then

$$L_q = \frac{\left(\frac{5}{8}\right)^2}{2\left(1 - \frac{5}{8}\right)} = \frac{25}{48} \text{ customer}$$

$$W_q = \frac{L_q}{\lambda} = \frac{\frac{25}{48}}{5} = \frac{5}{48} \text{ hour}$$

In this $M/D/1/GD/\infty/\infty$ system, a typical customer will spend only half as much time in line as in an $M/M/1/GD/\infty/\infty$ queuing system with identical arrival and service rates. As this example shows, even if mean service times are not decreased, a decrease in the variability of service times can substantially reduce queue size and customer waiting time.

PROBLEMS

Group A

1 An average of 20 cars per hour arrive at the drive-in window of a fast-food restaurant. If each car's service time is 2 minutes, how many cars (on the average) will be waiting in line? Assume exponential interarrival times.

2 Using the fact that $L_s = \frac{\lambda}{\mu}$, demonstrate that for an $M/G/1/GD/\infty/\infty$ queuing system, the probability that the server is busy is $\rho = \frac{\lambda}{\mu}$.

3 An average of 40 cars per hour arrive to be painted at a single-server GM painting facility. 95% of the cars require 1 minute to paint; 5% must be painted twice and require 2.5 minutes to paint. Assume that interarrival times are exponential.

a On the average, how long does a car wait before being painted?

b If cars never had to be repainted, how would your answer to part (a) change?

Group B

4 Consider an $M/G/1/GD/\infty/\infty$ queuing system in which an average of 10 arrivals occur each hour. Suppose that each customer's service time follows an Erlang distribution, with rate parameter 1 customer per minute and shape parameter 4.

a Find the expected number of customers waiting in line.

b Find the expected time that a customer will spend in the system.

c What fraction of the time will the server be idle?

5 Consider an $M/G/1/GD/\infty/\infty$ queuing system in which interarrival times are exponentially distributed with parameter λ and service times have a probability density function $s(t)$. Let $\mathbf{X}_i$ be the number of customers present an instant after the ith customer completes service.

a Explain why $\mathbf{X}_1, \mathbf{X}_2, \ldots, \mathbf{X}_k, \ldots$ is a Markov chain.

b Explain why $P_{ij} = P(\mathbf{X}_{k+1} = j|\mathbf{X}_k = i)$ is zero for $j < i - 1$.

c Explain why for $i > 0$, $P_{i,i-1} =$ (probability that no arrival occurs during a service time); $P_{ii} =$ (probability that one arrival occurs during a service time); and for j $\geq i$, $P_{ij} =$ (probability that $j - i + 1$ arrivals occur during a service time).

d Explain why, for $j \geq i - 1$ and $i > 0$,

$$P_{ij} = \int_0^\infty \frac{s(x)e^{-\lambda x}(\lambda x)^{j-i+1}}{(j - i + 1)!} \, dx$$

Hint: The probability that a service time is between x and $x + \Delta x$ is $\Delta x s(x)$. Given that the service time equals x, the probability that $j - i + 1$ arrivals will occur during the service time is

$$\frac{e^{-\lambda x}(\lambda x)^{j-i+1}}{(j - i + 1)!}$$

20.9 Finite Source Models: The Machine Repair Model

With the exception of the $M/M/1/GD/c/\infty$ model, all the models we have studied have displayed arrival rates that were independent of the state of the system. As discussed previously, there are two situations where the assumption of the state-independent arrival rates may be invalid:

1 If customers do not want to buck long lines, the arrival rate may be a decreasing function of the number of people present in the queuing system. For an illustration of this situation, see Problems 4 and 5 at the end of this section.

2 If arrivals to a system are drawn from a small population, the arrival rate may greatly depend on the state of the system. For example, if a bank has only 10 depositors, then at an instant when all depositors are in the bank, the arrival rate must be zero, while if fewer than 10 people are in bank, the arrival rate will be positive.

Models in which arrivals are drawn from a small population are called **finite source models.** We now analyze an important finite source model known as the *machine repair* (or *machine interference*) model.

In the machine repair problem, the system consists of K machines and R repair people. At any instant in time, a particular machine is in either good or bad condition. The length of time that a machine remains in good condition follows an exponential distribution with rate λ. Whenever a machine breaks down, the machine is sent to a repair center consisting of R repair people. The repair center services the broken machines as if they were arriving at an $M/M/R/GD/\infty/\infty$ system.

Thus, if $j \leq R$ machines are in bad condition, a machine that has just broken will immediately be assigned for repair; if $j > R$ machines are broken, $j - R$ machines will be waiting in a single line for a repair worker to become idle. The time it takes to complete repairs on a broken machine is assumed exponential with rate μ (or mean repair time is $\frac{1}{\mu}$). Once a machine is repaired, it returns to good condition and is again susceptible to breakdown. The machine repair model may be modeled as a birth–death process, where the state j at any time is the number of machines in bad condition. Using the Kendall–Lee notation, the model just described may be expressed as an $M/M/R/GD/K/K$ model. The first K indicates that at any time, no more than K customers (or machines) may be present, and the second K indicates that arrivals are drawn from a finite source of size K.

Table 8 exhibits the interpretation of each state for a machine repair model having $K = 5$ and $R = 2$ ($G =$ machine in good condition; $B =$ broken machine). To find the birth–death parameters for the machine repair model (see Figure 22), note that a birth cor-

TABLE 8
Possible States in a Machine Repair Problem When $K = 5$ and $R = 2$

State	No. of Good Machines	Repair Queue	No. of Repair Workers Busy
0	$G\,G\,G\,G\,G$		0
1	$G\,G\,G\,G$		1
2	$G\,G\,G$		2
3	$G\,G$	B	2
4	G	$B\,B$	2
5		$B\,B\,B$	2

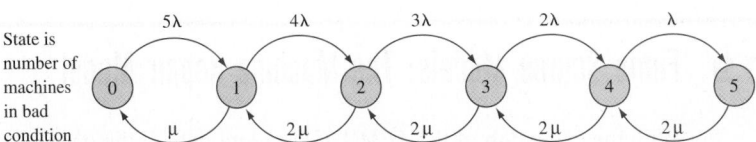

State is number of machines in bad condition

responds to a machine breaking down and a death corresponds to a machine having just been repaired. To figure out the birth rate in state j, we must determine the rate at which machines break down when the state of the system is j. When the state is j, there are $K - j$ machines in good condition. Since each machine breaks down at rate λ, the total rate at which breakdowns occur when the state is j is

$$\lambda_j = \underbrace{\lambda + \lambda + \cdots + \lambda}_{(K-j)\lambda\text{'s}} = (K - j)\lambda$$

To determine the death rate for the machine repair model, we proceed as we did in our discussion of the $M/M/s/GD/\infty/\infty$ queuing model. When the state is j, min (j, R) repair people will be busy. Since each occupied repair worker completes repairs at rate μ, the death rate μ_j is given by

$$\mu_j = j\mu \qquad (j = 0, 1, \ldots, R)$$
$$\mu_j = R\mu \qquad (j = R + 1, R + 2, \ldots, K)$$

If we define $\rho = \frac{\lambda}{\mu}$, an application of (16)–(18) yields the following steady-state probability distribution:

$$\pi_j = \binom{K}{j} \rho^j \pi_0 \qquad (j = 0, 1, \ldots, R)$$

$$= \frac{\binom{K}{j} \rho^j j! \pi_0}{R! R^{j-R}} \qquad (j = R + 1, R + 2, \ldots, K)$$

(52)

In (52),

$$\binom{K}{j} = \frac{K!}{j!(K - j)!}$$

where $0! = 1$, and for $n \geq 1$, $n! = n(n - 1) \cdots (2)(1)$. To use (52), begin by finding π_0 from the fact that $\pi_0 + \pi_1 + \cdots + \pi_k = 1$. Using the steady-state probabilities in (52), we can determine the following quantities of interest:

L = expected number of broken machines

L_q = expected number of machines waiting for service

W = average time a machine spends broken (down time)

W_q = average time a machine spends waiting for service

Unfortunately, there are no simple formulas for L, L_q, W, and W_q. The best we can do is to express these quantities in terms of the π_j's:

$$L = \sum_{j=0}^{j=K} j\pi_j \tag{53}$$

$$L_q = \sum_{j=R}^{j=K} (j - R)\pi_j \tag{54}$$

We can now use (28) and (29) to obtain W and W_q. Since the arrival rate is state-dependent, the average number of arrivals per unit time is given by $\bar{\lambda}$, where

$$\bar{\lambda} = \sum_{j=0}^{j=K} \pi_j\lambda_j = \sum_{j=0}^{j=K} \lambda(K - j)\pi_j = \lambda(K - L) \tag{55}$$

If (28) is applied to the machines being repaired and to those machines awaiting repairs, we obtain

$$W = \frac{L}{\bar{\lambda}} \tag{56}$$

Applying (29) to the machines awaiting repair, we obtain

$$W_q = \frac{L_q}{\bar{\lambda}} \tag{57}$$

The following example illustrates the use of these formulas.

EXAMPLE 12 **Patrol Cars**

The Gotham Township Police Department has 5 patrol cars. A patrol car breaks down and requires service once every 30 days. The police department has two repair workers, each of whom takes an average of 3 days to repair a car. Breakdown times and repair times are exponential.

1 Determine the average number of police cars in good condition.

2 Find the average down time for a police car that needs repairs.

3 Find the fraction of the time a particular repair worker is idle.

Solution This is a machine repair problem with $K = 5$, $R = 2$, $\lambda = \frac{1}{30}$ car per day, and $\mu = \frac{1}{3}$ car per day. Then

$$\rho = \frac{\frac{1}{30}}{\frac{1}{3}} = \frac{1}{10}$$

From (52),

$$\pi_1 = \binom{5}{1}\left(\frac{1}{10}\right)\pi_0 = .5\pi_0$$

$$\pi_2 = \binom{5}{2}\left(\frac{1}{10}\right)^2\pi_0 = .1\pi_0$$

$$\pi_3 = \binom{5}{3}\left(\frac{1}{10}\right)^3\frac{3!}{2!2}\pi_0 = .015\pi_0 \tag{58}$$

$$\pi_4 = \binom{5}{4}\left(\frac{1}{10}\right)^4\frac{4!}{2!(2)^2}\pi_0 = .0015\pi_0$$

$$\pi_5 = \binom{5}{5}\left(\frac{1}{10}\right)^5\frac{5!}{2!(2)^3}\pi_0 = .000075\pi_0$$

Then $\pi_0(1 + .5 + .1 + .015 + .0015 + .000075) = 1$, or $\pi_0 = .619$. Now (58) yields $\pi_1 = .310$, $\pi_2 = .062$, $\pi_3 = .009$, $\pi_4 = .001$, and $\pi_5 = 0$.

1 The expected number of cars in good condition is $K - L$, which is given by

$$K - \sum_{j=0}^{j=5} j\pi_j = 5 - [0(.619) + 1(.310) + 2(.062) + 3(.009) + 4(.001) + 5(0)]$$

$$= 5 - .465 = 4.535 \text{ cars in good condition}$$

2 We seek $W = \frac{L}{\bar{\lambda}}$. From (55),

$$\bar{\lambda} = \sum_{j=0}^{j=5} \lambda(5 - j)\pi_j = \frac{1}{30}(5\pi_0 + 4\pi_1 + 3\pi_2 + 2\pi_3 + \pi_4 + 0\pi_5)$$

$$= \frac{1}{30}[5(.619) + 4(.310) + 3(.062) + 2(.009) + 1(.001) + 0(0)]$$

$$= 0.151 \text{ car per day}$$

or

$$\bar{\lambda} = \lambda(K - L) = \frac{4.535}{30} = 0.151 \text{ car per day}$$

Since $L = 0.465$ car, we find that $W = \frac{0.465}{.0151} = 3.08$ days.

3 The fraction of the time that a particular repair worker will be idle is $\pi_0 + 0.5\pi_1 = .619 + .5(.310) = .774$.

If there were three repair people, the fraction of the time that a particular server would be idle would be $\pi_0 + (\frac{2}{3})\pi_1 + (\frac{1}{3})\pi_2$, and for a repair staff of R people, the probability that a particular server would be idle is given by

$$\pi_0 + \frac{(R - 1)\pi_1}{R} + \frac{(R - 2)\pi_2}{R} + \cdots + \frac{\pi_{R-1}}{R}$$

A Spreadsheet for the Machine Repair Problem

Machrep.xls

Figure 23 (file Machrep.xls) gives a spreadsheet template for the machine repair model. In cell B2, we input λ; in cell C2, μ; in cell D2, the number of repairers; and in cell F2,

FIGURE 23

	A	B	C	D	E	F	G	H
1	MACHINE	LAMBDA?	MU?	R?	RO	K?		
2	REPAIR	0.03333333	0.33333333	3	0.1	5		
3	MODEL	L	LS	LQ	W	WS	WQ	
4		0.45494681	0.45450532	0.0004415	3.00291412	3	0.002914124	
5								
6								
7								
8								
9								
10								
11								
12	STATE	LAMBDA(J)	MU(J)	CJ	PROB	#IN QUEUE	COLA*COLE	COLE*COL
13	0	0.16666667	0	1	0.62085236	0	0	0
14	1	0.13333333	0.33333333	0.5	0.31042618	0	0.310426181	0
15	2	0.1	0.66666667	0.1	0.06208524	0	0.124170472	0
16	3	0.06666667	1	0.01	0.00620852	0	0.018625571	0
17	4	0.03333333	1	0.00066667	0.0004139	1	0.001655606	0.0004139
18	5	0	1	2.2222E-05	1.3797E-05	2	6.89836E-05	2.7593E-05
19	6	0	1	0	0	3	0	0
20	7	0	1	0	0	4	0	0
21	8	0	1	0	0	5	0	0

the number of machines. In row 4, L, L_q, L_s, W, W_q, and W_s are computed. L_s equals the expected number of machines (in the steady state) being repaired and W_s equals the expected time that a broken machine spends being repaired. In column E, the steady-state probabilities are computed. We are assuming that $K \leq 1,000$. In Figure 23, we have input the information for Example 12.

Using LINGO for Machine Repair Model Computations

The LINGO function $@\mathbf{PFS}(K*\lambda/\mu,R,K)$ will yield L, the expected number (in the steady state) of machines in bad condition. The FS stands for Finite Source. Thus, for Example 12, $@\mathbf{PFS}(5*(1/30)/(1/3),2,5)$ will yield .465.

PROBLEMS

Group A

1 A laundromat has 5 washing machines. A typical machine breaks down once every 5 days. A repairer can repair a machine in an average of 2.5 days. Currently, three repairers are on duty. The owner of the laundromat has the option of replacing them with a superworker, who can repair a machine in an average of $\frac{5}{6}$ day. The salary of the superworker equals the pay of the three regular employees. Breakdown and service times are exponential. Should the laundromat replace the three repairers with the superworker?

2 My dog just had 3 frisky puppies who jump in and out of their whelping box. A puppy spends an average of 10 minutes (exponentially distributed) in the whelping box before jumping out. Once out of the box, a puppy spends an average of 15 minutes (exponentially distributed) before jumping back into the box.

  a At any given time, what is the probability that more puppies will be out of the box than will be in the box?

  b On the average, how many puppies will be in the box?

Group B

3[†] Gotham City has 10,000 streetlights. City investigators have determined that at any given time, an average of 1,000 lights are burned out. A streetlight burns out after an average of 100 days of use. The city has hired Mafia, Inc., to replace burned-out lamps. Mafia, Inc.'s contract states that the company is supposed to replace a burned-out street lamp in an average of 7 days. Do you think that Mafia, Inc. is living up to the contract?

4 This problem illustrates balking. The Oryo Cookie Ice Cream Shop in Dunkirk Square has three competitors. Since

[†]Based on Kolesar (1979).

people don't like to wait in long lines for ice cream, the arrival rate to the Oryo Cookie Ice Cream Shop depends on the number of people in the shop. More specifically, while $j \leq 4$ customers are present in the Oryo shop, customers arrive at a rate of $(20 - 5j)$ customers per hour. If more than 4 people are in the Oryo shop, the arrival rate is zero. For each customer, revenues less raw material costs are 50¢. Each server is paid $3 per hour. A server can serve an average of 10 customers per hour. To maximize expected profits (revenues less raw material and labor costs), how many servers should Oryo hire? Assume that interarrival and service times are exponential.

5 Suppose that interarrival times to a single-server system are exponential, but when n customers are present, there is a probability $\frac{n}{n+1}$ that an arrival will balk and leave the system before entering service. Also assume exponential service times.

  a Find the probability distribution of the number of people present in the steady state.

  b Find the expected number of people present in the steady state. (*Hint:* The fact that

$$e^x = 1 + x + \frac{x^2}{2!} + \frac{x^3}{3!} + \cdots$$

may be useful.)

6 For the machine repair model, show that $W = K/\bar{\lambda} - (1/\lambda)$.

7 (Requires use of a spreadsheet or LINGO) The machine repair model may often be used to approximate the behavior of a computer's CPU (central processing unit). Suppose that 20 terminals (assumed to always be busy) feed the CPU. After the CPU responds to a user, he or she takes an average of 80 seconds before sending another request to the CPU (this is called the *think time*). The CPU takes an average of

2 seconds to respond to any request. On the average, how long will a user have to wait before the CPU acts on his or her request? How will your answer change if there are 30 terminals? 40 terminals? Of course, you must make appropriate assumptions about exponentiality to answer this question.

8 Allbest airlines has 100 planes. Planes break down an average of twice a year and take one week to fix. Assuming the times between breakdowns and repairs are exponential, how many repairmen are needed to ensure that there is at least a 95% chance that 90 or more planes are available? (*Hint:* Use a one-way data table.)

9 An army has 200 tanks. Tanks need maintenance 10 times per year, and maintenance takes an average of 2 days. The army would like to have an average of at least 180 tanks working. How many repairmen are needed? Assume exponential interarrival and service times. (*Hint:* Use a one-way data table.)

Group C

10 Bectol, Inc. is building a dam. A total of 10 million cu ft of dirt is needed to construct the dam. A bulldozer is used to collect dirt for the dam. Then the dirt is moved via dumpers to the dam site. Only one bulldozer is available, and it rents for $100 per hour. Bectol can rent, at $40 per hour, as many dumpers as desired. Each dumper can hold 1,000 cu ft of dirt. It takes an average of 12 minutes for the bulldozer to load a dumper with dirt, and each dumper an average of five minutes to deliver the dirt to the dam and return to the bulldozer. Making appropriate assumptions about exponentiality, determine how Bectol can minimize the total expected cost of moving the dirt needed to build the dam. (*Hint:* There is a machine repair problem somewhere!)

20.10 Exponential Queues in Series and Open Queuing Networks

In the queuing models that we have studied so far, a customer's entire service time is spent with a single server. In many situations (such as the production of an item on an assembly line), the customer's service is not complete until the customer has been served by more than one server (see, for example, Figure 24).

Upon entering the system in Figure 24, the arrival undergoes stage 1 service (after waiting in line if all stage 1 servers are busy on arrival). After completing stage 1 service, the customer waits for and undergoes stage 2 service. This process continues until the customer completes stage k service. A system like Figure 24 is called a **k-stage series** (or tandem) **queuing system.** A remarkable theorem due to Jackson (1957) is as follows (see Heyman and Sobel (1984) for a proof).

> **THEOREM 4**
>
> If (1) interarrival times for a series queuing system are exponential with rate λ, (2) service times for each stage i server are exponential, and (3) each stage has an infinite-capacity waiting room, then interarrival times for arrivals to each stage of the queuing system are exponential with rate λ.

FIGURE **24**
Exponential Queues in Series

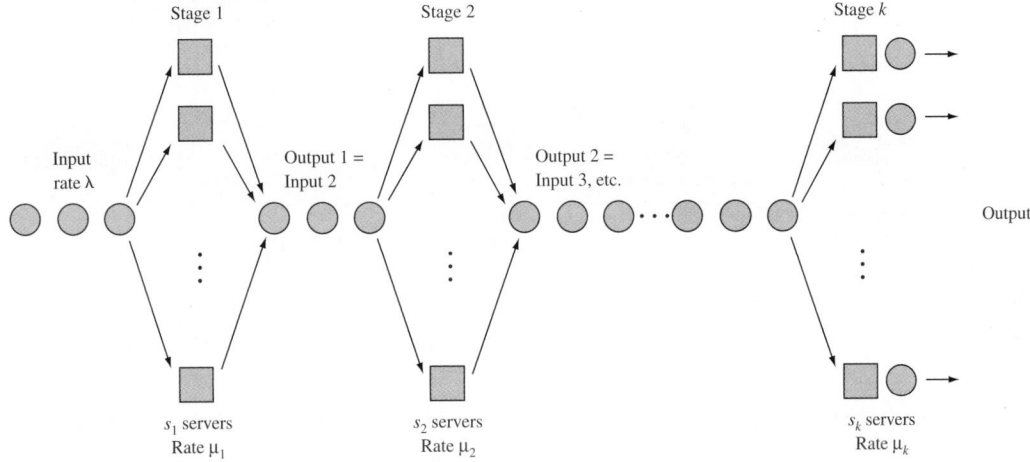

For this result to be valid, each stage must have sufficient capacity to service a stream of arrivals that arrives at rate λ; otherwise, the queue will "blow up" at the stage with insufficient capacity. From our discussion of the $M/M/s/GD/\infty/\infty$ queuing system in Section 20.6, we see that each stage will have sufficient capacity to handle an arrival stream of rate λ if and only if, for $j = 1, 2, \ldots, k$, $\lambda > s_j\mu_j$. If $\lambda < s_j\mu_j$, Jackson's result implies that stage j of the system in Figure 24 may be analyzed as an $M/M/s_j/GD/\infty/\infty$ system with exponential interarrival times having rate λ and exponential service times with a mean service time of $\frac{1}{\mu_j}$. The usefulness of Jackson's result is illustrated by the following example.

EXAMPLE 13 | **Auto Assembly**

The last two things that are done to a car before its manufacture is complete are installing the engine and putting on the tires. An average of 54 cars per hour arrive requiring these two tasks. One worker is available to install the engine and can service an average of 60 cars per hour. After the engine is installed, the car goes to the tire station and waits for its tires to be attached. Three workers serve at the tire station. Each works on one car at a time and can put tires on a car in an average of 3 minutes. Both interarrival times and service times are exponential.

1 Determine the mean queue length at each work station.

2 Determine the total expected time that a car spends waiting for service.

Solution This is a series queuing system with $\lambda = 54$ cars per hour, $s_1 = 1$, $\mu_1 = 60$ cars per hour, $s_2 = 3$, and $\mu_2 = 20$ cars per hour (see Figure 25). Since $\lambda < \mu_1$ and $\lambda < 3\mu_2$, neither queue will "blow up," and Jackson's theorem is applicable. For stage 1 (engine), $\rho = \frac{54}{60} = .90$. Then (27) yields

$$L_q \text{ (for engine)} = \left(\frac{\rho^2}{1 - \rho}\right) = \left[\frac{(.90)^2}{1 - .90}\right] = 8.1 \text{ cars}$$

Now (32) yields

$$W_q \text{ (for engine)} = \frac{L_q}{\lambda} = \frac{8.1}{54} = 0.15 \text{ hour}$$

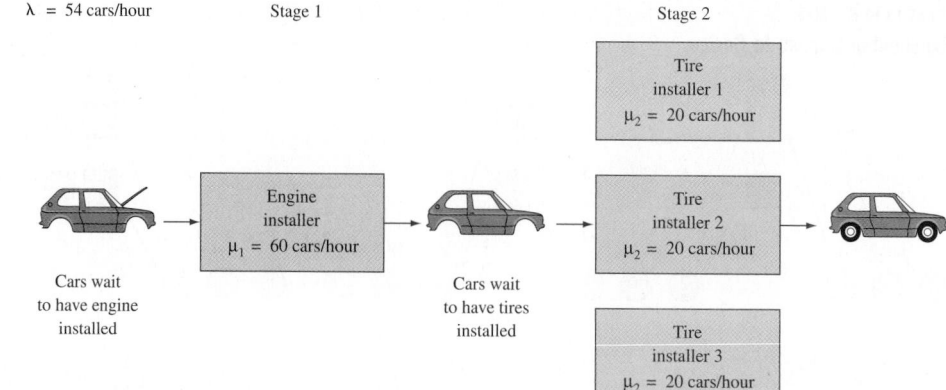

$\lambda = 54$ cars/hour

Stage 2

FIGURE 25
Series Queuing System
for Automobile

For stage 2 (tires), $\rho = \frac{54}{3(20)} = .90$. Table 6 yields $P(j \geq 3) = .83$. Now (41) yields

$$L_q \text{ (for tires)} = \frac{.83(.90)}{1 - .90} = 7.47 \text{ cars}$$

Then

$$W_q \text{ (for tires)} = \frac{L_q}{\lambda} = \frac{7.47}{54} = 0.138 \text{ hour}$$

Thus, the total expected time a car spends waiting for engine installation and tires is $0.15 + 0.138 = 0.288$ hour.

Open Queuing Networks

We now describe **open queuing networks,** a generalization of queues in series. As in Figure 24, assume that station j consists of s_j exponential servers, each operating at rate μ_j. Customers are assumed to arrive at station j from outside the queuing system at rate r_j. These interarrival times are assumed to be exponentially distributed. Once completing service at station i, a customer joins the queue at station j with probability p_{ij} and completes service with probability

$$1 - \sum_{j=1}^{j=k} p_{ij}$$

Define λ_j, the rate at which customers arrive at station j (this includes arrivals at station j from outside the system *and* from other stations). $\lambda_1, \lambda_2, \ldots, \lambda_k$ can be found by solving the following system of linear equations:

$$\lambda_j = r_j + \sum_{i=1}^{i=k} p_{ij}\lambda_i \qquad (j = 1, 2, \ldots, k)$$

This follows, because a fraction p_{ij} of the λ_i arrivals to station i will next go to station j. Suppose $s_i\mu_j > \lambda_j$ holds for all stations. Then it can be shown that the probability distribution of the number of customers present at station j and the expected number of customers present at station j can be found by treating station j as an $M/M/s_j/GD/\infty/\infty$ system with arrival rate λ_j and service rate μ_j. If for some j, $s_j\mu_j \leq \lambda_j$, then no steady-state distribution of

CHAPTER **20** Queuing Theory

customers exists. Remarkably, the numbers of customers present at each station are independent random variables. That is, knowledge of the number of people at all stations other than station j tells us nothing about the distribution of the number of people at station j! This result does not hold, however, if either interarrival or service times are not exponential.

To find L, the expected number of customers in the queuing system, simply add up the expected number of customers present at each station. To find W, the average time a customer spends in the system, simply apply the formula $L = \lambda W$ to the entire system. Here, $\lambda = r_1 + r_2 + \cdots + r_k$, because this represents the average number of customers per unit time arriving at the system. The following example illustrates the analysis of open queuing networks.

EXAMPLE 14   Open Queuing Network Example

Consider two servers. An average of 8 customers per hour arrive from outside at server 1, and an average of 17 customers per hour arrive from outside at server 2. Interarrival times are exponential. Server 1 can serve at an exponential rate of 20 customers per hour, and server 2 can serve at an exponential rate of 30 customers per hour. After completing service at server 1, half of the customers leave the system, and half go to server 2. After completing service at server 2, $\frac{3}{4}$ of the customers complete service, and $\frac{1}{4}$ return to server 1.

1   What fraction of the time is server 1 idle?

2   Find the expected number of customers at each server.

3   Find the average time a customer spends in the system.

4   How would the answers to parts (1)–(3) change if server 2 could serve only an average of 20 customers per hour?

Solution   We have an open queuing network with $r_1 = 8$ customers/hour and $r_2 = 17$ customers/hour. Also, $p_{12} = .5$, $p_{21} = .25$, and $p_{11} = p_{22} = 0$. We can find λ_1 and λ_2 by solving $\lambda_1 = 8 + .25\lambda_2$ and $\lambda_2 = 17 + .5\lambda_1$. This yields $\lambda_1 = 14$ customers/hour and $\lambda_2 = 24$ customers/hour.

1   Server 1 may be treated as an $M/M/1/GD/\infty/\infty$ system with $\lambda = 14$ customers/hour and $\mu = 20$ customers/hour. Then $\pi_0 = 1 - \rho = 1 - .7 = .3$. Thus, server 1 is idle 30% of the time.

2   From (26), we find L at server $1 = \frac{14}{20-14} = \frac{7}{3}$ and L at server $2 = \frac{24}{30-24} = 4$. Thus, an average of $4 + \frac{7}{3} = \frac{19}{3}$ customers will be present in the system.

3 $W = \frac{L}{\lambda}$, where $\lambda = 8 + 17 = 25$ customers/hour. Thus,

$$W = \frac{\left(\dfrac{19}{3}\right)}{25} = \frac{19}{75} \text{ hour}$$

4   In this case, $s_2\mu_2 = 20 < \lambda_2$, so no steady state exists.

Network Models of Data Communication Networks

Queuing networks are commonly used to model data communication networks. The queuing models enable us to determine the typical delay faced by transmitted data and also to design the network. Our discussion is based on Tannenbaum (1981). See file Compnetwork.xls.

Compnetwork.xls

Consider a data communication network with 5 nodes (*A*, *B*, *C*, *D*, *E*). Suppose each data packet transmitted consists of 800 bits, and the number of packets per second that must be transmitted between each pair of nodes is as shown in Figure 26.

For example, an average of 5 packets per second must be sent from node *A* to node *B*. Packets are not always transmitted over the most direct route. Suppose the routings used to transmit each type of message are as shown in Figure 27.

For example, all messages that must go from *A* to *D* are transmitted via the route *A–B–D*. Each arc or route connecting two nodes has a capacity measured in thousands of bits per second. For example, an arc with 16,000 bits/second of capacity can "serve" 16,000/800 = 20 packets/second. Each arc's capacity in thousands of bits per second is given in Figure 28. We are interested, of course, in the expected delay for a packet. Also, if total network capacity is limited, it is important to determine the capacity on each arc that will minimize the expected delay for a packet. The usual way to approach this problem is to treat each arc as if it were an independent *M/M*/1 queue and determine the expected time spent by each packet transmitted through that arc by the formula

	A	B	C	D	E	F	G
23	Packets/second		A	B	C	D	E
24		A	0	5	4	1	7
25		B	5	0	6	3	2
26		C	4	6	0	3	3
27		D	1	3	3	0	3
28		E	7	2	3	3	0

FIGURE 26

	A	B	C	D	E	F	G
30			A	B	C	D	E
31	Route used	A	-	AB	ABC	ABD	AE
32		B	BA	-	BC	BD	BDE
33		C	CBA	CB	-	CD	CDE
34		D	DBA	DB	DC	-	DCE
35		E	EA	EDB	EDC	ECD	-

FIGURE 27

	B	C	D	E	F
4	Line	Packets per second	Capacity (000) bits per second	Service Rate in Packets per second	W in seconds
5	AB	10	20	25	0.066667
6	AE	7	20	25	0.055556
7	BC	10	15	18.75	0.114286
8	BD	6	10	12.5	0.153846
9	CD	9	10	12.5	0.285714
10	CE	3	10	12.5	0.105263
11	DE	5	10	12.5	0.133333
12	BA	10	20	25	0.066667
13	EA	7	20	25	0.055556
14	CB	10	15	18.75	0.114286
15	DB	6	10	12.5	0.153846
16	DC	9	10	12.5	0.285714
17	EC	3	10	12.5	0.105263
18	ED	5	10	12.5	0.133333

FIGURE 28

$$W = \frac{1}{\mu - \lambda}$$

To illustrate, consider arc AB. Packets that are to be transmitted from A to B, A to C and A to D will use this arc. This is a total of $5 + 4 + 1 = 10$ packets per second. Suppose arc AB has a capacity of 20,000 bits per second. Then for arc AB, $\mu = 20{,}000/800 = 25$ packets per second, and $\lambda = 10$ packets per second. Then

$$W = \frac{1}{25 - 10} = .06667 \text{ second}$$

In rows 5–18 of Figure 28, we compute W for each arc in the communications network. Note that the network is assumed symmetric (that is, AB arrival rate and capacity equals BA arrival rate and capacity), so rows 12–18 are just copies of rows 5–11.

To determine the average delay faced by a packet, we use the following formula:

$$\text{Average delay per packet} = \frac{\sum_{\text{all arcs}} (\text{Arc arrival rate}) * (\text{expected time spent in arc})}{\text{Total number of arrivals}}$$

In Figure 29, we computed the average delay per packet in cell C20 with the formula

$$=\text{SUMPRODUCT(C5:C18,F5:F18)/SUM(C24:G28)}$$

Thus, average delay per packet is .18 second per packet.

	B	C
19		
20	Mean	0.180416
21	Time in system	
22	seconds	

FIGURE 29

	B	C	D	E	F	G
1		800				
2		bits/packet				
3						
4	Line	Packets per second	Capacity (000) bits per second	Service Rate in Packets per second	W in seconds	Diff
5	AB	10	18.31869	22.89836	0.077529	12.89836
6	AE	7	14.23287	17.79109	0.092669	10.79109
7	BC	10	18.31662	22.89577	0.077545	12.89577
8	BD	6	12.79577	15.99471	0.100053	9.994709
9	CD	9	16.98872	21.2359	0.081727	12.2359
10	CE	3	8.052237	10.0653	0.141537	7.065296
11	DE	5	11.2951	14.11888	0.109663	9.118877
12	BA	10	18.31869	22.89836	0.077529	12.89836
13	EA	7	14.23287	17.79109	0.092669	10.79109
14	CB	10	18.31662	22.89577	0.077545	12.89577
15	DB	6	12.79577	15.99471	0.100053	9.994709
16	DC	9	16.98872	21.2359	0.081727	12.2359
17	EC	3	8.052237	10.0653	0.141537	7.065296
18	ED	5	11.2951	14.11888	0.109663	9.118877
19		Total cap	200			
20	Mean	0.121843				

FIGURE 30

Suppose we had only 200,000 bits/second of total capacity to allocate to the network. How should we allocate capacity to minimize the expected delay per packet? See the sheet Optimization in file Compnetwork.xls (Figure 30). In G5:G18, we compute Service rate (in packets/second) − Arrival rate (in packets/second). We constrain this to be at least .01 so that a steady-state exists. Then our Solver window is as shown in Figure 31.

We choose capacities D5:D11 (remember that D12:D18 are just copies of D5:D11) to minimize expected system time (C20). We ensure that each arc's service rate exceeds its arrival rate (G5:G11≥.01), each capacity is nonnegative (D5:D11≥0), and total capacity is at most 200,000 (D19≤200). We find that we can reduce expected time in the system for a packet to .1218 second.

Of course, we are assuming a static routing, in which arrival rates to each node do not vary with the state of the network. In reality, many sophisticated dynamic routing schemes have been developed. A dynamic routing scheme would realize, for example, that if arc AB is congested and arc AD is relatively free, we should send messages directly from A to D instead of via route A–B–D.

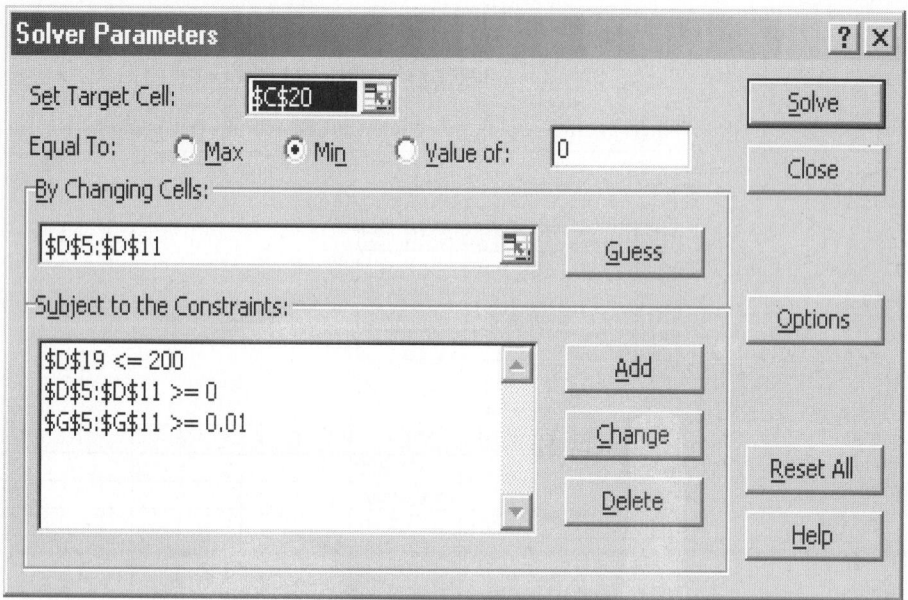

FIGURE 31

PROBLEMS

Group A

1 A Social Security Administration branch is considering the following two options for processing applications for social security cards:

Option 1 Three clerks process applications in parallel from a single queue. Each clerk fills out the form for the application in the presence of the applicant. Processing time is exponential with a mean of 15 minutes. Interarrival times are exponential.

Option 2 Each applicant first fills out an application without the clerk's help. The time to accomplish this is exponentially distributed, with a mean of 65 minutes. When the applicant has filled out the form, he or she joins a single line

to wait for one of the three clerks to check the form. It takes a clerk an average of 4 minutes (exponentially distributed) to review an application.

The interarrival time of applicants is exponential, and an average of 4.8 applicants arrive each hour. Which option will get applicants out of the office more quickly?

2 Consider an automobile assembly line in which each car undergoes two types of service: painting, then engine installation. Each hour, an average of 22.4 unpainted chassis arrive at the assembly line. It takes an average of 2.4 minutes to paint a car and an average of 3.75 minutes to install an engine. The assembly line has one painter and two engine installers. Assume that interarrival times and service times are exponential.

a On the average, how many painted cars without completely installed engines will be in the facility?

b On the average, how long will a painted car have to wait before installation of its engine begins?

3 Consider the following queuing systems:

System 1 An average of 40 customers arrive each hour; interarrival times are exponential. Customers must complete two types of service before leaving the system. The first server takes an average of 30 seconds (exponentially distributed) to perform type 1 service. After waiting in line, each customer obtains type 2 service (exponentially distributed with a mean of 1 minute) from a single server. After completing type 2 service, a customer leaves the system.

System 2 The arrival process for system 2 is identical to the interarrival process for system 1. In system 2, a customer must complete only one type of service. Service time averages 1.5 minutes and is exponentially distributed. Two servers are available.

In which system does a typical customer spend less time?

4 An average of 120 students arrive each hour (interarrival times are exponential) at State College's Registrar's Office to change their course registrations. To complete this process, a person must pass through three stations. Each station consists of a single server. Service times at each station are exponential, with the following mean times: station 1, 20 seconds; station 2, 15 seconds; station 3, 12 seconds. On the average, how many students will be present in the registrar's office for changing courses?

5 An average of 10 jobs per hour arrive at a job shop. Interarrival times of jobs are exponentially distributed. It takes an average of $\frac{10}{3}$ minutes (exponentially distributed) to complete a job. Unfortunately, $\frac{1}{3}$ of all completed jobs need to be reworked. Thus, with probability $\frac{1}{3}$, a completed job must wait in line to be reworked. In the steady state, how many jobs would one expect to find in the job shop? What would the answer be if it took an average of 5 minutes to finish a job?

6 Consider a queuing system consisting of three stations in series. Each station consists of a single server, who can process an average of 20 jobs per hour (processing times at each station are exponential). An average of 10 jobs per hour arrive (interarrival times are exponential) at station 1. When a job completes service at station 2, there is a .1 chance that it will return to station 1 and a .9 chance that it will move on to station 3. When a job completes service at station 3, there is a .2 chance that it will return to station 2 and a .8 chance that it will leave the system. All jobs completing service at station 1 immediately move on to station 2.

a Determine the fraction of time each server is busy.

b Determine the expected number of jobs in the system.

c Determine the average time a job spends in the system.

7 Before completing production, a product must pass through three stages of production. On the average, a new product begins at stage 1 every 6 minutes. The average time it takes to process the product at each stage is as follows: stage 1, 3 minutes; stage 2, 2 minutes; stage 3, 1 minute. After finishing at stage 3, the product is inspected (assume this takes no time). Ten percent of the final products are found to have a defective part and must return to stage 1 and go through the entire system again. After completing stage 3, 20% of the final products are found to be defective. They must return to stage 2 and pass through 2 and 3 again. On the average, how many jobs are in the system? Assume that all interarrival times and service times are exponential and that each stage consists of a single server.

8 A data communication network consists of three nodes, A, B, and C. Each packet transmitted contains 500 bits of information. The number of packets per second to be transmitted between each pair of nodes is as follows:

$$
\begin{array}{c}
\quad\quad A \quad B \quad C \\
\begin{array}{c} A \\ B \\ C \end{array}
\begin{bmatrix}
0 & 4 & 3 \\
4 & 0 & 6 \\
3 & 6 & 0
\end{bmatrix}
\end{array}
$$

The routing used for each pair of nodes is as follows:

$$
\begin{array}{c}
\quad\quad A \quad\quad B \quad\quad C \\
\begin{array}{c} A \\ B \\ C \end{array}
\begin{bmatrix}
- & ACB & AC \\
BCA & - & BAC \\
CA & CAB & -
\end{bmatrix}
\end{array}
$$

Assume that the capacities (in thousands of bits per second) for each arc are as follows:

Arc	Capacity
AB	12
AC	13
BC	15
BA	12
CA	13
CB	15

a Compute the expected delay for a packet.

b If a total of 75,000 bits/second of capacity is available, how should it be allocated?

9[†] Jobs arrive to a file server consisting of a CPU and two disks (disk 1 and disk 2). Currently there are six clients, and an average of three jobs per second arrive. Each visit to the CPU takes an average of .01 second, each visit to disk 1 takes an average of .02 second, and each visit to disk 2 takes an average of .03 second. An entering job first visits the CPU. After each visit to the CPU, with probability 7/16 the job next visits disk 1, with probability 8/16 the job next visits disk 2, and with probability 1/16 the job is completed. After visiting disk 1 or 2, the job immediately returns to the CPU.

a On the average, how many times does a job visit the CPU? How about disk 1? How about disk 2?

b On the average, how long does a job spend in the CPU? How long in disk 1? How long in disk 2? How long in the system?

10 Suppose the file server in Problem 9 now has 8 clients. Answer the questions in Problem 9.

[†]Based on Jain (1991).

11 Suppose we install a cache for disk 2. This will increase the mean time taken for a CPU visit by 30% and the mean time for a visit to disk 2 by 10%. On the other hand, the cache for disk 2 ensures that half the time the job was going to go to disk 2 the job will actually stay at the CPU and be processed there. Does the cache improve the operation of the system?

12 Suppose we eliminate disk 2. What will happen to the system response time? In this problem, you may assume that all requests that leave the CPU go to disk 1.

20.11 The *M/G/s/GD/s/∞* System (Blocked Customers Cleared)

In many queuing systems, an arrival who finds all servers occupied is, for all practical purposes, lost to the system. For example, a person who calls an airline for a reservation and gets a busy signal will probably call another airline. Or suppose that someone calls in a fire alarm and no engines are available; the fire will then burn out of control. Thus, in some sense, a request for a fire engine that occurs when no engines are available may be considered lost to the system. If arrivals who find all servers occupied leave the system, we call the system a **blocked customers cleared,** or BCC, system. Assuming that interarrival times are exponential, such a system may be modeled as an *M/G/s/GD/s/∞* system.

For an *M/G/s/GD/s/∞* system, *L, W, L_q,* and *W_q* are of limited interest. For example, since a queue can never occur, $L_q = W_q = 0$. If we let $\frac{1}{\mu}$ be the mean service time and λ be the arrival rate, then $W = W_s = \frac{1}{\mu}$.

In most BCC systems, primary interest is focused on the fraction of all arrivals who are turned away. Since arrivals are turned away only when *s* customers are present, a fraction π_s of all arrivals will be turned away. Hence, an average of $\lambda \pi_s$ arrivals per unit time will be lost to the system. Since an average of $\lambda(1 - \pi_s)$ arrivals per unit time will actually enter the system, we may conclude that

$$L = L_s = \frac{\lambda(1 - \pi_s)}{\mu}$$

For an *M/G/s/GD/s/∞* system, it can be shown that π_s depends on the service time distribution only through its mean ($\frac{1}{\mu}$). This fact is known as **Erlang's loss formula.** In other words, any *M/G/s/GD/s/∞* system with an arrival rate λ and mean service time of $\frac{1}{\mu}$ will have the same value of π_s. If we define $\rho = \frac{\lambda}{\mu}$, then for a given value of *s*, the value of π_s can be found from Figure 32. Simply read the value of ρ on the *x*-axis. Then the *y*-value on the *s*-server curve that corresponds to ρ will equal π_s. The following example illustrates the use of Figure 32.

EXAMPLE 15   **Ambulance Calls**

An average of 20 ambulance calls per hour are received by Gotham City Hospital. An ambulance requires an average of 20 minutes to pick up a patient and take the patient to the hospital. The ambulance is then available to pick up another patient. How many ambulances should the hospital have to ensure that there is at most a 1% probability of not being able to respond immediately to an ambulance call? Assume that interarrival times are exponentially distributed.

Solution   We are given that $\lambda = 20$ calls per hour, and $\frac{1}{\mu} = \frac{1}{3}$ hour. Thus, $\rho = \frac{\lambda}{\mu} = \frac{20}{3} = 6.67$. For $\rho = 6.67$, we seek the smallest value of *s* for which π_s is .01 or smaller. From Figure 32, we see that for $s = 13$, $\pi_s = .011$; and for $s = 14$, $\pi_s = .005$. Thus, the hospital needs 14 ambulances to meet its desired service standards.

FIGURE 32
Loss Probabilities for *M*/*G*/*s*/*GD*/*s*/∞ Queuing System

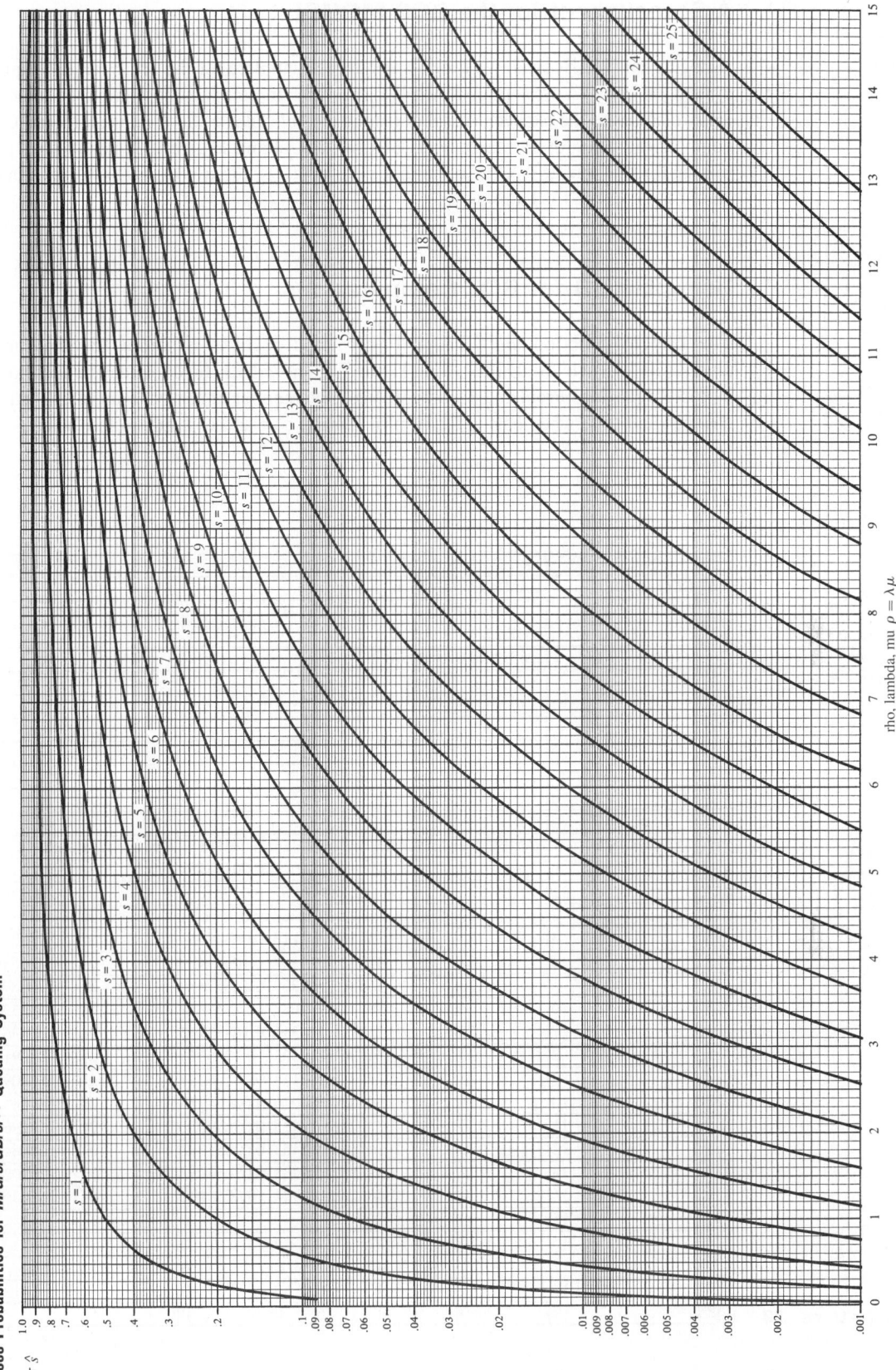

π ̂$_s$

rho, lambda, mu $\rho = \lambda\mu$

Source: Reprinted by permission of the publisher from *Introduction to Queuing Theory* by Robert B. Cooper. p. 316. Copyright©1980 by Elsevier Science Publishing Co., Inc.

A Spreadsheet for the BCC MODEL

Bcc.xls

In Figure 33 (file Bcc.xls) we give a spreadsheet template for the $M/G/s/GD/s/\infty$ queuing system. In cell B2, we input λ; in cell C2, μ; and in cell D2, the number of servers. In B4, we compute the expected number (in the steady state) of busy servers. In cell C4, we compute the value of π_s tabulated in Figure 32. Column E gives the steady-state probabilities for this model. We are assuming $s \leq 1,000$. In Figure 33, we have input the values of λ, μ, and s for Example 15.

Using LINGO for BCC Computations

The LINGO function $@\mathbf{PEL}(\lambda/\mu,s)$ will yield π_s. For Example 15, the function $@\mathbf{PEL}(20/3,13)$ yields .010627, as in Figure 32. The $@\mathbf{PEL}$ function may be used to solve a problem (such as Problem 6) where we seek the number of servers minimizing expected cost per-unit time when cost is the sum of service cost and cost due to lost business.

	A	B	C	D	E	F	G
1	BCC MODEL	LAMBDA?	MU?	s?			
2		20	3	14			
3		L OR LS	PI(s)				
4		6.63320534	0.0050192				
5							
6							
7							
8							
9							
10							
11							
12	STATE	LAMBDA(J)	MU(J)	CJ	PROB	#IN QUEUE	COLA*COLE
13	0	20	0	1	0.00127738	0	0
14	1	20	3	6.66666667	0.00851587	0	0.008515872
15	2	20	6	22.2222222	0.02838624	0	0.056772479
16	3	20	9	49.382716	0.06308053	0	0.189241596
17	4	20	12	82.3045267	0.10513422	0	0.420536879
18	5	20	15	109.739369	0.14017896	0	0.700894799
19	6	20	18	121.932632	0.1557544	0	0.934526398
20	7	20	21	116.126316	0.14833752	0	1.038362665
21	8	20	24	96.7719303	0.1236146	0	0.988916824
22	9	20	27	71.6829114	0.09156637	0	0.824097353
23	10	20	30	47.7886076	0.06104425	0	0.610442484
24	11	20	33	28.9627925	0.03699651	0	0.406961656
25	12	20	36	16.0904403	0.02055362	0	0.246643428
26	13	20	39	8.25150783	0.01054032	0	0.137024127
27	14	0	42	3.92928944	0.0050192	0	0.070268783
28	15	0	42	0	0	1	0
29	16	0	42	0	0	2	0
30	17	0	42	0	0	3	0
31	18	0	42	0	0	4	0
32	19	0	42	0	0	5	0

FIGURE 33

PROBLEMS

Group A

1 Suppose that a fire department receives an average of 24 requests for fire engines each hour. Each request causes a fire engine to be unavailable for an average of 20 minutes. To have at most a 1% chance of being unable to respond to a request, how many fire engines should the fire department have?

2 A telephone order sales company must determine how many telephone operators are needed to staff the phones during the 9-to-5 shift. It is estimated that an average of 480 calls are received during this period and that the average call lasts for 6 minutes. If the company wants to have at most 1 chance in 100 of a caller receiving a busy signal, how many operators should be hired for the 9-to-5 shift? What assumption does the answer require?

3 In Example 15, suppose the hospital had 10 ambulances. On the average, how many ambulances would be en route or returning from a call?

4 A phone system is said to receive 1 Erlang of usage per hour if callers keep lines busy for an average of 3,600 seconds per hour. Suppose a phone system receives 2 Erlangs of usage per hour. If you want only 1% of all calls blocked, how many phone lines do you need?[†]

5 (Requires the use of a spreadsheet or LINGO) At the peak usage time, an average of 200 people per hour attempt to log on the Jade Vax. The average length of time somebody spends on the Vax is 20 minutes. If the Indiana University Computing Service wants to ensure that during peak usage only 1% of all users receive an "All ports busy" message, how many ports should the Jade Vax have?

6 (Requires the use of a spreadsheet or LINGO) US Airlines receives an average of 500 calls per hour from customers who want to make a reservation (time between calls follows an exponential distribution). It takes an average of 3 minutes to handle each call. Each customer who buys a ticket contributes $100 to US Airlines profit. It costs $15 per hour to staff a telephone line. Any customer who receives a busy signal will purchase a ticket on another airline. How many telephone lines should US Airlines have?

Group B

7 On the average, 26 patrons per year come to the I.U. library to borrow the *I Ching* (assume that interarrival times are exponential). Borrowers who find the book unavailable leave and never return. A borrower keeps a copy of the *I Ching* for an average of 4 weeks.

    a If the library has only one copy, what is the expected number of borrowers who will come to borrow the *I Ching* each year and find that the book is not available?

    b Suppose that each person who comes to borrow the *I Ching* and is unable to is considered to cost the library $1 in goodwill. A copy of the *I Ching* lasts two years and costs $11. A thief has just stolen the library's only copy. To minimize the sum of purchasing and goodwill costs over the next two years, how many copies of the *I Ching* should be purchased?

8[‡] A company's warehouse can store up to 4 units of a good. Each month, an average of 10 orders for the good are received. The times between the receipt of successive orders are exponentially distributed. When an item is used to fill an order, a replacement item is immediately ordered, and it takes an average of one month for a replacement item to arrive. If no items are on hand when an order is received, the order is lost. What fraction of all orders will be lost due to shortages? (*Hint:* Let the storage space for each item be a server and think about what it means for a server to be busy. Then come up with an appropriate definition of "service" time.)

[†]Based on Green (1987).

[‡]Based on Karush (1957).

20.12 How to Tell Whether Interarrival Times and Service Times Are Exponential[§]

How can we determine whether the actual data are consistent with the assumption of exponential interarrival times and service times? Suppose, for example, that interarrival times of $t_1, t_2, \ldots, t_n$ have been observed. It can be shown that a reasonable estimate of the arrival rate λ is given by

$$\hat{\lambda} = \sum_{i=1}^{n} t_i^{\frac{i=n}{n}}$$

[§]This section covers topics that may be omitted with no loss of continuity.

For example, if $t_1 = 20$, $t_2 = 30$, $t_3 = 40$, and $t_4 = 50$, we have seen 4 arrivals in 140 time units, or an average of 1 arrival per 35 time units. In this case, our estimate of the arrival rate $\hat{\lambda}$ is given by

$$\hat{\lambda} = \frac{4}{20 + 30 + 40 + 50} = \frac{1}{35}$$

customer per unit time. Given $\hat{\lambda}$, we can try to determine whether $t_1, t_2, \ldots, t_n$ are consistent with the assumption that interarrival times are governed by an exponential distribution with rate $\hat{\lambda}$ and density $\hat{\lambda}e^{-\hat{\lambda}t}$. The easiest way to test this conjecture is by using a chi-square goodness-of-fit test to determine whether it is reasonable to conclude that t_1, $t_2, \ldots, t_n$ represent a random sample from a random variable with a given density function $f(t)$. A Kolmogorov–Smirnov test may also be used (see Law and Kelton (1990)).

To begin, we break up the set of possible interarrival times into k categories. Under the assumption that $f(t)$ does govern interarrival times, we determine the number of the t_i's that we would expect to fall into category i. We call this number e_i. Then we count up how many of the observed t_i's actually were in category i. We call this number o_i. Next, we use the following formula to compute the observed value of the chi-square statistic, written $\chi^2(\text{obs})$:

$$\chi^2(\text{obs}) = \sum_{i=1}^{i=k} \frac{(o_i - e_i)^2}{e_i}$$

The value of $\chi^2(\text{obs})$ follows a chi-square distribution, with $k - 2$ degrees of freedom. Important percentile points of the chi-square distribution are tabulated in Table 9.

If $\chi^2(\text{obs})$ is small, it is reasonable to assume that the t_i's are samples from a random variable with density function $f(t)$. (After all, a perfect fit would have $o_i = e_i$ for $i = 1$, $2, \ldots, k$, resulting in a χ^2 value of zero.) If $\chi^2(\text{obs})$ is large, it is reasonable to assume that the t_i's do not represent a random sample from a random variable with density $f(t)$.

More formally, we are interested in testing the following hypotheses:

H_0: $t_1, t_2, \ldots, t_n$ is a random sample from a random variable with density $f(t)$

H_a: $t_1, t_2, \ldots, t_n$ is not a random sample from a random variable with density function $f(t)$

Given a value of α (the desired Type I error), we accept H_0 if $\chi^2(\text{obs}) \leq \chi^2_{k-r-1}(\alpha)$ and accept H_a if $\chi^2(\text{obs}) > \chi^2_{k-r-1}(\alpha)$. From Table 9, we obtain $\chi^2_{k-r-1}(\alpha)$ which represents the point in the χ^2_{k-r-1} table that has an area α to the right of it. Here, r is the number of parameters that must be estimated to specify the interarrival time distribution. To find $\chi^2_{k-r-1}(\alpha)$ in Excel, we simply enter the formula CHINV(Alpha, $k-r-1$). Thus, if interarrival times are exponential, $r = 1$, and if interarrival times follow a normal distribution or an Erlang distribution, $r = 2$. When choosing the boundaries for the k categories, it is desirable to ensure that each e_i is at least 5, $k \leq 30$, and the e_i's be kept as equal as possible. Example 16 illustrates the use of the chi-square test.

EXAMPLE 16   **Interarrival Times: Exponential or Not Exponential?**

The following interarrival times (in minutes) have been observed: 0.01, 0.07, 0.03, 0.08, 0.04, 0.10, 0.05, 0.10, 0.11, 1.17, 1.50, 0.93, 0.54, 0.19, 0.22, 0.36, 0.27, 0.46, 0.51, 0.11, 0.56, 0.72, 0.29, 0.04, 0.73. Does it seem reasonable to conclude that these observations come from an exponential distribution?

Solution   There are 25 observations with $\sum_{i=1}^{i=25} t_i = 9.19$. Thus, $\bar{\lambda} = \frac{25}{9.19} = 2.72$ arrivals per minute. We now test whether or not our data are consistent with an exponential random variable

TABLE 9

Percentiles of Chi-Square Distribution

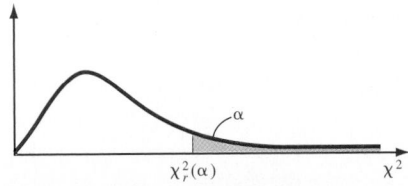

d.f. v	.990	.950	.900	.500	α .100	.050	.025	.010	.005
1	.0002	.004	.02	.45	2.71	3.84	5.02	6.63	7.88
2	.02	.10	.21	1.39	4.61	5.99	7.38	9.21	10.60
3	.11	.35	.58	2.37	6.25	7.81	9.35	11.34	12.84
4	.30	.71	1.06	3.36	7.78†	9.49	11.14	13.28	14.86
5	.55	1.15	1.61	4.35	9.24	11.07	12.83	15.09	16.75
6	.87	1.64	2.20	5.35	10.64	12.59	14.45	16.81	18.55
7	1.24	2.17	2.83	6.35	12.02	14.07	16.01	18.48	20.28
8	1.65	2.73	3.49	7.34	13.36	15.51	17.53	20.09	21.95
9	2.09	3.33	4.17	8.34	14.68	16.92	19.02	21.67	23.59
10	2.56	3.94	4.87	9.34	15.99	18.31	20.48	23.21	25.19
11	3.05	4.57	5.58	10.34	17.28	19.68	21.92	24.72	26.76
12	3.57	5.23	6.30	11.34	18.55	21.03	23.34	26.22	28.30
13	4.11	5.89	7.04	12.34	19.81	22.36	24.74	27.69	29.82
14	4.66	6.57	7.79	13.34	21.06	23.68	26.12	29.14	31.32
15	5.23	7.26	8.55	14.34	22.31	25.00	27.49	30.58	32.80
16	5.81	7.96	9.31	15.34	23.54	26.30	28.85	32.00	34.27
17	6.41	8.67	10.09	16.34	24.77	27.59	30.19	33.41	35.72
18	7.01	9.39	10.86	17.34	25.99	28.87	31.53	34.81	37.16
19	7.63	10.12	11.65	18.34	27.20	30.14	32.85	36.19	38.58
20	8.26	10.85	12.44	19.34	28.41	31.41	34.17	37.57	40.00
21	8.90	11.59	13.24	20.34	29.62	32.67	35.48	38.93	41.40
22	9.54	12.34	14.04	21.34	30.81	33.92	36.78	40.29	42.80
23	10.20	13.09	14.85	22.34	32.01	35.17	38.08	41.64	44.18
24	10.86	13.85	15.66	23.34	33.20	36.42	39.36	42.98	45.56
25	11.52	14.61	16.47	24.34	34.38	37.65	40.65	44.31	46.93
26	12.20	15.38	17.29	25.34	35.56	38.89	41.92	45.64	48.29
27	12.88	16.15	18.11	26.34	36.74	40.11	43.19	46.96	49.64
28	13.56	16.93	18.94	27.34	37.92	41.34	44.46	48.28	50.99
29	14.26	17.71	19.77	28.34	39.09	42.56	45.72	49.59	52.34
30	14.95	18.49	20.60	29.34	40.26	43.77	46.98	50.89	53.67
40	22.16	26.51	29.05	39.34	51.81	55.76	59.34	63.69	66.77
50	29.71	34.76	37.69	49.33	63.17	67.50	71.42	76.15	79.49
60	37.48	43.19	46.46	59.33	74.40	79.08	83.30	88.38	91.95
70	45.44	51.74	55.33	69.33	85.53	90.53	95.02	100.43	104.21
80	53.54	60.39	64.28	79.33	96.58	101.88	106.63	112.33	116.32
90	61.75	69.13	73.29	89.33	107.57	113.15	118.14	124.12	128.30
100	70.06	77.93	82.36	99.33	118.50	124.34	129.56	135.81	140.17

Source: Richard A. Johnson and Dean W. Wichern, *Applied Multivariate Statistical Analysis,* © 1982, p. 583. Reprinted by permission of Prentice Hall, Inc., Englewood Cliffs, New Jersey.

†*Note:* For example, $P(\chi_4^2 > 7.78) = .10$.

(call it **A**) having a density $f(t) = 2.72e^{-2.72t}$. We choose five categories so as to ensure that the probability that an observation from **A** falls into each of the five categories is .20. This yields $e_i = 25(.20) = 5$ for each category. To set the category boundaries, we need to determine the cumulative distribution function, $F(t)$, for **A**:

$$F(t) = P(\mathbf{A} \leq t) = \int_0^t 2.72e^{-2.72s}\, ds = 1 - e^{-2.72t}$$

Then we choose the categories to be as follows:

Category 1 $0 \leq t < m_1$ minutes

Category 2 $m_1 \leq t < m_2$ minutes

Category 3 $m_2 \leq t < m_3$ minutes

Category 4 $m_3 \leq t < m_4$ minutes

Category 5 $m_4 \leq t$ minutes

where $F(m_1) = .20$, $F(m_2) = .40$, $F(m_3) = .60$, and $F(m_4) = .80$.

Since $F(t) = 1 - e^{-2.72t}$, we see that for any number p, the value of t satisfying $F(t) = p$ may be found as follows:

$$1 - e^{-2.72t} = p$$
$$1 - p = e^{-2.72t}$$

Taking logarithms (to base e) of both sides yields

$$t = \frac{\ln(1 - p)}{-2.72}$$

$$m_1 = \frac{\ln .80}{-2.72} = 0.08$$

$$m_2 = \frac{\ln .60}{-2.72} = 0.19$$

$$m_3 = \frac{\ln .40}{-2.72} = 0.34$$

$$m_4 = \frac{\ln .20}{-2.72} = 0.59$$

Hence, our categories are as follows:

Category 1 $0 \leq t < 0.08$ minute

Category 2 $0.08 \leq t < 0.19$ minute

Category 3 $0.19 \leq t < 0.34$ minute

Category 4 $0.34 \leq t < 0.59$ minute

Category 5 $0.59 \leq t$

After classifying the data into these categories, we find that $o_1 = 6$, $o_2 = 5$, $o_3 = 4$, $o_4 = 5$, and $o_5 = 5$. By the construction of our categories, $e_1 = e_2 = e_3 = e_4 = e_5 = .20(25) = 5$. We now compute $\chi^2(\text{obs})$:

$$\chi^2(\text{obs}) = \frac{(6-5)^2}{5} + \frac{(5-5)^2}{5} + \frac{(4-5)^2}{5} + \frac{(5-5)^2}{5} + \frac{(5-5)^2}{5}$$
$$= .20 + 0 + .20 + 0 + 0 = .40$$

We arbitrarily choose $\alpha = .05$. Since we are trying to fit an exponential distribution to interarrival times, $r = 1$. Then $\chi^2_3(.05) = 7.81$, and we see that for $\alpha = .05$, we can accept the hypothesis that the observed interarrival times come from an exponential distribution with $\lambda = 2.72$ arrivals per minute.

Alternatively, we could have found the cutoff point for the chi-square test with the formula

$$=\text{CHINV}(.05,3)$$

This formula yields the value 7.81.

To test whether service times are exponentially distributed, simply apply the preceding approach to observed service times $s_1, s_2, \ldots, s_n$. Begin by obtaining an estimate (call it $\hat{\mu}$) of the actual service rate μ from

$$\hat{\mu} = \frac{n}{s_1 + s_2 + \cdots + s_n}$$

Then use the chi-square test to test whether or not it is reasonable to assume that the observed service times are observations from an exponential distribution with density $\hat{\mu}e^{-\hat{\mu}t}$.

PROBLEM

Group A

1 A travel agency wants to determine if the length of customers' phone calls can be adequately modeled by an exponential distribution. Last week, the agency recorded the length of all phone calls and obtained the following results (in seconds): 4, 6, 5, 8, 9, 10, 12, 8, 16, 20, 24, 27, 33, 37, 43, 50, 58, 68, 70, 78, 88, 100, 120, 130. Do these data indicate that the length of phone calls to the travel agency is governed by an exponential distribution?

20.13 Closed Queuing Networks

For manufacturing units attempting to implement just-in-time manufacturing, it makes sense to maintain a constant level of work in process. For a busy computer network, it may be convenient to assume that as soon as a job leaves the system, another job arrives to replace it. Such manufacturing and computer systems, where there is a constant number of jobs present, may be modeled as **closed queuing networks.** Recall that in an open queuing network, the numbers of jobs at each server were independent random variables. Since the number of jobs in the system is always constant, the distribution of jobs at different servers cannot be independent. We now discuss **Buzen's algorithm,** which can be used to determine steady-state probabilities for closed queuing networks.

We let P_{ij} be the probability that a job will go to server j after completing service at station i. Let P be the matrix whose $(i-j)$th entry is P_{ij}. We assume that service times at server j follow an exponential distribution with parameter μ_j. The system has s servers, and at all times, exactly N jobs are present. We let n_i be the number of jobs present at server i. Then the state of the system at any given time can be defined by an n-dimensional vector $\mathbf{n} = (n_1, n_2, \ldots, n_s)$. The set of possible states is given by $S_N = \{\mathbf{n}$ such that all $n_i \geq 0$ and $n_1 + n_2 + \cdots + n_s = N\}$.

Let λ_j equal the arrival rate to server j. Since there are no external arrivals, we may set all $r_j = 0$ and obtain the values of the λ_j's from the equation used in the open network situation. That is,

$$\lambda_j = \sum_{i=1}^{i=s} \lambda_i P_{ij} \qquad j = 1, 2, \ldots, s \tag{59}$$

Since jobs never leave the system, for each i, $\sum_{j=1}^{j=s} P_{ij} = 1$. This fact causes equation (59) to have no unique solution. Fortunately, it turns out that we can use any solution to (59) to help us get steady-state probabilities. If we define

$$\rho_i = \frac{\lambda_i}{\mu_i}$$

then we determine, for any state n, its steady-state probability $\Pi_N(\mathbf{n})$ from the following equation:

$$\Pi_N(\mathbf{n}) = \frac{\rho_1^{n_1} \, \rho_2^{n_2} \cdots \rho_n^{n_s}}{G(N)} \tag{60}$$

Here, $G(N) = \sum_{\mathbf{n} \in S_N} \rho_1^{n_1} \, \rho_2^{n_2} \cdots \rho_s^{n_s}$.

Buzen's algorithm gives us an efficient way to determine (in a spreadsheet) $G(N)$. Once we have the steady-state probability distribution, we can easily determine other measures of effectiveness, such as expected queue length at each server and expected time a job spends during each visit to a server, fraction of time a server is busy, and the throughput for each server (jobs per second processed by each server).

To obtain $G(N)$, we recursively compute the quantities $C_i(k)$, for $i = 1, 2, \ldots, s$ and $k = 0, 1, \ldots, N$. We initialize the recursion with $C_1(k) = \rho_1^k$, $k = 0, 1, \ldots, N$ and $C_i(0) = 1$, $i = 1, 2, \ldots, s$. For other values of k and i, we build up the values of $C_i(k)$ recursively via the following relationship:

$$C_i(k) = C_{i-1}(k) + \rho_i C_i(k - 1)$$

Then it can be shown that $G(N) = C_s(N)$. We illustrate the use of Buzen's algorithm with the following example.[†]

EXAMPLE 17   Flexible Manufacturing System

Consider a flexible manufacturing system in which 10 parts are always in process. Each part requires two operations. Each part begins by having operation 1 done at machine 1. Then, with probability .75 the part has operation 2 processed on machine 2, and with probability .25 the part has operation 2 processed on machine 3. Once a part completes operation 2, the part leaves the system and is immediately replaced by another part. We are given the following machine rates (the time for each operation is exponentially distributed): $\mu_1 = .25$ minute, $\mu_2 = .48$ minute, and $\mu_3 = .08$ minute.

a   Find the probability distribution of the number of parts at each machine.

b   Find the expected number of parts present at each machine.

c   What fraction of the time is each machine busy?

d   How many parts per minute are completed by each machine?

Buzen.xls   **Solution**   Our work is in file Buzen.xls. To begin, we need to compute one solution to the equations (59) defining λ_1, λ_2, and λ_3. We must solve

$$\lambda_1 = \lambda_2 + \lambda_3$$
$$\lambda_2 = .75\lambda_1$$
$$\lambda_3 = .25\lambda_1$$

[†]From Kao (1996).

There are an infinite number of solutions to this system. Arbitrarily choosing $\lambda_1 = 1$ yields the solution $\lambda_2 = .75$ and $\lambda_3 = .25$. In cells G8:I8, we compute $\rho_i = \frac{\lambda_i}{\mu_i}$. In G10:G20, we compute $C_1(k) = \rho_1^k$, $k = 0, 1, \ldots, 10$, and in G10:I10, we enter $C_i(0) = 1$, $i = 1, 2, 3$. Copying from H11 to H11:I20 the formula

$$=\text{G11+H\$8*H10}$$

implements the recursion $C_i(k) = C_{i-1}(k) + \rho_i C_i(k-1)$. Then we can find G(10) = 7,231,883 from the value of $C_3(10)$ in cell I20. See Figure 34.

We can now generate all possible system states efficiently by starting with $n_1 = 0$ and listing those states in order of increasing values of n_2. Then we increase n_1 to 1 and list all states in increasing values of n_2, etc. Once we have $n_1 = 10$, we will have listed all states. (See Figure 35.) To efficiently generate all possible states, we copy down from C25 the formula

$$=\text{IF(D25=0,B25+1,B25)}$$

This formula increments n_1 by 1 if $n_3 = 0$ (which is the same as having $n_2 = 10 - n_1$). Otherwise, the formula keeps n_1 constant.

Then we copy down from D25 the formula

$$=\text{IF(B25-B24=1,0,C24+1)}$$

This formula makes $n_2 = 0$ if we have just increased the value of n_1; otherwise, the formula increments the value of n_2 by 1.

Finally, from E25, we copy down the formula

$$=\text{10-B24-C24}$$

This ensures that $n_3 = 10 - n_1 - n_2$.

In E24:E89 we use (60) to compute the steady-state probability for each state by copying from E24 to E25:E89 the formula

$$=\text{(\$G\$8\textasciicircum B24)*(\$H\$8\textasciicircum C24)*(\$I\$8\textasciicircum D24)/\$I\$20}$$

Part (a)   Next, we answer part (a) by determining the probability distribution of the number of parts at each machine. We use the SUMIF function and a one-way data table to accomplish this goal. To begin, compute in H24 the probability of 0 parts at machine 1 with the formula

$$=\text{SUMIF(\$B\$24:\$B\$89,I23,E24:E89)}$$

This formula adds up every number in column D (which contains state probabilities) for the rows in which column B (which is parts at machine 1) has a 0 entry. See Figure 36.

	F	G	H	I
7	Mui	0.25	0.48	0.08
8	phoi	4	1.5625	3.125
9			1	2	3
10	0	1	1	1
11	1	4	5.5625	8.6875
12	2	16	24.6914063	51.83984
13	3	64	102.580322	264.5798
14	4	256	416.281754	1243.094
15	5	1024	1674.44024	5559.108
16	6	4096	6712.31287	24084.53
17	7	16384	26871.9889	102136.1
18	8	65536	107523.483	426698.9
19	9	262144	430149.442	1763583
20	10	1048576	1720684.5	7231883

FIGURE 34

	B	C	D	E
23	Parts at 1	Parts at 2	Parts at 3	Probability
24	0	0	10	0.01228143
25	0	1	9	0.00614071
26	0	2	8	0.00307036
27	0	3	7	0.00153518
28	0	4	6	0.00076759
29	0	5	5	0.00038379
30	0	6	4	0.0001919
31	0	7	3	9.5949E-05
32	0	8	2	4.7974E-05
33	0	9	1	2.3987E-05
34	0	10	0	1.1994E-05
35	1	0	9	0.01572023
36	1	1	8	0.00786011
37	1	2	7	0.00393006
38	1	3	6	0.00196503
39	1	4	5	0.00098251
40	1	5	4	0.00049126
41	1	6	3	0.00024563
42	1	7	2	0.00012281
43	1	8	1	6.1407E-05
44	1	9	0	3.0704E-05
45	2	0	8	0.02012189
46	2	1	7	0.01006094
47	2	2	6	0.00503047
48	2	3	5	0.00251524
49	2	4	4	0.00125762
50	2	5	3	0.00062881
51	2	6	2	0.0003144
52	2	7	1	0.0001572
53	2	8	0	7.8601E-05
54	3	0	7	0.02575602
55	3	1	6	0.01287801
56	3	2	5	0.006439
57	3	3	4	0.0032195
58	3	4	3	0.00160975
59	3	5	2	0.00080488
60	3	6	1	0.00040244
61	3	7	0	0.00020122
62	4	0	6	0.0329677
63	4	1	5	0.01648385
64	4	2	4	0.00824193
65	4	3	3	0.00412096
66	4	4	2	0.00206048
67	4	5	1	0.00103024
68	4	6	0	0.00051512
69	5	0	5	0.04219866
70	5	1	4	0.02109933
71	5	2	3	0.01054967
72	5	3	2	0.00527483
73	5	4	1	0.00263742
74	5	5	0	0.00131871
75	6	0	4	0.05401429
76	6	1	3	0.02700714
77	6	2	2	0.01350357
78	6	3	1	0.00675179
79	6	4	0	0.00337589
80	7	0	3	0.06913829
81	7	1	2	0.03456914
82	7	2	1	0.01728457
83	7	3	0	0.00864229
84	8	0	2	0.08849701
85	8	1	1	0.0442485
86	8	2	0	0.02212425
87	9	0	1	0.11327617
88	9	1	0	0.05663809
89	10	0	0	0.1449935

FIGURE 35

Selecting the table range G24:H35 and column input cell I23 enables us to loop through and compute the steady-state probabilities for each number of parts at machine 1. In a similar fashion, we obtain the following steady-state probability distributions for machines 2 and 3. See Figure 37.

Part (b) The mean number of parts present at machine 1 may be computed as $\sum_{i=0}^{i=10} i *$ (Probability of i parts at machine 1). In cell K31, we compute the mean number of parts at machine 1 with the formula

$$=\text{SUMPRODUCT(G25:G35,H25:H35)}$$

In a similar fashion, we compute the mean number of parts at machines 2 and 3 in cells K32 and K33. See Figure 38. Note that machine 1 is clearly the bottleneck.

Part (c) To compute the probability that each machine is busy, we just subtract from 1 the probability that each machine has 0 parts. These computations are done in L31:L33. We find that machine 1 is busy 97% of the time, machine 2 38% of the time, and machine 3 76% of the time.

	G	H	I
21			
22			Parts
23		Prob	0
24	Machine 1 parts	0.02455086	
25	0	0.02455086	
26	1	0.03140975	
27	2	0.04016518	
28	3	0.05131082	
29	4	0.06542029	
30	5	0.08307862	
31	6	0.10465268	
32	7	0.12963429	
33	8	0.15486976	
34	9	0.16991426	
35	10	0.1449935	

FIGURE 36

	G	H
37	Machine 2 Parts	0.61896518
38	0	0.61896518
39	1	0.23698584
40	2	0.09017388
41	3	0.03402481
42	4	0.01269126
43	5	0.00465769
44	6	0.00166949
45	7	0.00057718
46	8	0.00018798
47	9	5.4691E-05
48	10	1.1994E-05

	G	H
50	Machine 3 Parts	0.23793036
51	0	0.23793036
52	1	0.18587372
53	2	0.14519511
54	3	0.1133962
55	4	0.08851582
56	5	0.06900306
57	6	0.0536088
58	7	0.0412822
59	8	0.03105236
60	9	0.02186094
61	10	0.01228143

FIGURE 37

	J	K	L	M
29				
30	Mean Number	Mean Number	Prob busy	Completions per second
31	Machine 1	6.696224299	0.97544914	0.243862285
32	Machine 2	0.609634749	0.38103482	0.182896714
33	Machine 3	2.694140952	0.76206964	0.060965571

FIGURE 38

Part (d) To compute the mean number of service completions per minute by each machine, we simply multiply the probability that a machine is busy by the machine's service rate. These computations are done in M31:M33. We find that machine 1 on average completes .24 part/minute, machine 2 .18 part/minute, and machine 3 .06 part/minute.

PROBLEMS

Group A

1 Jobs arrive to a file server consisting of a CPU and two disks (disk 1 and disk 2). With probability 13/20, a job goes from CPU to disk 1, and with probability 6/20, a job goes from CPU to disk 2. With probability 1/20, a job is finished after its CPU operation and is immediately replaced by another job. There are always 3 jobs in the system. The mean time to complete the CPU operation is .039 second. The mean time to complete the disk 1 operation is .18 second, and the mean time to complete the disk 2 operation is .26 second.

a Determine the steady-state distribution of the number of jobs at each part of the system.

b What is the average number of jobs at CPU? Disk 1? Disk 2?

c What is the probability that CPU is busy? Disk 1? Disk 2?

d What is the average number of jobs completed per second by CPU? Disk 1? Disk 2?

2 A manufacturing process always has 8 parts in process. A part must successfully complete two steps (step 1 and step 2) to be completed. A single machine performs step 1 and can process an average of 8 parts per minute. A single machine performs step 2 and can process 11 parts per minute. Unfortunately, step 2 is not totally reliable. (Step 1 is totally reliable, however.) Each time a part is sent through step 2, there is a 10% chance that step 2 must be repeated.

a Find the steady-state distribution of parts at each machine.

b Find the average number of parts at each machine.

c Find the probability that each machine is busy.

d Find the number of parts per minute successfully completing service at each machine.

20.14 An Approximation for the *G/G/m* Queuing System

In most situations, interarrival times follow an exponential random variable. (See Denardo (1982) for an explanation of this fact.) Often, however, service times do not follow an exponential distribution. When interarrival times and service times each follow a nonexponential random variable, we call the queuing system a *G/G/m* system. The first *G* indicates that interarrival times always follow the same (but not necessarily exponential) random variable, while the second *G* indicates that service times always follow the same (but not necessarily exponential) random variable. For these situations, the templates discussed in the previous sections of this chapter are not valid. Fortunately, the Allen–Cunneen approximation (see Tanner (1995)) often gives a good approximation to *L*, *W*,

L_q, and W_q for $G/G/m$ systems. The file ggm.xls contains a spreadsheet implementation of the Allen–Cunneen approximation. The user need only input the following information:

- The average number of arrivals per unit time (λ) in cell B3.
- The average rate at which customers can be serviced (μ) in cell B4.
- The number of servers (s) in cell B5.
- The squared coefficient of variation—(variance of interarrival times)/(mean interarrival time)^2—of interarrival times in cell B6.
- The squared coefficient of variation—(variance of service times)/(mean service time)^2—of service times in cell B7.

The squared coefficient of variation for interarrival or service times can easily be estimated with the Excel functions =AVERAGE and =VARP. Recall that the exponential random variable has the property that variance = mean^2. Thus, the squared coefficient of variation for exponential interarrival or service times will equal 1, and the amount by which the squared coefficient of variation for interarrival or service times differs from 1 indicates the degree of departure from exponentiality. The Allen–Cunneen approximation is exact if interarrival times and service times are exponential. Extensive testing by Tanner indicates that in a wide variety of situations, the values of L, W, L_q, and W_q obtained by the approximation are within 10% of their true values. Here is an illustration of the Allen–Cunneen approximation.

EXAMPLE 18   NBD Bank

The NBD Bank branch in Bloomington, Indiana has 6 tellers. At peak times, an average of 4.8 customers per minute arrive at the bank. It takes a teller an average of 1 minute to serve a customer. The squared coefficient of variation for both interarrival times and service times is .5. Estimate the average time a customer will have to wait before seeing a teller. On average, how many customers will be present?

Solution After inputting the relevant information in cells B3 through B7 (see Figure 39), we find that on average, a customer will wait .216 minute for a teller. On average, 5.83 customers will be present in the bank. Congestion seems to be well under control. This favorable outcome is largely due to the low squared coefficient of variation for both interarrival and service times. For example, if both squared coefficients of variation were 4, then W_q would be 1.73 minutes, an 800% increase.

	A	B	C	D
1		G/G/m Template		
2		Allen-Cunneen Approximation		
3	Lambda	4.8		
4	Mu	1		
5	s	6		
6	CV arrive	0.5		
7	CV service	0.5		
8	u	4.8		
9	ro	0.8		
10	R(s,mu)	0.82322		
11	E$_C$(s,mu)	0.517772		
12	W$_q$	0.215738		
13	L$_q$	1.035544		
14	W	1.215738		
15	L	5.835544		

FIGURE 39

PROBLEMS

Group A

Problems 1–4 refer to Example 18.

1 NBD believes the congestion level is satisfactory if the average number of customers in line equals the number of servers. For the information given in the example, what is the maximum arrival rate that can be satisfactorily handled with 6 servers?

2 Show how the average time a customer must wait for a teller depends on the number of servers.

3 Using a two-way data table, determine how changes in the squared coefficient of variation for interarrival and service times affect the average number of customers in the NBD branch.

4 Suppose a teller costs $30 per hour. Suppose the bank values a customer's time at NBD at c per hour. Show how variations in c affect the number of tellers that NBD should use.

5 Southbest Airlines has an average of 230 customers per hour arriving at a ticket counter where 8 agents are working. Each agent can serve an average of 30 customers per hour. The squared coefficient of variation for the interarrival times is 1.5 and 2 for the service times.

   a On average, how many customers will be present at the ticket counter?

   b On average, how long will a customer have to wait for an agent?

20.15 Priority Queuing Models[†]

There are many situations in which customers are not served on a first come, first served (FCFS) basis. In Section 20.1, we also discussed the service in random order (SIRO) and last come, first served (LCFS) queue disciplines. Let $\mathbf{W}_{FCFS}$, $\mathbf{W}_{SIRO}$, and $\mathbf{W}_{LCFS}$ be the random variables representing a customer's waiting time in queuing systems under the disciplines FCFS, SIRO, and LCFS, respectively. It can be shown that

$$E(\mathbf{W}_{FCFS}) = E(\mathbf{W}_{SIRO}) = E(\mathbf{W}_{LCFS})$$

Thus, the average time (steady-state) that a customer spends in the system does not depend on which of these three queue disciplines is chosen. It can also be shown that

$$\text{var } \mathbf{W}_{FCFS} < \text{var } \mathbf{W}_{SIRO} < \text{var } \mathbf{W}_{LCFS} \tag{61}$$

Since a large variance is usually associated with a random variable that has a relatively large chance of assuming extreme values, (61) indicates that relatively large waiting times are most likely to occur with an LCFS discipline and least likely to occur with an FCFS discipline. This is reasonable, because in an LCFS system, a customer can get lucky and immediately enter service but can also be bumped to the end of a long line. In FCFS, however, the customer cannot be bumped to the end of a long line, so a very long wait is relatively unlikely.

In many organizations, the order in which customers are served depends on the customer's "type." For example, hospital emergency rooms usually serve seriously ill patients before they serve nonemergency patients. Also, in many computer systems, longer jobs do not enter service until all shorter jobs in the queue have been completed. Models in which a customer's type determines the order in which customers undergo service are called **priority queuing models.**

The following scenario encompasses many priority queuing models (including all the models discussed in this section). Assume there are n types of customers (labeled type 1, type 2, . . . , type n). The interarrival times of type i customers are exponentially distributed with rate λ_i. Interarrival times of different customer types are assumed to be independent. The service time of a type i customer is described by a random variable $\mathbf{S}_i$ (not necessarily exponential). We assume that lower-numbered customer types have priority over higher-numbered customer types.

[†]This section covers topics that may be omitted with no loss of continuity.

Nonpreemptive Priority Models

We begin by considering nonpreemptive priority models. In a nonpreemptive model, a customer's service cannot be interrupted. After each service completion, the next customer to enter service is chosen by giving priority to lower-numbered customer types (with ties broken on an FCFS basis). For example, if $n = 3$ and three type 2 and four type 3 customers are present, the next customer to enter service would be the type 2 customer who was the first of that type to arrive.

In the Kendall–Lee notation, a nonpreemptive priority model is indicated by labeling the fourth characteristic as NPRP. To indicate multiple customer types, we subscript the first two characteristics with i's. Thus, $M_i/G_i/\cdots$ would represent a situation in which the interarrival times for the ith customer type are exponential and the service times for the ith customer type have a general distribution. In what follows, we let

W_{qk} = expected steady-state waiting time in line spent by a type k customer

W_k = expected steady-state time in the system spent by a type k customer

L_{qk} = expected steady-state number of type k customers waiting in line

L_k = expected steady-state number of type k customers in the system

The $M_i/G_i/1$/NPRP/∞/∞ Model

Our first results concern the single-server, nonpreemptive $M_i/G_i/1$/NPRP/∞/∞ system. Define $\rho_i = \frac{\lambda_i}{\mu_i}$, $a_0 = 0$, and $a_k = \sum_{i=1}^{i=k}\rho_i$. We assume[†] that

$$\sum_{i=1}^{i=n} \frac{\lambda_i}{\mu_i} < 1$$

Then

$$
\begin{aligned}
W_{qk} &= \frac{\sum_{k=1}^{k=n} \lambda_k E(\mathbf{S}_k^2)/2}{(1 - a_{k-1})(1 - a_k)} \\
L_{qk} &= \lambda_k W_{qk} \\
W_k &= W_{qk} + \frac{1}{\mu_k} \\
L_k &= \lambda_k W_k
\end{aligned}
$$

(62)

The following example illustrates the use of (62).

EXAMPLE 19   **Copying Priority**

A copying facility gives shorter jobs priority over long jobs. Interarrival times for each type of job are exponential, and an average of 12 short jobs and 6 long jobs arrive each hour. Let type 1 job = short job and type 2 job = long job. Then we are given that

$$E(\mathbf{S}_1) = 2 \text{ minutes} \qquad E(\mathbf{S}_1^2) = 6 \text{ minutes}^2 = \frac{1}{600} \text{ hour}^2$$

$$E(\mathbf{S}_2) = 4 \text{ minutes} \qquad E(\mathbf{S}_2^2) = 18 \text{ minutes}^2 = \frac{1}{200} \text{ hour}^2$$

Determine the average length of time each type of job spends in the copying facility.

[†]If this condition does not hold, then for one or more customer types, no steady-state waiting time will exist.

Solution We are given that $\lambda_1 = 12$ jobs per hour, $\lambda_2 = 6$ jobs per hour, and $\mu_1 = 30$ jobs per hour, and $\mu_2 = 15$ jobs per hour. Then $\rho_1 = \frac{12}{30} = .4$ and $\rho_2 = \frac{6}{15} = .4$. Since $\rho_1 + \rho_2 < 1$, a steady state will exist. Now $a_0 = 0$, $a_1 = .4$, and $a_2 = .4 + .4 = .8$. Equations (62) now yield

$$W_{q1} = \frac{\dfrac{12\left(\dfrac{1}{600}\right)}{2} + \dfrac{6\left(\dfrac{1}{200}\right)}{2}}{(1-0)(1-.4)} = \frac{\dfrac{30}{1,200}}{.6} = 0.042 \text{ hour}$$

$$W_{q2} = \frac{\dfrac{12\left(\dfrac{1}{600}\right)}{2} + \dfrac{6\left(\dfrac{1}{200}\right)}{2}}{(1-.4)(1-.8)} = \frac{\dfrac{30}{1,200}}{.12} = 0.208 \text{ hour}$$

Also,

$$W_1 = W_{q1} + \frac{1}{\mu_1} = 0.042 + 0.033 = 0.075 \text{ hour}$$

$$W_2 = W_{q2} + \frac{1}{\mu_2} = 0.208 + 0.067 = 0.275 \text{ hour}$$

Thus, as expected, the long jobs spend much more time in the copying facility than the short jobs do.

The $M_i/G_i/1/NPRP/\infty/\infty$ Model with Customer-Dependent Waiting Costs

Consider a single-server, nonpreemptive priority system in which a cost c_k is charged for each unit of time that a type k customer spends in the system. If we want to minimize the expected cost incurred per unit time (in the steady state), what priority ordering should be placed on the customer types? Suppose the n customer types are numbered such that

$$c_1\mu_1 \geq c_2\mu_2 \geq \cdots \geq c_n\mu_n \tag{63}$$

Then expected cost is minimized by giving the highest priority to type 1 customers, the second-highest priority to type 2 customers, and so forth, and the lowest priority to type n customers. To see why this priority ordering is reasonable, observe that when a type k customer is being served, cost leaves the system at a rate $c_k\mu_k$. Thus, cost can be minimized by giving the highest priority to customer types with the largest values of $c_k\mu_k$.

As a special case of this result, suppose we want to minimize L, the expected number of jobs in the system. Let $c_1 = c_2 = \cdots = c_n = 1$. Then at any time, the cost per unit time is equal to the number of customers in the system. Thus, the expected cost per unit time will equal L. Now (63) becomes

$$\mu_1 \geq \mu_2 \geq \cdots \geq \mu_n \qquad \text{or} \qquad \frac{1}{\mu_1} \leq \frac{1}{\mu_2} \leq \cdots \leq \frac{1}{\mu_n}$$

Thus, we may conclude that the expected number of jobs in the system will be minimized if the highest priority is given to the customer types with the shortest mean service time. This priority discipline is known as the *shortest processing time* (SPT) discipline.

The $M_i/M/s/NPRP/\infty/\infty$ Model

To obtain tractable analytic results for multiserver priority systems, we must assume that each customer type has exponentially distributed service times with a mean of $\frac{1}{\mu}$, and that type i customers have interarrival times that are exponentially distributed with rate λ_i. Such a system with s servers is denoted by the notation $M_i/M/s/NPRP/\infty/\infty$. For this model,

$$W_{qk} = \frac{P(j \ge s)}{s\mu(1 - a_{k-1})(1 - a_k)} \tag{64}$$

In (64),

$$a_k = \sum_{i=1}^{i=k} \frac{\lambda_i}{s\mu} \qquad (k \ge 1)$$

$a_0 = 0$, and $P(j \ge s)$ is obtained from Table 6 for an s-server system having

$$\rho = \frac{\lambda_1 + \lambda_2 + \cdots + \lambda_n}{s\mu}$$

Example 20 illustrates the use of (64).

EXAMPLE 20   **Police Response**

Gotham Township has 5 police cars. The police department receives two types of calls: emergency (type 1) and nonemergency (type 2) calls. Interarrival times for each type of call are exponentially distributed, with an average of 10 emergency and 20 nonemergency calls being received each hour. Each type of call has an exponential service time, with a mean of 8 minutes (assume that, on the average, 6 of the 8 minutes is the travel time from the police station to the call and back to the station). Emergency calls are given priority over nonemergency calls. On the average, how much time will elapse between the placement of a nonemergency call and the arrival of a police car?

Solution   We are given that $s = 5$, $\lambda_1 = 10$ calls per hour, $\lambda_2 = 20$ calls per hour, $\mu = 7.5$ calls per hour, $\rho = \frac{10+20}{5(7.5)} = .80$, $a_0 = 0$, $a_1 = \frac{10}{37.5} = .267$, and $a_2 = \frac{10+20}{37.5} = .80$. From Table 6, with $s = 5$ and $\rho = .80$, $P(j \ge 5) = .55$. Then (64) yields

$$W_{q2} = \frac{.55}{5(7.5)(1 - .267)(1 - .80)} = \frac{.55}{5.50} = 0.10 \text{ hour} = 6 \text{ minutes}$$

The average time between the placement of a nonemergency call and the arrival of the car is $W_{q2} + (\frac{1}{2})$ (total travel time per call) $= 6 + 3 = 9$ minutes.

Preemptive Priorities

We close our discussion of priority queuing systems by discussing a single-server **preemptive queuing system.** In a preemptive queuing system, a lower-priority customer (say, a type i customer) can be bumped from service whenever a higher-priority customer arrives. Once no higher-priority customers are present, the bumped type i customer reenters service. In a **preemptive resume model,** a customer's service continues from the point at which it was interrupted. In a **preemptive repeat model,** a customer begins service anew each time he or she reenters service. Of course, if service times are exponentially distributed, the resume and repeat disciplines are identical. (Why?) In the Kendall–Lee notation, we denote a preemptive queuing system by labeling the fourth characteristic PRP. We now consider a single-server $M_i/M/1/PRP/\infty/\infty$ system in which the service time of each customer is exponential with mean $\frac{1}{\mu}$ and the interarrival times for the ith customer type are exponentially distributed with rate λ_i. Then

$$W_k = \frac{\dfrac{1}{\mu}}{(1 - a_{k-1})(1 - a_k)} \tag{65}$$

where $a_0 = 0$ and

$$a_k = \sum_{i=1}^{i=k} \frac{\lambda_i}{\mu}$$

For obvious reasons, preemptive disciplines are rarely used if the customers are people. Preemptive disciplines are sometimes used, however, for "customers" like computer jobs. The following example illustrates the use of (65).

EXAMPLE 21 **University Computer System**

On the Podunk U computer system, faculty jobs (type 1) always preempt student jobs (type 2). The length of each type of job follows an exponential distribution, with mean 30 seconds. Each hour, an average of 10 faculty and 50 student jobs are submitted. What is the average length of time between the submission and completion of a student's computer job? Assume that interarrival times are exponential.

Solution We are given that $\mu = 2$ jobs per minute, $\lambda_1 = \frac{1}{6}$ job per minute, and $\lambda_2 = \frac{5}{6}$ job per minute. Then

$$a_0 = 0, \qquad a_1 = \frac{\frac{1}{6}}{2} = \frac{1}{12}, \qquad a_2 = \frac{1}{12} + \frac{\frac{5}{6}}{2} = \frac{1}{2}$$

Equation (65) yields

$$W_2 = \frac{\dfrac{1}{2}}{\left(1 - \dfrac{1}{12}\right)\left(1 - \dfrac{1}{2}\right)} = \frac{12}{11} \text{ minutes} = 1.09 \text{ minutes}$$

An average of 1.09 minutes will elapse between the time a student submits a job and the time the job is completed.

PROBLEMS

Group A

1 English professor Jacob Bright has one typist, who types for 8 hours per day. He submits three types of jobs to the typist: tests, research papers, and class handouts. The information in Table 10 is available. Professor Bright has told the typist that tests have priority over research papers, and research papers have priority over class handouts. Assuming a nonpreemptive system, determine the expected time that Professor Bright will have to wait before each type of job is completed.

TABLE 10

Type of Job	Frequency (number per day)	$E(S_i)$ (hours)	$E(S_i^2)$ (hours)^2
Test	2	1	2
Research paper	0.5	4	20
Class handout	5	0.5	0.50

2 Suppose a supermarket uses a system in which all customers wait in a single line for the first available cashier. Assume that the service time for a customer who purchases k items is exponentially distributed, with mean k seconds. Also, a customer who purchases k items feels that the cost of waiting in line for 1 minute is $\frac{\$1}{k}$. If customers can be assigned priorities, what priority assignment will minimize the expected waiting cost incurred by the supermarket's customers? Why would a customer's waiting cost per minute be a decreasing function of k?

3 Four doctors work in a hospital emergency room that handles three types of patients. The time a doctor spends with each type of patient is exponentially distributed, with a mean of 15 minutes. Interarrival times for each customer type are exponential, with the average number of arrivals per hour for each patient type being as follows: type 1, 3 patients; type 2, 5 patients; type 3, 3 patients. Assume that type 1 patients have the highest priority, and type 3 patients have the lowest priority (no preemption is allowed). What is the average length of time that each type of patient must wait before seeing a doctor?

4 Consider a computer system to which two types of computer jobs are submitted. The mean time to run each type of job is $\frac{1}{\mu}$. The interarrival times for each type of job are exponential, with an average of λ_i type i jobs arriving each hour. Consider the following three situations.

a Type 1 jobs have priority over type 2 jobs, and preemption is allowed.

b Type 1 jobs have priority over type 2 jobs, and no preemption is allowed.

c All jobs are serviced on a FCFS basis.

Under which system are type 1 jobs best off? Worst off? Answer the same questions for type 2 jobs.

20.16 Transient Behavior of Queuing Systems

Throughout the chapter, we have assumed that the arrival rate, service rate, and number of servers have stayed constant over time. This allows us to talk reasonably about the existence of a steady state. In many situations, the arrival rate, service rate, and number of servers may vary. Here are some examples.

■ A fast-food restaurant is likely to experience a much larger arrival rate during the time from noon to 1:30 P.M. than during other hours of the day. Also, the number of servers (in a restaurant with parallel servers) will vary during the day, with more servers available during the busier periods.

■ Since most heart attacks occur during the morning, a coronary care unit will experience more arrivals during the morning.

■ Most voters vote either before or after work, so a polling place will be less busy during the middle of the day.

When the parameters defining the queuing system vary over time, we say that the system is **nonstationary.** Consider, for example, a fast-food restaurant that opens at 10 A.M. and closes at 6 P.M. We are interested in the probability distribution of the number of customers present at all times between 10 A.M. and closing. We call these probability distributions **transient probabilities.** For example, if we want to determine the probability that at least 10 customers are present, this probability will surely be larger at 12:30 P.M. than at 3 P.M.

We now assume that at time t, interarrival times are exponential with rate $\lambda(t)$. Also, $s(t)$ servers are available at time t, with service times exponential with rate $\mu(t)$. We assume that the maximum number of customers present at any time is given by N. To determine transient probabilities, we choose a small length of time Δt and assume at most one event (an arrival or service completion) can occur during an interval of length Δt. We assume that k customers are currently present at time t, and that

- The probability of an arrival during an interval of length Δt is $\lambda(t)*(\Delta t)$.
- The probability of more than one arrival during a time interval of length Δt is $o(\Delta t)$.
- Arrivals during different intervals are independent.
- The probability of a service completion during an interval of length Δt is given by $\min(s(t), k)*\mu t \Delta t$.
- The probability of more than one service completion during a time interval of length Δt is $o(\Delta t)$.

When arrivals are governed by the first three assumptions, we say that arrivals follow a **nonhomogeneous Poisson process.** Our assumptions imply that, given the arrival rate and service rate, the expected number of arrivals and/or service completions during the next Δt will match what we expect. The source of the error in our approximation is the fact that at least two events can occur during a length of time Δt. The probability of this occurring is $o(\Delta t)$, so if we make Δt small enough, our approximation should not cause large errors in computing transient probabilities.

We now define $P_i(t)$ to be the probability that i customers are present at time t. We will assume (although this is not necessary) that the system is initially empty, so $P_0(0) = 1$ and for $i > 0$, $P_0(i) = 0$. Then, given knowledge of $P_i(t)$, we may compute $P_i(t + \Delta t)$ as follows:

$$P_0(t + \Delta t) = (1 - \lambda(t)\Delta t)P_0(t) + \mu(t)\Delta t P_1(t)$$
$$P_i(t + \Delta t) = \lambda(t)\Delta t P_{i-1}(t) + (1 - \lambda(t)\Delta t - \min(s(t), i)\mu(t)\Delta t)P_i(t) + \min(s(t), i + 1)\mu(t)\Delta t P_{i+1}(t), \quad N - 1 \geq i \geq 1$$
$$P_N(t + \Delta t) = \lambda(t)\Delta t P_{N-1}(t) + (1 - \min(s(t), N)\mu(t))\Delta t P_N(t)$$

As previously stated, these equations are based on the assumption that if the state at time t is i, then during the next Δt, the probability of an arrival is $\lambda(t) \Delta t$, and the probability of a service completion is $\min(s(t), i)\mu(t)\Delta t$. The first equation then follows after observing that being in state 0 at time $t + \Delta t$ can only happen if we were in state 1 at time t and had a service completion during the next Δt or were in state 0 at time t and had no arrival during the next Δt. The second equation follows after observing that for $N - 1 \geq i \geq 1$, we can only be in state i at time $t + \Delta t$ if one of the following occurs.

- We were in state $i - 1$ at time t and had an arrival during the next Δt.
- We were in state $i + 1$ at time t and had a service completion during the next Δt.
- We were in state i at time t, and no arrival or service completion occurred during the next Δt.

The final equation follows after observing that to be in state N at time $t + \Delta t$, one of the following must occur:

- We were in state $N - 1$ at time t, and an arrival occurs during the next Δt.
- We were in state N at time t, and no service completion occurs during the next Δt.

The following example shows how we can use our approximations to determine transient probabilities for a nonstationary queuing system.

EXAMPLE 22 **Lunchtime Rush**

A small fast-food restaurant is trying to model the lunchtime rush. The restaurant opens at 11 A.M., and all customers wait in one line to have their orders filled. The arrival rate per hour at different times is as shown in Table 11. Arrivals follow a nonhomogeneous

TABLE 11

Time	Hourly Arrival Rate
11–11.30 A.M.	30
11:30 A.M.–noon	40
Noon–12:30 P.M.	50
12:30 P.M.–1 P.M.	60
1 P.M.–1:30 P.M.	35
1:30 P.M.–2 P.M.	25

Poisson process. The restaurant can serve an average of 50 people per hour. Service times are exponential. Management wants to model the probability distribution of customers from 11 A.M. through 2 P.M.

a At 12:30 P.M., estimate the average number of people in line or in service.

b At 11:30 A.M., estimate the average number of people in line or in service.

restaurant.xls **Solution** Our work is in the file restaurant.xls. (See Figure 40.) We use 5-second time increments and proceed as follows:

Step 1 In E4, we compute the probability of a service completion in 5 seconds by multiplying the hourly service rate by $\Delta t = 1/720$.

Step 2 In column A, we use the Excel DATA FILL command to generate times ranging in 5-second increments from 0 to 10,800 (2 P.M.).

Step 3 By copying from B11 to B11:B2171 the formula

$$A11/3600$$

we convert the time in seconds to hours.

Step 4 By copying from C11 to C12:C2171 the formula

$$=\text{VLOOKUP}(B11,\$G\$2:\$H\$7,2)/720$$

we look up the hourly arrival rate for the current time and convert it to a 5-second arrival rate by multiplying the hourly arrival rate by $\Delta t = 1/720$. Note that the arrival rate is highly nonstationary. This fact will greatly affect the system's level of congestion.

Step 5 We assume that a maximum of $N = 30$ customers will be present. Therefore, we need 31 columns to compute the probability of 0, 1, . . . 30 people being present at each time. At time 0, we assume that the restaurant is empty, so the probability that 0 people are present equals 1. For i at least 1, there is a 0 probability of i people being present. These probabilities are entered in row 11 of columns D–AH.

Step 6 In cell D12, we compute the probability that nobody is in the system at time 5 seconds with the formula

$$=(1\text{-}C11)*D11+\text{sprob}*E11$$

This formula implements the first of our approximating equations.

Step 7 By copying from cell E12 to E12:AG12 the formula

$$=\$C11*D11+(1\text{-}\$C11\text{-sprob})*E11+\text{sprob}*F11$$

we compute the probability that 1, 2, . . . , 29 people are present after 5 seconds. This formula implements our second approximating equation.

FIGURE 40

	A	B	C	D	E	F	G	H
1								
2							0	30
3				srate	50		0.5	40
4				sprob	0.069444		1	50
5							1.5	60
6							2	35
7							2.5	25
8								
9				0	1	2	3	4
10	Time	Hour	Arrival Prob	Prob 0	Prob 1	Prob 2	Prob 3	Prob 4
11	0	0	0.04166667	1	0	0	0	0
12	5	0.001389	0.04166667	0.958333	0.041667	0	0	0
13	10	0.002778	0.04166667	0.921296	0.076968	0.001736	0	0
14	15	0.004167	0.04166667	0.888254	0.106924	0.00475	7.23E-05	0
15	20	0.005556	0.04166667	0.858669	0.132384	0.008683	0.000262	3.01E-06
16	25	0.006944	0.04166667	0.832084	0.154055	0.013252	0.000595	1.36E-05
17	30	0.008333	0.04166667	0.808112	0.172528	0.01824	0.001082	3.69E-05
18	35	0.009722	0.04166667	0.786422	0.188297	0.023477	0.001724	7.79E-05
19	40	0.011111	0.04166667	0.766731	0.201773	0.028834	0.002516	0.000141
20	45	0.0125	0.04166667	0.748795	0.213303	0.034212	0.003448	0.000231

Step 8   In cell AH12, we compute the probability that 30 people are present after 5 seconds with the formula

$$=(1\text{-sprob})*AH11+C11*AG11$$

This implements the third approximating equation.

Step 9   Select the cell range D12:AH12 and position the cursor over the crosshair in the lower right-hand corner of cell AH12. Now double-clicking the left mouse button will copy the formulas in D12:AH12 down to match the number of rows in column C. Thus, we have now completed our computation of the probability distribution of customers from 11 A.M. to 2 P.M. See Figure 41.

Part (a)   In cell K5, we compute the expected number of customers present at 12:30 P.M. (note that row 1091 has time 1.5 hours or 5,400 seconds) with the formula

$$=SUMPRODUCT(\$D\$9:\$AH\$9,D1091:AH1091)$$

We find that an average of 6.25 customers will be present at 12:30 P.M.

FIGURE 41

	D	E	F	G	H	I	J	K	L	M	AH
2				0	30						
3	srate	50		0.5	40		minutes				
4	sprob	0.069444		1	50						
5				1.5	60		Mean #	6.253338	Mean#	1.430961	
6				2	35		12:30		11:30		
7				2.5	25						
8											
9	0	1	2	3	4	5	6	7	8	9	30
10	Prob 0	Prob 1	Prob 2	Prob 3	Prob 4	Prob 5	Prob 6	Prob 7	Prob 8	Prob 9	Prob 30
11	1	0	0	0	0	0	0	0	0	0	0
12	0.958333	0.041667	0	0	0	0	0	0	0	0	0
13	0.921296	0.076968	0.001736	0	0	0	0	0	0	0	0
14	0.888254	0.106924	0.00475	7.23E-05	0	0	0	0	0	0	0

Part (b) Note that row 371 is time 11:30 A.M. In cell M5, the formula

$$=\text{SUMPRODUCT(D9:AH9,D371:AH371)}$$

shows that an average of only 1.43 customers are expected to be present at 11:30.

PROBLEMS

Group A

1[†] A single machine is used between 8 A.M. and 4 P.M. to perform EKGs (electrocardiograms). There are 3 waiting spaces, and any arrival finding no available waiting space is lost to the system. The arrival rate per hour at time t ($t = 0$ is 8 A.M., and $t = 8$ is 4 P.M.) is given by

$$\lambda(t) = 9.24 - 1.584 \cos\left(\frac{\pi t}{1.51}\right) + 7.897 \sin\left(\frac{\pi t}{3.02}\right)$$

$$- 10.434 \cos\left(\frac{\pi t}{4.53}\right) + 4.293 \cos\left(\frac{\pi t}{6.04}\right)$$

Assume that service times are exponential and an average of 7 EKGs can be completed per hour. Also assume that arrivals follow a nonhomogeneous Poisson process. Determine how the probability that an arriving patient is lost to the system varies during the day.

2 The polls are open in Gotham City from 11 A.M. to 6 P.M. The city has 3 voting machines. It takes an average

of 1.5 minutes (exponentially distributed) for a voter to complete voting. The arrival rate of voters throughout the day is as shown in Table 12. What is the probability that all voting will be completed by 6:30 P.M.?

TABLE 12

Time	Hourly Arrival Rate
11 A.M.–noon	80
Noon–1 P.M.	125
1 P.M.–2 P.M.	110
2 P.M.–3 P.M.	90
3 P.M.–4 P.M.	80
4 P.M.–5 P.M.	70
5 P.M.–6 P.M.	100

[†]Based on Kao (1996).

SUMMARY Exponential Distribution

A random variable X has an exponential distribution with parameter λ if the density of X is given by

$$f(t) = \lambda e^{-\lambda t} \qquad (t \geq 0)$$

Then

$$E(X) = \frac{1}{\lambda} \qquad \text{and} \qquad \text{var } X = \frac{1}{\lambda^2}$$

The exponential distribution has the no-memory property. This means, for instance, that if interarrival times are exponentially distributed with rate or parameter λ, then no matter how long it has been since the last arrival, there is a probability $\lambda \Delta t$ that an arrival will occur during the next Δt time units.

Interarrival times are exponential with parameter λ if and only if the number of arrivals to occur in an interval of length t follows a Poisson distribution with parameter λt. The mass function for a Poisson distribution with parameter λ is given by

$$P(N = n) = \frac{e^{-\lambda}\lambda^n}{n!} \qquad (n = 0, 1, 2, \ldots)$$

Erlang Distribution

If interarrival or service times are not exponential, an Erlang random variable can often be used to model them. If $\mathbf{T}$ is an Erlang random variable with rate parameter R and shape parameter k, the density of $\mathbf{T}$ is given by

$$f(t) = \frac{R(Rt)^{k-1}e^{-Rt}}{(k-1)!} \qquad (t \geq 0)$$

and

$$E(\mathbf{T}) = \frac{k}{R} \qquad \text{and} \qquad \text{var } \mathbf{T} = \frac{k}{R^2}$$

Birth–Death Processes

For a birth-death process, the steady-state probability (π_j) or fraction of the time that the process spends in state j can be found from the following flow balance equations:

$$
\begin{aligned}
(j = 0) &\qquad \pi_0\lambda_0 = \pi_1\mu_1 \\
(j = 1) &\qquad (\lambda_1 + \mu_1)\pi_1 = \lambda_0\pi_0 + \mu_2\pi_2 \\
(j = 2) &\qquad (\lambda_2 + \mu_2)\pi_2 = \lambda_1\pi_1 + \mu_3\pi_3 \\
&\qquad \vdots \\
(j\text{th equation}) &\qquad (\lambda_j + \mu_j)\pi_j = \lambda_{j-1}\pi_{j-1} + \mu_{j+1}\pi_{j+1}
\end{aligned}
$$

The jth flow balance equation states that the expected number of transitions per unit time out of state j = (expected number of transitions per unit time into state j). The solution to the balance equations is found from

$$\pi_j = \pi_0 \frac{\lambda_0\lambda_1 \cdots \lambda_{j-1}}{\mu_1\mu_2 \cdots \mu_j} \qquad (j = 1, 2, \ldots)$$

and the fact that $\pi_0 + \pi_1 + \cdots = 1$.

Notation for Characteristics of Queuing Systems

π_j = steady-state probability that j customers are in system

L = expected number of customers in system

L_q = expected number of customers in line (queue)

L_s = expected number of customers in service

W = expected time a customer spends in system

W_q = expected time a customer spends waiting in line

W_s = expected time a customer spends in service

λ = average number of customers per unit time

μ = average number of service completions per unit time (service rate)

$\rho = \dfrac{\lambda}{s\mu}$ = traffic intensity

The *M/M/1/GD/∞/∞* Model

If $\rho \geq 1$, no steady state exists. For $\rho < 1$,

$$\pi_j = \rho^j (1 - \rho) \qquad (j = 0, 1, 2, \ldots)$$

$$L = \frac{\lambda}{\mu - \lambda}$$

$$L_q = \frac{\lambda^2}{\mu(\mu - \lambda)}$$

$$L_s = \rho$$

$$W = \frac{1}{\mu - \lambda}$$

$$W_q = \frac{\lambda}{\mu(\mu - \lambda)}$$

$$W_s = \frac{1}{\mu}$$

(The last three formulas were obtained from the L, L_q, and L_s formulas via the relation $L = \lambda W$.)

The *M/M/1/GD/c/∞* Model

If $\lambda \neq \mu$,

$$\pi_0 = \frac{1 - \rho}{1 - \rho^{c+1}}$$

$$\pi_j = \rho^j \pi_0 \qquad (j = 1, 2, \ldots, c)$$

$$\pi_j = 0 \qquad (j = c + 1, c + 2, \ldots)$$

$$L = \frac{\rho[1 - (c + 1)\rho^c + c\rho^{c+1}]}{(1 - \rho^{c+1})(1 - \rho)}$$

If $\lambda = \mu$,

$$\pi_j = \frac{1}{c + 1} \qquad (j = 0, 1, \ldots, c)$$

$$L = \frac{c}{2}$$

For all values of λ and μ,

$$L_s = 1 - \pi_0$$

$$L_q = L - L_s$$

$$W = \frac{L}{\lambda(1 - \pi_c)}$$

$$W_q = \frac{L_q}{\lambda(1 - \pi_c)}$$

$$W_s = \frac{1}{\mu}$$

The *M/M/s/GD/∞/∞* Model

For $\rho \geq 1$, no steady state exists. For $\rho < 1$,

$$\pi_0 = \frac{1}{\displaystyle\sum_{i=0}^{i=s-1} \frac{(s\rho)^i}{i!} + \frac{(s\rho)^s}{s!(1 - \rho)}}$$

$$\pi_j = \frac{(s\rho)^j \, \pi_0}{j!} \qquad (j = 1, 2, \ldots, s)$$

$$\pi_j = \frac{(s\rho)^j \, \pi_0}{s! s^{j-s}} \qquad (j = s, s + 1, s + 2, \ldots)$$

$$P(j \geq s) = \frac{(s\rho)^s \, \pi_0}{s!(1 - \rho)} \qquad \text{(tabulated in Table 6)}$$

$$L_q = \frac{P(j \geq s)\rho}{1 - \rho}$$

$$W_q = \frac{P(j \geq s)}{s\mu - \lambda}$$

$$L_s = \frac{\lambda}{\mu}$$

$$W_s = \frac{1}{\mu}$$

$$L = L_q + \frac{\lambda}{\mu}$$

$$W = \frac{L}{\lambda}$$

The *M/G/∞/GD/∞/∞* Model

$$L = L_s = \frac{\lambda}{\mu}$$

$$W = W_s = \frac{1}{\mu}$$

$$W_q = L_q = 0$$

The *M/G/1/GD/∞/∞* Model

$$\sigma^2 = \text{variance of service time distribution}$$

$$L_q = \frac{\lambda^2 \sigma^2 + \rho^2}{2(1 - \rho)}$$

$$L = L_q + \rho$$

$$L_s = \lambda \left(\frac{1}{\mu}\right)$$

$$W_q = \frac{L_q}{\lambda}$$

$$W = W_q + \frac{1}{\mu}$$

$$W_s = \frac{1}{\mu}$$

$$\pi_0 = 1 - \rho$$

Machine Repair ($M/M/R/GD/K/K$) Model

$$\rho = \frac{\lambda}{\mu}$$

L = expected number of broken machines

L_q = expected number of machines waiting for service

W = average time a machine spends broken

W_q = average time a machine spends waiting for service

π_j = steady-state probability that j machines are broken

λ = rate at which machine breaks down

μ = rate at which machine is repaired

Also,

$$\pi_j = \binom{K}{j} \rho^j \pi_0 \qquad (j = 0, 1, \ldots, R)$$

$$= \frac{\binom{K}{j} \rho^j j! \pi_0}{R! R^{j-R}} \qquad (j = R + 1, R + 2, \ldots, K)$$

$$L = \sum_{j=0}^{j=K} j \pi_j$$

$$L_q = \sum_{j=R}^{j=K} (j - R) \pi_j$$

$$\bar{\lambda} = \sum_{j=0}^{j=K} \pi_j \lambda_j = \sum_{j=0}^{j=K} \lambda(K - j)\pi_j = \lambda(K - L)$$

$$W = \frac{L}{\bar{\lambda}}$$

$$W_q = \frac{L_q}{\bar{\lambda}}$$

Exponential Queues in Series

If a steady state exists and if (1) interarrival times for a series queuing system are exponential with rate λ; (2) service times for each stage i server are exponential; and (3) each stage has an infinite-capacity waiting room, then interarrival times for arrivals to each stage of the queuing system are exponential with rate λ.

The *M/G/s/GD/s/∞* Model

A fraction π_s of all customers are lost to the system, and π_s depends only on the arrival rate λ and on the mean $\frac{1}{\mu}$ of the service time. Figure 21 can be used to find π_s.

What to Do If Interarrival or Service Times Are Not Exponential

A chi-square test may be used to determine if the actual data indicate that interarrival or service times are exponential. If interarrival and/or service times are not exponential, then L, L_q, W, and W_q may be approximated by Allen–Cunneen formula.

For many queuing systems, there is no formula or table that can be used to compute the system's operating characteristics. In this case, we must resort to simulation (see Chapters 21 and 22).

Closed Queuing Network

Manufacturing and computer systems in which there is a constant number of jobs present may be modeled as **closed queuing networks.**

We let P_{ij} be the probability that a job will go to server j after completing service at station i. Let P be the matrix whose $(i - j)$th entry is P_{ij}. We assume that service times at server j follow an exponential distribution with parameter μ_j. The system has s servers, and at all times, exactly N jobs are present. We let n_i be the number of jobs present at server i. Then the state of the system at any given time can be defined by an n-dimensional vector $\mathbf{n} = (n_1, n_2, \ldots, n_s)$. The set of possible states is given by $S_N = \{\mathbf{n}$ such that all $n_i \geq 0$ and $n_1 + n_2 + \cdots + n_s = N\}$.

Let λ_j equal the arrival rate to server j. Since there are no external arrivals, we may set all $r_j = 0$ and obtain the values of the λ_j's from the equation used in the open network situation. That is,

$$\lambda_j = \sum_{i=1}^{i=s} \lambda_i P_{ij} \qquad j = 1, 2, \ldots, s$$

Since jobs never leave the system, for each i, $\sum_{j=1}^{j=s} P_{ij} = 1$. This fact causes the above equation to have no unique solution. Fortunately, it turns out that we can use any solution to help us get steady-state probabilities. If we define

$$\rho_i = \frac{\lambda_i}{\mu_i}$$

then we determine, for any state n, its steady-state probability $\Pi_N(\mathbf{n})$ from the following equation:

$$\Pi_N(\mathbf{n}) = \frac{\rho_1^{n_1} \rho_2^{n_2} \cdots \rho_n^{n_s}}{G(N)}$$

Here, $G(N) = \sum_{\mathbf{n} \in S_N} \rho_1^{n_1} \rho_2^{n_2} \cdots \rho_s^{n_s}$.

Buzen's algorithm gives us an efficient way to determine (in a spreadsheet) $G(N)$. Once we have the steady-state probability distribution, we can easily determine other measures of effectiveness, such as expected queue length at each server and expected time a job spends during each visit to a server, fraction of time a server is busy, and the throughput for each server (jobs per second processed by each server).

To obtain $G(N)$, we recursively compute the quantities $C_i(k)$ for $i = 1, 2, \ldots, s$ and $k = 0, 1, \ldots, N$. We initialize the recursion with $C_1(k) = \rho_1^k$, $k = 0, 1, \ldots, N$ and $C_i(0) = 1$, $i = 1, 2, \ldots, s$. For other values of k and i, we build up the values of $C_i(k)$ recursively via the following relationship:

$$C_i(k) = C_{i-1}(k) + \rho_i C_i(k - 1)$$

Then it can be shown that $G(N) = C_s(N)$.

An Approximation for the *G/G/m* Queuing System

In most situations, interarrival times follow an exponential random variable. Often, however, service times do not follow an exponential distribution. When interarrival times and service times each follow a nonexponential random variable, we call the queuing system a *G/G/m* system. For these situations, the templates discussed in the previous sections of this chapter are not valid. Fortunately, the Allen–Cunneen approximation often gives a good approximation to L, W, L_q, and W_q for *G/G/m* systems. The file ggm.xls contains a spreadsheet implementation of the Allen–Cunneen approximation. The user need only input the following information:

ggm.xls

- The average number of arrivals per unit time (λ) in cell B3.
- The average rate at which customers can be serviced (μ) in cell B4.
- The number of servers (s) in cell B5.
- The squared coefficient of variation—(variance of interarrival times)/(mean interarrival time)^2—of interarrival times in cell B6.
- The squared coefficient of variation—(variance of service times)/(mean service time)^2—of service times in cell B7.

The Allen–Cunneen approximation is exact if interarrival times and service times are exponential. Extensive testing by Tanner indicates that in a wide variety of situations, the values of L, W, L_q, and W_q obtained by the approximation are within 10% of their true values.

Transient Behavior of Queuing Systems

We define $P_i(t)$ to be the probability that i customers are present at time t. We then assume (although this is not necessary) that the system is initially empty, so $P_0(0) = 1$ and, for $i > 0$, $P_0(i) = 0$. Then, given knowledge of $P_i(t)$, we may compute $P_i(t + \Delta t)$ as follows:

$$P_0(t + \Delta t) = (1 - \lambda(t)\Delta t)P_0(t) + \mu(t)\Delta t P_1(t)$$
$$P_i(t + \Delta t) = \lambda(t)\Delta t P_{i-1}(t) + (1 - \lambda(t)\Delta t - \min(s(t), i)\mu(t)\Delta t)P_i(t) + \min(s(t),$$
$$i + 1)\mu(t)\Delta t P_{i+1}(t), N - 1 \geq i \geq 1$$
$$P_N(t + \Delta t) = \lambda(t)\Delta t P_{N-1}(t) + (1 - \min(s(t), N)\mu(t))\Delta t P_N(t)$$

These equations are based on the assumption that, if the state at time t is i, then during the next Δt, the probability of an arrival is $\lambda(t) \Delta t$, and the probability of a service completion is $\min(s(t), i)\mu(t)\Delta t$.

REVIEW PROBLEMS

Group A

1 Buses arrive at the downtown bus stop and leave for the mall stop. Past experience indicates that 20% of the time, the interval between buses is 20 minutes; 40% of the time, the interval is 40 minutes; and 40% of the time, the interval is 2 hours. If I have just arrived at the downtown bus stop, how long, on the average, should I expect to wait for a bus?

2 Registration at State University proceeds as follows: Upon entering the registration hall, the students first wait in line to register for classes. A single clerk handles registration for classes, and it takes the clerk an average of 2 minutes to handle a student's registration. Next, the student must wait in line to pay fees. A single clerk handles the payment of fees. The clerk takes an average of 2 minutes to process a student's fees. Then the student leaves the registration building. An average of 15 students per hour arrive at the registration hall.

    a If interarrival and service times are exponential, what is the expected time a student spends in the registration hall?

    b What is the probability that during the next 5 minutes, exactly 2 students will enter the registration hall?

    c Without any further information, what is the probability that during the next 3 minutes, no student will arrive at the fee clerk's desk?

    d Suppose the registration system is changed so that a student can register for classes and pay fees at the same station. If the service time at this single station follows an Erlang distribution with rate parameter 1.5 per minute and shape parameter 2, what is the expected time a student spends waiting in line?

3 At the Smalltown post office, patrons wait in a single line for the first open window. An average of 100 patrons per hour enter the post office, and each window can serve an average of 45 patrons per hour. The post office estimates a cost of 10¢ for each minute a patron waits in line and believes that it costs $20 per hour to keep a window open. Interarrival times and service times are exponential.

    a To minimize the total expected hourly cost, how many windows should be open?

    b If the post office's goal is to ensure that at most 5% of all patrons will spend more than 5 minutes in line, how many windows should be open?

4 Each year, an average of 500 people pass the New York state bar exam and enter the legal profession. On the average, a lawyer practices law in New York State for 35 years. Twenty years from now, how many lawyers would you expect there to be in New York State?

5 There are 5 students and one keg of beer at a wild and crazy campus party. The time to draw a glass of beer follows an exponential distribution, with an average time of 2 minutes. The time to drink a beer also follows an exponential distribution, with a mean of 18 minutes. After finishing a beer, each student immediately goes back to get another beer.

    a On the average, how long does a student wait in line for a beer?

    b What fraction of the time is the keg not in use?

    c If the keg holds 500 glasses of beer, how long, on the average, will it take to finish the keg?

6 The manager of a large group of employees must decide if she needs another photocopying machine. The cost of a machine is $40 per 8-hour day whether or not the machine is in use. An average of 4 people per hour need to use the copying machine. Each person uses the copier for an average of 10 minutes. Interarrival times and copying times are exponentially distributed. Employees are paid $8 per hour, and we assume that a waiting cost is incurred when a worker is waiting in line or is using the copying machine. How many copying machines should be rented?

7 An automated car wash will wash a car in 10 minutes. Arrivals occur an average of 15 minutes apart (exponentially distributed).

    a On the average, how many cars are waiting in line for a wash?

    b If the car wash could be speeded up, what wash time would reduce the average waiting time to 5 minutes?

8 The Newcoat Painting Company has for some time been experiencing high demand for its automobile repainting service. Since it has had to turn away business, management is concerned that the limited space available to store cars awaiting painting has cost lost revenue. A small vacant lot next to the painting facility has recently been made available for lease on a long-term basis at a cost of $10 per day. Management believes that each lost customer costs $20 in profit. Current demand is estimated to be 21 cars per day with exponential interarrival times (including those turned away), and the facility can service at an exponential rate of 24 cars per day. Cars are processed on an FCFS basis. Waiting space is now limited to 9 cars but can be increased to 20 cars with the lease of the vacant lot. Newcoat wants to determine whether the vacant lot should be leased. Management also wants to know the expected daily lost profit due to turning away customers if the lot is leased. Only one car can be painted at a time.

9 At an exclusive restaurant, there is only one table and waiting space for only one other group; others that arrive when the waiting space is filled are turned away. The arrival rate follows an exponential distribution with a rate of one group per hour. It takes the average group 1 hour (exponentially distributed) to be served and eat the meal. What is the average time that a group spends waiting for a table?

10 The owner of an exclusive restaurant has two tables but only one waiter. If the second table is occupied, the owner waits on that table himself. Service times are exponentially distributed with mean 1 hour, and the time between arrivals is exponentially distributed with mean 1.5 hours. When the restaurant is full, people must wait outside in line.

a What percentage of the time is the owner waiting on a table?

b If the owner wants to spend at most 10% of his time waiting on tables, what is the maximum arrival rate that can be tolerated?

11 Ships arrive at a port facility at an average rate of 2 ships every 3 days. On the average, it takes a single crew 1 day to unload a ship. Assume that interarrival and service times are exponential. The shipping company owns the port facility as well as the ships using that facility. It is estimated to cost the company $1,000 per day that each ship spends in port. The crew servicing the ships consists of 100 workers, who are each paid an average of $30 per day. A consultant has recommended that the shipping company hire an additional 40 workers and split the employees into two equal-sized crews of 70 each. This would give each crew an unloading or loading time averaging $\frac{3}{2}$ days. Which crew arrangement would you recommend to the company?

12 An average of 40 jobs per day arrive at a factory. The time between arrivals of jobs is exponentially distributed. The factory can process an average of 42 jobs per day, and the time to process a job is exponentially distributed.

a What is the probability that exactly 180 jobs arrive at the factory during a 5-day period?

b On the average, how long does it take before a job is completed (measured from the time the job arrives at the factory)?

c What fraction of the time is the factory idle?

d What is the probability that work on a job will begin within 2 days of its arrival at the factory?

13 A printing shop receives an average of 1 order per day. The average length of time required to complete an order is .5 day. At any time, the print shop can work on at most one job.

a On the average, how many jobs are present in the print shop?

b On the average, how long will a person who places an order have to wait until it is finished?

c What is the probability that an order will be finished within 2 days of its arrival?

Group B

14 The mail order firm of L. L. Pea receives an average of 200 calls per hour (times between calls are exponentially distributed). It takes an L. L. Pea operator an average of 3 minutes to handle a call. If a caller gets a busy signal, L. L. Pea assumes that he or she will call Seas Beginning (a competing mail order house), and L. L. Pea will lose an average of $30 in profit. The cost of keeping a phone line open is $9 per hour. How many operators should L. L. Pea have on duty?

15 Each hour, an average of 3 type 1 and 3 type 2 customers arrive at a single-server station. Interarrival times for each customer type are exponential and independent. The average service time for a type 1 customer is 6 minutes, and the average service time for a type 2 customer is 3 minutes (all service times are exponentially distributed). Consider the following three service arrangements:

Arrangement 1 All customers wait in a single line and are served on an FCFS basis.

Arrangement 2 Type 1 customers are given nonpreemptive priority over type 2 customers.

Arrangement 3 Type 2 customers are given nonpreemptive priority over type 1 customers.

Which arrangement will result in the smallest average per-customer waiting time? Which arrangement will result in the largest average per-customer waiting time?

16 Podunk University Operations Research Department has two phone lines. An average of 30 people per hour try to call the OR Department, and the average length of a phone call is 1 minute. If a person attempts to call when both lines are busy, he or she hangs up and is lost to the system. Assume that the time between people attempting to call and service times is exponential.

a What fraction of the time will both lines be free? What fraction of the time will both lines be busy? What fraction of the time will exactly one line be free?

b On the average, how many lines will be busy?

c On the average, how many callers will hang up each hour?

17[†] Smalltown has two ambulances. Ambulance 1 is based at the local college, and ambulance 2 is based downtown. If a request for an ambulance comes from the college, the college-based ambulance is sent if it is available. Otherwise, the downtown-based ambulance is sent (if available). If no ambulance is available, the call is assumed to be lost to the system. If a request for an ambulance comes from anywhere else in the town, the downtown-based ambulance is sent if it is available. Otherwise, the college-based ambulance is sent if available. If no ambulance is available, the call is considered lost to the system. The time between calls is exponentially distributed. An average of 3 calls per hour are received from the college, and an average of 4 calls per hour are received from the rest of the town. The average time (exponentially distributed) it takes an ambulance to respond to a call and be ready to respond to another call is shown in Table 13.

a What fraction of the time is the downtown ambulance busy?

b What fraction of the time is the college ambulance busy?

c What fraction of all calls will be lost to the system?

d On the average, who waits longer for an ambulance, a college student or a town person?

18 An average of 10 people per hour arrive (interarrival times are exponential) intending to swim laps at the local

TABLE 13

Ambulance Comes From	Ambulance Goes to	
	College	Noncollege
College	4 minutes	7 minutes
Downtown	5 minutes	4 minutes

[†]Based on Carter (1972).

YMCA. Each intends to swim an average of 30 minutes. The YMCA has three lanes open for lap swimming. If one swimmer is in a lane, he or she swims up and down the right side of the lane. If two swimmers are in a lane, each swims up and down one side of the lane. Swimmers always join the lane with the fewest number of swimmers. If all three lanes are occupied by two swimmers, a prospective swimmer becomes disgusted and goes running.

a What fraction of the time will 3 people be swimming laps?

b On the average, how many people are swimming laps in the pool?

c How many lanes does the YMCA need to allot to lap swimming to ensure that at most 5% of all prospective swimmers will become disgusted and go running?

19[†] (Requires use of a spreadsheet) An average of 140 people per year apply for public housing in Boston. An average of 20 housing units per year become available. During a given year, there is a 10% chance that a family on the waiting list will find private housing and remove themselves from the list. Assume that all relevant random variables are exponentially distributed.

a On the average, how many families will be on the waiting list?

b On the average, how much time will a family spend on the list before obtaining housing (either public or private)? For the last question, remember that $L = \lambda W$!

[†]Based on Kaplan (1986)

REFERENCES

The following books contain excellent discussions of queuing theory at an intermediate level:

Cooper, R. *Introduction to Queuing Theory,* 2d ed. New York: North-Holland, 1981.

Gross, D., and C. Harris. *Fundamentals of Queuing Theory,* 3d ed. New York: Wiley, 1997.

Karlin, S., and H. Taylor. *A First Course in Stochastic Processes,* 2d ed. Orlando, Fla.: Academic Press, 1975.

Lee, A. *Applied Queuing Theory.* New York: St. Martin's Press, 1966.

Ross, S. *Applied Probability Models with Optimization Applications.* San Francisco, Calif.: Holden-Day, 1970.

Saaty, T. *Elements of Queuing Theory with Applications.* New York: Dover, 1983.

The following three books contain excellent discussions of queuing theory at a more advanced level:

Heyman, D., and M. Sobel. *Stochastic Models in Operations Research,* vol. 1. New York: McGraw-Hill, 1984.

Kao, E. *An Introduction to Stochastic Processes.* Belmont, Cal.: Duxbury, 1997.

Kleinrock, L. *Queuing Systems,* vols. 1 and 2. New York: Wiley, 1975.

For an excellent applications-oriented study of queuing theory we recommend:

Hall, R. *Queuing Methods for Service and Manufacturing.* Englewood Cliffs, N.J.: Prentice Hall, 1991.

Tanner, M. *Practical Queuing Analysis.* New York: McGraw-Hill, 1995.

Brigham, G. "On a Congestion Problem in an Aircraft Factory," *Operations Research* 3(1955):412–428.

Buzen, J. P. "Computational Algorithms for Closed Queuing Networks with Exponential Servers," *Communications of the ACM* 16:9(1973):527–531.

Carter, G., J. Chaiken, and E. Ignall. "Response Areas for Two Emergency Units," *Operations Research* 20(1972):571–594.

Denardo, E. *Dynamic Programming: Theory and Applications.* Englewood Cliffs, N.J.: Prentice Hall, 1982. Explains why interarrival times are often exponential.

Erickson, W. "Management Science and the Gas Shortage," *Interfaces* 4(1973):47–51.

Feller, W. *An Introduction to Probability Theory and Its Applications,* 2d ed., vol. 1. New York: Wiley, 1957. Proves that the exponential distribution is the only continuous random variable with the no-memory property.

Gilliam, R. "An Application of Queuing Theory to Airport Passenger Security Screening," *Interfaces* 9(1979): 117–123.

Green, J. "Managing a Telephone System Demands Skill," *The Office* (November 1987):144–145.

Hillier, F., and O. Yu. *Queuing Tables and Graphs.* New York: North-Holland, 1981.

Jackson, J. "Networks of Waiting Lines," *Operations Research* 5(1957):518–521.

Jain, R. *The Art of Computer System Performance Analysis.* New York: Wiley, 1991.

Kaplan, E. "Tenant Assignment Models," *Operations Research* 34(no. 6, 1986):833–843.

Karmarker, U. "Lot-Sizing and Lead-Time Performance in a Manufacturing Cell," *Interfaces* 15(no.2, 1985):1–9.

Karush, W. "A Queuing Model for an Inventory Problem," *Operations Research* 5(1957):693–703.

Kendall, D. "Some Problems in the Theory of Queues," *Journal of the Royal Statistical Society,* Series B, 13(1951):151–185.

Kolesar, P. "A Quick and Dirty Response to the Quick and Dirty Crowd: Particularly to Jack Byrd's 'The Value of Queuing Theory,' " *Interfaces* 9(1979):77–82.

Law, A., and W. Kelton. *Simulation Modeling and Analysis.* New York: McGraw-Hill, 1990. Discusses simulation of queuing systems and fitting random variables to actual interarrival and service time data.

Tannenbaum, A. *Computer Networks.* Englewood Cliffs, N.J.: Prentice Hall, 1981.

Vogel, M. "Queuing Theory Applied to Machine Manning," *Interfaces* 9(1979):1–8.

21

Simulation

Simulation is a very powerful and widely used management science technique for the analysis and study of complex systems. In previous chapters, we were concerned with the formulation of models that could be solved analytically. In almost all of those models, our goal was to determine optimal solutions. However, because of complexity, stochastic relations, and so on, not all real-world problems can be represented adequately in the model forms of the previous chapters. Attempts to use analytical models for such systems usually require so many simplifying assumptions that the solutions are likely to be inferior or inadequate for implementation. Often, in such instances, the only alternative form of modeling and analysis available to the decision maker is simulation.

Simulation may be defined as a technique that imitates the operation of a real-world system as it evolves over time. This is normally done by developing a simulation model. A *simulation model* usually takes the form of a set of assumptions about the operation of the system, expressed as mathematical or logical relations between the objects of interest in the system. In contrast to the exact mathematical solutions available with most analytical models, the simulation process involves executing or running the model through time, usually on a computer, to generate representative samples of the measures of performance. In this respect, simulation may be seen as a sampling experiment on the real system, with the results being sample points. For example, to obtain the best estimate of the mean of the measure of performance, we average the sample results. Clearly, the more sample points we generate, the better our estimate will be. However, other factors, such as the starting conditions of the simulation, the length of the period being simulated, and the accuracy of the model itself, all have a bearing on how good our final estimate will be. We discuss such issues later in the chapter.

As with most other techniques, simulation has its advantages and disadvantages. The major advantage of simulation is that simulation theory is relatively straightforward. In general, simulation methods are easier to apply than analytical methods. Whereas analytical models may require us to make many simplifying assumptions, simulation models have few such restrictions, thereby allowing much greater flexibility in representing the real system. Once a model is built, it can be used repeatedly to analyze different policies, parameters, or designs. For example, if a business firm has a simulation model of its inventory system, various inventory policies can be tried on the model rather than taking the chance of experimenting on the real-world system. However, it must be emphasized that simulation is not an optimizing technique. It is most often used to analyze "what if" types of questions. Optimization with simulation is possible, but it is usually a slow process. Simulation can also be costly. However, with the development of special-purpose simulation languages, decreasing computational cost, and advances in simulation methodologies, the problem of cost is becoming less important.

In this chapter, we focus our attention on simulation models and the simulation technique. We present several examples of simulation models and explore such concepts as random numbers, time flow mechanisms, Monte Carlo sampling, simulation languages, and statistical issues in simulation.

21.1 Basic Terminology

We begin our discussion by presenting some of the terminology used in simulation. In most simulation studies, we are concerned with the simulation of some system. Thus, in order to model a system, we must understand the concept of the system. Among the many different ways of defining a system, the most appropriate definition for simulation problems is the one proposed by Schmidt and Taylor (1970).

DEFINITION ■ A **system** is a collection of entities that act and interact toward the accomplishment of some logical end. ■

In practice, however, this definition generally tends to be more flexible. The exact description of the system usually depends on the objectives of the simulation study. For example, what may be a system for a particular study may be only a subset of the overall system for another.

Systems generally tend to be dynamic—their status changes over time. To describe this status, we use the concept of the state of a system.

DEFINITION ■ The **state** of a system is the collection of variables necessary to describe the status of the system at any given time. ■

As an example of a system, let us consider a bank. Here, the system consists of the servers and the customers waiting in line or being served. As customers arrive or depart, the status of the system changes. To describe these changes in status, we require a set of variables called the **state variables.** For example, the number of busy servers, the number of customers in the bank, the arrival time of the next customer, and the departure time of the customers in service together describe every possible change in the status of the bank. Thus, these variables could be used as the state variables for this system. In a system, an object of interest is called an **entity,** and any properties of an entity are called **attributes.** For example, the bank's customers may be described as the entities, and the characteristics of the customers (such as the occupation of a customer) may be defined as the attributes.

Systems may be classified as discrete or continuous.

DEFINITION ■ A **discrete system** is one in which the state variables change only at discrete or countable points in time. ■

A bank is an example of a discrete system, since the state variables change only when a customer arrives or when a customer finishes being served and departs. These changes take place at discrete points in time.

DEFINITION ■ A **continuous system** is one in which the state variables change continuously over time. ■

A chemical process is an example of a continuous system. Here, the status of the system is changing continuously over time. Such systems are usually modeled using differential equations. We do not discuss any continuous systems in this chapter.

There are two types of simulation models: static and dynamic.

We usually refer to a static simulation as a **Monte Carlo simulation.**

Within these two classifications, a simulation may be deterministic or stochastic. A **deterministic simulation model** is one that contains no random variables; a **stochastic simulation model** contains one or more random variables. Discrete and continuous simulation models are similar to discrete and continuous systems. In this chapter, we concentrate mainly on discrete stochastic models. Such models are called *discrete-event* simulation models. Discrete-event simulation concerns the modeling of a stochastic system as it evolves over time by a representation in which state variables change only at discrete points in time.

21.2 An Example of a Discrete-Event Simulation

Before we proceed to the details of simulation modeling, it will be useful to work through a simple simulation example to illustrate some of the basic concepts in discrete-event simulation. The model we have chosen as our initial example is a single-server queuing system. Customers arrive into this system from some population and either go into service immediately if the server is idle or join a waiting line (queue) if the server is busy. Examples of this kind of a system are a one-person barber shop, a small grocery store with only one checkout counter, and a single ticket counter at an airline terminal.

The same model was studied in Chapter 20 in connection with queuing theory. In that chapter, we used an analytical model to determine the various operating characteristics of the system. However, we had to make several restrictive assumptions to use queuing theory. In particular, when we studied an $M/M/1$ system, we had to assume that both interarrival times and service times were exponentially distributed. In many situations, these assumptions may not be appropriate. For example, arrivals at an airline counter generally tend to occur in bunches, because of such factors as the arrivals of shuttle buses and connecting flights. For such a system, an empirical distribution of arrival times must be used, which implies that the analytical model from queuing theory is no longer feasible. With simulation, any distribution of interarrival times and service times may be used, thereby giving much more flexibility to the solution process.

To simulate a queuing system, we first have to describe it. For this single-server system, we assume that arrivals are drawn from an infinite calling population. There is unlimited waiting room capacity, and customers will be served in the order of their arrival— that is, on a first come, first served (FCFS) basis. We further assume that arrivals occur one at a time in a random fashion, with the distribution of interarrival times as specified in Table 1. All arrivals are eventually served, with the distribution of service times shown in Table 2. Service times are also assumed to be random. After service, all customers return to the calling population. This queuing system can be represented as shown in Figure 1.

Before dealing with the details of the simulation itself, we must define the state of this system and understand the concepts of events and clock time within a simulation. For this

TABLE **1**		TABLE **2**	
Interarrival Time Distribution		Service Time Distribution	
Interarrival Time (minutes)	Probability	Service Time (minutes)	Probability
1	.20	1	.35
2	.30	2	.40
3	.35	3	.25
4	.15		

example, we use the following variables to define the state of the system: (1) the number of customers in the system; (2) the status of the server—that is, whether the server is busy or idle; and (3) the time of the next arrival.

Closely associated with the state of the system is the concept of an event. An **event** is defined as a situation that causes the state of the system to change instantaneously. In the single-server queuing model, there are only two possible events that can change the state of the system: an arrival into the system and a departure from the system at the completion of service. In the simulation, these events will be scheduled to take place at certain points in time. All the information about them is maintained in a list called the **event list.** Within this list, we keep track of the type of events scheduled and, more important, the time at which these events are scheduled to take place. Time in a simulation is maintained using a variable called the **clock time.** The concept of clock time will become clearer as we work through the example.

We begin this simulation with an empty system and arbitrarily assume that our first event, an arrival, takes place at clock time 0. This arrival finds the server idle and enters service immediately. Arrivals at other points in time may find the server either idle or busy. If the server is idle, the customer enters service. If the server is busy, the customer joins the waiting line. These actions can be summarized as shown in Figure 2.

Next, we schedule the departure time of the first customer. This is done by randomly generating a service time from the service time distribution (described later in the chapter) and setting the departure time as

$$\text{Departure time} = \text{clock time now} + \text{generated service time} \tag{1}$$

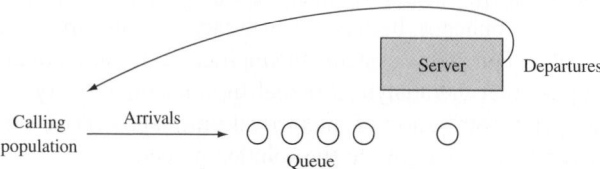

FIGURE 1
Single-Server Queuing System

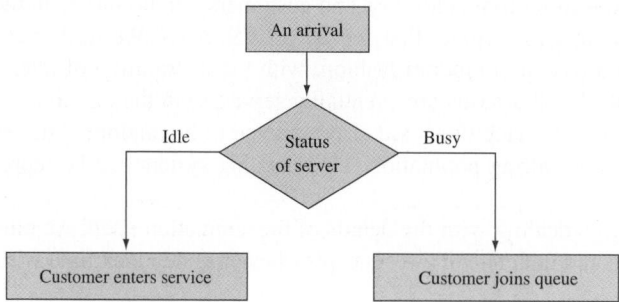

FIGURE 2
Flowchart for an Arrival

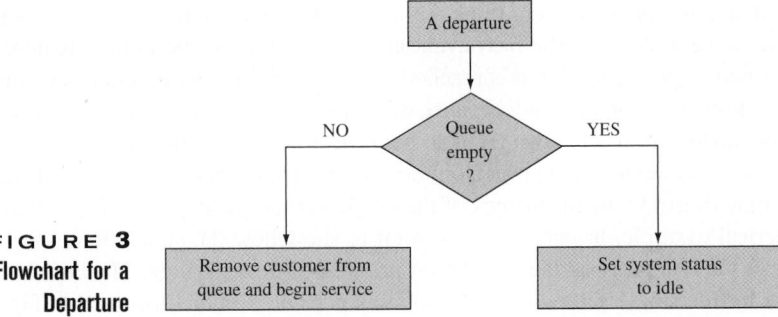

Also, we now schedule the next arrival into the system by randomly generating an interarrival time from the interarrival time distribution and setting the arrival time as

$$\text{Arrival time} = \text{clock time now} + \text{generated interarrival time} \qquad (2)$$

If, for example, we have generated a service time of 2 minutes, then the departure time for the first customer will be set at clock time 2. Similarly, if we have generated an interarrival time of 1 minute, the next arrival will be scheduled for clock time 1.

Both these events and their scheduled times are maintained on the event list. Once we have completed all the necessary actions for the first arrival, we scan the event list to determine the next scheduled event and its time. If the next event is determined to be an arrival, we move the clock time to the scheduled time of the arrival and go through the preceding sequence of actions for an arrival. If the next event is a departure, we move the clock time to the time of the departure and process a departure. For a departure, we check whether the length of the waiting line is greater than zero. If it is, we remove the first customer from the queue and begin service on this customer by setting a departure time using Equation (1). If no one is waiting, we set the status of the system to idle. These departure actions are summarized in Figure 3.

This approach of simulation is called the **next-event time-advance mechanism,** because of the way the clock time is updated. We advance the simulation clock to the time of the most imminent event—that is, the first event in the event list. Since the state variables change only at event times, we skip over the periods of inactivity between the events by jumping from event to event. As we move from event to event, we carry out the appropriate actions for each event, including any scheduling of future events. We continue in this manner until some prespecified stopping condition is satisfied. However, the procedure requires that at any point in the simulation, we have an arrival and a departure scheduled for the future. Thus, a future arrival is always scheduled when processing a new arrival into the system. A departure time, on the other hand, can only be scheduled when a customer is brought into service. Thus, if the system is idle, no departures can be scheduled. In such instances, the usual practice is to schedule a dummy departure by setting the departure time equal to a very large number—say, 9,999 (or larger if the clock time is likely to exceed 9,999). This way, our two events will consist of a real arrival and a dummy departure.

The jump to the next event in the next-event mechanism may be a large one or a small one; that is, the jumps in this method are variable in size. We contrast this approach with the **fixed-increment time-advance method.** With this method, we advance the simulation clock in increments of Δt time units, where Δt is some appropriate time unit, usually 1 time unit. After each update of the clock, we check to determine whether any event is scheduled to take place at the current clock time. If an event is scheduled, we carry out the appropriate actions for the event. If none is scheduled, or if we have completed all the required actions for the current time, we update the simulation clock by Δt units and repeat the process. As with the next-event approach, we continue in this manner until the prespecified stopping condition is reached. The fixed-increment time-advance mechanism is often simpler to comprehend, because of its fixed steps in time. For most models, however, the next-event mechanism tends to be more efficient computationally. Consequently, we use only the next-event approach in developing the models for the rest of the chapter.

We now illustrate the mechanics of the single-server queuing system simulation, using a numerical example. In particular, we want to show how the simulation model is represented in the computer as the simulation progresses through time. The entire simulation process for the single-server queuing model is presented in the flowchart in Figure 4. All the blocks in this flowchart are numbered for easy reference. For simplicity, we assume that both the interarrival times (*ITs*) and the service times (*STs*) have already been generated for the first few customers from the given probability distributions in Tables 1 and 2. These times are shown in Table 3, from which we can see that the time between the first and the second arrival is 2 time units, the time between the second and the third arrival is also 2 time units, and so on. Similarly, the service time for the first customer is 3 time units, *ST* for the second customer is also 3 time units, and so on.

To demonstrate the simulation model, we need to define several variables:

$$TM = \text{clock time of the simulation}$$
$$AT = \text{scheduled time of the next arrival}$$
$$DT = \text{scheduled time of the next departure}$$
$$SS = \text{status of the server } (1 = \text{busy, } 0 = \text{idle})$$
$$WL = \text{length of the waiting line}$$
$$MX = \text{length (in time units) of a simulation run}$$

Having taken care of these preliminaries, we now begin the simulation by initializing all the variables (block 1 in Figure 4). Since the first arrival is assumed to take place at time 0, we set $AT = 0$. We also assume that the system is empty at time 0, so we set $SS = 0$, $WL = 0$, and $DT = 9,999$. (Note that DT must be greater than MX). This implies that our list of events now consists of two scheduled events: an arrival at time 0 and a dummy departure at time 9,999. This completes the initialization process and gives us the computer representation of the simulation shown in Table 4.

We are now ready for our first action in the simulation: searching through the event list to determine the first event (block 2). Since our simulation consists of only two events, we simply determine the next event by comparing AT and DT. (In other simulations, we might have more than two events, so we would have to have an efficient system of searching through the event list.) An arrival is indicated by $AT < DT$, a departure by $DT < AT$. At this point, $AT = 0$ is less than $DT = 9,999$, indicating that an arrival will take place next. We label this event 1 and update the clock time, TM, to the time of event 1 (block 3). That is, we set $TM = 0$.

The arrival at time 0 finds the system empty, indicated by the fact that $SS = 0$ (block 4). Consequently, the customer enters service immediately. For this part of the simulation, we first set $SS = 1$ to signify that the server is now busy (block 6). We next generate a service time (block 7) and set the departure time for this customer (block 8). From Table 3, we see that ST for customer 1 is 3. Since $TM = 0$ at this point, we set $DT = 3$ for the first customer. In other words, customer 1 will depart from the system at clock time 3. Finally, to complete all the actions of processing an arrival, we schedule the next arrival into the system by generating an interarrival time, IT (block 9), and setting the time of this arrival using the equation $AT = TM + IT$ (block 10). Since $IT = 2$, we set $AT = 2$. That is, the second arrival will take place at clock time 2. At the end of event 1, our computer representation of the simulation will be as shown in Table 4.

FIGURE **4**

Flowchart for Simulation Model for Single-Server Queuing System

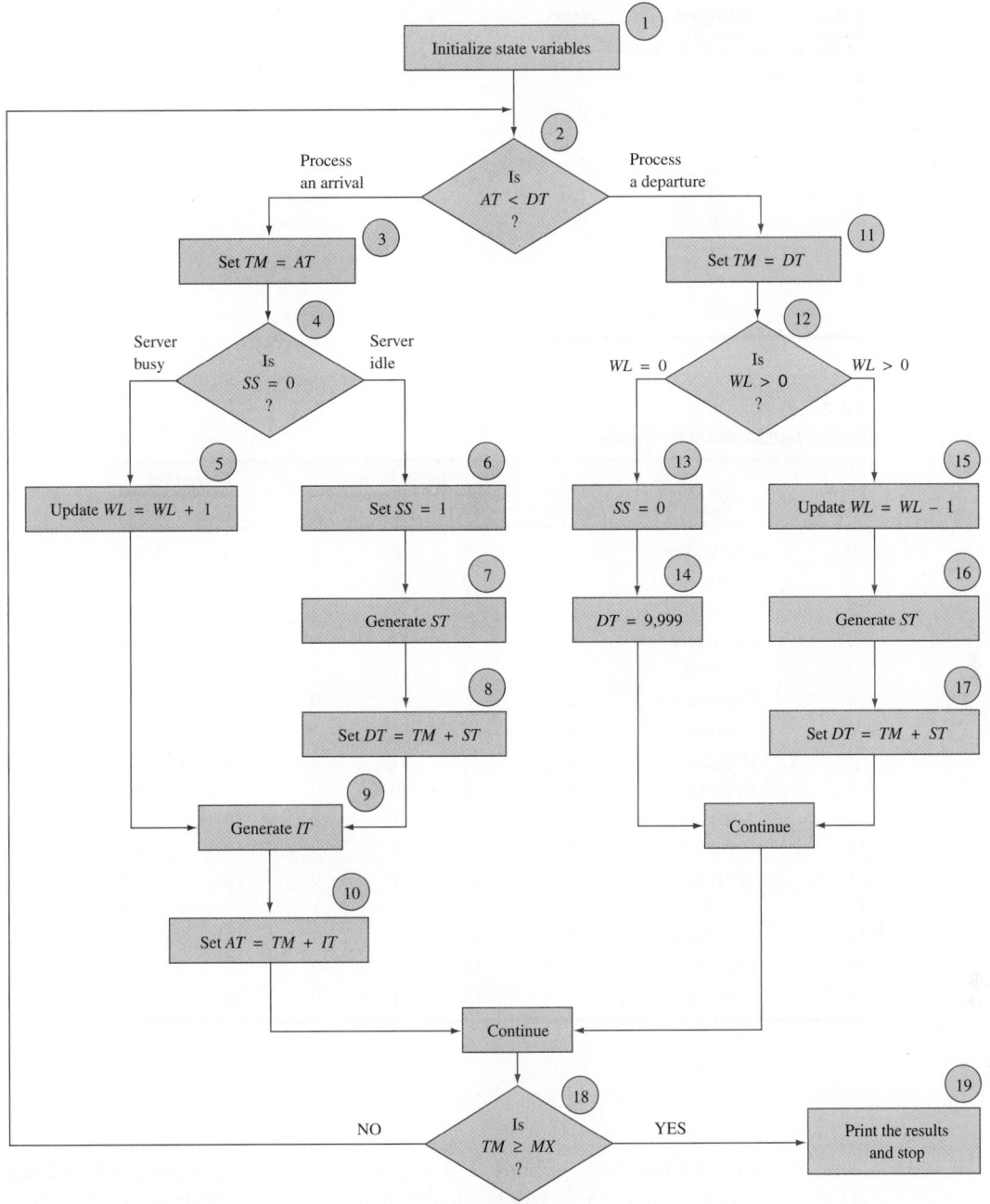

At this stage of the simulation, we proceed to block 18 to determine whether the clock time, *TM*, has exceeded the specified time length of simulation, *MX*. If it has, we print out the results (block 19) and stop the execution of the simulation model. If it has not, we continue with the simulation. We call this the termination process. We execute this process at the end of each event. However, for this example, we assume that *MX* is a large number. Consequently, from here on, we will not discuss the termination process.

TABLE 3
Generated Interarrival and Service Times

Customer Number	Interarrival Time (*IT*)	Service Time (*ST*)
1	—	3
2	2	3
3	2	2
4	3	1
5	4	1
6	2	2
7	1	1
8	3	2
9	3	—

TABLE 4

Computer Representation of the Simulation

End of Event	Type of Event	Customer Number	System Variables			Event List	
			TM	SS	WL	AT	DT
0	Initialization	—	0	0	0	0	9,999
1	Arrival	1	0	1	0	2	3
2	Arrival	2	2	1	1	4	3
3	Departure	1	3	1	0	4	6
4	Arrival	3	4	1	1	7	6
5	Departure	2	6	1	0	7	8
6	Arrival	4	7	1	1	11	8
7	Departure	3	8	1	0	11	9
8	Departure	4	9	0	0	11	9,999
9	Arrival	5	11	1	0	13	12
10	Departure	5	12	0	0	13	9,999
11	Arrival	6	13	1	0	14	15
12	Arrival	7	14	1	1	17	15
13	Departure	6	15	1	0	17	16
14	Departure	7	16	0	0	17	9,999
15	Arrival	8	17	1	0	20	19

At this point, we loop back to block 2 to determine the next event. Since $AT = 2$ and $DT = 3$, the next event, event 2, will be an arrival at time 2. Having determined the next event, we now advance the simulation to the time of this arrival by updating TM to 2.

The arrival at time 2 finds the server busy, so we put this customer in the waiting line by updating WL from 0 to 1 (block 5). Since the present event is an arrival, we now schedule the next arrival into the system. Given that $IT = 2$ for arrival 3, the next arrival takes place at clock time 4. This completes all the necessary actions for event 2. We again loop back to block 2 to determine the next event. From the computer representation of the system in Table 4, we see that at this point (end of event 2), $DT = 3$ is less than $AT = 4$. This implies that the next event, event 3, will be a departure at clock time 3. We advance the clock to the time of this departure; that is, we update TM to 3 (block 11).

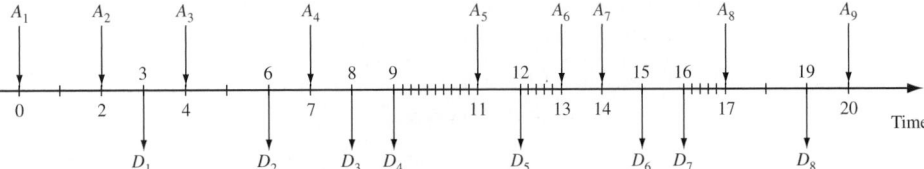

FIGURE 5
Time Continuum Representation of Single-Server Simulation

At time 3, we process the first departure from the system. With the departure, the server now becomes idle. We check the status of the waiting line to see whether there are any customers waiting for service (block 12). Since $WL = 1$, we have one customer waiting. We remove this customer from the waiting line, set $WL = 0$ (block 15), and bring this customer into service by generating a service time, ST (block 16), and setting the departure time using the relation $DT = TM + ST$ (block 17). From Table 3, we see that for customer 2, $ST = 3$. Since $TM = 3$, we set $DT = 6$. We have now completed all the actions for event 3, giving us the computer representation shown in Table 4.

From here on, we leave it to the reader to work through the logic of the simulation for the rest of the events in this example. Table 4 shows the status of the simulation at the end of each of these events. Note that at the end of events 8, 10, and 14 (all departures), the system becomes idle. During the sequence of actions for these events, we set $SS = 0$ (block 13) and $DT = 9,999$ (block 14). In each case, the system stays idle until an arrival takes place. This simulation is summarized in the time continuum diagram in Figure 5. Here, the A's represent the arrivals and the D's the departures. Note that the hatched areas, such as the one between times 9 and 11, signify that the system is idle.

This simple example illustrates some of the basic concepts in simulation and the way in which simulation can be used to analyze a particular problem. Although this model is not likely to be used to evaluate many situations of importance, it has provided us with a specific example and, more important, has introduced a variety of key simulation concepts. In the rest of the chapter, we analyze some of these simulation concepts in more detail. No mention was made in the example of the collection of statistics, but procedures can be easily incorporated into the model to determine the measures of performance of this system. For example, we could expand the flowchart to calculate and print the mean waiting time, the mean number in the waiting line, and the proportion of idle time. We discuss statistical issues in detail later in the chapter.

21.3 Random Numbers and Monte Carlo Simulation

In our queuing simulation example, we saw that the underlying movement through time is achieved in the simulation by generating the interarrival and the service times from the specified probability distributions. In fact, all event times are determined either directly or indirectly by these generated service and interarrival times. The procedure of generating these times from the given probability distributions is known as sampling from probability distributions, or random variate generation, or **Monte Carlo sampling.** In this section, we present and discuss several different methods of sampling from discrete distributions. We initially demonstrate the technique using a roulette wheel and then expand it by carrying out the sampling using random numbers.

The principle of sampling from discrete distributions is based on the frequency interpretation of probability. That is, in the long run, we would like the outcomes to occur with the frequencies specified by the probabilities in the distribution. For example, if we consider the service time distribution in Table 2, we would like, in the long run, to generate

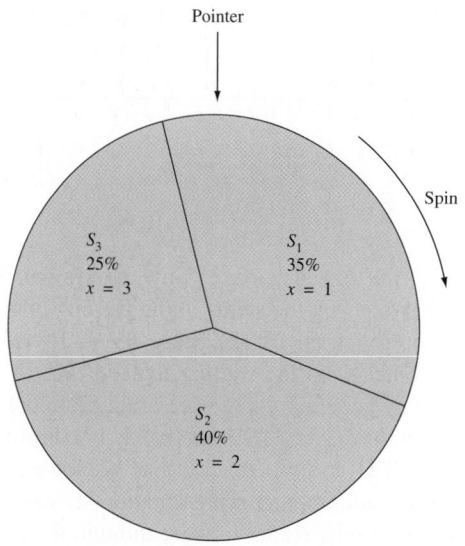

Pointer

Spin

S_3
25%
$x = 3$

S_1
35%
$x = 1$

S_2
40%
$x = 2$

FIGURE 6
Segmentation of
Roulette Wheel

a service time of 1 minute 35% of the time, a service time of 2 minutes 40% of the time, and a service time of 3 minutes 25% of the time. In addition to obtaining the right frequencies, the sampling procedure should be independent; that is, each generated service time should be independent of the service times that precede it and follow it.

To achieve these two properties using a roulette wheel, we first partition the wheel into three segments, each proportional in area to a probability in the distribution (see Figure 6). For example, the first segment (say, S_1) is allocated 35% of the area of the roulette wheel. This area corresponds to the probability of .35 and the service time of 1 minute. The second segment, S_2, covers 40% of the area and corresponds to the probability of .40 and the service time of 2 minutes. Finally, the third segment, S_3, is allocated the remaining 25% of the area, corresponding to the probability .25 and the service time of 3 minutes. If we now spin the roulette wheel and the pointer falls in segment S_1, it means that we have generated a service time of 1 minute; in segment S_2, 2 minutes; and in segment S_3, 3 minutes. If the roulette wheel is fair, as we assume, then in the long run, (1) we will generate the service times with approximately the same frequency as specified in the distribution, and (2) the results of each spin will be independent of the results that precede and follow it.

We now expand on this technique by using numbers for segmentation instead of areas. We assume that the roulette wheel has 100 numbers on it, ranging from 00 to 99, inclusive. We further assume that the segmentation is such that each number has the same probability, .01, of showing up. Using this method of segmentation, we allocate 35 numbers (say, from 00 to 34) to the service time of 1 minute. Since each number has a probability .01 of showing up, the 35 numbers together are equivalent to a probability of .35. Similarly, if we allocate the numbers from 35 to 74 to the service time of 2 minutes, and the numbers from 75 to 99 to the service time of 3 minutes, we achieve the desired probabilities. As before, we spin the roulette wheel to generate the service times, but with this method, the numbers directly determine the service times. In other words, if we generate a number between 00 and 34, we set the service time equal to 1 minute; between 35 and 74, to 2 minutes; and between 75 and 99, to 3 minutes.

This procedure of segmentation and using a roulette wheel is equivalent to generating integer random numbers between 00 and 99. This follows from the fact that each random number in a sequence (in this case from 00 to 99) has an equal probability (in this case, .01) of showing up, and each random number is independent of the numbers that precede and follow it. If we now had a procedure for generating the 100 random numbers between 00 and 99, then instead of spinning a roulette wheel to obtain a service time, we could use a generated random number. Technically, a random number, R_i, is defined as an independent random sample drawn from a continuous uniform distribution whose probability density function (pdf) is given by

$$f(x) = \begin{cases} 1 & 0 \le x \le 1 \\ 0 & \text{otherwise} \end{cases}$$

Thus, each random number will be uniformly distributed over the range between 0 and 1. Because of this, these random numbers are usually referred to as $U(0, 1)$ random numbers, or simply as uniform random numbers.

Random Number Generators

Uniform random numbers can be generated in many different ways. Since our interest in random numbers is for use within simulations, we need to be able to generate them on a computer. This is done using mathematical functions called **random number generators.**

Most random number generators use some form of a congruential relationship. Examples of such generators include the linear congruential generator, the multiplicative generator, and the mixed generator. The linear congruential generator is by far the most widely used. In fact, most built-in random number functions on computer systems use this generator. With this method, we produce a sequence of integers $x_1, x_2, x_3, \ldots$ between 0 and $m - 1$ according to the following recursive relation:

$$x_{i+1} = (ax_i + c) \text{ modulo } m \qquad (i = 0, 1, 2, \ldots)$$

The initial value of x_0 is called the seed, a is the constant multiplier, c is the increment, and m is the modulus. These four variables are called the parameters of the generator. Using this relation, the value of x_{i+1} equals the remainder from the division of $ax_i + c$ by m. The random number between 0 and 1 is then generated using the equation

$$R_i = \frac{x_i}{m} \qquad (i = 1, 2, 3, \ldots)$$

For example, if $x_0 = 35$, $a = 13$, $c = 65$, and $m = 100$, the algorithm works as follows:

Iteration 0 Set $x_0 = 35$, $a = 13$, $c = 65$, and $m = 100$.

Iteration 1 Compute

$$x_1 = (ax_0 + c) \text{ modulo } m$$
$$= [13(35) + 65] \text{ modulo } 100$$
$$= 20$$

Deliver

$$R_1 = \frac{x_1}{m}$$
$$= \frac{20}{100}$$
$$= 0.20$$

Iteration 2 Compute

$$x_2 = (ax_1 + c) \text{ modulo } m$$
$$= [13(20) + 65] \text{ modulo } 100$$
$$= 25$$

Deliver

$$R_2 = \frac{x_2}{m}$$
$$= \frac{25}{100}$$
$$= 0.25$$

Iteration 3 Compute

$$x_3 = (ax_2 + c) \text{ modulo } m$$
$$\vdots$$

and so on.

Each random number generated using this method will be a decimal number between 0 and 1. Note that although it is possible to generate a 0, a random number cannot equal 1. Random numbers generated using congruential methods are called **pseudorandom numbers.** They are not true random numbers in the technical sense, because they are completely determined once the recurrence relation is defined and the parameters of the generator are specified. However, by carefully selecting the values of a, c, m, and x_0, the pseudorandom numbers can be made to meet all the statistical properties of true random numbers. In addition to the statistical properties, random number generators must have several other important characteristics if they are to be used efficiently within computer simulations. (1) The routine must be fast; (2) the routine should not require a lot of core storage; (3) the random numbers should be replicable; and (4) the routine should have a sufficiently long cycle—that is, we should be able to generate a long sequence without repetition of the random numbers.

There is one important point worth mentioning at this stage: Most programming languages have built-in library functions that provide random (or pseudorandom) numbers directly. Therefore, most users need only know the library function for a particular system. In some systems, a user may have to specify a value for the seed, x_0, but it is unlikely that a user would have to develop or design a random number generator. However, for more information on random numbers and random number generators, the interested reader may consult Banks and Carson (1984), Knuth (1998), or Law and Kelton (1991).

Computer Generation of Random Numbers

We now take the method of Monte Carlo sampling a stage further and develop a procedure using random numbers generated on a computer. The idea is to transform the $U(0, 1)$ random numbers into integer random numbers between 00 and 99 and then to use these integer random numbers to achieve the segmentation by numbers. The transformation is a relatively straightforward procedure. If the $(0, 1)$ random numbers are multiplied by 100, they will be uniformly distributed over the range from 0 to 100. Then, if the fractional part of the number is dropped, the result will be integers from 00 to 99, all equally likely. For example, if we had generated the random number 0.72365, multiplying it by 100 gives

TABLE 5
Two-Digit Integer Random Numbers

69	56	30	32	66	79	55	24	80	35	10	98
92	92	88	82	13	04	86	31	13	23	44	93
13	42	51	16	17	29	62	08	59	41	47	72
25	96	58	14	68	15	18	99	13	05	03	83
34	78	50	89	98	93	70	11	49	01	9	35
64	43	71	48	36	78	53	67	37	57	25	17
84	59	68	45	12	53	68	38	18	60	02	82
31	28	52	89	27	35	34	74	96	93	45	63
21	17	71	55	32	74	20	68	44	34	53	68
91	84	39	25	20	83	60	62	99	61	32	98
55	86	18	93	51	77	68	37	69	02	85	60
43	16	20	42	82	17	41	50	54	21	25	43
40	98	71	03	68	05	37	02	86	17	38	99
42	37	72	33	72	43	51	60	17	94	51	39
18	06	28	75	69	80	33	69	12	25	53	36
13	20	42	92	57	08	24	06	41	12	89	95
58	18	98	89	08	60	89	93	58	13	29	34
63	68	69	62	07	49	95	48	20	03	71	90
92	54	29	31	80	28	48	45	92	71	31	33
84	11	57	64	93	69	86	22	23	84	38	60
33	24	65	76	87	95	98	47	00	71	31	97
53	08	80	85	73	13	25	35	22	82	26	43
02	19	61	38	00	21	42	79	31	70	00	17
22	81	43	44	78	88	30	31	15	63	09	99
38	25	32	92	11	55	18	52	47	30	43	87
04	61	82	18	82	75	12	19	44	87	77	93
06	54	51	64	81	98	63	47	57	52	74	56
51	51	00	41	78	84	42	79	06	82	58	53
99	93	87	86	83	79	16	33	53	34	40	32
29	12	64	73	38	08	49	32	53	33	91	90
31	78	93	25	37	51	68	40	34	47	83	76
81	69	27	35	71	12	69	78	96	93	35	96
26	73	28	81	38	09	55	10	27	29	52	46
92	29	08	15	73	26	33	05	89	08	26	99
00	86	32	46	80	22	97	19	99	95	53	20
39	25	07	41	74	71	01	64	23	69	74	95
38	86	41	38	71	91	75	54	65	73	47	86
41	74	68	21	74	89	43	19	98	74	09	50
63	53	45	07	47	15	58	75	88	51	88	99
00	54	86	59	77	09	54	55	99	15	67	63
01	38	88	03	71	88	72	39	76	45	11	07
38	05	53	31	18	11	26	65	61	77	19	03
34	43	19	12	35	02	09	86	69	90	53	50
23	41	56	34	77	30	50	02	34	68	49	16
57	24	80	69	51	81	83	05	19	45	30	20
93	86	08	08	99	62	75	97	29	51	68	96
16	10	38	33	32	25	34	66	72	17	51	97
75	28	35	14	01	00	98	51	74	10	79	30
53	38	65	32	78	77	64	11	31	06	73	47
91	90	95	95	66	80	10	90	51	24	81	06

TABLE 6
Cumulative Distribution Function
and Random Number Ranges for Interarrival Times

Interarrival Time (minutes)	Probability	Cumulative Probability	Random Number Ranges
1	.20	.20	00–19
2	.30	.50	20–49
3	.35	.85	50–84
4	.15	1.00	85–99

us 72.365. Truncating the decimal portion of the number will leave us with the integer random number 72. On the computer, we achieve this transformation by first generating a $U(0, 1)$ random number. Next, we multiply it by 100. Finally, we store the product using an integer variable; this final stage will truncate the decimal portion of the number. This procedure will give us integer random numbers between 00 and 99. Table 5 lists some integer pseudorandom numbers obtained using this procedure. (These random numbers will be used in several examples later in the chapter.)

We now formalize this procedure and use it to generate random variates for a discrete random variable. The procedure consists of two steps: (1) We develop the cumulative probability distribution (cdf) for the given random variable, and (2) we use the cdf to allocate the integer random numbers directly to the various values of the random variable. To illustrate this procedure, we use the distribution of interarrival times from the queuing example of Section 21.2 (see Table 1). If we develop the cdf for this distribution, we get the probabilities shown in Table 6. The first interarrival time of 1 minute occurs with a probability of .20. Thus, we need to allocate 20 random numbers to this outcome. If we assign the 20 numbers from 00 to 19, we utilize the decimal random number range from 0 to 0.19999. Note that the upper end of this range lies just below the cumulative probability of .20. For the interarrival time of 2 minutes, we allocate 30 random numbers. If we assign the integer numbers from 20 to 49, we notice that this covers the decimal random number range from 0.20 to 0.49999. As before, the upper end of this range lies just below the cumulative probability of .50, but the lower end coincides with the previous cumulative probability of .20. If we now allocate the integer random numbers from 50 to 84 to the interarrival time of 3 minutes, we notice that these numbers are obtained from the decimal random number range from 0.50 (the same as the cumulative probability associated with an interarrival time of 2 minutes) to 0.84999, which is a fraction smaller than .85. Finally, the same analyses apply to the interarrival time of 4 minutes. In other words, the cumulative probability distribution enables us to allocate the integer random number ranges directly. Once these ranges have been specified for a given distribution, all we must do to obtain the value of a random variable is generate an integer random number and match it against the random number allocations. For example, if the random number had turned out to be 35, this would translate to an interarrival time of 2 minutes. Similarly, the random number 67 would translate to an interarrival time of 3 minutes, and so on. We now demonstrate these concepts in an example of a Monte Carlo simulation.

21.4 An Example of Monte Carlo Simulation

In this section, we use a Monte Carlo simulation to simulate a news vendor problem (see Chapter 16).

EXAMPLE 1   **Pierre's Bakery**

Pierre's Bakery bakes and sells french bread. Each morning, the bakery satisfies the demand for the day using freshly baked bread. Pierre's can bake the bread only in batches of a dozen loaves each. Each loaf costs 25¢ to make. For simplicity, we assume that the total daily demand for bread also occurs in multiples of 12. Past data have shown that this demand ranges from 36 to 96 loaves per day. A loaf sells for 40¢, and any bread left over at the end of the day is sold to a charitable kitchen for a salvage price of 10¢/loaf. If demand exceeds supply, we assume that there is a lost-profit cost of 15¢/loaf (because of loss of goodwill, loss of customers to competitors, and so on). The bakery records show that the daily demand can be categorized into three types: high, average, and low. These demands occur with probabilities of .30, .45, and .25, respectively. The distribution of the demand by categories is given in Table 7. Pierre's would like to determine the optimal number of loaves to bake each day to maximize profit (revenues + salvage revenues − cost of bread − cost of lost profits).

Solution   To solve this problem by simulation, we require a number of different policies to evaluate. Here, we define a policy as the number of loaves to bake each day. Each given policy is then evaluated over a fixed period of time to determine its profit margin. The policy that gives the highest profit is selected as the best policy.

In the simulation process, we first develop a procedure for generating the demand for the day:

Step 1   Determine the type of demand—that is, whether the demand for the day is high, average, or low. To do this, calculate the cdf for this distribution and set up the random number assignments (see Table 8). Then, to determine the type of demand, all we have to do is to generate a two-digit random number and match it against the random number allocations in this table.

Step 2   Generate the actual demand for the day from the appropriate demand distribution. The cdf and the random number allocations for the distribution of each of the three demand types are presented in Table 9. Then, to generate a demand, we simply generate an integer random number and match it against the appropriate random number assignments. For example, if our demand type was "average" in step 1, the random number 80 would translate into a demand of 72. Similarly, if the type of demand was "high" in step 1, the random number 9 would translate into a demand of 48.

The simulation process for this problem is relatively simple. For each day, we generate a demand for the day. Then we evaluate the various costs for a given policy. Suppose, for example, that the policy is to bake 60 loaves each day. If the demand for a particular

TABLE 7
Demand Distribution by Demand Categories

| Demand | Demand Probability Distribution | | |
	High	Average	Low
36	.05	.10	.15
48	.10	.20	.25
60	.25	.30	.35
72	.30	.25	.15
84	.20	.10	.05
96	.10	.05	.05

TABLE 8
Distribution of Demand Type

Type of Demand	Probability	Cumulative Distribution	Random Number Ranges
High	.30	.30	00–29
Average	.45	.75	30–74
Low	.25	1.00	75–99

TABLE 9

Distribution by Demand Type

Demand	Cumulative Distribution			Random Number Ranges		
	High	Average	Low	High	Average	Low
36	.05	.10	.15	00–04	00–09	00–14
48	.15	.30	.40	05–14	10–29	15–39
60	.40	.60	.75	15–39	30–59	40–74
72	.70	.85	.90	40–69	60–84	75–89
84	.90	.95	.95	70–89	85–94	90–94
96	1.00	1.00	1.00	90–99	95–99	95–99

day turns out to be 72, we have 60(0.40) = $24.00 in revenues, 60(0.25) = $15.00 in production costs, and 12(0.15) = $1.80 in lost-profit costs (because of the shortfall of 12 loaves). This gives us a net profit of 24.00 − 15.00 − 1.80 = $7.20 for that day.

Using this procedure, we calculate a profit margin for each day in the simulation. To evaluate a policy, we run the simulation for a fixed number of days for the given policy. At the end of the simulation, we average the profit margins over the set number of days to obtain the expected profit margin per day for the policy. Note that the procedure in this simulation is different from the queuing simulation, in that the present simulation does not evolve over time in the same way. Here, each day is an independent simulation. Such simulations are commonly referred to as **Monte Carlo simulations.**

To illustrate this procedure, we present in Table 10 a manual simulation for the first 15 days for a policy where we bake 60 loaves per day. From this table, the demand for both day 1 and day 2 turns out to be 60 loaves. (Random numbers used in this example were obtained from Table 5.) This demand generates a revenue of $24.00 for each of these days. Since the 60 loaves cost $15.00 to bake, our profit margin for each of the first 2 days is $9.00. On day 3, the demand is 72, giving us a shortfall of 12 loaves. As shown in the table, the profit margin for day 3 is $7.20 (24.00 − 15.00 − 1.80). On day 4, we generate a demand of 48. Since our policy is to bake 60 loaves, we will have 12 loaves left over. The 48 loaves sold give us revenues of only $19.20. However, the 12 loaves left over provide an additional $1.20 in salvage revenue, yielding a profit of $5.40 (19.20 + 1.20 − 15.00) for day 4.

If we now complete the manual simulation for the period of 15 days, the total profit earned during this time comes to $97.20. This gives us an average daily profit figure of $\frac{97.20}{15}$ = $6.48. However, this cannot be accepted as the final profit margin for this policy. The simulation results over this short a period are likely to be highly dependent on the sequence of random numbers generated, so they cannot be accepted as statistically valid. The simulation would have to be carried out over a long period of time before the profit margin could be accepted as truly representative. These statistical issues are discussed later. In the meantime, we have evaluated several different policies for this problem using

TABLE **10**
Simulation Table for Baking 60 Loaves per Day

Day	Random No. for Demand Type	Type of Demand	Random No. for Demand	Demand	Revenue	Lost Profit	Salvage Revenue	Profit
1	69	Average	56	60	$24.00	—	—	$9.00
2	30	Average	32	60	$24.00	—	—	$9.00
3	66	Average	79	72	$24.00	$1.80	—	$7.20
4	55	Average	24	48	$19.20	—	$1.20	$5.40
5	80	Low	35	48	$19.20	—	$1.20	$5.40
6	10	High	98	96	$24.00	$5.40	—	$3.60
7	92	Low	88	72	$24.00	$1.80	—	$7.20
8	82	Low	17	48	$19.20	—	$1.20	$5.40
9	04	High	86	84	$24.00	$3.60	—	$5.40
10	31	Average	13	48	$19.20	—	$1.20	$5.40
11	23	High	44	72	$24.00	$1.80	—	$7.20
12	93	Low	13	36	$14.40	—	$2.40	$1.80
13	42	Average	51	60	$24.00	—	—	$9.00
14	16	High	17	60	$24.00	—	—	$9.00
15	29	High	62	72	$24.00	$1.80	—	$7.20

TABLE **11**

Evaluation of Policies

Policy	No. of Loaves Baked Daily	Average Daily Profit	
		Exact	Simulation
A	36	$1.273	$1.273
B	48	$4.347	$4.349
C	60	$6.435	$6.436
D	72	$6.917	$6.915
E	84	$6.102	$6.104
F	96	$4.653	$4.642

a simulation model on a computer. The results of these policies are presented in Table 11. We see that the best policy for Pierre's Bakery is to bake 72 loaves each day. This table also compares the results from the simulation with the exact solution for each policy. We can see that simulation does a remarkable job of converging to the right solution. The closeness of the two solutions is not totally unexpected, since we ran the simulation model for 10,000 days for each policy.

PROBLEMS

Group A

Use the random numbers in Table 5 to solve the following problems.

1 Simulate the single-server queuing system described in Section 21.2 for the first 25 departures from the system to develop an estimate for the expected time in the waiting line. Is this a reasonable estimate? Explain.

2 Perform the simulation for Pierre's Bakery for 25 more days (days 16 through 40) for policy C in Table 11. Compare the answer with the results in the table.

3 Consider the simplest form of craps. In this game, we roll a pair of dice. If we roll a 7 or an 11 on the first throw, we win right away. If we roll a 2 or a 3 or a 12, we lose right away. Any other total (that is, 4, 5, 6, 8, 9, or 10) gives us a second chance. In this part of the game, we keep rolling the dice until we get either a 7 or the total rolled on the first throw. If we get a 7, we lose. If we roll the same total as on the first throw, we win. Assuming that the dice are fair, develop a simulation experiment to determine what percentage of the time we win.

4 Tankers arrive at an oil port with the distribution of interarrival times shown in Table 12. The port has two terminals, A and B. Terminal B is newer and therefore more efficient than terminal A. The time it takes to unload a tanker depends on the tanker's size. A supertanker takes 4 days to unload at terminal A and 3 days at terminal B. A midsize tanker takes 3 days at terminal A and 2 days at terminal B. The small tankers take 2 days at terminal A and 1 day at terminal B. Arriving tankers form a single waiting line in the port area until a terminal becomes available for service. Service is given on an FCFS basis. The type of tankers and the frequency with which they visit this port is given by the distribution in Table 13. Develop a simulation model for this port. Compute such statistics as the average number of

TABLE 12

Interarrival Times (days)	Probability
1	.20
2	.25
3	.35
4	.15
5	.05

TABLE 13

Type of Tanker	Probability
Supertanker	.40
Midsize tanker	.35
Small tanker	.25

tankers in port, the average number of days in port for a tanker, and the percentage of idle time for each of the terminals. (*Hint:* Use the flowchart in Figure 4 and modify it for a multiserver queuing system.)

21.5 Simulations with Continuous Random Variables

The simulation examples presented thus far used only discrete probability distributions for the random variables. However, in many simulations, it is more realistic and practical to use continuous random variables. In this section, we present and discuss several procedures for generating random variates from continuous distributions. The basic principle is very similar to the discrete case. As in the discrete method, we first generate a $U(0, 1)$ random number and then transform it into a random variate from the specified distribution. The process for carrying out the transformation, however, is quite different from the discrete case.

There are many different methods for generating continuous random variates. The selection of a particular algorithm will depend on the distribution from which we want to generate, taking into account such factors as the exactness of the random variables, the computational and storage efficiencies, and the complexity of the algorithm. The two most commonly used algorithms are the inverse transformation method (ITM) and the acceptance–rejection method (ARM). Between these two methods, it is possible to generate random variables from almost all of the most frequently used distributions. We present a detailed description of both these algorithms, along with several examples for each method. In addition to this, we present two methods for generating random variables from the normal distribution.

Inverse Transformation Method

The inverse transformation method is generally used for distributions whose cumulative distribution function can be obtained in closed form. Examples include the exponential, the uniform, the triangular, and the Weibull distributions. For distributions whose cdf does not exist in closed form, it may be possible to use some numerical method, such as a power-series expansion, within the algorithm to evaluate the cdf. However, this is likely to complicate the procedure to such an extent that it may be more efficient to use a different algorithm to generate the random variates. The ITM is relatively easy to describe and execute. It consists of the following three steps:

Step 1 Given a probability density function $f(x)$ for a random variable $\mathbf{X}$, obtain the cumulative distribution function $F(x)$ as

$$F(x) = \int_{-\infty}^{x} f(t)dt$$

Step 2 Generate a random number r.

Step 3 Set $F(x) = r$ and solve for x. The variable x is then a random variate from the distribution whose pdf is given by $f(x)$.

We now describe the mechanics of the algorithm using an example. For this, we consider the distribution given by the function

$$f(x) = \begin{cases} \dfrac{x}{2} & 0 \le x \le 2 \\ 0 & \text{otherwise} \end{cases}$$

A function of this type is called a **ramp function.** It can be represented graphically as shown in Figure 7. The area under the curve, $f(x) = \frac{x}{2}$, represents the probability of the occurrence of the random variable $\mathbf{X}$. We assume that in this case, $\mathbf{X}$ represents the service times of a bank teller. To obtain random variates from this distribution using the inverse transformation method, we first compute the cdf as

$$F(x) = \int_{0}^{x} \frac{t}{2} \, dt$$

$$= \frac{x^2}{4}$$

This cdf is represented formally by the function

$$F(x) = \begin{cases} 0 & x < 0 \\ \dfrac{x^2}{4} & 0 \le x \le 2 \\ 1 & x \ge 2 \end{cases}$$

Next, in step 2, we generate a random number r. Finally, in step 3, we set $F(x) = r$ and solve for x.

$$\frac{x^2}{4} = r$$

$$x = \pm 2\sqrt{r}$$

Since the service times are defined only for positive values of x, a service time of $x = -2\sqrt{r}$ is not feasible. This leaves us with $x = 2\sqrt{r}$ as the solution for x. This equation is called a **random variate generator** or a **process generator.** Thus, to obtain a service time, we first generate a random number and then transform it using the preceding equation. Each execution of the equation will give us one service time from the given distribution. For instance, if a random number $r = 0.64$ is obtained, a service time of $x = 2\sqrt{0.64} = 1.6$ will be generated.

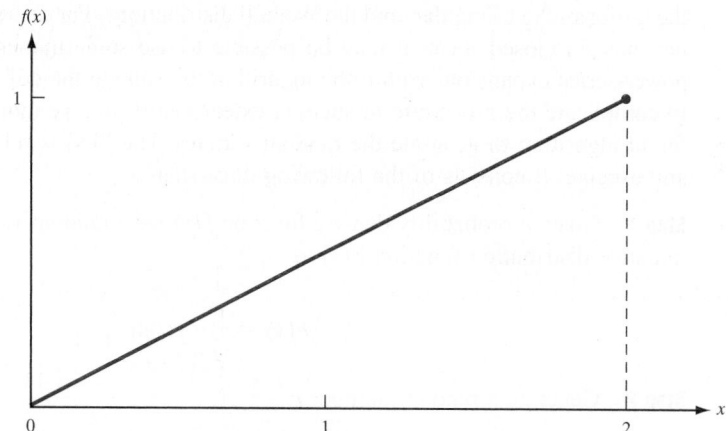

FIGURE 7
The pdf of a Ramp Function

Graphically, the inverse transformation method can be represented as shown in Figure 8. We see from this graph that the range of values for the random variable (that is, $0 \le x \le 2$) coincides with the cumulative probabilities, $0 \le F(x) \le 1.0$. In other words, for any value of $F(x)$ over the interval $[0, 1]$, there exists a corresponding value of the random variable, given by x. Since a random number is also defined in the range between 0 and 1, this implies that a random number can be translated directly into a corresponding value of x using the relation $r = F(x)$. The solution for x in terms of r is known as taking the inverse of $F(x)$, denoted by $x = F^{-1}(r)$—hence the name *inverse transformation*. Note that if r is equal to 0, we will generate a random variate equal to 0, the smallest possible value of x. Similarly, if we generate a random number equal to 1, it will be transformed to 2, the largest possible value of x.

To show that the ITM generates numbers with the same distribution as x, consider the fact that for any two numbers x_1 and x_2, the probability $P(x_1 \le \mathbf{X} \le x_2) = F(x_2) - F(x_1)$. Then what we have to show is that the probability that the generated value of $\mathbf{X}$ lies between x_1 and x_2 is also the same. From Figure 8, we see that the generated value of $\mathbf{X}$ will be between x_1 and x_2 if and only if the chosen random number is between $r_1 = F(x_1)$ and $r_2 = F(x_2)$. Thus, the probability that the generated value of $\mathbf{X}$ is between x_1 and x_2 is also $F(x_2) - F(x_1)$. This shows that the ITM does indeed generate numbers with the same distribution as $\mathbf{X}$.

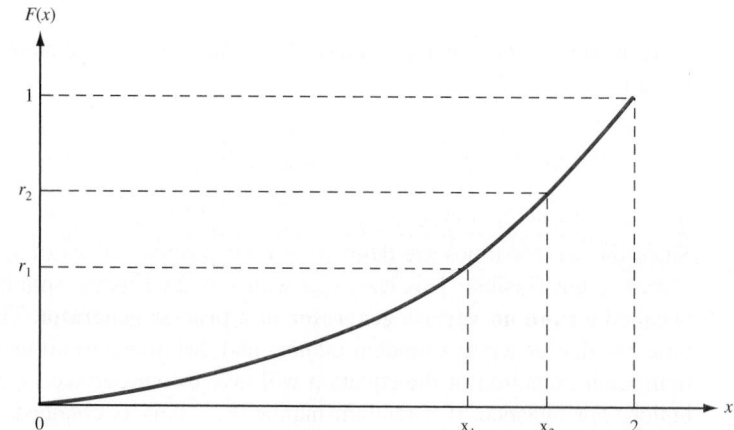

FIGURE 8
Graphical Representation of Inverse Transformation Method

As this example shows, the major advantage of the inverse transformation method is its simplicity and ease of application. However, as mentioned earlier, we must be able to determine $F(x)$ in closed form for the desired distribution before we can use the method efficiently. Also, in this example, we see that we need exactly one random number to produce one random variable. Other methods, such as the acceptance–rejection method, may require several random numbers to generate a single value of **X**. The following three examples illustrate the application of the ITM.

EXAMPLE 2   **The Exponential Distribution**

As mentioned in Chapter 20, the exponential distribution has important applications in the mathematical representation of queuing systems. The pdf of the exponential distribution is given by

$$f(x) = \begin{cases} \lambda e^{-\lambda x} & x \geq 0, \lambda > 0 \\ 0 & \text{otherwise} \end{cases}$$

Use the inverse transformation method to generate observations from an exponential distribution.

Solution   In step 1, we compute the cdf. This is given by

$$F(x) = \begin{cases} 0 & x < 0 \\ 1 - e^{-\lambda x} & x \geq 0 \end{cases}$$

Next, we generate a random number r and set $F(x) = r$ to solve for x. This gives us

$$1 - e^{-\lambda x} = r$$

Rearranging to

$$e^{-\lambda x} = 1 - r$$

and taking the natural logarithm of both sides, we have

$$-\lambda x = \ln(1 - r)$$

Finally, solving for x gives the solution

$$x = -\frac{1}{\lambda} \ln(1 - r)$$

To simplify our computations, we can replace $(1 - r)$ with r. Since r is a random number, $(1 - r)$ will also be a random number. This means that we have not changed anything except the way we are writing the $U(0,1)$ random number. Thus, our process generator for the exponential distribution will now be

$$x = -\frac{1}{\lambda} \ln r$$

For instance, $r = \frac{1}{e}$ yields $x = \frac{1}{\lambda}$, and $r = 1$ yields $x = 0$.

EXAMPLE 3   **The Uniform Distribution**

Consider a random variable **X** that is uniformly distributed on the interval $[a, b]$. The pdf of this distribution is given by the function

$$f(x) = \begin{cases} \dfrac{1}{b - a} & a \leq x \leq b \\ 0 & \text{otherwise} \end{cases}$$

Use the ITM to generate observations from this random variable.

Solution The cdf of this distribution is given by

$$F(x) = \begin{cases} 0 & x < a \\ \dfrac{x - a}{b - a} & a \le x \le b \\ 1 & x > b \end{cases}$$

To use the ITM to generate observations from a uniform distribution, we first generate a random number r and then set $F(x) = r$ to solve for x. This gives

$$\frac{x - a}{b - a} = r$$

Solving for x yields

$$x = a + (b - a)r$$

as the process generator for the uniform distribution. For example, $r = \frac{1}{2}$ yields $x = \frac{a+b}{2}$, $r = 1$ yields $x = b$, $r = 0$ yields $x = a$, and so on.

EXAMPLE 4 | **The Triangular Distribution**

Consider a random variable **X** whose pdf is given by

$$f(x) = \begin{cases} \frac{1}{2}(x - 2) & 2 \le x \le 3 \\ \frac{1}{2}(2 - \frac{x}{3}) & 3 \le x \le 6 \\ 0 & \text{otherwise} \end{cases}$$

Use the ITM to generate observations from the distribution. This distribution, called a *triangular* distribution, is represented graphically in Figure 9. It has the endpoints [2, 6], and its mode is at 3. We can see that 25% of the area under the curve lies in the range of x from 2 to 3, and the other 75% lies in the range from 3 to 6. In other words, 25% of the values of the random variable **X** lie between 2 and 3, and the other 75% fall between 3 and 6. The triangular distribution has important applications in simulation. It is often used to represent activities for which there are few or no data. (For a detailed account of this distribution, see Banks and Carson (1984) or Law and Kelton (1991).)

Solution The cdf of this triangular distribution is given by the function

$$F(x) = \begin{cases} 0 & x < 2 \\ \frac{1}{4}(x - 2)^2 & 2 \le x \le 3 \\ -\frac{1}{12}(x^2 - 12x + 24) & 3 \le x \le 6 \\ 1 & \text{otherwise} \end{cases}$$

For simplicity, we redefine $F(x) = (\frac{1}{4})(x - 2)^2$, for $2 \le x \le 3$, as $F_1(x)$, and $F(x) = (-\frac{1}{12})(x^2 - 12x + 24)$, for $3 \le x \le 6$, as $F_2(x)$.

This cdf can be represented graphically as shown in Figure 10. Note that at $x = 3$, $F(3) = 0.25$. This implies that the function $F_1(x)$ covers the first 25% of the range of the cdf, and $F_2(x)$ applies over the remaining 75% of the range. Since we now have two separate functions representing the cdf, the ITM has to be modified to account for these two functions, their ranges, and the distribution of the ranges. As far as the ITM goes, the distribution of the ranges is the most important. This distribution is achieved by using the random number from step 2. In other words, if $r < 0.25$, we use the function $F_1(x) = (\frac{1}{4})(x - 2)^2$ in step 3. Otherwise, we use $F_2(x) = (-\frac{1}{12})(x^2 - 12x + 24)$. Since $r < 0.25$ for 25% of the time and $r \ge 0.25$ for the other 75%, we achieve the desired distribution. In

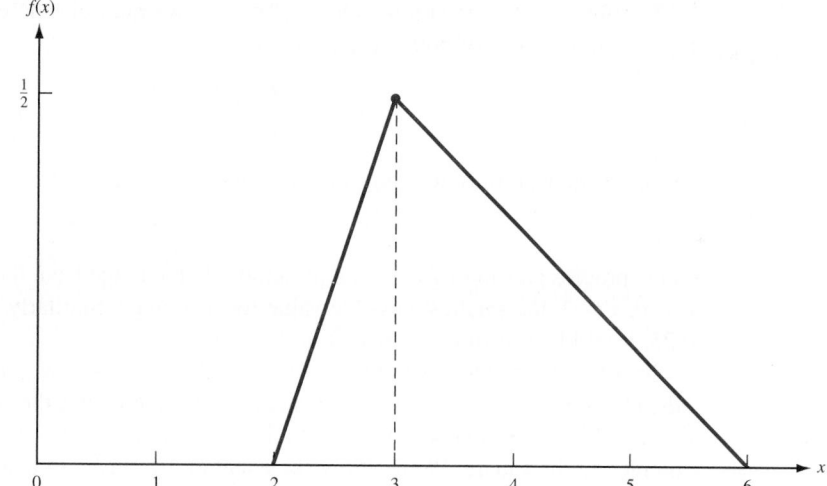

FIGURE 9
Density Function for a
Triangular Distribution

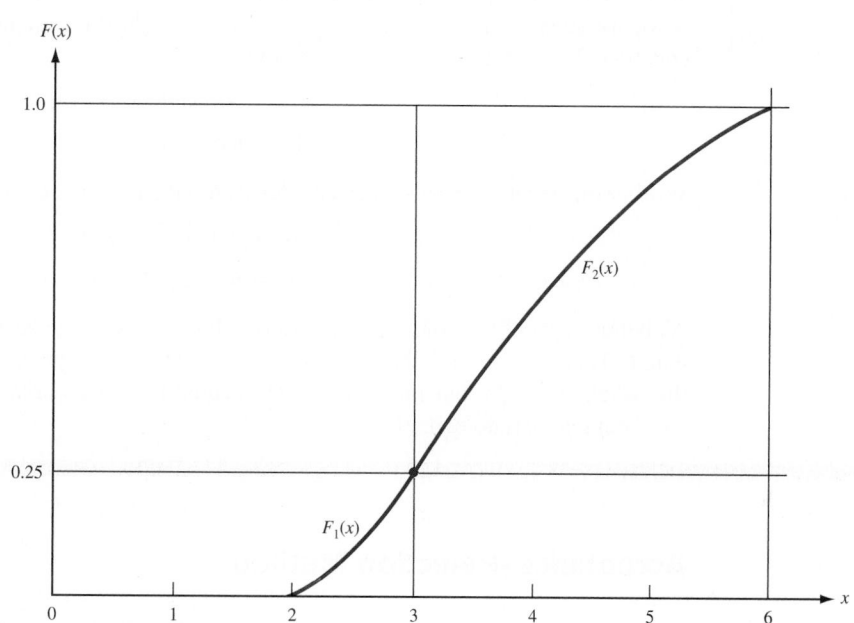

FIGURE 10
The cdf of a Triangular
Distribution

either case, we set the function $F_1(x)$ or $F_2(x)$ equal to r and solve for x. That is, we solve one of the following equations:

$$(\tfrac{1}{4})(x - 2)^2 = r \qquad \text{for } 0 \le r < 0.25$$
$$(-\tfrac{1}{12})(x^2 - 12x + 24) = r \qquad \text{for } 0.25 \le r \le 1.0$$

x will then be our random variable of interest.

As the graph in Figure 10 shows, a random number between 0 and 0.25 will be transformed into a value of x between 2 and 3. Similarly, if $r \ge 0.25$, it will be transformed into a value of x between 3 and 6.

To solve the first equation, $(\frac{1}{4})(x - 2)^2 = r$, we multiply the equation by 4 and then take the square root of both sides. This gives us

$$x - 2 = \pm\sqrt{4r}$$
$$x = 2 \pm 2\sqrt{r}$$

Since x is defined only for values greater than 2, $x = 2 - 2\sqrt{r}$ is infeasible, leaving

$$x = 2 + 2\sqrt{r}$$

as the process generator for a random number in the range from 0 to 0.25. Note that when $r = 0$, $x = 2$, the smallest possible value for this range. Similarly, when we generate $r = 0.25$, it will be transformed to $x = 3$.

To solve the second equation, $(-\frac{1}{12})(x^2 - 12x + 24) = r$, we can use one of two methods: (1) employing the quadratic formula or (2) completing the square. (See Banks and Carson (1984) for details of the quadratic formula method.) Here, we use the method of completing the square. We multiply the equation by -12 and rearrange the terms to get

$$x^2 - 12x = -24 - 12r$$

To complete the square, we first divide the x term's coefficient by 2. This gives us -6. Next, we square this value to get 36. Finally, we add this resultant to both sides of the equation. This leaves us with the equation

$$x^2 - 12x + 36 = 12 - 12r$$
$$(x - 6)^2 = 12 - 12r$$

Writing the equation in this form enables us to take the square root of both sides. That is,

$$x - 6 = \pm\sqrt{12 - 12r}$$
$$x = 6 \pm 2\sqrt{3 - 3r}$$

As before, part of the solution is infeasible. In this case, x is feasible only for values less than 6. Thus, we use only the equation $x = 6 - 2\sqrt{3 - 3r}$ as our process generator. Note that when $r = 0.25$, our random variate is equal to 3. Similarly, when $r = 1$, we generate a random variate equal to 6.

Acceptance–Rejection Method

There are several important distributions, including the Erlang (used in queuing models) and the beta (used in PERT), whose cumulative distribution functions do not exist in closed form. For these distributions, we must resort to other methods of generating random variates, one of which is the acceptance–rejection method (ARM). This method is generally used for distributions whose domains are defined over finite intervals. Thus, given a distribution whose pdf, $f(x)$, is defined over the interval $a \leq x \leq b$, the algorithm consists of the following steps.

Step 1   Select a constant M such that M is the largest value of $f(x)$ over the interval $[a, b]$.

Step 2   Generate two random numbers, r_1 and r_2.

Step 3   Compute $x^* = a + (b - a)r_1$. (This ensures that each member of $[a, b]$ has an equal chance to be chosen as x^*.)

Step 4   Evaluate the function $f(x)$ at the point x^*. Let this be $f(x^*)$.

Step 5   If

$$r_2 \leq \frac{f(x^*)}{M}$$

deliver x^* as a random variate from the distribution whose pdf is $f(x)$. Otherwise, reject x^* and go back to step 2.

Note that the algorithm continues looping back to step 2 until a random variate is accepted. This may take several iterations. For this reason, the algorithm can be relatively inefficient. The efficiency, however, is highly dependent on the shape of the distribution. There are several ways by which the method can be made more efficient. One of these is to use a function in step 1 instead of a constant. See Fishman (1978) or Law and Kelton (1991) for details of the algorithm.

We now illustrate the details of the algorithm using a ramp function. Consider a random variable **X** whose pdf is given by the function

$$f(x) = \begin{cases} 2x & 0 \leq x \leq 1 \\ 0 & \text{otherwise} \end{cases}$$

In step 1 of the ARM, it is generally useful to graph the pdf. Since our objective is to obtain the largest value of $f(x)$ over the domain of the function, graphing will enable us to determine the value of M simply by inspection. The graph of the pdf is shown in Figure 11. We see that the largest value of $f(x)$ occurs at $x = 1$ and is equal to 2. In other words, we set $M = 2$ in step 1. Next, we generate two random numbers, r_1 and r_2. In step 3, we transform the first random number, r_1, into an **X** value, x^*, using the relationship $x^* = a + (b - a)r_1$. This step is simply a procedure for randomly generating a value of the random variable **X**. Given that we are using a random number r_1 to determine x^*, every value over the interval $[a, b]$ has an equal probability of showing up. Note that if $r_1 = 0$, x^* will be equal to a, the left endpoint of the domain. Similarly, if $r_1 = 1$, x^* will be equal to b, the right endpoint of the domain. Since $a = 0$ and $b = 1$ for this distribution, it follows that $x^* = r_1$. This value of x^* now becomes our potential random variate for the current iteration. In steps 4 and 5, we have to determine whether to accept or reject x^*. We first evaluate the function $f(x)$ at $x = x^*$ to obtain $f(x^*)$ and then compute $\frac{f(x^*)}{M}$. If $r_2 \leq \frac{f(x^*)}{M}$, we accept x^*. Otherwise, we reject x^*. Substituting $x^* = r_1$ in $f(x)$ gives us $f(x^*) = 2r_1$. Since $M = 2$, the term $\frac{f(x^*)}{M}$ reduces to r_1. Given this, our decision rule (in step 5) for accepting x^* simplifies to a comparison of r_2 and r_1. If $r_2 \leq r_1$, we accept x^* as our random variate. Otherwise, we go back to step 2 and repeat the process until we obtain an acceptance.

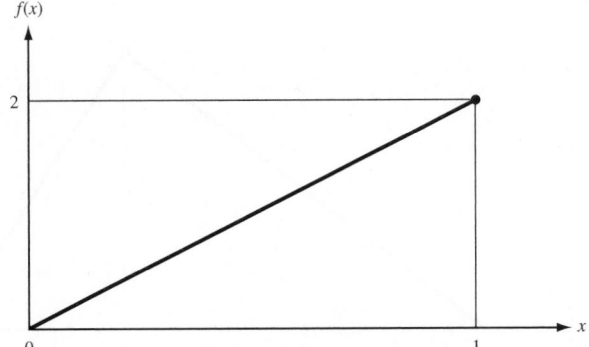

FIGURE 11
The pdf of a Ramp Function

For example, if $r_1 = 0.7$ and $r_2 = 0.6$, we choose $x = 0.7$, and if $r_1 = 0.7$ and $r_2 = 0.8$, no value of the random variable is generated. For this problem, exactly half the random variates generated in step 3 will be rejected in step 5.

EXAMPLE 5   The Acceptance–Rejection Method

Use the acceptance–rejection method to generate random variates from a triangular distribution whose pdf is given by

$$f(x) = \begin{cases} -\frac{1}{6} + \frac{x}{12} & 2 \leq x \leq 6 \\ \frac{4}{3} - \frac{x}{6} & 6 \leq x \leq 8 \end{cases}$$

Solution   For simplicity, we redefine $f(x) = -\frac{1}{6} + \frac{x}{12}$ as $f_1(x)$, and $f(x) = \frac{4}{3} - \frac{x}{6}$ as $f_2(x)$. This distribution is represented graphically in Figure 12.

Since this distribution is defined over two intervals, we must modify steps 4 and 5 of the acceptance–rejection method to account for these ranges. The first three steps of the algorithm, however, stay the same as before. That is, step 1 determines M, step 2 generates r_1 and r_2, and step 3 transforms r_1 into a value x^* of $\mathbf{X}$.

From the graph of the pdf in Figure 12, it is clear that $M = \frac{1}{3}$. This distribution has the endpoints [2, 8], which implies that $a = 2$ and $b = 8$. If we now substitute these endpoints in step 3, the x^* values are generated by the equation $x^* = 2 + 6r_1$. Then we see that if r_1 is between 0 and $\frac{2}{3}$, x^* will lie in the range from 2 to 6. If $r_1 > \frac{2}{3}$, x^* will lie in the interval [6, 8]. To account for this, we make our first modification in step 4. If x^* lies between 2 and 6, then in step 4, we use the function $f_1(x)$ to evaluate $f(x^*)$. Otherwise, we use $f_2(x)$ to compute $f(x^*)$. Step 4 now can be summarized as follows: If $2 \leq x^* \leq 6$,

$$f(x^*) = f_1(x^*)$$
$$= -\frac{1}{6} + \frac{x^*}{12}$$
$$= \frac{r_1}{2}$$

If $6 \leq x^* \leq 8$,

$$f(x^*) = f_2(x^*)$$
$$= \frac{4}{3} - \frac{x^*}{6}$$
$$= 1 - r_1$$

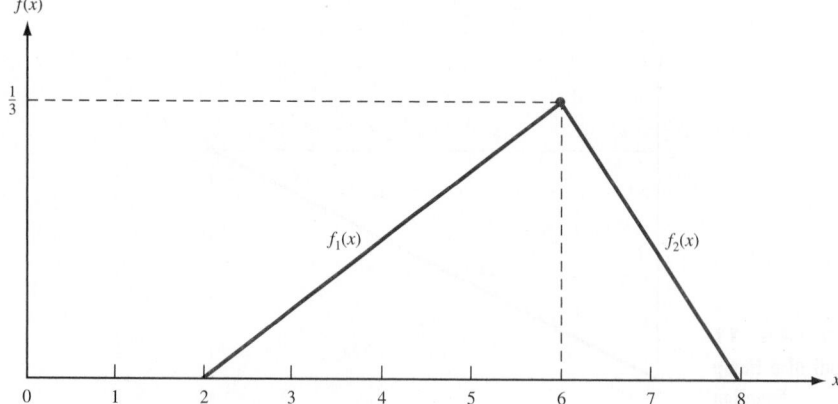

FIGURE 12
The pdf of
Triangular Distribution
of Example 5

The next step in the algorithm is either to accept or to reject the current value of x^*. We accept x^* if the condition $r_2 \leq \frac{f(x^*)}{M}$ is satisfied. However, following step 4, we need to evaluate this condition over the two intervals by substituting the appropriate function, $f(x^*)$, into the relation. In other words, step 5 for this distribution will now be as follows: For $2 \leq x^* \leq 6$, we accept x^* if $r_2 \leq \frac{f_1(x^*)}{M}$—that is, if $r_2 \leq \frac{3r_1}{2}$. For $6 \leq x^* \leq 8$, we accept x^* if $r_2 \leq \frac{f_2(x^*)}{M}$—that is, if $r_2 \leq 3(1 - r_1)$. If x^* is rejected, we go back to step 2 and repeat the process.

As before, some of the x^* values will be rejected. In this case, also, the probability of accepting a random variate is .5. That is, one half of all random variates generated in step 3 will, in the long run, be rejected in step 5.

We now give an intuitive justification of the validity of the ARM. In particular, we want to show that the ARM does generate observations from the given random variable $\mathbf{X}$. For any number x the ARM should yield $P(x \leq \mathbf{X} \leq x + \Delta) = f(x)\Delta$. Now the probability that the ARM generates an observation between x and $x + \Delta$ is given by

$$\sum_{i=1}^{i=\infty} \text{(probability first } i - 1 \text{ iterations yield no value and}$$
$$i\text{th iteration yields a value between } x \text{ and } x + \Delta)$$

$$= \sum_{i=1}^{i=\infty} \left(1 - \frac{1}{M(b - a)}\right)^{i-1} \frac{f(x)\Delta}{M(b - a)}$$

$$= \frac{f(x)\Delta}{M(b - a)} \left(\frac{1}{1 - (1 - 1/(M(b - a)))}\right) = f(x)\Delta$$

where we have used the fact that on any ARM iteration, there is a probability $1/M$ $(b - a)$ that a value of the random variable will be generated (see Problem 6), and that for $c < 1$,

$$\sum_{i=1}^{i=\infty} c^{i-1} = \frac{1}{1 - c}$$

Direct and Convolution Methods for the Normal Distribution

Because of the importance of the normal distribution, considerable attention has been paid to generating normal random variates. This has resulted in many different algorithms for the normal distribution. Both the inverse transformation method and the acceptance–rejection method are inappropriate for the normal distribution, because (1) the cdf does not exist in closed form and (2) the distribution is not defined over a finite interval. Although it is possible to use numerical methods within the ITM and to truncate the distribution for the acceptance–rejection method, other methods tend to be much more efficient. In this section, we describe two such methods—first, an algorithm based on convolution techniques, and then a direct transformation algorithm that produces two standard normal variates with mean 0 and variance 1.

The Convolution Algorithm

In the convolution algorithm, we make direct use of the Central Limit Theorem. The Central Limit Theorem states that the sum $\mathbf{Y}$ of n independent and identically distributed random variables (say, $\mathbf{Y}_1, \mathbf{Y}_2, \ldots, \mathbf{Y}_n$, each with mean μ and finite variance σ^2) is approximately normally distributed with mean $n\mu$ and variance $n\sigma^2$. If we now apply this to $U(0, 1)$ random variables, $\mathbf{R}_1, \mathbf{R}_2, \ldots, \mathbf{R}_n$, with mean $\mu = 0.5$ and $\sigma^2 = \frac{1}{12}$, it follows that

$$\mathbf{Z} = \frac{\sum_{i=1}^{n} \mathbf{R}_i - 0.5n}{\left(\frac{n}{12}\right)^{1/2}}$$

is approximately normal with mean 0 and variance 1. We would expect this approximation to work better as the value of n increases. However, most simulation literature suggests using a value of $n = 12$. Using 12 not only seems adequate but, more important, has the advantage that it simplifies the computational procedure. If we now substitute $n = 12$ into the preceding equation, the process generator simplifies to

$$\mathbf{Z} = \sum_{i=1}^{12} \mathbf{R}_i - 6$$

This equation avoids a square root and a division, both of which are relatively time-consuming routines on a computer.

If we want to generate a normal variate $\mathbf{X}$ with mean μ and variance σ^2, we first generate $\mathbf{Z}$ using this process generator and then transform it using the relation $\mathbf{X} = \mu + \sigma\mathbf{Z}$. Note that this convolution is unique to the normal distribution and cannot be extended to other distributions. Several other distributions do, of course, lend themselves to convolution methods. For example, we can generate random variates from an Erlang distribution with shape parameter k and rate parameter $k\lambda$, using the fact that an Erlang random variable can be obtained by the sum of k iid exponential random variables, each with parameter $k\lambda$.

The Direct Method

The direct method for the normal distribution was developed by Box and Muller (1958). Although it is not as efficient as some of the newer techniques, it is easy to apply and execute. The algorithm generates two $U(0, 1)$ random numbers, r_1 and r_2, and then transforms them into two normal random variates, each with mean 0 and variance 1, using the direct transformations

$$\mathbf{Z}_1 = (-2 \ln r_1)^{1/2} \sin 2\pi r_2$$
$$\mathbf{Z}_2 = (-2 \ln r_1)^{1/2} \cos 2\pi r_2$$

As in the convolution method, it is easy to transform these standardized normal variates into normal variates $\mathbf{X}_1$ and $\mathbf{X}_2$ from the distribution with mean μ and variance σ^2, using the equations

$$\mathbf{X}_1 = \mu + \sigma\mathbf{Z}_1$$
$$\mathbf{X}_2 = \mu + \sigma\mathbf{Z}_2$$

The direct method produces exact normal random variates, whereas the convolution method gives us only approximate normal random variates. For this reason, the direct method is much more commonly used. For details of these and other normal algorithms, see Fishman (1978) or Law and Kelton (1991).

PROBLEMS

Group A

1 Consider a continuous random variable with the following pdf:

$$f(x) = \begin{cases} \frac{1}{2} & 0 \le x \le 1 \\ \frac{3}{4} - \frac{x}{4} & 1 \le x \le 3 \end{cases}$$

Develop a process generator for these breakdown times using the inverse transformation method and the acceptance–rejection method.

2 A job shop manager wants to develop a simulation model to help schedule jobs through the shop. He has evaluated the completion times for all the different types of jobs. For one particular job, the times to completion can be represented by the following triangular distribution:

$$f(x) = \begin{cases} \frac{x}{8} - \frac{1}{4} & 2 \le x \le 4 \\ \frac{10}{24} - \frac{x}{24} & 4 \le x \le 10 \end{cases}$$

Develop a process generator for this distribution using the inverse transformation method.

3 Given the continuous triangular distribution in Figure 13, develop a process generator using the inverse transformation method.

Group B

4 For Problem 2, develop a computer program for the process generator. Generate 100 random variates and compare the mean and variance of this sample against the theoretical mean and variance of this distribution. Now repeat the experiment for the following numbers of random variates: 250; 500; 1,000; and 5,000. From these experiments, what can be said about the process generator?

5 A machine operator processes two types of jobs, A and B, during the course of the day. Analyses of past data show that 40% of all jobs are type A jobs, and 60% are type B jobs. Type A jobs have completion times that can be represented by an Erlang distribution with rate parameter 5 and shape parameter 2. Completion times of type B jobs can be represented by the following triangular distribution:

$$f(x) = \begin{cases} \dfrac{x - 1}{12} & 1 \le x \le 4 \\ \dfrac{9 - x}{20} & 4 \le x \le 9 \end{cases}$$

If jobs arrive at an exponential rate of 10 per hour, develop a simulation model to calculate the percentage of idle time of the operator and the average number of jobs in the line waiting to be processed.

6 Show that on any iteration of the acceptance–rejection method, there is a probability $\frac{1}{M(b-a)}$ that a value of the random variable is generated.

7 We all hate to bring small change to the store. Using random numbers, we can eliminate the need for change and give the store and the customer a fair shake.

    a Suppose you buy something that costs $.20. How could you use random numbers (built into the cash register system) to decide whether you should pay $1.00 or nothing? This eliminates the need for change!

    b If you bought something for $9.60, how would you use random numbers to eliminate the need for change?

    c In the long run, why is this method fair to both the store and the customer?

FIGURE 13

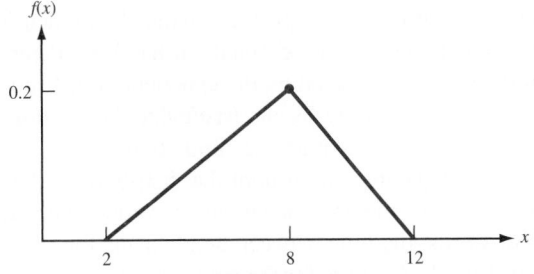

21.6 An Example of a Stochastic Simulation

We now present an example of a simulation using some of the concepts covered in Section 21.5. We consider the case of Cabot, Inc., a large mail order firm in Chicago. Orders arrive into the warehouse via telephones. At present, Cabot maintains 10 operators at work 24 hours a day. The operators take the orders and feed them directly into a central computer, using terminals. Each operator has one terminal. At present, the company has a total of 11 terminals. That is, if all terminals are working, there will be 1 spare terminal.

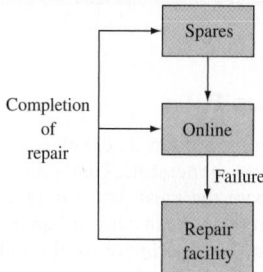

FIGURE 14
Flow of Terminals

A terminal that is online may break down. If that happens, the terminal is removed from the workstation and is replaced with a spare, if one is available. If none is available, the operator must wait until one becomes available. During this time, this operator does not take any orders. The broken terminal is sent to the workshop, where the company has one repair channel allocated to repairing terminals. At the completion of a repair, the terminal either acts as a spare or goes directly into service if an operator is waiting for a terminal. The flow of terminals in the system is shown in Figure 14.

The Cabot managers believe that the terminal system needs evaluation, because the downtime of operators due to broken terminals has been excessive. They feel that the problem can be solved by the purchase of some additional terminals for the spares pool. Accountants have determined that a new terminal will cost a total of $75 per week in such costs as investment cost, capital cost, maintenance, and insurance. It has also been estimated that the cost of terminal downtime, in terms of delays, lost orders, and so on, is $1,000 per week. Given this information, the Cabot managers would like to determine how many additional terminals they should purchase.

This model is a version of the machine repair problem (see Section 20.9). In such models (if both the breakdown and the repair times can be represented by the exponential distribution), it is easy to find an analytical solution to the problem using birth–death processes. However, in analyzing the historical data for the terminals, it has been determined that although the breakdown times can be represented by the exponential distribution, the repair times can be adequately represented only by the triangular distribution. This implies that analytical methods cannot be used; we must use simulation.

To simulate this system, we first require the parameters of both the distributions. For the breakdown time distribution, the data show that the breakdown rate is exponential and equal to 1 per week per terminal. In other words, the time between breakdowns for a terminal is exponential with a mean equal to 1 week. Analysis for the repair times (measured in weeks) shows that this distribution can be represented by the triangular distribution

$$f(x) = \begin{cases} -10 + 400x & 0.025 \leq x \leq 0.075 \\ 50 - 400x & 0.075 \leq x \leq 0.125 \end{cases}$$

which has a mean of 0.075 week. That is, the repair staff can, on the average, repair 13.33 terminals per week. We represent the repair time distribution graphically in Figure 15.

To find the optimum number of terminals for the system, we must balance the cost of the additional terminals against the increased revenues (because of reduced downtime costs) generated as a result of the increase in the number of terminals. In the simulation, we increase the number of terminals in the system, n, from the present total of 11 in increments of 1. For this fixed value of n, we then run our simulation model to estimate the net revenue. Net revenue here is defined as the difference between the increase in revenues due to the additional terminals and the cost of these additional terminals. We keep

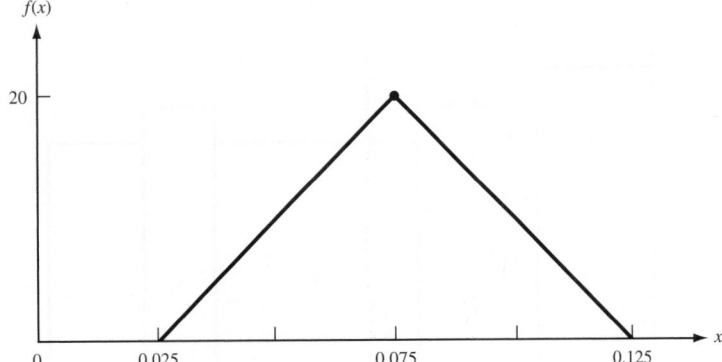

FIGURE 15
The pdf of the Repair Times

on adding terminals until the net revenue position reaches a peak. Thus, our primary objective in the simulation is to determine the net revenue for a fixed number of terminals.

To calculate the net revenue, we first compute the average number of online terminals, EL_n (or equivalently, the average number of downtime terminals, ED_n), for a fixed number of terminals in the system, n. In fact, ED_n is simply equal to $n - EL_n$. Once we have a value for EL_n, we can compute the expected weekly downtime costs, given by $1,000(10 - EL_n)$. Then the increase in revenues as a result of increasing the number of terminals from 11 to n is $1,000(EL_n - EL_{11})$.

Mathematically, we compute EL_n as

$$EL_n = \frac{\int_0^T N(t)dt}{T} = \frac{\sum_{i=1}^m A_i}{T}$$

where

T = length of the simulation

$N(t)$ = number of terminals online at time t $\qquad (0 \le t \le T)$

A_i = area of rectangle under $N(t)$ between e_{i-1} and e_i
(where e_i is the time of the ith event)

m = number of events that occur in the interval $[0, T]$

This computation is illustrated in Figure 16 for $n = 10$. In this example, we start with 10 terminals online at time 0. Between time 0 and time e_1, the time of the first event, the total online time for all the terminals is given by $10e_1$, since each terminal is online for a period of e_1 time units. Similarly, the total online time between events 1 and 2 is 9 $(e_2 - e_1)$, given that the breakdown at time e_1 leaves us with only 9 working terminals between time e_1 and time e_2. If we now run this simulation over T time units and sum up the areas $A_1, A_2, A_3, \ldots$, we can get an estimate for EL_{10} by dividing this sum by T. This statistic is called a **time-average statistic.** As long as the simulation is run for a sufficiently long period of time, our estimate for EL_{10} should be fairly close to the actual.

In this simulation, we would like to set up the process in such a way that it will be possible to collect the statistics to compute the areas $A_1, A_2, A_3, \ldots$. That is, as we move from event to event, we would like to keep track of at least the number of terminals online between the events and the time between events. To do this, we first define the state of the system as the number of terminals in the repair facility. From this definition, it follows that the only time the state of the system will change is when there is either a breakdown or a completion of a repair. This implies that there are two events in this simulation: breakdown and completion of repairs.

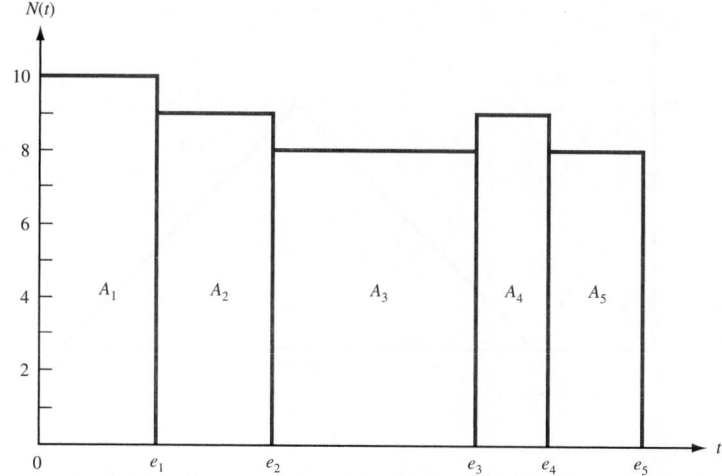

FIGURE **16**
Computation of EL_{10}

To set up the simulation, our first task is to determine the process generators for both the breakdown and the repair times. Since both these distributions have cdf's in closed form, we use the ITM to develop the process generators. For the exponential distribution, the process generator is simply

$$x = -\log r$$

In the case of the repair times, applying the ITM gives us

$$x = 0.025 + \sqrt{0.005r} \qquad (0 \leq r \leq 0.5)$$

and

$$x = 0.125 - \sqrt{0.005(1 - r)} \qquad (0.5 \leq r \leq 1.0)$$

as the process generators.

Within this experiment, we run several different simulations, one for each different value of n. Since n at present equals 11, we begin the experiment with this number and increase n until the net revenues reach a peak. For each n, we start the simulation in the state where there are no terminals in the repair facility. In this state, all 10 operators are online and any remaining terminals are in the spares pool.

Our first action in the simulation is to schedule the first series of events, the breakdown times for the terminals presently online. We do this in the usual way, by generating an exponential random variate for each online terminal from the breakdown distribution and setting the time of breakdown by adding this generated time to the current clock time, which is zero. Having scheduled these events, we next determine the first event, the first breakdown, by searching through the current event list. We then move the simulation clock to the time of this event and process this breakdown.

To process a breakdown, we take two separate series of actions: (1) Determine whether a spare is available. If one is available, bring the spare into service and schedule the breakdown time for this terminal. If none is available, update the back-order position. (2) Determine whether the repair staff is idle. If so, start the repair on the broken terminal by generating a random variate from the service times distribution and scheduling the completion time of the repair. If the repair staff is busy, place the broken terminal in the repair queue. Having completed these two series of actions, we now update all the statistical counters. These actions are summarized in the system flow diagram in Figure 17. We

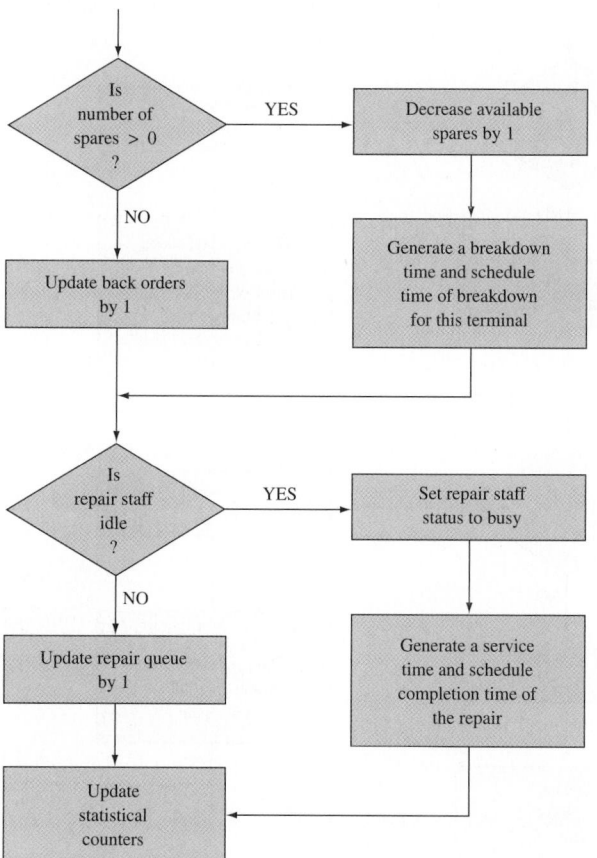

FIGURE 17
Flowchart for a
Breakdown

proceed with the simulation by determining the next event and moving the clock to the time of this event. If the next event is another breakdown, we repeat the preceding series of actions. Otherwise, we process a completion of a repair.

To process the completion of a repair, we also undertake two series of actions. (1) At the completion of a repair, we have an additional working terminal, so we determine whether the terminal goes directly to an operator or to the spares pool. If a back order exists, we bring the terminal directly into service and schedule the time of the breakdown for this terminal in the usual manner. If no operator is waiting for a terminal, the terminal goes into the spares pool. (2) We check the repair queue to see whether any terminals are waiting to be repaired. If the queue is greater than zero, we bring the first terminal from the queue into repair and schedule the time of the completion of this repair. Otherwise, we set the status of the repair staff to idle. Finally, at the completion of these actions, we update all the statistical counters. This part of the simulation is summarized in Figure 18.

We proceed with the simulation (for a given n) by moving from event to event until the termination time T. At this time, we calculate all the relevant measures of performance from the statistical counters. Our key measure is the net revenue for the current value of n. If this revenue is greater than the revenue for a system with $n - 1$ terminals, we increase the value of n by 1 and repeat the simulation with $n + 1$ terminals in the system.

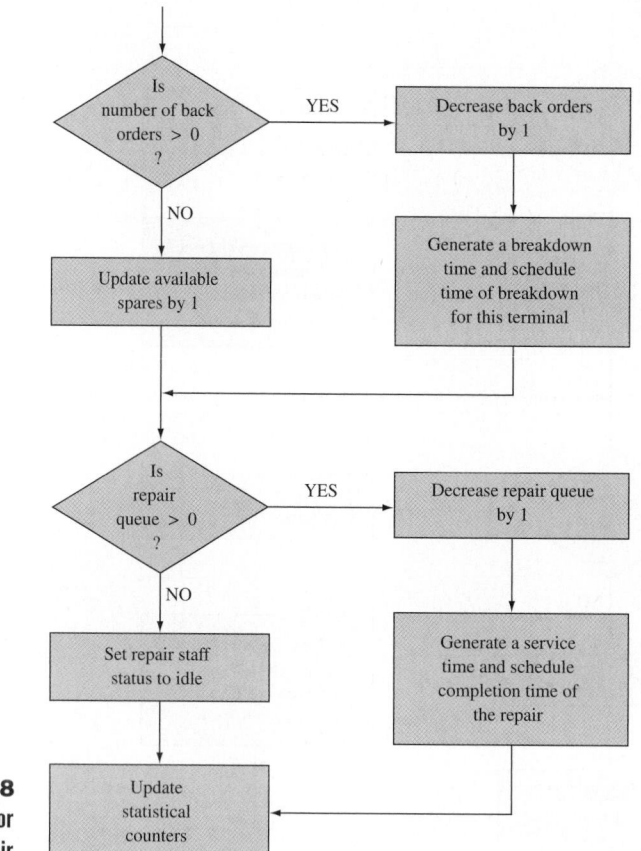

FIGURE **18**
Flowchart for
Completion of a Repair

Otherwise, the net revenue has reached a peak. If this is the case, we stop the experiment and accept $n - 1$ terminals as the optimal number of terminals to have in the system. This simulation experiment is summarized in Figure 19. For this experiment, we assume that the maximum number of terminals we can have is 25. (Note that the variable *REVO* in the flowchart is revenue for the system with $n - 1$ terminals.)

For this problem, we ran a complete experiment, whose results are summarized in Table 14. In this table, we show the overall effect on the net revenues as we increase the number of terminals from 11. For example, when we increase n from 11 to 12, the expected number of online terminals increases from 9.362 to 9.641, for a net increase of 0.279. This results in an increase of $279 in revenues per week at a cost of $75, giving us a net revenue increase of $204 per week. Similarly, if we increase the number of terminals from 11 to 13, we have a net increase of $289. The net increase peaks with 14 terminals in the system. This is further highlighted by the graph in Figure 20.

The simulation outlined in this example can be used to analyze other policy options that management may have. For example, instead of purchasing additional terminals, Cabot could hire a second repair worker or choose a preventive maintenance program for the terminals. Alternatively, the company might prefer a combination of these policies. The simulation model provides a very flexible mechanism for evaluating alternative policies.

FIGURE **19**
Flowchart for Terminal Simulation

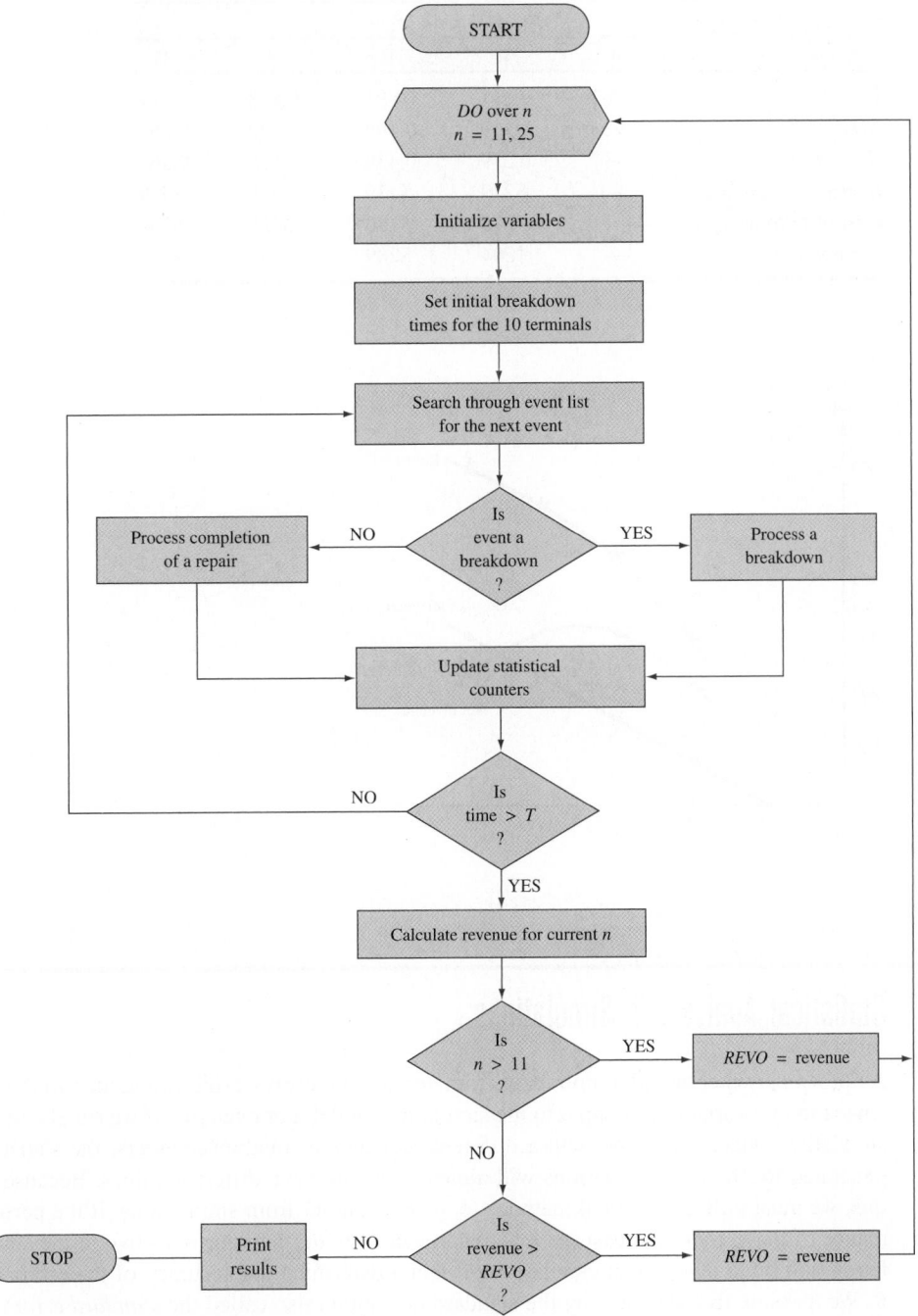

TABLE 14
Simulation Results for the Terminal System

	No. of Terminals (n)				
	11	12	13	14	15
EL_n	9.362	9.641	9.801	9.878	9.931
ED_n	0.638	0.359	0.199	0.122	0.069
$EL_n - EL_{11}$	—	0.279	0.439	0.516	0.569
Increase in revenue	—	$279	$439	$516	$569
Cost of terminals	—	$75	$150	$225	$300
Net revenue	—	$204	$289	$291	$269

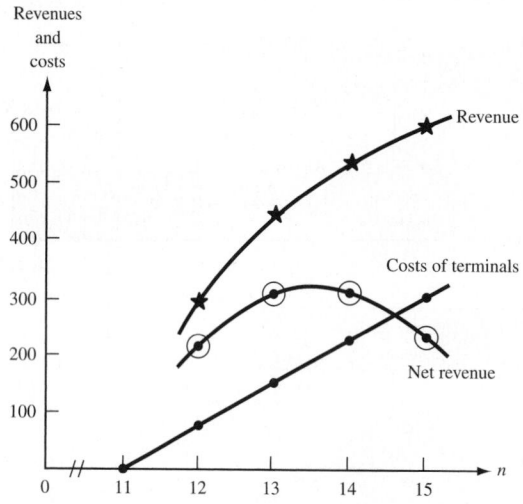

FIGURE 20
Revenue and Costs for Terminal System

21.7 Statistical Analysis in Simulations

As previously mentioned, output data from simulation always exhibit random variability, since random variables are input to the simulation model. For example, if we run the same simulation twice, each time with a different sequence of random numbers, the statistics generated for the two simulations will almost certainly have different values. Because of this, we must utilize statistical methods to analyze output from simulations. If the performance of the system is measured by a parameter (say, θ), then our objective in the simulation will be to develop an estimate $\hat{\theta}$ of θ, and determine the accuracy of the estimator $\hat{\theta}$. We measure this accuracy by the standard deviation (also called the *standard error*) of $\hat{\theta}$. The overall measure of variability is generally stated in the form of a confidence interval at a given level of confidence. Thus, the purpose of the statistical analysis is to estimate this confidence interval.

Determination of the confidence intervals in simulation is complicated by the fact that output data are rarely, if ever, independent. That is, the data are autocorrelated. For example, in a queuing simulation, the waiting time of a customer often depends on the prior customers. Similarly, in an inventory simulation, the models are usually set up such that the beginning inventory on a given day is the ending inventory from the previous day, thereby creating a correlation. This means that the classical methods of statistics, which assume independence, are not directly applicable to the analysis of simulation output data. Thus, we must modify the statistical methods to make proper inferences from simulation data.

In addition to the problem of autocorrelation, we may have a second problem, in that the specification of the initial conditions of the system at time 0 may influence the output data. For example, suppose that in the queuing simulation from Section 9.2, the arrival and the service distributions are such that the average waiting time per customer exceeds 15 minutes. In other words, the system is heavily congested. If we were to begin this simulation with no one in the system, the first few customers would have either zero or very small waiting times. These initial waiting times are highly dependent on the starting conditions and may therefore not be representative of the steady-state behavior of the system. This initial period of time before a simulation reaches steady state is called the **transient period** or **warmup period.**

There are two ways of overcoming the problems associated with the transient period. The first approach is to use a set of initial conditions that are representative of the system in steady state. However, in many simulations, it may be difficult to set such initial conditions. This is particularly true of queuing simulations. The alternative approach is to let the simulation run for a while and discard the initial part of the simulation. With this approach, we are assuming that the initial part of the simulation warms the model up to an equilibrium state. Since we do not collect any statistics during the warmup stage, we can reduce much of the initialization bias. Unfortunately, there are no easy ways to assess how much initial data to delete to reduce the initialization bias to negligible levels. Since each simulation model is different, it is up to the analyst to determine when the transient period ends. Although this is a difficult process, there are some general guidelines one can use. For these and other details of this topic, see Law and Kelton (1991).

Simulation Types

For the purpose of analyzing output data, we generally categorize simulations into one of two types: terminating simulations and steady-state simulations. A **terminating simulation** is one that runs for a duration of time T_E, where E is a specified event (or events) that stops the simulation. The event E may be a specified time, in which case the simulation runs for a fixed amount of time. Or, if it is a specified condition, the length of the simulation will be a random variable. A **steady-state simulation** is one that runs over a long period of time; that is, the length of the simulation goes to "infinity."

Often, the type of model determines which type of output analysis is appropriate for a particular simulation. For example, in the simulation of a bank, we would most likely use a terminating simulation, since the bank physically closes every evening, giving us an appropriate terminating event. When simulating a computer system, a steady-state simulation may be more appropriate, since most large computer systems do not shut down except in cases of breakdowns or maintenance. However, the system or model may not always be the best indicator of which simulation would be the most appropriate. It is quite possible to use the terminating simulation approach for systems more suited to steady-state simulations, and vice versa. In this section, we provide a detailed description of the statistical analysis associated with terminating simulations. The analysis for steady-state simulations is much more involved. For details of the latter, see Banks and Carson (1984) or Law and Kelton (1991).

Suppose we make n independent replications using a terminating simulation approach. If each of the n simulations is started with the same initial conditions and is executed using a different sequence of random numbers, then each simulation can be treated as an independent replication. For simplicity, we assume that there is only a single measure of performance, represented by the variable X. Thus, X_j is the estimator of the measure of performance from the jth replication. Then, given the conditions of the replications, the sequence $X_1, X_2, \ldots, X_n$ will be iid random variables. With these iid random variables, we can use classical statistical analysis to construct a $100(1 - \alpha)\%$ confidence interval for $\theta = E(X)$ as follows:

$$\bar{X} \pm t_{(\alpha/2, n-1)} \sqrt{\frac{S^2}{n}}$$

where

$$\bar{X} = \sum_{i=1}^{n} \frac{X_i}{n}$$

$$S^2 = \sum_{i=1}^{n} \frac{(X_i - \bar{X})^2}{n - 1}$$

and $t_{(\alpha, n-1)}$ is the number such that for a t-distribution with $n - 1$ degrees of freedom,

$$P(t_{n-1} \geq t_{(\alpha, n-1)}) = \alpha$$

(see Table 13 in Chapter 24). This probability can also be computed in Excel with the formula

$$=\text{TINV}(2*\text{alpha}, \text{degrees of freedom})$$

The overall mean $\bar{X}$ is simply the average of the X-values computed over the n samples and can be used as the best estimate of the measure of performance. The quantity S^2 is the sample variance.

To illustrate the terminating simulation approach, we use an example from the Cabot, Inc. case. For this illustration, we assume there are 11 terminals in the system, and we perform only 10 independent terminating runs of the simulation model. The terminating event, E, is a fixed time. That is, all 10 simulations are run for the same length of time. The results from these runs are shown in Table 15. The overall average for these 10 runs

TABLE 15
Sample Averages of Number of OnLine Terminals from 10 Replications

Run Number	X_j
1	9.252
2	9.273
3	9.413
4	9.198
5	9.532
6	9.355
7	9.155
8	9.558
9	9.310
10	9.269

for the expected number of terminals online turns out to be 9.331. (Compare this average with the result in Table 14 of 9.362, which was obtained using the steady-state approach.) If we now calculate the sample variance, we find $S^2 = 0.018$. Since $t_{(.025,9)} = 2.26$, we obtain

$$9.331 \pm 2.26 \sqrt{\frac{0.0180}{10}} = 9.331 \pm 0.096$$

as the 95% confidence interval for this sample.

We could have also computed $t_{(.025,9)}$ in Excel with the formula

$$=\text{TINV}(.05,9)$$

This yields 2.26, which is consistent with the table in Chapter 24.

The length of the confidence interval will, of course, depend on how good our sample results are. If this confidence interval is unacceptable, we can reduce its length by either increasing the number of terminating replications or the length of each simulation. For example, if we increase the number of runs from 10 to 20, we improve the results on two fronts. First, the overall average (9.359) approaches the result from the steady-state simulation; second, the confidence interval length decreases from 0.192 to 0.058. As we saw in this example, the terminating simulation approach offers a relatively easy method for analyzing output data. However, it must be emphasized that other methods for analyzing simulation data may be more efficient for a given problem. For a detailed treatment of this topic, see Banks and Carson (1984) or Law and Kelton (1991).

21.8 Simulation Languages

One of the most important aspects of a simulation study is the computer programming. Writing the computer code for a complex simulation model is often a difficult and arduous task. Because of this, several special-purpose computer simulation languages have been developed to simplify the programming. In this section, we describe several of the best-known and most readily available simulation languages, including GPSS, GASP IV, and SLAM.

Most simulation languages use one of two different modeling approaches or orientations: event scheduling or process interaction. As we have seen, in the event-scheduling approach, we model the system by identifying its characteristic events and writing routines to describe the state changes that take place at the time of each event. The simulation evolves over time by updating the clock to the next scheduled event and making whatever changes are necessary to the system and the statistics by executing the routines. In the process-interaction approach, we model the system as a series of activities that an entity (or a customer) must undertake as it passes through the system. For example, in a queuing simulation, the activities for an entity consist of arriving, waiting in line, getting service, and departing from the system. Thus, using the process-interaction approach, we model these activities instead of events. When programming in a general-purpose language such as FORTRAN or BASIC, we generally use the event-scheduling approach. GPSS uses the process-interaction approach. SLAM allows the modeler to use either approach or even a mixture of the two, whichever is the most appropriate for the model being analyzed.

Of the general-purpose languages, FORTRAN is the most commonly used in simulation. In fact, several simulation languages, including GASP IV and SLAM, use a FORTRAN base. Generally, simulation programs in FORTRAN are written as a series of subroutines, one for each major function of the simulation process. This is particularly true of the FORTRAN-based simulation languages. For example, in GASP IV, there are approximately 30 FORTRAN subroutines and functions. These include a time-advance routine, random variate generation routines, routines to manage the future events list, routines to collect statistics, and so on. To use GASP IV, we must provide a main program, an initialization routine, and the event routines. For the rest of the program, we use the GASP routines. Because of these prewritten routines, GASP IV provides a great deal of programming flexibility. For more details of this language, see Pritsker (1974).

GPSS, in contrast to GASP, is a highly structured special-purpose language. It was developed by IBM. GPSS does not require writing a program in the usual sense. The language is made up of about 40 standard statements or blocks. Building a GPSS model then consists of combining these sets of blocks into a flow diagram so that it represents the path an entity takes as it passes through the system. For example, for a single-server queuing system, the statements are of the form GENERATE (arrive in the system), QUEUE (join the waiting line), DEPART (leave the queue to enter service), ADVANCE (advance the clock to account for the service time), RELEASE (release the service facility at the end of service), and TERMINATE (leave the system). The simulation program is then compiled from these statements of the flow diagram. GPSS was designed for relatively easy simulation of queuing systems. However, because of its structure, it is not as flexible as GASP IV, especially for the nonqueuing type of simulations. For a more detailed description of GPSS, see Schriber (1974).

SLAM was developed by Pritsker and Pegden (1979). It allows us to develop simulation models as network models, discrete-event models, continuous models, or any combination of these. The discrete-event orientation is an extension of GASP IV. The network representation can be thought of as a pictorial representation of a system through which entities flow. In this respect, the structure of SLAM is similar to that of GPSS. Once the network model of the system has been developed, it is translated into a set of SLAM program statements for execution on the computer.

In Chapter 22, we will show how to use the powerful, user-friendly Process Model package that is included on this book's CD-ROM.

The decision of which language to use is one of the most important that a modeler or an analyst must make in performing a simulation study. The simulation languages offer several advantages. The most important of these is that the special-purpose languages provide a natural framework for simulation modeling and most of the features needed in programming a simulation model. However, this must be balanced against the fact that the general-purpose languages allow greater programming flexibility, and that languages like FORTRAN and BASIC are much more widely used and available.

21.9   The Simulation Process

In this chapter, we have considered several simulation models and presented a number of key simulation concepts. We now discuss the process for a complete simulation study and present a systematic approach of carrying out a simulation. A simulation study normally consists of several distinct stages. These are presented in Figure 21. However, not all simulation studies consist of all these stages or follow the order stated here. On the other hand, there may even be considerable overlap between some of these stages.

The initial stage of any scientific study, including a simulation project, requires an explicit *statement of the objectives* of the study. This should include the questions to be answered, the hypothesis to be tested, and the alternatives to be considered. Without a clear understanding and description of the problem, the chances of successful completion and implementation are greatly diminished. Also in this step, we address issues such as the

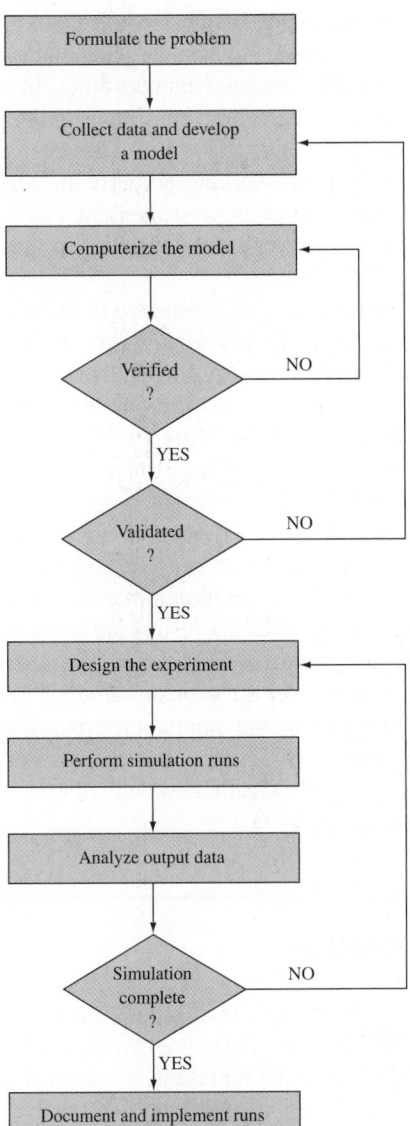

FIGURE 21
Steps in a Simulation Study

performance criteria, the model parameters, and the identification and definition of the state variables. It is, of course, very likely that the initial formulation of the problem will undergo many modifications as the study proceeds and as we learn more about the situation being studied. Nevertheless, a clear initial statement of the objectives is essential.

The next stage is the *development* of the model and the *collection of data.* The development of the model is probably the most difficult and critical part of a simulation study. Here, we try to represent the essential features of the systems under study by mathematical or logical relations. There are few firm rules to guide an analyst on how to go about this process. In many ways, this is as much an art as a science. However, most experts agree that the best approach is to start with a simple model and make it more detailed and complex as one learns more about the system.

Having developed the model, we next put it into a form in which it can be analyzed on the computer. This usually involves *developing a computer program* for the model. One of the key decisions here is the choice of the language. As noted earlier, the special-purpose languages require less programming than the general-purpose languages but are less flexible and tend to require longer computer running times. In either case, the programming part of the study is likely to be a time-consuming process, since simulation programs tend to be long and complex. Once the program has been developed and debugged, we determine whether the program is working properly. In other words, is the program doing what it is supposed to do? This process is called the *verification* step and is usually difficult, since for most simulations, we will not have any results with which to compare the computer output.

If we are satisfied with the program, we now move to the *validation* stage. This is another critical part of a simulation study. In this step, we validate the model to determine whether it realistically represents the system being analyzed and whether the results from the model will be reliable. As with the verification stage, this is generally a difficult process. Each model presents a different challenge. However, there are some general guidelines that one can follow. For more on these procedures, see Law and Kelton (1991) or Shannon (1979). If we are satisfied at this stage with the performance of the model, we can use the model to conduct the experiments to answer the questions at hand. The data generated by the simulation experiments must be collected, processed, and analyzed. The results are analyzed not only as the solution to the model but also in terms of statistical reliability and validity. Finally, after the results are processed and analyzed, a decision must be made whether to perform any additional experiments.

The primary emphasis in this chapter has been on sampling procedures and model construction. As a result, many topics of the simulation process are either not covered or treated only briefly. However, these are important issues in simulation, and the reader interested in using simulation should consult Law and Kelton (1991), Shannon (1979), Banks and Carson (1984), or Ross (1996).

SUMMARY

Introduction to Simulation

Simulation may be defined as a technique that imitates the operation of a real-world system as it evolves over a period of time. There are two types of simulation models: static and dynamic. A **static simulation model** represents a system at a particular point in time. A **dynamic simulation model** represents a system as it evolves over time. Simulations can be deterministic or stochastic. A **deterministic simulation** contains no random variables, whereas a **stochastic simulation** contains one or more random variables. Finally, simulations may be represented by either discrete or continuous models. A **discrete simulation** is one in which the state variables change only at discrete points in time. In a **continuous simulation,** the state variables change continuously over time. In this chapter, we have dealt only with discrete stochastic models. Such models are called **discrete-event simulation models.**

The Simulation Process

The simulation process consists of several distinct stages. Each study may be somewhat different, but in general, we use the following framework:

1   Formulate the problem.

2   Collect data and develop a model.

3   Computerize the model.

4   Verify the computer model.

5   Validate the simulation model.

6   Design the experiment.

7   Perform the simulation runs.

8   Document and implement.

Generating Random Variables

Random variables are represented using probability distributions. The procedure for generating random variables from given probability distributions is known as random variate generation or **Monte Carlo sampling.** The principle of sampling is based on the frequency interpretation of probability and requires a steady stream of random numbers. We generate random numbers for this procedure using congruential methods. The most commonly used of the congruential methods is the **linear congruential method.** Random numbers generated from a linear congruential generator use the following relation:

$$x_{i+1} = (ax_i + c) \text{ modulo } m \qquad (i = 0, 1, 2, \ldots)$$

This gives us the remainder from the division of $(ax_i + c)$ by m. The random numbers are delivered using the relation

$$R_i = \frac{x_i}{m} \qquad (i = 1, 2, 3, \ldots)$$

For discrete distributions, Monte Carlo sampling is achieved by allocating ranges of random numbers according to the probabilities in the distribution. For continuous distributions, we generate random variates using one of several algorithms, including the **inverse transformation method** and the **acceptance–rejection method.** The inverse transformation method requires a cdf in closed form and consists of the following steps.

Step 1   Given a probability density function $f(x)$, develop the cumulative distribution function as

$$F(x) = \int_{-\infty}^{x} f(t)dt$$

Step 2   Generate a random number r.

Step 3   Set $F(x) = r$ and solve for x. The variable x is then a random variate from the distribution whose pdf is given by $f(x)$.

The acceptance–rejection method requires the pdf to be defined over a finite interval. Thus, given a probability density function $f(x)$ over the interval $a \leq x \leq b$, we execute the acceptance–rejection algorithm as follows.

Step 1   Select a constant M such that M is the largest value of $f(x)$ over the interval $[a, b]$.

Step 2   Generate two random numbers, r_1 and r_2.

Step 3   Compute $x^* = a + (b - a)r_1$.

Step 4   Evaluate the function $f(x)$ at the point x^*. Let this be $f(x^*)$.

Step 5   If $r_2 < \frac{f(x^*)}{M}$, deliver x^* as a random variate. Otherwise, reject x^* and go back to step 2.

Between these two methods, it is possible to generate random variates from almost all of the commonly used distributions. The one exception is the normal distribution. For the normal distribution, we generate random variates directly by transforming the random numbers r_1 and r_2 into standardized normal variates, $\mathbf{Z}_1$ and $\mathbf{Z}_2$, using the relations

$$\mathbf{Z}_1 = (-2 \ln r_1)^{1/2} \sin 2\pi r_2$$
$$\mathbf{Z}_2 = (-2 \ln r_1)^{1/2} \cos 2\pi r_2$$

Types of Simulations

In discrete-event simulations, we generally simulate using the next-event time-advance approach. In this procedure, the simulation evolves over time by updating the clock to the next scheduled event and taking whatever actions are necessary for each event. The events are scheduled by generating random variates from probability distributions. Data from a simulation can be analyzed using either a terminating simulation approach or a steady-state simulation approach. In terminating simulations, we make n independent replications of the model, using the same initial conditions but running each replication with a different sequence of random numbers. If the measure of performance is represented by the variable X, this approach gives us the estimators $X_1, X_2, \ldots, X_n$ from the n replications. These estimators are used to develop a $100(1 - \alpha)\%$ confidence interval as

$$\overline{X} \pm t_{(\alpha/2, n-1)} \sqrt{\frac{S^2}{n}}$$

for a fixed value of n.

Simulation gives us the flexibility to study systems that are too complex for analytical methods. However, it must be put into proper perspective. Simulation models are time consuming and costly to construct and run. Additionally, the results may not be very precise and are often hard to validate. Simulation can be a powerful tool, but only if it is used properly.

REVIEW PROBLEMS

Group A

1 Use the linear congruential generator to obtain a sequence of 10 random numbers, given that $a = 17$, $c = 43$, $m = 100$, and $x_0 = 31$.

2 A news vendor sells newspapers and tries to maximize profits. The number of papers sold each day is a random variable. However, analysis of the past month's data shows the distribution of daily demand in Table 16. A paper costs the vendor 20¢. The vendor sells the paper for 30¢. Any unsold papers are returned to the publisher for a credit of 10¢. Any unsatisfied demand is estimated to cost 10¢ in goodwill and lost profit. If the policy is to order a quantity equal to the preceding day's demand, determine the average daily profit of the news vendor by simulating this system. Assume that the demand for day 0 is equal to 32.

3 An airport hotel has 100 rooms. On any given night, it takes up to 105 reservations, because of the possibility of no-shows. Past records indicate that the number of daily reservations is uniformly distributed over the integer range [96, 105]. That is, each integer number in this range has an

TABLE 16

Demand per Day	Probability
30	.05
31	.15
32	.22
33	.38
34	.14
35	.06

equal probability, .1, of showing up. The no-shows are represented by the distribution in Table 17. Develop a simulation model to find the following measures of performance of this booking system: the expected number of rooms used per night and the percentage of nights when more than 100 rooms are claimed.

TABLE 17	
Number of No-Shows	Probability
0	.10
1	.20
2	.25
3	.30
4	.10
5	.05

TABLE 18	
Interarrival Time (minutes)	Probability
1	.20
2	.25
3	.40
4	.10
5	.05

TABLE 21	
Length of Appointment (minutes)	Probability
24	.10
27	.20
30	.40
33	.15
36	.10
39	.05

TABLE 22	
Daily Demand (units)	Probability
12	.05
13	.15
14	.25
15	.35
16	.15
17	.05

4 The university library has one copying machine for the students to use. Students arrive at the machine with the distribution of interarrival times shown in Table 18. The time to make a copy is uniformly distributed over the range [16, 25] seconds. Analysis of past data has shown that the number of copies a student makes during a visit has the distribution in Table 19. The librarian feels that under the present system, the lines in front of the copying machine are too long and that the time a student spends in the system (waiting time + service time) is excessive. Develop a simulation model to estimate the average length of the waiting line and the expected waiting time in the system.

5 A salesperson in a large bicycle shop is paid a bonus if he sells more than 4 bicycles a day. The probability of selling more than 4 bicycles a day is only .40. If the number of bicycles sold is greater than 4, the distribution of sales is as shown in Table 20. The shop has four different models of bicycles. The amount of the bonus paid out varies by type. The bonus for model A is $10; 40% of the bicycles sold are of this type. Model B accounts for 35% of the sales and pays a bonus of $15. Model C has a bonus rating of $20 and makes up 20% of the sales. Finally, model D pays a bonus of $25 for each sale but accounts for only 5% of the sales. Develop a simulation model to calculate the bonus a salesperson can expect in a day.

6 A heart specialist schedules 16 patients each day, 1 every 30 minutes, starting at 9 A.M. Patients are expected to arrive for their appointments at the scheduled times. However, past experience shows that 10% of all patients arrive 15 minutes early, 25% arrive 5 minutes early, 50% arrive exactly on time, 10% arrive 10 minutes late, and 5% arrive 15 minutes late. The time the specialist spends with a patient varies, depending on the type of problem. Analysis of past data shows that the length of an appointment has the

distribution in Table 21. Develop a simulation model to calculate the average length of the doctor's day.

Group B

7 Suppose we are considering the selection of the reorder point, R, of a (Q, R) inventory policy. With this policy, we order up to Q when the inventory level falls to R or less. The probability distribution of daily demand is given in Table 22. The lead time is also a random variable and has the distribution in Table 23. We assume that the "order up to" quantity for each order stays the same at 100. Our interest here is to determine the value of the reorder point, R, that minimizes the total variable inventory cost. This variable cost is the sum of the expected inventory carrying cost, the expected ordering cost, and the expected stockout cost. All stockouts are backlogged. That is, a customer waits until an item is available. Inventory carrying cost is estimated to be 20¢/unit/day and is charged on the units in inventory at the end of a day. A stockout costs $1 for every unit short. The cost of ordering is $10 per order. Orders arrive at the beginning of a day. Develop a simulation model to simulate this inventory system to find the best value of R.

8 A large car dealership in Bloomington, Indiana, employs five salespeople. All salespeople work on commission; they are paid a percentage of the profits from the cars they sell. The dealership has three types of cars: luxury, midsize, and subcompact. Data from the past few years show that the car sales per week per salesperson have the distribution in Table 24. If the car sold is a subcompact, a salesperson is given a commission of $250. For a midsize car, the commission is either $400 or $500, depending on the model sold. On the

TABLE 19	
Number of Copies	Probability
6	.20
7	.25
8	.35
9	.15
10	.05

TABLE 20	
No. of Bicycles Sold	Probability
5	.35
6	.45
7	.15
8	.05

TABLE 23	
Lead Time (days)	Probability
1	.20
2	.30
3	.35
4	.15

TABLE 24	
No. of Cars Sold	Probability
0	.10
1	.15
2	.20
3	.25
4	.20
5	.10

TABLE 25			TABLE 26	
Type of Car Sold	Probability		Length of Queue (q)	Probability of Reneging
Subcompact	.40		$6 \leq q \leq 8$	.20
Midsize	.35		$9 \leq q \leq 10$	.40
Luxury	.25		$11 \leq q \leq 14$	.60
			$q > 14$	.80

midsize cars, a commission of $400 is paid out 40% of the time, and $500 is paid out the other 60% of the time. For a luxury car, commission is paid out according to three separate rates: $1,000 with a probability of 35%, $1,500 with a probability of 40%, and $2,000 with a probability of 25%. If the distribution of type of cars sold is as shown in Table 25, what is the average commission for a salesperson in a week?

9 Consider a bank with 4 tellers. Customers arrive at an exponential rate of 60 per hour. A customer goes directly into service if a teller is idle. Otherwise, the arrival joins a waiting line. There is only one waiting line for all the tellers. If an arrival finds the line too long, he or she may decide to leave immediately (reneging). The probability of a customer reneging is shown in Table 26. If a customer joins the waiting line, we assume that he or she will stay in the system until served. Each teller serves at the same service rate. Service times are uniformly distributed over the range [3, 5]. Develop a simulation model to find the following measures of performance for this system: (1) the expected time a customer spends in the system, (2) the percentage of customers who renege, and (3) the percentage of idle time for each teller.

10 Jobs arrive at a workshop, which has two work centers (A and B) in series, at an exponential rate of 5 per hour. Each job requires processing at both these work centers, first on A and then on B. Jobs waiting to be processed at each center can wait in line; the line in front of work center A has unlimited space, and the line in front of center B has space for only 4 jobs at a time. If this space reaches its capacity, jobs cannot leave center A. In other words, center A stops processing until space becomes available in front of B. The processing time for a job at center A is uniformly distributed over the range [6, 10]. The processing time for a job at center B is represented by the following triangular distribution:

$$f(x) = \begin{cases} \frac{1}{4}(x-1) & 1 \leq x \leq 3 \\ \frac{1}{4}(5-x) & 3 \leq x \leq 5 \end{cases}$$

Develop a simulation model of this system to determine the following measures of performance: (1) the expected number of jobs in the workshop at any given time, (2) the percentage of time center A is shut down because of shortage of queuing space in front of center B, and (3) the expected completion time of a job.

REFERENCES

There are several outstanding books on simulation. For a beginning book, we recommend Watson (1981); for an intermediate approach, Banks and Carson (1984); and for a more advanced treatment, Law and Kelton (1991).

Banks, J., and J. Carson. *Discrete-Event System Simulation.* Englewood Cliffs, N.J.: Prentice Hall, 1984.

Box, G., and M. Muller. "A Note on the Generation of Random Normal Deviates," *Annals of Mathematical Statistics* 29(1958):610–611.

Fishman, G. *Principles of Discrete Event Simulation.* New York: Wiley, 1978.

Fishman, G. *Monte Carlo: Concepts, Algorithms and Applications.* Berlin: Springer-Verlag, 1996.

Kelton, D., Sadowski, R., and Sadowski, S. *Simulation with Arena.* New York: McGraw-Hill, 2001.

Knuth, D.W. *The Art of Computer Programming: II. Seminumerical Algorithms.* Reading, Mass.: Addison-Wesley, 1998.

Law, A.M., and W. Kelton. *Simulation Modeling and Analysis.* New York: McGraw-Hill, 1991.

Pritsker, A. *The GASP IV Simulation Language.* New York: Wiley, 1974.

Pritsker, A., and C. Pegden. *Introduction to Simulation and SLAM.* New York: Wiley, 1979.

Ross, S. *Simulation.* San Francisco: Academic Press, 1996.

Schmidt, J.W., and R.E. Taylor. *Simulation and Analysis of Industrial Systems.* Homewood, Ill.: Irwin, 1970.

Schriber, T. *Simulation Using GPSS.* New York: Wiley, 1974.

Shannon, R. E. *Systems Simulation: The Art and Science.* Englewood Cliffs, N.J.: Prentice Hall, 1979.

Watson, H. *Computer Simulation in Business.* New York: Wiley, 1981.

Kelly, J. "A New Interpretation of Information Rate," *Bell System Technical Journal* 35(1956):917–926.

Marcus, A. "The Magellan Fund and Market Efficiency," *Journal of Portfolio Management* Fall (1990):85–88.

Morrison, D., and R. Wheat. "Pulling the Goalie Revisited," *Interfaces* 16(no. 6, 1984):28–34.

Simulation with Process Model

In Chapter 21, we learned how to build simulation models of many different situations. In this chapter, we will explain how the powerful, user-friendly simulation package Process Model can be used to simulate queuing systems.

22.1 Simulating an *M/M/*1 Queuing System

After installing Process Model from the book's CD-ROM, you can start Process Model by selecting Start Programs Process Model 4. You will see the screen shown in Figure 1, where some key icons have been labeled.

MM1.igx

It is simple to simulate an *M/M/*1 queuing system having $\lambda = 10$ arrivals/hour and $\mu = 15$ customers/hour. See file MM1.igx. Assume that these are calls for directory assistance.

Step 1 Click on one of the arrival icons (a person or a phone) and drag the icon to the blank part of the screen (called the Layout portion). We have chosen to use the phone icon. Your screen should look like Figure 2.

Step 2 Select the Process rectangle and drag it right over the arrival icon. Click on it and drag it to the right. You will now have a double-arrowed connection between the arrival icon and the Process rectangle. The double-arrowed icon indicates the arrival of entities into the system. Later we will tell Process Model that interarrival times are exponential with mean 6 minutes. After Taking Calls is typed within the Process rectangle, the Layout window looks as shown in Figure 3.

Step 3 Choose one of the server icons to represent a telephone operator (say, the person with the computer) and drag this icon to the Layout window above the Take Calls Process rectangle. Then type the word "operator" to indicate a phone operator. Next, click on the Connector Line tool in the Toolbox and place the cursor over the operator. We then click once and drag a connection down to the Take Calls activity. This indicates that the operator can take calls. The Layout window should now look as shown in Figure 4.

Step 4 Next, tell Process Model to make interarrival times exponential. Process Model works off the mean interarrival time or service time, not the arrival or service rates. Process Model supports many distributions, including the triangular, normal, and Erlang random variables. For now, we will use the exponential distribution. Since the average time between arrivals is 6 minutes, we will model the interarrival times as E(6). (E stands for exponential.) To enter the interarrival time distribution, click on the double arrow connecting Call to Take Calls and fill in the dialog box as shown in Figure 5. Entering Periodic and E(6) ensures that interarrival times will be generated over and over as independent exponential random variables with mean 6.

FIGURE 1

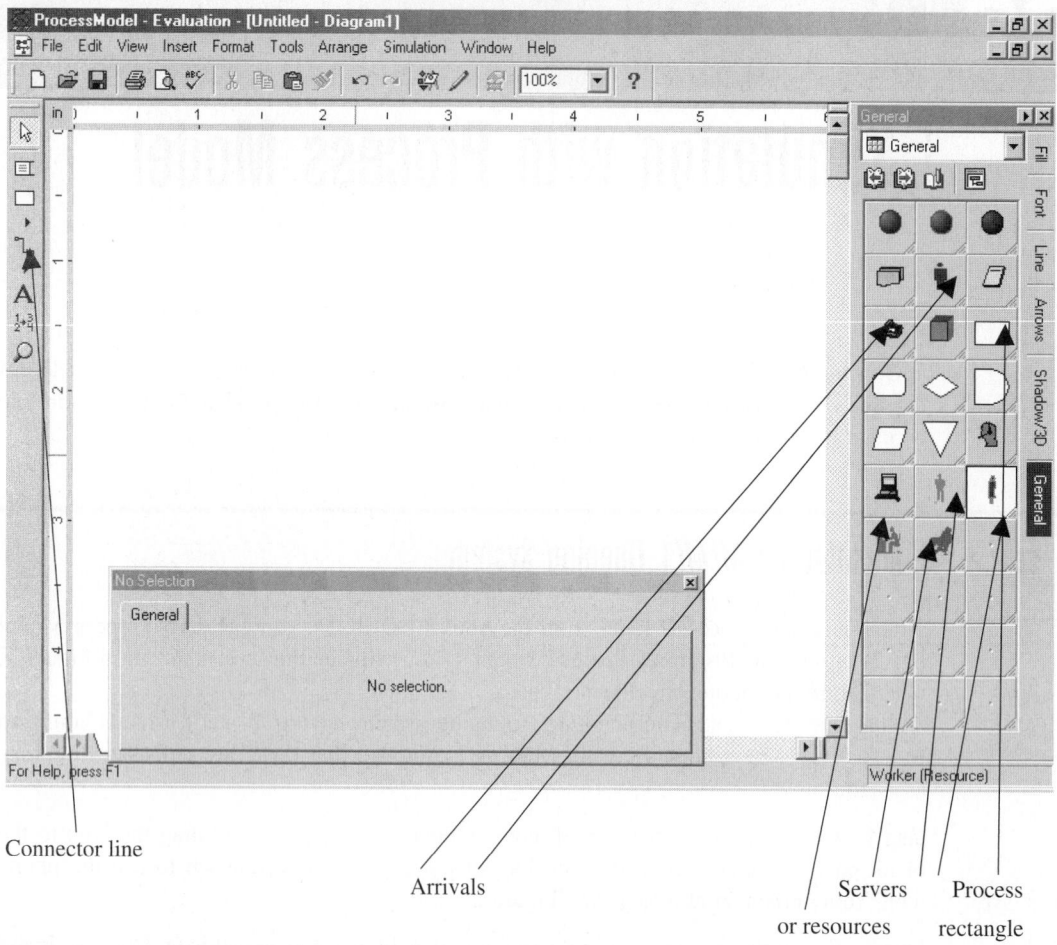

Connector line

Arrivals

Servers
or resources

Process
rectangle

Step 5 We now need to tell Process Model that service times are exponential with mean 4. To do this, click on the Take Calls Process rectangle and fill in the dialog box as shown in Figure 6.

Step 6 We have now completed the model setup. Select File Save As and save the model. (All models have the suffix .igx.)

Step 7 To run the simulation, select Simulation and then Options and fill in the dialog box as shown in Figure 7.

We have chosen to run the system for 4,000 hours. Choosing a Warmup length of 1 hour, the first hour of running the simulation will not be used in the collection of statistics. To start the simulation, choose Simulation Save and Simulation. As the simulation progresses, telephone calls moving through the flowchart illustrate the flow of calls through the process. Resources or servers will show a green light when the resource is being utilized and a blue light when idle. Counters above and to the left of each activity represent the number of calls waiting to be processed. The speed of the simulation can be

FIGURE 2

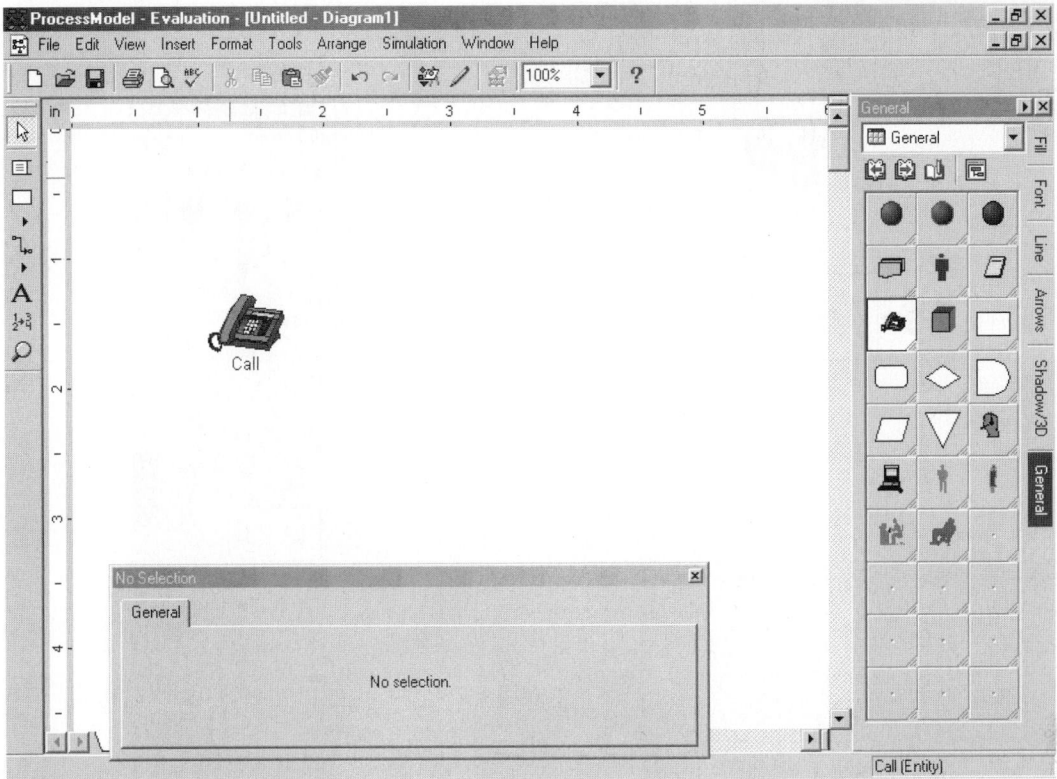

FIGURE 3

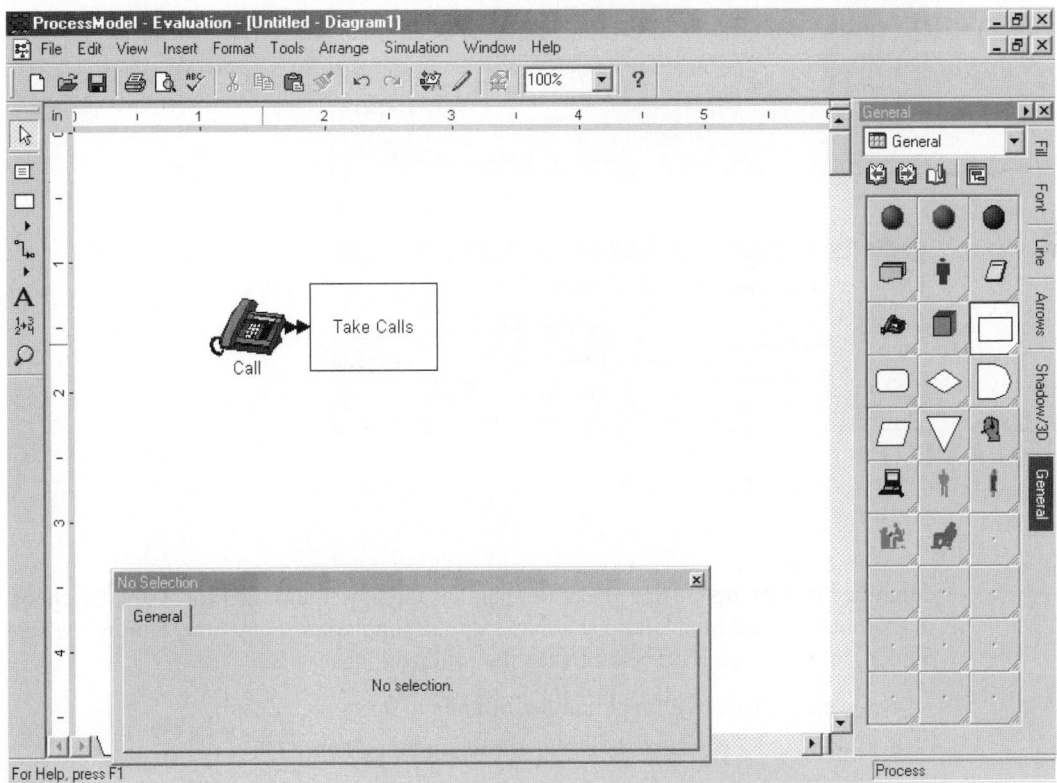

FIGURE 4

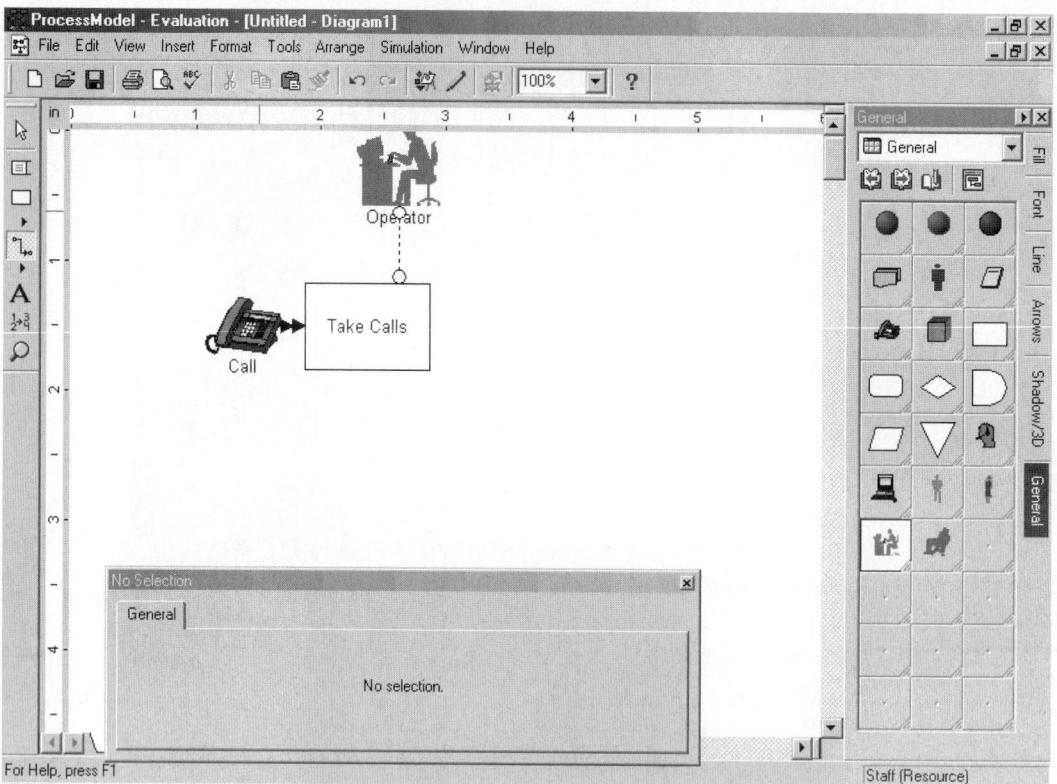

FIGURE 5

FIGURE 6

controlled by moving the Speed Control bar, left for slower and right for faster. By choosing Simulation End Simulation, you may stop the simulation at any time. During the simulation, an on-screen scoreboard tracks the following quantities:

- Quantity Processed (total number of units to leave the system)
- Cycle Time (average time a unit spends in the system)

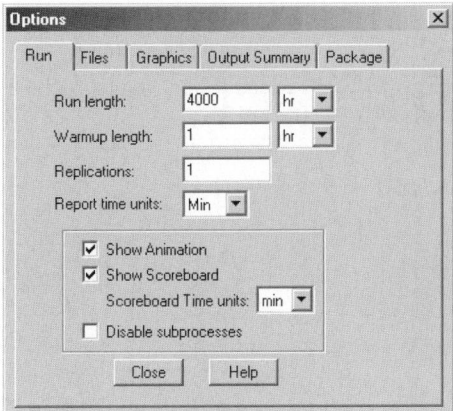

FIGURE 7

- Value Added Time (time a unit spends in service)
- Cost Per Unit (if costs are associated with the resources, the cost incurred per unit serviced is computed)

After completing the simulation, you are asked if you want to view the output. If so, you will see an output similar to Figure 8, and the output may be saved as a text file. Figure 8 includes comments (in **boldface**) to explain the key portions of the output.

If we treat this output as representative of the system's steady state, we have the following parameter estimates:

- $\Pi_0 = .3363$
- $\Pi_0 + \Pi_1 = .5634$
- $W_s = 4.01$ minutes
- $W_q = 7.71$ minutes
- $W = 11.72$ minutes

For an $M/M/1$ system, we can compute the steady-state values of these quantities exactly. Since $\rho = 10/15 = .667$, we find from Equation (24) of Chapter 20 that $\Pi_0 = 1 - .667 = .333$. From (25) of Chapter 20, we find that $\Pi_1 = .667(1 - .667) = .222$. Thus, $\Pi_0 + \Pi_1 = .555$. Clearly $W_s = 4$ minutes. From Formula (31) of Chapter 20, $W = \frac{1}{15 - 10} = .2$ hour $= 12$ minutes. Then $W_q = 12 - 4 = 8$ minutes. Note that the simulation yields very close agreement with the steady-state estimates.

22.2 Simulating an *M/M/2* System

MM2.igx

Let us modify the previous example by changing the number of operators to 2 and ensuring that up to 2 operators can be working on calls at the same time. See file MM2.igx. To change the number of operators to 2, click on the resource and fill in the dialog box as shown in Figure 9.

To ensure that two operators can work on calls at the same time click on the Process rectangle Take Calls and modify it as in Figure 10.

After saving this file as MM2.igx and running it for 1,000 hours, we obtain the output shown in Figure 11. (Boldface comments explain key portions.)

```
-------------------------------------------------------------------------
---------
General Report
Output from C:\Program Files\ProcessModel 4\mm1.mod
Date: Aug/13/2002   Time: 07:50:42 AM
-------------------------------------------------------------------------
---------
Scenario        : Normal Run
Replication     : 1 of 1
Warmup Time     : 1 hr
Simulation Time : 3986.90
-------------------------------------------------------------------------
---------


ACTIVITIES

                                            Average
Activity          Scheduled        Total    Minutes   Average
Maximum    Current
Name              Hours Capacity  Entries  Per Entry  Contents
Contents  Contents  % Util
------------    ---------  --------  -------  ---------  --------  -----
---  --------  ------
Take Call in Q   3985.90          999   39581       7.71      1.27
21         1     0.13
Take Call        3985.90            1   39581       4.01      0.66
1          1    66.37
```

In the nearly 3,986 hours for which data was collected, almost exactly 10 arrivals per hour occurred. The total time spent by a call in queue (waiting) averaged 7.71 minutes, and total service time averaged 4.01 minutes.

ACTIVITY STATES BY PERCENTAGE (Multiple Capacity)

```
                                 %
Activity       Scheduled    %  Partially    %
Name           Hours    Empty  Occupied   Full
------------   ---------  -----  ---------  ----
Take Call inQ   3985.90  56.34    43.66    0.00
```

The queue for calls was empty 56.34% of the time.

ACTIVITY STATES BY PERCENTAGE (Single Capacity)

```
Activity  Scheduled      %       %       %        %
Name       Hours    Operation  Idle  Waiting  Blocked
---------  ---------  ---------  -----  -------  -------
Take Call   3985.90     66.37   33.63    0.00     0.00
```

FIGURE 8   **The operator was idle 33.63% of the time.**

```
RESOURCES

                                     Average
                           Number    Minutes
Resource         Scheduled Of Times  Per
Name      Units  Hours     Used      Usage   % Util
--------  -----  --------- --------  ------- ------
Staff       1    3985.90    39581     4.01    66.37
```

Staff was busy 66.37% of the time, and average staff usage per call processed was 4.01 minutes.

```
RESOURCE STATES BY PERCENTAGE

Resource  Scheduled     %      %     %
Name      Hours     In Use  Idle  Down
--------  --------- ------  ----- ----
Staff      3985.90   66.37  33.63  0.00
```

Staff was busy 66.37% of the time.
ENTITY SUMMARY (Times in Scoreboard time units)

```
                     Average   Average
                     Cycle     VA
Entity      Qty      Time      Time      Average
Name      Processed  (Minutes) (Minutes) Cost
------    ---------  --------- --------- -------
Call        39580     11.72     4.01      1.33
```

FIGURE 8
(Continued) **Each call spent an average of 11.72 minutes in the system.**

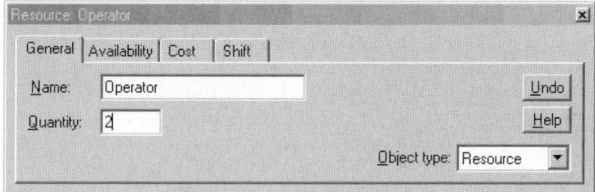

FIGURE 9

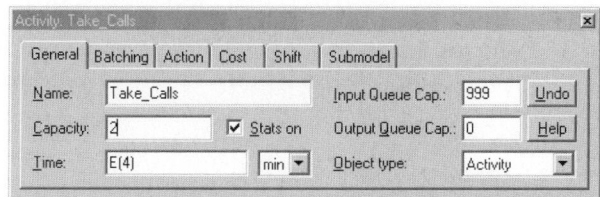

FIGURE 10

```
--------------------------------------------------------------------------
---------
General Report
Output from C:\Program Files\ProcessModel 4\second.mod
Date: Aug/13/2002   Time: 03:08:49 PM
--------------------------------------------------------------------------
---------
Scenario       : Normal Run
Replication    : 1 of 1
Warmup Time    : 1 hr
Simulation Time : 4001 hr
--------------------------------------------------------------------------
---------
```

ACTIVITIES

Activity Name	Scheduled Hours	Capacity	Total Entries	Average Minutes Per Entry	Average Contents	Maximum Contents	Current Contents	% Util
Take Calls inQ	4000	999	39968	0.51	0.08	7	0	0.01

We see that an average of almost exactly 10 arrivals per hour have been observed.

Activity Name	Scheduled Hours	Capacity	Total Entries	Average Minutes Per Entry	Average Contents	Maximum Contents	Current Contents	% Util
Take Calls	4000	2	39970	4.03	0.67	2	0	33.60

ACTIVITY STATES BY PERCENTAGE (Multiple Capacity)

Activity Name	Scheduled Hours	% Empty	% Partially Occupied	% Full
Take Calls inQ	4000	94.34	5.66	0.00
Take Calls	4000	49.58	33.64	16.78

We see that 5.66% of the time, people are waiting. Note from the *M/M/s* template below that probability of people waiting = probability >2 people present = 1 - .5 - .333 - .111 = .0566.

RESOURCES

Resource Name	Units	Scheduled Hours	Number Of Times Used	Average Minutes Per Usage	% Util
Operator.1	1	4000	19971	4.03	33.60
Operator.2	1	4000	19999	4.03	33.60
Operator	2	8000	39970	4.03	33.60

Each operator's mean service time is 4.03 minutes (compared to the 4 minutes we input).

FIGURE 11

RESOURCE STATES BY PERCENTAGE

Resource Name	Scheduled Hours	% In Use	% Idle	% Down
Operator.1	4000	33.60	66.40	0.00
Operator.2	4000	33.60	66.40	0.00
Operator	8000	33.60	66.40	0.00

Each operator was busy 33.60% of the time. Note from the steady-state probabilities in the *M/M/s* template that the probability that a server is busy is .5*(Prob. 1 person present) + Prob(>=2 people present) = .5*(.333) + .167 = .333.
ENTITY SUMMARY (Times in Scoreboard time units)

Entity Name	Qty Processed	Average Cycle Time (Minutes)	Average VA Time (Minutes)	Average Cost
Call	39970	4.55	4.03	0.00

The average time a call spends in the system is 4.55 minutes. From the *M/M/s* spreadsheet, we find that *W* = 4.5 minutes.

VARIABLES

Variable Name Average Value	Total Changes	Average Minutes Per Change	Minimum Value	Maximum Value	Current Value
Avg BVA Time Entity 0	1	0.00	0	0	0
Avg BVA Time Call 0	39971	6.00	0	0	0

	A	B	C	D	E	F	G
1	M/M/s/GD	LAMBDA?	MU?	s?	RO		
2		10	15	2	0.33333333		
3		L	LS	LQ	W	WS	WQ
4		0.75	0.66666667	0.08333333	0.075	0.066666667	0.008333333
5	STATE	P(j>=s)					
6		1	0.16666667				
7	P(Wq>t)	t?	P(W>t)				
8	9.57313E-60	6.70521931	3.1296E-44				
9							
10							
11							
12	STATE	LAMBDA(J)	MU(J)	CJ	PROB	#IN QUEUE	COLA*COLE
13	0	10	0	1	0.5	0	0
14	1	10	15	0.66666667	0.33333333	0	0.333333333
15	2	10	30	0.22222222	0.11111111	0	0.222222222

FIGURE 11
(Continued)

22.3 Simulating a Series System

In this section, we use Process Model to simulate a series queuing system. We will take the case of the auto assembly line (Example 13 of Section 20.10).

EXAMPLE 1   Auto Assembly

The last two things that are done to a car before its manufacture is complete are installing the engine and putting on the tires. An average of 54 cars per hour arrive requiring these

FIGURE 12

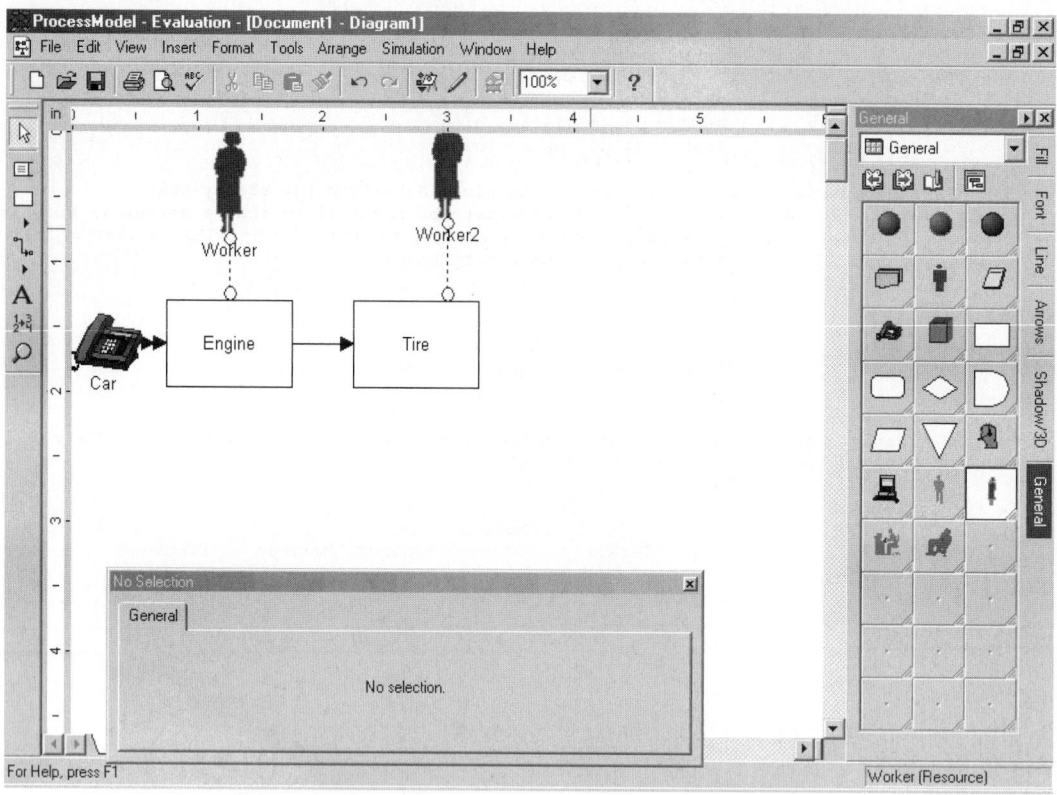

two tasks. One worker is available to install the engine and can service an average of 60 cars per hour. After the engine is installed, the car goes to the tire station and waits for its tires to be attached. Three workers serve at the tire station. Each works on one car at a time and can put tires on a car in an average of 3 minutes. Assume that interarrival times and service times are exponential. Simulate this system for 400 hours.

Solution

Carassembly.igx

See file Carassembly.igx. The key to creating a queuing network with Process Model is to build the diagram one service center at a time. We begin by creating the arrivals as in Section 22.1. Then we create the engine production center as in Section 22.1. Then we drag the Process rectangle over the Engine production center and pull it to the right to create the Tire production center. Of course, we must change the number of servers at the Tire production center to 3 (and also change the capacity of the tire operation to 3). We must enter service times of E(1) for the Engine production center and E(3) for the Tire production center. For the interarrival times, we must enter E(1.11), since there is an arrival an average of every 60/54 = 1.1 minutes. The flowchart looks like Figure 12.

Note that clicking on the arrow connecting the Engine and Tire rectangles results in the dialog box shown in Figure 13. This indicates that 100% of all cars completing Engine installation are sent to the Tire station. We also adjusted the Move time from 1 minute to 0 minute. The Move time indicates how many minutes are needed to move from the Engine to the Tire station. Suppose, for example 70% of the jobs completing Engine installation are sent to other stations (such as Final inspection), and 30% are sent on to the Tire station. We would model this by changing the percentage on the arrow joining Engine and Final inspection to 70%. The percentage going from Engine to Tire would automatically adjust to 30%. We will see how this works in the next example.

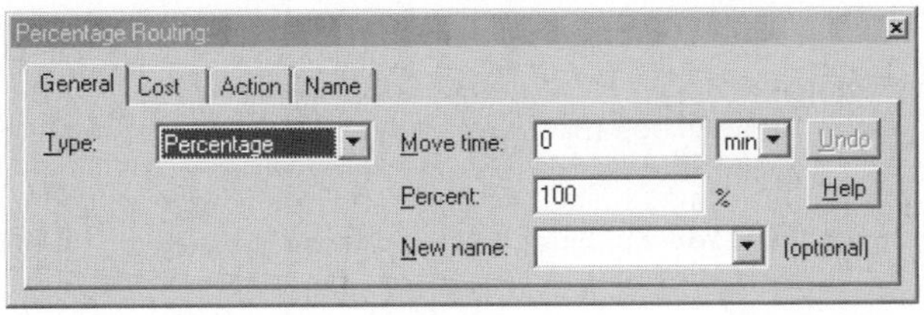

After running the simulation for 400 hours (with a 1-hour warmup period), we obtained the results shown in Figure 14 (explanatory comments in boldface).

```
----------------------------------------------------------------------
---------
General Report
Output from C:\Program Files\ProcessModel 4\carassembly.mod
Date: Aug/15/2002   Time: 10:14:13 AM
----------------------------------------------------------------------
---------
Scenario        : Normal Run
Replication     : 1 of 1
Warmup Time     : 1 hr
Simulation Time : 404.88
----------------------------------------------------------------------
---------


ACTIVITIES

                                           Average
Activity     Scheduled             Total   Minutes   Average   Maximum
Current
Name           Hours  Capacity  Entries  Per Entry  Contents  Contents
Contents  % Util
----------  ---------  --------  -------  ---------  --------  --------
--------  ------
Engine inQ    403.88       999    21741       8.32      7.46        54
1    0.75
Engine        403.88         1    21741       1.00      0.89         1
1   89.82
Tire inQ      403.88      1000    21743       8.00      7.18        46
0    0.72
Tire          403.88         3    21746       3.02      2.71         3
3   90.45
```
An average of 21,741/403.88 = 53.83 arrivals per hour were observed.

```
ACTIVITY STATES BY PERCENTAGE (Multiple Capacity)

                                %
Activity     Scheduled     %  Partially      %
Name           Hours   Empty  Occupied     Full
----------  ---------  -----  ---------   -----
Engine inQ    403.88   19.24     80.76     0.00
Tire inQ      403.88   25.74     74.26     0.00
Tire          403.88    2.37     15.20    82.42
```
19% of the time, no cars are waiting for engine installation. 26% of the time, no cars are waiting for tire installation. 82.4% of the time, all tire installers are busy; 2.37% of the time, no tire installers are busy; and 15.2% of the time, some (but not all) tire installers are busy.

FIGURE 14

ACTIVITY STATES BY PERCENTAGE (Single Capacity)

Activity Name	Scheduled Hours	% Operation	% Idle	% Waiting	% Blocked
Engine	403.88	89.82	10.18	0.00	0.00

The engine installer is busy 89.8% of the time. (In the steady state, she should be busy 90% of the time according to $\Pi_0 = \rho$.

RESOURCES

Resource Name	Units	Scheduled Hours	Number Of Times Used	Average Minutes Per Usage	% Util
Worker	1	403.88	21741	1.00	89.82
Worker2.1	1	403.88	7250	3.02	90.45
Worker2.2	1	403.88	7258	3.02	90.46
Worker2.3	1	403.88	7238	3.02	90.45
Worker2	3	1211.64	21746	3.02	90.45

Mean service time at the Engine station is 1 minute. At the Tire station, mean service time is 3.02 minutes.
RESOURCE STATES BY PERCENTAGE

Resource Name	Scheduled Hours	% In Use	% Idle	% Down
Worker	403.88	89.82	10.18	0.00
Worker2.1	403.88	90.45	9.55	0.00
Worker2.2	403.88	90.46	9.54	0.00
Worker2.3	403.88	90.45	9.55	0.00
Worker2	1211.64	90.45	9.55	0.00

Tire workers appear to be busy around 90.5% of the time, and the engine installer is busy 89.9% of the time.

ENTITY SUMMARY (Times in Scoreboard time units)

Entity Name	Qty Processed	Average Cycle Time (Minutes)	Average VA Time (Minutes)	Average Cost
Car	21743	20.36	4.02	0.00

Average time in the system is 20.4 minutes. In our discussion of Example 13 of Chapter 20, we found total time (in steady state) in the system to equal mean engine service time + mean tire installation time + mean time waiting for engine + mean time waiting for tires = 1 + 3 + 60(.15) + 60(.138) = 21.4 minutes.

VARIABLES

Variable Name Average Value	Total Changes	Average Minutes Per Change	Minimum Value	Maximum Value	Current Value
Avg BVA Time Entity 0	1	0.00	0	0	0
Avg BVA Time Car 0	21744	1.11	0	0	0

FIGURE 14
(Continued)

The Effect of a Finite Buffer

Suppose we have only enough space for two cars to wait for tire installation. This is called a *buffer* of size 2. We assume that if there are two cars waiting for tire installation, the engine installation center must shut down until there is room to "store" a car waiting for tire installation. To model this, change the Input Capacity in the Tire Activity dialog box to 2. (See Figure 15.) Rerunning the simulation, we now find the average time for a car in the system is nearly 6 hours! Clearly, we need more storage space. Even with a buffer of size 10, total time in the system is increased by around 50%.

22.4 Simulating Open Queuing Networks

In this section, we show how to use Process Model to simulate open queuing networks. To illustrate, we simulate Example 14 from Section 20.10.

EXAMPLE 2   Open Queuing Network

An open queuing network consists of two servers: server 1 and server 2. An average of 8 customers per hour arrive from outside at server 1. An average of 17 customers per hour arrive from outside at server 2. Interarrival times are exponential. Server 1 can serve at an exponential rate of 20 customers per hour, and server 2 can serve at an exponential rate of 30 customers per hour. After completing service at server 1, half the customers leave the system and half go to server 2. After completing service at server 2, 75% of the customers complete service and 25% return to server 1. Simulate this system for 400 hours.

Open.igx   Solution See file Open.igx. To begin, we need to create two arrival entities: one representing external arrivals to server 1 and one representing external arrivals to server 2. After creating Process rectangles for server 1 and server 2, we use the Connector tool to create a link from server 1 to server 2, a link from server 2 to server 1, and a link from Server 1 and 2 to exit the system. For server 1, arrivals are E(7.5), and for server 2, arrivals are E(3.53). Service time for server 1 is E(3), while service time for server 2 is E(2). After clicking on the link from server 1 to server 2, we make sure that the routing percentage is 50% (this is the default). After clicking on the link from server 2 to server 1, we change the routing percentage to 25%. Note that the routing percentages on the exit links automatically adjust so that the total routing percentage leaving a service center is 100%. As an example, the dialog box on the link leaving server 2 and going to server 1 should be filled in

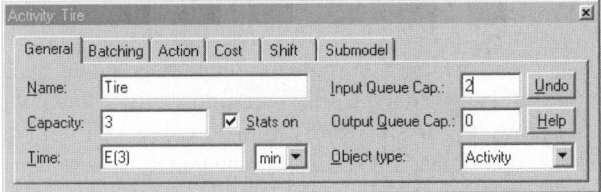

FIGURE 15

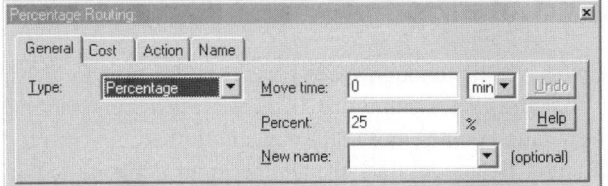

FIGURE 16

FIGURE 17

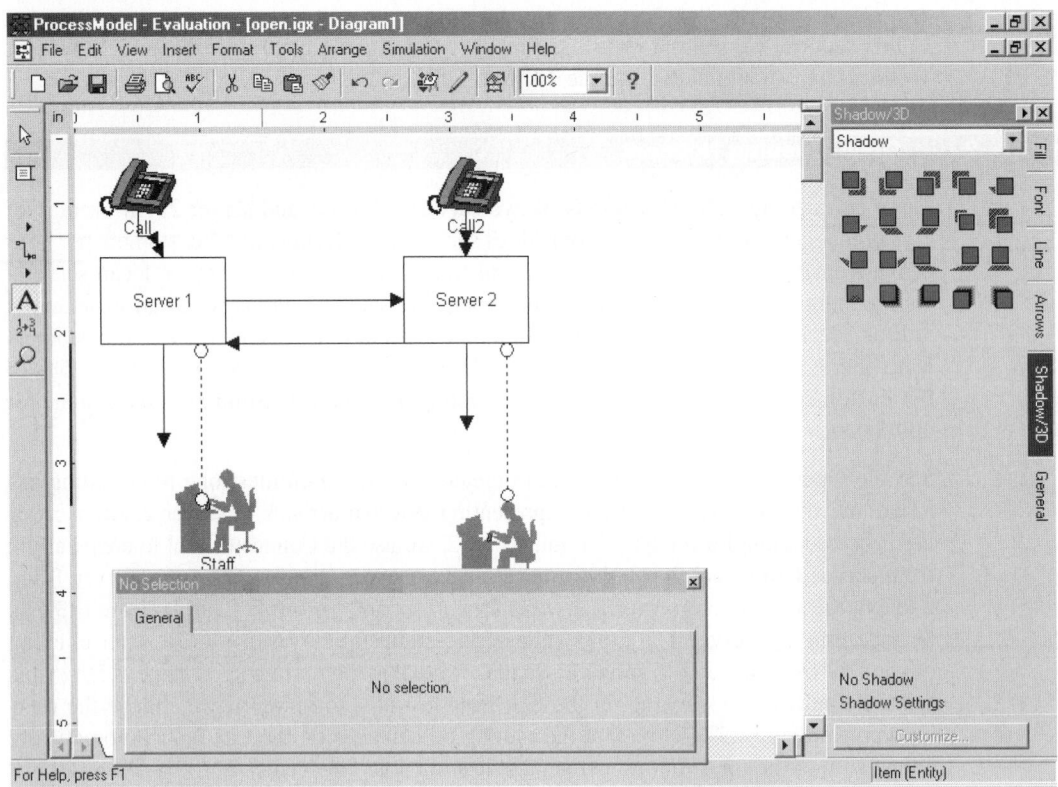

as shown in Figure 16. There, the Move time from server 2 to server 1 is made equal to 0, and we ensured that 25% of all customers completing server 2 go instantly to server 1. This implies that 75% of all customers completing server 2 instantly leave the system. (Click on the arc leaving server 2 if you do not believe this.) Note that the default Move times must always be adjusted from 1 minute unless you want a move time of 1 minute.

The flowchart is shown in Figure 17. Note that as the simulation runs, some calls move between the servers, and some exit the system. Seeing this movement really makes the concept of an open queuing network come alive. The simulation output is shown in Figure 18 (with explanatory comments in boldface).

```
-----------------------------------------------------------------------
---------
General Report
Output from C:\Program Files\ProcessModel 4\open.mod
Date: Aug/15/2002   Time: 10:17:11 AM
-----------------------------------------------------------------------
---------
Scenario        : Normal Run
Replication     : 1 of 1
Warmup Time     : 4 hr
Simulation Time : 404 hr
-----------------------------------------------------------------------
---------


ACTIVITIES

                                              Average
Activity         Scheduled         Total      Minutes    Average
Maximum     Current
Name              Hours  Capacity  Entries  Per Entry  Contents
Contents  Contents  % Util
------------   ---------  --------  -------  ---------  --------  ------
--   --------  ------
Server 1 inQ        400       999     5485       6.69      1.52
14          3    0.15
Server 1            400         1     5482       3.02      0.69
1           1   69.07
Server 2 inQ        400       999     9627       7.69      3.08
32          0    0.31
Server 2            400         1     9628       2.00      0.80
1           0   80.28
```

Note: Server 1 processes around 14 calls per hour, while server 2 processes around 24 calls per hour. This agrees with the results we obtained in our analysis of Example 14 in Chapter 20.

```
ACTIVITY STATES BY PERCENTAGE (Multiple Capacity)

                                  %
Activity      Scheduled      %  Partially    %
Name            Hours    Empty  Occupied   Full
------------  ---------  -----  ---------  ----
Server 1 inQ       400  45.19      54.81   0.00
Server 2 inQ       400  35.44      64.56   0.00
```

45% of the time, no jobs are waiting at server 1. 35% of the time, no jobs are waiting at server 2.

```
ACTIVITY STATES BY PERCENTAGE (Single Capacity)

Activity  Scheduled        %      %        %        %
Name        Hours   Operation   Idle  Waiting  Blocked
--------  ---------  ---------  -----  -------  -------
Server 1       400      69.07  30.93     0.00     0.00
Server 2       400      80.28  19.72     0.00     0.00
```

Server 1 is busy 69.1% of the time, while server 2 is busy 80.3% of the time.

FIGURE 18

Resource Name	Units	Scheduled Hours	Number Of Times Used	Average Minutes Per Usage	% Util
Staff	1	400	6787	2.83	80.04
Staff2	1	400	8323	1.99	69.32

RESOURCE STATES BY PERCENTAGE

Resource Name	Scheduled Hours	% In Use	% Idle	% Down
Staff	400	80.04	19.96	0.00
Staff2	400	69.32	30.68	0.00

ENTITY SUMMARY (Times in Scoreboard time units)

Entity Name	Qty Processed	Average Cycle Time (Minutes)	Average VA Time (Minutes)	Average Cost
Call	3114	16.30	4.57	0.00
Call2	6885	13.92	3.13	0.00

Calls that first arrive from outside to server 1 spend an average of 16.3 minutes in the system and 4.57 minutes in service. Calls that first arrive from outside to server 2 spend an average of 13.92 minutes in the system and 3.13 minutes in service.

VARIABLES

Variable Name Average Value	Total Changes	Average Minutes Per Change	Minimum Value	Maximum Value	Current Value
Avg BVA Time Entity 0	1	0.00	0	0	0
Avg BVA Time Call 0	3115	7.70	0	0	0
Avg BVA Time Call2 0	6886	3.48	0	0	0

FIGURE 18
(Continued)

22.5 Simulating Erlang Service Times

As we saw in Chapter 20, service times often do not follow an exponential distribution. The Erlang distribution is often used to model nonexponential service times. An Erlang distribution can be defined by a mean and a shape parameter k. The shape parameter must be an integer. It can be shown that

$$\text{Standard deviation of Erlang} = \frac{\text{mean}}{\sqrt{k}}$$

Therefore, if we know the mean and standard deviation of the service times, we may determine an appropriate value of k. The syntax for generating Erlang service times in Process Model is ER(Mean, k). We now illustrate how to simulate a queuing system with Erlang service times.

EXAMPLE 3   **Walk-in Clinic**

A walk-in hospital clinic has four doctors. An average of 12 patients per hour arrive at the clinic (interarrival times are assumed to be exponential). A doctor can see an average of 4 patients per hour, with the standard deviation of service times being 8.66 minutes. Simulate the operation of this clinic for 1,000 hours.

Doctor.igx   **Solution**   See file Doctor.igx. The flowchart is shown in Figure 19. We must remember to adjust the number of doctors to 4 and the See Doctor capacity to 4. By clicking on the arrow connecting Customer and See Doctor, we can input the interarrival times as E(5). We know that the mean service time is 15 minutes. To estimate the shape parameter k, we solve $\frac{15}{\sqrt{k}} = 8.66$ minutes. This yields $k = 3$. We now fill in the See Doctor dialog box as shown in Figure 20. This ensures that up to 4 patients can be seen at once and that service times will follow an Erlang random variable with mean 15 minutes and shape parameter 3. (This implies a standard deviation of 8.66 minutes.)

The output is shown in Figure 21, with explanatory comments in boldface.

FIGURE 19

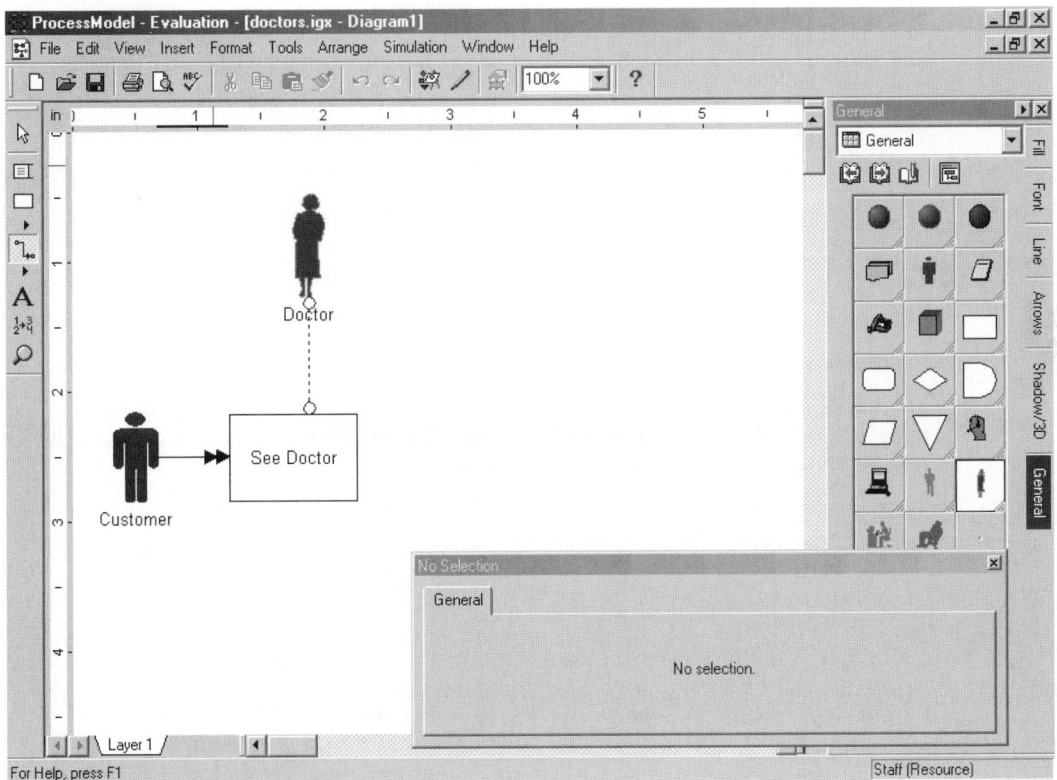

FIGURE 20

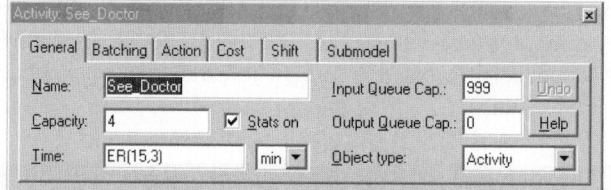

```
General Report
Output from C:\Program Files\ProcessModel 4\doctors.mod
Date: Aug/14/2002   Time: 10:29:27 AM
-----------------------------------------------------------------------
---------
Scenario        : Normal Run
Replication     : 1 of 1
Simulation Time : 1000 hr
-----------------------------------------------------------------------
---------


ACTIVITIES

                                                 Average
Activity          Scheduled          Total       Minutes    Average
Maximum    Current
Name              Hours    Capacity  Entries  Per Entry  Contents
Contents  Contents  % Util
--------------    ---------  --------  -------  ---------  --------  ----
----  --------  ------
See Doctor inQ    1000       999       11964       5.37       1.07
23         0     0.11
See Doctor        1000         4       11964      15.07       3.00
4          2    75.12
```

This implies that average time waiting for doctor is 5.37 minutes and average service time is 15.07 minutes.

```
ACTIVITY STATES BY PERCENTAGE (Multiple Capacity)

                                         %
Activity          Scheduled    %    Partially     %
Name              Hours      Empty   Occupied    Full
--------------    ---------  -----  ---------   -----
See Doctor inQ    1000       64.88    35.12      0.00
See Doctor        1000        3.65    45.95     50.40
```
35% of all patients will have to wait for a doctor. All doctors are busy 50% of the time, and all doctors are idle 4% of the time. Between 1 and 3 doctors are busy 46% of the time.

```
RESOURCES

                                 Average
                        Number   Minutes
Resource            Scheduled  Of Times    Per
Name       Units     Hours      Used     Usage   % Util
--------   -----   ---------  --------  -------  ------
Doctor.1     1        1000       2983     15.11   75.13
Doctor.2     1        1000       3025     14.90   75.12
Doctor.3     1        1000       2976     15.14   75.12
Doctor.4     1        1000       2980     15.12   75.12
Doctor       4        4000      11964     15.07   75.12
```
The average service time varies from a low of 14.9 minutes for

FIGURE 21
Doctor 1 to a high of 15.14 minutes for Doctor 3.

RESOURCE STATES BY PERCENTAGE

Resource Name	Scheduled Hours	% In Use	% Idle	% Down
Doctor.1	1000	75.13	24.87	0.00
Doctor.2	1000	75.12	24.88	0.00
Doctor.3	1000	75.12	24.88	0.00
Doctor.4	1000	75.12	24.88	0.00
Doctor	4000	75.12	24.88	0.00

Each doctor is busy about 75% of the time. This is reasonable because (12 patients/hour)*(15 minutes/patient)= 180 minutes/hour of work arrives for doctors, and doctors have 240 minutes per hour to work, so they should be busy 75% of the time.

ENTITY SUMMARY   (Times in Scoreboard time units)

Entity Name	Qty Processed	Average Cycle Time (Minutes)	Average VA Time (Minutes)	Average Cost
Customer	11962	20.45	15.07	0.00

On average, a patient spends a total of 20.45 minutes in the system. (Send me to this clinic!!)

VARIABLES

Variable Name Average Value	Total Changes	Average Minutes Per Change	Minimum Value	Maximum Value	Current Value
Avg BVA Time Entity 0	1	0.00	0	0	0
Avg BVA Time Customer 0	11963	5.01	0	0	0

22.6 What Else Can Process Model Do?

Our discussion of Process Model has only scratched the surface of its capabilities. Other modeling features include the following.

- **Bulk arrivals and services.** At a restaurant, people often arrive in groups. This arrival pattern is called *bulk arrivals.* Consider an amusement park ride seating 40 people. The attendant waits until 40 people are present and then runs the ride. This service mechanism is known as *bulk service.*

- **Reneging.** Perhaps people hang up when calling an 800 number if they are put on hold more than 5 minutes. Process Model can accommodate such balking or reneging behavior.

- **Variation in arrival pattern.** At a restaurant or bank, the arrival rate varies substantially over the course of a day (or a whole week). Variable arrival rate patterns can easily be simulated with Process Model.

- **Variation in number of servers.** During the day, workers take breaks and go to lunch. Also, many companies vary the number of servers during the day. Process Model can easily accommodate variation in service capacity.

- **Priorities.** In an emergency room, more seriously ill patients are given priority over earlier arriving, less ill patients. Process Model can handle complex priority mechanisms.

For more details on these and other features of Process Model, consult the online manual.

REVIEW PROBLEMS

Group A

1 At a manufacturing assembly line, 30 jobs arrive per hour. Each job must pass through two production stages: stage 1 and stage 2. Stage 1 takes an average of 1 minute to complete, and 1 worker is available to perform stage 1. After completing stage 1, the job immediately passes to stage 2. Stage 2 takes an average of 2 minutes to complete, and 2 workers are available to work on stage 2. After completing stage 2, each job is inspected. Inspection takes an average of 3 minutes, and 3 workers are available to perform inspection. After inspection, 10% of the jobs must be returned to stage 1, and they then repeat both stages 1 and 2. After inspection, 20% of all jobs return to stage 2 and repeat stage 2. Assume that interarrival times and service times are exponential.

a What is the average time a job spends in the system from arrival to completion?

b What percentage of the time is each worker busy?

2 The United Airlines security station for Terminal C in Indianapolis has 3 X-ray machines. During the busy early morning hours, an average of 400 passengers per hour arrive at Terminal C (with exponential interarrival times). Each X-ray machine can handle an average of 150 passengers per hour (with exponential service times for X-ray machines).

After going through security, 90% of the customers are free to go to their flight, but 10% must be "wanded." Three people are available to do the wanding. Wanding requires a mean of 4 minutes, with a standard deviation of 2 minutes.

a How long does it take the average passenger to pass through security?

b If there were no wanding, how long would it take the average passenger to pass through security?

c Which would improve the situation more: adding an X-ray machine or adding an additional person to perform wanding?

3 Consider an emergency room. An average of 10 patients arrive per hour (interarrival times are exponential). Upon entering, the patient fills out a form. Assume that this always takes 5 minutes. Then each patient is processed by one of two registration clerks. This takes an average of 7 minutes (exponentially distributed). Then each patient walks 2 minutes to a waiting room and waits for one of 4 doctors. The time a doctor takes to see a patient averages 20 minutes, with a standard deviation of 10 minutes.

a On the average, how long does a patient spend in the emergency room?

b On the average, how much of this time is spent waiting for a doctor?

c What percentage of the time is each doctor busy?

4 The Indiana University Credit Union has 4 tellers working. It takes an average of 3 minutes (exponentially distributed) to serve a customer. Assume that an average of 60 customers per hour arrive at the Credit Union (interarrival times are exponential).

a How long do customers have to wait for a teller?

b What percentage of the time is a teller busy?

5 A pharmacist has to fill an average of 15 orders per hour (interarrival times are exponentially distributed). 80% of the orders are relatively simple and take 2 minutes to fill. 20% of the orders take 10 minutes to fill.

a What percentage of the time is the pharmacist busy?

b On average, how long does it take to get a prescription filled?

6 Solve Problem 5 if the service times followed a normal distribution with mean 3 minutes and standard deviation .5 minute. Use the syntax N(3,.5) to generate service times.

7 At Indiana Pacer games, 10,000 fans must enter through 10 checkpoints in the hour before each game (interarrival times are exponential). It takes exactly 3 seconds to have a ticket processed. How long does an average ticketholder spend from arrival to passing through the checkpoint?

8 Since September 11, 2001, each Pacer ticketholder's clothing and handbags are searched. Assume that this takes exactly 10 seconds and occurs right after the ticketholder passes through the checkpoint. Four people are available at each checkpoint to do the searching. How long does the average ticketholder spend from arrival to passing through the checkpoint?

Simulation with the Excel Add-in @Risk

Many simulations, particularly those involving financial applications can easily be performed with the Excel add-in @Risk. @Risk makes it easy to generate random variables. For example, to generate a standard normal random variable in a cell, just enter the formula =RISKNORMAL(0,1). If you want to run 10,000 iterations of a spreadsheet, just tell @Risk to run 10,000 iterations. Then @Risk provides a complete statistical or graphical summary of the results. In this chapter, we will see how @Risk can be used to simulate a wide variety of situations, ranging from the NPV of a new project to the probability of winning at craps.

23.1 Introduction to @Risk: The News Vendor Problem

@Risk is used to model situations where decisions are to be made under uncertainty. Here is an easy example. See the @Risk crib sheet in Appendix 1.

EXAMPLE 1   **Ordering Calendars**

Our bookstore must determine how many 2005 nature calendars to order in August 2004. It costs $2.00 to order each calendar, and we sell each calendar for $4.50. After January 1, 2005, leftover calendars are returned for $.75. Our best guess is that the number of calendars demanded is governed by the following probabilities.

Demand	Probability
100	.3
150	.2
200	.3
250	.15
300	.05

How many calendars should we order?

Solution   The final result is in file Newsdiscrete.xls. See Figure 1.

Newsdiscrete.xls

Step 1   Enter parameter values in C3:C5.

Step 2   It can be shown that ordering an amount equal to one of the possible demands for calendars always maximizes expected profit. For now, we enter a trial order quantity of 200 calendars in cell C1.

	A	B	C	D	E	F	G
1	Order quantity		100				
2	Quantity demanded		100				
3	Sales price		$4.50				
4	Salvage value		$0.75			demand	prob
5	Purchase price		$2.00			100	0.3
6						150	0.2
7	Full price revenue	$450.00				200	0.3
8	Salvage revenue	$0.00				250	0.15
9	Costs	$200.00				300	0.05
10	Profit	$250.00					

FIGURE 1

Step 3 To tell @Risk to generate demand according to the above probabilities, type in C2 the formula

$$=RISKDISCRETE(F5:F9,G5:G9)$$

This generates a demand for calendars of 100 30% of the time, 150 20% of the time, etc. Essentially, for each iteration, @Risk generates a random number between 0 and 1. Then random numbers <.3 yield a demand of 100, random numbers ≥.3 and <.5 yield a demand of 150, random numbers ≥.5 and <.8 yield a demand of 200, random numbers ≥.8 and <.95 yield a demand of 250, and random numbers ≥.95 yield a demand of 300. Of course, successive random numbers generated by @Risk are independent of each other.

This demand could also have been generated with the formula

$$=RISKDISCRETE(\{100,150,200,250,300\},\{.3,.2,.3,.15,.05\})$$

In either format, the demands are listed first, followed by the probabilities. To see the spreadsheet recalculate when you hit F9, select Simulation Settings (the third icon from left) and choose from the Sampling tab Recalculation, and then choose Monte Carlo. Approximately 30% of the time a demand of 100 will occur, around 20% of the time a demand of 150 will occur, etc. If you change Sampling Recalculation to True EV, the mean of the random variable (172.5) will appear. If you change Sampling Recalculation to Expected Value, the value of the random variable nearest to the mean (in this case 150) will occur. We recommend always leaving Sampling Type on Latin Hypercube, because it is much more accurate than Monte Carlo. To illustrate how Latin Hypercube sampling works, suppose we told @Risk to sample from a normal distribution with mean 100 and standard deviation 15. The 5th, 10th, . . . , 95th percentile of a standard normal distribution can be found (using the NORMSINV function) to equal the values shown in Figure 2.

Suppose we want to simulate 100 values of a normal random variable with mean 100 and standard deviation 15. Then @Risk will ensure that 5 are less than or equal to 75.33, 5 are between 75.33 and 80.78, etc. Thus, the simulation will yield a very accurate representation of the random variable's distribution. In particular, our simulated means, variances, and other statistics will be much more accurate than if we used the Monte Carlo simulation. With Monte Carlo, 8 of 100 generated values could be <75.33, 3 out of 100 generated values between 75.33 and 80.78, etc.

Step 4 In cell B7, compute full-price revenue with the formula

$$=C3*MIN(C1,C2)$$

This ensures that we sell at full price the minimum of quantity ordered and quantity demanded.

Step 5 In B8, compute salvage revenue with the formula

$$=C4*IF(C1>C2,(C1-C2),0)$$

	D	E
5	Percentile	Value
6	0.05	75.3272
7	0.1	80.77672
8	0.15	84.4535
9	0.2	87.37568
10	0.25	89.88266
11	0.3	92.13399
12	0.35	94.22019
13	0.4	96.19979
14	0.45	98.11508
15	0.5	100
16	0.55	101.8849
17	0.6	103.8002
18	0.65	105.7798
19	0.7	107.866
20	0.75	110.1173
21	0.8	112.6243
22	0.85	115.5465
23	0.9	119.2233
24	0.95	124.6728

FIGURE **2**

This ensures that the number left over is (number ordered) − (number demanded)—as long as that is >0.

Step 6 In B9, compute ordering costs with the formula

$$=C1*C5$$

Step 7 In cell B10, compute profit with the formula

$$=B7+B8-B9$$

We now want to compute profit for each possible order quantity (100, 150, 200, 250, or 300). The RISKSIMTABLE function makes this easy to do.

FIGURE **3**

	D	E	F	G	H	I	J	K	L
13									
14		Name	Workbook	Worksheet	Cell	Sim#	Minimum	Mean	Maximum
15	Output 1	Profit	newsdiscrete.xl	Sheet1	B10	1	250	250	250
16	Output 1	Profit	newsdiscrete.xl	Sheet1	B10	2	187.5	318.75	375
17	Output 1	Profit	newsdiscrete.xl	Sheet1	B10	3	125	350	500
18	Output 1	Profit	newsdiscrete.xl	Sheet1	B10	4	62.5	325	625
19	Output 1	Profit	newsdiscrete.xl	Sheet1	B10	5	0	271.875	750
20	Input 1	Order quant	newsdiscrete.xl	Sheet1	C1	1	100	100	100
21	Input 1	Order quant	newsdiscrete.xl	Sheet1	C1	2	150	150	150
22	Input 1	Order quant	newsdiscrete.xl	Sheet1	C1	3	200	200	200
23	Input 1	Order quant	newsdiscrete.xl	Sheet1	C1	4	250	250	250
24	Input 1	Order quant	newsdiscrete.xl	Sheet1	C1	5	300	300	300
25	Input 2	Quantity der	newsdiscrete.xl	Sheet1	C2	1	100	172.5	300
26	Input 2	Quantity der	newsdiscrete.xl	Sheet1	C2	2	100	172.5	300
27	Input 2	Quantity der	newsdiscrete.xl	Sheet1	C2	3	100	172.5	300
28	Input 2	Quantity der	newsdiscrete.xl	Sheet1	C2	4	100	172.5	300
29	Input 2	Quantity der	newsdiscrete.xl	Sheet1	C2	5	100	172.5	300

	D	E	F	G	H	I
37						
38	Name	Profit	Profit	Profit	Profit	Profit
39	Description	Output (Sim	Output (Sim#2)	Output (Sim	Output (Sim	Output (Sim
40	Cell	B10	B10	B10	B10	B10
41	Minimum	250	187.5	125	62.5	0
42	Maximum	250	375	500	625	750
43	Mean	250	318.75	350	325	271.875
44	Std Deviatic	0	85.96629	163.5405	208.8956	225.6981
45	Variance	0	7390.203	26745.5	43637.39	50939.61
46	Skewness	Error!	-0.8715626	-0.397861	3.47E-02	0.2893988
47	Kurtosis	Error!	1.758383	1.42927	1.627334	2.06803
48	Errors Calcu	0	0	0	0	0
49	Mode	250	375	500	62.5	0
50	5% Perc	250	187.5	125	62.5	0
51	10% Perc	250	187.5	125	62.5	0
52	15% Perc	250	187.5	125	62.5	0
53	20% Perc	250	187.5	125	62.5	0
54	25% Perc	250	187.5	125	62.5	0
55	30% Perc	250	187.5	125	62.5	0
56	35% Perc	250	375	312.5	250	187.5
57	40% Perc	250	375	312.5	250	187.5
58	45% Perc	250	375	312.5	250	187.5
59	50% Perc	250	375	312.5	250	187.5
60	55% Perc	250	375	500	437.5	375
61	60% Perc	250	375	500	437.5	375
62	65% Perc	250	375	500	437.5	375
63	70% Perc	250	375	500	437.5	375
64	75% Perc	250	375	500	437.5	375
65	80% Perc	250	375	500	437.5	375
66	85% Perc	250	375	500	625	562.5
67	90% Perc	250	375	500	625	562.5
68	95% Perc	250	375	500	625	562.5
69	Filter Minimum					
70	Filter Maximum					
71	Type (1 or 2)					
72	# Values Fil	0	0	0	0	0
73	Scenario #1	>75%	>75%	>75%	>75%	>75%
74	Scenario #2	<25%	<25%	<25%	<25%	<25%
75	Scenario #3	>90%	>90%	>90%	>90%	>90%
76	Target #1 (V	400	400	400	400	400
77	Target #1 (F	100%	100%	50%	50%	80%
78	Target #2 (V	250	375	500	625	750
79	Target #2 (F	99%	99%	99%	99%	99%
80	Target #3 (V	360	360	360	360	360
81	Target #3 (F	100%	30%	50%	50%	50%

FIGURE 4

Step 8 In cell C1, enter the possible order quantities (100, 150, 200, 250, 300) with the formula

$$=\text{RISKSIMTABLE}(\{100,150,200,250,300\})$$

We could also have entered this RISKSIMTABLE function with the formula

$$=\text{RISKSIMTABLE}(F5:F9)$$

Note that if we obtain the arguments of an @Risk function such as RISKSIMTABLE or RISKDISCRETE by pointing to a different cell, we need to omit the { and } brackets.

On the first simulation, @Risk will put 100 in this cell and run the desired number of iterations. On the second simulation, @Risk will put 150 in this cell and run the desired number of iterations. Finally, on the fifth simulation, @Risk will put 300 in this cell and run the desired number of iterations.

Step 9   With the cursor in B10, select B10 as an output cell by selecting the single arrow icon. Note that the phrase RiskOutput() + appears before our Profit formula, indicating that Profit is an output cell. We could have entered this phrase instead of using the icon.

Step 10   Select the Simulations Settings icon. From the Iteration tab, select 1,000 iterations and 5 simulations. From Sampling tab, choose Latin Hypercube from the Sampling option. This will cause @Risk to recalculate demand and profit 1,000 times for each of the five order quantities. In general, if you have a RISKSIMTABLE in your spreadsheet, the number of simulations should equal the number of values in the RISKSIMTABLE. If you do not use a RISKSIMTABLE, leave Simulations at 1.

Step 11   Select the Run Simulation icon shown here. After running the simulation, you will see the summary statistics shown in Figure 3. The first simulation is for 100 calendars ordered, the second for 150 calendars ordered, etc.

To obtain detailed statistics, select Insert Detailed Statistics and obtain Figure 4. To paste the statistics into the spreadsheet, right click on Results and then select Copy. Click on the X icon and choose Paste to insert the results into the original spreadsheet.

Interpretation of Statistical Output   Figures 3 and 4 show that average profit for 1,000 trials when 200 calendars are ordered (for example) is $350.00. From Figure 4, the standard deviation for 1,000 trials is $163.54. It appears that ordering 200 calendars maximizes expected profit, but a case can be made for ordering 150 calendars. For 10% less expected profit, we can cut risk in half. The decision depends on the store's degree of aversion to risk.

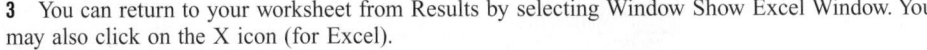

REMARKS   **1**   The RISKSIMTABLE function uses the same set of random numbers to generate demand for each simulation. Thus, for each order quantity, the profit keys off the same set of demands.
2   You can return to the Results at any time by selecting the Results icon.
3   You can return to your worksheet from Results by selecting Window Show Excel Window. You may also click on the X icon (for Excel).

Finding a Confidence Interval for Expected Profit

If we ran 1,000 more trials in Example 1, @Risk would generate a different set of profits[†], and we would get a different estimate of average profit. So no simulation gives average profit exactly. How accurate is the estimate of average profit @Risk gives?

From Section 21.9, we can be 95% sure that average or expected profit for 200 calendars is between

$$\text{(Mean profit)} \pm t_{(.025,\ 199)} \text{ mean standard error}$$

Using the Excel formula TINV(.05,199) = 1.97, we find $t_{(.025,\ 199)} = 1.97$. Here,

$$\text{Mean standard error} = \frac{\text{standard deviation}}{\sqrt{\text{iterations}}} = \frac{163.54}{\sqrt{1,000}} = 5.17$$

[†]When the seed is set (from Simulation Settings Sampling) to 0, each time you run a simulation you will obtain different results. Other possible seed values are integers between 1 and 32,767. Whenever a nonzero seed is chosen, the same values for the input cells and output cells will occur. For example, if we choose a seed value of 10, then each time we run the simulation we will obtain exactly the same results. We often choose a seed of 1. If you also choose a seed of 1, your statistical output should exactly match ours.

Thus, we are 95% sure that expected profit is between $350 \pm 1.97(163.54)/\sqrt{1,000}$, or $339.81 and $360.19.

To be 95% confident of estimating the mean within $1, how many iterations are needed? The required number of iterations must satisfy

$$\frac{1.97(163.54)}{\sqrt{\text{iterations}}} = 1 \quad \text{or} \quad \text{iterations} = 322.17^2 = 103,796$$

To achieve a precise estimate of expected profit requires many iterations!

Modeling Normal Demand with the RISKNORMAL Function

In Example 1, the assumption of discrete demand is unrealistic. Let's suppose demand is normally distributed, with a mean of 200 and a standard deviation of 30. Then we are 68% sure that demand is between 170 and 230, 95% sure between 140 and 270, etc. To model normal demand, simply change cell C2's formula to

$$=\text{RISKNORMAL}(200,30)$$

Normalsim.xls

(See file Normalsim.xls.) This implies (for example), by the well-known rule of thumb, that 68% of the time demand will be between 170 and 230, 95% of the time between 140 and 260, and 99.7% of the time between 110 and 290.

@Risk generates a normal random variable by the inverse transformation method. First, we generate a random number that is equally likely to be any value between 0 and 1. Suppose we generate .6. Then the generated value of the normal random variable will be the 60th percentile of the random variable ($=\text{NORMINV}(.6,200,30)$).

With normal demand, any order quantity is reasonable, because demand may assume any value. We will still try the same set of order quantities, however. After running the simulation and selecting Insert Detailed Statistics, we obtain the output shown in Figure 5.

Figure 5 shows that ordering 200 calendars yields a higher mean profit than ordering 100, 150, 250, or 300 calendars. Plotting the expected profit for each order quantity yields the graph shown in Figure 6.

Under the assumption that profit is a unimodal function of order quantity (which is indeed correct), Figure 6 shows that expected profit is maximized by ordering between 150 and 250 calendars. Another RISKSIMTABLE (with values 160, 170, 180, 190, 200, 210, 220, 230, 240, 250) would help zero in on the actual best order quantity (which turns out to be 213 calendars).

REMARK   To preclude the demand for calendars being a fraction, you could change the formula in cell C2 to

$$=\text{ROUND}(\text{RISKNORMAL}(200,30),0)$$

Then each demand generated by the RISKNORMAL function will be rounded to the nearest integer.

Finding Targets and Percentiles

At the bottom of the Detailed Statistics output, we may enter targets as values or percentages. Enter a value and @Risk tells you for what fraction of iterations the output cell was less than or equal to target. For example, we entered 400 under value and found that the profit for ordering 200 calendars was less than or equal to $400 18.7% of the time. We entered 34% under percentage and found that 34% of the time, profit was less than or equal to $453.44. We entered 99% under percentage and found that 99% of the time,

	D	E	F	G	H	I
76						
77	Name	Profit	Profit	Profit	Profit	Profit
78	Description	Output (Sim	Output (Sim#2)	Output (Sim	Output (Sim	Output (Sim
79	Cell	B10	B10	B10	B10	B10
80	Minimum	238.7325	176.2325	113.7325	51.23248	-11.26752
81	Maximum	250	375	500	625	722.9146
82	Mean	249.9887	372.7469	455.0987	435.2473	374.9326
83	Std Deviatic	0.3563102	13.68735	65.79792	107.9455	112.4699
84	Variance	0.1269569	187.3436	4329.366	11652.22	12649.48
85	Skewness	-31.52797	-8.285047	-1.647313	-0.230145	-1.48E-02
86	Kurtosis	996.006	85.60149	5.455792	2.672569	2.967134
87	Errors Calcu	0	0	0	0	0
88	Mode	250	375	500	625	390.5545
89	5% Perc	250	375	314.0754	251.5754	189.0754
90	10% Perc	250	375	355.4924	292.9924	230.4924
91	15% Perc	250	375	382.9496	320.4496	257.9496
92	20% Perc	250	375	405.072	342.572	280.072
93	25% Perc	250	375	423.7888	361.2888	298.7888
94	30% Perc	250	375	440.8098	378.3098	315.8098
95	35% Perc	250	375	456.6077	394.1077	331.6077
96	40% Perc	250	375	471.358	408.858	346.358
97	45% Perc	250	375	485.665	423.165	360.665
98	50% Perc	250	375	499.9558	437.4558	374.9558
99	55% Perc	250	375	500	451.3856	388.8856
100	60% Perc	250	375	500	465.8172	403.3172
101	65% Perc	250	375	500	480.6487	418.1487
102	70% Perc	250	375	500	496.4514	433.9514
103	75% Perc	250	375	500	513.1733	450.6732
104	80% Perc	250	375	500	532.158	469.658
105	85% Perc	250	375	500	553.6349	491.135
106	90% Perc	250	375	500	581.4592	518.9592
107	95% Perc	250	375	500	621.5056	559.0056
108	Filter Minimum					
109	Filter Maximum					
110	Type (1 or 2)					
111	# Values Fil	0	0	0	0	0
112	Scenario #1	>75%	>75%	>75%	>75%	>75%
113	Scenario #2	<25%	<25%	<25%	<25%	<25%
114	Scenario #3	>90%	>90%	>90%	>90%	>90%
115	Target #1 (V	400	400	400	400	400
116	Target #1 (F	100%	100%	18.70%	36.90%	58.82%
117	Target #2 (V	250	375	500	625	634.03864
118	Target #2 (F	99%	99%	99%	99%	99%
119	Target #3 (V	360	360	360	360	360
120	Target #3 (F	100%	3.60%	10.70%	24.60%	44.72%
121	Target #4 (Value)			453.44424		
122	Target #4 (Perc%)			34%		

FIGURE 5

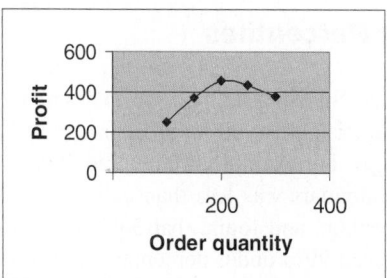

FIGURE 6

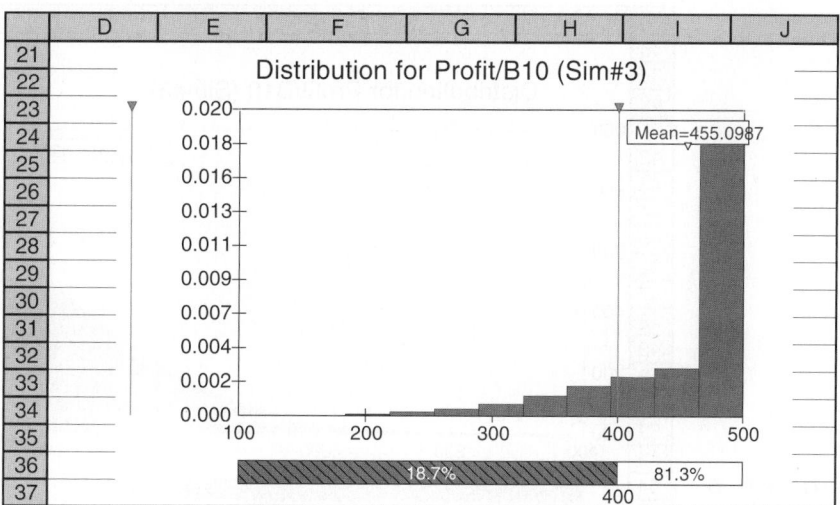

FIGURE 7

profit was less than or equal to $500. We entered $360 under value and found that profit was less than or equal to $360 10.7% of the time.

Creating Graphs with @Risk

To create a histogram of possible profits in Example 1, go to the Results menu and right click on the output cell Profit from the Explorer style list. Then choose Histogram and the third simulation (for 200 calendars ordered) to obtain a histogram similar to Figure 7. By moving the sliders at the bottom of the graph, you may zero in on the probability of any range of values. For example, there is an 18.7% chance that profit will be $400 or less. To paste any graph into Excel, right click on the graph and select Copy. You may also copy a graph into Excel by selecting Graph in Excel Option.

If we right click on a selected graph, we may change it to a **cumulative ascending graph.** See Figure 8.

Figure 8 gives the probability that profit is less than or equal to the x-value. Thus, there is around a 19% chance that profit is ≤$400.

By right clicking on a histogram or cumulative ascending graph and selecting Format, we can obtain a **cumulative descending graph.** (See Figure 9.) In a cumulative descending graph, the y-coordinate is the probability that profit exceeds the x-coordinate. For example, there is approximately an 81% probability that profit will exceed $400.

Using the Report Settings Option

 You may also create graphs and statistical reports directly with the Report Settings option. By choosing the Report Settings icon, any output may be sent directly to the current workbook or a new workbook. For example, see Figure 10. Checking the dialog box as shown there, and choosing Generate Reports Now, would place the output that has been generated in a new workbook.

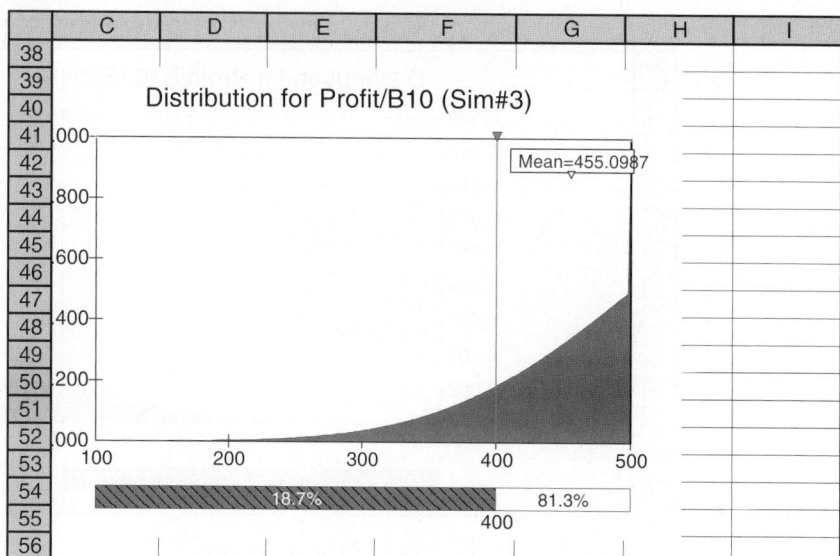

FIGURE 8
Cumulative Ascending
Graph

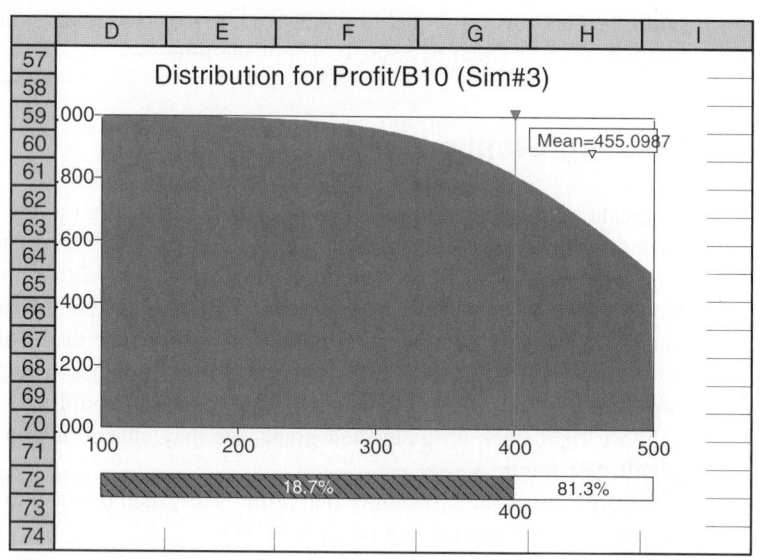

FIGURE 9
Cumulative Descending
Graph

Using @Risk Statistics

Instead of generating large reports, you may just want your spreadsheet to show the mean and standard deviation (and possibly other statistics) of your output cells. @Risk 4.5 contains statistical functions that accomplish this goal. To see how this works, enter in cell F12 the formula

$$=RISKMEAN(\$B\$10,F11)$$

and copy this formula to G12:J12. In F12, this keeps track of the mean of the first simulation. In G12, it keeps track of the mean for the second simulation, etc.

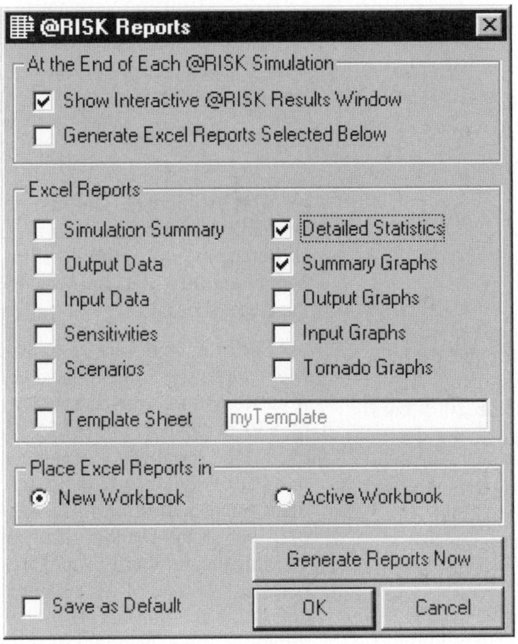

FIGURE 10

For example, in cell F12 we entered the formula

$$=\text{RISKSTDDEV}(\$B\$10,F11)$$

Then we copied this formula from F12 to G12:J12. This formula keeps track of the standard deviation from each order quantity. The results are as follows:

	E	F	G	H	I	J
10						
11		1	2	3	4	5
12	mean	$249.99	$372.75	$455.10	$435.25	$374.93
13	sigma	$0.36	$13.69	$65.80	$107.95	$112.47

For example, in the third simulation, for which we ordered 200 calendars, the mean profit for 1,000 iterations was $455.10, with a standard deviation of $65.80.

PROBLEMS

Group A

1 Explain why expected profit must be maximized by ordering a quantity equal to some possible demand for calendars. (*Hint:* If this is not the case, then some order quantity, such as 190 calendars, must maximize expected profit. If ordering 190 calendars maximizes expected profit, then it must yield a higher expected profit than an order size

of 150. But then an order of 200 calendars must also yield a larger expected profit than 190 calendars. This contradicts the assumed optimality of ordering 190 calendars!)

2 In August 2004, a car dealer is trying to determine how many 2005 cars should be ordered. Each car ordered in

TABLE 1

No. of Cars Demanded	Probability
20	.30
25	.15
30	.15
35	.20
40	.20

August 2004 costs $10,000. The demand for the dealer's 2005 models has the probability distribution shown in Table 1. Each car sells for $15,000. If demand for 2005 cars exceeds the number of cars ordered in August, the dealer must reorder at a cost of $12,000 per car. Excess cars may be disposed of at $9,000 per car. Use simulation to determine how many cars should be ordered in August. For your optimal order quantity, find a 95% confidence interval for expected profit.

3 Suppose that the bookstore in Example 1 receives no money for the first 50 excess calendars returned, but still receives $.75 for each subsequent calendar returned. Does this change the optimal order quantity?

4 A TSB (Tax Saver Benefit plan) allows you to put money into an account at the beginning of the calendar year to use for medical expenses. This amount is not subject to federal tax (hence the phrase TSB). As you pay medical expenses during the year, you are reimbursed by the administrator of the TSB, until the TSB account is exhausted. The catch is, however, that any money left in the TSB at the end of the year is lost to you. You estimate that it is equally likely that your medical expenses for next year will be $3,000, $4,000, $5,000, $6,000, or $7,000. Your federal income tax rate is 40%. Assume your annual salary is $50,000.

a How much should you put in a TSB? Consider both expected disposable income and the standard deviation of disposable income in your answer. (*Hint:* Your simu-

lation will indicate that two options have nearly the same expected disposable income.)

b Does your annual salary influence the correct decision?

Group B

5 For Problem 2, suppose that the demand for cars is normally distributed with $\mu = 40$ and $\sigma = 7$. Use simulation to determine an optimal order quantity. For your optimal order quantity, determine a 95% confidence interval for expected profit.

6 Six months before its annual convention, the American Medical Association must determine how many rooms to reserve. At this time, the AMA can reserve rooms at a cost of $50 per room. The AMA must pay the $50 room cost even if the room is not occupied. The AMA believes that the number of doctors attending the convention will be normally distributed, with a mean of 5,000 and a standard deviation of 1,000. If the number of people attending the convention exceeds the number of rooms reserved, extra rooms must be reserved at a cost of $80 per room. Use simulation to determine the number of rooms that should be reserved to minimize the expected cost to the AMA.

7 A ticket from Indianapolis to Orlando on Deleast Airlines sells for $150. The plane can hold 100 people. It costs $8,000 to fly an empty plane. The airline incurs variable costs of $30 (food and fuel) for each person on the plane. If the flight is overbooked, anyone who cannot get a seat receives $300 in compensation. On the average, 95% of all people who have a reservation show up for the flight. To maximize expected profit, how many reservations for the flight should be taken by Deleast? (*Hint:* The @Risk function RISKBINOMIAL can be used to simulate the number of passengers who show up. If the number of reservations taken is in cell A2, then the formula

$$=\text{RISKBINOMIAL}(A2,.95)$$

will generate the number of customers who actually show up for a flight!)

23.2 Modeling Cash Flows from a New Product

In this section, we will show how GM and Eli Lilly model the cash flows from new products. We begin by discussing the important triangular random variable.

The Triangular Random Variable

Managers often analyze in terms of best case, worst case, and most likely outcome. They often fail to realize that any value between the best and worst cases may occur. The triangular random variable can help.

Suppose we want to model first-year market share for a new product. We feel that the worst case is 20%, the most likely share is 40%, and the best case is 70%. We will model year 1 market share with a **triangular random variable.** See Figure 11. Basically, @Risk generates year 1 market share by making the likelihood of a given share proportional to

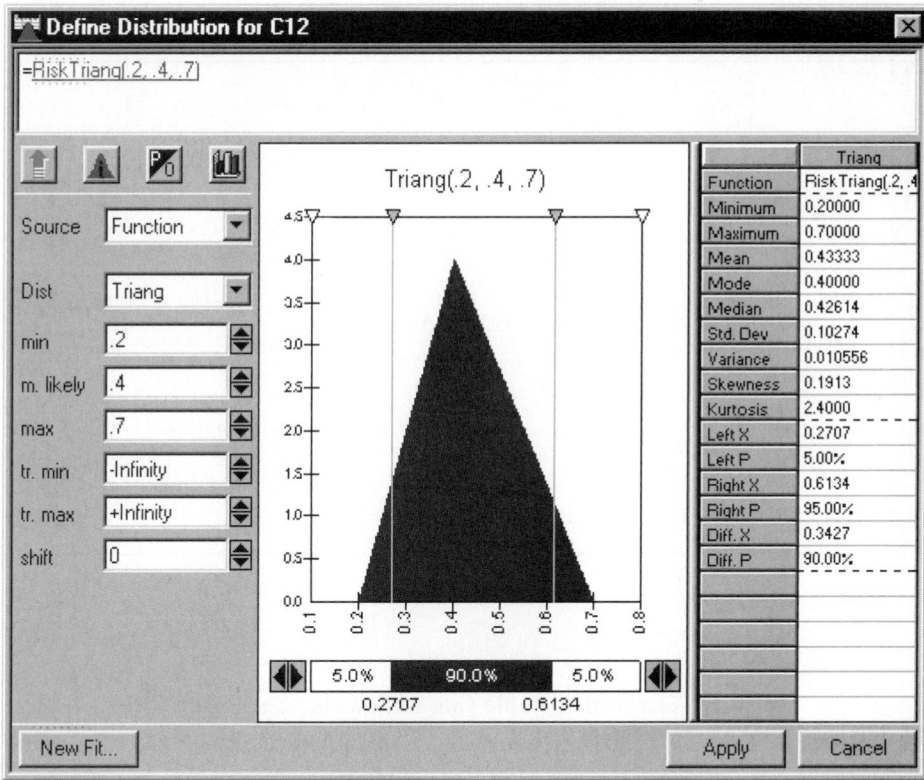

FIGURE 11

the height of the triangle in Figure 11. Thus, a 40% year 1 market share is most likely; all simulated market shares will be between 20% and 70%. A 30% market share occurs half as often as a year 1 40% market share, etc. The maximum height of the triangle is 4, because that makes the total area under the triangle equal to 1. The probability of market share being in a given range is equal to the area in that range under the triangle. For example, the chance of market share being at most 40% is .5*(4)*(.4 − .2) = .4 or 40%. To display this distribution, choose the Define Distributions icon and select the triangular random variable. Enter min = 0.2, m. likely = 0.4, and max = 0.7. You will see the picture in Figure 11.

EXAMPLE 2 General Motors

GM is trying to estimate the cash flows from a new car that will sell for 5 years. During the current year (year 0), a fixed development cost of $1.4 billion is incurred. This cost is depreciated on a straight-line basis over the next 5 years. Year 1 unit sales of the new model are assumed to follow a triangular random variable with worst case of 100,000 units, most likely case of 150,000 units, and best case of 170,000 units. Sales during years 2–5 are assumed to "decay" at the same rate each year. This annual decay rate is assumed to follow a triangular random variable with best case of 5%, most likely case of 8%, and worst case of 10%. Each year, a car sells for $15,000. During year 1, each car sold incurs a variable cost of $10,000. Due to increased labor costs, the variable cost of producing the car increases 4% a year. The tax rate is 40%, and cash flows are discounted at 15% a year. (Assume all cash flows occur at the end of the year.)

a Estimate the mean NPV of the cash flows from the new car.

b What fraction of the time will the new model add value to GM?

FIGURE **12**

	C	D	E	F	G	H	I
2							
3							
4	tax rate	0.4					
5	cost growth	0.04					
6	discount rate	0.15					
7	decay rate	0.072255268					
8							
9		Time					
10		0	1	2	3	4	5
11	Cost	1.40E+09					
12	Unit Sales		144227.4769	133806.2819	124138.0731	115168.4433	106846.9166
13	Price		$ 15,000.00	$ 15,000.00	$ 15,000.00	$ 15,000.00	$ 15,000.00
14	Unit cost		$ 10,000.00	$ 10,400.00	$ 10,816.00	$ 11,248.64	$ 11,698.59
15	Revenues		$ 2,163,412,154.19	$2,007,094,228.55	$ 1,862,071,096.57	$ 1,727,526,649.90	$ 1,602,703,748.32
16	Variable Cost		$ 1,442,274,769.46	$1,391,585,331.79	$ 1,342,677,398.70	$ 1,295,488,358.34	$ 1,249,957,799.41
17	Depreciation		$ 280,000,000.00	$ 280,000,000.00	$ 280,000,000.00	$ 280,000,000.00	$ 280,000,000.00
18	Before tax profit		$ 441,137,384.73	$ 335,508,896.76	$ 239,393,697.87	$ 152,038,291.56	$ 72,745,948.91
19	After tax profit		$ 264,682,430.84	$ 201,305,338.05	$ 143,636,218.72	$ 91,222,974.93	$ 43,647,569.34
20	Cash flow	-1400000000	$ 544,682,430.84	$ 481,305,338.05	$ 423,636,218.72	$ 371,222,974.93	$ 323,647,569.34
21							
22	npv cash flows	$77,633,524.27					
23							

Solution

Gmcashflow.xls

Our work is in file Gmcashflow.xls. (See Figure 12.)

Recall that in years 1–5, cash flow = after-tax profit + depreciation.

Step 1   In cell B11, enter the fixed cost of 1.4e9. In cell B20, enter the year 0 cash flow with the formula

$$=-B11$$

Step 2   In cell E12, compute year 1 unit sales with the formula

$$=RISKTRIANG(100000,150000,170000)$$

The syntax of the RISKTRIANG function requires that the lowest value of the random variable be entered first, followed by the most likely value, followed by the largest value.

Step 3   In cell D7, simulate the decay rate with the formula

$$=RISKTRIANG(0.05,0.08,0.1)$$

Step 4   In cells F12:I12, compute unit sales for years 2–5 by copying from F12 to G12:I12 the formula

$$=(1-decay\_rate)*E12$$

The cell D7 has been named decay_rate.

Step 5   Enter in E13:I13 the unit price of $15,000.

Step 6   In cell E14, enter the year 1 variable cost of $10,000. Then in cells F14:I14, compute the variable cost for years 1–5 by copying from F14 to G14:I14 the formula

$$=E14*(1+\$D\$5)$$

Step 7   Copying from E15 to F15:I15 the formula

$$=E13*E12$$

computes the sales revenue for each year.

	H	I
29		gmcashflowdecay.xls
30	Name	npv cash flows / Time
31	Description	Output
32	Cell	Sheet1!D22
33	Minimum	-2.19E+08
34	Maximum	2.55E+08
35	Mean	4.31E+07
36	Std Deviation	9.92E+07
37	Variance	9.84E+15
38	Skewness	-0.3451601
39	Kurtosis	2.396719
40	Errors Calculated	0
41	Mode	6.10E+07
42	5% Perc	-1.35E+08
43	10% Perc	-1.03E+08
44	15% Perc	-7.06E+07
45	20% Perc	-4.68E+07
46	25% Perc	-3.02E+07
47	30% Perc	-8040124
48	35% Perc	9848326
49	40% Perc	2.64E+07
50	45% Perc	4.13E+07
51	50% Perc	5.76E+07
52	55% Perc	6.90E+07
53	60% Perc	8.02E+07
54	65% Perc	9.26E+07
55	70% Perc	1.04E+08
56	75% Perc	1.18E+08
57	80% Perc	1.32E+08
58	85% Perc	1.49E+08
59	90% Perc	1.66E+08
60	95% Perc	1.90E+08
61	Filter Minimum	

	L	M	N
34			
35	95% CI		
36	for Mean		
37	NPV		
38	Lower	3.69E+07	43-2(99)/sqrt(1000)
39	Upper	4.94E+07	43+2(99)/sqrt(1000)

FIGURE 13

Step 8 Copying from E16 to F16:I16 the formula

$$=E14*E12$$

computes the variable cost for each year.

Step 9 In cells E17:I17, compute the depreciation for each of years 1–5 by copying from E17 to F17:I17 the formula

$$=\$D\$11/5$$

Step 10 By copying from E18 to F18:I18 the formula

$$=E15-E16-E17$$

we determine before-tax profit for years 1–5.

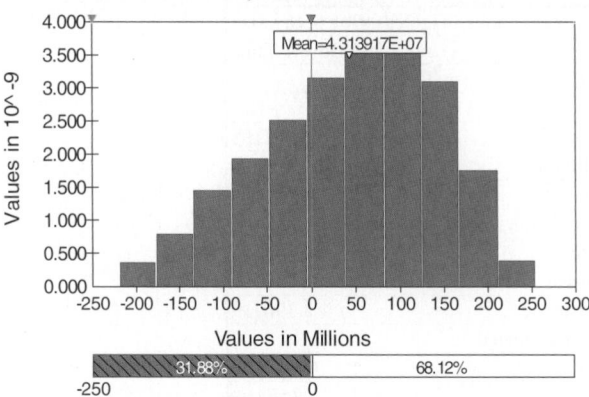

Distribution for npv cash flows / Time/D22

FIGURE 14

Step 11 By copying from E19 to F19:I19 the formula

$$=(1\text{-tax\_rate})*E18$$

we determine after-tax profit for years 1–5.

Step 12 By copying from E20 to F20:I20 the formula

$$=E19+E17$$

we add each year's depreciation to its after-tax profit to compute the year's cash flow.

Step 13 Assuming end-of-year cash flows, the formula

$$=NPV(0.15,D20:I20)$$

in cell D22 computes the NPV of all cash flows.

Step 14 After making cell D22 an output cell and running 1,000 iterations, we obtain the statistical output shown in Figure 13 and the graphical output in Figure 14.

From Figure 13, the mean NPV of cash flows (or risk-adjusted NPV) is $43 million. We are 95% certain that mean NPV is between $37 million and $49 million. Figure 14 shows that there is a 32% chance the project will have cash flows with a negative NPV (thereby reducing the company's value) and a 68% chance that cash flows will have a positive NPV.

The Lilly Model

In the car business, a new model virtually always has reduced sales every year. A new drug, however, sees increased sales in the first few years, followed by reduced sales. To model this form of the product life cycle, we must incorporate the following sources of uncertainty. (Note that we assume that total number of years for which the drug is sold is known).

- Number of years for which unit sales increase
- Average annual percentage increase in sales during the sales-increase portion of the sales period
- Average annual percentage decrease in sales during the sales-decrease portion of the sales period

FIGURE **15**

	B	C	D	E	F	G	H	I	J	K	L	M	N
1													
2			Growth then decay										
3		length of growth	5										
4		tax rate	0.4										
5		cost growth	0.04										
6		discount rate	0.15										
7		growth rate	0.055313219										
8		decay rate	0.117781276										
9			Time										
10			0	1	2	3	4	5	6	7	8	9	10
11		Cost	1.60E+09										
12		Unit Sales		1.12E+05	1.18E+05	1.25E+05	1.32E+05	1.39E+05	1.47E+05	1.29E+05	1.14E+05	1.01E+05	8.88E+04
13		Price		1.50E+04	1.50E+04	1.50E+04	1.50E+04	1.50E+04	1.50E+04	1.50E+04	1.50E+04	1.50E+04	1.50E+04
14		Unit cost		1.00E+04	1.04E+04	1.08E+04	1.12E+04	1.17E+04	1.22E+04	1.27E+04	1.32E+04	1.37E+04	1.42E+04
15		Revenues		1.68E+09	1.77E+09	1.87E+09	1.98E+09	2.08E+09	2.20E+09	1.94E+09	1.71E+09	1.51E+09	1.33E+09
16		Variable Cost		1.12E+09	1.23E+09	1.35E+09	1.48E+09	1.63E+09	1.78E+09	1.64E+09	1.50E+09	1.38E+09	1.26E+09
17		Depreciation		1.60E+08	1.60E+08	1.60E+08	1.60E+08	1.60E+08	1.60E+08	1.60E+08	1.60E+08	1.60E+08	1.60E+08
18		Before tax profit		4.00E+08	3.84E+08	3.62E+08	3.34E+08	2.99E+08	2.56E+08	1.44E+08	5.01E+07	-2.77E+07	-9.19E+07
19		After tax profit		2.40E+08	2.30E+08	2.17E+08	2.00E+08	1.79E+08	1.53E+08	8.62E+07	3.01E+07	-1.66E+07	-5.51E+07
20		Cash flow	-1600000000	4.00E+08	3.90E+08	3.77E+08	3.60E+08	3.39E+08	3.13E+08	2.46E+08	1.90E+08	1.43E+08	1.05E+08
21													
22		npv cash flows	($290,597,621.28)										

Lillygrowth.xls

Example 3 shows how to model this type of product life cycle. See file Lillygrowth.xls and Figure 15.

EXAMPLE 3   **Eli Lilly**

Lilly is producing a new drug that will be sold for 10 years. Year 1 unit sales are assumed to follow a triangular random variable with worst case 100,000 units, most likely case 150,000, and best case 170,000. The year 0 fixed cost of developing the drug is $1.6 billion, to be depreciated on a 10-year straight-line basis. Sales are equally likely to increase for 3, 4, 5, or 6 years, with the average percentage increase during those years following a triangular random variable with worst case 5%, most likely case 8%, and best case 10%. During the remainder of the 10-year sales life of the drug, unit sales will decrease at a rate governed by a triangular random variable having best case 8%, most likely case 12%, and worst case 18%. During each year, a unit of the drug sells for $15,000. Year 1 variable cost of producing a unit of the drug is $10,000. The unit variable cost of producing the drug increases at 4% a year.

a Estimate the mean NPV of the drug's cash flows.

b What is the probability that the drug will add value to Lilly?

c What source of uncertainty is the most important driver of the drug's NPV?

Solution After dragging our formulas to create years 6–10 and changing the depreciation in row 17 to be over a 10-year period, we simulate random variables in D3 (length of sales increase), D7 (annual percentage rate of sales increase), and D8 (annual percentage rate of sales decrease) with the following formulas

$$\text{Cell D3: } = \text{RISKDUNIFORM}(\{3,4,5,6\})$$

The RISKDUNIFORM variable is a discrete random variable that assigns equal probability to each listed value.

$$\text{Cell D7: } = \text{RISKTRIANG}(0.05, 0.08, 0.1)$$
$$\text{Cell D8: } = \text{RISKTRIANG}(0.08, 0.12, 0.18)$$

In cell E12, we generate year 1 units sales with the formula

$$=RISKTRIANG(100000,150000,170000)$$

Copying from F12 to G12:N12 the formula

$$=IF(F10 \leq length\_of\_growth+1,E12*(1+growth\_rate),E12*(1-decay\_rate))$$

generates unit sales for years 2–10. Note that our formula increases annual sales by the growth rate for length-of-growth years and decreases annual sales by decay rate during later years. (D3 is named length_of_growth, D7 is named growth_rate, and D8 is named decay_rate.)

We used Autoconvergence to determine the number of iterations for @Risk to run. Under Simulation Settings, selecting Iterations Auto and a change of 1% ensures that @Risk will keep running iterations until, during the last 100 iterations, the mean, standard deviation, and selected other statistics change by 1% or less. In this example, @Risk ran 1,800 iterations, yielding the results in Figure 16. There was an estimated mean of −$29 million and a 54% chance of negative NPV. Right clicking on NPV from the Explorer interface yields the histogram in Figure 17. The histogram shows a 53% chance that the drug will decrease Lilly's NPV.

For part (c), use a **tornado graph** to determine the key drivers of NPV. To obtain a tornado graph, you must have selected the Collect All Outputs box from the Simulation Settings Sampling dialog box. (Unless you want a tornado graph, it is probably best to uncheck that box. Checking that box adds a column to your output for each @Risk function in the model, and this can clutter up the output.) Right click on NPV in the Explorer interface and select Tornado Graph. We can obtain a correlation and/or regression tornado graph as shown in Figures 18 and 19.

Each bar of the correlation tornado graph (Figure 18) gives the correlation of the @Risk random variable with NPV. For example,

- Year 1 unit sales has a .98 correlation with NPV.
- Annual growth rate has a .14 correlation with NPV.

In short, the uncertainty about year 1 unit sales is very important for determining NPV, but other random variables could probably be replaced by their mean without changing the distribution of NPV by much.

For each @Risk random variable, the regression tornado graph (Figure 19) computes the *standardized regression coefficient* for the @Risk random variable when we try to predict NPV from all @Risk random variables in the spreadsheet. A standardized regression coefficient tells us (after adjusting for other variables in the equation) the number of standard deviations by which NPV changes when the given @Risk random variable changes by one standard deviation. For example,

- A one standard deviation change in year 1 unit sales will (ceteris paribus) change NPV by .98 standard deviation.
- A one standard deviation change in annual growth rate will increase NPV by .15 standard deviation (ceteris paribus).

Again it is clear that the uncertainty for year 1 sales is really all that matters here; other random variables may as well be replaced by their means.

	C	D	E
24			
25			
26		Name	npv cash fl
27		Description	Output
28		Cell	D22
29		Minimum	-3.54E+08
30		Maximum	2.37E+08
31		Mean	-2.86E+07
32		Std Deviation	1.23E+08
33		Variance	1.52E+16
34		Skewness	-0.34653
35		Kurtosis	2.440396
36		Errors Calculated	0.00E+00
37		Mode	2.33E+07
38		5% Perc	-2.52E+08
39		10% Perc	-2.06E+08
40		15% Perc	-1.71E+08
41		20% Perc	-1.41E+08
42		25% Perc	-1.14E+08
43		30% Perc	-9.07E+07
44		35% Perc	-6.99E+07
45		40% Perc	-5.01E+07
46		45% Perc	-3.21E+07
47		50% Perc	-1.31E+07
48		55% Perc	4.01E+06
49		60% Perc	1.94E+07
50		65% Perc	3.26E+07
51		70% Perc	4.98E+07
52		75% Perc	6.48E+07
53		80% Perc	8.02E+07
54		85% Perc	9.80E+07
55		90% Perc	1.23E+08
56		95% Perc	1.56E+08
57		Filter Minimum	
58		Filter Maximum	
59		Type (1 or 2)	
60		# Values Filtered	0
61		Scenario #1	>75%
62		Scenario #2	<25%
63		Scenario #3	>90%
64		Target #1 (Value)	0
65		Target #1 (Perc%)	53.73%

FIGURE 16

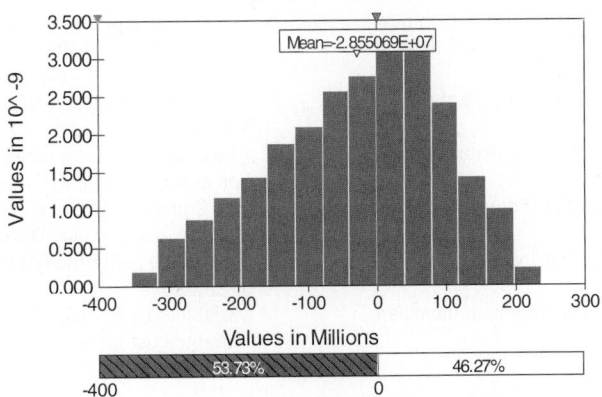

Distribution for npv cash flows / Time/D22

FIGURE 17

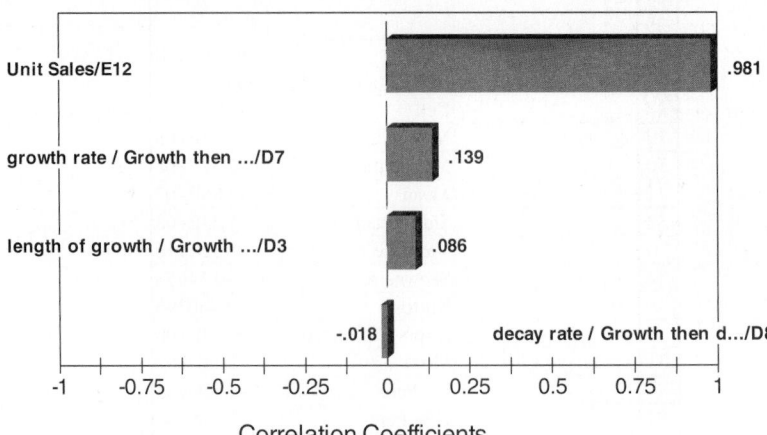

Correlations for npv cash flows / Time/D22

Unit Sales/E12	.981
growth rate / Growth then .../D7	.139
length of growth / Growth .../D3	.086
decay rate / Growth then d.../D8	-.018

FIGURE 18

Correlation Coefficients

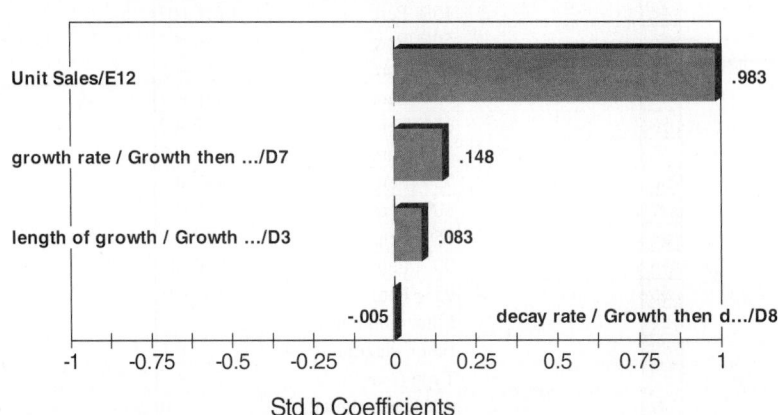

Regression Sensitivity for npv cash flows/
Time/D22

Unit Sales/E12	.983
growth rate / Growth then .../D7	.148
length of growth / Growth .../D3	.083
decay rate / Growth then d.../D8	-.005

FIGURE 19

Std b Coefficients

PROBLEMS

Group A

1 Dord Motors is considering whether to introduce a new model: the Racer. The profitability of the Racer will depend on the following factors:

- Fixed cost of developing Racer: Equally likely to be $3 billion or $5 billion.
- Sales: Year 1 sales will be normally distributed with $\mu = 200{,}000$ and $\sigma = 50{,}000$.
  Year 2 sales will be normally distributed with $\mu =$ year 1 sales and $\sigma = 50{,}000$.
  Year 3 sales will be normally distributed with $\mu =$ year 2 sales and $\sigma = 50{,}000$.
  For example, if year 1 sales = 180,000, then the mean for year 2 sales will be 180,000.

- Price: Year 1 price = $13,000
  Year 2 price = 1.05*{(year 1 price) + $30*(% by which year 1 sales exceed expected year 1 sales)}
  The 1.05 is the result of inflation!
  Year 3 price = 1.05*{(year 2 price) + $30*(% by which year 2 sales exceed expected year 2 sales)}
  For example, if year 1 sales = 180,000, then year 2 price = 1.05*{13,000 + 30(−10)} = $13,335
- Variable cost per car: During year 1, the variable cost per car is equally likely to be $5,000, $6,000, $7,000, or $8,000.
  Variable cost for year 2 = 1.05*(year 1 variable cost)
  Variable cost for year 3 = 1.05*(year 2 variable cost)

TABLE 2

Year	1	2	3
GNP	3%	5%	4%
INF	4%	7%	3%

TABLE 3

Number of Competitors	Probability
0	.50
1	.30
2	.10
3	.10

TABLE 4

	Year 1	Year 2	Year 3
Sales price	$15,000	$16,000	$17,000
Variable cost	$12,000	$13,000	$14,000

TABLE 5

Time Abandoned	Value Received
End of year 1	$3,000
End of year 2	$2,600
End of year 3	$1,900
End of year 4	$900

Your goal is to estimate the NPV of the new car during its first three years. Assume that cash flows are discounted at 10%; that is, $1 received now is equivalent to $1.10 received a year from now.

a Simulate 400 iterations and estimate the mean and standard deviation of the NPV the first three years of sales.

b I am 95% sure that the expected NPV of this project is between _____ and _____.

c Use the Target option to determine a 95% confidence interval for the actual NPV of the Racer during its first three years of production.

d Use a tornado graph to analyze which factors are most influential in determining the NPV of the Racer.

2 Trucko produces the Goatco truck. The company wants information about the discounted profits earned during the next three years. During a given year, the total number of trucks sold in the United States is 500,000 + 50,000*GNP − 40,000*INF, where

GNP = % increase in GNP during year

INF = % increase in Consumer Price Index during year

Value Line has made the predictions given in Table 2 for the increase in GNP and INF during the next three years.

In the past, 95% of Value Line's GNP predictions have been accurate within 6% of the actual GNP increase, and 95% of Value Line's INF predictions have been accurate within 5% of the actual inflation increase.

At the beginning of each year, a number of competitors may enter the trucking business. At the beginning of a year, the probability that a certain number of competitors will enter the trucking business is given in Table 3.

Before competitors join the industry at the beginning of year 1, there are two competitors. During a year that begins (after competitors have entered the business, but before any have left) with c competitors, Goatco will have a market share given by $.5*(.9)^c$. At the end of each year, there is a 20% chance that each competitor will leave the industry.

The sales price of the truck and production cost per truck are given in Table 4.

a Simulate 500 times the next three years of Truckco's profit. Estimate the mean and variance of the discounted three-year profits (use a discount rate of 10%).

b Do the same if during each year there is a 50% chance that each competitor leaves the industry.

(*Hint:* You can model the number of firms leaving the industry in a given period with the RISKBINOMIAL function. For example, if the number of competitors in the industry is in cell A8, then the number of firms leaving the industry during a period can be modeled with the statement =RISKBINOMIAL(A8,.20). Just remember that the RISKBINOMIAL function is not defined if its first argument equals 0.)

Group B

3 You have the opportunity to buy a project that yields at the end of years 1–5 the following (random) cash flows:

End of year 1 cash flow is normal with mean 1,000 and standard deviation 200.

For $t > 1$, end of year t cash flow is normal with Mean = actual end of year $(t − 1)$ cash flow and Standard deviation = .2*(mean of year t cash flow).

a Assuming cash flows are discounted at 10%, determine the expected NPV (in time 0 dollars) of the cash flows of this project.

b Suppose we are given the following option: At the end of year 1, 2, 3, or 4, we may give up our right to future cash flows. In return for doing this, we receive the *abandonment value* given in Table 5.

Assume that we make the abandonment decision as follows: We abandon if and only if the expected NPV of the cash flows from the remaining years is smaller than the abandonment value. For example, suppose end of year 1 cash flow is $900. At this point in time, our best guess is that cash flows from years 2–5 will also be $900. Thus, we would abandon the project at the end of year 1 if $3,000 exceeded the NPV of receiving $900 for four straight years. Otherwise, we would continue. What is the expected value of the abandonment option?

4 Mattel is developing a new Madonna doll. Managers have made the following assumptions.

It is equally likely that the doll will sell for two, four, six, eight, or ten years.

At the beginning of year 1, the potential market for the doll is 1 million. The potential market grows by an average of 5% per year. They are 95% sure that the growth in the potential market during any year will be between 3% and 7%.

They believe their share of the potential market during year 1 will be at worst 20%, most likely 40%, and at best 50%. All values between 20% and 50% are possible.

The variable cost of producing a doll during year 1 is equally likely to be $4 or $6.

The sales price of the doll during year 1 will be $10.

Each year, the sales price and variable cost of producing the doll will increase by 5%.

The fixed cost of developing the doll (incurred in year 0) is equally likely to be $4, $8, or $12 million.

At time 0, there is one competitor in the market. During each year that begins with four or fewer competitors, there is a 20% chance that a new competitor will enter the market.

To determine year t unit sales (for $t > 1$), proceed as follows. Suppose that at the end of year $t - 1$, x competitors were present. Then assume that during year t, a fraction $.9 - .1*x$ of loyal customers (last year's purchasers) will buy a doll during the next year and a fraction $.2 - .04*x$ of people currently in the market who did not purchase a doll last year will purchase a doll from the company this year. We now generate a prediction for year t unit sales. Of course, this prediction will not be precise. We assume that it is sure to be accurate within 15%, however.

Cash flows are discounted at 10% per year.

a Estimate the expected NPV (in time 0 dollars) of this project.

b You are 95% sure the expected NPV of this project is between _____ and _____.

c You are 95% sure that the actual NPV of the project is between _____ and _____.

d What two factors does the tornado diagram indicate are key drivers of the project's profitability?

5 GM is thinking of marketing a new car, the Batmobile. It is equally likely that the car will take 1, 2, or 3 years to develop. This may be modeled by a RISKDUNIFORM random variable. A RISKDUNIFORM function is equally likely to assume any of the values listed in the cell.

Development cost is assumed equally split over development time. The best case is development cost of

TABLE 6

Years	Probability
4	.1
5	.3
6	.4
7	.2

$300 million, the most likely case is $800 million, and the worst case is $1.7 billion.

The product will begin sales during the year after development concludes. The number of years the car will be sold is assumed to be governed by the probability distribution in Table 6.

The size of the market during the first year of sales is unknown, but the worst case is a market size of 100,000, the most likely case is 145,000, and the best case is 165,000. Annual growth in market size is unknown, but is assumed to have a worst case of 1% per year, a most likely case of 6% a year, and a best case of 8% per year.

First-year market share is unknown, but the worst case is a 30% market share, the most likely case is 45%, and the best case is 50%. After the first year of sales, market share will fluctuate. On average, next year's share will equal this year's share. We are 95% sure that next year's market share will be within 40% of this year's market share.

During the first year of sales, price is unknown, with a worst-case price of $16,000, a most likely price of $17,500, and a best-case price of $18,000. Each year, price increases by 5%.

During the first year of sales, the best-case estimate for the cost of producing a car is $11,000, the most likely cost is $13,000, and the worst-case cost is $14,500. Each year, variable cost increases by 5%.

The discount rate for this project is 15%.

a You are 95% sure that mean NPV for this project is between _____ and _____.

b What is the probability that the project will add value to the company?

c What are the key drivers of the project's success?

d Construct a graph that illustrates the range of possible NPVs that might be generated by this project.

23.3 Project Scheduling Models

In Chapter 7, we used linear programming to determine the length of time needed to complete a project. We also learned how to identify critical activities, where an activity is critical if increasing its activity time by a small amount increases the length of time needed to complete the project by the same amount. Our discussion there required the assumption that all activity times are known with certainty. In reality, these times are usually uncertain. Of course, this implies that the length of time needed to complete the project is also uncertain. It also implies that for each activity, there is a *probability* (not necessarily equal to 0 or 1) that the activity is critical.

To illustrate, suppose that activities *A* and *B* can begin immediately. Activity *C* can then begin as soon as activities *A* and *B* are both completed, and the project is completed as soon as activity *C* is completed. Activity *C* is clearly on the critical path, but what about *A* and *B*? Let's say that the *expected* activity times of *A* and *B* are 10 and 12. If we use these expected times and ignore any uncertainty about the actual times—that is, if we proceed as we did in Chapter 7—then activity *B* is definitely a critical activity. However, suppose there is some positive probability that *A* can have duration 12 and *B* can have duration 11. Under this scenario, *A* is a critical activity. Therefore, we cannot say in advance which of the activities, *A* or *B*, will be critical. However, by using simulation we can see how *likely* it is that each of these activities is critical. We can also see how long the entire project is likely to take. We illustrate with the following example.

EXAMPLE 4 | **Construction Project with Uncertain Activity Times**

Tom Lingley, an independent contractor, has agreed to build a new room on an existing house. He plans to begin work on Monday morning, June 1. The main question is when he will complete his work, given that he works only on weekdays. The owner of the house is particularly hopeful that the room will be ready by Saturday, June 27, that is, in 20 or fewer working days. The work proceeds in stages, labeled A through J, as summarized in Table 7. Three of these activities, E, F, and G, will be done by separate independent subcontractors. The *expected* durations of the activities (in days) are shown in the table. However, these are only best guesses. Lingley knows that the *actual* activities times can vary because of unexpected delays, worker illnesses, and so on. He would like to use computer simulation to see (1) how long the project is likely to take, (2) how likely it is that the project will be completed by the deadline, and (3) which activities are likely to be critical.

Solution

We first need to choose distributions for the uncertain activity times. Then, given any randomly generated activity times, we will illustrate a method for calculating the length of the project and identifying the activities on the critical path.

The Pert Distribution As always, there are several reasonable candidate probability distributions we could use for the random activity times. Here we illustrate a distribution that

TABLE 7
Activity Time Data

Description	Index	Predecessors	Expected Duration
Prepare foundation	A	None	4
Put up frame	B	A	4
Order custom windows	C	None	11
Erect outside walls	D	B	3
Do electrical wiring	E	D	4
Do plumbing	F	D	3
Put in ductwork	G	D	4
Hang drywall	H	E, F, G	3
Install windows	I	B, C	1
Paint and clean up	J	H	2

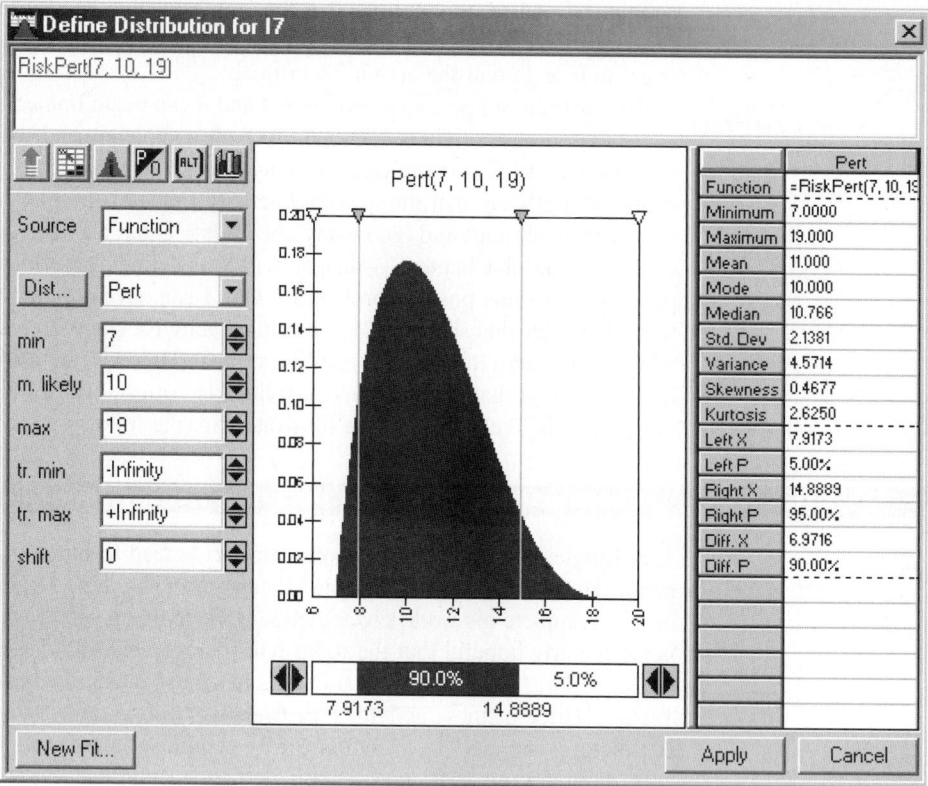

FIGURE 20
Pert Distribution

has become popular in project scheduling, called the *Pert distribution.*[†] As shown in Figure 20, it is a "rounded" version of the triangular distribution that is specified by three parameters: a minimum value, a most likely value, and a maximum value. The distribution in the figure uses the values 7, 10, and 19 for these three values, which implies a mean of 11. We will use this distribution for activity C. Similarly, for the other activities, we choose parameters for the Pert distribution that lead to the means in Table 7. In reality, it would be done the other way around. The contractor would estimate the minimum, most likely, and maximum parameters for the various activities, and the means would follow from these.

Developing the Simulation Model   The key to the model is representing the project network in activity-on-arc form, as in Figure 21, and then finding E_j for each j, where E_j is the earliest time we can get to node j. When the nodes are numbered so that all arcs go from lower-numbered nodes to higher-numbered nodes, we can calculate the E_j's iteratively, starting with $E_1 = 0$, with the equation

$$E_j = \max(E_i + t_{ij}) \tag{1}$$

Here, the maximum is taken over all arcs leading into node j, and t_{ij} is the activity time on such an arc. Then E_n is the time to complete the project, where n is the index of the finish node. This will make it very easy to calculate the project length.

[†]It is named after the acronym PERT (Program Review and Evaluation Technique) that is synonymous with project scheduling in an uncertain environment.

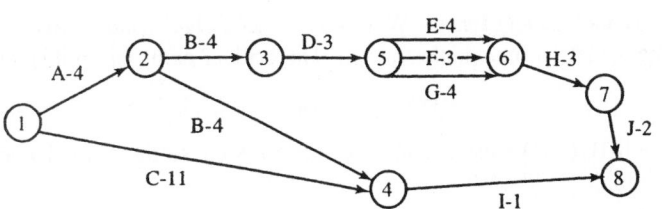

FIGURE 21
Project Network for Room-Building Project

FIGURE 22
Project Scheduling Simulation Model

	A	B	C	D	E	F	G	H	I	J
1	Room construction project									
2										
3	Data on activity network					Parameters of PERT distributions				
4	Activity	Code	Numeric index	Predecessors	Min	Most likely	Max	Implied mean	Duration	Duration+
5	Prepare foundation	A	1	None	1.5	3.5	8.5	4	2.158	2.159
6	Put up frame	B	2	A	3	4	5	4	4.513	4.513
7	Order custom windows	C	3	None	7	10	19	11	9.572	9.572
8	Erect outside walls	D	4	B	2	2.5	6	3	3.322	3.322
9	Do electrical wiring	E	5	D	3	3.5	7	4	3.282	3.282
10	Do plumbing	F	6	D	2	2.5	6	3	2.377	2.377
11	Put in duct work	G	7	D	2	4	6	4	4.668	4.668
12	Hang dry wall	H	8	E,F,G	2.5	3	3.5	3	3.197	3.197
13	Install windows	I	9	B,C	0.5	1	1.5	1	1.384	1.384
14	Paint and clean up	J	10	H	1.5	2	2.5	2	1.677	1.677
15										
16	Index of activity to increase		1							
17										
18	Event times									
19		Node	Event time	Event time+						
20		1	0	0						
21		2	2.158	2.159						
22		3	6.671	6.672						
23		4	9.572	9.572						
24		5	9.993	9.994						
25		6	14.661	14.662						
26		7	17.858	17.859						
27		8	19.536	19.537						
28										
29	Increase in project time?		1							
30										

We also need a method for identifying the critical activities for any given activity times. By definition, an activity is critical if a small increase in its activity time causes the project time to increase. Therefore, we will keep track of two sets of activity times and associated project times. The first uses the simulated activity times. The second adds a small amount, such as 0.001 day, to a "selected" activity's time. By using the RISKSIMTABLE function with a list as long as the number of activities, we can make each activity the "selected" activity in this method. The spreadsheet model appears in Figure 22, and the details are as follows. (See the Projectsim.xls file.)

Projectsim.xls

Inputs Enter the parameters of the Pert activity time distributions in the shaded cells and the implied means next to them. As discussed above, we actually chose the minimum, most likely, and maximum values while in @Risk's Model window to achieve the means in Table 7. Note that some of these distributions are symmetric about the most likely value, whereas others are skewed.

Activity Times Generate random activity times in column I by entering the formula

$$=RISKPERT(E5,F5,G5)$$

in cell I5 and copying it down.

Augmented Activity Times We want to successively add a small amount to each activity's time to determine whether it is on the critical path. To do this, enter the formula

$$=\text{RISKSIMTABLE}(\{1, 2, 3, 4, 5, 6, 7, 8, 9, 10\})$$

in cell B16. (We use a list of length 10 because there are 10 activities.) Then enter the formula

$$=\text{I5}+\text{IF}(\text{Index}=\text{C5},0.001,0)$$

in cell J5 and copy it down. (Here, Index is the range name of cell B16.) For example, if we are checking whether activity D (the 4th activity) is critical, the Index cell will be 4, and we will run a simulation where activity D's time is augmented by 0.001 and the other activity times are unchanged.

Event Times We want to use Equation (1) to calculate the node event times in the range B20:B27. There is no quick way to enter the required formulas. (We see no way of using Copy and Paste.) We need to use the project network as a guide for each node. Begin by entering 0 in cell B20. Then enter the appropriate formulas in the other cells. For example, the formulas in cells B22, B23, and B27 are

$$=\text{B21}+\text{I6}$$
$$=\text{MAX}(\text{B20}+\text{I7},\text{B21}+\text{I6})$$

and

$$=\text{RISKOUTPUT}()+\text{MAX}(\text{B23}+\text{I13},\text{B26}+\text{I14})$$

To understand these, note that node 3 has only one arc leading into it, and this arc originates at node 2. No MAX is required for this node's equation. In contrast, node 4 has two arcs leading into it, from nodes 1 and 2, so a MAX is required. Similarly, node 8 requires a MAX, because it has two arcs leading into it. Also, it is the finish node, so we designate its event time cell as an @Risk output cell—it contains the time to complete the project.

Augmented Event Times Copy the formulas in the range B20:B27 to the range C20:C27 to calculate the event times when the selected activity's time is augmented by 0.001.

Project Time Increases? To check whether the selected activity's increased activity time increases the project time, enter the formula

$$=\text{RISKOUTPUT}()+\text{IF}(\text{C27}>\text{B17},1,0)$$

If this calculates to 1, then the selected activity is critical for these particular activity times. Otherwise, it is not. Note that this cell is also designated as an @Risk output cell.

Using @Risk We set the number of iterations to 1,000 and the number of simulations to 10 (one for each activity that we want to check for being critical). After running @Risk, we request the histogram of project times in Figure 23. In Chapter 7, when the activity times were not considered random, the project time was 20 days. Now it varies from a low of 15.89 days to a high of 25.50 days, with an average of 20.42 days.[†] Although the 5th and 95th percentiles appear in the figure, it might be more interesting (and depressing) to Tom Lingley to see the probabilities of various project times being exceeded. For example, we entered 20 in the Left X box next to the histogram. The Left P value implies that there is about a 57% chance that the project will not be completed within 20 days.

[†]It can be shown mathematically that the expected project time is *always* greater than when the expected activity times are used to calculate the project time, as we did in Chapter 7. In other words, an assumption of certainty always leads to an underestimation of the true expected project time.

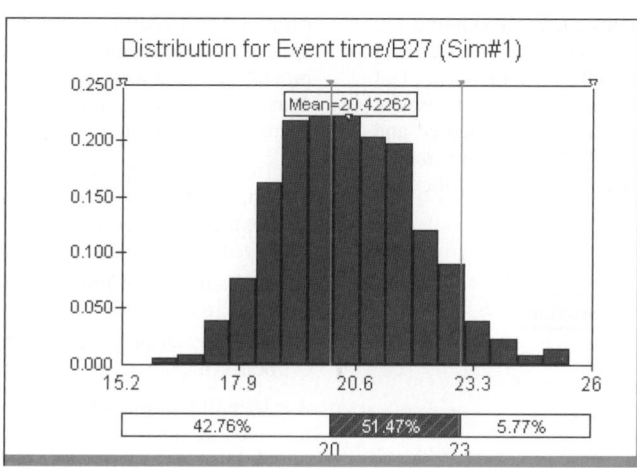

FIGURE 23
Histogram of Project Completion Time

Name	Event time
Cell	B27 Output (Sim#1)
Minimum	15.89254
Mean	20.42262
Maximum	25.50042
Std Dev	1.602013
Variance	2.566447
Skewness	0.292087
Kurtosis	2.924874
Mode	20.30279
Left X	20
Left P	42.75847%
Right X	23
Right P	94.2309%
Diff. X	3
Diff. P	51.47242%
5th Perc.	17.9489
95th Perc.	23.11927
#Errors	0
Filter Min	
Filter Max	
#Filtered	0

FIGURE 24
Probabilities of Activities Being Critical

	Name	Cell	Sim#	Minimum	Mean	Maximum
Output 2	Increase in project time? / Event time	B29	1	0	0.998	1
Output 2	Increase in project time? / Event time	B29	2	0	0.998	1
Output 2	Increase in project time? / Event time	B29	3	0	0.002	1
Output 2	Increase in project time? / Event time	B29	4	0	0.998	1
Output 2	Increase in project time? / Event time	B29	5	0	0.446	1
Output 2	Increase in project time? / Event time	B29	6	0	0.063	1
Output 2	Increase in project time? / Event time	B29	7	0	0.491	1
Output 2	Increase in project time? / Event time	B29	8	0	0.998	1
Output 2	Increase in project time? / Event time	B29	9	0	0.002	1
Output 2	Increase in project time? / Event time	B29	10	0	0.998	1

Similarly, the values in the Right X and Right P boxes imply that the chance of the project lasting longer than 23 days is slightly greater than 5%. This is certainly not good news for Lingley, and he might have to resort to the crashing we discussed in Chapter 8.

The summary measures for the B29 output cell appear in Figure 24. Each "simulation" in this output represents one selected activity being increased slightly. The Mean column indicates the fraction of iterations where the project time increases as a result of the selected activity's time increase. Hence, it represents the probability that this activity is critical. For example, the first activity (A) is always critical, the third activity (C) is never critical, and the fifth activity (E) is critical about 45% of the time. More specifically, we see that the critical path always includes activities A, B, D, H, J, and one of the three "parallel" activities E, F, and G.

PROBLEMS

Group A

1 The city of Bloomington is about to build a new water treatment plant. Once the plant is designed (D), we can select the site (S), the building contractor (C), and the operating personnel (P). Once the site is selected, we can erect the building (B). We can order the water treatment machine (W) and prepare the operations manual (M) only after the contractor is selected. We can begin training (T) the operators when both the operations manual and operating personnel selection are completed. When the treatment plant and the building are finished, we can install the treatment machine (I). Once the treatment machine is installed and operators are trained, we can obtain an operating license (L). The estimated mean and standard deviation of the time

TABLE 8

	Mean	Standard Deviation
Activity D	6	1.5
Activity S	2	3.0
Activity C	4	1.0
Activity P	3	1.0
Activity B	24	6.0
Activity W	14	4.0
Activity M	3	0.4
Activity T	4	1.0
Activity I	6	1.0
Activity L	3	6.0

TABLE 9

	Predecessors	Mean Time	Standard Deviation
Activity A: Hire workers	—	4	0.6
Activity B: Dig big hole	A	9	2.5
Activity C: Pour foundation	B	5	1.0
Activity D: Destroy room	A	7	2.0
Activity E: Build main structure	C	10	1.5

TABLE 10

	Predecessors	Mean Time	Standard Deviation
Activity A: Obtain funding	—	6	0.6
Activity B: Design building	A	8	1.3
Activity C: Prepare site	A	2	0.2
Activity D: Lay foundation	B, C	2	0.3
Activity E: Erect walls and roof	D	3	1.0
Activity F: Finish exterior	E	3	0.6
Activity G: Finish interior	D	7	1.5
Activity H: Landscape grounds	F, G	5	1.2

(in months) needed to complete each activity are given in Table 8. Use simulation to estimate the probability that the project will be completed in (a) under 50 days and (b) more than 55 days. Also estimate the probabilities that B, I, and T are critical activities.

2 To complete an addition to the Business Building, the activities in Table 9 need to be completed (all times are in months). The project is completed once Room 111 has been destroyed and the main structure has been built.

a Estimate the probability that it will take at least 3 years to complete the addition.

b For each activity, estimate the probability that it will be a critical activity.

3 To build Indiana University's new law building, the activities in Table 10 must be completed (all times are in months).

a Estimate the probability that the project will take less than 30 months to complete.

b Estimate the probability that the project will take more than 3 years to complete.

c For each of the activities A, B, C, and G, estimate the probability that it is a critical activity.

23.4 Reliability and Warranty Modeling

In today's high-tech world, it is very important to be able to compute the probability that a system made up of machines will work for a desired amount of time. The subject of estimating the distribution of machine failure times and the distribution of time to failure of a system is known as **reliability theory.**

Distribution of Machine Life

We assume the length of time (call it **X**) until failure of a machine is a continuous random variable having a distribution function $F(t) = P(\mathbf{X} \leq t)$ and a density function $f(t)$. Thus, for small Δt, the probability that a machine will fail between time t and $t + \Delta t$ is approximately $f(t)\Delta t$. The **failure rate** of a machine at time t [call it $r(t)$] is defined to be $(1/\Delta t)$ times the probability that the machine will fail between time t and time $t + \Delta t$, given that the machine has not failed by time t. Thus,

$$r(t) = \left(\frac{1}{\Delta t}\right) \text{Prob}(\mathbf{X} \text{ is between } t \text{ and } t + \Delta t | \mathbf{X} > t) = \frac{\Delta t f(t)}{\Delta t(1 - F(t))} = \frac{f(t)}{(1 - F(t))}$$

If $r(t)$ is an increasing function of t, the machine is said to have an **increasing failure rate (IFR)**. If $r(t)$ is a decreasing function of t, the machine is said to have a **decreasing failure rate (DFR).**

Consider an exponential distribution which has $f(t) = \lambda e^{-\lambda t}$ and $F(t) = 1 - e^{-\lambda t}$. Then we find that

$$r(t) = \frac{\lambda e^{-\lambda t}}{e^{-\lambda t}} = \lambda$$

Thus, a machine whose lifetime follows an exponential random variable has **constant failure rate.** This is analogous to the no-memory property of the exponential distribution discussed in Chapter 20.

The random variable that is most frequently used to model the time till failure of a machine is the **Weibull random variable.** The Weibull random variable has the following density and distribution functions:

$$f(t) = \frac{\alpha x^{\alpha-1}}{\beta^{\epsilon}} e^{-(t/\beta)^{\epsilon}}$$
$$F(t) = 1 - e^{(-t/\beta)^{\alpha}}$$

It can be shown that if $\beta < 1$, the Weibull random variable exhibits DFR, and if $\beta > 1$, the Weibull random variable exhibits IFR. The @Risk function RISKWEIBULL(alpha, beta) will generate an observation for a Weibull random variable having parameters α and β. If you input the mean and variance of observed machine times to failure into cells D4 and D5, respectively, of workbook Weibest.xls, the workbook computes the unique values of α and β that yield the observed mean and variance of times to failure. For example, we see in Figure 25 that if the mean time to machine failure were 12 months and the standard deviation were 6 months, then a Weibull with $\alpha = 2.2$ and $\beta = 13.55$ would yield the desired mean and variance.

Weibest.xls

Common Types of Machine Combinations

Three common types of machine combinations are as follows:

- **A series system.** A series system functions only as long as each machine functions. See Figure 26(a).
- **A parallel system.** A parallel system functions as long as at least one machine functions. See Figure 26(b).

	A	B	C	D	E	F	G
1		**Estimating Weibull**					
2		**Distribution Parameters**					
3							
4		Mean time to failure		12			
5		Variance of time to Failure		36			
6		Second Moment of failure time		180			
7		Second moment/(mean)^2		1.25		Beta	13.54976
8		Alpha				Alpha	2.2

FIGURE 25

(a) Series system
All *n* must work.

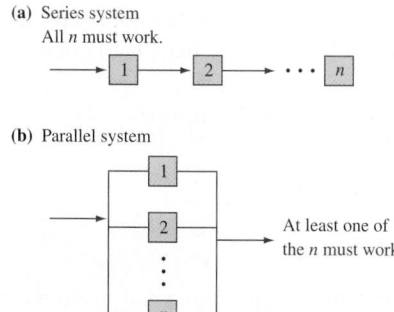

(b) Parallel system

At least one of
the *n* must work.

FIGURE 26

■ **A *k* out of *n* system.** A *k* out of *n* system consists of *n* machines and is considered working as long as *k* machines are working.

Of course, by combining these types, a very complex system may be modeled. We now show how to use @Risk to model the probability that a machine system will last a desired amount of time.

EXAMPLE 5 **Hubble Telescope**

Assume that the Hubble telescope contains four large mirrors. The time (in months) until a mirror fails follows a Weibull random variable with $\alpha = 25$ and $\beta = 50$.

a For certain types of pictures to be useful, all mirrors must be working. What is the probability that the telescope can produce these types of pictures for at least 5 years?

b Certain types of pictures can be taken as long as at least one mirror is working. What is the probability that these pictures can be taken for at least 7 years?

c Certain types of pictures can be taken as long as at least two mirrors are working. What is the probability that these pictures can be taken for at least 6 years?

Solution See file Reliability.xls.

Reliability.xls

Step 1 We begin by generating the length of time until each mirror fails in C3:C6 by copying from C3 to C4:C6 the formula

$$=RISKWEIBULL(25,50)$$

FIGURE 27

	A	B	C
1			
2	**Hubble Telescope**		
3		Mirror 1	49.30487
4		Mirror 2	30.19602
5		Mirror 3	38.99237
6		Mirror 4	37.64995
7			
8		Time all 4 work	30.19602
9		Time till last one fails	49.30487
10		Last time 2 are working	38.99237

	F	G	H	I
10	Name	Time all 4 v	Time till las	Last time 2
11	Description	Output	Output	Output
12	Cell	C8	C9	C10
13	Minimum	3.733206	31.46502	27.71785
14	Maximum	66.0223	101.8234	80.49436
15	Mean	35.63382	64.18716	54.32821
16	Std Deviati	10.08293	9.306231	8.747266
17	Variance	101.6655	86.60596	76.51466
18	Skewness	-0.104045	0.121514	-2.73E-02
19	Kurtosis	2.737432	3.290648	2.910271
20	Errors Calc	0	0	0
21	Mode	34.15707	62.93507	58.45681
22	5% Perc	18.52796	49.15086	39.91564
23	10% Perc	22.40516	52.22929	42.77759
24	15% Perc	24.85496	54.84017	44.97655
25	20% Perc	26.80984	56.5674	46.99073
26	25% Perc	28.67021	57.84864	48.3152
27	30% Perc	30.25738	59.51218	49.89228
28	35% Perc	31.90257	60.52841	50.94405
29	40% Perc	33.26531	61.73524	52.26505
30	45% Perc	34.4916	62.83329	53.25808
31	50% Perc	35.7727	63.89499	54.54087
32	55% Perc	37.04685	65.06183	55.4865
33	60% Perc	38.58305	66.16101	56.72353
34	65% Perc	39.88355	67.578	58.00496
35	70% Perc	41.17931	68.97778	58.9762
36	75% Perc	42.88946	70.39309	60.089
37	80% Perc	44.47398	71.75684	61.61526
38	85% Perc	46.13106	73.66335	63.45758
39	90% Perc	48.46651	76.06507	65.64239
40	95% Perc	51.94818	79.72974	68.19598
41	Filter Minimum			
42	Filter Maximum			
43	Filter Type			
44	# Values F	0	0	0
45	Scenario #	>75%	>75%	>75%
46	Scenario #	<25%	<25%	<25%
47	Scenario #	>90%	>90%	>90%
48	Target #1 (	60	84	72
49	Target #1 (	99.54%	98.29%	98.00%

FIGURE 28

Step 2   Part (a) is a series system. We can take the desired pictures until the first mirror fails. The first mirror fails at the smallest of the four mirror failure times. Thus, the length of time for which the first type of picture can be taken is computed in cell C8 with the formula

$$=MIN(C3:C6)$$

Step 3   Part (b) is a parallel system. We can take the desired pictures until the time the last mirror fails. We compute the time the last mirror fails in cell C9 with the formula

$$=MAX(C3:C6)$$

Step 4   Part (c) is a 2 out of 4 system. We can take the desired pictures until the time of the third mirror failure. The time of the third mirror failure is the second largest of the failure times. We compute the time of the third mirror failing in cell C10 with the formula

$$=LARGE(C3:C6,2)$$

This formula computes the second largest of the mirror failure times. Of course, this is the time the third mirror fails. See Figure 27.

Step 5   We now select cells C8:C10 as output cells and run 1,000 iterations. After using targets with the Detailed Statistics output, we obtain the results in Figure 28.

We find in part (a) that there is a 99.54% chance that all four mirrors will fail in 60 months or less, and only a .46% chance that all four mirrors will work for at least 60 months. In part (b), we find that there is a 98.29% chance that all four mirrors will fail within 7 years, and only a 1.71% chance that all four mirrors will be working for at least 7 years. In part (c), we find that there is a 98% chance that two or more mirrors will be working for 72 months or less, and only a 2% chance that two or more mirrors will be working for at least 72 months.

Estimating Warranty Expenses

If we know the distribution of the time till failure of a purchased product, @Risk makes it a simple matter to estimate the distribution of warranty costs associated with a product. The idea is illustrated in the following example.

EXAMPLE 6   **Refrigerator Failure**

The time until first failure of a refrigerator (in years) follows a Weibull random variable with $\alpha = 6.7$ and $\beta = 8.57$. If a refrigerator fails within 5 years, we must replace it with a new refrigerator costing $500. If the replacement refrigerator fails within 5 years, we must also replace that refrigerator with a new one costing $500. Thus, the warranty stays in force until a refrigerator lasts at least 5 years. Estimate the average warranty cost incurred with the sale of a new refrigerator. (Do not worry about discounting costs.)

Solution   See file Refrigerator.xls. We enter the length of time a refrigerator lasts in cell C6 with
Refrigerator.xls   the formula

$$=RISKWEIBULL(6.7,8.57)$$

	A	B	C	D	E	F	G
1							
2		**Refrigerator**					
3		**Warranty**					
4							
5		Number	Lasts	Cost			
6		1	8.113087	0			
7		2	6.91762	0			.027^5
8		3	7.233594	0			1.43489E-08
9		4	8.776642	0			
10		5	7.120917	0			
11			Total cost	0			
12							

FIGURE 29

We are not sure how many replacement refrigerators we might have to provide for the customer. By selecting the Define Distributions icon when we are in cell C6, we can move the sliders on the Weibull density function and determine the probability that we will have to replace a given refrigerator. We find that there is only a 2.7% chance that a refrigerator will have to be replaced. Then the chance that at least 5 refrigerators will have to be replaced is $(.027)^5 = .000014$. Thus, generating only 5 refrigerator lifetimes should give us an accurate estimate of total cost. We therefore copy the RISKWEIBULL formula from C6 to C7:C10. See Figure 29.

In cell D6, we compute the cost associated with a sold refrigerator with the formula

$$=IF(C6<5,500,0)$$

In cells D7:D10, we compute the cost (if any) associated with any replacement refrigerators by copying from D7 to D8:D10 the formula

$$=IF(AND(D6>0,C7<5),500,0)$$

This formula picks up the cost of a replacement if and only if the previous refrigerator failed and the current refrigerator lasts less than 5 years.

In cell D11, we compute total cost with the formula

$$=SUM(D6:D10)$$

After running 1,000 iterations and making cell D11 an output cell (see below), we find the mean warranty cost per refrigerator to be $14.50. Note that maximum cost was $1,000, so on at least one iteration, two refrigerators needed to be replaced.

	F	G	H	I	J	K	L	M
11								
12		Name	Workbook	Worksheet	Cell	Minimum	Mean	Maximum
13	Output 1	Total cost / Cos	refrigerator	Sheet1	D11	0	**14.5**	1000

PROBLEMS

Group A

Assume that the lifetimes of all machines described follow a Weibull random variable.

1 Suppose an auto engine consists of 12 components in series. The mean lifetime of each component is 5 years, with a standard deviation of 2 years.

    a What is the probability that the engine will work for at least 2 years?

    b If the engine were a parallel system, what is the probability that the engine would work for at least 10 years?

    c If at least 8 engine components need to work for the engine to work, what is the probability that the engine will work for at least 7 years?

2 An aircraft engine lasts an average of 5 years, with a standard deviation of 3 years before it needs to be replaced. Consider a plane with 4 new engines. On the average, how long will it be until an engine needs to be replaced?

3 A one-mile length of street has 5 street lights, equally spaced. The mean lifetime of a street light is 3 years, with a standard deviation of 1 year. Assume that all 5 lights have just been replaced. The street is considered too dark if at least one part of the street has no light working within .5 mile. On the average, how long will it be until the street is considered too dark?

4 In the refrigerator example, suppose the warranty works as follows. If a refrigerator fails at any time within 5 years of purchase, we give the consumer a prorated refund on the $500 purchase price. For example, if the refrigerator fails after 4 years, we pay the customer $100. If the refrigerator fails after 3 years, we pay the customer $200. Estimate our expected warranty expense per refrigerator sold.

5 The time to failure of a TV picture tube averages 5 years, with a standard deviation of 3 years. It costs an average of $250 to repair or replace a TV picture tube. Determine fair prices for a 3-year, 4-year, or 5-year warranty.

23.5 The RISKGENERAL Function

What if a continuous random variable (such as market share) does not appear to follow a normal or triangular distribution? We can model it with the **RISKGENERAL** function.

EXAMPLE 7   **RISKGENERAL Distribution**

Suppose that market shares between 0% and 60% are possible. A 45% share is most likely. There are five market-share levels for which we feel comfortable about comparing the relative likelihoods (see Table 11).

From the table, a market share of 45% is 8 times as likely as 10%; 20% and 55% are equally likely, etc. This distribution cannot be triangular, because then 20% would be (20/45) as likely as the peak of 45%. In fact, 20% is .75 as likely as 45%. See Figure 30 and file Riskgeneral.xls for our analysis.

Riskgeneral.xls

To model market share, enter the formula

$$=\text{RISKGENERAL}(0,60,\{10,20,45,50,55\},\{1,6,8,7,6\})$$

TABLE 11

Market Share	Relative Likelihood
10%	1
20%	6
45%	8
50%	7
55%	6

	B	C	D	E	F	G
1	**EXAMPLE OF**					
2	**RISKGENERAL**					
3	**DISTRIBUTION**					
4						
5			Minimum	0		
6			Maximum	60		
7			**Specified Points**			
8			10	1		
9			20	6		
10			45	8		
11			50	7		
12			55	6		
13	35.75	=RISKGENERAL(0,60,{10,20,45,50,55},{1,6,8,7,6})				

FIGURE **30**

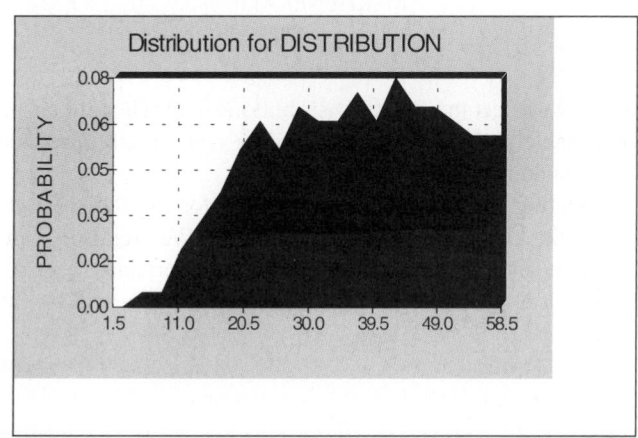

FIGURE **31**

	C	D
29	Share	Likelihood
30	0	0
31	10	1
32	20	6
33	45	8
34	50	7
35	55	6
36	60	0

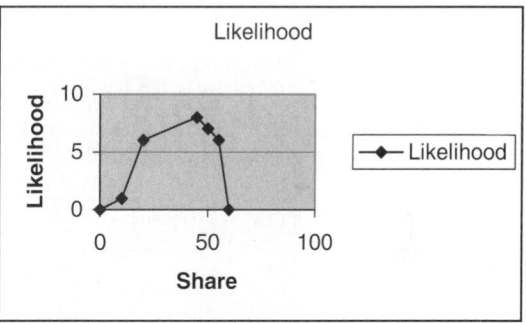

FIGURE **32**

The syntax of RISKGENERAL is as follows.

- Begin with the smallest and largest possible values.
- Then enclose in {} the numbers for which you feel you can compare relative likelihoods.
- Finally, enclose in {} the relative likelihoods of the numbers you have previously listed.

Running this in @Risk yields the output in Figure 31. Note that 20 is 6/8 as likely as 45; 10 is 1/8 as likely as 45; 50 is 7/8 as likely as 45; 55 is 6/8 as likely as 45, etc. In between the given points, the density function changes at a linear rate. Thus, 30 would have a likelihood of

$$6 + \frac{(30 - 20)*(8 - 6)}{(45 - 20)} = 6.8$$

Basically what @Risk has done is to take the curve constructed by connecting (with straight lines) the points $(0, 0)$, $(10,1), \ldots, (55,6)$, $(60,0)$. @Risk rescales the height of this curve so that the area under it equals 1, and then randomly selects points based on the height of the curve. Thus, a share around 45 is 8/6 as likely as a share around 20, etc. Figure 32 illustrates this idea.

REMARK For the spreadsheet in Figure 30, the syntax

$$=\text{RISKGENERAL}(0,60,D8:D12,E8:E12)$$

is also acceptable.

Suppose we select the Define Distributions icon. Then we choose the RISKGENERAL random variable and select Apply. Now we can directly insert the RISKGENERAL (or any other) random variable into a cell.

After entering the appropriate parameters for the RISKGENERAL random variable, we will see the histogram shown in Figure 33. We are also given statistical information, such as the mean and variance, for the random variable. If we select Apply, the formula defining the desired RISKGENERAL random variable will be entered into the cell.

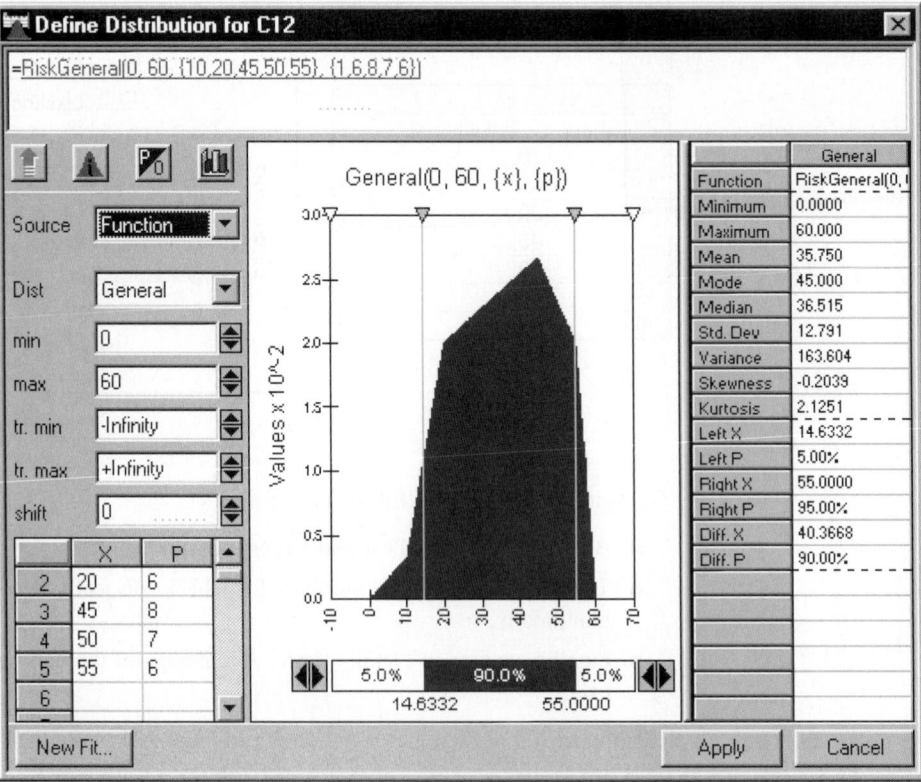

FIGURE 33

23.6 The RISKCUMULATIVE Random Variable

With the RISKGENERAL function, we estimated the relative likelihood of a random variable taking on various values. With the RISKCUMULATIVE function, we estimate the cumulative probability that the random variable is less than or equal to several given values. The RISKCUMULATIVE function can be used to approximate the cumulative distribution function for any continuous random variable.

| EXAMPLE 8 | RISKCUMULATIVE |

A large auto company's net income for North American operations (NAO) for the next year may be between 0 and $10 billion. The auto company estimates there is a 10% chance that net income will be less than or equal to $1 billion, a 70% chance that net income will be less than or equal to $5 billion, and a 90% chance that net income will be less than or equal to $9 billion. Use @Risk to simulate NAO's net income for the next year.

	A	B	C	D	E	F	G	H
1	Cumulative distribution							
2								
3	Min	0						
4	Max	10		4.2				
5	x	P(X<=x)	Slope	4.2	RiskCumul(B3,B4,A6:A8,B6:B8)			
6	1	0.1	0.1					
7	5	0.7	0.15					
8	9	0.9	0.05		Name	P(X<=x)		
9	>9		0.1		Description	Output		
10					Cell	D5		
11					Minimum =	4.89E-03		
12					Maximum	9.999967		
13					Mean =	4.199986		
14					Std Deviati	2.773699		
15					Variance =	7.693407		
16					Skewness	0.589373		
17					Kurtosis =	2.285831		
18					Errors Calc	0		
19					Mode =	3.43314		
20					5% Perc =	0.497997		
21					10% Perc :	0.999338	10%ile is 1!	
22					15% Perc :	1.333212		
23					20% Perc :	1.665637		
24					25% Perc :	1.996866		
25					30% Perc :	2.332803		
26					35% Perc :	2.664376		
27					40% Perc :	2.996635		
28					45% Perc :	3.330816		
29					50% Perc :	3.663554		
30					55% Perc :	3.995894		
31					60% Perc :	4.33135		
32					65% Perc :	4.664128		
33					70% Perc :	4.997442	70%ile is 5!	
34					75% Perc :	5.995409		
35					80% Perc :	6.993743		
36					85% Perc :	7.99109		
37					90% Perc :	8.989162	90%ile is near 9	
38					95% Perc :	9.499336		

FIGURE 34

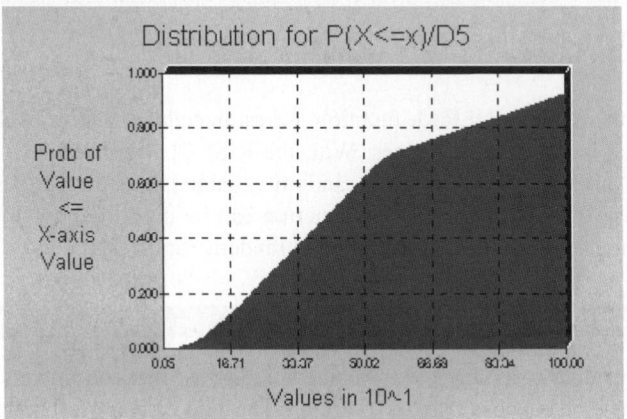

FIGURE 35

Our work is in the file Cumulative.xls. See Figure 34. The RISKCUMULATIVE function takes as inputs (in order) the following quantities:

- The smallest value assumed by the random variable
- The largest value assumed by the random variable
- Intermediate values assumed by the random variable
- For each intermediate value, the cumulative probability that the random variable is less than or equal to the intermediate value

In cell D5, we enter the following formula to simulate NAO's annual net income:

$$=\text{RISKCUMUL(B3,B4,A6:A8,B6:B8)}$$

We could have also used the following formula in cell D4:

$$=\text{RISKCUMUL}(0,10,\{1,5,9\},\{0.1,0.7,0.9\})$$

@Risk will now ensure that

- For net income x between 0 and \$1 billion, the cumulative probability that net income is less than or equal to x rises with a slope equal to $\frac{.1 - 0}{1 - 0} = .1$.
- For net income x between \$1 billion and \$5 billion, the cumulative probability that net income is less than or equal to x rises with a slope equal to $\frac{.7 - .1}{5 - 1} = .15$.
- For net income x between \$5 billion and \$9 billion, the cumulative probability that net income is less than or equal to x rises with a slope equal to $\frac{.9 - .7}{9 - 5} = .05$.
- For net income x greater than \$9 billion, the cumulative probability that net income is less than or equal to x rises with a slope equal to $\frac{1 - .9}{10 - 9} = .10$.

After running 1,600 iterations we found the output in Figure 34. Note that the 10th percentile of the random variable is near 1, the 70th percentile is near 5, and the 90th percentile is near 9. Figure 35 displays a cumulative ascending graph of net income. Note that (as described previously) the slope of the graph is relatively constant between 0 and 1, between 1 and 5, between 5 and 9, and between 9 and 10.

23.7 The RISKTRIGEN Random Variable

When we use the RISKTRIANG function, we are assuming we know the absolute worst and absolute best case that can occur. Many companies, such as Eli Lilly, prefer to use a triangular random variable in which the worst case and best case are defined by a percentile of the random variable. For example, at Eli Lilly the 10th percentile of demand, most likely demand, and 90th percentile of demand often define forecasts. The following example shows how to use the RISKTRIGEN function to model uncertainty.

EXAMPLE 9 **RISKTRIGEN**

Eli Lilly believes there is a 10% chance that its new drug Niagara's market share will be 25% or less, a 10% chance that market share will be 70% or more, and the most likely market share is 40%. Use @Risk to model the market share for Niagara.

Solution Our work is in the file Risktrigen.xls. See Figure 36. In B7, we just entered the formula

Risktrigen.xls

$$=\text{RISKTRIGEN(B3,B4,B5,10,90)}$$

	A	B
1	trigen function	
2		
3	10%ile	0.25
4	Most likely	0.4
5	90 %ile	0.7
6		
7	share	0.464537

FIGURE 36

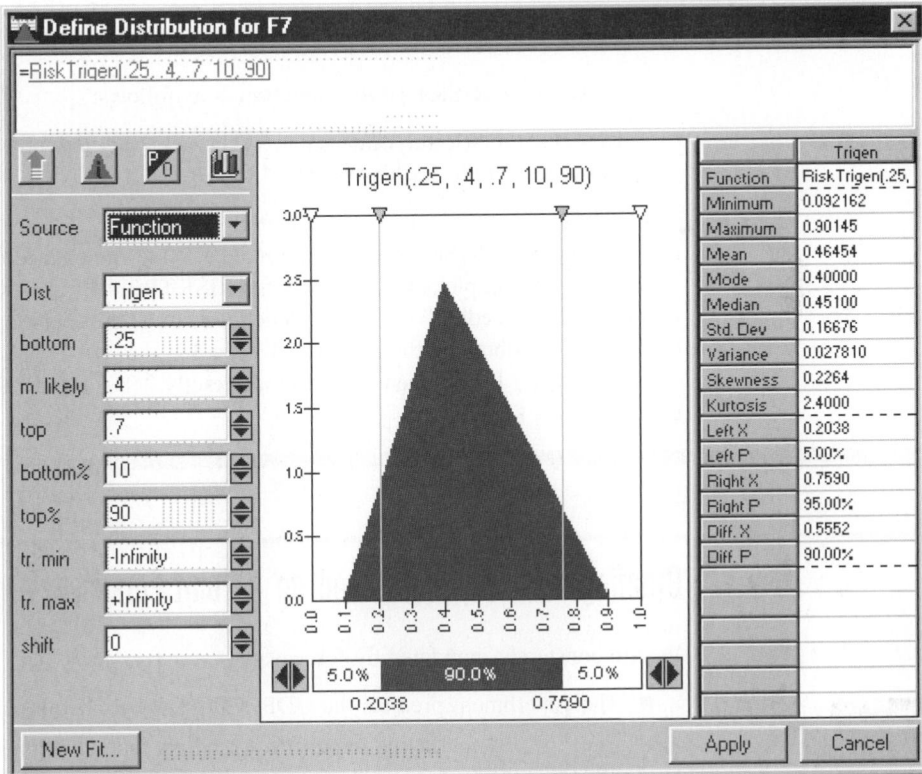

FIGURE 37

FIGURE 38

	C	D
34	Name	
35	Description	Output
36	Cell	[trigen.xls]$
37	Minimum =	9.73E-02
38	Maximum	0.886495
39	Mean =	0.464533
40	Std Deviat	0.166746
41	Variance =	2.78E-02
42	Skewness	0.22598
43	Kurtosis =	2.398804
44	Errors Calc	0
45	Mode =	0.401881
46	5% Perc =	0.203626
47	10% Perc :	0.249634
48	15% Perc :	0.285171
49	20% Perc :	0.315337
50	25% Perc :	0.341713
51	30% Perc :	0.365485
52	35% Perc :	0.387192
53	40% Perc :	0.407964
54	45% Perc :	0.428942
55	50% Perc :	0.450952
56	55% Perc :	0.473918
57	60% Perc :	0.498488
58	65% Perc :	0.524349
59	70% Perc :	0.552427
60	75% Perc :	0.58265
61	80% Perc :	0.61619
62	85% Perc :	0.654338
63	90% Perc :	0.699825
64	95% Perc :	0.758373

The syntax of the RISKTRIGEN function is as follows:

=RISKTRIGEN(lower value, most likely value, higher value, percentile for lower value, percentile for higher value)

In Figure 37, we show the density function for the market share. Note that @Risk picks the worst case for RISKTRIGEN (around 10%), so the chance of a market share below 25% is .10. @Risk picks the best case for RISKTRIGEN (around 89%), so the probability of a share exceeding 70% is .10. When we ran 1,600 iterations, with cell B7 being the output cell, we obtained the output in Figure 38.

Note that the 10th percentile is almost exactly 25%, and the 90th percentile is almost exactly 70%.

23.8 Creating a Distribution Based on a Point Forecast

We are constantly inundated by forecasts:

- The government predicts the GDP will grow by 4% during the next year.
- The Eli Lilly marketing department predicts that demand for a given drug will be 400,000,000 d.o.t. (days of therapy) during the next year.

- A Wall Street guru predicts that the Dow will go up 20% during the next 12 months.
- The bookmakers forecast that the Pacers will beat the Rockets by 6 in the opening game of the 2005 NBA season.

Although the forecasts may be the best available, they are almost sure to be incorrect. For example, the bookmakers' prediction that the Pacers will win by 6 points is incorrect unless the Pacers win by exactly 6 points. In short, any single-valued (or *point*) forecast implies a distribution for the quantity being forecasted. How can we find a random variable that correctly models the uncertainty inherent in the point forecast? The key to putting a distribution around a point forecast is to have some historical data about the accuracy of past forecasts of the quantity of interest. For example, with regard to our forecast for the Dow, we might have the forecast made in January of each of the past 10 years for the percentage change in the Dow and the actual change in the Dow for each of those years. We begin by seeing if past forecasts exhibit any bias. For each past forecast, we determine (actual value)/(forecast value). Then we average these ratios. If our forecasts are unbiased, this average should be around 1. Any significant deviation from 1 would indicate a significant bias.[†] For example, if the average of actual/forecast is 2, the actual results tend to be around twice our forecast. To correct for this bias, we should automatically double our forecast. If the average of actual/forecast is .5, the actual results tend to be around half our forecast; to eliminate bias, we should automatically halve our forecast. Once we have eliminated forecast bias, we look at the standard deviation of the percentage errors of the unbiased forecast. We use the following @Risk random variable to model the quantity being forecast.

RISKNORMAL(unbiased forecast, (percentage standard deviation of unbiased forecasts)*(unbiased forecast))

EXAMPLE 10   Drug Forecast

Drugforecast.xls

The file Drugforecast.xls contains actual and forecast sales (in millions of d.o.t.) for the years 1995–2002. See Figure 39. The forecast for 2003 is that 60 million d.o.t. will be sold. How would you model actual sales of the drug for 2003?

Solution   **Step 1**   In cells F5:F12, check for bias by computing actual sales/forecast sales for each year. To do this, copy from F5 to F6:F12 the formula

$$=D5/E5$$

Step 2   In cell F2, compute the bias of the original forecasts by averaging each year's actual/forecast sales.

$$=AVERAGE(F5:F12)$$

We find that actual sales tend to come in 8% under forecast.

Step 3   In G5:G12, correct past biased forecasts by multiplying them by .92. Simply copy from G5 to G6:G12 the formula

$$=\$F\$2*E5$$

[†]To see if the bias is significantly different from 1, compute

$$\frac{\text{Average of (actual)/(forecast)} - 1}{\text{Standard deviation of actual/forecast}}$$

If this exceeds $t_{(\alpha/2, n-1)}$ then there is significant bias. We usually choose $\alpha = .05$.

FIGURE 39

	C	D	E	F	G	H	I
1						mean	std dev
2			mean	0.918031		1	0.113753
3							
4	Year	Actual Sales	Forecast	A/F	Unbiased forecast	%age error	
5	1995	17	22	0.772727	20.19668	84%	
6	1996	59	61	0.967213	55.9999	105%	
7	1997	46	51	0.901961	46.81959	98%	
8	1998	85	86	0.988372	78.95067	108%	
9	1999	98	103	0.951456	94.5572	104%	
10	2000	94	118	0.79661	108.3277	87%	
11	2001	24	22	1.090909	20.19668	119%	
12	2002	14	16	0.875	14.6885	95%	

FIGURE 40

	E	F
14		
15	Mean 2003	55.08187
16	Sigma 2003	6.2657

Step 4 In H5:H12, compute each year's percentage error for the unbiased forecast. Copy from H5 to H6:H12 the formula

$$=D5/G5$$

Step 5 In cell I2, compute the standard deviation of the percentage errors with the formula

$$=STDEV(H5:H12)$$

We find that the standard deviation of past unbiased forecasts has been around 11% of the unbiased forecast. We now model the 2003 sales of the drug (in millions of d.o.t.) with the formula

$$=RISKNORMAL(60*(.918), (60*.918)*.114) \quad \text{or} \quad RISKNORMAL(55.08,6.27)$$

See Figure 40.

23.9 Forecasting the Income of a Major Corporation

In many large corporations, different parts of a company make forecasts for quarterly net income. An analyst in the CEO's office pulls together the individual predictions to forecast the entire company's net income. In this section, we show an easy way to pool forecasts from different portions of a company and create a probabilistic forecast for the entire company.

So far, we have usually assumed that @Risk functions in different cells are independent. For example, the value of a RISKNORMAL(0,1) in cell A6 has no effect on the value of a RISKNORMAL(0,1) in any other cell. In many situations, however, variables of interest might be correlated. For example, a weak yen will lower the price of a Japanese car in the United States and hurt GM market share. Since higher price incentives increase market share, GM market share may also be negatively correlated with car price. Also, net income of NAO (North American operations) is often correlated with net income in Europe. The following example shows how to model correlations with @Risk. Recall that the correlation between two random variables must lie between -1 and $+1$.

- Correlation near $+1$ implies a strong positive linear relationship.
- Correlation near -1 implies a strong negative linear relationship.
- Correlation near $+.5$ implies a moderate positive linear relationship.
- Correlation near $-.5$ implies a moderate negative linear relationship.
- Correlation near 0 implies a weak linear relationship.

EXAMPLE 11   Forecasting GM Net Income

Corrinc.xls

Suppose GM CEO Rick Waggoner has received the following forecast for quarterly net income (in billions of dollars) for Europe, NAO, Latin America, and Asia. See Figure 41 and file Corrinc.xls.

For example, we believe Latin American income will be on average $.4 billion. Based on past forecast records, the standard deviation of forecast errors is 25%, so the standard deviation of net income is $.1 billion. We assume that actual income will follow a normal distribution. Historically, net income in different parts of the world has been correlated. Suppose the correlations are as given in B10:F13. Latin America and Europe are most correlated, and Asia and NAO are least correlated. What is the probability that total net income will exceed $4 billion?

Solution

To correlate the net incomes of the different regions, we use the RISKCORRMAT function. The syntax is as follows:

= Actual @Risk formula, RISKCORRMAT(correlation matrix, relevant column of matrix)

where

Correlation matrix: cells where correlations between variables are located

Relevant column: column of correlation matrix that gives correlations for this cell

Actual @Risk formula: distribution of the random variable

	A	B	C	D	E	F	G
1	Net Income Consolidation						
2	with correlation					Goal is 4 billion!	
3			Mean	Std. Dev	Actual		
4	1	LA	0.4	0.1	0.449011	0.521472	
5	2	NAO	2	0.4	1.256578	1.264837	
6	3	Europe	1.1	0.3	1.14203	0.994558	
7	4	Asia	0.8	0.3	0.685143	0.707549	
8				Total!!	3.532761	3.488417	
9							
10		Correlations	LA	NAO	Europe	Asia	
11		LA	1	0.6	0.7	0.5	
12		NAO	0.6	1	0.6	0.4	
13		Europe	0.7	0.6	1	0.5	
14		Asia	0.5	0.4	0.5	1	
15							
16							

FIGURE 41

	B	C	D	E	F
54	Scenario #3 =	>90%		36% chance we fail	
55	**Target #1 (Value)**	4		to meet target	
56	**Target #1 (Perc%)**	**35.72%** ◀			
57					

	B	C	D
17	Name	Total!! / Actual	
18	Description	Output	
19	Cell	E8	
20	Minimum =	1.858541	
21	Maximum =	6.71191	
22	Mean =	4.300031	
23	Std Deviation =	0.895158	
24	Variance =	0.801308	
25	Skewness =	-5.82E-02	
26	Kurtosis =	2.894021	
27	Errors Calculated	0	
28	Mode =	4.470891	
29	5% Perc =	2.756473	
30	10% Perc =	3.186955	
31	15% Perc =	3.364678	
32	20% Perc =	3.554199	
33	25% Perc =	3.715597	
34	30% Perc =	3.854618	
35	35% Perc =	3.96633	
36	40% Perc =	4.080534	
37	45% Perc =	4.173182	
38	50% Perc =	4.306374	
39	55% Perc =	4.413318	
40	60% Perc =	4.530555	
41	65% Perc =	4.632649	
42	70% Perc =	4.7776	
43	75% Perc =	4.907873	
44	80% Perc =	5.04496	
45	85% Perc =	5.216321	
46	90% Perc =	5.456462	
47	95% Perc =	5.758535	

FIGURE 42

Step 1 Generate actual Latin American income in cell E4 with the formula

$$=RISKNORMAL(C4,D4,RISKCORRMAT(\$C\$11:\$F\$14,A4))$$

This ensures that the correlation of Latin American income with other incomes is created according to the first column of C11:F14. Also, Latin American income will be normally distributed, with a mean of \$.4 billion and standard deviation of \$.1 billion.

Step 2 Copying the formula in E4 to E5:E7 (respectively) generates the net income in each region and tells @Risk to use the correlations in C11:F14.

Step 3 In cell E8, compute total income with the formula

$$=SUM(E4:E7)$$

Step 4 Cell E8 has been made the output cell. We find from Targets (value of 4) that there is a 36% chance of not meeting the \$4 billion target. Also, the standard deviation of net income is \$895 million. See Figure 42.

	B	C	D
15	Name	Total!! / Actual	
16	Description	Output	
17	Cell	E8	
18	Minimum =	2.174825	
19	Maximum =	6.290998	
20	Mean =	4.299921	
21	Std Deviation =	0.605397	
22	Variance =	0.366506	

	B	C	D	E	F
53	Target #1 (Value)=	4			
54	Target #1 (Perc%)=	30.76%		31% chance we fail to meet target	
55					
56					
57					
58					

FIGURE 43

FIGURE 44

	B	C	D	E	F	G	H	I	J	K	L
5											
6	Name	Total!! / Ac	LA / Actual	NAO / Actu	Europe / A	Asia / Actual					
7	Description	Output	Normal(C4	Normal(C5	Normal(C6	Normal(C7,D7)					
8	Iteration#	E8	E4	E5	E6	E7		LA	NAO	Europe	Asia
9	1	4.804644	0.478546	2.196594	1.351783	0.777721	LA	1			
10	2	4.132098	0.441263	1.699526	1.184871	0.806438	NAO	0.591262	1		
11	3	6.129157	0.496915	2.453791	1.91255	1.265901	Europe	0.702735	0.587704	1	
12	4	6.54744	0.57896	2.424948	1.968532	1.574999	Asia	0.498132	0.399115	0.496651	1
13	5	3.057065	0.319965	1.517732	0.968105	0.251263					
14	6	5.324339	0.488499	2.292126	1.084479	1.459235					
907	899	4.735623	0.469691	2.19903	1.466369	0.600534					
908	900	4.901974	0.507751	2.242637	1.004801	1.146786					

What If Net Incomes Are Not Correlated?

Nocorrinc.xls

In workbook Nocorrinc.xls, we ran the simulation of Example 10, assuming that the net incomes in different regions were independent (that is, had 0 correlation). The results appear in Figure 43. Note that the absence of correlation has reduced the standard deviation to $600 million and our chance of not meeting our $4 billion income target. This is because if the incomes of all the regions are independent, then it is likely that a high income in one region will be cancelled out by a low income in another region. If the incomes of the regions are positively correlated, these correlations reduce the diversification or hedging effect.

Checking the Correlations

We can check that @Risk actually did correctly correlate net incomes. Make sure to check Collect Distribution Samples when you run the simulation. Once you have run the simulation, select the Data option from the Results menu. The results of each iteration will appear in the bottom half of the screen. You can Edit Copy Paste this data to a blank worksheet. See Figure 44. Now check the correlations between each region's net income with Data Analysis Tools Correlation. Select Data Analysis Tools Correlations and fill in the

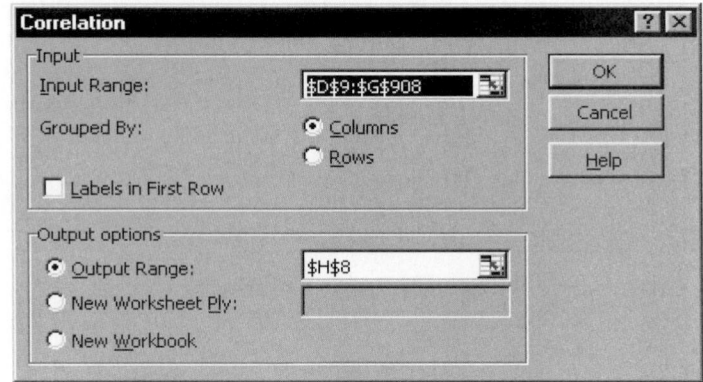

FIGURE 45

dialog box as in Figure 45. Note that the correlations between the net incomes are virtually identical to what we entered in the spreadsheet.

23.10 Using Data to Obtain Inputs for New Product Simulations

Many companies use subjective estimates to obtain inputs for new product simulations. For example, market size may be subjectively modeled as a triangular random variable, with the marketing department coming to a consensus on best-case, worst-case, and most likely scenarios. In many situations, however, past data may be used to obtain estimates of key variables. We now discuss how past data on similar products or projects can be used to model share, price, volume, and cost uncertainty. The utility of any model will depend on the type of data available.

The Scenario Approach to Modeling Volume Uncertainty

When trying to model volume of sales for a new product in the auto and drug industries, it is common to look for similar products sold in the past. We often have knowledge of the following:

- Accuracy of forecasts for year 1 sales volume
- Data on how sales change after the first year

Volume.xls

Consider Figure 46—data on actual and forecast year 1 sales for seven similar products. See file Volume.xls. For example, for product 1, actual year 1 sales were 80,000; the forecast for year 1 was 44,396. The percentage change in sales from year to year for the seven products is given in Figure 47.

For example, product 1 sales went up 43% during the second year, 33% during the third year, etc.

Suppose we forecast year 1 sales to be 90,000 units. How can we model the uncertain volume in product sales?

Step 1 From cell D11 (formula =AVERAGE(D4:D10)) of Figure 46, we see that past forecasts for year 1 sales of similar products have overforecast the actual sales by 36.3%.

FIGURE 46

	B	C	D	E	F
3	Actual	Forecast	Actual/For ecast	Unbiased forecast	%age error
4	80000	44396	1.8019641	60516.733	1.3219484
5	100000	99209	1.0079731	135233.01	0.7394644
6	120000	94808	1.265716	129233.95	0.9285486
7	150000	96813	1.5493787	131966.99	1.1366479
8	180000	172862	1.0412931	235630.31	0.7639085
9	200000	108770	1.8387423	148265.72	1.3489295
10	55000	53052	1.0367187	72315.832	0.7605527
11		mean	1.3631123	stdev	0.2677479

FIGURE 47

	A	B	C	D	E	F	G	H	I	J
13	Scenario	Year 2	Year 3	Year 4	Year 5	Year 6	Year 7	Year 8	Year 9	Year 10
14	1	1.43	1.33	0.93	0.75	0.57	0.40	0.37	0.38	0.24
15	2	1.39	1.13	0.96	0.59	0.49	0.45	0.46	0.40	0.24
16	3	1.30	1.38	0.98	0.84	0.80	0.65	0.57	0.48	0.35
17	4	1.47	1.49	1.36	1.15	1.20	1.15	0.93	0.99	0.71
18	5	1.23	1.06	0.73	0.45	0.39	0.31	0.28	0.23	0.15
19	6	1.26	1.22	1.08	0.79	0.77	0.70	0.60	0.60	0.49
20	7	1.30	1.02	0.84	0.62	0.45	0.32	0.27	0.24	0.22

Step 2   Therefore, we can create unbiased forecasts in column E by copying the formula

$$=\$D\$11*C4$$

from E4 to E5:E10.

Step 3   In column F, we compute the percentage error of our unbiased forecasts. In cell F4, we compute the percentage error for product 1 with the formula

$$=B4/E4$$

Copying this formula from F4 to F5:F10 generates percentage errors for the other products.

Step 4   In cell F11, we compute the standard deviation (26.7%) of these percentage errors with the formula

$$=STDEV(F4:F10)$$

We are now ready to model 10 years of sales for the new product. To generate year 1 sales, we model year 1 sales to be normally distributed, with a mean of 1.36*90,000 and a standard deviation of .267*(90,000*1.267). To model sales for years 2–10, we use @Risk to randomly choose one of the seven volume-change patterns (or **scenarios**) from Figure 47. Then we use the chosen scenario to generate sales growth for years 2–10.

Step 5   In cell G4, we choose a scenario with the formula

$$=RISKDUNIFORM(A14:A20)$$

This formula gives a 1/7 chance of choosing each scenario.

FIGURE 48

	G	H	I	J	K	L	M	N	O	P	Q
1	Year 1 Forecast		90000								
2		Year									
3	Scenario	1	2	3	4	5	6	7	8	9	10
4	4	102588.9	151164	225922	306360.9	351610	420801.1	484511.5	451618.5	445300.1	314821.9

Step 6 In H4, we generate year 1 sales with the formula

$$=RISKNORMAL(I1*D11,(I1*D11)*F11)$$

This implies that

Mean year 1 sales = (biased forecast)(factor to correct for bias)

(Standard deviation year 1 sales) = (unbiased forecast for year 1 sales)*(standard deviation of errors as percentage of unbiased forecast)

Step 7 In cell I4, we generate year 2 sales with the formula

$$=H4*VLOOKUP(\$G\$4,\$A\$14:\$J\$20,I3)$$

This formula takes year 1 generated sales and multiplies it by the year 2 growth factor for the chosen scenario. Copying this formula to I4:Q4 generates sales for years 2–10. See Figure 48.

Modeling Statistical Relationships with One Independent Variable

Suppose we want to model the dependence of a variable Y on a single independent variable X. We proceed as follows.

Step 1 Try to find the straight line, power curve, and exponential curve that best fit the data. The easiest way to do this is to plot the points with Excel and use the Trend Curve feature.

- The straight line is of the form $Y = a + bX$.
- The power function is of the form $Y = ax^b$.
- The exponential function is of the form $Y = ae^{bX}$.

Step 2 For each curve and each data point, compute the percentage error

$$\frac{\text{Actual value of } Y - \text{predicted value of } Y}{\text{Predicted value of } Y}$$

Step 3 For each curve, compute mean absolute percentage error (MAPE) by averaging the absolute percentage errors.

Step 4 Choose the curve that yields the lowest MAPE as the best fit.

Step 5 Does at least one of the three curves appear to have some predictive value? Check the plot for this, or look at the p-value from the regression; it should be $\leq .15$. If so, model the uncertainty associated with the relationship between X and Y as follows:

- If the straight line is the best fit, then model Y as

    =RISKNORMAL(prediction, standard deviation of actual (not percentage) errors)

- If the power curve or the exponential curve is the best fit, then model Y as

    =RISKNORMAL(prediction, prediction*(standard deviation of percentage errors))

EXAMPLE 12   **Modeling the Cost of Building Capacity**

We are not sure of the cost of building capacity for a new drug, but we believe that costs will run around 50% more (in real terms) than for the drug Zozac. Table 12 gives data on the costs incurred when capacity was built for Zozac.

For example, when 110,000 units of capacity for Zozac were built, the cost was $654,000 (in today's dollars). How would you model the uncertain cost of building capacity for the new product?

Capacity.xls   **Solution**   See the file Capacity.xls.

Step 1 To begin, we plot the best-fitting straight line, power curve, and exponential curve. To do this, use Chart Wizard (X-Y option 1) and click on points till they turn gold. Next, choose the desired curve and select R-SQ and the Equation option. We obtain the graphs in Figures 49–51.

Step 2 In C3:E8 (see Figure 52), we compute the predictions for each curve. In C3:C8, we compute the straight-line predictions by copying from C3 to C3:C8 the formula

$$=5.0623*A3+77.516$$

In D3:D8, we compute the power curve prediction by copying from D3 to D3:D8 the formula

$$=13.483*A3\char`^0.8229$$

In E3:E8, we compute the exponential curve predictions by copying from E3 to E3:E8 the formula

$$=164.52*EXP(0.0114*A3)$$

Step 3 In F3:H8, we use

$$\frac{\text{Actual value of } Y - \text{predicted value of } Y}{\text{Predicted value of } Y}$$

TABLE 12

Capacity (thousands)	Cost ($ thousands)
20	156
50	350
80	490
110	654
140	760
160	890

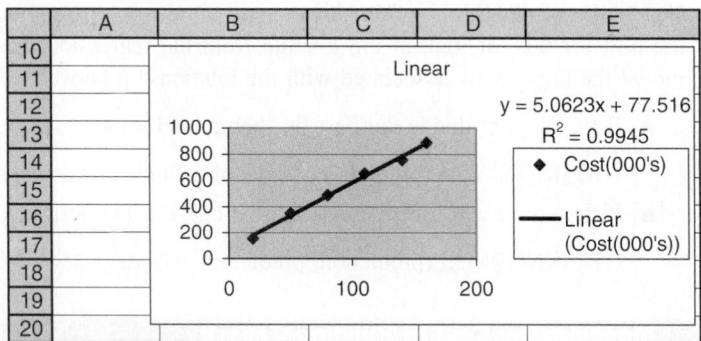

FIGURE 49

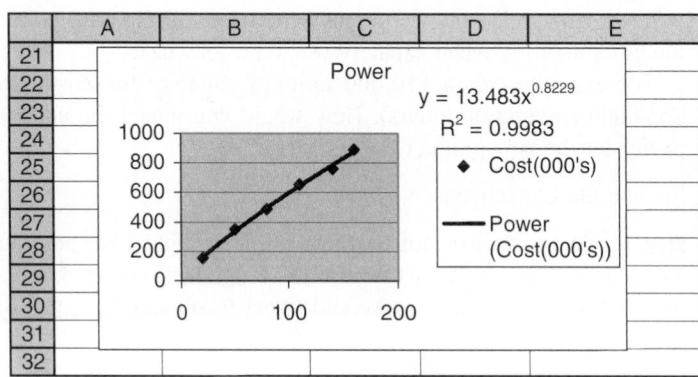

FIGURE 50

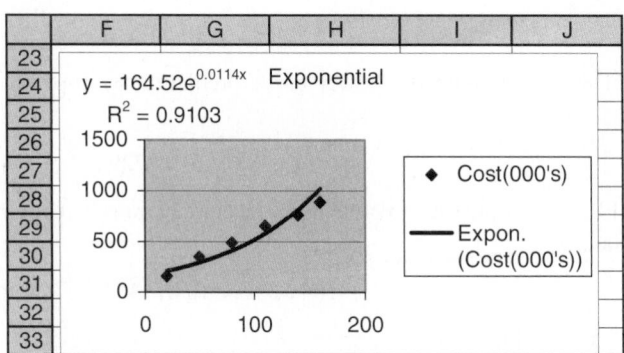

FIGURE 51

	A	B	C	D	E
1	Capacity Cost Modeling				
2	Capacity(0	Cost(000's)	Linear Prediction	Power Prediction	Exponential Prediction
3	20	156	178.762	158.6369	206.6511577
4	50	350	330.631	337.1855	290.9152953
5	80	490	482.5	496.4086	409.5390027
6	110	654	634.369	645.132	576.5327482
7	140	760	786.238	786.7474	811.6199132
8	160	890	887.484	878.1261	1019.463863

FIGURE 52

	F	G	H	I	J	K
1						
2	%age Error Linear	%age Error Power	%age Error Exponential	APE Linear	APE Power	APE Exponential
3	-0.127331	-0.016622	-0.2451046	0.127331	0.016622	0.24510464
4	0.058582	0.038004	0.20309934	0.058582	0.038004	0.20309934
5	0.015544	-0.01291	0.19646724	0.015544	0.01291	0.19646724
6	0.030946	0.013746	0.13436748	0.030946	0.013746	0.13436748
7	-0.033372	-0.033997	-0.0636011	0.033372	0.033997	0.06360109
8	0.002835	0.013522	-0.1269921	0.002835	0.013522	0.12699211
9	St dev	0.026132		0.044768	0.021467	0.16160532
10				MAPE		

FIGURE 53

to compute the percentage error for each model. (See Figure 53.) To do this, simply copy the formula

$$=(\$B3-C3)/C3$$

from F3 to F3:H8.

Step 4   In I3:K9, we compute the MAPE for each equation. We begin by computing the absolute percentage error for each point and each curve by copying the formula

$$=ABS(F3)$$

from I3:K8.

Next we compute the MAPE for each equation by copying the formula

$$=AVERAGE(I3:I8)$$

from I9:K9.

Step 5   We find that the power curve (see J9) has the lowest MAPE. Therefore, we model the cost of adding capacity with a power curve. By entering in G9 the formula

$$=STDEV(G3:G8)$$

we find 2.6% to be the standard deviation of the percentage errors for the power curve. We now model the cost of adding capacity for the new product with the formula

$$=1.5*RISKNORMAL(13.483*(Capacity)^\wedge.8229,.026*13.483*(Capacity)^\wedge.8229)$$

That is, our best guess for the cost of adding capacity has a mean equal to the power curve forecast and a standard deviation equal to 2.6% of our forecast.

EXAMPLE 13   **Bidding on a Construction Project**

We are bidding against a competitor for a construction project and want to model her bid. In the past, her bid has been closely related to our (estimated) cost of completing the project. See file Biddata.xls and Figure 54.

Biddata.xls

Figures 55–57 give the best fitting linear, power, and exponential curves.

As in Example 12, we compute predictions and MAPEs for each curve (see Figure 58). The linear curve has the smallest MAPE. Computing the actual errors for the linear curve's predictions (in column F) and their standard deviation, we find a standard deviation of .94. Therefore, we model our competitor's bid as

$$=RISKNORMAL(1.489*(Our\ cost) - 1.7893, .94)$$

	A	B	C	D	E	F
1	(All numbers in 000's)					
2	Our cost	Comp1 bid	Linear prediction	Power prediction	Exponential prediction	Actual Linear Error
3	10	13	13.1027	13.35697	16.3795084	-0.1027
4	14	20	19.0587	19.07213	19.5315493	0.9413
5	16	22	22.0367	21.96795	21.3282198	-0.0367
6	18	25	25.0147	24.88511	23.2901627	-0.0147
7	30	44	42.8827	42.73548	39.4893521	1.1173
8	25	34	35.4377	35.23444	31.6909474	-1.4377
9	38	56	54.7947	54.88668	56.1502464	1.2053
10	44	63	63.7287	64.10133	73.114819	-0.7287
11	24	33	33.9487	33.74424	30.3267775	-0.9487
12					stdev	0.94189151

FIGURE 54

FIGURE 55

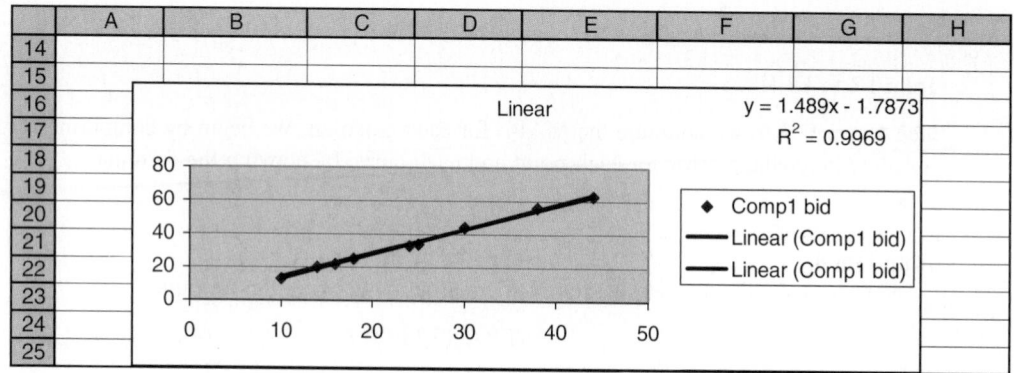

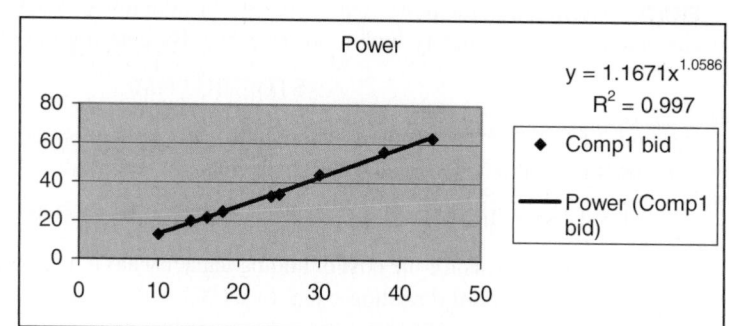

FIGURE 56

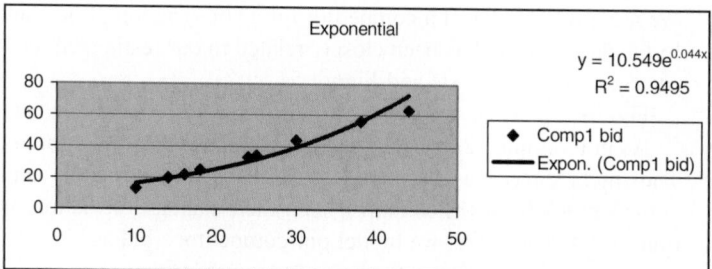

FIGURE 57

	G	H	I	J	K	L
2	Linear %age error	Power %age error	Exponential %age error	Linear abs %age error	Power abs %age error	Exponential %age error
3	-0.00784	-0.02673	-0.20633	0.007838	0.026726	0.206325
4	0.04939	0.048651	0.023984	0.04939	0.048651	0.023984
5	-0.00167	0.001459	0.031497	0.001665	0.001459	0.031497
6	-0.00059	0.004617	0.073415	0.000588	0.004617	0.073415
7	0.026055	0.029589	0.114224	0.026055	0.029589	0.114224
8	-0.04057	-0.03504	0.072862	0.04057	0.035035	0.072862
9	0.021997	0.020284	-0.00268	0.021997	0.020284	0.002676
10	-0.01143	-0.01718	-0.13834	0.011434	0.017181	0.138342
11	-0.02795	-0.02206	0.088147	0.027945	0.022055	0.088147
12				0.020831	0.022844	0.083497
13				MAPE		

FIGURE 58

EXAMPLE 14   **The Effects of New Competition on Price**

For similar products, the year after the first competitor comes in has historically shown a significant price drop. Figure 59 contains data on this situation.

For example, for the first product, a competitor entered in year 1. During year 2, a 22% price drop was observed, after allowing for a normal inflationary increase of 5% during the second year. Model the effect on price the year after the first competitor enters the market. See file Pricedata.xls.

Pricedata.xls

Solution   Figures 60–62 give the best-fitting linear, power, and exponential curves. The extremely low R^2 values imply that the year of entry has little or no effect on the price drop the year after the first competitor comes in. Therefore, we model price drop as a RISKNORMAL function, using the mean and standard deviation found in D14 and D15. If a competitor enters during year t, we would model the year $t + 1$ price with the formula

$$=1.05*(\text{year } t \text{ price})*\text{RISKNORMAL}(.803,.0366)$$

Note: $.803 = 1 - .197$.

	B	C	D
3	Year competitor enters	Share drop next year	Price drop next year
4		1	35	22
5		1	33	21
6		2	20	17
7		3	15	15
8		3	13	19
9		4	14	24
10		5	10	15
11		6	9	22
12		5	11	25
13		4	13	17
14			Mean	19.7
15			Std Dev	3.622461

FIGURE 59

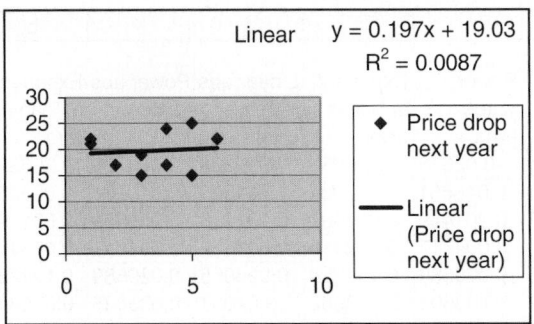

FIGURE **60**

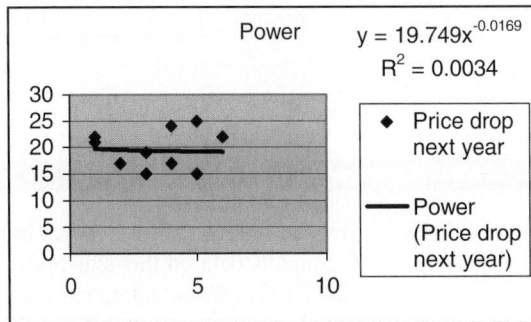

FIGURE **61**

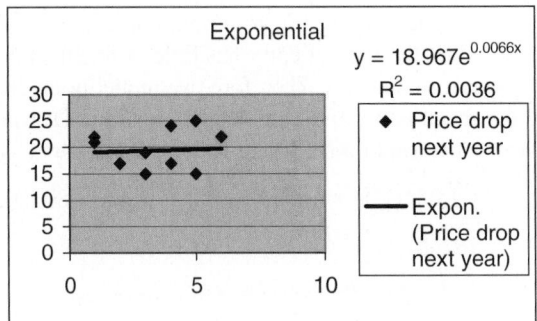

FIGURE **62**

Here, the assumption is that the market drop during a year is normally distributed. To check this, we could compute the skewness (with the SKEW function) and kurtosis (with the KURT function) of the data. If both the skewness and kurtosis are near 0, the market drop is probably normally distributed. An alternate approach to modeling the drop in price is to use the formula RISKDUNIFORM(D4:D13). This ensures that the drop in price is equally likely to assume one of the observed values. This approach has the advantage of not automatically assuming normality. The disadvantage, however, is that using the RISKDUNIFORM function implies that only 10 values of price drop are possible.

PROBLEMS

Group A

1 You are considering developing a new product. Forecast year 1 sales are for 80,000 units, and the year 1 price is $4.00 per unit. In file Simidata.xls you are given data on seven similar products from the past. (See Figure 63.)

For example,

For product 1, actual sales were 92.26% of forecast sales.

Year 2 price (in real dollars) was 76.7% of year 1 price.

Year 2 demand was 30.7% more than year 1 demand.

Product 1 only sold for 6 years.

The risk-adjusted discount rate is 11% per year. We assume that the price index will climb 5% per year.

We are unsure about the fixed cost of developing the product. It is equally likely to be $50,000 or $150,000. We are also unsure about the year 1 variable cost of producing it. It is equally likely to be $1, $1.50, or $2. After year 1, variable cost will climb by 5% per year. It costs $3 to build one unit of annual capacity.

a Assuming 80,000 units of annual capacity, estimate the 10-year risk-adjusted NPV of this product.

b What capacity level do you recommend?

2 You are trying to estimate NPV of profit for a new computer product, which you are confident will sell for ten years. You are given the following information.

The hurdle rate is 15%. Assume end of year for profits.

The total cost of developing the product will be spread equally over the product's life. Total development cost will be between $2 billion and $11 billion. There is a 25% chance that total fixed cost is $3 billion or less, a 50% chance of $6 billion or less, and a 75% chance of $9 billion or less.

The total year 1 market size (in terms of annual unit sales) is unknown but is believed to be between 0 and 600 million units. Unit sales of 100 million and 500 million are equally likely. Unit sales of 200 million and 400 million are equally likely, and are 4 times as likely as sales of 100

FIGURE 63

	A B		C	D	E	F	G	H
6	**Other Products**	1	2	3	4	5	6	7
7	Forecast Year 1	10000	15000	20000	25000	30000	18000	12000
8	Actual Year 1	9226	18544	20147	24093	27517	21670	12345
9	Year 1 Price	$ 10.00	$ 11.00	$ 12.00	$ 9.00	$ 8.00	$ 7.00	$ 9.00
10	Year 2 Price %age change	0 .7671376	1.01958	1.157148	0.799233	0.66222	0.96338	1.108995
11	Year 3 Price %age change	0.77301544	0.916731	0.629211	0.763033	0.785409	1.12807	0.770459
12	Year 4 Price %age change	0.93641094	1.07326	0.704279	1.032535	1.214266	0.607159	0.635232
13	Year 5 Price %age change	0.62486148	0.838744	0.730323	1.128628	1.222691	0.915762	0.709054
14	Year 6 Price %age change	0 .9909713	0.836317	1.178683	0.835511	1.186193	0.98035	0.870983
15	Year 7 Price %age change	0	1.154532	0.778286	1.008012	1.155539	0.83953	1.044165
16	Year 8 Price %age change	0	1.002691	0.991726	0.70686	0.871703	0	0.789561
17	Year 9 Price %age change	0	0.866046	0.933498	0	0.748535	0	0.800709
18	Year 10 Price %age change	0	0	1.137376	0	0.804221	0	0.963354
19	Year 2 % change in demand	1.30771172	1.257895	1.30467	1.326283	1.371715	1.203553	1.246827
20	Year 3 %age change change in demand	1.35463816	1.176022	1.439216	1.681935	1.186842	1.105765	0.705081
21	Year 4 % change change in demand	0.89116031	1.156565	0.940743	1.060037	0.953028	1.280444	0.927083
22	Year 5 %age change change in demand	0.62	0.728722	0.956427	0.744345	0.711915	0.710494	0.536207
23	Year 6 % change change in demand	0.53	0.529757	0.571999	0.6862	0.613999	0.572403	0.393907
24	Year 7 %age change change in demand	0	0.298447	0.193459	0.375018	0.432	0.269806	0.561965
25	Year 8 %age change change in demand	0	0.305065	0.314531	0.298697	0.294049	0	0.288539
26	Year 9 %age change change in demand	0	0.2	0.25	0	0.2	0	0.22
27	Year 10 %age change change in demand	0	0	0.15	0	0.16	0	0.12

TABLE 13

Year Competitor Entered	Drop in Share
2	21
3	17
4	15
5	13
6	12
2	20
4	16
5	12
6	11
7	10
8	9
8	10
10	9
12	8

TABLE 14

Predicted	Actual
40,000	37,000
50,000	42,000
60,000	56,000
70,000	67,000
80,000	75,000

million. Sales of 300 million are 5 times as likely as sales of 100 million. Each year, market growth is expected to average 5%, and during each year we are 95% sure that market growth will be between 3% and 7%.

Our most likely year 1 market share is 30%. There is a 5% chance that our market share will be less than or equal to 10% and a 5% chance that our market share will be more than 40%. A triangular distribution appears to be reasonable for market share. In later years, we expect market share, on average, to equal the previous year's share, but there is a 95% chance that market share could change by up to 20% of its current value.

The year 1 price charged for each unit follows a triangular random variable, with the most likely value $50, worst case $45, and best case $60. Each year, unit price will increase 5%.

The year 1 unit variable cost of production follows a triangular random variable with worst case $30, best case $20, and most likely case $24. Each year, variable costs will increase 5%.

a You are 95% sure that the mean NPV of the project is between _____ and _____. Run 1,600 iterations.

b What is the chance that this project will meet its hurdle rate?

c What are key drivers of the project's profitability?

3 You are trying to model what fraction of market share a new drug will lose the year a competitor comes in. Table 13 gives information for similar drugs. For example, competition for one drug entered the market 2 years after our drug, and we then lost 21% of our market share. How would you model the effect of competition on our product sales?

Group B

4 You own a small biotech firm. Eli Daisy wants to buy the rights to a potential cancer drug you are developing.

There is no way you could sell the product yourself. It will cost you $350,000 (payable at end of year 0) to develop the drug. Here's what Daisy has offered. At the end of years 1–8, Daisy will pay you 10% of the sales revenue for the drug, up to a maximum of $700,000. You discount cash flows at 20% per year. Each year, the drug sells for $20 per unit. You believe that the drug will sell 50,000 units during year 1. Table 14 shows your forecasts and actual year 1 sales for similar products in the past.

The pattern of sales for similar products is as follows. For a certain number of years, sales increase by a given percentage. Then, for all remaining years, sales decrease by a given percentage. You believe there is a 20% chance that sales will increase for 2 years, a 50% chance for 3 years, and a 30% chance for 4 years.

The percentage increase during the first path of the product life cycle will be between 2% and 20% per year. There is one chance in four that the annual percentage increase during this part of the product life cycle will be 5% or less; one chance in two of 15% or less, and three chances in 4 of 18% or less.

The annual percentage decrease during the remaining portion of the product life cycle will be between 2% and 10%. A 6% annual decrease is four times as likely as an 8% annual decrease. A 4% annual decrease is twice as likely as an 8% annual decrease.

Based on this information, would you take the deal? Explain your answer. What is the single most important driver of the deal's NPV?

5 You are trying to evaluate the profitability of a new drug produced by Eli Lilly. The drug will be sold during the years 2005–2010.

Development cost will be charged on September 10, 2004. The development cost will be between $.5 million and $5 million. A development cost of $2 million is four times as likely as a development cost of $1 million. A development cost of $4 million is twice as likely as a development cost of $1 million.

Unit sales during 2005 will be between 80,000 and 240,000. There is a 25% chance that 2005 unit sales will be less than or equal to 100,000 units, a 50% chance that they will be less than or equal to 140,000 units, and a 75% chance that they will be less than or equal to 200,000 units.

After year 1, sales will decay at a constant annual rate. For similar products, the decay rates have been 5%, 6%, 8%, 9%, 10%, 4%, 3%, and 8%.

Each year, you will charge $45 for the product.

Each year's variable production cost will depend on the number of units sold. For a drug with similar cost structure,

FIGURE **64**

	F	G
6	000's	000's
7	Units produced	Cost
8	40	813.323
9	50	999.459
10	60	1230.911
11	70	1399.077
12	80	1592.645
13	90	1812.399
14	100	2013.139
15	110	2709.943
16	120	3405.542
17	130	4096.212
18	140	4815.177
19	150	5516.294
20	160	6200.432
21	170	6914.829
22	180	7613.689
23	190	8320.617
24	200	9012.181

FIGURE **65**

	K	L
11	Actual	Forecast
12	20000	26000
13	30000	35000
14	10000	14000
15	40000	48000
16	50000	62000

(*Hint:* In modeling annual variable cost, you need not do a MAPE. From the proper plot, the appropriate model should be clear.)

6 GM is considering producing a new car. GM's current net income for each of the next 6 years is assumed to be $100 million, which we assume to be received on June 30 of the years 2004–2009. The tax rate is 40%. Assume that no other GM projects involve depreciation. The fixed cost of developing the new car will be between $20 million and $40 million, with a most likely value of $25 million. The entire fixed cost is incurred on June 30, 2004 and is depreciated on a straight-line basis during the years 2005–2009. All future cash flows are received midyear. The car is assumed to be sold during the years 2005–2009. Forecast for 2005 unit sales is 15,000. Past forecasts and actual sales during the first year of similar models are as shown in Figure 65.

During the years 2006–2009, sales are assumed to decay at the same rate each year. This rate will be between 5% and 20%, with a 12% decay rate twice as likely as an 8% decay rate and a 16% decay rate three times as likely as an 8% decay rate. During 2005, the car will sell for $13,000. The price will increase by the same percentage each year, with 1%, 2%, and 3% price increases being equally likely. During 2005, variable costs are $11,000. During 2006–2009, variable costs will increase by the same percentage, with increases of 2%, 4%, and 6% being equally likely. Discount cash flows at 15%. Should GM produce the car?

the variable cost as a function of units produced was as shown in Figure 64. For example, during year 1, 40,000 units were produced, and the cost was $813,323. (See file Sim3data.xls.)

Assume that cash flows are discounted at 10% and cash flows for years 2005–2010 may be considered to be received midyear (June 30).

a After running 900 iterations, you are 95% sure that actual NPV earned by the drug (in 09/10/04 dollars) is between _____ and _____.

b What is the key driver of the drug's NPV?

23.11 Simulation and Bidding

In situations in which you must bid against competitors, simulation can often be used to determine an appropriate bid. Usually you do not know what a competitor will bid, but you may have an idea about the range of bids a competitor may choose. In this section, we show how to use simulation to determine a bid that maximizes your expected profit. First, we briefly discuss generating observations from a uniformly distributed random variable.

Uniform Random Variables

A random variable is said to be uniformly distributed on the closed interval [a, b] (written U(a, b)) if the random variable is equally likely to assume any value between a and b inclusive. To generate samples from a U(a, b) random variable, enter the formula

$$=RISKUNIFORM(a,b)$$

into a cell.

We now show how to use simulation to determine a bid that maximizes expected profit.

EXAMPLE 15   Bidding

You are going to make a bid on a construction project. You believe it will cost you $10,000 to complete the project. Four competitors are going to bid against you. Based on past history, you believe that each competitor's bid is equally likely to be any value between your cost of completing the project and triple your cost of completing the project. You also believe that each competitor's bid is independent of the other competitors' bids. What bid maximizes your expected profit?

Solution

Bid.xls

In our solution, all amounts will be in thousands of dollars. The statement of the problem implies that each competitor's bid is U(10, 30), and the bids of the competitors are independent. Our simulation is shown in Figure 66 (file Bid.xls). We proceed as follows.

Step 1   In cell C3, we enter the cost of the project.

Step 2   In cell C4, we enter ten possible bids (11, 12, 13, 14, 15, 16, 17, 18, 19, and 20) with the formula

$$=RISKSIMTABLE(\{11,12,13,14,15,16,17,18,19,20\})$$

	A	B	C	D	E	F
1	Bidding Example					
2						
3	My cost(thousands)		10			
4	My bid(thousands)		11			
5	Competitor 1 Bid		15.23686647			
6	Competitor 2 Bid		24.37239289			
7	Competitor 3 Bid		23.26008201			
8	Competitor 4 Bid		25.59600472			
9	Profit(thousands)		1			
10						
11		Cell	Name	Minimum	Mean	Maximum
12						
13		C9	(Sim#1) Profit(thousa...	0	0.81	1
14		C9	(Sim#2) Profit(thousa...	0	1.3	2
15		C9	(Sim#3) Profit(thousa...	0	1.575	3
16		C9	(Sim#4) Profit(thousa...	0	1.66	4
17		C9	(Sim#5) Profit(thousa...	0	1.5625	5
18		C9	(Sim#6) Profit(thousa...	0	1.41	6
19		C9	(Sim#7) Profit(thousa...	0	1.155	7
20		C9	(Sim#8) Profit(thousa...	0	0.96	8
21		C9	(Sim#9) Profit(thousa...	0	0.7425	9
22		C9	(Sim#10) Profit(thous...	0	0.55	10

FIGURE 66
Bidding Simulation

Step 3   In C5, we generate the bid of the first competitor by entering the formula

$$=\text{RISKUNIFORM(C\$3,3*C\$3)}$$

Copying this formula to the range C6:C8 generates the bids of the other three competitors. (Why does this ensure that their bids are independent?)

Step 4   In cell C9, we compute the actual profit for this trial by entering the formula

$$=\text{IF(C4}\leq=\text{MIN(C5:C8),C4-C3,0)}$$

This ensures that if we win the bid (C4≤MIN(C5:C8)), then our profit equals our bid less the project cost of $10,000; if we don't win the bid (C4>MIN(C5:C8)), then we earn no profit. This statement assumes that we win all ties, but the chance of a tie bid is negligible (why?), so this really does not matter. To see how things work, hit the recalculation (F9) button and see how the cells of the spreadsheet change.

Step 5   To determine the bid that maximizes expected profit, we ran 400 iterations of this spreadsheet for each bid with @Risk. From Figure 66, it appears that a bid between $13,000 and $15,000 will maximize expected profit (with an expected profit of $1,660).

Step 6   To zero in on the bid that maximizes expected profit, we replaced the formula in cell C4 with

$$=\text{RISKSIMTABLE(\{13.2,13.4,13.6,13.8,14,14.2,14.4,14.6,14.8\})}$$

One hundred iterations of this spreadsheet indicate that a bid of around $14,200 maximizes expected profit (an expected profit of around $1,800 is earned).

PROBLEMS

Group A

1   If the number of competitors in Example 15 were to double, how would the optimal bid change?

2   If the average bid for each competitor stayed the same, but their bids exhibited less variability, would the optimal bid increase or decrease? To study this question, assume that each competitor's bid follows each of the following random variables:

   a $U(15, 25)$

   b $U(18, 22)$

3   Warren Millken is attempting to take over Biotech Corporation. The worth of Biotech depends on the success or failure of several drugs under development. Warren does not know the actual (per share) worth of Biotech, but the current owners of Biotech do know the actual worth of the company. Warren assumes that Biotech's actual worth is equally likely to be between $0 and $100 per share. Biotech will accept Warren's offer if it exceeds the true worth of the company. For example, if the current owners think Biotech is worth $40 per share and Warren bids $50 per share, they will accept the bid. If the current owners accept Warren's bid, then Warren's corporate strengths immediately increase Biotech's market value by 50%. How much should Warren bid?

23.12 Playing Craps with @Risk

Craps is a very complex game. With @Risk, it is easy to estimate the probability of winning at craps.

EXAMPLE 1 6   **Craps**

In the game of craps, a player tosses two dice. If the first toss yields a 2, 3, or 12, the player loses. If the player rolls a 7 or 11 on the first toss, he or she wins. Otherwise, the player continues tossing the dice until he or she either matches the number thrown on the first roll (called the *point*) or tosses a 7. Rolling the point before rolling a 7 wins. Rolling a 7 before the point loses. By complex calculations, it can be shown that a player wins at craps 49.3% of the time. Use @Risk to verify this.

Solution

The key observation is that we do not know how many rolls the game will take. Suppose the game does not end on the first toss. The least likely points to be made are 4 and 10 which have probability $3/36 = 1/12$ of being made. Therefore, after the first toss, there is at least a $(1/12) +$ probability of $7 = (1/12) + (1/6) = (1/4)$ chance that the game will end on each toss. Thus, the chance of the game continuing on each toss is at most $(3/4)$. After (say) 50 tosses, the probability that the game is still going on is at most $.75^{49} = 7$ in 10,000,000. Therefore, we can cut off the game after 50 tosses and not worry about the (fewer than 1 in a million) games that go on beyond 50 tosses. After each dice roll, we keep track of the game status:

$$0 = \text{game lost}$$
$$1 = \text{game won}$$
$$2 = \text{game still going}$$

Craps.xls

The output cell will keep track of the status of the game after the 50th toss. A 1 will indicate a win, and a 0 will indicate a loss. The work is in the file Craps.xls. See Figure 67.

Step 1   In B2, we use the RISKDUNIFORM function (discrete uniform random variable) to generate the roll of the dice on the first toss with the formula

$$=\text{RISKDUNIFORM(\$AD\$9:\$AD\$14)}$$

The RISKDUNIFORM function ensures that each of its arguments is equally likely. Therefore, each die has an equal $(1/6)$ chance of yielding a 1, 2, 3, 4, 5, or 6.

FIGURE **67**

	A	B	C	D	E	F	G	H	AX	AY
1	TOSS#	1	2	3	4	5	6	7	49	50
2	Die Toss 1	3.5	3.5	3.5	3.5	3.5	3.5	3.5	3.5	3.5
3	Die Toss 2	3.5	3.5	3.5	3.5	3.5	3.5	3.5	3.5	3.5
4	Total	7	7	7	7	7	7	7	7	7
5	GAME STATUS	1	1	1	1	1	1	1	1	1
6	0=LOSS	WIN??	1							
7	1=WIN									
8	2=STILL GOING	95% CI								
9	LOWER									
10	UPPER									

Copying this formula to the range B2:AY3 generates both dice rolls for 30 tosses. Note that we have hidden rolls 8–28.

Step 2   In B4:AY4, we compute the total dice roll on all 30 rolls by copying from B4 to C4:AY4 the formula

$$=SUM(B4:C4)$$

Step 3   In cell B5, we determine the game status after the first roll with the formula

$$=IF(OR(B4=2,B4=3,B4=12),0,IF(OR(B4=7,B4=11),1,2))$$

Note that a 2, 3, or 12 will result in a loss, a 7 or 11 will result in a win, and any other roll will result in the game continuing.

Step 4   In cell C5, we compute the status of the game after the second roll with the formula

$$=IF(OR(B5=0,B5=1),B5,IF(C4=\$B4,1,IF(C4=7,0,2)))$$

Note that if the game ended on the first roll, we maintain the status of the game. If we make our point, we record a win with a 1. If we roll a 7, we record a loss. Otherwise, the game is still going.

Copying this formula from C5 to D5:AY5 records the game status after rolls 2–50. The game result is in AY5, which we copy to C6 so that we can easily see it. After running 4,000 iterations with output cell C6, we obtain a 48.3% chance of winning. With 10,000 iterations, we usually obtain a probability very close to 49.3%.

23.13   Simulating the NBA Finals

Finals.xls

The Indiana Pacers came within two plays (one questionable foul call on Dale Davis in game 6 and Travis Best missing a shot in game 4) of winning the 2000 NBA championship. Before the series, what was the probability that the Lakers would win the series? From the Sagarin ratings (found at http://www.kiva.net/~jsagarin/), we found that the Lakers are around 4 points better than the Pacers. The home team has a 3-point edge, and games play out according to a normal distribution, with mean equal to our prediction and a standard deviation of 12 points. Past history shows that the Sagarin forecasts exhibit no bias. In the file Finals.xls and Figure 68, we simulate the 2000 NBA Finals. Recall that the Lakers were at home during games 1, 2, 6, and 7, while the Pacers were at home during games 3–5. (Note: We always make a series go 7 games, because we do not know when it will actually end.) If the Lakers win at least 4 of the 7 games, they win the series, which is indicated by a 1 in cell I14. We have named the cells in D2:D5 with the range names given in C2:C5.

Step 1   In G5:G11, we generate our forecast for each game by copying the formula

$$=IF(F5=``LA",HE+LA-IND,-HE+LA-IND)$$

from G5 to G6:G11.

Step 2   In H5:H11, we generate the Lakers' margin of victory in each game as normally distributed with a standard deviation of 12 and mean given in column G. Just copy from H5 to H6:H11 the formula

$$=RISKNORMAL(G5,STDEV)$$

FIGURE **68**

	B	C	D	E	F	G	H	I
1	NBA Finals 2000							
2		IND	5					
3		LA	9					
4		HE	3	Game	Home	LA Forecast	LA Margin	LA Win
5		STDEV	12	1	LA	7	1.312640391	1
6				2	LA	7	0.911581437	1
7				3	IND	1	-0.423239113	0
8				4	IND	1	22.02026177	1
9				5	IND	1	-13.825966	0
10				6	LA	7	-5.963644407	0
11				7	LA	7	-2.505744158	0
12							LA total wins	3
13								
14							LA Wins series?	0
15								
16								
17								
18		Name	NPV	LA Wins series? / LA Win				
19		Description	Output	Output				
20		Cell	[]Sheet1!F{	[finals.xls]Sheet1!I14				
21		Minimum =	0	0				
22		Maximum =	0	1				
23		Mean =	0	0.796875				

Step 3 In I5:I11, we determine if the Lakers won the game by copying from I5 to I6:I11 the formula

$$=IF(H6>0,1,0)$$

Step 4 In cell I12, we compute the total number of Lakers wins in the series with the formula

$$=SUM(I5:I11)$$

Step 5 Note that if the Lakers win at least 4 games, they win the series. In cell I14, we determine if the Lakers win the series with the formula

$$=IF(I12>=4,1,0)$$

From the @Risk output, we find that the Lakers had an 80% chance to win the series. The bookmakers had L.A. as a 7-1 favorite, which means (after taking out a 10% profit) they believed that the Lakers had around a 90% chance to win.

REVIEW PROBLEMS

Group A

1 The New York Knicks and the Chicago Bulls are ready for the best-of-seven NBA Eastern finals. The two teams are evenly matched, but the home team wins 60% of the games between the two teams. The sequence of home and away games is to be chosen by the Knicks. The Knicks have the home edge and will be the home team for four of the seven scheduled games. They have the following choices (home team is listed for each game):

Sequence 1: NY, NY, CHIC, CHIC, NY, CHIC, NY
Sequence 2: NY, NY, CHIC, CHIC, CHIC, NY, NY

Use simulation to show that either sequence gives the Knicks the same chance of winning the series.

2[†] You currently have $100. Each week, you can invest any amount of money you currently have in a risky investment. With probability .4, the amount you invest is tripled (e.g., if you invest $100, you increase your asset

[†]Based on Kelly (1956).

position by \$300), and with probability .6, the amount you invest is lost. Consider the following investment strategies:

(1) Each week, invest 10% of your money.

(2) Each week, invest 30% of your money.

(3) Each week, invest 50% of your money.

Simulate 100 weeks of each strategy 50 times. Which strategy appears to be best? In general, if you can multiply your investment by M with probability p and lose your investment with probability q, you should invest a fraction $\frac{p(M-1)-q}{M-1}$ of your money each week. This strategy maximizes (for a favorable game) the expected growth rate of your fortune and is known as the **Kelly criterion.**

3[†] The Magellan mutual fund has beaten the Standard and Poor's 500 during 11 of the last 13 years. People use this as an argument that you can "beat the market." Here's another way to look at it that shows that Magellan's beating the market 11 out of 13 times is not unusual. Consider 50 mutual funds, each of which has a 50% chance of beating the market during a given year. Use simulation to estimate the probability that over a 13-year period the "best" of the 50 mutual funds will beat the market for at least 11 out of 13 years. This probability turns out to exceed 40%, which means that the best mutual fund's beating the market 11 out of 13 years is not an unusual occurrence!

4 You have made it to the final round of "Let's Make a Deal." You know that there is \$1 million behind either door 1, door 2, or door 3. It is equally likely that the prize is behind any of the three doors. The two doors without a prize have nothing behind them. You randomly choose door 2. Before you see whether the prize is behind door 2, Monty chooses to open a door that has no prize behind it. For the sake of definiteness, suppose that before door 2 is opened, Monty reveals that there is no prize behind door 3. You now have the opportunity to switch and choose door 1. Should you switch?

Use a spreadsheet to simulate this situation 400 times. For each "trial" use an @Risk function to generate the door behind which the prize lies. Then use another @Risk function to generate the door that Monty will open. Assume that Monty plays as follows: Monty knows where the prize is and will open an empty door, but he cannot open door 2. If the prize is really behind door 2, Monty is equally likely to open door 1 or door 3. If the prize is really behind door 1, Monty must open door 3. If the prize is really behind door 3, Monty must open door 1.

5 Star-crossed soap-opera lovers Noah and Julia have had a big argument. Julia's sister Maria wants Noah and Julia to make up, so she has told them both to go to the romantic gazebo at 1 P.M. Unfortunately, Noah and Julia are not punctual. Each is equally likely to show up at the gazebo any time between 1 and 2 P.M. Assuming that each will stay for 20 minutes, what is the probability that they will meet? You can model the arrival of each person using a RISKUNIFORM random variable. For example, RISKUNIFORM(1,2) is equally likely to choose any number between 1 and 2 (including the endpoints 1 and 2).

6 The game of Chuck-a-Luck is played as follows: You pick a number between 1 and 6 and toss three dice. If your number does not appear, you lose \$1. If your number appears x times, you win \$$x$. On the average, how much money will you win or lose on each play of the game?

7 I toss a die several times until the total number of spots I have seen is at least 13. What is the most likely total that will occur?

Group B

8[‡] When the team is behind late in the game, a hockey coach usually waits until there is one minute left before pulling the goalie. Actually, coaches should pull their goalies much sooner. Suppose that if both teams are at full strength, each team scores an average of .05 goal per minute. Also suppose that if you pull your goalie, you score an average of .08 goal per minute, while your opponent scores an average of .12 goal per minute. Suppose you are one goal behind with five minutes left in the game. Consider the following two strategies:

Strategy 1: Pull your goalie if you are behind at any point in the last five minutes of the game; put him back in if you tie the score or go ahead.

Strategy 2: Pull your goalie if you are behind at any point in the last minute of the game; put him back in if you tie the score or go ahead.

Which strategy maximizes your chance of winning or tying the game? Simulate the game using ten-second increments of time. Use the RISKBINOMIAL function to determine whether a team scores a goal in a given ten-second segment. It is acceptable to do this because the probability of scoring two or more goals in a ten-second period is near 0.

9 Suppose we toss an ordinary die 5 times. A 4-straight occurs if exactly 4 (not 5) of our rolls are consecutive integers. For example, if we roll 1, 2, 3, 4, 6 we have a 4-straight. Also 3, 4, 5, 6, 1, 1 is a 4-straight. However 2, 3, 4, 5, 6 is not a 4-straight. After running 4,000 iterations, you are 95% sure that the chance of tossing a 4-straight is between _____ and _____.

10 Buffie the Vampire Slayer is going to Las Vegas to relax. She is going to play the following game of blackjack. She throws a pair of dice until the cumulative total of her tosses is at least 4. If her total is 8 or more, she loses. Assuming that Buffie has not yet lost, the croupier (Spike) tosses the dice until his total is at least 4. If Spike's total is 8 or more, then Buffie (assuming she did not total 8 or more) wins. Otherwise, we compare Spike's and Buffie's totals. The high total wins, with a tie going to Spike. After running 900 iterations, you are 95% sure Buffie's chance of winning the game is between _____ and _____.

11 Wheaties is producing cereals with five different sets of trading cards:

- Rock stars
- NBA stars
- Baseball stars
- Hockey stars
- Football stars

[†]Based on Marcus (1990).

[‡]Based on Morrison and Wheat (1984).

Each box contains one set of trading cards, and you do not know which set is in a box until you open it.

a On the average, how many boxes are needed to obtain all five sets of trading cards?

b You are 95% sure that between _____ and _____ boxes of Wheaties must be purchased to obtain all five sets of trading cards.

REFERENCES

Kelly, J. "A New Interpretation of Information Rate," *Bell System Technical Journal* 35(1956):917–926.

Marcus, A. "The Magellan Fund and Market Efficiency," *Journal of Portfolio Management* (1990, Fall):85–88.

Morrison, D., and R. Wheat. "Pulling the Goalie Revisited," *Interfaces* 16(no. 6, 1984):28–34.

Forecasting Models

In previous chapters, we have often blindly substituted numbers into problems without considering where the numbers came from. For example, in the Giapetto LP (Example 1 in Chapter 3), we assumed that the variable cost of producing a train was $21. In reality, we would have to estimate the cost of producing a train. This can be done using the method of *simple linear regression,* explained in Section 24.6.

In Chapters 15 and 16, we used inventory theory to determine production quantities and reorder points. To use the models of Chapters 15 and 16, we need to be able to forecast the demand for a product. In Sections 24.1–24.5, we discuss extrapolation and smoothing methods that can be used to forecast future demand for a product.

As another example of how we can use "good forecasts," suppose that we want to use the queuing models of Chapter 20 to determine how the number of tellers at a bank should vary with the day of the week and the time of day. To tackle this problem, we need to determine how the rate at which customers enter the bank depends on the time of day and the day of the week. For example, if we knew that over half the bank's customers arrived during the lunch hour (noon to 1 P.M.), that would have a significant effect on the optimal staffing policy.

In this chapter, we discuss two important types of forecasting methods: extrapolation methods and causal forecasting methods. In Sections 24.1–24.5, we discuss *extrapolation methods,* which are used to forecast future values of a time series from past values of a time series. To illustrate, consider Lowland Appliance Company's monthly sales of TVs, compact disc players (CDs), and air conditioners (ACs) for the last 24 months, given in Table 1. In an extrapolation forecasting method, it is assumed that past patterns and trends in sales will continue in future months. Thus, past data on appliance sales (and no other information) are used to generate forecasts for appliance sales during future months. Extrapolation methods (unlike the causal forecasting methods described in Sections 24.6–24.8) don't take into account what "caused" past data; they simply assume that past trends and patterns will continue in the future.

Causal forecasting methods attempt to forecast future values of a variable (called the dependent variable) by using past data to estimate the relationship between the dependent variable and one or more independent variables. For example, Lowland might try to forecast future monthly sales of air conditioners by using past data to determine how air conditioner sales are related to independent variables such as price, advertising, and the month of the year. Causal forecasting methods will be discussed in Sections 24.6–24.8.

24.1 Moving-Average Forecasting Methods

Let $x_1, x_2, \ldots, x_t, \ldots$ be observed values of a time series, where x_t is the value of the time series observed during period t. One of the most commonly used forecasting

TABLE **1**
Lowland Appliance Sales

Month	TV Sales	CD Sales	AC Sales	Month	TV Sales	CD Sales	AC Sales
1	30	40	13	13	38	79	36
2	32	47	7	14	30	82	21
3	30	50	23	15	35	80	47
4	39	49	32	16	30	85	81
5	33	56	58	17	34	94	112
6	34	53	60	18	40	89	139
7	34	55	90	19	36	96	230
8	38	63	93	20	32	100	201
9	36	68	63	21	40	100	122
10	39	65	39	22	36	105	84
11	30	72	37	23	40	108	74
12	36	69	29	24	34	110	62

methods is the moving-average method. We define $f_{t,1}$ to be the forecast period for period $t + 1$ made after observing x_t. For the moving-average method,

$$f_{t,1} = \text{average of the last } N \text{ observations}$$
$$= \text{average of } x_t, x_{t-1}, x_{t-2}, \ldots, x_{t-N+1}$$

where N is a given parameter.

To illustrate the moving-average method, we choose $N = 3$ and use the moving-average method to forecast TV sales for the first six months of data in Table 1. The resulting computations are given in Table 2. For months 1–3, we have not yet observed three months of data, so (for $N = 3$) we cannot develop a moving-average forecast for sales for these months. For month 4, we find our forecast, $f_{3,1} = \frac{30+32+30}{3} = 30.67$. For month 5, our forecast is $f_{4,1} = \frac{32+30+39}{3} = 33.67$. For month 6, our forecast is $f_{5,1} = \frac{30+39+33}{3} = 34$.

Note that from one period to the next, our forecast "moves" by replacing the "oldest" observation in the average by the most recent observation.

Choice of *N*

How should we choose N, the number of periods used to compute the moving average? To answer this question, we need to define a measure of forecast accuracy. We will use the **mean absolute deviation** (MAD) as our measure of forecast accuracy. Before defining the MAD, we need to define the concept of a **forecast error.** Given a forecast for x_t, we define e_t to be the error in our forecast for x_t, to be given by

$$e_t = x_t - (\text{forecast for } x_t)$$

From Table 2, we find $e_4 = 39 - 30.67 = 8.33$, $e_5 = 33 - 33.67 = -0.67$, and $e_6 = 34 - 34 = 0$. The MAD is simply the average of the absolute values of all the e_t's. Thus, for periods 1–6, our moving-average forecast yields a MAD given by

$$\text{MAD} = \frac{|e_4| + |e_5| + |e_6|}{3} = \frac{8.33 + 0.67 + 0}{3} = 3$$

Thus, on the average, our forecasts for TV sales are off by 3 TVs per month.

TABLE 2
Moving-Average Forecasts ($N = 3$)
for TV Sales

Month	Actual Sales	Predicted Sales
1	30	—
2	32	—
3	30	—
4	39	$\frac{30 + 32 + 30}{3}$
5	33	$\frac{32 + 30 + 39}{3}$
6	34	$\frac{30 + 39 + 33}{3}$

We are trying to forecast next month's TV sales as an average of the last N months' actual sales. What value of N will minimize our mean absolute error (obtained by averaging the actual error incurred during each month)? We will try $N = 1, 2, \ldots, 12$.

We begin with an explanation of the Excel OFFSET function. This function lets you pick out a cell range relative to a given location in the spreadsheet. The syntax of the OFFSET function is as follows:

OFFSET(reference, rows, columns, height, width)

- **Reference** is the cell from which you base the row and column references.
- **Rows** helps locate the upper left-hand corner of the OFFSET range. Rows is measured by number of rows up or down (up is negative, and down is positive) from the cell reference.
- **Columns** helps locate the upper left-hand corner of the OFFSET range. Columns is measured by number of columns left or right (left is negative, and right is positive) from the cell reference.
- **Height** is the number of rows in the selected range.
- **Width** is the number of columns in the selected range.

Offsetexample.xls

File Offsetexample.xls contains some examples of how the OFFSET function works. See Figure 1. The nice thing about the OFFSET function is that it can be copied like any formula. The next section will show the true power of the OFFSET function.

Tvsales.xls

Our work is in file Tvsales.xls. We begin creating a forecast in month 13, because that is the first month in which 12 months of historical data are available. See Figures 2 and 3.

Step 1 By copying from C17 to C18:C28 the formula

$$=\text{AVERAGE(OFFSET(B17,-\$D\$3,0,\$D\$3,1))}$$

obtain the average of the last D3 months of data.

- B17 ensures that we define our range relative to the cell directly to the left of the cell where the formula is entered.
- -$D$3 ensures that our range begins D3 rows above the row where the formula is entered.
- The 0 ensures that the OFFSET range will always remain in column B.
- $D$3 ensures that we average the last D3 observations.
- 1 ensures that the OFFSET range includes a single column.

FIGURE **1**

	A	B	C	D	E	F	G	H	I	J	K
1											
2											
3		Offset examples									
4											
5											
6		1	2	3	4			1	2	3	4
7		5	6	7	8			5	6	7	8
8		9	10	11	12			9	10	11	12
9											
10	=SUM(OFFSET(B7,-1,1,2,1))	**8**					=SUM(OFFSET(H6,0,1,3,2))	**39**			
11											
12											
13											
14		1	2	3	4						
15		5	6	7	8						
16		9	10	11	12						
17											
18	=SUM(OFFSET(E16,-2,-3,2,3))	**24**									
19											

	A	B
4	Month	TV Sales Actual
5	1.00	30
6	2.00	32
7	3.00	30
8	4.00	39
9	5.00	33
10	6.00	34
11	7.00	34
12	8.00	38
13	9.00	36
14	10.00	39
15	11.00	30
16	12.00	36
17	13.00	38
18	14.00	30
19	15.00	35
20	16.00	30
21	17.00	34
22	18.00	40
23	19.00	36
24	20.00	32
25	21.00	40
26	22.00	36
27	23.00	40
28	24.00	34

FIGURE **2**

	A	B	C	D	E	F	G	H
2				# OF PERIODS				
3				1				
4	Month	TV Sales Actual	Moving average forecast	Abs error	MAD	5		
5	1.00	30						
6	2.00	32						
7	3.00	30						
8	4.00	39					# of periods	5
9	5.00	33					1	5
10	6.00	34					2	3.666666667
11	7.00	34					3	3.361111111
12	8.00	38					4	3.333333333
13	9.00	36					5	3.016666667
14	10.00	39					6	3.111111111
15	11.00	30					7	3.226190476
16	12.00	36					8	3.21875
17	13.00	38	36	2			9	3.055555556
18	14.00	30	38	8			10	3.083333333
19	15.00	35	30	5			11	3.045454545
20	16.00	30	35	5			12	3.111111111
21	17.00	34	30	4			Min	3.016666667
22	18.00	40	34	6			best #	5
23	19.00	36	40	4				
24	20.00	32	36	4				
25	21.00	40	32	8				
26	22.00	36	40	4				
27	23.00	40	36	4				
28	24.00	34	40	6				

FIGURE 3

Step 2 By copying from D17 to D18:D28 the formula

$$=ABS(B17-C17)$$

compute the absolute value of the error in each month's forecast (based on a D3-month moving average).

Step 3 In cell F4, compute the average of the absolute errors (often called the MAD) with the formula

$$=AVERAGE(D17:D28)$$

Step 4 Enter the trial number of periods for the moving average (1–12) in G9:G20, and in cell H8, enter the MAD with the formula

$$=F4$$

Step 5 After selecting the table range G8:H20 and choosing a one-way data table with the column input cell of D3, we find that a 5-period moving average yields the smallest MAD (3.02).

Step 6 We obtain the minimum MAD in cell H21 with the formula

$$=MIN(H9:H20)$$

Step 7 Entering in cell H22 the formula

$$=MATCH(H21,H9:H21,0)$$

gives the number of periods (5) yielding the smallest MAD.

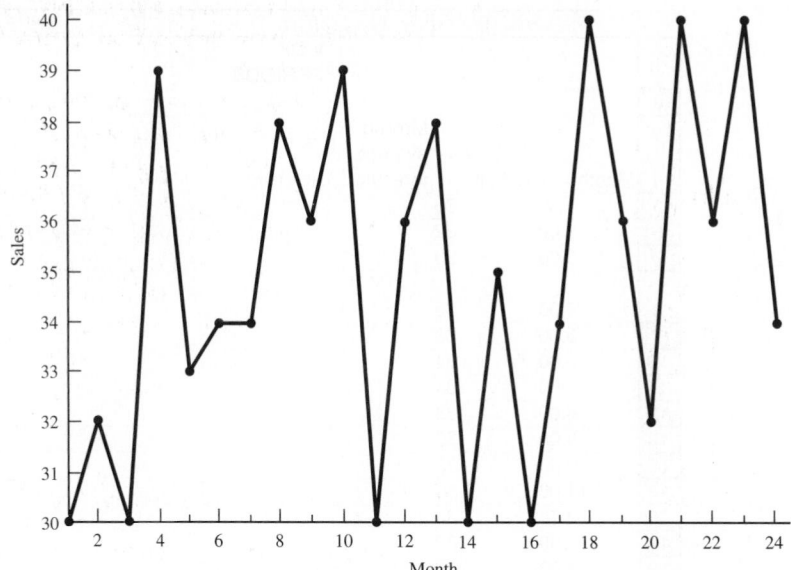

FIGURE 4
TV Sales

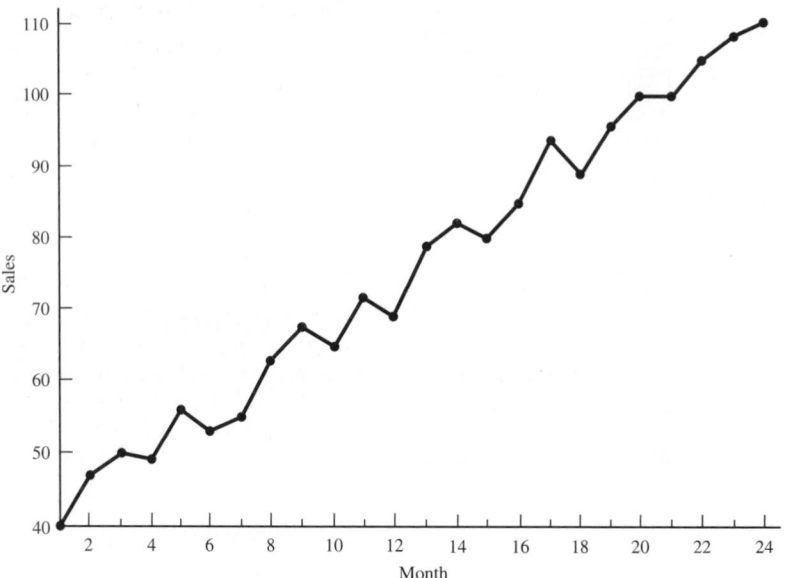

FIGURE 5
CD Player Sales

Moving-average forecasts perform well for a time series that fluctuates about a constant **base level.** From Figure 4, it appears that monthly TV sales fluctuate about a base level of 35. More formally, moving-average forecasts work well if

$$x_t = b + \varepsilon_t \tag{1}$$

where b is the base level for the series and ε_t is the random fluctuation in period t about the base level.

From Figures 5 and 6, we see that sales of CD players and air conditioners are not well described in Equation (1). From Figure 5, we see that there is an upward **trend** in CD player sales, so they do not fluctuate about a base level. From Figure 6, we find that air conditioner sales exhibit **seasonality:** The peaks and valleys of the series repeat at regular 12-month intervals. Figure 6 also shows that air conditioner sales exhibit an upward

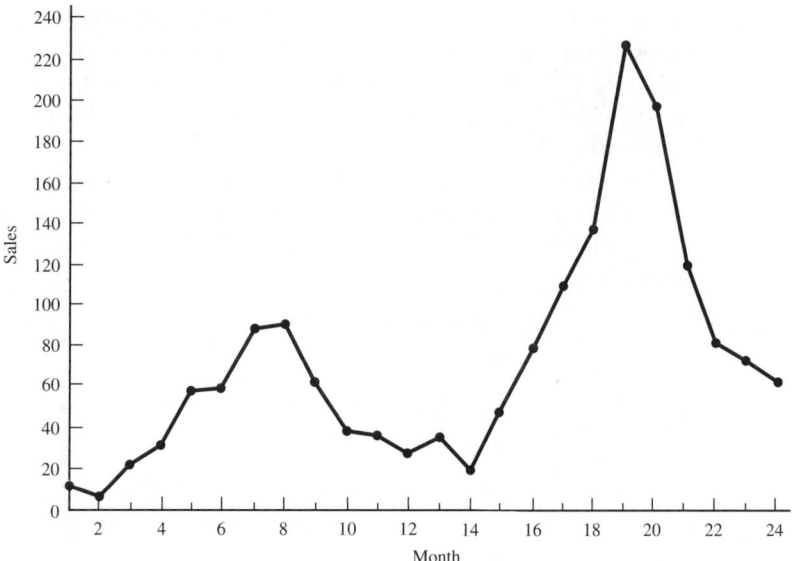

FIGURE 6
Air Conditioner Sales

trend. In situations where trend and/or seasonality are present, the moving-average method usually yields poor forecasts. To close this section, we note that in addition to trend and seasonality, a time series may exhibit **cyclic behavior.** For example, auto sales often follow the business cycle of the national economy. Cyclic behavior is much more irregular than a seasonal pattern and is often hard to detect.

24.2 Simple Exponential Smoothing

If a time series fluctuates about a base level, **simple exponential smoothing** may be used to obtain good forecasts for future values of the series. To describe simple exponential smoothing, let A_t = smoothed average of a time series after observing x_t. After observing x_t, A_t is the forecast for the value of the time series during any future period. The key equation in simple exponential smoothing is

$$A_t = \alpha x_t + (1 - \alpha)A_{t-1} \tag{2}$$

In (2), α is a **smoothing constant** that satisfies $0 < \alpha < 1$. To initialize the forecasting procedure, we must have (before observing x_1) a value for A_0. Usually, we let A_0 be the observed value for the period immediately preceding period 1. As with moving-average forecasts, we let $f_{t,k}$ be the forecast for x_{t+k} made at the end of period t. Then

$$A_t = f_{t,k} \tag{3}$$

Assuming that we are trying to forecast one period ahead, our error for predicting x_t (written again as e_t) is given by

$$e_t = x_t - f_{t-1,1} = x_t - A_{t-1} \tag{4}$$

To understand (2) better, we use (4) to rewrite (2) as

$$A_t = A_{t-1} + \alpha(x_t - A_{t-1}) = A_{t-1} + \alpha e_t$$

Thus, our new forecast $A_t = f_t, 1$ is equal to our old forecast (A_{t-1}) plus a fraction of our period t error (e_t). This implies that if we "overpredict" x_t, we lower our forecast, and if we "underpredict" x_t, we raise our forecast. For larger values of the smoothing constant α, more weight is given to the most recent observation (see Remark 3 at the end of the section).

We illustrate simple exponential smoothing (with $\alpha = 0.1$) for the first six months of TV sales. The results are given in Table 3. We assume that 32 TVs were sold last month, so we initialize the procedure with $A_0 = 32$. Here are some illustrations of the computations:

$$A_t = 0.1x_1 + 0.9A_0 = 0.1(30) + 0.9(32) = 31.8$$
$$f_{0,1} = A_0 = 32$$
$$e_1 = x_1 - A_0 = 30 - 32 = -2$$
$$f_{1,1} = A_1 = 31.8$$
$$e_2 = x_2 - A_1 = 32 - 31.8 = 0.2$$
$$A_2 = 0.1x_2 + 0.9A_1 = 0.1(32) + 0.9(31.8) = 31.82$$

For months 1–6, the MAD of our forecast is given by

$$\text{MAD} = \frac{|-2| + |0.2| + |-1.82| + |7.36| + |0.63| + |1.56|}{6}$$

$$= 2.26$$

For the entire 24-month period, we can determine (using a one-way data table) the value of α yielding the lowest MAD. The results are given in Table 4. It appears that a value of α between 0.20 and 0.30 yields the lowest MAD.

REMARKS **1** Since $\alpha < 1$, exponential smoothing "smooths out" variations in a time series by not giving total weight to the last observation.
2 If $\alpha = \frac{2}{N+1}$, simple exponential smoothing (with smoothing parameter α) and an N-period moving-average forecast will both yield similar forecasts. For example, $\alpha = 0.33$ is roughly equivalent to a five-period moving average.
3 To see why we call the method *exponential* smoothing, consider (2) for $t - 1$:

$$A_{t-1} = \alpha x_{t-1} + (1 - \alpha)A_{t-2} \tag{5}$$

Substituting (5) into (2) yields

$$A_t = \alpha x_t + (1 - \alpha)[\alpha x_{t-1} + (1 - \alpha)A_{t-2}]$$
$$= \alpha x_t + \alpha(1 - \alpha)x_{t-1} + (1 - \alpha)^2 A_{t-2} \tag{6}$$

Note that

$$A_{t-2} = \alpha x_{t-2} + (1 - \alpha)A_{t-3} \tag{7}$$

TABLE 3
Simple Exponential Smoothing for TV Sales ($\alpha = .1$)

Month	Actual Sales	Forecast	A_t	e_t
1	30	32	31.8	−2.00
2	32	31.8	31.82	0.20
3	30	31.82	31.64	−1.82
4	39	31.64	32.37	7.36
5	33	32.37	32.44	0.63
6	34	32.44	32.60	1.56

TABLE 4
MAD for TV Sales

α	MAD
0.05	3.20
0.10	3.04
0.15	2.94
0.20	2.89
0.25	2.88
0.30	2.90
0.35	2.94
0.40	2.98
0.45	3.05
0.50	3.13

Substituting (7) into (6) yields

$$A_t = \alpha x_t + \alpha(1 - \alpha)x_{t-1} + \alpha(1 - \alpha)^2 x_{t-2} + (1 - \alpha)^3 A_{t-3}$$

Repeating this process yields

$$A_t = \alpha x_t + \alpha(1 - \alpha)x_{t-1} + \alpha(1 - \alpha)^2 x_{t-2} + \cdots + \alpha(1 - \alpha)^k x_{t-k} + \cdots \tag{8}$$

Since $\alpha + \alpha(1 - \alpha) + \alpha(1 - \alpha)^2 + \cdots = 1$, (8) shows that if we go back an "infinite" number of periods, our current smoothed average is a weighted average of all past observations. The weight given to the observation from k periods in the past declines exponentially (by a factor of $1 - \alpha$). The larger the value of α, the more weight is given to the most recent observations. For example, for $\alpha = 0.2$, the three most recent observations have 49% of the weight (20%, 16%, and 13%), whereas for $\alpha = 0.5$, the three most recent observations have 88% of the weight (50%, 25%, and 13%).

4 In practice, α is usually chosen to equal 0.10, 0.30, or 0.50. If the value of α that minimizes the MAD exceeds 0.5, then trend, seasonality, or cyclical variation is probably present, and simple exponential smoothing is not a recommended forecast technique. In such cases, better forecasts will probably be provided by either Holt's method (exponential smoothing with trend, discussed in Section 24.3) or Winter's method (exponential smoothing with trend and seasonality, discussed in Section 24.4).

5 Even if a time series is not fluctuating about a constant base level, simple exponential smoothing may still provide good forecasts. If $x_t = m_t + \varepsilon_t$ and $m_t = m_{t-1} + \delta_t$, where ε_t and δ_t are independent error terms each having mean 0, then simple exponential smoothing will provide good forecasts. This implies that if the mean demand (m_t) for a product is randomly shifting over time, simple exponential smoothing will still provide good forecasts of product demand.

24.3 Holt's Method: Exponential Smoothing with Trend

If we believe that a time series exhibits a linear trend (and no seasonality), **Holt's method** often yields good forecasts. At the end of the tth period, Holt's method yields an estimate of the base level (L_t) and the per-period trend (T_t) of the series. For example, suppose that $L_{20} = 20$ and $T_{20} = 2$. This means that after observing x_{20}, we believe that the base level of the series is 20 and that the base level is increasing by two units per period. Thus, five periods from now, we estimate that the base level of the series will equal 30.

After observing x_t, equations (9) and (10) are used to update the base and trend estimates. α and β are smoothing constants, each between 0 and 1.

$$L_t = \alpha x_t + (1 - \alpha)(L_{t-1} + T_{t-1}) \tag{9}$$

$$T_t = \beta(L_t - L_{t-1}) + (1 - \beta)T_{t-1} \tag{10}$$

To compute L_t, we take a weighted average of the following two quantities:

1 x_t, which is an estimate of the period t base level from the current period

2 $L_{t-1} + T_{t-1}$, which is an estimate of the period t base level based on previous data

To compute T_t, we take a weighted average of the following two quantities:

1 An estimate of trend from the current period given by the increase in the smoothed base from period $t - 1$ to period t

2 T_{t-1}, which is our previous estimate of the trend

As before, we define $f_{t,k}$ to be the forecast for x_{t+k} made at the end of period t. Then

$$f_{t,k} = L_t + kT_t \tag{11}$$

To initialize Holt's method, we need an initial estimate (call it L_0) of the base and an initial estimate (call it T_0) of the trend. We might set T_0 equal to the average monthly increase in the time series during the previous year, and we might set L_0 equal to last month's observation.

From Figure 5, it is clear that CD player sales exhibit an upward trend, but no obvious seasonal pattern is present. Therefore, Holt's method should yield good forecasts. Let's assume that CD player sales during each of the last 12 months are given by 4, 6, 8, 10, 14, 18, 20, 22, 24, 28, 31, and 34. Then

$$T_0 = \frac{(6 - 4) + (8 - 6) + (10 - 8) + \cdots + (34 - 31)}{11}$$

$$= \frac{34 - 4}{11} = 2.73$$

We then estimate $L_0 = 34$.

Applying the Holt method to the first six months of sales (using $\alpha = 0.30$ and $\beta = 0.10$) we obtain the results shown in Table 5. Here are some illustrations of the calculations:

$$L_1 = 0.30x_1 + 0.70(L_0 + T_0) = 0.3(40) + 0.7(34 + 2.73) = 37.71$$
$$T_1 = 0.1(L_1 - L_0) + 0.9T_0 = 0.1(37.71 - 34) + 0.9(2.73) = 2.83$$
$$f_{1,1} = L_1 + T_1 = 37.71 + 2.83 = 40.54$$
$$e_2 = x_2 - f_{1,1} = 47 - 40.54 = 6.46$$

TABLE 5
Holt's Method for CD Player Sales ($\alpha = 0.30$, $\beta = 0.10$)

Month	Sales	L_t	T_t	$f_{t-1,1}$ ($L_{t-1} + T_{t-1}$)	e_t ($x_t - f_{t-1,1}$)
1	40	37.71	2.83	36.73	3.27
2	47	42.48	3.02	40.54	6.46
3	50	46.85	3.16	45.50	4.50
4	49	49.70	3.13	50.01	−1.01
5	56	53.78	3.22	52.83	3.17
6	53	55.80	3.10	57.00	−4.00

For the first six months of CD player sales, we find

$$\text{MAD} = \frac{3.27 + 6.46 + 4.5 + 1.01 + 3.17 + 4.00}{6} = 3.74$$

For the entire 24-month period, we find $\text{MAD} = 2.85$.

As one more illustration of Equation (11), suppose that we want to make a forecast at the end of month 6 for month 10 CD player sales. From (11), we find that $f_{6,4} = L_6 + 4T_6 = 55.80 + 4(3.10) = 68.2$. By trying various combinations of α and β, we could find the values of α and β that minimize the MAD. If these values are not both less than 0.5, then seasonality or cyclical behavior is probably present, and another forecasting method should be used.

In summary, Holt's method will provide good forecasts for a series with a linear trend. Such a series may be modeled as $x_t = a + bt + \varepsilon_t$, where

$$a = \text{base level at beginning of period 1}$$

$$b = \text{per-period trend}$$

$$\varepsilon_t = \text{error term for period } t$$

A multiplicative version of Holt's method (see Problem 15) can be used to generate good forecasts for a series of the form $x_t = ab^t\varepsilon_t$. Here, the value of b represents the percentage growth in the base level of the series during each period. Thus, $b = 1.1$ implies that the base level of the series is increasing by 10% per period. In this model, ε_t is a random error factor with a mean of 1.

A Spreadsheet Implementation of the Holt Method

Holt.xls

Figure 7 (obtained from the file Holt.xls) contains an implementation of the Holt method. In columns B and C, we have typed in the 24 months of CD player sales obtained from Table 1. In cells D4 and E4, we have input L_0 and T_0. Trial values of alpha and beta appear in cells E2 and F2. In cell D5, we compute L_1 by inputting the formula =E\$2*C5+(1−E\$2)*(D4+E4). In cell E5, we compute T_1 by inputting the formula =F\$2*(D5−D4)+(1−F\$2)*E4. In cell F5, we compute $f_{0,1}$ from the formula =D4+E4. In cell G5, we compute e_1 from the formula =C5−F5. In cell H5, we compute $|e_1|$ from the formula =ABS(G5). Copying the formulas from the range C5:H5 to the range C5:H28 completes the implementation of the Holt method. The formula =AVERAGE(H5:H28) in cell G2 computes the MAD (2.85) for the 24 months.

We can use an Excel two-way data table to determine values of α and β that yield a small MAD. We input possible values for α in a cell range B31:B39 and values for β in the cell range C30:K30. We input a formula to compute the MAD (=G2) into cell B30. Invoking the DATATABLE command, we choose the table range B30:K39. Then we select cell E2 as the column input cell and cell F2 as the row input cell. This causes the values in B1:B39 to be input into E2 and the values in C30:K30 to be input into cell F2. After selecting OK, for each combination of α and β in the table Excel computes the MAD. We see that of the combinations listed, $\alpha = .10$ and $\beta = .40$ yields the lowest MAD (2.70). If we wanted to obtain an even lower MAD, we could explore values of α and β near .10 and .40, respectively, by creating another data table. By the way, F9 will recalculate the last data table you have created in your spreadsheet.

FIGURE 7
Holt's Method

	B	C	D	E	F	G	H	I	J	K
1		HOLT	METHOD	ALPHA	BETA	MAD=				
2				0.3	0.1	2.84686937				
3	MONTH	CD SALES	Lt	Tt	f(t-1,1)	et	letl			
4	0		34	2.73						
5	1	40	37.711	2.8281	36.73	3.27	3.27			
6	2	47	42.47737	3.021927	40.5391	6.4609	6.4609			
7	3	50	46.8495079	3.15694809	45.499297	4.500703	4.500703			
8	4	49	49.7045192	3.12675441	50.006456	-1.006456	1.00645599			
9	5	56	53.7818915	3.2218162	52.8312736	3.1687264	3.1687264			
10	6	53	55.8025954	3.10170497	57.0037077	-4.0037077	4.00370772			
11	7	55	57.7330103	2.98457596	58.9043004	-3.9043004	3.90430038			
12	8	63	61.4023104	3.05304837	60.7175862	2.28241378	2.28241378			
13	9	68	65.5187511	3.15938761	64.4553587	3.54464127	3.54464127			
14	10	65	67.5746971	3.04904345	68.6781387	-3.6781387	3.67813872			
15	11	72	71.0366184	3.09033123	70.6237406	1.37625945	1.37625945			
16	12	69	72.5888647	2.93652274	74.1269496	-5.1269496	5.12694962			
17	13	79	76.5677712	3.04076112	75.5253875	3.47461252	3.47461252			
18	14	82	80.3259726	3.11250515	79.6085324	2.39146765	2.39146765			
19	15	80	82.4069345	3.00935081	83.4384778	-3.4384778	3.4384778			
20	16	85	85.2913997	2.99686226	85.4162853	-0.4162853	0.41628527			
21	17	94	90.0017834	3.1682144	88.2882619	5.71173805	5.71173805			
22	18	89	91.9189984	3.04311447	93.1699978	-4.1699978	4.16999776			
23	19	96	95.273479	3.07425108	94.9621129	1.0378871	1.0378871			
24	20	100	98.8434111	3.12381918	98.3477301	1.65226989	1.65226989			
25	21	100	101.377061	3.06480227	101.96723	-1.9672303	1.96723025			
26	22	105	104.609304	3.08154636	104.441863	0.55813656	0.55813656			
27	23	108	107.783596	3.09082084	107.690851	0.30914923	0.30914923			
28	24	110	110.612091	3.06458835	110.874416	-0.8744164	0.87441638			
29					BETA					
30	2.84686937	0.1	0.2	0.3	0.4	0.5	0.6	0.7	0.8	0.9
31	0.1	2.85549229	2.79917857	2.73752333	2.70190155	2.74959079	2.79516165	2.83287197	2.86298631	2.92313769
32	0.2	2.76620734	2.73280037	2.76089108	2.78733151	2.84068253	2.92107922	2.97420106	2.98283649	2.99339222
33	0.3	2.84686937	2.87216928	2.90902939	2.95265285	2.98980036	3.04270174	3.13130108	3.21889032	3.2935572
34	0.4	2.96069374	3.0030147	3.05465632	3.110774	3.17038284	3.23467306	3.30196817	3.35141111	3.38544721
35	0.5	3.0707409	3.1289783	3.19166282	3.25002095	3.31027756	3.36404349	3.42489473	3.50405565	3.58298443
36	0.6	3.19398831	3.26198014	3.33253292	3.39877183	3.48348934	3.58574625	3.68496609	3.78162671	3.87646977
37	0.7	3.31493308	3.39345284	3.47449093	3.59818192	3.71792621	3.83504256	3.95075646	4.06889148	4.21031682
38	0.8	3.43163015	3.52755966	3.66950469	3.80654215	3.940503	4.07309251	4.26812614	4.4709749	4.67844324
39	0.9	3.53843635	3.69071551	3.844996	4.02099412	4.2282208	4.45607616	4.68692747	4.9520137	5.24579003

24.4 Winter's Method: Exponential Smoothing with Seasonality

The appropriately named **Winter's method** is used to forecast time series for which trend and seasonality are present. As previously mentioned, Figure 6 shows that air conditioner sales exhibit an upward trend and seasonality, so Winter's method is a logical candidate for forecasting these sales.

To describe Winter's method, we require two definitions. Let c = the number of periods in the length of the seasonal pattern ($c = 4$ for quarterly data, and $c = 12$ for monthly data). Let s_t be an estimate of a seasonal multiplicative factor for month t, obtained after observing x_t. For instance, suppose month 7 is July and $s_7 = 2$. Then after observing month 7's air conditioner sales, we believe that July air conditioner sales will (all other things being equal) be twice the sales expected during an average month. If month 24 is December, and $s_{24} = 0.4$, then after observing month 24 sales, we predict that December air conditioner sales will be 40% of the expected sales during an average month. In what follows, L_t and T_t have the same meaning as they did in Holt's method. Each period, L_t, T_t, and s_t are updated (in that order) by using Equations (12)–(14). Again, α, β, and γ are smoothing constants, each of which is between 0 and 1.

$$L_t = \alpha \frac{x_t}{s_{t-c}} + (1 - \alpha)(L_{t-1} + T_{t-1}) \tag{12}$$

$$T_t = \beta(L_t - L_{t-1}) + (1 - \beta)T_{t-1} \tag{13}$$

$$s_t = \gamma \frac{x_t}{L_t} + (1 - \gamma)s_{t-c} \tag{14}$$

Equation (12) updates the estimate of the series base by taking a weighted average of the following two quantities:

1   $L_{t-1} + T_{t-1}$, which is our base level estimate before observing x_t

2   The deseasonalized observation $\dfrac{x_t}{s_{t-c}}$, which is an estimate of the base obtained from the current period

Equation (13) is identical to the T_t equation (10) used to update trend in the Holt method.

Equation (14) updates the estimate of month t's seasonality by taking a weighted average of the following two quantities:

1   Our most recent estimate of month t's seasonality (s_{t-c})

2   $\dfrac{x_t}{L_t}$, which is an estimate of month t's seasonality, obtained from the current month

At the end of period t, the forecast ($f_{t,k}$) for month $t + k$ is given by

$$f_{t,k} = (L_t + kT_t)s_{t+k-c} \tag{15}$$

Thus, to forecast the value of the series during period $t + k$, we multiply our estimate of the period $t + k$ base ($L_t + kT_t$) by our most recent estimate of month ($t + k$)'s seasonality factor (s_{t+k-c}).

Initialization of Winter's Method

To obtain good forecasts with Winter's method, we must obtain good initial estimates of base, trend, and all seasonal factors. Let

$$L_0 = \text{estimate of base at beginning of month 1}$$
$$T_0 = \text{estimate of trend at beginning of month 1}$$
$$s_{-11} = \text{estimate of January seasonal factor at beginning of month 1} \tag{16}$$
$$s_{-10} = \text{estimate of February seasonal factor at beginning of month 1}$$
$$\vdots$$
$$s_0 = \text{estimate of December seasonal factor at beginning of month 1}$$

A variety of methods are available to estimate the parameters in (16). We choose a simple method that requires two years of data. Suppose that the last two years of sales (by month) were as follows:

$$\text{Year } -2: \quad 4, 3, 10, 14, 25, 26, 38, 40, 28, 17, 16, 13$$
$$\text{Year } -1: \quad 9, 6, 18, 27, 48, 50, 75, 77, 52, 33, 31, 24$$
$$\text{Total sales during year } - 2 = 234$$
$$\text{Total sales during year } - 1 = 450$$

We estimate T_0 by

$$T_0 = \frac{(\text{Avg. monthly sales during year } -1) - (\text{Avg. monthly sales during year } -2)}{12}$$

or

$$T_0 = \frac{\frac{450}{12} - \frac{234}{12}}{12} = 1.5$$

To estimate L_0, we first determine the average monthly demand during year $-1(\frac{450}{12})$. This estimates the base at the middle of year -1 (month 6.5 of year -1). To bring this estimate to the end of month 12 of year -1, we add $(12 - 6.5)T_0 = 5.5T_0$. Thus, our estimate of $L_0 = 37.5 + 5.5(1.5) = 45.75$.

To estimate the seasonality factor for a given month (say, January $= s_{-11}$), we take an estimate of January seasonality for year -2 and year -1 and average them. In year -2, average monthly demand was $\frac{234}{12} = 19.5$; in January of year -2, 4 air conditioners were sold. Therefore,

$$\text{Year } -2 \text{ estimate of January seasonality} = \frac{4}{19.5} = 0.205$$

Similarly,

$$\text{Year } -1 \text{ estimate of January seasonality} = \frac{9}{37.5} = 0.240$$

Finally, we obtain $s_{-11} = \frac{0.205 + 0.24}{2} = 0.22$. In similar fashion, we obtain

$$s_{-10} = 0.16, \quad s_{-9} = 0.50, \quad s_{-8} = 0.72, \quad s_{-7} = 1.28, \quad s_{-6} = 1.33,$$
$$s_{-5} = 1.97, \quad s_{-4} = 2.05, \quad s_{-3} = 1.41, \quad s_{-2} = 0.88, \quad s_{-1} = 0.82, \quad s_0 = 0.65$$

As a check, initial seasonal factor estimates should average to 1.

Before showing how (12)–(14) are used, we demonstrate how to use (15) for forecasting. At the beginning of month 1, our forecast for month 1 air conditioner sales is

$$f_{0,1} = (L_0 + T_0)s_{0+1-12} = (45.75 + 1.5)0.22 = 10.40$$

At the beginning of month 1, our forecast for month 7 air conditioner sales is

$$f_{0,7} = (L_0 + 7T_0)s_{0+7-12} = (45.75 + 7(1.5))1.97 = 110.81$$

For $\alpha = 0.5$, $\beta = 0.4$, $\gamma = 0.6$, applying Winter's method to the first 12 months of air conditioner sales data yields the results in Table 6.

We illustrate the computations by computing L_1, T_1, and s_1.

$$L_1 = 0.5\left(\frac{x_1}{s_{-11}}\right) + 0.5(L_0 + T_0) = 0.5\left(\frac{13}{0.22}\right) + 0.5(45.75 + 1.5) = 53.17$$

$$T_1 = 0.4(L_1 - L_0) + 0.6T_0 = 0.4(53.17 - 45.75) + 0.6(1.5) = 3.87$$

$$s_1 = 0.6\left(\frac{x_1}{L_1}\right) + 0.4s_{-11} = 0.6\left(\frac{13}{53.17}\right) + 0.4(0.22) = 0.23$$

TABLE 6
Winter's Method for Air Conditioners ($\alpha = 0.5$, $\beta = 0.4$, $\gamma = 0.6$)

Month	Sales	L_t	T_t	s_t	$f_{t-1,1}$	Error
1	13	53.17	3.87	0.23	10.40	2.60
2	7	50.39	1.21	0.15	9.13	−2.13
3	23	48.80	0.09	0.48	25.80	−2.80
4	32	46.67	−0.80	0.70	35.20	−3.20
5	58	45.59	−0.91	1.28	58.71	−0.71
6	60	44.90	−0.82	1.33	59.42	0.58
7	90	44.88	−0.50	1.99	86.82	3.18
8	93	44.87	−0.30	2.06	90.97	2.03
9	63	44.62	−0.28	1.41	62.84	0.16
10	39	44.33	−0.29	0.88	39.02	−0.02
11	37	44.58	−0.07	0.83	36.12	0.88
12	29	44.56	−0.05	0.65	28.93	0.07

Thus, at the end of month 1, our forecast for (say) month 7 air conditioner sales is $f_{1,6} = (L_1 + 6T_1)s_{1+6-12} = (53.17 + 6(3.87))1.97 = 150.49$. Our forecast for month 7 at the end of month 1 exceeds the forecast for month 7 made at the beginning of month 1, because month 1 sales were higher than predicted.

For all 24 months of data, spreadsheet calculations show that MAD = 10.48.

REMARKS **1** Since Winter's method uses three smoothing constants, it is quite a chore to find the combination of α, β, and γ values that yields the smallest MAD. The use of a spreadsheet to do Winter's method is discussed in Review Problem 3. The Excel Solver can aid in finding good values of α, β, and γ. Just use Solver to find parameter values that minimize MAD.
2 Although the values of α and β that minimize MAD should not exceed 0.5 (as in the Holt method), it is not uncommon for the best value of γ to exceed 0.5. This is because for monthly data, each monthly seasonal factor is updated during only $\frac{1}{12}$ of all periods. Since the seasonality factors are updated so infrequently, we may need to give more weight to each observation, so $\gamma > 0.5$ is not out of the question.
3 Figure 8 shows how well forecasts of air conditioner sales (for $\alpha = 0.5$, $\beta = 0.4$, and $\gamma = 0.6$) compare to actual air conditioner sales. The agreement between predicted and actual sales is quite good except during months 15 and 17. During these months, our forecasts are much too high. Perhaps new salespeople were hired during these two months, causing sales to be less than anticipated.

Forecasting Accuracy

For any forecasting model in which forecast errors are normally distributed, we may use MAD to estimate s_e = standard deviation of our forecast errors. The relationship between MAD and s_e is given in Formula (17).

$$s_e = 1.25 \text{ MAD} \tag{17}$$

Assuming that errors are normally distributed, we know that approximately 68% of our predictions should be within s_e of the actual value, and approximately 95% of our predictions should be within $2s_e$ of the actual value. Thus, for our air conditioner sales predictions, we find that $s_e = 1.25(10.48) = 13.10$. So we would expect that for about $0.68(24) = 16$ of 24 months, our predictions for sales would be off by at most 13.10 air conditioners, and for $0.95(24) = 23$ of 24 months, our predictions would be off by at most $2(13.10) = 26.2$ air conditioners. Actually, our predictions for air conditioner sales are accurate within 13.10 during 17 months and accurate within 26.2 during 22 months.

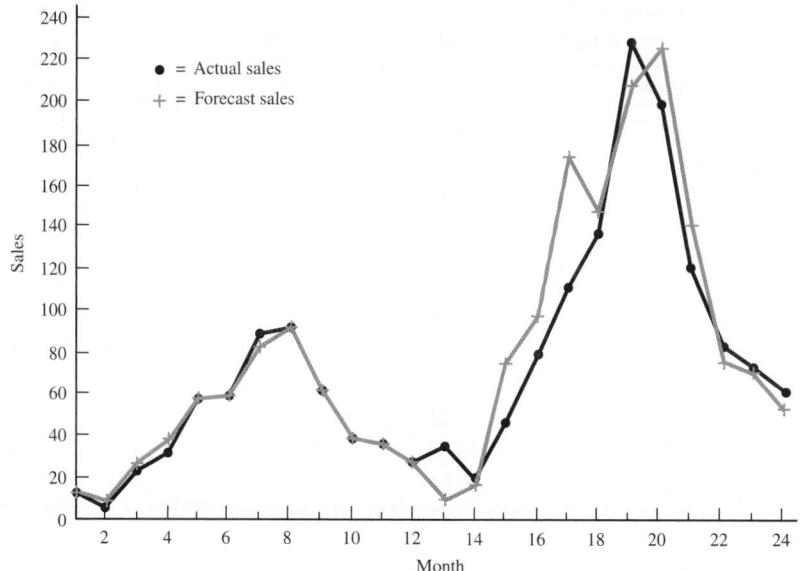

FIGURE 8
Air Conditioner Sales Predictions

We note that in most situations where a forecast is required, knowing something about the probable accuracy of the forecast is almost as important as the actual forecast. Thus, this short subsection is very important!

PROBLEMS

Group A

1 Simple exponential smoothing (with $\alpha = 0.2$) is being used to forecast monthly beer sales at Gordon's Liquor Store. After observing April's demand, the predicted demand for May is 4,000 cans of beer.

a At the beginning of May, what is the prediction for July's beer sales?

b Actual demand during May and June is as follows: May, 4,500 cans of beer; June, 3,500 cans of beer. After observing June's demand, what is the forecast for July's demand?

c The demand during May and June averages out to $\frac{4,500+3,500}{2} = 4,000$ cans per month. This is the same as the forecast for monthly sales before we observed the May and June data. Yet after observing the May and June demands for beer, our forecast for July demand has decreased from what it was at the end of April. Why?

2 We are predicting quarterly sales for soda at Gordon's Liquor Store using Winter's method. We are given the following information:

  Seasonality factors: fall = 0.8   spring = 1.2
  winter = 0.7   summer = 1.3
  Current base estimate = 400 cases per quarter
  Current trend estimate = 40 cases per quarter
  $\alpha = 0.2$ $\beta = 0.3$ $\gamma = 0.5$

Now sales of 650 cases during the summer quarter are observed.

a Use this information to update the estimates of base, trend, and seasonality.

b After observing the summer demand, forecast demand for the fall quarter and the winter quarter.

3 We are using Winter's method and monthly data to forecast the GDP. (All numbers are in billions of dollars.) At the end of January 2005, $L_t = 600$ and $T_t = 5$. We are given the following seasonalities: January, 0.80; February, 0.85; December, 1.2. During February 2005, the GDP is at a level of 630. At the end of February what is the forecast for the December 2005 level of the GDP? Use $\alpha = \beta = \gamma = 0.5$.

4 We are using the Holt method to predict monthly VCR sales at Highland Appliance. At the end of October, 2005, $L_t = 200$ and $T_t = 10$. During November, 2005, 230 VCRs are sold. At the end of November, MAD = 25, and we are 95% sure that VCR sales for December, 2005 will be between _____ and _____. Use $\alpha = \beta = 0.5$.

5 We are using simple exponential smoothing to predict monthly electric shaver sales at Hook's Drug Store. At the end of October 2006, our forecast for December 2006 sales was 40. In November 50 shavers were sold, and during December 45 shavers were sold. Suppose $\alpha = 0.50$. At the end of December, 2006, what is our prediction for the total number of shavers that will be sold during March and April of 2007?

6 We are using simple exponential smoothing to predict monthly auto sales at Bloomington Ford. The company believes that sales do not exhibit trend or seasonality, so simple exponential smoothing has yielded satisfactory forecasts for the most part. Each March, however, Bloomington Ford has observed that sales tend to exceed the simple exponential smoothing forecast (A_{Feb}) by 200. Suppose that at the end of February 2004, $A_t = 600$. During March 2004, 900 cars are sold.

a Using $\alpha = 0.3$, determine (at the end of March 2004) a forecast for April 2004 car sales.

b Assume that at the end of March, MAD = 60. We are 95% sure that April sales will be between _____ and _____.

7 The University Credit Union is open Monday through Saturday. Winter's method is being used (with $\alpha = \beta = \gamma = 0.5$) to predict the number of customers entering the bank each day. After incorporating the arrivals of October 16, $L_t = 200$ customers, $T_t = 1$ customer, and the seasonalities are as follows: Monday, 0.90; Tuesday, 0.70; Wednesday, 0.80; Thursday, 1.1; Friday, 1.2; Saturday, 1.3. For example, this means that on a typical Monday, the number of customers is 90% of the number of customers entering the bank on an average day. On Tuesday, October 17, 182 customers enter the bank. At the close of business on October 17, make a prediction for the number of customers to enter the bank on October 25.

8 The Holt method (exponential smoothing with trend and without seasonality) is being used to forecast weekly car sales at TOD Ford. Currently, the base is estimated to be 50 cars per week, and the trend is estimated to be 6 cars per week. During the current week, 30 cars are sold. After observing the current week's sales, forecast the number of cars to be sold during the week that begins three weeks after the conclusion of the current week. Use $\alpha = \beta = 0.3$.

9 Winter's method (with $\alpha = 0.2$, $\beta = 0.1$, and $\gamma = 0.5$) is being used to forecast the number of customers served each day by Last National Bank. The bank is open Monday through Friday. At present, the following seasonalities have been estimated: Monday, 0.80; Tuesday, 0.90; Wednesday, 0.95; Thursday, 1.10; Friday, 1.25. A seasonality of 0.80 for Monday means that on a Monday, the number of customers served by the bank tends to be 80% of average. Currently, the base is estimated to be 20 customers, and the trend is estimated to equal 1 customer. After observing that on Monday 30 customers are served by the bank, predict the number of customers to be served by the bank on Wednesday.

10 We have been assigned to forecast the number of aircraft engines ordered each month by Engine Company. At the end of February, the forecast is that 100 engines will be ordered during April. During March, 120 engines are ordered.

a Using $\alpha = 0.3$, determine (at the end of March) a forecast for the number of orders placed during April. Answer the same question for May.

b Suppose at the end of March, MAD = 16. At the end of March, we are 68% sure that April orders will be between _____ and _____.

11 Winter's method is being used to forecast quarterly U.S. retail sales (in billions of dollars). At the end of the first

quarter, $L_t = 300$, $T_t = 30$, and the seasonal indexes are as follows: quarter 1, 0.90; quarter 2, 0.95; quarter 3, 0.95; quarter 4, 1.20. During the second quarter, retail sales are $360 billion. Assume $\alpha = 0.2$, $\beta = 0.4$, and $\gamma = 0.5$.

a At the end of the second quarter, develop a forecast for retail sales during the fourth quarter of the year.

b At the end of the second quarter, develop a forecast for retail sales during the second quarter of the following year.

Group B

12 Simple exponential smoothing with $\alpha = 0.3$ is being used to predict sales of radios at Lowland Appliance. Predictions are made on a monthly basis. After observing August radio sales, the forecast for September is 100 radios.

a During September, 120 radios are sold. After observing September sales, what is the prediction for October radio sales? For November radio sales?

b It turns out that June sales were recorded as 10 radios. Actually, however, 100 radios were sold in June. After correcting for this error, what would be the prediction for October radio sales?

13 In our discussion of Winter's method, a monthly seasonality of (say) 0.80 for January means that during January, air conditioner sales are expected to be 80% of the sales during an average month. An alternative approach to modeling seasonality is to let the seasonality factor for each month represent how far above average air conditioner sales will be during the current month. For instance, if $s_{Jan} = -50$, then air conditioner sales during January are expected to be 50 less than air conditioner sales during an average month. If $s_{July} = 90$, then air conditioner sales during July are expected to be 90 more than air conditioner sales during an average month. Let

s_t = the seasonality for month t after month t demand is observed

L_t = the estimate of base after month t demand is observed

T_t = the estimate of trend after month t demand is observed

Then the Winter's method equations given in the text are modified to be as follows (* indicates multiplication):

$$L_t = \alpha * (I) + (1 - \alpha) * (L_{t-1} + T_{t-1})$$
$$T_t = \beta * (L_t - L_{t-1}) + (1 - \beta) * T_{t-1}$$
$$s_t = \gamma * (II) + (1 - \gamma) * s_{t-12}$$

a What should I and II be?

b Suppose that month 13 is a January, $L_{12} = 30$, $T_{12} = -3$, $s_1 = -50$, and $s_2 = -20$. Let $\alpha = \gamma = \beta = 0.5$. Suppose 12 air conditioners are sold during month 13. At the end of month 13, what is the prediction for air conditioner sales during month 14?

14 Winter's method assumes a multiplicative seasonality but an additive trend. For example, a trend of 5 means that the base will increase by 5 units per period. Suppose there is actually a multiplicative trend. Then (ignoring seasonality) if the current estimate of the base is 50 and the current estimate of the trend is 1.2, we would predict demand to increase by 20% per period. Ignoring seasonality, we would thus forecast the next period's demand to be 50(1.2) and forecast the demand two periods in the future to be $50(1.2)^2$.

If we want to use a multiplicative trend in Winter's method, we should use the following equations:

$$L_t = \alpha * \left(\frac{x_t}{s_{t-c}}\right) + (1 - \alpha) * (I)$$

$$T_t = \beta * (II) + (1 - \beta) * T_{t-1}$$

$$s_t = \gamma * \left(\frac{x_t}{L_t}\right) + (1 - \gamma)* s_{t-12}$$

a Determine what I and II should be.

b Suppose we are working with monthly data and month 12 is a December, month 13 a January, and so on. Also suppose that $L_{12} = 100$, $T_{12} = 1.2$, $s_1 = 0.90$, $s_2 = 0.70$, and $s_3 = 0.95$. Suppose $x_{13} = 200$. At the end of month 13, what is the prediction for x_{15}? Assume $\alpha = \beta = \gamma = 0.5$.

15 Holt's method assumes an additive trend. For example, a trend of 5 means that the base will increase by 5 units per period. Suppose there is actually a multiplicative trend. Thus, if the current estimate of the base is 50 and the current estimate of the trend is 1.2, we would predict demand to increase by 20% per period. So we would forecast the next period's demand to be 50(1.2) and forecast the demand two periods in the future to be $50(1.2)^2$. If we want to use a multiplicative trend in Holt's method, we should use the following equations:

$$L_t = \alpha * (x_t) + (1 - \alpha) * (I)$$

$$T_t = \beta * (II) + (1 - \beta) * T_{t-1}$$

a Determine what I and II should be.

b Suppose we are working with monthly data and month 12 is a December, month 13 a January, and so on. Also suppose that $L_{12} = 100$ and $T_{12} = 1.2$. Suppose $x_{13} = 200$. At the end of month 13, what is the prediction for x_{15}? Assume $\alpha = \beta = 0.5$.

16 A version of simple exponential smoothing can be used to predict the outcome of sporting events. To illustrate, consider pro football. We first assume that all games are played on a neutral field. Before each day of play, we assume that each team has a rating. For example, if the Bears' rating is +10 and the Bengals' rating is +6, we would predict the Bears to beat the Bengals by $10 - 6 = 4$ points. Suppose the Bears play the Bengals and win by 20 points. For this observation, we "underpredicted" the Bears' performance by $20 - 4 = 16$ points. The best α for pro football is 0.10. After the game, we therefore increase the Bears' rating by $16(0.1) = 1.6$ and decrease the Bengals' rating by 1.6 points. In a rematch, the Bears would be favored by $(10 + 1.6) - (6 - 1.6) = 7.2$ points.

a How does this approach relate to the equation $A_t = A_{t-1} + \alpha(e_t)$?

b Suppose the home-field advantage in pro football is 3 points; that is, home teams tend to outscore visiting teams by an average of 3 points a game. How could the home-field advantage be incorporated into this system?

c How could we determine the best α for pro football?

d How might we determine ratings for each team at the beginning of the season?

e Suppose we tried to apply the above method to predict pro football (16-game schedule), college football (11-game schedule), college basketball (30-game schedule), and pro basketball (82-game schedule). Which sport would have the smallest optimal α? Which sport would have the largest optimal α?

f Why would this approach probably yield poor forecasts for major league baseball?

24.5 Ad Hoc Forecasting

Suppose we want to determine how many tellers a bank must have working each day to provide adequate service. In order to use the queuing models of Chapter 20 to answer this question, we need to be able to predict the number of customers who will enter the bank each day. The bank manager believes that the month of the year and the day of the week influence the number of customers entering the bank. (The bank is open Monday through Saturday, except for holidays.) Can we develop a simple forecasting model to help the bank predict the number of customers who will enter each day?

The number of customers entering the bank each day during the last year is given in Table 7. We have used 1 = Monday, 2 = Tuesday, . . . , 6 = Saturday, and 7 = Sunday to denote the days of the week. A "Y" in the AH column means that the day is the day after the bank was closed for a holiday.

Let x_t = number of customers entering the bank on day t. We postulate that $x_t = B \times DW_t \times M_t \times \varepsilon_t$, where

B = base level of customer traffic corresponding to an average day

DW_t = day of the week factor corresponding to the day of the week on which day t falls

M_t = month factor corresponding to the month during which day t occurs

ε_t = random error term whose average value equals 1

TABLE 7
Arrivals to Bank

Month	Day M	Day W	Customer	AH	Forecast
1	1	1			
1	2	2	431	Y	399.13
1	3	3	271		415.88
1	4	4	362		416.51
1	5	5	696		560.10
1	6	6	315		356.32
1	7	7			
1	8	1	330		493.98
1	9	2	352		399.13
1	10	3	606		415.88
1	11	4	550		416.51
1	12	5	626		560.10
1	13	6	392		356.32
1	14	7			
1	15	1	540		493.98
1	16	2	474		399.13
1	17	3	457		415.88
1	18	4	401		416.51
1	19	5	691		560.10
1	20	6	388		356.32
1	21	7			
1	22	1	533		493.98
1	23	2	384		399.13
1	24	3	360		415.88
1	25	4	515		416.51
1	26	5	325		560.10
1	27	6	412		356.32
1	28	7			
1	29	1	592		493.98
1	30	2	366		399.13
1	31	3	512		415.88
2	1	4	476		425.33
2	2	5	531		571.97
2	3	6	303		363.87
2	4	7			
2	5	1	474		504.45
2	6	2	255		407.58
2	7	3	282		424.69
2	8	4	321		425.33
2	9	5	416		571.97
2	10	6	257		363.87
2	11	7			
2	12	1	638		504.45
2	13	2	506		407.58
2	14	3	420		424.69
2	15	4	459		425.33
2	16	5	515		571.97

(Continued)

TABLE 7
(Continued)

Month	Day M	Day W	Customer	AH	Forecast
2	17	6	501		363.87
2	18	7			
2	19	1	556		504.45
2	20	2	510		407.58
2	21	3	436		424.69
2	22	4	512		425.33
2	23	5	547		571.97
2	24	6	319		363.87
2	25	7			
2	26	1	637		504.45
2	27	2	474		407.58
2	28	3	487		424.69
2	29	4	402		425.33
3	1	5	778		574.26
3	2	6	374		365.32
3	3	7			
3	4	1	544		506.46
3	5	2	485		409.21
3	6	3	361		426.39
3	7	4	315		427.03
3	8	5	423		574.26
3	9	6	357		365.32
3	10	7			
3	11	1	649		506.46
3	12	2	351		409.21
3	13	3	405		426.39
3	14	4	404		427.03
3	15	5	483		574.26
3	16	6	411		365.32
3	17	7			
3	18	1	309		506.46
3	19	2	453		409.21
3	20	3	515		426.39
3	21	4	380		427.03
3	22	5	426		574.26
3	23	6	427		365.32
3	24	7			
3	25	1	489		506.46
3	26	2	341		409.21
3	27	3	471		426.39
3	28	4	517		427.03
3	29	5	647		574.26
3	30	6	415		365.32
3	31	7			
4	1	1	363		483.02
4	2	2	337		390.27
4	3	3	314		406.65

(Continued)

TABLE 7
(Continued)

Month	Day M	Day W	Customer	AH	Forecast
4	4	4	465		407.26
4	5	5	584		547.67
4	6	6	313		348.41
4	7	7			
4	8	1	376		483.02
4	9	2	292		390.27
4	10	3	484		406.65
4	11	4	227		407.26
4	12	5	496		547.67
4	13	6	395		348.41
4	14	7			
4	15	1	625		483.02
4	16	2	430		390.27
4	17	3	454		406.65
4	18	4	372		407.26
4	19	5	455		547.67
4	20	6	253		348.41
4	21	7			
4	22	1	432		483.02
4	23	2	469		390.27
4	24	3	392		406.65
4	25	4	467		407.26
4	26	5	684		547.67
4	27	6	349		348.41
4	28	7			
4	29	1	750		483.02
4	30	2	409		390.27
5	1	3	348		373.31
5	2	4	230		373.88
5	3	5	630		502.78
5	4	6	358		319.85
5	5	7			
5	6	1	269		443.43
5	7	2	107		358.27
5	8	3	360		373.31
5	9	4	208		373.88
5	10	5	547		502.78
5	11	6	325		319.85
5	12	7			
5	13	1	473		443.43
5	14	2	337		358.27
5	15	3	317		373.31
5	16	4	341		373.88
5	17	5	338		502.78
5	18	6	369		319.85
5	19	7			
5	20	1	618		443.43

(Continued)

TABLE **7**
(Continued)

Month	Day M	Day W	Customer	AH	Forecast
5	21	2	458		358.27
5	22	3	457		373.31
5	23	4	572		373.88
5	24	5	668		502.78
5	25	6	318		319.85
5	26	7			
5	27	1	300		443.43
5	28	2	469		358.27
5	29	3	434		373.31
5	30	4	419		373.88
5	31	5			
6	1	6	432	Y	354.08
6	2	7			
6	3	1	463		490.89
6	4	2	457		396.62
6	5	3	273		413.27
6	6	4	327		413.90
6	7	5	554		556.60
6	8	6	256		354.08
6	9	7			
6	10	1	465		490.89
6	11	2	479		396.62
6	12	3	437		413.27
6	13	4	585		413.90
6	14	5	616		556.60
6	15	6	318		354.08
6	16	7			
6	17	1	724		490.89
6	18	2	390		396.62
6	19	3	550		413.27
6	20	4	266		413.90
6	21	5	410		556.60
6	22	6	303		354.08
6	23	7			
6	24	1	514		490.89
6	25	2	353		396.62
6	26	3	397		413.27
6	27	4	539		413.90
6	28	5	411		556.60
6	29	6	413		354.08
6	30	7			
7	1	1	583		484.44
7	2	2	477		391.42
7	3	3	410		407.85
7	4	4			
7	5	5	615	Y	549.29
7	6	6	288		349.44

(Continued)

TABLE 7
(Continued)

Month	Day M	Day W	Customer	AH	Forecast
7	7	7			
7	8	1	478		484.44
7	9	2	298		391.42
7	10	3	253		407.85
7	11	4	366		408.46
7	12	5	410		549.29
7	13	6	270		349.44
7	14	7			
7	15	1	541		484.44
7	16	2	331		391.42
7	17	3	318		407.85
7	18	4	441		408.46
7	19	5	651		549.29
7	20	6	300		349.44
7	21	7			
7	22	1	608		484.44
7	23	2	401		391.42
7	24	3	390		407.85
7	25	4	391		408.46
7	26	5	619		549.29
7	27	6	391		349.44
7	28	7			
7	29	1	413		484.44
7	30	2	474		391.42
7	31	3	503		407.85
8	1	4	267		418.33
8	2	5	619		562.56
8	3	6	370		357.88
8	4	7			
8	5	1	406		496.15
8	6	2	432		400.87
8	7	3	333		417.70
8	8	4	327		418.33
8	9	5	647		562.56
8	10	6	407		357.88
8	11	7			
8	12	1	396		496.15
8	13	2	664		400.87
8	14	3	508		417.70
8	15	4	519		418.33
8	16	5	555		562.56
8	17	6	365		357.88
8	18	7			
8	19	1	492		496.15
8	20	2	420		400.87
8	21	3	360		417.70
8	22	4	469		418.33

(Continued)

TABLE 7
(Continued)

Month	Day M	Day W	Customer	AH	Forecast
8	23	5	488		562.56
8	24	6	326		357.88
8	25	7			
8	26	1	465		496.15
8	27	2	384		400.87
8	28	3	280		417.70
8	29	4	292		418.33
8	30	5	649		562.56
8	31	6	493		357.88
9	1	7			
9	2	1			
9	3	2	459	Y	391.76
9	4	3	353		408.21
9	5	4	287		408.82
9	6	5	471		549.77
9	7	6	266		349.74
9	8	7			
9	9	1	505		484.87
9	10	2	528		391.76
9	11	3	342		408.21
9	12	4	551		408.82
9	13	5	525		549.77
9	14	6	304		349.74
9	15	7			
9	16	1	479		484.87
9	17	2	258		391.76
9	18	3	263		408.21
9	19	4	450		408.82
9	20	5	540		549.77
9	21	6	297		349.74
9	22	7			
9	23	1	399		484.87
9	24	2	264		391.76
9	25	3	479		408.21
9	26	4	459		408.82
9	27	5	915		549.77
9	28	6	247		349.74
9	29	7			
9	30	1	725		484.87
10	1	2	197		390.39
10	2	3	326		406.78
10	3	4	374		407.39
10	4	5	477		547.85
10	5	6	367		348.52
10	6	7			
10	7	1	317		483.17
10	8	2	205		390.39

(Continued)

TABLE 7
(Continued)

Month	Day M	Day W	Customer	AH	Forecast
10	9	3	519		406.78
10	10	4	483		407.39
10	11	5	489		547.85
10	12	6	345		348.52
10	13	7			
10	14	1	660		483.17
10	15	2	262		390.39
10	16	3	395		406.78
10	17	4	522		407.39
10	18	5	582		547.85
10	19	6	335		348.52
10	20	7			
10	21	1	503		483.17
10	22	2	396		390.39
10	23	3	548		406.78
10	24	4	471		407.39
10	25	5	528		547.85
10	26	6	344		348.52
10	27	7			
10	28	1	419		483.17
10	29	2	429		390.39
10	30	3	609		406.78
10	31	4	519		407.39
11	1	5	674		596.31
11	2	6	352		379.35
11	3	7			
11	4	1	360		525.91
11	5	2	500		424.92
11	6	3	339		442.76
11	7	4	326		443.43
11	8	5	459		596.31
11	9	6	255		379.35
11	10	7			
11	11	1	432		525.91
11	12	2	527		424.92
11	13	3	394		442.76
11	14	4	424		443.43
11	15	5	388		596.31
11	16	6	356		379.35
11	17	7			
11	18	1	635		525.91
11	19	2	309		424.92
11	20	3	613		442.76
11	21	4	580		443.43
11	22	5	627		596.31
11	23	6	514		379.35
11	24	7			

(Continued)

TABLE **7**
(Continued)

Month	Day M	Day W	Customer	AH	Forecast
11	25	1	686		525.91
11	26	2	452		424.92
11	27	3	384		442.76
11	28	4			
11	29	5	701	Y	596.31
11	30	6	425		379.35
12	1	7			
12	2	1	291		510.06
12	3	2	407		412.12
12	4	3	458		429.42
12	5	4	243		430.06
12	6	5	449		578.34
12	7	6	315		367.91
12	8	7			
12	9	1	633		510.06
12	10	2	429		412.12
12	11	3	375		429.42
12	12	4	540		430.06
12	13	5	615		578.34
12	14	6	455		367.91
12	15	7			
12	16	1	385		510.06
12	17	2	472		412.12
12	18	3	576		429.42
12	19	4	321		430.06
12	20	5	679		578.34
12	22	7			
12	23	1	407		510.06
12	24	2	328		412.12
12	25	3			
12	26	4	491	Y	430.06
12	27	5	586		578.34
12	28	6	367		367.91
12	29	7			
12	30	1	707		510.06
12	31	2	400		412.12

To begin, we estimate B = average number of arrivals per day the bank is open = 438.33. We illustrate the estimation of the DW_t by

$$DW_t \text{ for Monday} = \frac{\text{average number of arrivals on Mondays bank is open}}{B}$$

$$= \frac{492.07}{438.33} = 1.122$$

Similarly, we find

$$DW_t \text{ for Tuesday } = 0.907$$
$$DW_t \text{ for Wednesday } = 0.945$$
$$DW_t \text{ for Thursday } = 0.947$$
$$DW_t \text{ for Friday } = 1.273$$
$$DW_t \text{ for Saturday } = 0.809$$

To estimate M_t (say, for May), we write

$$M_t \text{ for May } = \frac{\text{average number of arrivals on May day for which bank is open}}{B}$$

$$= \frac{395}{438.33} = 0.901$$

In a similar fashion, we find the M_t for the remaining months:

$$M_t \text{ for January } = 1.004$$
$$M_t \text{ for February } = 1.025$$
$$M_t \text{ for March } = 1.029$$
$$M_t \text{ for April } = 0.982$$
$$M_t \text{ for June } = 0.998$$
$$M_t \text{ for July } = 0.984$$
$$M_t \text{ for August } = 1.008$$
$$M_t \text{ for September } = 0.985$$
$$M_t \text{ for October } = 0.982$$
$$M_t \text{ for November } = 1.069$$
$$M_t \text{ for December } = 1.037$$

To illustrate how the forecasts in Table 7 were generated, consider how we would generate a forecast for the number of customers to enter the bank on Thursday, February 1, of the current year. Assuming ε_t equals its average value of 1, we would forecast $B \times$ (DW_t for Thursday) $\times$ (M_t for February) $= 438.33(0.947)(1.025) = 425.48$ customers would enter. (The difference from the printout value shown in the table is due to rounding of DW_t and M_t values.) To forecast customer arrivals for a future day (say, Saturday, February 8, of next year), we would obtain $B \times$ (DW_t for Saturday) $\times$ (M_t for February) $= 438.33\ (0.809)(1.025) = 363.47$ customers.

For the data given in Table 7, our simple model yielded a MAD of 79.1. If this method were used to generate forecasts for the coming year, however, the MAD would probably exceed 79.1. This is because we have fit our parameters to past data; there is no guarantee that future data will "know" that they should follow the same pattern as past data. We have also neglected to consider whether or not an upward trend in the data is present (see Problem 3).

Suppose the bank manager observes that on the day after a holiday, bank traffic is much higher than the model predicts. The data in Table 8 indicate that this is indeed the case. How can we use this information to obtain more accurate customer forecasts for days after holidays? From Table 8, we find that the average value of Actual/Forecast for days after a holiday is 1.15. Thus, for any day after a holiday, we obtain a new forecast simply by multiplying our previous forecast by 1.15.

TABLE 8
Bank Traffic on Day after Holiday

Day after Holiday	Actual	Forecast (rounded)	Actual/Forecast
January 2	431	399	1.08
June 1	432	354	1.22
July 5	615	549	1.12
September 3	459	392	1.17
November 29	701	596	1.18
December 26	491	430	1.14

PROBLEMS

Group A

1 Suppose the bank is a college credit union and that on days when the college's professors get paid, bank traffic is much higher than usual. Assuming that college professors are paid on the first weekday of each month, how could we incorporate this fact into the forecasting procedure described in this section?

2 Suppose again that the bank is a college credit union, but now the staff gets paid every other Friday. Again, bank traffic is much higher than usual on staff paydays. How could we incorporate this fact into the forecasting procedure described in this section?

3 Suppose that the number of customers entering the bank is growing at around 20% per year. How could we incorporate this fact into the forecasting procedure described in this section?

24.6 Simple Linear Regression

Often, we try to predict the value of one variable (called the **dependent variable**) from the value of another variable (the **independent variable**). Some examples follow:

Dependent Variable	**Independent Variable**
Sales of product	Price of product
Automobile sales	Interest rate
Total production cost	Units produced

If the dependent variable and the independent variable are related in a linear fashion, simple linear regression can be used to estimate this relationship. In Section 24.7, we will discuss how to estimate nonlinear relationships.

To illustrate simple linear regression, let's recall the Giapetto problem (Example 1 in Chapter 3). To set up this problem, we need to determine the cost of producing a soldier and the cost of producing a train. Let's suppose that we want to determine the cost of producing a train. To estimate this cost, we have observed for ten weeks the number of trains produced each week and the total cost of producing those trains. This information is given in Table 9.

The data from Table 9 are plotted in Figure 9. Observe that there appears to be a strong linear relationship between x_i (number of trains produced during week i) and y_i (cost of producing trains made during week i). The line plotted in Figure 9 appears, in a way to be made precise later, to come close to capturing the linear relationship between units produced and production cost. We will soon see how this line was chosen.

TABLE 9
Weekly Cost Data on Trains

Week	Trains Produced	Cost of Producing Trains
1	10	$257.40
2	20	$601.60
3	30	$782.00
4	40	$765.40
5	45	$895.50
6	50	$1,133.00
7	60	$1,152.80
8	55	$1,132.70
9	70	$1,459.20
10	40	$970.10

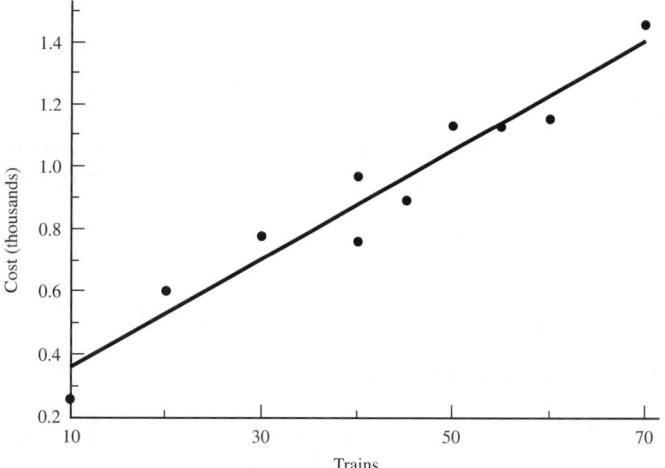

FIGURE 9
Scatterplot of Cost of Producing Trains

To begin, we model the linear relationship between x_i and y_i by the following equation:

$$y_i = \beta_0 + \beta_1 x_i + \varepsilon_i$$

where ε_i is an error term representing the fact that in a week during which x_i trains are produced, the production cost might not always equal $\beta_0 + \beta_1 x_i$. If $\varepsilon_i > 0$, the cost of producing x_i trains during week i will exceed $\beta_0 + \beta_1 x_i$, whereas if $\varepsilon_i < 0$, the cost of producing x_i trains during week i will be less than $\beta_0 + \beta_1 x_i$. However, we expect ε_i to average out to 0, so the expected cost during a week in which x_i trains are produced is $\beta_0 + \beta_1 x_i$.

The true values of β_0 and β_1 are unknown. Suppose we estimate β_0 using $\hat{\beta}_0$ and estimate β_1 using $\hat{\beta}_1$. Then our prediction for y_i (since the average value of $\varepsilon_i = 0$) is given by $\hat{y}_i = \hat{\beta}_0 + \hat{\beta}_1 x_i$.

Suppose we have data points of the form $(x_1, y_1), (x_2, y_2), \ldots, (x_n, y_n)$. How should we choose values of $\hat{\beta}_0$ and $\hat{\beta}_1$ that yield good estimates of β_0 and β_1? We select values of $\hat{\beta}_0$ and $\hat{\beta}_1$ that make our predictions $\hat{y}_i = \hat{\beta}_0 + \hat{\beta}_1 x_i$ close to the actual data points (x_i, y_i). To formalize this idea, define e_i = error or residual for data point i = (actual cost y_i) − (predicted cost $\hat{y}_i$) = $y_i - \hat{\beta}_0 - \hat{\beta}_1 x_i$. We now choose $\hat{\beta}_0$ and $\hat{\beta}_1$ to minimize

$$F(\hat{\beta}_0, \hat{\beta}_1) = \sum e_i^2 = \sum (y_i - \hat{\beta}_0 - \hat{\beta}_1 x_i)^2$$

The values $\hat{\beta}_0$ and $\hat{\beta}_1$ minimizing $F(\hat{\beta}_0, \hat{\beta}_1)$ are called the **least squares estimates** of β_0 and β_1. As described in Example 19 in Chapter 11, we find $\hat{\beta}_0$ and $\hat{\beta}_1$ by setting

$$\frac{\partial F}{\partial \hat{\beta}_0} = \frac{\partial F}{\partial \hat{\beta}_1} = 0$$

The resulting values of $\hat{\beta}_0$ and $\hat{\beta}_1$ are given by

$$\hat{\beta}_1 = \frac{\sum(x_i - \bar{x})(y_i - \bar{y})}{\sum(x_i - \bar{x})^2} \qquad \hat{\beta}_0 = \bar{y} - \hat{\beta}_1\bar{x} \qquad \text{(18)}$$

where $\bar{x}$ = average value of all x_i's and $\bar{y}$ = average value of all y_i's.

We call $\hat{y}_i = \hat{\beta}_0 + \hat{\beta}_1 x_i$ the **least squares regression line.** Essentially, if the least squares line fits the points well (in a sense to be made more precise later), we will use $\hat{\beta}_0 + \hat{\beta}_1 x_i$ as our prediction for y_1.

Usually, the least squares line is determined by computer. Excel, Minitab, and many other popular packages will provide $\hat{\beta}_0$ and $\hat{\beta}_1$. For the sake of completeness, however, the computations needed to determine $\hat{\beta}_0$ and $\hat{\beta}_1$ for the data in Table 9 are given in Table 10, where we have used

$$\bar{x} = \frac{\sum x_i}{10} = 42 \qquad \text{and} \qquad \bar{y} = \frac{\sum y_i}{10} = 914.97$$

From Table 10 (which can easily be implemented on a spreadsheet), we find that $\sum(x_i - \bar{x})(y_i - \bar{y}) = 53,756.6$ and $\sum(x_i - \bar{x})^2 = 3,010$. From (18), we now find that

$$\hat{\beta}_1 = \frac{53,756.6}{3,010} = 17.86 \qquad \text{and} \qquad \hat{\beta}_0 = 914.97 - (17.86)42 = 164.88$$

Our least squares line is $\hat{y} = 164.88 + 17.86x$. Thus, we estimate that each extra train incurs a variable cost of $\hat{\beta}_1 = \$17.86$.

Our predictions and errors for all ten weeks are given in Table 11. To illustrate the computations, consider the first point (10,257.4). The predicted cost is $\hat{y}_1 = 164.88 + 17.86(10) = 343.5$, and the error is given by $e_1 = 257.4 - 343.5 = -86.1$.

Every least squares line has two properties:

1 It passes through the point $(\bar{x}, \bar{y})$. Thus, during a week in which Giapetto produced $\bar{x} = 42$ trains, we would predict that these trains would cost $914.97 to produce.

TABLE 10

Computation of $\hat{\beta}_0$ and $\hat{\beta}_1$ for Train Cost Data

x_i	y_i	$x_i - \bar{x}$	$y_i - \bar{y}$	$(x_i - \bar{x})(y_i - \bar{y})$	$(x_i - \bar{x})^2$
10	257.4	−32	−657.57	21,042.24	1,024
20	601.6	−22	−313.37	6,894.14	484
30	782.0	−12	−132.97	1,595.64	144
40	765.4	−2	−149.57	299.14	4
45	895.5	3	−19.47	−58.41	9
50	1,133.0	8	218.03	1,744.24	64
60	1,152.8	18	237.83	4,280.94	324
55	1,132.7	13	217.73	2,830.49	169
70	1,459.2	28	544.23	15,238.44	784
40	970.1	−2	55.13	−110.26	4

TABLE 11
Computations of Errors

x_i	y_i	$\hat{y}_i$	e_i
10	257.4	343.5	−86.1
20	601.6	522.1	79.5
30	782.0	700.7	81.3
40	765.4	879.3	−113.9
45	895.5	968.5	−73.0
50	1,133.0	1,057.8	75.2
60	1,152.8	1,236.4	−83.6
55	1,132.7	1,147.1	−14.4
70	1,459.2	1,415	44.2
40	970.1	879.3	90.8

2 $\sum e_i = 0$. The least squares line "splits" the data points, in the sense that the sum of the vertical distances from points above the least squares line to the least squares line equals the sum of the vertical distances from points below the least squares line to the least squares line.

How Good a Fit?

How do we determine how well the least squares line fits our data points? To answer this question, we need to discuss three components of variation: **sum of squares total** (SST), **sum of squares error** (SSE), and **sum of squares regression** (SSR). Sum of squares total is given by SST $= \sum(y_i - \bar{y})^2$. SST measures the total variation of y_i about its mean $\bar{y}$. Sum of squares error is given by SSE $= \sum(y_i - \hat{y}_i)^2 = \sum e_i^2$. If the least squares line passes through all the data points, SSE $= 0$. Thus, a small SSE would indicate that the least squares line fits the data well. We define sum of squares regression to be SSR $= \sum(\hat{y}_i - \bar{y})^2$. It can be shown that

$$SST = SSR + SSE \tag{19}$$

Note that SST is a function only of the values of y. For a good fit, SSE will be small, so (19) shows that SSR will be large for a good fit. More formally, we may define the **coefficient of determination** (R^2) for y by

$$R^2 = \frac{SSR}{SST} = \text{percentage of variation in } y \text{ explained by } x$$

Equivalently, (19) allows us to write

$$1 - R^2 = \frac{SSE}{SST} = \text{percentage of variation in } y \text{ not explained by } x$$

From computer output, we find that SST $= 1,021,762$ and SSE $= 61,705$. Then (19) yields SSR $=$ SST $-$ SSE $= 960,057$. Thus, we find that $R^2 = \frac{960,057}{1,021,762} = 0.94$. This means that the number of trains produced during a week explains 94% of the variation in the weekly cost of producing trains. All other factors combined can explain at most 6% of the variation in weekly cost, so we can be quite sure that the linear relationship between x and y is strong.

A measure of the linear association between x and y is the **sample linear correlation** r_{xy}. A sample correlation near $+1$ indicates a strong positive linear relationship between

x and y; a sample correlation near -1 indicates a strong negative linear relationship between x and y; and a sample correlation near 0 indicates a weak linear relationship between x and y.

By the way, if $\hat{\beta}_1 \geq 0$, then r_{xy} equals $+\sqrt{R^2}$, whereas if $\hat{\beta}_1 \leq 0$, the sample correlation between x and y is given by $-\sqrt{R^2}$. Thus, in our cost example, $r_{xy} = \sqrt{0.94} = 0.97$, indicating a strong linear relationship between x and y.

Forecasting Accuracy

A measure of the accuracy of predictions derived from regression is given by the **standard error of the estimate** (s_e). If we let n = number of observations, s_e is given by

$$s_e = \sqrt{\frac{SSE}{n-2}}$$

For our example,

$$s_e = \sqrt{\frac{61,705}{10-2}} = 87.8$$

It is usually true[†] that approximately 68% of the values of y will be within s_e of the predicted value $\hat{y}$, and 95% of the values of y will be within $2s_e$ of the predicted value $\hat{y}$. In the current example, we expect that 68% of our cost estimates will be within \$87.80 of the true cost, and 95% will be within \$175.60. In actuality, for 80% of our data points, actual cost is within s_e of the predicted cost, and for 100% of our data points, actual cost is within $2s_e$ of the predicted cost.

Any observation for which y is not within $2s_e$ of $\hat{y}$ is called an **outlier.** Outliers represent unusual data points and should be carefully examined. Of course, if an outlier is the result of a data entry error, it should be corrected. If an outlier is in some way uncharacteristic of the remaining data points, it may be better to omit the outlier and re-estimate the least squares line. Since all the errors are smaller than $2s_e$ in absolute value, there are no outliers in our cost example.

t-Tests in Regression

Using a **t-test,** we can test the significance of a linear relationship. To test H_0: $\beta_1 = 0$ (no significant linear relationship between x and y) against H_a: $\beta_1 \neq 0$ (significant linear relationship between x and y) at a level of significance α, we compute the t-statistic given by

$$t = \frac{\hat{\beta}_1}{\text{StdErr}(\hat{\beta}_1)}$$

[†]Actually, approximately 68% of the points should be within

$$s_e \sqrt{1 + \frac{1}{n} + \frac{(x - \bar{x})^2}{\sum(x_i - \bar{x})^2}}$$

of $\hat{y}$, and 95% of the points should be within

$$2s_e \sqrt{1 + \frac{1}{n} + \frac{(x - \bar{x})^2}{\sum(x_i - \bar{x})^2}}$$

of $\hat{y}$.

TABLE **12**
Percentage Points of the *t*-Distribution†

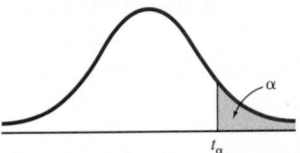

df	a = 0.1	a = 0.05	a = 0.025	a = 0.01	a = 0.005
1	3.078	6.314	12.706	31.821	63.657
2	1.886	2.920	4.303	6.965	9.925
3	1.638	2.353	3.182	4.541	5.841
4	1.533	2.132	2.776	3.747	4.604
5	1.476	2.015	2.571	3.365	4.032
6	1.440	1.943	2.447	3.143	3.707
7	1.415	1.895	2.365	2.998	3.499
8	1.397	1.860	2.306	2.896	3.355
9	1.383	1.833	2.262	2.821	3.250
10	1.372	1.812	2.228	2.764	3.169
11	1.363	1.796	2.201	2.718	3.106
12	1.356	1.782	2.179	2.681	3.055
13	1.350	1.771	2.160	2.650	3.012
14	1.345	1.761	2.145	2.624	2.977
15	1.341	1.753	2.131	2.602	2.947
16	1.337	1.746	2.120	2.583	2.921
17	1.333	1.740	2.110	2.567	2.898
18	1.330	1.734	2.101	2.552	2.878
19	1.328	1.729	2.093	2.539	2.861
20	1.325	1.725	2.086	2.528	2.845
21	1.323	1.721	2.080	2.518	2.831
22	1.321	1.717	2.074	2.508	2.819
23	1.319	1.714	2.069	2.500	2.807
24	1.318	1.711	2.064	2.492	2.797
25	1.316	1.708	2.060	2.485	2.787
26	1.315	1.706	2.056	2.479	2.779
27	1.314	1.703	2.052	2.473	2.771
28	1.313	1.701	2.048	2.467	2.763
29	1.311	1.699	2.045	2.462	2.756
30	1.310	1.697	2.042	2.457	2.750
40	1.303	1.684	2.021	2.423	2.704
60	1.296	1.671	2.000	2.390	2.660
120	1.289	1.658	1.980	2.358	2.617
240	1.285	1.651	1.970	2.342	2.596
inf.	1.282	1.645	1.960	2.326	2.576

†Computed by P. J. Hildebrand. Reprinted with permission of PWS-KENT Publishing Company.

StdErr($\hat{\beta}_1$) measures the uncertainty in our estimate of β_1; it can usually be found on a computer printout. We reject H_0 if $|t| \geq t_{(\alpha/2, n-2)}$, where $t_{(\alpha/2, n-2)}$ is obtained from Table 12. For our cost example, StdErr($\hat{\beta}_1$) = 1.6 (found from a computer printout), so $t = \frac{17.86}{1.6} = 11.16$. Using $\alpha = 0.05$, we find $t_{(.025, 8)} = 2.306$, so we reject H_0 and again conclude that there is a strong linear relationship between x and y.

Assumptions Underlying the Simple Linear Regression Model

Statistical analysis of the simple linear regression model requires that the following assumptions hold.

Assumption 1

The variance of the error term should not depend on the value of the independent variable x. This assumption is called **homoscedasticity.** If the variance of the error term depends on x, then we say that **heteroscedasticity** is present. To see whether the homoscedasticity assumption is satisfied, we plot the errors on the y-axis and the value of x on the x-axis. Figure 10 illustrates a situation where the homoscedasticity assumption is satisfied; the figure indicates no tendency for the size of the errors to depend on x. In Figure 11, however, the magnitude of the errors tends to increase as x increases. This is an example of heteroscedasticity. Using $\ln y$ or $y^{1/2}$ as the dependent variable will often eliminate heteroscedasticity.

Assumption 2

Errors are normally distributed. This assumption is not of vital importance, so we will not discuss it further.

Assumption 3

The errors should be independent. This assumption is often violated when data are collected (as in our example) over time. Independence of the errors implies that knowing the value of one error should tell us nothing about the value of the next (or any other) error.

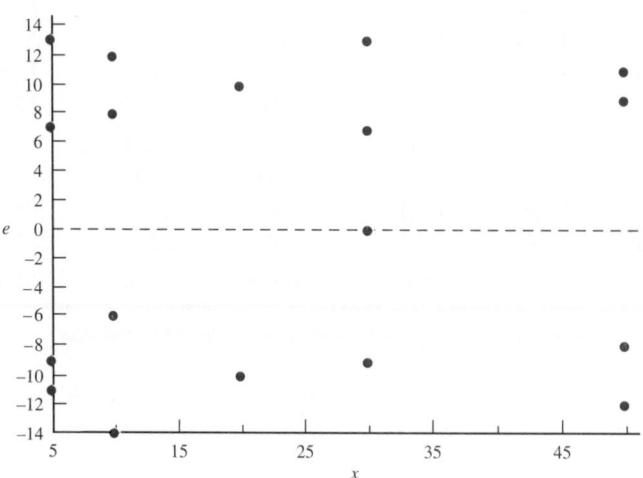

FIGURE 10
Homoscedasticity

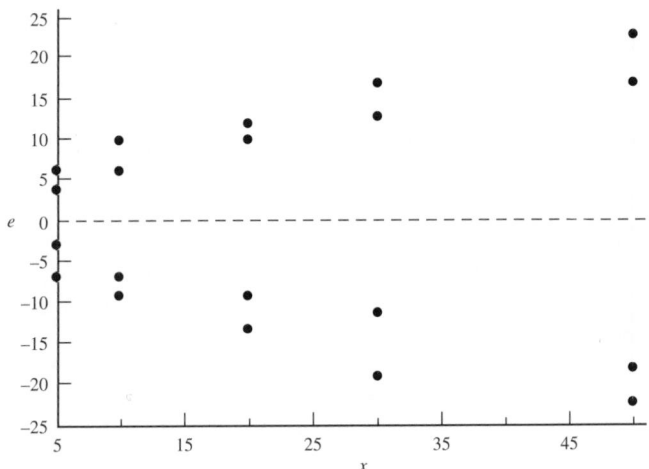

FIGURE 11
Heteroscedasticity

The validity of this assumption can be checked by plotting the errors in time-series sequence. In Figure 12, we find that the errors had the following signs: $+ + + + + + -$ $- - - - -$. This sequence of errors exhibits the following pattern: a positive error (corresponding to underprediction of the actual value of y) is usually followed by another positive error, and a negative error (corresponding to overprediction of the actual value of y) is usually followed by another negative error. This pattern indicates that successive errors are not independent; it is referred to as **positive autocorrelation.** In other words, positive autocorrelation indicates that successive errors have a positive linear relationship and are not linearly independent. If the sequence of errors in time sequence resembles Figure 13, we have **negative autocorrelation.** Here, the sequence of errors is $+ - + - + - + -$ $+ - + -$. This indicates that a positive error tends to be followed by a negative error, and vice versa. The conclusion is that successive errors have a negative linear relationship and are not independent. In Figure 14, we have the following sequence of errors: $+ + -$ $+ + - + - + + + -$. Here, no obvious pattern is present, and the independence assumption appears to be satisfied. Observe that the errors "average out" to 0, so we would expect about half our errors to be positive and half to be negative. Thus, if there is no pattern in the errors, we would expect the errors to change sign about half the time. This observation enables us to formalize the preceding discussion as follows.

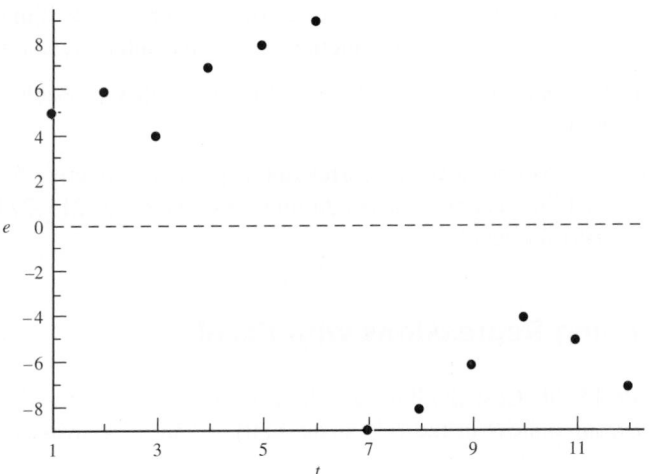

FIGURE 12
Positive Autocorrelation

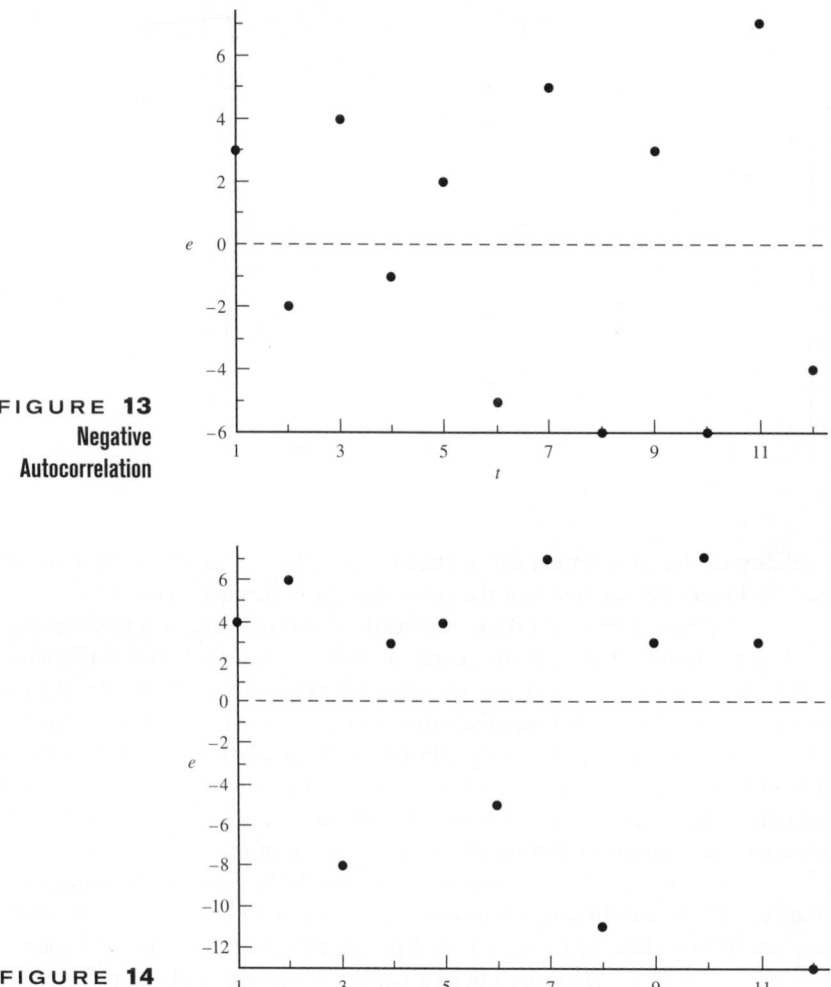

FIGURE **13**
Negative
Autocorrelation

FIGURE **14**
No Autocorrelation

1   If the errors change sign very rarely (much less than half the time), they probably violate the independence assumption, and positive autocorrelation is probably present.

2   If the errors change sign very often (much more than half the time), they probably violate the independence assumption, and negative autocorrelation is probably present.

3   If the errors change sign about half the time, they probably satisfy the independence assumption.

If positive or negative autocorrelation is present, correcting for the autocorrelation will often result in much more accurate forecasts. See pages 215–221 of Pindyck and Rubinfeld (1989) for details.

Running Regressions with Excel

Cost.xls

Figure 15 (file Cost.xls) illustrates how to run a regression with Excel. We have input the data from Table 9 in the cell range A2:B11 and then invoked the Tools Data Analysis

	A	B	C	D	E	F	G	H
1	trains	cost	yihatr	e				
2	10	257.4	9.32060032	248.0794	COST	REGRESSION		
3	20	601.6	18.6412006	582.958799	EXAMPLE			
4	30	782	27.961801	754.038199				
5	40	765.4	37.2824013	728.117599				
6	45	895.5	41.9427014	853.557299				
7	50	1133	46.6030016	1086.397				
8	60	1152.8	55.9236019	1096.8764				
9	55	1132.7	51.2633018	1081.4367				
10	70	1459.2	65.2442022	1393.9558				
11	40	970.1	37.2824013	932.817599				
12								
13								
14								
15		SUMMARY OUTPUT						
16								
17		*Regression Statistics*						
18		Multiple R	0.9693343					
19		R Square	0.9396089					
20		Adjusted R Squ	0.93206					
21		Standard Error	87.824643					
22		Observations	10					
23								
24		ANOVA						
25			*df*	*SS*	*MS*	*F*	*Significance F*	
26		Regression	1	960057.16	960057.16	124.4698887	3.72837E-06	
27		Residual	8	61705.344	7713.168			
28		Total	9	1021762.5				
29								
30			*Coefficients*	*Standard Err*	*t Stat*	*P-value*	*Lower 95%*	*Upper 95%*
31		Intercept	164.87791	72.743329	2.2665708	0.05317264	-2.8686199	332.62443
32		trains	17.859336	1.6007855	11.156607	3.72837E-06	14.16791514	21.550756
33								
34								
35								
36								

FIGURE 15

Regression Command.[†] Fill in the dialog box as shown in Figure 16. The Y range B1:B11 contains the name of the dependent variable and the values of the dependent variable. The X range A1:A11 contains the name of the independent variable and the values of the independent variable. Since the first row of the X and Y ranges include labels, we checked the Labels box. We checked cell B15 as the upper left-hand corner of the Output Range. We did not check the Residuals box. If we had, we would have obtained the predicted value and residual for each observation. Figure 15 shows the results of the regression.

Let's examine what the important numbers in the output mean. (We omit discussion of the portions of the output that are irrelevant to our discussion of regression.)

R Square   This is $r^2 = .939609$.

Multiple R   This is the square root of r^2, with the sign of Multiple R being the same as the slope of the regression line.

Standard Error   This is $s_e = 87.82$.

Observations   This is the number of data points (10).

[†]If the Analysis Tool Pak does not show up when you select Tools Data, go to Tools Add Ins and check the Analysis Tool Pak and Analysis Tool Pak Vba boxes.

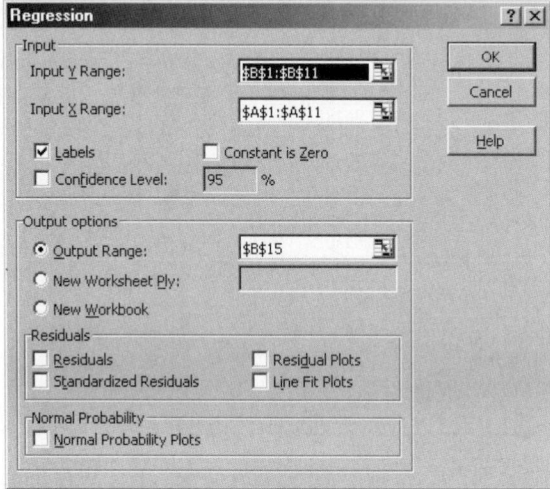

FIGURE 16

SS column The Regression entry (96,057.16) is SSR. The Residual entry (61,705.34) is SSE. The Total entry (1,021,762.5) is SST.

Coefficients column The Intercept entry (164.88) gives the value of $\hat{\beta}_0 = 164.88$, and the trains entry (17.86) gives the value of $\hat{\beta}_1 = 17.86$.

t stat This gives the observed t-statistic (coefficient/standard error) for the Intercept and the Trains variable.

Standard Error column The Intercept entry gives the standard error $\hat{\beta}_0 = 72.74$, and the Trains entry gives the standard error $\hat{\beta}_1 = 1.60$. The coefficient entry divided by the standard error entry yields the t-statistic for the intercept or slope (tabulated in the next column).

P-value For the intercept and slope, this gives Probability($|t_{n-2}| \geq$ |Observed t-statistic|). If, for example, the p-value for Trains is less than α, we reject H_0: $\beta_1 = 0$; otherwise, we accept $\beta_1 = 0$. For $\alpha = .05$, we reject $\beta_1 = 0$. For p-value $= .05$, it is borderline whether or not to accept the hypothesis that $\beta_0 = 0$.

In cell C2, we obtain $\hat{y}_1$ by inputting the formula =D$14+A2*C$20. In cell D2, we obtain e_1 by inputting the formula =B2−C2. Copying from the range C2:D2 to C2:D11 creates predictions and errors for all observations.

Obtaining a Scatterplot with Excel

To obtain a scatterplot with Excel, let the range where your independent variable is be the X range. Then let the range where your dependent variable is be the Y range. Then select X-Y Graph.

24.7 Fitting Nonlinear Relationships

Often, a plot of points of the form (x_i, y_i) indicates that y is not a linear function of x. In such cases, however, the plot may indicate that there is a nonlinear relationship between

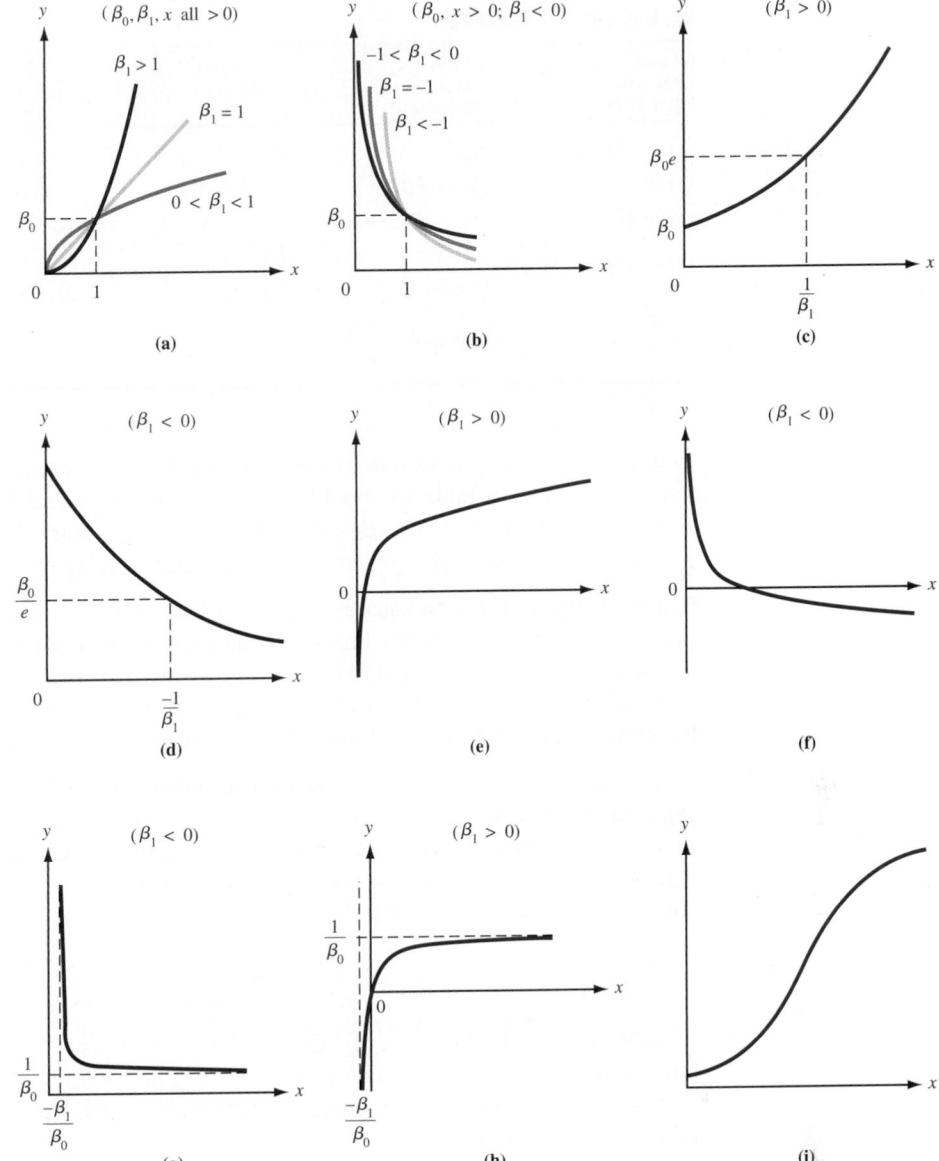

FIGURE 17
Graphs of Linearizable
Functions†

†Reprinted by permission from C. Daniel and F. Wood, Fitting Functions to Data, Copyright 1980, John Wiley and Sons.

x and y. For example, if the plot of the (x_i, y_i) looks like any of parts (a)–(i) of Figure 17, a nonlinear relationship between x and y is indicated.

The following procedure may be used to estimate a nonlinear relationship.

Step 1 Plot the points and find which part of Figure 17 best fits the data. For illustrative purposes, suppose the data look like part (c).

Step 2 The second column of Table 13 gives the functional relationship between x and y. For part (c), this would be $y = \beta_0 \exp(\beta_1 x)$.

Step 3 Transform each data point according to the rules in the third column of Table 13. Thus, if part (c) of the figure is relevant, we transform each value of y into $\ln y$ and trans-

TABLE 13

How to Fit a Nonlinear Relationship

If Graph Looks Like Figure 17 Part	We Have the Functional Relationship	Transform (x_i, y_i) into	Estimate of Functional Relationship
(a) or (b)	$y = \beta_0 x^{\beta_1}$	$(\ln x_i, \ln y_i)$	$\hat{y} = \exp(\hat{\beta} + s_e^2/2)x^{\hat{\beta_1}}$
(c) or (d)	$y = \beta_0 \exp(\beta_1 x)$	$(x_i, \ln y_i)$	$\hat{y} = \exp(\hat{\beta}_0 + \hat{\beta}_1 x + s_e^2/2)$
(e) or (f)	$y = \beta_0 + \beta_1(\ln x)$	$(\ln x_i, y_i)$	$\hat{y} = \hat{\beta}_0 + \hat{\beta}_1 (\ln x)$
(g) or (h)	$y = \dfrac{x}{\beta_0 x + \beta_1}$	$\left(\dfrac{1}{x_i}, \dfrac{1}{y_i}\right)$	$\hat{y} = \dfrac{x}{\hat{\beta}_0 x + \hat{\beta}_1}$
(i)	$y = \exp\left(\beta_0 + \dfrac{\beta_1}{x}\right)$	$\left(\dfrac{1}{x_i}, \ln y_i\right)$	$\hat{y} = \exp(\hat{\beta}_0 + \dfrac{\hat{\beta}_1}{x} + s_e^2/2)$

form each value of x into x. Given the relationship in the second column of Table 13, the transformed data in Table 13 should, if plotted, indicate a straight-line relationship. For part (c), for example, if $y = \beta_0 \exp(\beta_1 x)$, then taking natural logarithms of both sides yields $\ln y = \ln(\beta_0) + \beta_1 x$, so there is indeed a linear relationship between x and $\ln y$.

Step 4 Estimate the least squares regression line for the transformed data. If $\hat{\beta}_0$ is the intercept of the least squares line (for transformed data), $\hat{\beta}_1$ is the slope of the least squares line (for transformed data), and s_e is the standard error of the regression estimate, then we read the estimated relationship from the final column of Table 13. Thus, if part (c) were relevant, we would estimate that $\hat{y} = \exp(\hat{\beta}_0 + \hat{\beta}_1 x + s_e^2/2)$.

To illustrate the idea, suppose we want to predict future VCR sales for an appliance store. Sales for the last 24 months are given in Table 14 and are plotted in Figure 18 (where each dot indicates actual sales). We will use $x =$ number of the month as the independent variable. Figure 18 indicates an S-shaped relationship between $x =$ number of the month and $y =$ sales during the month (like part (i) of Figure 17). So according to Table 13,

$$y = \exp\left(\beta_0 + \frac{\beta_1}{x}\right)$$

Following the third column of Table 13, we now estimate the least squares regression line for the points $(\frac{1}{1}, \ln 23)$, $(\frac{1}{2}, \ln 156)$, ..., $(\frac{1}{24}, \ln 3{,}495)$. We find $\hat{\beta}_0 = 8.387$, $s_e = .276$, and $\hat{\beta}_1 = -5.788$. From the last column of Table 13, we obtain the estimated relationship between x and y:

$$\hat{y} = \exp\left(8.387 + .5(.276)^2 - \frac{5.788}{x}\right)$$

$$= \exp\left(8.425 - \frac{5.788}{x}\right)$$

To illustrate how this formula could be used to predict future sales, suppose we want to predict VCR sales during month 26. For $x = 26$, we would forecast that

$$y = \exp\left(8.425 - \frac{5.788}{26}\right) = 3{,}649.3 \text{ VCRs}$$

would be sold.

REMARKS 1 For the regression on the points $(\frac{1}{1}, \ln 23)$, $(\frac{1}{2}, \ln 156)$, ..., $R^2 = 0.95$. This means that 95% of the variation in $\ln y$ is explained by variation in $\frac{1}{x}$. Unfortunately, this does not tell us anything about how accurate our predictions of actual sales (y) are likely to be. For this, we compute the

TABLE **14**

VCR Sales

Month	Sale of VCRs
1	23
2	156
3	330
4	482
5	1,209
6	1,756
7	2,000
8	2,512
9	2,366
10	2,942
11	2,872
12	2,937
13	3,136
14	3,241
15	3,149
16	3,524
17	3,542
18	3,312
19	3,547
20	3,376
21	3,375
22	3,403
23	3,697
24	3,495

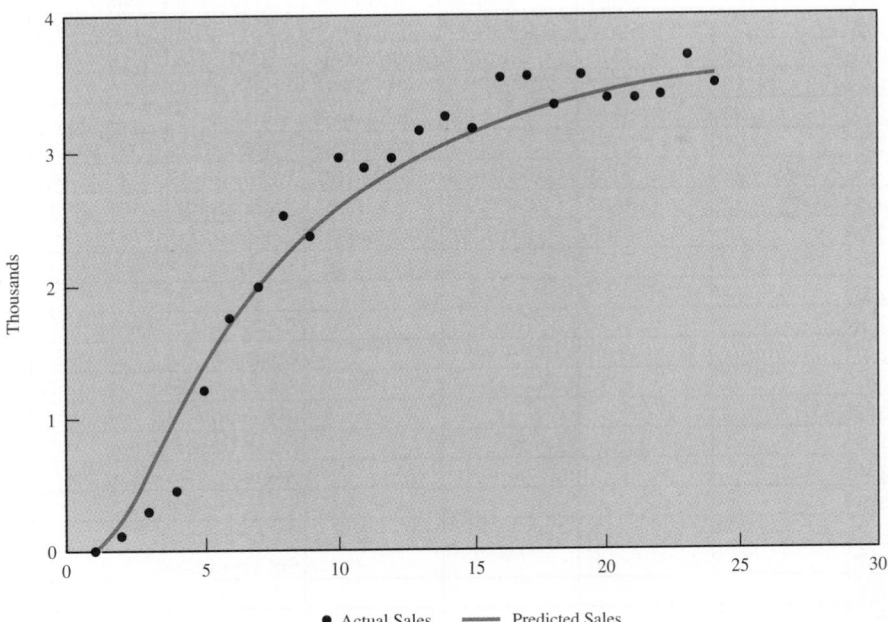

FIGURE **18**
VCR Sales

predicted sales for each month, and e_i = (actual month i sales) − (predicted month i sales). By averaging $|e_i|$ for the 24 months, we find the MAD of our predictions to be 170.3. By applying (17), we may estimate the standard deviation of our forecasts to be 1.25(170.3) = 212.88. Thus, 95% of the time, we would expect our predictions for VCR sales to be accurate within 2(212.88) = 425.76 VCRs.

2 If we had mistakenly tried to fit a straight line to this data, we would have obtained $s_e = 546$, so fitting the S-shaped curve has greatly improved our forecasts.

Using a Spreadsheet to Fit a Nonlinear Relationship

VCR.xls

Figure 19 (file VCR.xls) shows how we can use a spreadsheet to fit a curve to the data in Table 14. We input the data in the cell range A2:B26. In columns C and D, we create the

FIGURE 19

A	A	B	C	D	E	F	G	H
1		MAD=	170.2927		VCR	EXAMPLE		
2	MONTH	SALES	1/MONTH	LNSALES	PREDICT	ERROR	ABSERR	
3	1	23	1	3.1354942	13.969974	9.0300255	9.0300255	
4	2	156	0.5	5.049856	252.37307	-96.37307	96.373067	
5	3	330	0.3333333	5.7990927	662.20451	-332.2045	332.20451	
6	4	482	0.25	6.1779441	1072.6714	-590.6714	590.6714	
7	5	1209	0.2	7.0975489	1432.6874	-223.6874	223.6874	
8	6	1756	0.1666667	7.4707938	1737.5658	18.434166	18.434166	
9	7	2000	0.1428571	7.6009025	1994.303	5.6969922	5.6969922	
10	8	2512	0.125	7.8288345	2211.4573	300.54274	300.54274	
11	9	2366	0.1111111	7.768956	2396.5747	-30.57475	30.574747	
12	10	2942	0.1	7.9868449	2555.765	386.23499	386.23499	
13	11	2872	0.0909091	7.9627639	2693.8455	178.15445	178.15445	
14	12	2937	0.0833333	7.9851439	2814.5945	122.4055	122.4055	
15	13	3136	0.0769231	8.0507034	2920.9846	215.01543	215.01543	
16	14	3241	0.0714286	8.0836372	3015.3711	225.62887	225.62887	
17	15	3149	0.0666667	8.0548402	3099.6364	49.363637	49.363637	
18	16	3524	0.0625	8.167352	3175.2979	348.70211	348.70211	
19	17	3542	0.0588235	8.1724468	3243.5904	298.40965	298.40965	
20	18	3312	0.0555556	8.1053075	3305.5268	6.47315	6.47315	
21	19	3547	0.0526316	8.1738575	3361.9455	185.05446	185.05446	
22	20	3376	0.05	8.1244469	3413.5452	-37.54522	37.545218	
23	21	3375	0.047619	8.1241506	3460.9127	-85.91273	85.912726	
24	22	3403	0.0454545	8.1324127	3504.5442	-101.5442	101.54423	
25	23	3697	0.0434783	8.215277	3544.8619	152.13809	152.13809	
26	24	3495	0.0416667	8.1590887	3582.2271	-87.22711	87.227114	
27								
28								
29						Regression Output:		
30					Constant			8.3867886
31					Std Err of Y Est			0.2761082
32					R Squared			0.9527748
33					No. of Observations			24
34					Degrees of Freedom			22
35								
36					X Coefficient(s)		-5.787996	
37					Std Err of Coef.		0.2747317	
38								
39								
40								
41								

transformed variables 1/MONTH and LNSALES. In C3, we input the formula 1/A3. In D3, we input the formula =LN(B3). Copying from the range C3:D3 to C3:D26 yields the transformed values of x_i and y_i. We now run a regression with the X range C3:C26 and the Y range D3:D26. The results of this regression are used to predict VCR sales in each month. In cell E3, we enter the formula =EXP(C\$47+C\$48/A3+.5(C\$37^2)) to generate a forecast for month 1 VCR sales. In cell F3, we determine e_1 with the formula =B3−E3. In cell G3, we determine $|e_1|$ with the formula =ABS(F3). Copying from the range E3:G3 to the range E3:G26 generates forecasts and errors for all 24 months. In cell C1, we compute the MAD for all 24 months with the formula =AVERAGE(G3:G26).

Utilizing the Excel Trend Curve

The Excel Trend Curve makes it easy to fit an equation to a set of data. After creating an X-Y scatterplot, click on the points in the graph until the points turn gold. Then select Chart Add Trendline. See Figure 20.

- Choosing Linear yields the straight line that best fits the points.
- Choose Logarithmic if the scatterplot looks like (e) or (f) in Figure 17. Then Excel yields the best-fitting equation of the form $y = \beta_0 + \beta_1(\ln(x))$.
- Choose Power if the scatterplot looks like (a) or (b) in Figure 17. Then Excel yields the best-fitting equation of the form $y = \beta_0 x^{\beta_1}$.
- Choose Exponential if the scatterplot looks like (c) or (d) in Figure 17. Then Excel yields the best-fitting equation of the form $y = \beta_0 e^{\beta_1 x}$.
- Choosing Polynomial of order n ($n = 1, 2, 3, 4, 5,$ or 6) yields the best-fitting equation of the form $y = \beta_0 + \beta_1 x + \beta_2 x^2 + \cdots + \beta_n x^n$

Before having Trend Curve fit the curve, select Options and select Display Equation on Chart and Display R^2 on Chart. The R^2 value displayed is the R^2 associated with the linear regression based on the transformed (x_i, y_i) listed in the third column of Table 13. For the Linear, Polynomial, and Logarithmic options, choosing Intercept = 0 will set $\beta_0 = 0$.

Figure 21 shows the results obtained from Trend Curve when applied to find the best-fitting straight line for the data in worksheet Cost.xls. Note that the R^2 and equation estimates match those we obtained from the Analysis Tool Pak.

Cost.xls

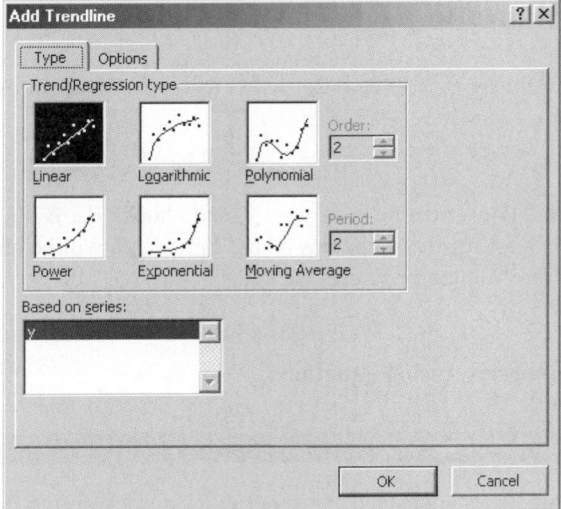

FIGURE 20

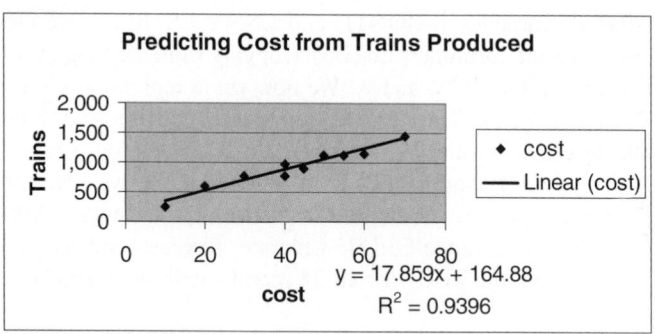

FIGURE 21

24.8 Multiple Regression

In many situations, more than one independent variable may be useful in predicting the value of a dependent variable. We then use **multiple regression.** For example, to predict the monthly sales for a national fast-food chicken chain, we might consider using the following independent variables: national income, price of chicken, dollars spent on advertising during the current month, and dollars spent on advertising during the previous month.

Suppose we are using k independent variables to predict the dependent variable y and we have n data points of the form $(y_i, x_{1i}, x_{2i}, \ldots, x_{ki})$, where x_{ji} = value of jth independent variable for ith data point and y_i = value of dependent variable for ith data point. In multiple regression, we model the relationship between y and the k independent variables by

$$y_i = \beta_0 + \beta_1 x_{1i} + \beta_2 x_{2i} + \cdots + \beta_k x_{ki} + \varepsilon_i$$

where ε_i is an error term with mean 0, representing the fact that the actual value of y_i may not equal $\beta_0 + \beta_1 x_{1i} + \beta_2 x_{2i} + \cdots + \beta_k x_{ki}$. β_j may be thought of as the increase in y if the value of the jth independent variable is increased by 1 and all other independent variables are held constant. Thus, β_j is analogous to $\frac{\partial y}{\partial x_j}$, where x_j is the jth independent variable.

Estimation of the β_i's

Suppose we estimate β_i ($i = 0, 1, 2, \ldots, k$) using $\hat{\beta}_i$. Then our prediction or estimate for y_i is given by

$$\hat{y}_i = \hat{\beta}_0 + \hat{\beta}_1 x_{1i} + \hat{\beta}_2 x_{2i} + \cdots + \hat{\beta}_k x_{ki}$$

As in Section 24.6, we define $e_i = y_i - \hat{y}_i$ and choose $\hat{\beta}_0, \hat{\beta}_1, \ldots, \hat{\beta}_k$ to minimize Σe_i^2. Usually, these least squares estimates of $\beta_0, \beta_1, \ldots, \beta_k$ will be obtained from a computer package such as Minitab or Excel. We call

$$\hat{y}_i = \hat{\beta}_0 + \hat{\beta}_1 x_{1i} + \hat{\beta}_2 x_{2i} + \cdots + \hat{\beta}_k x_{ki}$$

the **least squares regression equation.**

EXAMPLE 1   **Truck Maintenance**

We want to predict maintenance expense (y) for a truck during the current year, from the independent variables x_1 = miles driven (in thousands) during the current year and x_2 =

TABLE 15

Truck Maintenance Data

y	x_1	x_2
$832	6	8
$733	7	7
$647	9	6
$553	11	5
$467	13	4
$373	15	3
$283	17	2
$189	18	1
$96	19	0

TABLE 16

Computer Output for Example 1

Variable	Coefficient	Standard Error	t-Value
Constant	17.73846	31.0271	0.57171
x_1	4.061538	1.56742	2.59123
x_2	98.50769	2.756428	35.73744

Standard error of estimate = 2.106157

age of the truck (in years) at the beginning of the current year. We are given the information in Table 15.

Solution Computer output for this example is given in Table 16. Reading down the Coefficient column and rounding to two decimal places, we obtain $\hat{\beta}_0 = 17.74$, $\hat{\beta}_1 = 4.06$, and $\hat{\beta}_2 = 98.51$. Thus, we would predict annual maintenance cost for a truck from

$$\hat{y} = 17.74 + 4.06x_1 + 98.51x_2 \tag{20}$$

For a five-year-old truck that is driven 10,000 miles during a year, we predict annual maintenance costs by $17.74 + 4.06(10) + 98.51(5) = \$550.89$.

From (20), we conclude that (holding the age of the truck constant) driving an extra thousand miles during a year increases annual maintenance costs by $\hat{\beta}_1 = \$4.06$, and that an increase of one year in the age of the truck (holding miles driven constant) increases annual maintenance costs by $\hat{\beta}_2 = \$98.51$.

Goodness of Fit Revisited

For multiple regression, we define SSR, SSE, and SST as we did in Section 24.6. We also find that $R^2 = \frac{\text{SSR}}{\text{SST}}$ = percentage of variation in y explained by the k independent variables and $1 - R^2$ = percentage of variation in y not explained by the k independent variables. If we define the standard error of the estimate as

$$s_e = \sqrt{\frac{\text{SSE}}{(n - k - 1)}}$$

then (as in Section 24.6) we expect approximately 68% of the y-values to be within s_e of $\hat{y}$ and approximately 95% of the y-values to be within $2s_e$ of $\hat{y}$. We have already seen from Table 16 that $s_e = 2.106$. Thus, 95% of the time, we expect our predictions for annual truck maintenance expenditures to be accurate within \$4.21.

Hypothesis Testing

If we have included independent variables $x_1, x_2, \ldots, x_k$ in a multiple regression, we often want to test

H_0: $\beta_i = 0$ (x_i does not have a significant effect on y when the other independent variables are included in the regression equation)

against

H_a: $\beta_i \neq 0$ (x_i does have a significant effect on y when the other independent variables are included in the regression equation)

To test these hypotheses, we compute

$$t = \frac{\hat{\beta}_i}{\text{StdErr}(\hat{\beta}_i)}$$

where $\text{StdErr}(\hat{\beta}_i)$ measures the amount of uncertainty present in our estimate of β_i. $\text{StdErr}(\hat{\beta}_i)$ (and often the t-statistic) is read from computer output. At a level of significance α, we reject H_0 if $|t| > t_{(\alpha/2, n-k-1)}$. From Table 16, we find that (t for x_1) = 2.59 and (t for x_2) = 35.74. Suppose $\alpha = 0.05$. Since $t_{(.025, 9-2-1)} = 2.447$, we reject H_0 for each independent variable and conclude that both miles driven and age of the truck have a significant effect on annual maintenance cost.

Usually, variables included in a regression equation should have significant t-statistics. ($\alpha = 0.10$ or $\alpha = 0.05$ are commonly used levels of significance in regression analysis.) If an independent variable has an insignificant t-statistic, we usually remove the independent variable from the equation and obtain new least squares estimates. To illustrate this idea, suppose we have the data shown in Table 17 on sales at the Bloomington Happy Chicken Restaurant during the last 20 years (file Chicken.xls). (POP = population within 10 miles of Happy Chicken Restaurant, AD = thousands of dollars spent on advertising during the current year, LAGAD = thousands of dollars spent on advertising during the previous year, and SALES = sales in thousands of dollars.)

We attempt to estimate the model

$$\text{SALES} = \beta_0 + \beta_1 \text{YEAR} + \beta_2 \text{POP} + \beta_3 \text{AD} + \beta_4 \text{LAGAD} + \varepsilon$$

We are using YEAR as an independent variable in the hopes of picking up a possible upward trend in sales. LAGAD is used as an independent variable because we believe that last year's advertising might affect this year's sales. We obtain the following estimated regression equation (t-statistics for each independent variable are in parentheses):

$$\widehat{\text{SALES}} = 10{,}951.51 + 169.51\ \text{YEAR} - .059\ \text{POP} + 122.38\ \text{AD} + 276.93\ \text{LAGAD}$$
$$\qquad\qquad\qquad (1.91)\qquad\quad (-.70)\qquad (13.84)\qquad\quad (28.92)$$

(21)

We cannot use the first year of data, since LAGAD is undefined. Since $t_{(.05, 19-4-1)} = 1.761$, we find that all independent variables except for POP are significant for $\alpha = 0.10$. Thus, YEAR, AD, and LAGAD appear to have a significant effect on sales. After

TABLE **17**
Happy Chicken Sales Data

Year	POP	AD	LAGAD	Sales
1	96,020	30	—	13,000
2	102,558	20	30	15,713
3	101,792	15	20	12,937
4	104,347	25	15	12,872
5	106,180	30	25	16,227
6	106,562	15	30	15,388
7	105,209	25	15	13,180
8	109,185	35	25	17,199
9	109,976	40	35	20,674
10	110,659	20	40	20,350
11	111,844	25	20	14,444
12	111,576	35	25	17,530
13	113,784	5	35	16,711
14	112,482	12	5	9,715
15	116,487	16	12	12,248
16	117,316	21	16	13,856
17	117,830	22	21	15,285
18	118,148	24	22	15,620
19	118,481	26	24	17,158
20	121,069	28	26	17,800

dropping the insignificant variable POP from the equation, we obtain the following estimated regression equation:

$$\widehat{\text{SALES}} = 5150.94 + 108.58 \text{ YEAR} + 121.59 \text{ AD} + 274.30 \text{ LAGAD}$$
$$(8.29) \qquad (14.10) \qquad (31.72)$$

All independent variables are significant. Also, we find $R^2 = 0.99$ and $s_e = 309$. Thus, we are reasonably satisfied with this equation and expect that 95% of the time, our prediction for sales will be within \$618,000 of actual sales.

Choosing the Best Regression Equation

How can we choose between several regression equations having different sets of independent variables? We usually want to choose the equation with the lowest value of s_e, since that will yield the most accurate forecasts. We also want the t-statistics for all variables in the equation to be significant. These two objectives may conflict, in which case it is difficult to determine the "best" equation. If the available computer printout contains the C_p statistic, then the regression chosen should have a C_p value close to (number of independent variables in the equation) + 1. For example, if a regression with three independent variables has $C_p = 80$, we can be sure that it is not a "good" regression. Actually, if a regression has C_p much larger than p, it means that at least one important variable has been omitted from the regression. (See Daniel and Wood (1980) for a discussion of the C_p statistic.)

Multicollinearity

If an estimated regression equation contains two or more independent variables that exhibit a strong linear relationship, we say that **multicollinearity** is present. A strong linear relationship between some of the independent variables may make the computer's estimates of the β_i's unreliable. In certain circumstances, multicollinearity can even cause a variable that should have β_i positive to have $\hat{\beta}_i$ substantially less than 0. The Happy Chicken example illustrates multicollinearity. We began with both YEAR and POP as independent variables, and as POP and YEAR both increase over time, we would expect a strong positive linear relationship to exist between them. Indeed, the correlation between YEAR and POP is 0.98. To see that the estimates of β_{YEAR} and β_{POP} are unreliable, note that in (21), $\hat{\beta}_{\text{POP}} < 0$, indicating that an increase in the number of customers near Happy Chicken decreases sales. This anomaly is due to multicollinearity. The strong linear relationship between YEAR and POP makes it difficult for the computer to estimate β_{POP} and β_{YEAR} accurately. After we drop POP from the estimated equation, the multicollinearity problem disappears, because there is no strong linear relationship between any of the remaining independent variables.

By the way, if an *exact* linear relationship exists between two or more independent variables, there are an infinite number of combinations of the $\hat{\beta}_i$'s which will minimize the sum of the squared errors, and most computer packages will print an error message. For example, if we let $x_1 = $ U.S. consumer expenditure during a year, $x_2 = $ U.S. investment during a year, $x_3 = $ U.S. government expenditure during a year, and $x_4 = $ U.S. national income during a year, it is well known that $x_4 = x_1 + x_2 + x_3$. In this case, we cannot use x_1, x_2, x_3, and x_4 as independent variables; at least one should be dropped from the equation.

Dummy Variables

Often, a nonquantitative or qualitative independent variable may influence the dependent variable. Some examples are as follows:

Dependent Variable	Categorical Independent Variable
Salary of employee	Race of employee
Consumer expenditures during year	Whether it is a wartime or a peacetime year
Customers entering bank on a given day	Day of the week
Air conditioner sales during a given month	Month of the year

In each of these situations, the independent variable does not assume a numerical value, but it may be classified into one of c categories. To illustrate, for monthly air conditioner sales, $c = 12$, and for the consumer expenditure example, $c = 2$.

Let the possible values of the categorical variable be listed as value 1, value 2, ..., value c. To model the effect of a categorical variable on a dependent variable, we define $c - 1$ **dummy variables** as follows:

$x_1 = 1$ if observation takes on value 1 of categorical variable

$x_1 = 0$ otherwise

$x_2 = 1$ if observation takes on value 2 of categorical variable

$x_2 = 0$ otherwise

$$\vdots$$

$x_{c-1} = 1$ if observation takes on value $c - 1$ of categorical variable

$x_{c-1} = 0$ otherwise

TABLE 18
Customer Traffic at University Credit Union

Day Number	Day of Week	University Payday?	Customers Entering Credit Union
1	Monday	No	515
12	Tuesday	No	360
18	Wednesday	Yes	548
23	Wednesday	No	386
24	Thursday	No	440
46	Monday	Yes	687
48	Wednesday	No	350
52	Tuesday	No	430
54	Thursday	No	370
55	Friday	No	496
70	Friday	No	506
81	Monday	No	509
89	Thursday	Yes	508
104	Thursday	No	396
106	Monday	No	600
108	Wednesday	No	266
122	Tuesday	No	360
130	Friday	Yes	521
152	Tuesday	No	398

We now include $x_1, x_2, \ldots, x_{c-1}$ (along with any other relevant independent variables) in the estimated regression equation.

To illustrate the use of dummy variables, suppose we are trying to predict the number of customers to enter the University Credit Union each day. The bank manager believes that bank traffic is influenced by the day of the week (the credit union is open Monday through Friday) and by whether or not the day is a payday for university employees. We are given the number of people to enter the bank during 18 randomly chosen days. (Day 1 is the present, day 6 is a week from now, and so on.) The relevant information is given in Table 18.

In this situation, there are two categorical variables of interest: the day of the week ($c = 5$) and whether or not a day is a payday ($c = 2$). For the day of the week, we define value 1 = Monday, value 2 = Tuesday, value 3 = Wednesday, value 4 = Thursday, and value 5 = Friday. Then we let

$$x_1 = 1 \quad \text{if day is a Monday} \qquad x_1 = 0 \quad \text{otherwise}$$
$$x_2 = 1 \quad \text{if day is a Tuesday} \qquad x_2 = 0 \quad \text{otherwise}$$
$$x_3 = 1 \quad \text{if day is a Wednesday} \qquad x_3 = 0 \quad \text{otherwise}$$
$$x_4 = 1 \quad \text{if day is a Thursday} \qquad x_4 = 0 \quad \text{otherwise}$$

For whether or not a day is a payday, we define $c - 1 = 1$ dummy variables. Letting value 1 = payday and value 2 = not a payday, we define

$$x_5 = 1 \quad \text{if day is a payday} \quad x_5 = 0 \quad \text{otherwise}$$

To account for a possible trend in the number of customers, we include $T =$ number of the day as an independent variable. To illustrate how this information would be coded on the computer, we "code" the last two observations:

T	x_1	x_2	x_3	x_4	x_5	Customers
130	0	0	0	0	1	521
152	0	1	0	0	0	398

We obtain the following estimated regression equation (t-statistics in parentheses):

$$\hat{y} = 496.1 - 0.36T + 71.1x_1 - 78.5x_2 - 122.5x_3 - 74.8x_4 + 127.1x_5$$
$$(-1.29) \quad (1.87) \quad (-2.04) \quad (-3.17) \quad (-1.99) \quad (4.43)$$

For $\alpha = 0.10$, we find that all independent variables except T are significant. After eliminating T from the estimated equation, we obtain the following equation:

$$\hat{y} = 466.2 + 80.4x_1 - 79.2x_2 - 109.8x_3 - 68.8x_4 + 124.3x_5$$
$$(2.1) \quad (-2.0) \quad -(2.86) \quad (-1.79) \quad (4.24)$$

(22)

For $\alpha = 0.10$, all independent variables are significant, so this equation appears satisfactory. We also find that $R^2 = 0.85$ and $s_e = 48.8$. Thus, on 95% of all days, our prediction for the number of customers entering the credit union should be accurate within $2(48.8) = 97.6$ customers.

Interpretation of Coefficients of Dummy Variables

How do we interpret the coefficients of dummy variables? To illustrate, let's determine how whether or not a day is a payday affects credit union traffic. On a payday, $x_5 = 1$, and we predict that $466.2 + 80.4x_1 - 79.2x_2 - 109.8x_3 - 68.8x_4 + 124.3$ customers will enter the credit union. On a day that is not a payday, $x_5 = 0$, and we predict that $466.2 + 80.4x_1 - 79.2x_2 - 109.8x_3 - 68.8x_4$ customers will enter the credit union. Subtracting, we find that on a payday, we predict (all other things being equal) that $\hat{\beta}_5 = 124.3$ more customers will enter the credit union than on a day that is not a payday.

To see how the day of the week influences credit union traffic, we note that (22) yields a different prediction for each day of the week. For Monday, $x_1 = 1$, $x_2 = x_3 = x_4 = 0$, and we obtain $\hat{y} = 466.2 + 80.4 + 124.3x_5 = 546.6 + 124.3x_5$. For Tuesday, $x_2 = 1$, $x_1 = x_3 = x_4 = 0$, and we obtain $\hat{y} = 466.2 - 79.2 + 124.3x_5 = 387 + 124.3x_5$. For Wednesday, $x_3 = 1$, $x_2 = x_1 = x_4 = 0$, and we obtain $\hat{y} = 466.2 - 109.8 + 124.3x_5 = 356.4 + 124.3x_5$. For Thursday, $x_4 = 1$, $x_1 = x_3 = x_2 = 0$, and we obtain $\hat{y} = 466.2 - 68.8 + 124.3x_5 = 397.4 + 124.3x_5$. For Friday, $x_1 = x_2 = x_3 = x_4 = 0$, and we obtain $\hat{y} = 466.2 + 124.3x_5$. Thus, we find that (all other things being equal) credit union traffic is heaviest on Mondays, next heaviest on Fridays, third heaviest on Thursdays, fourth heaviest on Tuesdays, and lightest on Wednesdays.

Multiplicative Models

Often, we believe that there is a relationship of the following form:

$$Y = \beta_0 x_1^{\beta_1} x_2^{\beta_2} \cdots x_k^{\beta_k}$$

(23)

To estimate such a relationship, simply take logarithms of both sides of (23). This yields

$$\ln Y = \ln \beta_0 + \beta_1 (\ln x_1) + \beta_2 (\ln x_2) + \cdots + \beta_k (\ln x_k)$$

Thus, to estimate (23), we run a multiple regression with the dependent variable being $\ln Y$ and the independent variables being $\ln x_1, \ln x_2, \ldots, \ln x_k$. To illustrate the idea, suppose we want to determine how the annual operating costs of an insurance company depend on the number of home insurance and car insurance policies that have been written.

TABLE **19**
Insurance Company Branch Data

Branch	Annual Operating Cost	Number of Home Insurance Policies	Number of Car Insurance Policies
1	$124,000	400	1,200
2	$71,000	350	360
3	$136,000	600	800
4	$219,000	800	1,800
5	$230,000	900	1,600
6	$75,000	200	1,000
7	$56,000	120	900
8	$110,000	340	1,100
9	$120,000	490	900
10	$144,000	700	800

Branch.xls

Table 19 gives relevant information for ten branches of the insurance company (file Branch.xls).

To fit the model $Y = \beta_0 x_1^{\beta_1} x_2^{\beta_2}$, where Y = annual operating cost, x_1 = home insurance policies, and x_2 = car insurance policies, we would input into the computer points of the form (ln 124,000, ln 400, ln 1,200), and so on. The least squares estimates obtained are

$$\text{Constant term estimate} = 5.339$$
$$\text{Estimate for } \beta_1 = 0.583$$
$$\text{Estimate for } \beta_2 = 0.409$$

This regression yields an R^2 of 0.998, indicating a very good fit. The constant term estimate is an estimate of ln β_0, so our actual estimate of β_0 is $e^{5.339} = 208.3$, and we estimate that $Y = 208.3 x_1^{0.583} x_2^{0.409}$. To illustrate the use of this equation, we predict annual operating cost for an insurance branch writing 500 home policies and 1,200 car policies. For this branch, we obtain $Y = 208.3(500)^{0.583}(1,200)^{0.409} = \$141,767$.

Heteroscedasticity and Autocorrelation in Multiple Regression

By plotting the errors in time-series sequence, we may check (as described in Section 24.6) to see whether the errors from a multiple regression are independent. If autocorrelation is present and the errors do not appear to be independent, then correcting for autocorrelation will usually yield better forecasts.

By plotting the errors (on the y-axis) against the predicted value of y (on the x-axis), we can determine whether homoscedasticity or heteroscedasticity is present. If homoscedasticity is present, the plot should show no obvious pattern (that is, the plot should resemble Figure 10), whereas if heteroscedasticity is present, the plot should show an obvious pattern indicating that the errors somehow depend on the predicted value of y (perhaps as in Figure 11). If heteroscedasticity is present, the t-tests described in this section are invalid.

Implementing Multiple Regression on a Spreadsheet

In Figures 22 and 23 (file Credit.xls) we have run the regression for the data in Table 18. In the cell range A3:A21, we input the customer count for each day, and in the cell range B3:B21, the number of each day. In the cell range H3:H21, we input the day of the week for each observation (1 = Monday, . . . , 5 = Friday). In the cell range G3:G21, a dummy variable indicates whether each day is a payday. We then used =IF statements to create the dummy variables for the day of the week. In cell C3, we input the formula =IF(H3=1,1,0). This places a 1 in C3, indicating the first observation is on a Monday. In cell D3, we enter the formula =IF(H3=2,1,0); in cell E3, =IF(H3=3,1,0); and in cell F3 =IF(H3=4,1,0). Copying from the range C3:F3 to the range C3:F21 generates the values of the dummy variables for all observations. To run the regression, select the Y range of A2:A21 and the X range of B2:G21. The regression output has the following interpretation:

Intercept   This is $\hat{\beta}_0 = 496.0857$.

Standard Error   This is $s_e = 48.84517$.

R Square   This is $R^2 = .845927$. This means that together, all the independent variables in the regression explain 84.6% of the variation in the number of customers arriving daily.

Observations   This is the number of data points (19).

Total df   This is the degrees of freedom ($n - k - 1 = 19 - 6 - 1$) used for the t-test of H_0: $\beta_i = 0$ against H_1: $\beta \neq 0$.

Coefficients   For each independent variable, this column yields the coefficient of the independent variable in the least squares equation. For example, $\hat{\beta}_T = -0.36222$.

Standard Error   For each independent variable, this row yields StdErr $\hat{\beta}_i$. For example, StdErr $\hat{\beta}_T = 0.279852$. The X Coefficient divided by the Std Err of Coef. yields the t-statistic for testing H_0: $\beta_i = 0$ against H_1: $\beta_i \neq 0$.

t Stat   This gives the observed t-statistic (coefficient/standard error) for the Intercept and all independent variables. For example, the t-statistic for Monday is 1.87.

Standard Error column   The Intercept entry gives the standard error $\hat{\beta}_0 = 37.66$, and the coefficient entries give the standard error for each independent variable. For example,

	A	B	C	D	E	F	G	H
1		CREDIT	UNION	EXAMPLE				
2	CUSTOMER	DAY#	MON	TUES	WED	THUR	PAYDAY?	DAYWK
3	515	1	1	0	0	0	0	1
4	360	12	0	1	0	0	0	2
5	548	18	0	0	1	0	1	3
6	386	23	0	0	1	0	0	3
7	440	24	0	0	0	1	0	4
8	687	46	1	0	0	0	1	1
9	350	48	0	0	1	0	0	3
10	430	52	0	1	0	0	0	2
11	370	54	0	0	0	1	0	4
12	496	55	0	0	0	0	0	5
13	506	70	0	0	0	0	0	5
14	509	81	1	0	0	0	0	1
15	508	89	0	0	0	1	1	4
16	396	104	0	0	0	1	0	4
17	600	106	1	0	0	0	0	1
18	266	108	0	0	1	0	0	3
19	360	122	0	1	0	0	0	2
20	521	130	0	0	0	0	1	5
21	398	152	0	1	0	0	0	2

FIGURE 22
Credit Union Example

	B	C	D	E	F	G
22	SUMMARY OUTPUT					
23						
24	*Regression Statistics*					
25	Multiple R	0.9197432				
26	R Square	0.8459275				
27	Adjusted R Square	0.7478813				
28	Standard Error	51.017121				
29	Observations	19				
30						
31	SUMMARY OUTPUT					
32						
33	*Regression Statistics*					
34	Multiple R	0.9197432				
35	R Square	0.8459275				
36	Adjusted R Square	0.7688912				
37	Standard Error	48.845174				
38	Observations	19				
39						
40	ANOVA					
41		*df*	*SS*	*MS*	*F*	*Significance*
42	Regression	6	157192.73	26198.789	10.980899	0.0002823
43	Residual	12	28630.212	2385.851		
44	Total	18	185822.95			
45						
46		*Coefficients*	*Standard Err*	*t Stat*	*P-value*	*Lower 95%*
47	Intercept	496.08565	37.660289	13.172646	1.7E-08	414.03093
48	DAY#	-0.362218	0.2798522	-1.294318	0.2199096	-0.971963
49	MON	71.076945	38.076099	1.8667076	0.0865578	-11.88375
50	TUES	-78.47826	38.509376	-2.0379	0.0642282	-162.383
51	WED-	122.5236	38.6517	-3.16994	0.0080705	-206.7384
52	THUR	-74.82254	37.669971	-1.986265	0.070328	-156.8984
53	PAYDAY?	127.10854	28.682168	4.4316224	0.0008188	64.615462
54						

FIGURE 23

standard error $\hat{\beta}_{\text{Monday}} = 38.07$. The coefficient entry divided by the standard error entry yields the *t*-statistic for the intercept or slope (tabulated in the next column).

P-value For the intercept and each independent variable in a regression with k independent variables, this gives Probability($|t_{n-k-1}| \geq$ |Observed *t*-statistic|). If, for example, the *p*-value for Wednesday is less than α, we reject H_0: $\beta_{\text{Wednesday}} = 0$; otherwise, we accept $\beta_{\text{Wednesday}} = 0$. For $\alpha = .05$, we reject $\beta_{\text{Wednesday}} = 0$.

The Data Analysis Regression Tool can handle a maximum of 15 independent variables. The data for the independent variables must be in adjacent columns.

PROBLEMS

Group A

1 For the years 1961–1970, the annual return on General Motors stock and the return on the Standard and Poor's market index were as given in Table 20 (file Beta.xls).

a Let Y = return on General Motors stock during a year and X = return on Standard and Poor's index during a year. Financial theory suggests that $Y = \beta_0 + \beta_1 X + \varepsilon$, where β_1 is called the **beta** for General Motors. Give an interpretation for the beta of a stock (in this case, General Motors), and use the data in Table 20 to estimate the beta for General Motors.

b Does the Standard and Poor's index appear to have a significant effect (for $\alpha = 0.05$) on the return on General Motors stock?

c What percentage of the variation in the return on General Motors Stock is explained by variation in the Standard and Poor's index?

d What percentage of the variation in the return on General Motors stock is unexplained by variation in Standard and Poor's index?

e During a year in which the Standard and Poor's in-

TABLE **20**

Year	Return on General Motors Stock	Return on Standard and Poor's Index
1961	12%	21%
1962	2%	−3%
1963	38%	15%
1964	26%	20%
1965	18%	12%
1966	−10%	0%
1967	0%	10%
1968	9%	10%
1969	−2%	2%
1970	−1%	−15%

dex increased by 15%, what would we predict for the return on General Motors stock?

2 We are trying to determine the number of labor hours required to produce a unit of a product. We are given the information in Table 21 (file Learn.xls). For example, the 2nd unit produced required 517 labor hours, and the 600th unit produced required 34 labor hours.

a Try to determine a relationship between the number of units already produced and the labor hours needed to produce the next unit. Why is this relationship called the **learning curve?**

b How many labor hours would be needed to produce the 800th unit?

c We are 95% sure that the prediction in part (b) is accurate within _____ hours.

3 Quarterly sales for a department store over a six-year period are given in Table 22 (file Sales.xls).

a Use multiple regression to develop a model that can

TABLE 21

Cumulative Production	Labor Hours Needed for Last Unit
1	715
2	517
10	239
20	174
40	126
60	104
100	82
150	68
200	59
300	47
500	37
600	34

TABLE 22

Year	Quarter	Sales (millions)
1984	1	$50,147
1984	2	$49,325
1984	3	$57,048
1984	4	$76,781
1985	1	$48,617
1985	2	$50,898
1985	3	$58,517
1985	4	$77,691
1986	1	$50,862
1986	2	$53,028
1986	3	$58,849
1986	4	$79,660
1987	1	$51,640
1987	2	$54,119
1987	3	$65,681
1987	4	$85,175
1988	1	$56,405
1988	2	$60,031
1988	3	$71,486
1988	4	$92,183
1989	1	$60,800
1989	2	$64,900
1989	3	$76,997
1989	4	$103,337

be used to predict future quarterly sales. (*Hint:* Use dummy variables and an independent variable for the number of the quarter (quarter 1, quarter 2, . . . , quarter 24).

b Letting Y_t = sales during quarter number t, discuss how to fit the following model to the data in Table 22:

$$Y_t = \beta_0 \beta_1^t \beta_2^{x_2} \beta_3^{x_3} \beta_4^{x_4}$$

where $x_2 = 1$ if t is a first quarter, $x_3 = 1$ if t is a second quarter, and $x_4 = 1$ if t is a fourth quarter. (*Hint:* Take logarithms of both sides.)

c Interpret the answer to part (b).

d Which model appears to yield better predictions for sales?

4 To determine how price influences sales, a company changed the price of a product over a 20-week period. The price charged each week and the number of units sold are given in Table 23 (file Price.xls). Develop a model to relate sales to price.

5 Confederate Express Service is attempting to determine how its shipping costs for a month depend on the number of units shipped during a month. For the last 15 months, the number of units shipped and total shipping cost are given in Table 24 (file Ship.xls).

TABLE 23

Price	Units Sold
$1	1,145
$2	788
$3	617
$4	394
$5	275
$6	319
$7	289
$8	241
$9	259
$10	176
$11	179
$12	232
$13	183
$14	181
$15	222
$16	212
$17	186
$18	110
$19	183
$20	172

TABLE 24

Month	Units Shipped	Total Shipping Cost
1	300	$1,060
2	400	$1,380
3	500	$1,640
4	200	$740
5	300	$1,060
6	350	$1,190
7	460	$1,520
8	480	$1,580
9	120	$540
10	760	$2,420
11	580	$2,200
12	340	$1,470
13	120	$790
14	100	$720
15	500	$1,960

a Determine a relationship between units shipped and monthly shipping cost.

b Plot the errors for the predictions in order of time sequence. Is there any unusual pattern?

c We have been told that there was a trucking strike during months 11–15, and we believe that this may have influenced shipping costs. How could the answer to part (a) be modified to account for the effects of the strike?

After accounting for the effects of the strike, does the unusual pattern in part (b) disappear?

Group B

6 In Example 1, we ran a regression with only x_1 (miles driven) as an independent variable. We found the coefficient of x_1 in this regression to be -51.68. This appears to indicate (contrary to what we would expect) that increasing the miles driven will lead to decreased maintenance costs. Explain this result. (*Hint:* Estimate the correlation between x_1 and x_2.)

7 Suppose we are trying to fit a curve to data, and part (i) of Figure 17 is relevant. Explain why the points of the form $(\frac{1}{x_i}, \ln y_i)$ should, when plotted, indicate a straight-line relationship.

8 Consider the regression in which we estimated cost of running an insurance company as a function of the number of home and car insurance policies. If there were a 1% increase in the number of car insurance policies, by what percentage would we predict that total costs would increase?

9 In the example in which we predicted the number of customers to enter the credit union, suppose that we had used five (instead of four) dummy variables to represent the days of the week. What problem would have arisen?

SUMMARY   Moving-Average Forecasts

$$f_{t,1} = \text{average of last } N \text{ observations}$$
$$e_t = x_t - (\text{prediction for } x_t)$$
$$\text{MAD} = \text{average value of } |e_t|$$

Choose N to minimize MAD.

Simple Exponential Smoothing

$$A_t = \text{smoothed average at end of period } t$$
$$= f_{t,k} = \text{forecast for period } t + k \text{ made at end of period } t$$
$$A_t = \alpha x_t + (1 - \alpha)A_{t-1}$$

Choose α to minimize MAD.

Holt's Method

Holt's method is used when trend is present, but there is no seasonality.

$$L_t = \text{estimate of base at end of period } t$$
$$T_t = \text{estimate of per-period trend at end of period } t$$
$$L_t = \alpha x_t + (1 - \alpha)(L_{t-1} + T_{t-1})$$
$$T_t = \beta(L_t - L_{t-1}) + (1 - \beta)T_{t-1}$$
$$f_{t,k} = L_t + kT_t$$

Winter's Method

Winter's method is used when we believe that trend and seasonality may be present.

$$s_t = \text{estimate for month } t \text{ seasonal factor at the end of month } t$$
$$L_t = \frac{\alpha x_t}{s_{t-c}} + (1 - \alpha)(L_{t-1} + T_{t-1})$$
$$T_t = \beta(L_t - L_{t-1}) + (1 - \beta)T_{t-1}$$
$$s_t = \frac{\gamma x_t}{L_t} + (1 - \gamma)s_{t-c}$$
$$f_{t,k} = (L_t + kT_t)s_{t+k-c}$$

For all extrapolation methods, we expect 68% of our predictions to be within $s_e = 1.25$ MAD of the actual value and 95% of our predictions to be within $2s_e$ of the actual value.

Simple Linear Regression

Given data points $(x_1, y_1), \ldots, (x_n, y_n)$, we estimate a linear relationship between x and y by $\hat{y} = \hat{\beta}_0 + \hat{\beta}_1 x$, where

$$\hat{\beta}_1 = \frac{\sum(x_i - \bar{x})(y_i - \bar{y})}{\sum(x_i - \bar{x})^2} \quad \text{and} \quad \hat{\beta}_0 = \bar{y} - \hat{\beta}_1 \bar{x}$$

$$R^2 = \frac{\text{SSR}}{\text{SST}} = \text{percentage of variation in } y \text{ explained by } x$$

$$r_{xy} = \text{sample linear correlation between } x \text{ and } y$$
$$(r_{xy} \text{ indicates the strength of the linear relationship between } x \text{ and } y)$$

$$s_e = \sqrt{\frac{\text{SSE}}{n - 2}}$$

We expect 68% of our predictions to be within s_e of the actual value and 95% of our predictions to be within $2s_e$ of the actual value.

A t-statistic exceeding $t_{(\alpha/2, n-2)}$ in absolute value is evidence (at level of significance α) that there is a significant linear relationship between x and y.

Fitting a Nonlinear Relationship

Step 1 Plot the points and find the part of Figure 17 that best fits the data.

Step 2 The second column of Table 13 gives the functional relationship between x and y.

Step 3 Transform each data point according to the rules in the third column of Table 13.

Step 4 Estimate the least squares regression line for the transformed data. If $\hat{\beta}_0$ is the intercept of the least squares line (for transformed data) and $\hat{\beta}_1$ is the slope of the least squares line (for transformed data), then we read the estimated relationship from the final column of Table 13.

Multiple Regression

Multiple regression is used when more than one independent variable is needed to predict y.

R^2 = percentage of variation in y explained by the independent variables

Reject H_0: $\beta_i = 0$ at a level of significance α if $(t$ for $x_i) \geq t_{(\alpha/2, n-k-1)}$, where k is the number of independent variables being used to predict y.

If there is a strong linear relationship between two or more independent variables, then $\hat{\beta}_i$ may be an unreliable estimate of β_i. In such cases, we say that **multicollinearity** is present.

If a nonquantitative or qualitative independent variable (such as the day of the week or the month of the year) is believed to influence a dependent variable, **dummy variables** may be used to model the effect of the qualitative independent variable on the dependent variable. If the qualitative variable can assume c values, use only $c - 1$ dummy variables.

REVIEW PROBLEMS

Group A

1 Table 25 gives data concerning pork sales (file Pork.xls). Price is in dollars per hundred lb sold, quantity sold is in billions of pounds, per-capita income is in dollars, U.S. population is in millions, and GNP is in billions of dollars.

a Use this data to develop a regression equation that could be used to predict the quantity of pork sold during future periods. Is autocorrelation, heteroscedasticity, or multicollinearity a problem?

b Suppose that during each of the next two quarters, price = $45, U.S. population = 240, GNP = 2,620, and per-capita income = $10,000. Predict the quantity of pork sold during each of the next two quarters.

c 68% of the time, we expect our prediction for pork sales to be accurate within _____.

d Use Winter's method to develop a forecast for pork sales during the next two quarters. (Use the first two years to initialize.)

2 We are to predict sales for a motel chain based on the information in Table 26 (file Motel.xls).

a Use this data and multiple regression to make predictions for the motel chain's sales during the next four quarters. Assume that advertising during each of the next four quarters is $50,000.

TABLE 25

Quarter	Year	Price of Pork	Quantity Sold	Per-Capita Income	U.S. Population	GNP
1	1975	39.35	30.44	8,255	212	1,549
2	1975	46.11	29.23	8,671	213	1,589
3	1975	58.83	25.12	8,583	214	1,629
4	1975	52.2	28.35	8,649	215	1,669
1	1976	47.99	28.95	8,775	216	1,718
2	1976	49.19	27.83	8,812	217	1,768
3	1976	43.88	29.53	8,884	218	1,818
4	1976	34.25	35.9	8,967	219	1,868
1	1977	39.08	32.94	9,036	220	1,918
2	1977	40.87	31.86	9,125	221	1,978
3	1977	43.85	30.74	9,280	222	2,038
4	1977	41.38	34.99	9,399	223	2,098
1	1978	47.44	32.43	9,487	224	2,148
2	1978	47.84	32.65	9,530	225	2,218
3	1978	48.52	31.58	9,622	226	2,288
4	1978	50.05	35.40	9,732	227	2,338
1	1979	51.98	33.98	9,813	228	2,398
2	1979	48.04	37.58	9,778	229	2,448
3	1979	38.52	38.59	9,809	230	2,478
4	1979	36.39	43.47	9,867	231	2,508
1	1980	36.31	41.24	9,958	232	2,539
2	1980	31.18	43.00	9,805	235	2,598
3	1980	46.23	37.57	9,882	235	2,598

TABLE 26

Quarter	Potential Customers (thousands)	Advertising (thousands of dollars)	Season	Sales (millions)
1	100	30	Winter	1,200
2	105	20	Spring	880
3	111	15	Summer	1,800
4	117	40	Fall	1,050
5	122	10	Winter	1,700
6	128	50	Spring	350
7	135	5	Summer	2,500
8	142	40	Fall	760
9	149	20	Winter	2,300
10	156	10	Spring	1,000
11	164	60	Summer	1,570
12	172	5	Fall	2,430
13	181	35	Winter	1,320
14	190	15	Spring	1,400
15	200	70	Summer	1,890
16	210	25	Fall	3,200
17	221	30	Winter	2,200
18	232	60	Spring	1,440
19	243	80	Summer	4,000
20	264	60	Fall	4,100

TABLE **27**

Year	Jan.	Feb.	Mar.	Apr.	May	June	July	Aug.	Sept.	Oct.	Nov.	Dec.
1965	38	44	53	49	54	57	51	58	48	44	42	37
1966	42	43	53	49	49	40	40	36	29	31	26	23
1967	29	32	41	44	49	47	46	47	43	45	34	31
1968	35	43	46	46	43	41	44	47	41	40	32	32
1969	34	40	43	42	43	44	39	40	33	32	31	28
1970	34	29	36	42	43	44	44	48	45	44	40	37
1971	45	49	62	62	58	59	64	62	50	52	50	44
1972	51	56	60	65	64	63	63	72	61	65	51	47

b Use the Holt method to make forecasts for the motel chain's sales during the next four quarters.

c Use simple exponential smoothing to make predictions for the motel chain's sales during the next four quarters.

d Use Winter's method to determine predictions for the motel chain's sales during the next four quarters.

e Which forecasts would be expected to be the most reliable? (*Hint:* Use advertising, lagged by one period, as an independent variable.)

3 Table 27 gives the following data for monthly U.S. housing sales (in thousands of houses) for 1965–1972.

a Use the years 1965–1966 to initialize the parameters for Winter's method. Then find values of α, β, and γ that yield a MAD (for 1967–1972) of less than 3.5. (*Hint:* It may be necessary to use $\alpha > 0.5$.)

b We would expect 68% of our forecasts to be accurate within _____ and 95% of our forecasts to be accurate within _____.

c Check to see whether the data are consistent with the answer to part (b).

d Although we have not discussed autocorrelation for smoothing methods, good forecasts derived from smoothing methods should exhibit no autocorrelation. Do the forecast errors for this problem exhibit autocorrelation?

e It has been stated that if only trend and seasonality are important factors, then α should be at most 0.5. Explain why this problem required $\alpha > 0.5$.

f At the end of December 1972, what is the forecast for housing sales during the first three months of 1973?

Note: This assignment is a snap on a spreadsheet. The spreadsheet might be set up as in Table 28. In B14, enter $=A\$3*A14/D2+(1-A\$3)*(B13+C13)$. Insert analogous formulas in C14, D14, E14. Remember that the forecast must be made before "seeing" A14. In F14, enter $=A14-E14$. In G14, enter $=ABS(F14)$. Copy from B14:G14 to ??. To compute MAD, average the absolute errors for each month (rows 14–85).

4 Using x as the independent variable and y as the dependent variable, find the least squares line for the following three data points:

x	y
1	2
4	5
7	2

TABLE **28**

Row	A	B	C	D	E	F	G
1	SALES	BASE	TREND	SEASON	FORE	ERR	ABSERR
2	ALPHA	BETA	GAMMA	S-11			
3	.1	.2	.3	S-10			
.							
.							
.							
13		LO	TO	SO			
14	29						
15	32						
.							
.							
85	47						

TABLE 29

x	y
0	100
1	130
2	170
3	200
4	260
5	300
6	305
7	330
8	380

TABLE 30

Humidi	Temp	Press	Hard
40	1	0	148
60	1	0	209
50	1	0	177
70	1	0	208
80	1	0	262
60	1	1	248
65	1	1	253
70	1	1	263
35	1	1	184
45	1	1	220
70	0	0	129
28	0	0	53
49	0	0	98
89	0	0	170
90	0	0	172
34	0	1	80
56	0	1	90
77	0	1	151
23	0	1	58
56	0	1	107

5 We are trying to predict the number of uses of automatic bank teller machines as a function of time. The data are given in Table 29. Here, x = number of years after 1980 and y = number of monthly uses of ATMs (in millions) during the given year. The estimated regression equation is $\hat{y} = 102.3 + 34.8x$. We are given that SST = 74,100, SSE = 1,298, and StdErr($\hat{\beta_1}$) = 1.76.

a Test H_0: $\beta_1 = 0$ against H_a: $\beta_1 \neq 0$ for $\alpha = 0.05$. Interpret the result.

b Find the correlation between x and y.

c Is the 1987 entry an outlier?

d If present trends continue, what is the approximate probability that during 1990, more than 470 million ATM transactions per month will occur? (*Hint:* Use the fact that the errors are normally distributed.)

6 Carboco puts metal coatings on jet propeller blades. The harder the coating, the higher the quality of the coating. The coating is shot onto the blade using pressurized gas contained in an F-gun. Carboco can control the temperature and gas pressure in the F-gun and can also control the room humidity. To see how gas pressure, temperature, and humidity influence hardness, Carboco engineers have run a regression for which the dependent variable is

HARD = hardness of a coating

and the independent variables are

HUMIDI = room humidity

TEMP = 1 if temperature level is high

= 0 if temperature level is low

PRESS = 1 if gas pressure is high

= 0 if gas pressure is low

T*P = product of TEMP and PRESS

We assume that temperature and gas pressure have only two possible levels, low and high. The relevant data are given in Table 30 (file Temp.xls).

a Ignoring considerations of heteroscedasticity, multicollinearity, and autocorrelation, which equation should be used to predict hardness? Explain.

b What combination of gas pressure and temperature setting will maximize hardness?

c Explain how changing temperature from a low level to a high level affects hardness. Be specific!

7 We have been assigned to determine how the total weekly production cost for Widgetco depends on the number of widgets produced during the week. The following model has been proposed:

$$Y = \beta_0 + \beta_1 X + \beta_2 X^2 + \beta_3 X^3 + \varepsilon$$

where X = number of widgets produced during the week and Y = total production cost for the week. For 15 weeks of data, we found that SSR = 215,475 and SST = 229,228. For this model, we obtain the following estimated regression equation (t-statistics for each coefficient are in parentheses):

$$\hat{y} = -29.7 + 19.8X - 0.39X^2 + 0.005X^3$$
$$(0.78) \quad (0.62) \quad (1.25)$$

a For $\alpha = 0.10$, test H_0: $\beta_i = 0$ against H_a: $\beta_i \neq 0$ ($i = 1, 2, 3$).

b Determine R^2 for this model. How can the high R^2 value be reconciled with the answer to part (a)?

8 Let Y_t = sales during month t (in thousands of dollars) for a photography studio (SALES in Table 31) and P_t = price charged for portraits during month t (PRICE). Use a computer to fit the following model to the data in Table 31 (file Portrait.xls):

$$Y_t = \beta_0 + \beta_1 Y_{t-1} + \beta_2 P_t + \varepsilon_t$$

Thus, last month's sales and the current month's price are independent variables.

a If the price of a portrait during month 21 is $10, what would we predict for month 21's sales?

b Does there appear to be a problem with autocorrelation, heteroscedasticity, or multicollinearity?

TABLE **31**

Month	Sales	Price
1	400	5
2	1,042	4
3	1,129	8
4	1,110	6
5	1,336	6
6	1,363	10
7	1,177	9
8	603	8
9	582	12
10	697	9
11	586	8
12	673	9
13	546	10
14	334	11
15	27	8
16	76	9
17	298	10
18	746	6
19	962	7
20	907	8

TABLE **32**

Year	GNP
1975	1,060
1976	1,170
1977	1,305
1978	1,455
1979	1,630
1980	1,800
1981	2,000
1982	2,220
1983	2,450
1984	2,730

b When the regression on transformed data is done, we find that $\hat{\beta}_0 = 6.86$ and $\hat{\beta}_1 = 0.105$. What is the prediction for 1985 GNP?

Group B

10 Suppose the true relationship between Y and time t is given by

$$Y = \beta_0 e^{\beta_1 t} + \varepsilon$$

where $\beta_1 > 0$. If we try to fit our usual linear model $Y = \beta_0 + \beta_1 t + \varepsilon$ to the data, are we likely to encounter autocorrelation? Heteroscedasticity? Multicollinearity?

9 The U.S. GNP during the years 1975–1984 is given in Table 32 in billions of dollars (file GNP.xls).

a Plot x = years after 1974 against GNP, and use the plot to describe how to fit a curve that could be used to predict GNP during future years.

REFERENCES

For further information on extrapolation methods, we recommend:

Hax, A., and D. Candea. *Production and Inventory Management.* Englewood Cliffs, N.J.: Prentice Hall, 1984.

Makridakis, S., and S. Wheelwright. *Forecasting: Methods and Applications.* New York: Wiley, 1986.

For further information on regression, we recommend:

Chatterjee, S., and B. Price. *Regression Analysis by Example.* New York: Wiley, 1990.

Daniel, C., and F. Wood. *Fitting Equations to Data.* New York: Wiley, 1980.

Montgomery, D., and E. Peck. *Introduction to Linear Regression Analysis.* New York: Wiley, 1991.

Pindyck, R., and D. Rubinfeld. *Econometric Models and Economic Forecasts.* New York: McGraw-Hill, 1989.

@Risk Crib Sheet

@Risk Icons

Once you are familiar with the function of the @Risk icons, you will find @Risk easy to learn. Here is a description of the icons.

Opening an @Risk Simulation

This icon allows you to open up a saved @Risk simulation. I do not recommend saving simulations. Instead, I paste results into a spreadsheet.

Saving an @Risk Simulation

This icon allows you to save an @Risk simulation, including data and simulation settings.

Simulation Settings

This icon allows you to control the settings for the simulation. Clicking on this icon activates the dialog box shown in Figure 1. There follows a description of what each of the tabs can do.

Iterations Tab

Various options are associated with the Iterations tab.

#Iterations   #Iterations is how many times you want @Risk to recalculate the spreadsheet. For example, choosing 100 iterations means that 100 values of your output cells will be tabulated.

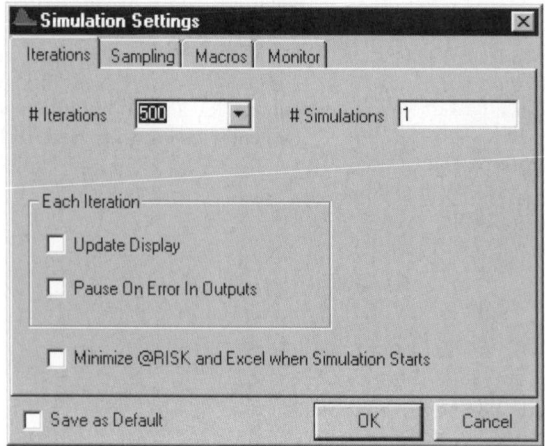

FIGURE 1

#Simulations   Leave this at 1 unless you have a =RISKSIMTABLE functon in the spread-sheet. In this case, choose #Simulations to equal the number of values in SIMTABLE. For example, if we have the formula =RISKSIMTABLE({100,150,200,250,300}) in cell A1, set #Simulations to 5. The first simulation will place 100 in A1, the second simulation will place 150 in A1, and the fifth simulation will place 300 in A1. #Iterations will be run for each simulation.

Pause on Error   Checking this box causes @Risk to pause if an error occurs in any cell during the simulation. @Risk will highlight the cells where the error occurs.

Update Display   Checking this box causes @Risk to show the results of each iteration on the screen. This is nice, but it slows things down.

See Figure 2. The Sampling tab options are as follows.

Sampling Type   While a little slower, Latin Hypercube sampling is much more accurate than Monte Carlo sampling. To illustrate, Latin Hypercube guarantees for a given cell that 5% of observations will come from the bottom 5th percentile of the actual random vari-able, 5% will come from the top 5th percentile of the actual random variable, etc. If we choose Monte Carlo sampling, 8% of our observations may come from the bottom 5% of the actual distribution, when in reality only 5% of observations should do so. When sim-ulating financial derivatives, it is crucial to use Latin Hypercube.

Standard Recalc   If you choose Expected Value, you obtain the expected value of the ran-dom variable unless the random variable is discrete. Then you obtain the possible value of the random variable that is closest to the random variable's expected value. For in-stance, for a statement

$$=RISKDISCRETE(\{1,2,\},\{.6,.4\})$$

the expected value is $1(.6) + 2(.4) = 1.4$, so Expected Value enters a 1.

If you choose the Monte Carlo option, *when you hit F9, all the random cells will re-calculate. This makes it much easier to understand and debug the spreadsheet.* Thus, with Monte Carlo selected,

$$=RISKDISCRETE(\{1,2,\},\{.6,.4\})$$

will return a 1 60% of the time and a 2 40% of the time.

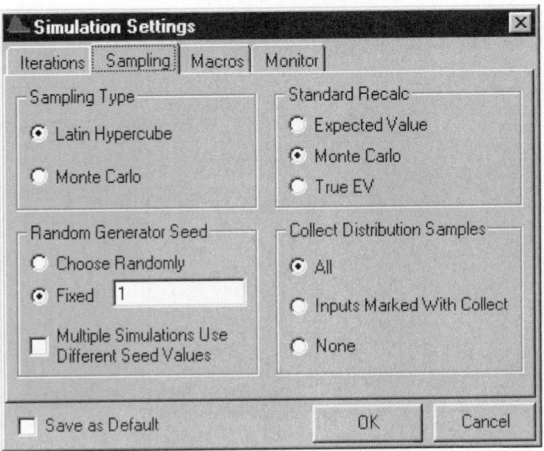

FIGURE 2

If you choose the True EV option, then the actual expected value of the random variable will be returned. Thus,

$$=\text{RISKDISCRETE}(\{1,2,\},\{.6,.4\})$$

will yield a 1.4.

Collecting Distribution Samples   Check All if you want to get Tornado Graphs, Scenario Analysis, or Extract Data. Also check this box if you want statistics on cells generated by @Risk functions. You can always check this box if you like, but if you have many @Risk functions in your spreadsheet, checking the box will slow down the simulation. Checking Inputs Marked With Collect will collect data on a subset of your risk functions marked with Riskcollect.

Random Number Generator Seed   When the seed is set to 0, each time you run a simulation, you will obtain different results. Other possible seed values are integers between 1 and 32,767. Whenever a nonzero seed is chosen, the same values for the input cells and output cells will occur. For example, if we choose a seed value of 10, each time we run the simulation, we will obtain exactly the same results.

Autoconvergence

Under #Iterations, you may select Auto. See Figure 3. You may then select a percentage such as 1%. Then @Risk keeps running until during the last 100 iterations, the mean and standard deviation change by at most 1%. This can be a lot of iterations! I prefer to choose the number of iterations myself by setting

$$\frac{2s}{\sqrt{n}}$$

equal to the desired level of accuracy for the output cell's mean. Here, $s =$ standard deviation of output cell for a trial simulation (say, 400 iterations). For example, if a trial simulation yields $s = 100$ and I want to be 95% sure that I am estimating the population mean within 10, I need

$$\frac{2(100)}{\sqrt{n}} = 10$$

or $n = 400$.

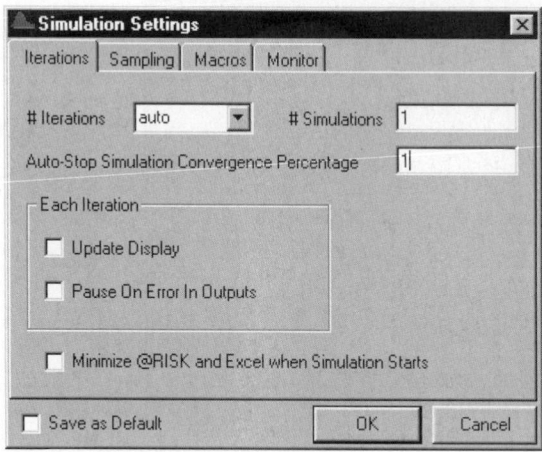

FIGURE 3

Macro Tab

See Figure 4. The Macro tab enables @Risk to run a macro before or after each iteration of a simulation. For example, checking After Each Iteration's Recalc and entering Macro1 after evaluating each ouput cell would result in the following sequence of events:

- Compute @Risk functions and calculate output cells.
- Run Macro1.
- Compute @Risk functions and calculate output cells, etc.

Select Output Cells

This icon enables you to select an output cell or cells for which @Risk will create statistics. Simply select a range of cells and click on the icon to select the range as output cells. You may select as many ranges as you desire.

List Input and Output Cells

This icon lists all output cells. Also listed are cells containing @Risk functions. These are called input cells. From this list, you can change the names of output cells or delete output cells.

Run Simulation

This icon starts the simulation. The status of the simulation is shown in the lower left-hand corner of your screen. Hitting the Escape key allows you to terminate the simulation.

Show Results

This icon allows you to see results. There are two windows:

- Summary Results, containing Minimum, Mean, and Maximum for all input and output cells.
- Simulation Statistics, containing more detailed statistics.

Clicking the Hide icon will send you back to your worksheet. To paste your statistics into your worksheet, simply select a window and Edit Copy Paste it into the worksheet.

Define Distribution Icon

This icon allows you to see the mass function or density function for any random variable. You may also use this icon directly to enter any @Risk formula into a cell.

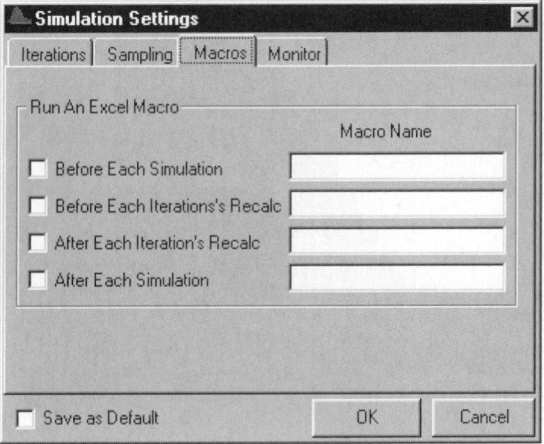

FIGURE 4

Graphing

To obtain a graph, right click on the cell from the Explorer interface. Then choose the type of graph desired. To copy the graph into Excel, right click on the graph and select Copy or Graph in Excel.

A histogram gives the fraction of iterations assuming different values. The histogram in Figure 5 was generated for a cell containing the formula

$$=RISKNORMAL(100,15)$$

The histogram indicates that the input cell was bell-shaped and that the most common values of the input cell were around 100.

For a cumulative ascending graph (Figure 6), the *y*-axis gives the fraction of iterations yielding a value $\leq$ the value on the *x*-axis. Thus, about 50% of all iterations in this case yielded a value $\leq$ 100.

For a cumulative descending graph, the *y*-axis gives the fraction of iterations yielding a value $\geq$ the value on the *x*-axis. In Figure 7, this input cell exceeded 85 about 84% of the time.

An area graph replaces bars with smooth areas. A fitted curve smooths out the variation in bar heights before creating an area graph.

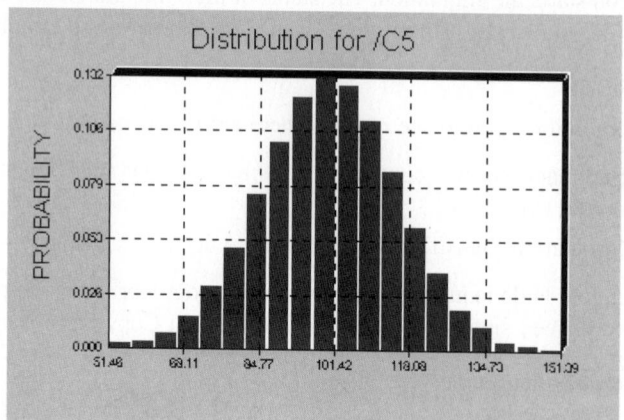

FIGURE 5
Histogram

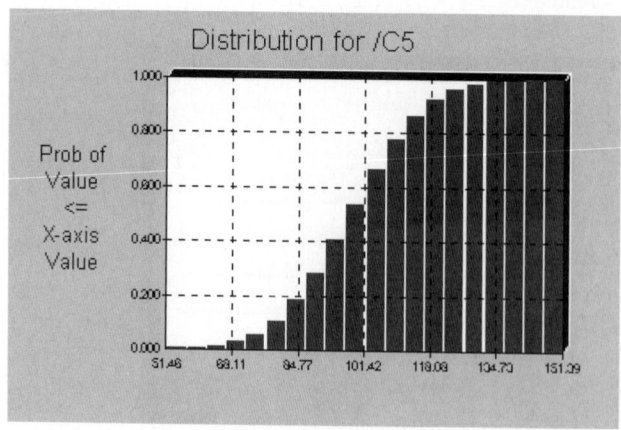

FIGURE 6
Cumulative Ascending Graph

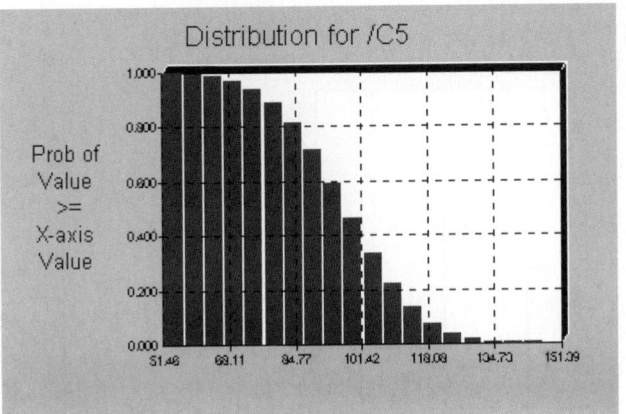

FIGURE 7
Cumulative Descending Graph

Targets

At the bottom of the Simulation Statistics window is a Target option. You may enter a Value or Percentile, and @Risk fills in the one you left out. For

$$=RISKNORMAL(100,15)$$

we obtained the following results:

```
Target #1 (Value) =      85
Target #1 (Perc%) =      15.87%
Target #2 (Value) =      130
Target #2 (Perc%) =      97.75%
Target #3 (Value) =      114.9159
Target #3 (Perc%) =      84%
```

- We entered Target#1(Value) of 85, and @Risk reported that the cell was ≤85 15.87% of the time.

- We entered Target#2(Value) of 130, and @Risk reported that the cell was ≤130 97.75% of the time.

- We entered Target#3(Perc%) of 84%, and @Risk reported that 84% of the time, the cell was ≤114.92.

Extracting Data

Sometimes you may want to see the values of @Risk functions and output cells that @Risk created on the iterations run. If so, check Collect Distribution Samples under Simulation Settings and then click on Data in the Results window. You can then Edit Copy Paste the data to your spreadsheet and subject it to further analysis.

Sensitivity

If you want a Tornado Graph, right click on the output cell and select Tornado Graph. This also requires that you check Collect Distribution Samples. You may choose either a Correlation or a Regression graph. Tornado graphs let you know which input cells have the largest influence on your output cell(s).

@Risk Functions

We now illustrate some of the most useful @Risk functions.

The RISKDISCRETE Function

This generates a discrete random variable that takes on a finite number of values with known probabilities. See Figure 8. First, enter the possible values of the random variable

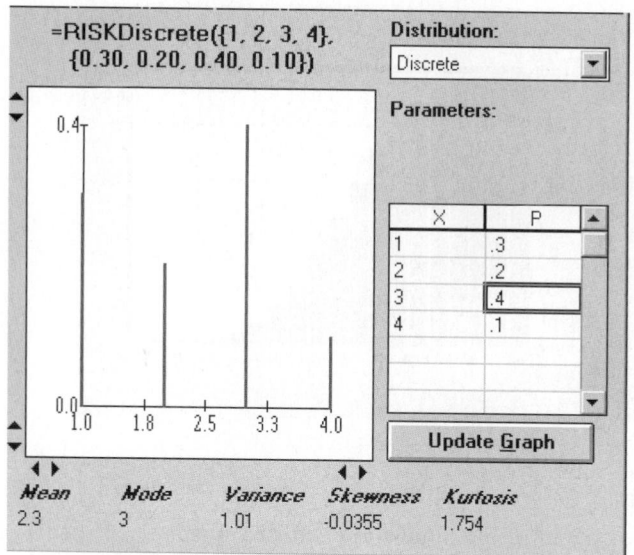

=RISKDiscrete({1, 2, 3, 4}, {0.30, 0.20, 0.40, 0.10})

Distribution:
Discrete

Parameters:

X	P
1	.3
2	.2
3	.4
4	.1

Update **G**raph

Mean	Mode	Variance	Skewness	Kurtosis
2.3	3	1.01	-0.0355	1.754

FIGURE 8

and then the probability for each value. Thus, =RISKDISCRETE({1,2,3,4},{.3,.2,.4,.1}) would generate 1 30% of the time, 2 20% of the time, 3 40% of the time, and 4 10% of the time.

If the values and probabilities were entered in A2:B5, we could have entered this random variable with formula

$$=RISKDISCRETE(A2:A5, B2:B5)$$

The RISKSIMTABLE Function

Suppose we enter

$$=RISKSIMTABLE(\{100,150,200,250,300\})$$

in cell A5, and #Iterations is 100. If we change #Simulations to 5, then on the first simulation, 100 iterations are run with 100 in cell A5. On the second simulation, 100 iterations are run with 150 in cell A5. Finally, on the fifth simulation, 100 iterations are run with 300 in cell A5. If the five arguments for the =RISKSIMTABLE function were in B1:B5, we could have also entered the =RISKSIMTABLE function as

$$=RISKSIMTABLE(B1:B5)$$

The RISKDUNIFORM Function

See Figure 9. We use the RISKDUNIFORM function when a random variable assumes several equally likely values. Thus,

$$=RISKDUNIFORM(\{1,2,3,4\})$$

is equally likely to generate 1, 2, 3, or 4. If 1, 2, 3, 4 were entered in A1:A4, then we could have entered

$$=RISKDUNIFORM(A1:A4)$$

The RISKBINOMIAL Function

See Figure 10. Use the =RISKBINOMIAL function when you have repeated independent trials, each having the same probability of success. For example, if there are 5 competi-

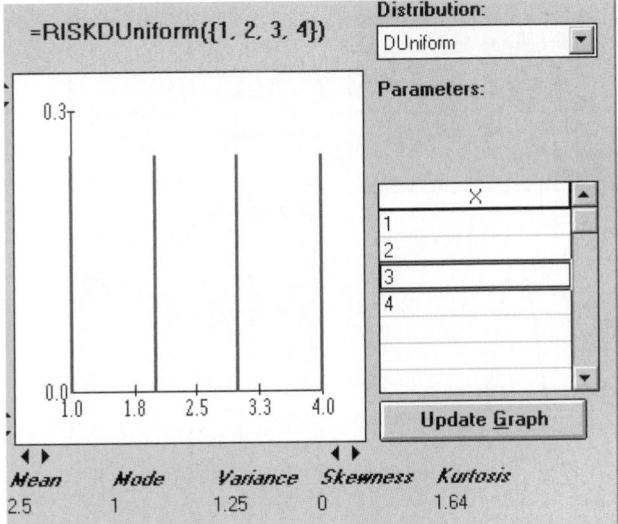

=RISKDUniform({1, 2, 3, 4})

Distribution:
DUniform

Parameters:

X
1
2
3
4

Update **G**raph

Mean	Mode	Variance	Skewness	Kurtosis
2.5	1	1.25	0	1.64

FIGURE 9

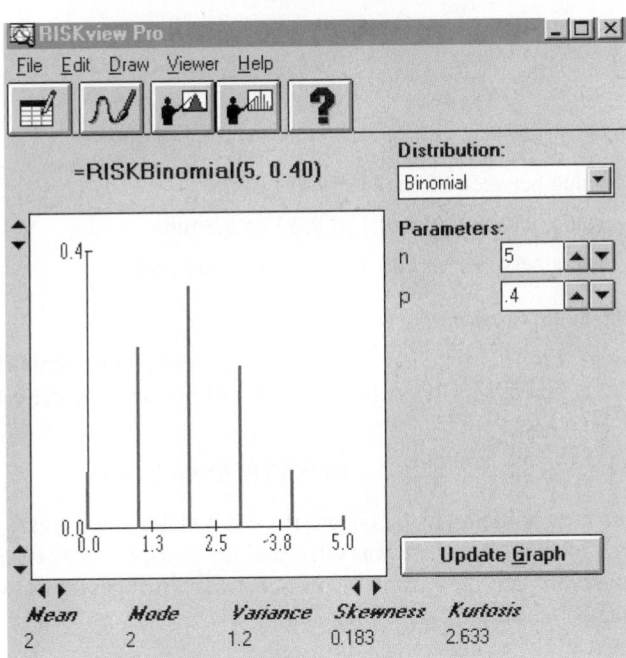

RISKview Pro

File Edit Draw Viewer Help

=RISKBinomial(5, 0.40)

Distribution:
Binomial

Parameters:

n   5

p   .4

Update **G**raph

Mean	Mode	Variance	Skewness	Kurtosis
2	2	1.2	0.183	2.633

FIGURE 10

tors who might enter an industry this year, each competitor has a 40% chance of entering, and entrants are independent, then we could model this situation with the formula

=RISKBINOMIAL(5,.4)

The RISKNORMAL Function

See Figure 11. Use this function to model a continuous, symmetric (or bell-shaped) random variable. The formula

=RISKNORMAL(100,15)

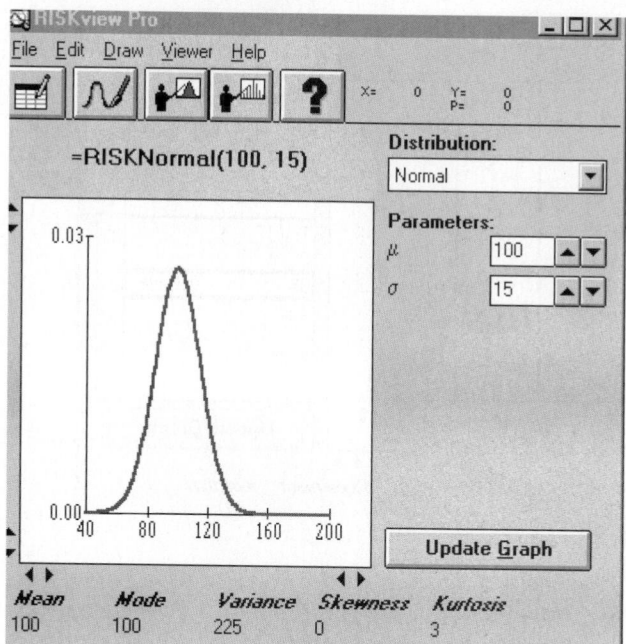

FIGURE 11

will yield

- a value between 85 and 115 68% of the time
- a value between 70 and 130 95% of the time
- a value between 55 and 145 99.7% of the time

The RISKTRIANG Function

See Figure 12. This function enables us to model a nonsymmetrical continuous random variable. It generalizes the well-known idea of best-case, worst-case, and most likely scenarios. For example,

$$=RISKTRIANG(.2,.4,.8)$$

could be used to model market share if we felt that the worst-case market share was 20%, the most likely market share was 40%, and the best-case market share was 80%. Note that the probability that the market share is between 30% and 40% would be the area under this triangle between .3 and .4. The entire triangle has an area of 1. This fact determines the height of the triangle.

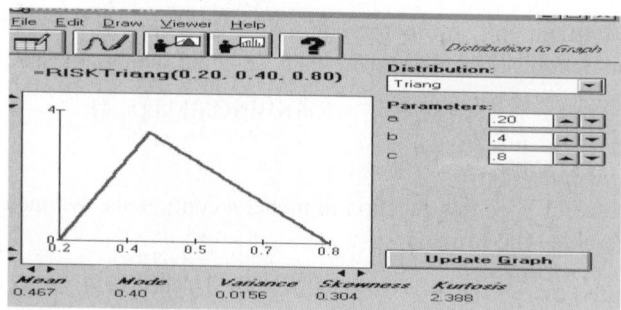

FIGURE 12

The RISKTRIGEN Function

See Figure 13. Sometimes we want to use a triangular random variable, but we are not sure of the absolute best and worst possibilities. We may believe that there is a 10% chance that market share will be less than or equal to 30%, that the most likely share is 40%, and that there is a 10% chance that share will exceed 75%. The RISKTRIGEN function is used in this situation. The formula

$$=RISKTRIGEN(.3,.4,.75,10,90)$$

would be appropriate for this situation. Then @Risk draws a triangle that yields

- A 10% chance that market share is less than or equal to 30%. This requires a worst possible market share of around 20%.
- A most likely market share of 40%.
- A 10% chance that market share is greater than or equal to 75%. This requires a best possible market share of around 95%.

Again, the probability of a market share between 20% and 50% is just the area under the triangle between 20% and 50%.

The RISKUNIFORM Function

See Figure 14. Suppose a competitor's bid is equally likely to be anywhere between 10 and 30 thousand dollars. This can be modeled by a uniform random variable with the formula

$$=RISKUNIFORM(10,30)$$

Again, this function makes any bid between 10 and 30 thousand dollars equally likely. The probability of a bid between 15 and 28 thousand would be the area of the rectangle bounded by $x = 15$ and $x = 28$. This would equal $(28 - 15)(.05) = .65$.

The RISKGENERAL Function

What if a continuous random variable does not appear to follow a normal or a triangular distribution? We can model it with the =RISKGENERAL function.

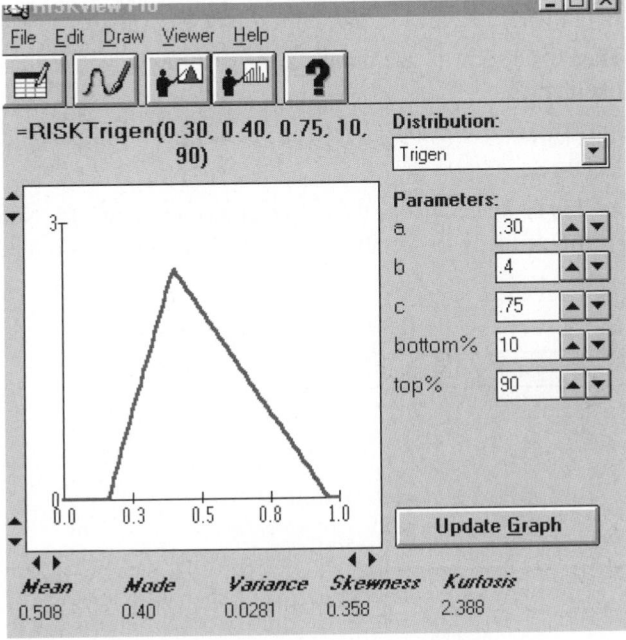

FIGURE 13

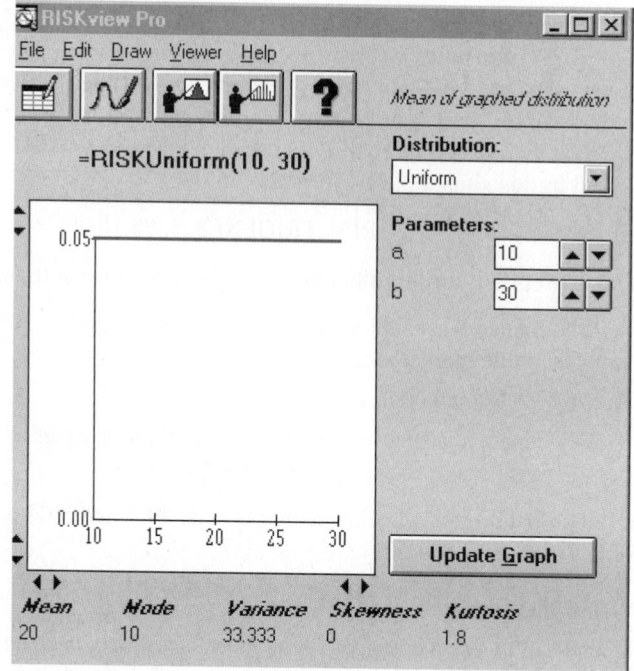

FIGURE 14

Suppose that a market share of between 0 and 60% is possible, and a 45% share is most likely. There are five market-share levels for which we feel comfortable about comparing relative likelihood. (See Table 1.) Thus, a market share of 45% is 8 times as likely as 10%; 20% and 55% are equally likely; etc. Note that this distribution cannot be triangular, because then 20% would be (20/45) as likely as peak of 45%, and 20% would be .75 as likely as 45%. To model this, enter the formula

$$=RISKGENERAL(0,60,\{10,20,45,50,55\},\{1,6,8,7,6\})$$

The syntax of RISKGENERAL is as follows:

- Begin with the smallest and largest possible values.
- Then enclose in {} the numbers for which you feel you can compare relative likelihoods.
- Finally, enclose in {} the relative likelihoods of the numbers you have previously listed.

Running this in @Risk yields the output shown in Figure 15. Note that 20 is 6/8 likely as 45; 10 is 1/8 as likely as 45; 50 is 7/8 as likely as 45; 55 is 6/8 as likely as 45; etc. In be-

TABLE 1

Market Share	Relative Likelihood
10%	1
20%	6
45%	8
50%	7
55%	6

FIGURE **15**

	A	B	C	D	E	F	G
1		**EXAMPLE OF**					
2		**RISKGENERAL**					
3		**DISTRIBUTION**					
4							
5				Minimum	0		
6				Maximum	60		
7				**Specified Points**			
8				10	1		
9				20	6		
10				45	8		
11				50	7		
12				55	6		
13		45.28889	=RISKGENERAL(0,60,{10,20,45,50,55},{1,6,8,7,6})				

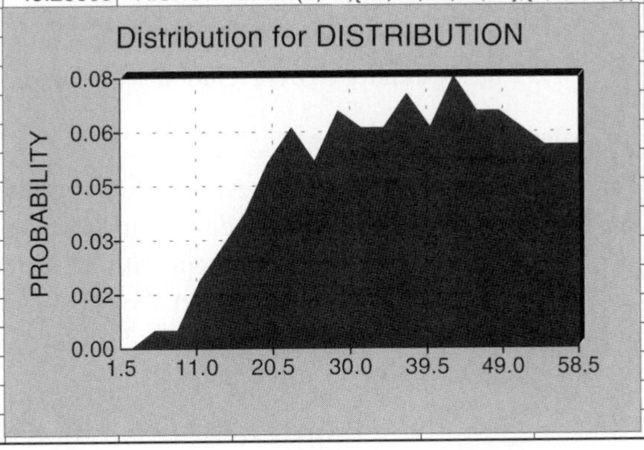

Distribution for DISTRIBUTION

FIGURE **15**

tween the given points, the density function changes at a linear rate. Thus, 30 would have a likelihood of

$$6 + \frac{(30 - 20)*(8 - 6)}{(45 - 20)} = 6.8$$

Modeling Correlations

Suppose we have three normal random variables, each having mean 0 and standard deviation 1, correlated as follows:

- Variable 1 and variable 2 have .7 correlation.
- Variable 1 and variable 3 have a .8 correlation.
- Variable 2 and variable 3 have a .75 correlation.

To model this correlation structure, we use the =RISKCORRMAT command. Simply enter your correlation matrix somewhere in the worksheet. In Figure 16, we chose C27:E29.

	B	C	D	E	F	G	H
25							
26							
27		1	0.7	0.8			
28		0.7	1	0.75			
29		0.8	0.75	1			
30							
31	1	Variable 1	1.793028	risknormal(0,1,riskcorrmat(c27:e29,1))			
32	2	Variable 2	-0.449129	risknormal(0,1,riskcorrmat(c27:e29,2))			
33	3	Variable 3	-0.521328	risknormal(0,1,riskcorrmat(c27:e29,3))			

FIGURE **16**

For each variable, type in front of the variable's actual distribution the syntax

=Actual Risk Function, RISKCORRMAT(Matrix, i)

Here, Matrix (C27:E29 in this case) indicates where the correlation matrix resides, and *i* is the column of the correlation matrix that contains the correlations for variable *i*. Thus, for variable 1, the correlations come from the first column of the correlation matrix.

If you run a simulation and extract the data for cells D31:D33, you will find that

- Each cell has a mean of around 0 and a standard deviation around 1.
- Each cell follows a normal distribution.
- D31 has around a .7 correlation with D32.
- D31 has around a .8 correlation with D33.
- D32 has around a .75 correlation with D33.

Truncating Random Variables

Suppose you believe that market share for a product is approximately normally distributed, with mean .6 and standard deviation .1. This random variable could exceed 1 or be negative, which would be inconsistent with the fact that market share must be between 0 and 1. To resolve this, you may enter the random variable from the Define Distribution icon as shown in Figure 17.

You could also type in formula

=RISKNORMAL(.6,.1,RISKTRUNCATE(0,1))

Then @Risk generates a normal random variable with mean .6 and standard deviation .1. If the random variable assumes a value between 0 (the lower truncation value) and 1 (the upper truncation value), that value is retained. Otherwise, another value is generated. The truncation values must be within 5 standard deviations of the mean.

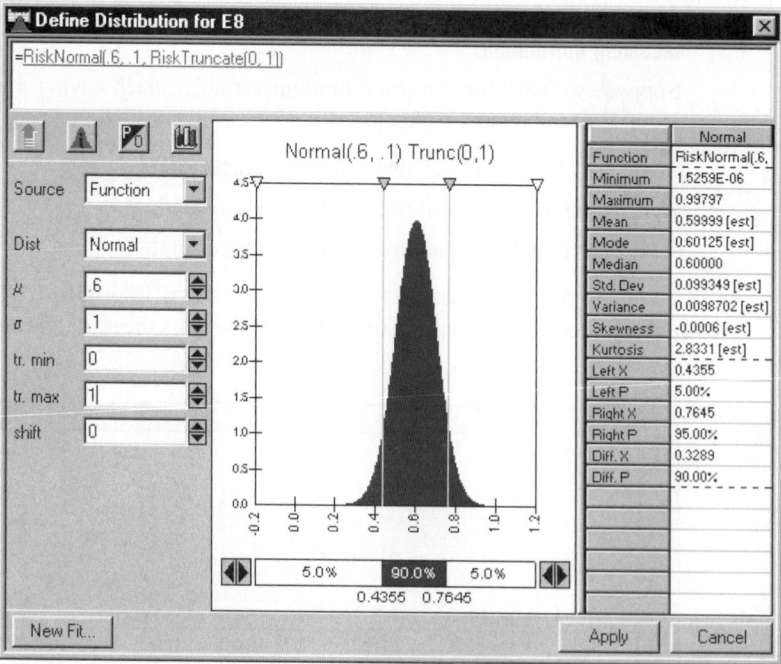

FIGURE 17

The RISKPERT Function

This function is similar to the RISKTRIANG function. The RISKPERT function is used to model the duration of projects. For example,

$$=RISKPERT(5,10,20)$$

would be used to model the duration of an activity that always takes at least 5 days, never takes more than 20 days, and is most likely to take 10 days. Whereas RISKTRIANG has a piecewise linear density function, the RISKPERT density has no linear segments. It is a special case of a Beta random variable.

Common Error Message

The error message "Invalid number of arguments" means that an incorrect syntax has been used with an @Risk function. For example, =RISKDUNIFORM({A1:A7}) may have been used instead of =RISKDUNIFORM(A1:A7).

Cases

Jeffrey B. Goldberg

UNIVERSITY OF ARIZONA

Help, I'm Not Getting Any Younger!

Profile of a university professor:

- 45-year-old formerly athletic male
- 215 pounds
- 71 inches tall
- Exercises no more than once a week. Walks 0.5 miles daily to and from the car while carrying a 10-pound briefcase.
- Family history of adult diabetes.

I need help! My diet is terrible, and I have been gaining weight and feeling more tired.

I heard that Professor George Dantzig of Stanford once used linear programming to construct a diet. It would be great if *you* could tell me what to eat during each day. So, because I'm a firm believer in mathematical models, I want you to use linear programming to determine a reasonable diet for me to eat during a week. It is your job to collect data for use in the model.

I have the following requirements for the diet:

- I like variety. You cannot prescribe a diet in which I eat just one food during the entire week (like 10 boxes of Total cereal). I would like to eat at least 15 different foods during the week.
- You have to give me something from each of the four basic food groups (dairy, fruit and vegetable, meat, and grains)—not Mcfood, frozen food, pizza food, or food on a stick.
- I like nutrition. You cannot prescribe a diet that does not meet minimum daily requirements for essential minerals and vitamins. You cannot prescribe a diet in which I gain a lot of weight. I could stand to lose a few pounds.
- I hate Brussels sprouts, sweet potatoes, pears, and organ meats such as liver and kidney.
- Forget about any canned fruits or vegetables. Yuck.
- I do not eat any pork or pork products.
- I am not a big fan of frozen dinners, no matter how nutritious or convenient they are.
- I don't drink milk with any meal except breakfast.
- I work for the university, so I have a limited budget for food. Try to keep costs less than $100 per week (the lower the better).

- I might consider taking vitamin pills to get nutritional requirements, but I would rather eat food.

Key Questions

- What should I eat at each meal?
- If I allowed less variety, would your recommendation change?
- If I allowed more than $100 per week, would your recommendation change? How?
- What key minerals and vitamins constrain the solution?

Solar Energy for Your Home

As our ability to extract and process fossil fuels decreases, many people are looking to renewable resources to meet their energy needs. In particular, solar energy is becoming an advanced technology that has economic promise. In areas with large solar insolations, there can be enough energy to power an entire home. The amount of solar energy reaching the earth each year is many times greater than worldwide energy demand; it varies, of course, with location, time of day, and season. Sunlight is also a widespread resource and can be captured from virtually anywhere on earth.

There are two categories of home solar systems: passive and active. In a passive system, the solar energy heats a material that is used in a productive manner. For example, in Arizona it is common to use a passive solar system to heat swimming pools and the water used in the home. Every building has some of its heating requirements met by solar energy. Sunlight passing through windows is a source of heat, and the value of passive solar heating is enhanced by proper building insulation. A well-insulated building requires less energy for heating; thus, much of the heating load can be met by passive solar features. Optimum passive solar design begins with the layout of a building lot; a house must be oriented so that it can take full advantage of available solar energy.

Active systems are more complex and generally involve converting the solar energy to electrical energy. Photovoltaic (PV) cells use the energy of the sun to produce electricity. They produce none of the greenhouse or acid gas emissions that are commonly associated with the use of fossil fuels to generate electricity. The main barrier to increased use of this technology is cost. A common semiconducting material used in PV cells is single crystal silicon. Single crystal silicon cells are generally the most efficient type of PV cells, converting as much as 23% of incoming solar energy into electricity. The main problem with them is their production cost. Polycrystalline silicon cells are less expensive to manufacture but less efficient than single crystal cells (15% to 17%). Thin films (0.001–0.002 mm thick) of amorphous or uncrystallized silicon are another PV alternative. These thin films are inexpensive and may be easily deposited on materials such as glass and metal, thus lending themselves to mass production. Amorphous silicon thin-film PV cells are widely used in commercial electronics, powering watches and calculators. These cells, however, are not especially efficient—12% in the lab, 7% for commercial cells—and they degrade with time, losing as much as 50% of their efficiency with exposure to sunlight.

Solar power is an intermittent source of electricity. If PV cells are your only source, then the storage of electricity may be necessary. Electricity for a home can be stored in batteries, which can be expensive. Also, to generate sufficient electricity, you need a large area of collectors on your roof or somewhere on your property. The amount of solar energy captured depends on the surface area of the collectors and their conversion efficiency.

A solar energy system can often be looked at as a conservation system. Figure 1 depicts one way to look at the daily flow of energy.

Your job is to design an active solar system for a home in your area. For the analysis, you will have to collect data on:

- system cost and efficiency,
- daily solar insolation in your area (usually measured in watts/meter^2; this information can be found locally where weather data are stored and collected), and
- typical daily power requirements for a home in your area.

The costs of the system generally include a fixed component and variable components that depend on the total area of the PV collectors, the type of material used in the collector (usually only material is chosen), and the amount of battery storage needed. Your analysis should cover at least 6 months of data (12 months would be better, because you would like your design to be appropriate for the entire year). You should assume that *all* energy requirements for the home will

FIGURE 1
Daily Energy Flow

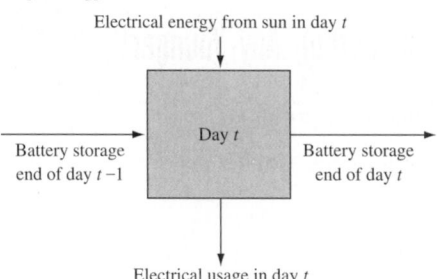

be met by this system (no natural gas will be used for heating or cooking, for example).

Your design should include the following:

- the area of the PV collectors and the amount of battery storage that you need,
- an estimate of the cost of the system (you may include any tax advantages that accrue from the purchase of solar energy systems),
- a profile of the battery storage levels at the end of each day for a six-month period,
- an estimate of cost savings (or loss) over buying your electrical power from the local utility company.

CASE 3

Golf-Sport: Managing Operations

Golf-Sport is a small-sized company that produces high-quality components for people who build their own golf clubs and prebuilt sets of clubs. There are five components—steel shafts, graphite shafts, forged iron heads, metal wood heads, and metal wood heads with titanium inserts—made in three plants—Chandler, Glendale, and Tucson—in the Golf-Sport system. Each plant can produce any of the components, although each plant has a different set of individual constraints and unit costs. These constraints cover labor and packaging machine time (the machine is used by all components); the specific values for each component–plant combination are given in Tables 1–3. Note that even though the components are identical in the three plants, different production processes are used, and therefore the products use different amounts of resources in different plants.

Besides component sales, the company takes the components and manufactures sets of golf clubs. Each set requires 13 shafts, 10 iron heads, and 3 wood

TABLE 1
Product-Resource Constraints: Chandler

Products	Resources		
	Labor (Minutes/Unit)	Packing (Minutes/Unit)	Advertising ($/Unit)
Steel shafts	1	4	1.0
Graphite shafts	1.5	4	1.5
Forged iron heads	1.5	5	1.1
Metal wood heads	3	6	1.5
Titanium insert heads	4	6	1.9
Monthly availability (minutes)	12,000	20,000	—

TABLE 2
Product-Resource Constraints: Glendale

Products	Resources		
	Labor (Minutes/Unit)	Packing (Minutes/Unit)	Advertising ($/Unit)
Steel shafts	3.5	7	1.1
Graphite shafts	3.5	7	1.1
Forged iron heads	4.5	8	1.1
Metal wood heads	4.5	9	1.2
Titanium insert heads	5.0	7	1.9
Monthly availability (minutes)	15,000	40,000	—

TABLE 3
Product-Resource Constraints: Tucson

Products	Resources		
	Labor (Minutes/Unit)	Packing (Minutes/Unit)	Advertising ($/Unit)
Steel shafts	3	7.5	1.3
Graphite shafts	3.5	7.5	1.3
Forged iron heads	4	8.5	1.3
Metal wood heads	4.5	9.5	1.3
Titanium insert heads	5.5	8.0	1.9
Monthly availability (minutes)	22,000	35,000	—

heads. All of the shafts in a set must be the same type (steel or graphite), and all of the wood heads must be the same type (metal or metal with inserts). Assembly times for the sets at each plant are shown in Table 4.

Each plant of Golf-Sport has a retail outlet to sell components and sets, and the specific plant is the only supplier for its retail outlet. The minimum and maximum amounts of demand for each plant–product pair are given in Table 5. Note that, although the minimums must be satisfied, you do not need to satisfy demand up to the maximum amount.

This planning problem is for two months. The costs in Table 6 increase by 12% for the second month, and production times are stationary. Inventory costs are based on end-of-period inventory for each product set and cost out at 8% of the cost values in Table 6. Table 7 lists the revenue generated by each product. Initially, there is no inventory.

The corporation controls the capital available for expenses; the cash requirements for each product are given in the last column of Tables 1–3. There is a total of $20,000 available for advertising for the entire system during each month, and any money not spent in a month is not available the next month. The corporation also controls graphite. Each shaft requires 4 ounces of graphite; a total of 1,000 pounds is available for each of the two months.

Your job is to determine a recommendation for the company. A recommendation must include a plan for production and sales. In addition, you should also address the following sensitivity-analysis issues in your recommendation:

- If you could get more graphite or advertising cash, how much would you like, how would you use it, and what would you be willing to pay?

- At what site(s) would you like to add extra packing machine hours, assembly hours, and/or extra labor hours? How much would you be willing to pay per hour and how many extra hours would you like?

- Marketing is trying to get Golf-Sport to consider an advertising program that promises a 50% increase in their maximum demand. Can we handle this with the current system or do we need more resources? How much more is the production going to cost if we take on the additional demand?

TABLE 4

Plant	Time (Minutes per set)	Total Time Available (Minutes)
Chandler	65	5,500
Glendale	60	5,000
Tucson	65	6,000

TABLE 6
Material, Production, and Assembly Costs ($) per Part or Set

Products	Plants		
	Chandler	Glendale	Tucson
Steel shafts	6	5	7
Graphite shafts	19	18	20
Forged iron heads	4	5	5
Metal wood heads	10	11	12
Titanium insert heads	26	24	27
Set: Steel, metal	178	175	180
Set: Steel, insert	228	220	240
Set: Graphite, metal	350	360	370
Set: Graphite, insert	420	435	450

TABLE 5
Minimum and Maximum Product Demand per Month

Products	Store (or Plant)		
	Chandler	Glendale	Tucson
Steel shafts	[0, 2,000]	[0, 2,000]	[0, 2,000]
Graphite shafts	[100, 2,000]	[100, 2,000]	[50, 2,000]
Forged iron heads	[200, 2,000]	[200, 2,000]	[100, 2,000]
Metal wood heads	[30, 2,000]	[30, 2,000]	[15, 2,000]
Titanium insert heads	[100, 2,000]	[100, 2,000]	[100, 2,000]
Set: Steel, metal	[0, 200]	[0, 200]	[0, 200]
Set: Steel, insert	[0, 100]	[0, 100]	[0, 100]
Set: Graphite, metal	[0, 300]	[0, 300]	[0, 300]
Set: Graphite, insert	[0, 400]	[0, 400]	[0, 400]

TABLE 7
Revenue per Part or Set ($)

Products	Plants		
	Chandler	Glendale	Tucson
Steel shafts	10	10	12
Graphite shafts	25	25	30
Forged iron heads	8	8	10
Metal wood heads	18	18	22
Titanium insert heads	40	40	45
Set: Steel, metal	290	290	310
Set: Steel, insert	380	380	420
Set: Graphite, metal	560	560	640
Set: Graphite, insert	650	650	720

CASE 4

Vision Corporation: Production Planning and Shipping

Vision is a large company that produces video-capturing devices for military applications such as missiles, long-range cameras, and aerial drones. Four different types of cameras (differing mainly by lens type) are made in the three plants in the system. Each plant can produce any of the four camera types, although each plant has its own individual constraints and unit costs. These constraints cover labor and machining restrictions, and the specific values are given in Tables 8–10. Note that even though the products are identical in the three plants, different production processes are used and thus the products use different amounts of resources in different plants. The corporation controls the material that goes into the lenses; the material requirements for each product are given in the last column of Tables 8–10. A total of 3,500

TABLE 8
Product-Resource Constraints: Plant 1

Products	Resources		
	Labor (Hours/Unit)	Machine (Hours/Unit)	Material (Lb./Unit)
Small	3	8	1.0
Medium	3	8.5	1.1
Large	4	9	1.2
Precision	4	9	1.3
Total available	6,000	10,000	—

TABLE 9
Product-Resource Constraints: Plant 2

Products	Resources		
	Labor (Hours/Unit)	Machine (Hours/Unit)	Material (Lb./Unit)
Small	3.5	7	1.1
Medium	3.5	7	1.0
Large	4.5	8	1.1
Precision	4.5	9	1.4
Total available	5,000	12,500	—

TABLE 10
Product-Resource Constraints: Plant 3

Products	Resources		
	Labor (Hours/Unit)	Machine (Hours/Unit)	Material (Lb./Unit)
Small	3	7.5	1.1
Medium	3.5	7.5	1.1
Large	4	8.5	1.3
Precision	4.5	8.5	1.3
Total available	3,000	6,000	—

pounds of material is available for the entire system during the planning period.

Transport has 3 major customers (RAYco, HONco, and MMco) for its products. The maximum sales for each customer–product pair is given in Table 11. Product sales prices are given in Table 12, and the shipping costs from each plant to each customer are detailed in Table 13. Table 14 contains the production costs for each product–plant pair.

All shipping from plants 1 and 2 that goes to RAYco or HONco must go through a special inspection. These units are sent to a central site, inspected, and then sent to their destination. The capacity of this special inspection site is 1,500 pieces.

Your job is to determine a recommendation for the company. A recommendation must include a plan for production and shipping as well as the cost and revenue generated from each plant. In addition, you should address the following potential issues in your recommendation:

- If you could get more material, how much would you like? How would you use it? What would you be willing to pay?

- If you could get more inspection capacity, how much would you like? How would you use it? What would you be willing to pay?

TABLE 11

Maximum Product Sales ($) per Unit

Products	Customers		
	RAYco	HONco	MMco
Small	200	400	200
Medium	300	300	400
Large	500	200	300
Precision	200	400	300

TABLE 12

Product Sales Price ($) per Unit

Products	Customers		
	RAYco	HONco	MMco
Small	17	16	16
Medium	18	18	17
Large	22	22	23
Precision	29	26	27

TABLE 13

Shipping Costs ($) per Unit

Plant	Customers		
	RAYco	HONco	MMco
1	1.0	1.6	1.1
2	1.2	1.5	1.0
3	1.4	1.5	1.3

TABLE 14

Production Costs ($) per Unit

Products	Plant		
	1	2	3
Small	14	13	14
Medium	16	17	15
Large	18	20	19
Precision	26	24	23

- At what plant(s) would you like to add extra machine hours? How much would you be willing to pay per hour? How many extra hours would you like?

- Marketing is trying to get RAYco to consider a 50% increase in its demand. Can we handle this with the current system or do we need more resources? How much more money can we make if we take on the additional demand?

Material Handling in a General Mail-Handling Facility[†]

For more than 200 years, the United States Postal Service (USPS) has delivered mail across the country. Daily delivery goes to some 137 million households; in 2001, the USPS processed and delivered more than 207 billion pieces of mail to a delivery network that grew by 1.7 million new addresses. Clearly, the USPS is the largest material handler (in terms of pieces) in the world. Statistics recorded by Pricewaterhouse Coopers show that 94% of first-class mail destined for next-day delivery received overnight service—and this was a record performance for a second straight year. Despite the high volume, the USPS managed to cut costs by $900 million in 2001 while maintaining record service performance and high levels of customer satisfaction.

To process mail quickly, one must use advanced mechanization. Mail-sorting machines can process 10 letters per second (we are long past the days of hand sorting in front of a large set of post boxes). Sorting using the zip+4 standard can result in a mail sort down to an individual carrier's walk sequence, which saves significant carrier time.

Five major operations can be performed on each letter, and each operation has its own machine:

Automatic facer and canceller (AFC) This machine cancels the stamp and orients all of the letters so that the stamp is in the upper-right corner. This machine also separates mail into one of three streams—automation, mechanization, or manual.

Letter sorting machine (LSM) This machine is semi-automated and helps human operators sort mail. The operator reads the address and then types in a destination code. The machine then routes the letter to the appropriate bin.

Optical character reader (OCR) This machine reads handwritten or typed addresses and then prints a machine-readable barcode on the envelope.

Barcode sorter (BCS) This machine reads the barcode on the letter (either printed by the OCR or by the sender's equipment) and then sorts it to a bin.

[†]Based on work done jointly with Ron Askin and Sanjay Jagdale, 1994.

Delivery barcode sorter (DBCS) This machine does a two-pass sort that uses barcodes with the zip+4 code and sorts down almost to a walk sequence. When using the DBCS, the result is such that a carrier requires little to no processing at the carrier station to deliver the mail.

Items known as *flats* (for example, magazines and 8.5 × 11–inch envelopes) also are processed through the system. A flat-sorting machine (FSM) is used in a semiautomated process. An operator loads a piece onto the machine and keys in a code based on the flat's address; the machine then routes the piece to an appropriate bin.

A letter that enters a general mail facility (GMF) follows a routing that depends on the machine readability of the address and the presence of an existing barcode. Although many routes are possible, the major ones are given in Figure 2. Once letters go through the AFC, they are stored in cardboard or plastic trays that hold approximately 400 letters. The letters are moved in these trays throughout the facility, and each machine sorts the mail into different trays.

As part of quality improvement, the post office is always looking for ways to cut costs while expediting mail processing. To meet goals for overnight delivery and three-day cross-country delivery, letters that arrive by 6 p.m. from box pickup must be processed that night and be on planes or trucks for the next destination. Because the sorting and character-reading machines cost millions of dollars each, increasing machine use and saving the purchase of even one machine in a GMF is a significant achievement.

One of the keys to faster processing and increasing utilization is an effective material-handling and data system. Each tray has a barcode that describes the salient characteristics of its mail. When a sorting machine is ready for operation (say, for example, that we are going to sort down to a group of 10 zip codes), a call goes out to bring all trays with appropriate mail to the appropriate BCS machine. The data system must (1) know where those trays are located, (2) go and get them, (3) bring the trays to the machine-input area, and (4) exit the area. The faster this can be done, the better.

The network for our GMF is given in Figure 3. Each machine has an input and output point. For example, nodes 1 and 28 are the inputs and output, respectively, for AFC 1. For this application, the material-handling system is an overhead monorail. Carriers that hold one tray circulate around the system to pick up and deliver trays; they rest in the parking lot when not in use. The arcs in the diagram are the links of the monorail; all links are one-way. The dotted lines on Figure 3 represent links that are above the machine level and offer shortcuts across the facility. The facility also contains switches (nodes 29, 18, and 32) that allow carriers to change directions. Node 34 is the link to the shipping dock, and all trays enter and exit the system at this point.

The carriers travel at approximately 1 mile per hour, and there must be 15 feet between carriers on the same link. For the purpose of this study, assume that there is a bypass at each node so a carrier can pass other carriers that are stopped for loading and unloading operations. Also assume that the switches operate quickly relative to the speed of the vehicles so that collisions do not occur and the switch capacity is not constraining. Figure 3 is drawn approximately to scale. The facility is approximately 220 feet long by 160 feet wide. At 1 mile per hour, it takes a carrier approximately 2.5 minutes to run the length of the facility (from node 14 to node 1, for example).

Table 15 contains the tray movement loads for the peak hour. Each load has an origin node, a destination node, and the number of trays that must be moved. Each load is a leg in the route for a particular tray of mail. The system must have capacity to move an empty carrier from the parking lot to the origin, load the tray, move the tray to the destination, unload the tray, and then return to the parking lot. All carriers are dispatched from the parking lot because this simplifies the logic of the scheduler. By capacity, there must be a sufficient number of vehicles and the capacity on each link between nodes cannot be overloaded. As-

FIGURE 2

Processing Routes for Letters

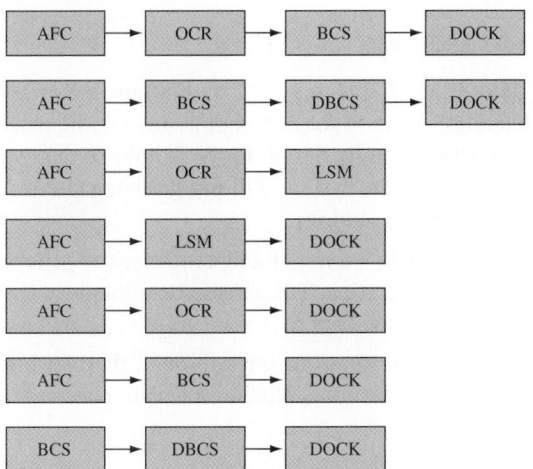

FIGURE **3**

General Mail Facility: Track Layout

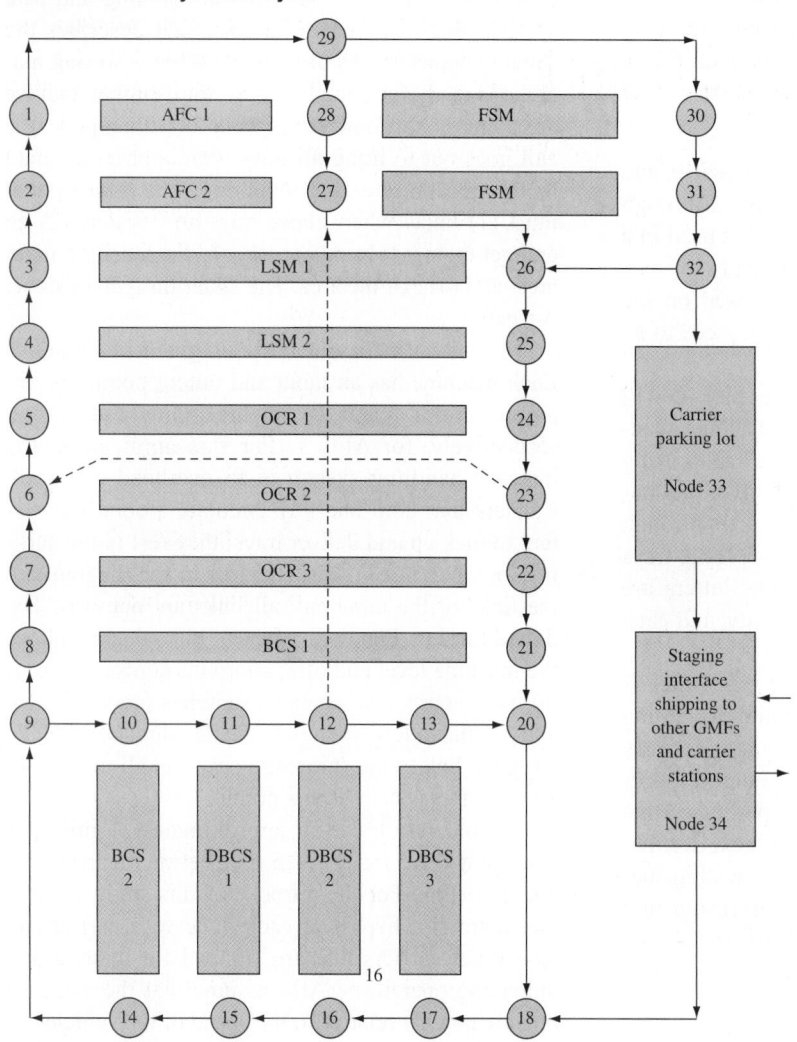

TABLE 15

Load Data for the Peak Hour

Load Number	Origin Node	Destination Node	Number of Trays
1	33	1	15
2	28	4	20
3	22	14	30
4	10	33	30
5	24	8	15
6	21	33	30
7	24	4	5
8	25	33	15
9	27	17	15
10	13	33	40

sume that it takes one minute to perform a loading or unloading operation.

Your jobs are to:

- Determine if this system has enough material-handling capacity for moving the trays in the peak hour (we generally design for peak hour so that we are sure that the system will not get bogged down when demand is high).

- Suggest where we might add extra track to relieve capacity congestion. This should be minimized because track cost is high.

- Determine the flows of trays through the network during the peak hour. Which routes are chosen for each load?

- Determine places in the network that are risky—that is, if a link goes down, machines can be cut off from the material-handling system.

- Estimate the total carrier travel distance during the peak hour.

- Investigate the effect of reducing the intercarrier spacing. You need space between carriers to prevent carrier collisions. If we put better sensors on the front of the carriers, they can stop more quickly and we can have less spacing.

Some tips:

- This is a difficult problem. Be patient and try not to become discouraged.

- Do not forget the empty carrier movements to the origin and from the destination.

- Compute the capacity on each arc. Initially, assume a single lane. To increase capacity, consider multiple lanes between nodes or consider adding arcs to give more paths between origin nodes and destination nodes.

- A precise formulation of this problem can be larger than most problems you have seen. You should not undertake to solve a large-sized formulation unless you have software that can handle large problems. Some approximate formulations are more manageable, but they still can require hundreds of variables and constraints.

CASE 6

Selecting Corporate Training Programs[†]

Introduction

Training has become a major cost of doing business. A 1995 survey of all U.S. businesses with 100 or more employees revealed that approximately $52 billion was being spent on training; it has been estimated that $90 to $100 billion is being spent for training overall. Developing strategies to implement cross-training is a current topic in the operations research (OR) literature. Management consultants advocate aggressive education and professional development to remain competitive in the global and local markets. Employees now expect job and skill growth to be a major component of their duties.

[†]Based on work done jointly with John V. Farr, and David A. Thomas at USMA, 1995.

Increased training costs have occurred for many reasons. Employees view training in the form of formal degrees and documented technical skills as important for job security. Technology is changing at a rapid pace. It has been claimed that high schools and universities are not producing the skills needed by industry, so industry must train and reeducate recent graduates. For high school graduates, this may include training in technology-based skills; for college graduates, this may include developing nontechnical skills such as leadership, communications, interpersonal relations, and ethics.

Problem Environment

For a corporation, the primary purpose of training is to ensure that employees have the key skills needed to effectively manage and operate the business. There are many options for providing training. For example, to train staff members in computer skills, a corporation may use any of the following strategies:

- hiring an outside consultant to develop and present an on-site training course,

- using corporate personnel to develop and present an on-site training course,

- purchasing a training course and having employees use it for self-study,

- contracting with a local college or university to provide training, or

- sending employees to an off-site training seminar.

The above possibilities are for a single skill. The purpose of many training programs, however, is to give employees a broad set of skills. Often the skill sets of two or more programs partially overlap. When this happens, the corporation must choose the set of programs that give employees the required skills for their jobs and the appropriate employees for each training program. In any case, training decisions made in an ad-hoc "pay-as-you-go" manner will be inefficient and generally result in additional expense.

To give the decision problem structure, the following assumptions are appropriate:

- We have a known study period—for example, the next 3 or the next 5 years—over which we need to plan training. The study period should fit with the overall business strategy and enable accurate estimates for training needs and available resources.

- There is a known set of skills that employees need. Among others, these may include technical, interpersonal, communication, and management skills.

TABLE 16
Skills List

No.	Skill	No.	Skill
1	New employee orientation	22	Stress management
2	Performance appraisals	23	Computer programming
3	Personal computer apps	24	Diversity
4	Leadership	25	Data processing/MIS
5	Sexual harassment	26	Planning
6	Team building	27	Public speaking and presentation
7	Safety	28	Strategic planning
8	Hiring and selection process	29	Writing skills
9	New equipment operation	30	Negotiating skills
10	Training the trainer	31	Finance
11	Product knowledge	32	Marketing
12	Decision making	33	Substance abuse
13	Listening skills	34	Ethics
14	Time management	35	Outplacement and retirement
15	Conducting meetings	36	Creativity
16	Quality imiprovements	37	Purchasing
17	Delegation skills	38	Smoking cessation
18	Problem solving	39	Financial and business literacy
19	Goal setting	40	Reengineering
20	Managing change	41	Foreign language
21	Motivation		

TABLE 17
Salary and Skills Required for Each Job Classification

Person	Salary ($)	Skills 1–20
Senior Manager	250,000	0 1 0 1 0 0 0 0 0 0 0 1 1 1 0 1 0 1 1 0
Project Manager	200,000	0 1 0 0 1 0 0 0 0 0 1 1 1 1 1 0 0 1 0 0
Professional	150,000	1 1 0 0 1 1 1 0 1 0 0 0 1 0 1 0 0 0 0 1
Sales	150,000	1 1 0 0 0 0 0 0 1 1 0 0 1 0 0 0 1 0 0 1
Technician	100,000	1 1 1 0 1 0 0 1 0 0 0 0 0 0 0 0 0 0 0 1
Administrative Assistant	80,000	1 1 1 0 1 0 0 0 0 0 0 0 0 0 0 0 0 0 0 0

Person	Skills 21–41
Senior Manager	1 1 1 1 0 1 1 0 1 0 1 1 0 1 0 1 1 1 0 0 0
Project Manager	0 1 1 1 0 1 1 1 1 0 1 0 1 0 0 1 1 1 1 0 1
Professional	0 0 0 0 1 1 0 0 0 1 0 0 1 0 1 0 0 1 1 0 1
Sales	0 0 0 0 1 1 0 0 0 1 0 0 1 0 1 0 0 1 1 1 1
Technician	0 0 0 0 0 0 0 0 0 0 0 0 0 0 0 0 0 1 1 1 1
Administrative Assistant	0 0 0 0 0 0 0 0 1 0 0 0 0 0 0 0 1 1 1 0

TABLE 18

Enrollment Cost and Skills of Each Program

Program	Enrollment Cost ($)	Skills 1–20
Program 1	500	1 1 0 1 0 0 0 0 0 0 1 1 1 1 1 0 1 0 0 0
Program 2	300	0 0 0 0 1 1 1 1 0 0 0 0 0 0 0 0 0 0 0 0
Program 3	500	0 0 0 0 0 1 0 0 1 0 0 0 0 0 0 1 0 0 0 1
Program 4	575	0 1 0 1 0 0 0 0 1 1 0 0 1 0 0 0 1 0 0 0
Program 5	800	0 1 1 1 0 0 0 0 0 1 0 0 1 0 0 0 1 0 0 0
Program 6	400	0 0 0 0 0 0 0 1 1 0 0 0 0 0 0 0 0 0 0 1
Program 7	200	0 0 0 0 1 1 0 0 0 0 0 0 1 0 0 0 0 0 0 1
Program 8	1,000	0 0 0 0 0 0 0 0 0 0 1 1 1 1 1 1 0 0 1 0
Program 9	200	1 1 1 1 0 0 0 0 0 0 0 0 0 0 1 1 0 0 0 0
Program 10	500	0 0 0 1 0 0 0 0 0 0 0 1 0 0 0 0 0 0 0 0
Program 11	700	0 0 0 0 0 0 0 0 0 1 0 1 0 0 0 0 0 0 0 0
Program 12	600	0 0 0 0 0 0 0 0 0 0 1 0 0 1 0 0 0 0 0 0
Program 13	400	0 0 0 0 0 0 0 0 0 0 0 1 1 0 0 0 1 1 1 0
Program 14	900	1 0 0 0 1 1 0 0 0 0 0 0 0 0 0 1 0 1 0 0
Program 15	700	1 0 1 0 1 0 1 1 1 0 0 0 0 0 0 0 0 0 1 0

Program	Skills 21–41
Program 1	1 0 0 0 0 0 0 0 0 0 0 1 0 0 0 0 0 0 0 0 0
Program 2	0 0 0 0 0 1 0 0 0 0 0 0 0 0 0 0 0 0 1 1 0
Program 3	1 1 1 1 1 1 1 0 1 0 1 0 0 0 0 1 0 0 0 0 0
Program 4	0 0 0 0 0 0 0 1 0 0 0 0 0 0 0 1 0 0 0 0 1
Program 5	0 0 0 1 0 1 0 0 0 0 0 1 0 0 0 0 0 1 1 0
Program 6	0 0 0 0 0 0 0 0 1 0 0 1 1 0 0 1 1 0 0 0 0
Program 7	0 1 0 0 0 0 0 0 0 0 0 1 0 1 0 0 0 0 0 0 0
Program 8	1 1 1 1 0 0 1 0 1 1 0 0 0 0 0 0 1 0 0 0 0
Program 9	0 0 0 1 1 1 0 0 0 0 0 0 0 0 0 0 0 0 1 1 0
Program 10	0 0 0 0 0 0 0 0 1 1 0 0 0 0 0 1 1 0 0 0
Program 11	0 1 1 0 0 0 1 0 1 1 0 1 0 0 1 1 0 0 0 0 0
Program 12	0 0 0 0 0 0 0 1 0 1 1 0 0 0 0 0 0 1 1 1 0
Program 13	1 0 1 0 1 0 0 0 0 0 1 0 1 0 0 0 0 0 0 0 0
Program 14	0 0 0 0 0 0 0 0 0 0 0 0 0 0 1 1 0 0 0 0 0 1
Program 15	0 0 0 0 0 0 1 0 0 0 0 0 0 0 0 0 1 0 0 1

- Employees are divided into classes. In each class, we have estimates for (1) the number of employees, (2) the employee hourly wage, (3) the number of employees that require each particular skill, and (4) the maximum time available for training employees in each class during the study period.

- There is a list of training programs. For each program, we assume we have (1) the set of skills taught, (2) the cost, (3) the development time, (4) the completion time for an employee, and (5) the maximum number of employees who can participate per decision cycle.

- Training is equally effective for all people, thus we are concerned with which programs to offer and which employees in each class to assign to each program. If we know the quality of the training for individual skills for individual classes, then we can relax this assumption.

Potential Corporate Setting

Your job is to develop models to aid businesses and corporations in determining the appropriate training programs to use. The type of model and issues often depend on the size of the corporation and the potential

TABLE 19
Interfering Programs

Program Number (Days Long)	Programs that Interfere			
Program 1 (2)	3	5	8	
Program 2 (2)	3	7	10	
Program 3 (4)	1	2	12	
Program 4 (3)	6	7	14	
Program 5 (2)	1	9	12	
Program 6 (3)	4	7	11	14
Program 7 (5)	2	4	6	
Program 8 (2)	1	10	13	
Program 9 (3)	5	15		
Program 10 (3)	2	8		
Program 11 (2)	6	12	15	
Program 12 (4)	3	5	11	
Program 13 (3)	8	14		
Program 14 (4)	4	6	13	
Program 15 (3)	9	11		

uses of the models. For large corporations, there are many employees in each class, so it is not necessary to model and schedule down to the individual employee. Concentrate instead on the assignment of classifications to programs and ignore the assignment of specific individuals to programs. Also, sufficient resources exist to develop internal training programs, hence you should consider program development costs as well as employee costs (lost work time, travel, lodging, meals, course materials) in the objective. A large corporation can use the model to plan the development of courses. This will help determine (1) program-development costs so that in-house programs are cost-effective and (2) appropriate programs for each employee classification so that, on average, there is sufficient time to complete the assigned programs within the available time.

For small businesses, the focus is often different. Typically, these companies do not develop in-house programs because they do not train enough employees to justify development costs. Because the number of employees is small, it is important to model down to the employee level and schedule employees so that both training and job tasks can be completed.

Your OR consulting firm has been hired to design the training program for a small company. There are no in-house classes, and vendors provide all training. The company has determined 41 skills that are important for its employees; these are listed in Table 16. There are six employees; the salary level and skills required for each person are given in Table 17. You can assume that there are 250 working days per year. There

are 15 programs available for use; Table 18 contains the cost per person and the skills covered for each program. In Table 18, a 1 in the row for program p and the column for skill s implies that program p contains skill s. Table 19 lists the programs that conflict in time with other programs (for example, programs 3, 5, and 8 conflict with program 1). An employee cannot take two programs simultaneously. It is company policy that each employee is limited to 15 days for training per year.

Key Questions

Your job is to develop a recommendation for the company for addressing its training needs. In particular, you should address the following key questions:

- Which training programs should we be using? What is the assignment of personnel to those programs?
- Identify programs with heavy use that may justify the development of an in-house course. How much would you be willing to pay for that development if you could use the program for the next three years?
- We have the opportunity to negotiate prices for programs. Which programs would you suggest are candidates for negotiation?
- What skills are especially expensive for us to cover? If we were to develop our own programs, what skills should be covered?
- Would your recommendation change if we allowed more days of training per year?

CASE 7
BestChip: Expansion Strategy

BestChip (BC) is a large nationwide corporation that produces low-fat snack products for an expanding market (pun intended). Basically, BC takes materials (corn, wheat, and potatoes) and turns them into two types of snacks: chips (regular and green onion) and party mix (one variety). BC is expanding into the western United States and is considering sites for locating production facilities.

BC currently has eight candidate sites. Table 20 shows the sites' purchase prices and the purchase and shipping cost per ton of each material to each site.

The purchase cost represents the yearly amortized cost of opening and operating the site (exclusive of

TABLE 20
Site Information and Material Shipping Cost

Site Location	Purchase Cost ($/Year)	Material Shipping Cost ($/Ton)		
		Corn	Wheat	Potato
Yuma, AZ	125,000	10	5	16
Fresno, CA	130,000	12	8	11
Tucson, AZ	140,000	9	10	15
Pomona, CA	160,000	11	7	14
Santa Fe, NM	150,000	8	14	10
Flagstaff, AZ	170,000	10	12	11
Las Vegas, NV	155,000	13	12	9
St. George, UT	115,000	14	15	8

TABLE 21
Demand Information

Company	Location	Demand		
		Regular	Green Onion	Party Mix
Jones	Salt Lake City	1,300	900	1,700
YZCO	Albuquerque	1,400	1,100	1,700
Square Q	Phoenix	1,200	800	1,800
AJ Stores	San Diego	1,900	1,200	2,200
Sun Quest	Los Angeles	1,900	1,400	2,300
Harm's Path	Tucson	1,500	1,000	1,400

shipping costs). Each site may produce as many as 20,000 tons of product per year.

BC has six major customers, and all demand is shipped by truck from the plant to the customer warehouse. The shipping cost depends on the tonnage and distance and comes to $0.15 per ton-mile. The customers, their location, and their yearly demand in tons for each product are listed in Table 21. You must meet demand.

The makeup of the products does not depend on the production plant. Table 22 gives the product-ingredient mix data. The company requires that we consolidate our business, so we cannot locate plants in more than two states.

For this analysis, ignore the differences in property and income tax rates between the states (this is usually critical, but it gets us far afield of the key issue of math programming). In addition, many critical factors actually determine locations; for example, the method of financing the site purchase will also be a major factor in the decision—but we will ignore that also.

Your job is to determine how we should expand into the west and develop alternatives. Questions you should answer include:

TABLE 22
Product-Ingredient Mix

Product	Ingredient		
	Corn	Wheat	Potato
Regular chips	70	20	10
Green onion chips	30	15	55
Party mix	20	50	30

- What sites should be selected? How should the customers be served?

- If gasoline gets more expensive and our trucking costs change, then how is the recommendation affected?

- If rail freight costs for material shipping increase, then how is the recommendation affected?

Please consider other sensitivity-analysis issues that you feel might be important for management's decision-making process.

CASE 8

Emergency Vehicle Location in Springfield

You are the logistics manager for the Springfield Fire Department. You are to develop a recommendation for providing emergency service to Springfield. The department's resources include engine trucks, ladder trucks, and paramedic vehicles. The budget suggests a total of 15 vehicles are fundable in the coming year. Currently, seven engines, three ladder trucks, and five paramedic vehicles are in operation. This system runs 24 hours per day.

The city has been divided into 10 zones (see Figure 4). The map is drawn to scale. For each zone, the department has estimated the number of fire calls, the number of false alarms, and the number of medical calls per 12-hour day. These data are listed in Table 23. Currently, there are five fire stations in the city; these are listed in Figure 4. Each existing station costs $20,000 per year to operate. The yearly costs for each potential station (including the amortized cost of construction) are also listed in Table 23. Each station can hold two vehicles at most.

For fire calls, an engine and a ladder truck must respond. For medical calls, a paramedic vehicle always responds and an engine also goes when one is available and closer than the nearest paramedic vehicle. On average, fire calls take 2.5 hours, false alarms 10 minutes, and medical calls 45 minutes.

FIGURE 4
Map of the City (17 miles by 11 miles)

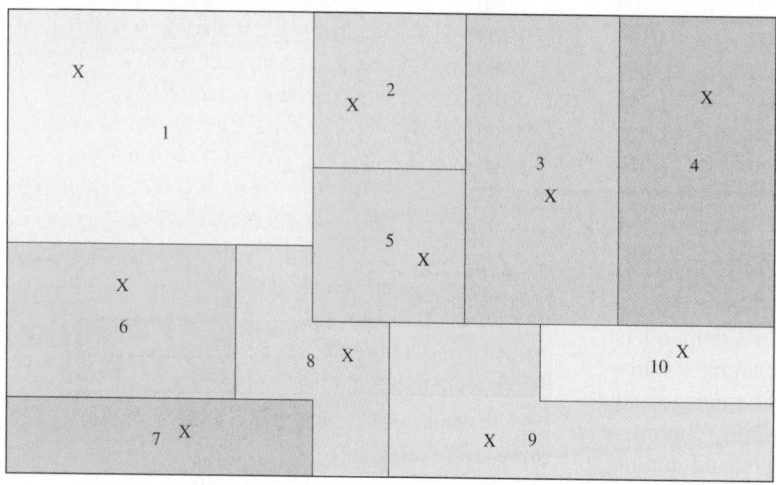

X marks the spot of the existing sites and the possible sites in each zone

TABLE 23
Demand Information per Year

Zone	Fire Calls	False Alarms	Medical Calls	Base Cost ($/Year)
1	100	200	1000	40,000
2	50	100	450	Existing
3	75	100	600	35,000
4	120	75	1300	50,000
5	150	100	1400	Existing
6	300	150	1000	50,000
7	200	100	800	Existing
8	250	175	1000	Existing
9	100	25	900	Existing
10	75	50	650	35,000

When vehicles are dispatched on a call, the closest idle vehicle is dispatched first. If no vehicles are idle, then the call must be sent to a private provider; these responses cost the city $5,000 per medical call, $15,000 per fire call, and $200 per false alarm. There is no queueing. The street network is largely rectangular, and the fire department estimates that the cost per mile for travel is $1.50 per mile for engines and ladders and $0.75 per mile for paramedic units.

Your job is to design a system for the fire department. The questions that should be considered are these:

- What sites should be selected and how should the vehicles be distributed?

- If travel gets more expensive, how is the recommendation affected?

- If the cost of using the private provider increases, then how should the system be changed?

- Is all of this equipment needed to serve the public?

- How much more demand can be handled with the full complement of vehicles?

Your write-up should include a description of your models and any assumptions made in model formulation. You will have to make simplifying assumptions, because this problem has details that may be difficult to model. There are *many* ways to model parts of this system, and you can use different approaches to answer different questions. You may use Excel or LINDO, or you may use heuristics. Your call will depend on your modeling approaches.

Hint: This case is less specific and has vague components; simplify as a first approach and then get more complex. If you try to include everything, you will become frustrated because this does not fit any standard modeling paradigm.

CASE 9

System Design: Project Management[†]

System Design (SD) is a small corporation that contracts to manage systems and industrial engineering projects. In this case it must manage the design and construction of a power plant's data-processing and data-collection system. SD's role in the project is to

[†]This material is expanded from a homework problem in *Applied Mathematical Programming* by Wayne Winston.

hire subcontractors, ensure each task is completed within specification, determine how much labor to assign to each task, and generally ensure the project's success.

SD is really a subcontractor within the larger project of building the power plant. Table 24 details SD's plant-construction and data-system-design tasks. SD is directly in charge of tasks 2, 6, 7, 10, 14, 15, and 19. The remaining tasks in Table 24 interact with those in SD's charge. Assume that the remaining tasks (1) will start whenever their predecessor tasks are complete and (2) will finish exactly after their duration.

To shorten the seven SD tasks, you must pay additional labor and overhead costs. Table 25 lists the functions you can use to compute the cost of changing each task duration to a new value. (*Note:* t_j is the original duration of task j; d_j is the minimum duration of task j.) The table also lists the lowest possible task-duration value. You may not increase the duration of any task.

The revenue that SD obtains from the project depends not only on its tasks but also on when the total project is completed. The project is due at day 900, and SD receives the contract price of $600,000 if the project is done then. If the project is finished x days early, then SD receives a total of $600,000 + $15,000x^{0.7}$ in revenue. If SD finishes x days late, then it receives $600,000 - $20,000x^{1.4}$ in revenue.

Expediting tasks can be profitable and necessary to meet deadlines, although employees do not really like it. Task completion quality is a function of the task completion time, and we would like to have a high quality. This may conflict with our objective of maximizing profit. Because quality affects future revenues, it is difficult to estimate the dollar impact of poor quality. If t_j is the original duration of task j and x_j is the expedited duration time for task j, then quality, measured on a scale of 0–100 (with 100 being the best), can be represented by the function

$$100 - \min [100, (t_j - x_j)^{2.2}]$$

This is only the quality for a task. It is unclear how one might quantify the quality of the project.

Your job is to determine how we should proceed with our tasks. Your analysis should answer some of the following questions:

- What tasks are critical to project completion? What tasks will you expedite?

- How are you measuring system quality, and how does your recommendation measure up relative to that objective?

TABLE **24**
Task Information

Task No.	Task Name	Task Duration (Days)	Immediate Predecessor Tasks
1	Preliminary system description	40	—
2	Develop specifications	100	1
3	Client approval	50	2
4	Develop input-output summary	60	2
5	Develop alarm list	40	4
6	Develop log formats	40	3, 5
7	Software definition	35	3
8	Hardware requirements	35	3
9	Finalize input-output summary	60	5, 6
10	Analysis performance calculations	70	9
11	Automatic turbine startup analysis	65	9
12	Boiler guides analysis	30	9
13	Fabricate and ship	200	10, 11, 12
14	Software preparation	80	7, 10, 11
15	Install and check	130	13, 14
16	Termination and wiring lists	30	9
17	Schematic wiring lists	60	16
18	Pulling terminals and cables	60	15, 17
19	Operational test	125	18
20	First firing	1	19

TABLE **25**
Expediting Costs ($1,000) and Limits

| Task No. | Duration (Days) | | Cost to Decrease t_j by x Days |
	Current t_j	Minimum d_j	
2	100	70	$1.5x^{1.8}$
6	40	20	$2x^{2.0}$
7	35	20	$1x^{2.0}$
10	70	40	$1.8x^{1.9}$
14	80	60	$1.9x^{1.6}$
15	130	120	$0.95x^{2.7}$
19	125	80	$0.9x^{2.9}$

- How much money do you make on the project?
- What will you give the decision maker to help with the decision?
- If we can move minimum duration days to lower values, then which values would you like to reduce?
- If you could control additional tasks by paying more money, which ones would you like to take, and how much would you be willing to pay for control?

CASE **10**

Modular Design for the Help-You Company

The Help-You Company is in the business of manufacturing first-aid kits for cars, hikers, campers, sports teams, and scouting groups. The company is located in Tucson, Arizona, and all materials must be sent to Tucson and then shipped to customers' warehouses. The company has done extensive market surveys of its customers; Table 26 shows its estimates for the demand for its kits in the coming year.

Each kit contains the individual items shown in Table 27, which are listed with their base sizes in pounds. Help-You, for example, can buy packs of Acetaminophen extra-strength caplets. Each pack contains 12 tablets and costs Help-You $1.50.

Help-You buys the individual items and then assembles kits based on the requirements for each part in each kit as shown in Table 28.

These are minimum requirements in that the customer expects at least the listed quantity of each item in each specific kit. For example, in the kit for campers, there must be at least four blankets and at least three cold units (six cold packs) as well as the other items.

There are two strategies available for assembly of the kits:

- In *direct assembly,* the exact requirements are put into each kit.

- In *modular assembly,* one or more standard modules are developed that can be assembled and combined into a kit with enough modules so that the

minimum requirements for each item are met. A graphic of the approach is detailed in Figure 5.

If you design a module, for example, that has two units of Band-Aids (as well as the other items) and place three of these modules in a scouting kit, then the kit will have 3*2 = 6 units of Band-Aids; this will meet the requirement of four units of Band-Aids for scouting kits. In this example, there is an "overage" of two extra units of Band-Aids that costs

$$2 * \$1 \text{ per unit} = \$2$$

per each scout kit demanded. Also, the total unit content in a module cannot weigh more than 15 pounds (an assembly requirement).

Direct assembly meets requirements exactly, although it usually has higher labor costs than modular assembly. For storing inventory, modules are easier to use because they tend to be smaller than kits.

Develop a strategy for modular assembly. The key costs of the modular system are the overages that occur. Your strategy must include the following:

- the number of modules you are designing (the more modules you have, the closer you can match requirements exactly, although the higher the costs for assembly and inventory);

- the unit content of each module designed; and

- an estimate for the total number of each module required.

Your analysis should consider issues such as:

- the trade-off between the number of different modules designed and the total overage cost (you do not need to try more than five different modules—why?);

TABLE 27
Item Cost and Base Size

Item	Cost ($)	Base Unit	Base Unit Weight (Lb.)
Adhesive Band-Aids	1	10 per pack	0.20
Ace bandages	2	1 bandage	0.20
Flares	4	3 per pack	1.00
Blankets	15	1 blanket	2.00
Adhesive tape	2.50	1 roll	0.40
Cold packs	4	2 per pack	0.80
Sunburn cream	3.50	1 tube	0.40
Antiseptic cream	2	1 tube	0.50
Acetaminophen extra-strength caplets	1.50	12 tablets	0.30
Rubber gloves	1.50	3 pairs	0.20

TABLE **28**

Item to Kit Requirements (Base Units)

Kit Item	Cars	Hikers	Campers	Sports Teams	Scouting Groups
Ace bandages	1	2	4	12	6
Band-Aids	0	2	4	4	4
Flares	2	1	1	0	2
Blankets	1	1	4	2	3
Adhesive tape	2	2	3	6	4
Cold packs	2	2	3	6	3
Sunburn cream	1	2	4	4	5
Antiseptic cream	1	2	3	2	4
Acetaminophen caplets	1	2	4	6	6
Rubber gloves	1	1	2	10	5

FIGURE **5**

Modular Assembly

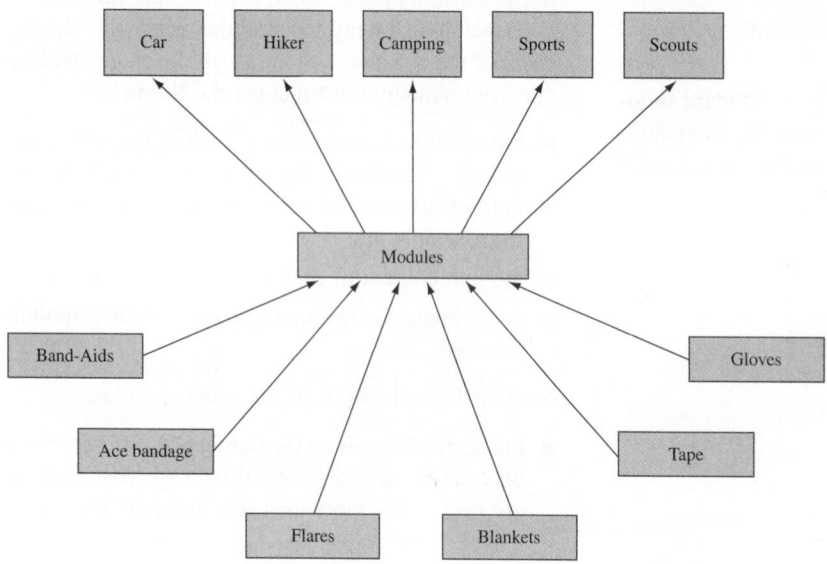

- the sensitivity of your solution to the kit demand estimates (for example, what happens to your recommendations if the number of scouting kits sold changes by 20%?);

- the sensitivity of your recommendations to the unit cost values;

- the sensitivity of your recommendations to the weight limit on the size of each module; and

- discussion about your confidence that you have the optimal solution in light of what you have covered concerning convex functions and sets.

CASE **11**

Brite Power: Capacity Expansion

Brite Power is a small power provider in the Finger Lakes region of New York state. Because of California's power shortage in summer 2001, Brite Power's board of directors has decided to commission a study to ensure that the company has sufficient power until 2020. The study requires a time horizon of 16 years; the first decisions can be implemented in January 2005. Even though operations plans for power companies are important, this study is at a higher level.

Our concern is with power capacity during the year and not with day-to-day or hour-to-hour power-usage fluctuations.

By the start of 2005, Brite will have plants with 60 megawatts' worth of production capacity. Estimates of demand for power from the company for the years 2005 to 2020 have been made and are listed in Table 29.

The economics of a coal-fired plant run according to a *power law*—that is, the cost of a new plant in constant dollars (sometimes called "year 0 dollars") follows the following estimation rule:

cost of plant with capacity K =

$$\left[\frac{\text{capacity K}}{\text{capacity of base size plant}}\right]^{0.8} * \text{cost of the base size plant}$$

For this analysis, the base plant's production capacity is 5 megawatts; its cost is $18 million. The company estimates inflation at 4% per year for the duration of the time horizon. The company uses a discount rate of 12% per year; this assumes that actual dollars are used in the analysis [1 actual dollar in year 1 is equivalent to $1/(1.12) = .89$ dollars now in year 0].

The time required for constructing a new plant is two years. The project requires 65% of the cost at the start of the first year; the remaining 35% is spent at the start of the second year. If Brite Power starts a new

plant in 2007, for example, then 65% of the costs occur in 2007 and 35% in 2008; the plant then comes online and can be used to satisfy demand in 2009. With this lead time, it is clear that Brite Power needs to do advanced planning.

Brite Power can build plants with 5-, 10-, 15-, and 20-megawatt capacities. If the company invests now in research for new technologies ($3 million per year for 5 years), then it can reduce the exponent in the power model from 0.8 to 0.65.

Besides building new capacity, Brite Power must operate plants; operations costs are based on the amount of capacity used. If demand in year t is D_t megawatts and total capacity in year t is C_t megawatts, then the operations cost in constant dollars for the year is:

$$\left(\frac{C_t}{D_t}\right)^{0.5} * D_t * \$400,000$$

By and large, the company must satisfy all demand in the year, although there are opportunities to buy 3 megawatts per year from neighboring power companies at $600,000 per megawatt (in constant dollars).

Key Questions

- The Brite Board would like to know when to augment capacity. How big should the expansions be? When should they start?

- What are the actual dollars spent over the time horizon to acquire and satisfy demand? What is the discounted value of these expenditures?

- Should the company invest in research to lower the power law coefficient?

- If demand estimates are increased or decreased, then how would the plan change?

- If the $400,000 value in the operations cost changes, then how would the plan change?

TABLE 29

Year	Demand (Megawatts)	Year	Demand (Megawatts)
2005	54	2013	87
2006	58	2014	87
2007	63	2015	90
2008	63	2016	90
2009	69	2017	100
2010	75	2018	110
2011	77	2019	110
2012	77	2020	120

Answers to Selected Problems

Chapter 2

SECTION 2.1

1 a $-A = \begin{bmatrix} -1 & -2 & -3 \\ -4 & -5 & -6 \\ -7 & -8 & -9 \end{bmatrix}$

b $3A = \begin{bmatrix} 3 & 6 & 9 \\ 12 & 15 & 18 \\ 21 & 24 & 27 \end{bmatrix}$

c $A + 2B$ is undefined.

d $A^T = \begin{bmatrix} 1 & 4 & 7 \\ 2 & 5 & 8 \\ 3 & 6 & 9 \end{bmatrix}$

e $B^T = \begin{bmatrix} 1 & 0 & 1 \\ 2 & -1 & 2 \end{bmatrix}$

f $AB = \begin{bmatrix} 4 & 6 \\ 10 & 15 \\ 16 & 24 \end{bmatrix}$

g BA is undefined.

2 $\begin{bmatrix} y_1 \\ y_2 \\ y_3 \end{bmatrix} = \begin{bmatrix} 0.50 & 0 & 0.10 \\ 0.30 & 0.70 & 0.30 \\ 0.20 & 0.30 & 0.60 \end{bmatrix}\begin{bmatrix} x_1 \\ x_2 \\ x_3 \end{bmatrix}$

SECTION 2.2

1 $\begin{bmatrix} 1 & -1 \\ 2 & 1 \\ 1 & 3 \end{bmatrix}\begin{bmatrix} x_1 \\ x_2 \end{bmatrix} = \begin{bmatrix} 4 \\ 6 \\ 8 \end{bmatrix}$ or $\begin{bmatrix} 1 & -1 & 4 \\ 2 & 1 & 6 \\ 1 & 3 & 8 \end{bmatrix}$

SECTION 2.3

1 No solution.

2 Infinite number of solutions of the form $x_1 = 2 - 2k$, $x_2 = 2 + k$, $x_3 = k$.

3 $x_1 = 2$, $x_2 = -1$.

SECTION 2.4

1 Linearly dependent.

2 Linearly independent.

SECTION 2.5

2 $A^{-1} = \begin{bmatrix} \frac{1}{2} & \frac{1}{2} & -\frac{1}{2} \\ -1 & -2 & 3 \\ \frac{1}{2} & -\frac{1}{2} & \frac{1}{2} \end{bmatrix}$

3 A^{-1} does not exist.

8 a $\frac{1}{100}B^{-1}$.

SECTION 2.6

2 30.

REVIEW PROBLEMS

1 Infinite number of solutions of the form $x_1 = k - 1$, $x_2 = 3 - k$, $x_3 = k$.

3 $\begin{bmatrix} U_{t+1} \\ T_{t+1} \end{bmatrix} = \begin{bmatrix} 0.75 & 0 \\ 0.20 & 0.90 \end{bmatrix}\begin{bmatrix} U_t \\ T_t \end{bmatrix}$

4 $x_1 = 0$, $x_2 = 1$.

13 Linearly independent.

14 Linearly dependent.

15 a Only if a, b, c, and d are all nonzero will rank $A = 4$. Thus, A^{-1} exists if and only if all of a, b, c, and d are nonzero.

b Applying the Gauss-Jordan method, we find if a, b, c, and d are all nonzero,

$$A^{-1} = \begin{bmatrix} \frac{1}{a} & 0 & 0 & 0 \\ 0 & \frac{1}{b} & 0 & 0 \\ 0 & 0 & \frac{1}{c} & 0 \\ 0 & 0 & 0 & \frac{1}{d} \end{bmatrix}$$

18 -4.

Chapter 3

SECTION 3.1

1 $\max z = 30x_1 + 100x_2$

s.t. $\quad x_1 + \quad x_2 \leq 7 \qquad$ **(Land constraint)**

$\quad\quad 4x_1 + 10x_2 \leq 40 \qquad$ **(Labor constraint)**

$\quad\quad 10x_1 \quad\quad\quad \geq 30 \qquad$ **(Government constraint)**

$\quad\quad x_1, x_2 \geq 0$

2 No, the government constraint is not satisfied.

b No, the labor constraint is not satisfied.

c No, $x_2 \geq 0$ is not satisfied.

SECTION 3.2

1 $z = \$370$, $x_1 = 3$, $x_2 = 2.8$.

3 $z = \$14$, $x_1 = 3$, $x_2 = 2$.

4 a We want to make x_1 larger and x_2 smaller, so we move down and to the right.

b We want to make x_1 smaller and x_2 larger, so we move up and to the left.

b We want to make both x_1 and x_2 smaller, so we move down and to the left.

SECTION 3.3

1 No feasible solution.

2 Alternative optimal solutions.

3 Unbounded LP.

SECTION 3.4

1 For $i = 1, 2, 3$, let x_i = tons of processed factory i waste. Then the appropriate LP is

$\min z = 15x_1 + 10x_2 + 20x_3$

s.t. $\quad 0.10x_1 + 0.20x_2 + 0.40x_3 \geq 30 \qquad$ **(Pollutant 1)**

$\quad\quad 0.45x_1 + 0.25x_2 + 0.30x_3 \geq 40 \qquad$ **(Pollutant 2)**

$\quad\quad\quad\quad\quad\quad x_1, x_2, x_3 \geq 0$

It is doubtful that the processing cost is proportional to the amount of waste processed. For example, processing 10 tons of waste is probably not 10 times as costly as processing 1 ton of waste. The Divisibility and Certainty Assumptions seem reasonable.

SECTION 3.5

1 Let x_1 = number of full-time employees (FTE) who start work on Sunday, x_2 = number of FTE who start work on Monday, . . . , x_7 = number of FTE who start work on Saturday; x_8 = number of part-time employees (PTE) who start work on Sunday, . . . , x_{14} = number of PTE who start work on Saturday. Then the appropriate LP is

$\min z = 15(8)(5)(x_1 + x_2 + \cdots + x_7)$

$\quad\quad + 10(4)(5)(x_8 + x_9 + \cdots + x_{14})$

s.t. $8(x_1 + x_4 + x_5 + x_6 + x_7) + 4(x_8 + x_{11} + x_{12}$

$\quad\quad + x_{13} + x_{14}) \geq 88 \qquad$ **(Sunday)**

$8(x_1 + x_2 + x_5 + x_6 + x_7) + 4(x_8 + x_9 + x_{12}$

$\quad\quad + x_{13} + x_{14}) \geq 136 \qquad$ **(Monday)**

$8(x_1 + x_2 + x_3 + x_6 + x_7) + 4(x_8 + x_9 + x_{10}$

$\quad\quad + x_{13} + x_{14}) \geq 104 \qquad$ **(Tuesday)**

$8(x_1 + x_2 + x_3 + x_4 + x_7) + 4(x_8 + x_9 + x_{10}$

$\quad\quad + x_{11} + x_{14}) \geq 120 \qquad$ **(Wednesday)**

$8(x_1 + x_2 + x_3 + x_4 + x_5) + 4(x_8 + x_9 + x_{10}$

$\quad\quad + x_{11} + x_{12}) \geq 152 \qquad$ **(Thursday)**

$8(x_2 + x_3 + x_4 + x_5 + x_6) + 4(x_9 + x_{10} + x_{11}$

$\quad\quad + x_{12} + x_{13}) \geq 112 \qquad$ **(Friday)**

$8(x_3 + x_4 + x_5 + x_6 + x_7) + 4(x_{10} + x_{11} + x_{12}$

$\quad\quad + x_{13} + x_{14}) \geq 128 \qquad$ **(Saturday)**

$20(x_8 + x_9 + x_{10} + x_{11} + x_{12} + x_{13} + x_{14})$

$\leq 0.25(136 + 104 + 120 + 152 + 112 + 128$

$\quad + 88)$

(The last constraint ensures that part-time labor will fulfill at most 25% of all labor requirements)

$$\text{All variables} \geq 0$$

3 Let x_1 = number of employees who start work on Sunday and work five days, x_2 = number of employees who start work on Monday and work five days, . . . , x_7 = number of employees who start work on Saturday and work five days. Also let o_1 = number of employees who start work on Sunday and work six days, . . . , o_7 = number of employees who start work on Saturday and work six days. Then the appropriate LP is

$\min z = 250(x_1 + x_2 + \cdots + x_7)$

$\quad\quad + 312(o_1 + o_2 + \cdots + o_7)$

s.t. $\quad x_1 + x_4 + x_5 + x_6 + x_7 + o_1 + o_3 + o_4$

$\quad\quad + o_5 + o_6 + o_7 \geq 11 \qquad$ **(Sunday)**

$\quad x_1 + x_2 + x_5 + x_6 + x_7 + o_1 + o_2 + o_4$

$\quad\quad + o_5 + o_6 + o_7 \geq 17 \qquad$ **(Monday)**

$\quad x_1 + x_2 + x_3 + x_6 + x_7 + o_1 + o_2 + o_3$

$\quad\quad + o_5 + o_6 + o_7 \geq 13 \qquad$ **(Tuesday)**

$\quad x_1 + x_2 + x_3 + x_4 + x_7 + o_1 + o_2 + o_3$

$\quad\quad + o_4 + o_6 + o_7 \geq 15 \qquad$ **(Wednesday)**

$\quad x_1 + x_2 + x_3 + x_4 + x_5 + o_1 + o_2 + o_3$

$\quad\quad + o_4 + o_5 + o_7 \geq 19 \qquad$ **(Thursday)**

$\quad x_2 + x_3 + x_4 + x_5 + x_6 + o_1 + o_2 + o_3$

$\quad\quad + o_4 + o_5 + o_6 \geq 14 \qquad$ **(Friday)**

$\quad x_3 + x_4 + x_5 + x_6 + x_7 + o_2 + o_3 + o_4$

$\quad\quad + o_5 + o_6 + o_7 \geq 16 \qquad$ **(Saturday)**

$$\text{All variables} \geq 0$$

SECTION 3.6

2 NPV of investment 1

$$= -6 - \frac{5}{1.1} + \frac{7}{(1.1)^2} + \frac{9}{(1.1)^3} = \$2.00.$$

NPV of investment 2

$$= -8 - \frac{3}{1.1} + \frac{9}{(1.1)^2} + \frac{7}{(1.1)^3} = \$1.97.$$

Let $x_1 =$ fraction of investment 1 that is undertaken and $x_2 =$ fraction of investment 2 that is undertaken. If we measure NPV in thousands of dollars, we want to solve the following LP:

$$\max z = 2x_1 + 1.97x_2$$

$$\text{s.t.} \quad 6x_1 + 8x_2 \leq 10$$
$$5x_1 + 3x_2 \leq 7$$
$$x_1 \qquad \leq 1$$
$$x_2 \leq 1$$

All variables ≥ 0

The optimal solution to this LP is $x_1 = 1$, $x_2 = 0.5$, $z = \$2,985$.

SECTION 3.7

1 $z = \$2,500$, $x_1 = 50$, $x_2 = 100$.

SECTION 3.8

1 Let ingredient 1 = sugar, ingredient 2 = nuts, ingredient 3 = chocolate, candy 1 = Slugger, and candy 2 = Easy Out. Let $x_{ij} =$ ounces of ingredient i used to make candy j. (All variables are in ounces.) The appropriate LP is

$$\max z = 25(x_{12} + x_{22} + x_{32}) + 20(x_{11} + x_{21} + x_{31})$$

$$\text{s.t.} \quad x_{11} + x_{12} \leq 100 \quad \textbf{(Sugar constraint)}$$
$$x_{21} + x_{22} \leq 20 \quad \textbf{(Nuts constraint)}$$
$$x_{31} + x_{32} \leq 30 \quad \textbf{(Chocolate constraint)}$$
$$x_{22} \geq 0.2(x_{12} + x_{22} + x_{32})$$
$$x_{21} \leq 0.1(x_{11} + x_{21} + x_{31})$$
$$x_{31} \geq 0.1(x_{11} + x_{21} + x_{31})$$

All variables ≥ 0

SECTION 3.9

1 Let $x_1 =$ hours of process 1 run per week

$x_2 =$ hours of process 2 run per week

$x_3 =$ hours of process 3 run per week

$g_2 =$ barrels of gas 2 sold per week

$o_1 =$ barrels of oil 1 purchased per week

$o_2 =$ barrels of oil 2 purchased per week

$$\max z = 9(2x_1) + 10g_2 + 24(2x_3) - 5x_1 - 4x_2$$

$$- x_3 - 2o_1 - 3o_2$$

$$\text{s.t.} \quad o_1 = 2x_1 + x_2$$
$$o_2 = 3x_1 + 3x_2 + 2x_3$$
$$o_1 \leq 200$$
$$o_2 \leq 300$$
$$g_2 + 3x_3 = x_1 + 3x_2 \quad \textbf{(Gas 2 production)}$$
$$x_1 + x_2 + x_3 \leq 100 \quad \textbf{(100 hours per week}$$
$$\textbf{of cracker time)}$$

All variables ≥ 0

5 Let $A =$ total number of units of A produced

$B =$ total number of units of B produced

$CS =$ total number of units of C produced (and sold)

$AS =$ units of A sold

$BS =$ units of B sold

$$\max z = 10AS + 56BS + 100CS$$

$$\text{s.t.} \quad A + 2B + 3C \leq 40$$
$$A = AS + 2B$$
$$B = BS + CS$$

All variables ≥ 0

SECTION 3.10

1 Let $x_t =$ production during month t and $i_t =$ inventory at end of month t.

$$\min z = 5x_1 + 8x_2 + 4x_3 + 7x_4$$
$$+ 2i_1 + 2i_2 + 2i_3 + 2i_4 - 6i_4$$

$$\text{s.t.} \quad i_1 = x_1 - 50$$
$$i_2 = i_1 + x_2 - 65$$
$$i_3 = i_2 + x_3 - 100$$
$$i_4 = i_3 + x_4 - 70$$

All variables ≥ 0

SECTION 3.11

3 Let $A =$ dollars invested in A, $B =$ dollars invested in B, $c_0 =$ leftover cash at time 0, $c_1 =$ leftover cash at time 1, and $c_2 =$ leftover cash at time 2. Then a correct formulation is

$$\max z = c_2 + 1.9B$$

$$\text{s.t.} \quad A + c_0 = 10,000$$

(Time 0 available = time 0 invested)

$$0.2A + c_0 = B + c_1$$

(Time 1 available = time 1 invested)

$$1.5A + c_1 = c_2$$

(Time 2 available = time 2 invested)

All variables ≥ 0

The optimal solution to this LP is $B = c_0 = \$10,000$, $A = c_1 = c_2 = 0$, and $z = \$19,000$. Notice that it is optimal to wait for the "good" investment (B) even though leftover cash earns no interest.

SECTION 3.12

2 Let JAN1 = number of computers rented at beginning of January for one month, and so on. Also define IJAN = number of computers available to meet January demand, and so on. The appropriate LP is

$$\min z = 200(\text{JAN1} + \text{FEB1} + \text{MAR1} + \text{APR1}$$
$$+ \text{MAY1} + \text{JUN1}) + 350(\text{JAN2} + \text{FEB2}$$
$$+ \text{MAR2} + \text{APR2} + \text{MAY2} + \text{JUN2})$$
$$+ 450(\text{JAN3} + \text{FEB3} + \text{MAR3} + \text{APR3})$$
$$+ \text{MAY3} + \text{JUN3}) - 150\text{MAY3}$$
$$- 300\,\text{JUN3} - 175\text{JUN2}$$

s.t. $\text{IJAN} = \text{JAN1} + \text{JAN2} + \text{JAN3}$

$\text{IFEB} = \text{IJAN} - \text{JAN1} + \text{FEB1} + \text{FEB2} + \text{FEB3}$

$\text{IMAR} = \text{IFEB} - \text{JAN2} - \text{FEB1} + \text{MAR1}$
$\qquad + \text{MAR2} + \text{MAR3}$

$\text{IAPR} = \text{IMAR} - \text{FEB2} - \text{MAR1} - \text{JAN3}$
$\qquad + \text{APR1} + \text{APR2} + \text{APR3}$

$\text{IMAY} = \text{IAPR} - \text{FEB3} - \text{MAR2} - \text{APR1}$
$\qquad + \text{MAY1} + \text{MAY2} + \text{MAY3}$

$\text{IJUN} = \text{IMAY} - \text{MAR3} - \text{APR2} - \text{MAY1}$
$\qquad + \text{JUN1} + \text{JUN2} + \text{JUN3}$

$\text{IJAN} \geq 9$

$\text{IFEB} \geq 5$

$\text{IMAR} \geq 7$

$\text{IAPR} \geq 9$

$\text{IMAY} \geq 10$

$\text{IJUN} \geq 5$

All variables ≥ 0

REVIEW PROBLEMS

2 Let x_1 = number of chocolate cakes baked and x_2 = number of vanilla cakes baked. Then we must solve

$$\max z = x_1 + \tfrac{1}{2}x_2$$
$$\text{s.t.} \quad \tfrac{1}{3}x_1 + \tfrac{2}{3}x_2 \leq 8$$
$$4x_1 + x_2 \leq 30$$
$$x_1, x_2 \geq 0$$

The optimal solution is $z = \frac{\$69}{7}$, $x_1 = \frac{36}{7}$, $x_2 = \frac{66}{7}$.

8 Let x_1 = acres of farm 1 devoted to corn, x_2 = acres of farm 1 devoted to wheat, x_3 = acres of farm 2 devoted to corn, x_4 acres of farm 2 devoted to wheat. Then a correct formulation is

$$\min z = 100x_1 + 90x_2 + 120x_3 + 80x_4$$

s.t. $x_1 + x_2 \qquad\qquad \leq 100$

(Farm 1 land)

$\qquad\qquad x_3 + x_4 \leq 100$

(Farm 2 land)

$500x_1 + 650x_3 \geq 7,000$

(Corn requirement)

$400x_2 + 350x_4 \geq 11,000$

(Wheat requirement)

$$x_1, x_2, x_3, x_4 \geq 0$$

9 Let x_1 = units of process 1, x_2 = units of process 2, and x_3 = modeling hours hired. Then a correct formulation is

$$\max z = 5(3x_1 + 5\,x_2) - 3(x_1 + 2x_2)$$
$$- 2(2x_1 + 3x_2) - 100x_3$$

s.t. $x_1 + 2x_2 \leq 20,000$ **(Limited labor)**

$2x_1 + 3x_2 \leq 35,000$ **(Limited chemicals)**

$3x_1 + 5x_2 = 1,000 + 200x_3$

(Perfume production = perfume demands)

$$x_1, x_2, x_3 \geq 0$$

17 Let OT = number of tables made of oak, OC = number of chairs made of oak, PT = number of tables made of pine, and PC = number of chairs made of pine. Then the correct formulation is

$$\max z = 40(OT + PT) + 15(OC + PC)$$

s.t. $17(OT) + 5(OC) \leq 150$

(Use at most 150 board ft of oak)

$30PT + 13PC \leq 210$

(Use at most 210 board ft of pine)

$$OT, OC, PT, PC \geq 0$$

18 Let school 1 = Cooley High, and school 2 = Walt Whitman High. Let M_{ij} = number of minority students who live in district i who will attend school j, and let NM_{ij} = number of nonminority students who live in district i who will attend school j. Then the correct LP is

$$\min z = (M_{11} + NM_{11}) + 2(M_{12} + NM_{12})$$
$$+ 2(M_{21} + NM_{21}) + (M_{22} + NM_{22})$$
$$+ (M_{31} + NM_{31}) + (M_{32} + NM_{32})$$

s.t. $M_{11} + M_{12} = 50$

$M_{21} + M_{22} = 50$

$M_{31} + M_{32} = 100$

$NM_{11} + NM_{12} = 200$

$NM_{21} + NM_{22} = 250$

$NM_{31} + NM_{32} = 150$

For school 1, we obtain the following blending constraints:

$$0.20 \leq \frac{M_{11} + M_{21} + M_{31}}{M_{11} + M_{21} + M_{31} + NM_{11} + NM_{21} + NM_{31}}$$
$$\leq 0.30$$

This yields the following two LP constraints:

$$0.8M_{11} + 0.8M_{21} + 0.8M_{31} - 0.2NM_{11}$$
$$- 0.2NM_{21} - 0.2NM_{31>} \geq 0$$
$$0.7M_{11} + 0.7M_{21} + 0.7M_{32} - 0.3NM_{11}$$
$$- 0.3NM_{21} - 0.3NM_{31} \leq 0$$

For school 2, we obtain the following blending constraints:

$$0.20 \leq \frac{M_{12} + M_{22} + M_{32}}{M_{12} + M_{22} + M_{32} + NM_{12} + NM_{22} + NM_{32}}$$
$$\leq 0.30$$

This yields the following two LP constraints:

$$0.8M_{12} + 0.8M_{22} + 0.8M_{32} - 0.20NM_{12}$$
$$- 0.20NM_{22} - 0.20NM_{32} \geq 0$$
$$0.7M_{12} + 0.7M_{22} + 0.7M_{32} - 0.30NM_{12}$$
$$- 0.30NM_{22} - 0.30NM_{32} \leq 0$$

We must also ensure that each school has between 300 and 500 students. Thus, we also need the following constraints:

$$300 \leq M_{11} + NM_{11} + M_{21} + NM_{21} + M_{31} + NM_{31}$$
$$\leq 500$$
$$300 \leq M_{12} + NM_{12} + M_{22} + NM_{22} + M_{32} + NM_{32}$$
$$\leq 500$$

To complete the formulation, add the sign restrictions that all variables are ≥ 0.

47 For $i < j$, let X_{ij} = number of workers who get off days i and j of week (day 1 = Sunday, day 2 = Monday, . . . , day 7 = Saturday).

$$\max z = X_{12} + X_{17} + X_{23} + X_{34} + X_{45} + X_{56} + X_{67}$$
$$\text{s.t.} \quad X_{17} + X_{27} + X_{37} + X_{47} + X_{57} + X_{67} = 2$$
(Saturday constraint)
$$X_{12} + X_{13} + X_{14} + X_{15} + X_{16} + X_{17} = 12$$
(Sunday constraint)
$$X_{12} + X_{23} + X_{24} + X_{25} + X_{26} + X_{27} = 12$$
(Monday constraint)
$$X_{13} + X_{23} + X_{34} + X_{35} + X_{36} + X_{37} = 6$$
(Tuesday constraint)
$$X_{14} + X_{24} + X_{34} + X_{45} + X_{46} + X_{47} = 5$$
(Wednesday constraint)
$$X_{15} + X_{25} + X_{35} + X_{45} + X_{56} + X_{57} = 14$$
(Thursday constraint)

$$X_{16} + X_{26} + X_{36} + X_{46} + X_{56} + X_{67} = 9$$
(Friday constraint)

All variables ≥ 0

49 Let X_{ij} = money invested at beginning of month i for a period of j months. After noting that for each month (money invested) + (bills paid) = (money available), we obtain the following formulation:

$$\max z = 1.08X_{14} + 1.03X_{23} + 1.01X_{32} + 1.001X_{41}$$
$$\text{s.t.} \quad X_{11} + X_{12} + X_{13} + X_{14} + 600 = 400 + 400$$
(Month 1)
$$X_{21} + X_{22} + X_{23} + 500 = 1.001X_{11} + 800$$
(Month 2)
$$X_{31} + X_{32} + 500 = 1.01X_{12} + 1.001X_{21} + 300$$
(Month 3)
$$X_{41} + 250 = 1.03X_{13} + 1.01X_{22} + 1.001X_{31} + 300$$
(Month 4)

All variables ≥ 0

53 Let T_1 = number of type 1 turkeys purchased

T_2 = number of type 2 turkeys purchased

D_1 = pounds of dark meat used in cutlet 1

W_1 = pounds of white meat used in cutlet 1

D_2 = pounds of dark meat used in cutlet 2

W_2 = pounds of white meat used in cutlet 2

Then the appropriate formulation is

$$\max z = 4(W_1 + D_1) + 3(W_2 + D_2) - 10T_1 - 8T_2$$

s.t. $W_1 + D_1 \leq 50$	**(Cutlet 1 demand)**
$W_2 + D_2 \leq 30$	**(Cutlet 2 demand)**
$W_1 + W_2 \leq 5T_1 + 3T_2$	**(Don't use more white meat than you have)**
$D_1 + D_2 \leq 2T_1 + 3T_2$	**(Don't use more dark meat than you have)**

$$W_1/(W_1 + D_1) \geq 0.7 \text{ or } 0.3W_1 \geq 0.7D_1$$
$$W_2/(W_2 + D_2) \geq 0.6 \text{ or } 0.4W_2 \geq 0.6D_2$$
$$T_1, T_2, D_1, W_1, D_2, W_2 \geq 0$$

Chapter 4

SECTION 4.1

1 $\max z = 3x_1 + 2x_2$

$$\text{s.t.} \quad 2x_1 + x_2 + s_1 \qquad\qquad = 100$$
$$x_1 + x_2 \qquad + s_2 \qquad = 80$$
$$x_1 \qquad\qquad\qquad + s_3 = 40$$

3 $\min z = 3x_1 + x_2$

s.t. $\quad x_1 \qquad - e_1 \qquad = 3$

$\qquad x_1 + x_2 \qquad + s_2 = 4$

$\qquad 2x_1 - x_2 \qquad = 3$

SECTION **4.4**

1 From Figure 2 of Chapter 3, we find the extreme points of the feasible region.

Point	Basic Variables
$H = (0, 0)$	$s_1 = 100, s_2 = 80, s_3 = 40$
$E = (40, 0)$	$x_1 = 40, \quad s_1 = 20, s_2 = 40$
$F = (40, 20)$	$x_1 = 40, \quad x_2 = 20, s_2 = 20$
$G = (20, 60)$	$x_1 = 20, \quad x_2 = 60, s_3 = 20$
$D = (0, 80)$	$x_2 = 80, \quad s_1 = 20, s_3 = 40$

SECTION **4.5**

1 $z = 180, x_1 = 20, x_2 = 60$.

2 $z = \frac{32}{3}, x_1 = \frac{10}{3}, x_2 = \frac{4}{3}$.

SECTION **4.6**

1 $z = -5, x_1 = 0, x_2 = 5$.

SECTION **4.7**

2 Solution 1: $z = 6, x_1 = 0, x_2 = 1$; solution 2: $z = 6$, $x_1 = \frac{56}{17}, x_2 = \frac{45}{17}$. By averaging these two solutions, we obtain solution 3: $z = 6, x_1 = \frac{28}{17}, x_2 = \frac{31}{17}$.

SECTION **4.8**

1 $x_1 = 4,999, x_2 = 5,000$ has $z = 10,000$.

SECTION **4.10**

1 **a** Both very small numbers (for example, 0.000003) and large numbers (for example, 3,000,000) appear in the problem.

b Let $x_1 =$ units of product i produced (in millions). If we measure our profit in millions of dollars, the LP becomes

$\max = 6x_1 + 4x_2 + 3x_3$

s.t. $\quad 4x_1 + 3x_2 + 2x_3 \le 3 \qquad$ **(Million labor hours)**

$\qquad 3x_1 + 2x_2 + \quad x_3 \le 2 \qquad$ **(lb of pollution)**

$\qquad x_1, x_2, x_3 \ge 0$

SECTION **4.11**

1 $z = 16, x_1 = x_2 = 2$. The point where all three constraints are binding ($x_1 = x_2 = 2$) corresponds to the following three sets of basic variables:

Set $1 = \{x_1, x_2, s_1\}$

Set $2 = \{x_1, x_2, s_2\}$

Set $3 = \{x_1, x_2, s_3\}$

SECTIONS **4.12** AND **4.13**

1 $z = 1, x_1 = x_2 = 0, x_3 = 1$.

4 Infeasible LP.

SECTION **4.14**

1 Let $i_t = i_t' - i_t''$ be the inventory position at the end of month t. For each constraint in the original problem, replace i_t by $i_t' - i_t''$. Also add the sign restrictions $i_t' \ge 0$ and $i_t'' \ge 0$. To ensure that demand is met by the end of quarter 4, add constraint $i_4'' = 0$. Replace the terms involving i_t in the objective function by

$(100i_1' + 100i_1'' + 100i_2' + 100i_2''$

$\qquad\qquad + 100i_3' + 110i_3'' + 100i_4' + 110i_4'')$

2 $z = 5, x_1 = 1, x_2 = 3$.

SECTION **4.16**

2 Let $x_i =$ number of lots purchased from supplier i. The appropriate LP is

$\min z = 10s_1^- + 6s_2^- + 4s_3^- + s_4^+$

s.t. $\quad 60x_1 + 50x_2 + 40x_3 + s_1^- - s_1^+ = 5,000$

$\qquad\qquad\qquad\qquad\qquad\qquad$ **(Excellent chips)**

$\qquad 20x_1 + 35x_2 + 20x_3 + s_2^- - s_2^+ = 3,000$

$\qquad\qquad\qquad\qquad\qquad\qquad$ **(Good chips)**

$\qquad 20x_1 + 15x_2 + 40x_3 + s_3^- - s_3^+ = 1,000$

$\qquad\qquad\qquad\qquad\qquad\qquad$ **(Mediocre chips)**

$\qquad 400x_1 + 300x_2 + 250x_3 + s_4^- - s_4^+ = 28,000$

$\qquad\qquad\qquad\qquad\qquad\qquad$ **(Budget constraint)**

$\qquad\qquad\qquad$ All variables ≥ 0

REVIEW PROBLEMS

4 Unbounded LP.

5 $z = -6, x_1 = 0, x_2 = 3$.

6 Infeasible LP.

8 $z = 12, x_1 = x_2 = 2$.

10 Four types of furniture.

15 $z = \frac{17}{2}, x_2 = \frac{3}{2}, x_4 = \frac{1}{2}$.

17 **a** $-c \ge 0$ and $b \ge 0$.

b $b \ge 0$ and $c = 0$. Also need $a_2 > 0$ and/or $a_3 > 0$ to ensure that when x_1 is pivoted in, a feasible solution results. If only $a_3 > 0$, then we need b to be strictly positive.

c $-c < 0$, $a_2 \leq 0$, $a_3 \leq 0$ ensures that x_1 can be made arbitrarily large and z will become arbitrarily large.

20 Let c_t = net number of drivers hired at the beginning of the year t. Then $c_t = h_t - f_t$, where h_t = number of drivers hired at beginning of year t, and f_t = number of drivers fired at beginning of year t. Also let d_t = number of drivers after drivers have been hired or fired at beginning of year t. Then a correct formulation is (cost in thousands of dollars)

$$\min z = 10(d_1 + d_2 + d_3 + d_4 + d_5)$$
$$+ 2(f_1 + f_2 + f_3 + f_4 + f_5)$$
$$+ 4(h_1 + h_2 + h_3 + h_4 + h_5)$$

s.t. $\quad d_1 = 50 + h_1 - f_1$
$$d_2 = d_1 + h_2 - f_2$$
$$d_3 = d_2 + h_3 - f_3$$
$$d_4 = d_3 + h_4 - f_4$$
$$d_5 = d_4 + h_5 - f_5$$
$$d_1 \geq 60,\ d_2 \geq 70,\ d_3 \geq 50,\ d_4 \geq 65,$$
$$d_5 \geq 75$$

All variables ≥ 0

26 Let

R_t = robots available during quarter t
 (after robots are bought or sold for the quarter)

B_t = robots bought during quarter t

S_t = robots sold during quarter t

I'_t = cars in inventory at end of quarter t

C_t = cars produced during quarter t

I''_t = backlogged demand for cars at end of quarter t

Then a correct formulation is

$$\min z = 500(R_1 + R_2 + R_3 + R_4)$$
$$+ 200(I'_1 + I'_2 + I'_3 + I'_4)$$
$$+ 5{,}000(B_1 + B_2 + B_3 + B_4)$$
$$- 3{,}000(S_1 + S_2 + S_3 + S_4)$$
$$+ 300(I''_1 + I''_2 + I''_3 + I''_4)$$

s.t. $\quad R_1 = 2 + B_1 - S_1$
$$R_2 = R_1 + B_2 - S_2$$
$$R_3 = R_2 + B_3 - S_3$$
$$R_4 = R_3 + B_4 - S_4$$
$$I'_1 - I''_1 = C_1 - 600$$
$$I'_2 - I''_2 = I'_1 - I''_1 + C_2 - 800$$
$$I'_3 - I''_3 = I'_2 - I''_2 + C_3 - 500$$
$$I'_4 - I''_4 = I'_3 - I''_3 + C_4 - 400$$
$$R_4 \geq 2$$
$$C_1 \leq 200R_1$$
$$C_2 \leq 200R_2$$

$$C_3 \leq 200R_3$$
$$C_4 \leq 200R_4,\ I''_4 = 0$$
$$B_1, B_2, B_3, B_4 \leq 2$$

All variables ≥ 0

Chapter 5

SECTION 5.1

1 Decision variables remain the same. New z-value is $210.

4 **a** $\frac{50}{3} \leq c_1 \leq 350$.

c $4{,}000{,}000 \leq \text{HIW} \leq 84{,}000{,}000$; $x_1 = 3.6 + 0.15\Delta$, $x_2 = 1.4 - 0.025\Delta$.

f $310,000.

SECTION 5.2

1 **a** $3,875.

b Decision variables remain the same. New z-value is $3,750.

c Solution remains the same.

3 **a** Still 90¢.

b 95¢.

c 95¢.

d Still 90¢.

e 82.5¢.

f 30¢ or less.

g 22.5¢ or less.

SECTION 5.3

3 2.5¢.

4 $2.

5 Buy raw material, because it will reduce cost by $6.67.

SECTION 5.4

3 See Figures 1–4.

FIGURE 1

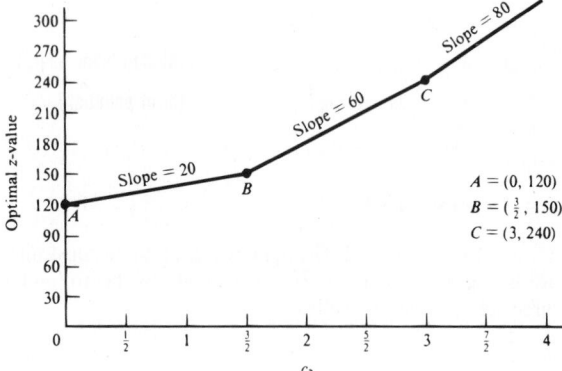

$A = (0, 120)$
$B = (\frac{1}{2}, 150)$
$C = (3, 240)$

FIGURE 2

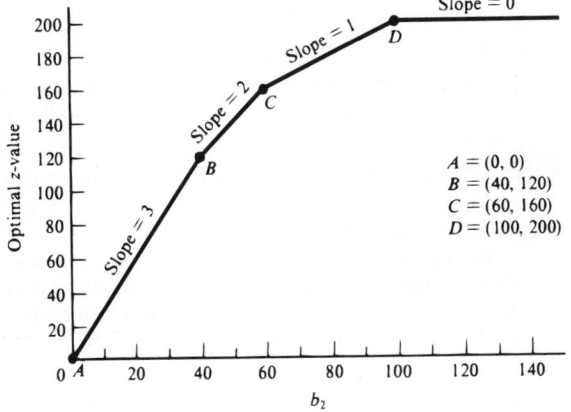

$A = (0, 0)$
$B = (40, 120)$
$C = (60, 160)$
$D = (100, 200)$

FIGURE 3

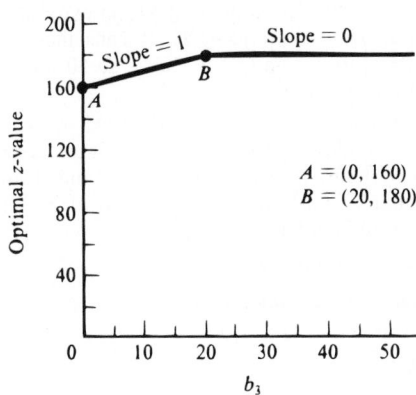

$A = (0, 160)$
$B = (20, 180)$

FIGURE 4

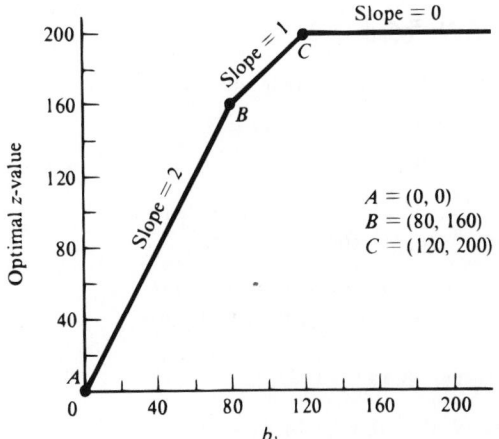

$A = (0, 0)$
$B = (80, 160)$
$C = (120, 200)$

1 **a** $1,046,667.

b Yes.

c $33.33.

b $333.33.

7 **a** Decision variables remain unchanged. New z-value is $1,815,000.

b Pay $0 for an additional 100 board ft of lumber. Pay $1,350 for an additional 100 hours of labor.

c $1,310,000.

d $1,665,000.

Chapter 6

SECTION 6.1

1 Decision variables remain the same. New z-value is $210.

SECTION 6.2

1 $z \quad + 4s_1 + 5s_2 = 28$

$x_1 + s_1 + s_2 = 6$

$x_2 + s_1 + 2s_2 = 10$

SECTION 6.3

3 $x_1 = 2$, $x_2 = 0$, $x_3 = 8$, $z = 280$ (same as original solution).

5 Home computer tables should not be produced.

6 **a** Profit for candy bar $1 \leq 6¢$. If type 1 candy bar earns 7¢ profit, new optimal solution is $z = \$3.50$, $x_1 = 50$, $x_2 = x_3 = 0$.

b $5¢ \leq$ candy bar 2 profit $\leq 15¢$. If candy bar 2 profit is 13¢, decision variables remain the same, but profit is now $4.50.

c $\frac{100}{3} \leq$ sugar ≤ 100.

d $z = \$3.40$, $x_1 = 0$, $x_2 = 20$, $x_3 = 40$. If 30 oz of sugar is available, current basis is no longer optimal, and problem must be solved again.

e Make type 1 candy bars.

f Make type 4 candy bars.

8 **a** $16,667.67 \leq$ comedy cost $\leq \$350,000.

b 4 million $\leq$ HIW ≤ 84 million. For 40 million HIW exposures, new optimal solution is $x_1 = 5.4$, $x_2 = 1.1$, $z = \$350,000.

c Advertise on news program.

SECTION 6.4

1 Yes.

2 No.

4 Yes.

1 min $w = y_1 + 3y_2 + 4y_3$

s.t. $-y_1 + y_2 + y_3 \geq 2$

$y_1 + y_2 - 2y_3 \geq 1$

$y_1, y_2, y_3 \geq 0$

2 max $z = 4x_1 + x_2 + 3x_3$

s.t. $2x_1 + x_2 + x_3 \leq 1$

$x_1 + x_2 + 2x_3 \leq -1$

$x_1, x_2, x_3 \geq 0$

3 min $w = 5y_1 + 7y_2 + 6y_3 + 4y_4$

s.t. $y_1 + 2y_2 \qquad + y_4 \geq 4$

$y_1 + y_2 + 2y_3 \qquad = -1$

$y_3 + y_4 = 2$

$y_1, y_2 \geq 0; y_3 \leq 0; y_4$ urs

4 max $z = 6x_1 + 8x_2$

s.t. $x_1 + x_2 \leq 4$

$2x_1 - x_2 \leq 2$

$2x_2 = -1$

$x_1 \leq 0; x_2$ urs

SECTION **6.7**

1 a min $w = 100y_1 + 80y_2 + 40y_3$

s.t. $2y_1 + y_2 + y_3 \geq 3$

$y_1 + y_2 \qquad \geq 2$

$y_1, y_2, y_3 \geq 0$

b and c $y_1 = 1, y_2 = 1, y_3 = 0, w = 180$. Observe that this solution has a w-value that equals the optimal primal z-value. Since this solution is dual feasible, it must be optimal (by Lemma 2) for the dual.

2 a min $w = 3y_1 + 2y_2 + y_3$

s.t. $y_1 + \qquad y_3 \geq -2$

$y_1 + y_2 \qquad \geq -1$

$y_1 + y_2 + y_3 \geq 1$

b $y_1 = $ coefficient of s_1 in optimal row $0 = 0$

$y_2 = -($coefficient of e_2 in optimal row 0$)$

$= -1$

$y_3 = $ coefficient of a_3 in optimal row $0 - M$

$= 2$

Optimal w-value $= 0$.

9 Dual is max $w = 28y_1 + 24y_2$

s.t. $7y_1 + 2y_2 \leq 50$

$2y_1 + 12y_2 \leq 100$

$y_1, y_2 \geq 0$

Optimal dual solution is $w = \$320{,}000, y_1 = 5, y_2 = 7.5$.

SECTION **6.8**

2 b New z-value $= \$3.40$.

c New z-value $= \$2.60$.

d Since current basis is no longer optimal, the current shadow prices cannot be used to determine the new z-value.

5 b Skilled labor shadow price $= 0$, unskilled labor shadow price $= 0$, raw material shadow price $= 15$, and product 2 constraint shadow price $= -5$.

We would be willing to pay $\$0$ for an additional hour of either type of labor. We would pay up to $\$15$ for an extra unit of raw material. Reducing the product 2 marketing requirement by 1 unit will save the company $\$5$.

c $\Delta b_3 = 5$, so new z-value $= 435 + 5(15) = \$510$.

d Since shadow price of each labor constraint is zero, the optimal z-value remains unchanged.

e For a 5-unit requirement, $\Delta b_4 = 2$. Thus, new z-value $= 435 + 2(-5) = \$425$. For a 2-unit requirement, $\Delta b_4 = -1$. Thus, new z-value $= 435 + (-1)(-5) = \$440$.

6 a If purchased at the given price of $\$1$, an extra unit of raw material increases profits by $\$2.50$. Thus, the firm would be willing to pay up to $1 + 2.5 = \$3.50$ for an extra unit of raw material.

b Both labor constraints are nonbinding. All we can say is that if an additional hour of skilled labor were available at $\$3$/hour, we would not buy it, and if an additional hour of unskilled labor were available at $\$2$/hour, we would not buy it.

7 a New z-value $= \$380{,}000$.

b New z-value $= \$290{,}000$.

SECTION **6.9**

1 The current basis is no longer optimal. We should make computer tables, because they sell for $\$35$ each and use only $\$30$ worth of resources.

2 a Current basis remains optimal if type 1 profit ≤ 6¢.

SECTION **6.10**

1 a min $w = 600y_1 + 400y_2 + 500y_3$

s.t. $4y_1 + y_2 + 3y_3 \geq 6$

$9y_1 + y_2 + 4y_3 \geq 10$

$7y_1 + 3y_2 + 2y_3 \geq 9$

$10y_1 + 40y_2 + y_3 \geq 20$

$y_1, y_2, y_3, y_4 \geq 0$

b $w = \frac{2{,}800}{3}, y_1 = \frac{22}{5}, y_2 = \frac{2}{15}, y_3 = 0$.

SECTION **6.11**

1 $z = -9, x_1 = 0, x_2 = 14, x_3 = 9$.

2 a The current solution is still optimal.

b The LP is now infeasible.

c The new optimal solution is $z = 10$, $x_1 = 1$, $x_2 = 4$.

SECTION 6.12

4 Only HPER is inefficient.

REVIEW PROBLEMS

1 **a** min $w = 6y_1 + 3y_2 + 10y_3$

s.t. $y_1 + y_2 + 2y_3 \geq 4$

$2y_1 - y_2 + y_3 \geq 1$

y_1 urs; $y_2 \leq 0$; $y_3 \geq 0$

Optimal dual solution is $w = \frac{58}{3}$, $y_1 = -\frac{2}{3}$, $y_2 = 0$, $y_3 = \frac{7}{3}$.

b $9 \leq b_3 \leq 12$. If $b_3 = 11$, the new optimal solution is $z = \frac{65}{3}$, $x_1 = \frac{16}{3}$, $x_2 = \frac{1}{3}$.

2 $c_1 \geq \frac{1}{2}$.

3 **a** min $w = 6y_1 + 8y_2 + 2y_3$

s.t. $y_1 + 6y_2 \geq 5$

$y_1 + y_3 \geq 1$

$y_1 + y_2 + y_3 \geq 2$

$y_1, y_2, y_3 \geq 0$

Optimal dual solution is $w = 9$, $y_1 = 0$, $y_2 = \frac{5}{6}$, $y_3 = \frac{7}{6}$.

b $0 \leq c_1 \leq 6$.

c $c_2 \leq \frac{7}{6}$.

4 **a** New z-value $= 32{,}540 + 10(88) = \$33{,}420$. Decision variables remain the same.

b Can't tell, since allowable increase is < 1.

c $\$0$.

d $32{,}540 + (-2)(-20) = \$32{,}580$.

e Produce jeeps.

8 **a** New z-value $= \$266.20$.

b New z-value $= \$270.70$. Decision variables remain the same.

c $\$12.60$.

d $20¢$.

e Produce product 3.

17 $z = -16$, $x_1 = 8$, $x_2 = 0$.

20 Optimal primal solution: $z = 13$, $x_1 = 1$, $x_2 = x_3 = 0$, $x_4 = 2$. Optimal dual solution: $w = 13$, $y_1 = 1$, $y_2 = 1$.

21 **a** $c_1 \geq 3$.

b $c_2 \leq \frac{4}{3}$.

c $0 \leq b_1 \leq 9$.

d $b_2 \geq 10$.

28 LP 2 optimal solution: $z = 550$, $x_1 = 0.5$, $x_2 = 5$. Optimal solution to dual of LP 2: $w = 550$, $y_1 = y_2 = \frac{100}{3}$.

36 $b_2 \geq 3$.

Chapter 7

SECTION 7.1

1

	CUSTOMER 1	CUSTOMER 2	CUSTOMER 3	SUPPLY
Warehouse 1	15	35	25	40
Warehouse 2	10	50	40	30
Shortage	90	80	110	20
DEMAND	30	30	30	

3

								SUPPLY
1-RT	7	8	9	10	11	12	0	200
1-OT	11	12	13	14	15	16	0	100
2-RT	M	7	8	9	10	11	0	200
2-OT	M	11	12	13	14	15	0	100
3-RT	M	M	7	8	9	10	0	200
3-OT	M	M	11	12	13	14	0	100
4-RT	M	M	M	7	8	9	0	200
4-OT	M	M	M	11	12	13	0	100
5-RT	M	M	M	M	7	8	0	200
5-OT	M	M	M	M	11	12	0	100
6-RT	M	M	M	M	M	7	0	200
6-OT	M	M	M	M	M	11	0	100
	200	260	240	340	190	150	420	

5

	MONTH 1	MONTH 2	DUMMY	SUPPLY
Daisy	800	720	0	5
Laroach	710	750	0	5
DEMAND	3	4	3	

7 This is a maximization problem, so number in each cell is a revenue, not a cost.

	CLIFF	BLAKE	ALEXIS	SUPPLY
Site 1	1000	900	1100	100,000
Site 2	2000	2200	1900	100,000
Dummy	0	0	0	40,000
DEMAND	80,000	80,000	80,000	

12 **a** Replace the M's by incorporating a backlogging cost. For example, month 3 regular production can be used to meet month 1 demand at a cost of $400 + 2(30) = \$460$.

b Add a supply point called "lost sales," with cost of shipping a unit to any month's demand being $450. Supply of "lost sales" supply point should equal total demand. Then adjust dummy demand point's demand to rebalance the problem.

c A shipment from month 1 production to month 4 demand should have a cost of M.

d For each month, add a month i subcontracting supply point, with a supply of 10 and a cost that is $40 more than the cost for the corresponding month i regular supply point. Then adjust the demand at the dummy demand point so that the problem is balanced.

SECTIONS 7.2 AND 7.3

The optimal solution to Problem 1 of Section 7.1 is to ship 10 units from warehouse 1 to customer 2, 30 units from warehouse 1 to customer 3, and 30 units from warehouse 2 to customer 1.

The optimal solution to Problem 5 of Section 7.1 is to buy 4 gallons from Daisy in month 2 and 3 gallons from Laroach in month 1.

In Problem 7 of Section 7.1, Cliff gets 20,000 acres at site 1 and 20,000 acres at site 2. Blake gets 80,000 acres at site 2. Alexis gets 80,000 acres at site 1.

SECTION 7.4

2 Current basis remains optimal if $c_{34} \leq 7$.

4 New optimal solution is $x_{12} = 12$, $x_{13} = 23$, $x_{21} = 45$, $x_{23} = 5$, $x_{32} = 8$, $x_{34} = 30$, and $z = 1{,}020 - 2(3) - 2(10) = 994$.

SECTION 7.5

1 Person 1 does job 2, person 2 does job 1, person 3 does no job, person 4 does job 4, and person 5 does job 3.

8 **a** Company 1 does route 1, company 2 does route 2, company 3 does route 3, and company 4 does route 4.

b Company 3 does routes 3 and 1, company 2 does routes 2 and 4.

SECTION 7.6

1 **a**

	L.A.	DETROIT	ATLANTA	HOUSTON	TAMPA	DUMMY	
L.A.	0	140	100	90	225	0	5,100
Detroit	145	0	111	110	119	0	6,900
Atlanta	105	115	0	113	78	0	4,000
Houston	89	109	121	0	M	0	4,000
Tampa	210	117	82	M	0	0	4,000
	4,000	4,000	4,000	6,400	5,500	100	

b

	L.A.	DETROIT	ATLANTA	HOUSTON	TAMPA	DUMMY	
L.A.	0	M	100	90	225	0	5,100
Detroit	M	0	111	110	119	0	6,900
Atlanta	105	115	0	113	78	0	4,000
Houston	89	109	121	0	M	0	4,000
Tampa	210	117	82	M	0	0	4,000
	4,000	4,000	4,000	6,400	5,500	100	

c

	L.A.	DETROIT	ATLANTA	HOUSTON	TAMPA	DUMMY	
L.A.	0	140	100	90	225	0	5,100
Detroit	145	0	111	110	119	0	6,900
Atlanta	105	115	0	113	78	0	4,000
Houston	89	109	121	0	5	0	4,000
Tampa	210	117	82	5	0	0	4,000
	4,000	4,000	4,000	6,400	5,500	100	

REVIEW PROBLEMS

3 Meet January demand with 30 units of January production. Meet February demand with 5 units of January production, 10 units of February production, and 15 units of March production. Meet March demand with 20 units of March production.

4 Maid 1 does the bathroom, maid 2 straightens up, maid 3 does the kitchen, maid 4 gets the day off, and maid 5 vacuums.

7 Shipping 1 unit from W_i to W_j means one white student from district i goes to school in district j. Shipping 1 unit from B_i to B_j means one black student from district i goes to school in district j. The costs of M ensure that shipments from W_i to B_j or B_i to W_j cannot occur (table on next page).

8 Optimal solution is $z = 1,580$, $x_{11} = 40$, $x_{12} = 10$, $x_{13} = 10$, $x_{22} = 50$, $x_{32} = 10$, $x_{34} = 30$.

13 Optimal solution is $z = 98$, $x_{13} = 5$, $x_{21} = 3$, $x_{24} = 7$, $x_{32} = 3$, $x_{33} = 7$, $x_{34} = 5$.

25 Sell painting 1 to customer 1, painting 2 to customer 2, painting 3 to customer 3, and painting 4 to customer 4.

	W_1	B_1	W_2	B_2	W_3	B_3	SUPPLY
W_1	0	M	3	M	5	M	210
B_1	M	0	M	3	M	5	120
W_2	3	M	0	M	4	M	210
B_2	M	3	M	0	M	4	30
W_3	5	M	4	M	0	M	180
B_3	M	5	M	4	M	0	150
DEMAND	200	100	200	100	200	100	

Chapter 8

SECTION 8.2

2 1–2–5 (length 14).

3

	NODE 2	NODE 3	NODE 4	NODE 5	SUPPLY
Node 1	2	8	M	M	1
Node 2	0	5	4	12	1
Node 3	M	0	6	M	1
Node 4	M	M	0	10	1
DEMAND	1	1	1	1	

M = large number to prevent shipping a unit through a nonexistent arc.

5 Replace the car at times 2, 4, and 6. Total cost = $14,400.

SECTION 8.3

1 max $z = x_0$

s.t. $x_{so,1} \le 6, x_{so,2} \le 2, x_{12} \le 1, x_{32} \le 3,$

$x_{13} \le 3, x_{3,si} \le 2, x_{24} \le 7, x_{4,si} \le 7$

$x_0 = x_{so,1} + x_{so,2}$ **(Node so)**

$x_{so,1} = x_{13} + x_{12}$ **(Node 1)**

$x_{12} + x_{32} + x_{so,2} = x_{24}$ **(Node 2)**

$x_{13} = x_{32} + x_{3,si}$ **(Node 3)**

$x_{24} = x_{4,si}$ **(Node 4)**

$x_{3,si} + x_{4,si} = x_0$ **(Node si)**

All variables ≥ 0

Maximum flow = 6. Cut associated with $V' = \{2, 3, 4, si\}$ has capacity 6.

2 max $z = x_0$

s.t. $x_{so,1} \leq 2, x_{12} \leq 4, x_{1,si} \leq 3, x_{2,si} \leq 2,$

$x_{23} \leq 1, x_{3,si} \leq 2, x_{so,3} \leq 1$

$x_0 = x_{so,1} + x_{so,3}$ **(Node so)**

$x_{so,1} = x_{1,si} + x_{12}$ **(Node 1)**

$x_{12} = x_{23} + x_{2,si}$ **(Node 2)**

$x_{23} + x_{so,3} = x_{3,si}$ **(Node 3)**

$x_{1,si} + x_{2,si} + x_{3,si} = x_0$ **(Node si)**

All variables ≥ 0

Maximum flow = 3. Cut associated with $V' = \{1, 2, 3, si\}$ has capacity 3.

6 See Figure 5. An arc of capacity 1 goes from each package type node to each truck node. If maximum flow = 21, all packages can be delivered.

FIGURE 5

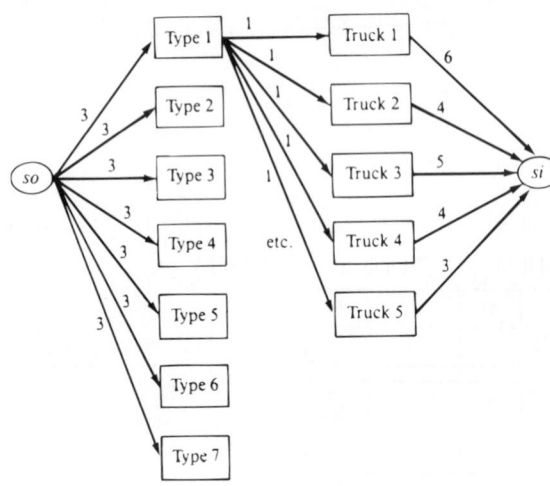

7 See Figure 6. If maximum flow = 4, then all jobs can be completed.

FIGURE 6

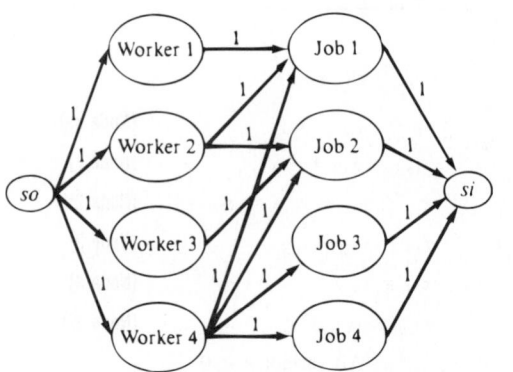

SECTION 8.4

4 a See Figure 7.

FIGURE 7

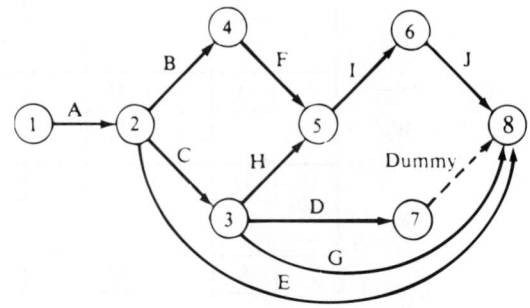

b Critical path is A–C–G (project duration is 14 days).
c Start project by June 13.
d min $z = x_8 - x_1$

s.t. $x_2 \geq x_1 + 3$

$x_3 \geq x_2 + 6$

$x_4 \geq x_2 + 2$

$x_5 \geq x_4 + 3$

$x_5 \geq x_3 + 1$

$x_6 \geq x_5 + 1.5$

$x_8 \geq x_6 + 2$

$x_8 \geq x_7$ $(x_7 \geq x_3 + 2)$

$x_8 \geq x_2 + 3$

$x_8 \geq x_3 + 5$

All variables urs

5 a See Figure 8. A–B–E–F–G and A–B–C–G are critical paths. Duration of project is 26 days.

FIGURE 8

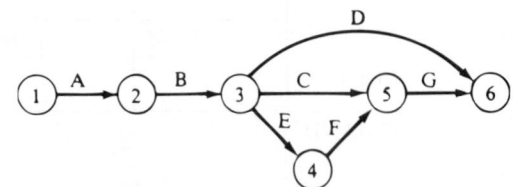

Activity	Total Float	Free Float
A	0	0
B	0	0
C	0	0
D	8	8
E	0	0
F	0	0
G	0	0

b $\min z = 30A + 15B + 20C + 40D$
$$+ 20E + 30F + 40G$$

s.t. $x_2 \geq x_1 + 5 - A$

$x_3 \geq x_2 + 8 - B$

$x_4 \geq x_3 + 4 - E$

$x_5 \geq x_3 + 10 - C$

$x_5 \geq x_4 + 6 - F$

$x_6 \geq x_3 + 5 - D$

$x_6 \geq x_5 + 3 - G$

$x_6 - x_1 \leq 20$

$A \leq 2, B \leq 3, C \leq 1, D \leq 2, E \leq 2,$

$F \leq 3, G \leq 1$

$A, B, C, D, E, F, G \geq 0$

All other variables urs

8 b From the LP output, we find the critical path 1–2–3–4–5–6. This implies that activities A, B, E, F, and G are critical. (Since 1–2–3–5–6 is also a critical path, activity C is also a critical activity, but the LP does not give us this information.)

SECTION 8.5

1 $\min z = 4x_{12} + 3x_{24} + 2x_{46} + 3x_{13}$
$$+ 3x_{35} + 2x_{25} + 2x_{56}$$

s.t. $x_{12} + x_{13} = 1$ **(Node 1)**

$x_{12} = x_{24} + x_{25}$ **(Node 2)**

$x_{13} = x_{35}$ **(Node 3)**

$x_{24} = x_{46}$ **(Node 4)**

$x_{25} = x_{56}$ **(Node 5)**

$x_{46} + x_{56} = 1$ **(Node 6)**

$x_{ij} \geq 0$

If $x_{ij} = 1$, the shortest path from node 1 to node 6 contains arc (i, j); if $x_{ij} = 0$, the shortest path from node 1 to node 6 does not contain arc (i, j).

4 a See Figure 9. All arcs have infinite capacity.

FIGURE 9

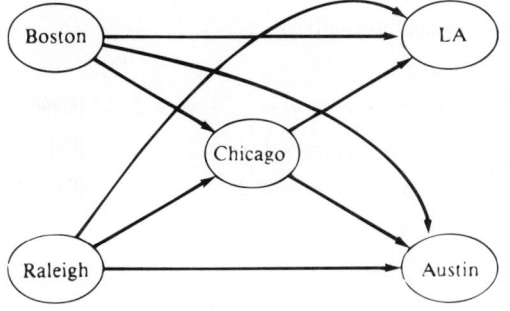

Arc	Shipping Cost
Bos.–Chic.	$800 + 80 = \$ 880$
Bos.–Aus.	$800 + 220 = \$1,020$
Bos.–L.A.	$800 + 280 = \$1,080$
Ral.–Chic.	$900 + 100 = \$1,000$
Ral.–Aus.	$900 + 140 = \$1,040$
Ral.–L.A.	$900 + 170 = \$1,070$
Chic.–Aus.	$\$40$
Chic.–L.A.	$\$50$

Problem is balanced, so no dummy point is needed.

City	Net Outflow
Boston	400
Raleigh	300
Chicago	0
L.A.	−400
Austin	−300

5

Arc	Unit Cost
S.D.–Dal.	$\$420$
S.D.–Hous.	$\$100$
L.A.–Dal.	$\$300$
L.A.–Hous.	$\$110$
S.D.–Dummy	$\$0$
L.A.–Dummy	$\$0$
Dal.–Chic.	$700 + 550 = \$1,250$
Dal.–N.Y.	$700 + 450 = \$1,150$
Hous.–Chic.	$900 + 530 = \$1,430$
Hous.–N.Y.	$900 + 470 = \$1,370$

City	Net Outflow (100,000 barrels/day)
San Diego	5
L.A.	4
Dallas	0
Houston	0
Chicago	−4
N.Y.	−3
Dummy	−2

SECTION 8.6

2 The MST consists of the arcs (1, 3), (3, 5), (3, 4), and (3, 2). Total length of MST is 15.

SECTION 8.7

1 c $z = 8, x_{12} = x_{25} = x_{56} = 1, x_{13} = x_{24} = x_{35} = x_{46} = 0$.

3 $z = 590, x_{12} = 20, x_{24} = 20, x_{34} = 2, x_{35} = 2, x_{13} = 12, x_{23} = 0, x_{25} = 0, x_{45} = 0$.

1 a N.Y.–St. Louis–Phoenix–L.A. uses 2,450 gallons of fuel.

2 a See Figure 10.

FIGURE 10

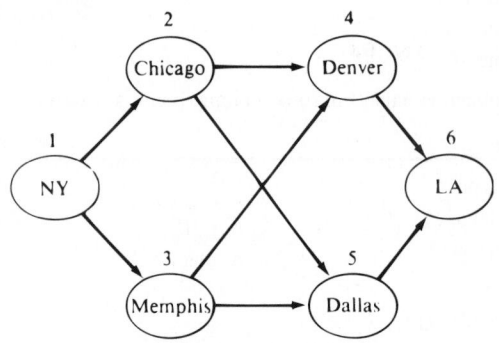

$$\max z = x_0$$

$$\text{s.t.} \quad x_{12} + x_{13} = x_0 \quad \textbf{(Node 1)}$$
$$x_{12} = x_{24} + x_{25} \quad \textbf{(Node 2)}$$
$$x_{13} = x_{34} = x_{35} \quad \textbf{(Node 3)}$$
$$x_{24} + x_{34} = x_{46} \quad \textbf{(Node 4)}$$
$$x_{25} + x_{35} = x_{56} \quad \textbf{(Node 5)}$$
$$x_{46} + x_{56} = x_0 \quad \textbf{(Node 6)}$$
$$x_{12} \le 500, \ x_{13} \le 400, \ x_{24} \le 300,$$
$$x_{25} \le 250, \ x_{34} \le 200, \ x_{35} \le 150,$$
$$x_{46} \le 400, \ x_{56} \le 350$$

$$\text{All variables} \ge 0$$

5 The MST consists of the following arcs: N.Y.–Clev., N.Y.–Nash., Nash.–Dal., Dal.–St.L., Dal.–Pho., Pho.–L.A., and S.L.C.–L.A. Total length of the MST is 4,300.

Chapter 9

SECTION 9.2

1 Let $x_1 = \begin{cases} 1 & \text{if player } i \text{ starts} \\ 0 & \text{otherwise} \end{cases}$

Then the appropriate IP is

$$\max z = 3x_1 + 2x_2 + 2x_3 + x_4 + 3x_5 + 3x_6 + x_7$$

$$\text{s.t.} \quad x_1 + x_3 + x_5 + x_7 \ge 4 \quad \textbf{(Guards)}$$
$$x_3 + x_4 + x_5 + x_6 + x_7 \ge 2 \quad \textbf{(Forwards)}$$
$$x_2 + x_4 + x_6 \ge 1 \quad \textbf{(Center)}$$
$$x_1 + x_2 + x_3 + x_4 + x_5 + x_6 + x_7 = 5$$
$$3x_1 + 2x_2 + 2x_3 + x_4 + 3x_5 + 3x_6 + 3x_7 \ge 10 \quad \textbf{(Ballhandling)}$$

$$3x_1 + x_2 + 3x_3 + 3x_4 + 3x_5 + x_6 + 2x_7 \ge 10 \quad \textbf{(Shooting)}$$

$$x_1 + 3x_2 + 2x_3 + 3x_4 + 3x_5 + 2x_6 + 2x_7 \ge 10 \quad \textbf{(Rebounding)}$$

$$x_6 + x_3 \le 1$$
$$-x_4 - x_5 + 2 \le 2y \quad \textbf{(If } x_1 > 0, \text{ then } x_4 + x_5 \ge 2\textbf{)}$$
$$x_1 \le 2(1 - y)$$
$$x_2 + x_3 \ge 1$$

$$x_1, x_2, \dots, x_7, y \text{ all } 0\text{–}1 \text{ variables}$$

3 Let $x_1 = $ units of product 1 produced

$x_2 = $ units of product 2 produced

$$y_i = \begin{cases} 1 & \text{if any product } i \text{ is ordered} \\ 0 & \text{otherwise} \end{cases}$$

Then the appropriate IP is

$$\max z = 2x_1 + 5x_2 - 10y_1 - 20y_2$$

$$\text{s.t.} \quad 3x_1 + 6x_2 \le 120$$
$$x_1 \le 40y_1$$
$$x_2 \le 20y_2$$
$$x_1, x_2 \ge 0; \ y_1, y_2 = 0 \text{ or } 1$$

6 $y_1 = \begin{cases} 1 & \text{if calculus is taken} \\ 0 & \text{otherwise} \end{cases}$

$y_2 = \begin{cases} 1 & \text{if operations research is taken} \\ 0 & \text{otherwise} \end{cases}$

$y_3 = \begin{cases} 1 & \text{if data structure is taken} \\ 0 & \text{otherwise} \end{cases}$

$y_4 = \begin{cases} 1 & \text{if business statistics is taken} \\ 0 & \text{otherwise} \end{cases}$

$y_5 = \begin{cases} 1 & \text{if computer simulation is taken} \\ 0 & \text{otherwise} \end{cases}$

$y_6 = \begin{cases} 1 & \text{if introduction to computer} \\ & \text{programming is taken} \\ 0 & \text{otherwise} \end{cases}$

$y_7 = \begin{cases} 1 & \text{if forecasting is taken} \\ 0 & \text{otherwise} \end{cases}$

Then the appropriate IP is

$$\min z = y_1 + y_2 + y_3 + y_4 + y_5 + y_6 + y_7$$

$$\text{s.t.} \quad y_1 + y_2 + y_3 + y_4 + y_7 \ge 2 \quad \textbf{(Math)}$$
$$y_2 + y_4 + y_5 + y_7 \ge 2 \quad \textbf{(OR)}$$
$$y_3 + y_5 + y_6 \ge 2 \quad \textbf{(Computers)}$$
$$y_4 \le y_1$$
$$y_5 \le y_6$$

$$y_3 \le y_6$$
$$y_7 \le y_4$$
$$y_1, y_2, \ldots, y_7 = 0 \text{ or } 1$$

10 Add the constraints $x + y - 3 \le Mz$, $2x + 5y - 12 \le M(1 - z)$, $z = 0$ or 1, where M is a large positive number.

11 Add the constraints $y - 3 \le Mz$, $3 - x \le (1 - z)M$, $z = 0$ or 1, where M is a large positive number.

13 Let $x_i =$ number of workers employed on line i

$$y_i = \begin{cases} 1 & \text{if line } i \text{ is used} \\ 0 & \text{otherwise} \end{cases}$$

Then the appropriate IP is

$$\min z = 1{,}000y_1 + 2{,}000y_2 + 500x_1 + 900x_2$$

$$\text{s.t.} \quad 20x_1 + 50x_2 \ge 120$$
$$30x_1 + 35x_2 \ge 150$$
$$40x_1 + 45x_2 \ge 200$$
$$x_1 \le 7y_1$$
$$x_2 \le 7y_2$$
$$x_1, x_2 \ge 0;\ y_1, y_2 = 0 \text{ or } 1$$

14 a Let $x_i = \begin{cases} 1 & \text{if disk } i \text{ is used} \\ 0 & \text{otherwise} \end{cases}$

Then the appropriate IP is

$$\min z = 3x_1 + 5x_2 + x_3 + 2x_4 + x_5 + 4x_6$$
$$\qquad + 3x_7 + x_8 + 2x_9 + 2x_{10}$$

$$\text{s.t.} \quad x_1 + x_2 + x_4 + x_5 + x_8 + x_9 \ge 1 \qquad \textbf{(File 1)}$$
$$x_1 + x_3 \ge 1 \qquad \textbf{(File 2)}$$
$$x_2 + x_5 + x_7 + x_{10} \ge 1 \qquad \textbf{(File 3)}$$
$$x_3 + x_6 + x_8 \ge 1 \qquad \textbf{(File 4)}$$
$$x_1 + x_2 + x_4 + x_6 + x_7 + x_9 + x_{10} \ge 1 \qquad \textbf{(File 5)}$$
$$x_i = 0 \text{ or } 1 \quad (i = 1, 2, \ldots, 10)$$

b Add the constraints $1 - x_2 \le 2y$, $x_3 + x_5 \le 2(1 - y)$, $y = 0$ or 1 (need $M = 2$, because $x_3 + x_5 = 2$ is possible). We could also have added the constraints $x_2 \ge x_3$ and $x_2 \ge x_5$.

SECTION 9.3

1 $z = 20$, $x_1 = 4$, $x_2 = 0$.

2 $z = \$400{,}000$. $(x_1 = 6, x_2 = 1)$ and $(x_1 = 4, x_2 = 2)$ are both optimal solutions.

4 $z = 8$, $x_1 = 2$, $x_2 = 0$.

SECTION 9.4

1 $z = 5.6$, $x_1 = 1.2$, $x_2 = 2$.

SECTION 9.5

2 $\max z = 60x_1 + 48x_2 + 14x_3 + 31x_4 + 10x_5$

s.t. $800x_1 + 600x_2 + 300x_3 + 400x_4 + 200x_5 \le 1{,}100$

$$x_i = 0 \text{ or } 1$$

Optimal solution is $z = 79$, $x_2 = x_4 = 1$, $x_1 = x_3 = x_5 = 0$.

SECTION 9.6

1 Do jobs in the following order: job 2, job 1, job 3, and job 4. Total delay is 20 minutes.

2 LFR—LFP—LP—LR—LFR has a total cost of $330.

8 Warehouse 1 to factory 1, warehouse 2 to factory 3, warehouse 3 to factory 4, warehouse 4 to factory 2, warehouse 5 to factory 5 has a total cost of $35,000.

SECTION 9.7

1 $z = 4$, $x_1 = x_2 = x_4 = x_5 = 1$, $x_3 = 0$.

2 $z = 3$, $x_1 = x_3 = 1$, $x_2 = 0$.

SECTION 9.8

1 $z = 110$, $x_1 = 4$, $x_2 = 3$.

REVIEW PROBLEMS

3 Let $z_i = \begin{cases} 1 & \text{if gymnast } i \text{ enters both events} \\ 0 & \text{otherwise} \end{cases}$

$$x_i = \begin{cases} 1 & \text{if gymnast } i \text{ enters only balance beam} \\ 0 & \text{otherwise} \end{cases}$$

$$y_i = \begin{cases} 1 & \text{if gymnast } i \text{ enters only floor exercises} \\ 0 & \text{otherwise} \end{cases}$$

Then the appropriate IP is

$$\max z = 16.7z_1 + 17.7z_2 + \cdots + 17.7z_6 + 8.8x_1$$
$$\qquad + 9.4x_2 + \cdots + 9.1x_6 + 7.9y_1$$
$$\qquad + 8.3y_2 + \cdots + 8.6y_6$$

$$\text{s.t.} \quad z_1 + z_2 + \cdots + z_6 = 3$$
$$x_1 + x_2 + \cdots + x_6 = 1$$
$$y_1 + y_2 + \cdots + y_6 = 1$$
$$x_1 + y_1 + z_1 \le 1$$
$$x_2 + y_2 + z_2 \le 1$$
$$x_3 + y_3 + z_3 \le 1$$
$$x_4 + y_4 + z_4 \le 1$$
$$x_5 + y_5 + z_5 \le 1$$
$$x_6 + y_6 + z_6 \le 1$$
$$\text{All variables 0 or 1}$$

4 Let $x_{ij} = \begin{cases} 1 & \text{if students from district } i \\ & \text{are sent to school } j \\ 0 & \text{otherwise} \end{cases}$

Then the appropriate IP is

$\min z = 110x_{11} + 220x_{12} + 37.5x_{21} + 127.5x_{22}$

$\qquad + 80x_{31} + 80x_{32} + 117x_{41} + 36x_{42}$

$\qquad + 135x_{51} + 54x_{52}$

s.t. $110x_{11} + 75x_{21} + 100x_{31} + 90x_{41} + 90x_{51} \geq 150$
$\qquad\qquad\qquad\qquad$ **(School 1 $\geq$ 150 students)**

$\qquad 110x_{12} + 75x_{22} + 100x_{32} + 90x_{42} + 90x_{52} \geq 150$
$\qquad\qquad\qquad\qquad$ **(School 2 $\geq$ 150 students)**

$0.20 \leq \dfrac{30x_{11} + 5x_{21} + 10x_{31} + 40x_{41} + 30x_{51}}{110x_{11} + 75x_{21} + 100x_{31} + 90x_{41} + 90x_{51}},$

$\quad$ or $0 \leq 8x_{11} - 10x_{21} - 10x_{31} + 22x_{41} + 12x_{51}$

$0.20 \leq \dfrac{30x_{12} + 5x_{22} + 10x_{32} + 40x_{42} + 30x_{52}}{110x_{12} + 75x_{22} + 100x_{32} + 90x_{42} + 90x_{52}},$

$\quad$ or $0 \leq 8x_{12} - 10x_{22} - 10x_{32} + 22x_{42} + 12x_{52}$

$x_{11} + x_{12} = 1$

$x_{21} + x_{22} = 1$

$x_{31} + x_{32} = 1$

$x_{41} + x_{42} = 1$

$x_{51} + x_{52} = 1$

$\qquad\qquad$ All variables 0 or 1

5 Let $x_1 = \begin{cases} 1 & \text{if RS is signed} \\ 0 & \text{otherwise} \end{cases}$

$\qquad\quad x_2 = \begin{cases} 1 & \text{if BS is signed} \\ 0 & \text{otherwise} \end{cases}$

$\qquad\quad x_3 = \begin{cases} 1 & \text{if DE is signed} \\ 0 & \text{otherwise} \end{cases}$

$\qquad\quad x_4 = \begin{cases} 1 & \text{if ST is signed} \\ 0 & \text{otherwise} \end{cases}$

$\qquad\quad x_5 = \begin{cases} 1 & \text{if TS is signed} \\ 0 & \text{otherwise} \end{cases}$

Then the appropriate IP is

$\qquad \max z = 6x_1 + 5x_2 + 3x_3 + 3x_4 + 2x_5$

$\qquad$ s.t. $6x_1 + 4x_2 + 3x_3 + 2x_4 + 2x_5 \leq 12$

$\qquad\qquad\qquad x_2 + x_3 + x_4 \qquad\quad \leq 2$

$\qquad\qquad x_1 + x_2 + x_3 + \qquad x_5 \leq 2$

$\qquad\qquad x_1 + x_2 \qquad\qquad\qquad\quad \leq 1$

$\qquad\qquad$ All variables 0 or 1

10 Use two 20¢ coins, one 50¢ coin, and one 1¢ coin.

13 Infeasible.

26 Let

$x_{it} = \begin{cases} 1 & \text{if building } i \text{ is started during year } t \\ 0 & \text{otherwise} \end{cases}$

Then the appropriate IP (in thousands) is

$\max z = 100x_{11} + 50x_{12} + 60x_{21} + 30x_{22} + 40x_{31}$

s.t. $30x_{11} + 20x_{21} + 20x_{31} \leq 60$ $\qquad$ **(Year 1 workers)**

$\qquad 30(x_{11} + x_{12}) + 20(x_{21} + x_{22})$

$\qquad\qquad + 20(x_{31} + x_{32}) \leq 60$ $\qquad$ **(Year 2 workers)**

$\qquad 30(x_{12} + x_{13}) + 20(x_{22} + x_{23})$

$\qquad\qquad + 20(x_{31} + x_{32} + x_{33}) \leq 60$ $\qquad$ **(Year 3 workers)**

$\qquad 30(x_{13} + x_{14}) + 20(x_{23} + x_{24})$

$\qquad\qquad + 20(x_{32} + x_{33} + x_{34}) \leq 60$ $\qquad$ **(Year 4 workers)**

$\left. \begin{array}{l} x_{11} + x_{21} + x_{31} \leq 1 \\ x_{12} + x_{22} + x_{32} \leq 1 \\ x_{13} + x_{23} + x_{33} \leq 1 \end{array} \right\}$ **(No more than one building begins during each year)**

$\left. \begin{array}{l} x_{11} + x_{12} + x_{13} + x_{14} \leq 1 \\ x_{21} + x_{22} + x_{23} + x_{24} \leq 1 \\ x_{31} + x_{32} + x_{33} + x_{34} \leq 1 \end{array} \right\}$ **(Each building is started at most once)**

$\qquad x_{21} + x_{22} + x_{23} = 1$ $\quad$ **(Building 2 is finished by end of year 4)**

$\qquad\qquad$ All variables 0 or 1

27 Let $y_i = \begin{cases} 1 & \text{if truck } i \text{ is used} \\ 0 & \text{otherwise} \end{cases}$

$\qquad\quad x_{ij} = \begin{cases} 1 & \text{if truck } i \text{ is used to deliver to grocer } j \\ 0 & \text{otherwise} \end{cases}$

Then the appropriate IP is

$\min z = 45y_1 + 50y_2 + 55y_3 + 60y_4$

s.t. $100x_{11} + 200x_{12} + 300x_{13} + 500x_{14} + 800x_{15}$
$\qquad\qquad\qquad\qquad\qquad\qquad \leq 400y_1$

$\qquad 100x_{21} + 200x_{22} + 300x_{23} + 500x_{24} + 800x_{25}$
$\qquad\qquad\qquad\qquad\qquad\qquad \leq 500y_2$

$\qquad 100x_{31} + 200x_{32} + 300x_{33} + 500x_{34} + 800x_{35}$
$\qquad\qquad\qquad\qquad\qquad\qquad \leq 600y_3$

$\qquad 100x_{41} + 200x_{42} + 300x_{43} + 500x_{44} + 800x_{45}$
$\qquad\qquad\qquad\qquad\qquad\qquad \leq 1{,}100y_4$

$\qquad x_{11} + x_{21} + x_{31} + x_{41} = 1$

$\qquad x_{12} + x_{22} + x_{32} + x_{42} = 1$

$\qquad x_{13} + x_{23} + x_{33} + x_{43} = 1$

$\qquad x_{14} + x_{24} + x_{34} + x_{44} = 1$

$\qquad x_{15} + x_{25} + x_{35} + x_{45} = 1$

$\qquad\qquad$ All variables 0 or 1

Chapter 10

SECTIONS 10.1 AND 10.2

1 $z = 11, x_1 = 3, x_2 = 0, x_3 = 2$.

2 $z = 10, x_1 = 2, x_2 = 2$.

SECTION 10.3

2 Let x_i = number of 15-ft boards cut according to combination i, where

Combination	3-Ft Boards	5-Ft Boards	8-Ft Boards
1	0	1	1
2	2	0	1
3	0	3	0
4	1	2	0
5	3	1	0
6	5	0	0

Then we want to solve

$$\min z = x_1 + x_2 + x_3 + x_4 + x_5 + x_6$$
$$\text{s.t.} \quad 2x_2 + x_4 + 3x_5 + 5x_6 \geq 10$$
$$x_1 + 3x_3 + 2x_4 + x_5 \geq 20$$
$$x_1 + x_2 \geq 15$$
$$x_i \geq 0$$

The optimal solution is $z = \frac{55}{3}, x_1 = 10, x_2 = 5, x_3 = \frac{10}{3}$.

SECTION 10.4

1 $z = 40, x_1 = 3, x_2 = 2, x_3 = 3$.

3 $z = 15, x_1 = 0, x_2 = 0, x_3 = 3$.

SECTION 10.5

1 $z = 29.5, x_1 = 2, x_2 = 0.5, x_3 = 4, x_4 = x_5 = 0$.

3 $z = \frac{81}{13}, x_1 = \frac{12}{13}, x_2 = \frac{11}{13}$.

SECTION 10.6

1 $\mathbf{y}^1 = \mathbf{x}^1 = [\frac{3}{8} \ \frac{1}{4} \ \frac{3}{8}]$.

REVIEW PROBLEMS

1 $z = 9, x_1 = x_3 = 0, x_2 = 3$.

3 $z = -12, x_2 = 2, x_4 = 10, x_1 = x_3 = 0$.

5 Maximum profit is $540. Optimal production levels are

Product 1 at plant 1 $= \frac{5}{3}$ units

Product 1 at plant 2 $= \frac{100}{3}$ units

Product 2 at plant 1 $= \frac{290}{9}$ units

Product 2 at plant 2 $= 0$ units

Chapter 11

SECTION 11.1

1 3.

3 a $-xe^{-x} + e^{-x}$.

e $\frac{3}{x}$.

5 $\ln(1 + x) = x - \frac{x^2}{2} + \frac{x^3}{3p^3}$ for some p between 1 and $1 + x$.

7 a k.

b The maximum size of the market (as measured in sales per year).

9 The machine time is the better buy.

SECTION 11.2

1 a Let S = soap opera ads and F = football ads. Then we want to solve the following LP:

$$\min z = 50S + 100F$$
$$\text{s.t.} \quad 5S^{1/2} + 17F^{1/2} \geq 40 \quad \textbf{(Men)}$$
$$20S^{1/2} + 7F^{1/2} \geq 60 \quad \textbf{(Women)}$$
$$S, F \geq 0$$

b Since doubling S does not double the contribution of S to the constraints, we are violating the Proportionality Assumption. Additivity is not violated.

5 $a = b = c = 20$.

SECTION 11.3

1 Convex.

2 Neither convex nor concave.

5 Concave.

8 Concave.

SECTION 11.4

1 If fixed cost is $5,000, spend $10,000 on advertising. If fixed cost is $20,000, don't spend any money on advertising.

2 Without tax, produce 12.25 units; with tax, produce 12 units.

5 $z = 1, x = 1$.

SECTION 11.5

2 After four iterations, the interval of uncertainty is $[-0.42, 0.17)$.

SECTION 11.6

1 $x = \dfrac{x_1 + x_2 + \cdots + x_n}{n}$,

$y = \dfrac{y_1 + y_2 + \cdots + y_n}{n}$.

3 $q_1 = 98.5$, $q_2 = 1$.

6 $(0, 0)$ is a saddle point. $(\frac{3}{2}, \frac{3}{2})$ and $(\frac{3}{2}, -\frac{3}{2})$ are each a local minimum.

SECTION 11.7

3 Successive points are $(\frac{1}{2}, \frac{3}{4})$, $(\frac{3}{4}, \frac{3}{4})$, $(\frac{3}{4}, \frac{7}{8})$.

SECTION 11.8

2 $L = K = \frac{10}{3}$; produce 10 machines.

4 $x_1 = \dfrac{900}{13}$, $x_2 = \dfrac{400}{13}$, $\lambda = \dfrac{13^{1/2}}{2} - 1$. An extra dollar spent on promotion would increase profit by approximately $\$\left(\dfrac{13^{1/2}}{2} - 1\right)$.

SECTION 11.9

1 Capacity = 27.5 kwh. Peak price = \$65. Off-peak price = \$20.

2 $z = 2^{1/2}$, $x_1 = \dfrac{2^{1/2}}{2}$, $x_2 = -\dfrac{2^{1/2}}{2}$.

6 $z = \frac{1}{2}$, $x_1 = \frac{1}{2}$, $x_2 = \frac{3}{2}$.

SECTION 11.10

1 $\min z = 0.09x_1^2 + 0.04x_2^2 + 0.01x_3^2$
$+ 0.012x_1x_2 - 0.008x_1x_3 + 0.010x_2x_3$

s.t. $\quad x_2 - x_3 \geq 0$

$x_1 + x_2 + x_3 = 100$

$x_1, x_2, x_3 \geq 0$

4 $p_1 = \$292.81$, $p_2 = \$158.33$. Pay no money for an additional hour of labor. Pay up to (approximately) \$53.81 for another chip.

SECTION 11.11

1 Using grid points 0, 0.5, 1, 1.5, and 2 for x_1 and grid points 0, 0.5, 1, 1.5, 2, and 2.5 for x_2, we obtain the following approximating problem:

$\min z = 0.25\delta_{12} + \delta_{13} + 2.25\delta_{14} + 4\delta_{15} + 0.25\delta_{22}$
$+ \delta_{23} + 2.25\delta_{24} + 4\delta_{25} + 6.25\delta_{26}$

s.t. $\quad 0.25\delta_{12} + \delta_{13} + 2.25\delta_{14} + 4\delta_{15} + 2(0.25\delta_{22}$
$+ \delta_{23} + 2.25\delta_{24} + 4\delta_{25} + 6.25\delta_{26}) \leq 4$

$0.25\delta_{12} + \delta_{13} + 2.25\delta_{14} + 4\delta_{15} + 0.25\delta_{22}$
$+ \delta_{23} + 2.25\delta_{24} + 4\delta_{25} + 6.25\delta_{26} \leq 6$

$\delta_{11} + \delta_{12} + \delta_{13} + \delta_{14} + \delta_{15} = 1$

$\delta_{21} + \delta_{22} + \delta_{23} + \delta_{24} + \delta_{25} + \delta_{26} = 1$

All variables nonnegative

Adjacency Assumption

SECTION 11.12

1 $\mathbf{x}^1 = [0 \quad 1]$ and $\mathbf{x}^2 = [\frac{1}{3} \quad \frac{5}{6}]$.

SECTION 11.13

2 First we attempt to maximize output by solving

$\max z = 20x_{11} + 12x_{12} + 10x_{13} + 12x_{21} + 15x_{22}$
$+ 9x_{23} + 6x_{31} + 5x_{32} + 10x_{33}$

s.t. $\quad x_{11} + x_{12} + x_{13} = 1$

$x_{21} + x_{22} + x_{23} = 1$

$x_{31} + x_{32} + x_{33} = 1$

$x_{11} + x_{21} + x_{31} = 1$

$x_{12} + x_{22} + x_{32} = 1$

$x_{13} + x_{23} + x_{33} = 1$

All $x_{ij} \geq 0$

Here x_{ij} fraction of day worker i spends working on product j. This LP has optimal solution $x_{11} = x_{22} = x_{33} = 1$, with Output = 45 and Happiness = $6 + 5 + 8 = 19$. Thus, point $(45, 19)$ is on a tradeoff curve. Now add constraint

$6x_{11} + 8x_{12} + 10x_{13} + 6x_{21} + 5x_{22} + 9x_{23} + 9x_{31}$
$+ 10x_{32} + 8x_{33} \geq \text{HAPP}$

where HAPP = 20, 25, and 26. (HAPP cannot exceed 26.) This yields four points on the tradeoff curve in Figure 11. Value of Output is optimal z-value for each LP.

FIGURE 11

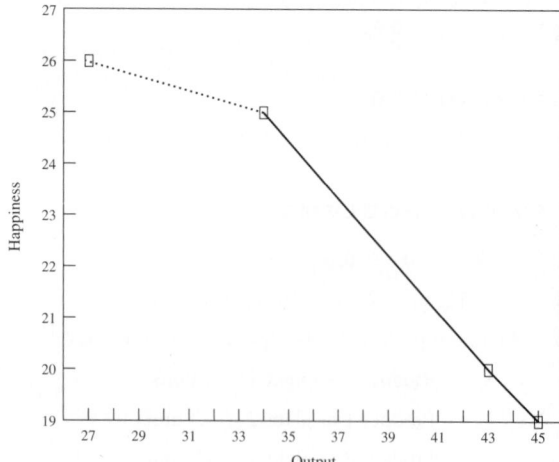

2 Locate the store at point 7. In general, locate store at arithmetic mean of the location of all customers.

3 **a** Use $\frac{93}{8}$ units of raw material, sell $\frac{93}{4}$ units of product 1, and sell $\frac{15}{2}$ units of product 2.
c Pay slightly less than $5 for an extra unit of raw material.

5 [1.18, 1.63).

7 Produce 20 units during each of the three months.

18 Locate the store at point 5; in general, locate store at the median of the customer's locations.

Chapter 12

SECTION 12.1

1 $\frac{e^{10} - 1}{2}$.

3 $Ih - \frac{dh}{2}$.

SECTION 12.2

1 $2y(2y + y^2) - 3y + 2(y^2 - y)$.

SECTION 12.3

1 **a** $\frac{2}{9}$.
c No.
e $\frac{1}{4}$.

SECTION 12.4

1 $\frac{2}{3}$.

3 .001.

SECTION 12.5

1 **a** Let $\mathbf{S}$ = number sold. Then $E(\mathbf{S}) = \frac{290}{3}$, and var $\mathbf{S} = \frac{200}{9}$.

3 **a** $F(a) = 0$ for $a \leq 0$, $F(a) = 1 - e^{-a}$ for $a \geq 0$.
b $E(\mathbf{X}) = 1$ var $\mathbf{X} = 1$.
c $e^{-1} - e^{-2}$.

SECTION 12.6

2 .8749.

REVIEW PROBLEMS

3 $\frac{25}{2}$.

5 **a** $E(\mathbf{X}) = 85$; var $\mathbf{X} = 9{,}000$.
b $P\left(\mathbf{Z} \geq \frac{91 - 85}{(9{,}000)^{1/2}}\right) = .476$.

7 **a** $\frac{1}{6}$.
b .004996.

Chapter 13

SECTION 13.1

1 Maximin decision; small campaign. Maximax decision: large campaign. Minimax regret decision: large campaign.

2 Maximin decision; don't build. Maximax decision: build. Maximax regret decision: build. Expected value decision: build.

5 Maximin decision: $6,000, $8,000, or $11,000 bid. Maximax decision: $11,000 bid. Minimax regret decision: $11,000 bid. Expected value decision: $11,000 bid.

SECTION 13.2

1 **a** Risk-averse.
b Prefer L_1; risk premium for $L_2 = \$339$.

2 **a** Risk-seeking.
b Prefer L_2; risk premium for $L_2 = -\$235$.

6 **b** $1,900.

7 Take statistics course.

13 L_1 is preferred.

SECTION 13.4

1 Hire the geologist. If she gives a favorable report, drill; if she gives an unfavorable report, don't drill. Expected profit is $180,000; EVSI = $20,000; EVPI = $55,000.

4 Market without testing. Expected profit = $16,000; EVSI = $3,800; EVPI = $14,000.

9 Play daringly during the first game. If he wins the first game, play conservatively during the second game. If he loses the first game, play daringly during the second game. If tied after two games, play daringly during the third game.

12 **a** Buy the gold now.
b Wait for Congress and (if possible) buy the gold later.

SECTION 13.5

2 Hire the geologist. If he predicts an earthquake, build at Roy Rogers; if he predicts no earthquake, build at Diablo. Expected total cost = $13,900,000. EVSI = $1,100,000; EVPI = $2,000,000.

4 Hire the firm. If it predicts a hit, air the show. If it predicts a flop, don't air the show. Expected profit = $35,000; EVSI = $50,000; EVPI = $75,000.

SECTION 13.6

1 **c** National's utility function is of the form $.3u_1(x_1) + .5u_2(x_2) + .2u_1(x_1)u_2(x_2)$.

6 d $k_3 > 0$.

1 a Professor 2 should receive the bigger raise.
   c Pairwise comparison matrix is consistent.

REVIEW PROBLEMS

1 a Invest in money market fund.
   b Invest in gold.
   c Money market fund has a maximum regret of $500.
   d All investments have the same expected return.

2 Because of the risk-averse nature of the utility function, invest in the least risky investment (the money market fund).

5 a Invest all money in stocks.
   b Indifferent between hiring and not hiring forecaster. Expected final asset position = $1,160,000; EVSI = $10,000; EVPI = $20,000.

12 If all potential litterers are risk-averse, then raising the fine will result in the larger decrease in littering.

Chapter 14

SECTION **14.1**

1 Value to row player = 2. Row player plays row 1, and column player plays column 1.

2 Value to row player = 6. Row player plays row 2, and column player plays column 1 or column 3.

SECTION **14.2**

1 Value to row player = $\frac{4}{3}$. Row player's optimal strategy is $(\frac{2}{3}, \frac{1}{3})$ and column player's optimal strategy is $(\frac{2}{3}, \frac{1}{3}, 0)$.

8 State's optimal strategy is, with probability $\frac{1}{2}$, play A first and B second; with probability $\frac{1}{2}$, play B first and A second. Ivy's optimal strategy is, with probability $\frac{1}{2}$, play X first and Y second; with probability $\frac{1}{2}$, play X second and Y first. Value of game to State = $\frac{1}{2}$.

SECTION **14.3**

1 a

Gunner	Soldier 1	2	3	4	5
Spot A	1	1	0	0	0
Spot B	0	1	1	0	0
Spot C	0	0	1	1	0
Spot D	0	0	0	1	1

   b Columns 2 and 4 are dominated.
   c Expected value to the gunner = $\frac{1}{3}$.
   d Always firing at A.

e Gunner's LP is

$$\max z = v$$
$$\text{s.t.} \quad v \le x_1$$
$$v \le x_1 + x_2$$
$$v \le x_2 + x_3$$
$$v \le x_3 + x_4$$
$$v \le x_4$$
$$x_1 + x_2 + x_3 + x_4 = 1$$
$$x_1, x_2, x_3, x_4 \ge 0; v \text{ urs}$$

Soldier's LP is

$$\min w$$
$$\text{s.t.} \quad w \ge y_1 + y_2$$
$$w \ge y_2 + y_3$$
$$w \ge y_3 + y_4$$
$$w \ge y_4 + y_5$$
$$\text{s.t.} \quad y_1 + y_2 + y_3 + y_4 + y_5 = 1$$
$$y_1, y_2, y_3, y_4, y_5 \ge 0; w \text{ urs}$$

3 Value to row player = $\frac{5}{2}$. Row player's optimal strategy is $(\frac{1}{2}, \frac{1}{2})$, and column player's optimal strategy is $(\frac{3}{4}, \frac{1}{4}, 0)$.

SECTION **14.4**

1 $(9, -1)$ is an equilibrium point.

3 This is a Prisoner's Dilemma game, with the equilibrium point occurring where each borough opposes the other borough's bond issues. Reward is $0 to each borough.

SECTION **14.7**

2 Core consists of the point $(25, 25, 25, 25)$.

3 a $v(\{\ \}) = \$0; v(\{1\}) = v(\{2\}) = v(\{3\}) = -\$2;$ $v(\{1, 2\}) = v(\{2, 3\}) = v(\{1, 3\}) = \$2; v(\{1, 2, 3\}) = \$3.$

   b and c The Shapley value gives $1 to each player. The core is ($1, $1, $1).

7 Assuming that the runway costs $1/ft, the Shapley value recommends the following fees per landing: type 1, $20; type 2, $\frac{240}{7}$; type 3, $\frac{440}{7}$; type 4, $\frac{1,140}{7}$.

REVIEW PROBLEMS

1 Both stores will be located at point B, and the two firms will each have 26 customers.

3 b Value to row player = $-\frac{5}{11}$. For each player, the optimal strategy is $(\frac{4}{11}, \frac{4}{11}, \frac{3}{11})$.

6 a $v(\{\ \}) = v(\{49\}) = v(\{50\}) = v(\{1\}) = v(\{1, 49\}) = 0; v(\{1, 50\}) = v(\{49, 50\}) = v(\{1, 49, 50\}) = 1.$

   b Core consists of point $(0, 0, 1)$.

   c Shapley value gives $\frac{1}{6}$ to player 1, $\frac{1}{6}$ to player 2, and $\frac{2}{3}$ to player 3.

Chapter 15

SECTION 15.2

1 **a** 4,000 gallons.
b 12 orders per year.
c One month.
e For a 2-week lead time, reorder point $= \frac{48,000}{26} = 1,846.15$ gallons. For a 10-week lead time, reorder point $= 1,230.77$ gallons.

3 **a** Send out $\frac{30}{7.07} = 4.24$ trucks per hour.
b Send out $\frac{30}{5} = 6$ trucks per hour.

12 **b** Six trainees in each program.
c Run 4.5 programs per year.
d 2.25 trainees.

SECTION 15.3

1 Order 300 boxes per year. Place 3.2 orders per year.
5 Order 100 thermometers.

SECTION 15.4

2 Optimal run size $= 692.82$. Do 34.64 runs per year.

SECTION 15.5

2 Whenever dealer is 10 cars short, an order for 50 cars should be placed. Maximum shortage will be 10 cars.

SECTION 15.6

1 Demand is too lumpy to justify using EOQ.

REVIEW PROBLEMS

1 **a** 268.33 desks.
b 22.36 orders per year.
c $2(22.36)(300) = \$13,416$.
d For 1-week lead time, reorder point $= 115.38$ desks; for 5-week lead time, reorder point $= 40.93$ desks.
e Order 342.05 desks 17.54 times per year.

3 The EOQ for the lowest price is optimal. Thus, 417.79 cameras should be ordered.

Chapter 16

SECTION 16.2

1 **a** $q = 6$.
b $q = 2$.
c $E(q)$ is not a convex function of q.

SECTION 16.3

1 Order 35 cars.
3 q^* will decrease.
5 **a** Order 60 cells.

SECTION 16.4

2 \$775,000, using $F(.25) = .60$.
3 107.5 trees, using $F(.25) = .60$.

SECTION 16.5

1 Locate at $\dfrac{2^{1/2}}{2}$.

SECTION 16.6

1 Order quantity $= 45.61$, reorder point $= 32.6$, safety stock $= 12.6$.

2 **a** Reorder print $= 57.9$, safety stock $= 17.9$.
b Reorder print $= 60.60$, safety stock $= 20.60$.

SECTION 16.7

1

SLM_1	Reorder Point
80%	11.06
90%	16.46
95%	20.24
99%	26.30

For $SLM_2 = 0.5$ stockout per year, the reorder point is 32.06.

3 **a** $SLM_1 = 98.75\%$.
b $r = 20$ is the smallest reorder point with SLM_1 exceeding 95%.
c Need a reorder point of at least 30 units.

SECTION 16.8

1 $R = 0.23$ years and $S = 3,241$.

SECTION 16.9

1 Type A items: 1 and 2; type B items: 3 through 6; type C items: 7 through 10. See Figure 12.

FIGURE 12

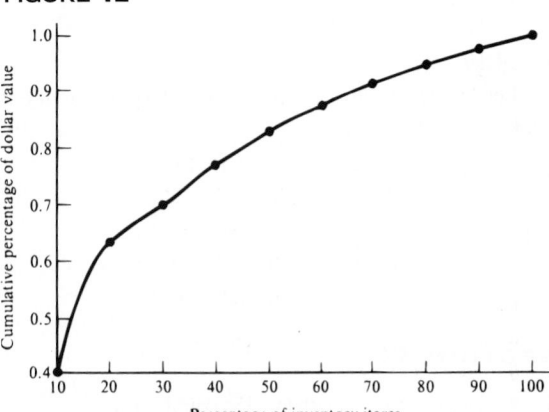

1 **a** See Figure 13.

FIGURE 13

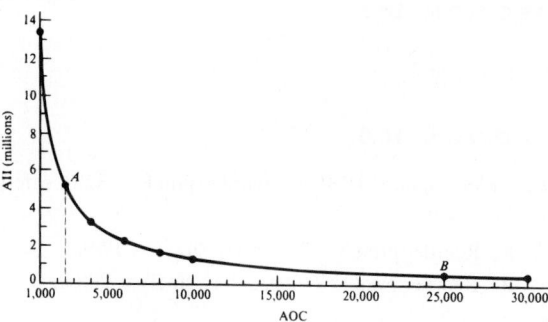

REVIEW PROBLEMS

1 **a** Bake 40 dozen cookies.
  b Bake 58.6 dozen cookies.
  c Bake 55 dozen cookies.

3 **a** Order quantity = 707.11, shortage cost = $12.41.
  b Order quantity = 707.11. Assuming a penalty for lost sales of $8 - 5 = 3, shortage cost = $9.12.
  c Reorder point of zero will do the job.

Chapter 17

SECTION 17.2

1

	Sunny	Cloudy
Sunny	.90	.10
Cloudy	.20	.80

2 State

$$
\begin{array}{c|ccccc}
 & 0 & 1 & 2 & 3 & 4 \\\hline
0 & 0 & 0 & \frac{1}{3} & \frac{1}{3} & \frac{1}{3} \\
1 & 0 & 0 & \frac{1}{3} & \frac{1}{3} & \frac{1}{3} \\
2 & \frac{1}{3} & \frac{1}{3} & \frac{1}{3} & 0 & 0 \\
3 & 0 & \frac{1}{3} & \frac{1}{3} & \frac{1}{3} & 0 \\
4 & 0 & 0 & \frac{1}{3} & \frac{1}{3} & \frac{1}{3}
\end{array}
$$

4 Let *SC* denote that yesterday was sunny and today is cloudy, and so on.

$$
\begin{array}{c|cccc}
 & SS & SC & CS & CC \\\hline
SS & .95 & .05 & 0 & 0 \\
SC & 0 & 0 & .40 & .60 \\
CS & .70 & .30 & 0 & 0 \\
CC & 0 & 0 & .20 & .80
\end{array}
$$

SECTION 17.3

1 **a** Urban, .651; suburban, .258; rural, .091.
  b 31.5%.

SECTION 17.4

1 2.

2 Yes.

3 **a** State 4.
  b States 1, 2, 3, 5, and 6.
  c {1, 3, 5} and {2, 6}.

4 P_1 is ergodic; P_2 is not ergodic.

SECTION 17.5

1 Urban, $\frac{38}{183}$; suburban, $\frac{90}{183}$; rural, $\frac{55}{183}$.

3 **a** State 1, $\frac{3}{5}$; state 2, $\frac{2}{5}$.

4 Replace a fair car.

7 Expected price of stock 1 = $16.67; expected price of stock 2 = $16.00.

SECTION 17.6

1 **a** $1.11 + 0.99 + 0.99 + 0.88 = 3.97$.
  b .748.

2 $1 + 0.80 + 18 = 19.80$ years.

11 **a** 14.3%.
  b Reducing warranty period will save $715,000 - $392,500 = $322,500.

SECTION 17.7

1 7,778 freshmen, 7,469 sophomores, 8,057 juniors, and 7,162 seniors.

2 Each working adult must contribute $2,000 more.

REVIEW PROBLEMS

1 An average of $\frac{260}{3}$ tools per day will be produced.

2 **a** .815.
  b Annual profit without warranty = $(3,000)$(total market size)$(\frac{1}{4})$. Annual profit with warranty = $(2,700)$(total market size)$(\frac{4}{9})$. Profit with warranty is larger.

3 Value of star = $4,400,000; value of starter = $3,199,000; value of substitute = $1,600,000.

4 **a** Each year, 1,638,270 new books, 1,474,443 once-used books, 1,179,554 twice-used books, and 707,733 thrice-used books will be sold.

Chapter 18

SECTION 18.1

1 Begin by picking up 4 matches. On each successive turn, pick up 5 − (number of matches picked up by opponent on last turn).

2 The players began with $16.25, $8.75, and $5.00.

SECTION 18.2

1 1–3–5–8–10, 1–4–6–9–10, and 1–4–5–8–10 are all shortest paths from node 1 to node 10 (each has length 11). The path 3–5–8–10 is the shortest path from node 3 to node 10 (this path has length 7).

3 Bloomington–Indianapolis–Dayton–Toledo–Cleveland takes 8 hours.

SECTION 18.3

1 Produce no units during month 1, 3 units during month 2, no units during month 3, and 4 units during month 4. Total cost is $15.00

2 Month 1, 200 radios; month 2, 600 radios; month 3, no radios. Total cost is $8,950.

3 a Produce 1 unit. Cost associated with arc is $4.50.

SECTION 18.4

1 Site 1, $1 million; site 2, $2 million; site 3, $1 million. Total revenue is $24 million.

2 Obtain a benefit of 10 with two type 1 items or two type 2 and one type 3 item.

3 a See Figure 14.

FIGURE 14

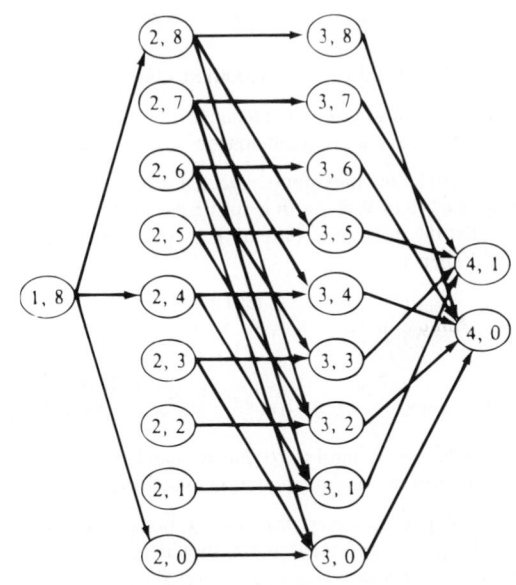

5 One type 2 item and one type 3 item yield a benefit of 75.

SECTION 18.5

2 Trade in car whenever it is two years old (at times 2, 4, and 6). Net cost is $14,400.

SECTION 18.6

1 Let $f_t(i)$ be the maximum expected net profit earned during years $t, t + 1, \ldots, 10$, given that Sunco has i barrels of reserves at the beginning of year t. Then

$$f_{10}(i) = \max_x \{xp_{10} - c(x)\}$$

where x must satisfy $0 \leq x \leq i$. For $t \leq 9$,

$$f_t(i) = \max_x \{xp_t - c(x) + f_{t+1}(i + b_t - x)\} \quad \textbf{(1)}$$

where $0 \leq x \leq i$. We use Equation (1) to work backward until $f_1(i_0)$ is determined. If discounting is allowed, let $\beta =$ the discount factor. Then we redefine $f_t(i)$ to be measured in terms of year t dollars. Then we replace (1) with (1′):

$$f_t(i) = \max_x \{xp_t - c(x) + \beta f_{t+1}(i + b_t - x)\} \quad \textbf{(1′)}$$

where $0 \leq x \leq b$.

2 b Let $f_t(d)$ be the maximum utility that can be earned during years $t, t + 1, \ldots, 10$, given that d dollars are available at the beginning of year t (including year t income). During year 10, it makes sense to consume all available money (after all, there is no future). Thus, $f_{10}(d) = \ln d$. For $t \leq 9$,

$$f_t(d) = \max_c \{\ln c + f_{t+1}(1.1(d - c) + i)\}$$

where $0 \leq c \leq d$. We work backward from the $f_{10}(\cdot)$'s to $f_1(D)$.

5 French, 1 hour; English, no hours; statistics, 3 hours. There is a .711 chance of passing at least one course.

7 Define $f_t(w)$ to be the maximum net profit (revenues less costs) obtained from the steer during weeks $t, t + 1, \ldots,$ 10, given that the steer weighs w pounds at the beginning of week t. Now

$$f_{10}(w) = \max_p \{10g(w, p) - c(p)\}$$

where $0 \leq p$. Then for $t \leq 9$,

$$f_t(w) = \max_p \{-c(p) + f_{t+1}(g(w, p))\}$$

Farmer Jones should work backward until $f_1(w_0)$ has been computed.

8 Define $f_t(i, d)$ to be the maximum number of loyal customers at the end of month 12, given that there are i loyal customers at the beginning of month t and d dollars available to spend on advertising during months $t, t + 1, \ldots, 12$. If there is only one month left, all available funds should be spent during that month. This yields

$$f_{12}(i, d) = (1 - p(d))i + (N - i)q(d)$$

For $t \leq 11$,

$$f_t(i, d) = \max_x \{f_{t+1}[(1 - p(x))i + (N - i)q(x), d - x]\}$$

where $0 \leq x \leq d$. We work backward until $f_1(i_0, D)$ has been determined.

10 Let $f_t(i_t, x_{t-1})$ be the minimum cost incurred during months $t, t + 1, \ldots, 12$, given that inventory at the beginning of month t is i_t, and production during month $t - 1$ was x_{t-1}. Then

$$f_{12}(i_{12}, x_{11}) = \min_{x_{12}} \{c_{12}x_{12} + 5|x_{12} - x_{11}| + h_{12}(i_{12} + x_{12} - d_{12})\}$$

where x_{12} must satisfy $x_{12} \geq 0$ and $i + x_{12} \geq d_{12}$. For $t \leq 11$,

$$f_t(i_t, x_{t-1})$$
$$= \min_{x_t} \{c_t x_t + 5|x_t - x_{t-1}| + h_t(i + x_t - d_t)$$
$$+ f_{t+1}(i_t + x_t - d_t, x_t)\}$$

where x_t must satisfy $x_t \geq 0$ and $i_t + x_t \geq d_t$. We work backward until $f_1(20, 20)$ has been computed.

SECTION 18.7

1 If initial inventory is 200 units, only modification is to produce 200 fewer units during period 1.

2 The Wagner–Whitin and Silver–Meal methods both yield the following production schedule: period 1, 90 units; period 4, 230 units. Total cost is $176.

REVIEW PROBLEMS

1 Shortest path from node 1 to node 10 is 1–4–8–10. Shortest path from node 2 to node 10 is 2–5–8–10.

2 Month 2, 1 unit; month 3, 4 units. Total cost is $12.

4 For 6 flights, the airline earns $540 with 3 Miami, 2 L.A. and 1 N.Y. flight; or 3 Miami and 3 L.A. flights. For four flights, the airline earns $375 with 2 Miami and 2 L.A. flights.

5 Without the 20¢ piece, use one 50¢, one 25¢, one 10¢, one 5¢, and one 1¢ piece. With the 20¢ piece, use one 50¢, two 20¢, and one 1¢ piece.

7 a Let $f_t(w)$ be the minimum cost incurred in meeting demands for the years $t, t + 1, \ldots, 5$, given that (before hiring and firing for year t) w workers are available.

h_t = workers hired at beginning of year t

d_t = workers fired at beginning of year t

w_t = workers required during year t

Then

$$f_t(w) = \min_{h_t, d_t} \{10{,}000h_t + 20{,}000d_t$$
$$+ 30{,}000(w + h_t - d_t) + f_{t+1}(h_t + .9(w - d_t))$$

where h_t and d_t must satisfy $0 \leq h_t$, $0 \leq d_t \leq w$, and $w + h_t - d_t \geq w_t$.

Chapter 19

SECTION 19.1

1 Two gallons of milk to each store.

2 $2 million to investment 1, $0 to investment 2, and $2 million to investment 3.

SECTION 19.2

1 Produce 3 units during period 1, no units during period 2, and 1 unit during period 3.

2 At any time, produce the number of units needed to bring the period's stock level (before the period's demand is met) to 2 units.

SECTION 19.3

1 Ulanowsky should play boldly during the first game. If he wins the first game, then he should play conservatively in the second game. If he loses the first game, then he should play boldly in the second game. If there is a tie-breaking game, Ulanowsky should play boldly. His chance of winning the match is .537.

2 On the first toss, Dickie should bet $2. If he loses, he is wiped out, but if he wins, he bets $1 on the second toss. If he wins on the second toss, he stops. If he loses on the second toss, he bets $2 on the third toss.

SECTION 19.4

2 Let $f_t(d)$ be the maximum expected asset position of the firm at the end of year 10, given that at the beginning of year t, the firm has d dollars in assets. Then

$$f_{10}(d) = \max_i \left\{ p \sum_y q_y(d + i + y) + (1 - p)\sum_y q_y(d - i + y) \right\}$$

where $0 \leq i \leq d$. For $t \leq 9$,

$$f_t(d) = \max_i \left\{ p \sum_y q_y f_{t+1}(d + i + y) + (1 - p)\sum_y q_y f_{t+1}(d - i + y) \right\}$$

We work backward until $f_1(10{,}000)$ has been computed.

5 We should always do maintenance on a running machine and always repair a broken machine.

6 Let $f_t(p)$ be the maximum expected revenue earned from selling a share of Wivco stock during days $t, t + 1, \ldots, 30$, given that the price of a share at the beginning of day t is p dollars. Then

$$f_{30}(p) = p \qquad \textbf{(Sell stock)}$$

and for $t \leq 29$,

$$f_t(p) = \max \begin{cases} p & \textbf{(Sell stock)} \\ \sum_x q(x)f_{t+1}((1 + x/100)p) & \textbf{(Keep stock)} \end{cases}$$

We work backward until $f_1(10)$ has been determined, and we continue until the optimal action is to sell the stock.

7 Sara should not accept the first cat, but any later cat that is the best Sara has seen so far should be accepted. The probability that Sara will get her preferred cat is $\frac{11}{24}$.

8 Define $f_t(i)$ to be the minimum net expected cost incurred during periods $t, t + 1, \ldots, 100$, given that the inventory level is i at the beginning of period t. Then

$$f_{100}(i) = \min_x \left\{ \sum_{d \leq i+x} q_d(i + x - d - rd) + \sum_{d > i+x} q_d(p(d - i - x) - r(i + x)) + c(x) \right\}$$

where $x \geq 0$. For $t \leq 99$,

$$f_t(i) = \min_x \left\{ \sum_{d \leq i+x} q_d(i + x - d - rd) \right.$$
$$+ \sum_{d > i+x} (p(d - i - x) - r(i + x))q_d + c(x)$$
$$+ \sum_{d \leq i+x} q_d f_{t+1}(i + x - d)$$
$$\left. + \sum_{d > i+x} q_d f_{t+1}(0) \right\}$$

where $x \geq 0$. We work backward until $f_1(0)$ has been determined.

10 Define $f_t(d)$ to be the maximum expected number of units sold in markets $t, t + 1, \ldots, T$, given that d dollars are available to spend on these markets. Then

$$f_T(d) = c_T k_T p_T(d)$$

and for $t \leq T - 1$,

$$f_t(d) = \max_x \{c_t k_t p_t(x) + f_{t+1}(d - x)\}$$

where $0 \leq x \leq d$. We work backward until $f_1(D)$ has been computed.

SECTION 19.5

1 a

Period's Beginning Inventory	Period's Production Level
0	4
1	3
2	0
3	0

2 a Always charge 11% interest rate on loans.

REVIEW PROBLEMS

1 Assign 2 sales reps to district 1, 1 sales rep to district 2, and 2 sales reps to district 3.

2 a Produce 2 units during period 1. During period 2, produce 1 or 2 units if beginning inventory is 0; if beginning inventory for period 2 is 1 or 2 units, produce no units during period 2. During period 3, produce 1 unit if beginning inventory is 0; otherwise, produce no units during period 3.

4 Let $f_t(b)$ be the maximum discounted net benefit earned during years $t, t + 1, \ldots, 2039$, given that b barrels of oil are available at the beginning of year t. Then

$$f_t(b) = \max_{d,x} \{u(x) - d + \beta p(d) f_{t+1}(b - x$$
$$+ 500,000) + \beta(1 - p(d)) f_{t+1}(b - x)\}$$

where $0 \leq d$ and $0 \leq x \leq b$. We work backward until $f_{2004}(B)$ is determined and then compute the optimal consumption strategy.

5 a Try to answer the first two questions and then stop. The expected amount of money won is $9,000.

Chapter 20

SECTION 20.2

1 $\frac{6}{7}$.

2 $\frac{555}{11}$ minutes.

4 a $\frac{2e^{-2}}{3} = .09$.

  b $1 - e^{-2} - 2e^{-2} = .594$.

SECTION 20.3

2 b $\frac{121}{144}$.

  c $\frac{1}{144}$.

SECTION 20.4

1 a $\frac{5}{6}$.

  b $\frac{25}{6}$ passengers.

  c $\frac{1}{2}$ minute.

3 a Unchanged.

  b Cut in half.

  c Unchanged.

8 Two checkpoints.

10 b 1 taxi.

  c $120 per hour.

SECTION 20.5

1 a 1.75 customers per hour.
  b $\frac{114}{130}$.

6 Barber 1's average hourly revenue = $53.23. Barber 2's average hourly revenue = $40.00.

SECTION 20.6

1 Two registers.

3 a Finance, $\frac{1}{5}$ day; Marketing, $\frac{1}{10}$ day.
  b 0.078 day.
  d .07.

4 4 servers.

13 $\frac{3}{4}$ mechanic.

14 System 1 is more efficient.

SECTION 20.7

1 5,200 members.

3 200 firms. The probability that there are at least 3,200 firms is zero.

SECTION 20.8

1 $\frac{2}{3}$ car.

4 a $\frac{5}{6}$ customer.

  b 9 minutes.

  c $\frac{1}{3}$.

SECTION 20.9

1 Superworker is better.

2 a $\frac{81}{125}$.

  b $\frac{6}{5}$ puppies.

SECTION 20.10

2 a 2.73 cars.

  b 0.06 hour.

4 $\frac{11}{3}$ students.

SECTION 20.11

1 15 fire engines.

7 b 2 copies.

SECTION 20.12

1 Using four categories (each having $e_i = 6$), we accept the hypothesis that the length of a telephone call is exponential with mean $\frac{1,024}{24}$ seconds.

SECTION 20.15

1 Tests spend an average of $\frac{19}{8}$ hours in the system, research papers spend an average of $\frac{27}{4}$ hours in the system, and class handouts spend an average of $\frac{23}{2}$ hours in the system.

2 Highest priority to $k = 1$ customers, next highest priority to $k = 2$ customers, and so on.

REVIEW PROBLEMS

2 a 8 minutes.

  b $\dfrac{(1.25)^2 e^{-1.25}}{2\,!} = .22.$

  c $e^{-0.75} = .47.$

3 a 3 windows.

  b 3 windows.

6 2 copiers

7 a $\frac{2}{3}$ car.

  b 8.2 minutes.

8 Rent the vacant lot. Then expected daily lost profit is $21(20)(0.008) = \$3.36$.

9 $\frac{1}{2}$ hour.

11 Have one crew of 100 workers.

16 a Both lines are free $\frac{8}{13}$ of time, one line is free $\frac{4}{13}$ of the time, and both lines are busy $\frac{1}{13}$ of the time.

  b $\frac{6}{13}$ line.

  c $30(\frac{1}{13}) = \frac{30}{13}$ callers per hour.

Chapter 21

SECTION 21.4

1 Approximately 0.76 minute. The answer may vary according to the random numbers used in the computations.

3 See Figure 15. Variables used in the model:

  VALUE = face value from the roll of the dice

  WINS = total number of wins up to current simulation

  LOSSES = total number of losses up to current simulation

  POINT = face value from the first roll

  PWON = proportion of wins

FIGURE **15**

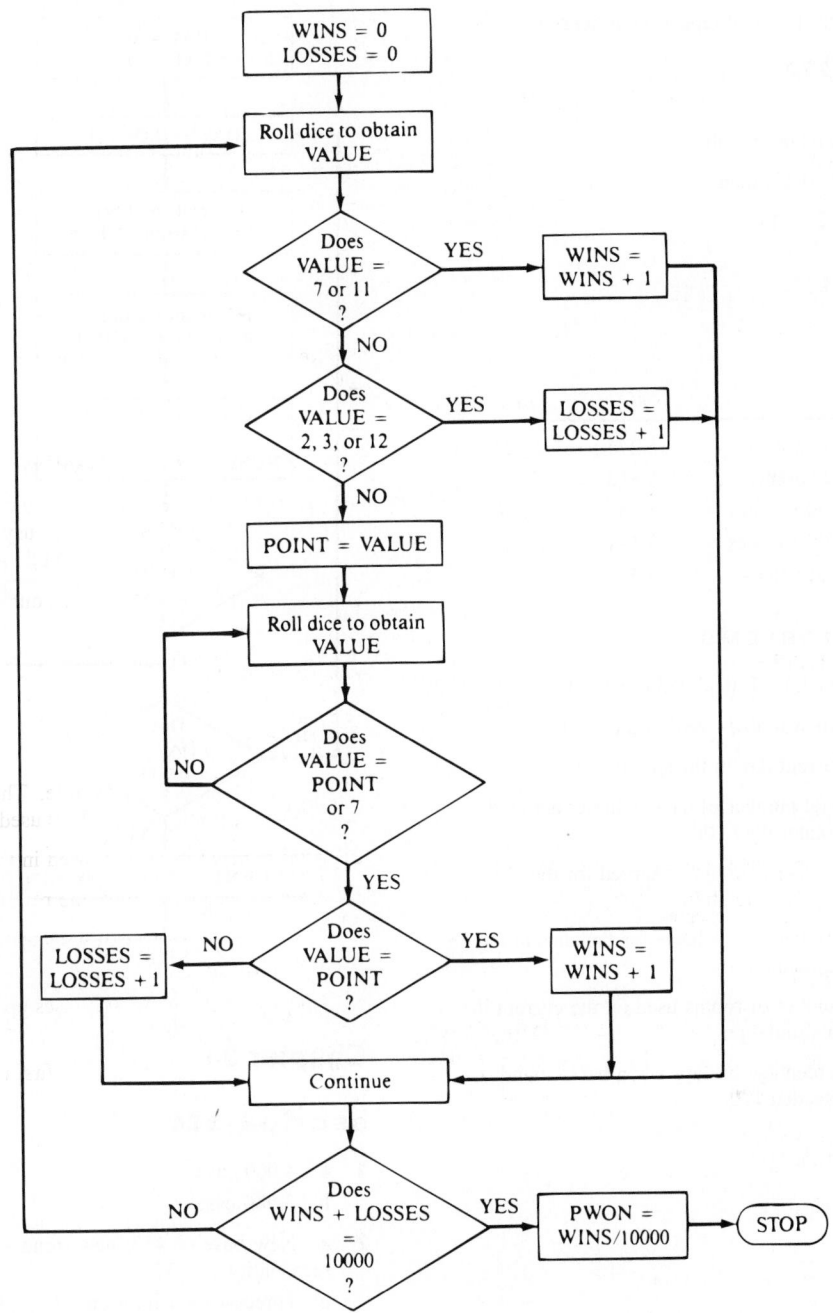

SECTION **21.5**

1 ITM: Generate a random number r.

    If ($r \leq 0.5$) then

      $x = 2r$

    else

      $x = 3 - 2\sqrt{2 - 2r}$

    endif

ARM: Generate two random numbers, r_1 and r_2.

    Set $x^* = 3r_1$

    If ($x^* \leq 1$) then

      Accept x^* as the random variate

    else

      If ($r_2 \leq \frac{3}{2}(1 - r_1)$) then

        Accept x^* as the random variate

            else

                Reject x* and repeat the process

            endif

        endif

2 Generate a random number *r*.

    If $(r \leq 0.25)$ then

        $x = 2 + 4\sqrt{r}$

    else

        $x = 10 - 4\sqrt{3 - 3r}$

    endif

4

	E(x)	var x
After 250 variates	5.274	2.823
After 500 variates	5.318	2.755
After 1,000 variates	5.364	2.856
After 5,000 variates	5.344	2.887
Theoretical values	5.333	2.889

REVIEW PROBLEMS

1 0.70, 0.33, 0.04, 0.11, 0.30, 0.53, 0.44, 0.91, 0.90, 0.73.

3 See Figure 16. Variables used in the model:

    DAY = current day in the simulation

    EXC = total number of days with net demand
          greater than 100

    NRES = number of rooms reserved for the current
           day in simulation

    NNSH = number of no-shows for the current day in
           simulation

    NUSD = number of rooms used on the current day
           in simulation

    PERCENT = percentage of days when net demand
              exeeded 100

FIGURE **16**

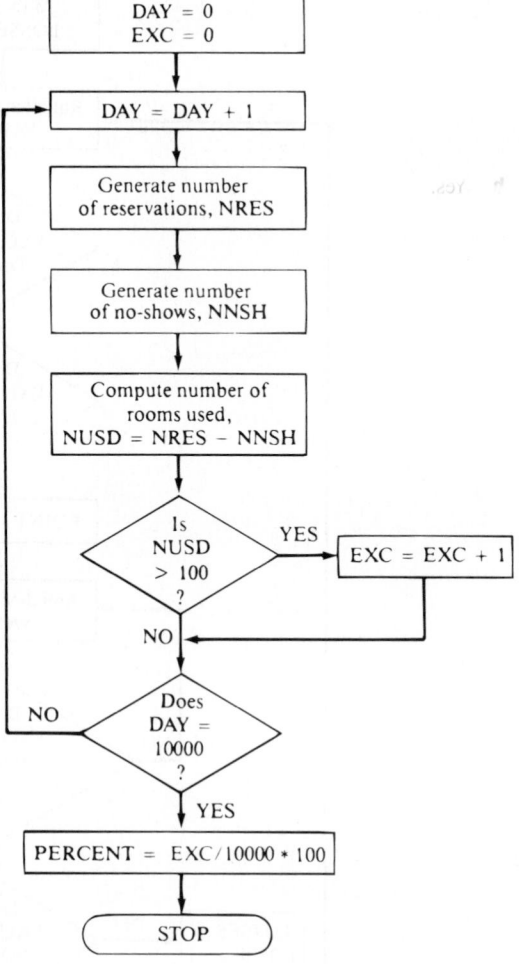

Chapter 24

SECTION 24.4

1 **a** 4,000 cases.

   b 3,980 cases.

2 **a** New base = 452; new trend = 43.6; new summer
       seasonality = 1.37.

   b Forecast for winter quarter = 377.44.

4 95% sure that December sales will be between 172.5
and 297.5.

12 **a** 106.

   b 115.26.

SECTION 24.5

1 For each professor's payday, compute

$$\frac{\text{Actual customers}}{\text{Forecast customers}}$$

Average these ratios. Suppose we obtain 1.3. Then to obtain forecast for a day on which professors are paid, compute a forecast by our basic method and multiply this forecast by 1.3.

SECTION 24.8

1 a Estimate beta = 0.88.
   b Yes.
   c 45%.
   e 16.1%.

3 a $\widehat{\text{SALES}} = 52{,}900 + 912.5T - 9{,}859Q_1 - 8{,}467Q_2 + 20{,}129Q_4$, where T = quarter number and Q_i = dummy variable for quarter i.

   d The part (b) model has smaller standard error than the part (a) model. Thus, the part (b) model will yield a better forecast.

4 $\widehat{\text{SALES}} = e^7 \, \text{PRICE}^{-0.67}$.

5 b Indicates positive autocorrelation.

Index